S0-BOI-235

MATHEMATICAL SYMBOLS

$=$	is equal to
$\neq$	is not equal to
$\approx$	is approximately equal to
$\propto$	is proportional to
$>$	is greater than
$\geq$	is greater than or equal to
$\gg$	is much greater than
$<$	is less than
$\leq$	is less than or equal to
$\ll$	is much less than
$\pm$	plus or minus
$\mp$	minus or plus
x_{av} or $\bar{x}$	average value of x
Δx	change in x ($x_f - x_i$)
$\lvert x \rvert$	absolute value of x
Σ	sum of
$\rightarrow 0$	approaches 0
∞	infinity

USEFUL MATHEMATICAL FORMULAS

Area of a triangle	$\frac{1}{2}bh$
Area of a circle	πr^2
Circumference of a circle	$2\pi r$
Surface area of a sphere	$4\pi r^2$
Volume of a sphere	$\frac{4}{3}\pi r^3$
Pythagorean theorem	$r^2 = x^2 + y^2$
Quadratic formula	if $ax^2 + bx + c = 0$, then $x = \frac{-b \pm \sqrt{b^2 - 4ac}}{2a}$

EXPONENTS AND LOGARITHMS

$x^n x^m = x^{n+m}$

$x^{-n} = \frac{1}{x^n}$ $\qquad \ln(xy) = \ln x + \ln y$

$\frac{x^n}{x^m} = x^{n-m}$ $\qquad \ln\left(\frac{x}{y}\right) = \ln x - \ln y$

$(xy)^n = x^n y^n$ $\qquad \ln x^n = n \ln x$

$(x^n)^m = x^{nm}$

VALUES OF SOME USEFUL NUMBERS

$\pi = 3.14159\ldots$	$\log 2 = 0.30103\ldots$	$\sqrt{2} = 1.41421\ldots$
$e = 2.71828\ldots$	$\ln 2 = 0.69315\ldots$	$\sqrt{3} = 1.73205\ldots$

TRIGONOMETRIC RELATIONSHIPS

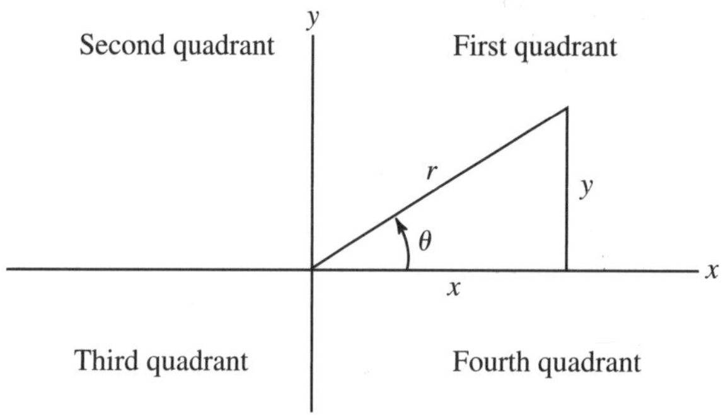

Definitions of Trigonometric Functions

$\sin\theta = \frac{y}{r} \qquad \cos\theta = \frac{x}{r} \qquad \tan\theta = \frac{\sin\theta}{\cos\theta} = \frac{y}{x}$

Trigonometric Functions of Important Angles

$\theta°$(rad)	$\sin\theta$	$\cos\theta$	$\tan\theta$
$0°(0)$	0	1	0
$30°(\pi/6)$	0.500	$\sqrt{3}/2 \approx 0.866$	$\sqrt{3}/3 \approx 0.577$
$45°(\pi/4)$	$\sqrt{2}/2 \approx 0.707$	$\sqrt{2}/2 \approx 0.707$	1.00
$60°(\pi/3)$	$\sqrt{3}/2 \approx 0.866$	0.500	$\sqrt{3} \approx 1.73$
$90°(\pi/2)$	1	0	∞

Trigonometric Identities

$\sin(\theta + \phi) = \sin\theta\cos\phi + \sin\phi\cos\theta$

$\cos(\theta + \phi) = \cos\theta\cos\phi - \sin\theta\sin\phi$

$\sin 2\theta = 2\sin\theta\cos\theta$

VECTOR MANIPULATIONS

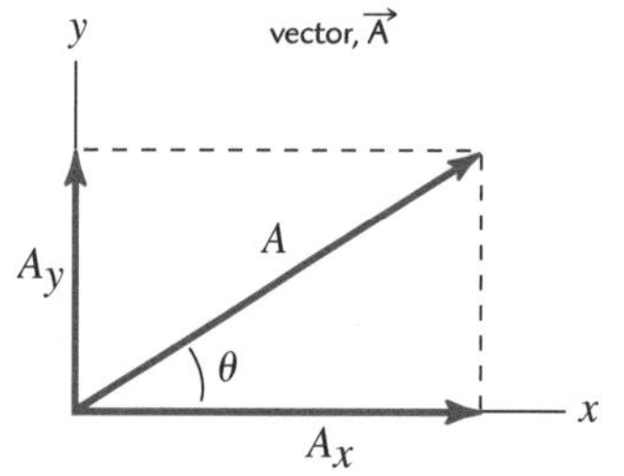

Magnitude and direction	$\longleftrightarrow$	***x* and *y* components**
A, θ	$\longrightarrow$	$A_x = A\cos\theta$ $A_y = A\sin\theta$
$A = \sqrt{A_x{}^2 + A_y{}^2}$ $\theta = \tan^{-1}\left(\frac{A_y}{A_x}\right)$	$\longleftarrow$	A_x, A_y

CONVERSION FACTORS

Mass

$1\ \text{kg} = 10^3\ \text{g}$
$1\ \text{g} = 10^{-3}\ \text{kg}$
$1\ \text{u} = 1.66 \times 10^{-24}\ \text{g} = 1.66 \times 10^{-27}\ \text{kg}$
1 slug = 14.6 kg
1 metric ton = 1000 kg

Length

$1\ \text{Å} = 10^{-10}\ \text{m}$
$1\ \text{nm} = 10^{-9}\ \text{m}$
$1\ \text{cm} = 10^{-2}\ \text{m} = 0.394\ \text{in.}$
1 yd = 3 ft
$1\ \text{m} = 10^{-3}\ \text{km} = 3.281\ \text{ft} = 39.4\ \text{in.}$
$1\ \text{km} = 10^3\ \text{m} = 0.621\ \text{mi}$
$1\ \text{in.} = 2.54\ \text{cm} = 2.54 \times 10^{-2}\ \text{m}$
1 ft = 0.305 m = 30.5 cm
1 mi = 5280 ft = 1609 m = 1.609 km
1 ly (light year) $= 9.46 \times 10^{12}$ km
1 pc (parsec) $= 3.09 \times 10^{13}$ km

Area

$1\ \text{cm}^2 = 10^{-4}\ \text{m}^2 = 0.1550\ \text{in.}^2 = 1.08 \times 10^{-3}\ \text{ft}^2$
$1\ \text{m}^2 = 10^4\ \text{cm}^2 = 10.76\ \text{ft}^2 = 1550\ \text{in.}^2$
$1\ \text{in.}^2 = 6.94 \times 10^{-3}\ \text{ft}^2 = 6.45\ \text{cm}^2 = 6.45 \times 10^{-4}\ \text{m}^2$
$1\ \text{ft}^2 = 144\ \text{in.}^2 = 9.29 \times 10^{-2}\ \text{m}^2 = 929\ \text{cm}^2$

Volume

$1\ \text{cm}^3 = 10^{-6}\ \text{m}^3 = 3.35 \times 10^{-5}\ \text{ft}^3 = 6.10 \times 10^{-2}\ \text{in.}^3$
$1\ \text{m}^3 = 10^6\ \text{cm}^3 = 10^3\ \text{L} = 35.3\ \text{ft}^3 = 6.10 \times 10^4\ \text{in.}^3 = 264\ \text{gal}$
$1\ \text{liter} = 10^3\ \text{cm}^3 = 10^{-3}\ \text{m}^3 = 1.056\ \text{qt} = 0.264\ \text{gal}$
$1\ \text{in.}^3 = 5.79 \times 10^{-4}\ \text{ft}^3 = 16.4\ \text{cm}^3 = 1.64 \times 10^{-5}\ \text{m}^3$
$1\ \text{ft}^3 = 1728\ \text{in.}^3 = 7.48\ \text{gal} = 0.0283\ \text{m}^3 = 28.3\ \text{L}$
$1\ \text{qt} = 2\ \text{pt} = 946\ \text{cm}^3 = 0.946\ \text{L}$
$1\ \text{gal} = 4\ \text{qt} = 231\ \text{in.}^3 = 0.134\ \text{ft}^3 = 3.785\ \text{L}$

Time

1 h = 60 min = 3600 s
$1\ \text{day} = 24\ \text{h} = 1440\ \text{min} = 8.64 \times 10^4\ \text{s}$
$1\ \text{y} = 365\ \text{days} = 8.76 \times 10^3\ \text{h} = 5.26 \times 10^5\ \text{min} = 3.16 \times 10^7\ \text{s}$

Speed

1 m/s = 3.60 km/h = 3.28 ft/s = 2.24 mi/h
1 km/h = 0.278 m/s = 0.621 mi/h = 0.911 ft/s
1 ft/s = 0.682 mi/h = 0.305 m/s = 1.10 km/h
1 mi/h = 1.467 ft/s = 1.609 km/h = 0.447 m/s
60 mi/h = 88 ft/s

Force

1 N = 0.225 lb
1 lb = 4.45 N
Equivalent weight of a mass of 1 kg on Earth's surface = 2.2 lb = 9.8 N
$1\ \text{dyne} = 10^{-5}\ \text{N} = 2.25 \times 10^{-6}\ \text{lb}$

Pressure

$1\ \text{Pa} = 1\ \text{N/m}^2 = 1.45 \times 10^{-4}\ \text{lb/in.}^2 = 7.5 \times 10^{-3}\ \text{mm Hg}$
$1\ \text{mm Hg} = 133\ \text{Pa} = 0.02\ \text{lb/in.}^2 = 1\ \text{torr}$
$1\ \text{atm} = 14.7\ \text{lb/in.}^2 = 101.3\ \text{kPa} = 30\ \text{in. Hg} = 760\ \text{mm Hg}$
$1\ \text{lb/in.}^2 = 6.89\ \text{kPa}$
$1\ \text{bar} = 10^5\ \text{Pa} = 100\ \text{kPa}$
$1\ \text{millibar} = 10^2\ \text{Pa}$

Energy

$1\ \text{J} = 0.738\ \text{ft} \cdot \text{lb} = 0.239\ \text{cal} = 9.48 \times 10^{-4}\ \text{Btu} = 6.24 \times 10^{18}\ \text{eV}$
1 kcal = 4186 J = 3.968 Btu
$1\text{Btu} = 1055\ \text{J} = 778\ \text{ft} \cdot \text{lb} = 0.252\ \text{kcal}$
$1\ \text{cal} = 4.186\ \text{J} = 3.97 \times 10^{-3}\ \text{Btu} = 3.09\ \text{ft} \cdot \text{lb}$
$1\ \text{ft} \cdot \text{lb} = 1.36\ \text{J} = 1.29 \times 10^{-3}\ \text{Btu}$
$1\ \text{eV} = 1.60 \times 10^{-19}\ \text{J}$
$1\ \text{kWh} = 3.6 \times 10^6\ \text{J}$
$1\ \text{erg} = 10^{-7}\ \text{J} = 7.38 \times 10^{-6}\ \text{ft} \cdot \text{lb}$

Power

$1\ \text{W} = 1\ \text{J/s} = 0.738\ \text{ft} \cdot \text{lb/s} = 1.34 \times 10^{-3}\ \text{hp} = 3.41\ \text{Btu/h}$
$1\ \text{ft} \cdot \text{lb/s} = 1.36\ \text{W} = 1.82 \times 10^{-3}\ \text{hp}$
$1\ \text{hp} = 550\ \text{ft} \cdot \text{lb/s} = 745.7\ \text{W} = 2545\ \text{Btu/h}$

Mass–Energy Equivalents

$1\ \text{u} = 1.66 \times 10^{-27}\ \text{kg} \leftrightarrow 931.5\ \text{MeV}$
$1\ \text{electron mass} = 9.11 \times 10^{-31}\ \text{kg} = 5.49 \times 10^{-4}\ \text{u} \leftrightarrow 0.511\ \text{MeV}$
$1\ \text{proton mass} = 1.673 \times 10^{-27}\ \text{kg} = 1.007\,267\ \text{u} \leftrightarrow 938.28\ \text{MeV}$
$1\ \text{neutron mass} = 1.675 \times 10^{-27}\ \text{kg} = 1.008\,665\ \text{u} \leftrightarrow 939.57\ \text{MeV}$

Temperature

$T_F = \frac{9}{5}T_C + 32$
$T_C = \frac{5}{9}(T_F - 32)$
$T_K = T_C + 273.15$

Angle

$1\ \text{rad} = 57.3°$

$1° = 0.0175$ rad	$60° = \pi/3$ rad
$15° = \pi/12$ rad	$90° = \pi/2$ rad
$30° = \pi/6$ rad	$180° = \pi$ rad
$45° = \pi/4$ rad	$360° = 2\pi$ rad

$1\ \text{rev/min} = (\pi/30)\ \text{rad/s} = 0.1047\ \text{rad/s}$

APPENDICES IN THE TEXT

Appendix A	Basic Mathematical Tools	p. APP-1
Appendix B	Typical Values	p. APP-11
Appendix C	Planetary Data	p. APP-12
Appendix D	Electrical Circuit Elements	p. APP-13
Appendix E	Periodic Table of the Elements	p. APP-14
Appendix F	Properties of Selected Isotopes	p. APP-15

Physics

AP* Edition

Physics

AP* Edition

Third Edition

James S. Walker

Western Washington University

PEARSON

Prentice Hall

Upper Saddle River, New Jersey 07458

Library of Congress Cataloging-in-Publication Data

WALKER, JAMES S.
Physics / James S. Walker.—3rd ed.

p. cm.
Includes index.
ISBN 0-13-153631-1
1. Physics—Textbooks. I. Title.
QC23.2.W35 2007
530—dc22

2005032634

Senior Editor: Erik Fahlgren
Editor-in-Chief, Science: Dan Kaveney
Media Editor: Michael J. Richards
Associate Editor: Christian Botting
Production Editor: Debra A. Wechsler
Executive Managing Editor: Kathleen Schiaparelli
Assistant Managing Editor: Beth Sweeten
Managing Editor, Science and Math Media: Nicole M. Jackson
Media Production Editor: Aaron Reid
Editor-in-Chief, Development: Carol Trueheart
Development Editor: David Chelton
Manufacturing Manager: Alexis Heydt-Long
Manufacturing Buyer: Alan Fischer
Director of Creative Services: Paul Belfanti
Creative Director: Juan R. López
Art Director: Jonathan Boylan
Interior and Cover Design: Koala Bear Design
Senior Managing Editor, AV Production and Management: Patricia Burns
Manager, Production Technologies: Matthew Haas
Managing Editor, Art Management: Abigail Bass
AV Production Editor: Xiaohong Zhu
Illustrations: Precision Graphics
Director, Image Resource Center: Melinda Reo
Manager, Rights and Permissions: Zina Arabia
Interior Image Specialist: Beth Brenzel
Cover Image Specialist: Karen Sanatar
Photo Researcher: Truitt & Marshall
Image Permission Coordinator: Richard Rodrigues
Editorial Assistant: Jessica Berta
Cover Image: Satellite image of Hurricane Fran off the coast of Florida: Stocktrek/CORBIS; barred spiral galaxy NGC1300: NASA, ESA, and the Hubble Heritage Team (STSci/AURA); interior of the Ouessant Island Lighthouse: Guy Thouvenin/Robert Harding World Imagery; person holding a magnifying glass up to oak tree: Cosmo Condina/Photographers Choice/Getty Images, Inc.; three rows of organ pipes: Kenneth Garrett/National Geographic/Getty Images, Inc.; opera singer in a Viking helmet; C Squared Studios/Photodisc Green/Getty Images, Inc.; magnet covered in iron fillings: Steve Taylor/Stone/Getty Images, Inc.; MRI scan of human head and shoulders (Digital Composite): David Job/Stone/Getty Images, Inc.

PEARSON Prentice Hall

© 2007, 2004, 2002 by Pearson Education, Inc.
Pearson Prentice Hall
Pearson Education, Inc.
Upper Saddle River, NJ 07458

All rights reserved. No part of this book may be reproduced in any form or by any means, without permission in writing from the publisher.

Pearson Prentice Hall™ is a trademark of Pearson Education, Inc.

*AP, Pre-AP, Advanced Placement, and Advanced Placement Program are registered trademarks of The College Entrance Examination Board, which was not involved in the production of, and does not endorse, this product.

Printed in the United States of America

10 9 8 7 6 5 4 3 2

ISBN 0-13-196067-9

Pearson Education LTD., *London*
Pearson Education Australia PTY, Limited, *Sydney*
Pearson Education Singapore, Pte. Ltd
Pearson Education North Asia Ltd, *Hong Kong*
Pearson Education Canada, Ltd., *Toronto*
Pearson Educación de Mexico, S.A. de C.V.
Pearson Education—Japan, *Tokyo*
Pearson Education Malaysia, Pte. Ltd

About the Author

James S. Walker James Walker obtained his Ph.D. in theoretical physics from the University of Washington in 1978. He subsequently served as a post-doc at the University of Pennsylvania, the Massachusetts Institute of Technology, and the University of California at San Diego before joining the physics faculty at Washington State University in 1983. Professor Walker's research interests include statistical mechanics, critical phenomena, and chaos. His many publications on the application of renormalization-group theory to systems ranging from absorbed monolayers to binary-fluid mixtures have appeared in *Physical Review, Physical Review Letters, Physica,* and a host of other publications. He has also participated in observations on the summit of Mauna Kea, looking for evidence of extra-solar planets.

Jim Walker likes to work with students at all levels, from judging elementary school science fairs to writing research papers with graduate students, and has taught introductory physics for many years. His enjoyment of this course and his empathy for students have earned him a reputation as an innovative, enthusiastic, and effective teacher. Jim's educational publications include "Reappearing Phases" (*Scientific American,* May 1987) as well as articles in the *American Journal of Physics* and *The Physics Teacher*. In recognition of his contributions to the teaching of physics, Jim was named Boeing Distinguished Professor of Science and Mathematics Education for 2001–2003.

When he is not writing, conducting research, teaching, or developing new classroom demonstrations and pedagogical materials, Jim enjoys amateur astronomy, bird watching, photography, juggling, unicycling, boogie boarding, and kayaking. Jim is also an avid jazz pianist and organist. He has served as ballpark organist for several Class A minor league baseball teams, including minor league affiliates of the Seattle Mariners and San Francisco Giants.

About the Cover

The photographs on the front and back covers of this book are a reminder that physics determines the behavior of everything in the universe—from the particles that make up atoms, to the tissues of the human body, to galaxies and the multitude of stars they contain. Grouped in pairs that share a common physical theme, these images touch on just a few of the many real-world applications of physics discussed in this text.

Magnet and MRI Image

A horseshoe magnet attracts iron filings to its poles, where its magnetic field is most intense. A much more powerful magnet is responsible for the striking image of a human head produced by magnetic resonance imaging (MRI). These and other aspects of magnetism are covered in Chapter 22.

Organ Pipes and Opera Singer

A grand organ contains thousands of pipes that allow it to produce a wide range of sounds. The human windpipe, though more limited in size and range, can still give the organ a run for its money. Turn to Chapter 14 for a detailed discussion of sound.

Lighthouse Lens and Magnifying Glass

The massive lens used by a lighthouse to guide ships safely to port operates on the same physical principles as the hand-held magnifying glass. Optics and optical instruments are the topics of Chapters 26 and 27.

Hurricane and Galaxy

The spiral structure of a hurricane bears a strong resemblance to that of a rotating galaxy, even though the forces at play in a hurricane (air pressure) are very different from those that act in a galaxy (gravity). The physics of rotation is studied in Chapters 10 and 11. Fluids are covered in Chapter 15, and gravity is presented in Chapter 12.

As you study the material in this book, your understanding of physics will deepen. You will begin to make the connection between magnets and MRI, organ pipes and windpipes, lighthouses and magnifying glasses, hurricanes and galaxies. At the same time, your appreciation for the world around you will increase as you come to recognize the fundamental physical principles on which all of our lives are based.

This book is dedicated to my parents,
Ivan and Janet Walker, and to my wife Betsy.

Brief Contents

1 Introduction to Physics 1

PART I MECHANICS

2 One-Dimensional Kinematics 17
3 Vectors in Physics 55
4 Two-Dimensional Kinematics 79
5 Newton's Laws of Motion 107
6 Applications of Newton's Laws 141
7 Work and Kinetic Energy 179
8 Potential Energy and Conservation of Energy 204
9 Linear Momentum and Collisions 240
10 Rotational Kinematics and Energy 281
11 Rotational Dynamics and Static Equilibrium 315
12 Gravity 358
13 Oscillations About Equilibrium 394
14 Waves and Sound 430
15 Fluids 476

PART II THERMAL PHYSICS

16 Temperature and Heat 514
17 Phases and Phase Changes 548
18 The Laws of Thermodynamics 585

PART III ELECTROMAGNETISM

19 Electric Charges, Forces, and Fields 625
20 Electric Potential and Electric Potential Energy 662
21 Electric Current and Direct-Current Circuits 695
22 Magnetism 732
23 Magnetic Flux and Faraday's Law of Induction 768
24 Alternating-Current Circuits 803

PART IV LIGHT AND OPTICS

25 Electromagnetic Waves 837
26 Geometrical Optics 870
27 Optical Instruments 909
28 Physical Optics: Interference and Diffraction 937

PART V MODERN PHYSICS

29 Relativity 970
30 Quantum Physics 1003
31 Atomic Physics 1034
32 Nuclear Physics and Nuclear Radiation 1071

Contents

Applications in the Text xxiii

Preface: To the AP* Teacher xxvii

Preface: To the AP* Student xli

Guide to Features of the Text xlii

1 Introduction to Physics 1

1–1 Physics and the Laws of Nature 2
1–2 Units of Length, Mass, and Time 2
1–3 Dimensional Analysis 4
1–4 Significant Figures 5
1–5 Converting Units 8
1–6 Order-of-Magnitude Calculations 10
1–7 Scalars and Vectors 11
1–8 Problem Solving in Physics 11
Chapter Summary 13
Conceptual Questions 14
Conceptual Exercises 14
Problems 14

PART I MECHANICS

2 One-Dimensional Kinematics 17

2–1 Position, Distance, and Displacement 18
2–2 Average Speed and Velocity 19
2–3 Instantaneous Velocity 23
2–4 Acceleration 25
2–5 Motion with Constant Acceleration 29
2–6 Applications of the Equations of Motion 35
2–7 Freely Falling Objects 37
Chapter Summary 43
Problem-Solving Summary 45
Conceptual Questions 46
Conceptual Exercises 46
Problems 47

3 Vectors in Physics 55

3–1 Scalars Versus Vectors 56
3–2 The Components of a Vector 56
3–3 Adding and Subtracting Vectors 61
3–4 Unit Vectors 64
3–5 Position, Displacement, Velocity, and Acceleration Vectors 65
3–6 Relative Motion 69
Chapter Summary 72
Problem-Solving Summary 73
Conceptual Questions 73
Conceptual Exercises 74
Problems 74

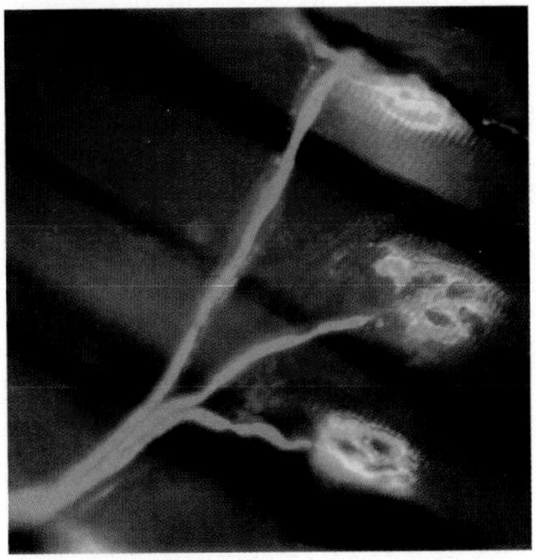

Can we measure the speed of a nerve impulse? (Page 16)

How long does it take for the chameleon to catch the bug? (Page 50)

How does a cyclist's velocity vector change as she navigates this hill? (Page 68)

Where do the lava bombs land? (Page 87)

4 Two-Dimensional Kinematics 79

4–1 Motion in Two Dimensions 80
4–2 Projectile Motion: Basic Equations 83
4–3 Zero Launch Angle 84
4–4 General Launch Angle 89
4–5 Projectile Motion: Key Characteristics 93
Chapter Summary 97
Problem-Solving Summary 99
Conceptual Questions 99
Conceptual Exercises 100
Problems 101

5 Newton's Laws of Motion 107

5–1 Force and Mass 108
5–2 Newton's First Law of Motion 108
5–3 Newton's Second Law of Motion 110
5–4 Newton's Third Law of Motion 118
5–5 The Vector Nature of Forces: Forces in Two Dimensions 121
5–6 Weight 124
5–7 Normal Forces 128
Chapter Summary 132
Problem-Solving Summary 133
Conceptual Questions 133
Conceptual Exercises 135
Problems 136

Why does water fly off this dog's back? (Page 134)

6 Applications of Newton's Laws 141

6–1 Frictional Forces 142
6–2 Strings and Springs 150
6–3 Translational Equilibrium 154
6–4 Connected Objects 158
6–5 Circular Motion 162
Chapter Summary 167
Problem-Solving Summary 168
Conceptual Questions 168
Conceptual Exercises 169
Problems 170

7 Work and Kinetic Energy 179

7–1 Work Done by a Constant Force 180
7–2 Kinetic Energy and the Work-Energy Theorem 186
7–3 Work Done by a Variable Force 190
7–4 Power 195
Chapter Summary 197
Problem-Solving Summary 198
Conceptual Discussion Questions 199
Conceptual Exercises 199
Problems 200

Does the sole of this shoe experience static or kinetic friction when it is in contact with the ground? (Page 143)

8 Potential Energy and Conservation of Energy 204

8–1 Conservative and Nonconservative Forces 205
8–2 Potential Energy and the Work Done by Conservative Forces 208
8–3 Conservation of Mechanical Energy 214
8–4 Work Done by Nonconservative Forces 222
8–5 Potential Energy Curves and Equipotentials 227
Chapter Summary 230
Problem-Solving Summary 231
Conceptual Questions 231
Conceptual Exercises 232
Problems 233

9 Linear Momentum and Collisions 240

9–1 Linear Momentum 241
9–2 Momentum and Newton's Second Law 243
9–3 Impulse 244
9–4 Conservation of Linear Momentum 248
9–5 Inelastic Collisions 253
9–6 Elastic Collisions 259
9–7 Center of Mass 264
*9–8 Systems with Changing Mass: Rocket Propulsion 269
Chapter Summary 271
Problem-Solving Summary 273
Conceptual Questions 273
Conceptual Exercises 274
Problems 275

10 Rotational Kinematics and Energy 281

10–1 Angular Position, Velocity, and Acceleration 282
10–2 Rotational Kinematics 286
10–3 Connections Between Linear and Rotational Quantities 289
10–4 Rolling Motion 294
10–5 Rotational Kinetic Energy and the Moment of Inertia 295
10–6 Conservation of Energy 299
Chapter Summary 304
Problem-Solving Summary 306
Conceptual Questions 307
Conceptual Exercises 307
Problems 308

11 Rotational Dynamics and Static Equilibrium 315

11–1 Torque 316
11–2 Torque and Angular Acceleration 319
11–3 Zero Torque and Static Equilibrium 323
11–4 Center of Mass and Balance 330
11–5 Dynamic Applications of Torque 333
11–6 Angular Momentum 335
11–7 Conservation of Angular Momentum 338
11–8 Rotational Work and Power 343
*11–9 The Vector Nature of Rotational Motion 344

How many energy conversions can you identify? (Page 231)

Is this collision more nearly elastic or inelastic? (Page 253)

How does the "tuck" affect the speed of the diver's spin? (Page 338)

Why doesn't this satellite fall to the ground? (Page 372)

Chapter Summary 346
Problem-Solving Summary 348
Conceptual Questions 348
Conceptual Exercises 348
Problems 350

12 Gravity 358

12–1 Newton's Law of Universal Gravitation 359
12–2 Gravitational Attraction of Spherical Bodies 362
12–3 Kepler's Laws of Orbital Motion 367
12–4 Gravitational Potential Energy 374
12–5 Energy Conservation 377
*12–6 Tides 384
Chapter Summary 386
Problem-Solving Summary 388
Conceptual Questions 388
Conceptual Exercises 388
Problems 389

What happened to the kinetic energy of this jumper? (Page 408)

13 Oscillations About Equilibrium 394

13–1 Periodic Motion 395
13–2 Simple Harmonic Motion 396
13–3 Connections Between Uniform Circular Motion and Simple Harmonic Motion 399
13–4 The Period of a Mass on a Spring 405
13–5 Energy Conservation in Oscillatory Motion 409
13–6 The Pendulum 412
13–7 Damped Oscillations 418
13–8 Driven Oscillations and Resonance 419
Chapter Summary 421
Problem-Solving Summary 423
Conceptual Questions 423
Conceptual Exercises 424
Problems 425

How many trumpeting elephants are needed to match the call of a whale? (Page 440)

14 Waves and Sound 430

14–1 Types of Waves 431
14–2 Waves on a String 433
*14–3 Harmonic Wave Functions 436
14–4 Sound Waves 437
14–5 Sound Intensity 441
14–6 The Doppler Effect 446
14–7 Superposition and Interference 452
14–8 Standing Waves 455
14–9 Beats 463
Chapter Summary 465
Problem-Solving Summary 468
Conceptual Questions 468
Conceptual Exercises 468
Problems 469

15 Fluids 476

15–1 Density 477
15–2 Pressure 477
15–3 Static Equilibrium in Fluids: Pressure and Depth 481
15–4 Archimedes' Principle and Buoyancy 486
15–5 Applications of Archimedes' Principle 488
15–6 Fluid Flow and Continuity 493
15–7 Bernoulli's Equation 495
15–8 Applications of Bernoulli's Equation 498
*15–9 Viscosity and Surface Tension 501
Chapter Summary 504
Problem-Solving Summary 506
Conceptual Questions 506
Conceptual Exercises 507
Problems 508

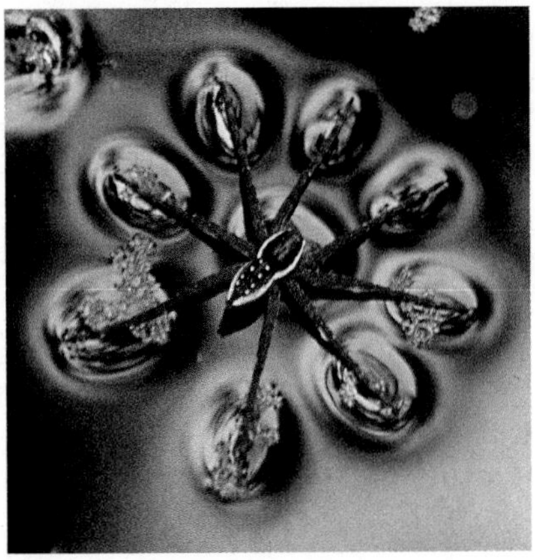

How does this spider use surface tension to survive? (Page 503)

PART II THERMAL PHYSICS

16 Temperature and Heat 514

16–1 Temperature and the Zeroth Law of Thermodynamics 515
16–2 Temperature Scales 516
16–3 Thermal Expansion 520
16–4 Heat and Mechanical Work 526
16–5 Specific Heats 528
16–6 Conduction, Convection, and Radiation 531
Chapter Summary 539
Problem-Solving Summary 541
Conceptual Questions 541
Conceptual Exercises 542
Problems 543

Why would this lizard stand on two legs instead of four? (Page 532)

17 Phases and Phase Changes 548

17–1 Ideal Gases 549
17–2 Kinetic Theory 555
17–3 Solids and Elastic Deformation 560
17–4 Phase Equilibrium and Evaporation 565
17–5 Latent Heats 571
17–6 Phase Changes and Energy Conservation 574
Chapter Summary 576
Problem-Solving Summary 578
Conceptual Questions 578
Conceptual Exercises 579
Problems 579

What causes this lenticular cloud to "park" above the mountain? (Page 600)

18 The Laws of Thermodynamics 585

18–1 The Zeroth Law of Thermodynamics 586
18–2 The First Law of Thermodynamics 586
18–3 Thermal Processes 588
18–4 Specific Heats for an Ideal Gas: Constant Pressure, Constant Volume 596
18–5 The Second Law of Thermodynamics 600

What role does electricity play in pollination? (Page 660)

18–6 Heat Engines and the Carnot Cycle 600
18–7 Refrigerators, Air Conditioners, and Heat Pumps 604
18–8 Entropy 608
18–9 Order, Disorder, and Entropy 612
18–10 The Third Law of Thermodynamics 614
Chapter Summary 615
Problem-Solving Summary 617
Conceptual Questions 617
Conceptual Exercises 618
Problems 619

PART III ELECTROMAGNETISM

19 Electric Charges, Forces, and Fields 625
19–1 Electric Charge 626
19–2 Insulators and Conductors 629
19–3 Coulomb's Law 630
19–4 The Electric Field 637
19–5 Electric Field Lines 643
19–6 Shielding and Charging by Induction 646
19–7 Electric Flux and Gauss's Law 649
Chapter Summary 653
Problem-Solving Summary 654
Conceptual Questions 655
Conceptual Exercises 655
Problems 657

How can this musician play a song without touching his instrument? (Page 683)

20 Electric Potential and Electric Potential Energy 662
20–1 Electric Potential Energy and the Electric Potential 663
20–2 Energy Conservation 666
20–3 The Electric Potential of Point Charges 669
20–4 Equipotential Surfaces and the Electric Field 673
20–5 Capacitors and Dielectrics 677
20–6 Electrical Energy Storage 683
Chapter Summary 685
Problem-Solving Summary 687
Conceptual Questions 687
Conceptual Exercises 688
Problems 689

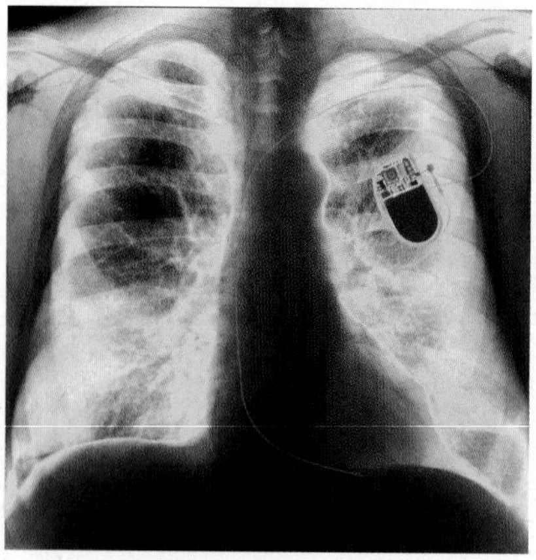

How does an RC circuit keep time for an unsteady heart? (Page 720)

21 Electric Current and Direct-Current Circuits 695
21–1 Electric Current 696
21–2 Resistance and Ohm's Law 699
21–3 Energy and Power in Electric Circuits 702
21–4 Resistors in Series and Parallel 705
21–5 Kirchhoff's Rules 711
21–6 Circuits Containing Capacitors 714
21–7 *RC* Circuits 717
*21–8 Ammeters and Voltmeters 720
Chapter Summary 721

Conceptual Questions 963
Conceptual Exercises 964
Problems 964

PART V MODERN PHYSICS

29 Relativity 970

29–1 The Postulates of Special Relativity 971
29–2 The Relativity of Time and Time Dilation 972
29–3 The Relativity of Length and Length Contraction 978
29–4 The Relativistic Addition of Velocities 981
29–5 Relativistic Momentum 983
29–6 Relativistic Energy and $E = mc^2$ 986
29–7 The Relativistic Universe 991
29–8 General Relativity 991
Chapter Summary 995
Problem-Solving Summary 997
Conceptual Questions 998
Conceptual Exercises 998
Problems 998

30 Quantum Physics 1003

30–1 Blackbody Radiation and Planck's Hypothesis of Quantized Energy 1004
30–2 Photons and the Photoelectric Effect 1007
30–3 The Mass and Momentum of a Photon 1013
30–4 Photon Scattering and the Compton Effect 1014
30–5 The de Broglie Hypothesis and Wave-Particle Duality 1017
30–6 The Heisenberg Uncertainty Principle 1021
30–7 Quantum Tunneling 1025
Chapter Summary 1026
Problem-Solving Summary 1028
Conceptual Questions 1028
Conceptual Exercises 1029
Problems 1030

31 Atomic Physics 1034

31–1 Early Models of the Atom 1035
31–2 The Spectrum of Atomic Hydrogen 1036
31–3 Bohr's Model of the Hydrogen Atom 1039
31–4 de Broglie Waves and the Bohr Model 1046
31–5 The Quantum Mechanical Hydrogen Atom 1047
31–6 Multielectron Atoms and the Periodic Table 1050
31–7 Atomic Radiation 1055
Chapter Summary 1064
Problem-Solving Summary 1066
Conceptual Questions 1067
Conceptual Exercises 1067
Problems 1067

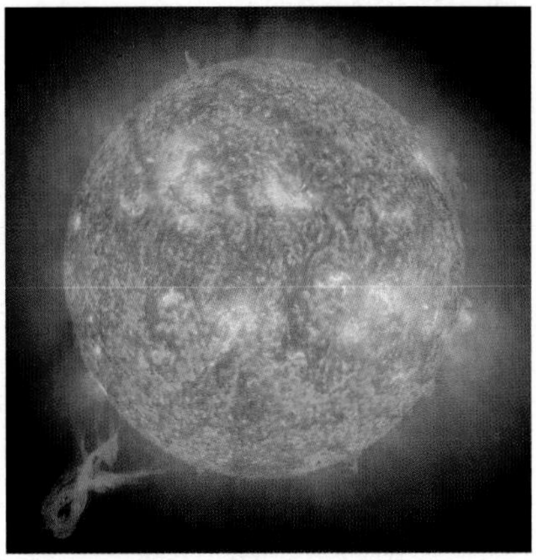

How does the Sun convert mass to energy? (Page 987)

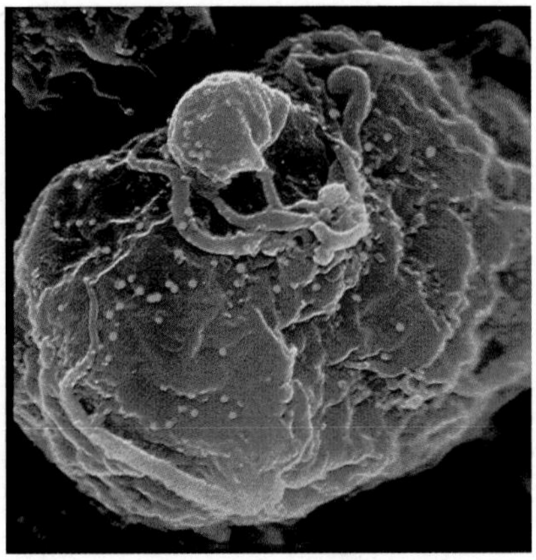

What role does quantum physics play in understanding how an electron microscope works? (Page 1020)

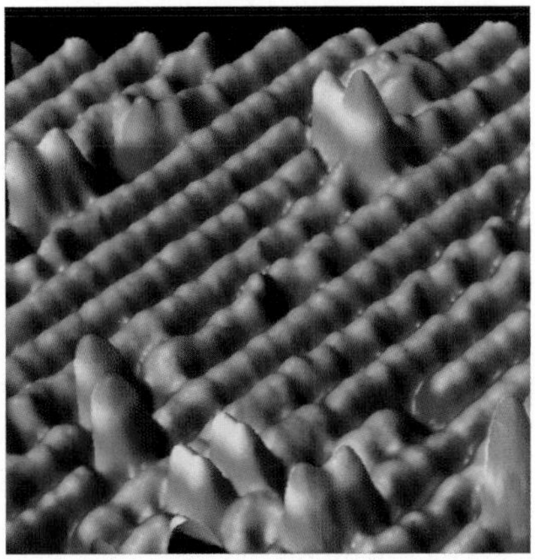

How does quantum physics allow us to map the "terrain" of a surface at the atomic level? (Page 1026)

How did Alba become the world's first fluorescent rabbit? (Page 1034)

32 Nuclear Physics and Nuclear Radiation 1071

32–1 The Constituents and Structure of Nuclei 1072
32–2 Radioactivity 1076
32–3 Half-Life and Radioactive Dating 1083
32–4 Nuclear Binding Energy 1089
32–5 Nuclear Fission 1090
32–6 Nuclear Fusion 1093
32–7 Practical Applications of Nuclear Physics 1095
32–8 Elementary Particles 1099
32–9 Unified Forces and Cosmology 1102
Chapter Summary 1103
Problem-Solving Summary 1105
Conceptual Questions 1106
Conceptual Exercises 1106
Problems 1107

What is responsible for the variety of color in this display? (Page 1055)

Appendices

Appendix A
Basic Mathematical Tools APP-1

Appendix B
Typical Values APP-11

Appendix C
Planetary Data APP-12

Appendix D
Elements of Electrical Circuits APP-13

Appendix E
Periodic Table of the Elements APP-14

Appendix F
Properties of Selected Isotopes APP-15

Answers to Your Turn Problems ANS-1

Answers to Odd-Numbered Conceptual Questions ANS-5

Answers to Odd-Numbered Conceptual Exercises ANS-13

Answers to Odd-Numbered Problems ANS-23

Photo Credits P-1

Index I-1

Problem-Solving Summary 723
Conceptual Questions 724
Conceptual Exercises 724
Problems 726

22 Magnetism 732

22–1 The Magnetic Field 733
22–2 The Magnetic Force on Moving Charges 735
22–3 The Motion of Charged Particles in a Magnetic Field 739
22–4 The Magnetic Force Exerted on a Current-Carrying Wire 744
22–5 Loops of Current and Magnetic Torque 746
22–6 Electric Currents, Magnetic Fields, and Ampère's Law 748
22–7 Current Loops and Solenoids 752
22–8 Magnetism in Matter 755
Chapter Summary 757
Problem-Solving Summary 760
Conceptual Questions 760
Conceptual Exercises 760
Problems 762

23 Magnetic Flux and Faraday's Law of Induction 768

23–1 Induced Electromotive Force 769
23–2 Magnetic Flux 770
23–3 Faraday's Law of Induction 772
23–4 Lenz's Law 775
23–5 Mechanical Work and Electrical Energy 778
23–6 Generators and Motors 781
23–7 Inductance 784
23–8 *RL* Circuits 787
23–9 Energy Stored in a Magnetic Field 788
23–10 Transformers 790
Chapter Summary 793
Problem-Solving Summary 795
Conceptual Questions 795
Conceptual Exercises 796
Problems 797

24 Alternating-Current Circuits 803

24–1 Alternating Voltages and Currents 804
24–2 Capacitors in AC Circuits 809
24–3 *RC* Circuits 812
24–4 Inductors in AC Circuits 817
24–5 *RLC* Circuits 820
24–6 Resonance in Electrical Circuits 824
Chapter Summary 829
Problem-Solving Summary 831
Conceptual Questions 831
Conceptual Exercises 831
Problems 832

How does the Earth's magnetic field affect the northern lights? (Page 744)

How is the motion of water through this dam converted to electrical energy? (Page 782)

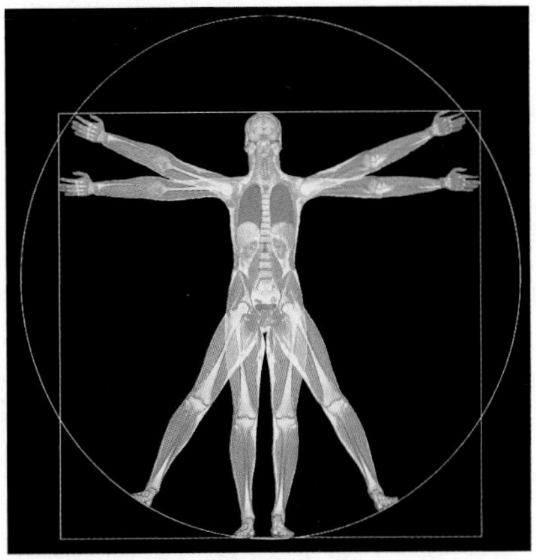

How can an alternating-current electrical circuit make magnetic resonance imaging possible? (Page 803)

Why does a daisy look different to a bee? (Page 837)

Why do these dew drops form an image of the flower? (Page 892)

How can a butterfly produce brilliant colors without the use of pigments? (Page 937)

PART IV LIGHT AND OPTICS

25 Electromagnetic Waves 837

25–1 The Production of Electromagnetic Waves 838
25–2 The Propagation of Electromagnetic Waves 841
25–3 The Electromagnetic Spectrum 845
25–4 Energy and Momentum in Electromagnetic Waves 849
25–5 Polarization 853
Chapter Summary 860
Problem-Solving Summary 862
Conceptual Questions 863
Conceptual Exercises 863
Problems 864

26 Geometrical Optics 870

26–1 The Reflection of Light 871
26–2 Forming Images with a Plane Mirror 872
26–3 Spherical Mirrors 875
26–4 Ray Tracing and the Mirror Equation 877
26–5 The Refraction of Light 884
26–6 Ray Tracing for Lenses 891
26–7 The Thin-Lens Equation 894
26–8 Dispersion and the Rainbow 896
Chapter Summary 898
Problem-Solving Summary 900
Conceptual Questions 900
Conceptual Exercises 901
Problems 902

27 Optical Instruments 909

27–1 The Human Eye and the Camera 910
27–2 Lenses in Combination and Corrective Optics 913
27–3 The Magnifying Glass 919
27–4 The Compound Microscope 923
27–5 Telescopes 924
27–6 Lens Aberrations 927
Chapter Summary 928
Problem-Solving Summary 930
Conceptual Questions 930
Conceptual Exercises 930
Problems 931

28 Physical Optics: Interference and Diffraction 937

28–1 Superposition and Interference 938
28–2 Young's Two-Slit Experiment 940
28–3 Interference in Reflected Waves 944
28–4 Diffraction 950
28–5 Resolution 954
28–6 Diffraction Gratings 957
Chapter Summary 961
Problem-Solving Summary 963

Applications in the Text

Note: This list includes applied topics that receive significant discussion in the chapter text or a worked Example, as well as topics that are touched on in end-of-chapter Conceptual Questions, Conceptual Exercises, and Problems. Topics of particular relevance to the life sciences or medicine are marked **BIO**.

Chapter 1
Estimation: How many raindrops in a storm 10
The strike of a Mantis shrimp 15 **BIO**
Mosquito courtship 16 **BIO**

Chapter 2
Takeoff distance for an airliner 34
The stopping distance of a car 35
Calculating the speed of a lava bomb 40

Chapter 3
Determining the height of a cliff 58
Crossing a river 71
The watch that won the Longitude Prize 75

Chapter 4
The parabolic trajectory of projectiles 87
Golf on the Moon 94
How an archerfish hunts 96 **BIO**
Volcanoes on Io 103
Punkin Chunkin 103
Landing rovers on Mars 104

Chapter 5
How walking affects your height 113 **BIO**
Astronaut jet packs 116
Stopping an airplane with Foamcrete 116
Simulating weightlessness 128 **BIO**

Chapter 6
Antilock braking systems 149
Setting a broken leg with traction 151 **BIO**
Skids and banked roadways 164
Centrifuges and ultracentrifuges 166 **BIO**
Human-powered centrifuge 174 **BIO**

Chapter 7
Human power output and flight 195 **BIO**
The reentry of Skylab 200
Human-powered flight 201 **BIO**
Power output of the human brain 203 **BIO**

Chapter 8
Converting food energy to mechanical energy 211 **BIO**
The wing of the hawkmoth 234 **BIO**
Nasal strips 237
The jump of a flea 238 **BIO**

Chapter 9
The force between a ball and a bat 245
The ballistocardiograph 251 **BIO**
Heartbeat detectors 251 **BIO**
Stellar explosions 252
The ballistic pendulum 256
Analyzing a traffic accident 257
The center of mass of the arm 265 **BIO**
An exploding rocket 269
The Saturn V rocket 270

Chapter 10
The operation of a CD 290
Types of centrifuges 291
The microhematocrit centrifuge 292 **BIO**
Moment of inertia of the Earth 298
Dental drills, the world's fastest turbines 309

Chapter 11
Applying the brakes 326
Forces required for structural stability 328
An arm in a sling 328 **BIO**
Hurricanes and tornadoes 339
The angular speed of a pulsar 341
The precession of the Earth 345
Gyroscopes in navigation and space 346

Chapter 12
The dependence of gravity on altitude 363
Weighing the Earth 365
The internal structure of the Earth and Moon 366
The Sun and Mercury 371
Geosynchronous satellites 372
The Global Positioning System (GPS) 373
Maneuvering spacecraft 373
The impact of meteorites 378
Planetary atmospheres 382
Black holes 384
Gravitational lensing 384
Tides 385
Tidal locking 385
The Roche limit and Saturn's rings 386

Chapter 13
Measuring the mass of a "weightless" astronaut 406
The pendulum clock and pulsilogium 413
Adjusting a grandfather clock 415
Walking speed 417 **BIO**
Resonance: radio tuners 420
Resonance: spider webs 420
Resonance: bridges 420
A cricket thermometer 428 **BIO**

Chapter 14
Calculating distance to a lightning strike 439
Ultrasonic sounds in nature 440
Ultrasonic scans 440 **BIO**
Shock wave lithotripsy 440 **BIO**
Infrasonic communication in elephants and whales 440
Infrasound produced by meteors 441
Echolocation 442 **BIO**
Human perception of sound intensity 444 **BIO**
Radar guns 451
Measuring the speed of blood flow 451 **BIO**
The red shift of distant galaxies 451
Connecting speakers in phase 454
Active noise reduction 455
The shape of a piano 458
Human perception of pitch 458 **BIO**
Frets on a guitar 459
Human hearing and the ear canal 460 **BIO**
Organ pipes 462

Chapter 15
Walking on lily pads 478
Pressure at the wreck of the *Titanic* 481
The barometer 482
The hydraulic lift 485
Measuring the body's density 488 **BIO**
Measuring body fat 489 **BIO**
The swim bladder in fish 491 **BIO**
Diving sea mammals 491 **BIO**
Maximum load indicator on a ship 491
The tip of the iceberg 492
Hoses and nozzles 494
The lift produced by an airplane wing 498
The force on a roof in a windstorm 498
Ventilation in a prairie dog burrow 499 **BIO**
Blood speed in the pulmonary artery 502 **BIO**
Breathing, alveoli, and premature birth 504 **BIO**

Chapter 16
Bimetallic strips 522
Antiscalding devices 522
Thermal expansion joints 522
Floating icebergs 525
The ecology of lakes 526 **BIO**
Bursting water pipes 526
Water and the climate 529
Insulated windows 534
Countercurrent exchange 535 **BIO**
Cold hands and feet 535 **BIO**
Convection on the Earth and Sun 535
Using color to measure temperature 536
Temperatures of the stars 536
Thermos bottles and the Dewar 538

Chapter 17
Take a deep breath 550 **BIO**
Stretching a bone 561 **BIO**
The autoclave 567 **BIO**
The pressure cooker 567
Adding salt to boiling water 567
Ice melting under pressure 568
Frost wedging and frost heaving 568
Biological antifreeze 569 **BIO**
Cooling the body with evaporation 569 **BIO**
Stability of planetary atmospheres 570
Homemade ice cream 574

Chapter 18
Diesel engines 595
Using adiabatic heating to start a fire 595
Sonoluminescence 595
Rain shadows 599
The steam engine 600
Refrigerators 605
Air conditioners 606
Heat pumps 606
Heat death of the universe 613
Entropy and life 613 **BIO**

Chapter 19
Bacterial infection from endoscopic surgery 629 **BIO**
Photocopiers and laser printers 630
Electrodialysis for water purification 639
Electric fish 642 **BIO**
Electrical shark repellent 643 **BIO**
Television screens and ink-jet printers 646
Electrical shielding 647
Lightning rods and Saint Elmo's fire 648
Electrostatic precipitation 648

Chapter 20
The electrocardiograph 676 **BIO**
The electroencephalograph 677 **BIO**
Computer keyboards 682
The Theremin—a musical instrument you play without touching 682
The electronic flash 684
The defibrillator 684 **BIO**
Capacitor hazards 685 **BIO**
Automatic external defibrillator 692 **BIO**

Chapter 21
The bolometer 701
Thermistors and fever thermometers 702
Superconductors and high-temperature superconductivity 702
"Battery check" meters 704
Three-way light bulbs 707
"Touch-sensitive" lamps 715
Delay circuits in windshield wipers and turn signals 719
Pacemakers 720 **BIO**

Chapter 22
Refrigerator magnets 734
The Earth's magnetic field 734
The electromagnetic flowmeter 740 **BIO**
The mass spectrometer 742
The aurora borealis and aurora australis 744
The galvanometer 748

MRI instruments 755 BIO
Magnetic reed switches in pacemakers 755 BIO
Magnetite in living organisms 756 BIO
Magnetism and the brain 756 BIO
Magnetic levitation 757

Chapter 23
Dynamic microphones and seismographs 773
Electric guitar pickups 773
Magnetic disk drives and credit card readers 774
T coils and induction loops 774
Magnetic braking and speedometers 777
Induction stove 778
Magnetic antitheft devices 781
Tracking the movement of insects 781 BIO
Tracking the motion of the human eye 781 BIO
Electric generators 781
Electric motors 783
Energy recovery technology in cars 784
High-voltage electric power transmission 792

Chapter 24
Electric shock hazard 807 BIO
Polarized plugs and grounded plugs 808
Ground fault circuit interrupter 808
Light dimmers 819
Tuning a radio or television 827
Metal detectors 827
Persistence of vision 831 BIO

Chapter 25
Radio and television communications 840
Doppler radar 844
Nexrad 844
Infrared receptors in pit vipers 847 BIO
Biological effects of ultraviolet light 848 BIO
Irradiated food 848 BIO
Photoelastic stress analysis 857
Liquid crystal displays (LCDs) 858
Navigating using polarized light from the sky 858 BIO
Why the sky is blue 859
How Polaroid sunglasses cut reflected glare 859

Chapter 26
Micromirror devices and digital movie projection 872
Corner reflectors and the Earth-Moon distance 875
Parabolic mirrors 877
Apparent depth 887
Mirages 887
Porro prisms in binoculars 890
Optical fibers and endoscopes 890 BIO
Underwater vision 894
The rainbow 897

Chapter 27
Optical properties of the human eye 910 BIO
Speed and aperture settings on a camera 912
Extended vision: Correcting nearsightedness 915 BIO
Intracorneal rings 917 BIO
Radial keratotomy 917 BIO
Correcting farsightedness 918 BIO
Keratometers 919 BIO
Achromatic lenses 928

Chapter 28
Newton's rings 946
Soap bubbles and oil slicks 947
Nonreflective coating 949
Reading the information on a CD 950
Pointillism and painting 957 BIO
Color television images 957 BIO
Acousto-optic modulation 959
X-ray diffraction and the structure of DNA 959
Grating spectroscopes 960
Measuring the red shift of a quasar 960
Iridescence in nature 961 BIO

Chapter 29
Nuclear power—converting mass to energy 986
The energy of the Sun 987
Positron-emission tomography 988 BIO
Gravitational lensing 993
Black holes 993
The search for gravity waves 995

Chapter 30
Measuring the temperature of a star 1005
Dark-adapted vision 1009 BIO
Photocells 1012
Solar energy panels 1013
Sailing on a beam of light 1014
Optical tweezers 1014
Electron microscopes 1020
Scanning tunneling microscopy 1025
Owl vision/Human vision 1031 BIO

Chapter 31
Medical X-ray tubes 1056 BIO
Computerized axial tomography 1058 BIO
Helium-neon laser 1059
Laser eye surgery 1060 BIO
Photodynamic therapy 1061 BIO
Holography 1061
Fluorescent light bulbs 1062
Applications of fluorescence in forensics 1063 BIO
Detecting scorpions at night 1063 BIO
The GFP Bunny 1064 BIO

Chapter 32
Smoke detector 1079
Dating the Iceman 1088
Nuclear reactors 1093
Powering the Sun: the proton-proton cycle 1094
Manmade fusion 1095
Radiation and cells 1095 BIO
Radioactive tracers 1097 BIO
Positron-emission tomography 1097 BIO
Magnetic resonance imaging (MRI) 1098 BIO

Preface: To the AP* Teacher

Teaching Advanced Placement* Physics can be a most challenging—and rewarding—experience. Students enter the course with a wide range of backgrounds, interests, and skills and we, the teachers, strive not only to convey the basic concepts and fundamental laws of physics, but also to give students an appreciation of its relevance and appeal.

I wrote this book to help with that task. It incorporates a number of unique and innovative pedagogical features that evolved from years of teaching experience. The materials have been tested extensively in the classroom and in focus groups, and refined based on comments from students and teachers who used the first and second editions of the text. The enthusiastic response I received from users of the first two editions was both flattering and motivating. The third edition has been improved in response to this feedback, with updated homework Problems and a new section of Conceptual Exercises designed to be used with classroom response systems and online homework systems.

Learning Tools in the Text

A key goal of this text is to help students make the connection between a conceptual understanding of physics and the various skills necessary to solve quantitative problems. One of the chief means to that end is the replacement of traditional "textbook" Examples with an integrated system of learning tools: fully worked *Examples with Solutions in Two-Column Format, Active Examples, Conceptual Checkpoints,* and *Exercises.* Each of these tools performs some of the functions of a standard Example, but each is specialized to meet the needs of students at a particular point in the development of a chapter.

These needs are not always the same. Sometimes students require a detailed explanation of how to tackle a particular problem; at other times, they must be allowed to take an active role and work out the details for themselves. Sometimes it is important for them to perform calculations and concentrate on numerical precision; at other times it is more fruitful for them to explore a key idea in a conceptual context. And sometimes, all that is required is practice using a new equation or definition.

A good teacher can sense when students need a patient, step-by-step exposition and when they need only minimal reinforcement; when they need to focus on concepts and when they need an opportunity to practice their quantitative skills.

This text attempts to emulate the teaching style of successful instructors by providing the right tool at the right place and the right time.

Worked Examples with Solutions in Two-Column Format

Examples model the most complete and detailed method of solving a particular type of problem. The Examples in this text are presented in a format that focuses on the basic strategies and thought processes involved in problem solving. The aim of this approach is to help students first visualize the situation, devise a *strategy* to be followed, and then implement a clear *step-by-step solution* to the problem. This focus on the intimate relationship between conceptual insights and problem-solving techniques encourages students to view the ability to solve problems as a logical outgrowth of conceptual understanding rather than a kind of parlor trick.

Each Example has the same basic structure:

- **Picture the Problem** This first step discusses how the physical situation can be represented visually and what such a representation can tell us about how to analyze and solve the problem. At this step, always accompanied by a figure, we set up a coordinate system where appropriate, label important quantities, and indicate which values are known.

- **Strategy** Closely linked with this visualization process is the formulation of a *Strategy* to be followed in solving the problem. The strategy addresses the commonly asked question, "How do I get started?" by providing a clear overview of the problem and helping students to identify the relevant physical principles. It then guides the student in using known relationships to map a step-by-step path to the solution.
- **Solution in Two Column Format** In the step-by-step *Solution* of the problem, each of the steps is presented with a prose statement in the left-hand column and the corresponding mathematical implementation in the right-hand column. Each step clearly translates the idea described in words into the appropriate equations.

  When reviewing an Example, note that the left-hand column gives the flow of ideas used in the solution; the right-hand column gives the mathematical calculations that were carried out.
- **Insight** Each Example wraps up with an *Insight*—a comment regarding the solution just obtained. Some Insights deal with possible alternative solution techniques, others with new ideas suggested by the results.
- **Practice Problem** Following the Insight is a *Practice Problem,* which gives the student a chance to practice the type of calculation just presented. The Practice Problems, always accompanied by their answers, provide students with a valuable check on their understanding of the material. Finally, each Example ends with a reference to some related end-of-chapter Problems to allow students to test their skills further.

Active Examples

Active Examples serve as a bridge between the fully worked Examples, in which every detail is fully discussed and every step is given, and the homework Problems, where no help is given at all. In an Active Example, students take an active role in solving the problem by thinking through the logic of the steps described on the left and checking their answers on the right. Students often find it useful to practice problem solving by covering one column of an Active Example with a sheet of paper and filling in the covered steps as they refer to the other column. Follow-up questions, called *Your Turns,* ask students to look at the problem in a slightly different way. Answers to Your Turns are provided at the end of the book. Working through Active Examples will make students better prepared to tackle homework problems on their own.

Conceptual Checkpoints

Conceptual Checkpoints help students sharpen their insight into key physical principles. A typical Conceptual Checkpoint presents a thought-provoking question that can be answered by logical reasoning based on physical concepts rather than by numerical calculations. These questions, which can be just as challenging as any numerical problem and just as educational, are presented in multiple-choice format to help focus the student's thinking. The statement of the question is followed by a detailed discussion and analysis in the section titled *Reasoning and Discussion,* and the *Answer* is given at the end of the checkpoint for quick and easy reference.

Exercises

Exercises present brief calculations designed to illustrate the application of important new relationships, without the expenditure of time and space required by a fully worked Example. Exercises generally give students an opportunity to practice the use of a new equation, become familiar with the units of a new physical quantity, and get a feeling for typical magnitudes.

Problem-Solving Notes

In addition to the in-text elements just described, each chapter includes a number of *Problem-Solving Notes* in the margin. These practical hints are designed to highlight useful problem-solving methods while helping students avoid common pitfalls and misconceptions.

PROBLEM-SOLVING NOTE

End-of-Chapter Learning Tools

The end-of-chapter material in this text also includes a number of innovations, along with refinements of more familiar elements.

Chapter Summary

Each chapter concludes with a *Chapter Summary* presented in an easy-to-use outline style. Key concepts, equations, and important figures are organized by topic for convenient reference.

Problem-Solving Summary

A unique feature of this text is the *Problem-Solving Summary* at the end of the chapter. This summary addresses common sources of misconceptions in problem solving, and gives specific references to Examples and Active Examples illustrating the correct procedures. Organized by type of problem, each entry in the *Problem-Solving Summary* relates a specific type of calculation to the relevant physical concepts.

Conceptual Questions

The homework for each chapter begins with a section of *Conceptual Questions.* Answers to the odd-numbered Questions can be found in the back of the book, so that students can check their reasoning and conclusions. Answers to even-numbered Conceptual Questions are available in the *Instructor's Solutions Manual.*

Conceptual Exercises

An important improvement to the third edition is a section of *Conceptual Exercises* containing multiple-choice and ranking questions. These questions have been carefully selected and written for maximum effectiveness when used with classroom-response systems (clickers). They can also be graded with online homework systems, and hence can be integrated seamlessly into weekly homework assignments. Answers to the odd-numbered Exercises can be found in the back of the book. Answers to even-numbered Conceptual Exercises are available in the Instructor's Solutions Manual.

Numerical and Integrated Homework Problems

A collection of numerical and *Integrated Problems* (**IP**) are presented at the end of each chapter. Note that a number of Problems are given for each section of the chapter. In addition, a section titled *General Problems* presents a variety of Problems that use material from two or more sections within the chapter, or refer to material covered in earlier chapters.

The difficulty of Problems is ranked using one, two, or three red dots. The most straightforward Problems are labeled with a single dot (•), Problems involving several steps and more detailed reasoning are labeled with two dots (••), and Problems of a more challenging nature are indicated with three dots (•••).

Problems of special biological or medical relevance are indicated with the symbol **BIO**.

Problems throughout the homework, labeled with the symbol **IP**, *integrate a conceptual question with a numerical problem.* Problems of this type, which stress the importance of reasoning from basic principles, show how conceptual insight and numerical calculation go hand in hand in physics. They afford students the opportunity to express their understanding of physics in both words and mathematics. Approximately 20 percent of the end-of-chapter Problems are Integrated Problems.

INTERACTIVE FIGURE

Finally, a new section of *Interactive Problems* has been added to the homework. In these Problems, an Example, Active Example, Conceptual Checkpoint, or Figure "comes alive" as the computer animates the corresponding physical system. For easy reference, these figures are denoted by an Interactive Figure icon. I have chosen the figures to be made "active," and have written the scripts from which the animations have been developed. These animations are designed to be flexible in their application—they can be used in lecture, as a "virtual lab," or as a component of a homework assignment. By giving direct visual feedback to the student, they help to reinforce what is being learned and to provide an additional pathway of understanding.

Scope and Organization

Table of Contents

As you will notice from the Table of Contents (pages xiii–xxii), the presentation of physics in this text follows the standard practice for introductory courses, with only a few well-motivated refinements.

First, note that Chapter 3 is devoted to vectors and their application to physics. This material could be presented in an Appendix, but my experience has been that students benefit greatly from a full discussion of vectors early in the course. Most students have seen vectors and trigonometric functions before, but rarely from the point of view of physics. Thus, including vectors in the text sends a message that this is important material, and it gives students an opportunity to brush up on their math skills.

Note also that additional time is given to some of the more fundamental aspects of physics, such as Newton's laws and energy. Presenting such material in two chapters gives the student a better opportunity to assimilate and master these crucial topics, which form the foundation for so much of what follows. Given the time constraints we all face in the classroom, the distribution of material presented in this text is designed to give time and emphasis where most appropriate. Sections considered optional are marked with an asterisk.

REAL-WORLD PHYSICS

REAL-WORLD PHYSICS: BIO

Real-World Physics

Since physics applies to everything in nature, it is only reasonable to point out applications of physics that students may encounter in the real world. Each chapter presents a number of discussions focusing on "Real-World Physics." Those of general interest are designated by a globe icon in the margin. Applications that pertain more specifically to biology and medicine are indicated by a CAT scan icon in the margin. The inclusion of a generous selection of such topics should help to make the course material more interesting and relevant to all students, including the many whose career orientation is toward the life sciences. A full list of the real-world applications in the text begins on page xxiii.

The Illustration Program

Drawings

Physics is a highly visual subject, and many physics concepts are best conveyed by graphic means. Figures do far more than illustrate a physics text—often, they bear the main burden of the exposition. Accordingly, great attention has been paid to the figures in this book so as to achieve the optimal balance of realism and stylization, with the primary emphasis always on the clarity of the analysis.

As mentioned previously, every Example in the text (as well as most of the Active Examples and many of the Conceptual Checkpoints) is accompanied by a figure. Every Example includes a section, "Picture the Problem," that concentrates on visual representation of the physical situation described in the problem, as well as the use of that representation in devising a problem-solving strategy. Great emphasis has been placed on how to draw a free-body diagram (see page 113), and all such diagrams in Examples and end-of-chapter Problems have been drawn to scale using the data given.

Color has been used consistently throughout the text to reinforce concepts and make the diagrams easier for students to understand. Thus, force vectors are always red, velocity vectors always green, and so on. Similarly, wave fronts, adiabats, heat flow, work output, and many other elements have each been assigned a characteristic color. While the student has not been asked to learn the color scheme consciously, there can be little question that such consistency is pedagogically helpful on a subliminal level.

Companion Photographs

One of the most fundamental ways in which we learn is by comparing and contrasting. This principle is exploited in the way photographs are arranged on the cover of the third edition and used throughout the text. Many photos are presented in groups of two or three that contrast opposing physical principles or illustrate a single concept in a variety of contexts. Grouping carefully chosen photographs in this way helps students to see the universality of physics.

In conclusion, it is my hope that this text will engage the imagination and interest of the students who use it, and open to them a world governed by the fundamental laws and principles of physics. Those of us who devote ourselves to studying and teaching physics know that it is a subject of great beauty and power. This book strives to share some of that appreciation with others.

Resources

The third edition is supplemented by a Media and Print Ancillary package developed to address the needs of both students and instructors.

For the Teacher

Instructor's Solutions Manual (Vol. I: 0-13-185110-1; Vol. II: 0-13-153632-X)
Kenneth L. Menningen (University of Wisconsin–Stevens Point)
Contains detailed, worked solutions to every Problem in the text, all solved using the step-by-step problem-solving strategy of the in-chapter Examples (*Picture the Problem, Strategy,* two-column *Solutions,* and *Insight*). The ISM also contains answers to the even-numbered Conceptual Questions and Exercises. Electronic versions of the ISM are available on the *Instructor Resource Center on CD-ROM.* Available to qualified adopters.

Instructor's Resource Manual with Instructor Notes on ConcepTest Questions (0-13-153633-8)
This manual has two parts. The first part, prepared by Katherine Whatley and Judith Beck (both of University of North Carolina–Asheville), contains sample syllabi, lecture outlines, notes, demonstration suggestions, readings, and additional references and resources. The second part, prepared by Cornelius Bennhold and Gerald Feldman (both of The George Washington University) contains an overview of the development and implementation of ConcepTests, as well as instructor notes for each ConcepTest found on the *Instructor Resource Center on CD-ROM.* Available to qualified adopters.

Test Item File (0-13-185112-8)
Revised by Delena Bell Gatch (Georgia Southern University), this test bank contains approximately 3000 multiple choice, short answer, and true/false questions, many conceptual in nature. All are referenced to the corresponding text section and ranked by level of difficulty. The *Test Item File* is also available in electronic format (see description of *Instructor Resource Center on CD-ROM*). Available to qualified adopters.

Transparency Pack (0-13-153645-1)
Includes approximately 400 full-color transparencies of illustrations and tables from the text. Available to qualified adopters.

Instructor Resource Center on CD-ROM (0-13-153635-4)
This multiple CD-ROM set provides virtually every electronic asset you'll need in and out of the classroom. Though you can navigate freely through the CDs to find the resources you want, the software allows you to browse and search through the catalog of assets. The CD-ROMs are organized by chapter and include all text illustrations and tables from *Physics, Third Edition* in JPEG and PowerPoint formats. Instructors can preview, sequence, and play back images, as well as perform keyword searches, add lecture notes, and incorporate their own digital resources. The IRC/CDs also contain *TestGenerator*, a powerful dual-platform, fully networkable software program for creating tests ranging from short quizzes to long exams. Questions from the third edition *Test Item File*, including randomized versions, are supplied, and professors can use the Question Editor to modify existing questions or create new questions. The IRC/CDs also contain the instructor's version of *Physlet Physics*, the new *Interactive Figures* also on the *Companion Website* and in *PHGradeAssist*, chapter-by-chapter lecture outlines in PowerPoint, *ConcepTest* "Clicker" Questions in PowerPoint, chapter-by-chapter files of all numbered equations, the eleven *Physics You Can See* demonstration videos, and electronic versions of the *Test Item File*, the *Instructor's Solutions Manual*, the *Instructor's Resource Manual*, a list of key equations for each chapter, and the end-of-chapter Questions, Exercises, and Problems from *Physics, Third Edition*. Available to qualified adopters.

The following optional resources are available to qualified college adopters.

"Physics You Can See" Video Demonstrations (0-205-12393-7)
Each segment, 2–5 minutes long, demonstrates a classical physics experiment. Eleven segments are included, such as "Coin & Feather" (acceleration due to gravity), "Monkey & Gun" (rate of vertical free fall), "Swivel Hips" (force pairs), and "Collapse a Can" (atmospheric pressure).

Peer Instruction: A User's Manual (0-13-565441-6)
Eric Mazur (Harvard University)
A manual with ready-to-use resources for an innovative new approach to teaching introductory physics, developed by a well-known physicist and leader in physics education.

Just-In-Time Teaching: Blending Active Learning with Web Technology (0-13-085034-9)
Gregor Novak (Air Force Academy), Andrew Gavrin (IUPUI), Wolfgang Christian (Davidson College), and *Evelyn Patterson (Air Force Academy)*
This resource book for educators discusses *Just-in-Time Teaching (JiTT)*—a teaching and learning methodology designed to engage students by using feedback from pre-class web assignments to adjust classroom lessons. This allows students to receive rapid response to the specific questions and problems they are having instead of more generic lectures that may or may not address what students actually need help with. Many teachers have found that this process also encourages students to take more control of the learning process and become active and interested learners.

Physlets®: Teaching Physics with Interactive Curricular Material (0-13-029341-5)
Wolfgang Christian and *Mario Belloni (Davidson College)*
This manual/CD package shows physics instructors—both web novices and Java savvy programmers alike—how to author their *own* interactive curricular material using Physlets—Java applets written for physics pedagogy that can be embedded directly into html documents and that can interact with the user. It demonstrates the use of Physlets in conjunction with JavaScript to deliver *a wide variety* of web-based interactive physics activities, and provides examples of Physlets created for classroom demonstrations, traditional and *Just-in-Time Teaching* homework problems, pre- and post-laboratory exercises, and Interactive Engagement activities. More than just a technical how-to book, the manual gives instructors some ideas about the new *possibilities* that Physlets offer, and is designed to make the transition to using Physlets *quick and easy*.

Course and Homework Management Tools

Blackboard

Blackboard™ is a comprehensive and flexible e-Learning software platform that delivers a course management system, customizable institution-wide portals, online communities, and an advanced architecture that allows for Web-based integration of multiple administrative systems.

Features:
- Course Management System including progress tracking, class and student management, grade book, communication, assignments, and reporting tools.
- Testing programs with the capability for instructors to create online quizzes and tests, and automatically grade and track results. All tests can be entered into the grade book for easy course management.

CourseCompass

Powered by **Blackboard**
With the highest level of service, support, and training available today, CourseCompass combines tested, quality online course resources with easy-to-use online course management tools. CourseCompass is designed to address the individual needs of instructors, who will be able to create an online course without any special technical skills or training.

Features:
- Higher flexibility—Professors can adapt *Physics, Third Edition* content to match their own teaching goals.
- Assessment, customization, class administration, and communication tools.
- Point-and-click access—Prentice Hall's vast educational resources are available to instructors at a click of the mouse.
- A nationally hosted and fully supported system that relieves individuals and institutions of the burdens of trouble-shooting and maintenance.

WebCT

WebCT offers a powerful set of tools that enables you to create practical Web-based educational programs—the ideal resources to enhance a campus course or to construct one entirely online. The WebCT shell and tools, integrated with the *Physics, Third Edition* content, results in a versatile, course-enhancing teaching and learning system.

Features:
- Page tracking, progress tracking, class and student management, grade book, communication, calendar, reporting tools, and more.

- Communication tools, including chat rooms, bulletin boards, private e-mail, and WhiteBoard.
- Testing tools that help create and administer timed online quizzes and tests, and automatically grade and track all results.

Course Management Content

Along with all the material from the Companion Website (www.prenhall.com/walker), our course management cartridges include additional materials only available through Blackboard, CourseCompass, or WebCT, and include:

- Test bank questions, converted from our TestGen *Test Item File.*
- Resources from the *Instructor Resource Center on CD.*

PHGradeAssist

PHGradeAssist is a timesaving homework and assessment system that lets you:

- Create, edit, and deliver student assignments ranging from self-study sessions, to homework and quizzes, to proctored exams;
- Select, edit, and organize text-specific questions and question banks to fit your course;
- Create, edit, and organize your own questions and question banks;
- Specify policies such as time limits, number of retakes, and feedback for each assignment;
- Track student and class performance in an online gradebook; and
- Export grades to common spreadsheet programs.

Content includes new Conceptual Exercises, end-of-chapter Problems with significantly enhanced hints and feedback as well as links to appropriate eBook sections, the Interactive Figures and Interactive Problems from the textbook, gradable questions to *Physlet® Physics* Problems, MCAT Study Guide quizzes from Kaplan Test Prep and Admissions, and Warm-Ups and Puzzles from the *Just-in-Time Teaching* method.

WebAssign *http://webassign.com*

WebAssign's homework delivery service allows you to create assignments from a database of end-of-chapter Conceptual Questions and Problems from *Physics, Third Edition* or write and customize your own.

LON-CAPA *http://www.lon-capa.org*

The Learning Online Network with a Computer Assisted Personalized Approach (LON-CAPA) is an integrated system for online learning and assessment. It consists of a learning content authoring and management system allowing new and existing content to be used flexibly, a course management system, an individualized homework and automatic grading system.

CAPA

With Computer Assisted Personalized Approach (CAPA), an instructor can create and/or assemble personalized assignments, quizzes, and examinations.

End-of-chapter Problems from the third edition of *Physics* are available in CAPA and LON-CAPA formats.

For the Student

Preparing for the AP* Physics B Exam with Physics, Third Edition (ISBN 0-13-220862-8)

Connie Wells (Pembroke Hill School)

This print review book is a resource for students or teachers, and is directly linked to Walker's third edition. Features of this test preparation guide for AP* Physics B include:

- Description of the AP* exam and unique approaches to success
- Correlation of each chapter of Walker's text to the AP* Syllabus
- Ten recommended homework AP*-practice problems from each chapter

- Sample multiple choice and free response questions for each chapter, with answers and explanations
- How to prepare for laboratory questions and a complete sample
- Insight notes in each chapter that point the student to tips and techniques that may improve accuracy and performance on the exam
- A full-length sample AP* exam with answers and explanations

Student Study Guide with Selected Solutions (Volume I: 0-13-153646-X; Volume II: 0-13-236963-X)
David Reid (University of Chicago)
The print study guide provides the following for each chapter:

- Objectives
- Warm-Up Questions from the *Just-in-Time Teaching* method by Gregor Novak and Andrew Gavrin (Indiana University–Purdue University, Indianapolis)
- Chapter Review with two-column Examples and integrated quizzes
- Reference Tools & Resources (equation summaries, important tips, and tools)
- Puzzle Questions (also from Novak & Gavrin's *JiTT* method)
- Selected Solutions for several end-of-chapter questions and problems

Student Pocket Companion (0-13-153634-6)
Biman Das (SUNY Potsdam)
This easy-to-carry 5″ × 7″ paperback contains a summary of the entire text, including all key concepts and equations, as well as tips and hints.

Companion Website with Grade Tracker
Our Website (*http://www.prenhall.com/walker*), which hosts contributions from leaders in physics education research, provides students with a variety of interactive explorations of each chapter's topics, easily accommodating differences in learning styles. Student tools provided on the Website include the new *Interactive Figures,* written by James Walker, where an Example, Active Example, Conceptual Checkpoint, or Figure from the book "comes alive" as the corresponding physical system is animated; a subset of the *Conceptual Exercises* from the third edition text in electronic format; *Warm Ups, Puzzles,* and *"What Is Physics Good For?" Applications* by Gregor Novak and Andy Gavrin (Indiana University–Purdue University, Indianapolis); multiple-choice *Practice Questions; Physlet® Problems,* with links to Java applets that check the student's understanding of key concepts; *MCAT Study Guide* questions by Kaplan Test Prep and Admissions, and a full electronic version of the text as an *eBook.* Instructor tools include online grading capabilities and a Syllabus Manager.

The following optional resources are available to qualified adopters

Physlet® Physics (0-13-101969-4)
Wolfgang Christian and Mario Belloni (Davidson College)
This text and CD-ROM package provides ready-to-run interactive Physlet-based curricular material to teachers and students. Physlets, award-winning Java applets written by the authors, have been widely adopted by the physics teaching community. This book provides the first class-tested collection of ready to run Physlet-based material that is easy to assign (like an end-of-chapter problem of a textbook) and easy to use (material is on a CD and in the book). A web-server or an internet connection is not required. The book contains over 800 Physlet-based activities that span all of introductory physics. An *Instructor's Guide* is available to qualified adopters.

Ranking Task Exercises in Physics: Student Edition (0-13-144851-X)
Thomas L. O'Kuma (Lee College), David P. Maloney (Indiana University–Purdue University, Fort Wayne), and Curtis J. Hieggelke (Joliet Junior College)
Ranking Tasks are an innovative type of conceptual exercise that asks students to make comparative judgments about a set of variations on a particular physical situation. This text is a unique resource for physics instructors who are looking for tools to incorporate more conceptual analysis in their courses. This supplement contains over 200 Ranking Task Exercises that cover all classical physics topics.

E&M TIPERs: Electricity & Magnetism Tasks Inspired by Physics Education Research (0-13-185499-2)
Curtis J. Hieggelke (Joliet Junior College), David P. Maloney (Indiana University Purdue University Fort Wayne), Stephen E. Kanim (New Mexico State University), and Thomas L. O'Kuma (Lee College)

E&M TIPERs is a comprehensive set of conceptual exercises for electricity and magnetism based on the results of education research into how students learn physics. This workbook contains over 300 tasks in eleven different task formats.

Tutorials in Introductory Physics and Homework Package (0-13-097069-7)
Lillian C. McDermott, Peter S. Shaffer, Physics Education Group (all of the University of Washington)
This landmark book presents a series of physics tutorials designed by a leading physics education research group. Emphasizing the development of concepts and scientific reasoning skills, the tutorials focus on the specific conceptual and reasoning difficulties that students tend to encounter. The tutorials cover a range of topics in Mechanics, E & M, and Waves & Optics.

Mathematics for College Physics (0-13-141427-5)
Biman Das (State University of New York-Potsdam)
A supplementary text for the introductory physics course, this text is ideal for students who lack necessary mathematical tools. It not only shows them clearly how mathematics is directly applied to physics, but discusses math anxiety in general and how to overcome it. Instead of a rigorous development of the concepts of mathematics (as is found in a typical math book), the text describes the various mathematical concepts and tools (including algebra, trigonometry, geometry, vectors, probability, and statistics) and their direct use in solving physics problems.

Interactive Physics Player Workbook, Second Edition (0-13-067108-8)
Cindy Schwarz (Vassar College), John Ertel (Naval Academy), MSC Software
This interactive workbook and hybrid CD-ROM package is designed to help students visualize and work with specific physics problems through simulations created with Interactive Physics files. Forty problems of varying degrees of difficulty require students to make predictions, change variables, run, and visualize motion on the computer. The accompanying workbook/study guide provides instructions, physics review, hints, and questions. The accompanying hybrid CD-ROM contains everything students need to run the simulations.

Acknowledgments

I would like to express my gratitude to my colleagues at Washington State University and Western Washington University, as well as others in the physics community for their contributions to this project.

My thanks are due also to many people at Prentice Hall who encouraged and aided me through the process of making this book a reality: Daniel Schiller, whose expert guidance was invaluable in developing the first edition of the book; Erik Fahlgren, whose experience with the first two editions helped shape the third edition; Jessica Berta, for her attention to detail and coordination of the reviewing program; Michael Richards, Christian Botting, Karen Bosch, and Nicole Jackson for their dedicated work on the supplements and media program; and David Chelton, for his invaluable input in developing the third edition.

In addition, I am grateful for the work of Debra Wechsler for choreographing the production process; and Jon Boylan for coordinating the interior and cover design. The patience, dedication, and skill of these people are much appreciated.

I owe a great debt to my students over the years. My experience with them in the classroom provided the motivation and the inspiration for this book.

Reviewers

We are grateful to the following instructors for their thoughtful comments on the manuscript of this text and careful checking of proofs.

Reviewers of the Third Edition

Alice M Hawthorne Allen
Virginia Tech
Bradley C. Antanaitis
Lafayette College
Michael Arnett
Kirkwood Community College
Robert W. Arts
Pikeville College
David Balogh
Fresno City College
David T. Bannon
Oregon State University
Anand Batra
Howard University
S. M. Bhagat
University of Maryland
James D. Borgardt
Juniata College
David Branning
Trinity College
Thomas B. Cobb
Bowling Green State University
David Craig
LeMoyne College
David Curott
University of North Alabama
Duane Deardorff
University of North Carolina at Chapel Hill
Edward Derringh
Wentworth Institute of Technology
Martha Dickinson
Maine Maritime Academy
David Elmore
Purdue University
David Faust
Mt. Hood Community College
Joseph Gallant
Kent State University, Trumbull
Thor Garber
Pensacola Junior College
David Gerdes
University of Michigan
John D. Gieringer
Alvernia College
Karen Gipson
Grand Valley State University
Steven Hagen
University of Florida
John Hopkins
The Pennsylvania State University
Adam Johnston
Weber State University
Gordon O. Johnson
Washington State University
Sanford Kern
Colorado State University
Ilkka Koskelo
San Francisco State University
Ian M. Lindevald
Truman State University
Jeff Loats
Fort Lewis College
Daniel Ludwigsen
Kettering University
Lorin Matthews
Baylor University
William McNairy
Duke University
Kenneth L. Menningen
University of Wisconsin — Stevens Point
Anatoly Miroshnichenko
University of Toledo
Wouter Montfrooij
University of Missouri
Gary Morris
Valparaiso University
Robert Oman
University of South Florida
Michael Ottinger
Missouri Western State College
Francis Pichanick
University of Massachusetts, Amherst
Jerry Polson
Southeastern Oklahoma State University
David Procopio
Mohawk Valley Community College
Michele Rallis
Ohio State University
Prabha Ramakrishnan
North Carolina State University
John F. Reading
Texas A&M University
John A. Rochowicz, Jr.
Alvernia College
Lawrence G. Rowan
University of North Carolina at Chapel Hill
Misa T. Saros
Viterbo University
C. Gregory Seab
University of New Orleans
George Strobel
University of Georgia
Carey E. Stronach
Virginia State University
Irina Struganova
Barry University
Daniel Stump
Michigan State University
John A. Underwood
Austin Community College, Rio Grande

Previous Editions

Daniel Akerib
Case Western Reserve University
Richard Akerib
Queens College
Barbara S. Andereck
Ohio Wesleyan University
Eva Andrei
Rutgers University
Rama Bansil
Boston University
Paul Beale
University of Colorado-Boulder
Mike Berger
Indiana University
David Berman
University of Iowa
James P. Boyle
Western Connecticut State University
Jeff Braun
University of Evansville
Matthew E. Briggs
University of Wisconsin-Madison
Jack Brockway
State University of New York-Oswego
Neal Cason
University of Notre Dame
Lattie Collins
Eastern Tennessee State University

James Cook
Middle Tennessee State University
Stephen Cotanch
North Carolina State University
David Curott
University of North Alabama
William Dabby
Edison Community College
Robert Davie
St. Petersburg Junior College
Steven Davis
University of Arkansas-Little Rock
N. E. Davison
University of Manitoba
Anthony DiStefano
University of Scranton
David C. Doughty, Jr.
Christopher Newport University
F. Eugene Dunnam
University of Florida
John J. Dykla
Loyola University-Chicago
Eldon Eckard
Bainbridge College
Donald Elliott
Carroll College
Robert Endorf
University of Cincinnati
Raymond Enzweiler
Northern Kentucky University
John Erdei
University of Dayton
Frank Ferrone
Drexel University
John Flaherty
Yuba College
Curt W. Foltz
Clarion University
Lewis Ford
Texas A&M University
Armin Fuchs
Florida Atlantic University
Asim Gangopadhyaya
Loyola University-Chicago
Barry Gilbert
Rhode Island College
Fred Gittes
Washington State University
Michael Graf
Boston College
William Gregg
Louisiana State University
Rainer Grobe
Illinois State University
Mitchell Haeri
Saddleback College
Parameswar Hari
California State University-Fresno
Xiaochun He
Georgia State University
Timothy G. Heil
University of Georgia
J. Erik Hendrickson
University of Wisconsin-Eau Claire
Scott Holmstrom
University of Tulsa
Manuel A. Huerta
University of Miami
Zafar Ismail
Daemen College
Nadejda Kaltcheva
University of Wisconsin-Oshkosh
William Karstens
Saint Michael's College
Dana Klinck
Hillsborough Community College
Laird Kramer
Florida International University
R. Gary Layton
Northern Arizona University
Kevin M. Lee
University of Nebraska-Lincoln
Michael Lieber
University of Arkansas
Mark Lindsay
University of Louisville
Hilliard Macomber
University of Northern Iowa
Trecia Markes
University of Nebraska-Kearny
Joseph Mills
University of Houston
Paul Morris
Abilene Christian University
David Moyle
Clemson University
Ashok Muthukrishnan
Texas A&M University
K. W. Nicholson
Central Alabama Community College
Larry Owens
College of the Sequoias
A. Ray Penner
Malaspina University
Robert Piserchio
San Diego State University
Anthony Pitucco
Pima Community College
William Pollard
Valdosta State University
Robert Pompi
Binghamton University
Earl Prohofsky
Purdue University
Jia Quan
Pasadena City College
David Raffaelle
Glendale Community College
Michael Ram
State University of New York-Buffalo
Rex Ramsier
University of Akron
Lawrence B. Rees
Brigham Young University
M. Anthony Reynolds
Embry Riddle University
Dennis Rioux
University of Wisconsin-Oshkosh
Bob Rogers
San Francisco State University
Gaylon Ross
University of Central Arkansas
Gerald Royce
Mary Washington College
Wolfgang Rueckner
Harvard University
Mats Selen
University of Illinois
Bartlett Sheinberg
Houston Community College
Peter Shull
Oklahoma State University
Christopher Sirola
Tri-County Technical College
Daniel Smith
South Carolina State University
Leigh M. Smith
University of Cincinnati
Soren Sorensen
University of Tennessee-Knoxville
Mark W. Sprague
East Carolina University
Leo Takahashi
Penn State University-Beaver
Harold Taylor
Richard Stockton College
Frederick Thomas
Sinclair Community College
Jack Tuszynski
University of Alberta
Lorin Vant Hull
University of Houston
Karl Vogler
Northern Kentucky University
Desmond Walsh
Memorial University of Newfoundland
Toby Ward
College of Lake County

Richard Webb
Pacific Union College
Lawrence Weinstein
Old Dominion University
Jeremiah Williams
Illinois Wesleyan University
Linda Winkler
Moorhead State University
Lowell Wood
University of Houston
Robert Wood
University of Georgia
Jeffrey L. Wragg
College of Charleston

Student Reviewers

We wish to thank the following students at Boston University and California State University-Chico for providing helpful feedback during the development of the third edition of this text. Their comments offered us valuable insight into the student experience.

Joel Amato
Sabrina Carrie
Suprna Chandra
Kara Coffee
Philip Dagostino
Andrew D. Fisher
Shadi Miri Ghomizadea
Colleen Hanlon
Tamara Jones
Ryan Morrison
Julie Palakovich
Suraj Parekh
Peter Ploewski
Darren B. Robertson
Chris Simons
ATTENDEES: Lynda Klein, California State University-Chico

We would also like to thank the following students at the University of Houston, Washington State University, and North Carolina State University for providing helpful feedback via review or focus group for the first two editions of this text:

Ali Ahmed
Max Aquino
Margaret Baker
Tynisa Bennett
Joshua Carmichael
Tyler Cumby
Rebecca Currell
Jonas Hauptmann
Parker Havron
Jamie Helms
Robert Hubbard
Bryce Lewis
Michelle Lim
Candida Mejia
Roderick Minogue
Hang Nguyen
Mary Nguyen
Scott Parsons
Tiffany Switzer
Steven Taylor
Monique Thomas
Khang Tran
Michael Vasquez
Jerod Williams
Nathan Witwer
Alexander Wood
Melissa Wright

Preface: To the AP* Student

As a student preparing to take an AP* algebra-based physics course, you are probably aware that physics applies to absolutely everything in the natural world, from raindrops and people to galaxies and atoms. Because physics is so wide-ranging and comprehensive, it can sometimes seem a bit overwhelming. This text, which reflects nearly two decades of classroom experience, is designed to help you deal with a large body of information and develop a working understanding of the basic concepts in physics. Now in its third edition, it incorporates many refinements that have come directly from interacting with students using the first two editions. As a result of these interactions, I am confident that as you develop a deeper understanding of physics, you will also enrich your experience of the world in which you live.

Now, I must admit that I like physics, and so I may be a bit biased in this respect. Still, the reason I teach and continue to study physics is that I enjoy the insight it gives into the physical world. I can't help but notice—and enjoy—aspects of physics all around me each and every day. As I always tell my students on the first day of class, I would like to share some of this enjoyment and delight in the natural world with you. It is for this reason that I undertook the task of writing this book.

To assist you in the process of studying physics, this text incorporates a number of learning aids, including *Two-Column Examples, Active Examples, and Conceptual Checkpoints.* These and other elements work together in a unified way to enhance your understanding of physics on both a conceptual and a quantitative level—they have been developed to give you the benefit of what we know about how students learn physics, and to incorporate strategies that have proven successful to students over the years. The pages that follow will introduce these elements to you, describe the purpose of each, and explain how they can help you.

As you progress through the text, you will encounter many interesting and intriguing applications of physics drawn from the world around you. Some of these, such as magnetically levitated trains or the satellite-based Global Positioning System that enables you to determine your position anywhere on Earth to within a few feet, are primarily technological in nature. Others focus on explaining familiar or not-so-familiar phenomena, such as why the Moon has no atmosphere, how sweating cools the body, or why flying saucer-shaped clouds often hover over mountain peaks even when the sky is clear. Still others, such as countercurrent heat exchange in animals and humans or the use of sound waves to destroy kidney stones, are of particular relevance to students of biology and the other life sciences.

In many cases, you may find the applications to be a bit surprising. Did you know, for example, that you are shorter at the end of the day than when you first get up in the morning? (This is discussed in Chapter 5.) That an instrument called the ballistocardiograph can detect the presence of a person hiding in a truck, just by registering the minute recoil from the beating of the stowaway's heart? (This is discussed in Chapter 9.) That if you hum next to a spider's web at just the right pitch you can cause a resonance effect that sends the spider into a tizzy? (This is discussed in Chapter 13.) That powerful magnets can exploit the phenomenon of diamagnetism to levitate living creatures? (This is discussed in Chapter 22.)

Writing this textbook was a rewarding experience for me. I hope using it will prove equally rewarding to you, and that it will inspire an interest in and appreciation of physics that will last a lifetime.

James S. Walker
jamesswalker@physics.wwu.edu

In-Chapter Learning Tools

Because one-size-fits-all examples do not sufficiently address the needs of students, the text employs a variety of pedagogical elements, each used where it can contribute most to developing conceptual insight and problem-solving skills.

Worked Examples with Solutions in Two-Column Format

The structure of the worked Example models good problem-solving habits while giving students greater insight into the physical principles and concepts behind the problem.

Problem Statement

Describes the physical situation to be analyzed, what is known, and what is to be found.

Picture the Problem

Reminds students to sketch the situation and helps them visualize the processes taking place, identify and label important quantities, and set up coordinate axes.

Strategy

Here, students learn how to analyze the problem, identify the key physical principles at work, and map a plan for obtaining the solution.

Solution

The two-column format helps students translate the language of problem-solving steps into the corresponding mathematical expressions.

Insight

Often, solving a problem leads to new insights. This step points out interesting or significant features of the problem, the solution process, or the result.

Practice Problem

Students are given an opportunity to test their skills on a problem similar to the one just worked. For immediate reinforcement, the answer is provided after the Practice Problem. Related homework problems also direct students to end-of-chapter Problems that are similar to the Example.

EXAMPLE 8–4 Compressed Energy and the Jump of a Flea

When a force of 120.0 N is applied to a certain spring, it causes a stretch of 2.25 cm. What is the potential energy of this spring when it is compressed by 3.50 cm?

Picture the Problem

The top sketch shows the spring stretched 2.25 cm by the force $F_1 = 120.0$ N. The lower sketch shows the same spring compressed by a second force, F_2, which causes a compression of 3.50 cm.

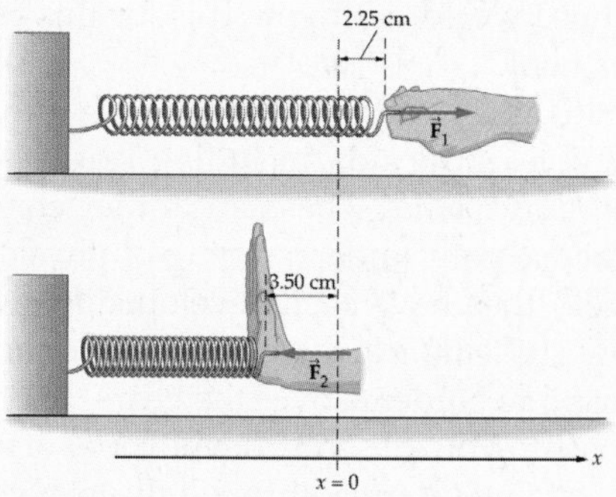

Strategy

From the first piece of information—a certain force causes a certain stretch—we can calculate the force constant using $F = kx$. Once we know k, we find the potential energy with $U = \frac{1}{2}kx^2$.

Solution

1. Solve $F = kx$ for the spring constant, k:

$$F = kx$$

$$k = \frac{F}{x} = \frac{120.0 \text{ N}}{0.0225 \text{ m}} = 5330 \text{ N/m}$$

2. Substitute $k = 5330$ N/m and $x = -0.0350$ m into the potential energy expression, $U = \frac{1}{2}kx^2$:

$$U = \tfrac{1}{2}kx^2 = \tfrac{1}{2}(5330 \text{ N/m})(-0.0350 \text{ m})^2 = 3.26 \text{ J}$$

Insight

Though this example deals with ideal springs, the same basic physics applies to many other real-world situations. A case in point is the jump of a flea, in which a flea can propel itself up to 100 times its body length. The physics behind this feat is the slow accumulation of energy in a "springy" strip of resilin in the coxa of the leg, as shown in the accompanying sketches, and the sudden release of this energy at a later time. Specifically, as the flea's muscles flex the leg, the resilin strip in the coxa is stretched, storing the work done by the muscles in the form of potential energy, $U = \frac{1}{2}kx^2$. Later, when a trigger mechanism unlocks the flexed leg, the energy stored in the resilin is released explosively—rapidly extending the leg and propelling the flea upward. See Problem 77 for a calculation using the force constant of resilin.

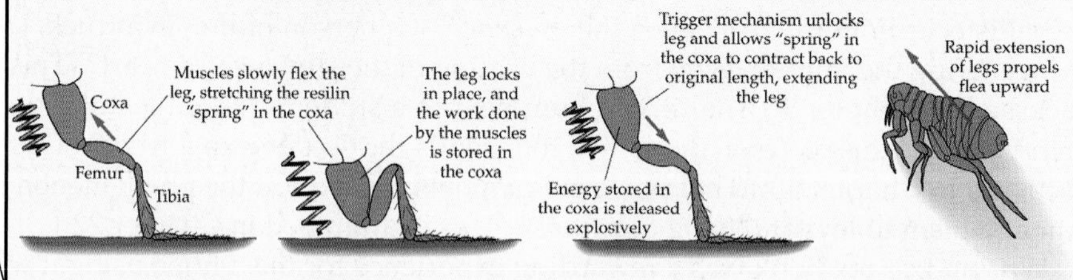

Practice Problem

What stretch is necessary for the spring in this Example to have a potential energy of 5.00 J? **[Answer:** 4.33 cm]

Some related homework problems: Problem 9, Problem 13

Active Examples

Active Examples require students to play an active role. These examples act as a bridge between the worked Examples in the chapter, in which the solution process is fully spelled out, and the end-of-chapter Problems, where students receive no guidance at all. **Your Turn** Problems ask students to look at the situation in a slightly different way.

ACTIVE EXAMPLE 6–2 Nasal Strips

An increasingly popular device for improving air flow through nasal passages is the nasal strip, which consists of two flat, polyester springs enclosed by an adhesive tape covering. Measurements show that a nasal strip can exert an outward force of 0.22 N on the nose, causing it to expand by 3.5 mm. **(a)** Treating the nose as an ideal spring, find its force constant in newtons per meter. **(b)** How much force would be required to expand the nose by 4.0 mm?

REAL-WORLD PHYSICS: BIO

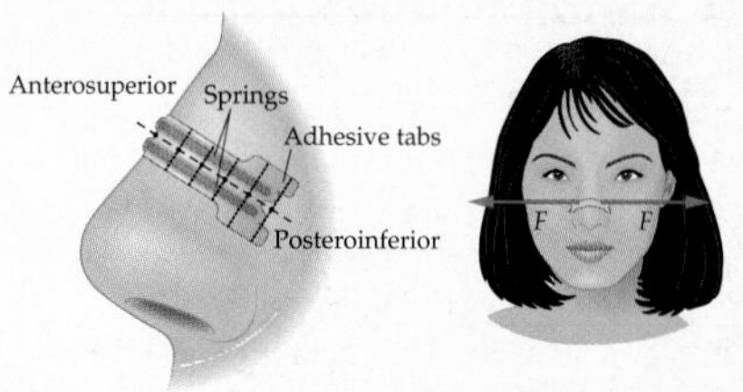

Conceptual Checkpoints

Conceptual Checkpoints test students' ability to make inferences based on chapter concepts. These qualitative questions are presented in multiple-choice format to guide student response. "Reasoning and Discussion" clearly explains the physical principles behind the answer, while recognizing and addressing students' possible misconceptions.

CONCEPTUAL CHECKPOINT 5–1 Tightening a Hammer

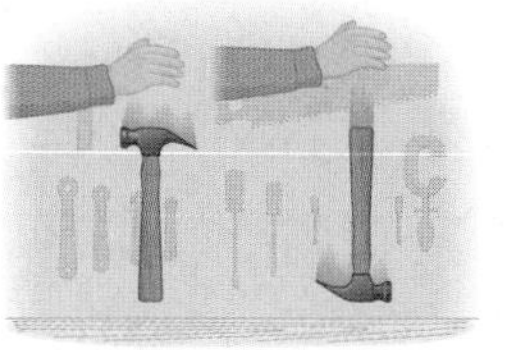

The metal head of a hammer is loose. To tighten it, you drop the hammer down onto a table. Should you **(a)** drop the hammer with the handle end down, **(b)** drop the hammer with the head end down, or **(c)** do you get the same result either way?

Reasoning and Discussion

It might seem that since the same hammer hits against the same table in either case, there shouldn't be a difference. Actually, there is.

In case **(a)**, the handle of the hammer comes to rest when it hits the table, but the head continues downward until a force acts on it to bring it to rest. The force that acts on it is supplied by the handle, which results in the head being wedged more tightly onto the handle. Since the metal head is heavy, the force wedging it onto the handle is great. In case **(b)** the head of the hammer comes to rest, but the handle continues to move until a force brings it to rest. The handle is lighter than the head, however; thus the force acting on it is less, resulting in less tightening.

Answer:

(a) Drop the hammer with the handle end down.

Exercises

When a more elaborate device is not needed, simple Exercises give students practice in using a new formula or relationships introduced in the text.

EXERCISE 12–3

Use $M_E = gR_E^2/G$ to calculate the mass of the Earth.

Solution

Substituting numerical values, we find

$$M_E = \frac{gR_E^2}{G} = \frac{(9.81\ \mathrm{m/s^2})(6.37 \times 10^6\ \mathrm{m})^2}{6.67 \times 10^{-11}\ \mathrm{N \cdot m^2/kg^2}} = 5.97 \times 10^{24}\ \mathrm{kg}$$

End-of-Chapter Learning Tools

Chapter Summary

The Chapter Summary is organized by section and topic. It provides an easy reference for students to review key terms, concepts, figures, and equations from the chapter.

Chapter Summary

11–1 Torque

A force applied so as to cause an angular acceleration is said to exert a torque, τ.

Tangential Force
A force is tangential if it is tangent to a circle centered on the axis of rotation.

Torque Due to a Tangential Force
A tangential force F applied at a distance r from the axis of rotation produces a torque

$$\tau = rF \qquad 11\text{–}1$$

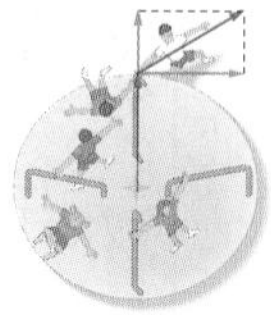

Torque for a General Force
A force exerted at an angle θ with respect to the radial direction, and applied at a distance r from the axis of rotation, produces the torque

$$\tau = rF \sin\theta \qquad 11\text{–}2$$

11–2 Torque and Angular Acceleration

A single torque applied to an object g

Newton's Second Law for Rotation
The connection between torque and a

$\sum\tau$

In this expression, I is the moment of
the angular acceleration about this ax

Rotational/Translational Analogies
Torque is analogous to force, the mon
the angular acceleration is analogous

Problem-Solving Summary

Type of Problem	Relevant Physical Concepts	Related Examples
Find the torque exerted on a system.	The torque exerted by a tangential force a distance r from the axis of rotation is $\tau = rF$. If the force is at an angle θ to the radial direction, the torque is $\tau = rF \sin\theta$.	Example 11–1
Determine the angular acceleration of a system.	First, calculate the torque exerted on the system. Next, find the angular acceleration using Newton's second law as applied to rotation, namely, $\tau = I\alpha$.	Examples 11–2, 11–3
Find the forces required for static equilibrium.	Static equilibrium requires that both the net force and the net torque acting on a system be zero.	Examples 11–4, 11–5, 11–6 Active Examples 11–1, 11–2, 11–3
Find the final angular momentum of a system.	A torque changes the angular momentum L of a system with time as follows; $\tau = \Delta L/\Delta t$. If no net torque acts on a system, its angular momentum is conserved.	Examples 11–8, 11–9 Active Examples 11–4, 11–5

Problem-Solving Summary

The Problem-Solving Summary presents an inventory of the types of problems encountered in the chapter. It also provides the relevant physical concepts and related chapter Examples for each.

Conceptual Questions

A separate section of qualitative, conceptual questions allows students to test their understanding of chapter principles. Answers to odd-numbered Conceptual Questions are located in the back of the book.

Conceptual Questions

(Answers to odd-numbered Conceptual Questions can be found in the back of the book.)

1. Two forces produce the same torque. Does it follow that they have the same magnitude? Explain.
2. A car pitches down in front when the brakes are applied sharply. Explain this observation in terms of torques.
3. A tightrope walker uses a long pole to aid in balancing. Why?
4. When a motorcycle accelerates rapidly from a stop it sometimes "pops a wheelie"; that is, its front wheel may lift off the ground. Explain this behavior in terms of torques.
5. Give an example of a system in which the net torque is zero but the net force is nonzero.
9. Give two everyday examples of objects that are in static equilibrium.
10. Can an object have zero translational acceleration and, at the same time, have nonzero angular acceleration? If your answer is no, explain why not. If your answer is yes, give a specific example.
11. Stars form when a large rotating cloud of gas collapses. What happens to the angular speed of the gas cloud as it collapses?
12. What purpose does the tail rotor on a helicopter serve?
13. Is it possible to change the angular momentum of an object

Conceptual Exercises

The Third Edition incorporates a new section of conceptual problems referred to as Conceptual Exercises. These Exercises consist of multiple-choice and ranking questions that can be used with classroom-response systems to enhance conceptual understanding. In addition, they can be graded with computer-grading systems, and hence can be incorporated seemlessly into homework assignments. Answers to odd-numbered Conceptual Exercises are given in the back of the book.

Conceptual Exercises

(Answers to odd-numbered Conceptual Exercises can be found in the back of the book.)

1. A given car battery is rated as 250 amp-hours. Does this rating indicate the amount of energy, power, charge, voltage, or current that can be drawn from the battery? Explain.
2. The four conducting cylinders shown in **Figure 21–26** are all made of the same material, though they differ in length and/or diameter. They are connected to four different batteries, which supply the necessary voltages to give the circuits the same current, I. Rank the four voltages, V_1, V_2, V_3, and V_4, in order of increasing value. Indicate ties where appropriate.
3. A conducting wire is quadrupled in length and tripled in diameter. **(a)** Does its resistance increase, decrease, or stay the same? Explain. **(b)** By what factor does its resistance change?
4. The resistivity of tungsten increases with temperature. When a light containing a tungsten filament heats up, does its power consumption increase, decrease, or stay the same? Explain.
5. **Figure 21–27** shows a plot of current versus voltage for two different materials, A and B. Which of these materials satisfies Ohm's law? Explain.

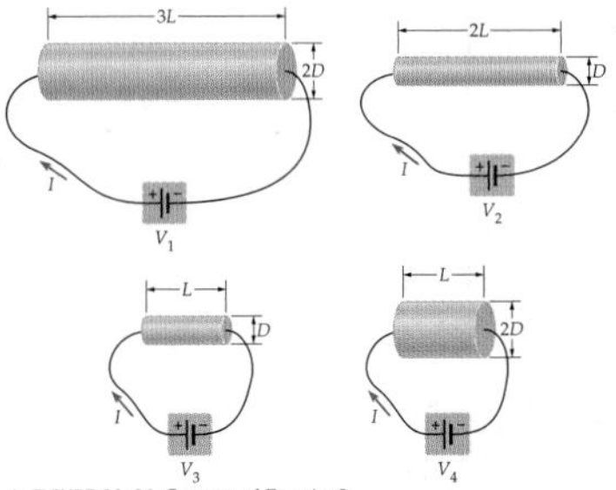

▲ **FIGURE 21–26** Conceptual Exercise 2

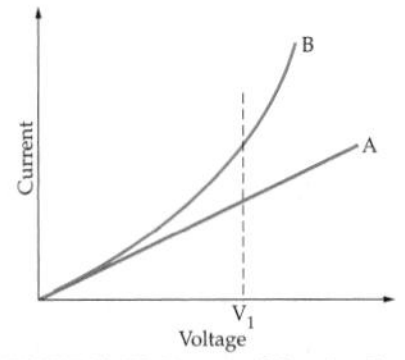

▲ **FIGURE 21–27** Conceptual Exercises 5 and 8

5. **Figure 21–27** shows a plot of current versus voltage for two different materials, A and B. Which of these materials satisfies Ohm's law? Explain.

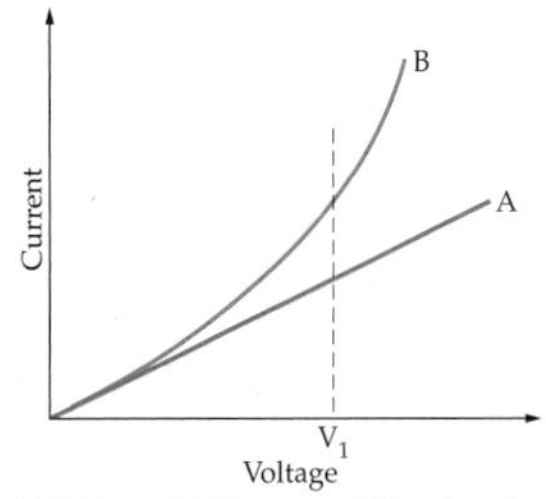

▲ **FIGURE 21–27** Conceptual Exercises 5 and 8

Integrated Problems

Each set of end-of-chapter Problems contains Integrated Problems, marked **IP**, that combine qualitative and quantitative parts. These Problems, which stress the importance of reasoning from principles, show how conceptual insight and numerical calculation go hand-in-hand in physics.

15. •• **IP** A white-crowned sparrow flying horizontally with a speed of 1.80 m/s folds its wings and begins to drop in free fall. **(a)** How far does the sparrow fall after traveling a horizontal distance of 0.500 m? **(b)** If the sparrow's initial speed is increased, does the distance of fall increase, decrease, or stay the same?

Interactive Problems

A new section of homework Problems deals with selected physical situations from the text that "come alive" via computer animation. The animations can be used in lecture or as a "virtual lab". They are also designed to be incorporated into homework assignments and to be graded with on-line homework systems.

83. •• ***Referring to Example 4–6*** Suppose the ball is dropped from a height of 4.50 m, and that the dolphin jumps with the same speed of 12.0 m/s. **(a)** What is the launch angle of the dolphin? **(b)** At what height does the dolphin catch the ball? **(c)** What is the minimum initial speed the dolphin must have to be able to catch the ball before it hits the water?

84. •• ***Referring to Example 4–6*** In Example 4–6, the dolphin catches the ball while it is still moving upward. **(a)** What initial speed does the dolphin need if it is to be moving horizontally when it catches the ball? **(b)** For the case in part (a), at what height is the ball caught? **(c)** Referring again to part (a), find the speed of both the dolphin and the ball just before the ball is caught.

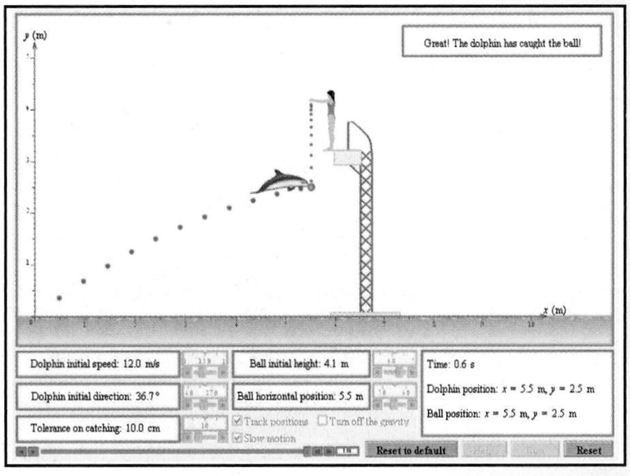

Physics

AP* Edition

Introduction to Physics

1–1 Physics and the Laws of Nature 2
1–2 Units of Length, Mass, and Time 2
1–3 Dimensional Analysis 4
1–4 Significant Figures 5
1–5 Converting Units 8
1–6 Order-of-Magnitude Calculations 10
1–7 Scalars and Vectors 11
1–8 Problem Solving in Physics 11

Physics is a quantitative science, based on careful measurements of quantities such as mass, length, and time. In the measurement shown here, a baby elephant is found to have a mass of approximately 425 kilograms, corresponding to a weight of about 935 pounds. Measurements of length and time indicate that the elephant's height is 1.25 meters, and its age is eleven months.

The goal of physics is to gain a deeper understanding of the world in which we live. For example, the laws of physics allow us to predict the behavior of everything from rockets sent to the Moon, to integrated chips in computers, to lasers used to perform eye surgery. In short, everything in nature—from atoms and subatomic particles to solar systems and galaxies—obeys the laws of physics.

As we begin our study of physics, it is useful to consider a range of issues that underlies everything to follow. One of the most fundamental of these is the system of units we use when we measure such things as the mass of an object, its length, and the time between two events. Other equally important issues include methods for handling numerical calculations and basic conventions of mathematical notation. By the end of the chapter we will have developed a common "language" of physics that will be used throughout this book and probably in any science that you study.

1–1 Physics and the Laws of Nature

Physics is the study of the fundamental laws of nature, which, simply put, are the laws that underlie all physical phenomena in the universe. Remarkably, we have found that these laws can be expressed in terms of mathematical equations. As a result, it is possible to make precise, quantitative comparisons between the predictions of theory—derived from the mathematical form of the laws—and the observations of experiments. Physics, then, is a science rooted equally firmly in theory and experiment, and, as physicists make new observations, they constantly test and—if necessary—refine the present theories.

What makes physics particularly fascinating is the fact that it relates to everything in the universe. There is a great beauty in the vision that physics brings to our view of the universe; namely, that all the complexity and variety that we see in the world around us, and in the universe as a whole, are manifestations of a few fundamental laws and principles. That we can discover and apply these basic laws of nature is both astounding and exhilarating.

For those not familiar with the subject, physics may seem to be little more than a confusing mass of formulas. Sometimes, in fact, these formulas can be the trees that block the view of the forest. For a physicist, however, the many formulas of physics are simply different ways of expressing a few fundamental ideas. It is the forest—the basic laws and principles of physical phenomena in nature—that is the focus of this text.

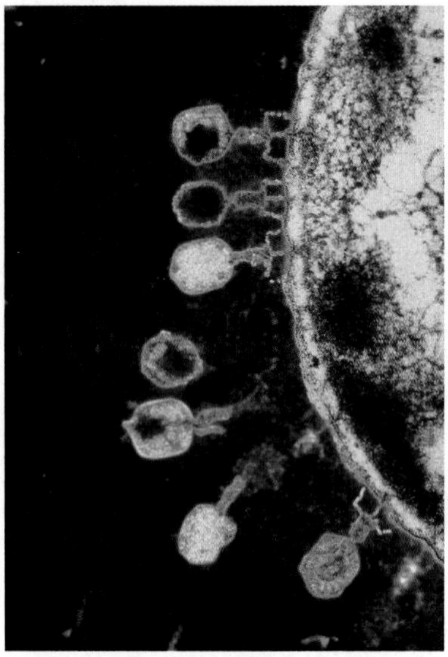

▲ The size of these viruses, seen here attacking a bacterial cell, is about 10^{-7} m.

1–2 Units of Length, Mass, and Time

To make quantitative comparisons between the laws of physics and our experience of the natural world, certain basic physical quantities must be measured. The most common of these quantities are **length** (L), **mass** (M), and **time** (T). In fact, in the next several chapters these are the only quantities that arise. Later in the text, additional quantities, such as temperature and electric current, will be introduced as needed.

We begin by defining the units in which each of these quantities is measured. Once the units are defined, the values obtained in specific measurements can be expressed as multiples of them. For example, our unit of length is the **meter** (m). It follows, then, that a person who is 1.94 m tall has a height 1.94 times this unit of length. Similar comments apply to the unit of mass, the **kilogram,** and the unit of time, the **second.**

The detailed system of units used in this book was established in 1960 at the Eleventh General Conference of Weights and Measures in Paris, France, and goes by the name Système International d'Unités, or SI for short. Thus, when we refer to **SI units,** we mean units of meters (m), kilograms (kg), and seconds (s). Taking the first letter from each of these units leads to an alternate name that is often used—the **mks system.**

In the remainder of this section we define each of the SI units.

▲ The diameter of this typical galaxy is about 10^{21} m. (How many viruses would it take to span the galaxy?)

Length

Early units of length were often associated with the human body. For example, the Egyptians defined the cubit to be the distance from the elbow to the tip of the middle finger. Similarly, the foot was at one time defined to be the length of the royal foot of King Louis XIV. As colorful as these units may be, they are not particularly reproducible—at least not to great precision.

In 1793 the French Academy of Sciences, seeking a more objective and reproducible standard, decided to define a unit of length equal to one ten-millionth the distance from the North Pole to the equator. This new unit was named the metre (from the Greek *metron* for "measure"). The preferred spelling in the United States is *meter*. This definition was widely accepted, and in 1799 a "standard" meter was produced. It consisted of a platinum-iridium alloy rod with two marks on it one meter apart.

TABLE 1–1 Typical Distances

Distance from Earth to the nearest large galaxy (the Andromeda galaxy, M31)	2×10^{22} m
Diameter of our galaxy (the Milky Way)	8×10^{20} m
Distance from Earth to the nearest star (other than the sun)	4×10^{16} m
One light year	9.46×10^{15} m
Average radius of Pluto's orbit	6×10^{12} m
Distance from Earth to the Sun	1.5×10^{11} m
Radius of Earth	6.37×10^{6} m
Length of a football field	10^{2} m
Height of a person	2 m
Diameter of a CD	0.12 m
Diameter of the aorta	0.018 m
Diameter of a period in a sentence	5×10^{-4} m
Diameter of a red blood cell	8×10^{-6} m
Diameter of the hydrogen atom	10^{-10} m
Diameter of a proton	2×10^{-15} m

Since 1983 we have used an even more precise definition of the meter, based on the speed of light in a vacuum. In particular:

> One meter is defined to be the distance traveled by light in a vacuum in 1/299,792,458 of a second.

No matter how its definition is refined, however, a meter is still about 3.28 feet, which is roughly 10 percent longer than a yard. A list of typical lengths is given in Table 1–1.

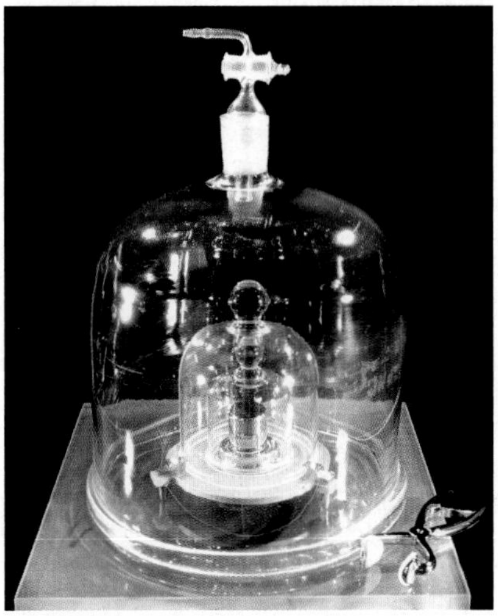

▲ The standard kilogram, a cylinder of platinum and iridium 0.039 m in height and diameter, is kept under carefully controlled conditions in Sèvres, France. Exact replicas are maintained in other laboratories around the world.

Mass

In SI units, mass is measured in kilograms. Unlike the meter, the kilogram is not based on any natural physical quantity. By convention, the kilogram has been defined as follows:

> The kilogram, by definition, is the mass of a particular platinum-iridium alloy cylinder at the International Bureau of Weights and Standards in Sèvres, France.

To put the kilogram in everyday terms, a quart of milk has a mass slightly less than 1 kilogram. Additional masses, in kilograms, are given in Table 1–2.

Note that we do not define the kilogram to be the *weight* of the platinum-iridium cylinder. In fact, weight and mass are quite different quantities, even though they are often confused in everyday language. Mass is an intrinsic, unchanging property of an object. Weight, in contrast, is a measure of the gravitational force acting on an object, which can vary depending on the object's location. For example, if you are fortunate enough to travel to Mars someday, you will find that your weight is less than on Earth, though your mass is unchanged. The force of gravity will be discussed in detail in Chapter 12.

TABLE 1–2 Typical Masses

Galaxy (Milky Way)	4×10^{41} kg
Sun	2×10^{30} kg
Earth	5.97×10^{24} kg
Space shuttle	2×10^{6} kg
Elephant	5400 kg
Automobile	1200 kg
Human	70 kg
Baseball	0.15 kg
Honeybee	1.5×10^{-4} kg
Red blood cell	10^{-13} kg
Bacterium	10^{-15} kg
Hydrogen atom	1.67×10^{-27} kg
Electron	9.11×10^{-31} kg

Time

Nature has provided us with a fairly accurate timepiece in the revolving Earth. In fact, prior to 1956 the mean solar day was defined to consist of 24 hours, with 60 minutes per hour, and 60 seconds per minute, for a total of $(24)(60)(60) = 84{,}400$ seconds. Even the rotation of the Earth is not completely regular, however.

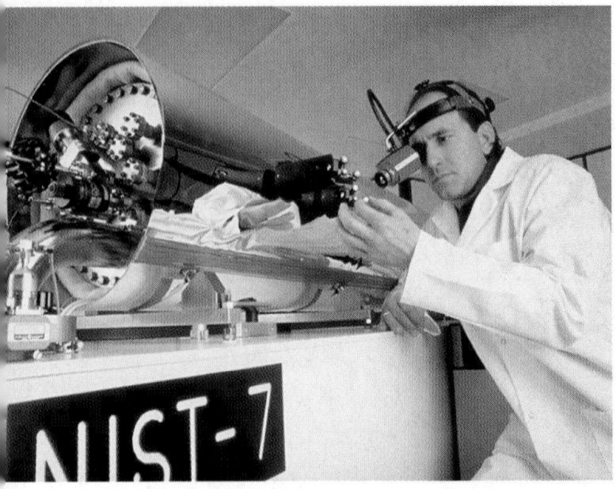

▲ This atomic clock, which keeps time on the basis of radiation from cesium atoms, is accurate to about three millionths of a second per year. (How long would it take for it to gain or lose an hour?)

© 2000 Sidney Harris

TABLE 1–3 Typical Times

Age of the universe	5×10^{17} s
Age of the Earth	1.3×10^{17} s
Existence of human species	6×10^{13} s
Human lifetime	2×10^{9} s
One year	3×10^{7} s
One day	8.6×10^{4} s
Time between heartbeats	0.8 s
Human reaction time	0.1 s
One cycle of a high-pitched sound wave	5×10^{-5} s
One cycle of an AM radio wave	10^{-6} s
One cycle of a visible light wave	2×10^{-15} s

Today, the most accurate timekeepers known are "atomic clocks," which are based on characteristic frequencies of radiation emitted by certain atoms. These clocks have typical accuracies of about 1 second in 300,000 years. The atomic clock used for defining the second operates with cesium-133 atoms. In particular, the second is defined as follows:

> One second is defined to be the time it takes for radiation from a cesium-133 atom to complete 9,192,631,770 cycles of oscillation.

A range of characteristic time intervals is given in Table 1–3.

The nation's time and frequency standard is determined by a *cesium fountain atomic clock* developed at the National Institute of Standards and Technology (NIST) in Boulder, Colorado. The fountain atomic clock, designated NIST-F1, produces a "fountain" of cesium atoms that are projected upward in a vacuum to a height of about a meter. It takes roughly a second for the atoms to rise and fall through this height (as we shall see in the next chapter), and during this relatively long period of time the frequency of their oscillation can be measured with great precision. In fact, the NIST-F1 will gain or lose no more than one second in every 20 million years of operation.

Atomic clocks are almost commonplace these days. For example, the satellites that participate in the Global Positioning System (GPS) actually carry atomic clocks with them in orbit. This allows them to make the precision time measurements that are needed for an equally precise determination of position and speed. Similarly, the "atomic clocks" that are advertised for use in the home, while not atomic in their operation, nonetheless get their time from radio signals sent out from the atomic clocks at NIST in Boulder. You can access the official U. S. time on your computer by going to http://time.gov on the Web.

Other Systems of Units and Standard Prefixes

Although SI units are used throughout most of this book and are used almost exclusively in scientific research and in industry, we will occasionally refer to other systems that you may encounter from time to time.

For example, a system of units similar to the mks system, though comprised of smaller units, is the **cgs system,** which stands for centimeter (cm), gram (g), and second (s). In addition, the British engineering system is often encountered in everyday usage in the United States. Its basic units are the slug for mass, the foot (ft) for length, and the second (s) for time.

Finally, multiples of the basic units are common no matter which system is used. Standard prefixes are used to designate common multiples in powers of ten. For example, the prefix *kilo* means one thousand, or, equivalently, 10^3. Thus, 1 kilogram is 10^3 grams, and 1 kilometer is 10^3 meters. Similarly, *milli* is the prefix for one thousandth, or 10^{-3}. Thus, a millimeter is 10^{-3} meter, and so on. The most common prefixes are listed in Table 1–4.

EXERCISE 1–1

a. A minivan sells for 33,200 dollars. Express the price of the minivan in kilodollars and megadollars.

b. A typical *E. coli* bacterium is about 5 micrometers (or microns) in length. Give this length in millimeters and kilometers.

Solution

a. 33.2 kilodollars, 0.0332 megadollars.

b. 0.005 mm, 0.000000005 km.

1–3 Dimensional Analysis

In physics, when we speak of the **dimension** of a physical quantity, we refer to the *type* of quantity in question, regardless of the units used in the measurement. For example, a distance measured in cubits and another distance measured in light-years both have the same dimension—length. The same is true of compound units such as velocity, which has the dimensions of length per unit time (length/time). A velocity measured in miles per hour has the same dimensions—length/time—as one measured in inches per century.

Now, any valid formula in physics must be **dimensionally consistent;** that is, each term in the equation must have the same dimensions. It simply doesn't make sense to add a distance to a time, for example, any more than it makes sense to add apples and oranges. They are different things.

To check the dimensional consistency of an equation, it is convenient to introduce a special notation for the dimension of a quantity. We will use square brackets, [ ], for this purpose. Thus, if x represents a distance, which has dimensions of length [L], we write this as $x = [\mathrm{L}]$. Similarly, a velocity, v, has dimensions of length per time [T]; thus we write $v = [\mathrm{L}]/[\mathrm{T}]$ to indicate its dimensions. Acceleration, a, which is the change in velocity per time, has the dimensions $a = ([\mathrm{L}]/[\mathrm{T}])/[\mathrm{T}] = [\mathrm{L}]/[\mathrm{T}^2]$. The dimensions of some common physical quantities are summarized in Table 1–5.

Let's use this notation to check the dimensional consistency of a simple equation. Consider the following formula:

$$x = x_0 + vt$$

In this equation, x and x_0 represent distances, v is a velocity, and t is time. Writing out the dimensions of each term, we have

$$[\mathrm{L}] = [\mathrm{L}] + \frac{[\mathrm{L}]}{[\mathrm{T}]}[\mathrm{T}]$$

It might seem at first that the last term has different dimensions than the other two. However, dimensions obey the same rules of algebra as other quantities. Thus the dimensions of time cancel in the last term:

$$[\mathrm{L}] = [\mathrm{L}] + \frac{[\mathrm{L}]}{[\cancel{\mathrm{T}}]}[\cancel{\mathrm{T}}] = [\mathrm{L}] + [\mathrm{L}]$$

As a result, we see that each term in this formula has the same dimensions. This type of calculation with dimensions is referred to as **dimensional analysis.**

EXERCISE 1–2

Show that $x = x_0 + v_0 t + \frac{1}{2}at^2$ is dimensionally consistent. The quantities x and x_0 are distances, v_0 is a velocity, and a is an acceleration.

Solution

Using the dimensions given in Table 1–5, we have

$$[\mathrm{L}] = [\mathrm{L}] + \frac{[\mathrm{L}]}{[\cancel{\mathrm{T}}]}[\cancel{\mathrm{T}}] + \frac{[\mathrm{L}]}{[\cancel{\mathrm{T}^2}]}[\cancel{\mathrm{T}^2}] = [\mathrm{L}] + [\mathrm{L}] + [\mathrm{L}]$$

Note that $\frac{1}{2}$ is ignored in this analysis because it has no dimensions.

Later in this text you will derive your own formulas from time to time. As you do so, it is helpful to check dimensional consistency at each step of the derivation. If at any time the dimensions don't agree, you will know that a mistake has been made, and you can go back and look for it. If the dimensions check, however, it's not a guarantee the formula is correct—after all, dimensionless factors, like 1/2 or 2, don't show up in a dimensional check.

TABLE 1–4 Common Prefixes

Power	Prefix	Abbreviation
10^{15}	peta	P
10^{12}	tera	T
10^{9}	giga	G
10^{6}	mega	M
10^{3}	kilo	k
10^{2}	hecto	h
10^{1}	deka	da
10^{-1}	deci	d
10^{-2}	centi	c
10^{-3}	milli	m
10^{-6}	micro	μ
10^{-9}	nano	n
10^{-12}	pico	p
10^{-15}	femto	f

TABLE 1–5 Dimensions of Some Common Physical Quantities

Quantity	Dimension
Distance	[L]
Area	[L^2]
Volume	[L^3]
Velocity	[L]/[T]
Acceleration	[L]/[T^2]
Energy	[M][L^2]/[T^2]

1–4 Significant Figures

When a mass, a length, or a time is measured in a scientific experiment, the result is known only to within a certain accuracy. The inaccuracy or uncertainty can be caused by a number of factors, ranging from limitations of the measuring device itself to limitations associated with the senses and the skill of the person performing the experiment. In any case, the fact that observed values of experimental quantities have inherent uncertainties should always be kept in mind when performing calculations with those values.

Suppose, for example, that you want to determine the walking speed of your pet tortoise. To do so, you measure the time, t, it takes for the tortoise to walk a distance, d, and then you calculate the quotient, d/t. When you measure the distance with a ruler, which has one tick mark per millimeter, you find that $d = 21.2$ cm, with the precise value of the digit in the second decimal place uncertain. Defining the number of **significant figures** in a physical quantity to be equal to the number of digits in it that are known with certainty, we say that d is known to *three* significant figures.

Similarly, you measure the time with an old pocket watch, and as best you can determine it, $t = 8.5$ s, with the second decimal place uncertain. Note that t is known to only *two* significant figures. If we were to make this measurement with a digital watch, with a readout giving the time to 1/100 of a second, the accuracy of the result would still be limited by the finite reaction time of the experimenter. The reaction time would have to be predetermined in a separate experiment. (See Problem 71 in Chapter 2 for a simple way to determine your reaction time.)

Returning to the problem at hand, we would now like to calculate the speed of the tortoise. Using the above values for d and t and a calculator with eight digits in its display, we find $(21.2 \text{ cm})/(8.5 \text{ s}) = 2.4941176$ cm/s. Clearly, such an accurate value for the speed is unjustified, considering the limitations of our measurements. After all, we can't expect to measure quantities to two and three significant figures and from them obtain results with eight significant figures. In general, the number of significant figures that result when we multiply or divide physical quantities is given by the following rule of thumb:

> The number of significant figures after multiplication or division is equal to the number of significant figures in the *least* accurately known quantity.

▲ Every measurement has some degree of uncertainty associated with it. How precise would you expect this measurement to be?

In our speed calculation, for example, we know the distance to three significant figures, but the time to only two significant figures. As a result, the speed should be given with just two significant figures, $d/t = (21.2 \text{ cm})/(8.5 \text{ s}) = 2.5$ cm/s. Note that we didn't just keep the first two digits in 2.4941176 cm/s and drop the rest. Instead, we "rounded up"; that is, because the first digit to be dropped (9 in this case) is greater than or equal to 5, we increase the previous digit (4 in this case) by 1. Thus, 2.5 cm/s is our best estimate for the tortoise's speed.

EXAMPLE 1–1 It's the Tortoise by a Hare

A tortoise races a rabbit by walking with a constant speed of 2.51 cm/s for 12.23 s. How much distance does the tortoise cover?

Picture the Problem

The race between the rabbit and the tortoise is shown in our sketch. The rabbit pauses to eat a carrot while the tortoise walks with a constant speed.

Strategy

The distance covered by the tortoise is the speed of the tortoise multiplied by the time during which it walks.

Solution

1. Multiply the speed by the time to find the distance d:

$$d = (\text{speed})(\text{time})$$
$$= (2.51 \text{ cm/s})(12.23 \text{ s}) = 30.7 \text{ cm}$$

Insight

Notice that if we simply multiply 2.51 cm/s by 12.23 s, we obtain 30.6973 cm. We don't give all of these digits in our answer, however. In particular, because the quantity that is known with the least accuracy (the speed) has only three significant figures, we give a result with three significant figures. Note, in addition, that the third digit in our answer has been rounded up from 6 to 7.

Practice Problem

How long does it take for the tortoise to walk 17 cm? [**Answer:** $t = (17\text{ cm})/(2.51\text{ cm/s}) = 6.8\text{ s}$]

Some related homework problems: Problem 11, Problem 15

Note that the distance of 17 cm in the Practice Problem has only two significant figures because we don't know the digits to the right of the decimal place. If the distance were given as 17.0 cm, on the other hand, it would have three significant figures.

When physical quantities are added or subtracted, we use a slightly different rule of thumb. In this case, the rule involves the number of decimal places in each of the terms:

> The number of decimal places after addition or subtraction is equal to the smallest number of decimal places in any of the individual terms.

Thus, if you make a time measurement of 16.74 s, and then a subsequent time measurement of 5.1 s, the total time of the two measurements should be given as 21.8 s, rather than 21.84 s.

▲ The finish of the 100-meter race at the 1996 Atlanta Olympics. This official timing photo shows Donovan Bailey setting a new world record of 9.84 s. (If the timing had been accurate to only tenths of a second—as would probably have been the case before electronic devices came into use—how many runners would have shared the winning time? How many would have shared the second-place and third-place times?)

EXERCISE 1–3

You and a friend pick some raspberries. Your flat weighs 12.7 lb, and your friend's weighs 7.25 lb. What is the combined weight of the raspberries?

Solution

Just adding the two numbers gives 19.95 lb. According to our rule of thumb, however, the final result must have only a single decimal place (corresponding to the term with the smallest number of decimal places). Rounding off to one place, then, gives 20.0 lb as the acceptable result.

Scientific Notation

The number of significant figures in a given quantity may be ambiguous due to the presence of zeros at the beginning or end of the number. For example, if a distance is stated to be 2500 m, the two zeros could be significant figures, or they could be zeros that simply show where the decimal point is located. If the two zeros are significant figures, the uncertainty in the distance is roughly a meter; if they are not significant figures, however, the uncertainty is about 100 m.

To remove this type of ambiguity, we can write the distance in **scientific notation**—that is, as a number of order unity times an appropriate power of ten. Thus, in this example, we would express the distance as 2.5×10^3 m if there are only two significant figures, or as 2.500×10^3 m to indicate four significant figures. Likewise, a time given as 0.000036 s has only two significant figures—the preceding zeros only serve to fix the decimal point. If this quantity were known to three significant figures, we would write it as 3.60×10^{-5} s to remove any ambiguity. See Appendix A for a more detailed discussion of scientific notation.

EXERCISE 1–4

How many significant figures are there in **(a)** 21.00, **(b)** 21, **(c)** 2.1×10^{-2}, **(d)** 2.10×10^{-3}?

Solution

(a) 4, **(b)** 2, **(c)** 2, **(d)** 3

Round-Off Error

Finally, even if you perform all your calculations to the same number of significant figures as in the text, you may occasionally obtain an answer that differs in its last digit from that given in the book. In most cases this is not an issue as far as understanding the physics is concerned—usually it is due to **round-off error.**

Round-off error occurs when numerical results are rounded off at different times during a calculation. To see how this works, let's consider a simple example. Suppose you are shopping for knickknacks, and you buy one item for \$2.21, plus 8 percent sales tax. The total price is \$2.3868, or, rounded off to the nearest penny, \$2.39. Later, you buy another item for \$1.35. With tax this becomes \$1.458 or, again to the nearest penny, \$1.46. The total expenditure for these two items is \$2.39 + \$1.46 = \$3.85.

Now, let's do the rounding off in a different way. Suppose you buy both items at the same time for a total before-tax price of \$2.21 + \$1.35 = \$3.56. Adding in the 8 percent tax gives \$3.8448, which rounds off to \$3.84, one penny different from the previous amount. This same type of discrepancy can occur in physics problems. In general, it's a good idea to keep one extra digit throughout your calculations whenever possible, rounding off only the final result. But while this practice can help to reduce the likelihood of round-off error, there is no way to avoid it in every situation.

1–5 Converting Units

It is often convenient to convert from one set of units to another. For example, suppose you would like to convert 316 ft to its equivalent in meters. Looking at the conversion factors on the inside front cover of the text, we see that

$$1 \text{ m} = 3.281 \text{ ft} \qquad \textbf{1–1}$$

Equivalently,

$$\frac{1 \text{ m}}{3.281 \text{ ft}} = 1 \qquad \textbf{1–2}$$

Now, to make the conversion, we simply multiply 316 ft by this expression, which is equivalent to multiplying by 1:

$$(316 \cancel{\text{ft}})\left(\frac{1 \text{ m}}{3.281 \cancel{\text{ft}}}\right) = 96.3 \text{ m}$$

▲ From this sign, you can calculate factors for converting miles to kilometers and vice versa. (Why do you think the conversion factors seem to vary for different destinations?)

Note that the conversion factor is written in this particular way, as 1 m divided by 3.281 ft, so that the units of feet cancel out, leaving the final result in the desired units of meters.

Of course, we can just as easily convert from meters to feet if we use the reciprocal of this conversion factor—which is also equal to 1:

$$1 = \frac{3.281 \text{ ft}}{1 \text{ m}}$$

For example, a distance of 26.4 m is converted to feet by canceling out the units of meters, as follows:

$$(26.4 \cancel{\text{m}})\left(\frac{3.281 \text{ ft}}{1 \cancel{\text{m}}}\right) = 86.6 \text{ ft}$$

Thus, we see that converting units is as easy as multiplying by 1—because that's really what you're doing.

EXAMPLE 1–2 A High-Volume Warehouse

A warehouse is 20.0 yards long, 10.0 yards wide, and 15.0 ft high. What is its volume in SI units?

Picture the Problem

In our sketch we picture the warehouse, and indicate the relevant lengths for each of its dimensions.

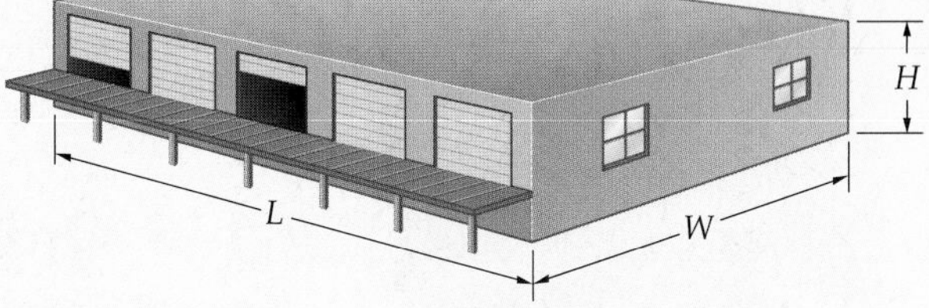

Strategy

We begin by converting the length, width, and height of the warehouse to meters. Once this is done, the volume in SI units is simply the product of the three dimensions.

Solution

1. Convert the length of the warehouse to meters: $L = (20.0 \text{ yard})\left(\frac{3 \text{ ft}}{1 \text{ yard}}\right)\left(\frac{1 \text{ m}}{3.281 \text{ ft}}\right) = 18.3 \text{ m}$

2. Convert the width to meters: $W = (10.0 \text{ yard})\left(\frac{3 \text{ ft}}{1 \text{ yard}}\right)\left(\frac{1 \text{ m}}{3.281 \text{ ft}}\right) = 9.14 \text{ m}$

3. Convert the height to meters: $H = (15.0 \text{ ft})\left(\frac{1 \text{ m}}{3.281 \text{ ft}}\right) = 4.57 \text{ m}$

4. Calculate the volume of the warehouse: $V = L \times W \times H = (18.3 \text{ m})(9.14 \text{ m})(4.57 \text{ m}) = 764 \text{ m}^3$

Insight

We would say, then, that the warehouse has a volume of 764 cubic meters—the same as 764 cubical boxes that are 1 m on a side.

Practice Problem

What is the volume of the warehouse if its length is one-hundredth of a mile, and the other dimensions are unchanged? [**Answer:** $V = 672 \text{ m}^3$]

Some related homework problems: Problem 17, Problem 18

Finally, the same procedure can be applied to conversions involving any number of units. For instance, if you walk at 3.00 mi/h, how fast is that in m/s? In this case we need the following additional conversion factors:

$$1 \text{ mi} = 5{,}280 \text{ ft} \qquad 1 \text{ h} = 3{,}600 \text{ s}$$

With these factors at hand, we carry out the conversion as follows:

$$(3.00 \cancel{\text{mi}}/\cancel{\text{h}})\left(\frac{5{,}280 \cancel{\text{ft}}}{1 \cancel{\text{mi}}}\right)\left(\frac{1 \text{ m}}{3.281 \cancel{\text{ft}}}\right)\left(\frac{1 \cancel{\text{h}}}{3{,}600 \text{ s}}\right) = 1.34 \text{ m/s}$$

Note that in each conversion factor the numerator is equal to the denominator. In addition, each conversion factor is written in such a way that the unwanted units cancel, leaving just meters per second in our final result.

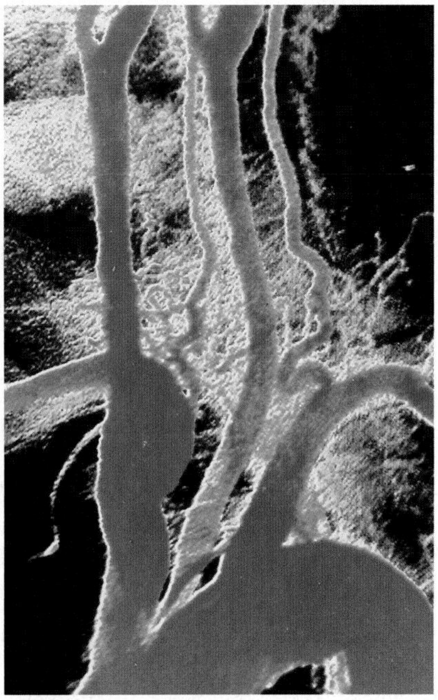

▲ Major blood vessels branch from the aorta (bottom), the artery that receives blood directly from the heart.

ACTIVE EXAMPLE 1–1 Find the Speed of Blood

Blood in the human aorta can attain speeds of 35.0 cm/s. How fast is this in **(a)** ft/s and **(b)** mi/h?

Solution *(Test your understanding by performing the calculations indicated in each step.)*

Part (a)

1. Convert centimeters to meters and then to feet: 1.15 ft/s

Part (b)

2. First, convert centimeters to miles: 2.17×10^{-4} mi/s

3. Next, convert seconds to hours: 0.783 mi/h

continued on next page

▲ Enrico Fermi (1901–1954) was renowned for his ability to pose and solve interesting order-of-magnitude problems. A winner of the 1938 Nobel Prize in physics, Fermi would ask his classes to obtain order-of-magnitude estimates for questions such as "How many piano tuners are there in Chicago?" or "How much is a tire worn down during one revolution?" Estimation questions like these are known to physicists today as "Fermi Problems."

continued from previous page

Insight

Of course, the conversions in part (b) can be carried out in a single calculation if desired.

Your Turn

Find the speed of blood in units of km/h.

(Answers to **Your Turn** *problems are given in the back of the book.)*

1–6 Order-of-Magnitude Calculations

An **order-of-magnitude** calculation is a rough "ballpark" estimate designed to be accurate to within a factor of about 10. One purpose of such a calculation is to give a quick idea of what order of magnitude should be expected from a complete, detailed calculation. If an order-of-magnitude calculation indicates that a distance should be on the order of 10^4 m, for example, and your calculator gives an answer on the order of 10^7 m, then there is an error somewhere that needs to be resolved.

For example, suppose you would like to estimate the speed of a cliff diver on entering the water. First, the cliff may be 20 or 30 feet high; thus in SI units we would say that the order of magnitude of the cliff's height is 10 m—certainly not 1 m or 10^2 m. Next, the diver hits the water something like a second later—certainly not 0.1 s later nor 10 s later. Thus, a reasonable order-of-magnitude estimate of the diver's speed is $10 \text{ m}/1 \text{ s} = 10 \text{ m/s}$, or roughly 20 mi/h. If you do a detailed calculation and your answer is on the order of 10^4 m/s, you probably entered one of your numbers incorrectly.

Another reason for doing an order-of-magnitude calculation is to get a feeling for what size numbers we are talking about in situations where a precise count is not possible. This is illustrated in the following Example.

EXAMPLE 1–3 Estimation: How Many Raindrops in a Storm

A thunderstorm drops half an inch (~ 0.01 m) of rain on Washington D.C., which covers an area of about 70 square miles ($\sim 10^8 \text{ m}^2$). Estimate the number of raindrops that fell during the storm.

Picture the Problem

Our sketch shows an area $A = 10^8 \text{ m}^2$ covered to a depth $d = 0.01$ m by rainwater from the storm. Each drop of rain is approximated by a small sphere with a diameter of 4 mm.

Strategy

To find the number of raindrops, we first calculate the volume of water required to cover 10^8 m^2 to a depth of 0.01 m. Next, we calculate the volume of an individual drop of rain, recalling that the volume of a sphere of radius r is $4\pi r^3/3$. We estimate the diameter a raindrop to be about 4 mm. Finally, dividing the volume of a drop into the volume of water that fell during the storm gives the number of drops.

Solution

1. Calculate the order of magnitude of the volume of water, V_{water}, that fell during the storm:

$$V_{\text{water}} = Ad = (10^8 \text{ m}^2)(0.01 \text{ m}) \approx 10^6 \text{ m}^3$$

2. Calculate the order of magnitude of the volume of a drop of rain, V_{drop}. Note that if the diameter of a drop is 4 mm, its radius is $r = 2 \text{ mm} = 0.002$ m:

$$V_{\text{drop}} = \frac{4}{3}\pi r^3 \approx \frac{4}{3}\pi(0.002 \text{ m})^3 \approx 10^{-8} \text{ m}^3$$

3. Divide V_{drop} into V_{water} to find the order of magnitude of the number of drops that fell during the storm:

$$\textit{number of raindrops} \approx \frac{V_{\text{water}}}{V_{\text{drop}}} \approx \frac{10^6 \text{ m}^3}{10^{-8} \text{ m}^3} = 10^{14}$$

Insight

Thus the number of raindrops in this one small storm is roughly 100,000 times greater than the current population of the Earth.

Practice Problem

If a storm pelts Washington D.C. with 10^{15} raindrops, how many inches of rain fall on the city? [**Answer:** About 5 inches.]

Some related homework problems: Problem 33, Problem 35

Appendix B provides a number of interesting "typical values" for length, mass, speed, acceleration, and many other quantities. You may find these to be of use in making your own order-of-magnitude estimates.

1–7 Scalars and Vectors

Physical quantities are sometimes defined solely in terms of a number and the corresponding unit, like the volume of a room or the temperature of the air it contains. Other quantities require both a numerical value *and* a direction. For example, suppose a car is traveling at a rate of 25 m/s in a direction that is due north. Both pieces of information—the rate of travel (25 m/s) and the direction (north)—are required to fully specify the motion of the car. The rate of travel is given the name **speed;** the rate of travel combined with the direction is referred to as the **velocity.**

In general, quantities that are specified by a numerical value only are referred to as **scalars;** quantities that require both a numerical value and a direction are called **vectors:**

- A scalar is a numerical value, expressed in terms of appropriate units. An example would be the temperature of a room or the speed of a car.
- A vector is a mathematical quantity with both a numerical value and a direction. An example would be the velocity of a car.

All the physical quantities discussed in this text are either vectors or scalars. The properties of numbers (scalars) are well known, but the properties of vectors are sometimes less well known—though no less important. For this reason, you will find that Chapter 3 is devoted entirely to a discussion of vectors in two and three dimensions and, more specifically, to how they are used in physics.

The rather straightforward special case of vectors in one dimension is discussed in Chapter 2. There, we see that the direction of a velocity vector, for example, can only be to the left or to the right, up or down, and so on. That is, only two choices are available for the direction of a vector in one dimension. This is illustrated in **Figure 1-1**, where we see two cars, each traveling with a speed of 25 m/s. We also see that the cars are traveling in opposite directions, with car 1 moving to the right and car 2 moving to the left. We indicate the direction of travel with a plus sign for motion to the right, and a negative sign for motion to the left. Thus, the velocity of car 1 is written, $v_1 = +25\ \mathrm{m/s}$, and the velocity of car 2 is $v_2 = -25\ \mathrm{m/s}$. The speed of each car is the absolute value, or **magnitude,** of the velocity; that is, speed $= |v_1| = |v_2| = 25\ \mathrm{m/s}$.

Whenever we deal with one-dimensional vectors, we shall indicate their direction with the appropriate sign. Many examples are found in Chapter 2 and, again, in later chapters where the simplicity of one dimension can again be applied.

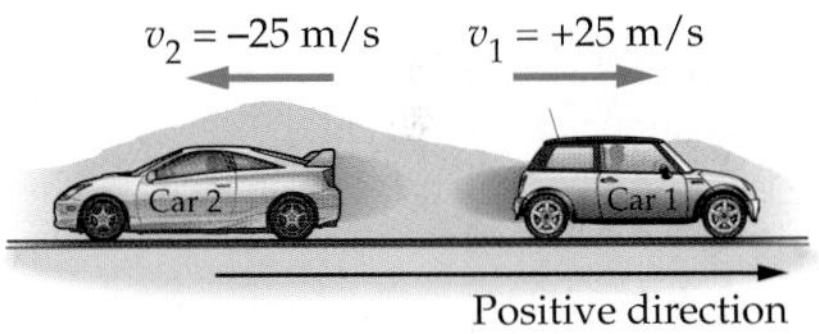

▲ **FIGURE 1–1 Velocity vectors in one dimension**

The two cars shown in this figure have equal speeds of 25 m/s, but are traveling in opposite directions. To indicate the direction of travel, we first choose a positive direction (to the right in this case), and then give appropriate signs to the velocity of each car. Notice that car 1 moves to the right, and hence its velocity is positive, $v_1 = +25\ \mathrm{m/s}$; car 2 has a negative velocity, $v_2 = -25\ \mathrm{m/s}$, because it moves to the left.

1–8 Problem Solving in Physics

Physics is a lot like swimming—you have to learn by doing. You could read a book on swimming and memorize every word in it, but when you jump into a pool the first time you are going to have problems. Similarly, you could read this book carefully, memorizing every formula in it, but when you finish, you still haven't learned physics. To learn physics, you have to go beyond passive reading; you have to interact with physics and experience it by doing problems.

In this section we present a general overview of problem solving in physics. The suggestions given below, which apply to problems in all areas of physics, should help to develop a systematic approach.

We should emphasize at the outset that there is no recipe for solving problems in physics—it is a creative activity. In fact, the opportunity to be creative is one of the attractions of physics. The following suggestions, then, are not intended as a rigid set of steps that must be followed like the steps in a computer program.

Rather, they provide a general guideline that experienced problem solvers find to be effective.

- **Read the problem carefully** Before you can solve a problem, you need to know exactly what information it gives and what it asks you to determine. Some information is given explicitly, as when a problem states that a person has a mass of 70 kg. Other information is implicit; for example, saying that a ball is dropped from rest means that its initial speed is zero. Clearly, a *careful* reading is the essential first step in problem solving.
- **Sketch the system** This may seem like a step you can skip—but don't. A sketch helps you to acquire a physical feeling for the system. It also provides an opportunity to label those quantities that are known and those that are to be determined. All Examples in this text begin with a sketch of the system, accompanied by a brief description in a section labeled "Picture the Problem."
- **Visualize the physical process** Try to visualize what is happening in the system as if you were watching it in a movie. Your sketch should help. This step ties in closely with the next step.
- **Strategize** This may be the most difficult, but at the same time the most creative, part of the problem-solving process. From your sketch and visualization, try to identify the physical processes at work in the system. Ask yourself what concepts or principles are involved in this situation. Then, develop a strategy—a game plan—for solving the problem. All Examples in this book have a "Strategy" spelled out before the solution begins.
- **Identify appropriate equations** Once a strategy has been developed, find the specific equations that are needed to carry it out.
- **Solve the equations** Use basic algebra to solve the equations identified in the previous step. Work with symbols such as x or y for the most part, substituting numerical values near the end of the calculations. Working with symbols will make it easier to go back over a problem to locate and identify mistakes, if there are any, and to explore limits and special cases.
- **Check your answer** Once you have an answer, check to see if it makes sense: (i) Does it have the correct dimensions? (ii) Is the numerical value reasonable?
- **Explore limits/special cases** Getting the correct answer is nice, but it's not all there is to physics. You can learn a great deal about physics and about the connection between physics and mathematics by checking various limits of your answer. For example, if you have two masses in your system, m_1 and m_2, what happens in the special case that $m_1 = 0$ or $m_1 = m_2$? Check to see whether your answer and your physical intuition agree.

The **Examples** in this text are designed to deepen your understanding of physics and at the same time develop your problem-solving skills. They all have the same basic structure: Problem Statement; Picture the Problem; Strategy; Solution, presenting the flow of ideas and the mathematics side-by-side in a two-column format; Insight; and a Practice Problem related to the one just solved. As you work through the Examples in the chapters to come, notice how the basic problem-solving guidelines outlined above are implemented in a consistent way.

In addition to the Examples, this text contains a new and innovative type of worked-out problem called the **Active Example,** the first one of which appears on page 9. The purpose of Active Examples is to encourage active participation in the solution of a problem and, in so doing, to act as a "bridge" between Examples—where each and every detail is worked out—and homework problems—where you are completely on your own. An analogy would be to think of Examples as like a tricycle, with no balancing required; homework problems as like a bicycle, where balancing is initially difficult to master; and Active Examples as like a bicycle with training wheels that give just enough help to prevent a fall. When you work through an Active Example, keep in mind that the work you are doing as you progress step-by-step through the problem is just the kind of work you'll be doing later in your homework assignments.

Finally, it is tempting to look for shortcuts when doing a problem—to look for a formula that seems to fit and some numbers to plug into it. It may seem harder to think ahead, to be systematic as you solve the problem, and then to think back over what you have done at the end of the problem. The extra effort is worth it, however, because by doing these things you will develop powerful problem-solving skills that can be applied to unexpected problems you may encounter on exams—and in life in general.

Chapter Summary

1–1 Physics and the Laws of Nature

Physics is based on a small number of fundamental laws and principles.

1–2 Units of Length, Mass, and Time

Length
One meter is defined as the distance traveled by light in a vacuum in 1/299,792,458 second.

Mass
One kilogram is the mass of a metal cylinder kept at the International Bureau of Weights and Standards.

Time
One second is the time required for a particular type of radiation from cesium-133 to undergo 9,192,631,770 oscillations.

1–3 Dimensional Analysis

Dimension
The dimension of a quantity is the type of quantity it is; for example, length [L], mass [M], or time [T].

Dimensional Consistency
An equation is dimensionally consistent if each term in it has the same dimensions. All valid physical equations are dimensionally consistent.

Dimensional Analysis
A calculation based on the dimensional consistency of an equation.

1–4 Significant Figures

Significant Figures
The number of digits reliably known, excluding digits that simply indicate the decimal place. For example, 3.45 and 0.0000345 both have three significant figures.

Round-off Error
Discrepancies caused by rounding off numbers in intermediate results.

1–5 Converting Units

Multiply by the ratio of two units to convert from one to another. As an example, to convert 3.5 m to feet, you multiply by the factor (1 ft/0.3048 m).

1–6 Order-of-Magnitude Calculations

A ballpark estimate designed to be accurate to within the nearest power of ten.

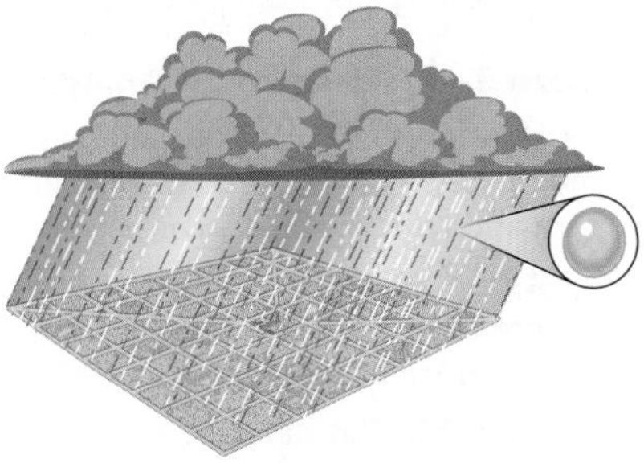

1–7 Scalars and Vectors

A physical quantity that can be represented by a numerical value only is called a scalar. Quantities that require a direction in addition to the numerical value are called vectors.

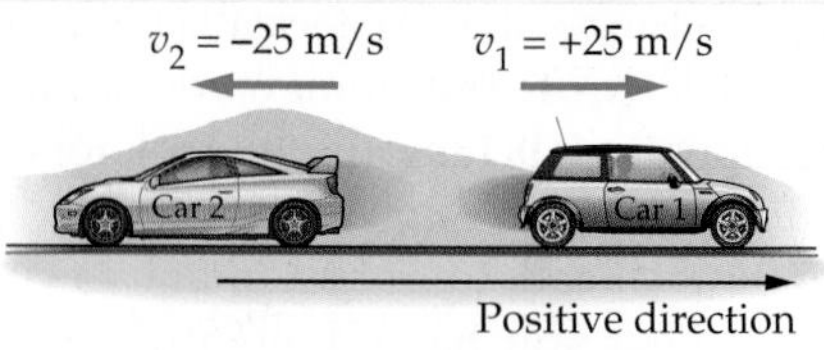

1–8 Problem Solving in Physics

A good general approach to problem solving is as follows: read; sketch; visualize; strategize; identify equations; solve; check; explore limits.

Conceptual Questions

(Answers to odd-numbered Conceptual Questions can be found in the back of the book.)

1. Can dimensional analysis determine whether the area of a circle is πr^2 or $2\pi r^2$? Explain.
2. If a distance d has units of meters, and a time T has units of seconds, does the quantity $T + d$ make sense physically? What about the quantity d/T? Explain in both cases.
3. Is it possible for two quantities to **(a)** have the same units but different dimensions or **(b)** have the same dimensions but different units? Explain.
4. Give an order-of-magnitude estimate for the time in seconds of the following: **(a)** a year; **(b)** a baseball game; **(c)** a heartbeat; **(d)** the age of the Earth; **(e)** the age of a person.
5. Give an order-of-magnitude estimate for the length in meters of the following: **(a)** a person; **(b)** a fly; **(c)** a car; **(d)** a 747 airplane; **(e)** an interstate freeway stretching coast-to-coast.

Conceptual Exercises

(Answers to odd-numbered Conceptual Exercises can be found in the back of the book.)

1. Which of the following equations are dimensionally consistent? **(a)** $x = vt$, **(b)** $x = \frac{1}{2}at^2$, **(c)** $t = (2x/a)^{1/2}$.
2. Which of the following equations are dimensionally consistent? **(a)** $v = at$, **(b)** $v = \frac{1}{2}at^2$, **(c)** $t = a/v$, **(d)** $v^2 = 2ax$.
3. Which of the following quantities have the dimensions of a distance? **(a)** vt, **(b)** $\frac{1}{2}at^2$, **(c)** $2at$, **(d)** v^2/a.
4. Which of the following quantities have the dimensions of a speed? **(a)** $\frac{1}{2}at^2$, **(b)** at, **(c)** $(2x/a)^{1/2}$, **(d)** $(2ax)^{1/2}$.
5. Which of the following quantities have the dimensions of an acceleration? **(a)** xt^2, **(b)** v^2/x, **(c)** x/t^2, **(d)** v/t.

Problems

Note: **IP** *denotes an integrated conceptual/quantitative problem.* **BIO** *identifies problems of biological or medical interest. Red bullets (•, ••, •••) are used to indicate the level of difficulty of each problem.*

Section 1–2 Units of Length, Mass, and Time

1. • **Spiderman** The movie *Spiderman* brought in \$114,000,000 in its opening weekend. Express this amount in **(a)** gigadollars and **(b)** teradollars.
2. • **BIO The Thickness of Hair** A human hair has a thickness of about 70 μm. What is this in **(a)** meters and **(b)** kilometers?
3. • The speed of light in a vacuum is approximately 0.3 Gm/s. Express the speed of light in meters per second.
4. • **A Fast Computer** IBM has a computer it calls the Blue Gene/L that can do 136.8 teracalculations per second. How many calculations can it do in a microsecond?

Section 1–3 Dimensional Analysis

5. • Velocity is related to acceleration and distance by the following expression, $v^2 = 2ax^p$. Find the power p that makes this equation dimensionally consistent.
6. • Acceleration is related to distance and time by the following expression, $a = 2xt^p$. Find the power p that makes this equation dimensionally consistent.
7. • Show that the equation $v = v_0 + at$ is dimensionally consistent. Note that v and v_0 are velocities and that a is an acceleration.
8. •• Newton's second law (to be discussed in Chapter 5) states that acceleration is proportional to the force acting on an object and is inversely proportional to the object's mass. What are the dimensions of force?
9. •• The time T required for one complete oscillation of a mass m on a spring of force constant k is

$$T = 2\pi\sqrt{\frac{m}{k}}$$

Find the dimensions k must have for this equation to be dimensionally correct.

Section 1–4 Significant Figures

10. • The first several digits of π are known to be $\pi = 3.14159265358979\ldots$. What is π to **(a)** three significant figures, **(b)** five significant figures, and **(c)** seven significant figures?
11. • The speed of light to five significant figures is 2.9979×10^8 m/s. What is the speed of light to three significant figures?
12. • A parking lot is 134.3 m long and 37.66 m wide. What is the perimeter of the lot?
13. • On a fishing trip you catch a 2.45-lb bass, a 10.1-lb rock cod, and a 16.13-lb salmon. What is the total weight of your catch?

14. •• How many significant figures are there in **(a)** 0.000054, **(b)** 3.001×10^5?

15. •• What is the area of a circle of radius **(a)** 14.37 m and **(b)** 3.8 m?

Section 1–5 Converting Units

16. • BIO **Mantis Shrimp** Peacock mantis shrimps (*Odontodactylus scyllarus*) feed largely on snails. They shatter the shells of their prey by delivering a sharp blow with their front legs, which have been observed to reach peak speeds of 23 m/s. What is this speed in **(a)** feet per second and **(b)** miles per hour?

17. • **(a)** The largest building in the world by volume is the Boeing 747 plant in Everett, Washington. It measures approximately 631 m long, 707 yards wide, and 110 ft high. What is its volume in cubic feet? **(b)** Convert your result from part (a) to cubic meters.

18. • The Ark of the Covenant is described as a chest of acacia wood 2.5 cubits in length and 1.5 cubits in width and height. Given that a cubit is equivalent to 17.7 in., find the volume of the ark in cubic feet.

19. • How long does it take for radiation from a cesium-133 atom to complete 1 million cycles?

20. • **Angel Falls** Water going over Angel Falls, in Venezuela, the world's highest waterfall, drops through a distance of 3212 ft. What is this distance in km?

21. • An electronic advertising sign repeats a message every 6 seconds, day and night, for a week. How many times did the message appear on the sign?

22. • BIO **Blue Whales** The blue whale (*Balaenoptera musculus*) is thought to be the largest animal ever to inhabit the Earth. The longest recorded blue whale had a length of 108 ft. What is this length in meters?

23. • **The Star of Africa** The Star of Africa, a diamond in the royal scepter of the British crown jewels, has a mass of 530.2 carats, where 1 carat = 0.20 g. Given that 1 kg has an approximate weight of 2.21 lb, what is the weight of this diamond in pounds?

24. • IP Many highways have a speed limit of 55 mi/h. **(a)** Is this speed greater than, less than, or equal to 55 km/h? Explain. **(b)** Find the speed limit in km/h that corresponds to 55 mi/h.

25. • What is the speed in miles per hour of a beam of light traveling at 3.00×10^8 m/s?

26. • BIO **Woodpecker Impact** When red-headed woodpeckers (*Melanerpes erythrocephalus*) strike the trunk of a tree, they can experience an acceleration ten times greater than the acceleration of gravity, or about 98.1 m/s^2. What is this acceleration in ft/s^2?

27. •• **A Jiffy** The American physical chemist Gilbert Newton Lewis (1875–1946) proposed a unit of time called the "jiffy." According to Lewis, 1 jiffy = the time it takes light to travel one centimeter. **(a)** If you perform a task in a jiffy, how long has it taken in seconds? **(b)** How many jiffys are in one minute? (Use the fact that the speed of light is approximately 2.9979×10^8 m/s.)

28. •• **The Mutchkin and the Noggin** **(a)** A mutchkin is a Scottish unit of liquid measure equal to 0.42 L. How many mutchkins are required to fill a container that measures one foot on a side? **(b)** A noggin is a volume equal to 0.28 mutchkin. What is the conversion factor between noggins and gallons?

29. •• Suppose 1.0 cubic meter of oil is spilled into the ocean. Find the area of the resulting slick, assuming that it is one molecule thick, and that each molecule occupies a cube 0.50 μm on a side.

30. •• IP **(a)** A standard sheet of paper measures 8 1/2 by 11 inches. Find the area of one such sheet of paper in m^2. **(b)** A second sheet of paper is half as long and half as wide as the one described in part (a). By what factor is its area less than the area found in part (a)?

31. •• BIO **Squid Nerve Impulses** Nerve impulses in giant axons of the squid can travel with a speed of 20.0 m/s. How fast is this in **(a)** ft/s and **(b)** mi/h?

32. •• The acceleration of gravity is approximately 9.81 m/s^2 (depending on your location). What is the acceleration of gravity in feet per second squared?

Section 1–6 Order-of-Magnitude Calculations

33. • Give a ballpark estimate of the number of seats in a typical major league ballpark.

Shea Stadium, in New York. How many fans can it hold? (Problem 33)

34. • Milk is often sold by the gallon in plastic containers. **(a)** Estimate the number of gallons of milk that are purchased in the United States each year. **(b)** What approximate weight of plastic does this represent?

35. •• New York is roughly 3000 miles from Seattle. When it is 10:00 A.M. in Seattle, it is 1:00 P.M. in New York. Using this information, estimate **(a)** the rotational speed of the surface of Earth, **(b)** the circumference of Earth, and **(c)** the radius of Earth.

36. •• You've just won the $1 million cash lottery, and you go to pick up the prize. What is the approximate weight of the cash if you request payment in **(a)** quarters or **(b)** dollar bills?

General Problems

37. • BIO **Photosynthesis** The light that plants absorb to perform photosynthesis has a wavelength that peaks near 675 nm. Express this distance in **(a)** millimeters and **(b)** inches.

38. • **Glacial Speed** On June 9, 1983, the lower part of the Variegated Glacier in Alaska was observed to be moving at a rate of 210 feet per day. What is this speed in meters per second?

Alaska's Variegated Glacier (Problem 38)

39. •• BIO **Mosquito Courtship** Male mosquitoes in the mood for mating find female mosquitoes of their own species by listening for the characteristic "buzzing" frequency of the female's wing beats. This frequency is about 605 wing beats per second. **(a)** How many wing beats occur in one minute? **(b)** How many cycles of oscillation does the radiation from a cesium-133 atom complete during one mosquito wing beat?

40. •• **Ten and Ten** When Coast Guard pararescue jumpers leap from a helicopter to save a person in the water, they like to jump when the helicopter is flying "ten and ten," which means it is 10 feet above the water and moving forward with a speed of 10 knots. What is "ten and ten" in SI units? (A knot is one nautical mile per hour, where a nautical mile is 1.852 km.)

41. •• IP A Porsche sports car can accelerate at 12 m/s^2. **(a)** Is this acceleration greater than, less than, or equal to 12 ft/s^2? Explain. **(b)** Determine the acceleration of a Porsche in ft/s^2. **(c)** Determine its acceleration in km/h^2.

42. •• BIO **Human Nerve Fibers** Type A nerve fibers in humans can conduct nerve impulses at speeds up to 140 m/s. **(a)** How fast are the nerve impulses in miles per hour? **(b)** How far (in meters) can the impulses travel in 5.0 ms?

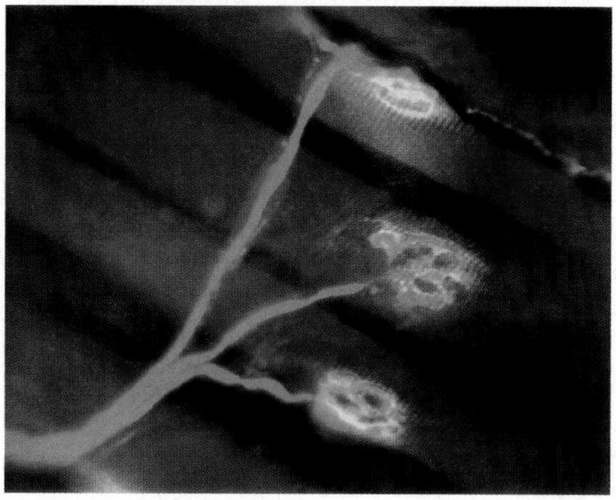

The impulses in these nerve axons, which carry commands to the skeletal muscle fibers in the background, travel at speeds of up to 140 m/s. (Problem 42)

43. •• BIO **Brain Growth** The mass of a newborn baby's brain has been found to increase by about 1.6 mg per minute. **(a)** How much does the brain's mass increase in one day? **(b)** How long does it take for the brain's mass to increase by 0.0055 kg?

44. •• BIO **Cricket Chirps** Crickets are well known for chirping at a rate that depends on temperature—some people even use a pet cricket as a thermometer. The snowy tree cricket (*Oecanthus fultoni* Walker) has been found to chirp according to the following formula:

$$N = \text{number of chirps per 13.0 seconds}$$
$$= T - 40.0$$

In this expression, T is the temperature in degrees Fahrenheit. Suppose a snowy cricket is chirping when the temperature is 65.0 degrees Fahrenheit. How many oscillations does the radiation from a cesium-133 atom complete between successive chirps?

45. •• **The Huygens Probe** NASA's Cassini mission to Saturn released a probe on December 25, 2004, that landed on the Saturnian moon Titan on January 14, 2005. The probe, which was named Huygens, was released with a gentle relative speed of 31 cm/s. As Huygens moved away from the main spacecraft, it rotated at a rate of seven revolutions per minute. **(a)** How many revolutions had Huygens completed when it was 150 yards from the mother ship? **(b)** How far did Huygens move away from the mother ship during each revolution? Give your answer in feet.

46. ••• Acceleration is related to velocity and time by the following expression, $a = v^p t^q$. Find the powers p and q that make this equation dimensionally consistent.

47. ••• The period T of a simple pendulum is the amount of time required for it to undergo one complete oscillation. If the length of the pendulum is L and the acceleration of gravity is g, then T is given by

$$T = 2\pi L^p g^q$$

Find the powers p and q required for dimensional consistency.

48. ••• Driving along a crowded freeway, you notice that it takes a time t to go from one mile marker to the next. When you increase your speed by 7.8 mi/h, the time to go one mile decreases by 14 s. What was your original speed?

One-Dimensional Kinematics

2-1 Position, Distance, and Displacement 18
2-2 Average Speed and Velocity 19
2-3 Instantaneous Velocity 23
2-4 Acceleration 25
2-5 Motion with Constant Acceleration 29
2-6 Applications of the Equations of Motion 35
2-7 Freely Falling Objects 37

These sprinters, crossing the finish line of the 100-m dash, illustrate one-dimensional motion. At the moment shown in the photograph the runners are moving with constant velocity at a speed of approximately 10 m/s.

We begin our study of physics with **mechanics,** the area of physics perhaps most apparent to us in our everyday lives. Every time you raise an arm, stand up or sit down, throw a ball or open a door, your actions are governed by the laws of mechanics. Basically, mechanics is the study of how objects move, how they respond to external forces, and how other factors, such as size, mass, and mass distribution, affect their motion. This is a lot to cover, and we certainly won't try to tackle it all in one chapter.

Furthermore, in this chapter we treat all physical objects as *point particles;* that is, we consider all the mass of the object to be concentrated at a single point. This is a common practice in physics. For example, if you are interested in calculating the time it takes the Earth to complete a revolution about the Sun, it is reasonable to consider the Earth and the Sun as simple particles. In later chapters, we extend our studies to increasingly realistic situations, involving motion in more than one dimension and physical objects with shape and size.

2–1 Position, Distance, and Displacement

The first step in describing the motion of a particle is to set up a **coordinate system** that defines its position. An example of a coordinate system in one dimension is shown in **Figure 2–1**. This is simply an x axis, with an origin (where $x = 0$) and an arrow indicating the positive direction—the direction in which x increases. In setting up a coordinate system, we are free to choose the origin and the positive direction as we like, but once we make a choice we must be consistent with it throughout any calculations that follow.

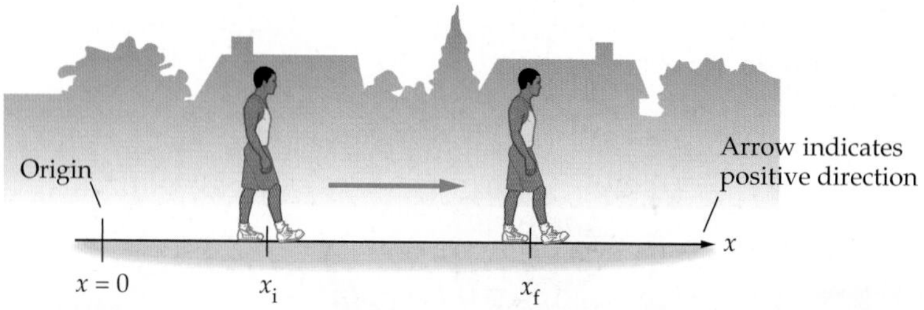

▶ **FIGURE 2–1 A one-dimensional coordinate system**

You are free to choose the origin and positive direction as you like, but once your choice is made, stick with it.

The particle in Figure 2–1 is a person who has moved to the right from an initial position, x_i, to a final position, x_f. Because the positive direction is to the right, it follows that x_f is greater than x_i; that is, $x_f > x_i$.

Now that we've seen how to set up a coordinate system, let's use one to investigate the situation sketched in **Figure 2–2**. Suppose that you leave your house, drive to the grocery store, and then return home. The **distance** you've covered in your trip is 8.6 mi. In general, distance is defined as follows:

Definition: Distance

distance = total length of travel

SI unit: meter, m

Using SI units, the distance in this case is

$$8.6 \text{ mi} = (8.6 \text{ mi})\left(\frac{1609 \text{ m}}{1 \text{ mi}}\right) = 1.4 \times 10^4 \text{ m}$$

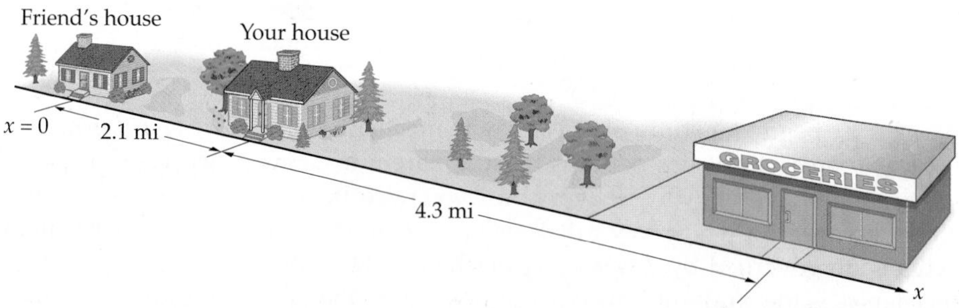

▶ **FIGURE 2–2 One-dimensional coordinates**

The locations of your house, your friend's house, and the grocery store in terms of a one-dimensional coordinate system.

In a car, the distance traveled is indicated by the odometer. Note that distance is always positive and, because it has no direction associated with it, it is a scalar, as discussed in Chapter 1.

Another useful way to characterize a particle's motion is in terms of the **displacement,** Δx, which is simply the change in position.

Definition: Displacement, Δx

displacement = change in position = final position − initial position

$$\text{displacement} = \Delta x = x_f - x_i \qquad \text{2–1}$$

SI unit: meter, m

Notice that we use the delta notation, Δx, as a convenient shorthand for the quantity $x_f - x_i$. (See Appendix A for a complete discussion of delta notation.) Also, note that Δx can be positive (if the final position is to the right of the initial position, $x_f > x_i$), negative (if the final position is to the left of the initial position, $x_f < x_i$), or zero (if the final and initial positions are the same, $x_f = x_i$). In fact, the displacement is a one-dimensional vector, as defined in Chapter 1, and its direction (right or left) is given by its sign (positive or negative, respectively).

The SI units of displacement are meters—the same as for distance—but displacement and distance are really quite different. For example, in the round trip from your house to the grocery store and back the distance traveled is 8.6 mi, whereas the displacement is zero because $x_f = 2.1\text{ mi} = x_i$. Suppose, instead, that you go from your house to the grocery store and then to your friend's house. On this trip the distance is 10.7 mi, but the displacement is

$$\Delta x = x_f - x_i = (0) - (2.1\text{ mi}) = -2.1\text{ mi}$$

As mentioned in the previous paragraph, the minus sign means your displacement is in the negative direction, that is, to the left.

ACTIVE EXAMPLE 2–1 Find the Distance and Displacement

Calculate **(a)** the distance and **(b)** the displacement for a trip from your friend's house to the grocery store and then to your house.

Solution *(Test your understanding by performing the calculations indicated in each step.)*

Part (a)

1. Add the distances for the various parts of the total trip:	$2.1\text{ mi} + 4.3\text{ mi} + 4.3\text{ mi} = 10.7\text{ mi}$

Part (b)

2. Determine the initial position for the trip, using Figure 2–2:	$x_i = 0$
3. Determine the final position for the trip, using Figure 2–2:	$x_f = 2.1\text{ mi}$
4. Subtract x_i from x_f to find the displacement	$\Delta x = 2.1\text{ mi}$

Your Turn

Suppose we choose the origin in Figure 2–2 to be at your house, rather than at your friend's house. In this case, find **(a)** the distance and **(b)** the displacement for the trip from your friend's house to the grocery store and then to your house.

(Answers to **Your Turn** *problems are given in the back of the book.)*

2–2 Average Speed and Velocity

The next step in describing motion is to consider how rapidly an object moves. For example, how long does it take for a Randy Johnson fastball to reach home plate? How far does an orbiting space shuttle travel in one hour? How fast do your eyelids move when you blink? These are examples of some of the most basic questions regarding motion, and in this section we learn how to answer them.

The simplest way to characterize the rate of motion is with the **average speed:**

$$\text{average speed} = \frac{\text{distance}}{\text{elapsed time}} \qquad \text{2–2}$$

The dimensions of average speed are distance per time, or in SI units—meters per second, m/s. Both distance and elapsed time are positive; thus average speed is always positive.

EXAMPLE 2–1 The Kingfisher Takes a Plunge

A kingfisher is a bird that catches fish by plunging into water from a height of several meters. If a kingfisher dives from a height of 7.0 m with an average speed of 4.00 m/s, how long does it take for it to reach the water?

Picture the Problem

As shown in the sketch, the kingfisher moves in a straight line through a vertical distance of 7.0 m. The average speed of the bird is 4.00 m/s.

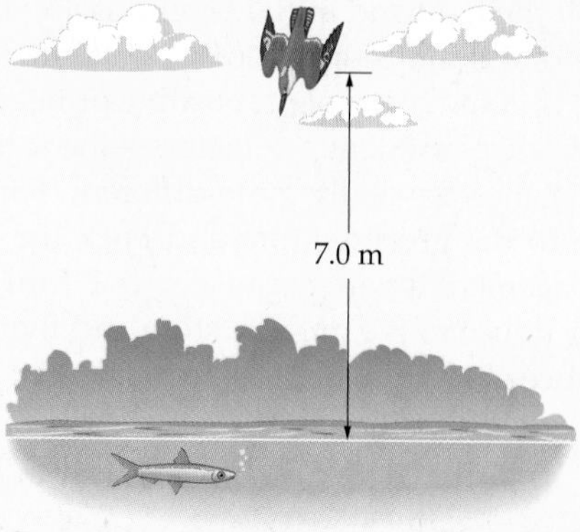

Strategy

By rearranging Equation 2–2 we can solve for the elapsed time.

Solution

1. Rearrange Equation 2–2 to solve for elapsed time:
$$\text{elapsed time} = \frac{\text{distance}}{\text{average speed}}$$

2. Substitute numerical values to find the time:
$$\text{elapsed time} = \frac{7.0\ \cancel{\text{m}}}{4.00\ \cancel{\text{m}}/\text{s}} = \frac{7.0}{4.00}\text{s} = 1.8\ \text{s}$$

Insight

Note that Equation 2–2 is not just a formula for calculating the average speed. It relates speed, time, and distance. Given any two of these quantities, Equation 2–2 can be used to find the third.

Practice Problem

A kingfisher dives with an average speed of 4.6 m/s for 1.4 s. What was the height of the dive? [**Answer:** distance = (average speed)(elapsed time) = (4.6 m/s)(1.4 s) = 6.4 m]

Some related homework problems: Problem 11, Problem 13

Next, we calculate the average speed for a trip consisting of two parts of equal length, each traveled with a different speed.

CONCEPTUAL CHECKPOINT 2–1 Average Speed

You drive 4.00 mi at 30.0 mi/h and then another 4.00 mi at 50.0 mi/h. Is your average speed for the 8.00-mi trip **(a)** greater than 40.0 mi/h, **(b)** equal to 40.0 mi/h, or **(c)** less than 40.0 mi/h?

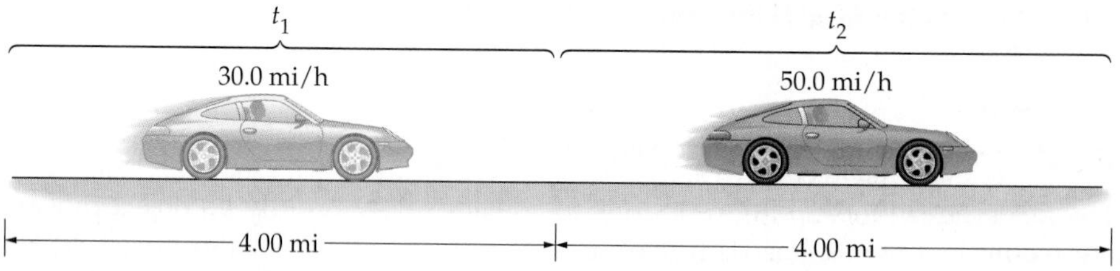

Reasoning and Discussion

At first glance it might seem that the average speed is definitely 40.0 mi/h. On further reflection, however, it is clear that it takes more time to travel 4.00 mi at 30.0 mi/h than it does to travel 4.00 mi at 50.0 mi/h. Therefore, you will be traveling at the lower speed for a greater period of time, and hence your average speed will be *less* than 40.0 mi/h—that is, closer to 30.0 mi/h than to 50.0 mi/h.

Answer

(c) The average speed is less than 40.0 mi/h.

To confirm the conclusion of the Conceptual Checkpoint, we simply apply the definition of average speed to find its value for this trip. We already know that the distance traveled is 8.00 mi; what we need now is the elapsed time. On the first 4.00 mi the time is

$$t_1 = \frac{4.00 \text{ mi}}{30.0 \text{ mi/h}} = (4.00/30.0) \text{ h}$$

The time required to cover the second 4.00 mi is

$$t_2 = \frac{4.00 \text{ mi}}{50.0 \text{ mi/h}} = (4.00/50.0) \text{ h}$$

Therefore, the elapsed time for the entire trip is

$$t_1 + t_2 = (4.00/30.0) \text{ h} + (4.00/50.0) \text{ h} = 0.213 \text{ h}$$

This gives the following average speed:

$$\text{average speed} = \frac{8.00 \text{ mi}}{0.213 \text{ h}} = 37.6 \text{ mi/h} < 40.0 \text{ mi/h}$$

Note that a "guess" will never give a detailed result like 37.6 mi/h; a systematic, step-by-step calculation is required.

In many situations, there is a quantity that is even more useful than the average speed. It is the **average velocity,** v_{av}, and it is defined as displacement per time:

Definition: Average velocity, v_{av}

$$\text{average velocity} = \frac{\text{displacement}}{\text{elapsed time}}$$

$$v_{av} = \frac{\Delta x}{\Delta t} = \frac{x_f - x_i}{t_f - t_i} \qquad \text{2–3}$$

SI unit: meter per second, m/s

Not only does the average velocity tell us, on average, how fast something is moving, it also tells us the *direction* the object is moving. For example, if an object moves in the positive direction, then $x_f > x_i$, and $v_{av} > 0$. On the other hand, if an object moves in the negative direction, it follows that $x_f < x_i$, and $v_{av} < 0$. As with displacement, the average velocity is a one-dimensional vector, and its direction is given by its sign. Average velocity gives more information than average speed; hence it is used more frequently in physics.

In the next Example, pay close attention to the positive and negative signs.

EXAMPLE 2–2 Sprint Training

An athlete sprints 50.0 m in 8.00 s, stops, and then walks slowly back to the starting line in 40.0 s. If the "sprint direction" is taken to be positive, what is **(a)** the average sprint velocity, **(b)** the average walking velocity, and **(c)** the average velocity for the complete round trip?

Picture the Problem

In our sketch we set up a coordinate system with the sprint going in the positive x direction, as described in the problem. For convenience, we choose the origin to be at the starting line. The finish line, then, is at $x = 50.0$ m.

Strategy

In each part of the problem we are asked for the average velocity and we are given information for times and distances. All that is needed, then, is to determine $\Delta x = x_f - x_i$ and $\Delta t = t_f - t_i$ in each case and apply Equation 2–3.

continued on next page

continued from previous page

Solution

Part (a)

1. Apply Equation 2–3 to the sprint, with $x_f = 50.0$ m, $x_i = 0$, $t_f = 8.00$ s, and $t_i = 0$:

$$v_{av} = \frac{\Delta x}{\Delta t} = \frac{x_f - x_i}{t_f - t_i} = \frac{50.0\text{ m} - 0}{8.00\text{ s} - 0} = \frac{50.0}{8.00}\text{ m/s} = 6.25\text{ m/s}$$

Part (b)

2. Apply Equation 2–3 to the walk. In this case, $x_f = 0$, $x_i = 50.0$ m, $t_f = 48.0$ s, and $t_i = 8.00$ s:

$$v_{av} = \frac{x_f - x_i}{t_f - t_i} = \frac{0 - 50.0\text{ m}}{48.0\text{ s} - 8.00\text{ s}} = -\frac{50.0}{40.0}\text{ m/s} = -1.25\text{ m/s}$$

Part (c)

3. For the round trip, $x_f = x_i = 0$; thus $\Delta x = 0$:

$$v_{av} = \frac{\Delta x}{\Delta t} = \frac{0}{48.0\text{ s}} = 0$$

Insight

Note that the sign of the velocities in parts **(a)** and **(b)** indicates the direction of motion; positive for motion to the right, negative for motion to the left. Also, notice that the average *speed* for the entire 100.0 m trip (100.0 m/48.0 s = 2.08 m/s) is nonzero, even though the average velocity vanishes.

Practice Problem

If the average velocity during the walk is −1.50 m/s, how long does it take the athlete to walk back to the starting line? [**Answer:** $\Delta t = \Delta x/v_{av} = (-50.0\text{ m})/(-1.50\text{ m/s}) = 33.3\text{ s}$]

Some related homework problems: Problem 7, Problem 15, Problem 16

Graphical Interpretation of Average Velocity

It is often useful to "visualize" a particle's motion by sketching its position as a function of time. For example, consider a particle moving back and forth along the x axis, as shown in **Figure 2–3**. In this plot, we have indicated the position of a particle at a variety of times.

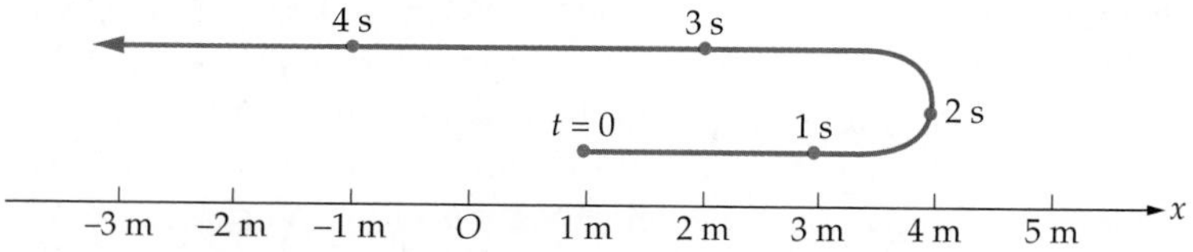

▲ **FIGURE 2–3 One-dimensional motion along the x axis**
The particle moves to the right for $0 \le t \le 2$ s, and to the left for $t > 2$ s. When the particle turns around at $t = 2$ s, we draw its path slightly above the path drawn for $t = 0$ to $t = 2$ s. This is simply for clarity—the particle is actually on the x axis at all times.

This way of keeping track of a particle's position and the corresponding time is a bit messy, though, so let's replot the same information with a different type of graph. In **Figure 2–4** we again plot the motion shown in Figure 2–3, but this time with the vertical axis representing the position, x, and the horizontal axis representing time, t. An **x-versus-t graph** like this makes it considerably easier to visualize a particle's motion.

An x-versus-t plot also leads to a particularly useful interpretation of average velocity. To see how, suppose you would like to know the average velocity of the particle in Figures 2–3 and 2–4 from $t = 0$ to $t = 3$ s. From our definition of average velocity in Equation 2–3, we know that $v_{av} = \Delta x/\Delta t = (2\text{ m} - 1\text{ m})/(3\text{ s} - 0) = +0.3\text{ m/s}$. To relate this to the x-versus-t plot, draw a straight line connecting the position at $t = 0$ (call this point A) and the position at $t = 3$ s (point B). The result is shown in **Figure 2–5 (a)**.

The slope of the straight line from A to B is equal to the rise over the run, which in this case is $\Delta x/\Delta t$. But $\Delta x/\Delta t$ is the average velocity. Thus we see that:

- The slope of a line connecting two points on an x-versus-t plot is equal to the average velocity during that time interval.

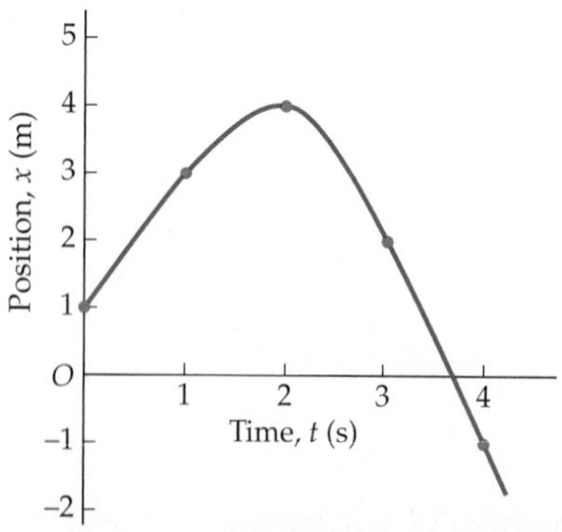

▲ **FIGURE 2–4 Motion along the x axis represented with an x-versus-t graph**
The motion shown here is the same as that indicated in Figure 2–3.

As an additional example, let's calculate the average velocity between times $t = 2$ s and $t = 3$ s in Figure 2–4. A line connecting the corresponding points is shown in **Figure 2–5 (b)**.

The first thing we notice about this line is that it has a negative slope; thus $v_{av} < 0$ and the particle is moving to the left. We also note that it is inclined more steeply than the line in Figure 2–5(a), hence the magnitude of its slope is greater. In fact, if we calculate the slope of this line we find that $v_{av} = -2$ m/s for this time interval.

Thus, connecting points on an x-versus-t plot gives an immediate "feeling" for the average velocity over a given time interval. This type of graphical analysis will be particularly useful in the next section.

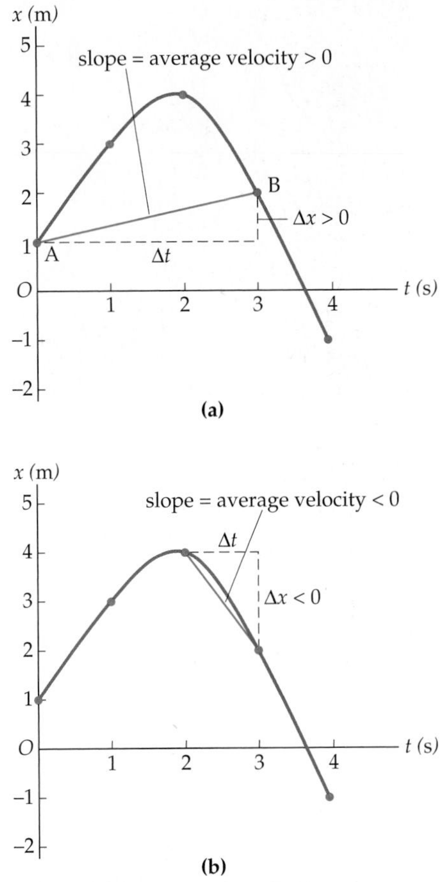

▲ FIGURE 2–5 Average velocity on an *x*-versus-*t* graph

(a) The average velocity between $t = 0$ (A) and $t = 3$ s (B) is equal to the slope of the straight line connecting A and B. **(b)** The slope of this straight line is the average velocity between $t = 2$ s and $t = 3$ s. Note that the average velocity is negative, indicating motion to the left.

2–3 Instantaneous Velocity

Though average velocity is a useful way to characterize motion, it can miss a lot. For example, suppose you travel by car on a long, straight highway, covering 92 mi in 2.0 hours. Your average velocity is 46 mi/h. Even so, there may have been only a few times during the trip when you were actually driving at 46 mi/h. You may have sped along at 65 mi/h during most of the time, except when you stopped to have a bite to eat at a roadside diner, during which time your average velocity was zero.

To have a more accurate representation of your trip, you should average your velocity over shorter periods of time. If you calculate your average velocity every 15 minutes, you have a better picture of what the trip was like. An even better, more realistic picture of the trip is obtained if you calculate the average velocity every minute or every second. Ideally, when dealing with the motion of any particle, it is desirable to know the velocity of the particle at each instant of time.

This idea of a velocity corresponding to an instant of time is just what is meant by the **instantaneous velocity.** Mathematically, we define the instantaneous velocity as follows:

Definition: Instantaneous Velocity, *v*

$$v = \lim_{\Delta t \to 0} \frac{\Delta x}{\Delta t} \qquad \text{2–4}$$

SI unit: meter per second, m/s

In this expression the notation $\lim_{\Delta t \to 0}$ means "evaluate the average velocity, $\Delta x/\Delta t$, over shorter and shorter time intervals, approaching zero in the limit." Note that the instantaneous velocity can be positive, negative, or zero, just like the average velocity—and just like the average velocity, the instantaneous velocity is a one-dimensional vector. The magnitude of the instantaneous velocity is called the **instantaneous speed.** In a car, the speedometer gives a reading of the vehicle's instantaneous speed.

As Δt becomes smaller, Δx becomes smaller as well, but the ratio $\Delta x/\Delta t$ approaches a constant value. To see how this works, consider first the simple case of a particle moving with a constant velocity of $+1$ m/s. If the particle starts at $x = 0$ at $t = 0$, then its position at $t = 1$ s is $x = 1$ m, its position at $t = 2$ s is $x = 2$ m, and so on. Plotting this motion in an x-versus-t plot gives a straight line, as shown in **Figure 2–6**.

Now, suppose we want to find the instantaneous velocity at $t = 3$ s. To do so, we calculate the average velocity over small intervals of time centered at 3 s, and let the time intervals become arbitrarily small, as shown in the Figure. Since x-versus-t is a straight line, it is clear that $\Delta x/\Delta t = \Delta x_1/\Delta t_1$, no matter how small the time interval Δt. As Δt becomes smaller, so does Δx, but the ratio $\Delta x/\Delta t$ is simply the slope of the line, 1 m/s. Thus, the instantaneous velocity at $t = 3$ s is 1 m/s.

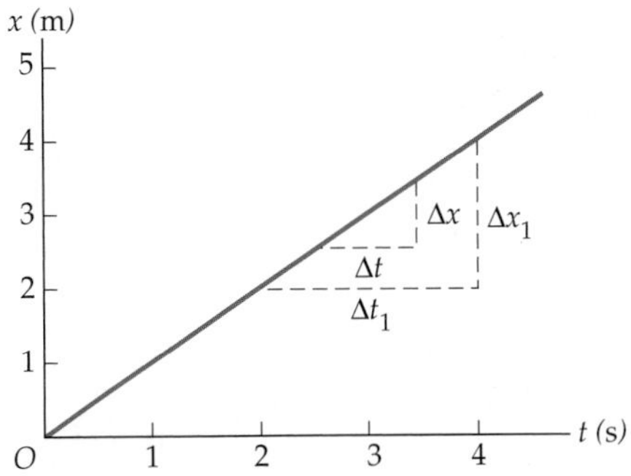

▲ FIGURE 2–6 Constant velocity on an *x*-versus-*t* graph

Motion with a constant velocity of $+1$ m/s. The motion begins at $x = 0$ at time $t = 0$. The slope $\Delta x_1/\Delta t_1$ is equal to $(2\text{ m})/(2\text{ s}) = 1$ m/s. Because x-versus-t is a straight line, the slope $\Delta x/\Delta t$ is also equal to 1 m/s for any value of Δt.

▲ A speedometer indicates the instantaneous speed of a car. Note that the speedometer gives no information about the *direction* of motion. Thus, the speedometer is truly a "speed meter," not a velocity meter.

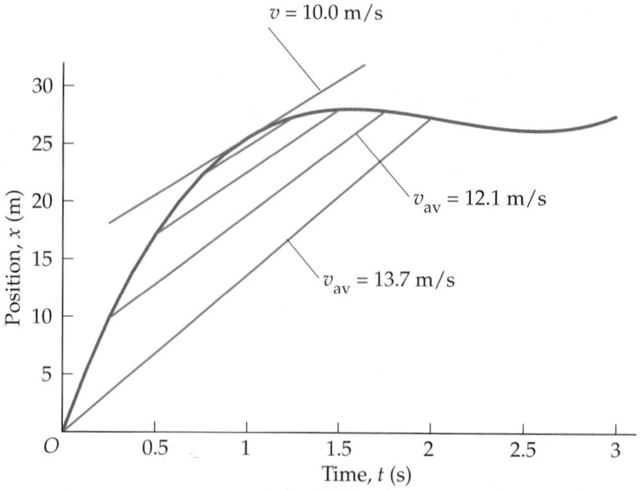

▲ FIGURE 2–7 Instantaneous velocity
An x-versus-t plot for motion with variable velocity. The instantaneous velocity at $t = 1$ s is equal to the slope of the tangent line at that time. The average velocity for a small time interval centered on $t = 1$ s approaches the instantaneous velocity at $t = 1$ s as the time interval goes to zero.

Of course, in this example the instantaneous velocity is 1 m/s for any instant of time, not just $t = 3$ s. Therefore:

- When velocity is constant, the average velocity over any time interval is equal to the instantaneous velocity at any time.

TABLE 2–1
x-versus-*t* Values for Figure 2–7

t (s)	x (m)
0	0
0.25	9.85
0.50	17.2
0.75	22.3
1.00	25.6
1.25	27.4
1.50	28.1
1.75	28.0
2.00	27.4

In general, a particle's velocity varies with time, and the x-versus-t plot is not a straight line. An example is shown in **Figure 2–7**, with the corresponding numerical values of x and t given in Table 2–1.

In this case, what is the instantaneous velocity at, say, $t = 1.00$ s? As a first approximation, let's calculate the average velocity for the time interval from $t_i = 0$ to $t_f = 2.00$ s. Note that this time interval is centered at $t = 1.00$ s. From Table 2–1 we see that $x_i = 0$, and $x_f = 27.4$ m, thus $v_{av} = 13.7$ m/s. The corresponding straight line connecting these two points is the lowest straight line in Figure 2–7.

The next three lines, in upward progression, refer to time intervals from 0.250 s to 1.75 s, 0.500 s to 1.50 s, and 0.750 s to 1.25 s, respectively. The corresponding average velocities, given in Table 2–2, are 12.1 m/s, 10.9 m/s, and 10.2 m/s. Table 2–2 also gives results for even smaller time intervals. In particular, for the interval from 0.900 s to 1.10 s the average velocity is 10.0 m/s. Smaller intervals also give 10.0 m/s. Thus, we can conclude that the instantaneous velocity at $t = 1.00$ s is $v = 10.0$ m/s.

The uppermost straight line in Figure 2–7 is the tangent line to the x-versus-t curve at the time $t = 1.00$ s; that is, it is the line that touches the curve at just a single point. Its slope is 10.0 m/s. Clearly, the average-velocity lines have slopes

TABLE 2–2 Calculating the Instantaneous Velocity at $t = 1$ s.

t_i(s)	t_f(s)	Δt(s)	x_i(m)	x_f(m)	Δx(m)	$v_{av} = \Delta x/\Delta t$(m/s)
0	2.00	2.00	0	27.4	27.4	13.7
0.250	1.75	1.50	9.85	28.0	18.2	12.1
0.500	1.50	1.00	17.2	28.1	10.9	10.9
0.750	1.25	0.50	22.3	27.4	5.10	10.2
0.900	1.10	0.20	24.5	26.5	2.00	10.0
0.950	1.05	0.10	25.1	26.1	1.00	10.0

that approach the slope of the tangent line as the time intervals become smaller. This is an example of the following general result:

- The instantaneous velocity at a given time is equal to the slope of the tangent line at that point on an x-versus-t graph.

Thus, a visual inspection of an x-versus-t graph gives information not only about the location of a particle, but also about its velocity.

CONCEPTUAL CHECKPOINT 2–2 Instantaneous Velocity

Referring to Figure 2–7, is the instantaneous velocity at $t = 0.500$ s **(a)** greater than, **(b)** less than, or **(c)** the same as the instantaneous velocity at $t = 1.00$ s?

Reasoning and Discussion

From the x-versus-t graph in Figure 2–7 it is clear that the slope of a tangent line drawn at $t = 0.500$ s is greater than the slope of the tangent line at $t = 1.00$ s. It follows that the particle's velocity at 0.500 s is greater than its velocity at 1.00 s.

Answer

(a) The instantaneous velocity is greater at $t = 0.500$ s.

In the remainder of the book, when we say velocity it is to be understood that we mean *instantaneous* velocity. If we want to refer to the average velocity, we will specifically say average velocity.

Graphical Interpretation of Average and Instantaneous Velocity

Let's summarize the graphical interpretations of average and instantaneous velocity on an x-versus-t graph:

- Average velocity is the slope of the straight line connecting two points corresponding to a given time interval.
- Instantaneous velocity is the slope of the tangent line at a given instant of time.

These relations are illustrated in **Figure 2–8**.

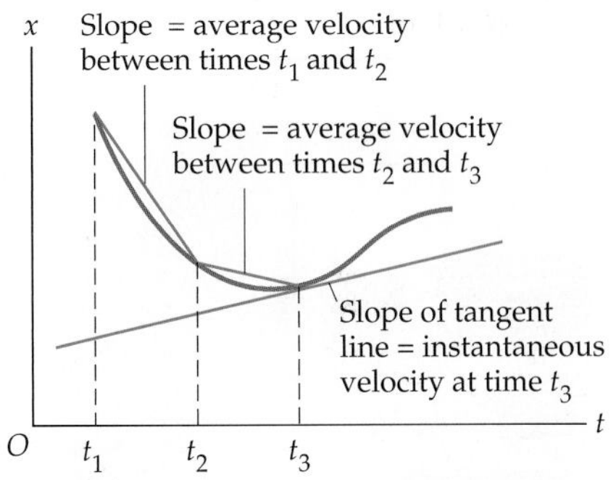

▲ **FIGURE 2–8 Graphical interpretation of average and instantaneous velocity**
Average velocities correspond to the slope of straight-line segments connecting different points on an x-versus-t graph. Instantaneous velocities are given by the slope of the tangent line at a given time.

2–4 Acceleration

Just as velocity is the rate of change of *displacement* with time, **acceleration** is the rate of change of *velocity* with time. Thus, an object accelerates whenever its velocity *changes,* no matter what the change—it accelerates when its velocity increases, it accelerates when its velocity decreases. Of all the concepts discussed in this chapter, perhaps none is more central to physics than acceleration. Galileo, for example, showed that falling bodies move with constant acceleration. Newton showed that acceleration and force are directly related, as we shall see in Chapter 5. Thus, it is particularly important to have a clear, complete understanding of acceleration before leaving this chapter.

We begin, then, with the definition of **average acceleration:**

Definition: Average Acceleration, a_{av}

$$a_{av} = \frac{\Delta v}{\Delta t} = \frac{v_f - v_i}{t_f - t_i} \qquad \text{2–5}$$

SI unit: meter per second per second, m/s^2

Note that the dimensions of average acceleration are the dimensions of velocity per time, or (meters per second) per second:

$$\frac{\text{meters per second}}{\text{second}} = \frac{m/s}{s} = \frac{m}{s^2}$$

This is generally expressed as meters per second squared. For example, the acceleration of gravity on the Earth's surface is approximately 9.81 m/s^2, which means

TABLE 2–3 Typical Accelerations (m/s^2)	
Ultracentrifuge	3×10^6
Bullet fired from a rifle	4.4×10^5
Batted baseball	3×10^4
Click beetle righting itself	400
Acceleration required to deploy airbags	60
Bungee jump	30
High jump	15
Acceleration of gravity on Earth	9.81
Emergency stop in a car	8
Airplane during takeoff	5
An elevator	3
Acceleration of gravity on the Moon	1.62

▲ The space shuttle *Discovery* accelerates upward on the initial phase of its journey into orbit. During this time the astronauts on board the shuttle experience an approximately linear acceleration that may be as great as 20 m/s^2.

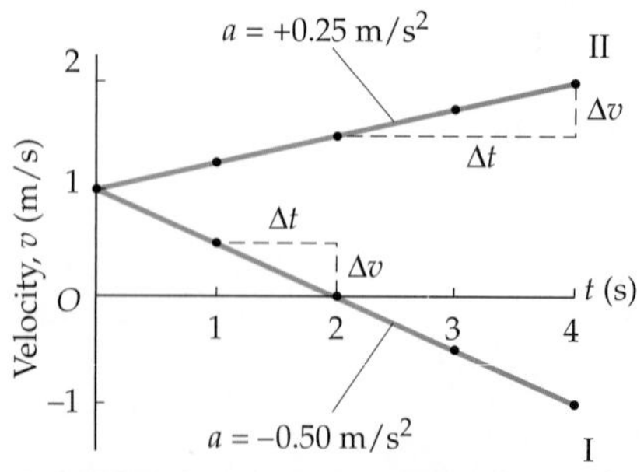

▲ **FIGURE 2–9 *v*-versus-*t* plots for motion with constant acceleration**
Curve I represents the movement of a particle with constant acceleration $a = -0.50\ m/s^2$. Curve II represents the motion of a particle with constant acceleration $a = +0.25\ m/s^2$.

that the velocity of a falling object changes by 9.81 meters per second (m/s) every second (s). In addition, we see that the average acceleration can be positive, negative, or zero. In fact, it is a one-dimensional vector, just like displacement, average velocity, and instantaneous velocity. Typical magnitudes of acceleration are given in Table 2–3.

EXERCISE 2–1

a. Saab advertises a car that goes from 0 to 60.0 mi/h in 6.2 s. What is the average acceleration of this car?

b. An airplane has an average acceleration of 5.6 m/s^2 during takeoff. How long does it take for the plane to reach a speed of 150 mi/h?

Solution

a. average acceleration $= a_{av} = (60.0\ mi/h)/(6.2\ s)$

$= (26.8\ m/s)/(6.2\ s) = 4.3\ m/s^2$

b. $\Delta t = \Delta v/a_{av} = (150\ mi/h)/(5.6\ m/s^2) = (67.0\ m/s)/(5.6\ m/s^2) = 12\ s$

Next, just as we considered the limit of smaller and smaller time intervals to find an instantaneous velocity, we can do the same to define an **instantaneous acceleration:**

Definition: Instantaneous Acceleration, *a*

$$a = \lim_{\Delta t \to 0} \frac{\Delta v}{\Delta t} \qquad \text{2–6}$$

SI unit: meter per second per second, m/s^2

As you might expect, the instantaneous acceleration is a one-dimensional vector, just like the average acceleration, and its direction is given by its sign. For simplicity, when we say acceleration in the future we are referring to the instantaneous acceleration.

One final note before we go on to some examples. If the acceleration is constant, it has the same value at all times. Therefore:

- When acceleration is constant, the instantaneous and average accelerations are the same.

We shall make use of this fact when we return to the special case of constant acceleration in the next section.

Graphical Interpretation of Acceleration

To see how acceleration can be interpreted graphically, suppose that a particle has a constant acceleration of $-0.50\ m/s^2$. This means that the velocity of the particle *decreases* by 0.50 m/s each second. Thus, if its velocity is 1.0 m/s at $t = 0$, then at $t = 1$ s its velocity is 0.50 m/s, at $t = 2$ s its velocity is 0, at $t = 3$ s its velocity is -0.50 m/s, and so on. This is illustrated by curve I in **Figure 2–9**, where we see that a plot of v-versus-t results in a straight line with a negative slope. Curve II in Figure 2–9 has a positive slope, corresponding to a constant acceleration of $+0.25\ m/s^2$. Thus, in terms of a v-versus-t plot, a constant acceleration results in a straight line with a slope equal to the acceleration.

CONCEPTUAL CHECKPOINT 2–3 Speed as a Function of Time

The speed of a particle with the *v*-versus-*t* graph shown by curve II in Figure 2–9 increases steadily with time. Consider, instead, a particle whose *v*-versus-*t* graph is given by curve I in Figure 2–9. As a function of time, does the speed of this particle **(a)** increase, **(b)** decrease, or **(c)** decrease and then increase?

Reasoning and Discussion

Recall that speed is the *magnitude* of velocity. In curve I of Figure 2–9 the speed starts out at 1.0 m/s, then *decreases* to 0 at $t = 2$ s. After $t = 2$ s the speed *increases* again. For example, at $t = 3$ s the speed is 0.50 m/s, and at $t = 4$ s the speed is 1 m/s.

Did you realize that the particle represented by curve I in Figure 2–9 changes direction at $t = 2$ s? It certainly does. Before $t = 2$ s the particle moves in the positive direction; after $t = 2$ s it moves in the negative direction. At precisely $t = 2$ s the particle is momentarily at rest. However, regardless of whether the particle is moving in the positive direction, moving in the negative direction, or instantaneously at rest, it still has the same constant acceleration. Acceleration has to do only with the way the velocity is *changing* at a given moment.

Answer:

(c) The speed decreases and then increases.

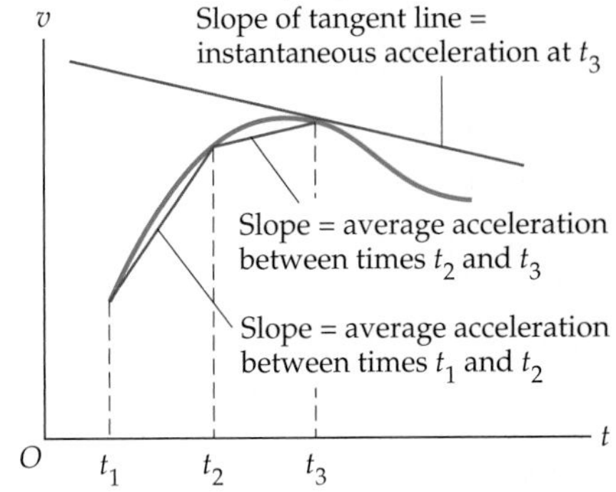

▲ **FIGURE 2–10 Graphical interpretation of average and instantaneous acceleration**

Average accelerations correspond to the slope of straight-line segments connecting different points on an v-versus-t graph. Instantaneous accelerations are given by the slope of the tangent line at a given time.

The graphical interpretations for velocity presented in Figure 2–8 apply equally well to acceleration, with just one small change: Instead of an x-versus-t graph, we use a v-versus-t graph, as in **Figure 2–10**. Thus, the average acceleration in a v-versus-t plot is the slope of a straight line connecting points corresponding to two different times. Similarly, the instantaneous acceleration is the slope of the tangent line at a particular time.

EXAMPLE 2–3 An Accelerating Train

A train moving in a straight line with an initial velocity of 0.50 m/s accelerates at 2.0 m/s² for 2.0 seconds, coasts with zero acceleration for 3.0 seconds, and then accelerates at −1.5 m/s² for 1.0 second. **(a)** What is the final velocity of the train? **(b)** What is the average acceleration of the train?

Picture the Problem

We begin by sketching a v-versus-t plot for the train. The basic idea is that each interval of constant acceleration is represented by a straight line of the appropriate slope. Therefore, we draw a straight line with the slope 2.0 m/s² from $t = 0$ to $t = 2.0$ s, a line with zero slope from $t = 2.0$ s to $t = 5.0$ s, and a line with the slope −1.5 m/s² from $t = 5.0$ s to $t = 6.0$ s. The line connecting the initial and final points determines the average acceleration.

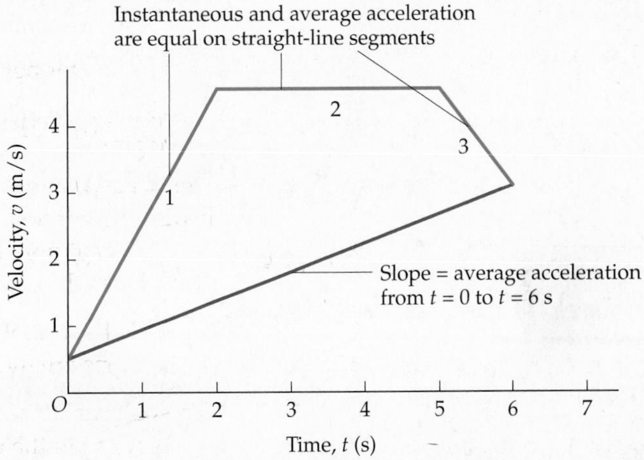

Strategy

During each period of constant acceleration the change in velocity is $\Delta v = a_{av}\Delta t = a\Delta t$.

a. Adding the individual changes in velocity gives the total change, $\Delta v = v_f - v_i$. Since v_i is known, this expression can be solved for the final velocity, v_f.

b. The average acceleration can be calculated using Equation 2–5, $a_{av} = \Delta v/\Delta t$. Note that Δv has been obtained in part (a), and that the total time interval is $\Delta t = 6.0$ s, as is clear from the graph.

Solution

Part (a)

1. Find the change in velocity during each of the three periods of constant acceleration:

$$\Delta v_1 = a_1\Delta t_1 = (2.0\ \text{m/s}^2)(2.0\ \text{s}) = 4.0\ \text{m/s}$$
$$\Delta v_2 = a_2\Delta t_2 = (0)(3.0\ \text{s}) = 0$$
$$\Delta v_3 = a_3\Delta t_3 = (-1.5\ \text{m/s}^2)(1.0\ \text{s}) = -1.5\ \text{m/s}$$

2. Sum the change in velocity for each period to obtain the total Δv:

$$\Delta v = \Delta v_1 + \Delta v_2 + \Delta v_3$$
$$= 4.0\ \text{m/s} + 0 - 1.5\ \text{m/s} = 2.5\ \text{m/s}$$

3. Use Δv to find v_f, recalling that $v_i = 0.50$ m/s:

$$\Delta v = v_f - v_i$$
$$v_f = \Delta v + v_i = 2.5\ \text{m/s} + 0.50\ \text{m/s} = 3.0\ \text{m/s}$$

Part (b)

4. The average acceleration is $\Delta v/\Delta t$:

$$a_{av} = \frac{\Delta v}{\Delta t} = \frac{2.5\ \text{m/s}}{6.0\ \text{s}} = 0.42\ \text{m/s}^2$$

continued on next page

continued from previous page

Insight

Note that the average acceleration for these six seconds is not simply the average of the individual accelerations; $2.0\ \mathrm{m/s^2}$, $0\ \mathrm{m/s^2}$, and $-1.5\ \mathrm{m/s^2}$. The reason is that different amounts of time are spent with each acceleration. In addition, the average acceleration can be found graphically, as indicated in the v-versus-t sketch on the previous page. Specifically, the graph shows that Δv is 2.5 m/s for the time interval from $t = 0$ to $t = 6.0$ s.

Practice Problem

What is the average acceleration of the train between $t = 2.0$ s and $t = 6.0$ s? [**Answer:** $a_{av} = \Delta v/\Delta t = (3.0\ \mathrm{m/s} - 4.5\ \mathrm{m/s})/(6.0\ \mathrm{s} - 2.0\ \mathrm{s}) = -0.38\ \mathrm{m/s^2}$]

Some related homework problems: Problem 32, Problem 34

In one dimension, nonzero velocities and accelerations are either positive or negative, depending on whether they point in the positive or negative direction of the coordinate system chosen. Thus, the velocity and acceleration of an object may have the same or opposite signs. (Of course, in two or three dimensions the relationship between velocity and acceleration can be much more varied, as we shall see in the next several chapters.) This leads to the following two possibilities:

- When the velocity and acceleration of an object have the same sign, the speed of the object increases. In this case, the velocity and acceleration point in the same direction.
- When the velocity and acceleration of an object have opposite signs, the speed of the object decreases. In this case, the velocity and acceleration point in opposite directions.

These two possibilities are illustrated in **Figure 2–11**. Notice that when a particle's speed increases, it means either that its velocity becomes more positive, as in **Figure 2–11 (a)**, or more negative, as in **Figure 2–11 (d)**. In either case, it is the magnitude of the velocity—the speed—that increases.

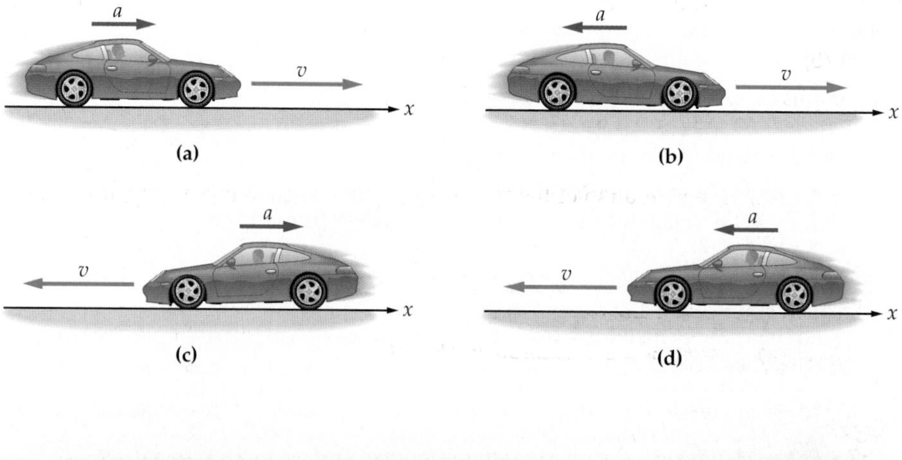

▶ **FIGURE 2–11 Cars accelerating or decelerating**

A car's speed increases when its velocity and acceleration point in the same direction, as in cases **(a)** and **(d)**. When the velocity and acceleration point in opposite directions, as in cases **(b)** and **(c)**, the car's speed decreases.

▶ The winner of this race was traveling at a speed of 313.91 mph at the end of the quarter-mile course. Since the winning time was just 4.607 s, the *average* acceleration during this race was approximately three times the acceleration of gravity (Section 2–7).

When a particle's speed decreases, it is often said to be *decelerating*. A common misconception is that deceleration implies a negative acceleration. This is not true. Deceleration can be caused by a positive or a negative acceleration, depending on the direction of the initial velocity. For example, the car in **Figure 2–11 (b)** has a positive velocity and a negative acceleration, while the car in **Figure 2–11 (c)** has a negative velocity and a positive acceleration. In both cases, the speed of the car decreases. Again, all that is required for deceleration in one dimension is that the velocity and acceleration have *opposite signs*; that is, they must point in *opposite directions*, as in parts (b) and (c) of Figure 2–11.

Velocity versus time plots for the four situations shown in Figure 2–11 are presented in **Figure 2–12**. In each of the four plots in Figure 2–12 we assume constant acceleration. Be sure to understand clearly the connection between the v-versus-t plots in Figure 2–12, and the corresponding physical motions indicated in Figure 2–11.

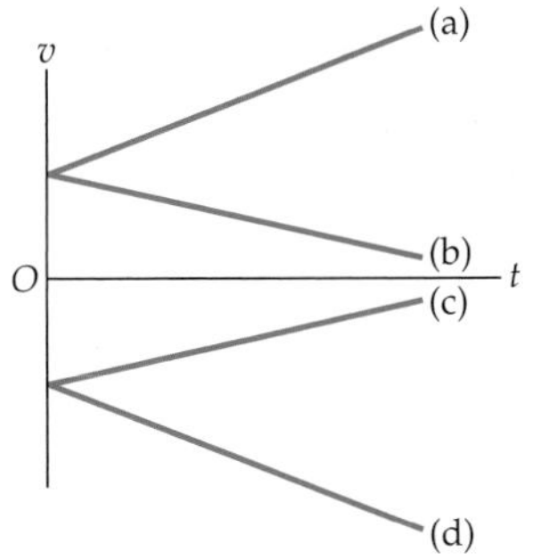

▲ **FIGURE 2–12 *v*-versus-*t* plots with constant acceleration**

Four plots of v versus t corresponding to the four situations shown in Figure 2–11. Note that the speed increases in cases **(a)** and **(d)**, but decreases in cases **(b)** and **(c)**.

EXAMPLE 2–4 The Ferry Docks

A ferry makes a short run between two docks; one in Anacortes, Washington, the other on Guemes Island. As the ferry approaches Guemes Island (traveling in the positive x direction), its speed is 7.4 m/s. **(a)** If the ferry slows to a stop in 12.3 s, what is its average acceleration? **(b)** As the ferry returns to the Anacortes dock, its speed is 7.3 m/s. If it comes to rest in 13.1 s, what is its average acceleration?

Picture the Problem

Our sketch shows the locations of the two docks and the positive direction indicated in the problem. Note that the distance between docks is not given, nor is it needed.

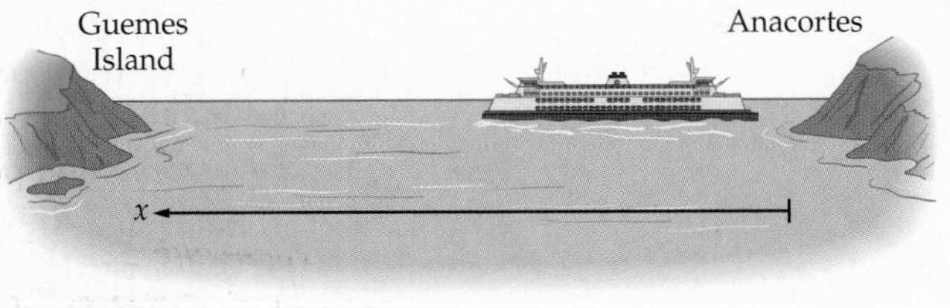

Strategy

We are given the initial and final velocities (the ferry comes to a stop in each case, so its final speed is zero) and the relevant times. Therefore, we can find the average acceleration using $a_{av} = \Delta v/\Delta t$, being careful to get the signs right.

Solution

Part (a)

1. Calculate the average acceleration, noting that $v_i = 7.4\ \text{m/s}$ and $v_f = 0$:

$$a_{av} = \frac{\Delta v}{\Delta t} = \frac{v_f - v_i}{\Delta t} = \frac{0 - 7.4\ \text{m/s}}{12.3\ \text{s}} = -0.60\ \text{m/s}^2$$

Part (b)

2. In this case, $v_i = -7.3\ \text{m/s}$ and $v_f = 0$:

$$a_{av} = \frac{\Delta v}{\Delta t} = \frac{v_f - v_i}{\Delta t} = \frac{0 - (-7.3\ \text{m/s})}{13.1\ \text{s}} = 0.56\ \text{m/s}^2$$

Insight

In each case, the acceleration of the ferry is opposite in sign to its velocity; therefore the ferry decelerates.

Practice Problem

When the ferry leaves Guemes Island, its speed increases from 0 to 5.8 m/s in 9.25 s. What is its average acceleration? [**Answer:** $a_{av} = -0.63\ \text{m/s}^2$]

Some related homework problems: Problem 30, Problem 31

2–5 Motion with Constant Acceleration

In this section, we derive equations describing the motion of particles moving with **constant acceleration.** These "equations of motion" can be used to describe a wide range of everyday phenomena. For example, in an idealized world with no air resistance, falling bodies have constant acceleration.

As mentioned in the previous section, if a particle has constant acceleration—that is, the same acceleration at every instant of time—then its instantaneous acceleration, a, is equal to its average acceleration, a_{av}. Recalling the definition of average acceleration, Equation 2–5, we have

$$a_{av} = \frac{v_f - v_i}{t_f - t_i} = a$$

where the initial and final times may be chosen arbitrarily. For example, let $t_i = 0$ for the initial time, and let $v_i = v_0$ denote the velocity at time zero. For the final time and velocity we drop the subscripts to simplify notation; thus we let $t_f = t$ and $v_f = v$. With these identifications we have

$$a_{av} = \frac{v - v_0}{t - 0} = a$$

Therefore,

$$v - v_0 = a(t - 0) = at$$

or

Constant-Acceleration Equation of Motion: Velocity as a Function of Time

$$v = v_0 + at \qquad \text{2–7}$$

Note that Equation 2–7 describes a straight line on a v-versus-t plot. The line crosses the velocity axis at the value v_0 and has a slope a, in agreement with the graphical interpretations discussed in the previous section. For example, in curve I of Figure 2–9, the equation of motion is $v = v_0 + at = (1\ \text{m/s}) + (-0.5\ \text{m/s}^2)t$. Also, note that $(-0.5\ \text{m/s}^2)t$ has the units $(\text{m/s}^2)(\text{s}) = \text{m/s}$; thus each term in Equation 2–7 has the same dimensions (as it must to be a valid physical equation).

EXERCISE 2–2

A ball is thrown straight upward with an initial velocity of +8.2 m/s. If the acceleration of the ball is $-9.81\ \text{m/s}^2$, what is its velocity after

a. 0.50 s, and **b.** 1.0 s?

Solution

a. Substituting $t = 0.50$ s in Equation 2–7 yields

$$v = 8.2\ \text{m/s} + (-9.81\ \text{m/s}^2)(0.50\ \text{s}) = 3.3\ \text{m/s}$$

b. Similarly, using $t = 1.0$ s in Equation 2–7 gives

$$v = 8.2\ \text{m/s} + (-9.81\ \text{m/s}^2)(1.0\ \text{s}) = -1.6\ \text{m/s}$$

Next, how far does a particle move in a given time if its acceleration is constant? To answer this question, recall the definition of average velocity:

$$v_{av} = \frac{\Delta x}{\Delta t} = \frac{x_f - x_i}{t_f - t_i}$$

Using the same identifications given previously for initial and final times, and letting $x_i = x_0$ and $x_f = x$, we have

$$v_{av} = \frac{x - x_0}{t - 0}$$

Thus,

$$x - x_0 = v_{av}(t - 0) = v_{av}t$$

or

$$x = x_0 + v_{av}t \qquad \text{2–8}$$

Now, Equation 2–8 is fine as it is. In fact, it applies whether the acceleration is constant or not. A more useful expression, for the case of constant acceleration, is obtained by writing v_{av} in terms of the initial and final velocities. This can be done by referring to **Figure 2–13 (a)**. Here the velocity changes linearly (since a is

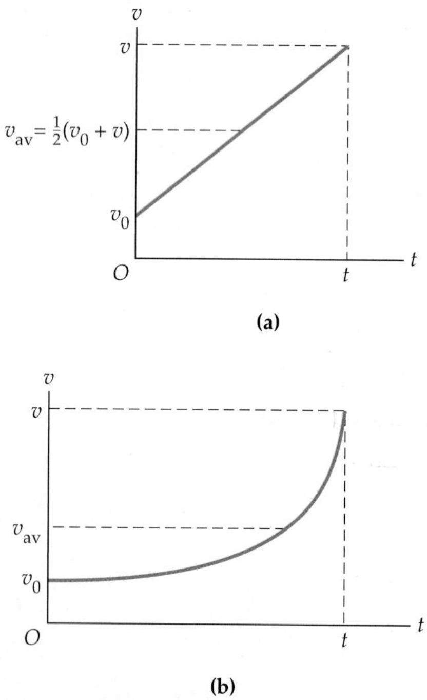

▲ **FIGURE 2–13 The average velocity** **(a)** When acceleration is constant, the velocity varies linearly with time. As a result, the average velocity, v_{av}, is simply the average of the initial velocity, v_0, and the final velocity, v. **(b)** The velocity curve for nonconstant acceleration is nonlinear. In this case, the average velocity is no longer midway between the initial and final velocities.

constant) from v_0 at $t = 0$ to v at some later time t. The average velocity during this period of time is simply the average of the initial and final velocities; that is, the sum of the two velocities divided by two:

Constant-Acceleration Equation of Motion: Average Velocity

$$v_{av} = \tfrac{1}{2}(v_0 + v) \qquad 2\text{–}9$$

The average velocity is indicated in the figure. Note that if the acceleration is not constant, as in **Figure 2–13 (b)**, this simple averaging of initial and final velocities is no longer valid.

Substituting the expression for v_{av} from Equation 2–9 into Equation 2–8 yields

Constant-Acceleration Equation of Motion: Position as a Function of Time

$$x = x_0 + \tfrac{1}{2}(v_0 + v)t \qquad 2\text{–}10$$

This equation, like Equation 2–7, is valid *only* for constant acceleration. The utility of Equations 2–7 and 2–10 is illustrated in the next Example.

PROBLEM-SOLVING NOTE

"Coordinate" the Problem

The first step in solving a physics problem is to produce a simple sketch of the system. Your sketch should include a coordinate system, along with an origin and a positive direction. Next, you should identify quantities that are given in the problem, such as initial position, initial velocity, acceleration, and so on. These preliminaries will help in producing a mathematical representation of the problem.

EXAMPLE 2–5 Full Speed Ahead

A boat moves slowly inside a marina (so as not to leave a wake) with a constant speed of 1.50 m/s. As soon as it passes the breakwater, leaving the marina, it throttles up and accelerates at 2.40 m/s^2. **(a)** How fast is the boat moving after accelerating for 5.00 s? **(b)** How far has the boat traveled in this time?

Picture the Problem

In our sketch we choose the origin to be at the breakwater, and the positive x direction to be the direction of motion. With this choice the initial position is $x_0 = 0$, and the initial velocity is $v_0 = 1.50$ m/s.

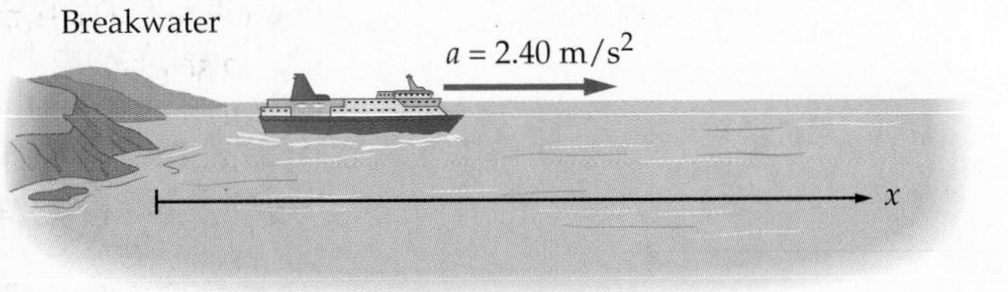

Strategy

The acceleration is constant, so we can use Equations 2–7 to 2–10. In part (a) we want to relate velocity to time, so we use Equation 2–7, $v = v_0 + at$. In part (b) our knowledge of the initial and final velocities allows us to relate position to time using Equation 2–10, $x = x_0 + \tfrac{1}{2}(v_0 + v)t$.

Solution

Part (a)

1. Use Equation 2–7 with $v_0 = 1.50$ m/s, and $a = 2.40$ m/s^2:

$$v = v_0 + at = 1.50\text{ m/s} + (2.40\text{ m/s}^2)(5.00\text{ s})$$
$$= 1.50\text{ m/s} + 12.0\text{ m/s} = 13.5\text{ m/s}$$

Part (b)

2. Apply Equation 2–10, using the result for v obtained in part (a):

$$x = x_0 + \tfrac{1}{2}(v_0 + v)t$$
$$= 0 + \tfrac{1}{2}(1.50\text{ m/s} + 13.5\text{ m/s})(5.00\text{ s})$$
$$= (7.50\text{ m/s})(5.00\text{ s}) = 37.5\text{ m}$$

Insight

Since the boat has a constant acceleration between $t = 0$ and $t = 5.00$ s, its velocity-versus-time curve is linear during this time interval. As a result, the average velocity for these 5.00 seconds is the average of the initial and final velocities, $v_{av} = \tfrac{1}{2}(1.50\text{ m/s} + 13.5\text{ m/s}) = 7.50$ m/s. Multiplying the average velocity by the time, 5.00 s, gives the distance traveled—which is exactly what Equation 2–10 does in Step 2.

Practice Problem

At what time is the boat's speed equal to 10.0 m/s? [**Answer:** $t = 3.54$ s]

Some related homework problems: Problem 42, Problem 43

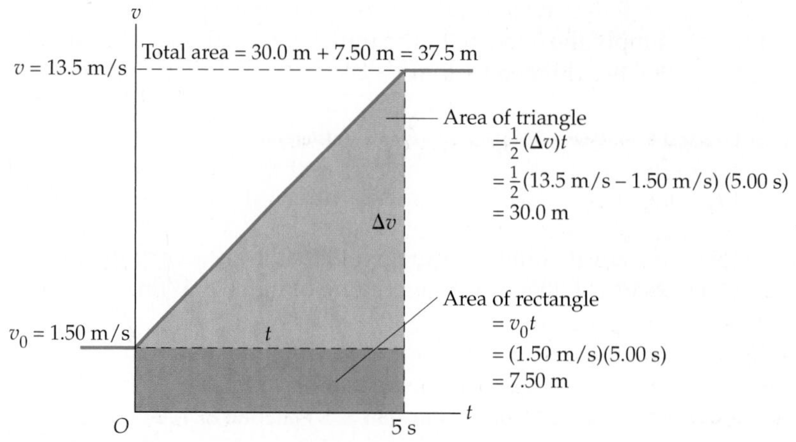

▶ **FIGURE 2–14 Velocity versus time for the boat in Example 2–5**
The distance traveled by the boat between $t = 0$ and $t = 5.00$ s is equal to the corresponding area under the velocity curve.

The velocity of the boat in Example 2–5 is plotted as a function of time in **Figure 2–14**, with the acceleration starting at time $t = 0$ and ending at $t = 5.00$ s. We will now show that the *distance* traveled by the boat from $t = 0$ to $t = 5.00$ s *is equal to the corresponding area under the velocity-versus-time curve.* This is a general result, valid for any velocity curve and any time interval:

- The distance traveled by an object from a time t_1 to a time t_2 is equal to the area under the velocity curve between those two times.

In this case, the area is the sum of the areas of a rectangle and a triangle. The rectangle has a base of 5.00 s and a height of 1.50 m/s, which gives an area of (5.00 s)(1.50 m/s) = 7.50 m. Similarly, the triangle has a base of 5.00 s and a height of (13.5 m/s − 1.50 m/s) = 12.0 m/s, for an area of $\frac{1}{2}$(5.00 s)(12.0 m/s) = 30.0 m. Clearly, the total area is 37.5 m, just as found in Example 2–5.

Staying with Example 2–5 for a moment, let's repeat the calculation of part (b), only this time for the general case. First, we use the final velocity from part (a), calculated with $v = v_0 + at$, in the expression for the average velocity, $v_{av} = \frac{1}{2}(v_0 + v)$. Symbolically, this gives the following:

$$\tfrac{1}{2}(v_0 + v) = \tfrac{1}{2}[v_0 + (v_0 + at)] = v_0 + \tfrac{1}{2}at \qquad \textit{(constant acceleration)}$$

Next, we substitute this result into Equation 2–10, which yields

$$x = x_0 + \tfrac{1}{2}(v_0 + v)t = x_0 + \left(v_0 + \tfrac{1}{2}at\right)t$$

Multiplying through by t gives the following result:

Constant-Acceleration Equation of Motion: Position as a Function of Time

$$x = x_0 + v_0t + \tfrac{1}{2}at^2 \qquad \textbf{2–11}$$

Here we have an expression for position versus time that is explicitly in terms of the acceleration, a.

Note that each term in Equation 2–11 has the same dimensions, as they must. For example, the velocity term, v_0t, has the units (m/s)(s) = m. Similarly, the acceleration term, $\frac{1}{2}at^2$, has the units $(m/s^2)(s^2) = m$.

EXERCISE 2–3

Repeat part (b) of Example 2–5 using Equation 2–11.

Solution

$$x = x_0 + v_0t + \tfrac{1}{2}at^2 = 0 + (1.50\ \text{m/s})(5.00\ \text{s}) + \tfrac{1}{2}(2.40\ \text{m/s}^2)(5.00\ \text{s})^2 = 37.5\ \text{m}$$

The next Example gives further insight into the physical meaning of Equation 2–11.

EXAMPLE 2–6 Put the Pedal to the Metal

A drag racer starts from rest and accelerates at 7.40 m/s^2. How far has it traveled in **(a)** 1.00 s, **(b)** 2.00 s, **(c)** 3.00 s?

Picture the Problem

We set up a coordinate system in which the drag racer starts at the origin and accelerates in the positive x direction. With this choice, it follows that $x_0 = 0$ and $a = +7.40\ \mathrm{m/s^2}$. Also, since the racer starts from rest, its initial velocity is zero, $v_0 = 0$. Incidentally, the positions of the racer in the sketch have been drawn to scale.

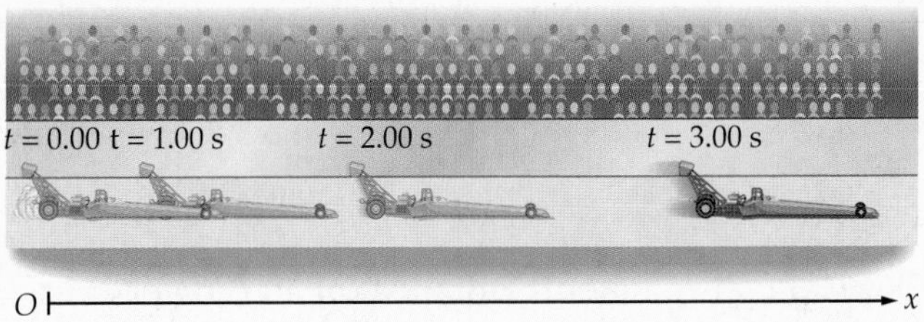

Strategy

Since this problem gives the acceleration, which is constant, and asks for a relationship between position and time, we use Equation 2–11.

Solution

Part (a)

1. Evaluate Equation 2–11 with $a = 7.40\ \mathrm{m/s^2}$ and $t = 1.00$ s:

$$x = x_0 + v_0t + \tfrac{1}{2}at^2 = 0 + 0 + \tfrac{1}{2}at^2 = \tfrac{1}{2}at^2$$

$$x = \tfrac{1}{2}(7.40\ \mathrm{m/s^2})(1.00\ \mathrm{s})^2 = 3.70\ \mathrm{m}$$

Part (b)

2. From the calculation in part (a), Equation 2–11 reduces to $x = \tfrac{1}{2}at^2$ in this situation. Evaluate $x = \tfrac{1}{2}at^2$ at $t = 2.00$ s:

$$x = \tfrac{1}{2}at^2$$

$$= \tfrac{1}{2}(7.40\ \mathrm{m/s^2})(2.00\ \mathrm{s})^2 = 14.8\ \mathrm{m} = 4(3.70\ \mathrm{m})$$

Part (c)

3. Repeat with $t = 3.00$ s:

$$x = \tfrac{1}{2}at^2$$

$$= \tfrac{1}{2}(7.40\ \mathrm{m/s^2})(3.00\ \mathrm{s})^2 = 33.3\ \mathrm{m} = 9(3.70\ \mathrm{m})$$

Insight

This Example illustrates one of the key features of accelerated motion—position does not change uniformly with time when an object accelerates. In this case, the distance traveled in the first two seconds is 4 times the distance traveled in the first second, and the distance traveled in the first three seconds is 9 times the distance traveled in the first second. This kind of behavior is a direct result of the fact that x depends on t^2 when the acceleration is nonzero.

Practice Problem

In one second the racer travels 3.70 m. How long does it take for the racer to travel 2(3.70 m) = 7.40 m? [**Answer:** $t = \sqrt{2}$ s = 1.41 s]

Some related homework problems: Problem 44, Problem 59

Figure 2–15 shows a graph of x versus t for Example 2–6. Notice the parabolic shape of the x-versus-t curve, which is due to the $\tfrac{1}{2}at^2$ term, and is characteristic of constant acceleration. In particular, if acceleration is positive ($a > 0$), then a plot of x-versus-t curves upward; if acceleration is negative ($a < 0$), a plot of x-versus-t curves downward. The greater the magnitude of a, the greater the curvature. In contrast, if a particle moves with constant velocity ($a = 0$) the t^2 dependence vanishes, and the x-versus-t plot is a straight line.

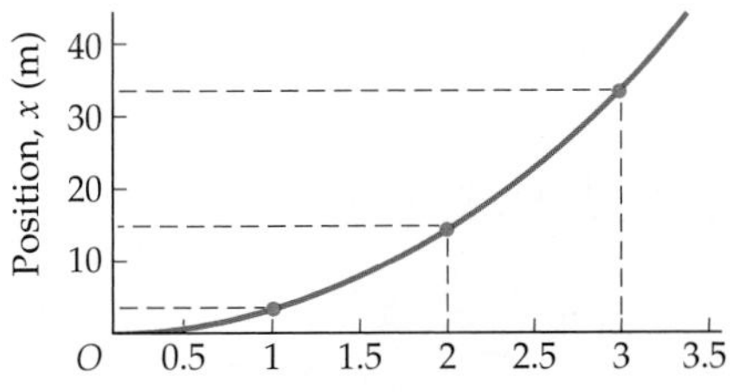

▲ **FIGURE 2–15 Position versus time for Example 2–6**
The upward-curving, parabolic shape of this x-versus-t plot indicates a positive, constant acceleration. The dots on the curve show the position of the drag racer in Example 2–6 at the times 1.00 s, 2.00 s, and 3.00 s.

Our final equation of motion with constant acceleration relates velocity to position. We start by solving for the time, t, in Equation 2–7:

$$v = v_0 + at \quad \text{or} \quad t = \frac{v - v_0}{a}$$

Next, we substitute this result into Equation 2–10, thus eliminating t:

$$x = x_0 + \tfrac{1}{2}(v_0 + v)t = x_0 + \tfrac{1}{2}(v_0 + v)\left(\frac{v - v_0}{a}\right)$$

Noting that $(v_0 + v)(v - v_0) = v_0v - v_0{}^2 + v^2 - vv_0 = v^2 - v_0{}^2$, we have

$$x = x_0 + \frac{v^2 - v_0{}^2}{2a}$$

Finally, a straightforward rearrangement of terms yields

Constant-Acceleration Equation of Motion: Velocity in Terms of Displacement

$$v^2 = v_0^2 + 2a(x - x_0) = v_0^2 + 2a\Delta x \qquad \text{2–12}$$

This equation allows us to relate the velocity at one position to the velocity at another position, without knowing how much time is involved. The next Example shows how Equation 2–12 can be used.

EXAMPLE 2–7 Takeoff Distance for an Airliner

REAL-WORLD PHYSICS

Jets at JFK International Airport accelerate from rest at one end of a runway, and must attain takeoff speed before reaching the other end of the runway. **(a)** Plane A has acceleration a and takeoff speed v_{to}. What is the minimum length of runway, Δx_A, required for this plane? Give a symbolic answer. **(b)** Plane B has the same acceleration as plane A, but requires twice the takeoff speed. Find Δx_B and compare with Δx_A. **(c)** Find the minimum runway length for plane A if $a = 2.20\text{ m/s}^2$ and $v_{to} = 95.0\text{ m/s}$. (These values are typical for a 747 jetliner.)

Picture the Problem

In our sketch, we choose the positive x direction to be the direction of motion. With this choice, it follows that the acceleration of the plane is positive, $a = +2.20\text{ m/s}^2$. Similarly, the takeoff velocity is positive as well, $v_{to} = +95.0\text{ m/s}$.

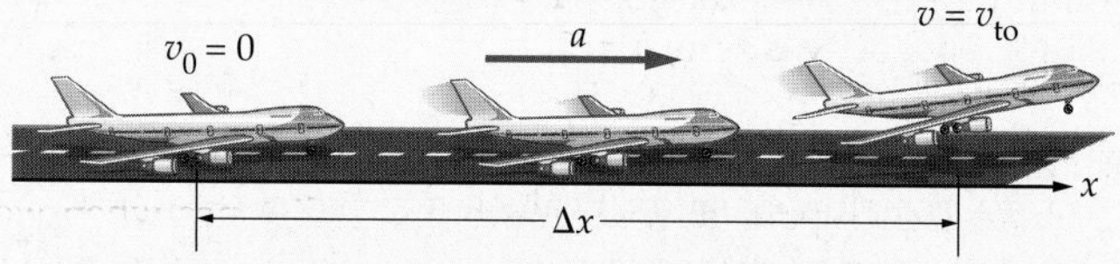

Strategy

From the sketch it is clear that we want to express Δx, the distance the plane travels in attaining takeoff speed, in terms of the acceleration, a, and the takeoff speed, v_{to}. Equation 2–12, which relates distance to velocity, allows us to do this.

Solution

Part (a)

1. Solve Equation 2–12 for Δx. To find Δx_A, set $v_0 = 0$ and $v = v_{to}$:

$$\Delta x = \frac{v^2 - v_0^2}{2a} \qquad \Delta x_A = \frac{v_{to}^2}{2a}$$

Part (b)

2. To find Δx_B, simply change v_{to} to $2v_{to}$ in part (a):

$$\Delta x_B = \frac{(2v_{to})^2}{2a} = \frac{4v_{to}^2}{2a} = 4\Delta x_A$$

Part (c)

3. Substitute numerical values into the result found in part (a):

$$\Delta x_A = \frac{v_{to}^2}{2a} = \frac{(95.0\text{ m/s})^2}{2(2.20\text{ m/s}^2)} = 2050\text{ m}$$

Insight

For purposes of comparison, the shortest runway at JFK International Airport is 04R/22L, which has a length of 2560 m.

This Example illustrates the fact that there are many advantages to obtaining symbolic results before substituting numerical values. In this case, we find that the take-off distance is proportional to v^2; hence, we conclude immediately that doubling v results in a fourfold increase of Δx.

Practice Problem

Find the minimum acceleration needed for a takeoff speed of $v_{to} = (95.0\text{ m/s})/2 = 47.5\text{ m/s}$ on a runway of length $\Delta x = (2050\text{ m})/4 = 513\text{ m}$. [**Answer:** $a = v_{to}^2/2\Delta x = 2.20\text{ m/s}^2$]

Some related homework problems: Problem 50, Problem 52

Finally, all of our constant-acceleration equations of motion are collected for easy reference in Table 2–4.

TABLE 2–4 Constant-Acceleration Equations of Motion

Variables related	Equation	Number
velocity, time, acceleration	$v = v_0 + at$	2–7
initial, final, and average velocity	$v_{av} = \frac{1}{2}(v_0 + v)$	2–9
position, time, velocity	$x = x_0 + \frac{1}{2}(v_0 + v)t$	2–10
position, time, acceleration	$x = x_0 + v_0 t + \frac{1}{2}at^2$	2–11
velocity, position, acceleration	$v^2 = v_0^2 + 2a(x - x_0) = v_0^2 + 2a\Delta x$	2–12

2–6 Applications of the Equations of Motion

We devote this section to a variety of examples further illustrating the use of the constant-acceleration equations of motion. In our first Example, we consider the distance and time needed to brake a vehicle to a complete stop.

EXAMPLE 2–8 Hit the Brakes!

A park ranger driving on a back country road suddenly sees a deer "frozen" in the headlights. The ranger, who is driving at 11.4 m/s, immediately applies the brakes and slows with an acceleration of 3.80 m/s^2. **(a)** If the deer is 20.0 m from the ranger's vehicle when the brakes are applied, how close does the ranger come to hitting the deer? **(b)** How much time is needed for the ranger's vehicle to stop?

Picture the Problem

We choose the positive x direction to be the direction of motion. With this choice it follows that $v_0 = +11.4$ m/s. In addition, the fact that the ranger's vehicle is slowing down means its acceleration points in the *opposite* direction to that of the velocity [see Figure 2–11 (b) and (c)]. Therefore, the vehicle's acceleration is $a = -3.80$ m/s^2. Finally, when the vehicle comes to rest its velocity is zero, $v = 0$.

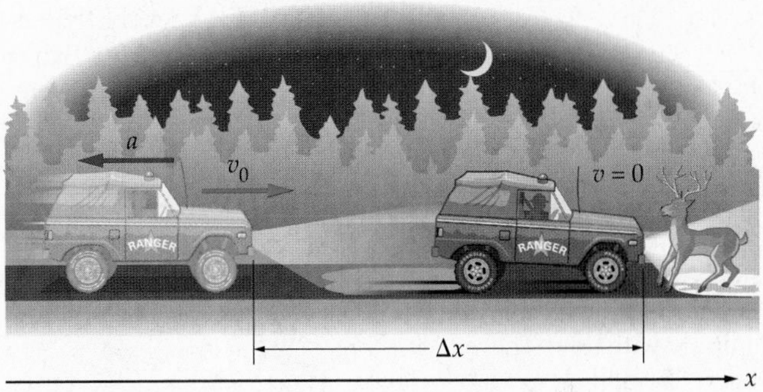

Strategy

The acceleration is constant, so we can use the equations listed in Table 2–4. In part **(a)** we want to find a distance when we know the velocity and acceleration, so we use a rearranged version of Equation 2–12. In part **(b)** we want to find a time when we know the velocity and acceleration, so we use a rearranged version of Equation 2–7.

Solution

Part (a)

1. Solve Equation 2–12 for Δx: $$\Delta x = \frac{v^2 - v_0^{\,2}}{2a}$$

2. Set $v = 0$, and substitute numerical values: $$\Delta x = -\frac{v_0^{\,2}}{2a} = -\frac{(11.4\text{ m/s})^2}{2(-3.80\text{ m/s}^2)} = 17.1\text{ m}$$

3. Subtract Δx from 20.0 m to find the distance between the stopped vehicle and the deer: $$20.0\text{ m} - 17.1\text{ m} = 2.9\text{ m}$$

Part (b)

4. Set $v = 0$ in Equation 2–7 and solve for t: $$v = v_0 + at = 0$$ $$t = -\frac{v_0}{a} = -\frac{11.4\text{ m/s}}{(-3.80\text{ m/s}^2)} = 3.00\text{ s}$$

Insight

Note the difference in the way t and Δx depend on the initial speed. If the initial speed is doubled, for example, the time needed to stop also doubles, but the distance needed to stop increases by a factor of four. This is one reason why speed on the highway has such a great influence on safety.

Practice Problem

Show that using $t = 3.00$ s in Equation 2–11 results in the same distance needed to stop. [**Answer:** $x = x_0 + v_0t + \frac{1}{2}at^2 = 0 + (11.4\text{ m/s})(3.00\text{ s}) + \frac{1}{2}(-3.80\text{ m/s}^2)(3.00\text{ s})^2 = 17.1$ m, as expected.]

Some related homework problems: Problem 52, Problem 53

In Example 2–8, we calculated the distance necessary for a vehicle to come to a complete stop. But how does v vary with distance as the vehicle slows down? The next Conceptual Checkpoint deals with this topic.

REAL-WORLD PHYSICS
The stopping distance of a car

CONCEPTUAL CHECKPOINT 2–4 Stopping Distance

The ranger in Example 2–8 brakes for 17.1 m. After braking for only half that distance, $\frac{1}{2}(17.1\text{ m}) = 8.55$ m, is the ranger's speed **(a)** equal to $\frac{1}{2}v_0$, **(b)** greater than $\frac{1}{2}v_0$, or **(c)** less than $\frac{1}{2}v_0$?

Reasoning and Discussion

As pointed out in the Insight for Example 2–8, the fact that the stopping distance, Δx, depends on v_0^2 means that this distance increases by a factor of four when the speed is doubled. For example, the stopping distance with an initial speed of v_0 is four times the stopping distance when the initial speed is $v_0/2$.

To apply this observation to the ranger, suppose that the stopping distance with an initial speed of v_0 is Δx. It follows that the stopping distance for an initial speed of $v_0/2$ is $\Delta x/4$. This means that as the ranger slows from v_0 to 0, it takes a distance $\Delta x/4$ to slow from $v_0/2$ to 0, and the remaining distance, $3\Delta x/4$, to slow from v_0 to $v_0/2$. Thus, at the halfway point the ranger has not yet slowed to half of the initial velocity—the speed at this point is greater than $v_0/2$.

Answer:

(b) The ranger's speed is greater than $\frac{1}{2}v_0$.

Clearly, v does not decrease uniformly with distance. A plot showing v as a function of x for Example 2–8 is shown in **Figure 2–16**. As we can see from the graph, v changes more in the second half of the braking distance than in the first half.

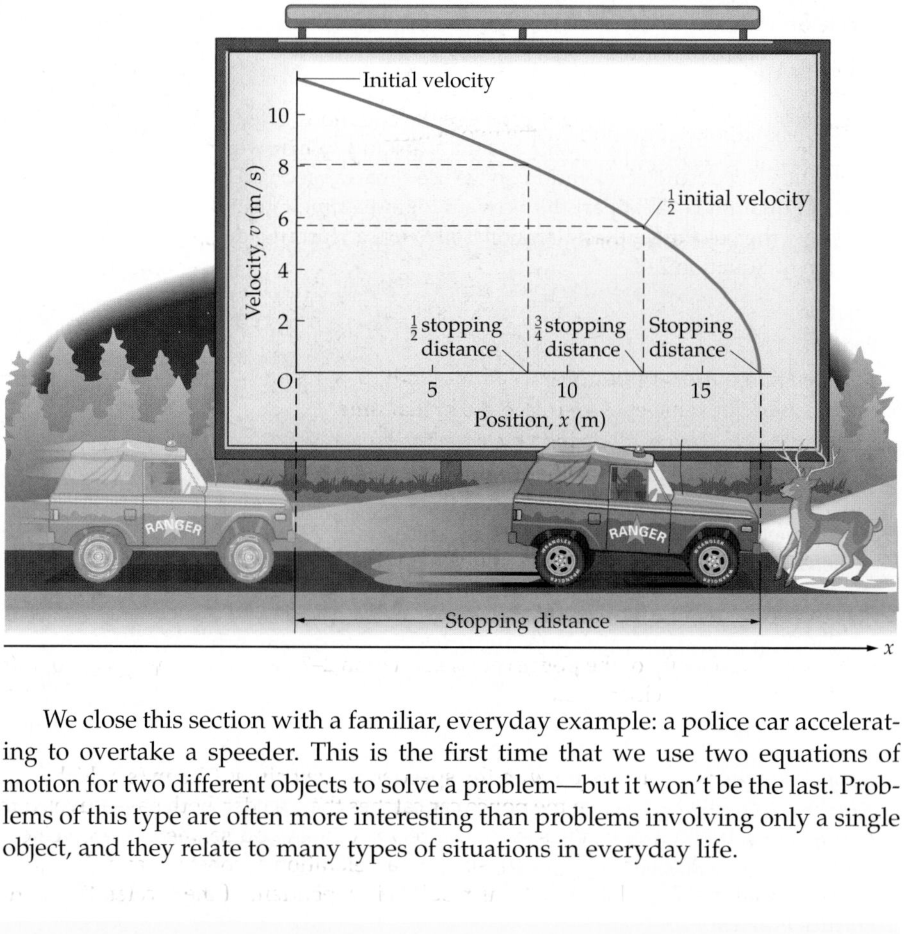

▶ **FIGURE 2–16 Velocity as a function of position for the ranger in Example 2–8**
The ranger's vehicle in Example 2–8 comes to rest with constant acceleration, which means that its velocity decreases uniformly with time. The velocity *does not* decrease uniformly with distance, however. In particular, note how rapidly the velocity decreases in the final one-quarter of the stopping distance.

PROBLEM-SOLVING NOTE

Strategize

Before attempting to solve a problem, it is a good idea to have some sort of plan, or "strategy," for how to proceed. It may be as simple as saying, "The problem asks me to relate velocity and time, therefore I will use Equation 2–7." In other cases the strategy is a bit more involved. Producing effective strategies is one of the most challenging—and creative—aspects of problem solving.

We close this section with a familiar, everyday example: a police car accelerating to overtake a speeder. This is the first time that we use two equations of motion for two different objects to solve a problem—but it won't be the last. Problems of this type are often more interesting than problems involving only a single object, and they relate to many types of situations in everyday life.

EXAMPLE 2–9 Catching a Speeder

A speeder doing 40.0 mi/h (about 17.9 m/s) in a 25 mi/h zone approaches a parked police car. The instant the speeder passes the police car, the police begin their pursuit. If the speeder maintains a constant velocity, and the police car accelerates with a constant acceleration of 4.50 m/s^2, **(a)** how long does it take for the police car to catch the speeder, **(b)** how far have the two cars traveled in this time, and **(c)** what is the velocity of the police car when it catches the speeder?

Picture the Problem

Our sketch shows the two cars at the moment the speeder passes the resting police car. At this instant, which we take to be $t = 0$, both the speeder and the police car are at the origin, $x = 0$. In addition, we choose the positive x direction to be the direction of motion; therefore, the speeder's initial velocity is given by $v_s = +17.9$ m/s, and the police car's initial velocity is zero. The speeder's acceleration is zero, but the police car has an acceleration given by $a = +4.50$ m/s^2. Finally, our plot shows the linear x-versus-t plot for the speeder, and the parabolic x-versus-t plot for the police car.

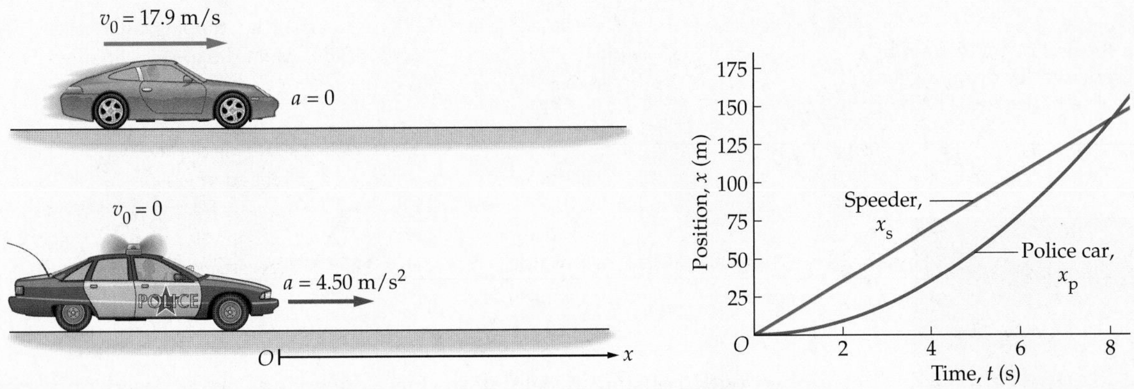

INTERACTIVE FIGURE

Strategy

To solve this problem, first write down a position-versus-time equation for the police car, x_p, and a separate equation for the speeder, x_s. Next, we find the time it takes the police car to catch the speeder by setting $x_p = x_s$ and solving the resulting equation for t. Once the catch time is determined, it is straightforward to calculate the distance traveled and the velocity of the police car.

Solution

Part (a)

1. Write equations of motion for the two vehicles. For the police car, $v_0 = 0$, and $a = 4.50\ \text{m/s}^2$. For the speeder, $v_0 = 17.9\ \text{m/s} = v_s$, and $a = 0$:

$$x_p = \tfrac{1}{2}at^2$$
$$x_s = v_s t$$

2. Set $x_p = x_s$ and solve for the time:

$$\tfrac{1}{2}at^2 = v_s t \quad \text{or} \quad \left(\tfrac{1}{2}at - v_s\right)t = 0$$

$$\text{two solutions: } t = 0 \quad \text{or} \quad t = \frac{2v_s}{a}$$

3. Clearly, $t = 0$ corresponds to the initial conditions, because both vehicles started at $x = 0$ at that time. The time of interest is obtained by substituting numerical values into the other solution:

$$t = \frac{2v_s}{a} = \frac{2(17.9\ \text{m/s})}{4.50\ \text{m/s}^2} = 7.96\ \text{s}$$

Part (b)

4. Substitute $t = 7.96$ s into the equations of motion for x_p and x_s. Note that $x_p = x_s$, as expected:

$$x_p = \tfrac{1}{2}at^2 = \tfrac{1}{2}(4.50\ \text{m/s}^2)(7.96\ \text{s})^2 = 143\ \text{m}$$
$$x_s = v_s t = (17.9\ \text{m/s})(7.96\ \text{s}) = 143\ \text{m}$$

Part (c)

5. To find the velocity of the police car use Equation 2–7, which relates velocity to time:

$$v_p = v_0 + at = 0 + (4.50\ \text{m/s}^2)(7.96\ \text{s}) = 35.8\ \text{m/s}$$

Insight

When the police car catches up with the speeder, its velocity is 35.8 m/s, which is exactly twice the velocity of the speeder. A coincidence? Not at all. When the police car catches the speeder, both have traveled the same distance (142 m) in the same time (7.96 s), therefore they have the same average velocity. Of course, the average velocity of the speeder is simply v_s. The average velocity of the police car is $\frac{1}{2}(v_0 + v)$, since its acceleration is constant, and thus $\frac{1}{2}(v_0 + v) = v_s$. Since $v_0 = 0$ for the police car, it follows that $v = 2v_s$. Notice that this result is independent of the acceleration of the police car, as we show in the following Practice Problem.

Practice Problem

Repeat this Example for the case where the acceleration of the police car is $a = 3.20\ \text{m/s}^2$. **[Answer: (a)** $t = 11.2$ s, **(b)** $x_p = x_s = 200$ m, **(c)** $v_p = 35.8$ m/s.]

Some related homework problems: Problem 49, Problem 60

2–7 Freely Falling Objects

The most famous example of motion with constant acceleration is **free fall**—the motion of an object falling freely under the influence of gravity. It was Galileo (1564–1642) who first showed, with his own experiments, that falling bodies

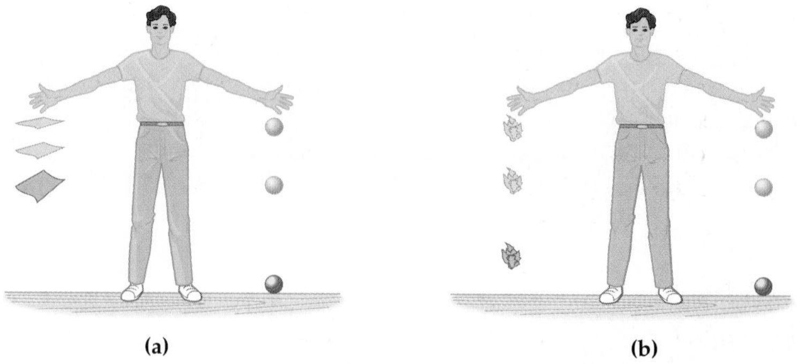

▶ **FIGURE 2–17 Free fall and air resistance**
(a) Dropping a sheet of paper and a rubber ball compared with **(b)** dropping a wadded-up sheet of paper and a ball.

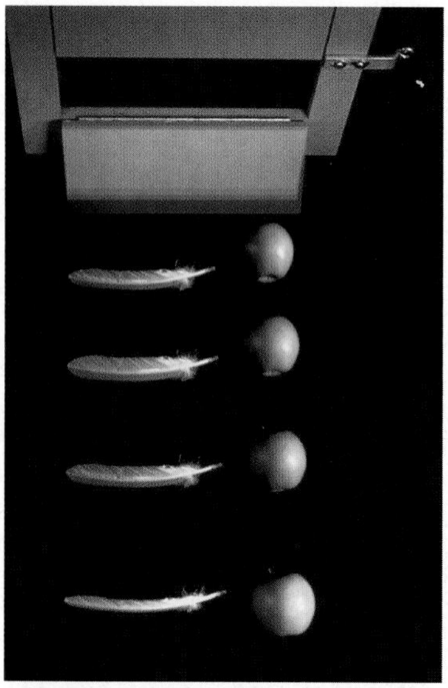

▲ In the absence of air resistance, all bodies fall with the same acceleration, regardless of their mass.

▲ Whether she is on the way up, at the peak of her flight, or on the way down, this girl is in free fall, accelerating downward with the acceleration of gravity. Only when she is in contact with the blanket does her acceleration change.

TABLE 2–5 Values of g at Different Locations on Earth (m/s^2)

Location	Latitude	g
North Pole	90° N	9.832
Oslo, Norway	60° N	9.819
Hong Kong	30° N	9.793
Quito, Ecuador	0°	9.780

move with constant acceleration. His conclusions were based on experiments done by rolling balls down inclines of various steepness. By using an incline, Galileo was able to reduce the acceleration of the balls, thus producing motion slow enough to be timed with the rather crude instruments available.

Galileo also pointed out that objects of different weight fall with the *same* constant acceleration—provided air resistance is small enough to be ignored. Whether he dropped objects from the Leaning Tower of Pisa to demonstrate this fact, as legend has it, will probably never be known for certain, but we do know that he performed extensive experiments to support his claim.

Today it is easy to verify Galileo's assertion by dropping objects in a vacuum chamber, where the effects of air resistance are essentially removed. In a standard classroom demonstration, a feather and a coin are dropped in a vacuum, and both fall at the same rate. In 1971, a novel version of this experiment was carried out on the Moon by astronaut David Scott. In the near perfect vacuum on the Moon's surface he dropped a feather and a hammer and showed a worldwide television audience that they fell to the ground in the same time.

To illustrate the effect of air resistance in everyday terms, consider dropping a sheet of paper and a rubber ball (**Figure 2–17**). The paper drifts slowly to the ground, taking much longer to fall than the ball. Now, wad the sheet of paper into a tight ball and repeat the experiment. This time the ball of paper and the rubber ball reach the ground in nearly the same time. What was different in the two experiments? Clearly, when the sheet of paper was wadded into a ball, the effect of air resistance on it was greatly reduced, so that both objects fell almost as they would in a vacuum.

Before considering a few examples, let's first discuss exactly what is meant by "free fall." To begin, the word *free* in free fall means free from any effects other than gravity. For example, in free fall we assume that an object's motion is not influenced by any form of friction or air resistance.

- Free fall is the motion of an object subject *only* to the influence of gravity.

Though free fall is an idealization—which does not apply to many real-world situations—it is still a useful approximation in many other cases. In the following examples we assume that the motion may be considered as free fall.

Next, it should be realized that the word *fall* in free fall does not mean the object is necessarily moving downward. By free fall, we mean *any* motion under the influence of gravity alone. If you drop a ball, it is in free fall. If you throw a ball upward or downward, it is in free fall as soon as it leaves your hand.

- An object is in free fall as soon as it is released, whether it is dropped from rest, thrown downward, or thrown upward.

Finally, the acceleration produced by gravity on the Earth's surface (sometimes called the gravitational strength) is denoted with the symbol g. As a shorthand name, we will frequently refer to g simply as "the acceleration due to gravity." In fact, as we shall see in Chapter 12, the value of g varies according to one's location on the surface of the earth, as well as one's altitude above it. Table 2–5 shows how g varies with latitude.

In all the calculations that follow in this book, we shall use $g = 9.81\ \mathrm{m/s^2}$ for the acceleration due to gravity. Note, in particular, that g always stands for $+9.81\ \mathrm{m/s^2}$, never $-9.81\ \mathrm{m/s^2}$. For example, if we choose a coordinate system with the positive direction upward, the acceleration in free fall is $a = -g$. If the positive direction is downward, then free-fall acceleration is $a = g$.

With these comments, we are ready to explore a variety of free-fall examples.

EXAMPLE 2–10 Do the Cannonball!

A person steps off the end of a 3.00-m-high diving board and drops to the water below. **(a)** How long does it take for the person to reach the water? **(b)** What is the person's speed on entering the water?

Picture the Problem

In our sketch we choose the origin to be at the height of the diving board, and we let the positive direction be downward. With these choices, $x_0 = 0$, $a = g$, and the water is at $x = 3.00$ m. Of course, $v_0 = 0$ since the person simply steps off the board.

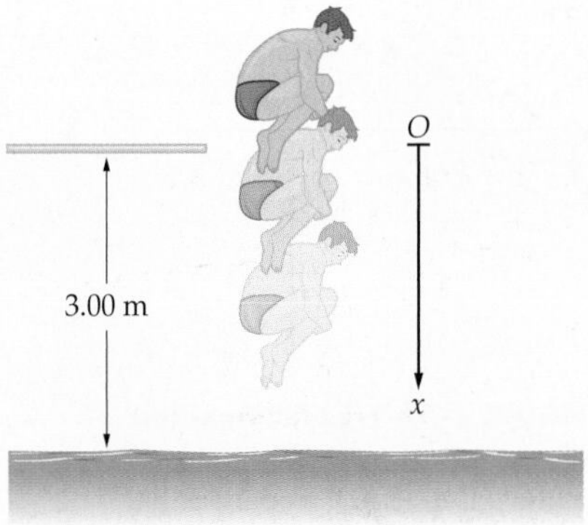

Strategy

We can neglect air resistance in this case and model the motion as free fall. This means we can assume a constant acceleration equal to g and use the equations of motion in Table 2–4. For part **(a)** we want to find the time of fall when we know the distance and acceleration, so we use Equation 2–11. For part **(b)** we can relate velocity to time by using Equation 2–7, or we can relate velocity to position by using Equation 2–12. We will implement both approaches and show that the results are the same.

Solution

Part (a)

1. Write Equation 2–11 with $x_0 = 0$, $v_0 = 0$, and $a = g$:

$$x = x_0 + v_0 t + \tfrac{1}{2}at^2 = 0 + 0 + \tfrac{1}{2}gt^2 = \tfrac{1}{2}gt^2$$

2. Solve for the time, t, and set $x = 3.00$ m:

$$t = \sqrt{\frac{2x}{g}} = \sqrt{\frac{2(3.00\ \mathrm{m})}{9.81\ \mathrm{m/s^2}}} = 0.782\ \mathrm{s}$$

Part (b)

3. Use the time found in part (a) in Equation 2–7:

$$v = v_0 + gt = 0 + (9.81\ \mathrm{m/s^2})(0.782\ \mathrm{s}) = 7.67\ \mathrm{m/s}$$

4. We can also find the velocity without knowing the time by using Equation 2–12 with $\Delta x = 3.00$ m:

$$v^2 = v_0{}^2 + 2a\Delta x = 0 + 2g\Delta x$$

$$v = \sqrt{2g\Delta x} = \sqrt{2(9.81\ \mathrm{m/s^2})(3.00\ \mathrm{m})} = 7.67\ \mathrm{m/s}$$

Insight

Let's put these results in more common, everyday units. If you step off a diving board 9.84 ft (3.00 m) above the water, you enter the water with a speed of 17.2 mi/h (7.67 m/s).

Practice Problem

What is your speed on entering the water if you step off a 10.0-m diving tower? [**Answer:** $v = \sqrt{2(9.81\ \mathrm{m/s^2})(10.0\ \mathrm{m})} = 14.0\ \mathrm{m/s} = 31.4\ \mathrm{mi/h}$]

Some related homework problems: Problem 65, Problem 74

The special case of free fall from rest occurs so frequently, and in so many different contexts, that it deserves special attention. If we take x_0 to be zero, and positive to be downward, then position as a function of time is $x = x_0 + v_0 t + \tfrac{1}{2}gt^2 = 0 + 0 + \tfrac{1}{2}gt^2$, or

$$x = \tfrac{1}{2}gt^2 \qquad 2\text{–}13$$

Similarly, velocity as a function of time is

$$v = gt \qquad 2\text{–}14$$

and velocity as a function of position is

$$v = \sqrt{2gx} \qquad 2\text{–}15$$

PROBLEM-SOLVING NOTE

Check Your Solution

Once you have a solution to a problem, check to see whether it makes sense. First, make sure the units are correct; m/s for speed, $\mathrm{m/s^2}$ for acceleration, and so on. Second, check the numerical value of your answer. If you are solving for the speed of a diver dropping from a 3.0-m diving board and you get an unreasonable value like 200 m/s (≈450 mi/h), chances are good that you've made a mistake.

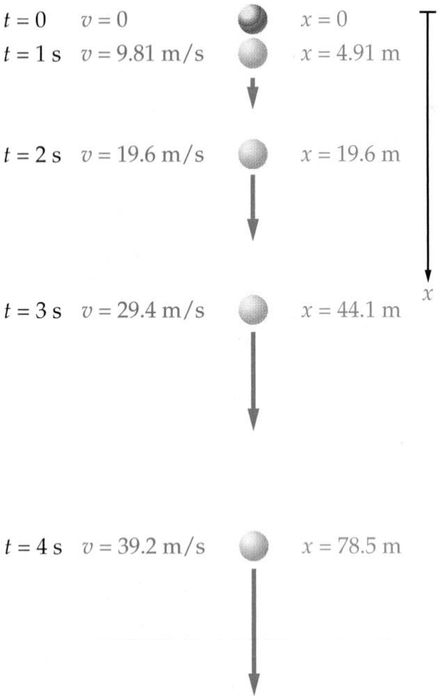

▲ **FIGURE 2–18 Free fall from rest**
Position and velocity are shown as functions of time. It is apparent that velocity depends linearly on t, whereas position depends on t^2.

The behavior of these functions is illustrated in **Figure 2–18**. Note that position increases with time squared, whereas velocity increases linearly with time.

Next we consider two objects dropped from rest, one after the other, and discuss how their separation varies with time.

CONCEPTUAL CHECKPOINT 2–5 **Free-fall Separation**

You drop a rock from a bridge to the river below. When the rock has fallen 4 m, you drop a second rock. As the rocks continue their free fall, does their separation (a) increase, (b) decrease, or (c) stay the same?

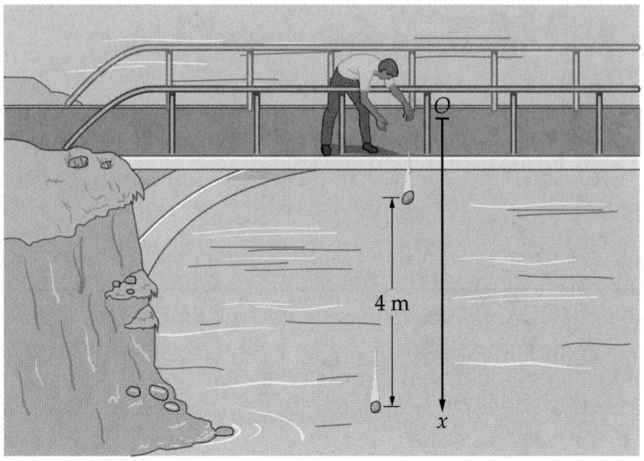

Reasoning and Discussion
It might seem that since both rocks are in free fall, their separation remains the same. This is not so. The rock that has a head start always has a greater velocity than the later one; thus it covers a greater distance in any interval of time. As a result, the separation between the rocks increases.

Answer
(a) The separation between the rocks increases.

An erupting volcano shooting out fountains of lava is an impressive sight. In the next Example we show how a simple timing experiment can determine the initial velocity of the erupting lava.

EXAMPLE 2–11 **Bombs Away: Calculating the Speed of a Lava Bomb**

REAL-WORLD PHYSICS A volcano shoots out blobs of molten lava, called lava bombs, from its summit. A geologist observing the eruption uses a stopwatch to time the flight of a particular lava bomb that is projected straight upward. If the time for it to rise and fall back to its launch height is 4.75 s, and its acceleration is 9.81 m/s^2 downward, what is its initial speed?

Picture the Problem
Our sketch shows a coordinate system with upward as the positive x direction. For clarity, we offset the upward and downward trajectories slightly. In addition, we choose $t = 0$ to be the time at which the lava bomb is launched. With these choices it follows that $x_0 = 0$ and the acceleration is $a = -g = -9.81\ \mathrm{m/s^2}$. The initial speed to be determined is v_0.

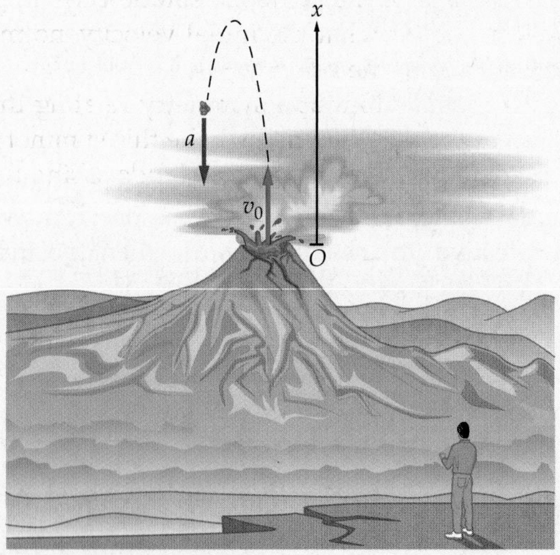

Strategy
Once again, we can neglect air resistance and model the motion of the lava bombs as free fall—this time with an initial upward velocity. We know that the lava bomb starts at $x = 0$ at the time $t = 0$ and returns to $x = 0$ at the time $t = 4.75$ s. This means that we know the bomb's position, time, and acceleration ($a = -g$), from which we would like to determine the initial velocity. A reasonable approach is to use Equation 2–11 and solve it for the one unknown it contains, v_0.

Solution

1. Write out $x = x_0 + v_0t + \frac{1}{2}at^2$ with $x_0 = 0$ and $a = -g$. Factor out a time, t, from the two remaining terms:
$$x = x_0 + v_0t + \tfrac{1}{2}at^2 = v_0t - \tfrac{1}{2}gt^2 = \left(v_0 - \tfrac{1}{2}gt\right)t$$

2. Set x equal to zero, since this is the position of the lava bomb at $t = 0$ and $t = 4.75$ s:
$$x = \left(v_0 - \tfrac{1}{2}gt\right)t = 0 \quad \text{two solutions:}$$
(i) $t = 0$
(ii) $v_0 - \frac{1}{2}gt = 0$

3. The first solution is simply the initial condition; that is, $x = 0$ at $t = 0$. Solve the second solution for the initial speed:
$$v_0 - \tfrac{1}{2}gt = 0 \quad \text{or} \quad v_0 = \tfrac{1}{2}gt$$

4. Substitute numerical values for g and the time the lava bomb lands:
$$v_0 = \tfrac{1}{2}gt = \tfrac{1}{2}(9.81\ \text{m/s}^2)(4.75\ \text{s}) = 23.3\ \text{m/s}$$

Insight

A geologist can determine a lava bomb's initial speed by simply observing its flight time. Knowing the lava bomb's initial speed can help geologists determine how severe a volcanic eruption will be, and how dangerous it might be to people in the surrounding area.

Practice Problem

A second lava bomb is projected straight upward with an initial speed of 25 m/s. How long is it in the air? [**Answer:** $t = 5.1$ s]

Some related homework problems: Problem 67, Problem 77

What is the speed of a lava bomb when it returns to Earth; that is, when it returns to the same level from which it was launched? Physical intuition might suggest that, in the absence of air resistance, it should be the same as the initial speed. To show that this hypothesis is indeed correct, write out Equation 2–7 for this case:

$$v = v_0 - gt$$

Substituting numerical values, we find

$$v = v_0 - gt = 23.3\ \text{m/s} - (9.81\ \text{m/s}^2)(4.75\ \text{s}) = -23.3\ \text{m/s}$$

Thus, the velocity of the lava when it lands is just the negative of the velocity it had when launched upward. Or put another way, when the lava lands, it has the same speed as when it was launched; it's just traveling in the opposite direction.

It is instructive to verify this result symbolically. Recall from Example 2–11 that $v_0 = \frac{1}{2}gt$, where t is the time the bomb lands. Substituting this result into Equation 2–7 we find

$$v = \tfrac{1}{2}gt - gt = -\tfrac{1}{2}gt = -v_0$$

The advantage of the symbolic solution lies in showing that the result is not a fluke—no matter what the initial velocity, no matter what the acceleration, the bomb lands with the velocity $-v_0$.

These results hint at a symmetry relating the motion on the way up to the motion on the way down. To make this symmetry more apparent, we first solve for the time when the lava bomb lands. Using the result $v_0 = \frac{1}{2}gt$ from Example 2–11, we find

$$t = \frac{2v_0}{g} \qquad \textit{(time of landing)}$$

Next, we find the time when the velocity of the lava is zero, which is at its highest point. Setting $v = 0$ in Equation 2–7, we have $v = v_0 - gt = 0$, or

$$t = \frac{v_0}{g} \qquad \textit{(time when } v = 0\textit{)}$$

Note that this is exactly half the time required for the lava to make the round trip. Thus, the velocity of the lava is zero and the height of the lava is greatest exactly halfway between launch and landing.

▲ In the absence of air resistance, these lava bombs from the Kilauea volcano on the big island of Hawaii would strike the water with the same speed they had when they were blasted into the air.

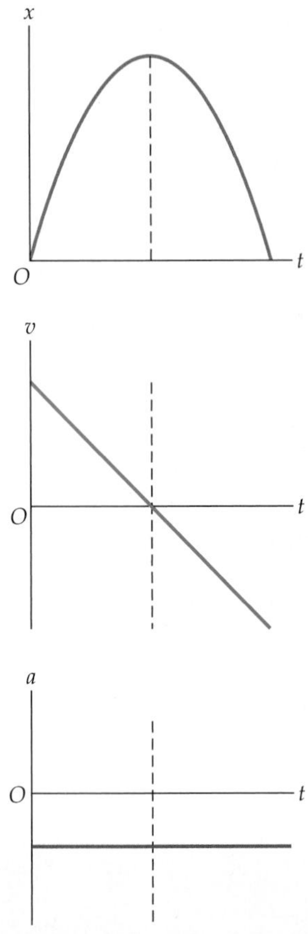

▲ **FIGURE 2–20 Position, velocity, and acceleration of a lava bomb as functions of time**

The fact that x versus t is curved indicates an acceleration; the downward curvature shows that the acceleration is negative. This is also clear from v versus t, which has a negative slope. The constant slope of the straight line in the v-versus-t plot indicates a constant acceleration, as shown in the a-versus-t plot.

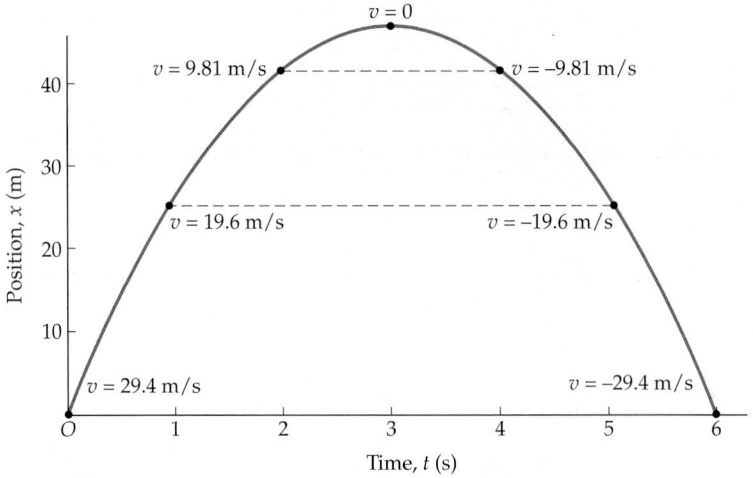

▲ **FIGURE 2–19 Position and velocity of a lava bomb**

This lava bomb is in the air for 6 seconds. Note the symmetry about the midpoint of the bomb's flight.

This symmetry is illustrated in **Figure 2–19**. In this case we consider a lava bomb that is in the air for 6.00 s, moving without air resistance. Note that at $t = 3.00$ s the lava is at its highest point and its velocity is zero. At times equally spaced before and after $t = 3.00$ s, the lava is at the same height, has the same speed, but is moving in opposite directions. As a result of this symmetry, a movie of the lava bomb's flight would look the same whether run forward or in reverse.

Figure 2–20 shows the time dependence of position, velocity, and acceleration for an object in free fall without air resistance after being thrown upward. As soon as the object is released, it begins to accelerate downward—as indicated by the negative slope of the velocity-versus-time plot—though it isn't necessarily moving downward. For example, if you throw a ball *upward* it begins to accelerate *downward* the moment it leaves your hand. It continues moving upward, however, until its speed diminishes to zero. Since gravity is causing the downward acceleration, and gravity doesn't turn off just because the ball's velocity goes through zero, the ball continues to accelerate downward even when it is momentarily at rest.

Similarly, in the next Example we consider a sandbag that falls from an ascending hot-air balloon. This means that before the bag is in free fall it was moving upward—just like a ball thrown upward. And just like the ball, the sandbag continues moving upward for a brief time before momentarily stopping and then moving downward.

EXAMPLE 2–12 Look Out Below! A Sand Bag in Free Fall

A hot-air balloon is rising straight upward with a constant speed of 6.5 m/s. When the basket of the balloon is 20.0 m above the ground, a bag of sand tied to the basket comes loose. **(a)** How long is the bag of sand in the air before it hits the ground? **(b)** What is the greatest height of the bag of sand during its fall to the ground?

Picture the Problem

We choose the origin to be at ground level and positive to be upward. This means that, for the bag, we have $x_0 = 20.0$ m, $v_0 = 6.5$ m/s, and $a = -g$. Our sketch also shows snapshots of the balloon and bag of sand at three different times, starting at $t = 0$ when the bag comes loose. Note that the bag is moving upward with the balloon at the time it comes loose. It therefore continues to move upward for a short time after it separates from the basket, exactly as if it had been thrown upward.

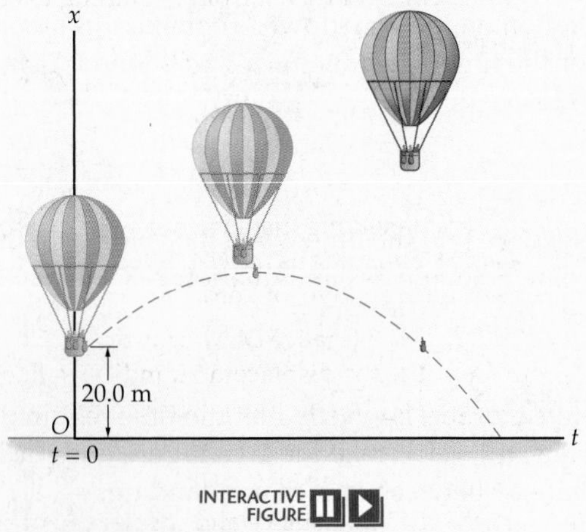

Strategy

The effects of air resistance on the sandbag can be ignored. As a result, we can use the equations in Table 2–4 with a constant acceleration $a = -g$.

In part (a) we want to relate position and time—knowing the initial position and initial velocity—so we use Equation 2–11. To find the time the bag hits the ground, we set $x = 0$ and solve for t.

For part (b) we have no expression that gives the maximum height of a particle—so we will have to come up with something on our own. We can start with the fact that $v = 0$ at the greatest height, since it is there the bag momentarily stops as it changes direction. Therefore, we can find the time t when $v = 0$ by using Equation 2–7, and then substitute t into Equation 2–11 to find x_{max}.

Solution

Part (a)

1. Apply Equation 2–11 to the bag of sand, where x_0 and v_0 have the values given. Set $x = 0$:

$$x = x_0 + v_0t - \tfrac{1}{2}gt^2 = 0$$

2. Note that we have a quadratic equation for t in the form $At^2 + Bt + C = 0$, where $A = -\frac{1}{2}g$, $B = v_0$, and $C = x_0$. Solve this equation for t. The positive solution, 2.78 s, applies to this problem: (Quadratic equations and their solutions are discussed in Appendix A. In general, one can expect two solutions to a quadratic equation.)

$$t = \frac{-v_0 \pm \sqrt{v_0{}^2 - 4\left(-\frac{1}{2}g\right)(x_0)}}{2\left(-\frac{1}{2}g\right)}$$

$$= \frac{-(6.5\text{ m/s}) \pm \sqrt{(6.5\text{ m/s})^2 + 2(9.81\text{ m/s}^2)(20.0\text{ m})}}{(-9.81\text{ m/s}^2)}$$

$$= \frac{-(6.5\text{ m/s}) \pm 20.8\text{ m/s}}{(-9.81\text{ m/s}^2)} = 2.78\text{ s}, -1.46\text{ s}$$

Part (b)

3. Apply Equation 2–7 to the bag of sand, then find the time when the velocity equals zero:

$$v = v_0 + at = v_0 - gt$$

$$v_0 - gt = 0 \quad \text{or} \quad t = \frac{v_0}{g} = \frac{6.5\text{ m/s}}{9.81\text{ m/s}^2} = 0.66\text{ s}$$

4. Use $t = 0.66$ s in Equation 2–11 to find the maximum height:

$$x_{max} = 20.0\text{ m} + (6.5\text{ m/s})(0.66\text{ s}) - \tfrac{1}{2}(9.81\text{ m/s}^2)(0.66\text{ s})^2$$

$$= 22\text{ m}$$

Insight

The positive solution to the quadratic equation is certainly the one that applies here, but the negative solution is not completely without meaning. What physical meaning might it have? Well, if the balloon had been *descending* with a speed of 6.5 m/s, instead of rising, then the time for the bag to reach the ground would have been 1.46 s. Try it! Let $v_0 = -6.5$ m/s and repeat the calculation given in part (a).

Practice Problem

What is the velocity of the bag of sand just before it hits the ground? [**Answer:** $v = v_0 - gt = (6.5\text{ m/s}) - (9.81\text{ m/s}^2)(2.78\text{ s}) = -20.8\text{ m/s}$; the minus sign indicates the bag is moving downward.]

Some related homework problems: Problem 81, Problem 97

Chapter Summary

2–1 Distance and Displacement

Distance

Total length of travel, from beginning to end. The distance is always positive.

Displacement

Displacement, Δx, is the change in position:

$$\Delta x = x_f - x_i \qquad \text{2–1}$$

When calculating displacement, it is important to remember that it is always *final* position minus *initial* position—never the other way. Displacement can be positive, negative, or zero.

Positive and Negative Displacement

The *sign* of the displacement indicates the *direction* of motion. For example, suppose we choose the positive direction to be to the right. Then $\Delta x > 0$ means motion to the right, and $\Delta x < 0$ means motion to the left.

Units

The SI unit of distance and displacement is the meter, m.

2–2 Average Speed and Velocity

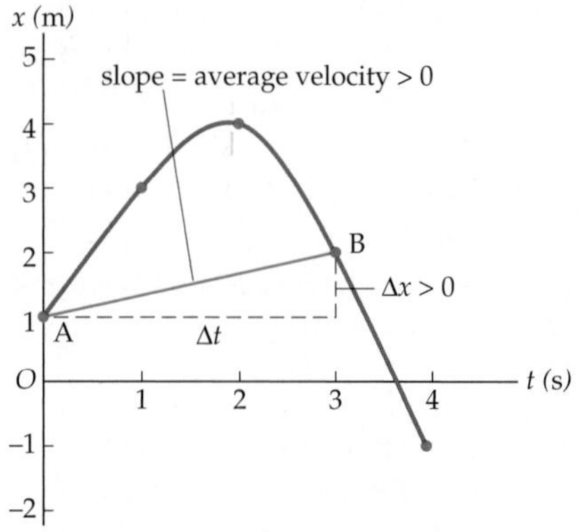

Average Speed
Average speed is *distance* divided by elapsed time:

$$\text{average speed} = \text{distance/time} \quad 2\text{–}2$$

Average speed is never negative.

Average Velocity
Average velocity, v_{av}, is *displacement* divided by time:

$$v_{av} = \frac{\Delta x}{\Delta t} = \frac{x_f - x_i}{t_f - t_i} \quad 2\text{–}3$$

Average velocity is positive if motion is in the positive direction, and negative if motion is in the negative direction.

Graphical Interpretation of Velocity
In an x-versus-t plot, the average velocity is the slope of a line connecting two points.

Units
The SI unit of speed and velocity is meters per second, m/s.

2–3 Instantaneous Velocity

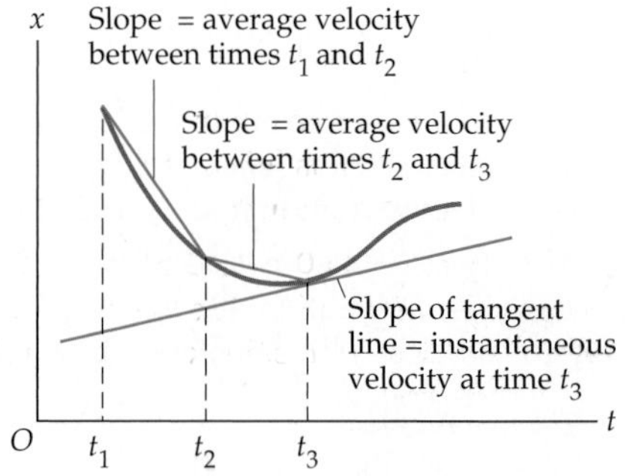

The velocity at an instant of time is the limit of the average velocity over shorter and shorter time intervals:

$$v = \lim_{\Delta t \to 0} \frac{\Delta x}{\Delta t} \quad 2\text{–}4$$

Instantaneous velocity can be positive, negative, or zero, with the sign indicating the direction of motion.

Constant Velocity
When velocity is constant, the instantaneous velocity is equal to the average velocity.

Graphical Interpretation
In an x-versus-t plot, the instantaneous velocity at a given time is equal to the slope of the tangent line at that time.

2–4 Acceleration

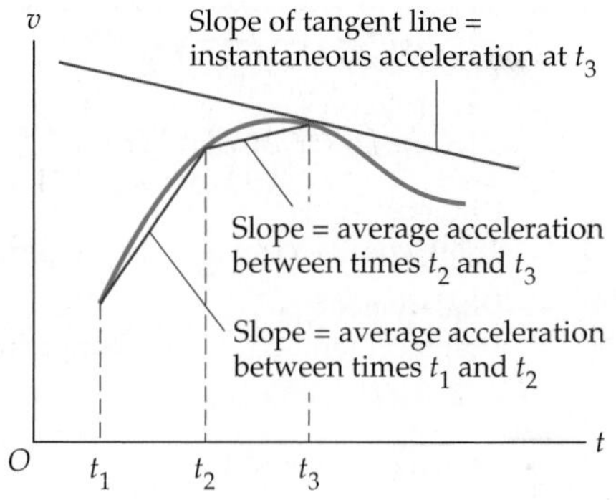

Average Acceleration
Average acceleration is the change in velocity divided by the change in time:

$$a_{av} = \frac{\Delta v}{\Delta t} = \frac{v_f - v_i}{t_f - t_i} \quad 2\text{–}5$$

Average acceleration is positive if $v_f > v_i$, is negative if $v_f < v_i$, and is zero if $v_f = v_i$.

Instantaneous Acceleration
Instantaneous acceleration is the limit of the average acceleration as the time interval goes to zero:

$$a = \lim_{\Delta t \to 0} \frac{\Delta v}{\Delta t} \quad 2\text{–}6$$

Instantaneous acceleration can be positive, negative, or zero, depending on whether the velocity is becoming more positive, more negative, or is staying the same. Knowing the sign of the acceleration *does not* tell you whether an object is speeding up or slowing down, and it *does not* give the direction of motion.

Constant Acceleration
When acceleration is constant, the instantaneous acceleration is equal to the average acceleration.

Deceleration
An object whose speed is decreasing is said to be decelerating. Deceleration occurs whenever the velocity and acceleration have opposite signs.

Graphical Interpretation
In a v-versus-t plot, the instantaneous acceleration is equal to the slope of the tangent line at a given time.

Units
The SI unit of acceleration is meters per second per second, or m/s^2.

2–5 Motion with Constant Acceleration

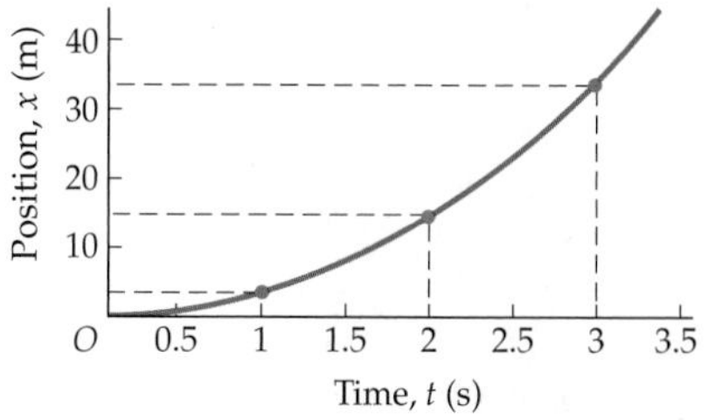

Several different "equations of motion" describe particles moving with constant acceleration. Each equation relates a different set of variables:

Velocity as a Function of Time $$v = v_0 + at \quad (2\text{–}7)$$

Initial, Final, and Average Velocity $$v_{av} = \tfrac{1}{2}(v_0 + v) \quad (2\text{–}9)$$

Position as a Function of Time and Velocity $$x = x_0 + \tfrac{1}{2}(v_0 + v)t \quad (2\text{–}10)$$

Position as a Function of Time and Acceleration $$x = x_0 + v_0 t + \tfrac{1}{2}at^2 \quad (2\text{–}11)$$

Velocity as a Function of Position $$v^2 = v_0^2 + 2a(x - x_0) = v_0^2 + 2a\Delta x \quad (2\text{–}12)$$

2–7 Freely Falling Objects

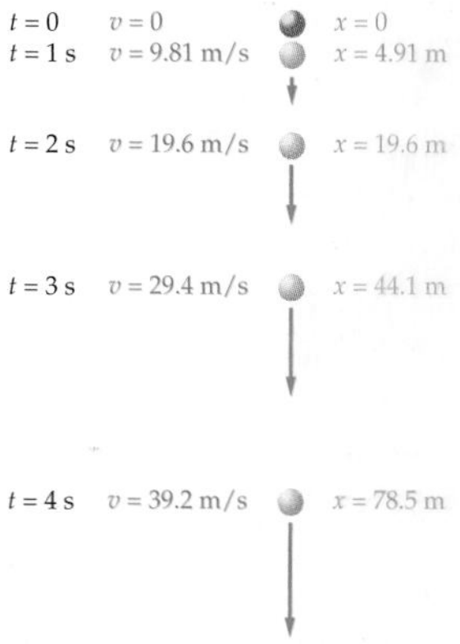

Objects in free fall move under the influence of gravity alone. An object is in free fall as soon as it is released, whether it is thrown upward, thrown downward, or released from rest.

Acceleration Due to Gravity
The acceleration due to gravity on the earth's surface varies slightly from place to place. In this book we shall define the acceleration of gravity to have the following magnitude:

$$g = 9.81\ m/s^2$$

Note that g is always a positive quantity. If we choose the positive direction of our coordinate system to be downward (in the direction of the acceleration of gravity), it follows that the acceleration of an object in free fall is $a = +g$. On the other hand, if we choose our positive direction to be upward, the acceleration of a freely falling object is in the negative direction; hence $a = -g$.

Problem-Solving Summary

Type of Calculation	Relevant Physical Concepts	Related Examples
Relate velocity to time.	In motion with uniform acceleration a, the velocity changes with time as $v = v_0 + at$ (Equation 2–7).	Examples 2–5, 2–8, 2–9, 2–10, 2–11, 2–12
Relate velocity to position.	If an object with an initial velocity v_0 accelerates with a uniform acceleration a for a distance Δx, the final velocity, v, is given by $v^2 = v_0^2 + 2a\Delta x$ (Equation 2–12)	Examples 2–7, 2–8, 2–10
Relate position to time.	The position of an object moving with constant acceleration a varies with time as follows: $x = x_0 + \frac{1}{2}(v_0 + v)t$ (Equation 2–10) or equivalently $x = x_0 + v_0 t + \frac{1}{2}at^2$ (Equation 2–11).	Examples 2–5, 2–6, 2–9, 2–10, 2–11, 2–12

Conceptual Questions

(Answers to odd-numbered Conceptual Questions can be found in the back of the book.)

(The effects of air resistance are to be ignored in this chapter.)

1. You and your dog go for a walk to a nearby park. On the way, your dog takes many short side trips to chase squirrels, examine fire hydrants, and so on. When you arrive at the park, do you and your dog have the same displacement? Have you traveled the same distance? Explain.
2. Does an odometer in a car measure distance or displacement? Explain.
3. Can you drive your car in such a way that the distance it covers is **(a)** greater than, **(b)** equal to, or **(c)** less than the magnitude of its displacement? In each case, give an example if your answer is yes, explain why not if your answer is no.
4. An astronaut orbits Earth in the space shuttle. In one complete orbit, is the magnitude of the displacement the same as the distance traveled? Explain.
5. After a tennis match the players dash to the net to congratulate one another. If they both run with a speed of 3 m/s, are their velocities equal? Explain.
6. Does a speedometer measure speed or velocity? Explain.
7. Is it possible for a car to circle a race track with constant velocity? Can it do so with constant speed? Explain.
8. Friends tell you that on a recent trip their average velocity was +20 m/s. Is it possible that their instantaneous velocity was negative at any time during the trip? Explain.
9. For what kind of motion are the instantaneous and average velocities equal?
10. If the position of an object is zero, does its speed have to be zero? Explain.
11. Assume that the brakes in your car create a constant deceleration, regardless of how fast you are going. If you double your driving speed, how does this affect **(a)** the time required to come to a stop, and **(b)** the distance needed to stop?
12. The velocity of an object is zero at a given instant of time. **(a)** Is it possible for the object's acceleration to be zero at this time? Explain. **(b)** Is it possible for the object's acceleration to be nonzero at this time? Explain.
13. If the velocity of an object is nonzero, can its acceleration be zero? Give an example if your answer is yes, explain why not if your answer is no.
14. Is it possible for an object to have zero average velocity over a given interval of time, yet still be accelerating during the interval? Give an example if your answer is yes, explain why not if your answer is no.
15. A batter hits a pop fly straight up. **(a)** Is the acceleration of the ball on the way up different from its acceleration on the way down? **(b)** Is the acceleration of the ball at the top of its flight different from its acceleration just before it lands?
16. A person on a trampoline bounces straight upward with an initial speed of 4.5 m/s. What is the person's speed when she returns to her initial height?
17. After winning a baseball game, one player drops a glove, while another tosses a glove into the air. How do the accelerations of the two gloves compare?
18. A volcano shoots a lava bomb straight upward. Does the displacement of the lava bomb depend on **(a)** your choice of origin for your coordinate system, or **(b)** your choice of a positive direction? Explain in each case.

Conceptual Exercises

(Answers to odd-numbered Conceptual Exercises can be found in the back of the book.)

(The effects of air resistance are to be ignored in this chapter.)

1. The position-versus-time plot of a boat positioning itself next to a dock is shown in **Figure 2–21**. Rank the six points indicated in the plot in order of increasing value of the velocity v, starting with the most negative. Indicate a tie with an equal sign.

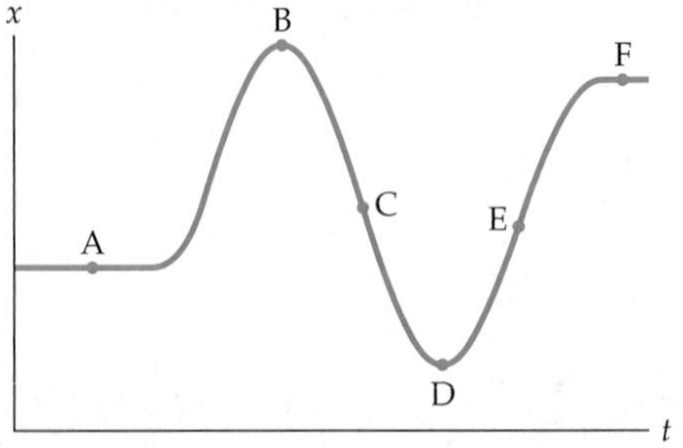

▲ **FIGURE 2–21** Conceptual Exercise 1

2. You drive in a straight line at 15 m/s for 10 kilometers, then at 25 m/s for another 10 kilometers. Is your average speed for the entire trip **(a)** 20 m/s, **(b)** more than 20 m/s, or **(c)** less than 20 m/s?
3. You drive in a straight line at 15 m/s for 10 minutes, then at 25 m/s for another 10 minutes. Is your average speed for the entire trip **(a)** 20 m/s, **(b)** more than 20 m/s, or **(c)** less than 20 m/s?
4. Two bows shoot arrows with the same initial speed. The string in bow A must be pulled back farther when shooting its arrow than the string in bow B. Is the acceleration of the arrow shot by bow A greater than, less than, or equal to the acceleration of the arrow shot by bow B?
5. A ball is released at the point $x = 2$ m on an inclined plane with a nonzero initial velocity. After being released, the ball moves with constant acceleration. The acceleration and initial velocity of the ball is described by one of the following four cases: case 1, $a > 0$, $v_0 > 0$; case 2, $a > 0$, $v_0 < 0$; case 3, $a < 0$, $v_0 > 0$; case 4, $a < 0$, $v_0 < 0$. **(a)** In which of these cases will the ball definitely pass $x = 0$ at some later time? **(b)** In which of these cases is more information needed to determine whether the ball will cross $x = 0$? **(c)** In which of these cases will the ball come to rest momentarily at some time during its motion?
6. A carpenter on the roof of a building accidentally drops her hammer. As the hammer falls, it passes two windows of equal height, as shown in **Figure 2–22 (a)**. Is the *increase* in speed of the hammer as it drops past window 1 greater than, less than, or equal to the *increase* in speed as it drops past window 2?

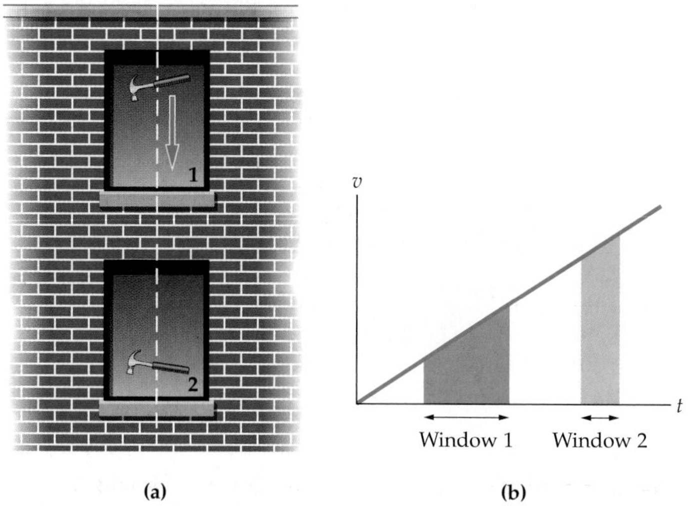

▲ **FIGURE 2–22** Conceptual Exercises 6 and 7

7. **Figure 2–22 (b)** shows a v-versus-t plot for the hammer dropped by the carpenter in Conceptual Exercise 6. Notice that the times when the hammer passes the two windows are indicated by shaded areas. Is the shaded area corresponding to window 1 greater than, less than, or equal to the shaded area corresponding to window 2?

8. At the edge of a roof you throw ball A upward with an initial speed of v_0, and then throw ball B downward with the same initial speed. Just before the balls hit the ground, which of the following is true: **(a)** the speed of ball A is greater than the speed of ball B, **(b)** the speed of ball A is equal to the speed of ball B, or **(c)** the speed of ball A is less than the speed of ball B?

9. At the edge of a roof you drop ball A from rest, and then throw ball B downward with an initial velocity of v_0. Is the increase in speed just before the balls land **(a)** more for ball A, **(b)** more for ball B, or **(c)** the same for each ball?

10. Suppose the two balls described in Conceptual Exercise 9 are released at the same time, with ball A dropped from rest and ball B thrown downward with an initial speed v_0. Identify which of the 5 plots shown in **Figure 2–23** corresponds to **(a)** ball A and **(b)** ball B.

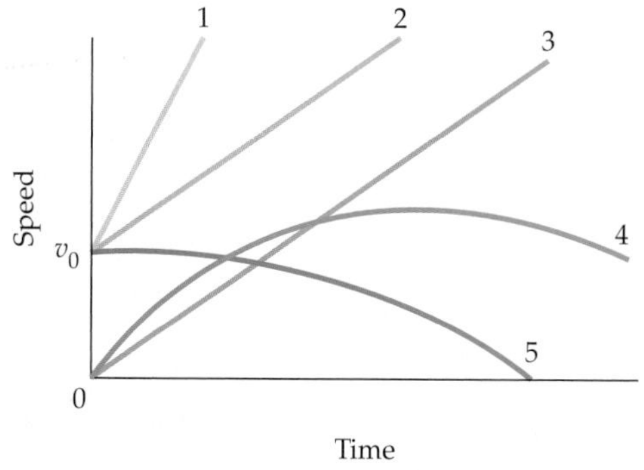

▲ **FIGURE 2–23** Conceptual Exercise 10

11. A ball is thrown straight upward with an initial speed v_0. When it reaches the top of its flight at height h, a second ball is thrown straight upward with the same initial speed. Do the balls cross paths **(a)** at height $\frac{1}{2}h$, **(b)** above $\frac{1}{2}h$, or **(c)** below $\frac{1}{2}h$?

Problems

Note: **IP** *denotes an integrated conceptual/quantitative problem.* **BIO** *identifies problems of biological or medical interest. Red bullets (•, ••, •••) are used to indicate the level of difficulty of each problem.*

(The effects of air resistance are to be ignored in this chapter.)

Section 2–1 Position, Distance, and Displacement

1. • Referring to **Figure 2–24**, you walk from your home to the library, then to the park. **(a)** What is the distance traveled? **(b)** What is your displacement?

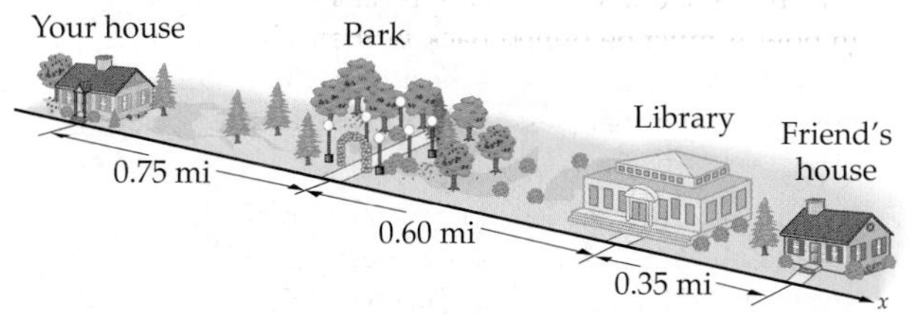

▲ **FIGURE 2–24** Problems 1 and 4

2. • The two tennis players shown in **Figure 2–25** walk to the net to congratulate one another. **(a)** Find the distance traveled and the displacement of player A. **(b)** Repeat for player B.

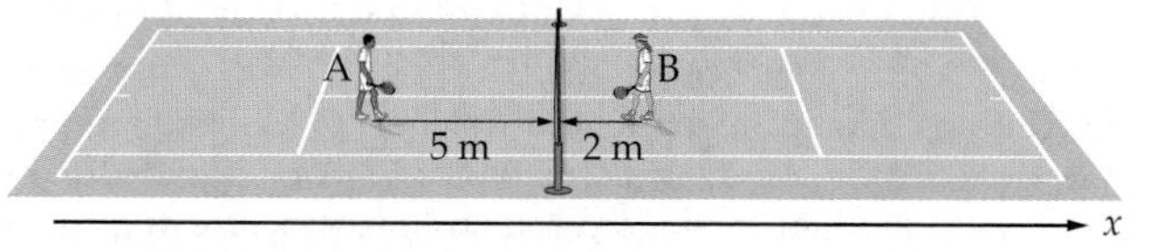

▲ **FIGURE 2–25** Problem 2

3. • The golfer in **Figure 2–26** sinks the ball in two putts, as shown. What is **(a)** the distance traveled by the ball, and **(b)** the displacement of the ball?

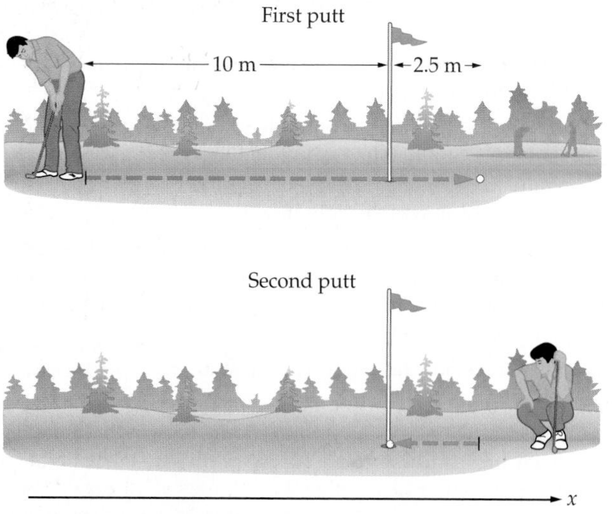

▲ **FIGURE 2–26** Problem 3

4. • In Figure 2–24, you walk from the park to your friend's house, then back to your house. What is your **(a)** distance traveled, and **(b)** displacement?

5. • A jogger runs on the track shown in **Figure 2–27**. Neglecting the curvature of the corners, **(a)** what is the distance traveled and the displacement in running from point A to point B? **(b)** Find the distance and displacement for a complete circuit of the track.

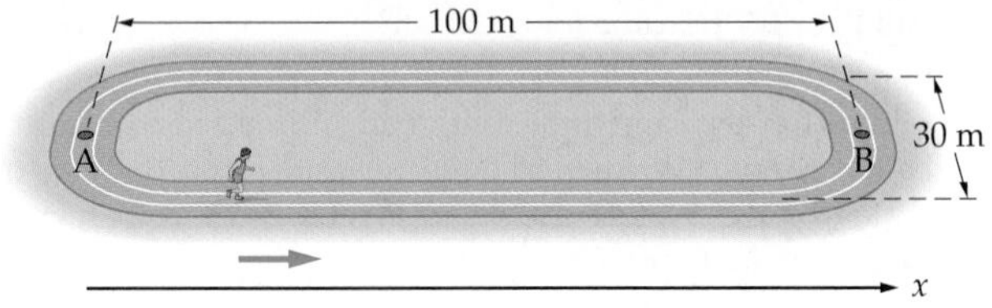

▲ **FIGURE 2–27** Problem 5

6. •• **IP** A child rides a pony on a circular track whose radius is 4.5 m. **(a)** Find the distance traveled and the displacement after the child has gone halfway around the track. **(b)** Does the distance traveled increase, decrease, or stay the same when the child completes one circuit of the track? Explain. **(c)** Does the displacement increase, decrease, or stay the same when the child completes one circuit of the track? Explain. **(d)** Find the distance and displacement after a complete circuit of the track.

Section 2–2 Average Speed and Velocity

7. • Joseph DeLoach of the United States set an Olympic record in 1988 for the 200-meter dash with a time of 19.75 seconds. What was his average speed? Give your answer in meters per second and miles per hour.

8. • In 1992 Zhuang Yong of China set a women's Olympic record in the 100-meter freestyle swim with a time of 54.64 seconds. What was her average speed in m/s and mi/h?

9. • **BIO** Kangaroos have been clocked at speeds of 65 km/h. **(a)** How far can a kangaroo hop in 3.2 minutes at this speed? **(b)** How long will it take a kangaroo to hop 0.25 km at this speed?

10. • **Rubber Ducks** A severe storm on January 10, 1992, caused a cargo ship near the Aleutian Islands to spill 29,000 rubber ducks and other bath toys into the ocean. Ten months later hundreds of rubber ducks began to appear along the shoreline near Sitka, Alaska, roughly 1600 miles away. What was the approximate average speed of the ocean current that carried the ducks to shore in **(a)** m/s and **(b)** mi/h? (Rubber ducks from the same spill began to appear on the coast of Maine in July 2003.)

11. • Radio waves travel at the speed of light, approximately 186,000 miles per second. How long does it take for a radio message to travel from the Earth to the Moon and back? (See the inside back cover for the necessary data.)

12. • It was a dark and stormy night, when suddenly you saw a flash of lightning. Three-and-a-half seconds later you heard the thunder. Given that the speed of sound in air is about 340 m/s, how far away was the lightning bolt?

13. • **BIO Nerve Impulses** The human nervous system can propagate nerve impulses at about 10^2 m/s. Estimate the time it takes for a nerve impulse generated when your finger touches a hot object to travel to your brain.

14. • Estimate how fast your hair grows in miles per hour.

15. •• A finch rides on the back of a Galapagos tortoise, which walks at the stately pace of 0.060 m/s. After 1.2 minutes the finch tires of the tortoise's slow pace, and takes flight in the same direction for another 1.2 minutes at 12 m/s. What was the average speed of the finch for this 2.4-minute interval?

16. •• You jog at 9.5 km/h for 8.0 km, then you jump into a car and drive an additional 16 km. With what average speed must you drive your car if your average speed for the entire 24 km is to be 22 km/h?

17. •• A dog runs back and forth between its two owners, who are walking toward one another (**Figure 2–28**). The dog starts running when the owners are 10.0 m apart. If the dog runs with a speed of 3.0 m/s, and the owners each walk with a speed of 1.3 m/s, how far has the dog traveled when the owners meet?

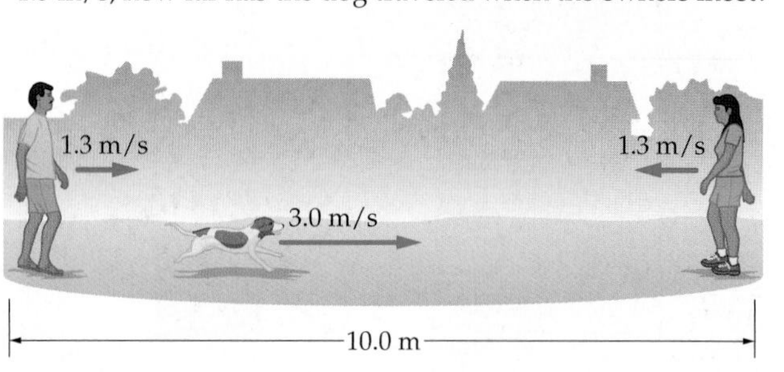

▲ **FIGURE 2–28** Problem 17

18. •• **IP** You drive in a straight line at 20.0 m/s for 10.0 minutes, then at 30.0 m/s for another 10.0 minutes. **(a)** Is your average speed 25.0 m/s, more than 25.0 m/s, less than 25.0 m/s? Explain. **(b)** Verify your answer to part (a) by calculating the average speed.

19. •• In heavy rush-hour traffic you drive in a straight line at 12 m/s for 1.5 minutes, then you have to stop for 3.5 minutes, and finally you drive at 15 m/s for another 2.5 minutes. **(a)** Plot a position-versus-time graph for this motion. Your plot should extend from $t = 0$ to $t = 7.5$ minutes. **(b)** Use your plot from part (a) to calculate the average velocity between $t = 0$ and $t = 7.5$ minutes.

20. •• **IP** You drive in a straight line at 20.0 m/s for 10.0 miles, then at 30.0 m/s for another 10.0 miles. **(a)** Is your average speed 25.0 m/s, more than 25.0 m/s, less than 25.0 m/s? Explain. **(b)** Verify your answer to part (a) by calculating the average speed.

21. •• **IP** An expectant father paces back and forth, producing the position-versus-time graph shown in **Figure 2–29**. Without performing a calculation, indicate whether the father's velocity is positive, negative, or zero on each of the following segments of the graph: **(a)** A, **(b)** B, **(c)** C, and **(d)** D. Calculate the numerical value of the father's velocity for the segments **(e)** A, **(f)** B, **(g)** C, and **(h)** D, and show that your results verify your answers to parts (a)–(d).

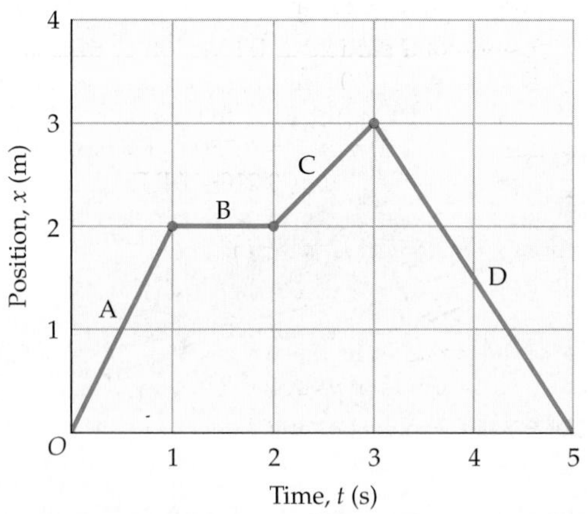

▲ **FIGURE 2–29** Problem 21

22. •• The position of a particle as a function of time is given by $x = (-5\text{ m/s})t + (3\text{ m/s}^2)t^2$. **(a)** Plot x versus t for $t = 0$ to $t = 2$ s. **(b)** Find the average velocity of the particle from $t = 0$ to $t = 1$ s. **(c)** Find the average speed from $t = 0$ to $t = 1$ s.

23. •• The position of a particle as a function of time is given by $x = (6\text{ m/s})t + (-2\text{ m/s}^2)t^2$. **(a)** Plot x versus t for $t = 0$ to $t = 2$ s. **(b)** Find the average velocity of the particle from $t = 0$ to $t = 1$ s. **(c)** Find the average speed from $t = 0$ to $t = 1$ s.

24. •• **IP** A tennis player moves back and forth along the baseline while waiting for her opponent to serve, producing the position-versus-time graph shown in **Figure 2–30**. **(a)** Without performing a calculation, indicate on which of the segments of the graph, A, B, or C, the player has the greatest speed. Calculate the player's speed for **(b)** segment A, **(c)** segment B, and **(d)** segment C, and show that your results verify your answers to part (a).

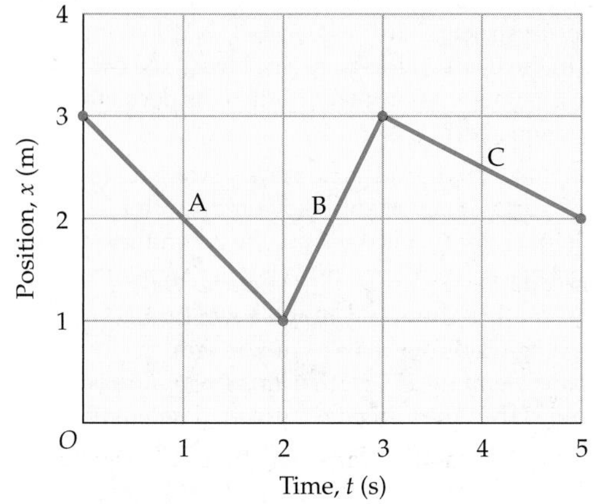

▲ **FIGURE 2–30** Problem 24

25. ••• On your wedding day you leave for the church 30.0 minutes before the ceremony is to begin, which should be plenty of time since the church is only 10.0 miles away. On the way, however, you have to make an unanticipated stop for construction work on the road. As a result, your average speed for the first 15 minutes is only 5.0 mi/h. What average speed do you need for the rest of the trip to get you to the church on time?

Section 2–3 Instantaneous Velocity

26. •• The position of a particle as a function of time is given by $x = (2.0\ \text{m/s})t + (-3.0\ \text{m/s}^3)t^3$. **(a)** Plot x versus t for time from $t = 0$ to $t = 1.0$ s. **(b)** Find the average velocity of the particle from $t = 0.35$ s to $t = 0.45$ s. **(c)** Find the average velocity from $t = 0.39$ s to $t = 0.41$ s. **(d)** Do you expect the instantaneous velocity at $t = 0.40$ s to be closer to 0.54 m/s, 0.56 m/s, or 0.58 m/s? Explain.

27. •• The position of a particle as a function of time is given by $x = (-2.00\ \text{m/s})t + (3.00\ \text{m/s}^3)t^3$. **(a)** Plot x versus t for time from $t = 0$ to $t = 1.00$ s. **(b)** Find the average velocity of the particle from $t = 0.150$ s to $t = 0.250$ s. **(c)** Find the average velocity from $t = 0.190$ s to $t = 0.210$ s. **(d)** Do you expect the instantaneous velocity at $t = 0.200$ s to be closer to -1.62 m/s, -1.64 m/s, or -1.66 m/s? Explain.

Section 2–4 Acceleration

28. • A 747 airliner reaches its takeoff speed of 173 mi/h in 35.2 s. What is the magnitude of its average acceleration?

29. • At the starting gun, a runner accelerates at 1.9 m/s^2 for 5.2 s. The runner's acceleration is zero for the rest of the race. What is the speed of the runner **(a)** at $t = 2.0$ s, and **(b)** at the end of the race?

30. • A jet makes a landing traveling due east with a speed of 115 m/s. If the jet comes to rest in 13.0 s, what is the magnitude and direction of its average acceleration?

31. • A car is traveling due north at 18.1 m/s. Find the velocity of the car after 7.50 s if its acceleration is **(a)** 1.30 m/s^2 due north, or **(b)** 1.15 m/s^2 due south.

32. •• A motorcycle moves according to the velocity-versus-time graph shown in **Figure 2–31**. Find the average acceleration of the motorcycle during each of the following segments of the motion: **(a)** A, **(b)** B, and **(c)** C.

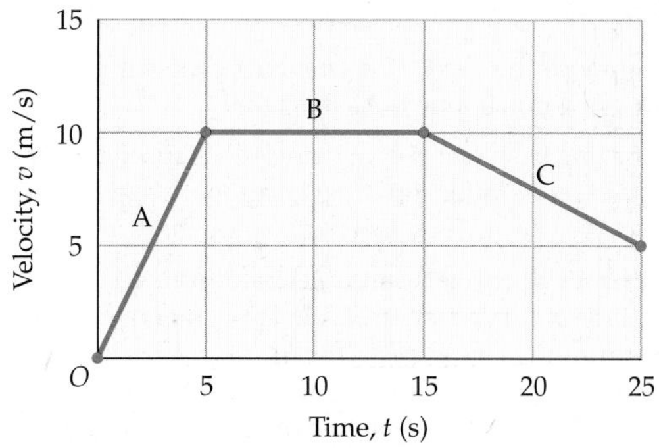

▲ **FIGURE 2–31** Problem 32

33. •• A person on horseback moves according to the velocity-versus-time graph shown in **Figure 2–32**. Find the displacement of the person for each of the following segments of the motion: **(a)** A, **(b)** B, and **(c)** C.

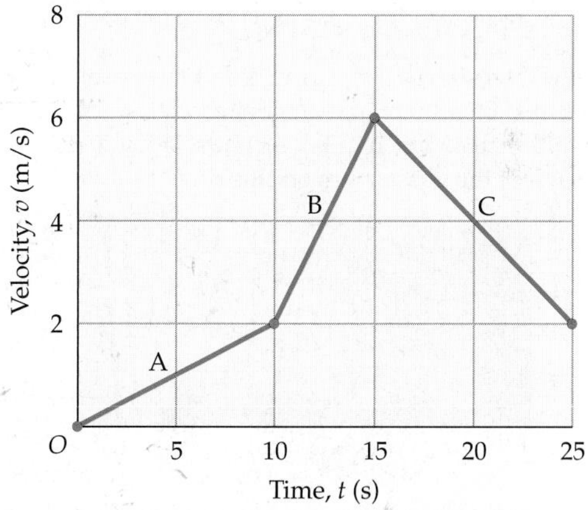

▲ **FIGURE 2–32** Problem 33

34. •• Running with an initial velocity of +11 m/s, a horse has an average acceleration of -1.81 m/s^2. How long does it take for the horse to decrease its velocity to +6.5 m/s?

35. •• **IP** Assume that the brakes in your car create a constant deceleration of 4.2 m/s^2 regardless of how fast you are driving. If you double your driving speed from 16 m/s to 32 m/s, **(a)** does the time required to come to a stop increase by a factor of two or a factor of four? Explain. Verify your answer to part (a) by calculating the stopping times for initial speeds of **(b)** 16 m/s and **(c)** 32 m/s.

36. •• **IP** In the previous problem, **(a)** does the distance needed to stop increase by a factor of two or a factor of four? Explain. Verify your answer to part (a) by calculating the stopping distances for initial speeds of **(b)** 16 m/s and **(c)** 32 m/s.

37. •• As a train accelerates away from a station, it reaches a speed of 4.7 m/s in 5.0 s. If the train's acceleration remains constant, what is its speed after an additional 6.0 s has elapsed?

38. •• A particle has an acceleration of +6.24 m/s^2 for 0.300 s. At the end of this time the particle's velocity is +9.31 m/s. What was the particle's initial velocity?

Section 2–5 Motion with Constant Acceleration

39. • Landing with a speed of 81.9 m/s, and traveling due south, a jet comes to rest in 949 m. Assuming the jet slows with constant acceleration, find the magnitude and direction of its acceleration.

40. • When you see a traffic light turn red, you apply the brakes until you come to a stop. If your initial speed was 12 m/s, and you were heading due west, what was your average velocity during braking? Assume constant deceleration.

41. •• Suppose the car in the previous problem comes to rest in 35 m. How much time does this take?

42. •• Starting from rest, a boat increases its speed to 4.12 m/s with constant acceleration. **(a)** What is the boat's average speed? **(b)** If it takes the boat 4.77 s to reach this speed, how far has it traveled?

43. •• BIO IP A cheetah can accelerate from rest to 25.0 m/s in 6.22 s. Assuming constant acceleration, **(a)** how far has the cheetah run in this time? **(b)** After sprinting for just 3.11 s, is the cheetah's speed 12.5 m/s, more than 12.5 m/s, or less than 12.5 m/s? Explain. **(c)** What is the cheetah's average speed for the first 3.11 s of its sprint? For the second 3.11 s of its sprint? **(d)** Calculate the distance covered by the cheetah in the first 3.11 s and the second 3.11 s.

Section 2–6 Applications of the Equations of Motion

44. • A child slides down a hill on a toboggan with an acceleration of 1.8 m/s^2. If she starts at rest, how far has she traveled in **(a)** 1.0 s, **(b)** 2.0 s, and **(c)** 3.0 s?

45. • **The Detonator** On a ride called the Detonator at Worlds of Fun in Kansas City, passengers accelerate straight downward from rest to 45 mi/h in 2.2 seconds. What is the average acceleration of the passengers on this ride?

The Detonator (Problem 45)

46. • **Air Bags** Air bags are designed to deploy in 10 ms. Estimate the acceleration of the front surface of the bag as it expands. Express your answer in terms of the acceleration of gravity g.

47. • **Jules Verne** In his novel *From the Earth to the Moon* (1866), Jules Verne describes a spaceship that is blasted out of a cannon, called the *Columbiad*, with a speed of 12,000 yards/s. The *Columbiad* is 900 ft long, but part of it is packed with powder, so the spaceship accelerates over a distance of only 700 ft. Estimate the acceleration experienced by the occupants of the spaceship during launch. Give your answer in m/s^2. (Verne realized that the "travelers would . . . encounter a violent recoil," but he probably didn't know that people generally lose consciousness if they experience accelerations greater than about $7g \sim 70$ m/s^2.)

48. •• BIO **Bacterial Motion** Approximately 0.1% of the bacteria in an adult human's intestines are *Escherichia coli*. These bacteria have been observed to move with speeds up to 15 μm/s and maximum accelerations of 166 μm/s^2. Suppose an *E. coli* bacterium in your intestines starts at rest and accelerates at 156 μm/s^2. How much **(a)** time and **(b)** distance are required for the bacterium to reach a speed of 12 μm/s?

49. •• Two cars drive on a straight highway. At time $t = 0$, car 1 passes mile marker 0 traveling due east with a speed of 20.0 m/s. At the same time, car 2 is 1.0 km east of mile marker 0 traveling at 30.0 m/s due west. Car 1 is speeding up with an acceleration of magnitude 2.5 m/s^2, and car 2 is slowing down with an acceleration of magnitude 3.2 m/s^2. **(a)** Write x-versus-t equations of motion for both cars, taking east as the positive direction. **(b)** At what time do the cars pass next to one another?

50. •• **A Meteorite Strikes** On October 9, 1992, a 27-pound meteorite struck a car in Peekskill, NY, leaving a dent 22 cm deep in the trunk. If the meteorite struck the car with a speed of 130 m/s, what was the magnitude of its deceleration, assuming it to be constant?

51. •• A rocket blasts off and moves straight upward from the launch pad with constant acceleration. After 3.0 s the rocket is at a height of 77 m. **(a)** What are the magnitude and direction of the rocket's acceleration? **(b)** What is its speed at this time?

52. •• IP You are driving through town at 12.0 m/s when suddenly a ball rolls out in front of you. You apply the brakes and begin decelerating at 3.5 m/s^2. **(a)** How far do you travel before stopping? **(b)** When you have traveled only half the distance in part (a), is your speed 6.0 m/s, greater than 6.0 m/s, or less than 6.0 m/s? Support your answer with a calculation.

53. •• IP You are driving through town at 16 m/s when suddenly a car backs out of a driveway in front of you. You apply the brakes and begin decelerating at 3.2 m/s^2. **(a)** How much time does it take to stop? **(b)** After braking half the time found in part (a), is your speed 8.0 m/s, greater than 8.0 m/s, or less than 8.0 m/s? Support your answer with a calculation. **(c)** If the car backing out was initially 55 m in front of you, what is the maximum reaction time you can have before hitting the brakes and still avoid hitting the car?

54. •• IP BIO **A Tongue's Acceleration** When a chameleon captures an insect, its tongue can extend 16 cm in 0.10 s. **(a)** Find the magnitude of the tongue's acceleration, assuming it to be constant. **(b)** In the first 0.050 s, does the tongue extend 8.0 cm, more than 8.0 cm, or less than 8.0 cm? Support your conclusion with a calculation.

It's not polite to reach! (Problem 54)

55. •• IP Coasting due west on your bicycle at 8.4 m/s, you encounter a sandy patch of road 7.2 m across. When you leave the sandy patch your speed has been reduced by 2.0 m/s to 6.4 m/s. **(a)** Assuming the sand causes a constant acceleration, what was the bicycle's acceleration in the sandy patch? Give both magnitude and direction. **(b)** How long did it take to cross the sandy patch? **(c)** Suppose you enter the sandy patch with a speed of only 5.4 m/s. Is your final speed in this case 3.4 m/s, more than 3.4 m/s, or less than 3.4 m/s? Explain.

56. •• **BIO Surviving a Large Deceleration** On July 13, 1977, while on a test drive at Britain's Silverstone racetrack, the throttle on David Purley's car stuck wide open. The resulting crash subjected Purley to the greatest "g-force" ever survived by a human—he decelerated from 173 km/h to zero in a distance of only about 0.66 m. Calculate the magnitude of the acceleration experienced by Purley (assuming it to be constant), and express your answer in units of the acceleration of gravity, $g = 9.81\ \mathrm{m/s^2}$.

57. •• **IP** A boat is cruising in a straight line at a constant speed of 2.6 m/s when it is shifted into neutral. After coasting 12 m the engine is engaged again, and the boat resumes cruising at the reduced constant speed of 1.6 m/s. Assuming constant acceleration while coasting, **(a)** how long did it take for the boat to coast the 12 m? **(b)** What was the boat's acceleration while it was coasting? **(c)** When the boat had coasted for 6.0 m, was its speed 2.1 m/s, more than 2.1 m/s, or less than 2.1 m/s? Explain.

58. •• A model rocket rises with constant acceleration to a height of 3.2 m, at which point its speed is 26.0 m/s. **(a)** How much time does it take for the rocket to reach this height? **(b)** What was the magnitude of the rocket's acceleration? **(c)** Find the height and speed of the rocket 0.10 s after launch.

59. •• The infamous chicken is dashing toward home plate with a speed of 5.8 m/s when he decides to hit the dirt. The chicken slides for 1.1 s, just reaching the plate as he stops (safe, of course). **(a)** What is the magnitude and direction of the chicken's acceleration? **(b)** How far did the chicken slide?

60. •• A bicyclist is finishing his repair of a flat tire when a friend rides by with a constant speed of 3.5 m/s. Two seconds later the bicyclist hops on his bike and accelerates at $2.4\ \mathrm{m/s^2}$ until he catches his friend. **(a)** How much time does it take until he catches his friend? **(b)** How far has he traveled in this time? **(c)** What is his speed when he catches up?

61. •• A car in stop-and-go traffic starts at rest, moves forward 13 m in 8.0 s, then comes to rest again. The velocity-versus-time plot for this car is given in **Figure 2–33**. What distance does the car cover in **(a)** the first 4.0 seconds of its motion and **(b)** the last 2.0 seconds of its motion? **(c)** What is the constant speed V that characterizes the middle portion of its motion?

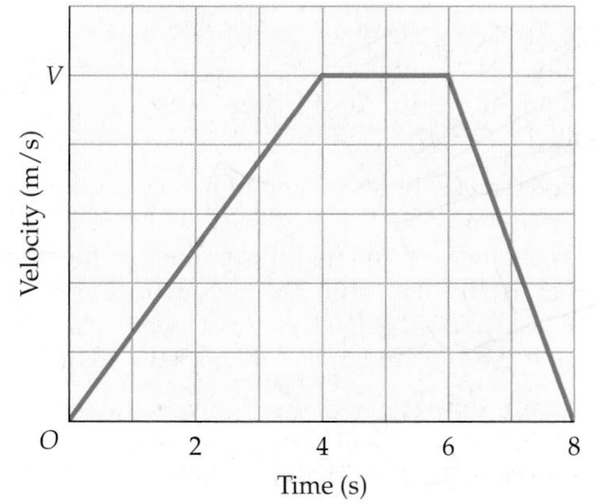

▲ **FIGURE 2–33** Problem 61

62. ••• A car and a truck are heading directly toward one another on a straight and narrow street, but they avoid a head-on collision by simultaneously applying their brakes at $t = 0$. The resulting velocity versus time graphs are shown in **Figure 2–34**. What is the separation between the car and the truck when they have come to rest, given that at $t = 0$ the car is at $x = 15$ m and the truck is at $x = -35$ m. (Note that this information determines which line in the graph corresponds to which vehicle.)

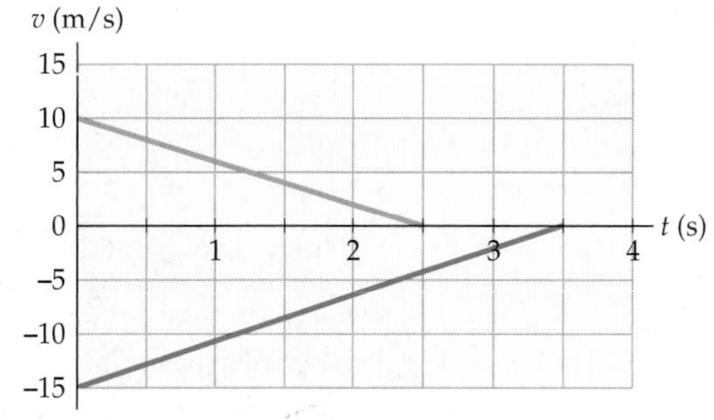

▲ **FIGURE 2–34** Problem 62

63. ••• In a physics lab, students measure the time it takes a small cart to slide a distance of 1.00 m on a smooth track inclined at an angle θ above the horizontal. Their results are given in the following table.

θ	10.0°	20.0°	30.0°
time, s	1.08	0.770	0.640

(a) Find the magnitude of the cart's acceleration for each angle. **(b)** Show that your results for part (a) are in close agreement with the formula, $a = g \sin \theta$. (We will derive this formula in Chapter 5.)

Section 2–7 Freely Falling Objects

64. • Legend has it that Isaac Newton was hit on the head by a falling apple, thus triggering his thoughts on gravity. Assuming the story to be true, estimate the speed of the apple when it struck Newton.

65. • The cartoon below shows a car in free fall. Is the statement made in the cartoon accurate? Justify your answer.

"IT GOES FROM ZERO TO SIXTY IN ABOUT THREE SECONDS."

(Problems 65 and 66)

66. • Referring to the cartoon in Problem 65, how long would it take for the car to go from 0 to 30 mi/h?

67. • **Jordan's Jump** Michael Jordan's vertical leap is reported to be 48 inches. What is his takeoff speed? Give your answer in meters per second.

68. • **BIO** Gulls are often observed dropping clams and other shellfish from a height to the rocks below, as a means of opening the shells. If a seagull drops a shell from rest at a height of 14 m, how fast is the shell moving when it hits the rocks?

69. • A volcano launches a lava bomb straight upward with an initial speed of 28 m/s. Taking upward to be the positive direction, find the speed and direction of motion of the lava bomb **(a)** 2.0 seconds and **(b)** 3.0 seconds after it is launched.

70. • **An Extraterrestrial Volcano** The first active volcano observed outside the Earth was discovered in 1979 on Io, one of the moons of Jupiter. The volcano was observed to be ejecting material to a height of about 2.00×10^5 m. Given that the acceleration of gravity on Io is 1.80 m/s^2, find the initial velocity of the ejected material.

71. • **BIO Measure Your Reaction Time** Here's something you *can* try at home—an experiment to measure your reaction time. Have a friend hold a ruler by one end, letting the other end hang down vertically. At the lower end, hold your thumb and index finger on either side of the ruler, ready to grip it. Have your friend release the ruler without warning. Catch it as quickly as you can. If you catch the ruler 5.2 cm (~2 inches) from the lower end, what is your reaction time?

How fast are your reactions? (Problem 71)

72. •• Bill steps off a 3.0-m-high diving board and drops to the water below. At the same time, Ted jumps upward with a speed of 4.2 m/s from a 1.0-m-high diving board. Choosing the origin to be at the water's surface, and upward to be the positive x direction, write x-versus-t equations of motion for both Bill and Ted.

73. •• Repeat the previous problem, this time with the origin 3.0 m above the water, and with downward as the positive x direction.

74. •• On a hot summer day in the state of Washington while kayaking, I saw several swimmers jump from a railroad bridge into the Snohomish River below. The swimmers stepped off the bridge, and I estimated that they hit the water 1.5 s later. **(a)** How high was the bridge? **(b)** How fast were the swimmers moving when they hit the water? **(c)** What would the swimmer's drop time be if the bridge were twice as high?

75. •• **Highest Water Fountain** The world's highest fountain of water is located, appropriately enough, in Fountain Hills, Arizona. The fountain rises to a height of 560 ft (5 feet higher than the Washington Monument). **(a)** What is the initial speed of the water? **(b)** How long does it take for water to reach the top of the fountain?

76. •• Wrongly called for a foul, an angry basketball player throws the ball straight down to the floor. If the ball bounces straight up and returns to the floor 2.8 s after first striking it, what was the ball's greatest height above the floor?

77. •• To celebrate a victory, a pitcher throws her glove straight upward with an initial speed of 6.0 m/s. **(a)** How long does it take for the glove to return to the pitcher? **(b)** How long does it take for the glove to reach its maximum height?

78. •• **IP** Standing at the edge of a cliff 32.5 m high, you drop a ball. Later, you throw a second ball downward with an initial speed of 11.0 m/s. **(a)** Which ball has the greater *increase* in speed when it reaches the base of the cliff, or do both balls speed up by the same amount? **(b)** Verify your answer to part (a) with a calculation.

79. •• You shoot an arrow into the air. Two seconds later (2.00 s) the arrow has gone straight upward to a height of 30.0 m above its launch point. **(a)** What was the arrow's initial speed? **(b)** How long did it take for the arrow to first reach a height of 15.0 m above its launch point?

80. •• While riding on an elevator descending with a constant speed of 3.0 m/s, you accidentally drop a book from under your arm. **(a)** How long does it take for the book to reach the elevator floor, 1.2 m below your arm? **(b)** What is the book's speed relative to you when it hits the elevator floor?

81. •• A hot-air balloon is descending at a rate of 2.0 m/s when a passenger drops a camera. If the camera is 45 m above the ground when it is dropped, **(a)** how long does it take for the camera to reach the ground, and **(b)** what is its velocity just before it lands? Let upward be the positive direction for this problem.

82. •• **IP** Standing side by side, you and a friend step off a bridge at different times and fall for 1.6 s to the water below. Your friend goes first, and you follow after she has dropped a distance of 2.0 m. **(a)** When your friend hits the water, is the separation between the two of you 2.0 m, less than 2.0 m, or more than 2.0 m? **(b)** Verify your answer to part (a) with a calculation.

83. •• A model rocket blasts off and moves upward with an acceleration of 12 m/s^2 until it reaches a height of 26 m, at which point its engine shuts off and it continues its flight in free fall. **(a)** What is the maximum height attained by the rocket? **(b)** What is the speed of the rocket just before it hits the ground? **(c)** What is the total duration of the rocket's flight?

84. ••• **Hitting the "High Striker"** A young woman at a carnival steps up to the "high striker," a popular test of strength where the contestant hits one end of a lever with a mallet, propelling a small metal plug upward toward a bell. She gives the mallet a mighty swing and sends the plug to the top of the striker, where it rings the bell. **Figure 2–35** shows the corresponding position-versus-time plot for the plug. Using the information given in the plot, answer the following questions: **(a)** What is the average speed of the plug during its upward journey? **(b)** By how much does the speed of the plug decrease during its upward journey? **(c)** What is the initial speed of the plug? (Assume the plug to be in free fall during its upward motion, with no effects of air resistance or friction.)

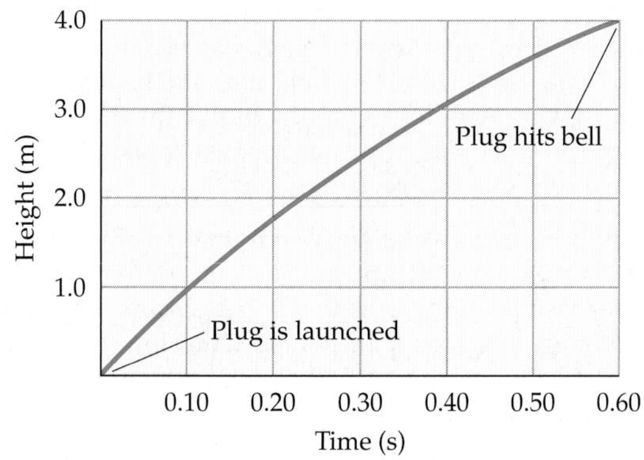

▲ **FIGURE 2–35** Problem 84

85. ••• While sitting on a tree branch 10.0 m above the ground, you drop a chestnut. When the chestnut has fallen 2.5 m, you throw a second chestnut straight down. What initial speed must you give the second chestnut if they are both to reach the ground at the same time?

General Problems

86. • In a well-known Jules Verne novel, Phileas Fogg travels around the world in 80 days. What was Mr. Fogg's approximate average speed during his adventure?

87. • An astronaut on the Moon drops a rock straight downward from a height of 1.25 m. If the acceleration of gravity on the Moon is 1.62 m/s^2, what is the speed of the rock just before it lands?

88. • You jump from the top of a boulder to the ground 1.5 m below. Estimate your deceleration on landing.

89. • **A Supersonic Waterfall** Geologists have learned of periods in the past when the Strait of Gibraltar closed off, and the Mediterranean Sea dried out and become a desert. Later, when the strait reopened, a massive saltwater waterfall was created. According to geologists, the water in this waterfall was supersonic; that is, it fell with speeds in excess of the speed of sound. Ignoring air resistance, what is the minimum height necessary to create a supersonic waterfall? (The speed of sound may be taken to be 340 m/s.)

90. •• Astronauts on a distant planet throw a rock straight upward and record its motion with a video camera. After digitizing their video, they are able to produce the graph of height, y, versus time, t, shown in **Figure 2–36.** **(a)** What is the acceleration of gravity on this planet? **(b)** What was the initial speed of the rock?

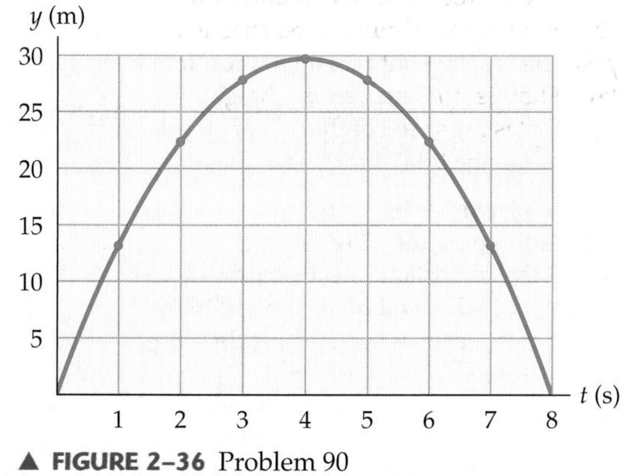

▲ **FIGURE 2–36** Problem 90

91. •• **Bam!** The first word spoken on the surface of the Moon after Apollo 15 landed on July 30, 1971, was "Bam!" The reason for this was that the rocket engine was shut off a bit early, when the lander was still 4.3 ft above the lunar surface and moving downward with a speed of 0.50 ft/s. From that point on the lander descended in lunar free fall. **(a)** What was the impact speed when the lander touched down? Give your answer in feet per second, the units used by the astronauts. **(b)** How long did it take for the lander to drop the last 4.3 ft? (The acceleration of gravity on the Moon is 1.62 m/s^2.)

92. •• **Drop Tower** NASA operates a 2.2-second drop tower at the Glenn Research Center in Cleveland, Ohio. At this facility, experimental packages are dropped from the top of the tower, on the 8th floor of the building. During their 2.2 seconds of free fall, experiments experience a microgravity environment similar to that of a spacecraft in orbit. **(a)** What is the drop distance of a 2.2-s tower? **(b)** How fast are the experiments traveling when they hit the air bags at the bottom of the tower? **(c)** If the experimental package comes to rest over a distance of 0.75 m upon hitting the air bags, what is the average stopping acceleration?

93. •• **IP** A youngster bounces straight up and down on a trampoline. Suppose she doubles her initial speed from 2.0 m/s to 4.0 m/s. **(a)** By what factor does her time in the air increase? **(b)** By what factor does her maximum height increase? **(c)** Verify your answers to parts (a) and (b) with an explicit calculation.

94. •• At the 18th green of the U. S. Open you need to make a 20.5 ft putt to win the tournament. When you hit the ball, giving it an initial speed of 1.57 m/s, it stops 6.00 ft short of the hole. **(a)** Assuming the deceleration caused by the grass is constant, what should the initial speed have been to just make the putt? **(b)** What initial speed do you need to make the remaining 6.00-ft putt?

95. •• **IP** A popular entertainment at some carnivals is the blanket toss (see photo, p. 38). **(a)** If a person is thrown to a maximum height of 28.0 ft above the blanket, how long does she spend in the air? **(b)** Is the amount of time the person is above a height of 14.0 ft more than, less than, or equal to the amount of time the person is below a height of 14.0 ft? Explain. **(c)** Verify your answer to part (b) with a calculation.

96. •• Referring to Conceptual Checkpoint 2–5, find the separation between the rocks at **(a)** $t = 1.0$ s, **(b)** $t = 2.0$ s, and **(c)** $t = 3.0$ s, where time is measured from the instant the second rock is dropped. **(d)** Verify that the separation increases linearly with time.

97. •• **IP** A glaucous-winged gull, ascending straight upward at 5.20 m/s, drops a shell when it is 12.5 m above the ground. **(a)** What is the magnitude and direction of the shell's acceleration just after it is released? **(b)** Find the maximum height above the ground reached by the shell. **(c)** How long does it take for the shell to reach the ground? **(d)** What is the speed of the shell at this time?

98. •• A doctor, preparing to give a patient an injection, squirts a small amount of liquid straight upward from a syringe. If the liquid emerges with a speed of 1.5 m/s, **(a)** how long does it take for it to return to the level of the syringe? **(b)** What is the maximum height of the liquid above the syringe?

99. •• A hot-air balloon has just lifted off and is rising at the constant rate of 2.0 m/s. Suddenly one of the passengers realizes she has left her camera on the ground. A friend picks it up and tosses it straight upward with an initial speed of 13 m/s. If the passenger is 2.5 m above her friend when the camera is tossed, how high is she when the camera reaches her?

100. ••• In the previous problem, what is the minimum initial speed of the camera if it is to just reach the passenger? (Hint: When the camera is thrown with its minimum speed, its speed on reaching the passenger is the same as the speed of the passenger.)

101. ••• **Old Faithful** Watching Old Faithful erupt, you notice that it takes a time t for water to emerge from the base of the geyser and reach its maximum height. **(a)** What is the height of the geyser, and **(b)** what is the initial speed of the water? Evaluate your expressions for **(c)** the height and **(d)** the initial speed for a measured time of 1.65 s.

102. ••• **IP** A ball is thrown upward with an initial speed v_0. When it reaches the top of its flight, at a height h, a second ball is thrown upward with the same initial velocity. **(a)** Sketch an x-versus-t plot for each ball. **(b)** From your graph, decide whether the balls cross paths at $h/2$, above $h/2$, or below $h/2$. **(c)** Find the height where the paths cross.

103. ••• Weights are tied to each end of a 20.0-cm string. You hold one weight in your hand and let the other hang vertically a height h above the floor. When you release the weight in your hand, the two weights strike the ground one after the other with audible thuds. Find the value of h for which the time between release and the first thud is equal to the time between the first thud and the second thud.

104. ••• A ball, dropped from rest, covers three-quarters of the distance to the ground in the last second of its fall. **(a)** From what height was the ball dropped? **(b)** What was the total time of fall?

105. ••• A stalactite on the roof of a cave drips water at a steady rate to a pool 4.0 m below. As one drop of water hits the pool, a second drop is in the air, and a third is just detaching from the stalactite. **(a)** What is the position and velocity of the second drop when the first drop hits the pool? **(b)** How many drops per minute fall into the pool?

106. ••• You drop a ski glove from a height h onto fresh snow, and it sinks to a depth d before coming to rest. **(a)** In terms of g and h, what is the speed of the glove when it reaches the snow? **(b)** What are the magnitude and direction of the glove's acceleration as it moves through the snow, assuming it to be constant? Give your answer in terms of g, h, and d.

107. ••• To find the height of an overhead power line, you throw a ball straight upward. The ball passes the line on the way up after 0.75 s, and passes it again on the way down 1.5 s after it was tossed. What is the height of the power line and the initial speed of the ball?

108. ••• Suppose the first rock in Conceptual Checkpoint 2–5 drops through a height h before the second rock is released from rest. Show that the separation between the rocks, S, is given by the following expression:

$$S = h + \left(\sqrt{2gh}\right)t$$

In this result, the time t is measured from the time the second rock is dropped.

109. ••• An arrow is fired with a speed of 20.0 m/s at a block of Styrofoam resting on a smooth surface. The arrow penetrates a certain distance into the block before coming to rest relative to it. During this process the arrow's deceleration has a magnitude of 1550 m/s^2 and the block's acceleration has a magnitude of 450 m/s^2. **(a)** How long does it take for the arrow to stop moving with respect to the block? **(b)** What is the common speed of the arrow and block when this happens? **(c)** How far into the block does the arrow penetrate?

110. ••• Sitting in a second-story apartment, a physicist notices a ball moving straight upward just outside her window. The ball is visible for 0.25 s as it moves a distance of 1.05 m from the bottom to the top of the window. **(a)** How long does it take before the ball reappears? **(b)** What is the greatest height of the ball above the top of the window?

111. ••• **The Quadratic Formula from Kinematics** In this problem we show how the kinematic equations of motion can be used to derive the quadratic formula. First, consider an object with an initial position x_0, an initial velocity v_0, and an acceleration a. To find the time when this object reaches the position $x = 0$ we can use the quadratic formula, or apply the following two-step procedure: **(a)** Use Equation 2–12 to show that the velocity of the object when it reaches $x = 0$ is given by $v = \pm\sqrt{v_0{}^2 - 2ax_0}$. **(b)** Use Equation 2–7 to show that the time corresponding to the velocity found in part (a) is $t = \dfrac{-v_0 \pm \sqrt{v_0{}^2 - 2ax_0}}{a}$. **(c)** To complete our derivation, show that the result of part (b) is the same as applying the quadratic formula to $x = x_0 + v_0t + \frac{1}{2}at^2 = 0$.

Interactive Problems

112. •• ***Referring to Example 2–9*** Suppose the speeder (red car) is traveling with a constant speed of 25 m/s, and that the maximum acceleration of the police car (blue car) is 3.8 m/s^2. If the police car is to start from rest and catch the speeder in 15 s or less, what is the maximum head-start distance the speeder can have? Measure time from the moment the police car starts.

113. •• ***Referring to Example 2–9*** The speeder passes the position of the police car with a constant speed of 15 m/s. The police car immediately starts from rest and pursues the speeder with constant acceleration. What acceleration must the police car have if it is to catch the speeder in 7.0 s? Measure time from the moment the police car starts.

114. •• **IP** ***Referring to Example 2–12*** **(a)** In Example 2–12, the bag of sand is released at 20.0 m and reaches a maximum height of 22 m. If the bag had been released at 30.0 m instead, with everything else remaining the same, would its maximum height be 32 m, greater than 32 m, or less than 32 m? **(b)** Find the speed of the bag just before it lands when it is released from 30.0 m.

115. •• ***Referring to Example 2–12*** Suppose the balloon is descending with a constant speed of 4.2 m/s when the bag of sand comes loose at a height of 35 m. **(a)** How long is the bag in the air? **(b)** What is the speed of the bag when it is 15 m above the ground?

3 Vectors in Physics

3–1	Scalars Versus Vectors	56
3–2	The Components of a Vector	56
3–3	Adding and Subtracting Vectors	61
3–4	Unit Vectors	64
3–5	Position, Displacement, Velocity, and Acceleration Vectors	65
3–6	Relative Motion	69

The points of the compass have long been used as a framework for indicating directions. In physics, we more frequently use x and y rather than N, S, E, and W. Either way, specifying a direction as well as a magnitude is essential to defining one of the physicist's basic tools, the vector.

Of all the mathematical tools used in this book, perhaps none is more important than the vector. In the next chapter, for example, we use vectors to extend our study of motion from one dimension to two dimensions. More generally, vectors are *indispensable* when a physical quantity has a direction associated with it. Suppose, for example, that a pilot wants to fly from Denver to Dallas. If the air is still, the pilot can simply head the plane toward the destination. If there is a wind blowing from west to east, however, the pilot must use vectors to determine the correct heading so that the plane and its passengers will arrive in Dallas and not Little Rock.

In this chapter we discuss what a vector is, how it differs from a scalar, and how it can represent a physical quantity. We also show how to find the components of a vector and how to add and subtract vectors. All of these techniques are used time and again throughout the book. Other useful aspects of vectors, such as how to multiply them, will be presented in later chapters when the need arises.

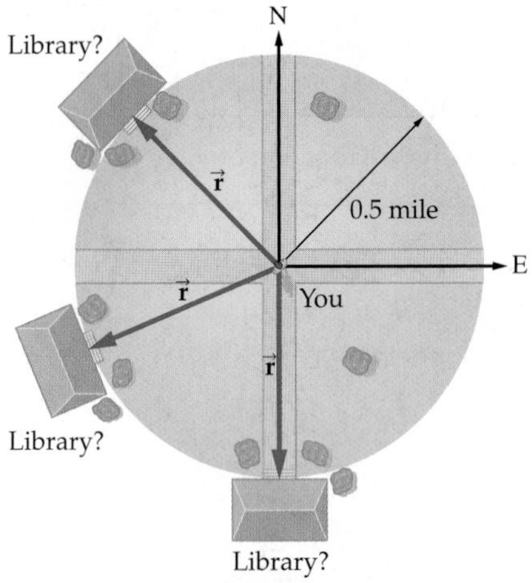

▲ **FIGURE 3–1 Distance and direction**
If you know only that the library is 0.5 mi from you, it could lie anywhere on a circle of radius 0.5 mi. If, instead, you are told the library is 0.5 mi northwest, you know its precise location.

▲ The information given by this sign includes both a distance and a direction for each city. In effect, the sign defines a displacement vector for each of these destinations.

3–1 Scalars Versus Vectors

Numbers can represent many quantities in physics. For example, a numerical value, together with the appropriate units, can specify the volume of a container, the temperature of the air, or the time of an event. In physics, a number with its units is referred to as a **scalar:**

- A scalar is a number with units. It can be positive, negative, or zero.

Sometimes, however, a scalar isn't enough to adequately describe a physical quantity—in many cases, a direction is needed as well. For example, suppose you're walking in an unfamiliar city and you want directions to the library. You ask a passerby, "Do you know where the library is?" If the person replies "Yes," and walks on, he hasn't been too helpful. If he says, "Yes, it is half a mile from here," that is more helpful, but you still don't know where it is. The library could be anywhere on a circle of radius one-half mile, as shown in **Figure 3–1**. To pin down the location, you need a reply such as, "Yes, the library is half a mile northwest of here." With both a distance *and* a direction, you know the location of the library.

Thus, if you walk northwest for half a mile you arrive at the library, as indicated by the upper left arrow in Figure 3–1. The arrow points in the direction traveled, and its **magnitude,** 0.5 mi in this case, represents the distance covered. In general, a quantity that is specified by both a *magnitude* and a *direction* is represented by a **vector:**

- A vector is a mathematical quantity with both a direction and a magnitude.

In the example of walking to the library, the vector corresponding to the trip is the displacement vector. Other examples of vector quantities are the velocity and the acceleration of an object. For example, the magnitude of a velocity vector is its speed, and its direction is the direction of motion, as we shall see later in this chapter.

When we indicate a vector on a diagram or a sketch, we draw an arrow, as in Figure 3–1. To indicate a vector with a written symbol, we use **boldface** for the vector itself, with a small arrow above it to remind us of its vector nature, and *italic* for its magnitude. Thus, for example, the upper-left vector in Figure 3–1 is designated by the symbol $\vec{\mathbf{r}}$, and its magnitude is $r = 0.5$ mi. (When we represent a vector in a graph, we sometimes label it with the corresponding boldface symbol, and sometimes with the appropriate magnitude.) It is common in handwritten material to draw a small arrow over the vector's symbol, which is very similar to the way vectors are represented in this text.

3–2 The Components of a Vector

When we discussed directions for finding a library in the previous section, we pointed out that knowing the magnitude and direction angle—0.5 mi northwest—gives its precise location. We left out one key element in actually *getting* to the library, however. In most cities it would not be possible to simply walk in a straight line for 0.5 mi directly to the library, since to do so would take you through buildings where there are no doors, through people's backyards, and through all kinds of other obstructions. In fact, if the city streets are laid out along north–south and east–west directions, you might instead walk west for a certain distance, then turn and proceed north an equal distance until you reach the library, as illustrated in **Figure 3–2**. What you have just done is "resolved" the displacement vector $\vec{\mathbf{r}}$ between you and the library into east–west and north–south "components."

In general, to find the components of a vector we need to set up a coordinate system. In two dimensions we choose an origin, *O*, and a positive direction for both the *x* and the *y* axes, as in **Figure 3–3**. If the system were three-dimensional, we would also indicate a *z* axis.

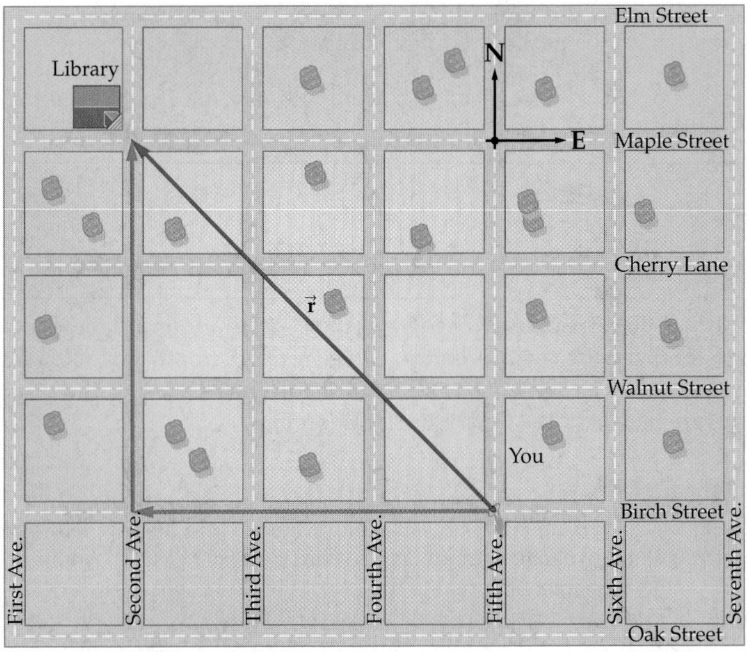

◀ **FIGURE 3–2 A walk along city streets to the library**

By taking the indicated path, we have "resolved" the vector $\vec{\mathbf{r}}$ into east–west and north–south components.

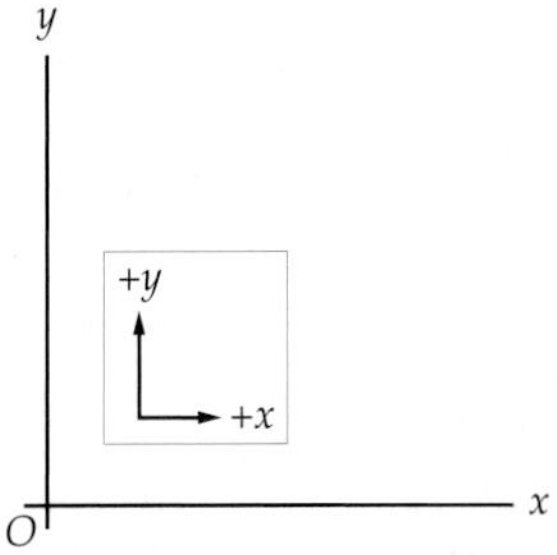

▲ **FIGURE 3–3 A two-dimensional coordinate system**

The positive x and y directions are indicated in this shorthand form.

Now, a vector is defined by its magnitude (indicated by the length of the arrow representing the vector) and its direction. For example, suppose an ant leaves its nest at the origin and, after foraging for some time, is at the location given by the vector $\vec{\mathbf{r}}$ in **Figure 3–4 (a)**. This vector has a magnitude $r = 1.50$ m and points in a direction $\theta = 25.0°$ above the x axis. Equivalently, $\vec{\mathbf{r}}$ can be defined by saying that it extends a distance r_x in the x direction and a distance r_y in the y direction, as shown in **Figure 3–4 (b)**. The quantities r_x and r_y are referred to as the x and y **scalar components** of the vector $\vec{\mathbf{r}}$.

We can find r_x and r_y by using standard trigonometric relations, as summarized in the Problem-Solving Note on this page. Referring to Figure 3–4 (b), we see that

$$r_x = r\cos 25.0° = (1.50\text{ m})(0.906) = 1.36\text{ m}$$

and

$$r_y = r\sin 25.0° = (1.50\text{ m})(0.423) = 0.634\text{ m}$$

Thus, we can say that the ant's final displacement is equivalent to what it would be if the ant had simply walked 1.36 m in the x direction and then 0.634 m in the y direction.

To show the equivalence of these two ways of describing a vector, let's start with the components of $\vec{\mathbf{r}}$, as determined previously, and use them to calculate the magnitude r and the angle θ. First, note that r_x, r_y, and r form a right triangle with r as the hypotenuse. Thus, we can use the Pythagorean theorem (Appendix A) to find r in terms of r_x and r_y. This gives

$$r = \sqrt{r_x^2 + r_y^2} = \sqrt{(1.36\text{ m})^2 + (0.634\text{ m})^2} = \sqrt{2.25\text{ m}^2} = 1.50\text{ m}$$

PROBLEM-SOLVING NOTE

A Vector and Its Components

Given the magnitude and direction of a vector, find its components:

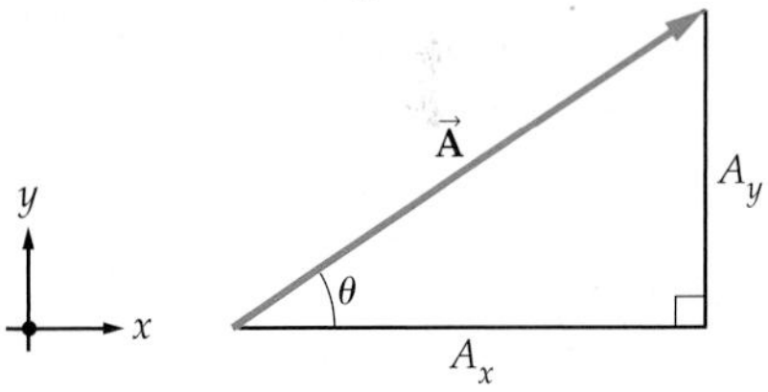

$$A_x = A\cos\theta$$
$$A_y = A\sin\theta$$

Given the components of a vector, find its magnitude and direction:

$$A = \sqrt{A_x^2 + A_y^2}$$
$$\theta = \tan^{-1}\frac{A_y}{A_x}$$

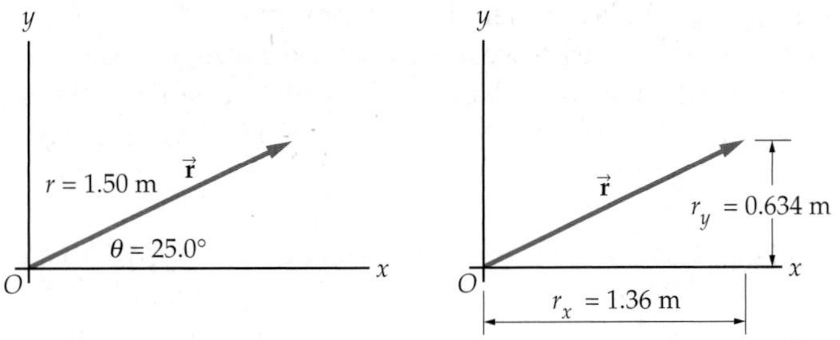

(a) **(b)**

◀ **FIGURE 3–4 A vector and its scalar components**

(a) The vector $\vec{\mathbf{r}}$ is defined by its length ($r = 1.50$ m) and its direction angle ($\theta = 25.0°$) measured counterclockwise from the positive x axis. **(b)** Alternatively, the vector $\vec{\mathbf{r}}$ can be defined by its x component, $r_x = 1.36$ m, and its y component, $r_y = 0.634$ m.

as expected. Second, we can use any two sides of the triangle to obtain the angle θ, as shown in the next three calculations:

$$\theta = \sin^{-1}\left(\frac{0.634 \text{ m}}{1.50 \text{ m}}\right) = \sin^{-1}(0.423) = 25.0°$$

$$\theta = \cos^{-1}\left(\frac{1.36 \text{ m}}{1.50 \text{ m}}\right) = \cos^{-1}(0.907) = 25.0°$$

$$\theta = \tan^{-1}\left(\frac{0.634 \text{ m}}{1.36 \text{ m}}\right) = \tan^{-1}(0.466) = 25.0°$$

In some situations we know a vector's magnitude and direction; in other cases we are given the vector's components. You will find it useful to be able to convert quickly and easily from one description of a vector to the other using trigonometric functions and the Pythagorean theorem.

EXAMPLE 3–1 Determining the Height of a Cliff

REAL-WORLD PHYSICS In the Jules Verne novel *Mysterious Island*, Captain Cyrus Harding wants to find the height of a cliff. He stands with his back to the base of the cliff, then marches straight away from it for 5.00×10^2 ft. At this point he lies on the ground and measures the angle from the horizontal to the top of the cliff. If the angle is 34.0°, **(a)** how high is the cliff? **(b)** What is the straight-line distance from Captain Harding to the top of the cliff?

Picture the Problem

Our sketch shows Cyrus Harding making his measurement of the angle, $\theta = 34.0°$, to the top of the cliff. The relevant triangle for this problem is also indicated. Note that the opposite side of the triangle is the height of the cliff, h; the adjacent side is the distance from the base of the cliff to Harding, $b = 5.00 \times 10^2$ ft; and finally, the hypotenuse is the distance, d, from Harding to the top of the cliff.

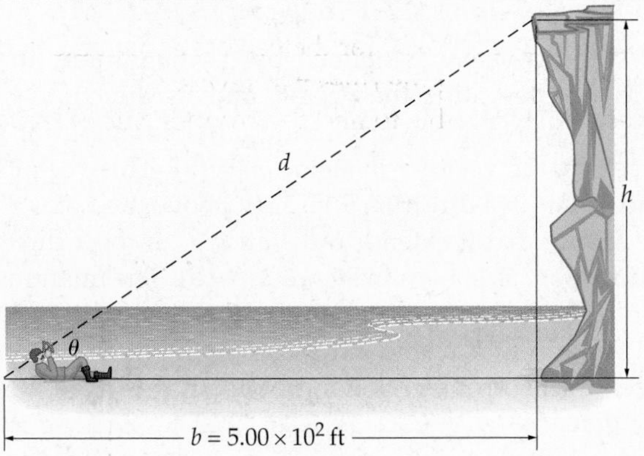

Strategy

The tangent of θ is the height of the triangle divided by the base: $\tan\theta = h/b$. Since we know both θ and the base, we can find the height using this relation. Similarly, the distance from Harding to the top of the cliff can be obtained by solving $\cos\theta = b/d$ for d.

Solution

Part (a)

1. Use $\tan\theta = h/b$ to solve for the height of the cliff, h: $h = b\tan\theta = (5.00 \times 10^2 \text{ ft})\tan 34.0° = 337 \text{ ft}$

Part (b)

2. Similarly, use $\cos\theta = b/d$ to solve for the distance d from Captain Harding to the top of the cliff: $d = \dfrac{b}{\cos\theta} = \dfrac{5.00 \times 10^2 \text{ ft}}{\cos 34.0°} = 603 \text{ ft}$

Insight

An alternative way to solve part **(b)** is to use the Pythagorean theorem:

$$d = \sqrt{h^2 + b^2} = \sqrt{(337 \text{ ft})^2 + (5.00 \times 10^2 \text{ ft})^2} = 603 \text{ ft}$$

Thus, if we let $\vec{\mathbf{r}}$ denote the vector from Cyrus Harding to the top of the cliff, as shown here, its magnitude is 603 ft and its direction is 34.0° above the x axis. Alternatively, the x component of $\vec{\mathbf{r}}$ is 5.00×10^2 ft and its y component is 337 ft.

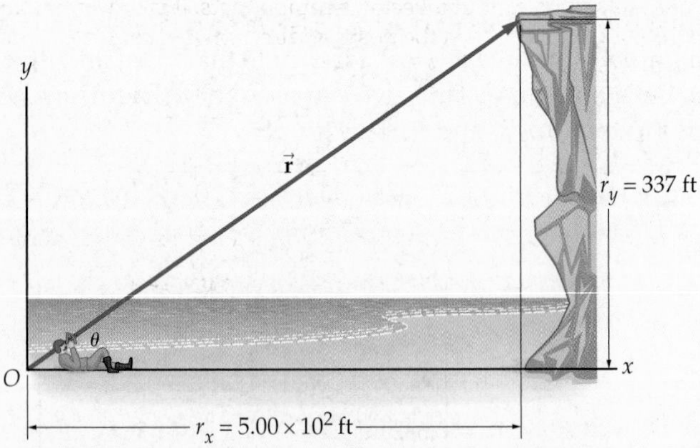

Practice Problem

What angle would Cyrus Harding have found if he had walked 6.00×10^2 ft from the cliff to make his measurement? [**Answer:** $\theta = 29.3°$]

Some related homework problems: Problem 1, Problem 13

EXERCISE 3–1

a. Find A_x and A_y for the vector $\vec{\mathbf{A}}$ with magnitude and direction given by $A = 3.5$ m and $\theta = 66°$, respectively.

b. Find B and θ for the vector $\vec{\mathbf{B}}$ with components $B_x = 75.5$ m and $B_y = 6.20$ m.

Solution

a. $A_x = 1.4$ m, $A_y = 3.2$ m

b. $B = 75.8$ m, $\theta = 4.69°$

Next, how do you determine the correct signs for the x and y components of a vector? This can be done by considering the right triangle formed by A_x, A_y, and $\vec{\mathbf{A}}$, as shown in **Figure 3–5**. To determine the sign of A_x, start at the tail of the vector and move along the x axis toward the right angle. If you are moving in the positive x direction, then A_x is positive ($A_x > 0$); if you are moving in the negative x direction, then A_x is negative ($A_x < 0$). For the y component, start at the right angle and move toward the tip of the arrow. A_y is positive or negative depending on whether you are moving in the positive or negative y direction.

For example, consider the vector shown in **Figure 3–6 (a)**. In this case, $A_x > 0$ and $A_y < 0$, as indicated in the figure. Similarly, the signs of A_x and A_y are given in **Figures 3–6 (b, c, d)** for the vectors shown there. Be sure to verify each of these cases by applying the rules just given. As we continue our study of physics, it is important to be able to find the components of a vector *and* to assign to them the correct signs.

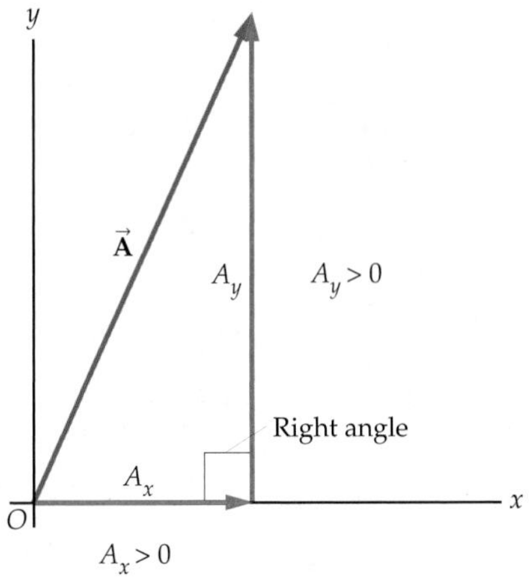

▲ **FIGURE 3–5 A vector whose x and y components are positive**

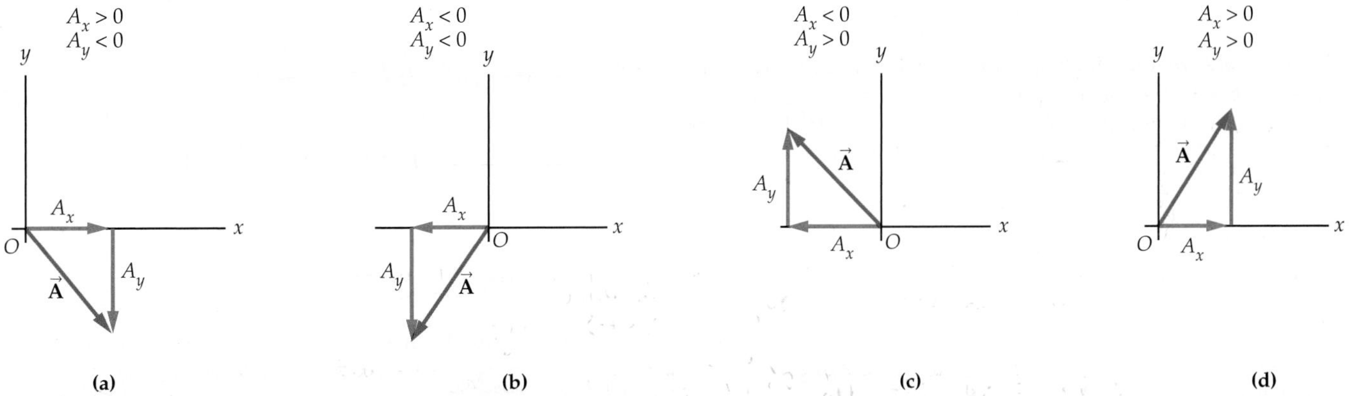

▲ **FIGURE 3–6 Examples of vectors with components of different signs**

To determine the signs of a vector's components, it is only necessary to observe the direction in which they point. For example, in part (a) the x component points in the positive direction; hence $A_x > 0$. Similarly, the y component in part (a) points in the negative y direction; therefore $A_y < 0$.

EXERCISE 3–2

The vector $\vec{\mathbf{A}}$ has a magnitude of 7.25 m. Find its components for direction angles of

a. $\theta = 5.00°$,

b. $\theta = 125°$,

c. $\theta = 245°$, and

d. $\theta = 335°$.

Solution

a. $A_x = 7.22$ m, $A_y = 0.632$ m

b. $A_x = -4.16$ m, $A_y = 5.94$ m

c. $A_x = -3.06$ m, $A_y = -6.57$ m

d. $A_x = 6.57$ m, $A_y = -3.06$ m

Be careful when using your calculator to determine the direction angle, θ, because you may need to add 180° to get the correct angle, as measured counterclockwise from the positive x axis. For example, if $A_x = -0.50$ m and $A_y = 1.0$ m, your calculator will give the following result:

$$\theta = \tan^{-1}\left(\frac{1.0\text{ m}}{-0.50\text{ m}}\right) = \tan^{-1}(-2.0) = -63°$$

Does this angle correspond to the specified vector? The way to check is to sketch $\vec{\mathbf{A}}$. When you do, your drawing is similar to Figure 3–6 (c), and thus the direction angle of $\vec{\mathbf{A}}$ should be between 90° and 180°. To obtain the correct angle, add 180° to the calculator's result,

$$\theta = -63° + 180° = 117°$$

This, in fact, is the direction angle for the vector $\vec{\mathbf{A}}$.

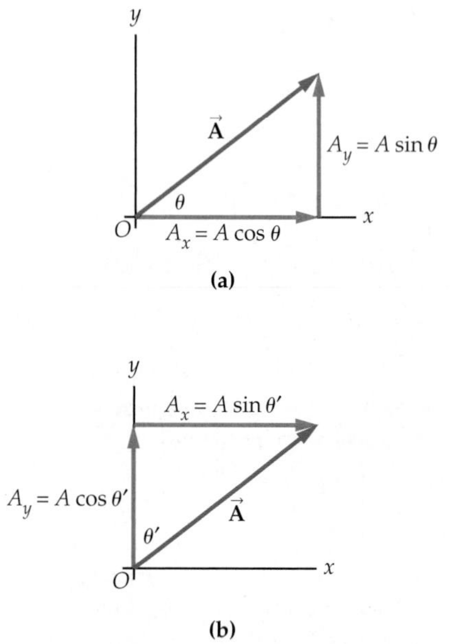

▲ **FIGURE 3–7 Vector direction angles**
Vector $\vec{\mathbf{A}}$ and its components in terms of **(a)** the angle relative to the x axis and **(b)** the angle relative to the y axis.

EXERCISE 3–3

The vector $\vec{\mathbf{B}}$ has components $B_x = -2.10$ m and $B_y = -1.70$ m. Find the direction angle, θ, for this vector.

Solution

$\tan^{-1}[(-1.70\text{ m})/(-2.10\text{ m})] = \tan^{-1}(1.70/2.10) = 39.0°, \theta = 39.0 + 180° = 219°$

Finally, in many situations the direction of a vector $\vec{\mathbf{A}}$ is given by the angle θ, measured relative to the x axis, as in **Figure 3–7 (a)**. In these cases we know that

$$A_x = A\cos\theta$$

and

$$A_y = A\sin\theta$$

On the other hand, we are sometimes given the angle between the vector and the y axis, as in **Figure 3–7 (b)**. If we call this angle θ', then it follows that

$$A_x = A\sin\theta'$$

and

$$A_y = A\cos\theta'$$

These two seemingly different results are actually in complete agreement. Note that $\theta + \theta' = 90°$, or $\theta' = 90° - \theta$. If we use the trigonometric identities given in Appendix A, we find

$$A_x = A\sin\theta' = A\sin(90° - \theta) = A\cos\theta$$

and

$$A_y = A\cos\theta' = A\cos(90° - \theta) = A\sin\theta$$

EXERCISE 3–4

If a vector's direction angle relative to the x axis is 35°, then its direction angle relative to the y axis is 55°. Find the components of a vector $\vec{\mathbf{A}}$ of magnitude 5.2 m in terms of

a. its direction relative to the x axis, and
b. its direction relative to the y axis.

Solution

a. $A_x = (5.2\text{ m})\cos 35° = 4.3\text{ m}, A_y = (5.2\text{ m})\sin 35° = 3.0\text{ m}$
b. $A_x = (5.2\text{ m})\sin 55° = 4.3\text{ m}, A_y = (5.2\text{ m})\cos 55° = 3.0\text{ m}$

3–3 Adding and Subtracting Vectors

One important reason for determining the components of a vector is that they are useful in adding and subtracting vectors. In this section we begin by defining vector addition graphically, and then show how the same addition can be performed more concisely and accurately with components.

Adding Vectors Graphically

One day you open an old chest in the attic and find a treasure map inside. To locate the treasure, the map says that you must "Go to the sycamore tree in the backyard, march 5 paces north, then 3 paces east." If these two displacements are represented by the vectors $\vec{\mathbf{A}}$ and $\vec{\mathbf{B}}$ in **Figure 3–8**, the total displacement from the tree to the treasure is given by the vector $\vec{\mathbf{C}}$. We say that $\vec{\mathbf{C}}$ is the *vector sum* of $\vec{\mathbf{A}}$ and $\vec{\mathbf{B}}$; that is, $\vec{\mathbf{C}} = \vec{\mathbf{A}} + \vec{\mathbf{B}}$. In general, vectors are added graphically according to the following rule:

- To add the vectors $\vec{\mathbf{A}}$ and $\vec{\mathbf{B}}$, place the tail of $\vec{\mathbf{B}}$, at the head of $\vec{\mathbf{A}}$. The sum, $\vec{\mathbf{C}} = \vec{\mathbf{A}} + \vec{\mathbf{B}}$., is the vector extending from the tail of $\vec{\mathbf{A}}$ to the head of $\vec{\mathbf{B}}$.

If the instructions to find the treasure were a bit more complicated—5 paces north, 3 paces east, then 4 paces southeast, for example—the path from the sycamore tree to the treasure would be like that shown in **Figure 3–9**. In this case, the total displacement, $\vec{\mathbf{D}}$, is the sum of the three vectors $\vec{\mathbf{A}}$, $\vec{\mathbf{B}}$, and $\vec{\mathbf{C}}$; that is, $\vec{\mathbf{D}} = \vec{\mathbf{A}} + \vec{\mathbf{B}} + \vec{\mathbf{C}}$. It follows that to add more than two vectors, we just keep placing the vectors head-to-tail, head-to-tail, and then draw a vector from the tail of the first vector to the head of the last vector, as in Figure 3–9.

In order to place a given pair of vectors head-to-tail, it may be necessary to move the corresponding arrows. This is fine, as long as you don't change their length or their direction. After all, a vector is defined by its length and direction; if these are unchanged, so is the vector.

- A vector is defined by its magnitude and direction, regardless of its location.

▲ To a good approximation, these snow geese are all moving in the same direction with the same speed. As a result, their velocity vectors are equal, even though their positions are different.

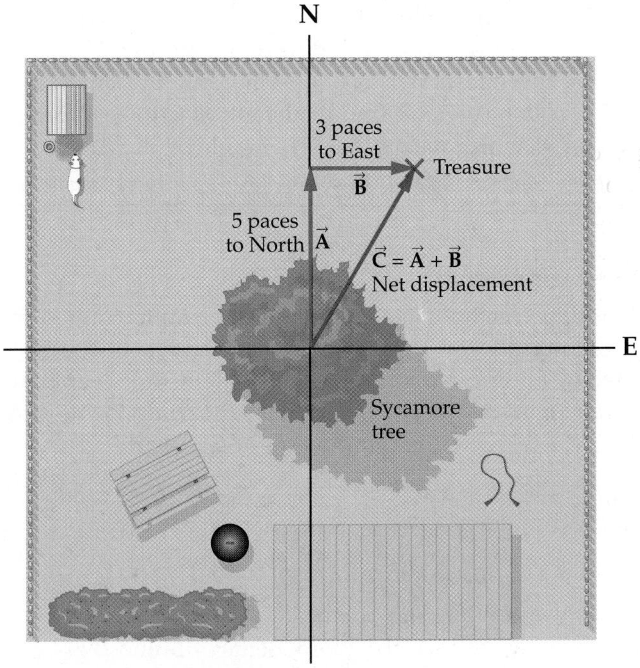

▲ **FIGURE 3–8 The sum of two vectors**

To go from the sycamore tree to the treasure, one must first go 5 paces north ($\vec{\mathbf{A}}$), and then 3 paces east ($\vec{\mathbf{B}}$). The net displacement from the tree to the treasure is $\vec{\mathbf{C}} = \vec{\mathbf{A}} + \vec{\mathbf{B}}$.

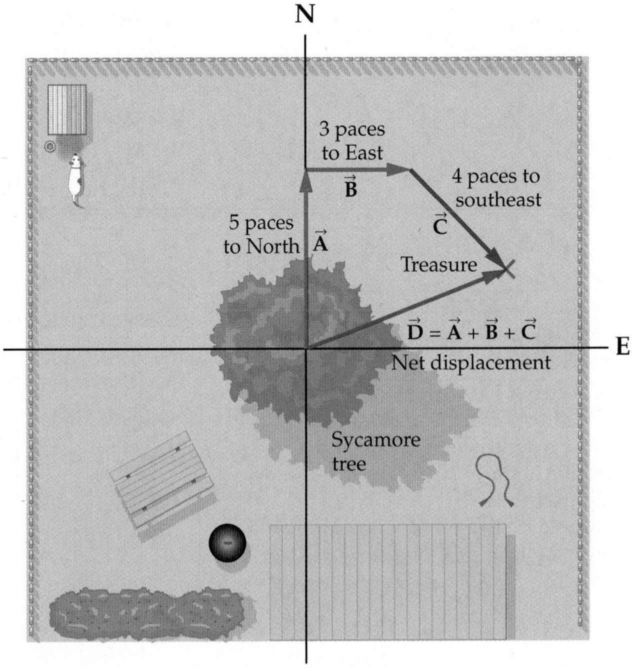

▲ **FIGURE 3–9 Adding several vectors**

Searching for a treasure that is 5 paces north ($\vec{\mathbf{A}}$), 3 paces east ($\vec{\mathbf{B}}$), and 4 paces southeast ($\vec{\mathbf{C}}$) of the sycamore tree. The net displacement from the tree to the treasure is $\vec{\mathbf{D}} = \vec{\mathbf{A}} + \vec{\mathbf{B}} + \vec{\mathbf{C}}$.

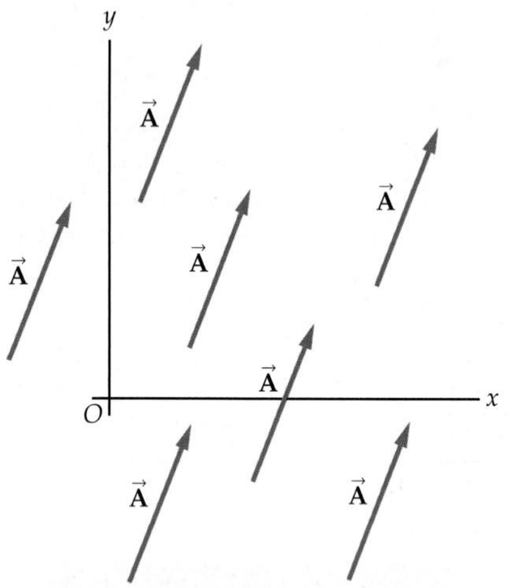

▲ **FIGURE 3–10 Identical vectors $\vec{A}$ at different locations**

A vector is defined by its direction and length; its location is immaterial.

For example, in **Figure 3–10** all of the vectors are the same, even though they are at different locations on the graph.

As an example of moving vectors, consider two vectors, $\vec{A}$ and $\vec{B}$, and their vector sum, $\vec{C}$,

$$\vec{C} = \vec{A} + \vec{B}$$

as illustrated in **Figure 3–11 (a)**. By moving the arrow representing $\vec{B}$ so that its tail is at the origin, and moving the arrow for $\vec{A}$ so that its tail is at the head of $\vec{B}$, we obtain the construction shown in **Figure 3–11 (b)**. From this graph we see that $\vec{C}$, which is $\vec{A} + \vec{B}$, is also equal to $\vec{B} + \vec{A}$:

$$\vec{C} = \vec{A} + \vec{B} = \vec{B} + \vec{A}$$

That is, the sum of vectors is independent of the order in which the vectors are added.

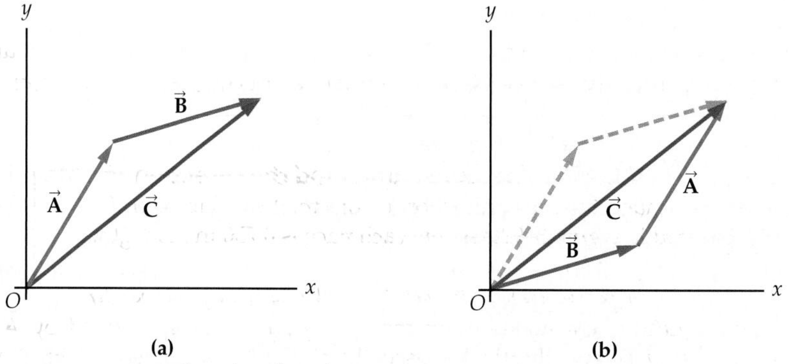

▲ **FIGURE 3–11 $\vec{A} + \vec{B} = \vec{B} + \vec{A}$**

The vector $\vec{C}$ is equal to **(a)** $\vec{A} + \vec{B}$ and **(b)** $\vec{B} + \vec{A}$. Note also that $\vec{C}$ is the diagonal of the parallelogram formed by the vectors $\vec{A}$ and $\vec{B}$. For this reason, this method of vector addition is referred to as the "parallelogram method."

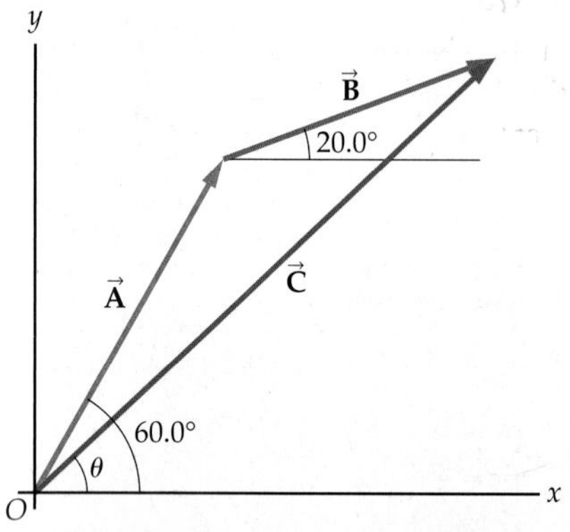

▲ **FIGURE 3–12 Graphical addition of vectors**

The vector $\vec{A}$ has a magnitude of 5.00 m and a direction angle of 60.0°; the vector $\vec{B}$ has a magnitude of 4.00 m and a direction angle of 20.0°. The magnitude and direction of $\vec{C} = \vec{A} + \vec{B}$ can be measured on the graph with a ruler and a protractor.

Now, suppose that $\vec{A}$ has a magnitude of 5.00 m and a direction angle of 60.0° above the x axis, and that $\vec{B}$ has a magnitude of 4.00 m and a direction angle of 20.0° above the x axis. These two vectors and their sum, $\vec{C}$, are shown in **Figure 3–12**. The question is: What is the length and direction angle of $\vec{C}$?

A graphical way to answer this question is to simply measure the length and direction of $\vec{C}$ in Figure 3–12. With a ruler, we find the length of $\vec{C}$ to be approximately 1.75 times the length of $\vec{A}$, which means that $\vec{C}$ is roughly 1.75 (5.00 m) = 8.75 m. Similarly, with a protractor we measure the angle θ to be about 45.0° above the x axis.

Adding Vectors Using Components

The graphical method of adding vectors yields approximate results, limited by the accuracy with which the vectors can be drawn and measured. In contrast, exact results can be obtained by adding $\vec{A}$ and $\vec{B}$ in terms of their components. To see how this is done, consider **Figure 3–13 (a)**, which shows the components of $\vec{A}$ and $\vec{B}$, and **Figure 3–13 (b)**, which shows the components of $\vec{C}$. Clearly,

$$C_x = A_x + B_x$$

and

$$C_y = A_y + B_y$$

Thus, to add vectors, you simply add the components.

Returning to our example in Figure 3–12, the components of $\vec{A}$ and $\vec{B}$ are

$$A_x = (5.00\text{ m}) \cos 60.0° = 2.50\text{ m} \quad A_y = (5.00\text{ m}) \sin 60.0° = 4.33\text{ m}$$

and

$$B_x = (4.00\text{ m}) \cos 20.0° = 3.76\text{ m} \quad B_y = (4.00\text{ m}) \sin 20.0° = 1.37\text{ m}$$

Adding component by component yields the components of $\vec{C} = \vec{A} + \vec{B}$:

$$C_x = A_x + B_x = 2.50 \text{ m} + 3.76 \text{ m} = 6.26 \text{ m}$$

and

$$C_y = A_y + B_y = 4.33 \text{ m} + 1.37 \text{ m} = 5.70 \text{ m}$$

With these results, we can now find *precise* values for C, the magnitude of vector $\vec{C}$, and its direction angle θ. In particular,

$$C = \sqrt{C_x^2 + C_y^2} = \sqrt{(6.26 \text{ m})^2 + (5.70 \text{ m})^2} = \sqrt{71.7 \text{ m}^2} = 8.47 \text{ m}$$

and

$$\theta = \tan^{-1}\left(\frac{C_y}{C_x}\right) = \tan^{-1}\left(\frac{5.70 \text{ m}}{6.26 \text{ m}}\right) = \tan^{-1}(0.911) = 42.3°$$

Note that these exact values are in rough agreement with the approximate results found by graphical addition.

In the future, we will always add vectors using components—graphical addition is useful primarily as a rough check on the results obtained with components.

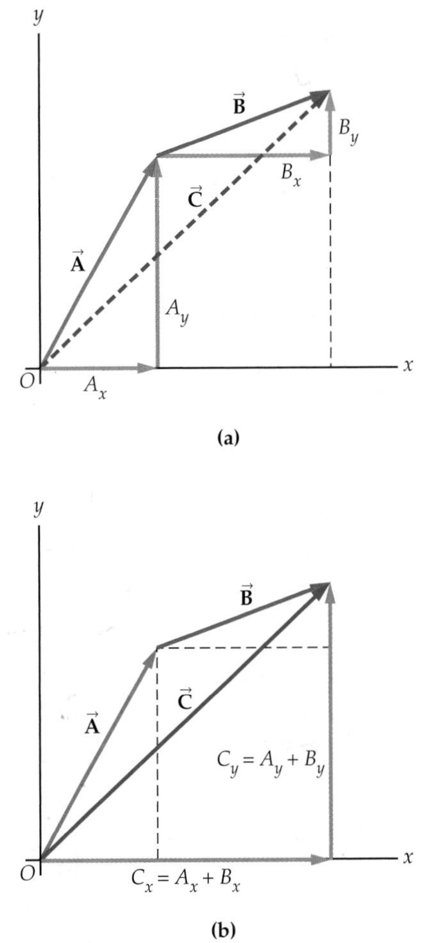

▲ **FIGURE 3–13 Component addition of vectors**

(a) The x and y components of $\vec{A}$ and $\vec{B}$. **(b)** The x and y components of $\vec{C}$. Notice that $C_x = A_x + B_x$ and $C_y = A_y + B_y$

ACTIVE EXAMPLE 3–1 Treasure Hunt: Find the Direction and Magnitude

What are the magnitude and direction of the total displacement for the treasure hunt illustrated in Figure 3–9? Assume each pace is 0.750 m in length.

Solution *(Test your understanding by performing the calculations indicated in each step.)*

To define a convenient notation, let the first 5 paces be represented by $\vec{A}$, the next 3 paces by $\vec{B}$, and the final 4 paces by $\vec{C}$. The total displacement, then, is $\vec{D} = \vec{A} + \vec{B} + \vec{C}$.

1. Find the components of $\vec{A}$:	$A_x = 0, A_y = 3.75$ m
2. Find the components of $\vec{B}$:	$B_x = 2.25$ m, $B_y = 0$
3. Find the components of $\vec{C}$:	$C_x = 2.12$ m, $C_y = -2.12$ m
4. Sum the components of $\vec{A}$, $\vec{B}$, and $\vec{C}$ to find the components of $\vec{D}$:	$D_x = 4.37$ m, $D_y = 1.63$ m
5. Determine D and θ:	$D = 4.66$ m, $\theta = 20.5°$

Your Turn

If the length of each pace is decreased by a factor of two, to 0.375 m, by what factors do you expect D and θ to change? Verify your answers with a numerical calculation.

(Answers to **Your Turn** *problems are given in the back of the book.)*

Subtracting Vectors

Next, how do we subtract vectors? Suppose, for example, that we would like to determine the vector $\vec{D}$, where

$$\vec{D} = \vec{A} - \vec{B}$$

and $\vec{A}$ and $\vec{B}$ are the vectors shown in Figure 3–12. To find $\vec{D}$, we start by rewriting it as follows:

$$\vec{D} = \vec{A} + (-\vec{B})$$

That is, $\vec{D}$ is the sum of $\vec{A}$ and $-\vec{B}$. Now the negative of a vector has a very simple graphical interpretation:

- The negative of a vector is represented by an arrow of the same length as the original vector, but pointing in the opposite direction. That is, multiplying a vector by minus one *reverses its direction*.

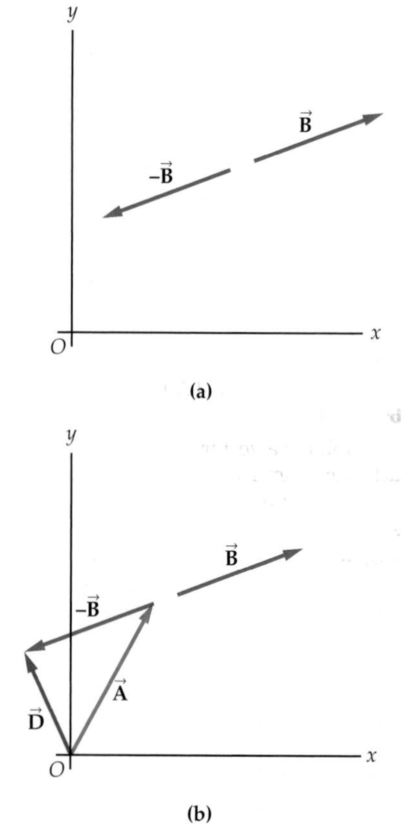

▲ **FIGURE 3–14 Vector subtraction**
(a) The vector $\vec{B}$ and its negative $-\vec{B}$.
(b) A vector construction for $\vec{D} = \vec{A} - \vec{B}$.

For example, the vectors $\vec{B}$ and $-\vec{B}$ are indicated in **Figure 3–14 (a)**. Thus, to subtract $\vec{B}$ from $\vec{A}$, simply reverse the direction of $\vec{B}$ and add it to $\vec{A}$, as indicated in **Figure 3–14 (b)**.

In terms of components, you subtract vectors by simply subtracting the components. For example, if

$$\vec{D} = \vec{A} - \vec{B}$$

then

$$D_x = A_x - B_x$$

and

$$D_y = A_y - B_y$$

Once the components of $\vec{D}$ are found, its magnitude and direction angle can be calculated as usual.

EXERCISE 3–5

a. For the vectors given in Figure 3–12, find the components of $\vec{D} = \vec{A} - \vec{B}$.
b. Find D and θ and compare with the vector $\vec{D}$ shown in Figure 3–14 (b).

Solution

a. $D_x = -1.26$ m, $D_y = 2.96$ m
b. $D = 3.22$ m, $\theta = -66.9° + 180° = 113°$. In Figure 3–14 (b) we see that $\vec{D}$ is shorter than $\vec{B}$, which has a magnitude of 4.00 m, and its direction angle is somewhat greater than 90°, in agreement with our numerical results.

3–4 Unit Vectors

Unit vectors provide a convenient way of expressing an arbitrary vector in terms of its components, as we shall see. But first, let's define what we mean by a unit vector. In particular, the unit vectors $\hat{\mathbf{x}}$ and $\hat{\mathbf{y}}$ are defined to be dimensionless vectors of unit magnitude pointing in the positive x and y directions, respectively:

- The x unit vector, $\hat{\mathbf{x}}$, is a dimensionless vector of unit length pointing in the positive x direction.
- The y unit vector, $\hat{\mathbf{y}}$, is a dimensionless vector of unit length pointing in the positive y direction.

Figure 3–15 shows $\hat{\mathbf{x}}$ and $\hat{\mathbf{y}}$ on a two-dimensional coordinate system. Since unit vectors have no physical dimensions—like mass, length, or time—they are used to specify direction only.

▲ **FIGURE 3–15 Unit vectors**
The unit vectors $\hat{\mathbf{x}}$ and $\hat{\mathbf{y}}$ point in the positive x and y directions, respectively.

Multiplying Unit Vectors by Scalars

To see the utility of unit vectors, consider the effect of multiplying a vector by a scalar. For example, multiplying a vector by 3 increases its magnitude by a factor of 3, but does not change its direction, as shown in **Figure 3–16**. Multiplying by -3 increases the magnitude by a factor of 3 *and* reverses the direction of the vector. This is also shown in Figure 3–16. In the case of unit vectors—which have a magnitude of 1 and are dimensionless—multiplication by a scalar results in a vector with the same magnitude and dimensions as the scalar.

For example, if a vector $\vec{A}$ has the scalar components $A_x = 5$ m and $A_y = 3$ m, we can write it as

$$\vec{A} = (5\text{ m})\hat{\mathbf{x}} + (3\text{ m})\hat{\mathbf{y}}$$

We refer to the quantities (5 m) $\hat{\mathbf{x}}$ and (3 m) $\hat{\mathbf{y}}$ as the x and y **vector components** of the vector $\vec{A}$. In general, an arbitrary two-dimensional vector $\vec{A}$ can always be written as the sum of a vector component in the x direction and a vector component in the y direction:

$$\vec{\mathbf{A}} = A_x\hat{\mathbf{x}} + A_y\hat{\mathbf{y}}$$

This is illustrated in **Figure 3–17 (a)**. An equivalent way of representing the vector components of a vector is illustrated in **Figure 3–17 (b)**. In this case we see that the vector components are the *projection* of a vector onto the x and y axes. The sign of the vector components is positive if they point in the positive x or y direction, and negative if they point in the negative x or y direction. This is how vector components will generally be shown in later chapters.

Finally, note that vector addition and subtraction is straightforward with unit vector notation:

$$\vec{\mathbf{C}} = \vec{\mathbf{A}} + \vec{\mathbf{B}} = (A_x + B_x)\hat{\mathbf{x}} + (A_y + B_y)\hat{\mathbf{y}}$$

and

$$\vec{\mathbf{D}} = \vec{\mathbf{A}} - \vec{\mathbf{B}} = (A_x - B_x)\hat{\mathbf{x}} + (A_y - B_y)\hat{\mathbf{y}}$$

Clearly, unit vectors provide a useful way to keep track of the x and y components of a vector.

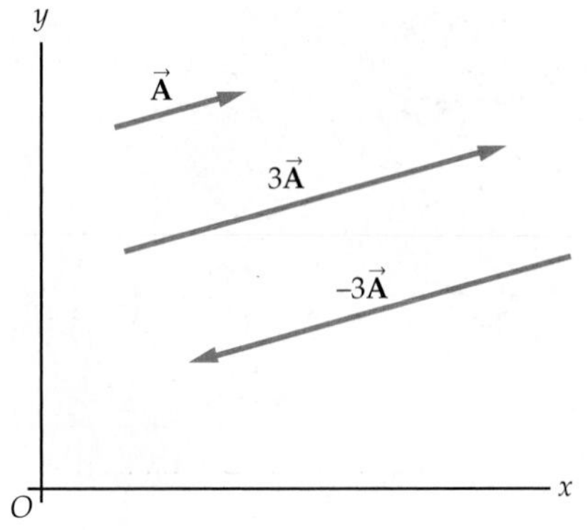

▲ **FIGURE 3–16 Multiplying a vector by a scalar**
Multiplying a vector by a positive scalar different from 1 will change the length of the vector but leave its direction the same. If the vector is multiplied by a negative scalar its direction is reversed.

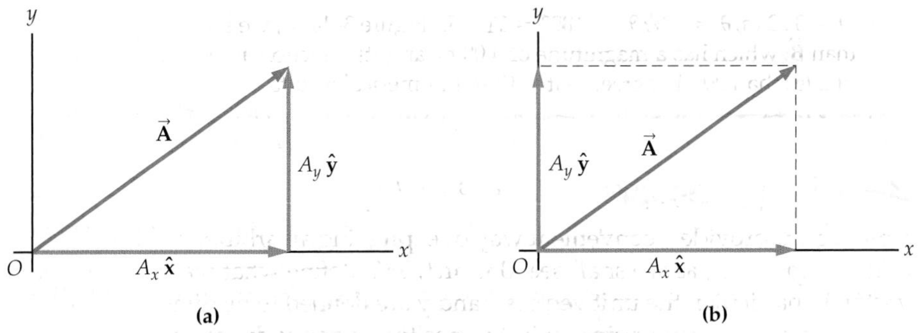

◀ **FIGURE 3–17 Vector components**
(a) A vector $\vec{\mathbf{A}}$ can be written in terms of unit vectors as $\vec{\mathbf{A}} = A_x\hat{\mathbf{x}} + A_y\hat{\mathbf{y}}$. **(b)** Vector components can be thought of as the projection of the vector onto the x and y axes. This method of representing vector components will be used frequently in subsequent chapters.

3–5 Position, Displacement, Velocity, and Acceleration Vectors

In Chapter 2 we discussed four different one-dimensional vectors: position, displacement, velocity, and acceleration. Each of these quantities had a direction associated with it, indicated by its sign; positive meant in the positive direction, negative meant in the negative direction. Now we consider these vectors again, this time in two dimensions, where the possibilities for direction are not so limited.

Position Vectors

To begin, imagine a two-dimensional coordinate system, as in **Figure 3–18**. Position is indicated by a vector from the origin to the location in question. We refer to the position vector as $\vec{\mathbf{r}}$; its units are meters, m.

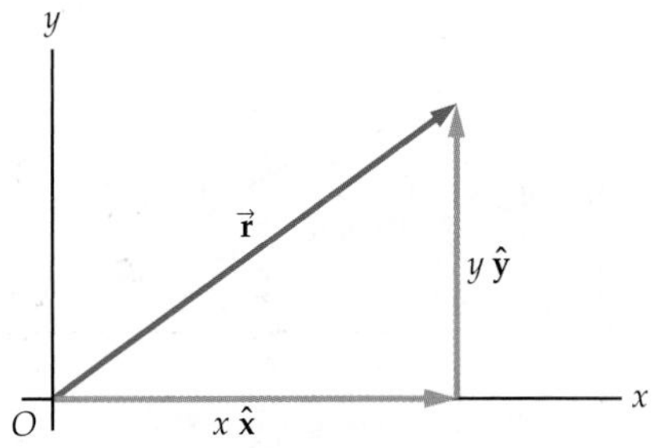

▲ **FIGURE 3–18 Position vector**
The position vector $\vec{\mathbf{r}}$ points from the origin to the current location of an object. The x and y vector components of $\vec{\mathbf{r}}$ are $x\hat{\mathbf{x}}$ and $y\hat{\mathbf{y}}$, respectively.

Definition: Position Vector, $\vec{\mathbf{r}}$

position vector $= \vec{\mathbf{r}}$ 3–1

SI unit: meter, m

In terms of unit vectors, the position vector is simply $\vec{\mathbf{r}} = x\hat{\mathbf{x}} + y\hat{\mathbf{y}}$.

▲ A map can be used to determine the direction and magnitude of the displacement vector from your initial position to your destination.

Now, suppose that initially you are at the location indicated by the position vector $\vec{\mathbf{r}}_i$, and that later you are at the final position represented by the position vector $\vec{\mathbf{r}}_f$. Your displacement vector, $\Delta\vec{\mathbf{r}}$, is the change in position:

Definition: Displacement Vector, $\Delta\vec{\mathbf{r}}$

$$\Delta\vec{\mathbf{r}} = \vec{\mathbf{r}}_f - \vec{\mathbf{r}}_i \qquad \text{3–2}$$

SI unit: meter, m

Rearranging this definition slightly, we see that

$$\vec{\mathbf{r}}_f = \vec{\mathbf{r}}_i + \Delta\vec{\mathbf{r}}$$

That is, the final position is equal to the initial position plus the change in position. This is illustrated in **Figure 3–19**, where we see that $\Delta\vec{\mathbf{r}}$ extends from the head of $\vec{\mathbf{r}}_i$ to the head of $\vec{\mathbf{r}}_f$.

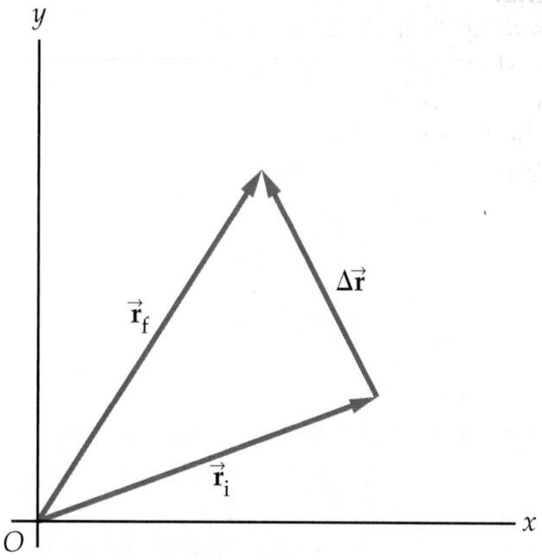

▲ **FIGURE 3–19 Displacement vector**
The displacement vector $\Delta\vec{\mathbf{r}}$ is the change in position. It points from the head of the initial position vector $\vec{\mathbf{r}}$ to the head of the final position vector $\vec{\mathbf{r}}_f$. Thus $\vec{\mathbf{r}}_f = \vec{\mathbf{r}}_i + \Delta\vec{\mathbf{r}}$ or $\Delta\vec{\mathbf{r}} = \vec{\mathbf{r}}_f - \vec{\mathbf{r}}_i$.

Velocity Vectors

Next, the average velocity vector is defined as the displacement vector $\Delta\vec{\mathbf{r}}$ divided by the elapsed time Δt.

Definition: Average Velocity Vector, $\vec{\mathbf{v}}_{av}$

$$\vec{\mathbf{v}}_{av} = \frac{\Delta\vec{\mathbf{r}}}{\Delta t} \qquad \text{3–3}$$

SI unit: meter per second, m/s

Since $\Delta\vec{\mathbf{r}}$ is a vector, it follows that $\vec{\mathbf{v}}_{av}$ is also a vector; it is the vector $\Delta\vec{\mathbf{r}}$ times the scalar $(1/\Delta t)$. Thus $\vec{\mathbf{v}}_{av}$ is parallel to $\Delta\vec{\mathbf{r}}$ and has the units m/s.

EXERCISE 3–6

A dragonfly is observed initially at the position $\vec{\mathbf{r}}_i = (2.00\text{ m})\hat{\mathbf{x}} + (3.50\text{ m})\hat{\mathbf{y}}$. Three seconds later it is at the position $\vec{\mathbf{r}}_f = (-3.00\text{ m})\hat{\mathbf{x}} + (5.50\text{ m})\hat{\mathbf{y}}$. What was the dragonfly's average velocity during this time?

Solution

$$\vec{\mathbf{v}}_{av} = (\vec{\mathbf{r}}_f - \vec{\mathbf{r}}_i)/\Delta t = [(-5.00\text{ m})\hat{\mathbf{x}} + (2.00\text{ m})\hat{\mathbf{y}}]/(3.00\text{ s})$$
$$= (-1.67\text{ m/s})\hat{\mathbf{x}} + (0.667\text{ m/s})\hat{\mathbf{y}}$$

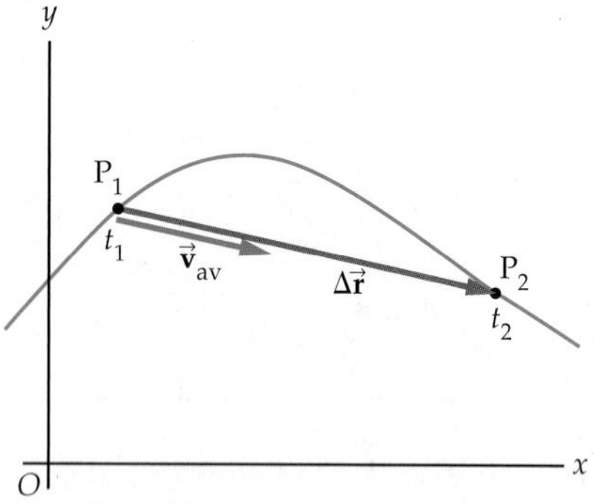

▲ **FIGURE 3–20 Average velocity vector**
The average velocity $\vec{\mathbf{v}}_{av}$ points in the same direction as $\Delta\vec{\mathbf{r}}$.

To help visualize $\vec{\mathbf{v}}_{av}$, imagine a particle moving in two dimensions along the blue path shown in **Figure 3–20**. If the particle is at point P_1 at time t_1, and at P_2 at time t_2, its displacement is indicated by the vector $\Delta\vec{\mathbf{r}}$. The average velocity is parallel to $\Delta\vec{\mathbf{r}}$, as indicated in Figure 3–20. It makes sense physically that $\vec{\mathbf{v}}_{av}$ is parallel to $\Delta\vec{\mathbf{r}}$; after all, on *average* you have moved in the direction of $\Delta\vec{\mathbf{r}}$ during the time from t_1 to t_2. To put it another way, a particle that starts at P_1 at the time t_1 and moves with the velocity $\vec{\mathbf{v}}_{av}$ until the time t_2 will arrive in precisely the same location as the particle that follows the blue path.

By considering smaller and smaller time intervals, as in **Figure 3–21**, it is possible to calculate the instantaneous velocity vector:

Definition: Instantaneous Velocity Vector, $\vec{\mathbf{v}}$

$$\vec{\mathbf{v}} = \lim_{\Delta t \to 0} \frac{\Delta\vec{\mathbf{r}}}{\Delta t} \qquad \text{3–4}$$

SI unit: meter per second, m/s

As can be seen in Figure 3–21, the instantaneous velocity at a given time is tangential to the path of the particle at that time. In addition, the magnitude of the velocity vector is the speed of the particle. Thus, the instantaneous velocity vector tells you both how fast a particle is moving and in what direction.

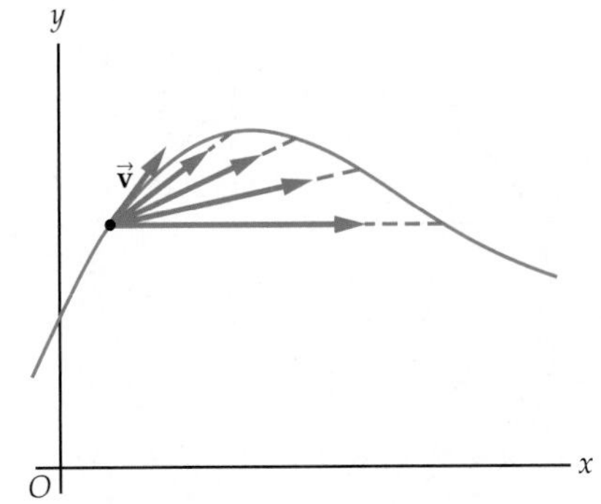

▲ **FIGURE 3–21 Instantaneous velocity vector**

The instantaneous velocity vector $\vec{\mathbf{v}}$ is obtained by calculating the average velocity vector over smaller and smaller time intervals. In the limit of vanishingly small time intervals, the average velocities approach the instantaneous velocity. Notice that the instantaneous velocity vector points in the direction of motion at any given time.

EXERCISE 3–7

Find the speed and direction of motion for a rainbow trout whose velocity is $\vec{\mathbf{v}} = (3.7\text{ m/s})\hat{\mathbf{x}} + (-1.3\text{ m/s})\hat{\mathbf{y}}$.

Solution

$$\text{speed} = v = \sqrt{(3.7\text{ m/s})^2 + (-1.3\text{ m/s})^2} = 3.9\text{ m/s}, \theta = \tan^{-1}\left(\frac{-1.3\text{ m/s}}{3.7\text{ m/s}}\right) = -19°,$$

that is, 19° below the x axis.

Acceleration Vectors

Finally, the average acceleration vector over an interval of time, Δt, is defined as the change in the velocity vector, $\Delta\vec{\mathbf{v}}$, divided by the scalar Δt.

Definition: Average Acceleration Vector, $\vec{\mathbf{a}}_{av}$

$$\vec{\mathbf{a}}_{av} = \frac{\Delta\vec{\mathbf{v}}}{\Delta t} \qquad 3\text{–}5$$

SI unit: meter per second per second, m/s^2

An example is given in **Figure 3–22**, where we show the initial and final velocity vectors corresponding to two different times. Since the change in velocity is defined as

$$\Delta\vec{\mathbf{v}} = \vec{\mathbf{v}}_f - \vec{\mathbf{v}}_i$$

it follows that

$$\vec{\mathbf{v}}_f = \vec{\mathbf{v}}_i + \Delta\vec{\mathbf{v}}$$

as indicated in Figure 3–22. Thus, $\Delta\vec{\mathbf{v}}$ is the vector extending from the head of $\vec{\mathbf{v}}_i$ to the head of $\vec{\mathbf{v}}_f$, just as $\Delta\vec{\mathbf{r}}$ extends from the head of $\vec{\mathbf{r}}_i$ to the head of $\vec{\mathbf{r}}_f$ in Figure 3–19. The direction of $\vec{\mathbf{a}}_{av}$ is the direction of $\Delta\vec{\mathbf{v}}$, as shown in Figure 3–22 (b).

Can an object accelerate if its speed is constant? Absolutely—if its direction changes. Consider a car driving with a constant speed on a circular track, as shown

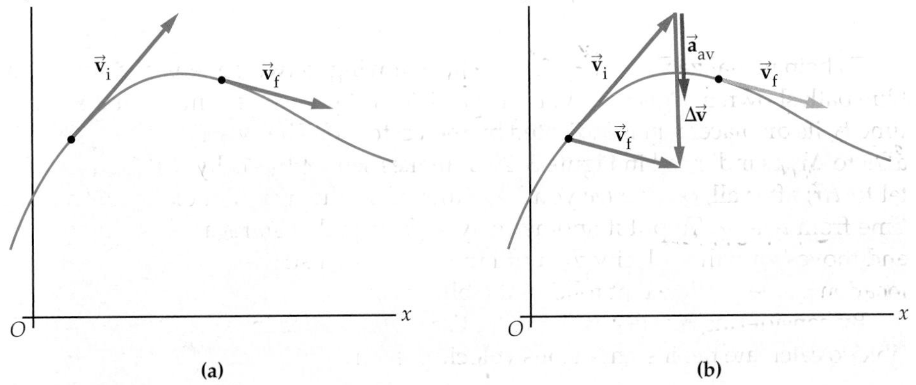

▲ **FIGURE 3–22 Average acceleration vector**

(a) As a particle moves along the blue path its velocity changes in magnitude and direction. At the time t_i the velocity is $\vec{\mathbf{v}}_i$; at the time t_f the velocity is $\vec{\mathbf{v}}_f$. **(b)** The average acceleration vector $\vec{\mathbf{a}}_{av} = \Delta\vec{\mathbf{v}}/\Delta t$ points in the direction of the change in velocity vector $\Delta\vec{\mathbf{v}}$. We obtain $\Delta\vec{\mathbf{v}}$ by moving $\vec{\mathbf{v}}_f$ so that its tail coincides with the tail of $\vec{\mathbf{v}}_i$, and then drawing the arrow that connects the head of $\vec{\mathbf{v}}_i$ to the head of $\vec{\mathbf{v}}_f$. Note that $\vec{\mathbf{a}}_{av}$ need not point in the direction of motion, and in general it doesn't.

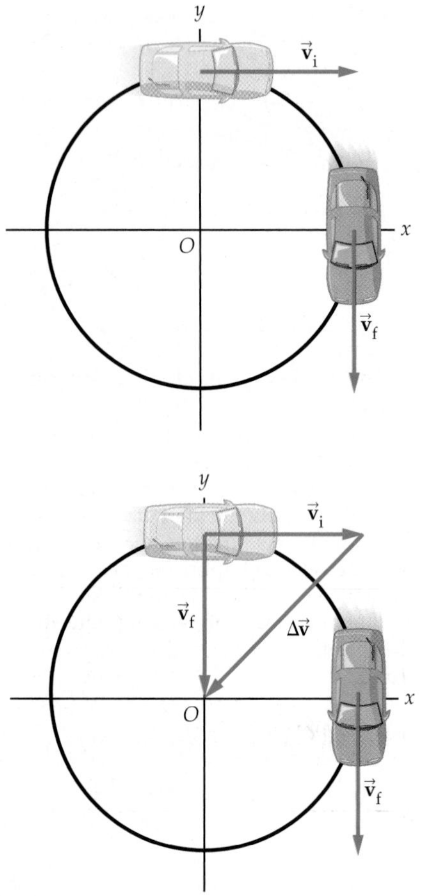

▲ **FIGURE 3–23 Average acceleration for a car traveling in a circle with constant speed**

Although the speed of this car never changes, it is still accelerating—due to the change in its direction of motion. For the time interval depicted, the car's average acceleration is in the direction of $\Delta\vec{v}$, which is toward the center of the circle. (As we shall see in Chapter 6, the car's acceleration is always toward the center of the circle.)

in **Figure 3–23**. Suppose that the initial velocity of the car is $\vec{v}_i = (12\text{ m/s})\hat{x}$, and that 10.0 s later its final velocity is $\vec{v}_f = (-12\text{ m/s})\hat{y}$. Note that the speed is 12 m/s in each case, but the velocity is different because the *direction* has changed. Calculating the average acceleration, we find a nonzero acceleration:

$$\vec{a}_{av} = \frac{\Delta\vec{v}}{\Delta t} = \frac{\vec{v}_f - \vec{v}_i}{10.0\text{ s}}$$

$$= \frac{(-12\text{ m/s})\hat{y} - (12\text{ m/s})\hat{x}}{10.0\text{ s}} = (-1.2\text{ m/s}^2)\hat{x} + (-1.2\text{ m/s}^2)\hat{y}$$

Thus, a change in direction is just as important as a change in speed in producing an acceleration. We shall study circular motion in detail in Chapter 6.

Finally, by going to infinitesimally small time intervals, $\Delta t \to 0$, we can define the instantaneous acceleration:

Definition: Instantaneous Acceleration Vector, $\vec{a}$

$$\vec{a} = \lim_{\Delta t \to 0} \frac{\Delta\vec{v}}{\Delta t} \qquad 3\text{–}6$$

SI unit: meter per second per second, m/s^2

ACTIVE EXAMPLE 3–2 Find the Average Acceleration

A car is traveling northwest at 9.00 m/s. Eight seconds later it has rounded a corner and is now heading north at 15.0 m/s. What is the magnitude and direction of its average acceleration during those 8.00 seconds?

Let the positive x direction be east, and the positive y direction be north.

Solution *(Test your understanding by performing the calculations indicated in each step.)*

1. Write out $\vec{v}_i$:	$\vec{v}_i = (-6.36\text{ m/s})\hat{x} + (6.36\text{ m/s})\hat{y}$
2. Write out $\vec{v}_f$:	$\vec{v}_f = (15.0\text{ m/s})\hat{y}$
3. Calculate $\Delta\vec{v}$:	$\Delta\vec{v} = (6.36\text{ m/s})\hat{x} + (8.64\text{ m/s})\hat{y}$
4. Find $\vec{a}_{av}$:	$\vec{a}_{av} = (0.795\text{ m/s}^2)\hat{x} + (1.08\text{ m/s}^2)\hat{y}$
5. Determine a_{av} and θ:	$a_{av} = 1.34\text{ m/s}^2$, $\theta = 53.6°$ north of east

Your Turn

Find the magnitude and direction of the average acceleration if the same change in velocity occurs in 4.00 s rather than 8.00 s.

(Answers to **Your Turn** *problems are given in the back of the book.)*

▶ The velocities of these cyclists change in both magnitude and direction as they slow to negotiate a series of sharp curves and then speed up again. Both kinds of velocity change involve an acceleration.

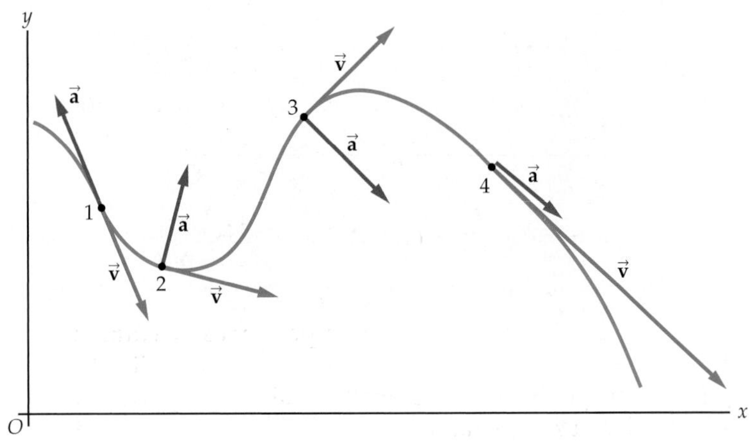

◀ **FIGURE 3–24 Velocity and acceleration vectors for a particle moving along a winding path**

The acceleration of a particle need not point in the direction of motion. At point (1) the particle is slowing down, at (2) it is turning to the left, at (3) it is turning to the right, and, finally, at point (4) it is speeding up.

Note carefully the following critical distinctions between the velocity vector and the acceleration vector:

- The velocity vector, $\vec{v}$, is always in the direction of a particle's motion.
- The acceleration vector, $\vec{a}$, can point in directions other than the direction of motion, and in general it does.

An example of a particle's motion, showing the velocity and acceleration vectors at various times, is presented in **Figure 3–24**.

Note that in all cases the velocity is tangential to the motion, though the acceleration points in various directions. When the acceleration is perpendicular to the velocity of an object, as at points (2) and (3) in Figure 3–24, its speed remains constant while its direction of motion changes. At points (1) and (4) in Figure 3–24 the acceleration is antiparallel (opposite) or parallel to the velocity of the object, respectively. In such cases, the direction of motion remains the same while the speed changes. Throughout the next chapter we shall see further examples of motion in which the velocity and acceleration are in different directions.

3–6 Relative Motion

A good example of the use of vectors is in the description of relative motion. Suppose, for example, that you are standing on the ground as a train goes by at 15.0 m/s, as shown in **Figure 3–25**. Inside the train, a free-riding passenger is walking in the forward direction at 1.2 m/s relative to the train. How fast is the passenger moving relative to you? Clearly, the answer is $1.2\ \text{m/s} + 15.0\ \text{m/s} = 16.2\ \text{m/s}$. What if the passenger had been walking with the same speed, but toward the back of the train? In this case, you would see the passenger going by with a speed of $-1.2\ \text{m/s} + 15.0\ \text{m/s} = 13.8\ \text{m/s}$.

Let's generalize these results. Call the velocity of the *t*rain relative to the *g*round $\vec{v}_{tg}$, the velocity of the *p*assenger relative to the *t*rain $\vec{v}_{pt}$, and the velocity of the *p*assenger relative to the *g*round $\vec{v}_{pg}$. As we saw in the previous paragraph, the velocity of the *p*assenger relative to the *g*round is

$$\vec{v}_{pg} = \vec{v}_{pt} + \vec{v}_{tg} \qquad \text{3–7}$$

This vector addition is illustrated in **Figure 3–26** for the two cases we discussed.

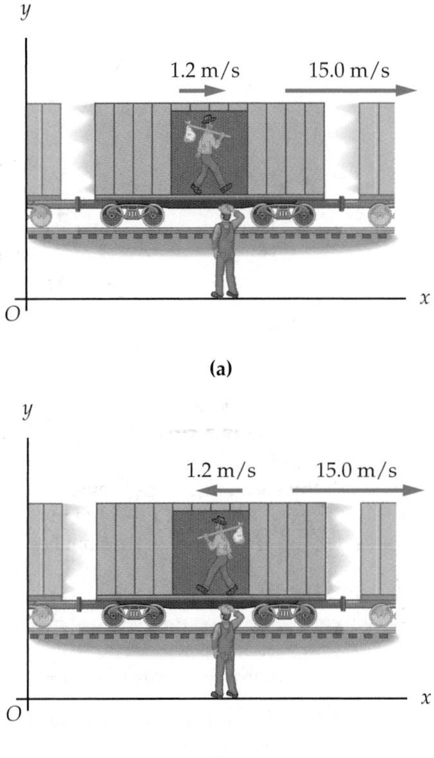

▲ **FIGURE 3–25 Relative velocity of a passenger on a train with respect to a person on the ground**

(a) The passenger walks toward the front of the train. **(b)** The passenger walks toward the rear of the train.

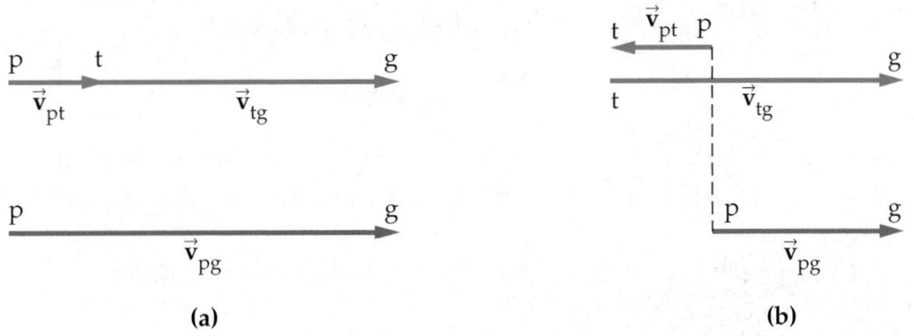

◀ **FIGURE 3–26 Adding velocity vectors**

Vector addition to find the velocity of the passenger with respect to the ground for **(a)** Figure 3-25 (a) and **(b)** Figure 3-25 (b).

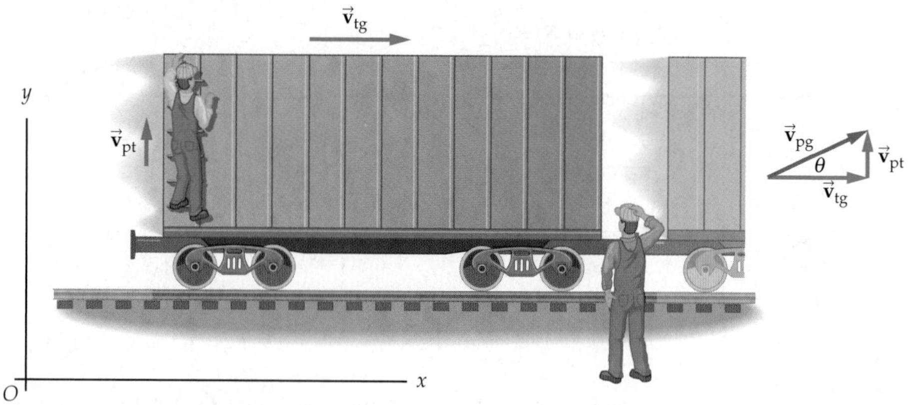

▲ **FIGURE 3–27 Relative velocity in two dimensions**

A person climbs up a ladder on a moving train with velocity $\vec{v}_{pt}$ relative to the train. If the train moves relative to the ground with a velocity $\vec{v}_{tg}$, the velocity of the person on the train relative to the ground is $\vec{v}_{pg} = \vec{v}_{pt} + \vec{v}_{tg}$.

Though this example dealt with one-dimensional motion, Equation 3–7 is valid for velocity vectors pointing in arbitrary directions. For example, instead of walking on the car's floor, the passenger might be climbing a ladder to the roof of the car, as in **Figure 3–27**. In this case $\vec{v}_{pt}$ is vertical, $\vec{v}_{tg}$ is horizontal, and $\vec{v}_{pg}$ is simply the vector sum $\vec{v}_{pt} + \vec{v}_{tg}$.

EXERCISE 3–8

Suppose the passenger in Figure 3–27 is climbing a vertical ladder with a speed of 0.20 m/s, and the train is slowly coasting forward at 0.70 m/s. Find the speed and direction of the passenger relative to the ground.

Solution

$$\vec{v}_{pg} = (0.70\text{ m/s})\hat{x} + (0.20\text{ m/s})\hat{y};\text{ thus}$$

$$v_{pg} = \sqrt{(0.70\text{ m/s})^2 + (0.20\text{ m/s})^2} = 0.73\text{ m/s}, \theta = \tan^{-1}(0.20/0.70) = 16°$$

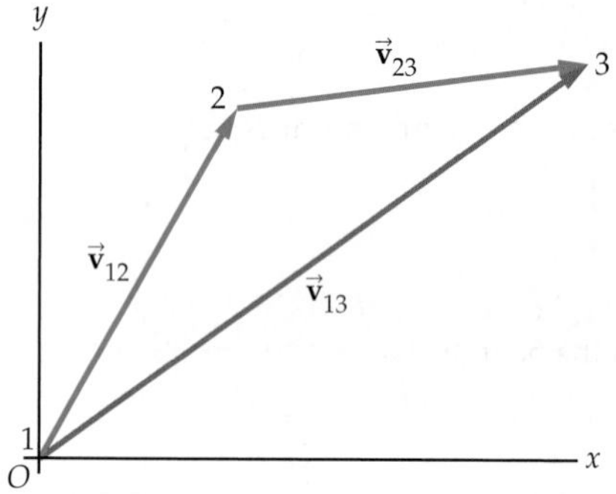

▲ **FIGURE 3–28 Vector addition used to determine relative velocity**

When we add the velocity vectors $\vec{v}_{12}$ and $\vec{v}_{23}$ the subscript 2 disappears, and we are left with $\vec{v}_{13}$.

Note that the subscripts in Equation 3–7 follow a definite pattern. On the left-hand side of the equation we have the subscripts pg. On the right-hand side we have two sets of subscripts, pt and tg; note that a pair of t's has been inserted between the p and the g. This pattern always holds for any relative motion problem, though the subscripts will be different when referring to different objects. Thus, we can say quite generally that

$$\vec{v}_{13} = \vec{v}_{12} + \vec{v}_{23} \quad \textbf{3–8}$$

where, in the train example, we can identify 1 as the *passenger*, 2 as the *train*, and 3 as the *ground*.

The vector addition in Equation 3–8 is shown in **Figure 3–28**. For convenience in seeing how the subscripts are ordered in the equation, we have labeled the tail of each vector with its first subscript and the head of each vector with its second subscript.

One final note about velocities and their subscripts: Reversing the subscripts reverses the velocity. This is indicated in **Figure 3–29**, where we see that

$$\vec{v}_{ba} = -\vec{v}_{ab}$$

▲ **FIGURE 3–29 Reversing the subscripts of a velocity reverses the corresponding velocity vector**

Physically, what we are saying is that if you are riding in a car due *north* at 20 m/s relative to the ground, then the ground, relative to you, is moving due *south* at 20 m/s.

Let's apply these results to a two-dimensional example.

EXAMPLE 3–2 Crossing a River

REAL-WORLD PHYSICS

You are riding in a boat whose speed relative to the water is 6.1 m/s. The boat points at an angle of 25° upstream on a river flowing at 1.4 m/s. What is your velocity relative to the ground?

Picture the Problem

We choose the x axis to be perpendicular to the river, and the y axis to point upstream. With these choices the velocity of the boat relative to the water is 25° above the x axis. In addition, the velocity of the water relative to the ground has a magnitude of 1.4 m/s and points in the negative y direction.

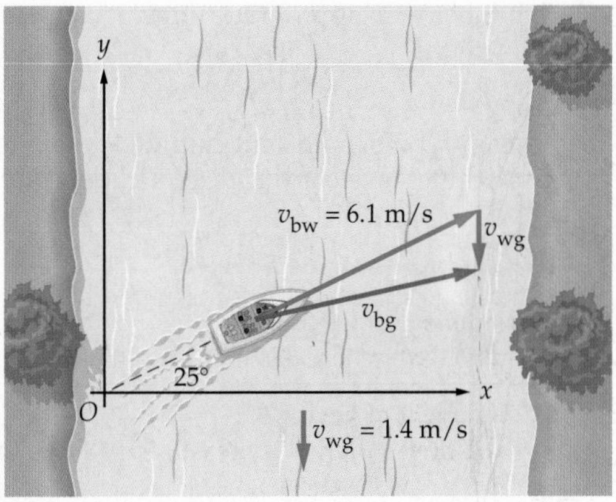

Strategy

If the water were still, the boat would move in the direction in which it is pointed. With the water flowing downstream, as shown, the boat will move in a direction closer to the x axis. To find the velocity of the boat we use $\vec{\mathbf{v}}_{13} = \vec{\mathbf{v}}_{12} + \vec{\mathbf{v}}_{23}$ with 1 referring to the boat (b), 2 referring to the water (w), and 3 referring to the ground (g).

INTERACTIVE FIGURE

Solution

1. Rewrite $\vec{\mathbf{v}}_{13} = \vec{\mathbf{v}}_{12} + \vec{\mathbf{v}}_{23}$ with $1 \rightarrow$ b, $2 \rightarrow$ w, and $3 \rightarrow$ g: $\vec{\mathbf{v}}_{bg} = \vec{\mathbf{v}}_{bw} + \vec{\mathbf{v}}_{wg}$

2. From our sketch we see that the water flows at 1.4 m/s in the negative y direction relative to the ground: $\vec{\mathbf{v}}_{wg} = (-1.4\text{ m/s})\hat{\mathbf{y}}$

3. The velocity of the boat relative to the water is given in the problem statement: $\vec{\mathbf{v}}_{bw} = (6.1\text{ m/s})\cos 25°\,\hat{\mathbf{x}} + (6.1\text{ m/s})\sin 25°\,\hat{\mathbf{y}} = (5.5\text{ m/s})\hat{\mathbf{x}} + (2.6\text{ m/s})\hat{\mathbf{y}}$

4. Carry out the vector sum in step 1 to find $\vec{\mathbf{v}}_{bg}$: $\vec{\mathbf{v}}_{bg} = (5.5\text{ m/s})\hat{\mathbf{x}} + (2.6\text{ m/s} - 1.4\text{ m/s})\hat{\mathbf{y}} = (5.5\text{ m/s})\hat{\mathbf{x}} + (1.2\text{ m/s})\hat{\mathbf{y}}$

Insight

Note that the speed of the boat relative to the ground is $\sqrt{(5.5\text{ m/s})^2 + (1.2\text{ m/s})^2} = 5.6\text{ m/s}$, and the direction angle is $\theta = \tan^{-1}(1.2/5.5) = 12°$ upstream.

Practice Problem

Find the speed and direction of the boat relative to the ground if the river flows at 4.5 m/s. [**Answer:** $v_{bg} = 5.8\text{ m/s}$, $\theta = -19°$. In this case, a person on the ground sees the boat going slowly downstream, even though the boat itself points upstream.]

Some related homework problems: Problem 41, Problem 45, Problem 47

Suppose the problem had been to find the velocity of the boat relative to the water so that it goes straight across the river at 5.0 m/s. That is, we want to find $\vec{\mathbf{v}}_{bw}$ such that $\vec{\mathbf{v}}_{bg} = (5.0\text{ m/s})\hat{\mathbf{x}}$. One approach is to simply solve $\vec{\mathbf{v}}_{bg} = \vec{\mathbf{v}}_{bw} + \vec{\mathbf{v}}_{wg}$ for $\vec{\mathbf{v}}_{bw}$, which gives

$$\vec{\mathbf{v}}_{bw} = \vec{\mathbf{v}}_{bg} - \vec{\mathbf{v}}_{wg} \tag{3–9}$$

Another approach is to go back to our general relation, $\vec{\mathbf{v}}_{13} = \vec{\mathbf{v}}_{12} + \vec{\mathbf{v}}_{23}$ and choose 1 to be the *boat*, 2 to be the *ground*, and 3 to be the *water*. With these substitutions we find

$$\vec{\mathbf{v}}_{bw} = \vec{\mathbf{v}}_{bg} + \vec{\mathbf{v}}_{gw}$$

This is the same as Equation 3–9, since $\vec{\mathbf{v}}_{gw} = -\vec{\mathbf{v}}_{wg}$. In either case, the desired velocity of the boat relative to the water is

$$\vec{\mathbf{v}}_{bw} = (5.0\text{ m/s})\hat{\mathbf{x}} + (1.4\text{ m/s})\hat{\mathbf{y}}$$

which corresponds to a speed of 5.2 m/s and a direction angle of 16° upstream.

Chapter Summary

3–1 Scalars Versus Vectors

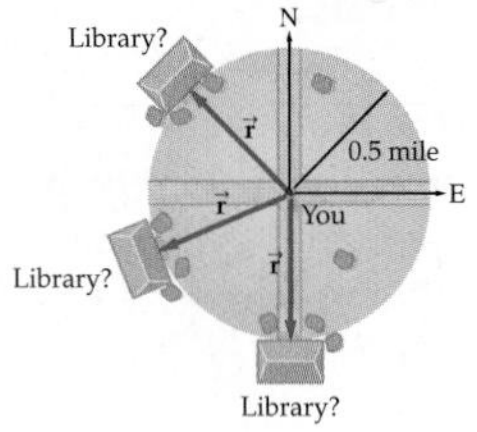

Scalar
A number with appropriate units. Examples of scalar quantities include time and length.

Vector
A quantity with both a magnitude and a direction. Examples include displacement, velocity, and acceleration.

3–2 The Components of a Vector

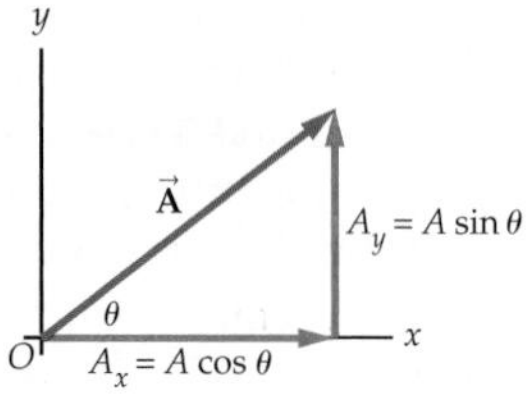

x Component of Vector $\vec{A}$
$A_x = A\cos\theta$, where θ is measured relative to the x axis.

y Component of Vector $\vec{A}$
$A_y = A\sin\theta$, where θ is measured relative to the x axis.

Sign of the Components
A_x is positive if $\vec{\mathbf{A}}$ points in the positive x direction, and negative if it points in the negative x direction. Similar remarks apply to A_y.

Magnitude of Vector $\vec{A}$
The magnitude of $\vec{\mathbf{A}}$ is $A = \sqrt{A_x^2 + A_y^2}$.

Direction Angle of Vector $\vec{A}$
The direction angle of $\vec{\mathbf{A}}$ is $\theta = \tan^{-1}(A_y/A_x)$, where θ is measured relative to the x axis.

3–3 Adding and Subtracting Vectors

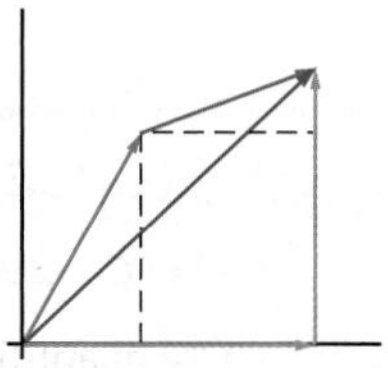

Graphical Method
To add $\vec{\mathbf{A}}$ and $\vec{\mathbf{B}}$, place them so that the tail of $\vec{\mathbf{B}}$ is at the head of $\vec{\mathbf{A}}$. The sum $\vec{\mathbf{C}} = \vec{\mathbf{A}} + \vec{\mathbf{B}}$ is the arrow from the tail of $\vec{\mathbf{A}}$ to the head of $\vec{\mathbf{B}}$. See Figure 3–8.

To find $\vec{\mathbf{A}} - \vec{\mathbf{B}}$, place $\vec{\mathbf{A}}$ and $-\vec{\mathbf{B}}$ head-to-tail and draw an arrow from the tail of $\vec{\mathbf{A}}$ to the head of $-\vec{\mathbf{B}}$. See Figure 3–14.

Component Method
If $\vec{\mathbf{C}} = \vec{\mathbf{A}} + \vec{\mathbf{B}}$, then $C_x = A_x + B_x$ and $C_y = A_y + B_y$. If $\vec{\mathbf{C}} = \vec{\mathbf{A}} - \vec{\mathbf{B}}$, then $C_x = A_x - B_x$ and $C_y = A_y - B_y$.

3–4 Unit Vectors

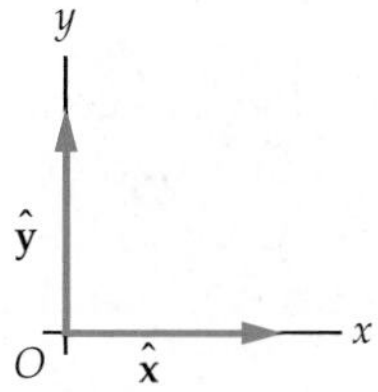

x Unit Vector
Written $\hat{\mathbf{x}}$, the x unit vector is a dimensionless vector of unit length in the positive x direction.

y Unit Vector
Written $\hat{\mathbf{y}}$, the y unit vector is a dimensionless vector of unit length in the positive y direction.

Vector Addition
$\vec{\mathbf{A}} + \vec{\mathbf{B}} = (A_x + B_x)\hat{\mathbf{x}} + (A_y + B_y)\hat{\mathbf{y}}$

3–5 Position, Displacement, Velocity, and Acceleration Vectors

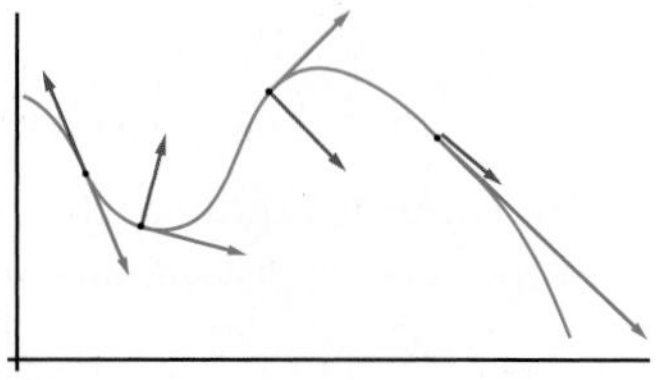

Position Vector
The position vector $\vec{\mathbf{r}}$ points from the origin to a particle's location.

Displacement Vector
The displacement vector $\Delta\vec{\mathbf{r}}$ is the change in position; $\Delta\vec{\mathbf{r}} = \vec{\mathbf{r}}_f - \vec{\mathbf{r}}_i$.

Velocity Vector
The velocity vector $\vec{\mathbf{v}}$ points in the direction of motion and has a magnitude equal to the speed.

Acceleration Vector
The acceleration vector $\vec{a}$ indicates how quickly and in what direction the velocity is changing. It need not point in the direction of motion.

3–6 Relative Motion

Velocity of Object 1 Relative to Object 3
$\vec{v}_{13} = \vec{v}_{12} + \vec{v}_{23}$ where object 2 can be anything.

Reversing the Subscripts on a Velocity
$\vec{v}_{12} = -\vec{v}_{21}$.

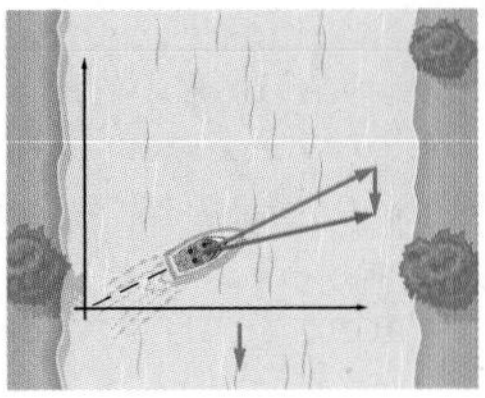

Problem-Solving Summary

Type of Problem	Relevant Physical Concepts	Related Examples
Add or subtract vectors.	Resolve the vectors into x and y components, then add or subtract the components.	Active Example 3–1 Exercise 3–5
Calculate the average velocity.	Divide the displacement, $\Delta\vec{r}$, by the elapsed time Δt.	Exercise 3–6
Calculate the average acceleration.	Divide the change in velocity, $\Delta\vec{v}$, by the elapsed time Δt.	Active Example 3–2
Find the relative velocity of object 1 with respect to object 3.	Use $\vec{v}_{13} = \vec{v}_{12} + \vec{v}_{23}$ with the appropriate choices for 1, 2, and 3.	Example 3–2 Exercise 3–8

Conceptual Questions

(Answers to odd-numbered Conceptual Questions can be found in the back of the book.)

1. For the following quantities, indicate which is a scalar and which is a vector: **(a)** the time it takes for you to run the 100-yard dash; **(b)** your displacement after running the 100-yard dash; **(c)** your average velocity while running; **(d)** your average speed while running.

2. Which, if any, of the vectors shown in **Figure 3–30** are equal?

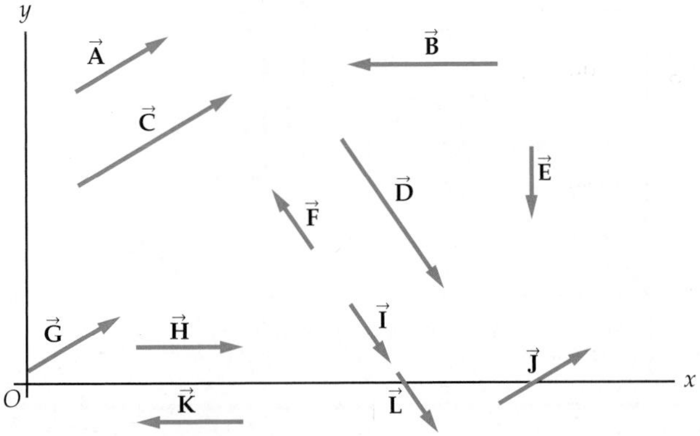

▲ **FIGURE 3–30** Conceptual Question 2

3. Given that $\vec{A} + \vec{B} = 0$, **(a)** how does the magnitude of $\vec{B}$ compare with the magnitude of $\vec{A}$? **(b)** How does the direction of $\vec{B}$ compare with the direction of $\vec{A}$?

4. Can a component of a vector be greater than the vector's magnitude?

5. Suppose that $\vec{A}$ and $\vec{B}$ have nonzero magnitude. Is it possible for $\vec{A} + \vec{B}$ to be zero?

6. Can a vector with zero magnitude have one or more components that are nonzero? Explain.

7. Given that $\vec{A} + \vec{B} = \vec{C}$, and that $A^2 + B^2 = C^2$, how are $\vec{A}$ and $\vec{B}$ oriented relative to one another?

8. Given that $\vec{A} + \vec{B} = \vec{C}$, and that $A + B = C$, how are $\vec{A}$ and $\vec{B}$ oriented relative to one another?

9. Given that $\vec{A} + \vec{B} = \vec{C}$, and that $A - B = C$, how are $\vec{A}$ and $\vec{B}$ oriented relative to one another?

10. Vector $\vec{A}$ has x and y components of equal magnitude. What can you say about the possible directions of $\vec{A}$?

11. The components of a vector $\vec{A}$ satisfy the relation $A_x = -A_y \neq 0$. What are the possible directions of $\vec{A}$?

12. Use a sketch to show that two vectors of unequal magnitude cannot add to zero, but that three vectors of unequal magnitude can.

13. Rain is falling vertically downward and you are running for shelter. To keep driest, should you hold your umbrella vertically, tilted forward, or tilted backward? Explain.

14. When sailing, the wind feels stronger when you sail upwind ("beating") than when you are sailing downwind ("running"). Explain.

Conceptual Exercises

(Answers to odd-numbered Conceptual Exercises can be found in the back of the book.)

1. Suppose that each component of a certain vector is doubled. **(a)** By what multiplicative factor does the magnitude of the vector change? **(b)** By what multiplicative factor does the direction angle of the vector change?
2. Rank the vectors in **Figure 3–31** in order of increasing magnitude.

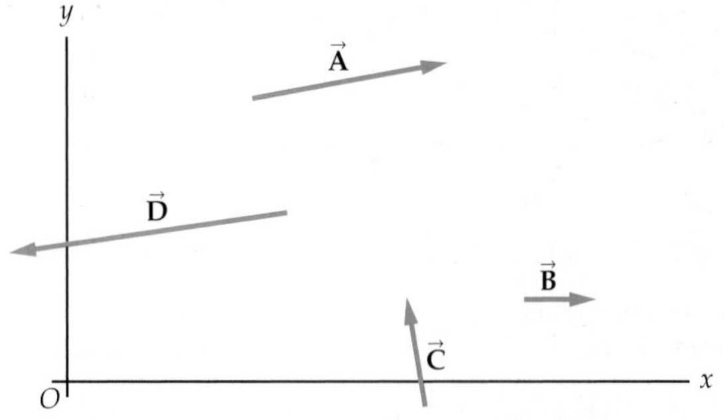

▲ **FIGURE 3–31** Conceptual Exercises 2, 3, and 4

3. Rank the vectors in Figure 3–31 in order of increasing value of their x component.
4. Rank the vectors in Figure 3–31 in order of increasing value of their y component.
5. Consider the vectors $\vec{\mathbf{A}} = (1.2\text{ m})\hat{\mathbf{x}}$ and $\vec{\mathbf{B}} = (-3.4\text{ m})\hat{\mathbf{x}}$. Which of these vectors has **(a)** the greater x component and **(b)** the greater magnitude?
6. Two vectors are defined as follows: $\vec{\mathbf{A}} = (-2.2\text{ m})\hat{\mathbf{x}}$ and $\vec{\mathbf{B}} = (1.4\text{ m})\hat{\mathbf{y}}$. **(a)** Is the magnitude of $1.4\,\vec{\mathbf{A}}$ greater than, less than, or equal to the magnitude of $2.2\,\vec{\mathbf{B}}$? **(b)** Is the x component of $1.4\,\vec{\mathbf{A}}$ greater than, less than, or the same as the y component of $2.2\,\vec{\mathbf{B}}$?
7. Consider the vectors $\vec{\mathbf{A}}$ and $\vec{\mathbf{B}}$ shown in **Figure 3–32**. Which of the other four vectors in the figure ($\vec{\mathbf{C}}$, $\vec{\mathbf{D}}$, $\vec{\mathbf{E}}$, and $\vec{\mathbf{F}}$) best represents the *direction* of **(a)** $\vec{\mathbf{A}} + \vec{\mathbf{B}}$, **(b)** $\vec{\mathbf{A}} - \vec{\mathbf{B}}$, and **(c)** $\vec{\mathbf{B}} - \vec{\mathbf{A}}$?

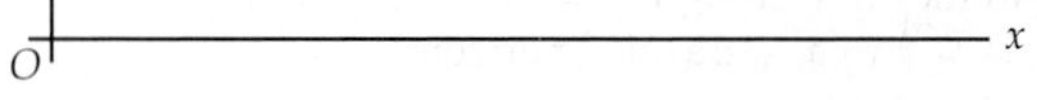

▲ **FIGURE 3–32** Conceptual Exercises 7 and 8

8. Refer to Figure 3–32 for the following questions: **(a)** Is the magnitude of $\vec{\mathbf{A}} + \vec{\mathbf{D}}$ greater than, less than, or equal to the magnitude of $\vec{\mathbf{A}} + \vec{\mathbf{E}}$? **(b)** Is the magnitude of $\vec{\mathbf{A}} + \vec{\mathbf{E}}$ greater than, less than, or equal to the magnitude of $\vec{\mathbf{A}} + \vec{\mathbf{F}}$?
9. The components of a vector $\vec{\mathbf{A}}$ satisfy $A_x < 0$ and $A_y < 0$. Is the direction angle of $\vec{\mathbf{A}}$ **(a)** between 0° and 90°, **(b)** between 90° and 180°, **(c)** between 180° and 270°, or **(d)** between 270° and 360°?
10. The components of a vector $\vec{\mathbf{B}}$ satisfy $B_x > 0$ and $B_y < 0$. Is the direction angle of $\vec{\mathbf{B}}$ **(a)** between 0° and 90°, **(b)** between 90° and 180°, **(c)** between 180° and 270°, or **(d)** between 270° and 360°?
11. The accompanying photo shows a KC-10A Extender using a boom to refuel an aircraft in flight. If the velocity of the KC-10A is 125 m/s due east relative to the ground, what is the velocity of the aircraft being refueled relative to **(a)** the ground, and **(b)** the KC-10A?

Air-to-air refueling. (Conceptual Exercise 11)

12. The blue curves shown in **Figure 3–33** display the motion of two different particles in the x-y plane. For each of the eight vectors in Figure 3–33, state whether it is **(a)** a position vector, **(b)** a velocity vector, or **(c)** an acceleration vector for the particles.

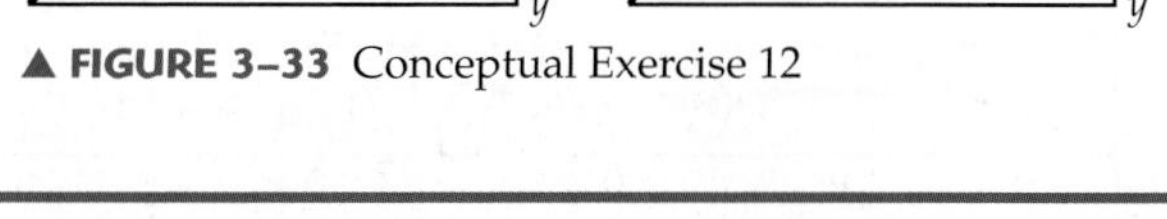

▲ **FIGURE 3–33** Conceptual Exercise 12

Problems

Note: **IP** *denotes an integrated conceptual/quantitative problem.* **BIO** *identifies problems of biological or medical interest. Red bullets (•, ••, •••) are used to indicate the level of difficulty of each problem.*

Section 3–2 The Components of a Vector

1. • The press box at a baseball park is 38.0 ft above the ground. A reporter in the press box looks at an angle of 15.0° below the horizontal to see second base. What is the horizontal distance from the press box to second base?
2. • You are driving up a long, inclined road. After 1.2 miles you notice that signs along the roadside indicate that your elevation has increased by 530 ft. **(a)** What is the angle of the road above the horizontal? **(b)** How far do you have to drive to gain an additional 150 ft of elevation?

3. • **A One Percent Grade** A road that rises 1 ft for every 100 ft traveled horizontally is said to have a 1 percent grade. Portions of the Lewiston grade, near Lewiston, Idaho, have a 6 percent grade. At what angle is this road inclined above the horizontal?

4. • Find the x and y components of a position vector, $\vec{\mathbf{r}}$, of magnitude $r = 75$ m, if its angle relative to the x axis is **(a)** 35.0° and **(b)** 65.0°.

5. • A baseball "diamond" (**Figure 3–34**) is a square with sides 90 ft in length. If the positive x axis points from home plate to first base, and the positive y axis points from home plate to third base, find the displacement vector of a base runner who has just hit **(a)** a double, **(b)** a triple, or **(c)** a home run.

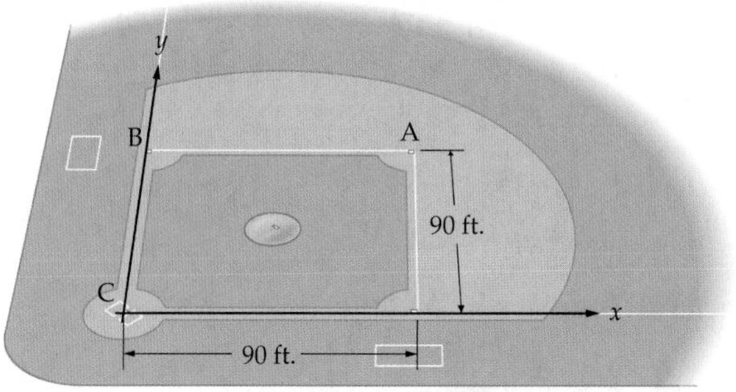

▲ **FIGURE 3–34** Problem 5

6. •• A lighthouse that rises 49 ft above the surface of the water sits on a rocky cliff that extends 19 ft from its base, as shown in **Figure 3–35**. A sailor on the deck of a ship sights the top of the lighthouse at an angle of 30.0° above the horizontal. If the sailor's eye level is 14 ft above the water, how far is the ship from the rocks?

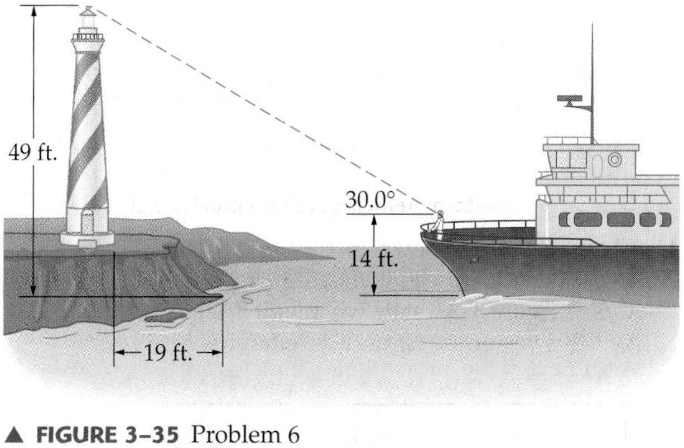

▲ **FIGURE 3–35** Problem 6

7. •• **H_2O** A water molecule is shown schematically in **Figure 3–36**. The distance from the center of the oxygen atom to the center of a hydrogen atom is 0.96 Å, and the angle between the hydrogen atoms is 104.5°. Find the center-to-center distance between the hydrogen atoms. (1 Å $= 10^{-10}$ m.)

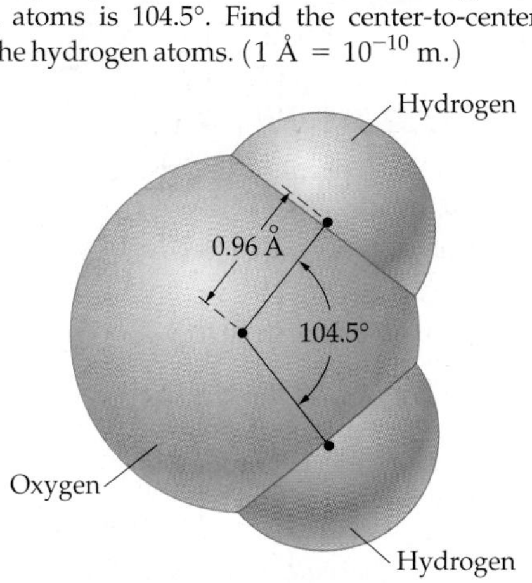

▲ **FIGURE 3–36** Problem 7

8. •• **IP** The x and y components of a vector $\vec{\mathbf{r}}$ are $r_x = 14$ m and $r_y = -9.5$ m, respectively. Find **(a)** the direction and **(b)** the magnitude of the vector $\vec{\mathbf{r}}$. **(c)** If both r_x and r_y are doubled, how do your answers to parts (a) and (b) change?

9. •• **IP The Longitude Problem** In 1755, John Harrison (1693–1776) completed his fourth precision chronometer, the H4, which eventually won the celebrated Longitude Prize. (For the human drama behind the Longitude Prize, see *Longitude*, by Dava Sobel.) When the minute hand of the H4 indicated 10 minutes past the hour, it extended 3.0 cm in the horizontal direction. **(a)** How long was the H4's minute hand? **(b)** At 10 minutes past the hour, was the extension of the minute hand in the vertical direction more than, less than, or equal to 3.0 cm? Explain. **(c)** Calculate the vertical extension of the minute hand at 10 minutes past the hour.

Not just a watch! The Harrison H4. (Problem 9)

10. •• You drive a car 660 ft to the east, then 370 ft to the north. **(a)** What is the magnitude of your displacement? **(b)** Using a sketch, estimate the direction of your displacement. **(c)** Verify your estimate in part (b) with a numerical calculation of the direction.

11. •• Vector $\vec{\mathbf{A}}$ has a magnitude of 50 units and points in the positive x direction. A second vector, $\vec{\mathbf{B}}$, has a magnitude of 120 units and points at an angle of 70° below the x axis. Which vector has **(a)** the greater x component, and **(b)** the greater y component?

12. •• A treasure map directs you to start at a palm tree and walk due north for 12.0 m. You are then to turn 90° and walk 25.0 m; then turn 90° again and walk 5.00 m. Give the distance from the palm tree, and the direction relative to north, for each of the four possible locations of the treasure.

13. •• A whale comes to the surface to breathe and then dives at an angle of 20.0° below the horizontal (**Figure 3–37**). If the whale continues in a straight line for 150 m, **(a)** how deep is it, and **(b)** how far has it traveled horizontally?

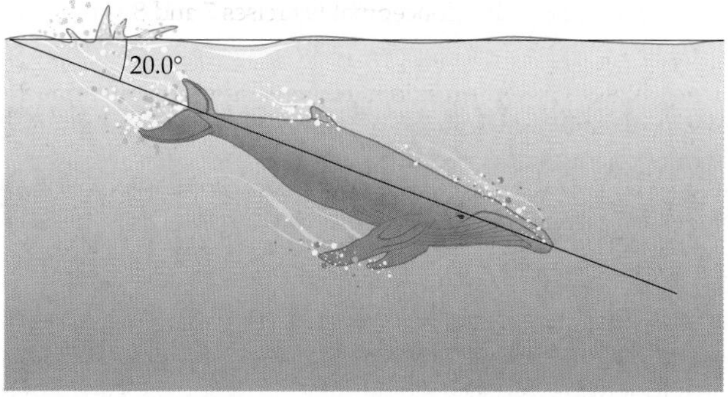

▲ **FIGURE 3–37** Problem 13

Section 3–3 Adding and Subtracting Vectors

14. • A vector $\vec{A}$ has a magnitude of 50.0 m and points in a direction 20.0° below the positive x axis. A second vector, $\vec{B}$, has a magnitude of 70.0 m and points in a direction 50.0° above the positive x axis. **(a)** Sketch the vectors $\vec{A}$, $\vec{B}$, and $\vec{C} = \vec{A} + \vec{B}$. **(b)** Using the component method of vector addition, find the magnitude and direction of the vector $\vec{C}$.

15. • An air traffic controller observes two airplanes approaching the airport. The displacement from the control tower to plane 1 is given by the vector $\vec{A}$, which has a magnitude of 220 km and points in a direction 32° north of west. The displacement from the control tower to plane 2 is given by the vector $\vec{B}$, which has a magnitude of 140 km and points 65° east of north. **(a)** Sketch the vectors $\vec{A}$, $-\vec{B}$, and $\vec{D} = \vec{A} - \vec{B}$. Notice that $\vec{D}$ is the displacement from plane 2 to plane 1. **(b)** Find the magnitude and direction of the vector $\vec{D}$.

16. • The initial velocity of a car, $\vec{v}_i$, is 45 km/h in the positive x direction. The final velocity of the car, $\vec{v}_f$, is 66 km/h in a direction that points 75° above the positive x axis. **(a)** Sketch the vectors $-\vec{v}_i$, $\vec{v}_f$, and $\Delta\vec{v} = \vec{v}_f - \vec{v}_i$. **(b)** Find the magnitude and direction of the change in velocity, $\Delta\vec{v}$.

17. •• Vector $\vec{A}$ points in the positive x direction and has a magnitude of 75 m. The vector $\vec{C} = \vec{A} + \vec{B}$ points in the positive y direction and has a magnitude of 95 m. **(a)** Sketch $\vec{A}$, $\vec{B}$, and $\vec{C}$. **(b)** Estimate the magnitude and direction of the vector $\vec{B}$. **(c)** Verify your estimate in part (b) with a numerical calculation.

18. •• Vector $\vec{A}$ points in the negative x direction and has a magnitude of 22 units. The vector $\vec{B}$ points in the positive y direction. **(a)** Find the magnitude of $\vec{B}$ if $\vec{A} + \vec{B}$ has a magnitude of 37 units. **(b)** Sketch $\vec{A}$ and $\vec{B}$.

19. •• Vector $\vec{A}$ points in the negative y direction and has a magnitude of 5 units. Vector $\vec{B}$ has twice the magnitude and points in the positive x direction. Find the direction and magnitude of **(a)** $\vec{A} + \vec{B}$, **(b)** $\vec{A} - \vec{B}$, and **(c)** $\vec{B} - \vec{A}$.

20. •• A basketball player runs down the court, following the path indicated by the vectors $\vec{A}$, $\vec{B}$, and $\vec{C}$ in **Figure 3–38**. The magnitudes of these three vectors are $A = 10.0$ m, $B = 20.0$ m, and $C = 7.0$ m. Find the magnitude and direction of the net displacement of the player using **(a)** the graphical method and **(b)** the component method of vector addition. Compare your results.

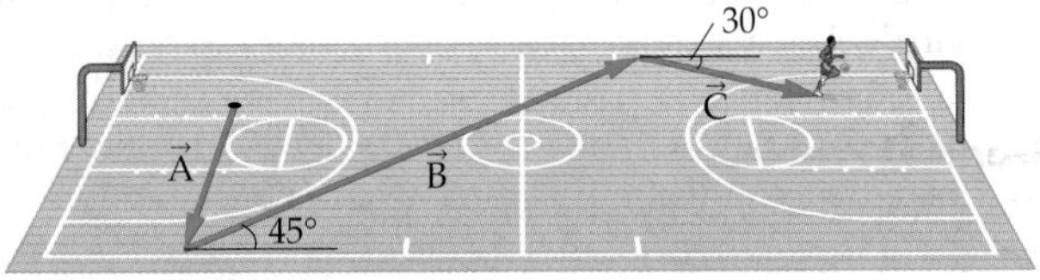

▲ **FIGURE 3–38** Problem 20

Section 3–4 Unit Vectors

21. • A particle undergoes a displacement $\Delta\vec{r}$ of magnitude 54 m in a direction 42° below the x axis. Express $\Delta\vec{r}$ in terms of the unit vectors $\hat{x}$ and $\hat{y}$.

22. • A vector has a magnitude of 2.5 m and points in a direction that is 140° counterclockwise from the x axis. Find the x and y components of this vector.

23. • A vector $\vec{A}$ has a length of 6.1 m and points in the negative x direction. Find **(a)** the x component and **(b)** the magnitude of the vector $-3.7\,\vec{A}$.

24. • The vector $-5.2\,\vec{A}$ has a magnitude of 44 m and points in the positive x direction. Find **(a)** the x component and **(b)** the magnitude of the vector $\vec{A}$.

25. • Find the direction and magnitude of the vectors.
(a) $\vec{A} = (5.0\text{ m})\hat{x} + (-2.0\text{ m})\hat{y}$,
(b) $\vec{B} = (-2.0\text{ m})\hat{x} + (5.0\text{ m})\hat{y}$, and
(c) $\vec{A} + \vec{B}$.

26. • Find the direction and magnitude of the vectors.
(a) $\vec{A} = (25\text{ m})\hat{x} + (-12\text{ m})\hat{y}$,
(b) $\vec{B} = (2.0\text{ m})\hat{x} + (15\text{ m})\hat{y}$, and
(c) $\vec{A} + \vec{B}$.

27. • For the vectors given in Problem 26, express **(a)** $\vec{A} - \vec{B}$, and **(b)** $\vec{B} - \vec{A}$ in unit vector notation.

28. • Express each of the vectors in **Figure 3–39** in unit vector notation.

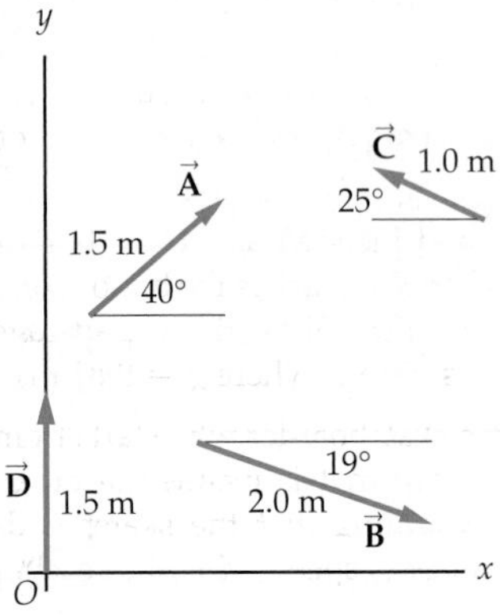

▲ **FIGURE 3–39** Problems 28 and 29

29. •• Referring to the vectors in Figure 3–39, express the sum $\vec{A} + \vec{B} + \vec{C}$ in unit vector notation.

Section 3–5 Position, Displacement, Velocity, and Acceleration Vectors

30. • **IP Moving the Knight** Two of the allowed chess moves for a knight are shown in **Figure 3–40**. **(a)** Is the magnitude of displacement 1 greater than, less than, or equal to the magnitude of displacement 2? Explain. **(b)** Find the magnitude and direction of the knight's displacement for each of the two moves. Assume that the checkerboard squares are 3.5 cm on a side.

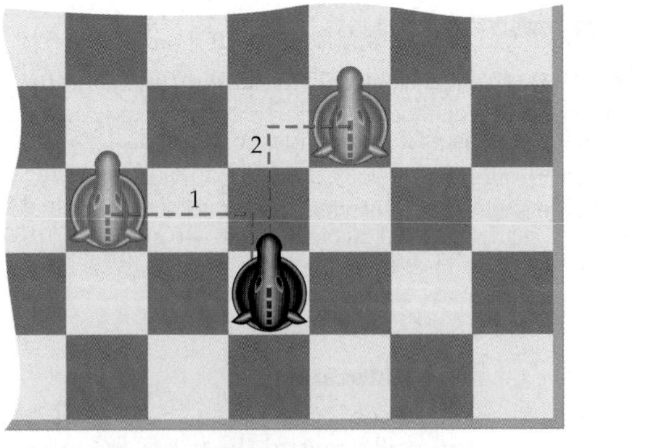

▲ **FIGURE 3–40** Problem 30

31. • **IP** In its daily prowl of the neighborhood, a cat makes a displacement of 120 m due north, followed by a 72-m displacement due west. **(a)** Find the magnitude and direction of the displacement required for the cat to return home. **(b)** If, instead, the cat had first prowled 72 m west and then 120 m north, how would this affect the displacement needed to bring it home? Explain.

32. • If the cat in Problem 31 takes 45 minutes to complete the 120-m displacement and 17 minutes to complete the 72-m displacement, what is the magnitude and direction of its average velocity during this 62-minute period of time?

33. • What is the direction and magnitude of your total displacement if you have traveled due west with a speed of 27 m/s for 130 s, then due south at 14 m/s for 62 s?

34. •• You drive a car 1500 ft to the east, then 2500 ft to the north. If the trip took 3.0 minutes, what was the direction and magnitude of your average velocity?

35. •• IP A jogger runs with a speed of 3.25 m/s in a direction 30.0° above the x axis. **(a)** Find the x and y components of the jogger's velocity. **(b)** How will the velocity components found in part (a) change if the jogger's speed is halved?

36. •• You throw a ball upward with an initial speed of 4.5 m/s. When it returns to your hand 0.92 s later, it has the same speed in the downward direction (assuming air resistance can be ignored). What was the average acceleration vector of the ball?

37. •• A skateboarder rolls from rest down an inclined ramp that is 15.0 m long and inclined above the horizontal at an angle of $\theta = 20.0°$. When she reaches the bottom of the ramp 3.00 s later her speed is 10.0 m/s. Show that the average acceleration of the skateboarder is $g \sin \theta$, where $g = 9.81 \text{ m/s}^2$.

38. •• Consider a skateboarder who starts from rest at the top of a 12.5 m-long ramp that is inclined at an angle of 15.0° to the horizontal. Assuming that the skateboarder's acceleration is $g \sin 15.0°$, find his speed when he reaches the bottom of the ramp in 3.00 s.

39. ••• IP **The Position of the Moon** Relative to the center of the Earth, the position of the Moon can be approximated by

$$\vec{r} = (3.84 \times 10^8 \text{ m})\{\cos[(2.46 \times 10^{-6} \text{ radians/s})t]\hat{x} + \sin[(2.46 \times 10^{-6} \text{ radians/s})t]\hat{y}\}$$

where t is measured in seconds. **(a)** Find the magnitude and direction of the Moon's average velocity between $t = 0$ and $t = 7.38$ days. (This time is one-quarter of the 29.5 days it takes the Moon to complete one orbit.) **(b)** Is the instantaneous speed of the Moon greater than, less than, or the same as the average speed found in part (a)? Explain.

40. ••• **The Velocity of the Moon** The velocity of the Moon relative to the center of the Earth can be approximated by

$$\vec{v} = (945 \text{ m/s})\{-\sin[(2.46 \times 10^{-6} \text{ radians/s})t]\hat{x} + \cos[(2.46 \times 10^{-6} \text{ radians/s})t]\hat{y}\}$$

where t is measured in seconds. To approximate the instantaneous acceleration of the Moon at $t = 0$, calculate the magnitude and direction of the average acceleration between the times **(a)** $t = 0$ and $t = 0.100$ days and **(b)** $t = 0$ and $t = 0.0100$ days. (The time required for the Moon to complete one orbit is 29.5 days.)

Section 3–6 Relative Motion

41. • As an airplane taxies on the runway with a speed of 16.5 m/s, a flight attendant walks toward the tail of the plane with a speed of 1.22 m/s. What is the flight attendant's speed relative to the ground?

42. • Referring to Example 3–2, find the time it takes for the boat to reach the opposite shore if the river is 25 m wide.

43. •• As you hurry to catch your flight at the local airport, you encounter a moving walkway that is 85 m long and has a speed of 2.2 m/s relative to the ground. If it takes you 68 s to cover 85 m when walking on the ground, how long will it take you to cover the same distance on the walkway? Assume that you walk with the same speed on the walkway as you do on the ground.

44. •• In Problem 43, how long would it take you to cover the 85 m length of the walkway if, once you get on the walkway, you immediately turn around and start walking in the opposite direction with a speed of 1.3 m/s relative to the walkway?

45. •• IP The pilot of an airplane wishes to fly due north, but there is a 65 km/h wind blowing toward the east. **(a)** In what direction should the pilot head her plane if its speed relative to the air is 340 km/h? **(b)** Draw a vector diagram that illustrates your result in part (a). **(c)** If the pilot decreases the air speed of the plane, but still wants to head due north, should the angle found in part (a) be increased or decreased?

46. •• A passenger walks from one side of a ferry to the other as it approaches a dock. If the passenger's velocity is 1.50 m/s due north relative to the ferry, and 4.50 m/s at an angle of 30.0° west of north relative to the water, what are the direction and magnitude of the ferry's velocity relative to the water?

47. •• You are riding on a Jet Ski at an angle of 35° upstream on a river flowing with a speed of 2.8 m/s. If your velocity relative to the ground is 9.5 m/s at an angle of 20.0° upstream, what is the speed of the Jet Ski relative to the water? (Note: Angles are measured relative to the x axis shown in Example 3–2.)

48. •• IP In Problem 47, suppose the Jet Ski is moving at a speed of 12 m/s relative to the water. **(a)** At what angle must you point the Jet Ski if your velocity relative to the ground is to be perpendicular to the shore of the river? **(b)** If you increase the speed of the Jet Ski relative to the water, does the angle in part (a) increase, decrease, or stay the same? Explain. (Note: Angles are measured relative to the x axis shown in Example 3–2.)

49. ••• IP Two people take identical Jet Skis across a river, traveling at the same speed relative to the water. Jet Ski A heads directly across the river and is carried downstream by the current before reaching the opposite shore. Jet Ski B travels in a direction that is 35° upstream and arrives at the opposite shore directly across from the starting point. **(a)** Which Jet Ski reaches the opposite shore in the least amount of time? **(b)** Confirm your answer to part (a) by finding the ratio of the time it takes for the two Jet Skis to cross the river. (Note: Angles are measured relative to the x axis shown in Example 3–2.)

General Problems

50. • You slide a box up a loading ramp that is 10.0 ft long. At the top of the ramp the box has risen a height of 3.00 ft. What is the angle of the ramp above the horizontal?

51. • Find the direction and magnitude of the vector $2\vec{A} + \vec{B}$, where $\vec{A} = (12.1 \text{ m})\hat{x}$ and $\vec{B} = (-32.2 \text{ m})\hat{y}$.

52. •• It is given that $\vec{A} - \vec{B} = (-51.4 \text{ m})\hat{x}$, $\vec{C} = (62.2 \text{ m})\hat{x}$, and $\vec{A} + \vec{B} + \vec{C} = (13.8 \text{ m})\hat{x}$. Find the vectors $\vec{A}$ and $\vec{B}$.

53. •• IP Two students perform an experiment with a train and a ball. Michelle rides on a flatcar pulled at 8.20 m/s by a train on a straight, horizontal track; Gary stands at rest on the ground near the tracks. When Michelle throws the ball with an initial angle of 55.0° above the horizontal, from her point of view, Gary sees the ball rise straight up and back down above a fixed point on the ground. **(a)** Did Michelle throw the ball toward the front of the train or toward the rear of the train? Explain. **(b)** What was the initial speed of Michelle's throw? **(c)** What was the initial speed of the ball as seen by Gary?

54. •• An off-roader explores the open desert in her Hummer. First she drives 25° west of north with a speed of 6.5 km/h for 15 minutes, then due east with a speed of 12 km/h for 7.5 minutes. She completes the final leg of her trip in 22 minutes. What is the direction and speed of travel on the final leg? (Assume her speed is constant on each leg, and that she returns to her starting point at the end of the final leg.)

55. •• Find the x, y, and z components of the vector $\vec{\mathbf{A}}$ shown in **Figure 3–41**, given that $A = 65$ m.

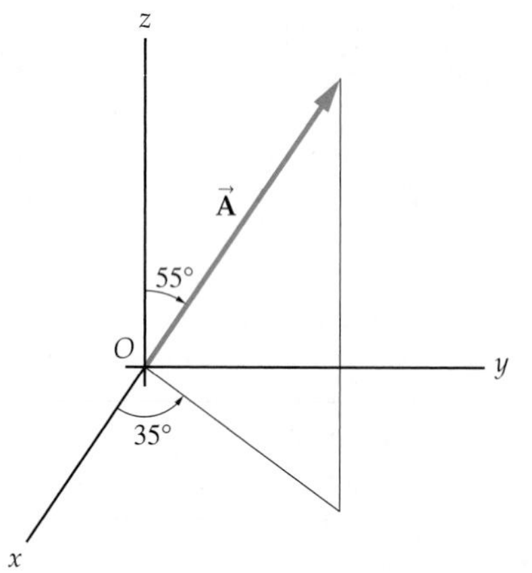

▲ **FIGURE 3–41** Problem 55

56. •• A football is thrown horizontally with an initial velocity of $(16.6\ \text{m/s})\hat{\mathbf{x}}$. Ignoring air resistance, the average acceleration of the football over any period of time is $(-9.81\ \text{m/s}^2)\hat{\mathbf{y}}$. **(a)** Find the velocity vector of the ball 1.75 s after it is thrown. **(b)** Find the magnitude and direction of the velocity at this time.

57. •• As a function of time, the velocity of the football described in Problem 56 can be written as $\vec{\mathbf{v}} = (16.6\ \text{m/s})\hat{\mathbf{x}} - [(9.81\ \text{m/s}^2)t]\hat{\mathbf{y}}$. Calculate the average acceleration vector of the football for the time periods **(a)** $t = 0$ to $t = 1.00$ s, **(b)** $t = 0$ to $t = 2.50$ s, and **(c)** $t = 0$ to $t = 5.00$ s. (If the acceleration of an object is constant, its average acceleration is the same for all time periods.)

58. •• Two airplanes taxi as they approach the terminal. Plane 1 taxies with a speed of 12 m/s due north. Plane 2 taxies with a speed of 7.5 m/s in a direction 20° north of west. **(a)** What are the direction and magnitude of the velocity of plane 1 relative to plane 2? **(b)** What are the direction and magnitude of the velocity of plane 2 relative to plane 1?

59. •• A shopper at the supermarket follows the path indicated by vectors $\vec{\mathbf{A}}$, $\vec{\mathbf{B}}$, $\vec{\mathbf{C}}$, and $\vec{\mathbf{D}}$ in **Figure 3–42**. Given that the vectors have the magnitudes $A = 51$ ft, $B = 45$ ft, $C = 35$ ft, and $D = 13$ ft, find the total displacement of the shopper using **(a)** the graphical method and **(b)** the component method of vector addition. Give the direction of the displacement relative to the direction of vector $\vec{\mathbf{A}}$.

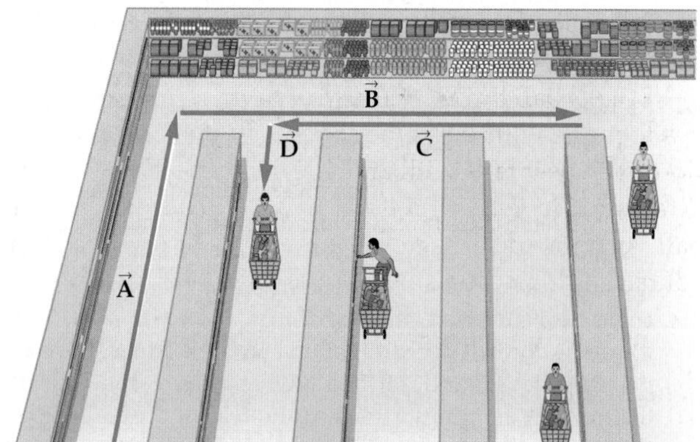

▲ **FIGURE 3–42** Problem 59

60. •• Initially, a particle is moving at 4.40 m/s at an angle of 37.5° above the horizontal. Two seconds later, its velocity is 6.25 m/s at an angle of 56.0° below the horizontal. What was the particle's average acceleration during these 2.00 seconds?

61. •• A passenger on a stopped bus notices that rain is falling vertically just outside the window. When the bus moves with constant velocity, the passenger observes that the falling raindrops are now making an angle of 15° with respect to the vertical. **(a)** What is the ratio of the speed of the raindrops to the speed of the bus? **(b)** Find the speed of the raindrops, given that the bus is moving with a speed of 18 m/s.

62. •• **A Big Clock** The clock that rings the bell known as Big Ben has an hour hand that is 9.0 feet long and a minute hand that is 14 feet long, where the distance is measured from the center of the clock to the tip of each hand. What is the tip-to-tip distance between these two hands when the clock reads 12 minutes after four o'clock?

63. •• **IP** Suppose we orient the x axis of a two-dimensional coordinate system along the beach at Waikiki. Waves approaching the beach have a velocity relative to the shore given by $\vec{\mathbf{v}}_{\text{ws}} = (1.3\ \text{m/s})\hat{\mathbf{y}}$. Surfers move more rapidly than the waves, but at an angle to the beach. The angle is chosen so that the surfers approach the shore with the same speed as the waves. **(a)** If a surfer has a speed of 7.2 m/s relative to the water, what is her direction of motion relative to the positive x axis? **(b)** What is the surfer's velocity relative to the wave? **(c)** If the surfer's speed is increased, will the angle in part (a) increase or decrease? Explain.

64. ••• **IP** Referring to Example 3–2, **(a)** what heading must the boat have if it is to land directly across the river from its starting point? **(b)** How much time is required for this trip if the river is 25.0 m wide? **(c)** Suppose the speed of the boat is increased, but it is still desired to land directly across from the starting point. Should the boat's heading be more upstream, more downstream, or the same as in part (a)? Explain.

65. ••• Vector $\vec{\mathbf{A}}$ points in the negative x direction. Vector $\vec{\mathbf{B}}$ points at an angle of 30.0° above the positive x axis. Vector $\vec{\mathbf{C}}$ has a magnitude of 15 m and points in a direction 40.0° below the positive x axis. Given that $\vec{\mathbf{A}} + \vec{\mathbf{B}} + \vec{\mathbf{C}} = 0$, find the magnitudes of $\vec{\mathbf{A}}$ and $\vec{\mathbf{B}}$.

66. ••• As two boats approach the marina, the velocity of boat 1 relative to boat 2 is 2.30 m/s in a direction 40.0° east of north. If boat 1 has a velocity that is 0.750 m/s due north, what is the velocity (magnitude and direction) of boat 2?

Interactive Problems

67. •• **IP** ***Referring to Example 3–2*** Suppose the speed of the boat relative to the water is 7.0 m/s. **(a)** At what angle to the x axis must the boat be headed if it is to land directly across the river from its starting position? **(b)** If the speed of the boat relative to the water is increased, will the angle needed to go directly across the river increase, decrease, or stay the same? Explain.

68. ••• ***Referring to Example 3–2*** Suppose the boat has a speed of 6.7 m/s relative to the water, and that the dock on the opposite shore of the river is at the location $x = 55$ m and $y = 28$ m relative to the starting point of the boat. **(a)** At what angle relative to the x axis must the boat be pointed in order to reach the other dock? **(b)** With the angle found in part (a), what is the speed of the boat relative to the ground?

4 Two-Dimensional Kinematics

4–1	Motion in Two Dimensions	80
4–2	Projectile Motion: Basic Equations	83
4–3	Zero Launch Angle	84
4–4	General Launch Angle	89
4–5	Projectile Motion: Key Characteristics	93

When you hear the word "projectile," you probably think of an artillery shell or perhaps a home run into the upper deck. But as we'll see in this chapter, the term applies to any object moving under the influence of gravity alone. For example, each of these juggling balls undergoes projectile motion as it moves from one hand to the other. In this chapter we will explore the physical laws that govern such motion, and will learn—among other things—that these balls follow a parabolic path.

We now extend our study of kinematics to motion in two dimensions. This allows us to consider a much wider range of physical phenomena observed in everyday life. Of particular interest is **projectile motion,** the motion of objects that are initially launched—or "projected"—and which then continue moving under the influence of gravity alone. Examples of projectile motion include balls thrown from one person to another, water spraying from a hose, salmon leaping over rapids, and divers jumping from the cliffs of Acapulco.

The main idea of this chapter is quite simple: Horizontal and vertical motions are independent. That's it. For example, a ball thrown horizontally with a speed v continues to move with the same speed v in the horizontal direction, even as it falls with an increasing speed in the vertical direction. Similarly, the time of fall is the same whether a ball is dropped from rest straight down, or thrown horizontally. Simply put, each motion continues as if the other motion were not present.

This chapter develops and applies the idea of independence of motion to many common physical systems.

4–1 Motion in Two Dimensions

In this section we develop equations of motion to describe objects moving in two dimensions. First, we consider motion with constant velocity, determining x and y as functions of time. Next, we investigate motion with constant acceleration. We show that the one-dimensional kinematic equations of Chapter 2 can be extended in a straightforward way to apply to two dimensions.

Constant Velocity

To begin, consider the simple situation shown in **Figure 4–1**. A turtle starts at the origin at $t = 0$ and moves with a constant speed $v_0 = 0.26\ \text{m/s}$ in a direction 25° above the x-axis. How far has the turtle moved in the x and y directions after 5.0 seconds?

First, note that the turtle moves in a straight line a distance

$$d = v_0 t = (0.26\ \text{m/s})(5.0\ \text{s}) = 1.3\ \text{m}$$

as indicated in Figure 4–1(a). From the definitions of sine and cosine given in the previous chapter, we see that

$$x = d \cos 25° = 1.2\ \text{m}$$
$$y = d \sin 25° = 0.55\ \text{m}$$

An alternative way to approach this problem is to treat the x and y motions separately. First, we determine the speed of the turtle in each direction. Referring to Figure 4–1(b), we see that the x component of velocity is

$$v_{0x} = v_0 \cos 25° = 0.24\ \text{m/s}$$

and the y component is

$$v_{0y} = v_0 \sin 25° = 0.11\ \text{m/s}$$

Next, we find the distance traveled by the turtle in the x and y directions by multiplying the speed in each direction by the time:

$$x = v_{0x}t = (0.24\ \text{m/s})(5.0\ \text{s}) = 1.2\ \text{m}$$

and

$$y = v_{0y}t = (0.11\ \text{m/s})(5.0\ \text{s}) = 0.55\ \text{m}$$

This is in agreement with our previous results. To summarize, we can think of the turtle's actual motion as a combination of separate x and y motions.

In general, the turtle might start at a position $x = x_0$ and $y = y_0$ at time $t = 0$. In this case, we have

$$x = x_0 + v_{0x}t \qquad \textbf{4–1}$$

and

$$y = y_0 + v_{0y}t \qquad \textbf{4–2}$$

as the x and y equations of motion.

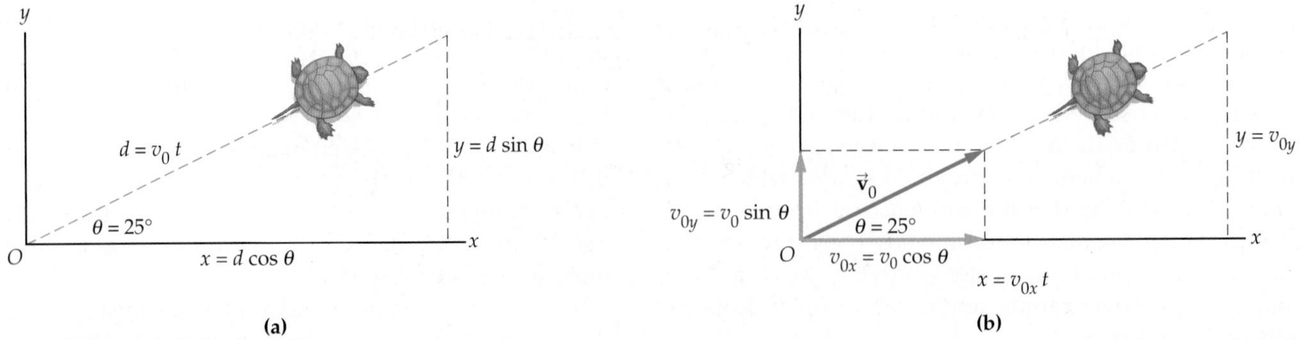

▲ **FIGURE 4–1 Constant velocity**
A turtle walks from the origin with a speed of $v_0 = 0.26\ \text{m/s}$. **(a)** In a time t the turtle moves through a straight-line distance of $d = v_0 t$; thus the x and y displacements are $x = d \cos \theta$, $y = d \sin \theta$. **(b)** Equivalently, the turtle's x and y components of velocity are $v_{0x} = v_0 \cos \theta$ and $v_{0y} = v_0 \sin \theta$; hence $x = v_{0x}t$ and $y = v_{0y}t$.

Compare these equations with Equation 2–11, $x = x_0 + v_0t + \frac{1}{2}at^2$, which gives position as a function of time in one dimension. When acceleration is zero, as it is for the turtle, Equation 2–11 reduces to $x = x_0 + v_0t$. Replacing v_0 with the x component of the velocity, v_{0x}, yields Equation 4–1. Similarly, replacing each x in Equation 4–1 with y converts it to Equation 4–2, the y equation of motion.

A situation illustrating the use of Equations 4–1 and 4–2 is given in Example 4–1.

EXAMPLE 4–1 The Eagle Descends

An eagle perched on a tree limb 19.5 m above the water spots a fish swimming near the surface. The eagle pushes off from the branch and descends toward the water. By adjusting its body in flight, the eagle maintains a constant speed of 3.10 m/s at an angle of 20.0° below the horizontal. **(a)** How long does it take for the eagle to reach the water? **(b)** How far has the eagle traveled in the horizontal direction when it reaches the water?

Picture the Problem

We set up our coordinate system so that the eagle starts at $x_0 = 0$ and $y_0 = h = 19.5$ m. The water level is $y = 0$. As indicated in our sketch, $v_{0x} = v_0 \cos\theta$ and $v_{0y} = -v_0 \sin\theta$, where $v_0 = 3.10$ m/s. Notice that both components of the eagle's velocity are constant, and therefore the equations of motion given in Equations 4–1 and 4–2 apply.

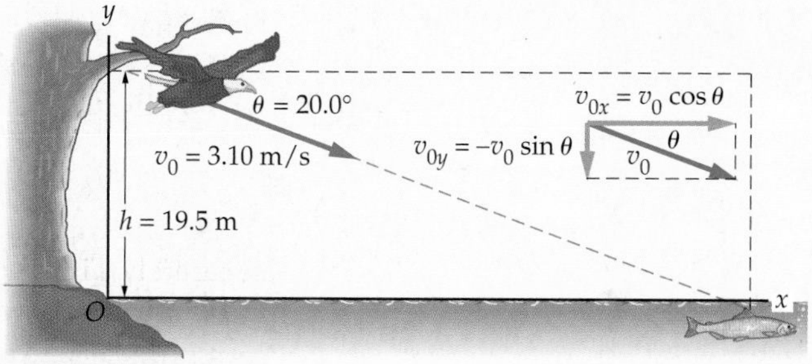

Strategy

As usual in such problems, it is best to treat the eagle's flight as a combination of separate x and y motions. Since we are given the constant speed of the eagle, and the angle at which it descends, we can find the x and y components of its velocity. We then use the y equation of motion, $y = y_0 + v_{0y}t$, to find the time t when the eagle reaches the water. Finally, we use this value of t in the x equation of motion, $x = x_0 + v_{0x}t$, to find the horizontal distance the bird travels.

Solution

Part (a)

1. Begin by determining v_{0x} and v_{0y}:

$$v_{0x} = v_0 \cos\theta = (3.10 \text{ m/s}) \cos 20.0° = 2.91 \text{ m/s}$$
$$v_{0y} = -v_0 \sin\theta = -(3.10 \text{ m/s}) \sin 20.0° = -1.06 \text{ m/s}$$

2. Now, set $y = 0$ in $y = y_0 + v_{0y}t$ and solve for t:

$$y = y_0 + v_{0y}t = h + v_{0y}t = 0$$
$$t = -\frac{h}{v_{0y}} = -\frac{19.5 \text{ m}}{(-1.06 \text{ m/s})} = 18.4 \text{ s}$$

Part (b)

3. Substitute $t = 18.4$ s into $x = x_0 + v_{0x}t$ to find x:

$$x = x_0 + v_{0x}t = 0 + (2.91 \text{ m/s})(18.4 \text{ s}) = 53.5 \text{ m}$$

Insight

Notice how the two minus signs in step 2 combine to give a positive time. One minus sign comes from setting $y = 0$, the other from the fact that v_{0y} is negative. No matter where we choose the origin, or what direction we choose to be positive, the time will always have the same value.

As mentioned in the problem statement, the eagle cannot travel in a straight line by simply dropping from the tree limb—it has to adjust its wings and tail to produce enough lift to balance the force of gravity. Airplanes do the same thing when they adjust their flight surfaces to make a smooth landing.

Practice Problem

What is the location of the eagle 2.00 s after it takes flight? [**Answer:** $x = 5.82$ m, $y = 17.4$ m.]

Some related homework problems: Problem 1, Problem 2

Constant Acceleration

To study motion with constant acceleration in two dimensions we repeat what was done in one dimension in Chapter 2, but with separate equations for both x and y. For example, to obtain x as a function of time we start with $x = x_0 + v_0t + \frac{1}{2}at^2$ (Equation 2–11), and replace both v_0 and a with the corresponding x components,

v_{0x} and a_x. This gives

$$x = x_0 + v_{0x}t + \tfrac{1}{2}a_xt^2 \qquad \text{4–3(a)}$$

To obtain y as a function of time, we write y in place of x in Equation 4–3(a):

$$y = y_0 + v_{0y}t + \tfrac{1}{2}a_yt^2 \qquad \text{4–3(b)}$$

These are the position versus time equations of motion for two dimensions. (In three dimensions we introduce a third coordinate direction and label it z. We would then simply replace x with z in Equation 4–3(a) to obtain z as a function of time.)

The same approach gives velocity as a function of time. Start with Equation 2–7, $v = v_0 + at$, and write it in terms of x and y components. This yields

$$v_x = v_{0x} + a_xt \qquad \text{4–4(a)}$$

$$v_y = v_{0y} + a_yt \qquad \text{4–4(b)}$$

Note that we simply repeat everything we did for one dimension, only now with separate equations for the x and y components.

Finally, we can write $v^2 = v_0{}^2 + 2a\Delta x$ in terms of components as well:

$$v_x{}^2 = v_{0x}{}^2 + 2a_x\Delta x \qquad \text{4–5(a)}$$

$$v_y{}^2 = v_{0y}{}^2 + 2a_y\Delta y \qquad \text{4–5(b)}$$

The following table summarizes our results:

TABLE 4–1 Constant-Acceleration Equations of Motion

Position as a function of time	Velocity as a function of time	Velocity as a function of position
$x = x_0 + v_{0x}t + \frac{1}{2}a_xt^2$	$v_x = v_{0x} + a_xt$	$v_x{}^2 = v_{0x}{}^2 + 2a_x\Delta x$
$y = y_0 + v_{0y}t + \frac{1}{2}a_yt^2$	$v_y = v_{0y} + a_yt$	$v_y{}^2 = v_{0y}{}^2 + 2a_y\Delta y$

These are the fundamental equations that will be used to obtain *all* of the results presented throughout the rest of this chapter. Though it may appear sometimes that we are writing new sets of equations for different special cases, the equations aren't new—what we are actually doing is simply writing these equations again, but with specific values substituted for the constants that appear in them.

EXAMPLE 4–2 A Hummer Accelerates

A hummingbird is flying in such a way that it is initially moving vertically with a speed of 4.6 m/s and accelerating horizontally at 11 m/s^2. Assuming the bird's acceleration remains constant for the time interval of interest, find the horizontal and vertical distance through which it moves in 0.55 s.

Picture the Problem

In our sketch we have placed the origin of a two-dimensional coordinate system at the location of the hummingbird at the initial time, $t = 0$. In addition, we have chosen the initial direction of motion to be in the positive y direction, and the direction of acceleration to be in the positive x direction. As a result, it follows that $x_0 = y_0 = 0$, $v_{0x} = 0$, $v_{0y} = 4.6$ m/s, $a_x = 11$ m/s^2, and $a_y = 0$. As the hummingbird moves upward, its x component of velocity increases, resulting in a curved path, as shown.

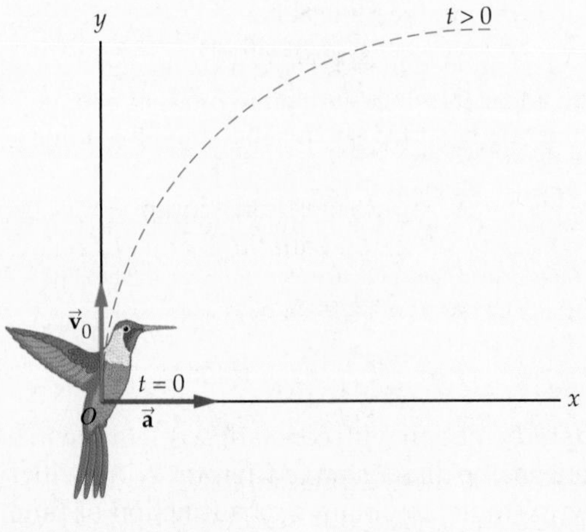

Strategy

Since we want to relate position and time, we find the horizontal position of the hummingbird using $x = x_0 + v_{0x}t + \frac{1}{2}a_xt^2$, and the vertical position using $y = y_0 + v_{0y}t + \frac{1}{2}a_yt^2$.

Solution

1. Use $x = x_0 + v_{0x}t + \frac{1}{2}a_x t^2$ to find x at $t = 0.55$ s: $x = x_0 + v_{0x}t + \frac{1}{2}a_x t^2 = 0 + 0 + \frac{1}{2}(11\ \text{m/s}^2)(0.55\ \text{s})^2 = 1.7\ \text{m}$

2. Use $y = y_0 + v_{0y}t + \frac{1}{2}a_y t^2$ to find y at $t = 0.55$ s: $y = y_0 + v_{0y}t + \frac{1}{2}a_y t^2 = 0 + (4.6\ \text{m/s})(0.55\ \text{s}) + 0 = 2.5\ \text{m}$

Insight

In 0.55 s the hummingbird moves 1.7 m horizontally and 2.5 m vertically. The horizontal position of the bird will eventually increase more rapidly with time than the vertical position, due to the t^2 dependence of x as compared with the t dependence of y. This results in a curved, parabolic path for the hummingbird, as shown in our sketch.

Practice Problem

How much time is required for the hummingbird to move 2.0 m horizontally from its initial position? [**Answer:** $t = 0.60$ s]

Some related homework problems: Problem 3, Problem 4

4–2 Projectile Motion: Basic Equations

We now apply the independence of horizontal and vertical motions to projectiles. Just what do we mean by a projectile? Well, a **projectile** is an object that is thrown, kicked, batted, or otherwise launched into motion and then allowed to follow a path determined solely by the influence of gravity. As you might expect, this covers a wide variety of physical systems.

In studying projectile motion we make the following assumptions:

- air resistance is ignored
- the acceleration due to gravity is constant, downward, and has a magnitude equal to $g = 9.81\ \text{m/s}^2$
- the Earth's rotation is ignored

Air resistance can be significant when a projectile moves with relatively high speed or if it encounters a strong wind. In many everyday situations, however, like tossing a ball to a friend or dropping a book, air resistance is relatively insignificant. As for the acceleration due to gravity, $g = 9.81\ \text{m/s}^2$, this value varies slightly from place to place on the Earth's surface and decreases with increasing altitude. In addition, the rotation of the Earth can be significant when considering projectiles that cover great distances. Little error is made in ignoring the variation of g or the rotation of the Earth, however, in the examples of projectile motion considered in this chapter.

Let's incorporate these assumptions into the equations of motion given in the previous section. Suppose, as in **Figure 4–2**, that the x axis is horizontal and the y axis is vertical, with the positive direction upward. Since downward is the negative direction, it follows that

$$a_y = -9.81\ \text{m/s}^2 = -g$$

Gravity causes no acceleration in the x direction. Thus, the x component of acceleration is zero:

$$a_x = 0$$

With these acceleration components substituted into the fundamental constant-acceleration equations of motion (Table 4–1) we find:

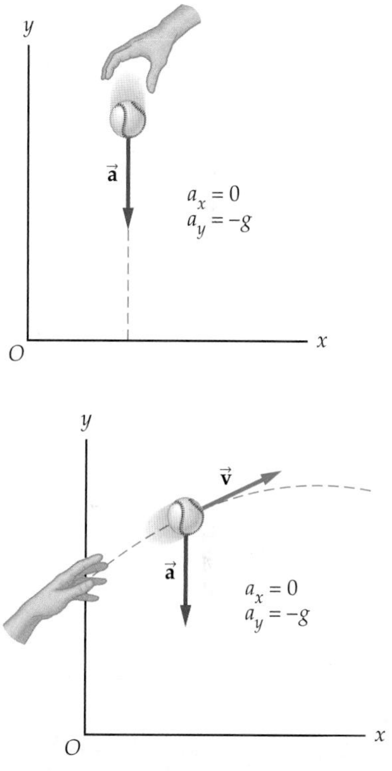

▲ **FIGURE 4–2 Acceleration in free fall**
All objects in free fall have acceleration components $a_x = 0$ and $a_y = -g$ when the coordinate system is chosen as shown here. This is true regardless of whether the object is dropped, thrown, kicked, or otherwise set into motion.

Projectile Motion ($a_x = 0$, $a_y = -g$)

$$x = x_0 + v_{0x}t \qquad v_x = v_{0x} \qquad v_x^2 = v_{0x}^2$$
$$y = y_0 + v_{0y}t - \tfrac{1}{2}gt^2 \qquad v_y = v_{0y} - gt \qquad v_y^2 = v_{0y}^2 - 2g\Delta y \qquad \text{4–6}$$

Note that in these expressions the positive y direction is upward and the quantity g is positive. All of our studies of *projectile motion* will use Equations 4–6 as our

PROBLEM-SOLVING NOTE

Acceleration of a Projectile

When the x axis is chosen to be horizontal and the y axis points vertically upward, it follows that the acceleration of an ideal projectile is $a_x = 0$ and $a_y = -g$.

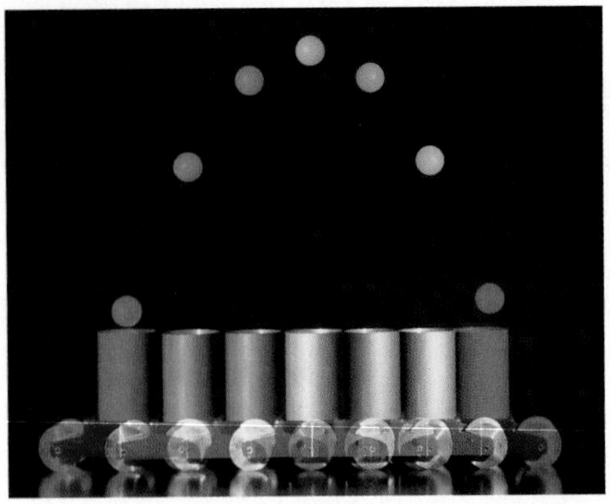

▲ In the multiple-exposure photo at left, a ball is projected upward from a moving cart. The ball retains its initial horizontal velocity; as a result, it follows a parabolic path and remains directly above the cart at all times. When the ball lands, it falls back into the cart, just as it would if the cart had been at rest. (In this sequence, the exposures were made at equal time intervals with light of different colors, making it easier to follow the relative motion of the ball and the cart.) In the photo at right, the pilot ejection seat of a jet fighter is being ground-tested. Here too, the horizontal and vertical motions are independent; thus, the test dummy is still almost directly above the cockpit from which it was ejected. Note, however, that air resistance is beginning to reduce the dummy's horizontal velocity. Eventually, it will fall far behind the speeding plane.

fundamental equations—again, special cases will simply correspond to substituting specific values for the constants.

A simple demonstration illustrates the independence of horizontal and vertical motions in projectile motion. First, while standing still, drop a rubber ball to the floor and catch it on the rebound. Note that the ball goes straight down, lands near your feet, and returns almost to the level of your hand in about a second.

Next, walk—or roller skate—with constant speed before dropping the ball, then observe its motion carefully. To you, its motion looks the same as before: It goes straight down, lands near your feet, bounces straight back up, and returns in about one second. This is illustrated in **Figure 4–3**. The fact that you were moving in the horizontal direction the whole time had no effect on the ball's vertical motion—the motions were independent.

▶ **FIGURE 4–3 Independence of vertical and horizontal motions**
When you drop a ball while walking, running, or skating with constant velocity, it appears to you to drop straight down from the point where you released it. To a person at rest, the ball follows a curved path that combines horizontal and vertical motions.

To an observer who sees you walking by, the ball follows a curved path, as shown. The precise shape of this curved path is determined in the next section.

4–3 Zero Launch Angle

A special case of some interest is a projectile launched horizontally, so that the angle between the initial velocity and the horizontal is $\theta = 0$. We devote this section to a brief look at this type of motion.

Equations of Motion

Suppose you are walking with a speed v_0 when you release a ball from a height h, as discussed in the previous section. If we choose ground level to be $y = 0$ and the release point to be directly above the origin, the initial position of the

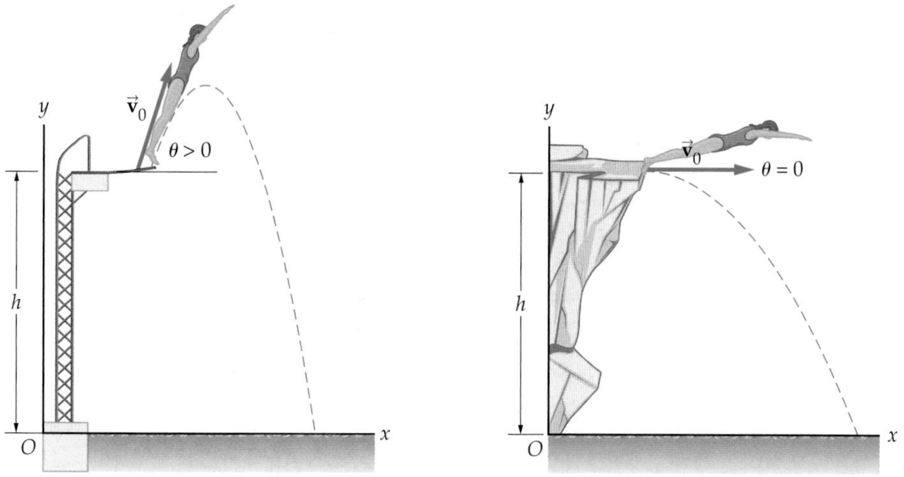

◀ **FIGURE 4–4 Launch angle of a projectile**

(a) A projectile launched at an angle above the horizontal, $\theta > 0$. A launch below the horizontal would correspond to $\theta < 0$. **(b)** A projectile launched horizontally, $\theta = 0$. In this section we consider $\theta = 0$. The next section deals with $\theta \neq 0$.

ball is given by

$$x_0 = 0$$

and

$$y_0 = h$$

This is illustrated in Figure 4–3.

The initial velocity is horizontal, corresponding to $\theta = 0$ in **Figure 4–4**. As a result, the x component of the initial velocity is simply the initial speed,

$$v_{0x} = v_0 \cos 0° = v_0$$

and the y component of the initial velocity is zero,

$$v_{0y} = v_0 \sin 0° = 0$$

Substituting these specific values into our fundamental equations for projectile motion (Equations 4–6) gives the following simplified results for zero launch angle ($\theta = 0$):

$$x = v_0 t \qquad v_x = v_0 = \text{constant} \qquad {v_x}^2 = {v_0}^2 = \text{constant}$$
$$y = h - \tfrac{1}{2}gt^2 \qquad v_y = -gt \qquad {v_y}^2 = -2g\Delta y \tag{4–7}$$

Note that the x component of velocity remains the same for all time and that the y component steadily decreases with time. As a result, x increases linearly with time, and y decreases with a t^2 dependence. Snapshots of this motion at equal time intervals are shown in **Figure 4–5**.

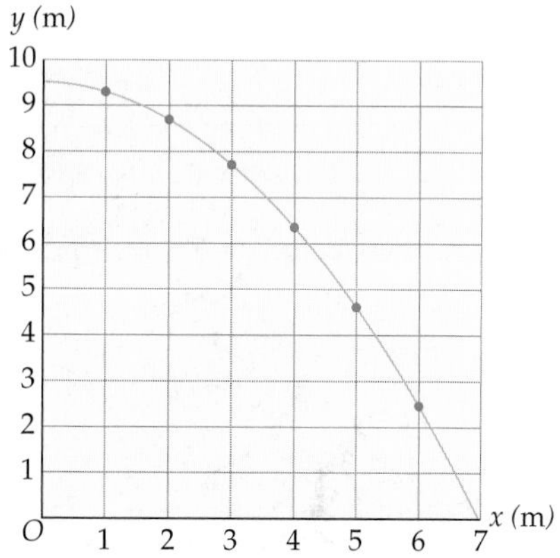

▲ **FIGURE 4–5 Trajectory of a projectile launched horizontally**

In this plot, the projectile was launched from a height of 9.5 m with an initial speed of 5.0 m/s. The positions shown in the plot correspond to the times $t = 0.20$ s, 0.40 s, 0.60 s, ... Note the uniform motion in the x direction, and the accelerated motion in the y direction.

PROBLEM-SOLVING NOTE

Identify Initial Conditions

The launch point of a projectile determines x_0 and y_0. The initial velocity of a projectile determines v_{0x} and v_{0y}.

EXAMPLE 4–3 Dropping a Ball

A person skateboarding with a constant speed of 1.30 m/s releases a ball from a height of 1.25 m above the ground. Given that $x_0 = 0$ and $y_0 = h = 1.25$ m, find x and y for **(a)** $t = 0.250$ s and **(b)** $t = 0.500$ s. **(c)** Find the velocity, speed, and direction of motion of the ball at $t = 0.500$ s.

Picture the Problem

The ball starts at $x_0 = 0$ and $y_0 = h = 1.25$ m. Its initial velocity is horizontal, therefore $v_{0x} = v_0 = 1.30$ m/s and $v_{0y} = 0$. In addition, it accelerates with the acceleration due to gravity in the negative y direction, $a_y = -g$, and moves with constant speed in the x direction, $a_x = 0$.

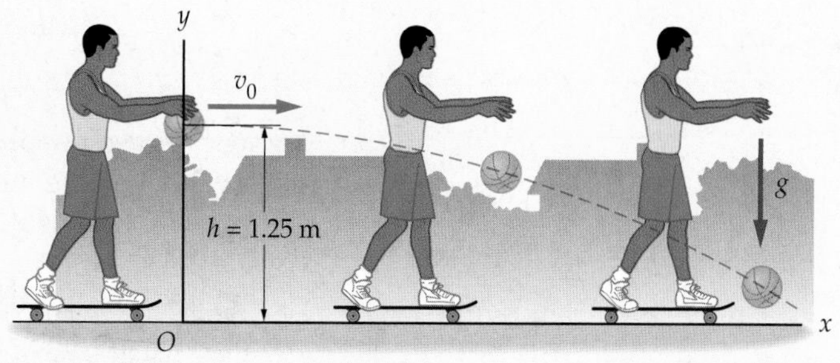

Strategy

The x and y positions are given by $x = v_0 t$ and $y = h - \tfrac{1}{2}gt^2$, respectively. We simply substitute time into these expressions. Similarly, the velocity components are $v_x = v_0$ and $v_y = -gt$.

continued on next page

continued from previous page

Solution

Part (a)

1. Substitute $t = 0.250$ s into the x and y equations of motion:

$$x = v_0 t = (1.30\ \text{m/s})(0.250\ \text{s}) = 0.325\ \text{m}$$
$$y = h - \tfrac{1}{2}gt^2$$
$$= 1.25\ \text{m} - \tfrac{1}{2}(9.81\ \text{m/s}^2)(0.250\ \text{s})^2 = 0.943\ \text{m}$$

Part (b)

2. Substitute $t = 0.500$ s into the x and y equations of motion: Note that the ball is only about an inch above the ground at this time:

$$x = v_0 t = (1.30\ \text{m/s})(0.500\ \text{s}) = 0.650\ \text{m}$$
$$y = h - \tfrac{1}{2}gt^2$$
$$= 1.25\ \text{m} - \tfrac{1}{2}(9.81\ \text{m/s}^2)(0.500\ \text{s})^2 = 0.0238\ \text{m}$$

Part (c)

3. First, calculate the x and y components of the velocity at $t = 0.500$ s using $v_x = v_0$ and $v_y = -gt$:

$$v_x = v_0 = 1.30\ \text{m/s}$$
$$v_y = -gt = -(9.81\ \text{m/s}^2)(0.500\ \text{s}) = -4.91\ \text{m/s}$$

4. Use these components to determine $\vec{\mathbf{v}}$, v, and θ:

$$\vec{\mathbf{v}} = (1.30\ \text{m/s})\hat{\mathbf{x}} + (-4.91\ \text{m/s})\hat{\mathbf{y}}$$
$$v = \sqrt{v_x^{\,2} + v_y^{\,2}}$$
$$= \sqrt{(1.30\ \text{m/s})^2 + (-4.91\ \text{m/s})^2} = 5.08\ \text{m/s}$$
$$\theta = \tan^{-1}\frac{v_y}{v_x} = \tan^{-1}\frac{(-4.91\ \text{m/s})}{1.30\ \text{m/s}} = -75.2^\circ$$

Insight

Note that the x position of the ball does not depend on the acceleration of gravity, g, and that its y position does not depend on the initial horizontal speed of the ball, v_0. For example, if the person is running when he drops the ball, the ball is moving faster in the horizontal direction, and it keeps up with the person when it is dropped. Its vertical motion doesn't change at all, however; it drops to the ground in exactly the same time and bounces back to the same height as before.

Practice Problem

How long does it take for the ball to land? [**Answer:** Referring to the results of part (b), it is clear that the time of landing is slightly greater than 0.500 s. Setting $y = 0$ gives a precise answer; $t = \sqrt{2h/g} = 0.505$ s.]

Some related homework problems: Problem 12, Problem 13, Problem 17

CONCEPTUAL CHECKPOINT 4–1 Compare Splashdown Speeds

Two youngsters dive off an overhang into a lake. Diver 1 drops straight down, diver 2 runs off the cliff with an initial horizontal speed v_0. Is the splashdown speed of diver 2 **(a)** greater than, **(b)** less than, or **(c)** equal to the splashdown speed of diver 1?

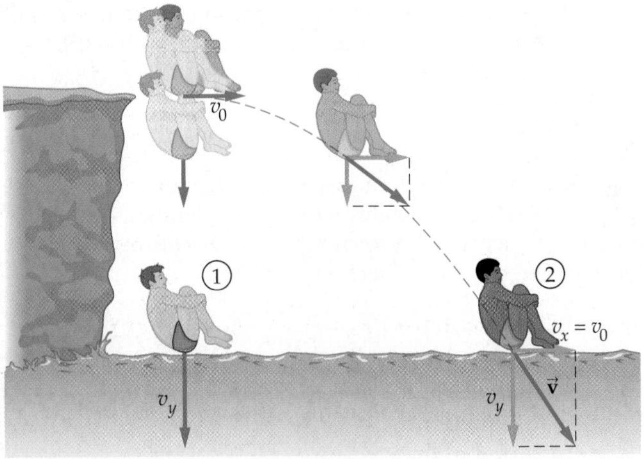

Reasoning and Discussion

Note that neither diver has an initial y component of velocity, and that they both fall with the same vertical acceleration—the acceleration due to gravity. Therefore, the two divers fall for the same amount of time, and their y components of velocity are the same at splashdown. Since diver 2 also has a nonzero x component of velocity, unlike diver 1, the speed of diver 2 is greater.

Answer:

(a) The speed of diver 2 is greater than that of diver 1.

Parabolic Path

Just what is the shape of the curved path followed by a projectile launched horizontally? This can be found by combining $x = v_0 t$ and $y = h - \frac{1}{2}gt^2$, which allows us to express y in terms of x. First, solve for time using the x equation. This gives

$$t = x/v_0$$

Next, substitute this result into the y equation to eliminate t:

$$y = h - \tfrac{1}{2}g\left(\frac{x}{v_0}\right)^2 = h - \left(\frac{g}{2v_0^2}\right)x^2 \qquad \textbf{4–8}$$

Note that y has the form

$$y = a + bx^2$$

where $a = h$ = constant and $b = -g/2v_0^2$ = constant. This is the equation of a parabola that curves downward, a characteristic shape in projectile motion.

REAL-WORLD PHYSICS

The parabolic trajectory of projectiles

◀ Lava bombs (left) and fountain jets (right) trace out parabolic paths, as is typical in projectile motion. The trajectories are only slightly altered by air resistance.

Landing Site

Where does a projectile land if it is launched horizontally with a speed v_0 from a height h?

The most direct way to answer this question is to set $y = 0$ in Equation 4–8, since $y = 0$ corresponds to ground level. This gives

$$0 = h - \left(\frac{g}{2v_0^2}\right)x^2$$

Solving for x yields the landing site:

$$x = v_0\sqrt{\frac{2h}{g}} \qquad \textbf{4–9}$$

Note that we have chosen the positive sign for the square root since the projectile was launched in the positive x direction, and hence lands at a positive value of x.

A useful alternative approach is to find the time of landing with the kinematic relations given in Equation 4–7, and then substitute this time into $x = v_0 t$. This approach is illustrated in the next Example.

PROBLEM-SOLVING NOTE

Use Independence of Motion

Projectile problems can be solved by breaking the problem into its x and y components, and then solving for the motion of each component separately.

EXAMPLE 4–4 Jumping a Crevasse

A mountain climber encounters a crevasse in an ice field. The opposite side of the crevasse is 2.75 m lower, and is separated horizontally by a distance of 4.10 m. To cross the crevasse, the climber gets a running start and jumps in the horizontal direction. **(a)** What is the minimum speed needed by the climber to safely cross the crevasse? If, instead, the climber's speed is 6.00 m/s, **(b)** where does the climber land, and **(c)** what is the climber's speed on landing?

Picture the Problem

The mountain climber jumps from $x_0 = 0$ and $y_0 = h = 2.75$ m. The landing site for part (a) is $x = w = 4.10$ m and $y = 0$. Note that the y position of the climber *decreases* by h, and therefore $\Delta y = -h = -2.75$ m. As for the initial velocity, we are given that $v_{0x} = v_0$ and $v_{0y} = 0$. Finally, with our choice of coordinates it follows that $a_x = 0$ and $a_y = -g$.

Strategy

We can model the climber as a projectile, and apply our equations for projectile motion with a horizontal launch.

a. From Equation 4–7 we have that $x = v_0 t$ and $y = h - \frac{1}{2}gt^2$. Setting $y = 0$ determines the time of landing. Using this time in the x equation gives the horizontal landing position in terms of the initial speed.

b. We can now use the relation from part (a) to find x in terms of $v_0 = 6.00$ m/s.

c. We already know v_x, since it remains constant, and we can calculate v_y using $v_y^2 = -2g\Delta y$ (Equation 4–7). With the velocity components known, we can use the Pythagorean theorem to find the speed.

Solution

Part (a)

1. Set $y = h - \frac{1}{2}gt^2$ equal to zero (landing condition) and solve for the corresponding time t:

$$y = h - \tfrac{1}{2}gt^2 = 0$$
$$t = \sqrt{\frac{2h}{g}}$$

2. Substitute this expression for t into the x equation of motion, $x = v_0 t$, and solve for the speed, v_0:

$$x = v_0 t = v_0\sqrt{\frac{2h}{g}} \quad \text{or} \quad v_0 = x\sqrt{\frac{g}{2h}}$$

3. Substitute numerical values in this expression:

$$v_0 = x\sqrt{\frac{g}{2h}} = (4.10\text{ m})\sqrt{\frac{9.81\text{ m/s}^2}{2(2.75\text{ m})}} = 5.48\text{ m/s}$$

Part (b)

4. Substitute $v_0 = 6.00$ m/s into the expression for x obtained in step 2, $x = v_0\sqrt{2h/g}$:

$$x = v_0\sqrt{\frac{2h}{g}} = (6.00\text{ m/s})\sqrt{\frac{2(2.75\text{ m})}{9.81\text{ m/s}^2}} = 4.49\text{ m}$$

Part (c)

5. Use the fact that the x component of velocity does not change to determine v_x, and use $v_y^2 = -2g\Delta y$ to determine v_y. For v_y, note that we choose the minus sign for the square root because the climber is moving downward:

$$v_x = v_0 = 6.00\text{ m/s}$$
$$v_y = \pm\sqrt{-2g\Delta y}$$
$$= -\sqrt{-2(9.81\text{ m/s}^2)(-2.75\text{ m})} = -7.35\text{ m/s}$$

6. Use the Pythagorean theorem to determine the speed:

$$v = \sqrt{v_x^2 + v_y^2}$$
$$= \sqrt{(6.00\text{ m/s})^2 + (-7.35\text{ m/s})^2} = 9.49\text{ m/s}$$

Insight

The minimum speed needed to safely cross the crevasse is 5.48 m/s. If the initial horizontal speed is 6.00 m/s, the climber will land 4.49 m − 4.10 m = 0.39 m beyond the edge of the crevasse with a speed of 9.49 m/s.

Practice Problem

(a) When the climber's speed is the minimum needed to cross the crevasse, $v_0 = 5.48$ m/s, how long is the climber in the air? **(b)** How long is the climber in the air when $v_0 = 6.00$ m/s? **[Answer: (a)** $t = x/v_0 = (4.10\text{ m})/(5.48\text{ m/s}) = 0.748$ s. **(c)** $t = x/v_0 = (4.49\text{ m})/(6.00\text{ m/s}) = 0.748$ s. The times are the same! The answer to both parts is simply the time needed to fall through a height h; $t = \sqrt{2h/g} = 0.748$ s.]

Some related homework problems: Problem 8, Problem 9, Problem 14

CONCEPTUAL CHECKPOINT 4–2 Minimum Speed

If the height h is increased in the previous example but the width w remains the same, does the minimum speed needed to cross the crevasse **(a)** increase, **(b)** decrease, or **(c)** stay the same?

Reasoning and Discussion

If the height is greater, the time of fall is also greater. Since the climber is in the air for a greater time, the horizontal distance covered for a given initial speed is also greater. Thus, if the width of the crevasse is the same, a lower initial speed allows for a safe crossing.

Answer

(b) The minimum speed decreases.

4–4 General Launch Angle

We now consider the more general case of a projectile launched at an arbitrary angle with respect to the horizontal. This means we can no longer use the simplifications associated with zero launch angle. As always, we return to our basic equations for projectile motion (Equations 4–6), and this time we simply let θ be nonzero.

Figure 4–6(a) shows a projectile launched with an initial speed v_0 at an angle θ above the horizontal. Since the projectile starts at the origin, the initial x and y positions are zero:

$$x_0 = y_0 = 0$$

The components of the initial velocity are determined as indicated in **Figure 4–6(b)**:

$$v_{0x} = v_0 \cos\theta$$

and

$$v_{0y} = v_0 \sin\theta$$

As a quick check, note that if $\theta = 0$, then $v_{0x} = v_0$ and $v_{0y} = 0$. Similarly, if $\theta = 90°$ we find $v_{0x} = 0$ and $v_{0y} = v_0$. These checks are depicted in **Figure 4–6(c)**.

Substituting these results into the basic equations for projectile motion yields the following results for a general launch angle:

$$\begin{aligned} x &= (v_0 \cos\theta)t & v_x &= v_0 \cos\theta & v_x^2 &= v_0^2 \cos^2\theta \\ y &= (v_0 \sin\theta)t - \tfrac{1}{2}gt^2 & v_y &= v_0 \sin\theta - gt & v_y^2 &= v_0^2 \sin^2\theta - 2g\Delta y \end{aligned} \qquad \text{4–10}$$

Note that these equations, which are valid for any launch angle, reduce to the simpler Equations 4–7 when we set $\theta = 0$ and $y_0 = h$. In the next two Exercises, we use Equations 4–10 to calculate a projectile's position and velocity for three equally spaced times.

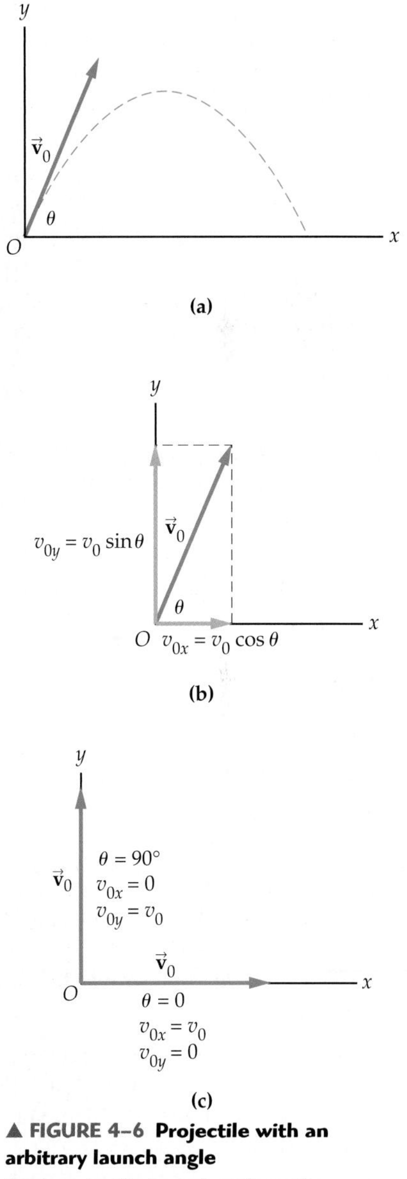

▲ **FIGURE 4–6 Projectile with an arbitrary launch angle**
(a) A projectile launched from the origin at an angle θ above the horizontal. **(b)** The x and y components of the initial velocity. **(c)** Velocity components in the limits $\theta = 0$ and $\theta = 90°$.

EXERCISE 4–1

A projectile is launched from the origin with an initial speed of 20.0 m/s at an angle of 35.0° above the horizontal. Find the x and y positions of the projectile at times **(a)** $t = 0.500$ s, **(b)** $t = 1.00$ s, and **(c)** $t = 1.50$ s.

Solution

a. $x = 8.19$ m, $y = 4.51$ m,

b. $x = 16.4$ m, $y = 6.57$ m,

c. $x = 24.6$ m, $y = 6.17$ m. Note that x increases steadily; y increases, then decreases.

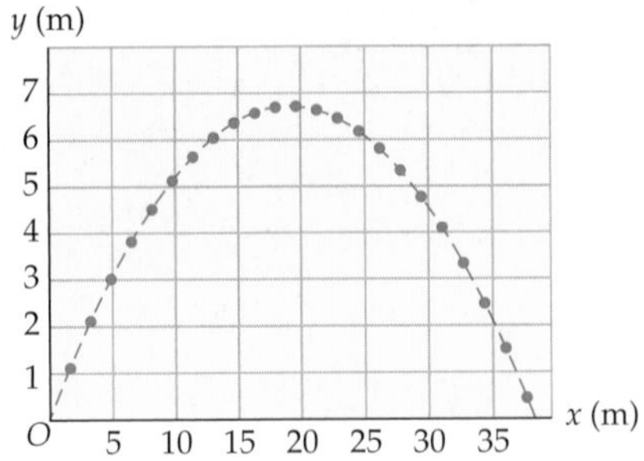

▲ **FIGURE 4–7 Snapshots of a trajectory**
This plot shows a projectile launched from the origin with an initial speed of 20.0 m/s at an angle of 35.0° above the horizontal. The positions shown in the plot correspond to the times $t = 0.1$ s, 0.2 s, 0.3 s, ... Red dots mark the positions considered in Exercises 4–1 and 4–2.

EXERCISE 4–2

Referring to Exercise 4–1, find the velocity of the projectile at times **(a)** $t = 0.500$ s, **(b)** $t = 1.00$ s, and **(c)** $t = 1.50$ s.

Solution

a. $\vec{v} = (16.4\ \text{m/s})\hat{x} + (6.57\ \text{m/s})\hat{y}$, **b.** $\vec{v} = (16.4\ \text{m/s})\hat{x} + (1.66\ \text{m/s})\hat{y}$,
c. $\vec{v} = (16.4\ \text{m/s})\hat{x} + (-3.24\ \text{m/s})\hat{y}$.

Figure 4–7 shows the projectile referred to in the previous Exercises for a series of times spaced by 0.10 s. Note that the points in Figure 4–7 are not evenly spaced in terms of position, even though they are evenly spaced in time. In fact, the points bunch closer together at the top of the trajectory, showing that a comparatively large fraction of the flight time is spent near the highest point. This is why it seems that a basketball player soaring toward a slam dunk, or a ballerina performing a grand jeté, is "hanging" in air.

▲ "Hanging" in air near the peak of a jump requires no special knack—in fact, it's an unavoidable consequence of the laws of physics. This phenomenon, which makes big leapers (such as deer and dancers) look particularly graceful, can also make life more dangerous for salmon fighting their way upstream to spawn.

EXAMPLE 4–5 A Rough Shot

Chipping from the rough, a golfer sends the ball over a 3.00 m-high tree that is 14.0 m away. The ball lands at the same level from which it was struck after traveling a horizontal distance of 17.8 m—on the green, of course. **(a)** If the ball left the club 54.0° above the horizontal and landed on the green 2.24 s later, what was its initial speed? **(b)** How high was the ball when it passed over the tree?

Picture the Problem

Our sketch shows the ball taking flight from the origin, $x_0 = y_0 = 0$, with a launch angle of 54.0°, and arcing over the tree. The individual points along the parabolic trajectory correspond to equal time intervals.

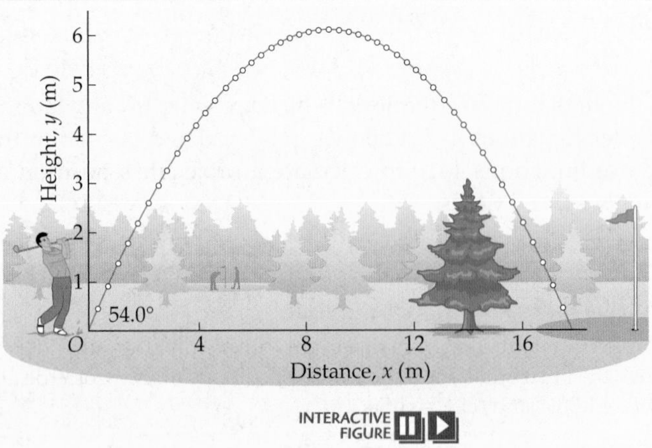

Strategy

a. Since the projectile moves with constant speed in the x direction, the x component of velocity is simply horizontal distance divided by time. Knowing v_x and θ, we can find v_0 from $v_x = v_0 \cos\theta$.

b. We can use $x = (v_0 \cos\theta)t$ to find the time when the ball is at $x = 14.0$ m. Substituting this time into $y = (v_0 \sin\theta)t - \frac{1}{2}gt^2$ gives the height.

Solution

Part (a)

1. Divide the horizontal distance, d, by the time of flight, t, to obtain v_x:

$$v_x = \frac{d}{t} = \frac{17.8\ \text{m}}{2.24\ \text{s}} = 7.95\ \text{m/s}$$

2. Use $v_x = v_0 \cos\theta$ to find v_0, the initial speed: $v_x = v_0 \cos\theta$ or $v_0 = \dfrac{v_x}{\cos\theta} = \dfrac{7.95\text{ m/s}}{\cos 54.0°} = 13.5\text{ m/s}$

Part (b)

3. Use $x = (v_0 \cos\theta)t$ to find the time when $x = 14.0$ m. Recall that $x_0 = 0$: $x = (v_0 \cos\theta)t$ or $t = \dfrac{x}{v_0 \cos\theta} = \dfrac{14.0\text{ m}}{7.95\text{ m/s}} = 1.76\text{ s}$

4. Evaluate $y = (v_0 \sin\theta)t - \frac{1}{2}gt^2$ at the time found in step 3. Recall that $y_0 = 0$:

$$y = (v_0 \sin\theta)t - \tfrac{1}{2}gt^2$$
$$= [(13.5\text{ m/s})\sin 54.0°](1.76\text{ s}) - \tfrac{1}{2}(9.81\text{ m/s}^2)(1.76\text{ s})^2$$
$$= 4.03\text{ m}$$

Insight

The ball clears the top of the tree by 1.03 m and lands on the green 0.48 s later. When it lands, its speed (in the absence of air resistance) is again 13.5 m/s—the same as when it was launched. This result will be verified in the next section.

Practice Problem

What are the speed and direction of the ball when it passes over the tree? [**Answer:** To find the ball's speed and direction, note that $v_x = 7.95$ m/s and $v_y = v_0 \sin\theta - gt = -6.34$ m/s. It follows that $v = \sqrt{v_x^2 + v_y^2} = 10.2$ m/s and $\theta = \tan^{-1}(v_y/v_x) = -38.6°$.]

Some related homework problems: Problem 25, Problem 33

ACTIVE EXAMPLE 4–1 An Elevated Green

A golfer hits a ball from the origin with an initial speed of 30.0 m/s at an angle of 50.0° above the horizontal. The ball lands on a green that is 5.00 m above the level where the ball was struck.

a. How long is the ball in the air?

b. How far has the ball traveled in the horizontal direction when it lands?

c. What is the speed and direction of motion of the ball just before it lands?

Solution *(Test your understanding by performing the calculations indicated in each step.)*

Part (a)

1. Let $y = (v_0 \sin\theta)t - \frac{1}{2}gt^2 = 5.00$ m and solve for t: $t = 0.229$ s, 4.46 s

2. When $t = 0.229$ s, the ball is moving upward; when $t = 4.46$ s, the ball is on the way down. Choose the later time: $t = 4.46$ s

Part (b)

3. Substitute $t = 4.46$ s into $x = (v_0 \cos\theta)t$: $x = 86.0$ m

Part (c)

4. Use $v_x = v_0 \cos\theta$ to calculate v_x: $v_x = 19.3$ m/s

5. Substitute $t = 4.46$ s into $v_y = v_0 \sin\theta - gt$ to find v_y: $v_y = -20.8$ m/s

6. Calculate v and θ: $v = 28.4$ m/s, $\theta = -47.1°$

Your Turn

How long is the ball in the air if the green is 5.00 m *below* the level where the ball was struck?

(Answers to **Your Turn** *problems are given in the back of the book.)*

The next Example presents a classic situation in which two projectiles collide. One projectile is launched from the origin, and thus its equations of motion are given by Equation 4–10. The second projectile is simply dropped from a height, which is a special case of the equations of motion in Equation 4–7 with $v_0 = 0$.

EXAMPLE 4–6 A Leap of Faith

A trained dolphin leaps from the water with an initial speed of 12.0 m/s. It jumps directly toward a ball held by the trainer a horizontal distance of 5.50 m away and a vertical distance of 4.10 m above the water. In the absence of gravity the dolphin would move in a straight line to the ball and catch it, but because of gravity the dolphin follows a parabolic path well below the ball's initial position, as shown in the sketch. If the trainer releases the ball the instant the dolphin leaves the water, show that the dolphin and the falling ball meet.

Picture the Problem

In our sketch we have the dolphin leaping from the water at the origin $x_0 = y_0 = 0$ with an angle above the horizontal given by $\theta = \tan^{-1}(h/d)$. The initial position of the ball is $x_0 = d = 5.50$ m and $y_0 = h = 4.10$ m, and its initial velocity is zero. The ball drops straight down with the acceleration of gravity, $a_y = -g$.

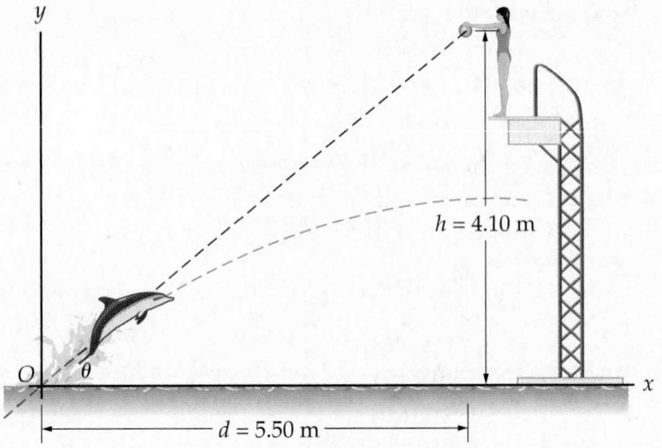

INTERACTIVE FIGURE

Strategy

We want to show that when the dolphin is at $x = d$, its height above the water is the same as the height of the ball above the water. To do this we first find the time when the dolphin is at $x = d$, then calculate y for the dolphin at this time. Next, we calculate y of the ball at the same time and then check to see if they are equal.

Since the ball drops from rest from a height h, its y equation of motion is $y = h - \frac{1}{2}gt^2$, as in Equation 4–7 in Section 4–3.

Solution

1. Calculate the angle at which the dolphin leaves the water:

$$\theta = \tan^{-1}\left(\frac{h}{d}\right) = \tan^{-1}\left(\frac{4.10\text{ m}}{5.50\text{ m}}\right) = 36.7°$$

2. Use this angle and the initial speed to find the time t when the x position of the dolphin, x_d, is equal to 5.50 m. The x equation of motion is $x_d = (v_0 \cos\theta)t$:

$$x_d = (v_0 \cos\theta)t = [(12.0\text{ m/s})\cos 36.7°]t = (9.62\text{ m/s})t$$
$$= 5.50\text{ m}$$
$$t = \frac{5.50\text{ m}}{9.62\text{ m/s}} = 0.572\text{ s}$$

3. Evaluate the y position of the dolphin, y_d, at $t = 0.572$ s. The y equation of motion is $y_d = (v_0 \sin\theta)t - \frac{1}{2}gt^2$:

$$y_d = (v_0 \sin\theta)t - \tfrac{1}{2}gt^2$$
$$= [(12.0\text{ m/s})\sin 36.7°](0.572\text{ s}) - \tfrac{1}{2}(9.81\text{ m/s}^2)(0.572\text{ s})^2$$
$$= 4.10\text{ m} - 1.60\text{ m} = 2.50\text{ m}$$

4. Finally, evaluate the y position of the ball, y_b, at $t = 0.572$ s. The ball's equation of motion is $y_b = h - \frac{1}{2}gt^2$:

$$y_b = h - \tfrac{1}{2}gt^2 = 4.10\text{ m} - \tfrac{1}{2}(9.81\text{ m/s}^2)(0.572\text{ s})^2$$
$$= 4.10\text{ m} - 1.60\text{ m} = 2.50\text{ m}$$

Insight

In the absence of gravity, both the dolphin and the ball would be at $x = 5.50$ m and $y = 4.10$ m at $t = 0.572$ s. Because of gravity, however, the dolphin and the ball fall below their zero-gravity positions—and by the same amount, 1.60 m. In fact, from the point of view of the dolphin, the ball is always at the same angle of 36.7° above the horizontal until it is caught.

This is shown in the accompanying plot, where the red dots show the position of the ball at 10 equally spaced times, and the blue dots show the position of the dolphin at the corresponding times. In addition, the dashed lines from the dolphin to the ball all make the same angle with the horizontal, 36.7°.

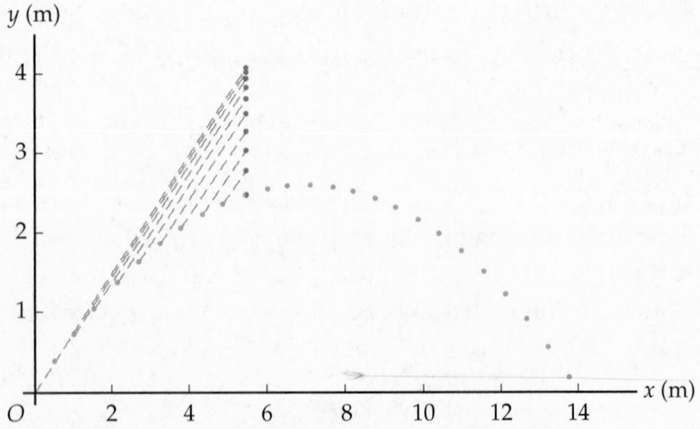

Practice Problem

At what height does the dolphin catch the ball if it leaves the water with an initial speed of 8.00 m/s?
[**Answer:** $y_d = y_b = 0.493$ m. If the dolphin's initial speed is less than 7.50 m/s, it hits the water before catching the ball.]

Some related homework problems: Problem 25, Problem 34

4–5 Projectile Motion: Key Characteristics

We conclude this chapter with a brief look at some additional characteristics of projectile motion that are both interesting and useful. In all cases our results follow as a direct consequence of the fundamental kinematic equations (Equation 4–10) describing projectile motion.

Range

The **range**, R, of a projectile is the horizontal distance it travels before landing. We consider the case shown in **Figure 4–8**, where the initial and final elevations are the same ($y = 0$). One way to obtain the range, then, is as follows: (i) Find the time when the projectile lands by setting $y = 0$ in the expression $y = (v_0 \sin\theta)t - \frac{1}{2}gt^2$; (ii) Substitute the time found in (i) into the x equation of motion.

Carrying out the first part of the calculation yields the following:

$$(v_0 \sin\theta)t - \tfrac{1}{2}gt^2 = 0 \quad \text{or} \quad (v_0 \sin\theta)t = \tfrac{1}{2}gt^2$$

Clearly, $t = 0$ is a solution to this equation—corresponding to the initial condition—but the solution we seek is a time that is greater than zero. We can find the desired time by dividing both sides of the equation by t. This gives

$$(v_0 \sin\theta) = \tfrac{1}{2}gt \quad \text{or} \quad t = \left(\frac{2v_0}{g}\right)\sin\theta \qquad \textbf{4–11}$$

This is the time when the projectile lands—also known as the time of flight.

Now, substitute this time into $x = (v_0 \cos\theta)t$ to find the value of x when the projectile lands:

$$x = (v_0 \cos\theta)t = (v_0 \cos\theta)\left(\frac{2v_0}{g}\right)\sin\theta = \left(\frac{2{v_0}^2}{g}\right)\sin\theta\cos\theta$$

This value of x is the range, R, thus

$$R = \left(\frac{2{v_0}^2}{g}\right)\sin\theta\cos\theta$$

Using the trigonometric identity $\sin 2\theta = 2\sin\theta\cos\theta$, as given in Appendix A, we can write this more compactly as follows:

$$R = \left(\frac{{v_0}^2}{g}\right)\sin 2\theta \qquad \textit{(same initial and final elevation)} \qquad \textbf{4–12}$$

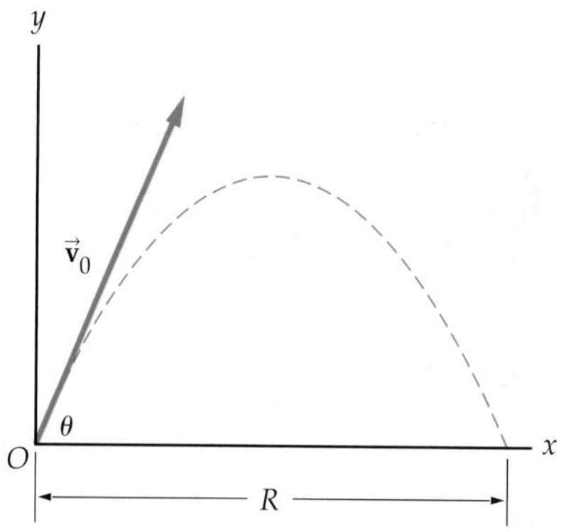

▲ **FIGURE 4–8 Range of a projectile**
The range R of a projectile is the horizontal distance it travels between its takeoff and landing positions.

PROBLEM-SOLVING NOTE

Use the Same Math Regardless of the Initial Conditions

Once an object is launched, its trajectory follows the kinematic equations of motion, regardless of specific differences in the initial conditions. Thus, our equations of motion can be used to derive any desired characteristic of projectile motion, including range, symmetry, and maximum height.

ACTIVE EXAMPLE 4–2 Find the Initial Speed

A football game begins with a kickoff in which the ball travels a horizontal distance of 45 yd and lands on the ground. If the ball was kicked at an angle of 40.0° above the horizontal, what was its initial speed?

Solution *(Test your understanding by performing the calculations indicated in each step.)*

1. Solve Equation 4–12 for the initial speed v_0: $\quad v_0 = \sqrt{gR/\sin 2\theta}$
2. Convert the range to meters: $\quad R = 41\text{ m}$
3. Substitute numerical values: $\quad v_0 = 20\text{ m/s}$

Insight

Note that we choose the positive square root in step 1 because we are interested only in the *speed* of the ball, which is always positive.

Your Turn

Suppose the initial speed of the ball is increased by 10 percent, to 22 m/s. By what percentage does the range increase?

(*Answers to* **Your Turn** *problems are given in the back of the book.*)

REAL-WORLD PHYSICS
Golf on the Moon

Note that R depends inversely on the acceleration of gravity, g—thus the smaller g, the larger the range. For example, a projectile launched on the Moon, where the acceleration of gravity is only about 1/6 that on Earth, travels about six times as far as it would on Earth. It was for this reason that astronaut Alan Shepard simply couldn't resist the temptation of bringing a golf club and ball with him on the third lunar landing mission in 1971. He ambled out onto the Fra Mauro Highlands and became the first person to hit a tee shot on the Moon. His distance was undoubtedly respectable—unfortunately, his ball landed in a sand trap.

Now, what launch angle gives the greatest range? From Equation 4–12 we see that R varies with angle as $\sin 2\theta$; thus R is largest when $\sin 2\theta$ is largest—that is, when $\sin 2\theta = 1$. Since $\sin 90° = 1$, it follows that $\theta = 45°$ gives the maximum range. Thus

$$R_{max} = \frac{v_0^2}{g} \qquad \text{4–13}$$

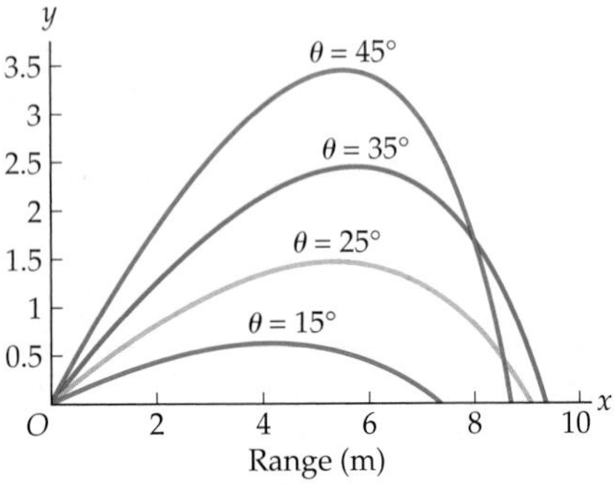

▲ **FIGURE 4–9 Projectiles with air resistance**

Projectiles with the same initial speed but different launch angles showing the effects of air resistance. Notice that the maximum range occurs for a launch angle less than 45°, and that the projectiles return to the ground at a steeper angle than the launch angle.

As expected, the range (Equation 4–12) and maximum range (Equation 4–13) depend strongly on the initial speed of the projectile—they are both proportional to v_0^2.

Note that these results are specifically for the case where a projectile lands at the same level from which it was launched. If a projectile lands at a higher level, for example, the launch angle that gives maximum range is greater than 45°, and if it lands at a lower level, the angle for maximum range is less than 45°.

Finally, the range given here applies only to the ideal case of no air resistance. In cases where air resistance is significant, as in the flight of a rapidly moving golf ball, for example, the overall range of the ball is reduced. In addition, the maximum range occurs for a launch angle less than 45° (**Figure 4–9**). The reason is that with a smaller launch angle the golf ball is in the air for less time, giving air resistance less time to affect its flight.

Symmetry in Projectile Motion

There are many striking symmetries in projectile motion, beginning with the graceful symmetry of the parabola itself. As a first example, recall that earlier in this section, in Equation 4–11, we found the time when a projectile lands:

$$t = \left(\frac{2v_0}{g}\right)\sin\theta$$

Now, by symmetry, the time it takes a projectile to reach its highest point (in the absence of air resistance) should be just half this time. After all, the projectile moves in the x direction with constant speed, and the highest point—by symmetry—occurs at $x = \frac{1}{2}R$.

This all seems reasonable, but is there another way to check? Well, at the highest point the projectile is moving horizontally, thus its y component of velocity is zero. Let's find the time when $v_y = 0$ and compare with the time to land:

$$v_y = v_{0y} - gt = v_0 \sin\theta - gt = 0$$
$$t = \left(\frac{v_0}{g}\right)\sin\theta \qquad \text{4–14}$$

As expected from symmetry, the time at the highest point is one-half the time at landing.

▲ To be successful, a juggler must master the behavior of projectile motion. Physicist Richard Feynman shows that just knowing the appropriate equations is not enough; one must also practice. In this sense, learning to juggle is similar to learning to solve physics problems.

There is another interesting symmetry concerning speed. Recall that when a projectile is launched its y component of velocity is $v_y = v_0 \sin\theta$. When the projectile lands, at time $t = (2v_0/g)\sin\theta$, its y component of velocity is

$$v_y = v_0 \sin\theta - gt = v_0 \sin\theta - g\left(\frac{2v_0}{g}\right)\sin\theta = -v_0 \sin\theta$$

This is exactly the opposite of the y component of the velocity when it was launched. Since the x component of velocity is always the same, it follows that when the projectile lands, its speed, $v = \sqrt{v_x^2 + v_y^2}$, is the same as when it was launched—as one might expect from symmetry.

The velocities are different, however, since the direction of motion is different at launch and landing. Even so, there is still a symmetry—the initial velocity is *above* the horizontal by the angle θ; the landing velocity is *below* the horizontal by the same angle θ.

So far, these results have referred to launching and landing, which both occur at $y = 0$. The same symmetry extends to any level, though. That is, at a given height the speed of a projectile is the same on the way up as on the way down. In addition, the angle of the velocity above the horizontal on the way up is the same as the angle below the horizontal on the way down. This is illustrated in **Figure 4–10** and in the next Conceptual Checkpoint.

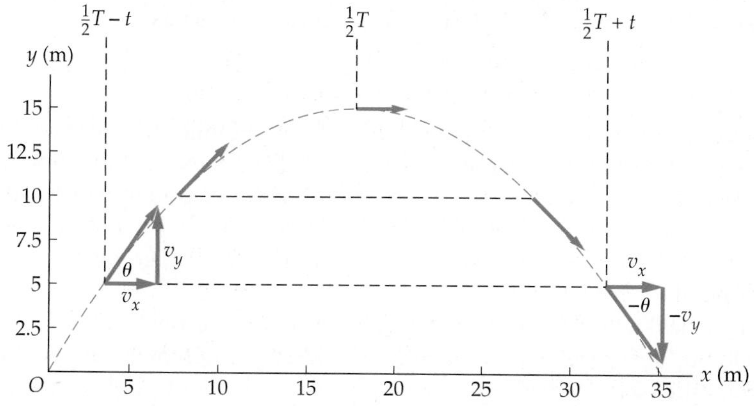

▲ **FIGURE 4–10 Velocity vectors for a projectile launched at the origin**
At a given height the speed (length of velocity vector) is the same on the way up as on the way down. The direction of motion on the way up is above the horizontal by the same amount that it is below the horizontal on the way down. In this case, the total time of flight is T, and the greatest height is reached at the time $T/2$. Notice that the speed is the same at the time $(T/2) - t$ as it is at the time $(T/2) + t$.

CONCEPTUAL CHECKPOINT 4–3 Compare Landing Speeds

You and a friend stand on a snow-covered roof. You both throw snowballs with the same initial speed, but in different directions. You throw your snowball downward, at 40° *below* the horizontal; your friend throws her snowball upward, at 40° *above* the horizontal. When the snowballs land on the ground, is the speed of your snowball **(a)** greater than, **(b)** less than, or **(c)** the same as the speed of your friend's snowball?

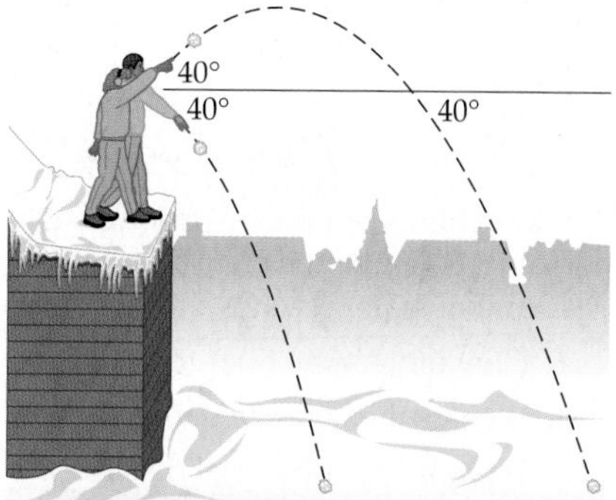

Reasoning and Discussion

One consequence of symmetry in projectile motion is that when your friend's snowball returns to the level of the throw, its speed will be the same as the initial speed. In addition, it will be moving downward, at 40° below the horizontal. From that point on its motion is the same as that of your snowball; thus it lands with the same speed.

What if you throw your snowball horizontally? Or suppose you throw it straight down? In either case, the final speed is unchanged! In fact, for a given initial speed, the speed on landing simply doesn't depend on the direction in which you throw the ball. This is shown in homework problems 29 and 66. We return to this point in Chapter 8 when we discuss potential energy and energy conservation.

Answer

(c) The snowballs have the same speed.

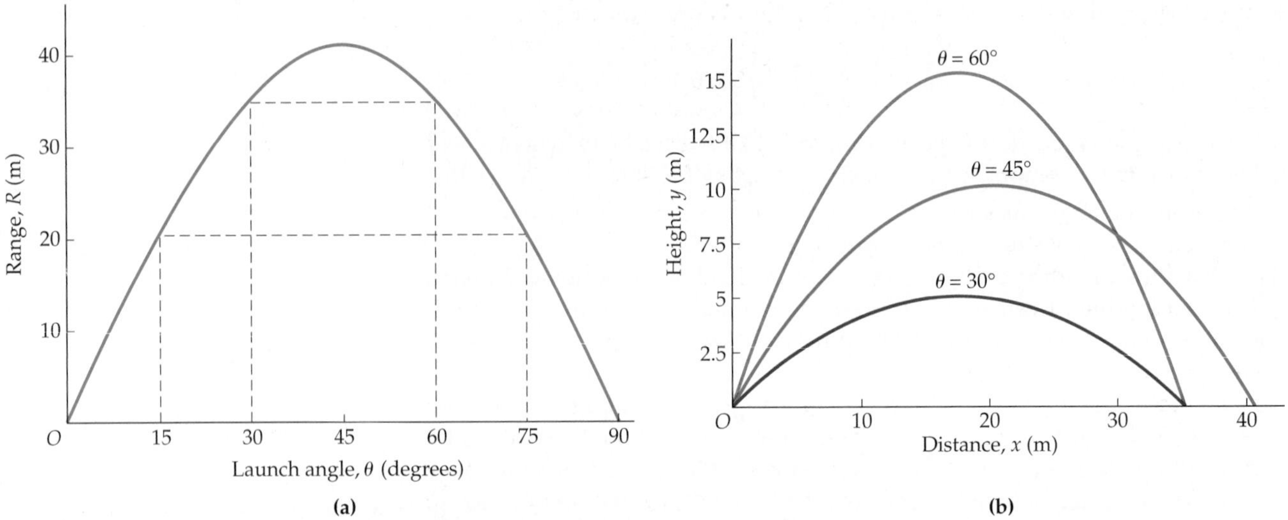

▲ **FIGURE 4–11 Range and launch angle in the absence of air resistance**

(a) A plot of range versus launch angle for a projectile launched with an initial speed of 20 m/s. Note that the maximum range occurs at $\theta = 45°$. Launch angles equally greater than or less than 45°, such as 30° and 60°, give the same range. **(b)** Trajectories of projectiles with initial speeds of 20 m/s and launch angles of 60°, 45°, and 30°. The projectiles with launch angles of 30° and 60° land at the same location.

▲ An archerfish would have trouble procuring its lunch without an instinctive grasp of projectile motion.

As our final example of symmetry, consider the range R. A plot of R versus launch angle θ is shown in **Figure 4–11** for $v_0 = 20$ m/s. Note that in the absence of air resistance, R is greatest at $\theta = 45°$, as pointed out previously. In addition, we can see from the figure that the range for angles equally above or below 45° is the same. For example, if air resistance is negligible, the range for $\theta = 30°$ is the same as the range for $\theta = 60°$.

Symmetries such as these are just some of the many reasons why physicists find physics to be "beautiful" and "aesthetically pleasing." Discovering such patterns and symmetries in nature is really what physics is all about. A physicist does not consider the beauty of projectile motion to be diminished by analyzing it in detail. Just the opposite—detailed analysis reveals deeper, more subtle, and sometimes unexpected levels of beauty.

Maximum Height

Let's follow up on an observation made earlier in this section, namely, that a projectile is at maximum height when its y component of velocity is zero. In fact, we will use this observation to determine the maximum height of an arbitrary projectile. This can be accomplished with the following two-step calculation: (i) Find the time when $v_y = 0$; (ii) Substitute this time into the y-versus-t equation of motion, $y = (v_0 \sin\theta)t - \frac{1}{2}gt^2$. This calculation is carried out in the next Example.

EXAMPLE 4–7 What a Shot!

The archerfish hunts by dislodging an unsuspecting insect from its resting place with a stream of water expelled from the fish's mouth. Suppose the archerfish squirts water with an initial speed of 2.30 m/s at an angle of 19.5° above the horizontal. When the stream of water reaches a beetle on a leaf at height h above the water's surface, it is moving horizontally.

a. How much time does the beetle have to react?

b. What is the height h of the beetle?

c. What is the horizontal distance d between the fish and the beetle when the water is launched?

Picture the Problem

Our sketch shows the fish squirting water from the origin, $x_0 = y_0 = 0$, and the beetle at $x = d$, $y = h$. The stream of water starts off with a speed $v_0 = 2.30$ m/s at an angle $\theta = 19.5°$ above the horizontal. Note that the water is moving horizontally when it reaches the beetle.

Strategy

a. Because the stream of water is moving horizontally when it reaches the beetle, it is at the top of its parabolic trajectory, as can be seen in Figure 4–10. This means that its y component of velocity is zero. Therefore, we can set $v_y = 0$ in $v_y = v_0 \sin\theta - gt$ and solve for the time t.

b. To find the maximum height of the stream of water, and of the beetle, we substitute the time found in part (a) into $y = (v_0 \sin\theta)t - \frac{1}{2}gt^2$.

c. Similarly, we can find the horizontal distance d by substituting the time from part (a) into $x = (v_0 \cos\theta)t$.

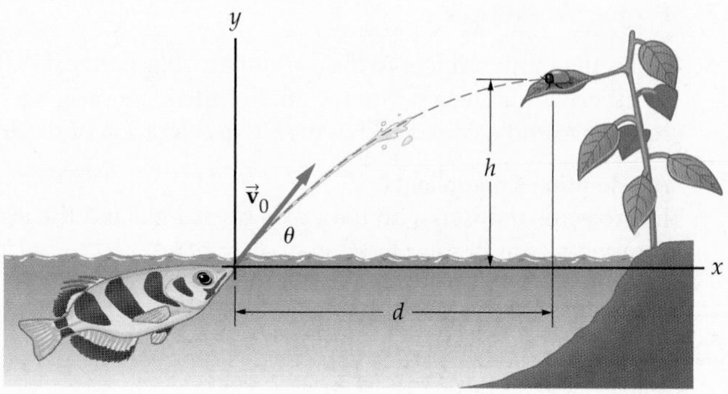

Solution

Part (a)

1. Set $v_y = v_0 \sin\theta - gt$ equal to zero and solve for the corresponding time t:

$$v_y = v_{0y} - gt = v_0 \sin\theta - gt = 0$$

$$t = \frac{v_0 \sin\theta}{g}$$

2. Substitute numerical values to determine the reaction time:

$$t = \frac{v_0 \sin\theta}{g} = \frac{(2.30\ \text{m/s}) \sin 19.5°}{9.81\ \text{m/s}^2} = 0.0783\ \text{s}$$

Part (b)

3. To calculate the height, we substitute $t = (v_0 \sin\theta)/g$ into $y = (v_0 \sin\theta)t - \frac{1}{2}gt^2$:

$$y = (v_0 \sin\theta)\left(\frac{v_0 \sin\theta}{g}\right) - \frac{1}{2}g\left(\frac{v_0 \sin\theta}{g}\right)^2 = \frac{(v_0 \sin\theta)^2}{2g}$$

4. Substitute numerical values to find the height h:

$$h = \frac{(v_0 \sin\theta)^2}{2g} = \frac{[(2.30\ \text{m/s}) \sin 19.5°]^2}{2(9.81\ \text{m/s}^2)} = 0.0300\ \text{m}$$

Part (c)

5. We can find the horizontal distance d using x as a function of time, $x = (v_0 \cos\theta)t$:

$$x = (v_0 \cos\theta)t$$
$$d = [(2.30\ \text{m/s}) \cos 19.5°](0.0783\ \text{s}) = 0.170\ \text{m}$$

Insight

To hit the beetle, the fish aims 19.5° above the horizontal. For comparison, note that the straight-line angle to the beetle is $\tan^{-1}(0.0300/0.170) = 10.0°$. Therefore, the fish cannot aim directly at its prey if it wants a meal.

Finally, note that by working symbolically in step 3 we have derived a general result for the maximum height of a projectile. In particular, we find $y_{max} = (v_0 \sin\theta)^2/2g$, a result that is valid for any launch speed and angle. As a check of our result, note that if we launch a projectile straight upward ($\theta = 90°$), the maximum height is $y_{max} = v_0{}^2/2g$. Comparing with the one-dimensional kinematics of Chapter 2, if an object is thrown straight upward with an initial speed v_0, and the object accelerates downward with the acceleration of gravity, $a = -g$, it comes to rest ($v = 0$) after covering a vertical distance Δy given by $0 = v_0{}^2 + 2(-g)\Delta y$. Solving for the distance yields $\Delta y = v_0{}^2/2g = y_{max}$. This is an example of the internal consistency that characterizes all of physics.

Practice Problem

How far does the stream of water go if it happens to miss the beetle? [**Answer:** By symmetry, the distance d is half the range. Thus the stream of water travels a distance $R = 2d = 0.340$ m.]

Some related homework problems: Problem 71, Problem 72

Chapter Summary

4–1 Motion in Two Dimensions

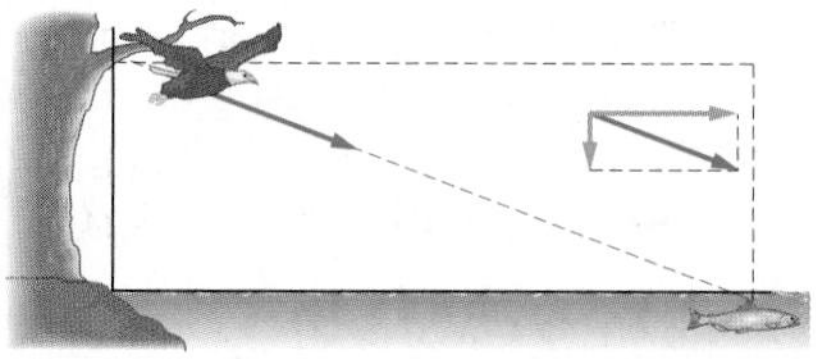

Independence of Motion

Components of motion in the x and y directions can be treated independently of one another. Thus, two-dimensional motion with constant acceleration is described by the same kinematic equations derived in Chapter 2, only now written in terms of x and y components.

4–2 Projectile Motion

Projectile motion refers to the path of an object after it is thrown, kicked, batted, or otherwise launched into the air. For the ideal case, we assume no air resistance and a constant downward acceleration of magnitude g.

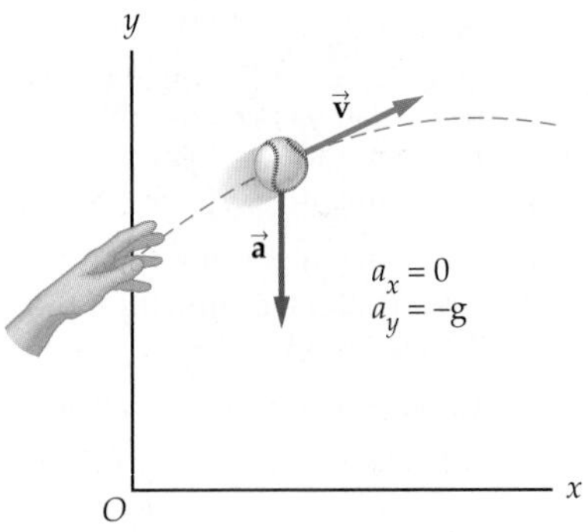

Acceleration Components
In projectile motion, with the x axis horizontal and the y axis upward, the components of the acceleration of gravity are

$$a_x = 0$$
$$a_y = -g$$

x and y as Functions of Time
The x and y equations of motion are

$$x = x_0 + v_{0x}t$$
$$y = y_0 + v_{0y}t - \tfrac{1}{2}gt^2 \qquad \textbf{4–6}$$

v_x and v_y as Functions of Time
The velocity components vary with time as follows:

$$v_x = v_{0x}$$
$$v_y = v_{0y} - gt \qquad \textbf{4–6}$$

v_x and v_y as Functions of Displacement
v_x and v_y vary with displacement as

$$v_x^2 = v_{0x}^2$$
$$v_y^2 = v_{0y}^2 - 2g\Delta y \qquad \textbf{4–6}$$

4–3 Zero Launch Angle

Equations of Motion
A projectile launched horizontally from $x_0 = 0$, $y_0 = h$ with an initial speed v_0 has the following equations of motion:

$$x = v_0 t \qquad v_x = v_0 \qquad v_x^2 = v_0^2$$
$$y = h - \tfrac{1}{2}gt^2 \qquad v_y = -gt \qquad v_y^2 = -2g\Delta y \qquad \textbf{4–7}$$

Parabolic Path
The path followed by a projectile launched horizontally with an initial speed v_0 is described by

$$y = h - \left(\frac{g}{2v_0^2}\right)x^2 \qquad \textbf{4–8}$$

This path is a parabola.

Landing Site
The landing site of a projectile launched horizontally is

$$x = v_0\sqrt{\frac{2h}{g}} \qquad \textbf{4–9}$$

In this expression, v_0 is the initial speed and h is the initial height. Note that this result is simply the speed in the x direction multiplied by the time of fall.

4–4 General Launch Angle

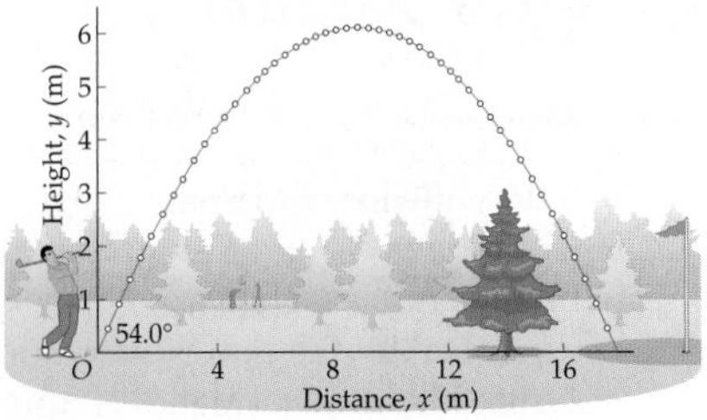

Launch from the Origin
The equations of motion for a launch from the origin with an initial speed v_0 at an angle of θ with respect to the horizontal are

$$x = (v_0 \cos\theta)t \qquad v_x = v_0\cos\theta \qquad v_x^2 = v_0^2\cos^2\theta$$
$$y = (v_0\sin\theta)t - \tfrac{1}{2}gt^2 \qquad v_y = v_0\sin\theta - gt \qquad v_y^2 = v_0^2\sin^2\theta - 2g\Delta y \qquad \textbf{4–10}$$

4–5 Projectile Motion: Key Characteristics

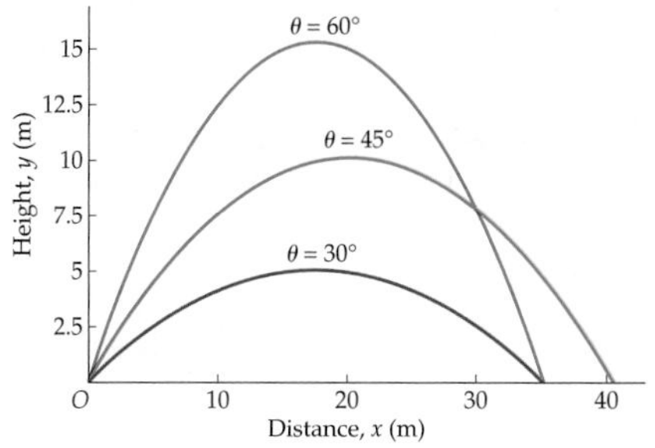

Range

The range of a projectile launched from the origin with an initial speed v_0 and a launch angle θ is

$$R = \left(\frac{{v_0}^2}{g}\right)\sin 2\theta \qquad 4\text{–}12$$

This expression applies only to projectiles that land at the same level from which they were launched.

Symmetry

Projectile motion exhibits many symmetries. For example, the speed of a projectile depends only on its height and not on whether it is moving upward or downward.

Maximum Height

The maximum height of a projectile above its launch site is

$$y_{max} = \frac{{v_0}^2 \sin^2\theta}{2g}$$

In this equation, v_0 is the initial speed and θ is the launch angle.

Problem-Solving Summary

Type of Problem	Relevant Physical Concepts	Related Examples
Study two-dimensional motion with constant acceleration.	Motion in the x direction is independent of motion in the y direction. This is the basis for the equations of motion given in Table 4–1. Note that these equations are the same as the kinematic equations of Chapter 2, only written in terms of x and y components.	Examples 4–1, 4–2
Find the location and velocity of a projectile launched horizontally.	When a projectile is launched horizontally with a speed v_0 its initial velocity components are $v_{0x} = v_0$ and $v_{0y} = 0$. Make these substitutions in the equations of projectile motion given in Equation 4–6.	Examples 4–3, 4–4 Conceptual Checkpoints 4–1, 4–2
Find the location and velocity of a projectile launched with an arbitrary launch angle.	If a projectile is launched at an angle θ, its initial velocity components are $v_{0x} = v_0 \cos\theta$ and $v_{0y} = v_0 \sin\theta$. Make these substitutions in the equations of projectile motion given in Equation 4–6.	Examples 4–5, 4–6, 4–7 Active Examples 4–1, 4–2

Conceptual Questions

(Answers to odd-numbered Conceptual Questions can be found in the back of the book.)

1. What is the acceleration of a projectile when it reaches its highest point? What is its acceleration just before and just after reaching this point?
2. A projectile is launched with an initial speed of v_0 at an angle θ above the horizontal. It lands at the same level from which it was launched. What was its average velocity between launch and landing? Explain.
3. A projectile is launched from level ground. When it lands, its direction of motion has rotated clockwise through 60°. What was the launch angle? Explain.
4. In a game of baseball, a player hits a high fly ball to the outfield. **(a)** Is there a point during the flight of the ball where its velocity is parallel to its acceleration? **(b)** Is there a point where the ball's velocity is perpendicular to its acceleration? Explain in each case.
5. A projectile is launched with an initial velocity of $\vec{\mathbf{v}} = (4\text{ m/s})\hat{\mathbf{x}} + (3\text{ m/s})\hat{\mathbf{y}}$. What is the velocity of the projectile when it reaches its highest point? Explain.
6. A projectile is launched from a level surface with an initial velocity of $\vec{\mathbf{v}} = (2\text{ m/s})\hat{\mathbf{x}} + (4\text{ m/s})\hat{\mathbf{y}}$. What is the velocity of the projectile just before it lands? Explain.
7. Do projectiles for which air resistance is nonnegligible, such as a bullet fired from a rifle, have maximum range when the launch angle is greater than, less than, or equal to 45°? Explain.
8. Two projectiles are launched from the same point at the same angle above the horizontal. Projectile 1 reaches a maximum height twice that of projectile 2. What is the ratio of the initial speed of projectile 1 to the initial speed of projectile 2? Explain.
9. A child rides on a pony walking with constant velocity. The boy leans over to one side and a scoop of ice cream falls from his ice cream cone. Describe the path of the scoop of ice cream as seen by **(a)** the child and **(b)** his parents standing on the ground nearby.
10. Driving down the highway, you find yourself behind a heavily loaded tomato truck. You follow close behind the truck, keeping the same speed. Suddenly a tomato falls from the back of

the truck. Will the tomato hit your car or land on the road, assuming you continue moving with the same speed and direction? Explain.

11. A projectile is launched from the origin of a coordinate system where the positive x axis points horizontally to the right and the positive y axis points vertically upward. What was the projectile's launch angle with respect to the x axis if, at its highest point, its direction of motion has rotated **(a)** clockwise through 50° or **(b)** counterclockwise through 30°? Explain.

Conceptual Exercises

(Answers to odd-numbered Conceptual Exercises can be found in the back of the book.)

1. As you walk briskly down the street, you toss a small ball into the air. If you want the ball to land in your hand when it comes back down, should you toss the ball **(a)** straight upward relative to yourself, **(b)** in a forward direction, or **(c)** in a backward direction? Explain.

2. A certain projectile is launched with an initial speed v_0. At its highest point its speed is $\frac{1}{2}v_0$. Was the launch angle of the projectile **(a)** 30°, **(b)** 45°, or **(c)** 60°?

3. Two divers run horizontally off the edge of a low cliff. Diver 2 runs with twice the speed of diver 1. When the divers hit the water, is the horizontal distance covered by diver 2 **(a)** the same as, **(b)** twice as much as, or **(c)** four times the horizontal distance covered by diver 1?

4. Two youngsters dive off an overhang into a lake. Diver 1 drops straight down, diver 2 runs off the cliff with an initial horizontal speed v_0. Is the splashdown speed of diver 2 **(a)** greater than, **(b)** less than, or **(c)** equal to the splashdown speed of diver 1?

5. Three projectiles (A, B, and C) are launched with the same initial speed but with different launch angles, as shown in **Figure 4–12**. List the projectiles in order of increasing **(a)** horizontal component of initial velocity and **(b)** time of flight. Indicate a tie with an equal sign.

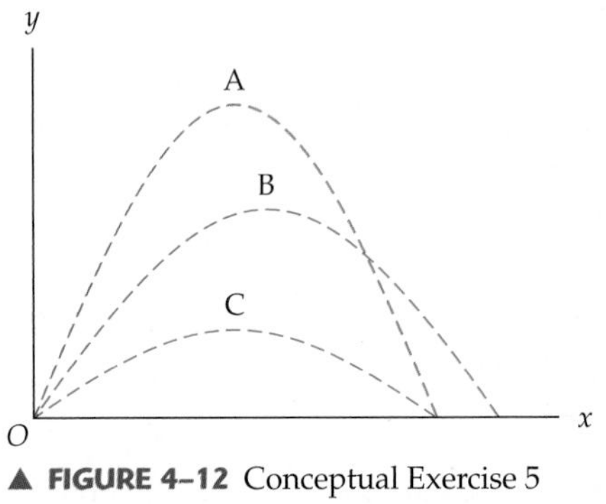

▲ **FIGURE 4–12** Conceptual Exercise 5

6. Three projectiles (A, B, and C) are launched with different initial speeds so that they reach the same maximum height, as shown in **Figure 4–13**. List the projectiles in order of increasing **(a)** initial speed and **(b)** time of flight. Indicate a tie with an equal sign.

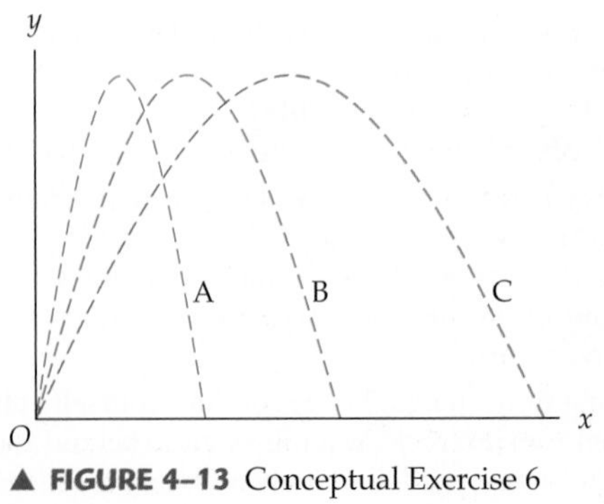

▲ **FIGURE 4–13** Conceptual Exercise 6

7. The penguin to the left in the accompanying photo is about to land an ice floe. Just before it lands, is its speed **(a)** greater than, **(b)** less than, or **(c)** the same as when it left the water? Explain.

This penguin behaves much like a projectile from the time it leaves the water until it touches down on the ice. (Conceptual Exercise 7)

8. A person flips a coin into the air and it lands on the ground a few feet away. If the person were to perform an identical coin flip on an elevator rising with constant speed, would the coin's time of flight be **(a)** greater than, **(b)** less than, or **(c)** the same as when the person was at rest?

9. Suppose the elevator in the previous Exercise is rising with constant acceleration, rather than constant velocity. In this case, would the coin's time of flight be **(a)** greater than, **(b)** less than, or **(c)** the same as when the person was at rest?

10. You throw a ball into the air with an initial speed of 10 m/s at an angle of 60° above the horizontal. The ball returns to the level from which it was thrown in the time T. Referring to **Figure 4-14**, which of the plots (1, 2, or 3) best represents the speed of the ball as a function of time? What is wrong with the other two plots?

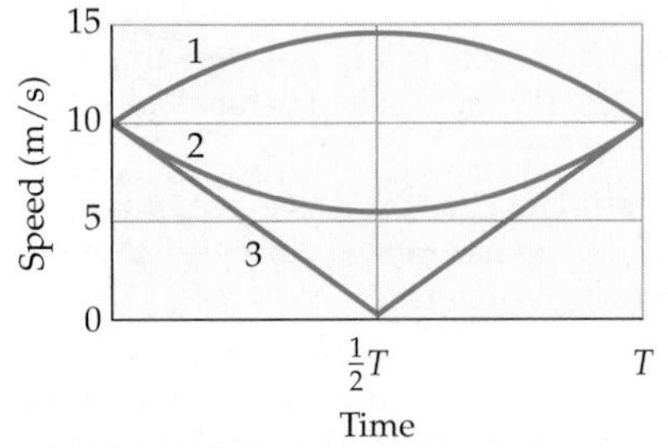

▲ **FIGURE 4–14** Conceptual Exercise 10

11. Child 1 throws a snowball horizontally from the top of a roof; child 2 throws a snowball straight down. Once in flight, is the acceleration of snowball 2 **(a)** greater than, **(b)** equal to, or **(c)** less than the acceleration of snowball 1?

Problems

Note: **IP** *denotes an integrated conceptual/quantitative problem.* **BIO** *identifies problems of biological or medical interest. Red bullets (•, ••, •••) are used to indicate the level of difficulty of each problem.*

Air resistance should be ignored in the problems for this chapter, unless specifically stated otherwise.

Section 4–1 Motion in Two Dimensions

1. • A sailboat runs before the wind with a constant speed of 3.2 m/s in a direction 42° north of west. How far **(a)** west and **(b)** north has the sailboat traveled in 25 min?

2. • As you walk to class with a constant speed of 1.60 m/s, you are moving in a direction that is 15.0° north of east. How much time does it take to change your displacement by **(a)** 20.0 m east or **(b)** 30.0 m north?

3. • Starting from rest, a car accelerates at $2.0\ \mathrm{m/s^2}$ up a hill that is inclined 5.5° above the horizontal. How far **(a)** horizontally and **(b)** vertically has the car traveled in 12 s?

4. •• **IP** A particle passes through the origin with a velocity of $(6.2\ \mathrm{m/s})\hat{\mathbf{y}}$. If the particle's acceleration is $(-4.4\ \mathrm{m/s^2})\hat{\mathbf{x}}$ **(a)** what are its x and y positions after 5.0 s? **(b)** What are v_x and v_y at this time? **(c)** Does the speed of this particle increase with time, decrease with time, or increase and then decrease? Explain.

5. •• An electron in a cathode-ray tube is traveling horizontally at $2.10 \times 10^9\ \mathrm{cm/s}$ when deflection plates give it an upward acceleration of $5.30 \times 10^{17}\ \mathrm{cm/s^2}$. **(a)** How long does it take for the electron to cover a horizontal distance of 6.20 cm? **(b)** What is its vertical displacement during this time?

6. •• Two canoeists start paddling at the same time and head toward a small island in a lake, as shown in **Figure 4–15**. Canoeist 1 paddles with a speed of 1.35 m/s at an angle of 45° north of east. Canoeist 2 starts on the opposite shore of the lake, a distance of 1.5 km due east of canoeist 1. **(a)** In what direction relative to north must canoeist 2 paddle to reach the island? **(b)** What speed must canoeist 2 have if the two canoes are to arrive at the island at the same time?

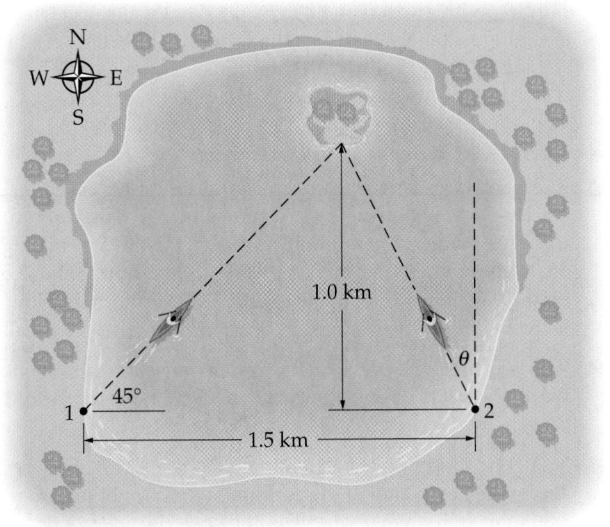

▲ **FIGURE 4–15** Problem 6

Section 4–3 Zero Launch Angle

7. • An archer shoots an arrow horizontally at a target 15 m away. The arrow is aimed directly at the center of the target, but it hits 52 cm lower. What was the initial speed of the arrow?

8. • **Victoria Falls** The great, gray-green, greasy Zambezi River flows over Victoria Falls in south central Africa. The falls are approximately 108 m high. If the river is flowing horizontally at 3.60 m/s just before going over the falls, what is the speed of the water when it hits the bottom? Assume the water is in free fall as it drops.

9. • A diver runs horizontally off the end of a diving board with an initial speed of 1.75 m/s. If the diving board is 3.00 m above the water, what is the diver's speed just before she enters the water?

10. • An astronaut on the planet Zircon tosses a rock horizontally with a speed of 6.95 m/s. The rock falls through a vertical distance of 1.40 m and lands a horizontal distance of 8.75 m from the astronaut. What is the acceleration of gravity on Zircon?

11. •• **IP Pitcher's Mounds** Pitcher's mounds are raised to compensate for the vertical drop of the ball as it travels a horizontal distance of 18 m to the catcher. **(a)** If a pitch is thrown horizontally with an initial speed of 32 m/s, how far does it drop by the time it reaches the catcher? **(b)** If the speed of the pitch is increased, does the drop distance increase, decrease, or stay the same? Explain. **(c)** If this baseball game were to be played on the Moon, would the drop distance increase, decrease, or stay the same? Explain.

12. •• Playing shortstop, you pick up a ground ball and throw it to second base. The ball is thrown horizontally, with a speed of 22 m/s, directly toward point A (**Figure 4–16**). When the ball reaches the second baseman 0.45 s later, it is caught at point B. **(a)** How far were you from the second baseman? **(b)** What is the distance of vertical drop, AB?

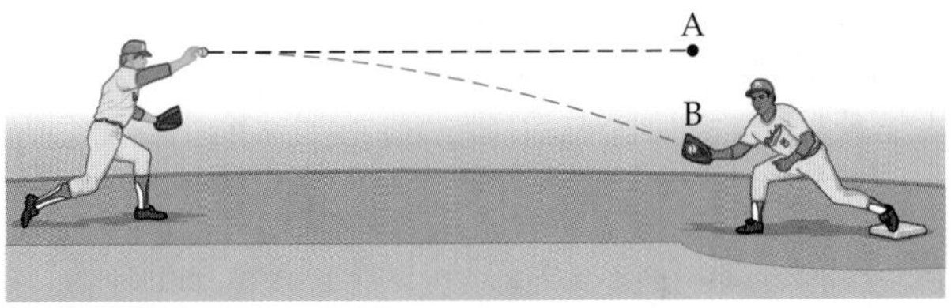

▲ **FIGURE 4–16** Problem 12

13. •• **IP** A crow is flying horizontally with a constant speed of 2.70 m/s when it releases a clam from its beak (**Figure 4–17**). The clam lands on the rocky beach 2.10 s later. Just before the clam lands, what is **(a)** its horizontal component of velocity, and **(b)** its vertical component of velocity? **(c)** How would your answers to parts (a) and (b) change if the speed of the crow were increased? Explain.

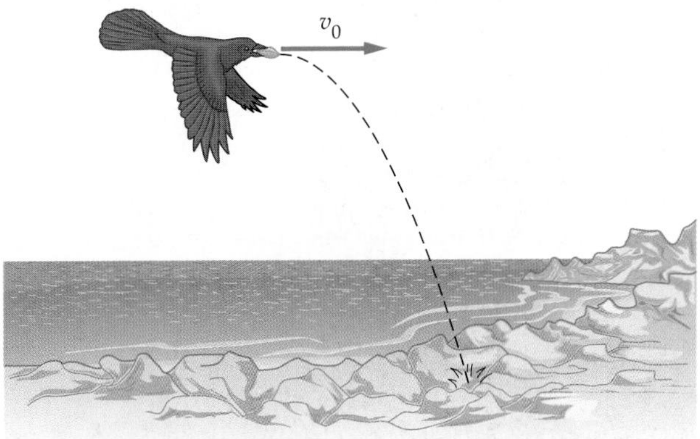

▲ **FIGURE 4–17** Problem 13

14. •• A mountain climber jumps a 2.8-m wide crevasse by leaping horizontally with a speed of 7.8 m/s. **(a)** If the climber's

direction of motion on landing is $-45°$, what is the height difference between the two sides of the crevasse? **(b)** Where does the climber land?

15. •• **IP** A white-crowned sparrow flying horizontally with a speed of 1.80 m/s folds its wings and begins to drop in free fall. **(a)** How far does the sparrow fall after traveling a horizontal distance of 0.500 m? **(b)** If the sparrow's initial speed is increased, does the distance of fall increase, decrease, or stay the same?

16. •• **Pumpkin Toss** In Denver, children bring their old jack-o-lanterns to the top of a tower and compete for accuracy in hitting a target on the ground (**Figure 4–18**). Suppose that the tower is 9.0 m high and that the bull's-eye is a horizontal distance of 3.5 m from the launch point. If the pumpkin is thrown horizontally, what is the launch speed needed to hit the bull's-eye?

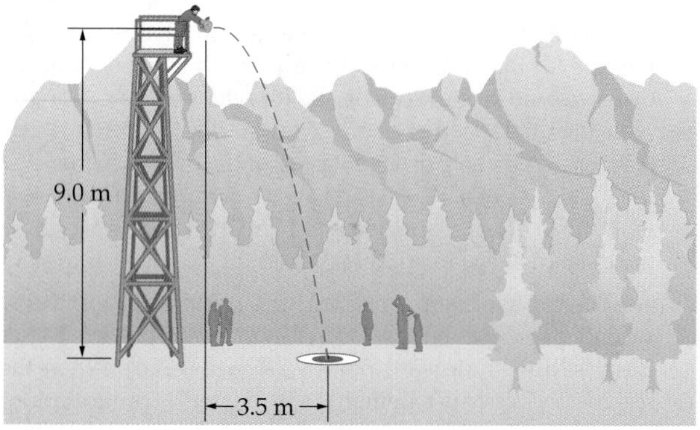

▲ **FIGURE 4–18** Problems 16 and 17

17. •• If, in the previous problem, a jack-o-lantern is given an initial horizontal speed of 3.3 m/s, what are the direction and magnitude of its velocity **(a)** 0.75 s later, and **(b)** just before it lands?

18. •• Fairgoers ride a Ferris wheel with a radius of 5.00 m (**Figure 4–19**). The wheel completes one revolution every 32.0 s. **(a)** What is the average speed of a rider on this Ferris wheel? **(b)** If a rider accidentally drops a stuffed animal at the top of the wheel, where does it land relative to the base of the ride? (Note: The bottom of the wheel is 1.75 m above the ground.)

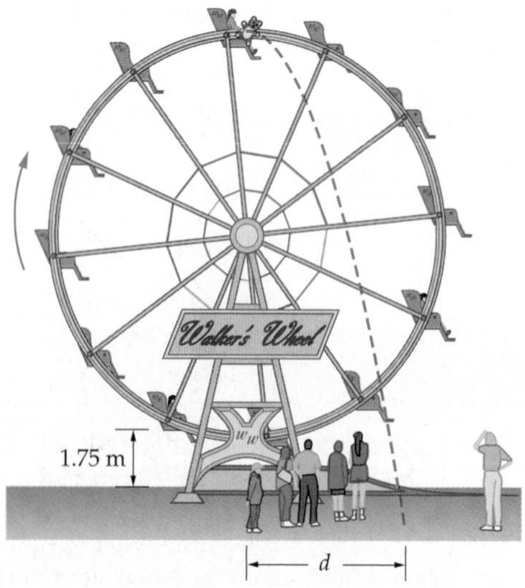

▲ **FIGURE 4–19** Problems 18 and 36

19. •• **IP** A swimmer runs horizontally off a diving board with a speed of 3.62 m/s and hits the water a horizontal distance of 1.68 m from the end of the board. **(a)** How high above the water was the diving board? **(b)** If the swimmer runs off the board with a reduced speed, does it take more, less, or the same time to reach the water?

20. •• **Baseball and the Washington Monument** On August 25, 1894, Chicago catcher William Schriver caught a baseball thrown from the top of the Washington Monument (555 ft, 898 steps). **(a)** If the ball was thrown horizontally with a speed of 5.00 m/s, where did it land? **(b)** What was the ball's speed and direction of motion when caught?

21. ••• A basketball is thrown horizontally with an initial speed of 4.20 m/s (**Figure 4–20**). A straight line drawn from the release point to the landing point makes an angle of 30.0° with the horizontal. What was the release height?

▲ **FIGURE 4–20** Problem 21

22. ••• **IP** A ball rolls off a table and falls 0.75 m to the floor, landing with a speed of 4.0 m/s. **(a)** What is the acceleration of the ball just before it strikes the ground? **(b)** What was the initial speed of the ball? **(c)** What initial speed must the ball have if it is to land with a speed of 5.0 m/s?

Section 4–4 General Launch Angle

23. • A second baseman tosses the ball to the first baseman, who catches it at the same level from which it was thrown. The throw is made with an initial speed of 17.0 m/s at an angle of 35.0° above the horizontal. **(a)** What is the horizontal component of the ball's velocity just before it is caught? **(b)** How long is the ball in the air?

24. • Referring to the previous problem, what is the y component of the ball's velocity and its direction of motion just before it is caught?

25. • A cork shoots out of a champagne bottle at an angle of 35.0° above the horizontal. If the cork travels a horizontal distance of 1.30 m in 1.25 s, what was its initial speed?

26. • A soccer ball is kicked with a speed of 9.50 m/s at an angle of 25.0° above the horizontal. If the ball lands at the same level from which it was kicked, how long was it in the air?

27. •• In a game of basketball, a forward makes a bounce pass to the center. The ball is thrown with an initial speed of 4.3 m/s at an angle of 15° below the horizontal. It is released 0.80 m above the floor. What horizontal distance does the ball cover before bouncing?

28. •• Repeat the previous problem for a bounce pass in which the ball is thrown 15° *above* the horizontal.

29. •• **IP** Snowballs are thrown with a speed of 13 m/s from a roof 7.0 m above the ground. Snowball A is thrown straight downward; snowball B is thrown in a direction 25° above the horizontal. **(a)** Is the landing speed of snowball A greater than, less than, or the same as the landing speed of snowball B? Explain. **(b)** Verify your answer to part (a) by calculating the landing speed of both snowballs.

30. •• In the previous problem, find the direction of motion of the two snowballs just before they land.

31. •• A golfer gives a ball a maximum initial speed of 30.0 m/s. **(a)** What is the longest possible hole-in-one for this golfer? Neglect any distance the ball might roll on the green and assume that the tee and the green are at the same level. **(b)** What is the minimum speed of the ball during this hole-in-one shot?

32. •• What is the highest tree the ball in the previous problem could clear on its way to the longest possible hole-in-one?

33. •• The "hang time" of a punt is measured to be 4.50 s. If the ball was kicked at an angle of 63.0° above the horizontal and was caught at the same level from which it was kicked, what was its initial speed?

34. •• In a friendly game of handball, you hit the ball essentially at ground level and send it toward the wall with a speed of 14 m/s at an angle of 34° above the horizontal. **(a)** How long does it take for the ball to reach the wall if it is 3.8 m away? **(b)** How high is the ball when it hits the wall?

35. •• **IP** In the previous problem, **(a)** what are the magnitude and direction of the ball's velocity when it strikes the wall? **(b)** Has the ball reached the highest point of its trajectory at this time? Explain.

36. •• A passenger on the Ferris wheel described in Problem 18 drops his keys when he is on the way up and at the 10 o'clock position. Where do the keys land relative to the base of the ride?

37. •• On a hot summer day, a young girl swings on a rope above the local swimming hole (**Figure 4–21**). When she lets go of the rope her initial velocity is 2.25 m/s at an angle of 35.0° above the horizontal. If she is in flight for 0.616 s, how high above the water was she when she let go of the rope?

▲ **FIGURE 4–21** Problem 37

38. ••• A certain projectile is launched with an initial speed v_0. At its highest point its speed is $v_0/4$. What was the launch angle?

Section 4–5 Projectile Motion: Key Characteristics

39. • **Punkin Chunkin** In Sussex County, Delaware, a post-Halloween tradition is "Punkin Chunkin," in which contestants build cannons, catapults, trebuchets, and other devices to launch pumpkins and compete for the greatest distance. Though hard to believe, pumpkins have been projected a distance of 4086 feet in this contest. What is the minimum initial speed needed for such a shot?

40. • A dolphin jumps with an initial velocity of 12.0 m/s at an angle of 40.0° above the horizontal. The dolphin passes through the center of a hoop before returning to the water. If the dolphin is moving horizontally when it goes through the hoop, how high above the water is the center of the hoop?

41. • A player passes a basketball to another player who catches it at the same level from which it was thrown. The initial speed of the ball is 7.1 m/s, and it travels a distance of 4.6 m. What was **(a)** the initial direction of the ball and **(b)** its time of flight?

42. • A golf ball is struck with a five iron on level ground. It lands 92.2 m away 4.30 s later. What was **(a)** the direction and **(b)** the magnitude of the initial velocity?

43. • **A Record Toss** Babe Didrikson holds the world record for the longest baseball throw (296 ft) by a woman. For the following questions, assume that the ball was thrown at an angle of 45.0° above the horizontal, that it traveled a horizontal distance of 296 ft, and was caught at the same level from which it was thrown. **(a)** What was the ball's initial speed? **(b)** How long was the ball in the air?

44. • In the photograph to the left on page 84, suppose the cart that launches the ball is 11 cm high. Estimate **(a)** the launch speed of the ball and **(b)** the time interval between successive stroboscopic exposures.

45. •• **IP** **Volcanoes on Io** Astronomers have discovered several volcanoes on Io, a moon of Jupiter. One of them, named Loki, ejects lava to a maximum height of 2.00×10^5 m. **(a)** What is the initial speed of the lava? (The acceleration of gravity on Io is 1.80 m/s^2.) **(b)** If this volcano were on Earth, would the maximum height of the ejected lava be greater than, less than, or the same as on Io? Explain.

A volcano on Io, the innermost moon of Jupiter, displays the characteristic features of projectile motion. (Problem 45)

46. •• **IP** A soccer ball is kicked with an initial speed of 10.2 m/s in a direction 25.0° above the horizontal. Find the magnitude and direction of its velocity **(a)** 0.250 s and **(b)** 0.500 s after being kicked. **(c)** Is the ball at its greatest height before or after 0.500 s? Explain.

47. •• A second soccer ball is kicked with the same initial speed as in Problem 46. After 0.750 s it is at its highest point. What was its initial direction of motion?

48. •• IP A golfer tees off on level ground, giving the ball an initial speed of 42.0 m/s and an initial direction of 35.0° above the horizontal. **(a)** How far from the golfer does the ball land? **(b)** The next golfer in the group hits a ball with the same initial speed but at an angle above the horizontal that is greater than 45.0°. If the second ball travels the same horizontal distance as the first ball, what was its initial direction of motion? Explain.

49. •• IP One of the most popular events at Highland games is the hay toss, where competitors use a pitchfork to throw a bale of hay over a raised bar. Suppose the initial velocity of a bale of hay is $\vec{v} = (1.12\ \text{m/s})\hat{x} + (8.85\ \text{m/s})\hat{y}$. **(a)** After what minimum time is its speed equal to 5.00 m/s? **(b)** How long after the hay is tossed is it moving in a direction that is 45.0° below the horizontal? **(c)** If the bale of hay is tossed with the same initial speed, only this time straight upward, will its time in the air increase, decrease, or stay the same? Explain.

General Problems

50. • A train moving with constant velocity travels 170 m north in 12 s and an undetermined distance to the west. The speed of the train is 32 m/s. **(a)** Find the direction of the train's motion relative to north. **(b)** How far west has the train traveled in this time?

51. • Referring to Example 4–2, find **(a)** the x component and **(b)** the y component of the hummingbird's velocity at the time $t = 0.55$ s. **(c)** What is the bird's direction of travel at this time, relative to the positive x axis?

52. • A racket ball is struck in such a way that it leaves the racket with a speed of 4.87 m/s in the horizontal direction. When the ball hits the court, it is a horizontal distance of 1.95 m from the racket. Find the height of the racket ball when it left the racket.

53. •• IP A hot-air balloon rises from the ground with a velocity of $(2.00\ \text{m/s})\hat{y}$. A champagne bottle is opened to celebrate takeoff, expelling the cork horizontally with a velocity of $(5.00\ \text{m/s})\hat{x}$ relative to the balloon. When opened, the bottle is 6.00 m above the ground. **(a)** What is the initial velocity of the cork, as seen by an observer on the ground? Give your answer in terms of the x and y unit vectors. **(b)** What is the speed of the cork, and its initial direction of motion, as seen by the same observer? **(c)** Determine the maximum height above the ground attained by the cork. **(d)** How long does the cork remain in the air?

54. •• Repeat the previous problem, this time assuming that the balloon is *descending* with a speed of 2.00 m/s.

55. •• IP A soccer ball is kicked from the ground with an initial speed of 14.0 m/s. After 0.275 s its speed is 12.9 m/s. **(a)** Give a strategy that will allow you to calculate the ball's initial direction of motion. **(b)** Use your strategy to find the initial direction.

56. •• A particle leaves the origin with an initial velocity $\vec{v} = (2.40\ \text{m/s})\hat{x}$, and moves with constant acceleration $\vec{a} = (-1.90\ \text{m/s}^2)\hat{x} + (3.20\ \text{m/s}^2)\hat{y}$. **(a)** How far does the particle move in the x direction before turning around? **(b)** What is the particle's velocity at this time? **(c)** Plot the particle's position at $t = 0.500$ s, 1.00 s, 1.50 s, and 2.00 s. Use these results to sketch position versus time for the particle.

57. •• When the dried-up seed pod of a scotch broom plant bursts open, it shoots out a seed with an initial velocity of 2.62 m/s at an angle of 60.5° above the horizontal. If the seed pod is 0.455 m above the ground, **(a)** how long does it take for the seed to land? **(b)** What horizontal distance does it cover during its flight?

58. •• Referring to Problem 57, a second seed shoots out from the pod with the same speed but with a direction of motion 30.0° below the horizontal. **(a)** How long does it take for the second seed to land? **(b)** What horizontal distance does it cover during its flight?

59. •• A shot-putter throws the shot with an initial speed of 11.2 m/s from a height of 5.00 ft above the ground. What is the range of the shot if the launch angle is **(a)** 20.0°, **(b)** 30.0°, or **(c)** 40.0°?

60. •• **Pararescue Jumpers** Coast Guard pararescue jumpers are trained to leap from helicopters into the sea to save boaters in distress. The rescuers like to step off their helicopter when it is "ten and ten", which means that it is *ten* feet above the water and moving forward, horizontally at *ten* knots. What is **(a)** the speed and **(b)** the direction of motion as a pararescuer enters the water following a ten and ten jump?

61. •• **Landing on Mars** The method by which the Mars exploration rovers, *Spirit* and *Opportunity*, landed on the surface of Mars in January of 2004 was quite elaborate. The rovers began their descent through the thin Martian atmosphere on a parachute until they reached an altitude of about 16.7 m. At that point a system of four air bags with six lobes each were inflated, retro rockets brought the craft to rest, and the rovers detached from their parachutes. After free-falling to the surface, with an acceleration of 3.72 m/s^2, and bouncing about a dozen times, the rovers deflated their air bags, righted themselves, and went about their explorations. On the first bounce from the surface (**Figure 4–22**), the rovers had an initial velocity of 9.92 m/s at an angle of 75.0° above the horizontal. **(a)** What was the maximum height of a rover between the first and second bounces? **(b)** How far did a rover travel in the horizontal direction between the first and second bounces?

▲ **FIGURE 4–22** Problem 61

62. •• A ball thrown straight upward returns to its original level in 2.50 s. A second ball is thrown at an angle of 40.0° above the horizontal. What is the initial speed of the second ball if it also returns to its original level in 2.50 s?

63. •• IP To decide who pays for lunch, a passenger on a moving train tosses a coin straight upward with an initial speed of 4.38 m/s and catches it again when it returns to its initial level. From the point of view of the passenger, then, the coin's initial velocity is $(4.38\ \text{m/s})\hat{y}$. The train's velocity relative to the ground is $(12.1\ \text{m/s})\hat{x}$. **(a)** What is the minimum speed of the coin relative to the ground during its flight? At what point in the coin's flight does this minimum speed occur? Explain. **(b)** Find the initial speed and direction of the coin as seen by an observer on the ground. **(c)** Use the expression for y_{max} derived in Example 4–7 to calculate the maximum height of the coin, as

seen by an observer on the ground. **(d)** Calculate the maximum height of the coin from the point of view of the passenger, who sees only one-dimensional motion.

64. •• **IP** A cannon is placed at the bottom of a cliff 61.5 m high. If the cannon is fired straight upward, the cannonball just reaches the top of the cliff. **(a)** What is the initial speed of the cannonball? **(b)** Suppose a second cannon is placed at the top of the cliff. This cannon is fired horizontally, giving its cannonballs the same initial speed found in part (a). Show that the range of this cannon is the same as the maximum range of the cannon at the base of the cliff. (Assume the ground at the base of the cliff is level, though the result is valid even if the ground is not level.)

65. • **Shot Put Record** The men's world record for the shot put, 23.12 m, was set by Randy Barnes of the United States on May 20, 1990. If the shot was launched from 6.00 ft above the ground at an initial angle of 42.0°, what was its initial speed?

66. •• Referring to Conceptual Checkpoint 4–3, suppose the two snowballs are thrown from an elevation of 15 m with an initial speed of 12 m/s. What is the speed of each ball when it is 5.0 m above the ground?

67. •• **IP** A hockey puck just clears the 2.00-m high boards on its way out of the rink. The base of the boards is 20.2 m from the point where the puck is launched. **(a)** Given the launch angle of the puck, θ, outline a strategy that you can use to find its initial speed, v_0. **(b)** Use your strategy to find v_0 for $\theta = 15.0°$.

68. •• Referring to Active Example 4–2, suppose the ball is punted from an initial height of 0.750 m. What is the initial speed of the ball in this case?

69. •• **A "Lob" Pass Versus a "Bullet"** A quarterback can throw a receiver a high, lazy "lob" pass or a low, quick "bullet" pass. These passes are indicated by curves 1 and 2, respectively, in **Figure 4–23**. **(a)** The lob pass is thrown with an initial speed of 21.5 m/s and its time of flight is 3.97 s. What is its launch angle? **(b)** The bullet pass is thrown with a launch angle of 25.0°. What is the initial speed of this pass? **(c)** What is the time of flight of the bullet pass?

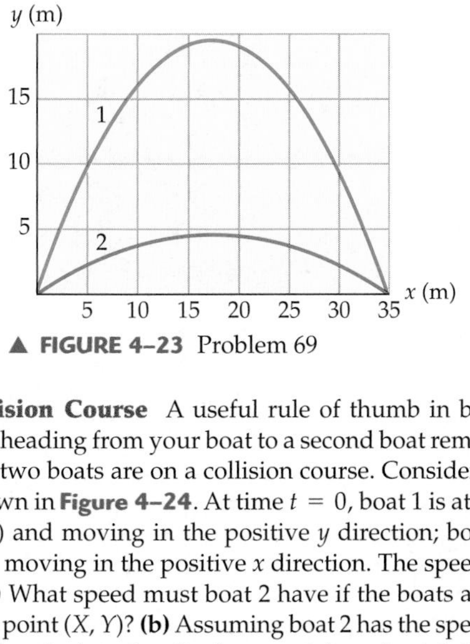

▲ **FIGURE 4–23** Problem 69

70. ••• **Collision Course** A useful rule of thumb in boating is that if the heading from your boat to a second boat remains constant, the two boats are on a collision course. Consider the two boats shown in **Figure 4–24**. At time $t = 0$, boat 1 is at the location $(X, 0)$ and moving in the positive y direction; boat 2 is at $(0, Y)$ and moving in the positive x direction. The speed of boat 1 is v_1. **(a)** What speed must boat 2 have if the boats are to collide at the point (X, Y)? **(b)** Assuming boat 2 has the speed found in part (a), calculate the displacement from boat 1 to boat 2, $\Delta\vec{r} = \vec{r}_2 - \vec{r}_1$. **(c)** Use your results from part (b) to show that $(\Delta r)_y/(\Delta r)_x = -Y/X$, independent of time. This shows that $\Delta\vec{r} = \vec{r}_2 - \vec{r}_1$, maintains a constant direction until the collision, as specified in the rule of thumb.

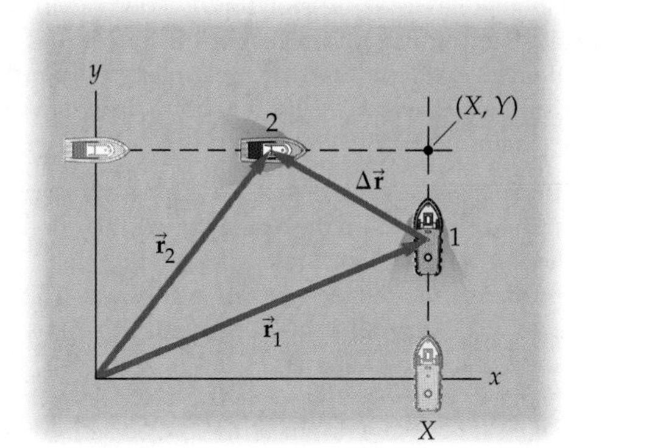

▲ **FIGURE 4–24** Problem 70

71. ••• As discussed in Example 4–7, the archerfish hunts by dislodging an unsuspecting insect from its resting place with a stream of water expelled from the fish's mouth. Suppose the archerfish squirts water with a speed of 2.00 m/s at an angle of 50.0° above the horizontal, and aims for a beetle on a leaf 3.00 cm above the water's surface. **(a)** At what horizontal distance from the beetle should the archerfish fire if it is to hit its target in the least time? **(b)** How much time will the beetle have to react?

72. ••• **(a)** What is the greatest horizontal distance from which the archerfish can hit the beetle, assuming the same squirt speed and direction as in Problem 71? **(b)** How much time does the beetle have to react in this case?

73. ••• Find the launch angle for which the range and maximum height of a projectile are the same.

74. ••• A mountain climber jumps a crevasse of width W by leaping horizontally with speed v_0. **(a)** If the height difference between the two sides of the crevasse is h, what is the minimum value of v_0 for the climber to land safely on the other side? **(b)** In this case, what is the climber's direction of motion on landing?

75. ••• Prove that the landing speed of a projectile is independent of launch angle for a given height of launch.

76. •• **Maximum Height and Range** Prove that the maximum height of a projectile, H, divided by the range of the projectile, R, satisfies the relation $H/R = \frac{1}{4}\tan\theta$.

77. •• **Landing on a Different Level** A projectile fired from $y = 0$ with initial speed v_0 and initial angle θ lands on a different level, $y = h$. Show that the time of flight of the projectile is

$$T = \tfrac{1}{2}T_0\left(1 + \sqrt{1 - \frac{h}{H}}\right)$$

where T_0 is the time of flight for $h = 0$ and H is the maximum height of the projectile.

78. ••• A mountain climber jumps a crevasse by leaping horizontally with speed v_0. If the climber's direction of motion on landing is θ below the horizontal, what is the height difference h between the two sides of the crevasse?

79. ••• **IP** Referring to Problem 63, suppose the initial velocity of the coin tossed by the passenger is $\vec{v} = (-2.25\text{ m/s})\hat{x} + (4.38\text{ m/s})\hat{y}$. The train's velocity relative to the ground is still $(12.1\text{ m/s})\hat{x}$. **(a)** What is the minimum speed of the coin relative to the ground during its flight? At what point in the coin's flight does this minimum speed occur? Explain. **(b)** Find the initial speed and direction of the coin as seen by an observer on the

ground. **(c)** Use the expression for y_{max} derived in Example 4–7 to calculate the maximum height of the coin, as seen by an observer on the ground. **(d)** Repeat part (c) from the point of view of the passenger. Verify that both observers calculate the same maximum height.

80. ••• **Projectiles: Coming or Going?** Most projectiles continually move farther from the origin during their flight, but this is not the case if the launch angle is greater than $\cos^{-1}\left(\frac{1}{3}\right) = 70.5°$. For example, the projectile shown in **Figure 4–25** has a launch angle of 75.0° and an initial speed of 10.1 m/s. During the portion of its motion shown in red, it is moving closer to the origin—it is moving away on the blue portions. Calculate the distance from the origin to the projectile **(a)** at the start of the red portion, **(b)** at the end of the red portion, and **(c)** just before the projectile lands. Notice that the distance for part (b) is the smallest of the three.

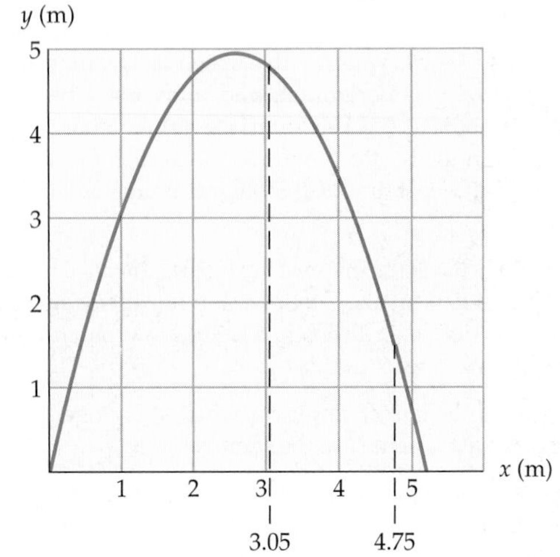

▲ **FIGURE 4–25** Problem 80

Interactive Homework Problems

81. •• ***Referring to Example 4–5*** **(a)** At what launch angle *greater* than 54.0° does the golf ball just barely miss the top of the tree in front of the green? Assume the ball has an initial speed of 13.5 m/s, and that the tree is 3.00-m high and is a horizontal distance of 14.0 m from the launch point. **(b)** Where does the ball land in the case described in part (a)? **(c)** At what launch angle *less* than 54.0° does the golf ball just barely miss the top of the tree in front of the green? **(d)** Where does the ball land in the case described in part (c)?

82. •• ***Referring to Example 4–5*** Suppose the golf ball is launched with a speed of 15.0 m/s at an angle of 57.5° above the horizontal, and that it lands on a green 3.50 m above the level where it was struck. **(a)** What horizontal distance does the ball cover during its flight? **(b)** What increase in initial speed would be needed to increase the horizontal distance in part (a) by 7.50 m? Assume everything else remains the same.

83. •• ***Referring to Example 4–6*** Suppose the ball is dropped at the horizontal distance of 5.50 m, but from a new height of 5.00 m. The dolphin jumps with the same speed of 12.0 m/s. **(a)** What launch angle must the dolphin have if it is to catch the ball? **(b)** At what height does the dolphin catch the ball in this case? **(c)** What is the minimum initial speed the dolphin must have to catch the ball before it hits the water?

84. •• **IP** ***Referring to Example 4–6*** Suppose we change the dolphin's launch angle to 45.0°, but everything else remains the same. Thus, the horizontal distance to the ball is 5.50 m, the drop height is 4.10 m, and the dolphin's launch speed is 12.0 m/s. **(a)** What is the vertical distance between the dolphin and the ball when the dolphin reaches the horizontal position of the ball? We refer to this as the "miss distance." **(b)** If the dolphin's launch speed is reduced, will the miss distance increase, decrease, or stay the same? **(c)** Find the miss distance for a launch speed of 10.0 m/s.

5 Newton's Laws of Motion

5–1	Force and Mass	108
5–2	Newton's First Law of Motion	108
5–3	Newton's Second Law of Motion	110
5–4	Newton's Third Law of Motion	118
5–5	The Vector Nature of Forces: Forces in Two Dimensions	121
5–6	Weight	124
5–7	Normal Forces	128

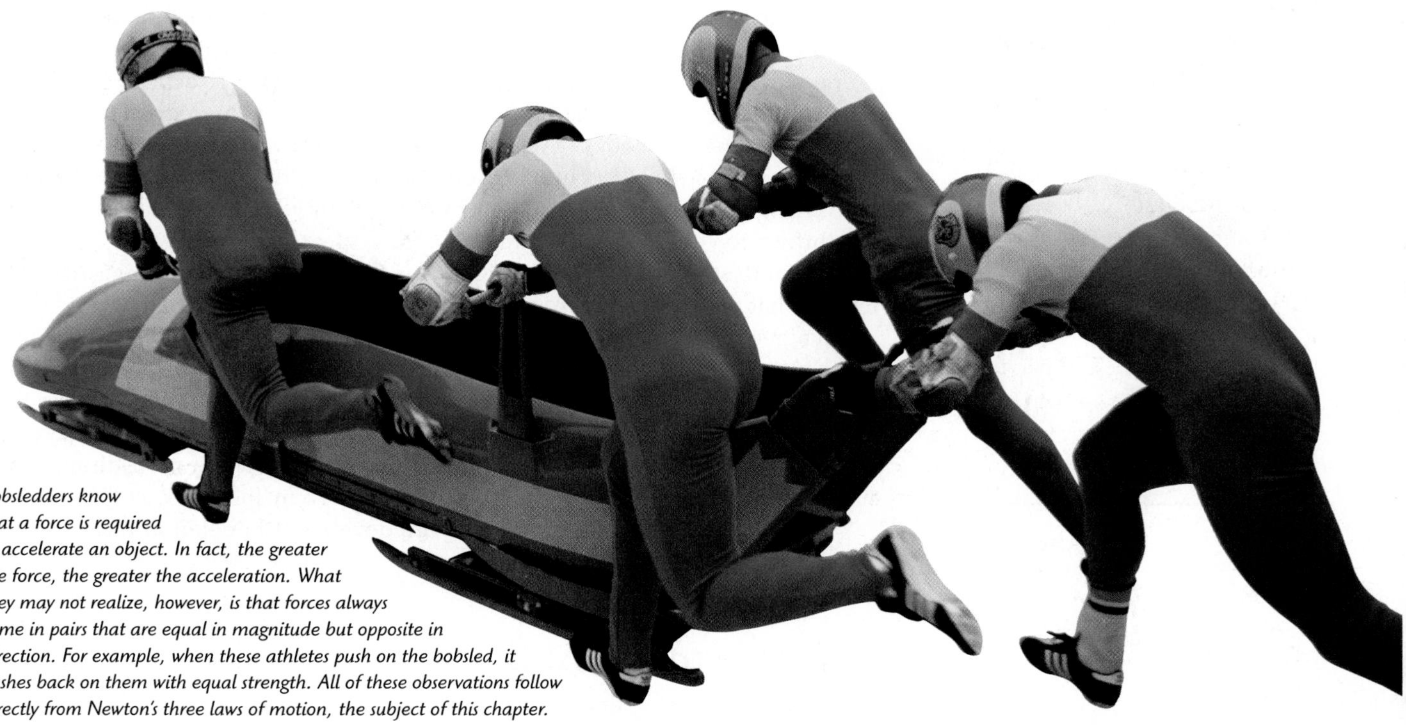

Bobsledders know that a force is required to accelerate an object. In fact, the greater the force, the greater the acceleration. What they may not realize, however, is that forces always come in pairs that are equal in magnitude but opposite in direction. For example, when these athletes push on the bobsled, it pushes back on them with equal strength. All of these observations follow directly from Newton's three laws of motion, the subject of this chapter.

We are all subject to Newton's laws of motion, whether we know it or not. You can't move your body, drive a car, or toss a ball in a way that violates his rules. In short, our very existence is constrained and regulated by these three fundamental statements concerning matter and its motion.

Yet Newton's laws are surprisingly simple, especially when you consider that they apply equally well to galaxies, planets, comets, and yes, even apples falling from trees. In this chapter we present the three laws of Newton, and we show how they can be applied to everyday situations. Using them, we go beyond a simple description of motion, as in kinematics, to a study of the *causes* of motion, referred to as **dynamics.**

With the advent of Newtonian dynamics in 1687, science finally became quantitative and predictive. Edmund Halley, inspired by Newton's laws, used them to predict the return of the comet that today bears his name. In all of recorded history, no one had ever before predicted the appearance of a comet; in fact, they were generally regarded as supernatural apparitions. Though Halley didn't live to see his comet's return, his correct prediction illustrated the power of Newton's laws in a most dramatic and memorable way.

Today, we still recognize Newton's laws as the indispensable foundation for all of physics. It would be nice to say that these laws are the complete story when it comes to analyzing motion, but that is not the case. In the early part of the last century, physicists discovered that Newton's laws must be modified for objects moving at speeds near that of light and for objects comparable in size to atoms. In the world of everyday experience, however, Newton's laws still reign supreme.

5–1 Force and Mass

A **force,** simply put, is a push or a pull. When you push on a box to slide it across the floor, for example, or pull on the handle of a wagon to give a child a ride, you are exerting a force. Similarly, when you hold this book in your hand, you exert an upward force to oppose the downward pull of gravity. If you set the book on a table, the table exerts the same upward force you exerted a moment before. Forces are truly all around us.

Now, when you push or pull on something, there are two quantities that characterize the force you are exerting. The first is the strength or **magnitude** of your force; the second is the **direction** in which you are pushing or pulling. Because a force is determined by both a magnitude and a direction, it is a vector. We consider the vector properties of forces in more detail in Section 5–5.

TABLE 5–1
Typical Masses in Kilograms (kg)

Earth	5.97×10^{24}
Space Shuttle	2,000,000
Blue whale (largest animal on Earth)	178,000
Whale shark (largest fish)	18,000
Elephant (largest land animal)	5400
Automobile	1200
Human (adult)	70
Gallon of milk	3.6
Quart of milk	0.9
Baseball	0.145
Honeybee	0.00015
Bacterium	10^{-15}

In general, an object has several forces acting on it at any given time. In the previous example, a book at rest on a table experiences a downward force due to gravity and an upward force due to the table. If you push the book across the table, it also experiences a horizontal force due to your push. The total, or net, force exerted on the book is the vector sum of the individual forces acting on it.

After the net force acting on an object, the second key ingredient in Newton's laws is the **mass** of an object, which is a measure of how difficult it is to change its velocity—to start an object moving if it is at rest, to bring it to rest if it is moving, or to change its direction of motion. For example, if you throw a baseball or catch one thrown to you, the force required is not too great. But if you want to start a car moving or to stop one that is coming at you, the force involved is much greater. It follows that the mass of a car is greater than the mass of a baseball.

In agreement with everyday usage, mass can also be thought of as a measure of the quantity of matter in an object. Thus, it is clear that the mass of an automobile, for example, is much greater than the mass of a baseball, but much less than the mass of the Earth. We measure mass in units of kilograms (kg), where one kilogram is defined as the mass of a standard cylinder of platinum-iridium, as discussed in Chapter 1. A list of typical masses is given in Table 5–1.

These properties of force and mass are developed in detail in the next three sections.

5–2 Newton's First Law of Motion

If you've ever stood in line at an airport, pushing your bags forward a few feet at a time, you know that as soon as you stop pushing the bags, they stop moving. Observations such as this often lead to the erroneous conclusion that a force is required for an object to move. In fact, according to Newton's first law of motion, a force is required only to *change* an object's motion.

What is missing in this analysis is the force of friction between the bags and the floor. When you stop pushing the bags, it is not true that they stop moving because they no longer have a force acting on them. On the contrary, there is a rather large *frictional force* between the bags and the floor. It is this force that causes the bags to come to rest.

To see how motion is affected by reducing friction, imagine that you slide on dirt into second base during a baseball game. You won't slide very far before stopping. On the other hand, if you slide with the same initial speed on a sheet of ice—where the friction is much less than on a ball field—you slide considerably farther. If you could reduce the friction more, you would slide even farther.

In the classroom, air tracks allow us to observe motion with practically no friction. An example of such a device is shown in **Figure 5–1**. Note that air is blown through small holes in the track, creating a cushion of air for a small "cart" to ride on. A cart placed at rest on a level track remains at rest—unless you push on it to get it started.

Once set in motion, the cart glides along with constant velocity—constant speed in a straight line—until it hits a bumper at the end of the track. The bumper

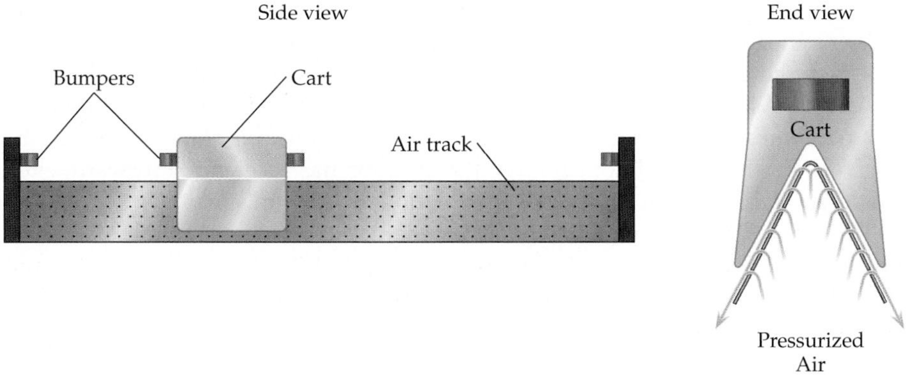

◀ **FIGURE 5–1 The air track**
An air track provides a cushion of air on which a cart can ride with virtually no friction.

exerts a force on the cart, causing it to change its direction of motion. After bouncing off the bumper, the cart again moves with constant velocity. If the track could be extended to infinite length, and could be made perfectly frictionless, the cart would simply keep moving with constant velocity forever.

Newton's first law of motion summarizes these observations in the following statements:

Newton's First Law

An object at rest remains at rest as long as no net force acts on it.

An object moving with constant velocity continues to move with the same speed and in the same direction as long as no net force acts on it.

▲ An air track provides a nearly frictionless environment for experiments involving linear motion.

Notice the recurring phrase, "no net force," in these statements. It is important to realize that this can mean one of two things: (i) no force acts on the object; or (ii) forces act on the object, but they sum to zero. We shall see examples of the second possibility later in this chapter and again in the next chapter.

Newton's first law, which was first enunciated by Galileo, is also known as the **law of inertia,** which is appropriate since the literal meaning of the word inertia is "laziness." Speaking loosely, we can say that matter is "lazy," in that it won't change its motion unless forced to do so. For example, if an object is at rest, it won't start moving on its own. If an object is already moving with constant velocity, it won't alter its speed *or* direction, unless a force causes the change. We call this property of matter its inertia

According to Newton's first law, being at rest and moving with constant velocity are actually equivalent. To see this, imagine two observers: one is in a train moving with constant velocity; the second is standing next to the tracks, at rest on the ground. The observer in the train places an ice cube on a dinner tray. From that person's point of view—that is, in that person's **frame of reference**—the ice cube has no net force acting on it and it is at rest on the tray. It obeys the first law. In the frame of reference of the observer on the ground, the ice cube has no net force on it and it moves with constant velocity. This also agrees with the first law. Thus Newton's first law holds for both observers: They both see an ice cube with zero net force moving with constant velocity—it's just that for the first observer the constant velocity happens to be zero.

In this example, we say that each observer is in an **inertial frame of reference;** that is, a frame of reference in which the law of inertia holds. In general, if one frame is an inertial frame of reference, then any frame of reference that moves with constant velocity relative to that frame is also an inertial frame of reference. Thus, if an object moves with constant velocity in one inertial frame, it is always possible to find another inertial frame in which the object is at rest. It is in this sense that there really isn't any difference between being at rest and moving with constant velocity. It's all relative—relative to the frame of reference the object is viewed from.

This gives us a more compact statement of the first law:

If the net force on an object is zero, its velocity is constant.

As an example of a frame of reference that is not inertial, imagine that the train carrying the first observer suddenly comes to a halt. From the point of view of that observer, there is still no net force on the ice cube. However, because of the rapid braking, the ice cube flies off the tray. In fact, the ice cube simply continues to move forward with the same constant velocity while the *train* comes to rest. To the observer on the train, it appears that the ice cube has accelerated forward, even though no force acts on it, which is in violation of Newton's first law.

In general, any frame that accelerates relative to an inertial frame is a noninertial frame. The surface of the Earth accelerates slightly, due to its rotational and orbital motions, but since the acceleration is so small, it may be considered an excellent approximation to an inertial frame of reference. Unless specifically stated otherwise, we will always consider the surface of the Earth to be an inertial frame.

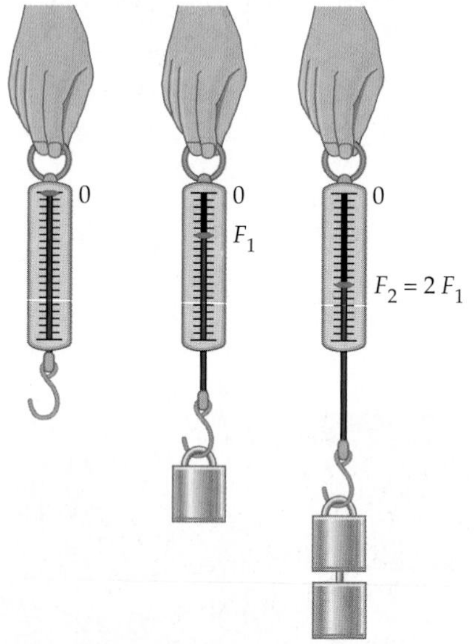

▲ **FIGURE 5–2 Calibrating a "force meter"**
With two weights, the force exerted by the scale is twice the force exerted when only a single weight is attached.

5–3 Newton's Second Law of Motion

To hold an object in your hand, you have to exert an upward force to oppose, or "balance," the force of gravity. If you suddenly remove your hand so that the only force acting on the object is gravity, it accelerates downward, as discussed in Chapter 2. This is one example of Newton's second law, which states, basically, that unbalanced forces cause accelerations.

To explore this in more detail, consider a spring scale of the type used to weigh fish. The scale gives a reading of the force, F, exerted by the spring contained within it. If we hang one weight from the scale, it gives a reading that we will call F_1. If two identical weights are attached, the scale reads $F_2 = 2F_1$, as indicated in **Figure 5–2**. With these two forces marked on the scale, we are ready to perform some force experiments.

First, attach the scale to an air-track cart, as in **Figure 5–3**. If we pull with a force F_1, we observe that the cart accelerates at the rate a_1. If we now pull with a force $F_2 = 2F_1$, the acceleration we observe is $a_2 = 2a_1$. Thus, the acceleration is proportional to the force—the greater the force, the greater the acceleration.

▶ **FIGURE 5–3 Acceleration is proportional to force**
The spring calibrated in Figure 5–2 is used to accelerate a mass on a "frictionless" air track. If the force is doubled, the acceleration is also doubled.

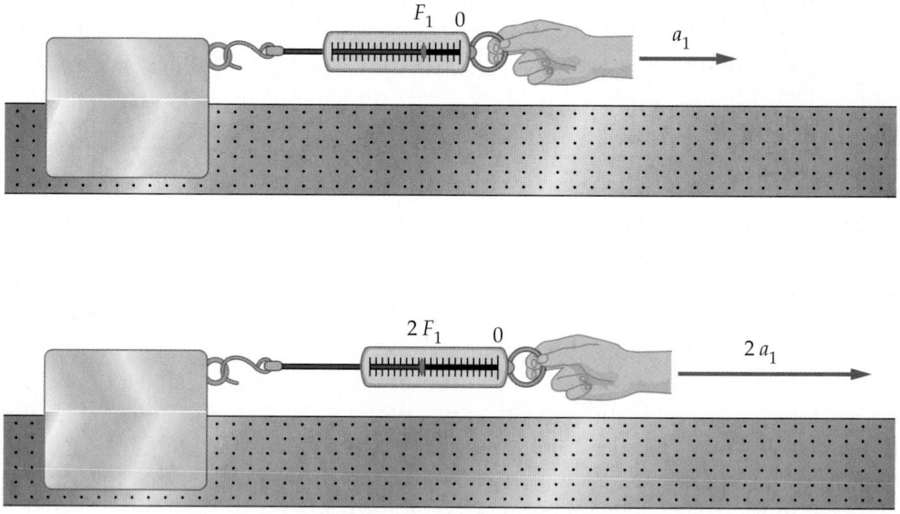

Second, instead of doubling the force, let's double the mass of the cart by connecting two together, as in **Figure 5–4**. In this case, if we pull with a force F_1 we find an acceleration equal to $\frac{1}{2}a_1$. Thus, the acceleration is inversely proportional to mass—the greater the mass, the less the acceleration.

Combining these results, we find that in this simple case—with just one force in just one direction—the acceleration is given by

$$a = \frac{F}{m}$$

Rearranging the equation yields the form of Newton's law that is perhaps best known, $F = ma$.

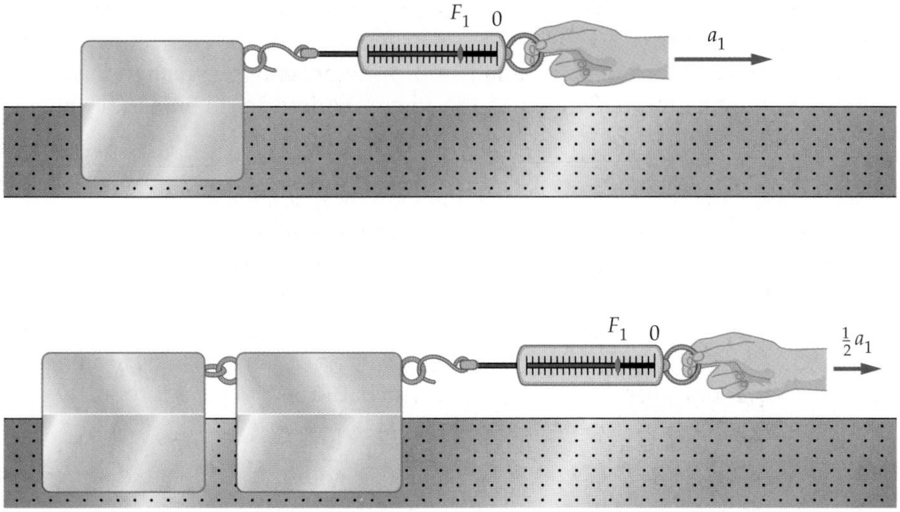

FIGURE 5–4 Acceleration is inversely proportional to mass
If the mass of an object is doubled but the force remains the same, the acceleration is halved.

In general, there may be several forces acting on a given mass, and these forces may be in different directions. Thus, we replace F with the sum of the force vectors acting on a mass:

$$\text{sum of force vectors} = \vec{\mathbf{F}}_{\text{net}} = \sum \vec{\mathbf{F}}$$

The notation, $\Sigma\vec{\mathbf{F}}$, which uses the Greek letter sigma (Σ), is read "sum $\vec{\mathbf{F}}$." Recalling that acceleration is also a vector, we arrive at the formal statement of Newton's second law of motion:

Newton's Second Law

$$\vec{\mathbf{a}} = \frac{\sum \vec{\mathbf{F}}}{m} \quad \text{or} \quad \sum \vec{\mathbf{F}} = m\vec{\mathbf{a}} \qquad 5\text{–}1$$

In words:

If an object of mass m is acted on by a net force $\Sigma\vec{\mathbf{F}}$, it will experience an acceleration $\vec{\mathbf{a}}$ that is equal to the net force divided by the mass. Because the net force is a vector, the acceleration is also a vector. In fact, the direction of an object's acceleration is the *same* as the direction of the net force acting on it.

One should note that Newton's laws cannot be derived from anything more basic. In fact, this is what we mean by a law of nature. The validity of Newton's laws, and all other laws of nature, comes directly from comparisons with experiment.

In terms of vector components, an equivalent statement of the second law is:

$$\sum F_x = ma_x \qquad \sum F_y = ma_y \qquad \sum F_z = ma_z \qquad 5\text{–}2$$

Note that Newton's second law holds independently for each coordinate direction. This component form of the second law is particularly useful when solving problems.

Let's pause for a moment to consider an important special case of the second law. Suppose an object has zero net force acting upon it. This may be because no forces act on it at all, or that it is acted on by forces whose vector sum is zero. In either case, we can state this mathematically as:

$$\sum \vec{\mathbf{F}} = 0$$

Now, according to Newton's second law, we conclude that the acceleration of this object must be zero:

$$\vec{\mathbf{a}} = \frac{\sum \vec{\mathbf{F}}}{m} = \frac{0}{m} = 0$$

But if an object's acceleration is zero, its velocity must be constant. In other words, if the net force on an object is zero, the object moves with constant velocity. This is

▲ Even though the tugboat exerts a large force on this ship, the ship's acceleration is small. This is because the acceleration of an object is inversely proportional to its mass, and the mass of the ship is enormous. The force exerted on the unfortunate hockey player is much smaller. The resulting acceleration is much larger, however, due to the relatively small mass of the player compared to that of the ship.

Newton's first law. Thus we see that Newton's first and second laws are consistent with one another.

Forces are measured in units called, appropriately enough, the **newton (N).** In particular, one newton is defined as the force required to give one kilogram of mass an acceleration of 1 m/s^2. Thus,

$$1\ \text{N} = (1\ \text{kg})(1\ \text{m/s}^2) = 1\ \text{kg}\cdot\text{m/s}^2 \qquad \text{5–3}$$

In everyday terms, a newton is roughly a quarter of a pound. Note that a force in newtons divided by a mass in kilograms has the units of acceleration:

$$\frac{1\ \text{N}}{1\ \text{kg}} = \frac{1\ \text{kg}\cdot\text{m/s}^2}{1\ \text{kg}} = 1\ \text{m/s}^2 \qquad \text{5–4}$$

Other common units for force are presented in Table 5–2. Typical forces and their magnitudes in newtons are listed in Table 5–3.

TABLE 5–2 Units of Mass, Acceleration, and Force

System of units	Mass	Acceleration	Force
SI	kilogram (kg)	m/s^2	newton (N)
cgs	gram (g)	cm/s^2	dyne (dyn)
British	slug	ft/s^2	pound (lb)

(*Note:* 1 N = 10^5 dyne = 0.225 lb.)

TABLE 5–3 Typical Forces in Newtons (N)

Main engines of space shuttle	31,000,000
Pulling force of locomotive	250,000
Thrust of jet engine	75,000
Force to accelerate a car	7000
Weight of adult human	700
Weight of an apple	1
Weight of a rose	0.1
Weight of an ant	0.001

EXERCISE 5–1

The net force acting on a Jaguar XK8 has a magnitude of 6800 N. If the car's acceleration is 3.8 m/s^2, what is its mass?

Solution

Since the net force and the acceleration are always in the same direction, we can replace the vectors in Equation 5–1 with magnitudes. Solving $\Sigma F = ma$ for the mass yields

$$m = \frac{\sum F}{a} = \frac{6800\ \text{N}}{3.8\ \text{m/s}^2} = 1800\ \text{kg}$$

The following Conceptual Checkpoint presents a situation in which both Newton's first and second laws play an important role.

CONCEPTUAL CHECKPOINT 5–1 Tightening a Hammer

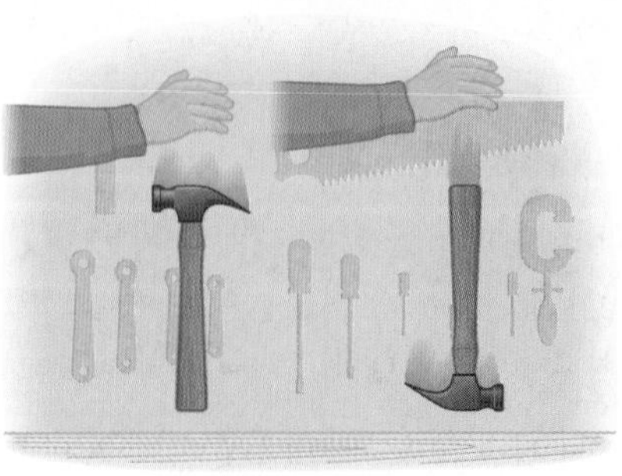

The metal head of a hammer is loose. To tighten it, you drop the hammer down onto a table. Should you **(a)** drop the hammer with the handle end down, **(b)** drop the hammer with the head end down, or **(c)** do you get the same result either way?

Reasoning and Discussion

It might seem that since the same hammer hits against the same table in either case, there shouldn't be a difference. Actually, there is.

In case **(a)**, the handle of the hammer comes to rest when it hits the table, but the head continues downward until a force acts on it to bring it to rest. The force that acts on it is supplied by the handle, which results in the head being wedged more tightly onto the handle. Since the metal head is heavy, the force wedging it onto the handle is great. In case **(b)** the head of the hammer comes to rest, but the handle continues to move until a force brings it to rest. The handle is lighter than the head, however; thus the force acting on it is less, resulting in less tightening.

Answer:

(a) Drop the hammer with the handle end down.

A similar effect occurs when you walk—with each step you take you tamp your head down onto your spine, as when dropping a hammer handle end down.

This causes you to grow shorter during the day! Try it. Measure your height first thing in the morning, then again before going to bed. If you're like many people, you'll find that you have shrunk by an inch or so during the day.

REAL-WORLD PHYSICS: BIO

How walking affects your height

Free-Body Diagrams

When solving problems involving forces and Newton's laws, it is essential to begin by making a sketch that indicates *each and every external force* acting on a given object. This type of sketch is referred to as a **free-body diagram.** If we are concerned only with nonrotational motion, as is the case in this and the next chapter, we treat the object of interest as a point particle and apply each of the forces acting on the object to that point, as **Figure 5–5** shows. Once the forces are drawn, we choose a coordinate system and resolve each force into components. At this point, Newton's second law can be applied to each coordinate direction separately.

PROBLEM-SOLVING NOTE

External Forces

External forces acting on an object fall into two main classes: (i) Forces at the point of contact with another object, and (ii) forces exerted by an external agent, such as gravity.

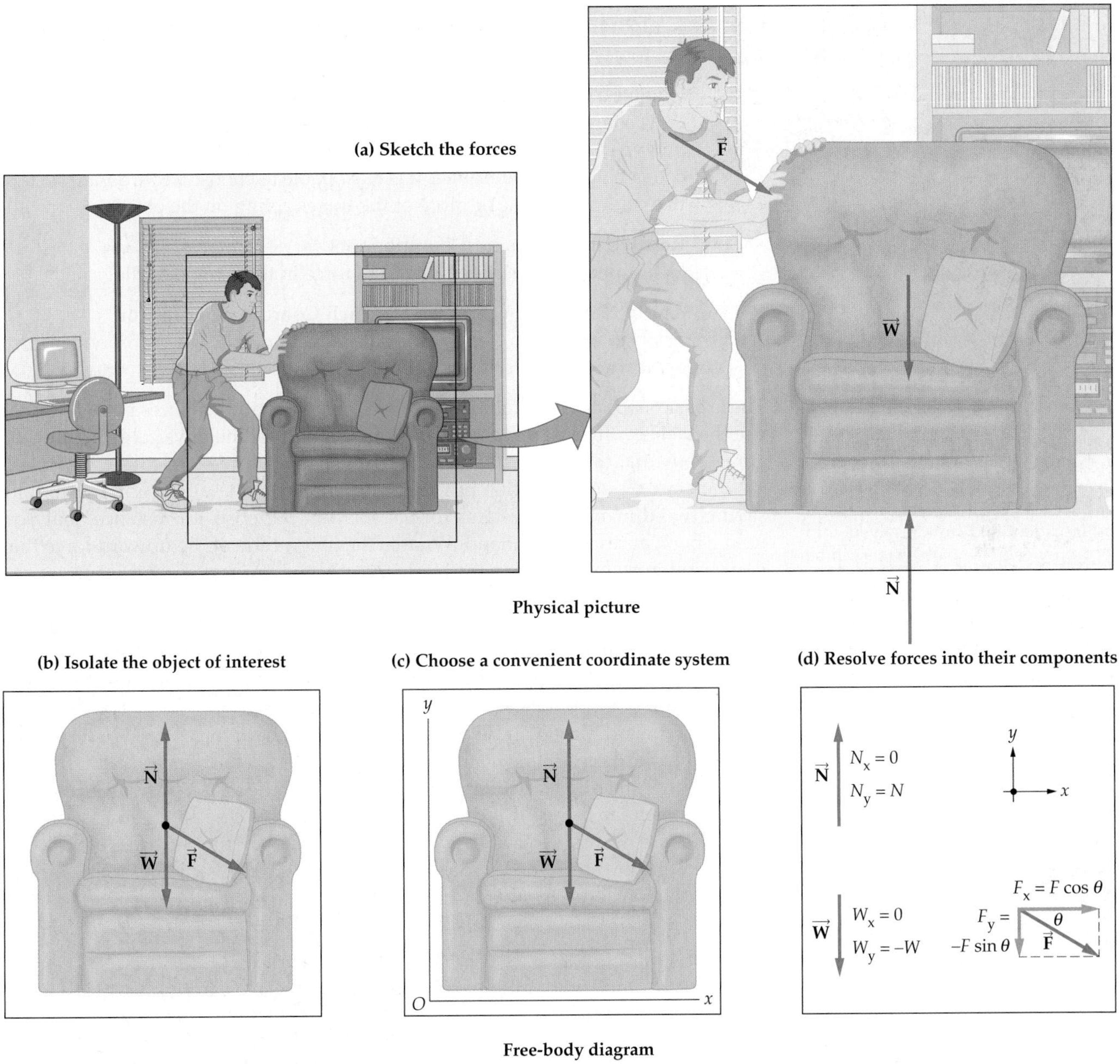

▲ **FIGURE 5–5 Constructing and using a free-body diagram**

The four basic steps in constructing and using a free-body diagram are illustrated in these sketches. **(a)** Sketch all of the external forces acting on an object of interest. Note that only forces acting *on* the object are shown; none of the forces exerted *by* the object are included. **(b)** Isolate the object and treat it as a point particle. **(c)** Choose a convenient coordinate system. This will often mean aligning a coordinate axis to coincide with the direction of one or more forces in the system. **(d)** Resolve each of the forces into components using the coordinate system of part (c).

For example, in Figure 5–5 there are three external forces acting on the chair. One is the force $\vec{\mathbf{F}}$ exerted by the person. In addition, gravity exerts a downward force, $\vec{\mathbf{W}}$, which is simply the weight of the chair. Finally, the floor exerts an upward force on the chair that prevents it from falling toward the center of the Earth. This force is referred to as the *normal force*, $\vec{\mathbf{N}}$, because it is perpendicular (that is, normal) to the surface of the floor. We will consider the weight and the normal force in greater detail in Sections 5–6 and 5–7, respectively.

We can summarize the steps involved in constructing a free-body diagram as follows:

Sketch the Forces
Identify and sketch all of the external forces acting on an object. Sketching the forces roughly to scale will help in estimating the direction and magnitude of the net force.

Isolate the Object of Interest
Replace the object with a point particle of the same mass. Apply each of the forces acting on the object to that point.

Choose a Convenient Coordinate System
Any coordinate system will work; however, if the object moves in a known direction, it is often convenient to pick that direction for one of the coordinate axes. Otherwise, it is reasonable to choose a coordinate system that aligns with one or more of the forces acting on the object.

Resolve the Forces into Components
Determine the components of each force in the free-body diagram.

Apply Newton's Second Law to Each Coordinate Direction
Analyze motion in each coordinate direction using the component form of Newton's second law, as given in Equation 5–2.

PROBLEM-SOLVING NOTE

Picture the Problem

In problems involving Newton's laws, it is important to begin with a free-body diagram and to identify all the external forces that act on an object. Once these forces are identified and resolved into their components, Newton's laws can be applied in a straightforward way. It is crucial, however, that only external forces acting on the object be included, and that none of the external forces be omitted.

These basic steps are illustrated in Figure 5–5. Note that the figures in this chapter use the labels "Physical picture" to indicate a sketch of the physical situation and "Free-body diagram" to indicate a free-body sketch.

We start by applying this procedure to a simple one-dimensional example, saving two-dimensional systems for Section 5–5. Suppose, for instance, that you hold a book at rest in your hand. What is the magnitude of the upward force that your hand must exert to keep the book at rest? From everyday experience, we expect that the upward force must be equal in magnitude to the weight of the book, but let's see how this result can be obtained directly from Newton's second law.

We begin with a sketch of the physical situation, as shown in **Figure 5–6 (a)**. The corresponding free-body diagram, in **Figure 5–6 (b)**, shows just the book, represented by a point, and the forces acting on it. Note that two forces act on the book: (i) the downward force of gravity, $\vec{\mathbf{W}}$, and (ii) the upward force, $\vec{\mathbf{F}}$, exerted by your hand. Only the forces acting *on* the book are included in the free-body diagram.

Now that the free-body diagram is drawn, we indicate a coordinate system so that the forces can be resolved into components. In this case all the forces are vertical. Thus we draw a y axis in the vertical direction in Figure 5–6 (b). Note that we have chosen upward to be the positive direction. With this choice, the y components of the forces are $F_y = F$ and $W_y = -W$. It follows that

$$\sum F_y = F - W$$

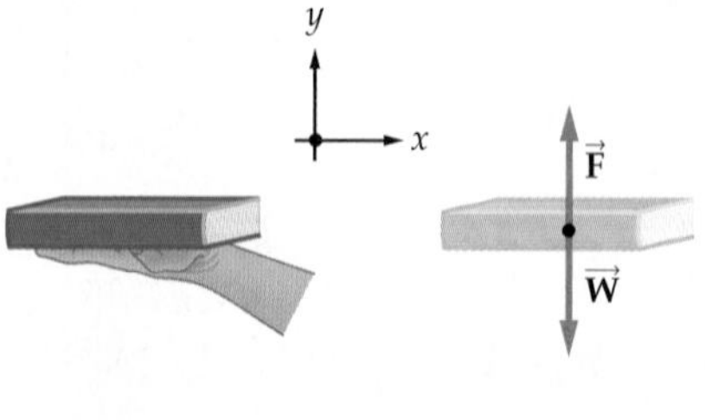

(a) Physical picture

(b) Free-body diagram

▲ **FIGURE 5–6 A book supported in a person's hand**

(a) The physical situation. **(b)** The free-body diagram for the book, showing the two external forces acting on it. We also indicate our choice for a coordinate system.

Using the y component of the second law ($\Sigma F_y = ma_y$) we find

$$F - W = ma_y$$

Since the book remains at rest, its acceleration is zero. Thus, $a_y = 0$, which gives

$$F - W = ma_y = 0 \quad \text{or} \quad F = W$$

as expected.

Next, we consider a situation where the net force acting on an object is nonzero, meaning that its acceleration is also nonzero.

EXAMPLE 5–1 Three Forces

Moe, Larry, and Curly push on a 752-kg boat that floats next to a dock. They each exert an 80.5-N force parallel to the dock. **(a)** What is the acceleration of the boat if they all push in the same direction? Give both direction and magnitude. **(b)** What is the magnitude and direction of the boat's acceleration if Larry and Curly push in the opposite direction to Moe's push?

Picture the Problem

In our sketch we indicate the three relevant forces acting on the boat: $\vec{\mathbf{F}}_M$, $\vec{\mathbf{F}}_L$, and $\vec{\mathbf{F}}_C$. Note that we have chosen the positive x direction to the right, in the direction that all three push for part (a). Therefore, all three forces have a positive x component in part (a). In part (b), however, the forces exerted by Larry and Curly have negative x components.

Physical pictures

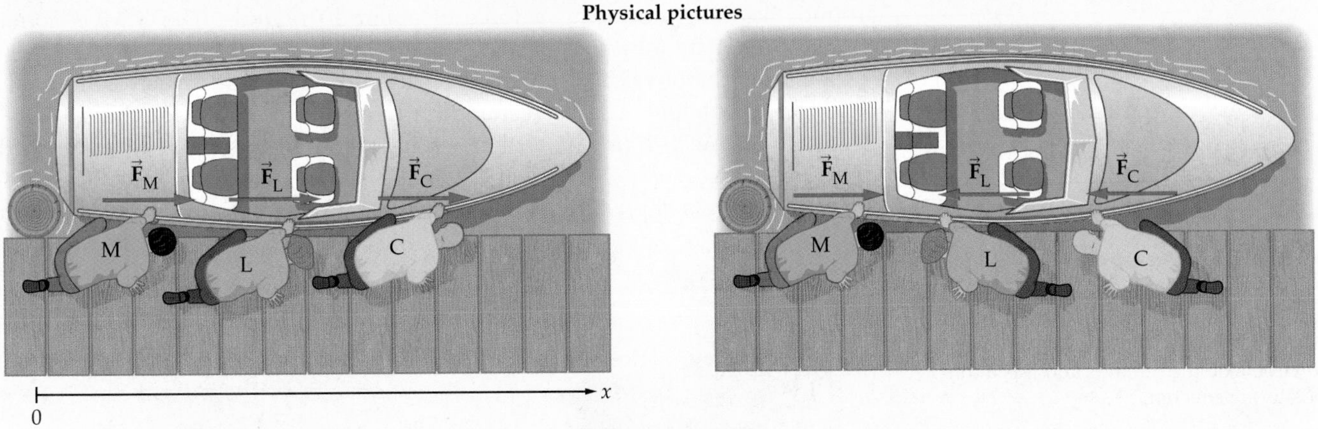

Free-body diagrams

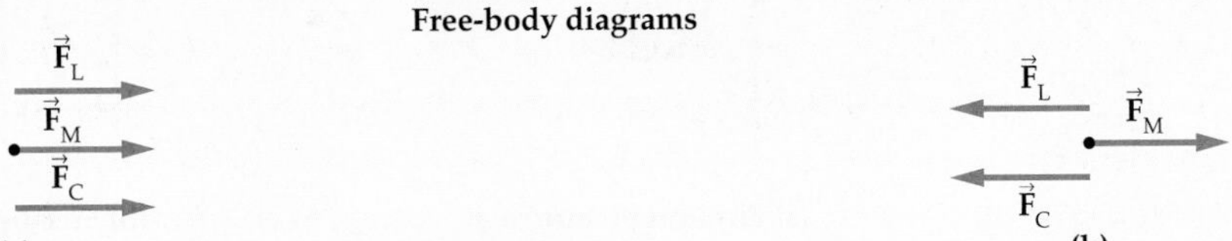

Strategy

Since we know the mass of the boat and the forces acting on it, we can find the acceleration using $\Sigma F_x = ma_x$. Even though this problem is one-dimensional, it is important to think of it in terms of vector components. For example, when we sum the x components of the forces, we are careful to use the appropriate signs—just as we always do when dealing with vectors.

Solution

Part (a)

1. Write out the x component for each of the three forces: $F_{M,x} = F_{L,x} = F_{C,x} = 80.5 \text{ N}$
2. Sum the x components of force and set equal to ma_x: $\sum F_x = F_{M,x} + F_{L,x} + F_{C,x} = 241.5 \text{ N} = ma_x$
3. Divide by the mass to find a_x. Since a_x is positive, the acceleration is to the right, as expected: $a_x = \frac{\sum F_x}{m} = \frac{241.5 \text{ N}}{752 \text{ kg}} = 0.321 \text{ m/s}^2$

Part (b)

4. Again, start by writing the x component for each force:
$$F_{M,x} = 80.5 \text{ N}$$
$$F_{L,x} = F_{C,x} = -80.5 \text{ N}$$
5. Sum the x components of force and set equal to ma_x:
$$\sum F_x = F_{M,x} + F_{L,x} + F_{C,x}$$
$$= 80.5 \text{ N} - 80.5 \text{ N} - 80.5 \text{ N} = -80.5 \text{ N} = ma_x$$
6. Solve for a_x. In this case a_x is negative, indicating an acceleration to the left: $a_x = \frac{\sum F_x}{m} = \frac{-80.5 \text{ N}}{752 \text{ kg}} = -0.107 \text{ m/s}^2$

Insight

The results of this Example are in agreement with everyday experience: three forces in the same direction cause more acceleration than three forces in opposing directions. The method of using vector components and being careful about their signs gives the expected results in a simple situation like this, and also works in more complicated situations where everyday experience may be of little help.

Practice Problem

If Moe, Larry, and Curly all push to the right with 85.0-N forces, and the boat accelerates at 0.530 m/s^2, what is its mass? [**Answer:** 481 kg]

Some related homework problems: Problem 1, Problem 3

In some problems, we are given information that allows us to calculate an object's acceleration using the kinematic equations of Chapters 2 and 4. Once the acceleration is known, the second law can be used to find the net force that caused the acceleration.

REAL-WORLD PHYSICS
Astronaut jet packs

For example, suppose that an astronaut uses a jet pack to push a satellite toward the space shuttle. These jet packs, which are known to NASA as Manned Maneuvering Units, or MMUs, are basically small "one-person rockets" strapped to the back of an astronaut's spacesuit. An MMU contains pressurized nitrogen gas that can be released through varying combinations of 24 nozzles spaced around the unit, producing a force of about 10 pounds. The MMUs contain enough propellant for a six-hour EVA (extra-vehicular activity).

We show the physical situation in **Figure 5–7 (a)**, where an astronaut pushes on a 655-kg satellite. The corresponding free-body diagram for the satellite is shown in **Figure 5–7 (b)**. Note that we have chosen the x axis to point in the direction of the push. Now, if the satellite starts at rest and moves 0.675 m after 5.00 seconds of pushing, what is the force, F, exerted on it by the astronaut?

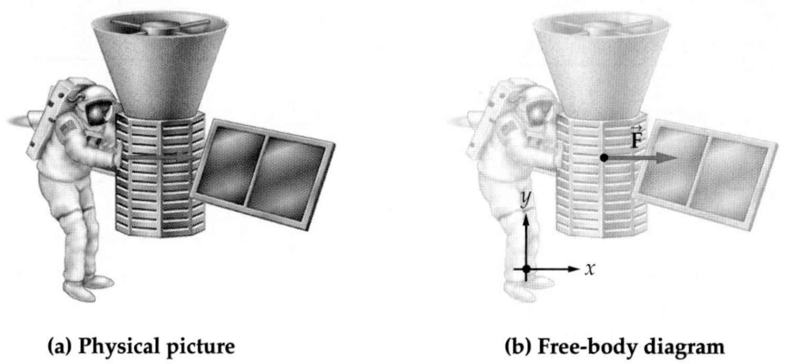

▶ **FIGURE 5–7 An astronaut using a jet pack to push a satellite**
(a) The physical situation. **(b)** The free-body diagram for the satellite. Only one force acts on the satellite, and it is in the positive x direction.

▲ A technician inspects the landing gear of an airliner in a test of Foamcrete, a solid paving material that is just soft enough to collapse under the weight of an airliner. A plane that has run off the runway will slow safely to a stop as its wheels plow through the crumbling Foamcrete.

Clearly, we would like to use Newton's second law (basically, $\vec{\mathbf{F}} = m\vec{\mathbf{a}}$) to find the force, but we know only the mass of the satellite, not its acceleration. We can find the acceleration, however, by assuming constant acceleration (after all, the force is constant) and using the kinematic equation relating position to time $x = x_0 + v_{0x}t + \frac{1}{2}a_xt^2$. We can choose the initial position of the satellite to be $x_0 = 0$, and we are given that it starts at rest, thus $v_{0x} = 0$. Hence,

$$x = \frac{1}{2}a_xt^2$$

Since we know the distance covered in a given time, we can solve for the acceleration:

$$a_x = \frac{2x}{t^2} = \frac{2(0.675\text{ m})}{(5.00\text{ s})^2} = 0.0540\text{ m/s}^2$$

Now that kinematics has provided the acceleration, we use the x component of the second law to find the force. Only one force acts on the satellite, and its x component is F; thus,

$$\sum F_x = F = ma_x$$
$$F = ma_x = (655\text{ kg})(0.0540\text{ m/s}^2) = 35.4\text{ N}$$

This force corresponds to a push of about 8 lb.

Another problem in which we use kinematics to find the acceleration is presented in the following Active Example.

ACTIVE EXAMPLE 5–1 The Force Exerted by Foamcrete

REAL-WORLD PHYSICS Foamcrete is a substance designed to stop an airplane that has run off the end of a runway, without causing injury to passengers. It is solid enough to support a car, but crumbles under the weight of a large airplane. By crumbling, it slows the plane to a safe stop. For example, suppose a 747 jetliner with a mass of 1.75×10^5 kg and an initial speed of 26.8 m/s is slowed to a stop in 122 m. What is the magnitude of the average retarding force $\vec{\mathbf{F}}$ exerted by the Foamcrete on the plane?

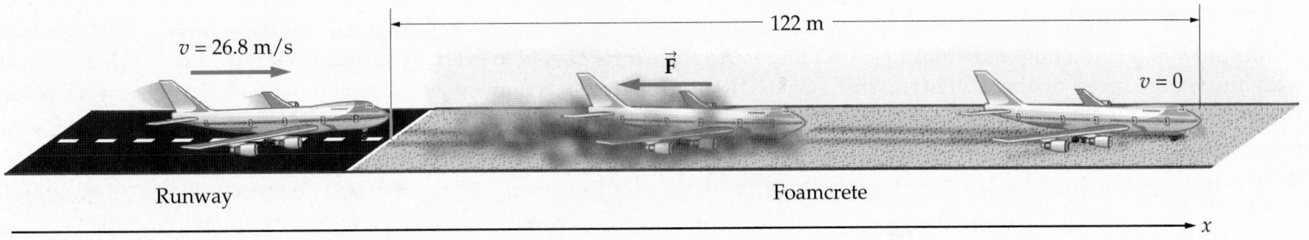

Solution *(Test your understanding by performing the calculations indicated in each step.)*

1. Use $v^2 = v_0{}^2 + 2a_x \Delta x$ to find the plane's average acceleration: $a_x = -2.94\ \text{m/s}^2$
2. Sum the forces in the x direction. Let F represent the magnitude of the force $\vec{\mathbf{F}}$: $\sum F_x = -F$
3. Set the sum of forces equal to mass times acceleration: $-F = ma_x$
4. Solve for the magnitude of the average force, F: $F = -ma_x = 5.15 \times 10^5\ \text{N}$

Insight
Though the plane moves in the positive direction, its acceleration, and the net force exerted on it, are in the negative direction. As a result, the plane's speed decreases with time.

Your Turn
Find the plane's stopping distance if the magnitude of the average force exerted by the Foamcrete is doubled.
(Answers to **Your Turn** *problems are given in the back of the book.)*

Note again the care we take with the signs. The plane's acceleration is negative, hence the net force acting on it, $\vec{\mathbf{F}}$, is in the negative x direction. On the other hand, the magnitude of the force, F, is positive, as is always the case for magnitudes.

Finally, we end this section with an estimation problem.

EXAMPLE 5–2 Pitch Man: Estimate the Force on the Ball

A pitcher throws a 0.15-kg baseball, accelerating it from rest to a speed of about 90 mi/h. Estimate the force exerted by the pitcher on the ball.

Picture the Problem
We choose the x axis to point in the direction of the pitch. Also indicated in the sketch is the distance over which the pitcher accelerates the ball, Δx. Since we are interested only in the pitch, and not in the subsequent motion of the ball, we ignore the effects of gravity.

Strategy
We know the mass, so we can find the force with $F_x = ma_x$ if we can estimate the acceleration. To find the acceleration, we start with the fact that $v_0 = 0$ and $v \approx 90$ mi/h. In addition, we can see from the sketch that a reasonable estimate for Δx is about 2.0 m. Combining these results with the kinematic equation $v^2 = v_0{}^2 + 2a_x \Delta x$ yields the acceleration, which we then use to find the force.

Solution

1. Starting with the fact that 60 mi/h = 1 mi/min, perform a rough back-of-the-envelope conversion of 90 mi/h to meters per second: $v \approx 90\ \text{mi/h} = \dfrac{1.5\ \text{mi}}{\text{min}} \approx \dfrac{2400\ \text{m}}{60\ \text{s}} = 40\ \text{m/s}$
2. Solve $v^2 = v_0{}^2 + 2a_x \Delta x$ for the acceleration, a_x. Use the estimates $\Delta x \approx 2.0$ m and $v \approx 40$ m/s: $a_x = \dfrac{v^2 - v_0{}^2}{2\,\Delta x} \approx \dfrac{(40\ \text{m/s})^2 - 0}{2(2.0\ \text{m})} = 400\ \text{m/s}^2$
3. Find the corresponding force with $F_x = ma_x$: $F_x = ma_x \approx (0.15\ \text{kg})(400\ \text{m/s}^2) = 60\ \text{N} \approx 10\ \text{lb}$

Insight
On the one hand, this is a sizable force, especially when you consider that the ball itself weighs only about 1/3 lb. Thus, the pitcher exerts a force on the ball that is about 30 times greater than the force exerted by Earth's gravity. It follows that ignoring gravity during the pitch is a reasonable approximation.

continued on next page

continued from previous page

On the other hand, you might say that 10 lb isn't that much force for a person to exert. That's true, but this force is being exerted with an average speed of about 20 m/s, which means that the pitcher is actually generating about 1.5 horsepower—a sizeable power output for a person. We will cover power in detail in Chapter 7, and relate it to human capabilities.

Practice Problem

What is the approximate speed of the pitch if the force exerted by the pitcher is $\frac{1}{2}(60\text{ N}) = 30\text{ N}$? [**Answer:** 30 m/s or 60 mi/h.]

Some related homework problems: Problem 4, Problem 6

Another way to find the acceleration is to estimate the amount of time it takes to make the pitch. However, since the pitch is delivered so quickly—about 1/10 s—estimating the time would be more difficult than estimating the distance Δx.

5–4 Newton's Third Law of Motion

Nature never produces just one force at a time; *forces always come in pairs*. In addition, the forces in a pair, which always act on *different objects*, are equal in magnitude and opposite in direction. This is Newton's third law of motion.

Newton's Third Law

For every force that acts on an object, there is a reaction force acting on a different object that is equal in magnitude and opposite in direction.

In a somewhat more specific form:

> If object 1 exerts a force $\vec{\mathbf{F}}$ on object 2, then object 2 exerts a force $-\vec{\mathbf{F}}$ on object 1.

This law, more commonly known by its abbreviated form, "for every action there is an equal and opposite reaction," completes Newton's laws of motion.

Figure 5–8 illustrates some action-reaction pairs. Notice that there is always a reaction force, whether the action force pushes on something hard to move, like a refrigerator, or on something that moves with no friction, like an air-track cart. In some cases, the reaction force tends to be overlooked, as when the Earth exerts a *downward* gravitational force on the space shuttle, and the shuttle exerts an equal and opposite *upward* gravitational force on the Earth. Still, the reaction force always exists.

Another important aspect of the third law is that the action-reaction forces always act on *different* objects. This, again, is illustrated in Figure 5–8. Thus, in drawing a free-body diagram, only one of the action-reaction pair of forces would be drawn for a given object. The other force in the pair would appear in the free-body diagram of a different object. As a result, *the two forces do not cancel.*

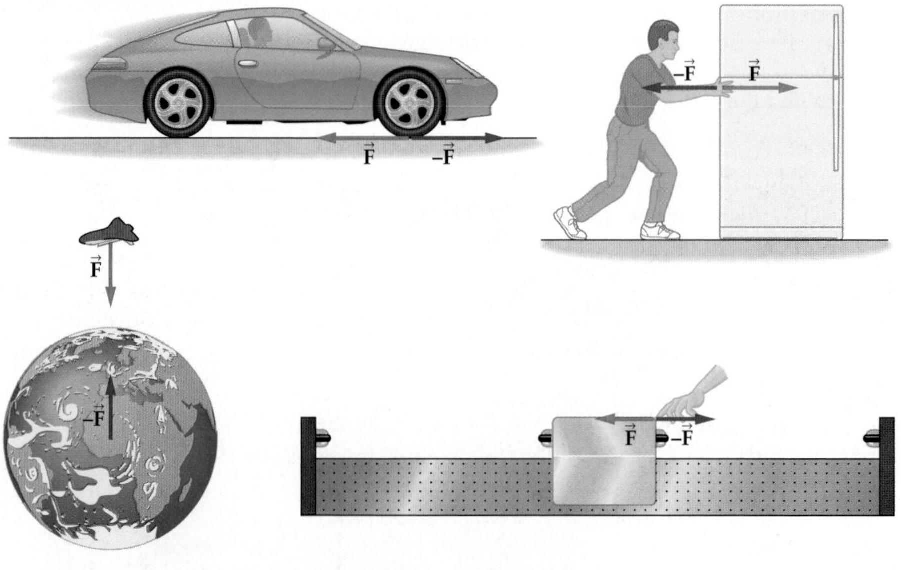

▶ **FIGURE 5–8 Examples of action-reaction force pairs**

For example, consider a car accelerating from rest, as in Figure 5–8. As the car's engine turns the wheels, the tires exert a force on the road. By the third law, the road exerts an equal and opposite force on the car's tires. It is this second force—which acts on the car through its tires—that propels the car forward. The force exerted by the tires on the road does not accelerate the car.

Since the action-reaction forces act on different objects, they generally produce different accelerations. This is the case in the next Example.

EXAMPLE 5–3 Tippy Canoe

Two groups of canoeists meet in the middle of a lake. After a brief visit, a person in canoe 1 pushes on canoe 2 with a force of 46 N to separate the canoes. If the mass of canoe 1 and its occupants is $m_1 = 150$ kg, and the mass of canoe 2 and its occupants is $m_2 = 250$ kg, **(a)** find the acceleration the push gives to each canoe. **(b)** What is the separation of the canoes after 1.2 s of pushing?

Picture the Problem

We have chosen the positive x direction to point from canoe 1 to canoe 2. With this choice, the force exerted on canoe 2 is $\vec{\mathbf{F}}_2 = (+46\text{ N})\hat{\mathbf{x}}$. By Newton's third law, the force exerted on the person in canoe 1, and thus on canoe 1 itself if the person is firmly seated, is $\vec{\mathbf{F}}_1 = (-46\text{ N})\hat{\mathbf{x}}$. For convenience, we have placed the origin at the point where the canoes touch.

Physical picture **Free-body diagrams**

Strategy

From Newton's third law, the force on canoe 1 is equal in magnitude to the force on canoe 2—the masses of the canoes are different, however, and therefore their accelerations are different as well. **(a)** We can find the acceleration of each canoe by solving $\Sigma F_x = ma_x$ for a_x. **(b)** The kinematic equation relating position to time, $x = x_0 + v_{0x}t + \frac{1}{2}a_xt^2$, can then be used to find the displacement of each canoe.

Solution

Part (a)

1. Use Newton's second law to find the acceleration of canoe 2: $$a_{2,x} = \frac{\sum F_{2,x}}{m_2} = \frac{46\text{ N}}{250\text{ kg}} = 0.18\text{ m/s}^2$$

2. Do the same calculation for canoe 1. Note that the acceleration of canoe 1 is in the negative direction: $$a_{1,x} = \frac{\sum F_{1,x}}{m_1} = \frac{-46\text{ N}}{150\text{ kg}} = -0.31\text{ m/s}^2$$

Part (b)

3. Use $x = x_0 + v_{0x}t + \frac{1}{2}a_xt^2$ to find the position of canoe 2 at $t = 1.2$ s. From the problem statement, we know the canoes start at the origin ($x_0 = 0$), and at rest ($v_{0x} = 0$): $$x_2 = \tfrac{1}{2}a_{2,x}t^2 = \tfrac{1}{2}(0.18\text{ m/s}^2)(1.2\text{ s})^2 = 0.13\text{ m}$$

4. Repeat the calculation for canoe 1: $$x_1 = \tfrac{1}{2}a_{1,x}t^2 = \tfrac{1}{2}(-0.31\text{ m/s}^2)(1.2\text{ s})^2 = -0.22\text{ m}$$

5. Subtract the two positions to find the separation of the canoes: $$x_2 - x_1 = 0.13\text{ m} - (-0.22\text{ m}) = 0.35\text{ m}$$

Insight

The same magnitude of force acts on each canoe, hence the lighter one has the greater acceleration and the greater displacement. If the heavier canoe were replaced by a large ship of great mass, both vessels would still accelerate as a result of the push. However, the acceleration of the large ship would be so small as to be practically imperceptible. In this case, it would appear as if only the canoe moved, whereas, in fact, both vessels move.

Practice Problem

If the mass of canoe 2 is increased, does its acceleration increase, decrease, or stay the same? Check your answer by calculating the acceleration for the case where canoe 2 is replaced by a 25,000-kg ship. [**Answer:** The acceleration will decrease. In this case, $a = 0.0018\text{ m/s}^2$.]

Some related homework problems: Problem 14, Problem 15

When objects are touching one another, the action-reaction forces are often referred to as **contact forces.** The behavior of contact forces is explored in the following Conceptual Checkpoint.

CONCEPTUAL CHECKPOINT 5–2 Contact Forces

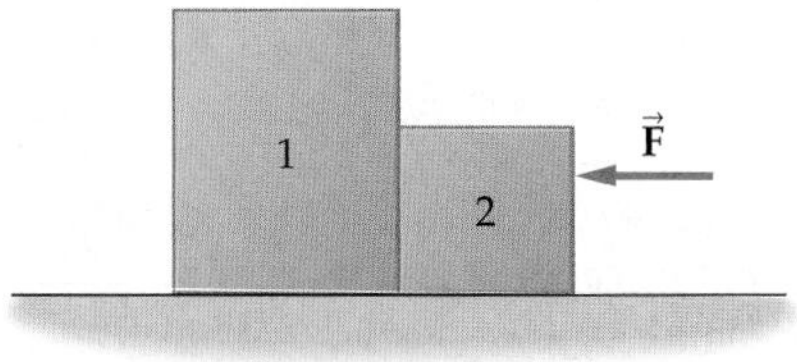

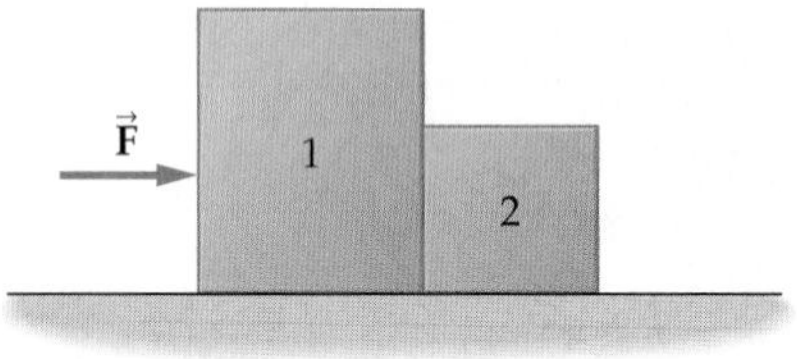

Two boxes—one large and heavy, the other small and light—rest on a smooth, level floor. You push with a force $\vec{F}$ on either the small box or the large box. Is the contact force between the two boxes **(a)** the same in either case, **(b)** larger when you push on the large box, or **(c)** larger when you push on the small box?

Reasoning and Discussion

Since the same force pushes on the boxes, you might think the force of contact is the same in both cases. It is not. What we can conclude, however, is that the boxes have the same acceleration in either case—the same net force acts on the same total mass, so the same acceleration, a, results.

To find the contact force between the boxes, we focus our attention on each box individually, and note that *Newton's second law must be satisfied for each of the boxes, just as it is for the entire two-box system.* For example, when the external force is applied to the small box, the only force acting on the *large* box (mass m_1) is the contact force; hence, the contact force must have a magnitude equal to m_1a. In the second case, the only force acting on the *small* box (mass m_2) is the contact force, and so the magnitude of the contact force is m_2a. Since m_1 is greater than m_2, it follows that the force of contact is larger when you push on the small box, m_1a, than when you push on the large box, m_2a.

To summarize, the contact force is larger when *it* must push the larger box.

Answer:

(c) The contact force is larger when you push on the small box.

In the next Example, we calculate a numerical value for the contact force in a system similar to that described in Conceptual Checkpoint 5–2. We also show explicitly that Newton's third law is required for a full analysis of this system.

EXAMPLE 5–4 When Push Comes to Shove

A box of mass $m_1 = 10.0$ kg rests on a smooth, horizontal floor next to a box of mass $m_2 = 5.00$ kg. If you push on box 1 with a horizontal force of magnitude $F = 20.0$ N, **(a)** what is the acceleration of the boxes? **(b)** What is the force of contact between the boxes?

Picture the Problem

We choose the x axis to be horizontal and pointing to the right. Thus, $\vec{F} = (20.0\text{ N})\hat{\mathbf{x}}$. The contact forces are labeled as follows: $\vec{F}_1$ is the contact force exerted on box 1; $\vec{F}_2$ is the contact force exerted on box 2. By Newton's third law, the contact forces have the same magnitude, f, but point in opposite directions. With our coordinate system, we have $\vec{F}_1 = -f\hat{\mathbf{x}}$ and $\vec{F}_2 = f\hat{\mathbf{x}}$.

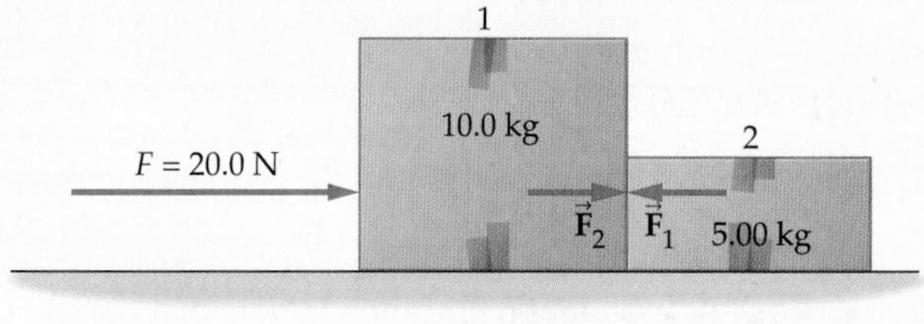

Physical picture

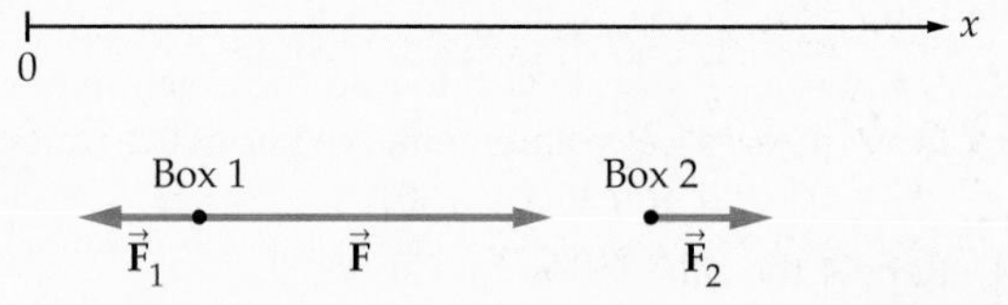

Free-body diagrams

INTERACTIVE FIGURE

Strategy

a. Since the two boxes are in contact, they have the same acceleration. We find this acceleration with Newton's second law; that is, we divide the net horizontal force by the total mass of the two boxes.

b. Now let's consider the system consisting solely of box 2. The mass in this case is 5.00 kg, and the only horizontal force acting on the system is $\vec{F}_2$. Thus, we can find f, the magnitude of $\vec{F}_2$, by requiring that box 2 have the acceleration found in part (a).

Solution

Part (a)

1. Find the net horizontal force acting on the two boxes. Note that $\vec{F}_1$ and $\vec{F}_2$ are equal in magnitude but opposite in direction. Hence, they sum to zero; $\vec{F}_1 + \vec{F}_2 = 0$:

$$\sum_{\substack{\text{both}\\\text{boxes}}} F_x = F = 20.0\text{ N}$$

2. Divide the net force by the total mass, $m_1 + m_2$, to find the acceleration of the boxes:

$$a_x = \frac{\sum F_x}{m_1 + m_2} = \frac{20.0\text{ N}}{(10.0\text{ kg} + 5.00\text{ kg})} = \frac{20.0\text{ N}}{15.0\text{ kg}} = 1.33\text{ m/s}^2$$

Part (b)

3. Find the net horizontal force acting on box 2, and set it equal to the mass of box 2 times its acceleration:

$$\sum_{\text{box2}} F_x = F_{2,x} = f = m_2 a_x$$

4. Determine the magnitude of the contact force, f, by substituting numerical values for m_2 and a_x:

$$f = m_2 a_x = (5.00\ \text{kg})(1.33\ \text{m/s}^2) = 6.67\ \text{N}$$

Insight

Since the net horizontal force acting on box 1 is $F - f = 20.0\ \text{N} - 6.67\ \text{N} = 13.3\ \text{N}$, it follows that its acceleration is $(13.3\ \text{N})/(10.0\ \text{kg}) = 1.33\ \text{m/s}^2$. Thus, as expected, box 1 and box 2 have precisely the same acceleration.

If box 2 were not present, the 20.0-N force acting on box 1 would give it an acceleration of $2.00\ \text{m/s}^2$. As it is, the contact force between the boxes slows box 1 so that its acceleration is less than $2.00\ \text{m/s}^2$, and accelerates box 2 so that its acceleration is greater than zero. The precise value of the contact force is simply the value that gives both boxes the same acceleration.

Practice Problem

Suppose the relative positions of the boxes are reversed, so that F pushes on the small box, as shown to the right. Calculate the contact force for this case, and show that the force is greater than 6.67 N, as expected from Conceptual Checkpoint 5-2. [**Answer:** The contact force in this case is 13.3 N, double its previous value. This follows because the box being pushed has twice the mass of the box that was pushed originally.]

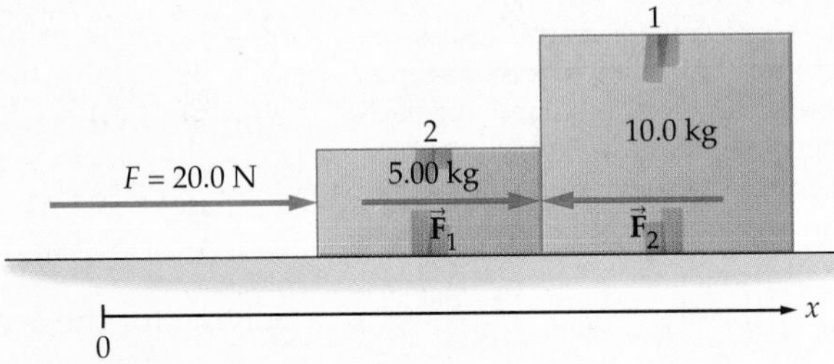

Some related homework problems: Problem 16, Problem 17

5–5 The Vector Nature of Forces: Forces in Two Dimensions

When we presented Newton's second law in Section 5–3, we said that an object's acceleration is equal to the net force acting on it divided by its mass. For example, if only a single force acts on an object, its acceleration is found to be in the same direction as the force. If more than one force acts on an object, experiments show that its acceleration is in the direction of the vector sum of the forces. Thus forces are indeed vectors, and they exhibit all the vector properties discussed in Chapter 3.

The mass of an object, on the other hand, is simply a positive number with no associated direction. It represents the amount of matter in an object.

As an example of the vector nature of forces, suppose two astronauts are using jet packs to push a 940-kg satellite toward the space shuttle, as shown in **Figure 5–9**. With the coordinate system indicated in the figure, astronaut 1 pushes in the positive x direction and astronaut 2 pushes in a direction 52° above the x axis. If astronaut 1 pushes with a force of magnitude $F_1 = 26$ N and astronaut 2 pushes with a force of magnitude $F_2 = 41$ N, what are the magnitude and direction of the satellite's acceleration?

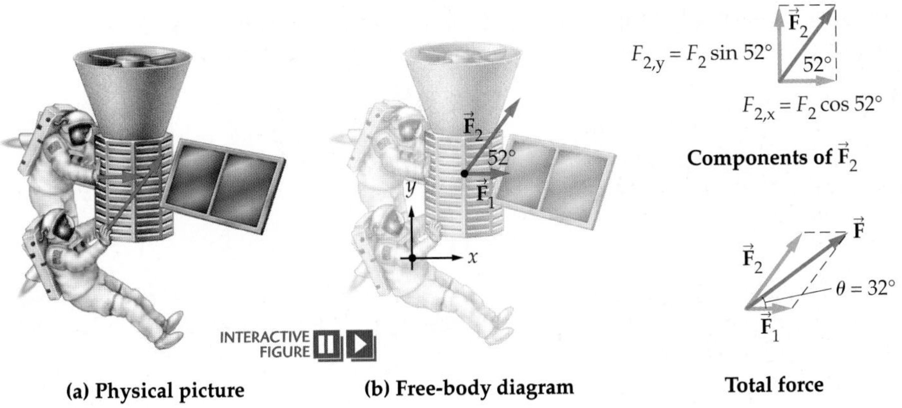

◀ FIGURE 5–9 Two astronauts pushing a satellite with forces that differ in magnitude and direction
The acceleration of the satellite can be found by calculating a_x and a_y separately, then combining these components to find a and θ.

The easiest way to solve a problem like this is to treat each coordinate direction independently of the other, just as we did many times when studying two-dimensional kinematics in Chapter 4. Thus, we first resolve each force into its x and y components. Referring to Figure 5–9, we see that for the x direction

$$F_{1,x} = F_1$$
$$F_{2,x} = F_2 \cos 52^\circ$$

For the y direction

$$F_{1,y} = 0$$
$$F_{2,y} = F_2 \sin 52^\circ$$

Next, we find the acceleration in the x direction by using the x component of Newton's second law:

$$\sum F_x = ma_x$$

Applied to this system, we have

$$\sum F_x = F_{1,x} + F_{2,x} = F_1 + F_2 \cos 52^\circ = 26\text{ N} + (41\text{ N})\cos 52^\circ = 51\text{ N}$$
$$= ma_x$$

Solving for the acceleration yields

$$a_x = \frac{\sum F_x}{m} = \frac{51\text{ N}}{940\text{ kg}} = 0.054\text{ m/s}^2$$

Similarly, in the y direction we start with

$$\sum F_y = ma_y$$

This gives

$$\sum F_y = F_{1,y} + F_{2,y} = 0 + F_2 \sin 52^\circ = (41\text{ N})\sin 52^\circ = 32\text{ N}$$
$$= ma_y$$

As a result, the y component of acceleration is:

$$a_y = \frac{\sum F_y}{m} = \frac{32\text{ N}}{940\text{ kg}} = 0.034\text{ m/s}^2$$

Thus, the satellite accelerates in both the x and the y directions. Its total acceleration has a magnitude of

$$a = \sqrt{a_x^2 + a_y^2} = \sqrt{(0.054\text{ m/s}^2)^2 + (0.034\text{ m/s}^2)^2} = 0.064\text{ m/s}^2$$

From Figure 5–9 we expect the total acceleration to be in a direction above the x axis but at an angle less than 52°. Straightforward calculation yields

$$\theta = \tan^{-1}\left(\frac{a_y}{a_x}\right) = \tan^{-1}\left(\frac{0.034\text{ m/s}^2}{0.054\text{ m/s}^2}\right) = \tan^{-1}(0.63) = 32^\circ$$

This is the same direction as the total force in Figure 5–9, as expected.

The following Example and Active Example give further practice with resolving force vectors and using Newton's second law in component form.

PROBLEM-SOLVING NOTE

Component–by–Component Application of Newton's Laws

Newton's laws can be applied to each coordinate direction independently of the others. Therefore, when drawing a free-body diagram, be sure to include a coordinate system. Once the forces are resolved into their x and y components, the second law can be solved for each component separately. Working in a component-by-component fashion is the systematic way of using Newton's laws.

EXAMPLE 5–5 Jack and Jill

Jack and Jill lift upward on a 1.3-kg pail of water, with Jack exerting a force $\vec{\mathbf{F}}_1$ of magnitude 7.0 N and Jill exerting a force $\vec{\mathbf{F}}_2$ of magnitude 11 N. Jill's force is exerted at an angle of 28° with the vertical, as shown below. At what angle θ with respect to the vertical should Jack exert his force if the pail is to accelerate straight upward?

Picture the Problem

Our physical picture and free-body diagram show the pail and the three forces acting on it, as well as the angles relative to the vertical. In the panels at the right, we show the x and y components of the forces $\vec{\mathbf{F}}_1$ and $\vec{\mathbf{F}}_2$. Notice, in particular, that $F_{1,x} = -F_1 \sin\theta$ and $F_{1,y} = F_1 \cos\theta$. Similarly, $F_{2,x} = F_2 \sin 28^\circ$ and $F_{2,y} = F_2 \cos 28^\circ$.

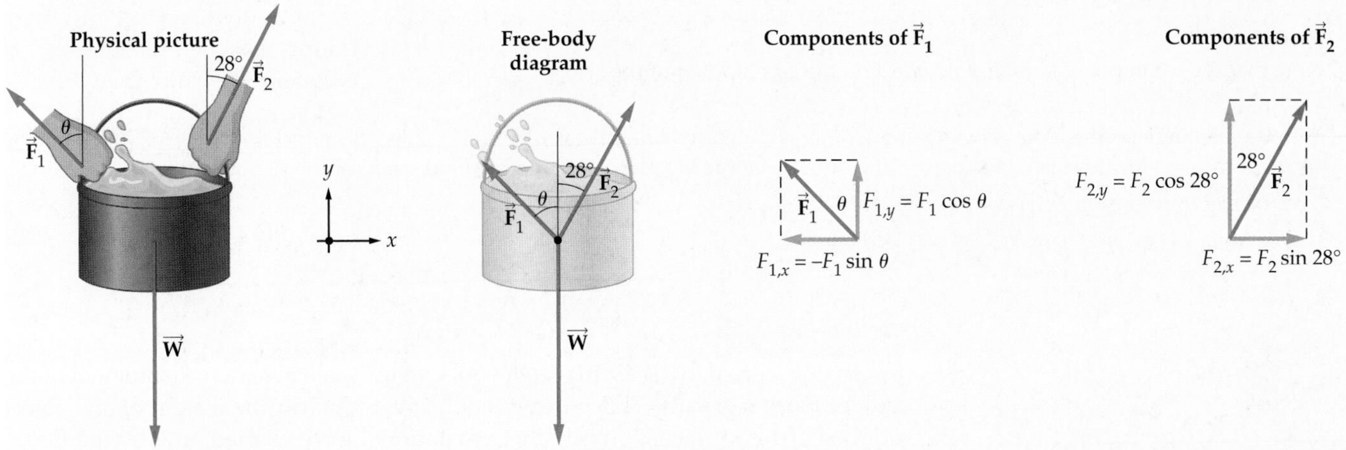

Strategy

We want the acceleration to be purely vertical. This means that the x component of acceleration must be zero, $a_x = 0$. For a_x to be zero it is necessary that the sum of forces in the x direction be zero, $\Sigma F_x = 0$. Since the x component of $\vec{\mathbf{F}}_1$ depends on the angle θ, the equation $\Sigma F_x = 0$ can be used to find θ.

Solution

1. Begin by writing out the x components of each force. Note that $\vec{\mathbf{W}}$ has no x component, and that the x component of $\vec{\mathbf{F}}_1$ points in the negative x direction:
$$F_{1,x} = -F_1 \sin \theta \qquad F_{2,x} = F_2 \sin 28° \qquad W_x = 0$$

2. Sum the x components of force and set equal to zero:
$$\sum F_x = -F_1 \sin \theta + F_2 \sin 28° + 0 = ma_x = 0 \quad \text{or}$$
Note that θ is the only unknown in this equation.
$$F_1 \sin \theta = F_2 \sin 28°$$

3. Solve for $\sin \theta$ and then for θ:
$$\sin \theta = \frac{F_2 \sin 28°}{F_1} = \frac{(11\ \text{N})\sin 28°}{7.0\ \text{N}} = 0.74$$
$$\theta = \sin^{-1}(0.74) = 48°$$

Insight

Note that only the y components of $\vec{\mathbf{F}}_1$ and $\vec{\mathbf{F}}_2$ contribute to the vertical acceleration of the pail. The x components of the applied forces only influence the horizontal motion, they have no effect on the vertical acceleration of the pail. In this case the horizontal components of the applied forces cancel, and hence the pail moves straight upward.

Practice Problem

At what angle must Jack exert his force for the pail to accelerate straight upward if **(a)** $\vec{\mathbf{F}}_2$ is at an angle of 19° with the vertical or **(b)** $\vec{\mathbf{F}}_2$ is at an angle of 35° with the vertical? **[Answer: (a)** 31°, **(b)** 64°**]**

Some related homework problems: Problem 22, Problem 26

ACTIVE EXAMPLE 5–2 Find the Speed of the Sled

A 4.60-kg sled is pulled across a smooth ice surface. The force acting on the sled is of magnitude 6.20 N and points in a direction 35.0° above the horizontal. If the sled starts at rest, how fast is it going after being pulled for 1.15 s?

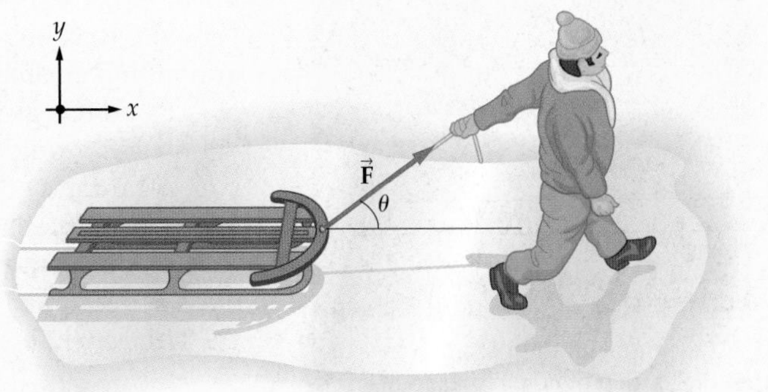

Solution *(Test your understanding by performing the calculations indicated in each step.)*

1. Find the x component of $\vec{\mathbf{F}}$: $F_x = 5.08\ \text{N}$
2. Apply Newton's second law to the x direction: $\sum F_x = F_x = ma_x$
3. Solve for the x component of acceleration: $a_x = 1.10\ \text{m/s}^2$
4. Use $v_x = v_{0x} + a_x t$ to find the speed of the sled: $v_x = 1.27\ \text{m/s}$

continued on next page

continued from previous page

Insight

Note that the y component of $\vec{\mathbf{F}}$ has no effect on the acceleration of the sled.

Your Turn

Suppose the angle of the force above the horizontal is decreased, and the sled is again pulled from rest for 1.15 s. **(a)** Is the final speed of the sled greater than, less than, or the same as before? Explain. **(b)** Find the final speed of the sled for the case $\theta = 25.0°$.

(*Answers to* **Your Turn** *problems are given in the back of the book.*)

5–6 Weight

When you step onto a scale to weigh yourself, the scale gives a measurement of the pull of Earth's gravity. This is your weight, W. Similarly, the weight of any object on the Earth's surface is simply the gravitational force exerted on it by the Earth.

- The weight, W, of an object on the Earth's surface is the gravitational force exerted on it by the Earth.

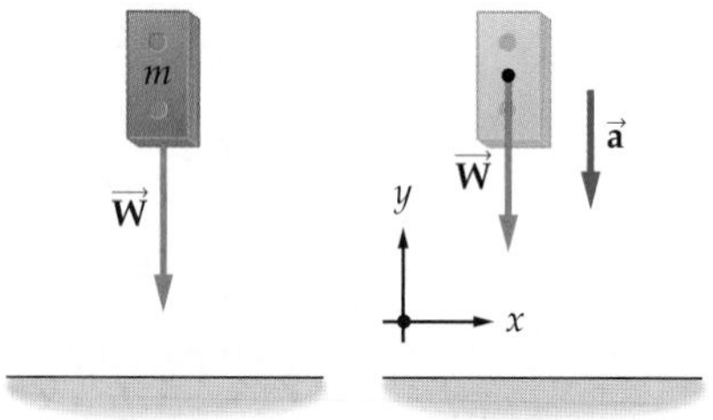

▲ **FIGURE 5–10 Weight and mass**
A brick of mass m has only one force acting on it in free fall—its weight, $\vec{\mathbf{W}}$. The resulting acceleration has a magnitude $a = g$; hence $W = mg$.

As we know from everyday experience, the greater the mass of an object, the greater its weight. For example, if you put a brick on a scale and weigh it, you might get a reading of 9.0 N. If you put a second, identical brick on the scale—which doubles the mass—you will find a weight of 2(9.0 N) = 18 N. Clearly, there must be a simple connection between weight, W, and mass, m.

To see exactly what this connection is, consider taking one of the bricks just mentioned and letting it drop in free fall. As indicated in **Figure 5–10**, the only force acting on the brick is its weight, W, which is downward. If we choose upward to be the positive direction, we have

$$\sum F_y = -W$$

In addition, we know from Chapter 2 that the brick moves downward with an acceleration of $g = 9.81\ \mathrm{m/s^2}$ regardless of its mass. Thus,

$$a_y = -g$$

Using these results in Newton's second law

$$\sum F_y = ma_y$$

we find

$$-W = -mg$$

Therefore, the weight of an object of mass m is $W = mg$:

Definition: Weight, *W*

$$W = mg \qquad \textbf{5–5}$$

SI unit: newton, N

▲ At the moment this picture was taken, the acceleration of both climbers was zero because the net force acting on them was zero. In particular, the upward forces exerted on the lower climber by the other climber and the ropes exactly cancel the downward force that gravity exerts on her.

Note that there is a clear distinction between weight and mass. Weight is a gravitational force, measured in newtons; mass is a measure of the inertia of an object, and it is given in kilograms. For example, if you were to travel to the Moon, your mass would not change—you would have the same amount of matter in you, regardless of your location. On the other hand, the gravitational force on the Moon's surface is less than the gravitational force on the Earth's surface. As a result, you would weigh less on the Moon than on the Earth, even though your mass is the same.

To be specific, on Earth an 81.0-kg person has a weight given by

$$W_{\mathrm{Earth}} = mg_{\mathrm{Earth}} = (81.0\ \mathrm{kg})(9.81\ \mathrm{m/s^2}) = 795\ \mathrm{N}$$

In contrast, the same person on the Moon, where the acceleration of gravity is $1.62\ \mathrm{m/s^2}$, weighs only

$$W_{\mathrm{Moon}} = mg_{\mathrm{Moon}} = (81.0\ \mathrm{kg})(1.62\ \mathrm{m/s^2}) = 131\ \mathrm{N}$$

This is roughly one-sixth the weight on Earth. If, sometime in the future, there is a Lunar Olympics, the Moon's low gravity would be a boon for pole-vaulters, gymnasts, and others.

Finally, since weight is a force—which is a vector quantity—it has both a magnitude and a direction. Its magnitude, of course, is mg, and its direction is simply the direction of gravitational acceleration. Thus, if $\vec{\mathbf{g}}$ denotes a vector of magnitude g, pointing in the direction of free-fall acceleration, the weight of an object can be written in vector form as follows:

$$\vec{\mathbf{W}} = m\vec{\mathbf{g}}$$

We use the weight vector and its magnitude, mg, in the next example.

EXAMPLE 5–6 Where's the Fire?

The fire alarm goes off, and a 97-kg fireman slides 3.0 m down a pole to the ground floor. Suppose the fireman starts from rest, slides with constant acceleration, and reaches the ground floor in 1.2 s. What was the upward force $\vec{\mathbf{F}}$ exerted by the pole on the fireman?

Picture the Problem

Our sketch shows the fireman sliding down the pole and the two forces acting on him: the upward force exerted by the pole, $\vec{\mathbf{F}}$, and the downward force of gravity, $\vec{\mathbf{W}}$. We choose the positive y direction to be upward, therefore $\vec{\mathbf{F}} = F\hat{\mathbf{y}}$ and $\vec{\mathbf{W}} = (-mg)\hat{\mathbf{y}}$. In addition, we choose $y = 0$ to be at ground level.

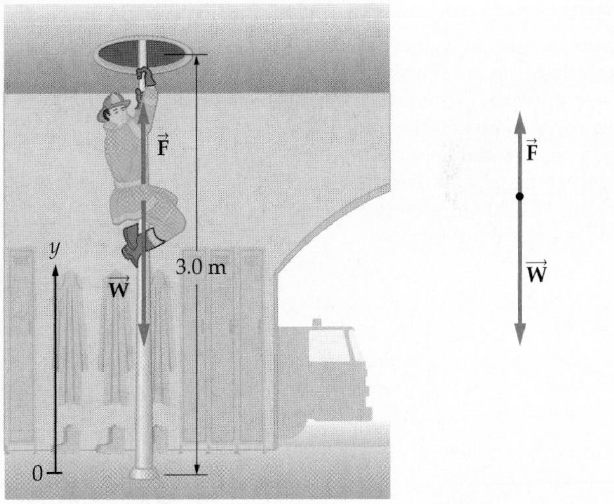

Physical picture **Free-body diagram**

Strategy

The basic idea in approaching this problem is to apply Newton's second law to the y direction; $\Sigma F_y = ma_y$. The acceleration is not given directly, but we can find it using the kinematic equation $y = y_0 + v_{0y}t + \frac{1}{2}a_y t^2$. Substituting the result for a_y into Newton's second law, along with $W_y = -W = -mg$, allows us to solve for the unknown force, $\vec{\mathbf{F}}$.

Solution

1. Solve $y = y_0 + v_{0y}t + \frac{1}{2}a_y t^2$ for a_y, using the fact that $v_{0y} = 0$:

$$y = y_0 + v_{0y}t + \tfrac{1}{2}a_y t^2 = y_0 + \tfrac{1}{2}a_y t^2$$
$$a_y = \frac{2(y - y_0)}{t^2}$$

2. Substitute $y = 0$, $y_0 = 3.0$ m, and $t = 1.2$ s to find the acceleration:

$$a_y = \frac{2(0 - 3.0\text{ m})}{(1.2\text{ s})^2} = -4.2\text{ m/s}^2$$

3. Sum the forces in the y direction:

$$\sum F_y = F - mg$$

4. Set the sum of the forces equal to mass times acceleration:

$$F - mg = ma_y$$

5. Solve for F, the y component of the force exerted by the pole. Use the result for F to write the force vector $\vec{\mathbf{F}}$:

$$F = mg + ma_y = m(g + a_y)$$
$$= (97\text{ kg})(9.81\text{ m/s}^2 - 4.2\text{ m/s}^2) = 540\text{ N}$$
$$\vec{\mathbf{F}} = (540\text{ N})\hat{\mathbf{y}}$$

Insight

How is it that the pole exerts a force on the fireman? Well, by wrapping his arms and legs around the pole as he slides, the fireman exerts a downward force on the pole. By Newton's third law, the pole exerts an upward force of equal magnitude on the fireman. These forces are due to friction, which we shall study in detail in Chapter 6.

Practice Problem

What is the fireman's acceleration if the force exerted on him by the pole is 650 N? [**Answer:** $a_y = -3.1\text{ m/s}^2$]

Some related homework problems: Problem 29, Problem 33

Apparent Weight

We have all had the experience of riding in an elevator and feeling either heavy or light, depending on its motion. For example, when an elevator moving downward comes to rest by accelerating upward, we feel heavier. On the other hand, we feel lighter when an elevator moving upward comes to rest by accelerating downward. In short, the motion of an elevator can give rise to an **apparent weight** that differs from our true weight. Why?

The reason is that our sensation of weight in this case is due to the force exerted on our feet by the floor of the elevator. If this force is greater than our weight, mg, we feel heavy; if it is less than mg, we feel light.

As an example, imagine you are in an elevator that is moving with an upward acceleration a, as indicated in **Figure 5–11**. Two forces act on you: (i) your weight, W, acting downward; and (ii) the upward normal force exerted on your feet by the floor of the elevator. Let's call the second force W_a, since it represents your apparent weight—that is, W_a is the force that pushes upward on your feet and gives you the sensation of your "weight" pushing down on the floor. We can find W_a by applying Newton's second law to the vertical direction.

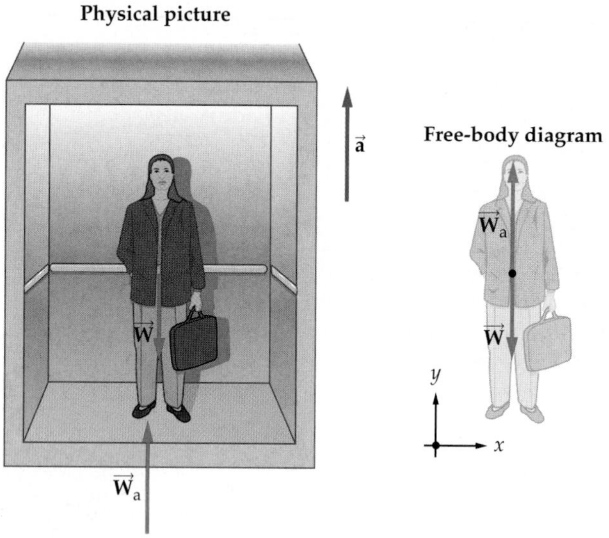

▶ **FIGURE 5–11 Apparent weight**
A person rides in an elevator that is accelerating upward. Because the acceleration is upward, the net force must also be upward. As a result, the force exerted on the person by the floor of the elevator, $\vec{W}_a$, must be greater than the person's weight, $\vec{W}$. This means that the person feels heavier than normal.

To be specific, the sum of the forces acting on you is

$$\sum F_y = W_a - W$$

By Newton's second law, this sum must equal ma_y. Since $a_y = a$, we find

$$W_a - W = ma$$

Solving for the apparent weight, W_a, yields

$$\begin{aligned} W_a &= W + ma \\ &= mg + ma = m(g + a) \end{aligned} \qquad 5\text{–}6$$

Note that W_a is greater than your weight, mg, and hence you feel heavier. In fact, your apparent weight is precisely what it would be if you were suddenly "transported" to a planet where the acceleration of gravity is $g + a$ instead of g.

On the other hand, if the elevator accelerates downward, so that $a_y = -a$, your apparent weight is found by simply replacing a with $-a$ in Equation 5–6:

$$\begin{aligned} W_a &= W - ma \\ &= mg - ma = m(g - a) \end{aligned} \qquad 5\text{–}7$$

In this case you feel lighter than usual.

We explore these results in the next Example, in which we consider weighing a fish on a scale. The reading on the scale is equal to the upward force it exerts on an object. Thus, the upward force exerted by the scale is the apparent weight, W_a.

EXAMPLE 5–7 How Much Does the Salmon Weigh?

As part of an attempt to combine physics and biology in the same class, an instructor asks students to weigh a 5.0-kg salmon by hanging it from a fish scale attached to the ceiling of an elevator. What is the apparent weight of the salmon, $\vec{\mathbf{W}}_a$, if the elevator **(a)** is at rest, **(b)** moves with an upward acceleration of 2.5 m/s^2, or **(c)** moves with a downward acceleration of 3.2 m/s^2?

Picture the Problem

The free-body diagram for the salmon shows the weight of the salmon, $\vec{\mathbf{W}}$, and the force exerted by the scale, $\vec{\mathbf{W}}_a$. Note that upward is the positive direction. Therefore, the y component of $\vec{\mathbf{W}}$ is $-W = -mg$ and the y component of $\vec{\mathbf{W}}_a$ is W_a.

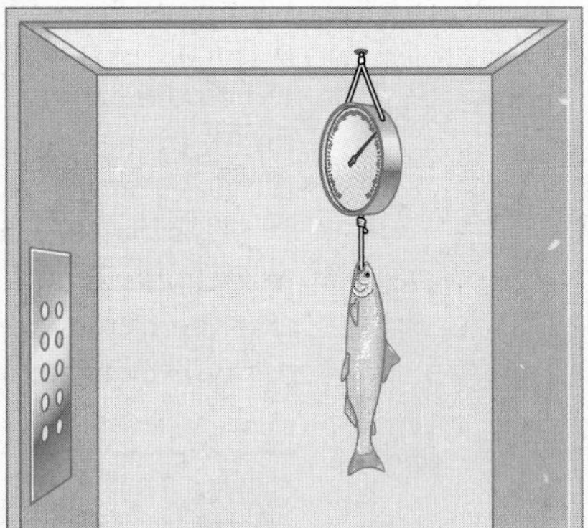

Physical picture

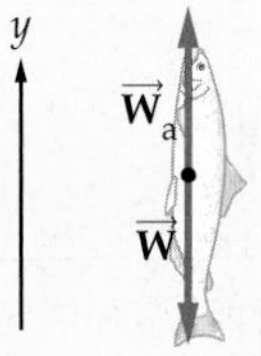

Free-body diagram

Strategy

We know the weight, $W = mg$, and the acceleration, a. To find the apparent weight, W_a, we use $\Sigma F_y = ma_y$. **(a)** Set $a_y = 0$. **(b)** Set $a_y = 2.5$ m/s^2. **(c)** Set $a_y = -3.2$ m/s^2.

Solution

Part (a)

1. Sum the y component of the forces and set equal to mass times the y component of acceleration, with $a_y = 0$:

$$\sum F_y = W_a - W = ma_y = 0$$

2. Solve for W_a, then write the vector $\vec{\mathbf{W}}_a$:

$$W_a = W = mg = (5.0\text{ kg})(9.81\text{ m/s}^2) = 49\text{ N}$$
$$\vec{\mathbf{W}}_a = (49\text{ N})\hat{\mathbf{y}}$$

Part (b)

3. Again, sum the forces and set equal to mass times acceleration, this time with $a_y = a = 2.5$ m/s^2:

$$\sum F_y = W_a - W = ma_y = ma$$

4. Solve for W_a, then write the vector $\vec{\mathbf{W}}_a$:

$$W_a = W + ma$$
$$= mg + ma = 49\text{ N} + (5.0\text{ kg})(2.5\text{ m/s}^2) = 62\text{ N}$$
$$\vec{\mathbf{W}}_a = (62\text{ N})\hat{\mathbf{y}}$$

Part (c)

5. Finally, sum the forces and set equal to mass times acceleration, with $a_y = -a = -3.2$ m/s^2:

$$\sum F_y = W_a - W = ma_y = -ma$$

6. Solve for W_a, then write the vector $\vec{\mathbf{W}}_a$:

$$W_a = W - ma$$
$$= mg - ma = 49\text{ N} - (5.0\text{ kg})(3.2\text{ m/s}^2) = 33\text{ N}$$
$$\vec{\mathbf{W}}_a = (33\text{ N})\hat{\mathbf{y}}$$

Insight

When the salmon is at rest, or moving with constant velocity, its acceleration is zero and the apparent weight is equal to the actual weight, mg. In part (b) the apparent weight is greater than the actual weight because the scale must exert an upward force capable not only of supporting the salmon, but of accelerating it upward as well. In part (c) the apparent weight is less than the actual weight. In this case the net force acting on the salmon is downward, and hence its acceleration is downward.

Practice Problem

What is the elevator's acceleration if the scale gives a reading of **(a)** 55 N or **(b)** 45 N? [**Answer:** **(a)** $a_y = 1.2$ m/s^2, **(b)** $a_y = -0.80$ m/s^2]

Some related homework problems: Problem 31, Problem 32

Let's return for a moment to Equation 5–7:

$$W_a = m(g - a)$$

This result indicates that a person feels lighter than normal when riding in an elevator with a downward acceleration a. In particular, if the elevator's downward acceleration is g—that is, if the elevator is in free fall—it follows that

REAL-WORLD PHYSICS: BIO
Simulating weightlessness

▲ Astronaut candidates pose for a floating class picture during weightlessness training aboard the "vomit comet."

$W_a = m(g - g) = 0$. Thus, a person feels "weightless" (zero apparent weight) in a freely falling elevator!

NASA uses this effect when training astronauts. Trainees are sent aloft in a KC-135 airplane affectionately known as the "vomit comet" (since many trainees experience nausea along with the weightlessness). To generate an experience of weightlessness, the plane flies on a parabolic path—the same path followed by a projectile in free fall. Each round of weightlessness lasts about half a minute, after which the plane pulls up to regain altitude and start the cycle again. On a typical flight, trainees experience about 40 cycles of weightlessness. Many scenes in the movie *Apollo 13* were shot in 30-second takes aboard the vomit comet.

This idea of free-fall weightlessness applies to more than just the vomit comet. In fact, astronauts in orbit experience weightlessness for the same reason—they and their craft are actually in free fall. As we shall see in detail in Chapter 12 (Gravity), orbital motion is just a special case of free fall.

CONCEPTUAL CHECKPOINT 5–3 Elevator Ride

If you ride in an elevator moving upward with constant speed, is your apparent weight **(a)** the same as, **(b)** greater than, or **(c)** less than mg?

Reasoning and Discussion

If the elevator is moving in a straight line with constant speed, its acceleration is zero. Now, if the acceleration is zero, the net force must also be zero. Hence, the upward force exerted by the floor of the elevator, W_a, must equal the downward force of gravity, mg. As a result, your apparent weight is equal to mg.

Note that this conclusion agrees with Equations 5–6 and 5–7, with $a = 0$.

Answer:

(a) Your apparent weight is the same as mg.

5–7 Normal Forces

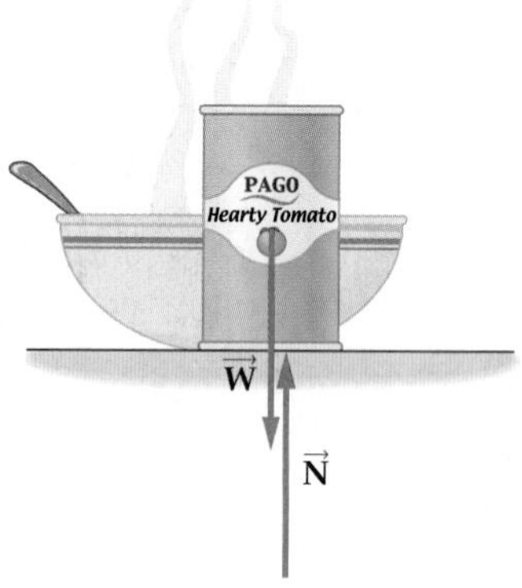

Physical picture

Free-body diagram

▲ **FIGURE 5–12 The normal force is perpendicular to a surface**

A can of soup rests on a kitchen counter, which exerts a normal (perpendicular) force, $\vec{N}$, to support it. In the special case shown here, the normal force is equal in magnitude to the weight, $W = mg$, and opposite in direction.

As you get ready for lunch, you take a can of soup from the cupboard and place it on the kitchen counter. The can is now at rest, which means that its acceleration is zero, so the net force acting on it is also zero. Thus, you know that the downward force of gravity is being opposed by an upward force exerted by the counter, as shown in **Figure 5–12**. As we have mentioned before, this force is referred to as the **normal force, $\vec{N}$**. The reason the force is called normal is that it is *perpendicular to the surface*, and in mathematical terms, normal simply means perpendicular.

The origin of the normal force is the interaction between atoms in a solid that act to maintain its shape. When the can of soup is placed on the countertop, for example, it causes an imperceptibly small compression of the surface of the counter. This is similar to compressing a spring, and just like a spring, the countertop exerts a force to oppose the compression. Therefore, the greater the weight placed on the countertop, the greater the normal force it exerts to oppose being compressed.

In the example of the soup can and the countertop, the magnitude of the normal force is equal to the weight of the can. This is a special case, however. In general, the normal force may be greater than or less than the weight of an object.

To see how this can come about, consider pulling a 12.0-kg suitcase across a smooth floor by exerting a force, $\vec{F}$, at an angle θ above the horizontal. The weight of the suitcase is $mg = (12.0\text{ kg})(9.81\text{ m/s}^2) = 118\text{ N}$. The normal force will have a magnitude less than this, however, because the force $\vec{F}$ has an upward component that supports part of the suitcase's weight. To be specific, suppose that $\vec{F}$ has a magnitude of 45.0 N and that $\theta = 20.0°$. What is the normal force exerted by the floor on the suitcase?

The situation is illustrated in **Figure 5–13**, where we show the three forces acting on the suitcase: (i) the weight of the suitcase, $\vec{W}$, (ii) the force $\vec{F}$, and (iii) the normal force, $\vec{N}$. We also indicate a typical coordinate system in the figure,

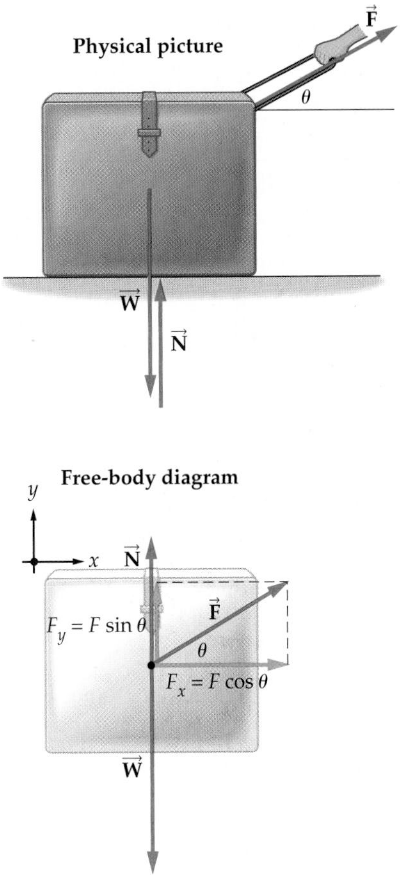

▶ **FIGURE 5–13 The normal force may differ from the weight**
A suitcase pulled across the floor by an applied force of magnitude F, directed at an angle θ above the horizontal. As a result of the upward component of $\vec{\mathbf{F}}$, the normal force $\vec{\mathbf{N}}$ will have a magnitude less than the weight of the suitcase.

with the x axis horizontal and the y axis vertical. Now, the key to solving a problem like this is to realize that since the suitcase does not move in the y direction, its y component of acceleration is zero; that is, $a_y = 0$. It follows, from Newton's second law, that the sum of the y components of force must also equal zero; that is, $\Sigma F_y = ma_y = 0$. Using this condition, we can solve for the one force that is unknown, $\vec{\mathbf{N}}$.

To find $\vec{\mathbf{N}}$, then, we start by writing out the y component of each force. For the weight we have $W_y = -mg = -118$ N; for the applied force, $\vec{\mathbf{F}}$, the y component is $F_y = F \sin 20.0° = (45.0\text{ N}) \sin 20.0° = 15.4$ N; finally, the y component of the normal force is $N_y = N$. Setting the sum of the y components of force equal to zero yields

$$\sum F_y = W_y + F_y + N_y = -mg + F\sin 20.0° + N = 0$$

Solving for N gives

$$N = mg - F\sin 20.0° = 118\text{ N} - 15.4\text{ N} = 103\text{ N}$$

In vector form,

$$\vec{\mathbf{N}} = \mathrm{N}_y\hat{\mathbf{y}} = (103\text{ N})\hat{\mathbf{y}}$$

Thus, as mentioned, the normal force has a magnitude less than $mg = 118$ N because the y component of $\vec{\mathbf{F}}$, $F_y = F \sin 20.0°$, supports part of the weight. In the following Example, however, the applied forces cause the normal force to be greater than the weight.

EXAMPLE 5–8 Ice Block

A 6.0-kg block of ice is acted on by two forces, $\vec{\mathbf{F}}_1$ and $\vec{\mathbf{F}}_2$, as shown in the diagram. If the magnitudes of the forces are $F_1 = 13$ N and $F_2 = 11$ N, find **(a)** the acceleration of the ice and **(b)** the normal force exerted on it by the table.

Picture the Problem

The sketch shows our choice of coordinate system, as well as all the forces acting on the block of ice. Note that $\vec{\mathbf{F}}_1$ has a positive x component and a negative y component; $\vec{\mathbf{F}}_2$ has negative x and y components. The weight and the normal force have only y components, therefore $W_x = 0$, $W_y = -W = -mg$, $N_x = 0$, and $N_y = N$.

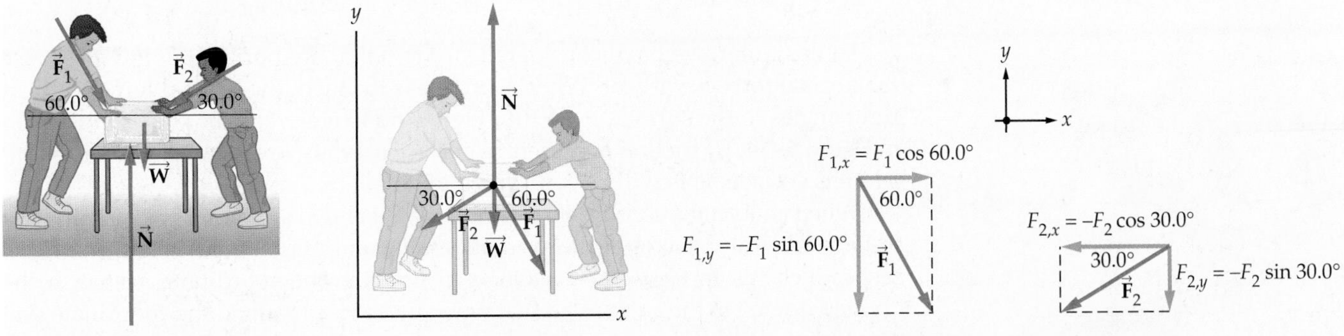

Physical picture — Free-body diagram — Components of $\vec{\mathbf{F}}_1$ (enlarged) — Components of $\vec{\mathbf{F}}_2$ (enlarged)

Strategy

The basic idea in this problem is to apply Newton's second law to the x and y directions separately. **(a)** The block can accelerate only in the horizontal direction; thus we find the acceleration by solving $\Sigma F_x = ma_x$ for a_x. **(b)** There is no motion in the y direction, and therefore the acceleration in the y direction is zero. Hence, we can find the normal force $\vec{\mathbf{N}}$ by setting $\Sigma F_y = ma_y = 0$.

continued on next page

continued from previous page

Solution

Part (a)

1. Write out the x component of each force:
$$F_{1,x} = F_1 \cos 60.0° = (13\ \text{N}) \cos 60.0° = 6.5\ \text{N}$$
$$F_{2,x} = -F_2 \cos 30.0° = -(11\ \text{N}) \cos 30.0° = -9.5\ \text{N}$$
$$N_x = 0 \qquad W_x = 0$$

2. Sum the x components of force:
$$\sum F_x = F_{1,x} + F_{2,x} + N_x + W_x$$
$$= 6.5\ \text{N} - 9.5\ \text{N} + 0 + 0 = -3.0\ \text{N}$$

3. Divide by the mass to obtain the acceleration:
$$a_x = \frac{\sum F_x}{m} = \frac{-3.0\ \text{N}}{6.0\ \text{kg}} = -0.50\ \text{m/s}^2$$
$$\vec{\mathbf{a}} = (-0.50\ \text{m/s}^2)\hat{\mathbf{x}}$$

Part (b)

4. Write out the y component of each force:
$$F_{1,y} = -F_1 \sin 60° = -(13\ \text{N}) \sin 60.0° = -11\ \text{N}$$

The only force we don't know is the normal. We represent its magnitude by N:
$$F_{2,y} = -F_2 \sin 30° = -(11\ \text{N}) \sin 30.0° = -5.5\ \text{N}$$
$$N_y = N \qquad W_y = -W = -mg$$

5. Sum the y components of force:
$$\sum F_y = F_{1,y} + F_{2,y} + N_y + W_y$$
$$= -11\ \text{N} - 5.5\ \text{N} + N - mg$$

6. Set this sum equal to 0 since the acceleration in the y direction is zero, and solve for N:
$$-11\ \text{N} - 5.5\ \text{N} + N - mg = 0$$
$$N = 11\ \text{N} + 5.5\ \text{N} + mg$$
$$= 11\ \text{N} + 5.5\ \text{N} + (6.0\ \text{kg})(9.81\ \text{m/s}^2) = 75\ \text{N}$$

7. Finally, we write the normal force in vector form:
$$\vec{\mathbf{N}} = (75\ \text{N})\hat{\mathbf{y}}$$

Insight

The block accelerates to the left, even though the force acting to the right, $\vec{\mathbf{F}}_1$, has a greater magnitude than the force acting to the left, $\vec{\mathbf{F}}_2$. This is because $\vec{\mathbf{F}}_2$ has the greater x component. Also, note that the normal force is greater in magnitude than the weight, $mg = 59$ N.

In general, the normal force exerted by a surface is just as large as is necessary to prevent motion of an object into the surface. If the required force is larger than the material can provide, the surface will break.

Practice Problem

At what angle must $\vec{\mathbf{F}}_2$ be applied if the block of ice is to have zero acceleration? [**Answer:** $a_x = 0$ implies $F_1 \cos 60.0° = F_2 \cos \theta$. Thus, $\theta = 54°$.]

Some related homework problems: Problem 37, Problem 43

▲ **FIGURE 5–14 An object on an inclined surface**

The normal force $\vec{\mathbf{N}}$ is always at right angles to the surface; hence, it is not always in the vertical direction.

To this point, we have considered surfaces that are horizontal, in which case the normal force is vertical. When a surface is inclined, the normal force is still at right angles to the surface, even though it is no longer vertical. This is illustrated in **Figure 5–14**. (If friction is present, a surface may also exert a force that is parallel to its surface. This will be considered in detail in Chapter 6.)

When choosing a coordinate system for an inclined surface, it is generally best to have the x and y axes of the system parallel and perpendicular to the surface, respectively, as in **Figure 5–15**. One can imagine the coordinate system to be "bolted down" to the surface, so that when the surface is tilted the coordinate system tilts along with it.

With this choice of coordinate system, there is no motion in the y direction, even on the inclined surface, and the normal force points in the positive y direction. Thus, the condition that determines the normal force is still $\Sigma F_y = ma_y = 0$, as before. In addition, if the object slides on the surface, its motion is purely in the x direction.

Finally, if the surface is inclined by an angle θ, note that the weight—which is still vertically downward—is at the same angle θ with respect to the negative y axis, as shown in Figure 5–15. As a result, the x and y components of the weight are

$$W_x = W \sin \theta = mg \sin \theta \qquad \textbf{5–8}$$

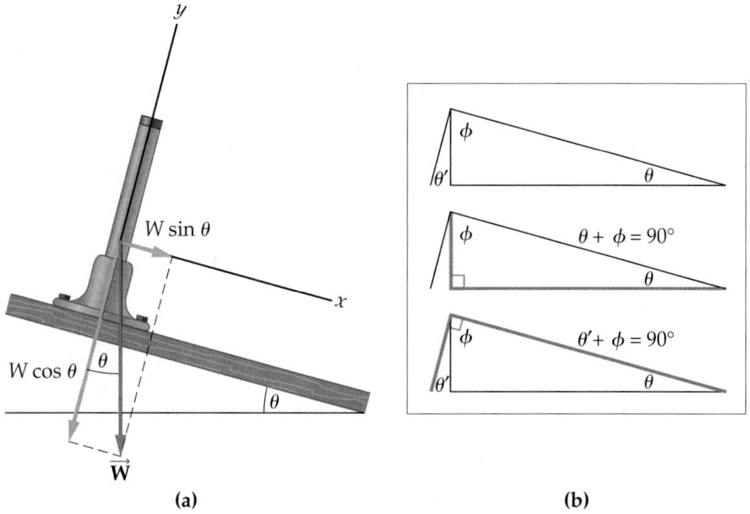

◀ FIGURE 5–15 Components of the weight on an inclined surface

Whenever a surface is tilted by an angle θ, the weight $\vec{\mathbf{W}}$ makes the same angle θ with respect to the negative y axis. This is proven at right, where we show that $\theta + \phi = 90°$, and that $\theta' + \phi = 90°$. From these results it follows that $\theta' = \theta$. The component of the weight perpendicular to the surface is $W_y = -W\cos\theta$; the component parallel to the surface is $W_x = W\sin\theta$.

and

$$W_y = -W\cos\theta = -mg\cos\theta \qquad \textbf{5–9}$$

Let's quickly check some special cases of these results. First, if $\theta = 0$ the surface is horizontal, and we find $W_x = 0$, $W_y = -mg$, as expected. Second, if $\theta = 90°$ the surface is vertical; therefore, the weight is parallel to the surface, pointing in the positive x direction. In this case, $W_x = mg$ and $W_y = 0$.

The next Example shows how to use the weight components to find the acceleration of an object on an inclined surface.

EXAMPLE 5–9 Toboggan to the Bottom

A child of mass m rides on a toboggan down a slick, ice-covered hill inclined at an angle θ with respect to the horizontal. **(a)** What is the acceleration of the child? **(b)** What is the normal force exerted on the child by the toboggan?

Picture the Problem

We choose the x axis to be parallel to the slope, with the positive direction pointing downhill. Similarly, we choose the y axis to be perpendicular to the slope, pointing up and to the right. With these choices, the x component of $\vec{\mathbf{W}}$ is positive, $W_x = W\sin\theta$, and its y component is negative, $W_y = -W\cos\theta$. Finally, the x component of the normal force is zero, $N_x = 0$, and its y component is positive, $N_y = N$.

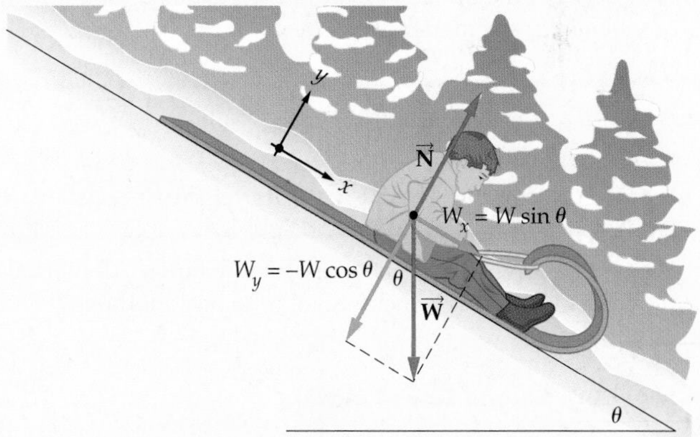

Strategy

Note that only two forces act on the child: (i) the weight, $\vec{\mathbf{W}}$ and (ii) the normal force, $\vec{\mathbf{N}}$. **(a)** We find the child's acceleration by solving $\Sigma F_x = ma_x$ for a_x. **(b)** Because there is no motion in the y direction, the y component of acceleration is zero. Therefore, we can find the normal force by setting $\Sigma F_y = ma_y = 0$.

Solution

Part (a)

1. Write out the x components of the forces acting on the child: $N_x = 0 \qquad W_x = W\sin\theta = mg\sin\theta$
2. Sum the x components of the forces and set equal to ma_x: $\sum F_x = N_x + W_x = mg\sin\theta = ma_x$
3. Divide by the mass m to find the acceleration in the x direction: $a_x = \dfrac{\sum F_x}{m} = \dfrac{mg\sin\theta}{m} = g\sin\theta$

continued on next page

continued from previous page

Part (b)

4. Write out the y components of the forces acting on the child: $N_y = N \quad W_y = -W\cos\theta = -mg\cos\theta$

5. Sum the y components of the forces and set the sum equal to zero, since $a_y = 0$: $\sum F_y = N_y + W_y = N - mg\cos\theta = ma_y = 0$

6. Solve for the magnitude of the normal force, N: $N - mg\cos\theta = 0$ or $N = mg\cos\theta$

Write the normal force in vector form: $\vec{\mathbf{N}} = (mg\cos\theta)\hat{\mathbf{y}}$

Insight

Note that for θ between 0 and 90° the acceleration of the child is *less* than the acceleration of gravity. This is because only a *component* of the weight is causing the acceleration.

Let's check some special cases of our general result, $a_x = g\sin\theta$. First, let $\theta = 0$. In this case, we find zero acceleration; $a_x = g\sin 0 = 0$. This makes sense because with $\theta = 0$ the hill is actually level, and we don't expect an acceleration. Second, let $\theta = 90°$. In this case, the hill is vertical, and the toboggan should drop straight down in free fall. This also agrees with our general result; $a_x = g\sin 90° = g$.

Practice Problem

What is the child's acceleration if its mass is doubled to $2m$? [**Answer:** The acceleration is still $a_x = g\sin\theta$. As in free fall, the acceleration produced by gravity is independent of mass.]

Some related homework problems: Problem 38, Problem 42

Chapter Summary

5–1 Force and Mass

Force

A push or a pull.

Mass

A measure of the difficulty in accelerating an object. Equivalently, a measure of the quantity of matter in an object.

5–2 Newton's First Law of Motion

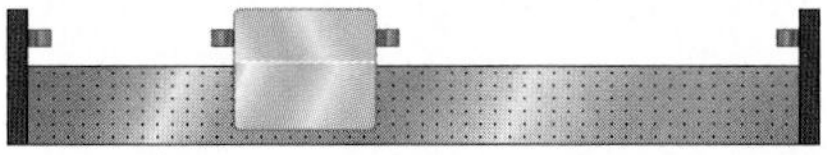

First Law (Law of Inertia)

If the net force on an object is zero, its velocity is constant.

Inertial Frame of Reference

Frame of reference in which the first law holds. All inertial frames of reference move with constant velocity relative to one another.

5–3 Newton's Second Law of Motion

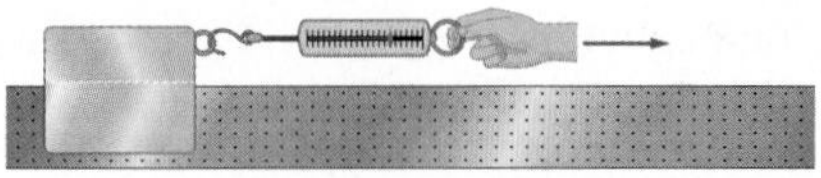

Second Law

An object of mass m has an acceleration $\vec{\mathbf{a}}$ given by the net force $\Sigma\vec{\mathbf{F}}$ divided by m. That is

$$\vec{\mathbf{a}} = \sum\vec{\mathbf{F}}/m \qquad 5\text{–}1$$

Component Form

$$a_x = \sum F_x/m \qquad a_y = \sum F_y/m \qquad a_z = \sum F_z/m \qquad 5\text{–}2$$

SI Unit: Newton (N)

$$1\text{ N} = 1\text{ kg}\cdot\text{m/s}^2 \qquad 5\text{–}3$$

Free-Body Diagram

A sketch showing all external forces acting on an object.

5–4 Newton's Third Law of Motion

Third Law

For every force that acts on an object, there is a reaction force acting on a different object that is equal in magnitude and opposite in direction.

Contact Forces

Action-reaction pair of forces produced by physical contact of two objects.

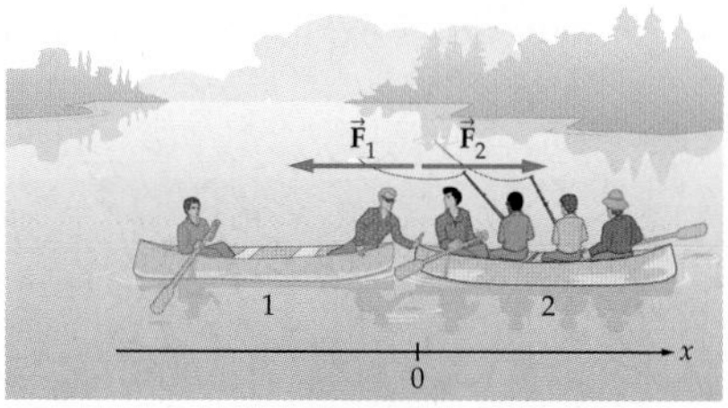

Physical picture

5–5 The Vector Nature of Forces: Forces in Two Dimensions

Forces are vectors.

Newton's second law can be applied to each component of force separately and independently.

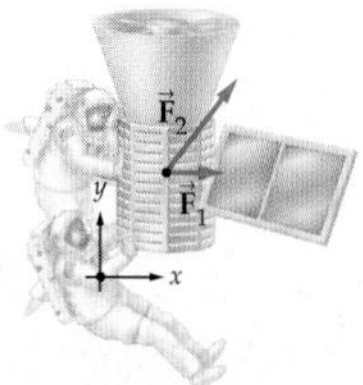

5–6 Weight

Gravitational force exerted *by the Earth on an object*.

On the surface of the Earth the weight, W, of an object of mass m has the magnitude

$$W = mg \qquad \text{5–5}$$

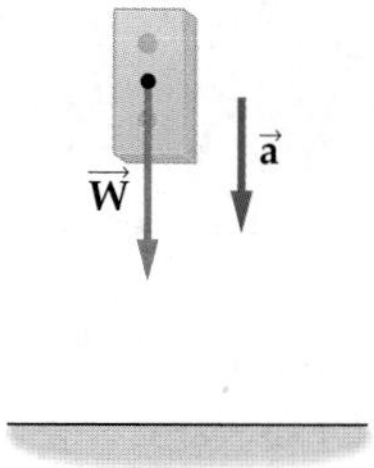

Apparent Weight

Force felt from contact with the floor or a scale in an accelerating system. For example, the sensation of feeling heavier or lighter in an accelerating elevator.

5–7 Normal Forces

Force exerted by a surface that is *perpendicular* to the surface.

The normal force is equal to the weight of an object only in special cases. In general, the normal force is greater than or less than the object's weight.

Problem-Solving Summary

Type of Calculation	Relevant Physical Concepts	Related Examples
Find the acceleration of an object.	Solve Newton's second law for each component of the acceleration; that is, $a_x = \Sigma F_x/m$ and $a_y = \Sigma F_y/m$.	Examples 5–1, 5–3, 5–4, 5–5, 5–8, 5–9 Active Examples 5–1, 5–2
Solve problems involving action-reaction forces.	Apply Newton's third law, being careful to note that the action-reaction forces act on different objects.	Examples 5–3, 5–4
Find the normal force exerted on an object.	Since there is no acceleration in the normal direction, set the sum of the normal components of force equal to zero.	Examples 5–8, 5–9

Conceptual Questions

(Answers to odd-numbered Conceptual Questions can be found in the back of the book.)

1. Driving down the road, you hit the brakes suddenly. As a result, your body moves toward the front of the car. Explain, using Newton's laws.

2. You've probably seen pictures of someone pulling a tablecloth out from under glasses, plates, and silverware set out for a formal dinner. Perhaps you've even tried it yourself. Using Newton's laws of motion, explain how this stunt works.

3. As you read this, you are most likely sitting quietly in a chair. Can you conclude, therefore, that you are at rest? Explain.

4. When a dog gets wet, it shakes its body from head to tail to shed the water. Explain, in terms of Newton's first law, why this works.

A dog uses the principle of inertia to shake water from its coat. (Conceptual Question 4)

5. A young girl slides down a rope. As she slides faster and faster she tightens her grip, increasing the force exerted on her by the rope. What happens when this force is equal in magnitude to her weight? Explain.

6. A drag-racing car accelerates forward because of the force exerted on it by the road. Why, then, does it need an engine? Explain.

7. A block of mass m hangs from a string attached to a ceiling, as shown in **Figure 5–16**. An identical string hangs down from the bottom of the block. Which string breaks if **(a)** the lower string is pulled with a slowly increasing force or **(b)** the lower string is jerked rapidly downward? Explain.

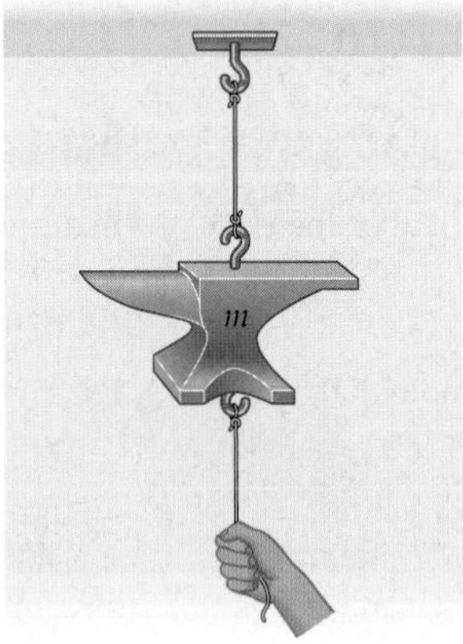

▲ **FIGURE 5–16** Conceptual Question 7

8. An astronaut on a space walk discovers that his jet pack no longer works, leaving him stranded 50 m from the spacecraft. If the jet pack is removable, explain how the astronaut can still use it to return to the ship.

9. Two untethered astronauts on a space walk decide to take a break and play catch with a baseball. Describe what happens as the game of catch progresses.

10. What are the action-reaction forces when a baseball bat hits a fast ball? What is the effect of each force?

11. In **Figure 5–17** Wilbur asks Mr. Ed, the talking horse, to pull a cart. Mr. Ed replies that he would like to, but the laws of nature just won't allow it. According to Newton's third law, he says, if he pulls on the wagon it pulls back on him with an equal force. Clearly, then, the net force is zero and the wagon will stay put. How should Wilbur answer the clever horse?

▲ **FIGURE 5–17** Conceptual Question 11

12. A whole brick has more mass than half a brick, thus the whole brick is harder to accelerate. Why doesn't a whole brick fall more slowly than half a brick? Explain.

13. The force exerted by gravity on a whole brick is greater than the force exerted by gravity on half a brick. Why, then, doesn't a whole brick fall faster than half a brick? Explain.

14. Is it possible for an object at rest to have only a single force acting on it? If your answer is yes, provide an example. If your answer is no, explain why not.

15. Is it possible for an object to be in motion and yet have zero net force acting on it? Explain.

16. A bird cage, with a parrot inside, hangs from a scale. The parrot decides to hop to a higher perch. What can you say about the reading on the scale **(a)** when the parrot jumps, **(b)** when the parrot is in the air, and **(c)** when the parrot lands on the second perch? Assume that the scale responds rapidly so that it gives an accurate reading at all times.

17. Suppose you jump from the cliffs of Acapulco and perform a perfect swan dive. As you fall, you exert an upward force on the Earth equal in magnitude to the downward force the Earth exerts on you. Why, then, does it seem that you are the one doing all the accelerating? Since the forces are the same, why aren't the accelerations?

18. A friend tells you that since his car is at rest, there are no forces acting on it. How would you reply?

19. Since all objects are "weightless" in orbit, how is it possible for an orbiting astronaut to tell if one object has more mass than another object? Explain.

20. To clean a rug, you can hang it from a clothesline and beat it with a tennis racket. Use Newton's laws to explain why beating the rug should have a cleansing effect.

21. If you step off a high board and drop to the water below, you plunge into the water without injury. On the other hand, if you were to drop the same distance onto solid ground, you might break a leg. Use Newton's laws to explain the difference.

22. A moving object is acted on by a net force. Give an example of a situation in which the object moves **(a)** in the same direction as the net force, **(b)** at right angles to the net force, or **(c)** in the opposite direction of the net force.

23. Is it possible for an object to be moving in one direction while the net force acting on it is in another direction? If your answer is yes, provide an example. If your answer is no, explain why not.

24. Since a bucket of water is "weightless" in space, would it hurt to kick the bucket? Explain.

25. In the movie *The Rocketeer*, a teenager discovers a jet-powered backpack in an old barn. The backpack allows him to fly at incredible speeds. In one scene, however, he uses the backpack to rapidly accelerate an old pickup truck that is being chased by "bad guys." He does this by bracing his arms against the cab of the pickup and firing the backpack, giving the truck the acceleration of a drag racer. Is the physics of this scene "Good," "Bad," or "Ugly?" Explain.

26. List three common objects that have a weight of approximately 1 N.

Conceptual Exercises

(Answers to odd-numbered Conceptual Exercises can be found in the back of the book.)

1. A small car collides with a large truck. **(a)** Is the force experienced by the car greater than, less than, or equal to the force experienced by the truck? Explain. **(b)** Is the acceleration experienced by the car greater than, less than, or equal to the acceleration experienced by the truck? Explain.

2. A skateboarder on a ramp is accelerated by a nonzero net force. For each of the following statements, state whether it is always true, never true, or sometimes true. **(a)** The skateboarder is moving in the direction of the net force. **(b)** The acceleration of the skateboarder is at right angles to the net force. **(c)** The acceleration of the skateboarder is in the same direction as the net force. **(d)** The skateboarder is instantaneously at rest.

3. Three objects, A, B, and C, have x and y components of velocity that vary with time as shown in **Figure 5–18**. What is the direction of the net force acting on each of these objects, as measured from the positive x axis. (All of the nonzero slopes have the same magnitude.)

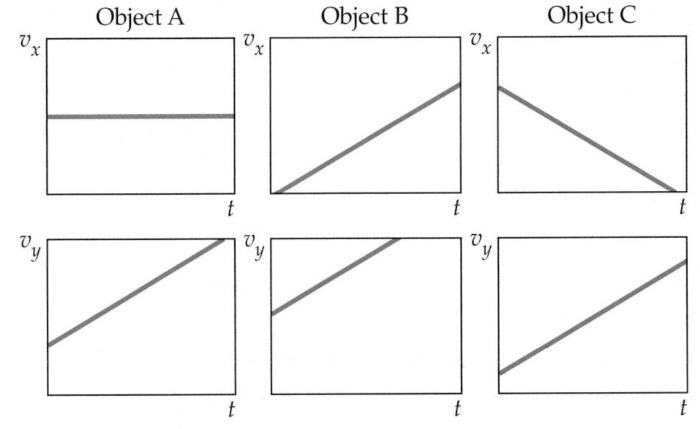

▲ **FIGURE 5–18** Conceptual Exercise 3

4. You drop two balls of equal diameter from the same height at the same time. Ball 1 is made of metal and has a greater mass than ball 2, which is made of wood. If the upward force due to air resistance is the same for both balls, does ball 1 reach the ground before, after, or at the same time as ball 2?

5. Riding in an elevator moving upward with constant speed, you begin a game of darts. Do you have to aim your darts higher, lower, or the same as when you play darts on solid ground?

6. Riding in an elevator moving with a constant upward acceleration, you begin a game of darts. Do you have to aim your darts higher, lower, or the same as when you play darts on solid ground?

7. Give the direction of the net force acting on each of the following objects. If the net force is zero, state "zero." **(a)** A car accelerating northward from a stoplight. **(b)** A car traveling southward and slowing down. **(c)** A car traveling westward with constant speed. **(d)** A skydiver parachuting downward with constant speed. **(e)** A baseball during its flight from pitcher to catcher (ignoring air resistance).

8. An object of mass m is initially at rest. After a force of magnitude F acts on it for a time T, the object has a speed v. Suppose the mass of the object is doubled, and the magnitude of the force acting on it is quadrupled. In terms of T, how long does it take for the object to accelerate from rest to a speed v now?

9. You jump out of an airplane and open your parachute after a brief period of free fall. To decelerate your fall, must the force exerted on you by the parachute be greater than, less than, or equal to your weight?

10. A hockey puck is acted on by one or more forces, as shown in **Figure 5–19**. Rank the four cases, A, B, C, and D, in order of the magnitude of the puck's acceleration, starting with the smallest. Indicate ties with an equal sign.

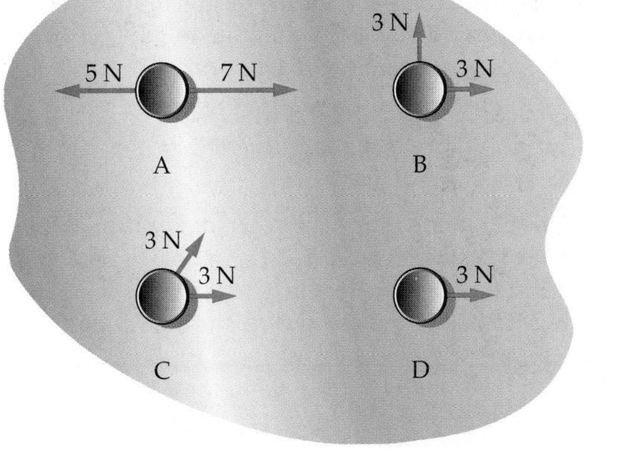

▲ **FIGURE 5–19** Conceptual Exercise 10

11. Each of the three identical hockey pucks shown in **Figure 5–20** is acted on by a 3-N force. Puck A moves with a speed of 7 m/s in a direction opposite to the force; puck B is instantaneously at rest; puck C moves with a speed of 7 m/s at right angles to the force. Rank the three pucks in order of the magnitude of their acceleration, starting with the smallest. Indicate ties with an equal sign.

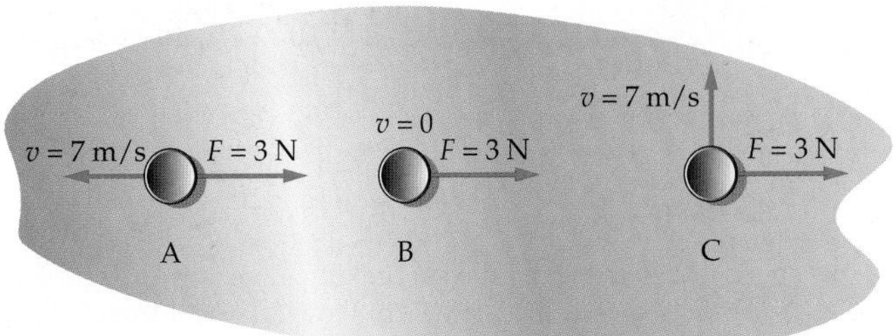

▲ **FIGURE 5–20** Conceptual Exercise 11

Problems

Note: **IP** *denotes an integrated conceptual/quantitative problem.* **BIO** *identifies problems of biological or medical interest.* Red bullets (•,•• ,•••) are used to indicate the level of difficulty of each problem.

Section 5–3 Newton's Second Law of Motion

1. • On a planet far, far away, an astronaut picks up a rock. The rock has a mass of 5.00 kg, and on this particular planet its weight is 40.0 N. If the astronaut exerts an upward force of 46.2 N on the rock, what is its acceleration?

2. • In a grocery store, you push a 10.3-kg shopping cart with a force of 12.1 N. If the cart starts at rest, how far does it move in 2.50 s?

3. • You are pulling your little sister on her sled across an icy (frictionless) surface. When you exert a constant horizontal force of 110 N, the sled has an acceleration of 2.5 m/s^2. If the sled has a mass of 7.0 kg, what is the mass of your little sister?

4. • A 0.53-kg billiard ball initially at rest is given a speed of 12 m/s during a time interval of 4.0 ms. What average force acted on the ball during this time?

5. • A 92-kg water skier floating in a lake is pulled from rest to a speed of 12 m/s in a distance of 25 m. What is the net force exerted on the skier, assuming his acceleration is constant?

6. •• **IP** A 42.0-kg parachutist lands moving straight downward with a speed of 3.85 m/s. **(a)** If the parachutist comes to rest with constant acceleration over a distance of 0.750 m, what force does the ground exert on her? **(b)** If the parachutist comes to rest over a shorter distance, is the force exerted by the ground greater than, less than, or the same as in part (a)? Explain.

7. •• **IP** In baseball, a pitcher can accelerate a 0.15-kg ball from rest to 98 mi/h in a distance of a 1.7 m. **(a)** What is the average force exerted on the ball during the pitch? **(b)** If the mass of the ball is increased, is the force required of the pitcher increased, decreased, or unchanged? Explain.

8. •• A major-league catcher gloves a 92 mi/h pitch and brings it to rest in 0.15 m. If the force exerted by the catcher is 803 N, what is the mass of the ball?

9. •• Driving home from school one day, you spot a ball rolling out into the street (**Figure 5–21**). You brake for 1.20 s, slowing your 950-kg car from 16.0 m/s to 9.50 m/s. **(a)** What was the average force exerted on your car during braking? **(b)** How far did you travel while braking?

▲ **FIGURE 5–21** Problem 9

10. •• **Stopping a 747** A 747 jetliner lands and begins to slow to a stop as it moves along the runway. If its mass is 3.50×10^5 kg, its speed is 27.0 m/s, and the net braking force is 4.30×10^5 N, **(a)** what is its speed 7.50 s later? **(b)** How far has it traveled in this time?

11. •• **IP** A drag racer crosses the finish line doing 212 mi/h and promptly deploys her drag chute (the small parachute used for braking). **(a)** What force must the drag chute exert on the 898-kg car to slow it to 45.0 mi/h in a distance of 185 m? **(b)** Describe the strategy you used to solve part (a).

Section 5–4 Newton's Third Law of Motion

12. • You hold a brick at rest in your hand. **(a)** How many forces act on the brick? **(b)** Identify these forces. **(c)** Are these forces equal in magnitude and opposite in direction? **(d)** Are these forces an action-reaction pair? Explain.

13. • Referring to Problem 12, you are now accelerating the brick upward. **(a)** How many forces act on the brick in this case? **(b)** Identify these forces. **(c)** Are these forces equal in magnitude and opposite in direction? **(d)** Are these forces an action-reaction pair? Explain.

14. •• On vacation, your 1300-kg car pulls a 540-kg trailer away from a stoplight with an acceleration of 1.90 m/s^2. **(a)** What is the net force exerted by the car on the trailer? **(b)** What force does the trailer exert on the car? **(c)** What is the net force acting on the car?

15. •• **IP** A 71-kg parent and a 19-kg child meet at the center of an ice rink. They place their hands together and push. **(a)** Is the force experienced by the child more than, less than, or the same as the force experienced by the parent? **(b)** Is the acceleration of the child more than, less than, or the same as the acceleration of the parent? Explain. **(c)** If the acceleration of the child is 2.6 m/s^2 in magnitude, what is the magnitude of the parent's acceleration?

16. •• A force of magnitude 7.50 N pushes three boxes with masses $m_1 = 1.30$ kg, $m_2 = 3.20$ kg, and $m_3 = 4.90$ kg, as shown in **Figure 5–22**. Find the magnitude of the contact force **(a)** between boxes 1 and 2, and **(b)** between boxes 2 and 3.

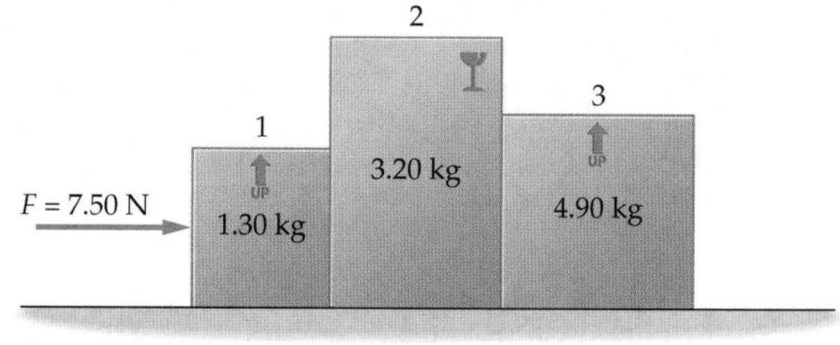

▲ **FIGURE 5–22** Problem 16

17. •• A force of magnitude 7.50 N pushes three boxes with masses $m_1 = 1.30$ kg, $m_2 = 3.20$ kg, and $m_3 = 4.90$ kg, as shown in **Figure 5–23**. Find the magnitude of the contact force **(a)** between boxes 1 and 2, and **(b)** between boxes 2 and 3.

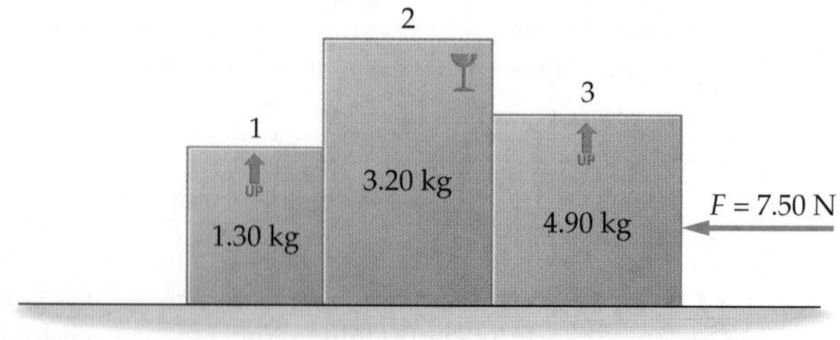

▲ **FIGURE 5–23** Problem 17

18. •• **IP** Two boxes sit side-by-side on a smooth horizontal surface. The lighter box has a mass of 5.2 kg, the heavier box has a mass of 7.4 kg. **(a)** Find the contact force between these boxes when a horizontal force of 5.0 N is applied to the light box. **(b)** If the 5.0-N force is applied to the heavy box instead, is the contact force between the boxes the same as, greater than, or less than the contact force in part (a)? Explain. **(c)** Verify your answer to part (b) by calculating the contact force in this case.

Section 5–5 The Vector Nature of Forces

19. • A farm tractor tows a 3900-kg trailer up a 16° incline with a steady speed of 3.0 m/s. What force does the tractor exert on the trailer? (Ignore friction.)

20. • A surfer "hangs ten," and accelerates down the sloping face of a wave. If the surfer's acceleration is 3.25 m/s^2 and friction can be ignored, what is the angle at which the face of the wave is inclined above the horizontal?

21. • A shopper pushes a 7.5-kg shopping cart up a 13° incline, as shown in **Figure 5–24**. Find the magnitude of the horizontal force, $\vec{F}$, needed to give the cart an acceleration of 1.41 m/s^2.

▲ **FIGURE 5–24** Problem 21

22. • Two crewmen pull a raft through a lock, as shown in **Figure 5–25**. One crewman pulls with a force of 130 N at an angle of 34° relative to the forward direction of the raft. The second crewman, on the opposite side of the lock, pulls at an angle of 45°. With what force should the second crewman pull so that the net force of the two crewmen is in the forward direction?

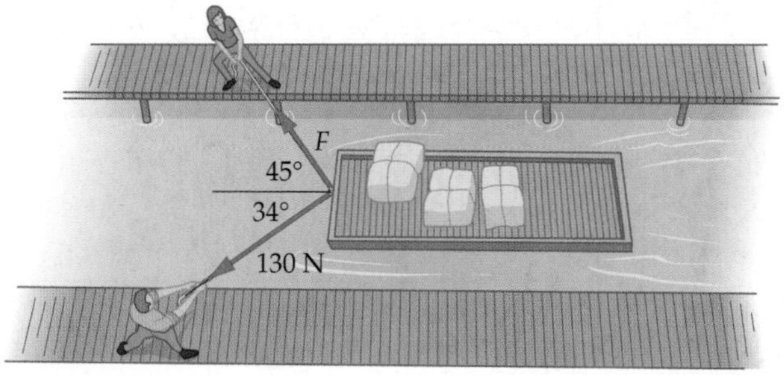

▲ **FIGURE 5–25** Problem 22

23. •• To give a 19-kg child a ride, two teenagers pull on a 3.7-kg sled with ropes, as indicated in **Figure 5–26**. Both teenagers pull with a force of 55 N at an angle of 35° relative to the forward direction, which is the direction of motion. In addition, the snow exerts a retarding force on the sled that points opposite to the direction of motion, and has a magnitude of 57 N. Find the acceleration of the sled and child.

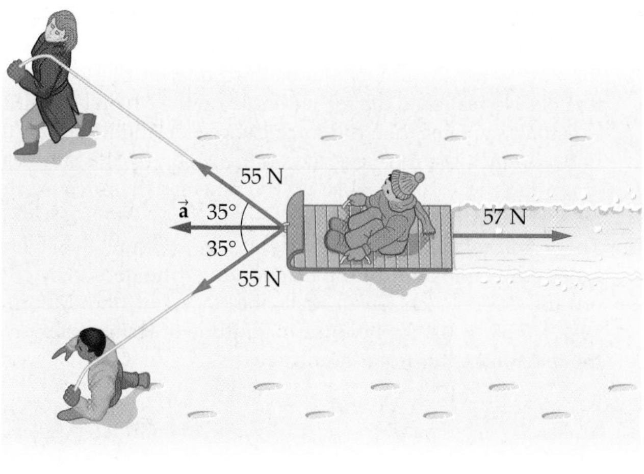

▲ **FIGURE 5–26** Problem 23

24. •• **IP** Before practicing his routine on the rings, a 67-kg gymnast stands motionless, with one hand grasping each ring and his feet touching the ground. Both arms slope upward at an angle of 24° above the horizontal. **(a)** If the force exerted by the rings on each arm has a magnitude of 290 N, and is directed along the length of the arm, what is the magnitude of the force exerted by the floor on his feet? **(b)** If the angle his arms make with the horizontal is greater that 24°, and everything else remains the same, is the force exerted by the floor on his feet greater than, less than, or the same as the value found in part (a)? Explain.

25. •• **IP** A 65-kg skier speeds down a trail, as shown in **Figure 5–27**. The surface is smooth and inclined at an angle of 22° with the horizontal. **(a)** Find the direction and magnitude of the net force acting on the skier. **(b)** Does the net force exerted on the skier increase, decrease, or stay the same as the slope becomes steeper? Explain.

▲ **FIGURE 5–27** Problems 25 and 38

26. •• An object acted on by three forces moves with constant velocity. One force acting on the object is in the positive x direction and has a magnitude of 6.5 N; a second force has a magnitude of 4.4 N and points in the negative y direction. Find the direction and magnitude of the third force acting on the object.

27. •• A train is traveling up a 3.9° incline at a speed of 3.15 m/s when the last car breaks free and begins to coast without friction. **(a)** How long does it take for the last car to come to rest momentarily? **(b)** How far did the last car travel before momentarily coming to rest?

28. •• **The Force Exerted on the Moon** Figure 5–28 shows the Earth, Moon, and Sun (not to scale) in their relative positions at the time when the Moon is in its third quarter phase. Though few people realize it, the force exerted on the Moon by the Sun is actually greater than the force exerted on the Moon by the Earth. In fact, the force exerted on the Moon by the Sun has a magnitude of $F_{SM} = 4.34 \times 10^{20}$ N, whereas the force exerted by the Earth has a magnitude of only $F_{EM} = 1.98 \times 10^{20}$ N. These forces are indicated to scale in Figure 5–28. Find **(a)** the direction and **(b)** the magnitude of the net force acting on the Moon. **(c)** Given that the mass of the Moon is $M_M = 7.35 \times 10^{22}$ kg, find the magnitude of its acceleration at the time of the third-quarter phase.

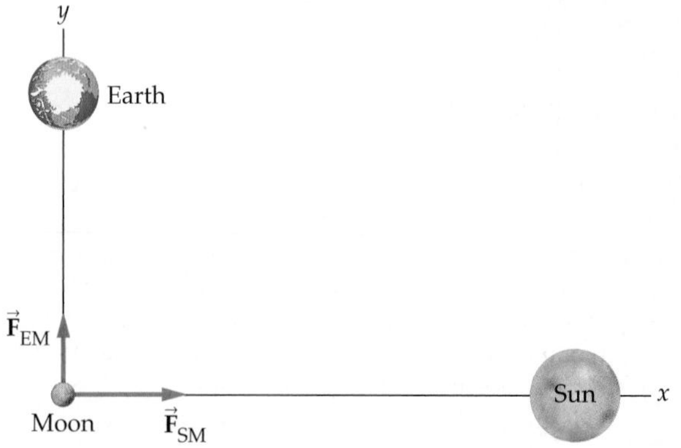

▲ **FIGURE 5–28** Problem 28

Section 5–6 Weight

29. • You pull upward on a stuffed suitcase with a force of 115 N, and it accelerates upward at 0.725 m/s^2. What is **(a)** the mass and **(b)** the weight of the suitcase?

30. • **BIO Brain Growth** A newborn baby's brain grows rapidly. In fact, it has been found to increase in mass by about 1.6 mg per minute. **(a)** How much does the brain's weight increase in one day? **(b)** How long does it take for the brain's weight to increase by 0.15 N?

31. • Suppose a rocket launches with an acceleration of 30.2 m/s^2. What is the apparent weight of an 88-kg astronaut aboard this rocket?

32. • At the bow of a ship on a stormy sea, a crewman conducts an experiment by standing on a bathroom scale. In calm waters, the scale reads 182 lb. During the storm, the crewman finds a maximum reading of 225 lb and a minimum reading of 138 lb. Find **(a)** the maximum upward acceleration and **(b)** the maximum downward acceleration experienced by the crewman.

33. •• **IP** As part of a physics experiment, you stand on a bathroom scale in an elevator. Though your normal weight is 610 N, the scale at the moment reads 730 N. **(a)** Is the acceleration of the elevator upward, downward, or zero? Explain. **(b)** Calculate the magnitude of the elevator's acceleration. **(c)** What, if anything, can you say about the velocity of the elevator? Explain.

34. •• When you weigh yourself on good old *terra firma* (solid ground), your weight is 141 lb. In an elevator your apparent weight is 122 lb. What are the direction and magnitude of the elevator's acceleration?

35. •• **IP BIO Flight of the Samara** A 1.21-g samara—the winged fruit of a maple tree—falls toward the ground with a constant speed of 1.1 m/s (**Figure 5–29**). **(a)** What is the force of air resistance exerted on the samara? **(b)** If the constant speed of descent is greater than 1.1 m/s, is the force of air resistance greater than, less than, or the same as in part (a)? Explain.

▲ **FIGURE 5–29** Problem 35

36. ••• When you lift a bowling ball with a force of 82 N, the ball accelerates upward with an acceleration a. If you lift with a force of 92 N, the ball's acceleration is $2a$. Find **(a)** the weight of the bowling ball, and **(b)** the acceleration a.

Section 5–7 Normal Forces

37. • A 23-kg suitcase is being pulled with constant speed by a handle that is at an angle of 25° above the horizontal. If the normal force exerted on the suitcase is 180 N, what is the force F applied to the handle?

38. • **(a)** Draw a free-body diagram for the skier in Problem 25. **(b)** Determine the normal force acting on the skier.

39. • A 9.3-kg child sits in a 3.7-kg high chair. **(a)** Draw a free-body diagram for the child, and find the normal force exerted by the chair on the child. **(b)** Draw a free-body diagram for the chair, and find the normal force exerted by the floor on the chair.

40. •• **Figure 5–30** shows the normal force as a function of the angle θ for the suitcase shown in Figure 5–13. Determine the magnitude of the force $\vec{F}$ for each of the three curves shown in Figure 5–30. Give your answer in terms of the weight of the suitcase, mg.

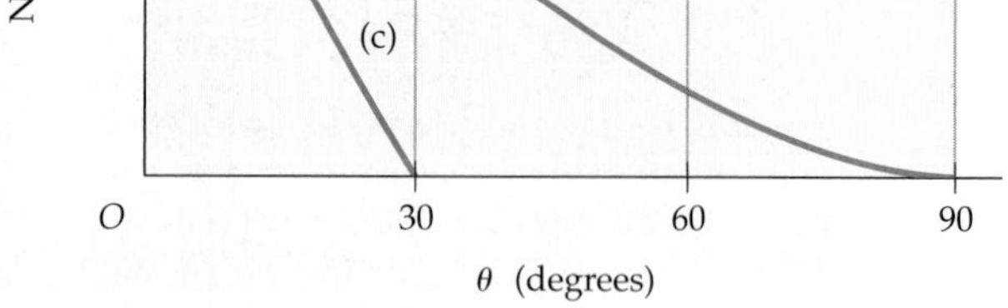

▲ **FIGURE 5–30** Problem 40

41. •• A 5.0-kg bag of potatoes sits on the bottom of a stationary shopping cart. **(a)** Sketch a free-body diagram for the bag of potatoes. **(b)** Now suppose the cart moves with a constant velocity. How does this affect your free-body diagram? Explain.

42. •• **IP (a)** Find the normal force exerted on a 2.7-kg book resting on a surface inclined at 32° above the horizontal. **(b)** If the angle of the incline is reduced, do you expect the normal force to increase, decrease, or stay the same? Explain.

43. •• **IP** A gardener mows a lawn with an old-fashioned push mower. The handle of the mower makes an angle of 32° with the surface of the lawn. **(a)** If a 209-N force is applied along the handle of the 18-kg mower, what is the normal force exerted by the lawn on the mower? **(b)** If the angle between the surface of the lawn and the handle of the mower is increased, does the normal force exerted by the lawn increase, decrease, or stay the same? Explain.

44. ••• An ant walks slowly away from the top of a bowling ball, as shown in **Figure 5–31**. If the ant starts to slip when the normal force on its feet drops below one-half its weight, at what angle θ does slipping begin?

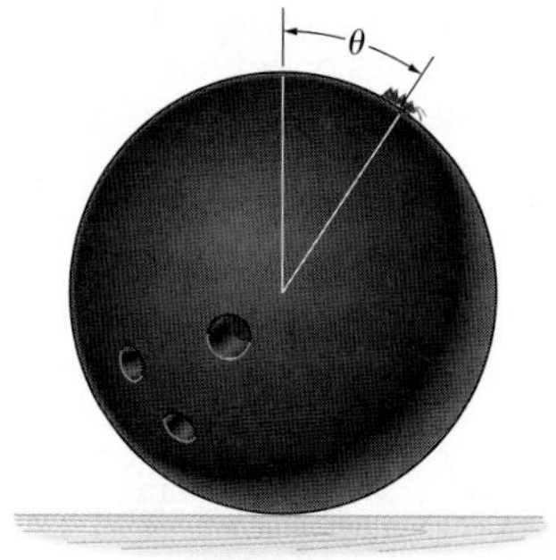

▲ **FIGURE 5–31** Problem 44

General Problems

45. • In a tennis serve, a 0.070-kg ball can be accelerated from rest to 35 m/s over a distance of 0.80 m. Find the magnitude of the average force exerted by the racket on the ball during the serve.

46. • A 49.5-kg swimmer with an initial speed of 1.25 m/s decides to coast until she comes to rest. If she slows with constant acceleration and stops after coasting 2.20 m, what was the force exerted on her by the water?

47. •• **IP The VASIMR Rocket** NASA plans to use a new type of rocket, a Variable Specific Impulse Magnetoplasma Rocket (VASIMR), on future missions. A VASIMR can produce 1200 N of thrust (force) when in operation. If a VASIMR has a mass of 2.2×10^5 kg, **(a)** what acceleration will it experience? Assume that the only force acting on the rocket is its own thrust, and that the mass of the rocket is constant. **(b)** Over what distance must the rocket accelerate from rest to achieve a speed of 9500 m/s? **(c)** When the rocket has covered one-quarter the acceleration distance found in part (b), is its average speed 1/2, 1/3, or 1/4 its average speed during the final three-quarters of the acceleration distance? Explain.

48. •• An object of mass $m = 5.95$ kg has an acceleration $\vec{\mathbf{a}} = (1.17\text{ m/s}^2)\hat{\mathbf{x}} + (-0.664\text{ m/s}^2)\hat{\mathbf{y}}$. Three forces act on this object: $\vec{\mathbf{F}}_1$, $\vec{\mathbf{F}}_2$, and $\vec{\mathbf{F}}_3$. Given that $\vec{\mathbf{F}}_1 = (3.22\text{ N})\hat{\mathbf{x}}$ and $\vec{\mathbf{F}}_2 = (-1.55\text{ N})\hat{\mathbf{x}} + (2.05\text{ N})\hat{\mathbf{y}}$, find $\vec{\mathbf{F}}_3$.

49. •• At the local grocery store, you push a 14.5-kg shopping cart. You stop for a moment to add a bag of dog food to your cart. With a force of 12.0 N, you now accelerate the cart from rest through a distance of 2.29 m in 3.00 s. What was the mass of the dog food?

50. •• **IP BIO The Force of Running** Biomechanical research has shown that when a 67-kg person is running, the force exerted on each foot as it strikes the ground can be as great as 2300 N. **(a)** What is the ratio of the force exerted on the foot by the ground to the person's body weight? **(b)** If the only forces acting on the person are (i) the force exerted by the ground and (ii) the person's weight, what is the magnitude and direction of the person's acceleration? **(c)** If the acceleration found in part (b) acts for 10.0 ms, what is the resulting change in the vertical component of the person's velocity?

51. •• **IP BIO Grasshopper Liftoff** To become airborne, a 2.0-g grasshopper requires a takeoff speed of 2.7 m/s. It acquires this speed by extending its hind legs through a distance of 3.7 cm. **(a)** What is the average acceleration of the grasshopper during takeoff? **(b)** Find the magnitude of the average net force exerted on the grasshopper by its hind legs during takeoff. **(c)** If the mass of the grasshopper increases, does the takeoff acceleration increase, decrease, or stay the same? **(d)** If the mass of the grasshopper increases, does the required takeoff force increase, decrease, or stay the same? Explain.

52. •• **Takeoff from an Aircraft Carrier** On an aircraft carrier, a jet can be catapulted from 0 to 155 mi/h in 2.00 s. If the average force exerted by the catapult is 9.35×10^5 N, what is the mass of the jet?

A jet takes off from the flight deck of an aircraft carrier. (Problem 52)

53. •• **IP** An archer shoots a 0.022-kg arrow at a target with a speed of 57 m/s. When it hits the target, it penetrates to a depth of 0.085 m. **(a)** What was the average force exerted by the target on the arrow? **(b)** If the mass of the arrow is doubled, and the force exerted by the target on the arrow remains the same, by what multiplicative factor does the penetration depth change? Explain.

54. •• An apple of mass $m = 0.13$ kg falls out of a tree from a height $h = 3.2$ m. **(a)** What is the magnitude of the force of gravity, mg, acting on the apple? **(b)** What is the apple's speed, v, just before it lands? **(c)** Show that the force of gravity times the height, mgh, is equal to $\frac{1}{2}mv^2$. (We shall investigate the significance of this result in Chapter 8.) Be sure to show that the dimensions are in agreement as well as the numerical values.

55. •• An apple of mass $m = 0.22$ kg falls from a tree and hits the ground with a speed of $v = 14$ m/s. **(a)** What is the magnitude of the force of gravity, mg, acting on the apple? **(b)** What is the time, t, required for the apple to reach the ground? **(c)** Show that the force of gravity times the time, mgt, is equal to mv. (We shall investigate the significance of this result in Chapter 9.) Be sure to show that the dimensions are in agreement as well as the numerical values.

56. •• **BIO The Fall of *T. rex*** Paleontologists estimate that if a *Tyrannosaurus rex* were to trip and fall, it would have experienced a force of approximately 260,000 N acting on its torso when it hit the ground. Assuming the torso has a mass of 3800 kg, **(a)** find the magnitude of the torso's upward acceleration as it comes to rest. (For comparison, humans lose consciousness with an acceleration of about $7g$.) **(b)** Assuming the torso is in free fall for a distance 1.46 m as it falls to the ground, how much time is required for the torso to come to rest once it contacts the ground?

57. •• **Deep Space I** The NASA spacecraft *Deep Space I* was shut down on December 18, 2001, following a three-year journey to the asteroid Braille and the comet Borrelly. This spacecraft used a solar-powered ion engine to produce 0.064 ounces of thrust (force) by stripping electrons from neon atoms and accelerating the resulting ions to 70,000 mi/h. The thrust was only as much as the weight of a couple sheets of paper, but the engine operated continuously for 16,000 hours. As a result, the speed of the spacecraft increased by 7900 mi/h. What was the mass of *Deep Space I*? (Assume that the mass of the neon gas is negligible.)

58. •• Your groceries are in a bag with paper handles. The handles will tear off if a force greater than 51.0 N is applied to them. What is the greatest mass of groceries that can be lifted safely with this bag, given that the bag is raised **(a)** with constant speed, or **(b)** with an acceleration of 1.35 m/s^2?

59. •• IP While waiting at the airport for your flight to leave, you observe some of the jets as they take off. With your watch you find that it takes about 35 seconds for a plane to go from rest to takeoff speed. In addition, you estimate that the distance required is about 1.5 km. **(a)** If the mass of a jet is 1.70×10^5 kg, what force is needed for takeoff? **(b)** Describe the strategy you used to solve part (a).

60. •• BIO **Gecko Feet** Researchers have found that a gecko's foot is covered with hundreds of thousands of small hairs (*setae*) that allow it to walk up walls and even across ceilings. A single foot pad, which has an area of 1.0 cm^2, can attach to a wall or ceiling with a force of 11 N. **(a)** How many 250-g geckos could be suspended from the ceiling by a single foot pad? **(b)** Estimate the force per square centimeter that your body exerts on the soles of your shoes, and compare with the 11 N/cm^2 of the sticky gecko foot.

A Tokay gecko (*Gekko gecko*) shows off its famous feet. (Problem 60)

61. •• Two boxes are at rest on a smooth, horizontal surface. The boxes are in contact with one another. If box 1 is pushed with a force of magnitude $F = 12.00$ N, the contact force between the boxes is 8.50 N; if, instead, box 2 is pushed with the force F, the contact force is 12.00 N − 8.50 N = 3.50 N. In either case, the boxes move together with an acceleration of 1.70 m/s^2. What is the mass of **(a)** box 1 and **(b)** box 2?

62. ••• IP Responding to an alarm, a 92-kg fireman slides down a pole to the ground floor, 3.2 m below. The fireman starts at rest and lands with a speed of 4.0 m/s. **(a)** Find the average force exerted on the fireman by the pole. **(b)** If the landing speed is half that in part (a), is the average force exerted on the fireman doubled? Explain. **(c)** Find the average force exerted on the fireman when the landing speed is 2.0 m/s.

63. ••• For a birthday gift, you and some friends take a hot-air balloon ride. One friend is late, so the balloon floats a couple of feet off the ground as you wait. Before this person arrives, the combined weight of the basket and people is 1220 kg, and the balloon is neutrally buoyant. When the late arrival climbs up into the basket, the balloon begins to accelerate downward at 0.56 m/s^2. What was the mass of the last person to climb aboard?

64. ••• A baseball of mass m and initial speed v strikes a catcher's mitt. If the mitt moves a distance Δx as it brings the ball to rest, what is the average force it exerts on the ball?

65. ••• When two people push in the same direction on an object of mass m they cause an acceleration of magnitude a_1. When the same people push in opposite directions, the acceleration of the object has a magnitude a_2. Determine the magnitude of the force exerted by each of the two people in terms of m, a_1, and a_2.

66. ••• BIO **Seatbelt Safety** Seatbelts provide two main advantages in a car accident: (i) they keep you from being thrown from the car, and (ii) they reduce the force that acts on you during the collision to survivable levels. The second benefit can be illustrated by comparing the net force exerted on the driver of a car in a head-on collision with and without a seatbelt. **(a)** A driver wearing a seatbelt decelerates at roughly the same rate as the car itself. Since modern cars have a "crumple zone" built into the front of the car, the car will decelerate over a distance of roughly 1.0 m. Find the net force acting on a 65-kg driver who is decelerated from 18 m/s to rest in a distance of 1.0 m. **(b)** A driver who does not wear a seatbelt continues to move forward with a speed of 18 m/s (due to inertia) until something solid is encountered. The driver now comes to rest in a much shorter distance—perhaps only a centimeter. Find the net force acting on a 65-kg driver who is decelerated from 18 m/s to rest in 1.0 cm.

67. ••• An air-track cart of mass $m_1 = 0.14$ kg is moving with a speed $v_0 = 1.3$ m/s to the right when it collides with a cart of mass $m_2 = 0.25$ kg that is at rest. Each cart has a wad of putty on its bumper, and hence they stick together as a result of their collision. Suppose the average contact force between the carts is $F = 1.5$ N during the collision. **(a)** What is the acceleration of cart 1? Give direction and magnitude. **(b)** What is the acceleration of cart 2? Give direction and magnitude. **(c)** How long does it take for both carts to have the same speed? (Once the carts have the same speed the collision is over and the contact force vanishes.) **(d)** What is the final speed of the carts, v_f? **(e)** Show that $m_1 v_0$ is equal to $(m_1 + m_2)v_f$. (We shall investigate the significance of this result in Chapter 9.)

Interactive Problems

68. •• IP ***Referring to Example 5–4*** Suppose we would like the contact force between the boxes to have a magnitude of 5.00 N, and that the only thing in the system we are allowed to change is the mass of box 2—the mass of box 1 is 10.0 kg and the applied force is 20.0 N. **(a)** Should the mass of box 2 be increased or decreased? Explain. **(b)** Find the mass of box 2 that results in a contact force of magnitude 5.00 N? **(c)** What is the acceleration of the boxes in this case?

69. •• ***Referring to Example 5–4*** Suppose the force of 20.0 N pushes on two boxes of unknown mass. We know, however, that the acceleration of the boxes is 1.20 m/s^2 and the contact force has a magnitude of 4.45 N. Find the mass of **(a)** box 1 and **(b)** box 2.

70. •• IP ***Referring to Figure 5–9*** Suppose the magnitude of $\vec{F}_2$ is increased from 41 N to 55 N, and that everything else in the system remains the same. **(a)** Do you expect the direction of the satellite's acceleration to be greater than, less than, or equal to 32°? Explain. Find **(b)** the direction and **(c)** the magnitude of the satellite's acceleration in this case.

71. •• IP ***Referring to Figure 5–9*** Suppose we would like the acceleration of the satellite to be at an angle of 25°, and that the only quantity we can change in the system is the magnitude of $\vec{F}_1$. **(a)** Should the magnitude of $\vec{F}_1$ be increased or decreased? Explain. **(b)** What is the magnitude of the satellite's acceleration in this case?

Applications of Newton's Laws

6–1 Frictional Forces 142
6–2 Strings and Springs 150
6–3 Translational Equilibrium 154
6–4 Connected Objects 158
6–5 Circular Motion 162

The climber in this photograph may not be thinking about Newton's laws, but they are involved in every aspect of his endeavor. He relies on the forces transmitted through the ropes to support his weight, and on the pulleys through which the ropes are threaded to give those forces the desired directions. His forward progress is made possible by the substantial frictional force between his hands and the rope. These are but a few of the many real-world applications of Newton's laws that are explored in this chapter.

Newton's laws of motion can be applied to an immense variety of systems, a sampling of which was discussed in Chapter 5. In this chapter we extend our discussion of Newton's laws by introducing new types of forces and by considering new classes of systems.

For example, we begin by considering the forces due to friction between two surfaces. As we shall see, the force of friction is different depending on whether the surfaces are in static contact, or are moving relative to one another—an important consideration in antilock braking systems. And though friction may seem like something that should be eliminated, we show that it is actually essential to life as we know it.

Next, we investigate the forces exerted by strings and springs, and show how these forces can safely suspend a mountain climber over a chasm, or cushion the ride of a locomotive. Finally, we consider the key role that force plays in making circular motion possible.

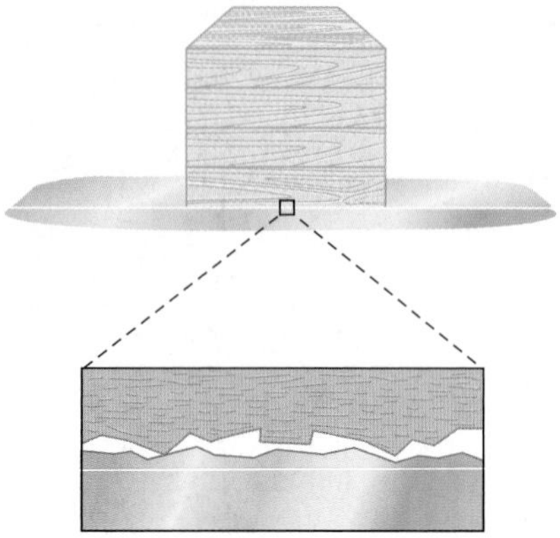

▲ **FIGURE 6–1 The origin of friction**
Even "smooth" surfaces have irregularities when viewed at the microscopic level. This type of roughness contributes to friction.

6–1 Frictional Forces

In Chapter 5 we always assumed that surfaces were smooth and that objects could slide without resistance to their motion. No surface is perfectly smooth, however. When viewed on the atomic level, even the "smoothest" surface is actually rough and jagged, as indicated in **Figure 6–1**. To slide one such surface across another requires a force large enough to overcome the resistance of microscopic hills and valleys bumping together. This is the origin of the force we call **friction.**

We often think of friction as something that should be reduced, or even eliminated if possible. For example, roughly 20 percent of the gasoline you buy does nothing but overcome friction within your car's engine. Clearly, reducing that friction would be most desirable.

On the other hand, friction can be helpful—even indispensable—in other situations. Suppose, for example, that you are standing still and then decide to begin walking forward. The force that accelerates you is the force of friction between your shoes and the ground. We simply couldn't walk or run without friction—it's hard enough when friction is merely reduced, as on an icy sidewalk. Similarly, starting or stopping a car, or even turning a corner, all require friction. Friction is an important and common feature of everyday life.

Since friction is caused by the random, microscopic irregularities of a surface, and since it is greatly affected by other factors such as the presence of lubricants, there is no simple "law of nature" for friction. There are, however, some very useful rules of thumb that give us rather accurate, approximate results for calculating frictional forces. In what follows, we describe these rules of thumb for the two types of friction most commonly used in this text—kinetic friction and static friction.

Kinetic Friction

As its name implies, kinetic friction is the friction encountered when surfaces slide against one another with a finite relative speed. The force generated by this friction, which will be designated with the symbol f_k, acts to oppose the sliding motion at the point of contact between the surfaces.

A series of simple experiments illustrates the main characteristics of kinetic friction. First, imagine attaching a spring scale to a rough object, like a brick, and pulling it across a table, as shown in **Figure 6–2**. If the brick moves with constant velocity, Newton's second law tells us that the net force on the brick must be zero. Hence, the force read on the scale, F, has the same magnitude as the force of kinetic friction, f_k. Now, if we repeat the experiment, but this time put a second brick on top of the first, we find that the force needed to pull the brick with constant velocity is doubled, to $2F$.

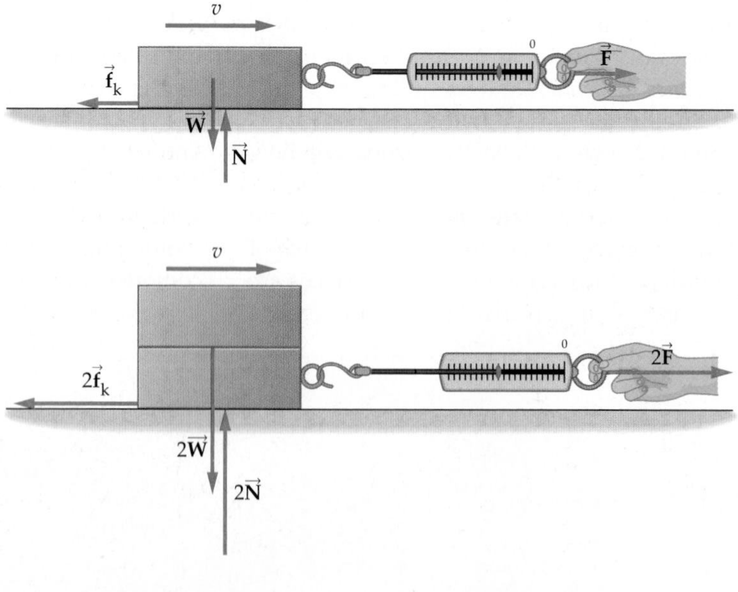

▶ **FIGURE 6–2 Kinetic friction and the normal force**
In the top part of the figure, a force F is required to pull the brick with constant speed v. Thus the force of kinetic friction is $f_k = F$. In the bottom part of the figure, the normal force has been doubled, and so has the force of kinetic friction, to $f_k = 2F$.

From this experiment we see that when we double the normal force—by stacking up two bricks, for example—the force of kinetic friction is also doubled. In general, the force of kinetic friction is found to be proportional to the magnitude of the normal force, N. Stated mathematically, this observation can be written as follows:

$$f_k = \mu_k N \qquad 6\text{–}1$$

The constant of proportionality, μ_k (pronounced "mew sub k"), is referred to as the **coefficient of kinetic friction.** In Figure 6–2 the normal force is equal to the weight of the bricks, but this is a special case. The normal force is greater than the weight if someone pushes down on the bricks, and this would cause more friction, or less than the weight if the bricks are placed on an incline. The former case is considered in several homework problems, and the latter case is considered in Examples 6–2 and 6–3.

Since f_k and N are both forces, and hence have the same units, we see that μ_k is a dimensionless number. The coefficient of kinetic friction is always positive, and typical values range between 0 and 1, as indicated in Table 6–1. The interpretation of μ_k is simple: If $\mu_k = 0.1$, for example, the force of kinetic friction is one-tenth of the normal force. Simply put, the greater μ_k the greater the friction; the smaller μ_k the smaller the friction.

TABLE 6–1 Typical Coefficients of Friction

Materials	Kinetic, μ_k	Static, μ_s
Rubber on concrete (dry)	0.80	1–4
Steel on steel	0.57	0.74
Glass on glass	0.40	0.94
Wood on leather	0.40	0.50
Copper on steel	0.36	0.53
Rubber on concrete (wet)	0.25	0.30
Steel on ice	0.06	0.10
Waxed ski on snow	0.05	0.10
Teflon on Teflon	0.04	0.04
Synovial joints in humans	0.003	0.01

As we know from everyday experience, the force of kinetic friction tends to oppose motion, as shown in Figure 6–2. Thus, $f_k = \mu_k N$ is not a vector equation, because N is perpendicular to the direction of motion. When doing calculations with the force of kinetic friction, we use $f_k = \mu_k N$ to find its magnitude, and we draw its direction so that it is opposite to the direction of motion.

There are two more friction experiments of particular interest. First, suppose that when we pull a brick, we initially pull it at the speed v, then later at the speed $2v$. What forces do we measure? It turns out that the force of kinetic friction is approximately the same in each case—it certainly does not double when we double the speed. Second, let's try standing the brick on end, so that it has a smaller area in contact with the table. If this smaller area is half the previous area, is the force halved? No, the force remains essentially the same, regardless of the area of contact.

We summarize these observations with the following three rules of thumb for kinetic friction:

Rules of Thumb for Kinetic Friction

The force of kinetic friction between two surfaces is:

1. Proportional to the magnitude of the normal force, N, between the surfaces,

$$f_k = \mu_k N$$

2. Independent of the relative speed of the surfaces.
3. Independent of the area of contact between the surfaces.

▲ Friction plays an important role in almost everything we do. Sometimes it is desirable to reduce friction; in other cases we want as much friction as possible. For example, it is more fun to ride on a water slide (upper) if the friction is low. Similarly, an engine operates more efficiently when it is oiled. When running, however, we need friction to help us speed up, slow down, and make turns. The sole of this running shoe (lower), like a car tire, is designed to maximize friction.

Again, these rules are useful and fairly accurate, though they are still only approximate. For simplicity, when we do calculations involving kinetic friction in this text, we will use these rules as if they were exact.

Before we show how to use f_k in calculations, we should make a comment regarding rule 3. This rule often seems rather surprising and counterintuitive. How is it that a larger area of contact doesn't produce a larger force? One way to think about this is to consider that when the area of contact is large, the normal force is spread out over a large area, giving a small force per area, F/A. As a result, the microscopic hills and valleys are not pressed too deeply against one another. On the other hand, if the area is small the normal force is concentrated in a small region, which presses the surfaces together more firmly, due to the large force per area. The net effect is roughly the same in either case.

Now, let's consider a commonly encountered situation in which kinetic friction plays a decisive role.

EXAMPLE 6–1 Pass the Salt—Please

Someone at the other end of the table asks you to pass the salt. Feeling quite dashing, you slide the 50.0-g salt shaker in their direction, giving it an initial speed of 1.15 m/s. If the shaker comes to rest with constant acceleration in 0.840 m, what is the coefficient of kinetic friction between the shaker and the table?

Picture the Problem

We choose the positive x direction to be the direction of motion, and the positive y direction to be upward. Two forces act in the y direction; the shaker's weight, $\vec{\mathbf{W}} = -W\hat{\mathbf{y}} = -mg\hat{\mathbf{y}}$, and the normal force, $\vec{\mathbf{N}} = N\hat{\mathbf{y}}$. Only one force acts in the x direction; the force of kinetic friction, $\vec{\mathbf{f}}_k = -\mu_k N\hat{\mathbf{x}}$. Note that the shaker moves through a distance of 0.840 m with an initial speed $v_{0x} = 1.15$ m/s.

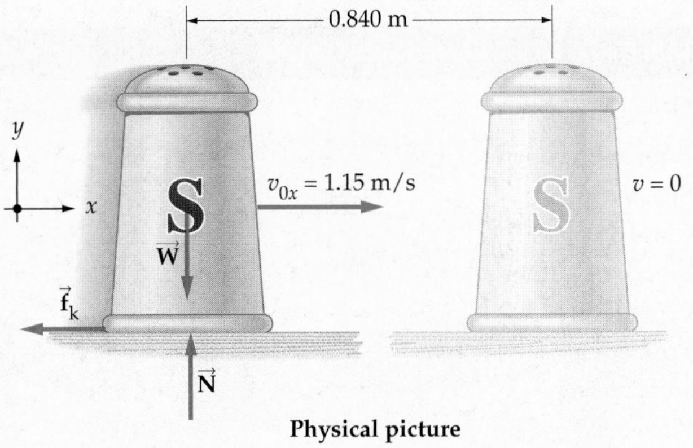

Physical picture

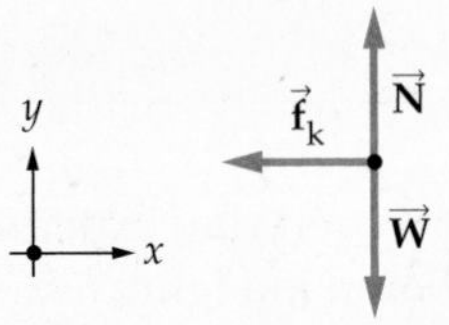

Free-body diagram

Strategy

Since the frictional force has a magnitude of $f_k = \mu_k N$, it follows that $\mu_k = f_k/N$. Therefore, we need to find the magnitudes of the frictional force, f_k, and the normal force, N.

To find f_k we set $\Sigma F_x = ma_x$, and find a_x with the kinematic equation $v_x^2 = v_{0x}^2 + 2a_x\Delta x$.

To find N we set $a_y = 0$ (since there is no motion in the y direction) and solve for N using $\Sigma F_y = ma_y = 0$.

Solution

1. Set $\Sigma F_x = ma_x$ to find f_k in terms of a_x:

$$\sum F_x = -f_k = ma_x \quad \text{or} \quad f_k = -ma_x$$

2. Determine a_x by using the kinematic equation relating velocity to position, $v_x^2 = v_{0x}^2 + 2a_x\Delta x$:

$$v_x^2 = v_{0x}^2 + 2a_x\Delta x$$

$$a_x = \frac{v_x^2 - v_{0x}^2}{2\Delta x} = \frac{0 - (1.15\text{ m/s})^2}{2(0.840\text{ m})} = -0.787\text{ m/s}^2$$

3. Set $\Sigma F_y = ma_y = 0$ to find the normal force, N:

$$\sum F_y = N + (-W) = ma_y = 0 \quad \text{or} \quad N = W = mg$$

4. Substitute $N = mg$ and $f_k = -ma_x$ (with $a_x = -0.787\text{ m/s}^2$) into $\mu_k = f_k/N$ to find μ_k:

$$\mu_k = \frac{f_k}{N} = \frac{-ma_x}{mg} = \frac{-a_x}{g} = \frac{-(-0.787\text{ m/s}^2)}{9.81\text{ m/s}^2} = 0.0802$$

Insight

Note that m canceled in line 4, so our final result is independent of the shaker's mass. For example, if we were to slide a shaker with twice the mass, but with the same initial speed, it would slide the same distance. It is unlikely this independence would have been apparent if we had worked the problem numerically rather than symbolically.

Practice Problem

Given the same initial speed and a coefficient of kinetic friction equal to 0.120, what is **(a)** the acceleration of the shaker, and **(b)** the distance it slides? **[Answer: (a)** $a_x = -1.18\text{ m/s}^2$, **(b)** 0.560 m]

Some related homework problems: Problem 1, Problem 14

In the next Example we consider a system that is inclined at an angle θ relative to the horizontal. As a result, the normal force responsible for the kinetic friction is less than the weight of the object. To be very clear about how we handle the force vectors in such a case, we begin by resolving each vector into its x and y components.

PROBLEM-SOLVING NOTE

Choice of Coordinate System: Incline

On an incline, align one axis (x) parallel to the surface, and the other axis (y) perpendicular to the surface. That way the motion is in the x direction. Since no motion occurs in the y direction, we know that $a_y = 0$.

EXAMPLE 6–2 Making a Big Splash

A trained sea lion slides from rest with constant acceleration down a 3.0-m-long ramp into a pool of water. If the ramp is inclined at an angle of 23° above the horizontal and the coefficient of kinetic friction between the sea lion and the ramp is 0.26, how long does it take for the sea lion to make a splash in the pool?

Picture the Problem

As is usual with inclined surfaces, we choose one axis to be parallel to the surface and the other to be perpendicular to it. In our sketch, the sea lion accelerates in the positive x direction ($a_x > 0$), having started from rest, $v_{0x} = 0$. We are free to choose the initial position of the sea lion to be $x_0 = 0$. There is no motion in the y direction, and therefore $a_y = 0$. Finally, we note from the free-body diagram that $\vec{\mathbf{N}} = N\hat{\mathbf{y}}$, $\vec{\mathbf{f}}_k = -\mu_k N\hat{\mathbf{x}}$, and $\vec{\mathbf{W}} = (mg \sin\theta)\hat{\mathbf{x}} + (-mg\cos\theta)\hat{\mathbf{y}}$.

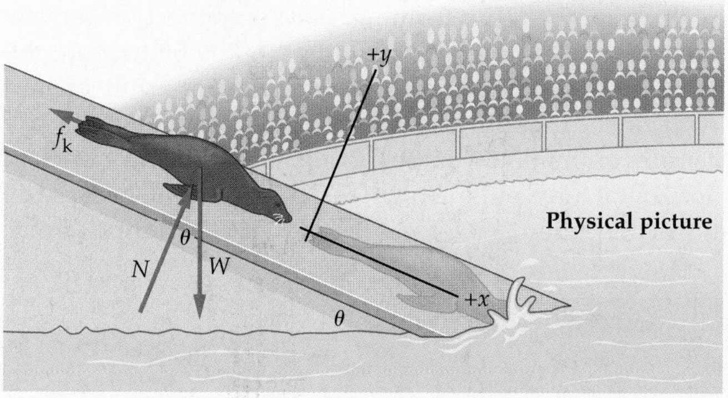

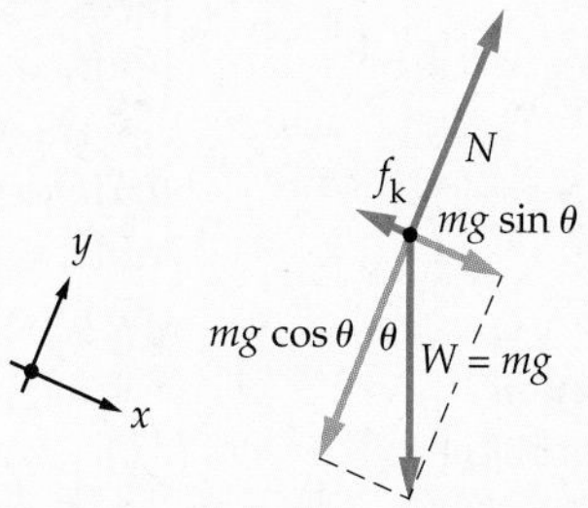

Strategy

We can use the kinematic equation relating position to time, $x = x_0 + v_{0x}t + \frac{1}{2}a_xt^2$, to find the time of the sea lion's slide. It will be necessary, however, to first determine the acceleration of the sea lion in the x direction, a_x.

To find a_x we apply Newton's second law to the sea lion. First, we can find N by setting $\Sigma F_y = ma_y$ equal to zero (since $a_y = 0$). It is important to start by finding N because we need it to find the force of kinetic friction, $f_k = \mu_k N$. Using f_k in the sum of forces in the x direction, $\Sigma F_x = ma_x$, allows us to solve for a_x and, finally, for the time.

Solution

1. We begin by resolving each of the three force vectors into x and y components:

$$N_x = 0 \qquad N_y = N$$
$$f_{k,x} = -f_k = -\mu_k N \qquad f_{k,y} = 0$$
$$W_x = mg\sin\theta \qquad W_y = -mg\cos\theta$$

2. Set $\Sigma F_y = ma_y = 0$ to find N:
We see that N is less than the weight, mg:

$$\sum F_y = N - mg\cos\theta = ma_y = 0$$
$$N = mg\cos\theta$$

3. Next, set $\Sigma F_x = ma_x$:
Note that the mass cancels in this equation:

$$\sum F_x = mg\sin\theta - \mu_k N$$
$$= mg\sin\theta - \mu_k mg\cos\theta = ma_x$$

4. Solve for the acceleration in the x direction, a_x:

$$a_x = g(\sin\theta - \mu_k\cos\theta)$$
$$= (9.81\ \mathrm{m/s^2})[\sin 23° - (0.26)\cos 23°] = 1.5\ \mathrm{m/s^2}$$

5. Use $x = x_0 + v_{0x}t + \frac{1}{2}a_xt^2$ to find the time when the sea lion reaches the bottom. We choose $x_0 = 0$, and we are given that $v_{0x} = 0$, hence we set $x = \frac{1}{2}a_xt^2 = 3.0$ m and solve for t:

$$x = \tfrac{1}{2}a_xt^2$$
$$t = \sqrt{\frac{2x}{a_x}} = \sqrt{\frac{2(3.0\ \mathrm{m})}{1.5\ \mathrm{m/s^2}}} = 2.0\ \mathrm{s}$$

continued on next page

continued from previous page

Insight

Note that we don't need the sea lion's mass to find the time. On the other hand, if we wanted the magnitude of the force of kinetic friction, $f_k = \mu_k N = \mu_k mg \cos\theta$, the mass would be needed.

It is useful to compare the sliding salt shaker in Example 6–1 with the sliding sea lion in this Example. In the case of the salt shaker, friction is the only force acting along the direction of motion (opposite to the direction of motion, in fact), and it brings the object to rest. Because of the slope on which the sea lion slides, however, it experiences both a component of its weight in the forward direction and the friction force opposite to the motion. Since the component of the weight is the larger of the two forces, the sea lion accelerates down the slope—friction only acts to slow its progress.

Practice Problem

How long would it take the sea lion to reach the water if there were no friction in this system? [**Answer:** 1.3 s]

Some related homework problems: Problem 8, Problem 56

Static Friction

Static friction tends to keep two surfaces from moving relative to one another. It, like kinetic friction, is due to the microscopic irregularities of surfaces that are in contact. In fact, static friction is typically stronger than kinetic friction because when surfaces are in static contact, their microscopic hills and valleys can nestle down deeply into one another, thus forming a strong connection between the surfaces that may even include molecular bonding. In kinetic friction, the surfaces bounce along relative to one another and don't become as firmly enmeshed.

As we did with kinetic friction, let's use the results of some simple experiments to determine the rules of thumb for static friction. We start with a brick at rest on a table, with no horizontal force pulling on it, as in **Figure 6–3**. Of course, in this case the force of static friction is zero; no force is needed to keep the brick from sliding.

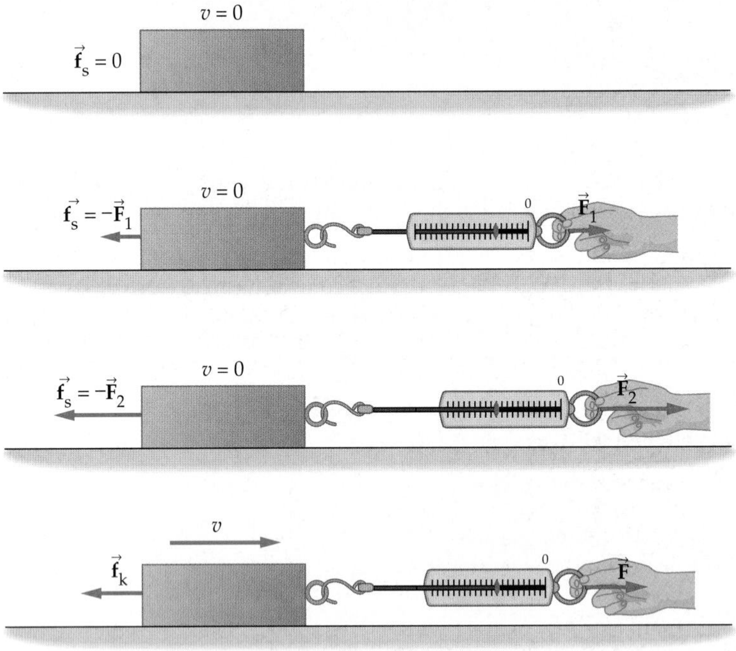

▶ **FIGURE 6–3 Static friction**
As the force applied to an object increases, so does the force of static friction—up to a certain point. Beyond this maximum value, static friction can no longer hold the object, and it begins to slide. Now kinetic friction takes over.

Next, attach a spring scale to the brick and pull with a small force of magnitude F_1, a force small enough that the brick doesn't move. Since the brick is still at rest, it follows that the force of static friction, f_s, is equal in magnitude to the applied force; that is, $f_s = F_1$. Now, increase the applied force to a new value, F_2, which is still small enough that the brick stays at rest. In this case, the force of static friction has also increased so that $f_s = F_2$. If we continue increasing the applied force, we eventually reach a value beyond which the brick starts to move and kinetic friction takes over, as shown in the figure. Thus, there is an upper limit to the force that can be exerted by static friction, and we call this upper limit $f_{s,max}$.

To summarize, the force of static friction, f_s, can have any value between zero and $f_{s,max}$. This can be written mathematically as follows:

$$0 \leq f_s \leq f_{s,max} \qquad 6\text{–}2$$

Imagine repeating the experiment, only now with a second brick on top of the first. This doubles the normal force and it also doubles the maximum force of static friction. Thus, the maximum force is proportional to the magnitude of the normal force, or

$$f_{s,max} = \mu_s N \qquad 6\text{–}3$$

The constant of proportionality is called μ_s (pronounced "mew sub s"), the **coefficient of static friction.** Note that μ_s, like μ_k, is dimensionless. Typical values are given in Table 6–1. In most cases, μ_s is greater than μ_k, indicating that the force of static friction is greater than the force of kinetic friction, as mentioned. In fact, it is not uncommon for μ_s to be greater than 1, as in the case of rubber in contact with dry concrete.

Finally, two additional comments regarding the nature of static friction: (i) Experiments show that static friction, like kinetic friction, is independent of the area of contact. (ii) The force of static friction is not in the direction of the normal force, thus $f_{s,max} = \mu_s N$ is not a vector relation. The direction of f_s is parallel to the surface of contact, and opposite to the direction the object would move if there were no friction.

These observations are summarized in the following rules of thumb:

Rules of Thumb for Static Friction

The force of static friction between two surfaces has the following properties:

1. It takes on any value between zero and the maximum possible force of static friction, $f_{s,max} = \mu_s N$:

$$0 \leq f_s \leq \mu_s N$$

2. It is independent of the area of contact between the surfaces.
3. It is parallel to the surface of contact, and in the direction that opposes relative motion.

▲ The coefficient of static friction between two surfaces depends on many factors, including whether the surfaces are dry or wet. On the desert floor of Death Valley, California, occasional rains can reduce the friction between rocks and the sandy ground to such an extent that strong winds can move the rocks over considerable distances. This results in linear "rock trails," which record the direction of the winds at different times.

Next, we consider a practical method of determining the coefficient of static friction. As with the last Example, we begin by resolving all relevant force vectors into their x and y components.

EXAMPLE 6–3 Slightly Tilted

A flatbed truck slowly tilts its bed upward to dispose of a 95.0-kg crate. For small angles of tilt the crate stays put, but when the tilt angle exceeds 23.2°, the crate begins to slide. What is the coefficient of static friction between the bed of the truck and the crate?

Picture the Problem

We align our coordinate system with the incline, and choose the positive x direction to point down the slope. Note that three forces act on the crate: the normal force, $\vec{\mathbf{N}} = N\hat{\mathbf{y}}$, the force of static friction, $\vec{\mathbf{f}}_s = -\mu_s N\hat{\mathbf{x}}$, and the weight, $\vec{\mathbf{W}} = (mg\sin\theta)\hat{\mathbf{x}} + (-mg\cos\theta)\hat{\mathbf{y}}$.

INTERACTIVE FIGURE

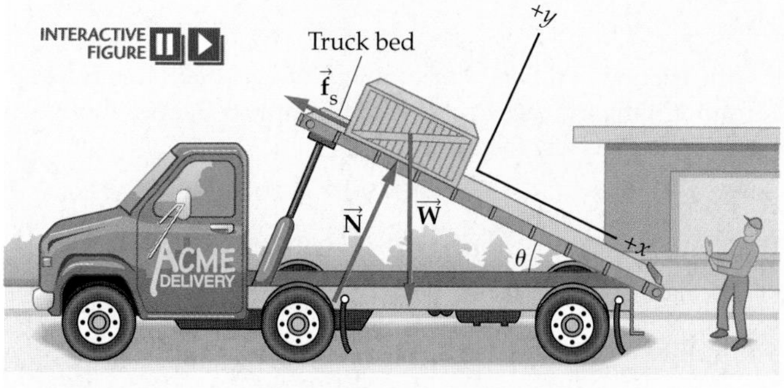

Physical picture

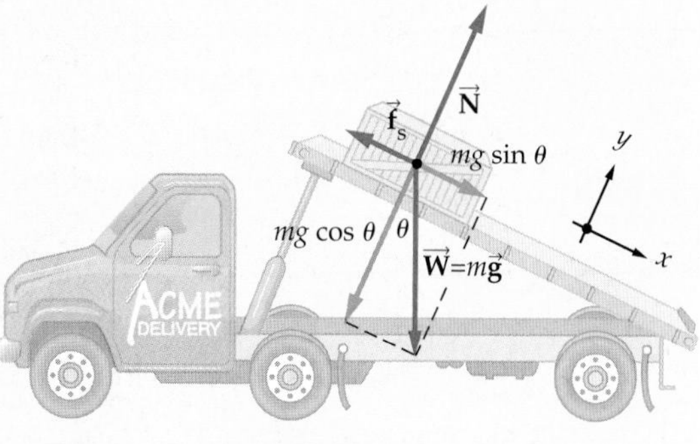

Free-body diagram

continued on next page

continued from previous page

Strategy

When the crate is on the verge of slipping, but has not yet slipped, its acceleration is zero in both the x and y directions. In addition, "verge of slipping" means that the magnitude of the static friction is at its maximum value, $f_s = f_{s,max} = \mu_s N$. Thus, we set $\Sigma F_y = ma_y = 0$ to find N, then use $\Sigma F_x = ma_x = 0$ to find μ_s.

Solution

1. Resolve the three force vectors acting on the crate into x and y components:

$$N_x = 0 \qquad N_y = N$$
$$f_{s,x} = -f_{s,max} = -\mu_s N \qquad f_{s,y} = 0$$
$$W_x = mg\sin\theta \qquad W_y = -mg\cos\theta$$

2. Set $\Sigma F_y = ma_y = 0$, since $a_y = 0$:

$$\sum F_y = N_y + f_{s,y} + W_y = N + 0 - mg\cos\theta = ma_y = 0$$

Solve for the normal force, N:

$$N = mg\cos\theta$$

3. Set $\Sigma F_x = ma_x = 0$, since the crate is at rest, and use the result for N obtained in step 2:

$$\sum F_x = N_x + f_{s,x} + W_x = ma_x = 0$$
$$= 0 - \mu_s N + mg\sin\theta$$
$$= 0 - \mu_s mg\cos\theta + mg\sin\theta$$

4. Solve the expression for the coefficient of static friction, μ_s:

$$\mu_s mg\cos\theta = mg\sin\theta$$
$$\mu_s = \frac{mg\sin\theta}{mg\cos\theta} = \tan\theta = \tan 23.2° = 0.429$$

Insight

In general, if an object is on the verge of slipping when the surface on which it rests is tilted at an angle θ_c, the coefficient of static friction between the object and the surface is $\mu_s = \tan\theta_c$. Note that this result is independent of the mass of the object. In particular, the critical angle for this crate is precisely the same whether it is filled with feathers or lead bricks.

Practice Problem

Find the magnitude of the force of static friction acting on the crate. [**Answer:** $f_{s,max} = \mu_s N = 367$ N]

Some related homework problems: Problem 9, Problem 68

Recall that static friction can have magnitudes less than its maximum possible value. This point is emphasized in the following Active Example.

ACTIVE EXAMPLE 6–1 The Force of Static Friction

In the previous Example, what is the magnitude of the force of static friction acting on the crate when the truck bed is tilted at an angle of 20.0°?

Solution *(Test your understanding by performing the calculations indicated in each step.)*

1. Sum the x components of force acting on the crate: $\sum F_x = 0 - f_s + mg\sin\theta$
2. Set this sum equal to zero (since $a_x = 0$) and solve for the magnitude of the static friction force, f_s: $f_s = mg\sin\theta$
3. Substitute numerical values, including $\theta = 20.0°$: $f_s = 319$ N

Insight

Notice that the force of static friction in this case has a magnitude (319 N) that is less than the value of 367 N found in the Practice Problem of Example 6–3, even though the coefficient of static friction is precisely the same.

Your Turn

At what tilt angle will the force of static friction have a magnitude of 225 N?

(Answers to **Your Turn** *problems are given in the back of the book.)*

Finally, friction often enters into problems dealing with vehicles with rolling wheels. In Conceptual Checkpoint 6–1, we consider which type of friction is appropriate in such cases.

CONCEPTUAL CHECKPOINT 6–1 Friction for Rolling Tires

A car drives with its tires rolling freely. Is the friction between the tires and the road (a) kinetic or (b) static?

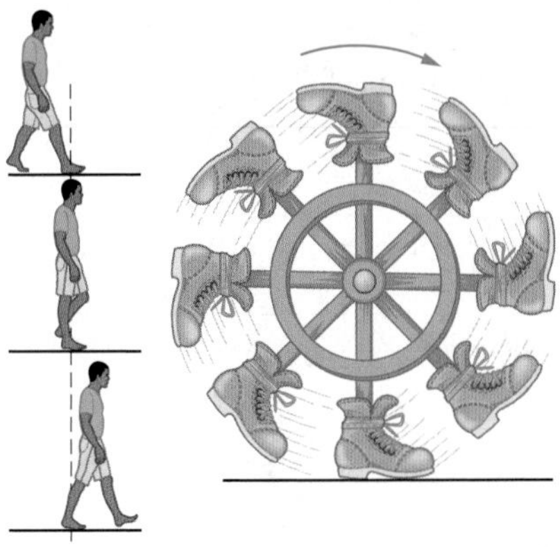

Reasoning and Discussion

A reasonable-sounding answer is that because the car is moving, the friction between its tires and the road must be kinetic friction—but this is not the case.

Actually, the friction is static because the bottom of the tire is in static contact with the road. To understand this, watch your feet as you walk. Even though you are moving, each foot is in static contact with the ground once you step down on it. Your foot doesn't move again until you lift it up and move it forward for the next step. A tire can be thought of as a succession of feet arranged in a circle, each of which is momentarily in static contact with the ground.

Answer:

(b) The friction between the tires and the road is static friction.

To summarize, if a car skids, the friction acting on it is kinetic; if its wheels are rolling, the friction is static. Since static friction is generally greater than kinetic friction, it follows that a car can be stopped in less distance if its wheels are rolling (static friction) than if its wheels are locked up (kinetic friction). This is the idea behind the antilock braking systems (ABS) that are available on many cars. When the brakes are applied in a car with ABS, an electronic rotation sensor at each wheel detects whether the wheel is about to start skidding. To prevent skidding, a small computer automatically begins to modulate the hydraulic pressure in the brake lines in short bursts, causing the brakes to release and then reapply in rapid succession. This allows the wheels to continue rotating, even in an emergency stop, and for static friction to determine the stopping distance. **Figure 6–4** shows a comparison of braking distances for cars with and without ABS. An added benefit of ABS is that a driver is better able to steer and control a braking car if its wheels are rotating.

REAL-WORLD PHYSICS
Antilock braking systems

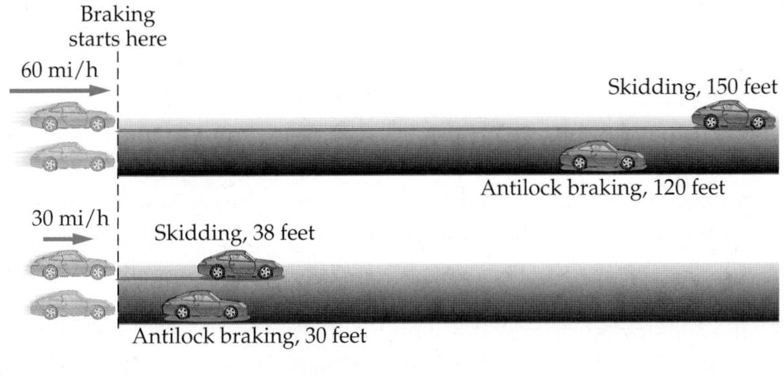

◀ **FIGURE 6–4 Stopping distance with and without ABS**

Antilock braking systems (ABS) allow a car to stop with static friction rather than kinetic friction—even in a case where a person slams on the brakes. As a result, the braking distance is reduced, due to the fact that μ_s is typically greater than μ_k. Professional drivers can beat the performance of ABS by carefully adjusting the force they apply to the brake pedal during a stop, but ABS provides essentially the same performance—within a few percent—for a person who simply pushes the brake pedal to the floor and holds it there.

▲ The angle that the sloping sides of a sand pile (left) make with the horizontal is determined by the coefficient of static friction between grains of sand, in much the same way that static friction determines the angle at which the crate in Example 6–3 begins to slide. The same basic mechanism determines the angle of the cone-shaped mass of rock debris at the base of a cliff, known as a talus slope (right).

6–2 Strings and Springs

A common way to exert a force on an object is to pull on it with a string, a rope, a cable, or a wire. Similarly, you can push or pull on an object if you attach it to a spring. In this section we discuss the basic features of strings and springs and how they transmit forces.

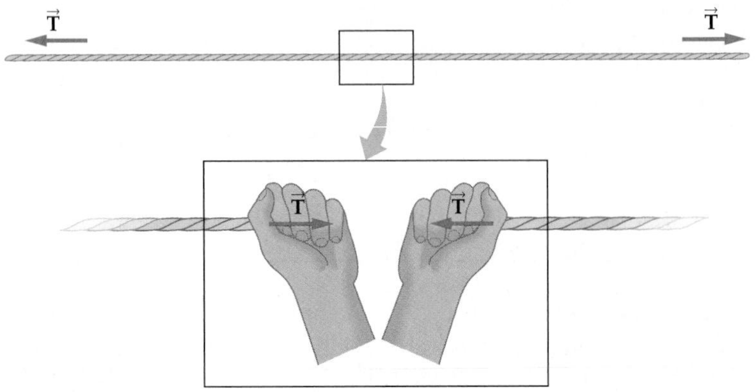

▲ **FIGURE 6–5 Tension in a string**

A string, pulled from either end, has a tension, T. If the string were to be cut at any point, the force required to hold the ends together is T.

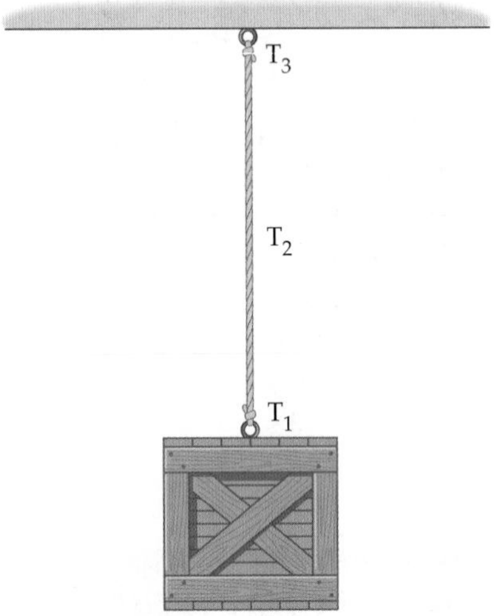

▲ **FIGURE 6–6 Tension in a heavy rope**

Because of the weight of the rope, the tension is noticeably different at points 1, 2, and 3. As the rope becomes lighter, however, the difference in tension decreases. In the limit of a rope of zero mass, the tension is the same throughout the rope.

Strings and Tension

Imagine picking up a light string and holding it with one end in each hand. If you pull to the right with your right hand with a force T and to the left with your left hand with a force T, the string becomes taut. In such a case, we say that there is a **tension** T in the string. To be more specific, if your friend were to cut the string at some point, the tension T is the force pulling the ends apart, as illustrated in **Figure 6–5**—that is, T is the force your friend would have to exert with each hand to hold the cut ends together. Note that at any given point, the tension pulls equally to the right and to the left.

As an example, consider a rope that is attached to the ceiling at one end, and to a box with a weight of 105 N at the other end, as shown in **Figure 6–6**. In addition, suppose the rope is uniform, and that it has a total weight of 2.00 N. What is the tension in the rope (i) where it attaches to the box, (ii) at its midpoint, and (iii) where it attaches to the ceiling?

First, the rope holds the box at rest; thus, the tension where the rope attaches to the box is simply the weight of the box, $T_1 = 105\text{ N}$. At the midpoint of the rope, the tension supports the weight of the box, plus the weight of half the rope. Thus, $T_2 = 105\text{ N} + \frac{1}{2}(2.00\text{ N}) = 106\text{ N}$. Similarly, at the ceiling the tension supports the box plus all of the rope, giving a tension of $T_3 = 107\text{ N}$. Note that the tension pulls down on the ceiling but pulls up on the box.

From this discussion, we can see that the tension in the rope changes slightly from top to bottom because of the mass of the rope. If the rope had less mass, the difference in tension between its two ends would also be less. In particular, if the rope's mass were to be vanishingly small, the difference in tension would vanish as well. In this text, we will assume that all ropes, strings, wires, and so on are practically massless—unless specifically stated otherwise—and, hence, that the tension is the same throughout their length.

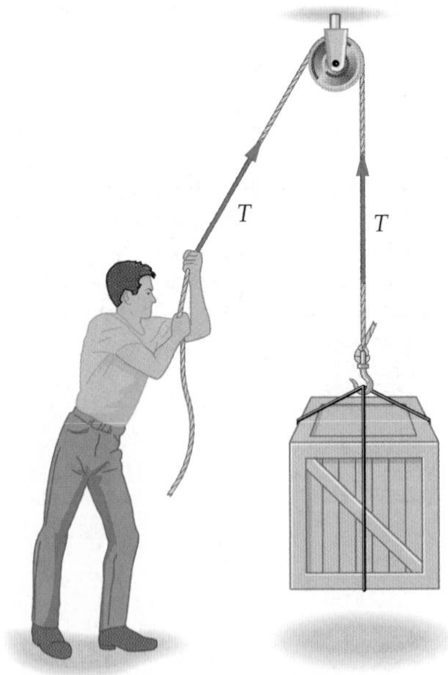

▲ **FIGURE 6–7 A pulley changes the direction of a tension**

In an ideal string, the tension has the same magnitude, T, throughout its length. A pulley can serve to redirect the string, however, so that the tension acts in a different direction.

Pulleys are often used to redirect a force exerted by a string, as indicated in **Figure 6–7**. In the ideal case, a pulley has no mass, and no friction in its bearings. Thus, *an ideal pulley simply changes the direction of the tension in a string, without changing its magnitude*. If a system contains more than one pulley, however, it is possible to arrange them in such a way as to "magnify a force," even if each pulley itself merely redirects the tension in a string. The traction device considered in the next Example shows one way this can be accomplished in a system that uses three ideal pulleys.

EXAMPLE 6–4 A Bad Break: Setting a Broken Leg with Traction

A traction device employing three pulleys is applied to a broken leg, as shown in the sketch. The middle pulley is attached to the sole of the foot, and a mass m supplies the tension in the ropes. Find the value of the mass m if the force exerted on the sole of the foot by the middle pulley is to be 165 N.

Picture the Problem

Our sketch shows the physical picture as well as the tension forces acting on the middle pulley. Notice that on the upper portion of the rope the tension is $\vec{\mathbf{T}}_1 = (T\cos 40.0°)\hat{\mathbf{x}} + (T\sin 40.0°)\hat{\mathbf{y}}$; on the lower portion it is $\vec{\mathbf{T}}_2 = (T\cos 40.0°)\hat{\mathbf{x}} + (-T\sin 40.0°)\hat{\mathbf{y}}$.

Strategy

We begin by noting that the rope supports the hanging mass m. As a result, the tension in the rope, T, must be equal in magnitude to the weight of the mass; $T = mg$.

Next, the pulleys simply change the direction of the tension without changing its magnitude. Therefore, the net force exerted on the sole of the foot is the sum of the tension T at 40.0° above the horizontal plus the tension T at 40.0° below the horizontal. We will calculate the net force component by component.

Once we calculate the net force acting on the foot, we set it equal to 165 N and solve for the tension T. Finally, we find the mass using the relation $T = mg$.

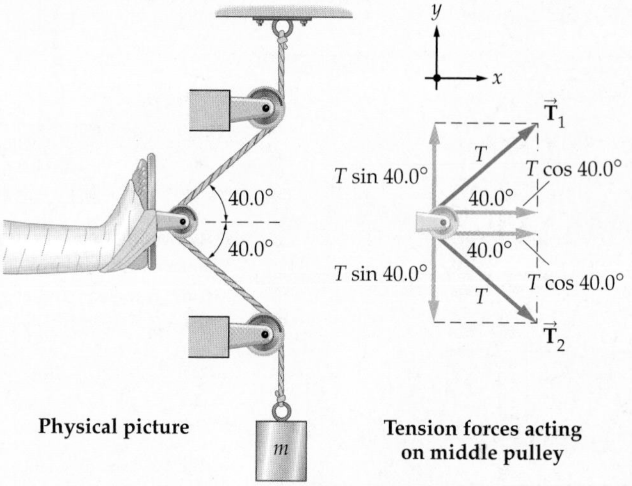

Physical picture

Tension forces acting on middle pulley

Solution

1. First, consider the tension that acts upward and to the right on the middle pulley. Resolve this tension into x and y components:

$$T_{1,x} = T\cos 40.0° \qquad T_{1,y} = T\sin 40.0°$$

2. Next, consider the tension that acts downward and to the right on the middle pulley. Resolve this tension into x and y components. Note the minus sign in the y component:

$$T_{2,x} = T\cos 40.0° \qquad T_{2,y} = -T\sin 40.0°$$

3. Sum the x and y components of force acting on the middle pulley. We see that the net force acts only in the x direction, as one might expect from symmetry:

$$\sum F_x = T\cos 40.0° + T\cos 40.0° = 2T\cos 40.0°$$
$$\sum F_y = T\sin 40.0° - T\sin 40.0° = 0$$

4. Step 3 shows that the net force acting on the middle pulley is $2T\cos 40.0°$. Set this force equal to 165 N and solve for T:

$$2T\cos 40.0° = 165\text{ N}$$
$$T = \frac{165\text{ N}}{2\cos 40.0°} = 108\text{ N}$$

5. Solve for the mass, m, using $T = mg$:

$$T = mg$$
$$m = \frac{T}{g} = \frac{108\text{ N}}{9.81\text{ m/s}^2} = 11.0\text{ kg}$$

Insight

As pointed out earlier, this pulley arrangement "magnifies the force" in the sense that a 108-N weight attached to the rope produces a 165-N force exerted on the foot by the middle pulley. Note that the tension in the rope always has the same value—$T = 108$ N—as expected with ideal pulleys, but because of the arrangement of the pulleys the force applied to the foot by the rope is $2T\cos 40.0° > T$.

In addition, notice that the force exerted on the foot by the middle pulley produces an opposing force in the leg that acts in the direction of the head (a cephalad force), as desired to set a broken leg and keep it straight as it heals.

Practice Problem

(a) Would the required mass m increase or decrease if the angles in this device were changed from 40.0° to 30.0°? **(b)** Find the mass m for an angle of 30.0°. [**Answer: (a)** The required mass m will decrease. **(b)** 9.71 kg]

Some related homework problems: Problem 18, Problem 21, Problem 31

CONCEPTUAL CHECKPOINT 6–2 Compare the Readings on the Scales

The scale at left reads 9.81 N. Is the reading of the scale at right **(a)** greater than 9.81 N, **(b)** equal to 9.81 N, or **(c)** less than 9.81 N?

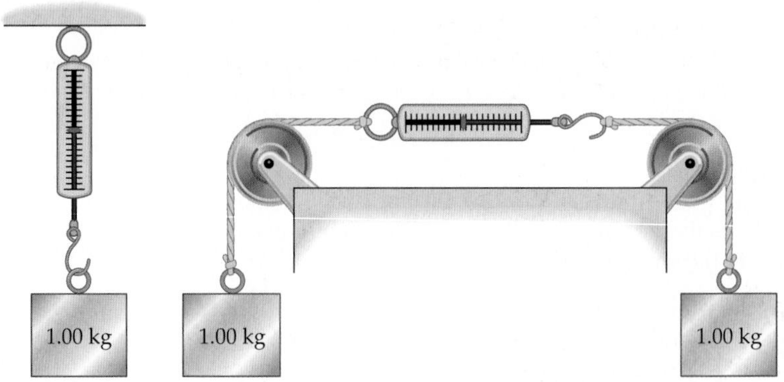

Reasoning and Discussion

Since a pulley simply changes the direction of the tension in a string without changing its magnitude, it is clear that the scale attached to the ceiling reads the same as the scale shown in the figure below.

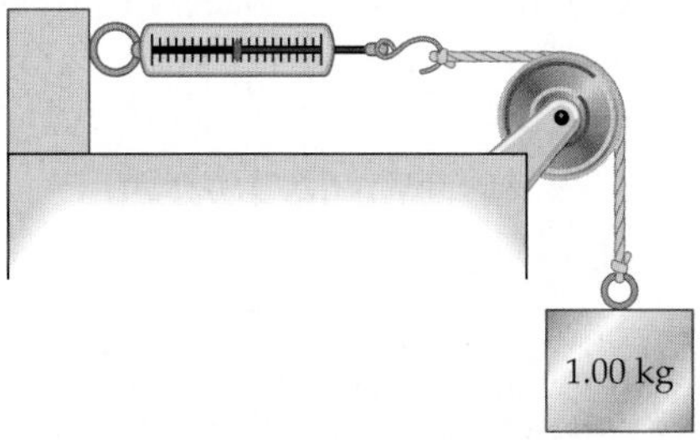

There is no difference, however, between attaching the top end of the scale to something rigid and attaching it to another 1.00-kg hanging mass. In either case, the fact that the scale is at rest means that a force of 9.81 N must be exerted to the left on the top of the scale to balance the 9.81-N force exerted on the lower end of the scale. As a result, the two scales read the same.

Answer:

(b) The reading of the scale at right is equal to 9.81 N.

Springs and Hooke's Law

Suppose you take a spring of length L, as shown in **Figure 6–8 (a)**, and attach it to a block. If you pull on the spring, causing it to stretch to a length $L + x$, the spring pulls on the block with a force of magnitude F. If you increase the length of the spring to $L + 2x$, the force exerted by the spring increases to $2F$. Similarly, if you compress the spring to a length $L - x$, the spring pushes on the block with a force of magnitude F, where F is the same force given previously. As you might expect, compression to a length $L - 2x$ results in a push of magnitude $2F$.

As a result of these experiments, we can say that a spring exerts a force that is proportional to the amount, x, by which it is stretched or compressed. Thus, if F is the magnitude of the spring force, we can say that

$$F = kx$$

In this expression, k is a constant of proportionality, referred to as the **force constant,** or, equivalently, as the spring constant. Since F has units of newtons and x has units of meters, it follows that k has units of newtons per meter, or N/m. The larger the value of k, the stiffer the spring.

To be more precise, consider the spring shown in **Figure 6–8 (b)**. Note that we have placed the origin of the x axis at the equilibrium length of the spring—that is, at the position of the spring when no force acts on it. Now, if we stretch the spring so that the end of the spring is at a positive value of $x (x > 0)$, we find that

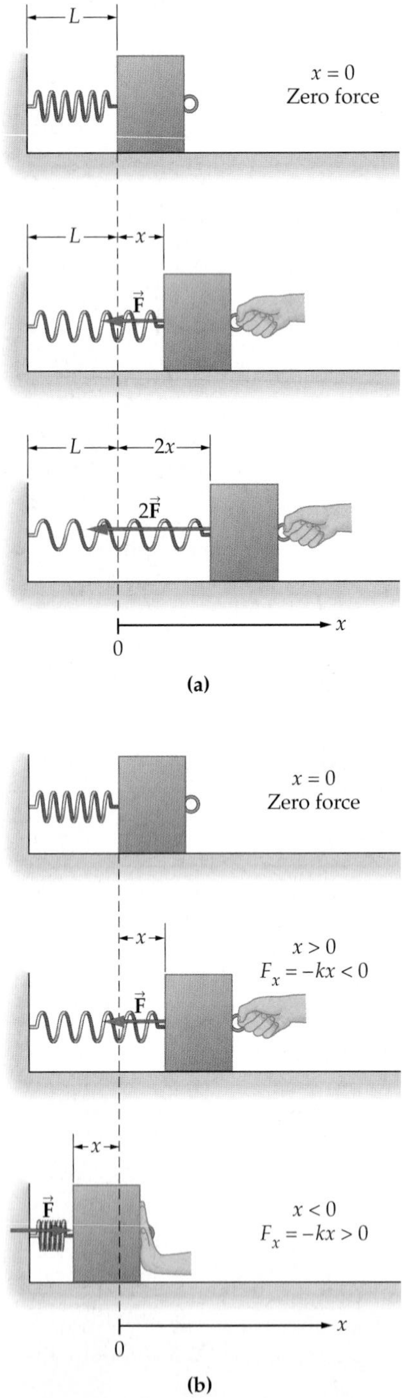

▲ **FIGURE 6–8 Spring Forces**
When dealing with a spring, it is convenient to choose the origin at the equilibrium (zero force) position. In the cases shown above, the force is strictly in the x direction, and is given by $F_x = -kx$. Note that the minus sign means that the force is opposite to the displacement; that is, the force is restoring.

the spring exerts a force of magnitude kx in the negative x direction. Thus, the spring force (which has only an x component) can be written as

$$F_x = -kx$$

Similarly, consider compressing the spring so that its end is at a negative value of $x(x < 0)$. In this case, the force exerted by the spring is of magnitude kx, and points in the positive x direction, as is shown in Figure 6–8 (b). Again, we can write the spring force as

$$F_x = -kx$$

To see that this is correct—that is, that F_x is positive in this case—recall that x is negative, which means that $(-x)$ is positive.

This result for the force of a spring is known as Hooke's law, after Robert Hooke (1635–1703). It is really just a good rule of thumb rather than a law of nature. Clearly, it can't work for any amount of stretching. For example, we know that if we stretch a spring far enough it will be permanently deformed, and will never return to its original length. Still, for small stretches or compressions, Hooke's law is quite accurate.

Rules of Thumb for Springs (Hooke's Law)

A spring stretched or compressed by the amount x from its equilibrium length exerts a force whose x component is given by

$F_x = -kx$ *(gives magnitude and direction)* **6–4**

If we are interested only in the magnitude of the force associated with a given stretch or compression, we use the somewhat simpler form of Hooke's law:

$F = kx$ *(gives magnitude only)* **6–5**

In this text, we consider only **ideal springs**—that is, springs that are massless, and that are assumed to obey Hooke's law exactly.

Since the stretch of a spring and the force it exerts are proportional, we can now see how a spring scale operates. In particular, pulling on the two ends of a scale stretches the spring inside it by an amount proportional to the applied force. Once the scale is calibrated—by stretching the spring with a known, or reference, force—we can use it to measure other unknown forces.

Finally, it is useful to note that Hooke's law, which we've introduced in the context of ideal springs, is particularly important in physics because it applies to so much more than just springs. For example, the forces that hold atoms together are often modeled by Hooke's law—that is, as "interatomic springs"—and these are the forces that are ultimately responsible for the normal force (Chapter 5), vibrations and oscillations (Chapter 13), wave motion (Chapter 14), and even the thermal expansion of solids (Chapter 16). And this just scratches the surface—Hooke's law comes up in one form or another in virtually every field of physics. In the following Active Example, we present a biomedical application of Hooke's law.

▲ Springs come in a variety of sizes and shapes. The large springs on a railroad car (top) are so stiff and heavy that you can't compress or stretch them by hand. Still, three of them are needed to smooth the ride of this car. In contrast, the delicate spiral spring inside a watch (bottom) flexes with even the slightest touch. It exerts enough force, however, to power the equally delicate mechanism of the watch.

ACTIVE EXAMPLE 6–2 Nasal Strips

An increasingly popular device for improving air flow through nasal passages is the nasal strip, which consists of two flat, polyester springs enclosed by an adhesive tape covering. Measurements show that a nasal strip can exert an outward force of 0.22 N on the nose, causing it to expand by 3.5 mm. **(a)** Treating the nose as an ideal spring, find its force constant in newtons per meter. **(b)** How much force would be required to expand the nose by 4.0 mm?

REAL-WORLD PHYSICS: BIO

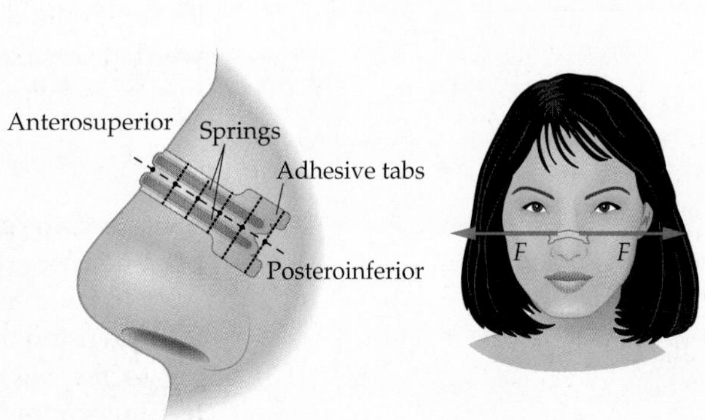

continued on next page

continued from previous page

Solution *(Test your understanding by performing the calculations indicated in each step.)*

Part (a)

1. Solve the magnitude form of Hooke's law, $F = kx$, for the force constant, k: $k = F/x$
2. Substitute numerical values for F and x: $k = 62\ \text{N/m}$

Part (b)

3. Use $F = kx$ to find the required force: $F = 0.25\ \text{N}$

Insight

Even though the human nose is certainly not an ideal spring, Hooke's law is still a useful way to model its behavior when dealing with forces and the stretches they cause.

Your Turn

Suppose a new nasal strip comes on the market that exerts an outward force of 0.32 N. What expansion of the nose will be caused by this strip?

(*Answers to* **Your Turn** *problems are given in the back of the book.*)

6–3 Translational Equilibrium

When we say that an object is in **translational equilibrium,** we mean that the net force acting on it is zero:

$$\sum \vec{\mathbf{F}} = 0 \qquad \textbf{6–6}$$

From Newton's second law, this is equivalent to saying that the object's acceleration is zero. In two-dimensional systems, translational equilibrium implies two independent conditions: $\Sigma F_x = 0$ and $\Sigma F_y = 0$. In one dimension, only one of these conditions will apply.

Later, in Chapters 10 and 11, we will study objects that have both rotational and linear motions. In such cases, rotational equilibrium will be as important as translation equilibrium. For now, however, when we say equilibrium, we simply mean translational equilibrium.

As a first example, consider the one-dimensional situation illustrated in **Figure 6–9**. Here we see a person lifting a bucket of water from a well by pulling down on a rope that passes over a pulley. If the bucket's mass is m, and it is rising with constant speed v, what is the tension T_1 in the rope attached to the bucket? In addition, what is the tension T_2 in the chain that supports the pulley?

To answer these questions, we first note that both the bucket and the pulley are in equilibrium; that is, they both have zero acceleration. As a result, the net force on each of them must be zero.

Let's start with the bucket. In Figure 6–9, we see that just two forces act on the bucket: (i) its weight $W = mg$ downward, and (ii) the tension in the rope, T_1 upward. If we take upward to be the positive direction, we can write $\Sigma F_y = 0$ for the bucket as follows:

$$T_1 - mg = 0$$

Therefore, the tension in the rope is $T_1 = mg$. Note that this is also the force the person must exert downward on the rope, as expected.

Next, we consider the pulley. In Figure 6–9, we see that three forces act on it: (i) the tension in the chain, T_2 upward, (ii) the tension in the part of the rope leading to the bucket, T_1 downward, and (iii) the tension in the part of the rope leading to the person, T_1 downward. Note that we don't include the weight of the pulley since we consider it to be ideal; that is, massless and frictionless. If we

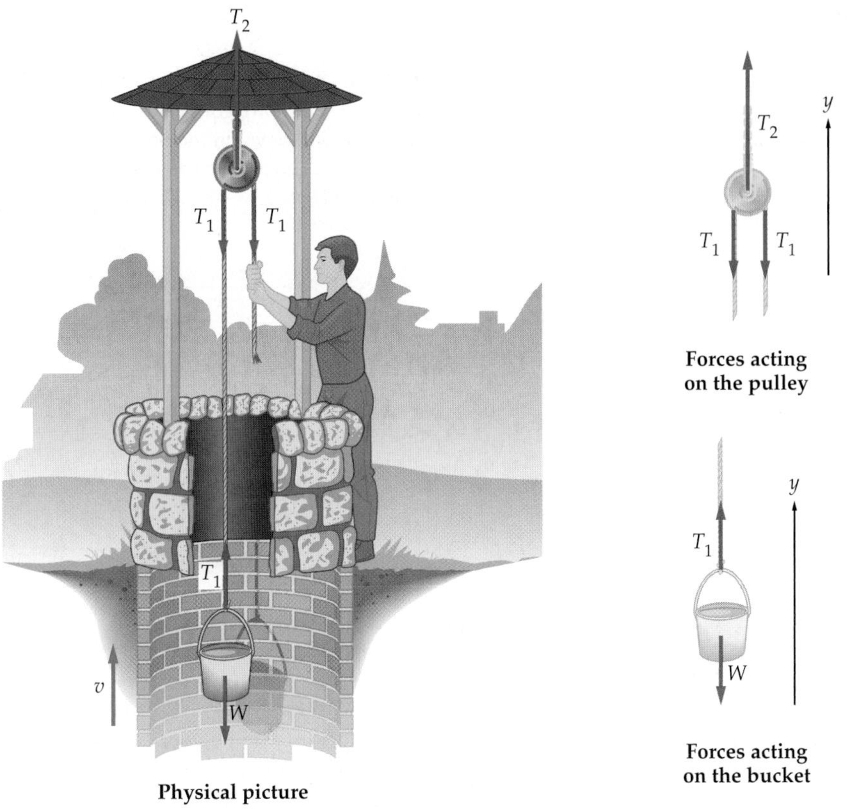

◀ **FIGURE 6–9 Raising a bucket**

A person lifts a bucket of water from the bottom of a well with a constant speed, v. Because the speed is constant, the net force acting on the bucket must be zero.

again take upward to be positive, the statement that the net force acting on the pulley is zero ($\Sigma F_y = 0$) can be written

$$T_2 - T_1 - T_1 = 0$$

It follows that the tension in the chain is $T_2 = 2T_1 = 2mg$, twice the weight of the bucket of water!

In the next Conceptual Checkpoint we consider a slight variation of this situation.

CONCEPTUAL CHECKPOINT 6–3 Comparing Tensions

A person hoists a bucket of water from a well and holds the rope, keeping the bucket at rest, as at left. A short time later, the person ties the rope to the bucket so that the rope holds the bucket in place, as at right. In this case, is the tension in the rope **(a)** greater than, **(b)** less than, or **(c)** equal to the tension in the first case?

continued on next page

continued from previous page

Reasoning and Discussion

In the first case (left), the only upward force exerted on the bucket is the tension in the rope. Since the bucket is at rest, the tension must be equal in magnitude to the weight of the bucket. In the second case (right), the two ends of the rope exert equal upward forces on the bucket, hence the tension in the rope is only half the weight of the bucket. To see this more clearly, imagine cutting the bucket in half so that each end of the rope supports half the weight, as indicated in the accompanying diagram.

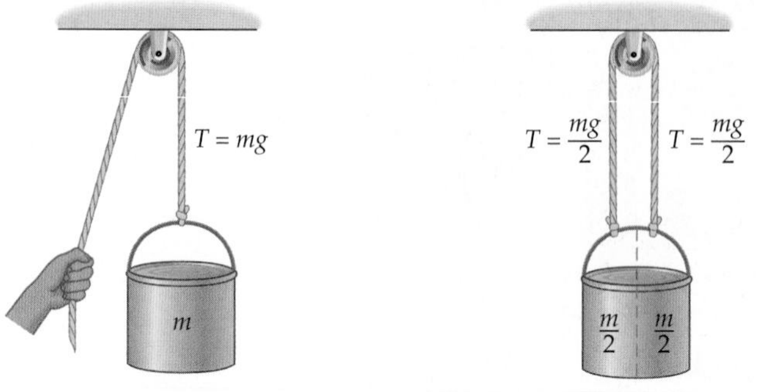

Answer:

(b) The tension in the second case is less than in the first.

In the next two Examples, we consider two-dimensional systems in which forces act at various angles with respect to one another. Hence, our first step is to resolve the relevant vectors into their x and y components. Following that, we apply the conditions for translational equilibrium, $\Sigma F_x = 0$ and $\Sigma F_y = 0$.

EXAMPLE 6–5 Suspended Vegetation

To hang a 6.20-kg pot of flowers, a gardener uses two wires—one attached horizontally to a wall, the other sloping upward at an angle of $\theta = 40.0°$ and attached to the ceiling. Find the tension in each wire.

Picture the Problem

We choose a typical coordinate system, with the positive x direction to the right and the positive y direction upward. With this choice, tension 1 is in the positive x direction, $\vec{\mathbf{T}}_1 = T_1\hat{\mathbf{x}}$, the weight is in the negative y direction, $\vec{\mathbf{W}} = -mg\hat{\mathbf{y}}$, and tension 2 has a negative x component and a positive y component, $\vec{\mathbf{T}}_2 = (-T_2 \cos\theta)\hat{\mathbf{x}} + (T_2 \sin\theta)\hat{\mathbf{y}}$.

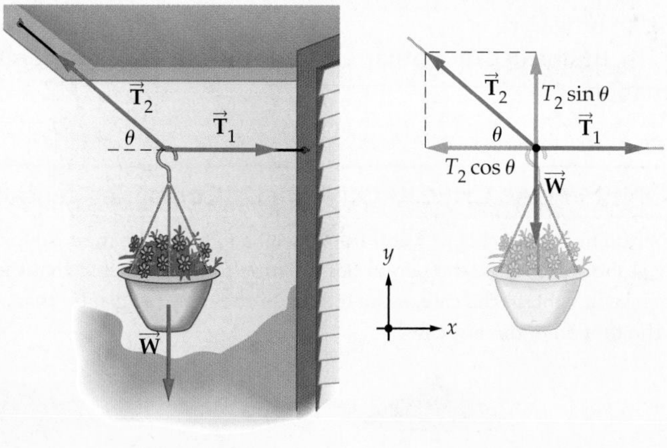

Physical picture

Free-body diagram

Strategy

The pot is at rest, and therefore the net force acting on it is zero. As a result, we can say that (i) $\Sigma F_x = 0$ and (ii) $\Sigma F_y = 0$. These two conditions allow us to determine the magnitude of the two tensions, T_1 and T_2.

Solution

1. First, resolve each of the forces acting on the pot into x and y components:

$$T_{1,x} = T_1 \qquad T_{1,y} = 0$$
$$T_{2,x} = -T_2 \cos\theta \qquad T_{2,y} = T_2 \sin\theta$$
$$W_x = 0 \qquad W_y = -mg$$

2. Now, set $\Sigma F_x = 0$. Note that this condition gives a relation between T_1 and T_2:

$$\sum F_x = T_{1,x} + T_{2,x} + W_x = T_1 + (-T_2 \cos\theta) + 0 = 0$$
$$T_1 = T_2 \cos\theta$$

3. Next, set $\Sigma F_y = 0$. This time, the resulting condition determines T_2 in terms of the weight, mg:

$$\sum F_y = T_{1,y} + T_{2,y} + W_y = 0 + T_2 \sin\theta + (-mg) = 0$$
$$T_2 \sin\theta = mg$$

4. Use the relation obtained in step 3 to find T_2: $$T_2 = \frac{mg}{\sin\theta} = \frac{(6.20\text{ kg})(9.81\text{ m/s}^2)}{\sin 40.0°} = 94.6\text{ N}$$

5. Finally, use the connection between the two tensions (obtained from $\Sigma F_x = 0$) to find T_1: $$T_1 = T_2\cos\theta = (94.6\text{ N})\cos 40.0° = 72.5\text{ N}$$

Insight

Notice that even though two wires suspend the pot, they both have tensions *greater* than the pot's weight, $mg = 60.8$ N. This is an important point for architects and engineers to consider when designing structures.

Practice Problem

Find T_1 and T_2 if the second wire slopes upward at the angle **(a)** $\theta = 20°$, **(b)** $\theta = 60.0°$, or **(c)** $\theta = 90.0°$.
[**Answer: (a)** $T_1 = 167$ N, $T_2 = 178$ N **(b)** $T_1 = 35.1$ N, $T_2 = 70.2$ N **(c)** $T_1 = 0$, $T_2 = mg = 60.8$ N]

Some related homework problems: Problem 29, Problem 32

ACTIVE EXAMPLE 6–3 The Forces in a Low-Tech Laundry

A 1.84-kg bag of clothespins hangs in the middle of a clothesline, causing it to sag by an angle $\theta = 3.50°$. Find the tension, T, in the clothesline.

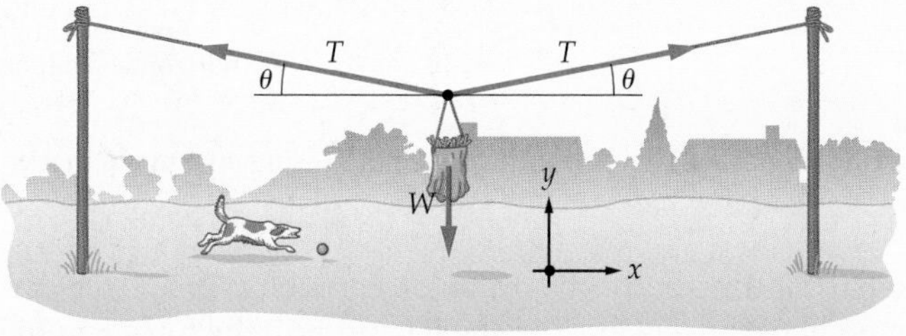

▲ Like the bag of clothespins in Active Example 6–3, this mountain climber is in static equilibrium. Since the ropes suspending the climber are nearly horizontal, the tension in them is significantly greater than the climber's weight.

Solution *(Test your understanding by performing the calculations indicated in each step.)*

1. Find the y component for each tension: $T_y = T\sin\theta$
2. Find the y component of the weight: $W_y = -mg$
3. Set $\Sigma F_y = 0$: $T\sin\theta + T\sin\theta - mg = 0$
4. Solve for T: $T = mg/(2\sin\theta) = 148$ N

Insight

Note that we only considered the y components of force in our calculation. This is because forces in the x direction automatically balance, due to the symmetry of the system.

Your Turn

At what sag angle, θ, will the tension in the clothesline have a magnitude of 175 N?

(Answers to **Your Turn** *problems are given in the back of the book.)*

At 148 N, the tension in the clothesline is quite large, especially when you consider that the weight of the clothespin bag itself is only 18.1 N. The reason for such a large value is that the vertical component of the two tensions is $2T\sin\theta$, which, for $\theta = 3.50°$, is $(0.122)T$. If $(0.122)T$ is to equal the weight of the bag, it is clear that T must be roughly eight times the bag's weight.

If you and a friend were to pull on the two ends of the clothesline, in an attempt to straighten it out, you would find that no matter how hard you pulled, the line would still sag. You may be able to reduce θ to quite a small value, but as you do so the corresponding tension increases rapidly. In principle, it would take an infinite force to completely straighten the line and reduce θ to zero.

On the other hand, if θ were 90°, so that the two halves of the clothesline were vertical, the tension would be $T = mg/(2 \sin 90°) = mg/2$. In this case, each side of the line supports half the weight of the bag, as expected.

6–4 Connected Objects

Interesting applications of Newton's laws arise when we consider accelerating objects that are tied together. Suppose, for example, that a force of magnitude F pulls two boxes—connected by a string—along a frictionless surface, as in **Figure 6–10**. In such a case, the string has a certain tension, T, and the two boxes have the same acceleration, a. Given the masses of the boxes and the applied force F, we would like to determine both the tension in the string and the acceleration of the boxes.

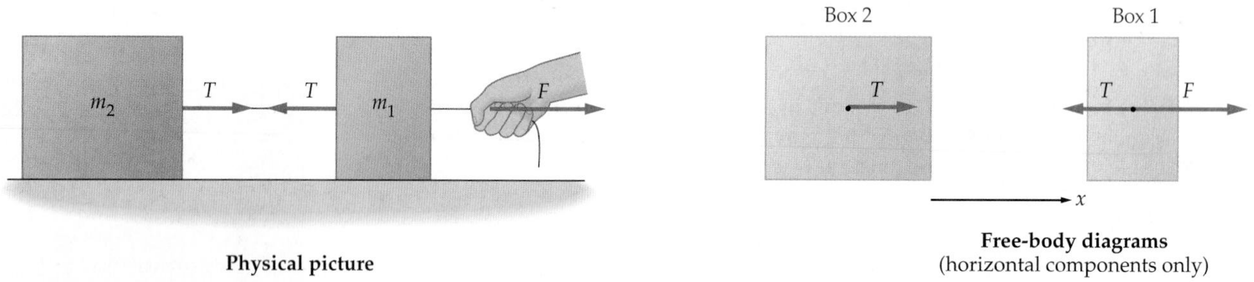

▲ **FIGURE 6–10 Two boxes connected by a string**
The string ensures that the two boxes have the same acceleration. This physical connection results in a mathematical connection, as shown in Equation 6–7. Note that in this case we treat each box as a separate system.

First, sketch the free-body diagram for each box. Box 1 has two horizontal forces acting on it: (i) the tension T to the left, and (ii) the force F to the right. Box 2 has only a single horizontal force, the tension T to the right. If we take the positive direction to be to the right, Newton's second law for the two boxes can be written as follows:

$$\begin{aligned} F - T &= m_1a_1 = m_1a \qquad \text{box 1} \\ T &= m_2a_2 = m_2a \qquad \text{box 2} \end{aligned} \tag{6–7}$$

Since the boxes have the same acceleration, a, we have set $a_1 = a_2 = a$.

Next, we can eliminate the tension T by adding the two equations:

$$\begin{aligned} F - T &= m_1a \\ T &= m_2a \\ \hline F &= (m_1 + m_2)a \end{aligned}$$

With this result, it is straightforward to solve for the acceleration in terms of the applied force F:

$$a = \frac{F}{m_1 + m_2} \tag{6–8}$$

Finally, substitute this expression for a into either of the second law equations to find the tension. The algebra is simpler if we use the equation for box 2. We find

$$T = m_2a = \left(\frac{m_2}{m_1 + m_2}\right)F \tag{6–9}$$

It is left as an exercise to show that the equation for box 1 gives the same expression for T.

A second way to approach this problem is to treat both boxes together as a single system with a mass $m_1 + m_2$, as shown in **Figure 6–11**. The only *external* horizontal force acting on this system is the applied force F—the two tension forces are now *internal* to the system, and internal forces are not included when applying Newton's second law. As a result, the horizontal acceleration is simply $F/(m_1 + m_2)$, as given

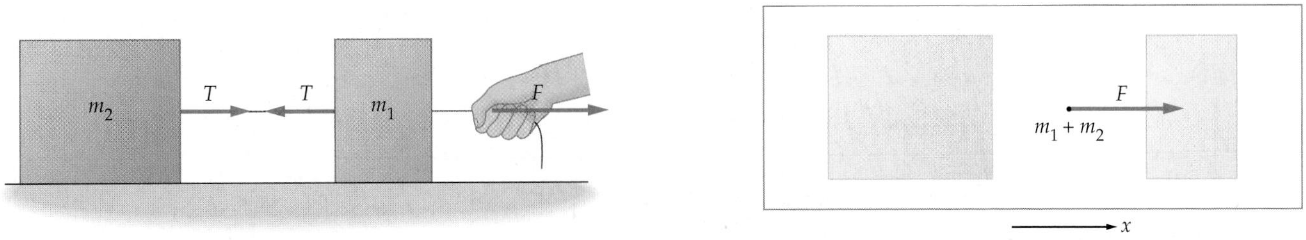

Physical picture

Free-body diagram (horizontal components only)

▲ **FIGURE 6–11 Two boxes, one system**

In this case we consider the two boxes together as a single system of mass $m_1 + m_2$. The only external horizontal force acting on this system is $\vec{\mathbf{F}}$; hence the horizontal acceleration of the system is $a = F/(m_1 + m_2)$, in agreement with Equation 6–8.

in Equation 6–8. This is certainly a quick way to find the acceleration a, but to find the tension T we must still use one of the relations given in Equation 6–7.

In general, we are always free to choose the "system" any way we like—we can choose any individual object, as when we considered box 1 and box 2 separately, or we can choose all the objects together. The important point is that Newton's second law is equally valid no matter what choice we make for the system, as long as we remember to include only forces *external* to *that system* in the corresponding free-body diagram.

CONCEPTUAL CHECKPOINT 6–4 Tension in the String

Two masses, m_1 and m_2, are connected by a string that passes over a pulley. Mass m_1 slides without friction on a horizontal tabletop, and mass m_2 falls vertically downward. Both masses move with a constant acceleration of magnitude a. Is the tension in the string **(a)** greater than, **(b)** equal to, or **(c)** less than m_2g?

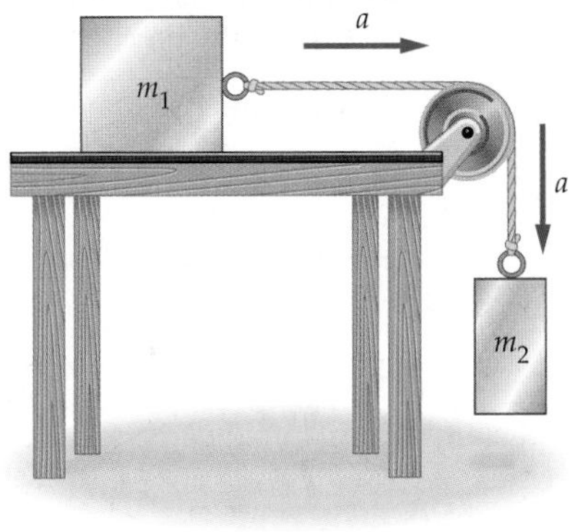

Reasoning and Discussion

First, note that m_2 accelerates downward, which means that the net force acting on it is downward. Only two forces act on m_2, however; the tension in the string (upward), and its weight (downward). Since the net force is downward, the tension in the string must be less than the weight, m_2g.

A common misconception is that since m_2 has to pull m_1 behind it, the tension in the string must be greater than m_2g. Certainly, attaching the string to m_1 has an effect on the tension. If the string were not attached, for example, its tension would be zero. Hence, m_2 pulling on m_1 increases the tension to a value greater than zero, though still less than m_2g.

Answer:

(c) The tension in the string is less than m_2g.

In the next Example, we verify the qualitative conclusions given in the Conceptual Checkpoint with a detailed calculation. But first, a note about choosing a coordinate system for a problem such as this. Rather than apply the same coordinate system to both masses, it is useful to take into consideration the fact that a pulley simply changes the direction of the tension in a string. With this in mind, we choose a set of axes that "follow the motion" of the string, so that both masses accelerate in the positive x direction with accelerations of equal magnitude. Example 6–6 illustrates the use of this type of coordinate system.

PROBLEM-SOLVING NOTE

Choice of Coordinate System: Connected Objects

If two objects are connected by a string passing over a pulley, let the coordinate system follow the direction of the string. With this choice, both objects have accelerations of the same magnitude and in the same coordinate direction.

EXAMPLE 6–6 Connected Blocks

A block of mass m_1 slides on a frictionless tabletop. It is connected to a string that passes over a pulley and suspends a mass m_2. Find the acceleration of the masses and the tension in the string.

Picture the Problem

Our coordinate system follows the motion of the string so that both masses move in the positive x direction. Since the masses are connected, their accelerations have the same magnitude. Thus, $a_{1,x} = a_{2,x} = a$. In addition, note that the tension, $\vec{\mathbf{T}}$, is in the positive x direction for mass 1, but in the negative x direction for mass 2. Its magnitude, T, is the same for each mass, however. Finally, the weight of mass 2, W_2, acts in the positive x direction, whereas the weight of mass 1 is offset by the normal force, N.

continued on next page

continued from previous page

INTERACTIVE FIGURE

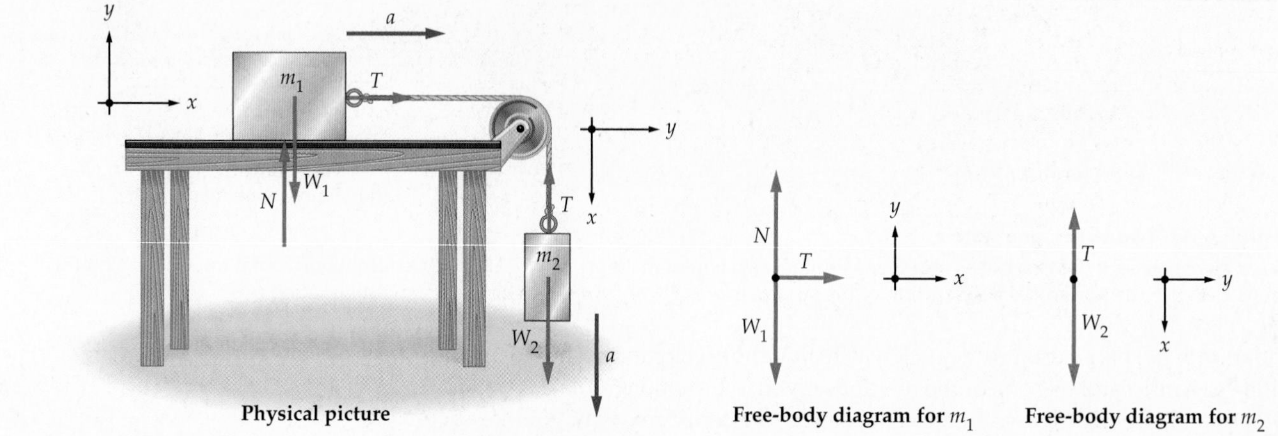

Physical picture **Free-body diagram for m_1** **Free-body diagram for m_2**

Strategy

Applying Newton's second law to the two masses yields the following relations: For mass 1, $\Sigma F_{1,x} = T = m_1 a_{1,x} = m_1 a$ and for mass 2, $\Sigma F_{2,x} = m_2 g - T = m_2 a_{2,x} = m_2 a$. These two equations can be solved for the two unknowns, a and T.

Solution

1. First, write $\Sigma F_{1,x} = m_1 a$. Note that the only force acting on m_1 in the x direction is T:

$$\sum F_{1,x} = T = m_1 a$$
$$T = m_1 a$$

2. Next, write $\Sigma F_{2,x} = m_2 a$. In this case, two forces act in the x direction; $W_2 = m_2 g$ (positive direction) and T (negative direction):

$$\sum F_{2,x} = m_2 g - T = m_2 a$$
$$m_2 g - T = m_2 a$$

3. Sum the two relations obtained to eliminate T:

$$T = m_1 a$$
$$m_2 g - T = m_2 a$$
$$m_2 g = (m_1 + m_2)a$$

4. Solve for a:

$$a = \left(\frac{m_2}{m_1 + m_2}\right)g$$

5. Substitute a into the first relation ($T = m_1 a$) to find T:

$$T = m_1 a = \left(\frac{m_1 m_2}{m_1 + m_2}\right)g$$

Insight

We could just as well have determined T using $m_2 g - T = m_2 a$, though the algebra is a bit messier. Also, note that $a = 0$ if $m_2 = 0$, and that $a = g$ if $m_1 = 0$, as expected. Similarly, $T = 0$ if either m_1 or m_2 is zero. This type of check, where you connect equations with physical situations, is one of the best ways to increase your understanding of physics.

Practice Problem

Find the tension for the case $m_1 = 1.50$ kg and $m_2 = 0.750$ kg, and compare the tension to $m_2 g$. [**Answer:** $a = 3.27\ \text{m/s}^2$, $T = 4.91\ \text{N} < m_2 g = 7.36\ \text{N}$]

Some related homework problems: Problem 36, Problem 40

Conceptual Checkpoint 6–4 shows that the tension in the string should be less than $m_2 g$. Let's rewrite our solution for T to show that this is indeed the case. From Example 6–6 we have

$$T = \left(\frac{m_1 m_2}{m_1 + m_2}\right)g = \left(\frac{m_1}{m_1 + m_2}\right)m_2 g$$

Since the ratio $m_1/(m_1 + m_2)$ is always less than 1 (as long as m_2 is nonzero), it follows that $T < m_2 g$, as expected.

We conclude this section with a classic system that can be used to measure the acceleration of gravity. It is referred to as Atwood's machine, and it is basically two blocks of different mass connected by a string that passes over a pulley. The resulting acceleration of the blocks is related to the acceleration of gravity by a relatively simple expression, which we derive in the following Example.

EXAMPLE 6–7 Atwood's Machine

Atwood's machine consists of two masses connected by a string that passes over a pulley, as shown below. Find the acceleration of the masses for general m_1 and m_2, and evaluate for the specific case $m_1 = 3.1$ kg, $m_2 = 4.4$ kg.

Picture the Problem

Our sketch shows Atwood's machine, along with our choice of coordinate directions for the two blocks. Note that both blocks accelerate in the positive x direction with accelerations of equal magnitude, a. From the free-body diagrams we can see that for mass 1 the weight is in the negative x direction and the tension is in the positive x direction. For mass 2, the tension is in the negative x direction and the weight is in the positive x direction. The tension has the same magnitude T for both masses, but their weights are different.

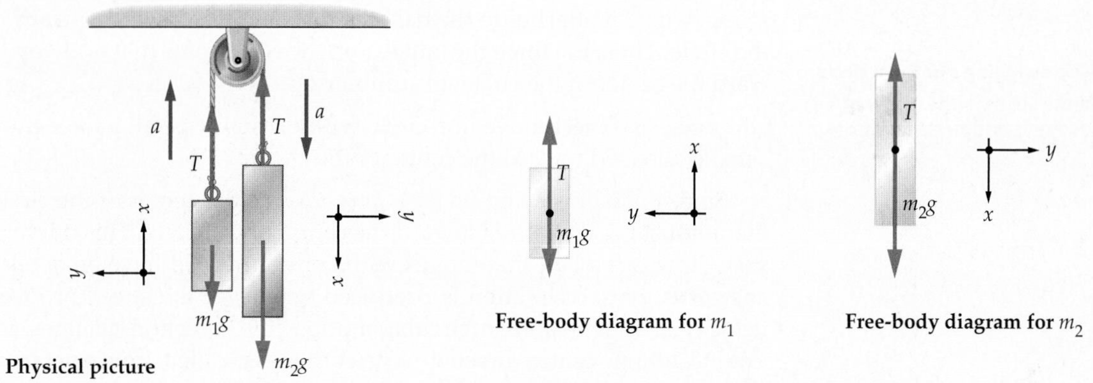

Strategy

To find the acceleration of the blocks, we follow the same strategy given in the previous Example. In particular, we start by applying Newton's second law to each block individually, using the fact that $a_{1,x} = a_{2,x} = a$. This gives two equations, both involving the tension T and the acceleration a. Eliminating T allows us to solve for the acceleration.

Solution

1. Begin by writing out the expression $\Sigma F_{1,x} = m_1 a$. Note that two forces act in the x direction; T (positive direction) and $m_1 g$ (negative direction):

$$\sum F_{1,x} = T - m_1 g = m_1 a$$

2. Next, write out $\Sigma F_{2,x} = m_2 a$. The two forces acting in the x direction in this case are $m_2 g$ (positive direction) and T (negative direction):

$$\sum F_{2,x} = m_2 g - T = m_2 a$$

3. Sum the two relations obtained above to eliminate T:

$$\begin{aligned} T - m_1 g &= m_1 a \\ m_2 g - T &= m_2 a \\ \hline (m_2 - m_1)g &= (m_1 + m_2)a \end{aligned}$$

4. Solve for a:

$$a = \left(\frac{m_2 - m_1}{m_1 + m_2}\right)g$$

5. To evaluate the acceleration, substitute numerical values for the masses and for g:

$$a = \left(\frac{m_2 - m_1}{m_1 + m_2}\right)g$$
$$= \left(\frac{4.4\text{ kg} - 3.1\text{ kg}}{3.1\text{ kg} + 4.4\text{ kg}}\right)(9.81\text{ m/s}^2) = 1.7\text{ m/s}^2$$

Insight

Since m_2 is greater than m_1, we find that the acceleration is positive, meaning that the masses accelerate in the positive x direction. On the other hand, if m_1 were greater than m_2, we would find that a is negative, indicating that the masses accelerate in the negative x direction. Finally, if $m_1 = m_2$ we have $a = 0$, as expected.

Practice Problem

If m_1 is increased by a small amount, does the acceleration of the blocks increase, decrease, or stay the same? Check your answer by evaluating the acceleration for $m_1 = 3.3$ kg. [**Answer:** If m_1 is increased only slightly, the acceleration will decrease. For $m_1 = 3.3$ kg, we find $a = 1.4\text{ m/s}^2$.]

Some related homework problems: Problem 40, Problem 42

▲ **FIGURE 6–12 Swinging a ball in a circle**
The tension in the string pulls outward on the person's hand and pulls inward on the ball.

6–5 Circular Motion

According to Newton's second law, if no force acts on an object, it will move with constant speed in a constant direction. A force is required to change the speed, the direction, or both. For example, if you drive a car with constant speed on a circular track, the direction of the car's motion changes continuously. A force must act on the car to cause this change in direction. We would like to know two things about a force that causes circular motion: what is its direction, and what is its magnitude?

First, let's consider the direction of the force. Imagine swinging a ball tied to a string in a circle about your head, as shown in **Figure 6–12**. As you swing the ball, you feel a tension in the string pulling outward. Of course, on the other end of the string, where it attaches to the ball, the tension pulls inward, toward the center of the circle. Thus, the force the ball experiences is a force that is always directed toward the center of the circle. In summary,

> To make an object move in a circle with constant speed, a force must act on it that is directed toward the center of the circle.

Since the ball is acted on by a *force* toward the center of the circle, it follows that it must be *accelerating* toward the center of the circle. This might seem odd at first: How can a ball that moves with constant speed have an acceleration? The answer is that acceleration is produced whenever the speed or direction of the velocity changes—and in circular motion, the direction changes continuously. The resulting center-directed acceleration is called **centripetal acceleration** (centripetal is from the Latin for "center seeking").

Let's calculate the magnitude of the centripetal acceleration, a_{cp}, for an object moving with a constant speed v in a circle of radius r. **Figure 6–13** shows the circular path of an object, with the center of the circle at the origin. To calculate the acceleration at the top of the circle, at point P, we first calculate the average acceleration from point 1 to point 2:

$$\vec{\mathbf{a}}_{av} = \frac{\Delta\vec{\mathbf{v}}}{\Delta t} = \frac{\vec{\mathbf{v}}_2 - \vec{\mathbf{v}}_1}{\Delta t} \qquad \textbf{6–10}$$

The instantaneous acceleration at P is the limit of $\vec{\mathbf{a}}_{av}$ as points 1 and 2 move closer to P.

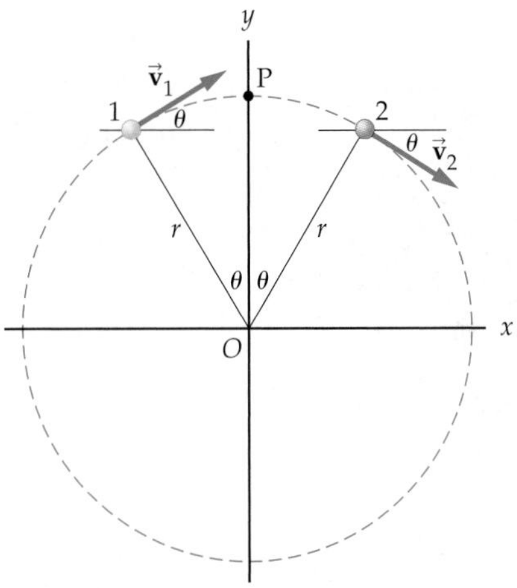

▲ **FIGURE 6–13 A particle moving in a circular path centered on the origin**
The speed of the particle is constant, but its velocity is constantly changing.

Referring to Figure 6–13, we see that $\vec{\mathbf{v}}_1$ is at an angle θ above the horizontal, and $\vec{\mathbf{v}}_2$ is at an angle θ below the horizontal. Both $\vec{\mathbf{v}}_1$ and $\vec{\mathbf{v}}_2$ have a magnitude v. Therefore, we can write the two velocities in vector form as follows:

$$\vec{\mathbf{v}}_1 = (v\cos\theta)\hat{\mathbf{x}} + (v\sin\theta)\hat{\mathbf{y}}$$
$$\vec{\mathbf{v}}_2 = (v\cos\theta)\hat{\mathbf{x}} + (-v\sin\theta)\hat{\mathbf{y}}$$

Substituting these results into $\vec{\mathbf{a}}_{av}$ gives

$$\vec{\mathbf{a}}_{av} = \frac{\vec{\mathbf{v}}_2 - \vec{\mathbf{v}}_1}{\Delta t} = \frac{-2v\sin\theta}{\Delta t}\hat{\mathbf{y}} \qquad \textbf{6–11}$$

Note that $\vec{\mathbf{a}}_{av}$ points in the negative y direction—which, at point P, is toward the center of the circle.

To complete the calculation, we need Δt, the time it takes the object to go from point 1 to point 2. Since the object's speed is v, and the distance from point 1 to point 2 is $d = r(2\theta)$ where θ is measured in radians (see Appendix A, page A-2 for a discussion of radians and degrees), we find

$$\Delta t = \frac{d}{v} = \frac{2r\theta}{v} \qquad \textbf{6–12}$$

TABLE 6–2
$\frac{\sin\theta}{\theta}$ for Values of θ Approaching Zero

θ, radians	$\frac{\sin\theta}{\theta}$
1.00	0.841
0.500	0.959
0.250	0.990
0.125	0.997
0.0625	0.999

Combining this result for Δt with the previous result for $\vec{\mathbf{a}}_{av}$ gives

$$\vec{\mathbf{a}}_{av} = \frac{-2v\sin\theta}{(2r\theta/v)}\hat{\mathbf{y}} = -\frac{v^2}{r}\left(\frac{\sin\theta}{\theta}\right)\hat{\mathbf{y}} \qquad \textbf{6–13}$$

To find $\vec{\mathbf{a}}$ at point P, we let points 1 and 2 approach P, which means letting θ go to zero. Table 6–2 shows that as θ goes to zero ($\theta \rightarrow 0$) the ratio $(\sin\theta)/\theta$ goes to 1:

$$\frac{\sin\theta}{\theta} \xrightarrow[\text{as } \theta \rightarrow 0]{} 1$$

Finally, then, the instantaneous acceleration at point P is

$$\vec{\mathbf{a}} = -\frac{v^2}{r}\hat{\mathbf{y}} = -a_{cp}\hat{\mathbf{y}} \quad \text{6–14}$$

As mentioned, the direction of the acceleration is toward the center of the circle, and now we see that its magnitude is

$$a_{cp} = \frac{v^2}{r} \quad \text{6–15}$$

We can summarize these results as follows:

- When an object moves in a circle of radius r with constant speed v, its centripetal acceleration is $a_{cp} = v^2/r$.
- A force must be applied to an object to give it circular motion. For an object of mass m, the net force acting on it must have a magnitude given by

$$f_{cp} = ma_{cp} = m\frac{v^2}{r} \quad \text{6–16}$$

and must be directed toward the center of the circle.

Note that the **centripetal force,** f_{cp}, can be produced in any number of ways. For example, f_{cp} might be the tension in a string, as in the example with the ball, or it might be due to friction between tires and the road, as when a car turns a corner. In addition, f_{cp} could be the force of gravity causing a satellite, or the Moon, to orbit the Earth. Thus, f_{cp} is a force that must be present to cause circular motion, but the specific cause of f_{cp} varies from system to system.

We now show how these results for centripetal force and centripetal acceleration can be applied in practice.

▲ The people enjoying this carnival ride are experiencing a centripetal acceleration of roughly 10 m/s^2 directed inward, toward the axis of rotation. The force needed to produce this acceleration, which keeps the riders moving in a circular path, is provided by the horizontal component of the tension in the chains.

PROBLEM-SOLVING NOTE

Choice of Coordinate System: Circular Motion

In circular motion, it is convenient to choose the coordinate system so that one axis points toward the center of the circle. Then, we know that the acceleration in that direction must be $a_{cp} = v^2/r$.

EXAMPLE 6–8 Rounding a Corner

A 1200-kg car rounds a corner of radius $r = 45$ m. If the coefficient of static friction between the tires and the road is $\mu_s = 0.82$, what is the greatest speed the car can have in the corner without skidding?

Picture the Problem

In the first sketch we show a bird's-eye view of the car as it moves along its circular path. The next sketch shows the car moving directly toward the observer. Note that we have chosen the positive x direction to point toward the center of the circular path, and the positive y axis to point vertically upward. We also indicate the three forces acting on the car: gravity, $\vec{\mathbf{W}} = -W\hat{\mathbf{y}} = -mg\hat{\mathbf{y}}$; the normal force, $\vec{\mathbf{N}} = N\hat{\mathbf{y}}$; and the force of static friction, $\vec{\mathbf{f}}_s = \mu_s N\hat{\mathbf{x}}$.

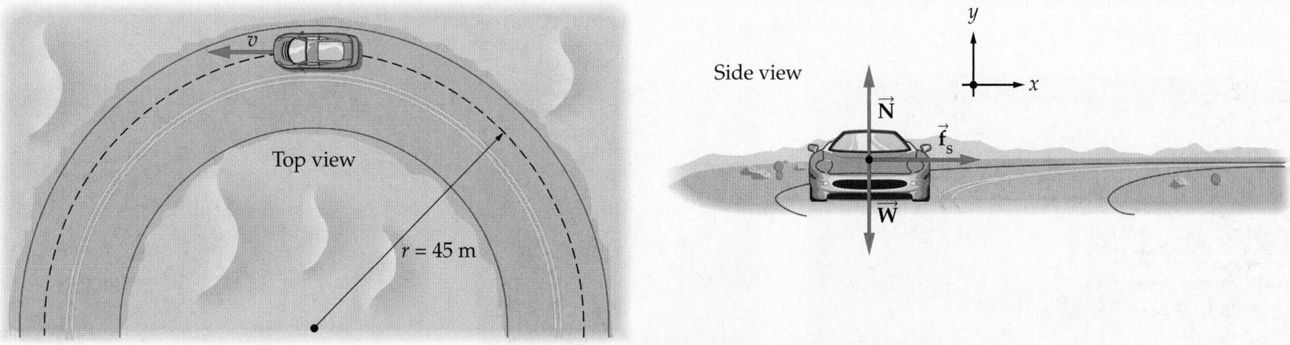

INTERACTIVE FIGURE

Strategy

In this system, the force of static friction provides the centripetal force required for the car to move in a circular path. That is why the force of friction is at right angles to the car's direction of motion; it is directed toward the center of the circle. In addition, the friction in this case is static because the car's tires are rolling without slipping—always making static contact with the ground. Finally, if the car moves faster, more centripetal force (i.e., more friction) is required. Thus, the greatest speed for the car corresponds to the maximum static friction, $f_s = \mu_s N$. Hence, if we set $\mu_s N$ equal to the centripetal force, $ma_{cp} = mv^2/r$, we can solve for v.

continued on next page

continued from previous page

Solution

1. Sum the x components of force to relate the force of static friction to the centripetal acceleration of the car:

$$\sum F_x = f_s = ma_x$$

Since the car moves in a circular path, with the center of the circle in the x direction, it follows that $a_x = a_{cp} = v^2/r$. Make this substitution, along with $f_s = \mu_s N$ for the force of static friction:

$$\mu_s N = ma_{cp} = m\frac{v^2}{r}$$

2. Next, set the sum of the y components of force equal to zero (since $a_y = 0$) to find the normal force, N:

$$\sum F_y = N - W = ma_y = 0$$

Solve for the normal force:

$$N = W = mg$$

3. Substitute $N = mg$ in step 1 and solve for v. Notice that the mass of the car cancels:

$$\mu_s mg = m\frac{v^2}{r}$$

$$v = \sqrt{\mu_s rg}$$

4. Substitute numerical values to determine v:

$$v = \sqrt{(0.82)(45\ \text{m})(9.81\ \text{m/s}^2)} = 19\ \text{m/s}$$

Insight

Note that the maximum speed is less if the radius is smaller (tighter corner) or if μ_s is smaller (slick road). The mass of the vehicle, however, is irrelevant. For example, the maximum speed is precisely the same for a motorcycle rounding this corner as it is for a large, heavily loaded truck.

Practice Problem

Suppose the situation described in this Example takes place on the Moon, where the acceleration of gravity is less than it is on Earth. If a lunar rover goes around this same corner, is its maximum speed greater than, less than, or the same as the result found in step 4? To check your answer, find the maximum speed for a lunar rover when it rounds a corner with $r = 45$ m and $\mu_s = 0.82$. (On the Moon, $g = 1.62\ \text{m/s}^2$.) [**Answer:** The maximum speed will be less. On the Moon we find $v = 7.7$ m/s.]

Some related homework problems: Problem 43, Problem 45, Problem 49

REAL-WORLD PHYSICS

Skids and banked roadways

If you try to round a corner too rapidly, you may experience a skid; that is, your car may begin to slide sideways across the road. A common bit of road wisdom is that you should turn in the direction of the skid to regain control—which, to most people, sounds counterintuitive. The advice is sound, however. Suppose, for example, that you are turning to the left and begin to skid to the right. If you turn more sharply to the left to try to correct for the skid, you simply reduce the turning radius of your car, r. The result is that the centripetal acceleration, v^2/r, becomes larger, and an even larger force would be required from the road to make the turn. The tendency to skid would therefore be increased. On the other hand, if you turn slightly to the right when you start to skid, you *increase* your turning radius and the centripetal acceleration decreases. In this case your car may stop skidding, and you can then regain control of your vehicle.

▲ The steeply banked track at the Talladega Speedway in Alabama (left) helps to keep the rapidly moving cars from skidding off along a tangential path. Even when there is no solid roadway, however, banking can still help—airplanes bank when making turns (center) to keep from "skidding" sideways. Banking is beneficial in another way as well. Occupants of cars on a banked roadway or of a banking airplane feel no sideways force when the banking angle is just right, so turns become a safer and more comfortable experience. For this reason, some trains use hydraulic suspension systems to bank when rounding corners (right), even though the tracks themselves are level.

You may also have noticed that many roads are tilted, or banked, when they round a corner. The same type of banking is observed on many automobile racetracks as well. Next time you drive around a banked curve, notice that the banking tilts you in toward the center of the circular path you are following. This is by design. On a banked curve, the normal force exerted by the road contributes to the required centripetal force. If the tilt angle is just right, the normal force provides all of the centripetal force so that the car can negotiate the curve even if there is no friction between its tires and the road. The next Example determines the optimum banking angle for a given speed and given radius of turn.

EXAMPLE 6–9 Bank on It

If a roadway is banked at the proper angle, a car can round a corner without any assistance from friction between the tires and the road. Find the appropriate banking angle for a 900-kg car traveling at 20.5 m/s in a turn of radius 85.0 m.

Picture the Problem

Note that we choose the positive y axis to point vertically upward and the positive x direction to point toward the center of the circular path. Since $\vec{\mathbf{N}}$ is perpendicular to the banked roadway, it is at an angle θ to the y axis. Therefore, $\vec{\mathbf{N}} = (N \sin\theta)\hat{\mathbf{x}} + (N\cos\theta)\hat{\mathbf{y}}$ and $\vec{\mathbf{W}} = -W\hat{\mathbf{y}} = -mg\hat{\mathbf{y}}$.

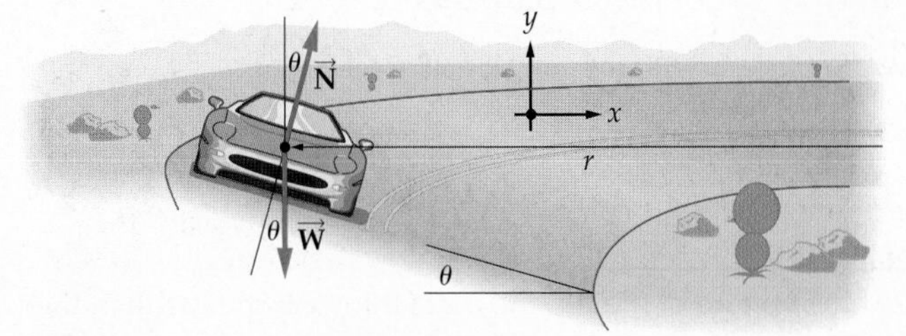

Strategy

In order for the car to move in a circular path, there must be a force acting on it in the positive x direction. Since the weight $\vec{\mathbf{W}}$ has no x component, it follows that the normal force $\vec{\mathbf{N}}$ must supply the needed centripetal force. Thus, we find N by setting $\Sigma F_y = ma_y = 0$, since there is no motion in the y direction. Then we use N in $\Sigma F_x = ma_x = mv^2/r$ to find the angle θ.

Solution

1. Start by determining N from the condition $\Sigma F_y = 0$:

$$\sum F_y = N\cos\theta - W = 0$$
$$N = \frac{W}{\cos\theta} = \frac{mg}{\cos\theta}$$

2. Next, set $\Sigma F_x = mv^2/r$:

$$\sum F_x = N\sin\theta$$
$$= ma_x = ma_{cp} = m\frac{v^2}{r}$$

3. Substitute $N = mg/\cos\theta$ (from $\Sigma F_y = 0$) and solve for θ, using the fact that $\sin\theta/\cos\theta = \tan\theta$. Notice that, once again, the mass of the car cancels:

$$N\sin\theta = \frac{mg}{\cos\theta}\sin\theta = m\frac{v^2}{r}$$
$$\tan\theta = \frac{v^2}{gr} \quad \text{or} \quad \theta = \tan^{-1}\left(\frac{v^2}{gr}\right)$$

4. Substitute numerical values to determine θ:

$$\theta = \tan^{-1}\left[\frac{(20.5\ \mathrm{m/s})^2}{(9.81\ \mathrm{m/s^2})(85.0\ \mathrm{m})}\right] = 26.7°$$

Insight

The symbolic result in Step 3 shows that the banking angle increases with increasing speed and decreasing radius of turn, as one would expect.

From the point of view of a passenger, the experience of rounding a properly banked corner is basically the same as riding on a level road—there are no "sideways forces" to make the turn uncomfortable. There is one small difference, however—the passenger feels heavier due to the increased normal force.

Practice Problem

A turn of radius 65 m is banked at 30.0°. What speed should a car have in order to make the turn with no assistance from friction? [**Answer:** $v = 19\ \mathrm{m/s}$]

Some related homework problems: Problem 46, Problem 92

If you've ever driven through a dip in the road, you know that you feel momentarily heavier near the bottom of the dip, just like a passenger in Example 6–9. This change in apparent weight is due to the approximately circular motion of the car, as we show next.

ACTIVE EXAMPLE 6–4 Find the Normal Force

While driving along a country lane with a constant speed of 17.0 m/s, you encounter a dip in the road. The dip can be approximated as a circular arc, with a radius of 65.0 m. What is the normal force exerted by a car seat on an 80.0-kg passenger when the car is at the bottom of the dip?

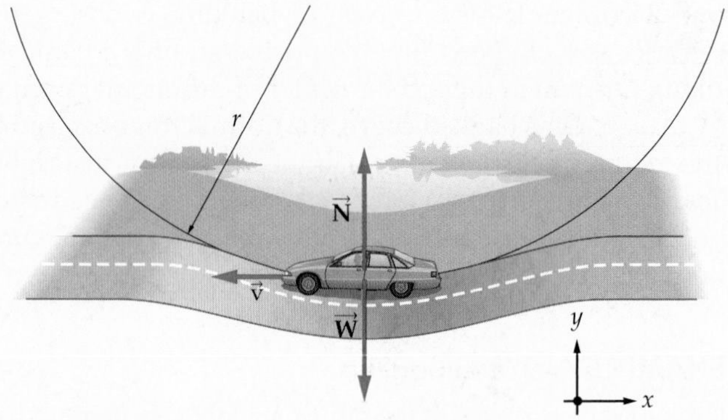

Solution *(Test your understanding by performing the calculations indicated in each step.)*

1. Write $\Sigma F_y = ma_y$ for the passenger: $N - mg = ma_y$
2. Replace a_y with the centripetal acceleration: $a_y = v^2/r$
3. Solve for N: $N = mg + mv^2/r$
4. Substitute numerical values: $N = 1140 \text{ N}$

Insight

At the bottom of the dip the normal force is greater than the weight of the passenger, since it must also supply the centripetal force. As a result, the passenger feels heavier than usual. In this case, the 80.0-kg passenger feels as if his mass has increased by 45 percent, to 116 kg!

The same physics applies to a jet pilot who pulls a plane out of a high-speed dive. In that case, the magnitude of the effect can be much larger, resulting in a decrease of blood flow to the brain and eventually to loss of consciousness. Here's a case where basic physics really can be a matter of life and death.

Your Turn

At what speed will the magnitude of the normal force be equal to 1250 N?

(Answers to **Your Turn** *problems are given in the back of the book.)*

A similar calculation can be applied to a car going over the top of a bump. In that case, circular motion results in a reduced apparent weight.

REAL-WORLD PHYSICS: BIO
Centrifuges and ultracentrifuges

Finally, we determine the acceleration produced in a **centrifuge,** a common device in biological and medical laboratories that uses large centripetal accelerations to perform such tasks as separating red and white blood cells from serum. A simplified top view of a centrifuge is shown in **Figure 6–14**.

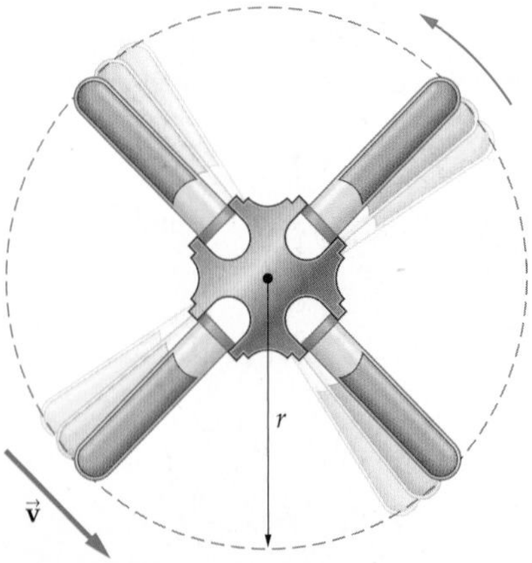

▲ **FIGURE 6–14 Simplified top view of a centrifuge in operation**

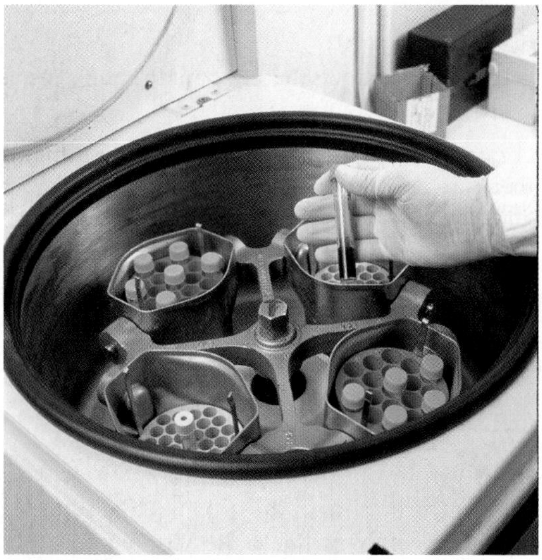

▲ A laboratory centrifuge of the kind commonly used to separate blood components.

EXERCISE 6–1

The centrifuge in Figure 6–14 rotates at a rate that gives the bottom of the test tube a linear speed of 89.3 m/s. If the bottom of the test tube is 8.50 cm from the axis of rotation, what is the centripetal acceleration experienced there?

Solution

Applying the relation $a_{cp} = v^2/r$ yields

$$a_{cp} = \frac{v^2}{r} = \frac{(89.3\ \text{m/s})^2}{0.0850\ \text{m}} = 93{,}800\ \text{m/s}^2 = 9560g$$

In this expression, g is the acceleration of gravity, 9.81 m/s^2.

Thus, a centrifuge can produce centripetal accelerations that are many thousand times greater than the acceleration of gravity. In fact, devices referred to as **ultracentrifuges** can produce accelerations as great as 1 million g. Even in the relatively modest case considered in Exercise 6–1, the forces involved in a centrifuge can be quite significant. For example, if the contents of the test tube have a mass of 12.0 g, the centripetal force that must be exerted by the bottom of the tube is $(0.0120\ \text{kg})(9560\ g) = 1130\ \text{N}$, or about 250 lbs!

Finally, an object moving in a circular path may increase or decrease its speed. In such a case, the object has both an acceleration tangential to its path that changes its speed, $\vec{a}_t$, plus a centripetal acceleration perpendicular to its path, $\vec{a}_{cp}$, that changes its direction of motion. Such a situation is illustrated in **Figure 6–15**. The total acceleration of the object is the vector sum of $\vec{a}_t$ and $\vec{a}_{cp}$. We will explore this case more fully in Chapter 10.

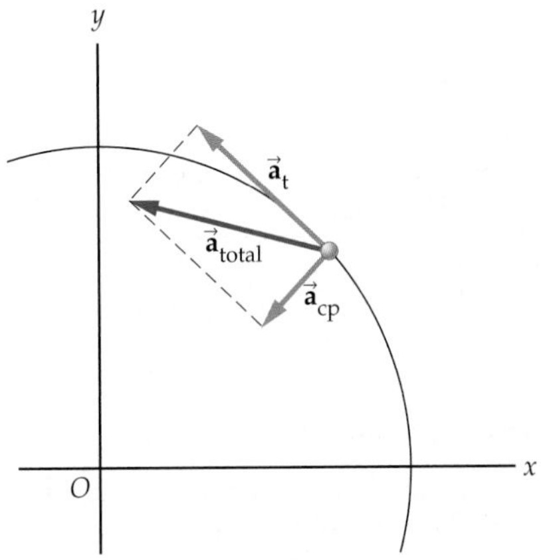

▲ **FIGURE 6–15 A particle moving in a circular path with tangential acceleration**
In this case, the particle's speed is increasing at the rate given by a_t.

Chapter Summary

6–1 Frictional Forces

Frictional forces are due to the microscopic roughness of surfaces in contact. As a rule of thumb, friction is independent of the area of contact and independent of the relative speed of the surfaces.

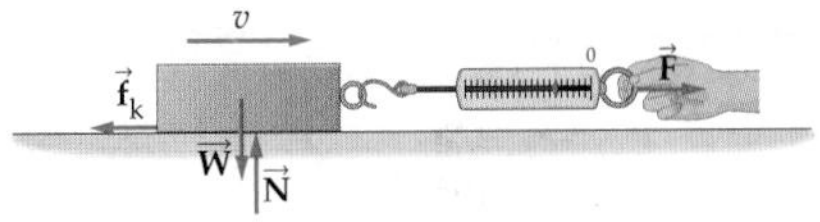

Kinetic Friction
Friction experienced by surfaces that are in contact and moving relative to one another. The force of kinetic friction is given by

$$f_k = \mu_k N \qquad \textbf{6–1}$$

In this expression, μ_k is the coefficient of kinetic friction and N is the magnitude of the normal force.

Static Friction
Friction experienced by surfaces that are in static contact. The maximum force of static friction is given by

$$f_{s,max} = \mu_s N \qquad \textbf{6–3}$$

In this expression, μ_s is the coefficient of static friction and N is the magnitude of the normal force. The force of static friction can have any magnitude between zero and its maximum value.

6–2 Strings and Springs

Strings and springs provide a common way of exerting forces on objects. Ideal strings and springs are massless.

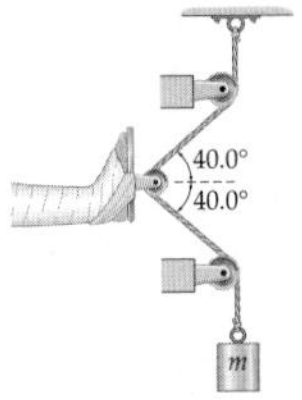

Tension
The force transmitted through a string. The tension is the same throughout the length of an ideal string.

Hooke's Law
The force exerted by an ideal spring stretched by the amount x is

$$F_x = -kx \qquad \textbf{6–4}$$

In words, the force exerted by a spring is proportional to the amount of stretch or compression, and is in the opposite direction.

6–3 Translational Equilibrium

An object is in translational equilibrium if the net force acting on it is zero. Equivalently, an object is in equilibrium if it has zero acceleration.

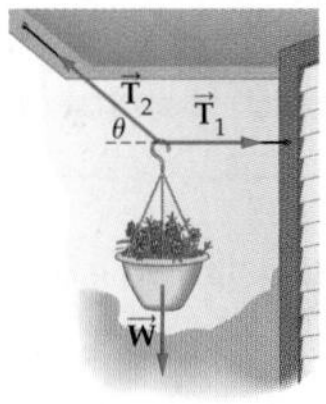

6–4 Connected Objects

Connected objects are linked physically, and hence they are linked mathematically as well. For example, objects connected by strings have the same magnitude of acceleration.

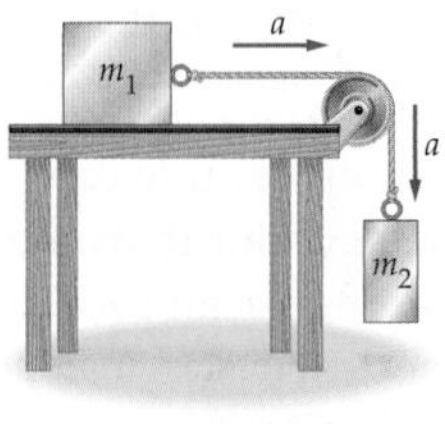

6–5 Circular Motion

An object moving with speed v in a circle of radius r has an acceleration of magnitude v^2/r directed toward the center of the circle: This is referred to as the centripetal acceleration, a_{cp}. If the object has a mass m, the force required for the circular motion is

$$f_{cp} = ma_{cp} = mv^2/r \qquad 6\text{–}16$$

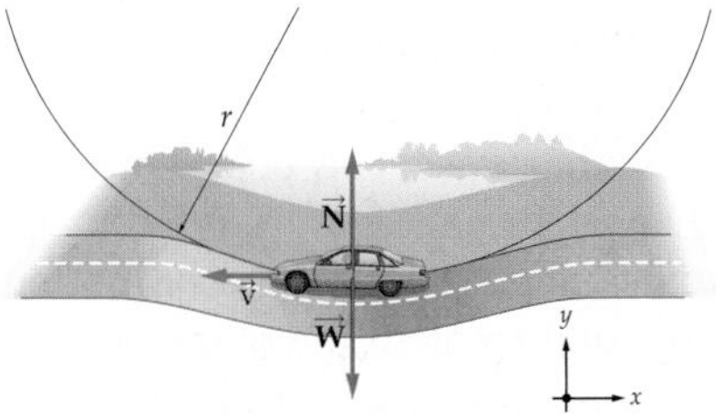

Problem-Solving Summary

Type of Calculation	Relevant Physical Concepts	Related Examples
Find the acceleration when kinetic friction is present.	First, find the magnitude of the normal force, N. The corresponding kinetic friction has a magnitude of $f_k = \mu_k N$ and points opposite to the direction of motion. Include this force with the others when applying Newton's second law.	Examples 6–1, 6–2
Solve problems involving static friction.	Start by finding the magnitude of the normal force, N. The corresponding static friction has a magnitude between zero and $\mu_s N$. Its direction opposes motion.	Example 6–3 Active Example 6–1
Find the acceleration and the tension for masses connected by a string.	Apply Newton's second law to each mass separately. This generates two equations, which can be solved for the two unknowns, a and T.	Examples 6–6, 6–7
Solve problems involving circular motion.	Set up the coordinate system so that one axis points to the center of the circle. When applying Newton's second law to that direction, set the acceleration equal to $a_{cp} = v^2/r$.	Examples 6–8, 6–9 Active Example 6–4

Conceptual Questions

(Answers to odd-numbered Conceptual Questions can be found in the back of the book.)

1. A clothesline always sags a little, even if nothing hangs from it. Explain.
2. In the *Jurassic Park* sequel, *The Lost World*, a man tries to keep a large vehicle from going over a cliff by connecting a cable from his Jeep to the vehicle. The man then puts the Jeep in gear and spins the rear wheels. Do you expect that spinning the tires will increase the force exerted by the Jeep on the vehicle? Why or why not?
3. When a traffic accident is investigated, it is common for the length of the skid marks to be measured. How could this information be used to estimate the initial speed of the vehicle that left the skid marks?
4. In a car with rear-wheel drive, the maximum acceleration is often less than the maximum deceleration. Why?
5. A train typically requires a much greater distance to come to rest, for a given initial speed, than does a car. Why?
6. Give some everyday examples of situations in which friction is beneficial.

7. At the local farm, you buy a flat of strawberries and place them on the backseat of the car. On the way home, you begin to brake as you approach a stop sign. At first the strawberries stay put, but as you brake a bit harder, they begin to slide off the seat. Explain.
8. It is possible to spin a bucket of water in a vertical circle and have none of the water spill when the bucket is upside down. How would you explain this to members of your family?
9. Water sprays off a rapidly turning bicycle wheel. Why?
10. Can an object be in equilibrium if it is moving? Explain.
11. In a dramatic circus act, a motorcyclist drives his bike around the inside of a vertical circle. How is this possible, considering that the motorcycle is upside down at the top of the circle?
12. The gravitational attraction of the Earth is only slightly less at the altitude of an orbiting spacecraft than it is on the Earth's surface. Why is it, then, that astronauts feel weightless?
13. A popular carnival ride has passengers stand with their backs against the inside wall of a cylinder. As the cylinder begins to spin, the passengers feel as if they are being pushed against the wall. Explain.
14. Referring to Question 13, after the cylinder reaches operating speed, the floor is lowered away, leaving the passengers "stuck" to the wall. Explain.
15. Your car is stuck on an icy side street. Some students on their way to class see your predicament and help out by sitting on the trunk of your car to increase its traction. Why does this help?
16. The parking brake on a car causes the rear wheels to lock up. What would be the likely consequence of applying the parking brake in a car that is in rapid motion? Note: Do *not* try this at home.
17. **BIO** The foot of your average gecko is covered with billions of tiny hair tips—called spatulae—that are made of keratin, the protein found in human hair. A subtle shift of the electron distribution in both the spatulae and the wall to which a gecko clings produces an adhesive force by means of the van der Waals interaction between molecules. Suppose a gecko uses its spatulae to cling to a vertical windowpane. If you were to describe this situation in terms of a coefficient of static friction, μ_s, what value would you assign to μ_s? Is this a sensible way to model the gecko's feat? Explain.
18. Discuss the physics involved in the spin cycle of a washing machine. In particular, how is circular motion related to the removal of water from the clothes?
19. The gas pedal and the brake pedal are capable of causing a car to accelerate. Can the steering wheel also produce an acceleration? Explain.
20. In the movie *2001: A Space Odyssey*, a rotating space station provides "artificial gravity" for its inhabitants. How does this work?

The rotating space station from the movie *2001: A Space Odyssey* (Conceptual Question 20)

21. When rounding a corner on a bicycle or a motorcycle, the driver leans inward, toward the center of the circle. Why?
22. In *Robin Hood: Prince of Thieves*, starring Kevin Costner, Robin swings between trees on a vine that is on fire. At the lowest point of his swing, the vine burns through and Robin begins to fall. The next shot, from high up in the trees, shows Robin falling straight downward. Would you rate the physics of this scene "Good," "Bad," or "Ugly?" Explain.

Conceptual Exercises

(Answers to odd-numbered Conceptual Exercises can be found in the back of the book.)

1. Referring to the hanging planter in Example 6–5, which of the three graphs (A, B, or C) in **Figure 6–16** shows an accurate plot of the tensions T_1 and T_2 as a function of the angle θ?

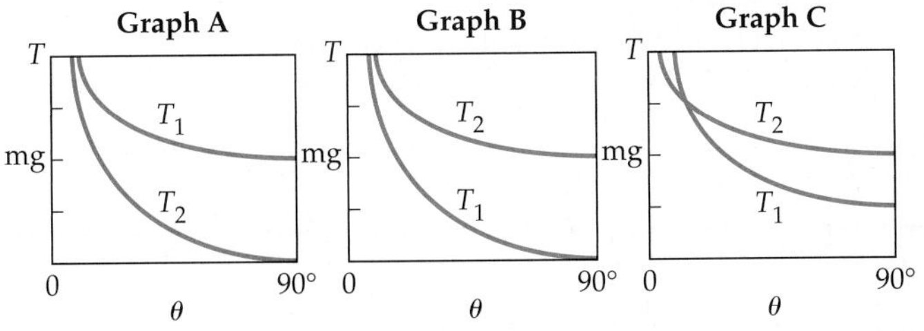

▲ **FIGURE 6–16** Conceptual Exercise 1

2. If you weigh yourself on a bathroom scale at the equator, is the reading you get greater than, less than, or equal to the reading you get if you weigh yourself at the North Pole?
3. In Example 6–6 (Connected Blocks), suppose m_1 and m_2 are both increased by a factor of 2. **(a)** Does the acceleration of the blocks increase, decrease, or stay the same? **(b)** Does the tension in the string increase, decrease, or stay the same?
4. Suppose m_1 and m_2 in Example 6–7 (Atwood's Machine) are both increased by 1 kg. Does the acceleration of the blocks increase, decrease, or stay the same?
5. The three identical boxes shown in **Figure 6–17** remain at rest on a rough, horizontal surface, even though they are acted on by two different forces, $\vec{F}_1$ and $\vec{F}_2$. All of the forces labeled $\vec{F}_1$ have the same magnitude; all of the forces labeled $\vec{F}_2$ are identical to one another. Rank the boxes in order of increasing magnitude of the force static friction between them and the surface. Indicate a tie when appropriate.

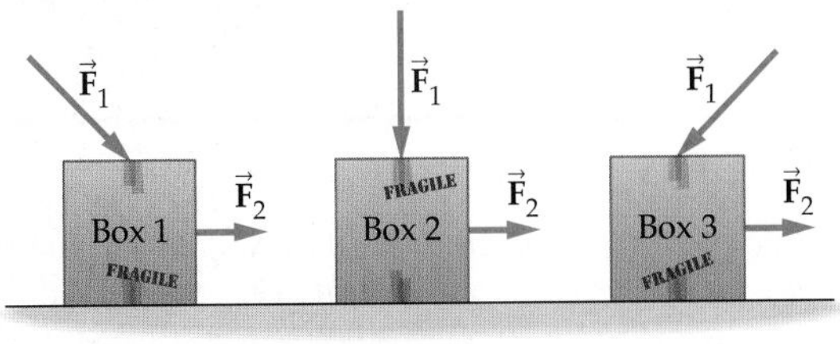

▲ **FIGURE 6–17** Conceptual Exercise 5

6. Suppose you stand on a bathroom scale and get a reading of 700 N. In principle, would the scale read more, less, or the same if the Earth did not rotate?
7. An object moves on a flat surface with an acceleration of constant magnitude. If the acceleration is always perpendicular to the object's direction of motion, **(a)** is the shape of the object's path circular, straight, or parabolic? **(b)** Does the object's velocity change during its motion? **(c)** Does its speed change?

8. You push two identical bricks across a tabletop with constant speed, v, as shown in **Figure 6–18**. In case 1, you place the bricks end to end; in case 2, you stack the bricks one on top of the other. Is the force of kinetic friction in case 1 greater than, less than, or equal to the force of kinetic friction in case 2?

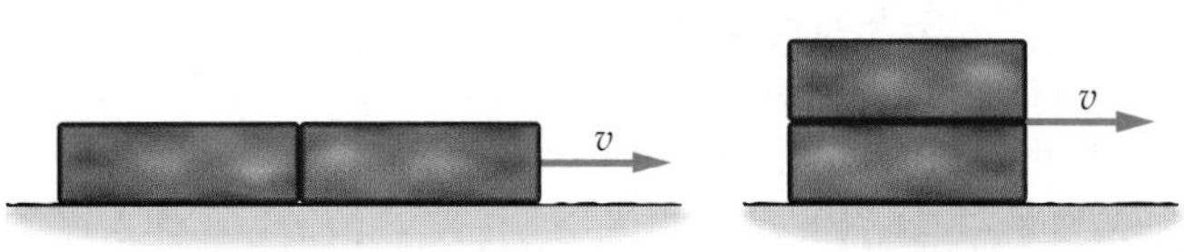

▲ **FIGURE 6–18** Conceptual Exercise 8

9. Two drivers traveling side-by-side at the same speed suddenly see a deer in the road ahead of them and begin braking. Driver 1 stops by locking up his brakes and screeching to a halt; driver 2 stops by applying her brakes just to the verge of locking, so that the wheels continue to turn until her car comes to a complete stop. All other factors being equal, is the stopping distance of driver 1 greater than, less than, or equal to the stopping distance of driver 2?

10. A car drives with constant speed on an elliptical track, as shown in **Figure 6–19**. Rank the points A, B, and C in order of increasing likelihood that the car might skid. Indicate a tie when appropriate.

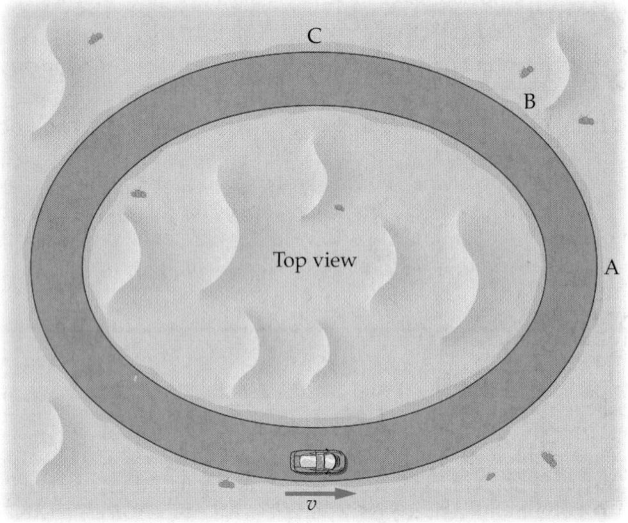

▲ **FIGURE 6–19** Conceptual Exercise 10

11. A car is driven with constant speed around a circular track. **(a)** Is its velocity constant? **(b)** Is the magnitude of its acceleration constant? **(c)** Is the direction of its acceleration constant?

12. A certain spring has a force constant k. **(a)** If this spring is cut in half, does the resulting half spring have a force constant that is greater than, less than, or equal to k? **(b)** If two of the original full-length springs are connected end to end, does the resulting double spring have a force constant that is greater than, less than, or equal to k?

13. **BIO Maneuvering a Jet** Humans lose consciousness if exposed to prolonged accelerations of more than about 7 g. This is of concern to jet fighter pilots, who may experience centripetal accelerations of this magnitude when making high-speed turns. Suppose we would like to decrease the centripetal acceleration of a jet. Rank the following changes in order of how effective they are in decreasing the centripetal acceleration, starting with the least effective: (A) decrease the turning radius by a factor of two; (B) decrease the speed by a factor of three; (C) increase the turning radius by a factor of four.

14. A puck attached to a string undergoes circular motion on an air table. If the string breaks at the point indicated in **Figure 6–20**, is the subsequent motion of the puck best described by path A, B, C, or D?

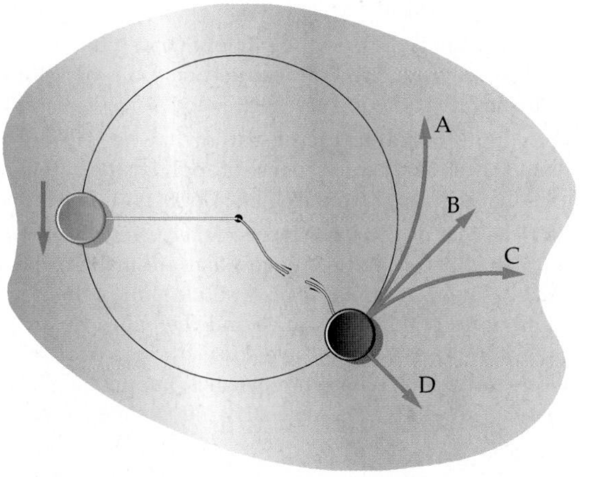

▲ **FIGURE 6–20** Conceptual Exercise 14

15. **BIO Gravitropism** As plants grow, they tend to align their stems and roots along the direction of the gravitational field. This tendency, which is related to differential concentrations of plant hormones known as auxins, is referred to as *gravitropism*. As an illustration of gravitropism, experiments show that seedlings placed in pots on the rim of a rotating turntable do not grow in the vertical direction. Do you expect their stems to tilt inward—toward the axis of rotation—or outward?

Problems

Note: **IP** *denotes an integrated conceptual/quantitative problem.* **BIO** *identifies problems of biological or medical interest. Red bullets (•, ••, •••) are used to indicate the level of difficulty of each problem.*

Section 6–1 Frictional Forces

1. • A baseball player slides into third base with an initial speed of 4.0 m/s. If the coefficient of kinetic friction between the player and the ground is 0.46, how far does the player slide before coming to rest?

2. • A child goes down a playground slide with an acceleration of 1.16 m/s^2. Find the coefficient of kinetic friction between the child and the slide if the slide is inclined at an angle of 31.0° below the horizontal.

3. • Hopping into your Porsche, you floor it and accelerate at 12 m/s^2 without spinning the tires. Determine the minimum coefficient of static friction between the tires and the road needed to make this possible.

4. • When you push a 1.80-kg book resting on a tabletop, it takes 2.25 N to start the book sliding. Once it is sliding, however, it takes only 1.50 N to keep the book moving with constant speed. What are the coefficients of static and kinetic friction between the book and the tabletop?

5. • In Problem 4, what is the frictional force exerted on the book when you push on it with a force of 0.75 N?

6. •• **IP** A tie of uniform width is laid out on a table, with a fraction of its length hanging over the edge. Initially, the tie is at rest. **(a)** If the fraction hanging from the table is increased, the tie eventually slides to the ground. Explain. **(b)** What is the coefficient of static friction between the tie and the table if the tie begins to slide when one-fourth of its length hangs over the edge?

7. •• To move a large crate across a rough floor, you push on it with a force F at an angle of 21° below the horizontal, as shown in **Figure 6–21**. Find the force necessary to start the crate moving, given that the mass of the crate is 32 kg and the coefficient of static friction between the crate and the floor is 0.57.

▲ **FIGURE 6–21** Problems 7, 8, and 91

8. •• In Problem 7, find the acceleration of the crate if the applied force is 330 N and the coefficient of kinetic friction is 0.45.

9. •• **IP** A 45-kg crate is placed on an inclined ramp. When the angle the ramp makes with the horizontal is increased to 23°, the crate begins to slide downward. **(a)** What is the coefficient of static friction between the crate and the ramp? **(b)** At what angle does the crate begin to slide if its mass is doubled?

10. •• **IP** A 95-kg sprinter wishes to accelerate from rest to a speed of 12 m/s in a distance of 20 m. **(a)** What coefficient of static friction is required between the sprinter's shoes and the track? **(b)** Explain the strategy used to find the answer to part (a).

11. •• A person places a cup of coffee on the roof of her car while she dashes back into the house for a forgotten item. When she returns to the car, she hops in and takes off with the coffee cup still on the roof. **(a)** If the coefficient of static friction between the coffee cup and the roof of the car is 0.24, what is the maximum acceleration the car can have without causing the cup to slide? Ignore the effects of air resistance. **(b)** What is the smallest amount of time in which the person can accelerate the car from rest to 15 m/s and still keep the coffee cup on the roof?

12. •• **IP Force Times Distance I** At the local hockey rink, a puck with a mass of 0.12 kg is given an initial speed of 5.3 m/s. **(a)** If the coefficient of kinetic friction between the ice and the puck is 0.11, what distance d does the puck slide before coming to rest? **(b)** If the mass of the puck is doubled, does the frictional force F exerted on the puck increase, decrease, or stay the same? Explain. **(c)** Does the stopping distance of the puck increase, decrease, or stay the same when its mass is doubled? Explain. **(d)** For the situation considered in part (a), show that $Fd = \frac{1}{2}mv^2$. (The significance of this result will be discussed in Chapter 7.)

13. •• **Force Times Distance II** A block of mass $m = 1.95$ kg slides with an initial speed $v_i = 4.33$ m/s on a smooth, horizontal surface. The block now encounters a rough patch with a coefficient of kinetic friction given by $\mu_k = 0.260$. The rough patch extends for a distance $d = 0.125$ m, after which the surface is again frictionless. **(a)** What is the acceleration of the block when it is in the rough patch? **(b)** What is the final speed, v_f, of the block when it exits the rough patch? **(c)** Show that $-Fd = -(\mu_k mg)d = \frac{1}{2}mv_f^2 - \frac{1}{2}mv_i^2$. (The significance of this result will be discussed in Chapter 7.)

14. ••• **IP** The coefficient of kinetic friction between the tires of your car and the roadway is μ. **(a)** If your initial speed is v and you lock your tires during braking, how far do you skid? Give your answer in terms of v, μ, and m, the mass of your car. **(b)** If you double your speed, what happens to the stopping distance? **(c)** What is the stopping distance for a truck with twice the mass of your car, assuming the same initial speed and coefficient of kinetic friction?

Section 6–2 Strings and Springs

15. • Pulling up on a rope, you lift a 4.25-kg bucket of water from a well with an acceleration of 1.80 m/s^2. What is the tension in the rope?

16. • When a 9.29-kg mass is placed on top of a vertical spring, the spring compresses 4.11 cm. Find the force constant of the spring.

17. • A 110-kg box is loaded into the trunk of a car. If the height of the car's bumper decreases by 13 cm, what is the force constant of its rear suspension?

18. • A 50.0-kg person takes a nap in a backyard hammock. Both ropes supporting the hammock are at an angle of 15.0° above the horizontal. Find the tension in the ropes.

19. • **IP** A backpack full of books weighing 52.0 N rests on a table in a physics laboratory classroom. A spring with a force constant of 150 N/m is attached to the backpack and pulled horizontally, as indicated in **Figure 6–22**. **(a)** If the spring is pulled until it stretches 2.00 cm and the pack remains at rest, what is the force of friction exerted on the backpack by the table? **(b)** Does your answer to part (a) change if the mass of the backpack is doubled? Explain.

▲ **FIGURE 6–22** Problems 19 and 20

20. • If the 52.0-N backpack in Problem 19 begins to slide when the spring ($k = 150$ N/m) stretches by 2.50 cm, what is the coefficient of static friction between the backpack and the table?

21. •• **IP** The equilibrium length of a certain spring with a force constant of $k = 250$ N/m is 0.18 m. **(a)** What is the magnitude of the force that is required to hold this spring at twice its equilibrium length? **(b)** Is the magnitude of the force required to keep the spring compressed to half its equilibrium length greater than, less than, or equal to the force found in part (a)? Explain.

22. •• **IP** Illinois Jones is being pulled from a snake pit with a rope that breaks if the tension in it exceeds 755 N. **(a)** If Illinois Jones has a mass of 70.0 kg and the snake pit is 3.40 m deep, what is the minimum time that is required to pull our intrepid explorer from the pit? **(b)** Explain why the rope breaks if Jones is pulled from the pit in less time than that calculated in part (a).

23. •• **IP** A spring with a force constant of 120 N/m is used to push a 0.27-kg block of wood against a wall, as shown in **Figure 6–23**. **(a)** Find the minimum compression of the spring needed to keep the block from falling, given that the coefficient of static friction between the block and the wall is 0.46. **(b)** Does your answer to part (a) change if the mass of the block of wood is doubled? Explain.

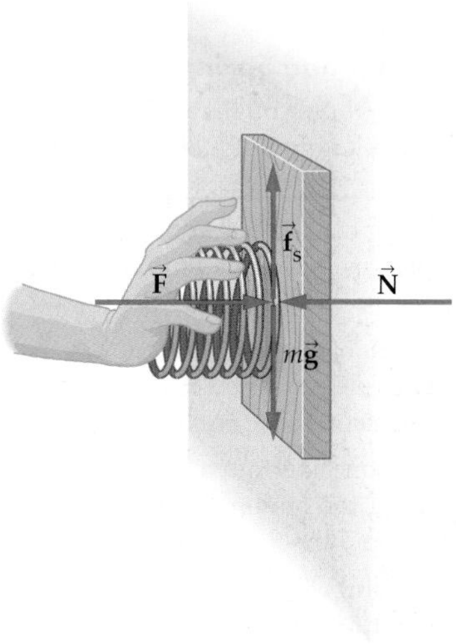

▲ **FIGURE 6–23** Problem 23

24. •• **IP** Your friend's 12.6-g graduation tassel hangs on a string from his rear-view mirror. **(a)** When he accelerates from a stoplight, the tassel deflects backward toward the rear of the car. Explain. **(b)** If the tassel hangs at an angle of 6.20° relative to the vertical, what is the acceleration of the car?

25. •• In Problem 24, **(a)** find the tension in the string holding the tassel. **(b)** At what angle to the vertical will the tension in the string be twice the weight of the tassel?

26. •• **IP** A picture hangs on the wall suspended by two strings, as shown in **Figure 6–24**. The tension in string 1 is 1.7 N. **(a)** Is the tension in string 2 greater than, less than, or equal to 1.7 N? Explain. **(b)** Verify your answer to part (a) by calculating the tension in string 2. **(c)** What is the weight of the picture?

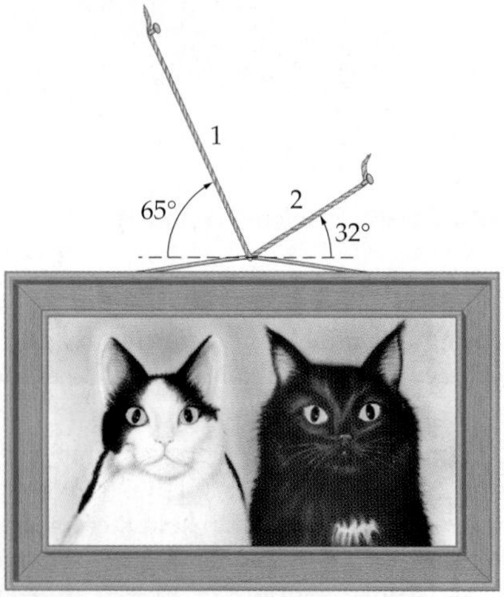

▲ **FIGURE 6–24** Problems 26 and 69

27. •• **Mechanical Advantage** The pulley system shown in **Figure 6–25** is used to lift a 52-kg crate. Note that one chain connects the upper pulley to the ceiling and a second chain connects the lower pulley to the crate. Assuming the masses of the chains, pulleys, and ropes are negligible, determine **(a)** the force $\vec{F}$ required to lift the crate with constant speed, **(b)** the tension in the upper chain, and **(c)** the tension in the lower chain.

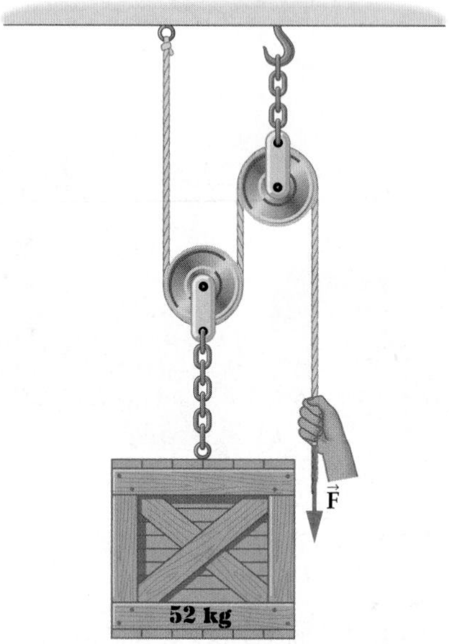

▲ **FIGURE 6–25** Problems 27 and 28

28. •• In Problem 27, determine **(a)** the force $\vec{F}$, **(b)** the tension in the upper chain, and **(c)** the tension in the lower chain, given that the crate is rising with an acceleration of 2.3 m/s^2.

Section 6–3 Translational Equilibrium

29. • Pulling the string on a bow back with a force of 27.5 lb, an archer prepares to shoot an arrow. If the archer pulls in the center of the string, and the angle between the two halves is 135°, what is the tension in the string?

30. • In **Figure 6–26** we see two blocks connected by a string and tied to a wall. The mass of the lower block is 1.0 kg; the mass of the upper block is 2.0 kg. Given that the angle of the incline is 31°, find the tensions in **(a)** the string connecting the two blocks and **(b)** the string that is tied to the wall.

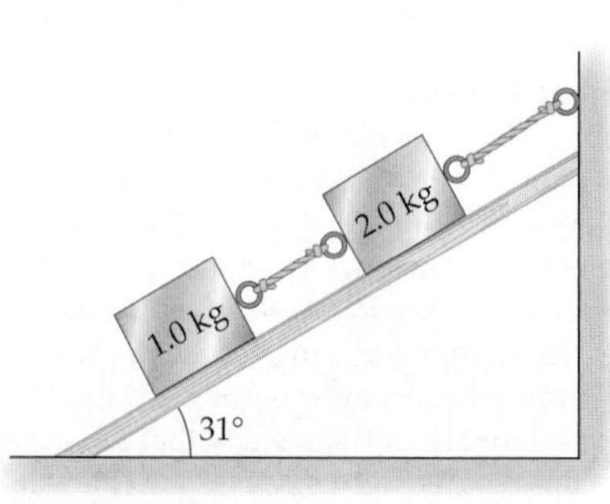

▲ **FIGURE 6–26** Problem 30

31. • **BIO Traction** After a skiing accident, your leg is in a cast and supported in a traction device, as shown in **Figure 6–27**. Find the magnitude of the force $\vec{F}$ exerted by the leg on the small pulley. (By Newton's third law, the small pulley exerts an equal and opposite force on the leg.) Let the mass m be 2.50 kg.

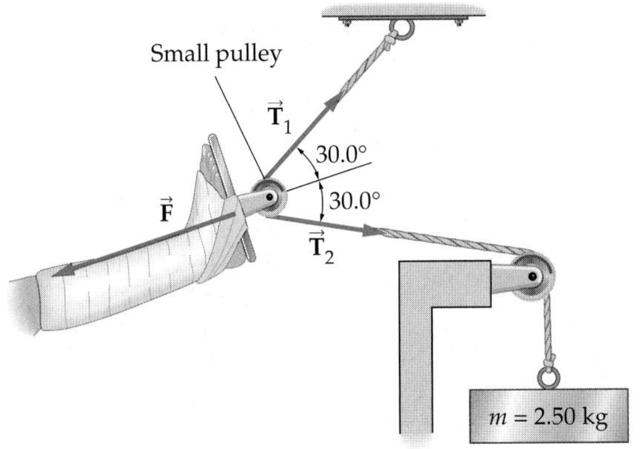

▲ **FIGURE 6–27** Problems 31 and 53

32. • Two blocks are connected by a string, as shown in **Figure 6–28**. The smooth inclined surface makes an angle of 42° with the horizontal, and the block on the incline has a mass of 6.7 kg. Find the mass of the hanging block that will cause the system to be in equilibrium. (The pulley is assumed to be ideal.)

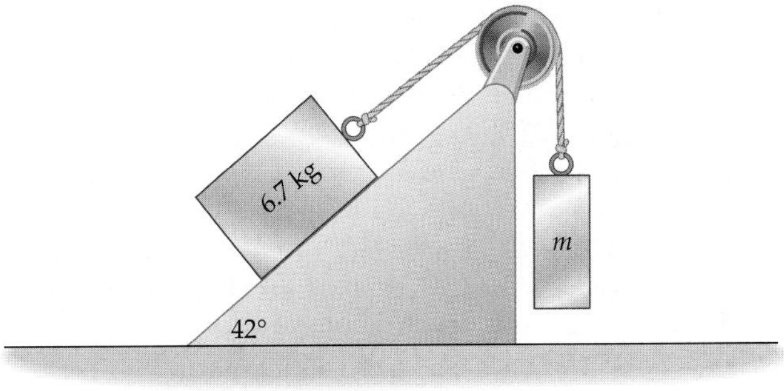

▲ **FIGURE 6–28** Problem 32

33. •• A 0.15-kg ball is placed in a shallow wedge with an opening angle of 120°, as shown in **Figure 6–29**. For each contact point between the wedge and the ball, determine the force exerted on the ball. Assume the system is frictionless.

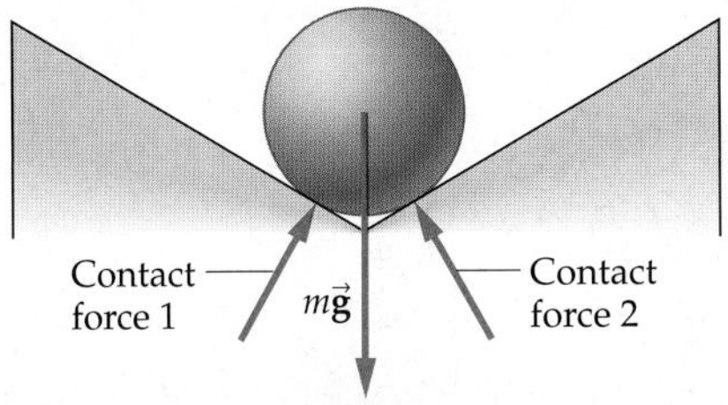

▲ **FIGURE 6–29** Problem 33

34. •• **IP** You want to nail a 1.6-kg board onto the wall of a barn. To position the board before nailing, you push it against the wall with a horizontal force $\vec{F}$ to keep it from sliding to the ground. (**Figure 6–30**) **(a)** If the coefficient of static friction between the board and the wall is 0.79, what is the least force you can apply and still hold the board in place? **(b)** What happens to the force of static friction if you push against the wall with a force greater than that found in part (a)?

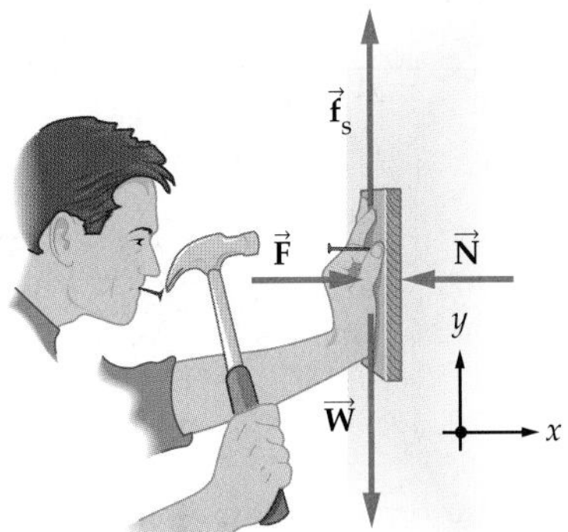

▲ **FIGURE 6–30** Problem 34

35. ••• **BIO The Russell Traction System** To immobilize a fractured femur (the thigh bone), doctors often utilize the Russell traction system illustrated in **Figure 6–31**. Notice that one force is applied directly to the knee, $\vec{F}_1$, while two other forces, $\vec{F}_2$ and $\vec{F}_3$, are applied to the foot. The latter two forces combine to give a force $\vec{F}_2 + \vec{F}_3$ that is transmitted through the lower leg to the knee. The result is that the knee experiences the total force $\vec{F}_{total} = \vec{F}_1 + \vec{F}_2 + \vec{F}_3$. The goal of this traction system is to have $\vec{F}_{total}$ directly in line with the fractured femur, at an angle of 20.0° above the horizontal. Find **(a)** the angle θ required to produce this alignment of $\vec{F}_{total}$ and **(b)** the magnitude of the force, $\vec{F}_{total}$ that is applied to the femur in this case. (Assume the pulleys are ideal.)

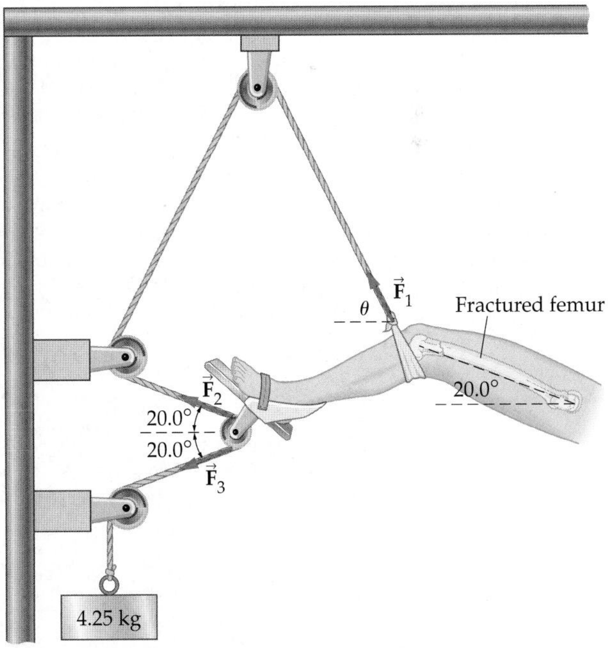

▲ **FIGURE 6–31** Problem 35

Section 6–4 Connected Objects

36. • Find the acceleration of the masses shown in **Figure 6–32**, given that m_1 = 1.0 kg, m_2 = 2.0 kg, and m_3 = 3.0 kg. Assume the table is frictionless and the masses move freely.

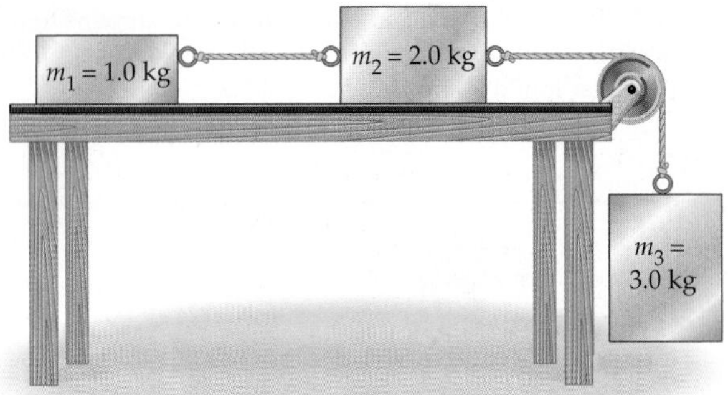

▲ **FIGURE 6–32** Problems 36, 39, and 88

37. • Two blocks are connected by a string, as shown in **Figure 6–33**. The smooth inclined surface makes an angle of 35° with the horizontal, and the block on the incline has a mass of 5.7 kg. The mass of the hanging block is $m = 3.2$ kg. Find **(a)** the direction and **(b)** the magnitude of the hanging block's acceleration.

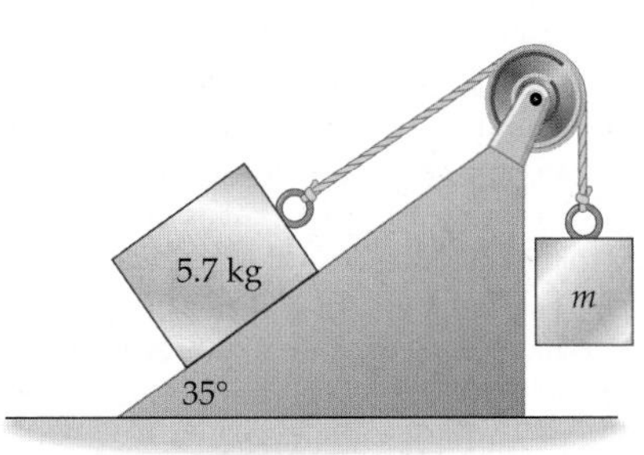

▲ **FIGURE 6–33** Problems 37 and 38

38. • Referring to Problem 37, find **(a)** the direction and **(b)** the magnitude of the hanging block's acceleration if its mass is $m = 4.2$ kg.

39. •• Referring to Figure 6-32, find the tension in the string connecting **(a)** m_1 and m_2 and **(b)** m_2 and m_3. Assume the table is frictionless and the masses move freely.

40. •• **IP** A 3.50-kg block on a smooth tabletop is attached by a string to a hanging block of mass 2.80 kg, as shown in **Figure 6–34**. The blocks are released from rest and allowed to move freely. **(a)** Is the tension in the string greater than, less than, or equal to the weight of the hanging mass? Find **(b)** the acceleration of the blocks and **(c)** the tension in the string.

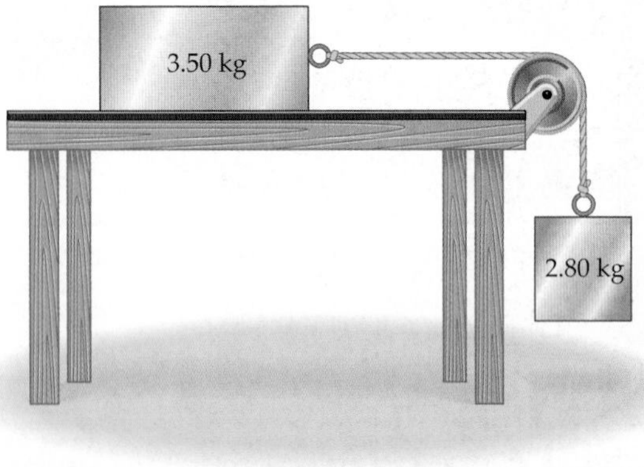

▲ **FIGURE 6–34** Problem 40

41. •• **IP** A 7.3-N force pulls horizontally on a 1.7-kg block that slides on a smooth horizontal surface. This block is connected by a horizontal string to a second block of mass $m_2 = 0.81$ kg on the same surface. **(a)** What is the acceleration of the blocks? **(b)** What is the tension in the string? **(c)** If the mass of block 1 is increased, does the tension in the string increase, decrease, or stay the same?

42. ••• **Buckets and a Pulley** Two buckets of sand hang from opposite ends of a rope that passes over an ideal pulley. One bucket is full and weighs 110 N; the other bucket is only partly filled and weighs 63 N. **(a)** Initially, you hold onto the lighter bucket to keep it from moving. What is the tension in the rope? **(b)** You release the lighter bucket and the heavier one descends. What is the tension in the rope now? **(c)** Eventually the heavier bucket lands and the two buckets come to rest. What is the tension in the rope now?

Section 6–5 Circular Motion

43. • When you take your 1200-kg car out for a spin, you go around a corner of radius 57 m with a speed of 15 m/s. The coefficient of static friction between the car and the road is 0.88. Assuming your car doesn't skid, what is the force exerted on it by static friction?

44. • Find the linear speed of the bottom of a test tube in a centrifuge if the centripetal acceleration there is 52,000 times the acceleration of gravity. The distance from the axis of rotation to the bottom of the test tube is 7.5 cm.

45. • **BIO A Human Centrifuge** To test the effects of high acceleration on the human body, the National Aeronautics and Space Administration (NASA) has constructed a large centrifuge at the Manned Spacecraft Center in Houston. In this device, astronauts are placed in a capsule that moves in a circular path with a radius of 15 m. If the astronauts in this centrifuge experience a centripetal acceleration 9.0 times that of gravity, what is the linear speed of the capsule?

46. • A car goes around a curve on a road that is banked at an angle of 31.5°. Even though the road is slick, the car will stay on the road without any friction between its tires and the road when its speed is 22.7 m/s. What is the radius of the curve?

47. •• Jill of the Jungle swings on a vine 6.5 m long. What is the tension in the vine if Jill, whose mass is 61 kg, is moving at 2.4 m/s when the vine is vertical?

48. •• **IP** In Problem 47, **(a)** how does the tension in the vine change if Jill's speed is doubled? Explain. **(b)** How does the tension change if her mass is doubled instead? Explain.

49. •• **IP (a)** As you ride on a Ferris wheel, your apparent weight is different at the top than at the bottom. Explain. **(b)** Calculate your apparent weight at the top and bottom of a Ferris wheel, given that the radius of the wheel is 7.2 m, it completes one revolution every 28 s, and your mass is 55 kg.

A Ferris Wheel (Problems 49 and 70)

50. •• Driving in your car with a constant speed of 12 m/s, you encounter a bump in the road that has a circular cross-section, as indicated in **Figure 6–35**. If the radius of curvature of the bump is 35 m, find the apparent weight of a 67-kg person in your car as you pass over the top of the bump.

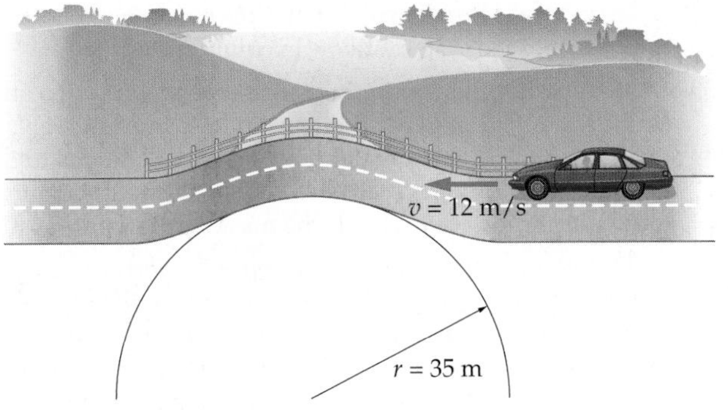

▲ **FIGURE 6–35** Problems 50 and 51

51. •• Referring to Problem 50, at what speed must you go over the bump if people in your car are to feel "weightless"?

52. •• **IP** You swing a 4.5-kg bucket of water in a vertical circle of radius 1.1 m. **(a)** What speed must the bucket have if it is to complete the circle without spilling any water? **(b)** How does your answer depend on the mass of the bucket?

General Problems

53. • **BIO** A skateboard accident leaves your leg in a cast and supported by a traction device, as in Figure 6–27. Find the mass m that must be attached to the rope if the net force exerted by the small pulley on the foot is to have a magnitude of 32 N.

54. • Find the centripetal acceleration at the top of a test tube in a centrifuge, given that the top is 4.2 cm from the axis of rotation and that its linear speed is 77 m/s.

55. • Find the coefficient of kinetic friction between a 3.85-kg block and the horizontal surface on which it rests if an 850-N/m spring must be stretched by 6.20 cm to pull it with constant speed. Assume that the spring pulls in the horizontal direction.

56. • A child goes down a playground slide that is inclined at an angle of 24.5° below the horizontal. Find the acceleration of the child given that the coefficient of kinetic friction between the child and the slide is 0.375.

57. • When a block is placed on top of a vertical spring, the spring compresses 3.15 cm. Find the mass of the block, given that the force constant of the spring is 1750 N/m.

58. • **BIO Nasal Strips** Many people use nasal strips, or external nasal dilator strips (ENDS), to alleviate a number of respiratory problems. The mechanism shown in **Figure 6–36** is used to measure the force constant of these strips. If a weight with a mass of 25 g is placed on the moveable platform, the ENDS is found to compress a distance of 28 mm. Find **(a)** the force constant of the ENDS (assuming it to be an ideal spring) in N/m and **(b)** the compression one would expect when a 32-g mass is placed on the moveable platform.

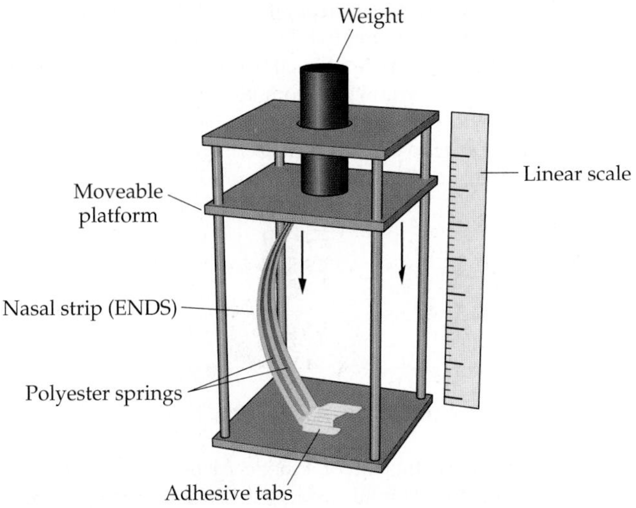

▲ **FIGURE 6–36** Problem 58

59. •• **The da Vinci Code** Leonardo da Vinci (1452–1519) is credited with being the first to perform quantitative experiments on friction, though his results weren't known until centuries later, due in part to the secret code (mirror writing) he used in his notebooks. Leonardo would place a block of wood on an inclined plane and measure the angle at which the block begins to slide. He reports that the coefficient of static friction was 0.25 in his experiments. At what angle did Leonardo's blocks begin to slide?

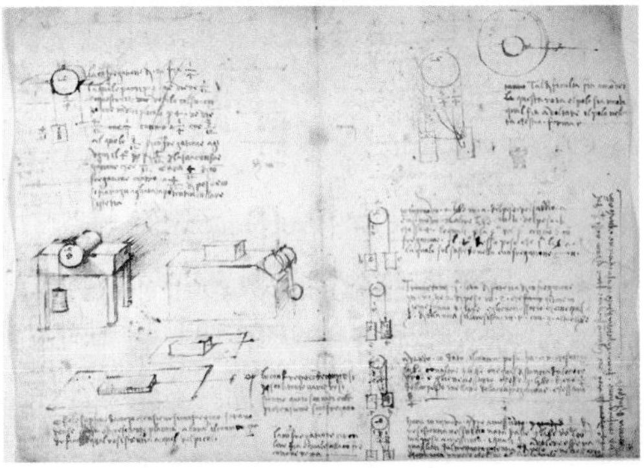

Sketches from the notebooks of Leonardo da Vinci showing experiments he performed on friction (Problem 59)

60. •• A force of 9.4 N pulls horizontally on a 1.1-kg block that slides on a rough, horizontal surface. This block is connected by a horizontal string to a second block of mass $m_2 = 1.92$ kg on the same surface. The coefficient of kinetic friction is $\mu_k = 0.24$ for both blocks. **(a)** What is the acceleration of the blocks? **(b)** What is the tension in the string?

61. •• You swing a 3.25-kg bucket of water in a vertical circle of radius 0.950 m. At the top of the circle the speed of the bucket is 3.23 m/s; at the bottom of the circle its speed is 6.91 m/s. Find the tension in the rope tied to the bucket at **(a)** the top and **(b)** the bottom of the circle.

62. •• A 12-g coin slides upward on a surface that is inclined at an angle of 15° above the horizontal. The coefficient of kinetic friction between the coin and the surface is 0.23; the coefficient of static friction is 0.31. Find the magnitude and direction of the force of friction **(a)** when the coin is sliding and **(b)** after it comes to rest.

63. •• In Problem 62, the angle of the incline is increased to 25°. Find the magnitude and direction of the force of friction when the coin is **(a)** sliding upward initially and **(b)** sliding back downward later.

64. •• A physics textbook weighing 22 N rests on a table. The coefficient of static friction between the book and the table is $\mu_s = 0.60$; the coefficient of kinetic friction is $\mu_k = 0.40$. You push horizontally on the book with a force that gradually increases from 0 to 15 N, and then slowly decreases to 5.0 N, as indicated in the following table. For each value of the applied force given in the table, give the magnitude of the force of friction and state whether the book is accelerating, decelerating, at rest, or moving with constant speed.

Applied force	Friction force	Motion
0		
5.0 N		
11 N		
15 N		
11 N		
8.0 N		
5.0 N		

65. •• A ball of mass m is placed in a wedge, as shown in **Figure 6–37**, in which the two walls meet at a right angle. Assuming the walls of the wedge are frictionless, determine the magnitude of **(a)** contact force 1 and **(b)** contact force 2.

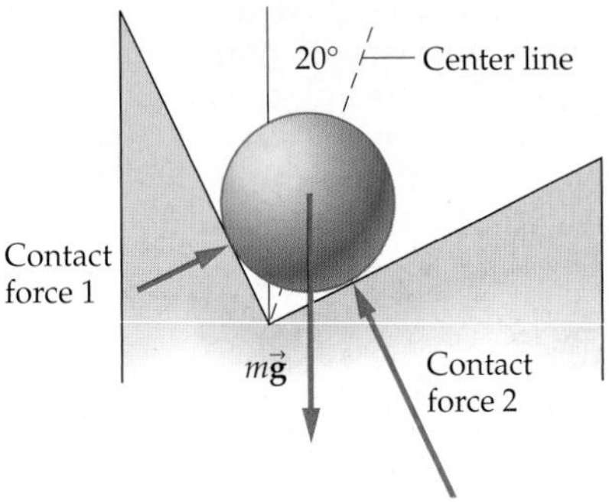

▲ **FIGURE 6–37** Problem 65

66. •• A 2.0-kg box rests on a plank that is inclined at an angle of 65° above the horizontal. The upper end of the box is attached to a spring with a force constant of 360 N/m, as shown in **Figure 6–38**. If the coefficient of static friction between the box and the plank is 0.22, what is the maximum amount the spring can be stretched and the box remain at rest?

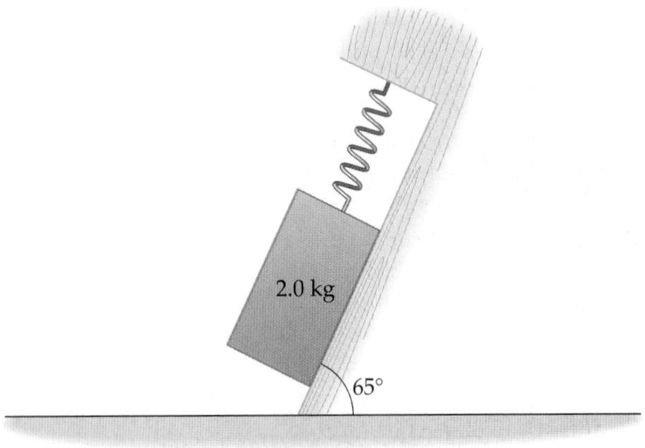

▲ **FIGURE 6–38** Problem 66

67. •• **IP** The blocks shown in **Figure 6–39** are at rest. **(a)** Find the frictional force exerted on block A given that the mass of block A is 8.50 kg, the mass of block B is 2.25 kg, and the coefficient of static friction between block A and the surface on which it rests is 0.320. **(b)** If the mass of block A is doubled, does the frictional force exerted on it increase, decrease, or stay the same? Explain.

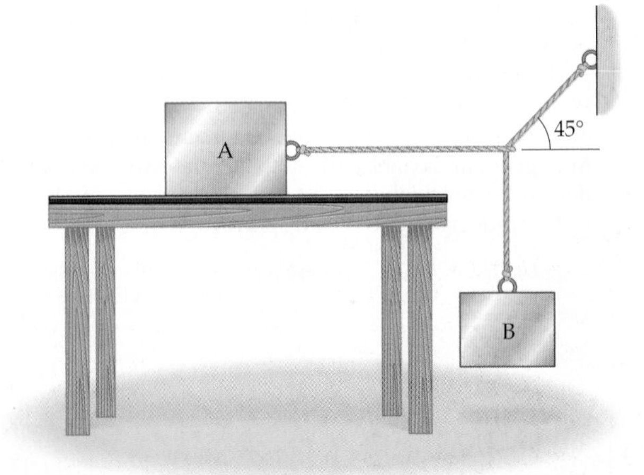

▲ **FIGURE 6–39** Problems 67 and 68

68. •• In part (a) of Problem 67, what is the maximum mass block B can have and the system still be in equilibrium?

69. •• **IP** A picture hangs on the wall suspended by two strings, as shown in Figure 6–24. The tension in string 2 is 1.7 N. **(a)** Is the tension in string 1 greater than, less than or equal to 1.7 N? Explain. **(b)** Verify your answer to part (a) by calculating the tension in string 1. **(c)** What is the mass of the picture?

70. •• **IP** Referring to Problem 49, suppose the Ferris wheel rotates fast enough to make you feel "weightless" at the top. **(a)** How many seconds does it take to complete one revolution in this case? **(b)** How does your answer to part (a) depend on your mass? Explain. **(c)** What is the direction and magnitude of your acceleration when you are at the bottom of the wheel? Assume that its rotational speed has remained constant.

71. •• **A Conical Pendulum** A 0.075-kg toy airplane is tied to the ceiling with a string. When the airplane's motor is started, it moves with a constant speed of 1.21 m/s in a horizontal circle of radius 0.44 m, as illustrated in **Figure 6–40**. Find **(a)** the angle the string makes with the vertical and **(b)** the tension in the string.

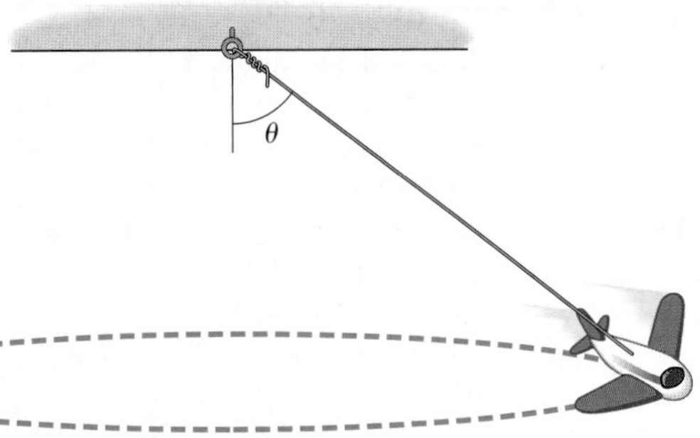

▲ **FIGURE 6–40** Problem 71

72. •• A tugboat tows a barge at constant speed with a 3500-kg cable, as shown in **Figure 6–41**. If the angle the cable makes with the horizontal where it attaches to the barge and the tugboat is 22°, find the force the cable exerts on the barge in the forward direction.

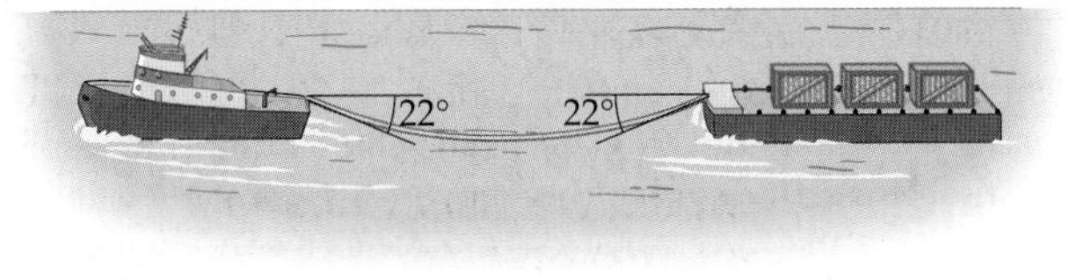

▲ **FIGURE 6–41** Problem 72

73. •• **IP** Two blocks, stacked one on top of the other, can move without friction on the horizontal surface shown in **Figure 6–42**. The surface between the two blocks is rough, however, with a coefficient of static friction equal to 0.47. **(a)** If a horizontal force F is applied to the 5.0-kg bottom block, what is the maximum value F can have before the 2.0-kg top block begins to slip? **(b)** If the mass of the top block is increased, does the maximum value of F increase, decrease, or stay the same? Explain.

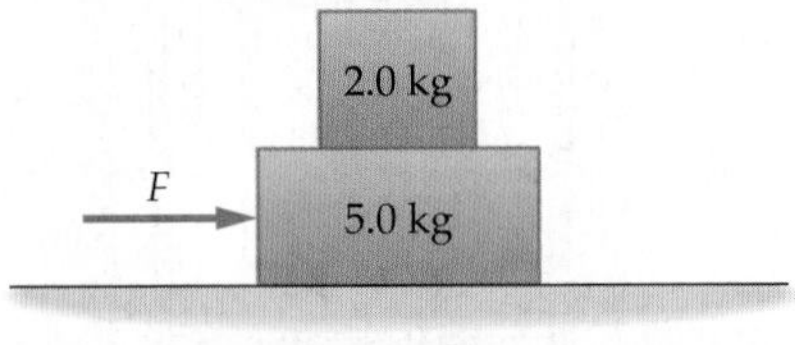

▲ **FIGURE 6–42** Problem 73

74. •• Find the coefficient of kinetic friction between a 4.5-kg block and the horizontal surface on which it rests if an 85 N/m

spring must be stretched by 2.1 cm to pull the block with constant speed. Assume the spring pulls in a direction 13° above the horizontal.

75. •• **IP** In a daring rescue by helicopter, two men with a combined mass of 172 kg are lifted to safety. **(a)** If the helicopter lifts the men straight up with constant acceleration, is the tension in the rescue cable greater than, less than, or equal to the combined weight of the men? Explain. **(b)** Determine the tension in the cable if the men are lifted with a constant acceleration of 1.10 m/s^2.

76. •• At the airport, you pull a 18-kg suitcase across the floor with a strap that is at an angle of 45° above the horizontal. Find **(a)** the normal force and **(b)** the tension in the strap, given that the suitcase moves with constant speed and that the coefficient of kinetic friction between the suitcase and the floor is 0.38.

77. •• **IP** A light spring with a force constant of 13 N/m is connected to a wall and to a 1.2-kg toy bulldozer, as shown in **Figure 6–43**. When the electric motor in the bulldozer is turned on, it stretches the spring for a distance of 0.45 m before its tread begins to slip on the floor. **(a)** Which coefficient of friction (static or kinetic) can be determined from this information? Explain. **(b)** What is the numerical value of this coefficient of friction?

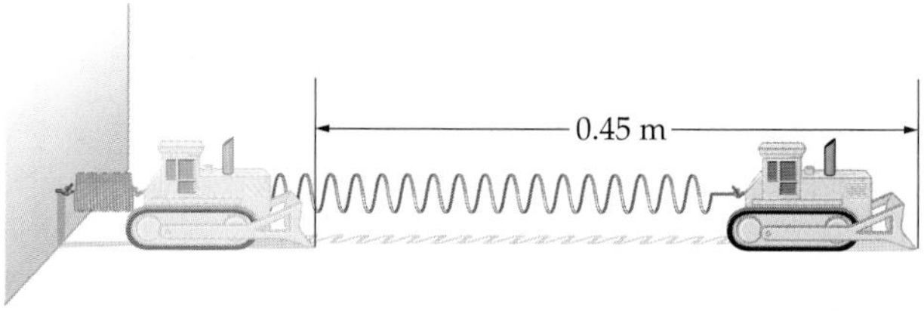

▲ **FIGURE 6–43** Problem 77

78. •• **IP** A 0.10-g spider hangs from the middle of the first thread of its future web. The thread makes an angle of 5.2° with the horizontal on both sides of the spider. **(a)** What is the tension in the thread? **(b)** If the angle made by the thread had been less than 5.2°, would its tension have been greater than, less than, or the same as in part (a)? Explain.

79. •• Find the acceleration the cart in **Figure 6–44** must have in order for the cereal box at the front of the cart not to fall. Assume that the coefficient of static friction between the cart and the box is 0.38.

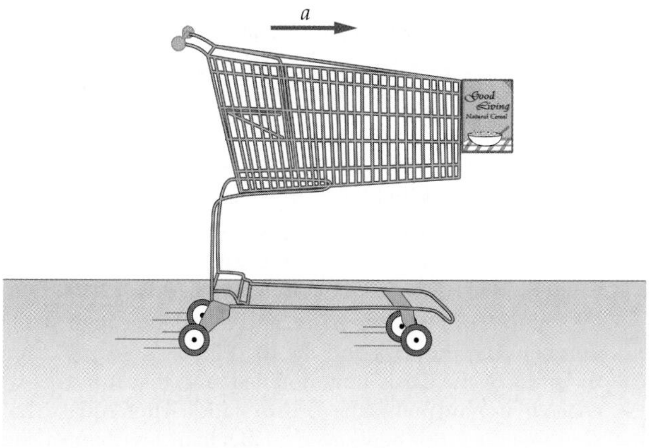

▲ **FIGURE 6–44** Problem 79

80. •• **IP Playing a Violin** The tension in a violin string is 2.7 N. When pushed down against the neck of the violin, the string makes an angle of 4.1° with the horizontal. **(a)** With what force must you push down on the string to bring it into contact with the neck? **(b)** If the angle were less than 4.1°, would the required force be greater than, less than, or the same as in part (a)? Explain.

81. •• **IP** A pair of fuzzy dice hangs from a string attached to your rear-view mirror. As you turn a corner with a radius of 95 m and a constant speed of 25 mi/h, what angle will the dice make with the vertical? Why is it unnecessary to give the mass of the dice?

82. •• Find the tension in each of the two ropes supporting a hammock if one is at an angle of 15° above the horizontal and the other is at an angle of 35° above the horizontal. The person sleeping in the hammock (unconcerned about tensions and ropes) has a mass of 64 kg.

83. •• As your plane circles an airport, it moves in a horizontal circle of radius 2100 m with a speed of 380 km/h. If the lift of the airplane's wings is perpendicular to the wings, at what angle should the plane be banked so that it doesn't tend to slip sideways?

84. •• **IP** A block with a mass of 3.1 kg is placed at rest on a surface inclined at an angle of 45° above the horizontal. The coefficient of static friction between the block and the surface is 0.50, and a force of magnitude F pushes upward on the block, parallel to the inclined surface. **(a)** The block will remain at rest only if F is greater than a minimum value, F_{min}, and less than a maximum value, F_{max}. Explain the reasons for this behavior. **(b)** Calculate F_{min}. **(c)** Calculate F_{max}.

85. •• A mountain climber of mass m hangs onto a rope to keep from sliding down a smooth, ice-covered slope (**Figure 6–45**). Find a formula for the tension in the rope when the slope is inclined at an angle θ above the horizontal. Check your results in the limits $\theta = 0$ and $\theta = 90°$.

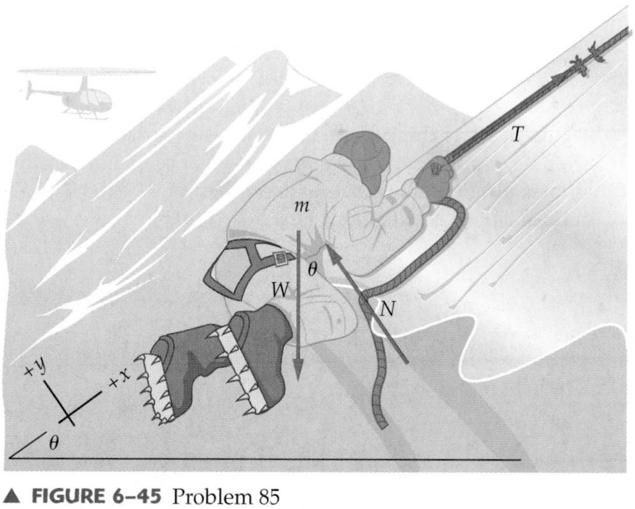

▲ **FIGURE 6–45** Problem 85

86. •• A child sits on a rotating merry-go-round, 2.3 m from its center. If the speed of the child is 2.2 m/s, what is the minimum coefficient of static friction between the child and the merry-go-round that will prevent the child from slipping?

87. ••• A wood block of mass m rests on a larger wood block of mass M that rests on a wooden table. The coefficients of static and kinetic friction between all surfaces are μ_s and μ_k, respectively. What is the minimum horizontal force, F, applied to the lower block that will cause it to slide out from under the upper block?

88. ••• Find the tension in each of the two strings shown in Figure 6–32 for general values of the masses. Your answer should be in terms of m_1, m_2, m_3, and g.

89. ••• The coefficient of static friction between a rope and the table on which it rests is μ_s. Find the fraction of the rope that can hang over the edge of the table before it begins to slip.

90. ••• A hockey puck of mass m is attached to a string that passes through a hole in the center of a table, as shown in **Figure 6–46**. The hockey puck moves in a circle of radius r. Tied to the other end of the string, and hanging vertically beneath the table, is a mass M. Assuming the tabletop is perfectly smooth, what speed must the hockey puck have if the mass M is to remain at rest?

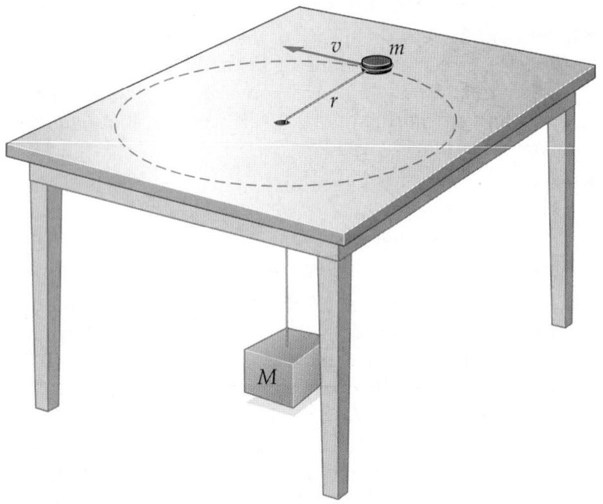

▲ **FIGURE 6–46** Problem 90

91. ••• **The Force Needed to Move a Crate** To move a crate of mass m across a rough floor, you push down on it at an angle θ, as shown in Figure 6–21 for the special case of $\theta = 21°$. **(a)** Find the force necessary to start the crate moving as a function of θ, given that the coefficient of static friction between the crate and the floor is μ_s. **(b)** Show that it is impossible to move the crate, no matter how great the force, if the coefficient of static friction is greater than or equal to $1/\tan\theta$.

92. ••• **IP** A popular ride at amusement parks is illustrated in **Figure 6–47**. In this ride, people sit in a swing that is suspended from a rotating arm. Riders are at a distance of 12 m from the axis of rotation and move with a speed of 25 mi/h. **(a)** Find the centripetal acceleration of the riders. **(b)** Find the angle θ the supporting wires make with the vertical. **(c)** If you observe a ride like that in Figure 6–47, or as shown in the photo on page 163, you will notice that all the swings are at the same angle θ to the vertical, regardless of the weight of the rider. Explain.

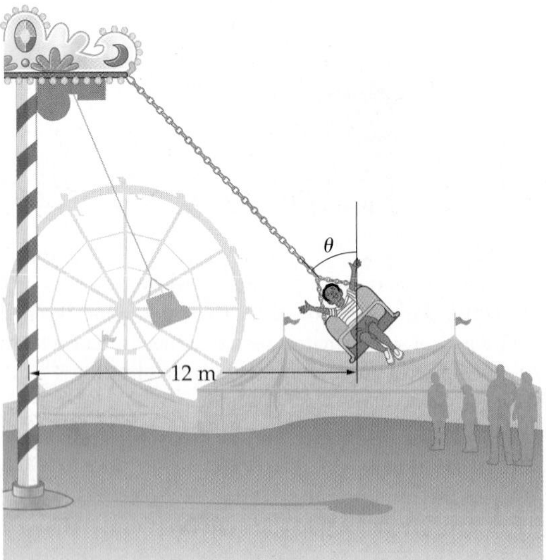

▲ **FIGURE 6–47** Problem 92

93. ••• **A Conveyor Belt** A box is placed on a conveyor belt that moves with a constant speed of 1.25 m/s. The coefficient of kinetic friction between the box and the belt is 0.780. **(a)** How much time does it take for the box to stop sliding relative to the belt? **(b)** How far does the box move in this time?

94. ••• You push a box along the floor against a constant force of friction. When you push with a horizontal force of 75 N, the acceleration of the box is 0.50 m/s^2; when you increase the force to 81 N, the acceleration is 0.75 m/s^2. Find **(a)** the mass of the box and **(b)** the coefficient of kinetic friction between the box and the floor.

95. ••• As part of a circus act, a person drives a motorcycle with constant speed v around the inside of a vertical track of radius r, as indicated in **Figure 6–48**. If the combined mass of the motorcycle and rider is m, find the normal force exerted on the motorcycle by the track at the points **(a)** A, **(b)** B, and **(c)** C.

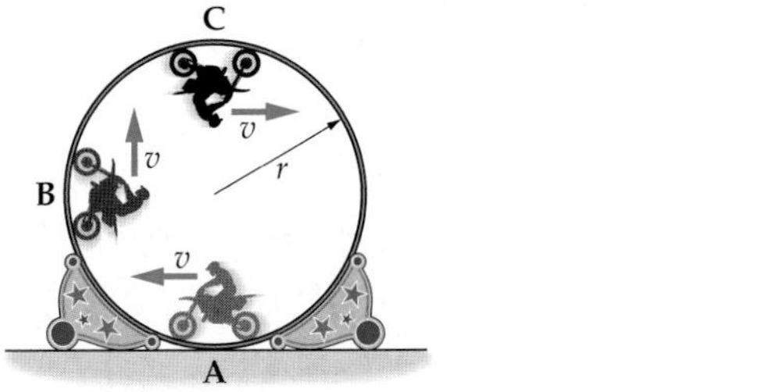

▲ **FIGURE 6–48** Problem 95

Interactive Problems

96. •• **IP** ***Referring to Example 6–3*** Suppose the coefficients of static and kinetic friction between the crate and the truck bed are 0.415 and 0.382, respectively. **(a)** Does the crate begin to slide at a tilt angle that is greater than, less than, or equal to 23.2°? **(b)** Verify your answer to part (a) by determining the angle at which the crate begins to slide. **(c)** Find the length of time it takes for the crate to slide a distance of 2.75 m when the tilt angle has the value found in part (b).

97. •• **IP** ***Referring to Example 6–3*** The crate begins to slide when the tilt angle is 17.5°. When the crate reaches the bottom of the flatbed, after sliding a distance of 2.75 m, its speed is 3.11 m/s. Find **(a)** the coefficient of static friction and **(b)** the coefficient of kinetic friction between the crate and the flatbed.

98. •• ***Referring to Example 6–6*** Suppose the mass on the frictionless tabletop has the value $m_1 = 2.45$ kg. **(a)** Find the value of m_2 that gives an acceleration of 2.85 m/s^2. **(b)** What is the corresponding tension, T, in the string? **(c)** Calculate the ratio T/m_2g and show that it is less than 1, as expected.

99. •• ***Referring to Example 6–8*** **(a)** At what speed will the force of static friction exerted on the car by the road be equal to half the weight of the car? The mass of the car is $m = 1200$ kg, the radius of the corner is $r = 45$ m, and the coefficient of static friction between the tires and the road is $\mu_s = 0.82$. **(b)** Suppose the mass of the car is now doubled, and that it moves with a speed that again makes the force of static friction equal to half the car's weight. Is this new speed greater than, less than, or equal to the speed in part (a)?

7 Work and Kinetic Energy

7–1	Work Done by a Constant Force	180
7–2	Kinetic Energy and the Work-Energy Theorem	186
7–3	Work Done by a Variable Force	190
7–4	Power	195

We all know intuitively that motion, energy, and work are somehow related. For example, the chemical energy stored in this pitcher's muscles enables him to do work on a baseball. This means, basically, that he exerts a force on it over a distance. The work done on the ball appears as kinetic energy—the energy of motion—and when the ball is caught, its kinetic energy can in turn do work on the catcher. In this chapter we'll give precise definitions of the concepts of work, kinetic energy, and power, and explore the physical relationships among them.

The concept of force is one of the foundations of physics, as we have seen in the previous two chapters. Equally fundamental, though less obvious, is the idea that a force times the displacement through which it acts is also an important physical quantity. We refer to this quantity as the *work* done by a force.

Now, we all know what work means in everyday life: We get up in the morning and go to work, or we "work up a sweat" as we hike a mountain trail. Later in the day we eat lunch, which gives us the "energy" to continue working or to continue our hike. In this chapter we give a precise physical definition of work, and show how it is related to another important physical quantity—the energy of motion, or *kinetic energy*. When these concepts are extended in the next chapter, we are led to the rather sweeping observation that the total amount of energy in the universe remains constant at all times.

7–1 Work Done by a Constant Force

In this section we define work—in the physics sense of the word—and apply our definition to a variety of physical situations. We start with the simplest case; namely, the work done when force and displacement are in the same direction. Later in the section we generalize our definition to include cases where the force and displacement are in arbitrary directions. We conclude with a discussion of the work done on an object when it is acted on by more than one force.

Force in the Direction of Displacement

When we push a shopping cart in a store or pull a suitcase through an airport, we do work. The greater the force, the greater the work; the greater the distance, the greater the work. These simple ideas form the basis for our definition of work.

To be specific, suppose we push a box with a constant force $\vec{\mathbf{F}}$, as shown in **Figure 7–1**. If we move the box *in the direction of* $\vec{\mathbf{F}}$ through a displacement $\vec{\mathbf{d}}$, the **work** W we have done is Fd:

Definition of Work, *W*, When a Constant Force Is in the Direction of Displacement

$$W = Fd \qquad \text{7–1}$$

SI unit: newton-meter $(\mathrm{N \cdot m})$ = joule, J

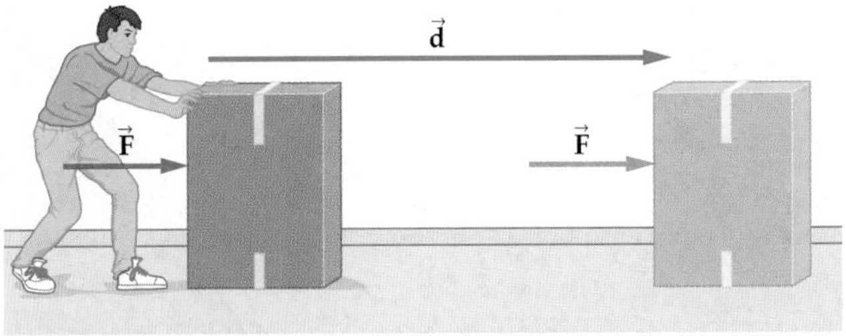

▶ **FIGURE 7–1 Work: constant force in the direction of motion**

A constant force $\vec{\mathbf{F}}$ pushes a box through a displacement $\vec{\mathbf{d}}$. In this special case, where the force and displacement are in the *same* direction, the work done on the box by the force is $W = Fd$.

Note that work is the product of two magnitudes, and hence it is a scalar. In addition, notice that a small force acting over a large distance gives the same work as a large force acting over a small distance. For example, $W = (1\ \mathrm{N})(400\ \mathrm{m}) = (400\ \mathrm{N})(1\ \mathrm{m})$.

The dimensions of work are newtons (force) times meters (distance), or $\mathrm{N \cdot m}$. This combination of dimensions is called the **joule** (rhymes with school, as commonly pronounced) in honor of James Prescott Joule (1818–1889), a dedicated physicist who is said to have conducted physics experiments even while on his honeymoon. We define a joule as follows:

Definition of the joule, J

$$1\ \text{joule} = 1\ \mathrm{J} = 1\ \mathrm{N \cdot m} = 1(\mathrm{kg \cdot m/s^2}) \cdot \mathrm{m} = 1\ \mathrm{kg \cdot m^2/s^2} \qquad \text{7–2}$$

To get a better feeling for work and the associated units, suppose you exert a force of 82.0 N on the box in Figure 7–1 and move it in the direction of the force through a distance of 3.00 m. The work you have done is

$$W = Fd = (82.0\ \mathrm{N})(3.00\ \mathrm{m}) = 246\ \mathrm{N \cdot m} = 246\ \mathrm{J}$$

Similarly, if you do 5.00 J of work to lift a book through a vertical distance of 0.750 m, the force you exerted on the book is

$$F = \frac{W}{d} = \frac{5.00\ \mathrm{J}}{0.750\ \mathrm{m}} = \frac{5.00\ \mathrm{N \cdot m}}{0.750\ \mathrm{m}} = 6.67\ \mathrm{N}$$

EXERCISE 7–1

One species of Darwin's finch, *Geospiza magnirostris*, can exert a force of 205 N with its beak as it cracks open a *Tribulus* seed case. If its beak moves through a distance of 0.40 cm during this operation, how much work does the finch do to get the seed?

Solution

$W = Fd = (205\ \text{N})(0.0040\ \text{m}) = 0.82\ \text{J}$

Just how much work is a joule, anyway? Well, you do one joule of work when you lift a gallon of milk through a height of about an inch, or lift an apple a meter. One joule of work lights a 100-watt lightbulb for 0.01 seconds or heats a glass of water 0.00125 degrees Celsius. Clearly, a joule is a modest amount of work in everyday terms. Additional examples of work are listed in Table 7–1.

TABLE 7–1 Typical Values of Work

Activity	Equivalent work (J)
Annual U. S. energy use	8×10^{19}
Mt. St. Helens eruption	10^{18}
Burning one gallon of gas	10^{8}
Human food intake/day	10^{7}
Melting an ice cube	10^{4}
Lighting a 100-W bulb for 1 minute	6000
Heartbeat	0.5
Turning page of a book	10^{-3}
Hop of a flea	10^{-7}
Breaking a bond in DNA	10^{-20}

EXAMPLE 7–1 Heading for the ER

An intern pushes a 72-kg patient on a 15-kg gurney, producing an acceleration of $0.60\ \text{m/s}^2$. How much work does the intern do by pushing the patient and gurney through a distance of 2.5 m? Assume the gurney moves without friction.

Picture the Problem

Our sketch shows the physical situation for this problem. Note that the force exerted by the intern is in the same direction as the displacement of the gurney; therefore, we know that $W = Fd$.

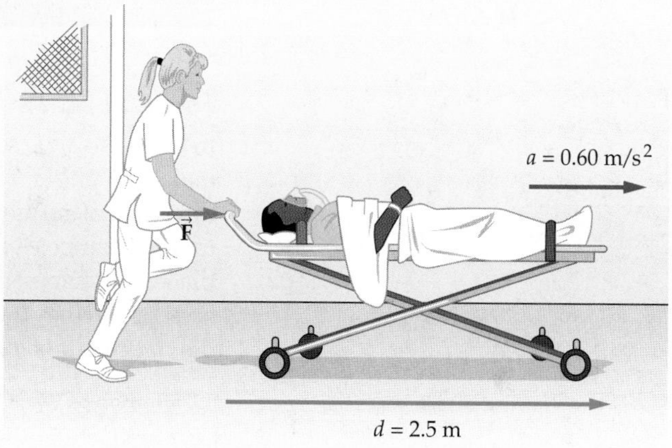

Strategy

We are not given the magnitude of the force F, so we cannot apply Equation 7–1 directly. However, we are given the mass and acceleration of the patient and gurney, from which we can calculate the force with $F = ma$. The work done by the intern is then $W = Fd$.

Solution

1. First, find the force F exerted by the intern: $F = ma = (72\ \text{kg} + 15\ \text{kg})(0.60\ \text{m/s}^2) = 52\ \text{N}$
2. The work done by the intern, W, is the force times the distance: $W = Fd = (52\ \text{N})(2.5\ \text{m}) = 130\ \text{J}$

Insight

You might wonder whether the work done by the intern depends on the speed of the gurney. The answer is no. The work done on an object, $W = Fd$, doesn't depend on whether the object moves through the distance d quickly or slowly. What does depend on the speed of the gurney is the *rate* at which work is done, as we discuss in detail in Section 7–4.

Practice Problem

If the total mass of the gurney plus patient is halved and the acceleration is doubled, does the work done by the intern increase, decrease, or remain the same? [**Answer:** The work remains the same.]

Some related homework problems: Problem 1, Problem 2

Before moving on, let's note an interesting point about our definition of work. It's clear from Equation 7–1 that *the work W is zero if the distance d is zero*—and this is true regardless of how great the force might be. For example, if you push against a solid wall you do no work on it, even though you may become tired from your efforts. Similarly, if you stand in one place holding a 50-pound suitcase in your hand, you do no work on the suitcase. The fact that we become tired when we push against a wall or hold a heavy object is due to the repeated contraction and expansion of individual cells within our muscles. Thus, even when we are "at rest," our muscles are doing mechanical work on the microscopic level.

▶ The weightlifter at right does more work in raising 150 kilograms above her head than Atlas, who is supporting the entire world. Why?

Force at an Angle to the Displacement

In **Figure 7–2** we see a person pulling a suitcase on a level surface with a strap that makes an angle θ with the horizontal—in this case the force is at an angle to the direction of motion. How do we calculate the work now? Well, instead of force times distance, we say that work is the *component* of force in the *direction* of displacement times the magnitude of the displacement. In Figure 7–2, the component of force in the direction of displacement is $F \cos \theta$ and the magnitude of the displacement is d. Therefore, the work is $F \cos \theta$ times d:

Definition of Work When the Angle Between a Constant Force and the Displacement Is θ

$$W = (F \cos \theta)d = Fd \cos \theta \qquad \textbf{7–3}$$

SI unit: joule, J

Of course, in the case where the force is in the direction of motion, the angle θ is zero; then $W = Fd \cos 0° = Fd \cdot 1 = Fd$, in agreement with Equation 7–1.

Equally interesting is a situation in which the force and the displacement are at right angles to one another. In this case $\theta = 90°$ and the work done by the force F is zero; $W = Fd \cos 90° = 0$.

This result leads naturally to an alternative way to think about the expression $W = Fd \cos \theta$. In **Figure 7–3** we show the displacement and the force for the suitcase

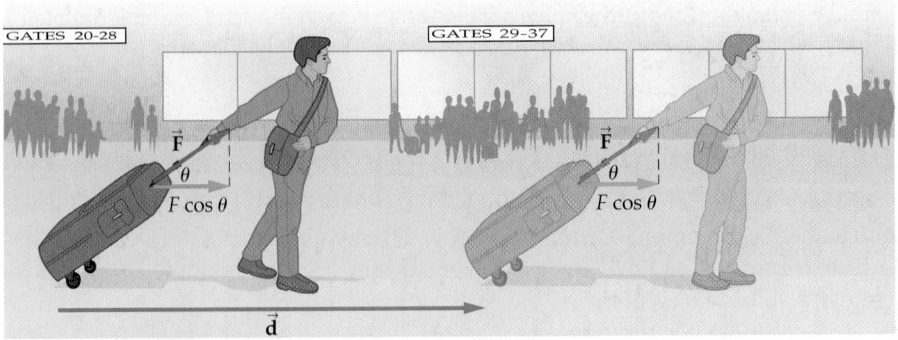

▶ **FIGURE 7–2 Work: force at an angle to direction of motion**

A person pulls a suitcase with a strap at an angle θ to the direction of motion. The component of force in the direction of motion is $F \cos \theta$, and the work done by the person is $W = (F \cos \theta)d$.

in Figure 7–2. Notice that the displacement is equivalent to a displacement in the direction of the force of magnitude ($d \cos \theta$) *plus* a displacement at right angles to the force of magnitude ($d \sin \theta$). Since the displacement at right angles to the force corresponds to zero work and the displacement in the direction of the force corresponds to a work $W = F(d \cos \theta)$, it follows that the work done in this case is $Fd \cos \theta$, as given in Equation 7–3. Thus, the work done by a force can be thought of in the following two *equivalent* ways:

(i) *Work is the component of force in the direction of displacement times the magnitude of the displacement.*

(ii) *Work is the component of displacement in the direction of the force times the magnitude of the force.*

In either of these interpretations, the mathematical expression for work is exactly the same, $W = Fd \cos \theta$, where θ is the angle between the force vector and the displacement vector when they are placed tail-to-tail. This definition of θ is illustrated in Figure 7–3.

Finally, we can also express work as the **dot product** between the vectors $\vec{\mathbf{F}}$ and $\vec{\mathbf{d}}$; that is, $W = \vec{\mathbf{F}} \cdot \vec{\mathbf{d}} = Fd \cos \theta$. Note that the dot product, which is always a scalar, is simply the magnitude of one vector times the magnitude of the second vector times the cosine of the angle between them. We discuss the dot product in greater detail in Appendix A.

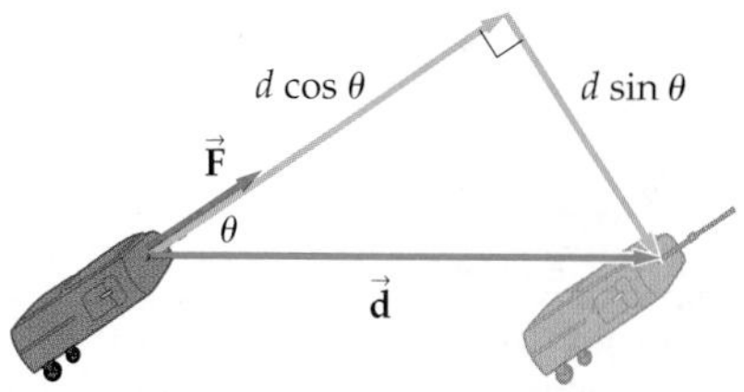

▲ **FIGURE 7–3 Force at an angle to direction of motion: another look**

The displacement of the suitcase in Figure 7–2 is equivalent to a displacement of magnitude $d \cos \theta$ in the direction of the force $\vec{\mathbf{F}}$ plus a displacement of magnitude $d \sin \theta$ perpendicular to the force. Only the displacement parallel to the force results in nonzero work, hence the total work done is $F(d \cos \theta)$ as expected.

EXAMPLE 7–2 Gravity Escape System

REAL-WORLD PHYSICS In a gravity escape system (GES), an enclosed lifeboat on a large ship is deployed by letting it slide down a ramp and then continuing in free fall to the water below. Suppose a 4970-kg lifeboat slides a distance of 5.00 m on a ramp, dropping through a vertical height of 2.50 m. How much work does gravity do on the boat?

Picture the Problem

From our sketch, we see that the force of gravity $m\vec{\mathbf{g}}$ and displacement $\vec{\mathbf{d}}$ are at an angle θ relative to one another when placed tail-to-tail, and that θ is also the angle the ramp makes with the vertical. In addition, we note that the vertical height of the ramp is $h = 2.50$ m and the length of the ramp is $d = 5.00$ m.

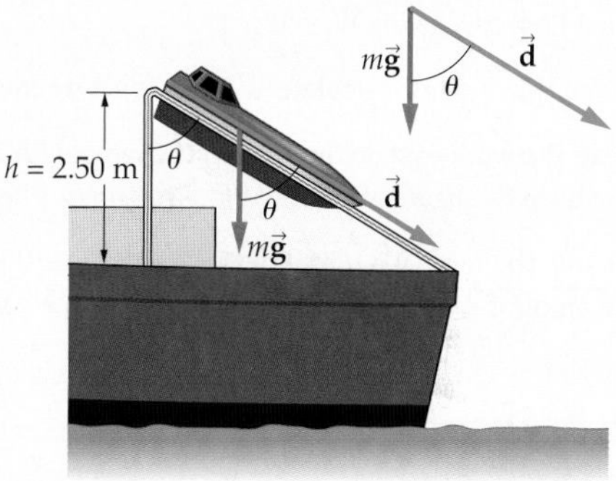

Strategy

By definition, the work done on the lifeboat by gravity is $W = Fd \cos \theta$, where $F = mg$, $d = 5.00$ m, and θ is the angle between $m\vec{\mathbf{g}}$ and $\vec{\mathbf{d}}$. We are not given θ in the problem statement, but from the right triangle that forms the ramp we see that $\cos \theta = h/d$. Once θ is determined from the geometry of our sketch, it is straightforward to calculate W.

Solution

1. First, find the component of $\vec{\mathbf{F}} = m\vec{\mathbf{g}}$ in the direction of motion:

$$F \cos \theta = (mg)\left(\frac{h}{d}\right)$$
$$= (4970 \text{ kg})(9.81 \text{ m/s}^2)\left(\frac{2.50 \text{ m}}{5.00 \text{ m}}\right) = 24{,}400 \text{ N}$$

2. Multiply by distance to find the work:

$$W = (F \cos \theta)d = (24{,}400 \text{ N})(5.00 \text{ m}) = 122{,}000 \text{ J}$$

3. Alternatively, cancel d algebraically before substituting numerical values:

$$W = Fd \cos \theta = (mg)(d)\left(\frac{h}{d}\right)$$
$$= mgh = (4970 \text{ kg})(9.81 \text{ m/s}^2)(2.50 \text{ m}) = 122{,}000 \text{ J}$$

Insight

The work is simply $W = mgh$, exactly the same as if the lifeboat had fallen straight down through the height h.

Notice that working the problem symbolically, as in step 3, results in two distinct advantages. First, it makes for a simpler expression for the work. Second, and more importantly, it shows that the distance d cancels; hence the work depends on the height h but not on d. Such a result is not apparent when we work solely with numbers, as in steps 1 and 2.

Practice Problem

Suppose the lifeboat slides halfway to the water, gets stuck for a moment, and then starts up again and continues to the end of the ramp. What is the work done by gravity in this case? [**Answer:** The work done by gravity is exactly the same, $W = mgh$, independent of how the boat moves down the ramp.]

Some related homework problems: Problem 8, Problem 9

Next, we present a Conceptual Checkpoint that compares the work required to move an object along two different paths.

CONCEPTUAL CHECKPOINT 7–1 Path Dependence of Work

You want to load a box into the back of a truck. One way is to lift it straight up through a height h, as shown, doing a work W_1. Alternatively, you can slide the box up a loading ramp a distance L, doing a work W_2. Assuming the box slides on the ramp without friction, which of the following is correct: **(a)** $W_1 < W_2$, **(b)** $W_1 = W_2$, **(c)** $W_1 > W_2$?

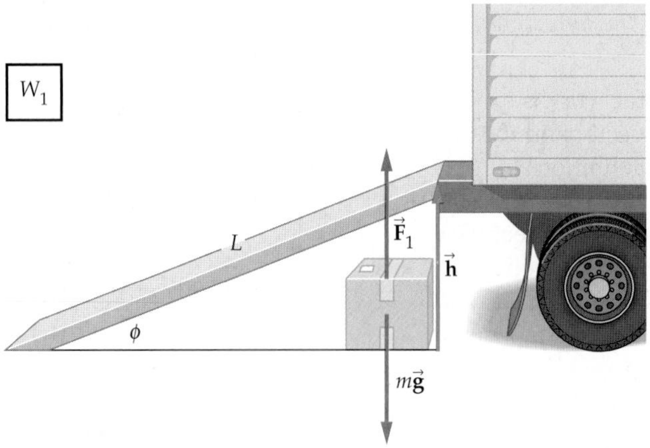

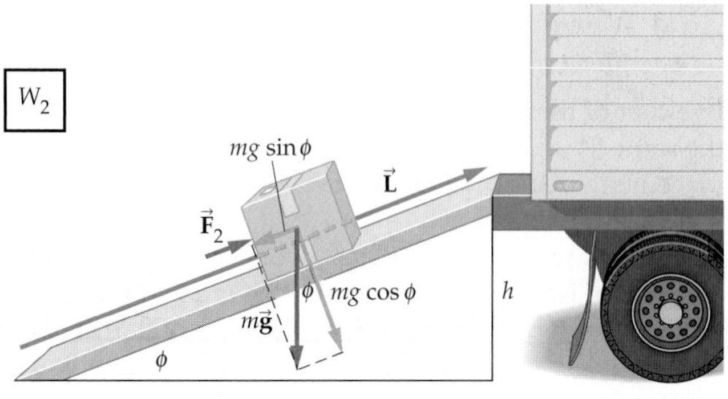

Reasoning and Discussion

You might think that W_2 is less than W_1, since the force needed to slide the box up the ramp, F_2, is less than the force needed to lift it straight up. On the other hand, the distance up the ramp, L, is greater than the vertical distance, h, so perhaps W_2 should be greater than W_1. In fact, these two effects cancel exactly, giving $W_1 = W_2$.

To see this, we first calculate W_1. The force needed to lift the box with constant speed is $F_1 = mg$, and the height is h, therefore $W_1 = mgh$.

Next, the work to slide the box up the ramp with constant speed is $W_2 = F_2L$, where F_2 is the force required to push against the tangential component of gravity. In the figure we see that $F_2 = mg \sin \phi$. The figure also shows that $\sin \phi = h/L$; thus $W_2 = (mg \sin \phi)L = (mg)(h/L)L = mgh = W_1$.

Clearly, the ramp is a useful device—it reduces the *force* required to move the box upward from $F_1 = mg$ to $F_2 = mg(h/L)$. Even so, it doesn't decrease the amount of *work* we need to do. As we have seen, the reduced force on the ramp is offset by the increased distance.

Answer:

(b) $W_1 = W_2$

Negative Work and Total Work

Work depends on the angle between the force, $\vec{F}$, and the displacement (or direction of motion), $\vec{d}$. This dependence gives rise to three distinct possibilities, as shown in **Figure 7–4**:

(i) *Work is positive if the force has a component in the direction of motion* $(-90° < \theta < 90°)$.
(ii) *Work is zero if the force has no component in the direction of motion* $(\theta = \pm 90°)$.
(iii) *Work is negative if the force has a component opposite to the direction of motion* $(90° < \theta < 270°)$.

Thus, whenever we calculate work, we must be careful about its sign and not just assume it to be positive.

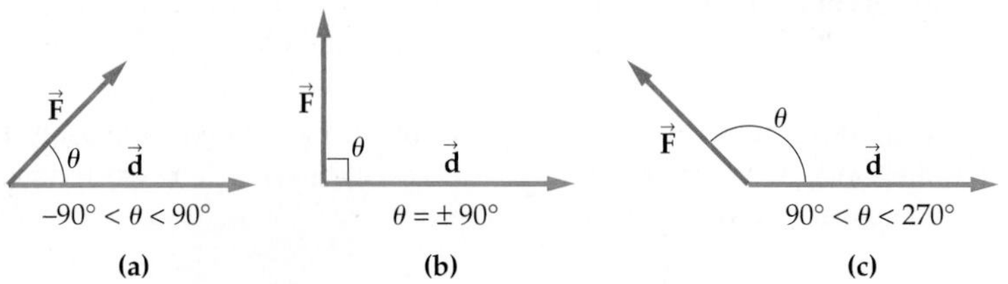

▶ **FIGURE 7–4 Positive, negative, and zero work**
Work is positive when the force is in the same general direction as the displacement and is negative if the force is generally opposite to the displacement. Zero work is done if the force is at right angles to the displacement.

When more than one force acts on an object, the total work is the sum of the work done by each force separately. Thus, if force $\vec{\mathbf{F}}_1$ does work W_1, force $\vec{\mathbf{F}}_2$ does work W_2, and so on, the total work is

$$W_{\text{total}} = W_1 + W_2 + W_3 + \cdots = \sum W_i \qquad \textbf{7–4}$$

Equivalently, the total work can be calculated by first performing a vector sum of all the forces acting on an object to obtain $\vec{\mathbf{F}}_{\text{total}}$ and then using our basic definition of work:

$$W_{\text{total}} = (F_{\text{total}} \cos \theta)d = F_{\text{total}}\, d \cos \theta \qquad \textbf{7–5}$$

where θ is the angle between $\vec{\mathbf{F}}_{\text{total}}$ and the displacement $\vec{\mathbf{d}}$. In the next two Examples we calculate the total work in each of these ways.

PROBLEM-SOLVING NOTE

Be Careful About the Angle θ

In calculating $W = Fd \cos \theta$ be sure that the angle you use in the cosine is the angle between the force and the displacement vectors when they are placed tail to tail. Sometimes θ may be used to label a different angle in a given problem. For example, θ is often used to label the angle of a slope, in which case it may have nothing to do with the angle between the force and the displacement. To summarize: Just because an angle is labeled θ doesn't mean it's automatically the correct angle to use in the work formula.

EXAMPLE 7–3 A Coasting Car I

A car of mass m coasts down a hill inclined at an angle ϕ below the horizontal. The car is acted on by three forces: (i) the normal force $\vec{\mathbf{N}}$ exerted by the road, (ii) a force due to air resistance, $\vec{\mathbf{F}}_{\text{air}}$, and (iii) the force of gravity, $m\vec{\mathbf{g}}$. Find the total work done on the car as it travels a distance d along the road.

Picture the Problem

Because ϕ is the angle the slope makes with the horizontal, it is also the angle between $m\vec{\mathbf{g}}$ and the downward normal direction, as was shown in Figure 5–15. It follows that the angle between $m\vec{\mathbf{g}}$ and the displacement $\vec{\mathbf{d}}$ is $\theta = 90° - \phi$. Our sketch also shows that the angle between $\vec{\mathbf{N}}$ and $\vec{\mathbf{d}}$ is $\theta = 90°$, and the angle between $\vec{\mathbf{F}}_{\text{air}}$ and $\vec{\mathbf{d}}$ is $\theta = 180°$.

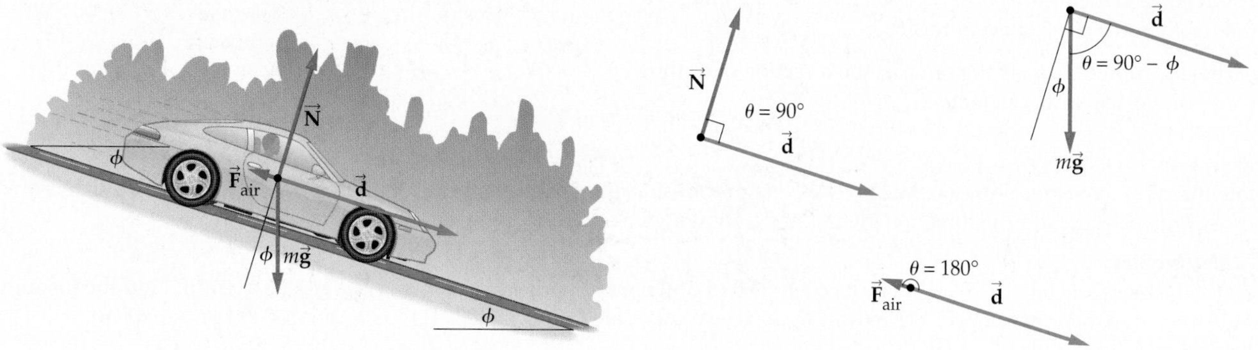

Strategy

For each force we calculate the work using $W = Fd \cos \theta$, where θ is the angle between that particular force and the displacement $\vec{\mathbf{d}}$. The total work is the sum of the work done by each of the three forces.

Solution

1. We start with the work done by the normal force, $\vec{\mathbf{N}}$. From the figure we see that $\theta = 90°$ for this force:

$$W_N = Nd \cos \theta = Nd \cos 90° = Nd(0) = 0$$

2. For the force of air resistance, $\theta = 180°$:

$$W_{\text{air}} = F_{\text{air}}d \cos 180° = F_{\text{air}}d(-1) = -F_{\text{air}}d$$

3. For gravity the angle θ is $\theta = 90° - \phi$, as indicated in the figure. Recall that $\cos(90° - \phi) = \sin \phi$ (see Appendix A):

$$W_{mg} = mgd \cos(90° - \phi) = mgd \sin \phi$$

4. The total work is the sum of the individual works:

$$W_{\text{total}} = W_N + W_{\text{air}} + W_{mg} = 0 - F_{\text{air}}d + mgd \sin \phi$$

Insight

The normal force is perpendicular to the motion of the car, and thus does no work. Air resistance points in a direction that opposes the motion, so it does negative work. On the other hand, gravity has a component in the direction of motion; therefore, its work is positive. The physical significance of positive, negative, and zero work will be discussed in detail in the next section.

Practice Problem

Calculate the total work done on a 1550-kg car as it coasts 20.4 m down a hill with $\phi = 5.00°$. Let the force due to air resistance be 15.0 N. [**Answer:** $W_{\text{total}} = W_N + W_{\text{air}} + W_{mg} = 0 - F_{\text{air}}d + mgd \sin \phi = 0 - 306\text{ J} + 2.70 \times 10^4\text{ J} = 2.67 \times 10^4\text{ J}$]

Some related homework problems: Problem 12, Problem 66

In the previous Example, we showed that the total work can be calculated by finding the work done by each force separately, and then summing the individual works. In the next Example, we take a different approach. We first sum the forces acting on the car to find F_{total}. Once the total force is determined, we calculate the total work using $W_{\text{total}} = F_{\text{total}} d \cos\theta$.

EXAMPLE 7–4 A Coasting Car II

Consider the car described in Example 7–3. Calculate the total work done on the car using $W_{\text{total}} = F_{\text{total}} d \cos\theta$.

Picture the Problem

First, we choose the x axis to point down the slope, and the y axis to be at right angles to the slope. With this choice, there is no acceleration in the y direction, which means that the total force in that direction must be zero. As a result, the total force acting on the car is in the x direction. The magnitude of the total force is $mg \sin\phi - F_{\text{air}}$, as can be seen in our sketch.

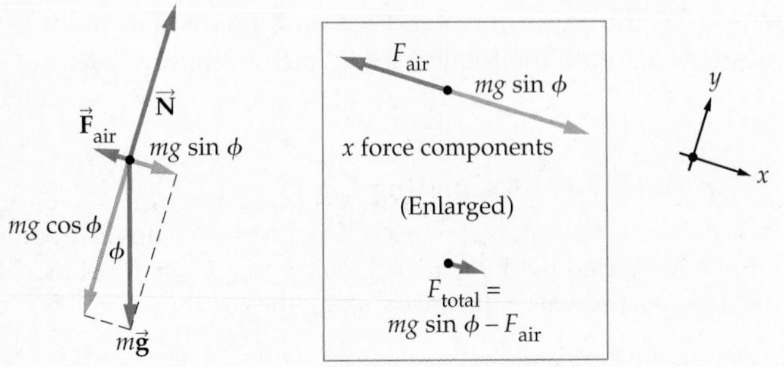

Strategy

We begin by finding the x component of each force vector and then summing them to find the total force acting on the car. As can be seen from the figure, the total force points in the positive x direction; that is, in the same direction as the displacement. Therefore, the angle θ in $W = F_{\text{total}} d \cos\theta$ is zero.

Solution

1. Referring to the figure above, we see that the *magnitude* of the total force is $mg \sin\phi$ minus F_{air}:

$$F_{\text{total}} = mg \sin\phi - F_{\text{air}}$$

2. The direction of $\vec{\mathbf{F}}_{\text{total}}$ is the same as the direction of $\vec{\mathbf{d}}$, thus $\theta = 0°$. We can now calculate W_{total}:

$$W_{\text{total}} = F_{\text{total}} d \cos\theta = (mg \sin\phi - F_{\text{air}}) d \cos 0°$$
$$= mgd \sin\phi - F_{\text{air}} d$$

Insight

Note that we were careful to calculate both the magnitude and the direction of the total force. The magnitude (which is always positive) gives F_{total} and the direction gives $\theta = 0°$, allowing us to use $W_{\text{total}} = F_{\text{total}} d \cos\theta$.

Practice Problem

The total work done on a 1620-kg car as it coasts 25.0 m down a hill with $\phi = 6.00°$ is $W_{\text{total}} = 3.75 \times 10^4$ J. Find the magnitude of the force due to air resistance. [**Answer:** $F_{\text{air}} d = -W_{\text{total}} + mgd \sin\phi = 4030$ J, thus $F_{\text{air}} = (4030 \text{ J})/d = 161$ N]

Some related homework problems: Problem 12, Problem 66

The full significance of positive versus negative work is seen in the next section, where we relate the work done on an object to the change in its speed.

7–2 Kinetic Energy and the Work-Energy Theorem

Suppose you drop an apple. As it falls, gravity does positive work on it, as indicated in **Figure 7–5**, and its speed increases. If you toss the apple upward, gravity does negative work, and the apple slows down. In general, whenever the total work done on an object is positive, its speed increases; when the total work is negative its speed decreases. In this section we derive an important result, the **work-energy theorem,** which makes this connection between work and change in speed precise.

To begin, consider an apple of mass m falling through the air, and suppose that two forces act on the apple—gravity, $m\vec{\mathbf{g}}$, and the average force of air resistance, $\vec{\mathbf{F}}_{\text{air}}$. The total force acting on the apple, $\vec{\mathbf{F}}_{\text{total}}$, gives the apple a constant downward acceleration of magnitude

$$a = F_{\text{total}}/m$$

Since the total force is downward and the motion is downward, the work done on the apple is positive.

◀ **FIGURE 7–5 Gravitational work**
The work done by gravity on an apple that moves downward is positive. If the apple is in free fall, this positive work will result in an increase in speed. On the other hand, the work done by gravity on an apple that moves upward is negative. If the apple is in free fall, the negative work done by gravity will result in a decrease of speed.

Now, suppose the initial speed of the apple is v_i, and that after falling a distance d its speed increases to v_f. The apple falls with constant acceleration a, hence constant-acceleration kinematics (Equation 2–12) gives

$$v_f^2 = v_i^2 + 2ad$$

or, with a slight rearrangement,

$$2ad = v_f^2 - v_i^2$$

Next, substitute $a = F_{total}/m$ into this equation:

$$2\left(\frac{F_{total}}{m}\right)d = v_f^2 - v_i^2$$

Multiplying both sides by m and dividing by 2 yields

$$F_{total}d = \tfrac{1}{2}mv_f^2 - \tfrac{1}{2}mv_i^2$$

where $F_{total}d$ is simply the total work done on the apple. Thus we find

$$W_{total} = \tfrac{1}{2}mv_f^2 - \tfrac{1}{2}mv_i^2$$

showing that total work is directly related to change in speed, as just mentioned. Note that $W_{total} > 0$ means $v_f > v_i$, $W_{total} < 0$ means $v_f < v_i$, and $W_{total} = 0$ implies that $v_f = v_i$.

The quantity $\tfrac{1}{2}mv^2$ in the equation for W_{total} has a special significance in physics, as we shall see. We call it the **kinetic energy,** K:

Definition of Kinetic Energy, *K*

$$K = \tfrac{1}{2}mv^2 \qquad \text{7–6}$$

SI unit: $kg \cdot m^2/s^2$ = joule, J

In general, the kinetic energy of an object is the energy due to its motion. We measure kinetic energy in joules, the same units as work, and both kinetic energy and work are scalars. Unlike work, however, kinetic energy is never negative. Instead, K is always greater than or equal to zero, independent of the direction of motion or the direction of any forces.

To get a feeling for typical values of kinetic energy, consider your kinetic energy when jogging. Assuming a mass of about 62 kg and a speed of 2.5 m/s, your kinetic energy is $K = \tfrac{1}{2}(62\text{ kg})(2.5\text{ m/s})^2 = 190\text{ J}$. Additional examples of kinetic energy are given in Table 7–2.

PROBLEM-SOLVING NOTE

Work Can Be Positive, Negative, or Zero

When you calculate work, be sure to keep track of whether it is positive or negative. The distinction is important, since positive work increases speed, whereas negative work decreases speed. Zero work, of course, has no effect on speed.

TABLE 7–2 Typical Kinetic Energies

Source	Approximate kinetic energy (joules)
Jet aircraft at 500 mph	10^9
Car at 60 mph	10^6
Home-run baseball	10^3
Person at walking speed	50
Housefly in flight	10^{-3}

EXERCISE 7–2

A truck moving at 15 m/s has a kinetic energy of 4.2×10^5 J. **(a)** What is the mass of the truck? **(b)** By what multiplicative factor does the kinetic energy of the truck increase if its speed is doubled?

Solution

(a) $K = \frac{1}{2}mv^2$; therefore $m = 2K/v^2 = 3700$ kg. **(b)** Kinetic energy depends on the speed squared, and hence doubling the speed increases the kinetic energy by a factor of four.

In terms of kinetic energy, the work-energy theorem can be stated as follows:

Work-Energy Theorem

The total work done on an object is equal to the change in its kinetic energy:

$$W_{\text{total}} = \Delta K = \tfrac{1}{2}mv_f^2 - \tfrac{1}{2}mv_i^2 \qquad 7\text{–}7$$

PROBLEM-SOLVING NOTE

Starts from Rest Means $v_i = 0$

A problem statement that uses a phrase like "starts from rest" or "is raised from rest" is telling you that $v_i = 0$.

Thus, the work-energy theorem says that when a force acts on an object over a distance—doing work on it—the result is a change in the speed of the object, and hence a change in its energy of motion. Equation 7–7 is the quantitative expression of this connection.

Finally, though we have derived the work-energy theorem for a force that is constant in direction and magnitude, it is valid for any force, as can be shown using the methods of calculus. In fact, the work-energy theorem is completely general, making it one of the more important and fundamental results in physics. It is also a very handy tool for problem solving, as we shall see many times throughout this text.

EXERCISE 7–3

How much work is required for a 74-kg sprinter to accelerate from rest to 2.2 m/s?

Solution

Since $v_i = 0$, we have $W = \frac{1}{2}mv_f^2 - \frac{1}{2}mv_i^2 = \frac{1}{2}mv_f^2 = \frac{1}{2}(74\text{ kg})(2.2\text{ m/s})^2 = 180$ J.

We now present a variety of Examples showing how the work-energy theorem is used in practical situations.

EXAMPLE 7–5 Hit the Books

A 4.10-kg box of books is lifted vertically from rest a distance of 1.60 m with a constant, upward applied force of 52.7 N. Find **(a)** the work done by the applied force, **(b)** the work done by gravity, and **(c)** the final speed of the box.

Picture the Problem

Our sketch shows that the direction of motion of the box is upward. In addition, we see that the applied force, $\vec{\mathbf{F}}_{\text{app}}$, is upward and the force of gravity, $m\vec{\mathbf{g}}$, is downward. Finally, the box is lifted from rest ($v_i = 0$) through a distance $\Delta y = 1.60$ m.

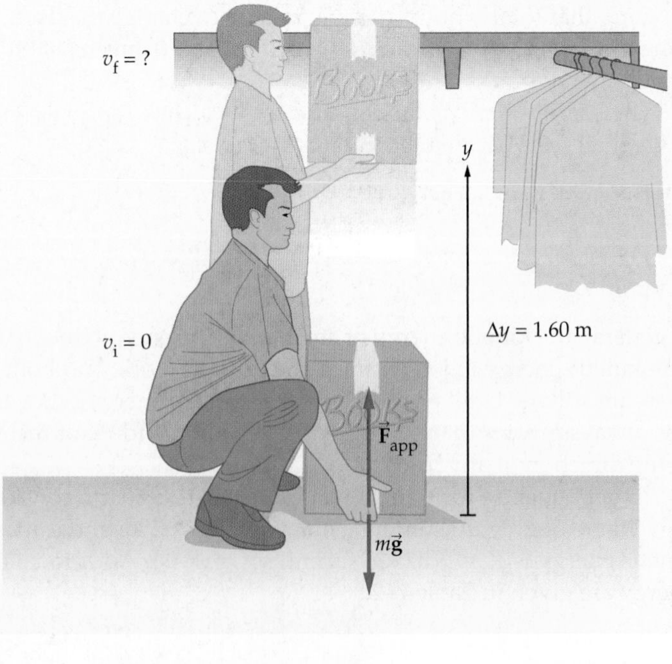

Strategy

The applied force is in the direction of motion, so the work it does, W_{app}, is positive. Gravity is opposite in direction to the motion, thus its work, W_g, is negative. The total work is the sum of W_{app} and W_g, and the final speed of the box is found by applying the work-energy theorem, $W_{\text{total}} = \Delta K$.

Solution

Part (a)

1. First we find the work done by the applied force. In this case, $\theta = 0°$ and the distance is $\Delta y = 1.60$ m:

$$W_{\text{app}} = F_{\text{app}} \cos 0° \, \Delta y = (52.7 \text{ N})(1)(1.60 \text{ m}) = 84.3 \text{ J}$$

Part (b)

2. Next, we calculate the work done by gravity. The distance is $\Delta y = 1.60$ m, as before, but now $\theta = 180°$:

$$W_g = mg \cos 180° \, \Delta y = (4.10 \text{ kg})(9.81 \text{ m/s}^2)(-1)(1.60 \text{ m}) = -64.4 \text{ J}$$

Part (c)

3. The total work done on the box, W_{total}, is the sum of W_{app} and W_g:

$$W_{\text{total}} = W_{\text{app}} + W_g = 84.3 \text{ J} - 64.4 \text{ J} = 19.9 \text{ J}$$

4. To find the final speed, v_f, we apply the work-energy theorem. Recall that the box started at rest, thus $v_i = 0$:

$$W_{\text{total}} = \tfrac{1}{2}mv_f^2 - \tfrac{1}{2}mv_i^2 = \tfrac{1}{2}mv_f^2$$

$$v_f = \sqrt{\frac{2W_{\text{total}}}{m}} = \sqrt{\frac{2(19.9 \text{ J})}{4.10 \text{ kg}}} = 3.12 \text{ m/s}$$

Insight

As a check on our result, we can find v_f in a completely different way. First, calculate the acceleration of the box with the result $a = (F_{\text{app}} - mg)/m = 3.04 \text{ m/s}^2$. Next, use this result in the kinematic equation $v^2 = v_0^2 + 2a\Delta y$. With $v_0 = 0$ and $\Delta y = 1.60$ m we find $v = 3.12$ m/s, in agreement with the results using the work-energy theorem.

Practice Problem

If the box is lifted only a quarter of the distance, is the final speed 1/8, 1/4, or 1/2 of the value found in step 4? Calculate v_f in this case as a check on your answer. [**Answer:** Since work depends linearly on Δy, and v_f depends on the square root of the work, it follows that the final speed is $\sqrt{1/4} = \frac{1}{2}$ the value in step 4. Letting $\Delta y = (1.60 \text{ m})/4 = 0.400$ m we find $v_f = \frac{1}{2}(3.12 \text{ m/s}) = 1.56$ m/s.]

Some related homework problems: Problem 15, Problem 18, Problem 19

In the previous Example the initial speed was zero. This is not always the case, of course. The next Example illustrates how to use the work-energy theorem when the initial velocity is nonzero.

EXAMPLE 7–6 Pulling a Sled

A boy exerts a force of 11.0 N at 29.0° above the horizontal on a 6.40-kg sled. Find the work done by the boy and the final speed of the sled after it moves 2.00 m, assuming the sled starts with an initial speed of 0.500 m/s and slides horizontally without friction.

Picture the Problem

Our sketch shows the direction of motion and the directions of each of the forces. Note that the normal force and the force due to gravity are vertical, whereas the displacement is horizontal. The force exerted by the boy has both a vertical component, $F \sin \theta$, and a horizontal component, $F \cos \theta$.

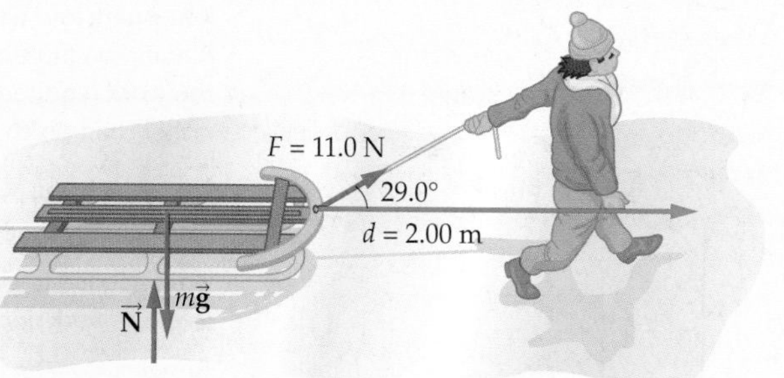

Strategy

The forces $\vec{N}$ and $m\vec{g}$ do no work because they are at right angles to the horizontal displacement. The force exerted by the boy, however, has a horizontal component that does positive work on the sled. Therefore, the total work is simply the work done by the boy. After calculating this work, we find v_f by applying the work-energy theorem with $v_i = 0.500$ m/s.

Solution

1. The work done by the boy is $(F \cos \theta)d$, where $\theta = 29.0°$. This is also the total work done on the sled:

$$W_{\text{boy}} = (F \cos \theta)d = (11.0 \text{ N})(\cos 29.0°)(2.00 \text{ m}) = 19.2 \text{ J} = W_{\text{total}}$$

2. Use the work-energy theorem to solve for the final speed:

$$W_{\text{total}} = \Delta K = \tfrac{1}{2}mv_f^2 - \tfrac{1}{2}mv_i^2$$

$$\tfrac{1}{2}mv_f^2 = W_{\text{total}} + \tfrac{1}{2}mv_i^2$$

$$v_f = \sqrt{\frac{2W_{\text{total}}}{m} + v_i^2}$$

continued on next page

3. Substitute numerical values to get the final answer:

$$v_f = \sqrt{\frac{2(19.2 \text{ J})}{6.40 \text{ kg}} + (0.500 \text{ m/s})^2}$$
$$= 2.50 \text{ m/s}$$

Insight

If the sled had started from rest, instead of with an initial speed of 0.500 m/s, would its final speed be 2.50 m/s − 0.500 m/s = 2.00 m/s?

No. If the initial speed is zero, then $v_f = \sqrt{\frac{2W_{total}}{m}} = \sqrt{\frac{2(19.2 \text{ J})}{6.40 \text{ kg}}} = 2.45$ m/s. Why don't the speeds add and subtract in a straightforward way? The reason is that the work-energy theorem depends on the *square* of the speeds rather than on v_i and v_f directly.

Practice Problem

Suppose the sled starts with a speed of 0.500 m/s and has a final speed of 2.50 m/s after the boy pulls it through a distance of 3.00 m. What force did the boy exert on the sled? [**Answer:** $F = W_{total}/(d \cos\theta) = \Delta K/(d \cos\theta) = 7.32$ N]

Some related homework problems: Problem 22, Problem 48

The final speeds in the previous Examples could have been found using Newton's laws and the constant-acceleration kinematics of Chapter 2, as indicated in the Insight following Example 7–5. The work-energy theorem provides an alternative method of calculation that is often much easier to apply than Newton's laws. We return to this point in Chapter 8.

CONCEPTUAL CHECKPOINT 7–2 Compare the Work

To accelerate a certain car from rest to the speed v requires the work W_1. The work needed to accelerate the car from v to $2v$ is W_2. Which of the following is correct: **(a)** $W_2 = W_1$, **(b)** $W_2 = 2W_1$, **(c)** $W_2 = 3W_1$, **(d)** $W_2 = 4W_1$?

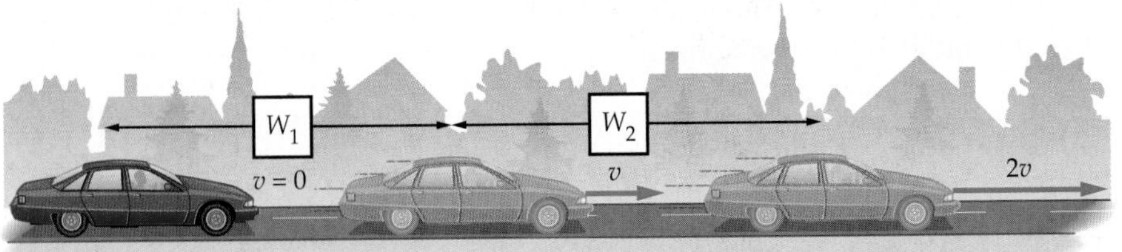

Reasoning and Discussion

A common mistake is to reason that since we increase the speed by the same amount in each case, the work required is the same. It is not, and the reason is that work depends on the speed squared rather than on the speed itself.

To see how this works, first calculate W_1, the work needed to go from rest to a speed v. From the work-energy theorem, with $v_i = 0$, and $v_f = v$, we find $W_1 = \frac{1}{2}mv_f^2 - \frac{1}{2}mv_i^2 = \frac{1}{2}mv^2$. Similarly, the work needed to go from rest, $v_i = 0$, to a speed $v_f = 2v$, is simply $\frac{1}{2}m(2v)^2 = 4\left(\frac{1}{2}mv^2\right) = 4W_1$. Therefore, the work needed to increase the speed from v to $2v$ is the difference: $W_2 = 4W_1 - W_1 = 3W_1$.

Answer:

(c) $W_2 = 3W_1$

PROBLEM-SOLVING NOTE

Be Careful About Linear Reasoning

Though some relations are linear—if you *double* the mass, you *double* the kinetic energy—others are not. For example, if you *double* the speed, you *quadruple* the kinetic energy. Be careful not to jump to conclusions based on linear reasoning.

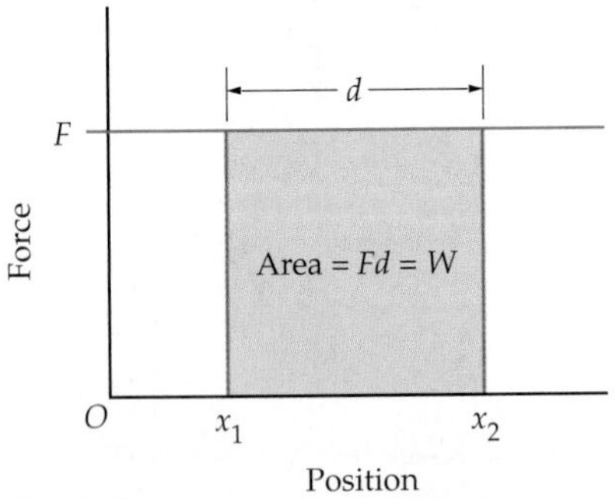

▲ **FIGURE 7–6 Graphical representation of the work done by a constant force**

A constant force F acting through a distance d does a work $W = Fd$. Note that Fd is also equal to the shaded area between the force line and the x axis.

7–3 Work Done by a Variable Force

Thus far we have calculated work only for constant forces, yet most forces in nature vary with position. For example, the force exerted by a spring depends on how far the spring is stretched, and the force of gravity between planets depends on their separation. In this section we show how to calculate the work for a force that varies with position.

First, let's review briefly the case of a constant force, and develop a graphical interpretation of work. **Figure 7–6** shows a constant force plotted versus position, x.

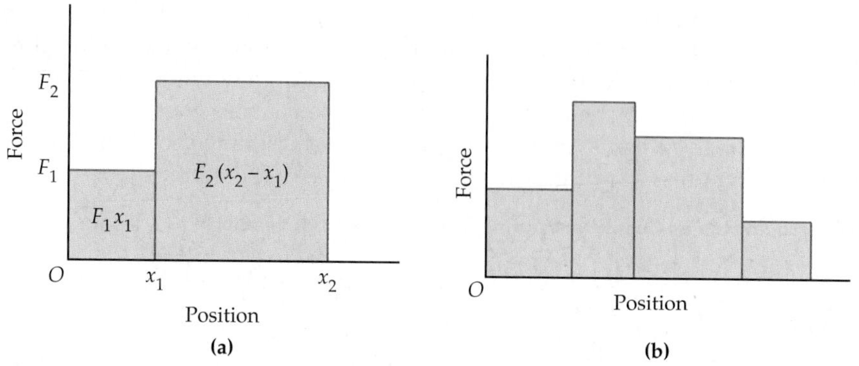

◀ **FIGURE 7–7 Work done by a nonconstant force**

(a) A force with a value F_1 from 0 to x_1 and a value F_2 from x_1 to x_2 does the work $W = F_1x_1 + F_2(x_2 - x_1)$. This is simply the area of the two shaded rectangles. **(b)** If a force takes on a number of different values, the work it does is still the total area between the force lines and the x axis, just as in part (a).

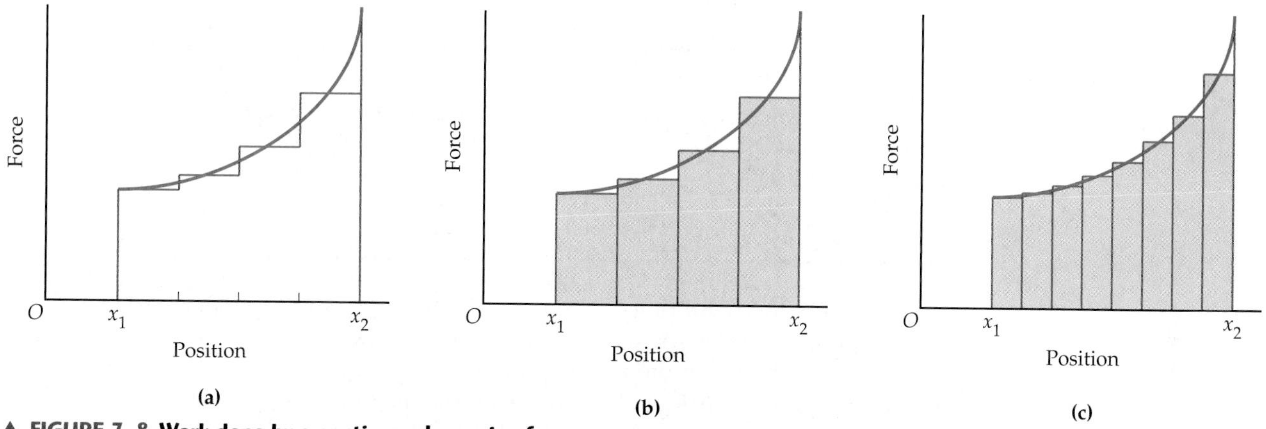

▲ **FIGURE 7–8 Work done by a continuously varying force**

(a) A continuously varying force can be approximated by a series of constant values that follow the shape of the curve. **(b)** The work done by the continuous force is approximately equal to the area of the small rectangles corresponding to the constant values of force shown in part (a). **(c)** In the limit of an infinite number of vanishingly small rectangles, we see that the work done by the force is equal to the area between the force curve and the x axis.

If the force acts in the positive x direction and moves an object a distance d, from x_1 to x_2, the work it does is $W = Fd = F(x_2 - x_1)$. Referring to the figure, we see that the work is equal to the shaded area[1] between the force line and the x axis.

Next, consider a force that has the value F_1 from $x = 0$ to $x = x_1$ and a different value F_2 from $x = x_1$ to $x = x_2$, as in **Figure 7–7 (a)**. The work in this case is the sum of the works done by F_1 and F_2. Therefore, $W = F_1x_1 + F_2(x_2 - x_1)$ which, again, is the area between the force lines and the x axis. Clearly, this type of calculation can be extended to a force with any number of different values, as indicated in **Figure 7–7 (b)**.

If a force varies continuously with position, we can approximate it with a series of constant values that follow the shape of the curve, as shown in **Figure 7–8 (a)**. It follows that the work done by the continuous force is approximately equal to the area of the corresponding rectangles, as **Figure 7–8 (b)** shows. The approximation can be made better by using more rectangles, as illustrated in **Figure 7–8 (c)**. In the limit of an infinite number of vanishingly small rectangles, the area of the rectangles becomes identical to the area under the force curve. Hence this area is the work done by the continuous force. To summarize:

> The work done by a force in moving an object from x_1 and x_2 is equal to the corresponding area between the force curve and the x axis.

A case of particular interest is that of a spring. Since the force exerted by a spring is given by $F_x = -kx$ (Section 6–2), it follows that the force we must exert to hold it at the position x is $+kx$. This is illustrated in **Figure 7–9**, where we also

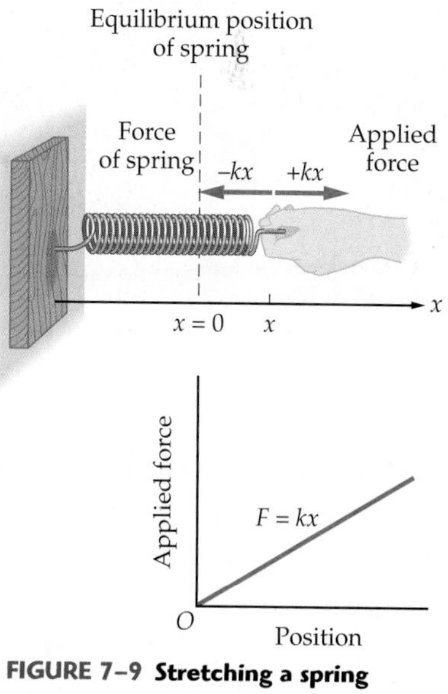

▲ **FIGURE 7–9 Stretching a spring**

The force we must exert on a spring to stretch it a distance x is $+kx$. Thus, applied force versus position for a spring is a straight line of slope k.

[1]Usually, area has the dimensions of (length) × (length), or length^2. In this case, however, the vertical axis is force and the horizontal axis is distance. As a result, the dimensions of area are (force) × (distance), which in SI units is $\text{N} \cdot \text{m} = \text{J}$.

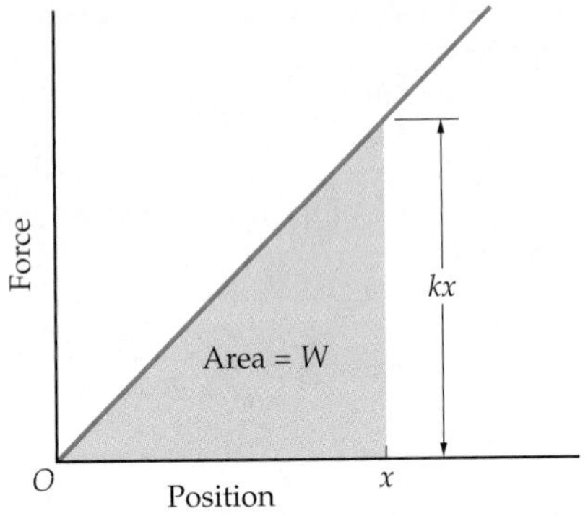

▲ **FIGURE 7–10 Work needed to stretch a spring a distance x**

The work done is equal to the shaded area, which is a right triangle. The area of the triangle is $\frac{1}{2}(x)(kx) = \frac{1}{2}kx^2$.

show that the corresponding force curve is a straight line extending from the origin. Therefore, the work we do in stretching a spring from $x = 0$ (equilibrium) to the general position x is the shaded, triangular area shown in **Figure 7–10**. This area is equal to $\frac{1}{2}$(base)(height), where in this case the base is x and the height is kx. As a result, the work is $\frac{1}{2}(x)(kx) = \frac{1}{2}kx^2$. Similar reasoning shows that the work needed to compress a spring a distance x is also $\frac{1}{2}kx^2$. Therefore,

Work to Stretch or Compress a Spring a Distance x from Equilibrium

$$W = \frac{1}{2}kx^2 \qquad \text{7–8}$$

SI unit: joule, J

We can get a feeling for the amount of work required to compress a typical spring in the following Exercise.

EXERCISE 7–4

The spring in a pinball launcher has a force constant of 405 N/m. How much work is required to compress the spring a distance of 3.00 cm?

Solution

$$W = \frac{1}{2}kx^2 = \frac{1}{2}(405\ \text{N/m})(0.0300\ \text{m})^2 = 0.182\ \text{J}$$

Note that the work done in compressing or expanding a spring varies with the second power of x, the displacement from equilibrium. The consequences of this dependence are explored throughout the rest of this section.

Before we consider a specific example, however, recall that the results for a spring apply to more than just the classic case of a helical coil of wire. In fact, any flexible structure satisfies the relations $F_x = -kx$ and $W = \frac{1}{2}kx^2$, given the appropriate value of the force constant, k, and small enough displacements, x. Several examples were mentioned in Section 6-2.

Here we consider an example from the field of nanotechnology; namely, the cantilevers used in **atomic-force microscopy** (AFM). As we show in Example 7–7, a typical atomic-force cantilever is basically a thin silicon bar about 250 μm in length, supported at one end like a diving board, with a sharp, hanging point at the other end. When the point is pulled across the surface of a material—like an old-fashioned phonograph needle in the groove of a record—individual atoms on the surface cause the point to move up and down, deflecting the cantilever. These deflections, which can be measured by reflecting a laser beam from the top of the cantilever, are then converted into an atomic-level picture of the surface, as shown in the accompanying photograph.

A typical force constant for an AFM cantilever is on the order of 1 N/m, much smaller than the 100–500 N/m force constant of a common lab spring. The implications of this are discussed in the following Example.

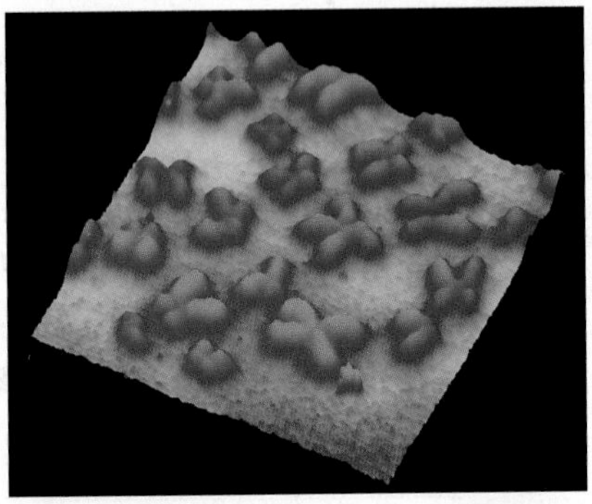

▲ Human chromosomes, as imaged by an atomic force microscope.

EXAMPLE 7–7 Flexing an AFM Cantilever

The work required to deflect a typical AFM cantilever by 0.10 nm is 1.2×10^{-20} J. **(a)** What is the force constant of the cantilever, treating it as an ideal spring? **(b)** How much work is required to increase the deflection of the cantilever from 0.10 nm to 0.20 nm?

Picture the Problem

The sketch on the left (on the next page) shows the cantilever and its sharp point being dragged across the surface of a material. In the sketch to the right, we show an exaggerated view of the cantilever's deflection, and indicate that it is equivalent to the stretch of an "effective" ideal spring with a force constant k.

Strategy

a. Given that $W = 1.2 \times 10^{-20}$ J for a deflection of $x = 0.10$ nm, we can find the effective force constant k using $W = \frac{1}{2}kx^2$.

b. To find the work required to deflect from $x = 0.10$ nm to $x = 0.20$ nm, $W_{1\to2}$, we calculate the work to deflect from $x = 0$ to $x = 0.20$ nm, $W_{0\to2}$, and then subtract the work needed to deflect from $x = 0$ to $x = 0.10$ nm, $W_{0\to1}$. (Note that we can *not* simply assume the work to go from $x = 0.10$ nm to $x = 0.20$ nm is the same as the work to go from $x = 0$ to $x = 0.10$ nm).

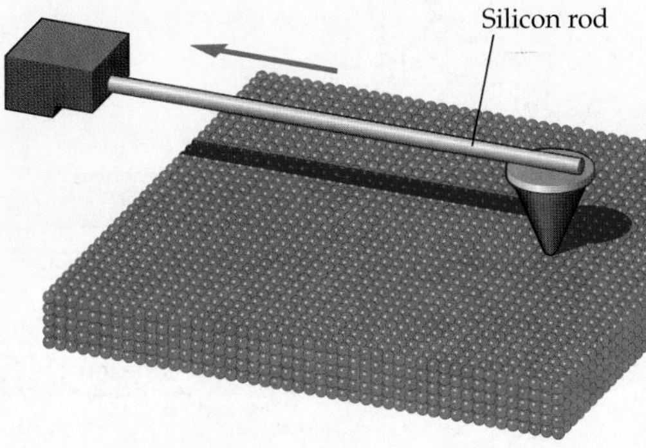

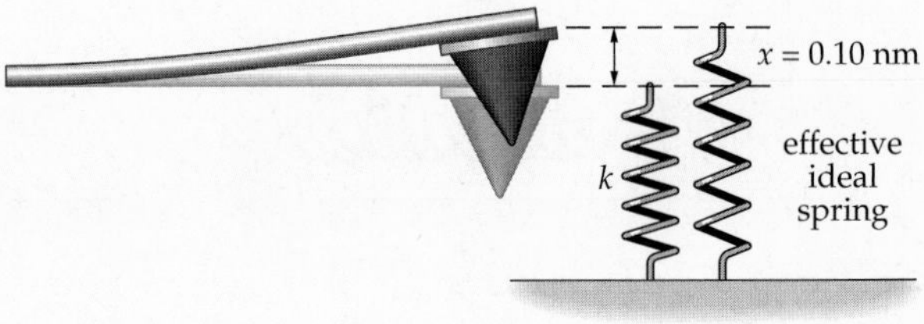

Solution

Part (a)

1. Solve $W = \frac{1}{2}kx^2$ for the force constant k:

$$k = \frac{2W}{x^2} = \frac{2(1.2 \times 10^{-20}\text{ J})}{(0.10 \times 10^{-9}\text{ m})^2} = 2.4\text{ N/m}$$

Part (b)

1. First, calculate the work needed to deflect the cantilever from $x = 0$ to $x = 0.20$ nm:

$$W_{0\to 2} = \tfrac{1}{2}kx^2$$
$$= \tfrac{1}{2}(2.4\text{ N/m})(0.2 \times 10^{-9}\text{ m})^2 = 4.8 \times 10^{-20}\text{ J}$$

2. Subtract from the above result the work to deflect from $x = 0$ to $x = 0.10$ nm, which the problem statement gives as 1.2×10^{-20} J:

$$W_{1\to 2} = W_{0\to 2} - W_{0\to 1}$$
$$= 4.8 \times 10^{-20}\text{ J} - 1.2 \times 10^{-20}\text{ J} = 3.6 \times 10^{-20}\text{ J}$$

Insight

Our results show that more energy is needed to deflect the cantilever the second 0.10 nm than to deflect it the first 0.10 nm. Why? The reason is that the force of the cantilever increases with distance; thus, the average force over the second 0.10 nm is greater than the average force over the first 0.10 nm. In fact, we can see from the adjacent figure that the average force between 0.10 nm and 0.20 nm (3.6 nN) is three times the average force between 0 and 0.10 nm (1.2 nN). It follows, then, that the work required for the second 0.10 nm is three times the work required for the first 0.10 nm.

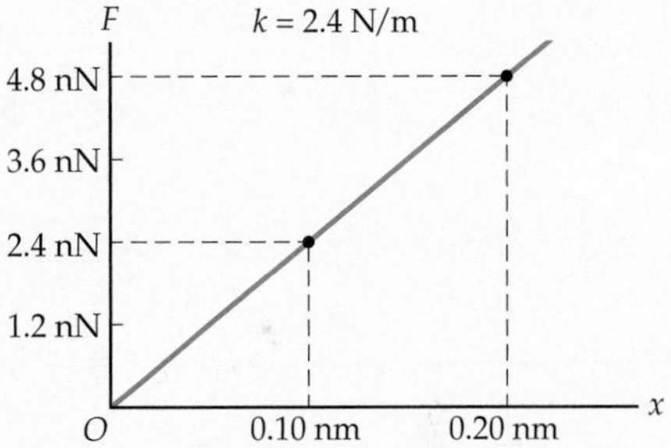

Practice Problem

A second cantilever has half the force constant of the cantilever in this Example. Is the work required to deflect the second cantilever by 0.20 nm greater than, less than, or equal to the work required to deflect the cantilever in this Example by 0.10 nm? [**Answer:** Halving the force constant halves the work, but doubling the deflection quadruples the work. The net effect is that the work increases by a factor of two, to 2.4×10^{-20} J.]

Some related homework problems: Problem 25, Problem 30

An equivalent way to calculate the work for a variable force is to multiply the average force, F_{av}, by the distance, d:

$$W = F_{av}d \qquad 7\text{–}9$$

For a spring that is stretched a distance x from equilibrium the force varies linearly from 0 to kx. Thus, the average force is $F_{av} = \frac{1}{2}kx$, as indicated in **Figure 7–11**. Therefore, the work is

$$W = \left(\tfrac{1}{2}kx\right)(x) = \tfrac{1}{2}kx^2$$

As expected, our result agrees with Equation 7–8.

Finally, when you stretch or compress a spring from its equilibrium position, the work you do is always positive. The work done *by* a spring, however, may be positive or negative, depending on the situation. For example, consider a block sliding to the right with an initial speed v_0 on a smooth, horizontal surface, as

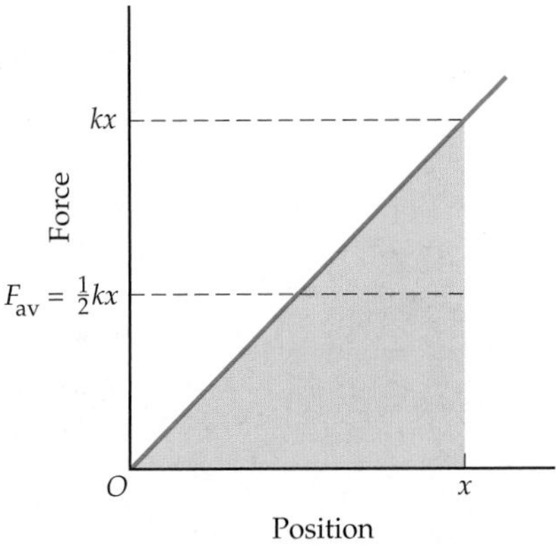

▲ **FIGURE 7–11 Work done in stretching a spring: average force**
The average force of a spring from $x = 0$ to x is $F_{av} = \frac{1}{2}kx$, and the work done is $W = F_{av}d = \left(\frac{1}{2}kx\right)(x) = \frac{1}{2}kx^2$.

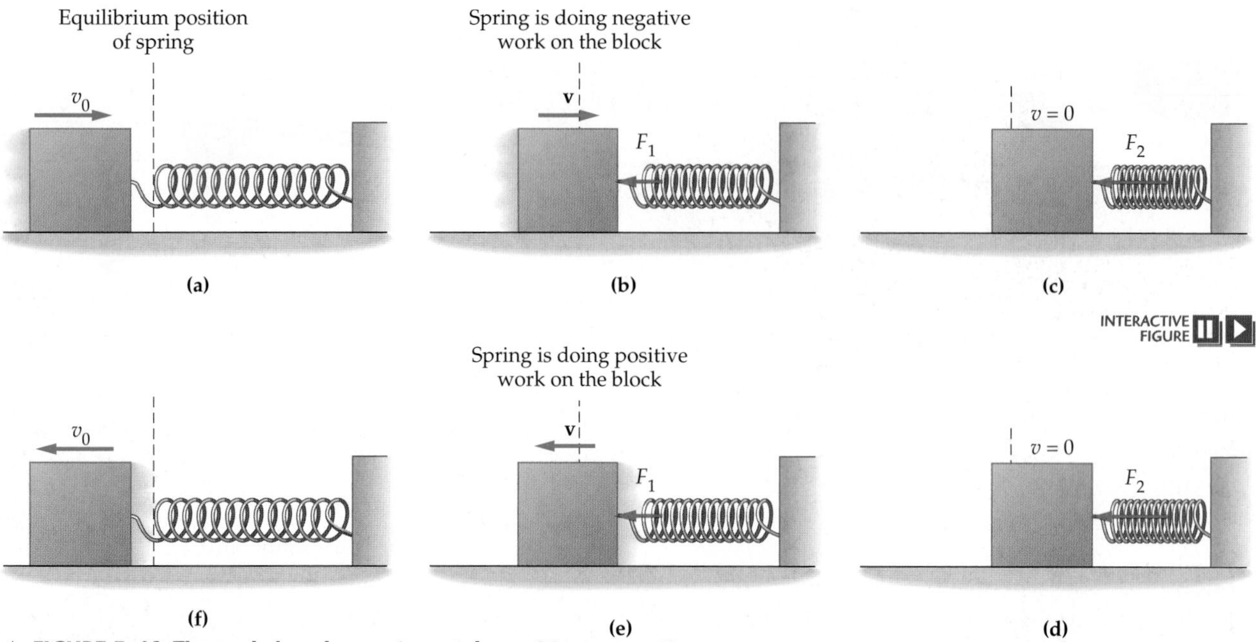

▲ **FIGURE 7–12 The work done by a spring can be positive or negative**
(a) A block slides to the right on a frictionless surface with a speed v_0 until it encounters a spring. **(b)** The spring now exerts a force to the left—opposite to the block's motion—and hence it does negative work on the block. This causes the block's speed to decrease. **(c)** The negative work done by the spring eventually is equal in magnitude to the block's initial kinetic energy, at which point the block comes to rest momentarily. As the spring expands, **(d)** and **(e)**, it does positive work on the block and increases its speed. **(f)** When the block leaves the spring its speed is again equal to v_0.

shown in **Figure 7–12 (a)**. When the block begins to compress the spring, as in **Figure 7–12 (b)**, the spring exerts a force on the block to the left—that is, opposite to the block's direction of motion. As a result, the spring does *negative* work on the block, which causes the block's speed to decrease. Eventually the negative work done by the spring, $W = -\frac{1}{2}kx^2$, is equal in magnitude to the initial kinetic energy of the block. At this point, **Figure 7–12 (c)**, the block comes to rest momentarily, and $W = \Delta K = K_f - K_i = 0 - K_i = -K_i = -\frac{1}{2}mv_0^2 = -\frac{1}{2}kx^2$. We apply this result in Active Example 7–1.

ACTIVE EXAMPLE 7–1 A Block Compresses a Spring

Suppose the block in Figure 7–12 (a) has a mass of 1.5 kg and moves with an initial speed of $v_0 = 2.2$ m/s. Find the compression of the spring, whose force constant is 475 N/m, when the block momentarily comes to rest.

Solution *(Test your understanding by performing the calculations indicated in each step.)*

1. Calculate the initial and final kinetic energies of the block: $K_i = 3.6$ J, $K_f = 0$
2. Calculate the change in kinetic energy of the block: $\Delta K = -3.6$ J
3. Set the negative work done by the spring equal to the change in kinetic energy of the block: $-\frac{1}{2}kx^2 = \Delta K = -3.6$ J
4. Solve for the compression, x, and substitute numerical values: $x = 0.12$ m

Insight

After the block comes to rest, the spring expands back to its equilibrium position, as shown in **Figures 7–12 (d)–(f)**. During this expansion the force exerted by the spring is in the same direction as the block's motion, and hence it does *positive* work in the amount $W = \frac{1}{2}kx^2$. As a result, the block leaves the spring with the same speed it had initially.

Your Turn

Find the compression of the spring for the case where the mass of the block is doubled to 3.0 kg.

(Answers to **Your Turn** *problems are given in the back of the book.)*

7–4 Power

Power is a measure of how *quickly* work is done. To be precise, suppose the work W is performed in the time t. The average power delivered during this time is defined as follows:

Definition of Average Power, *P*

$$P = \frac{W}{t} \qquad \text{7–10}$$

SI unit: J/s = watt, W

For simplicity of notation we drop the usual subscript av for an average quantity and simply understand that the power P refers to an average power unless stated otherwise.

Note that the dimensions of power are joules (work) per second (time). We define one joule per second to be a watt (W), after James Watt (1736–1819), the Scottish engineer and inventor who played a key role in the development of practical steam engines:

$$1 \text{ watt} = 1 \text{ W} = 1 \text{ J/s} \qquad \text{7–11}$$

Of course, the watt is the unit of power used to rate the output of lightbulbs. Another common unit of power is the horsepower (hp), which is used to rate the output of car engines. It is defined as follows:

$$1 \text{ horsepower} = 1 \text{ hp} = 746 \text{ W} \qquad \text{7–12}$$

Though it sounds like a horse should be able to produce one horsepower, in fact, a horse can generate only about 2/3 hp for sustained periods. The reason for the discrepancy is that when James Watt defined the horsepower—as a way to characterize the output of his steam engines—he purposely chose a unit that was overly generous to the horse, so that potential investors couldn't complain he was overstating the capability of his engines.

To get a feel for the magnitude of the watt and the horsepower, consider the power you might generate when walking up a flight of stairs. Suppose, for example, that an 80.0-kg person walks up a flight of stairs in 20.0 s, and that the altitude gain is 12.0 ft (3.66 m). Referring to Example 7–2 and Conceptual Checkpoint 7–1, we find that the work done by the person is $W = mgh = (80.0 \text{ kg})(9.81 \text{ m/s}^2)(3.66 \text{ m}) = 2870 \text{ J}$. To find the power, we simply divide by the time: $P = W/t = (2870 \text{ J})/(20.0 \text{ s}) = 144 \text{ W} = 0.193 \text{ hp}$. Thus, a leisurely stroll up the stairs requires about 1/5 hp or 150 W. Similarly, the power produced by a sprinter bolting out of the starting blocks is about 1 hp, and the greatest power most people can produce for sustained periods of time is roughly 1/3 to 1/2 hp. Further examples of power are given in Table 7–3.

TABLE 7–3
Typical Values of Power

Source	Approximate power (W)
Hoover Dam	1.34×10^9
Car moving at 40 mph	7×10^4
Home stove	1.2×10^4
Sunlight falling on one square meter	1380
Refrigerator	615
Television	200
Person walking up stairs	150
Human brain	20

REAL-WORLD PHYSICS: BIO
Human Power Output and Flight

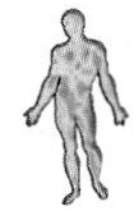

Human-powered flight is a feat just barely within our capabilities, since the most efficient human-powered airplanes require a steady power output of about 1/3 hp. On August 23, 1977, the *Gossamer Condor*, designed by Paul MacCready and flown by Bryan Allen, became the first human-powered airplane to complete a prescribed one-mile, figure-eight course and claim the Kremer Prize of £ 50,000. Allen, an accomplished bicycle racer, used bicycle-like pedals to spin the propeller. Controlling the slow-moving craft while pedaling at full power was no easy task. Allen also piloted the *Gossamer Albatross*, which, in 1979, became the first (and so far the only) human-powered aircraft to fly across the English Channel. This 22.25-mile flight—from Folkestone, England, to Cap Gris-Nez, France—took 2 hours 49 minutes and required a total energy output roughly equivalent to climbing to the top of the Empire State Building 10 times.

▲ The *Gossamer Albatross* on its record-breaking flight across the English Channel in 1979. On two occasions the aircraft actually touched the surface of the water, but the pilot was able to maintain control and complete the 22.25-mile flight.

Power output is also an important factor in the performance of a car. For example, suppose it takes a certain amount of work, W, to accelerate a car from 0 to 60 mi/h. If the average power provided by the engine is P, then according to Equation 7–10 the amount of time required to reach 60 mi/h is $t = W/P$. Clearly, the greater the power P, the less the time required to accelerate. Thus, in a loose way of speaking, we can say that the power of a car is a measure of "how fast it can go fast."

EXAMPLE 7–8 Passing Fancy

To pass a slow-moving truck, you want your fancy 1.30×10^3-kg car to accelerate from 13.4 m/s (30.0 mph) to 17.9 m/s (40.0 mph) in 3.00 s. What is the minimum power required for this pass?

Picture the Problem

Our sketch shows the car accelerating from an initial speed of $v_i = 13.4$ m/s to a final speed of $v_f = 17.9$ m/s. We assume the road is level, so that no work is done against gravity, and that friction and air resistance may be ignored.

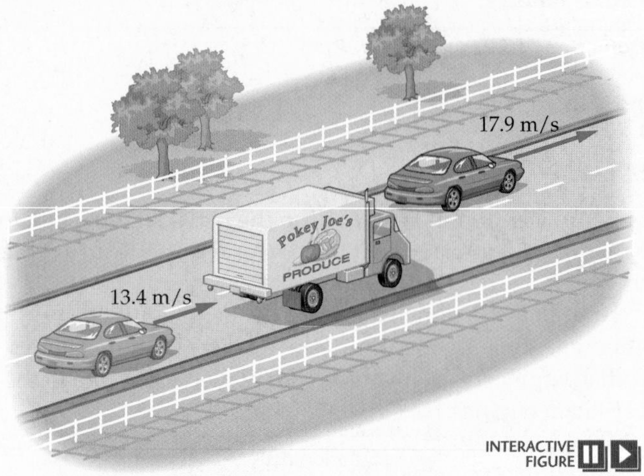

Strategy

Power is work divided by time, and work is equal to the change in kinetic energy as the car accelerates. We can determine the change in kinetic energy from the given mass of the car and its initial and final speeds. With this information at hand, we can determine the power with the relation $P = W/t = \Delta K/t$.

Solution

1. First, calculate the change in kinetic energy:

$$\Delta K = \tfrac{1}{2}mv_f^2 - \tfrac{1}{2}mv_i^2 = \tfrac{1}{2}(1.30 \times 10^3 \text{ kg})(17.9 \text{ m/s})^2 - \tfrac{1}{2}(1.30 \times 10^3 \text{ kg})(13.4 \text{ m/s})^2 = 9.16 \times 10^4 \text{ J}$$

2. Divide by time to find the minimum power. (The actual power would have to be greater to overcome frictional losses.):

$$P = \frac{W}{t} = \frac{\Delta K}{t} = \frac{9.16 \times 10^4 \text{ J}}{3.00 \text{ s}} = 3.05 \times 10^4 \text{ W} = 40.9 \text{ hp}$$

Insight

Suppose that your fancy car continues to produce the same 3.05×10^4 W of power as it accelerates from $v = 17.9$ m/s (40.0 mph) to $v = 22.4$ ms (50.0 mph). Is the time required more than, less than, or equal to 3.00 s? *Answer*: It will take more than 3.00 s. The reason is that ΔK is greater for a change in speed from 40.0 mph to 50.0 mph than for a change in speed from 30.0 mph to 40.0 mph, since K depends on speed squared. Since ΔK is greater, the time $t = \Delta K/P$ is also greater.

Practice Problem

Find the time required to accelerate from 40.0 mph to 50.0 mph with 3.05×10^4 W of power. [**Answer:** First, $\Delta K = 1.18 \times 10^5$ J. Second, $P = \Delta K/t$ can be solved for time to give $t = \Delta K/P$. Thus, $t = 3.87$ s.]

Some related homework problems: Problem 35, Problem 46

Finally, consider a system in which a car, or some other object, is moving with a constant speed v. For example, a car might be traveling uphill on a road inclined at an angle θ above the horizontal. To maintain a constant speed, the engine must exert a constant force F equal to the combined effects of friction, gravity, and air resistance, as indicated in **Figure 7–13**. Now, as the car travels a distance d, the work done by the engine is $W = Fd$, and the power it delivers is

$$P = \frac{W}{t} = \frac{Fd}{t}$$

Since the car has a constant speed, $v = d/t$, it follows that

$$P = \frac{Fd}{t} = F\left(\frac{d}{t}\right) = Fv \qquad \text{7–13}$$

Note that power is directly proportional to both the force and the speed. For example, suppose you push a heavy shopping cart with a force F. You produce twice as much power when you push at 2 m/s than when you push at 1 m/s, even though you are pushing no harder. It's just that the amount of work you do in a given time period is doubled.

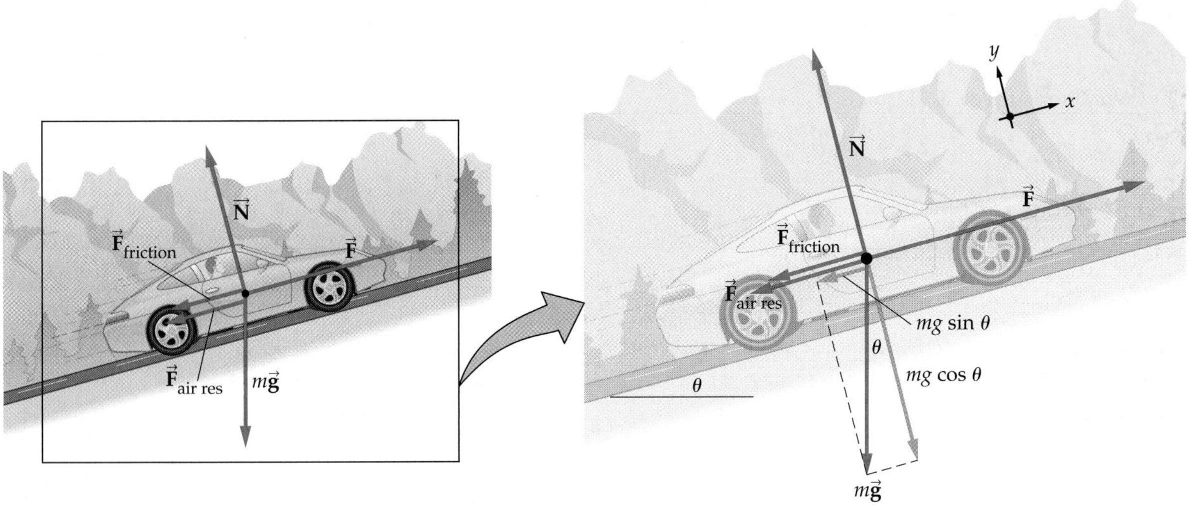

▲ **FIGURE 7–13 Driving up a hill**
A car traveling uphill at constant speed requires a constant force, F, of magnitude $mg \sin\theta + F_{\text{air res}} + F_{\text{friction}}$, applied in the direction of motion.

ACTIVE EXAMPLE 7–2 Find the Maximum Speed

It takes a force of 1280 N to keep a 1500-kg car moving with constant speed up a slope of 5.00°. If the engine delivers 50.0 hp to the drive wheels, what is the maximum speed of the car?

Solution *(Test your understanding by performing the calculations indicated in each step.)*

1. Convert the power of 50.0 hp to watts: $P = 3.73 \times 10^4\ \text{W}$
2. Solve Equation 7–13 for the speed v: $v = P/F$
3. Substitute numerical values for the power and force: $v = 29.1\ \text{m/s}$

Insight
Thus, the maximum speed of the car on this slope is approximately 65 mi/h.

Your Turn
How much power is required for a maximum speed of 32.0 m/s?

(Answers to **Your Turn** *problems are given in the back of the book.)*

Chapter Summary

7–1 Work Done by a Constant Force

A force exerted through a distance performs mechanical work.

Force in Direction of Motion
In this, the simplest case, work is force times distance:

$$W = Fd \qquad 7\text{–}1$$

Force at an Angle θ to Motion
Work is the component of force in the direction of motion, $F\cos\theta$, times distance, d:

$$W = (F\cos\theta)d = Fd\cos\theta \qquad 7\text{–}3$$

Negative and Total Work
Work is negative if the force opposes the motion; that is, if $\theta > 90°$. If more than one force does work, the total work is the sum of the works done by each force separately:

$$W_{\text{total}} = W_1 + W_2 + W_3 + \cdots \qquad 7\text{–}4$$

Equivalently, sum the forces first to find F_{total}, then

$$W_{\text{total}} = (F_{\text{total}} \cos \theta)d = F_{\text{total}}\, d \cos \theta \qquad \textbf{7–5}$$

Units
The SI unit of work and energy is the joule, J:

$$1\ \text{J} = \text{N} \cdot \text{m} \qquad \textbf{7–2}$$

7–2 Work-Energy Theorem and Kinetic Energy

Total work is equal to the change in kinetic energy:

$$W_{\text{total}} = \Delta K = \tfrac{1}{2}mv_f^2 - \tfrac{1}{2}mv_i^2 \qquad \textbf{7–7}$$

Note: To apply this theorem correctly, you must use the *total* work. Kinetic energy is one-half mass times speed squared:

$$K = \tfrac{1}{2}mv^2 \qquad \textbf{7–6}$$

It follows that kinetic energy is always positive or zero.

7–3 Work Done by a Variable Force

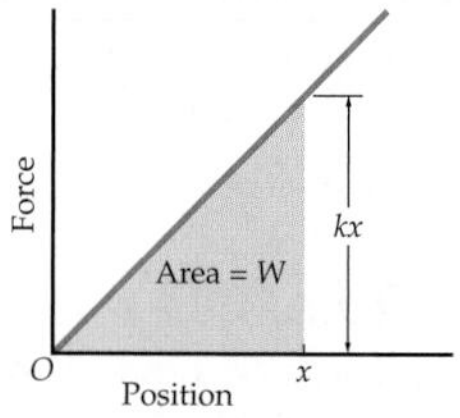

Work is equal to the area between the force curve and the displacement on the x axis. For the case of a spring force, the work to stretch or compress a distance x from equilibrium is

$$W = \tfrac{1}{2}kx^2 \qquad \textbf{7–8}$$

7–4 Power

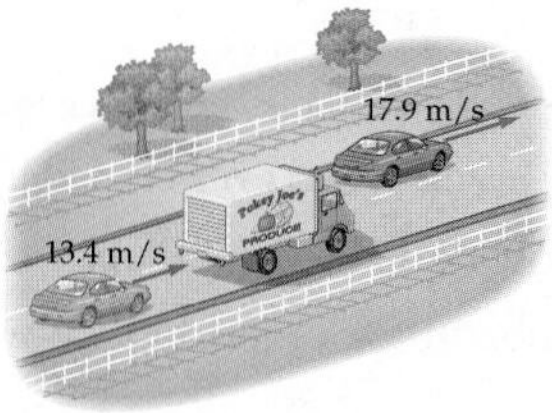

Average power is work divided by the time required to do the work:

$$P = \frac{W}{t} \qquad \textbf{7–10}$$

Equivalently, power is force times speed:

$$P = Fv \qquad \textbf{7–13}$$

Units
The SI unit of power is the watt, W:

$$1\ \text{W} = 1\ \text{J/s} \qquad \textbf{7–11}$$

$$746\ \text{W} = 1\ \text{hp} \qquad \textbf{7–12}$$

Problem-Solving Summary

Type of Calculation	Relevant Physical Concepts	Related Examples
Find the work done by a constant force.	Work is defined as force times displacement, $W = Fd$, when F is in the direction of motion. Use $W = (F \cos \theta)d$ when there is an angle θ between the force and the direction of motion.	Examples 7–1 through 7–6
Calculate the change in speed.	The change in kinetic energy is given by the work-energy theorem, $W_{\text{total}} = \Delta K$. From this, the change in speed can be found by recalling that $K = \frac{1}{2}mv^2$. Be sure W_{total} is the total work and that it has the correct sign.	Examples 7–5, 7–6
Calculate the power.	Find the work done, then divide by time: $P = W/t$. Alternatively, find the force, then multiply by the speed: $P = Fv$.	Example 7–8 Active Example 7–2

Conceptual Questions

(Answers to odd-numbered Conceptual Questions can be found in the back of the book.)

1. Is it possible to do work on an object that remains at rest?
2. A friend makes the statement, "Only the total force acting on an object can do work." Is this statement true or false? If it is true, state why; if it is false, give a counterexample.
3. A friend makes the statement, "A force that is always perpendicular to the velocity of a particle does no work on the particle." Is this statement true or false? If it is true, state why; if it is false, give a counterexample.
4. The net work done on a certain object is zero. What can you say about its speed?
5. To get out of bed in the morning, do you have to do work? Explain.
6. Give an example of a frictional force doing negative work.
7. Give an example of a frictional force doing positive work.
8. A ski boat moves with constant velocity. Is the net force acting on the boat doing work? Explain.
9. A package rests on the floor of an elevator that is rising with constant speed. The elevator exerts an upward normal force on the package, and hence does positive work on it. Why doesn't the kinetic energy of the package increase?
10. An object moves with constant velocity. Is it safe to conclude that no force acts on the object? Why, or why not?
11. Engine 1 does twice the work of engine 2. Is it correct to conclude that engine 1 produces twice as much power as engine 2? Explain.
12. Engine 1 produces twice the power of engine 2. Is it correct to conclude that engine 1 does twice as much work as engine 2? Explain.

Conceptual Exercises

(Answers to odd-numbered Conceptual Exercises can be found in the back of the book.)

1. The International Space Station orbits the Earth in an approximately circular orbit at a height of $h = 375$ km above the Earth's surface. In one complete orbit, is the work done by the Earth on the space station positive, negative, or zero?
2. A pendulum bob swings from point A to point B along the circular arc indicated in **Figure 7–14**. Indicate whether the work done on the bob by **(a)** gravity and **(b)** the string is positive, negative, or zero.

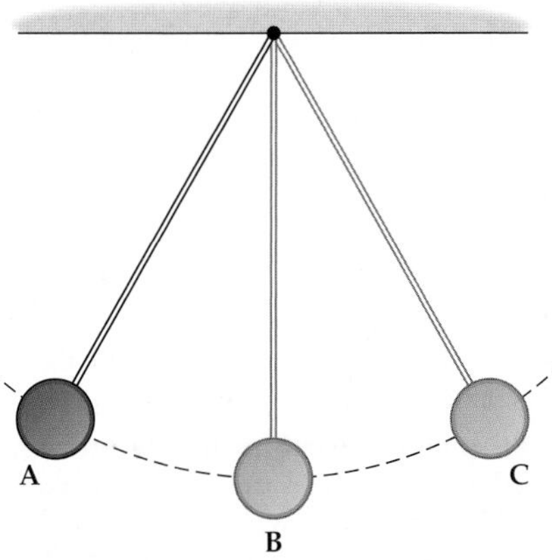

▲ **FIGURE 7–14** Conceptual Exercises 2 and 3

3. A pendulum bob swings from point B to point C along the circular arc indicated in Figure 7–14. Indicate whether the work done on the bob by **(a)** gravity and **(b)** the string is positive, negative, or zero.
4. A pitcher throws a ball at 90 mph and the catcher stops it in her glove. **(a)** Is the work done on the ball by the pitcher positive, negative, or zero? **(b)** Is the work done on the ball by the catcher positive, negative, or zero?
5. The work W_0 accelerates a car from 0 to 50 km/h. How much work is required to accelerate the car from 50 km/h to 150 km/h?
6. The work W_0 is required to accelerate a car from rest to the speed v_0. How much work is required to accelerate the car **(a)** from rest to the speed $v_0/2$ and **(b)** from $v_0/2$ to v_0?
7. As the three small sailboats shown in **Figure 7–15** drift next to a dock, because of wind and water currents, students pull on a line attached to the bow and exert forces of equal magnitude F.

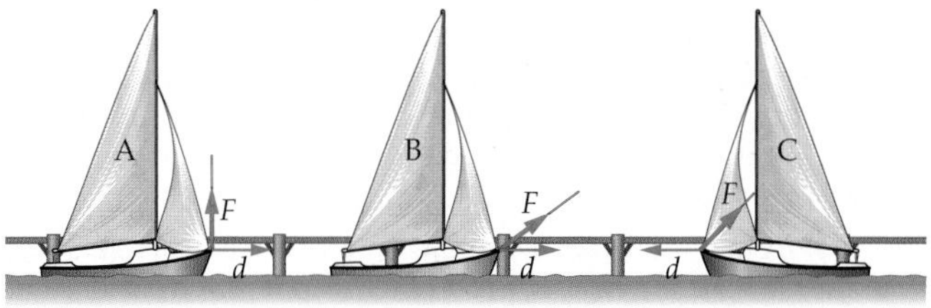

▲ **FIGURE 7–15** Conceptual Exercise 7

Each boat drifts through the same distance d. Rank the three boats in order of increasing work done on the boat by the force F. Indicate ties where appropriate.

8. Jogger A has a mass m and a speed v, jogger B has a mass $m/2$ and a speed $3v$, jogger C has a mass $3m$ and a speed $v/2$, and jogger D has a mass $4m$ and a speed $v/2$. Rank the joggers in order of increasing kinetic energy. Indicate ties where appropriate.
9. A youngster rides on a skateboard with a speed of 2 m/s. After a force acts on the youngster, her speed is 3 m/s. Was the work done by the force positive, negative, or zero?
10. A car is accelerated by a constant force, F. If the distance required to accelerate the car from rest to the speed v is Δx, what distance, $\Delta x'$, is required to accelerate the car from the speed v to the speed $2v$?
11. Car 1 has twice the mass of car 2, but they both have the same kinetic energy. If the speed of car 1 is v, what is the speed of car 2?
12. A block of mass m and speed v collides with a spring, compressing it a distance Δx. What is the compression of the spring if its force constant is increased by a factor of four?
13. A block of mass m and speed v collides with a spring, compressing it a distance Δx. What is the compression of the spring if the mass of the block is halved and its speed is doubled?
14. A work W_0 is required to stretch a certain spring 2 cm from its equilibrium position. **(a)** How much work is required to stretch the spring 1 cm from equilibrium? **(b)** Suppose the spring is already stretched 2 cm from equilibrium. How much additional work is required to stretch it to 3 cm from equilibrium?
15. Force F_1 does 5 J of work in 10 seconds, force F_2 does 3 J of work in 5 seconds, force F_3 does 6 J of work in 18 seconds, and force F_4 does 25 J of work in 125 seconds. Rank these forces in order of increasing power they produce. Indicate ties where appropriate.

Problems

Note: **IP** *denotes an integrated conceptual/quantitative problem.* **BIO** *identifies problems of biological or medical interest. Red bullets (•,••,•••) are used to indicate the level of difficulty of each problem.*

Section 7–1 Work Done by a Constant Force

1. • A farmhand pushes a 23-kg bale of hay 3.4 m across the floor of a barn. If she exerts a horizontal force of 86 N on the hay, how much work has she done?

2. • Children in a tree house lift a small dog in a basket 4.70 m up to their house. If it takes 201 J of work to do this, what is the combined mass of the dog and basket?

3. • Early one October, you go to a pumpkin patch to select your Halloween pumpkin. You lift the 3.2-kg pumpkin to a height of 1.2 m, then carry it 50.0 m (on level ground) to the check-out stand. **(a)** Calculate the work you do on the pumpkin as you lift it from the ground. **(b)** How much work do you do on the pumpkin as you carry it from the field?

4. • The coefficient of kinetic friction between a suitcase and the floor is 0.26. If the suitcase has a mass of 70.0 kg, how far can it be pushed across the level floor with 640 J of work?

5. •• You pick up a 3.4-kg can of paint from the ground and lift it to a height of 1.8 m. **(a)** How much work do you do on the can of paint? **(b)** You hold the can stationary for half a minute, waiting for a friend on a ladder to take it. How much work do you do during this time? **(c)** Your friend decides against the paint, so you lower it back to the ground. How much work do you do on the can as you lower it?

6. •• **IP** A tow rope, parallel to the water, pulls a water skier directly behind the boat with constant velocity for a distance of 65 m before the skier falls. The tension in the rope is 120 N. **(a)** Is the work done on the skier by the rope positive, negative, or zero? Explain. **(b)** Calculate the work done by the rope on the skier.

7. •• **IP** In the situation described in the previous problem, **(a)** is the work done on the boat by the rope positive, negative, or zero? Explain. **(b)** Calculate the work done by the rope on the boat.

8. •• A child pulls a friend in a little red wagon with constant speed. If the child pulls with a force of 16 N for 10.0 m, and the handle of the wagon is inclined at an angle of 25° above the horizontal, how much work does the child do on the wagon?

9. •• A 55-kg packing crate is pulled with constant speed across a rough floor with a rope that is at an angle of 40.0° above the horizontal. If the tension in the rope is 125 N, how much work is done on the crate to move it 5.0 m?

10. •• **IP** To clean a floor, a janitor pushes on a mop handle with a force of 50.0 N. **(a)** If the mop handle is at an angle of 55° above the horizontal, how much work is required to push the mop 0.50 m? **(b)** If the angle the mop handle makes with the horizontal is increased to 65°, does the work done by the janitor increase, decrease, or stay the same? Explain.

11. •• A small plane tows a glider at constant speed and altitude. If the plane does 2.00×10^5 J of work to tow the glider 145 m and the tension in the tow rope is 2560 N, what is the angle between the tow rope and the horizontal?

12. •• A young woman on a skateboard is pulled by a rope attached to a bicycle. The velocity of the skateboarder is $\vec{v} = (4.1\ \text{m/s})\hat{x}$ and the force exerted on her by the rope is $\vec{F} = (17\ \text{N})\hat{x} + (12\ \text{N})\hat{y}$. **(a)** Find the work done on the skateboarder by the rope in 25 seconds. **(b)** Assuming the velocity of the bike is the same as that of the skateboarder, find the work the rope does on the bicycle in 25 seconds.

13. •• To keep her dog from running away while she talks to a friend, Susan pulls gently on the dog's leash with a constant force given by $\vec{F} = (2.2\ \text{N})\hat{x} + (1.1\ \text{N})\hat{y}$. How much work does she do on the dog if its displacement is **(a)** $\vec{d} = (0.25\ \text{m})\hat{x}$, **(b)** $\vec{d} = (0.25\ \text{m})\hat{y}$, or **(c)** $\vec{d} = (-0.50\ \text{m})\hat{x} + (-0.25\ \text{m})\hat{y}$?

14. •• Water skiers often ride to one side of the center line of a boat, as shown in **Figure 7–16**. In this case, the ski boat is traveling at 15 m/s and the tension in the rope is 75 N. If the boat does 3500 J of work on the skier in 50.0 m, what is the angle θ between the tow rope and the center line of the boat?

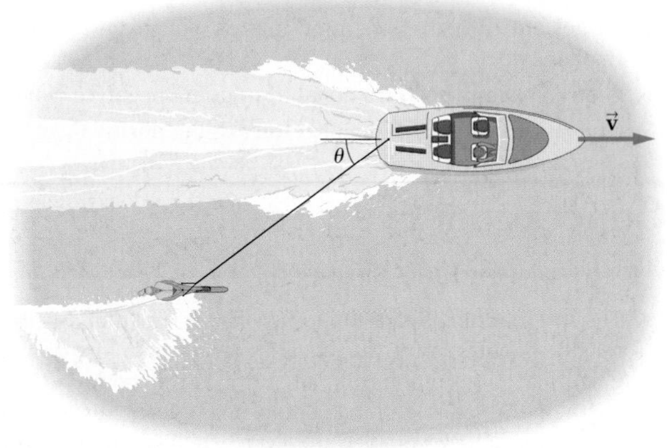

▲ **FIGURE 7–16** Problems 14 and 54

Section 7–2 Kinetic Energy and the Work-Energy Theorem

15. • How much work is needed for a 68-kg runner to accelerate from rest to 7.5 m/s?

16. • **Skylab's Reentry** When Skylab reentered the Earth's atmosphere on July 11, 1979, it broke into a myriad of pieces. One of the largest fragments was a 1770-kg lead-lined film vault, and it landed with an estimated speed of 120 m/s. What was the kinetic energy of the film vault when it landed?

17. • **IP** A 9.50-g bullet has a speed of 1.30 km/s. **(a)** What is its kinetic energy in joules? **(b)** What is the bullet's kinetic energy if its speed is halved? **(c)** If its speed is doubled?

18. •• **IP** A 0.14-kg pinecone falls 16 m to the ground, where it lands with a speed of 13 m/s. **(a)** With what speed would the pinecone have landed if there had been no air resistance? **(b)** Did air resistance do positive work, negative work, or zero work on the pinecone? Explain.

19. •• In the previous problem, **(a)** how much work was done on the pinecone by air resistance? **(b)** What was the average force of air resistance exerted on the pinecone?

20. •• At $t = 1.0$ s, a 0.40-kg object is falling with a speed of 6.0 m/s. At $t = 2.0$ s, it has a kinetic energy of 25 J. **(a)** What is the kinetic energy of the object at $t = 1.0$ s? **(b)** What is the speed of the object at $t = 2.0$ s? **(c)** How much work was done on the object between $t = 1.0$ s and $t = 2.0$ s?

21. •• After hitting a long fly ball that goes over the right fielder's head and lands in the outfield, the batter decides to keep going past second base and try for third base. The 62.0-kg player begins sliding 3.40 m from the base with a speed of 4.35 m/s. If the player comes to rest at third base, **(a)** how much work was done on the player by friction? **(b)** What was the coefficient of kinetic friction between the player and the ground?

22. •• **IP** A 1300-kg car coasts on a horizontal road with a speed of 18 m/s. After crossing an unpaved, sandy stretch of road 30.0 m

long, its speed decreases to 15 m/s. **(a)** Was the net work done on the car positive, negative, or zero? Explain. **(b)** Find the magnitude of the average net force on the car in the sandy section.

23. •• **IP** **(a)** In the previous problem, the car's speed decreased by 3.0 m/s as it coasted across a sandy section of road 30.0 m long. If the sandy portion of the road had been only 15.0 m long, would the car's speed have decreased by 1.5 m/s, more than 1.5 m/s, or less than 1.5 m/s? Explain. **(b)** Calculate the change in speed in this case.

24. •• A 65-kg bicyclist rides his 8.8-kg bicycle with a speed of 14 m/s. **(a)** How much work must be done by the brakes to bring the bike and rider to a stop? **(b)** How far does the bicycle travel if it takes 4.0 s to come to rest? **(c)** What is the magnitude of the braking force?

Section 7–3 Work Done by a Variable Force

25. • A spring with a force constant of 3.5×10^4 N/m is initially at its equilibrium length. **(a)** How much work must you do to stretch the spring 0.050 m? **(b)** How much work must you do to compress it 0.050 m?

26. • A 1.2-kg block is held against a spring of force constant 1.0×10^4 N/m, compressing it a distance of 0.15 m. How fast is the block moving after it is released and the spring pushes it away?

27. • Initially sliding with a speed of 2.2 m/s, a 1.8-kg block collides with a spring and compresses it 0.31 m before coming to rest. What is the force constant of the spring?

28. • The force shown in **Figure 7–17** moves an object from $x = 0$ to $x = 0.75$ m. **(a)** How much work is done by the force? **(b)** How much work is done by the force if the object moves from $x = 0.15$ m to $x = 0.60$ m?

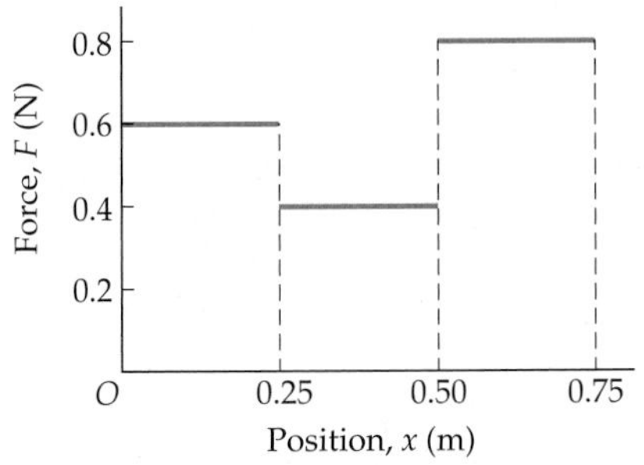

▲ **FIGURE 7–17** Problem 28

29. • An object is acted on by the force shown in **Figure 7–18**. What is the final position of the object if its initial position is $x = 0.40$ m and the work done on it is equal to **(a)** 0.21 J, or **(b)** −0.19 J?

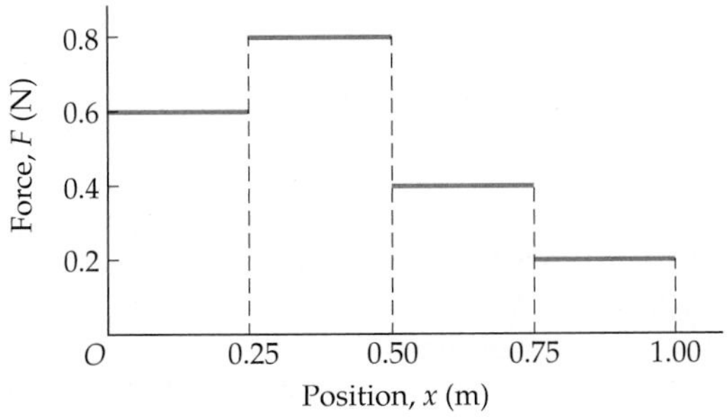

▲ **FIGURE 7–18** Problems 29 and 32

30. •• To compress spring 1 by 0.20 m takes 150 J of work. Stretching spring 2 by 0.30 m requires 210 J of work. Which spring is stiffer?

31. •• **IP** It takes 160 J of work to compress a certain spring 0.14 m. **(a)** What is the force constant of this spring? **(b)** To compress the spring an additional 0.14 m, does it take 160 J, more than 160 J, or less than 160 J? Verify your answer with a calculation.

32. •• The force shown in Figure 7–18 acts on a 1.7-kg object whose initial speed is 0.44 m/s and initial position is $x = 0.27$ m. **(a)** Find the speed of the object when it is at the location $x = 0.99$ m. **(b)** At what location would the object's speed be 0.32 m/s?

33. ••• A block is acted on by a force that varies as $(2.0 \times 10^4\ \text{N/m})x$ for $0 \le x \le 0.21$ m, and then remains constant at 4200 N for larger x. How much work does the force do on the block in moving it **(a)** from $x = 0$ to $x = 0.30$ m, or **(b)** from $x = 0.10$ m to $x = 0.40$ m?

Section 7–4 Power

34. • **BIO Climbing the Empire State Building** A new record for running the stairs of the Empire State Building was set on February 3, 2003. The 86 flights, with a total of 1576 steps, was run in 9 minutes and 33 seconds. If the height gain of each step was 0.20 m, and the mass of the runner was 70.0 kg, what was his average power output during the climb? Give your answer in both watts and horsepower.

35. • How many joules of energy are in a kilowatt-hour?

36. • Calculate the power output of a 1.3-g fly as it walks straight up a windowpane at 2.5 cm/s.

37. • An ice cube is placed in a microwave oven. Suppose the oven delivers 105 W of power to the ice cube and that it takes 32,200 J to melt it. How long does it take for the ice cube to melt?

38. • You raise a bucket of water from the bottom of a deep well. If your power output is 108 W, and the mass of the bucket and the water in it is 5.00 kg, with what speed can you raise the bucket? Ignore the weight of the rope.

39. •• In order to keep a leaking ship from sinking, it is necessary to pump 12.0 lb of water each second from below deck up a height of 2.00 m and over the side. What is the minimum horsepower motor that can be used to save the ship?

40. •• **IP** A kayaker paddles with a power output of 50.0 W to maintain a steady speed of 1.50 m/s. **(a)** Calculate the resistive force exerted by the water on the kayak. **(b)** If the kayaker doubles her power output, and the resistive force due to the water remains the same, by what factor does the kayaker's speed change?

41. •• **BIO Human-Powered Flight** Human-powered aircraft require a pilot to pedal, as in a bicycle, and produce a sustained power output of about 0.30 hp. The *Gossamer Albatross* flew across the English Channel on June 12, 1979, in 2 h 49 min. **(a)** How much energy did the pilot expend during the flight? **(b)** How many Snickers candy bars (280 Cal per bar) would the pilot have to consume to be "fueled up" for the flight? [Note: The nutritional calorie, 1 Cal, is equivalent to 1000 calories (1000 cal) as defined in physics. In addition, the conversion factor between calories and joules is as follows: 1 Cal = 1000 cal = 1 kcal = 4186 J.]

42. •• **IP** A grandfather clock is powered by the descent of a 4.15-kg weight. **(a)** If the weight descends through a distance of 0.720 m in 3.25 days, how much power does it deliver to the clock? **(b)** To increase the power delivered to the clock, should the time it takes for the mass to descend be increased or decreased? Explain.

43. •• **BIO The Power You Produce** Estimate the power you produce in running up a flight of stairs. Give your answer in horsepower.

44. ••• **IP** A certain car can accelerate from rest to the speed v in T seconds. If the power output of the car remains constant, **(a)** how long does it take for the car to accelerate from v to $2v$? **(b)** How fast is the car moving at $2T$ seconds after starting?

General Problems

45. • **BIO Muscle Cells** Biological muscle cells can be thought of as nanomotors that use the chemical energy of ATP to produce mechanical work. Measurements show that the active proteins within a muscle cell (such as myosin and actin) can produce a force of about 7.5 pN and displacements of 8.0 nm. How much work is done by such proteins?

46. • When you take a bite out of an apple, you do about 19 J of work. Estimate **(a)** the force and **(b)** the power produced by your jaw muscles during the bite.

47. • A Mountain bar has a mass of 0.045 kg and a calorie rating of 210 Cal. What speed would this candy bar have if its kinetic energy were equal to its metabolic energy. (See the note following Problem 41.)

48. • A small motor runs a lift that raises a load of bricks weighing 866 N to a height of 12.7 m in 26.2 s. Assuming that the bricks are lifted with constant speed, what is the minimum power the motor must produce?

49. • You push a 67-kg box across a floor where the coefficient of kinetic friction is $\mu_k = 0.55$. The force you exert is horizontal. **(a)** How much power is needed to push the box at a speed of 0.50 m/s? **(b)** How much work do you do if you push the box for 35 s?

50. • **BIO The Beating Heart** The average power output of the human heart is 1.33 watts. **(a)** How much energy does the heart produce in a day? **(b)** Compare the energy found in part (a) with the energy required to walk up a flight of stairs. Estimate the height a person could attain on a set of stairs using nothing more than the daily energy produced by the heart.

51. • **The Atmos Clock** The Atmos clock (the so-called perpetual motion clock) gets its name from the fact that it runs off pressure variations in the atmosphere, which drive a bellows containing a mixture of gas and liquid ethyl chloride. Because the power to drive these clocks is so limited, they must be very efficient. In fact, a single 60.0-W lightbulb could power 240 million Atmos clocks simultaneously. Find the amount of energy, in joules, required to run an Atmos clock for one day.

52. •• After a tornado, a 0.60-g straw was found embedded 2.6 cm into the trunk of a tree. If the average force exerted on the straw by the tree was 65 N, what was the speed of the straw when it hit the tree?

53. •• You throw a glove straight upward to celebrate a victory. Its initial kinetic energy is K and it reaches a maximum height h. What is the kinetic energy of the glove when it is at the height $h/2$?

54. •• The water skier in Figure 7–16 is at an angle of 35° with respect to the center line of the boat, and is being pulled at a constant speed of 14 m/s. If the tension in the tow rope is 90.0 N, **(a)** how much work does the rope do on the skier in 10.0 s? **(b)** How much work does the resistive force of water do on the skier in the same time?

55. •• **IP** A sled with a mass of 5.80 kg is pulled along the ground through a displacement given by $\vec{\mathbf{d}} = (4.55\text{ m})\hat{\mathbf{x}}$. (Let the x axis be horizontal and the y axis be vertical.) **(a)** How much work is done on the sled when the force acting on it is $\vec{\mathbf{F}} = (2.89\text{ N})\hat{\mathbf{x}} + (0.131\text{ N})\hat{\mathbf{y}}$? **(b)** How much work is done on the sled when the force acting on it is $\vec{\mathbf{F}} = (2.89\text{ N})\hat{\mathbf{x}} + (0.231\text{ N})\hat{\mathbf{y}}$? **(c)** If the mass of the sled is increased, does the work done by the forces in parts (a) and (b) increase, decrease, or stay the same? Explain.

56. •• **IP** A 0.16-kg apple falls from a branch 3.5 m above the ground. **(a)** Does the power delivered to the apple by gravity increase, decrease, or stay the same during the time the apple falls to the ground? Explain. Find the power delivered by gravity to the apple when the apple is **(b)** 2.5 m and **(c)** 1.5 m above the ground.

57. •• A juggling ball of mass m is thrown straight upward from an initial height h with an initial speed v_0. How much work has gravity done on the ball **(a)** when it reaches its greatest height, h_{max}, and **(b)** when it reaches ground level? **(c)** Find an expression for the kinetic energy of the ball as it lands.

58. •• The force shown in **Figure 7–19** acts on an object that moves along the x axis. How much work is done by the force as the object moves from **(a)** $x = 0$ to $x = 2.0$ m, **(b)** $x = 1.0$ m to $x = 4.0$ m, and **(c)** $x = 3.5$ m to $x = 1.2$ m?

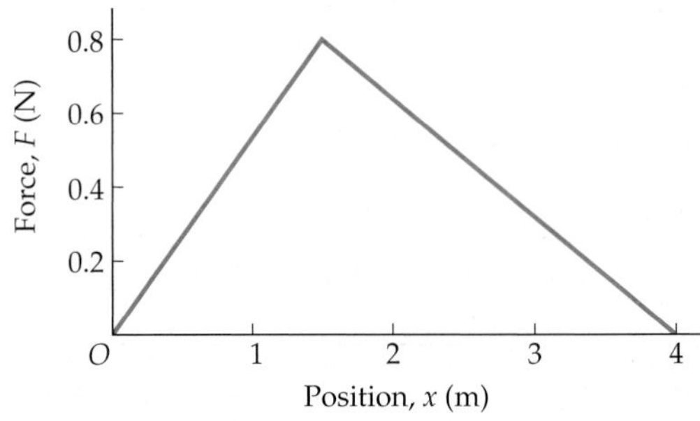

▲ **FIGURE 7–19** Problem 58

59. •• Calculate the power output of a 1.8-g spider as it walks up a windowpane at 2.3 cm/s. The spider walks on a path that is at 25° to the vertical, as illustrated in **Figure 7–20**.

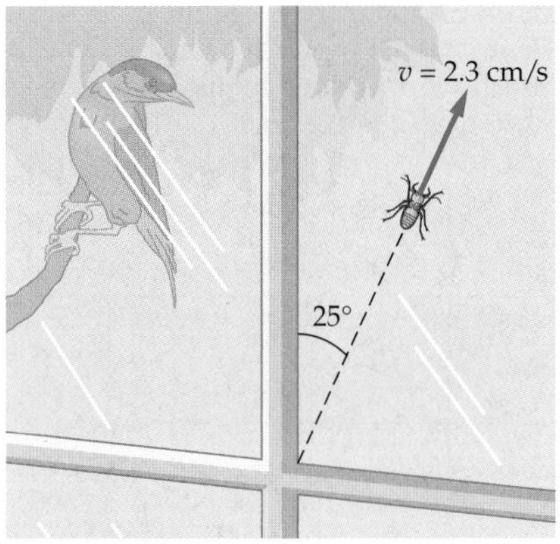

▲ **FIGURE 7–20** Problem 59

60. •• The motor of a ski boat produces a power of 36,600 W to maintain a constant speed of 14.0 m/s. To pull a water skier at the same constant speed, the motor must produce a power of 37,800 W. What is the tension in the rope pulling the skier?

61. •• **Cookie Power** To make a batch of cookies, you mix half a bag of chocolate chips into a bowl of cookie dough, exerting a 23-N force on the stirring spoon. Assume that your force is always in the direction of motion of the spoon. **(a)** What power is needed to move the spoon at a speed of 0.24 m/s? **(b)** How much work do you do if you stir the mixture for 1.5 min?

62. •• **IP** A pitcher accelerates a 0.14-kg hardball from rest to 42.5 m/s in 0.060 s. **(a)** How much work does the pitcher do on the ball? **(b)** What is the pitcher's power output during the pitch? **(c)** Suppose the ball reaches 42.5 m/s in less than 0.060 s. Is the power produced by the pitcher in this case more than, less than, or the same as the power found in part (b)? Explain.

63. •• **Catapult Launcher** A catapult launcher on an aircraft carrier accelerates a jet from rest to 72 m/s. The work done by the catapult during the launch is 7.6×10^7 J. **(a)** What is the mass of the jet? **(b)** If the jet is in contact with the catapult for 2.0 s, what is the power output of the catapult?

64. •• **BIO Brain Power** The human brain consumes about 22 W of power under normal conditions, though more power may be required during exams. **(a)** How long can one Snickers bar (see Problem 41) power the normally functioning brain? **(b)** At what rate must you lift a 3.6-kg container of milk (one gallon) if the power output of your arm is to be 22 W? **(c)** How long does it take to lift the milk container through a distance of 1.0 m at this rate?

65. •• **IP** A 1400-kg car delivers a constant 47 hp to the drive wheels. We assume the car is traveling on a level road and that all frictional forces may be ignored. **(a)** What is the acceleration of this car when its speed is 14 m/s? **(b)** If the speed of the car is doubled, does its acceleration increase, decrease, or stay the same? Explain. **(c)** Calculate the car's acceleration when its speed is 28 m/s.

66. •• **Meteorite** On October 9, 1992, a 27-pound meteorite struck a car in Peekskill, NY, creating a dent about 22 cm deep. If the initial speed of the meteorite was 550 m/s, what was the average force exerted on the meteorite by the car?

An interplanetary fender-bender (Problem 66)

67. ••• **BIO Powering a Pigeon** A pigeon in flight experiences a force of air resistance given approximately by $F = bv^2$, where v is the flight speed and b is a constant. **(a)** What are the units of the constant b? **(b)** What is the largest possible speed of the pigeon if its maximum power output is P? **(c)** By what factor does the largest possible speed increase if the maximum power is doubled?

68. ••• **Springs in Series** Two springs, with force constants k_1 and k_2, are connected in series, as shown in **Figure 7–21**. How much work is required to stretch this system a distance x from the equilibrium position?

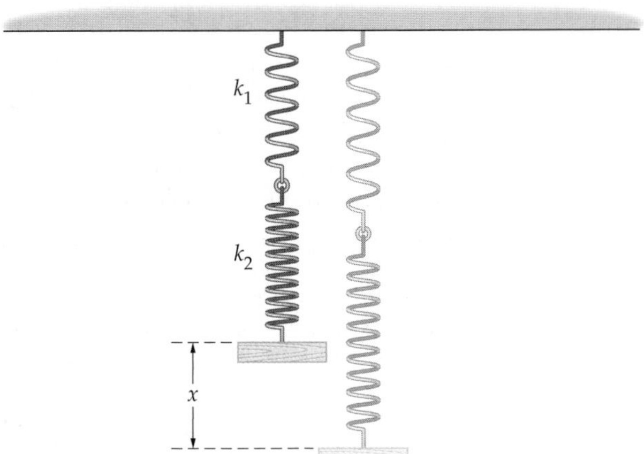

▲ **FIGURE 7–21** Problem 68

69. ••• **Springs in Parallel** Two springs, with force constants k_1 and k_2, are connected in parallel, as shown in **Figure 7–22**. How much work is required to stretch this system a distance x from the equilibrium position?

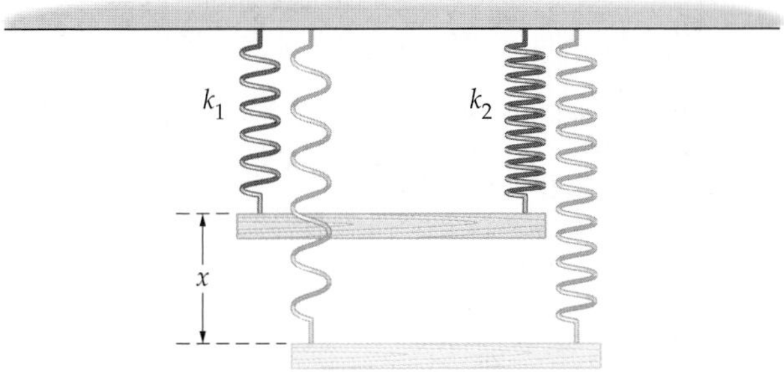

▲ **FIGURE 7–22** Problem 69

70. ••• A block rests on a horizontal frictionless surface. A string is attached to the block, and is pulled with a force of 45.0 N at an angle θ above the horizontal, as shown in **Figure 7–23**. After the block is pulled through a distance of 1.50 m, its speed is 2.60 m/s, and 50.0 J of work has been done on it. **(a)** What is the angle θ? **(b)** What is the mass of the block?

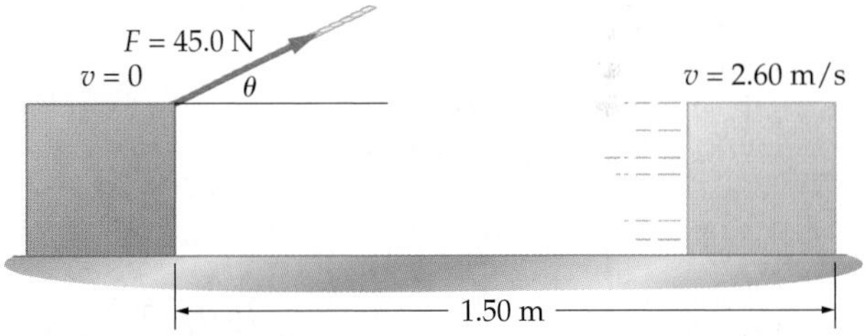

▲ **FIGURE 7–23** Problem 70

Interactive Problems

71. •• ***Referring to Figure 7–12*** Suppose the block has a mass of 1.4 kg and an initial speed of 0.62 m/s. **(a)** What force constant must the spring have if the maximum compression is to be 2.4 cm? **(b)** If the spring has the force constant found in part (a), find the maximum compression if the mass of the block is doubled *and* its initial speed is halved.

72. •• **IP** ***Referring to Figure 7–12*** In the situation shown in Figure 7–12 (d), a spring with a force constant of 750 N/m is compressed by 4.1 cm. **(a)** If the speed of the block in Figure 7–12 (f) is 0.88 m/s, what is its mass? **(b)** If the mass of the block is doubled, is the final speed greater than, less than, or equal to 0.44 m/s? **(c)** Find the final speed for the case described in part (b).

73. •• **IP** ***Referring to Example 7–8*** Suppose the car has a mass of 1400 kg and delivers 48 hp to the wheels. **(a)** How long does it take for the car to increase its speed from 15 m/s to 25 m/s? **(b)** Would the time required to increase the speed from 5.0 m/s to 15 m/s be greater than, less than, or equal to the time found in part (a)? **(c)** Determine the time required to accelerate from 5.0 m/s to 15 m/s.

74. •• **IP** ***Referring to Example 7–8*** A 1250-kg car accelerates from an initial speed of 15.0 m/s. After accelerating for 2.50 s, its speed is 17.0 m/s. **(a)** What is the minimum power delivered to the wheels by this car? **(b)** After accelerating for an additional 2.50 s, is the speed of the car greater than, less than, or equal to 19.0 m/s? Assume constant power output. **(c)** Calculate the speed of the car at the time described in part (b).

8 Potential Energy and Conservation of Energy

8–1 Conservative and Nonconservative Forces 205

8–2 Potential Energy and the Work Done by Conservative Forces 208

8–3 Conservation of Mechanical Energy 214

8–4 Work Done by Nonconservative Forces 222

8–5 Potential Energy Curves and Equipotentials 227

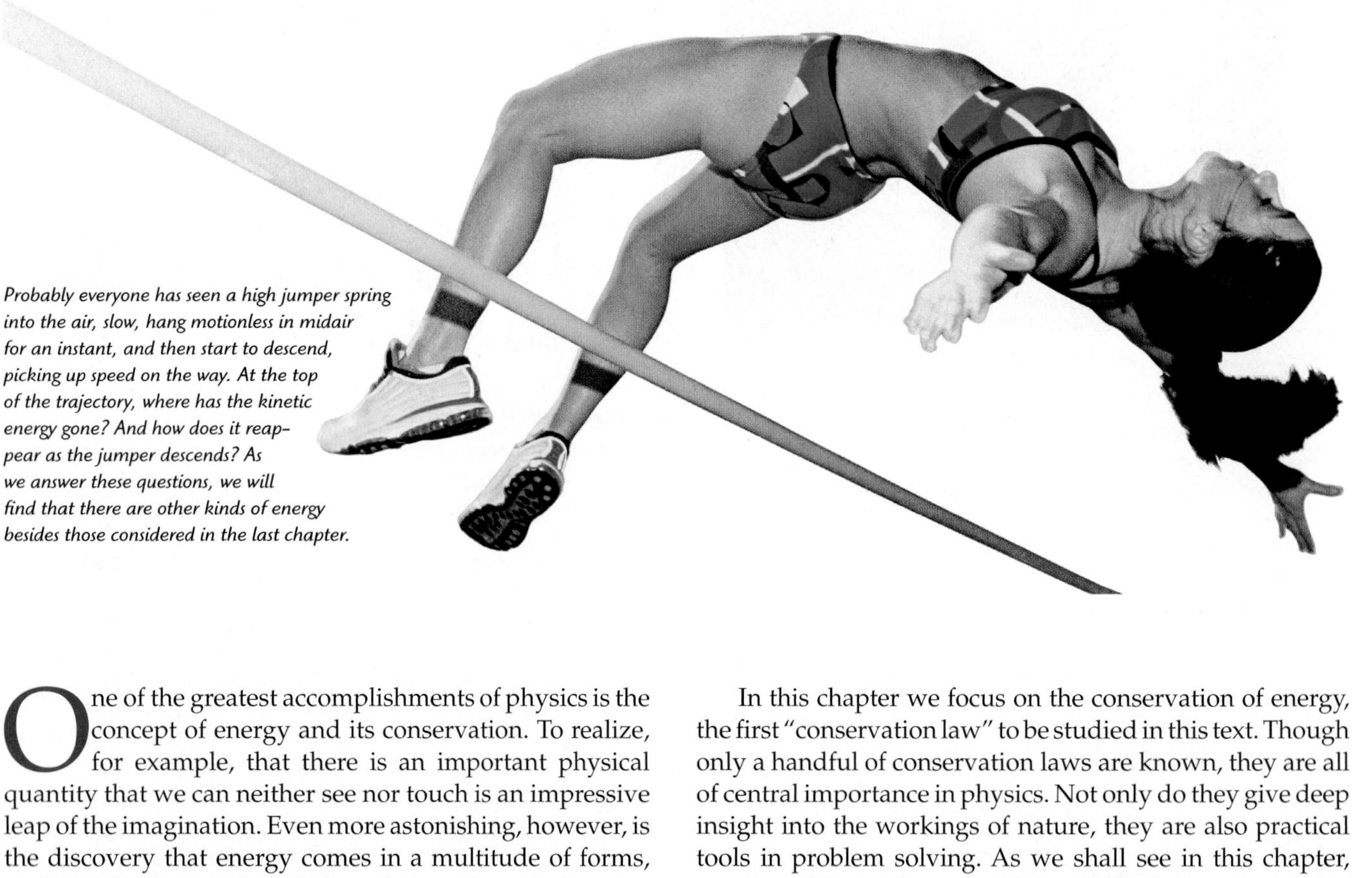

Probably everyone has seen a high jumper spring into the air, slow, hang motionless in midair for an instant, and then start to descend, picking up speed on the way. At the top of the trajectory, where has the kinetic energy gone? And how does it reappear as the jumper descends? As we answer these questions, we will find that there are other kinds of energy besides those considered in the last chapter.

One of the greatest accomplishments of physics is the concept of energy and its conservation. To realize, for example, that there is an important physical quantity that we can neither see nor touch is an impressive leap of the imagination. Even more astonishing, however, is the discovery that energy comes in a multitude of forms, and that the sum total of all these forms of energy is a constant. The universe, in short, has a certain amount of energy, and that energy simply ebbs and flows from one form to another, with the total amount remaining fixed.

In this chapter we focus on the conservation of energy, the first "conservation law" to be studied in this text. Though only a handful of conservation laws are known, they are all of central importance in physics. Not only do they give deep insight into the workings of nature, they are also practical tools in problem solving. As we shall see in this chapter, many problems that would be difficult to solve using Newton's laws can be solved with ease using the principle of energy conservation.

8–1 Conservative and Nonconservative Forces

In physics, we classify forces according to whether they are *conservative* or *nonconservative*. The key distinction is that when a **conservative force** acts, the work it does is stored in the form of energy that can be released at a later time. In this section, we sharpen this distinction and explore some examples of conservative and nonconservative forces.

Perhaps the simplest case of a conservative force is gravity. Imagine lifting a box of mass m from the floor to a height h, as in **Figure 8–1**. To lift the box with constant speed, the force you must exert against gravity is mg. Since the upward distance is h, the work you do on the box is $W = mgh$. If you now release the box and allow it to drop back to the floor, gravity does the same work, $W = mgh$, and in the process gives the box an equivalent amount of kinetic energy.

◀ **FIGURE 8–1 Work against gravity**
Lifting a box against gravity with constant speed takes a work mgh. When the box is released, gravity does the same work on the box as it falls. Gravity is a conservative force.

Contrast this with the force of kinetic friction, which is nonconservative. To slide a box of mass m across the floor with constant speed, as shown in **Figure 8–2**, you must exert a force of magnitude $\mu_k N = \mu_k mg$. After sliding the box a distance d, the work you have done is $W = \mu_k mgd$. In this case, when you release the box it simply stays put—friction does no work on it after you let go. Thus, the work done by a **nonconservative force** cannot be recovered later as kinetic energy; instead, it is converted to other forms of energy, such as a slight warming of the floor and box in our example.

The differences between conservative and nonconservative forces are even more apparent if we consider moving an object around a closed path. Consider, for example, the path shown in **Figure 8–3**. If we move a box of mass m along this path, the total work done by gravity is the sum of the work done on each segment of the path; that is $W_{total} = W_{AB} + W_{BC} + W_{CD} + W_{DA}$. The work done by gravity from A to B and from C to D is zero, since the force is at right angles to the displacement on these segments. Thus $W_{AB} = W_{CD} = 0$. On the segment from B to C, gravity does negative work (displacement and force are in opposite directions),

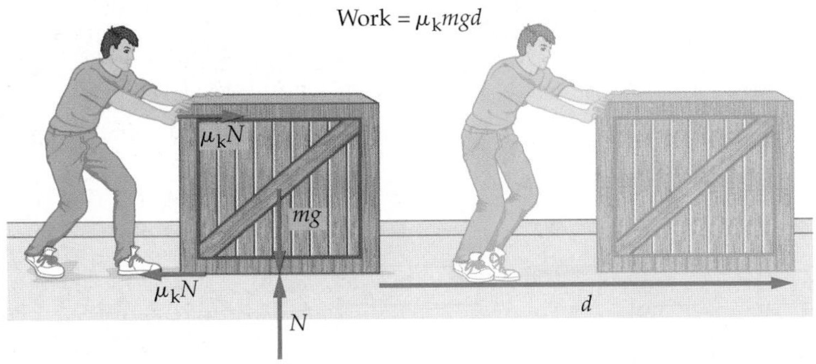

◀ **FIGURE 8–2 Work against friction**
Pushing a box with constant speed against friction takes a work $\mu_k mgd$. When the box is released, it quickly comes to rest and friction does no further work. Friction is a nonconservative force.

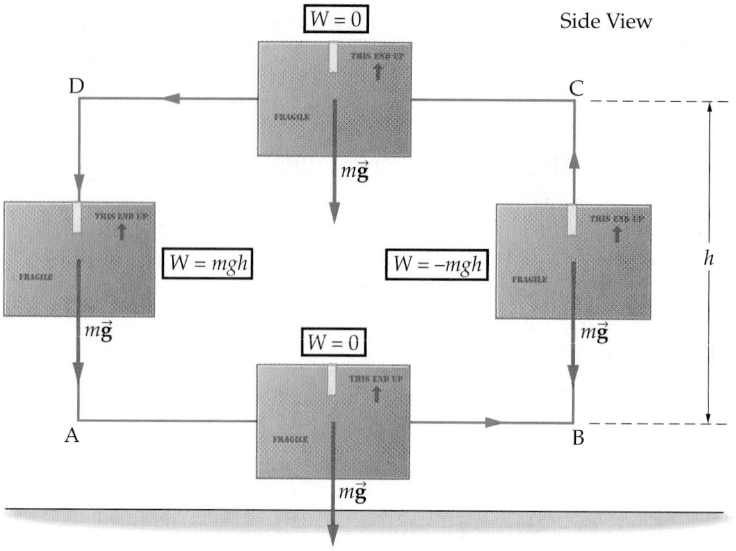

▶ **FIGURE 8–3 Work done by gravity on a closed path is zero**
Gravity does no work on the two horizontal segments of the path. On the two vertical segments, the amounts of work done are equal in magnitude but opposite in sign. Therefore, the total work done by gravity on this—or any—closed path is zero.

but it does positive work from D to A (displacement and force are in the same direction). Hence, $W_{BC} = -mgh$ and $W_{DA} = mgh$. As a result, the total work done by gravity is zero:

$$W_{total} = 0 + (-mgh) + 0 + mgh = 0$$

With friction, the results are quite different. If we push the box around the closed horizontal path shown in **Figure 8–4**, the total work done by friction does not vanish. In fact, friction does the negative work $W = -f_k d = -\mu_k mgd$ on each segment. Therefore, the total work done by kinetic friction is

$$W_{total} = (-\mu_k mgd) + (-\mu_k mgd) + (-\mu_k mgd) + (-\mu_k mgd) = -4\,\mu_k mgd$$

These results lead to the following definition of a conservative force:

Conservative Force: Definition 1
A conservative force is a force that does zero total work on any closed path.

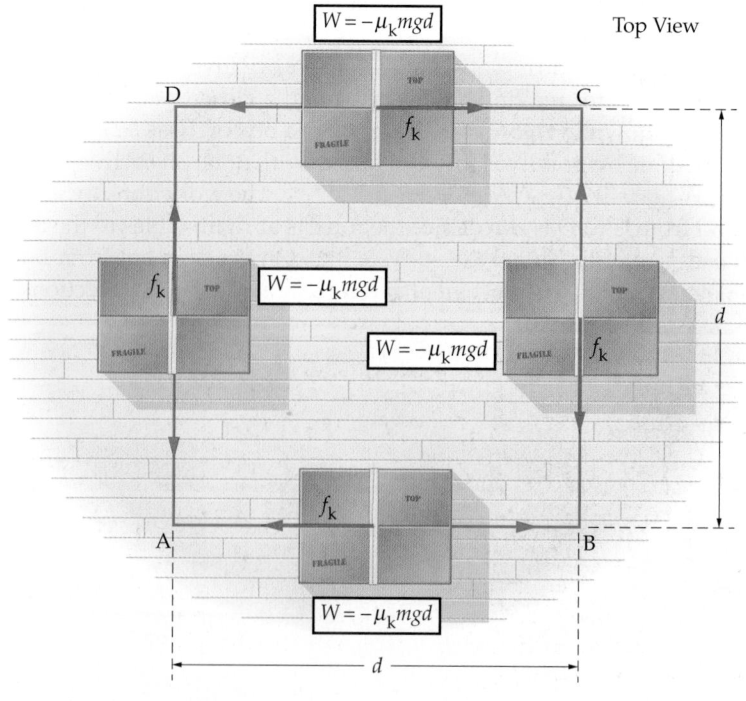

▶ **FIGURE 8–4 Work done by friction on a closed path is nonzero**
The work done by friction when an object moves through a distance d is $-\mu_k mgd$. Thus, the total work done by friction on a closed path is nonzero. In this case, it is equal to $-4\,\mu_k mgd$.

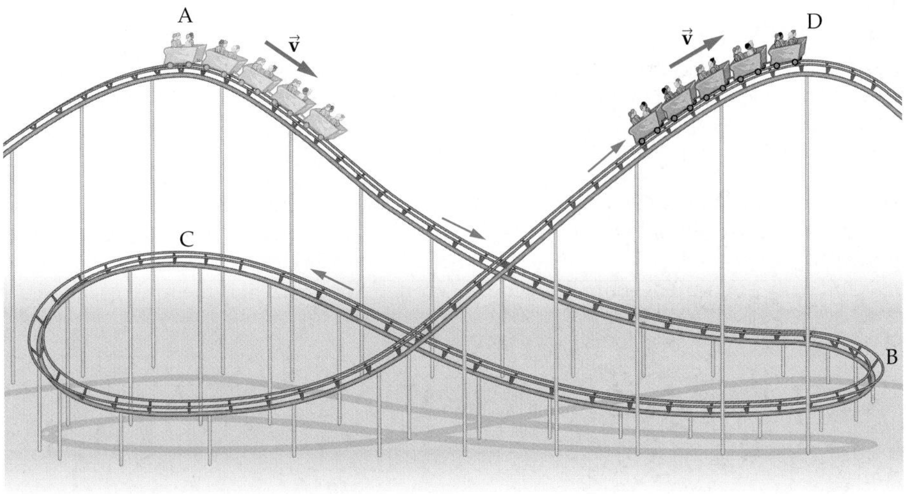

▲ **FIGURE 8–5 Gravity is a conservative force**
If frictional forces can be ignored, a roller coaster car will have the same speed at points A and D, since they are at the same height. Hence, after any complete circuit of the track the speed of the car returns to its initial value. It follows that the change in kinetic energy is zero for a complete circuit, and, therefore, the work done by gravity is also zero.

A roller coaster provides a good illustration of this definition. If a car on a roller coaster has a speed v at point A in **Figure 8–5**, it speeds up as it drops to point B, slows down as it approaches point C, and so on. When the car returns to its original height, at point D, it will again have the speed v, as long as friction and other nonconservative forces can be neglected. Similarly, if the car completes a circuit of the track and returns to point A, it will again have the speed v. Hence, a car's kinetic energy is unchanged ($\Delta K = 0$) after *any* complete circuit of the track. From the work energy theorem, $W_{total} = \Delta K$, it follows that the work done by gravity is zero for the closed path of the car, as expected for a conservative force.

This property of conservative forces has interesting consequences. For instance, consider the closed paths shown in **Figure 8–6**. On each of these paths, we know that the work done by a conservative force is zero. Thus, it follows from paths 1 and 2 that $W_{total} = W_1 + W_2 = 0$, or

$$W_2 = -W_1$$

Similarly, using paths 1 and 3 we have $W_{total} = W_1 + W_3 = 0$, or

$$W_3 = -W_1$$

As a result, we see that the work done on path 3 is the same as the work done on path 2:

$$W_3 = W_2$$

But paths 2 and 3 are arbitrary, as long as they start at point B and end at point A. This leads to an equivalent definition of a conservative force:

Conservative Force: Definition 2
If the work done by a force in going from an arbitrary point A to an arbitrary point B is *independent of the path* from A to B, the force is conservative.

This definition is given an explicit check in Example 8–1.

Table 8–1 summarizes the different kinds of conservative and nonconservative forces we have encountered thus far in this text.

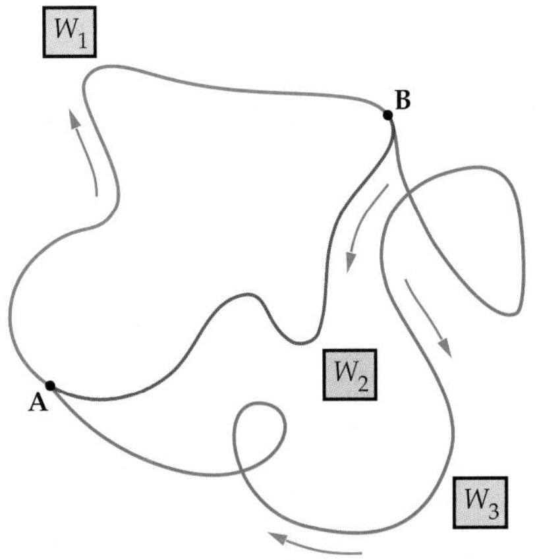

▲ **FIGURE 8–6 The work done by a conservative force is independent of path**
Considering paths 1 and 2, we see that $W_1 + W_2 = 0$, or $W_2 = -W_1$. From paths 1 and 3, however, we see that $W_1 + W_3 = 0$, or $W_3 = -W_1$. It follows, then, that $W_3 = W_2$, since they are both equal to $-W_1$; hence the work done in going from A to B is independent of the path.

TABLE 8–1
Conservative and Nonconservative Forces

Force	Section
Conservative forces	
Gravity	5–6
Spring force	6–2
Nonconservative forces	
Friction	6–1
Tension in a rope, cable, etc.	6–2
Forces exerted by a motor	7–4
Forces exerted by muscles	5–3

EXAMPLE 8–1 Different Paths, Different Forces

(a) A 4.57-kg box is moved with constant speed from A to B along the two paths shown at left below. Calculate the work done by gravity on each of these paths. **(b)** The same box is pushed across a floor from A to B along path 1 and path 2 at right below. If the coefficient of kinetic friction between the box and the surface is $\mu_k = 0.63$, how much work is done by friction along each path?

Picture the Problem

Part **(a)** of our sketch shows two different paths a box might be taken through in going from point A to point B. Path 1 is indicated by two red lines, indicating a vertical displacement of 1.0 m and a horizontal displacement of 3.0 m. Path 2, indicated in green, consists of two horizontal and two vertical displacements. In this case, we are interested in the work done by gravity. Part **(b)** shows the same basic paths—path 1 in orange and path 2 in purple—only this time on a rough floor. Here it is the force of kinetic friction that is of interest.

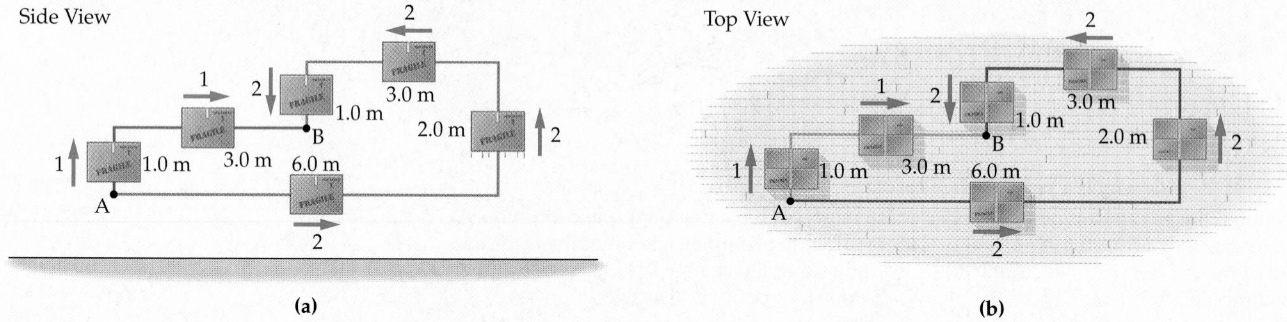

Strategy

To calculate the work for each path, we break it down into segments. Path 1 is made up of two segments, path 2 has four segments.

a. For gravity, the work is zero on horizontal segments. On vertical segments, the work done by gravity is positive when motion is downward and negative when motion is upward.

b. The work done by kinetic friction is negative on all segments of both paths.

Solution

Part (a)

1. Using $W = Fd = mgy$, calculate the work done by gravity along the two segments of path 1:

$$W_1 = -(4.57\ \text{kg})(9.81\ \text{m/s}^2)(1.0\ \text{m}) + 0 = -45\ \text{J}$$

2. In the same way, calculate the work done by gravity along the four segments of path 2:

$$W_2 = 0 - (4.57\ \text{kg})(9.81\ \text{m/s}^2)(2.0\ \text{m}) + 0 + (4.57\ \text{kg})(9.81\ \text{m/s}^2)(1.0\ \text{m}) = -45\ \text{J}$$

Part (b)

3. Using $F = \mu_k N$, calculate the work done by kinetic friction along the two segments of path 1:

$$W_1 = -(0.63)(4.57\ \text{kg})(9.81\ \text{m/s}^2)(1.0\ \text{m}) - (0.63)(4.57\ \text{kg})(9.81\ \text{m/s}^2)(3.0\ \text{m}) = -110\ \text{J}$$

4. Similarly, calculate the work done by kinetic friction along the four segments of path 2:

$$W_2 = -(0.63)(4.57\ \text{kg})(9.81\ \text{m/s}^2)(6.0\ \text{m}) - (0.63)(4.57\ \text{kg})(9.81\ \text{m/s}^2)(2.0\ \text{m}) - (0.63)(4.57\ \text{kg})(9.81\ \text{m/s}^2)(3.0\ \text{m}) - (0.63)(4.57\ \text{kg})(9.81\ \text{m/s}^2)(1.0\ \text{m}) = -340\ \text{J}$$

Insight

As expected, the conservative force of gravity gives the same work in going from A to B, regardless of the path. The work done by kinetic friction, however, is greater on the path of greater length.

Practice Problem

The work done by gravity when the box is moved from point B to a point C is 140 J. Is point C above or below point B? What is the vertical distance between points B and C? [**Answer:** Point C is 3.1 m below point B.]

Some related homework problems: Problem 1, Problem 2

8–2 Potential Energy and the Work Done by Conservative Forces

Work must be done to lift a bowling ball from the floor to a shelf. Once on the shelf, the bowling ball has zero kinetic energy, just as it did on the floor. Even so, the work done in lifting the ball has not been lost. If the ball is allowed to fall from

▲ Because gravity is a conservative force, the work done against gravity in lifting these logs (left) can, in principle, all be recovered. If the logs are released, for example, they will acquire an amount of kinetic energy exactly equal to the work done to lift them and to the gravitational potential energy that they gained in being lifted. Friction, by contrast, is a nonconservative force. Some of the work done by this spinning grindstone (right) goes into removing material from the object being ground, while the rest is transformed into sound energy and (especially) heat. Most of this work can never be recovered as kinetic energy.

the shelf, gravity does the same amount of work on it as you did to lift it in the first place. As a result, the work you did is "recovered" in the form of kinetic energy. Thus we say that when the ball is lifted to a new position, there is an increase in **potential energy,** U, and that this potential energy can be converted to kinetic energy when the ball falls.

In a sense, potential energy is a storage system for energy. When we increase the separation between the ball and the ground, the work we do is stored in the form of an increased potential energy. Not only that, but the storage system is perfect, in the sense that the energy is never lost, as long as the separation remains the same. The ball can rest on the shelf for a million years, and still, when it falls, it gains the same amount of kinetic energy.

Work done against friction, however, is not "stored" as potential energy. Instead, it is dissipated into other forms of energy such as heat or sound. The same is true of other nonconservative forces. Only conservative forces have the potential-energy storage system.

Before proceeding, we should point out an interesting difference between kinetic and potential energy. Kinetic energy is given by the expression $K = \frac{1}{2}mv^2$, no matter what force might be involved. On the other hand, each different conservative force has a different expression for its potential energy. To see how this comes about, we turn now to a precise definition of potential energy.

Potential Energy, *U*

When a conservative force does an amount of work W_c (the subscript c stands for conservative), the corresponding potential energy U is changed according to the following definition:

Definition of Potential Energy, *U*

$$W_c = U_i - U_f = -(U_f - U_i) = -\Delta U \qquad \text{8–1}$$

SI unit: joule, J

In words, the work done by a conservative force is equal to the negative of the change in potential energy. For example, when an object falls, gravity does *positive* work on it and its potential energy *decreases*. Similarly, when an object is lifted, gravity does *negative* work and the potential energy *increases*.

Note that since work is a scalar with units of joules, the same is true of potential energy. In addition, our definition determines only the *difference* in potential

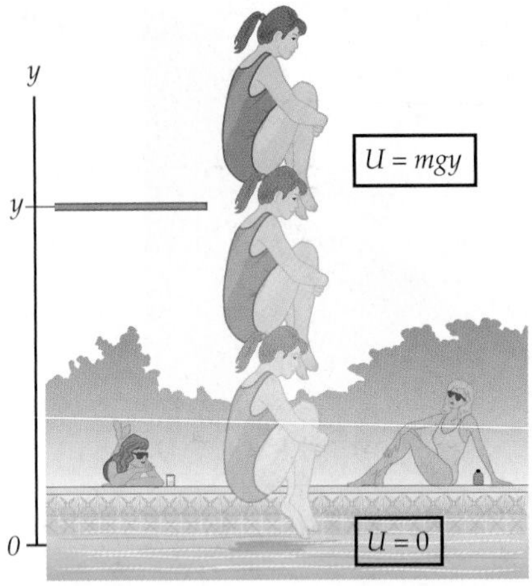

◀ **FIGURE 8–7 Gravitational Potential Energy**
A person drops from a diving board into a swimming pool. The diving board is at the height y, and the surface of the water is at $y = 0$. We choose the gravitational potential energy to be zero at $y = 0$; hence, the potential energy is mgy at the diving board.

energy between two points, not the actual value of the potential energy. Hence, we are free to choose the place where the potential energy is zero ($U = 0$) in much the same way we are free to choose the location of the origin in a coordinate system.

Gravity

Let's apply our definition of potential energy to the force of gravity near the Earth's surface. Suppose a person of mass m drops a distance y from a diving board into a pool, as shown in **Figure 8–7**. As the person drops, gravity does the work

$$W_c = mgy$$

Applying the definition given in Equation 8–1, the corresponding change in potential energy is

$$-\Delta U = U_i - U_f = W_c = mgy$$

In this expression, U_i is the potential energy when the diver is on the board, and U_f is the potential energy when the diver enters the water. Rearranging slightly, we have

$$U_i = mgy + U_f \qquad \text{8–2}$$

Note that U_i is greater than U_f.

As mentioned above, we are free to choose $U = 0$ anywhere we like; *only the difference* in U is important. For example, if you slip and fall to the ground, you hit with the same thud whether you fall in Denver (altitude 1 mile) or in Honolulu (at sea level). It's the difference in height that matters, not the height itself. (The acceleration of gravity does vary slightly with altitude, as we shall see, but the difference is small enough to be unimportant in this case.) The only point to be careful about when choosing a location for $U = 0$ is to be consistent with the choice once it is made.

In general, we choose $U = 0$ in a convenient location. In Figure 8–7, a reasonable place for $U = 0$ is the surface of the water, where $y = 0$; that is, $U_f = 0$. Then, Equation 8–2 becomes $U_i = mgy$. If we omit the subscript on U_i, letting U stand for the potential energy at the arbitrary height y, we have

Gravitational Potential Energy (Near Earth's Surface)

$$U = mgy \qquad \text{8–3}$$

Note that the gravitational potential energy depends only on the height, y, and is independent of horizontal position.

PROBLEM-SOLVING NOTE

Zero of Potential Energy

When working potential energy problems it is important to make a definite choice for the location where the potential energy is to be set equal to zero. Any location can be chosen, but once the choice is made, it must be used consistently.

▲ The $U = 0$ level for the gravitational potential energy of this system can be assigned to the point where the diver starts his dive, to the water level, or to any other level. Regardless of the choice, however, his kinetic energy when he strikes the water will be exactly equal to the difference in gravitational potential energy between his launch and splashdown points.

EXERCISE 8–1

Find the gravitational potential energy of a system consisting of a 65-kg person on a 3.0-m-high diving board. Let $U = 0$ be at water level.

Solution

Substituting $m = 65$ kg and $y = 3.0$ m in Equation 8–3 yields

$$U = mgy = (65\,\text{kg})(9.81\,\text{m/s}^2)(3.0\,\text{m}) = 1900\,\text{J}$$

The next Example considers the change in gravitational potential energy of a mountain climber, given different choices for the location of $U = 0$.

EXAMPLE 8–2 Pikes Peak or Bust

An 82.0-kg mountain climber is in the final stage of the ascent of 4301-m-high Pikes Peak. What is the change in gravitational potential energy as the climber gains the last 100.0 m of altitude? Let $U = 0$ be **(a)** at sea level or **(b)** at the top of the peak.

Picture the Problem

Our sketch shows the mountain climber and the last 100.0 m of altitude to be climbed. We choose a typical coordinate system, with the positive y axis upward and the positive x axis to the right.

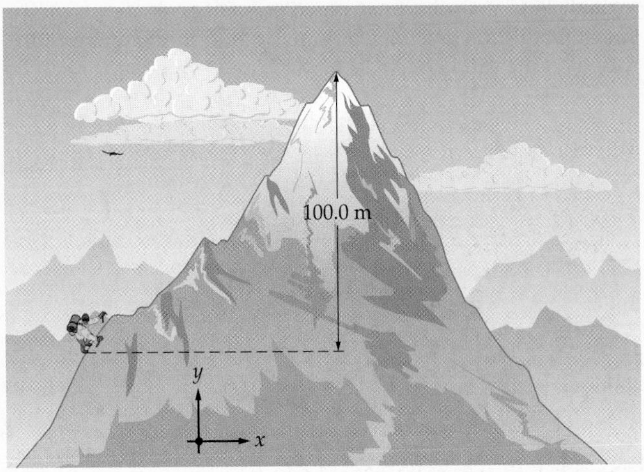

Strategy

The gravitational potential energy of the Earth-climber system depends only on the height y; the path followed in gaining the last 100.0 m of altitude is unimportant. The change in potential energy is $\Delta U = U_f - U_i = mgy_f - mgy_i$, where y_f is the altitude of the peak and y_i is 100.0 m less than y_f.

Solution

Part (a)

1. Calculate ΔU with $y_f = 4301$ m and $y_i = 4201$ m:

$$\Delta U = mgy_f - mgy_i$$
$$= (82.0 \text{ kg})(9.81 \text{ m/s}^2)(4301 \text{ m})$$
$$- (82.0 \text{ kg})(9.81 \text{ m/s}^2)(4201 \text{ m}) = 80{,}400 \text{ J}$$

Part (b)

2. Calculate ΔU with $y_f = 0$ and $y_i = -100.0$ m:

$$\Delta U = mgy_f - mgy_i$$
$$= (82.0 \text{ kg})(9.81 \text{ m/s}^2)(0)$$
$$- (82.0 \text{ kg})(9.81 \text{ m/s}^2)(-100.0 \text{ m}) = 80{,}400 \text{ J}$$

Insight

As expected, the *change* in gravitational potential energy does not depend on where we choose $U = 0$. Nor does it depend on the path taken between the initial and final points.

Practice Problem

Find the altitude of the climber for which the gravitational potential energy of the Earth-climber system is 1.00×10^5 J less than it is when the climber is at the summit. [**Answer:** 4180 m]

Some related homework problems: Problem 7, Problem 14

A single item of food can be converted into a surprisingly large amount of potential energy. This is shown for the case of a candy bar in Example 8–3.

EXAMPLE 8–3 Converting Food Energy to Mechanical Energy

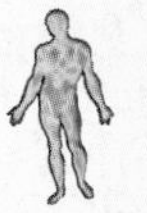

REAL-WORLD PHYSICS: BIO A candy bar called the Mountain Bar has a calorie content of 212 Cal = 212 kcal, which is equivalent to an energy of 8.87×10^5 J. If an 81.0-kg mountain climber eats a Mountain Bar and magically converts it all to potential energy, what gain of altitude would be possible?

Picture the Problem

We show the mountain climber eating the candy bar at a given level on the mountain, which we can take to be $y = 0$. The altitude gain, then, corresponds to $y = h$.

continued on next page

continued from previous page

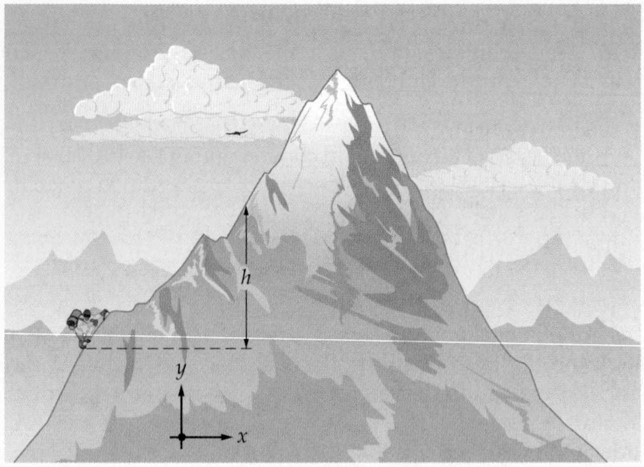

Strategy

The initial gravitational potential energy of the Earth-climber system is $U = 0$; the final potential energy is $U = mgh$. To find the altitude gain, set $U = mgh$ equal to the energy provided by the candy bar, 8.87×10^5 J, and solve for h.

Solution

1. Solve $U = mgh$ for h:

$$U = mgh$$
$$h = \frac{U}{mg}$$

2. Substitute numerical values, with $U = 8.87 \times 10^5$ J:

$$h = \frac{U}{mg} = \frac{8.87 \times 10^5 \text{ J}}{(81.0 \text{ kg})(9.81 \text{ m/s}^2)} = 1120 \text{ m}$$

Insight

This is more than two-thirds of a mile in elevation. Even if we take into account the fact that metabolic efficiency is only about 25%, the height would still be 280 m, or nearly two-tenths of a mile. It's remarkable just how much our bodies can do with so little.

Practice Problem

If the mass of the mountain climber is increased—by adding more items to the backpack, for example—does the possible elevation gain increase, decrease, or stay the same? Calculate the elevation gain for a climber with a mass of 91.0 kg. [**Answer:** The altitude gain will decrease. For $m = 91.0$ kg we find $h = 994$ m.]

Some related homework problems: Problem 7, Problem 14

▲ Because springs, and bungee cords, exert conservative forces, they can serve as energy storage devices. In this case, the stretched bungee cord is beginning to give up the energy it has stored, and to convert that potential energy into kinetic energy as the jumper is pulled rapidly skyward.

We have been careful *not* to say that the potential energy of the mountain climber—or any object—increases when its height increases. The reason is that the potential energy is a property of an entire system, not of its individual parts. The correct statement is that if an object is lifted, the potential energy of the Earth-object system is increased.

Springs

Consider a spring that is stretched from its equilibrium position a distance x. According to Equation 7–8, the work required to cause this stretch is $W = \frac{1}{2}kx^2$. Therefore, if the spring is released—and allowed to move from the stretched position back to the equilibrium position—it will do the same work, $\frac{1}{2}kx^2$. From our definition of potential energy, then, we see that

$$W_c = \tfrac{1}{2}kx^2 = U_i - U_f \qquad 8\text{–}4$$

Note that in this case U_f is the potential energy when the spring is at $x = 0$ (equilibrium position), and U_i is the potential energy when the spring is stretched by the amount x.

A convenient choice for $U = 0$ is the equilibrium position of the spring. With this choice we have $U_f = 0$, and Equation 8–4 becomes $U_i = \frac{1}{2}kx^2$. Omitting the subscript i, so that U represents the potential energy of the spring for an arbitrary amount of stretch x, we have

Potential Energy of a Spring

$$U = \tfrac{1}{2}kx^2 \qquad 8\text{–}5$$

Since U depends on x^2, which is positive even if x is negative, the potential energy of a spring is always greater than or equal to zero. Thus, a spring's potential energy increases whenever it is displaced from equilibrium.

EXERCISE 8–2

Find the potential energy of a spring with force constant $k = 680$ N/m if it is **(a)** stretched by 5.00 cm or **(b)** compressed by 7.00 cm.

Solution

Substituting $x = 0.0500$ m and $x = -0.0700$ m in Equation 8–5 yields

a. $U = \frac{1}{2}(680 \text{ N/m})(0.0500 \text{ m})^2 = 0.850 \text{ J}$
b. $U = \frac{1}{2}(680 \text{ N/m})(-0.0700 \text{ m})^2 = 1.67 \text{ J}$

Finally, comparing Equation 8–3 to Equation 8–5, we see that the potential energies for gravity and for a spring are given by different expressions. As mentioned, each conservative force has its own potential energy.

EXAMPLE 8–4 Compressed Energy and the Jump of a Flea

When a force of 120.0 N is applied to a certain spring, it causes a stretch of 2.25 cm. What is the potential energy of this spring when it is compressed by 3.50 cm?

Picture the Problem

The top sketch shows the spring stretched 2.25 cm by the force $F_1 = 120.0$ N. The lower sketch shows the same spring compressed by a second force, F_2, which causes a compression of 3.50 cm.

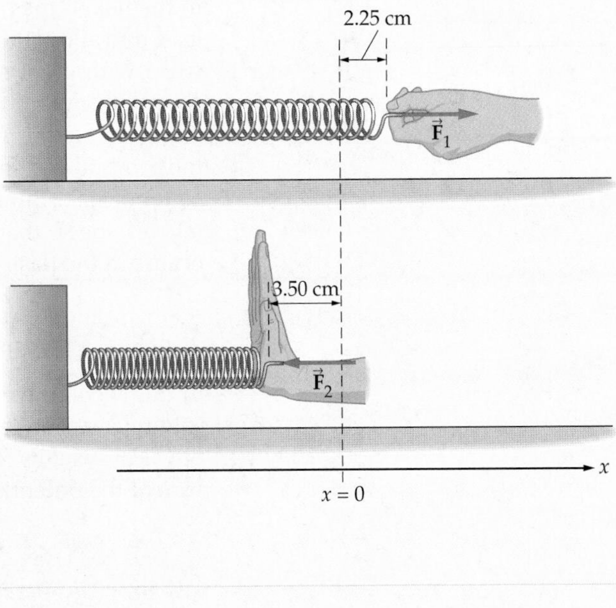

Strategy

From the first piece of information—a certain force causes a certain stretch—we can calculate the force constant using $F = kx$. Once we know k, we find the potential energy with $U = \frac{1}{2}kx^2$.

Solution

1. Solve $F = kx$ for the spring constant, k:

$$F = kx$$

$$k = \frac{F}{x} = \frac{120.0 \text{ N}}{0.0225 \text{ m}} = 5330 \text{ N/m}$$

2. Substitute $k = 5330$ N/m and $x = -0.0350$ m into the potential energy expression, $U = \frac{1}{2}kx^2$:

$$U = \tfrac{1}{2}kx^2 = \tfrac{1}{2}(5330 \text{ N/m})(-0.0350 \text{ m})^2 = 3.26 \text{ J}$$

continued on next page

continued from previous page

Insight

Though this example deals with ideal springs, the same basic physics applies to many other real-world situations. A case in point is the jump of a flea, in which a flea can propel itself up to 100 times its body length. The physics behind this feat is the slow accumulation of energy in a "springy" strip of resilin in the coxa of the leg, as shown in the accompanying sketches, and the sudden release of this energy at a later time. Specifically, as the flea's muscles flex the leg, the resilin strip in the coxa is stretched, storing the work done by the muscles in the form of potential energy, $U = \frac{1}{2}kx^2$. Later, when a trigger mechanism unlocks the flexed leg, the energy stored in the resilin is released explosively—rapidly extending the leg and propelling the flea upward. See Problem 77 for a calculation using the force constant of resilin.

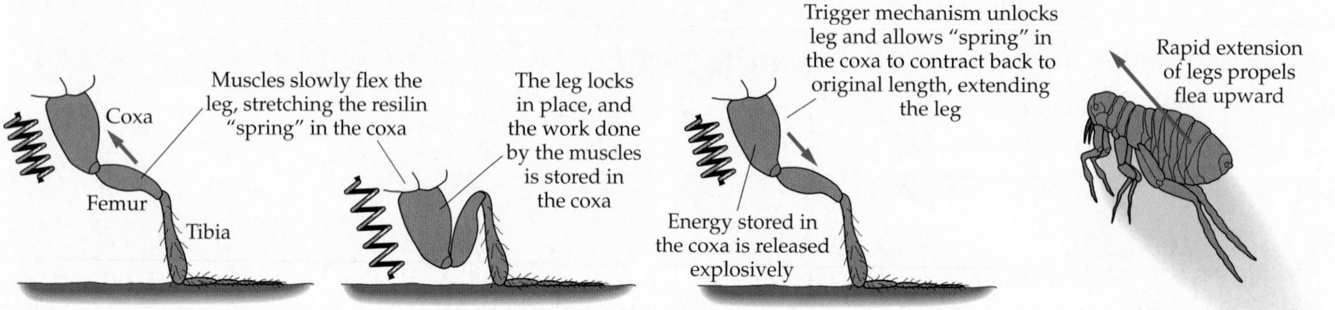

Practice Problem

What stretch is necessary for the spring in this Example to have a potential energy of 5.00 J? [**Answer:** 4.33 cm]

Some related homework problems: Problem 9, Problem 13

The jump of a flea is similar in many respects to the operation of a bow and arrow. In the latter case, the work done in slowly pulling the string back is stored in the flex of the bow. The string is held in place while aim is taken, and then released to allow the bow to return to its original shape. This propels the arrow forward with great speed—a speed many times faster than could be obtained by simply throwing the arrow with the same arm muscles that pulled back on the string. In fact, if the string returns to its original position in 1/1000th the time it took to pull the string back, the power it delivers to the arrow is magnified by a factor of 1000. Similarly, the spring-loaded jump of the flea gives it a much greater takeoff speed than if it relied solely on muscle power. An analogous process occurs in the flash unit of a camera, as we shall see in Chapter 20.

8–3 Conservation of Mechanical Energy

In this section, we show how potential energy can be used as a powerful tool in solving a variety of problems and in gaining greater insight into the workings of physical systems. To do so, we begin by defining the **mechanical energy,** E, as the sum of the potential and kinetic energies of an object:

$$E = U + K \qquad \textbf{8–6}$$

The significance of mechanical energy is that it is **conserved** in systems involving only conservative forces. By conserved, we mean that its value never changes; that is, E = constant. (In situations where nonconservative forces are involved, the mechanical energy can change, as when friction causes warming by converting mechanical energy to thermal energy. When *all* possible forms of energy are considered, energy is always found to be conserved.)

To show that E is conserved for conservative forces, we start with the work-energy theorem from Chapter 7:

$$W_{\text{total}} = \Delta K = K_{\text{f}} - K_{\text{i}}$$

Suppose for a moment that the system has only a single force and that the force is conservative. If this is the case, then the total work, W_{total}, is the work done by the conservative force, W_c:

$$W_{\text{total}} = W_c$$

From the definition of potential energy, we know that $W_c = -\Delta U = U_i - U_f$. Combining these results, we have

$$W_{\text{total}} = W_c$$
$$K_f - K_i = U_i - U_f$$

With a slight rearrangement we find

$$U_f + K_f = U_i + K_i$$

or

$$E_f = E_i$$

Since the initial and final points can be chosen arbitrarily, it follows that E is conserved:

$$E = \text{constant}$$

If the system has more than one conservative force, the only change to these results is to replace U with the sum of potential energies of all the forces.

To summarize:

Conservation of Mechanical Energy

In systems with conservative forces only, the mechanical energy E is conserved; that is, $E = U + K = \text{constant}$.

In terms of physical systems, conservation of mechanical energy means that energy can be converted between potential and kinetic forms, but that the sum remains the same. As an example, in the roller coaster shown in Figure 8–5, the gravitational potential energy decreases as the car approaches point B; as it does, the car's kinetic energy increases by the same amount. From a practical point of view, conservation of mechanical energy means that many physics problems can be solved by what amounts to simple bookkeeping.

For example, consider a key chain of mass m that is dropped to the floor from a height h, as illustrated in **Figure 8–8**. The question is, how fast are the keys moving just before they land? We know how to solve this problem using Newton's laws and kinematics, but now let's see how energy conservation can be used instead.

▲ A roller coaster (top) illustrates the conservation of mechanical energy. With every descent, gravitational potential energy is converted into kinetic energy; with every rise, kinetic energy is converted back into gravitational potential energy. If friction is neglected, the total mechanical energy of the car remains constant. The same principle is exploited at a pumped-storage facility, such as this one at the Mormon Flat Dam in Phoenix, Arizona (bottom). When surplus electrical power is available, it is used to pump water uphill into the reservoir. This process, in effect, stores electrical energy as gravitational potential energy. When power demand is high, the stored water is allowed to flow back downhill through the electrical generators in the dam, converting the gravitational energy to kinetic energy and the kinetic energy to electrical energy.

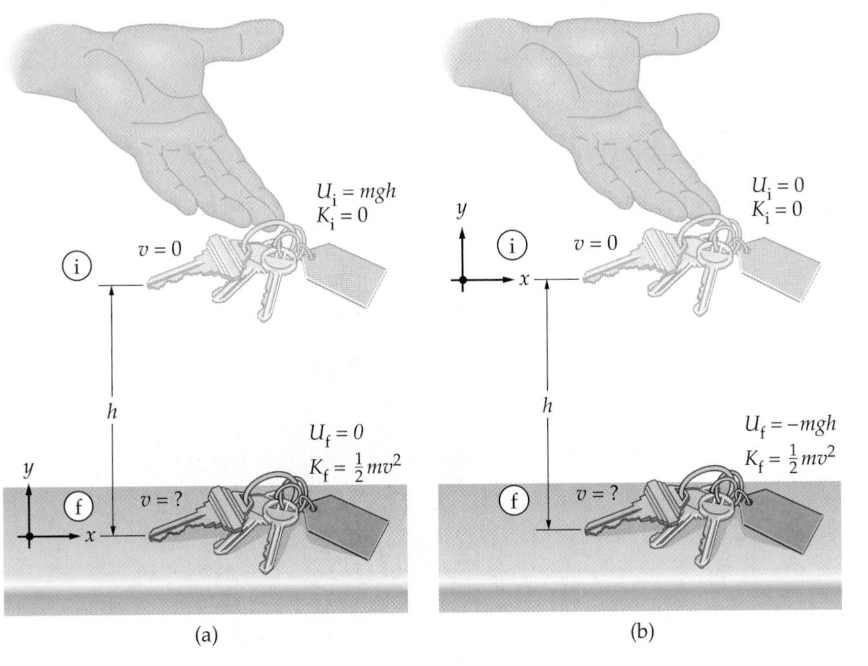

◀ **FIGURE 8–8 Solving a kinematics problem using conservation of energy**

(a) A set of keys falls to the floor. Ignoring frictional forces, we know that the mechanical energy at points i and f must be equal; $E_i = E_f$. Using this condition, we can find the speed of the keys just before they land. **(b)** The same physical situation as in part (a), except this time we have chosen $y = 0$ to be at the point where the keys are dropped. As before, we set $E_i = E_f$ to find the speed of the keys just before they land. The result is the same.

PROBLEM-SOLVING NOTE

Conservative Systems

A convenient approach to problems involving energy conservation is to first sketch the system, and then label the initial and final points with i and f, respectively. To apply energy conservation, write out the energy at these two points and set $E_i = E_f$.

First, note that the only force acting on the keys is gravity—ignoring air resistance, of course—and that gravity is a conservative force. As a result, we can say that $E = U + K$ is constant during the entire time the keys are falling. To solve the problem, then, we pick two points on the motion of the keys, say i and f in Figure 8–8, and we set the mechanical energy equal at these points:

$$E_i = E_f \qquad \textbf{8–7}$$

Writing this out in terms of potential and kinetic energies, we have

$$U_i + K_i = U_f + K_f \qquad \textbf{8–8}$$

This one equation—which is nothing but bookkeeping—can be used to solve for the one unknown, the final speed.

To be specific, in Figure 8–8 (a) we choose $y = 0$ at ground level, which means that $U_i = mgh$. In addition, the fact that the keys are released from rest means that $K_i = 0$. Similarly, at point f—just before hitting the ground—the energy is all kinetic, and the potential energy is zero; that is, $U_f = 0$, $K_f = \frac{1}{2}mv^2$. Substituting these values into Equation 8–8, we find

$$mgh + 0 = 0 + \tfrac{1}{2}mv^2$$

Canceling m and solving for v yields the same result we get with kinematics:

$$v = \sqrt{2gh}$$

Suppose, instead, that we had chosen $y = 0$ to be at the release point of the keys, as in Figure 8–8 (b), so that the keys land at $y = -h$. Now, when the keys are released, we have $U_i = 0$ and $K_i = 0$, and when they land $U_f = -mgh$ and $K_f = \frac{1}{2}mv^2$. Substituting these results in $U_i + K_i = U_f + K_f$ yields

$$0 + 0 = -mgh + \tfrac{1}{2}mv^2$$

Solving for v gives the same result:

$$v = \sqrt{2gh}$$

Thus, as expected, changing the zero level has no effect on the physical results.

EXAMPLE 8–5 Graduation Fling

At the end of a graduation ceremony, graduates fling their caps into the air. Suppose a 0.120-kg cap is thrown straight upward with an initial speed of 7.85 m/s, and that frictional forces can be ignored. **(a)** Use kinematics to find the speed of the cap when it is 1.18 m above the release point. **(b)** Show that the mechanical energy at the release point is the same as the mechanical energy 1.18 m above the release point.

Picture the Problem

In our sketch we choose $y = 0$ to be at the level where the cap is released with an initial speed of 7.85 m/s. In addition, note that we designate the release point as i (initial) and the point at which $y = 1.18$ m as f (final). It is the speed at point f that we wish to find.

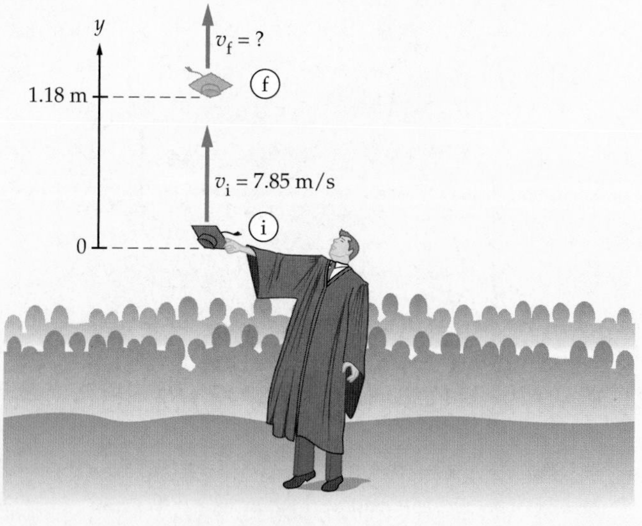

Strategy

a. The cap is in free fall, which justifies the use of constant-acceleration kinematics. Since we want to relate velocity to position, we use $v_y^2 = v_{0y}^2 + 2a_y\Delta y$ (Section 2–5). In this case, $v_{0y} = 7.85$ m/s, $\Delta y = 1.18$ m, and $a_y = -g$. Substituting these values gives v_y.

b. At each point we simply calculate $E = U + K$, with $U = mgy$ and $K = \frac{1}{2}mv^2$.

Solution

Part (a)

1. Use kinematics to solve for v_y:

$$v_y^2 = v_{0y}^2 + 2a_y \Delta y$$
$$v_y = \pm\sqrt{v_{0y}^2 + 2a_y \Delta y}$$

2. Substitute $v_{0y} = 7.85\ \text{m/s}$, $\Delta y = 1.18\ \text{m}$, and $a_y = -g$ to find v_y. Choose the plus sign, since we are interested only in the speed:

$$v_y = \sqrt{v_{0y}^2 + 2a_y \Delta y}$$
$$= \sqrt{(7.85\ \text{m/s})^2 + 2(-9.81\ \text{m/s}^2)(1.18\ \text{m})} = 6.20\ \text{m/s}$$

Part (b)

3. Calculate E_i. At this point $y_i = 0$ and $v_i = 7.85\ \text{m/s}$:

$$E_i = U_i + K_i = mgy_i + \tfrac{1}{2}mv_i^2$$
$$= 0 + \tfrac{1}{2}(0.120\ \text{kg})(7.85\ \text{m/s})^2 = 3.70\ \text{J}$$

4. Calculate E_f. At this point $y_f = 1.18\ \text{m}$ and $v_f = 6.20\ \text{m/s}$:

$$E_f = U_f + K_f = mgy_f + \tfrac{1}{2}mv_f^2$$
$$= (0.120\ \text{kg})(9.81\ \text{m/s}^2)(1.18\ \text{m}) + \tfrac{1}{2}(0.120\ \text{kg})(6.20\ \text{m/s})^2$$
$$= 1.39\ \text{J} + 2.31\ \text{J} = 3.70\ \text{J}$$

Insight

As expected, E_f is equal to E_i. In the remaining Examples in this section we turn this process around; we start with $E_f = E_i$, and use this relation to find a final speed or a final height. As we shall see, this procedure of using energy conservation is a more powerful approach—it actually makes the calculations simpler.

Practice Problem

Use energy conservation to find the height at which the speed of the cap is 5.00 m/s. [**Answer:** 1.87 m]

Some related homework problems: Problem 20, Problem 21, and Problem 23.

An interesting extension of this Example is shown in **Figure 8–9**. In this case, we are given that the speed of the cap is v_i at the height y_i, and we would like to know its speed v_f when it is at the height y_f.

To find v_f, we apply energy conservation to the points i and f:

$$U_i + K_i = U_f + K_f$$

Writing out U and K specifically for these two points yields the following:

$$mgy_i + \tfrac{1}{2}mv_i^2 = mgy_f + \tfrac{1}{2}mv_f^2$$

As before, we cancel m and solve for the unknown speed, v_f:

$$v_f = \sqrt{v_i^2 + 2g(y_i - y_f)}$$

This result is in agreement with the kinematic equation, $v_y^2 = v_{0y}^2 + 2a_y \Delta y$.

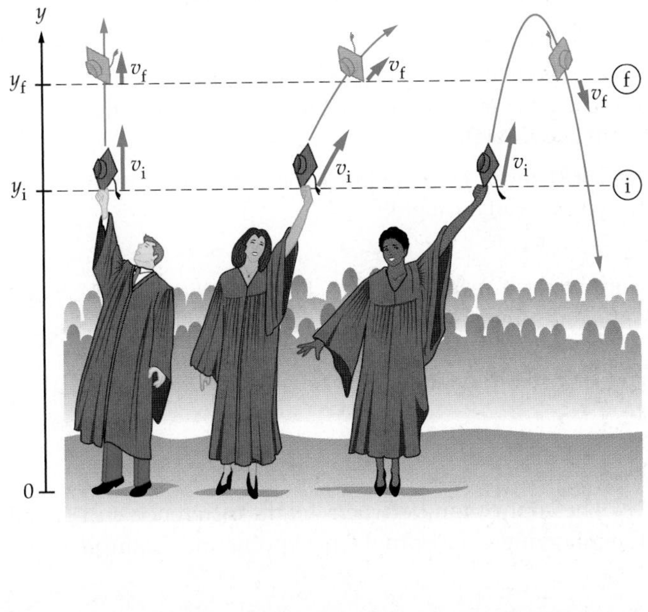

◀ **FIGURE 8–9 Speed is independent of path**

If the speed of the cap is v_i at the height y_i, its speed is v_f at the height y_f, independent of the path between the two heights. This assumes, of course, that frictional forces can be neglected.

Note that v_f depends only on y_i and y_f, not on the path connecting them. This is because conservative forces such as gravity do work that is path-independent. What this means physically is that the cap has the same speed v_f at the height y_f, whether it goes straight upward or follows some other trajectory, as in Figure 8–9. All that matters is the height difference.

EXAMPLE 8–6 Catching a Home Run

In the bottom of the ninth inning, a player hits a 0.15-kg baseball over the outfield fence. The ball leaves the bat with a speed of 36 m/s, and a fan in the bleachers catches it 7.2 m above the point where it was hit. Assuming frictional forces can be ignored, find **(a)** the kinetic energy of the ball when it is caught and **(b)** its speed when caught.

Picture the Problem

Our sketch shows the ball's trajectory. We label the hit point i and the catch point f. At point i we choose $y_i = 0$; at point f, then, $y_f = h = 7.2$ m. In addition, we are given that $v_i = 36$ m/s; v_f is to be determined.

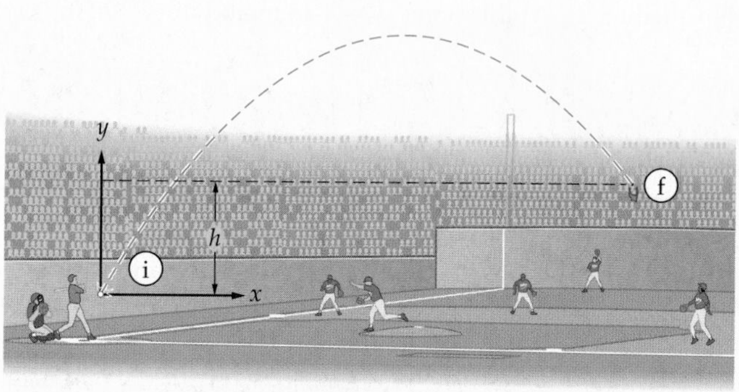

Strategy

a. Because frictional forces can be ignored, it follows that the initial mechanical energy is equal to the final mechanical energy; that is, $U_i + K_i = U_f + K_f$. Use this relation to find K_f.

b. Once K_f is determined, use $K_f = \frac{1}{2}mv_f^2$ to find v_f.

Solution

Part (a)

1. Begin by writing U and K for point i:

$$U_i = 0$$
$$K_i = \tfrac{1}{2}mv_i^2 = \tfrac{1}{2}(0.15\text{ kg})(36\text{ m/s})^2 = 97\text{ J}$$

2. Next, write U and K for point f:

$$U_f = mgh = (0.15\text{ kg})(9.81\text{ m/s}^2)(7.2\text{ m}) = 11\text{ J}$$
$$K_f = \tfrac{1}{2}mv_f^2$$

3. Set the total mechanical energy at point i, $E_i = U_i + K_i$, equal to the total mechanical energy at point f, $E_f = U_f + K_f$, and solve for K_f:

$$U_i + K_i = U_f + K_f$$
$$0 + 97\text{ J} = 11\text{ J} + K_f$$
$$K_f = 97\text{ J} - 11\text{ J} = 86\text{ J}$$

Part (b)

4. Use $K_f = \frac{1}{2}mv_f^2$ to find v_f:

$$K_f = \tfrac{1}{2}mv_f^2$$
$$v_f = \sqrt{\frac{2K_f}{m}} = \sqrt{\frac{2(86\text{ J})}{0.15\text{ kg}}} = 34\text{ m/s}$$

Insight

To find the ball's speed when it was caught, we need to know the height of point f, but we don't need to know any details about the ball's trajectory. For example, it is not necessary to know the angle at which the ball leaves the bat or its maximum height.

The histograms to the right show the values of U and K at the points i and f. Notice that the energy of the system is mostly kinetic at the time the ball is caught.

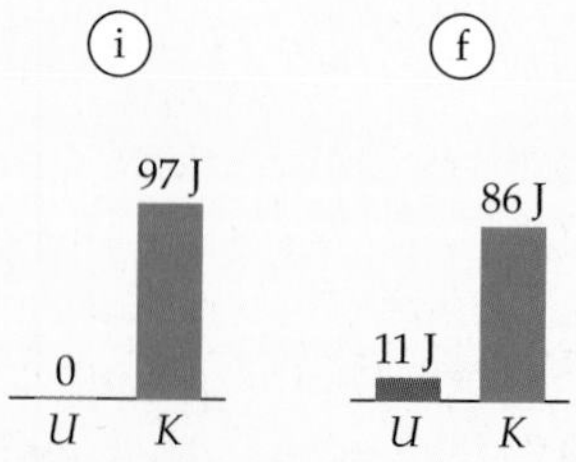

Practice Problem

If the mass of the ball were increased, would the catch speed be greater than, less than, or the same as the value we just found? [**Answer:** The same. U and K depend on mass in the same way, hence the mass cancels.]

Some related homework problems: Problem 19, Problem 20

The connection between height difference and speeds is explored further in the following Conceptual Checkpoint and Example.

CONCEPTUAL CHECKPOINT 8–1 Compare the Final Speeds

Swimmers at a water park can enter a pool using one of two frictionless slides of equal height. Slide 1 approaches the water with a uniform slope; slide 2 dips rapidly at first, then levels out. Is the speed v_2 at the bottom of slide 2 **(a)** greater than, **(b)** less than, or **(c)** the same as the speed v_1 at the bottom of slide 1?

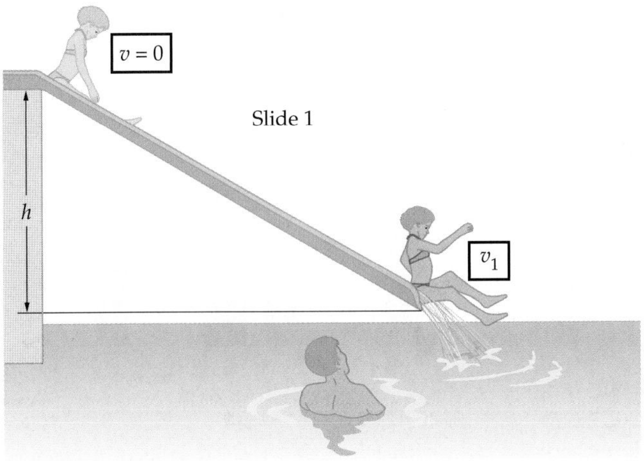

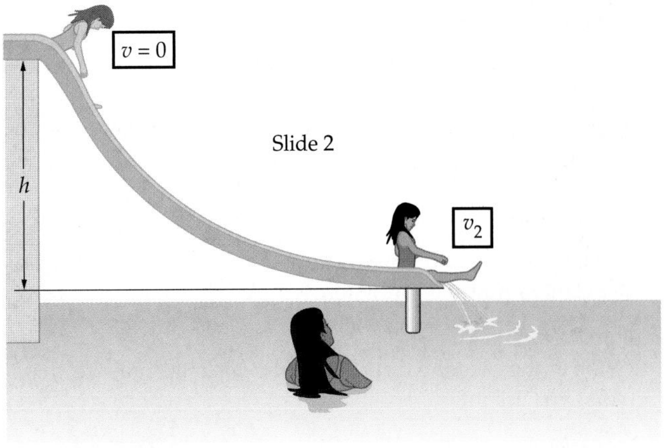

Reasoning and Discussion

In both cases, the same amount of potential energy, *mgh*, is converted to kinetic energy. Since the conversion of gravitational potential energy to kinetic energy is the *only* energy transaction taking place, it follows that the speed is the same for each slide.

Interestingly, although the final speeds are the same, the time required to reach the water is less for slide 2. The reason is that swimmer 2 reaches a high speed early and maintains it, whereas the speed of swimmer 1 increases slowly and steadily.

Answer:

(c) The speeds are the same.

EXAMPLE 8–7 Skateboard Exit Ramp

A 55-kg skateboarder enters a ramp moving horizontally with a speed of 6.5 m/s and leaves the ramp moving vertically with a speed of 4.1 m/s. Find the height of the ramp, assuming no energy loss to frictional forces.

Picture the Problem

We choose $y = 0$ to be the level of the bottom of the ramp, thus the gravitational potential energy is zero there. Point i indicates the skateboarder entering the ramp with a speed of 6.5 m/s; point f is the top of the ramp, where the speed is 4.1 m/s.

Strategy

To find h, simply set the initial energy, $E_i = U_i + K_i$, equal to the final energy, $E_f = U_f + K_f$.

Solution

1. Write expressions for U_i and K_i: $U_i = mg \cdot 0 = 0 \quad K_i = \frac{1}{2}mv_i^2$
2. Write expressions for U_f and K_f: $U_f = mgh \quad K_f = \frac{1}{2}mv_f^2$
3. Set the total mechanical energy at point i, $E_i = U_i + K_i$, equal to the total mechanical energy at point f, $E_f = U_f + K_f$: $U_i + K_i = U_f + K_f$
   $0 + \frac{1}{2}mv_i^2 = mgh + \frac{1}{2}mv_f^2$

continued on next page

continued from previous page

4. Solve for h. Note that m cancels:

$$mgh = \frac{1}{2}mv_i^2 - \frac{1}{2}mv_f^2$$

$$h = \frac{v_i^2 - v_f^2}{2g}$$

5. Substitute numerical values:

$$h = \frac{(6.5\ \mathrm{m/s})^2 - (4.1\ \mathrm{m/s})^2}{2(9.81\ \mathrm{m/s^2})} = 1.3\ \mathrm{m}$$

Insight

Note that our value for h is independent of the shape of the ramp—it is equally valid for one with the shape shown here, or one that simply inclines upward at a constant angle. In addition, the height does not depend on the person's mass, as we see in step 4.

The histograms to the right show U and K to scale at the points i and f, as well as at the maximum height where $K = 0$.

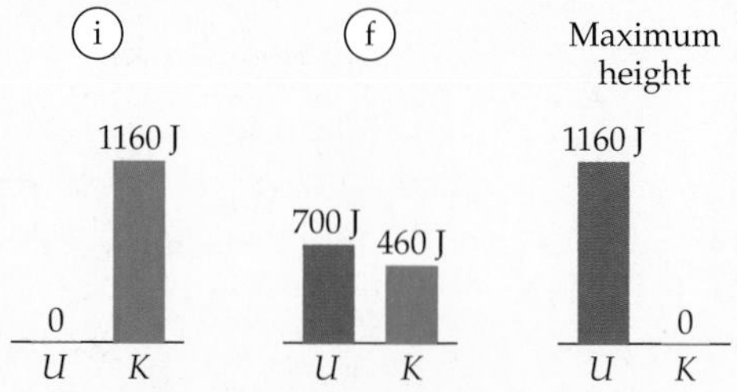

Practice Problem

What is the skateboarder's maximum height above the bottom of the ramp? [**Answer:** 2.2 m]

Some related homework problems: Problem 19, Problem 23

▲ Does the shape of the slide matter? (See Conceptual Checkpoint 8–1.)

It is interesting to express the equation in line 3 from Example 8–7 in words. First, the left side of the equation is the initial kinetic energy of the skateboarder, $\frac{1}{2}mv_i^2$. This is the initial energy content of the system. At point f the system still has the same amount of energy, only now part of it, mgh, is in the form of gravitational potential energy. The remainder is the final kinetic energy, $\frac{1}{2}mv_f^2$.

Conceptual Checkpoint 8–2 considers the effect of a slight change in the initial speed of an object.

CONCEPTUAL CHECKPOINT 8–2 What is the Final Speed?

A snowboarder coasts on a smooth track that rises from one level to another. If the snowboarder's initial speed is 4 m/s, the snowboarder just makes it to the upper level and comes to rest. With a slightly greater initial speed of 5 m/s, the snowboarder is still moving to the right on the upper level. Is the snowboarder's final speed in this case **(a)** 1 m/s, **(b)** 2 m/s, or **(c)** 3 m/s?

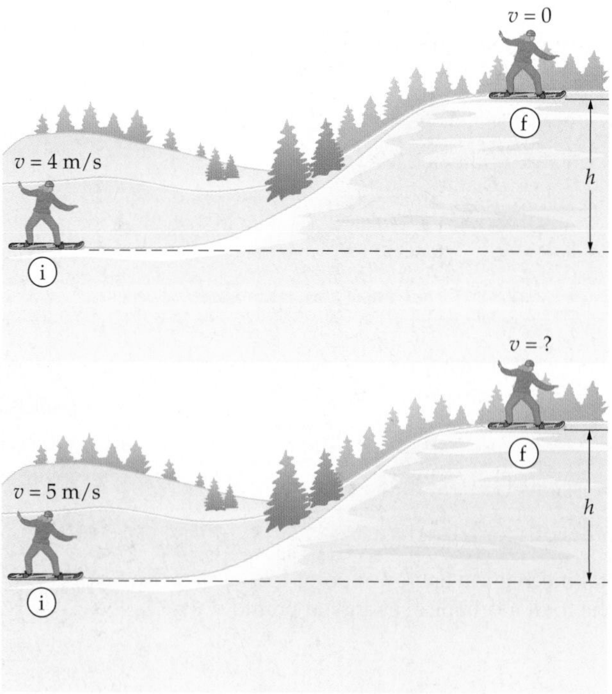

Reasoning and Discussion

A plausible-sounding answer is that since the initial speed is greater by 1 m/s in the second case, the final speed should be greater by 1 m/s as well. Therefore, the answer should be 0 + 1 m/s = 1 m/s. This is incorrect, however.

As surprising as it may seem, an increase in the initial speed from 4 m/s to 5 m/s results in an increase in the final speed from 0 to 3 m/s. This is due to the fact that kinetic energy depends on v^2 rather than v; thus, it is the difference in v^2 that counts. In this case, the initial value of v^2 increases from 16 m^2/s^2 to 25 m^2/s^2, for a total increase of 25 m^2/s^2 – 16 m^2/s^2 = 9 m^2/s^2. The final value of v^2 must increase by the same amount, 9 $m^2/s^2 = (3\text{ m/s})^2$. As a result, the final speed is 3 m/s.

Answer:

(c) The final speed of the snowboarder in the second case is 3 m/s.

Let's check the results of the previous Conceptual Checkpoint with a specific numerical example. Suppose the snowboarder has a mass of 74.0 kg. It follows that in the first case the initial kinetic energy is $K_i = \frac{1}{2}(74.0\text{ kg})(4.00\text{ m/s})^2 = 592\text{ J}$. At the top of the hill all of this kinetic energy is converted to gravitational potential energy, mgh.

In the second case, the initial speed of the snowboarder is 5.00 m/s; thus, the initial kinetic energy is $K_i = \frac{1}{2}(74.0\text{ kg})(5.00\text{ m/s})^2 = 925\text{ J}$. When the snowboarder reaches the top of the hill, 592 J of this kinetic energy is converted to gravitational potential energy, leaving the snowboarder with a final kinetic energy of 925 J – 592 J = 333 J. The corresponding speed is given by

$$\frac{1}{2}mv^2 = 333\text{ J}$$

$$v = \sqrt{\frac{2(333\text{ J})}{m}} = \sqrt{\frac{2(333\text{ J})}{74.0\text{ kg}}} = \sqrt{9.00\text{ m}^2/\text{s}^2} = 3.00\text{ m/s}$$

Thus, as expected, the snowboarder in the second case has a final speed of 3.00 m/s.

We conclude this section with two Examples involving springs.

EXAMPLE 8–8 Spring Time

A 1.70-kg block slides on a horizontal, frictionless surface until it encounters a spring with a force constant of 955 N/m. The block comes to rest after compressing the spring a distance of 4.60 cm. Find the initial speed of the block. (Ignore air resistance and any energy lost when the block initially contacts the spring.)

Picture the Problem

Point i refers to times before the block makes contact with the spring, which means the block has a speed v and the end of the spring is at $x = 0$. Point f refers to the time when the block has come to rest, and the spring is compressed to $x = -d = -4.60$ cm.

We can choose the center of the block to be the $y = 0$ level. With this choice, the gravitational potential energy of the system is zero at all times.

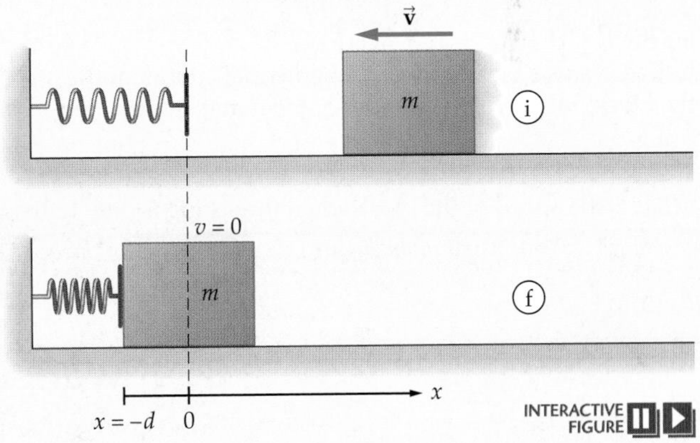

Strategy

Set E_i equal to E_f to find the one unknown, v. Note that the initial energy, E_i, is the kinetic energy of the block before it reaches the spring. The final energy, E_f, is the potential energy of the compressed spring.

Solution

1. Write expressions for U_i and K_i. For U, we consider only the potential energy of the spring, $U = \frac{1}{2}kx^2$:
$$U_i = \tfrac{1}{2}k \cdot 0^2 = 0 \quad K_i = \tfrac{1}{2}mv^2$$
2. Do the same for U_f and K_f:
$$U_f = \tfrac{1}{2}k(-d)^2 = \tfrac{1}{2}kd^2 \quad K_f = \tfrac{1}{2}m \cdot 0^2 = 0$$
3. Set the initial mechanical energy, $E_i = U_i + K_i$, equal to the final mechanical energy, $E_f = U_f + K_f$, and solve for v:
$$U_i + K_i = U_f + K_f$$
$$0 + \tfrac{1}{2}mv^2 = \tfrac{1}{2}kd^2 + 0$$
$$v = d\sqrt{\frac{k}{m}}$$
4. Substitute numerical values:
$$v = d\sqrt{\frac{k}{m}} = (0.0460\text{ m})\sqrt{\frac{955\text{ N/m}}{1.70\text{ kg}}} = 1.09\text{ m/s}$$

continued on next page

continued from previous page

Insight

After the block comes to rest, the spring expands again, converting its potential energy back into the kinetic energy of the block. When the block leaves the spring, moving to the right, its speed is once again 1.09 m/s.

Practice Problem

What is the compression distance, d, if the block's initial speed is 0.500 m/s? [**Answer:** 2.11 cm]

Some related homework problems: Problem 22, Problem 24

ACTIVE EXAMPLE 8–1 Find the Speed of the Block

Suppose the spring and block in Example 8–8 are oriented vertically, as shown here. Initially, the spring is compressed 4.60 cm and the block is at rest. When the block is released, it accelerates upward. Find the speed of the block when the spring has returned to its equilibrium position.

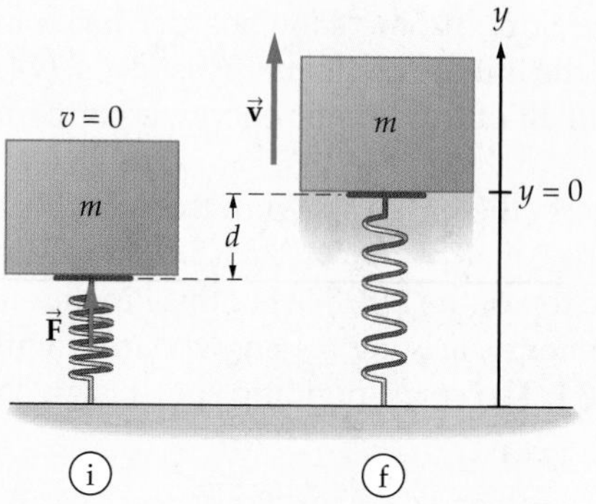

Solution *(Test your understanding by performing the calculations indicated in each step.)*

1. Write an expression for the initial mechanical energy E_i: $\quad E_i = U_i + K_i = -mgd + \frac{1}{2}kd^2 + 0$
2. Write an expression for the final mechanical energy E_f: $\quad E_f = U_f + K_f = 0 + 0 + \frac{1}{2}mv^2$
3. Set E_i equal to E_f and solve for v: $\quad -mgd + \frac{1}{2}kd^2 = \frac{1}{2}mv^2$
   $v = \sqrt{kd^2/m - 2gd}$
4. Substitute numerical values: $\quad v = 0.535 \text{ m/s}$

Insight

In this system, part of the initial potential energy of the spring ($\frac{1}{2}kd^2$) goes into increasing the gravitational potential energy of the block (mgd). The remainder of the initial energy, $\frac{1}{2}kd^2 - mgd$, is converted into the block's kinetic energy.

Your Turn

What is the speed of the block when the spring is only halfway back to its equilibrium position?

(Answers to **Your Turn** *problems are given in the back of the book.)*

8–4 Work Done by Nonconservative Forces

Nonconservative forces change the amount of mechanical energy in a system. They might decrease the mechanical energy by converting it to thermal energy, or increase it by converting muscular work to kinetic or potential energy. In some systems, both types of processes occur at the same time.

To see the connection between the work done by a nonconservative force, W_{nc}, and the mechanical energy, E, we return once more to the work-energy theorem, which says that the *total* work is equal to the change in kinetic energy:

$$W_{total} = \Delta K$$

Suppose, for instance, that a system has one conservative and one nonconservative force. In this case, the total work is the sum of the conservative work W_c and the nonconservative work W_{nc}:

$$W_{total} = W_c + W_{nc}$$

Recalling that conservative work is related to the change in potential energy by the definition given in Equation 8–1, $W_c = -\Delta U$, we have

$$W_{\text{total}} = -\Delta U + W_{\text{nc}} = \Delta K$$

Solving this relation for the nonconservative work yields

$$W_{\text{nc}} = \Delta U + \Delta K$$

Finally, since the total mechanical energy is $E = U + K$, it follows that the *change* in mechanical energy is $\Delta E = \Delta U + \Delta K$. As a result, the nonconservative work is simply the change in mechanical energy:

$$W_{\text{nc}} = \Delta E = E_f - E_i \qquad \textbf{8–9}$$

If more than one nonconservative force acts, we simply add the nonconservative work done by each such force to obtain W_{nc}.

At this point it may be useful to collect the three "working relationships" that have been introduced in the last two chapters:

$$W_{\text{total}} = \Delta K$$
$$W_c = -\Delta U$$
$$W_{\text{nc}} = \Delta E \qquad \textbf{8–10}$$

Note that positive nonconservative work increases the total mechanical energy of a system, while negative nonconservative work decreases the mechanical energy—and converts it to other forms. In the next Example, for instance, part of the initial mechanical energy of a leaf is converted to heat and other forms of energy by air resistance as it falls to the ground.

PROBLEM-SOLVING NOTE

Nonconservative Systems

Start by sketching the system and labeling the initial and final points with i and f, respectively. The initial and final mechanical energies are related to the nonconservative work by $W_{\text{nc}} = E_f - E_i$.

EXAMPLE 8–9 A Leaf Falls in the Forest: Find the Nonconservative Work

Deep in the forest, a 17.0-g leaf falls from a tree and drops straight to the ground. If its initial height was 5.30 m and its speed on landing was 1.3 m/s, how much nonconservative work was done on the leaf?

Picture the Problem

The leaf drops from rest at a height $y = h = 5.30$ m and lands with a speed $v = 1.3$ m/s at $y = 0$. These two points are labeled i and f, respectively.

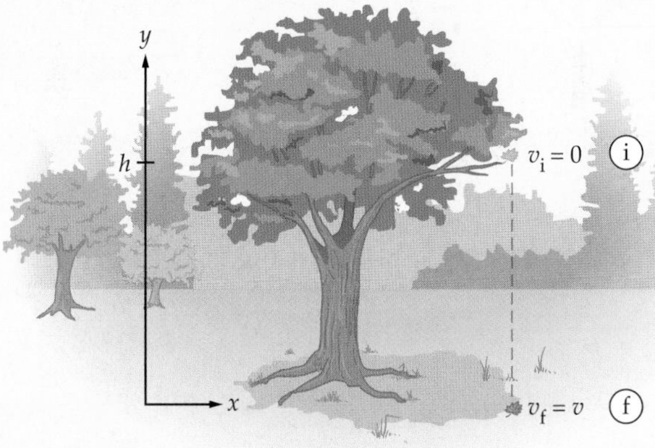

Strategy

To begin, calculate the initial mechanical energy, E_i, and the final mechanical energy, E_f. Once these energies have been determined, the nonconservative work is $W_{\text{nc}} = \Delta E = E_f - E_i$.

Solution

1. Evaluate U_i, K_i, and E_i:

$$U_i = mgh = (0.0170 \text{ kg})(9.81 \text{ m/s}^2)(5.30 \text{ m}) = 0.884 \text{ J}$$
$$K_i = \tfrac{1}{2}m \cdot 0^2 = 0$$
$$E_i = U_i + K_i = 0.884 \text{ J}$$

2. Next, evaluate U_f, K_f, and E_f.

$$U_f = mg \cdot 0 = 0$$
$$K_f = \tfrac{1}{2}mv^2 = \tfrac{1}{2}(0.0170 \text{ kg})(1.3 \text{ m/s})^2 = 0.014 \text{ J}$$
$$E_f = U_f + K_f = 0.014 \text{ J}$$

3. Use $W_{\text{nc}} = \Delta E$ to find the nonconservative work:

$$W_{\text{nc}} = \Delta E = E_f - E_i = 0.014 \text{ J} - 0.884 \text{ J} = -0.87 \text{ J}$$

continued on next page

continued from previous page

Insight

Note that most of the initial mechanical energy is dissipated as the leaf falls. This is indicated in the histograms to the right. The small amount that remains (only about 1.6%) appears as the kinetic energy of the leaf just before it lands. If a cherry had fallen from the tree, it would have struck the ground with a considerably greater speed—perhaps five times the speed of the leaf. In that case, the percentage of the initial potential energy remaining as kinetic energy would have been $5^2 = 25$ times greater than the percentage retained by the leaf.

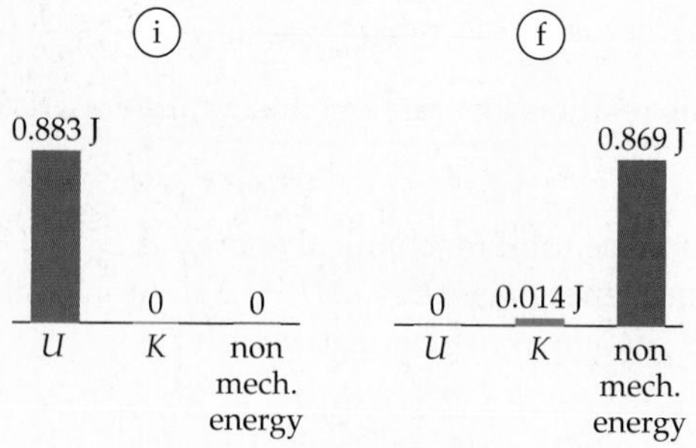

Practice Problem

What was the average nonconservative force exerted on the leaf as it fell? [**Answer:** $W_{nc} = -Fh$, $F = -W_{nc}/h = 0.16$ N, upward]

Some related homework problems: Problem 30, Problem 31

In the following Active Example, we use a knowledge of the nonconservative work to find the depth at which a diver comes to rest.

ACTIVE EXAMPLE 8–2 Find the Diver's Depth

A 95.0-kg diver steps off a diving board and drops into the water 3.00 m below. At some depth d below the water's surface, the diver comes to rest. If the nonconservative work done on the diver is $W_{nc} = -5120$ J, what is the depth, d?

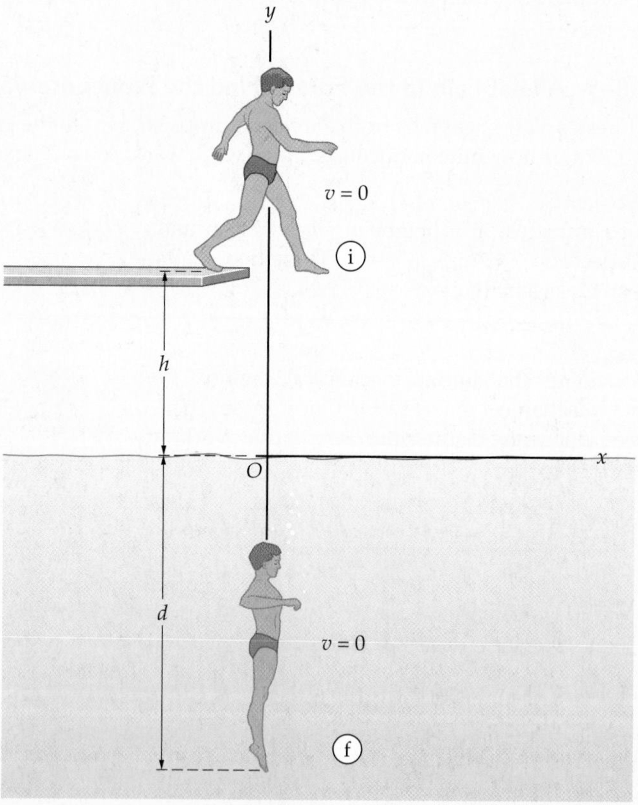

Solution *(Test your understanding by performing the calculations indicated in each step.)*

1. Write the initial mechanical energy, E_i: $E_i = mgh + 0 = mgh$
2. Write the final mechanical energy, E_f: $E_f = mg(-d) + 0 = -mgd$
3. Set W_{nc} equal to ΔE: $W_{nc} = \Delta E = E_f - E_i = -mgd - mgh$

4. Solve for d: $d = -(W_{nc} + mgh)/mg$

5. Substitute numerical values: $d = 2.49$ m

Insight

Another way to write step 3 is $E_f = E_i + W_{nc}$. In words, this equation says that the final mechanical energy is the initial mechanical energy plus the nonconservative work done on the system. In this case, $W_{nc} < 0$; hence the final mechanical energy is less than the initial mechanical energy.

Your Turn

Suppose the diver descends to a depth of 3.50 m. How much nonconservative work is done in this case?

(Answers to **Your Turn** *problems are given in the back of the book.)*

We now present a Conceptual Checkpoint that further examines the relationship between nonconservative work and distance.

CONCEPTUAL CHECKPOINT 8–3 Judging a Putt

A golfer badly misjudges a putt, sending the ball only one-quarter of the distance to the hole. The original putt gave the ball an initial speed of v_0. If the force of resistance due to the grass is constant, would an initial speed of (a) $2v_0$, (b) $3v_0$, or (c) $4v_0$ be needed to get the ball to the hole from its original position?

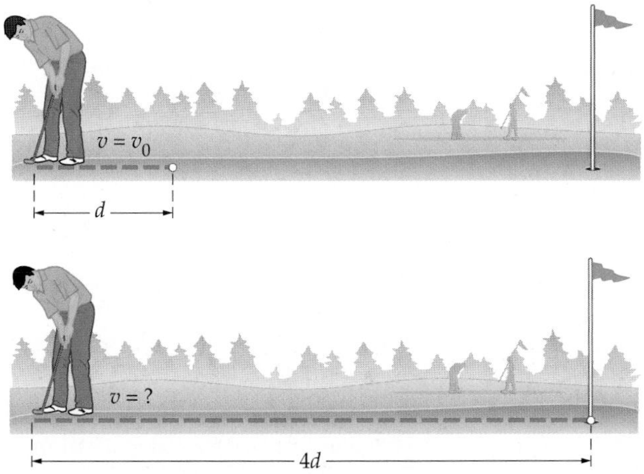

Reasoning and Discussion

In the original putt, the ball started with a kinetic energy of $\frac{1}{2}mv_0^2$ and came to rest in the distance d. The kinetic energy was dissipated by the nonconservative force due to grass resistance, F, which does the work $W_{nc} = -Fd$. Since the change in mechanical energy is $\Delta E = 0 - \frac{1}{2}mv_0^2 = -\frac{1}{2}mv_0^2$, it follows from $W_{nc} = \Delta E$ that $Fd = \frac{1}{2}mv_0^2$. Therefore, to go four times the distance, $4d$, we need to give the ball four times as much kinetic energy. Noting that kinetic energy is proportional to v^2, we see that the initial speed need only be doubled.

Answer:

(a) The initial speed should be doubled to $2v_0$.

A common example of a nonconservative force is kinetic friction. In the next Example, we show how to include the effects of friction in a system that also includes kinetic energy and gravitational potential energy.

EXAMPLE 8–10 Landing with a Thud

A block of mass $m_1 = 2.40$ kg is connected to a second block of mass $m_2 = 1.80$ kg, as shown here. When the blocks are released from rest, they move through a distance $d = 0.500$ m, at which point m_2 hits the floor. Given that the coefficient of kinetic friction between m_1 and the horizontal surface is $\mu_k = 0.450$, find the speed of the blocks just before m_2 lands.

Picture the Problem

We choose $y = 0$ to be at floor level; therefore, the gravitational potential energy of m_2 is zero when it lands. The potential energy of m_1 doesn't change during this process; it is always m_1gh. Thus, it isn't necessary to know the value of h. Note that we label the beginning and ending points with i and f, respectively.

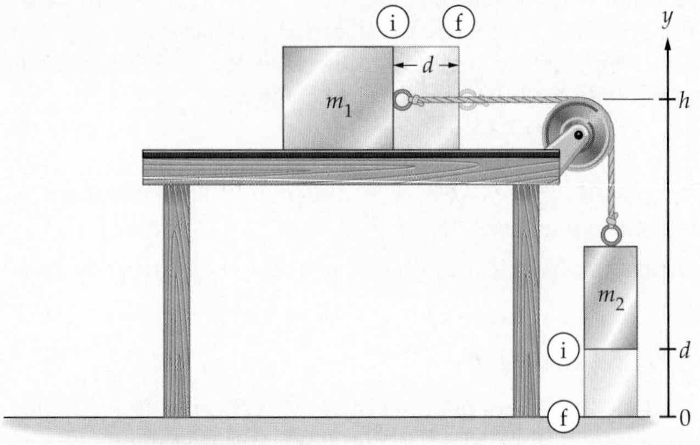

INTERACTIVE FIGURE

Strategy

Since a nonconservative force (friction) is doing work in this system, we use $W_{nc} = \Delta E = E_f - E_i$. Thus, we must calculate not only the mechanical energies, E_i and E_f, but also the nonconservative work, W_{nc}. Note that E_f can be written in terms of the unknown speed of the blocks just before m_2 lands. Therefore, we can set W_{nc} equal to ΔE and solve for the final speed.

Solution

1. Evaluate U_i, K_i, and E_i. Be sure to include contributions from both masses:

$$U_i = m_1gh + m_2gd$$
$$K_i = \tfrac{1}{2}m_1 \cdot 0^2 + \tfrac{1}{2}m_2 \cdot 0^2 = 0$$
$$E_i = U_i + K_i = m_1gh + m_2gd$$

2. Next, evaluate U_f, K_f, and E_f. Note that E_f depends on the unknown speed, v:

$$U_f = m_1gh + 0$$
$$K_f = \tfrac{1}{2}m_1v^2 + \tfrac{1}{2}m_2v^2$$
$$E_f = U_f + K_f = m_1gh + \tfrac{1}{2}m_1v^2 + \tfrac{1}{2}m_2v^2$$

3. Calculate the nonconservative work, W_{nc}. Recall that the force of friction is $f_k = \mu_k N = \mu_k m_1 g$, and that it points opposite to the displacement of distance d:

$$W_{nc} = -f_k d = -\mu_k m_1 g d$$

4. Set W_{nc} equal to $\Delta E = E_f - E_i$. Notice that m_1gh cancels because it occurs in both E_i and E_f.

$$W_{nc} = E_f - E_i$$
$$-\mu_k m_1 g d = \tfrac{1}{2}m_1v^2 + \tfrac{1}{2}m_2v^2 - m_2gd$$

5. Solve for v:

$$v = \sqrt{\frac{2(m_2 - \mu_k m_1)gd}{m_1 + m_2}}$$

6. Substitute numerical values:

$$v = \sqrt{\frac{2[1.80\text{ kg} - (0.450)(2.40\text{ kg})](9.81\text{ m/s}^2)(0.500\text{ m})}{1.80\text{ kg} + 2.40\text{ kg}}}$$
$$= 1.30\text{ m/s}$$

Insight

Note that step 4 can be rearranged as follows: $\tfrac{1}{2}m_1v^2 + \tfrac{1}{2}m_2v^2 = m_2gd - \mu_k m_1 g d$. Translating this to words, we can say that the final kinetic energy of the blocks is equal to the initial gravitational potential energy of m_2, minus the energy dissipated by friction.

Practice Problem

Find the coefficient of kinetic friction if the final speed of the blocks is 0.950 m/s. **[Answer:** $\mu_k = 0.589$**]**

Some related homework problems: Problem 35, Problem 36

Finally, we present an Active Example for the common situation of a system in which two different nonconservative forces do work.

ACTIVE EXAMPLE 8–3 Marathon Man: Find the Height of the Hill

An 80.0-kg jogger starts from rest and runs uphill into a stiff breeze. At the top of the hill the jogger has done the work $W_{nc1} = +1.80 \times 10^4$ J, air resistance has done the work $W_{nc2} = -4420$ J, and the jogger's speed is 3.50 m/s. Find the height of the hill.

Solution *(Test your understanding by performing the calculations indicated in each step.)*

1. Write the initial mechanical energy, E_i: $E_i = U_i + K_i = 0 + 0 = 0$
2. Write the final mechanical energy, E_f: $E_f = U_f + K_f = mgh + \frac{1}{2}mv^2$
3. Set W_{nc} equal to ΔE: $W_{nc} = \Delta E = mgh + \frac{1}{2}mv^2$
4. Use $W_{nc} = \Delta E$ to solve for h: $h = (W_{nc} - \frac{1}{2}mv^2)/mg$
5. Calculate the total nonconservative work: $W_{nc} = W_{nc1} + W_{nc2} = 13{,}600$ J
6. Substitute numerical values to determine h: $h = 16.7$ m

Insight

As usual when dealing with energy calculations, our final result is independent of the shape of the hill.

Your Turn

Suppose the jogger's mass had been 90.0 kg rather than 80.0 kg. What would be the height of the hill in this case?

(Answers to **Your Turn** *problems are given in the back of the book.)*

▲ Highways that descend steeply are often provided with escape ramps that enable truck drivers whose brakes fail to bring their rigs to a safe stop. These ramps provide a perfect illustration of the conservation of energy. From a physics point of view, the driver's problem is to get rid of an enormous amount of kinetic energy in the safest possible way. The ramps run uphill, so some of the kinetic energy is simply converted back into gravitational potential energy (just as in a roller coaster). In addition, the ramps are typically surfaced with sand or gravel, allowing much of the initial kinetic energy to be dissipated by friction into other forms of energy, such as sound and heat.

8–5 Potential Energy Curves and Equipotentials

Figure 8–10 shows a metal ball rolling on a roller coaster-like track. Initially the ball is at rest at point A. Since the height at A is $y = h$, the ball's initial mechanical energy is $E_0 = mgh$. If friction and other nonconservative forces can be ignored, the ball's mechanical energy remains fixed at E_0 throughout its motion. Thus,

$$E = U + K = E_0$$

As the ball moves, its potential energy falls and rises in the same way as the track. After all, the gravitational potential energy, $U = mgy$, is directly proportional

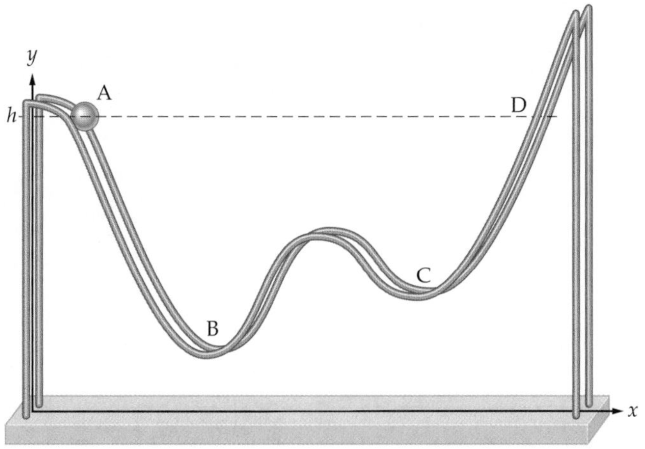

◀ **FIGURE 8–10 A ball rolling on a frictionless track**

The ball starts at A, where $y = h$, with zero speed. Its greatest speed occurs at B. At D, where $y = h$ again, its speed returns to zero.

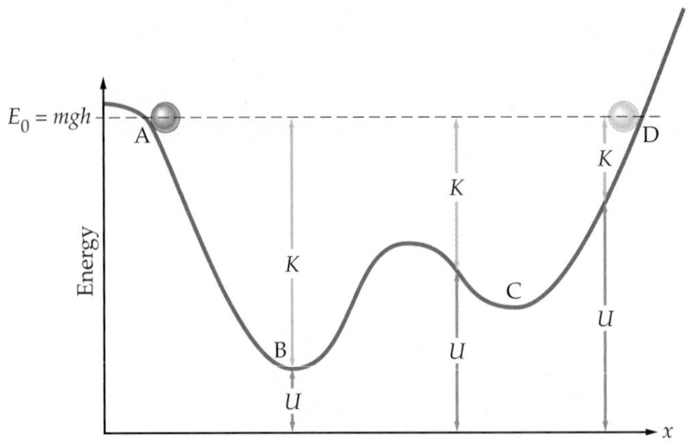

▶ **FIGURE 8–11 Gravitational potential energy versus position for the track shown in Figure 8–10**

The shape of the potential energy curve is exactly the same as the shape of the track. In this case, the total mechanical energy is fixed at its initial value, $E_0 = U + K = mgh$. Because the height of the curve is U, by definition, it follows that K is the distance from the curve up to the dashed line at $E_0 = mgh$. Note that K is largest at B. In addition, K vanishes at A and D, which are turning points of the motion.

to the height of the track, y. In a sense, then, the track itself represents a graph of the corresponding potential energy.

This is shown explicitly in **Figure 8–11**, where we plot energy on the vertical axis and x on the horizontal axis. The potential energy U looks just like the track in Figure 8–10. In addition, we plot a horizontal line at the value E_0, indicating the constant energy of the ball. Since the potential energy plus the kinetic energy must always add up to E_0, it follows that K is the amount of energy from the potential energy curve up to the horizontal line at E_0. This is also shown in Figure 8–11.

Examining an energy plot like Figure 8–11 gives a great deal of information about the motion of an object. For example, at point B the potential energy has its lowest value, and thus the kinetic energy is greatest there. At point C the potential energy has increased, indicating a corresponding decrease in kinetic energy. As the ball continues to the right, the potential energy increases until, at point D, it is again equal to the total energy, E_0. At this point the kinetic energy is zero, and the ball comes to rest momentarily. It then "turns around" and begins to move to the left, eventually returning to point A where it again stops, changes direction, and begins a new cycle. Points A and D, then, are referred to as **turning points** of the motion.

Turning points are also seen in the motion of a mass on a spring, as indicated in **Figure 8–12**. Figure 8–12 (a) shows a mass pulled to the position $x = A$, and released from rest; Figure 8–12 (b) shows the potential energy of the system, $U = \frac{1}{2}kx^2$. Starting the system this way gives it an initial energy $E_0 = \frac{1}{2}kA^2$, shown

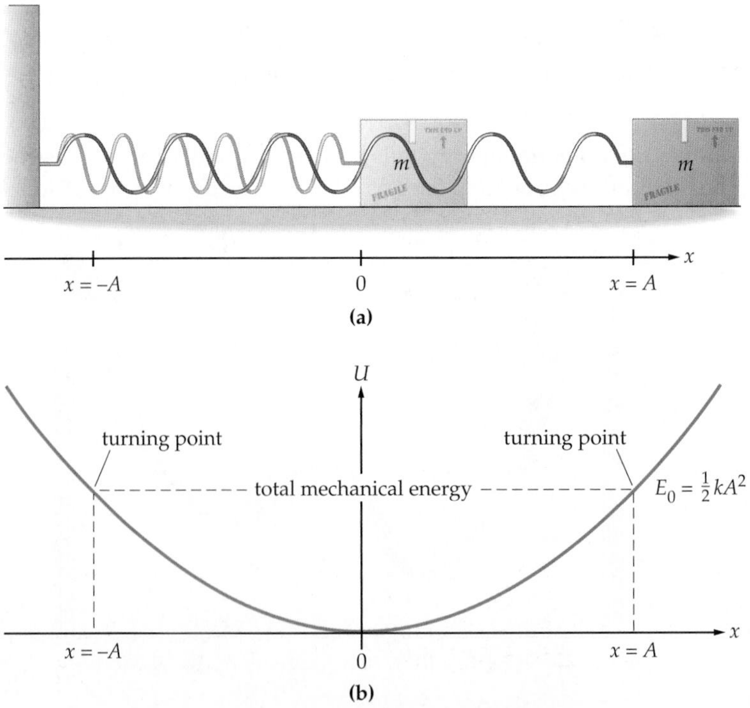

▶ **FIGURE 8–12 A mass on a spring**

(a) A spring is stretched by an amount A, giving it a potential energy of $U = \frac{1}{2}kA^2$. **(b)** The potential energy curve, $U = \frac{1}{2}kx^2$, for the spring in (a). Because the mass starts at rest, its initial mechanical energy is $E_0 = \frac{1}{2}kA^2$. The mass oscillates between $x = A$ and $x = -A$.

by the horizontal line in Figure 8–12 (b). As the mass moves to the left, its speed increases, reaching a maximum where the potential energy is lowest, at $x = 0$. If no nonconservative forces act, the mass continues to $x = -A$, where it stops momentarily before returning to $x = A$. This type of **oscillatory motion** will be studied in detail in Chapter 13.

The next Example uses a potential-energy curve to find the speed of an object at a given value of x.

EXAMPLE 8–11 A Potential Problem

A 1.60-kg object in a conservative system moves along the x axis, where the potential energy is as shown. A physical example would be a bead sliding on a wire with the shape of the potential energy curve. If the object's speed at $x = 0$ is 2.30 m/s, what is its speed at $x = 2.00$ m?

Picture the Problem

The plot shows U as a function of x. The values of U at $x = 0$ and $x = 2.00$ m are 9.35 J and 4.15 J, respectively. It follows that the object's speed at $x = 2.00$ m will be greater than its speed at $x = 0$.

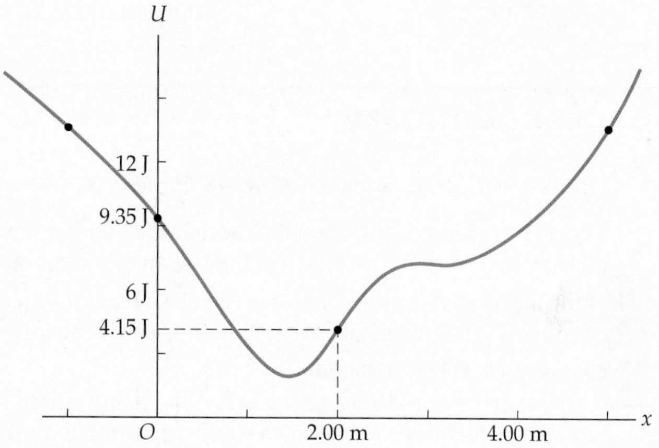

Strategy

Since mechanical energy is conserved, we know that the total energy at $x = 0$ $(U_i + K_i)$ is equal to the total energy at $x = 2.00$ m $(U_f + K_f)$.

The problem statement gives U_i, and since we also know the speed at $x = 0$ we can use $K = \frac{1}{2}mv^2$ to calculate the corresponding kinetic energy, K_i. At $x = 2.00$ m we know the potential energy, U_f; hence we can use $U_i + K_i = U_f + K_f$ to solve for K_f. Once the final kinetic energy is known, it is possible to solve for the final speed by once again using $K = \frac{1}{2}mv^2$.

Solution

1. Evaluate U_i, K_i, and E_i at $x = 0$:

$$U_i = 9.35\text{ J}$$
$$K_i = \tfrac{1}{2}mv_i^2 = \tfrac{1}{2}(1.60\text{ kg})(2.30\text{ m/s})^2 = 4.23\text{ J}$$
$$E_i = U_i + K_i = 9.35\text{ J} + 4.23\text{ J} = 13.58\text{ J}$$

2. Write expressions for U_f, K_f, and E_f at $x = 2.00$ m:

$$U_f = 4.15\text{ J}$$
$$K_f = \tfrac{1}{2}mv_f^2$$
$$E_f = U_f + K_f = 4.15\text{ J} + \tfrac{1}{2}mv_f^2$$

3. Set E_f equal to E_i and solve for v_f:

$$4.15\text{ J} + \tfrac{1}{2}mv_f^2 = 13.58\text{ J}$$

Solve for v_f:

$$v_f = \sqrt{\frac{2(13.58\text{ J} - 4.15\text{ J})}{m}}$$

4. Substitute the numerical value of the object's mass:

$$v_f = \sqrt{\frac{2(13.58\text{ J} - 4.15\text{ J})}{1.60\text{ kg}}} = 3.43\text{ m/s}$$

Insight

As we see in step 1, the total mechanical energy of the system is 13.58 J. This means that turning points for this object occur at values of x where $U = 13.58$ J.

Practice Problem

Using the graph provided, estimate the location of the turning points for this object. [**Answer:** $x = -1.00$ m and $x = 5.00$ m]

Some related homework problems: Problem 44, Problem 45

Oscillatory motion between turning points is also observed in molecules. If the energy of oscillation is relatively small, as is usual at room temperature, the atoms in a molecule simply vibrate back and forth—like masses connected by a spring. As long as no energy is gained or lost, the molecular oscillations continue unchanged. On the other hand, if the energy of the molecule is increased by heating, or some other mechanism, the molecule will eventually dissociate—fly apart—as the atoms move to infinite separation.

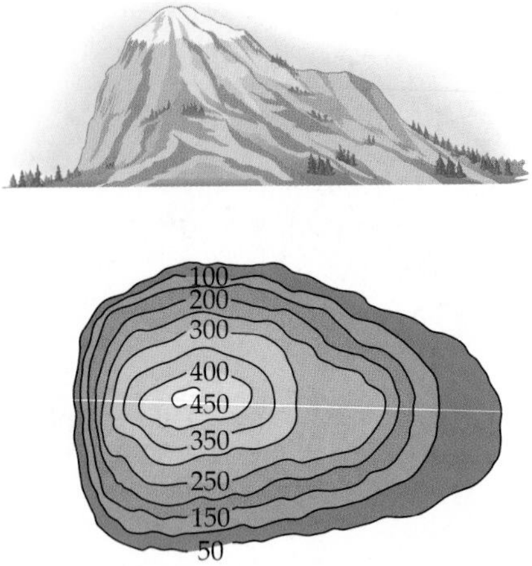

◀ **FIGURE 8–13 A contour map**
The map shows equal-altitude contour lines for a hill (top) that is very steep on one side (left) but slopes more gently on the other (right). Note that the contour lines are packed close together where the terrain is steep.

In some cases, a two-dimensional plot of potential energy contours is useful. For instance, **Figure 8–13** shows a contour map of a hill. Each contour corresponds to a given altitude and, hence, to a given value of the gravitational potential energy. In general, lines corresponding to constant values of potential energy are called **equipotentials.** Since the altitude changes by equal amounts from one contour to the next, it follows that when gravitational equipotentials are packed close together, the corresponding terrain is steep. On the other hand, when the equipotentials are widely spaced, the ground is nearly flat, since a large horizontal distance is required for a given change in altitude. We shall see similar plots with similar interpretations when we study electric potential energy in Chapter 21.

Chapter Summary

8–1 Conservative and Nonconservative Forces

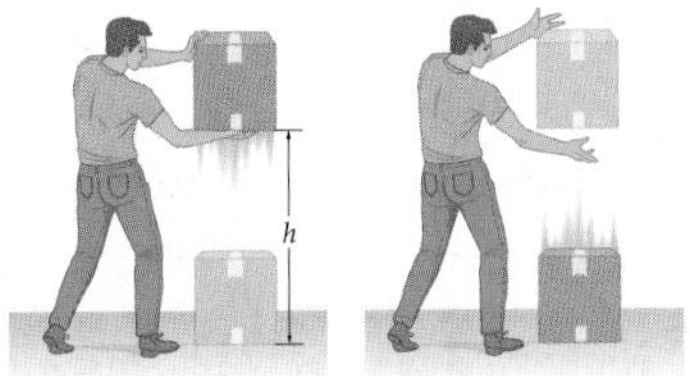

Conservative forces conserve the mechanical energy of a system. Thus, in a conservative system the total mechanical energy remains constant.

Nonconservative forces convert mechanical energy into other forms of energy, or convert other forms of energy into mechanical energy.

Conservative Force, Definition
A conservative force does zero total work on any closed path. In addition, the work done by a conservative force in going from point A to point B is *independent of the path* from A to B.

Examples of Conservative Forces
Gravity, spring.

Nonconservative Force, Definition
The work done by a nonconservative force on a closed path is nonzero. The work is also path dependent.

Examples of Nonconservative Forces
Friction, air resistance, tension in ropes and cables, forces exerted by muscles and motors.

8–2 Potential Energy and the Work Done by Conservative Forces

Potential energy, U, can "store" energy in a system. Energy in the form of potential energy can be converted to kinetic or other forms of energy.

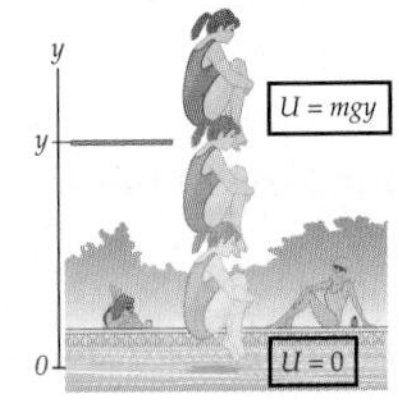

Potential Energy, Definition
The work done by a conservative force is minus the change in potential energy:

$$W_c = -\Delta U = U_i + U_f \qquad \textbf{8–1}$$

Zero Level
Any location can be chosen for $U = 0$. Once the choice is made, however, it must be used consistently.

Gravity
Choosing $y = 0$ to be the zero level near Earth's surface, $U = mgy$. **8–3**

Spring
Choosing $x = 0$ (the equilibrium position) to be the zero level, $U = \frac{1}{2}kx^2$. **8–5**

8–3 Conservation of Mechanical Energy

Mechanical energy, E, is conserved in systems with conservative forces only.

Mechanical Energy, Definition
Mechanical energy is the sum of the potential and kinetic energies of a system:

$$E = U + K \qquad \textbf{8–6}$$

8–4 Work Done by Nonconservative Forces

Nonconservative forces can change the mechanical energy of a system.

Change in Mechanical Energy
The work done by a nonconservative force is equal to the change in the mechanical energy of a system:

$$W_{nc} = \Delta E = E_f - E_i \qquad \textbf{8–9}$$

8–5 Potential Energy Curves and Equipotentials

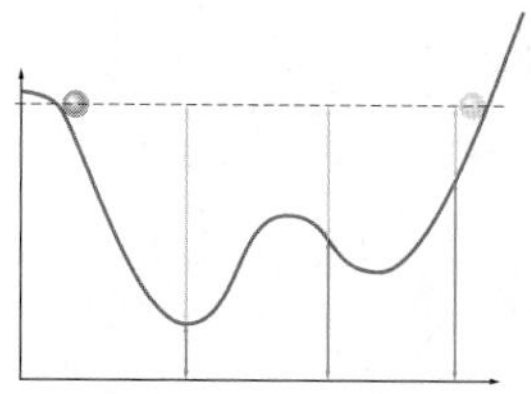

A potential energy curve plots U as a function of position.

An equipotential plot shows contours corresponding to constant values of U.

Turning Points
Turning points occur where an object stops momentarily before reversing direction. At turning points the kinetic energy is zero.

Oscillatory Motion
An object moving back and forth between two turning points is said to have oscillatory motion.

Problem-Solving Summary

Type of Calculation	Relevant Physical Concepts	Related Examples
Calculate the gravitational or spring potential energy.	The potential energy for gravity is $U = mgy$; the potential energy for a spring is $U = \frac{1}{2}kx^2$.	Examples 8–2, 8–3, 8–4
Apply energy conservation in a system involving gravity.	Choose a horizontal level for $y = 0$, then use $U = mgy$.	Examples 8–6, 8–7 Active Example 8–1
Apply energy conservation in a system involving a spring.	Use $U = \frac{1}{2}kx^2$, where x measures the expansion or compression of the spring from its equilibrium position.	Example 8–8, Active Example 8–1
Find the nonconservative work done on a system.	Calculate the initial energy, E_i, and the final energy, E_f. Then use $W_{nc} = \Delta E = E_f - E_i$.	Examples 8–9, 8–10, Active Examples 8–2, 8–3

Conceptual Questions

(Answers to odd-numbered Conceptual Questions can be found in the back of the book.)

1. Is it possible for the kinetic energy of an object to be negative? Is it possible for the gravitational potential energy of an object to be negative? Explain.
2. An avalanche occurs when a mass of snow slides down a steep mountain slope. Discuss the energy conversions responsible for water vapor rising to form clouds, falling as snow on a mountain, and then sliding down a slope as an avalanche.
3. If the stretch of a spring is doubled, the force it exerts is also doubled. By what factor does the spring's potential energy increase?
4. When a mass is placed on top of a vertical spring, the spring compresses and the mass moves downward. Analyze this system in terms of its mechanical energy.
5. If a spring is stretched so far that it is permanently deformed, its force is no longer conservative. Why?
6. An object is thrown upward to a person on a roof. At what point is the object's kinetic energy at maximum? At what point is the potential energy of the system at maximum? At what locations do these energies have their minimum values?
7. It is a law of nature that the total energy of the universe is conserved. What do politicians mean, then, when they urge "energy conservation"?
8. Discuss the various energy conversions that occur when a person performs a pole vault. Include as many conversions as you can, and consider times before, during, and after the actual vault itself.

How many energy conversions can you identify? (Conceptual Question 8)

9. Discuss the nature of the work done by the equipment shown in this photo. What types of forces are involved?

Conservative or nonconservative? (Conceptual Question 9)

10. A toy frog consists of a suction cup and a spring. When the suction cup is pressed against a smooth surface the frog is held down. When the suction cup lets go, the frog leaps into the air. Discuss the behavior of the frog in terms of energy conversions.
11. If the force on an object is zero, does that mean the potential energy of the system is zero? If the potential energy of a system is zero, is the force zero?
12. When a ball is thrown upward, its mechanical energy, $E = mgy + \frac{1}{2}mv^2$, is constant with time if air resistance can be ignored. How does E vary with time if air resistance cannot be ignored?
13. Discuss the various energy conversions that occur when a person performs a swan dive from a diving board. Include as many conversions as you can, and consider times before, during, and after the actual dive itself.
14. When a ball is thrown upward, it spends the same amount of time on the way up as on the way down—as long as air resistance can be ignored. If air resistance is taken into account, is the time on the way down the same as, greater than, or less than the time on the way up? Explain.

Conceptual Exercises

(Answers to odd-numbered Conceptual Exercises can be found in the back of the book.)

1. The work done by a conservative force is indicated in **Figure 8–14** for a variety of different paths connecting the points A and B. What is the work done by this force **(a)** on path 1 and **(b)** on path 2?

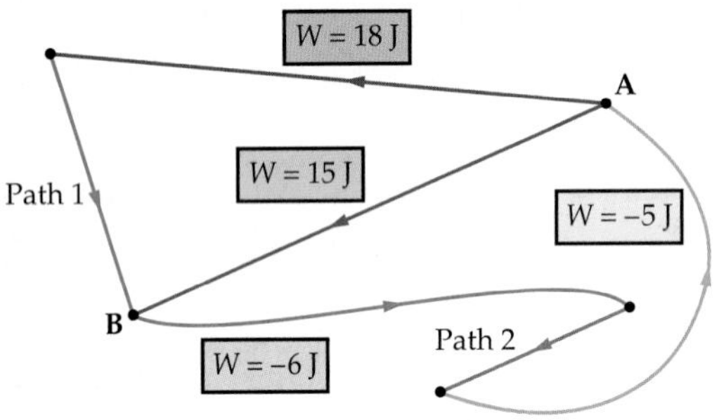

▲ **FIGURE 8–14** Conceptual Exercise 1

2. Taking a leap of faith, a bungee jumper steps off a platform and falls until the cord brings her to rest. Suppose you analyze this system by choosing $y = 0$ at the platform level, and your friend chooses $y = 0$ at the ground level. **(a)** Is the jumper's initial potential energy in your calculation greater than, less than, or equal to the same quantity in your friend's calculation? **(b)** Is the change in the jumper's potential energy in your calculation greater than, less than, or equal to the same quantity in your friend's calculation?
3. Ball 1 is thrown to the ground; ball 2 is dropped to the ground from rest. Assuming the balls have the same mass and are released from the same height, is the change in gravitational potential energy of ball 1 greater than, less than, or equal to the change in gravitational potential energy of ball 2?
4. You throw a ball upward and let it fall to the ground. Your friend drops an identical ball straight down to the ground from the same height. Is the change in kinetic energy of your ball greater than, less than, or equal to the change in kinetic energy of your friend's ball?
5. A mass is attached to the bottom of a vertical spring. This causes the spring to stretch and the mass to move downward. **(a)** Does the potential energy of the spring increase, decrease, or stay the same during this process? **(b)** Does the gravitational potential energy of the Earth-mass system increase, decrease, or stay the same during this process?
6. You and a friend both solve a problem involving a skier going down a slope. When comparing solutions, you notice that your choice for the $y = 0$ level is different than the $y = 0$ level chosen by your friend. Will your answers agree or disagree on the following quantities: **(a)** The skier's potential energy; **(b)** the skier's change in potential energy; **(c)** the skier's kinetic energy?
7. Suppose the situation described in Conceptual Checkpoint 8–2 is repeated on the fictional planet Epsilon, where the acceleration due to gravity is less than it is on the Earth. **(a)** Would the height of a hill on Epsilon that causes a reduction in speed from 4 m/s to 0 be greater than, less than, or the same as the height of the corresponding hill on Earth? **(b)** If the initial speed at the bottom of the hill on Epsilon is 5 m/s, will the final speed at the top of the hill be greater than, less than, or equal to 3 m/s?
8. A particle moves under the influence of a conservative force. At point A the particle has a kinetic energy of 12 J; at point B the particle is momentarily at rest, and the potential energy of the system is 25 J; at point C the potential energy of the system is 5 J. **(a)** What is the potential energy of the system when the particle is at point A? **(b)** What is the kinetic energy of the particle at point C?
9. Three balls are thrown upward with the same initial speed v_0, but at different angles relative to the horizontal, as shown in **Figure 8–15**. Ignoring air resistance, indicate which of the

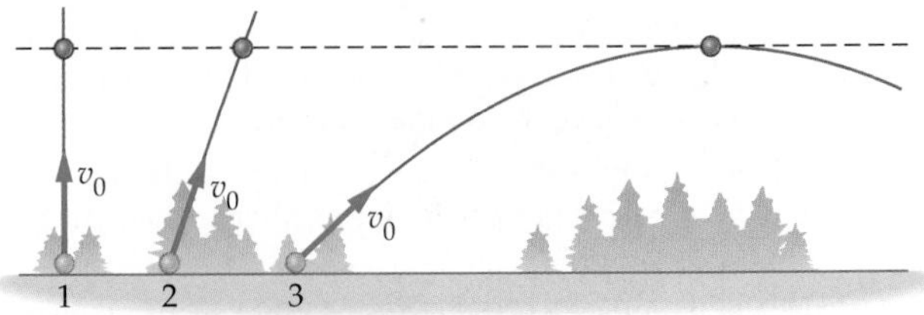

▲ **FIGURE 8–15** Conceptual Exercise 9

following statements is correct: At the dashed level, **(1)** ball 3 has the lowest speed; **(2)** ball 1 has the lowest speed; **(3)** all three balls have the same speed; **(4)** the speed of the balls depends on their mass.

10. When a ball of mass m is dropped from a height h, its kinetic energy just before landing is K. Now, suppose a second ball of mass $2m$ is dropped from a height $h/2$. Just before ball 2 lands, is its kinetic energy $4K$, $2K$, K, $K/2$ or $K/4$?

11. You coast up a hill on your bicycle with decreasing speed. Your friend pedals up the hill with constant speed. **(a)** Ignoring friction, does the mechanical energy of your system (you-bike-Earth) increase, decrease, or stay the same? **(b)** Does the mechanical energy of your friend's system increase, decrease, or stay the same?

12. You ride your bicycle down a hill, maintaining a constant speed the entire time. **(a)** As you ride, does the gravitational potential energy of the Earth-bike-rider system increase, decrease, or stay the same? **(b)** Does the kinetic energy of you and your bike increase, decrease, or stay the same? **(c)** Has the mechanical energy of this system increased, decreased, or remained the same?

13. A leaf falls to the ground with constant speed. Is $U_i + K_i$ greater than, less than, or the same as $U_f + K_f$ for this system?

14. On reentry, the space shuttle's protective heat tiles become extremely hot. Is the mechanical energy of the shuttle-Earth system when the shuttle lands greater than, less than, or the same as when it is in orbit?

15. Consider the two-block system shown in Example 8–10. **(a)** As block 2 descends through the distance d, does its mechanical energy increase, decrease, or stay the same? **(b)** Is the nonconservative work done on block 2 by the tension in the rope positive, negative, or zero?

16. For an object moving along the x axis, the potential energy of the frictionless system is shown in **Figure 8–16**. Suppose the object is released from rest at the point A. Rank the other points in the figure in increasing order of the object's speed. Indicate ties where appropriate.

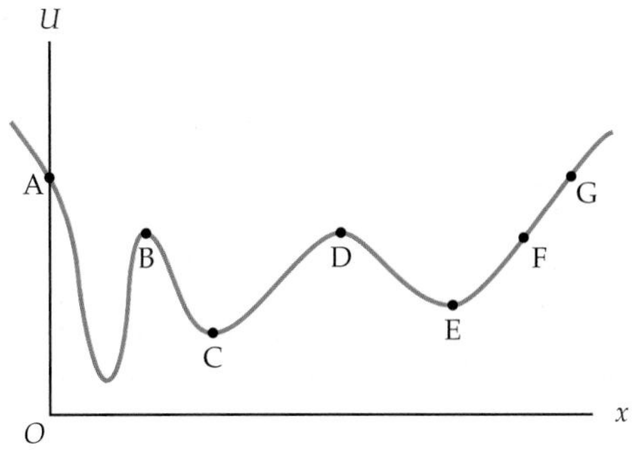

▲ **FIGURE 8–16** Conceptual Exercises 16 and 17

17. Referring to Conceptual Exercise 16, suppose the object is released from rest at a point halfway between the points F and G. Rank the other points in the figure in increasing order of the object's speed, if the object can reach that point. Indicate ties where appropriate.

Problems

Note: **IP** *denotes an integrated conceptual/quantitative problem.* **BIO** *identifies problems of biological or medical interest. Red bullets (•,••,•••) are used to indicate the level of difficulty of each problem.*

Section 8–1 Conservative and Nonconservative Forces

1. • Calculate the work done by gravity as a 3.2-kg object is moved from point A to point B in **Figure 8–17** along paths 1, 2, and 3.

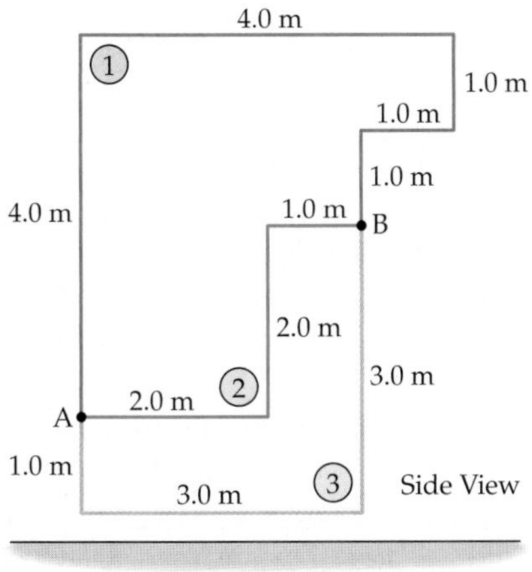

▲ **FIGURE 8–17** Problem 1

2. • Calculate the work done by friction as a 3.2-kg box is slid along a floor from point A to point B in **Figure 8–18** along paths 1, 2, and 3. Assume that the coefficient of kinetic friction between the box and the floor is 0.21.

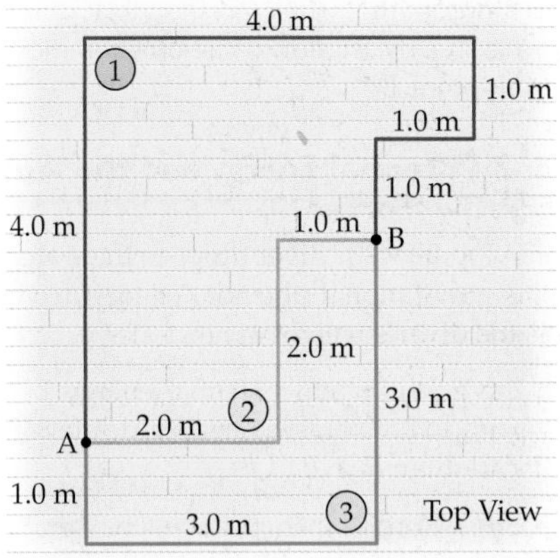

▲ **FIGURE 8–18** Problem 2

3. • **IP** A 4.1-kg block is attached to a spring with a force constant of 550 N/m, as shown in **Figure 8–19**. **(a)** Find the work done by the spring on the block as the block moves from A to B along paths 1 and 2. **(b)** How do your results depend on the mass of the block? Specifically, if you increase the mass, does the work done by the spring increase, decrease, or stay the same? (Assume the system is frictionless.)

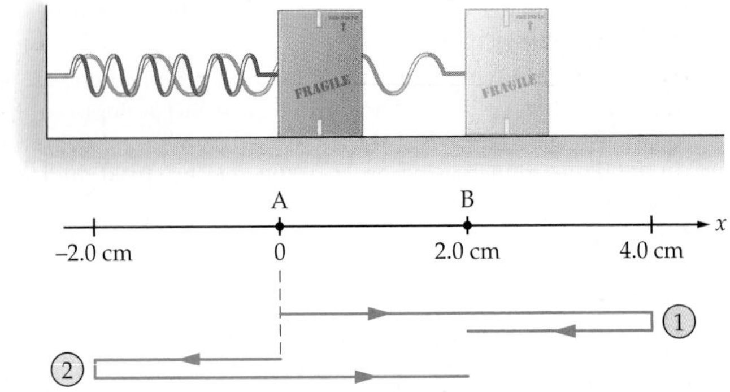

▲ **FIGURE 8–19** Problems 3 and 5

4. • **IP** **(a)** Calculate the work done by gravity as a 5.2-kg object is moved from A to B in **Figure 8–20** along paths 1 and 2. **(b)** How do your results depend on the mass of the block? Specifically, if you increase the mass, does the work done by gravity increase, decrease, or stay the same?

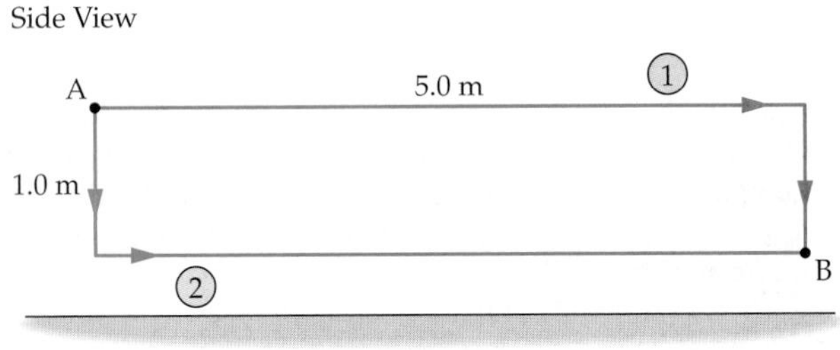

▲ **FIGURE 8–20** Problem 4

5. •• In the system shown in Figure 8–19, suppose the block has a mass of 2.7 kg, the spring has a force constant of 480 N/m, and the coefficient of kinetic friction between the block and the floor is 0.16. **(a)** Find the work done on the block by the spring and by friction as the block is moved from point A to point B along path 2. **(b)** Find the work done on the block by the spring and by friction if the block is moved directly from point A to point B.

Section 8–2 Potential Energy and the Work Done by Conservative Forces

6. • As an Acapulco cliff diver drops to the water from a height of 46 m, his gravitational potential energy decreases by 25,000 J. What is the diver's weight in newtons?

7. • Find the gravitational potential energy of an 83-kg person standing atop Mt. Everest at an altitude of 8848 m. Use sea level as the location for $y = 0$.

8. • **Jeopardy!** Contestants on the game show *Jeopardy!* depress spring-loaded buttons to "buzz in" and provide the question corresponding to the revealed answer. The force constant on these buttons is about 120 N/m. Estimate the amount of energy it takes—at a minimum—to buzz in.

9. •• **BIO The Wing of the Hawkmoth** Experiments performed on the wing of a hawkmoth (*Manduca sexta*) show that it deflects by a distance of $x = 4.8$ mm when a force of magnitude $F = 3.0$ mN is applied at the tip, as indicated in **Figure 8–21**. Treating the wing as an ideal spring, find **(a)** the force constant of the wing and **(b)** the energy stored in the wing when it is deflected. **(c)** What force must be applied to the tip of the wing to store twice the energy found in part (b)?

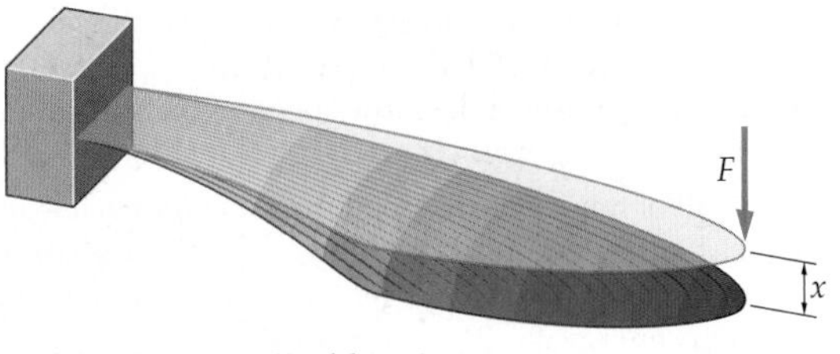

▲ **FIGURE 8–21** Problem 9

10. •• **IP** A vertical spring stores 0.962 J in spring potential energy when a 3.0-kg mass is suspended from it. **(a)** By what multiplicative factor does the spring potential energy change if the mass attached to the spring is doubled? **(b)** Verify your answer to part (a) by calculating the spring potential energy when a 6.0-kg mass is attached to the spring.

11. •• Pushing on the pump of a soap dispenser compresses a small spring. When the spring is compressed 0.50 cm, its potential energy is 0.0025 J. **(a)** What is the force constant of the spring? **(b)** What compression is required for the spring potential energy to equal 0.0084 J?

12. •• A force of 4.7 N is required to stretch a certain spring by 1.3 cm. **(a)** How far must this spring be stretched for its potential energy to be 0.020 J? **(b)** How much stretch is required for the spring potential energy to be 0.080 J?

13. •• **IP** The work required to stretch a certain spring from an elongation of 4.00 cm to an elongation of 5.00 cm is 30.5 J. **(a)** Is the work required to increase the elongation of the spring from 5.00 cm to 6.00 cm greater than, less than, or equal to 30.5 J? Explain. **(b)** Verify your answer to part (a) by calculating the required work.

14. •• A 0.33-kg pendulum bob is attached to a string 1.2 m long. What is the change in the gravitational potential energy of the system as the bob swings from point A to point B in **Figure 8–22**?

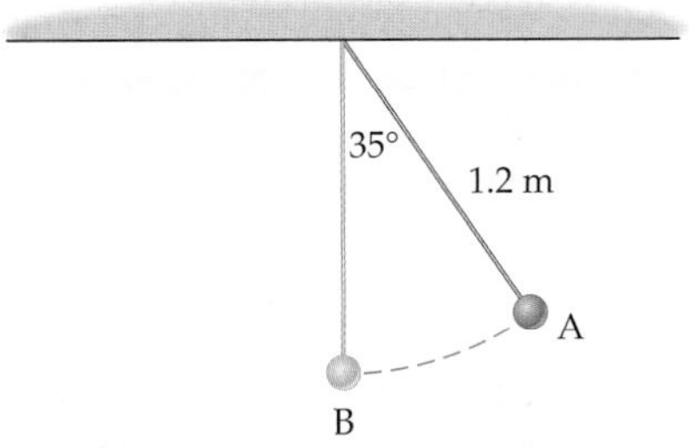

▲ **FIGURE 8–22** Problems 14, 24, 25, and 58

Section 8–3 Conservation of Mechanical Energy

15. • At an amusement park, a swimmer uses a water slide to enter the main pool. If the swimmer starts at rest, slides without friction, and descends through a vertical height of 2.61 m, what is her speed at the bottom of the slide?

16. • In the previous problem, find the swimmer's speed at the bottom of the slide if she starts with an initial speed of 0.840 m/s.

17. • **IP** A player passes a 0.600-kg basketball downcourt for a fast break. The ball leaves the player's hands with a speed of 8.30 m/s and slows down to 7.10 m/s at its highest point. **(a)** Ignoring air resistance, how high above the release point is the ball when it is at its maximum height? **(b)** How would doubling the ball's mass affect the result in part (a)? Explain.

18. •• **IP** In a tennis match, a player wins a point by hitting the ball sharply to the ground on the opponent's side of the net. **(a)** If

the ball bounces upward from the ground with a speed of 16 m/s, and is caught by a fan in the stands with a speed of 12 m/s, how high above the court is the fan? Ignore air resistance. **(b)** Explain why it is not necessary to know the mass of the tennis ball.

19. •• A 0.21-kg apple falls from a tree to the ground, 4.0 m below. Ignoring air resistance, determine the apple's kinetic energy, K, the gravitational potential energy of the system, U, and total mechanical energy of the system, E, when the apple's height above the ground is **(a)** 4.0 m, **(b)** 3.0 m, **(c)** 2.0 m, **(d)** 1.0 m, and **(e)** 0 m. Take ground level to be $y = 0$.

20. •• **IP** A 2.7-kg block slides with a speed of 1.4 m/s on a frictionless horizontal surface until it encounters a spring. **(a)** If the block compresses the spring 4.8 cm before coming to rest, what is the force constant of the spring? **(b)** What initial speed should the block have to compress the spring by 1.2 cm?

21. •• A 0.26-kg rock is thrown vertically upward from the top of a cliff that is 32 m high. When it hits the ground at the base of the cliff, the rock has a speed of 29 m/s. Assuming that air resistance can be ignored, find **(a)** the initial speed of the rock and **(b)** the greatest height of the rock as measured from the base of the cliff.

22. •• A 1.40-kg block slides with a speed of 0.950 m/s on a frictionless horizontal surface until it encounters a spring with a force constant of 734 N/m. The block comes to rest after compressing the spring 4.15 cm. Find the spring potential energy, U, the kinetic energy of the block, K, and the total mechanical energy of the system, E, for compressions of **(a)** 0 cm, **(b)** 1.00 cm, **(c)** 2.00 cm, **(d)** 3.00 cm, and **(e)** 4.00 cm.

23. •• A 5.00-kg rock is dropped and allowed to fall freely. Find the initial kinetic energy, the final kinetic energy, and the change in kinetic energy for **(a)** the first 2.00 m of fall and **(b)** the second 2.00 m of fall.

24. •• **IP** Suppose the pendulum bob in Figure 8–22 has a mass of 0.33 kg and is moving to the right at point B with a speed of 2.4 m/s. Air resistance is negligible. **(a)** What is the change in the system's gravitational potential energy when the bob reaches point A? **(b)** What is the speed of the bob at point A? **(c)** If the mass of the bob is increased, does your answer to part (a) increase, decrease, or stay the same? Explain. **(d)** If the mass of the bob is increased, does your answer to part (b) increase, decrease, or stay the same? Explain.

25. •• **IP** In the previous problem, **(a)** what is the bob's kinetic energy at point B? **(b)** At some point the bob will come to rest momentarily. Without doing an additional calculation, determine the change in the system's gravitational potential energy between point B and the point where the bob comes to rest. **(c)** Find the maximum angle the string makes with the vertical as the bob swings back and forth. Ignore air resistance.

26. ••• The two masses in the Atwood's machine shown in **Figure 8–23** are initially at rest at the same height. After they are released, the large mass, m_2, falls through a height h and hits the floor, and the small mass, m_1, rises through a height h. **(a)** Find the speed of the masses just before m_2 lands, giving your answer in terms of m_1, m_2, g, and h. Assume the ropes and pulley have negligible mass and that friction can be ignored. **(b)** Evaluate your answer to part (a) for the case $h = 1.2$ m, $m_1 = 3.7$ kg, and $m_2 = 4.1$ kg.

27. ••• In the previous problem, suppose the masses have an initial speed of 0.20 m/s, and that m_2 is moving upward. How high does m_2 rise above its initial position before momentarily coming to rest, given that $m_1 = 3.7$ kg and $m_2 = 4.1$ kg?

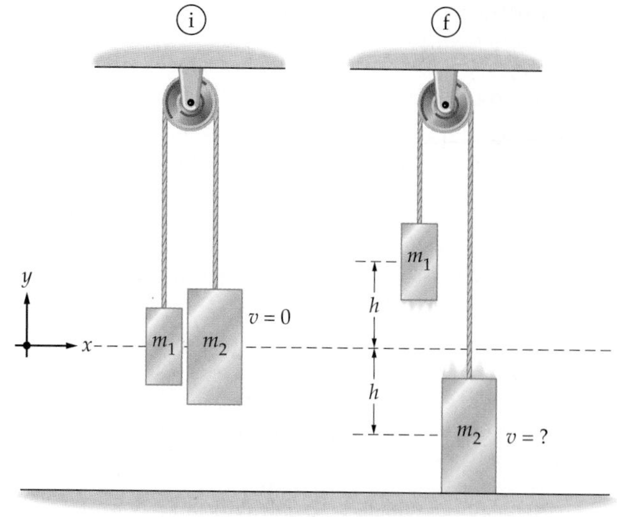

▲ **FIGURE 8–23** Problems 26, 27, 66, and 87

Section 8–4 Work Done by Nonconservative Forces

28. • Catching a wave, a 72-kg surfer starts with a speed of 1.3 m/s, drops through a height of 1.75 m, and ends with a speed of 8.2 m/s. How much nonconservative work was done on the surfer?

29. • At a playground, a 19-kg child plays on a slide that drops through a height of 2.1 m. The child starts at rest at the top of the slide. On the way down, the slide does a nonconservative work of −361 J on the child. What is the child's speed at the bottom of the slide?

30. • Starting at rest at the edge of a swimming pool, a 72.0-kg athlete swims along the surface of the water and reaches a speed of 1.20 m/s by doing the work $W_{nc1} = +161$ J. Find the nonconservative work, W_{nc2}, done by the water on the athlete.

31. • A 17,000-kg airplane lands with a speed of 82 m/s on a stationary aircraft carrier deck that is 115 m long. Find the work done by nonconservative forces in stopping the plane.

32. • **IP** The driver of a 1100-kg car moving at 17 m/s brakes quickly to 12 m/s when he spots a local garage sale. **(a)** Find the change in the car's kinetic energy. **(b)** Explain where the "missing" kinetic energy has gone.

33. •• Suppose the system in Example 8–10 starts with m_2 moving downward with a speed of 1.3 m/s. What speed do the masses have just before m_2 lands?

34. •• A 42.0-kg seal at an amusement park slides from rest down a ramp into the pool below. The top of the ramp is 1.75 m higher than the surface of the water and the ramp is inclined at an angle of 35.0° above the horizontal. If the seal reaches the water with a speed of 4.40 m/s, what is **(a)** the work done by kinetic friction and **(b)** the coefficient of kinetic friction between the seal and the ramp?

35. •• A 1.9-kg rock is released from rest at the surface of a pond 1.8 m deep. As the rock falls, a constant upward force of 4.6 N is exerted on it by water resistance. Calculate the nonconservative work, W_{nc}, done by water resistance on the rock, the gravitational potential energy of the system, U, the kinetic energy of the rock, K, and the total mechanical energy of the system, E, when the depth of the rock below the water's surface is **(a)** 0 m, **(b)** 0.50 m, and **(c)** 1.0 m. Let $y = 0$ be at the bottom of the pond.

36. •• A 1250-kg car drives up a hill that is 16.2 m high. During the drive, two nonconservative forces do work on the car: (i) the force of friction, and (ii) the force generated by the car's engine. The work done by friction is -3.11×10^5 J; the work done by the engine is $+6.44 \times 10^5$ J. Find the change in the car's kinetic energy from the bottom of the hill to the top of the hill.

37. •• **IP** An 81.0-kg in-line skater does +3420 J of nonconservative work by pushing against the ground with his skates. In addition, friction does −715 J of nonconservative work on the skater. The skater's initial and final speeds are 2.50 m/s and 1.22 m/s, respectively. **(a)** Has the skater gone uphill, downhill, or remained at the same level? Explain. **(b)** Calculate the change in height of the skater.

38. •• In Example 8–10, suppose the two masses start from rest and are moving with a speed of 2.05 m/s just before m_2 hits the floor. **(a)** If the coefficient of kinetic friction is $\mu_k = 0.350$, what is the distance of travel, d, for the masses? **(b)** How much conservative work was done on this system? **(c)** How much nonconservative work was done on this system? **(d)** Verify the three work relations given in Equation 8–10.

39. •• **IP** A 15,800-kg truck is moving at 12.0 m/s when it starts down a 6.00° incline in the Canadian Rockies. At the start of the descent the driver notices that the altitude is 1630 m. When she reaches an altitude of 1440 m, her speed is 29.0 m/s. Find the change in **(a)** the gravitational potential energy of the system and **(b)** the truck's kinetic energy. **(c)** Is the total mechanical energy of the system conserved? Explain.

40. ••• A 1.80-kg block slides on a rough horizontal surface. The block hits a spring with a speed of 2.00 m/s and compresses it a distance of 11.0 cm before coming to rest. If the coefficient of kinetic friction between the block and the surface is $\mu_k = 0.560$, what is the force constant of the spring?

Section 8–5 Potential Energy Curves and Equipotentials

41. • **Figure 8–24** shows a potential energy curve as a function of x. In qualitative terms, describe the subsequent motion of an object that starts at rest at point A.

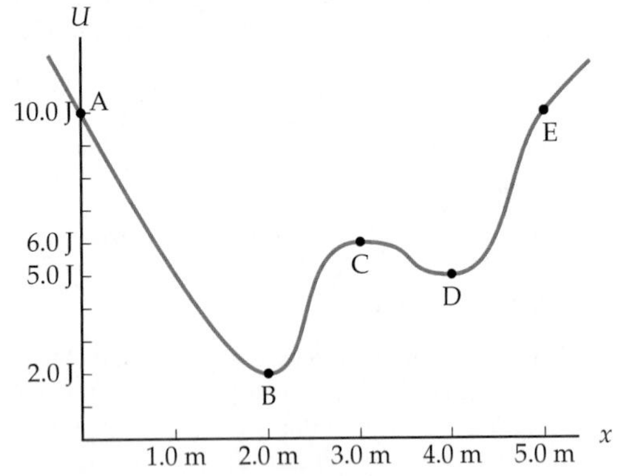

▲ **FIGURE 8–24** Problems 41, 42, 43, and 46

42. • An object moves along the x axis, subject to the potential energy shown in Figure 8–24. The object has a mass of 1.1 kg and starts at rest at point A. **(a)** What is the object's speed at point B? **(b)** At point C? **(c)** At point D? **(d)** What are the turning points for this object?

43. • A 1.44-kg object moves along the x axis, subject to the potential energy shown in Figure 8–24. If the object's speed at point C is 1.25 m/s, what are the approximate locations of its turning points?

44. • A 23-kg child swings back and forth on a swing suspended by 2.0-m-long ropes. Plot the gravitational potential energy of this system as a function of the angle the ropes make with the vertical, assuming the potential energy is zero when the ropes are vertical. Consider angles up to 90° on either side of the vertical.

45. •• Find the turning-point angles in the previous problem if the child has a speed of 0.89 m/s when the ropes are vertical. Indicate the turning points on a plot of the system's potential energy.

46. •• The potential energy of a particle moving along the x axis is shown in Figure 8–24. When the particle is at $x = 1.0$ m it has 3.6 J of kinetic energy. Give approximate answers to the following questions. **(a)** What is the total mechanical energy of the system? **(b)** What is the smallest value of x the particle can reach? **(c)** What is the largest value of x the particle can reach?

47. •• A block of mass $m = 0.95$ kg is connected to a spring of force constant $k = 775$ N/m on a smooth, horizontal surface. **(a)** Plot the potential energy of the spring from $x = -5.00$ cm to $x = 5.00$ cm. **(b)** Determine the turning points of the block if its speed at $x = 0$ is 1.3 m/s.

48. •• A ball of mass $m = 0.75$ kg is thrown straight upward with an initial speed of 8.9 m/s. **(a)** Plot the gravitational potential energy of the block from its launch height, $y = 0$, to the height $y = 5.0$ m. Let $U = 0$ correspond to $y = 0$. **(b)** Determine the turning point (maximum height) of this mass.

49. ••• Two blocks, each of mass m, are connected on a frictionless horizontal table by a spring of force constant k and equilibrium length L. Find the maximum and minimum separation between the two blocks in terms of their maximum speed, v_{max}, relative to the table. (The two blocks always move in opposite directions as they oscillate back and forth about a fixed position.)

General Problems

50. •• **IP** A sled slides without friction down a small, ice-covered hill. If the sled starts from rest at the top of the hill, its speed at the bottom is 7.50 m/s. **(a)** On a second run, the sled starts with a speed of 1.50 m/s at the top. When it reaches the bottom of the hill, is its speed 9.00 m/s, more than 9.00 m/s, or less than 9.00 m/s? Explain. **(b)** Find the speed of the sled at the bottom of the hill after the second run.

51. •• In the previous problem, what is the height of the hill?

52. •• A 61-kg skier encounters a dip in the snow's surface that has a circular cross section with a radius of curvature of 12 m. If the skier's speed at point A in **Figure 8–25** is 8.0 m/s, what is the normal force exerted by the snow on the skier at point B? Ignore frictional forces.

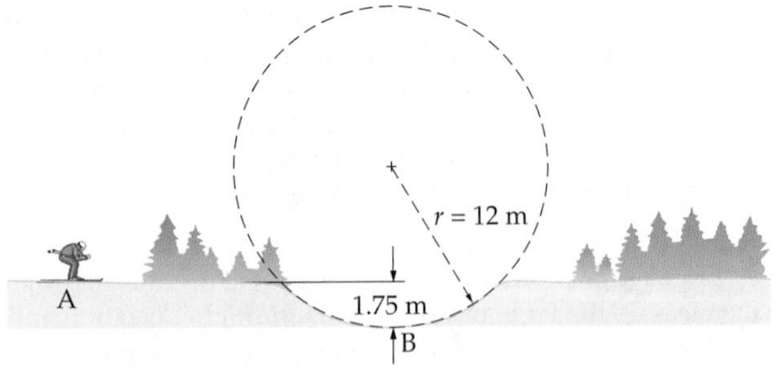

▲ **FIGURE 8–25** Problem 52

53. •• The spring in a clothespin is compressed 0.40 cm, storing energy in the form of spring potential energy. If the spring is compressed twice as far, the spring potential energy increases by 0.0046 J. What is the force constant, k, for this spring?

54. •• In a circus act, a 66-kg trapeze artist starts from rest with the 4.5-m trapeze rope horizontal. What is the tension in the rope when it is vertical?

55. •• **Running Shoes** The soles of a popular make of running shoe have a force constant of 2.0×10^5 N/m. Treat the soles as ideal springs for the following questions. **(a)** If a 62-kg person stands in a pair of these shoes, with her weight distributed equally on both feet, how much does she compress the soles? **(b)** How much energy is stored in the soles of her shoes when she's standing?

56. •• **Nasal Strips** The force required to flex a nasal strip and apply it to the nose is 0.25 N; the energy stored in the strip when flexed is 0.0022 J. Assume the strip to be an ideal spring for the following questions. Find **(a)** the distance through which the strip is flexed and **(b)** the force constant of the strip.

57. •• **IP** A pendulum bob with a mass of 0.13 kg is attached to a string with a length of 0.95 m. We choose the potential energy to be zero when the string makes an angle of 90° with the vertical. **(a)** Find the potential energy of this system when the string makes an angle of 45° with the vertical. **(b)** Is the magnitude of the change in potential energy from an angle of 90° to 45° greater than, less than, or the same as the magnitude of the change from 45° to 0°? Explain. **(c)** Calculate the potential energy of the system when the string is vertical.

58. •• Suppose the pendulum bob in Figure 8–22 has a mass of 0.25 kg. **(a)** How much work does gravity do on the bob as it moves from point A to point B? **(b)** From point B to point A? **(c)** How much work does the string do on the bob as it moves from point A to point B? **(d)** From point B to point A?

59. •• An 1865-kg airplane starts at rest on an airport runway at sea level. **(a)** What is the change in mechanical energy of the airplane if it climbs to a cruising altitude of 2420 m and maintains a constant speed of 96.5 m/s? **(b)** What cruising speed would the plane need at this altitude if its increase in kinetic energy is to be equal to its increase in potential energy?

60. •• **IP** At the local playground a child on a swing has a speed of 2.12 m/s when the swing is at its lowest point. **(a)** To what maximum vertical height does the child rise, assuming he sits still and "coasts"? Ignore air resistance. **(b)** How do your results change if the initial speed of the child is halved?

61. •• The water slide shown in **Figure 8–26** ends at a height of 1.50 m above the pool. If the person starts from rest at point A and lands in the water at point B, what is the height h of the water slide? (Assume the water slide is frictionless.)

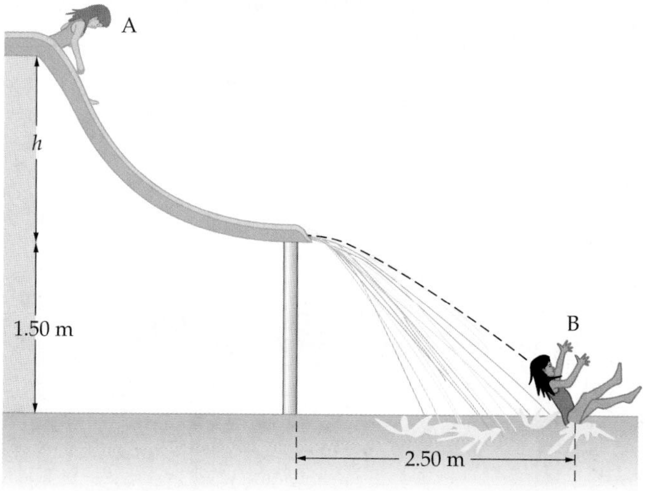

▲ **FIGURE 8–26** Problems 61 and 62

62. •• If the height of the water slide in Figure 8–26 is $h = 3.2$ m, and the person's initial speed at point A is 0.54 m/s, what is the new horizontal distance between the base of the slide and the splashdown point of the person?

63. •• **IP** A person is to be released from rest on a swing pulled away from the vertical by an angle of 20.0°. The two frayed ropes of the swing are 2.75 m long, and will break if the tension in either of them exceeds 355 N. **(a)** What is the maximum weight the person can have and not break the ropes? **(b)** If the person is released at an angle greater than 20.0°, does the maximum weight increase, decrease, or stay the same? Explain.

64. •• **IP** A car is coasting without friction toward a hill of height h and radius of curvature r. **(a)** What initial speed, v_0, will result in the car's wheels just losing contact with the roadway as the car crests the hill? **(b)** What happens if the initial speed of the car is greater than the value found in part (a)?

65. •• A skateboarder starts at point A in **Figure 8–27** and rises to a height of 2.64 m above the top of the ramp at point B. What was the skateboarder's initial speed at point A?

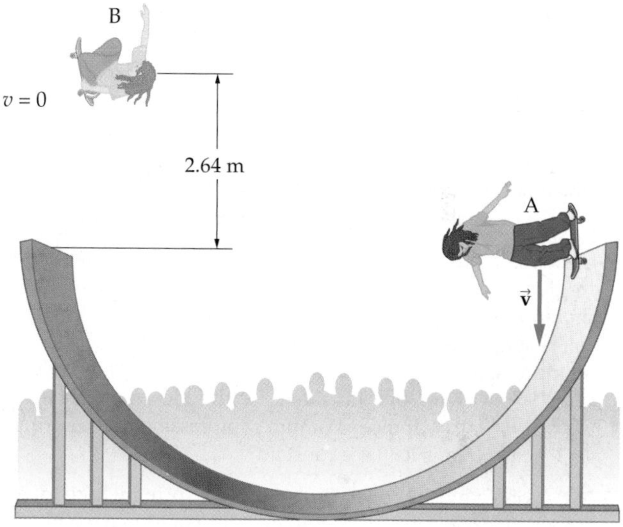

▲ **FIGURE 8–27** Problems 65 and 70

66. •• In the Atwood's machine of Problem 26, the mass m_2 remains at rest once it hits the floor, but the mass m_1 continues moving upward. How much higher does m_1 go after m_2 has landed? Give your answer for the case $h = 1.2$ m, $m_1 = 3.7$ kg, and $m_2 = 4.1$ kg.

67. •• An 8.70-kg block slides with an initial speed of 1.66 m/s up a ramp inclined at an angle of 27.4° with the horizontal. The coefficient of kinetic friction between the block and the ramp is 0.62. Use energy conservation to find the distance the block slides before coming to rest.

68. •• Repeat the previous problem for the case of an 8.70-kg block sliding down the ramp, with an initial speed of 1.66 m/s.

69. •• Jeff of the Jungle swings on a 7.6-m vine that initially makes an angle of 37° with the vertical. If Jeff starts at rest and has a mass of 78 kg, what is the tension in the vine at the lowest point of the swing?

70. •• A skateboard track has the form of a circular arc with a 4.00 m radius, extending to an angle of 90.0° relative to the vertical on either side of the lowest point, as shown in Figure 8–27. A 57.0-kg skateboarder starts from rest at the top of the circular arc. What is the normal force exerted on the skateboarder at the bottom of the circular arc?

71. •• **IP** A 2.2-kg mass is attached to a spring with a force constant of 560 N/m on a smooth horizontal surface. The mass is pulled

to one side, released, and allowed to oscillate back and forth. **(a)** Find the maximum compression and expansion of the spring if the maximum speed of the mass is 1.7 m/s. **(b)** By what multiplicative factor do your answers to part (a) change if the force constant of the spring is doubled?

72. •• A 1.9-kg block slides down a frictionless ramp, as shown in **Figure 8–28**. The top of the ramp is 1.5 m above the ground; the bottom of the ramp is 0.25 m above the ground. The block leaves the ramp moving horizontally, and lands a horizontal distance d away. Find the distance d.

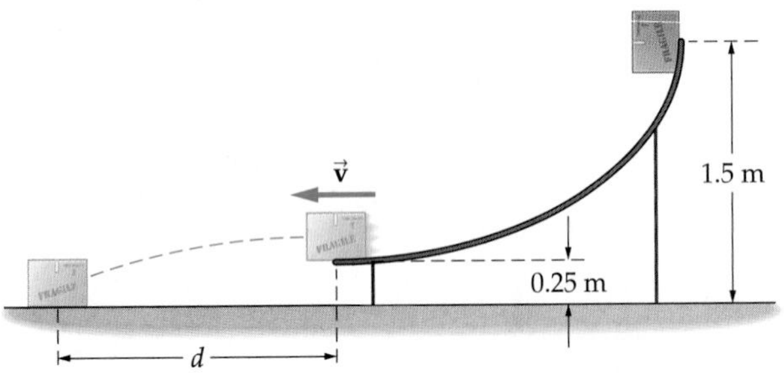

▲ **FIGURE 8–28** Problems 72 and 73

73. •• Suppose the ramp in Figure 8–28 is not frictionless. Find the distance d for the case in which friction on the ramp does −9.7 J of work on the block before it becomes airborne.

74. •• Two blocks, each with a mass $m = 0.25$ kg, can slide without friction on a horizontal surface. Initially, block 1 is in motion with a speed $v = 1.2$ m/s; block 2 is at rest. When block 1 collides with block 2, a spring bumper on block 1 is compressed. Maximum compression of the spring occurs when the two blocks move with the same speed, $v/2 = 0.60$ m/s. If the maximum compression of the spring is 1.8 cm, what is its force constant?

75. •• A 1.8-kg block is pushed up against a spring whose force constant is 650 N/m, compressing it 5.1 cm. When the block is released it moves horizontally without friction until it launches off the edge of a table 0.53 m above the ground. Find the horizontal distance, d, covered by the block during its fall to the ground.

76. •• BIO **Compressing the Ground** A running track at Harvard University uses a surface with a force constant of 2.5×10^5 N/m. This surface is compressed slightly every time a runner's foot lands on it. The force exerted by the foot, according to the Saucony shoe company, has a magnitude of 2700 N for a typical runner. Treating the track's surface as an ideal spring, find **(a)** the amount of compression caused by a foot hitting the track and **(b)** the energy stored briefly in the track every time a foot lands.

77. •• BIO **A Flea's Jump** The resilin in the upper leg (coxa) of a flea has a force constant of about 26 N/m, and when the flea cocks its jumping legs, the resilin in each leg is stretched by approximately 0.10 mm. Given that the flea has a mass of 0.50 mg, and that two legs are used in a jump, estimate the maximum height a flea can attain by using the energy stored in the resilin. (Assume the resilin to be an ideal spring.)

78. ••• IP A trapeze artist of mass m swings on a rope of length L. Initially, the trapeze artist is at rest and the rope makes an angle θ with the vertical. **(a)** Find the tension in the rope when it is vertical. **(b)** Explain why your result for part (a) depends on L in the way it does.

79. ••• IP **Tension at the Bottom** A ball of mass m is attached to a string of length L and released from rest at the point A in **Figure 8–29**. **(a)** Show that the tension in the string when the ball reaches point B is $3\,mg$, independent of the length l. **(b)** Give a detailed physical explanation for the fact that the tension at point B is independent of the length l.

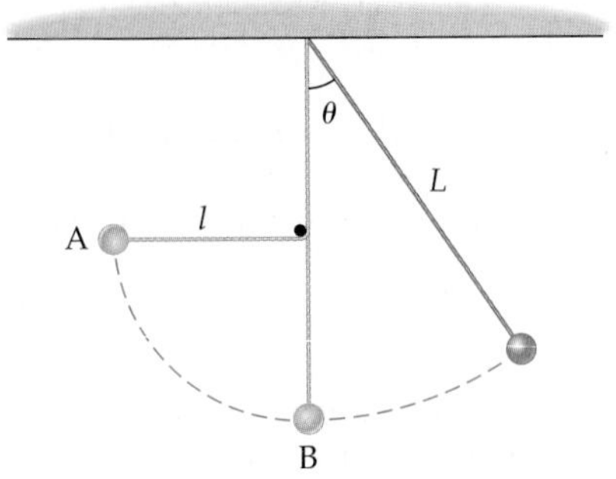

▲ **FIGURE 8–29** Problems 79 and 80

80. ••• IP In Figure 8–29, suppose that $L = 0.652$ m and $l = 0.325$ m. **(a)** Find the maximum angle the string makes with the vertical when the mass is released from rest at point A and swings as far to the right as it can. **(b)** At the point found in part (a), find the height of the mass above point B. Explain the physical significance of your result. **(c)** Give the angle of part (a) as a general expression in terms of L and l.

81. ••• An ice cube is placed on top of an overturned spherical bowl of radius r, as indicated in **Figure 8–30**. If the ice cube slides downward from rest at the top of the bowl, at what angle θ does it separate from the bowl? In other words, at what angle does the normal force between the ice cube and the bowl go to zero?

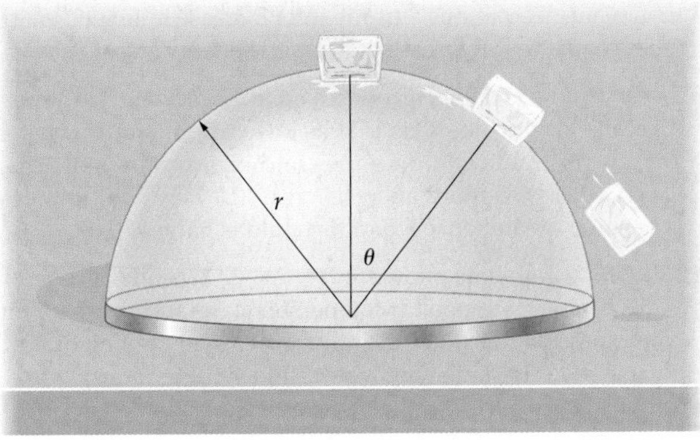

▲ **FIGURE 8–30** Problem 81

82. ••• IP The two blocks shown in **Figure 8–31** are moving with an initial speed v. **(a)** If the system is frictionless, find the

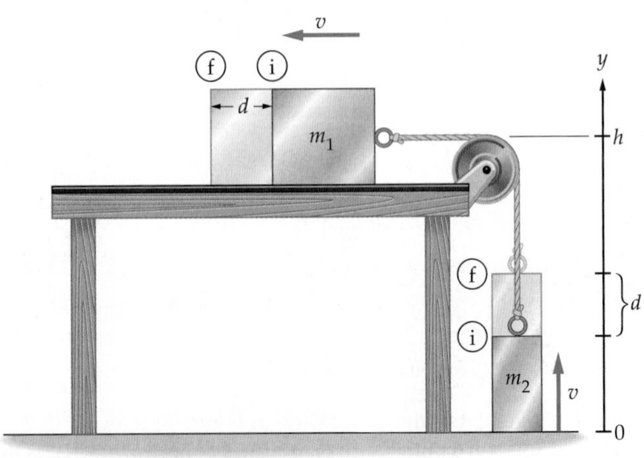

▲ **FIGURE 8–31** Problems 82 and 83

distance d the blocks travel before coming to rest. (Let $U = 0$ correspond to the initial position of block 2.) **(b)** Is the work done on block 2 by the rope positive, negative, or zero? Explain. **(c)** Calculate the work done on block 2 by the rope.

83. ••• **IP** Consider the system shown in Figure 8–31. **(a)** What initial speed v is required if the blocks $m_1 = 2.4$ kg and $m_2 = 1.1$ kg are to travel a distance $d = 6.5$ cm before coming to rest? Assume the coefficient of kinetic friction between m_1 and the tabletop is $\mu_k = 0.25$. **(b)** Is the work done on m_2 by the rope positive, negative, or zero? Explain. **(c)** Calculate the work done on m_2 by the rope.

84. ••• **IP Loop-the-Loop (a)** A block of mass m slides from rest on a frictionless loop-the-loop track, as shown in **Figure 8–32**. What is the minimum release height, h, required for the block to maintain contact with the track at all times? Give your answer in terms of the radius of the loop, r. **(b)** Explain why the release height obtained in part (a) is independent of the block's mass.

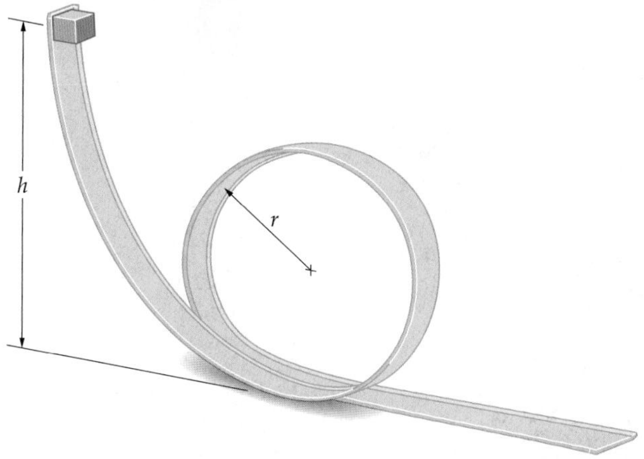

▲ **FIGURE 8–32** Problem 84

85. ••• **Figure 8–33** shows a 1.50-kg block at rest on a ramp of height h. When the block is released, it slides without friction to the bottom of the ramp, and then continues across a surface that is frictionless except for a rough patch of width 10.0 cm that has a coefficient of kinetic friction $\mu_k = 0.640$. Find h such that the block's speed after crossing the rough patch is 3.50 m/s.

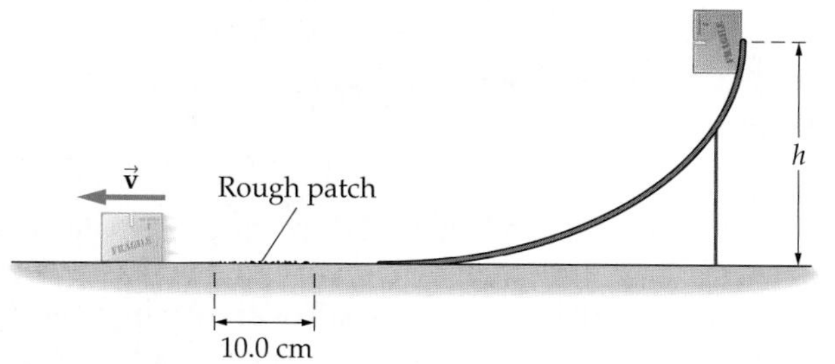

▲ **FIGURE 8–33** Problem 85

86. ••• In **Figure 8–34** a 1.2-kg block is held at rest against a spring with a force constant $k = 730$ N/m. Initially, the spring is compressed a distance d. When the block is released, it slides across a surface that is frictionless except for a rough patch of width 5.0 cm that has a coefficient of kinetic friction $\mu_k = 0.44$. Find d such that the block's speed after crossing the rough patch is 2.3 m/s.

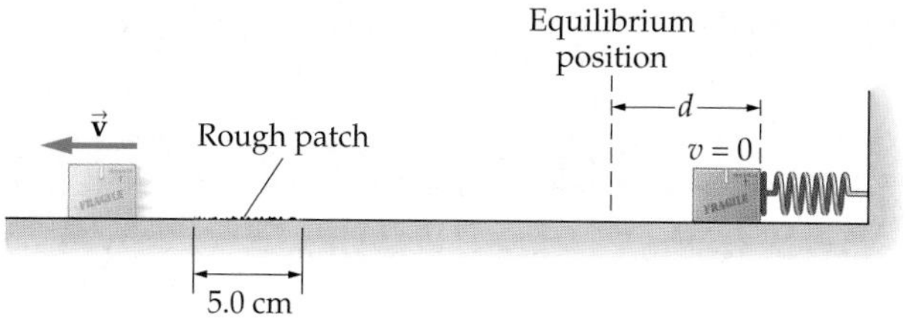

▲ **FIGURE 8–34** Problem 86

87. ••• **IP Using Work and Energy to Calculate Tension** Consider the Atwood's machine shown in Figure 8–23, with $h = 1.2$ m, $m_1 = 3.7$ kg, and $m_2 = 4.1$ kg. In this problem, we show how to calculate the tension in the rope using energy and work, rather than Newton's laws. **(a)** Is the change in mechanical energy for block 2 as it drops through the height h positive, negative, or zero? Explain. **(b)** Use energy conservation applied to the entire system to calculate the change in mechanical energy for block 2 as it drops through the height h. **(c)** Use your answer to part (b), and the known drop height, to find the magnitude of the tension in the rope.

Interactive Problems

88. •• ***Referring to Example 8–8*** Suppose the block is released from rest with the spring compressed 5.00 cm. The mass of the block is 1.70 kg and the force constant of the spring is 955 N/m. **(a)** What is the speed of the block when the spring expands to a compression of only 2.50 cm? **(b)** What is the speed of the block after it leaves the spring?

89. •• **IP** ***Referring to Example 8–8*** Consider a spring with a force constant of 955 N/m. **(a)** Suppose the mass of the block is 1.70 kg, but its initial speed can be varied. What initial speed is required to give a maximum spring compression of 4.00 cm? **(b)** Suppose the initial speed of the block is 1.09 m/s, but its mass can be varied. What mass is required to give a maximum spring compression of 4.00 cm?

90. •• ***Referring to Example 8–10*** Suppose we would like the landing speed of block 2 to be increased to 1.50 m/s. **(a)** Should the coefficient of kinetic friction between block 1 and the tabletop be increased or decreased? **(b)** Find the required coefficient of kinetic friction for a landing speed of 1.50 m/s. Note that $m_1 = 2.40$ kg, $m_2 = 1.80$ kg, and $d = 0.500$ m.

91. ••• **IP** ***Referring to Example 8–10*** Consider the case where $m_1 = 2.40$ kg, $m_2 = 1.80$ kg, $d = 0.500$ m, and $\mu_k = 0.450$. **(a)** Does the mechanical energy of block 2 increase, decrease, or stay the same as it moves from point i to point f? **(b)** Find the change in mechanical energy for block 2 as it moves from point i to point f. **(c)** What is the nonconservative work done on block 2 by the tension in the rope? **(d)** What is the tension in the rope? **(e)** What is the nonconservative work done on block 1 by the tension in the rope?

Linear Momentum and Collisions

9–1	Linear Momentum	241
9–2	Momentum and Newton's Second Law	243
9–3	Impulse	244
9–4	Conservation of Linear Momentum	248
9–5	Inelastic Collisions	253
9–6	Elastic Collisions	259
9–7	Center of Mass	264
*9–8	Systems with Changing Mass: Rocket Propulsion	269

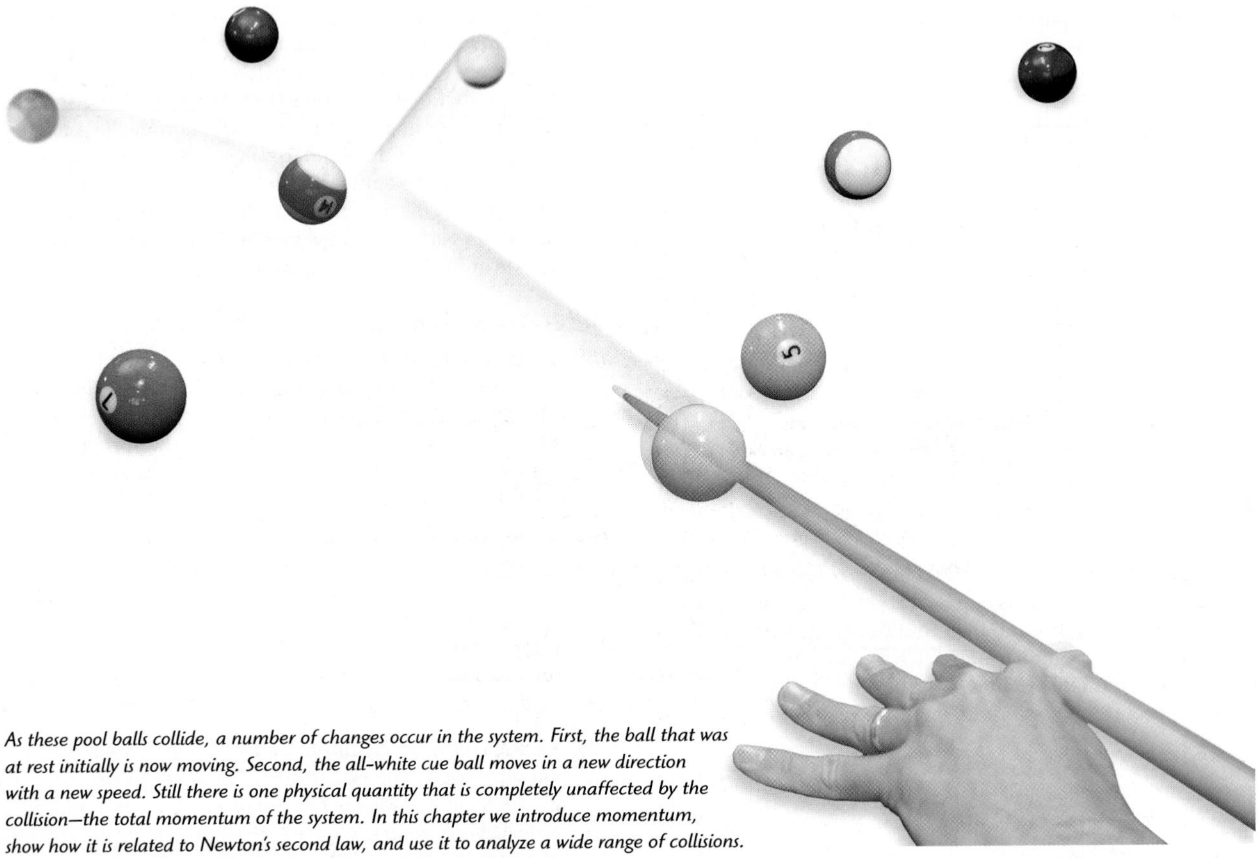

As these pool balls collide, a number of changes occur in the system. First, the ball that was at rest initially is now moving. Second, the all-white cue ball moves in a new direction with a new speed. Still there is one physical quantity that is completely unaffected by the collision—the total momentum of the system. In this chapter we introduce momentum, show how it is related to Newton's second law, and use it to analyze a wide range of collisions.

Conservation laws play a central role in physics. In this chapter we introduce the concept of *momentum* and show that it, like energy, is a conserved quantity. Nothing we can do—in fact, nothing that can occur in nature—can change the total energy or the total momentum of the universe.

As with conservation of energy, we shall see that the conservation of momentum provides a powerful way of approaching a variety of problems that would be extremely difficult to solve using Newton's laws directly. In particular, problems involving the collision of two or more objects—such as a baseball bat striking a ball or one car bumping into another at an intersection—are especially well suited to a momentum approach. Finally, we introduce the concept of the *center of mass* and show that it allows us to extend many of the results that have been obtained for point particles to systems involving more realistic objects.

9–1 Linear Momentum

Imagine for a moment that you are sitting at rest on a skateboard that can roll without friction on a smooth surface. If you catch a heavy, slow-moving ball tossed to you by a friend, you begin to move. If, on the other hand, your friend tosses you a light, yet fast-moving ball, the net effect may be the same—that is, catching a lightweight ball moving fast enough will cause you to move with the same speed as when you caught the heavy ball.

In physics, the previous observations are made precise by defining a quantity called the **linear momentum,** $\vec{\mathbf{p}}$, which is defined as the product of the mass m and velocity $\vec{\mathbf{v}}$ of an object:

Definition of Linear Momentum, $\vec{\mathbf{p}}$

$$\vec{\mathbf{p}} = m\vec{\mathbf{v}} \qquad \text{9–1}$$

SI unit: kg · m/s

In our example, if the heavy ball has twice the mass of the light ball but the light ball has twice the speed of the heavy ball, the momenta of the two balls are equal in magnitude. We can see from Equation 9–1 that the units of linear momentum are simply the units of mass times the units of velocity: kg · m/s. There is no special shorthand name given to this combination of units.

It is important to note that a constant *linear* momentum $\vec{\mathbf{p}}$ is the momentum of an object of mass m that is *moving in a straight line* with a velocity $\vec{\mathbf{v}}$. In Chapter 11 we introduce a similar quantity to describe the momentum of an object that rotates. This momentum will be referred to as the *angular momentum*. In general, when we simply say momentum, we are referring to the linear momentum $\vec{\mathbf{p}}$. We will always specify angular momentum when referring to the momentum associated with rotation.

Because the velocity $\vec{\mathbf{v}}$ is a vector with both a magnitude and a direction, so too is the momentum, $\vec{\mathbf{p}} = m\vec{\mathbf{v}}$. The next exercise gives some feeling for the *magnitude* of the momentum, $p = mv$, for everyday objects.

EXERCISE 9–1

(a) A 1180-kg car drives along a city street at 30.0 miles per hour (13.4 m/s). What is the magnitude of the car's momentum? **(b)** A major-league pitcher can give a 0.142-kg baseball a speed of 101 mi/h (45.1 m/s). Find the magnitude of the baseball's momentum.

Solution

a. Using $p = mv$, we find

$$p_c = m_c v_c = (1180 \text{ kg})(13.4 \text{ m/s}) = 15{,}800 \text{ kg} \cdot \text{m/s}$$

b. Similarly,

$$p_b = m_b v_b = (0.142 \text{ kg})(45.1 \text{ m/s}) = 6.40 \text{ kg} \cdot \text{m/s}$$

As an illustration of the vector nature of momentum, consider the situations shown in **Figures 9–1 (a)** and **(b)**. In Figure 9–1 (a), a 0.10-kg beanbag bear is dropped to the floor, where it hits with a speed of 4.0 m/s and sticks. In Figure 9–1 (b) a 0.10-kg rubber ball also hits the floor with a speed of 4.0 m/s, but in this case the ball bounces upward off the floor. Assuming an ideal rubber ball, its initial upward speed is 4.0 m/s. Now the question in each case is, "What is the change in momentum?"

To approach the problem systematically, we introduce a coordinate system as shown in Figure 9–1. With this choice, we can see that neither object has momentum in the x direction; thus we need only consider the y component of momentum, p_y. The problem, therefore, is one-dimensional; still, we must be careful about the sign of p_y.

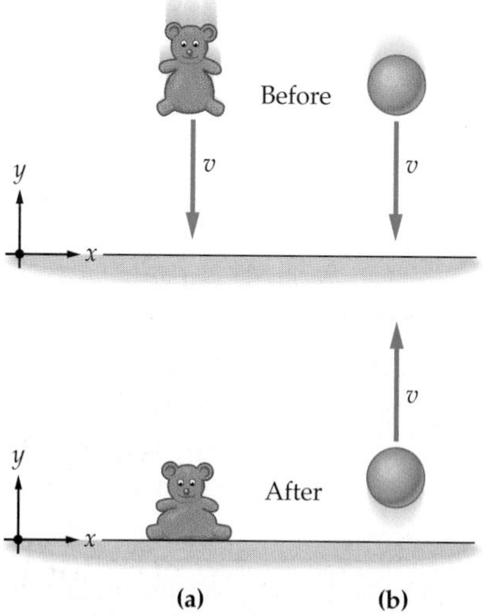

▲ **FIGURE 9–1 Change in momentum**
A beanbag bear and a rubber ball, with the same mass m and the same downward speed v, hit the floor. **(a)** The beanbag bear comes to rest on hitting the floor. Its change in momentum is mv upward. **(b)** The rubber ball bounces upward with a speed v. Its change in momentum is $2mv$ upward.

We begin with the beanbag. Just before hitting the floor, it moves downward (that is, in the negative y direction) with a speed of $v = 4.0\ \mathrm{m/s}$. Letting m stand for the mass of the beanbag, we find that the initial momentum is

$$p_{y,\mathrm{i}} = m(-v)$$

After landing on the floor, the beanbag is at rest; hence, its final momentum is zero:

$$p_{y,\mathrm{f}} = m(0) = 0$$

Therefore the change in momentum is

$$\begin{aligned} \Delta p_y &= p_{y,\mathrm{f}} - p_{y,\mathrm{i}} = 0 - m(-v) = mv \\ &= (0.10\ \mathrm{kg})(4.0\ \mathrm{m/s}) = 0.40\ \mathrm{kg\cdot m/s} \end{aligned}$$

Note that the change in momentum is positive—that is, in the upward direction. This makes sense because, before the bag landed, it had a negative (downward) momentum in the y direction. In order to increase the momentum from a negative value to zero, it is necessary to add a positive (upward) momentum.

Next, consider the rubber ball in Figure 9–1 (b). Before bouncing, its momentum is

$$p_{y,\mathrm{i}} = m(-v)$$

the same as for the beanbag. After bouncing, when the ball is moving in the upward (positive) direction, its momentum is

$$p_{y,\mathrm{f}} = mv$$

As a result, the change in momentum for the rubber ball is

$$\begin{aligned} \Delta p_y &= p_{y,\mathrm{f}} - p_{y,\mathrm{i}} = mv - m(-v) = 2mv \\ &= 2(0.10\ \mathrm{kg})(4.0\ \mathrm{m/s}) = 0.80\ \mathrm{kg\cdot m/s} \end{aligned}$$

This is *twice* the change in momentum of the beanbag! The reason is that in this case, the momentum in the y direction must first be increased from $-mv$ to 0, then increased again from 0 to mv. For the beanbag, the change was merely from $-mv$ to 0.

Note how important it is to be careful about the vector nature of the momentum and to use the correct sign for p_y. Otherwise, we might have concluded—erroneously—that the rubber ball had zero change in momentum, since the *magnitude* of its momentum was unchanged by the bounce. In fact, its momentum does change due to the change in its *direction* of motion.

One additional point: Since momentum is a vector, the total momentum of a system of objects is the *vector* sum of the momenta of all the objects. That is,

$$\vec{\mathbf{p}}_{\mathrm{total}} = \vec{\mathbf{p}}_1 + \vec{\mathbf{p}}_2 + \vec{\mathbf{p}}_3 + \cdots \qquad 9\text{–}2$$

This is illustrated for the case of three objects in the following Example.

PROBLEM-SOLVING NOTE

Coordinate Systems

Be sure to draw a coordinate system for momentum problems, even if the problem is only one-dimensional. It is important to use the coordinate system to assign the correct sign to velocities and momenta in the system.

EXAMPLE 9–1 Duck, Duck, Goose: Adding Momenta

At a city park, a person throws some bread into a duck pond. Two 4.00-kg ducks and a 9.00-kg goose paddle rapidly toward the bread, as shown in our sketch. If the ducks swim at 1.10 m/s, and the goose swims with a speed of 1.30 m/s, find the magnitude and direction of the total momentum of the three birds.

Picture the Problem

In our sketch we place the origin where the bread floats on the water. Note that duck 1 swims in the positive x direction, duck 2 swims in the negative y direction, and the goose swims in the positive y direction. Therefore, $\vec{\mathbf{p}}_{\mathrm{d1}} = m_{\mathrm{d}}v_{\mathrm{d}}\hat{\mathbf{x}}$, $\vec{\mathbf{p}}_{\mathrm{d2}} = -m_{\mathrm{d}}v_{\mathrm{d}}\hat{\mathbf{y}}$, and $\vec{\mathbf{p}}_{\mathrm{g}} = m_{\mathrm{g}}v_{\mathrm{g}}\hat{\mathbf{y}}$, where $v_{\mathrm{d}} = 1.10\ \mathrm{m/s}$, $m_{\mathrm{d}} = 4.00\ \mathrm{kg}$, $v_{\mathrm{g}} = 1.30\ \mathrm{m/s}$ and $m_{\mathrm{g}} = 9.00\ \mathrm{kg}$. The total momentum, $\vec{\mathbf{p}}_{\mathrm{total}}$, points at an angle θ relative to the positive x axis.

Strategy

Write the momentum of each bird as a vector, using unit vectors in the x and y directions. Next, sum these vectors component by component to find the total momentum. Finally, use the components of the total momentum to calculate its magnitude and direction.

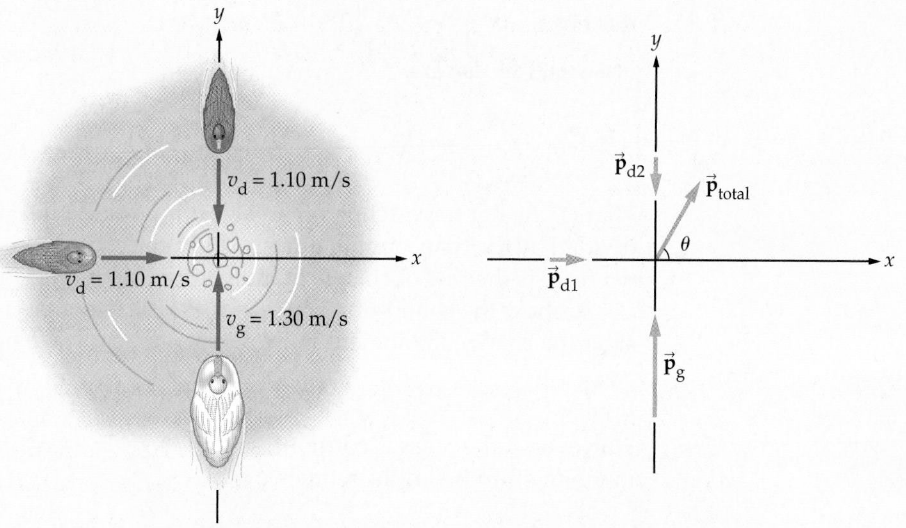

Solution

1. Use x and y unit vectors to express the momentum of each bird in vector form:

$$\vec{p}_{d1} = m_d v_d \hat{x} = (4.00\ \text{kg})(1.10\ \text{m/s})\hat{x} = (4.40\ \text{kg}\cdot\text{m/s})\hat{x}$$
$$\vec{p}_{d2} = -m_d v_d \hat{y} = -(4.00\ \text{kg})(1.10\ \text{m/s})\hat{y} = -(4.40\ \text{kg}\cdot\text{m/s})\hat{y}$$
$$\vec{p}_{g} = m_g v_g \hat{y} = (9.00\ \text{kg})(1.30\ \text{m/s})\hat{y} = (11.7\ \text{kg}\cdot\text{m/s})\hat{y}$$

2. Sum the momentum vectors to obtain the total momentum:

$$\vec{p}_{total} = \vec{p}_{d1} + \vec{p}_{d2} + \vec{p}_g = (4.40\ \text{kg}\cdot\text{m/s})\hat{x} + [-4.40\ \text{kg}\cdot\text{m/s} + 11.7\ \text{kg}\cdot\text{m/s}]\hat{y} = (4.40\ \text{kg}\cdot\text{m/s})\hat{x} + (7.30\ \text{kg}\cdot\text{m/s})\hat{y}$$

3. Calculate the magnitude of the total momentum:

$$p_{total} = \sqrt{p_{total,x}^2 + p_{total,y}^2} = \sqrt{(4.40\ \text{kg}\cdot\text{m/s})^2 + (7.30\ \text{kg}\cdot\text{m/s})^2} = 8.52\ \text{kg}\cdot\text{m/s}$$

4. Calculate the direction of the total momentum:

$$\theta = \tan^{-1}\left(\frac{p_{total,y}}{p_{total,x}}\right) = \tan^{-1}\left(\frac{7.30\ \text{kg}\cdot\text{m/s}}{4.40\ \text{kg}\cdot\text{m/s}}\right) = 58.9^\circ$$

Insight

Note that the momentum of each bird depends only on its mass and velocity; it is independent of the bird's location. In addition, we observe that the magnitude of the total momentum is less than the sum of the magnitudes of each bird's momentum individually. This is generally the case when dealing with vector addition—the only exception is when all vectors point in the same direction.

Practice Problem

Should the speed of the goose be increased or decreased if the total momentum of the three birds is to point in the positive x direction? Verify your answer by calculating the required speed. [**Answer:** The goose's speed must be decreased. Setting the momentum of the goose equal to minus the momentum of duck 2 yields $v_g = 0.489$ m/s.]

Some related homework problems: Problem 1, Problem 2, Problem 3

9–2 Momentum and Newton's Second Law

In Chapter 5 we introduced Newton's second law:

$$\sum \vec{F} = m\vec{a}$$

As mentioned, this expression is valid only for objects that have constant mass. The more general law, which holds even if the mass changes, is expressed in terms of momentum. In fact, Newton's original statement of the second law was in just this form:

Newton's Second Law

$$\sum \vec{\mathbf{F}} = \frac{\Delta \vec{\mathbf{p}}}{\Delta t} \qquad 9\text{–}3$$

That is, the net force acting on an object is equal to the change in its momentum divided by the time interval during which the change occurs—in other words, the net force is the rate of change of momentum with time.

To show the connection between these two statements of the second law, consider the change in momentum, $\Delta\vec{\mathbf{p}}$. Since $\vec{\mathbf{p}} = m\vec{\mathbf{v}}$, we have

$$\Delta\vec{\mathbf{p}} = \vec{\mathbf{p}}_\text{f} - \vec{\mathbf{p}}_\text{i} = m_\text{f}\vec{\mathbf{v}}_\text{f} - m_\text{i}\vec{\mathbf{v}}_\text{i}$$

However, if the mass is constant, so that $m_\text{f} = m_\text{i} = m$, it follows that the change in momentum is simply m times $\Delta\vec{\mathbf{v}}$:

$$\Delta\vec{\mathbf{p}} = m_\text{f}\vec{\mathbf{v}}_\text{f} - m_\text{i}\vec{\mathbf{v}}_\text{i} = m(\vec{\mathbf{v}}_\text{f} - \vec{\mathbf{v}}_\text{i}) = m\Delta\vec{\mathbf{v}}$$

As a result, Newton's second law, for objects of constant mass, can be written as follows:

$$\sum \vec{\mathbf{F}} = \frac{\Delta \vec{\mathbf{p}}}{\Delta t} = m\frac{\Delta \vec{\mathbf{v}}}{\Delta t}$$

Finally, recall that acceleration is the rate of change of velocity with time,

$$\vec{\mathbf{a}} = \frac{\Delta \vec{\mathbf{v}}}{\Delta t}$$

Therefore, we can write Equation 9–3 as

$$\sum \vec{\mathbf{F}} = \frac{\Delta \vec{\mathbf{p}}}{\Delta t} = m\vec{\mathbf{a}} \qquad 9\text{–}4$$

Hence, the two statements are equivalent if the mass is constant.

It should be noted, however, that $\Sigma\vec{\mathbf{F}} = \Delta\vec{\mathbf{p}}/\Delta t$ is the general form of Newton's second law, and that it is valid no matter how the mass may vary. In the remainder of this chapter we use this form of the second law to investigate the connections between forces and changes in momentum.

9–3 Impulse

The pitcher delivers a fastball, the batter takes a swing, and with a crack of the bat the ball that was approaching home plate at 95.0 mi/h is now heading toward the pitcher at 115 mi/h. In the language of physics, we say that the bat has delivered an **impulse,** $\vec{\mathbf{I}}$, to the ball.

During the brief time the ball and bat are in contact—perhaps as little as a thousandth of a second—the force between them rises rapidly to a large value, as shown in **Figure 9–2**, then falls back to zero as the ball takes flight. It would be almost impossible, of course, to describe every detail of the way the force varies with time. Instead, we focus on the average force exerted by the bat, $\vec{\mathbf{F}}_\text{av}$, which is also shown in Figure 9–2. The impulse, then, is defined to be $\vec{\mathbf{F}}_\text{av}$ times the length of time, Δt, that the ball and bat are in contact, which is simply the area under the force-versus-time curve:

Definition of Impulse, $\vec{\mathbf{I}}$

$$\vec{\mathbf{I}} = \vec{\mathbf{F}}_\text{av}\,\Delta t \qquad 9\text{–}5$$

SI unit: N · s = kg · m/s

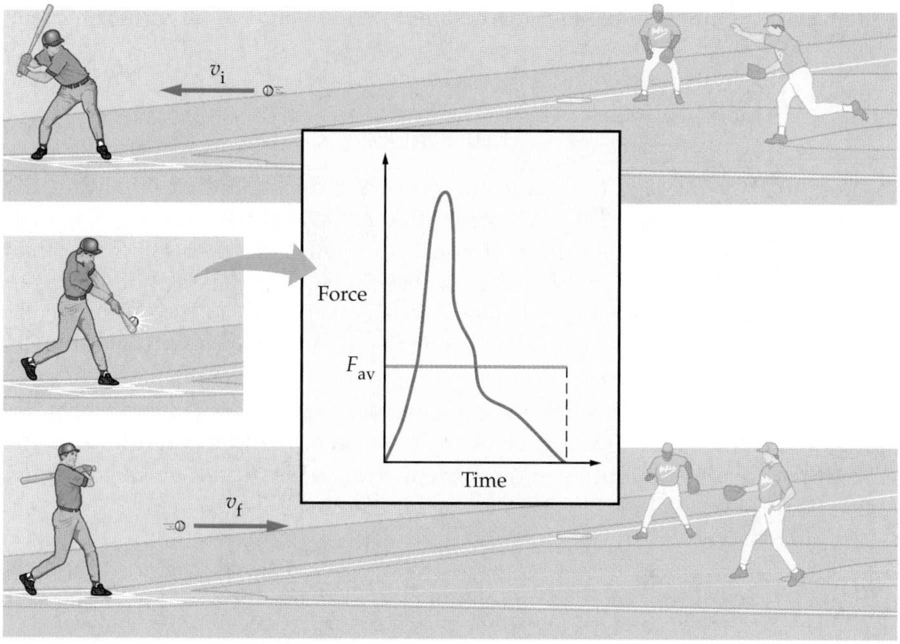

◀ **FIGURE 9–2 The average force during a collision**

The force between two objects that collide, as when a bat hits a baseball, rises rapidly to very large values, then drops again to zero in a matter of milliseconds. Rather than try to describe the complex behavior of the force, we focus on its average value, F_{av}. Note that the area under the F_{av} rectangle is the same as the area under the actual force curve.

Note that impulse is a vector and that it points in the same direction as the average force. In addition, its units are $N \cdot s = (kg \cdot m/s^2) \cdot s = kg \cdot m/s$, the same as the units of momentum.

It is no accident that impulse and momentum have the same units. In fact, rearranging Newton's second law, Equation 9–3, we see that the average force times Δt is simply the change in momentum of the ball due to the bat:

$$\vec{\mathbf{F}}_{av} = \frac{\Delta\vec{\mathbf{p}}}{\Delta t}$$

$$\vec{\mathbf{F}}_{av}\,\Delta t = \Delta\vec{\mathbf{p}}$$

Hence, in general, impulse is just the change in momentum:

Momentum-Impulse Theorem

$$\vec{\mathbf{I}} = \vec{\mathbf{F}}_{av}\,\Delta t = \Delta\vec{\mathbf{p}} \qquad 9\text{–}6$$

For instance, if we know the impulse delivered to an object—that is, its change in momentum—and the time interval during which the change occurs, we can find the average force that caused the impulse.

As an example, let's calculate the impulse given to the baseball considered at the beginning of this section, as well as the average force between the ball and the bat. First, set up a coordinate system with the positive x axis pointing from home plate toward the pitcher's mound, as indicated in **Figure 9–3**. If the ball's mass is 0.145 kg, its initial momentum—which is in the negative x direction—is

$$\vec{\mathbf{p}}_i = -mv_i\hat{\mathbf{x}} = -(0.145\text{ kg})(95.0\text{ mi/h})\left(\frac{0.447\text{ m/s}}{1\text{ mi/h}}\right)\hat{\mathbf{x}} = -(6.16\text{ kg}\cdot\text{m/s})\hat{\mathbf{x}}$$

Immediately after the hit, the ball's final momentum is in the positive x direction:

$$\vec{\mathbf{p}}_f = mv_f\hat{\mathbf{x}} = (0.145\text{ kg})(115\text{ mi/h})\left(\frac{0.447\text{ m/s}}{1\text{ mi/h}}\right)\hat{\mathbf{x}} = (7.45\text{ kg}\cdot\text{m/s})\hat{\mathbf{x}}$$

The impulse, then, is

$$\vec{\mathbf{I}} = \Delta\vec{\mathbf{p}} = \vec{\mathbf{p}}_f - \vec{\mathbf{p}}_i = [7.45\text{ kg}\cdot\text{m/s} - (-6.16\text{ kg}\cdot\text{m/s})]\hat{\mathbf{x}} = (13.6\text{ kg}\cdot\text{m/s})\hat{\mathbf{x}}$$

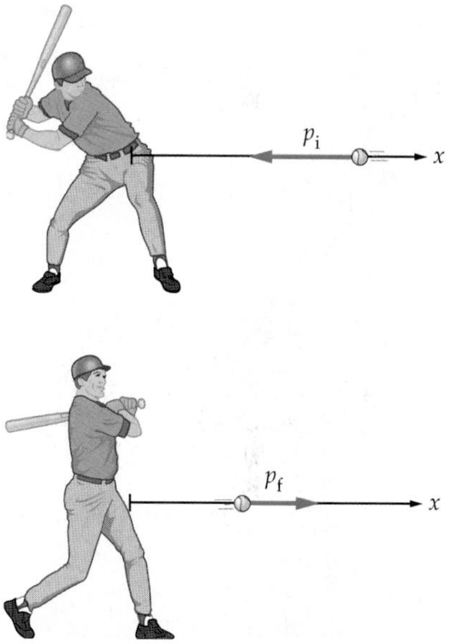

▲ **FIGURE 9–3 Hitting a baseball**

A batter hits a ball, sending it back toward the pitcher's mound. The impulse delivered to the ball by the bat changes its the ball's momentum from $-p_i\hat{\mathbf{x}}$ to $p_f\hat{\mathbf{x}}$.

REAL-WORLD PHYSICS

The force between a ball and a bat

If the ball and bat are in contact for 1.20 ms $= 1.20 \times 10^{-3}$ s, a typical time, the average force is

$$\vec{\mathbf{F}}_{av} = \frac{\Delta\vec{\mathbf{p}}}{\Delta t} = \frac{\vec{\mathbf{I}}}{\Delta t} = \frac{(13.6\ \text{kg}\cdot\text{m/s})\hat{\mathbf{x}}}{1.20 \times 10^{-3}\ \text{s}} = (1.13 \times 10^{4}\ \text{N})\hat{\mathbf{x}}$$

Note that the average force is in the positive x direction; that is, toward the pitcher, as expected. In addition, the magnitude of the average force is remarkably large. In everyday units, the force between the ball and the bat is more than 2500 pounds! This explains why the ball is observed in high-speed photographs to deform significantly during a hit—the force is so large that, for an instant, it partially flattens the ball. Finally, notice that the weight of the ball, which is only about 0.3 lb, is completely negligible compared to the forces involved during the hit.

In problems that are strictly one-dimensional, we can drop the vector notation when dealing with impulse. However, we must still be careful about the signs of the various quantities in the system. This is illustrated in the following Active Example.

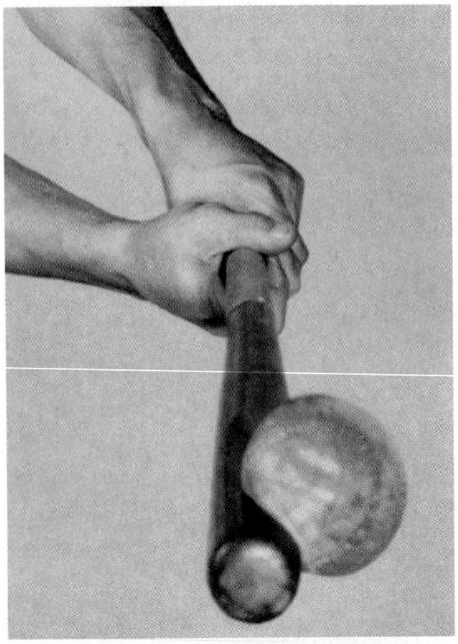

▲ When a softball is hit by a bat (top), an enormous force (thousands of newtons) acts for a very short period of time—perhaps only a few ms. During this time, the ball is dramatically deformed by the impact. To keep the same thing from happening to a pole vaulter, who must fall nearly 20 feet after clearing the bar (bottom), a deeply padded landing area is provided. The change in the pole vaulter's momentum as he is brought to a stop, $mv = F\Delta t$, is the same whether he lands on a mat or on concrete. However, the padding is very yielding, greatly prolonging the time Δt during which he is in contact with the mat. The corresponding force on the vaulter is thus markedly decreased.

ACTIVE EXAMPLE 9–1 Find the Final Speed of the Ball

A 0.144-kg baseball is moving toward home plate with a speed of 43.0 m/s when it is bunted (hit softly). The bat exerts an average force of 6.50×10^3 N on the ball for 1.30 ms. The average force is directed toward the pitcher, which we take to be the positive x direction. What is the final speed of the ball?

Solution *(Test your understanding by performing the calculations indicated in each step.)*

1. Relate change in momentum to impulse (Equation 9–5): $\Delta p = p_f - p_i = I = F_{av}\,\Delta t$
2. Solve for the final momentum: $p_f = F_{av}\,\Delta t + p_i$
3. Calculate the initial momentum: $p_i = -6.19\ \text{kg}\cdot\text{m/s}$
4. Calculate the impulse: $I = F_{av}\,\Delta t = 8.45\ \text{kg}\cdot\text{m/s}$
5. Use these results to find the final momentum: $p_f = 2.26\ \text{kg}\cdot\text{m/s}$
6. Divide by the mass to find the final velocity: $v_f = p_f/m = 15.7\ \text{m/s}$

Insight

With our choice of coordinate system, we see that the initial momentum of the ball was in the negative x direction. The impulse applied to the ball, however, resulted in a final momentum (and velocity) in the positive x direction.

Your Turn

Suppose the bat is in contact with the ball for 2.60 ms rather than 1.30 ms. What is the final speed of the ball in this case?

(Answers to **Your Turn** *problems are given in the back of the book.)*

We saw in Section 9–1 that the change in momentum is different for an object that hits something and sticks compared with an object that hits and bounces off. This means that the impulse, and hence the force, is different in the two cases. We explore this in the following Conceptual Checkpoint.

CONCEPTUAL CHECKPOINT 9–1 Rain Versus Hail

A person stands under an umbrella during a rain shower. A few minutes later the raindrops turn to hail—though the number of "drops" hitting the umbrella per time and their speed remains the same. Is the force required to hold the umbrella in the hail **(a)** the same as, **(b)** more than, or **(c)** less than the force required in the rain?

▲ Most bats can take off simply by dropping from their perch on a branch or the ceiling of a cave, but vampire bats like this one must leap from the ground to become airborne. They do so by rocking forward onto their front limbs and then pushing off, using the extremely strong pectoral muscles that are also their main source of power in flight. Pushing downward on the ground, a bat experiences an upward reaction force exerted on it by the ground, with a corresponding impulse sufficient to propel it upward a considerable distance. In fact, a vampire bat can launch itself 1 m or more into the air in a mere 30 ms.

Reasoning and Discussion

When raindrops strike the umbrella, they tend to splatter and run off; when hailstones hit the umbrella, they bounce back upward. As a result, the change in momentum is greater for the hail—just as the change in momentum is greater for a rubber ball bouncing off the floor than it is for a beanbag landing on the floor. Hence, the impulse and the force are greater with hail.

Answer:

(b) The force is greater in the hail.

We conclude this section with an additional calculation involving impulse.

EXAMPLE 9–2 Jumping for Joy

After winning a prize on a game show, a 72-kg contestant jumps for joy. **(a)** If the jump results in an upward speed of 2.1 m/s, what is the impulse experienced by the contestant? **(b)** Before the jump, the floor exerts an upward force of mg on the contestant. What additional average upward force does the floor exert if the contestant pushes down on it for 0.36 s during the jump?

Picture the Problem

Our sketch shows that the contestant's motion is purely one-dimensional, with a final speed of 2.1 m/s in the positive vertical direction. Note that we have chosen the positive y direction to be upward, therefore $\vec{\mathbf{v}}_i = 0$ and $\vec{\mathbf{v}}_f = (2.1 \text{ m/s})\hat{\mathbf{y}}$.

Strategy

a. From the momentum-impulse theorem, we know that impulse is equal to the change in momentum. We are given the initial and final velocities of the contestant, and his mass as well; hence the change in momentum, $\Delta\vec{\mathbf{p}}$, can be calculated using the definition of momentum, $\vec{\mathbf{p}} = m\vec{\mathbf{v}}$.

b. The average value of the additional force exerted on the contestant by the floor is $\Delta\vec{\mathbf{p}}/\Delta t$, where Δt is given as 0.36 s and $\Delta\vec{\mathbf{p}}$ is calculated in part (a).

Solution

Part (a)

1. Write an expression for the impulse, noting that $\vec{\mathbf{v}}_i = 0$: $\vec{\mathbf{I}} = \Delta\vec{\mathbf{p}} = \vec{\mathbf{p}}_f - \vec{\mathbf{p}}_i = m\vec{\mathbf{v}}_f$

2. Substitute numerical values: $\vec{\mathbf{I}} = m\vec{\mathbf{v}}_f = (72 \text{ kg})(2.1 \text{ m/s})\hat{\mathbf{y}} = (150 \text{ kg} \cdot \text{m/s})\hat{\mathbf{y}}$

continued on next page

continued from previous page

Part (b)

3. Express the average force in terms of the impulse $\vec{\mathbf{I}}$ and the time interval Δt:

$$\vec{\mathbf{F}}_{\text{av}} = \frac{\vec{\mathbf{I}}}{\Delta t} = \frac{(150\ \text{kg}\cdot\text{m/s})\hat{\mathbf{y}}}{0.36\ \text{s}} = (420\ \text{kg}\cdot\text{m/s}^2)\hat{\mathbf{y}} = (420\ \text{N})\hat{\mathbf{y}}$$

Insight

The magnitude of the additional average force exerted by the floor is rather large; in fact, 420 N is approximately 95 lb, or about 60% of the contestant's weight of 160 lb. Thus, the total upward force exerted by the floor is mg + 420 N = 710 N + 420 N, which is about 250 lb. The contestant, of course, exerts the same force downward. Fortunately, the contestant only needs to exert that force for a third of a second.

When the contestant lands, an impulse is required to bring him to rest. If he lands with stiff legs, the impulse occurs in a short time, resulting in a large force delivered to the knees—with possible harmful effects. If he bends his legs on landing, on the other hand, the time duration is significantly increased, and the force applied to the contestant is correspondingly reduced.

Practice Problem

Suppose the contestant lands with a speed of 2.1 m/s and comes to rest in 0.25 s. What is the magnitude of the average force exerted by the floor during landing? [**Answer:** mg + 600 N ~ 290 lb]

Some related homework problems: Problem 8, Problem 9

9–4 Conservation of Linear Momentum

In this section we turn to perhaps the most significant aspect of linear momentum—the fact that it is a conserved quantity. In this respect, it plays a fundamental role in physics similar to that of energy. We shall also see that momentum conservation leads to calculational simplifications, making it of great practical significance.

First, recall that the net force acting on an object is equal to the rate of change of its momentum

$$\sum \vec{\mathbf{F}} = \frac{\Delta \vec{\mathbf{p}}}{\Delta t}$$

Rearranging this expression, we find that the change in momentum during a time interval Δt is

$$\Delta \vec{\mathbf{p}} = \left(\sum \vec{\mathbf{F}}\right)\Delta t \qquad \text{9–7}$$

Clearly, then, if the net force acting on an object is zero,

$$\sum \vec{\mathbf{F}} = 0$$

its change in momentum is also zero

$$\Delta \vec{\mathbf{p}} = \left(\sum \vec{\mathbf{F}}\right)\Delta t = 0$$

Writing the change of momentum in terms of its initial and final values, we have

$$\Delta \vec{\mathbf{p}} = \vec{\mathbf{p}}_{\text{f}} - \vec{\mathbf{p}}_{\text{i}} = 0$$

or

$$\vec{\mathbf{p}}_{\text{f}} = \vec{\mathbf{p}}_{\text{i}} \qquad \text{9–8}$$

Since the momentum does not change in this case, we say that it is **conserved.** To summarize:

Conservation of Momentum

If the net force acting on an object is zero, its momentum is conserved; that is, $\vec{\mathbf{p}}_{\text{f}} = \vec{\mathbf{p}}_{\text{i}}$

Note that in some cases the force may be zero in one direction and nonzero in another. For example, an object in free fall has a nonzero y component of force, $F_y \neq 0$, but no force in the x direction, $F_x = 0$. As a result, the object's y component of momentum changes with time while its x component of momentum remains constant. Therefore, in applying momentum conservation, we must remember that both the force and the momentum are vector quantities and that the momentum conservation principle applies separately to each coordinate direction.

Thus far, our discussion has referred to the forces acting on a single object. Next, we consider a system composed of more than one object.

Internal Versus External Forces

The net force acting on a system of objects is the sum of forces applied from outside the system (external forces, $\vec{\mathbf{F}}_{\text{ext}}$) and forces acting between objects within the system (internal forces, $\vec{\mathbf{F}}_{\text{int}}$). Thus, we can write

$$\vec{\mathbf{F}}_{\text{net}} = \sum \vec{\mathbf{F}} = \sum \vec{\mathbf{F}}_{\text{ext}} + \sum \vec{\mathbf{F}}_{\text{int}}$$

As we shall see, internal and external forces play very *different* roles in terms of how they affect the momentum of a system.

To illustrate the distinction, consider the case of two canoes floating at rest next to one another on a lake, as described in Example 5–3 and shown in **Figure 9–4**. In this case, let's consider the "system" to be the two canoes and the people inside them. When a person in canoe 1 pushes on canoe 2, a force $\vec{\mathbf{F}}_2$ is exerted on canoe 2. By Newton's third law, an equal and opposite force, $\vec{\mathbf{F}}_1 = -\vec{\mathbf{F}}_2$, is exerted on the person in canoe 1. Note that $\vec{\mathbf{F}}_1$ and $\vec{\mathbf{F}}_2$ are internal forces, since they act between objects in the system. In addition, note that they sum to zero:

$$\vec{\mathbf{F}} + \vec{\mathbf{F}}_2 = (-\vec{\mathbf{F}}_2) + \vec{\mathbf{F}}_2 = 0$$

◀ **FIGURE 9–4 Separating two canoes**
A system comprised of two canoes and their occupants. The forces $\vec{\mathbf{F}}_1$ and $\vec{\mathbf{F}}_2$ are internal to the system. They sum to zero.

This is a special case, of course, but it demonstrates the following general principles:

- Internal forces, like all forces, always occur in action-reaction pairs.
- Because the forces in action-reaction pairs are equal and opposite—due to Newton's third law—internal forces must *always* sum to zero:

$$\sum \vec{\mathbf{F}}_{\text{int}} = 0$$

The fact that internal forces always cancel means that the net force acting on a system of objects is simply the sum of the *external* forces acting on it:

$$\vec{\mathbf{F}}_{\text{net}} = \sum \vec{\mathbf{F}}_{\text{ext}} + \sum \vec{\mathbf{F}}_{\text{int}} = \sum \vec{\mathbf{F}}_{\text{ext}}$$

The external forces, on the other hand, may or may not sum to zero—it all depends on the particular situation. For example, if the system consists of the two canoes in Figure 9–4, the external forces are the weights of the people and the canoes acting downward, and the upward, normal force exerted by the water to keep the canoes afloat. These forces sum to zero, and there is no acceleration in the vertical direction. In the next few sections we consider a variety of systems in which the external forces either sum to zero, or are so small that they can be ignored. Later, in Section 9–7, we consider situations where the external forces do not sum to zero and hence must be taken into account.

▲ If the astronaut in this photo pushes on the satellite, the satellite exerts an equal but opposite force on him, in accordance with Newton's third law. If we are calculating the change in the astronaut's momentum, we must take this force into account. However, if we define the system to be the astronaut *and* the satellite, the forces between them are internal to the system. Whatever effect they may have on the astronaut or the satellite individually, they do not affect the momentum of the system as a whole. Therefore, whether a particular force counts as internal or external depends entirely on where we draw the boundaries of the system.

Finally, how do external and internal forces affect the momentum of a system? To see the connection, first note that Newton's second law gives the change in the net momentum for a given time interval Δt:

$$\Delta \vec{\mathbf{p}}_{\text{net}} = \vec{\mathbf{F}}_{\text{net}}\,\Delta t$$

Because the internal forces cancel, however, the change in the net momentum is directly related to the net *external* force:

$$\Delta \vec{\mathbf{p}}_{\text{net}} = \left(\sum \vec{\mathbf{F}}_{\text{ext}}\right)\Delta t \qquad 9\text{–}9$$

Therefore, the key distinction between internal and external forces is the following:

Conservation of Momentum for a System of Objects

- *Internal* forces have absolutely no effect on the net momentum of a system.
- If the *net external* force acting on a system is zero, its net momentum is conserved. That is,

$$\vec{\mathbf{p}}_{1,\text{f}} + \vec{\mathbf{p}}_{2,\text{f}} + \vec{\mathbf{p}}_{3,\text{f}} + \cdots = \vec{\mathbf{p}}_{1,\text{i}} + \vec{\mathbf{p}}_{2,\text{i}} + \vec{\mathbf{p}}_{3,\text{i}} + \cdots$$

It is important to note that these statements apply only to the *net* momentum of a system, not to the momentum of each individual object. For example, suppose a system consists of two objects, 1 and 2, and that the net external force acting on the system is zero. As a result, the net momentum must remain constant:

$$\vec{\mathbf{p}}_{\text{net}} = \vec{\mathbf{p}}_1 + \vec{\mathbf{p}}_2 = \text{constant}$$

PROBLEM-SOLVING NOTE

Internal versus External Forces

It is important to keep in mind that internal forces cannot change the momentum of a system—only a net external force can do that.

This does not mean, however, that $\vec{\mathbf{p}}_1$ is constant or that $\vec{\mathbf{p}}_2$ is constant. All we can say is that the *sum* of $\vec{\mathbf{p}}_1$ and $\vec{\mathbf{p}}_2$ does not change.

As a specific example, consider the case of the two canoes floating on a lake, as described previously. Initially the momentum of the system is zero, because the canoes are at rest. After a person pushes the canoes apart, they are both moving, and hence both have nonzero momentum. Thus, the momentum of each canoe has changed. On the other hand, because the net external force acting on the system is zero, the sum of the canoes' momenta must still vanish. We show this in the next Example.

EXAMPLE 9–3 Tippy Canoe: Comparing Velocity and Momentum

Two groups of canoeists meet in the middle of a lake. After a brief visit, a person in canoe 1 pushes on canoe 2 with a force of 46 N to separate the canoes. If the mass of canoe 1 and its occupants is 130 kg, and the mass of canoe 2 and its occupants is 250 kg, find the momentum of each canoe after 1.20 s of pushing.

Picture the Problem

We choose the positive x direction to point from canoe 1 to canoe 2. With this choice, the force exerted on canoe 2 is $\vec{\mathbf{F}}_2 = (46\text{ N})\hat{\mathbf{x}}$ and the force exerted on canoe 1 is $\vec{\mathbf{F}}_1 = (-46\text{ N})\hat{\mathbf{x}}$.

Strategy

First, we find the acceleration of each canoe using $a_x = F_x/m$. Next, we use $v_x = v_{0x} + a_x t$ to find the velocity at time t. Note that the canoes start at rest, hence $v_{0x} = 0$. Finally, the momentum can be calculated using $p_x = mv_x$.

Solution

1. Use Newton's second law to find the acceleration of canoe 2:
$$a_{2,x} = \frac{\sum F_{2,x}}{m_2} = \frac{46\ \text{N}}{250\ \text{kg}} = 0.18\ \text{m/s}^2$$

2. Do the same calculation for canoe 1. Note that the acceleration of canoe 1 is in the negative direction:
$$a_{1,x} = \frac{\sum F_{1,x}}{m_1} = \frac{-46\ \text{N}}{130\ \text{kg}} = -0.35\ \text{m/s}^2$$

3. Calculate the velocity of each canoe at $t = 1.20$ s:
$$v_{1,x} = a_{1,x}t = (-0.35\ \text{m/s}^2)(1.20\ \text{s}) = -0.42\ \text{m/s}$$
$$v_{2,x} = a_{2,x}t = (0.18\ \text{m/s}^2)(1.20\ \text{s}) = 0.22\ \text{m/s}$$

4. Calculate the momentum of each canoe at $t = 1.20$ s:
$$p_{1,x} = m_1 v_{1,x} = (130\ \text{kg})(-0.42\ \text{m/s}) = -55\ \text{kg} \cdot \text{m/s}$$
$$p_{2,x} = m_2 v_{2,x} = (250\ \text{kg})(0.22\ \text{m/s}) = 55\ \text{kg} \cdot \text{m/s}$$

Insight

Note that the sum of the momenta of the two canoes is zero. This is just what one would expect: The canoes start at rest with zero momentum, there is zero net external force acting on the system, hence the final momentum must also be zero. The final velocities *do not* add to zero; it is momentum ($m\vec{\mathbf{v}}$) that is conserved, not velocity ($\vec{\mathbf{v}}$).

Finally, we solved this problem using one-dimensional kinematics so that we could clearly see the distinction between velocity and momentum. An alternative way to calculate the final momentum of each canoe is to use $\Delta\vec{\mathbf{p}} = \vec{\mathbf{p}}_\text{f} - \vec{\mathbf{p}}_\text{i} = \vec{\mathbf{F}}\Delta t$. For canoe 1 we have $\vec{\mathbf{p}}_{1,\text{f}} = \vec{\mathbf{F}}_1\,\Delta t + \vec{\mathbf{p}}_{1,\text{i}} = (-46\ \text{N})\hat{\mathbf{x}}(1.20\ \text{s}) + 0 = (-55\ \text{kg} \cdot \text{m/s})\hat{\mathbf{x}}$, in agreement with our results above. A similar calculation yields $\vec{\mathbf{p}}_{2,\text{f}} = (55\ \text{kg} \cdot \text{m/s})\hat{\mathbf{x}}$ for canoe 2.

Practice Problem

What are the final momenta if the canoes are pushed apart with a force of 56 N? [**Answer:** $p_{1,x} = -67\ \text{kg} \cdot \text{m/s}$, $p_{2,x} = 67\ \text{kg} \cdot \text{m/s}$]

Some related homework problems: Problem 16, Problem 17

In a situation like that described in Example 9–3, the person in canoe 1 pushes canoe 2 away. At the same time, canoe 1 begins to move in the opposite direction. This is referred to as **recoil.** It is essentially the same as the recoil one experiences when firing a gun or when turning on a strong stream of water.

A particularly interesting example of recoil involves the human body. Perhaps you have noticed, when resting quietly in a rocking or reclining chair, that the chair wobbles back and forth slightly about once a second. The reason for this movement is that each time your heart pumps blood in one direction (from the atria to the ventricles, then to the aorta and pulmonary arteries, and so on) your body recoils in the opposite direction. Because the recoil depends on the force exerted by your heart on the blood and the volume of blood expelled from the heart with each beat, it is possible to gain valuable medical information regarding the health of your heart by analyzing the recoil it produces.

REAL-WORLD PHYSICS: BIO
The ballistocardiograph

The medical instrument that employs the physical principle of recoil is called the *ballistocardiograph*. It is a completely noninvasive technology that simply requires the patient to sit comfortably in a chair fitted with sensitive force sensors under the seat and behind the back. Sophisticated bathroom scales also utilize this technology. A ballistocardiographic (BCG) scale detects the recoil vibrations of the body as a person stands on the scale. This allows the BCG scale to display not only the person's body weight but his or her heart rate as well.

A more dramatic application of heartbeat recoil is currently being used at the Riverbend Maximum Security Institution in Tennessee. The only successful breakout from this prison occurred when four inmates hid in a secret compartment in a delivery truck that was leaving the facility. The institution now uses a heartbeat recoil detector that would have foiled this escape. Vehicles leaving the prison must stop at a checkpoint where a small motion detector is attached to it with a suction cup. Any persons hidden in the vehicle will reveal their presence by the very beating of their hearts. These heartbeat detectors have proved to be 100 percent effective, even though the recoil of the heart may displace a large truck by only a few millionths of an inch. Similar systems are being used at other high-security installations and border crossings.

REAL-WORLD PHYSICS: BIO
Heartbeat detectors

CONCEPTUAL CHECKPOINT 9–2 Momentum Versus Kinetic Energy

In Example 9–3, the final momentum of the system (consisting of the two canoes and their occupants) is equal to the initial momentum of the system. Is the final kinetic energy (a) equal to, (b) less than, or (c) greater than the initial kinetic energy?

Reasoning and Discussion

The final momentum of the two canoes is zero because one canoe has a positive momentum and the other has a negative momentum of the same magnitude. The two momenta, then, sum to zero. Kinetic energy, which is $\frac{1}{2}mv^2$, cannot be negative, hence no such cancellation is possible. Both canoes have positive kinetic energies, hence, the final kinetic energy is greater than the initial kinetic energy, which is zero.

Where does the increase in kinetic energy come from? It comes from the muscular work done by the person who pushes the canoes apart.

Answer:

(c) K_f is greater than K_i.

REAL-WORLD PHYSICS
Stellar explosions

A special case of some interest is the universe. Since there is nothing external to the universe—by definition—it follows that the net external force acting on it is zero. Therefore, its net momentum is conserved. No matter what happens—a comet collides with the Earth, a star explodes and becomes a supernova, a black hole swallows part of a galaxy—the total momentum of the universe simply cannot change. A particularly vivid illustration of momentum conservation in our own galaxy is provided by the exploding star Eta Carinae. As can be seen in the Hubble Space Telescope photograph, jets of material are moving away from the star in opposite directions, just like the canoes moving apart from one another in Example 9–3.

Conservation of momentum also applies to the more everyday situation described in the next Active Example.

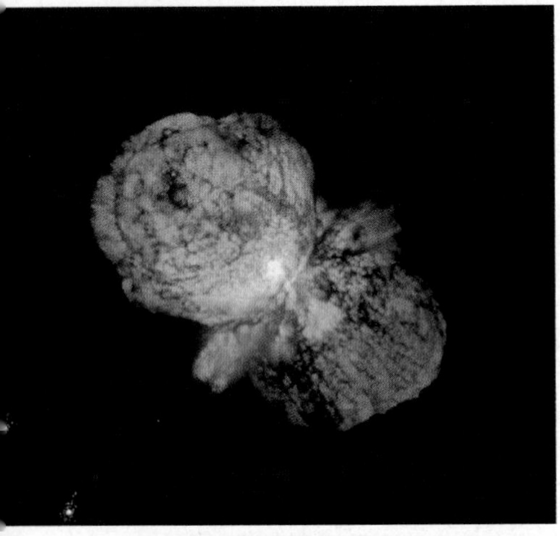

▲ This Hubble Space Telescope photograph shows the aftermath of a violent explosion of the star Eta Carinae. The explosion, which was observed on Earth in 1841 and briefly made Eta Carinae the second brightest star in the sky, produced two bright lobes of matter spewing outward in opposite directions. In this photograph, these lobes have expanded to about the size of our solar system. The momentum of the star before the explosion must be the same as the total momentum of the star and the bright lobes after the explosion. Since the lobes are roughly symmetric and move in opposite directions, their net momentum is essentially zero. Thus, we conclude that the momentum of the star itself was virtually unchanged by the explosion.

ACTIVE EXAMPLE 9–2 Find the Velocity of the Bee

A honeybee with a mass of 0.150 g lands on one end of a floating 4.75-g popsicle stick. After sitting at rest for a moment, it runs toward the other end with a velocity $\vec{\mathbf{v}}_b$ relative to the still water. The stick moves in the opposite direction with a speed of 0.120 cm/s. What is the velocity of the bee? (Let the direction of the bee's motion be the positive x direction.)

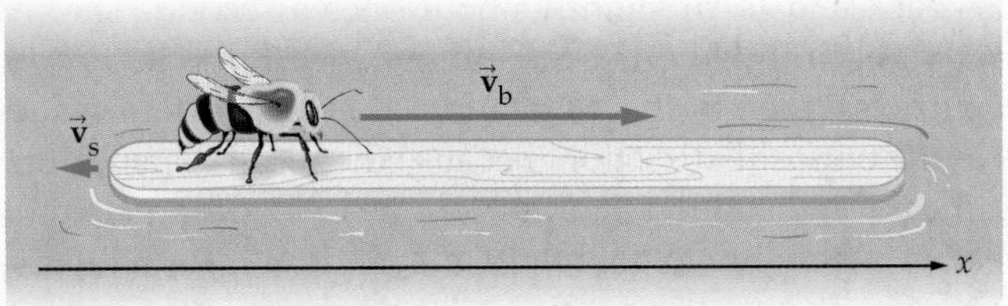

Solution *(Test your understanding by performing the calculations indicated in each step.)*

1. Set the total momentum of the system equal to zero:	$\vec{\mathbf{p}}_b + \vec{\mathbf{p}}_s = 0$
2. Solve for the momentum of the bee:	$\vec{\mathbf{p}}_b = -\vec{\mathbf{p}}_s = m_b v_b \hat{\mathbf{x}}$
3. Calculate the momentum of the stick:	$\vec{\mathbf{p}}_s = -m_s v_s \hat{\mathbf{x}} = (-0.570\ \mathrm{g \cdot cm/s})\hat{\mathbf{x}}$
4. Calculate the momentum of the bee:	$\vec{\mathbf{p}}_b = m_b v_b \hat{\mathbf{x}} = -\vec{\mathbf{p}}_s = (0.570\ \mathrm{g \cdot cm/s})\hat{\mathbf{x}}$
5. Divide by the bee's mass to find its velocity:	$\vec{\mathbf{v}}_b = \vec{\mathbf{p}}_b / m_b = (3.80\ \mathrm{cm/s})\hat{\mathbf{x}}$

Insight

Because only internal forces are at work while the bee walks on the stick, the system's total momentum must remain zero.

Your Turn

Suppose the mass of the popsicle stick is 9.50 g rather than 4.75 g. What is the bee's velocity in this case?

*(Answers to **Your Turn** problems are given in the back of the book.)*

9–5 Inelastic Collisions

We now turn our attention to **collisions.** By a collision we mean a situation in which two objects strike one another, and in which the net external force is either zero or negligibly small. For example, if two train cars roll along on a level track and hit one another, this is a collision. In this case, the net external force—the weight downward and the normal force exerted by the tracks upward—is zero. As a result, the momentum of the two-car system is conserved.

Another example of a collision is a baseball being struck by a bat. In this case, the external forces are not zero because the weight of the ball is not balanced by any other force. However, as we have seen in Section 9–3, the forces exerted during the hit are much larger than the weight of the ball or the bat. Hence, to a good approximation, we may neglect the external forces (the weight of the ball and bat) in this case, and say that the momentum of the ball-bat system is conserved.

Now it may seem surprising at first, but the fact that the momentum of a system is conserved during a collision does not necessarily mean that the system's kinetic energy is conserved. In fact most, or even all, of a system's kinetic energy may be converted to other forms during a collision while, at the same time, not one bit of momentum is lost. This shall be explored in detail in this section.

In general, collisions are categorized according to what happens to the kinetic energy of the system. There are two possibilities. After a collision, the final kinetic energy, K_f, is either equal to the initial kinetic energy, K_i, or it is not. If $K_f = K_i$, the collision is said to be **elastic.** We shall consider elastic collisions in the next section.

On the other hand, the kinetic energy may change during a collision. Usually it decreases due to losses associated with sound, heat, and deformation. Sometimes it increases, if the collision sets off an explosion, for instance. In any event, collisions in which the kinetic energy is not conserved are referred to as **inelastic:**

Inelastic Collisions

In an inelastic collision, the momentum of a system is conserved,

$$\vec{\mathbf{p}}_f = \vec{\mathbf{p}}_i$$

but its kinetic energy is not,

$$K_f \neq K_i$$

Finally, in the special case where objects stick together after the collision, we say that the collision is **completely inelastic.**

Completely Inelastic Collisions

When objects stick together after colliding, the collision is completely inelastic.

◀ In both elastic and inelastic collisions, momentum is conserved. The same is not true of kinetic energy, however. In the largely inelastic collision at left, much of the hockey players' initial kinetic energy is transformed into work: rearranging the players' anatomies and shattering the glass of the rink. In the highly elastic collision at right, the ball rebounds with very little diminution of its kinetic energy (though a little energy is lost as sound and heat).

In a completely inelastic collision, the maximum amount of kinetic energy is lost. If the total momentum of the system is zero, this means that all of the kinetic energy is lost. For systems with nonzero total momentum, however, some kinetic energy will remain after the collision—still, the amount lost is the maximum permitted by momentum conservation.

Inelastic Collisions in One Dimension

Consider a system of two identical train cars of mass m on a smooth, level track. One car is at rest initially while the other moves toward it with a speed v_0, as shown in **Figure 9–5**. When the cars collide, the coupling mechanism latches, causing the cars to stick together and move as a unit. What is the speed of the cars after the collision?

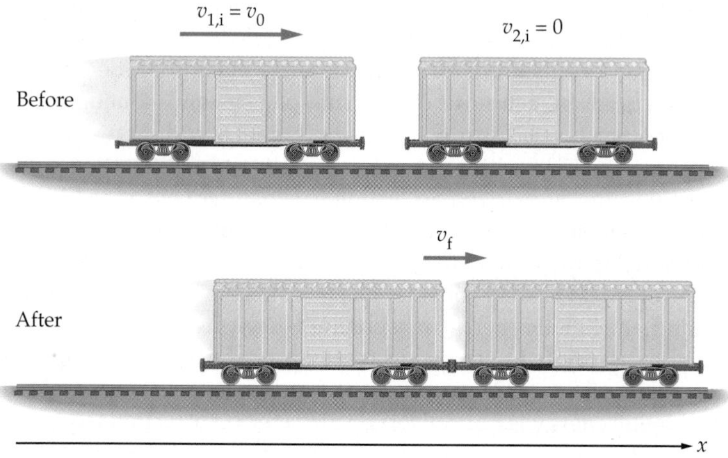

▶ **FIGURE 9–5 Railroad cars collide and stick together**

A moving train car collides with an identical car that is stationary. After the collision, the cars stick together and move with the same speed.

To answer this question, we begin by considering the general case that applies to any completely inelastic collision, and then we look at the specific case of the two train cars. In general, suppose that two masses, m_1 and m_2, have initial velocities $v_{1,i}$ and $v_{2,i}$ respectively. The initial momentum of the system is

$$p_i = m_1 v_{1,i} + m_2 v_{2,i}$$

After the collision, the objects move together with a common velocity v_f. Therefore, the final momentum is

$$p_f = (m_1 + m_2)v_f$$

Equating the initial and final momenta yields $m_1 v_{1,i} + m_2 v_{2,i} = (m_1 + m_2)v_f$, or

$$v_f = \frac{m_1 v_{1,i} + m_2 v_{2,i}}{m_1 + m_2} \qquad \textbf{9–10}$$

We can apply this general result to the case of the two railroad cars by noting that $m_1 = m_2 = m$, $v_{1,i} = v_0$, and $v_{2,i} = 0$. Thus, the final velocity is

$$v_f = \frac{mv_0 + m \cdot 0}{m + m} = \frac{m}{2m}v_0 = \frac{1}{2}v_0 \qquad \textbf{9–11}$$

As you might have guessed, the final speed is one-half the initial speed.

EXERCISE 9–2

A 1200-kg car moving at 2.5 m/s is struck in the rear by a 2600-kg truck moving at 6.2 m/s. If the vehicles stick together after the collision, what is their speed immediately after colliding? (Assume that external forces may be ignored.)

Solution

Applying Equation 9–10 with $m_1 = 1200$ kg, $v_{1,i} = 2.5$ m/s, $m_2 = 2600$ kg and $v_{2,i} = 6.2$ m/s yields $v_f = 5.0$ m/s.

During the collision of the railroad cars, some of the initial kinetic energy is converted to other forms. Some propagates away as sound, some is converted to heat, some creates permanent deformations in the metal of the latching mechanism. The precise amount of kinetic energy that is lost is addressed in the following Conceptual Checkpoint.

CONCEPTUAL CHECKPOINT 9–3 How Much Kinetic Energy Is Lost?

A railroad car of mass m and speed v collides and sticks to an identical railroad car that is initially at rest. After the collision, is the kinetic energy of the system **(a)** 1/2, **(b)** 1/3, or **(c)** 1/4 of its initial kinetic energy?

Reasoning and Discussion

Before the collision, the kinetic energy of the system is

$$K_i = \frac{1}{2}mv^2$$

After the collision, the mass doubles and the speed is halved. Hence, the final kinetic energy is

$$K_f = \frac{1}{2}(2m)\left(\frac{v}{2}\right)^2 = \frac{1}{2}\left(\frac{1}{2}mv^2\right) = \frac{1}{2}K_i$$

Therefore, one-half of the initial kinetic energy is converted to other forms of energy.

An equivalent way to arrive at this conclusion is to express the kinetic energy in terms of the momentum, $p = mv$:

$$K = \frac{1}{2}mv^2 = \frac{1}{2}\left(\frac{m^2v^2}{m}\right) = \frac{p^2}{2m}$$

Since the momentum is the same before and after the collision, the fact that the mass doubles means the kinetic energy is halved.

Answer:

(a) The final kinetic energy is one-half the initial kinetic energy.

Note that we know the precise amount of kinetic energy that was lost, even though we don't know just how much went into sound, how much went into heat, and so on. It is not necessary to know all of those details to determine how much kinetic energy was lost.

We also know how much momentum was lost—none.

PROBLEM-SOLVING NOTE

Momentum versus Energy Conservation

Be sure to distinguish between momentum conservation and energy conservation. A common error is to assume that kinetic energy is conserved just because the momentum is conserved.

EXAMPLE 9–4 Goal-Line Stand

On a touchdown attempt, a 95.0-kg running back runs toward the end zone at 3.75 m/s. A 111-kg linebacker moving at 4.10 m/s meets the runner in a head-on collision. If the two players stick together, **(a)** what is their velocity immediately after the collision? **(b)** What are the initial and final kinetic energies of the system?

Picture the Problem

In our sketch, we let subscript 1 refer to the red-and-gray running back, who carries the ball, and subscript 2 refer to the blue-and-gold linebacker, who will make the tackle. The direction of the running back's initial motion is taken to be in the positive x direction. Therefore, the initial velocities of the players are $\vec{v}_1 = (3.75\text{ m/s})\hat{x}$ and $\vec{v}_2 = (-4.10\text{ m/s})\hat{x}$.

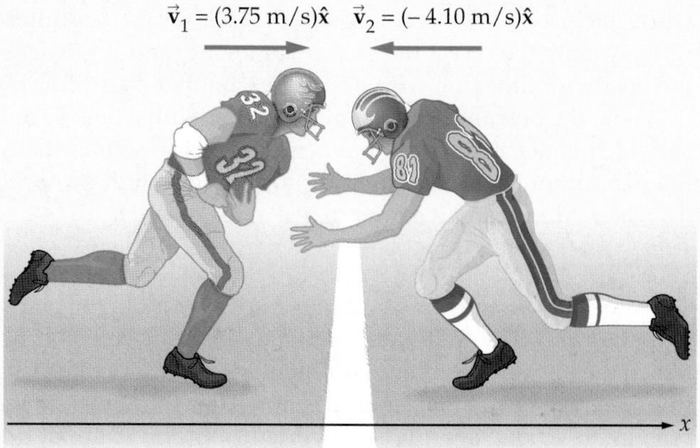

Strategy

a. The final velocity can be found by applying momentum conservation to the system consisting of the two players. Initially, the players have momenta in opposite directions. After the collision, the players move together with a combined mass $m_1 + m_2$ and a velocity $\vec{v}_f$.

continued on next page

continued from previous page

b. The kinetic energies can be found by applying $\frac{1}{2}mv^2$ to the players individually to obtain the initial kinetic energy, and then to their combined motion for the final kinetic energy.

Solution

Part (a)

1. Set the initial momentum equal to the final momentum: $m_1\vec{\mathbf{v}}_1 + m_2\vec{\mathbf{v}}_2 = (m_1 + m_2)\vec{\mathbf{v}}_f$

2. Solve for the final velocity and substitute numerical values, being careful to use the appropriate signs:

$$\vec{\mathbf{v}}_f = \frac{m_1\vec{\mathbf{v}}_1 + m_2\vec{\mathbf{v}}_2}{m_1 + m_2}$$
$$= \frac{(95.0\ \mathrm{kg})(3.75\ \mathrm{m/s})\hat{\mathbf{x}} + (111\ \mathrm{kg})(-4.10\ \mathrm{m/s})\hat{\mathbf{x}}}{95.0\ \mathrm{kg} + 111\ \mathrm{kg}}$$
$$= (-0.480\ \mathrm{m/s})\hat{\mathbf{x}}$$

Part (b)

3. Calculate the initial kinetic energy of the two players:

$$K_i = \tfrac{1}{2}m_1{v_1}^2 + \tfrac{1}{2}m_2{v_2}^2$$
$$= \tfrac{1}{2}(95.0\ \mathrm{kg})(3.75\ \mathrm{m/s})^2 + \tfrac{1}{2}(111\ \mathrm{kg})(-4.10\ \mathrm{m/s})^2$$
$$= 1600\ \mathrm{J}$$

4. Calculate the final kinetic energy of the players, noting that they both move with the same velocity after the collision:

$$K_f = \tfrac{1}{2}(m_1 + m_2){v_f}^2$$
$$= \tfrac{1}{2}(95.0\ \mathrm{kg} + 111\ \mathrm{kg})(-0.480\ \mathrm{m/s})^2 = 23.7\ \mathrm{J}$$

Insight

After the collision, the two players are moving in the negative direction; that is, away from the end zone. This is because the linebacker had more negative momentum than the running back had positive momentum. As for the kinetic energy, of the original 1600 J, only 23.7 J is left after the collision. This means that over 98% of the original kinetic energy is converted to other forms. Even so, *none* of the momentum is lost.

Practice Problem

If the final speed of the two players is to be zero, should the speed of the running back be increased or decreased? Check your answer by calculating the required speed for the running back. [**Answer**: The running back's speed should be increased to 4.79 m/s.]

Some related homework problems: Problem 23, Problem 30

EXAMPLE 9–5 Ballistic Pendulum

REAL-WORLD PHYSICS

In a ballistic pendulum, an object of mass m is fired with an initial speed v_0 at the bob of a pendulum. The bob has a mass M, and is suspended by a rod of negligible mass. After the collision, the object and the bob stick together and swing through an arc, eventually gaining a height h. Find the height h in terms of m, M, v_0, and g.

Picture the Problem

Our sketch shows the physical setup of a ballistic pendulum. Initially, only the object of mass m is moving, and it moves in the positive x direction with a speed v_0. Immediately after the collision, the bob and object move together with a new speed, v_f, which is determined by momentum conservation. Finally, the pendulum continues to swing to the right until its speed decreases to zero and it comes to rest at the height h.

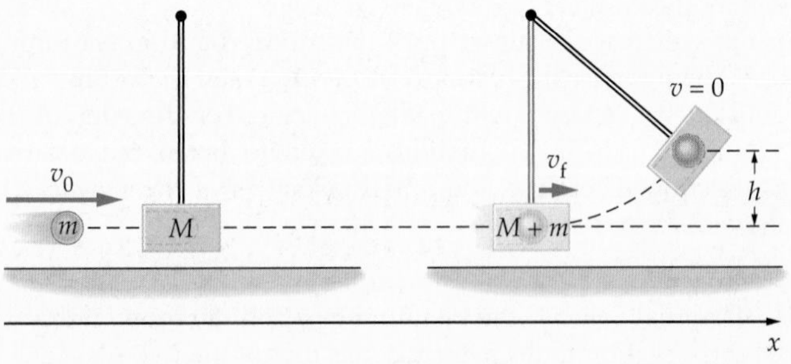

Strategy

There are two distinct physical processes at work in the ballistic pendulum. The first is a completely inelastic collision between the bob and the object. Momentum is conserved during this collision, but kinetic energy is not. After the collision, the remaining kinetic energy is converted into gravitational potential energy, which determines how high the bob and object will rise.

INTERACTIVE FIGURE

Solution

1. Set the momentum just before the bob–object collision equal to the momentum just after the collision. Let v_f be the speed just after the collision:

$$mv_0 = (M + m)v_f$$

2. Solve for the speed just after the collision, v_f:

$$v_f = \left(\frac{m}{M + m}\right)v_0$$

3. Calculate the kinetic energy just after the collision:

$$K_f = \frac{1}{2}(M + m)v_f^2 = \frac{1}{2}(M + m)\left(\frac{m}{M + m}\right)^2 v_0^2$$

$$= \frac{1}{2}mv_0^2\left(\frac{m}{M + m}\right)$$

4. Set the kinetic energy after the collision equal to the gravitational potential energy at the height h:

$$\frac{1}{2}mv_0^2\left(\frac{m}{M + m}\right) = (M + m)gh$$

5. Solve for the height, h:

$$h = \left(\frac{m}{M + m}\right)^2\left(\frac{v_0^2}{2g}\right)$$

Insight

A ballistic pendulum is often used to measure the speed of a rapidly moving object, such as a bullet. If a bullet were shot straight up, it would rise to the height $v_0^2/2g$, which can be thousands of feet. On the other hand, if a bullet of mass m is fired into a ballistic pendulum, in which M is much greater than m, the bullet reaches only a small fraction of this height. Thus, the ballistic pendulum makes for a more convenient and practical measurement.

Practice Problem

A 7.00-g bullet is fired into a ballistic pendulum whose bob has a mass of 0.950 kg. If the bob rises to a height of 0.220 m, what was the initial speed of the bullet? [**Answer:** $v_0 = 284$ m/s. If this bullet were fired straight up, it would rise 4.11 km $\approx$ 13,000 ft in the absence of air resistance.]

Some related homework problems: Problem 27, Problem 28

Inelastic Collisions in Two Dimensions

Next we consider collisions in two dimensions, where we must conserve the momentum component by component. To do this, we set up a coordinate system and resolve the initial momentum into x and y components. Next, we demand that the final momentum have precisely the same x and y components as the initial momentum. That is,

$$p_{x,i} = p_{x,f}$$

and

$$p_{y,i} = p_{y,f}$$

The following Example shows how to carry out such a calculation in a practical situation.

PROBLEM-SOLVING NOTE

Sketch the System Before and After the Collision

In problems involving collisions, it is useful to draw the system before and after the collision. Be sure to label the relevant masses, velocities, and angles.

EXAMPLE 9–6 Bad Intersection: Analyzing a Traffic Accident

REAL-WORLD PHYSICS

A car with a mass of 950 kg and a speed of 16 m/s approaches an intersection, as shown on the next page. A 1300-kg minivan traveling at 21 m/s is heading for the same intersection. The car and minivan collide and stick together. Find the speed and direction of the wrecked vehicles just after the collision, assuming external forces can be ignored.

Picture the Problem

In our sketch, we align the x and y axes with the crossing streets. With this choice, $\vec{\mathbf{v}}_1$ (the car's velocity) is in the positive x direction, and $\vec{\mathbf{v}}_2$ (the minivan's velocity) is in the positive y direction. In addition, the problem statement indicates that

continued on next page

continued from previous page

$m_1 = 950$ kg and $m_2 = 1300$ kg. After the collision, the two vehicles move together (as a unit) with a speed v_f in a direction θ with respect to the positive x axis.

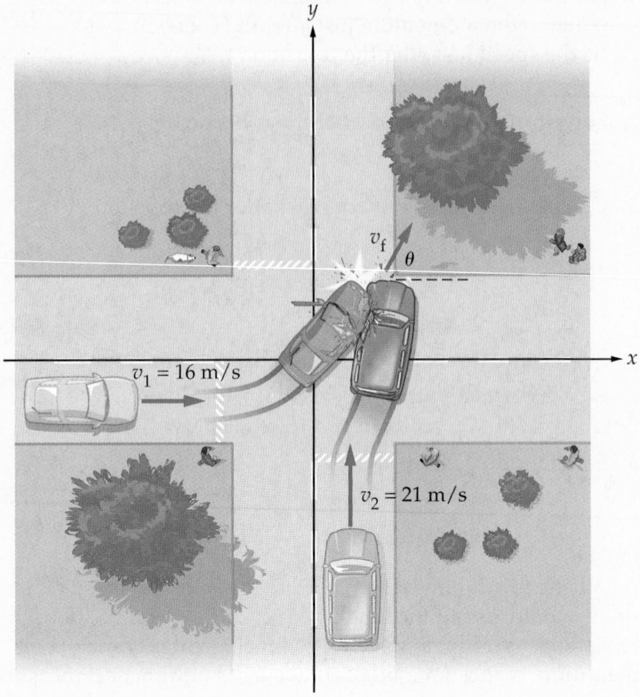

Strategy

Because external forces can be ignored, the total momentum of the system must be conserved during the collision. This is really two conditions: (i) the x component of momentum is conserved, and (ii) the y component of momentum is conserved. These two conditions determine the two unknowns: the final speed, v_f, and the final direction, θ.

Solution

1. Set the initial x component of momentum equal to the final x component of momentum:

$$m_1v_1 = (m_1 + m_2)v_f \cos\theta$$

2. Do the same for the y component of momentum:

$$m_2v_2 = (m_1 + m_2)v_f \sin\theta$$

3. Divide the y momentum equation by the x momentum equation. This eliminates v_f, giving an equation involving θ alone:

$$\frac{m_2v_2}{m_1v_1} = \frac{(m_1 + m_2)v_f \sin\theta}{(m_1 + m_2)v_f \cos\theta} = \frac{\sin\theta}{\cos\theta} = \tan\theta$$

4. Solve for θ:

$$\theta = \tan^{-1}\left(\frac{m_2v_2}{m_1v_1}\right) = \tan^{-1}\left[\frac{(1300\text{ kg})(21\text{ m/s})}{(950\text{ kg})(16\text{ m/s})}\right]$$
$$= \tan^{-1}(1.8) = 61°$$

5. The final speed can be found using either the x or the y momentum equation. Here we use the x equation:

$$v_f = \frac{m_1v_1}{(m_1 + m_2)\cos\theta}$$
$$= \frac{(950\text{ kg})(16\text{ m/s})}{(950\text{ kg} + 1300\text{ kg})\cos 61°} = 14\text{ m/s}$$

Insight

As a check, you should verify that the y momentum equation gives the same value for v_f.

When a collision occurs in the real world, a traffic-accident investigation team will measure skid marks at the scene of the crash and use this information—along with some basic physics—to determine the initial speeds and directions of the vehicles. This information is often presented in court, where it can lead to a clear identification of the driver at fault.

Practice Problem

Suppose the speed and direction immediately after the collision are known to be $v_f = 12.5$ m/s and $\theta = 42°$, respectively. Find the initial speed of each car. [**Answer:** $v_1 = 22$ m/s, $v_2 = 14$ m/s]

Some related homework problems: Problem 24, Problem 25

9–6 Elastic Collisions

In this section we consider collisions in which both momentum and kinetic energy are conserved. As mentioned in the previous section, such collisions are referred to as elastic:

Elastic Collisions

In an elastic collision, momentum and kinetic energy are conserved. That is,

$$\vec{\mathbf{p}}_{\mathrm{f}} = \vec{\mathbf{p}}_{\mathrm{i}}$$

and

$$K_{\mathrm{f}} = K_{\mathrm{i}}$$

Most collisions in everyday life are rather poor approximations to being elastic—usually there is a significant amount of energy converted to other forms. However, the collision of objects that bounce off one another with little deformation—like billiard balls, for example—provides a reasonably good approximation to an elastic collision. In the subatomic world, on the other hand, elastic collisions are common. Elastic collisions, then, are not merely an ideal that is approached but never attained—they are constantly taking place in nature.

Elastic Collisions in One Dimension

Consider a head-on collision of two carts on an air track, as pictured in **Figure 9–6**. The carts are provided with bumpers that give an elastic bounce when the carts collide. Let's suppose that initially cart 1 is moving to the right with a speed v_0 toward cart 2, which is at rest. If the masses of the carts are m_1 and m_2, respectively, then momentum conservation can be written as follows:

$$m_1 v_0 = m_1 v_{1,\mathrm{f}} + m_2 v_{2,\mathrm{f}}$$

In this expression, $v_{1,\mathrm{f}}$ and $v_{2,\mathrm{f}}$ are the final velocities of the two carts. Note that we say velocities, not speeds, since it is possible for cart 1 to reverse direction, in which case $v_{1,\mathrm{f}}$ would be negative.

Next, the fact that this is an elastic collision means the final velocities must also satisfy energy conservation:

$$\frac{1}{2} m_1 v_0^2 = \frac{1}{2} m_1 v_{1,\mathrm{f}}^2 + \frac{1}{2} m_2 v_{2,\mathrm{f}}^2$$

Thus, we now have two equations for the two unknowns, $v_{1,\mathrm{f}}$ and $v_{2,\mathrm{f}}$. Straightforward—though messy—algebra yields the following results:

$$v_{1,\mathrm{f}} = \left(\frac{m_1 - m_2}{m_1 + m_2}\right) v_0$$

$$v_{2,\mathrm{f}} = \left(\frac{2m_1}{m_1 + m_2}\right) v_0 \qquad \text{9–12}$$

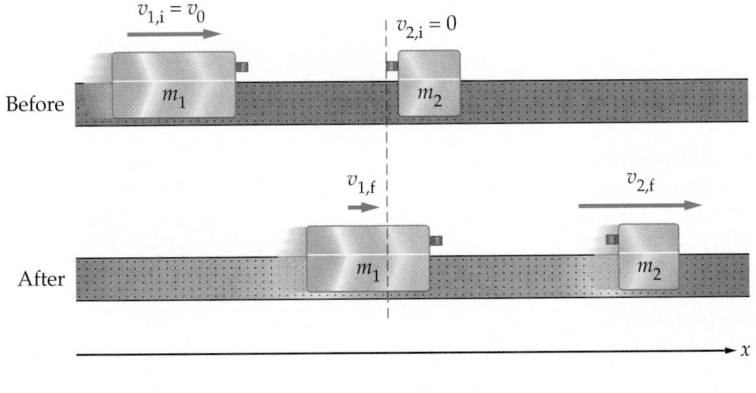

◀ **FIGURE 9–6 An elastic collision between two air carts**

In the case pictured, $v_{1,\mathrm{f}}$ is to the right (positive), which means that m_1 is greater than m_2. In fact, we have chosen $m_1 = 2m_2$ for this plot; therefore, $v_{1,\mathrm{f}} = v_0/3$ and $v_{2,\mathrm{f}} = 4v_0/3$ as given by Equations 9–12. If m_1 were less than m_2, cart 1 would bounce back toward the left, meaning that $v_{1,\mathrm{f}}$ would be negative.

Note that the final velocity of cart 1 can be positive, negative, or zero, depending on whether m_1 is greater than, less than, or equal to m_2, respectively. The final velocity of cart 2, however, is always positive.

EXERCISE 9–3

At an amusement park, a 96.0-kg bumper car moving with a speed of 1.24 m/s bounces elastically off a 135-kg bumper car at rest. Find the final velocities of the cars.

Solution

Using Equations 9–12, we find the final velocities to be $v_{1,\text{f}} = -0.209\text{ m/s}$ and $v_{2,\text{f}} = 1.03\text{ m/s}$. Note that the direction of travel of car 1 has been reversed.

Let's check a few special cases of our results. First, consider the case where the two carts have equal masses, $m_1 = m_2 = m$. Substituting into Equations 9–12, we find

$$v_{1,\text{f}} = \left(\frac{m - m}{m + m}\right)v_0 = 0$$

and

$$v_{2,\text{f}} = \left(\frac{2m}{m + m}\right)v_0 = v_0$$

Thus, after the collision, the cart that was moving with velocity v_0 is now at rest, and the cart that was at rest is now moving with velocity v_0. In effect, the carts have "exchanged" velocities. This case is illustrated in **Figure 9–7 (a)**.

Next, suppose that m_2 is much greater than m_1, or, equivalently, that m_1 approaches zero. Returning to Equations 9–12, and setting $m_1 = 0$, we find

$$v_{1,\text{f}} = \left(\frac{0 - m_2}{0 + m_2}\right)v_0 = \left(\frac{-m_2}{m_2}\right)v_0 = -v_0$$

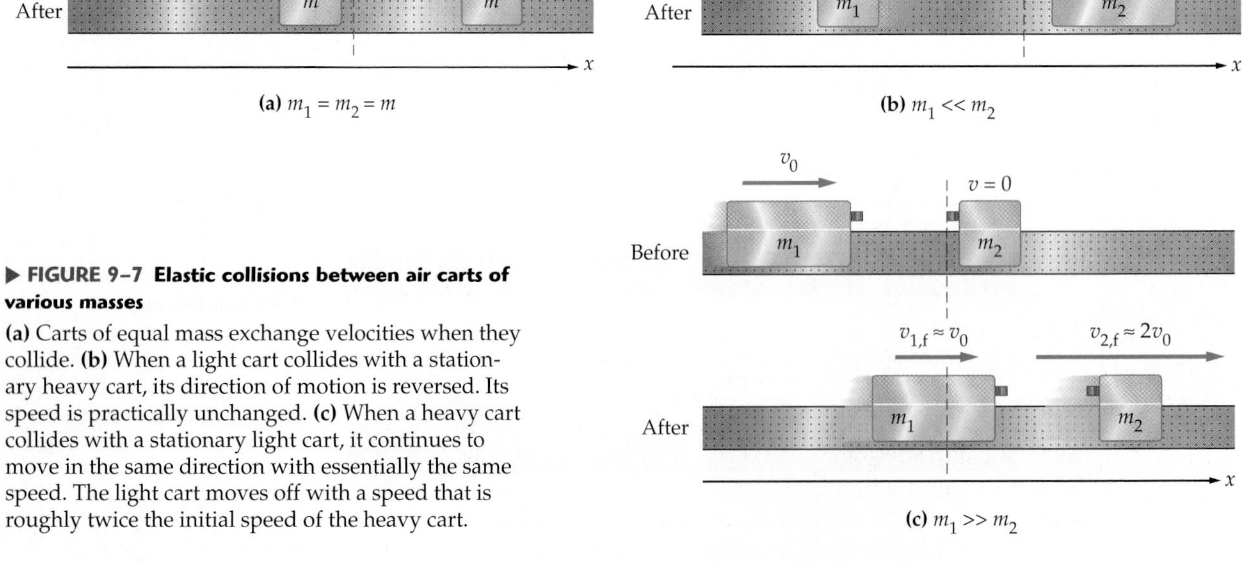

▶ **FIGURE 9–7 Elastic collisions between air carts of various masses**

(a) Carts of equal mass exchange velocities when they collide. **(b)** When a light cart collides with a stationary heavy cart, its direction of motion is reversed. Its speed is practically unchanged. **(c)** When a heavy cart collides with a stationary light cart, it continues to move in the same direction with essentially the same speed. The light cart moves off with a speed that is roughly twice the initial speed of the heavy cart.

and

$$v_{2,f} = \frac{2 \cdot 0}{0 + m_2} v_0 = 0$$

Physically, we interpret these results as follows: A very light cart collides with a heavy cart that is at rest. The heavy cart hardly budges, but the light cart is reflected, heading *backward* (remember the minus sign in $-v_0$) with the same speed it had initially. For example, if you throw a ball against a wall, the wall is the very heavy object and the ball is the light object. The ball bounces back with the same speed it had initially (assuming an ideal elastic collision). We show a case in which m_1 is much less than m_2 in **Figure 9–7 (b)**.

Finally, what happens when m_1 is much greater than m_2? To check this limit we can set m_2 equal to zero. We consider the results in the following Conceptual Checkpoint.

CONCEPTUAL CHECKPOINT 9–4 Speed After a Collision

A hoverfly is happily maintaining a fixed position about 10 ft above the ground when an elephant charges out of the bush and collides with it. The fly bounces elastically off the forehead of the elephant. If the initial speed of the elephant is v_0, is the speed of the fly after the collision equal to **(a)** v_0, **(b)** $1.5v_0$, or **(c)** $2v_0$?

Reasoning and Discussion

We can use Equations 9–12 to find the final speeds of the fly and the elephant. First, let m_1 be the mass of the elephant, and m_2, be the mass of the fly. Clearly, m_2 is vanishingly small compared with m_1, hence we can evaluate Equations 9–12 in the limit $m_2 \to 0$. This yields

$$v_{1,f} = \left(\frac{m_1 - m_2}{m_1 + m_2}\right) v_0 \xrightarrow[m_2 \to 0]{} \left(\frac{m_1}{m_1}\right) v_0 = v_0$$

and

$$v_{2,f} = \left(\frac{2m_1}{m_1 + m_2}\right) v_0 \xrightarrow[m_2 \to 0]{} \left(\frac{2m_1}{m_1}\right) v_0 = 2v_0$$

As expected, the speed of the elephant is unaffected. The fly, however, rebounds with twice the speed of the elephant. **Figure 9–7 (c)** illustrates this case with air carts.

Answer:

(c) The speed of the fly is $2v_0$.

Note that after the collision the fly is separating from the elephant with the speed $2v_0 - v_0 = v_0$. Before the collision the elephant was approaching the fly with the same speed, v_0. This is a special case of the following general result:

> The speed of separation after a head-on elastic collision is always equal to the speed of approach before the collision.

The proof of this statement is the subject of Problem 77.

▲ The apparatus shown here illustrates some of the basic features of elastic collisions between objects of equal mass. The device consists of five identical metal balls suspended by strings. When the end ball is pulled out to the side and then released so as to fall back and strike the second ball, it creates a rapid succession of elastic collisions among the balls. In each collision, one ball comes to rest while the next one begins to move with the original speed, just as with the air carts in Figure 9–7 (a). When the collisions reach the other end of the apparatus, the last ball swings out to the same height from which the first ball was released.

If two balls are pulled out and released, two balls swing out at the other side, and so on. To see why this must be so, imagine that the two balls swing in with a speed v and a single ball swings out at the other side with a speed v'. What value must v' have (a) to conserve momentum, and (b) to conserve kinetic energy? Since the required speed is $v' = 2v$ for (a) and $v' = \sqrt{2}v$ for (b), it follows that it is not possible to conserve both momentum and kinetic energy with two balls swinging in and one ball swinging out.

Elastic Collisions in Two Dimensions

In a two-dimensional elastic collision, if we are given the final speed and direction of one of the objects, we can find the speed and direction of the other object using energy conservation and momentum conservation. For example, consider the collision of two 7.00-kg curling stones, as depicted in **Figure 9–8**. One stone is at rest initially, the other approaches with a speed $v_{1,i} = 1.50$ m/s. The collision is not head-on, and after the collision, stone 1 moves with a speed of $v_{1,f} = 0.610$ m/s in a direction 66.0° away from the initial line of motion. What is the speed and direction of stone 2?

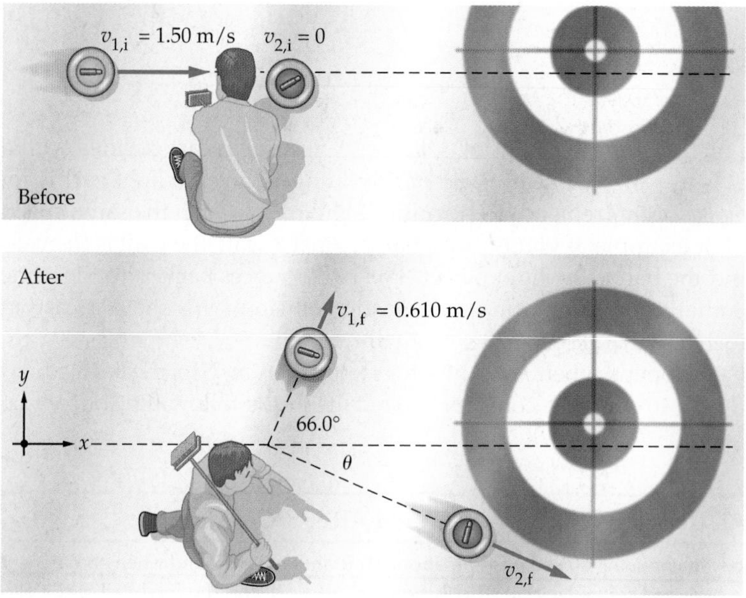

▶ **FIGURE 9–8 Two curling stones undergo an elastic collision**
The speed of curling stone 2 after this collision can be determined using energy conservation; its direction of motion can be found using momentum conservation in either the x or the y direction.

PROBLEM-SOLVING NOTE

Kinetic Energy in Elastic Collisions

Remember that in elastic collisions, by definition, the kinetic energy is conserved.

First, let's find the speed of stone 2. The easiest way to do this is to simply require that the final kinetic energy be equal to the initial kinetic energy. Initially, the kinetic energy is

$$K_i = \tfrac{1}{2}m_1 v_{1,i}^2 = \tfrac{1}{2}(7.00\text{ kg})(1.50\text{ m/s})^2 = 7.88\text{ J}$$

After the collision stone 1 has a speed of 0.610 m/s and stone 2 has the speed $v_{2,f}$. Hence, the final kinetic energy is

$$K_f = \tfrac{1}{2}m_1 v_{1,f}^2 + \tfrac{1}{2}m_2 v_{2,f}^2 = \tfrac{1}{2}(7.00\text{ kg})(0.610\text{ m/s})^2 + \tfrac{1}{2}m_2 v_{2,f}^2$$
$$= 1.30\text{ J} + \tfrac{1}{2}m_2 v_{2,f}^2 = K_i$$

Solving for the speed of stone 2, we find

$$v_{2,f} = 1.37\text{ m/s}$$

Next, we can find the direction of motion of stone 2 by requiring that the momentum be conserved. For example, initially there is no momentum in the y direction. This must be true after the collision as well. Hence, we have the following condition:

$$0 = m_1 v_{1,f}\sin 66.0^\circ - m_2 v_{2,f}\sin\theta$$

Solving for the angle θ we find

$$\theta = 24.0^\circ$$

As a final check, compare the initial and final x components of momentum. Initially, we have

$$p_{x,i} = m_1 v_{1,i} = (7.00\text{ kg})(1.50\text{ m/s}) = 10.5\text{ kg}\cdot\text{m/s}$$

Following the collision, the x component of momentum is

$$p_{x,f} = m_1 v_{1,f}\cos 66.0^\circ + m_2 v_{2,f}\cos 24.0^\circ$$
$$= (7.00\text{ kg})(0.610\text{ m/s})\cos 66.0^\circ + (7.00\text{ kg})(1.37\text{ m/s})\cos 24.0^\circ$$
$$= 10.5\text{ kg}\cdot\text{m/s}$$

As expected, the momentum is unchanged.

EXAMPLE 9–7 Two Fruits in Two Dimensions: Analyzing an Elastic Collision

Two astronauts on opposite ends of a spaceship are comparing lunches. One has an apple, the other has an orange. They decide to trade. Astronaut 1 tosses the 0.130-kg apple toward astronaut 2 with a speed of 1.11 m/s. The 0.160-kg orange is tossed from astronaut 2 to astronaut 1 with a speed of 1.21 m/s. Unfortunately, the fruits collide, sending the orange off with a speed of 1.16 m/s at an angle of 42.0° with respect to its original direction of motion. Find the final speed and direction of the apple, assuming an elastic collision. Give the apple's direction relative to its original direction of motion.

Picture the Problem

In our sketch we refer to the apple as object 1 and to the orange as object 2. We also choose the positive x direction to be in the initial direction of motion of the apple. We shall describe the "Before" and "After" sketches separately:

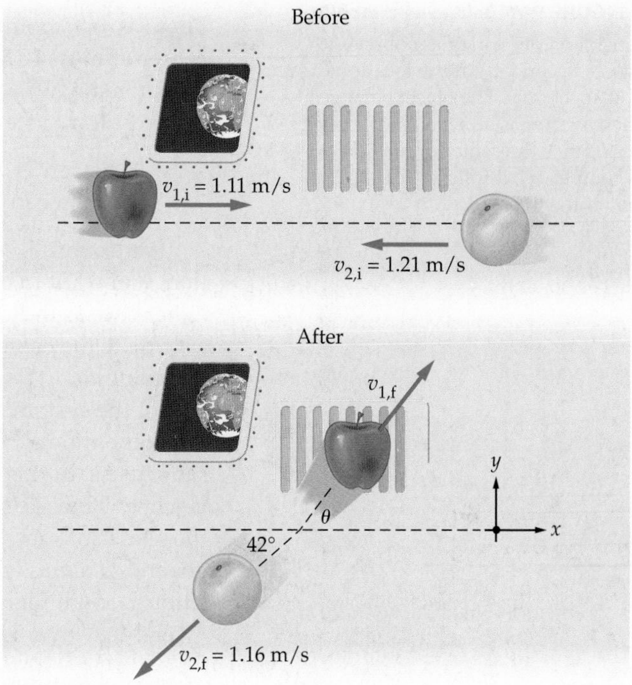

Before

Initially, the apple moves in the positive x direction with a speed of 1.11 m/s, and the orange moves in the negative x direction with a speed of 1.21 m/s. There is no momentum in the y direction before the collision.

After

After the collision, the orange moves with a speed of 1.16 m/s in a direction 42° below the negative x axis. As a result, the orange now has momentum in the negative y direction. To cancel this y momentum, the apple must move in a direction that is above the positive x axis, as indicated in the sketch.

Strategy

As described in the text, we first find the speed of the apple by demanding that the initial and final kinetic energies be the same. Next, we find the angle θ by conserving momentum in either the x or the y direction—the results are the same whichever direction is chosen.

Solution

1. Calculate the initial kinetic energy of the system:

$$K_i = \tfrac{1}{2}m_1 v_{1,i}^2 + \tfrac{1}{2}m_2 v_{2,i}^2$$
$$= \tfrac{1}{2}(0.130\text{ kg})(1.11\text{ m/s})^2 + \tfrac{1}{2}(0.160\text{ kg})(1.21\text{ m/s})^2$$
$$= 0.197\text{ J}$$

2. Calculate the final kinetic energy of the system in terms of $v_{1,f}$:

$$K_f = \tfrac{1}{2}m_1 v_{1,f}^2 + \tfrac{1}{2}m_2 v_{2,f}^2$$
$$= \tfrac{1}{2}(0.130\text{ kg})v_{1,f}^2 + \tfrac{1}{2}(0.160\text{ kg})(1.16\text{ m/s})^2$$
$$= \tfrac{1}{2}(0.130\text{ kg})v_{1,f}^2 + 0.108\text{ J}$$

3. Set $K_f = K_i$ to find $v_{1,f}$:

$$v_{1,f} = \sqrt{\frac{2(0.197\text{ J} - 0.108\text{ J})}{0.130\text{ kg}}} = 1.17\text{ m/s}$$

4. Set the final y component of momentum equal to zero to determine the angle, θ:

$$0 = m_1 v_{1,f}\sin\theta - m_2 v_{2,f}\sin 42.0°$$

Solve for $\sin\theta$:

$$\sin\theta = \frac{m_2 v_{2,f}\sin 42.0°}{m_1 v_{1,f}}$$

5. Substitute numerical values:

$$\sin\theta = \frac{(0.160\text{ kg})(1.16\text{ m/s})\sin 42.0°}{(0.130\text{ kg})(1.17\text{ m/s})} = 0.817$$
$$\theta = \sin^{-1}(0.817) = 54.8°$$

Insight

The x momentum equation gives the same value for θ, as expected.

Practice Problem

Suppose that after the collision the apple moves in the positive y direction with a speed of 1.27 m/s. What is the final speed and direction of the orange in this case? [**Answer:** The orange moves with a speed of 1.07 m/s in a direction of 74.7° below the negative x axis.

Some related homework problems: Problem 33, Problem 81

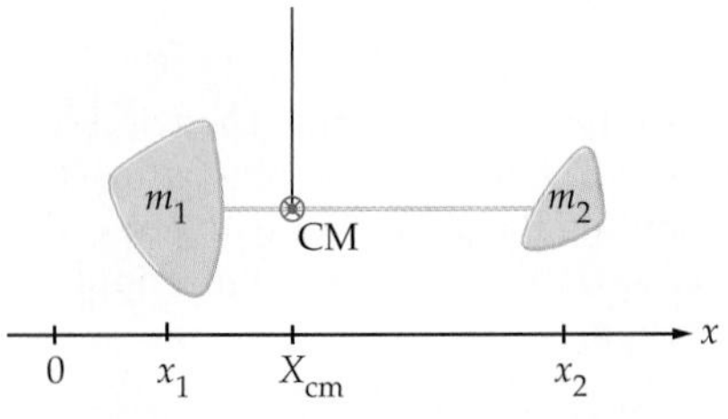

▲ **FIGURE 9–9 Balancing a mobile**
Consider a portion of a mobile with masses m_1 and m_2 at the locations x_1 and x_2, respectively. The object balances when a string is attached at the center of mass. Since the center of mass is closer to m_1 than to m_2, it follows that m_1 is greater than m_2.

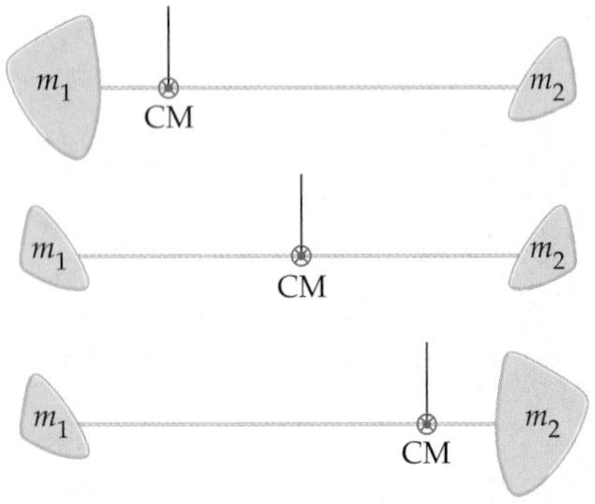

▲ **FIGURE 9–10 The center of mass of two objects**
The center of mass is closest to the larger mass, or equidistant between the masses if they are equal.

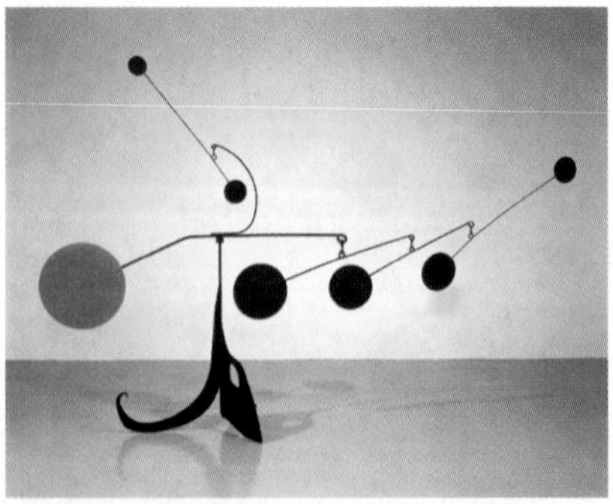

▲ Mobiles like *Myxomatose* by Alexander Calder illustrate the concept of center of mass with artistic flair. Each arm of the mobile is in balance because it is suspended at its center of mass.

9–7 Center of Mass

In this section we introduce the concept of the center of mass. We begin by defining its location for a given system of masses. Next we consider the motion of the center of mass and show how it is related to the net external force acting on the system. As we shall see, the center of mass plays a key role in the analysis of collisions.

Location of the Center of Mass

There is one point in any system of objects that has special significance—the **center of mass (CM).** One of the reasons the center of mass is so special is the fact that, in many ways, a system behaves as if all of its mass were concentrated there. As a result, a system can be balanced at its center of mass:

> The center of mass of a system of masses is the point where the system can be balanced in a uniform gravitational field.

For example, suppose you are making a mobile. At one stage in its construction, you want to balance a light rod with objects of mass m_1 and m_2 connected to either end, as indicated in **Figure 9–9**. To make the rod balance, you should attach a string to the center of mass of the system, just as if all its mass were concentrated at that point. In a sense, you can think of the center of mass as the "average" location of the system's mass.

To be more specific, suppose the two objects connected to the rod have the same mass. In this case the center of mass is at the midpoint of the rod, since this is where it balances. On the other hand, if one object has more mass than the other, the center of mass is closer to the heavier object, as indicated in **Figure 9–10**. In general, if a mass m_1 is on the x axis at the position x_1, and a mass m_2 is at the position x_2, as in Figure 9–9, the location of the center of mass, X_{cm}, is defined as the "weighted" average of the two positions:

Center of Mass for Two Objects

$$X_{cm} = \frac{m_1x_1 + m_2x_2}{m_1 + m_2} = \frac{m_1x_1 + m_2x_2}{M} \qquad \text{9–13}$$

Note that we have used $M = m_1 + m_2$ for the total mass of the two objects, and that the two positions, x_1 and x_2, are multiplied—or weighted—by their respective masses.

To see that this definition of X_{cm} agrees with our expectations, consider first the case where the masses are equal; $m_1 = m_2 = m$. In this case, $M = m_1 + m_2 = 2m$, and $X_{cm} = (mx_1 + mx_2)/2m = \frac{1}{2}(x_1 + x_2)$. Thus, as expected, if two masses are equal, their center of mass is halfway between them. On the other hand, if m_1 is significantly greater than m_2, it follows that $M = m_1 + m_2 \sim m_1$ and $m_1x_1 + m_2x_2 \sim m_1x_1$, since m_2 can be ignored in comparison to m_1. As a result, we find that $X_{cm} \sim m_1x_1/m_1 = x_1$; that is, the center of mass is essentially at the location of the extremely heavy mass, m_1. In general, as one mass becomes larger than the other, the center of mass moves closer to the larger mass.

EXERCISE 9–4

Suppose the masses in Figure 9–9 are separated by 0.500 m, and that $m_1 = 0.260$ kg and $m_2 = 0.170$ kg. What is the distance from m_1 to the center of mass of the system?

Solution

Letting $x_1 = 0$ and $x_2 = 0.500$ m in Figure 9–9, we have

$$X_{cm} = \frac{m_1x_1 + m_2x_2}{m_1 + m_2} = \frac{(0.260\text{ kg})\cdot 0 + (0.170\text{ kg})(0.500\text{ m})}{0.260\text{ kg} + 0.170\text{ kg}} = 0.198\text{ m}$$

Thus, the center of mass is closer to m_1 (the larger mass) than to m_2.

To extend the definition of X_{cm} to more general situations, first consider a system that contains many objects, not just two. In that case, X_{cm} is the sum of m

times x for each object, divided by the total mass of the system, M. If, in addition, the objects in the system are not in a line, but are distributed in two dimensions, the center of mass will have both an x coordinate, X_{cm}, and a y coordinate, Y_{cm}. As one would expect, Y_{cm} is simply the sum of m times y for each object, divided by M. Thus, the x coordinate of the center of mass is

***X* Coordinate of the Center of Mass**

$$X_{cm} = \frac{m_1x_1 + m_2x_2 + \cdots}{m_1 + m_2 + \cdots} = \frac{\sum mx}{M} \qquad 9\text{–}14$$

Similarly, the y coordinate of the center of mass is

***Y* Coordinate of the Center of Mass**

$$Y_{cm} = \frac{m_1y_1 + m_2y_2 + \cdots}{m_1 + m_2 + \cdots} = \frac{\sum my}{M} \qquad 9\text{–}15$$

In systems with a continuous, uniform distribution of mass, the center of mass is at the geometric center of the object, as illustrated in **Figure 9–11**. Note that it is common for the center of mass to be located in a position where no mass exists, as in a life preserver, where the center of mass is precisely in the center of the hole.

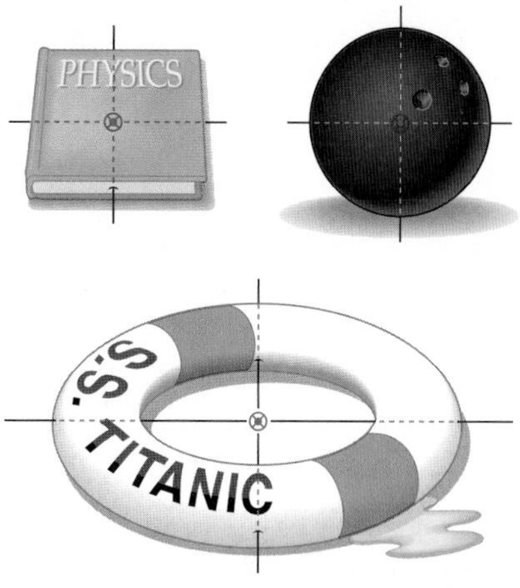

▲ **FIGURE 9–11 Locating the center of mass**
In an object of continuous, uniform mass distribution, the center of mass is located at the geometric center of the object. In some cases, this means that the center of mass is not located within the object.

EXAMPLE 9–8 Center of Mass of the Arm

REAL-WORLD PHYSICS: BIO

A person's arm is held with the upper arm vertical, the lower arm and hand horizontal. Find the center of mass of the arm in this configuration, given the following data: The upper arm has a mass of 2.5 kg and a center of mass 0.18 m above the elbow; the lower arm has a mass of 1.6 kg and a center of mass 0.12 m to the right of the elbow; the hand has a mass of 0.64 kg and a center of mass 0.40 m to the right of the elbow.

Picture the Problem
We place the origin at the elbow, with the x and y axes pointing along the lower and upper arms, respectively. The center of mass of each of the three parts of the arm is indicated by an x; the center of mass of the entire arm is at the point labeled CM.

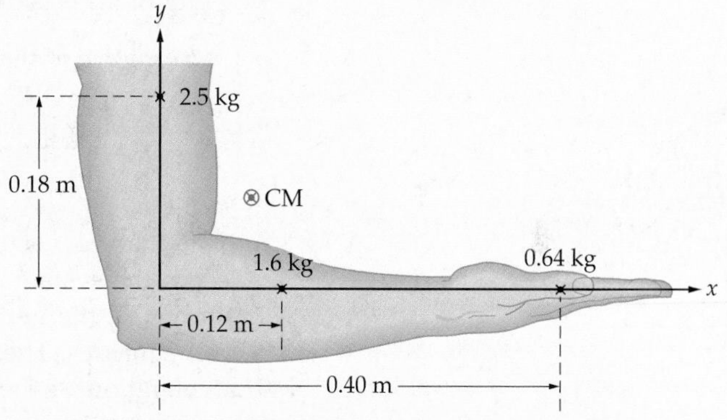

Strategy
Using the information given in the problem statement, we can treat the arm as a system of three point masses placed as follows: 2.5 kg at (0, 0.18 m); 1.6 kg at (0.12 m, 0); 0.64 kg at (0.40 m, 0). We substitute these masses and locations into Equations 9–14 and 9–15 to find the x and y coordinates of the center of mass, respectively.

Solution

1. Calculate the x coordinate of the center of mass:

$$X_{cm} = \frac{(2.5\text{ kg})(0) + (1.6\text{ kg})(0.12\text{ m}) + (0.64\text{ kg})(0.40\text{ m})}{2.5\text{ kg} + 1.6\text{ kg} + 0.64\text{ kg}} = 0.095\text{ m}$$

2. Do the same calculation for the y coordinate of the center of mass:

$$Y_{cm} = \frac{(2.5\text{ kg})(0.18\text{ m}) + (1.6\text{ kg})(0) + (0.64\text{ kg})(0)}{2.5\text{ kg} + 1.6\text{ kg} + 0.64\text{ kg}} = 0.095\text{ m}$$

Insight
As is often the case, the center of mass of an arm held in this position is in a location where no mass exists—you might say the center of mass is having an out-of-body experience. This effect can sometimes be put to good use, as when the center of mass of a high jumper passes beneath the horizontal bar while the body passes above it. See Conceptual Question 18 for a photo of this technique in action, in the famous "Fosbury flop."

Practice Problem
Suppose a 1.2-kg ball is placed in the center of the hand; that is, 0.40 m to the right of the elbow. **(a)** Does X_{cm} increase, decrease, or stay the same? **(b)** Does Y_{cm} increase, decrease, or stay the same? **(c)** Check your answers to parts (a) and (b) by finding the center of mass of the arm-ball system. [**Answer:** (a) increases; (b) decreases; (c) $X_{cm} = 0.16$ m, $Y_{cm} = 0.076$ m]

Some related homework problems: Problem 42, Problem 44

Motion of the Center of Mass

Another reason the center of mass is of such importance is that its motion often displays a remarkable simplicity when compared with the motion of other parts of a system. To analyze this motion, we consider both the velocity and the acceleration of the center of mass. Each of these quantities is defined in complete analogy with the definition of the center of mass itself.

For example, to find the velocity of the center of mass, we first multiply the mass of each object in a system, m, by its velocity, $\vec{\mathbf{v}}$, to give $m_1\vec{\mathbf{v}}_1$, $m_2\vec{\mathbf{v}}_2$, and so on. Next, we add all these products together, $m_1\vec{\mathbf{v}}_1 + m_2\vec{\mathbf{v}}_2 + \cdots$, and divide by the total mass, $M = m_1 + m_2 + \cdots$. The result, by definition, is the velocity of the center of mass, $\vec{\mathbf{V}}_{\text{cm}}$:

Velocity of the Center of Mass

$$\vec{\mathbf{V}}_{\text{cm}} = \frac{m_1\vec{\mathbf{v}}_1 + m_2\vec{\mathbf{v}}_2 + \cdots}{m_1 + m_2 + \cdots} = \frac{\sum m\vec{\mathbf{v}}}{M} \qquad \text{9–16}$$

Comparing with Equation 9–14, we see that $\vec{\mathbf{V}}_{\text{cm}}$ is the same as X_{cm} with each position x replaced with a velocity vector $\vec{\mathbf{v}}$. In addition, note that the total mass of the system, M, times the velocity of the center of mass, $\vec{\mathbf{V}}_{\text{cm}}$, is simply the total momentum of the system:

$$M\vec{\mathbf{V}}_{\text{cm}} = m_1\vec{\mathbf{v}}_1 + m_2\vec{\mathbf{v}}_2 + \cdots = \vec{\mathbf{p}}_1 + \vec{\mathbf{p}}_2 + \cdots = \vec{\mathbf{p}}_{\text{total}}$$

To gain more information on how the center of mass moves, we next consider its acceleration, $\vec{\mathbf{A}}_{\text{cm}}$. As expected by analogy with $\vec{\mathbf{V}}_{\text{cm}}$, the acceleration of the center of mass is defined as follows:

Acceleration of the Center of Mass

$$\vec{\mathbf{A}}_{\text{cm}} = \frac{m_1\vec{\mathbf{a}}_1 + m_2\vec{\mathbf{a}}_2 + \cdots}{m_1 + m_2 + \cdots} = \frac{\sum m\vec{\mathbf{a}}}{M} \qquad \text{9–17}$$

Note that the vector $\vec{\mathbf{A}}_{\text{cm}}$ contains terms like $m_1\vec{\mathbf{a}}_1$, $m_2\vec{\mathbf{a}}_2$, and so on, for each object in the system. From Newton's second law, however, we know that $m_1\vec{\mathbf{a}}_1$ is simply $\vec{\mathbf{F}}_1$, the net force acting on mass 1. The same conclusion applies to each of the masses. Therefore, we find that the total mass of the system, M, times the acceleration of the center of mass, $\vec{\mathbf{A}}_{\text{cm}}$, is simply the total force acting on the system:

$$M\vec{\mathbf{A}}_{\text{cm}} = m_1\vec{\mathbf{a}}_1 + m_2\vec{\mathbf{a}}_2 + \cdots = \vec{\mathbf{F}}_1 + \vec{\mathbf{F}}_2 + \cdots = \vec{\mathbf{F}}_{\text{total}}$$

Recall, however, that the total force acting on a system is the same as the net external force, $\vec{\mathbf{F}}_{\text{net,ext}}$, since the internal forces cancel. Therefore, $M\vec{\mathbf{A}}_{\text{cm}}$ is the net external force acting on the system:

Newton's Second Law for a System of Particles

$$M\vec{\mathbf{A}}_{\text{cm}} = \vec{\mathbf{F}}_{\text{net,ext}} \qquad \text{9–18}$$

Zero Net External Force For systems in which $\vec{\mathbf{F}}_{\text{net,ext}}$ is zero, it follows that the acceleration of the center of mass is zero. Hence, if the center of mass is initially at rest, it remains at rest. Similarly, if the center of mass is moving initially, it continues to move with the same velocity. For example, in a collision between two air-track carts, the velocity of each cart changes as a result of the collision. The velocity of the center of mass of the two carts, however, is the same before and after the collision. We explore cases in which $\vec{\mathbf{F}}_{\text{net,ext}} = 0$ in the following Example and Active Example.

EXAMPLE 9–9 Crash of the Air Carts

An air cart of mass m and speed v_0 moves toward a second, identical air cart that is at rest. When the carts collide they stick together and move as one. Find the velocity of the center of mass of this system **(a)** before and **(b)** after the carts collide.

Picture the Problem

We choose the positive x direction to be the direction of motion of the incoming cart, whose initial speed is v_0. Note that the carts have wads of putty on their bumpers; this ensures that they stick together when they collide and thereafter move as a unit. Their final speed is v_f.

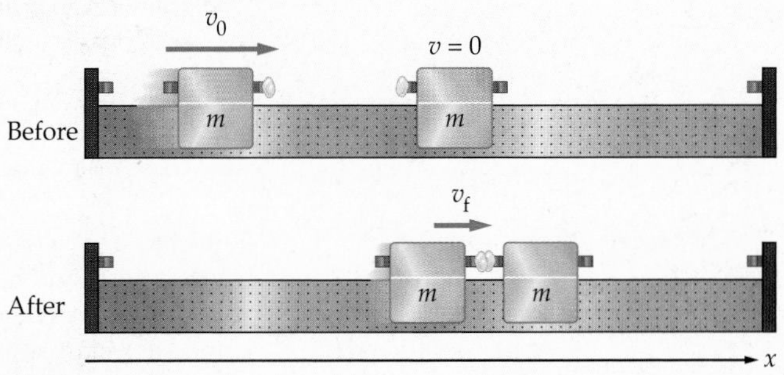

INTERACTIVE FIGURE

Strategy

a. We can find the velocity of the center of mass by applying Equation 9–16 to the case of just two masses; $\vec{\mathbf{V}}_{cm} = (m_1\vec{\mathbf{v}}_1 + m_2\vec{\mathbf{v}}_2)/M$. In this case, $\vec{\mathbf{v}}_1 = v_0\hat{\mathbf{x}}$, $\vec{\mathbf{v}}_2 = 0$, and $m_1 = m_2 = m$.

b. After the collision the two masses have the same velocity, $\vec{\mathbf{v}}_f = v_f\hat{\mathbf{x}}$, which is given by momentum conservation (Equations 9–10 and 9–11). Hence, $\vec{\mathbf{V}}_{cm} = (m_1\vec{\mathbf{v}}_f + m_2\vec{\mathbf{v}}_f)/M$.

Solution

Part (a)

1. Use $\vec{\mathbf{V}}_{cm} = (m_1\vec{\mathbf{v}}_1 + m_2\vec{\mathbf{v}}_2)/M$ to find the velocity of the center of mass before the collision:

$$\vec{\mathbf{V}}_{cm} = \frac{(m_1\vec{\mathbf{v}}_1 + m_2\vec{\mathbf{v}}_2)}{m_1 + m_2} = \frac{(mv_0\hat{\mathbf{x}} + m \cdot 0)}{m + m} = \tfrac{1}{2}v_0\hat{\mathbf{x}}$$

Part (b)

2. Use momentum conservation in the x direction to find the speed of the carts after the collision:

$$mv_0 = mv_f + mv_f$$
$$v_f = \tfrac{1}{2}v_0$$

3. Calculate the velocity of the center of mass of the two carts after the collision:

$$\vec{\mathbf{V}}_{cm} = \frac{(m_1\vec{\mathbf{v}}_1 + m_2\vec{\mathbf{v}}_2)}{m_1 + m_2} = \frac{(mv_f\hat{\mathbf{x}} + mv_f\hat{\mathbf{x}})}{m + m} = v_f\hat{\mathbf{x}} = \tfrac{1}{2}v_0\hat{\mathbf{x}}$$

Insight

As expected, the velocity of each cart changes when they collide. On the other hand, the velocity of the center of mass is completely unaffected by the collision. This is illustrated to the right, where we show a sequence of equal-time snapshots of the system just before and just after the collision. First, we note that the incoming cart moves two distance units for every time interval until it collides with the second cart. From that point on, the two carts are locked together, and move one distance unit per time interval. In contrast, the center of mass (CM), which is centered between the two equal-mass carts, progresses uniformly throughout the sequence, advancing one unit of distance for each time interval.

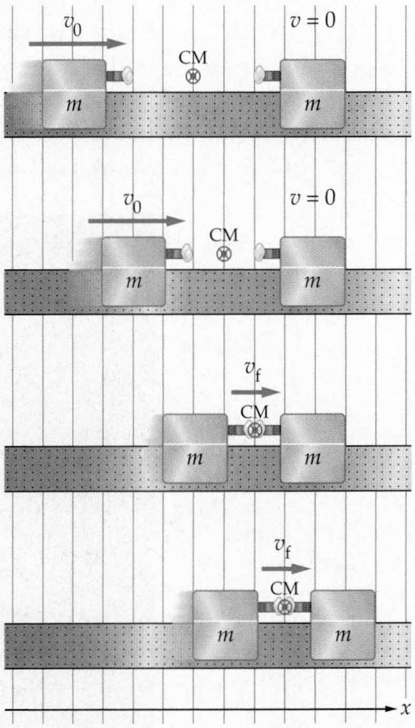

Practice Problem

If the mass of the cart that is moving initially is doubled to $2m$, does the velocity of the center of mass increase, decrease, or stay the same? Verify your answer by calculating the velocity of the center of mass in this case. [**Answer:** The velocity of the center of mass increases. We find that $\vec{\mathbf{V}}_{cm} = (2v_0/3)\hat{\mathbf{x}}$, both before and after the collision.

Some related homework problems: Problem 49, Problem 69

ACTIVE EXAMPLE 9–3 Find the Velocity of the Center of Mass

In Active Example 9–2 we found that as a 0.150-g bee runs with a speed of 3.80 cm/s in one direction, the 4.75-g popsicle stick on which it floats moves with a speed of 0.120 cm/s in the opposite direction. Find the velocity of the center of mass of the bee and the stick.

Solution *(Test your understanding by performing the calculations indicated in each step.)*

1. Write the velocity of the bee: $\vec{\mathbf{v}}_b = (3.80 \text{ cm/s})\hat{\mathbf{x}}$
2. Write the velocity of the stick: $\vec{\mathbf{v}}_s = (-0.120 \text{ cm/s})\hat{\mathbf{x}}$
3. Use these velocities to calculate $\vec{\mathbf{V}}_{cm}$: $\vec{\mathbf{V}}_{cm} = (m_b\vec{\mathbf{v}}_b + m_s\vec{\mathbf{v}}_s)/(m_b + m_s) = 0$

Insight

$\vec{\mathbf{V}}_{cm}$ is zero, and hence the center of mass stays at rest as the bee and the stick move. This is as expected, since the net external force is zero for this system, and the bee and stick started at rest initially.

Your Turn

If the bee increases its speed, will the velocity of the center of mass be nonzero?

(*Answers to* **Your Turn** *problems are given in the back of the book.*)

Nonzero Net External Force Recall that Newton's second law, as expressed in Equation 9–18, states that the acceleration of the center of mass is related to the net external force as follows:

$$M\vec{\mathbf{A}}_{cm} = \vec{\mathbf{F}}_{net,ext}$$

This is completely analogous to the relationship between the acceleration of an object of mass m and the net force $\vec{\mathbf{F}}_{net}$ applied to it:

$$m\vec{\mathbf{a}} = \vec{\mathbf{F}}_{net}$$

Therefore, when $\vec{\mathbf{F}}_{net,ext}$ is nonzero, we can conclude the following:

> The center of mass of a system accelerates precisely as if it were a point particle of mass M acted on by the force $\vec{\mathbf{F}}_{net,ext}$.

For this reason, the motion of the center of mass can be quite simple compared to the motion of its constituent parts. For example, a hammer tossed into the air with a rotation is shown in **Figure 9–12**. The motion of one part of the hammer, the tip of the handle, let's say, follows a complicated path in space. On the other hand, the path of the center of mass is a simple parabola, precisely the same path that a point mass would follow.

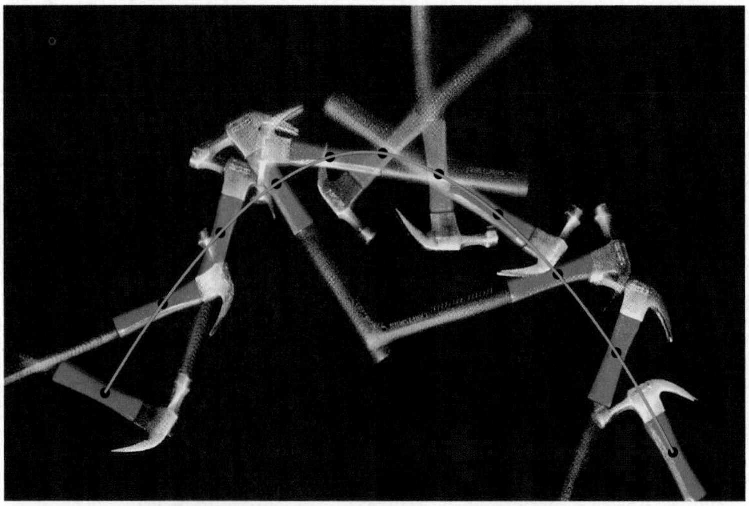

▶ **FIGURE 9–12 Motion of the center of mass**

At first glance the motion of this thrown hammer seems complex. However, if you draw a curve through the successive positions of the center of mass (marked with a black dot), you will find that it follows a parabolic path, just as if the hammer were a projectile with all of its mass concentrated at that one spot.

Similarly, consider a fireworks rocket launched into the sky, as illustrated in **Figure 9–13**. The center of mass of the rocket follows a parabolic path, ignoring air resistance. At some point in its path it explodes into numerous individual pieces. The explosion is due to internal forces, however, which must therefore sum to zero. Hence, the net external force acting on the pieces of the rocket is the same before, during, and after the explosion. As a result, the center of mass has a constant downward acceleration and continues to follow the original parabolic path. It is only when an additional external force acts on the system, as when one of the pieces of the rocket hits the ground, that the path of the center of mass changes.

REAL-WORLD PHYSICS
An Exploding Rocket

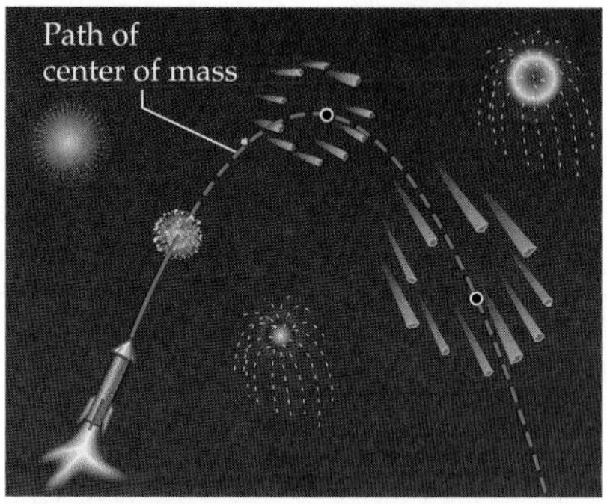

▲ **FIGURE 9–13 Center of mass of an exploding rocket**

A fireworks rocket follows a parabolic path, ignoring air resistance, until it explodes. After exploding, its center of mass continues on the same parabolic path until some of the fragments start to land.

To see how to apply $M\vec{\mathbf{A}}_{\text{cm}} = \vec{\mathbf{F}}_{\text{net,ext}}$, consider the system shown in **Figure 9–14**. Here we see a box of mass m_1, inside of which is a ball of mass m_2 suspended from a light string. The entire system rests on a scale reading its weight. The scale exerts an upward force on the box of magnitude F_s. Initially, of course, $F_s = (m_1 + m_2)g$.

Now, suppose the string breaks, allowing the ball to fall with constant acceleration g toward the bottom of the box. What is the reading on the scale while the ball falls? We can guess that the answer should be simply $m_1 g$, the weight of the box alone, but let's analyze the problem from the point of view of the center of mass.

Taking upward as the positive direction, the net external force acting on the box and the ball is

$$F_{\text{net,ext}} = F_s - m_1 g - m_2 g$$

The acceleration of the center of mass is

$$A_{\text{cm}} = \frac{m_1 \cdot 0 - m_2 g}{M} = -\frac{m_2}{M}g$$

Setting $MA_{\text{cm}} = F_{\text{net,ext}}$ yields

$$MA_{\text{cm}} = M\left(-\frac{m_2}{M}\right)g = -m_2 g = F_{\text{net,ext}} = F_s - m_1 g - m_2 g$$

Finally, canceling the term $-m_2 g$ and solving for the weight read by the scale, F_s, we find, as expected, that

$$F_s = m_1 g$$

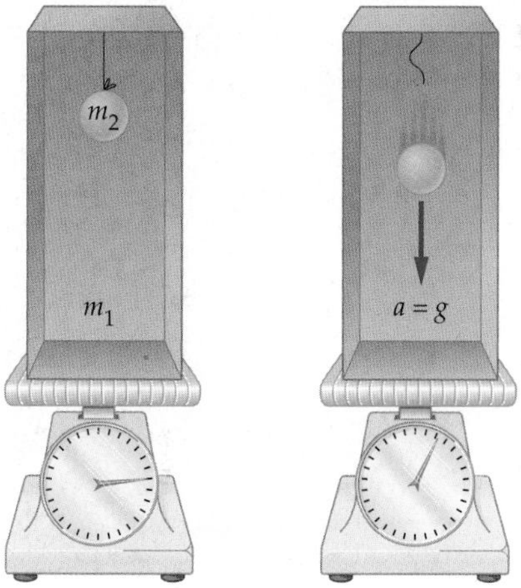

▲ **FIGURE 9–14 Weight and acceleration of the center of mass**

A box with a ball suspended from a string is weighed on a scale. The scale reads the weight of the box and the ball. When the string breaks and the ball falls with the acceleration of gravity, the scale reads only the weight of the box.

*9–8 Systems with Changing Mass: Rocket Propulsion

We close this chapter by considering systems in which the mass can change. A rocket, for example, changes its mass as its engines operate because it ejects part of the fuel as it burns. The burning process is produced by internal forces, hence the total momentum of the rocket and its fuel remains constant.

Consider, then, a rocket in outer space, far from any large, massive objects. When the rocket's engine is fired, it expels a certain mass of fuel out the back with a speed v. If the mass of the ejected fuel is Δm, then the momentum of the ejected fuel has a magnitude equal to $(\Delta m)v$. Since the total momentum of the system must still be zero, the rocket acquires an equivalent amount of momentum in the forward direction. Hence, the momentum increase of the rocket is

$$\Delta p = (\Delta m)v$$

If the mass of fuel Δm is ejected in the time Δt, the force exerted on the rocket is the change in its momentum divided by the time interval (Equation 9–3); that is

$$F = \frac{\Delta p}{\Delta t} = \left(\frac{\Delta m}{\Delta t}\right)v$$

The force exerted on the rocket by the ejected fuel is referred to as the **thrust.** Thus, the thrust of a rocket is

Thrust

$$\text{thrust} = \left(\frac{\Delta m}{\Delta t}\right)v \qquad \text{9–19}$$

SI unit: newton, N

REAL-WORLD PHYSICS
Saturn V rocket

By $\Delta m/\Delta t$, we simply mean the amount of mass per time coming out of the rocket. For example, on the Saturn V rocket, the one used on the manned missions to the Moon, the main engines eject fuel at the rate of 13,800 kg/s with a speed of 2440 m/s. As a result, the thrust produced by these engines is

$$\text{thrust} = \left(\frac{\Delta m}{\Delta t}\right)v = (13{,}800\ \text{kg/s})(2440\ \text{m/s}) = 33.7 \times 10^6\ \text{N}$$

Since this is about 7.60 million pounds, and the weight of the rocket at liftoff is only 6.30 million pounds $= 28.0 \times 10^6$ N, the thrust is sufficient to launch the rocket and give it an upward acceleration. In fact, the initial net force acting on the rocket is

$$F_{\text{net}} = \text{thrust} - mg = 33.7 \times 10^6\ \text{N} - 28.0 \times 10^6\ \text{N} = 5.7 \times 10^6\ \text{N}$$

The rocket's initial weight is $W = mg = 28.0 \times 10^6$ N, and hence its initial mass is $m = W/g = 2.85 \times 10^6$ kg. Therefore, the rocket lifts off with an upward acceleration of

$$a = \frac{F_{\text{net}}}{m} = \frac{5.7 \times 10^6\ \text{N}}{2.85 \times 10^6\ \text{kg}} = 2.0\ \text{m/s}^2 \approx 0.20\,g$$

This is a rather gentle acceleration. The gentleness lasts only a matter of seconds, however, since the decreasing mass of the rocket results in an increasing acceleration.

▲ A rocket (top) makes use of the principle of conservation of momentum: mass (the products of explosive burning of fuel) is ejected at high speed in one direction, causing the rocket to move in the opposite direction. The same method of propulsion has evolved in octopi (bottom) and some other animals. When danger threatens and a quick escape is needed, powerful muscles contract to create a jet of water that propels the animal to safety. (A smokescreen of ink provides additional security.)

EXERCISE 9–5

The ascent stage of the lunar lander was designed to produce 15,500 N of thrust at liftoff. If the speed of the ejected fuel is 2500 m/s, what is the rate at which the fuel must be burned?

Solution

The rate of fuel consumption is

$$\frac{\Delta m}{\Delta t} = \frac{\text{thrust}}{v} = \frac{15{,}500\ \text{N}}{2500\ \text{m/s}} = 6.2\ \text{kg/s}$$

A common question regarding rockets is: "How can a rocket accelerate in outer space when it has nothing to push against?" The answer is that rockets, in effect, push against their own fuel. The situation is similar to firing a gun. When a bullet is ejected by the internal combustion of the gunpowder, the person firing the gun feels a recoil. If the person were in space, or standing on a frictionless surface, the recoil would give him or her a speed in the direction opposite to the bullet. The burning of a rocket engine provides a continuous recoil, almost as if the rocket were firing a steady stream of bullets out the back.

Chapter Summary

9–1 Linear Momentum

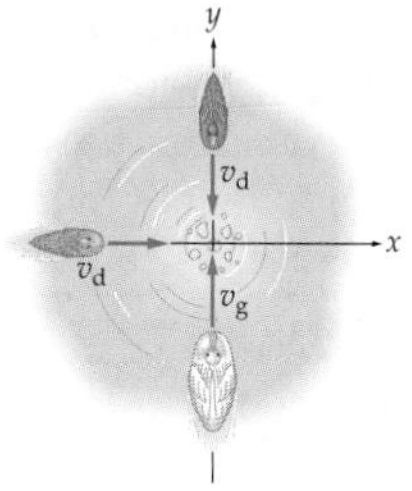

The linear momentum of an object of mass m moving with velocity $\vec{\mathbf{v}}$ is

$$\vec{\mathbf{p}} = m\vec{\mathbf{v}} \qquad 9\text{–}1$$

Momentum is a vector

Linear momentum is a vector, pointing in the same direction as the velocity vector, $\vec{\mathbf{v}}$.

Momentum of a system of objects

In a system of several objects, the total linear momentum is the vector sum of the individual momenta:

$$\vec{\mathbf{p}}_{\text{total}} = \vec{\mathbf{p}}_1 + \vec{\mathbf{p}}_2 + \vec{\mathbf{p}}_3 + \cdots \qquad 9\text{–}2$$

9–2 Newton's Second Law

In terms of momentum, Newton's second law is

$$\sum \vec{\mathbf{F}} = \frac{\Delta \vec{\mathbf{p}}}{\Delta t} \qquad 9\text{–}3$$

That is, the net force acting on an object is equal to the rate of change of its momentum.

Constant mass

For cases in which the mass is constant, Newton's second law reduces to the familiar form

$$\sum \vec{\mathbf{F}} = m\vec{\mathbf{a}} \qquad 9\text{–}4$$

9–3 Impulse

The impulse delivered to an object by an average force $\vec{\mathbf{F}}_{\text{av}}$ acting for a time Δt is

$$\vec{\mathbf{I}} = \vec{\mathbf{F}}_{\text{av}}\,\Delta t \qquad 9\text{–}5$$

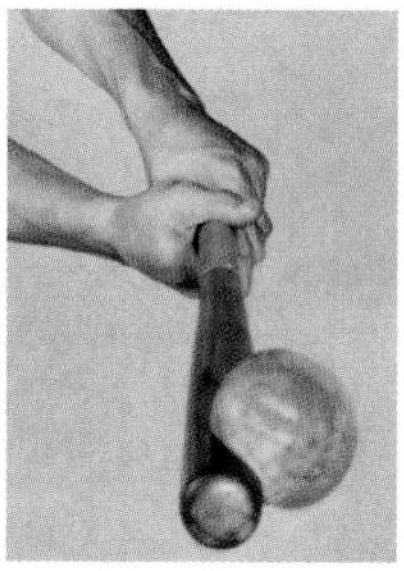

Impulse is a vector

Impulse is a vector, proportional to the force vector.

Impulse and momentum

By Newton's second law, the impulse delivered to an object is equal to the change in its momentum:

$$\vec{\mathbf{I}} = \vec{\mathbf{F}}_{\text{av}}\,\Delta t = \Delta\vec{\mathbf{p}} \qquad 9\text{–}6$$

Magnitude of the impulse and force

Since an impulse is often delivered in a very short time interval, the average force can be large.

9–4 Conservation of Linear Momentum

The momentum of an object is conserved (remains constant) if the net force acting on it is zero.

Internal/external forces

In a system of objects, internal forces always sum to zero. The net force acting on a system of objects, then, is the sum of the external forces.

Conservation of momentum in a system

In a system of objects, the net momentum is conserved if the net external force acting on the system is zero.

9–5 Inelastic Collisions

In collisions, we assume that external forces either sum to zero or are small enough to be ignored. Hence, momentum is conserved in all collisions.

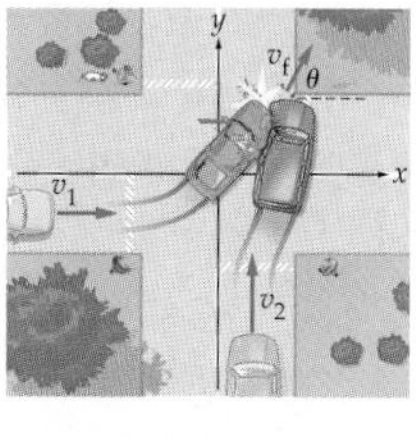

Inelastic collision

In an inelastic collision, the final kinetic energy is different from the initial kinetic energy. The kinetic energy is usually less after a collision, but it can also be more than the initial kinetic energy.

Completely inelastic collision
A collision in which objects hit and stick together is referred to as completely inelastic.

Collisions in one dimension
A one-dimensional collision occurs along a line, which we can choose to be the x axis. After the collision, the x component of momentum is equal to the x component of momentum before the collision; that is, the x component of momentum is conserved.

If two objects, of mass m_1 and m_2 and with initial velocities $v_{1,i}$ and $v_{2,i}$, collide and stick, the final velocity is

$$v_f = \frac{m_1 v_{1,i} + m_2 v_{2,i}}{m_1 + m_2} \quad \text{9–10}$$

Collisions in two dimensions
In a two-dimensional collision, there are two separate momentum relations to be satisfied: (i) the x component of momentum is conserved, and (ii) the y component of momentum is conserved.

9–6 Elastic Collisions

In collisions, we assume that external forces either sum to zero or are small enough to be ignored. Hence, momentum is conserved in all collisions.

Elastic collision
In an elastic collision, the final kinetic energy is equal to the initial kinetic energy.

Collisions in one dimension
In an elastic collision in one dimension where mass m_1 is moving with an initial velocity v_0, and mass m_2 is initially at rest, the velocities of the masses after the collision are:

$$v_{1,f} = \left(\frac{m_1 - m_2}{m_1 + m_2}\right) v_0$$

and

$$v_{2,f} = \left(\frac{2m_1}{m_1 + m_2}\right) v_0 \quad \text{9–12}$$

Collisions in two dimensions
In elastic collisions in two dimensions, three separate conditions are satisfied: (i) kinetic energy is conserved, (ii) the x component of momentum is conserved, and (iii) the y component of momentum is conserved.

9–7 Center of Mass

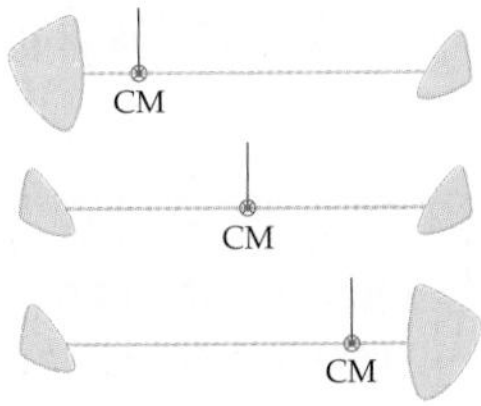

The location of the center of mass of a two-dimensional system of objects is defined as follows:

$$X_{cm} = \frac{m_1 x_1 + m_2 x_2 + \cdots}{m_1 + m_2 + \cdots} = \frac{\sum mx}{M} \quad \text{9–14}$$

and

$$Y_{cm} = \frac{m_1 y_1 + m_2 y_2 + \cdots}{m_1 + m_2 + \cdots} = \frac{\sum my}{M} \quad \text{9–15}$$

Motion of the center of mass
The velocity of the center of mass is

$$\vec{\mathbf{V}}_{cm} = \frac{m_1 \vec{\mathbf{v}}_1 + m_2 \vec{\mathbf{v}}_2 + \cdots}{m_1 + m_2 + \cdots} = \frac{\sum m\vec{\mathbf{v}}}{M} \quad \text{9–16}$$

Note that $M\vec{\mathbf{V}}_{cm} = m_1\vec{\mathbf{v}}_1 + m_2\vec{\mathbf{v}}_2 + \cdots = \vec{\mathbf{p}}_{total}$. If a system's momentum is conserved, its center of mass has constant velocity. Similarly, the acceleration of the center of mass is

$$\vec{\mathbf{A}}_{cm} = \frac{m_1 \vec{\mathbf{a}}_1 + m_2 \vec{\mathbf{a}}_2 + \cdots}{m_1 + m_2 + \cdots} = \frac{\sum m\vec{\mathbf{a}}}{M} \quad \text{9–17}$$

Note that $M\vec{\mathbf{A}}_{cm} = m_1\vec{\mathbf{a}}_1 + m_2\vec{\mathbf{a}}_2 + \cdots =$ (net external force). That is,

$$M\vec{\mathbf{A}}_{cm} = \vec{\mathbf{F}}_{net,ext} \qquad \textbf{9–18}$$

The center of mass accelerates as if the net external force acted on a single object of mass $M = m_1 + m_2 + \cdots$.

*9–8 Systems with Changing Mass: Rocket Propulsion

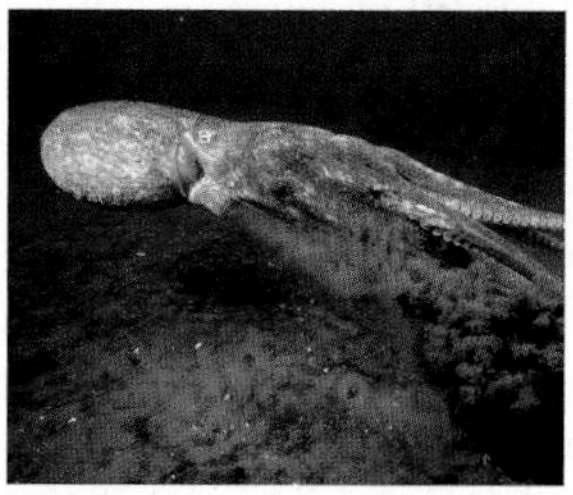

The mass of a rocket changes because its engines expel fuel when they are fired. If fuel is expelled with the speed v and at the rate $\Delta m/\Delta t$, the thrust experienced by the rocket is

$$\text{thrust} = \left(\frac{\Delta m}{\Delta t}\right)v \qquad \textbf{9–19}$$

Problem-Solving Summary

Type of Calculation	Relevant Physical Concepts	Related Examples
Calculate the momentum of a system.	Each object in a system has a momentum of magnitude mv that points in the direction of its velocity vector. The total momentum is the vector sum of the individual momenta.	Example 9–1
Relate force and time to the impulse.	The impulse acting on a system is the average force, F_{av}, times the time interval, Δt.	Example 9–2 Active Example 9–1
Apply momentum conservation.	Momentum is conserved when the net external force acting on a system is zero.	Examples 9–3, 9–4, 9–5, 9–6, 9–7 Active Example 9–2
Find the center of mass.	The location of the center of mass is given by Equations 9–14 and 9–15.	Example 9–8
Determine the motion of the center of mass.	The center of mass moves the same as if it were a point particle of mass M (the total mass of the system) acted on by the net external force, $\vec{\mathbf{F}}_{net,ext}$.	Example 9–9 Active Example 9–3

Conceptual Questions

(Answers to odd-numbered Conceptual Questions can be found in the back of the book.)

1. If you drop your keys, their momentum increases as they fall. Why is the momentum of the keys not conserved? Does this mean that the momentum of the universe increases as the keys fall? Explain.
2. By what factor does an object's kinetic energy change if its speed is doubled? By what factor does its momentum change?
3. A system of particles is known to have zero kinetic energy. What can you say about the momentum of the system?
4. A system of particles is known to have zero momentum. Does it follow that the kinetic energy of the system is also zero? Explain.
5. On a calm day you connect an electric fan to a battery on your sailboat and generate a breeze. Can the wind produced by the fan be used to power the sailboat? Explain.
6. In the previous question, can you use the wind generated by the fan to move a boat that has no sail? Explain why or why not.
7. Crash statistics show that it is safer to be riding in a heavy car in an accident than in a light car. Explain in terms of physical principles.
8. **(a)** As you approach a stoplight, you apply the brakes and bring your car to rest. What happened to your car's initial momentum? **(b)** When the light turns green, you accelerate until you reach cruising speed. What force was responsible for increasing your car's momentum?
9. An object at rest on a frictionless surface is struck by a second object. Is it possible for both objects to be at rest after the collision? Explain.
10. In the previous question, is it possible for one of the two objects to be at rest after the collision? Explain.
11. **(a)** Can two objects on a horizontal frictionless surface have a collision in which all the initial kinetic energy of the system is lost? Explain and give a specific example if your answer is yes. **(b)** Can two such objects have a collision in which all the initial momentum of the system is lost? Explain, and give a specific example if your answer is yes.
12. Two cars collide at an intersection. If the cars do not stick together, can we conclude that their collision was elastic? Explain.
13. At the instant a bullet is fired from a gun, the bullet and the gun have equal and opposite momenta. Which object—the bullet or the gun—has the greater kinetic energy? Explain. How does your answer apply to the observation that it is safe to hold a gun while it is fired, whereas the bullet is deadly?
14. An hourglass is turned over, and the sand is allowed to pour from the upper half of the glass to the lower half. If the hourglass is resting on a scale, and the total mass of the hourglass and sand is M, describe the reading on the scale as the sand runs to the bottom.

15. In the classic movie *The Spirit of St. Louis*, Jimmy Stewart portrays Charles Lindbergh on his history-making transatlantic flight. Lindbergh is concerned about the weight of his fuel-laden airplane. As he flies over Newfoundland he notices a fly on the dashboard. Speaking to the fly, he wonders aloud, "Does the plane weigh less if you fly inside it as it's flying? Now that's an interesting question." What do you think?

16. A tall, slender drinking glass with a thin base is initially empty. **(a)** Where is the center of mass of the glass? **(b)** Suppose the glass is now filled slowly with water until it is completely full. Describe the position and motion of the center of mass during the filling process.

17. Lifting one foot into the air, you balance on the other foot. What can you say about the location of your center of mass?

18. In the "Fosbury flop" method of high jumping, named for the track and field star Dick Fosbury, an athlete's center of mass may pass under the bar while the athlete's body passes over the bar. Explain how this is possible.

The "Fosbury flop." (Conceptual Question 18)

Conceptual Exercises

(Answers to odd-numbered Conceptual Exercises can be found in the back of the book.)

1. Object A has a mass m, object B has a mass $2m$, and object C has a mass $m/2$. Rank these objects in order of increasing kinetic energy, given that they all have the same momentum. Indicate ties where appropriate.

2. Object A has a mass m, object B has a mass $4m$, and object C has a mass $m/4$. Rank these objects in order of increasing momentum, given that they all have the same kinetic energy. Indicate ties where appropriate.

3. A block of wood is struck by a bullet. Is the block more likely to be knocked over if the bullet is metal and embeds itself in the wood, or if the bullet is rubber and bounces off the wood?

4. Your car rolls slowly in a parking lot and bangs into the metal base of a light pole. In terms of safety, is it better for your collision with the light pole to be elastic, inelastic, or is the safety risk the same for either case?

5. A net force of 200 N acts on a 100-kg boulder, and a force of the same magnitude acts on a 100-g pebble. Is the change of the boulder's momentum in one second greater than, less than, or equal to the change of the pebble's momentum in the same time period?

6. Referring to the previous question, is the change in the boulder's speed in one second greater than, less than, or equal to the change in speed of the pebble in the same time period?

7. A friend tosses a ball of mass m to you with a speed v. When you catch the ball, you feel a noticeable sting in your hand. If you now catch a ball of mass $2m$ and speed $v/2$, is the sting you feel greater than, less than, or the same as that felt when you caught the first ball?

8. Suppose you throw a rubber ball at a charging elephant (not a good idea). When the ball bounces back toward you, is its speed greater than, less than, or equal to the speed with which you threw it?

9. Force A has a magnitude F and acts for the time Δt, force B has a magnitude $2F$ and acts for the time $\Delta t/3$, force C has a magnitude $5F$ and acts for the time $\Delta t/10$, and force D has a magnitude $10F$ and acts for the time $\Delta t/100$. Rank these forces in order of increasing impulse. Indicate ties where appropriate.

10. A stalactite in a cave has drops of water falling from it to the cave floor below. The drops are equally spaced in time and come in rapid succession, so that at any given moment there are many drops in midair. Is the center of mass of the midair drops higher than, lower than, or at the halfway distance between the tip of the stalactite and the cave floor?

11. A pencil standing upright on its eraser end falls over and lands on a table. As the pencil falls, its eraser does not slip. The following questions refer to the contact force exerted on the pencil by the table. Let the positive x direction be in the direction the pencil falls, and the positive y direction be vertically upward. **(a)** During the pencil's fall, is the x component of the contact force positive, negative, or zero? **(b)** Is the y component of the contact force greater than, less than, or equal to the weight of the pencil?

12. A juggler performs a series of tricks with three bowling balls while standing on a bathroom scale. Is the average reading of the scale greater than, less than, or equal to the weight of the juggler plus the weight of the three balls?

13. **Figure 9-15** shows a block of mass $2m$ at rest on a horizontal, frictionless table. Attached to this block by a string that passes over a pulley is a second block, with a mass m. The initial position of the center of mass of the blocks is indicated by the point i. The blocks are now released and allowed to accelerate; a short time later their center of mass is at the point f. Did the center of mass follow the red path, the green path, or the blue path?

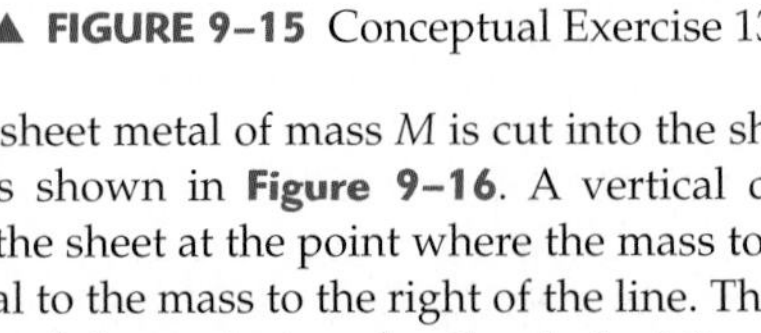

▲ **FIGURE 9-15** Conceptual Exercise 13

14. A piece of sheet metal of mass M is cut into the shape of a right triangle, as shown in **Figure 9-16**. A vertical dashed line is drawn on the sheet at the point where the mass to the left of the line is equal to the mass to the right of the line. The sheet is now placed on a fulcrum just under the dashed line and released from rest. Does the metal sheet remain level, tip to the left, or tip to the right?

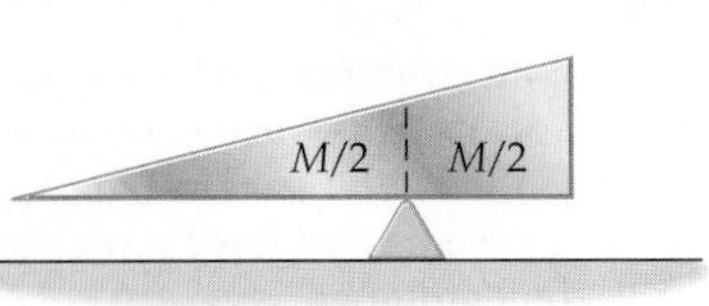

▲ **FIGURE 9-16** Conceptual Exercise 14

Problems

Note: **IP** *denotes an integrated conceptual/quantitative problem.* **BIO** *identifies problems of biological or medical interest. Red bullets (•, ••, •••) are used to indicate the level of difficulty of each problem.*

Section 9–1 Linear Momentum

1. • Referring to Exercise 9–1, what speed must the baseball have if its momentum is to be equal in magnitude to that of the car? Give your result in miles per hour.

2. • Find the total momentum of the birds in Example 9–1 if the goose reverses direction.

3. •• A 20.0-kg dog is running northward at 2.50 m/s, while a 5.00-kg cat is running eastward at 3.00 m/s. Their 70.0-kg owner has the same momentum as the two pets taken together. Find the direction and magnitude of the owner's velocity.

4. •• **IP** Two air-track carts move toward one another on an air track. Cart 1 has a mass of 0.35 kg and a speed of 1.2 m/s. Cart 2 has a mass of 0.61 kg. **(a)** What speed must cart 2 have if the total momentum of the system is to be zero? **(b)** Since the momentum of the system is zero, does it follow that the kinetic energy of the system is also zero? **(c)** Verify your answer to part (b) by calculating the system's kinetic energy.

5. •• A 0.150-kg baseball is dropped from rest. If the magnitude of the baseball's momentum is 0.780 kg · m/s just before it lands on the ground, from what height was it dropped?

6. •• **IP** A 220-g ball falls vertically downward, hitting the floor with a speed of 2.5 m/s and rebounding upward with a speed of 2.0 m/s. **(a)** Find the magnitude of the change in the ball's momentum. **(b)** Find the change in the magnitude of the ball's momentum. **(c)** Which of the two quantities calculated in parts (a) and (b) is more directly related to the net force acting on the ball during its collision with the floor? Explain.

7. ••• Object 1 has a mass m_1 and a velocity $\vec{\mathbf{v}}_1 = (2.80\ \text{m/s})\hat{\mathbf{x}}$. Object 2 has a mass m_2 and a velocity $\vec{\mathbf{v}}_2 = (3.10\ \text{m/s})\hat{\mathbf{y}}$. The total momentum of these two objects has a magnitude of 17.6 kg · m/s and points in a direction 66.5° above the positive x axis. Find m_1 and m_2.

Section 9–3 Impulse

8. • Find the magnitude of the impulse delivered to a soccer ball when a player kicks it with a force of 1450 N. Assume that the player's foot is in contact with the ball for 5.80×10^{-3} s.

9. • In a typical golf swing, the club is in contact with the ball for about 0.0010 s. If the 45-g ball acquires a speed of 67 m/s, estimate the magnitude of the force exerted by the club on the ball.

10. • A 0.50-kg croquet ball is initially at rest on the grass. When the ball is struck by a mallet, the average force exerted on it is 230 N. If the ball's speed after being struck is 3.2 m/s, how long was the mallet in contact with the ball?

11. • When spiking a volleyball, a player changes the velocity of the ball from 4.2 m/s to −24 m/s along a certain direction. If the impulse delivered to the ball by the player is −9.3 kg · m/s, what is the mass of the volleyball?

12. •• **IP** A 15.0-g marble is dropped from rest onto the floor 1.44 m below. **(a)** If the marble bounces straight upward to a height of 0.640 m, what is the magnitude and direction of the impulse delivered to the marble by the floor? **(b)** If the marble had bounced to a greater height, would the impulse delivered to it have been greater or less than the impulse found in part (a)? Explain.

13. •• To make a bounce pass, a player throws a 0.60-kg basketball toward the floor. The ball hits the floor with a speed of 5.4 m/s at an angle of 65° to the vertical. If the ball rebounds with the same speed and angle, what was the impulse delivered to it by the floor?

14. •• **IP** A 0.14-kg baseball moves toward home plate with a velocity $\vec{\mathbf{v}}_i = (-36\ \text{m/s})\hat{\mathbf{x}}$. After striking the bat, the ball moves vertically upward with a velocity $\vec{\mathbf{v}}_f = (18\ \text{m/s})\hat{\mathbf{y}}$. **(a)** Find the direction and magnitude of the impulse delivered to the ball by the bat. Assume that the ball and bat are in contact for 1.5 ms. **(b)** How would your answer to part (a) change if the mass of the ball were doubled? **(c)** How would your answer to part (a) change if the mass of the bat were doubled instead?

15. •• A player bounces a 0.43-kg soccer ball off her head, changing the velocity of the ball from $\vec{\mathbf{v}}_i = (8.8\ \text{m/s})\hat{\mathbf{x}} + (-2.3\ \text{m/s})\hat{\mathbf{y}}$ to $\vec{\mathbf{v}}_f = (5.2\ \text{m/s})\hat{\mathbf{x}} + (3.7\ \text{m/s})\hat{\mathbf{y}}$. If the ball is in contact with the player's head for 6.7 ms, what are **(a)** the direction and **(b)** the magnitude of the impulse delivered to the ball?

Section 9–4 Conservation of Momentum

16. • In a situation similar to Example 9–3, suppose the speeds of the two canoes after they are pushed apart are 0.56 m/s for canoe 1 and 0.41 m/s for canoe 2. If the mass of canoe 1 is 320 kg, what is the mass of canoe 2?

17. • Two ice skaters stand at rest in the center of an ice rink. When they push off against one another the 45-kg skater acquires a speed of 0.62 m/s. If the speed of the other skater is 0.89 m/s, what is this skater's mass?

18. • Suppose the bee in Active Example 9–2 has a mass of 0.175 g. If the bee walks with a speed of 1.41 cm/s relative to the still water, what is the speed of the 4.75-g stick relative to the water?

19. •• An object initially at rest breaks into two pieces as the result of an explosion. One piece has twice the kinetic energy of the other piece. What is the ratio of the masses of the two pieces? Which piece has the larger mass?

20. •• A 92-kg astronaut and a 1200-kg satellite are at rest relative to the space shuttle. The astronaut pushes on the satellite, giving it a speed of 0.14 m/s directly away from the shuttle. Seven-and-a-half seconds later the astronaut comes into contact with the shuttle. What was the initial distance from the shuttle to the astronaut?

21. •• **IP** An 85-kg lumberjack stands at one end of a 380-kg floating log, as shown in **Figure 9–17**. Both the log and the lumberjack are at rest initially. **(a)** If the lumberjack now trots toward the other end of the log with a speed of 2.7 m/s relative to the log, what is the lumberjack's speed relative to the shore? Ignore friction between the log and the water. **(b)** If the mass of the log had been greater, would the lumberjack's speed relative to the shore be greater than, less than, or the same as in part (a)? Explain. **(c)** Check your answer to part (b) by calculating the lumberjack's speed relative to the shore for the case of a 450-kg log.

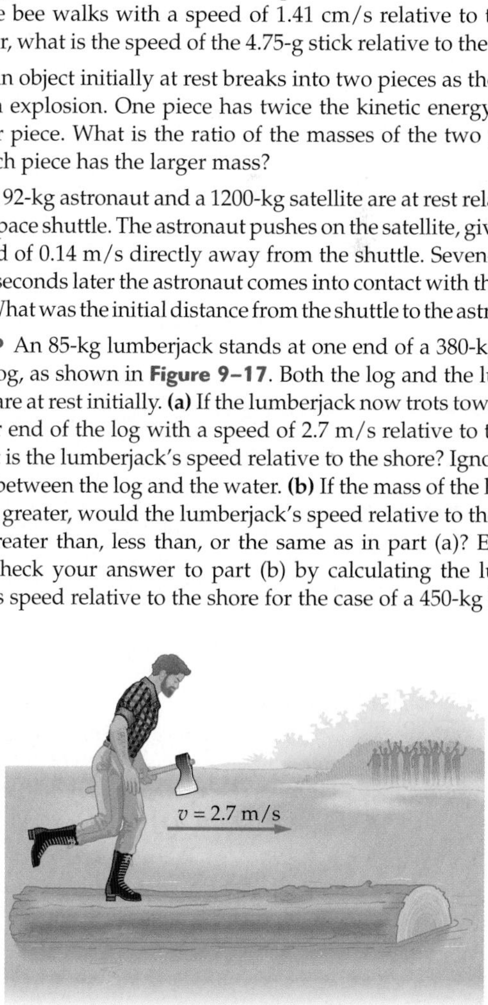

▲ **FIGURE 9–17** Problem 21

22. ••• A plate drops onto a smooth floor and shatters into three pieces of equal mass. Two of the pieces go off with equal speeds v at right angles to one another. Find the speed and direction of the third piece.

Section 9–5 Inelastic Collisions

23. • A cart of mass m moves with a speed v on a frictionless air track and collides with an identical cart that is stationary. If the two carts stick together after the collision, what is the final kinetic energy of the system?

24. • Suppose the car in Example 9–6 has an initial speed of 20.0 m/s and that the direction of the wreckage after the collision is 40.0° above the x axis. Find the initial speed of the minivan and the final speed of the wreckage.

25. • Two 78.0-kg hockey players skating at 5.25 m/s collide and stick together. If the angle between their initial directions was 115°, what is their speed after the collision?

26. •• **IP** **(a)** Referring to Exercise 9–2, is the final kinetic energy of the car and truck together greater than, less than, or equal to the sum of the initial kinetic energies of the car and truck separately? Explain. **(b)** Verify your answer to part (a) by calculating the initial and final kinetic energies of the system.

27. •• **IP** A bullet with a mass of 4.0 g and a speed of 650 m/s is fired at a block of wood with a mass of 0.095 kg. The block rests on a frictionless surface, and is thin enough that the bullet passes completely through it. Immediately after the bullet exits the block, the speed of the block is 23 m/s. **(a)** What is the speed of the bullet when it exits the block? **(b)** Is the final kinetic energy of this system equal to, less than, or greater than the initial kinetic energy? Explain. **(c)** Verify your answer to part (b) by calculating the initial and final kinetic energies of the system.

28. •• **IP** A 0.440-kg block of wood hangs from the ceiling by a string, and a 0.0720-kg wad of putty is thrown straight upward, striking the bottom of the block with a speed of 5.74 m/s. The wad of putty sticks to the block. **(a)** Is the mechanical energy of this system conserved? **(b)** How high does the putty-block system rise above the original position of the block?

29. •• A 0.430-kg block is attached to a horizontal spring that is at its equilibrium length, and whose force constant is 20.0 N/m. The block rests on a frictionless surface. A 0.0500-kg wad of putty is thrown horizontally at the block, hitting it with a speed of 2.30 m/s and sticking. How far does the putty-block system compress the spring?

30. ••• Two objects moving with a speed v travel in opposite directions in a straight line. The objects stick together when they collide, and move with a speed of $v/4$ after the collision. **(a)** What is the ratio of the final kinetic energy of the system to the initial kinetic energy? **(b)** What is the ratio of the mass of the more massive object to the mass of the less massive object?

Section 9–6 Elastic Collisions

31. • A 722-kg car stopped at an intersection is rear-ended by a 1620-kg truck moving with a speed of 14.5 m/s. If the car was in neutral and its brakes were off, so that the collision is approximately elastic, find the final speed of both vehicles after the collision.

32. • The collision between a hammer and a nail can be considered to be approximately elastic. Calculate the kinetic energy acquired by a 12-g nail when it is struck by a 550-g hammer moving with an initial speed of 4.5 m/s.

33. •• In the apple-orange collision in Example 9–7, suppose the final velocity of the orange is 1.03 m/s in the negative y direction. What is the final speed and direction of the apple in this case?

34. •• **Moderating a Neutron** In a nuclear reactor, neutrons released by nuclear fission must be slowed down before they can trigger additional reactions in other nuclei. To see what sort of material is most effective in slowing (or moderating) a neutron, calculate the ratio of a neutron's final kinetic energy to its initial kinetic energy, K_f/K_i, for a head-on elastic collision with each of the following stationary target particles. (*Note:* The mass of a neutron is $m = 1.009$ u, where the atomic mass unit, u, is defined as follows: $1 \text{ u} = 1.66 \times 10^{-27}$ kg.) **(a)** An electron ($M = 5.49 \times 10^{-4}$ u). **(b)** A proton ($M = 1.007$ u). **(c)** The nucleus of a lead atom ($M = 207.2$ u).

35. •• **IP** A charging bull elephant with a mass of 5230 kg comes directly toward you with a speed of 4.45 m/s. You toss a 0.150-kg rubber ball at the elephant with a speed of 7.91 m/s. **(a)** When the ball bounces back toward you, what is its speed? **(b)** How do you account for the fact that the ball's kinetic energy has increased?

36. •• The three air carts shown in **Figure 9–18** have masses, reading from left to right, of $4m$, $2m$, and m, respectively. The most massive cart has an initial speed of v_0; the other two carts are at rest initially. All carts are equipped with spring bumpers that give elastic collisions. **(a)** Find the final speed of each cart. **(b)** Verify that the final kinetic energy of the system is equal to the initial kinetic energy. (Assume the air track is long enough to accommodate all collisions.)

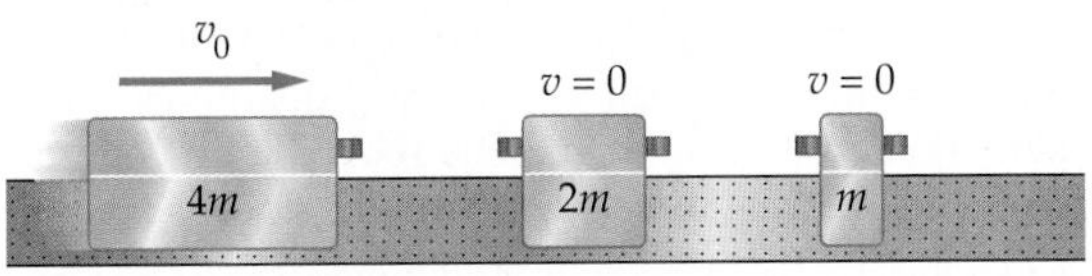

▲ **FIGURE 9–18** Problem 36

37. •• In this problem we show that when one ball is pulled to the left in the photo on page 261, only a single ball recoils to the right—under ideal elastic-collision conditions. To begin, suppose that each ball has a mass m, and that the ball coming in from the left strikes the other balls with a speed v_0. Now, consider the hypothetical case of two balls recoiling to the right. Determine the speed the two recoiling balls must have in order to satisfy **(a)** momentum conservation and **(b)** energy conservation. Since these speeds are not the same, it follows that momentum and energy cannot be conserved simultaneously with a recoil of two balls.

Section 9–7 Center of Mass

38. • Find the x coordinate of the center of mass of the bricks shown in **Figure 9–19**.

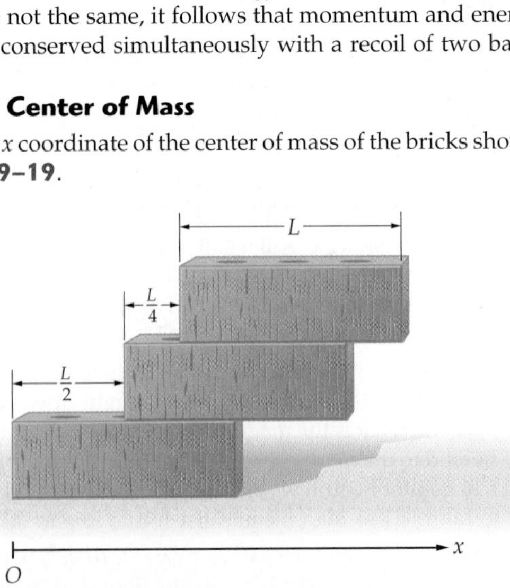

▲ **FIGURE 9–19** Problem 38

39. • You are holding a shopping basket at the grocery store with two 0.55-kg cartons of cereal at the left end of the basket. The basket is 0.75-m long. Where should you place a 1.8-kg half gallon of milk, relative to the left end of the basket, so that the center of mass of your groceries is at the center of the basket?

40. • **Earth–Moon Center of Mass** The Earth has a mass of 5.98×10^{24} kg, the Moon has a mass of 7.35×10^{22} kg, and their center-to-center distance is 3.85×10^8 m. How far from the center of the Earth is the Earth–Moon center of mass? Is the Earth–Moon center of mass above or below the surface of the Earth? By what distance? (As the Earth and Moon orbit one another, their centers orbit about their common center of mass.)

41. •• A cardboard box is in the shape of a cube with each side of length L. If the top of the box is missing, where is the center of mass of the open box? Give your answer relative to the geometric center of the box.

42. •• The location of the center of mass of the partially eaten, 12-inch diameter pizza shown in **Figure 9–20** is $X_{cm} = -1.4$ in. and $Y_{cm} = -1.4$ in. Assuming each quadrant of the pizza to be the same, find the center of mass of the uneaten pizza above the x axis (that is, the portion of the pizza in the second quadrant).

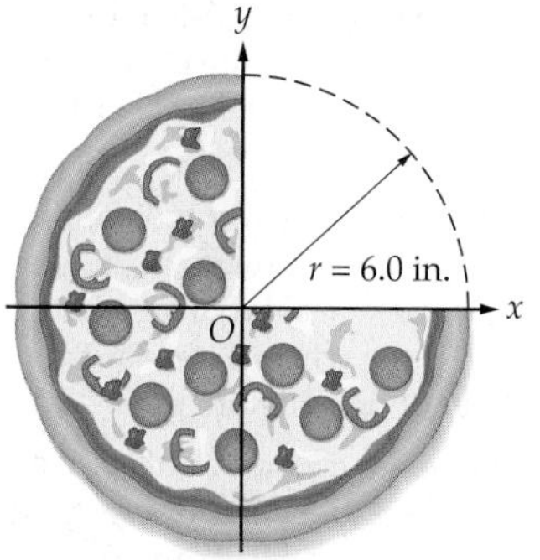

▲ **FIGURE 9–20** Problem 42

43. •• **The Center of Mass of Sulfur Dioxide** Sulfur dioxide (SO_2) consists of two oxygen atoms (each of mass 16 u, where u is defined in Problem 34) and a single sulfur atom (of mass 32 u). The center-to-center distance between the sulfur atom and either of the oxygen atoms is 0.143 nm, and the angle formed by the three atoms is 120°, as shown in **Figure 9–21**. Find the x and y coordinates of the center of mass of this molecule.

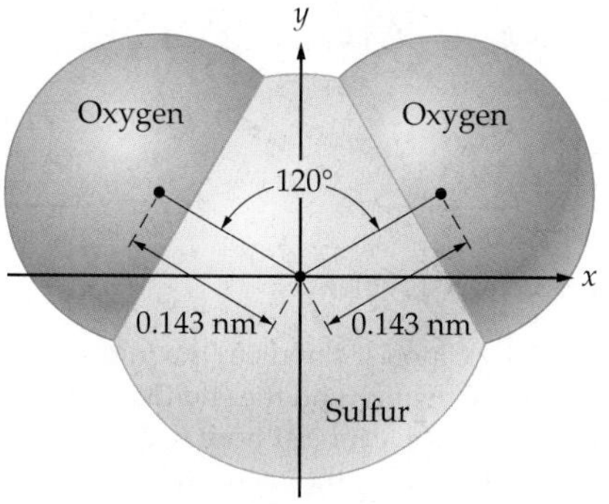

▲ **FIGURE 9–21** Problem 43

44. •• **IP** Three uniform meter sticks, each of mass m, are placed on the floor as follows: stick 1 lies along the y axis from $y = 0$ to $y = 1.0$ m, stick 2 lies along the x axis from $x = 0$ to $x = 1.0$ m, stick 3 lies along the x axis from $x = 1.0$ m to $x = 2.0$ m. **(a)** Find the location of the center of mass of the meter sticks. **(b)** How would the location of the center of mass be affected if the mass of the meter sticks were doubled?

45. •• A 0.721-kg rope 2.00 meters long lies on a floor. You grasp one end of the rope and begin lifting it upward with a constant speed of 0.810 m/s. Find the position and velocity of the rope's center of mass from the time you begin lifting the rope to the time the last piece of rope lifts off the floor. Plot your results. (Assume the rope occupies negligible volume directly below the point where it is being lifted.)

46. •• Repeat the previous problem, this time lowering the rope onto a floor instead of lifting it.

47. •• Consider the system shown in **Figure 9–22**. Assume that after the string breaks the ball falls through the liquid with constant speed. If the mass of the bucket *and* the liquid is 1.20 kg, and the mass of the ball is 0.150 kg, what is the reading on the scale **(a)** before and **(b)** after the string breaks?

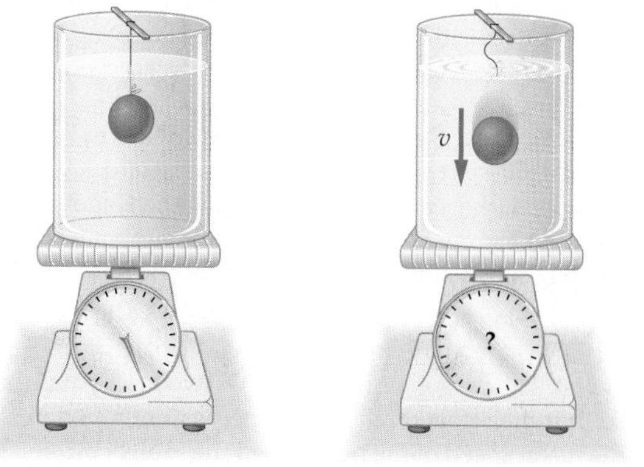

▲ **FIGURE 9–22** Problems 47 and 65

48. ••• A metal block of mass m is attached to the ceiling by a spring. Connected to the bottom of this block is a string that supports a second block of the same mass m, as shown in **Figure 9–23**. The string connecting the two blocks is now cut. **(a)** What is the net force acting on the two-block system immediately after the string is cut? **(b)** What is the acceleration of the center of mass of the two-block system immediately after the string is cut?

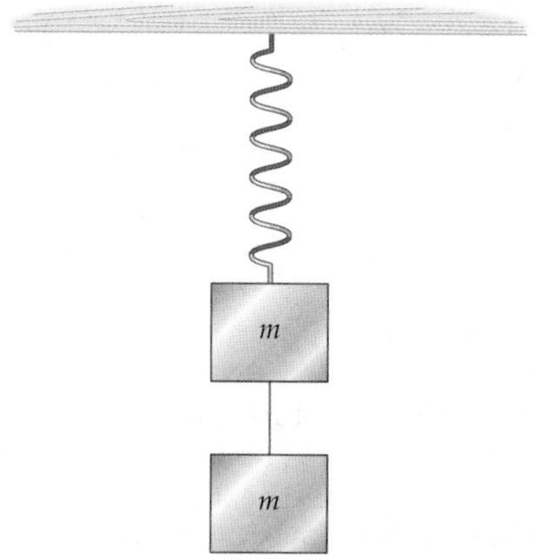

▲ **FIGURE 9–23** Problem 48

*Section 9–8 Systems with Changing Mass: Rocket Propulsion

49. • **Helicopter Thrust** During a rescue operation, a 5300-kg helicopter hovers above a fixed point. The helicopter blades send air downward with a speed of 62 m/s. What mass of air must pass through the blades every second to produce enough thrust for the helicopter to hover?

The powerful downdraft from this helicopter's blades creates a circular wave pattern in the water below. The thrust resulting from this downdraft is sufficient to support the weight of the helicopter. (Problem 49)

50. • **Rocks for a Rocket Engine** A child sits in a wagon with a pile of 0.50-kg rocks. If she can throw each rock with a speed of 11 m/s relative to the ground, causing the wagon to move, how many rocks must she throw per minute to maintain a constant average speed against a 3.4-N force of friction?

51. • A 57.0-kg person holding two 0.850-kg bricks stands on a 2.10-kg skateboard. Initially, the skateboard and the person are at rest. The person now throws the two bricks at the same time so that their speed relative to the person is 18.0 m/s. What is the recoil speed of the person and the skateboard relative to the ground, assuming the skateboard moves without friction?

52. •• In the previous problem, calculate the final speed of the person and the skateboard relative to the ground if the person throws the bricks one at a time. Assume that each brick is thrown with a speed of 18.0 m/s relative to the person.

53. •• A 0.540-kg bucket rests on a scale. Into this bucket you pour sand at the constant rate of 56.0 g/s. If the sand lands in the bucket with a speed of 3.20 m/s, **(a)** what is the reading of the scale when there is 0.750 kg of sand in the bucket? **(b)** What is the weight of the bucket and the 0.750 kg of sand?

54. •• **IP** Holding a long rope by its upper end, you lower it onto a scale. The rope has a mass of 0.13 kg per meter of length, and is lowered onto the scale at the constant rate of 1.4 m/s. **(a)** Calculate the thrust exerted by the rope as it lands on the scale. **(b)** At the instant when the amount of rope at rest on the scale has a weight of 2.5 N, does the scale read 2.5 N, more than 2.5 N, or less than 2.5 N? Explain. **(c)** Check your answer to part (b) by calculating the reading on the scale at this time.

General Problems

55. • A 72.5-kg tourist climbs the stairs to the top of the Washington Monument, which is 555 ft high. How far does the Earth move in the opposite direction as the tourist climbs?

56. •• A car moving with an initial speed v collides with a second stationary car that is one-half as massive. After the collision the first car moves in the same direction as before with a speed $v/3$. **(a)** Find the final speed of the second car. **(b)** Is this collision elastic or inelastic?

57. •• A 1.25-kg block of wood sits at the edge of a table, 0.752 m above the floor. A 0.0105-kg bullet moving horizontally with a speed of 725 m/s embeds itself within the block. What horizontal distance does the block cover before hitting the ground?

58. •• **IP** The carton of eggs shown in **Figure 9–24** is filled with a dozen eggs, each of mass m. Initially, the center of mass of the eggs is at the center of the carton. **(a)** Does the location of the center of mass of the eggs change more if egg 1 is removed or if egg 2 is removed? Explain. **(b)** Find the center of mass of the eggs when egg 1 is removed. **(c)** Find the center of mass of the eggs if egg 2 is removed instead.

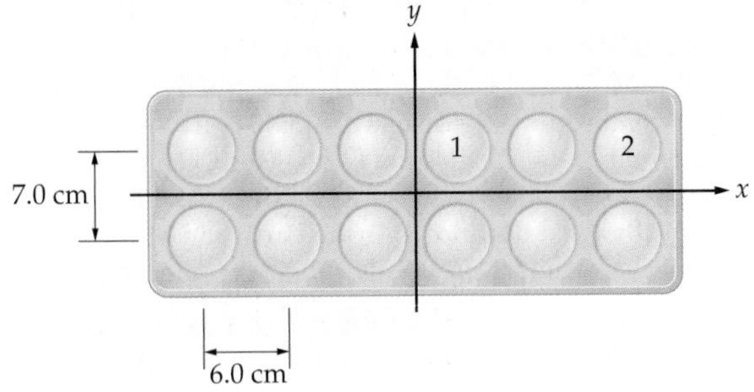

▲ **FIGURE 9–24** Problem 58

59. •• **The Force of a Storm** During a severe storm in Palm Beach, FL, on January 2, 1999, 31 inches of rain fell in a period of nine hours. Assuming that the raindrops hit the ground with a speed of 10 m/s, estimate the average upward force exerted by one square meter of ground to stop the falling raindrops during the storm. (Note: One cubic meter of water has a mass of 1000 kg.)

60. •• An apple that weighs 2.2 N falls vertically downward from rest for 1.3 s. **(a)** What is the change in the apple's momentum per second? **(b)** What is the total change in its momentum during the 1.3-second fall?

61. •• To balance a 35.5-kg automobile tire and wheel, a mechanic must place a 50.2-g lead weight 25.0 cm from the center of the wheel. When the wheel is balanced, its center of mass is exactly at the center of the wheel. How far from the center of the wheel was its center of mass before the lead weight was added?

62. •• A hoop of mass M and radius R rests on a smooth, level surface. The inside of the hoop has ridges on either side, so that it forms a track on which a ball can roll, as indicated in **Figure 9–25**. If a ball of mass $2M$ and radius $r = R/4$ is released as shown, the system rocks back and forth until it comes to rest with the ball at the bottom of the hoop. When the ball comes to rest, what is the x coordinate of its center?

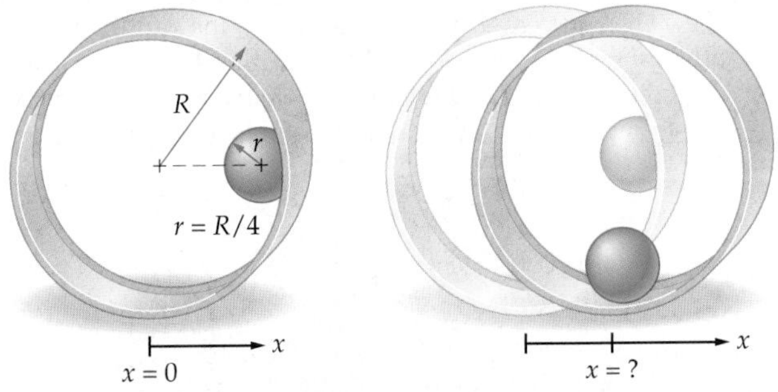

▲ **FIGURE 9–25** Problem 62

63. •• **IP** A 63-kg canoeist stands in the middle of her 22-kg canoe. The canoe is 3.0 m long, and the end that is closest to land is 2.5 m from the shore. The canoeist now walks toward the shore until she comes to the end of the canoe. **(a)** When the canoeist stops at the end of her canoe, is her distance from the shore equal to, greater than, or less than 2.5 m? Explain. **(b)** Verify your answer to part (a) by calculating the distance from the canoeist to shore.

64. •• In the previous problem, suppose the canoeist is 3.4 m from shore when she reaches the end of her canoe. What is the canoe's mass?

65. •• Referring to Problem 47, find the reading on the scale **(a)** before and **(b)** after the string breaks, assuming the ball falls through the liquid with an acceleration equal to 0.250 g.

66. •• A young hockey player stands at rest on the ice holding a 1.1-kg helmet. The player tosses the helmet with a speed of 6.2 m/s in a direction 13° above the horizontal, and recoils with a speed of 0.25 m/s. Find the mass of the hockey player.

67. •• Suppose the air carts in Example 9–9 are both moving to the right initially. The cart to the left has a mass m and an initial speed v_0; the cart to the right has an initial speed $v_0/2$. If the center of mass of this system moves to the right with a speed $2v_0/3$, what is the mass of the cart on the right?

68. •• A long, uniform rope with a mass of 0.135 kg per meter lies on the ground. You grab one end of the rope and lift it at the constant rate of 1.13 m/s. Calculate the upward force you must exert at the moment when the top end of the rope is 0.525 m above the ground.

69. •• **The Center of Mass of Water** Find the center of mass of a water molecule, referring to **Figure 9–26** for the relevant angles and distances. The mass of a hydrogen atom is 1.0 u, and the mass of an oxygen atom is 16 u, where u is the atomic mass unit (see Problem 34). Use the center of the oxygen atom as the origin of your coordinate system.

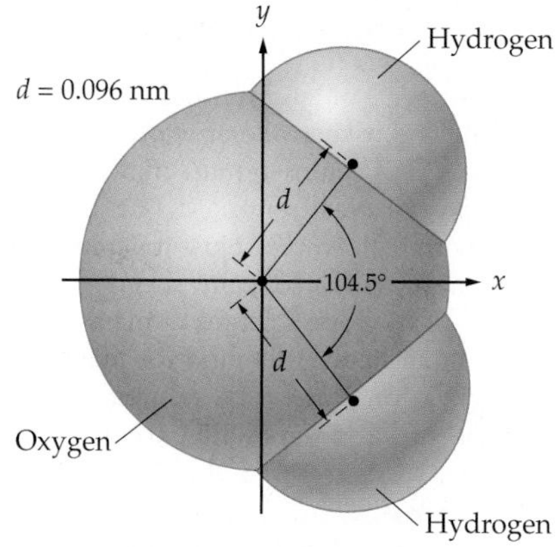

▲ **FIGURE 9–26** Problem 69

70. •• The three air carts shown in **Figure 9–27** have masses, reading from left to right, of m, $2m$, and $4m$, respectively. Initially, the cart on the right is at rest, whereas the other two carts are moving to the right with a speed v_0. All carts are equipped with putty bumpers that give completely inelastic collisions. **(a)** Find the final speed of the carts. **(b)** Calculate the ratio of the final kinetic energy of the system to the initial kinetic energy.

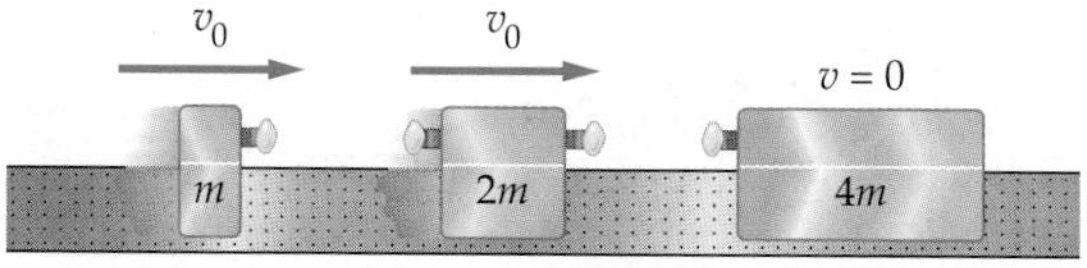

▲ **FIGURE 9–27** Problem 70

71. •• **IP** A fireworks rocket is launched vertically into the night sky with an initial speed of 44.2 m/s. The rocket coasts after being launched, then explodes and breaks into two pieces of equal mass 2.50 s later. **(a)** If each piece follows a trajectory that is initially at 45.0° to the vertical, what was their speed immediately after the explosion? **(b)** What is the velocity of the rocket's center of mass before and after the explosion? **(c)** What is the acceleration of the rocket's center of mass before and after the explosion?

72. •• **IP** The total momentum of two cars approaching an intersection is $\vec{\mathbf{p}}_{\text{total}} = (15{,}000 \text{ kg}\cdot\text{m/s})\hat{\mathbf{x}} + (2100 \text{ kg}\cdot\text{m/s})\hat{\mathbf{y}}$. **(a)** If the momentum of car 1 is $\vec{\mathbf{p}}_1 = (11{,}000 \text{ kg}\cdot\text{m/s})\hat{\mathbf{x}} + (-370 \text{ kg}\cdot\text{m/s})\hat{\mathbf{y}}$, what is the momentum of car 2? **(b)** Does your answer to part (a) depend on which car is closer to the intersection? Explain.

73. •• **Unlimited Overhang** Four identical textbooks, each of length L, are stacked near the edge of a table, as shown in **Figure 9–28**. The books are stacked in such a way that the distance they overhang the edge of the table, d, is maximized. Find the maximum overhang distance d in terms of L. In particular, show that $d > L$; that is, the top book is completely to the right of the table edge. (In principle, the overhang distance d can be made as large as desired simply by increasing the number of books in the stack.)

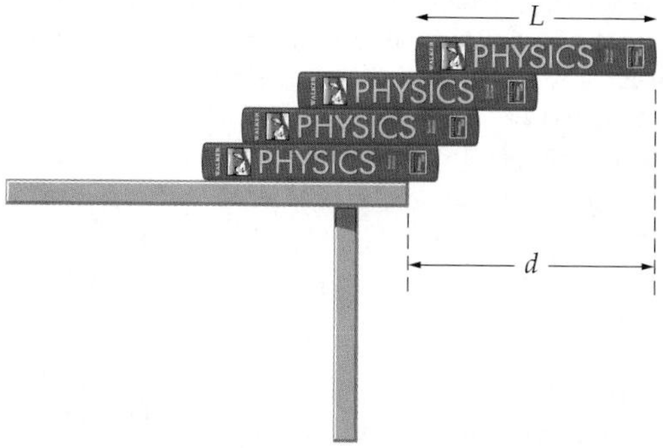

▲ **FIGURE 9–28** Problem 73

74. ••• Consider a one-dimensional, head-on elastic collision. One object has a mass m_1 and an initial velocity v_1; the other has a mass m_2 and an initial velocity v_2. Use momentum conservation and energy conservation to show that the final velocities of the two masses are

$$v_{1,f} = \left(\frac{m_1 - m_2}{m_1 + m_2}\right)v_1 + \left(\frac{2m_2}{m_1 + m_2}\right)v_2$$

$$v_{2,f} = \left(\frac{2m_1}{m_1 + m_2}\right)v_1 + \left(\frac{m_2 - m_1}{m_1 + m_2}\right)v_2$$

75. ••• Two air carts of mass $m_1 = 0.84$ kg and $m_2 = 0.42$ kg are placed on a frictionless track. Cart 1 is at rest initially, and has a spring bumper with a force constant of 610 N/m. Cart 2 has a flat metal surface for a bumper, and moves toward the bumper of the stationary cart with an initial speed $v = 0.68$ m/s. **(a)** What is the speed of the two carts at the moment when their speeds are equal? **(b)** How much energy is stored in the spring bumper when the carts have the same speed? **(c)** What is the final speed of the carts after the collision?

76. ••• **Golden Earrings and the Golden Ratio** A popular earring design features a circular piece of gold of diameter D with a circular cutout of diameter d, as shown in **Figure 9–29**. If this earring is to balance at the point P, show that the diameters must satisfy the condition $D = \phi d$, where $\phi = (1 + \sqrt{5})/2 = 1.61803\ldots$ is the famous "golden ratio."

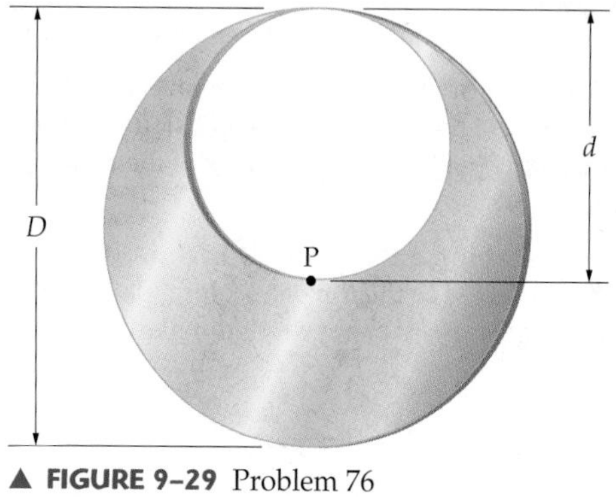

▲ **FIGURE 9–29** Problem 76

77. ••• Two objects with masses m_1 and m_2 and initial velocities $v_{1,i}$ and $v_{2,i}$ move along a straight line and collide elastically. Assuming that the objects move along the same straight line after the collision, show that their relative velocities are unchanged; that is, show that $v_{1,i} - v_{2,i} = v_{2,f} - v_{1,f}$. (You can use the results given in Problem 74.)

78. ••• **Amplified Rebound Height** Two small rubber balls are dropped from rest at a height h above a hard floor. When the balls are released, the lighter ball (with mass m) is directly above the heavier ball (with mass M). Assume the heavier ball reaches the floor first and bounces elastically; thus, when the balls collide, the ball of mass M is moving upward with a speed v and the ball of mass m is moving downward with essentially the same speed. In terms of h, find the height to which the ball of mass m rises after the collision. (Use the results given in Problem 74, and assume the balls collide at ground level.)

79. ••• **Gravitational Slingshot** Many spacecraft make use of the "gravitational slingshot" effect, in which a close encounter with a planet results in a significant increase in the spacecraft's speed. The basic features of the effect are shown in **Figure 9–30**, where we see the spacecraft moving to the left with an initial speed v_i, the planet moving to the right with a speed u, and the spacecraft departing from the planet after the encounter with a final speed v_f to the right. **(a)** Treating this interaction as a one-dimensional elastic collision, and using the result of Problem 77, determine the final speed of the spacecraft. **(b)** In the special case $v_i = u$, show that the spacecraft's final kinetic energy is 9 times greater than its initial kinetic energy.

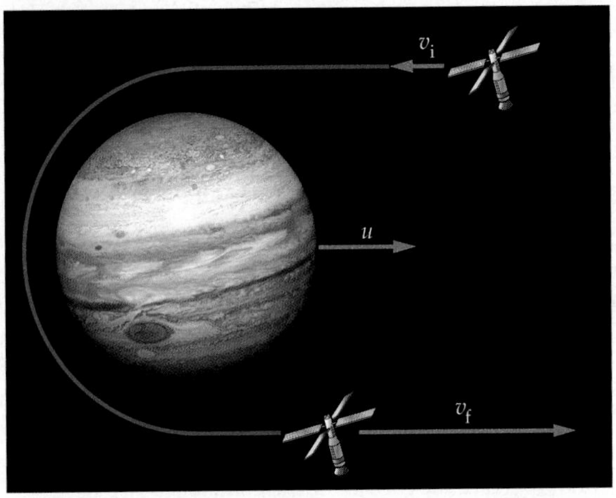

▲ **FIGURE 9–30** Problem 79

80. ••• On a cold winter morning, a child sits on a sled resting on smooth ice. When the 9.50-kg sled is pulled with a horizontal force of 40.0 N, it begins to move with an acceleration of 2.32 m/s^2. The 21.0-kg child accelerates too, but with a smaller acceleration than that of the sled. Thus, the child moves forward relative to the ice, but slides backward relative to the sled. Find the acceleration of the child relative to the ice.

81. ••• An object of mass m undergoes an elastic collision with an identical object that is at rest. The collision is not head-on. Show that the angle between the velocities of the two objects after the collision is 90°.

82. ••• **IP Weighing a Block on an Incline** A wedge of mass m_1 is firmly attached to the top of a scale, as shown in **Figure 9–31**. The inclined surface of the wedge makes an angle θ with the horizontal. Now, a block of mass m_2 is placed on the inclined surface of the wedge and allowed to accelerate without friction down the slope. **(a)** Show that the reading on the scale while the block slides is

$$(m_1 + m_2 \cos^2 \theta)g$$

(b) Explain why the reading on the scale is less than $(m_1 + m_2)g$. **(c)** Show that the expression in part (a) gives the expected results for $\theta = 0$ and $\theta = 90°$.

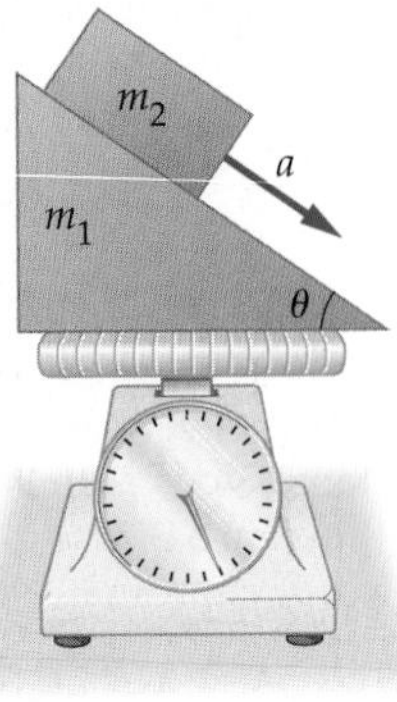

▲ **FIGURE 9–31** Problem 82

83. ••• **IP** A uniform rope of length L and mass M rests on a table. **(a)** If you lift one end of the rope upward with a constant *speed*, v, show that the rope's center of mass moves upward with constant *acceleration*. **(b)** Next, suppose you hold the rope suspended in air, with its lower end just touching the table. If you now lower the rope with a constant speed, v, onto the table, is the acceleration of the rope's center of mass upward or downward? Explain your answer. **(c)** Find the magnitude and direction of the acceleration of the rope's center of mass for the case described in part (b). Compare with part (a).

Interactive Problems

84. •• ***Referring to Example 9–5*** Suppose a bullet of mass $m = 6.75$ g is fired into a ballistic pendulum whose bob has a mass of $M = 0.675$ kg. **(a)** If the bob rises to a height of 0.128 m, what was the initial speed of the bullet? **(b)** What was the speed of the bullet-bob combination immediately after the collision takes place?

85. •• ***Referring to Example 9–5*** A bullet with a mass $m = 8.10$ g and an initial speed $v_0 = 320$ m/s is fired into a ballistic pendulum. What mass must the bob have if the bullet-bob combination is to rise to a maximum height of 0.125 m after the collision?

86. •• ***Referring to Example 9–9*** Suppose that cart 1 has a mass of 3.00 kg and an initial speed of 0.250 m/s. Cart 2 has a mass of 1.00 kg and is at rest initially. **(a)** What is the final speed of the carts? **(b)** How much kinetic energy is lost as a result of the collision?

87. •• ***Referring to Example 9–9*** Suppose the two carts have equal masses and are both moving to the right before the collision. The initial speed of the cart 1 (on the left) is v_0 and the initial speed of the cart 2 (on the right) is $v_0/2$. **(a)** What is the speed of the center of mass of this system? **(b)** What percentage of the initial kinetic energy is lost as a result of the collision? **(c)** Suppose the collision is elastic. What are the final speeds of the two carts in this case?

10 Rotational Kinematics and Energy

10–1 Angular Position, Velocity, and Acceleration 282

10–2 Rotational Kinematics 286

10–3 Connections Between Linear and Rotational Quantities 289

10–4 Rolling Motion 294

10–5 Rotational Kinetic Energy and the Moment of Inertia 295

10–6 Conservation of Energy 299

Can you imagine life without rotating objects: vehicles without wheels, machinery without gears, carnivals without merry-go-rounds? The people on this roller coaster certainly know that rotational motion is very different from motion on a straight, linear stretch of track. In this chapter we show that the motion of rotating objects, such as a roller coaster executing a loop-the-loop, can be analyzed using many of the same methods that we applied earlier to linear motion.

It is certainly no exaggeration to say that rotation is a part of everyday life. After all, we live on a planet that rotates about its axis once a day and that revolves about the Sun once a year. The apparent motion of the Sun across the sky, for example, is actually the result of the Earth's rotational motion. In addition, engines that power cars and trucks have moving parts that rotate quite rapidly, as do CDs, CD-ROMs, and DVDs, not to mention the tumbling, rotating molecules in the air we breathe. Thus, a study of rotation yields results that apply to a great variety of natural phenomena.

In this chapter, then, we study various aspects of rotational motion. As we do, we shall make extensive use of the close analogies that exist between rotational and linear motion. In fact, many of the results derived in earlier chapters can be applied to rotation by simply replacing linear quantities with their rotational counterparts.

10–1 Angular Position, Velocity, and Acceleration

To describe the motion of an object moving in a straight line, it is useful to establish a coordinate system with a definite origin and positive direction. In terms of this coordinate system we can measure the object's position, velocity, and acceleration.

Similarly, to describe rotational motion, we define "angular" quantities that are analogous to the linear position, velocity, and acceleration. These angular quantities form the basis of our study of rotation. We begin by defining the most basic angular quantity—the angular position.

Angular Position, θ

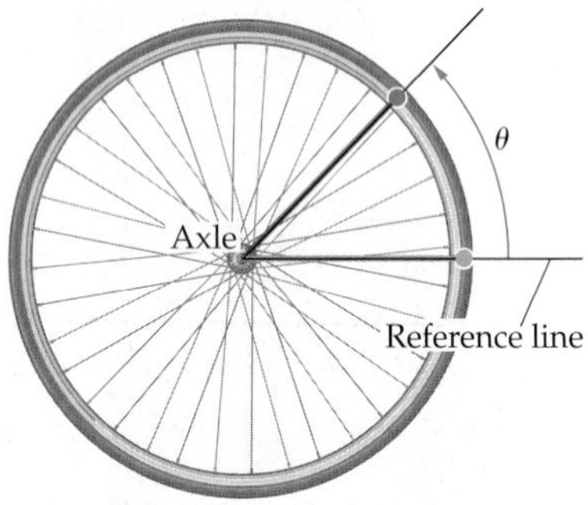

▲ **FIGURE 10–1 Angular position**

The angular position, θ, of a spot of paint on a bicycle wheel. The reference line, where $\theta = 0$, is drawn horizontal here but can be chosen in any direction.

Consider a bicycle wheel that is free to rotate about its axle, as shown in **Figure 10–1**. We say that the axle is the **axis of rotation** for the wheel. As the wheel rotates, each and every point on it moves in a circular path centered on the axis of rotation.

Now, suppose there is a small spot of red paint on the tire, and we want to describe the rotational motion of the spot. The **angular position** of the spot is defined to be the angle, θ, that a line from the axle to the spot makes with a reference line, as indicated in Figure 10–1.

Definition of Angular Position, θ

$$\theta = \text{angle measured from reference line} \qquad \text{10–1}$$

SI unit: radian (rad), which is dimensionless

The reference line simply defines $\theta = 0$; it is analogous to the origin in a linear coordinate system. The reference line begins at the axis of rotation, and may be chosen to point in any direction—just as an origin may be placed anywhere along a coordinate axis. Once chosen, however, the reference line must be used consistently.

Note that the spot of paint in Figure 10–1 is rotated counterclockwise from the reference line by the angle θ. By convention, we say that this angle is positive. Similarly, clockwise rotations correspond to negative angles.

Sign Convention for Angular Position

By convention:

$\theta > 0$ counterclockwise rotation from reference line
$\theta < 0$ clockwise rotation from reference line

▲ Rotational motion is everywhere in our universe, on every scale of length and time. A galaxy like the one at left may take millions of years to complete a single rotation about its center, while the skater in the middle photo spins several times in a second. The bacterium at right moves in a corkscrew path by rapidly twirling its flagella (the fine projections at either end of the cell) like whips.

Now that we have established a reference line (for $\theta = 0$), and a positive direction of rotation (counterclockwise), we must choose units in which to measure angles. Common units are degrees (°) and revolutions (rev), where one revolution—that is, going completely around a circle—corresponds to 360°:

$$1 \text{ rev} = 360°$$

The most convenient units for scientific calculations, however, are radians. A **radian** (rad) is defined as follows:

> A radian is the angle for which the arc length on a circle of radius r is equal to the radius of the circle.

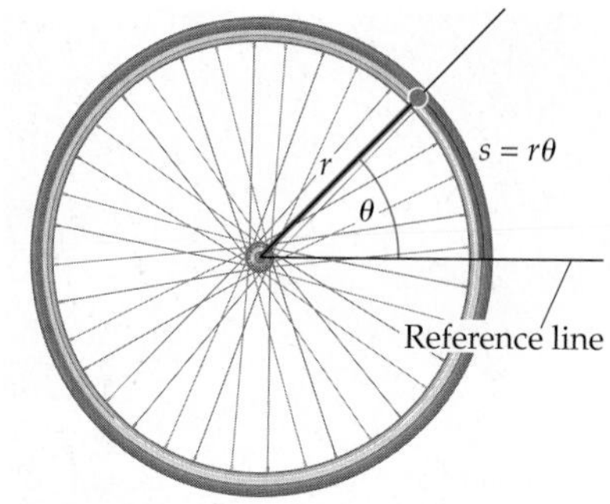

▲ **FIGURE 10–2 Arc length**
The arc length, s, from the reference line to the spot of paint is given by $s = r\theta$ if the angular position θ is measured in radians.

This definition is useful because it establishes a particularly simple relationship between an angle measured in radians and the corresponding arc length, as illustrated in **Figure 10–2**. For example, it follows from our definition that for an angle of one radian, the arc length s is equal to the radius: $s = r$. Similarly, an angle of two radians corresponds to an arc length of two radii, $s = 2r$, and so on. Thus, the arc length s for an arbitrary angle θ measured in radians is given by the following relation:

$$s = r\theta \qquad 10\text{–}2$$

This simple and straightforward relation does not hold for degrees or revolutions—additional conversion factors would be needed.

PROBLEM-SOLVING NOTE

Radians

Remember to measure angles in radians when using the relation $s = r\theta$.

In one complete revolution, the arc length is the circumference of a circle, $C = 2\pi r$. Comparing with $s = r\theta$, we see that a complete revolution corresponds to 2π radians:

$$1 \text{ rev} = 360° = 2\pi \text{ rad}$$

Equivalently,

$$1 \text{ rad} = \frac{360°}{2\pi} = 57.3°$$

One final note on the units for angles: Radians, as well as degrees and revolutions, are dimensionless. In the relation $s = r\theta$, for example, the arc length and the radius both have SI units of meters. For the equation to be dimensionally consistent, it is necessary that θ have no dimensions. Still, if an angle θ is, let's say, three radians, we will write it as $\theta = 3$ rad to remind us of the angular units being used.

Angular Velocity, ω

As the bicycle wheel in Figure 10–1 rotates, the angular position of the spot of red paint changes. This is illustrated in **Figure 10–3**. The **angular displacement** of the spot, $\Delta\theta$, is

$$\Delta\theta = \theta_f - \theta_i$$

If we divide the angular displacement by the time, Δt, during which the displacement occurs, the result is the **average angular velocity,** ω_{av}.

Definition of Average Angular Velocity, ω_{av}

$$\omega_{av} = \frac{\Delta\theta}{\Delta t} \qquad 10\text{–}3$$

SI unit: radian per second (rad/s) = s^{-1}

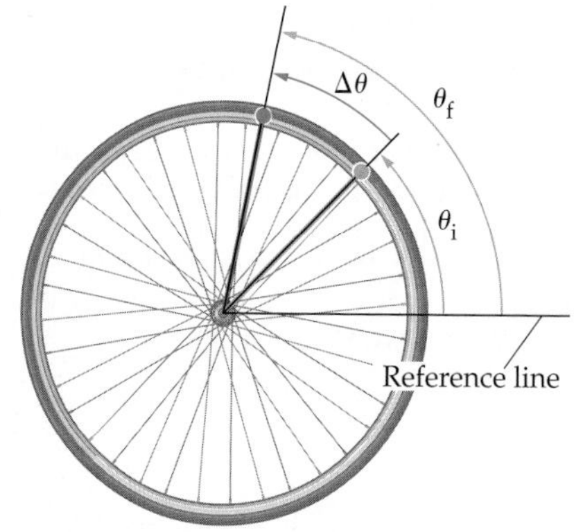

▲ **FIGURE 10–3 Angular displacement**
As the wheel rotates, the spot of paint undergoes an angular displacement, $\Delta\theta = \theta_f - \theta_i$.

This is analogous to the definition of the average linear velocity $v_{av} = \Delta x/\Delta t$. Note that the units of linear velocity are m/s, whereas the units of angular velocity are rad/s.

▲ Star trails provide a clear illustration of the relationship between angle, arc, and radius in circular motion. The stars, of course, do not actually move like this, but because of the Earth's rotation they appear to follow circular paths across the sky each night, with Polaris, the North Star, very near the axis of rotation. This photo was made by opening the camera shutter for an extended period of time. Notice that each star moves through the same angle in the course of the exposure. However, the farther a star is from the axis of rotation, the longer the arc it traces out in a given period of time. (Can you estimate the length of the exposure?)

In addition to the average angular velocity, we can define an **instantaneous angular velocity** as the limit of ω_{av} as the time interval, Δt, approaches zero. The instantaneous angular velocity, then, is

Definition of Instantaneous Angular Velocity, ω

$$\omega = \lim_{\Delta t \to 0} \frac{\Delta \theta}{\Delta t} \qquad \text{10–4}$$

SI unit: $\text{rad/s} = \text{s}^{-1}$

Generally, we shall refer to the instantaneous angular velocity simply as the angular velocity.

Note that we call ω the angular velocity, not the angular speed. The reason is that ω can be positive or negative, depending on the sense of rotation. For example, if the red paint spot rotates in the counterclockwise sense, the angular position, θ, increases. As a result, $\Delta\theta$ is positive and therefore, so is ω. Similarly, clockwise rotation corresponds to a negative $\Delta\theta$ and hence a negative ω.

Sign Convention for Angular Velocity

By convention:

$\omega > 0$ counterclockwise rotation

$\omega < 0$ clockwise rotation

The sign convention for angular velocity is illustrated in **Figure 10–4**. In analogy with linear motion, the sign of ω indicates the *direction* of the angular velocity *vector*, as we shall see in detail in Chapter 11. Similarly, the magnitude of the angular velocity is the **angular speed,** just as in the one-dimensional case.

In Exercise 10–1 we utilize the definitions and conversion factors presented so far in this section.

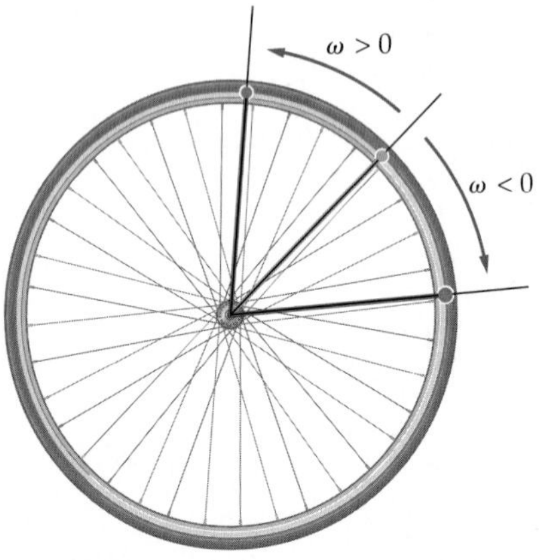

▲ **FIGURE 10–4 Angular speed and velocity**

Counterclockwise rotation is defined to correspond to a positive angular velocity, ω. Similarly, clockwise rotation corresponds to a negative angular velocity. The magnitude of the angular velocity is referred to as the angular speed.

EXERCISE 10–1

(a) An old phonograph record rotates clockwise at $33\frac{1}{3}$ rpm (revolutions per minute). What is its angular velocity in rad/s? **(b)** If a CD rotates at 22.0 rad/s, what is its angular speed in rpm?

Solution

a. Convert from rpm to rad/s, and note that clockwise rotation corresponds to a negative angular velocity:

$$\omega = -33\tfrac{1}{3}\,\text{rpm} = \left(-33\tfrac{1}{3}\frac{\text{rev}}{\text{min}}\right)\left(\frac{2\pi\,\text{rad}}{1\,\text{rev}}\right)\left(\frac{1\,\text{min}}{60\,\text{s}}\right) = -3.49\,\text{rad/s}$$

b. Converting angular speed from rad/s to rpm gives

$$\omega = \left(22.0\frac{\text{rad}}{\text{s}}\right)\left(\frac{1\,\text{rev}}{2\pi\,\text{rad}}\right)\left(\frac{60\,\text{s}}{1\,\text{min}}\right) = 210\frac{\text{rev}}{\text{min}} = 210\,\text{rpm}$$

Note that the same symbol, ω, is used for both angular velocity and angular speed in Exercise 10–1. Which quantity is meant in a given situation will be clear from the context in which it is used.

As a simple application of angular velocity, consider the following question: An object rotates with a constant angular velocity, ω. How much time, T, is required for it to complete one full revolution?

To solve this problem, note that since ω is constant, the instantaneous angular velocity is equal to the average angular velocity. That is,

$$\omega = \omega_{av} = \frac{\Delta\theta}{\Delta t}$$

In one revolution, we know that $\Delta\theta = 2\pi$ and $\Delta t = T$. Therefore,

$$\omega = \frac{\Delta\theta}{\Delta t} = \frac{2\pi}{T}$$

Finally, solving for T we find

$$T = \frac{2\pi}{\omega}$$

The time to complete one revolution, T, is referred to as the **period.**

Definition of Period, *T*

$$T = \frac{2\pi}{\omega} \qquad \text{10–5}$$

SI unit: second, s

EXERCISE 10–2

Find the period of a record that is rotating at 45 rpm.

Solution

To apply $T = 2\pi/\omega$ we must first express ω in terms of rad/s:

$$45 \text{ rpm} = \left(45\frac{\text{rev}}{\text{min}}\right)\left(\frac{2\pi \text{ rad}}{1 \text{ rev}}\right)\left(\frac{1 \text{ min}}{60 \text{ s}}\right) = 4.7 \text{ rad/s}$$

Now we can calculate the period:

$$T = \frac{2\pi}{\omega} = \frac{2\pi \text{ rad}}{4.7 \text{ rad/s}} = 1.3 \text{ s}$$

Angular Acceleration, α

If the angular velocity of the rotating bicycle wheel increases or decreases with time, we say that the wheel experiences an **angular acceleration,** α. The average angular acceleration is the change in angular velocity, $\Delta\omega$, in a given interval of time, Δt:

Definition of Average Angular Acceleration, α_{av}

$$\alpha_{av} = \frac{\Delta\omega}{\Delta t} \qquad \text{10–6}$$

SI unit: radian per second per second $(\text{rad/s}^2) = \text{s}^{-2}$

Note that the SI units of α are rad/s^2, which, since rad is dimensionless, is simply s^{-2}.

As expected, the instantaneous angular acceleration is the limit of α_{av} as the time interval, Δt, approaches zero:

Definition of Instantaneous Angular Acceleration, α

$$\alpha = \lim_{\Delta t \to 0} \frac{\Delta\omega}{\Delta t} \qquad \text{10–7}$$

SI unit: rad/s^2 = s^{-2}

When referring to the instantaneous angular acceleration, we will usually just say angular acceleration.

The sign of the angular acceleration is determined by whether the change in angular velocity is positive or negative. For example, if ω is becoming more positive, so that ω_f is greater than ω_i, it follows that α is positive. Similarly, if ω is becoming more negative, so that ω_f is less than ω_i, it follows that α is negative. Therefore, if ω and α have the same sign, the speed of rotation is increasing. If ω and α have opposite signs, the speed of rotation is decreasing. This is illustrated in **Figure 10–5**.

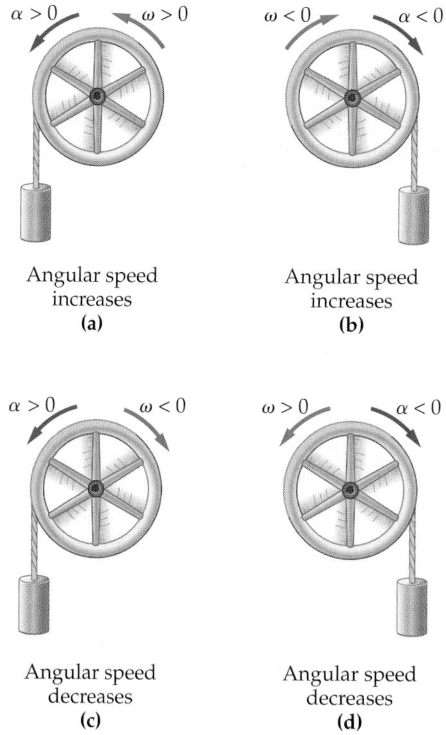

▲ **FIGURE 10–5 Angular acceleration and angular speed**

When angular velocity and acceleration have the same sign, as in **(a)** and **(b)**, the angular speed increases. When angular velocity and angular acceleration have opposite signs, as in **(c)** and **(d)**, the angular speed decreases.

EXERCISE 10–3

As the wind dies, a windmill that was rotating at 2.1 rad/s begins to slow down with a constant angular acceleration of 0.45 rad/s^2. How long does it take for the windmill to come to a complete stop?

Solution

If we choose the initial angular velocity to be positive, the angular acceleration is negative, corresponding to a deceleration. Hence, Equation 10–6 gives

$$\Delta t = \frac{\Delta\omega}{\alpha_{av}} = \frac{\omega_f - \omega_i}{\alpha} = \frac{0 - 2.1\ \text{rad/s}}{-0.45\ \text{rad/s}^2} = 4.7\ \text{s}$$

10–2 Rotational Kinematics

Just as the kinematics of Chapter 2 described linear motion, rotational kinematics describes rotational motion. In this section, as in Chapter 2, we concentrate on the important special case of constant acceleration.

As an example of a system with constant angular acceleration, consider the pulley shown in **Figure 10–6**. Wrapped around the circumference of the pulley is a string, with a mass attached to its free end. When the mass is released, the pulley begins to rotate—slowly at first, but then faster and faster. As we shall see in Chapter 11, the pulley is accelerating with constant angular acceleration.

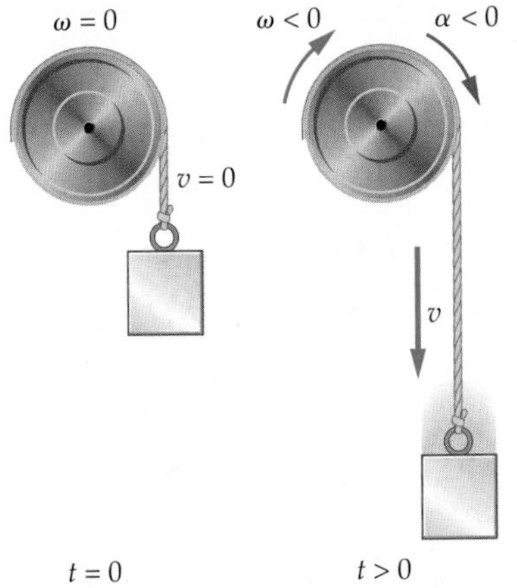

▲ FIGURE 10–6 A pulley with constant angular acceleration

A mass is attached to a string wrapped around a pulley. As the mass falls, it causes the pulley to increase its angular speed with a constant angular acceleration.

Since α is constant, it follows that the average and instantaneous angular accelerations are equal. Hence,

$$\alpha = \alpha_{av} = \frac{\Delta\omega}{\Delta t}$$

Suppose the pulley starts with the initial angular velocity ω_0 at time $t = 0$, and that at the later time t its angular velocity is ω. Substituting these values into the preceding expression for α yields

$$\alpha = \frac{\Delta\omega}{\Delta t} = \frac{\omega - \omega_0}{t - 0} = \frac{\omega - \omega_0}{t}$$

Rearranging, we see that the angular velocity, ω, varies with time as follows:

$$\omega = \omega_0 + \alpha t \qquad \textbf{10–8}$$

EXERCISE 10–4

If the angular velocity of the pulley in Figure 10–6 is −8.4 rad/s at a given time, and its angular acceleration is −2.8 rad/s^2, what is the angular velocity of the pulley 1.5 s later?

Solution

The angular velocity, ω, is found by applying Equation 10–8:

$$\omega = \omega_0 + \alpha t = -8.4\ \text{rad/s} + (-2.8\ \text{rad/s}^2)(1.5\ \text{s}) = -12.6\ \text{rad/s}$$

Note that the angular speed has increased, as expected, since ω and α have the same sign.

Note the close analogy between Equation 10–8 for angular velocity and the corresponding relation for linear velocity, Equation 2–7:

$$v = v_0 + at$$

Clearly, the equation for angular velocity can be obtained from our previous equation for linear velocity by replacing v with ω and replacing a with α. This type of analogy between linear and angular quantities can be most useful both in deriving angular equations—by starting with linear equations and using analogies—and in obtaining a better physical understanding of angular systems. Several linear-to-angular analogs are listed in the adjacent table.

Linear Quantity	Angular Quantity
x	θ
v	ω
a	α

Using these analogies, we can rewrite all the kinematic equations in Chapter 2 in angular form. The following table gives both the linear kinematic equations and their angular counterparts.

Linear Equation (a = constant)		Angular Equation (α = constant)	
$v = v_0 + at$	2–7	$\omega = \omega_0 + \alpha t$	10–8
$x = x_0 + \frac{1}{2}(v_0 + v)t$	2–10	$\theta = \theta_0 + \frac{1}{2}(\omega_0 + \omega)t$	10–9
$x = x_0 + v_0 t + \frac{1}{2}at^2$	2–11	$\theta = \theta_0 + \omega_0 t + \frac{1}{2}\alpha t^2$	10–10
$v^2 = v_0{}^2 + 2a(x - x_0)$	2–12	$\omega^2 = \omega_0{}^2 + 2\alpha(\theta - \theta_0)$	10–11

In solving kinematic problems involving rotation, we apply these angular equations in the same way that the linear equations were applied in Chapter 2. In a sense, then, this material is a review—since the mathematics is essentially the same. The only difference comes in the physical interpretation of the results. We will emphasize the rotational interpretations throughout the chapter.

PROBLEM-SOLVING NOTE

Rotational Kinematics

Using analogies between linear and angular quantities often helps when solving problems involving rotational kinematics.

EXAMPLE 10–1 Thrown for a Curve

To throw a curve ball, a pitcher gives the ball an initial angular speed of 36.0 rad/s. When the catcher gloves the ball 0.595 s later, its angular speed has decreased (due to air resistance) to 34.2 rad/s. **(a)** What is the ball's angular acceleration, assuming it to be constant? **(b)** How many revolutions does the ball make before being caught?

Picture the Problem

We choose the ball's initial direction of rotation to be positive. As a result, the angular acceleration will be negative. We can also identify the initial angular velocity to be $\omega_0 = 36.0$ rad/s, and the final angular velocity to be $\omega = 34.2$ rad/s.

Strategy

The problem states that the angular acceleration of the ball is constant. It follows that Equations 10–8 to 10–11 apply to its rotation.

a. To relate angular velocity to time, we use $\omega = \omega_0 + \alpha t$. This can be solved for α.

b. To relate angle to time we use $\theta = \theta_0 + \omega_0 t + \frac{1}{2}\alpha t^2$. The angular displacement of the ball is $\theta - \theta_0$.

Solution

Part (a)

1. Solve $\omega = \omega_0 + \alpha t$ for the angular acceleration, α:

$$\omega = \omega_0 + \alpha t$$
$$\alpha = \frac{\omega - \omega_0}{t}$$

2. Substitute numerical values to find α:

$$\alpha = \frac{\omega - \omega_0}{t} = \frac{34.2 \text{ rad/s} - 36.0 \text{ rad/s}}{0.595 \text{ s}} = -3.03 \text{ rad/s}^2$$

Part (b)

3. Use $\theta = \theta_0 + \omega_0 t + \frac{1}{2}\alpha t^2$ to calculate the angular displacement of the ball:

$$\theta - \theta_0 = \omega_0 t + \tfrac{1}{2}\alpha t^2 = (36.0 \text{ rad/s})(0.595 \text{ s}) + \tfrac{1}{2}(-3.03 \text{ rad/s}^2)(0.595 \text{ s})^2 = 20.9 \text{ rad}$$

4. Convert the angular displacement to revolutions:

$$\theta - \theta_0 = 20.9 \text{ rad} = 20.9 \text{ rad}\left(\frac{1 \text{ rev}}{2\pi \text{ rad}}\right) = 3.33 \text{ rev}$$

continued on next page

continued from previous page

Insight

The ball rotates through three-and-one-third revolutions during its time in flight.

An alternative method of solution is to use the kinematic relation given in Equation 10–9. This procedure yields $\theta - \theta_0 = \frac{1}{2}(\omega_0 + \omega)t = 20.9$ rad, in agreement with our previous result.

Practice Problem

(a) What is the angular velocity of the ball 0.500 s after it is thrown? **(b)** What is the ball's angular velocity after it completes its first full revolution? [**Answer:** (a) Use $\omega = \omega_0 + \alpha t$ to find $\omega = 34.5$ rad/s. (b) Use $\omega^2 = \omega_0^2 + 2\alpha(\theta - \theta_0)$ to find $\omega = 35.5$ rad/s.]

Some related homework problems: Problem 17, Problem 20

EXAMPLE 10–2 Wheel of Misfortune

On a certain game show, contestants spin a wheel when it is their turn. One contestant gives the wheel an initial angular speed of 3.40 rad/s. It then rotates through one-and-one-quarter revolutions and comes to rest on the BANKRUPT space. **(a)** Find the angular acceleration of the wheel, assuming it to be constant. **(b)** How long does it take for the wheel to come to rest?

Picture the Problem

We choose the initial angular velocity to be positive, $\omega_0 = +3.40$ rad/s, and indicate it with a counterclockwise rotation in our sketch. Since the wheel slows to a stop, the angular acceleration must be negative; that is, in the clockwise direction. After a rotation of 1.25 rev the wheel will read BANKRUPT.

Strategy

As in Example 10–1, we can use the kinematic equations for constant angular acceleration, Equations 10–8 to 10–11.

a. To begin, we are given the initial angular velocity, $\omega_0 = +3.40$ rad/s, the final angular velocity, $\omega = 0$ (the wheel comes to rest), and the angular displacement, $\theta - \theta_0 = 1.25$ rev. We can find the angular acceleration using $\omega^2 = \omega_0^2 + 2\alpha(\theta - \theta_0)$.

b. Knowing the angular velocity and acceleration, we can find the time with $\omega = \omega_0 + \alpha t$.

Solution

Part (a)

1. Solve $\omega^2 = \omega_0^2 + 2\alpha(\theta - \theta_0)$ for the angular acceleration, α:

$$\omega^2 = \omega_0^2 + 2\alpha(\theta - \theta_0)$$
$$\alpha = \frac{\omega^2 - \omega_0^2}{2(\theta - \theta_0)}$$

2. Convert $\theta - \theta_0 = 1.25$ rev to radians:

$$\theta - \theta_0 = 1.25 \text{ rev} = 1.25 \text{ rev}\left(\frac{2\pi \text{ rad}}{1 \text{ rev}}\right) = 7.85 \text{ rad}$$

3. Substitute numerical values to find α:

$$\alpha = \frac{\omega^2 - \omega_0^2}{2(\theta - \theta_0)} = \frac{0 - (3.40 \text{ rad/s})^2}{2(7.85 \text{ rad})} = -0.736 \text{ rad/s}^2$$

Part (b)

4. Solve $\omega = \omega_0 + \alpha t$ for the time, t:

$$\omega = \omega_0 + \alpha t$$
$$t = \frac{\omega - \omega_0}{\alpha}$$

5. Substitute numerical values to find t:

$$t = \frac{\omega - \omega_0}{\alpha} = \frac{0 - 3.40 \text{ rad/s}}{(-0.736 \text{ rad/s}^2)} = 4.62 \text{ s}$$

Insight

Note that it was not necessary to define a reference line; that is, a direction for $\theta = 0$. All we need to know is the angular displacement, $\theta - \theta_0$, not the individual angles θ and θ_0. Finally, notice that we can also solve Equation 10–9 for the time in part (b), which yields $t = 2(\theta - \theta_0)/(\omega_0 + \omega) = 4.62$ s, as expected.

Practice Problem

What is the angular speed of the wheel after one complete revolution? [**Answer:** $\omega = 1.52$ rad/s]

Some related homework problems: Problem 16, Problem 18

Finally, we consider a pulley that is rotating in such a way that initially it is lifting a mass with speed v. Gravity acting on the mass causes it and the pulley to slow and momentarily come to rest.

ACTIVE EXAMPLE 10–1 Find the Time To Rest

A pulley rotating in the counterclockwise direction is attached to a mass suspended from a string. The mass causes the pulley's angular velocity to decrease with a constant angular acceleration $\alpha = -2.10 \text{ rad/s}^2$. **(a)** If the pulley's initial angular velocity is $\omega_0 = 5.40$ rad/s, how long does it take for the pulley to come to rest? **(b)** Through what angle does the pulley turn during this time?

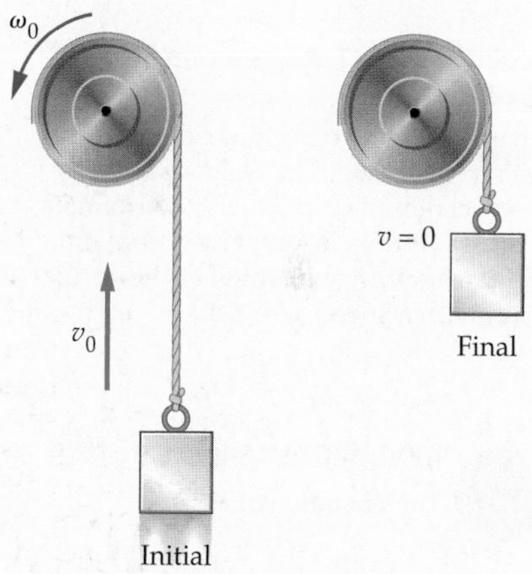

Solution *(Test your understanding by performing the calculations indicated in each step.)*

1. **(a)** Relate angular velocity to time: $\omega = \omega_0 + \alpha t$
2. Solve for the time, t: $t = (\omega - \omega_0)/\alpha$
3. Substitute numerical values: $t = 2.57$ s
4. **(b)** Use $\theta = \theta_0 + \omega_0 t + \frac{1}{2}\alpha t^2$ to solve for $\theta - \theta_0$: $\theta - \theta_0 = \omega_0 t + \frac{1}{2}\alpha t^2 = 6.94$ rad
5. Alternatively, use $\omega^2 = \omega_0{}^2 + 2\alpha(\theta - \theta_0)$: $\theta - \theta_0 = (\omega^2 - \omega_0{}^2)/2\alpha = 6.94$ rad

Insight

After the pulley comes to rest, it immediately begins to rotate in the clockwise direction as the mass falls. The pulley's angular *acceleration* is constant—it has the same value before the pulley stops, when it stops, and after it begins rotating in the opposite direction. This is analogous to a projectile thrown straight upward, where the linear velocity starts out positive, goes to zero, then changes sign, all while the linear acceleration remains constant in the negative direction.

Your Turn

Find the angular displacement of the pulley at the time when its angular velocity is half its initial value.

(Answers to **Your Turn** *problems are given in the back of the book.)*

10–3 Connections Between Linear and Rotational Quantities

At a local county fair a child rides on a merry-go-round. The ride completes one circuit every $T = 7.50$ s. Therefore, the angular velocity of the child, from Equation 10–5, is

$$\omega = \frac{2\pi}{T} = \frac{2\pi}{7.50 \text{ s}} = 0.838 \text{ rad/s}$$

The path followed by the child is circular, with the center of the circle at the axis of rotation of the merry-go-round. In addition, at any instant of time the child is moving in a direction that is *tangential* to the circular path, as **Figure 10–7** shows. What is the tangential speed, v_t, of the child? In other words, what is the speed of the wind in the child's face?

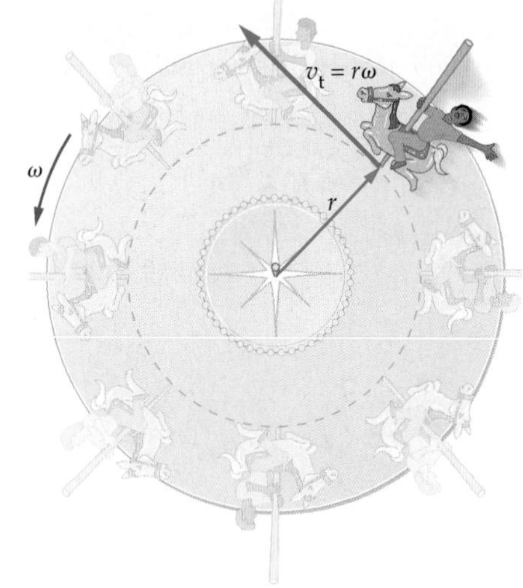

▲ **FIGURE 10–7 Angular and linear speed**
Overhead view of a child riding on a merry-go-round. The child's path is a circle centered on the merry-go-round's axis of rotation. At any given time the child is moving tangential to the circular path with a speed $v_t = r\omega$.

We can find the child's tangential speed by dividing the circumference of the circular path, $2\pi r$, by the time required to complete one circuit, T. Thus,

$$v_t = \frac{2\pi r}{T} = r\left(\frac{2\pi}{T}\right)$$

Because $2\pi/T$ is simply ω, we can express the tangential speed as follows:

Tangential Speed of a Rotating Object

$$v_t = r\omega \qquad 10\text{–}12$$

SI unit: m/s

Note that ω must be given in rad/s for this relation to be valid.

In the case of the merry-go-round, if the radius of the child's circular path is $r = 4.25$ m, the tangential speed is $v_t = r\omega = (4.25\text{ m})(0.838\text{ rad/s}) = 3.56\text{ m/s}$. When it is clear that we are referring to the tangential speed, we will often drop the subscript t, and simply write $v = r\omega$.

An interesting application of the relation between linear and angular speeds is provided in the operation of a compact disk (CD). As you know, a CD is played by shining a laser beam onto the disk, and then converting the pattern of reflected light into a pattern of sound waves. For proper operation, however, the linear speed of the disk where the laser beam shines on it must be maintained at the constant value of 1.25 m/s. As the CD is played, the laser beam scans the disk in a spiral track from near the center outward to the rim. In order to maintain the required linear speed, the angular speed of the disk must decrease as the beam scans outward. The required angular speeds are determined in the following Exercise.

REAL-WORLD PHYSICS
The operation of a CD

EXERCISE 10–5

Find the angular speed a CD must have to give a linear speed of 1.25 m/s when the laser beam shines on the disk **(a)** 2.50 cm and **(b)** 6.00 cm from its center.

Solution

a. Using $v = 1.25$ m/s and $r = 0.0250$ m in Equation 10–12, we find

$$\omega = \frac{v}{r} = \frac{1.25\text{ m/s}}{0.0250\text{ m}} = 50.0\text{ rad/s} = 477\text{ rpm}$$

b. Similarly, with $r = 0.0600$ m we find

$$\omega = \frac{v}{r} = \frac{1.25\text{ m/s}}{0.0600\text{ m}} = 20.8\text{ rad/s} = 199\text{ rpm}$$

Thus, a CD slows from about 500 rpm to roughly 200 rpm as it plays.

How do the angular and tangential speed of an object vary from one point to another? We explore this question in the following Conceptual Checkpoint.

CONCEPTUAL CHECKPOINT 10–1 Compare the Speeds

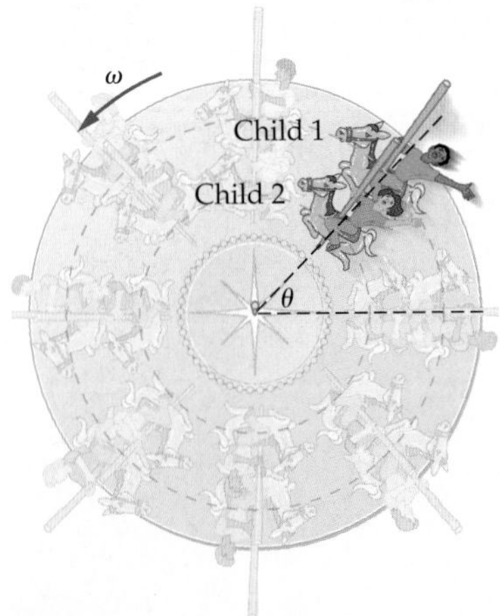

Two children ride on a merry-go-round, with child 1 at a greater distance from the axis of rotation than child 2. Is the angular speed of child 1 **(a)** greater than, **(b)** less than, or **(c)** the same as the angular speed of child 2?

Reasoning and Discussion

At any given time, the angle θ for child 1 is the same as the angle for child 2, as shown. Therefore, when the angle for child 1 has gone through 2π, for example, so has the angle for child 2. As a result, they have the same angular speed. In fact, *each and every point on the merry-go-round has exactly the same angular speed.*

The tangential speeds are different, however. Child 1 has the greater tangential speed since he travels around a larger circle in the same time that child 2 travels around a smaller circle. This is in agreement with the relation $v = r\omega$, since the radius to child 1 is greater than the radius to child 2. That is, $v_1 = r_1\omega > v_2 = r_2\omega$.

Answer:

(c) The angular speeds are the same.

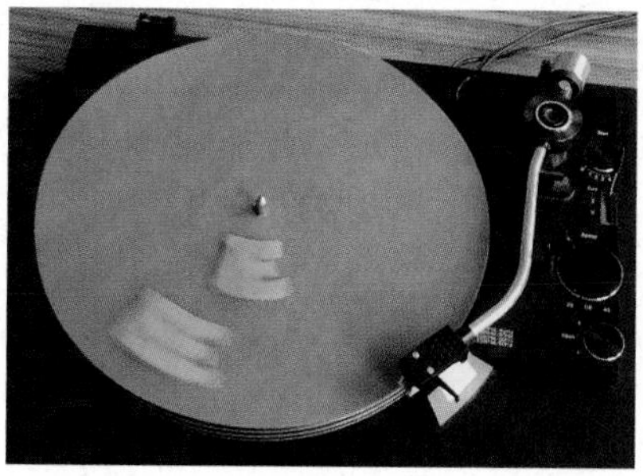

◀ In the photo at left, two identical plastic letter "E"s have been placed on a rotating turntable at different distances from the axis of rotation. The stretching and blurring of the image of the outermost letter clearly shows that it is moving faster than the letter closer to the axis. Similarly, the boy near the rim of this playground merry-go-round is moving faster than the girl near the hub.

Because the children on the merry-go-round move in a circular path, they experience a centripetal acceleration, a_{cp} (Section 6–5). The centripetal acceleration is always directed toward the axis of rotation and has a magnitude given by

$$a_{cp} = \frac{v^2}{r}$$

Note that the speed v in this expression is the tangential speed, $v = v_t = r\omega$, and therefore the centripetal acceleration in terms of ω is

$$a_{cp} = \frac{(r\omega)^2}{r}$$

Canceling one power of r, we have

Centripetal Acceleration of a Rotating Object

$$a_{cp} = r\omega^2 \qquad 10\text{–}13$$

SI unit: m/s^2

If the radius of a child's circular path on the merry-go-round is 4.25 m, and the angular speed of the ride is 0.838 rad/s, the centripetal acceleration of the child is $a_{cp} = r\omega^2 = (4.25\ m)(0.838\ rad/s)^2 = 2.98\ m/s^2$.

Though the centripetal acceleration of a merry-go-round is typically only a fraction of the acceleration of gravity, rotating devices referred to as **centrifuges** can produce centripetal accelerations many times greater than gravity. For example, the world's most powerful research centrifuge, operated by the U. S. Army Corps of Engineers, can subject 2.2-ton payloads to accelerations as high as 350g (350 times greater than the acceleration of gravity.) This centrifuge is used to study earthquake engineering and dam erosion. The Air Force uses centrifuges to subject prospective jet pilots to the accelerations they will experience during rapid flight maneuvers, and in the future NASA may even use a human-powered centrifuge for gravity studies aboard the International Space Station.

REAL-WORLD PHYSICS
The centrifuge

▲ The large centrifuge shown at left, at the Gagarin Cosmonaut Training Center, is used to train Russian cosmonauts for space missions. This device, which rotates at 36 rpm, can produce a centripetal acceleration of over 290 m/s^2, 30 times the acceleration of gravity. The device at right is a microhematocrit centrifuge, used to separate blood cells from plasma. The volume of red blood cells in a given quantity of whole blood is a major factor in determining the oxygen-carrying capacity of the blood, an important clinical indicator.

The centrifuges most commonly encountered in everyday life are those found in virtually every medical laboratory in the world. These devices, which can produce centripetal accelerations in excess of 13,000 g, are used to separate blood cells from blood plasma. They do this by speeding up the natural tendency of cells to settle out of plasma from days to minutes. The ratio of the packed cell volume to the total blood volume gives the *hematocrit value*, which is a useful clinical indicator of blood quality. In the next Example we consider the operation of a *microhematocrit centrifuge*, which measures the hematocrit value of a small (micro) sample of blood.

REAL-WORLD PHYSICS: BIO
The microhematocrit centrifuge

EXAMPLE 10–3 The Microhematocrit

REAL-WORLD PHYSICS: BIO

In a microhematocrit centrifuge, small samples of blood are placed in heparinized capillary tubes (heparin is an anticoagulant). The tubes are rotated at 11,500 rpm, with the bottom of the tubes 9.07 cm from the axis of rotation. **(a)** Find the linear speed of the bottom of the tubes. **(b)** What is the centripetal acceleration at the bottom of the tubes?

Picture the Problem

Our sketch shows a top view of the centrifuge, with the capillary tubes rotating at 11,500 rpm. Notice that the bottom of the tubes move in a circular path of radius 9.07 cm.

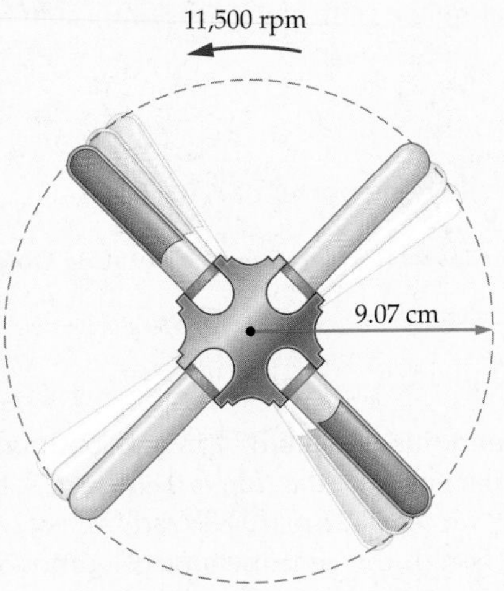

Strategy

a. Linear and angular speeds are related by $v = r\omega$. Once we convert the angular speed to rad/s we can use this relation to determine v.

b. The centripetal acceleration is $a_{cp} = r\omega^2$. Using ω from part (a) yields the desired result.

Solution

Part (a)

1. Convert the angular speed, ω, to radians per second:

$$\omega = (11{,}500 \text{ rev/min})\left(\frac{2\pi \text{ rad}}{1 \text{ rev}}\right)\left(\frac{1 \text{ min}}{60 \text{ s}}\right) = 1.20 \times 10^3 \text{ rad/s}$$

2. Use $v = r\omega$ to calculate the linear speed:

$$v = r\omega = (0.0907 \text{ m})(1.20 \times 10^3 \text{ rad/s}) = 109 \text{ m/s}$$

Part (b)

3. Calculate the centripetal acceleration using $a_{cp} = r\omega^2$:

$$a_{cp} = r\omega^2 = (0.0907 \text{ m})(1.20 \times 10^3 \text{ rad/s})^2 = 131{,}000 \text{ m/s}^2$$

4. As a check, calculate the centripetal acceleration using $a_{cp} = v^2/r$:

$$a_{cp} = \frac{v^2}{r} = \frac{(109 \text{ m/s})^2}{0.0907 \text{ m}} = 131{,}000 \text{ m/s}^2$$

Insight

Note that every point on a tube has the same angular speed. As a result, points near the top of a tube have smaller linear speeds and centripetal accelerations than do points near the bottom of a tube. In this case, the bottom of the tubes experience a centripetal acceleration about 13,400 times greater than the acceleration of gravity on the surface of the Earth; that is, $a_{cp} = 131{,}000 \text{ m/s}^2 = 13{,}400\, g$.

Practice Problem

What angular speed must this centrifuge have if the centripetal acceleration at the bottom of the tubes is to be $98{,}100 \text{ m/s}^2$ ($\approx 10{,}000\, g$)? [**Answer:** $\omega = \sqrt{a_{cp}/r} = 1040 \text{ rad/s} = 9930 \text{ rpm}$]

Some related homework problems: Problem 29, Problem 32

If the angular speed of the merry-go-round in Conceptual Checkpoint 10–1 changes, the tangential speed of the children changes as well. It follows, then, that the children will experience a tangential acceleration, a_t. We can determine a_t by considering the relation $v_t = r\omega$. If ω changes by the amount $\Delta\omega$, with r remaining constant, the corresponding change in tangential speed is

$$\Delta v_t = r\Delta\omega$$

If this change in ω occurs in the time Δt, the tangential acceleration is

$$a_t = \frac{\Delta v_t}{\Delta t} = r\frac{\Delta\omega}{\Delta t}$$

Since $\Delta\omega/\Delta t$ is the angular acceleration, α, we find that

Tangential Acceleration of a Rotating Object

$$a_t = r\alpha \qquad \textbf{10–14}$$

SI unit: m/s^2

As with the tangential speed, we will often drop the subscript t in a_t when no confusion will arise.

In general, the children on the merry-go-round may experience both tangential and centripetal accelerations at the same time. Recall that a_t is due to a changing tangential speed, and that a_{cp} is caused by a changing direction of motion, even if the tangential speed remains constant. To summarize:

Tangential Versus Centripetal Acceleration

$a_t = r\alpha$	due to changing angular speed
$a_{cp} = r\omega^2$	due to changing direction of motion

As the names suggest, the tangential acceleration is always tangential to an object's path; the centripetal acceleration is always perpendicular to its path.

In cases in which both the centripetal and tangential accelerations are present, the total acceleration is the vector sum of the two, as indicated in **Figure 10–8**. Note that $\vec{a}_t$ and $\vec{a}_{cp}$ are at right angles to one another, and hence the magnitude of the total acceleration is given by the Pythagorean theorem:

$$a = \sqrt{a_t^2 + a_{cp}^2}$$

The direction of the total acceleration, measured relative to the tangential direction, is

$$\phi = \tan^{-1}\left(\frac{a_{cp}}{a_t}\right)$$

This angle is shown in Figure 10–8.

In the next Active Example, we consider an object that is rotating with a constant angular acceleration, α. In this case, the tangential acceleration, $a_t = r\alpha$, is constant in magnitude. On the other hand, the centripetal acceleration, $a_{cp} = r\omega^2$, changes with time since the angular speed changes.

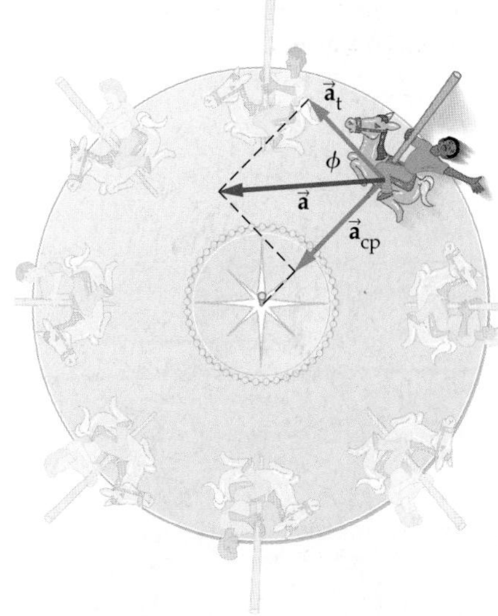

▲ **FIGURE 10–8 Centripetal and tangential acceleration**

If the angular speed of the merry-go-round is increased, the child will experience two accelerations: (i) a tangential acceleration, $\vec{a}_t$, and (ii) a centripetal acceleration, $\vec{a}_{cp}$. The child's total acceleration, $\vec{a}$, is the vector sum of $\vec{a}_t$ and $\vec{a}_{cp}$.

ACTIVE EXAMPLE 10–2 Find the Acceleration

Suppose the centrifuge in Example 10–3 is starting up with a constant angular acceleration of 95.0 rad/s². **(a)** What is the magnitude of the centripetal, tangential, and total acceleration of the bottom of a tube when the angular speed is 8.00 rad/s? **(b)** What angle does the total acceleration make with the direction of motion?

continued on next page

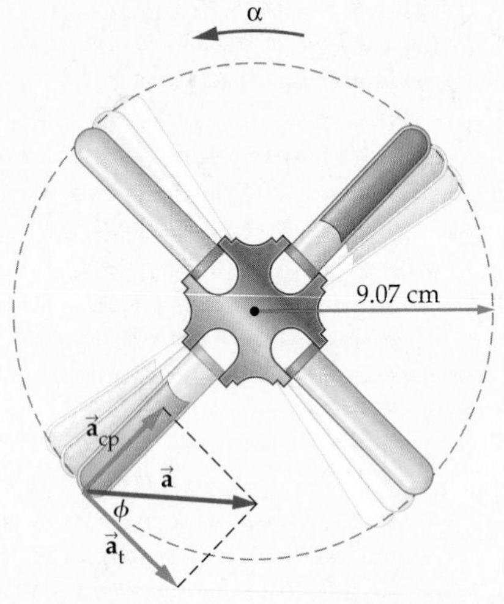

continued from previous page

Solution *(Test your understanding by performing the calculations indicated in each step.)*

Part (a)

1. Calculate the centripetal acceleration: $a_{cp} = r\omega^2 = 5.80\ \mathrm{m/s^2}$
2. Calculate the tangential acceleration: $a_t = r\alpha = 8.62\ \mathrm{m/s^2}$
3. Find the magnitude of the total acceleration: $a = \sqrt{a_{cp}^2 + a_t^2} = 10.4\ \mathrm{m/s^2}$

Part (b)

4. Find the angle ϕ for the total acceleration: $\phi = \tan^{-1}(a_{cp}/a_t) = 33.9°$

Insight

Note that all points on a tube have the same angular speed. In addition, all points have the same angular acceleration. In contrast, different points have different centripetal and tangential accelerations, due to their dependence on the distance r from the axis of rotation.

Your Turn

Find the magnitude and direction of the total acceleration of a point *halfway* between the top and bottom of a tube.

(Answers to **Your Turn** *problems are given in the back of the book.)*

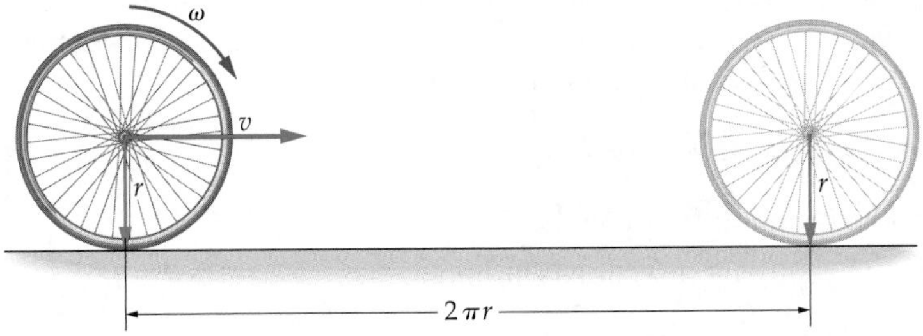

FIGURE 10–9 Rolling without slipping

A wheel of radius r rolling without slipping. During one complete revolution, the center of the wheel moves forward through a distance $2\pi r$.

10–4 Rolling Motion

We began this chapter with a bicycle wheel rotating about its axle. In that case, the axle was at rest and every point on the wheel, such as the spot of red paint, moved in a circular path about the axle. We would like to consider a different situation now. Suppose the bicycle wheel is rolling freely, as indicated in **Figure 10–9**, with no slipping between the tire and the ground. The wheel still rotates about the axle, but the axle itself is moving in a straight line. As a result, the motion of the wheel is a combination of both rotational motion and linear (or **translational**) motion.

To see the connection between the wheel's rotational and translational motions, we show one full rotation of the wheel in Figure 10–9. During this rotation, the axle translates forward through a distance equal to the circumference of the wheel, $2\pi r$. Because the time required for one rotation is the period, T, the translational speed of the axle is

$$v = \frac{2\pi r}{T}$$

Recalling that $\omega = 2\pi/T$, we find

$$v = r\omega = v_t \qquad \text{10–15}$$

Hence, the translational speed of the axle is equal to the tangential speed of a point on the rim of a wheel spinning with angular speed ω.

A rolling object, then, combines rotational motion with angular speed ω, and translational motion with linear speed $v = r\omega$, where r is the radius of the object. Let's consider these two motions one at a time. First, in **Figure 10–10 (a)** we show pure rotational motion with angular speed ω. In this case, the axle is at rest, and

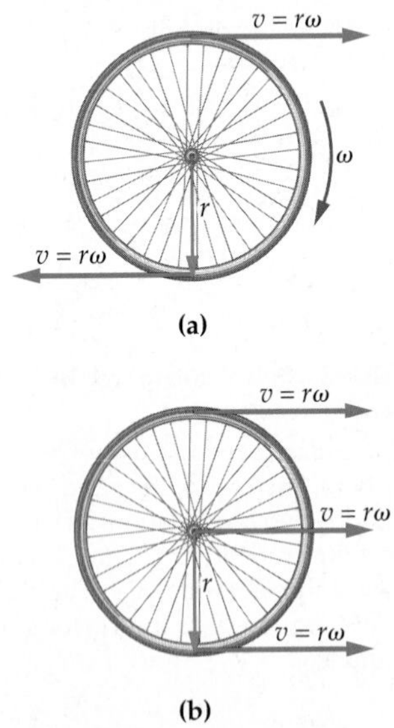

▲ **FIGURE 10–10 Rotational and translational motions of a wheel**

(a) In pure rotational motion, the velocities at the top and bottom of the wheel are in opposite directions. **(b)** In pure translational motion, each point on the wheel moves with the same speed in the same direction.

points at the top and bottom of the wheel have tangential velocities that are equal in magnitude, $v = r\omega$, but point in opposite directions.

Next, we consider translational motion with speed $v = r\omega$. This is illustrated in **Figure 10–10 (b)**, where we see that each point on the wheel moves in the same direction with the same speed. If this were the only motion the wheel had, it would be skidding across the ground, instead of rolling without slipping.

Finally, we combine these two motions by simply adding the velocity vectors in Figures 10–10 (a) and (b). The result is shown in **Figure 10–11**. At the top of the wheel the two velocity vectors are in the same direction, so they sum to give a speed of $2r\omega$. At the axle, the velocity vectors sum to give a speed $r\omega$. Finally, at the bottom of the wheel, the velocity vectors from rotation and translation have equal magnitude, but are in opposite directions. As a result, these velocities cancel, giving a speed of zero where the wheel is in contact with the ground.

The fact that the bottom of the wheel is instantaneously at rest, so that it is in static contact with the ground, is precisely what is meant by "rolling without slipping." Thus, a wheel that rolls without slipping is just like the situation when you are walking—even though your body as a whole moves forward, the soles of your shoes are momentarily at rest every time you place them on the ground. This point was discussed in detail in Conceptual Checkpoint 6–1.

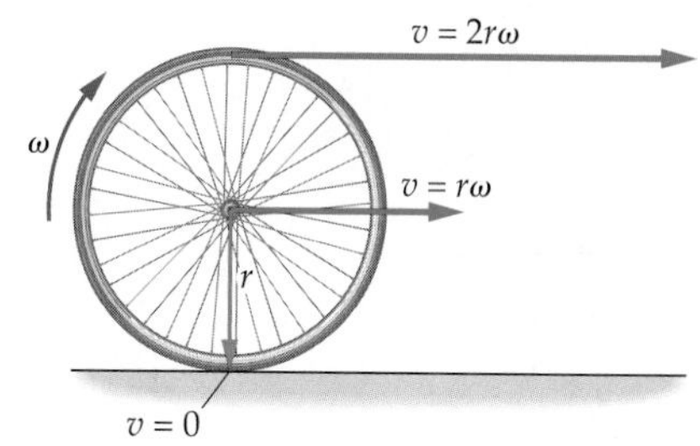

▲ **FIGURE 10–11 Velocities in rolling motion**

In a wheel that rolls without slipping, the point in contact with the ground is instantaneously at rest. The center of the wheel moves forward with the speed $v = r\omega$, and the top of the wheel moves forward with twice that speed, $v = 2r\omega$.

▲ This photograph of a rolling wheel gives a visual indication of the speed of its various parts. The bottom of the wheel is at rest at any instant, so the image there is sharp. The top of the wheel has the greatest speed, and the image there shows the most blurring. (Compare Figure 10–11.)

EXERCISE 10–6

A car with tires of radius 32 cm drives on the highway at 55 mph. **(a)** What is the angular speed of the tires? **(b)** What is the linear speed of the top of the tires?

Solution

a. Using Equation 10–15 we find

$$\omega = \frac{v}{r} = \frac{(55\text{ mph})\left(\dfrac{0.447\text{ m/s}}{1\text{ mph}}\right)}{0.32\text{ m}} = 77\text{ rad/s}$$

This is about 12 revolutions per second.

b. The top of the tires have a speed of $2v = 110$ mph.

10–5 Rotational Kinetic Energy and the Moment of Inertia

An object in motion has kinetic energy, whether that motion is translational, rotational, or a combination of the two. In translational motion, for example, the kinetic energy of a mass m moving with a speed v is $K = \frac{1}{2}mv^2$. We cannot use this expression for a rotating object, however, because the speed v of each particle within a rotating object varies with its distance r from the axis of rotation, as we have seen in Equation 10–12. Thus, there is no unique value of v for an entire rotating object. On the other hand, there *is* a unique value of ω, the angular speed, that applies to all particles in the object.

To see how the kinetic energy of a rotating object depends on its angular speed, we start with a particularly simple system consisting of a rod of length r and negligible mass rotating about one end with an angular speed ω. Attached to the other end of the rod is a point mass m, as **Figure 10–12** shows. To find the kinetic energy of the mass, recall that its linear speed is $v = r\omega$ (Equation 10–12). Therefore, the translational kinetic energy of the mass m is

$$K = \tfrac{1}{2}mv^2 = \tfrac{1}{2}m(r\omega)^2 = \tfrac{1}{2}(mr^2)\omega^2 \qquad \textbf{10–16}$$

Notice that the kinetic energy of the mass depends not only on the angular speed squared (analogous to the way the translational kinetic energy depends on the linear speed squared), but also on the radius squared—that is, the kinetic energy depends on the *distribution* of mass in the rotating object. To be specific, mass near

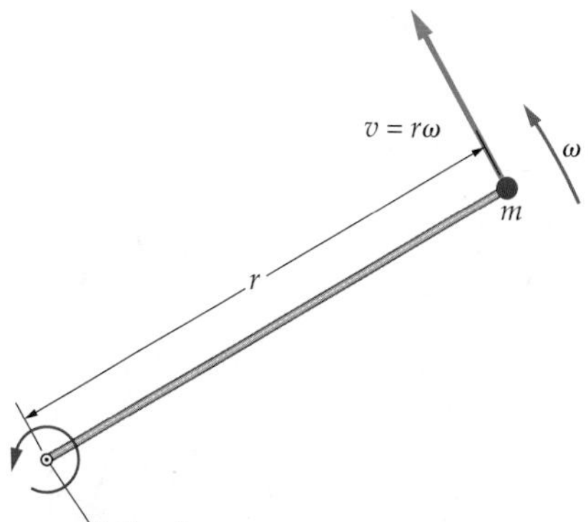

▲ **FIGURE 10–12 Kinetic energy of a rotating object**

As this rod rotates about the axis of rotation with an angular speed ω, the mass has a speed of $v = r\omega$. It follows that the kinetic energy of the mass is $K = \frac{1}{2}mv^2 = \frac{1}{2}(mr^2)\omega^2$.

the axis of rotation contributes little to the kinetic energy since its speed ($v = r\omega$) is small. On the other hand, the farther a mass is from the axis of rotation the greater its speed v for a given angular velocity, and thus the greater its kinetic energy.

You have probably noticed that the kinetic energy in Equation 10–16 is similar in form to the translational kinetic energy. Instead of $\frac{1}{2}(m)v^2$, we now have $\frac{1}{2}(mr^2)\omega^2$. Clearly, then, the quantity mr^2 plays the role of the mass for the rotating object. This "rotational mass" is given a special name in physics: the **moment of inertia,** I. Thus, in general, the kinetic energy of an object rotating with an angular speed ω can be written as:

Rotational Kinetic Energy

$$K = \frac{1}{2}I\omega^2 \qquad \text{10–17}$$

SI unit: J

The greater the moment of inertia—which some books call the rotational inertia—the greater an object's rotational kinetic energy. As we have just seen, in the special case of a point mass m a distance r from the axis of rotation, the moment of inertia is simply $I = mr^2$.

We now show how to find the moment of inertia for an object of arbitrary, fixed shape, as in **Figure 10–13**. Suppose, for example, that this object rotates about the axis indicated in the figure with an angular speed ω. To calculate the kinetic energy of the object, we first imagine dividing it into a collection of small mass elements, m_i. We then calculate the kinetic energy of each element and sum over all elements. This extends to a large number of mass elements what we did for the single mass m.

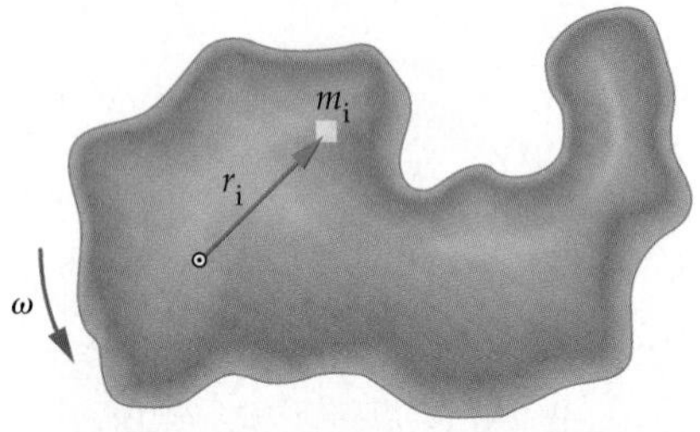

▲ **FIGURE 10–13 Kinetic energy of a rotating object of arbitrary shape**
To calculate the kinetic energy of an object of arbitrary shape as it rotates about an axis with angular speed ω, imagine dividing it into small mass elements, m_i. The total kinetic energy of the object is the sum of the kinetic energies of all the mass elements.

Following this plan, the total kinetic energy of an arbitrary rotating object is

$$K = \sum\left(\tfrac{1}{2}m_i v_i^2\right)$$

In this expression, m_i is the mass of one of the small mass elements and v_i is its speed. If m_i is at the radius r_i from the axis of rotation, as indicated in Figure 10–13, its speed is $v_i = r_i\omega$. Note that it is not necessary to write a separate angular speed, ω_i, for each element, because *all* mass elements of the object have exactly the same angular speed, ω. Therefore,

$$K = \sum\left(\tfrac{1}{2}m_i r_i^2\omega^2\right) = \tfrac{1}{2}\left(\sum m_i r_i^2\right)\omega^2$$

Now, in analogy with our results for the single mass, we can define the moment of inertia, I, as follows:

Definition of Moment of Inertia, *I*

$$I = \sum m_i r_i^2 \qquad \text{10–18}$$

SI unit: $kg \cdot m^2$

The precise value of I for a given object depends on its distribution of mass. A simple example of this dependence is given in the following Exercise.

EXERCISE 10–7

Use the general definition of the moment of inertia, as given in Equation 10–18, to find the moment of inertia for the dumbbell-shaped object shown in **Figure 10–14**. Note that the axis of rotation goes through the center of the object and points out of the page. In addition, assume that the masses may be treated as point masses.

Solution

Referring to Figure 10–14, we see that $m_1 = m_2 = m$ and $r_1 = r_2 = r$. Therefore, the moment of inertia is

$$I = \sum m_i r_i^2 = m_1 r_1^2 + m_2 r_2^2 = mr^2 + mr^2 = 2mr^2$$

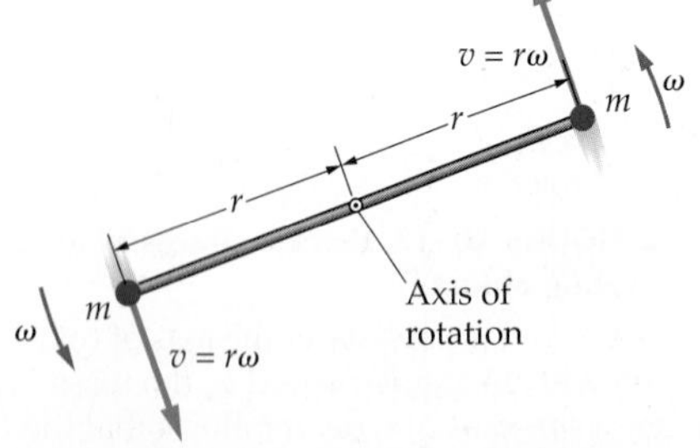

▲ **FIGURE 10–14 A dumbbell-shaped object rotating about its center.**

The connection between rotational kinetic energy and the moment of inertia is explored in more detail in the following Example.

EXAMPLE 10–4 Nose to the Grindstone

A grindstone with a radius of 0.610 m is being used to sharpen an ax. If the linear speed of the stone relative to the ax is 1.50 m/s, and the stone's rotational kinetic energy is 13.0 J, what is its moment of inertia?

Picture the Problem

Our sketch shows the grindstone spinning with an angular speed ω, which is not given in the problem statement. We do know, however, that the linear speed of the grindstone at its rim is $v = 1.50$ m/s and that its radius is $r = 0.610$ m. At this rate of rotation, the stone has a kinetic energy of 13.0 J.

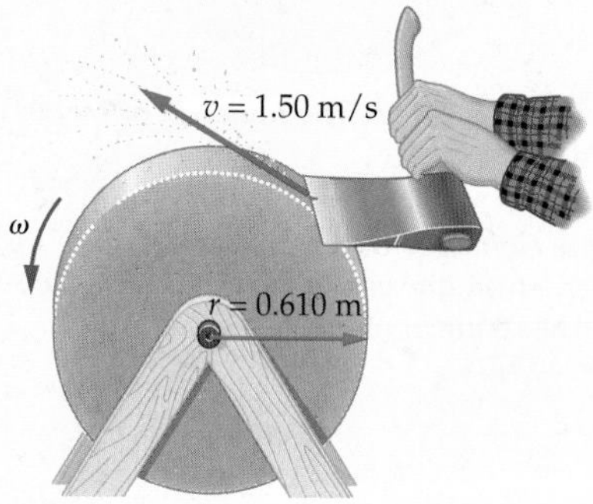

Strategy

Recall that rotational kinetic energy and moment of inertia are related by $K = \frac{1}{2}I\omega^2$; thus $I = 2K/\omega^2$. We are not given ω, but we can find it from the connection between linear and angular speed, $v = r\omega$. Thus, we begin by finding ω. We then use ω, along with the kinetic energy K, to find I.

Solution

1. Find the angular speed of the grindstone:
$$\omega = \frac{v}{r} = \frac{1.50\ \text{m/s}}{0.610\ \text{m}} = 2.46\ \text{rad/s}$$

2. Solve for the moment of inertia in terms of kinetic energy:
$$K = \tfrac{1}{2}I\omega^2, \quad \text{or} \quad I = \frac{2K}{\omega^2}$$

3. Substitute numerical values for K and ω:
$$I = \frac{2K}{\omega^2} = \frac{2(13.0\ \text{J})}{(2.46\ \text{rad/s})^2} = 4.30\ \text{J}\cdot\text{s}^2 = 4.30\ \text{kg}\cdot\text{m}^2$$

Insight

In this calculation we found I by relating it to the rotational kinetic energy of the grindstone. Later in the section we show how to calculate the moment of inertia of a disk, given its radius and mass.

Practice Problem

When the ax is pressed firmly against the grindstone for sharpening, the angular speed of the grindstone decreases. If the rotational kinetic energy of the grindstone is cut in half to 6.50 J, what is its angular speed? [**Answer:** The moment of inertia is unchanged; it depends only on the size, shape, and mass of the grindstone. Hence, $\omega = \sqrt{2K/I} = 1.74$ rad/s.]

Some related homework problems: Problem 47, Problem 48

We return now to the dependence of the moment of inertia on the particular shape, or mass distribution, of an object. Suppose, for example, that a mass M is formed into the shape of a *hoop* of radius R. In addition, consider the case where the axis of rotation is perpendicular to the plane of the hoop and passes through its center, as shown in **Figure 10–15**. This is similar to a bicycle wheel rotating about its axle, if one ignores the spokes. In terms of small mass elements, we can write the moment of inertia as

$$I = \sum m_i r_i^2$$

Each mass element of the hoop, however, is at the same radius R from the axis of rotation; that is, $r_i = R$. Hence, the moment of inertia in this case is

$$I = \sum m_i r_i^2 = \sum m_i R^2 = \left(\sum m_i\right)R^2$$

Clearly, the sum of all the elementary masses is simply the total mass of the hoop, $\Sigma m_i = M$. Therefore, the moment of inertia of a hoop of mass M and radius R is

$$I = MR^2 \text{ (hoop)}$$

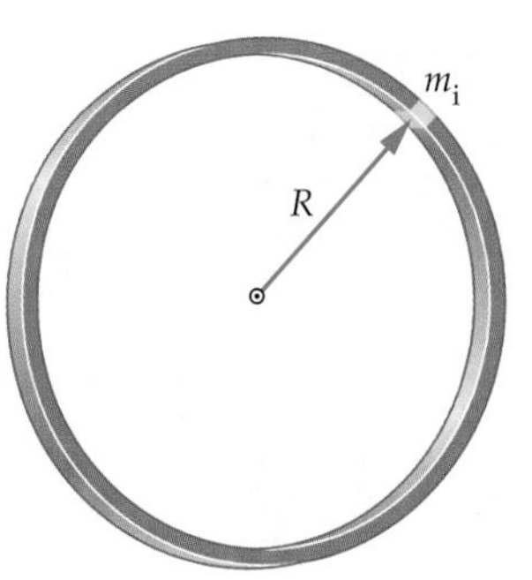

▲ **FIGURE 10–15 The moment of inertia of a hoop**
Consider a hoop of mass M and radius R. Each small mass element is at the same distance, R, from the center of the hoop. The moment of inertia in this case is $I = MR^2$.

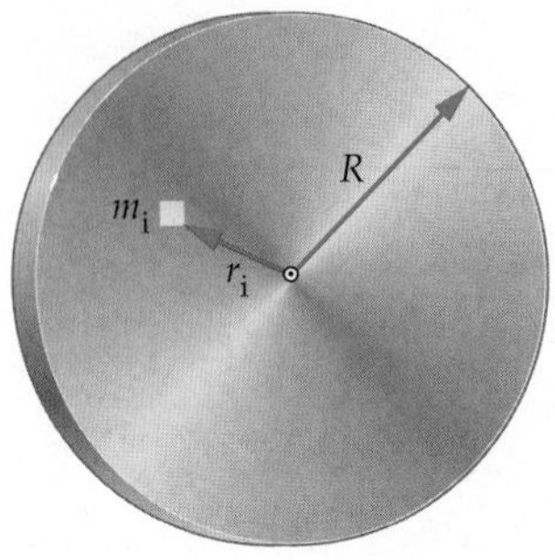

▲ **FIGURE 10–16 The moment of inertia of a disk**

Consider a disk of mass M and radius R. Mass elements for the disk are at distances from the center ranging from 0 to R. The moment of inertia in this case is $I = \frac{1}{2}MR^2$.

In contrast, if the same mass, M, is formed into a uniform *disk* of the same radius, R, the moment of inertia is different. To see this, note that it is no longer true that $r_i = R$ for all mass elements. In fact, most of the mass elements are closer to the axis of rotation than was the case for the hoop, as indicated in **Figure 10–16**. Thus, since the r_i are generally less than R, the moment of inertia will be smaller for the disk than for the hoop. A detailed calculation, summing over all mass elements, yields the following result:

$$I = \tfrac{1}{2}MR^2 \text{ (disk)}$$

As expected, I is less for the disk than for the hoop.

EXERCISE 10–8

If the grindstone in Example 10–4 is a uniform disk, what is its mass?

Solution

Applying the preceding equation yields

$$M = \frac{2I}{R^2} = \frac{2(4.30 \text{ kg}\cdot\text{m}^2)}{(0.610 \text{ m})^2} = 23.1 \text{ kg}$$

Thus, the grindstone has a weight of roughly 51 lb.

Table 10–1 collects moments of inertia for a variety of objects. Note that in all cases the moment of inertia is of the form $I = (\text{constant})MR^2$. It is only the constant in front of MR^2 that changes from one object to another.

Note also that objects of the same general shape but with different mass distributions—such as solid and hollow spheres—have different moments of inertia. In particular, a hollow sphere has a larger I than a solid sphere of the same mass, for the same reason that a hoop's moment of inertia is greater than a disk's—more of its mass is at a greater distance from the axis of rotation. Thus, I is a measure of both the shape *and* the mass distribution of an object.

REAL-WORLD PHYSICS
Moment of inertia of the Earth

Consider, for example, the moment of inertia of the Earth. If the Earth were a uniform sphere of mass M_E and radius R_E, its moment of inertia would be $\frac{2}{5}M_ER_E{}^2 = 0.4M_ER_E{}^2$. In fact, the Earth's moment of inertia is only $0.331M_ER_E{}^2$,

TABLE 10–1 Moments of Inertia for Uniform, Rigid Objects of Various Shapes and Total Mass M

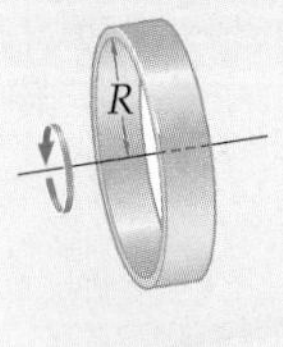	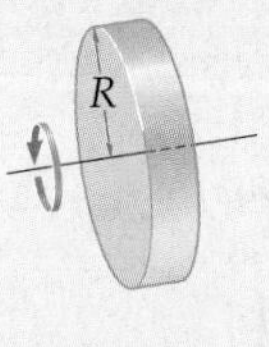	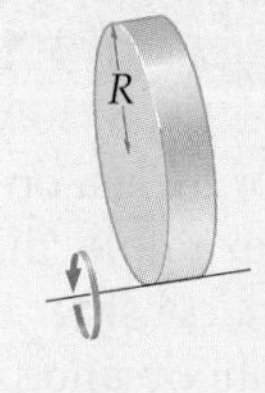	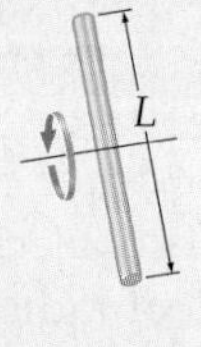	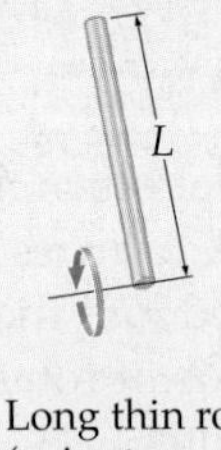
Hoop or cylindrical shell $I = MR^2$	Disk or solid cylinder $I = \frac{1}{2}MR^2$	Disk or solid cylinder (axis at rim) $I = \frac{3}{2}MR^2$	Long thin rod (axis through midpoint) $I = \frac{1}{12}ML^2$	Long thin rod (axis at one end) $I = \frac{1}{3}ML^2$
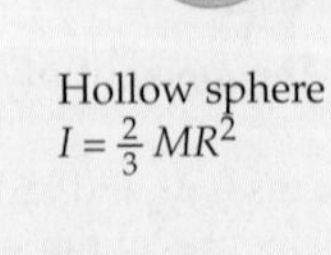	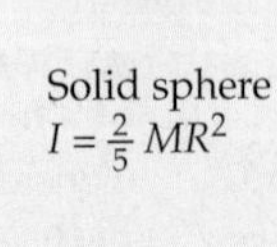	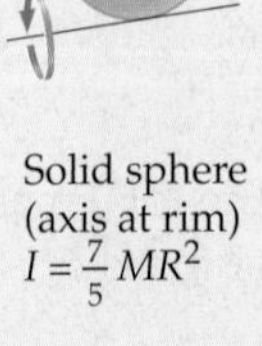	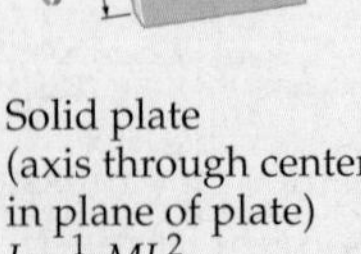	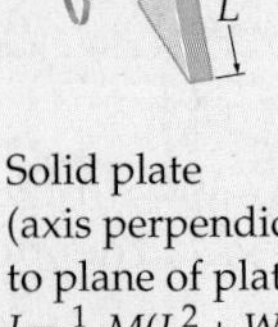
Hollow sphere $I = \frac{2}{3}MR^2$	Solid sphere $I = \frac{2}{5}MR^2$	Solid sphere (axis at rim) $I = \frac{7}{5}MR^2$	Solid plate (axis through center, in plane of plate) $I = \frac{1}{12}ML^2$	Solid plate (axis perpendicular to plane of plate) $I = \frac{1}{12}M(L^2 + W^2)$

considerably less than for a uniform sphere. This is due to the fact that the Earth is not homogeneous, but instead has a dense inner core surrounded by a less dense outer core and an even less dense mantle. The resulting concentration of mass near its axis of rotation gives the Earth a much smaller moment of inertia than it would have if its mass were uniformly distributed.

On the other hand, if the polar ice caps were to melt and release their water into the oceans, the Earth's moment of inertia would increase. This is because mass that had been near the axis of rotation (in the polar ice) would now be distributed more or less uniformly around the Earth (in the oceans). With more of the Earth's mass at greater distances from the axis of rotation, the moment of inertia would increase. If such an event were to occur, not only would the moment of inertia increase, but the length of the day would increase as well. We shall discuss the reasons for this in the next chapter.

The moment of inertia of an object also depends on the location and orientation of the axis of rotation. If the axis of rotation is moved, all of the r_i change, leading to a different result for I. This is investigated for the dumbbell system in the following Conceptual Checkpoint.

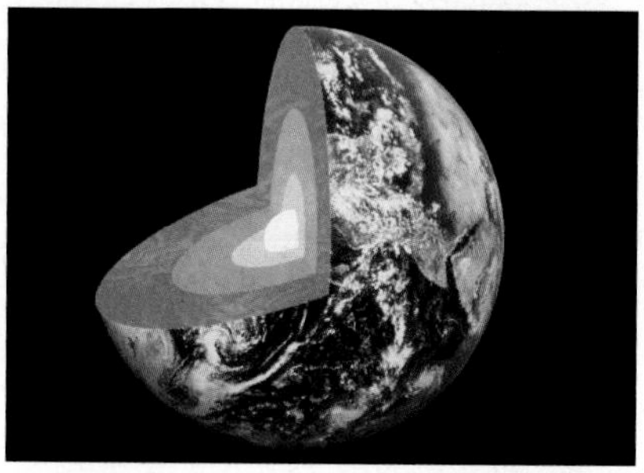

▲ The distribution of mass in the Earth is not uniform. Dense materials, like iron and nickel, have concentrated near the center, while less dense materials, like silicon and aluminum, have risen to the surface. This concentration of mass near the axis of rotation lowers the Earth's moment of inertia.

CONCEPTUAL CHECKPOINT 10–2 Compare the Moments of Inertia

If the dumbbell-shaped object in Figure 10–14 is rotated about one end, is its moment of inertia **(a)** more than, **(b)** less than, or **(c)** the same as the moment of inertia about its center? As before, assume that the masses can be treated as point masses.

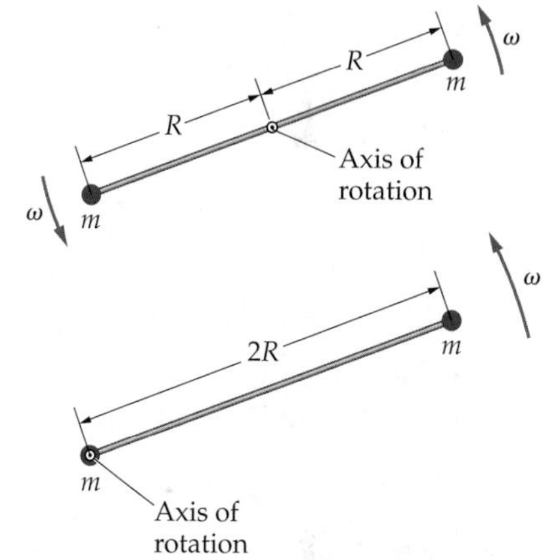

Reasoning and Discussion

As we saw in Exercise 10–7, the moment of inertia about the center of the dumbbell is $I = 2mR^2$. When the axis is at one end, that mass is at the radius $r = 0$, and the other mass is at $r = 2R$. Therefore, the moment of inertia is

$$I = \sum m_i r_i^2 = m \cdot 0 + m(2R)^2 = 4mR^2$$

Thus, the moment of inertia doubles when the axis of rotation is moved from the center to one end.

The reason I increases is that the moment of inertia depends on the radius squared. Hence, even small increases in r can cause significant increases in I. By moving the axis to one end, the radius to the other mass is increased to its greatest possible value. As a result, I increases.

Answer:

(a) The moment of inertia is greater about one end than about the center.

Finally, we summarize in the accompanying table the similarities between the translational kinetic energy, $K = \frac{1}{2}mv^2$, and the rotational kinetic energy, $K = \frac{1}{2}I\omega^2$. As expected, we see that the linear speed, v, has been replaced with the angular speed, ω. In addition, note that the mass m has been replaced with the moment of inertia I.

Linear Quantity	Angular Quantity
v	ω
m	I
$\frac{1}{2}mv^2$	$\frac{1}{2}I\omega^2$

As suggested by these analogies, the moment of inertia I plays the same role in rotational motion that mass plays in translational motion. For example, the larger I the more resistant an object is to any change in its angular velocity—an object with a large I is difficult to start rotating, and once it is rotating, it is difficult to stop. We shall see further applications of this analogy in the next chapter when we consider angular momentum.

10–6 Conservation of Energy

In this section, we consider the mechanical energy of objects that roll without slipping, and show how to apply energy conservation to such systems. In addition, we consider objects that rotate as a string or rope unwinds: for example, a pulley with a string wrapped around its circumference, or a yo-yo with a string wrapped around its axle. As long as the unwinding process and the rolling motion occur without slipping, the two situations are basically the same—at least as far as energy considerations are concerned.

To apply energy conservation to rolling objects, we first need to determine the kinetic energy of rolling motion. In Section 10–4 we saw that rolling motion is a combination of rotation and translation. It follows, then, that the kinetic energy of a rolling object is simply the sum of its translational kinetic energy, $\frac{1}{2}mv^2$, and its rotational kinetic energy, $\frac{1}{2}I\omega^2$:

PROBLEM-SOLVING NOTE

Energy Conservation with Rotational Motion

When applying energy conservation to a system with rotational motion, be sure to include the rotational kinetic energy, $\frac{1}{2}I\omega^2$.

Kinetic Energy of Rolling Motion

$$K = \tfrac{1}{2}mv^2 + \tfrac{1}{2}I\omega^2 \qquad \text{10–19}$$

Note that I in this expression is the moment of inertia about the center of the rolling object.

We can simplify the expression for the kinetic energy of a rolling object by using the fact that linear and angular speeds are related. In fact, recall that $v = r\omega$ (Equation 10–12), which can be rewritten as $\omega = v/r$. Substituting this into our expression for the rolling kinetic energy yields

Kinetic Energy of Rolling Motion: Alternative Form

$$K = \tfrac{1}{2}mv^2 + \tfrac{1}{2}I\left(\frac{v}{r}\right)^2 = \tfrac{1}{2}mv^2\left(1 + \frac{I}{mr^2}\right) \qquad \text{10–20}$$

Since $I = (\text{constant})mr^2$, the last term in Equation 10–20 is a constant that depends on the shape and mass distribution of the rolling object.

A special case of some interest is the point particle. In this case, by definition, all of the mass is at a single point. Therefore, $r = 0$, and hence $I = 0$. Substituting $I = 0$ in either Equation 10–19 or Equation 10–20 yields $K = \frac{1}{2}mv^2$, as expected.

Next, we apply Equations 10–19 and 10–20 to a disk that rolls with no slipping.

EXAMPLE 10–5 Like a Rolling Disk

A 1.20-kg disk with a radius of 10.0 cm rolls without slipping. If the linear speed of the disk is 1.41 m/s, find **(a)** the translational kinetic energy, **(b)** the rotational kinetic energy, and **(c)** the total kinetic energy of the disk.

Picture the Problem

Because the disk rolls without slipping, the angular speed and the linear speed are related by $v = r\omega$. Note that the linear speed is $v = 1.41$ m/s and the radius is $r = 10.0$ cm. Finally, we are given that the mass of the disk is 1.20 kg.

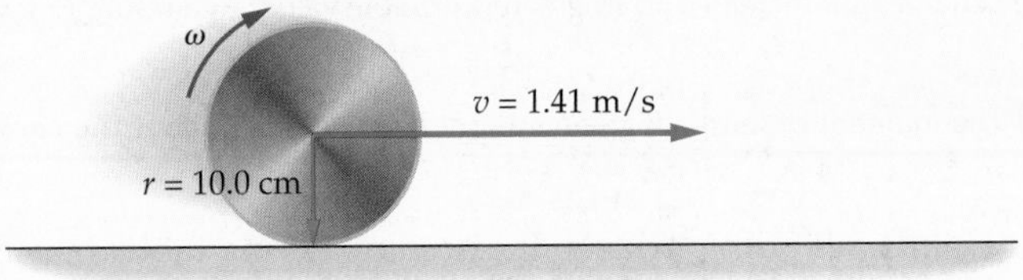

Strategy

We calculate each contribution to the kinetic energy separately. The linear kinetic energy, of course, is simply $\frac{1}{2}mv^2$. For the rotational kinetic energy, $\frac{1}{2}I\omega^2$, we must use the fact that the moment of inertia for a disk is $I = \frac{1}{2}mr^2$. Finally, since the disk rolls without slipping, its angular speed is $\omega = v/r$.

Solution

Part (a)

1. Calculate the translational kinetic energy, $\frac{1}{2}mv^2$:

$$\tfrac{1}{2}mv^2 = \tfrac{1}{2}(1.20\text{ kg})(1.41\text{ m/s})^2 = 1.19\text{ J}$$

Part (b)

2. Calculate the rotational kinetic energy symbolically, using $I = \frac{1}{2}mr^2$ and $\omega = v/r$:

$$\tfrac{1}{2}I\omega^2 = \tfrac{1}{2}\left(\tfrac{1}{2}mr^2\right)\left(\frac{v}{r}\right)^2 = \tfrac{1}{2}\left(\tfrac{1}{2}mv^2\right)$$

3. Substitute the numerical value for $\frac{1}{2}mv^2$ (the translational kinetic energy) obtained in step 1:

$$\tfrac{1}{2}I\omega^2 = \tfrac{1}{2}(1.19\text{ J}) = 0.595\text{ J}$$

Part (c)

4. Sum the kinetic energies obtained in parts (a) and (b):

$$K = 1.19\text{ J} + 0.595\text{ J} = 1.79\text{ J}$$

5. Note that the same result is obtained using Equation 10–20:

$$K = \tfrac{1}{2}mv^2\left(1 + \frac{I}{mr^2}\right) = \tfrac{1}{2}mv^2\left(1 + \frac{1}{2}\right)$$
$$= \tfrac{3}{2}\left(\tfrac{1}{2}mv^2\right) = \tfrac{3}{2}(1.19\text{ J}) = 1.79\text{ J}$$

Insight

The symbolic result in step 2 shows that the rotational kinetic energy of a uniform disk rolling without slipping is precisely one-half the disk's translational kinetic energy. Thus, 2/3 of the disk's kinetic energy is translational, 1/3 rotational. This result is independent of the disk's radius, as we can see by the cancellation of the radius r in step 2.

To understand this cancellation, note that a larger disk has a larger moment of inertia, since it has mass farther from the axis of rotation. On the other hand, the larger disk also has a smaller angular speed, since the angular speed is inversely proportional to the radius: $\omega = v/r$. These two effects cancel, giving the same rotational kinetic energy for uniform disks of any radius—provided their linear speed is the same.

Practice Problem

Repeat this problem for the case of a rolling, hollow sphere. [**Answer:** (a) 1.19 J, (b) 0.793 J, (c) 1.98 J]

Some related homework problems: Problem 49, Problem 52

CONCEPTUAL CHECKPOINT 10–3 Compare Kinetic Energies

A solid sphere and a hollow sphere of the same mass and radius roll without slipping at the same speed. Is the kinetic energy of the solid sphere **(a)** more than, **(b)** less than, or **(c)** the same as the kinetic energy of the hollow sphere?

Reasoning and Discussion

Both spheres have the same translational kinetic energy since they have the same mass and speed. The rotational kinetic energy, however, is proportional to the moment of inertia. Since the hollow sphere has the greater moment of inertia, it has the greater kinetic energy.

Answer:

(b) The solid sphere has less kinetic energy than the hollow sphere.

Now that we can calculate the kinetic energy of rolling motion, we show how to apply it to energy conservation. For example, consider an object of mass m, radius r, and moment of inertia I at the top of a ramp, as shown in **Figure 10–17**. The object is released from rest and allowed to roll to the bottom, a vertical height h below the starting point. What is the object's speed on reaching the bottom?

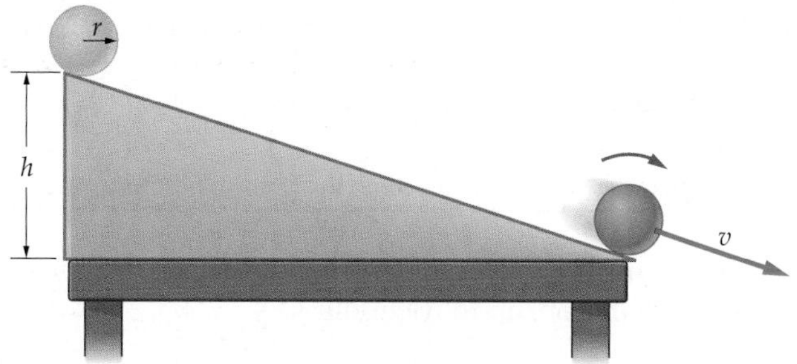

◀ FIGURE 10–17 An object rolls down an incline

An object starts at rest at the top of an inclined plane and rolls without slipping to the bottom. The speed of the object at the bottom depends on its moment of inertia—a larger moment of inertia results in a lower speed.

The simplest way to solve this problem is to use energy conservation. To do so, we set the initial mechanical energy at the top (i) equal to the final mechanical energy at the bottom (f). That is,

$$K_i + U_i = K_f + U_f$$

Since we are dealing with rolling motion, the kinetic energy is

$$K = \tfrac{1}{2}mv^2\left(1 + \frac{I}{mr^2}\right)$$

The potential energy is simply that due to the uniform gravitational field. Therefore,

$$U = mgy$$

With $y = h$ at the top of the ramp and the object starting at rest, we have

$$K_i + U_i = 0 + mgh = mgh$$

Similarly, with $y = 0$ at the bottom of the ramp and the object rolling with a speed v, we find

$$K_f + U_f = \frac{1}{2}mv^2\left(1 + \frac{I}{mr^2}\right) + 0 = \frac{1}{2}mv^2\left(1 + \frac{I}{mr^2}\right)$$

Setting the initial and final energies equal yields

$$mgh = \frac{1}{2}mv^2\left(1 + \frac{I}{mr^2}\right)$$

Solving for v, we find

$$v = \sqrt{\frac{2gh}{1 + \frac{I}{mr^2}}}$$

Let's quickly check one special case: namely, $I = 0$. With this substitution, we find

$$v = \sqrt{2gh}$$

This is the speed an object would have after falling straight down with no rotation through a distance h. Thus, setting $I = 0$ means there is no rotational kinetic energy, and hence the result is the same as for a point particle. As I becomes larger, the speed at the bottom of the ramp is smaller.

CONCEPTUAL CHECKPOINT 10–4 Which Object Wins the Race?

INTERACTIVE FIGURE

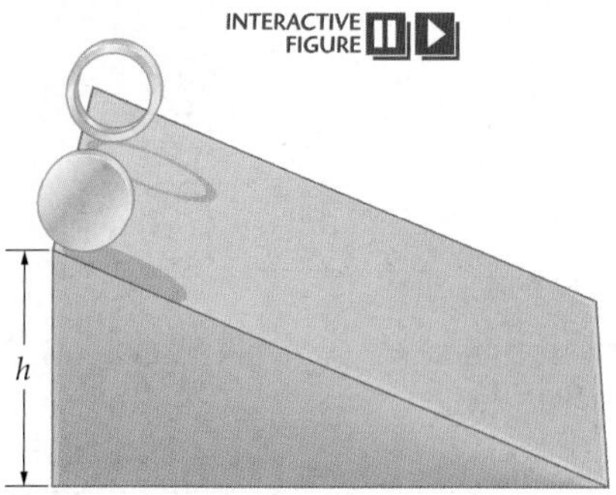

A disk and a hoop of the same mass and radius are released at the same time at the top of an inclined plane. Does the disk reach the bottom of the plane **(a)** before, **(b)** after, or **(c)** at the same time as the hoop?

Reasoning and Discussion

As we have just seen, the larger the moment of inertia, I, the smaller the speed, v. Hence the object with the larger moment of inertia (the hoop in this case) loses the race to the bottom, because its speed is less than the speed of the disk at any given height.

Another way to think about this is to recall that both objects have the same mechanical energy to begin with, namely, mgh. For the hoop, more of this initial potential energy goes into rotational kinetic energy, since the hoop has the larger moment of inertia; therefore, less energy is left for translational motion. As a result, the hoop moves more slowly and loses the race.

Answer:

(a) The disk wins the race by reaching the bottom before the hoop.

In the next Conceptual Checkpoint, we consider the effects of a surface that changes from nonslip to frictionless.

CONCEPTUAL CHECKPOINT 10–5 Compare Heights

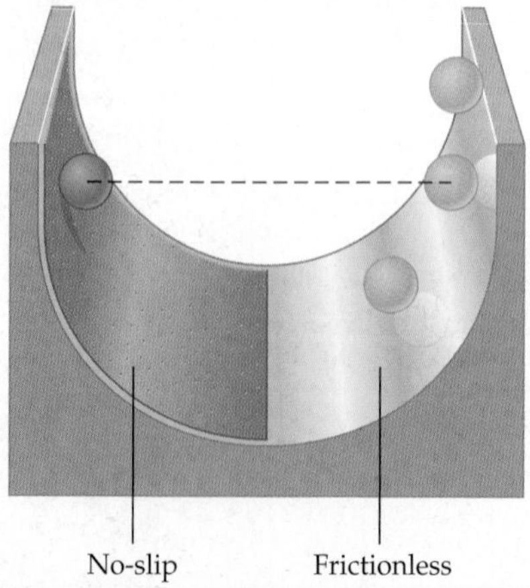

A ball is released from rest on a no-slip surface, as shown. After reaching its lowest point, the ball begins to rise again, this time on a frictionless surface. When the ball reaches its maximum height on the frictionless surface, is it **(a)** at a greater height, **(b)** at a lesser height, or **(c)** at the same height as when it was released?

Reasoning and Discussion

As the ball descends on the no-slip surface, it begins to rotate, increasing its angular speed until it reaches the lowest point of the surface. When it begins to rise again, there is no friction to slow the rotational motion; thus, the ball continues to rotate with the same angular speed it had at its lowest point. Therefore, some of the ball's initial gravitational potential energy remains in the form of rotational kinetic energy. As a result, less energy is available to be converted back into gravitational potential energy, and the height is less.

Answer:

(b) The height on the frictionless side is less.

We can also apply energy conservation to the case of a pulley, or similar object, with a string that winds or unwinds without slipping. In such cases, the relation $v = r\omega$ is valid and we can follow the same methods applied to an object that rolls without slipping.

EXAMPLE 10–6 Spinning Wheel

A block of mass m is attached to a string that is wrapped around the circumference of a wheel of radius R and moment of inertia I. The wheel rotates freely about its axis and the string wraps around its circumference without slipping. Initially the wheel rotates with an angular speed ω, causing the block to rise with a linear speed v. To what height does the block rise before coming to rest? Give a symbolic answer.

Picture the Problem

Note in our sketch that we choose the origin of the y axis to be at the initial height of the block. The positive y direction, as usual, is chosen to be upward. When the block comes to rest, then, it is at the height $y = h > 0$, where h is to be determined from the initial speed of the block and the properties of the wheel.

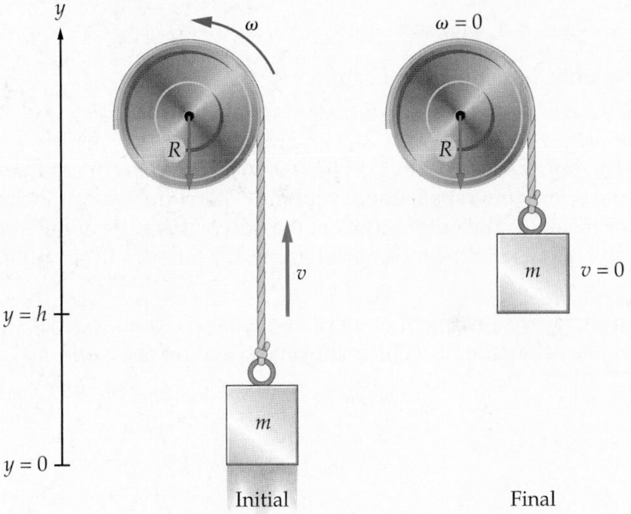

Strategy

The problem statement gives two key pieces of information. First, the string wraps onto the disk without slipping; therefore, $v = R\omega$. Second, the wheel rotates freely, which means that the mechanical energy of the system is conserved. Thus, at the height h the initial kinetic energy of the system has been converted to gravitational potential energy. This condition can be used to find h.

Before we continue, note that the mechanical energy of the system includes the following contributions: (i) linear kinetic energy for the block, (ii) rotational kinetic energy for the wheel, and (iii) gravitational potential energy for the block. We do not include the gravitational potential energy of the wheel because its height does not change.

Solution

1. Write an expression for the initial mechanical energy of the system, E_i, including all three contributions mentioned in the Strategy:

$$E_i = \tfrac{1}{2}mv^2 + \tfrac{1}{2}I\omega^2 + mgy$$
$$= \tfrac{1}{2}mv^2 + \tfrac{1}{2}I\left(\frac{v}{R}\right)^2 + 0$$

2. Write an expression for the final mechanical energy of the system, E_f:

$$E_f = \tfrac{1}{2}mv^2 + \tfrac{1}{2}I\omega^2 + mgy$$
$$= 0 + 0 + mgh$$

3. Set the initial and final mechanical energies equal to one another, $E_i = E_f$:

$$E_i = \tfrac{1}{2}mv^2 + \tfrac{1}{2}I\left(\frac{v}{R}\right)^2 = \tfrac{1}{2}mv^2\left(1 + \frac{I}{mR^2}\right)$$
$$= E_f = mgh$$

4. Solve for the height, h:

$$h = \left(\frac{v^2}{2g}\right)\left(1 + \frac{I}{mR^2}\right)$$

Insight

If the block were moving upward with speed v on its own—not attached to anything—it would rise to the height $h = v^2/2g$. We recover this result if $I = 0$, since in that case it is as if the wheel were not there. If the wheel is there, and I is nonzero, the block rises to a height that is *greater* than $v^2/2g$. The reason is that the wheel has kinetic energy, in addition to the kinetic energy of the block, and the sum of these kinetic energies must be converted to gravitational potential energy before the block and the wheel stop moving.

Practice Problem

Suppose the wheel is a disk with a mass equal to the mass m of the block. Find an expression for the height h in this case. [**Answer:** The moment of inertia of the wheel is $I = \tfrac{1}{2}mR^2$. Therefore, $h = (3/2)(v^2/2g)$.]

Some related homework problems: Problem 55, Problem 58, Problem 61

The situation with a yo-yo is similar, as we see in the next Active Example.

ACTIVE EXAMPLE 10–3 Find the Yo-Yo's Speed

Yo-Yo man releases a yo-yo from rest and allows it to drop, as he keeps the top end of the string stationary. The mass of the yo-yo is 0.056 kg, its moment of inertia is $2.9 \times 10^{-5}\ \mathrm{kg \cdot m^2}$, and the radius, r, of the axle the string wraps around is 0.0064 m. What is the linear speed, v, of the yo-yo after it has dropped through a height $h = 0.50$ m?

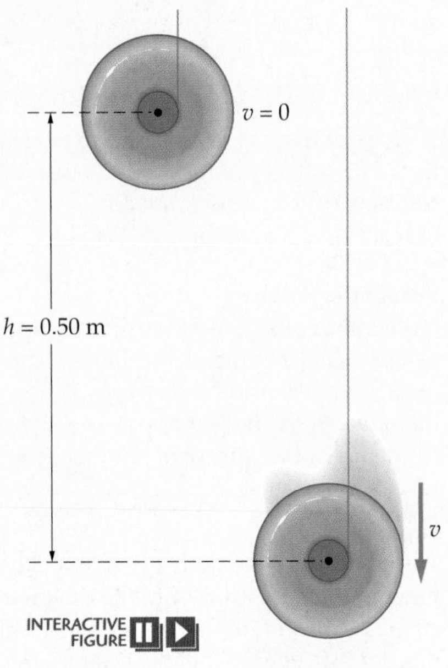

Solution *(Test your understanding by performing the calculations indicated in each step.)*

1. Write the initial energy of the system: $E_i = mgh$
2. Write the final energy of the system: $E_f = \frac{1}{2}mv^2(1 + I/mr^2)$
3. Set $E_f = E_i$ and solve for v: $v = \sqrt{2gh/(1 + I/mr^2)}$
4. Substitute numerical values: $v = 0.85$ m/s

Insight

The linear speed of the yo-yo is $v = r\omega$, where r is the radius of the axle from which the string unwraps without slipping. Therefore, the r in the term I/mr^2 is the radius of the axle. The outer radius of the yo-yo affects its moment of inertia, but since I is given to us in the problem statement, the outer radius is not pertinent.

Your Turn

If the yo-yo's moment of inertia is increased, does its final speed increase, decrease, or stay the same? Calculate the final speed for the case $I = 3.9 \times 10^{-5}\ \mathrm{kg \cdot m^2}$.

(Answers to **Your Turn** *problems are given in the back of the book.)*

Chapter Summary

10–1 Angular Position, Velocity, and Acceleration

To describe rotational motion, rotational analogs of position, velocity, and acceleration are defined.

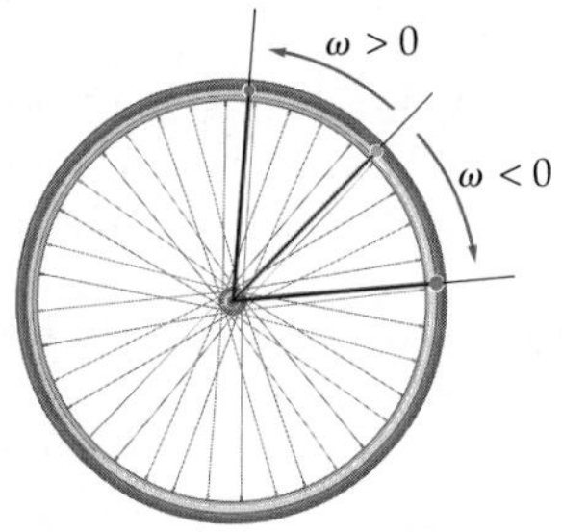

Angular Position

Angular position, θ, is the angle measured from an arbitrary reference line.

$$\theta \text{ (in radians)} = \text{arc length/radius} = s/r \qquad \text{10–2}$$

Angular Velocity

Angular velocity, ω, is the rate of change of angular position. The average angular velocity is

$$\omega_{av} = \frac{\Delta\theta}{\Delta t} \qquad \text{10–3}$$

The instantaneous angular velocity is the limit of ω_{av} as Δt approaches zero:

$$\omega = \lim_{\Delta t \to 0} \frac{\Delta\theta}{\Delta t} \qquad \text{10–4}$$

Angular Acceleration

Angular acceleration, α, is the rate of change of angular velocity. The average angular acceleration is

$$\alpha_{av} = \frac{\Delta\omega}{\Delta t} \qquad \text{10–6}$$

The instantaneous angular acceleration is the limit of α_{av} as Δt approaches zero:

$$\alpha = \lim_{\Delta t \to 0} \frac{\Delta\omega}{\Delta t} \qquad \text{10–7}$$

Period of Rotation

The period, T, is the time required to complete one full rotation. If the angular velocity is constant, T is related to ω as follows:

$$T = \frac{2\pi}{\omega} \qquad 10\text{–}5$$

Sign Convention

Counterclockwise rotations are positive; clockwise rotations are negative.

10–2 Rotational Kinematics

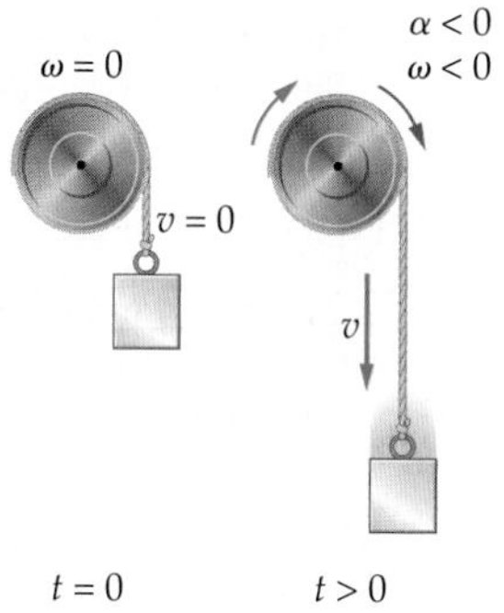

Rotational kinematics is the description of angular motion, in the same way that linear kinematics describes linear motion. In both cases, we assume constant acceleration.

Linear–Angular Analogs

Rotational kinematics is related to linear kinematics by the following linear–angular analogies:

Linear Quantity	Angular Quantity
x	θ
v	ω
a	α

Kinematic Equations (Constant Acceleration)

The equations of rotational kinematics are the same as the equations of linear kinematics, with the substitutions indicated by the linear-angular analogies:

Linear Equation		Angular Equation	
$v = v_0 + at$	2–7	$\omega = \omega_0 + \alpha t$	10–8
$x = x_0 + \frac{1}{2}(v_0 + v)t$	2–10	$\theta = \theta_0 + \frac{1}{2}(\omega_0 + \omega)t$	10–9
$x = x_0 + v_0 t + \frac{1}{2}at^2$	2–11	$\theta = \theta_0 + \omega_0 t + \frac{1}{2}\alpha t^2$	10–10
$v^2 = v_0{}^2 + 2a(x - x_0)$	2–12	$\omega^2 = \omega_0{}^2 + 2\alpha(\theta - \theta_0)$	10–11

10–3 Connections Between Linear and Rotational Quantities

A point on a rotating object follows a circular path. At any instant of time, the point is moving in a direction tangential to the circle, with a linear speed and acceleration. The linear speed and acceleration are related to the angular speed and acceleration.

Tangential Speed

The tangential speed, v_t, of a point on a rotating object is

$$v_t = r\omega \qquad 10\text{–}12$$

Centripetal Acceleration

The centripetal acceleration, a_{cp}, of a point on a rotating object is

$$a_{cp} = r\omega^2 \qquad 10\text{–}13$$

Centripetal acceleration is due to a change in direction of motion.

Tangential Acceleration

The tangential acceleration, a_t, of a point on a rotating object is

$$a_t = r\alpha \qquad 10\text{–}14$$

Tangential acceleration is due to a change in speed.

Total Acceleration

The total acceleration of a rotating object is the vector sum of its tangential and centripetal accelerations.

10–4 Rolling Motion

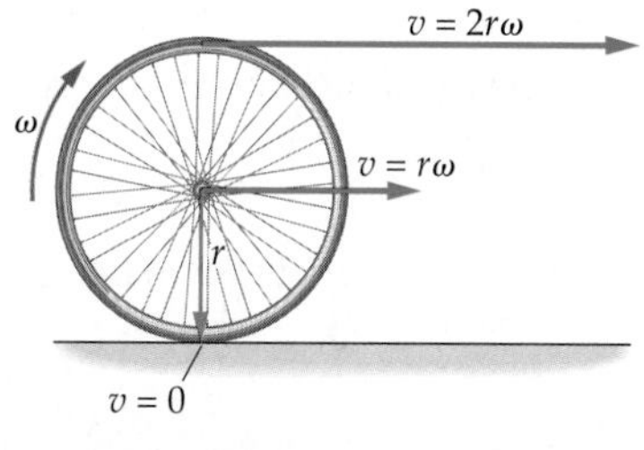

Rolling motion is a combination of translational and rotational motions. An object of radius r, rolling without slipping, translates with linear speed v and rotates with angular speed

$$\omega = v/r \qquad 10\text{–}15$$

10–5 Rotational Kinetic Energy and the Moment of Inertia

Rotating objects have kinetic energy, just as objects in linear motion have kinetic energy.

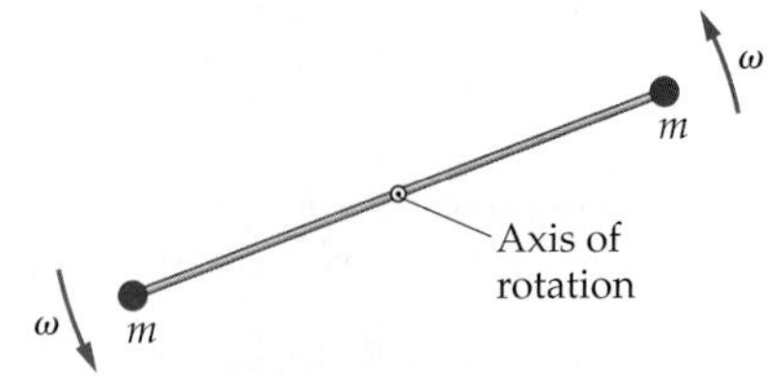

Rotational Kinetic Energy

The kinetic energy of a rotating object is

$$K = \tfrac{1}{2}I\omega^2 \qquad \textbf{10–17}$$

The quantity I is the moment of inertia.

Moment of Inertia, Discrete Masses

The moment of inertia, I, of a collection of masses, m_i, at distances r_i from the axis of rotation is

$$I = \sum m_i r_i^2 \qquad \textbf{10–18}$$

Moment of Inertia, Continuous Distribution of Mass

In a continuous object, the moment of inertia is calculated by dividing the object into a collection of small mass elements and summing $m_i r_i^2$ for each element. Results for a variety of continuous objects are collected in Table 10–1 on p. 298.

Linear–Angular Analog

The moment of inertia is the rotational analog to mass in linear systems. In particular, an object with a large moment of inertia is hard to start rotating and hard to stop rotating.

10–6 Energy Conservation

Energy conservation can be applied to a variety of rotational systems in the same way that it is applied to translational systems.

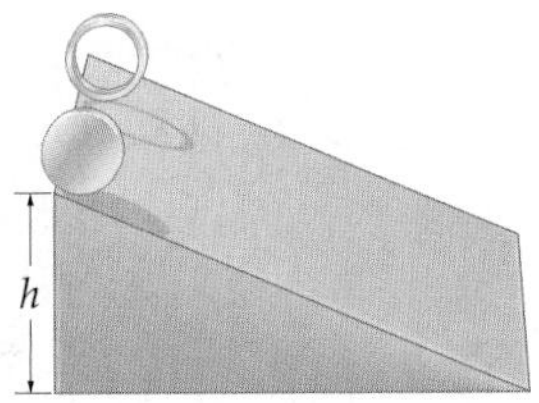

Kinetic Energy of Rolling Motion

The kinetic energy of an object that rolls without slipping is

$$K = \tfrac{1}{2}mv^2 + \tfrac{1}{2}I\omega^2 \qquad \textbf{10–19}$$

Since rolling without slipping implies that $\omega = v/r$, the kinetic energy can be written as follows:

$$K = \tfrac{1}{2}mv^2 + \tfrac{1}{2}I\left(\frac{v}{r}\right)^2 = \tfrac{1}{2}mv^2\left(1 + \frac{I}{mr^2}\right) \qquad \textbf{10–20}$$

Energy Conservation

Conservation of mechanical energy is a statement that the initial kinetic plus potential energy is equal to the final kinetic plus potential energy: $K_i + U_i = K_f + U_f$. By taking into account both rotational and translational kinetic energy, energy conservation can be applied in the same way as was done for linear systems.

Problem-Solving Summary

Type of Problem	Relevant Physical Concepts	Related Examples
Apply rotational kinematics with constant angular acceleration.	Rotational kinematics is completely analogous to the linear kinematics studied in Chapter 2. Angular problems are solved in the same way as the corresponding linear problems.	Example 10–1, Example 10–2 Active Example 10–1
Relate linear and angular motion.	Linear speed and angular speed are related by $v = r\omega$. Similarly, linear and angular acceleration are related by $a = r\alpha$. The centripetal acceleration of an object in circular motion is $a_{cp} = r\omega^2$.	Example 10–3 Active Example 10–2
Find the rotational kinetic energy of an object.	Rotational kinetic energy is given by $K = \frac{1}{2}I\omega^2$. The moment of inertia, I, plays the same role in rotational motion as the mass in linear motion.	Example 10–4, Example 10–5
Apply energy conservation to a rotational system.	To use energy conservation in a system with rotational motion, it is necessary to include the kinetic energy of rotation as one of the forms of energy.	Example 10–6 Active Example 10–3

Conceptual Questions

(Answers to odd-numbered Conceptual Questions can be found in the back of the book.)

1. A rigid object rotates about a fixed axis. Do all points on the object have the same angular speed? Do all points on the object have the same linear speed? Explain.
2. Can you drive your car in such a way that your tangential acceleration is zero while at the same time your centripetal acceleration is nonzero? Give an example if your answer is yes, state why not if your answer is no.
3. Can you drive your car in such a way that your tangential acceleration is nonzero while at the same time your centripetal acceleration is zero? Give an example if your answer is yes, state why not if your answer is no.
4. The fact that the Earth rotates gives people in New York a linear speed of about 750 mi/h. Where should you stand on the Earth to have the smallest possible linear speed?
5. At the local carnival you and a friend decide to take a ride on the Ferris wheel. As the wheel rotates with a constant angular speed, your friend poses the following questions: **(a)** Is my linear velocity constant? **(b)** Is my linear speed constant? **(c)** Is the magnitude of my centripetal acceleration constant? **(d)** Is the direction of my centripetal acceleration constant? What is your answer to each of these questions?
6. Why should changing the axis of rotation of an object change its moment of inertia, given that its shape and mass remain the same?
7. Give a common, everyday example for each of the following: **(a)** An object that has zero rotational kinetic energy but nonzero translational kinetic energy. **(b)** An object that has zero translation kinetic energy but nonzero rotational kinetic energy. **(c)** An object that has nonzero rotational and translational kinetic energies.
8. Two spheres have identical radii and masses. How might you tell which of these spheres is hollow and which is solid?
9. At the grocery store you pick up a can of beef broth and a can of chunky beef stew. The cans are identical in diameter and weight. Rolling both of them down the aisle with the same initial speed, you notice that the can of chunky stew rolls much farther than the can of broth. Why?
10. Suppose we change the race shown in Conceptual Checkpoint 10–4 so that a hoop of radius R and mass M races a hoop of radius R and mass $2M$. **(a)** Does the hoop with mass M finish before, after, or at the same time as the hoop with mass $2M$? Explain. **(b)** How would your answer to part (a) change if the hoops had different radii? Explain.

Conceptual Exercises

(Answers to odd-numbered Conceptual Exercises can be found in the back of the book.)

1. An object at rest begins to rotate with a constant angular acceleration. If this object rotates through an angle θ in the time t, through what angle did it rotate in the time $t/2$?
2. An object at rest begins to rotate with a constant angular acceleration. If the angular speed of the object is ω after the time t, what was its angular speed at the time $t/2$?
3. Two children, Jason and Betsy, ride on the same merry-go-round. Jason is a distance R from the axis of rotation; Betsy is a distance $2R$ from the axis. Is the rotational period of Jason greater than, less than, or equal to the rotational period of Betsy?
4. Referring to the previous Exercise, what is **(a)** the ratio of Jason's angular speed to Betsy's angular speed, **(b)** the ratio of Jason's linear speed to Betsy's linear speed, and **(c)** the ratio of Jason's centripetal acceleration to Betsy's centripetal acceleration?
5. **A Tall Building** The world's tallest building is the Taipei 101 Tower in Taiwan, which rises to a height of 508 m (1667 ft). When standing on the top floor of the building, is your angular speed due to the Earth's rotation greater than, less than, or the same as when you stand on the ground floor?
6. When you stand on the observation deck of the Empire State Building in New York, is your linear speed due to the Earth's rotation greater than, less than, or the same as when you were waiting for the elevators on the ground floor?
7. As you drive down the highway, the top of your tires are moving with a speed v. What is the reading on your speedometer?
8. The minute and hour hands of a clock have a common axis of rotation and equal mass. The minute hand is long, thin, and uniform; the hour hand is short, thick, and uniform. Is the moment of inertia of the minute hand greater than, less than, or equal to the moment of inertia of the hour hand?
9. Tons of dust and small particles rain down onto the Earth from space everyday. As a result, does the Earth's moment of inertia increase, decrease, or stay the same?
10. Place two quarters on a table with their rims touching, as shown in **Figure 10–18**. While holding one quarter fixed, roll the other one—without slipping—around the circumference of the fixed quarter until it has completed one round trip. How many revolutions has the rolling quarter made about its center?

▲ **FIGURE 10–18** Conceptual Exercise 10

11. **Hard-Boiled versus Raw Eggs** One way to tell whether an egg is raw or hard boiled—without cracking it open—is to place it on a kitchen counter and give it a spin. If you do this to two eggs, one raw the other hard boiled, you will find that one spins considerably longer than the other. Is the raw egg the one that spins a long time, or the one that stops spinning in a short time?
12. When the Hoover Dam was completed and the reservoir behind it filled with water, did the moment of inertia of the Earth increase, decrease, or stay the same?
13. Suppose a bicycle wheel is rotated about an axis through its rim and parallel to its axle. Is its moment of inertia greater than, less than, or the same as when it rotates about its axle?

14. The L-shaped object in **Figure 10–19** can be rotated in one of the following three ways: case 1, rotation about the x axis; case 2, rotation about the y axis; and case 3, rotation about the z axis (which passes through the origin perpendicular to the plane of the figure.) Rank these three cases in order of increasing moment of inertia. Indicate ties where appropriate.

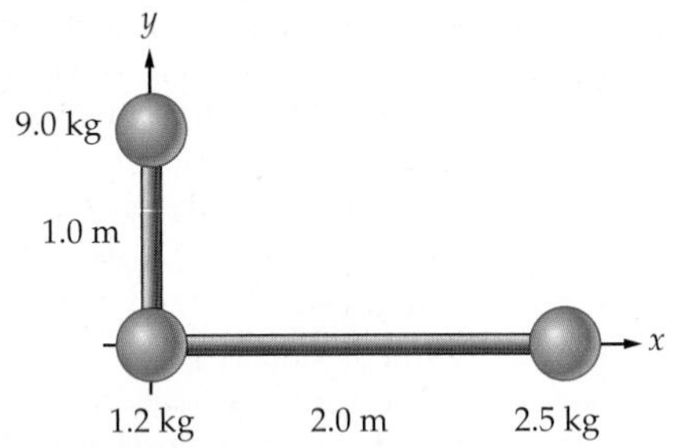

▲ **FIGURE 10–19** Conceptual Exercise 14

15. Consider the physical situation shown in Conceptual Checkpoint 10–5. Suppose this time a ball is released from rest on the frictionless surface. When the ball comes to rest on the no-slip surface, is its height greater than, less than, or the same as the height from which it was released?

16. Suppose we change the race shown in Conceptual Checkpoint 10–4 to a race between three different disks. Let disk 1 have a mass M and a radius R, disk 2 have a mass M and a radius $2R$, and disk 3 have a mass $2M$ and a radius R. Rank the three disks in the order in which they finish the race. Indicate ties where appropriate.

17. The object shown in **Figure 10–20** can be rotated in three different ways: case 1, rotation about the x axis; case 2, rotation about the y axis; and case 3, rotation about the z axis. Rank these three cases in order of increasing moment of inertia. Indicate ties where appropriate.

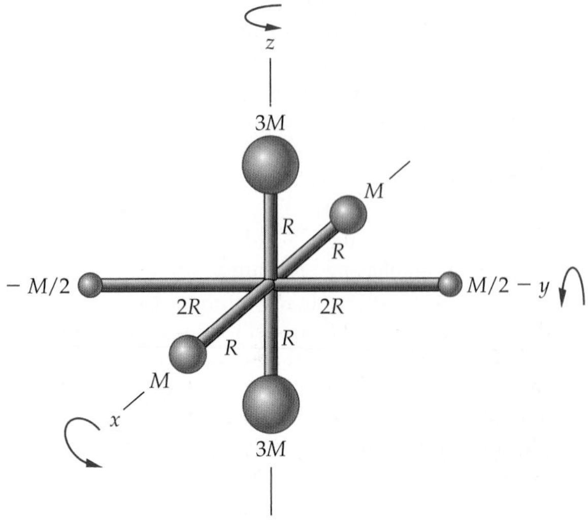

▲ **FIGURE 10–20** Conceptual Exercise 17

Problems

Note: **IP** *denotes an integrated conceptual/quantitative problem.* **BIO** *identifies problems of biological or medical interest. Red bullets (•, ••, •••) are used to indicate the level of difficulty of each problem.*

Section 10–1 Angular Position, Velocity, and Acceleration

1. • The following angles are given in degrees. Convert them to radians: 30°, 45°, 90°, 180°.

2. • The following angles are given in radians. Convert them to degrees: $\pi/6$, 0.70π, 1.5π, 5π.

3. • Find the angular speed of **(a)** the minute hand and **(b)** the hour hand of the famous clock in London, England, that rings the bell known as Big Ben.

4. • Express the angular velocity of the second hand on a clock in the following units: **(a)** rev/hr and **(b)** deg/min and **(c)** rad/s.

5. • Rank the following in order of increasing angular speed: an automobile tire rotating at 2.00×10^3 deg/s, an electric drill rotating at 400.0 rev/min, and an airplane propeller rotating at 40.0 rad/s.

6. • A spot of paint on a bicycle tire moves in a circular path of radius 0.33 m. When the spot has traveled a linear distance of 1.75 m, through what angle has the tire rotated? Give your answer in radians.

7. • What is the angular speed (in rev/min) of the Earth as it orbits about the Sun?

8. • Find the angular speed of the Earth as it spins about its axis. Give your result in rad/s.

9. • **The Crab Nebula** One of the most studied objects in the night sky is the Crab nebula, the remains of a supernova explosion observed by the Chinese in 1054. In 1968 it was discovered that a pulsar—a rapidly rotating neutron star that emits a pulse of radio waves with each revolution—lies near the center of the Crab nebula. The period of this pulsar is 33 ms. What is the angular speed (in rad/s) of the Crab nebula pulsar?

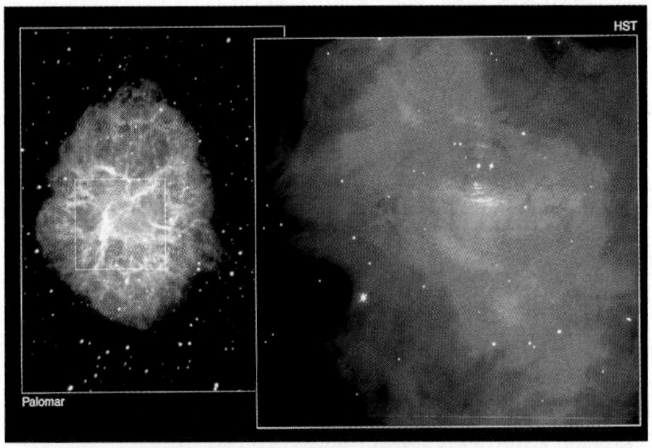

The photo at left is a true-color visible light image of the Crab nebula. In the false-color breakout, the pulsar can be seen as the left member of the pair of stars just above the center of the frame. (Problems 9 and 90)

10. •• **IP** A 3.5-inch floppy disk in a computer rotates with a period of 2.00×10^{-1} s. What is **(a)** the angular speed of the disk and **(b)** the linear speed of a point on the rim of the disk? **(c)** Does a point near the center of the disk have an angular speed that is greater than, less than, or the same as the angular speed found in part (a)? Explain. (Note: A 3.5-inch floppy disk is 3.5 inches in diameter.)

11. •• **IP** The angle an airplane propeller makes with the horizontal as a function of time is given by $\theta = (125\text{ rad/s})t +$

$(42.5\ \text{rad/s}^2)t^2$. **(a)** Estimate the instantaneous angular velocity at $t = 0.00$ s by calculating the average angular velocity from $t = 0.00$ s to $t = 0.010$ s. **(b)** Estimate the instantaneous angular velocity at $t = 1.000$ s by calculating the average angular velocity from $t = 1.000$ s to $t = 1.010$ s. **(c)** Estimate the instantaneous angular velocity at $t = 2.000$ s by calculating the average angular velocity from $t = 2.000$ s to $t = 2.010$ s. **(d)** Based on your results from parts (a), (b), and (c), is the angular acceleration of the propeller positive, negative, or zero? Explain. **(e)** Calculate the average angular acceleration from $t = 0.00$ s to $t = 1.00$ s and from $t = 1.00$ s to $t = 2.00$ s.

Section 10–2 Rotational Kinematics

12. • In Active Example 10–1, how long does it take before the angular velocity of the pulley is equal to -5.0 rad/s?

13. • In Example 10–2, through what angle has the wheel turned when its angular speed is 2.25 rad/s?

14. • The angular speed of a propeller on a boat increases with constant acceleration from 12 rad/s to 26 rad/s in 2.5 revolutions. What is the acceleration of the propeller?

15. • The angular speed of a propeller on a boat increases with constant acceleration from 12 rad/s to 26 rad/s in 2.5 seconds. Through what angle did the propeller turn during this time?

16. •• After fixing a flat tire on a bicycle you give the wheel a spin. **(a)** If its initial angular speed was 6.35 rad/s and it rotated 14.2 revolutions before coming to rest, what was its average angular acceleration? **(b)** For what length of time did the wheel rotate?

17. •• **IP** A ceiling fan is rotating at 0.90 rev/s. When turned off, it slows uniformly to a stop in 2.2 min. **(a)** How many revolutions does the fan make in this time? **(b)** Using the result from part (a), find the number of revolutions the fan must make for its speed to decrease from 0.90 rev/s to 0.45 rev/s.

18. •• A discus thrower starts from rest and begins to rotate with a constant angular acceleration of $2.2\ \text{rad/s}^2$. **(a)** How many revolutions does it take for the discus thrower's angular speed to reach 6.3 rad/s? **(b)** How much time does this take?

19. •• **Half Time** At 3:00 the hour hand and the minute hand of a clock point in directions that are 90.0° apart. What is the first time after 3:00 that the angle between the two hands has decreased by half to 45.0°?

When the little hand is on the 3 and the big hand is on the 12 . . . . (Problem 19)

20. •• **BIO** A centrifuge is a common laboratory instrument that separates components of differing densities in solution. This is accomplished by spinning a sample around in a circle with a large angular speed. Suppose that after a centrifuge in a medical laboratory is turned off, it continues to rotate with a constant angular deceleration for 10.2 s before coming to rest. **(a)** If its initial angular speed was 3850 rpm, what is the magnitude of its angular deceleration? **(b)** How many revolutions did the centrifuge complete after being turned off?

21. •• **The Slowing Earth** The Earth's rate of rotation is constantly decreasing, causing the day to increase in duration. In the year 2006 the Earth takes about 0.840 s longer to complete 365 revolutions than it did in the year 1906. What is the average angular acceleration of the Earth? Give your answer in rad/s^2.

22. •• **IP** A compact disk (CD) speeds up uniformly from rest to 310 rpm in 3.0 s. **(a)** Describe a strategy that allows you to calculate the number of revolutions the CD makes in this time. **(b)** Use your strategy to find the number of revolutions.

23. •• When a carpenter shuts off his circular saw, the 10.0-inch diameter blade slows from 4830 rpm to zero in 2.00 s. **(a)** What is the angular acceleration of the blade? **(b)** What is the distance traveled by a point on the rim of the blade during the deceleration? **(c)** What is the magnitude of the net displacement of a point on the rim of the blade during the deceleration?

24. •• **The World's Fastest Turbine** The drill used by most dentists today is powered by a small air-turbine that can operate at angular speeds of 350,000 rpm. These drills, along with ultrasonic dental drills, are the fastest turbines in the world—far exceeding the angular speeds of jet engines. Suppose a drill starts at rest and comes up to operating speed in 2.1 s. **(a)** Find the angular acceleration produced by the drill, assuming it to be constant. **(b)** How many revolutions does the drill bit make as it comes up to speed?

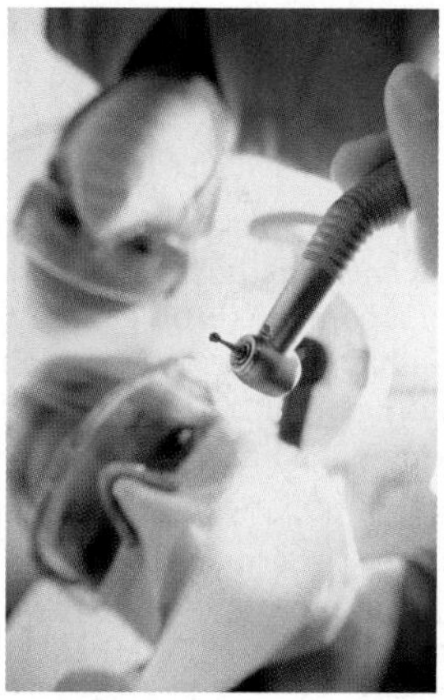

An air-turbine dentist drill—faster than a jet engine. (Problem 24)

Section 10–3 Connections Between Linear and Rotational Quantities

25. • The hour hand on a certain clock is 8.2 cm long. Find the tangential speed of the tip of this hand.

26. • Two children ride on the merry-go-round shown in Conceptual Checkpoint 10–1. Child 1 is 2.0 m from the axis of rotation, and child 2 is 1.5 m from the axis. If the merry-go-round completes one revolution every 4.5 s, find **(a)** the angular speed and **(b)** the linear speed of each child.

27. • The outer edge of a rotating Frisbee with a diameter of 29 cm has a linear speed of 3.4 m/s. What is the angular speed of the Frisbee?

28. • A carousel at the local carnival rotates once every 45 seconds. **(a)** What is the linear speed of an outer horse on the carousel, which is 2.75 m from the axis of rotation? **(b)** What is the linear speed of an inner horse that is 1.75 m from the axis of rotation?

29. •• **IP** Jeff of the Jungle swings on a vine that is 7.20 m long (**Figure 10–21**). At the bottom of the swing, just before hitting the tree, Jeff's linear speed is 8.50 m/s. **(a)** Find Jeff's angular

▲ **FIGURE 10–21** Problems 29 and 30

speed at this time. **(b)** What centripetal acceleration does Jeff experience at the bottom of his swing? **(c)** What exerts the force that is responsible for Jeff's centripetal acceleration?

30\. •• Suppose, in Problem 29, that at some point in his swing Jeff of the Jungle has an angular speed of 0.850 rad/s and an angular acceleration of 0.620 rad/s^2. Find the magnitude of his centripetal, tangential, and total accelerations, and the angle his total acceleration makes with respect to the tangential direction of motion.

31\. •• A compact disk, which has a diameter of 12.0 cm, speeds up uniformly from zero to 4.00 rev/s in 3.00 s. What is the tangential acceleration of a point on the outer rim of the disk at the moment when its angular speed is **(a)** 2.00 rev/s and **(b)** 3.00 rev/s?

32\. •• **IP** When a compact disk with a 12.0-cm diameter is rotating at 5.25 rad/s, what is **(a)** the linear speed and **(b)** the centripetal acceleration of a point on its outer rim? **(c)** Consider a point on the CD that is halfway between its center and its outer rim. Without repeating all of the calculations required for parts (a) and (b), determine the linear speed and the centripetal acceleration of this point.

33\. •• **IP** As Tony the fisherman reels in a "big one," he turns the spool on his fishing reel at the rate of 3.0 complete revolutions every second (**Figure 10–22**). **(a)** If the radius of the reel is 3.7 cm, what is the linear speed of the fishing line as it is reeled in? **(b)** How would your answer to part (a) change if the radius of the reel were doubled?

▲ **FIGURE 10–22** Problem 33

34\. •• A Ferris wheel with a radius of 9.5 m rotates at a constant rate, completing one revolution every 32 s. Find the direction and magnitude of a passenger's acceleration when **(a)** at the top and **(b)** at the bottom of the wheel.

35\. •• Suppose the Ferris wheel in the previous problem begins to decelerate at the rate of 0.22 rad/s^2 when the passenger is at the top of the wheel. Find the direction and magnitude of the passenger's acceleration at that time.

36\. •• **IP** A person swings a 0.52-kg tether ball tied to a 4.5-m rope in an approximately horizontal circle. **(a)** If the maximum tension the rope can withstand before breaking is 11 N, what is the maximum angular speed of the ball? **(b)** If the rope is shortened, does the maximum angular speed found in part (a) increase, decrease, or stay the same? Explain.

37\. •• To polish a filling, a dentist attaches a sanding disk with a radius of 3.20 mm to the drill. **(a)** When the drill is operated at 2.15×10^4 rad/s, what is the tangential speed of the rim of the disk? **(b)** What period of rotation must the disk have if the tangential speed of its rim is to be 275 m/s?

38\. •• In the previous problem, suppose the disk has an angular acceleration of 232 rad/s^2 when its angular speed is 640 rad/s. Find both the tangential and centripetal acceleration of a point on the rim of the disk.

39\. •• **The Bohr Atom** The Bohr model of the hydrogen atom pictures the electron as a tiny particle moving in a circular orbit about a stationary proton. In the lowest-energy orbit the distance from the proton to the electron is 5.29×10^{-11} m, and the linear speed of the electron is 2.18×10^6 m/s. **(a)** What is the angular speed of the electron? **(b)** How many orbits about the proton does it make each second? **(c)** What is the electron's centripetal acceleration?

40\. ••• A wheel of radius R starts from rest and accelerates with a constant angular acceleration α about a fixed axis. At what time t will the centripetal and tangential accelerations of a point on the rim have the same magnitude?

Section 10–4 Rolling Motion

41\. • The tires on a car have a radius of 31 cm. What is the angular speed of these tires when the car is driven at 12 m/s?

42\. • A child pedals a tricycle, giving the driving wheel an angular speed of 0.373 rev/s (**Figure 10–23**). If the radius of the wheel is 0.260 m, what is the child's linear speed?

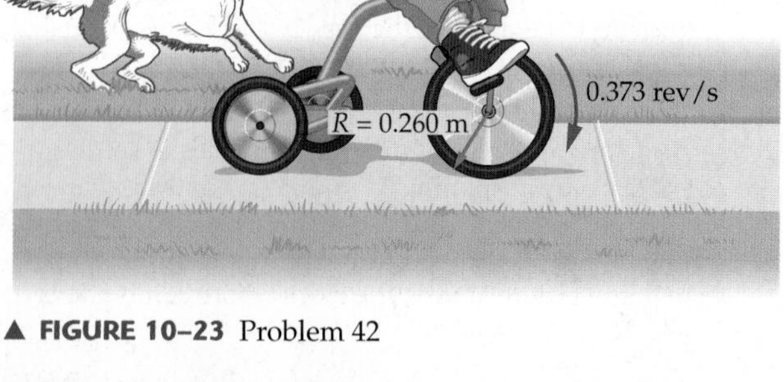

▲ **FIGURE 10–23** Problem 42

43. • A soccer ball, which has a circumference of 70.0 cm, rolls 12.0 yards in 3.45 s. What was the average angular speed of the ball during this time?

44. •• As you drive down the road at 17 m/s, you press on the gas pedal and speed up with a uniform acceleration of 1.12 m/s^2 for 0.65 s. If the tires on your car have a radius of 33 cm, what is their angular displacement during this period of acceleration?

45. •• **IP** A bicycle coasts downhill and accelerates from rest to a linear speed of 8.90 m/s in 12.2 s. **(a)** If the bicycle's tires have a radius of 36.0 cm, what is their angular acceleration? **(b)** If the radius of the tires had been smaller, would their angular acceleration be greater than or less than the result found in part (a)?

Section 10–5 Rotational Kinetic Energy and the Moment of Inertia

46. • The moment of inertia of a 0.98-kg bicycle wheel rotating about its center is $0.13 \text{ kg} \cdot \text{m}^2$. What is the radius of this wheel, assuming the weight of the spokes can be ignored?

47. • What is the kinetic energy of the grindstone in Example 10–4 if it completes one revolution every 4.20 s?

48. • An electric fan spinning with an angular speed of 12 rad/s has a kinetic energy of 4.1 J. What is the moment of inertia of the fan?

49. • Repeat Example 10–5 for the case of a rolling hoop of the same mass and radius.

50. •• **IP** A 13-g CD with a radius of 6.0 cm rotates with an angular speed of 32 rad/s. **(a)** What is its kinetic energy? **(b)** What angular speed must the CD have if its kinetic energy is to be doubled?

51. •• When a pitcher throws a curve ball, the ball is given a fairly rapid spin. If a 0.15-kg baseball with a radius of 3.7 cm is thrown with a linear speed of 46 m/s and an angular speed of 41 rad/s, how much of its kinetic energy is translational and how much is rotational? Assume the ball is a uniform, solid sphere.

52. •• **IP** A basketball rolls along the floor with a constant linear speed v. **(a)** Find the fraction of its total kinetic energy that is in the form of rotational kinetic energy about the center of the ball. **(b)** If the linear speed of the ball is doubled to $2v$, does your answer to part (a) increase, decrease, or stay the same? Explain.

53. •• Find the rate at which the rotational kinetic energy of the Earth is decreasing. The Earth has a moment of inertia of $0.331 M_E R_E^2$, where $R_E = 6.38 \times 10^6$ m and $M_E = 5.97 \times 10^{24}$ kg, and its rotational period increases by 2.3 ms with each passing century. Give your answer in watts.

54. •• A lawn mower has a flat, rod-shaped steel blade that rotates about its center. The mass of the blade is 0.58 kg and its length is 0.56 m. **(a)** What is the rotational energy of the blade at its operating angular speed of 3500 rpm? **(b)** If all of the rotational kinetic energy of the blade could be converted to gravitational potential energy, to what height would the blade rise?

Section 10–6 Conservation of Energy

55. • Suppose the block in Example 10–6 has a mass of 2.1 kg and an initial upward speed of 0.33 m/s. Find the moment of inertia of the wheel if its radius is 8.0 cm and the block rises to a height of 7.4 cm before momentarily coming to rest.

56. • Through what height must the yo-yo in Active Example 10–3 fall for its linear speed to be 0.65 m/s?

57. •• Calculate the speeds of **(a)** the disk and **(b)** the hoop at the bottom of the inclined plane in Conceptual Checkpoint 10–4 if the height of the incline is 0.82 m.

58. •• **IP Atwood's Machine** The two masses ($m_1 = 5.0$ kg and $m_2 = 3.0$ kg) in the Atwood's machine shown in **Figure 10–24** are released from rest, with m_1 at a height of 0.75 m above the floor. When m_1 hits the ground its speed is 1.8 m/s. Assuming that the pulley is a uniform disk with a radius of 12 cm, **(a)** outline a strategy that allows you to find the mass of the pulley. **(b)** Implement the strategy given in part (a) and determine the pulley's mass.

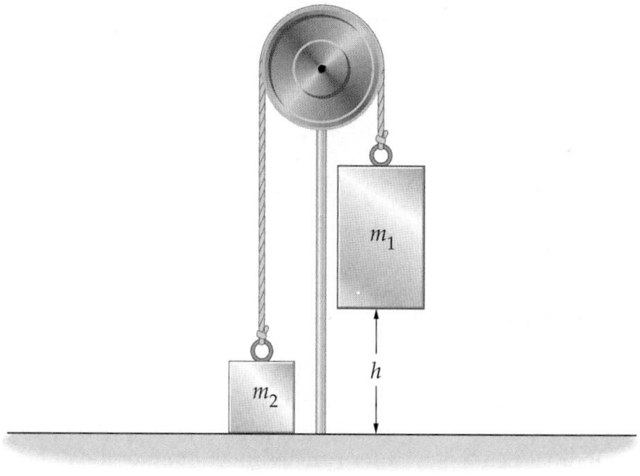

▲ **FIGURE 10–24** Problem 58

59. •• In Conceptual Checkpoint 10–5, assume the ball is a solid sphere of radius 2.8 cm and mass 0.10 kg. If the ball is released from rest at a height of 0.75 m above the bottom of the track on the no-slip side, **(a)** what is its angular speed when it is on the frictionless side of the track? **(b)** How high does the ball rise on the frictionless side?

60. •• **IP** After you pick up a spare, your bowling ball rolls without slipping back toward the ball rack with a linear speed of 2.85 m/s (**Figure 10–25**). To reach the rack, the ball rolls up a ramp that rises through a vertical distance of 0.53 m. **(a)** What is the linear speed of the ball when it reaches the top of the ramp? **(b)** If the radius of the ball were increased, would the speed found in part (a) increase, decrease, or stay the same? Explain.

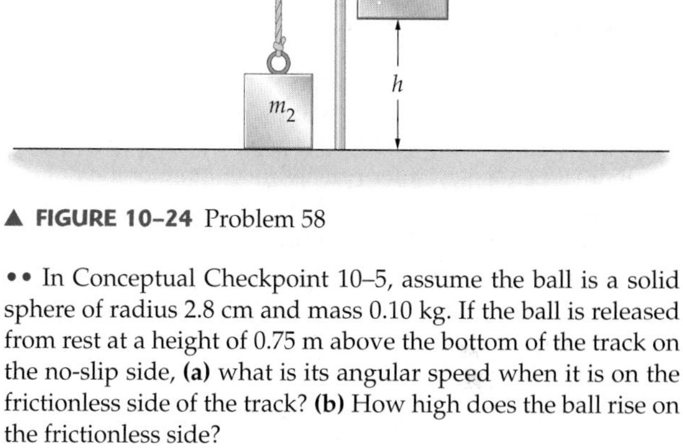

▲ **FIGURE 10–25** Problem 60

61. •• **IP** A 1.3-kg block is tied to a string that is wrapped around the rim of a pulley of radius 7.2 cm. The block is released from rest. **(a)** Assuming the pulley is a uniform disk with a mass of 0.31 kg, find the speed of the block after it has fallen through a height of 0.50 m. **(b)** If a small lead weight is attached near the rim of the pulley and this experiment is repeated, will the speed of the block increase, decrease, or stay the same? Explain.

62. •• After doing some exercises on the floor, you are lying on your back with one leg pointing straight up. If you allow your leg to fall freely until it hits the floor (**Figure 10–26**), what is the tangential speed of your foot just before it lands? Assume the leg can be treated as a uniform rod 0.95-m long that pivots freely about the hip.

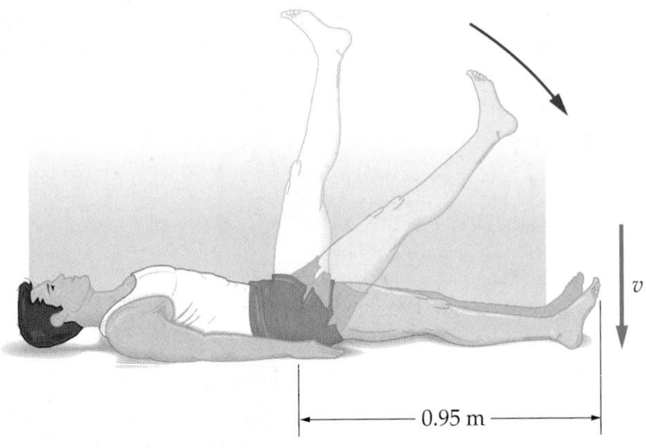

▲ **FIGURE 10–26** Problem 62

63\. ••• A 2.0-kg solid cylinder (radius = 0.10 m, length = 0.50 m) is released from rest at the top of a ramp and allowed to roll without slipping. The ramp is 0.75 m high and 5.0 m long. When the cylinder reaches the bottom of the ramp, what is **(a)** its total kinetic energy, **(b)** its rotational kinetic energy, and **(c)** its translational kinetic energy?

64\. ••• A 2.0-kg solid sphere (radius = 0.10 m) is released from rest at the top of a ramp and allowed to roll without slipping. The ramp is 0.75 m high and 5.3 m long. When the sphere reaches the bottom of the ramp, what is **(a)** its total kinetic energy, **(b)** its rotational kinetic energy, and **(c)** its translational kinetic energy?

General Problems

65\. • **Weightless on the Equator** In Quito, Ecuador, near the equator, you weigh about half a pound less than in Barrow, Alaska, near the pole. Find the rotational period of the Earth that would make you feel weightless at the equator. (With this rotational period, your centripetal acceleration would be equal to the acceleration due to gravity, g.)

66\. • BIO **Human-Powered Centrifuge** One method NASA is considering to provide astronauts on long space flights with exercise and "gravitational forces" simultaneously is the human-powered centrifuge. This device is basically a rotating, circular platform on which two astronauts lie supine along a diameter, head-to-head at the center, with their feet at opposite rims, as shown the accompanying photo. The radius of the platform is 6.25 ft. What angular speed (in rpm) must the platform have to give a centripetal acceleration of 5.0 gs at the rim?

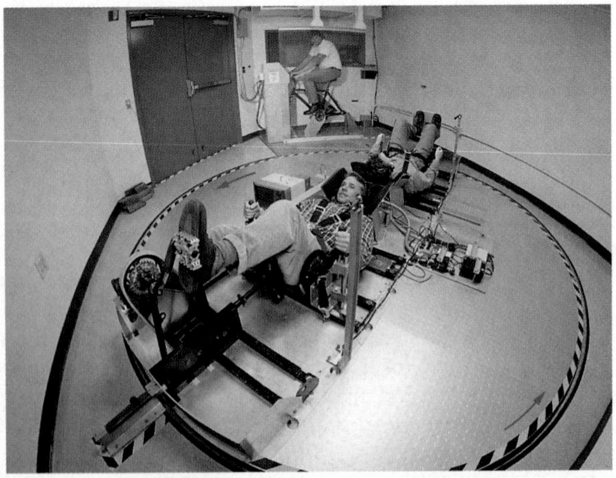

Human-powered centrifuge (Problem 66)

67\. • A diver completes $2\frac{1}{2}$ somersaults during a 2.1-s dive. What was the diver's average angular speed during the dive?

68\. • What linear speed must a 0.050-kg hula hoop have if its total kinetic energy is to be 0.10 J? Assume the hoop rolls on the ground without slipping.

69\. • BIO **Losing Consciousness** A pilot performing a horizontal turn will lose consciousness if she experiences a centripetal acceleration greater than 7.00 times the acceleration of gravity. What is the minimum radius turn she can make without losing consciousness if her plane is flying with a constant speed of 245 m/s?

70\. •• The accompanying double-exposure photograph illustrates a method for determining the speed of a BB. The circular disk in the upper part of the photo rotates with a constant angular speed of 50.4 revolutions per second. A single white radial line drawn on the disk is seen in two locations in the double exposure. Below the disk are two bright images of a BB taken during the two exposures. Use the information given here and in the photo to estimate the speed of the BB.

Speeding BB and spinning wheel. (Problems 70 and 71)

71\. •• Referring to the previous problem, **(a)** estimate the linear speed of a point on the rim of the rotating disk. **(b)** By comparing the arc length between the two white lines to the distance covered by the BB, estimate the speed of the BB. **(c)** What radius must the disk have for the linear speed of a point on its rim to be the same as the speed of the BB? **(d)** Suppose a 1.0-g lump of putty is stuck to the rim of the disk. What centripetal force is required to hold the putty in place?

72\. •• IP **When the Hands Align** A mathematically inclined friend e-mails you the following instructions: "Meet me in the cafeteria the first time after 2:00 P.M. today that the hands of a clock point in the same direction." **(a)** Is the desired meeting time before, after, or equal to 2:10 P.M.? Explain. **(b)** Is the desired meeting time before, after, or equal to 2:15 P.M.? Explain. **(c)** When should you meet your friend?

73\. •• IP A diver runs horizontally off the end of a diving tower 3.0 m above the surface of the water with an initial speed of 2.8 m/s. During her fall she rotates with an average angular speed of 2.4 rad/s. **(a)** How many revolutions has she made when she hits the water? **(b)** How does your answer to part (a) depend on the diver's initial speed? Explain.

74\. •• IP A potter's wheel of radius 6.5 cm rotates with a period of 0.55 s. What is **(a)** the linear speed and **(b)** the centripetal acceleration of a small lump of clay on the rim of the wheel? **(c)** How do your answers to parts (a) and (b) change if the period of rotation is doubled?

75\. •• IP **Playing a CD** The record in an old-fashioned record player always rotates at the same angular speed. With CDs, the situation is different. For a CD to play properly, the point on the

CD where the laser beam shines must have a linear speed $v_t = 1.25$ m/s, as indicated in **Figure 10–27**. **(a)** As the CD plays from the center outward, does its angular speed increase, decrease, or stay the same? Explain. **(b)** Find the angular speed of a CD when the laser beam is 2.50 cm from its center. **(c)** Repeat part (b) for the laser beam 6.00 cm from the center. **(d)** If the CD plays for 66.5 min, and the laser beam moves from 2.50 cm to 6.00 cm during this time, what is the CD's average angular acceleration?

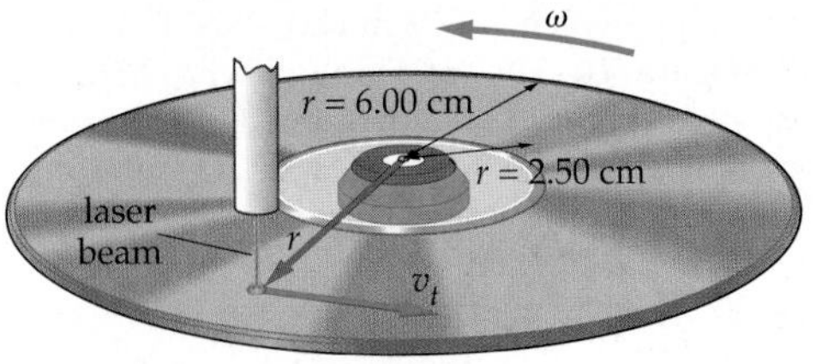

▲ **FIGURE 10–27** Problem 75

76. •• **BIO Roller Pigeons** Pigeons are bred to display a number of interesting characteristics. One breed of pigeon, the "roller," is remarkable for the fact that it does a number of backward somersaults as it drops straight down toward the ground. Suppose a roller pigeon drops from rest and free falls downward for a distance of 14 m. If the pigeon somersaults at the rate of 12 rad/s, how many revolutions has it completed by the end of its fall?

77. •• As a marble with a diameter of 1.8 cm rolls down an incline, its center moves with a linear acceleration of 3.2 m/s^2. **(a)** What is the angular acceleration of the marble? **(b)** What is the angular speed of the marble after it rolls for 1.5 s from rest?

78. •• A rubber ball with a radius of 3.2 cm rolls along the horizontal surface of a table with a constant linear speed v. When the ball rolls off the edge of the table, it falls 0.66 m to the floor below. If the ball completes 0.37 revolutions during its fall, what was its linear speed, v?

79. •• A college campus features a large fountain surrounded by a circular pool. Two students start at the northernmost point of the pool and walk slowly around it in opposite directions. **(a)** If the angular speed of the student walking in the clockwise direction (as viewed from above) is 0.042 rad/s and the angular speed of the other student is 0.027 rad/s, how long does it take before they meet? **(b)** At what angle, measured clockwise from due north, do the students meet? **(c)** If the difference in linear speed between the students is 0.23 m/s, what is the radius of the fountain?

80. •• **IP** A yo-yo moves downward until it reaches the end of its string, where it "sleeps." As it sleeps—that is, spins in place—its angular speed decreases from 35 rad/s to 25 rad/s. During this time it completes 120 revolutions. **(a)** How long did it take for the yo-yo to slow from 35 rad/s to 25 rad/s? **(b)** How long does it take for the yo-yo to slow from 25 rad/s to 15 rad/s? Assume a constant angular acceleration as the yo-yo sleeps.

81. •• **IP (a)** An automobile with tires of radius 32 cm accelerates from 0 to 45 mph in 9.1 s. Find the angular acceleration of the tires. **(b)** How does your answer to part (a) change if the radius of the tires is halved?

82. •• **IP** In Problems 63 and 64 we considered a cylinder and a solid sphere, respectively, rolling down a ramp. **(a)** Which object do you expect to have the greater speed at the bottom of the ramp? **(b)** Verify your answer to part (a) by calculating the speed of the cylinder and of the sphere when they reach the bottom of the ramp.

83. •• A centrifuge (Problem 20) with an angular speed of 6050 rpm produces a maximum centripetal acceleration equal to 6840 g (that is, 6840 times the acceleration of gravity). **(a)** What is the diameter of this centrifuge? **(b)** What force must the bottom of the sample holder exert on a 15.0-g sample under these conditions?

84. •• **A Yo-Yo with a Brain** Yomega ("The yo-yo with a brain") is constructed with a clever clutch mechanism in its axle that allows it to rotate freely and "sleep" when its angular speed is greater than a certain critical value. When the yo-yo's angular speed falls below this value, the clutch engages, causing the yo-yo to climb the string to the user's hand. If the moment of inertia of the yo-yo is 7.4×10^{-5} kg · m^2, its mass is 0.11 kg, and the string is 1.0 m long, what is the smallest angular speed that will allow the yo-yo to return to the user's hand?

A brain or just a clutch? (Problem 84)

85. •• The rotor in a centrifuge has an initial angular speed of 430 rad/s. After 8.2 s of constant angular acceleration its angular speed has increased to 550 rad/s. During this time, what was **(a)** the angular acceleration of the rotor and **(b)** the angle through which it turned?

86. •• **BIO** A honey bee has two pairs of wings that can beat 250 times a second. Estimate **(a)** the maximum angular speed of the wings and **(b)** the maximum linear speed of a wing tip.

87. •• The Sun, with Earth in tow, orbits about the center of the Milky Way galaxy at a speed of 137 miles per second, completing one revolution every 240 million years. **(a)** Find the angular speed of the Sun relative to the center of the Milky Way. **(b)** Find the distance from the Sun to the center of the Milky Way.

88. •• A person walks into a room and switches on the ceiling fan. The fan accelerates with constant angular acceleration for 15 s until it reaches its operating angular speed of 1.7 rotations/s—after that its speed remains constant as long as the switch is "on." The person stays in the room for a short time; then, 5.0 minutes after turning the fan on, she switches it off again and leaves the room. The fan now decelerates with constant angular acceleration, taking 2.4 minutes to come to rest. What is the total number of revolutions made by the fan, from the time it was turned on until the time it stopped?

89. •• **BIO Preventing Bone Loss in Space** When astronauts return from prolonged space flights, they often suffer from bone loss, resulting in brittle bones that may take weeks for their bodies to rebuild. One solution may be to expose astronauts to periods of substantial "g forces" in a centrifuge carried aboard their spaceship. To test this approach, NASA conducted a study in which four people spent 22 hours each in a compartment attached to the end of a 28-foot arm that rotated with an angular speed of 10.0 rpm. **(a)** What centripetal acceleration did these volunteers experience? Express your answer in terms of g. **(b)** What was their linear speed?

90. ••• **Angular Acceleration of the Crab Nebula** The pulsar in the Crab nebula (Problem 9) was created by a supernova explosion that was observed on Earth in A.D. 1054. Its current period of rotation (33.0 ms) is observed to be increasing by

1.26×10^{-5} seconds per year. **(a)** What is the angular acceleration of the pulsar in rad/s^2? **(b)** Assuming the angular acceleration of the pulsar to be constant, how many years will it take for the pulsar to slow to a stop? **(c)** Under the same assumption, what was the period of the pulsar when it was created?

91. ••• A thin, uniform rod of length L and mass M is pivoted about one end, as shown in **Figure 10–28**. The rod is released from rest in a horizontal position, and allowed to swing downward without friction or air resistance. When the rod is vertical, what is **(a)** its angular speed ω and **(b)** the tangential speed v_t of its free end?

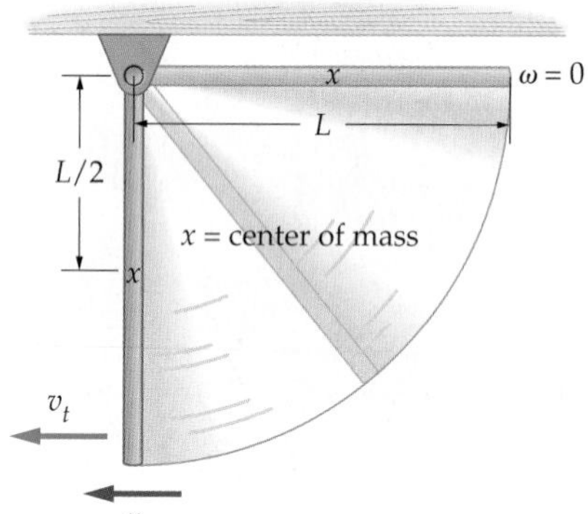

▲ **FIGURE 10–28** Problem 91

92. ••• **Center of Percussion** In the previous problem, suppose a small metal ball of mass $m = 2M$ is attached to the rod a distance d from the pivot. The rod and ball are released from rest in the horizontal position. **(a)** Show that when the rod reaches the vertical position, the speed of its tip is

$$v_t = \sqrt{3gL}\sqrt{\frac{1 + 4(d/L)}{1 + 6(d/L)^2}}$$

(b) At what finite value of d/L is the speed of the rod the same as it is for $d = 0$? (This value of d/L is the **center of percussion,** or "sweet spot," of the rod.)

93. ••• A wooden plank rests on two soup cans laid on their sides. Each can has a diameter of 6.5 cm, and the plank is 3.0 m long. Initially, one can is placed 1.0 m inward from either end of the plank, as **Figure 10–29** shows. The plank is now pulled 1.0 m to the right, and the cans roll without slipping. **(a)** How far does the center of each can move? **(b)** How many rotations does each can make?

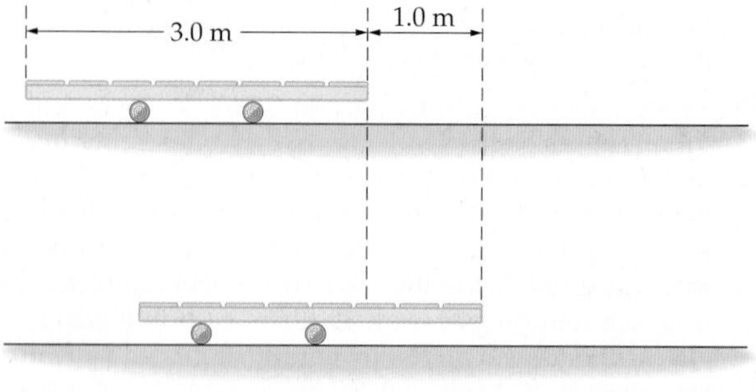

▲ **FIGURE 10–29** Problem 93

94. ••• A person rides on a 12-m diameter Ferris wheel that rotates at the constant rate of 8.1 rpm. Calculate the magnitude and direction of the force that the seat exerts on a 65-kg person when he is **(a)** at the top of the wheel, **(b)** at the bottom of the wheel, and **(c)** halfway up the wheel.

95. ••• **IP** A solid sphere with a diameter of 0.17 m is released from rest; it then rolls without slipping down a ramp, dropping through a vertical height of 0.61 m. The ball leaves the bottom of the ramp, which is 1.22 m above the floor, moving horizontally (**Figure 10–30**). **(a)** Through what horizontal distance d does the ball move before landing? **(b)** How many revolutions does the ball make during its fall? **(c)** If the ramp were to be made frictionless, would the distance d increase, decrease, or stay the same? Explain.

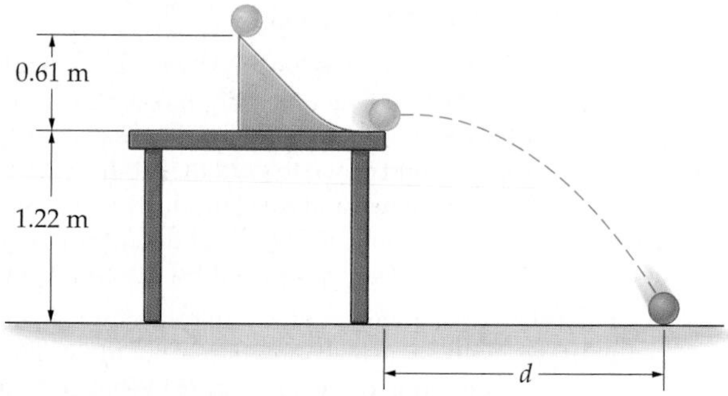

▲ **FIGURE 10–30** Problem 95

Interactive Problems

96. •• ***Referring to Conceptual Checkpoint 10–4*** Suppose we race a disk and a hollow spherical shell, like a basketball. The spherical shell has a mass M and a radius R; the disk has a mass $2M$ and a radius $2R$. **(a)** Which object wins the race? If the two objects are released at rest, and the height of the ramp is $h = 0.75$ m, find the speed of **(b)** the disk and **(c)** the spherical shell when they reach the bottom of the ramp.

97. •• ***Referring to Conceptual Checkpoint 10–4*** Consider a race between the following three objects: object 1, a disk; object 2, a solid sphere; object 3, a hollow spherical shell. All objects have the same mass and radius. **(a)** Rank the three objects in the order in which they finish the race. Indicate a tie where appropriate. **(b)** Rank the objects in order of increasing kinetic energy at the bottom of the ramp. Indicate a tie where appropriate.

98. •• ***Referring to Active Example 10–3*** **(a)** Suppose the radius of the axle the string wraps around is increased. Does the speed of the yo-yo after falling through a given height increase, decrease, or stay the same? **(b)** Find the speed of the yo-yo after falling from rest through a height $h = 0.50$ m if the radius of the axle is 0.0075 m. Everything else in Active Example 10-3 remains the same.

99. •• ***Referring to Active Example 10–3*** Suppose we use a new yo-yo that has the same mass as the original yo-yo and an axle of the same radius. The new yo-yo has a different mass distribution—most of its mass is concentrated near the rim. **(a)** Is the moment of inertia of the new yo-yo greater than, less than, or the same as that of the original yo-yo? **(b)** Find the moment of inertia of the new yo-yo if its speed after dropping from rest through a height $h = 0.50$ m is $v = 0.64$ m/s.

11 Rotational Dynamics and Static Equilibrium

11–1	Torque	316
11–2	Torque and Angular Acceleration	319
11–3	Zero Torque and Static Equilibrium	323
11–4	Center of Mass and Balance	330
11–5	Dynamic Applications of Torque	333
11–6	Angular Momentum	335
11–7	Conservation of Angular Momentum	338
11–8	Rotational Work and Power	343
*11–9	The Vector Nature of Rotational Motion	344

Equilibrium isn't just a matter of forces that add to zero—it also depends on where the forces are applied. These acrobats can maintain this precarious position by applying well-chosen forces at precise locations. Not only do they experience zero net force, but the net value of a new physical quantity—the torque—is also zero. In this chapter, we introduce the torque, show how it determines equilibrium, and discuss what happens when the net torque acting on a system is not zero.

In the previous chapter we learned how to describe uniformly accelerated rotational motion, but we did not discuss how a given angular acceleration is caused by a given force. The connection between forces and angular acceleration is the focus of this chapter.

We begin by defining a quantity that is the rotational equivalent of force. This quantity is called the *torque*. Although torque may not be as familiar a term as force, your muscles are exerting torques on your body at this very moment. In fact, every time you raise an arm, extend a finger, or stretch a leg, you exert torques to carry out these motions. Thus, our ability to move from place to place, or to hold our body still, is intimately related to our ability to exert precisely controlled torques on our limbs.

We also introduce the notion of *angular momentum* in this chapter and show that it is related to torque in essentially the same way that linear momentum is related to force. As a result, it follows that angular momentum is conserved when the net external torque acting on a system is zero. Thus, conservation of angular momentum joins conservation of energy and conservation of linear momentum as one of the fundamental principles on which all physics is based.

▲ The long handle of this wrench enables the user to produce a large torque without having to exert a very great force.

11–1 Torque

Suppose you want to loosen a nut by rotating it counterclockwise with a wrench. If you have ever used a wrench in this way, you probably know that the nut is more likely to turn if you apply your force as far from the nut as possible, as indicated in **Figure 11–1 (a)**. Applying a force near the nut would not be very effective—you could still get the nut to turn, but it would require considerably more effort! Similarly, it is much easier to open a revolving door if you push far from the axis of rotation, as indicated in **Figure 11–1 (b)**. Clearly, then, the tendency for a force to cause a rotation increases with the distance, r, from the axis of rotation to the force. As a result, it is useful to define a quantity called the **torque**, τ, that takes into account both the magnitude of the force, F, *and* the distance from the axis of rotation, r:

Definition of Torque, τ, for a Tangential Force

$$\tau = rF \qquad 11\text{–}1$$

SI unit: N · m

Note that the torque increases with both the force and the distance.

▶ **FIGURE 11–1 Applying a torque**
(a) When a wrench is used to loosen a nut, less force is required if it is applied far from the nut. **(b)** Similarly, less force is required to open a revolving door if it is applied far from the axis of rotation.

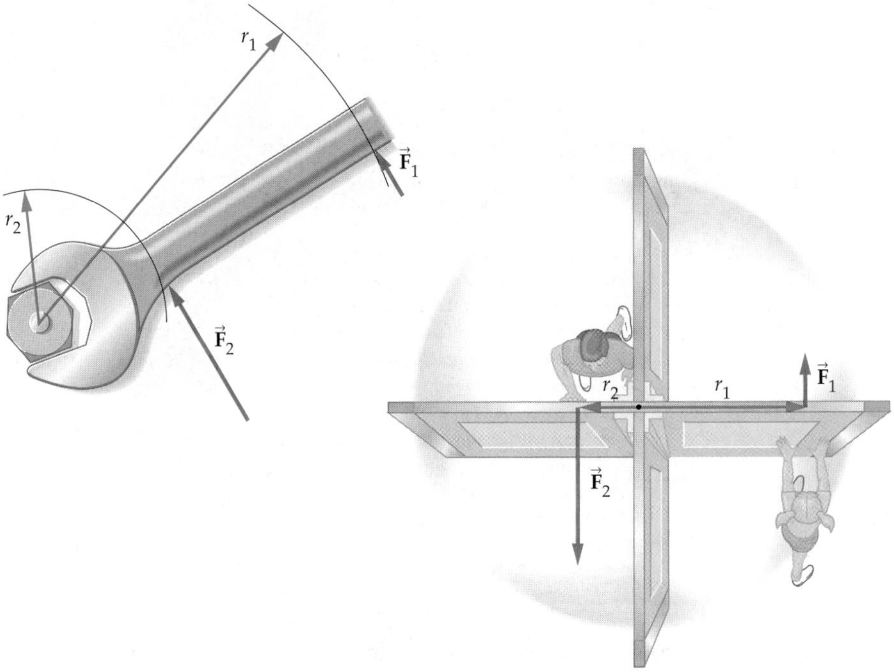

Equation 11–1 is valid only when the applied force is *tangential* to a circle of radius r centered on the axis of rotation, as indicated in Figure 11–1. The more general case is considered later in this section. First, we use Equation 11–1 to determine how much force is needed to open a swinging door, depending on where we apply the force.

EXERCISE 11–1

To open the door in Figure 11–1 (b) a tangential force F is applied at a distance r from the axis of rotation. If the minimum torque required to open the door is 3.1 N · m, what force must be applied if r is **(a)** 0.94 m or **(b)** 0.35 m?

Solution

(a) Setting $\tau = r_1F_1 = 3.1\ \text{N}\cdot\text{m}$, we find that the required force is

$$F_1 = \frac{\tau}{r_1} = \frac{3.1\ \text{N}\cdot\text{m}}{0.94\ \text{m}} = 3.3\ \text{N}$$

PROBLEM-SOLVING NOTES

The units of torque

Note that the units of torque are $N \cdot$ m, the same as the units of work. Though their units are the same, torque, τ, and work, W, represent different physical quantities and should not be confused with one another.

(b) Repeat the calculation, this time with $r_2 = 0.35$ m:

$$F_2 = \frac{\tau}{r_2} = \frac{3.1\ \text{N}\cdot\text{m}}{0.35\ \text{m}} = 8.9\ \text{N}$$

As expected, the required force is greater when it is applied closer to the hinges.

To this point we have considered tangential forces only. What happens if you exert a force in a direction that is not tangential? Suppose, for example, that you pull on a playground merry-go-round in a direction that is radial—that is, along a line that extends through the axis of rotation—as in **Figure 11–2 (a)**. In this case, your force has no tendency to cause a rotation. Instead, the axle of the merry-go-round simply exerts an equal and opposite force, and the merry-go-round remains at rest. Similarly, if you were to push or pull in a radial direction on a swinging door it would not rotate. We conclude that a *radial force produces zero torque*.

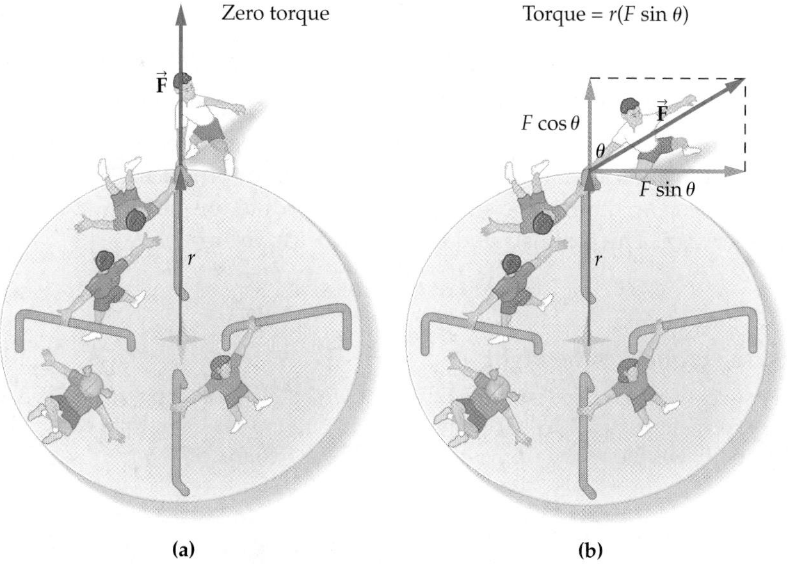

◀ **FIGURE 11–2 Only the tangential component of a force causes a torque**
(a) A radial force causes no rotation. In this case, the force $\vec{F}$ is opposed by an equal and opposite force exerted by the axle of the merry-go-round. The merry-go-round does not rotate. **(b)** A force applied at an angle θ with respect to the radial direction. The radial component of this force, $F \cos\theta$, causes no rotation; the tangential component, $F \sin\theta$, can cause a rotation.

On the other hand, what if your force is at an angle θ relative to a radial line, as shown in **Figure 11–2 (b)**? To analyze this case, we first resolve the force vector $\vec{F}$ into radial and tangential components. Referring to the figure, we see that the radial component has a magnitude of $F \cos\theta$, and the tangential component has a magnitude of $F \sin\theta$. Because it is the tangential component alone that causes rotation, we define the torque to have a magnitude of $r(F \sin\theta)$. That is,

General Definition of Torque, τ

$$\tau = r(F \sin\theta) \qquad \text{11–2}$$

SI units: N · m

(More generally, the torque can be defined as the **cross product** between the vectors $\vec{r}$ and $\vec{F}$; that is, $\vec{\tau} = \vec{r} \times \vec{F}$. The cross product is discussed in detail in Appendix A.)

As a quick check, note that a radial force corresponds to $\theta = 0$. In this case, $\tau = r(F \sin 0) = 0$, as expected. If the force is tangential, however, it follows that $\theta = \pi/2$. This gives $\tau = r(F \sin \pi/2) = rF$, in agreement with Equation 11–1.

An equivalent way to define the torque is in terms of the **moment arm,** $r_\perp$. The idea here is to extend a line through the force vector, as in **Figure 11–3**, and then draw a second line from the axis of rotation perpendicular to the line of the force. The perpendicular distance from the axis of rotation to the line of the force is defined to be $r_\perp$. From the figure, we see that

$$r_\perp = r \sin\theta$$

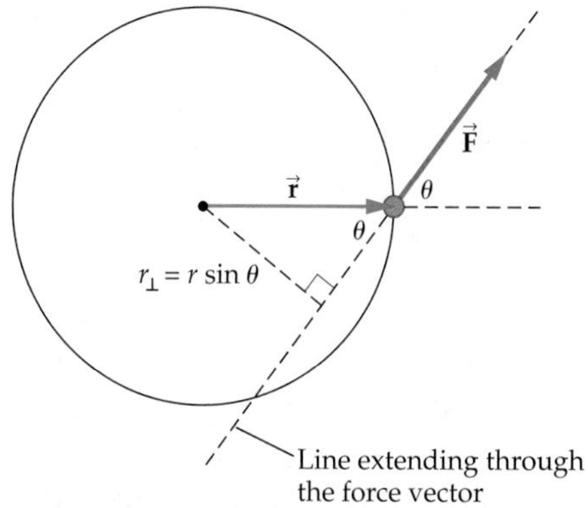

▲ **FIGURE 11–3 The moment arm**
To find the moment arm, $r_\perp$, for a given force, first extend a line through the force vector. This line is sometimes referred to as the "line of action." Next, drop a perpendicular line from the axis of rotation to the line of the force. The perpendicular distance is $r_\perp = r \sin\theta$.

▲ The net torque on the wheel of this ship is the sum of the torques exerted by the two helmsmen. At the moment pictured, they are both exerting negative torques on the wheel, causing it to rotate in the clockwise direction. This will turn the boat to its left—or, in nautical terms, to port.

In addition, we note that a simple rearrangement of the torque expression in Equation 11–2 yields

$$\tau = r(F \sin\theta) = (r \sin\theta)F$$

Thus, the torque can be written as the moment arm times the force:

$$\tau = r_{\perp}F \quad \text{11–3}$$

Just as a force applied to an object gives rise to a linear acceleration, a torque applied to an object gives rise to an angular acceleration. For example, if a torque acts on an object at rest, the object will begin to rotate; if a torque acts on a rotating object, the object's angular velocity will change. In fact, the greater the torque applied to an object, the greater its angular acceleration, as we shall see in the next section. For this reason, the sign of the torque is determined by the same convention used in Section 10–1 for angular acceleration:

Sign Convention for Torque

By convention, if a torque τ acts alone, then

$\tau > 0$ if the torque causes a counterclockwise angular acceleration

$\tau < 0$ if the torque causes a clockwise angular acceleration

In a system with more than one torque, the sign of each torque is determined by the type of angular acceleration *it alone* would produce. The net torque acting on the system, then, is the sum of each individual torque, taking into account the proper sign. This is illustrated in the following Example.

PROBLEM-SOLVING NOTES

The Sign of Torques

The sign of a torque is determined by the direction of rotation it would cause if it were the only torque acting in the system.

EXAMPLE 11–1 Torques to the Left and Torques to the Right

Two helmsmen, in disagreement about which way to turn a ship, exert the forces shown below on a ship's wheel. The wheel has a radius of 0.74 m, and the two forces have the magnitudes $F_1 = 72$ N and $F_2 = 58$ N. Find **(a)** the torque caused by $\vec{F}_1$ and **(b)** the torque caused by $\vec{F}_2$. **(c)** In which direction does the wheel turn as a result of these two forces?

Picture the Problem

Our sketch shows that both forces are applied at the distance $r = 0.74$ m from the axis of rotation. However, $F_1 = 72$ N is at an angle of 50.0° relative to the radial direction, whereas $F_2 = 58$ N is tangential, which means that its angle relative to the radial direction is 90.0°.

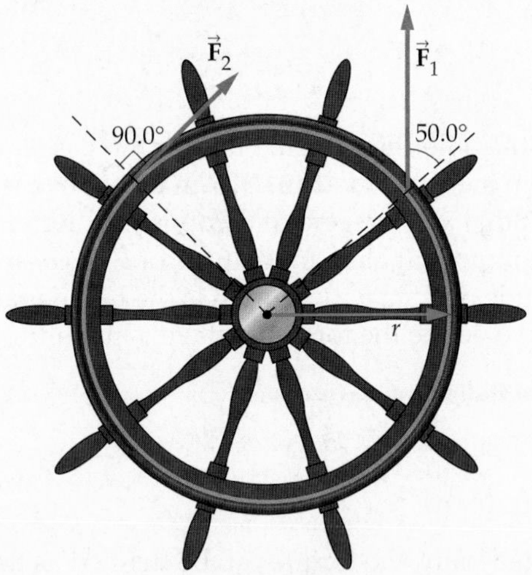

Strategy

For each force, we find the magnitude of the corresponding torque, using $\tau = rF\sin\theta$. As for the signs of the torques, we must consider the angular acceleration each force alone would cause. $\vec{F}_1$ acting alone would cause the wheel to accelerate counterclockwise, hence its torque is positive. $\vec{F}_2$ would accelerate the wheel clockwise if it acted alone, hence its torque is negative. If the sum of the two torques is positive, the wheel accelerates counterclockwise; if the sum of the two torques is negative, the wheel accelerates clockwise.

Solution

Part (a)

1. Use Equation 11–2 to calculate the torque due to $\vec{F}_1$. Recall that this torque is positive: $\tau_1 = rF_1 \sin 50.0° = (0.74\text{ m})(72\text{ N})\sin 50.0° = 41\text{ N}\cdot\text{m}$

Part (b)

2. Similarly, calculate the torque due to $\vec{F}_2$. Recall that this torque is negative: $\tau_2 = -rF_2 \sin 90.0° = -(0.74\text{ m})(58\text{ N}) = -43\text{ N}\cdot\text{m}$

Part (c)

3. Sum the torques from parts **(a)** and **(b)** to find the net torque: $\tau_{\text{net}} = \tau_1 + \tau_2 = 41\text{ N}\cdot\text{m} - 43\text{ N}\cdot\text{m} = -2\text{ N}\cdot\text{m}$

Insight

Because the net torque is negative, the wheel accelerates clockwise. Thus, even though $\vec{F}_2$ is the smaller force, it has the greater effect in determining the wheel's direction of acceleration. This is because $\vec{F}_2$ is applied tangentially, whereas $\vec{F}_1$ is applied in a direction that is partially radial.

Practice Problem

What magnitude of $\vec{F}_2$ would yield zero net torque on the wheel? [**Answer:** $F_2 = 55$ N]

Some related homework problems: Problem 1, Problem 3

11–2 Torque and Angular Acceleration

In the previous section we indicated that a torque causes a change in the rotational motion of an object. To be more precise, a single torque, τ, acting on an object causes the object to have an angular acceleration, α. In this section we develop the specific relationship between τ and α.

Consider, for example, a small object of mass m connected to an axis of rotation by a light rod of length r, as in **Figure 11–4**. If a tangential force of magnitude F is applied to the mass, it will move with an acceleration given by Newton's second law:

$$a = \frac{F}{m}$$

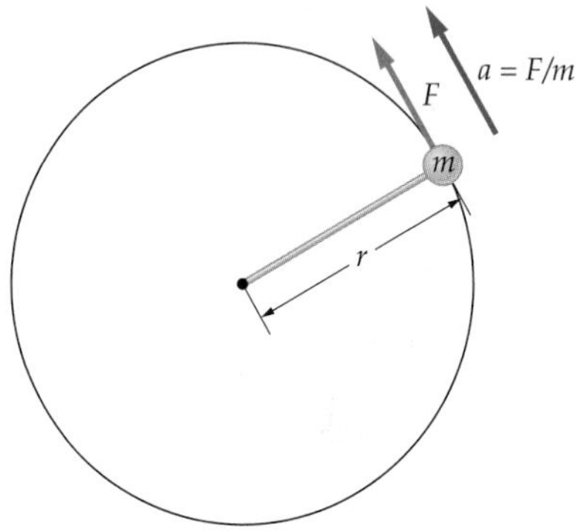

▲ **FIGURE 11–4 Torque and angular acceleration**

A tangential force F applied to a mass m gives it a linear acceleration of magnitude $a = F/m$. The corresponding angular acceleration is $\alpha = \tau/I$, where $\tau = rF$ and $I = mr^2$.

From Equation 10–14, we know that the linear and angular accelerations are related by

$$\alpha = \frac{a}{r}$$

Combining these results yields the following expression for the angular acceleration:

$$\alpha = \frac{a}{r} = \frac{F}{mr}$$

Finally, multiplying both numerator and denominator by r gives

$$\alpha = \left(\frac{r}{r}\right)\frac{F}{mr} = \frac{rF}{mr^2}$$

Now this last result is rather interesting, since the numerator and denominator have simple interpretations. First, the numerator is the torque, $\tau = rF$, for the case of a tangential force (Equation 11–1). Second, the denominator is the moment of inertia of a single mass m rotating at a radius r; that is, $I = mr^2$. Therefore, we find that

$$\alpha = \frac{rF}{mr^2} = \frac{\tau}{I}$$

or, rewriting slightly,

$$\tau = I\alpha$$

Thus, once we calculate the torque, as described in the previous section, we can find the angular acceleration of a system using $\tau = I\alpha$. Notice that the angular acceleration is directly proportional to the torque, and inversely proportional to the moment of inertia—that is, a large moment of inertia means a small angular acceleration.

Now, the relationship $\tau = I\alpha$ was derived for the special case of a tangential force and a single mass rotating at a radius r. However, the result is completely general. For example, in a system with more than one torque, the relation $\tau = I\alpha$

is replaced with $\tau_{net} = \Sigma\tau = I\alpha$, where τ_{net} is the net torque acting on the system. This gives us the *rotational* version of Newton's second law:

Newton's Second Law for Rotational Motion

$$\sum \tau = I\alpha \qquad 11\text{–}4$$

If only a single torque acts on a system, we will simply write $\tau = I\alpha$.

EXERCISE 11–2

A light rope wrapped around a disk-shaped pulley is pulled tangentially with a force of 0.53 N. Find the angular acceleration of the pulley, given that its mass is 1.3 kg and its radius is 0.11 m.

Solution

The torque applied to the disk is

$$\tau = rF = (0.11\text{ m})(0.53\text{ N}) = 5.8 \times 10^{-2}\text{ N}\cdot\text{m}$$

Since the pulley is a disk, its moment of inertia is given by

$$I = \tfrac{1}{2}mr^2 = \tfrac{1}{2}(1.3\text{ kg})(0.11\text{ m})^2 = 7.9 \times 10^{-3}\text{ kg}\cdot\text{m}^2$$

Thus, the angular acceleration of the pulley is

$$\alpha = \frac{\tau}{I} = \frac{5.8 \times 10^{-2}\text{ N}\cdot\text{m}}{7.9 \times 10^{-3}\text{ kg}\cdot\text{m}^2} = 7.3\text{ rad/s}^2$$

It is easy to remember the rotational version of Newton's second law, $\Sigma\tau = I\alpha$, by using analogies between rotational and linear quantities. We have already seen that I is the analog of m, and that α is the analog of a. Similarly, τ, which causes an angular acceleration, is the analog of F, which causes a linear acceleration. To summarize:

Linear Quantity	Angular Quantity
m	I
a	α
F	τ

Thus, just as $\Sigma F = ma$ describes linear motion, $\Sigma\tau = I\alpha$ describes rotational motion.

EXAMPLE 11–2 A Fish Takes the Line

A fisherman is dozing when a fish takes the line and pulls it with a tension T. The spool of the fishing reel is at rest initially and rotates without friction (since the fisherman left the drag off) as the fish pulls for a time t. If the radius of the spool is R, and its moment of inertia is I, find **(a)** the angular displacement of the spool, **(b)** the length of line pulled from the spool, and **(c)** the final angular speed of the spool.

Picture the Problem

Our sketch shows the fishing line being pulled tangentially from the spool with a tension T. Because the radius of the spool is R, the torque produced by the line is $\tau = RT$. Also note that as the spool rotates through an angle $\Delta\theta$, the line moves through a linear distance $\Delta x = R\,\Delta\theta$. Finally, the spool starts at rest, hence $\omega_0 = 0$.

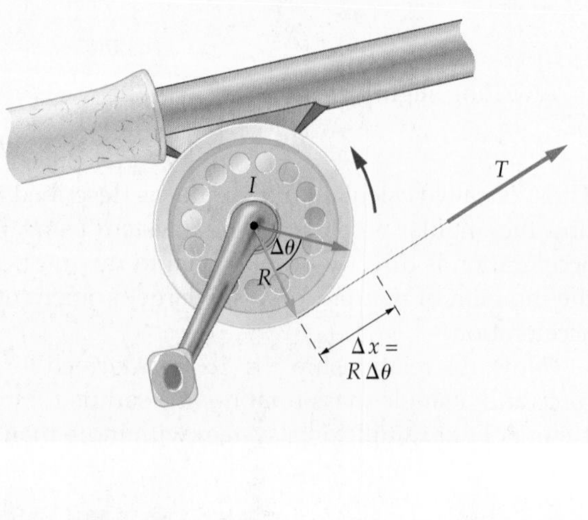

Strategy

This is basically an angular kinematics problem, as in Chapter 10, but in this case we must first calculate the angular acceleration using $\alpha = \tau/I$. Once α is known, we can find the angular displacement, $\Delta\theta$, using $\theta = \theta_0 + \omega_0 t + \tfrac{1}{2}\alpha t^2$. Similarly, we can find the angular speed of the spool, ω, using $\omega = \omega_0 + \alpha t$.

Solution

1. Calculate the torque acting on the spool. Note that $\theta = 90°$, since the pull is tangential. The radius is $r = R$, and the force applied to the reel is the tension in the line, T:

$$\tau = rF \sin\theta = RT \sin 90° = RT$$

2. Using the result just obtained for the torque, find the angular acceleration of the reel:

$$\alpha = \frac{\tau}{I} = \frac{RT}{I}$$

Part (a)

3. Calculate the angular displacement $\Delta\theta = \theta - \theta_0$:

$$\Delta\theta = \theta - \theta_0 = \omega_0 t + \tfrac{1}{2}\alpha t^2 = \tfrac{1}{2}\alpha t^2 = \left(\frac{RT}{2I}\right)t^2$$

Part (b)

4. Calculate the length of line pulled from the spool with $\Delta x = R\,\Delta\theta$:

$$\Delta x = R\,\Delta\theta = \left(\frac{R^2T}{2I}\right)t^2$$

Part (c)

5. Use $\omega = \omega_0 + \alpha t$ to find the final angular speed:

$$\omega = \omega_0 + \alpha t = \left(\frac{RT}{I}\right)t$$

Insight

Note that the final angular speed can also be obtained from the kinematic equation relating angular speed and angular distance; $\omega^2 = \omega_0{}^2 + 2\alpha\,\Delta\theta = 0 + 2(RT/I)(RT/2I)t^2 = (RT/I)^2\,t^2$.

This calculation also applies to other situations in which a "line" is pulled from a "reel." Examples include telephone line or sewing thread pulled from a spool.

Practice Problem

How fast is the line moving at time t? [**Answer:** $v = R\omega = (R^2T/I)t$]

Some related homework problems: Problem 7, Problem 14

CONCEPTUAL CHECKPOINT 11–1 Which Block Lands First?

The rotating systems shown below differ only in that the two spherical movable masses are positioned either far from the axis of rotation (left), or near the axis of rotation (right). If the hanging blocks are released simultaneously from rest, is it observed that **(a)** the block at left lands first, **(b)** the block at right lands first, or **(c)** both blocks land at the same time?

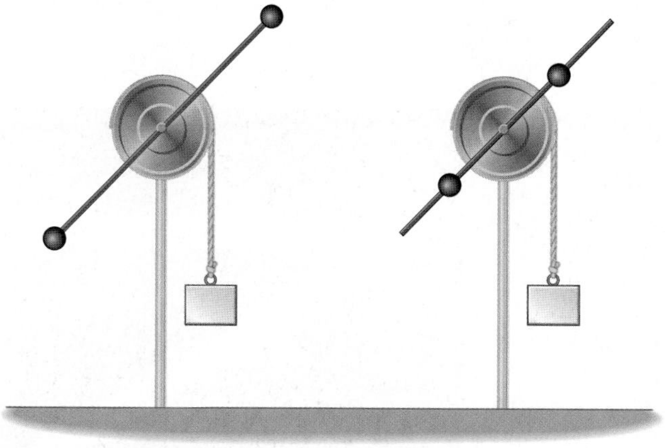

Reasoning and Discussion

The net external torque, supplied by the hanging blocks, is the same for each of these systems. However, the moment of inertia of the system at right is less than that of the system at left because the movable masses are closer to the axis of rotation. Since the angular acceleration is inversely proportional to the moment of inertia ($\alpha = \tau_{net}/I$), the system at right has the greater angular acceleration, and it wins the race.

Answer:

(b) The block at right lands first.

EXAMPLE 11–3 Drop It

A person holds his outstretched arm at rest in a horizontal position. The mass of the arm is m and its length is 0.740 m. When the person releases his arm, allowing it to drop freely, it begins to rotate about the shoulder joint. Find **(a)** the initial angular acceleration of the arm, and **(b)** the initial linear acceleration of the man's hand. (Hint: In calculating the torque, assume the mass of the arm is concentrated at its midpoint. In calculating the angular acceleration, use the moment of inertia of a uniform rod of length L about one end; $I = \frac{1}{3}mL^2$.)

Picture the Problem

The arm is initially horizontal and at rest. When released, it rotates downward about the shoulder joint. The force of gravity, mg, acts at a distance of $(0.740\text{ m})/2 = 0.370\text{ m}$ from the shoulder.

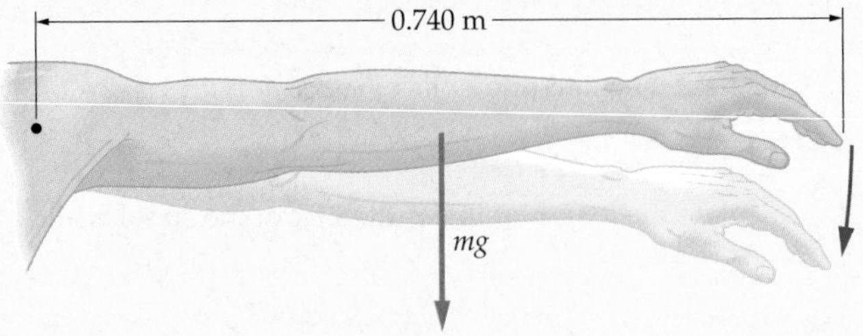

Strategy

The angular acceleration, α, can be found using $\tau = I\alpha$. In this case, the initial torque is $\tau = mg(L/2)$, where $L = 0.740$ m, and the moment of inertia is $I = \frac{1}{3}mL^2$.

Once the initial angular acceleration is found, the corresponding linear acceleration is obtained from $a = r\alpha$.

Solution

Part (a)

1. Use $\tau = I\alpha$ to find the angular acceleration, α:

$$\alpha = \frac{\tau}{I}$$

2. Write expressions for the initial torque, τ, and the moment of inertia, I:

$$\tau = mg\frac{L}{2}$$

$$I = \frac{1}{3}mL^2$$

3. Substitute τ and I into the expression for the angular acceleration. Note that the mass of the arm cancels:

$$\alpha = \frac{\tau}{I} = \frac{mgL/2}{mL^2/3} = \frac{3g}{2L}$$

4. Substitute numerical values:

$$\alpha = \frac{3g}{2L} = \frac{3(9.81\text{ m/s}^2)}{2(0.740\text{ m})} = 19.9\text{ rad/s}^2$$

Part (b)

5. Use $a = r\alpha$ to calculate the linear acceleration at the man's hand, a distance $r = L$ from the shoulder:

$$a = L\alpha = L\left(\frac{3g}{2L}\right) = \frac{3}{2}g = 14.7\text{ m/s}^2$$

Insight

Note that the linear acceleration of the hand is 1.50 times greater than the acceleration of gravity, regardless of the mass of the arm. This can be demonstrated with the following simple experiment: Hold your arm straight out with a pen resting on your hand. Now, relax your deltoid muscles, and let your arm rotate freely downward about your shoulder joint. Notice that as your arm falls downward, your hand moves more rapidly than the pen, which appears to "lift off" your hand. The pen drops with the acceleration of gravity, which is clearly less than the acceleration of the hand. This effect can be seen clearly in the adjacent photo.

▲ As a rod of length L rotates freely about one end, points farther from the axle than $2L/3$ have an acceleration greater than g (see the Practice Problem for Example 11–3). Thus, the rod falls out from under the last two dice.

Practice Problem

At what distance from the shoulder is the initial linear acceleration of the arm equal to the acceleration of gravity?
[**Answer:** Set $a = r\alpha$ equal to g. This gives $r = 2L/3 = 0.493$ m.]

Some related homework problems: Problem 10, Problem 12

11–3 Zero Torque and Static Equilibrium

The parents of a young boy are supporting him on a long, lightweight plank, as illustrated in **Figure 11–5**. If the mass of the child is m, the upward forces exerted by the parents must sum to mg; that is,

$$F_1 + F_2 = mg$$

This condition ensures that the net force acting on the plank is zero. It *does not*, however, guarantee that the plank remains at rest.

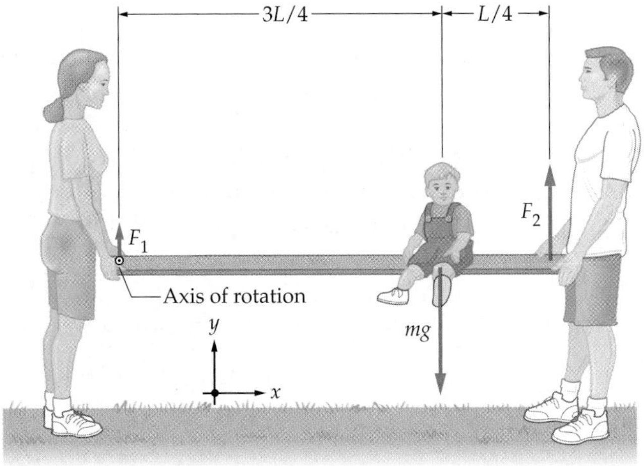

◀ **FIGURE 11–5 Forces required for static equilibrium**

Two parents support a child on a lightweight plank of length L. For the calculation described in the text, we choose the axis of rotation to be the left end of the plank.

To see why, imagine for a moment that the parent on the right lets go of the plank and that the parent on the left increases her force until it is equal to the weight of the child. In this case, $F_1 = mg$ and $F_2 = 0$, which clearly satisfies the force equation we have just written. Since the right end of the plank is no longer supported, however, it drops toward the ground while the left end rises. In other words, the plank rotates in a clockwise sense. For the plank to remain completely at rest, with no translation or rotation, we must impose the following *two* conditions: First, the net force acting on the plank must be zero, so that there is no translational acceleration. Second, the net torque acting on the plank must also be zero, so that there is no rotational acceleration. If both of these conditions are met, an extended object, like the plank, will remain at rest if it starts at rest. To summarize:

Conditions for Static Equilibrium

For an extended object to be in static equilibrium, the following two conditions must be met:

(i) The net force acting on the object must be zero,

$$\sum F_x = 0, \qquad \sum F_y = 0 \qquad \textbf{11–5}$$

(ii) The net torque acting on the object must be zero,

$$\sum \tau = 0 \qquad \textbf{11–6}$$

Note that these two conditions are independent of one another; that is, satisfying one does *not* guarantee that the other is satisfied.

Let's apply these conditions to the plank that supports the child. First, we consider the forces acting on the plank, with upward chosen as the positive direction, as in Figure 11–5. Setting the net force equal to zero yields

$$F_1 + F_2 - mg = 0$$

Clearly, this agrees with the force equation we wrote down earlier.

Next, we apply the torque condition. To do so, we must first choose an axis of rotation. For example, we might take the left end of the plank to be the axis, as in Figure 11–5. With this choice, we see that the force F_1 exerts zero torque, since it acts directly through the axis of rotation. On the other hand, F_2 acts at the far end of the plank, a distance L from the axis. In addition, F_2 would cause a counterclockwise (positive) rotation if it acted alone, as we can see in Figure 11–5. Therefore, the torque due to F_2 is

$$\tau_2 = F_2 L$$

Finally, the weight of the child, mg, acts at a distance of $3L/4$ from the axis, and would cause a clockwise (negative) rotation if it acted alone. Hence, its torque is negative:

$$\tau_{mg} = -mg\left(\tfrac{3}{4}L\right)$$

Setting the net torque equal to zero, then, yields the following condition:

$$F_2 L - mg\left(\tfrac{3}{4}L\right) = 0$$

This torque condition, along with the force condition in $F_1 + F_2 - mg = 0$, can be used to determine the two unknowns, F_1 and F_2. For example, we can begin by canceling L in the torque equation to find F_2:

$$F_2 = \tfrac{3}{4}mg$$

Substituting this result into the force condition gives

$$F_1 + \tfrac{3}{4}mg - mg = 0$$

Therefore, F_1 is

$$F_1 = \tfrac{1}{4}mg$$

These two forces support the plank, *and* keep it from rotating. As one might expect, the force nearest the child is greatest.

Our choice of the left end of the plank as the axis of rotation was completely arbitrary. In fact, if an object is in static equilibrium, the net torque acting on it is zero, regardless of the location of the axis of rotation. Hence, we are free to choose an axis of rotation that is most convenient for a given problem. In general, it is useful to pick the axis to be at the location of one of the unknown forces. This eliminates that force from the torque condition, and simplifies the remaining algebra. We consider an alternative choice for the axis of rotation in the following Active Example.

PROBLEM-SOLVING NOTES

Axis of Rotation

Any point in a system may be used as the axis of rotation when calculating torque. It is generally best, however, to choose an axis that gives zero torque for at least one of the unknown forces in the system. Such a choice simplifies the algebra needed to solve for the forces.

ACTIVE EXAMPLE 11–1 Find the Forces: Axis on the Right

A child of mass m is supported on a light plank by his parents, who exert the forces F_1 and F_2 as indicated. Find the forces required to keep the plank in static equilibrium. Use the right end of the plank as the axis of rotation.

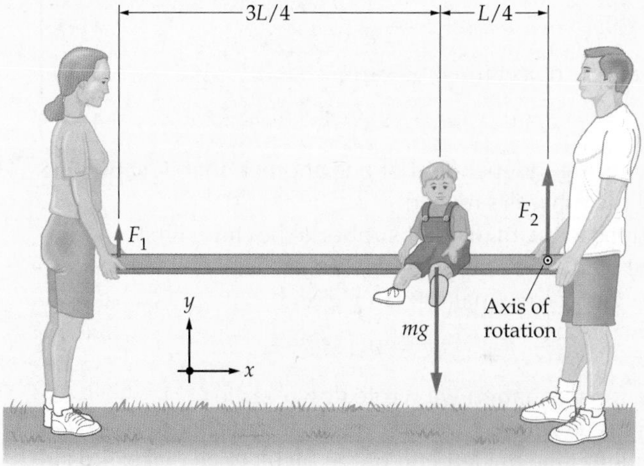

Solution *(Test your understanding by performing the calculations indicated in each step.)*

1. Set the net force acting on the plank equal to zero: $F_1 + F_2 - mg = 0$
2. Set the net torque acting on the plank equal to zero: $-F_1(L) + mg\left(\frac{1}{4}L\right) = 0$
3. Note that the torque condition involves only one of the two unknowns, F_1. Use this condition to solve for F_1: $F_1 = \frac{1}{4}mg$
4. Substitute F_1 into the force condition to solve for F_2: $F_2 = mg - \frac{1}{4}mg = \frac{3}{4}mg$

Insight

As expected, the results are identical to those obtained previously. Note that in this case the torque produced by the child would cause a counterclockwise rotation, hence it is positive. Thus, the magnitude *and* sign of the torque produced by a given force depend on the location chosen for the axis of rotation.

Your Turn

Suppose the child moves to a new position, with the result that the force exerted by the father is reduced to 0.60 mg. Did the child move to the left or to the right? How far did the child move?

(Answers to **Your Turn** *problems are given in the back of the book.)*

A third choice for the axis of rotation is considered in Problem 18. As expected, all three choices give the same results.

In the next Example, we show that the forces supporting a person or other object sometimes act in different directions. To emphasize the direction of the forces, we solve the Example in terms of the components of the relevant forces.

EXAMPLE 11–4 Taking the Plunge

A 5.00-m long diving board of negligible mass is supported by two pillars. One pillar is at the left end of the diving board, as shown below; the other is 1.50 m away. Find the forces exerted by the pillars when a 90.0-kg diver stands at the far end of the board.

Picture the Problem

We choose upward to be the positive direction for the forces. When calculating torques, we use the left end of the diving board as the axis of rotation. Note that $\vec{\mathbf{F}}_2$ would cause a counterclockwise rotation if it acted alone, so its torque is positive. On the other hand, $m\vec{\mathbf{g}}$ would cause a clockwise rotation, so its torque is negative. Finally, $\vec{\mathbf{F}}_2$ acts at a distance d from the axis of rotation, and $m\vec{\mathbf{g}}$ acts at a distance L.

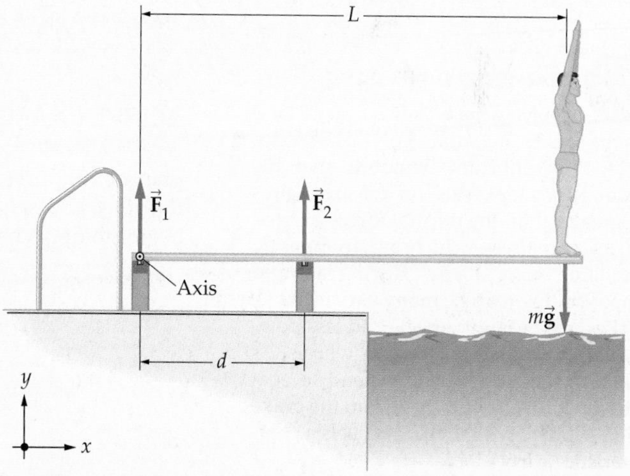

Strategy

As usual in static equilibrium problems, we use the conditions of (i) zero net force and (ii) zero net torque to determine the unknown forces, $\vec{\mathbf{F}}_1$ and $\vec{\mathbf{F}}_2$. In this system all forces act in the positive or negative y direction; thus we need only set the net y component of force equal to zero.

Solution

1. Set the net y component of force acting on the diving board equal to zero: $\sum F_y = F_{1,y} + F_{2,y} - mg = 0$
2. Calculate the torque due to each force, using the left end of the board as the axis of rotation. Note that each force is at right angles to the radius and that $\vec{\mathbf{F}}_1$ goes directly through the axis of rotation:
$$\tau_1 = F_{1,y}(0) = 0$$
$$\tau_2 = F_{2,y}(d)$$
$$\tau_3 = -mg(L)$$
3. Set the net torque acting on the diving board equal to zero: $\sum \tau = F_{1,y}(0) + F_{2,y}(d) - mg(L) = 0$

continued on next page

continued from previous page

4. Solve the torque equation for the force $F_{2,y}$: $F_{2,y} = mg(L/d)$
$= (90.0\text{ kg})(9.81\text{ m/s}^2)(5.00\text{ m}/1.50\text{ m}) = 2940\text{ N}$

5. Use the force equation to determine $F_{1,y}$: $F_{1,y} = mg - F_{2,y}$
$= (90.0\text{ kg})(9.81\text{ m/s}^2) - 2940\text{ N} = -2060\text{ N}$

Insight

The first point to notice about our solution is that $F_{1,y}$ is negative, which means that $\vec{\mathbf{F}}_1$ is actually directed *downward*, as shown to the right. To see why, imagine for a moment that the board is no longer connected to the first pillar. In this case, the board would rotate clockwise about the second pillar, and the left end of the board would move upward. Thus, a downward force is required on the left end of the board to hold it in place.

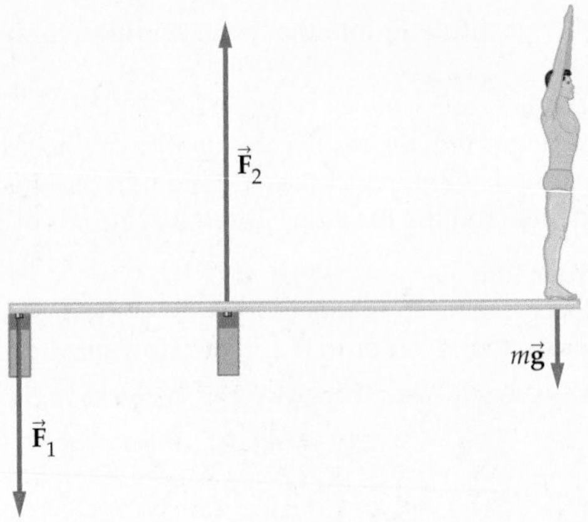

The second point is that both pillars exert forces with magnitudes that are considerably larger than the diver's weight, $mg = 883$ N. In particular, the first pillar must pull downward with a force of 2.33 mg, while the second pillar pushes upward with a force of $2.33mg + mg = 3.33mg$. This is not unusual. In fact, it is common for the forces in a structure, such as a bridge, a building, or the human body, to be much greater than the weight it supports.

Practice Problem

Find the forces exerted by the pillars when the diver is 1.00 m from the right end. [**Answer:** $F_{1,y} = -1470$ N, $F_{2,y} = 2350$ N]

Some related homework problems: Problem 20, Problem 26

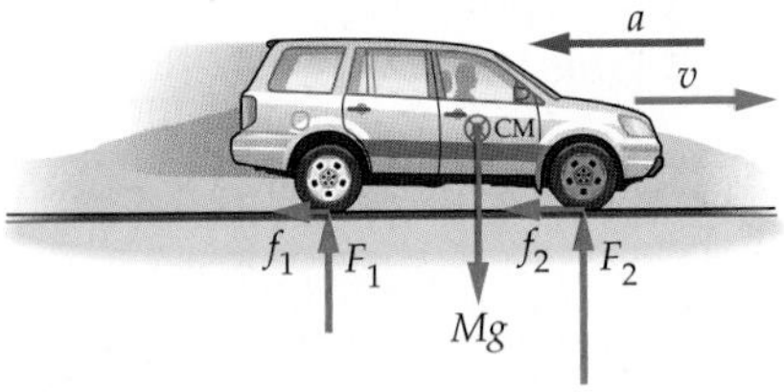

REAL-WORLD PHYSICS

Applying the brakes

As the brakes are applied on this SUV, rotational equilibrium demands that the normal forces exerted on the front tires be greater than the normal forces exerted on the rear tires—which is why braking cars are "nose down" during a rapid stop. For this reason, many cars use disk brakes for the front wheels and less powerful drum brakes for the rear wheels. As the disk brakes wear, they tend to coat the front wheels with dust from the brake pads, which give the front wheels a characteristic "dirty" look.

To this point we have ignored the mass of the plank holding the child and the diving board holding the swimmer, since they were described as lightweight. If we want to consider the torque exerted by an extended object of finite mass, however, we can simply treat it as if all its mass were concentrated at its center of mass, as was done in similar situations in Section 9–7. We consider such a system in the next Active Example.

ACTIVE EXAMPLE 11–2 Walking the Plank: Find the Mass

A cat walks along a uniform plank that is 4.00 m long and has a mass of 7.00 kg. The plank is supported by two sawhorses, one 0.440 m from the left end of the board and the other 1.50 m from its right end. When the cat reaches the right end, the plank just begins to tip. What is the mass of the cat?

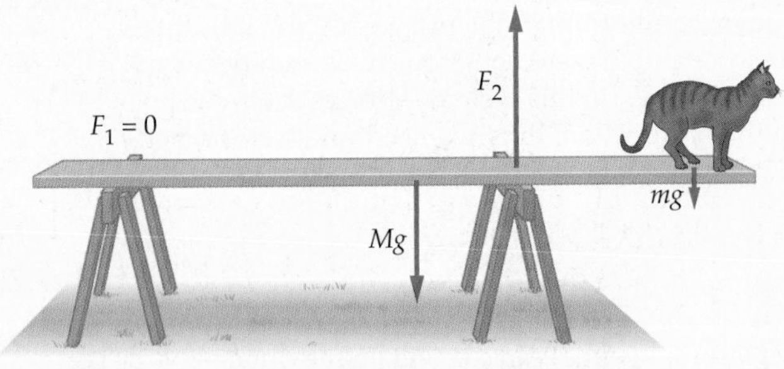

Solution *(Test your understanding by performing the calculations indicated in each step.)*

1. Since the board is just beginning to tip, there is no weight on the left sawhorse: $F_1 = 0$

2. Calculate the torque about the right sawhorse: $Mg(0.500\text{ m}) - mg(1.50\text{ m}) = 0$

3. Solve the torque equation for the mass of the cat, m: $m = 0.333M = 2.33\text{ kg}$

Insight

Note that we did not include a torque for the left sawhorse, since F_1 is zero. As an exercise, you might try repeating the calculation with the axis of rotation at the left sawhorse, or at the center of mass of the plank.

Your Turn

Write both the zero force and zero torque condition for the case where the axis of rotation is at the left sawhorse.

(Answers to **Your Turn** *problems are given in the back of the book.)*

Forces with both Vertical and Horizontal Components

Note that all of the previous examples have dealt with forces that point either directly upward or directly downward. We now consider a more general situation, where forces may have both vertical and horizontal components. For example, consider the wall-mounted lamp (sconce) shown in **Figure 11–6**. The sconce consists of a light curved rod that is bolted to the wall at its lower end. Suspended from the upper end of the rod, a horizontal distance H from the wall, is the lamp of mass m. The rod is also connected to the wall by a horizontal wire a vertical distance V above the bottom of the rod.

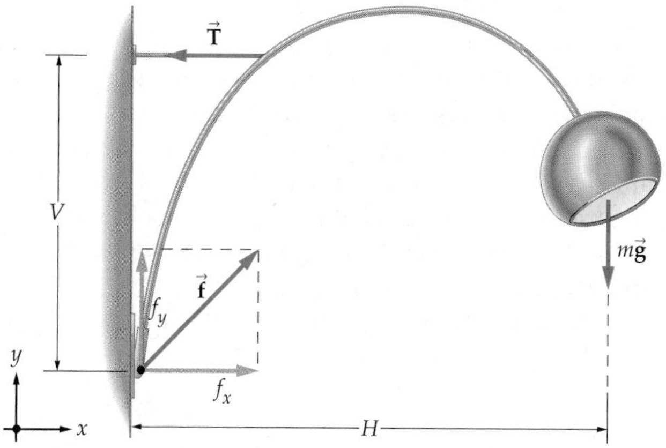

◀ **FIGURE 11–6 A lamp in static equilibrium**

A wall-mounted lamp of mass m is suspended from a light curved rod. The bottom of the rod is bolted to the wall. The rod is also connected to the wall by a horizontal wire a vertical distance V above the bottom of the rod.

Now, suppose we are designing this sconce to be placed in the lobby of a building on campus. To ensure its structural stability, we would like to know the tension T the wire must exert and the vertical and horizontal components of the force $\vec{f}$ that must be exerted by the bolt on the rod. This information will be important in deciding on the type of wire and bolt to be used in the construction.

To find these forces, we apply the same conditions as before: the net force and the net torque must be zero. In this case, however, forces may have both horizontal and vertical components. Thus, the condition of zero net force is really two separate conditions: (i) zero net force in the horizontal direction; and (ii) zero net force in the vertical direction. These two conditions plus (iii) zero net torque, allow for a full solution of the problem.

We begin with the torque condition. A convenient choice for the axis of rotation is the bottom end of the rod, since this eliminates one of the unknown forces ($\vec{f}$). With this choice we can readily calculate the torques acting on the rod by using the moment arm expression for the torque, $\tau = r_{\perp}F$ (Equation 11–3). We find

$$\sum \tau = T(V) - mg(H) = 0$$

This relation can be solved immediately for the tension, giving

$$T = mg(H/V)$$

Note that the tension is increased if the wire is connected closer to the bottom of the rod; that is, if V is reduced.

▲ The chains that support this sign maintain it in a state of translational and rotational equilibrium. The forces in the chains are most easily analyzed by resolving them into vertical and horizontal components and applying the conditions for equilibrium. In particular, the net vertical force, the net horizontal force, and the net torque must all be zero.

Next, we apply the force conditions. First, we sum the y components of all the forces and set the sum equal to zero:

$$\sum F_y = f_y - mg = 0$$

Thus, the vertical component of the force exerted by the bolt simply supports the weight of the lamp

$$f_y = mg$$

Finally, we sum the x components of the forces and set that sum equal to zero:

$$\sum F_x = f_x - T = 0$$

Clearly, the x component of the force exerted by the bolt is of the same magnitude as the tension, but it points in the opposite direction:

$$f_x = T = mg(H/V)$$

The bolt, then, pushes upward on the rod to support the lamp, and at the same time it pushes to the right to keep the rod from rotating.

For example, suppose the lamp in Figure 11–6 has a mass of 2.00 kg, and that $V = 12.0$ cm and $H = 15.0$ cm. In this case, we find the following forces:

$$T = mg(H/V) = (2.00\text{ kg})(9.81\text{ m/s}^2)(15.0\text{ cm})/(12.0\text{ cm}) = 24.5\text{ N}$$
$$f_x = T = 24.5\text{ N}$$
$$f_y = mg = (2.00\text{ kg})(9.81\text{ m/s}^2) = 19.6\text{ N}$$

REAL-WORLD PHYSICS
Forces required for structural stability

Note that f_x and T are greater than the weight, mg, of the lamp. Just as we found with the diving board in Example 11–4, the forces required of structural elements can be greater than the weight of the object to be supported—an important consideration when designing a structure like a bridge, an airplane, or a sconce. The same effect occurs in the human body. We find in Problem 19, for example, that the force exerted by the biceps to support a baseball in the hand is several times larger than the baseball's weight. Similar conclusions apply to muscles throughout the body.

In Example 11–5 we consider another system in which forces have both vertical and horizontal components.

EXAMPLE 11–5 Arm in a Sling

A hiker who has broken his forearm rigs a temporary sling using a cord stretching from his shoulder to his hand. The cord holds the forearm level and makes an angle of 40.0° with the horizontal where it attaches to the hand. Considering the forearm and hand to be uniform, with a total mass of 1.31 kg and a length of 0.300 m, find **(a)** the tension in the cord and **(b)** the horizontal and vertical components of the force, $\vec{\mathbf{f}}$, exerted by the humerus (the bone of the upper arm) on the radius and ulna (the bones of the forearm).

Picture the Problem

In our sketch, we use the typical conventions for the positive x and y directions. In addition, since the forearm and hand are assumed to be a uniform object, we indicate the weight mg as acting at its center. The length of the forearm and hand is $L = 0.300$ m. Finally, two other forces act on the forearm: (i) the tension in the cord, $\vec{\mathbf{T}}$, at an angle of 40.0° above the negative x axis, and (ii) the force $\vec{\mathbf{f}}$ exerted at the elbow joint.

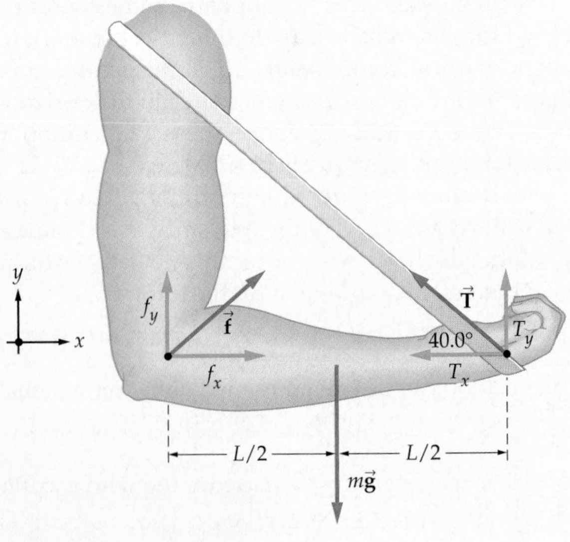

Strategy

In this system there are three unknowns: T, f_x, and f_y. These unknowns can be determined using the following three conditions: (i) net torque equals zero; (ii) net x component of force equals zero; and (iii) net y component of force equals zero.

We start with the torque condition, using the elbow joint as the axis of rotation. As we shall see, this choice of axis eliminates f, and gives a direct solution for the tension T. Next, we use T and the two force conditions to determine f_x and f_y.

Solution

Part (a)

1. Calculate the torque about the elbow joint. Note that f causes zero torque, mg causes a negative torque, and the vertical component of T causes a positive torque. The horizontal component of T produces no torque, since it is on a line with the axis:
$$\sum \tau = (T \sin 40.0°)L - mg(L/2) = 0$$

2. Solve the torque condition for the tension, T:
$$T = \frac{mg}{2 \sin 40.0°} = \frac{(1.31 \text{ kg})(9.81 \text{ m/s}^2)}{2 \sin 40.0°} = 10.0 \text{ N}$$

Part (b)

3. Set the sum of the x components of force equal to zero, and solve for f_x:
$$\sum F_x = f_x - T \cos 40.0° = 0$$
$$f_x = T \cos 40.0° = (10.0 \text{ N}) \cos 40.0° = 7.66 \text{ N}$$

4. Set the sum of the y components of force equal to zero, and solve for f_y:
$$\sum F_y = f_y - mg + T \sin 40.0° = 0$$
$$f_y = mg - T \sin 40.0°$$
$$= (1.31 \text{ kg})(9.81 \text{ m/s}^2) - (10.0 \text{ N}) \sin 40.0° = 6.42 \text{ N}$$

Insight

It is not necessary to determine T_x and T_y separately, since we know the direction of the cord. In particular, it is clear from our sketch that the components of $\vec{\mathbf{T}}$ are $T_x = -T \cos 40.0° = -7.66$ N and $T_y = T \sin 40.0° = 6.43$ N.

Did you notice that $\vec{\mathbf{f}}$ is at an angle of 40.0° with respect to the positive x axis, the same angle that $\vec{\mathbf{T}}$ makes with the negative x axis? The reason for this symmetry, of course, is that mg acts at the center of the forearm. If mg were to act closer to the elbow, for example, $\vec{\mathbf{f}}$ would make a larger angle with the horizontal, as we see in the following Practice Problem.

Practice Problem

Suppose the forearm and hand are nonuniform, and that the center of mass is located at a distance of $L/4$ from the elbow joint. What are T, f_x, and f_y in this case? [**Answer:** $T = 5.00$ N, $f_x = 3.83$ N, $f_y = 9.64$ N. In this case, $\vec{\mathbf{f}}$ makes an angle of 68.3° with the horizontal.]

Some related homework problems: Problem 27, Problem 79

ACTIVE EXAMPLE 11–3 Don't Walk Under the Ladder: Find the Forces

An 85-kg person stands on a lightweight ladder, as shown. The floor is rough; hence, it exerts both a normal force, f_1, and a frictional force, f_2, on the ladder. The wall, on the other hand, is frictionless; it exerts only a normal force, f_3. Using the dimensions given in the figure, find the magnitudes of f_1, f_2, and f_3.

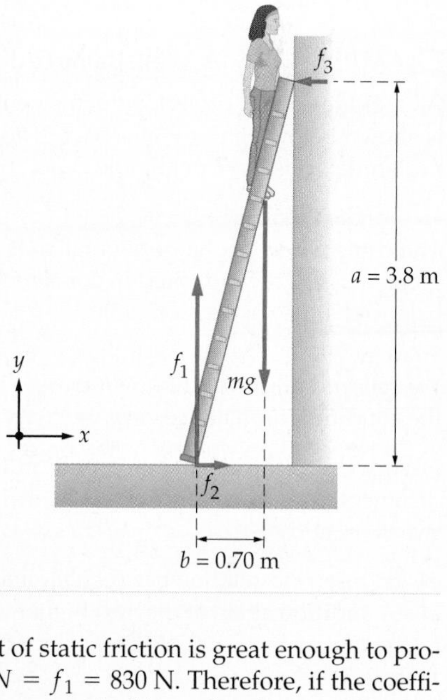

Solution *(Test your understanding by performing the calculations indicated in each step.)*

1. Set the net torque acting on the ladder equal to zero. Use the bottom of the ladder as the axis: $f_3(a) - mg(b) = 0$
2. Solve for f_3: $f_3 = mg(b/a) = 150$ N
3. Sum the x components of force and set equal to zero: $f_2 - f_3 = 0$
4. Solve for f_2: $f_2 = f_3 = 150$ N
5. Sum the y components of force and set equal to zero: $f_1 - mg = 0$
6. Solve for f_1: $f_1 = mg = 830$ N

Insight

If the floor is quite smooth, the ladder might slip—it depends on whether the coefficient of static friction is great enough to provide the needed force $f_2 = 150$ N. In this case, the normal force exerted by the floor is $N = f_1 = 830$ N. Therefore, if the coefficient of static friction is greater than 0.18 [since 0.18(830 N) = 150 N], the ladder will stay put. Ladders often have rubberized pads on the bottom in order to increase the static friction, and hence increase the safety of the ladder.

Your Turn

Write both the zero force and zero torque condition for the case where the axis of rotation is at the top of the ladder.

(Answers to **Your Turn** *problems are given in the back of the book.)*

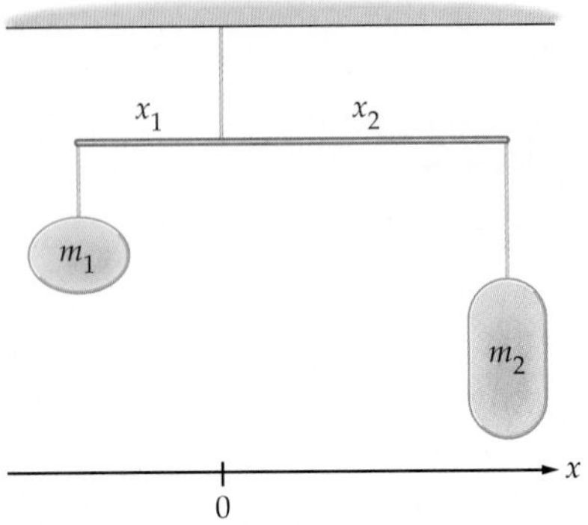

▲ **FIGURE 11–7 Zero torque and balance**

One section of a mobile. The rod is balanced when the net torque acting on it is zero. This is equivalent to having the center of mass directly under the suspension point.

11–4 Center of Mass and Balance

Suppose you decide to construct a mobile. To begin, you tie a thread to a light rod, as in **Figure 11–7**. Note that the rod extends a distance x_1 to the left of the thread and a distance x_2 to the right. At the left end of the rod you attach an object of mass m_1. What mass, m_2, should be attached to the right end if the rod is to be balanced?

From the discussions in the previous sections, it is clear that if the rod is to be in static equilibrium (balanced), the net torque acting on it must be zero. Taking the point where the thread is tied to the rod as the axis of rotation, this zero-torque condition can be written as:

$$m_1g(x_1) - m_2g(x_2) = 0$$

Canceling g and rearranging, we find

$$m_1x_1 = m_2x_2 \qquad \textbf{11–7}$$

This gives the following result for m_2:

$$m_2 = m_1(x_1/x_2)$$

For example, if $x_2 = 2x_1$, it follows that m_2 should be one-half of m_1.

Let's now consider a slightly different question: Where is the center of mass of m_1 and m_2? Choosing the origin of the x axis to be at the location of the thread, as indicated in Figure 11–7, we can use the definition of the center of mass, Equation 9–13, to find x_{cm}:

$$x_{cm} = \frac{m_1(-x_1) + m_2(x_2)}{m_1 + m_2} = -\left(\frac{m_1x_1 - m_2x_2}{m_1 + m_2}\right)$$

Referring to the zero-torque condition in Equation 11–7, we see that $m_1x_1 - m_2x_2 = 0$; hence the center of mass is at the origin:

$$x_{cm} = 0$$

This is precisely where the string is attached. We conclude, then, that the rod balances when the center of mass is directly below the point from which the rod is suspended. This is a general result.

Let's apply this result to the case of the mobile shown in the next Example.

EXAMPLE 11–6 A Well-Balanced Meal

As a grade-school project, students construct a mobile representing some of the major food groups. Their completed artwork is shown below. Find the masses m_1, m_2, and m_3 that are required for a perfectly balanced mobile. Assume the strings and the horizontal rods have negligible mass.

Picture the Problem

The dimensions of the horizontal rods, and the values of the given masses, are indicated in our sketch. Note that each rod is balanced at its suspension point.

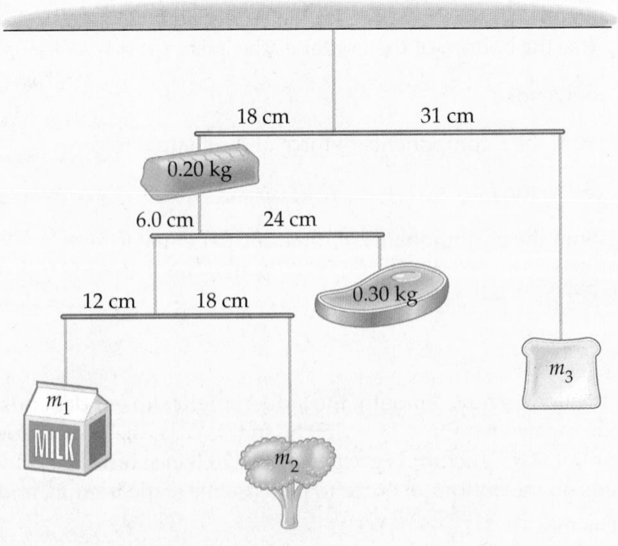

Strategy

We can find all three unknown masses by repeatedly applying the condition for balance, $m_1x_1 = m_2x_2$.

First, we apply the balance condition to m_1 and m_2, with the distances $x_1 = 12$ cm and $x_2 = 18$ cm. This gives a relation between m_1 and m_2.

To get a second relation between m_1 and m_2, we apply the balance condition again at the next higher level of the mobile. That is, the mass $(m_1 + m_2)$ at the distance 6.0 cm must balance the mass 0.30 kg at the distance 24 cm. These two conditions determine m_1 and m_2.

To find m_3 we again apply the balance condition, this time with the mass $(m_1 + m_2 + 0.30 \text{ kg} + 0.20 \text{ kg})$ at the distance 18 cm, and the mass m_3 at the distance 31 cm.

Solution

1. Apply the balance condition to m_1 and m_2:

$$m_1(12\text{ cm}) = m_2(18\text{ cm})$$
$$m_1 = (1.5)m_2$$

2. Apply the balance condition to the next level up in the mobile. Solve for the sum, $m_1 + m_2$:

$$(m_1 + m_2)(6.0\text{ cm}) = (0.30\text{ kg})(24\text{ cm})$$
$$m_1 + m_2 = \frac{(0.30\text{ kg})(24\text{ cm})}{6.0\text{ cm}} = 1.2\text{ kg}$$

3. Substitute $m_1 = (1.5)m_2$ into $m_1 + m_2 = 1.2$ kg to find m_2:

$$(1.5)m_2 + m_2 = (2.5)m_2 = 1.2\text{ kg}$$
$$m_2 = 1.2\text{ kg}/2.5 = 0.48\text{ kg}$$

4. Use $m_1 = (1.5)m_2$ to find m_1:

$$m_1 = (1.5)m_2 = (1.5)0.48\text{ kg} = 0.72\text{ kg}$$

5. Apply the balance condition to the top level of the mobile:

$$(0.72\text{ kg} + 0.48\text{ kg} + 0.30\text{ kg} + 0.20\text{ kg})(18\text{ cm}) = m_3(31\text{ cm})$$

6. Solve for m_3:

$$m_3 = \frac{(1.70\text{ kg})(18\text{ cm})}{31\text{ cm}} = 0.99\text{ kg}$$

Insight

With the values for m_1, m_2, and m_3 found above, the mobile balances at every level. In fact, the center of mass of the *entire* mobile is directly below the point where the uppermost string attaches to the ceiling.

Practice Problem

Find m_1, m_2, and m_3 if the 0.30 kg mass is replaced with a 0.40 kg mass. [**Answer:** $m_1 = 0.96$ kg, $m_2 = 0.64$ kg, $m_3 = 1.3$ kg]

Some related homework problems: Problem 38, Problem 40

In general, if you allow an arbitrarily shaped object to hang freely, its center of mass is directly below the suspension point. To see why, note that when the center of mass is directly below the suspension point, the torque due to gravity is zero, since the force of gravity extends right through the axis of rotation. This is shown in **Figure 11–8 (a)**. If the object is rotated slightly, as in **Figure 11–8 (b)**, the force of gravity is not in line with the axis of rotation—hence gravity produces a torque. This torque tends to rotate the object, bringing the center of mass back under the suspension point.

For example, suppose you cut a piece of wood into the shape of the continental United States, as shown in **Figure 11–9**, drill a small hole in it and hang it from

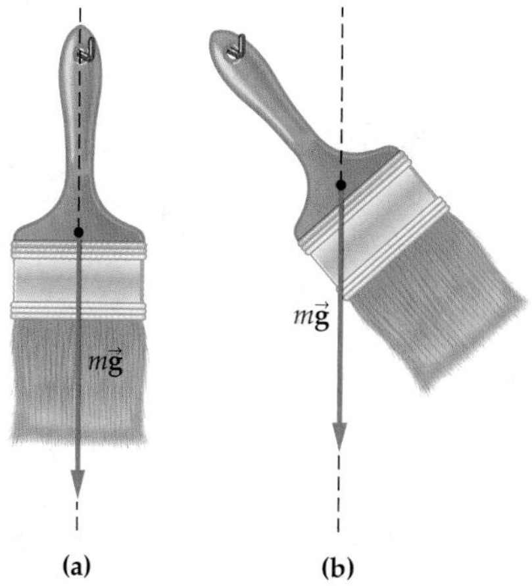

▲ **FIGURE 11–8 Equilibrium of a suspended object**

(a) If an object's center of mass is directly below the suspension point, its weight creates zero torque and the object is in equilibrium. **(b)** When an object is rotated, so that the center of mass is no longer directly below the suspension point, the object's weight creates a torque. The torque tends to rotate the object to bring the center of mass under the suspension point.

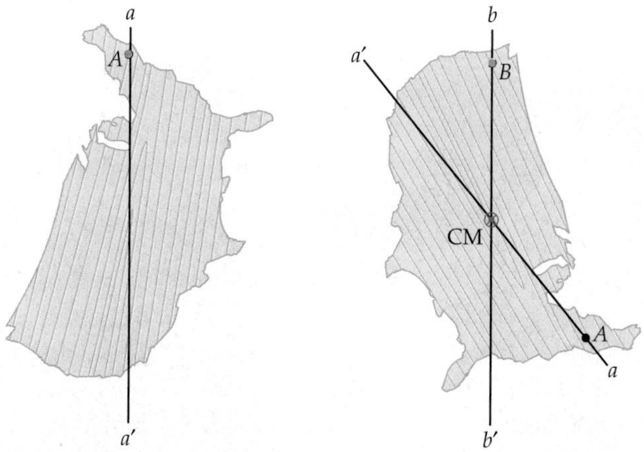

▲ **FIGURE 11–9 The geometric center of the United States**

To find the center of mass of an irregularly shaped object, such as a wooden model of the continental United States, suspend it from two or more points. The center of mass lies on a vertical line extending downward from the suspension point. The intersection of these vertical lines gives the precise location of the center of mass.

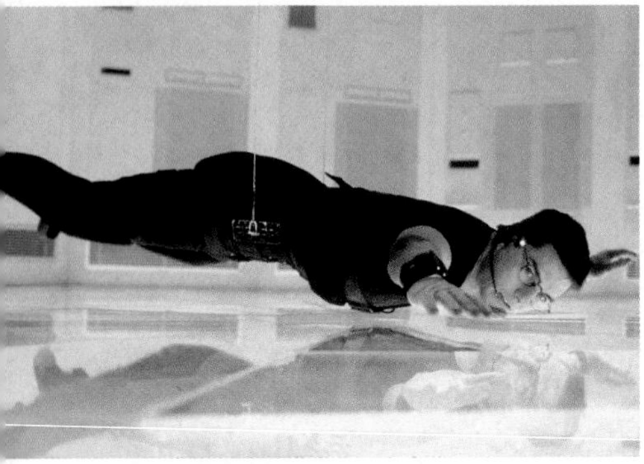

▲ In this scene from the movie *Mission Impossible*, Tom Cruise is attempting to download top-secret computer files without setting off the elaborate security system in the room. To accomplish this nearly impossible mission, he is suspended from the ceiling, since touching the floor would immediately give away his presence. To remain in equilibrium above the floor as he works, he must carefully adjust the position of his arms and legs to keep his center of mass directly below the suspension point.

the point A. The result is that the center of mass lies somewhere on the line aa'. Similarly, if a second hole is drilled at point B, we find that the center of mass lies somewhere on the line bb'. The only point that is on both the line aa' *and* the line bb' is the point CM, near Smith Center, Kansas, which marks the location of the center of mass.

CONCEPTUAL CHECKPOINT 11–2 Compare the Masses

A croquet mallet balances when suspended from its center of mass, as indicated in the drawing at left. If you cut the mallet in two at its center of mass, as in the drawing at right, how do the masses of the two pieces compare? **(a)** The masses are equal; **(b)** the piece with the head of the mallet has the greater mass; or **(c)** the piece with the head of the mallet has the smaller mass.

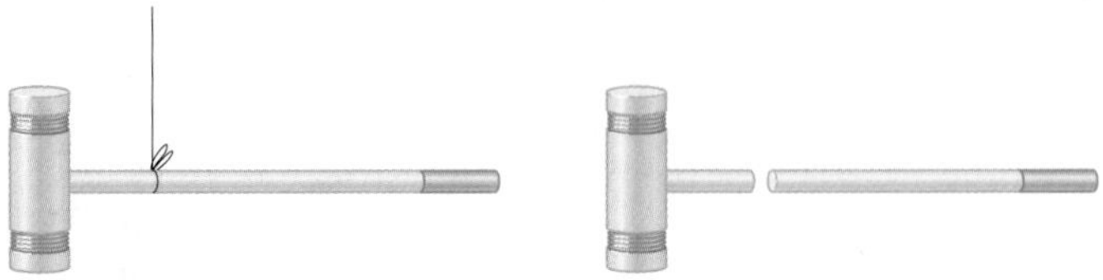

Reasoning and Discussion

The mallet balances because the torques due to the two pieces are of equal magnitude. The piece with the head of the mallet extends a smaller distance from the point of suspension than does the other piece, hence its mass must be greater; that is, a large mass at a small distance creates the same torque as a small mass at a large distance.

Answer:

(b) The piece with the head of the mallet has the greater mass.

Similar considerations apply to an object that is at rest on a surface, as opposed to being suspended from a point. In such a case, the object is in equilibrium as long as its center of mass is directly above the base on which it is supported. For example, when you stand upright with normal posture your feet provide a base of support, and your center of mass is above a point roughly halfway between your feet. If you lift your right foot from the floor—without changing your posture—you will begin to lose your balance and tip over. The reason is that your center of mass is no longer above the base of support, which is now your left foot. To balance on your left foot, you must lean slightly in that direction so as to position your center of mass directly above the foot. This principle applies to everything from a performer in a high-wire act to one of the "balancing rocks" that are a familiar sight in the desert southwest. In Problem 39 we apply this condition for stability to a stack of books on the edge of a table.

▶ (Left) Although it looks precarious, this rock in Arches National Park, Utah, has probably been balancing above the desert for many thousands of years. It will remain secure on its perch as long as its center of mass lies above its base of support. (Right) Although his knowledge may be based more on experience than on physics, the man on this ladder knows what he must do to keep from falling. By extending his leg backward as he leans forward, he keeps his center of mass safely positioned over the foot that supports him.

11–5 Dynamic Applications of Torque

In this section we focus on applications of Newton's second law for rotation. For example, consider a disk-shaped pulley of radius R and mass M with a string wrapped around its circumference, as in **Figure 11–10 (a)**. Hanging from the string is a mass m. When the mass is released, it accelerates downward and the pulley begins to rotate. If the pulley rotates without friction, and the string unwraps without slipping, what is the acceleration of the mass and the tension in the string?

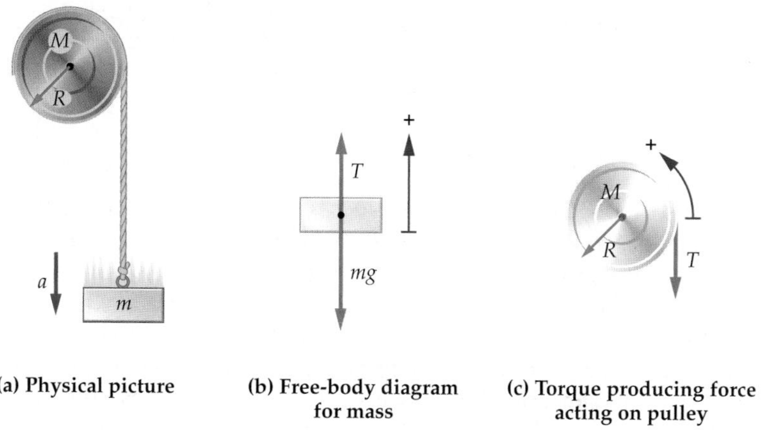

FIGURE 11–10 A mass suspended from a pulley

A mass m hangs from a string wrapped around the circumference of a disk-shaped pulley of radius R and mass M. When the mass is released, it accelerates downward. Positive directions of motion for the system are shown in parts (b) and (c). In part (c), the weight of the pulley acts downward at its center, and the axle exerts an upward force equal in magnitude to the weight of the pulley plus the tension in the string. Of the three forces acting on the pulley, only the tension in the string produces a torque about the axle.

At first it may seem that since the pulley rotates freely, the mass will simply fall with the acceleration of gravity. But remember, the pulley has a nonzero moment of inertia, $I > 0$, which means that it resists any change in its rotational motion. In order for the pulley to rotate, the string must pull downward on it. This means that the string also pulls upward on the mass m with a tension T. As a result, the net downward force on m is less than mg, and thus its acceleration is less than g.

To solve for the acceleration of the mass, we must apply Newton's second law to both the linear motion of the mass *and* the rotational motion of the pulley. The first step is to define a consistent choice of positive directions for the two motions. In Figure 11–10 (a) we note that when the pulley rotates counterclockwise, the mass moves upward. Thus, we choose counterclockwise to be positive for the pulley and upward to be positive for the mass.

With our positive directions established, we proceed to apply Newton's second law. Referring to the free-body diagram for the mass, shown in **Figure 11–10 (b)**, we see that

$$T - mg = ma \qquad \textbf{11–8}$$

Similarly, the free-body diagram for the pulley is shown in **Figure 11–10 (c)**. Note that the tension in the string, T, exerts a tangential force on the pulley at a distance R from the axis of rotation. This produces a torque of magnitude TR. Since the tension tends to cause a clockwise rotation, it follows that the torque is negative; thus, $\tau = -TR$. As a result, Newton's second law for the pulley gives

$$-TR = I\alpha \qquad \textbf{11–9}$$

Now, these two statements of Newton's second law are related by the fact that the string unwraps without slipping. As was discussed in Chapter 10, when a string unwraps without slipping, the angular and linear accelerations are related by

$$\alpha = \frac{a}{R}$$

Using this relation in Equation 11–9 we have

$$-TR = I\frac{a}{R}$$

or, dividing by R,

$$T = -I\frac{a}{R^2}$$

Substituting this result into Equation 11–8 yields

$$-I\frac{a}{R^2} - mg = ma$$

Finally, dividing by m and rearranging yields the acceleration, a:

$$a = -\frac{g}{\left(1 + \dfrac{I}{mR^2}\right)} \qquad \textbf{11–10}$$

Let's briefly check our solution for a. First, note that a is negative. This is to be expected, since the mass accelerates downward, which is the negative direction. Second, if the moment of inertia were zero, $I = 0$, or if the mass m were infinite, $m \to \infty$, the mass would fall with the acceleration of gravity, $a = -g$. When I is greater than zero and m is finite, however, the acceleration of the mass has a magnitude less than g. In fact, in the limit of an infinite moment of inertia, $I \to \infty$, the acceleration vanishes—the mass is simply unable to cause the pulley to rotate in this case.

The next Example presents another system in which Newton's laws are used to relate linear and rotational motions.

EXAMPLE 11–7 The Pulley Matters

A 0.31-kg cart on a horizontal air track is attached to a string. The string passes over a disk-shaped pulley of mass 0.080 kg and radius 0.012 m and is pulled vertically downward with a constant force of 1.1 N. Find **(a)** the tension in the string between the pulley and the cart and **(b)** the acceleration of the cart.

Picture the Problem

The system is shown below. We label the mass of the cart with M, the mass of the pulley with m, and the radius of the pulley with r. The applied downward force creates a tension $T_1 = 1.1$ N in the vertical portion of the string. The horizontal portion of the string, from the pulley to the cart, has a tension T_2. If the pulley had zero mass, these two tensions would be equal. In this case, however, T_2 will have a different value than T_1.

We also show the relevant forces acting on the pulley and the cart. The positive direction of rotation is counterclockwise, and the corresponding positive direction of motion for the cart is to the left.

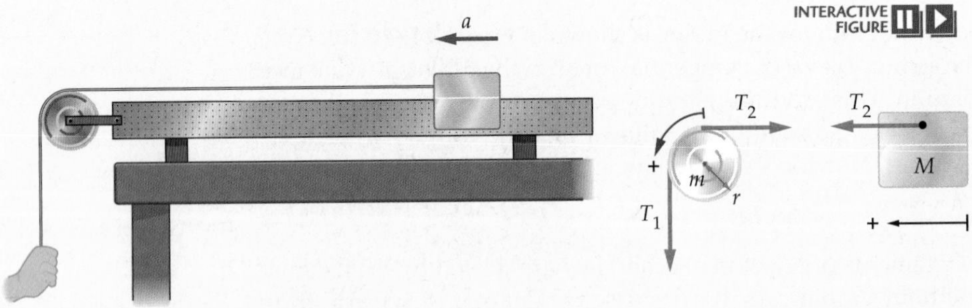

Physical picture

Relevant forces

Strategy

The two unknowns, T_2 and a, can be found by applying Newton's second law to both the pulley and the cart. This gives two equations for two unknowns.

In applying Newton's second law to the pulley, note that since the pulley is a disk, it follows that $I = \frac{1}{2}mr^2$. Also, since the string is not said to slip as it rotates the pulley, we can assume that the angular and linear accelerations are related by $\alpha = a/r$.

Solution

Part (a)

1. Apply Newton's second law to the cart: $T_2 = Ma$

2. Apply Newton's second law to the pulley. Note that T_1 causes a positive torque, and T_2 causes a negative torque. In addition, use the relation $\alpha = a/r$:

$$\Sigma\tau = I\alpha$$

$$rT_1 - rT_2 = \left(\tfrac{1}{2}mr^2\right)\left(\frac{a}{r}\right) = \tfrac{1}{2}mra$$

3. Use the cart equation, $T_2 = Ma$, to eliminate a in the pulley equation:

$$a = \frac{T_2}{M}$$

$$rT_1 - rT_2 = \tfrac{1}{2}mr\left(\frac{T_2}{M}\right)$$

4. Cancel r and solve for T_2:

$$T_2 = \frac{T_1}{1 + m/2M} = \frac{1.1\text{ N}}{1 + 0.080\text{ kg}/[2(0.31\text{ kg})]} = 0.97\text{ N}$$

Part (b)

5. Use $T_2 = Ma$ to find the acceleration:

$$a = \frac{T_2}{M} = \frac{0.97\text{ N}}{0.31\text{ kg}} = 3.1\text{ m/s}^2$$

Insight

Note that T_2 is less than T_1. As a result, the net torque acting on the pulley is in the counterclockwise direction, causing a rotation in that direction, as expected. If the mass of the pulley were zero ($m = 0$), the two tensions would be equal, and the acceleration of the cart would be $T_1/M = 3.5\text{ m/s}^2$.

Practice Problem

What applied force is necessary to give the cart an acceleration of 2.2 m/s^2?
[**Answer:** $T_1 = T_2(1 + m/2M) = (Ma)(1 + m/2M) = 0.77\text{ N}$]

Some related homework problems: Problem 44, Problem 45

11–6 Angular Momentum

When an object of mass m moves with a speed v in a straight line, we say that it has a linear momentum, $p = mv$. When the same object moves with an angular speed ω along the circumference of a circle of radius r, as in **Figure 11–11**, we say that it has an **angular momentum,** L. The magnitude of L is given by replacing m and v in the expression for p with their angular analogs I and ω (Section 10–5). Thus, we define the angular momentum as follows:

Definition of the Angular Momentum, *L*

$$L = I\omega \qquad \text{11–11}$$

SI unit: $\text{kg}\cdot\text{m}^2/\text{s}$

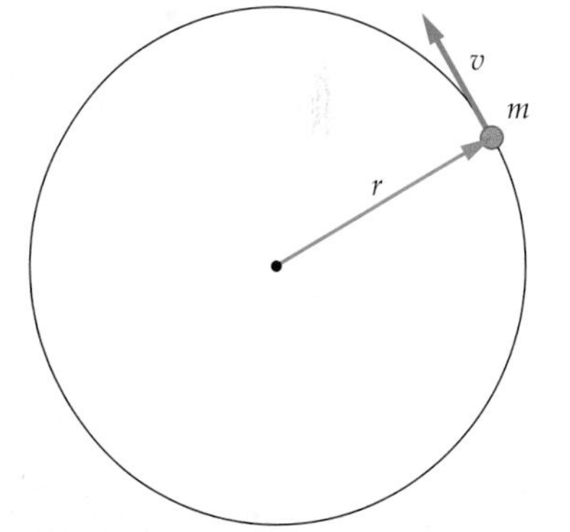

▲ **FIGURE 11–11 The angular momentum of circular motion**

A particle of mass m, moving in a circle of radius r with a speed v. This particle has an angular momentum of magnitude $L = rmv$.

This expression applies to any object undergoing angular motion, whether it is a point mass moving in a circle, as in Figure 11–11, or a rotating hoop, disk, or other object.

Returning for a moment to the case of a point mass m moving in a circle of radius r, recall that the moment of inertia in this case is $I = mr^2$ (Equation 10–18). In addition, the linear speed of the mass is $v = r\omega$ (Equation 10–12). Combining these results, we find

$$L = I\omega = (mr^2)(v/r) = rmv$$

Noting that mv is the linear momentum p, we find that the angular momentum of a point mass can be written in the following form:

$$L = rmv = rp \qquad \text{11–12}$$

It is important to recall that this expression applies specifically to a point particle moving along the circumference of a circle.

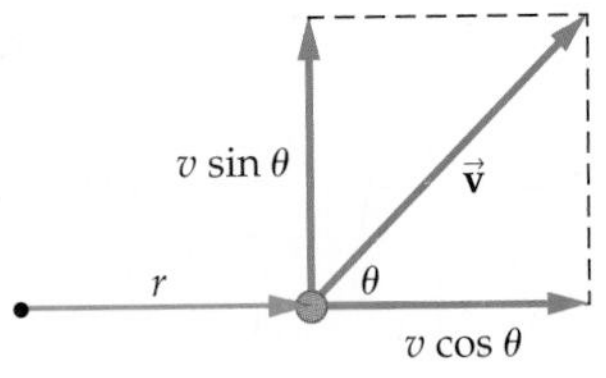

(a)

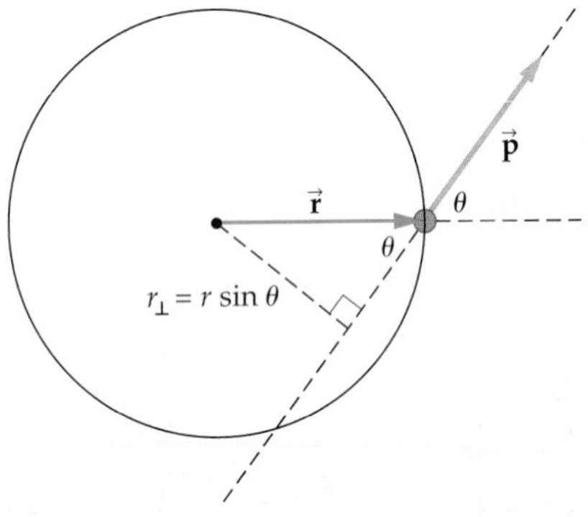

(b)

▲ **FIGURE 11–12 The angular momentum of nontangential motion**

(a) When a particle moves at an angle θ with respect to the radial direction, only the tangential component of velocity, $v \sin \theta$, contributes to the angular momentum. In the case shown here, the particle's angular momentum has a magnitude given by $L = rmv \sin \theta$. **(b)** The angular momentum of an object can also be defined in terms of the moment arm, $r_\perp$. Since $r_\perp = r \sin \theta$, it follows that $L = rmv \sin \theta = r_\perp mv$. Note the similarity between this figure and Figure 11–3.

More generally, a point object may be moving at an angle θ with respect to a radial line, as indicated in **Figure 11–12 (a)**. In this case, it is only the tangential component of the momentum, $p \sin \theta = mv \sin \theta$, that contributes to the angular momentum, just as the tangential component of the force, $F \sin \theta$, is all that contributes to the torque. Thus, the magnitude of the angular momentum for a point particle is defined as:

Angular Momentum, *L*, for a Point Particle

$$L = rp \sin \theta = rmv \sin \theta \qquad \text{11–13}$$

SI unit: $\text{kg} \cdot \text{m}^2/\text{s}$

Note that if the particle moves in a circular path the angle θ is 90° and the angular momentum is $L = rmv$, in agreement with Equation 11–12. On the other hand, if the object moves radially, so that $\theta = 0$, the angular momentum is zero; $L = rmv \sin 0 = 0$.

EXERCISE 11–3

Find the angular momentum of **(a)** a 0.13-kg Frisbee (considered to be a uniform disk of radius 7.5 cm) spinning with an angular speed of 1.15 rad/s, and **(b)** a 95-kg person running with a speed of 5.1 m/s on a circular track of radius 25 m.

Solution

a. Recalling that $I = \frac{1}{2}mR^2$ for a uniform disk (Table 10–1), we have

$$L = I\omega$$
$$= \left(\tfrac{1}{2}mR^2\right)\omega = \tfrac{1}{2}(0.13 \text{ kg})(0.075 \text{ m})^2(1.15 \text{ rad/s}) = 4.2 \times 10^{-4} \text{ kg} \cdot \text{m}^2/\text{s}$$

b. Treating the person as a particle of mass m, we find

$$L = rmv = (25 \text{ m})(95 \text{ kg})(5.1 \text{ m/s}) = 12{,}000 \text{ kg} \cdot \text{m}^2/\text{s}$$

An alternative definition of the angular momentum uses the moment arm, $r_\perp$, as was done for the torque in Equation 11–3. To apply this definition, start by extending a line through the momentum vector, $\vec{p}$, as in **Figure 11–12 (b)**. Next, draw a line from the axis of rotation perpendicular to the line through $\vec{p}$. The perpendicular distance from the axis of rotation to the line of $\vec{p}$ is the moment arm. From the figure we see that $r_\perp = r \sin \theta$. Hence, from Equation 11–13, the angular momentum is

$$L = r_\perp p = r_\perp mv$$

If an object moves in a circle of radius r, the moment arm is $r_\perp = r$ and the angular momentum reduces to our earlier result, $L = rp$.

CONCEPTUAL CHECKPOINT 11–3 Angular Momentum?

Does an object moving in a straight line have nonzero angular momentum **(a)** always, **(b)** sometimes, or **(c)** never?

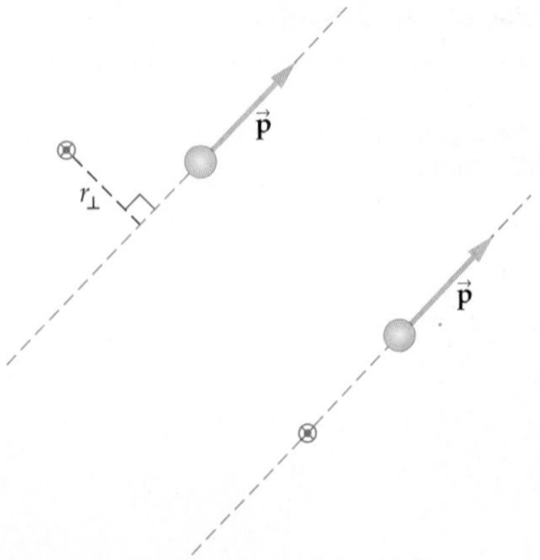

Reasoning and Discussion

The answer is sometimes, because it depends on the choice of the axis of rotation. If the axis of rotation is not on the line drawn through the momentum vector, as in the left sketch below, the moment arm is nonzero, and therefore $L = r_\perp p$ is also nonzero. If the axis of rotation is on the line of motion, as in the right sketch, the moment arm is zero; hence the linear momentum is radial and L vanishes.

Answer:

(b) An object moving in a straight line may or may not have angular momentum, depending on the location of the axis of rotation.

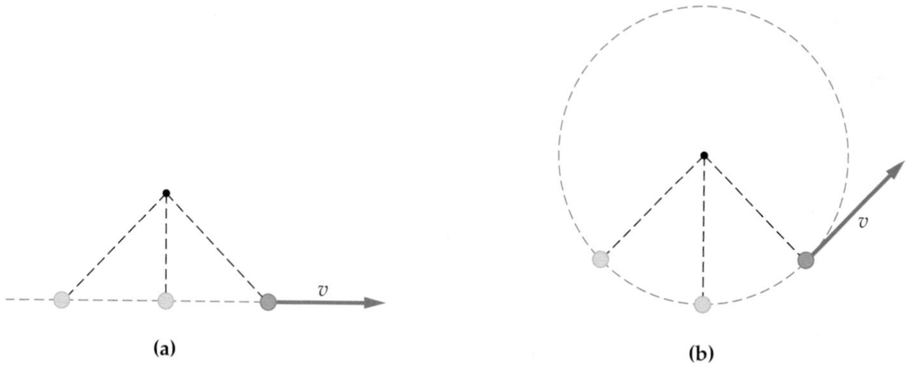

◀ **FIGURE 11–13 Angular momentum in linear and circular motion**

An object moving in **(a)** a straight line and **(b)** a circular path. In both cases, the angular position increases with time; hence, the angular momentum is positive.

Note that an object moving with a momentum p in a straight line that does not go through the axis of rotation has an *angular* position that changes with time. This is illustrated in **Figure 11–13 (a)**. It is for this reason that such an object is said to have an *angular* momentum.

The sign of L is determined by whether the angle to a given object is increasing or decreasing with time. For example, the object moving counterclockwise in a circular path in **Figure 11–13 (b)** has a positive angular momentum, since θ is increasing with time. Similarly, the object in Figure 11–13 (a) also has an angle θ that increases with time, hence its angular momentum is positive as well. On the other hand, if these objects were to have their direction of motion reversed, they would have angles that decrease with time and their angular momenta would be negative.

EXAMPLE 11–8 Jump On

Running with a speed of 4.10 m/s, a 21.2-kg child heads toward the rim of a merry-go-round, as shown. If the radius of the merry-go-round is 2.00 m, and the child moves in the direction shown, what is the child's angular momentum with respect to the center of the merry-go-round?

Picture the Problem

Our sketch shows the child approaching the rim of the merry-go-round at an angle of 135° relative to the radial direction. Note that the line of motion of the child does not go through the axis of the merry-go-round. As a result, the child has a nonzero initial angular momentum with respect to that axis of rotation.

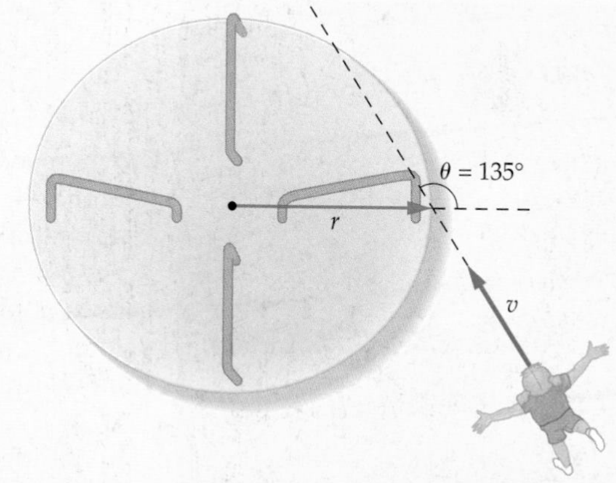

Strategy

The child's angular momentum can be found by applying $L = rmv \sin\theta$. In this case, we see from the sketch that $\theta = 135°$ and $r = 2.00$ m. Finally, the values of m and v are given in the problem statement.

Solution

1. Evaluate $L = rmv \sin\theta$:

$$L = rmv \sin\theta = (2.00 \text{ m})(21.2 \text{ kg})(4.10 \text{ m/s}) \sin 135° = 123 \text{ kg}\cdot\text{m}^2/\text{s}$$

Insight

When the child lands on the merry-go-round, she will transfer angular momentum to it causing the merry-go-round to rotate about its center. This will be discussed in more detail in the next section.

Practice Problem

For what angle relative to the radial line does the child have a maximum angular momentum? What is the angular momentum in this case? [**Answer:** $\theta = 90°$, for which $L = rmv = 174 \text{ kg}\cdot\text{m}^2/\text{s}$]

Some related homework problems: Problem 51, Problem 52, Problem 53

Next, we consider the rate of change of angular momentum with time. Since the moment of inertia is a constant—as long as the mass and shape of the object remain unchanged—the change in L in a time interval Δt is

$$\frac{\Delta L}{\Delta t} = I\frac{\Delta \omega}{\Delta t}$$

Recall, however, that $\Delta\omega/\Delta t$ is the angular acceleration, α. Therefore, we have

$$\frac{\Delta L}{\Delta t} = I\alpha$$

Since $I\alpha$ is the torque, it follows that Newton's second law for rotational motion can be written as

Newton's Second Law for Rotational Motion

$$\sum\tau = I\alpha = \frac{\Delta L}{\Delta t} \qquad \text{11–14}$$

Clearly, this is the rotational analog of $\Sigma F_x = ma_x = \Delta p_x/\Delta t$. Just as force can be expressed as the change in *linear* momentum in a given time interval, the torque can be expressed as the change in *angular* momentum in a time interval.

EXERCISE 11–4

In a light wind, a windmill experiences a constant torque of 255 N · m. If the windmill is initially at rest, what is its angular momentum 2.00 s later?

Solution

Solve Equation 11–14 for the change in angular momentum due to a single torque τ:

$$\Delta L = L_f - L_i = \left(\sum\tau\right)\Delta t = \tau\,\Delta t$$

Since the initial angular momentum of the windmill is zero, its final angular momentum is

$$L_f = \tau\,\Delta t = (255\ \text{N}\cdot\text{m})(2.00\ \text{s}) = 510\ \text{kg}\cdot\text{m}^2/\text{s}$$

▲ Once she has launched herself into space, this diver is essentially a projectile. However, the principle of conservation of angular momentum allows her to control the rotational part of her motion. By curling her body up into a tight "tuck," she decreases her moment of inertia, thereby increasing the speed of her spin. To slow down for an elegant entry into the water, she will extend her body, increasing her moment of inertia.

11–7 Conservation of Angular Momentum

When an ice skater goes into a spin and pulls her arms inward to speed up, she probably doesn't think about angular momentum. Neither does a diver who springs into the air and folds her body to speed her rotation. Most people, in fact, are not aware that the actions of these athletes are governed by the same basic laws of physics that cause a collapsing star to spin faster as it becomes a rapidly rotating pulsar. Yet in all these cases, as we shall see, **conservation of angular momentum** is at work.

To see the origin of angular momentum conservation, consider an object with an initial angular momentum L_i acted on by a single torque τ. After a period of time, Δt, the object's angular momentum changes in accordance with Newton's second law:

$$\tau = \frac{\Delta L}{\Delta t}$$

Solving for ΔL, we find

$$\Delta L = L_f - L_i = \tau\,\Delta t$$

Thus, the final angular momentum of the object is

$$L_f = L_i + \tau\,\Delta t$$

If the torque acting on the object is zero, $\tau = 0$, it follows that the initial and final angular momenta are equal—that is, the angular momentum is conserved:

$$L_f = L_i \qquad (\text{if } \tau = 0)$$

Angular momentum is also conserved in systems acted on by more than one torque, provided that the *net external torque* is zero. The reason that internal torques can be ignored is that, just as internal forces come in equal and opposite pairs that cancel, so too do internal torques. As a result, the internal torques in a system sum to zero, and the net torque acting on it is simply the net external torque. Thus, for a general system, angular momentum is conserved if $\tau_{\text{net, ext}}$ is zero:

Conservation of Angular Momentum

$$L_f = L_i \quad (\text{if } \tau_{\text{net, ext}} = 0) \qquad \text{11–15}$$

As an illustration of angular momentum conservation, we consider the case of a student rotating on a piano stool in the next Example. Notice how a change in moment of inertia results in a change in angular speed.

EXAMPLE 11–9 Going for a Spin

For a classroom demonstration, a student sits on a piano stool holding a sizable mass in each hand. Initially, the student holds his arms outstretched and spins about the axis of the stool with an angular speed of 3.74 rad/s. The moment of inertia in this case is 5.33 kg · m². While still spinning, the student pulls his arms in to his chest, reducing the moment of inertia to 1.60 kg · m². What is the student's angular speed now?

Picture the Problem

The initial and final configurations of the student are shown in our sketch. Clearly, the mass distribution in the final configuration, with the masses held closer to the axis of rotation, results in a smaller moment of inertia.

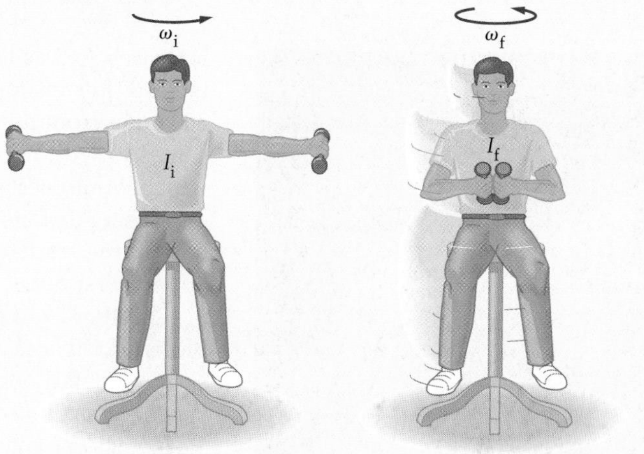

Strategy

Ignoring friction in the axis of the stool, since none was mentioned, we conclude that no external torques act on the system. As a result, the angular momentum is conserved. Therefore, setting the initial angular momentum, $L_i = I_i\omega_i$, equal to the final angular momentum, $L_f = I_f\omega_f$, yields the final angular speed.

Solution

1. Apply angular momentum conservation to this system:

$$L_i = L_f$$
$$I_i\omega_i = I_f\omega_f$$

2. Solve for the final angular speed, ω_f:

$$\omega_f = \left(\frac{I_i}{I_f}\right)\omega_i$$

3. Substitute numerical values:

$$\omega_f = \left(\frac{5.33\ \text{kg}\cdot\text{m}^2}{1.60\ \text{kg}\cdot\text{m}^2}\right)(3.74\ \text{rad/s}) = 12.5\ \text{rad/s}$$

Insight

Initially the student completes one revolution roughly every two seconds. After pulling the weights in, the student's rotation rate has increased to almost two revolutions a second—quite a dizzying pace. The same physics applies to a rotating diver or a spinning ice skater.

Practice Problem

What moment of inertia would be required to give a final spin rate of 10.0 rad/s? [**Answer:** $I_f = (\omega_i/\omega_f)I_i = 1.96\ \text{kg}\cdot\text{m}^2$]

Some related homework problems: Problem 55, Problem 64

REAL-WORLD PHYSICS
Hurricanes and tornadoes

An increasing angular speed, as experienced by the student in Example 11–9, can be observed in nature as well. For example, a hurricane draws circulating air in at ground level toward its "eye," where it then rises to an altitude of 10 miles or more. As air moves inward toward the axis of rotation, its angular speed

▲ This 1992 satellite photo of Hurricane Andrew (left), one of the most powerful hurricanes of recent decades, clearly suggests the rotating structure of the storm. The violence of the hurricane winds can be attributed in large part to conservation of angular momentum: as air is pushed inward toward the low pressure near the eye of the storm, its rotational velocity increases. The same principle, operating on a smaller scale, explains the tremendous destructive power of tornadoes. The tornado shown at right passed through downtown Miami on May 12, 1997.

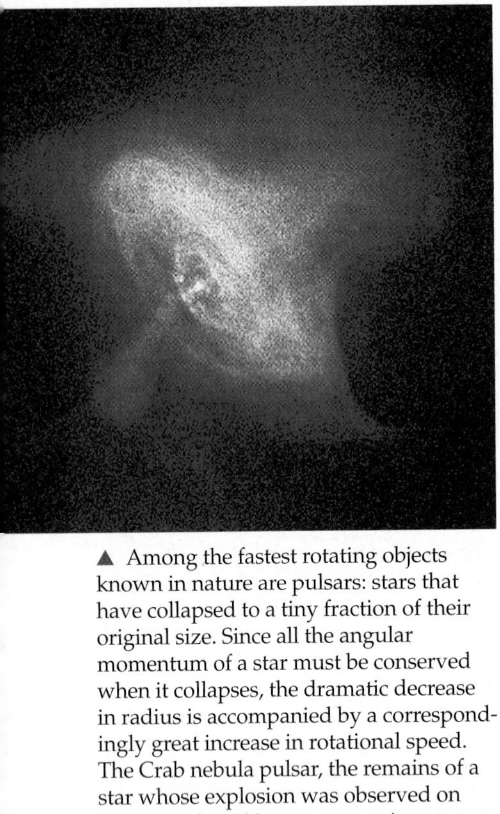

▲ Among the fastest rotating objects known in nature are pulsars: stars that have collapsed to a tiny fraction of their original size. Since all the angular momentum of a star must be conserved when it collapses, the dramatic decrease in radius is accompanied by a correspondingly great increase in rotational speed. The Crab nebula pulsar, the remains of a star whose explosion was observed on Earth nearly 1000 years ago, spins at about 30 rev/s. This X-ray photograph shows rings and jets of high-energy particles flying outward from the whirling neutron star at the center.

increases, just as the masses held by the student speed up when they are pulled inward. For example, if the wind has a speed of only 3.0 mph at a distance of 300 miles from the center of the hurricane, it would speed up to 150 mph when it comes to within 6.0 miles of the center. Of course, this analysis ignores friction, which would certainly decrease the wind speed. Still, the basic principle—that a decreasing distance from the axis of rotation implies an increasing speed—applies to both the student and the hurricane. Similar behavior is observed in tornadoes and waterspouts.

Another example of conservation of angular momentum occurs in stellar explosions. On occasion a star will explode, sending a portion of its material out into space. After the explosion, the star collapses to a fraction of its original size, speeding up its rotation in the process. If the mass of the star is greater than 1.44 times the mass of the Sun, the collapse can continue until a *neutron star* is formed, with a radius of only about 10 to 20 km. Neutron stars have incredibly high densities; in fact, if you could bring a teaspoonful of neutron star material to the Earth, it would weigh about 100 million tons! In addition, neutron stars produce powerful beams of X-rays and other radiation that sweep across the sky like a gigantic lighthouse beam as the star rotates. On the Earth we see pulses of radiation from these rotating beams, one for each revolution of the star. These "pulsating stars," or *pulsars*, typically have periods ranging from about 2 ms to nearly one second. The Crab nebula (see Problems 9 and 90 in Chapter 10) is a famous example of such a system. The dependence of angular speed on radius for a collapsing star is considered in Active Example 11–4.

ACTIVE EXAMPLE 11–4 A Stellar Performance: Find the Angular Speed

A star of radius $R = 2.3 \times 10^8$ m rotates with an angular speed $\omega = 2.4 \times 10^{-6}$ rad/s. If this star collapses to a radius of 20.0 km, find its final angular speed. (Treat the star as if it were a uniform sphere, and assume that no mass is lost as the star collapses.)

Solution *(Test your understanding by performing the calculations indicated in each step.)*

1. Apply conservation of angular momentum: $I_i\omega_i = I_f\omega_f$
2. Write expressions for the initial and final moments of inertia: $I_i = \frac{2}{5}MR_i^2$, and $I_f = \frac{2}{5}MR_f^2$
3. Solve for the final angular speed: $\omega_f = (I_i/I_f)\omega_i = (R_i^2/R_f^2)\omega_i$
4. Substitute numerical values: $\omega_f = 320$ rad/s

Insight
The final angular speed corresponds to a period of about 20 ms, a typical period for pulsars. Since 320 rad/s is roughly 3000 rpm, it follows that a pulsar, which has the mass of a star, rotates as fast as the engine in a racing car.

Your Turn
At what radius will the star's period of rotation be equal to 15 ms?
*(Answers to **Your Turn** problems are given in the back of the book.)*

REAL-WORLD PHYSICS
Angular speed of a pulsar

Note that if the student in Example 11–9 were to stretch his arms back out again, the resulting *increase* in the moment of inertia would cause a *decrease* in his angular speed. The same effect might apply to the Earth one day. For example, a melting of the polar ice caps would lead to an increase in the Earth's moment of inertia (as we saw in Chapter 10) and thus, by angular momentum conservation, the angular speed of the Earth would decrease. This would mean that more time would be required for the Earth to complete a revolution about its axis of rotation; that is, the day would lengthen.

Since angular momentum is conserved in the systems we have studied so far, it is natural to ask whether the energy is conserved as well. We consider this question in the next Conceptual Checkpoint.

CONCEPTUAL CHECKPOINT 11–4 Compare Kinetic Energies

A skater pulls in her arms, decreasing her moment of inertia by a factor of two, and doubling her angular speed. Is her final kinetic energy **(a)** equal to, **(b)** greater than, or **(c)** less than her initial kinetic energy?

Reasoning and Discussion
Let's calculate the initial and final kinetic energies, and compare them. The initial kinetic energy is

$$K_i = \tfrac{1}{2}I_i\omega_i^2$$

After pulling in her arms, the skater has half the moment of inertia and twice the angular speed. Hence, her final kinetic energy is

$$K_f = \tfrac{1}{2}I_f\omega_f^2 = \tfrac{1}{2}(I_i/2)(2\omega_i)^2 = 2\left(\tfrac{1}{2}I_i\omega_i^2\right) = 2K_i$$

Thus, the fact that K depends on the square of ω leads to an increase in the kinetic energy. The source of this additional energy is the work done by the muscles in the skater's arms as she pulls them in to her body.

Answer:
(b) The skater's kinetic energy increases.

Rotational Collisions

In the not-too-distant past, a person would play music by placing a record on a rotating turntable. Suppose, for example, that a turntable with a moment of inertia I_t is rotating freely with an initial angular speed ω_0. A record, with a moment of inertia I_r and initially at rest, is dropped straight down onto the rotating turntable, as in **Figure 11–14**. When the record lands, frictional forces between it and the turntable cause the record to speed up and the turntable to slow down, until they

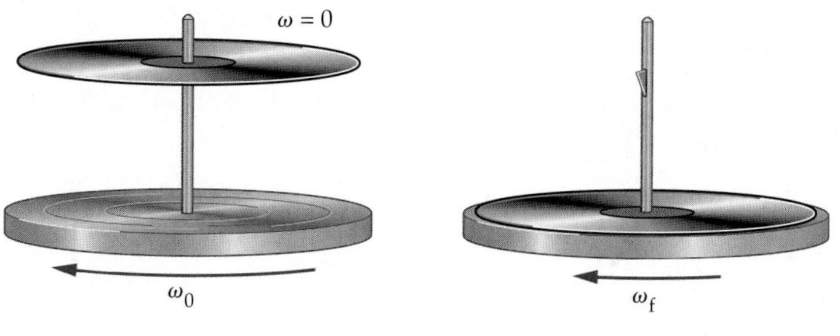

◀ **FIGURE 11–14 A rotational collision**
A nonrotating record dropped onto a rotating turntable is an example of a "rotational collision." Since only internal forces are involved during the collision, the final angular momentum is equal to the initial angular momentum.

both have the same angular speed. Since only internal forces are involved during this process, it follows that the system's angular momentum is conserved. We can think of this event, then, as a "rotational collision."

Before the collision, the angular momentum of the system is

$$L_i = I_t\omega_0$$

After the collision, when both the record and the turntable are rotating with the angular speed ω_f, the system's angular momentum is

$$L_f = I_t\omega_f + I_r\omega_f$$

Setting $L_f = L_i$ yields the final angular speed:

$$\omega_f = \left(\frac{I_t}{I_t + I_r}\right)\omega_0 \qquad \textbf{11–16}$$

Since this collision is completely inelastic, we expect the final kinetic energy to be less than the initial kinetic energy.

We conclude this section with a somewhat different example of a rotational collision. The physical principles involved are precisely the same, however.

ACTIVE EXAMPLE 11–5 Conserve Angular Momentum: Find the Angular Speed

A 34.0-kg child runs with a speed of 2.80 m/s tangential to the rim of a stationary merry-go-round. The merry-go-round has a moment of inertia of 512 $kg \cdot m^2$ and a radius of 2.31 m. When the child jumps onto the merry-go-round, the entire system begins to rotate. What is the angular speed of the system?

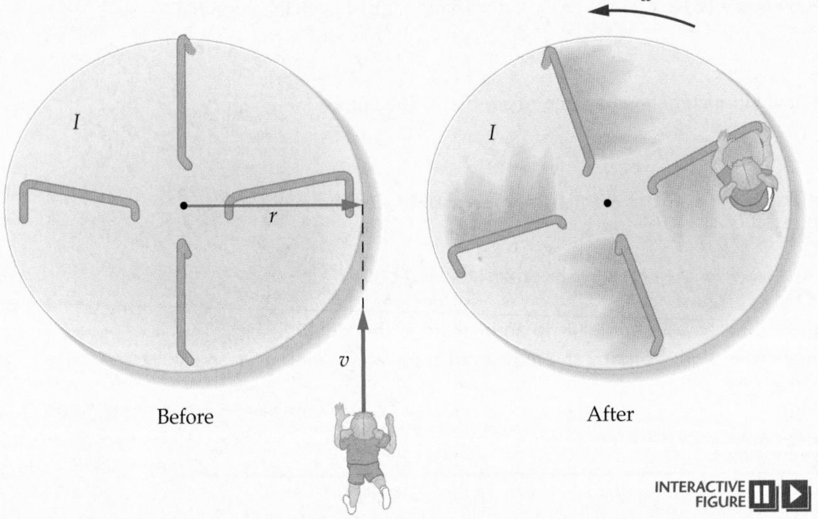

Solution *(Test your understanding by performing the calculations indicated in each step.)*

1. Write the initial angular momentum of the child:	$L_i = rmv$
2. Write the final angular momentum of the system:	$L_f = (I + mr^2)\omega$
3. Set $L_f = L_i$ and solve for the angular speed:	$\omega = rmv/(I + mr^2)$
4. Substitute numerical values:	$\omega = 0.317$ rad/s

Insight

If the moment of inertia of the merry-go-round had been zero, $I = 0$, the angular speed would be $\omega = v/r$. This means that the linear speed of the child, $r\omega = v$, is unchanged. If $I > 0$, however, the linear speed of the child is decreased. In this particular case, the child's linear speed after the collision is only $v = r\omega = 0.733$ m/s.

Your Turn

What initial speed does the child have if, after landing on the merry-go-round, it takes her 22.5 s to complete one revolution?

(Answers to **Your Turn** *problems are given in the back of the book.)*

The initial and final kinetic energies of the system in Active Example 11–5 are considered in Problem 58.

11–8 Rotational Work and Power

Just as a force acting through a distance performs work on an object, so too does a torque acting through an angular displacement. To see this, consider again the fishing line pulled from a reel. If the line is pulled with a force F for a distance Δx, as in **Figure 11–15**, the work done on the reel is

$$W = F\Delta x$$

Now, since the line is unwinding without slipping, it follows that the linear displacement of the line, Δx, is related to the angular displacement of the reel, $\Delta\theta$, by the following relation:

$$\Delta x = R\Delta\theta$$

In this equation, R is the radius of the reel, and $\Delta\theta$ is measured in radians. Thus, the work can be written as

$$W = F\Delta x = FR\Delta\theta$$

Finally, the torque exerted on the reel by the line is $\tau = RF$, and hence the work done on the reel is simply torque times angular displacement:

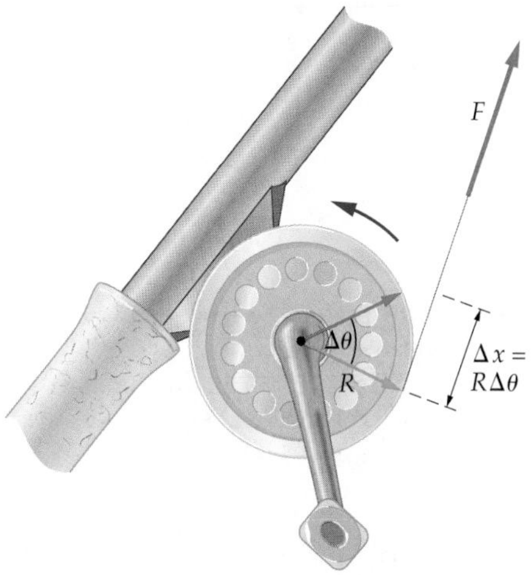

▲ **FIGURE 11–15 Rotational work**
A force F pulling a length of line Δx from a fishing reel does the work $W = F\,\Delta x$. In terms of torque and angular displacement, the work can be expressed as $W = \tau\,\Delta\theta$.

Work Done by Torque

$$W = \tau\Delta\theta \qquad 11\text{–}17$$

Note again the analogies between angular and linear quantities in $W = F\Delta x$ and $W = \tau\Delta\theta$. As usual, τ is the analog of F, and θ is the analog of x.

As we saw in Chapter 7, the net work done on an object is equal to the change in its kinetic energy. This is the work-energy theorem:

$$W = \Delta K = K_f - K_i \qquad 11\text{–}18$$

The work–energy theorem applies regardless of whether the work is done by a force acting through a distance or a torque acting through an angle.

Similarly, power is the amount of work done in a given time, regardless whether the work is done by a force or a torque. In the case of a torque, we have $W = \tau\Delta\theta$, and hence

Power Produced by a Torque

$$P = \frac{W}{\Delta t} = \tau\frac{\Delta\theta}{\Delta t} = \tau\omega \qquad 11\text{–}19$$

Again, the analogy is clear between $P = Fv$ for the linear case, and $P = \tau\omega$ for the rotational case.

EXERCISE 11–5

It takes a good deal of effort to make homemade ice cream. **(a)** If the torque required to turn the handle on an ice cream maker is 5.7 N·m, how much work is expended on each complete revolution of the handle? **(b)** How much power is required to turn the handle if each revolution is completed in 1.5 s?

Solution

a. Applying Equation 11–17 yields

$$W = \tau\Delta\theta = (5.7\ \text{N}\cdot\text{m})(2\pi\ \text{rad}) = 36\ \text{J}$$

b. Power is the work per time; that is,

$$P = W/\Delta t = (36\ \text{J})/(1.5\ \text{s}) = 24\ \text{W}$$

Equivalently, the angular speed of the handle is $\omega = (2\pi)/T = (2\pi)/(1.5\ \text{s}) = 4.2\ \text{rad/s}$, and therefore Equation 11–19 yields $P = \tau\omega = (5.7\ \text{N}\cdot\text{m})(4.2\ \text{rad/s}) = 24\ \text{W}$.

*11–9 The Vector Nature of Rotational Motion

We have mentioned many times that the angular velocity is a vector, and that we must be careful to use the proper sign for ω. But if the angular velocity is a vector, what is its direction?

To address this question, consider the rotating wheel shown in **Figure 11–16**. Each point on the rim of this wheel has a velocity vector pointing in a different direction in the plane of rotation. Since different parts of the wheel move in different directions, how can we assign a single direction to the angular velocity vector, $\vec{\omega}$? The answer is that there is only one direction that remains fixed as the wheel rotates; the direction of the axis of rotation. By definition, then, the angular velocity vector, $\vec{\omega}$, is taken to point along the axis of rotation.

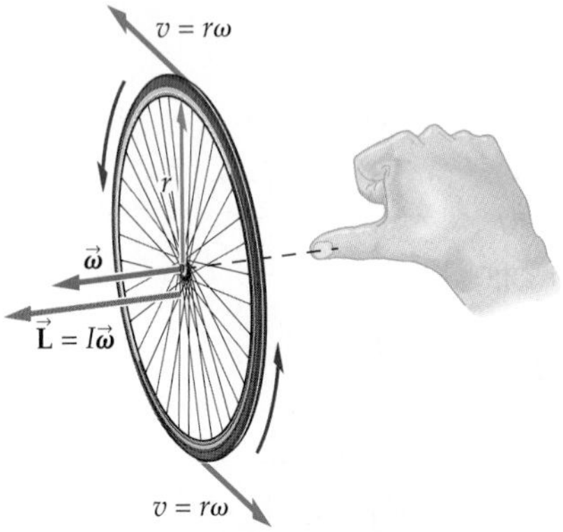

▲ **FIGURE 11–16 The right-hand rule for angular velocity**

The angular velocity, $\vec{\omega}$, of a rotating wheel points along the axis of rotation. Its direction is given by the right-hand rule.

Given that $\vec{\omega}$ points along the axis of rotation, we must still decide whether it points to the left or to the right in Figure 11–16. The convention we use for assigning the direction of $\vec{\omega}$ is referred to as the right-hand rule:

Right-Hand Rule for the Angular Velocity, $\vec{\omega}$
Curl the fingers of the right hand in the direction of rotation.
The thumb now points in the direction of the angular velocity, $\vec{\omega}$.

The right-hand rule for $\vec{\omega}$ is illustrated in Figure 11–16.

The same convention for direction applies to the angular momentum vector. First, recall that the angular momentum has a magnitude given by $L = I\omega$. Hence, we choose the direction of $\vec{L}$ to be the same as the direction of $\vec{\omega}$. That is

$$\vec{L} = I\vec{\omega} \qquad 11\text{–}20$$

The angular momentum vector is also illustrated in Figure 11–16.

Similarly, torque is a vector, and it too is defined to point along the axis of rotation. The right-hand rule for torque is similar to that for angular velocity:

Right-Hand Rule for Torque, $\vec{\tau}$
Curl the fingers of the right hand in the direction of rotation that this torque would cause if it acted alone.
The thumb now points in the direction of the torque vector, $\vec{\tau}$.

Examples of torque vectors are given in **Figure 11–17**.

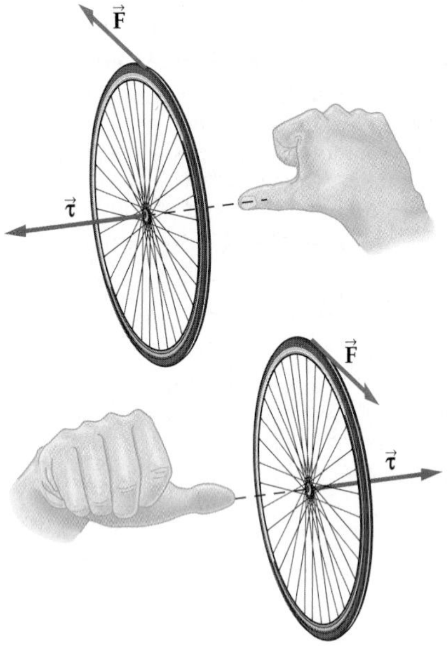

▲ **FIGURE 11–17 The right-hand rule for torque**

Examples of torque vectors obtained using the right-hand rule.

As an example of torque and angular momentum vectors, consider the spinning bicycle wheel shown in **Figure 11–18**. The angular momentum vector for the wheel points to the left, along the axis of rotation. If a person pushes on the rim of the wheel in the direction indicated, the resulting torque is also to the left, as shown in the figure. If this torque lasts for a time Δt, the angular momentum changes by the amount

$$\Delta\vec{L} = \vec{\tau}\,\Delta t$$

Adding $\Delta\vec{L}$ to the original angular momentum $\vec{L}_i$ yields the final angular momentum, $\vec{L}_f$, shown in Figure 11–18. Since $\vec{L}_f$ is in the same direction as $\vec{L}_i$, but with a greater magnitude, it follows that the wheel is spinning in the same direction as

▶ Children have always been fascinated by tops—but not only children. The physicists in the photo at right, Wolfgang Pauli and Niels Bohr, seem as delighted by a spinning top as any child. Their contributions to modern physics, discussed in Chapter 30, helped to show that subatomic particles, the ultimate constituents of matter, have a property (now referred to as "spin") that is in some ways analogous to the rotation of a top or a gyroscope.

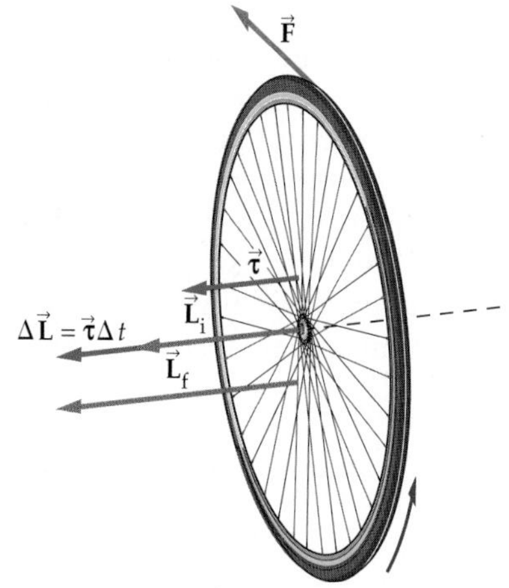

▲ **FIGURE 11–18 Torque and angular momentum vectors**

A tangential push on the spinning wheel in the direction shown causes a torque to the left. As a result, the angular momentum increases. Hence, the wheel spins faster, as expected.

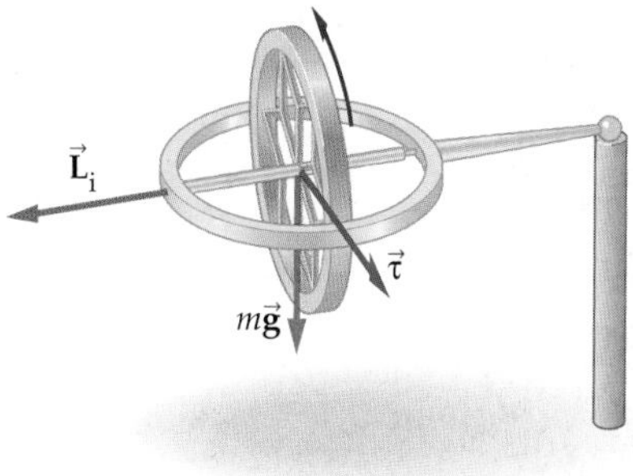

▲ **FIGURE 11–19 The torque exerted on a gyroscope**

A spinning gyroscope has an initial angular momentum to the left. The torque due to gravity is out of the page.

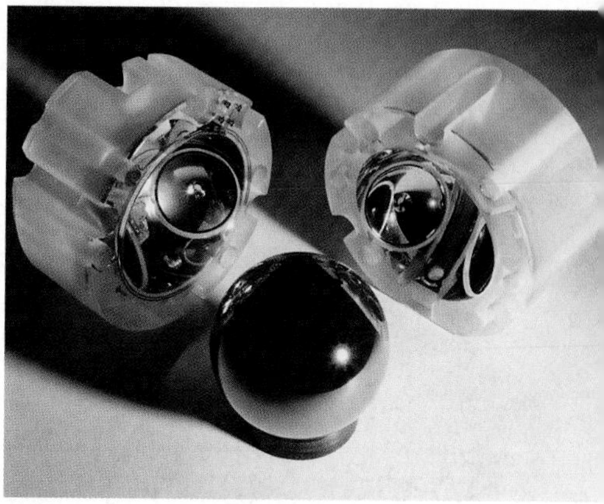

▲ The 1.5-inch fused quartz sphere shown here is no ordinary ball. In fact, it is the most perfect sphere ever manufactured. If the Earth were this smooth, the change in elevation from the deepest ocean trench to the highest mountain peak would be only 16 feet. Such precision is required because this sphere is designed to serve as the rotor for an extremely sensitive gyroscope. The device, a million times more sensitive than those used in the best inertial navigation systems, orbits the Earth as part of an experiment to test predictions of Einstein's theory of general relativity.

before, only faster. This is to be expected, considering the direction of the person's push on the wheel.

On the other hand, if a person pushes on the wheel in the opposite direction, the torque vector points to the right. As a result, $\Delta\vec{L}$ points to the right as well. When we add $\Delta\vec{L}$ and $\vec{L}_i$ to obtain the final angular momentum, $\vec{L}_f$, we find that it has the same direction as $\vec{L}_i$, but a smaller magnitude. Hence, we conclude that the wheel spins more slowly, as one would expect.

Finally, a case of considerable interest is when the torque and angular momentum vectors are at right angles to one another. The classic example of such a system is the **gyroscope.** To begin, consider a gyroscope whose axis of rotation is horizontal, as in **Figure 11–19**. If the gyroscope were to be released with no spin it would simply fall, rotating counterclockwise downward about its point of support. Curling the fingers of the right hand in the counterclockwise direction, we see that the thumb, and hence the torque due to gravity, points out of the page.

Next, imagine the gyroscope to be spinning rapidly—as would normally be the case—with its angular momentum pointing to the left in Figure 11–19. If the gyroscope is released now, it doesn't fall as before, even though the torque is the same. To see what happens instead, consider the change in angular momentum, $\Delta\vec{L}$, caused by the torque, $\vec{\tau}$, acting for a small interval of time. As shown in **Figure 11–20**, the small change, $\Delta\vec{L}$, is at right angles to $\vec{L}_i$; hence the final angular momentum, $\vec{L}_f$, is essentially the same length as $\vec{L}_i$, but pointing in a direction slightly out of the page. With each small interval of time, the angular momentum vector continues to change in direction so that, viewed from above as in Figure 11–20, the gyroscope as a whole rotates in a counterclockwise sense around its support point. This type of motion, where the axis of rotation changes direction with time, is referred to as **precession.**

Because of its spinning motion about its rotational axis, the Earth may be considered as one rather large gyroscope. Gravitational forces exerted on the Earth by the Sun and the Moon subject it to external torques that cause its rotational axis to precess. At the moment, the rotational axis of the Earth points toward Polaris, the "North Star," which remains almost fixed in position in time-lapse photographs while the other stars move in circular paths about it. In a few hundred years, however, Polaris will also move in a circular path in the sky because the Earth's axis of rotation will point in a different direction. After 26,000 years the Earth will complete one full cycle of precession, and Polaris will again be the pole star.

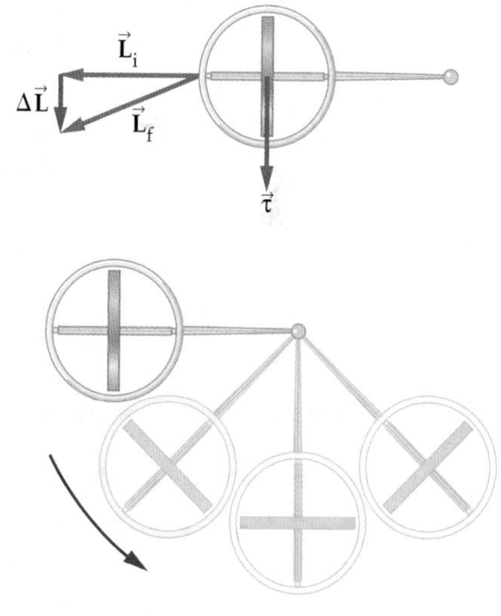

Top view

▲ **FIGURE 11–20 Precession of a gyroscope**

The gyroscope as viewed from above. In a time Δt the angular momentum changes by the amount $\Delta\vec{L} = \vec{\tau}\Delta t$. This causes the angular momentum vector, and hence the gyroscope as a whole, to rotate in a counterclockwise direction.

REAL-WORLD PHYSICS
Precession of the Earth

REAL-WORLD PHYSICS
Gyroscopes in navigation and space

On a smaller scale, gyroscopes are used in the navigational systems of a variety of vehicles. In such applications, the rapidly spinning wheel of a gyroscope is mounted on nearly frictionless bearings so that it is practically free from external torques. If no torque acts on the gyroscope, its angular momentum vector remains unchanged both in magnitude and—here is the important point—in direction. With the axis of its gyroscope always pointing in the same, known direction, it is possible for a vehicle to maintain a desired direction of motion relative to the gyroscope's reference direction. On the Hubble Space Telescope, for example, six gyroscopes are used for pointing and stability, though it can operate with only three working gyroscopes if necessary.

Chapter Summary

11–1 Torque

A force applied so as to cause an angular acceleration is said to exert a torque, τ.

Tangential Force
A force is tangential if it is tangent to a circle centered on the axis of rotation.

Torque Due to a Tangential Force
A tangential force F applied at a distance r from the axis of rotation produces a torque

$$\tau = rF \qquad 11\text{–}1$$

Torque for a General Force
A force exerted at an angle θ with respect to the radial direction, and applied at a distance r from the axis of rotation, produces the torque

$$\tau = rF \sin\theta \qquad 11\text{–}2$$

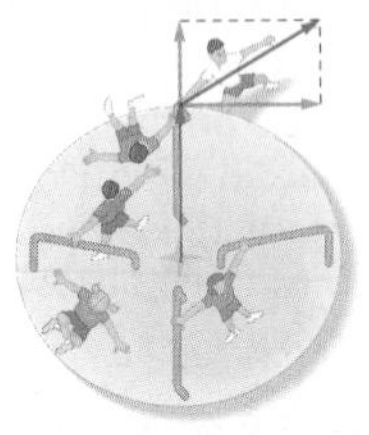

11–2 Torque and Angular Acceleration

A single torque applied to an object gives it an angular acceleration.

Newton's Second Law for Rotation
The connection between torque and angular acceleration is

$$\sum \tau = I\alpha \qquad 11\text{–}4$$

In this expression, I is the moment of inertia about the axis of rotation and α is the angular acceleration about this axis.

Rotational/Translational Analogies
Torque is analogous to force, the moment of inertia is analogous to mass, and the angular acceleration is analogous to linear acceleration. Therefore, the rotational analog of $F = ma$ is $\tau = I\alpha$.

11–3 Zero Torque and Static Equilibrium

The conditions for an object to be in static equilibrium are that the total force and the total torque acting on the object must be zero:

$$\sum F_x = 0, \quad \sum F_y = 0, \quad \sum \tau = 0$$

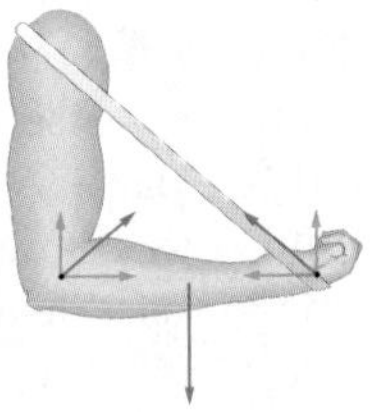

11–4 Center of Mass and Balance

An object balances when it is supported at its center of mass.

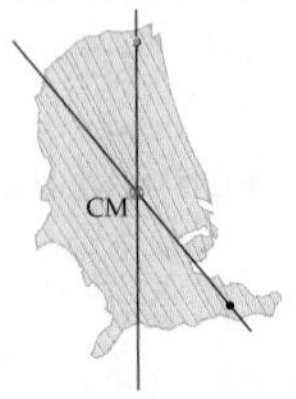

11–5 Dynamic Applications of Torque

Newton's second law can be applied to rotational systems in a way that is completely analogous to its application to linear systems.

Systems Involving Both Rotational and Linear Elements

In a system with both rotational and linear motion—such as a string passing over a pulley and attached to a mass—Newton's second law must be applied separately to the rotational and linear motions of the system. Connections between the two motions, such as $\alpha = a/r$, can be used to solve for all the accelerations in the system.

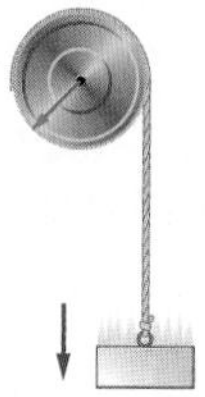

11–6 Angular Momentum

A moving object has angular momentum, as long as its direction of motion does not extend through the axis of rotation.

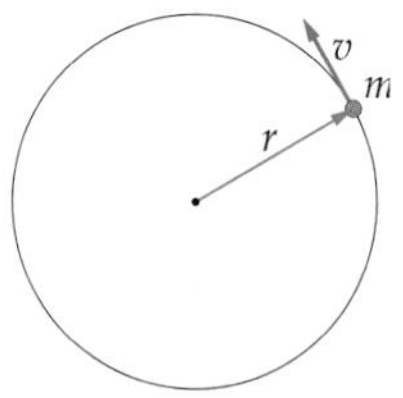

Angular Momentum and Angular Speed

Angular momentum can be expressed in terms of angular speed and the moment of inertia as follows:

$$L = I\omega \qquad \text{11–11}$$

This is the rotational analog of $p = mv$.

Tangential Motion

An object of mass m moving tangentially with a speed v at a distance r from the axis of rotation has an angular momentum, L, given by

$$L = rmv \qquad \text{11–12}$$

General Motion

If an object of mass m is a distance r from the axis of rotation and moves with a speed v at an angle θ with respect to the radial direction, its angular momentum is

$$L = rmv \sin\theta \qquad \text{11–13}$$

Newton's Second Law

Newton's second law can be expressed in terms of the rate of change of the angular momentum:

$$\sum \tau = I\alpha = \frac{\Delta L}{\Delta t} \qquad \text{11–14}$$

This is the rotational analog of $\Sigma F = \Delta p/\Delta t$.

11–7 Conservation of Angular Momentum

If the net external torque acting on a system is zero, its angular momentum is conserved:

$$L_f = L_i$$

Rotational Collisions

Systems in which two rotational objects come into contact can be thought of in terms of a "rotational collision." In such a case, the total angular momentum of the system is conserved.

11–8 Rotational Work and Power

A torque acting through an angle does work, just as does a force acting through a distance.

Work Done by a Torque

A torque τ acting through an angle $\Delta\theta$ does a work W given by

$$W = \tau\Delta\theta \qquad \text{11–17}$$

Work–Energy Theorem

The work–energy theorem is

$$W = \Delta K = K_f - K_i \qquad \text{11–18}$$

This theorem applies whether the work is done by a force or by a torque. In the linear case the kinetic energy is $\frac{1}{2}mv^2$; in the rotational case, the kinetic energy is $K = \frac{1}{2}I\omega^2$: (Equation 10–17).

*11–9 The Vector Nature of Rotational Motion

Rotational quantities have directions that point along the axis of rotation. The precise direction is given by the right-hand rule.

Right-Hand Rule
If the fingers of the *right hand* are curled in the direction of rotation, the thumb points in the direction of the rotational quantity in question. This rule applies to the angular velocity vector, $\vec{\omega}$, the angular acceleration vector, $\vec{\alpha}$, the angular momentum vector, $\vec{L}$, and the torque vector, $\vec{\tau}$.

Problem-Solving Summary

Type of Problem	Relevant Physical Concepts	Related Examples
Find the torque exerted on a system.	The torque exerted by a tangential force a distance r from the axis of rotation is $\tau = rF$. If the force is at an angle θ to the radial direction, the torque is $\tau = rF \sin \theta$.	Example 11–1
Determine the angular acceleration of a system.	First, calculate the torque exerted on the system. Next, find the angular acceleration using Newton's second law as applied to rotation, namely, $\tau = I\alpha$.	Examples 11–2, 11–3
Find the forces required for static equilibrium.	Static equilibrium requires that both the net force and the net torque acting on a system be zero.	Examples 11–4, 11–5, 11–6 Active Examples 11–1, 11–2, 11–3
Find the final angular momentum of a system.	A torque changes the angular momentum L of a system with time as follows; $\tau = \Delta L/\Delta t$. If no net torque acts on a system, its angular momentum is conserved.	Examples 11–8, 11–9 Active Examples 11–4, 11–5

Conceptual Questions

(Answers to odd-numbered Conceptual Questions can be found in the back of the book.)

1. Two forces produce the same torque. Does it follow that they have the same magnitude? Explain.
2. A car pitches down in front when the brakes are applied sharply. Explain this observation in terms of torques.
3. A tightrope walker uses a long pole to aid in balancing. Why?
4. When a motorcycle accelerates rapidly from a stop it sometimes "pops a wheelie"; that is, its front wheel may lift off the ground. Explain this behavior in terms of torques.
5. Give an example of a system in which the net torque is zero but the net force is nonzero.
6. Give an example of a system in which the net force is zero but the net torque is nonzero.
7. Is the normal force exerted by the ground the same for all four tires on your car? Explain.
8. Give two everyday examples of objects that are not in static equilibrium.
9. Give two everyday examples of objects that are in static equilibrium.
10. Can an object have zero translational acceleration and, at the same time, have nonzero angular acceleration? If your answer is no, explain why not. If your answer is yes, give a specific example.
11. Stars form when a large rotating cloud of gas collapses. What happens to the angular speed of the gas cloud as it collapses?
12. What purpose does the tail rotor on a helicopter serve?
13. Is it possible to change the angular momentum of an object without changing its linear momentum? If your answer is no, explain why not. If your answer is yes, give a specific example.
14. Suppose a diver springs into the air with no initial angular velocity. Can the diver begin to rotate by folding into a tucked position? Explain.

Conceptual Exercises

(Answers to odd-numbered Conceptual Exercises can be found in the back of the book.)

1. Consider the pulley-block systems shown in Conceptual Checkpoint 11–1. **(a)** Is the tension in the left-hand string greater than, less than, or equal to the weight of the mass attached to the string? Explain. **(b)** Is the tension in the left-hand string greater than, less than, or equal to the tension in the right-hand string? Explain.
2. Suppose a torque rotates your body about one of three different axes of rotation: case A, an axis through your spine; case B, an axis through your hips; case C, an axis through your ankles. Rank these three axes of rotation in increasing order of the angular acceleration produced by the torque.
3. The L-shaped object described in Problem 12 can be rotated in one of the following three ways: case A, about the x axis; case B, about the y axis; and case C, about the z axis (which passes through the origin perpendicular to the plane of the figure.) If the same torque τ is applied in each of these cases, rank them in increasing order of the resulting angular acceleration.

4. Suppose the person in Active Example 11–3 climbs higher on the ladder. As a result, is the ladder more likely, less likely, or equally likely to slip? Explain.

5. A uniform disk stands upright on its edge, and rests on a sheet of paper placed on a tabletop. If the paper is pulled horizontally to the right, as in **Figure 11–21**, **(a)** does the disk rotate clockwise or counterclockwise about its center? Explain. **(b)** Does the center of the disk move to the right, or to the left, or does it stay in the same location? Explain.

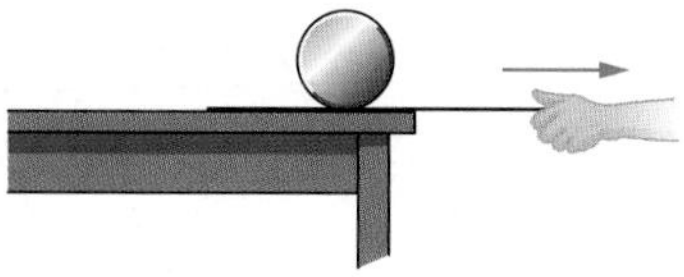

▲ **FIGURE 11–21** Conceptual Exercise 5

6. Consider the two rotating systems shown in **Figure 11–22**, each consisting of a mass m attached to a rod of negligible mass pivoted at one end. On the left, the mass is attached at the midpoint of the rod; to the right, it is attached to the free end of the rod. The rods are released from rest in the horizontal position at the same time. When the rod to the left reaches the vertical position, is the rod to the right (1) not yet vertical, (2) vertical, or (3) past vertical? Explain.

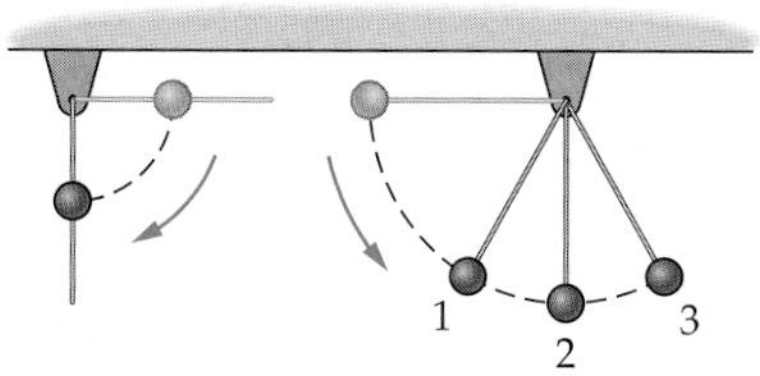

▲ **FIGURE 11–22** Conceptual Exercise 6

7. A disk and a hoop (bicycle wheel) of equal radius and mass each have a string wrapped around their circumferences. Hanging from the strings, halfway between the disk and the hoop, is a block of mass m, as shown in **Figure 11–23**. The disk and the hoop are free to rotate about their centers. When the block is allowed to fall, does it stay on the center line, move toward the right, or move toward the left? Explain.

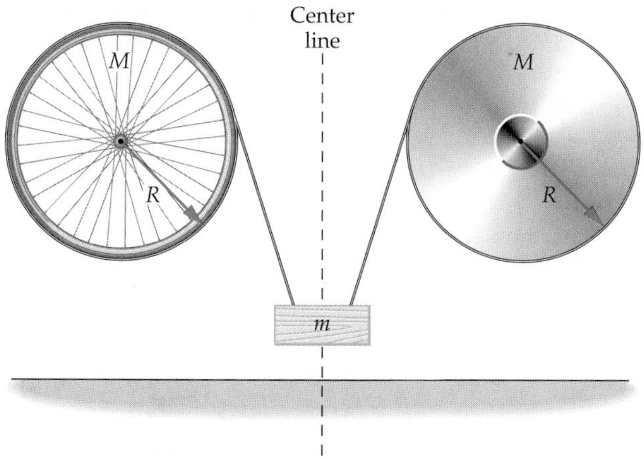

▲ **FIGURE 11–23** Conceptual Exercise 7

8. A motorcycle accelerates from rest, and both tires rotate without slipping. **(a)** Is the force exerted by the ground on the rear tire in the forward or backward direction? Explain. **(b)** Is the force exerted by the ground on the front tire in the forward or backward direction? Explain. **(c)** If the moment of inertia of the front tire is increased, will the motorcycle's acceleration increase, decrease, or remain the same? Explain.

9. A student rotates on a frictionless piano stool with his arms outstretched, a heavy weight in each hand. Suddenly he lets go of the weights, and they fall to the floor. As a result, does the student's angular speed increase, decrease, or stay the same? Explain.

10. A puck on a horizontal, frictionless surface is attached to a string that passes through a hole in the surface, as shown in **Figure 11–24**. As the puck rotates about the hole, the string is pulled downward, bringing the puck closer to the hole. During this process, does the puck's **(a)** linear speed, **(b)** angular speed, and **(c)** angular momentum increase, decrease, or remain the same? Explain in each case.

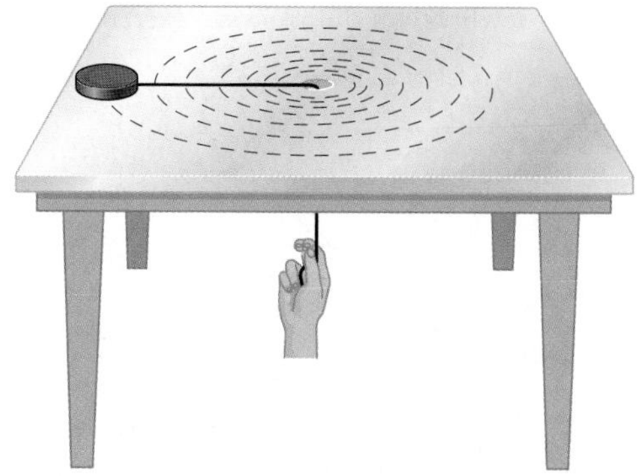

▲ **FIGURE 11–24** Conceptual Exercise 10 and Problem 78

11. A puck on a horizontal, frictionless surface is attached to a string that wraps around a pole of finite radius, as shown in **Figure 11–25**. **(a)** As the puck moves along the spiral path, does its speed increase, decrease, or stay the same? Explain. **(b)** Does its angular momentum increase, decrease, or stay the same? Explain.

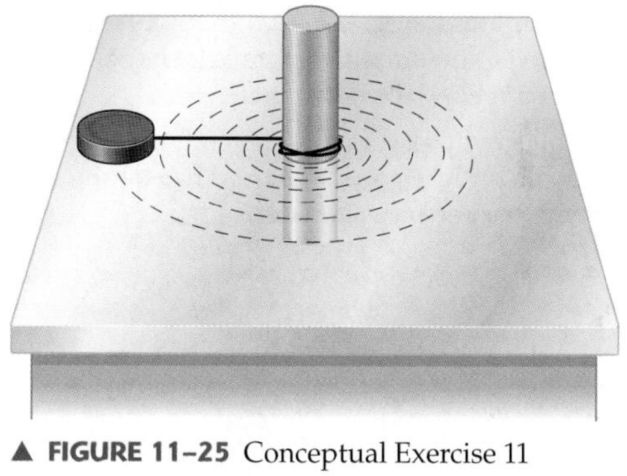

▲ **FIGURE 11–25** Conceptual Exercise 11

12. A beetle sits at the rim of a turntable that is at rest but is free to rotate about a vertical axis. Suppose the beetle now begins to walk around the perimeter of the turntable. Does the beetle move forward, backward, or does it remain in the same location relative to the ground? Answer for two different cases, **(a)** the turntable is much more massive than the beetle and **(b)** the turntable is massless.

13. A beetle sits near the rim of a turntable that is rotating without friction about a vertical axis. The beetle now begins to walk toward the center of the turntable. As a result, does the angular speed of the turntable increase, decrease, or stay the same? Explain.

14. Suppose the Earth were to magically expand, doubling its radius while keeping its mass the same. Would the length of the day increase, decrease, or stay the same? Explain.

15. Two spheres of equal mass and radius are rolling across the floor with the same speed. Sphere 1 is solid; sphere 2 is hollow. Is the work required to stop sphere 1 greater than, less than, or equal to the work required to stop sphere 2? Explain.

Problems

Note: **IP** *denotes an integrated conceptual/quantitative problem.* **BIO** *identifies problems of biological or medical interest. Red bullets (•, ••, •••) are used to indicate the level of difficulty of each problem.*

Section 11–1 Torque

1. • To tighten a spark plug, it is recommended that a torque of 15 N · m be applied. If a mechanic tightens the spark plug with a wrench that is 25 cm long, what is the minimum force necessary to create the desired torque?

2. • **Pulling a Weed** The gardening tool shown in **Figure 11–26** is used to pull weeds. If a 1.23-N · m torque is required to pull a given weed, what force did the weed exert on the tool?

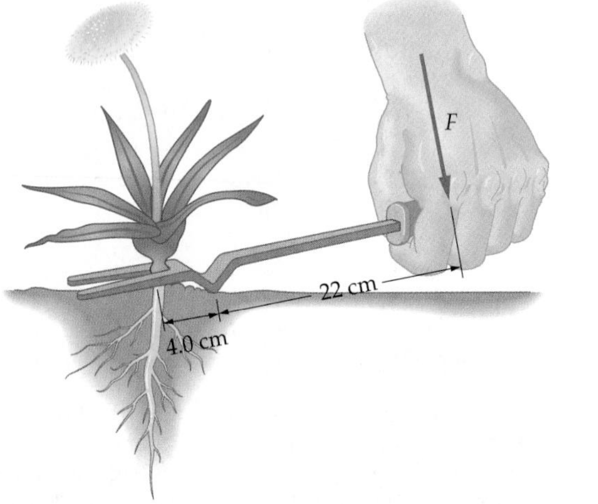

▲ **FIGURE 11–26** Problem 2

3. • A 1.51 kg bowling trophy is held at arm's length, a distance of 0.605 m from the shoulder joint. What torque does the trophy exert about the shoulder if the arm is **(a)** horizontal, or **(b)** at an angle of 22.5° below the horizontal?

4. • A person slowly lowers a 3.6-kg crab trap over the side of a dock, as shown in **Figure 11–27**. What torque does the trap exert about the person's shoulder?

▲ **FIGURE 11–27** Problem 4

5. •• **IP BIO Force to Hold a Baseball** A person holds a 1.42-N baseball in his hand, a distance of 34.0 cm from the elbow joint, as shown in **Figure 11–28**. The biceps, attached at a distance of 2.75 cm from the elbow, exerts an upward force of 12.6 N on the forearm. Consider the forearm and hand to be a uniform rod with a mass of 1.20 kg. **(a)** Calculate the net torque acting on the forearm and hand. Use the elbow joint as the axis of rotation. **(b)** If the net torque obtained in part (a) is nonzero, in which direction will the forearm and hand rotate? **(c)** Would the torque exerted on the forearm by the biceps increase or decrease if the biceps were attached farther from the elbow joint?

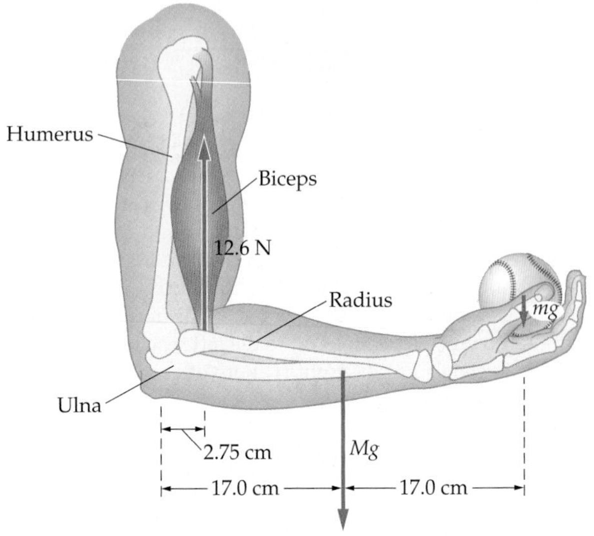

▲ **FIGURE 11–28** Problems 5 and 19

6. •• At the local playground, a 16-kg child sits on the end of a horizontal teeter-totter, 1.5 m from the pivot point. On the other side of the pivot an adult pushes straight down on the teeter-totter with a force of 95 N. In which direction does the teeter-totter rotate if the adult applies the force at a distance of **(a)** 3.0 m, **(b)** 2.5 m, or **(c)** 2.0 m from the pivot?

Section 11–2 Torque and Angular Acceleration

7. • A torque of 0.97 N · m is applied to a bicycle wheel of radius 35 cm and mass 0.75 kg. Treating the wheel as a hoop, find its angular acceleration.

8. • When a ceiling fan rotating with an angular speed of 2.65 rad/s is turned off, a frictional torque of 0.120 N · m slows it to a stop in 19.5 s. What is the moment of inertia of the fan?

9. • When the play button is pressed, a CD accelerates uniformly from rest to 450 rev/min in 3.0 revolutions. If the CD has a radius of 6.0 cm and a mass of 17 g, what is the torque exerted on it?

10. •• A person holds a ladder horizontally at its center. Treating the ladder as a uniform rod of length 3.15 m and mass 8.22 kg, find the torque the person must exert on the ladder to give it an angular acceleration of 0.302 rad/s^2.

11. •• **IP** A wheel on a game show is given an initial angular speed of 1.22 rad/s. It comes to rest after rotating through 0.75 of a turn. **(a)** Find the average torque exerted on the wheel given that it is a disk of radius 0.71 m and mass 6.4 kg. **(b)** If the mass of the wheel is doubled and its radius is halved, will the angle through which it rotates before coming to rest increase, decrease, or stay the same? Explain. (Assume that the average torque exerted on the wheel is unchanged.)

12. •• The L-shaped object in **Figure 11–29** consists of three masses connected by light rods. What torque must be applied to this object to give it an angular acceleration of 1.20 rad/s^2 if it is rotated about **(a)** the x axis, **(b)** the y axis, or **(c)** the z axis (which is through the origin and perpendicular to the page)?

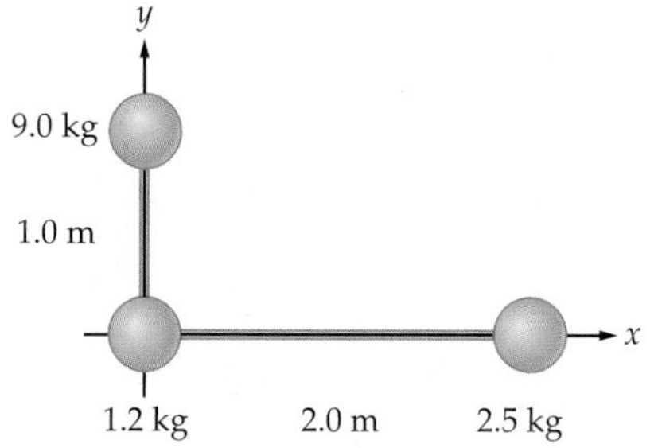

▲ **FIGURE 11–29** Problems 12 and 73 and Conceptual Exercise 3

13. •• **IP** A torque of 13 N · m is applied to the rectangular object shown in **Figure 11–30**. The torque can act about the x axis, the y axis, or the z axis, which passes through the origin and points out of the page. **(a)** In which case does the object experience the greatest angular acceleration? The least angular acceleration? Explain. Find the angular acceleration when the torque acts about **(b)** the x axis, **(c)** the y axis, and **(d)** the z axis.

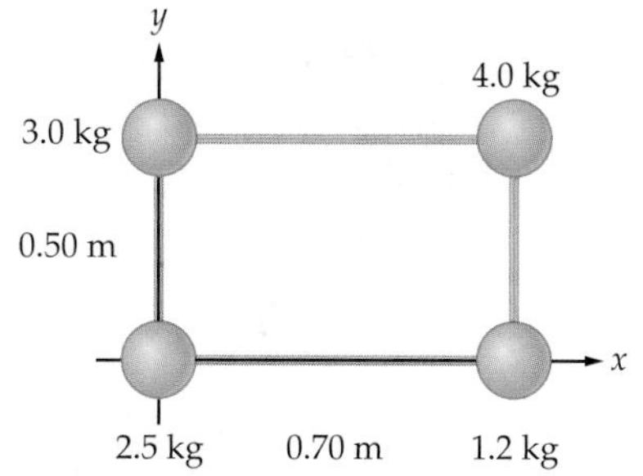

▲ **FIGURE 11–30** Problems 13 and 74

14. •• A fish takes the bait and pulls on the line with a force of 2.1 N. The fishing reel, which rotates without friction, is a cylinder of radius 0.055 m and mass 0.84 kg. **(a)** What is the angular acceleration of the fishing reel? **(b)** How much line does the fish pull from the reel in 0.25 s?

15. •• Repeat the previous problem, only now assume the reel has a friction clutch that exerts a restraining torque of 0.047 N · m.

Section 11–3 Zero Torque and Static Equilibrium

16. • A string that passes over a pulley has a 0.301-kg mass attached to one end and a 0.635-kg mass attached to the other end. The pulley, which is a disk of radius 9.50 cm, has friction in its axle. What is the magnitude of the frictional torque that must be exerted by the axle if the system is to be in static equilibrium?

17. • To loosen the lid on a jar of jam 8.9 cm in diameter, a torque of 8.5 N · m must be applied to the circumference of the lid. If a jar wrench whose handle extends 15 cm from the center of the jar is attached to the lid, what is the minimum force required to open the jar?

18. • Consider the system in Active Example 11–1, this time with the axis of rotation at the location of the child. Write out both the condition for zero net force and the condition for zero net torque. Solve for the two forces.

19. •• **IP BIO** Referring to the person holding a baseball in Problem 5, suppose the biceps exert just enough upward force to keep the system in static equilibrium. **(a)** Is the force exerted by the biceps more than, less than, or equal to the combined weight of the forearm, hand, and baseball? Explain. **(b)** Determine the force exerted by the biceps.

20. •• **IP BIO A Person's Center of Mass** To determine the location of her center of mass, a physics student lies on a lightweight plank supported by two scales 2.50 m apart, as indicated in **Figure 11–31**. If the left scale reads 290 N, and the right scale reads 122 N, find **(a)** the student's mass and **(b)** the distance from the student's head to her center of mass.

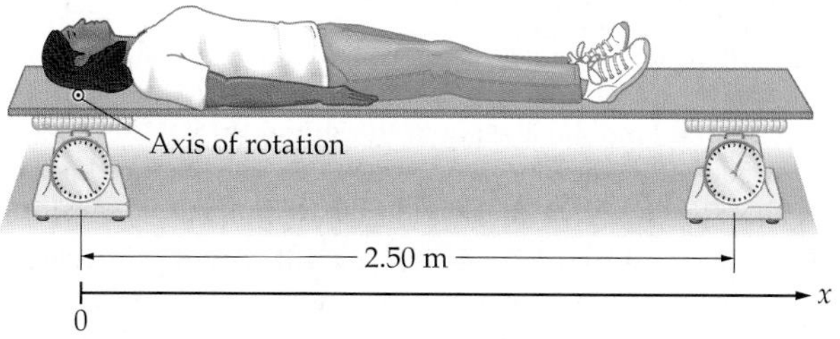

▲ **FIGURE 11–31** Problem 20

21. •• **Triceratops** A set of fossilized triceratops footprints discovered in Texas show that the front and rear feet were 3.2 m apart, as shown in **Figure 11–32**. The rear footprints were observed to be twice as deep as the front footprints. Assuming that the rear feet pressed down on the ground with twice the force exerted by the front feet, find the horizontal distance from the rear feet to the triceratops's center of mass.

▲ **FIGURE 11–32** Problem 21

22. •• **IP** A school yard teeter-totter with a total length of 5.2 m and a mass of 36 kg is pivoted at its center. A 18-kg child sits on one end of the teeter-totter. **(a)** Where should a parent push vertically downward with a force of 210 N in order to hold the teeter-totter level? **(b)** Where should the parent push with a force of 310 N? **(c)** How would your answers to parts (a) and (b) change if the mass of the teeter-totter were doubled? Explain.

23. •• A 0.122-kg remote control 23.0 cm long rests on a table, as shown in **Figure 11–33**, with a length L overhanging its edge. To operate the power button on this remote requires a force of 0.365 N. How far can the remote control extend beyond the

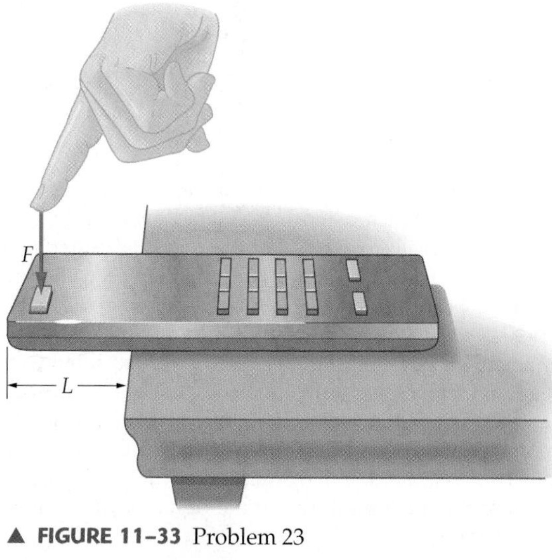

▲ **FIGURE 11–33** Problem 23

edge of the table and still not tip over when you press the power button? Assume the mass of the remote is distributed uniformly, and that the power button is 1.41 cm from the overhanging end of the remote.

24. •• **IP** A 0.11-kg meter stick is held perpendicular to a vertical wall by a 2.5-m string going from the wall to the far end of the stick. **(a)** Find the tension in the string. **(b)** If a shorter string is used, will its tension be greater than, less than, or the same as that found in part (a)? **(c)** Find the tension in a 2.0-m string.

25. •• Repeat Example 11–4, this time with a uniform diving board that weighs 225 N.

26. •• Babe Ruth steps to the plate and casually points to left center field to indicate the location of his next home run. The mighty Babe holds his bat across his shoulder, with one hand holding the small end of the bat. The bat is horizontal, and the distance from the small end of the bat to the shoulder is 22.5 cm. If the bat has a mass of 1.10 kg and has a center of mass that is 67.0 cm from the small end of the bat, find the magnitude and direction of the force exerted by **(a)** the hand and **(b)** the shoulder.

27. •• A uniform metal rod, with a mass of 3.1 kg and a length of 1.2 m, is attached to a wall by a hinge at its base. A horizontal wire bolted to the wall 0.51 m above the base of the rod holds the rod at an angle of 25° above the horizontal. The wire is attached to the top of the rod. **(a)** Find the tension in the wire. Find **(b)** the horizontal and **(c)** the vertical components of the force exerted on the rod by the hinge.

28. •• **IP** In the previous problem, suppose the wire is shortened, so that the rod now makes an angle of 35° with the horizontal. The wire is horizontal, as before. **(a)** Do you expect the tension in the wire to increase, decrease, or stay the same as a result of its new length? Explain. **(b)** Calculate the tension in the wire.

29. •• Repeat Active Example 11–3, this time with a uniform 7.2-kg ladder that is 4.0 m long.

30. •• A rigid, vertical rod of negligible mass is connected to the floor by a bolt through its lower end, as shown in **Figure 11–34**. The rod also has a wire connected between its top end and the floor. If a horizontal force F is applied at the midpoint of the rod, find **(a)** the tension in the wire, and **(b)** the horizontal and **(c)** the vertical components of force exerted by the bolt on the rod.

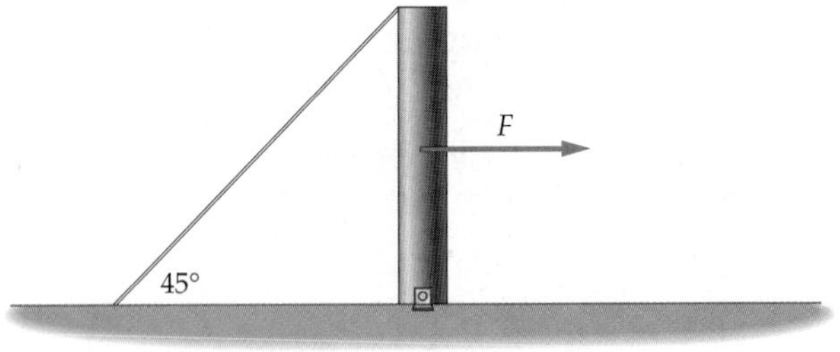

▲ **FIGURE 11–34** Problems 30, 96, and 97

31. •• **BIO Correcting Torsiversion** Torsiversion is a medical condition in which a tooth is rotated away from its normal position about the long axis of the root. For example, the first premolar in **Figure 11–35** is not only displaced slightly in the negative y direction, it is also rotated clockwise from its proper orientation. To correct this condition, an orthodontist might use an archwire and a bracket to apply both a force and a torque to the tooth. In the case shown, the forces F_1 and F_2 produce a net force in the positive y direction of 1.8 N and a torque about the center of the tooth (the origin of the coordinate system) of 0.0092 N · m. Find the magnitude of the two forces, given that the distances indicated in the figure are $d = 3.2$ mm and $D = 4.5$ mm.

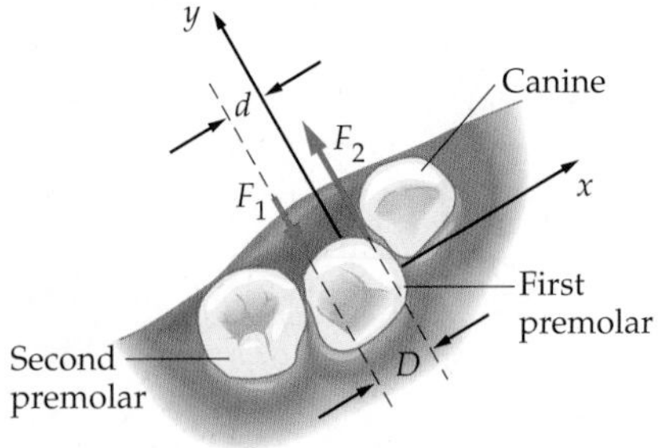

▲ **FIGURE 11–35** Problem 31

32. ••• **BIO Forces in the Foot Figure 11–36** shows the forces acting on a sprinter's foot just before she takes off at the start of the race. Find the magnitude of the force exerted on the heel by the Achilles tendon, F_H, and the magnitude of the force exerted on the foot at the ankle joint, F_J.

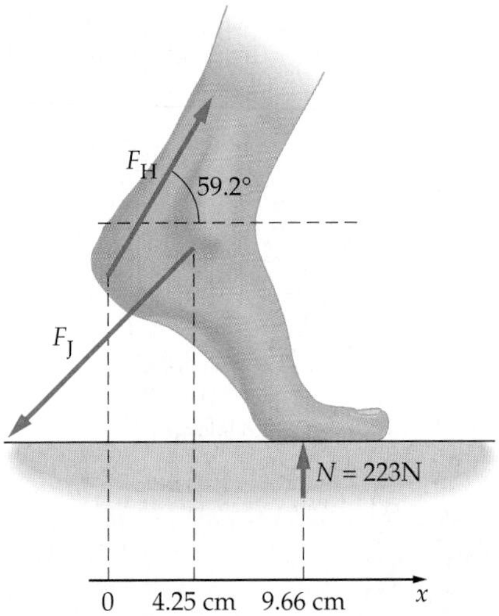

▲ **FIGURE 11–36** Problem 32

33. ••• A stick with a mass of 0.214 kg and a length of 0.436 m rests in contact with a bowling ball and a rough floor, as shown in **Figure 11–37**. The bowling ball has a diameter of 21.6 cm, and the angle the stick makes with the horizontal is 30.0°. You may assume there is no friction between the stick and the bowling ball, though friction with the floor must be taken into account. **(a)** Find the magnitude of the force exerted on the stick by the bowling ball. **(b)** Find the horizontal component of the force exerted on the stick by the floor. **(c)** Repeat part (b) for the vertical component of the force.

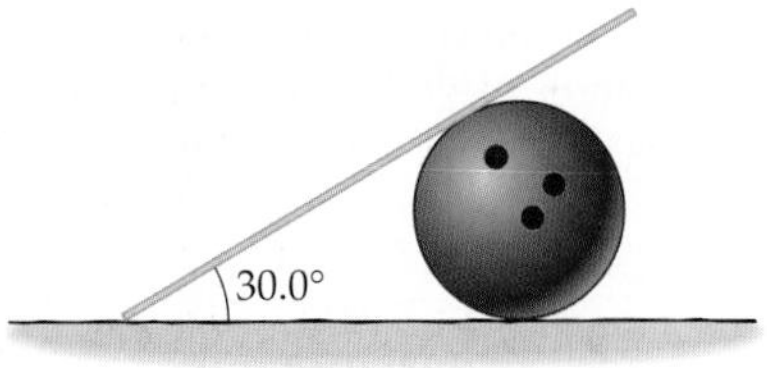

▲ **FIGURE 11–37** Problem 33

34. ••• **IP** A uniform crate with a mass of 16.2 kg rests on a floor with a coefficient of static friction equal to 0.571. The crate is a uniform cube with sides 1.21 m in length. **(a)** What horizontal force applied to the top of the crate will initiate tipping? **(b)** If the horizontal force is applied halfway to the top of the crate, it will begin to slip before it tips. Explain.

35. ••• In the previous problem, **(a)** what is the minimum height where the force F can be applied so that the crate begins to tip before sliding? **(b)** What is the magnitude of the force in this case?

Section 11–4 Center of Mass and Balance

36. • A hand-held shopping basket 62.0 cm long has a 1.81-kg carton of milk at one end, and a 0.722-kg box of cereal at the other end. Where should a 1.80-kg container of orange juice be placed so that the basket balances at its center?

37. • If the cat in Active Example 11–2 has a mass of 2.8 kg, how close to the right end of the two-by-four can it walk before the board begins to tip?

38. •• **IP** A 0.34-kg meter stick balances at its center. If a necklace is suspended from one end of the stick, the balance point moves 9.5 cm toward that end. **(a)** Is the mass of the necklace more than, less than, or the same as that of the meter stick? Explain. **(b)** Find the mass of the necklace.

39. •• **Maximum Overhang** Three identical, uniform books of length L are stacked one on top the other. Find the maximum overhang distance d in **Figure 11–38** such that the books do not fall over.

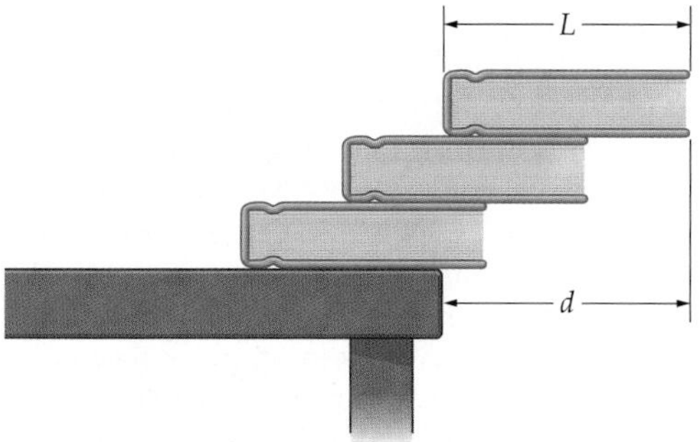

▲ **FIGURE 11–38** Problems 39 and 92

40. •• A baseball bat balances 71.1 cm from one end. If a 0.560-kg glove is attached to that end, the balance point moves 24.7 cm toward the glove. Find the mass of the bat.

Section 11–5 Dynamic Applications of Torque

41. •• A 2.85-kg bucket is attached to a disk-shaped pulley of radius 0.121 m and mass 0.742 kg. If the bucket is allowed to fall, **(a)** what is its linear acceleration? **(b)** What is the angular acceleration of the pulley? **(c)** How far does the bucket drop in 1.50 s?

42. •• **IP** In the previous problem, **(a)** is the tension in the rope greater than, less than, or equal to the weight of the bucket? Explain. **(b)** Calculate the tension in the rope.

43. •• A child exerts a tangential 40.0-N force on the rim of a disk-shaped merry-go-round with a radius of 2.40 m. If the merry-go-round starts at rest and acquires an angular speed of 0.0870 rev/s in 3.50 s, what is its mass?

44. •• **IP** You pull downward with a force of 25 N on a rope that passes over a disk-shaped pulley of mass 1.3 kg and radius 0.075 m. The other end of the rope is attached to a 0.67-kg mass. **(a)** Is the tension in the rope the same on both sides of the pulley? If not, which side has the largest tension? **(b)** Find the tension in the rope on both sides of the pulley.

45. •• Referring to the previous problem, find the linear acceleration of the 0.67-kg mass.

46. ••• A uniform meter stick of mass M has an empty paint can of mass m hanging from one end. The meter stick and the can balance at a point 20.0 cm from the end of the stick where the can is attached. When the balanced stick–can system is suspended from a scale, the reading on the scale is 2.54 N. Find the mass of **(a)** the meter stick and **(b)** the paint can.

47. ••• **Atwood's Machine** An Atwood's machine consists of two masses, m_1 and m_2, connected by a string that passes over a pulley. If the pulley is a disk of radius R and mass M, find the acceleration of the masses.

Section 11–6 Angular Momentum

48. • Calculate the angular momentum of the Earth about its own axis, due to its daily rotation. Assume that the Earth is a uniform sphere.

49. • A 0.015-kg record with a radius of 15 cm rotates with an angular speed of 33 1/3 rpm. Find the angular momentum of the record.

50. • In the previous problem, a 1.1-g fly lands on the rim of the record. What is the fly's angular momentum?

51. • Jogger 1 in **Figure 11–39** has a mass of 65.3 kg and runs in a straight line with a speed of 3.35 m/s. **(a)** What is the magnitude of the jogger's linear momentum? **(b)** What is the magnitude of the jogger's angular momentum with respect to the origin, O?

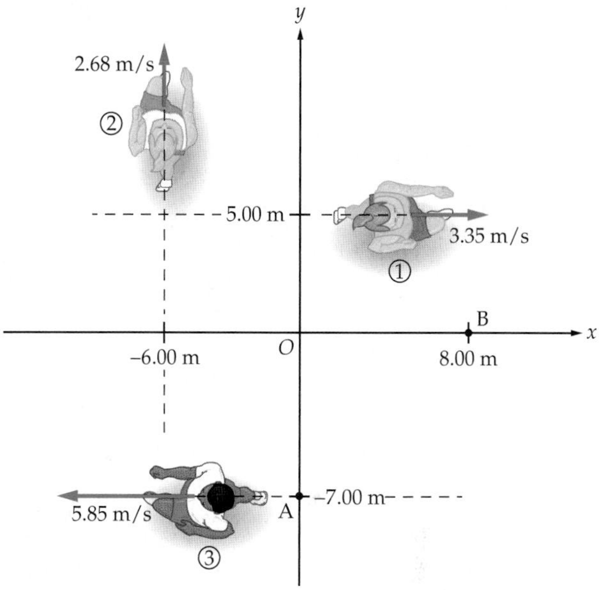

▲ **FIGURE 11–39** Problems 51, 52, and 53

52. • Repeat the previous problem for the case of jogger 2, whose speed is 2.68 m/s and whose mass is 58.2 kg.

53. •• **IP** Suppose jogger 3 in Figure 11–33 has a mass of 62.2 kg and a speed of 5.85 m/s. **(a)** Is the magnitude of the jogger's angular momentum greater with respect to point A or point B? Explain. **(b)** Is the magnitude of the jogger's angular momentum with respect to point B greater than, less than, or the same as it is with respect to the origin, O? Explain. **(c)** Calculate the magnitude of the jogger's angular momentum with respect to points A, B, and O.

54. •• A torque of 0.12 N · m is applied to an egg beater. **(a)** If the egg beater starts at rest, what is its angular momentum after 0.50 s? **(b)** If the moment of inertia of the egg beater is 2.5×10^{-3} kg · m^2, what is its angular speed after 0.50 s?

55. •• A windmill has an initial angular momentum of 8500 kg · m^2/s. The wind picks up, and 5.66 s later the windmill's angular momentum is 9700 kg · m^2/s. What was the torque acting on the windmill, assuming it was constant during this time?

56. •• Two gerbils run in place with a linear speed of 0.55 m/s on an exercise wheel that is shaped like a hoop. Find the angular momentum of the system if each gerbil has a mass of 0.22 kg and the exercise wheel has a radius of 9.5 cm and a mass of 5.0 g.

Section 11–7 Conservation of Angular Momentum

57. • As an ice skater begins a spin, his angular speed is 3.17 rad/s. After pulling in his arms, his angular speed increases to 5.66 rad/s. Find the ratio of the skater's final moment of inertia to his initial moment of inertia.

58. • Calculate both the initial and the final kinetic energy of the system described in Active Example 11–5.

59. • A diver tucks her body in midflight, decreasing her moment of inertia by a factor of two. By what factor does her angular speed change?

60. •• **IP** In the previous problem, **(a)** does the diver's kinetic energy increase, decrease, or stay the same? **(b)** Calculate the ratio of the final kinetic energy to the initial kinetic energy for the diver.

61. •• A disk-shaped merry-go-round of radius 2.63 m and mass 155 kg rotates freely with an angular speed of 0.641 rev/s. A 59.4-kg person running tangential to the rim of the merry-go-round at 3.41 m/s jumps onto its rim and holds on. Before jumping on the merry-go-round, the person was moving in the same direction as the merry-go-round's rim. What is the final angular speed of the merry-go-round?

62. •• **IP** In the previous problem, **(a)** does the kinetic energy of the system increase, decrease, or stay the same when the person jumps on the merry-go-round? **(b)** Calculate the initial and final kinetic energies for this system.

63. •• A student sits at rest on a piano stool that can rotate without friction. The moment of inertia of the student–stool system is 4.1 kg · m^2. A second student tosses a 1.5-kg mass with a speed of 2.7 m/s to the student on the stool, who catches it at a distance of 0.40 m from the axis of rotation. What is the resulting angular speed of the student and the stool?

64. •• **IP** Referring to the previous problem, **(a)** does the kinetic energy of the mass–student–stool system increase, decrease, or stay the same as the mass is caught? **(b)** Calculate the initial and final kinetic energy of the system.

65. •• **IP** A turntable with a moment of inertia of 5.4×10^{-3} kg · m^2 rotates freely with an angular speed of 33 1/3 rpm. Riding on the rim of the turntable, 15 cm from the center, is a cute, 36-g mouse. **(a)** If the mouse walks to the center of the turntable, will the turntable rotate faster, slower, or at the same rate? Explain. **(b)** Calculate the angular speed of the turntable when the mouse reaches the center.

66. •• A student on a piano stool rotates freely with an angular speed of 2.95 rev/s. The student holds a 1.25-kg mass in each outstretched arm, 0.759 m from the axis of rotation. The combined moment of inertia of the student and the stool, ignoring the two masses, is 5.43 kg · m^2, a value that remains constant. **(a)** As the student pulls his arms inward, his angular speed increases to 3.54 rev/s. How far are the masses from the axis of rotation at this time, considering the masses to be points? **(b)** Calculate the initial and final kinetic energy of the system.

67. ••• **Walking on a Merry-Go-Round** A child of mass m stands at rest near the rim of a stationary merry-go-round of radius R and moment of inertia I. The child now begins to walk around the circumference of the merry-go-round with a tangential speed v with respect to the merry-go-round's surface. **(a)** What is the child's speed with respect to the ground? Check your result in the limits **(b)** $I \to 0$ **(c)** and $I \to \infty$.

Section 11–8 Rotational Work and Power

68. • How much work must be done to accelerate a baton from rest to an angular speed of 7.4 rad/s about its center. Consider the baton to be a uniform rod of length 0.53 m and mass 0.44 kg.

69. • Turning a doorknob through 0.25 of a revolution requires 0.12 J of work. What is the torque required to turn the doorknob?

70. • A person exerts a tangential force of 36.1 N on the rim of a disk-shaped merry-go-round of radius 2.74 m and mass 167 kg. If the merry-go-round starts at rest, what is its angular speed after the person has rotated it through an angle of 32.5°?

71. • To prepare homemade ice cream, a crank must be turned with a torque of 3.8 N · m. How much work is required for each complete turn of the crank?

72. • **Power of a Dental Drill** A popular make of dental drill can operate at a speed of 42,500 rpm while producing a torque of 3.68 oz · in. What is the power output of this drill? Give your answer in watts.

73. •• The L-shaped object in Figure 11–29 consists of three masses connected by light rods. Find the work that must be done on this object to accelerate it from rest to an angular speed of 2.35 rad/s about **(a)** the x axis, **(b)** the y axis, and **(c)** the z axis (which is through the origin and perpendicular to the page).

74. •• The rectangular object in Figure 11–30 consists of four masses connected by light rods. What power must be applied to this object to accelerate it from rest to an angular speed of 2.5 rad/s in 6.4 s about **(a)** the x axis, **(b)** the y axis, and **(c)** the z axis (which is through the origin and perpendicular to the page)?

75. •• **IP** A circular saw blade accelerates from rest to an angular speed of 3620 rpm in 6.30 revolutions. **(a)** Find the torque exerted on the saw blade, assuming it is a disk of radius 15.2 cm and mass 0.755 kg. **(b)** Is the angular speed of the saw blade after 3.15 revolutions greater than, less than, or equal to 1810 rpm? Explain. **(c)** Find the angular speed of the blade after 3.15 revolutions.

General Problems

76. • After getting a drink of water, a hamster jumps onto an exercise wheel for a run. A few seconds later the hamster is running in place with a speed of 1.4 m/s. Find the work done by the hamster to get the exercise wheel moving, assuming it is a hoop of radius 0.13 m and mass 6.5 g.

77. •• A 47.0-kg uniform rod 4.25 m long is attached to a wall with a hinge at one end. The rod is held in a horizontal position by a wire attached to its other end. The wire makes an angle of 30.0° with the horizontal, and is bolted to the wall directly above the hinge. If the wire can withstand a maximum tension of 1450 N before breaking, how far from the wall can a 68.0-kg person sit without breaking the wire?

78. •• **IP** A puck attached to a string moves in a circular path on a frictionless surface, as shown in Figure 11–24. Initially, the speed of the puck is v and the radius of the circle is r. If the string passes through a hole in the surface, and is pulled downward until the radius of the circular path is $r/2$, **(a)** does the speed of the puck increase, decrease, or stay the same? **(b)** Calculate the final speed of the puck.

79. •• **BIO The Masseter Muscle** The masseter muscle, the principal muscle for chewing, is one of the strongest muscles for its size in the human body. It originates on the lower edge of the zygomatic arch (cheekbone) and inserts in the angle of the mandible. Referring to the lower diagram in **Figure 11–40**, where d = 7.60 cm and D = 10.85 cm, **(a)** find the torque produced about the axis of rotation by the masseter muscle. The force exerted by the masseter muscle is F_M = 455 N. **(b)** Find the biting force, F_B, exerted on the mandible by the upper teeth. Find **(c)** the horizontal and **(d)** the vertical component of the force F_J exerted on the mandible at the joint where it attaches to the skull. Assume that the mandible is in static equilibrium, and that upward is the positive vertical direction.

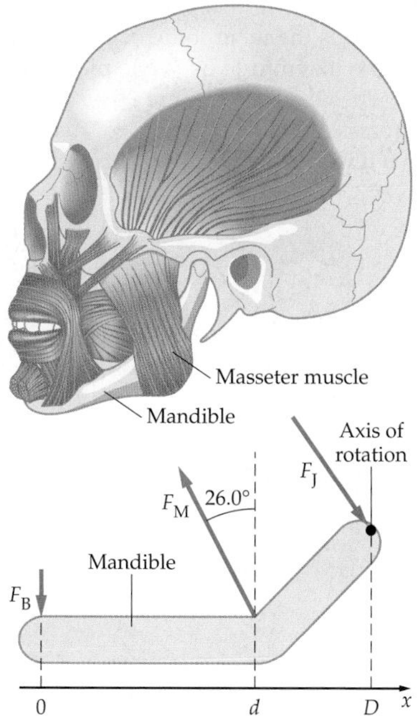

▲ **FIGURE 11–40** Problem 79

80. •• You are designing exercise equipment to operate as shown in **Figure 11–41**, where a person pulls upward on an elastic cord. The cord behaves like an ideal spring and has an unstretched length of 31 cm. If you would like the torque about the elbow joint to be 81 N · m in the position shown, what force constant, *k*, is required for the cord?

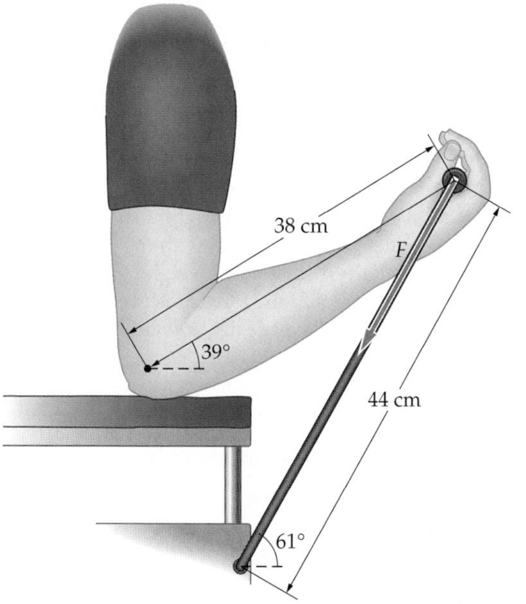

▲ **FIGURE 11–41** Problem 80

81. •• **Horsepower of a Car** Auto mechanics use the following formula to calculate the horsepower (HP) of a car engine:

$$\text{HP} = \text{Torque} \cdot \text{RPM}/C$$

In this expression, Torque is the torque produced by the engine in ft · lb, RPM is the angular speed of the engine in revolutions per minute, and *C* is a dimensionless constant. **(a)** Find the numerical value of *C*. **(b)** The Shelby Series 1 engine is advertised to generate 320 hp at 6500 rpm. What is the corresponding torque produced by this engine? Give your answer in ft · lbs.

82. •• **Balancing a *T. rex*** Paleontologists believe that *Tyrannosaurous rex* stood and walked with its spine almost horizontal, as indicated in **Figure 11–42**, and that its tail was held off the ground to balance its upper torso about the hip joint. Given that the total mass of *T. rex* was 5400 kg, and that the placement of the center of mass of the tail and the upper torso was as shown in Figure 11–42, find the mass of the tail required for balance.

▲ **FIGURE 11–42** Problem 82

83. •• **IP** You hold a uniform, 25-g pen horizontal with your thumb pushing down on one end and your index finger pushing upward 3.5 cm from your thumb. The pen is 14 cm long. **(a)** Which of these two forces is greater in magnitude? **(b)** Find the two forces.

84. •• In Active Example 11–3, suppose the ladder is uniform, 4.0 m long and weighs 60.0 N. Find the forces exerted on the ladder when the person is **(a)** halfway up the ladder and **(b)** three-fourths of the way up the ladder.

85. •• When you arrive at Duke's Dude Ranch, you are greeted by the large wooden sign shown in **Figure 11–43**. The left end of the sign is held in place by a bolt, the right end is tied to a rope that makes an angle of 20.0° with the horizontal. If the sign is uniform, 3.20 m long, and has a mass of 16.0 kg, what is **(a)** the tension in the rope, and **(b)** the horizontal and vertical components of the force, $\vec{F}$, exerted by the bolt?

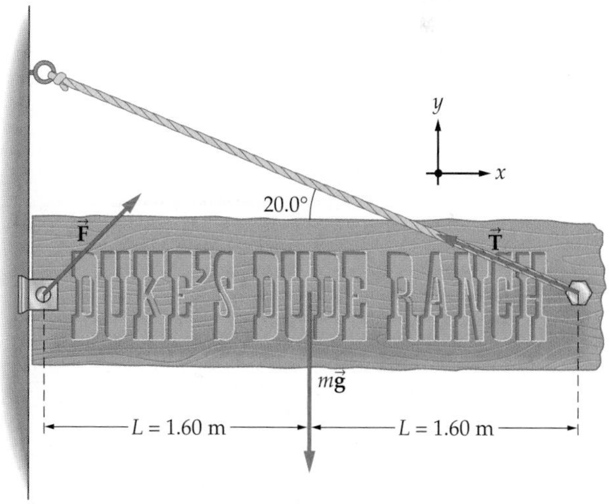

▲ **FIGURE 11–43** Problem 85

86. •• A 64.0-kg person stands on a lightweight diving board supported by two pillars, one at the end of the board, the other 1.10-m away. The pillar at the end of the board exerts a downward force of 828 N. **(a)** How far from that pillar is the person standing? **(b)** Find the force exerted by the second pillar.

87. •• In Example 11–4, find $\vec{F}_1$ and $\vec{F}_2$ as a function of the distance, *x*, of the swimmer from the left end of the diving board. Assume that the diving board is uniform and has a mass of 85 kg.

88. •• **Flats Versus Heels** A woman might wear a pair of flat shoes to work during the day, as in **Figure 11–44 (a)**, but a pair of high heels, **Figure 11–44 (b)**, when going out for the evening. Assume that each foot supports half her weight, $w = W/2 = 279$ N, and that the forces exerted by the floor on her feet occur at the points A and B in both figures. Find the forces F_A (point A) and F_B (point B) for **(a)** flat shoes and **(b)** high heels. **(c)** How have the high heels changed the weight distribution between the woman's heels and toes?

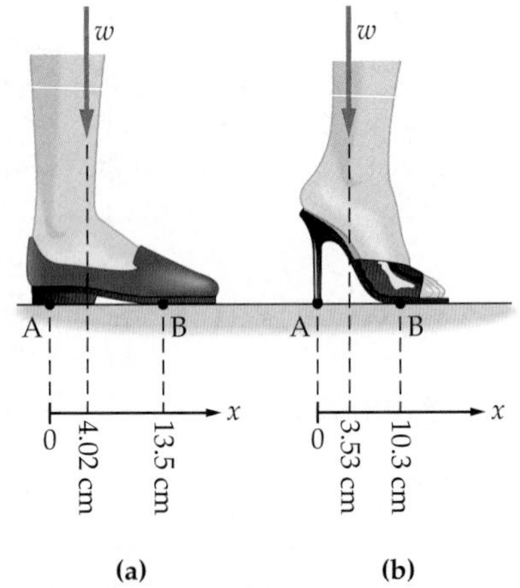

▲ **FIGURE 11–44** Problem 88

89. •• **BIO** A young girl sits at the edge of a dock by the bay, dipping her feet in the water. At the instant shown in **Figure 11–45**, she holds her lower leg stationary with her quadriceps muscle at an angle of 39° with respect to the horizontal. Use the information given in the figure, plus the fact that her lower leg has a mass of 3.4 kg, to determine the magnitude of the force, F_Q, exerted on the lower leg by the quadriceps.

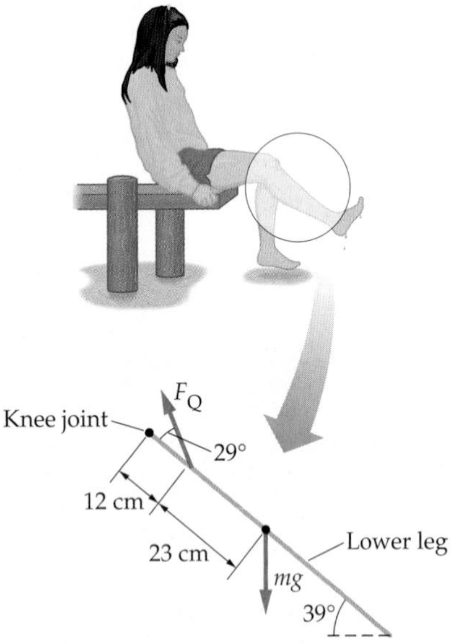

▲ **FIGURE 11–45** Problem 89

90. •• **BIO Deltoid Muscle** A crossing guard holds a STOP sign at arm's length, as shown in **Figure 11–46**. Her arm is horizontal, and we assume that the deltoid muscle is the only muscle supporting her arm. The weight of her upper arm is $W_u = 18$ N, the weight of her lower arm is $W_l = 11$ N, the weight of her hand is $W_h = 4.0$ N, and the weight of the sign is $W_s = 8.9$ N. The location where each of these forces acts on the arm is indicated in the figure. A force of magnitude f_d is exerted on the humerus by the deltoid, and the shoulder joint exerts a force on the humerus with horizontal and vertical components given by f_x and f_y, respectively. **(a)** Is the magnitude of f_d greater than, less than, or equal to the magnitude of f_x? Explain. Find **(b)** f_d, **(c)** f_x, and **(d)** f_y. (The weights in Figure 11–46 are drawn to scale; the unknown forces are to be determined. If a force is found to be negative, its direction is opposite to that shown.)

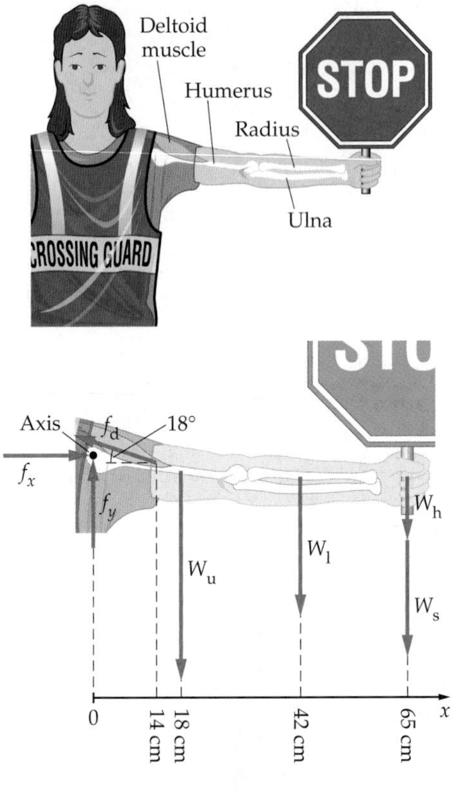

Free-Body Diagram of the Arm

▲ **FIGURE 11–46** Problem 90

91. •• **BIO Triceps** To determine the force a person's triceps muscle can exert, a doctor uses the procedure shown in **Figure 11–47**, where the patient pushes down with the palm of his hand on a force meter. Given that the weight of the lower arm is $Mg = 15.6$ N, and that the force meter reads $F = 89.0$ N, what is the force F_T exerted vertically upward by the triceps?

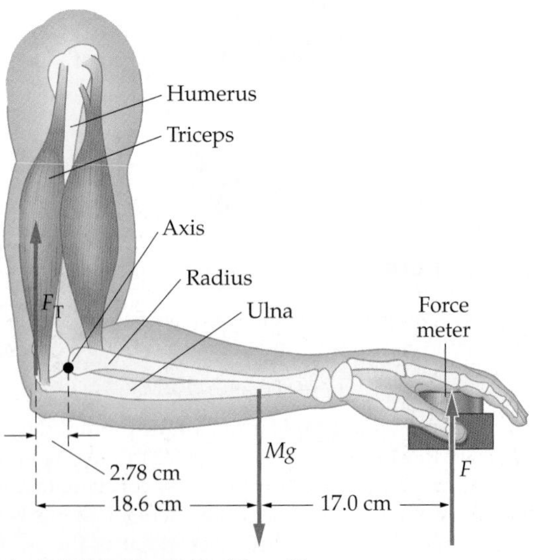

▲ **FIGURE 11–47** Problem 91

92. •• **IP** Suppose a fourth book, the same as the other three, is added to the stack of books shown in Figure 11–38. **(a)** What is the maximum overhang distance, d, in this case? **(b)** If the mass of each book is increased by the same amount, does your answer to part (a) increase, decrease, or stay the same? Explain.

93. •• **IP** Suppose partial melting of the polar ice caps increases the moment of inertia of the Earth from $0.331\ M_E R_E^2$ to $0.332\ M_E R_E^2$. **(a)** Would the length of a day (the time required for the Earth to complete one revolution about its axis) increase or decrease? Explain. **(b)** Calculate the change in the length of a day. Give your answer in seconds.

94. ••• A bicycle wheel of radius R and mass M is at rest against a step of height $3R/4$, as illustrated in **Figure 11–48**. Find the minimum horizontal force F that must be applied to the axle to make the wheel start to rise up over the step.

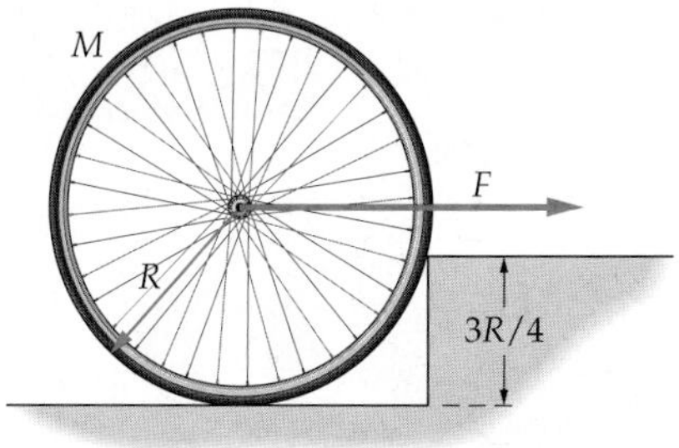

▲ **FIGURE 11–48** Problem 94

95. ••• A 0.105-kg yo-yo has an outer radius R that is 5.60 times greater than the radius r of its axle. The yo-yo is in equilibrium if a mass m is suspended from its outer edge, as shown in **Figure 11–49**. Find the tension in the two strings, T_1 and T_2, and the mass m.

▲ **FIGURE 11–49** Problem 95

96. ••• In Problem 30, assume that the rod has a mass of M and that its bottom end simply rests on the floor, held in place by static friction. If the coefficient of static friction is μ_s, find the maximum force F that can be applied to the rod at its midpoint before it slips.

97. ••• In the previous problem, suppose the rod has a mass of 2.3 kg and the coefficient of static friction is 1/7. **(a)** Find the greatest force F that can be applied at the midpoint of the rod without causing it to slip. **(b)** Show that if F is applied 1/8 of the way down from the top of the rod, it will never slip at all, no matter how large the force F.

98. ••• A cylinder of mass m and radius r has a string wrapped around its circumference. The upper end of the string is held fixed, and the cylinder is allowed to fall. Show that its linear acceleration is $(2/3)g$.

99. ••• Repeat the previous problem, replacing the cylinder with a solid sphere. Show that its linear acceleration is $(5/7)g$.

100. ••• A mass M is attached to a rope that passes over a disk-shaped pulley of mass m and radius r. The mass hangs to the left side of the pulley. On the right side of the pulley, the rope is pulled downward with a force F. Find **(a)** the acceleration of the mass, **(b)** the tension in the rope on the left side of the pulley, and **(c)** the tension in the rope on the right side of the pulley. **(d)** Check your results in the limits $m \to 0$ and $m \to \infty$.

101. ••• **Bricks in Equilibrium** Consider a system of four uniform bricks of length L stacked as shown in **Figure 11–50**. What is the maximum distance, x, that the middle bricks can be displaced outward before they begin to tip?

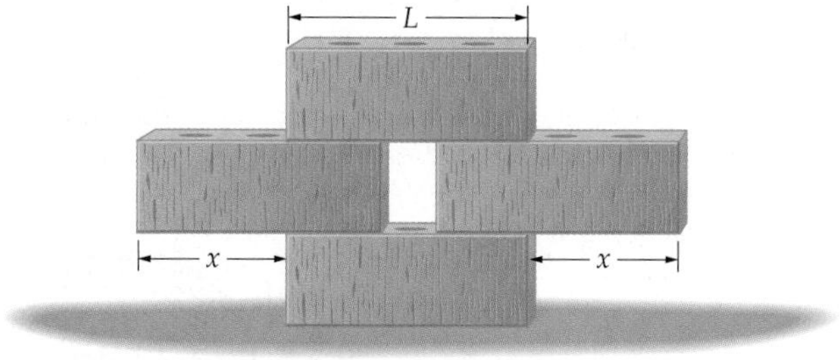

▲ **FIGURE 11–50** Problem 101

Interactive Problems

102. •• ***Referring to Example 11–7*** Suppose the mass of the pulley is doubled, to 0.160 kg, and that everything else in the system remains the same. **(a)** Do you expect the value of T_2 to increase, decrease, or stay the same? Explain. **(b)** Calculate the value of T_2 for this case.

103. •• ***Referring to Example 11–7*** Suppose the mass of the cart is doubled, to 0.62 kg, and that everything else in the system remains the same. **(a)** Do you expect the value of T_2 to increase, decrease, or stay the same? Explain. **(b)** Calculate the value of T_2 for this case.

104. •• ***Referring to Active Example 11–5*** Suppose the child runs with a different initial speed, but that everything else in the system remains the same. What initial speed does the child have if the angular speed of the system after the collision is 0.425 rad/s?

105. •• ***Referring to Active Example 11–5*** Suppose everything in the system is as described in Active Example 11–5 except that the child approaches the merry-go-round in a direction that is not tangential. Find the angle θ between the direction of motion and the outward radial direction (as in Example 11–8) that is required if the final angular speed of the system is to be 0.272 rad/s.

12 Gravity

12–1	Newton's Law of Universal Gravitation	359
12–2	Gravitational Attraction of Spherical Bodies	362
12–3	Kepler's Laws of Orbital Motion	367
12–4	Gravitational Potential Energy	374
12–5	Energy Conservation	377
*12–6	Tides	384

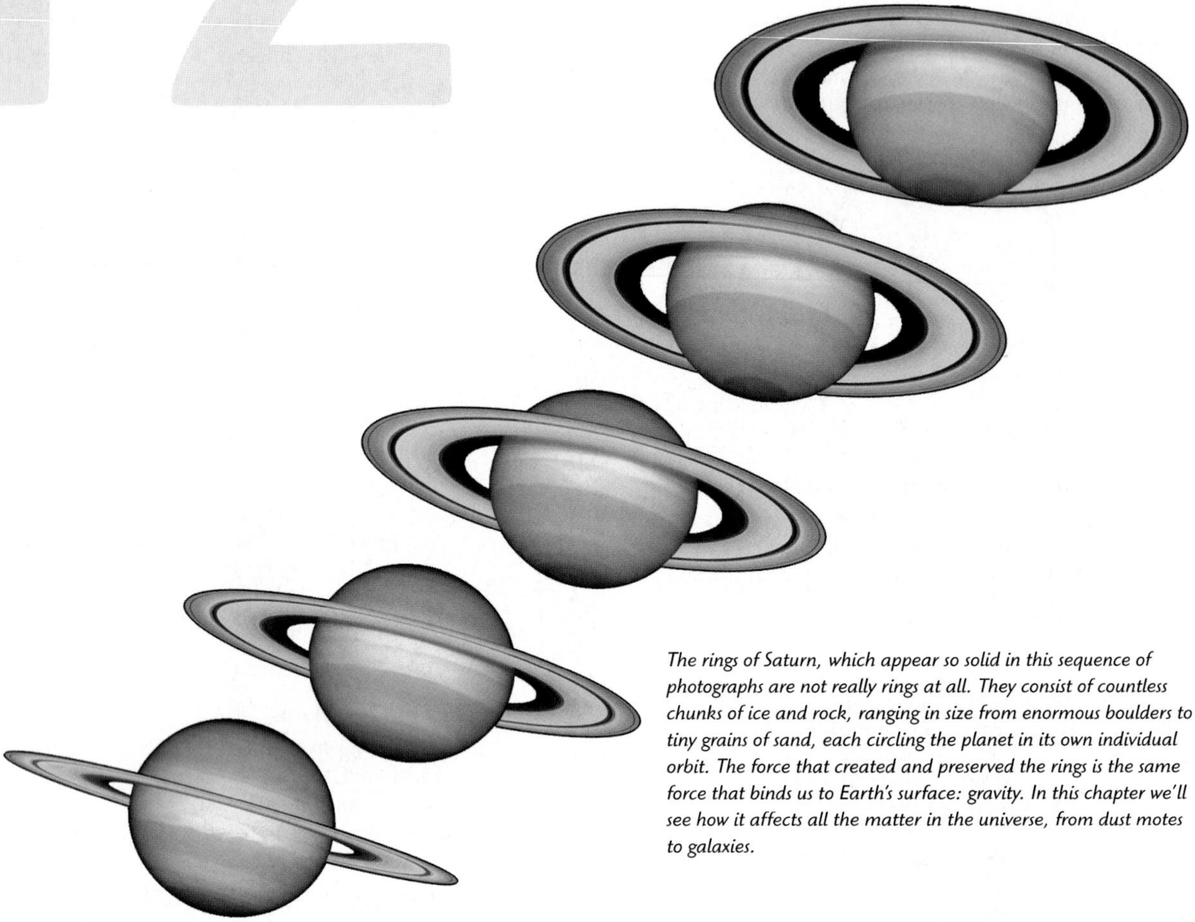

The rings of Saturn, which appear so solid in this sequence of photographs are not really rings at all. They consist of countless chunks of ice and rock, ranging in size from enormous boulders to tiny grains of sand, each circling the planet in its own individual orbit. The force that created and preserved the rings is the same force that binds us to Earth's surface: gravity. In this chapter we'll see how it affects all the matter in the universe, from dust motes to galaxies.

The study of gravity has always been a central theme in physics, from Galileo's early experiments on free fall in the seventeenth century, to Einstein's general theory of relativity in the early years of the twentieth century, and Stephen Hawking's work on black holes in recent years. Perhaps the grandest milestone in this endeavor, however, was the discovery by Newton of the **universal law of gravitation.** With just one simple equation to describe the force of gravity, Newton was able to determine the orbits of planets, moons, and comets, and to explain such earthly phenomena as the tides and the fall of an apple.

Before Newton's work, it was generally thought that the heavens were quite separate from the Earth, and that they obeyed their own "heavenly" laws. Newton showed, on the contrary, that the same law of gravity that operates on the surface of the Earth applies to the Moon and to other astronomical objects. As a result of Newton's efforts, physics expanded its realm of applicability to natural phenomena throughout the universe.

So successful was Newton's law of gravitation that Edmond Halley (1656–1742) was able to use it to predict the return of the comet that today bears his name. Though he did not live to see its return in 1758, the fact that the comet did reappear when predicted was an event unprecedented in human history. Roughly a hundred years later, Newton's theory of gravity scored an even more impressive success. Astronomers observing the planet Uranus noticed small deviations in its orbit, which they thought might be due to the gravitational tug of a previously unknown planet. Using Newton's law to calculate the predicted position of the new planet—now called Neptune—it was found on the very first night of observations, September 23, 1846. The fact that Neptune was precisely where the law of gravitation said it should be still stands as one of the most astounding triumphs in the history of science.

Today, Newton's law of gravitation is used to determine the orbits that take spacecraft from the Earth to various destinations within our solar system and beyond. Appropriately enough, spacecraft were even sent to view Halley's comet at close range in 1986. In addition, the law allows us to calculate with pinpoint accuracy the time of solar eclipses and other astronomical events in the distant past and remote future. This incredibly powerful and precise law of nature is the subject of this chapter.

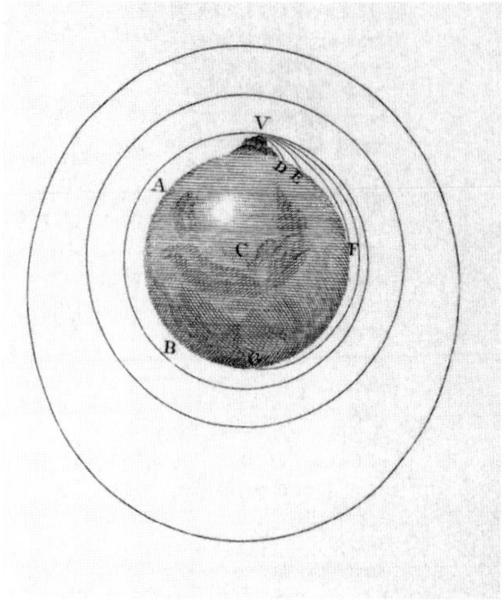

▲ In this illustration from his great work, the *Principia*, published in 1687, Newton presents a "thought experiment" to show the connection between free fall and orbital motion. Imagine throwing a projectile horizontally from the top of a mountain. The greater the initial speed of the projectile, the farther it travels in free fall before striking the ground. In the absence of air resistance, a great enough initial speed could result in the projectile circling the Earth and returning to its starting point. Thus, an object orbiting the Earth is actually in free fall—it simply has a large horizontal speed.

12–1 Newton's Law of Universal Gravitation

It's ironic, but the first fundamental force of nature to be recognized as such, **gravity,** is also the weakest of the fundamental forces. Still, it is the force most apparent to us in our everyday lives, and is the force responsible for the motion of the Moon, the Earth, and the planets. Yet the connection between falling objects on Earth and planets moving in their orbits was not known before Newton.

The flash of insight that came to Newton—whether it was due to seeing an apple fall to the ground or not—is simply this: The force causing an apple to accelerate downward is the same force causing the Moon to move in a circular path around the Earth. To put it another way, Newton was the first to realize that the Moon is *constantly falling* toward the Earth, though without ever getting closer to it, and that it falls for the same reason that an apple falls. This is illustrated in a classic drawing due to Newton, shown to the right.

To be specific, in the case of the apple the motion is linear as it accelerates downward toward the center of the Earth. In the case of the Moon the motion is circular with constant speed. As discussed in Section 6–5, an object in uniform circular motion accelerates toward the center of the circle. It follows, therefore, that the Moon *also* accelerates toward the center of the Earth. In fact, the force responsible for the Moon's centripetal acceleration is the Earth's gravitational attraction, the same force responsible for the fall of the apple.

To describe the force of gravity, Newton proposed the following simple law:

Newton's Law of Universal Gravitation

The force of gravity between any two point objects of mass m_1 and m_2 is attractive and of magnitude

$$F = G\frac{m_1 m_2}{r^2} \qquad \text{12–1}$$

In this expression, r is the distance between the masses, and G is a constant referred to as the **universal gravitation constant.** Its value is

$$G = 6.67 \times 10^{-11}\ \mathrm{N \cdot m^2/kg^2} \qquad \text{12–2}$$

The force is directed along the line connecting the masses, as indicated in **Figure 12–1**.

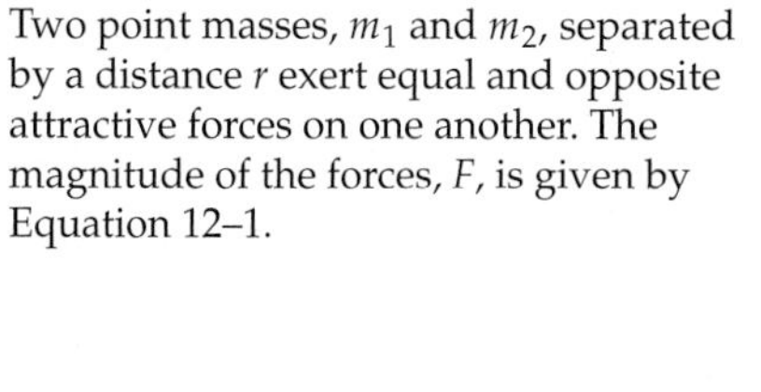

▲ **FIGURE 12–1 Gravitational force between point masses**

Two point masses, m_1 and m_2, separated by a distance r exert equal and opposite attractive forces on one another. The magnitude of the forces, F, is given by Equation 12–1.

Note that each mass experiences a force of the same magnitude, $F = Gm_1m_2/r^2$, but acting in opposite directions. That is, the force of gravity between two objects forms an action-reaction pair.

According to Newton's law, all objects in the universe attract all other objects in the universe by way of the gravitational interaction. It is in this sense that the force law is termed "universal." Thus, the net gravitational force acting on you is due not only to the planet on which you stand, which is certainly responsible for the majority of the net force, but also to people nearby, planets, and even stars in far-off galaxies. In short, everything in the universe "feels" everything else, thanks to gravity.

The fact that G is such a small number means that the force of gravity between objects of human proportions is imperceptibly small. This is shown in the following Exercise.

EXERCISE 12–1

A man takes his dog for a walk on a deserted beach. Treating people and dogs as point objects for the moment, find the force of gravity between the 105-kg man and his 11.2-kg dog when they are separated by a distance of **(a)** 1.00 m and **(b)** 10.0 m.

Solution

a. Substituting numerical values into Equation 12–1 yields

$$F = G\frac{m_1 m_2}{r^2} = (6.67 \times 10^{-11}\,\text{N}\cdot\text{m}^2/\text{kg}^2)\frac{(105\,\text{kg})(11.2\,\text{kg})}{(1.00\,\text{m})^2} = 7.84 \times 10^{-8}\,\text{N}$$

b. Repeating the calculation for $r = 10.0$ m gives

$$F = G\frac{m_1 m_2}{r^2} = (6.67 \times 10^{-11}\,\text{N}\cdot\text{m}^2/\text{kg}^2)\frac{(105\,\text{kg})(11.2\,\text{kg})}{(10.0\,\text{m})^2} = 7.84 \times 10^{-10}\,\text{N}$$

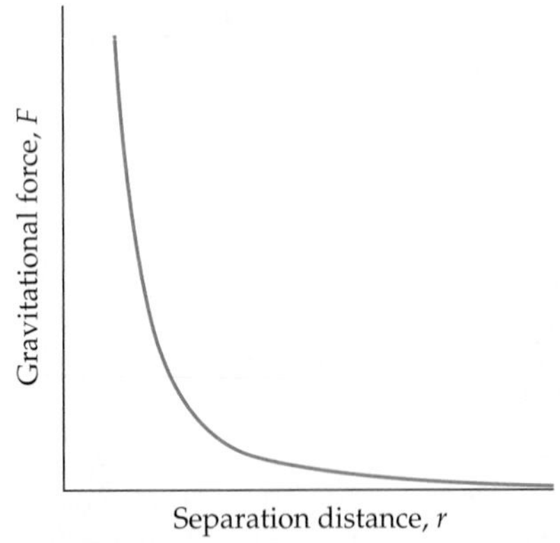

▲ **FIGURE 12–2 Dependence of the gravitational force on separation distance, *r***

Note that the gravitational force decreases rapidly with distance, but never completely vanishes. We say that gravity is a force of infinite range.

The forces found in Exercise 12–1 are imperceptibly small. In comparison, the force exerted by the Earth on the man is 1030 N and the force exerted on the dog is 110 N—these forces are several orders of magnitude greater than the force between the man and the dog. In general, gravitational forces are significant only when large masses, such as the Earth or the Moon, are involved.

Exercise 12–1 also illustrates how rapidly the force of gravity decreases with distance. In particular, since F varies as $1/r^2$, it is said to have an **inverse square dependence** on distance. Thus, for example, an increase in distance by a factor of 10 results in a decrease in the force by a factor of $10^2 = 100$. A plot of the force of gravity versus distance is given in **Figure 12–2**. Note that even though the force diminishes rapidly with distance, it never completely vanishes; thus, we say that gravity is a force of infinite range.

Note also that the force of gravity between two masses depends on the product of the masses, m_1 times m_2. With this type of dependence, it follows that if either mass is doubled, the force of gravity is doubled as well. This would not be the case, for example, if the force of gravity depended on the *sum* of the masses, $m_1 + m_2$.

Finally, if a given mass is acted on by gravitational interactions with a number of other masses, the net force acting on it is simply the vector sum of each of the forces individually. This property of gravity is referred to as **superposition**. As an example, superposition implies that the net gravitational force exerted on you at this moment is the vector sum of the force exerted by the Earth, plus the force exerted by the Moon, plus the force exerted by the Sun, and so on. The following Example illustrates superposition.

PROBLEM-SOLVING NOTE

Net Gravitational Force

To find the net gravitational force acting on an object, you should (i) resolve each of the forces acting on the object into components and (ii) add the forces component by component.

EXAMPLE 12–1 How Much Force Is With You?

As part of a daring rescue attempt, the *Millennium Eagle* passes between a pair of twin asteroids, as shown. If the mass of the spaceship is 2.50×10^7 kg and the mass of each asteroid is 3.50×10^{11} kg, find the net gravitational force exerted on the *Millennium Eagle* **(a)** when it is at location A and **(b)** when it is at location B. Treat the spaceship and the asteroids as if they were point objects.

Picture the Problem

Our sketch shows the spaceship as it follows a path between the twin asteroids. The relevant distances and masses are indicated, as are the two points of interest, A and B. Note that at location A the force $\vec{\mathbf{F}}_1$ points above the x axis at the angle θ_1 (to be determined); the force $\vec{\mathbf{F}}_2$ points below the x axis at the angle $\theta_2 = -\theta_1$, as can be seen by symmetry. At location B, the two forces act in opposite directions.

Strategy

To find the net gravitational force exerted on the spaceship, we first determine the magnitude of the force exerted on it by each asteroid. This is done by using Equation 12–1 and the distances given in our sketch. Next, we resolve these forces into x and y components. Finally, we sum the force components to find the net force.

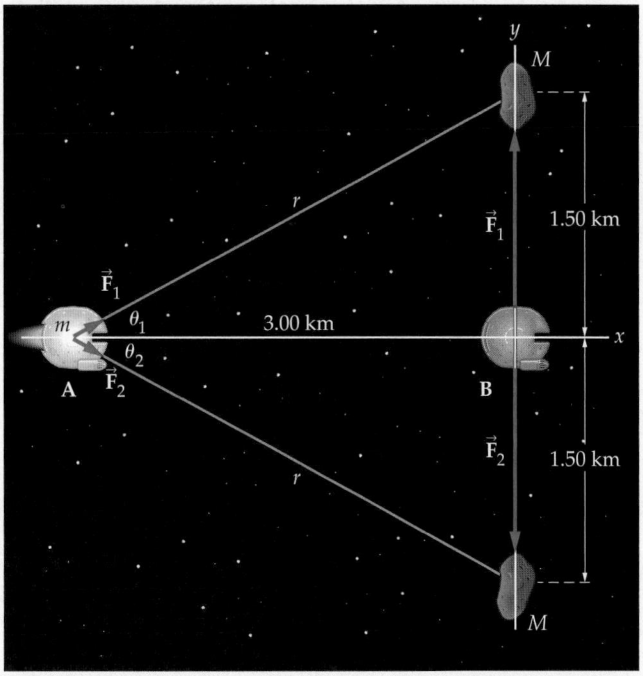

Solution

Part (a)

1. Use the Pythagorean theorem to find the distance r from point A to each asteroid. Also, refer to the sketch to find the angle between $\vec{F}_1$ and the x axis. The angle between $\vec{F}_2$ and the x axis has the same magnitude but the opposite sign:

$$r = \sqrt{(3.00 \times 10^3 \text{ m})^2 + (1.50 \times 10^3 \text{ m})^2} = 3350 \text{ m}$$

$$\theta_1 = \tan^{-1}\left(\frac{1.50 \times 10^3 \text{ m}}{3.00 \times 10^3 \text{ m}}\right) = \tan^{-1}(0.500) = 26.6°$$

$$\theta_2 = -\theta_1 = -26.6°$$

2. Use r and Equation 12–1 to calculate the magnitude of the forces $\vec{F}_1$ and $\vec{F}_2$ at point A:

$$F_1 = F_2 = G\frac{mM}{r^2}$$

$$= (6.67 \times 10^{-11} \text{ N}\cdot\text{m}^2/\text{kg}^2)\frac{(2.50 \times 10^7 \text{ kg})(3.50 \times 10^{11} \text{ kg})}{(3350 \text{ m})^2}$$

$$= 52.0 \text{ N}$$

3. Use the values of θ_1 and θ_2 found in step 1 to calculate the x and y components of $\vec{F}_1$ and $\vec{F}_2$:

$$F_{1,x} = F_1 \cos\theta_1 = (52.0 \text{ N}) \cos 26.6° = 46.5 \text{ N}$$

$$F_{1,y} = F_1 \sin\theta_1 = (52.0 \text{ N}) \sin 26.6° = 23.3 \text{ N}$$

$$F_{2,x} = F_2 \cos\theta_2 = (52.0 \text{ N}) \cos(-26.6°) = 46.5 \text{ N}$$

$$F_{2,y} = F_2 \sin\theta_2 = (52.0 \text{ N}) \sin(-26.6°) = -23.3 \text{ N}$$

4. Add the components of $\vec{F}_1$ and $\vec{F}_2$ to find the components of the net force, $\vec{F}$:

$$F_x = F_{1,x} + F_{2,x} = 93.0 \text{ N}$$

$$F_y = F_{1,y} + F_{2,y} = 0$$

Part (b)

5. Use Equation 12–1 to find the magnitude of the forces exerted on the spaceship by the asteroids at location B:

$$F_1 = F_2 = G\frac{mM}{r^2}$$

$$= (6.67 \times 10^{-11} \text{ N}\cdot\text{m}^2/\text{kg}^2)\frac{(2.50 \times 10^7 \text{ kg})(3.50 \times 10^{11} \text{ kg})}{(1.50 \times 10^3 \text{ m})^2}$$

$$= 259 \text{ N}$$

6. Use the fact that $\vec{F}_1$ and $\vec{F}_2$ have equal magnitudes and point in opposite directions to determine the net force, $\vec{F}$, acting on the spaceship:

$$\vec{F} = \vec{F}_1 + \vec{F}_2 = 0$$

continued on next page

continued from previous page

Insight

We find that the net force at location A is in the positive x direction, as one would expect by symmetry. At location B, where the force exerted by each asteroid is about 5 times greater than it is at location A, the *net* force is zero since the attractive forces exerted by the two asteroids are equal and opposite, and thus cancel. Note that the forces in our sketch have been drawn in correct proportion.

Rocket scientists often use the gravitational force between astronomical objects and spacecraft to accelerate the spacecraft and send them off to distant parts of the solar system. In fact, this gravitational attraction makes possible the "slingshot" effect discussed earlier in homework Problem 9–81.

Practice Problem

Find the net gravitational force acting on the spaceship when it is at the location $x = 5.00 \times 10^3$ m, $y = 0$. [**Answer:** 41.0 N in the negative x direction.]

Some related homework problems: Problem 8, Problem 10, Problem 11

12–2 Gravitational Attraction of Spherical Bodies

Newton's law of gravity applies to point objects. How, then, do we calculate the force of gravity for an object of finite size? In general, the approach is to divide the finite object into a collection of small mass elements, then use superposition and the methods of calculus to determine the net gravitational force. For an arbitrary shape, this calculation can be quite difficult. For objects with a uniform spherical shape, however, the final result is remarkably simple, as was shown by Newton.

Uniform Sphere

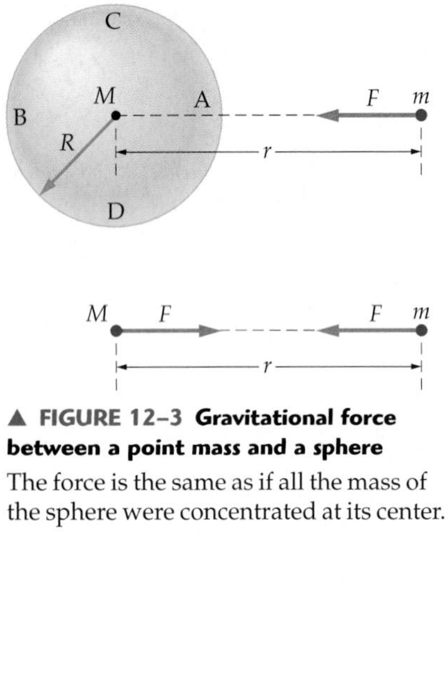

▲ **FIGURE 12–3 Gravitational force between a point mass and a sphere**
The force is the same as if all the mass of the sphere were concentrated at its center.

Consider a uniform sphere of radius R and mass M, as in **Figure 12–3**. A point object of mass m is brought near the sphere, though still outside it at a distance r from its center. The object experiences a relatively strong attraction from mass near the point A, and a weaker attraction from mass near point B. In both cases the force is along the line connecting the mass m and the center of the sphere; that is, along the x axis. In addition, mass at the points C and D exert a net force that is also along the x axis—just as in the case of the twin asteroids in Example 12–1. Thus, the symmetry of the sphere guarantees that the net force it exerts on m is directed toward the sphere's center. The magnitude of the force exerted by the sphere must be calculated with the methods of calculus—which Newton invented and then applied to this problem. As a result of his calculations, Newton was able to show that **the net force exerted by the sphere on the mass m is the same as if all the mass of the sphere were concentrated at its center.** That is, the force between the mass m and the sphere of mass M has a magnitude that is simply

$$F = G\frac{mM}{r^2} \qquad \text{12–3}$$

Let's apply this result to the case of a mass m on the surface of the Earth. If the mass of the Earth is M_E, and its radius is R_E, it follows that the force exerted on m by the Earth is

$$F = G\frac{mM_E}{{R_E}^2} = m\left(\frac{GM_E}{{R_E}^2}\right)$$

We also know, however, that the gravitational force experienced by a mass m on the Earth's surface is simply $F = mg$, where g is the acceleration due to gravity. Therefore, we see that

$$m\left(\frac{GM_E}{{R_E}^2}\right) = mg$$

or

$$g = \frac{GM_E}{{R_E}^2} = \frac{(6.67 \times 10^{-11}\ \text{N}\cdot\text{m}^2/\text{kg}^2)(5.97 \times 10^{24}\ \text{kg})}{(6.37 \times 10^6\ \text{m})^2} = 9.81\ \text{m/s}^2 \qquad \text{12–4}$$

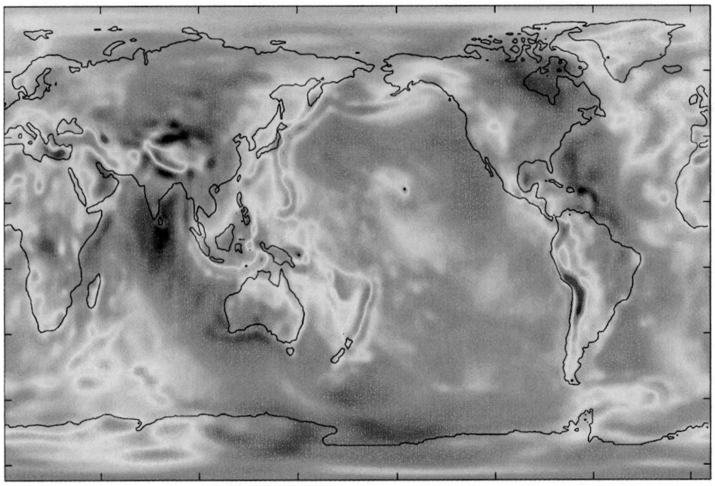

◀ This global model of the Earth's gravitational strength was constructed from a combination of surface gravity measurements and satellite tracking data. It shows how the acceleration of gravity varies from the value at an idealized "sea level" that takes into account the Earth's nonspherical shape. (The Earth is somewhat flattened at the poles—its radius is greatest at the equator.) Gravity is strongest in the red areas and weakest in the dark blue areas.

This result can be extended to objects above the Earth's surface, and hence farther from the center of the Earth, as we show in the next Example.

EXAMPLE 12–2 The Dependence of Gravity on Altitude

REAL-WORLD PHYSICS

If you climb to the top of Mt. Everest, you will be about 5.50 mi above sea level. What is the acceleration due to gravity at this altitude?

Picture the Problem

At the top of the mountain, your distance from the center of the Earth is $r = R_E + h$, where $h = 5.50$ mi is the altitude.

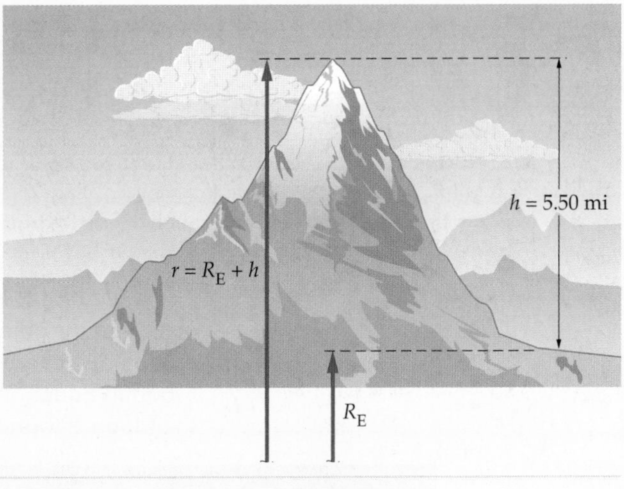

Strategy

First, use $F = GmM_E/r^2$ to find the force due to gravity on the mountaintop. Then, set $F = mg_h$ to find the acceleration g_h at the height h.

Solution

1. Calculate the force F due to gravity at a height h above the Earth's surface:

$$F = G\frac{mM_E}{(R_E + h)^2}$$

2. Set F equal to mg_h and solve for g_h:

$$F = G\frac{mM_E}{(R_E + h)^2} = mg_h$$

$$g_h = G\frac{M_E}{(R_E + h)^2}$$

3. Factor out $R_E{}^2$ from the denominator, and use the fact that $GM_E/R_E{}^2 = g$:

$$g_h = \left(\frac{GM_E}{R_E{}^2}\right)\frac{1}{\left(1 + \dfrac{h}{R_E}\right)^2} = \frac{g}{\left(1 + \dfrac{h}{R_E}\right)^2}$$

4. Substitute numerical values, with $h = 5.50$ mi $= (5.50\text{ mi})(1609\text{ m/mi}) = 8850$ m, and $R_E = 6.37 \times 10^6$ m:

$$g_h = \frac{g}{\left(1 + \dfrac{h}{R_E}\right)^2} = \frac{9.81\text{ m/s}^2}{\left(1 + \dfrac{8850\text{ m}}{6.37 \times 10^6\text{ m}}\right)^2} = 9.78\text{ m/s}^2$$

continued on next page

continued from previous page

Insight

As expected, the acceleration due to gravity is less as one moves farther from the center of the Earth. Thus, if you were to climb to the top of Mt. Everest, you would lose weight—not only because of the physical exertion required for the climb, but also because of the reduced gravity. In particular, a person with a mass of 60 kg (about 130 lbs) would lose about half a pound of weight just by standing on the summit of the mountain.

A plot of g_h as a function of h is shown in **Figure 12–4 (a)**. The plot indicates the altitude of Mt. Everest and the orbit of the space shuttle. **Figure 12–4 (b)** shows g_h out to the orbit of communications and weather satellites, which orbit at an altitude of roughly 22,300 mi.

Practice Problem

Find the acceleration due to gravity at the altitude of the space shuttle's orbit, 250 km above the Earth's surface.
[**Answer:** $g_h = 9.08\ \mathrm{m/s^2}$, a reduction of only 7.44% compared to the acceleration of gravity on the surface of the Earth.]

Some related homework problems: Problem 14, Problem 16

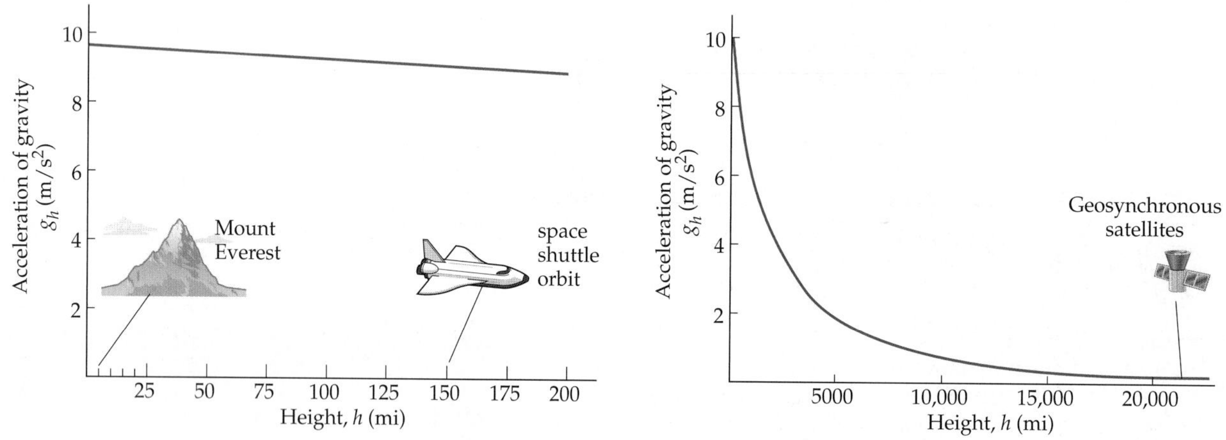

▲ **FIGURE 12–4 The acceleration due to gravity at a height *h* above the Earth's surface**

(a) In this plot, the peak of Mt. Everest is at about $h = 5.50$ mi, and the space shuttle orbit is at roughly $h = 150$ mi. **(b)** This shows the decrease in the acceleration of gravity from the surface of the Earth to an altitude of about 25,000 mi. The orbit of geosynchronous satellites—ones that orbit above a fixed point on the Earth—is at roughly $h = 22{,}300$ mi.

Equation 12–4 can be used to calculate the acceleration due to gravity on other objects in the solar system besides the Earth. For example, to calculate the acceleration due to gravity on the Moon, g_m, we simply use the mass and radius of the Moon in Equation 12–4. Once g_m is known, the weight of an object of mass m on the Moon is found by using $W_m = mg_m$.

◀ (Left) The weak lunar gravity permits astronauts, even encumbered by their massive space suits, to bound over the Moon's surface. The low gravitational pull, only about one-sixth that of Earth, is a consequence not only of the Moon's smaller size, but also of its lower average density. (Above) The *Sojourner* rover vehicle, brought to Mars by the Pathfinder mission of 1997, exploring the Martian surface. The gravity of Mars is greater than that of the Moon but still only about 38% that of Earth.

EXERCISE 12–2

a. Find the acceleration due to gravity on the surface of the Moon.

b. The lunar rover had a mass of 225 kg. What was its weight on the Earth and on the Moon? (Note, the mass of the Moon is $M_m = 7.35 \times 10^{22}$ kg and its radius is $R_m = 1.74 \times 10^6$ m.)

Solution

a. For the Moon, the acceleration due to gravity is

$$g_m = \frac{GM_m}{R_m^{\,2}} = \frac{(6.67 \times 10^{-11}\ \mathrm{N \cdot m^2/kg^2})(7.35 \times 10^{22}\ \mathrm{kg})}{(1.74 \times 10^6\ \mathrm{m})^2} = 1.62\ \mathrm{m/s^2}$$

This is about one-sixth the acceleration due to gravity on the Earth.

b. On the Earth, the rover's weight was

$$W = mg = (225\ \mathrm{kg})(9.81\ \mathrm{m/s^2}) = 2210\ \mathrm{N}$$

On the Moon, its weight was

$$W_m = mg_m = (225\ \mathrm{kg})(1.62\ \mathrm{m/s^2}) = 365\ \mathrm{N}$$

As expected, this is roughly one-sixth its Earth weight.

The replacement of a sphere with a point mass at its center can be applied to many physical systems. For example, the force of gravity between two spheres of finite size is the same as if *both* were replaced by point masses. Thus, the gravitational force between the Earth, with mass M_E, and the Moon, with mass M_m, is

$$F = G\frac{M_E M_m}{r^2}$$

The distance r in this expression is the center-to-center distance between the Earth and the Moon, as shown in **Figure 12–5**. It follows, then, that in many calculations involving the solar system, moons and planets can be treated as point objects.

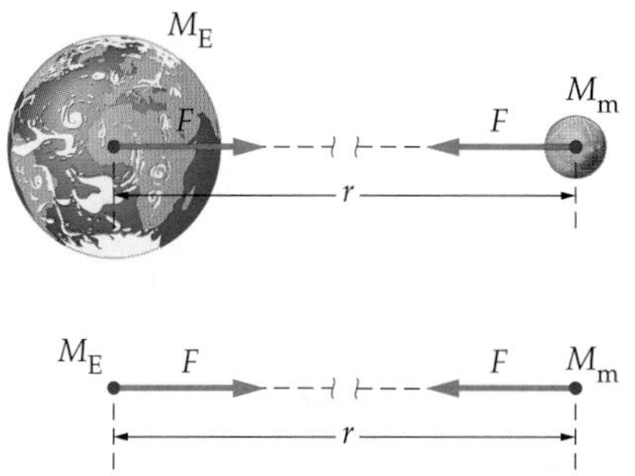

▲ **FIGURE 12–5 Gravitational force between the Earth and the Moon**

The force is the same as if both the Earth and the Moon were point masses. (The sizes of the Earth and Moon are in correct proportion in this figure, but the separation between the two should be much greater than that shown here. In reality, it is about 30 times the diameter of the Earth, and so would be about 2 ft on this scale.)

Weighing the Earth

The British physicist Henry Cavendish performed an experiment in 1798 that is often referred to as "weighing the Earth." What he did, in fact, was measure the value of the universal gravitation constant, G, that appears in Newton's law of gravity. As we have pointed out before, G is a very small number; hence a sensitive experiment is needed for its measurement. It is because of this experimental difficulty that G was not measured until more than 100 years after Newton published the law of gravitation.

In the Cavendish experiment, illustrated in **Figure 12–6**, two masses m are suspended from a thin thread. Near each suspended mass is a large stationary mass M, as shown. Each suspended mass is attracted by the force of gravity toward the large mass near it; hence the rod holding the suspended masses tends to rotate and twist the thread. The angle through which the thread twists can be measured by bouncing a beam of light from a mirror attached to the thread. If the force required to twist the thread through a given angle is known (from previous experiments), a measurement of the twist angle gives the magnitude of the force of gravity. Finally, knowing the masses m and M, and the distance between their centers, r, we can use Equation 12–1 to solve for G. Cavendish found $6.754 \times 10^{-11}\ \mathrm{N \cdot m^2/kg^2}$, in good agreement with the currently accepted value given in Equation 12–2.

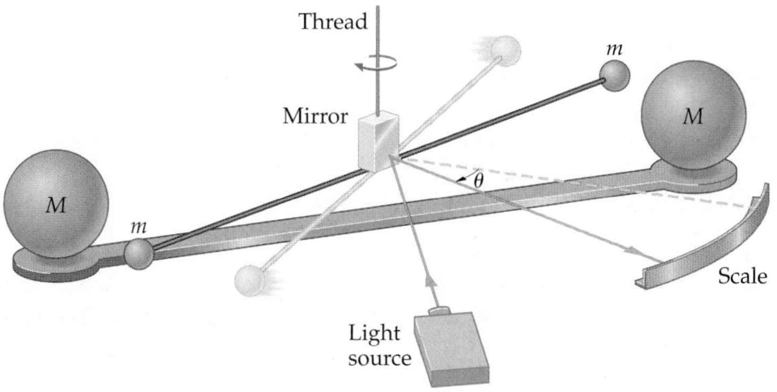

◀ **FIGURE 12–6 The Cavendish experiment**

The gravitational attraction between the masses m and M causes the rod and the suspending thread to twist. Measurement of the twist angle allows for a direct measurement of the gravitational force.

To see why Cavendish is said to have weighed the Earth, recall that the force of gravity on the surface of the Earth, mg, can be written as follows:

$$mg = G\frac{mM_E}{R_E^2}$$

Canceling m and solving for M_E yields

$$M_E = \frac{gR_E^2}{G} \qquad \text{12–5}$$

Before the Cavendish experiment, the quantities g and R_E were known from direct measurement, but G had yet to be determined. When Cavendish measured G, he didn't actually "weigh" the Earth, of course. Instead, he calculated its mass, M_E.

EXERCISE 12–3

Use $M_E = gR_E^2/G$ to calculate the mass of the Earth.

Solution

Substituting numerical values, we find

$$M_E = \frac{gR_E^2}{G} = \frac{(9.81\ \text{m/s}^2)(6.37 \times 10^6\ \text{m})^2}{6.67 \times 10^{-11}\ \text{N}\cdot\text{m}^2/\text{kg}^2} = 5.97 \times 10^{24}\ \text{kg}$$

As soon as Cavendish determined the mass of the Earth, geologists were able to use the result to calculate its average density; that is, its average mass per volume. Assuming a spherical Earth of radius R_E, its total volume is

$$V_E = \tfrac{4}{3}\pi R_E^3 = \tfrac{4}{3}\pi(6.37 \times 10^6\ \text{m})^3 = 1.08 \times 10^{21}\ \text{m}^3$$

Dividing this into the total mass yields the average density, ρ:

$$\rho = \frac{M_E}{V_E} = \frac{5.97 \times 10^{24}\ \text{kg}}{1.08 \times 10^{21}\ \text{m}^3} = 5530\ \text{kg/m}^3 = 5.53\ \text{g/cm}^3$$

This is an interesting result because typical rocks found near the surface of the Earth, such as granite, have a density of only about $3.00\ \text{g/cm}^3$. We conclude, then, that the interior of the Earth must have a greater density than its surface. In fact, by analyzing the propagation of seismic waves around the world, we now know that the Earth has a rather complex interior structure, including a solid inner core with a density of about $15.0\ \text{g/cm}^3$ (see Section 10–5).

REAL-WORLD PHYSICS
The internal structure of the Earth and the Moon

A similar calculation for the Moon yields an average density of about $3.33\ \text{g/cm}^3$, essentially the same as the density of the lunar rocks brought back during the Apollo program. Hence, it is likely that the Moon does not have an internal structure similar to that of the Earth.

Since G is a universal constant—with the same value everywhere in the universe—it can be used to calculate the mass of other bodies in the solar system as well. This is illustrated in the following Example.

EXAMPLE 12–3 Mars Attracts!

Find the mass of Mars, given that its radius is 3.39×10^6 m and that the acceleration of gravity on its surface is $3.73\ \text{m/s}^2$.

Picture the Problem

Our sketch shows an astronaut dropping a camera to the ground on the surface of Mars. If the acceleration of the camera is measured, we find $g_M = 3.73\ \text{m/s}^2$, where the subscript M refers to Mars. In addition, we indicate the radius of Mars in our sketch, where $R_M = 3.39 \times 10^6$ m.

Strategy

Since the acceleration of gravity is g_M on the surface of Mars, it follows that the force of gravity on an object of mass m is $F = mg_M$. This force is also given by Newton's law of gravity; that is, $F = GmM_M/R_M^2$. Setting these expressions for the force equal to one another yields the mass of Mars, M_M.

Solution

1. Set mg_M equal to GmM_M/R_M^2: $$mg_M = G\frac{mM_M}{R_M^2}$$

2. Cancel m and solve for the mass of Mars: $$M_M = \frac{g_M R_M^2}{G}$$

3. Substitute numerical values: $$M_M = \frac{(3.73\ \mathrm{m/s^2})(3.39 \times 10^6\ \mathrm{m})^2}{6.67 \times 10^{-11}\ \mathrm{N \cdot m^2/kg^2}} = 6.43 \times 10^{23}\ \mathrm{kg}$$

Insight

The important point here is that the universal gravitation constant, G, applies as well on Mars as on Earth, or any other object. Therefore, knowledge of the size and acceleration of gravity of an astronomical body is sufficient to determine its mass.

Practice Problem

If the radius of Mars were reduced to 3.00×10^6 m, with its mass remaining the same, would the acceleration of gravity on Mars increase, decrease, or stay the same? Check your answer by calculating the acceleration of gravity for this case. [**Answer:** The acceleration of gravity increases to $4.77\ \mathrm{m/s^2}$.]

Some related homework problems: Problem 19, Problem 20

12–3 Kepler's Laws of Orbital Motion

If you go outside each clear night and observe the position of Mars with respect to the stars, you will find that its apparent motion across the sky is rather complex. Instead of moving on a simple curved path, it occasionally reverses direction (this is known as *retrograde motion*). A few months later it reverses direction yet again and resumes its original direction of motion. Other planets exhibit similar odd behavior.

The Danish astronomer Tycho Brahe (1546–1601) followed the paths of the planets, and Mars in particular, for many years, even though the telescope had not yet been invented. He used, instead, an elaborate sighting device to plot the precise position of the planets. Brahe was joined in his work by Johannes Kepler (1571–1630) in 1600, and after Brahe's death, Kepler inherited his astronomical observations.

Kepler made good use of Brahe's life work, extracting from his carefully collected data the three laws of orbital motion we know today as Kepler's laws. These laws make it clear that the Sun and the planets do not orbit the Earth, as Ptolemy—the ancient Greek astronomer—claimed, but rather that the Earth, along with the other planets, orbit the Sun, as proposed by Copernicus (1473–1543).

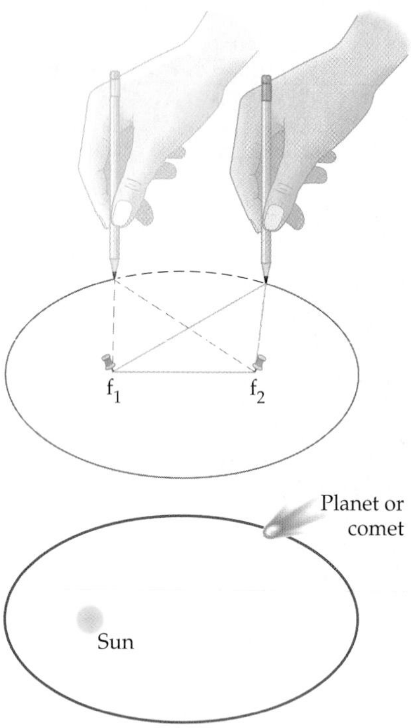

▲ **FIGURE 12–7 Drawing an ellipse**

To draw an ellipse, put two tacks in a piece of cardboard. The tacks define the "foci" of the ellipse. Now connect a length of string to the two tacks, and use a pencil and the string to sketch out a smooth closed curve, as shown. This closed curve is an ellipse. In a planetary orbit a planet follows an elliptical path, with the Sun at one focus. Nothing is at the other focus.

Why the planets obey Kepler's laws no one knew—not even Kepler—until Newton considered the problem decades after Kepler's death. Newton was able to show that each of Kepler's laws follows as a direct consequence of the universal law of gravitation. In the remainder of this section we consider Kepler's three laws one at a time, and point out the connection between them and the law of gravitation.

Kepler's First Law

Kepler tried long and hard to find a circular orbit around the Sun that would match Brahe's observations of Mars. After all, up to that time everyone from Ptolemy to Copernicus believed that celestial objects moved in circular paths of one sort or another. Though the orbit of Mars was exasperatingly close to being circular, the small differences between a circular path and the experimental observations just could not be ignored. Eventually, after a great deal of hard work and disappointment over the loss of circular orbits, Kepler discovered that Mars followed an orbit that was elliptical rather than circular. The same applied to the other planets. This observation became Kepler's first law:

Planets follow elliptical orbits, with the Sun at one focus of the ellipse.

This is a fine example of the scientific method in action. Though Kepler expected and wanted to find circular orbits, he would not allow himself to ignore the data. If Brahe's observations had not been so accurate, Kepler probably would have chalked up the small differences between the data and a circular orbit to error. As it was, he had to discard a treasured—but incorrect—theory, and move on to an unexpected, but ultimately correct, view of nature.

Kepler's first law is illustrated in **Figure 12–7**, along with a definition of an ellipse in terms of its two foci. In the case where the two foci merge, as in **Figure 12–8**, the ellipse reduces to a circle. Thus, a circular orbit *is* allowed by Kepler's first law, but only as a special case.

Newton was able to show that, because the force of gravity decreases with distance as $1/r^2$, closed orbits must have the form of ellipses or circles, as stated in Kepler's first law. He also showed that orbits that are not closed—say the orbit of a comet that passes by the Sun once and then leaves the solar system—are either parabolic or hyperbolic.

Kepler's Second Law

When Kepler plotted the position of a planet on its elliptical orbit, indicating at each position the time the planet was there, he made an interesting observation. First, draw a line from the Sun to a planet at a given time. Then a certain time later—perhaps a month—draw a line again from the Sun to the new position of the planet. The result is that the planet has "swept out" a wedge-shaped area, as indicated in **Figure 12–9 (a)**. If this procedure is repeated when the planet is on a different part of its orbit, another wedge-shaped area is generated. Kepler's observation was that the areas of these two wedges are equal:

As a planet moves in its orbit, it sweeps out an equal amount of area in an equal amount of time.

Kepler's second law follows from the fact that the force of gravity on a planet is directly toward the Sun. As a result, gravity exerts zero torque about the Sun, which means that the angular momentum of a planet in its orbit must be conserved. As Newton showed, conservation of angular momentum is equivalent to the equal-area law stated by Kepler.

▶ **FIGURE 12–8 The circle as a special case of the ellipse**

As the two foci of an ellipse approach one another, the ellipse becomes more circular. In the limit that the foci merge, the ellipse becomes a circle.

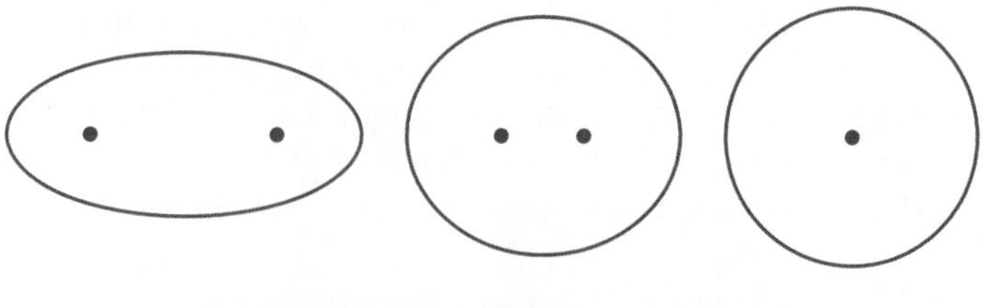

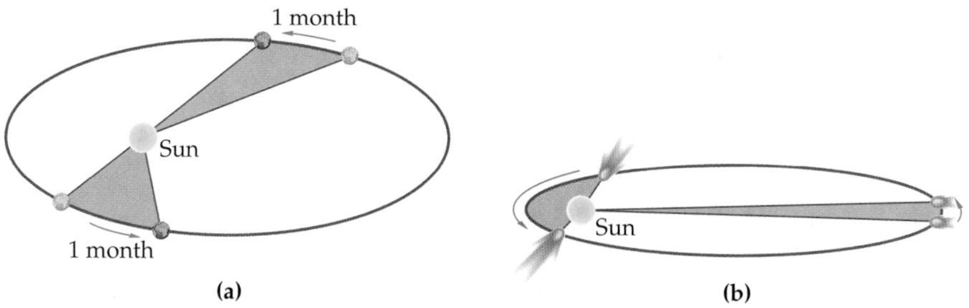

◀ **FIGURE 12–9 Kepler's second law**

(a) The second law states that a planet sweeps out equal areas in equal times. **(b)** In a highly elliptical orbit, the long, thin area is equal to the broad, fan-shaped area.

CONCEPTUAL CHECKPOINT 12–1 Compare Speeds

The Earth's orbit is slightly elliptical. In fact, the Earth is closer to the Sun during the northern hemisphere winter than it is during the summer. Is the speed of the Earth during winter (a) greater than, (b) less than, or (c) the same as its speed during summer?

Reasoning and Discussion

According to Kepler's second law, the area swept out by the Earth per month is the same in winter as it is in summer. In winter, however, the radius from the Sun to the Earth is less than it is in summer. Therefore, if this smaller radius is to sweep out the same area, the Earth must move more rapidly.

Answer:

(a) The speed of the Earth is greater during the winter.

Though we have stated the first two laws in terms of planets, they apply equally well to any object orbiting the Sun. For example, a comet might follow a highly elliptical orbit, as in **Figure 12–9 (b)**. When it is near the Sun, it moves very quickly, for the reason discussed in Conceptual Checkpoint 12–1, sweeping out a broad wedge-shaped area in a month's time. Later in its orbit, the comet is far from the Sun and moving slowly. In this case, the area it sweeps out in a month is a long, thin wedge. Still, the two wedges have equal areas.

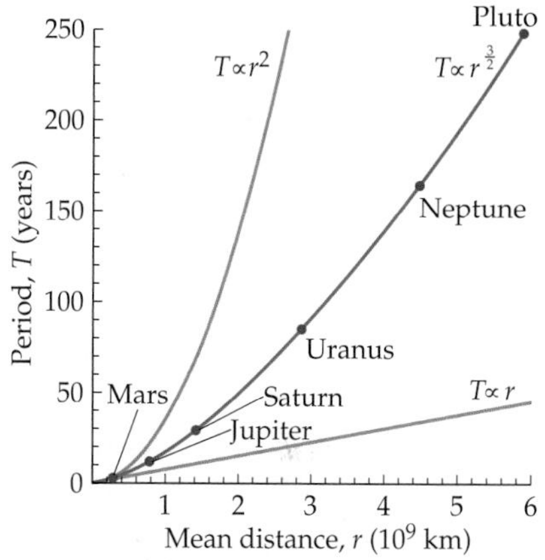

▲ **FIGURE 12–10 Kepler's third law and some near misses**

These plots represent three possible mathematical relationships between period of revolution, T (in years), and mean distance from the Sun, r (in kilometers). The lower curve shows $T = (\text{constant})r$; the upper curve is $T = (\text{constant})r^2$. The middle curve, which fits the data, is $T = (\text{constant})r^{3/2}$. This is Kepler's third law.

Kepler's Third Law

Finally, Kepler studied the relation between the mean distance of a planet from the Sun, r, and its period—that is, the time, T, it takes for the planet to complete one orbit. **Figure 12–10** shows a plot of period versus distance for the planets of the solar system. Kepler tried to "fit" these results to a simple dependence between T and r. If he tried a linear fit, that is T proportional to r, (the bottom curve in Figure 12–10), he found that the period did not increase rapidly enough with distance. On the other hand, if he tried T proportional to r^2, (the top curve in Figure 12–10), the period increased too rapidly. Splitting the difference, and trying T proportional to $r^{3/2}$, yields a good fit (the middle curve in Figure 12–10). This is Kepler's third law:

> The period, T, of a planet increases as its mean distance from the Sun, r, raised to the 3/2 power. That is,
>
> $$T = (\text{constant})r^{3/2} \quad \text{12–6}$$

It is straightforward to derive this result for the special case of a circular orbit. Consider, then, a planet orbiting the Sun at a distance r, as in **Figure 12–11**. Since the planet moves in a circular path, a centripetal force must act on it, as we saw in Section 6–5. In addition, this force must be directed toward the center of the circle; that is, toward the Sun. It is as if you were to swing a ball on the end of a string in a circle above your head, as in Figure 6–12 (p. 162). In order for the ball to move in a circular path, you have to exert a force on the ball toward the center of the circular path. This force is exerted through the string. In the case of a planet orbiting the Sun, the centripetal force is provided by the force of gravity between the Sun and the planet.

If the planet has a mass m, and the Sun has a mass M_s, the force of gravity between them is

$$F = G\frac{mM_s}{r^2}$$

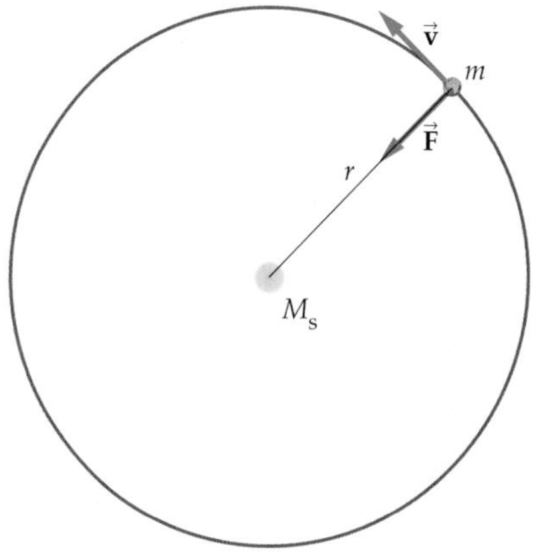

▲ **FIGURE 12–11 Centripetal force on a planet in orbit**

As a planet revolves about the Sun in a circular orbit of radius r, the force of gravity between it and the Sun, $F = GmM_s/r^2$, provides the required centripetal force.

Now, this force creates the centripetal acceleration of the planet, a_{cp}, which, according to Equation 6–15, is

$$a_{cp} = \frac{v^2}{r}$$

Thus, the centripetal force necessary for the planet to orbit is ma_{cp}:

$$F = ma_{cp} = m\frac{v^2}{r}$$

Since the speed of the planet, v, is the circumference of the orbit, $2\pi r$, divided by the time to complete an orbit, T, we have

$$F = m\frac{v^2}{r} = m\frac{(2\pi r/T)^2}{r} = \frac{4\pi^2 rm}{T^2}$$

Setting the centripetal force equal to the force of gravity yields

$$\frac{4\pi^2 rm}{T^2} = G\frac{mM_s}{r^2}$$

Eliminating m and rearranging, we find

$$T^2 = \frac{4\pi^2}{GM_s}r^3$$

or

$$T = \left(\frac{2\pi}{\sqrt{GM_s}}\right)r^{3/2} = (\text{constant})r^{3/2} \qquad \textbf{12–7}$$

As predicted by Kepler, T is proportional to $r^{3/2}$.

Deriving Kepler's third law by using Newton's law of gravitation has allowed us to calculate the constant that multiplies $r^{3/2}$. Note that the constant depends on the mass of the Sun; that is, *T depends on the mass being orbited*. It does not depend on the mass of the planet orbiting the Sun, however, as long as the planet's mass is much less than the mass of the Sun. As a result, Equation 12–7 applies equally to all the planets.

This result can also be applied to the case of a moon or a satellite (an artificial moon) orbiting a planet. To do so, we simply note that it is the planet that is being orbited, not the Sun. Hence, to apply Equation 12–7, we just replace the mass of the Sun, M_s, with the mass of the appropriate planet.

As an example, let's calculate the mass of Jupiter. One of the four moons of Jupiter discovered by Galileo is Io, which completes one orbit every 42 h 27 min = 1.53×10^5 s. Given that the average distance from the center of Jupiter to Io is

PROBLEM-SOLVING NOTE

The Mass in Kepler's Third Law

When applying Kepler's third law, recall that the mass in Equation 12–7, M_s, refers to the mass of the object being orbited. Thus, the third law can be applied to satellites of any object, as long as M_s is replaced by the orbited mass.

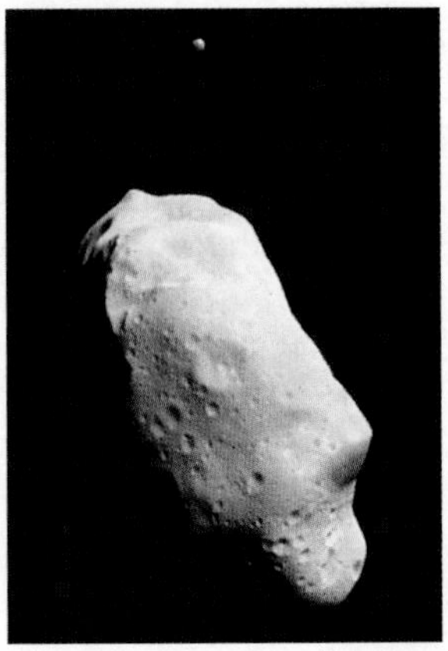

▶ Kepler's laws of orbital motion apply to planetary satellites as well as planets. Jupiter, the largest planet in the solar system, has at least 16 moons, all of which travel in elliptical orbits that obey Kepler's laws. (The moons in the photo at left, passing in front of Jupiter, are Io and Europa, two of the four largest Jovian satellites discovered by Galileo in 1609.) Even some asteroids have been found to have their own satellites. The large cratered object in the photo at right is Ida, an asteroid some 56 km long; its miniature companion at the top of the photo is Dactyl, about 1.5 km in diameter. Like all gravitationally bound bodies, Ida and Dactyl orbit their common center of mass.

4.22×10^8 m, we can find the mass of Jupiter as follows:

$$T = \left(\frac{2\pi}{\sqrt{GM_J}}\right)r^{3/2}$$

$$M_J = \frac{4\pi^2 r^3}{GT^2} = \frac{4\pi^2(4.22 \times 10^8\ \text{m})^3}{(6.67 \times 10^{-11}\ \text{N}\cdot\text{m}^2/\text{kg}^2)(1.53 \times 10^5\ \text{s})^2} = 1.90 \times 10^{27}\ \text{kg}$$

EXAMPLE 12–4 The Sun and Mercury

The Earth revolves around the Sun once a year at an average distance of 1.50×10^{11} m. **(a)** Use this information to calculate the mass of the Sun. **(b)** Find the period of revolution for the planet Mercury, whose average distance from the Sun is 5.79×10^{10} m.

Picture the Problem

Our sketch shows the orbits of Mercury, Venus, and the Earth in correct proportion. In addition, each of these orbits is slightly elliptical, though the deviation from circularity is too small for the eye to see. Finally, we indicate that the orbital radius for Mercury is 5.79×10^{10} m and the orbital radius for Earth is 1.50×10^{11} m.

INTERACTIVE FIGURE

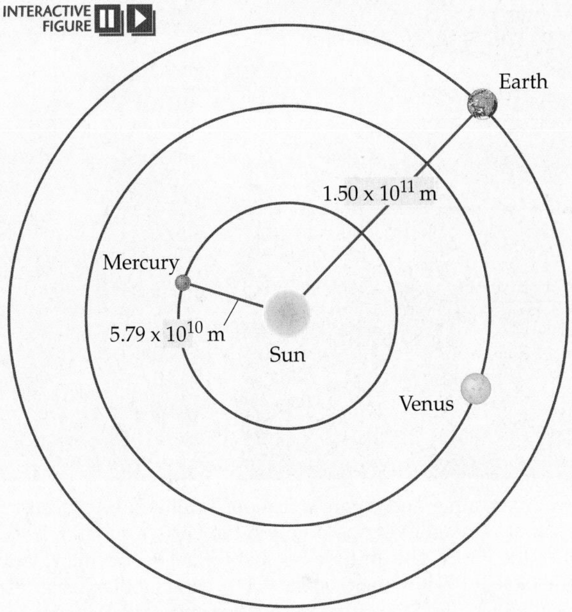

Strategy

a. To find the mass of the Sun, we solve Equation 12–7 for M_S. Note that the period $T = 1$ yr must be converted to seconds before we evaluate the formula.

b. The period of Mercury is found by substituting $r = 5.79 \times 10^{10}$ m in Equation 12–7.

Solution

Part (a)

1. Solve Equation 12–7 for the mass of the Sun:

$$T = \left(\frac{2\pi}{\sqrt{GM_S}}\right)r^{3/2}$$

$$M_S = \frac{4\pi^2 r^3}{GT^2}$$

2. Calculate the period of the Earth in seconds:

$$T = 1\ \text{yr}\left(\frac{365.24\ \text{days}}{1\ \text{yr}}\right)\left(\frac{24\ \text{hr}}{1\ \text{day}}\right)\left(\frac{3600\ \text{s}}{1\ \text{hr}}\right) = 3.16 \times 10^7\ \text{s}$$

3. Substitute numerical values in the expression for the mass of the Sun obtained in step 1:

$$M_S = \frac{4\pi^2 r^3}{GT^2}$$

$$= \frac{4\pi^2(1.50 \times 10^{11}\ \text{m})^3}{(6.67 \times 10^{-11}\ \text{N}\cdot\text{m}^2/\text{kg}^2)(3.16 \times 10^7\ \text{s})^2}$$

$$= 2.00 \times 10^{30}\ \text{kg}$$

Part (b)

4. Substitute $r = 5.79 \times 10^{10}$ m into Equation 12–7. In addition, use the mass of the Sun obtained in part (a):

$$T = \left(\frac{2\pi}{\sqrt{GM_S}}\right)r^{3/2}$$

$$= \left(\frac{2\pi}{\sqrt{(6.67 \times 10^{-11}\ \text{N}\cdot\text{m}^2/\text{kg}^2)(2.00 \times 10^{30}\ \text{kg})}}\right) \times (5.79 \times 10^{10}\ \text{m})^{3/2}$$

$$= 7.58 \times 10^6\ \text{s} = 0.240\ \text{y} = 87.7\ \text{days}$$

continued on next page

continued from previous page

Insight

In part (a), notice that the mass of the Sun is almost a million times more than the mass of the Earth, as determined in Exercise 12–3. In fact, the Sun accounts for 99.9% of all the mass in the solar system.

In part (b) we see that Mercury, with its smaller orbital radius, has a shorter year than the Earth.

Practice Problem

Venus orbits the Sun with a period of 1.94×10^7 s. What is its average distance from the Sun? [**Answer:** $r = 1.08 \times 10^{11}$ m]

Some related homework problems: Problem 23, Problem 27

▲ Many weather and communications satellites are placed in geosynchronous orbits that allow them to remain "stationary" in the sky—that is, fixed over one point on the Earth's equator. Because the Earth rotates, the period of such a satellite must exactly match that of the Earth. The altitude needed for such an orbit is about 36,000 km (see Active Example 12–1). Other satellites, such as those used in the Global Positioning System (GPS), the Hubble Space Telescope, and the American space shuttles, operate at much lower altitudes—typically just a few hundred miles. The photo at left shows the communications satellite Intelsat VI just prior to its capture by astronauts of the space shuttle *Endeavour*. A launch failure had left the satellite stranded in low orbit. The astronauts snared the satellite (right) and fitted it with a new engine that boosted it to its geosynchronous orbit, where it is still in operation today.

REAL-WORLD PHYSICS
Geosynchronous satellites

A *geosynchronous satellite* is one that orbits above the equator with a period equal to one day. From the Earth, such a satellite appears to be in the same location in the sky at all times, making it particularly useful for applications such as communications and weather forecasting. From Kepler's third law, we know that a satellite has a period of one day only if its orbital radius has a particular value. We determine this value in the following Active Example.

ACTIVE EXAMPLE 12–1 Find the Altitude of a Geosynchronous Satellite

Find the altitude above the Earth's surface where a satellite orbits with a period of one day ($R_E = 6.37 \times 10^6$ m, $M_E = 5.97 \times 10^{24}$ kg, $T = 1$ day $= 8.64 \times 10^4$ s).

Solution *(Test your understanding by performing the calculations indicated in each step.)*

1. Rewrite Equation 12–7, using the mass of the Earth in place of the mass of the Sun: $T = \left(2\pi/\sqrt{GM_E}\right)r^{3/2}$
2. Solve for the radius, r: $r = (T/2\pi)^{2/3}(GM_E)^{1/3}$
3. Substitute numerical values: $r = 4.22 \times 10^7$ m
4. Subtract the radius of the Earth to find the altitude: $r - R_E = 3.58 \times 10^7$ m

Insight

Thus, *all* geosynchronous satellites orbit 3.58×10^7 m $\approx$ 22,300 mi above our heads.

Your Turn

Find the altitude above the surface of the Moon where a "lunasynchronous" satellite would orbit. [Note: The length of a lunar day is one month (27.332 days), which is why we see only one side of the Moon.]

(Answers to **Your Turn** *problems are given in the back of the book.)*

REAL-WORLD PHYSICS

The Global Positioning System (GPS)

Not all spacecraft are placed in geosynchronous orbits, however. The U. S. space shuttle, for example, orbits at an altitude of about 150 mi. At that altitude, it takes less than an hour and a half to complete one orbit. The International Space Station, operational although still under construction, orbits at a similar altitude.

The 24 satellites of the Global Positioning System (GPS) are also in relatively low orbits. These satellites, which have an average altitude of 12,550 mi and orbit the Earth every 12 hours, are used to provide a precise determination of an observer's position anywhere on Earth. The operating principle of the GPS is illustrated in **Figure 12–12**. Imagine, for example, that satellite 2 emits a radio signal at a particular time (all GPS satellites carry atomic clocks on board). This signal travels away from the satellite with the speed of light (see Chapter 25) and is detected a short time later by an observer's GPS receiver. Multiplying the time delay by the speed of light gives the distance of the receiver from satellite 2. Thus, in our example, the observer must lie somewhere on the red circle in Figure 12–12. Similar time delay measurements for signals from satellite 11 show that the observer is also somewhere on the green circle; hence the observer is either at the point shown in Figure 12–12, or at the second intersection of the red and green circles on the other side of the planet. Measurements from satellite 6 can resolve the ambiguity and place the observer at the point shown in the figure. Measurements from additional satellites can even determine the observer's altitude. GPS receivers, which are used by hikers, boaters, and others who need to know their precise location, typically use signals from as many as 12 satellites. As currently operated, the GPS gives positions with a typical accuracy of 2 m to 10 m.

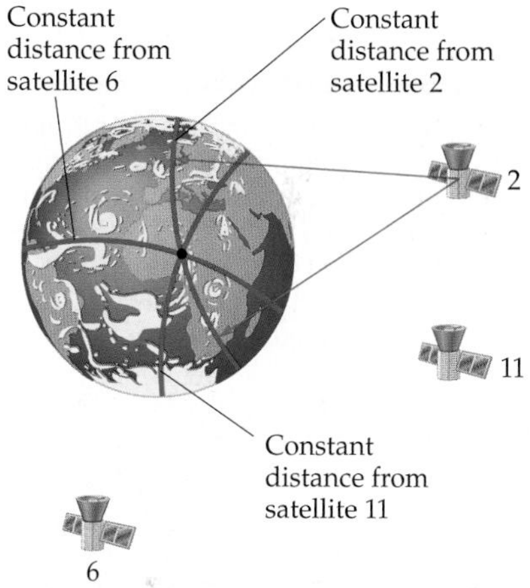

▲ **FIGURE 12–12 The Global Positioning System**

A system of 24 satellites in orbit about the Earth makes it possible to determine a person's location with great accuracy. Measuring the distance of a person from satellite 2 places the person somewhere on the red circle. Similar measurements using satellite 11 place the person's position somewhere on the green circle, and further measurements can pinpoint the person's location.

Orbital Maneuvers

We now show how Kepler's laws can give insight into maneuvering a satellite in orbit. Suppose, for example, that you are piloting a spacecraft in a circular orbit, and you would like to move to a lower circular orbit. As you might expect, you should begin by using your rockets to decrease your speed—that is, fire the rockets that point in the forward direction so that their thrust (Section 9–8) is opposite to your direction of motion. The result of firing the decelerating rockets at a given point A in your original orbit is shown in **Figure 12–13 (a)**. Note that your new orbit is not a circle, as desired, but rather an ellipse. To produce a circular orbit you can simply fire the decelerating rockets once again at point B, on the opposite side of the Earth from point A. The net result of these two firings is that you now move in a circular orbit of smaller radius.

REAL-WORLD PHYSICS

Maneuvering spacecraft

Similarly, to move to a larger orbit, you must fire your accelerating rockets twice. The first firing puts you into an elliptical orbit that moves farther from the Earth, as **Figure 12–13 (b)** shows. After the second firing you are again in a circular orbit. This simplest type of orbital transfer, requiring just two rocket burns, is referred to as a *Hohmann transfer*. The Hohmann transfer is the basic maneuver used to send spacecraft such as the Mars lander from Earth's orbit about the Sun to the orbit of Mars.

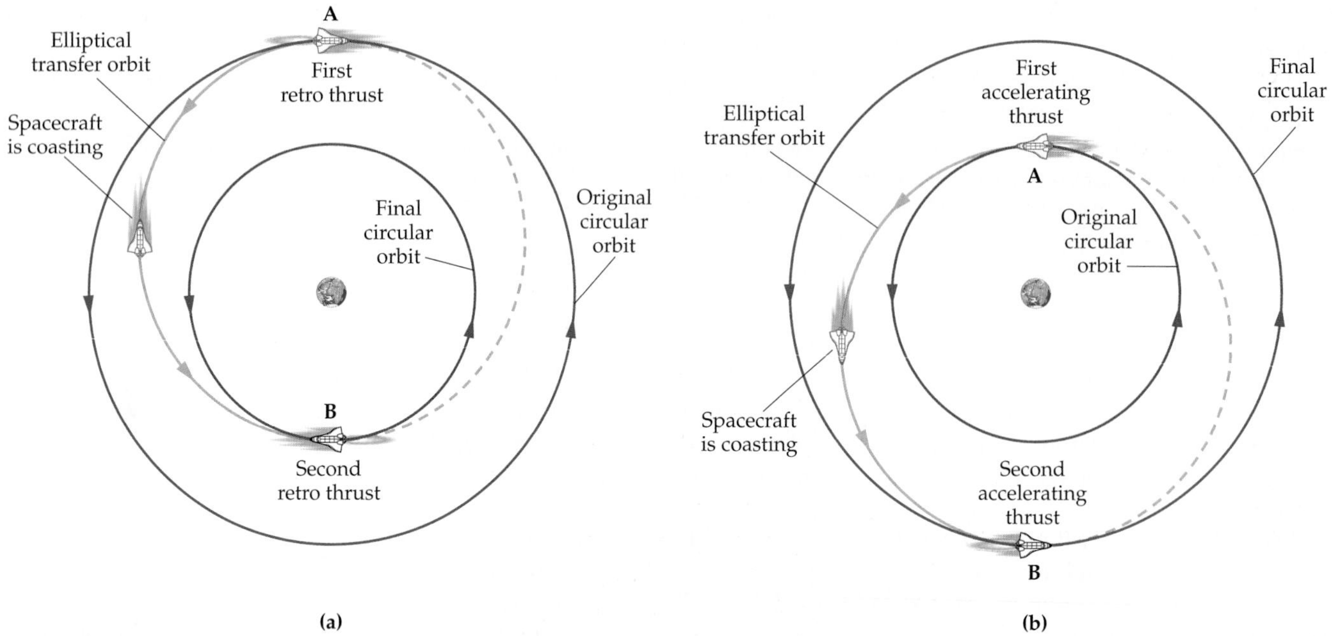

▲ **FIGURE 12–13 Orbital maneuvers**

(a) The radius of a satellite's orbit can be decreased by firing the decelerating rockets once at point A and again at point B. Between firings the satellite follows an elliptical orbit. The satellite speeds up as it falls inward toward the Earth during this maneuver. For this reason its final speed in the new circular orbit is greater than its speed in the original orbit, even though the decelerating rockets have slowed it down twice. **(b)** The radius of a satellite's orbit can be increased by firing the accelerating rockets once at point A and again at point B. Between firings the satellite follows an elliptical orbit. The satellite slows down as it moves farther from the Earth during this maneuver. For this reason its final speed in the new circular orbit is less than its speed in the original orbit, even though the accelerating rockets have sped it up twice.

CONCEPTUAL CHECKPOINT 12–2 **Which Rockets to Use?**

As you pilot your spacecraft in a circular orbit about the Earth, you notice the space station you want to dock with several miles ahead in the same orbit. To catch up with the space station, should you **(a)** fire your accelerating rockets or **(b)** fire your decelerating rockets?

Reasoning and Discussion

Since you want to catch up with something miles ahead, you must accelerate, right? Well, not in this case. Accelerating moves you into an elliptical orbit, as in Figure 12–13 (b), and with a second acceleration you can make your new orbit circular with a greater radius. Recall from Kepler's third law, however, that the larger the radius of an orbit the larger the period, as Equation 12–7 shows. Thus, on your new higher path you take longer to complete an orbit, so you fall farther behind the space station. The same is true even if you fire your rockets only once and stay on the elliptical orbit—it also has a longer period than the original orbit.

On the other hand, two decelerating burns will put you into a circular orbit of smaller radius, and thus smaller period. As a result, you complete an orbit in less time than before and catch up with the space station. After catching up, you can perform two accelerating burns to move you back into the original orbit to dock.

Answer:

(b) You should fire your decelerating rockets.

12–4 Gravitational Potential Energy

In Chapter 8 we saw that the principle of conservation of energy can be used to solve a number of problems that would be difficult to handle with a straightforward application of Newton's laws of mechanics. Before we can apply energy conservation to astronomical situations, however, we must know the gravitational potential energy for a spherical object such as the Earth. Now you may be wondering, "Don't we already know the potential energy of gravity?" Well, in fact, in Chapter 8 we said that the gravitational potential energy a distance h above the Earth's surface is $U = mgh$. As was mentioned at the time, however, this result is valid only near the Earth's surface, where we can say that the acceleration of gravity, g, is approximately constant.

As the distance from the Earth increases we know that g decreases, as was shown in Example 12–2. It follows that mgh cannot be valid for arbitrary h. Indeed, it can be shown that the gravitational potential energy of a system consisting of a mass m a distance r from the center of the Earth is

$$U = -G\frac{mM_E}{r} \qquad \text{12–8}$$

A plot of $U = -GmM_E/r$ is presented in **Figure 12–14**. Note that U approaches zero as r approaches infinity. This is a common convention in astronomical systems. In fact, since only *differences* in potential energy matter, as was mentioned in Chapter 8, the choice of the reference point $(U = 0)$ is completely arbitrary. When we considered systems that were near the Earth's surface, it was natural to let $U = 0$ at ground level. When we consider, instead, distances of astronomical scale, it is generally more convenient to choose the potential energy to be zero when objects are separated by an infinite distance.

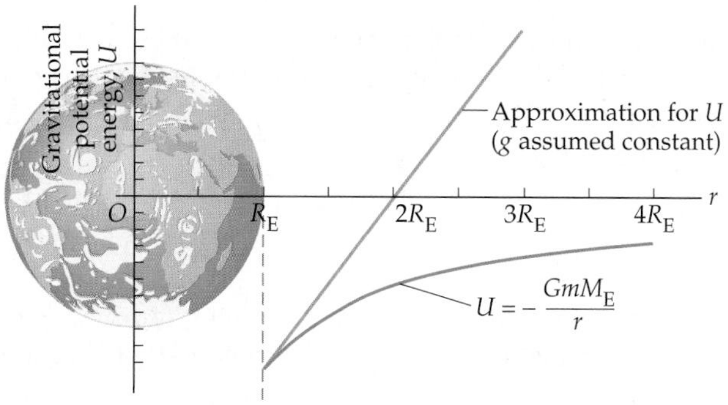

◀ **FIGURE 12–14 Gravitational potential energy as a function of the distance *r* from the center of the Earth**

The lower curve in this plot shows the gravitational potential energy, $U = -GmM_E/r$, for r greater than R_E. Near the Earth's surface, U is approximately linear, corresponding to the result $U = mgh$ given in Chapter 8.

EXERCISE 12–4

Use Equation 12–8 to find the gravitational potential energy of a 12.0-kg meteorite when it is **(a)** one Earth radius above the surface of the Earth, and **(b)** on the surface of the Earth.

Solution

a. In this case, the distance from the center of the Earth is $2R_E$, thus

$$U = -G\frac{mM_E}{2R_E}$$

$$= -(6.67 \times 10^{-11}\,\text{N}\cdot\text{m}^2/\text{kg}^2)\frac{(12.0\,\text{kg})(5.97 \times 10^{24}\,\text{kg})}{2(6.37 \times 10^{6}\,\text{m})} = -3.75 \times 10^{8}\,\text{J}$$

b. Now, the distance from the center of the Earth is R_E, therefore

$$U = -\frac{GmM_E}{R_E}$$

$$= -(6.67 \times 10^{-11}\,\text{N}\cdot\text{m}^2/\text{kg}^2)\frac{(12.0\,\text{kg})(5.97 \times 10^{24}\,\text{kg})}{6.37 \times 10^{6}\,\text{m}} = -7.50 \times 10^{8}\,\text{J}$$

Note that the potential energy in part (b) is twice what it was in part (a), since the distance from the center of the Earth to the meteorite has been halved.

At first glance, Equation 12–8 doesn't seem to bear any similarity to mgh, which we know to be valid near the surface of the Earth. Even so, there is a direct connection between these two expressions. Recall that when we say that the potential energy at a height h is mgh, what we mean is that when a mass m is raised from the ground to a height h, the potential energy of the system increases by the amount mgh. Let's calculate the corresponding difference in potential energy using Equation 12–8.

First, at a height h above the surface of the Earth we have $r = R_E + h$; hence the potential energy there is

$$U = -G\frac{mM_E}{R_E + h}$$

On the surface of the Earth, where $r = R_E$, we have

$$U = -G\frac{mM_E}{R_E}$$

The corresponding difference in potential energy is

$$\Delta U = \left(-G\frac{mM_E}{R_E + h}\right) - \left(-G\frac{mM_E}{R_E}\right)$$
$$= \left(-G\frac{mM_E}{R_E}\right)\left(\frac{1}{1 + h/R_E}\right) - \left(-G\frac{mM_E}{R_E}\right)$$

If h is much smaller than the radius of the Earth, it follows that h/R_E is a small number. In this case, we can apply the useful approximation $1/(1 + x) \approx 1 - x$ [see Figure A–5 (b) in Appendix A] to write $1/(1 + h/R_E) \approx 1 - h/R_E$. As a result, we have

$$\Delta U = \left(-G\frac{mM_E}{R_E}\right)(1 - h/R_E) - \left(-G\frac{mM_E}{R_E}\right) = m\left[\frac{GM_E}{R_E^2}\right]h$$

The term in square brackets should look familiar—according to Equation 12–4 it is simply g. Hence, the increase in potential energy at the height h is

$$\Delta U = mgh$$

as expected.

The straight line in Figure 12–14 corresponds to the potential energy mgh. Near the Earth's surface, it is clear that mgh and $-GmM_E/r$ are in close agreement. For larger r, however, the fact that gravity is getting weaker means that the potential energy does not continue rising as rapidly as it would if gravity were of constant strength.

An important distinction between the potential energy, U, and the gravitational force, $\vec{\mathbf{F}}$, is that the force is a vector, whereas the potential energy is a scalar—that is, U is simply a number. As a result:

> The total gravitational potential energy of a system of objects is the sum of the gravitational potential energies of each pair of objects separately.

Since U is not a vector, there are no x or y components to consider, as would be the case with a vector. Finally, the potential energy given in Equation 12–8 applies to a mass m and the Earth, with mass M_E. More generally, if two point masses, m_1 and m_2 are separated by a distance r, their gravitational potential energy is

Gravitational Potential Energy, *U*

$$U = -G\frac{m_1 m_2}{r} \qquad \text{12–9}$$

SI unit: joule, J

In the next Example we use this result, and the fact that U is a scalar, to find the total gravitational potential energy for a system of three point masses.

EXAMPLE 12–5 Simple Addition

Three masses are positioned as follows: $m_1 = 2.5$ kg is at the origin; $m_2 = 0.75$ kg is at $x = 0$, $y = 1.25$ m; and $m_3 = 0.75$ kg is at $x = 1.25$ m and $y = 1.25$ m. Find the total gravitational potential energy of this system.

Picture the Problem

The masses and their positions are shown in our sketch. The horizontal and vertical distances are $r = 1.25$ m; the diagonal distance is $\sqrt{2}r$.

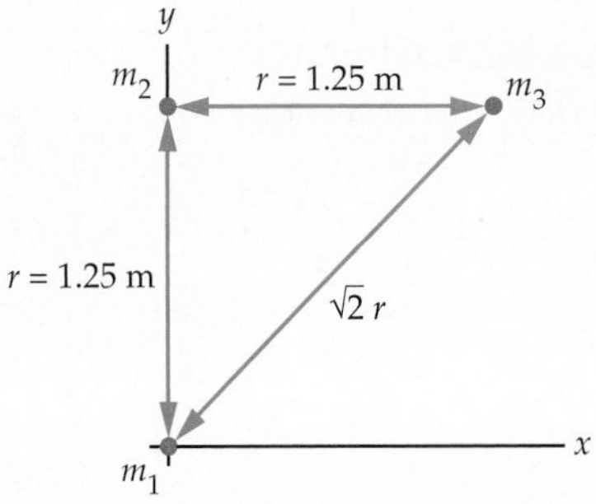

Strategy

The potential energy associated with each pair of masses is given by Equation 12–9. The total potential energy of the system is the sum of the potential energy for each of the three pairs of masses.

Solution

1. Use Equation 12–9 to calculate the potential energy for masses 1 and 2:

$$U_{12} = -G\frac{m_1 m_2}{r_{12}}$$
$$= -(6.67 \times 10^{-11}\,\text{N}\cdot\text{m}^2/\text{kg}^2)\frac{(2.5\text{ kg})(0.75\text{ kg})}{(1.25\text{ m})}$$
$$= -1.0 \times 10^{-10}\text{ J}$$

2. Similarly, calculate the potential energy for masses 2 and 3:

$$U_{23} = -G\frac{m_2 m_3}{r_{23}}$$
$$= -(6.67 \times 10^{-11}\,\text{N}\cdot\text{m}^2/\text{kg}^2)\frac{(0.75\text{ kg})(0.75\text{ kg})}{(1.25\text{ m})}$$
$$= -3.0 \times 10^{-11}\text{ J}$$

3. Do the same calculation for masses 1 and 3:

$$U_{13} = -G\frac{m_1 m_3}{r_{13}}$$
$$= -(6.67 \times 10^{-11}\,\text{N}\cdot\text{m}^2/\text{kg}^2)\frac{(2.5\text{ kg})(0.75\text{ kg})}{\sqrt{2}(1.25\text{ m})}$$
$$= -7.1 \times 10^{-11}\text{ J}$$

4. The total potential energy is the sum of the three contributions calculated above:

$$U_{\text{total}} = U_{12} + U_{23} + U_{13}$$
$$= -1.0 \times 10^{-10}\text{ J} - 3.0 \times 10^{-11}\text{ J} - 7.1 \times 10^{-11}\text{ J}$$
$$= -2.0 \times 10^{-10}\text{ J}$$

Insight

Note that the total gravitational potential energy of this system, $U_{\text{total}} = -2.0 \times 10^{-10}$ J, is less than it would be if the separation of the masses were to approach infinity, in which case $U_{\text{total}} = 0$. The implications of this change in potential energy, in terms of energy conservation, are considered in the next section.

Practice Problem

If the distance $r = 1.25$ m is reduced by a factor of two to $r = 0.625$ m, does the potential energy of the system increase, decrease, or stay the same? Verify your answer by calculating the potential energy in this case. [**Answer:** The potential energy decreases; that is, it becomes more negative. We find $U = 2(-2.0 \times 10^{-10}\text{ J})$].

Some related homework problems: Problem 36, Problem 37

12–5 Energy Conservation

Now that we know the gravitational potential energy, U, at an arbitrary distance from a spherical object, we can apply energy conservation to astronomical situations in the same way we applied it to systems near the Earth's surface in Chapter 8. To be specific, the mechanical energy, E, of an object of mass m a distance r from the Earth is

$$E = K + U = \tfrac{1}{2}mv^2 - G\frac{mM_E}{r} \qquad \text{12–10}$$

Using energy conservation—that is, setting the initial mechanical energy equal to the final mechanical energy—we can answer questions such as the following: Suppose that an asteroid has zero speed infinitely far from the Earth. If this asteroid were to fall directly toward the Earth, what speed would it have when it strikes the Earth's surface?

REAL-WORLD PHYSICS
The impact of meteorites

As you probably know, this is not an entirely academic question. Asteroids and comets, both large and small, have struck the Earth innumerable times during its history. In fact, a particularly large object appears to have struck the Earth on the Yucatan Peninsula in Mexico, near the town of Chicxulub, some 65 million years ago. Evidence suggests that this impact may have lead to the mass extinctions of the Cretaceous period, during which the dinosaurs disappeared from the Earth. Unfortunately, such events are not limited to the distant past. For example, as recently as 50,000 years ago, an iron asteroid tens of meters in diameter and shining 10,000 times brighter than the Sun (from atmospheric heating) slammed into the ground near Winslow, Arizona, forming the 1.2-km wide Barringer Meteor Crater. More recently yet, at sunrise on June 30, 1908, a relatively small stony asteroid streaked through the atmosphere and exploded at an altitude of several kilometers near the Tunguska River in Siberia. The energy released by the explosion was comparable to that of an H-bomb, and it flattened the forest for kilometers in all directions. One can only imagine the consequences if an event like this were to occur near a populated area. Finally, an uncomfortably close call occurred in the early evening of December 9, 1994, when an asteroid the size of a mountain passed the Earth at a distance only one-third the distance from the Earth to the Moon. Thus, though extremely unlikely, the scenarios depicted in movies such as *Armageddon* and *Deep Impact* are not completely unrealistic.

PROBLEM-SOLVING NOTE

Energy Conservation in Astronomical Systems

To apply conservation of energy to an object that moves far from the surface of a planet, one must use $U = -GmM/r$, where r is the distance from the center of the planet.

Returning to our original question, we can use energy conservation to determine the speed such an asteroid or comet might have when it hits the Earth. To begin, we assume the asteroid starts at rest, and hence its initial kinetic energy is zero, $K_i = 0$. In addition, the initial potential energy of the system, U_i, is also zero, since $U = -GmM_E/r$ approaches zero as r approaches infinity. As a result, the total initial mechanical energy of the asteroid-Earth system is zero: $E_i = K_i + U_i = 0$. Because gravity is a conservative force (as discussed in Section 8–1), the total mechanical energy remains constant as the asteroid falls toward the Earth. Thus,

▲ Bodies from space have struck the Earth countless times in the past and continue to do so on a regular basis. Most such objects are relatively small, ranging in size from grains of dust to fist-sized rocks, and burn up from friction as they pass through the atmosphere, creating the bright streaks that we know as meteors. But larger objects, including the occasional comet or asteroid, also cross our path from time to time, and some of these make it to the surface—often with very dramatic results. The crater at left, in Rotorua, New Zealand, must be of relatively recent origin (thousands rather than millions of years old), since erosion has not yet erased this scar on the Earth's surface.

The image at right is a false-color gravity anomaly map of the Chicxulub impact crater in Mexico. The object that struck here some 65 million years ago may have produced such far-reaching climatic disruption that the dinosaurs and many other species became extinct as a result. At the center of the crater the strength of gravity is lower than normal (blue) because of the presence of low-density rock: debris from the impact and sediments that have accumulated in the crater.

as the asteroid moves closer to the Earth and U becomes increasingly negative, the kinetic energy K must become increasingly positive so that their sum, $U + K$, is always zero.

We now set the initial energy equal to the final energy to determine the final speed, v_f. Recalling that the final distance r is the radius of the Earth, R_E, we have

$$E_i = E_f$$
$$0 = \tfrac{1}{2}mv_f^2 - G\frac{mM_E}{R_E}$$

Solving for the final speed yields

$$v_f = \sqrt{\frac{2GM_E}{R_E}} \qquad \textbf{12–11}$$

Substituting numerical values into this expression gives

$$v_f = \sqrt{\frac{2GM_E}{R_E}} = \sqrt{\frac{2(6.67 \times 10^{-11}\,\mathrm{N\cdot m^2/kg^2})(5.97 \times 10^{24}\,\mathrm{kg})}{6.37 \times 10^6\,\mathrm{m}}}$$
$$= 11{,}200\,\mathrm{m/s}\ (25{,}000\,\mathrm{mi/h}) \qquad \textbf{12–12}$$

Thus, a typical asteroid hits the Earth moving at about 7.0 mi/s—about 16 times faster than a rifle bullet! Note that this result is independent of the asteroid's mass.

To help visualize energy conservation in this system, we plot the gravitational potential energy U in **Figure 12–15**. Also indicated in the plot is the total energy, $E = 0$. Since $U + K$ must always equal zero, the value of K goes up as the value of U goes down. This is illustrated graphically in the figure with the help of several histogram bars.

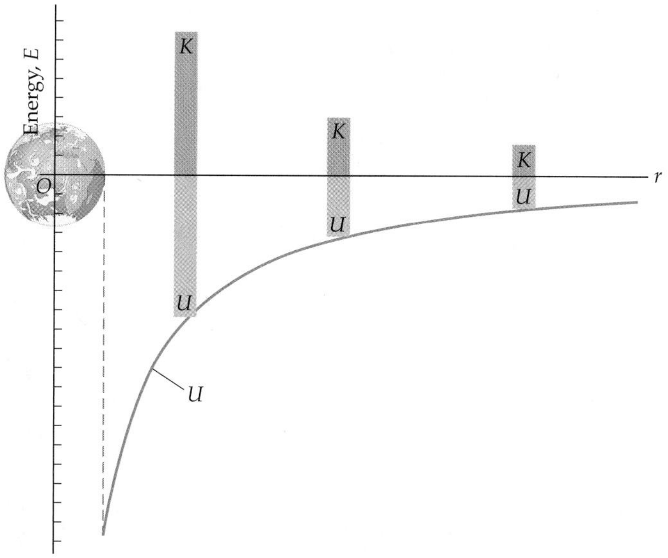

◀ **FIGURE 12–15 Potential and kinetic energies of an object falling toward Earth**
As an object with zero total energy moves closer to the Earth, its gravitational potential energy, U, becomes increasingly negative. In order for the total energy to remain zero, $E = U + K = 0$, it is necessary for the kinetic energy to become increasingly positive.

Another way to think about this is to imagine a smooth wooden or plastic surface constructed to have the same shape as the plot of U shown in Figure 12–15. An object placed on this surface has a gravitational potential energy proportional to the height of the surface above a given reference level. Thus, if a small block is allowed to slide without friction on the surface, it will move downhill and speed up as it drops lower in elevation. That is, its kinetic energy will increase as the potential energy of the system decreases. This is completely analogous to the behavior of an asteroid as it "falls" toward the Earth.

A somewhat more elaborate plot showing the same physics is presented in **Figure 12–16**. The two-dimensional surface in this case represents the potential energy function U as one moves away from the Earth in any direction. In particular,

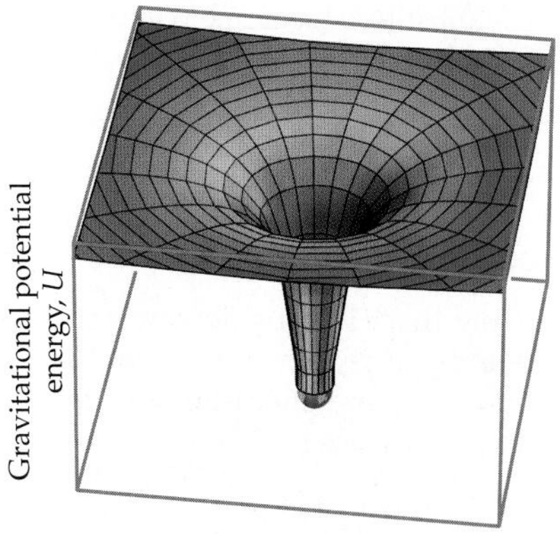

▲ **FIGURE 12–16 A gravitational potential "well"**
The illustration is a three-dimensional plot of the gravitational potential energy near an object such as the Earth. An object approaching the Earth speeds up as it "falls" into the gravitational potential well.

the dependence of U on distance r along any radial line in Figure 12–16 is the same as the shape of U versus r in Figure 12–15. Because the potential energy drops downward in such a plot, this type of situation is often referred to as a "potential well." If a marble is allowed to roll on such a surface, its motion is similar in many ways to the motion of an object near the Earth. In fact, if the marble is started with the right initial velocity, it will roll in a circular or elliptical "orbit" for a long time before falling into the center of the well. (Eventually, of course, the well does swallow up the marble. Though the retarding force of rolling friction is quite small, it still causes the marble to descend into a lower and lower orbit—just as air resistance causes a satellite to descend lower and lower into the Earth's atmosphere until it finally burns up.)

EXAMPLE 12–6 Armageddon Rendezvous

In the movie *Armageddon*, a crew of hard-boiled oil drillers rendezvous with a menacing asteroid just as it passes the orbit of the Moon on its way toward Earth. Assuming the asteroid starts at rest infinitely far from the Earth, as in the previous discussion, find its speed when it passes the Moon's orbit. Assume the Moon orbits at a distance of $60R_E$ from the center of the Earth and that its gravitational influence may be neglected.

Picture the Problem

Our sketch shows the Earth, the Moon, and the asteroid. The initial position of the asteroid is at infinity, where its speed is zero. For the purposes of this problem, its final position is at the Moon's orbit, where its speed is v_f. At this point, the asteroid is heading directly for the Earth.

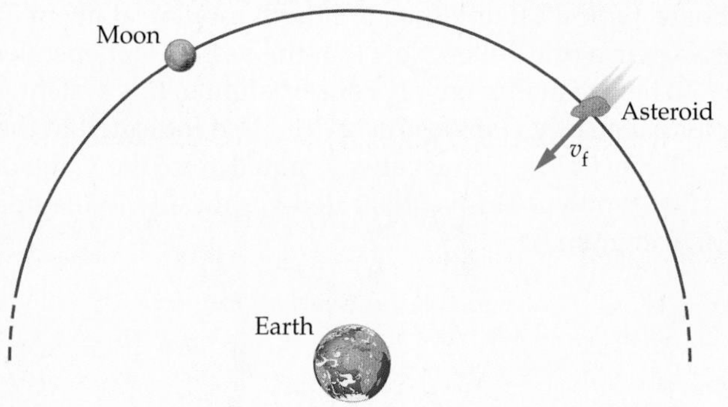

Strategy

The basic strategy is the same as that used to obtain the speed of an asteroid in Equation 12–12; namely, we set the initial energy equal to the final energy and solve for the final speed v_f. In this case, the final radius is $r = 60R_E$. As before, the initial energy is zero.

Solution

1. Set the initial energy of the system equal to its final energy:

$$E_i = E_f$$

$$0 = \tfrac{1}{2}mv_f^2 - G\frac{mM_E}{60R_E}$$

2. Solve for the final speed, v_f:

$$v_f = \sqrt{\frac{2GM_E}{60R_E}} = \frac{1}{\sqrt{60}}\left(\sqrt{\frac{2GM_E}{R_E}}\right)$$

3. Substitute the numerical value given in Equation 12–12 for the quantity in parentheses:

$$v_f = \frac{1}{\sqrt{60}}(11{,}200\ \text{m/s}) = 1450\ \text{m/s} \sim 3200\ \text{mi/h}$$

Insight

Note that the majority of the asteroid's increase in speed occurs after it passes the Moon. The reason for this can be seen in the accompanying plot of the gravitational potential energy, U.

Note that U drops downward more and more rapidly as the Earth is approached. Thus, while there is relatively little increase in K from infinite distance to $r = 60R_E$, there is a substantially larger increase in K from $r = 60R_E$ to $r = R_E$.

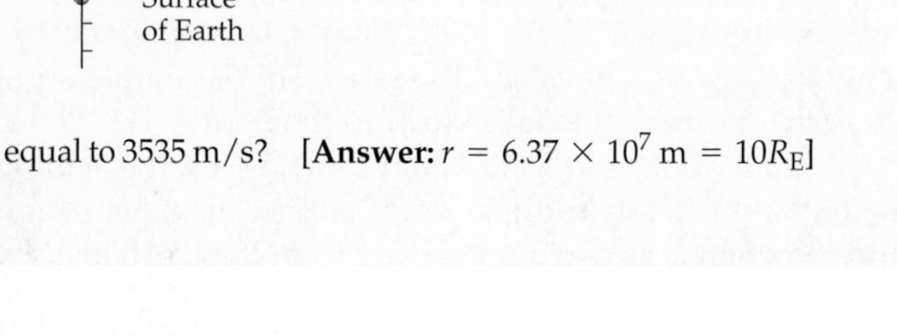

Practice Problem

At what distance from the center of the Earth is the asteroid's speed equal to 3535 m/s? [**Answer:** $r = 6.37 \times 10^7\ \text{m} = 10R_E$]

Some related homework problems: Problem 42, Problem 44

ACTIVE EXAMPLE 12–2 Find the Distance to a Satellite

A satellite in an elliptical orbit has a speed of 9.00 km/s when it is at its closest approach to the Earth (perigee). The satellite is 7.00×10^6 m from the center of the Earth at this time. When the satellite is at its greatest distance from the center of the Earth (apogee), its speed is 3.66 km/s. How far is the satellite from the center of the Earth at apogee? ($R_E = 6.37 \times 10^6$ m, $M_E = 5.97 \times 10^{24}$ kg)

Solution *(Test your understanding by performing the calculations indicated in each step.)*

1. Set the energy at perigee, E_1, equal to the energy at apogee, E_2:	$\frac{1}{2}mv_1^2 - GmM_E/r_1 = \frac{1}{2}mv_2^2 - GmM_E/r_2$
2. Solve for $1/r_2$:	$1/r_2 = 1/r_1 + (v_2^2 - v_1^2)/(2GM_E)$
3. Substitute numerical values:	$1/r_2 = 5.80 \times 10^{-8}\ \text{m}^{-1}$
4. Invert to obtain r_2:	$r_2 = 1.72 \times 10^7$ m

Insight
In this case, apogee is about 2.5 times farther from the center of the Earth than perigee.

Your Turn
What is the speed of the satellite when it is 8.75×10^6 m from the center of the Earth?
(Answers to **Your Turn** *problems are given in the back of the book.)*

▲ Comet Hale-Bopp, one of the largest and brightest comets to visit our celestial neighborhood in recent decades, photographed in April 1997. While most of the planets and planetary satellites in the solar system have roughly circular orbits, the orbits of many comets are highly elliptical. In accordance with Kepler's second law, these objects spend most of their time moving slowly through cold, distant regions of the solar system (often far beyond the orbit of Pluto). Their visits to the inner solar system are infrequent and relatively brief.

Escape Speed

Resisting the pull of Earth's gravity has always held a fascination for the human species, from Daedalus and Icarus with their wings of feathers and wax, to Leonardo da Vinci and his flying machine, to the Montgolfier brothers and their hot-air balloons. In his 1865 novel, *From the Earth to the Moon,* Jules Verne imagined launching a spacecraft to the Moon by firing it straight upward from a cannon. Not a bad idea—if you could survive the initial blast. Today, rockets are fired into space using the same basic idea, though they smooth out the initial blast by burning their engines over a period of several minutes.

Suppose, then, that you would like to launch a rocket of mass m with an initial speed sufficient not only to reach the Moon, but to allow it to escape the Earth altogether. If we refer to this speed as the **escape speed** for the Earth, v_e, the initial energy of the rocket is

$$E_i = K_i + U_i = \frac{1}{2}mv_e^2 - G\frac{mM_E}{R_E}$$

If the rocket just barely escapes the Earth, its speed will decrease to zero as its distance from the Earth approaches infinity. Therefore, the rocket's final kinetic energy is zero, as is the potential energy of the system, since $U = -GmM_E/r$ goes to zero as $r \rightarrow \infty$. It follows that

$$E_f = K_f + U_f = 0 - 0 = 0$$

Equating these energies yields

$$\frac{1}{2}mv_e^2 - G\frac{mM_E}{R_E} = 0$$

Therefore, the escape speed from the Earth is

$$v_e = \sqrt{\frac{2GM_E}{R_E}} = 11{,}200\ \text{m/s} \approx 25{,}000\ \text{mi/h} \qquad \textbf{12–13}$$

Note that the escape speed is precisely the same as the speed of the asteroid calculated in Equation 12–12. This is not surprising when you consider that an object launched from the Earth to infinity is just the reverse of an object falling from infinity to the Earth.

The result given in Equation 12–13 can be applied to other astronomical objects as well by simply replacing M_E and R_E with the appropriate mass and radius for that object.

EXERCISE 12–5

Calculate the escape speed for an object launched from the Moon.

Solution

For the Moon we use $M_m = 7.35 \times 10^{22}$ kg and $R_m = 1.74 \times 10^6$ m. With these values, the escape speed is

$$v_e = \sqrt{\frac{2GM_m}{R_m}} = \sqrt{\frac{2(6.67 \times 10^{-11}\ \text{N}\cdot\text{m}^2/\text{kg}^2)(7.35 \times 10^{22}\ \text{kg})}{1.74 \times 10^6\ \text{m}}}$$

$$= 2370\ \text{m/s}\ (5320\ \text{mi/h})$$

The relatively low escape speed of the Moon means that it is much easier to launch a rocket into space from the Moon than from the Earth. For example, the tiny lunar module that blasted off from the Moon to return the astronauts to Earth could not have come close to escaping from the Earth.

REAL-WORLD PHYSICS
Planetary atmospheres

Similarly, the Moon's low escape speed is the reason it has no atmosphere. Even if you could magically supply the Moon with an atmosphere, it would soon evaporate into space because the individual molecules in the air move with speeds great enough to escape. On the Earth, however, where the escape speed is much higher, gravity can prevent the rapidly moving molecules from moving off into space. Even so, light molecules, like hydrogen and helium, move faster for a given temperature than the heavier molecules like nitrogen and oxygen, as we shall see in Chapter 17. For this reason, the Earth's atmosphere contains virtually no hydrogen or helium. (In fact, helium was first discovered on the Sun, as we point out in Chapter 31; hence its name, which derives from the Greek word for the Sun, "helios.") Since a stable atmosphere is a likely requirement for the development of life, it follows that the escape speed is an important quantity when considering the possibility of life on other planets.

CONCEPTUAL CHECKPOINT 12–3 Compare Escape Speeds

Is the escape speed for a 10-N rocket **(a)** equal to, **(b)** less than, or **(c)** greater than the escape speed for a 10,000-N rocket?

Reasoning and Discussion

The derivation of the escape speed in Equation 12–13 shows that the mass of the rocket, m, cancels. Hence, the escape speed is the same for all objects, regardless of their mass. On the other hand, the kinetic energy required to give the 10,000-N rocket the escape speed is 1000 times greater than the kinetic energy required for the 10-N rocket.

Answer:

(a) Equal. The escape speed is independent of the mass that is escaping.

EXAMPLE 12–7 Half Escape

Suppose Jules Verne's cannon launches a rocket straight upward with an initial speed equal to one-half the escape speed. How far from the center of the Earth does this rocket travel before momentarily coming to rest? (Ignore air resistance in the Earth's atmosphere.)

Picture the Problem

Our sketch shows the rocket launched vertically from the Earth's surface with an initial speed equal to half the escape speed, $v_0 = \frac{1}{2}v_e$. The rocket moves radially away from the Earth until it comes to rest, $v = 0$, at a distance r from the center of the Earth.

Strategy

Since we ignore air resistance, the final energy of the rocket, E_f, must be equal to its initial energy, E_0. Setting these energies equal determines the point where the rocket comes to rest.

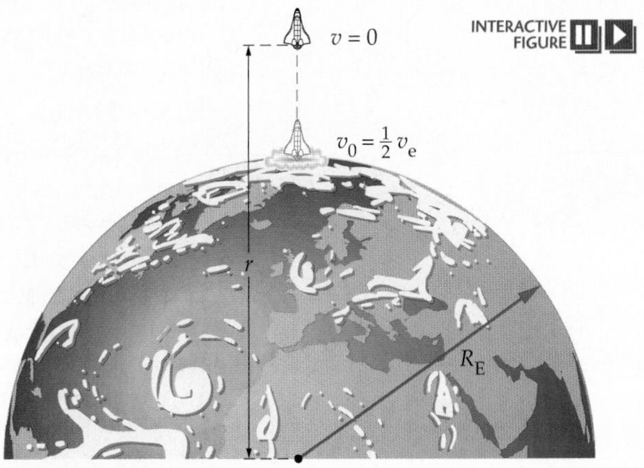

Solution

1. The initial speed, v_0, is one-half the escape speed. Use Equation 12–13 to write an expression for v_0:

$$v_0 = \tfrac{1}{2}v_e = \tfrac{1}{2}\sqrt{\frac{2GM_E}{R_E}} = \sqrt{\frac{GM_E}{2R_E}}$$

2. Write out the initial energy of the rocket, E_0:

$$E_0 = K_0 + U_0 = \tfrac{1}{2}mv_0^2 - \frac{GmM_E}{R_E}$$
$$= \tfrac{1}{2}m\left(\sqrt{\frac{GM_E}{2R_E}}\right)^2 - \frac{GmM_E}{R_E} = -\frac{3}{4}\frac{GmM_E}{R_E}$$

3. Write out the final energy of the rocket. Note that the rocket is a distance r from the center of the Earth when it comes to rest:

$$E_f = K_f + U_f = 0 - \frac{GmM_E}{r} = -\frac{GmM_E}{r}$$

4. Equate the initial and final energies:

$$-\frac{3}{4}\frac{GmM_E}{R_E} = -\frac{GmM_E}{r}$$

5. Solve the relation for r:

$$r = \tfrac{4}{3}R_E$$

Insight

An initial speed of v_e allows the rocket to go to infinity before stopping. If the rocket is launched with half that initial speed, however, it can only rise to a height of $4R_E/3 - R_E = R_E/3$ above the Earth's surface. Quite a dramatic difference.

Practice Problem

Find the rocket's maximum distance from the center of the Earth, r, if its launch speed is $3v_e/4$. [**Answer:** $r = 16R_E/7 = 2.29R_E$]

Some related homework problems: Problem 41, Problem 48

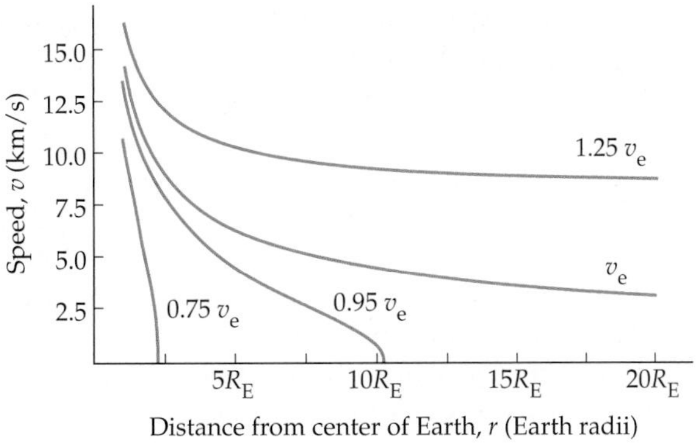

◀ **FIGURE 12–17 Speed of a rocket as a function of distance from the center of the Earth, r, for various vertical launch speeds**

The lower two curves show launch speeds that are less than the escape speed, v_e. In these cases the rocket comes to rest momentarily at a finite height above the Earth. The top curve corresponds to a launch speed greater than the escape speed—this rocket has a finite speed even at infinite distance. The final curve shows the speed of a rocket launched with the escape speed, v_e. It slows to zero speed as the distance approaches infinity.

A plot of the speed of a rocket as a function of its distance from the center of the Earth is presented in **Figure 12–17** for a variety of initial speeds. Note that when the initial speed is less than the escape speed, the rocket comes to rest momentarily at a finite distance, r. In particular, if the launch speed is $0.75v_e$, as in the

Practice Problem of Example 12–7, the rocket's maximum distance from the center of the Earth is $2.29R_E$.

Black Holes

As we can see from Equation 12–13, the escape speed of an object increases with increasing mass and decreasing radius. Thus, for example, if a massive star were to collapse to a relatively small size, its escape speed would become very large. According to Einstein's theory of general relativity, the escape speed of a compressed, massive star could even exceed the speed of light. In this case nothing—not even light—could escape from the star. For this reason, such objects are referred to as *black holes*. Anything entering a black hole would be making a one-way trip to an unknown destiny.

Since black holes cannot be seen directly, our evidence for their existence is indirect. However, we can predict that as matter is drawn toward a black hole it should become heated to the point where it would emit strong beams of X-rays before disappearing from view. X-ray beams matching these predictions have in fact been observed. These observations, coupled with a variety of others, give astronomers confidence that massive black holes reside at the core of many galaxies—including our own!

REAL-WORLD PHYSICS
Black holes and gravitational lensing

Finally, just as a black hole can bend a beam of light back on itself and prevent it from escaping, any massive object can bend light—at least a little. For example, light from distant stars is deflected as it passes by the Sun by 1.75 seconds of an arc (the size of a quarter at a distance of 1.8 miles). Light passing by an entire galaxy of stars or cluster of galaxies can be bent by significant amounts, however, as **Figure 12–18** indicates. This effect is referred to as *gravitational lensing*, since the galaxies act much like the lenses we will study in Chapter 26. Because of gravitational lensing, the images of very distant galaxies or quasars in deep-space astronomical photographs sometimes appear in duplicate, in quadruplicate, or even spread out into circular arcs.

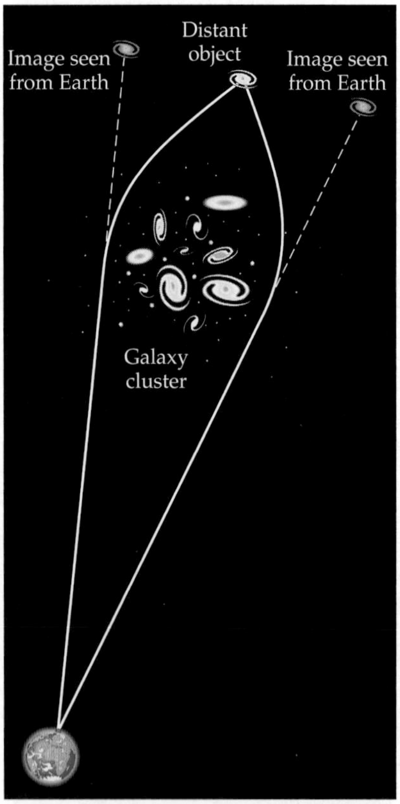

▲ **FIGURE 12–18 Gravitational lensing**
Astronomers often find that very distant objects seem to produce multiple images in their photographs. The cause is the gravitational attraction of intervening galaxies or clusters of galaxies, which are so massive that they can significantly bend the light from remote objects as it passes by them on its way to Earth.

*12–6 Tides

The reason for the ocean tides that rise and fall twice a day was a perplexing and enduring mystery until Newton introduced his law of universal gravitation. Even Galileo, who made so many advances in physics and astronomy, could not explain the tides. However, with the understanding that a force is required to cause an object to move in a circular path, and that the force of gravity becomes weaker with distance, it is possible to describe the tides in detail. In this section we show how it can be done. In addition, we extend the basic idea of tides to several related phenomena.

To begin, consider the idealized situation shown in **Figure 12–19 (a)**. Here we see an object of finite size (a moon or a planet, for example) orbiting a point mass. If all the mass of the object were concentrated at its center, the gravitational force exerted on it by the central mass would be precisely the amount needed to cause it to move in its circular path. Since the object is of finite size, however, the force exerted on various parts of it has different magnitudes. For example, points closer to the central mass experience a greater force than points farther away.

To see the effect of this variation in force, we use a dark red vector in Figure 12–19 (a) to indicate the force exerted by the central mass at three different points on the object. In addition, we use a light red vector to show the force that is required at each of these three points to cause a mass at that distance to orbit the central mass. Comparing these vectors, we see that the forces are identical at the center of the object—as expected. On the near side of the object, however, the force exerted by the central mass is larger than the force needed to hold the object in orbit, and on the far side the force due to the central mass is less than the force needed to hold the object in orbit. The result is that the near side of the object is pulled closer to the central mass and the far side tends to move farther from the

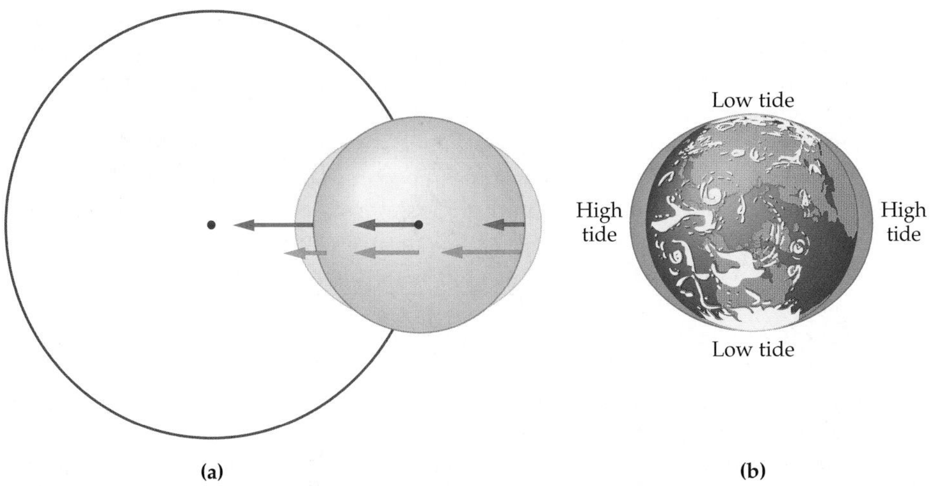

▲ **FIGURE 12–19 The reason for two tides a day**
(a) Tides are caused by a disparity between the gravitational force exerted at various points on a finite-sized object (dark red arrows) and the centripetal force needed for circular motion (light red arrows). Note that the gravitational force decreases with distance, as expected. On the other hand, the centripetal force required to keep an object moving in a circular path *increases* with distance. On the near side, therefore, the gravitational force is stronger than required, and the object is stretched inward. On the far side, the gravitational force is weaker than required, and the object stretches outward. **(b)** On the Earth, the water in the oceans responds more to the deforming effects of tides than do the solid rocks of the land. The result is two high tides and two low tides daily on opposite sides of the Earth.

central mass. This causes an egg-shaped deformation of the object, as indicated in Figure 12–19 (a).

REAL-WORLD PHYSICS
Tides

Any two objects orbiting one another cause deformations of this type. For example, the Earth causes a deformation in the Moon, and the Moon causes a similar deformation in the Earth. In **Figure 12–19 (b)** we show the Earth and the waters of its ocean deformed into an egg shape. Since the waters in the oceans can flow, they deform much more than the underlying rocky surface of the Earth. As a result, the water level relative to the surface of the Earth is greater at the *tidal bulges* shown in the figure. As the Earth rotates about its axis, a person at a given location will observe two high tides and two low tides each day. This is the basic mechanism of the tides on Earth, but the actual situation is complicated by the shape of the coastline at different locations and by the additional tidal effects due to the Sun.

The Moon has no oceans, of course, but the tidal bulges produced in it by the Earth are the reason we see only one side of the Moon. Specifically, the Earth exerts gravitational forces on the tidal bulges of the Moon, causing them to point directly toward the Earth. If the Moon were to rotate slightly away from this alignment, the forces exerted by the Earth would cause a torque that would return the Moon to the original alignment. The net result is that the Moon's period of rotation about its axis is equal to its period of revolution about the Earth. This effect, known as **tidal locking,** is common among the various moons in the solar system.

REAL-WORLD PHYSICS
Tidal locking

A particularly interesting example of tidal locking is provided by Jupiter's moon Io, a site of intense volcanism (see the photo on p. 103). Io follows an elliptical orbit around Jupiter, and its tidal deformation is larger when it is closer

◀ Tides on Earth are caused chiefly by the Moon's gravitational pull, though at full and new moon, when the Moon and Sun are aligned, the Sun's gravity can enhance the effect. In some places on Earth, such as the Bay of Fundy between Maine and Nova Scotia, local topographic conditions produce abnormally large tides.

to Jupiter than when it is farther away. As a result, each time Io orbits Jupiter it is squeezed into a greater deformation and then released. This continual flexing of Io causes its internal temperature to rise, just as a rubber ball gets warmer if you squeeze and release it in your hand over and over. It is this mechanism that is largely responsible for Io's ongoing volcanic activity.

REAL-WORLD PHYSICS
Roche limit and Saturn's rings

In extreme cases, tidal deformation can become so large that an object is literally torn apart. Since tidal deformation increases as a moon moves closer to the planet it orbits, there is a limiting orbital radius—known as the **Roche limit**—inside of which this breakup occurs. A most spectacular example of this effect can be seen in the rings of Saturn, all of which exist well within the Roche limit. The small chunks of ice and other materials that make up the rings may be the remains of a moon that moved too close to Saturn and was destroyed by tidal forces. On the other hand, they may represent material that tidal forces prevented from aggregating to form a moon in the first place. In either case, this dramatic debris field will now never coalesce to form a moon—tidal effects will not allow such a process to occur. Similar remarks apply to the smaller, much fainter rings that spacecraft have observed around Jupiter, Uranus, and Neptune.

Chapter Summary

12–1 Newton's Law of Universal Gravitation

The force of gravity between two point masses, m_1 and m_2, separated by a distance r is attractive and of magnitude

$$F = G\frac{m_1 m_2}{r^2} \qquad 12\text{–}1$$

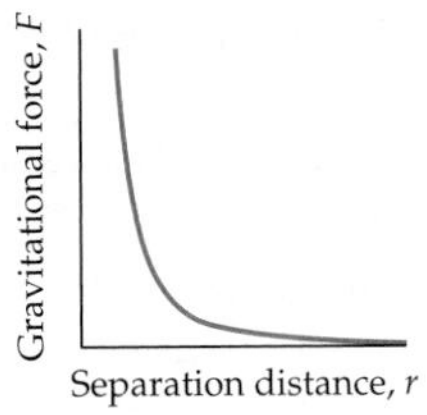

G is the universal gravitation constant,

$$G = 6.67 \times 10^{-11}\ \mathrm{N \cdot m^2/kg^2} \qquad 12\text{–}2$$

Gravity exerts an action-reaction pair of forces on m_1 and m_2; that is, the force exerted by gravity on m_1 is equal in magnitude but opposite in direction to the force exerted on m_2.

Inverse Square Dependence
The force of gravity decreases with distance, r, as $1/r^2$. This is referred to as an inverse square dependence.

Superposition
If more than one mass exerts a gravitational force on a given object, the net force is simply the vector sum of each force individually.

12–2 Gravitational Attraction of Spherical Bodies

In calculating gravitational forces, spherical objects can be replaced by point masses.

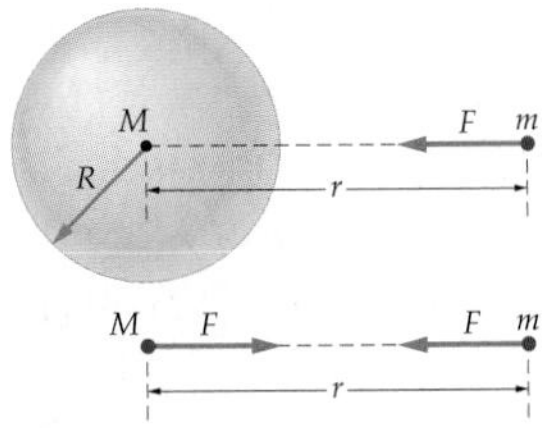

Uniform Sphere
If a mass m is outside a uniform sphere of mass M, the gravitational force between m and the sphere is equivalent to the force exerted by a point mass M located at the center of the sphere.

Acceleration of Gravity
Replacing the Earth with a point mass at its center, we find that the acceleration of gravity on the surface of the Earth is

$$g = \frac{GM_E}{R_E^{\,2}} \qquad 12\text{–}4$$

Weighing the Earth
Cavendish was the first to determine the value of the universal gravitation constant G by direct experiment. Knowing G allows one to calculate the mass of the Earth:

$$M_E = \frac{gR_E^{\,2}}{G} \qquad 12\text{–}5$$

12–3 Kepler's Laws of Orbital Motion

Kepler determined three laws that describe the motion of the planets in our solar system. Newton showed that Kepler's laws are a direct consequence of his law of universal gravitation.

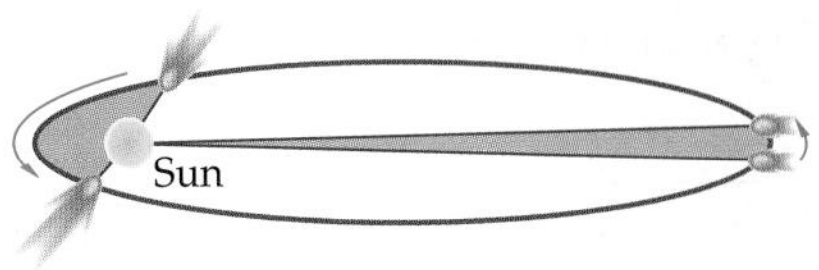

Kepler's First Law
The orbits of the planets are ellipses, with the Sun at one focus.

Kepler's Second Law
Planets sweep out equal area in equal time.

Kepler's Third Law
The period of a planet's orbit, T, is proportional to the 3/2 power of its average distance from the Sun, r:

$$T = \left(\frac{2\pi}{\sqrt{GM_s}}\right)r^{3/2} = (\text{constant})r^{3/2} \qquad \textbf{12–7}$$

12–4 Gravitational Potential Energy

The gravitational potential energy, U, between two point masses m_1 and m_2 separated by a distance r is

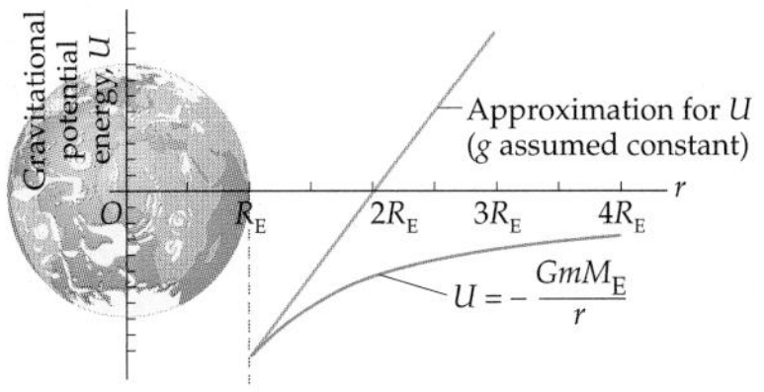

$$U = -G\frac{m_1 m_2}{r} \qquad \textbf{12–9}$$

Zero Level
The zero level of the gravitational potential energy between two point masses is chosen to be at infinite separation of the two masses.

***U* Is a Scalar**
The gravitational potential energy, U, is a scalar. Therefore, the total potential energy for a group of objects is simply the numerical sum of the potential energy associated with each pair of masses.

12–5 Energy Conservation

With the gravitational potential energy given in Section 12–4, energy conservation can be applied to astronomical situations.

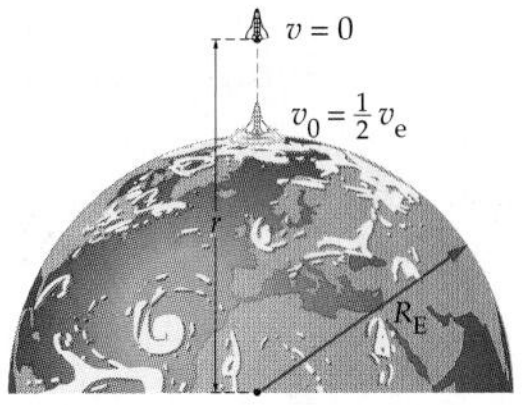

Total Mechanical Energy
An object with mass m, speed v, and at a distance r from the center of the Earth has a total energy given by

$$E = K + U = \tfrac{1}{2}mv^2 - \frac{GmM_E}{r} \qquad \textbf{12–10}$$

Escape Speed
An object launched from the surface of the Earth with the escape speed v_e can move infinitely far from the Earth. In the limit of infinite separation, the object slows to zero speed.

The escape speed for the Earth is given by

$$v_e = \sqrt{\frac{2GM_E}{R_E}} \qquad \textbf{12–13}$$

Its numerical value is 11,200 m/s = 25,000 mi/h. A similar expression can be applied to other astronomical bodies.

*12–6 Tides

Tides result from the variation of the gravitational force from one side of an astronomical object to the other side.

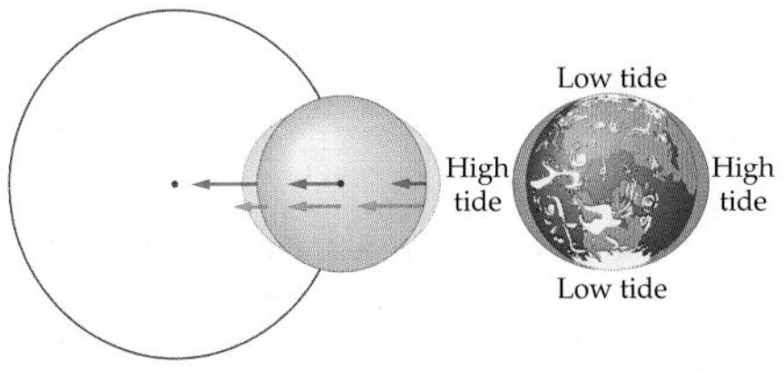

(a) (b)

Tidal Locking
Tidal locking occurs when one astronomical object always points its tidal bulge at the object it orbits.

Roche Limit
Tidal deformation increases as an astronomical object moves closer to the body it orbits. At the Roche limit, the tidal deformation is so great that it breaks the object into small pieces.

Problem-Solving Summary

Type of Problem	Relevant Physical Concepts	Related Examples
Find the force due to gravity.	The magnitude of the force is given by Newton's law of universal gravitation, $F = Gm_1m_2/r^2$. The direction of the force is attractive and along the line connecting m_1 and m_2. If more than one force is involved, the net force is the vector sum of the individual forces.	Examples 12–1, 12–2, 12–3
Relate the period of a planet to the radius of its orbit and the mass of the body it orbits.	Use Kepler's third law, $T = \left(2\pi/\sqrt{GM}\right)r^{3/2}$.	Example 12–4 Active Example 12–1
Determine the speed of an object at a particular location, given its initial speed and location.	Use energy conservation, with the gravitational potential energy given by $U = -Gm_1m_2/r$.	Examples 12–6, 12–7 Active Example 12–2

Conceptual Questions

(Answers to odd-numbered Conceptual Questions can be found in the back of the book.)

1. It is often said that astronauts in orbit experience weightlessness because they are beyond the pull of Earth's gravity. Is this statement correct? Explain.
2. When a person passes you on the street, you do not feel a gravitational tug. Explain.
3. Two objects experience a gravitational attraction. Give a reason why the gravitational force between them does not depend on the sum of their masses.
4. Imagine bringing the tips of your index fingers together. Each finger contains a certain finite mass, and the distance between them goes to zero as they come into contact. From the force law $F = Gm_1m_2/r^2$ one might conclude that the attractive force between the fingers is infinite, and, therefore, that your fingers must remain forever stuck together. What is wrong with this argument?
5. Does the radius vector of Mars sweep out the same amount of area per time as that of the Earth? Why or why not?
6. When a communications satellite is placed in a geosynchronous orbit above the equator, it remains fixed over a given point on the ground. Is it possible to put a satellite into an orbit so that it remains fixed above the North Pole? Explain.
7. The mass of Pluto on June 22, 1978, James Christy made the first observation of a moon orbiting Pluto. Until that time the mass of Pluto was not known, but with the discovery of its moon, Charon, its mass could be calculated with some accuracy. Explain.
8. Rockets are launched into space from Cape Canaveral in an easterly direction. Is there an advantage to launching to the east versus launching to the west? Explain.
9. One day in the future you may take a pleasure cruise to the Moon. While there you might climb a lunar mountain and throw a rock horizontally from its summit. If, in principle, you could throw the rock fast enough, it might end up hitting you in the back. Explain.
10. Apollo astronauts orbiting the Moon at low altitude noticed occasional changes in their orbit that they attributed to localized concentrations of mass below the lunar surface. Just what effect would such "mascons" have on their orbit?
11. If you light a candle on the space shuttle—which would not be a good idea—would it burn the same as on the Earth? Explain.
12. The force exerted by the Sun on the Moon is more than twice the force exerted by the Earth on the Moon. Should the Moon be thought of as orbiting the Earth or the Sun? Explain.
13. **The Path of the Moon** The Earth and Moon exert gravitational forces on one another as they orbit the Sun. As a result, the path they follow is not the simple circular orbit you would expect if either one orbited the Sun alone. Occasionally you will see a suggestion that the Moon follows a path like a sine wave centered on a circular path, as in the upper part of **Figure 12–20**. This is *incorrect*. The Moon's path is qualitatively like that shown in the lower part of **Figure 12–20**. Explain. (Refer to Conceptual Question 12.)

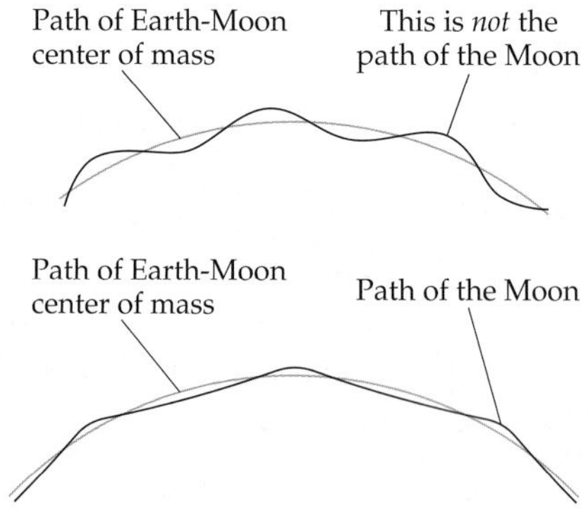

▲ **FIGURE 12–20** Conceptual Question 13

Conceptual Exercises

(Answers to odd-numbered Conceptual Exercises can be found in the back of the book.)

1. System A has masses m and m separated by a distance r; system B has masses m and $2m$ separated by a distance $2r$; system C has masses $2m$ and $3m$ separated by a distance $2r$; and system D has masses $4m$ and $5m$ separated by a distance $3r$. Rank these systems in order of increasing gravitational force.
2. You weigh yourself on a scale inside an airplane flying due east above the equator. If the airplane now turns around and heads due west with the same speed, will the reading on the scale increase, decrease, or stay the same? Explain.

3. Rank the three objects shown in **Figure 12–21** in increasing order of the net gravitational force the object experiences.

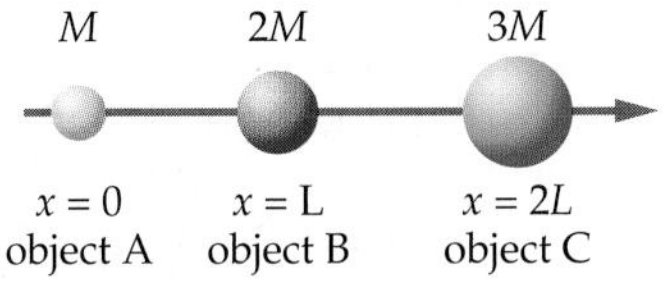

▲ **FIGURE 12–21** Conceptual Exercises 3 and 4

4. Referring to Figure 12–21, rank the three objects in increasing order of the initial acceleration it would experience if it alone were free to move.
5. When the Moon is in its new moon position (directly between the Earth and the Sun), is the net force exerted on it by the Sun and the Earth directed toward the Sun or directed toward the Earth? Explain. (Refer to Conceptual Question 13.)
6. **The Speed of the Earth** The orbital speed of the Earth is greatest around January 4 and least around July 4. Is the distance from the Earth to the Sun on January 4 greater than, less than, or the same as its distance from the Sun on July 4? Explain.
7. A satellite orbits the Earth in a circular orbit of radius r. At some point its rocket engine is fired in such a way that its speed increases rapidly by a small amount. As a result, does the **(a)** apogee distance and **(b)** perigee distance increase, decrease, or stay the same? Explain.
8. Repeat the previous problem, only this time with the rocket engine of the satellite fired in such a way as to slow the satellite.
9. **The Earth-Moon Distance Is Increasing** Laser reflectors left on the surface of the Moon by the Apollo astronauts show that the average distance from the Earth to the Moon is increasing at the rate of 3.8 cm per year. As a result, will the length of the month increase, decrease, or remain the same? Explain.
10. Is the amount of energy required to get a spacecraft from the Earth to the Moon greater than, less than, or the same as the energy required to get the same spacecraft from the Moon to the Earth? Explain.
11. A satellite goes through one complete orbit of the Earth. **(a)** Is the net work done on it by the Earth's gravitational force positive, negative, or zero? Explain. **(b)** Does your answer to part (a) depend on whether the orbit is circular or elliptical? Explain.
12. **The Crash of Skylab** Skylab, the largest spacecraft ever to fall back to the Earth, met its end on July 11, 1979. The cause of Skylab's crash was the friction it experienced in the upper reaches of the Earth's atmosphere. As the radius of Skylab's orbit decreased, did its speed increase, decrease, or stay the same? Explain.
13. Suppose the Earth were to suddenly shrink to half its current diameter, with its mass remaining constant. Would the escape speed of the Earth increase, decrease, or stay the same? Explain.
14. Is the energy required to launch a rocket vertically to a height h greater than, less than, or equal to the energy required to put the same rocket into orbit at the height h? Explain.

Problems

Note: **IP** *denotes an integrated conceptual/quantitative problem.* **BIO** *identifies problems of biological or medical interest. Red bullets (•, ••, •••) are used to indicate the level of difficulty of each problem. Astronomical data needed for problems in this chapter can be found on the inside back cover and in Appendix C. Ignore air resistance.*

Section 12–1 Newton's Law of Universal Gravitation

1. • In each hand you hold a 0.26-kg apple. What is the gravitational force exerted by each apple on the other when their separation is **(a)** 0.25 m and **(b)** 0.50 m?

2. • A 6.1-kg bowling ball and a 7.2-kg bowling ball rest on a rack 0.75 m apart. **(a)** What is the force of gravity exerted on each of the balls by the other ball? **(b)** At what separation is the force of gravity between the balls equal to 2.0×10^{-9} N?

3. • A communications satellite with a mass of 450 kg is in a circular orbit about the Earth. The radius of the orbit is 35,000 km as measured from the center of the Earth. Calculate **(a)** the weight of the satellite on the surface of the Earth and **(b)** the gravitational force exerted on the satellite by the Earth when it is in orbit.

4. • **The Attraction of Ceres** Ceres, the largest asteroid known, has a mass of roughly 8.7×10^{20} kg. If Ceres passes within 12,000 km of the spaceship in which you are traveling, what force does it exert on you? (Use an approximate value for your mass, and treat yourself and the asteroid as point objects.)

5. • In one hand you hold a 0.12-kg apple, in the other hand a 0.20-kg orange. The apple and orange are separated by 0.75 m. What is the magnitude of the force of gravity that **(a)** the orange exerts on the apple and **(b)** the apple exerts on the orange?

6. •• **IP** A spaceship of mass m travels from the Earth to the Moon along a line that passes through the center of the Earth and the center of the Moon. **(a)** At what distance from the center of the Earth is the force due to the Earth twice the magnitude of the force due to the Moon? **(b)** How does your answer to part (a) depend on the mass of the spaceship? Explain.

7. •• At new moon, the Earth, Moon, and Sun are in a line, as indicated in **Figure 12–22**. Find the direction and magnitude of the net gravitational force exerted on **(a)** the Earth, **(b)** the Moon, and **(c)** the Sun.

▲ **FIGURE 12–22** Problem 7

8. •• When the Earth, Moon, and Sun form a right triangle, with the Moon located at the right angle, as shown in **Figure 12–23**, the Moon is in its third quarter phase. (The Earth is viewed here from above its North Pole.) Find the magnitude and direction of the net force exerted on the Moon. Give the direction relative to the line connecting the Moon and the Sun.

▲ **FIGURE 12–23** Problems 8, 9, and 59

9. •• Repeat the previous problem, this time finding the magnitude and direction of the net force acting on the Sun. Give the direction relative to the line connecting the Sun and the Moon.

10. •• **IP** Three 6.75-kg masses are at the corners of an equilateral triangle and located in space far from any other masses. **(a)** If the sides of the triangle are 1.25 m long, find the magnitude of the net force exerted on each of the three masses. **(b)** How does

your answer to part (a) change if the sides of the triangle are doubled in length?

11. •• IP Four masses are positioned at the corners of a rectangle, as indicated in **Figure 12–24**. **(a)** Find the magnitude and direction of the net force acting on the 2.0-kg mass. **(b)** How do your answers to part (a) change (if at all) if all sides of the rectangle are doubled in length?

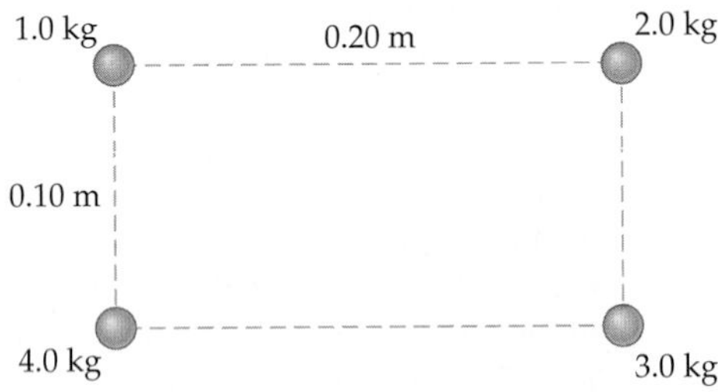

▲ **FIGURE 12–24** Problems 11 and 36

12. ••• Suppose that three astronomical objects (1, 2, and 3) are observed to lie on a line, and that the distance from object 1 to object 3 is D. Given that object 1 has four times the mass of object 3 and seven times the mass of object 2, find the distance between objects 1 and 2 for which the net force on object 2 is zero.

Section 12–2 Gravitational Attraction of Spherical Bodies

13. • Find the acceleration due to gravity on the surface of **(a)** Mercury, and **(b)** Venus.

14. • At what altitude above the Earth's surface is the acceleration due to gravity equal to $g/2$?

15. • Two 6.7-kg bowling balls, each with a radius of 0.11 m, are in contact with one another. What is the gravitational attraction between the bowling balls?

16. • What is the acceleration due to Earth's gravity at a distance from the center of the Earth equal to the orbital radius of the Moon?

17. • **Gravity on Titan** Titan is the largest moon of Saturn and the only moon in the solar system known to have a substantial atmosphere. Find the acceleration due to gravity on Titan's surface, given that its mass is 1.35×10^{23} kg and its radius is 2570 km.

18. •• IP At a certain distance from the center of the Earth, a 4.0-kg object has a weight of 2.0 N. **(a)** Find this distance. **(b)** If the object is released at this location and allowed to fall toward the Earth, what is its initial acceleration? **(c)** If the object is now moved twice as far from the Earth, by what factor does its weight change? Explain. **(d)** By what factor does its initial acceleration change? Explain.

19. •• The acceleration due to gravity on the Moon's surface is known to be about one-sixth the acceleration due to gravity on the Earth. Given that the radius of the Moon is roughly one-quarter that of the Earth, find the mass of the Moon in terms of the mass of the Earth.

20. •• IP **An Extraterrestrial Volcano** Several volcanoes have been observed erupting on the surface of Jupiter's closest Galilean moon, Io. Suppose that material ejected from one of these volcanoes reaches a height of 5.00 km after being projected straight upward with an initial speed of 134 m/s. Given that the radius of Io is 1820 km, **(a)** outline a strategy that allows you to calculate the mass of Io. **(b)** Use your strategy to calculate Io's mass.

21. •• IP **Verne's Trip to the Moon** In his novel *From the Earth to the Moon*, Jules Verne imagined that astronauts inside a spaceship would walk on the floor of the cabin when the force exerted on the ship by the Earth was greater than the force exerted by the Moon. When the force exerted by the Moon was greater, he thought the astronauts would walk on the ceiling of the cabin. **(a)** At what distance from the center of the Earth would the forces exerted on the spaceship by the Earth and the Moon be equal? **(b)** Explain why Verne's description of gravitational effects is incorrect.

22. ••• Consider an asteroid with a radius of 14 km and a mass of 3.35×10^{15} kg. Assume the asteroid is roughly spherical. **(a)** What is the acceleration due to gravity on the surface of the asteroid? **(b)** Suppose the asteroid spins about an axis through its center, like the Earth, with a rotational period T. What is the smallest value T can have before loose rocks on the asteroid's equator begin to fly off the surface?

Section 12–3 Kepler's Laws of Orbital Motion

23. • **Apollo Missions** On Apollo missions to the Moon, the command module orbited at an altitude of 110 km above the lunar surface. How long did it take for the command module to complete one orbit?

24. • Find the orbital speed of a satellite in a geosynchronous circular orbit 3.58×10^7 m above the surface of the Earth.

25. • **An Extrasolar Planet** In July of 1999 a planet was reported to be orbiting the Sun-like star Iota Horologii with a period of 320 days. Find the radius of the planet's orbit, assuming that Iota Horologii has the same mass as the Sun. (This planet is presumably similar to Jupiter, but it may have large, rocky moons that enjoy a relatively pleasant climate.)

26. • Phobos, one of the moons of Mars, orbits at a distance of 9378 km from the center of the red planet. What is the orbital period of Phobos?

27. • The largest moon in the solar system is Ganymede, a moon of Jupiter. Ganymede orbits at a distance of 1.07×10^9 m from the center of Jupiter with an orbital period of about 6.18×10^5 s. Using this information, find the mass of Jupiter.

28. •• IP **As Asteroid with Its Own Moon** The asteroid 243 Ida has its own small moon, Dactyl. (See the photo on page 370) **(a)** Outline a strategy to find the mass of 243 Ida, given that the orbital radius of Dactyl is 89 km and its period is 19 hr. **(b)** Use your strategy to calculate the mass of 243 Ida.

29. •• **GPS Satellites** GPS (Global Positioning System) satellites orbit at an altitude of 2.0×10^7 m. Find **(a)** the orbital period, and **(b)** the orbital speed of such a satellite.

30. •• IP Two satellites orbit the Earth, with satellite 1 at a greater altitude than satellite 2. **(a)** Which satellite has the greater orbital speed? Explain. **(b)** Calculate the orbital speed of a satellite that orbits at an altitude of one Earth radius above the surface of the Earth. **(c)** Calculate the orbital speed of a satellite that orbits at an altitude of two Earth radii above the surface of the Earth.

31. •• IP Calculate the orbital periods of satellites that orbit **(a)** one Earth radius above the surface of the Earth and **(b)** two Earth radii above the surface of the Earth. **(c)** How do your answers to parts (a) and (b) depend on the mass of the satellites? Explain. **(d)** How do your answers to parts (a) and (b) depend on the mass of the Earth? Explain.

32. •• IP The Martian moon Deimos has an orbital period that is greater than the other Martian moon, Phobos. Both moons have approximately circular orbits. **(a)** Is Deimos closer to or farther from Mars than Phobos? Explain. **(b)** Calculate the distance

from the center of Mars to Deimos given that its orbital period is 1.10×10^5 s.

33. ••• **Binary Stars** Centauri A and Centauri B are binary stars with a separation of 3.45×10^{12} m and an orbital period of 2.52×10^9 s. Assuming the two stars are equally massive (which is approximately the case), determine their mass.

34. ••• Find the speed of Centauri A and Centauri B, using the information given in the previous problem.

Section 12–4 Gravitational Potential Energy

35. • **Sputnik** The first artificial satellite to orbit the Earth was Sputnik I, launched October 4, 1957. The mass of Sputnik I was 83.5 kg, and its distances from the center of the Earth at apogee and perigee were 7330 km and 6610 km, respectively. Find the difference in gravitational potential energy for Sputnik I as it moved from apogee to perigee.

36. •• IP Consider the four masses shown in Figure 12–24. **(a)** Find the total gravitational potential energy of this system. **(b)** How does your answer to part (a) change if all the masses in the system are doubled? **(c)** How does your answer to part (a) change if, instead, all the sides of the rectangle are halved in length?

37. •• Calculate the gravitational potential energy of a 7.7-kg mass **(a)** on the surface of the Earth and **(b)** at an altitude of 350 km. **(c)** Take the difference between the results for parts (b) and (a), and compare with mgh, where $h = 350$ km.

38. •• Two 0.59-kg basketballs, each with a radius of 12 cm, are just touching. How much energy is required to change the separation between the centers of the basketballs to **(a)** 1.0 m and **(b)** 10.0 m? (Ignore any other gravitational interactions.)

39. •• Find the minimum kinetic energy needed for a 39,000-kg rocket to escape **(a)** the Moon or **(b)** the Earth.

Section 12–5 Energy Conservation

40. • Suppose one of the Global Positioning System satellites has a speed of 4.46 km/s at perigee and a speed of 3.64 km/s at apogee. If the distance from the center of the Earth to the satellite at perigee is 2.00×10^4 km, what is the corresponding distance at apogee?

41. • **Meteorites from Mars** Several meteorites found in Antarctica are believed to have come from Mars, including the famous ALH84001 meteorite that some believe contains fossils of ancient life on Mars. Meteorites from Mars are thought to get to Earth by being blasted off the Martian surface when a large object (such as an asteroid or a comet) crashes into the planet. What speed must a rock have to escape Mars?

The meteorite ALH84001 (left), dislodged from the Martian surface by a tremendous impact, drifted through space for millions of years before falling to Earth in Antarctica about 13,000 years ago. The electron micrograph at right shows tubular structures within the meteorite; some scientists think they are traces of primitive, bacteria-like organisms that may have lived on Mars billions of years ago. (Problem 41)

42. • Referring to Example 12–1, if the *Millennium Eagle* is at rest at point A, what is its speed at point B?

43. • What is the launch speed of a projectile that rises vertically above the Earth to an altitude equal to one Earth radius before coming to rest momentarily?

44. • A projectile launched vertically from the surface of the Moon rises to an altitude of 365 km. What was the projectile's initial speed?

45. • Find the escape velocity for **(a)** Mercury and **(b)** Venus.

46. •• IP **Halley's Comet** Halley's comet, which passes around the Sun every 76 years, has an elliptical orbit. When closest to the Sun (perihelion) it is at a distance of 8.823×10^{10} m and moves with a speed of 54.6 km/s. The greatest distance between Halley's comet and the Sun (aphelion) is 6.152×10^{12} m. **(a)** Is the speed of Halley's comet greater than or less than 54.6 km/s when it is at aphelion? Explain. **(b)** Calculate its speed at aphelion.

47. •• **The End of the Lunar Module** On Apollo Moon missions, the lunar module would blast off from the Moon's surface and dock with the command module in lunar orbit. After docking, the lunar module would be jettisoned and allowed to crash back onto the lunar surface. Seismometers placed on the Moon's surface by the astronauts would then pick up the resulting seismic waves. Find the impact speed of the lunar module, given that it is jettisoned from an orbit 110 km above the lunar surface moving with a speed of 1630 m/s.

48. •• If a projectile is launched vertically from the Earth with a speed equal to the escape speed, how high above the Earth's surface is it when its speed is half the escape speed?

49. •• Suppose a planet is discovered orbiting a distant star with 10 times the mass of the Earth and one-tenth its radius. How does the escape speed on this planet compare with that of the Earth?

50. •• A projectile is launched vertically from the surface of the Moon with an initial speed of 1250 m/s. At what altitude is the projectile's speed one-half its initial value?

51. •• To what radius would the Sun have to be contracted for its escape speed to equal the speed of light? (Black holes have escape speeds greater than the speed of light; hence we see no light from them.)

52. •• IP Two baseballs, each with a mass of 0.148 kg, are separated by a distance of 355 m in outer space, far from any other objects. **(a)** If the balls are released from rest, what speed do they have when their separation has decreased to 145 m? **(b)** Suppose the mass of the balls is doubled. Would the speed found in part (a) increase, decrease, or stay the same? Explain.

53. ••• On Earth, a person can jump vertically and rise to a height h. What is the radius of the largest spherical asteroid from which this person could escape by jumping straight upward? Assume that each cubic meter of the asteroid has a mass of 3500 kg.

*Section 12–6 Tides

54. •• As will be shown in Problem 55, the magnitude of the tidal force exerted on an object of mass m and length a is approximately $4GmMa/r^3$. In this expression, M is the mass of the body causing the tidal force and r is the distance from the center of m to the center of M. Suppose you are 1 million miles away from a black hole whose mass is a million times that of the Sun. **(a)** Estimate the tidal force exerted on your body by the black hole. **(b)** At what distance will the tidal force be approximately 10 times greater than your weight?

55. ••• A dumbbell has a mass m on either end of a rod of length $2a$. The center of the dumbbell is a distance r from the center of the Earth, and the dumbbell is aligned radially. If $r \gg a$, show that the difference in the gravitational force exerted on the two masses by the Earth is approximately $4GmM_{E}a/r^3$. (Note: The difference in force causes a tension in the rod connecting the masses. We refer to this as a *tidal force*.) [*Hint:* Use the fact that $1/(r-a)^2 - 1/(r+a)^2 \sim 4a/r^3$ for $r \gg a$.]

56. ••• Referring to the previous problem, suppose the rod connecting the two masses m is removed. In this case, the only force between the two masses is their mutual gravitational attraction. In addition, suppose the masses are spheres of radius a and mass $m = \frac{4}{3}\pi a^3 \rho$ that touch each other. (The Greek letter ρ stands for the density of the masses.) **(a)** Write an expression for the gravitational force between the masses m. **(b)** Find the distance from the center of the Earth, r, for which the gravitational force found in part (a) is equal to the tidal force found in Problem 55. This distance is known as the *Roche limit*. **(c)** Calculate the Roche limit for Saturn, assuming $\rho = 3330\ \text{kg/m}^3$. (The famous rings of Saturn are within the Roche limit for that planet. Thus, the innumerable small objects, composed mostly of ice, that make up the rings will never coalesce to form a moon.)

General Problems

57. • Consider a system consisting of three masses on the x axis. Mass $m_1 = 1.00$ kg is at $x = 1.00$ m; mass $m_2 = 2.00$ kg is at $x = 2.00$ m; and mass $m_3 = 3.00$ kg is at $x = 3.00$ m. What is the total gravitational potential energy of this system?

58. •• An astronaut exploring a distant solar system lands on an unnamed planet with a radius of 3660 km. When the astronaut jumps upward with an initial speed of 3.00 m/s, she rises to a height of 0.580 m. What is the mass of the planet?

59. •• **IP** When the Moon is in its third quarter phase, the Earth, Moon, and Sun form a right triangle, as shown in Figure 12–23. Calculate the magnitude of the force exerted on the Moon by **(a)** the Earth and **(b)** the Sun. **(c)** Does it make more sense to think of the Moon as orbiting the Sun, with a small effect due to the Earth, or as orbiting the Earth, with a small effect due to the Sun?

60. •• An equilateral triangle 10.0 m on a side has a 1.00-kg mass at one corner, a 2.00-kg mass at another corner, and a 3.00-kg mass at the third corner (**Figure 12–25**). Find the magnitude and direction of the net force acting on the 1.00-kg mass.

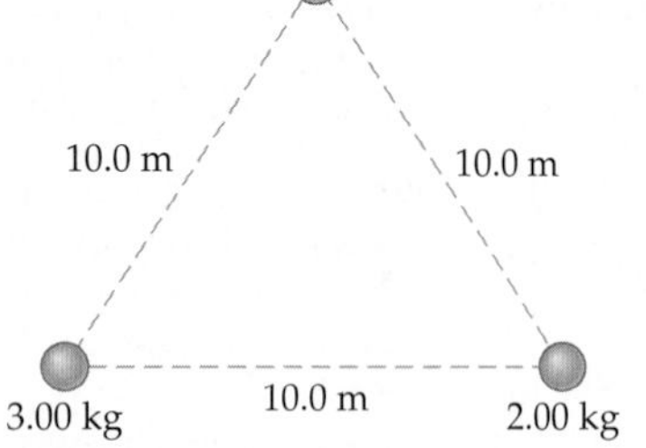

▲ **FIGURE 12–25** Problems 60 and 61

61. •• Suppose that each of the three masses in Figure 12–25 is replaced by a mass of 5.75 kg and radius 0.0716 m. If the balls are released from rest, what speed will they have when they collide at the center of the triangle? Ignore gravitational effects from any other objects.

62. •• **A Near Miss!** In the early morning hours of June 14, 2002, the Earth had a remarkably close encounter with an asteroid the size of a small city. The previously unknown asteroid, now designated 2002 MN, remained undetected until three days after it had passed the Earth. At its closest approach, the asteroid was 73,600 miles from the center of the Earth—about a third of the distance to the Moon. **(a)** Find the speed of the asteroid at closest approach, assuming its speed at infinite distance to be zero and considering only its interaction with the Earth. **(b)** Observations indicate the asteroid to have a diameter of about 2.0 km. Estimate the kinetic energy of the asteroid at closest approach, assuming it has an average density of $3.33\ \text{g/cm}^3$. (For comparison, a 1-megaton nuclear weapon releases about 5.6×10^{15} J of energy.)

63. •• **IP** Suppose a planet is discovered that has the same amount of mass in a given volume as the Earth, but has half its radius. **(a)** Is the acceleration due to gravity on this planet more than, less than, or the same as the acceleration due to gravity on the Earth? Explain. **(b)** Calculate the acceleration due to gravity on this planet.

64. •• **IP** Suppose a planet is discovered that has the same total mass as the Earth, but half its radius. **(a)** Is the acceleration due to gravity on this planet more than, less than, or the same as the acceleration due to gravity on the Earth? Explain. **(b)** Calculate the acceleration due to gravity on this planet.

65. •• Show that the speed of a satellite in a circular orbit a height h above the surface of the Earth is

$$v = \sqrt{\frac{GM_E}{R_E + h}}$$

66. •• In a binary star system, two stars orbit about their common center of mass, as shown in **Figure 12–26**. If $r_2 = 2r_1$, what is the ratio of the masses m_2/m_1 of the two stars?

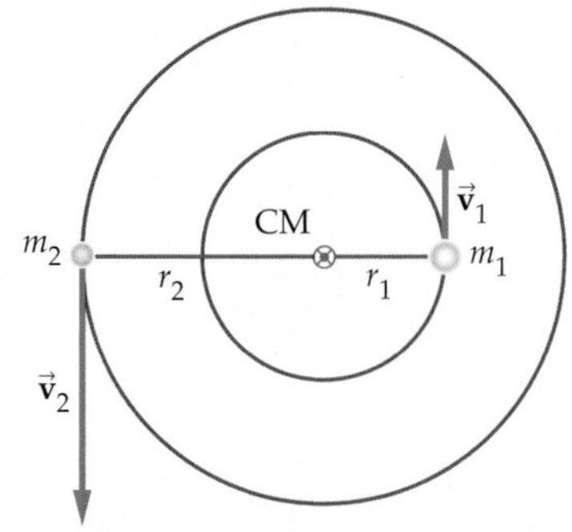

▲ **FIGURE 12–26** Problems 66 and 67

67. •• Find the orbital period of the binary star system described in the previous problem.

68. •• Using the results from Problem 46, find the angular momentum of Halley's comet **(a)** at perihelion and **(b)** at aphelion. (Take the mass of Halley's comet to be 9.8×10^{14} kg.)

69. •• In the not-too-distant future astronauts will travel to Mars to carry out scientific explorations. As part of their mission, it is likely that a "geosynchronous" satellite will be placed above a given point on the Martian equator to facilitate communications. At what altitude above the surface of Mars should such a satellite orbit? (Note: The Martian "day" is 24.6229 hours. Other relevant information can be found in Appendix C.)

70. •• **IP** A satellite is placed in Earth orbit 1000 miles higher than the altitude of a geosynchronous satellite. Referring to Active Example 12–1, we see that the altitude of the satellite is 23,300 mi. **(a)** Is the period of this satellite greater than or less than 24 hours? **(b)** As viewed from the surface of the Earth, does the satellite move eastward or westward? Explain. **(c)** Find the orbital period of this satellite.

71. •• Find the speed of the *Millennium Eagle* at point A in Example 12–1 if its speed at point B is 0.985 m/s.

72. •• Show that the force of gravity between the Moon and the Sun is always greater than the force of gravity between the Moon and the Earth.

73. •• The astronomical unit A.U. is defined as the mean distance from the Sun to the Earth ($1\text{ A.U.} = 1.50 \times 10^{11}\text{ m}$). Apply Kepler's third law (Equation 12–7) to the solar system, and show that it can be written as

$$T = Cr^{3/2}$$

In this expression, the period T is measured in years, the distance r is measured in astronomical units, and the constant C has a magnitude that you must determine.

74. •• **(a)** Find the kinetic energy of a 1520-kg satellite in a circular orbit about the Earth, given that the radius of the orbit is 12,600 miles. **(b)** How much energy is required to move this satellite to a circular orbit with a radius of 25,200 miles?

75. •• **IP Space Shuttle Orbit** On a typical mission, the space shuttle ($m = 2.00 \times 10^6$ kg) orbits at an altitude of 250 km above the Earth's surface. **(a)** Does the orbital speed of the shuttle depend on its mass? Explain. **(b)** Find the speed of the shuttle in its orbit. **(c)** How long does it take for the shuttle to complete one orbit of the Earth?

76. •• **Sampling Comet Wild 2** The spacecraft *Stardust* flew to within 147 miles of Comet Wild 2 on January 2, 2004, snapping pictures and collecting tiny specks of dust in its glistening halo to be returned to the Earth. The roughly spherical comet has a radius of 2.7 km, and the acceleration due to gravity on its surface is 0.00010 g. Find **(a)** the mass of the comet and **(b)** the speed needed to escape from its surface. (Note: It is easy for a person to jump upward with a speed of 3 m/s.)

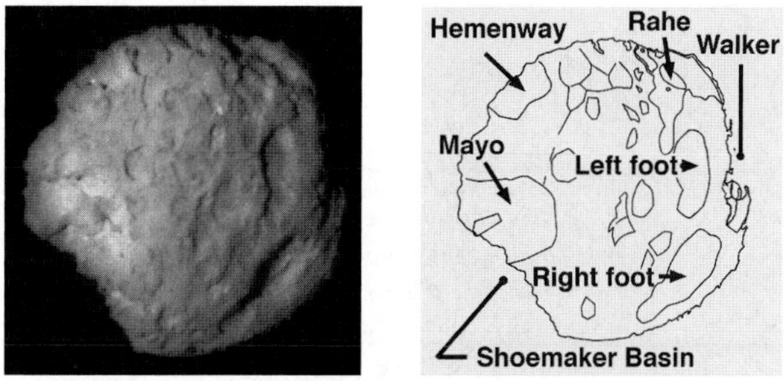

Comet Wild 2 and some of its surface features, including the Walker basin. (Problem 76)

77. ••• **IP** Consider an object of mass m orbiting the Earth at a radius r. **(a)** Find the speed of the object. **(b)** Show that the total mechanical energy of this object is equal to (-1) times its kinetic energy. **(c)** Does the result of part (b) apply to an object orbiting the Sun? Explain.

78. ••• In a binary star system two stars orbit about their common center of mass. Find the orbital period of such a system, given that the stars are separated by a distance d and have masses m and $2m$.

79. ••• Three identical stars, at the vertices of an equilateral triangle, orbit about their common center of mass (**Figure 12–27**). Find the period of this orbital motion in terms of the orbital radius, R, and the mass of each star, M.

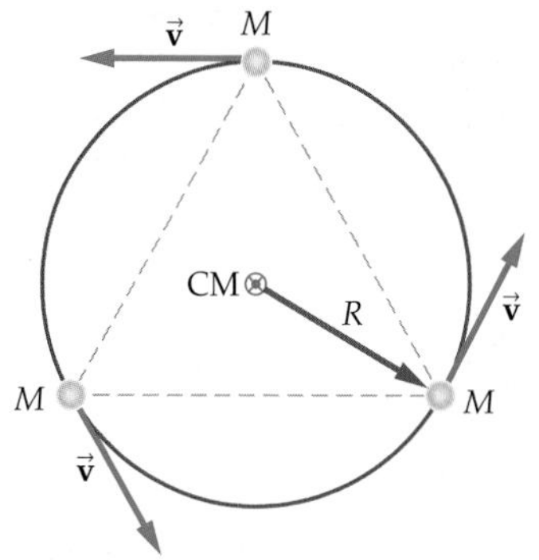

▲ **FIGURE 12–27** Problem 79

80. ••• Find an expression for the kinetic energy of a satellite of mass m in an orbit of radius r about a planet of mass M.

81. ••• Referring to Example 12–1, find the x component of the net force acting on the *Millennium Eagle* as a function of x. Plot your result, showing both negative and positive values of x.

82. ••• A satellite orbits the Earth in an elliptical orbit. At perigee its distance from the center of the Earth is 22,500 km and its speed is 4280 m/s. At apogee its distance from the center of the Earth is 24,100 km and its speed is 3990 m/s. Using this information, calculate the mass of the Earth.

Interactive Problems

83. •• ***Referring to Example 12–4*** Find the orbital radius that corresponds to a "year" of 150 days.

84. •• **IP** ***Referring to Example 12–4*** Suppose the mass of the Sun is suddenly doubled, but the Earth's orbital radius remains the same. **(a)** Would the length of an Earth year increase, decrease, or stay the same? **(b)** Find the length of a year for the case of a Sun with twice the mass. **(c)** Suppose the Sun retains its present mass, but the mass of the Earth is doubled instead. Would the length of the year increase, decrease, or stay the same?

85. •• **IP** ***Referring to Example 12–7*** **(a)** If the mass of the Earth were doubled, would the escape speed of a rocket increase, decrease, or stay the same? **(b)** Calculate the escape speed of a rocket for the case of an Earth with twice its present mass. **(c)** If the mass of the Earth retains its present value, but the mass of the rocket is doubled, does the escape speed increase, decrease, or stay the same?

86. •• **IP** ***Referring to Example 12–7*** Suppose the Earth is suddenly shrunk to half its present radius without losing any of its mass. **(a)** Would the escape speed of a rocket increase, decrease, or stay the same? **(b)** Find the escape speed for an Earth with half its present radius.

13 Oscillations About Equilibrium

13–1	Periodic Motion	395
13–2	Simple Harmonic Motion	396
13–3	Connections Between Uniform Circular Motion and Simple Harmonic Motion	399
13–4	The Period of a Mass on a Spring	405
13–5	Energy Conservation in Oscillatory Motion	409
13–6	The Pendulum	412
13–7	Damped Oscillations	418
13–8	Driven Oscillations and Resonance	419

In this era of atomic timekeepers and electronic digital readouts, a pendulum seems little more than a quaint reminder of the age of grandfather clocks. But pendulums played an important role in the development of physics, and analyzing the motion of a pendulum still provides insight into key physical principles. In this chapter we will explore the behavior of objects that swing, vibrate, or oscillate—and lay the foundations for understanding many natural phenomena, including sound.

In Chapter 11 we considered systems in static equilibrium. Such systems are seldom left undisturbed for very long, it seems, before they are displaced from equilibrium by a bump, a kick, or a nudge. When this happens to a system, it often results in **oscillations** back and forth from one side of the equilibrium position to the other.

The basic cause of oscillations is the fact that when an object is displaced from a position of stable equilibrium it experiences a *restoring force* that is directed back toward the equilibrium position. Thus, the restoring force accelerates the object in the direction of its initial, equilibrium position. When it reaches equilibrium the force acting on it is zero, but it doesn't come to rest. In moving back to equilibrium, it has gained speed and momentum, and hence its inertia carries it through the equilibrium position to the other side, where the restoring force is now in the opposite direction. The process repeats itself, leading to a series of oscillations.

Perhaps the most familiar oscillating system is the simple pendulum, like the ones that keep time in grandfather clocks. Modern digital wristwatches also use oscillators to keep time, but theirs are tiny quartz crystals. In fact, oscillating systems are found in nature over virtually all length scales, from water molecules that oscillate in a microwave oven, to planets that oscillate when struck by an asteroid, to the universe itself, which some think may oscillate in a series of "big bangs" followed by equally momentous "big crunches."

13–1 Periodic Motion

A motion that repeats itself over and over is referred to as **periodic motion.** The beating of your heart, the ticking of a clock, and the movement of a child on a swing are familiar examples. One of the key characteristics of a periodic system is the time required for the completion of one cycle of its repetitive motion. For example, the pendulum in a grandfather clock might take one second to swing from maximum displacement in one direction to maximum displacement in the opposite direction, or two seconds for a complete cycle of oscillation. In this case, we say that the **period,** T, of the pendulum is 2 s.

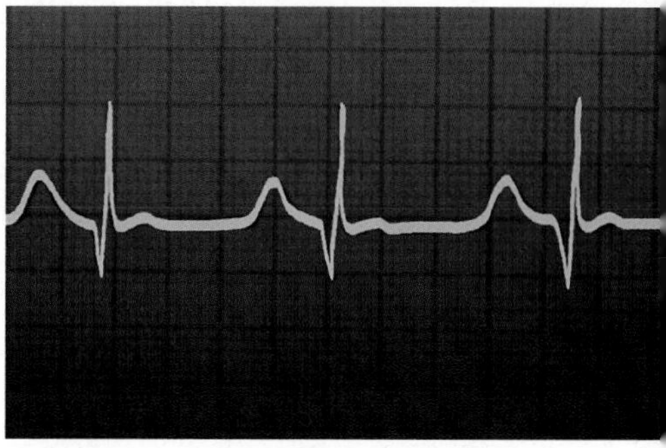

▲ Periodic phenomena are found everywhere in nature, from the movements of the heavenly bodies to the vibration of individual atoms. The trace of an electrocardiogram (ECG or EKG), as shown here, records the rhythmic electrical activity that accompanies the beating of our hearts.

Definition of Period, *T*

T = time required for one cycle of a periodic motion

SI unit: seconds/cycle = s

Note that a cycle (that is, an oscillation) is dimensionless.

Closely related to the period is another common measure of periodic motion, the **frequency,** f. The frequency of an oscillation is simply the number of oscillations per unit of time. Thus, f tells us how frequently, or rapidly, an oscillation takes place—the higher the frequency, the more rapid the oscillations. By definition, the frequency is simply the inverse of the period, T:

Definition of Frequency, *f*

$$f = \frac{1}{T} \qquad \text{13–1}$$

SI unit: cycle/second = 1/s = s^{-1}

Note that if the period of an oscillation, T, is very small, corresponding to rapid oscillations, the corresponding frequency, $f = 1/T$, will be large, as expected.

A special unit has been introduced for the measurement of frequency. It is the **hertz** (Hz), named for the German physicist Heinrich Hertz (1857–1894), in honor of his pioneering studies of radio waves. By definition, one Hz is one cycle per second:

$$1 \text{ Hz} = 1 \text{ cycle/second}$$

High frequencies are often measured in kilohertz (kHz), where 1 kHz = 10^3 Hz, or megahertz (MHz), where 1 MHz = 10^6 Hz.

EXERCISE 13–1

The processing "speed" of a computer refers to the number of binary operations it can perform in one second, so it is really a frequency. If the processor of a personal computer operates at 1.80 GHz, how much time is required for one processing cycle?

Solution

The frequency of the computer's processor is $f = 1.80$ GHz. Therefore, we can use Equation 13–1 to solve for the processing period:

$$T = \frac{1}{f} = \frac{1}{1.80 \text{ GHz}} = \frac{1}{1.80 \times 10^9 \text{ cycles/s}} = 5.56 \times 10^{-10} \text{ s}$$

The high frequency of the computer corresponds to a very small period—less than a billionth of a second to complete one operation. We next consider a situation in which the frequency is considerably smaller.

EXERCISE 13–2

A tennis ball is hit back and forth between two players warming up for a match. If it takes 2.31 s for the ball to go from one player to the other, what are the period and frequency of the ball's motion?

continued on next page

continued from previous page

Solution

The period of this motion is the time for the ball to complete one round trip from one player to the other and back. Therefore,

$$T = 2(2.31\ \text{s}) = 4.62\ \text{s}$$

$$f = \frac{1}{T} = \frac{1}{4.62\ \text{s}} = 0.216\ \text{Hz}$$

Notice that the period and frequency of periodic motion can vary over a remarkably large range. Table 13–1 gives a sampling of typical values of T and f.

TABLE 13–1 Typical Periods and Frequencies

System	Period (s)	Frequency (Hz)
Precession of the Earth	8.2×10^{11} (26,000 y)	1.2×10^{-12}
Hour hand of a clock	43,200 (12 h)	2.3×10^{-5}
Minute hand of a clock	3600	2.8×10^{-4}
Second hand of clock	60	0.017
Pendulum in grandfather clock	2.0	0.50
Human heartbeat	1.0	1.0
Lower range of human hearing	5.0×10^{-2}	20
Wing beat of housefly	5.0×10^{-3}	200
Upper range of human hearing	5.0×10^{-5}	20,000
Computer processor	5.6×10^{-10}	1.8×10^{9}

13–2 Simple Harmonic Motion

Periodic motion can take many forms, as illustrated by the tennis ball going back and forth between players in Exercise 13–2, or an expectant father pacing up and down in a hospital hallway. There is one type of periodic motion, however, that is of particular importance. It is referred to as **simple harmonic motion.**

A classic example of simple harmonic motion is provided by the oscillations of a mass attached to a spring. (See Section 6–2 for a discussion of ideal springs and the forces they exert.) To be specific, consider an air-track cart of mass m attached to a spring of force constant k, as in **Figure 13–1**. When the spring is at its equilibrium length—neither stretched nor compressed—the cart is at the position $x = 0$, where it will remain at rest if left undisturbed. If the cart is displaced from equilibrium by a distance x, however, the spring exerts a restoring force given by Hooke's law, $F = -kx$. In words:

> A spring exerts a restoring force whose magnitude is proportional to the distance it is displaced from equilibrium.

This direct proportionality between distance from equilibrium and force is the key feature of a mass–spring system that leads to simple harmonic motion. As for the direction of the spring force:

> The force exerted by a spring is opposite in direction to its displacement from equilibrium; this accounts for the minus sign in $F = -kx$. In general, a restoring force is one that *always* points toward the equilibrium position.

Now, suppose the cart is released from rest at the location $x = A$. As indicated in Figure 13–1, the spring exerts a force on the cart to the left, causing the cart to accelerate toward the equilibrium position. When the cart reaches $x = 0$, the net force acting on it is zero. Its speed is not zero at this point, however, and so it continues to move to the left. As the cart compresses the spring, it experiences a

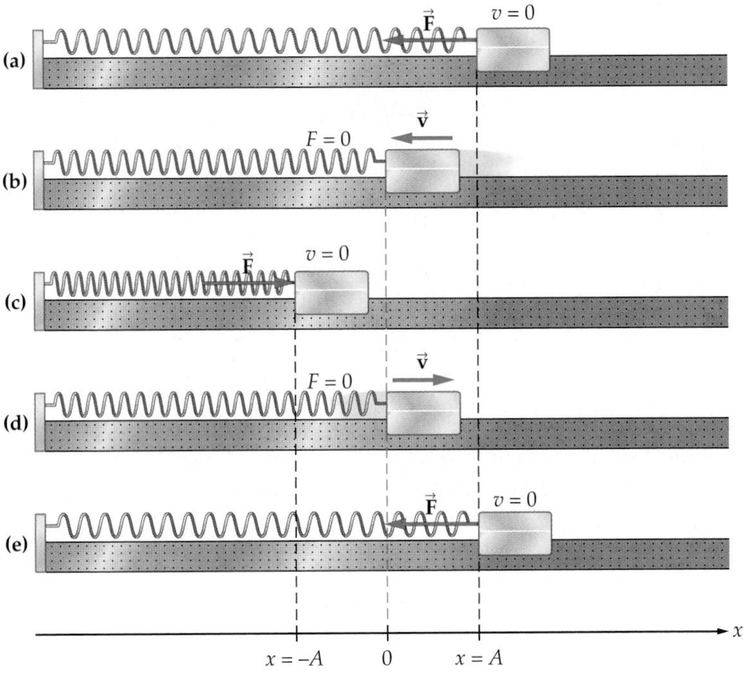

FIGURE 13–1 A mass attached to a spring undergoes simple harmonic motion about $x = 0$

(a) The mass is at its maximum positive value of x. Its velocity is zero, and the force on it points to the left with maximum magnitude. **(b)** The mass is at the equilibrium position of the spring. Here the speed has its maximum value, and the force exerted by the spring is zero. **(c)** The mass is at its maximum displacement in the negative x direction. The velocity is zero here, and the force points to the right with maximum magnitude. **(d)** The mass is at the equilibrium position of the spring, with zero force acting on it and maximum speed. **(e)** The mass has completed one cycle of its oscillation about $x = 0$.

force to the right, causing it to decelerate and finally come to rest at $x = -A$. The spring continues to exert a force to the right; thus, the cart immediately begins to move to the right until it comes to rest again at $x = A$, completing one oscillation in the time T.

If a pen is attached to the cart, it can trace its motion on a strip of paper moving with constant speed, as indicated in **Figure 13–2**. On this "strip chart" we obtain a record of the cart's motion as a function of time. As we see in Figure 13–2, the motion of the cart looks like a sine or a cosine function.

Mathematical analysis, using the methods of calculus, shows that this is indeed the case; that is, the position of the cart as a function of time can be represented by a sine or a cosine function. The reason that either function works can be seen by considering **Figure 13–3**. If we take $t = 0$ to be at point (1), for example, the position as a function of time starts at zero, just like a sine function; if we choose $t = 0$ to be at point (2), however, the position versus time starts at its maximum value, just like a cosine function. It is really the same mathematical function, differing only in the choice of starting point.

Returning to Figure 13–1, note that the position of the mass oscillates between $x = +A$ and $x = -A$. Since A represents the extreme displacement of the cart on either side of equilibrium, we refer to it as the **amplitude** of the motion. It follows

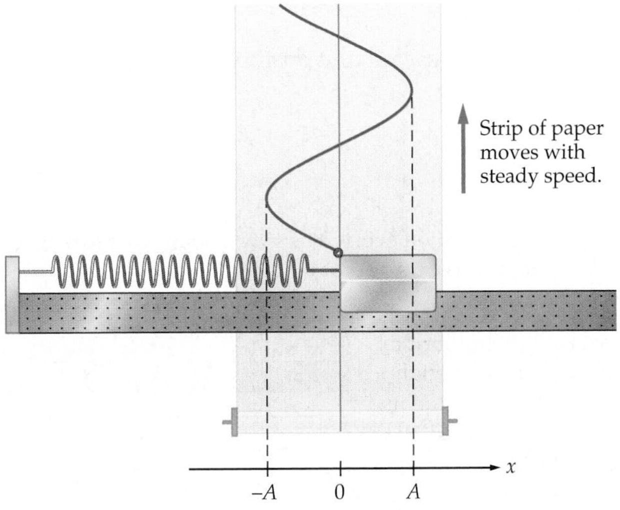

FIGURE 13–2 Displaying position versus time for simple harmonic motion

As an air-track cart oscillates about its equilibrium position, a pen attached to it traces its motion onto a moving sheet of paper. This produces a "strip chart," showing that the cart's motion has the shape of a sine or a cosine.

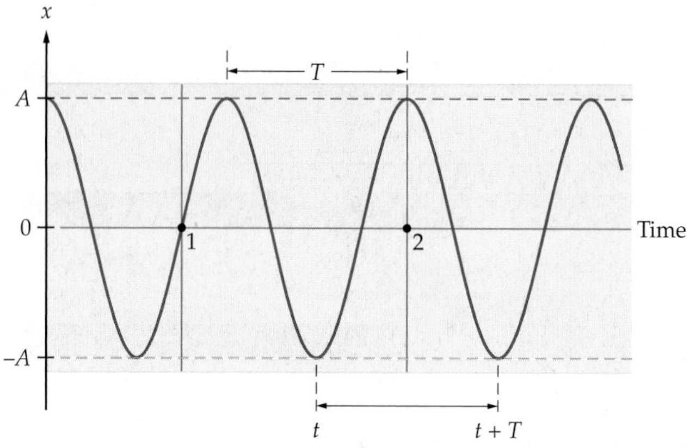

▶ **FIGURE 13–3 Simple harmonic motion as a sine or a cosine**

The strip chart from Figure 13–2. The cart oscillates back and forth from $x = +A$ to $x = -A$, completing one cycle in the time T. The function traced by the pen can be represented as a sine function if $t = 0$ is taken to be at point (1), where the function is equal to zero. The function can be represented by a cosine if $t = 0$ is taken to be at point (2), where the function has its maximum value.

that the amplitude is one-half the total range of motion. In addition, recall that the cart's motion repeats with a period T. As a result, the position of the cart is the same at the time $t + T$ as it is at the time t, as shown in Figure 13–3. Combining all these observations results in the following mathematical description of position versus time:

PROBLEM-SOLVING NOTE

Using Radians

When evaluating Equation 13–2, be sure you have your calculator set to "radians" mode.

Position Versus Time in Simple Harmonic Motion

$$x = A\cos\left(\frac{2\pi}{T}t\right) \qquad 13\text{–}2$$

SI unit: m

This type of dependence on time—as a sine or a cosine—is characteristic of simple harmonic motion. With this particular choice, the position at $t = 0$ is $x = A\cos(0) = A$; thus, Equation 13–2 describes an object that has its maximum displacement at $t = 0$, as in Figure 13–3.

To see how Equation 13–2 works, recall that the cosine oscillates between +1 and −1. Therefore, $x = A\cos(2\pi t/T)$ will oscillate between $+A$ and $-A$, just as in the strip chart. Next, consider what happens if we replace the time t with the time $t + T$. This gives

$$x = A\cos\left(\frac{2\pi}{T}(t + T)\right)$$

$$= A\cos\left(\frac{2\pi}{T}t + \frac{2\pi}{T}T\right) = A\cos\left(\frac{2\pi}{T}t + 2\pi\right)$$

Finally, using the fact that $\cos(\theta + 2\pi) = \cos\theta$ for any angle θ, we can rewrite the last expression as follows:

$$x = A\cos\left(\frac{2\pi}{T}t\right)$$

Therefore, as expected, the position at time $t + T$ is precisely the same as the position at time t.

EXAMPLE 13–1 Spring Time

An air-track cart attached to a spring completes one oscillation every 2.4 s. At $t = 0$ the cart is released from rest at a distance of 0.10 m from its equilibrium position. What is the position of the cart at **(a)** 0.30 s, **(b)** 0.60 s, **(c)** 2.7 s, and **(d)** 3.0 s?

Picture the Problem

In our sketch, we place the origin of the x-axis at the equilibrium position of the cart and the positive direction to point to the right. The cart is released from rest at $x = 0.10$ m, which means that its amplitude is $A = 0.10$ m. After it is released, the cart oscillates back and forth between $x = 0.10$ m and $x = -0.10$ m.

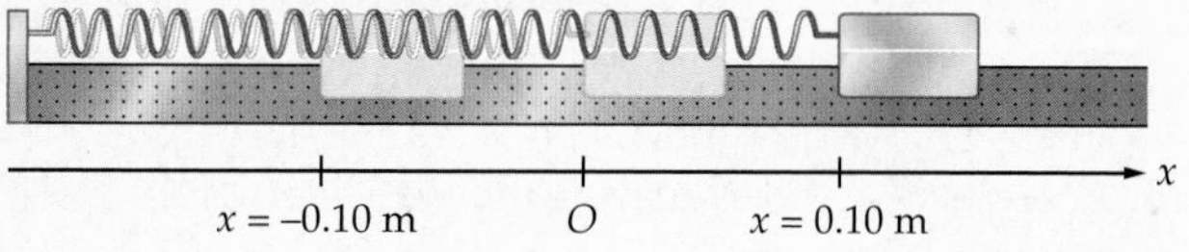

Strategy

Note that the period of oscillation, $T = 2.4$ s, is given in the problem statement. Thus, we can find the position of the cart by evaluating $x = A\cos(2\pi t/T)$ at the desired times, using $A = 0.10$ m. (Note: Remember to have your calculators set to radian mode when evaluating these cosine functions.)

Solution

Part (a)

1. Calculate x at the time $t = 0.30$ s:

$$x = A\cos\left(\frac{2\pi}{T}t\right) = (0.10\text{ m})\cos\left[\left(\frac{2\pi}{2.4\text{ s}}\right)(0.30\text{ s})\right]$$
$$= (0.10\text{ m})\cos(\pi/4) = 7.1\text{ cm}$$

Part (b)

2. Now, substitute $t = 0.60$ s:

$$x = A\cos\left(\frac{2\pi}{T}t\right) = (0.10\text{ m})\cos\left[\left(\frac{2\pi}{2.4\text{ s}}\right)(0.60\text{ s})\right]$$
$$= (0.10\text{ m})\cos(\pi/2) = 0$$

Part (c)

3. Repeat with $t = 2.7$ s:

$$x = A\cos\left(\frac{2\pi}{T}t\right) = (0.10\text{ m})\cos\left[\left(\frac{2\pi}{2.4\text{ s}}\right)(2.7\text{ s})\right]$$
$$= (0.10\text{ m})\cos(9\pi/4) = 7.1\text{ cm}$$

Part (d)

4. Repeat with $t = 3.0$ s:

$$x = A\cos\left(\frac{2\pi}{T}t\right) = (0.10\text{ m})\cos\left[\left(\frac{2\pi}{2.4\text{ s}}\right)(3.0\text{ s})\right]$$
$$= (0.10\text{ m})\cos(5\pi/2) = 0$$

Insight

Note that the results for parts (c) and (d) are the same as for parts (a) and (b), respectively. This is because the times in (c) and (d) are greater than the corresponding times in (a) and (b) by one period; that is, 2.7 s = 0.30 s + 2.4 s and 3.0 s = 0.60 s + 2.4 s.

Practice Problem

What is the first time the cart is at the position $x = -5.0$ cm? [**Answer:** $t = T/3 = 0.80$ s]

Some related homework problems: Problem 8, Problem 13, Problem 14

Though the motion of the air-track cart is strictly one-dimensional, it bears a close relationship to uniform circular motion. In the next section we explore this connection between simple harmonic motion and uniform circular motion in detail.

13–3 Connections Between Uniform Circular Motion and Simple Harmonic Motion

Imagine a turntable that rotates with a constant angular speed $\omega = 2\pi/T$, taking the time T to complete a revolution. At the rim of the turntable we place a small peg, as indicated in **Figure 13–4**. If we view the turntable from above, we see the peg undergoing uniform circular motion.

On the other hand, suppose we view the turntable from the side, so that the peg appears to move back and forth. Perhaps the easiest way to view this motion is to shine a light that casts a shadow of the peg on a screen, as shown in Figure 13–4. While the peg itself moves on a circular path, its shadow moves back and forth in a straight line.

To be specific, let the radius of the turntable be $r = A$, so that the shadow moves from $x = +A$ to $x = -A$. When the shadow is at $x = +A$, release a mass on a spring that is also at $x = +A$, so that the mass and the shadow start together. If we adjust the period of the mass so that it completes one oscillation in the same time T that the turntable completes one revolution, we find that the mass and the shadow move as one for all times. This is also illustrated in Figure 13–4. Since the mass undergoes simple harmonic motion, it follows that the shadow does so as well.

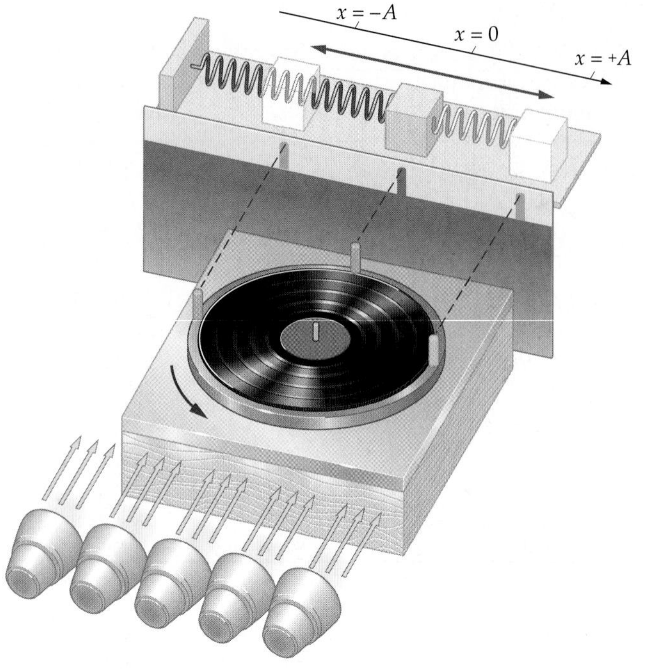

▶ **FIGURE 13–4 The relationship between uniform circular motion and simple harmonic motion**

A peg is placed at the rim of a turntable that rotates with constant angular velocity. Viewed from above, the peg exhibits uniform circular motion. If the peg is viewed from the side, however, it appears to move back and forth in a straight line, as we can see by shining a light to cast a shadow of the peg onto a screen. The shadow moves with simple harmonic motion. If we compare this motion with the behavior of a mass on a spring, moving with the same period as the turntable and an amplitude of motion equal to the radius of the turntable, we find that the mass and the shadow of the peg move together in simple harmonic motion.

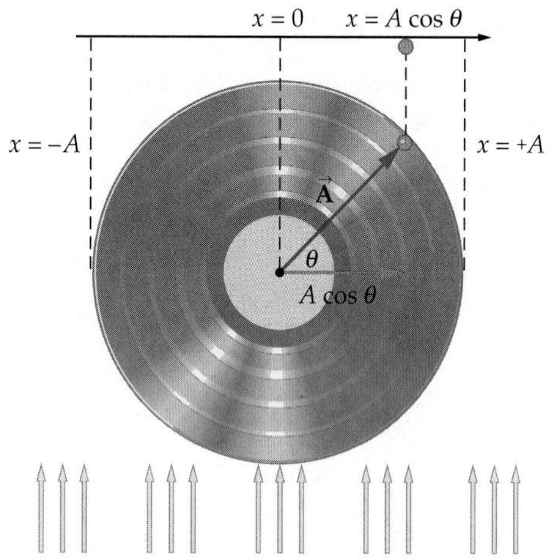

▲ **FIGURE 13–5 Position versus time in simple harmonic motion**

A peg rotates on the rim of a turntable of radius A. When the peg is at the angular position θ, its shadow is at $x = A\cos\theta$. Note that $A\cos\theta$ is also the x component of the radius vector $\vec{\mathbf{A}}$ from the center of the turntable to the peg.

We now use this connection, plus our knowledge of circular motion, to obtain detailed results for the position, velocity, and acceleration of a particle undergoing simple harmonic motion.

Position

In **Figure 13–5** we show the peg at the angular position θ, where θ is measured relative to the x-axis. If the peg starts at $\theta = 0$ at $t = 0$, and the turntable rotates with a constant angular speed ω, we know from Equation 10–10 that the angular position of the peg is simply

$$\theta = \omega t \qquad \text{13–3}$$

That is, the angular position increases linearly with time.

Now, imagine drawing a radius vector of length A to the position of the peg, as indicated in Figure 13–5. When we project the shadow of the peg onto the screen, the shadow is at the location $x = A\cos\theta$, which is the x component of the radius vector. Therefore, the position of the shadow as a function of time is

Position of the Shadow as a Function of Time

$$x = A\cos\theta = A\cos(\omega t) = A\cos\left(\frac{2\pi}{T}t\right) \qquad \text{13–4}$$

SI unit: m

Note that we have used Equation 13–3 to express θ in terms of the time, t. Clearly, Equations 13–4 and 13–2 are identical, so the shadow does indeed exhibit simple harmonic motion, just like a mass on a spring.

For notational simplicity, we will often write the position of a mass on a spring in the form $x = A\cos(\omega t)$, which is more compact than $x = A\cos(2\pi t/T)$. When referring to a rotating turntable, ω is called the angular speed; when referring to simple harmonic motion, or other periodic motion, we have a slightly different name for ω. In these situations, ω is called the **angular frequency:**

Definition of Angular Frequency, ω

$$\omega = 2\pi f = \frac{2\pi}{T} \qquad \text{13–5}$$

SI unit: $\text{rad/s} = \text{s}^{-1}$

Velocity

We can find the velocity of the shadow in the same way that we determined its position; first find the velocity of the peg, then take its x component. The result of this calculation will be the velocity as a function of time for simple harmonic motion.

To begin, recall that the velocity of an object in uniform circular motion of radius r has a magnitude equal to

$$v = r\omega$$

In addition, the velocity is tangential to the object's circular path, as indicated in **Figure 13–6 (a)**. Therefore, referring to the figure, we see that when the peg is at the angular position θ, the velocity vector makes an angle θ with the vertical. As a result, the x component of the velocity is $-v \sin \theta$. Combining these results, we find that the velocity of the peg, along the x-axis, is

$$v_x = -v \sin \theta = -r\omega \sin \theta$$

In what follows we shall drop the x subscript, since we know that the shadow and a mass on a spring move only along the x-axis. Recalling that $r = A$ and $\theta = \omega t$, we have

Velocity in Simple Harmonic Motion

$$v = -A\omega \sin(\omega t) \qquad \text{13–6}$$

SI unit: m/s

We plot x and v for simple harmonic motion in **Figure 13–6 (b)**. Note that when the displacement from equilibrium is a maximum, the velocity is zero. This is to be expected, since at $x = +A$ and $x = -A$ the object is momentarily at rest as it turns around. Not surprisingly, these points are referred to as **turning points** of the motion.

On the other hand, the speed is a maximum when the displacement from equilibrium is zero. Similarly, a mass on a spring is moving with its greatest speed as it goes through $x = 0$. From the expression $v = -A\omega \sin(\omega t)$, and the fact that the largest value of $\sin \theta$ is 1, we see that the maximum speed of the mass is

$$v_{max} = A\omega \qquad \text{13–7}$$

After the mass passes $x = 0$ it begins either to compress or to stretch the spring, and hence it slows down.

Acceleration

The acceleration of an object in uniform circular motion has a magnitude given by

$$a_{cp} = r\omega^2$$

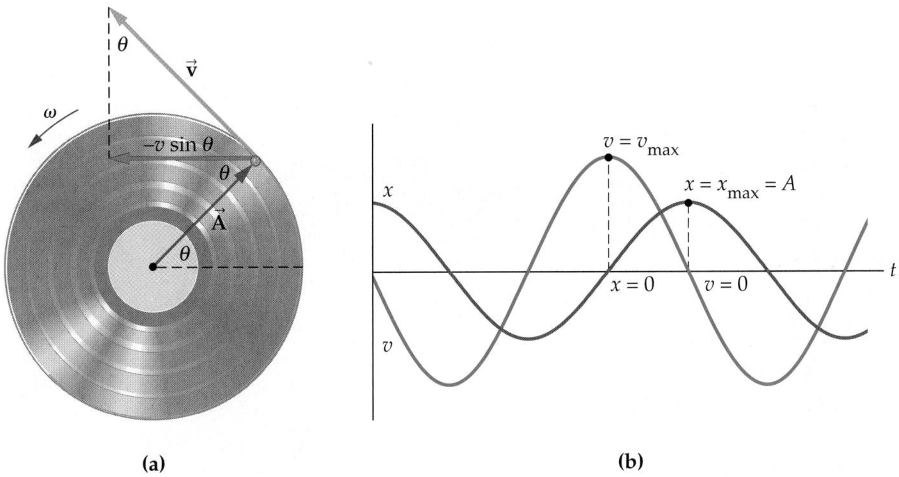

◀ **FIGURE 13–6 Velocity versus time in simple harmonic motion**
(a) The velocity of a peg rotating on the rim of a turntable is tangential to its circular path. As a result, when the peg is at the angle θ, its velocity makes an angle of θ with the vertical. The x component of the velocity, then, is $-v \sin \theta$. **(b)** Position, x, and velocity, v, as a function of time for simple harmonic motion. The speed is greatest when the object passes through equilibrium, $x = 0$. On the other hand, the speed is zero when the position is greatest; that is, at the turning points. Finally, note that as x moves in the negative direction, the velocity is negative. Similar remarks apply to the positive direction.

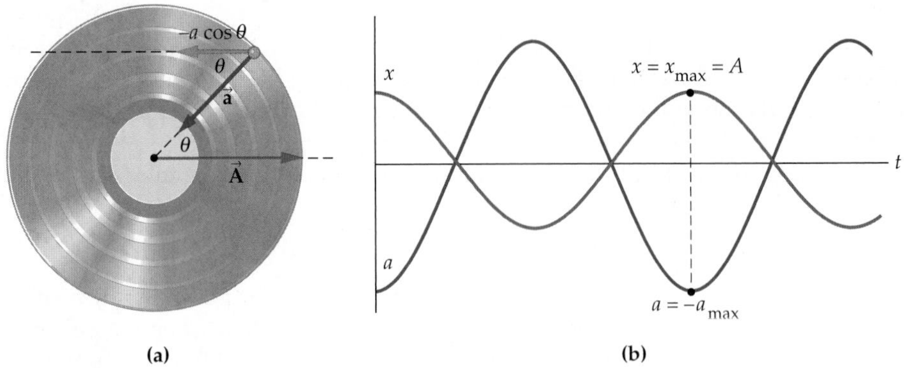

▶ **FIGURE 13–7 Acceleration versus time in simple harmonic motion**

(a) The acceleration of a peg on the rim of a uniformly rotating turntable is directed toward the center of the turntable. Hence, when the peg is at the angle θ, the acceleration makes an angle θ with the horizontal. The x component of the acceleration is $-a \cos \theta$. **(b)** Position, x, and acceleration, a, as a function of time for simple harmonic motion. Note that when the position has its greatest positive value, the acceleration has its greatest negative value.

In addition, the direction of the acceleration is toward the center of the circular path, as indicated in **Figure 13–7 (a)**. Thus, when the angular position of the peg is θ, the acceleration vector is at an angle θ below the x-axis, and its x component is $-a_{cp} \cos \theta$. Again setting $r = A$ and $\theta = \omega t$, we find

Acceleration in Simple Harmonic Motion

$$a = -A\omega^2 \cos(\omega t) \quad \text{13–8}$$

SI unit: m/s^2

PROBLEM-SOLVING NOTE

Be Sure to Use Radians

Note that Equations 13–4, 13–6, and 13–8 must all be evaluated in terms of radians

The position and acceleration for simple harmonic motion are plotted in **Figure 13–7 (b)**. Note that the acceleration and position vary with time in the same way but with opposite signs. That is, when the position has its maximum *positive* value, the acceleration has its maximum *negative* value, and so on. After all, the restoring force of the spring is opposite to the position, hence the acceleration, $a = F/m$, must also be opposite to the position. In fact, comparing Equations 13–4 and 13–8 we see that the acceleration can be written as:

$$a = -\omega^2 x$$

Finally, since the largest value of x is the amplitude A, we see that the maximum acceleration is of magnitude

$$a_{max} = A\omega^2 \quad \text{13–9}$$

We conclude this section with a few examples using position, velocity, and acceleration in simple harmonic motion.

EXAMPLE 13–2 Velocity and Acceleration

As in Example 13–1, an air-track cart attached to a spring completes one oscillation every 2.4 s. At $t = 0$ the cart is released from rest with the spring stretched 0.10 m from its equilibrium position. What are the velocity and acceleration of the cart at **(a)** 0.30 s and **(b)** 0.60 s?

Picture the Problem

Once again, we place the origin of the x-axis at the equilibrium position of the cart, with the positive direction pointing to the right. In addition, the cart is released from rest at $x = 0.10$ m, which means that it will have zero speed at $x = 0.10$ m and $x = -0.10$ m. Its speed will be a maximum at $x = 0$, however, which is also the point where the acceleration is zero.

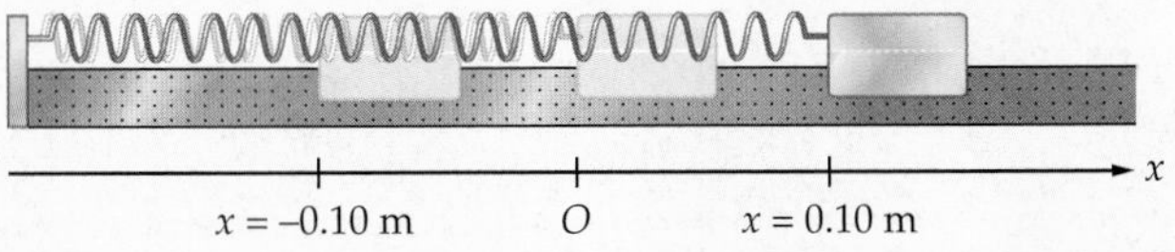

Strategy

After calculating the angular frequency, $\omega = 2\pi/T$, we simply substitute $t = 0.30$ s and $t = 0.60$ s into $v = -A\omega \sin(\omega t)$ and $a = -A\omega^2 \cos(\omega t)$. Remember to set your calculators to radian mode.

Solution

1. Calculate the angular frequency for this motion:

$$\omega = \frac{2\pi}{T} = \frac{2\pi}{(2.4\text{ s})} = 2.6\text{ rad/s}$$

Part (a)

2. Calculate v at the time $t = 0.30$ s. Express ω in terms of π—that is, $\omega = 2\pi/(2.4\text{ s})$. This isn't necessary, but it makes it easier to evaluate the sine and cosine functions. (See the discussion following the Example for additional details):

$$v = -A\omega\sin(\omega t)$$
$$= -(0.10\text{ m})(2.6\text{ rad/s})\sin\left[\left(\frac{2\pi}{2.4\text{ s}}\right)(0.30\text{ s})\right]$$
$$= -(26\text{ cm/s})\sin(\pi/4) = -18\text{ cm/s}$$

3. Similarly, calculate a at $t = 0.30$ s:

$$a = -A\omega^2\cos(\omega t)$$
$$= -(0.10\text{ m})(2.6\text{ rad/s})^2\cos\left[\left(\frac{2\pi}{2.4\text{ s}}\right)(0.30\text{ s})\right]$$
$$= -(68\text{ cm/s}^2)\cos(\pi/4) = -48\text{ cm/s}^2$$

Part (b)

4. Calculate v at the time $t = 0.60$ s:

$$v = -A\omega\sin(\omega t)$$
$$= -(0.10\text{ m})(2.6\text{ rad/s})\sin\left[\left(\frac{2\pi}{2.4\text{ s}}\right)(0.60\text{ s})\right]$$
$$= -(26\text{ cm/s})\sin(\pi/2) = -26\text{ cm/s}$$

5. Similarly, calculate a at $t = 0.60$ s:

$$a = -A\omega^2\cos(\omega t)$$
$$= -(0.10\text{ m})(2.6\text{ rad/s})^2\cos\left[\left(\frac{2\pi}{2.4\text{ s}}\right)(0.60\text{ s})\right]$$
$$= -(68\text{ cm/s}^2)\cos(\pi/2) = 0$$

Insight

Note that the cart speeds up from $t = 0.30$ s to $t = 0.60$ s; in fact, the maximum speed of the cart, $v_{max} = A\omega = 26$ cm/s, occurs at $t = 0.60\text{ s} = T/4$. Referring to Example 13–1, we see that this is precisely the time when the cart is at the equilibrium position, $x = 0$. As expected, the acceleration is zero at this time.

Practice Problem

What is the first time the velocity of the cart is +26 cm/s? [**Answer:** $v = +26$ cm/s at $t = 3T/4 = 1.8$ s]

Some related homework problems: Problem 19, Problem 20

In problems like the preceding Example, it is often useful to express the time t in terms of the period, T. For example, if the period is $T = 2.4$ s it follows that $t = 0.60$ s is $T/4$. Thus, the angular frequency times the time is

$$\omega t = \left(\frac{2\pi}{T}\right)\left(\frac{T}{4}\right) = \pi/2$$

This result was used in steps 4 and 5 in Example 13–2. Using $\omega t = \pi/2$ we find that the position of the cart is

$$x = A\cos(\omega t) = A\cos(\pi/2) = 0$$

Similarly, the velocity of the cart is

$$v = -A\omega\sin(\omega t) = -A\omega\sin(\pi/2) = -A\omega$$

and its acceleration is

$$a = -A\omega^2\cos(\omega t) = -A\omega^2\cos(\pi/2) = 0$$

When expressed in this way, it is clear why x and a are zero and why v has its maximum negative value.

PROBLEM-SOLVING NOTE

Expressing Time in Terms of the Period

When evaluating the expression $x = A\cos(2\pi t/T)$ at the time t, it is often helpful to express t in terms of the period T.

EXAMPLE 13–3 Turbulence!

On December 29, 1997, a United Airlines flight from Tokyo to Honolulu was hit with severe turbulence 31 minutes after takeoff. Data from the airplane's "black box" indicated the 747 moved up and down with an amplitude of 30.0 m and a maximum acceleration of 1.8 g. Treating the up-and-down motion of the plane as simple harmonic, find **(a)** the time required for one complete oscillation and **(b)** the plane's maximum vertical speed.

Picture the Problem

Our sketch shows the 747 airliner moving up and down with an amplitude of $A = 30.0$ m relative to its normal horizontal flight path. The period of motion, T, is the time required for one complete cycle of this up-and-down motion.

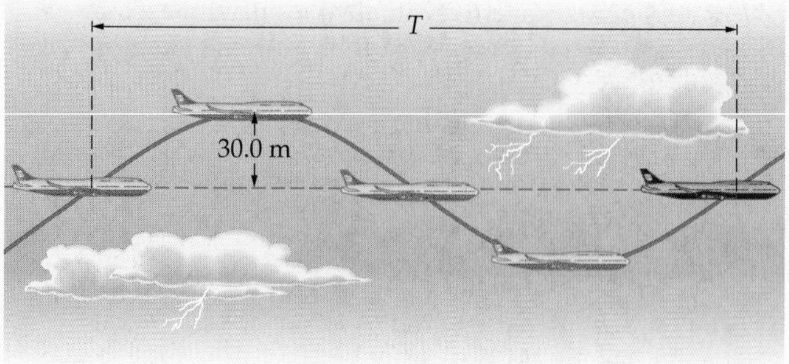

INTERACTIVE FIGURE

Strategy

We are given the maximum acceleration and the amplitude of motion. With these quantities, and the basic equations of simple harmonic motion, we can determine the other characteristics of the motion.

a. We know that the maximum acceleration of simple harmonic motion is $a_{max} = A\omega^2$ (Equation 13–9). This relation can be solved for ω in terms of the known quantities a_{max} and A. Next we rearrange $\omega = 2\pi/T$ to solve for the period of motion, T.

b. The maximum vertical speed is found using $v_{max} = A\omega$ (Equation 13–7).

Solution

Part (a)

1. Relate a_{max} to the angular frequency, ω: $a_{max} = A\omega^2$
2. Solve for ω, and express in terms of T: $\omega = \sqrt{a_{max}/A} = 2\pi/T$
3. Solve for T and substitute numerical values for g and A:

$$T = \frac{2\pi}{\sqrt{a_{max}/A}} = \frac{2\pi}{\sqrt{1.8g/A}} = \frac{2\pi}{\sqrt{1.8(9.81\text{ m/s}^2)/(30.0\text{ m})}} = 8.2\text{ s}$$

Part (b)

4. Calculate the maximum vertical speed using $v_{max} = A\omega = 2\pi A/T$:

$$v_{max} = A\omega = \frac{2\pi A}{T} = \frac{2\pi(30.0\text{ m})}{8.2\text{ s}} = 23\text{ m/s}$$

Insight

We don't expect the up-and-down motion of the plane to be exactly simple harmonic—after all, it's unlikely the plane's path has precisely the shape of a cosine function. Still, the approximation is reasonable. In fact, many systems are well approximated by simple harmonic motion, and hence the equations derived in this section are used widely in physics.

Notice that the maximum vertical speed of the passengers (23 m/s) is roughly 50 mi/h, and that this speed is first in the upward direction, and then 4.1 s later in the downward direction.

Practice Problem

What amplitude of motion would result in a maximum acceleration of $0.50g$, everything else remaining the same? [**Answer:** $A = 0.50\,g/\omega^2 = 0.50\,gT^2/4\pi^2 = 8.4$ m]

Some related homework problems: Problem 19, Problem 21

ACTIVE EXAMPLE 13–1 Bobbing for Apples: Find the Position, Velocity, and Acceleration

A Red Delicious apple floats in a barrel of water. If you lift the apple 2.00 cm above its floating level and release it, it bobs up and down with a period of $T = 0.750$ s. Assuming the motion is simple harmonic, find the position, velocity, and acceleration of the apple at the times **(a)** $T/4$ and **(b)** $T/2$.

Solution *(Test your understanding by performing the calculations indicated in each step.)*

1. Identify the amplitude of motion: $A = 2.00$ cm
2. Calculate the angular frequency: $\omega = 2\pi/T = 8.38$ rad/s

Part (a)

3. Evaluate $x = A\cos(\omega t)$ at $t = T/4$:	$x = A\cos(\pi/2) = 0$
4. Evaluate $v = -A\omega\sin(\omega t)$ at $t = T/4$:	$v = -A\omega\sin(\pi/2) = -A\omega = -16.8\text{ cm/s}$
5. Evaluate $a = -A\omega^2\cos(\omega t)$ at $t = T/4$:	$a = -A\omega^2\cos(\pi/2) = 0$

Part (b)

6. Evaluate $x = A\cos(\omega t)$ at $t = T/2$:	$x = A\cos(\pi) = -A = -2.00\text{ cm}$
7. Evaluate $v = -A\omega\sin(\omega t)$ at $t = T/2$:	$v = -A\omega\sin(\pi) = 0$
8. Evaluate $a = -A\omega^2\cos(\omega t)$ at $t = T/2$:	$a = -A\omega^2\cos(\pi) = A\omega^2 = 140\text{ cm/s}^2$

Insight

The acceleration in part (b) may seem rather large, but remember that $140\text{ cm/s}^2 = 1.40\text{ m/s}^2$, so the acceleration is only a fraction of the acceleration of gravity.

Your Turn

The maximum kinetic energy of this bobbing apple is 0.00388 J. What is its mass?

(Answers to **Your Turn** *problems are given in the back of the book.)*

13–4 The Period of a Mass on a Spring

In this section we show how the period of a mass on a spring is related to the mass, m, the force constant of the spring, k, and the amplitude of motion, A. As a first step, note that the net force acting on the mass at the position x is

$$F = -kx$$

Now, since $F = ma$, it follows that

$$ma = -kx$$

Substituting the time dependence of x and a, as given in Equations 13–4 and 13–8 in the previous section, we find

$$m[-A\omega^2\cos(\omega t)] = -k[A\cos(\omega t)]$$

Canceling $-A\cos(\omega t)$ from each side of the equation yields

$$\omega^2 = k/m$$

or

$$\omega = \sqrt{\frac{k}{m}} \qquad \text{13–10}$$

Finally, noting that $\omega = 2\pi/T$, it follows that the period of a mass on a spring is

Period of a Mass on a Spring

$$T = 2\pi\sqrt{\frac{m}{k}} \qquad \text{13–11}$$

SI unit: s

EXERCISE 13–3

When a 0.22-kg air-track cart is attached to a spring, it oscillates with a period of 0.84 s. What is the force constant for this spring?

Solution

From Equation 13–11 we find

$$k = 4\pi^2 m/T^2 = 12\text{ N/m}$$

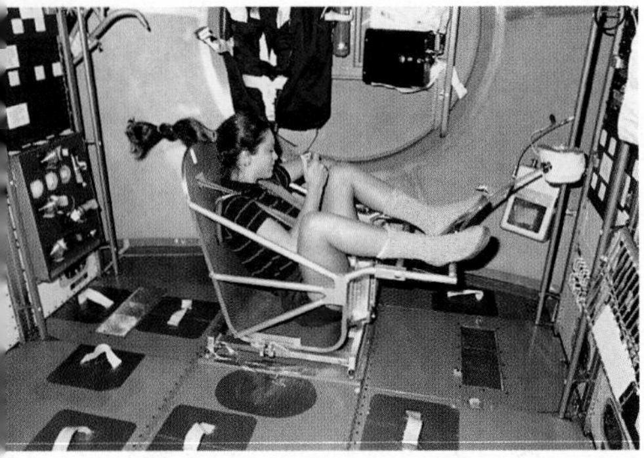

▲ Because astronauts are in free fall when orbiting the Earth, they behave as if they were "weightless." It is nevertheless possible to determine the mass of an astronaut by exploiting the properties of oscillatory motion. The chair into which astronaut Tamara Jernigan is strapped is attached to a spring. If the force constant of the spring is known, her mass can be determined simply by measuring the period with which she rocks back and forth. This instrument is known as a Body Mass Measurement Device (BMMD).

REAL-WORLD PHYSICS
Measuring the mass of a "weightless" astronaut

As one might expect, the period increases with the mass and decreases with the spring's force constant. For example, a larger mass has greater inertia, and hence it takes longer for the mass to move back and forth through an oscillation. On the other hand, a larger value of the force constant, k, indicates a stiffer spring. Clearly, a mass on a stiff spring completes an oscillation in less time than one on a soft, squishy spring.

The relationship between mass and period given in Equation 13–11 is used by NASA to measure the mass of astronauts in orbit. Recall that astronauts are in free fall as they orbit, as was discussed in Chapter 12, and therefore they are "weightless." As a result, they cannot simply step onto a bathroom scale to determine their mass. Thus, NASA has developed a device, known as the Body Mass Measurement Device (BMMD), to get around this problem. The BMMD is basically a spring attached to a chair, into which an astronaut is strapped. As the astronaut oscillates back and forth, the period of oscillation is measured. Knowing the force constant k and the period of oscillation, T, the astronaut's mass can be determined from Equation 13–11. The result is simply $m = kT^2/4\pi^2$. See Problem 60 for an application of this result.

Note that the period given in Equation 13–11 is independent of the amplitude, A, which canceled in the derivation of this equation. This might seem counterintuitive at first: Shouldn't it take more time for a mass to cover the greater distance implied by a larger amplitude? While it is true that a mass will cover a greater distance when the amplitude is increased, it is also true that a larger amplitude implies a larger force exerted by the spring. With a greater force acting on it, the mass moves more rapidly; in fact, the speed of the mass is increased just enough that it covers the greater distance in precisely the same time.

These relationships between the motion of a mass on a spring and the mass, the force constant, and the amplitude are summarized in **Figure 13–8**.

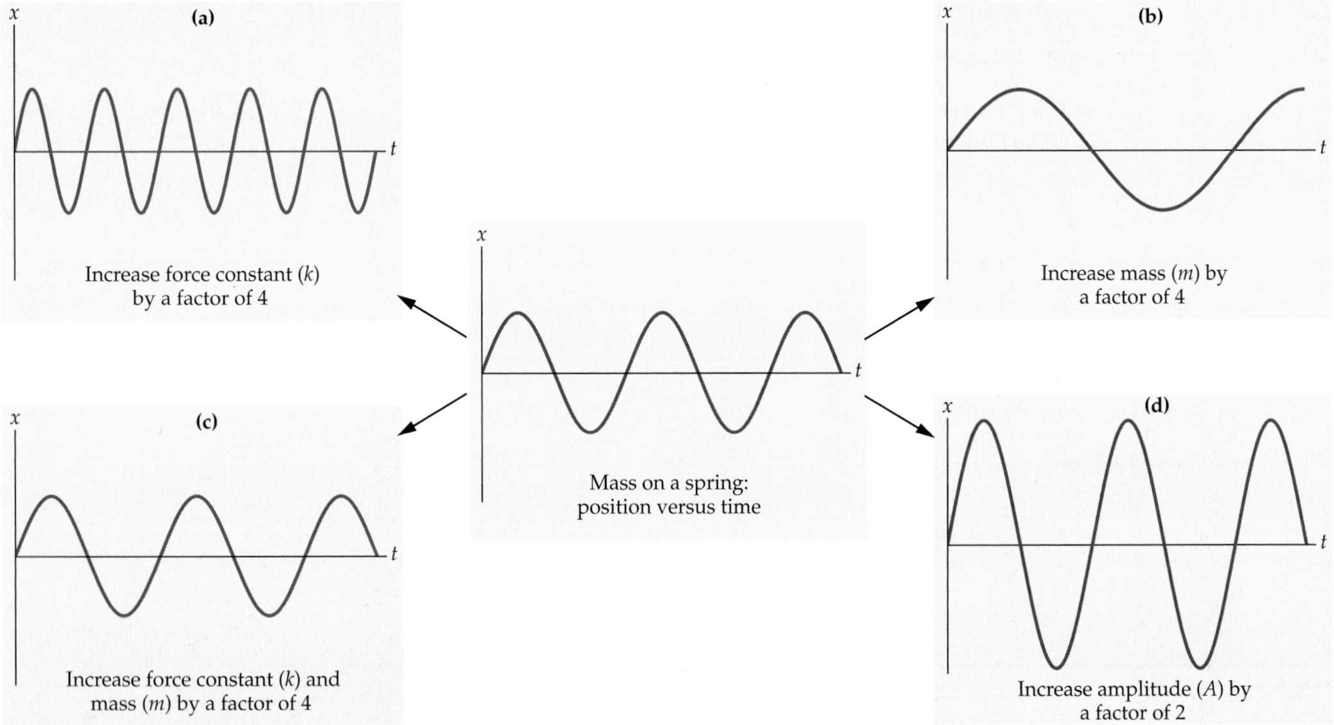

▲ **FIGURE 13–8 Factors affecting the motion of a mass on a spring**

The motion of a mass on a spring is determined by the force constant of the spring, k, the mass, m, and the amplitude, A. **(a)** Increasing the force constant causes the mass to oscillate with a greater frequency. **(b)** Increasing the mass lowers the frequency of oscillation. **(c)** If the force constant and the mass are both increased by the same factor, the effects described in parts (a) and (b) cancel, resulting in no change in the motion. **(d)** An increase in amplitude has no effect on the oscillation frequency. However, it will increase the maximum speed and maximum acceleration of the mass.

EXAMPLE 13–4 Spring into Motion

A 0.120-kg mass attached to a spring oscillates with an amplitude of 0.0750 m and a maximum speed of 0.524 m/s. Find **(a)** the force constant and **(b)** the period of motion.

Picture the Problem

Our sketch shows a mass oscillating about the equilibrium position of a spring, which we place at $x = 0$. The amplitude of the oscillations is 0.0750 m, and therefore the mass moves back and forth between $x = 0.0750$ m and $x = -0.0750$ m. The maximum speed of the mass, which occurs at $x = 0$, is $v_{max} = 0.524$ m/s.

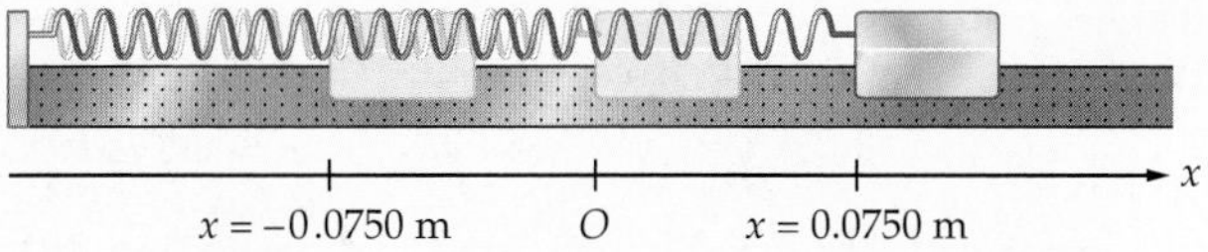

Strategy

a. To find the force constant, we first use the maximum speed, $v_{max} = A\omega$ (Equation 13–7), to determine the angular frequency ω. Once we know ω, we can obtain the force constant with $\omega = \sqrt{k/m}$ (Equation 13–10).

b. We can find the period from the angular frequency, using $\omega = 2\pi/T$. Alternatively, we can use the force constant and the mass in $T = 2\pi\sqrt{m/k}$ (Equation 13–11).

Solution

Part (a)

1. Calculate the angular frequency in terms of the maximum speed:

$$v_{max} = A\omega$$

$$\omega = \frac{v_{max}}{A} = \frac{0.524 \text{ m/s}}{0.0750 \text{ m}} = 6.99 \text{ rad/s}$$

2. Solve $\omega = \sqrt{k/m}$ for the force constant:

$$\omega = \sqrt{k/m}$$

$$k = m\omega^2 = (0.120 \text{ kg})(6.99 \text{ rad/s})^2 = 5.86 \text{ N/m}$$

Part (b)

3. Use $\omega = 2\pi/T$ to find the period:

$$\omega = 2\pi/T$$

$$T = \frac{2\pi}{\omega} = \frac{2\pi}{6.99 \text{ rad/s}} = 0.899 \text{ s}$$

4. Use $T = 2\pi\sqrt{m/k}$ to find the period:

$$T = 2\pi\sqrt{m/k} = 2\pi\sqrt{\frac{0.120 \text{ kg}}{5.86 \text{ N/m}}} = 0.899 \text{ s}$$

Insight

What would happen if we were to attach a larger mass to this same spring and release it with the same amplitude? The answer is that the period would increase, the angular frequency would decrease, and hence the maximum speed would also decrease.

Practice Problem

What is the maximum acceleration of the mass described in this Example? [**Answer:** $a_{max} = A\omega^2 = 3.66 \text{ m/s}^2$]

Some related homework problems: Problem 29, Problem 30

ACTIVE EXAMPLE 13–2 Mass on a Spring: Find the Force Constant and the Mass

When a 0.420-kg mass is attached to a spring, it oscillates with a period of 0.350 s. If, instead, a different mass, m_2, is attached to the same spring, it oscillates with a period of 0.700 s. Find **(a)** the force constant of the spring and **(b)** the mass m_2.

Solution *(Test your understanding by performing the calculations indicated in each step.)*

Part (a)

1. Let the initial mass and period be m_1 and T_1, respectively: $m_1 = 0.420$ kg, $T_1 = 0.350$ s

2. Use Equation 13–11 to write an expression for T_1: $T_1 = 2\pi\sqrt{m_1/k}$

3. Solve this expression for the force constant, k: $k = 4\pi^2 m_1/T_1^2 = 135$ N/m

continued on next page

continued from previous page

Part (b)

4. Write an expression for T_2: $T_2 = 2\pi\sqrt{m_2/k}$

5. Solve for m_2: $m_2 = kT_2^{\,2}/4\pi^2 = 1.68\text{ kg}$

Insight

In general, to double the period of a mass on a given spring—as in this case where it went from 0.350 s to 0.700 s—the mass must be increased by a factor of 4, in accordance with Equation 13–11. Therefore, an alternative way to find the second mass is $m_2 = 4(0.420\text{ kg}) = 1.68\text{ kg}$, in agreement with our result in step 5.

Your Turn

The maximum speed of the second mass is 0.787 m/s. What is its amplitude of motion?

(Answers to **Your Turn** *problems are given in the back of the book.)*

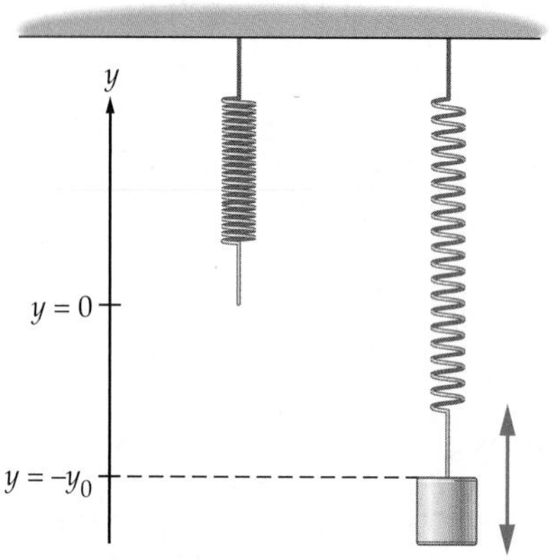

▲ **FIGURE 13–9 A mass on a vertical spring**

A mass stretches a vertical spring from its initial equilibrium at $y = 0$ to a new equilibrium at $y = -y_0 = -mg/k$. The mass executes simple harmonic motion about this new equilibrium.

A Vertical Spring

To this point we have considered only springs that are horizontal, and that are therefore unstretched at their equilibrium position. In many cases, however, we may wish to consider a vertical spring, as in **Figure 13–9**.

Now, when a mass m is attached to a vertical spring, it causes the spring to stretch. In fact, the vertical spring is in equilibrium when it exerts an upward force equal to the weight of the mass. That is, the spring stretches by an amount y_0 given by

$$ky_0 = mg$$

or

$$y_0 = mg/k \qquad \textbf{13–12}$$

Thus, a mass on a vertical spring oscillates about the equilibrium point $y = -y_0$. In all other respects the oscillations are the same as for a horizontal spring. In particular, the motion is simple harmonic, and the period is given by Equation 13–11.

▲ A ball attached to a vertical spring oscillates up and down with simple harmonic motion. If successive images taken at equal time intervals are displaced laterally, as in the sequence of photos at left, the ball appears to trace out a sinusoidal pattern (compare with Figure 13–2). The bungee jumper at right will oscillate in a similar fashion, though friction and air resistance will reduce the amplitude of his bounces.

EXAMPLE 13–5 It's a Stretch

A 0.260-kg mass is attached to a vertical spring. When the mass is put into motion, its period is 1.12 s. How much does the mass stretch the spring when it is at rest in its equilibrium position?

Picture the Problem

We choose the vertical axis to have its origin at the unstretched position of the spring. Once the mass is attached, the spring stretches to the position $y = -y_0$. The mass oscillates about this point with a period of 1.12 s.

Strategy

In order to find the stretch of the spring, $y_0 = mg/k$, we need to know the force constant, k. We can find k from the period of oscillation; that is, from $T = 2\pi\sqrt{m/k}$.

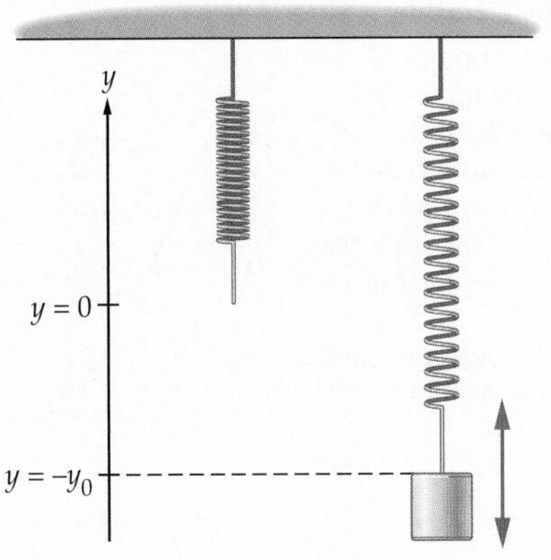

Solution

1. Use the period $T = 2\pi\sqrt{m/k}$ to solve for the force constant:

$$T = 2\pi\sqrt{m/k}$$

$$k = \frac{4\pi^2 m}{T^2} = \frac{4\pi^2(0.260\text{ kg})}{(1.12\text{ s})^2} = 8.18\text{ kg/s}^2 = 8.18\text{ N/m}$$

2. Set the magnitude of the spring force, ky_0, equal to mg to solve for y_0:

$$ky_0 = mg$$

$$y_0 = mg/k = \frac{(0.260\text{ kg})(9.81\text{ m/s}^2)}{8.18\text{ N/m}} = 0.312\text{ m}$$

Insight

Note that even though gravity determines the amount by which the spring stretches when at rest, it plays no role in determining the period of oscillation. Thus, one would observe the same period on the Moon, or even in orbit, as in the case of the Body Mass Measurement Device mentioned earlier in this section.

Practice Problem

A 0.170-kg mass stretches a vertical spring 0.250 m when at rest. What is its period when set into vertical motion? [**Answer:** $T = 1.00$ s]

Some related homework problems: Problem 31, Problem 34

CONCEPTUAL CHECKPOINT 13–1 **Compare Periods**

When a mass m is attached to a vertical spring with a force constant k, it oscillates with a period T. If the spring is cut in half and the same mass is attached to it, is the period of oscillation (a) greater than, (b) less than, or (c) equal to T?

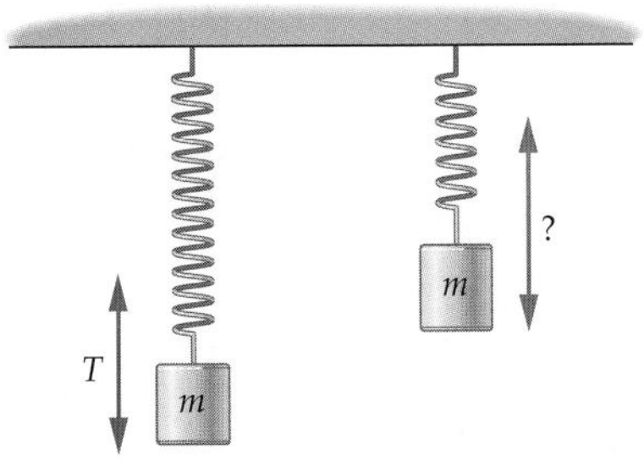

Reasoning and Discussion

The downward force exerted by the mass is of magnitude mg, as is the upward force exerted by the spring. Note that *each coil* of the spring experiences the *same force*, just as each point in a string experiences the same tension. Therefore, each coil elongates by the same amount, regardless of how many coils there are in a given spring. It follows, then, that the total elongation of the spring with half the number of coils is half the total elongation of the longer spring. Since half the elongation for the same applied force means a greater force constant, the half-spring has a larger value of k—it is stiffer. As a result, its period of oscillation is less than the period of the full spring.

Answer:

(b) The period for the half-spring is less than for the full spring.

13–5 Energy Conservation in Oscillatory Motion

In an ideal system with no friction or other nonconservative forces, the total energy is conserved. For example, the total energy E of a mass on a horizontal spring is the sum of its kinetic energy, $K = \frac{1}{2}mv^2$, and its potential energy, $U = \frac{1}{2}kx^2$. Therefore,

$$E = K + U = \tfrac{1}{2}mv^2 + \tfrac{1}{2}kx^2 \qquad 13\text{–}13$$

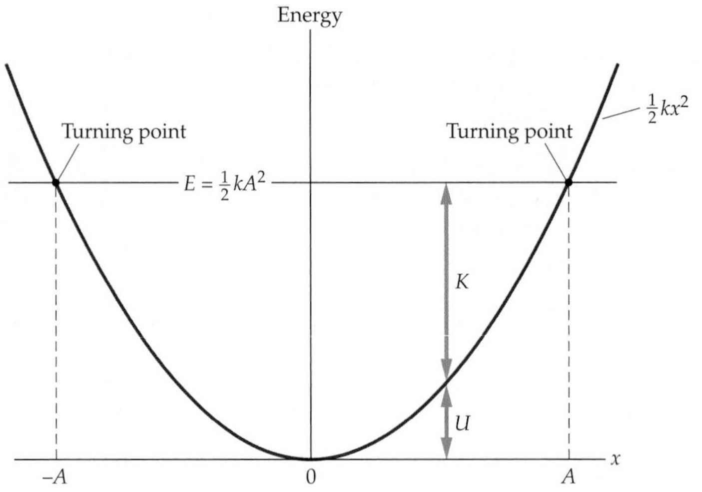

FIGURE 13–10 Energy as a function of position in simple harmonic motion
The parabola represents the potential energy, U, of the spring. The horizontal line shows the total energy, E, of the system, which is constant, and the distance from the parabola to the horizontal line is the kinetic energy, K. Note that the kinetic energy vanishes at the turning points, $x = A$ and $x = -A$. At these points the energy is purely potential, and thus the total energy of the system is $E = \frac{1}{2}kA^2$.

Since E remains the same throughout the motion, it follows that there is a continual tradeoff between kinetic and potential energy.

This energy tradeoff is illustrated in **Figure 13–10**, where the horizontal line represents the total energy of the system, E, and the parabolic curve is the spring's potential energy, U. At any given value of x, the sum of U and K must equal E; therefore, since U is the amount of energy from the axis to the parabola, K is the amount of energy from the parabola to the horizontal line. We can see, then, that the kinetic energy vanishes at the turning points, $x = +A$ and $x = -A$, as expected. On the other hand, the kinetic energy is greatest at $x = 0$ where the potential energy vanishes.

Since the mass oscillates back and forth with time, the kinetic and potential energies also change with time. For example, the potential energy is

$$U = \frac{1}{2}kx^2$$

Letting $x = A\cos(\omega t)$, we have

$$U = \frac{1}{2}kA^2\cos^2(\omega t) \qquad \textbf{13–14}$$

Clearly, the maximum value of U is

$$U_{max} = \frac{1}{2}kA^2$$

From Figure 13–10, we see that the maximum value of U is simply the total energy of the system, E. Therefore

$$E = U_{max} = \frac{1}{2}kA^2 \qquad \textbf{13–15}$$

This result leads to the following conclusion:

> In simple harmonic motion, the total energy is proportional to the square of the amplitude of motion.

Similarly, the kinetic energy of the mass is

$$K = \frac{1}{2}mv^2$$

Letting $v = -A\omega\sin(\omega t)$ yields

$$K = \frac{1}{2}mA^2\omega^2\sin^2(\omega t)$$

It follows that the maximum kinetic energy is

$$K_{max} = \frac{1}{2}mA^2\omega^2 \qquad \textbf{13–16}$$

As noted, the maximum kinetic energy occurs when the potential energy is zero; hence the maximum kinetic energy must equal the total energy, E. That is,

$$E = U + K = U_{\text{max}} + 0 = 0 + K_{\text{max}}$$

At first glance, Equation 13–16 doesn't seem to be the same as Equation 13–15. However, if we recall that $\omega^2 = k/m$ (Equation 13–10), we see that

$$K_{\text{max}} = \tfrac{1}{2}mA^2\omega^2 = \tfrac{1}{2}mA^2(k/m) = \tfrac{1}{2}kA^2$$

Therefore, $K_{\text{max}} = U_{\text{max}} = E$, as expected, and the kinetic energy as a function of time is

$$K = \tfrac{1}{2}kA^2 \sin^2(\omega t) \qquad \textbf{13–17}$$

K and U are plotted as functions of time in **Figure 13–11**. The horizontal line at the top is E, the sum of U and K at all times. This shows quite graphically the back-and-forth tradeoff of energy between kinetic and potential. Mathematically, we can see that the total energy E is constant as follows:

$$\begin{aligned} E = U + K &= \tfrac{1}{2}kA^2\cos^2(\omega t) + \tfrac{1}{2}kA^2\sin^2(\omega t) \\ &= \tfrac{1}{2}kA^2[\cos^2(\omega t) + \sin^2(\omega t)] = \tfrac{1}{2}kA^2 \end{aligned}$$

The last step follows from the fact that $\cos^2\theta + \sin^2\theta = 1$ for all θ.

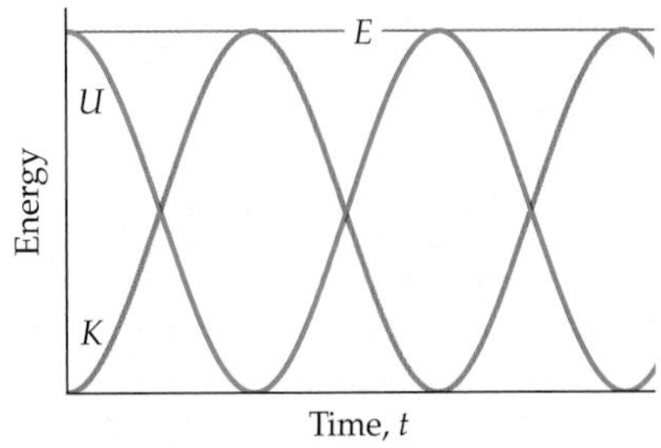

▲ **FIGURE 13–11 Energy as a function of time in simple harmonic motion**

The sum of the potential energy, U, and the kinetic energy, K, is equal to the (constant) total energy E at all times. Note that when one energy (U or K) has its maximum value, the other energy is zero.

PROBLEM-SOLVING NOTE

Maximum Potential and Kinetic Energy

The maximum potential energy of a mass–spring system is the same as the maximum kinetic energy of the mass. When the system has its maximum potential energy, the kinetic energy of the mass is zero; when the mass has its maximum kinetic energy, the potential energy of the system is zero.

EXAMPLE 13–6 Stop the Block

A 0.980-kg block slides on a frictionless, horizontal surface with a speed of 1.32 m/s. The block encounters an unstretched spring with a force constant of 245 N/m, as shown in the sketch. **(a)** How far is the spring compressed before the block comes to rest? **(b)** How long is the block in contact with the spring before it comes to rest?

Picture the Problem

As our sketch shows, the initial energy of the system is entirely kinetic; namely, the kinetic energy of the block with mass $m = 0.980$ kg and speed $v_0 = 1.32$ m/s. When the block momentarily comes to rest after compressing the spring by the amount A, its kinetic energy has been converted into the potential energy of the spring.

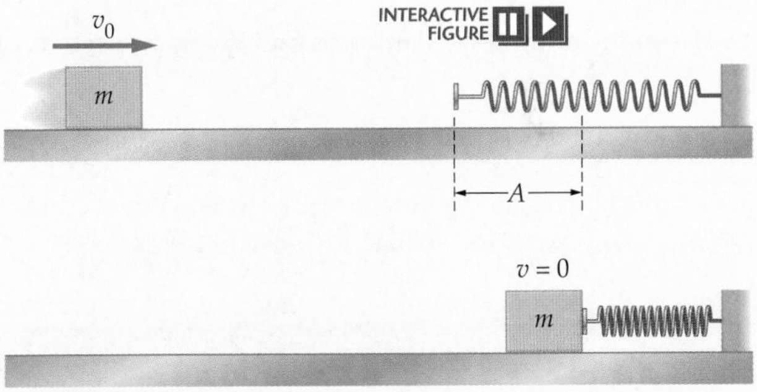

Strategy

a. We can find the compression, A, by using energy conservation. We set the initial kinetic energy of the block, $\tfrac{1}{2}mv_0^2$, equal to the spring potential energy, $\tfrac{1}{2}kA^2$, and solve for A.

b. If the mass were attached to the spring, it would complete one oscillation in the time $T = 2\pi\sqrt{m/k}$. In moving from the equilibrium position of the spring to maximum compression, the mass has undergone one-quarter of a cycle; thus the time is $T/4$.

Solution

Part (a)

1. Set the initial kinetic energy of the block equal to the spring potential energy: $\tfrac{1}{2}mv_0^2 = \tfrac{1}{2}kA^2$

2. Solve for A, the maximum compression: $A = v_0\sqrt{m/k} = (1.32\text{ m/s})\sqrt{\dfrac{0.980\text{ kg}}{245\text{ N/m}}} = 0.0835\text{ m}$

Part (b)

3. Calculate the period of one oscillation: $T = 2\pi\sqrt{m/k} = 2\pi\sqrt{\dfrac{0.980\text{ kg}}{245\text{ N/m}}} = 0.397\text{ s}$

4. Divide T by four, since the block has been in contact with the spring for one-quarter of an oscillation: $t = \tfrac{1}{4}T = \tfrac{1}{4}(0.397\text{ s}) = 0.0993\text{ s}$

continued on next page

continued from previous page

Insight

If the horizontal surface had been rough, some of the block's initial kinetic energy would have been converted to thermal energy. In this case, the maximum compression of the spring would be less than that just calculated.

Practice Problem

If the initial speed of the mass in this Example is increased, does the time required to bring it to rest increase, decrease, or stay the same? Check your answer by calculating the time for an initial speed of 1.50 m/s. [**Answer:** Increasing v increases the amplitude, A. The period is independent of amplitude, however. Thus, the time is the same, $t = 0.0993$ s.]

Some related homework problems: Problem 46, Problem 72

In the following Active Example, we consider a bullet striking a block that is attached to a spring. As the bullet embeds itself in the block, the completely inelastic bullet-block collision dissipates some of the initial kinetic energy into thermal energy. Only the kinetic energy remaining after the collision is available for compressing the spring.

ACTIVE EXAMPLE 13–3 Bullet–Block Collision: Find the Compression and Compression Time

A bullet of mass m embeds itself in a block of mass M, which is attached to a spring of force constant k. If the initial speed of the bullet is v_0, find **(a)** the maximum compression of the spring and **(b)** the time for the bullet-block system to come to rest. (See Example 9–5 for a similar system involving a pendulum.)

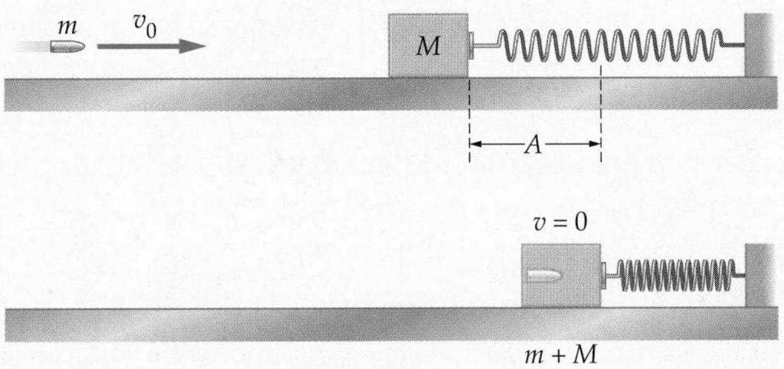

Solution *(Test your understanding by performing the calculations indicated in each step.)*

Part (a)

1. Use momentum conservation to find the final speed, v, of the bullet–block system: $mv_0 = (m + M)v$; $v = mv_0/(m + M)$
2. Find the kinetic energy of the bullet–block system after the collision: $\frac{1}{2}(m + M)v^2 = \frac{1}{2}m^2v_0^{\,2}/(m + M)$
3. Set this kinetic energy equal to $\frac{1}{2}kA^2$ to find the maximum compression of the spring: $A = mv_0/\sqrt{k(m + M)}$

Part (b)

4. As in the previous Example, the time to come to rest is one-quarter of a period: $t = T/4 = \frac{1}{2}\pi\sqrt{(m + M)/k}$

Insight

Note that the maximum compression depends directly on the initial speed of the bullet. Thus, a measurement of the compression could be used to determine the bullet's speed. On the other hand, the time for the bullet–block to come to rest is independent of the bullet's speed.

Your Turn

Suppose the force constant of the spring is quadrupled. By what factor does the amplitude of compression, A, change? By what factor does the time to come to rest, t, change?

(Answers to **Your Turn** *problems are given in the back of the book.)*

13–6 The Pendulum

One Sunday in 1583, as Galileo Galilei attended services in a cathedral in Pisa, Italy, he suddenly realized something interesting about the chandeliers hanging from the ceiling. Air currents circulating through the cathedral had set them in motion with small oscillations, and Galileo noticed that chandeliers of equal length oscillated with equal periods, even if their amplitudes were different. Indeed, as any given chandelier oscillated with decreasing amplitude its period remained constant. He verified this observation by timing the oscillations with his pulse!

Galileo was much struck by this observation, and after rushing home, he experimented with pendula constructed from different lengths of string and different weights. Continuing to use his pulse as a stopwatch, he observed that the period of a pendulum varies with its length, but is independent of the weight attached to the string. Thus, in one exhilarating afternoon, the young medical student discovered the key characteristics of a pendulum and launched himself on a new career in science. Later, he would go on to construct the first crude pendulum clock and a medical device, known as the pulsilogium, to measure a patient's pulse rate.

REAL-WORLD PHYSICS
The pendulum clock and pulsilogium

In modern terms, we would say that the chandeliers observed by Galileo were undergoing simple harmonic motion, as expected for small oscillations. As we know, the period of simple harmonic motion is independent of amplitude. The fact that the period is also independent of the mass is a special property of the pendulum, as we shall see next.

The Simple Pendulum

A simple pendulum consists of a mass m suspended by a light string or rod of length L. The pendulum has a stable equilibrium when the mass is directly below the suspension point, and oscillates about this position if displaced from it.

To understand the behavior of the pendulum, let's begin by considering the potential energy of the system. As shown in **Figure 13–12**, when the pendulum is at an angle θ with respect to the vertical, the mass m is above its lowest point by a vertical height $L(1 - \cos\theta)$. If we let the potential energy be zero at $\theta = 0$, the potential energy for general θ is

$$U = mgL(1 - \cos\theta) \qquad \textbf{13–18}$$

This function is plotted in **Figure 13–13**.

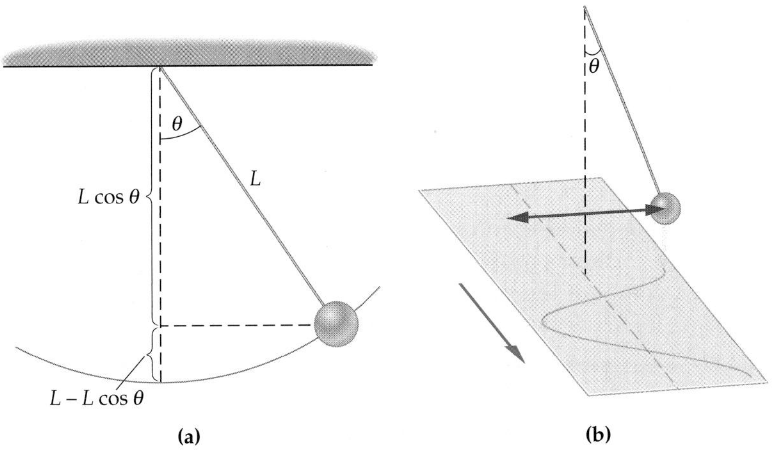

◀ **FIGURE 13–12 Motion of a pendulum**
(a) As a pendulum swings away from its equilibrium position, it rises a vertical distance $L - L\cos\theta = L(1 - \cos\theta)$. **(b)** As a pendulum bob swings back and forth, it leaks a trail of sand onto a moving sheet of paper, creating a "strip chart" similar to that produced by a mass on a spring in Figure 13–2. In particular, the angle of the pendulum with the vertical varies with time like a sine or a cosine.

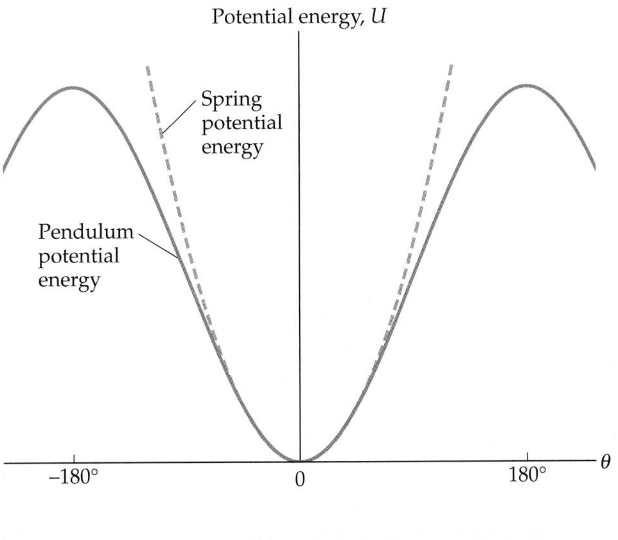

◀ **FIGURE 13–13 The potential energy of a simple pendulum**
As a simple pendulum swings away from the vertical by an angle θ, its potential energy increases, as indicated by the solid curve. Near $\theta = 0$ the potential energy of the pendulum is essentially the same as that of a mass on a spring (dashed curve). Therefore, when a pendulum oscillates with small displacements from the vertical, it exhibits simple harmonic motion—the same as a mass on a spring.

Note that the stable equilibrium of the pendulum corresponds to a minimum of the potential energy, as expected. Near this minimum, the shape of the potential energy curve is approximately the same as for a mass on a spring, as indicated in Figure 13–13. As a result, when a pendulum oscillates with small displacements from the vertical, its motion is virtually the same as the motion of a mass on a spring; that is, the pendulum exhibits simple harmonic motion.

Next, we consider the forces acting on the mass m. In **Figure 13–14** we show the force of gravity, $m\vec{g}$, and the tension force in the supporting string, $\vec{T}$. The tension acts in the radial direction and supplies the force needed to keep the mass moving along its circular path. The net tangential force acting on m, then, is simply the tangential component of its weight:

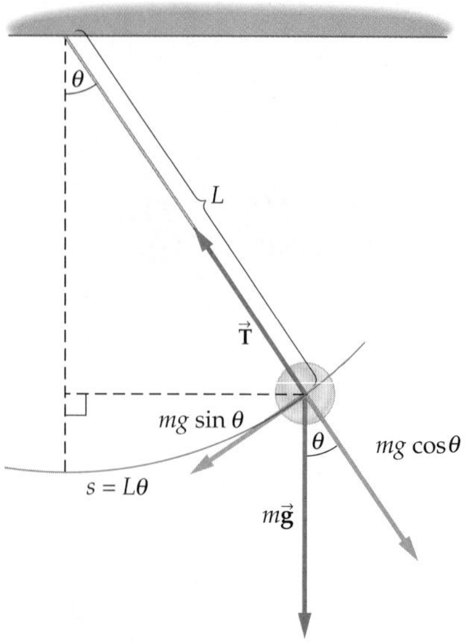

▲ **FIGURE 13–14 Forces acting on a pendulum bob**

When the bob is displaced by an angle θ from the vertical, the restoring force is the tangential component of the weight, $mg\sin\theta$.

$$F = mg\sin\theta$$

The direction of the net tangential force is always toward the equilibrium point. Thus F is a restoring force, as expected.

Now, for small angles θ (measured in *radians*), the sine of θ is approximately equal to the angle itself. That is,

$$\sin\theta \approx \theta$$

This is illustrated in **Figure 13–15**. Notice that there is little difference between $\sin\theta$ and θ for angles smaller than about $\pi/8$ rad $= 22.5$ degrees. In addition, we see from Figure 13–14 that the arc length displacement of the mass from equilibrium is

$$s = L\theta$$

Equivalently,

$$\theta = s/L$$

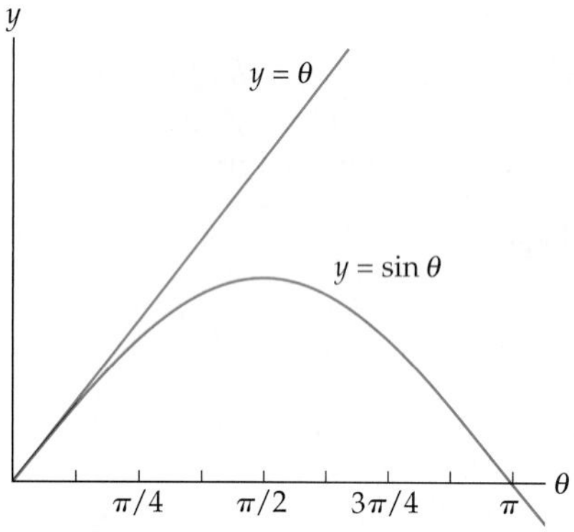

▲ **FIGURE 13–15 Relationship between sin θ and θ**

For small angles measured in radians, $\sin\theta$ is approximately equal to θ. Thus, when considering small oscillations of a pendulum, we can replace $\sin\theta$ with θ. (See also Appendix A.)

Therefore, if the mass m is displaced from equilibrium by a small arc length s, the force it experiences is restoring and of magnitude

$$F = mg\sin\theta \approx mg\theta = (mg/L)s \qquad \textbf{13–19}$$

Note that the restoring force is proportional to the displacement, just as expected for simple harmonic motion.

Let's compare the pendulum to a mass on a spring. In the latter case, the restoring force has a magnitude given by

$$F = kx$$

The restoring force acting on the pendulum has precisely the same form, if we let $x = s$ and

$$k = mg/L$$

Therefore, the period of a pendulum is simply the period of a mass on a spring, $T = 2\pi\sqrt{m/k}$, with k replaced by mg/L:

$$T = 2\pi\sqrt{\frac{m}{k}} = 2\pi\sqrt{\frac{m}{(mg/L)}}$$

Canceling the mass m, we find

Period of a Pendulum (small amplitude)

$$T = 2\pi\sqrt{\frac{L}{g}} \qquad \textbf{13–20}$$

SI unit: s

This is the classic formula for the period of a pendulum. Note that T depends on the length of the pendulum, L, and on the acceleration of gravity, g. It is independent, however, of the mass m and the amplitude A, as noted by Galileo.

The mass does not appear in the expression for the period of a pendulum, for the same reason that different masses free fall with the same acceleration. In particular, a large mass tends to move more slowly because of its large inertia; on the other hand, the larger a mass the greater the gravitational force acting on it. These two effects cancel in free fall, as well as in a pendulum.

EXERCISE 13–4

The pendulum in a grandfather clock is designed to take one second to swing in each direction; that is, 2.00 seconds for a complete period. Find the length of a pendulum with a period of 2.00 seconds.

Solution

Solve Equation 13–20 for L and substitute numerical values:

$$L = \frac{gT^2}{4\pi^2} = \frac{(9.81\ \text{m/s}^2)(2.00\ \text{s})^2}{4\pi^2} = 0.994\ \text{m}$$

CONCEPTUAL CHECKPOINT 13–2 Raise or lower the weight?

If you look carefully at a grandfather clock, you will notice that the weight at the bottom of the pendulum can be moved up or down by turning a small screw. Suppose you have a grandfather clock at home that runs slow. Should you turn the adjusting screw so as to **(a)** raise the weight or **(b)** lower the weight?

REAL-WORLD PHYSICS
Adjusting a grandfather clock

Reasoning and Discussion

To make the clock run faster, we want it to go from *tick* to *tick* more rapidly; in other words, we want the period of the pendulum to be decreased. From Equation 13–20 we can see that shortening the pendulum—that is, decreasing L—decreases the period. Hence, the weight should be raised, which effectively shortens the pendulum.

Answer:

(a) The weight should be raised. This shortens the period and makes the clock run faster.

EXAMPLE 13–7 Drop Time

A pendulum is constructed from a string 0.627 m long attached to a mass of 0.250 kg. When set in motion, the pendulum completes one oscillation every 1.59 s. If the pendulum is held at rest and the string is cut, how long will it take for the mass to fall through a distance of 1.00 m?

Picture the Problem

Note that the pendulum has a length $L = 0.627$ m and a period of oscillation $T = 1.59$ s. When the pendulum is held at rest in a vertical position, a pair of scissors is used to cut the string. The mass then falls straight downward with an acceleration g through a distance $y = 1.00$ m.

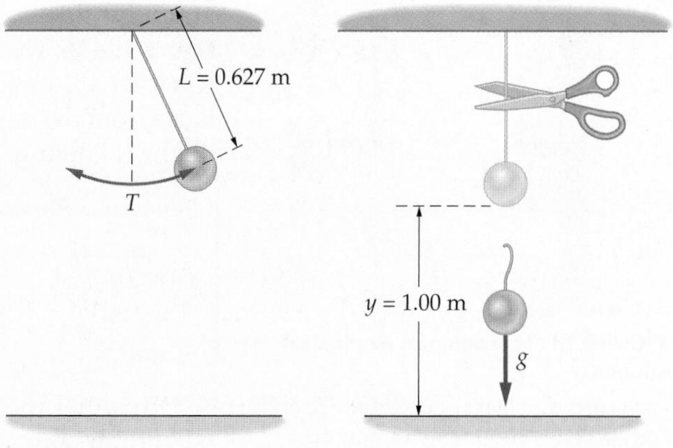

Strategy

At first it might seem that the period of oscillation and the time of fall are unrelated. Recall, however, that the period of a pendulum depends on both its length *and* the acceleration of gravity g *at the location of the pendulum.*

To solve this problem, then, we first use the period T to find the acceleration of gravity g. Once g is known, the time of fall is a straightforward kinematics problem (see Example 2–10).

continued on next page

continued from previous page

Solution

1. Use the formula for the period of a pendulum to solve for the acceleration of gravity: $T = 2\pi\sqrt{L/g}$ $\quad g = \dfrac{4\pi^2 L}{T^2}$

2. Substitute numerical values to find g: $g = \dfrac{4\pi^2 L}{T^2} = \dfrac{4\pi^2(0.627\text{ m})}{(1.59\text{ s})^2} = 9.79\text{ m/s}^2$

3. Use kinematics to solve for the time to drop from rest through a distance y: $y = \frac{1}{2}gt^2$ $\quad t = \sqrt{2y/g}$

4. Substitute $y = 1.00$ m and the value of g just found to find the time: $t = \sqrt{2y/g} = \sqrt{\dfrac{2(1.00\text{ m})}{9.79\text{ m/s}^2}} = 0.452\text{ s}$

Insight

The fact that the acceleration of gravity g varies from place to place on the Earth was mentioned in Chapter 2. In fact, "gravity maps," such as the one shown in the photo at the right, are valuable tools for geologists attempting to understand the underground properties of a given region. One instrument geologists use to make gravity maps is basically a very precise pendulum whose period can be accurately measured. Slight changes in the period from one location to another, or from one elevation to another, correspond to slight changes in g. More common in recent decades is an electronic "gravimeter" that uses a mass on a spring and relates the force of gravity to the stretch of the spring. These gravimeters can measure g with an accuracy of one part in 1000 million.

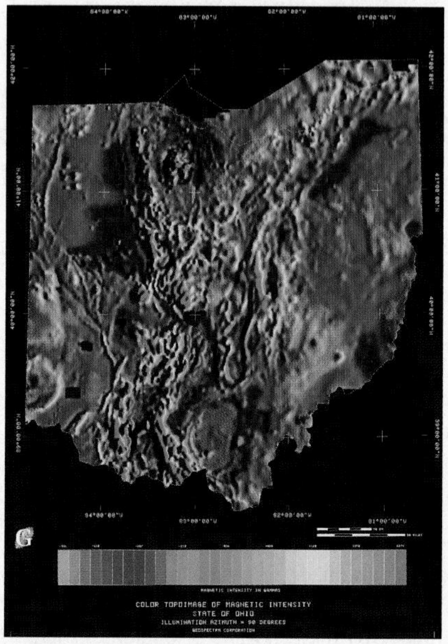

▲ A map of gravitational strength for the state of Ohio. The purple areas are those where the gravitational field is weakest. Areas where the field is strongest (red) represent regions where denser rocks lie near the surface.

Practice Problem

If a mass falls 1.00 m in a time of 0.451 s, what is **(a)** the acceleration of gravity and **(b)** the period of a pendulum of length 0.500 m at that location? [**Answer:** (a) 9.83 m/s^2, (b) 1.42 s]

Some related homework problems: Problem 49, Problem 51

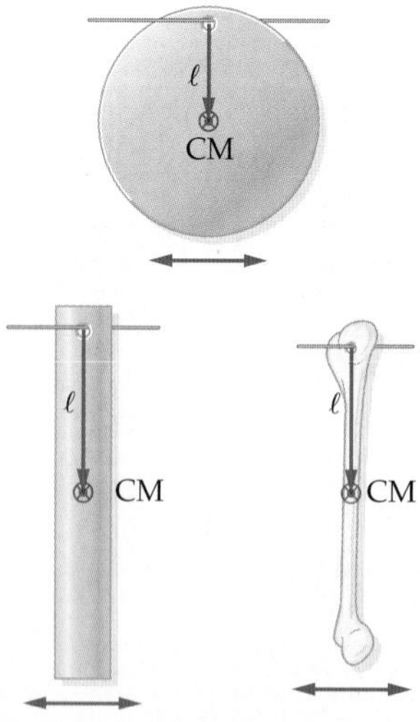

▲ **FIGURE 13–16 Examples of physical pendulums**

In each case, an object of definite size and shape oscillates about a given pivot point. The period of oscillation depends in detail on the location of the pivot point as well as on the distance ℓ from it to the center of mass, CM.

*The Physical Pendulum

In the ideal version of a simple pendulum, a bob of mass m swings back and forth a distance L from the suspension point. All of the pendulum's mass is assumed to be concentrated in the bob, which is treated as a point mass. On the other hand, a **physical pendulum** is one in which the mass is not concentrated at a point, but instead is distributed over a finite volume. Examples are shown in **Figure 13–16**. Detailed mathematical analysis shows that if the moment of inertia (Chapter 10) of a physical pendulum about its axis of rotation is I, and the distance from the axis to the center of mass is ℓ, the period of the pendulum is given by the following:

Period of a Physical Pendulum

$$T = 2\pi\sqrt{\frac{\ell}{g}}\left(\sqrt{\frac{I}{m\ell^2}}\right) \qquad \text{13–21}$$

SI unit: s

Note that the first part of the expression, $2\pi\sqrt{\ell/g}$, is the period of a simple pendulum with all its mass concentrated at the center of mass. The second factor, $\sqrt{I/m\ell^2}$, is a correction that takes into account the size and shape of the physical pendulum. Thus, writing the period in this form, rather than canceling one power of ℓ, makes for a convenient comparison with the simple pendulum.

To see how Equation 13–21 works in practice, we first apply it to a simple pendulum of mass m and length L. In this case, the moment of inertia is $I = mL^2$, and the distance to the center of mass is $\ell = L$. As a result, the period is

$$T = 2\pi\sqrt{\frac{\ell}{g}}\left(\sqrt{\frac{I}{m\ell^2}}\right) = 2\pi\sqrt{\frac{L}{g}}\left(\sqrt{\frac{mL^2}{mL^2}}\right) = 2\pi\sqrt{\frac{L}{g}}$$

Thus, as expected, Equation 13–21 also applies to a simple pendulum.

Next, we apply Equation 13–21 to a nontrivial physical pendulum; namely, your leg. When you walk, your leg rotates about the hip joint much like a uniform rod pivoted about one end, as indicated in **Figure 13–17**. Thus, if we approximate your leg as a uniform rod of length L, its period can be found using Equation 13–21. Recall from Chapter 10 (see Table 10–1) that the moment of inertia of a rod of length L about one end is

$$I = \tfrac{1}{3}mL^2$$

Similarly, the center of mass of your leg is essentially at the center of your leg; thus,

$$\ell = \tfrac{1}{2}L$$

Combining these results, we find that the period of your leg is roughly

$$T = 2\pi\sqrt{\frac{\ell}{g}}\left(\sqrt{\frac{I}{m\ell^2}}\right) = 2\pi\sqrt{\frac{\frac{1}{2}L}{g}}\left(\sqrt{\frac{\frac{1}{3}mL^2}{m\left(\frac{1}{2}L\right)^2}}\right) = 2\pi\sqrt{\frac{L}{g}}\left(\sqrt{\frac{2}{3}}\right)$$

Given that a typical human leg is about a meter long ($L = 1.0$ m), we find that its period of oscillation about the hip is approximately $T = 1.6$ s.

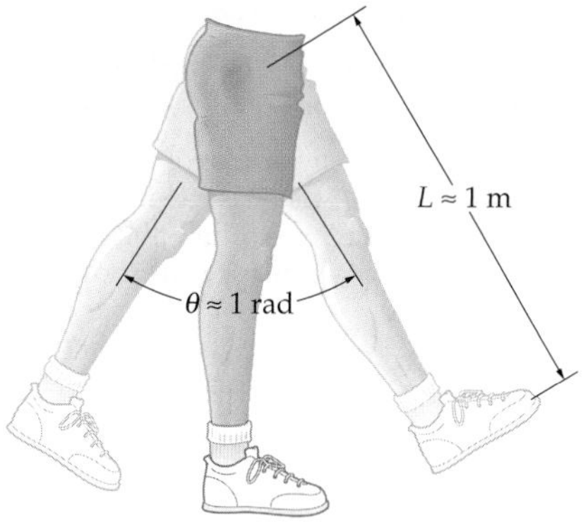

▲ **FIGURE 13–17 The leg as a physical pendulum**

As a person walks, each leg swings much like a physical pendulum. A reasonable approximation is to treat the leg as a uniform rod about 1 m in length.

REAL-WORLD PHYSICS: BIO

Walking speed

The significance of this result is that the natural walking pace of humans and other animals is largely controlled by the swinging motion of their legs as physical pendula. In fact, a great deal of research in animal locomotion has focused on precisely this type of analysis. In the case just considered, suppose that the leg in Figure 13–17 swings through an angle of roughly 1.0 radian, as the foot moves from behind the hip to in front of the hip for the next step. The arc length through which the foot moves is $s = r\theta \approx (1.0\text{ m})(1.0\text{ rad}) = 1.0$ m. Since it takes half a period ($T/2 = 0.80$ s) for the foot to move from behind the hip to in front of it, the average speed of the foot is roughly

$$v = \frac{d}{t} \approx \frac{1.0\text{ m}}{0.80\text{ s}} \approx 1.3\text{ m/s}$$

This is the typical speed of a person taking a brisk walk.

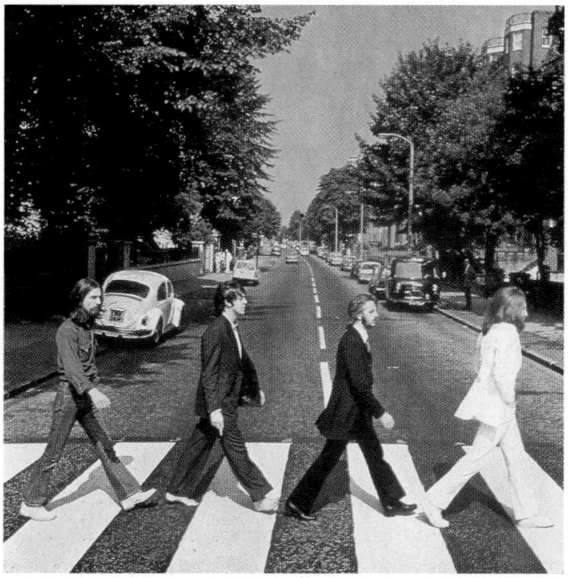

▼ When animals walk, the swinging movement of their legs can be approximated fairly well by treating the legs as physical pendulums. Such analysis has proved useful in analyzing the gaits of various creatures, from Beatles to elephants.

PROBLEM-SOLVING NOTE

The Period of a Physical Pendulum

When finding the period of a physical pendulum, recall that I is the moment of inertia about the pivot point and ℓ is the distance from the pivot point to the center of mass.

Returning to Equation 13–21, we can see that the smaller the moment of inertia I, the smaller the period. After all, an object with a small moment of inertia rotates quickly and easily. It therefore completes an oscillation in less time than an object with a larger moment of inertia. In the case of the human leg, the fact that its mass is distributed uniformly along its length—rather than concentrated at the far end—means that its moment of inertia and period are less than for a simple pendulum 1 m long.

ACTIVE EXAMPLE 13–4 Find the Period of Oscillation

A Christmas ornament is made from a hollow glass sphere of mass M and radius R. The ornament is suspended from a small hook near its surface. If the ornament is nudged slightly, what is its period of oscillation? (*Note*: The moment of inertia about the pivot point is $I = \frac{5}{3}MR^2$.)

Solution *(Test your understanding by performing the calculations indicated in each step.)*

1. Identify the distance from the axis to the center of mass: $\ell = R$
2. Substitute I and ℓ into Equation 13–21: $T = 2\pi\sqrt{\frac{R}{g}}\left(\sqrt{\frac{5}{3}}\right)$

Insight

The period is greater than that of a simple pendulum of length R.

Your Turn

Would you expect the period for a solid spherical ornament to be greater than, less than, or the same as for a hollow spherical ornament? Verify your conclusion by calculating T for a solid spherical ornament. Refer to Table 10–1 for the appropriate moment of inertia.

(Answers to **Your Turn** *problems are given in the back of the book.)*

13–7 Damped Oscillations

To this point we have restricted our considerations to oscillating systems in which no mechanical energy is gained or lost. In most physical systems, however, there is some loss of mechanical energy to friction, air resistance, or other nonconservative forces. As the mechanical energy of a system decreases, its amplitude of oscillation decreases as well, as expected from Equation 13–15. This type of motion is referred to as a **damped oscillation.**

In a typical situation, an oscillating mass may lose its mechanical energy to a force such as air resistance that is proportional to the speed of the mass and opposite in direction. The force in such a case can be written as:

$$\vec{\mathbf{F}} = -b\vec{\mathbf{v}}$$

The constant b is referred to as the **damping constant;** it is a measure of the strength of the damping force, and its SI units are kg/s.

If the damping constant is small, the system will continue to oscillate, but with a continuously decreasing amplitude. This type of motion, referred to as **underdamped,** is illustrated in **Figure 13–18 (a)**. In such cases, the amplitude decreases exponentially with time. Thus, if A_0 is the initial amplitude of an oscillating mass m, the amplitude at the time t is

$$A = A_0 e^{-bt/2m}$$

The exponential dependence of the amplitude is indicated by the dashed curve in the figure.

As the damping is increased, a point is reached where the system no longer oscillates, but instead simply relaxes back to the equilibrium position, as shown in **Figure 13–18 (b)**. A system with this type of behavior is said to be **critically damped.**

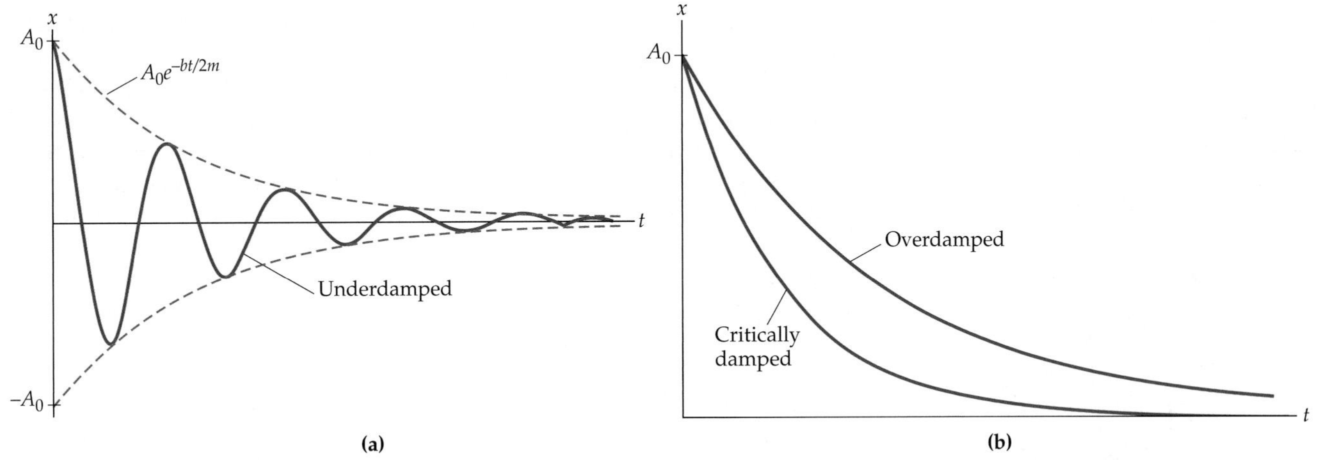

▲ **FIGURE 13–18 Damped oscillations**

In underdamped oscillations **(a)**, the position continues to oscillate as a function of time, but the amplitude of oscillation decreases exponentially. In critically damped and overdamped motion **(b)**, no oscillations occur. Instead, an object simply settles back to its equilibrium position without overshooting. Equilibrium is reached most rapidly in the critically damped case.

If the damping is increased beyond this point, the system is said to be **overdamped.** In this case, the system still returns to equilibrium without oscillating, but the time required is greater. This is also illustrated in the figure.

Some mechanical systems are designed to be near the condition for critical damping. If such a system is displaced from equilibrium, it will return to equilibrium, without oscillating, in the shortest possible time. For example, shock absorbers are designed so that a car that has just hit a bump will return to equilibrium quickly, without a lot of up-and-down oscillations.

13–8 Driven Oscillations and Resonance

In the previous section we considered the effects of removing energy from an oscillating system. It is also possible, however, to increase the energy of a system, or to replace the energy lost to various forms of friction. This can be done by applying an external force that does positive work.

Suppose, for example, that you hold the end of a string from which a small weight is suspended, as in **Figure 13–19**. If the weight is set in motion and you hold your hand still, it will soon stop oscillating. If you move your hand back and forth in a horizontal direction, however, you can keep the weight oscillating indefinitely. The motion of your hand is said to be "driving" the weight, leading to **driven oscillations.**

The response of the weight in this example depends on the frequency of your hand's back-and-forth motion, as you can readily verify for yourself. For instance, if you move your hand very slowly, the weight will simply track the motion of your hand. Similarly, if you oscillate your hand very rapidly, the weight will exhibit only small oscillations. Oscillating your hand at an intermediate frequency, however, can result in large amplitude oscillations for the weight.

Just what is an appropriate intermediate frequency? Well, to achieve a large response, your hand should drive the weight at the frequency at which it oscillates when not being driven. This is referred to as the **natural frequency,** f_0, of the system. For example, the natural frequency of a pendulum of length L is simply the inverse of its period:

$$f_0 = \frac{1}{T} = \frac{1}{2\pi}\sqrt{\frac{g}{L}}$$

Similarly, the natural frequency of a mass on a spring is

$$f_0 = \frac{1}{T} = \frac{1}{2\pi}\sqrt{\frac{k}{m}}$$

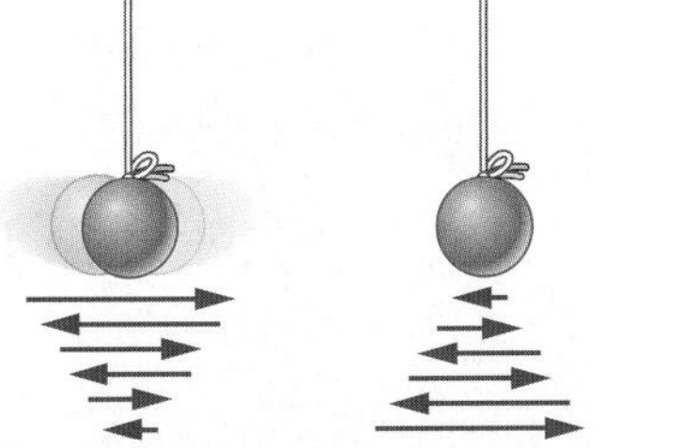

▲ **FIGURE 13–19 Driven oscillations**

If the support point of a pendulum is held still, its oscillations quickly die away. If the support point is oscillated back and forth, however, the pendulum will continue swinging. This is called "driving" the pendulum. If the driving frequency is close to the natural frequency of the pendulum—the frequency at which it oscillates when the support point is held still—its amplitude of motion can become quite large.

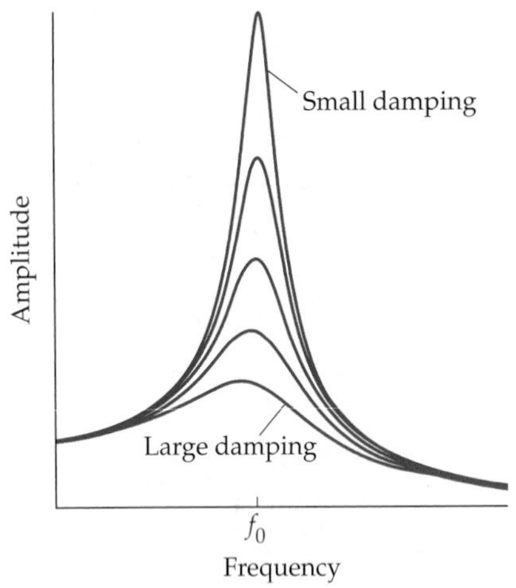

▲ **FIGURE 13–20 Resonance curves for various amounts of damping**

When the damping is small, the amplitude of oscillation can become very large for frequencies close to the natural frequency, f_0. When the damping is large, the amplitude has only a low, broad peak near the natural frequency.

In general, driving any system at a frequency near its natural frequency results in large oscillations.

As an example, **Figure 13–20** shows a plot of amplitude, A, versus driving frequency, f, for a mass on a spring. Note the large amplitude for frequencies near f_0. This type of large response, due to frequency matching, is known as **resonance,** and the curves shown in Figure 13–20 are referred to as **resonance curves.** The five curves shown in Figure 13–20 correspond to different amounts of damping, as indicated. As we can see, systems with small damping have a high, narrow peak on their resonance curve. This means that resonance is a large effect in these systems, and that it is very sensitive to frequency. On the other hand, systems with large damping have resonance peaks that are broad and low.

REAL-WORLD PHYSICS
Radio tuners and spiderwebs

Resonance plays an important role in a variety of physical systems, from a pendulum to atoms in a laser to a tuner in a radio or TV. As we shall see in Chapter 24, for example, adjusting the tuning knob in a radio changes the resonance frequency of the electrical circuit in the tuner. When its resonance frequency matches the frequency being broadcast by a station (101 MHz, perhaps), that station is picked up. To change stations, we simply change the resonance frequency of the tuner to the frequency of another station. A good tuner will have little damping, so stations that are even slightly "off resonance" will have small response, and hence will not be heard.

Mechanical examples of resonance are all around us as well. In fact, you might want to try the following experiment the next time you notice a spider in its web. Move close to the web and hum, starting with a low pitch. Slowly increase the pitch of your humming, and soon you will notice the spider react excitedly. You have hit the resonance frequency of its web, causing it to vibrate and making the spider think he has snagged a lunch. If you continue to increase your humming frequency, the spider quiets down again, because you have gone past the resonance. Each individual spider web you encounter will resonate at a specific, and different, frequency.

REAL-WORLD PHYSICS
Resonance: bridges

Man-made structures can show resonance effects as well. One of the most dramatic and famous examples is the collapse of Washington's Tacoma Narrows bridge in 1940. High winds through the narrows had often set the bridge into a gentle swaying motion, resulting in its being known by the affectionate nickname "Galloping Girdie." During one particular wind storm, however, the bridge experienced a resonance-like effect, and the amplitude of its swaying motion began to increase. Alarmed officials closed the bridge to traffic, and a short time later the swaying motion became so great that the bridge broke apart and fell into the waters below. Needless to say, bridges built since that time have been designed to prevent such catastrophic oscillations.

▲ Anyone who has ever pushed a child on a swing (left) knows that the timing of the pushes is critical. If they are synchronized with the natural frequency of the swing, the amplitude can increase rapidly. This phenomenon of resonance can have dangerous consequences. In 1940, the Tacoma Narrows bridge (right), completed only four months earlier, collapsed when high winds set the bridge swaying at one of its resonant frequencies.

Chapter Summary

13–1 Periodic Motion

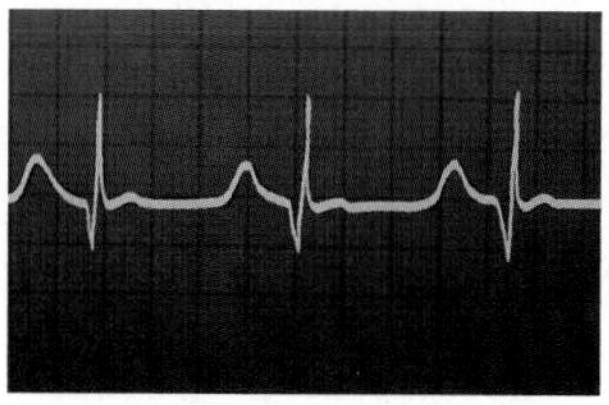

Periodic motion repeats after a definite length of time.

Period

The period, T, is the time required for a motion to repeat:

$$T = \text{time required for one cycle of a periodic motion}$$

Frequency

Frequency, f, is the number of oscillations per unit time.

Equivalently, f is the inverse of the period:

$$f = \frac{1}{T} \qquad \text{13–1}$$

Angular Frequency

Angular frequency, ω, is 2π times the frequency:

$$\omega = 2\pi f = \frac{2\pi}{T} \qquad \text{13–5}$$

Rapid motion corresponds to a short period and a large frequency.

13–2 Simple Harmonic Motion

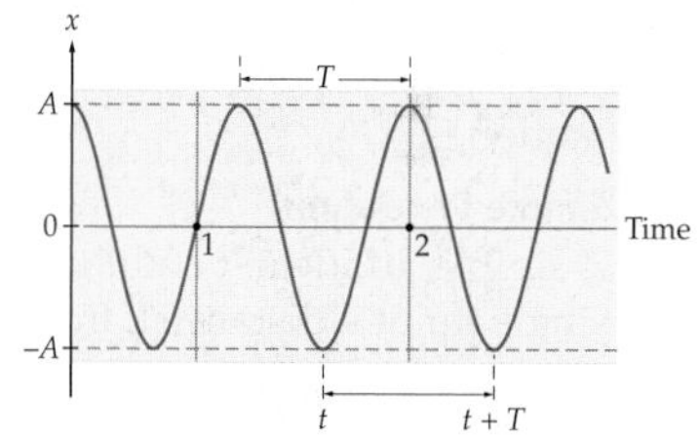

A particular type of periodic motion is simple harmonic motion. A classic example of simple harmonic motion is the oscillation of a mass attached to a spring.

Restoring Force

Simple harmonic motion occurs when the restoring force is proportional to the displacement from equilibrium.

Amplitude

The maximum displacement from equilibrium is referred to as the amplitude, A.

Position versus time

The position, x, of an object undergoing simple harmonic motion varies with time as $A\cos(\omega t)$:

$$x = A\cos\left(\frac{2\pi}{T}t\right) = A\cos(\omega t) \qquad \text{13–2, 13–4}$$

13–3 Connections Between Uniform Circular Motion and Simple Harmonic Motion

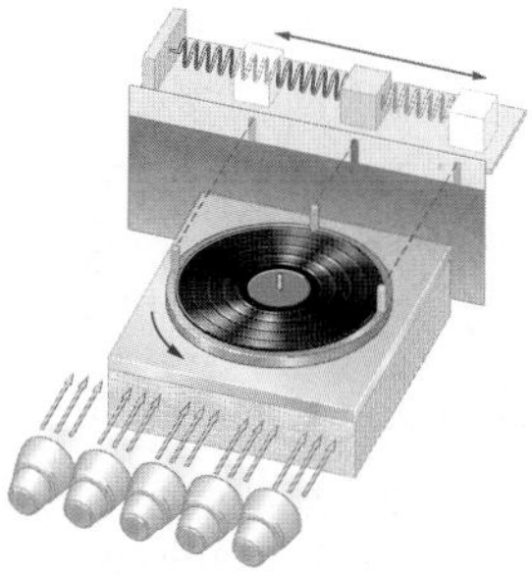

A close relationship exists between uniform circular motion and simple harmonic motion. In particular, circular motion viewed from the side—by projecting a shadow on a screen, for example—is simple harmonic.

Velocity

The velocity as a function of time in simple harmonic motion is

$$v = -A\omega\sin(\omega t) \qquad \text{13–6}$$

Acceleration

The acceleration as a function of time in simple harmonic motion is

$$a = -A\omega^2\cos(\omega t) \qquad \text{13–8}$$

Maximum Speed and Acceleration

The maximum speed of an object in simple harmonic motion is

$$v_{\text{max}} = A\omega \qquad \text{13–7}$$

Its maximum acceleration has a magnitude of

$$a_{\text{max}} = A\omega^2 \qquad \text{13–9}$$

13–4 The Period of a Mass on a Spring

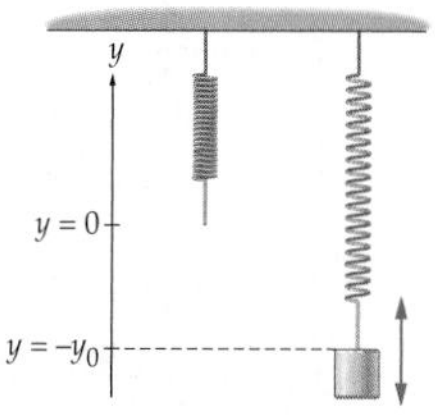

An important special case of simple harmonic motion is a mass on a spring

Period

The period of a mass m attached to a spring of force constant k is

$$T = 2\pi\sqrt{\frac{m}{k}} \qquad \text{13–11}$$

Vertical Spring
A mass attached to a vertical spring causes it to stretch to a new equilibrium position. Oscillations about this new equilibrium are simple harmonic, with a period given by Equation 13–11.

13–5 Energy Conservation in Oscillatory Motion

In an ideal oscillatory system, the total energy remains constant. This means that the kinetic and potential energy of the system vary with time in such a way that their sum remains fixed.

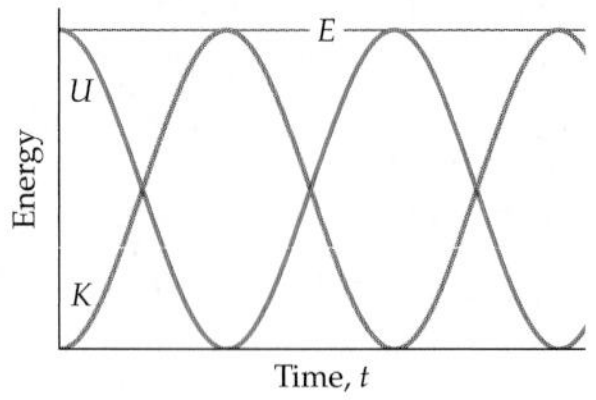

Total Energy
The total energy in simple harmonic motion is proportional to the amplitude, A, squared. For a mass on a spring, the total energy, E, is

$$E = \tfrac{1}{2}kA^2 \qquad \text{13–15}$$

Potential Energy as a Function of Time
For a mass on a spring, the potential energy varies with time as follows:

$$U = \tfrac{1}{2}kA^2 \cos^2(\omega t) \qquad \text{13–14}$$

Kinetic Energy as a Function of Time
Similarly, the kinetic energy as a function of time is

$$K = \tfrac{1}{2}kA^2 \sin^2(\omega t) \qquad \text{13–17}$$

13–6 The Pendulum

A pendulum oscillating with small amplitude also exhibits simple harmonic motion.

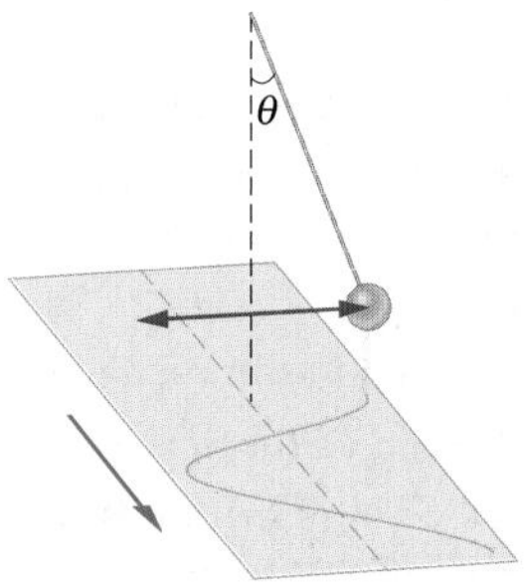

Simple Pendulum
A simple, or ideal, pendulum is one in which all the mass is concentrated at a single point a distance L from the suspension point.

Period of a Simple Pendulum
The period of a simple pendulum of length L is

$$T = 2\pi\sqrt{\frac{L}{g}} \qquad \text{13–20}$$

***Physical Pendulum**
In a physical pendulum, the mass is distributed throughout a finite volume. Thus, a physical pendulum has a definite shape, whereas a simple pendulum is characterized by a point mass.

***Period of a Physical Pendulum**
The period of a physical pendulum is

$$T = 2\pi\sqrt{\frac{\ell}{g}}\left(\sqrt{\frac{I}{m\ell^2}}\right) \qquad \text{13–21}$$

In this expression, I is the moment of inertia about the pivot point, and ℓ is the distance from the pivot point to the center of mass.

13–7 Damped Oscillations

Systems in which mechanical energy is lost to other forms, such as heat or sound, eventually come to rest at the equilibrium position. How they move as they come to rest depends on the amount of damping.

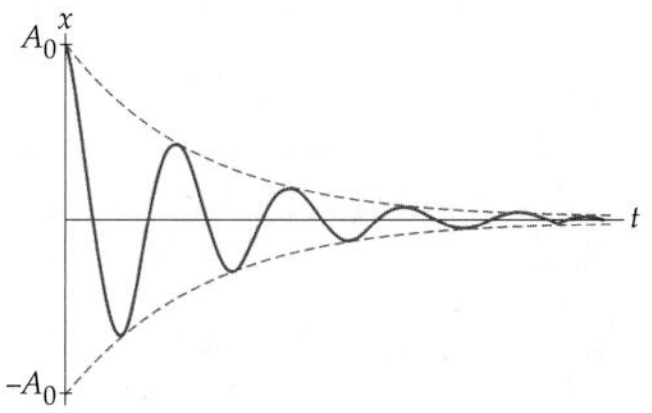

Underdamping
In an underdamped case, a system of mass m and damping constant b continues to oscillate as its amplitude steadily decreases with time. The decrease in amplitude is exponential:

$$A = A_0 e^{-bt/2m}$$

Critical Damping
In a system with critical damping, no oscillations occur. The system simply relaxes back to equilibrium in the least possible time.

Overdamping
An overdamped system also relaxes back to equilibrium with no oscillations. The relaxation occurs more slowly in this case than in critical damping.

13–8 Driven Oscillations and Resonance

If an oscillating system is driven by an external force, it is possible for energy to be added to the system. This added energy may simply replace energy lost to friction, or, in the case of resonance, it may result in oscillations of large amplitude and energy.

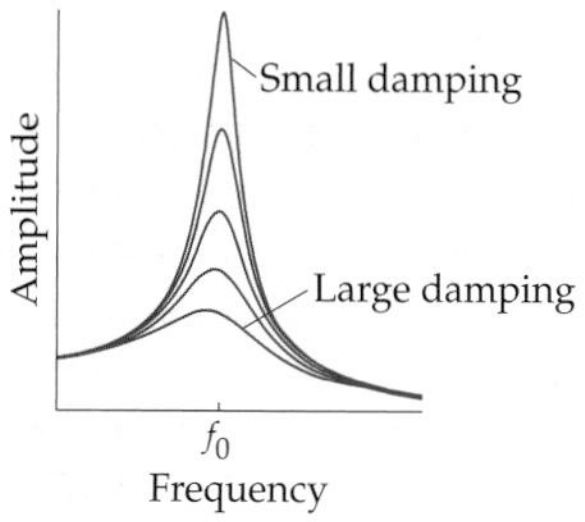

Natural Frequency
The natural frequency of an oscillating system is the frequency at which it oscillates when free from external disturbances.

Resonance
Resonance is the response of an oscillating system to a driving force of the appropriate frequency.

Problem-Solving Summary

Type of Problem	Relevant Physical Concepts	Related Examples
Find the position, velocity, and acceleration as a function of time for an object undergoing simple harmonic motion.	In simple harmonic motion, the position, velocity, and acceleration are all sinusoidal functions of time. In particular, $x = A\cos(\omega t)$, $v = -A\omega\sin(\omega t)$, and $a = -A\omega^2\cos(\omega t)$, where $\omega = 2\pi/T$.	Examples 13–1, 13–2 Active Example 13–1
Calculate the maximum speed and acceleration for simple harmonic motion.	The maximum speed is $v_{max} = A\omega$, and the maximum acceleration is $a_{max} = A\omega^2$.	Examples 13–3, 13–4
Relate the period of a mass on a spring to the mass and the force constant.	The period of a mass m attached to a spring of force constant k is $T = 2\pi\sqrt{m/k}$.	Examples 13–4, 13–5, 13–6 Active Examples 13–2, 13–3
Relate the period of a pendulum to its length and to the acceleration of gravity.	The period of a simple pendulum of length L is $T = 2\pi\sqrt{L/g}$. Note that the period is independent of the mass, but does depend on the acceleration of gravity.	Example 13–7
Find the period of a physical pendulum.	Identify the moment of inertia, I, about the pivot point and the distance from the pivot point to the center of mass, ℓ. Then use $T = 2\pi\sqrt{\ell/g}\left(\sqrt{I/m\ell^2}\right)$.	Active Example 13–4

Conceptual Questions

(Answers to odd-numbered Conceptual Questions can be found in the back of the book.)

1. A basketball player dribbles a ball with a steady period of T seconds. Is the motion of the ball periodic? Is it simple harmonic? Explain.
2. A person rides on a Ferris wheel that rotates with constant angular speed. If the Sun is directly overhead, does the person's shadow on the ground undergo periodic motion? Does it undergo simple harmonic motion? Explain.
3. An air-track cart bounces back and forth between the two ends of an air track. Is this motion periodic? Is it simple harmonic? Explain.
4. If a mass m and a mass $2m$ oscillate on identical springs with identical amplitudes, they both have the same maximum kinetic energy. How can this be? Shouldn't the larger mass have more kinetic energy? Explain.
5. An object oscillating with simple harmonic motion completes a cycle in a time T. If the object's amplitude is doubled, the time required for one cycle is still T, even though the object covers twice the distance. How can this be? Explain.
6. The position of an object undergoing simple harmonic motion is given by $x = A\cos(Bt)$. Explain the physical significance of the constants A and B. What is the frequency of this object's motion?
7. The velocity of an object undergoing simple harmonic motion is given by $v = -C\sin(Dt)$. Explain the physical significance of the constants C and D. What are the amplitude and period of this object's motion?
8. The pendulum bob in Figure 13–12 leaks sand onto the strip chart. What effect does this loss of sand have on the period of the pendulum? Explain.
9. Soldiers on the march are often ordered to break cadence in their step when crossing a bridge. Why is this a good idea?

Conceptual Exercises

(Answers to odd-numbered Conceptual Exercises can be found in the back of the book.)

1. A mass moves back and forth in simple harmonic motion with amplitude A and period T. **(a)** In terms of A, through what distance does the mass move in the time T? **(b)** Through what distance does it move in the time $5T/2$?

2. A mass moves back and forth in simple harmonic motion with amplitude A and period T. **(a)** In terms of T, how long does it take for the mass to move through a total distance of $2A$? **(b)** How long does it take for the mass to move through a total distance of $3A$?

3. A position-versus-time plot for an object undergoing simple harmonic motion is given in **Figure 13–21**. Rank the six points indicated in the figure in order of increasing **(a)** speed, **(b)** velocity, and **(c)** acceleration. Indicate a tie where necessary.

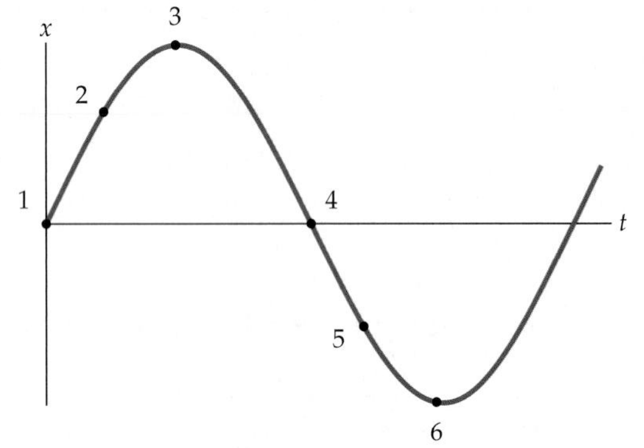

▲ **FIGURE 13–21** Conceptual Exercise 3

4. An object undergoes simple harmonic motion with a period T. In the time $3T/2$ the object moves through a total distance of $12D$. In terms of D, what is the object's amplitude of motion?

5. An object undergoing simple harmonic motion with a period T is at the position $x = 0$ at the time $t = 0$. At the time $t = 0.25T$ the position of the object is positive. State whether x is positive, negative, or zero at the following times: **(a)** $t = 1.5T$, **(b)** $t = 2T$, **(c)** $t = 2.25T$, and **(d)** $t = 6.75T$.

6. An object moves with simple harmonic motion. If the period of motion is doubled, by what multiplicative factor do the following quantities change: **(a)** angular frequency, **(b)** frequency, **(c)** maximum speed, **(d)** maximum acceleration, **(e)** total mechanical energy?

7. If the amplitude of a simple harmonic oscillator is doubled, by what multiplicative factor do the following quantities change: **(a)** angular frequency, **(b)** frequency, **(c)** period, **(d)** maximum speed, **(e)** maximum acceleration, **(f)** total mechanical energy?

8. A mass on a spring oscillates with simple harmonic motion of amplitude A about the equilibrium position $x = 0$. Its maximum speed is v_{max} and its maximum acceleration is a_{max}. **(a)** What is the speed of the mass at $x = 0$? **(b)** What is the acceleration of the mass at $x = 0$? **(c)** What is the speed of the mass at $x = A$? **(d)** What is the acceleration of the mass at $x = A$?

9. If a mass m is attached to a given spring, its period of oscillation is T. If two such springs are connected end to end and the same mass m is attached, is the resulting period of oscillation greater than, less than, or equal to T? Explain.

10. An old car with worn-out shock absorbers oscillates with a given frequency when it hits a speed bump. If the driver adds a couple of passengers to the car and hits another speed bump, is the car's frequency of oscillation greater than, less than, or equal to what it was before? Explain.

11. A mass m is suspended from the ceiling of an elevator by a spring of force constant k. When the elevator is at rest, the period of the mass is T. Does the period increase, decrease, or remain the same when the elevator **(a)** moves upward with constant speed or **(b)** moves upward with constant acceleration? Explain.

12. System A consists of a mass m attached to a spring with a force constant k; system B has a mass $2m$ attached to a spring with a force constant k; system C has a mass $3m$ attached to a spring with a force constant $6k$; and system D has a mass m attached to a spring with a force constant $4k$. Rank these systems in order of increasing period of oscillation.

13. The two blocks in **Figure 13–22** have the same mass. When set into oscillation, is the period of block 1 greater than, less than, or equal to the period of block 2? Explain.

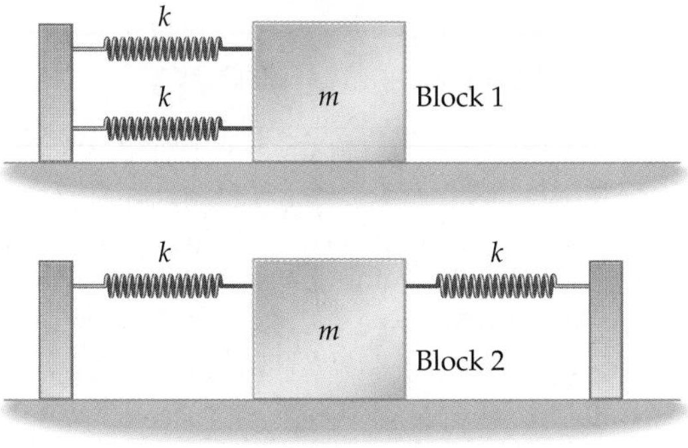

▲ **FIGURE 13–22** Conceptual Exercise 13, Problem 30

14. Metronomes, such as the penguin shown in the photo, are useful devices for music students. If it is desired to have the metronome tick with a greater frequency, should the penguin's bow tie be moved upward or downward? Explain.

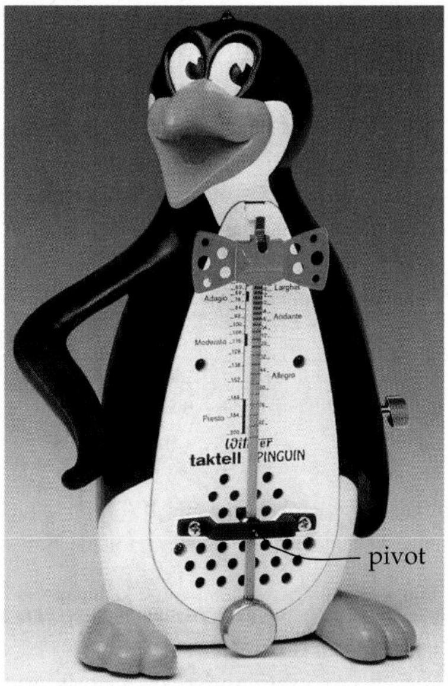

How do you like my tie? Question 14

15. A grandfather clock keeps correct time at sea level. If the clock is taken to the top of a nearby mountain, would you expect it to keep correct time, run slow, or run fast? Explain.

16. A pendulum of length L has a period T. How long must the pendulum be if its period is to be $2T$?

17. A pendulum of length L is suspended from the ceiling of an elevator. When the elevator is at rest, the period of the pendulum is T. Does the period increase, decrease, or remain the same when the elevator **(a)** moves upward with constant speed or **(b)** moves upward with constant acceleration? Explain.

Problems

Note: **IP** *denotes an integrated conceptual/quantitative problem.* **BIO** *identifies problems of biological or medical interest. Red bullets (•, ••, •••) are used to indicate the level of difficulty of each problem.*

Section 13–1 Periodic Motion

1. • A small cart on a 5.0-m long air track moves with a speed of 0.75 m/s. Bumpers at either end of the track cause the cart to reverse direction and maintain the same speed. Find the period and frequency of this motion.

2. • A person in a rocking chair completes 12 cycles in 21 s. What are the period and frequency of the rocking?

3. • While fishing for catfish, a fisherman suddenly notices that the bobber (a floating device) attached to his line is bobbing up and down with a frequency of 2.2 Hz. What is the period of the bobber's motion?

4. • If you dribble a basketball with a frequency of 1.77 Hz, how long does it take for you to complete 12 dribbles?

5. • You take your pulse and observe 74 heartbeats in a minute. What are the period and frequency of your heartbeat?

6. •• **IP** **(a)** Your heart beats with a frequency of 1.45 Hz. How many beats occur in a minute? **(b)** If the frequency of your heartbeat increases, will the number of beats in a minute increase, decrease, or stay the same? **(c)** How many beats occur in a minute if the frequency increases to 1.55 Hz?

7. •• You rev your car's engine to 2500 rpm (rev/min). **(a)** What is the period and frequency of the engine? **(b)** If you change the period of the engine to 0.034 s, how many rpms is it doing?

Section 13–2 Simple Harmonic Motion

8. • The position of a mass oscillating on a spring is given by $x = (3.2 \text{ cm}) \cos[2\pi t/(0.58 \text{ s})]$. **(a)** What is the period of this motion? **(b)** What is the first time the mass is at the position $x = 0$?

9. • The position of a mass oscillating on a spring is given by $x = (7.8 \text{ cm}) \cos[2\pi t/(0.68 \text{ s})]$. **(a)** What is the frequency of this motion? **(b)** When is the mass first at the position $x = -7.8$ cm?

10. •• A mass oscillates on a spring with a period of 0.83 s and an amplitude of 6.4 cm. Write an equation giving x as a function of time, assuming the mass starts at $x = A$ at time $t = 0$.

11. •• **IP** **Molecular Oscillations** An atom in a molecule oscillates about its equilibrium position with a frequency of 2.00×10^{14} Hz and a maximum displacement of 3.50 nm. **(a)** Write an expression giving x as a function of time for this atom, assuming that $x = A$ at $t = 0$. **(b)** If, instead, we assume that $x = 0$ at $t = 0$, would your expression for position versus time use a sine function or a cosine function? Explain.

12. •• A mass oscillates on a spring with a period T and an amplitude 0.48 cm. The mass is at the equilibrium position $x = 0$ at $t = 0$, and is moving in the positive direction. Where is the mass at the times **(a)** $t = T/8$, **(b)** $t = T/4$, **(c)** $t = T/2$ and **(d)** $t = 3T/4$? **(e)** Plot your results for parts (a) through (d) with the vertical axis representing position and the horizontal axis representing time.

13. •• The position of a mass on a spring is given by $x = (6.5 \text{ cm}) \cos[2\pi t/(0.88 \text{ s})]$. **(a)** What is the period, T, of this motion? **(b)** Where is the mass at $t = 0.25$ s? **(c)** Show that the mass is at the same location at 0.25 s + T seconds as it is at 0.25 s.

14. •• **IP** A mass attached to a spring oscillates with a period of 3.15 s. **(a)** If the mass starts from rest at $x = 0.0440$ m and time $t = 0$, where is it at time $t = 6.37$ s? **(b)** Is the mass moving in the positive or negative x direction at $t = 6.37$ s? Explain.

15. ••• An object moves with simple harmonic motion of period T and amplitude A. During one complete cycle, for what length of time is the position of the object greater than $A/2$?

16. ••• An object moves with simple harmonic motion of period T and amplitude A. During one complete cycle, for what length of time is the speed of the object greater than $v_{max}/2$?

17. ••• An object executing simple harmonic motion has a maximum speed v_{max} and a maximum acceleration a_{max}. Find **(a)** the amplitude and **(b)** the period of this motion. Express your answers in terms of v_{max} and a_{max}.

Section 13–3 Connections Between Uniform Circular Motion and Simple Harmonic Motion

18. • A ball rolls on a circular track of radius 0.62 m with a constant angular speed of 1.3 rad/s in the counterclockwise direction. If the angular position of the ball at $t = 0$ is $\theta = 0$, find the x component of the ball's position at the times 2.5 s, 5.0 s, and 7.5 s. Let $\theta = 0$ correspond to the positive x direction.

19. •• An object executing simple harmonic motion has a maximum speed of 4.3 m/s and a maximum acceleration of 0.65 m/s^2. Find **(a)** the amplitude and **(b)** the period of this motion.

20. • A child rocks back and forth on a porch swing with an amplitude of 0.204 m and a period of 2.80 s. Assuming the motion is approximately simple harmonic, find the child's maximum speed.

21. •• **IP** A 30.0-g goldfinch lands on a slender branch, where it oscillates up and down with simple harmonic motion of amplitude 0.0335 m and period 1.65 s. **(a)** What is the maximum acceleration of the finch? Express your answer as a fraction of the acceleration of gravity, g. **(b)** What is the maximum speed of the goldfinch? **(c)** At the time when the goldfinch experiences its maximum acceleration, is its speed a maximum or a minimum? Explain.

22. •• **BIO** **Tuning Forks in Neurology** Tuning forks are used in the diagnosis of nervous afflictions known as large-fiber polyneuropathies, which are often manifested in the form of reduced sensitivity to vibrations. Disorders that can result in this type of pathology include diabetes and nerve damage from exposure to heavy metals. The tuning fork in **Figure 13-23** has a frequency of 128 Hz. If the tips of the fork move with an amplitude of 1.25 mm, find **(a)** their maximum speed and **(b)** their maximum acceleration. Give your answer to part (b) as a multiple of g.

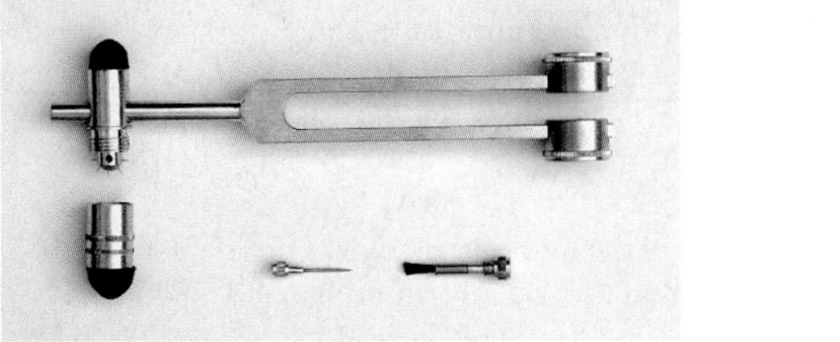

▲ **FIGURE 13-23** A Buck neurological hammer with tuning fork and Wartenburg pinwheel. (Problem 22)

23. •• A vibrating structural beam in a spacecraft can cause problems if the frequency of vibration is fairly high. Even if the amplitude of vibration is only a fraction of a millimeter, the acceleration of the beam can be several times greater than the acceleration due to gravity. As an example, find the maximum acceleration of a beam that vibrates with an amplitude of 0.25 mm at the rate of 110 vibrations per second. Give your answer as a multiple of g.

24. •• A peg on a turntable moves with a constant tangential speed of 0.67 m/s in a circle of radius 0.25 m. The peg casts a shadow on a wall. Find the following quantities related to the motion of the shadow: **(a)** the period, **(b)** the amplitude, **(c)** the maximum speed, and **(d)** the maximum magnitude of the acceleration.

25. •• The pistons in an internal combustion engine undergo a motion that is approximately simple harmonic. If the amplitude of motion is 3.5 cm, and the engine runs at 1700 rev/min, find **(a)** the maximum acceleration of the pistons and **(b)** their maximum speed.

26. •• A 0.64-kg air cart is attached to a spring and allowed to oscillate. If the displacement of the air cart from equilibrium is $x = (10.0\text{ cm})\cos[(2.00\text{ s}^{-1})t + \pi]$, find **(a)** the maximum kinetic energy of the cart and **(b)** the maximum force exerted on it by the spring.

27. •• **IP** A person rides on a mechanical bucking horse (see **Figure 13–24**) that oscillates up and down with simple harmonic motion. The period of the bucking is 0.74 s and the amplitude is slowly increasing. At a certain amplitude the rider must hang on to prevent separating from the mechanical horse. **(a)** Give a strategy that will allow you to calculate this amplitude. **(b)** Carry out your strategy and find the desired amplitude.

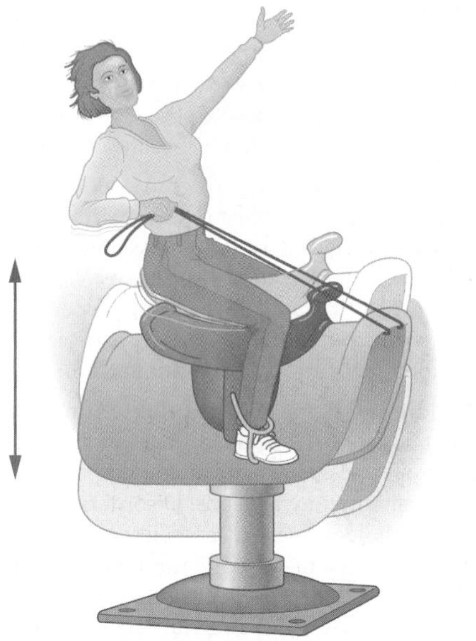

▲ **FIGURE 13–24** Problem 27

Section 13–4 The Period of a Mass on a Spring

28. • Show that the units of the quantity $\sqrt{k/m}$ are s^{-1}.

29. • A 0.42-kg mass attached to a spring undergoes simple harmonic motion with a period of 0.75 s. What is the force constant of the spring?

30. •• Find the periods of block 1 and block 2 in Figure 13–22, given that $k = 49.2$ N/m and $m = 1.25$ kg.

31. •• When a 0.50-kg mass is attached to a vertical spring, the spring stretches by 15 cm. How much mass must be attached to the spring to result in a 0.75-s period of oscillation?

32. •• A spring with a force constant of 65 N/m is attached to a 0.50-kg mass. Assuming that the amplitude of motion is 3.1 cm, determine the following quantities for this system: **(a)** ω, **(b)** v_{max}, **(c)** T.

33. •• Two people with a combined mass of 125 kg hop into an old car with worn-out shock absorbers. This causes the springs to compress by 8.00 cm. When the car hits a bump in the road, it oscillates up and down with a period of 1.65 s. Find **(a)** the total load supported by the springs and **(b)** the mass of the car.

34. •• A 0.85-kg mass attached to a vertical spring of force constant 150 N/m oscillates with a maximum speed of 0.35 m/s. Find the following quantities related to the motion of the mass: **(a)** the period, **(b)** the amplitude, **(c)** the maximum magnitude of the acceleration.

35. •• When a 0.213-kg mass is attached to a vertical spring, it causes the spring to stretch a distance d. If the mass is now displaced slightly from equilibrium, it is found to make 102 oscillations in 56.7 s. Find the stretch distance, d.

36. •• **IP** The springs of a 511-kg motorcycle have an effective force constant of 9130 N/m. **(a)** If a person sits on the motorcycle, does its period of oscillation increase, decrease, or stay the same? **(b)** By what percent and in what direction does the period of oscillation change when a 112-kg person rides the motorcycle?

37. ••• **IP** If a mass m is attached to a given spring, its period of oscillation is T. If two such springs are connected end to end, and the same mass m is attached, **(a)** is its period more than, less than, or the same as with a single spring? **(b)** Verify your answer to part (a) by calculating the new period, T', in terms of the old period T.

Section 13–5 Energy Conservation in Oscillatory Motion

38. • How much work is required to stretch a spring 0.133 m if its force constant is 9.67 N/m?

39. • A 0.321-kg mass is attached to a spring with a force constant of 12.3 N/m. If the mass is displaced 0.256 m from equilibrium and released, what is its speed when it is 0.128 m from equilibrium?

40. • Find the total mechanical energy of the system described in the previous problem.

41. •• A 1.8-kg mass attached to a spring oscillates with an amplitude of 7.1 cm and a frequency of 2.6 Hz. What is its energy of motion?

42. •• **IP** A 0.40-kg mass is attached to a spring with a force constant of 26 N/m and released from rest a distance of 3.2 cm from the equilibrium position of the spring. **(a)** Give a strategy that allows you to find the speed of the mass when it is halfway to the equilibrium position. **(b)** Use your strategy to find this speed.

43. •• **(a)** What is the maximum speed of the mass in the previous problem? **(b)** How far is the mass from the equilibrium position when its speed is half the maximum speed?

44. •• A bunch of grapes is placed in a spring scale at a supermarket. The grapes oscillate up and down with a period of 0.45 s, and the spring in the scale has a force constant of 650 N/m. What are **(a)** the mass and **(b)** the weight of the grapes?

45. •• What is the maximum speed of the grapes in the previous problem if their amplitude of oscillation is 2.3 cm?

46. •• **IP** A 0.505-kg block slides on a frictionless horizontal surface with a speed of 1.18 m/s. The block encounters an unstretched spring and compresses it 23.2 cm before coming to rest. **(a)** What is the force constant of this spring? **(b)** For what length of time is the block in contact with the spring before it comes to rest? **(c)** If the force constant of the spring is increased, does the time required to stop the block increase, decrease, or stay the same? Explain.

47. •• A 2.25-g bullet embeds itself in a 1.50-kg block, which is attached to a spring of force constant 785 N/m. If the maximum

compression of the spring is 5.88 cm, find **(a)** the initial speed of the bullet and **(b)** the time for the bullet-block system to come to rest.

Section 13–6 The Pendulum

48. • An observant fan at a baseball game notices that the radio commentators have lowered a microphone from their booth to just a few inches above the ground, as shown in **Figure 13–25**. The microphone is used to pick up sound from the field and from the fans. The fan also notices that the microphone is slowly swinging back and forth like a simple pendulum. Using her digital watch, she finds that 10 complete oscillations take 60.0 s. How high above the field is the radio booth? (Assume the microphone and its cord can be treated as a simple pendulum.)

▲ **FIGURE 13–25** Problem 48

49. • A simple pendulum of length 2.5 m makes 5.0 complete swings in 16 s. What is the acceleration of gravity at the location of the pendulum?

50. • **United Nations Pendulum** A large pendulum with a 200-lb gold-plated bob 12 inches in diameter is on display in the lobby of the United Nations building. The pendulum has a length of 75 ft. It is used to show the rotation of the Earth—for this reason it is referred to as a Foucault pendulum. What is the least amount of time it takes for the bob to swing from a position of maximum displacement to the equilibrium position of the pendulum? (Assume that the acceleration due to gravity is $g = 9.81\ \mathrm{m/s^2}$ at the UN building.)

51. • Find the length of a simple pendulum that has a period of 1.00 s. Assume that the acceleration of gravity is $g = 9.81\ \mathrm{m/s^2}$.

52. •• **IP** If the pendulum in the previous problem were to be taken to the Moon, where the acceleration of gravity is $g/6$, **(a)** would its period increase, decrease, or stay the same? **(b)** Check your result in part (a) by calculating the period of the pendulum on the Moon.

*53. •• A hula hoop hangs from a peg. Find the period of the hoop as it gently rocks back and forth on the peg. (For a hoop with axis at the rim $I = 2mR^2$, where R is the radius of the hoop.)

*54. •• A fireman tosses his 0.98-kg hat onto a peg, where it oscillates as a physical pendulum (**Figure 13–26**). If the center of mass of the hat is 8.4 cm from the pivot point, and its period of oscillation is 0.73 s, what is the moment of inertia of the hat about the pivot point?

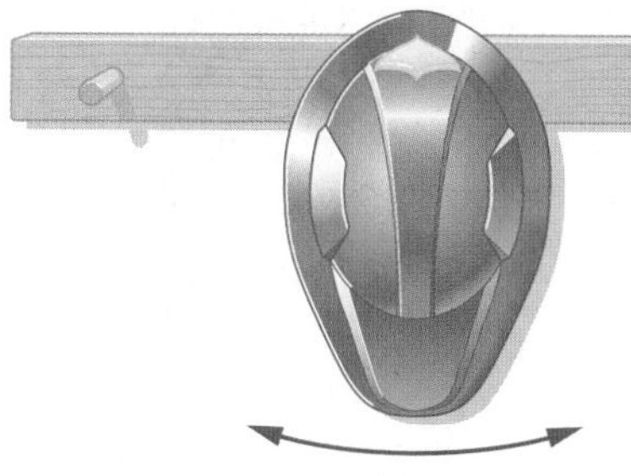

▲ **FIGURE 13–26** Problem 54

*55. •• **IP** Consider a meter stick that oscillates back and forth about a pivot point at one of its ends. **(a)** Is the period of a simple pendulum of length $L = 1.00$ m greater than, less than, or the same as the period of the meter stick? Explain. **(b)** Find the length L of a simple pendulum that has a period equal to the period of the meter stick.

*56. •• On the construction site for a new skyscraper, a uniform beam of steel is suspended from one end. If the beam swings back and forth with a period of 2.00 s, what is its length?

*57. •• **BIO (a)** Find the period of a child's leg as it swings about the hip joint. Assume the leg is 0.55 m long and can be treated as a uniform rod. **(b)** Estimate the child's walking speed.

58. ••• Suspended from the ceiling of an elevator is a simple pendulum of length L. What is the period of this pendulum if the elevator **(a)** accelerates upward with an acceleration a, or **(b)** accelerates downward with an acceleration whose magnitude is greater than zero but less than g? Give your answer in terms of L, g, and a.

General Problems

59. • A 1.6-kg mass is attached to a spring with a force constant of 59 N/m. If the mass is released with a speed of 0.25 m/s at a distance of 8.8 cm from the equilibrium position of the spring, what is its speed when it is halfway to the equilibrium position?

60. • **BIO Measuring an Astronaut's Mass** An astronaut uses a Body Mass Measurement Device (BMMD) to determine her mass. What is the astronaut's mass, given that the force constant of the BMMD is 2600 N/m and the period of oscillation is 0.85 s? (See the discussion on page 406 for more details on the BMMD.)

61. • A typical atom in a solid might oscillate with a frequency of 10^{12} Hz and an amplitude of 0.10 angstrom (10^{-11} m). Find the maximum acceleration of the atom and compare it with the acceleration of gravity.

62. • **Sunspot Observations** Sunspots vary in number as a function of time, exhibiting an approximately 11-year cycle. Galileo made the first European observations of sunspots in 1610, and daily observations were begun in Zurich in 1749. At the present time we are well into the 23rd observed cycle. What is the frequency of the sunspot cycle? Give your answer in Hz.

63. • **BIO Weighing a Bacterium** Scientist are using tiny, nanoscale cantilevers 4 micrometers long and 500 nanometers wide—essentially miniature diving boards—as a sensitive way to measure mass. The cantilevers oscillate up and down with a frequency that depends on the mass placed near the tip, and a laser beam is used to measure the frequency. A single *E. coli* bacterium was measured to have a mass of 665 femtograms = 6.65×10^{-16} kg with this device, as the cantilever oscillated with a frequency of 14.5 MHz. Treating the cantilever as an ideal, massless spring, find its effective force constant.

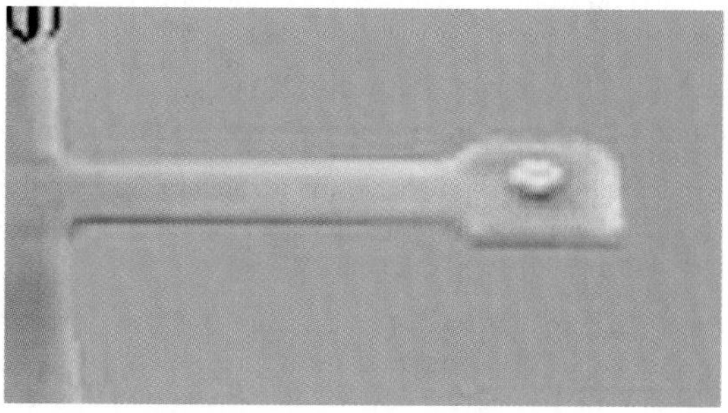

A silicon and silicon nitride cantilever with a 50-nanometer gold dot near its tip. (Problem 63)

64. •• The maximum speed of a 3.7-kg mass attached to a spring is 0.66 m/s, and the maximum force exerted on the mass is 11 N. **(a)** What is the amplitude of motion for this mass? **(b)** What is the force constant of the spring? **(c)** What is the frequency of this system?

65. •• The acceleration of a block attached to a spring is given by $a = -(0.302\ \text{m/s}^2)\cos([2.41\ \text{rad/s}]t)$. **(a)** What is the frequency of the block's motion? **(b)** What is the maximum speed of the block? **(c)** What is the amplitude of the block's motion?

66. •• **Helioseismology** In 1962, physicists at Cal Tech discovered that the surface of the Sun vibrates due to the violent nuclear reactions that roil within its core. This has lead to a new field of solar science known as helioseismology. A typical vibration of the Sun is shown in **Figure 13–27**; it has a period of 5.7 minutes. The blue patches in Figure 13–27 are moving outward; the red patches are moving inward. **(a)** Find the angular frequency of this vibration. **(b)** The maximum speed at which a patch of the surface moves during a vibration is 4.5 m/s. What is the amplitude of the vibration, assuming it to be simple harmonic motion.

▲ **FIGURE 13–27** A typical vibration pattern of the Sun. (Problem 66)

67. •• **IP** A 9.50-g bullet, moving horizontally with an initial speed v_0, embeds itself in a 1.45-kg pendulum bob that is initially at rest. The length of the pendulum is $L = 0.745$ m. After the collision, the pendulum swings to one side and comes to rest when it has gained a vertical height of 12.4 cm. **(a)** Is the kinetic energy of the bullet–bob system immediately after the collision greater than, less than, or the same as the kinetic energy of the system just before the collision? Explain. **(b)** Find the initial speed of the bullet. **(c)** How long does it take for the bullet-bob system to come to rest for the first time?

68. •• **BIO Spiderweb Oscillations** A 1.34-g spider oscillates on its web, which has a damping constant of 3.30×10^{-5} kg/s. How long does it take for the spider's amplitude of oscillation to decrease by 10.0%?

69. •• An object undergoes simple harmonic motion with a period T and amplitude A. In terms of T, how long does it take the object to travel from $x = A$ to $x = A/2$?

*70. •• Find the period of oscillation of a disk of mass 0.32 kg and radius 0.15 m if it is pivoted about a small hole drilled near its rim.

71. •• Calculate the ratio of the kinetic energy to the potential energy of a simple harmonic oscillator when its displacement is half its amplitude.

72. •• A 0.323-kg mass slides on a frictionless floor with a speed of 1.04 m/s. The mass strikes and compresses a spring with a force constant of 44.5 N/m. **(a)** How far does the mass travel after contacting the spring before it comes to rest? **(b)** How long does it take for the spring to stop the mass?

73. •• A large rectangular barge floating on a lake oscillates up and down with a period of 4.5 s. Find the damping constant for the barge, given that its mass is 2.44×10^5 kg and that its amplitude of oscillation decreases by a factor of 2.0 in 5.0 minutes.

74. •• **IP Figure 13–28** shows a displacement-versus-time graph of the periodic motion of a 3.8-kg mass on a spring. **(a)** Referring to the figure, do you expect the maximum speed of the mass to be greater than, less than, or equal to 0.50 m/s? Explain. **(b)** Calculate the maximum speed of the mass. **(c)** How much energy is stored in this system?

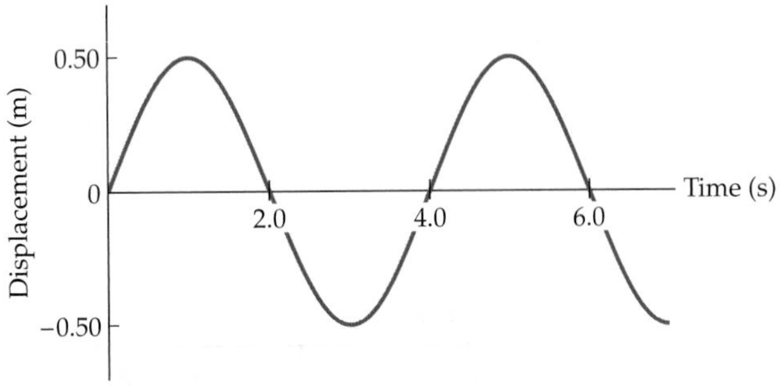

▲ **FIGURE 13–28** Problems 74 and 75

75. •• **IP** A 3.8-kg mass on a spring oscillates as shown in the displacement-versus-time graph in Figure 13–28. **(a)** Referring to the graph, at what times between $t = 0$ and $t = 6.0$ s does the mass experience a force of maximum magnitude? Explain. **(b)** Calculate the magnitude of the maximum force exerted on the mass. **(c)** At what times shown in the graph does the mass experience zero force? Explain. **(d)** How much force is exerted on the mass at the time $t = 0.50$ s?

76. •• **BIO A Cricket Thermometer, by Jiminy** The snowy tree cricket (*Oecanthus fultoni* Walker) chirps at a rate that is an accurate measure of the ambient temperature. Studies show that the cricket's chirping follows the expression $N = T - 39$, where N is the number of chirps in 13 seconds, and T is the numerical value of the temperature, in degrees Fahrenheit. **(a)** What is the frequency of the cricket's chirping (in Hz) when the temperature is 68 degrees Fahrenheit? **(b)** Suppose the temperature decreases uniformly from 75 degrees Fahrenheit to 62 degrees Fahrenheit in 12 minutes. How many chirps does the cricket produce in this time?

The snowy tree cricket. (Problem 76)

77. •• A 0.45-kg crow lands on a slender branch and bobs up and down with a period of 1.1 s. An eagle flies up to the same branch, scaring the crow away, and lands. The eagle now bobs up and down with a period of 3.6 s. Treating the branch as an ideal spring, find **(a)** the effective force constant of the branch and **(b)** the mass of the eagle.

78. ••• A mass m is connected to the bottom of a vertical spring whose force constant is k. Attached to the bottom of the mass is a string that is connected to a second mass m, as shown in **Figure 13-29**. Both masses are undergoing simple harmonic vertical motion of amplitude A. At the instant when the acceleration of the masses is a maximum in the upward direction the string breaks, allowing the lower mass to drop to the floor. Find the resulting amplitude of motion of the remaining mass.

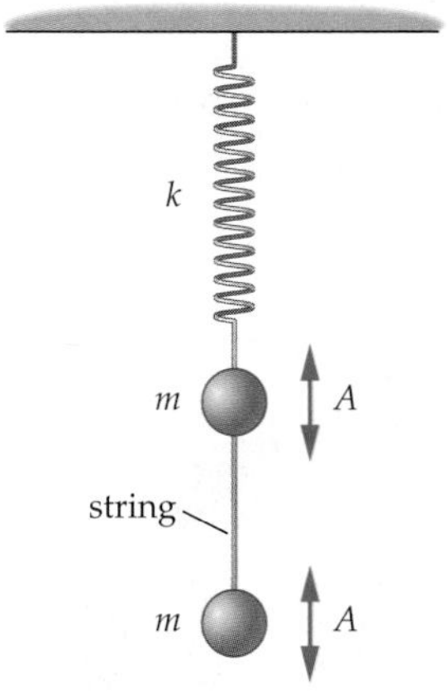

▲ **FIGURE 13-29** Problem 78

79. ••• **IP** Consider the pendulum shown in **Figure 13-30**. Note that the pendulum's string is stopped by a peg when the bob swings to the left, but moves freely when the bob swings to the right. **(a)** Is the period of this pendulum greater than, less than, or the same as the period of the same pendulum without the peg? **(b)** Calculate the period of this pendulum in terms of L and ℓ. **(c)** Evaluate your result for $L = 1.0$ m and $\ell = 0.25$ m.

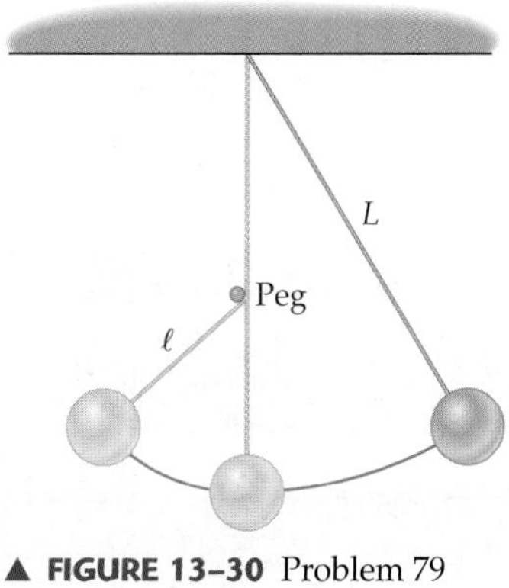

▲ **FIGURE 13-30** Problem 79

80. ••• When a mass m is attached to a vertical spring with a force constant k, it stretches the spring by the amount L. Calculate **(a)** the period of this mass and **(b)** the period of a simple pendulum of length L.

81. ••• An object undergoes simple harmonic motion of amplitude A and angular frequency ω about the equilibrium point $x = 0$. Use energy conservation to show that the speed of the object at the general position x is given by the following expression:

$$v = \omega\sqrt{A^2 - x^2}$$

*82. ••• A physical pendulum consists of a light rod of length L suspended in the middle. A large mass m_1 is attached to one end of the rod, and a lighter mass m_2 is attached to the other end, as illustrated in **Figure 13-31**. Find the period of oscillation for this pendulum.

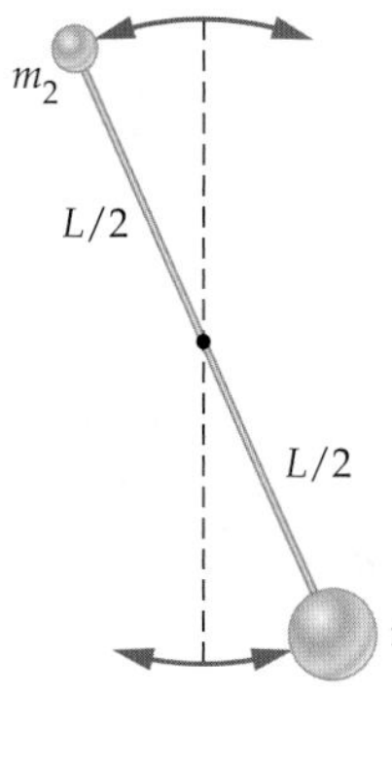

▲ **FIGURE 13-31** Problem 82

83. ••• **IP** A vertical hollow tube is connected to a speaker, which vibrates vertically with simple harmonic motion (**Figure 13-32**). The speaker operates with constant amplitude, A, but variable frequency, f. A slender pencil is placed inside the tube. **(a)** At low frequencies the pencil stays in contact with the speaker at all times; at higher frequencies the pencil begins to rattle. Explain the reason for this behavior. **(b)** Find an expression for the frequency at which rattling begins.

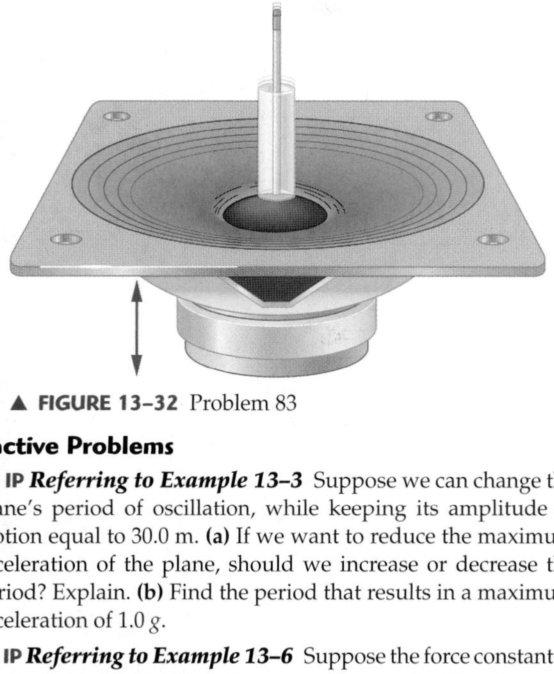

▲ **FIGURE 13-32** Problem 83

Interactive Problems

84. •• **IP** ***Referring to Example 13-3*** Suppose we can change the plane's period of oscillation, while keeping its amplitude of motion equal to 30.0 m. **(a)** If we want to reduce the maximum acceleration of the plane, should we increase or decrease the period? Explain. **(b)** Find the period that results in a maximum acceleration of 1.0 g.

85. •• **IP** ***Referring to Example 13-6*** Suppose the force constant of the spring is doubled, but the mass and speed of the block are still 0.980 kg and 1.32 m/s, respectively. **(a)** By what multiplicative factor do you expect the maximum compression of the spring to change? Explain. **(b)** Find the new maximum compression of the spring. **(c)** Find the time required for the mass to come to rest after contacting the spring.

86. •• **IP** ***Referring to Example 13-6*** **(a)** If the block's initial speed is increased, does the total time the block is in contact with the spring increase, decrease, or stay the same? **(b)** Find the total time of contact for $v_0 = 1.65$ m/s, $m = 0.980$ kg, and $k = 245$ N/m.

14 Waves and Sound

14–1	Types of Waves	431
14–2	Waves on a String	433
*14–3	Harmonic Wave Functions	436
14–4	Sound Waves	437
14–5	Sound Intensity	441
14–6	The Doppler Effect	446
14–7	Superposition and Interference	452
14–8	Standing Waves	455
14–9	Beats	463

Have you ever wondered why a grand piano has this somewhat peculiar shape? It's not just tradition—there's also a physical reason, having to do with the way vibrating strings produce sound. But to understand this and other aspects of sound, it is first necessary to learn about waves in general—for sound, as we shall see, is merely a particular kind of wave, though one that has a special importance in our lives.

In the last chapter, we studied the behavior of an oscillator. Here, we consider the behavior of a series of oscillators that are connected to one another. Connecting oscillators leads to an assortment of new phenomena, including waves on a string, water waves, and sound. In this chapter, we focus our attention on the behavior of such waves, and in particular on the way they propagate, their speed of propagation, and their interactions with one another. Later, in Chapter 25, we shall see that light is also a type of wave, and that it displays many of the same phenomena exhibited by the waves considered in this chapter.

14–1 Types of Waves

Consider a group of swings in a playground swing set. We know that each swing by itself behaves like a simple pendulum; that is, like an oscillator. Now, let's connect the swings to one another. To be specific, suppose we tie a rope from the seat of the first swing to its neighbor, and then another rope from the second swing to the third swing, and so on. When the swings are at rest—in equilibrium—the connecting ropes have no effect. If you now sit in the first swing and begin oscillating—thus "disturbing" the equilibrium—the connecting ropes cause the other swings along the line to start oscillating as well. You have created a traveling disturbance.

In general, a disturbance that propagates from one place to another is referred to as a **wave.** Waves propagate with well-defined speeds determined by the properties of the material through which they travel. In addition, waves carry energy. For example, part of the energy you put into sound waves when you speak is carried to the ears of others, where some of the sound energy is converted into electrical energy carried by nerve impulses to the brain which, in turn, creates the sensation of hearing.

It is important to distinguish between the motion of the wave itself and the motion of the individual particles that make up the wave. Common examples include the waves that propagate through a field of wheat. The individual wheat stalks sway back and forth as a wave passes, but they do not change their location. Similarly, a "wave" at a ball game may propagate around the stadium more quickly than a person can run, but the individual people making up the wave simply stand and sit in one place. From these simple examples it is clear that waves can come in a variety of types. We discuss some of the more common types in this section. In addition, we show how the speed of a wave is related to some of its basic properties.

▲ A wave can be viewed as a disturbance that propagates through space. Although the wave itself moves steadily in one direction, the particles that create the wave do not share in this motion. Instead, they oscillate back and forth about their equilibrium positions. The water in an ocean wave, for example, moves mainly up and down—as it passes, you bob up and down with it rather than being carried onto the shore. Similarly, the people in a human "wave" at a ballpark simply stand or raise their arms in place—they do not travel around the stadium.

Transverse Waves

Perhaps the easiest type of wave to visualize is a wave on a string, as illustrated in **Figure 14–1**. To generate such a wave, start by tying one end of a long string or rope to a wall. Pull on the free end with your hand, producing a tension in the string, and then move your hand up and down. As you do so, a wave will travel along the string toward the wall. In fact, if your hand moves up and down with simple harmonic motion, the wave on the string will have the shape of a sine or a cosine; we refer to such a wave as a **harmonic wave.**

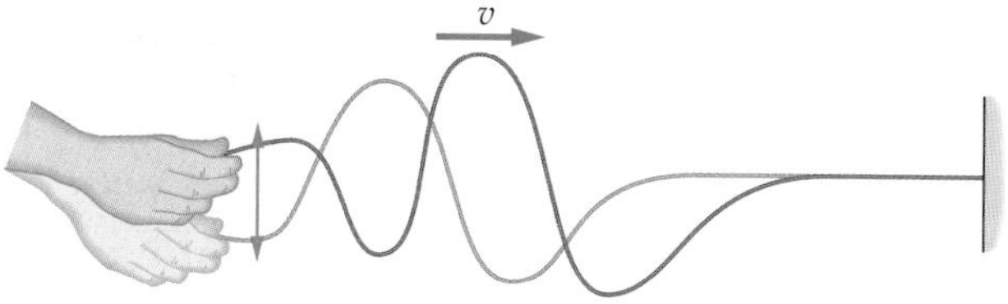

◀ **FIGURE 14–1 A wave on a string**
A wave on a string can be generated by vibrating one end of the string with an up-and-down motion.

Note that the wave travels in the horizontal direction, even though your hand oscillates vertically about one spot. In fact, if you look at any point on the string, it too moves vertically up and down, with no horizontal motion at all. This is shown in **Figure 14–2**, where we see the location of an individual point on a string as a wave travels past. Notice, in particular, that the displacement of particles in a string is at right angles to the direction of propagation of the wave. A wave with this property is called a **transverse wave:**

> In a transverse wave, the displacement of individual particles is at *right angles* to the direction of propagation of the wave.

Other examples of transverse waves include light and radio waves. These will be discussed in detail in Chapter 25.

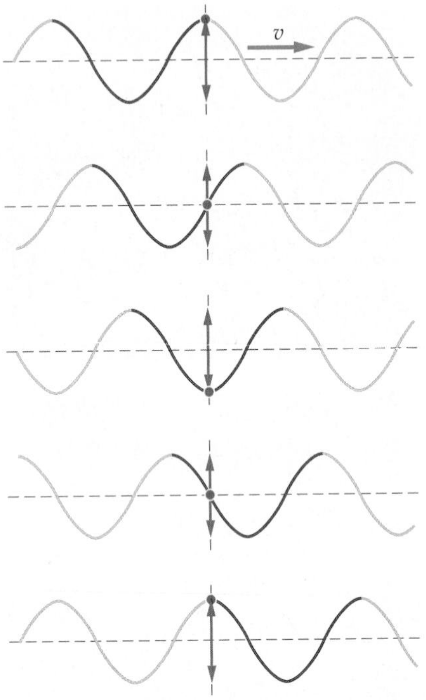

▲ **FIGURE 14–2 The motion of a wave on a string**

As a wave on a string moves horizontally, all points on the string vibrate in the vertical direction, as indicated by the blue arrow.

Longitudinal Waves

Longitudinal waves differ from transverse waves in the way that particles in the wave move. In particular, a longitudinal wave is defined as follows:

> In a longitudinal wave, the displacement of individual particles is parallel to the direction of propagation of the wave.

The classic example of a longitudinal wave is sound. When you speak, for example, the vibrations in your vocal cords create a series of compressions and expansions (rarefactions) in the air. The same kind of situation occurs with a loudspeaker, as illustrated in **Figure 14–3**. Here we see a speaker diaphragm vibrating horizontally with simple harmonic motion. As it moves to the right it compresses the air momentarily; as it moves to the left it rarefies the air. A series of compressions and rarefactions then travel horizontally away from the loudspeaker with the speed of sound.

Figure 14–3 also indicates the motion of an individual particle in the air as a sound wave passes. Note that the particle moves back and forth horizontally; that is, in the same direction as the propagation of the wave. The particle does not travel with the wave—each individual particle simply oscillates about a given position in space.

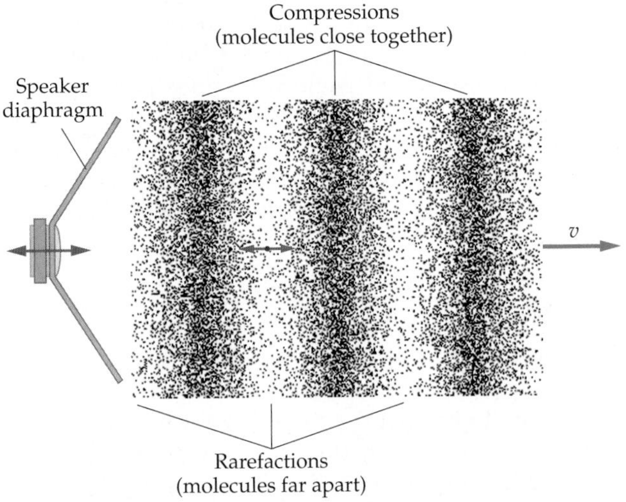

▲ **FIGURE 14–3 Sound produced by a speaker**

As the diaphragm of a speaker vibrates back and forth, it alternately compresses and rarefies the surrounding air. These regions of high and low density propagate away from the speaker with the speed of sound. Individual particles in the air oscillate back and forth about a given position, as indicated by the blue arrow.

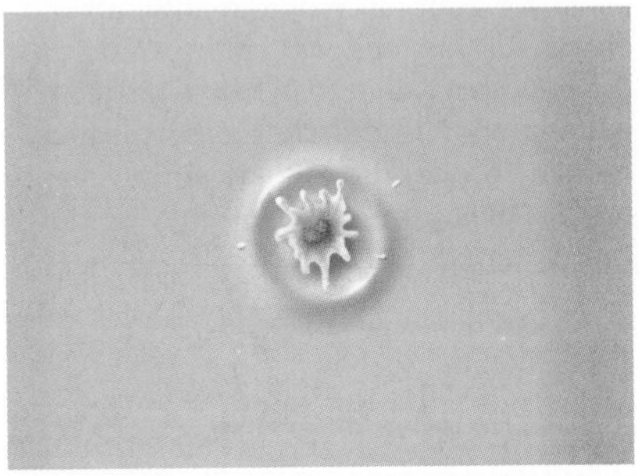

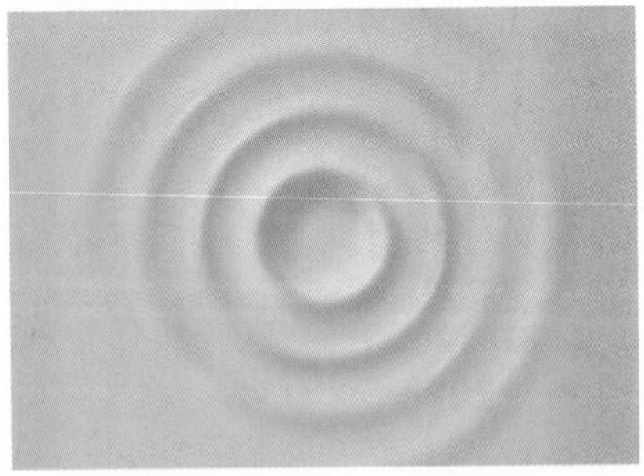

▲ **FIGURE 14–4 Water waves from a disturbance**

An isolated disturbance in a pool of water, caused by a pebble dropped into the water, creates waves that propagate symmetrically away from the disturbance. The crests and troughs form circles on the surface of the water as they move outward.

Water Waves

If a pebble is dropped into a pool of water, a series of concentric waves move away from the drop point. This is illustrated in **Figure 14–4**. To visualize the movement of the water as a wave travels by, place a small piece of cork into the water. As a wave passes, the motion of the cork will trace out the motion of the water itself, as indicated in **Figure 14–5**.

Notice that the cork moves in a roughly circular path, returning to approximately its starting point. Thus, each element of water moves both vertically and horizontally as the wave propagates by in the horizontal direction. In this sense, a water wave is a combination of both transverse and longitudinal waves. This makes the water wave more difficult to analyze. Hence, most of our results will refer to the simpler cases of purely transverse and purely longitudinal waves.

Wavelength, Frequency, and Speed

A simple wave can be thought of as a regular, rhythmic disturbance that propagates from one point to another, repeating itself both in *space* and in *time*. We now

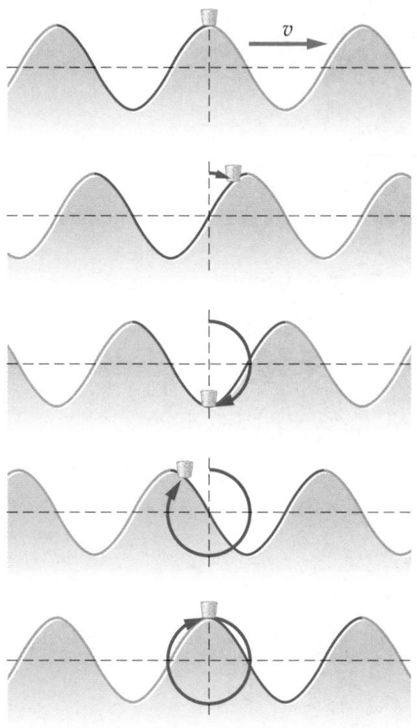

▶ **FIGURE 14–5 The motion of a water wave**
As a water wave passes a given point, a molecule (or a small piece of cork) moves in a roughly circular path. This means that the water molecules move both vertically and horizontally. In this sense, the water wave has characteristics of both transverse and longitudinal waves.

show that the repeat length and the repeat time of a wave are directly related to its speed of propagation.

We begin by considering the snapshots of a wave shown in **Figure 14–6**. Points on the wave corresponding to maximum upward displacement are referred to as **crests;** points corresponding to maximum downward displacement are called **troughs.** The distance from one crest to the next, or from one trough to the next, is the repeat length—or **wavelength,** λ—of the wave.

Definition of Wavelength, λ

λ = distance over which a wave repeats

SI unit: m

Similarly, the repeat time—or **period,** T—of a wave is the time required for one wavelength to pass a given point, as illustrated in Figure 14–6. Closely related to the period of a wave is its **frequency,** f, which, as with oscillations, is defined by the relation $f = 1/T$.

Combining these observations, we see that a wave travels a distance λ in the time T. Applying the definition of speed—distance divided by time—it follows that the speed of a wave is

Speed of a Wave

$$v = \frac{\text{distance}}{\text{time}} = \frac{\lambda}{T} = \lambda f \qquad \textbf{14–1}$$

SI unit: m/s

This result applies to all waves.

EXERCISE 14–1

Sound waves travel in air with a speed of 343 m/s. The lowest frequency sound we can hear is 20.0 Hz; the highest frequency is 20.0 kHz. Find the wavelength of sound for frequencies of 20.0 Hz and 20.0 kHz.

Solution

Solve Equation 14–1 for λ:

$$\lambda = \frac{v}{f} = \frac{343\text{ m/s}}{20.0\text{ s}^{-1}} = 17.2\text{ m}$$

$$\lambda = \frac{v}{f} = \frac{343\text{ m/s}}{20{,}000\text{ s}^{-1}} = 1.72\text{ cm}$$

14–2 Waves on a String

In this section we consider some of the basic properties of waves traveling on a string, a rope, a wire, or any similar linear medium.

The Speed of a Wave on a String

The speed of a wave is determined by the properties of the medium through which it propagates. In the case of a string of length L, there are two basic characteristics that determine the speed of a wave: (i) the tension in the string, and (ii) the mass of the string.

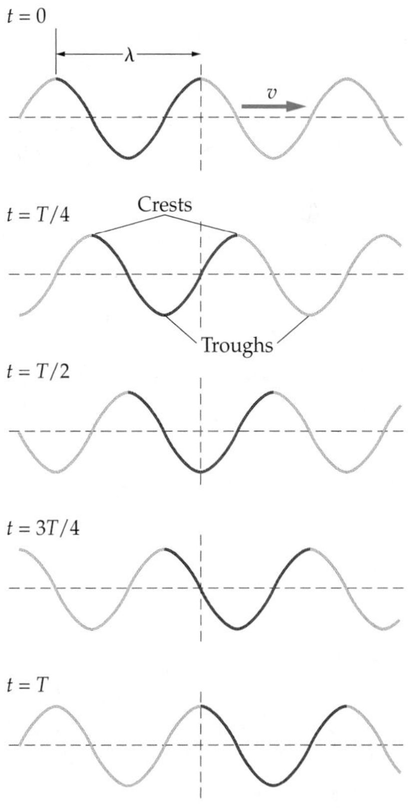

▲ **FIGURE 14–6 The speed of a wave**
A wave repeats over a distance equal to the wavelength, λ. The time necessary for a wave to move one wavelength is the period, T. Thus, the speed of a wave is $v = \lambda/T = \lambda f$.

Let's begin with the tension, which is the force F transmitted through the string (we will use F for the tension rather than T, to avoid confusion between the tension and the period). Clearly, there must be a tension in a string in order for it to propagate a wave. Imagine, for example, that a string lies on a smooth floor with both ends free. If you take one end into your hand and shake it, the portions of the string near your hand will oscillate slightly, but no wave will travel to the other end of the string. If someone else takes hold of the other end of the string and pulls enough to set up a tension, then any movement you make on your end will propagate to the other end. In fact, if the tension is increased—so that the string becomes less slack—waves will travel through the string more rapidly.

Next, we consider the mass m of the string. A heavy string responds slowly to a given disturbance because of its inertia. Thus, if you try sending a wave through a kite string or a large rope, both under the same tension, you will find that the wave in the rope travels more slowly. In general, the heavier a rope or string the slower the speed of waves in it. Of course, the total mass of a string doesn't really matter; a longer string has more mass, but its other properties are basically the same. What is important is the mass of the string per length. We give this quantity the label μ:

Definition of Mass per Length, μ

μ = mass per length = m/L

SI unit: kg/m

To summarize, we expect the speed v to increase with the tension F and decrease with the mass per length, μ. Assuming these are the only factors determining the speed of a wave on a string, we can obtain the dependence of v on F and μ using dimensional analysis (see Chapter 1, Section 3). First, we identify the dimensions of v, F, and μ:

$$[v] = \text{m/s}$$
$$[F] = \text{N} = \text{kg}\cdot\text{m/s}^2$$
$$[\mu] = \text{kg/m}$$

Next, we seek a combination of F and μ that has the dimensions of v; namely, m/s. Suppose, for example, that v depends on F to the power a and μ to the power b. Then, we have

$$v = F^a \mu^b$$

In terms of dimensions, this equation is

$$\text{m/s} = (\text{kg}\cdot\text{m/s}^2)^a(\text{kg/m})^b = \text{kg}^{a+b}\text{m}^{a-b}\text{s}^{-2a}$$

Comparing dimensions, we see that kg does not appear on the left side of the equation; therefore, we conclude that $a + b = 0$ so that kg does not appear on the right side of the equation. Hence, $a = -b$. Looking at the time dimension, s, we see that on the left we have s^{-1}; thus on the right side we must have $-2a = -1$, or $a = \frac{1}{2}$. It follows that $b = -a = -\frac{1}{2}$. This gives the following result:

Speed of a Wave on a String, v

$$v = \sqrt{\frac{F}{\mu}} \qquad \text{14–2}$$

SI unit: m/s

As expected, the speed increases with F and decreases with μ.

Dimensional analysis does not guarantee that this is the complete, final result; there could be a dimensionless factor like $\frac{1}{2}$ or 2π left unaccounted for. It turns out, however, that a complete analysis based on Newton's laws gives precisely the same result.

EXERCISE 14–2

A 5.0-m length of rope, with a mass of 0.52 kg, is pulled taut with a tension of 46 N. Find the speed of waves on the rope.

Solution

First, calculate the mass per length, μ:

$$\mu = m/L = (0.52 \text{ kg})/(5.0 \text{ m}) = 0.10 \text{ kg/m}$$

Now, substitute μ and F into Equation 14–2:

$$v = \sqrt{\frac{F}{\mu}} = \sqrt{\frac{46 \text{ N}}{0.10 \text{ kg/m}}} = 21 \text{ m/s}$$

PROBLEM-SOLVING NOTE

Mass versus Mass-per-Length

To find the mass of a string, multiply its mass per length, μ, by its length L. That is, $m = \mu L$.

EXAMPLE 14–1 A Wave on a Rope

A 12-m rope is pulled tight with a tension of 92 N. When one end of the rope is given a "thunk" it takes 0.45 s for the disturbance to propagate to the other end. What is the mass of the rope?

Picture the Problem

Our sketch shows a wave pulse traveling with a speed v from one end of the rope to the other, a distance of 12 m. The tension in the rope is 92 N, and the travel time of the pulse is 0.45 s.

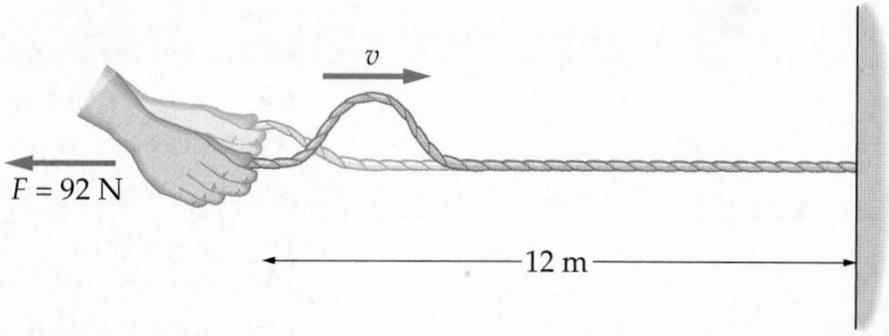

Strategy

We know that the speed of waves (disturbances) on a rope is determined by the tension and the mass per length. Thus, we first calculate the speed of the wave with the information given in the problem statement. Next, we solve for the mass per length, then multiply by the length to get the mass.

Solution

1. Calculate the speed of the wave: $v = \frac{d}{t} = \frac{12 \text{ m}}{0.45 \text{ s}} = 27 \text{ m/s}$

2. Use $v = \sqrt{F/\mu}$ to solve for the mass per length: $\mu = F/v^2$

3. Substitute numerical values for F and v: $\mu = \frac{F}{v^2} = \frac{92 \text{ N}}{(27 \text{ m/s})^2} = 0.13 \text{ kg/m}$

4. Multiply μ by $L = 12$ m to find the mass: $m = \mu L = (0.13 \text{ kg/m})(12 \text{ m}) = 1.6 \text{ kg}$

Insight

Note that the speed of a wave on this rope (about 60 mi/h) is comparable to the speed of a car on a highway. This speed could be increased even further by pulling harder on the rope, thus increasing its tension.

Practice Problem

If the tension in this rope is doubled, how long will it take for the thunk to travel from one end to the other? [**Answer:** In this case the wave speed is $v = 38$ m/s; hence the time is $t = 0.32$ s.]

Some related homework problems: Problem 9, Problem 10, Problem 11

In the following Conceptual Checkpoint, we consider the speed of a wave on a vertical rope of finite mass.

CONCEPTUAL CHECKPOINT 14–1 Speed of a Wave

A rope of length L and mass M hangs from a ceiling. If the bottom of the rope is given a gentle wiggle, a wave will travel to the top of the rope. As the wave travels upward does its speed **(a)** increase, **(b)** decrease, or **(c)** stay the same?

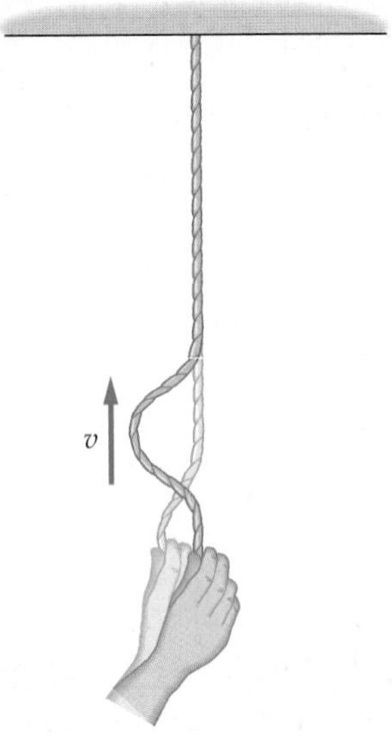

Reasoning and Discussion

The speed of the wave is determined by the tension in the rope and its mass per length. The mass per length is the same from bottom to top, but not the tension. In particular, the tension at any point in the rope is equal to the weight of rope below that point. Thus, the tension increases from almost zero near the bottom to essentially Mg near the top. Since the tension increases with height, so too does the speed, according to Equation 14–2.

Answer:

(a) The speed increases.

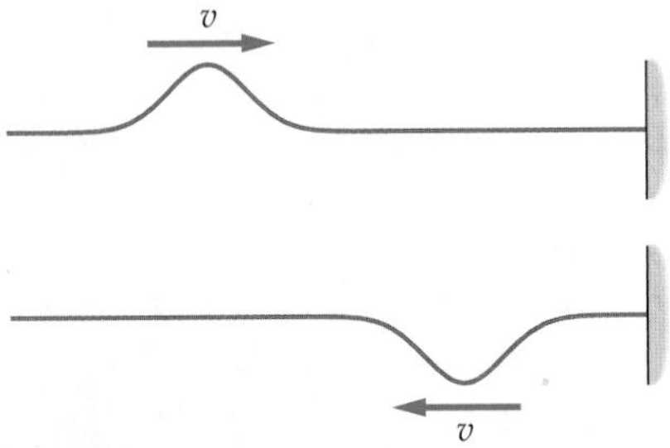

▲ **FIGURE 14–7 A reflected wave pulse: fixed end**

A wave pulse on a string is inverted when it reflects from an end that is tied down.

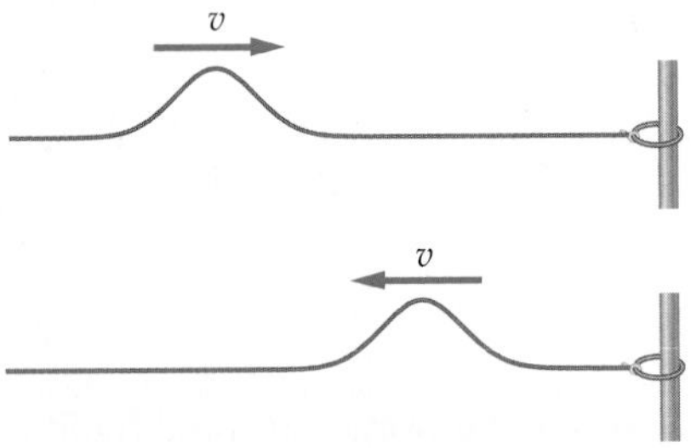

▲ **FIGURE 14–8 A reflected wave pulse: free end**

A wave pulse on a string whose end is free to move is reflected without inversion.

Reflections

Thus far we have discussed only the situation in which a wave travels along a string; but at some point the wave must reach the end of the string. What happens then? Clearly, we expect the wave to be reflected, but the precise way in which the reflection occurs needs to be considered.

Suppose, for example, that the far end of a string is anchored firmly into a wall, as shown in **Figure 14–7**. If you give a flick to your end of the string, you set up a wave "pulse" that travels toward the far end. When it reaches the end, it exerts an upward force on the wall, trying to pull it up into the pulse. Since the end is tied down, however, the wall exerts an equal and opposite downward force to keep the end at rest. Thus, the wall exerts a downward force on the string that is just the *opposite* of the upward force you exerted when you created the pulse. As a result, the reflection is an inverted, or upside-down, pulse, as indicated in Figure 14–7. We shall encounter this same type of inversion under reflection when we consider the reflection of light in Chapter 28.

Another way to tie off the end of the string is shown in **Figure 14–8**. In this case, the string is tied to a small ring that slides vertically with little friction on a vertical pole. In this way, the string still has a tension in it, since it pulls on the ring, but it is also free to move up and down.

Consider a pulse moving along such a string, as in Figure 14–8. When the pulse reaches the end, it lifts the ring upward and then lowers it back down. In fact, the pulse flicks the far end of the string in the *same* way that you flicked it when you created the pulse. Therefore, the far end of the string simply creates a new pulse, identical to the first, only traveling in the opposite direction. This is illustrated in the figure.

Thus, when waves reflect, they may or may not be inverted, depending on how the reflection occurs.

*14–3 Harmonic Wave Functions

If a wave is generated by oscillating one end of a string with simple harmonic motion, the waves will have the shape of a sine or a cosine. This is shown in **Figure 14–9**, where the y direction denotes the vertical displacement of the string, and $y = 0$ corresponds to the flat string with no wave present. In what follows, we consider the mathematical formula that describes y as a function of time, t, and position, x, for such a harmonic wave.

First, note that the harmonic wave in Figure 14–9 repeats when x increases by an amount equal to the wavelength, λ. Thus, the dependence of the wave on x must be of the form

$$y(x) = A\cos\left(\frac{2\pi}{\lambda}x\right) \qquad \textbf{14–3}$$

To see that this is the correct dependence, note that replacing x with $x + \lambda$ gives the same value for y:

$$y(x + \lambda) = A\cos\left[\frac{2\pi}{\lambda}(x + \lambda)\right] = A\cos\left(\frac{2\pi}{\lambda}x + 2\pi\right) = A\cos\left(\frac{2\pi}{\lambda}x\right) = y(x)$$

It follows that Equation 14–3 describes a vertical displacement that repeats with a wavelength λ, as desired for a wave.

This is only part of the "wave function," however, since we have not yet described the way the wave changes with time. This is illustrated in Figure 14–9, where we see a harmonic wave at time $t = 0$, $t = T/4$, $t = T/2$, $t = 3T/4$, and $t = T$. Note that the peak in the wave that was originally at $x = 0$ at $t = 0$ moves to $x = \lambda/4$, $x = \lambda/2$, $x = 3\lambda/4$, and $x = \lambda$ for the times just given. Thus, the position x of this peak can be written as follows:

$$x = \lambda\frac{t}{T}$$

Equivalently, we can say that the peak that was at $x = 0$ is now at the location given by

$$x - \lambda\frac{t}{T} = 0$$

Similarly, the peak that was originally at $x = \lambda$ at $t = 0$ is at the following position at the general time t:

$$x - \lambda\frac{t}{T} = \lambda$$

In general, if the position of a given point on a wave at $t = 0$ is $x(0)$, and its position at the time t is $x(t)$, the relation between these positions is $x(t) - \lambda t/T = x(0)$. Therefore, to take into account the time dependence of a wave, we replace $x = x(0)$ in Equation 14–3 with $x(0) = x - \lambda t/T$. This yields the harmonic wave function:

$$y(x, t) = A\cos\left[\frac{2\pi}{\lambda}\left(x - \lambda\frac{t}{T}\right)\right] = A\cos\left(\frac{2\pi}{\lambda}x - \frac{2\pi}{T}t\right) \qquad \textbf{14–4}$$

Note that the wave function, $y(x, t)$, depends on both time and position, and that the wave repeats whenever position increases by the wavelength, λ, or time increases by the period, T.

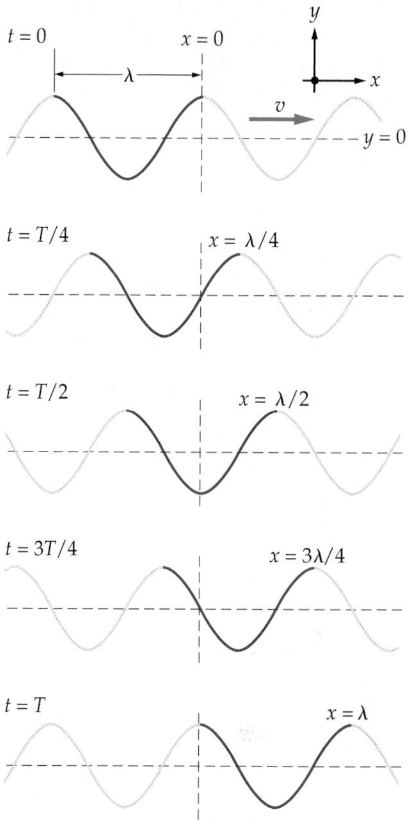

▲ **FIGURE 14–9 A harmonic wave moving to the right**
As a wave moves, the peak that was at $x = 0$ at time $t = 0$ moves to the position $x = \lambda t/T$ at the time t.

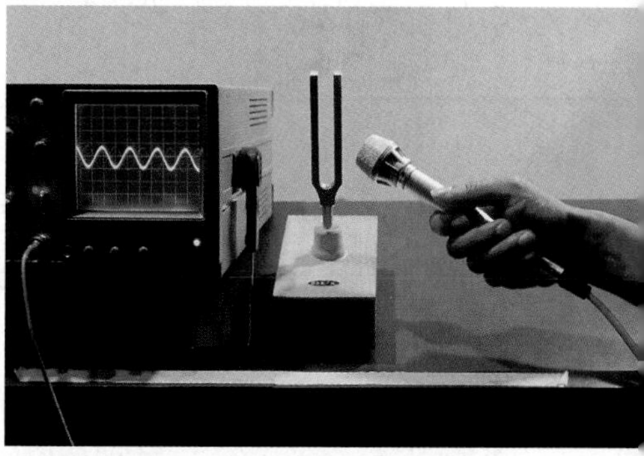

▲ An oscilloscope connected to a microphone can be used to display the wave form of a pure tone, created here by a tuning fork. The trace on the screen shows that the wave form is sinusoidal.

14–4 Sound Waves

The first thing we do when we come into this world is make a sound. It is many years later before we realize that sound is a wave propagating through the air at a speed of about 770 mi/h. More years are required to gain an understanding of the physics of a sound wave.

A useful mechanical model of a sound wave is provided by a Slinky. If we oscillate one end of a Slinky back and forth horizontally, as in **Figure 14–10**, we send out a longitudinal wave that also travels in the horizontal direction. The wave consists of regions where the coils of the Slinky are compressed alternating with regions where the coils are more widely spaced.

In close analogy with the Slinky model, a speaker produces sound waves by oscillating a diaphragm back and forth horizontally, as we saw in Figure 14–3. Just as with the Slinky, a wave travels away from the source horizontally. The wave consists of compressed regions alternating with rarefied regions.

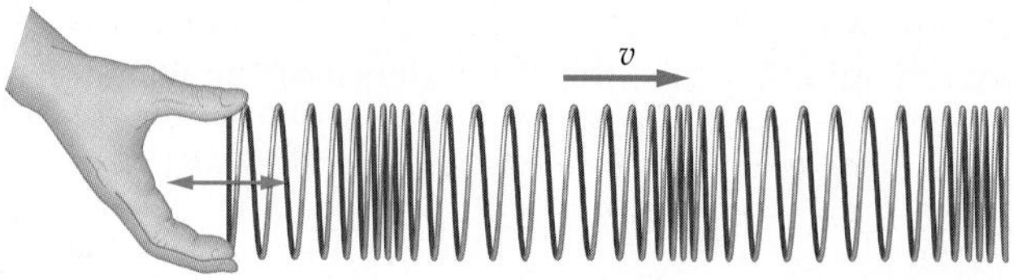

FIGURE 14–10 A wave on a Slinky

If one end of a Slinky is oscillated back and forth, a series of longitudinal waves are produced. These Slinky waves are analogous to sound waves.

At first glance, the sound wave seems very different from the wave on a string. In particular, the sound wave doesn't seem to have the nice, sinusoidal shape of a wave. Certainly, Figure 14–3 gives no hint of such a wavelike shape.

If we plot the appropriate quantities, however, the classic wave shape emerges. For example, in **Figure 14–11 (a)** we plot the rarefactions and compressions of a typical sound wave, while in **Figure 14–11 (b)** we plot the fluctuations in the density of the air versus x. Clearly, the density oscillates in a wavelike fashion. Similarly, **Figure 14–11 (c)** shows a plot of the fluctuations in the pressure of the air as a function of x. In regions where the density is high, the pressure is also high; and where the density is low, the pressure is low. Thus, pressure versus position again shows that a sound wave has the usual wave-like properties.

Just like the speed of a wave on a string, the speed of sound is determined by the properties of the medium through which it propagates. In air, under normal atmospheric pressure and temperature, the speed of sound is approximately the following:

Speed of Sound in Air (at room temperature, 20° C)

$$v = 343 \text{ m/s} \approx 770 \text{ mi/h}$$

SI unit: m/s

When we refer to the speed of sound in this text we will always assume the value is 343 m/s, unless stated specifically otherwise.

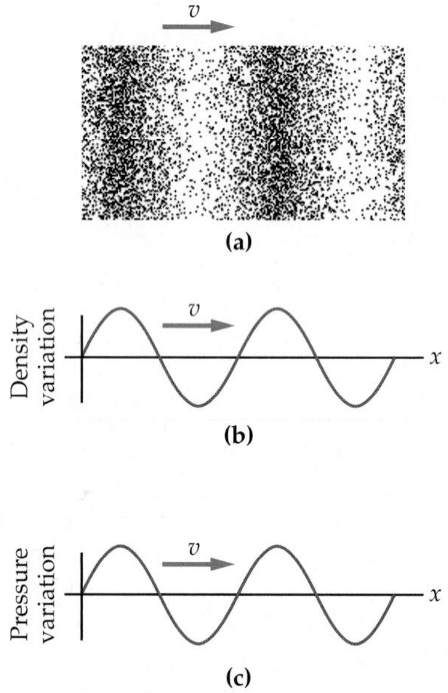

FIGURE 14–11 Wave properties of sound

A sound wave moving through the air **(a)** produces a wavelike disturbance in the **(b)** density and **(c)** pressure of the air.

As we shall see in Chapter 17, where we study the kinetic theory of gases, the speed of sound in air is directly related to the speed of the molecules themselves. Did you know, for example, that the air molecules colliding with your body at this moment have speeds that are essentially the speed of sound? As the air is heated the molecules will move faster, and hence the speed of sound also increases with temperature.

In a solid, the speed of sound is determined in part by how stiff the material is. The stiffer the material, the faster the sound wave, just as having more tension in a string causes a faster wave. Thus the speed of sound in plastic is rather large (2680 m/s), and in steel it is greater still (5960 m/s). Both speeds are much higher than the speed in air, which is certainly a "squishy" material in comparison. Table 14–1 gives a sampling of sound speed in a range of different materials.

TABLE 14–1
Speed of Sound in Various Materials

Material	Speed (m/s)
Aluminum	6420
Granite	6000
Steel	5960
Pyrex glass	5640
Copper	5010
Plastic	2680
Fresh water (20 °C)	1482
Fresh water (0 °C)	1402
Hydrogen (0 °C)	1284
Helium (0 °C)	965
Air (20 °C)	343
Air (0 °C)	331

CONCEPTUAL CHECKPOINT 14–2 How Far to the Lightning?

Five seconds after a brilliant flash of lightning, thunder shakes the house. Was the lightning **(a)** about a mile away, **(b)** much closer than a mile, or **(c)** much farther away than a mile?

Reasoning and Discussion

As mentioned, the speed of sound is 343 m/s, which is just over 1,000 ft/s. Thus, in five seconds sound travels slightly more than one mile. This gives rise to the following popular rule of thumb: The distance to a lightning strike (in miles) is the time for the thunder to arrive (in seconds) divided by 5.

Notice that we have neglected the travel time of light in our discussion. This is because light propagates with such a large speed (approximately 186,000 mi/s) that its travel time is about a million times less than that of sound.

Answer:

(a) The lightning was about a mile away.

EXAMPLE 14–2 Wishing Well

You drop a stone from rest into a well that is 7.35 m deep. How long does it take before you hear the splash?

Picture the Problem

Our sketch shows the well into which the stone is dropped. Notice that the depth of the well is $d = 7.35$ m. After the stone falls a distance d, the sound from the splash rises the same distance d before it is heard.

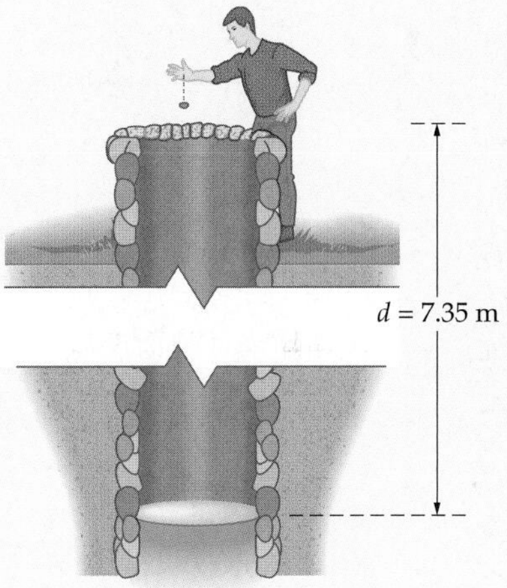

Strategy

The time until the splash is heard is the sum of (i) the time, t_1, for the stone to drop a distance d, and (ii) the time, t_2, for sound to travel a distance d.

For the time of drop, we use one-dimensional kinematics with an initial velocity $v = 0$, since the stone is dropped from rest, and an acceleration g. Therefore, the relationship between distance and time for the stone is $d = \frac{1}{2}gt_1^2$, with $g = 9.81\ \text{m/s}^2$.

For the sound wave, we use $d = vt_2$, with $v = 343$ m/s.

Solution

1. Calculate the time for the stone to drop:

$$d = \frac{1}{2}gt_1^2$$

$$t_1 = \sqrt{\frac{2d}{g}} = \sqrt{\frac{2(7.35\ \text{m})}{9.81\ \text{m/s}^2}} = 1.22\ \text{s}$$

2. Calculate the time for sound to travel a distance d:

$$d = vt_2$$

$$t_2 = \frac{d}{v} = \frac{7.35\ \text{m}}{343\ \text{m/s}} = 0.0214\ \text{s}$$

3. Sum the times found above:

$$t = t_1 + t_2 = 1.22\ \text{s} + 0.0214\ \text{s} = 1.24\ \text{s}$$

Insight

Note that the time of travel for the sound is quite small, only a couple hundredths of a second. It is still nonzero, however, and must be taken into account to obtain the correct total time.

In addition, notice that we use the same speed for a sound wave whether it is traveling horizontally, vertically upward, or vertically downward—its speed is independent of its direction of motion. As a result, the waves emanating from a source of sound propagate outward in a spherical pattern, with the wave crests forming concentric spheres around the source.

Practice Problem

You drop a stone into a well and hear the splash 1.47 s later. How deep is the well? [**Answer:** 10.2 m]

Some related homework problems: Problem 23, Problem 24

The Frequency of a Sound Wave

When we hear a sound, its frequency makes a great impression on us; in fact, the frequency determines the **pitch** of a sound. For example, the keys on a piano produce sound with frequencies ranging from 55 Hz for the key farthest to the left to 4187 Hz for the rightmost key. Similarly, as you hum a song you change

▶ Many animal species use sound waves with frequencies that are too high (ultrasonic) or too low (infrasonic) for human ears to detect. Bats, for example, navigate in the dark and locate their prey by means of a system of biological sonar. They emit a continuous stream of ultrasonic sounds and detect the echoes from objects around them. Blue whales, by contrast, communicate over long distances by means of infrasonic sounds.

the shape and size of your vocal chords slightly to change the frequency of the sound you produce.

The frequency range of human hearing extends well beyond the range of a piano, however. As a rule of thumb, humans can hear sounds between 20 Hz on the low-frequency end and 20,000 Hz on the high-frequency end. Sounds with frequencies above this range are referred to as **ultrasonic,** while those with frequencies lower than 20 Hz are classified as **infrasonic.** Though we are unable to hear ultrasound and infrasound, these frequencies occur commonly in nature, and are used in many technological applications as well.

REAL-WORLD PHYSICS
Ultrasonic sounds in nature

For example, bats and dolphins produce ultrasound almost continuously as they go about their daily lives. By listening to the echoes of their calls—that is, by using *echolocation*—they are able to navigate about their environment and detect their prey. As a defense mechanism, some of the insects that are preyed upon by bats have the ability to hear the ultrasonic frequency of a hunting bat and take evasive action. For instance, the praying mantis has a specialized ultrasound receptor on its abdomen that allows it to take cover in response to an approaching bat. More dramatically, certain moths fold their wings in flight and drop into a precipitous dive toward the ground when they hear a bat on the prowl.

REAL-WORLD PHYSICS: BIO
Medical applications of ultrasound: ultrasonic scans

Medical applications of ultrasound are also common. Perhaps the most familiar is the ultrasound scan that is used to image a fetus in the womb. By sending bursts of ultrasound into the body and measuring the time delay of the resulting echoes—the technological equivalent of echolocation—it is possible to map out the location of structures that lie hidden beneath the skin. In addition to imaging the interior of a body, ultrasound can also produce changes within the body that would otherwise require surgery. For example, in a technique called *shock wave lithotripsy* (SWL), an intense beam of ultrasound is concentrated onto a kidney stone that must be removed. After being hit with as many as 1000 to 3000 pulses of sound (at 23 joules per pulse), the stone is fractured into small pieces that the body can then eliminate on its own.

REAL-WORLD PHYSICS: BIO
Medical applications of ultrasound: shock wave lithotripsy

As for infrasound, it has been discovered in recent years that elephants can communicate with one another using sounds with frequencies as low as 15 Hz. In fact, it may be that *most* elephant communication is infrasonic. These sounds, which humans feel as vibration rather than hear as sound, can carry over an area of about thirty square kilometers on the dry African savanna. And elephants are not alone in this ability. Whales, such as the blue and the finback, produce powerful infrasonic calls as well. Since sound generally travels farther in water than in

REAL-WORLD PHYSICS
Infrasonic communication among animals

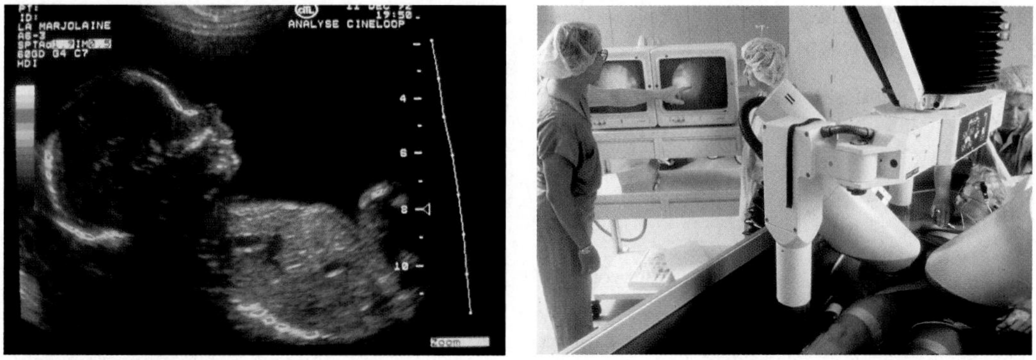

▲ Ultrasound is used in medicine both as an imaging medium and for therapeutic purposes. Ultrasound scans, or sonograms, are created by beaming ultrasonic pulses into the body and measuring the time required for the echoes to return. This technique is commonly used to evaluate heart function (echocardiograms) and to visualize the fetus in the uterus, as shown above (left). In shock wave lithotripsy (right), pulses of high-frequency sound waves are used to shatter kidney stones into fragments that can be passed in the urine.

air, the whale calls can be heard by others of their species over distances of thousands of kilometers.

One final example of infrasound is related to a dramatic event that occurred in southern New Mexico about a decade ago. At 12:47 in the afternoon of October 10, 1997, a meteor shining as bright as the full Moon streaked across the sky for a few brief moments. The event was observed not just visually, however, but with infrasound as well. An array of special microphones at the Los Alamos National Laboratory—originally designed to listen for clandestine nuclear weapons tests—heard the infrasonic boom created by the meteor. By tracking the sonic signals of such meteors it may be possible to recover fragments that manage to reach the ground. The Los Alamos detector is in constant operation, and it detects about 10 rather large objects (2 m or more in diameter) entering the Earth's atmosphere each year.

REAL-WORLD PHYSICS
Infrasound produced by meteors

It should be noted, in light of the wide range of frequencies observed in sound, that the speed of sound is the same for all frequencies. Thus, in the relation

$$v = \lambda f$$

the speed v remains fixed. For example, if the frequency of a wave is doubled, its wavelength is halved, so that the speed v stays the same. The fact that different frequencies travel with the same speed is evident when we listen to an orchestra in a large room. Different instruments are producing sounds of different frequencies, but we hear the sounds at the same time. Otherwise, listening to music from a distance would be quite a different and inharmonious experience.

14–5 Sound Intensity

The noise made by a jackhammer is much louder than the song of a sparrow. On this we can all agree. But how do we express such an observation physically? What physical quantity determines the loudness of a sound? We address these questions in this section, and we also present a quantitative scale by which loudness may be measured.

Intensity

The loudness of a sound is determined by its **intensity;** that is, by the amount of energy that passes through a given area in a given time. This is illustrated in **Figure 14–12**. If the energy E passes through the area A in the time t, the intensity, I, of the wave carrying the energy is

$$I = \frac{E}{At}$$

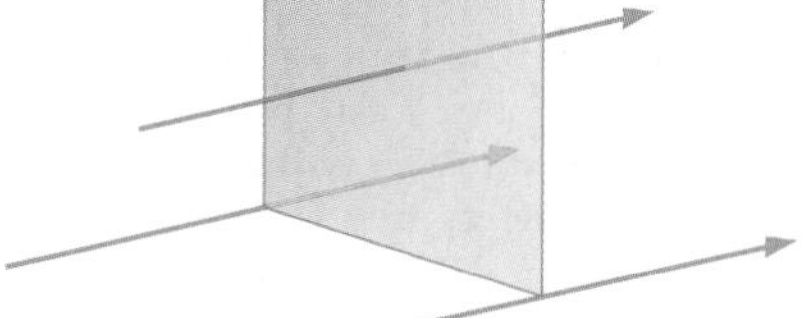

▲ FIGURE 14–12 Intensity of a wave
If a wave carries an energy E through an area A in the time t, the corresponding intensity is $I = E/At = P/A$, where $P = E/t$ is the power.

TABLE 14–2 Sound Intensities (W/m^2)

Loudest sound produced in laboratory	10^9
Saturn V rocket at 50 m	10^8
Rupture of the eardrum	10^4
Jet engine at 50 m	10
Threshold of pain	1
Rock concert	10^{-1}
Jackhammer at 1 m	10^{-3}
Heavy street traffic	10^{-5}
Conversation at 1 m	10^{-6}
Classroom	10^{-7}
Whisper at 1 m	10^{-10}
Normal breathing	10^{-11}
Threshold of hearing	10^{-12}

Recalling that power is energy per time, $P = E/t$, we can express the intensity as follows:

Definition of Intensity, *I*

$$I = \frac{P}{A} \qquad 14\text{–}5$$

SI unit: W/m^2

The units are those of power (watts, W) divided by area (meters squared, m^2).

Though we have introduced the concept of intensity in terms of sound, it applies to all types of waves. For example, the intensity of light from the Sun as it reaches the Earth's upper atmosphere is about 1380 W/m^2. If this intensity could be heard as sound, it would be painfully loud—roughly the equivalent of four jet airplanes taking off simultaneously. By comparison, the intensity of microwaves in a microwave oven is even greater, about 6000 W/m^2, whereas the intensity of a whisper is an incredibly tiny 10^{-10} W/m^2. A selection of representative intensities is given in Table 14–2.

EXERCISE 14–3

A loudspeaker puts out 0.15 W of sound through a square area 2.0 m on each side. What is the intensity of this sound?

Solution

Applying Equation 14–5, with $A = (2.0\text{ m})^2$, we find

$$I = \frac{P}{A} = \frac{0.15\text{ W}}{(2.0\text{ m})^2} = 0.038\text{ W/m}^2$$

When we listen to a source of sound, such as a person speaking or a radio playing a song, we notice that the loudness of the sound decreases as we move away from the source. This means that the intensity of the sound is also decreasing. The reason for this reduction in intensity is simply that the energy emitted per time by the source spreads out over a larger area—just as spreading a certain amount of jam over a larger piece of bread reduces the intensity of the taste.

In **Figure 14–13** we show a source of sound (a bat) and two observers (moths) listening at the distances r_1 and r_2. Notice that the waves emanating from the bat propagate outward spherically, with the wave crests forming a series of concentric spheres. Assuming no reflections of sound, and a power output by the bat equal to P, the intensity detected by the first moth is

$$I_1 = \frac{P}{4\pi r_1^{\,2}}$$

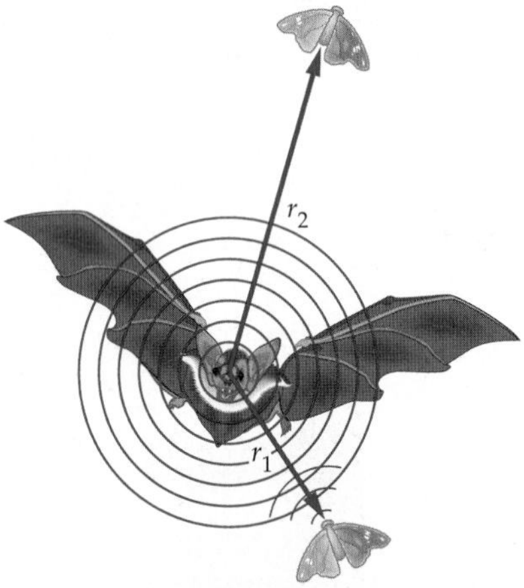

▲ **FIGURE 14–13 Echolocation**
Two moths, at distances r_1 and r_2, hear the sonar signals sent out by a bat. The intensity of the signal decreases with the square of the distance from the bat. The bat, in turn, hears the echoes sent back by the moths. It can then use the direction and intensity of the returning echoes to locate its prey.

In writing this expression, we have used the fact that the area of a sphere of radius r is $A = 4\pi r^2$. Similarly, the second moth hears the same sound with an intensity of

$$I_2 = \frac{P}{4\pi r_2^{\,2}}$$

The power P is the same in each case—it simply represents the amount of sound emitted by the bat. Solving for the intensity at moth 2 in terms of the intensity at moth 1 we find

$$I_2 = \left(\frac{r_1}{r_2}\right)^2 I_1 \qquad 14\text{–}6$$

In words, the intensity falls off with the square of the distance; doubling the distance reduces the intensity by a factor of 4.

To summarize, the intensity a distance r from a point source of power P is

Intensity with Distance from a Point Source

$$I = \frac{P}{4\pi r^2} \qquad 14\text{–}7$$

SI unit: W/m^2

This result assumes that no sound is reflected—which could increase the amount of energy passing through a given area—that no sound is absorbed, and that the sound propagates outward spherically. These assumptions are applied in the next Example.

PROBLEM-SOLVING NOTE

Intensity Variation with Distance

Suppose the intensity of a point source is I_1 at a distance r_1. This is enough information to find its intensity at any other distance. For example, to find the intensity I_2 at a distance r_2 we use the relation $I_2 = (r_1/r_2)^2 I_1$.

EXAMPLE 14–3 The Power of Song

Two people relaxing on a deck listen to a songbird sing. One person, only 1.00 m from the bird, hears the sound with an intensity of $2.80 \times 10^{-6}\ W/m^2$. **(a)** What intensity is heard by the second person, who is 4.25 m from the bird? Assume that no reflected sound is heard by either person. **(b)** What is the power output of the bird's song?

Picture the Problem

Our sketch shows the two observers, one at a distance of $r_1 = 1.00$ m from the bird, the other at a distance of $r_2 = 4.25$ m. The sound emitted by the bird is assumed to spread out spherically, with no reflections.

Strategy

a. The two intensities are related by Equation 14–6, with $r_1 = 1.00$ m and $r_2 = 4.25$ m.

b. The power output can be obtained from the definition of intensity, $I = P/A$. We can calculate P for either observer, noting that $A = 4\pi r^2$.

Solution

Part (a)

1. Substitute numerical values into Equation 14–6:

$$I_2 = \left(\frac{r_1}{r_2}\right)^2 I_1 = \left(\frac{1.00\ \text{m}}{4.25\ \text{m}}\right)^2 (2.80 \times 10^{-6}\ \text{W/m}^2)$$
$$= 1.55 \times 10^{-7}\ \text{W/m}^2$$

Part (b)

2. Solve $I = P/A$ for the power, P, using data for observer 1:

$$I_1 = P/A_1$$
$$P = I_1 A_1 = (2.80 \times 10^{-6}\ \text{W/m}^2)[4\pi(1.00\ \text{m})^2]$$
$$= 3.52 \times 10^{-5}\ \text{W}$$

3. As a check, repeat the calculation for observer 2:

$$I_2 = P/A_2$$
$$P = I_2 A_2 = (1.55 \times 10^{-7}\ \text{W/m}^2)[4\pi(4.25\ \text{m})^2]$$
$$= 3.52 \times 10^{-5}\ \text{W}$$

Insight

The intensity at observer 1 is $4.25^2 = 18.1$ times the intensity at observer 2. Even so, the bird only *seems* to be about 2.5 times louder to observer 1. The connection between intensity and perceived (subjective) loudness is discussed in detail later in this section.

Practice Problem

If the intensity at observer 2 were $7.40 \times 10^{-7}\ W/m^2$, how far would he be from the bird? [**Answer:** $r_2 = 1.95$ m]

Some related homework problems: Problem 28, Problem 33

ACTIVE EXAMPLE 14–1 The Big Hit: Find the Intensity

Ken Griffey, Jr. connects with a fast ball and sends it out of the park. A fan in the outfield bleachers, 140 m away, hears the hit with an intensity of $3.80 \times 10^{-7}\ W/m^2$. Assuming no reflected sounds, what is the intensity heard by the first-base umpire, 90 ft (27.4 m) away from home plate?

continued on next page

continued from previous page

Solution *(Test your understanding by performing the calculations indicated in each step.)*

1. Label the data given in the problem. Let the umpire be observer 1 and the fan be observer 2: $r_1 = 27.4\text{ m}$, $r_2 = 140\text{ m}$, $I_2 = 3.80 \times 10^{-7}\text{ W/m}^2$
2. Solve Equation 14–6 for I_1: $I_1 = (r_2/r_1)^2 I_2$
3. Substitute numerical values: $I_1 = 9.9 \times 10^{-6}\text{ W/m}^2$

Insight

For the fan, the sound from the hit is somewhat less intense than normal conversation. For the umpire it is comparable to the sound of a busy street.

Your Turn

Find the distance at which the sound of the hit has the intensity of a whisper. Refer to Table 14–2 for the necessary information.

(Answers to **Your Turn** *problems are given in the back of the book.)*

REAL-WORLD PHYSICS: BIO

Human perception of sound intensity

Human Perception of Sound

Hearing, like most of our senses, is incredibly versatile and sensitive. We can detect sounds that are about a million times fainter than a typical conversation, and listen to sounds that are a million times louder before experiencing pain. In addition, we are able to hear sounds over a wide range of frequencies, from 20 Hz to 20,000 Hz.

When detecting the faintest of sounds, our hearing is more sensitive than one would ever guess. For example, a faint sound, with an intensity of about 10^{-11} W/m^2, causes a displacement of molecules in the air of about 10^{-10} m. This displacement is roughly the diameter of an atom!

Equally interesting is the way we perceive the loudness of a sound. As an example, suppose you hear a sound of intensity I_1. Next, you listen to a second sound of intensity I_2, and this sound seems to be "twice as loud" as the first. If the two intensities are measured, it turns out that I_2 is about 10 times I_1. Similarly, a third sound, twice as loud as I_2, has an intensity I_3 that is 10 times greater than I_2. Thus, $I_2 = 10I_1$ and $I_3 = 10I_2 = 100I_1$.

Our perception of sound, then, is such that uniform increases in loudness correspond to intensities that increase by multiplicative factors. For this reason, a convenient scale to measure loudness depends on the logarithm of intensity, as we discuss next.

PROBLEM-SOLVING NOTE

Intensity versus Intensity Level

When reading a problem statement, be sure to note carefully whether it refers to the intensity, I, or to the intensity level, β. These two quantities have similar names but completely different meanings and units, as indicated in the following table:

Physical quantity	Physical meaning	Units
Intensity, I	Energy per time per area	W/m^2
Intensity level, β	A measure of relative loudness	dB

Intensity Level and Decibels

In the study of sound, loudness is measured by the **intensity level** of a wave. Designated by the symbol β, the intensity level is defined as follows:

Definition of Intensity Level, β

$$\beta = (10\text{ dB})\log(I/I_0) \qquad \textbf{14–8}$$

SI unit: decibel (dB), which is dimensionless

In this expression, log indicates the logarithm to the base 10, and I_0 is the intensity of the faintest sounds that can be heard. Experiments show this lowest detectable intensity to be

$$I_0 = 10^{-12}\text{ W/m}^2$$

Note that β is dimensionless; the only dimensions that enter into the definition are those of intensity, and they cancel in the logarithm. Still, just as with radians, it is convenient to label the values of intensity level with a name. The name we use—the bel—honors the work of Alexander Graham Bell (1847–1922), the inventor of the telephone. Since the bel is a fairly large unit, it is more common to measure β in units that are one-tenth of a bel. This unit is referred to as the **decibel,** and its abbreviation is dB.

To get a feeling for the decibel scale, let's start with the faintest sounds. If a sound has an intensity $I = I_0$, the corresponding intensity level is

$$\beta = (10 \text{ dB}) \log(I_0/I_0) = 10 \log(1) = 0$$

Increasing the intensity by a factor of 10 makes the sound seem twice as loud. In terms of decibels, we have

$$\beta = (10 \text{ dB}) \log(10I_0/I_0) = (10 \text{ dB}) \log(10) = 10 \text{ dB}$$

Going up in intensity by another factor of 10 doubles the loudness of the sound again, and yields

$$\beta = (10 \text{ dB}) \log(100I_0/I_0) = (10 \text{ dB}) \log(100) = 20 \text{ dB}$$

Thus, *the loudness of a sound doubles with each increase in intensity level of 10 dB.* The *smallest* increase in intensity level that can be detected by the human ear is about 1 dB.

The intensity of a number of independent sound sources is simply the sum of the individual intensities. We use this fact in the following Example.

PROBLEM-SOLVING NOTE

Calculating the Intensity Level

When determining the intensity level β, be sure to use the base 10 logarithm (log), as opposed to the "natural," or base e, logarithm (ln).

EXAMPLE 14–4 Pass the Pacifier

A crying child emits sound with an intensity of $8.0 \times 10^{-6}\ \text{W/m}^2$. Find **(a)** the intensity level in decibels for the child's sounds, and **(b)** the intensity level for this child and its twin, both crying with identical intensities.

Picture the Problem

We consider the crying sounds of either one or two children. Each child emits sound with an intensity $I = 8.0 \times 10^{-6}\ \text{W/m}^2$. If two children are crying together, the total intensity of their sound is $2I$.

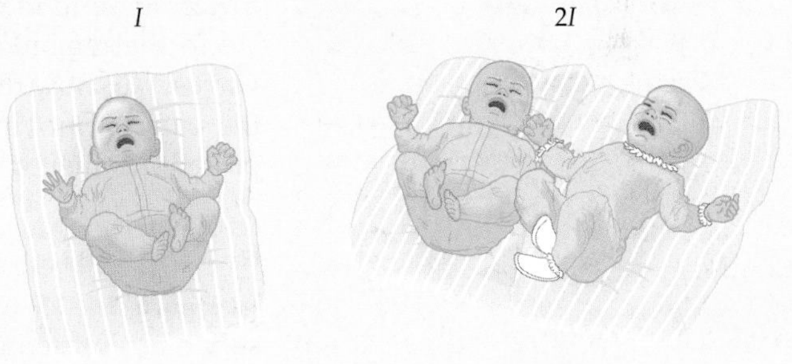

(a) **(b)**

Strategy

The intensity level, β, is obtained by applying Equation 14–8.

Solution

Part (a)

1. Calculate β for $I = 8.0 \times 10^{-6}\ \text{W/m}^2$:

$$\beta = (10 \text{ dB}) \log(I/I_0)$$
$$= (10 \text{ dB}) \log\left(\frac{8.0 \times 10^{-6}\ \text{W/m}^2}{10^{-12}\ \text{W/m}^2}\right) = (10 \text{ dB}) \log(8.0 \times 10^6)$$
$$= (10 \text{ dB}) \log(8.0) + (10 \text{ dB}) \log(10^6) = 69 \text{ dB}$$

Part (b)

2. Repeat the calculation with I replaced by $2I$:

$$\beta = (10 \text{ dB}) \log(2I/I_0)$$
$$= (10 \text{ dB}) \log(2) + (10 \text{ dB}) \log(I/I_0)$$
$$= 3.0 \text{ dB} + 69 \text{ dB} = 72 \text{ dB}$$

Insight

Note that the intensity level is increased by $(10 \text{ dB}) \log(2) = 3$ dB. This is a general rule: When the intensity is doubled, the intensity level, β, increases by 3 dB. Similarly, when the intensity is halved, β decreases by 3 dB.

Practice Problem

What is the intensity level of four identically crying quadruplets? [**Answer:** $\beta = 75$ dB]

Some related homework problems: Problem 30, Problem 31

Even though a change of 3 dB is relatively small—after all, a change of 10 dB is required to make a sound seem twice as loud—it still requires changing the intensity by a factor of two. For example, suppose a large nursery in a hospital has so many crying babies that the intensity level is 6 dB above the safe value, as determined by OSHA (Occupational Safety and Health Administration). To reduce the level by 6 dB it would be necessary to remove three-quarters of the children, leaving only one-quarter the original number. To our ears, however, the nursery will *sound* only 40 percent quieter!

Figure 14–14 shows the decibel scale with representative values indicated for a variety of common sounds.

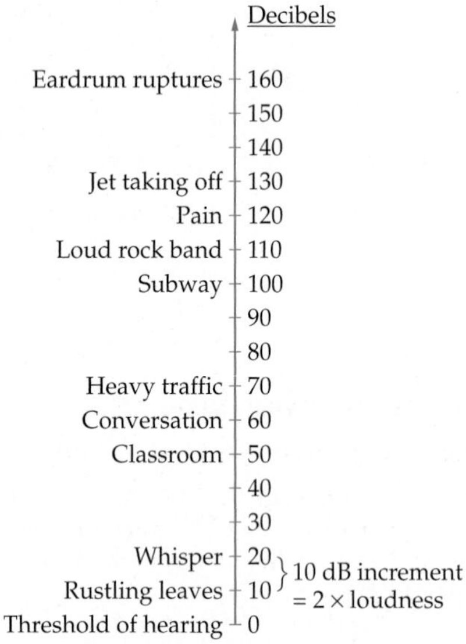

▲ **FIGURE 14–14 Representative intensity levels for common sounds**

14–6 The Doppler Effect

One of the most common physical phenomena involving sound is the change in pitch of a train whistle or a car horn as the vehicle moves past us. This change in pitch, due to the relative motion between a source of sound and the receiver, is called the **Doppler effect,** after the Austrian physicist Christian Doppler (1803–1853). If you listen carefully to the Doppler effect, you will notice that the pitch increases when the observer and the source are moving closer together, and decreases when the observer and source are separating.

One of the more fascinating aspects of the Doppler effect is the fact that it applies to all wave phenomena, not just to sound. In particular, the frequency of light is also Doppler-shifted when there is relative motion between the source and receiver. For light, this change in frequency means a change in color. In fact, most distant galaxies are observed to be red-shifted in the color of their light, which means they are moving away from the Earth. Some galaxies, however, are moving toward us, and their light shows a blue shift.

In the remainder of this section, we focus on the Doppler effect in sound waves. We show that the effect is different depending on whether the observer or the source is moving. Finally, both observer and source may be in motion, and we present results for such cases as well.

Moving Observer

In **Figure 14–15** we see a stationary source of sound in still air. The radiated sound is represented by the circular patterns of compressions moving away from the source with a speed v. The distance between the compressions is the wavelength, λ, and the frequency of the sound is f. As for any wave, these quantities are related by

$$v = \lambda f$$

For an observer moving toward the source with a speed u, as in Figure 14–15, the sound *appears* to have a higher speed, $v + u$ (though, of course, the speed of sound relative to the air is always the same). As a result, more compressions move past the observer in a given time than if the observer had been at rest. To the observer, then, the sound has a frequency, f', that is higher than the frequency of the source, f.

We can find the frequency f' by first noting that the wavelength of the sound does not change—it is still λ. The speed, however, has increased to $v' = v + u$. Thus, we can solve $v' = \lambda f'$ for the frequency, which yields

$$f' = \frac{v'}{\lambda} = \frac{v + u}{\lambda}$$

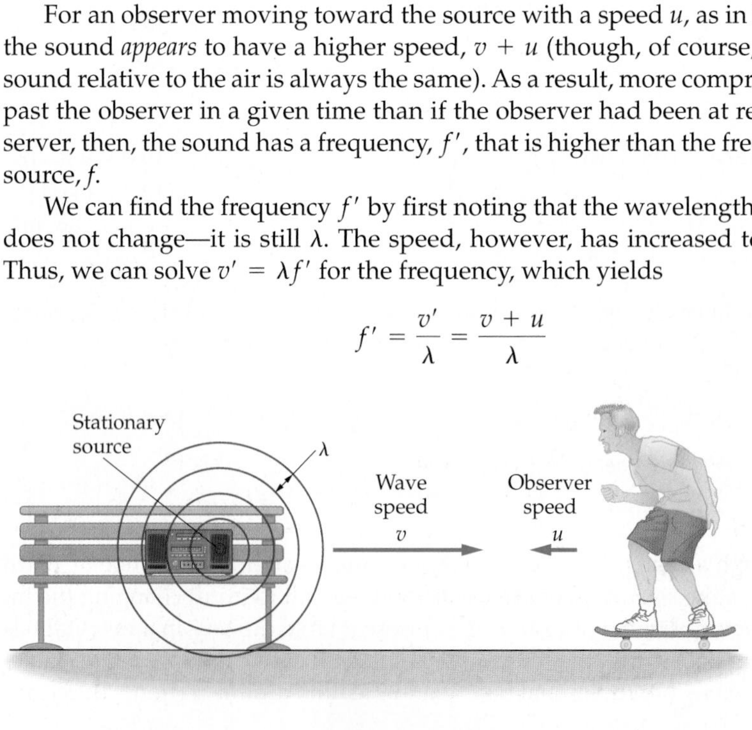

▶ **FIGURE 14–15 The Doppler effect: a moving observer**

Sound waves from a stationary source form concentric circles moving outward with a speed v. To the observer, who moves toward the source with a speed u, the waves are moving with a speed $v + u$.

Finally, we recall from Equation 14–1 that $\lambda = v/f$, and hence

$$f' = \frac{v + u}{(v/f)} = \left(\frac{v + u}{v}\right)f = (1 + u/v)f$$

Note that f' is greater than f. This is the Doppler effect.

If the observer had been moving away from the source with a speed u, the sound would *appear* to the observer to have the reduced speed $v' = v - u$. Repeating the calculation just given, we find that

$$f' = \frac{v'}{\lambda} = \frac{v - u}{\lambda} = (1 - u/v)f$$

In this case the Doppler effect results in f' being less than f.

Combining these results, we have

Doppler Effect for Moving Observer

$$f' = (1 \pm u/v)f \qquad \textbf{14–9}$$

SI unit: $1/\mathrm{s} = \mathrm{s}^{-1}$

In this expression u and v are speeds, and hence are always positive. The appropriate signs are obtained by using the *plus* sign when the observer moves *toward* the source, and the *minus* sign when the observer moves *away from* the source.

PROBLEM-SOLVING NOTE

Using the Correct Sign for the Doppler Effect

When an observer approaches a source, the frequency heard by the observer is greater than the frequency of the source; that is, $f' > f$. This means that we must use the plus sign in $f' = (1 \pm u/v)f$. Similarly, we must use the minus sign when the observer moves away from the source.

EXAMPLE 14–5 A Moving Observer

A street musician sounds the A string of his violin, producing a tone of 440 Hz. What frequency does a bicyclist hear as he **(a)** approaches and **(b)** recedes from the musician with a speed of 11.0 m/s?

Picture the Problem

The sketch shows a stationary source of sound and a moving observer. In part (a) the observer approaches the source with a speed $u = 11.0$ m/s; in part (b) the observer has passed the source and is moving away with the same speed.

Strategy

The frequency heard by the observer is given by Equation 14–9, with the plus sign for part (a) and the minus sign for part (b).

(a)

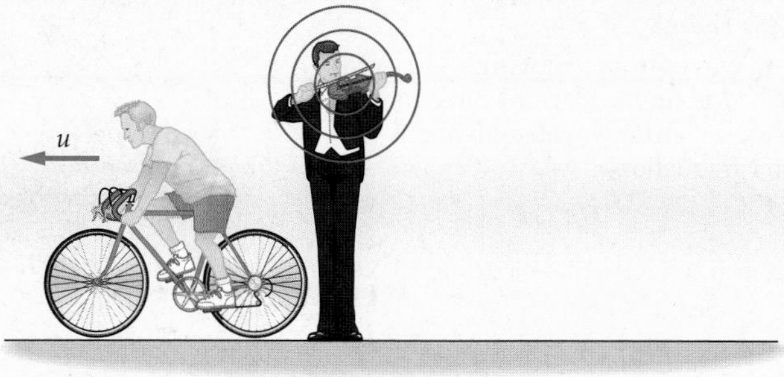

(b)

Solution

Part (a)

1. Apply Equation 14–9 with the plus sign and $u = 11.0$ m/s: $f' = (1 + u/v)f = \left(1 + \frac{11.0\ \mathrm{m/s}}{343\ \mathrm{m/s}}\right)(440\ \mathrm{Hz}) = 454\ \mathrm{Hz}$

Part (b)

2. Now use the minus sign in Equation 14–9: $f' = (1 - u/v)f = \left(1 - \frac{11.0\ \mathrm{m/s}}{343\ \mathrm{m/s}}\right)(440\ \mathrm{Hz}) = 426\ \mathrm{Hz}$

continued on next page

continued from previous page

Insight

As the bicyclist passes the musician, the observed frequency of sound decreases, giving a "wow" effect. The difference in frequency is about 1 semitone, the frequency difference between adjacent notes on the piano. See Table 14–3 on page 458 for semitones in the vicinity of middle C.

Practice Problem

If the bicyclist hears a frequency of 451 Hz when approaching the musician, what is his speed? [**Answer:** $u = 8.58 \text{ m/s}$]

Some related homework problems: Problem 37, Problem 40

Moving Source

With a stationary observer and a moving source, the Doppler effect is not due to the sound wave appearing to have a higher or lower speed, as when the observer moves. On the contrary, the speed of a wave is determined by the properties of the medium through which it propagates. Thus, once the source emits a sound wave, it travels through the medium with its characteristic speed v regardless of what the source is doing.

By way of analogy, consider a water wave. The speed of such waves is the same whether they are created by a rock dropped into the water or by a stick moved rapidly through the water. To take an extreme case, the waves coming to the beach from a slow-moving tugboat move with the same speed as the waves produced by a 100-mph speed boat. The same is true of sound waves.

Consider, then, a source moving toward an observer with a speed u, as shown in **Figure 14–16**. If the frequency of the source is f, it emits one compression every T seconds, where $T = 1/f$. Therefore, during one cycle of the wave a compression travels a distance vT while the source moves a distance uT. As a result, the next compression is emitted a distance $vT - uT$ behind the previous compression, as illustrated in **Figure 14–17**. This means that the wavelength in the forward direction is

$$\lambda' = vT - uT = (v - u)T$$

Now, the speed of the wave is still v, as mentioned, hence

$$v = \lambda' f'$$

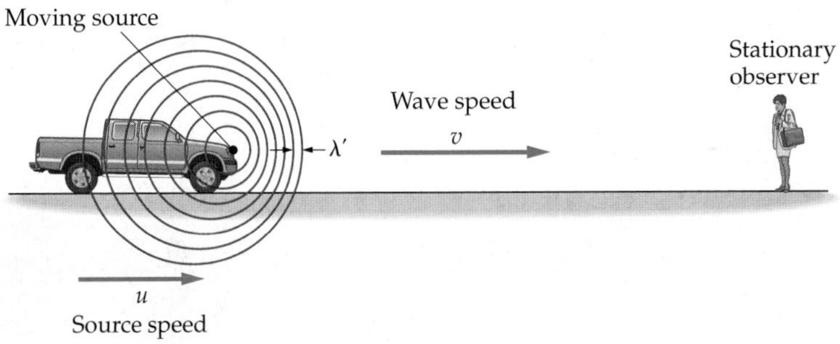

▶ **FIGURE 14–16 The Doppler effect: a moving source**

Sound waves from a moving source are bunched up in the forward direction, causing a shorter wavelength and a higher frequency.

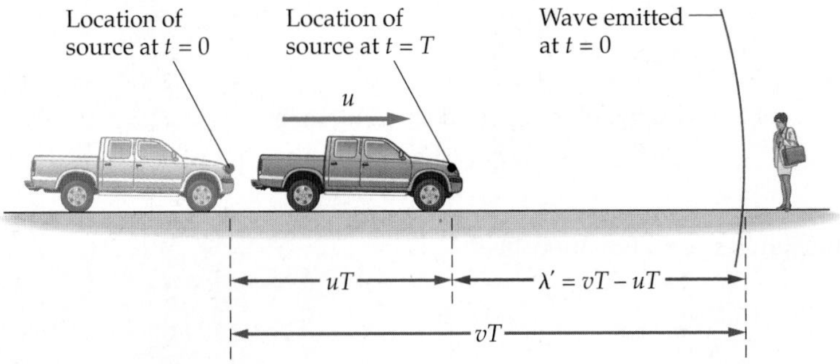

▶ **FIGURE 14–17 The Doppler-shifted wavelength**

During one period, T, the wave emitted at $t = 0$ moves through a distance vT. In the same time, the source moves toward the observer through the distance uT. At the time $t = T$ the next wave is emitted from the source; hence, the distance between the waves (the wavelength) is $\lambda' = vT - uT$.

Solving for the new frequency, f', we find

$$f' = \frac{v}{\lambda'} = \frac{v}{(v - u)T}$$

Finally, recalling that $T = 1/f$, we have

$$f' = \frac{v}{(v - u)(1/f)} = \frac{v}{v - u}f = \left(\frac{1}{1 - u/v}\right)f$$

Note that f' is greater than f, as expected.

In the reverse direction, the wavelength is increased by the amount uT. Thus,

$$\lambda' = vT + uT = (v + u)T$$

It follows that the Doppler-shifted frequency is

$$f' = \frac{v}{(v + u)T} = \left(\frac{1}{1 + u/v}\right)f$$

This is less than the source frequency, f.

Finally, we can combine these results to yield

Doppler Effect for Moving Source

$$f' = \left(\frac{1}{1 \mp u/v}\right)f \qquad \text{14–10}$$

SI unit: $1/\text{s} = \text{s}^{-1}$

As before, u and v are positive quantities. The *minus* sign is used when the source moves *toward* the observer, and the *plus* sign when the source moves *away from* the observer.

PROBLEM-SOLVING NOTE

Using Correct Signs

When a source approaches an observer, the frequency heard by the observer is greater than the frequency of the source; that is, $f' > f$. This means that we must use the minus sign in Equation 14–10, $f' = f/(1 - u/v)$, since this makes the denominator less than one. Similarly, use the plus sign when the source moves away from the observer.

EXAMPLE 14–6 Whistle Stop

A train sounds its whistle as it approaches a tunnel in a cliff. The whistle produces a tone of 650.0 Hz, and the train travels with a speed of 21.2 m/s. **(a)** Find the frequency heard by an observer standing near the tunnel entrance. **(b)** The sound from the whistle reflects from the cliff back to the engineer in the train. What frequency does the engineer hear?

Picture the Problem

The train moves with a speed $u = 21.2$ m/s and emits sound with a frequency $f = 650.0$ Hz. The observer near the tunnel hears the Doppler-shifted frequency f', and the engineer on the train hears the reflected sound at an even higher frequency f''.

INTERACTIVE FIGURE

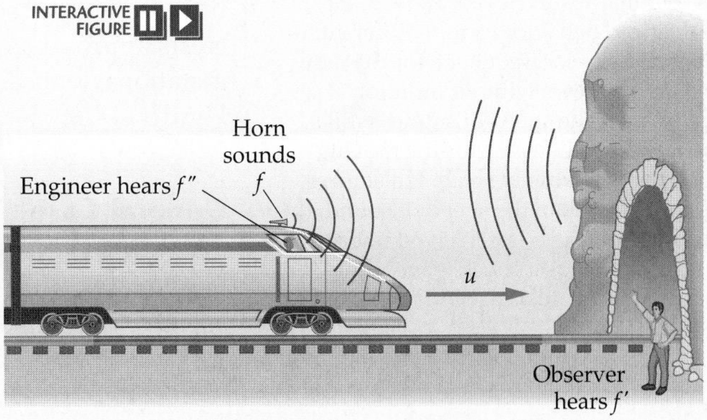

Strategy

Two Doppler shifts are involved in this problem. The first is due to the motion of the train toward the cliff. This shift causes the observer at the cliff to hear sound with a higher frequency f', given by Equation 14–10 with the minus sign. The reflected sound has the same frequency, f'.

The second shift is due to the engineer moving toward the reflected sound. Thus, the engineer hears a frequency f'' that is greater than f'. We find f'' using Equation 14–9 with the plus sign.

Solution

Part (a)

1. Use Equation 14–10, with the minus sign, to Doppler shift from f to f'.

$$f' = \left(\frac{1}{1 - u/v}\right)f = \left(\frac{1}{1 - \frac{21.2\text{ m/s}}{343\text{ m/s}}}\right)(650.0\text{ Hz})$$

$$= \left(\frac{1}{1 - 0.0618}\right)(650.0\text{ Hz}) = 693\text{ Hz}$$

continued on next page

continued from previous page

Part (b)

2. Now use Equation 14–9, with the plus sign, to Doppler shift from f' to f''.

$$f'' = (1 + u/v)f' = \left(1 + \frac{21.2\text{ m/s}}{343\text{ m/s}}\right)(693\text{ Hz})$$

$$= (1 + 0.0618)(693\text{ Hz}) = 736\text{ Hz}$$

Insight

Note that the reflected sound has the same frequency f' heard by the stationary observer, since all the cliff does is reverse the direction of motion of the sound heard at the cliff. Therefore, the cliff acts as a stationary source of sound at the frequency f'. The engineer's motion toward this stationary source results in the Doppler shift from f' to f''.

Practice Problem

If the stationary observer hears a frequency of 700.0 Hz, what is **(a)** the speed of the train and **(b)** the frequency heard by the engineer? [**Answer:** (a) $u = 24.5$ m/s, (b) $f'' = 750$ Hz]

Some related homework problems: Problem 35, Problem 85

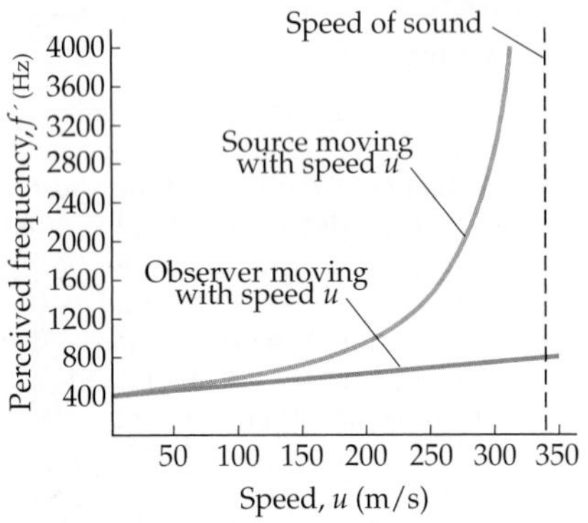

▲ **FIGURE 14–18 Doppler-shifted frequency versus speed for a 400-Hz sound source**

The upper curve corresponds to a moving source, the lower curve to a moving observer. Notice that while the two cases give similar results for low speed, the high-speed behavior is quite different. In fact, the Doppler frequency for the moving source grows without limit for speeds near the speed of sound, while the Doppler frequency for the moving observer is relatively small. If a source moves faster than the speed of sound, the sound it produces is perceived not as a pure tone, but as a shock wave, commonly referred to as a sonic boom.

Now that we have obtained the Doppler effect for both moving observers and moving sources, it is interesting to compare the results. **Figure 14–18** shows the Doppler-shifted frequency versus speed for a 400-Hz source of sound. The upper curve corresponds to a moving source, the lower curve to a moving observer. Notice that while the two cases give similar results for low speed, the high-speed behavior is quite different. In fact, the Doppler frequency for the moving source grows without limit for speeds near the speed of sound, while the Doppler frequency for the moving observer is relatively small.

These results can be understood both in terms of mathematics—by simply comparing Equations 14–9 and 14–10—and physically. In physical terms, recall that a moving observer encounters wave crests separated by the wavelength, as indicated in Figure 14–15. Doubling the speed of the observer simply reduces the time required to move from one crest to the next by a factor of 2, which doubles the frequency. Thus, in general, the frequency is proportional to the speed, as we see in the lower curve in Figure 14–18. In contrast, when the source moves, as in Figure 14–16, the wave crests become "bunched up" in the forward direction, since the source is almost keeping up with the propagating waves. As the speed of the source approaches the speed of sound, the separation between crests approaches zero. Consequently, the frequency with which the crests pass a stationary observer approaches infinity, as indicated by the upper curve in Figure 14–18.

General Case

The results derived earlier in this section can be combined to give the Doppler effect for situations in which both observer and source move. Letting u_s be the speed of the source, and u_o be the speed of the observer, we have

Doppler Effect for Moving Source and Observer

$$f' = \left(\frac{1 \pm u_o/v}{1 \mp u_s/v}\right)f \qquad \text{14–11}$$

SI unit: $1/\text{s} = \text{s}^{-1}$

As in the previous cases, u_s, u_o, and v are positive quantities. In the numerator, the plus sign corresponds to the case in which the observer moves in the direction of the source, whereas the minus sign indicates motion in the opposite direction. In the denominator, the minus sign corresponds to the case in which the source moves in the direction of the observer, whereas the plus sign indicates motion in the opposite direction.

EXERCISE 14–4

A car moving at 18 m/s sounds its 550-Hz horn. A bicyclist, traveling with a speed of 7.2 m/s, moves toward the approaching car. What frequency is heard by the bicyclist?

Solution

We use Equation 14–11 with $u_s = 18$ m/s and $u_o = 7.2$ m/s. Since the source and observer are approaching, we use the plus sign in the numerator and the minus sign in the denominator:

$$f' = \left(\frac{1 + u_o/v}{1 - u_s/v}\right)f = \left(\frac{1 + 7.2/343}{1 - 18/343}\right)(550\text{ Hz}) = 590\text{ Hz}$$

REAL-WORLD PHYSICS
Radar guns

The Doppler effect is used in an amazing variety of technological applications. Perhaps one of the most familiar of these is the "radar gun" which is used to measure the speed of a pitched baseball or a car breaking the speed limit. Though the radar gun uses radio waves rather than sound waves, the basic physical principle is the same—by measuring the Doppler-shifted frequency of waves reflected from an object, it is possible to determine its speed. Doppler radar, used in weather forecasting, applies this same technology to tracking the motion of precipitation caused by storm clouds.

REAL-WORLD PHYSICS: BIO
Measuring the speed of blood flow

In medicine, the Doppler shift is used to measure the speed of blood flow in an artery or in the heart itself. In this application, a beam of ultrasound is directed toward an artery in a patient. Some of the sound is reflected back by red blood cells moving through the artery. The reflected sound is detected, and its frequency is used to determine the speed of blood flow. If this information is color coded, with different colors indicating different speeds and directions of flow, an impressive image of blood flow can be constructed.

REAL-WORLD PHYSICS
Red shift of distant galaxies

Finally, the Doppler effect applies to the light of distant galaxies as well. For example, if a galaxy moves away from us—as most do—the light we observe from that galaxy has a lower frequency than if the galaxy were at rest relative to our galaxy, the Milky Way. Since red light has the lowest frequency of visible light, we refer to this reduction in frequency as a "red shift." Thus, by measuring the red shift of a galaxy, we can determine its speed relative to us. Such measurements show that the more distant a galaxy, the greater its speed relative to us—a result

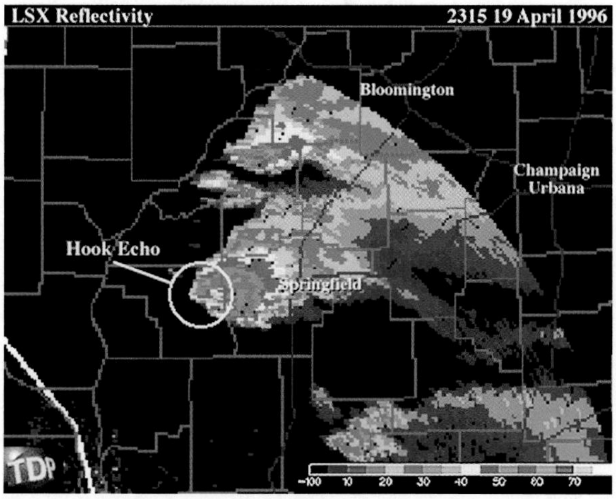

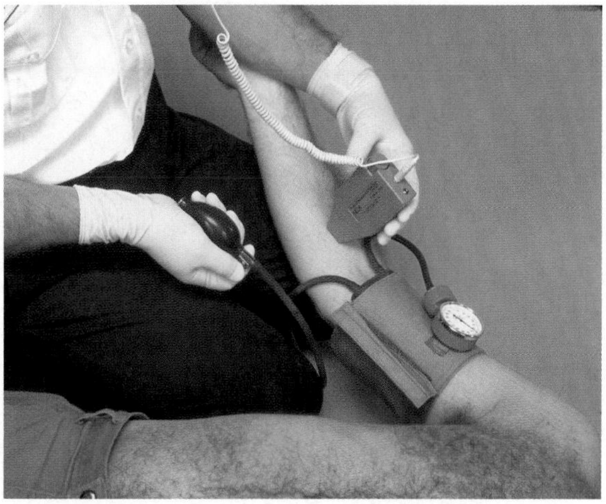

▲ Many familiar and not-so-familiar devices utilize the Doppler effect. Doppler radar, now widely used at airports and for weather forecasting, makes it possible to determine the speed and direction of winds in a distant storm by measuring the Doppler shift they produce—winds shift the radar frequency upward if they blow toward the source and downward if they blow away from the source. The image at left is the Doppler radar scan of a severe thunderstorm that struck the town of Ogden, Illinois, on April 19, 1996. Reddish colors indicate winds blowing toward the radar station, bluish colors indicate winds blowing away. The hook-shaped echo marked on the image is characteristic of tornadoes in the making. In the photo at right, a medical technician uses a Doppler blood flow meter instead of a stethoscope while measuring the blood pressure of a patient.

known as Hubble's law. This is just what one would expect from an explosion, or "Big Bang," in which rapidly moving pieces travel farther in a given amount of time. Thus the Doppler effect, and red-shift measurements, provide strong evidence for the Big Bang and an expanding universe.

14–7 Superposition and Interference

So far we have considered only a single wave at a time. In this section we turn our attention to the way waves combine when more than one is present. As we shall see, the behavior of waves is quite simple in this respect.

Superposition

The combination of two or more waves to form a resultant wave is referred to as **superposition.** When waves are of small amplitude, they superpose in the simplest of ways—they just add. For example, consider two waves on a string, as in **Figure 14–19**, described by the wave functions y_1 and y_2. If these two waves are present on the same string at the same time, the result is a wave given by

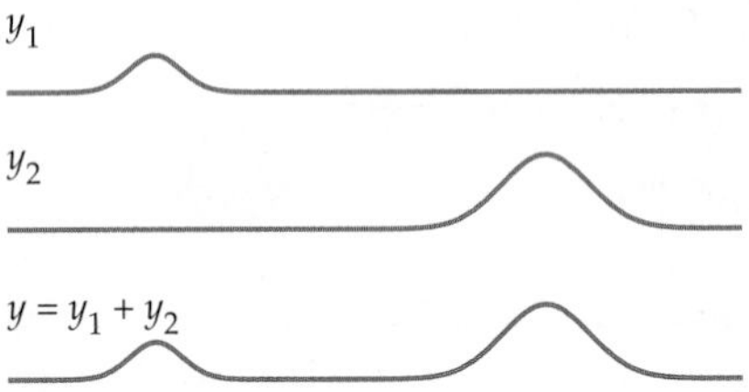

▲ **FIGURE 14–19 Wave superposition**
Waves of small amplitude superpose (that is, combine) by simple addition.

$$y = y_1 + y_2$$

To see how superposition works as a function of time, consider a string with two wave pulses on it, one traveling in each direction as shown in **Figure 14–20 (a)**. When the pulses arrive in the same region, they add, as stated. This is illustrated in Figure 14–20 (a). The question is, "What do the pulses look like after they have passed through one another? Does their interaction change them in any way?"

The answer is that the waves are completely unaffected by their interaction. This is also shown in Figure 14–20 (a). After the wave pulses pass through one another they continue on as if nothing had happened. It is like listening to an orchestra, where many different instruments are playing simultaneously, and their sounds are combining throughout the concert hall. Even so, you can still hear individual instruments, each making its own sound as if the others were not present.

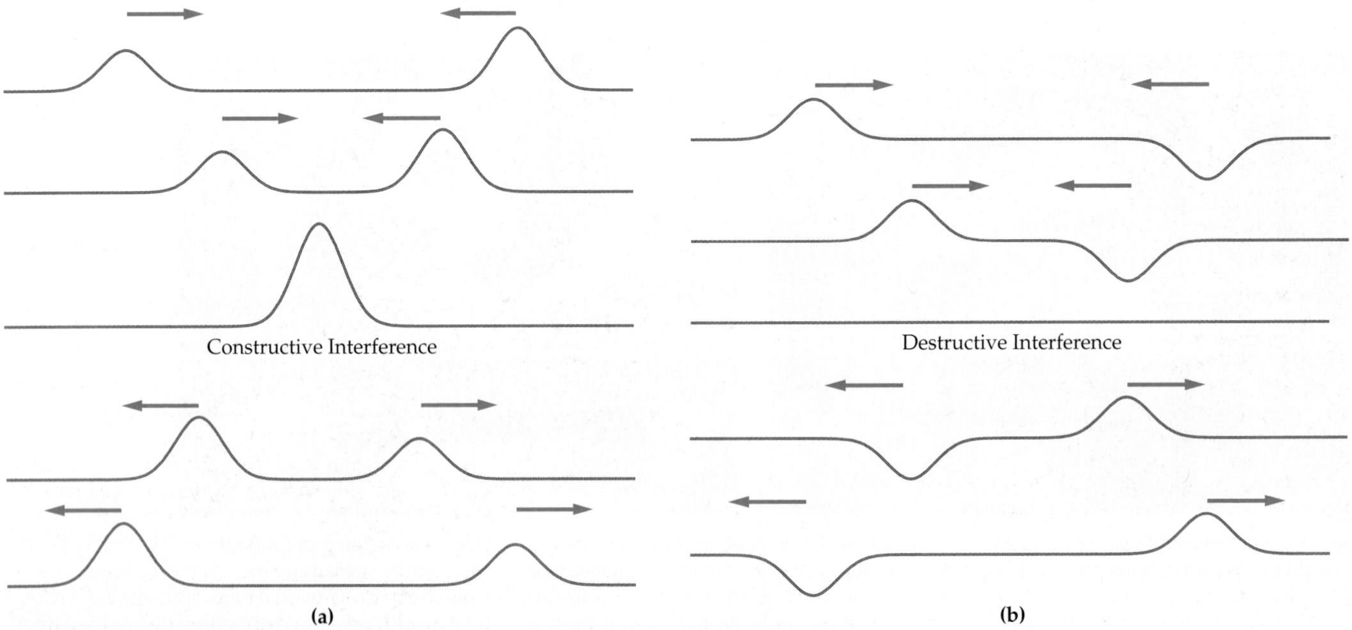

▲ **FIGURE 14–20 Interference**
Wave pulses superpose as they pass through one another. Afterward, the pulses continue on unchanged. In **(a)**, the pulses combine to give a larger amplitude. This is an example of *constructive interference*. When a positive pulse superposes with a negative pulse **(b)**, the result is *destructive interference*. In this case, with symmetrical pulses, there is one moment of complete cancellation.

CONCEPTUAL CHECKPOINT 14–3 Amplitude of a Resultant Wave

Since waves add, does the resultant wave y always have a greater amplitude than the individual waves y_1 and y_2?

Reasoning and Discussion

The wave y is the sum of y_1 and y_2, but remember that y_1 and y_2 are sometimes positive and sometimes negative. Thus, if y_1 is positive at a given time, for example, and y_2 is negative, the sum $y_1 + y_2$ can be zero or even negative. For example, if y_1 and y_2 both have the amplitude A, the amplitude of y can take any value from 0 to $2A$.

Answer:

No. The amplitude of y can be greater than, less than, or equal to the amplitudes of y_1 and y_2.

Interference

As simple as the principle of superposition is, it still leads to interesting consequences. For example, consider the wave pulses on a string shown in Figure 14–20 (a). When they combine, the resulting pulse has an amplitude equal to the sum of the amplitudes of the individual pulses. This is referred to as **constructive interference.**

On the other hand, two pulses like those in **Figure 14–20 (b)** may combine. When this happens, the positive displacement of one wave adds to the negative displacement of the other to create a net displacement of zero. That is, the pulses momentarily cancel one another. This is **destructive interference.**

It is important to note that the waves don't simply disappear when they experience destructive interference. For example, in Figure 14–20 (b) the wave pulses continue on unchanged after they interact. This makes sense from an energy point of view—after all, each wave carries energy, hence the waves, along with their energy, cannot simply vanish. In fact, when the string is flat in Figure 14–20 (b) it has its greatest transverse speed—just like a swing has its highest speed when it is in its equilibrium position. Therefore, the energy of the wave is still present at this instant of time—it is just in the form of the kinetic energy of the string.

It should also be noted that interference is not limited to waves on a string; all waves exhibit interference effects. In fact, you could say that interference is one of the key characteristics that define waves. In general, when waves combine, they form **interference patterns** that include regions of both constructive and destructive interference. An example is shown in **Figure 14–21**, where two circular waves are interfering. Note the regions of constructive interference separated by regions of destructive interference.

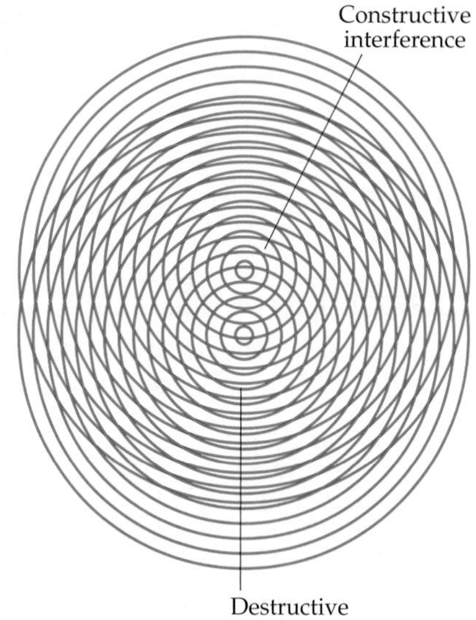

▲ **FIGURE 14–21 Interference of circular waves**

Interference pattern formed by the superposition of two sets of circular waves. The light radial "rays" are regions where crest meets crest and trough meets trough (constructive interference). The dark areas in between the light rays are regions where the crest of one wave overlaps the trough of another wave (destructive interference).

To understand the formation of such patterns, consider a system of two identical sources, as in **Figure 14–22**. Each source sends out waves consisting of alternating crests and troughs. We set up the system so that when one source emits a crest, the other emits a crest as well. Sources that are synchronized like this are said to be **in phase.**

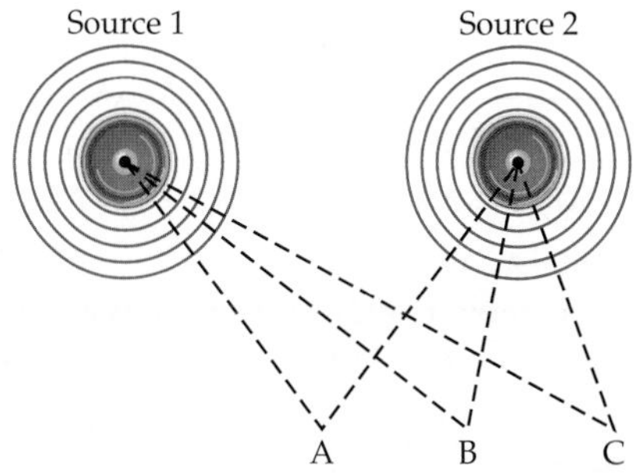

▲ **FIGURE 14–22 Interference with two sources**

Suppose the two sources emit waves in phase. At point A the distance to each source is the same, and, hence, crest meets crest. The result is constructive interference. At B the distance from source 1 is greater than that from source 2 by half a wavelength. The result is crest meeting trough and destructive interference. Finally, at C the distance from source 1 is one wavelength greater than the distance from source 2. Hence, we find constructive interference at C. If the sources had been opposite in phase, then A and C would be points of destructive interference, and B would be a point of constructive interference.

Now, at a point like A, the distance to each source is the same. Thus, if the wave from one source produces a crest at point A, so too does the wave from the other source. As a result, with crest combining with crest, the interference at A is constructive.

Next consider point B. At this location the wave from source 1 must travel a greater distance than the wave from source 2. If the extra distance is half a wavelength, it follows that when the wave from source 2 produces a crest at B the wave from source 1 produces a trough. As a result, the waves combine to give destructive interference at B. At point C, on the other hand, the distance from source 1 is one wavelength greater than the distance from source 2. Hence the waves are in phase again at C, with crest meeting crest for constructive interference.

In general, then, we can say that constructive and destructive interference occur under the following conditions for two sources that are in phase:

> Constructive interference occurs when the path length from the two sources differs by $0, \lambda, 2\lambda, 3\lambda, \ldots$.
>
> Destructive interference occurs when the path length from the two sources differs by $\lambda/2, 3\lambda/2, 5\lambda/2, \ldots$.

Other path length differences result in intermediate degrees of interference, between the extremes of destructive and constructive interference.

A specific example of interference patterns is provided by sound, using speakers that emit sound in phase with the same frequency. This situation is analogous to the two sources in Figure 14–22. As a result, constructive and destructive interference is to be expected, depending on the path length from each speaker. This is illustrated in the next Example.

EXAMPLE 14–7 Sound Off

Two speakers separated by a distance of 4.30 m emit sound of frequency 221 Hz. The speakers are in phase with one another. A person listens from a location 2.80 m directly in front of one of the speakers. Does the person hear constructive or destructive interference?

Picture the Problem

Our sketch shows the two speakers emitting sound in phase at the frequency $f = 221$ Hz. The speakers are separated by the distance $D = 4.30$ m, and the observer is a distance $d_1 = 2.80$ m directly in front of one of the speakers.

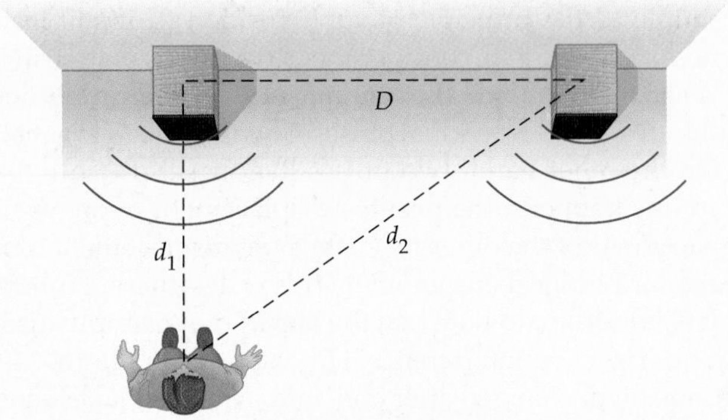

Strategy

The type of interference depends on whether the difference in path length, $d_2 - d_1$, is one or more wavelengths or an odd multiple of half a wavelength. Thus, we begin by calculating the wavelength, λ. Next, we find d_2, and compare the difference in path length to λ.

Solution

1. Calculate the wavelength of this sound, using $v = \lambda f$. As usual, let $v = 343$ m/s be the speed of sound:

$$\lambda = \frac{v}{f} = \frac{343\text{ m/s}}{221\text{ Hz}} = 1.55\text{ m}$$

2. Find the path length d_2:

$$d_2 = \sqrt{D^2 + d_1{}^2} = \sqrt{(4.30\text{ m})^2 + (2.80\text{ m})^2} = 5.13\text{ m}$$

3. Determine the difference in path length, $d_2 - d_1$:

$$d_2 - d_1 = 5.13\text{ m} - 2.80\text{ m} = 2.33\text{ m}$$

4. Divide λ into $d_2 - d_1$ to find the number of wavelengths that fit into the path difference:

$$\frac{d_2 - d_1}{\lambda} = \frac{2.33\text{ m}}{1.55\text{ m}} = 1.50$$

Insight

Since the path difference is $3\lambda/2$, we expect destructive interference. In the ideal case, the person would hear no sound. As a practical matter, some sound will be reflected from objects in the vicinity, resulting in a finite sound intensity.

Practice Problem

We know that 221 Hz gives destructive interference. What is the lowest frequency that gives constructive interference for the case described in this Example? [**Answer:** Set $\lambda = d_2 - d_1 = 2.33$ m. This gives $f = 147$ Hz.]

Some related homework problems: Problem 51, Problem 53

REAL-WORLD PHYSICS
Connecting speakers in phase

It is possible to connect a speaker with its wires reversed, which can result in a set of speakers that have **opposite phase.** In this case, as one speaker emits a compression the other sends out a rarefaction. When you set up a stereo system, it is important to be sure the wires are connected in a consistent fashion so that your speakers will be in phase.

If the two speakers in Figure 14–22 have opposite phase, for example, the conditions for constructive and destructive interference are changed, as are the interference patterns. For example, at point A, where the distances from the two speakers are the same, the wave from one speaker is a compression when the wave from the other speaker is a rarefaction. Thus, point A is now a point of destructive interference rather than constructive interference. In general, then, the conditions for constructive and destructive interference are simply reversed—a path difference of $0, \lambda, 2\lambda, \ldots$ results in destructive interference, a path difference of $\lambda/2, 3\lambda/2, \ldots$ results in constructive interference.

ACTIVE EXAMPLE 14–2 Opposite Phase Interference

The speakers shown to the right have opposite phase. They are separated by a distance of 5.20 m and emit sound with a frequency of 104 Hz. A person stands 3.00 m in front of the speakers and 1.30 m to one side of the center line between them. What type of interference occurs at the person's location?

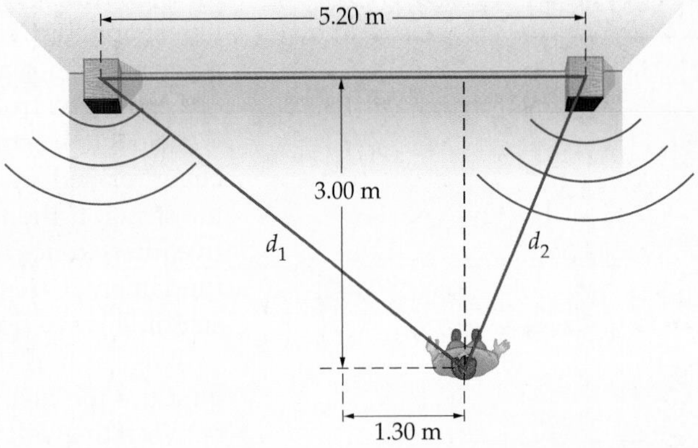

Solution *(Test your understanding by performing the calculations indicated in each step.)*

1. Calculate the wavelength: $\lambda = 3.30$ m
2. Find the path length d_1: $d_1 = 4.92$ m
3. Find the path length d_2: $d_2 = 3.27$ m
4. Calculate the path length difference, $d_1 - d_2$: $d_1 - d_2 = 1.65$ m
5. Divide the path length difference by the wavelength: $(d_1 - d_2)/\lambda = 0.500$

Insight

The path difference is half a wavelength, and the speakers have opposite phase. As a result, the person experiences constructive interference.

Your Turn

Find the next higher frequency for which constructive interference occurs at the person's location.

(Answers to **Your Turn** *problems are given in the back of the book.)*

Destructive interference can be used to reduce the intensity of noise in a variety of situations, such as a factory, a busy office, or even the cabin of an airplane. The process, referred to as Active Noise Reduction (ANR), begins with a microphone that picks up the noise to be reduced. The signal from the microphone is then reversed in phase and sent to a speaker. As a result, the speaker emits sound that is opposite in phase to the incoming noise—in effect, the speaker produces "anti-noise." In this way, the noise is *actively* canceled by destructive interference, rather than simply reduced by absorption. The effect when wearing a pair of ANR headphones can be as much as a 30-dB reduction in the intensity level of noise.

REAL-WORLD PHYSICS
Active noise reduction

14–8 Standing Waves

If you have ever plucked a guitar string, or blown across the mouth of a pop bottle to create a tone, you have generated **standing waves.** In general, a standing wave is one that oscillates with time, but remains fixed in its location. It is in this sense that the wave is said to be "standing."

In some respects, a standing wave can be considered as resulting from constructive interference of a wave with itself. As one might expect, then, standing waves occur only if specific conditions are satisfied. We explore these conditions in this section for two cases: (i) waves on a string and (ii) sound waves in a hollow, cylindrical structure.

Waves on a String

We begin by considering a string of length L that is tied down at both ends, as in **Figure 14–23 (a)**. If you pluck this string in the middle it vibrates as shown

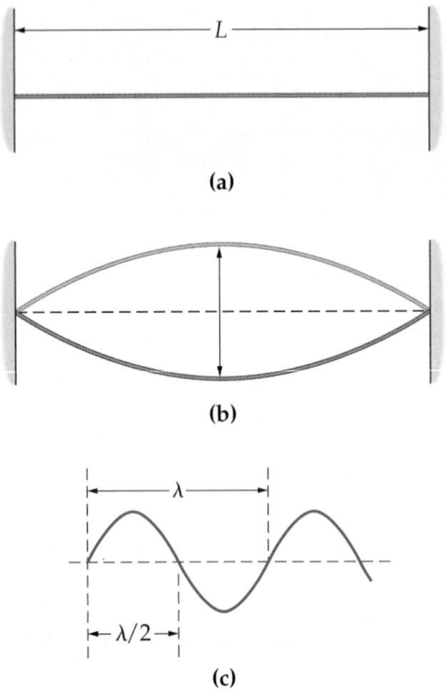

▲ **FIGURE 14–23 A standing wave**
(a) A string is tied down at both ends. **(b)** If the string is plucked in the middle, a standing wave results. This is the fundamental mode of oscillation of the string. **(c)** The fundamental consists of one-half (1/2) a wavelength between the two ends of the string. Hence, its wavelength is $2L$.

in **Figure 14–23 (b)**. This is referred to as the **fundamental mode** of vibration for this string or, also, as the **first harmonic.** Clearly, the string assumes a wave-like shape, but because of the boundary conditions—the ends tied down—the wave stays in place.

As is clear from **Figure14–23 (c)**, the fundamental mode corresponds to half a wavelength of a usual wave on a string. One can think of the fundamental as being formed by this wave reflecting back and forth between the walls holding the string. If the frequency is just right, the reflections combine to give constructive interference and the fundamental is formed; if the frequency differs from the fundamental frequency, the reflections result in destructive interference and a standing wave does not result.

We can find the frequency of the fundamental, or first harmonic, as follows: First use the fact that the wavelength of the first harmonic is twice the distance between the walls. Thus,

$$\lambda_1 = 2L$$

If the speed of waves on the string is v, it follows that the frequency of the first harmonic, f_1, is determined by $v = \lambda_1 f_1 = (2L)f_1$. Therefore,

$$f_1 = \frac{v}{\lambda_1} = \frac{v}{2L}$$

Note that the frequency of the first harmonic increases with the speed of the waves, and decreases as the string is lengthened.

The first harmonic is not the only standing wave that can exist on a string, however. In fact, there are an infinite number of standing wave modes—or **harmonics**—for any given string, with frequencies that are integer multiples of the first harmonic. To find the higher harmonics, note that the two ends of the string must remain fixed. Points on a standing wave that stay fixed are referred to as **nodes.** Halfway between any two nodes is a point on the wave that has a maximum displacement, as indicated in **Figure 14–24**. Such a point is called an **antinode.** Referring to Figure 14–24 (a), then, we see that the first harmonic consists of two nodes (N) and one antinode (A); the sequence is N-A-N.

The *second* harmonic can be constructed by including one more half wavelength in the standing wave, as in Figure 14–24 (b). This mode has the sequence N-A-N-A-N, and has one complete wavelength between the walls; that is, $\lambda_2 = L$. Therefore, the frequency of the second harmonic, f_2, is

$$f_2 = \frac{v}{\lambda_2} = \frac{v}{L} = 2f_1$$

Similarly, the *third* harmonic again includes one more half wavelength, as in Figure 14–24 (c). Now there are one-and-a-half wavelengths in the length L, and hence $(3/2)\lambda_3 = L$, or $\lambda_3 = 2L/3$. The corresponding third-harmonic frequency, f_3, is

$$f_3 = \frac{v}{\lambda_3} = \frac{v}{\frac{2}{3}L} = 3\frac{v}{2L} = 3f_1$$

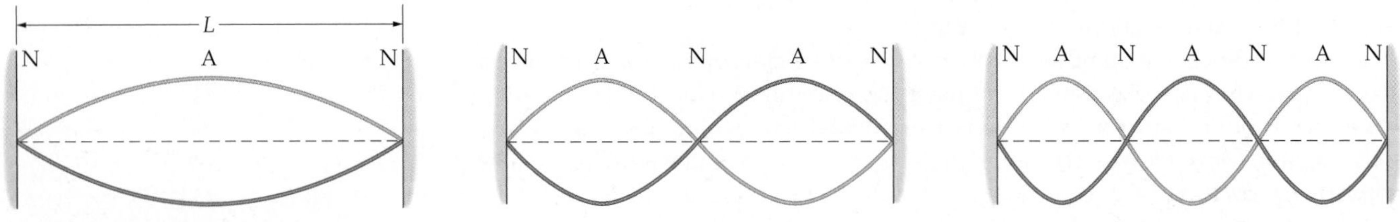

(a) First harmonic (fundamental) **(b) Second harmonic** **(c) Third harmonic**

▲ **FIGURE 14–24 Harmonics**
The first three harmonics for a string tied down at both ends. Note that an extra half wavelength is added to go from one harmonic to the next. **(a)** $\lambda/2 = L, \lambda = 2L$; **(b)** $\lambda = L$; **(c)** $3\lambda/2 = L, \lambda = 2L/3$.

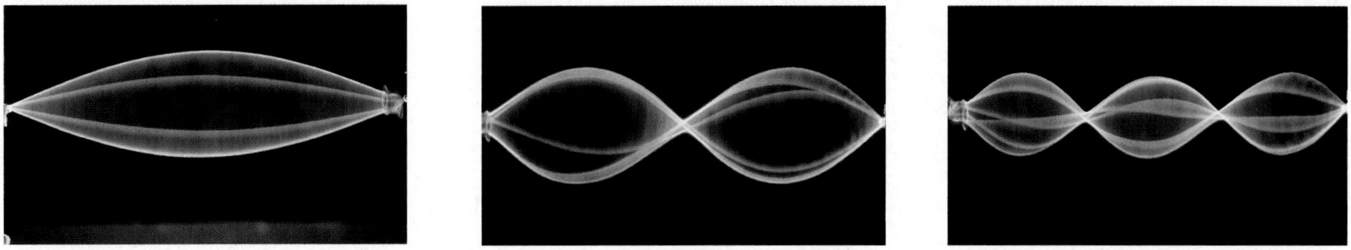

▲ The string in these multiflash photographs vibrates in one of three different standing wave patterns, each with its own characteristic frequency. The lowest frequency standing wave—the fundamental, or first harmonic—is shown in the photograph at left. In this case, there are only two nodes, one at each end of the string where it is tied down. If the length of the string is L, we see that the wavelength of the fundamental is twice this length, or $\lambda_1 = 2L$. Thus, if waves have a speed v on this string, their frequency is $f_1 = v/2L$. Higher harmonics are produced by adding one node at a time to the standing wave pattern. The second harmonic, shown in the middle photograph, has a node at either end and one in the middle. In this case the wavelength is $\lambda_2 = L$ and the frequency is $f_2 = v/L = 2f_1$. The photograph at right shows the third harmonic, where $\lambda_3 = 2L/3$ and $f_3 = v/(2L/3) = 3v/2L = 3f_1$. In general, the nth harmonic on a string tied down at both ends is $f_n = nf_1$.

Note that the frequencies of the harmonics are increasing in integer steps; that is, each harmonic has a frequency that is an integer multiple of the first-harmonic frequency. Clearly, then, the sequence of standing waves is characterized by the following:

Standing Waves on a String

First harmonic (fundamental) frequency and wavelength:

$$f_1 = \frac{v}{2L}$$

$$\lambda_1 = 2L \qquad \text{14–12}$$

Frequency and wavelength of the nth harmonic, with $n = 1, 2, 3, \ldots$:

$$f_n = nf_1 = n\frac{v}{2L}$$

$$\lambda_n = \lambda_1/n = 2L/n \qquad \text{14–13}$$

Note that the difference in frequency between any two successive harmonics is equal to the first-harmonic frequency, f_1, and that n represents the number of half wavelengths in the standing wave.

EXAMPLE 14–8 It's Fundamental

One of the harmonics on a string 1.30-m long has a frequency of 15.60 Hz. The next higher harmonic has a frequency of 23.40 Hz. Find **(a)** the fundamental frequency, and **(b)** the speed of waves on this string.

Picture the Problem

The problem statement does not tell us directly which two harmonics have the given frequencies. We do know, however, that they are *successive* harmonics of the string. For example, if one harmonic has one node between the two ends of the string, the next harmonic has two nodes. Our sketch illustrates this case, which turns out to be appropriate for this problem.

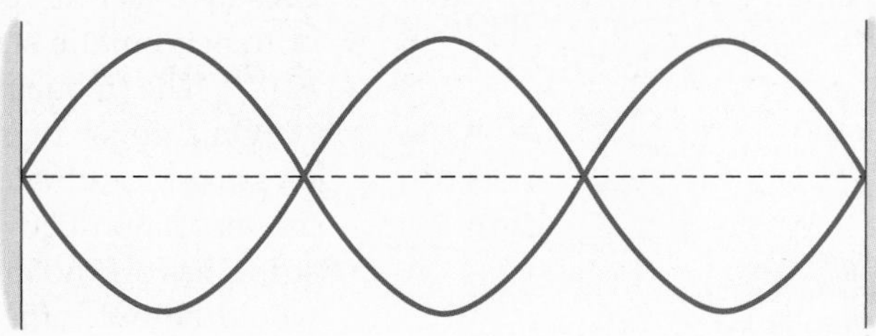

Strategy

a. We know from Equation 14–13 that the frequencies of successive harmonics increase by f_1. That is, $f_2 = f_1 + f_1 = 2f_1$, $f_3 = f_2 + f_1 = 3f_1$, $f_4 = f_3 + f_1 = 4f_1, \ldots$. Therefore, we can find the fundamental frequency, f_1, by taking the difference between the given frequencies.

b. Once the fundamental frequency is determined, we can find the speed of waves in the string from the relation $f_1 = v/2L$.

continued on next page

continued from previous page

Solution

Part (a)

1. The fundamental frequency is the difference between the two given frequencies: $f_1 = 23.40\text{ Hz} - 15.60\text{ Hz} = 7.80\text{ Hz}$

Part (b)

2. Solve $f_1 = v/2L$ for the speed, v: $f_1 = v/2L$, $v = 2Lf_1$

3. Substitute numerical values: $v = 2Lf_1 = 2(1.30\text{ m})(7.80\text{ Hz}) = 20.3\text{ m/s}$

Insight

Now that we know the fundamental frequency, we can identify the harmonics given in the problem statement. First, $15.6\text{ Hz} = 2(7.80\text{ Hz})$, so this is the second harmonic. The next mode, $23.4\text{ Hz} = 3(7.80\text{ Hz})$, is the third harmonic, as expected.

Practice Problem

Suppose the tension in this string is increased until the speed of the waves is 22.0 m/s. What are the frequencies of the first three harmonics in this case? [**Answer:** $f_1 = 8.46\text{ Hz}$, $f_2 = 16.9\text{ Hz}$, $f_3 = 25.4\text{ Hz}$]

Some related homework problems: Problem 60, Problem 61

When a guitar string is plucked or a piano string is struck, it vibrates primarily in its fundamental mode, with smaller contributions coming from the higher harmonics. It follows that notes of different pitch can be produced by using strings of different length. Recalling that the fundamental frequency for a string of length L is $f_1 = v/2L$, we see that long strings produce low frequencies and short strings produce high frequencies—all other variables remaining the same.

REAL-WORLD PHYSICS
The shape of a piano

This fact accounts for the general shape of a piano. Note that the strings shorten toward the right side of the piano, where the notes are of higher frequency. Similarly, a double bass is a larger instrument with longer strings than a violin, as one would expect by the different frequencies the instruments produce. To tune a stringed instrument, the tension in the strings is adjusted—since changing the length of the instrument is impractical. This in turn varies the speed v of waves on the string, and hence the fundamental frequency $f_1 = v/2L$ can be adjusted as desired.

REAL-WORLD PHYSICS: BIO
Human perception of pitch

The human ear responds to frequency in a rather interesting and unexpected way. In particular, frequencies that seem to increase by the same amount are in fact increasing by the same multiplicative factor. For example, if three frequencies, f_1, f_2, and f_3, sound equally spaced to our ears, you might think that f_2 is greater than f_1 by a certain amount, x, and that f_3 is greater than f_2 by the same amount. Mathematically, we would write this as $f_2 = f_1 + x$ and $f_3 = f_2 + x = f_1 + 2x$. In fact, when we measure the frequencies and compare, we find that f_2 is greater than f_1 by a multiplicative factor x, and that f_3 is greater than f_2 by the same factor; that is $f_2 = xf_1$ and $f_3 = xf_2 = x^2 f_1$.

For instance, middle C on the piano has a frequency of 261.7 Hz. If we move up one octave to the next C, the frequency is 523.3 Hz; going up one more octave, the next C is 1047 Hz. Note that with each octave the frequency doubles; that is, it goes up by a multiplicative factor of 2. Since there are 12 semitones in one octave of the chromatic scale, the frequency increase from one semitone to the next is $(2)^{1/12}$. The frequencies for a full chromatic octave are given in Table 14–3.

TABLE 14–3 Chromatic Musical Scale

Note	Frequency (Hz)
Middle C	261.7
$C^{\#}$ (C-sharp), D^{b} (D-flat)	277.2
D	293.7
$D^{\#}$, E^{b}	311.2
E	329.7
F	349.2
$F^{\#}$, G^{b}	370.0
G	392.0
$G^{\#}$, A^{b}	415.3
A	440.0
$A^{\#}$, B^{b}	466.2
B	493.9
C	523.3

On a guitar two full octaves and more can be produced on a single string by pressing the string down against frets to effectively change its length. Notice that the separation between frets is not uniform. In particular, suppose the unfretted string has a fundamental frequency of 250 Hz. Since one octave up on the scale would be twice the frequency, 500 Hz, the length of the string must be halved to produce that note. To go to the next octave, and double the frequency again to 1000 Hz, the string must be shortened by a factor of two again, to one quarter its

▲ Three factors determine the pitch of a vibrating string: mass per unit length, μ; tension, F; and length, L. In an instrument such as a guitar, the first of these factors is fixed once the strings are put on. (Note in the photos that the strings vary in thickness; other things being equal, the heavier the string, the lower the pitch.) The second factor, the tension, can be varied by means of pegs that the player uses to tune the instrument (left), adjusting the pitch of each "open" string to its correct value. The third factor, the length of the string, is the only one that the performer controls while playing. Pressing a string against one of the frets (right), changes its effective length—the length of string that is free to vibrate—and thus the note that is produced.

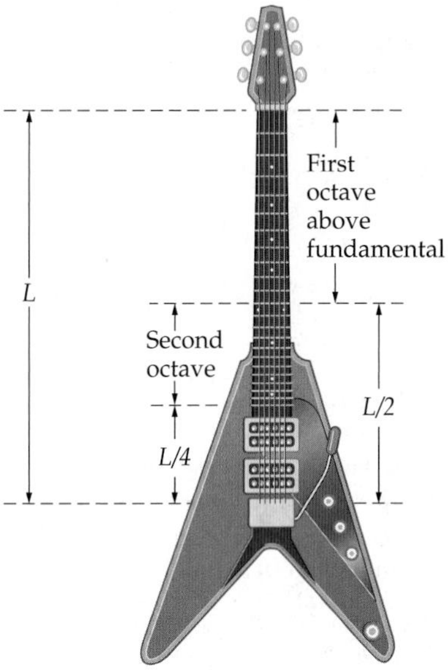

▲ **FIGURE 14–25 Frets on a guitar**

To go up one octave from the fundamental, the effective length of a guitar string must be halved. To increase one more octave, it is necessary to halve the length of the string again. Thus, the distance between frets is not uniform; they are more closely spaced near the base of the neck.

original length. This is illustrated in **Figure 14–25**. Since the distance between successive octaves is decreasing—in this case from $L/2$ to $L/4$—it follows that the spacing between frets must decrease as one goes to higher notes. As a result, the frets on a guitar are always more closely spaced as one moves toward the base of the neck.

REAL-WORLD PHYSICS
Frets on a guitar

Vibrating Columns of Air

If you blow across the open end of a pop bottle, as in **Figure 14–26**, you hear a tone of a certain frequency. If you pour some water into the bottle and repeat the experiment, the sound you hear has a higher frequency. In both cases you have excited the fundamental mode of the column of air within the bottle. When water was added to the bottle, however, the column of air was shortened, leading to a higher frequency—in the same way that a shortened string has a higher frequency.

Let's examine the situation more carefully. When you blow across the opening in the bottle, the result is a swirling movement of air that excites rarefactions and compressions, as illustrated in the figure. For this reason, the opening is an antinode (A) for sound waves. On the other hand, the bottom of the bottle is closed, preventing movement of the air; hence, it must be a node (N). Any standing wave in the bottle must have a node at the bottom and an antinode at the top.

The lowest frequency standing wave that is consistent with these conditions is shown in **Figure 14–27 (a)**. If we plot the density variation of the air for this wave, we see that one-quarter of a wavelength fits into the column of air in the bottle. Thus, if the length of the bottle is L, the first harmonic (fundamental) has a wavelength satisfying the following:

$$\tfrac{1}{4}\lambda = L$$
$$\lambda = 4L$$

The first-harmonic frequency, f_1, is given by

$$v = \lambda f_1$$

Solving for f_1 we find

$$f_1 = \frac{v}{\lambda} = \frac{v}{4L}$$

This is half the corresponding fundamental frequency for a wave on a string.

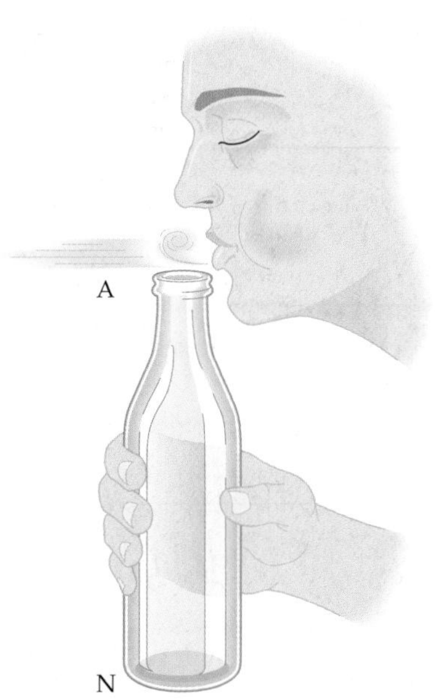

▲ **FIGURE 14–26 Exciting a standing wave**

When air is blown across the open top of a soda pop bottle, the turbulent air flow can cause an audible standing wave. The standing wave will have an antinode, A, at the top (where the air is moving) and a node, N, at the bottom (where the air cannot move).

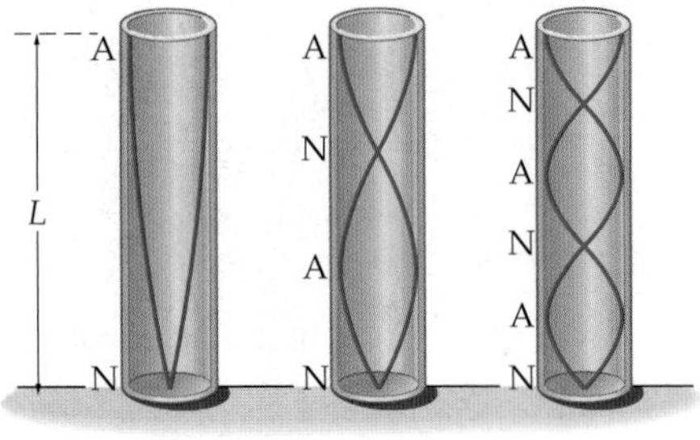

▲ **FIGURE 14–27 Standing waves in a pipe that is open at one end**

The first three harmonics for waves in a column of air of length L that is open at one end. **(a)** $\lambda/4 = L$, $\lambda = 4L$; **(b)** $3\lambda/4 = L$, $\lambda = 4L/3$; **(c)** $5\lambda/4 = L$, $\lambda = 4L/5$.

The next harmonic is produced by adding half a wavelength, just as in the case of the string. Thus, if the fundamental is represented by N-A, the second harmonic can be written as N-A-N-A. Since the distance from a node to an antinode is a quarter of a wavelength, we see that three-quarters (3/4) of a wavelength fits into the bottle for this mode. This is shown in **Figure 14–27 (b)**. Therefore, $3\lambda/4 = L$, and, hence,

$$\lambda = \tfrac{4}{3}L$$

As a result, the frequency is

$$\frac{v}{\lambda} = \frac{v}{\frac{4}{3}L} = 3\frac{v}{4L} = 3f_1$$

Notice that this is the *third* harmonic of the pipe, since its frequency is three times f_1.

Similarly, the next-higher harmonic is represented by N-A-N-A-N-A, as indicated in **Figure 14–27 (c)**. In this case, $5\lambda/4 = L$, and the frequency is

$$\frac{v}{\lambda} = \frac{v}{\frac{4}{5}L} = 5\frac{v}{4L} = 5f_1$$

This is the *fifth* harmonic of the pipe.

Clearly, the progression of harmonics for a column of air that is closed at one end and open at the other end is described by the following frequencies and wavelengths:

Standing Waves in a Column of Air Closed at One End

$$f_1 = \frac{v}{4L}$$

$$f_n = nf_1 = n\frac{v}{4L} \qquad n = 1, 3, 5, \ldots \qquad \textbf{14–14}$$

$$\lambda_n = \lambda_1/n = 4L/n$$

Note that only the odd harmonics are present in this case, as opposed to waves on a string, in which all integer harmonics occur.

REAL-WORLD PHYSICS: BIO
Human hearing and the ear canal

The human ear canal is an example of a column of air that is closed at one end (the eardrum) and open at the other end. Standing waves in the ear canal can lead to an increased sensitivity of hearing. This is illustrated in **Figure 14–28**, which

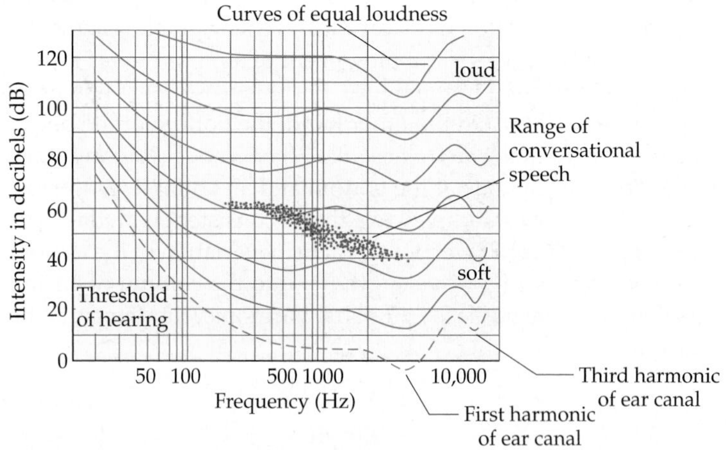

▲ **FIGURE 14–28 Human response to sound**

The human ear is more sensitive to some frequencies of sound than to others. For example, every point on a "curve of equal loudness" seems just as loud to us as any other point, even though the corresponding physical intensities may be quite different. To illustrate, note that the threshold of hearing is not equal to 0 dB for all frequencies. In fact, it is approximately 25 dB at 100 Hz, about 5 dB at 1000 Hz, and is even slightly negative near 3500 Hz. Thus, regions where the curves dip downward correspond to increased sensitivity of the ear—in fact, near 3500 Hz we can hear sounds that are a thousand times less intense than sounds at 100 Hz. The two most prominent dips occur near 3500 Hz and 11,000 Hz, corresponding to standing waves in the ear canal analogous to those shown in Figure 14–27 (a) and (b), respectively. (See Problem 59.)

shows "curves of equal loudness" as a function of frequency. Where these curves dip downward, sounds of lower intensity seem just as loud as sounds of higher intensity at other frequencies. The two prominent dips near 3500 Hz and 11,000 Hz are due to standing waves in the ear canal corresponding to Figures 14–27 (a) and (b), respectively.

EXAMPLE 14–9 Pop Music

An empty soda pop bottle is to be used as a musical instrument in a band. In order to be tuned properly, the fundamental frequency of the bottle must be 440.0 Hz. If the bottle is 26.0 cm tall, how high should it be filled with water to produce the desired frequency? Treat the bottle as a pipe that is closed at one end (the surface of the water) and open at the other end.

Picture the Problem

In our sketch, we label the height of the bottle with $H = 26.0$ cm, and the unknown height of water with h. Clearly, then, the length of the vibrating column of air is $L = H - h$.

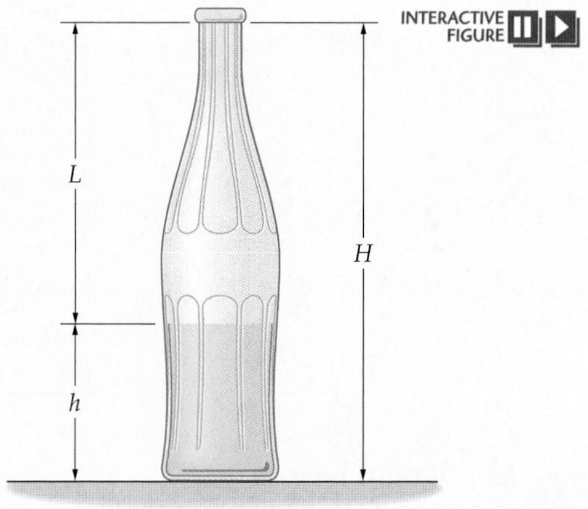

Strategy

Given the frequency of the fundamental ($f_1 = 440.0$ Hz) and the speed of sound in air ($v = 343$ m/s), we can use $f_1 = v/4L$ to solve for the length L of the air column. The height of water is then $h = H - L$.

Solution

1. Solve $f_1 = v/4L$ for the length L:

$$f_1 = v/4L$$
$$L = v/4f_1$$

2. Substitute numerical values:

$$L = \frac{v}{4f_1} = \frac{343\text{ m/s}}{4(440.0\text{ Hz})} = 0.195\text{ m}$$

3. Use $h = H - L$ to find the height of the water:

$$h = H - L = 0.260\text{ m} - 0.195\text{ m} = 0.065\text{ m} = 6.5\text{ cm}$$

Insight

If more water is added to the bottle, the air column will shorten and the fundamental frequency will become higher than 440.0 Hz.

Practice Problem

Calculate the fundamental frequency if the water level is increased to 7.00 cm. [**Answer:** $f_1 = 451$ Hz]

Some related homework problems: Problem 57, Problem 63

It is also possible to excite standing waves in columns of air that are open at both ends, as illustrated in **Figure 14–29**. In this case there is an antinode at each end of the column. Hence, the first harmonic, or fundamental, is A-N-A, as shown in Figure 14–29 (a). Note that half a wavelength fits into the pipe, thus

$$f_1 = \frac{v}{2L}$$

This is the same as the corresponding result for a wave on a string.

The next harmonic is A-N-A-N-A, which fits one complete wavelength in the pipe. This harmonic is shown in Figure 14–29 (b), and has the frequency

$$f_2 = \frac{v}{L} = 2f_1$$

This is the *second* harmonic of the pipe. The rest of the harmonics continue in exactly the same manner as for waves on a string, with all integer harmonics

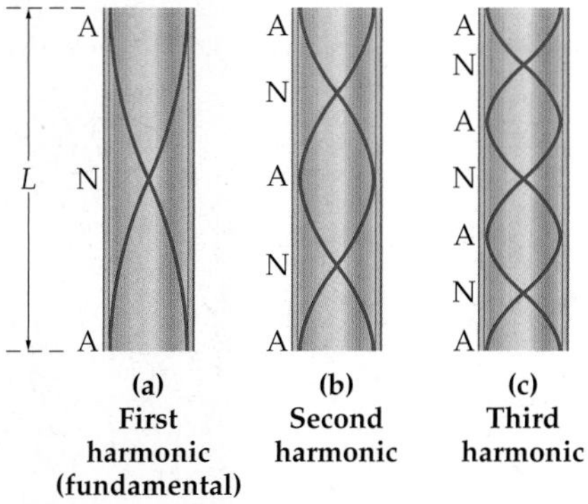

▲ **FIGURE 14–29 Standing waves in a pipe that is open at both ends**

The first three harmonics for waves in a column of air of length L that is open at both ends. **(a)** $\lambda/2 = L$, $\lambda = 2L$; **(b)** $\lambda = L$; **(c)** $3\lambda/2 = L$, $\lambda = 2L/3$.

present. Thus, the frequencies and wavelengths in a column of air open at both ends are as follows:

Standing Waves in a Column of Air Open at Both Ends

$$f_1 = \frac{v}{2L}$$

$$f_n = nf_1 = n\frac{v}{2L} \quad n = 1, 2, 3, \ldots \qquad \text{14–15}$$

$$\lambda_n = \lambda_1/n = 2L/n$$

CONCEPTUAL CHECKPOINT 14–4

If you fill your lungs with helium and speak, you sound something like Donald Duck. From this observation, we can conclude that the speed of sound in helium must be **(a)** less than, **(b)** the same as, or **(c)** greater than the speed of sound in air.

Reasoning and Discussion

When we speak with helium, our words are higher pitched. Looking at Equation 14–15, we see that for the frequency to increase, while the length of the vocal chords remains the same, the speed of sound must be higher.

Answer:

(c) The speed of sound is greater in helium than in air.

REAL-WORLD PHYSICS
Organ pipes

A pipe organ uses a variety of pipes of different length, with some being open at both ends, others open at one end only. When a key is pressed on the console of the organ, air is forced through a given pipe. By accurately adjusting the length of the pipe it can be given the desired tone. In addition, since open and closed pipes have different harmonic frequencies, they sound distinctly different to the ear, even if they have the same fundamental frequency. Thus, by judiciously choosing both the length and the type of a pipe, an organ can be given a range of different sounds, allowing it to mimic a trumpet, a trombone, a clarinet, and so on.

Standing waves have also been observed in the Sun. Like an enormous, low-frequency musical instrument, the Sun vibrates once roughly every five minutes, a result of the roiling nuclear reactions that take place within its core. One of the

▲ Blowing across the mouth of a bottle (Figure 14–26) sets the air column within the bottle vibrating, producing a tone. This principle is put to use in the pipe organ. A large organ may have hundreds of pipes of different lengths, some open at both ends and some at only one, affording the performer great control over the tonal quality of the sound produced, as well as its pitch.

goals of SOHO, the Solar and Heliospheric Observatory, is to study these solar vibrations in detail. By observing the variety of standing waves produced in the Sun, we can learn more about its internal structure and dynamics.

14–9 Beats

An interference pattern, such as that shown in Figure 14–21, is a snapshot at a given time, showing locations where constructive and destructive interference occur. It is an interference pattern in space. **Beats,** on the other hand, can be thought of as an interference pattern in time.

To be specific, imagine plucking two guitar strings that have slightly different frequencies. If you listen carefully, you notice that the sound produced by the strings is not constant in time. In fact, the intensity increases and decreases with a definite period. These fluctuations in intensity are the beats, and the frequency of successive maximum intensities is the **beat frequency.**

As an example, suppose two waves, with frequencies $f_1 = 1/T_1$ and $f_2 = 1/T_2$, interfere at a given, fixed location. At this location, each wave moves up and down with simple harmonic motion, as described by Equation 13–2. Applying this result to the vertical position, y, of each wave yields the following:

$$y_1 = A\cos\left(\frac{2\pi}{T_1}t\right) = A\cos(2\pi f_1 t)$$
$$y_2 = A\cos\left(\frac{2\pi}{T_2}t\right) = A\cos(2\pi f_2 t) \qquad \text{14–16}$$

These equations are plotted in **Figure 14–30 (a)**, with $A = 1$ m, and their superposition, $y_{total} = y_1 + y_2$, is shown in **Figure 14–30 (b)**.

Note that at the time $t = 0$, both y_1 and y_2 are equal to A; thus their superposition gives $2A$. Since the waves have different frequencies, however, they do not stay in phase. At a later time, t_1, we find that $y_1 = A$ and $y_2 = -A$; their superposition gives zero at this time. At a still later time, $t_2 = 2t_1$, the waves are again in phase and add to give $2A$. Thus, a person listening to these two waves hears a sound whose amplitude and loudness varies with time; that is, the person hears beats.

Superposing these waves mathematically, we find

$$\begin{aligned} y_{total} &= y_1 + y_2 \\ &= A\cos(2\pi f_1 t) + A\cos(2\pi f_2 t) \\ &= 2A\cos\left(2\pi\frac{f_1 - f_2}{2}t\right)\cos\left(2\pi\frac{f_1 + f_2}{2}t\right) \end{aligned} \qquad \text{14–17}$$

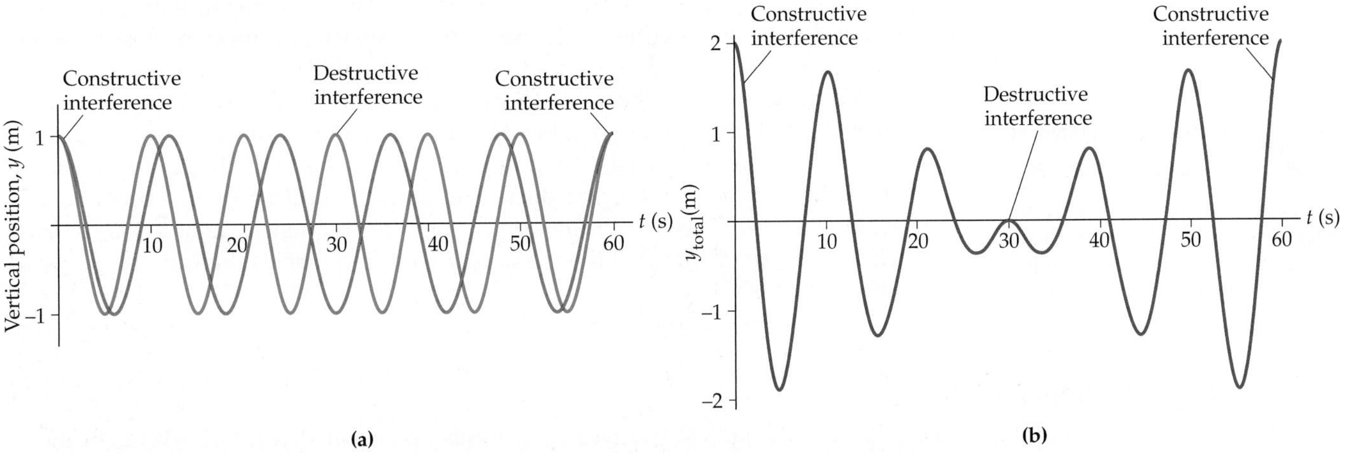

▲ **FIGURE 14–30 Interference of two waves with slightly different frequencies**

(a) A plot of the two waves, y_1 (blue) and y_2 (red), given in Equation 14–16. **(b)** The resultant wave y_{total} for the two waves shown in part (a). Note the alternately constructive and destructive interference leading to beats.

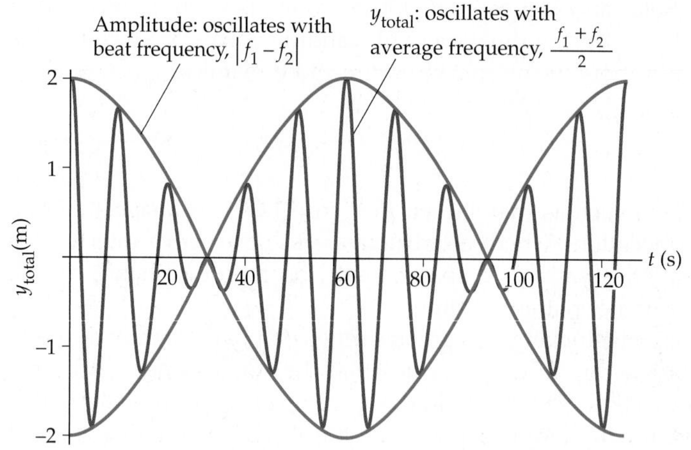

▶ FIGURE 14–31 Beats
Beats can be understood as oscillations at the average frequency, modulated by a slowly varying amplitude.

The final step in the expression follows from the trigonometric identities given in Appendix A. The first part of y_{total} is

$$2A\cos\left(2\pi\frac{f_1 - f_2}{2}t\right)$$

This gives the slowly varying amplitude of the beats, as indicated in **Figure 14–31**. Since a loud sound is heard whenever this term is equal to $-2A$ or $2A$—which happens twice during any given oscillation—the beat frequency is

Definition of Beat Frequency

$$f_{beat} = |f_1 - f_2| \qquad \textbf{14–18}$$

SI unit: $1/s = s^{-1}$

Finally, the rapid oscillations within each beat are due to the second part of y_{total}:

$$\cos\left(2\pi\frac{f_1 + f_2}{2}t\right)$$

Note that these oscillations have a frequency that is the average of the two input frequencies.

These results apply to any type of wave. In particular, if two sound waves produce beats, your ear will hear the average frequency with a loudness that varies with the beat frequency. For example, suppose the two guitar strings mentioned at the beginning of this section have the frequencies 438 Hz and 442 Hz. If you sound them simultaneously, you will hear the average frequency, 440 Hz, increasing and decreasing in loudness with a beat frequency of 4 Hz. This means that you will hear maximum loudness 4 times a second. If the frequencies are brought closer together, the beat frequency will be less and fewer maxima will be heard each second.

Clearly, beats can be used to tune a musical instrument to a desired frequency. To tune a guitar string to 440 Hz, for example, the string can be played simultaneously with a 440-Hz tuning fork. Listening to the beats, the tension in the string can be increased or decreased until the beat frequency becomes vanishingly small. This technique applies only to frequencies that are reasonably close to begin with, since the maximum beat frequencies the ear can detect are about 15 to 20 Hz.

PROBLEM-SOLVING NOTE

Calculating the Beat Frequency

The beat frequency of two waves is the *magnitude* of the difference in their frequencies. Thus, the beat frequency is always positive.

EXAMPLE 14–10 Getting a Tune-up

An experimental way to tune the soda pop bottle in Example 14–9 is to compare its frequency with that of a 440-Hz tuning fork. Initially, a beat frequency of 4 Hz is heard. As a small amount of water is added to that already present, the beat frequency increases steadily to 5 Hz. What were the initial and final frequencies of the bottle?

Picture the Problem

Our sketch shows the before and after situations for this problem. With the low water level the beat frequency is 4 Hz, with the higher level it is 5 Hz.

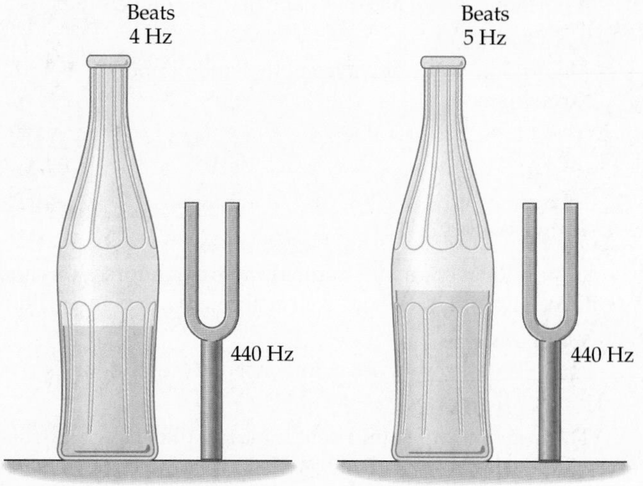

Strategy/Solution

The fact that the initial beat frequency is 4 Hz means the initial frequency of the bottle is either 336 Hz or 444 Hz.

As water is added, we know from Example 14–9 that the bottle's frequency will increase. We also know that the new beat frequency is 5 Hz, and hence the final frequency is either 335 Hz or 445 Hz. Only 445 Hz satisfies the condition that the frequency must have increased.

Therefore, the initial frequency is 444 Hz, and the final frequency is 445 Hz.

Insight

In this case, the initial frequency was too high. To tune the bottle properly, it is necessary to lower the water level.

Practice Problem

Suppose the initial beat frequency was 4 Hz and that adding a small amount of water caused the beat frequency to decrease steadily to 2 Hz. What were the initial and final frequencies in this case? [**Answer:** Initial frequency, 336 Hz; final frequency, 338 Hz]

Some related homework problems: Problem 68, Problem 70

Chapter Summary

14–1 Types of Waves

A wave is a propagating disturbance.

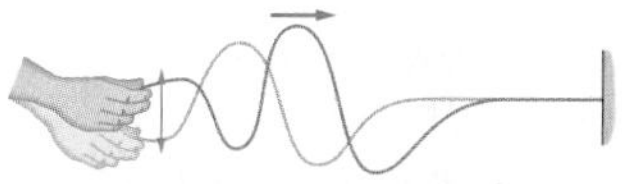

Transverse Waves and Longitudinal Waves

In a transverse wave individual particles move at right angles to the direction of wave propagation. In a longitudinal wave individual particles move in the same direction as the wave propagation.

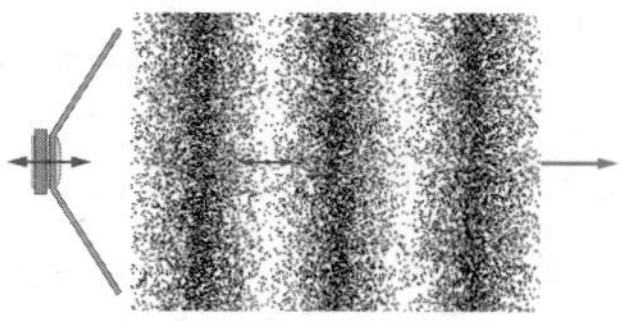

Wavelength, Frequency, and Speed

The wavelength, λ, frequency, f, and speed, v, of a wave are related by

$$v = \lambda f \quad \text{14–1}$$

14–2 Waves on a String

Transverse waves can propagate on a string held taut with a tension force, F.

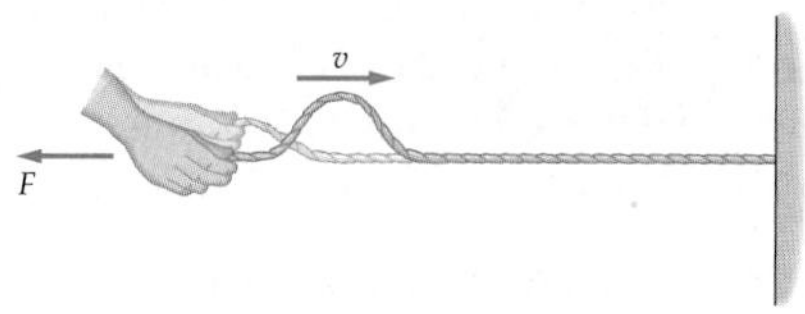

Mass per Length

The mass per length of a string is $\mu = m/L$

Speed of a Wave of a String

The speed of a wave on a string with a tension force F and a mass per length μ is

$$v = \sqrt{\frac{F}{\mu}} \quad \text{14–2}$$

Reflections

If the end of a string is fixed, the reflection of a wave is inverted. If the end of a string is free to move transversely, waves are reflected with no inversion.

*14–3 Harmonic Wave Functions

A harmonic wave has the shape of a sine or a cosine.

Wave Function
A harmonic wave of wavelength λ and period T is described by the following expression:

$$y(x, t) = A\cos\left(\frac{2\pi}{\lambda}x - \frac{2\pi}{T}t\right) \quad \text{14–4}$$

14–4 Sound Waves

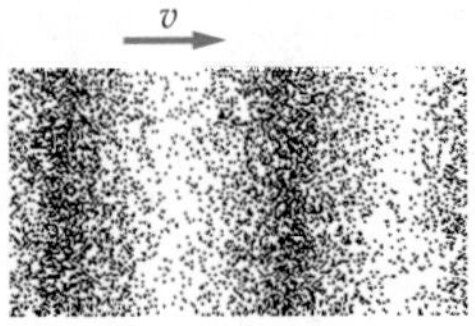

A sound wave is a longitudinal wave of compressions and rarefactions that can travel through the air, as well as through other gases, liquids, and solids.

Speed of Sound
The speed of sound in air, under typical conditions, is $v = 343$ m/s.

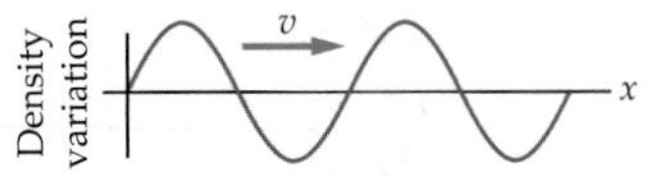

Frequency of Sound
The frequency of sound determines its pitch. High-pitched sounds have high frequencies; low-pitched sounds have low frequencies.

Human Hearing Range
Human hearing extends from 20 Hz to 20,000 Hz.

14–5 Sound Intensity

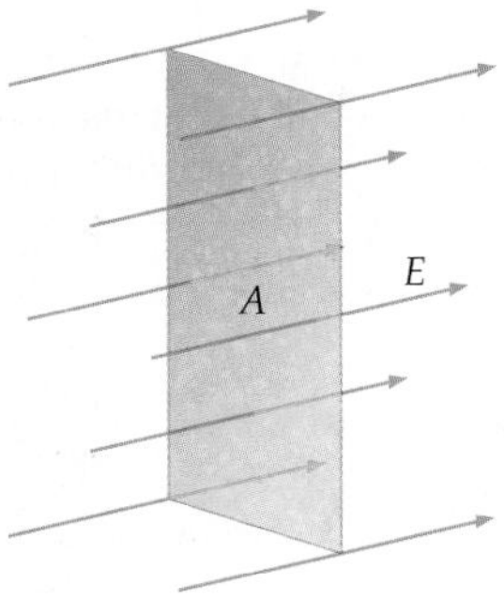

The loudness of a sound is determined by its intensity.

Intensity
Intensity, I, is a measure of the amount of energy per time that passes through a given area. Since energy per time is power, P, the intensity of a wave is

$$I = \frac{P}{A} \quad \text{14–5}$$

Point Source
If a point source emits sound with a power P, and there are no reflections, the intensity a distance r from the source is

$$I = \frac{P}{4\pi r^2} \quad \text{14–7}$$

Human Perception of Loudness
The intensity of a sound must be increased by a factor of 10 in order for it to seem twice as loud to our ears.

Intensity Level and Decibels
The intensity level, β, of a sound gives an indication of how loud it sounds to our ears. The intensity level is defined as follows:

$$\beta = 10\log(I/I_0) \quad \text{14–8}$$

The value of β is given in decibels.

14–6 The Doppler Effect

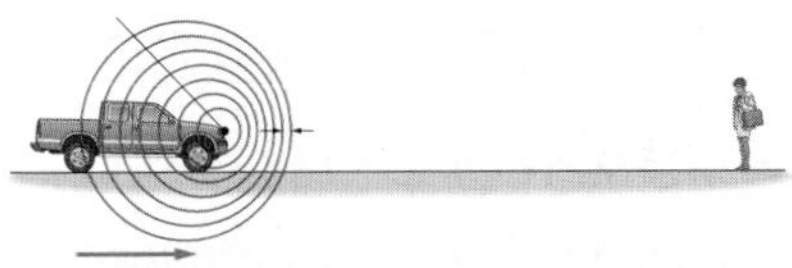

The change in frequency due to relative motion between a source and a receiver is called the Doppler effect.

Moving Observer
Suppose an observer is moving with a speed u relative to a stationary source. If the frequency of the source is f, and the speed of the waves is v, the frequency f' detected by the observer is

$$f' = (1 \pm u/v)f \quad \text{14–9}$$

The plus sign applies to the observer approaching the source, and the minus sign to the observer receding from the source.

Moving Source
If the source is moving with a speed u and the observer is at rest, the observed frequency is

$$f' = \left(\frac{1}{1 \mp u/v}\right)f \quad \text{14–10}$$

The minus sign applies to the source approaching the observer, and the plus sign to the source receding from the observer.

General Case

If the observer moves with a speed u_o and the source moves with a speed u_s, the Doppler effect gives

$$f' = \left(\frac{1 \pm u_o/v}{1 \mp u_s/v}\right)f \qquad \text{14–11}$$

The meaning of the plus and minus signs is the same as for the moving-observer and moving-source cases given above.

14–7 Superposition and Interference

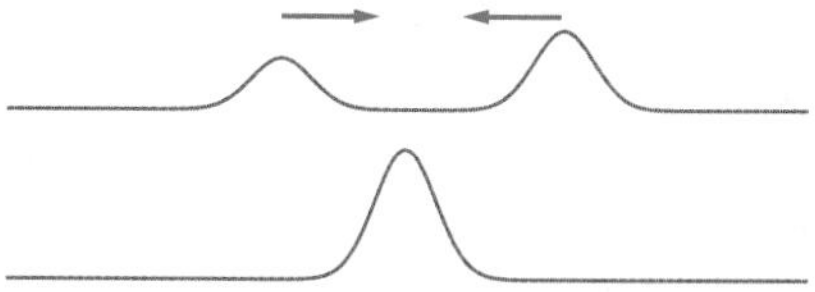

Waves can combine to give a variety of effects.

Superposition

When two or more waves occupy the same location at the same time they simply add, $y_{total} = y_1 + y_2$.

Constructive Interference

Waves that add to give a larger amplitude exhibit constructive interference.

Destructive Interference

Waves that add to give a smaller amplitude exhibit destructive interference.

Interference Patterns

Waves that overlap can create patterns of constructive and destructive interference. These are referred to as interference patterns.

In Phase/Opposite Phase

Two sources are in phase if they both emit crests at the same time.

Sources have opposite phase if one emits a crest at the same time the other emits a trough.

14–8 Standing Waves

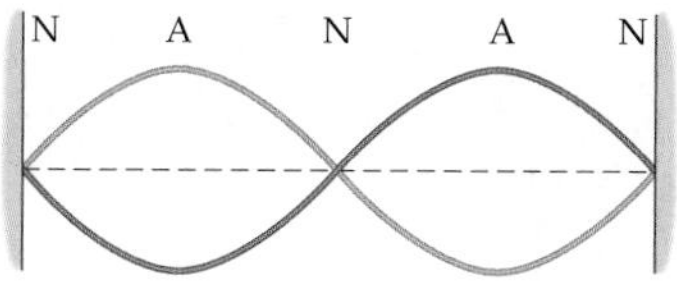

Standing waves oscillate in a fixed location.

Waves on a String

The fundamental, or first harmonic, corresponds to half a wavelength fitting into the length of the string. The fundamental for waves of speed v on a string of length L is

$$f_1 = \frac{v}{2L}$$
$$\lambda_1 = 2L \qquad \text{14–12}$$

The higher harmonics, with $n = 1, 2, 3, \ldots$ are described by

$$f_n = nf_1 = n(v/2L)$$
$$\lambda_n = \lambda_1/n = 2L/n \qquad \text{14–13}$$

Vibrating Columns of Air

The harmonics for a column of air closed at one end are

$$f_n = nf_1 = n(v/4L) \quad n = 1, 3, 5, \ldots$$
$$\lambda_n = \lambda_1/n = 4L/n \qquad \text{14–14}$$

The harmonics for a column of air open at both ends are

$$f_n = nf_1 = n(v/2L) \quad n = 1, 2, 3, \ldots$$
$$\lambda_n = \lambda_1/n = 2L/n \qquad \text{14–15}$$

In both of these expressions the speed of sound is v and the length of the column is L.

14–9 Beats

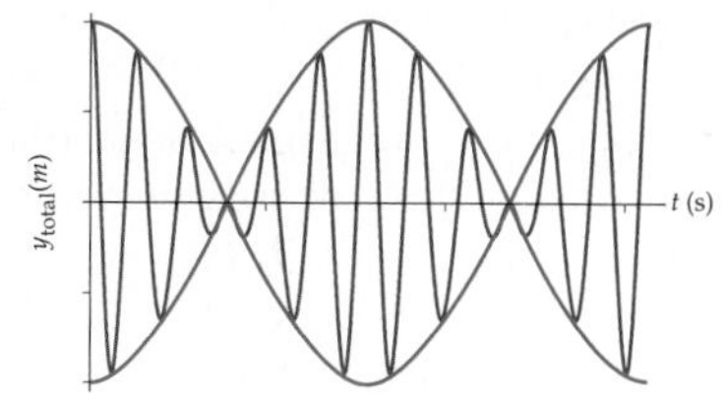

Beats occur when waves of slightly different frequencies interfere.

They can be thought of as interference patterns in time. To the ear, beats are perceived as an alternating loudness and softness to the sound.

Beat Frequency

If waves of frequencies f_1 and f_2 interfere, the beat frequency is

$$f_{beat} = |f_1 - f_2| \qquad \text{14–18}$$

Problem-Solving Summary

Type of Problem	Relevant Physical Concepts	Related Examples
Find the speed of a wave on a string, or relate the speed of a wave to the mass of a string.	The speed of a wave on a string is related to the tension in the string, F, and the string's mass per length, $\mu = m/L$, by the expression $v = \sqrt{F/\mu}$.	Example 14–1
Relate the intensity of a sound wave to its intensity level.	The intensity level of a sound wave, β, depends on the logarithm of the wave's intensity, I. The relation between β and I is $\beta = 10 \log(I/I_0)$, where $I_0 = 10^{-12}\ \mathrm{W/m^2}$.	Example 14–4
Calculate the Doppler shift for a moving source or observer.	If an observer and a source of sound with frequency f approach one another, the frequency heard by the observer is greater than f. If the source and observer recede from one another, the frequency heard by the observer is less than f.	Examples 14–5, 14–6
Calculate the beat frequency.	The beat frequency produced when sounds of frequency f_1 and f_2 are heard simultaneously is the magnitude of the difference in frequencies; $f_{\mathrm{beat}} = \lvert f_1 - f_2 \rvert$.	Example 14–10

Conceptual Questions

(Answers to odd-numbered Conceptual Questions can be found in the back of the book.)

1. A long nail has been driven halfway into the side of a barn. How should you hit the nail with a hammer to generate a longitudinal wave? How should you hit it to generate a transverse wave?
2. What type of wave is exhibited by "amber waves of grain?"
3. At a ball game, a "wave" circulating through the stands can be an exciting event. What type of wave (longitudinal or transverse) are we talking about? Is it possible to change the type of wave? Explain how people might move their bodies to accomplish this.
4. In a classic TV commercial, a group of cats feed from bowls of cat food that are lined up side by side. Initially there is one cat for each bowl. When an additional cat is added to the scene, it runs to a bowl at the end of the line and begins to eat. The cat that was there originally moves to the next bowl, displacing that cat, which moves to the next bowl, and so on down the line. What type of wave have the cats created? Explain.
5. Describe how the sound of a symphony played by an orchestra would be altered if the speed of sound depended on the frequency of sound.
6. A "radar gun" is often used to measure the speed of a major league pitch by reflecting a beam of radio waves off a moving ball. Describe how the Doppler effect can give the speed of the ball from a measurement of the frequency of the reflected beam.
7. When you drive a nail into a piece of wood, you hear a tone with each blow of the hammer. In fact, the tone increases in pitch as the nail is driven farther into the wood. Explain.
8. Explain the function of the sliding part of a trombone.
9. When you tune a violin string, what causes its frequency to change?
10. On a guitar, some strings are single wires, others are wrapped with another wire to increase the mass per length. Which type of string would you expect to be used for a low-frequency note? Explain.
11. As a string oscillates in its fundamental mode, there are times when it is completely flat. Is the energy of oscillation zero at these times? Explain.
12. On a rainy day, while driving your car, you notice that your windshield wipers are moving in synchrony with the wiper blades of the car in front of you. After several cycles, however your wipers and the wipers of the other car are moving opposite to one another. A short time later the wipers are synchronous again. What wave phenomena do the wipers illustrate? Explain.
13. To play a C major chord on the piano, you hit the C, E, and G keys simultaneously. When you do so, you hear no beats. Why? (Refer to Table 14–3.)

Conceptual Exercises

(Answers to odd-numbered Conceptual Exercises can be found in the back of the book.)

1. Consider a wave on a string with constant tension. If the frequency of the wave is doubled, by what multiplicative factor does **(a)** the wavelength and **(b)** the speed change reverse (a) and (b)?
2. Suppose you would like to double the speed of a wave on a string. By what multiplicative factor must you increase the tension?
3. Two strings are made of the same material and have equal tensions. String 1 is thick; string 2 is thin. Is the speed of waves on string 1 greater than, less than, or the same as the speed of waves on string 2? Explain.
4. Two strings are made of the same material and have waves of equal speed. String 1 is thick; string 2 is thin. Is the tension in string 1 greater than, less than, or the same as the tension in string 2? Explain.
5. A harmonic wave travels along a string. **(a)** At a point where the displacement of the string is greatest, is the kinetic energy of the string a maximum or a minimum? Explain. **(b)** At a point where the displacement of the string is zero, is the kinetic energy of the string a maximum or a minimum? Explain.

6. A harmonic wave travels along a string. **(a)** At a point where the displacement of the string is greatest, is the potential energy of the string a maximum or a minimum? Explain. **(b)** At a point where the displacement of the string is zero, is the potential energy of the string a maximum or a minimum? Explain.

7. The three waves, A, B, and C, shown in **Figure 14–32** propagate on strings with equal tensions and equal mass per length. Rank the waves in order of increasing **(a)** frequency, **(b)** wavelength, and **(c)** speed. Indicate ties where appropriate.

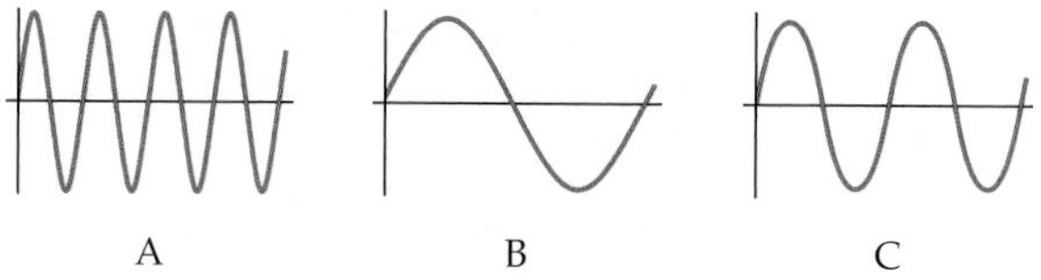

▲ **FIGURE 14–32** Conceptual Exercise 7

8. **Figure 14–33** shows a wave on a string moving to the right. For each of the points indicated on the string, A–F, state whether it is moving upward, moving downward, or is instantaneously at rest at the instant pictured.

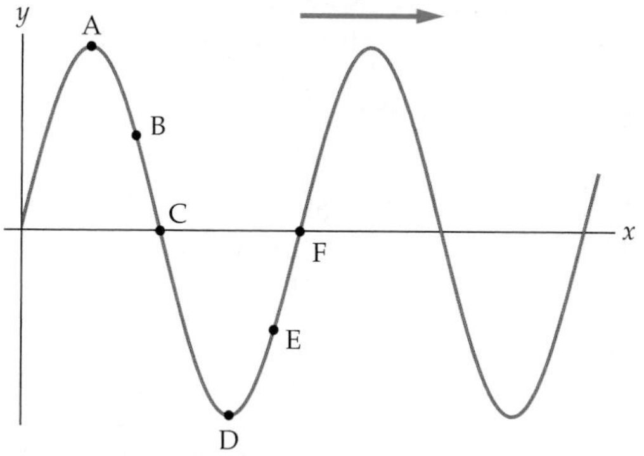

▲ **FIGURE 14–33** Conceptual Exercise 8

9. The vertical displacement of a wave on a string is described by the equation $y(x, t) = A\sin(Bx - Ct)$, in which A, B, and C are positive constants. **(a)** Does this wave propagate in the positive or negative x direction? Explain. **(b)** What is the wavelength of this wave? **(c)** What is the frequency of this wave? **(d)** What is the smallest positive value of x where the displacement of this wave is zero at $t = 0$?

10. The vertical displacement of a wave on a string is described by the equation $y(x, t) = A\sin(Bx + Ct)$, in which A, B, and C are positive constants. **(a)** Does this wave propagate in the positive or negative x direction? Explain. **(b)** What is the physical meaning of the constant A? **(c)** What is the speed of this wave? **(d)** What is the smallest positive time t for which the wave has zero displacement at the point $x = 0$?

11. If the distance to a point source of sound is doubled, by what multiplicative factor does the intensity change?

12. A horn produces sound with a frequency f_0. Let the frequency you hear when you are at rest and the horn moves toward you with a speed u be f_1; let the frequency you hear when the horn is at rest and you move toward it with a speed u be f_2. Is f_1 greater than, less than, or equal to f_2? Explain.

13. You are heading toward an island in your speedboat when you see a friend standing on shore at the base of a cliff. You sound the boat's horn to get your friend's attention. Let the wavelength of the sound produced by the horn be λ_1, the wavelength as heard by your friend be λ_2, and the wavelength of the echo as heard on the boat be λ_3. Rank these wavelengths in order of increasing length. Indicate ties where appropriate.

14. When you blow across the opening of a partially filled two-liter soda pop bottle you hear a tone. If you take a sip of the pop and blow across the opening again, does the tone you hear have a higher frequency, a lower frequency, or the same frequency as before? Explain.

15. A string is tied down at both ends. Some of the standing waves on this string have the following frequencies: 100 Hz, 200 Hz, 250 Hz, and 300 Hz. It is also known that there are no standing waves with frequencies between 250 Hz and 300 Hz. **(a)** What is the fundamental frequency of this string? **(b)** What is the frequency of the third harmonic?

16. When guitar strings A and B are plucked at the same time, a beat frequency of 2 Hz is heard. If string 1 is tightened, the beat frequency increases to 3 Hz. Which of the two strings had the lower frequency initially? Explain.

17. Is the beat frequency produced when a 245 Hz tone and a 240 Hz tone are played together greater than, less than, or equal to the beat frequency produced when a 140 Hz tone and a 145 Hz tone are played together. Explain.

18. You stand near the tracks as a train approaches with constant speed. The train is operating its horn continuously, and you listen carefully to the sound it makes. For each of the following properties of the sound, state whether it increases, decreases, or stays the same as the trains gets closer: **(a)** the intensity; **(b)** the frequency; **(c)** the wavelength; **(d)** the speed.

Problems

Note: **IP** *denotes an integrated conceptual/quantitative problem.* **BIO** *identifies problems of biological or medical interest. Red bullets (•, ••, •••) are used to indicate the level of difficulty of each problem. The speed of sound in air is assumed to be 343 m/s unless specifically stated otherwise.*

Section 14–1 Types of Waves

1. • A wave travels along a stretched horizontal rope. The vertical distance from crest to trough for this wave is 11 cm and the horizontal distance from crest to trough is 26 cm. What is **(a)** the wavelength and **(b)** the amplitude of this wave?

2. • A surfer floating beyond the breakers notes 14 waves per minute passing her position. If the wavelength of these waves is 34 m, what is their speed?

3. • The speed of surface waves in water decreases as the water becomes shallower. Suppose waves travel across the surface of a lake with a speed of 2.0 m/s and a wavelength of 1.5 m. When these waves move into a shallower part of the lake, their speed decreases to 1.6 m/s, though their frequency remains the same. Find the wavelength of the waves in the shallower water.

4. • **Tsunami** A tsunami traveling across deepwater can have a speed of 750 km/h and a wavelength of 310 km. What is the frequency of such a wave?

5. •• **IP** A 4.5-Hz wave with an amplitude of 12 cm and a wavelength of 27 cm travels along a stretched horizontal string. **(a)** How far does a given peak on the wave travel in a time interval of 0.50 s? **(b)** How far does a knot on the string travel in the same time interval? **(c)** How would your answers to parts (a) and (b) change if the amplitude of the wave were halved? Explain.

6. •• **Deepwater Waves** The speed of a deepwater wave with a wavelength λ is given approximately by $v = \sqrt{g\lambda/2\pi}$. Find the speed and frequency of a deepwater wave with a wavelength of 4.5 m.

7. •• **Shallow-Water Waves** In shallow water of depth d the speed of waves is approximately $v = \sqrt{gd}$. Find the speed and frequency of a wave with wavelength 0.75 cm in water that is 2.6 cm deep.

Section 14–2 Waves on a String

8. • Waves on a particular string travel with a speed of 16 m/s. By what factor should the tension in this string be changed to produce waves with a speed of 32 m/s?

9. •• A brother and sister try to communicate with a string tied between two tin cans (**Figure 14–34**). If the string is 9.5 m long, has a mass of 32 g, and is pulled taut with a tension of 8.6 N, how much time does it take for a wave to travel from one end of the string to the other?

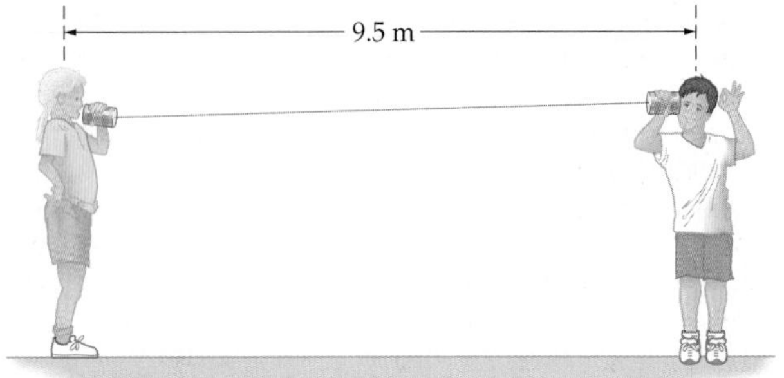

▲ **FIGURE 14–34** Problems 9 and 10

10. •• **IP** **(a)** Suppose the tension is increased in the previous problem. Does a wave take more, less, or the same time to travel from one end to the other? Calculate the time of travel for tensions of **(b)** 9.0 N and **(c)** 10.0 N.

11. •• **IP** A 7.3-m wire with a mass of 85 g is attached to the mast of a sailboat. If the wire is given a "thunk" at one end, it takes 0.94 s for the resulting wave to reach the other end. **(a)** What is the tension in the wire? **(b)** Would the tension found in part (a) be larger or smaller if the mass of the wire is greater than 85 g? **(c)** Calculate the tension for a 95-g wire.

12. •• Two steel guitar strings have the same length. String A has a diameter of 0.50 mm and is under 410.0 N of tension. String B has a diameter of 1.0 mm and is under a tension of 820.0 N. Find the ratio of the wave speeds, v_A/v_B, in these two strings.

13. ••• Use dimensional analysis to show how the speed v of a wave on a string of circular cross section depends on the tension in the string, T, the radius of the string, R, and its mass per volume, ρ.

*Section 14–3 Harmonic Wave Functions

14. • Write an expression for a harmonic wave with an amplitude of 0.16 m, a wavelength of 2.1 m, and a period of 1.8 s. The wave is transverse, travels to the right, and has a displacement of 0.16 m at $t = 0$ and $x = 0$.

15. • Write an expression for a transverse harmonic wave that has a wavelength of 2.8 m and propagates to the left with a speed of 13.3 m/s. The amplitude of the wave is 0.12 m, and its displacement at $t = 0$ and $x = 0$ is 0.12 m.

16. •• **IP** A wave on a string is described by the following equation:

$$y = (15 \text{ cm}) \cos\left(\frac{\pi}{5.0 \text{ cm}}x - \frac{\pi}{12 \text{ s}}t\right)$$

(a) What is the amplitude of this wave? **(b)** What is its wavelength? **(c)** What is its period? **(d)** What is its speed? **(e)** In which direction does the wave travel?

17. •• Consider the wave function given in the previous problem. Sketch this wave from $x = 0$ to $x = 10$ cm for the following times: **(a)** $t = 0$; **(b)** $t = 3.0$ s; **(c)** $t = 6.0$ s. **(d)** What is the least amount of time required for a given point on this wave to move from $y = 0$ to $y = 15$ cm? Verify your answer by referring to the sketches for parts (a), (b), and (c).

18. •• **IP** Four waves are described by the following equations, in which all distances are measured in centimeters and all times are measured in seconds:

$$y_A = 10 \cos(3x - 4t)$$
$$y_B = 10 \cos(5x + 4t)$$
$$y_C = 20 \cos(-10x + 60t)$$
$$y_D = 20 \cos(-4x - 20t)$$

(a) Which of these waves travel in the $+x$ direction? **(b)** Which of these waves travel in the $-x$ direction? **(c)** Which wave has the highest frequency? **(d)** Which wave has the greatest wavelength? **(e)** Which wave has the greatest speed?

Section 14–4 Sound Waves

19. • At Zion National Park a loud shout produces an echo 1.85 s later from a colorful sandstone cliff. How far away is the cliff?

20. • **BIO** **Dolphin Ultrasound** Dolphins of the open ocean are classified as Type II Odontocetes (toothed whales). These animals use ultrasonic "clicks" with a frequency of about 55 kHz to navigate and find prey. **(a)** Suppose a dolphin sends out a series of clicks that are reflected back from the bottom of the ocean 75 m below. How much time elapses before the dolphin hears the echoes of the clicks? (The speed of sound in seawater is approximately 1530 m/s.) **(b)** What is the wavelength of 55 kHz sound in the ocean?

21. • The lowest note on a piano is A, four octaves below the A given in Table 14–3. The highest note on a piano is a C, four octaves above middle C. Find the frequencies and wavelengths (in air) of these notes.

22. •• **IP** A sound wave in air has a frequency of 425 Hz. **(a)** What is its wavelength? **(b)** If the frequency of the sound is increased, does its wavelength increase, decrease, or stay the same? Explain. **(c)** Calculate the wavelength for a sound wave with a frequency of 475 Hz.

23. •• **IP** When you drop a rock into a well, you hear the splash 1.5 seconds later. **(a)** How deep is the well? **(b)** If the depth of the well were doubled, would the time required to hear the splash be greater than, less than, or equal to 3.0 seconds? Explain.

24. •• A rock is thrown downward into a well that is 8.80 m deep. If the splash is heard 1.20 seconds later, what was the initial speed of the rock?

Section 14–5 Sound Intensity

25. • The intensity level in a truck is 92 dB. What is the intensity of this sound?

26. • The distance to a point source is decreased by a factor of three. **(a)** By what multiplicative factor does the intensity increase? **(b)** By what additive amount does the intensity level increase?

27. • Sound 1 has an intensity of 38.0 W/m^2. Sound 2 has an intensity level that is 2.5 dB greater than the intensity level of sound 1. What is the intensity of sound 2?

28. •• A bird-watcher is hoping to add the white-throated sparrow to his "life list" of species. How far could he be from the bird described in Example 14–3 and still be able to hear it? Assume no reflections or absorption of the sparrow's sound.

29. •• Residents of Hawaii are warned of the approach of a tsunami by sirens mounted on the top of towers. Suppose a siren produces a sound that has an intensity level of 120 dB at a distance of 2.0 m. Treating the siren as a point source of sound, and ignoring reflections and absorption, find the intensity level heard by an observer at a distance of **(a)** 12 m and **(b)** 21 m from the siren. **(c)** How far away can the siren be heard?

30. •• In a pig-calling contest, a caller produces a sound with an intensity level of 110 dB. How many such callers would be required to reach the pain level of 120 dB?

31. •• **IP** Twenty violins playing simultaneously with the same intensity combine to give an intensity level of 82.5 dB. **(a)** What is the intensity level of each violin? **(b)** If the number of violins is increased to 40, will the combined intensity level be more than, less than, or equal to 165 dB? Explain.

32. •• **BIO The Human Eardrum** The radius of a typical human eardrum is about 4.0 mm. Find the energy per second received by an eardrum when it listens to sound that is **(a)** at the threshold of hearing and **(b)** at the threshold of pain.

33. ••• A point source of sound that emits uniformly in all directions is located in the middle of a large, open field. The intensity at Brittany's location directly north of the source is twice that at Phillip's position due east of the source. What is the distance between Brittany and Phillip if Brittany is 12.5 m from the source?

Section 14–6 The Doppler Effect

34. • A person with perfect pitch sits on a bus bench listening to the 450-Hz horn of an approaching car. If the person detects a frequency of 470 Hz, how fast is the car moving?

35. • A train moving with a speed of 31.8 m/s sounds a 136-Hz horn. What frequency is heard by an observer standing near the tracks as the train approaches?

36. • In the previous problem, suppose the stationary observer sounds a horn that is identical to the one on the train. What frequency is heard from this horn by a passenger in the train?

37. • **BIO** A bat moving with a speed of 3.25 m/s and emitting sound of 35.0 kHz approaches a moth at rest on a tree trunk. **(a)** What frequency is heard by the moth? **(b)** If the speed of the bat is increased, is the frequency heard by the moth higher or lower? **(c)** Calculate the frequency heard by the moth when the speed of the bat is 4.25 m/s.

38. • A motorcycle and a police car are moving toward one another. The police car emits sound with a frequency of 512 Hz and has a speed of 27.0 m/s. The motorcycle has a speed of 13.0 m/s. What frequency does the motorcyclist hear?

39. • In the previous problem, suppose that the motorcycle and the police car are moving in the same direction, with the motorcycle in the lead. What frequency does the motorcyclist hear in this case?

40. •• Hearing the siren of an approaching fire truck, you pull over to the side of the road and stop. As the truck approaches, you hear a tone of 460 Hz; as the truck recedes, you hear a tone of 410 Hz. How much time will it take for the truck to get from your position to the fire 5.0 km away, assuming it maintains a constant speed?

41. •• With what speed must you approach a source of sound to observe a 15% change in frequency?

42. •• **IP** A particular jet engine produces a tone of 495 Hz. Suppose that one jet is at rest on the tarmac while a second identical jet flies overhead at 82.5% of the speed of sound. The pilot of each jet listens to the sound produced by the engine of the other jet. **(a)** Which pilot hears a greater Doppler shift? Explain. **(b)** Calculate the frequency heard by the pilot in the moving jet. **(c)** Calculate the frequency heard by the pilot in the stationary jet.

43. •• **IP** Two bicycles approach one another, each traveling with a speed of 8.50 m/s. **(a)** If bicyclist A beeps a 315-Hz horn, what frequency is heard by bicyclist B? **(b)** Which of the following would cause the greater increase in the frequency heard by bicyclist B: (i) Bicyclist A speeds up by 1.50 m/s, or (ii) bicyclist B speeds up by 1.50 m/s? Explain.

44. •• A train on one track moves in the same direction as a second train on the adjacent track. The first train, which is ahead of the second train and moves with a speed of 35.8 m/s, blows a horn whose frequency is 124 Hz. If the frequency heard on the second train is 133 Hz, what is its speed?

45. •• Two cars traveling with the same speed move directly away from one another. One car sounds a horn whose frequency is 205 Hz and a person in the other car hears a frequency of 192 Hz. What is the speed of the cars?

46. ••• **The Bullet Train** The Shinkansen, the Japanese "bullet" train, runs at high speed from Tokyo to Nagoya. Riding on the Shinkansen, you notice that the frequency of a crossing signal changes markedly as you pass the crossing. As you approach the crossing, the frequency you hear is f; as you recede from the crossing the frequency you hear is $2f/3$. What is the speed of the train?

Section 14–7 Superposition and Interference

47. • Two wave pulses on a string approach one another at the time $t = 0$, as shown in **Figure 14–35**. Each pulse moves with a speed of 1.0 m/s. Make a careful sketch of the resultant wave at the times $t = 1.0$ s, 2.0 s, 2.5 s, 3.0 s, and 4.0 s, assuming that the superposition principle holds for these waves.

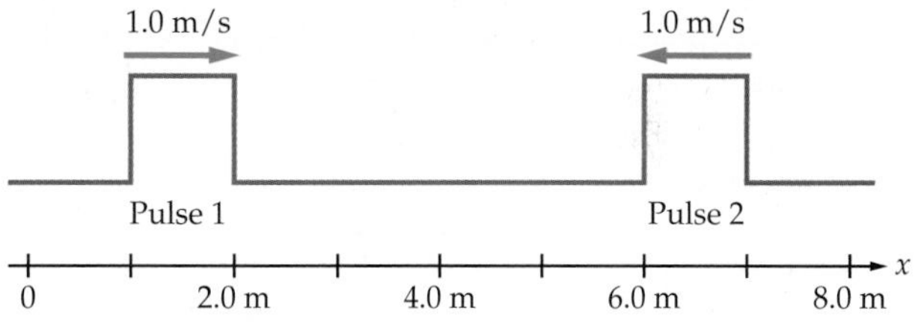

▲ **FIGURE 14–35** Problems 47 and 48

48. • Suppose pulse 2 in Problem 47 is inverted, so that it is a downward deflection of the string rather than an upward deflection. Repeat Problem 47 in this case.

49. • Two wave pulses on a string approach one another at the time $t = 0$, as shown in **Figure 14–36**. Each pulse moves with a speed of 1.0 m/s. Make a careful sketch of the resultant wave at the times $t = 1.0$ s, 2.0 s, 2.5 s, 3.0 s, and 4.0 s, assuming that the superposition principle holds for these waves.

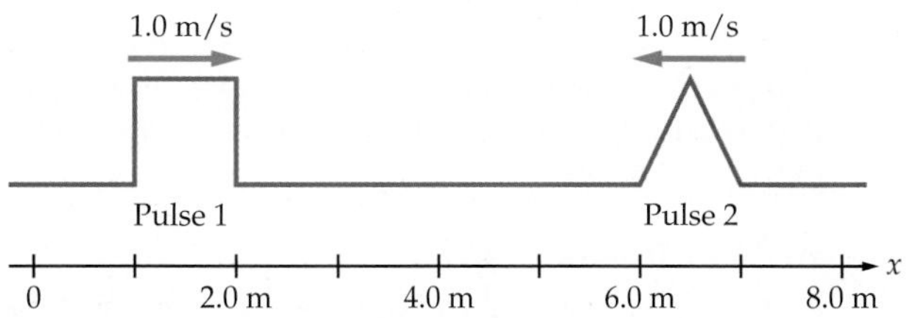

▲ **FIGURE 14–36** Problems 49 and 50

50. • Suppose pulse 2 in Problem 49 is inverted, so that it is a downward deflection of the string rather than an upward deflection. Repeat Problem 49 in this case.

51. •• A pair of in-phase stereo speakers is placed side by side, 0.85 m apart. You stand directly in front of one of the speakers, 1.1 m from the speaker. What is the lowest frequency that will produce constructive interference at your location?

52. •• **IP** Two violinists, one directly behind the other, play for a listener directly in front of them. Both violinists sound concert A (440 Hz). **(a)** What is the smallest separation between the violinists that will produce destructive interference for the listener? **(b)** Does this smallest separation increase or decrease if the violinists produce a note with a higher frequency? **(c)** Repeat part (a) for violinists who produce sounds of 540 Hz.

53. •• Two loudspeakers are placed at either end of a gymnasium, both pointing toward the center of the gym and equidistant from it. The speakers emit 256-Hz sound that is in phase. An observer at the center of the gym experiences constructive interference. How far toward either speaker must the observer walk to first experience destructive interference?

54. •• **IP (a)** In the previous problem, does the required distance increase, decrease, or stay the same if the frequency of the speakers is lowered? **(b)** Calculate the distance to the first position of destructive interference if the frequency emitted by the speakers is lowered to 242 Hz.

55. •• Two speakers with opposite phase are positioned 3.5 m apart, both pointing toward a wall 5.0 m in front of them (**Figure 14–37**). An observer standing against the wall midway between the speakers hears destructive interference. If the observer hears constructive interference after moving 0.84 m to one side along the wall, what is the frequency of the sound emitted by the speakers?

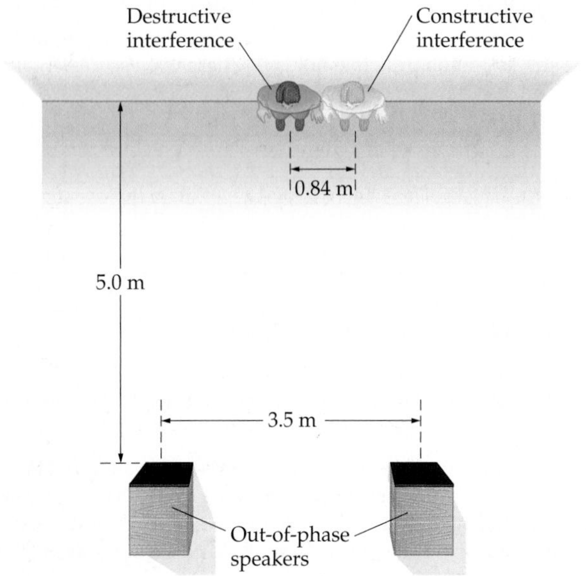

▲ **FIGURE 14–37** Problem 55

56. •• Suppose, in Example 14–7, that the speakers have opposite phase. What is the lowest frequency that gives destructive interference in this case?

Section 14–8 Standing Waves

57. • An organ pipe that is open at both ends is 3.5 m long. What is its fundamental frequency?

58. • A string 1.5 m long with a mass of 2.1 g is stretched between two fixed points with a tension of 95 N. Find the frequency of the fundamental on this string.

59. •• **IP BIO Standing Waves in the Human Ear** The human ear canal is much like an organ pipe that is closed at one end (at the tympanic membrane or eardrum) and open at the other (see **Figure 14–38**). A typical ear canal has a length of about 2.4 cm. **(a)** What is the fundamental frequency and wavelength of the ear canal? **(b)** Find the frequency and wavelength of the ear canal's third harmonic. (Recall that the third harmonic in this case is the standing wave with the second-lowest frequency.) **(c)** Suppose a person has an ear canal that is shorter than 2.4 cm. Is the fundamental frequency of that person's ear canal greater than, less than, or the same as the value found in part (a)? Explain. (Note that the frequencies found in parts (a) and (b) correspond closely to the frequencies of enhanced sensitivity in Figure 14–28.)

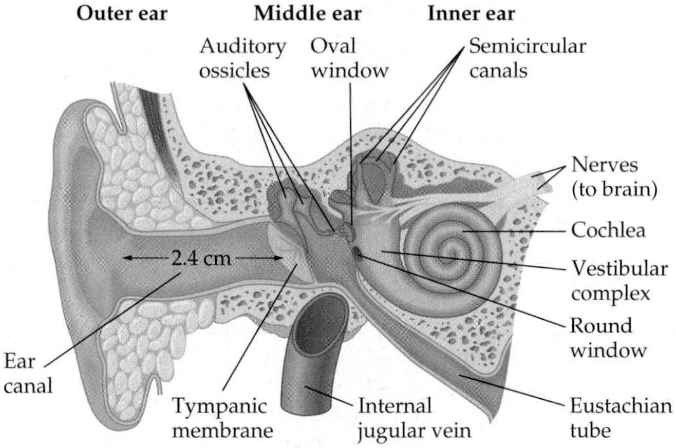

▲ **FIGURE 14–38** Problem 59

60. •• A guitar string 62 cm long vibrates with a standing wave that has three antinodes. **(a)** Which harmonic is this? **(b)** What is the wavelength of this wave?

61. •• **IP** A 12.5-g clothesline is stretched with a tension of 22.1 N between two poles 7.66 m apart. What is the frequency of **(a)** the fundamental and **(b)** the second harmonic? **(c)** If the tension in the clothesline is increased, do the frequencies in parts (a) and (b) increase, decrease, or stay the same? Explain.

62. •• **IP (a)** In the previous problem, will the frequencies increase, decrease, or stay the same if a more massive rope is used? **(b)** Repeat Problem 61 for a clothesline with a mass of 15.0 g.

63. •• The organ pipe in **Figure 14–39** is 2.75 m long. **(a)** What is the frequency of the standing wave shown in the pipe? **(b)** What is the fundamental frequency of this pipe?

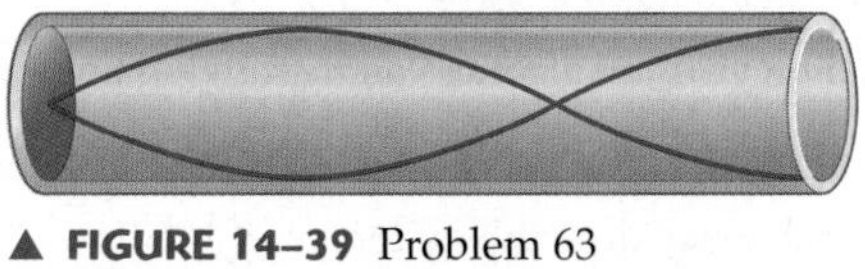

▲ **FIGURE 14–39** Problem 63

64. •• The frequency of the standing wave shown in **Figure 14–40** is 202 Hz. **(a)** What is the fundamental frequency of this pipe? **(b)** What is the length of the pipe?

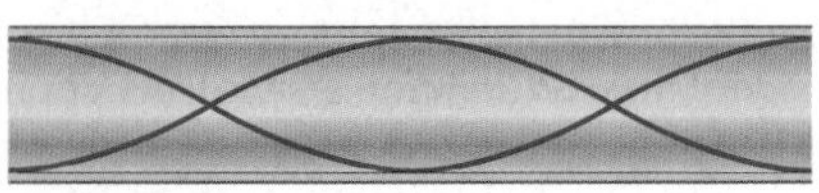

▲ **FIGURE 14–40** Problem 64

65. ••• An organ pipe open at both ends has a harmonic with a frequency of 440 Hz. The next higher harmonic in the pipe has a frequency of 528 Hz. Find **(a)** the frequency of the fundamental and **(b)** the length of the pipe.

Section 14–9 Beats

66. • Two tuning forks have frequencies of 275 Hz and 292 Hz. What is the beat frequency if both tuning forks are sounded simultaneously?

67. • **Tuning a Piano** To tune middle C on a piano, a tuner hits the key and at the same time sounds a 261-Hz tuning fork. If the tuner hears 3 beats per second, what are the possible frequencies of the piano key?

68. • Two musicians are comparing their clarinets. The first clarinet produces a tone that is known to be 441 Hz. When the two clarinets play together they produce eight beats every 2.00 seconds. If the second clarinet produces a higher pitched tone than the first clarinet, what is the second clarinet's frequency?

69. •• IP Two strings that are fixed at each end are identical, except that one is 0.560 cm longer than the other. Waves on these strings propagate with a speed of 34.2 m/s, and the fundamental frequency of the shorter string is 212 Hz. **(a)** What beat frequency is produced if each string is vibrating with its fundamental frequency? **(b)** Does the beat frequency in part (a) increase or decrease if the longer string is increased in length? **(c)** Repeat part (a), assuming that the longer string is 0.761 cm longer than the shorter string.

70. •• IP A tuning fork with a frequency of 320.0 Hz and a tuning fork of unknown frequency produce beats with a frequency of 4.5 Hz. If the frequency of the 320.0-Hz fork is lowered slightly by placing a bit of putty on one of its tines, the new beat frequency is 7.5 Hz. **(a)** Which tuning fork has the lower frequency? Explain. **(b)** What is the final frequency of the 320.0-Hz tuning fork? **(c)** What is the frequency of the other tuning fork?

71. •• Identical cellos are being tested. One is producing a fundamental frequency of 130.9 Hz on a string that is 1.25 m long and has a mass of 109 g. On the second cello the same string is fingered to reduce the length that can vibrate. If the beat frequency produced by these two strings is 4.33 Hz, what is the vibrating length of the second string?

72. ••• A friend in another city tells you that she has two organ pipes of different lengths, one open at both ends, the other open at one end only. In addition, she has determined that the beat frequency caused by the second-lowest frequency of each pipe is equal to the beat frequency caused by the third-lowest frequency of each pipe. Her challenge to you is to calculate the length of the organ pipe that is open at both ends, given that the length of the other pipe is 1.00 m.

General Problems

73. • Sitting peacefully in your living room one stormy day, you see a flash of lightning through the windows. Eight and a half seconds later thunder shakes the house. Estimate the distance from your house to the bolt of lightning.

74. • The fundamental of an organ pipe that is closed at one end and open at the other end is 261.6 Hz (middle C). The second harmonic of an organ pipe that is open at both ends has the same frequency. What are the lengths of these two pipes?

75. • **The Loudest Animal** The loudest sound produced by a living organism on Earth is made by the bowhead whale (*Balaena mysticetus*). These whales can produce a sound that, if heard in air at a distance of 3.00 m, would have an intensity level of 127 dB. This is roughly the equivalent of 5000 trumpeting elephants. How far away can you be from a 127 dB sound and still just barely hear it?(Assume a point source, and ignore reflections and absorption.)

76. • **Hearing a Good Hit** Physicist Robert Adair, once appointed the "official physicist to the National League" by the commissioner of baseball, believes that the crack-of-the-bat can tell an outfielder how well the ball has been hit. According to Adair, a good hit makes a sound of 510 Hz, while a poor hit produces a sound of 170 Hz. What is the difference in wavelength of these sounds?

77. • A standing wave of 603 Hz is produced on a string that is 1.33 m long and fixed on both ends. If the speed of waves on this string is 402 m/s, how many antinodes are there in the standing wave?

78. • BIO **Measuring Hearing Loss** To determine the amount of temporary hearing loss loud music can cause in humans, researchers studied a group of 20 adult females who were exposed to 110-dB music for 60 minutes. Eleven of the 20 subjects showed a 20.0-dB reduction in hearing sensitivity at 4000 Hz. What is the intensity corresponding to the threshold of hearing for these subjects?

79. •• BIO **Hearing a Pin Drop** The ability to hear a "pin drop" is the sign of sensitive hearing. Suppose a 0.55-g pin is dropped from a height of 28 cm, and that the pin emits sound for 1.5 s when it lands. Assuming all of the mechanical energy of the pin is converted to sound energy, and that the sound radiates uniformly in all directions, find the maximum distance from which a person can hear the pin drop. (This is the ideal maximum distance, but atmospheric absorption and other factors will make the actual maximum distance considerably smaller.)

80. •• A machine shop has 100 equally noisy machines that together produce an intensity level of 90 dB. If the intensity level must be reduced to 80 dB, how many machines must be turned off?

81. •• IP When you blow across the top of a soda pop bottle you hear a fundamental frequency of 206 Hz. Suppose the bottle is now filled with helium. **(a)** Does the fundamental frequency increase, decrease, or stay the same? Explain. **(b)** Find the new fundamental frequency. (Assume that the speed of sound in helium is three times that in air.)

82. •• **Speed of a Tsunami** Tsunamis can have wavelengths between 100 and 400 km. Since this is much greater than the average depth of the oceans (about 4.3 km), the ocean can be considered as shallow water for these waves. Using the speed of waves in shallow water of depth d given in Problem 7, find the typical speed for a tsunami. (*Note*: In the open ocean, tsunamis generally have an amplitude of less than a meter, allowing them to pass ships unnoticed. As they approach shore, however, the water depth decreases and the waves slow down. This can result in an increase of amplitude to as much as 37 m or more.)

83. •• Two trains with 124-Hz horns approach one another. The slower of the two trains has a speed of 22 m/s. What is the speed of the fast train if an observer standing near the tracks between the trains hears a beat frequency of 4.4 Hz?

84. •• IP Jim is speeding toward James Island with a speed of 24 m/s when he sees Betsy standing on shore at the base of a cliff (**Figure 14–41**). Jim sounds his 330-Hz horn. **(a)** What frequency does Betsy hear? **(b)** Jim can hear the echo of his horn

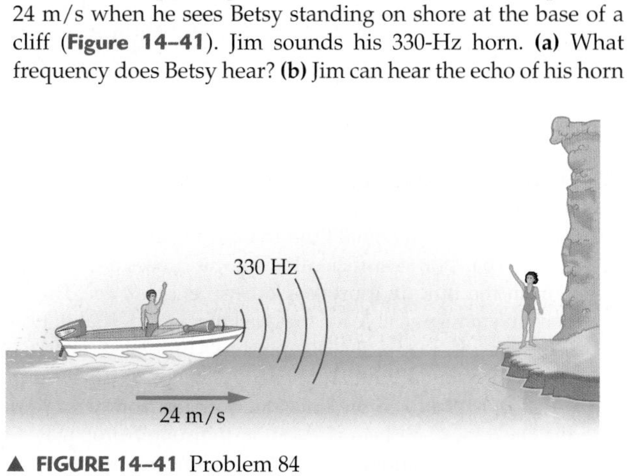

▲ **FIGURE 14–41** Problem 84

reflected back to him by the cliff. Is the frequency of this echo greater than or equal to the frequency heard by Betsy? Explain. **(c)** Calculate the frequency Jim hears in the echo from the cliff.

85. •• Two ships in a heavy fog are blowing their horns, both of which produce sound with a frequency of 175.0 Hz (**Figure 14–42**). One ship is at rest; the other moves on a straight line that passes through the one at rest. If people on the stationary ship hear a beat frequency of 3.5 Hz, what are the two possible speeds and directions of motion of the moving ship?

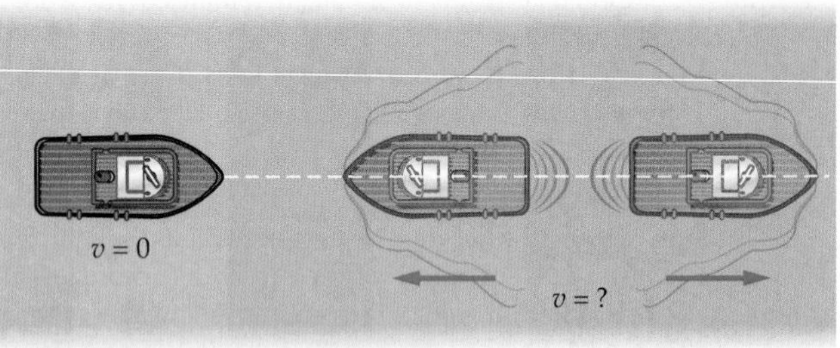

▲ **FIGURE 14–42** Problem 85

86. •• **IP Dinosaur Sounds** The long-crested, duck-billed dinosaur *Parasaurolophus walkeri*, which roamed the Earth 75 million years ago, possessed the largest crest of any duck bill. Many paleontologists believe the air passages in the dinosaur's crest acted like a bent organ pipe open at both ends and that it was used to produce sounds. **(a)** Suppose the nasal passages in a certain *P. walkeri* crest produce a bent tube 2.7 m long. What is the fundamental frequency of this tube, assuming the bend has no effect on the frequency? (For comparison, a typical human hearing range is 20 Hz to 20 kHz.) **(b)** Female *P. walkeri* may have had shorter crests than the male's. If this was the case, would the fundamental frequency of a female be greater than, less than, or the same as that of a male? Explain.

▲ **FIGURE 14–43** The long crest of *Parasaurolophus walkeri*. (Problem 86)

87. •• **BIO Cracking Your Knuckles** When you "crack" a knuckle, you cause the knuckle cavity to widen rapidly. This, in turn, allows the synovial fluid to expand into a larger volume. If this expansion is sufficiently rapid, it causes a gas bubble to form in the fluid in a process known as *cavitation*. This is the mechanism responsible for the cracking sound. (Cavitation can also cause pits in rapidly rotating ship's propellers.) If a "crack" produces a sound with an intensity level of 57 dB at your ear, which is 18 cm from the knuckle, how far from your knuckle can the "crack" be heard? Assume the sound propagates uniformly in all directions, with no reflections or absorption.

88. •• A steel guitar string has a tension T, length L, and diameter D. Give the multiplicative factor by which the fundamental frequency of the string changes under the following conditions: **(a)** The tension in the string is increased by a factor of 4. The diameter is D and the length is L. **(b)** The diameter of the string is increased by a factor of 3. The tension is T and the length is L **(c)** The length of the string is halved. The tension is T and the diameter is D.

89. •• A Slinky has a mass of 0.23 kg and negligible length. When it is stretched 1.5 m, it is found that transverse waves travel the length of the Slinky in 0.75 s. **(a)** What is the force constant, k, of the Slinky? **(b)** If the Slinky is stretched farther, will the time required for a wave to travel the length of the Slinky increase, decrease, or stay the same? Explain. **(c)** If the Slinky is stretched 3.0 m, how much time does it take a wave to travel the length of the Slinky? (The Slinky stretches like an ideal spring, and propagates transverse waves like a rope with variable tension.)

90. •• **IP BIO OSHA Noise Standards** OSHA, the Occupational Safety and Health Administration, has established standards for workplace exposure to noise. According to OSHA's Hearing Conservation Standard, the permissible noise exposure per day is 95.0 dB for 4 hours or 105 dB for 1 hour. Assuming the eardrum is 9.5 mm in diameter, find the energy absorbed by the eardrum **(a)** with 95.0 dB for 4 hours and **(b)** with 105 dB for 1 hour. **(c)** Is OSHA's safety standard simply a measure of the amount of energy absorbed by the ear? Explain.

91. •• **IP Thundersticks at Ball Games** "Thundersticks" are a popular noisemaking device at many sporting events. A typical thunderstick is a hollow plastic tube about 82 cm long and 8.5 cm in diameter. When two thundersticks are hit sharply together, they produce a copious amount of noise. **(a)** Which dimension, the length or diameter, is more important in determining the frequency of the sound emitted by the thundersticks? Explain. **(b)** Estimate the characteristic frequency of the thunderstick's sound. **(c)** Suppose a single pair of thundersticks produces sound with an intensity level of 95 dB. What is the intensity level of 1200 pairs of thundersticks clapping simultaneously?

92. •• An organ pipe 2.5 m long is open at one end and closed at the other end. What is the linear distance between a node and the adjacent antinode for the third harmonic in this pipe?

93. •• Two identical strings with the same tension vibrate at 631 Hz. If the tension in one of the strings is increased by 2.00% what is the resulting beat frequency?

94. •• **The Sound of a Black Hole** Astronomers using the Chandra X-ray Observatory have discovered that the Perseus Black Hole, some 250 million light years away, produces sound waves in the gaseous halo that surrounds it. The frequency of this sound is the same as the frequency of the 59th B-flat below the B-flat given in Table 14–3. How long does it take for this sound wave to complete one cycle? Give your answer in years.

95. •• **BIO The Love Song of the Midshipman Fish** When the plainfin midshipman fish (*Porichthys notatus*) migrates from deep Pacific waters to the west coast of North America each summer, the males (**Figure 14–44**) begin to sing their "love song," which some describe as sounding like a low-pitched motorboat. Houseboat residents and shore dwellers are kept awake for nights on end by the amorous fish. The love song consists of a single note, the second A flat below middle C. **(a)** If the speed of sound in seawater is 1531 m/s, what is the wavelength of the midshipman's song? **(b)** What is the wavelength of the sound after it emerges into the air? (Information on the musical scale is given in Table 14–3.)

▲ **FIGURE 14–44** Problem 95

96. ••• **IP** A rope of length L and mass M hangs vertically from a ceiling. The tension in the rope is only that due to its own weight. **(a)** Suppose a wave starts near the bottom of the rope and propagates upward. Does the speed of the wave increase, decrease, or stay the same as it moves up the rope? Explain. **(b)** Show that the speed of waves a height y above the bottom of the rope is $v = \sqrt{gy}$.

97. ••• Experiments on water waves show that the speed of waves in shallow water is independent of their wavelength (see Problem 7). Using this observation and dimensional analysis, determine how the speed v of shallow-water waves depends on the depth of the water, d, the mass per volume of water, ρ, and the acceleration of gravity, g.

98. ••• A deepwater wave of wavelength λ has a speed given approximately by $v = \sqrt{g\lambda/2\pi}$. Find an expression for the period of a deepwater wave in terms of its wavelength. (Note the similarity of your result to the period of a pendulum.)

99. ••• **Beats and Standing Waves** In Problem 53, suppose the observer walks toward one speaker with a speed of 1.35 m/s. **(a)** What frequency does the observer hear from each speaker? **(b)** What beat frequency does the observer hear? **(c)** How far must the observer walk to go from one point of constructive interference to the next? **(d)** How many times per second does the observer hear maximum loudness from the speakers? Compare your result with the beat frequency from part (b).

Interactive Problems

100. •• **IP** ***Referring to Example 14–6*** Suppose the engineer adjusts the speed of the train until the sound he hears reflected from the cliff is 775 Hz. The train's whistle still produces a tone of 650.0 Hz. **(a)** Is the new speed of the train greater than, less than, or equal to 21.2 m/s? Explain. **(b)** Find the new speed of the train.

101. •• ***Referring to Example 14–6*** Suppose the train is backing away from the cliff with a speed of 18.5 m/s and is sounding its 650.0-Hz whistle. **(a)** What is the frequency heard by the observer standing near the tunnel entrance? **(b)** What is the frequency heard by the engineer?

102. •• **IP** ***Referring to Example 14–9*** Suppose we add more water to the soda pop bottle. **(a)** Does the fundamental frequency increase, decrease, or stay the same? Explain. **(b)** Find the fundamental frequency if the height of water in the bottle is increased to 7.5 cm. The height of the bottle is still 26.0 cm.

103. •• **IP** ***Referring to Example 14–9*** The speed of sound increases slightly with temperature. **(a)** Does the fundamental frequency of the bottle increase, decrease, or stay the same as the air heats up on a warm day? Explain. **(b)** Find the fundamental frequency if the speed of sound in air increases to 348 m/s. Assume the bottle is 26.0 cm tall, and that it contains water to a depth of 6.5 cm.

15 Fluids

15–1	Density	477
15–2	Pressure	477
15–3	Static Equilibrium in Fluids: Pressure and Depth	481
15–4	Archimedes' Principle and Buoyancy	486
15–5	Applications of Archimedes' Principle	488
15–6	Fluid Flow and Continuity	493
15–7	Bernoulli's Equation	495
15–8	Applications of Bernoulli's Equation	498
*15–9	Viscosity and Surface Tension	501

Each of these common household liquids (from top to bottom: corn oil, water, shampoo, dish detergent, antifreeze, maple syrup) has a slightly different density, which determines its order in the vessel. In this chapter we'll see how density is related to buoyancy and other properties of fluids.

When we speak of *fluids* in physics, we refer to substances that can readily flow from place to place, and that take on the shape of a container rather than retain a shape of their own. Thus, when we use the term fluids, we are referring to both liquids and gases.

It is hard to think of a subject more relevant to our everyday lives than fluids. After all, we begin life as a fluid-filled cell suspended in a fluid. We live our independent lives immersed in a fluid that we breathe. In fact, fluids coursing through our circulatory system are literally the lifeblood of our existence. If it were not for the gases in our atmosphere and the liquid water on the Earth's surface, we could not exist.

In this chapter, we examine some of the fundamental physical principles that apply to fluids. All of these principles derive from the basic physics we have learned to this point. For example, straightforward considerations of force and weight lead to an understanding of buoyancy. Similarly, the work–energy theorem results in an understanding of how fluids behave when they flow. As such, fluids provide a wonderful opportunity for us to apply our knowledge of physics to a whole new array of interesting physical systems.

15–1 Density

The properties of a fluid can be hard to pin down, given that it can flow, change shape, and either split into smaller portions or combine into a larger system. Thus, one of the best ways to quantify a fluid is in terms of its **density.** Specifically, the density, ρ, of a material (fluid or not) is defined as the mass, M, per volume, V:

Definition of Density, ρ

$$\rho = M/V \qquad 15\text{–}1$$

SI unit: kg/m^3

The denser a material, the more mass it has in any given volume. Note, however, that the density of a substance is the same regardless of the total amount we have in a system.

To get a feel for densities in common substances, we start with water. For example, to fill a cubic container one meter on a side would take over 2000 pounds of water. More precisely, water has the following density:

$$\rho_w = \text{density of water} = 1000\ kg/m^3$$

A gallon (1 gallon $= 3.79\ L = 3.79 \times 10^{-3}\ m^3$) of water, then, has a mass of

$$M = \rho V = (1000\ kg/m^3)(3.79 \times 10^{-3}\ m^3) = 3.79\ kg$$

As a rule of thumb, a gallon of water weighs just over 8 pounds.

In comparison, the helium in a helium-filled balloon has a density of only about $0.179\ kg/m^3$, and the density of the air in your room is roughly $1.29\ kg/m^3$. On the higher end of the density scale, solid gold "weighs" in with a hefty $19{,}300\ kg/m^3$. Further examples of densities for common materials are given in Table 15–1.

TABLE 15–1
Densities of Common Substances

Substance	Density (kg/m^3)
Gold	19,300
Mercury	13,600
Lead	11,300
Silver	10,500
Iron	7860
Aluminum	2700
Ebony (wood)	1220
Ethylene glycol (antifreeze)	1114
Whole blood (37 °C)	1060
Seawater	1025
Freshwater	1000
Olive oil	920
Ice	917
Ethyl alcohol	806
Cherry (wood)	800
Balsa (wood)	120
Styrofoam	100
Oxygen	1.43
Air	1.29
Helium	0.179

CONCEPTUAL CHECKPOINT 15–1

One day you look in your refrigerator and find nothing but a dozen eggs (44 g each). A quick measurement shows that the inside of the refrigerator is 1.0 m by 0.60 m by 0.75 m. Is the weight of the *air* in your refrigerator **(a)** much less than, **(b)** about the same as, or **(c)** much more than the weight of the *eggs*?

Reasoning and Discussion

At first it might seem that the "thin air" in the refrigerator weighs practically nothing compared with a carton full of eggs. A brief calculation shows this is not the case. For the eggs, we have

$$m_{eggs} = 12(44\ g) = 0.53\ kg$$

For the air,

$$m_{air} = \rho V = (1.29\ kg/m^3)(1.0\ m \times 0.60\ m \times 0.75\ m) = 0.58\ kg$$

Thus, the air, with a mass of 0.58 kg (1.28 lb), actually weighs slightly more than the eggs, which have a mass of 0.53 kg (1.17 lb)!

Answer:

(b) The air and the eggs weigh about the same.

15–2 Pressure

If you have ever pushed a button, or pressed a key on a keyboard, you have applied pressure. Now, you might object that you simply exerted a force on the button, or the key, which is correct. That force is spread out over an area, however. For example, when you press a button, the tip of your finger contacts the button over a small but finite area. **Pressure,** P, is a measure of the amount of force, F, per area A:

Definition of Pressure, P

$$P = F/A \qquad 15\text{–}2$$

SI unit: N/m^2

PROBLEM-SOLVING NOTE

Pressure Is Force per Area

Remember that pressure is proportional to the applied force and *inversely* proportional to the area over which it acts.

Pressure is increased if the force applied to a given area is increased, or if a given force is applied to a smaller area. For example, if you press your finger against a balloon, not much happens—your finger causes a small indentation. On the other hand, if you push a needle against the balloon with the same force, you get an explosive pop. The difference is that the same force applied to the small area of a needle tip causes a large enough pressure to rupture the balloon.

EXAMPLE 15–1 Popping a Balloon

Find the pressure exerted on the skin of a balloon if you press with a force of 2.1 N using **(a)** your finger or **(b)** a needle. Assume the area of your fingertip is $1.0 \times 10^{-4}\ \mathrm{m}^2$, and the area of the needle tip is $2.5 \times 10^{-7}\ \mathrm{m}^2$. **(c)** Find the minimum force necessary to pop the balloon with the needle, given that the balloon pops with a pressure of $3.0 \times 10^5\ \mathrm{N/m^2}$.

Picture the Problem

Our sketch shows a balloon deformed by the press of a finger and of a needle. The difference is not the amount of force that is applied but the area over which it is applied. In the case of the needle, the force may be sufficient to pop the balloon.

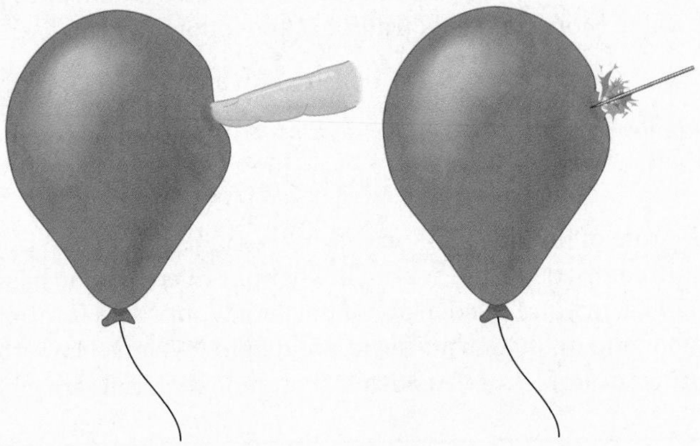

Strategy

(a), (b) The force F and the area A are given in the problem statement. We can find the pressure, then, by applying its definition, $P = F/A$.

(c) Rearrange $P = F/A$ to find the force corresponding to a given pressure and area.

Solution

Part (a)

1. Calculate the pressure exerted by the finger: $P = \dfrac{F}{A} = \dfrac{2.1\ \mathrm{N}}{1.0 \times 10^{-4}\ \mathrm{m}^2} = 2.1 \times 10^4\ \mathrm{N/m^2}$

Part (b)

2. Calculate the pressure exerted by the needle: $P = \dfrac{F}{A} = \dfrac{2.1\ \mathrm{N}}{2.5 \times 10^{-7}\ \mathrm{m}^2} = 8.4 \times 10^6\ \mathrm{N/m^2}$

Part (c)

3. Solve Equation 15–2 for the force: $F = PA$

4. Substitute numerical values: $F = (3.0 \times 10^5\ \mathrm{N/m^2})(2.5 \times 10^{-7}\ \mathrm{m}^2) = 0.075\ \mathrm{N}$

Insight

Note that the pressure exerted by the needle in part (b) is 400 times greater than the pressure due to the finger in part (a). This increase in pressure with decreasing area accounts for the sharp tips to be found on such disparate objects as nails, pens and pencils, and syringes.

Practice Problem

Find the area that a force of 2.1 N would have to act on to produce a pressure of $3.0 \times 10^5\ \mathrm{N/m^2}$. [**Answer:** $A = 7.0 \times 10^{-6}\ \mathrm{m}^2$]

Some related homework problems: Problem 8, Problem 9

REAL-WORLD PHYSICS: BIO

Walking on lily pads

An interesting example of force, area, and pressure in nature is provided by the family of small aquatic birds referred to as rails, and in particular by the gallinule. This bird has exceptionally long toes that are spread out over a large area. The result is that the weight of the bird causes only a relatively small pressure as it walks on the soft, muddy shorelines encountered in its habitat. In some species, the pressure exerted while walking is so small that the birds can actually walk across lily pads without sinking into the water.

Atmospheric Pressure and Gauge Pressure

We are all used to working under pressure—about 14.7 pounds per square inch to be precise. This is **atmospheric pressure,** P_{at}, a direct result of the weight of the air above us. In SI units, atmospheric pressure has the following value:

Atmospheric Pressure, P_{at}

$$P_{at} = 1.01 \times 10^5 \text{ N/m}^2 \qquad \text{15–3}$$

SI unit: N/m^2

A shorthand unit for N/m^2 is the **pascal** (Pa):

$$1 \text{ Pa} = 1 \text{ N/m}^2 \qquad \text{15–4}$$

The pascal honors the pioneering studies of fluids by the French scientist Blaise Pascal (1623–1662). Thus, atmospheric pressure can be written as

$$P_{at} = 101 \text{ kPa}$$

In British units, pressure is measured in pounds per square inch, and

$$P_{at} = 14.7 \text{ lb/in}^2$$

Finally, a common unit for atmospheric pressure in weather forecasting is the **bar,** defined as follows:

$$1 \text{ bar} = 10^5 \text{ Pa} \approx 1\, P_{at}$$

EXERCISE 15–1

Find the force exerted on the palm of your hand by atmospheric pressure. Assume your palm measures 0.080 m by 0.10 m.

Solution

Applying Equations 15–3 and 15–2, we find

$$F = P_{at}A = (1.01 \times 10^5 \text{ Pa})(0.080 \text{ m})(0.10 \text{ m}) = 810 \text{ N}$$

Thus, the atmosphere pushes on the palm of your hand with a force of approximately 180 pounds! Of course, it also pushes on the back of your hand with essentially the same force, but in the opposite direction.

Figure 15–1 illustrates the forces exerted on your hand by atmospheric pressure. If your hand is vertical, atmospheric pressure pushes to the right and to the

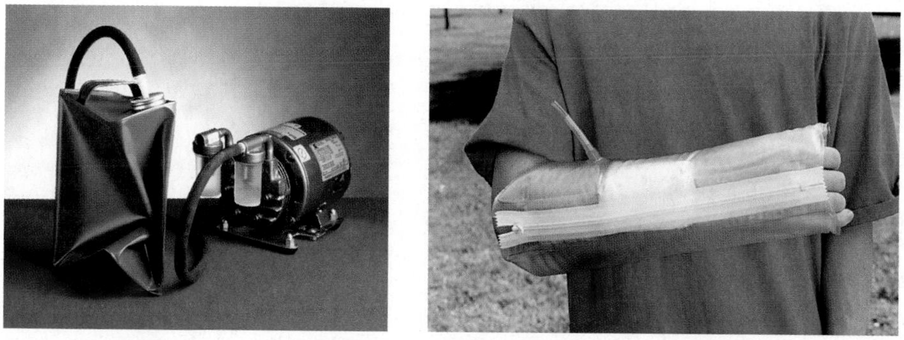

▲ The air around you exerts a force of about 14.7 pounds on every square inch of your body. Because this force is the same in all directions and is opposed by an equal pressure inside your body, you are generally unaware of it. However, when the air is pumped out of a sealed can (left), atmospheric pressure produces an inward force that is unopposed. The resulting collapse of the can vividly illustrates the pressure that is all around us. An air splint (right) utilizes the same principle of unequal pressure. A plastic sleeve is placed around an injured limb and inflated to a pressure greater than that of the atmosphere—and thus of the body's internal pressure. The increased external pressure retards bleeding from the injured area, and also tends to immobilize the limb in case it might be fractured. (Air splints are carried by hikers and others who might be in need of emergency treatment far from professional medical facilities. Before it is inflated, the air splint is about the size and weight of a credit card.)

▲ This bird exerts only a small pressure on the lily pad on which it walks because its weight is spread out over a large area by its long toes. Since the pressure is not enough to sink a lily pad, the bird can seemingly "walk on water."

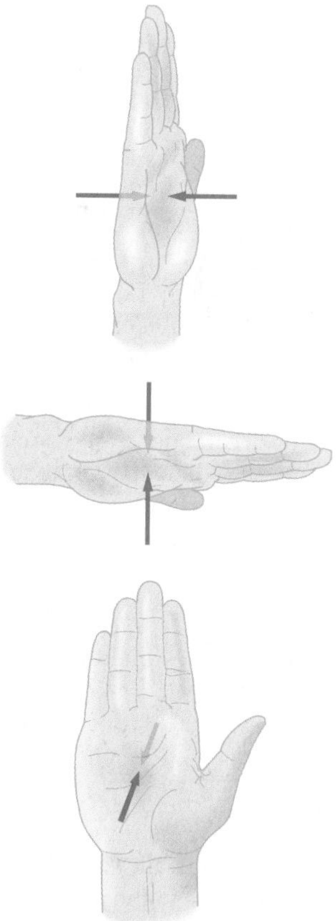

▲ **FIGURE 15–1 Pressure is the same in all directions**

The forces exerted on the two sides of a hand cancel, regardless of the hand's orientation. Hence, pressure acts equally in all directions.

left equally, so your hand feels zero net force. If your hand is horizontal, atmospheric pressure exerts upward and downward forces on your hand that are essentially the same in magnitude, again giving zero net force. This cancellation of forces occurs no matter what the orientation of your hand; thus, we can conclude the following:

> The pressure in a fluid acts equally in all directions, and acts at right angles to any surface.

In many cases we are interested in the difference between a given pressure and atmospheric pressure. For example, a flat tire does not have zero pressure in it; the pressure in the tire is atmospheric pressure. To inflate the tire to 241 kPa ($35\ \text{lb/in}^2$), the pressure inside the tire must be greater than atmospheric pressure by this amount; that is, $P = 241\ \text{kPa} + P_{at} = 342\ \text{kPa}$.

PROBLEM-SOLVING NOTE

Gauge Pressure

If a problem gives you the gauge pressure, recall that the actual pressure is the gauge pressure *plus* atmospheric pressure.

To deal with such situations, we introduce the **gauge pressure,** P_g, defined as follows:

$$P_g = P - P_{at} \qquad 15\text{–}5$$

It is the gauge pressure, then, that is determined by a tire gauge. Many problems in this chapter refer to the gauge pressure. Hence it must be remembered that the actual pressure in these cases is greater by the amount P_{at}.

EXAMPLE 15–2 Pressuring the Ball: Estimate the Gauge Pressure

Estimate the gauge pressure in a basketball by pushing down on it and noting the area of contact it makes with the surface on which it rests.

Picture the Problem

Our sketch shows the basketball both in its original state, and when a force F pushes downward on it. In the latter case, the ball flattens out on the bottom, forming a circular area of contact with the floor of diameter d.

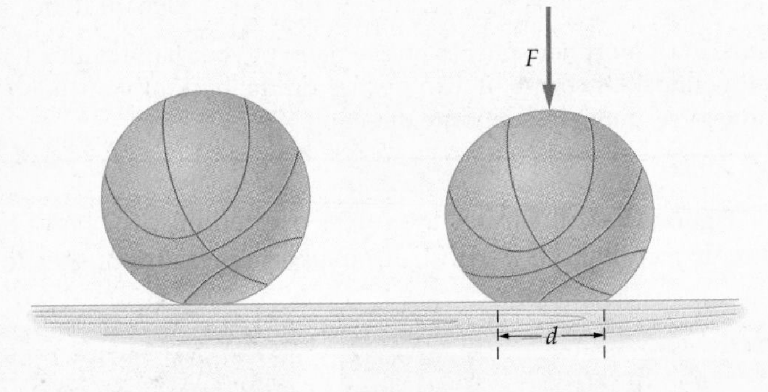

Strategy

To solve this problem, we have to make reasonable estimates of the force applied to the ball and the area of contact.

Suppose, for example, that we push down with a moderate force of 22 N (about 5 lb). The circular area of contact will probably have a diameter of about 2.0 centimeters. This can be verified by carrying out the experiment. Thus, given $F = 22\ \text{N}$ and $A = \pi(d/2)^2$ we can find the gauge pressure.

Solution

1. Using these estimates, calculate the gauge pressure, P_g:

$$P_g = \frac{F}{A} = \frac{22\ \text{N}}{\pi\left(\frac{0.020\ \text{m}}{2}\right)^2} = 7.0 \times 10^4\ \text{Pa}$$

Insight

Given that a pressure of one atmosphere, $P_{at} = 101\ \text{kPa} = 1.01 \times 10^5\ \text{Pa}$, corresponds to $14.7\ \text{lb/in}^2$, it follows that the gauge pressure of the ball, $P_g = 7.0 \times 10^4\ \text{Pa}$, corresponds to a pressure of roughly $10\ \text{lb/in}^2$. Thus, a basketball will typically have a gauge pressure in the neighborhood of $10\ \text{lb/in}^2$, and hence a total pressure inside the ball of about $25\ \text{lb/in}^2$.

Practice Problem

What is the diameter of the circular area of contact if a basketball with a $12\ \text{lb/in}^2$ gauge pressure is pushed down with a force of 44 N (about 10 lb)? [**Answer:** $d = 2.6\ \text{cm}$]

Some related homework problems: Problem 11, Problem 12

15–3 Static Equilibrium in Fluids: Pressure and Depth

Countless war movies have educated us on the perils of taking a submarine too deep. The hull creaks and groans, rivets start to pop, water begins to spray into the ship, and the captain keeps a close eye on the depth gauge. But what causes the pressure to increase as a submarine dives, and how much does it go up for a given increase in depth?

The answer to the first question is that the increased pressure is due to the added weight of water pressing on the submarine as it goes deeper. To see how this works, consider a cylindrical container filled to a height h with a fluid of density ρ, as in **Figure 15–2**. The top surface of the fluid is open to the atmosphere, with a pressure P_{at}. If the cross-sectional area of the container is A, the downward force exerted on the top surface by the atmosphere is

$$F_{top} = P_{at}A$$

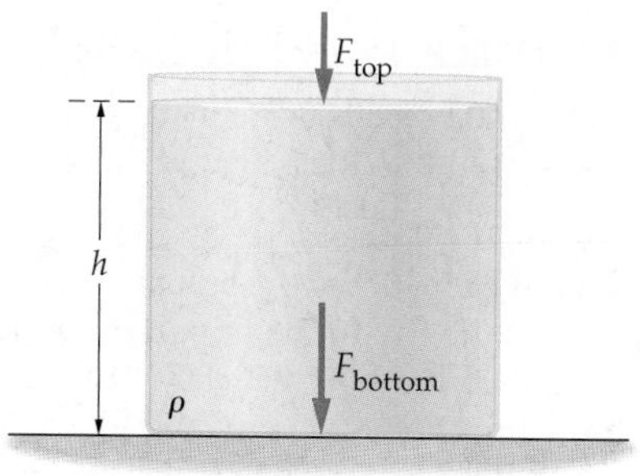

▲ **FIGURE 15–2 Pressure and the weight of a fluid**
The force pushing down on the bottom of the flask is greater than the force pushing down on the surface of the fluid. The difference in force is the weight of fluid in the flask.

Now, at the bottom of the container, the downward force is F_{top} *plus* the weight of the fluid. Recalling that $M = \rho V$, and that $V = hA$ for a cylinder of height h and area A, this weight is

$$W = Mg = \rho Vg = \rho(hA)g$$

Hence, we have

$$F_{bottom} = F_{top} + W = P_{at}A + \rho(hA)g$$

Finally, the pressure at the bottom is obtained by dividing F_{bottom} by the area A:

$$P_{bottom} = \frac{F_{bottom}}{A} = \frac{P_{at}A + \rho(hA)g}{A} = P_{at} + \rho gh$$

Of course, this relation holds not only for the bottom of the container, but for *any depth h below the surface*. Thus, the answer to the second question is that if the depth increases by the amount h, the pressure increases by the amount ρgh. At a depth h below the surface of a fluid, then, the pressure P is given by

$$P = P_{at} + \rho gh \qquad \textbf{15–6}$$

This expression holds for any fluid with constant density ρ and a pressure P_{at} at its upper surface.

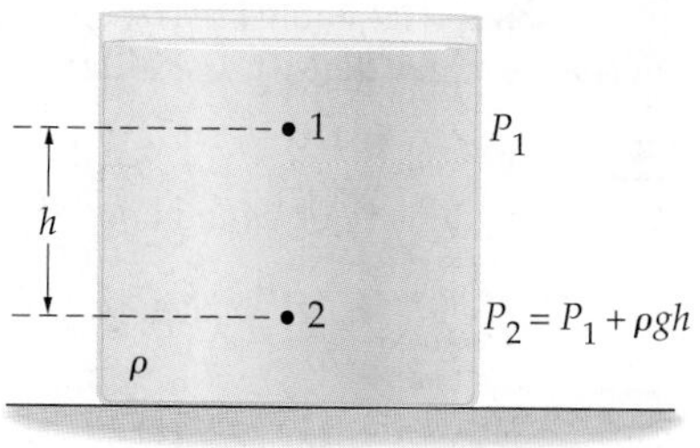

▲ **FIGURE 15–3 Pressure variation with depth**
If point 2 is deeper than point 1 by the amount h, its pressure is greater by the amount ρgh.

EXERCISE 15–2

The *Titanic* was found in 1985 lying on the bottom of the North Atlantic at a depth of 2.5 miles. What is the pressure at this depth?

Solution

Applying Equation 15–6 with $\rho = 1025\ \text{kg/m}^3$, we have

$$P = P_{at} + \rho gh = 1.01 \times 10^5\ \text{Pa} + (1025\ \text{kg/m}^3)(9.81\ \text{m/s}^2)\left(\frac{1609\ \text{m}}{1\ \text{mi}}\right)$$
$$= 4.1 \times 10^7\ \text{Pa}$$

This is about 400 atmospheres.

REAL-WORLD PHYSICS
Pressure at the wreck of the *Titanic*

The relation $P = P_{at} + \rho gh$ can be applied to any two points in a fluid. For example, if the pressure at one point is P_1, the pressure P_2 at a depth h below that point is the following:

Dependence of Pressure on Depth

$$P_2 = P_1 + \rho gh \qquad \textbf{15–7}$$

This relation is illustrated in **Figure 15–3**, and utilized in the next Example.

PROBLEM-SOLVING NOTE
Pressure Depends Only on Depth
The pressure in a static fluid depends only on the depth of the fluid. It is independent of the shape of the container.

EXAMPLE 15–3 Pressure and Depth

A cubical box 20.00 cm on a side is completely immersed in a fluid. At the top of the box the pressure is 105.0 kPa; at the bottom the pressure is 106.8 kPa. What is the density of the fluid?

Picture the Problem

Our sketch shows the box at an unknown depth d below the surface of the fluid. The important dimension for this problem is the height of the box, which is 20.00 cm. We are also given that the pressures at the top and bottom of the box are $P_1 = 105.0$ kPa and $P_2 = 106.8$ kPa, respectively.

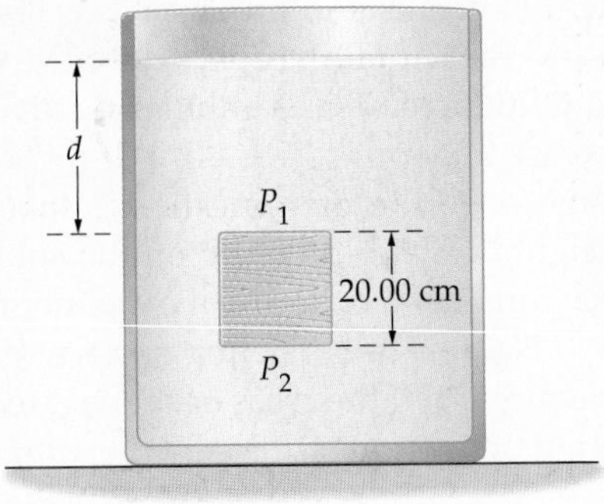

Strategy

The pressures at the top and bottom of the box are related by $P_2 = P_1 + \rho g h$. Since the pressures and the height of the box are given, this relation can be solved for the unknown density, ρ.

Solution

1. Solve $P_2 = P_1 + \rho g h$ for the density: $\rho = \dfrac{P_2 - P_1}{gh}$

2. Substitute numerical values: $\rho = \dfrac{1.068 \times 10^5 \text{ Pa} - 1.050 \times 10^5 \text{ Pa}}{(9.81 \text{ m/s}^2)(0.2000 \text{ m})} = 920 \text{ kg/m}^3$

Insight

Comparing with Table 15–1, it appears that the fluid in question may be olive oil. If the box had been immersed in water instead, with its greater density, the difference in pressure between the top and bottom of the box would have been greater as well.

Practice Problem

Given the density obtained above, what is the depth of fluid, d, at the top of the box? [**Answer:** $d = 0.44$ m]

Some related homework problems: Problem 13, Problem 18

CONCEPTUAL CHECKPOINT 15–2 The Size of Bubbles

One day, while swimming below the surface of the ocean, you let out a small bubble of air from your mouth. As the bubble rises toward the surface, does its diameter (a) increase, (b) decrease, or (c) stay the same?

Reasoning and Discussion

As the bubble rises, the pressure in the surrounding water decreases. This allows the air in the bubble to expand and occupy a larger volume.

Answer:

(a) The diameter of the bubble increases.

REAL-WORLD PHYSICS
The barometer

An interesting application of the variation of pressure with depth is the **barometer,** which can be used to measure atmospheric pressure. We consider here the simplest type of barometer, which was first proposed by Evangelista Torricelli (1608–1647) in 1643. First, fill a long glass tube—open at one end and closed at the other—with a fluid of density ρ. Next, invert the tube and place its open end below the surface of the same fluid in a bowl, as shown in **Figure 15–4**. Some of the fluid in the tube will flow into the bowl, leaving an empty space (vacuum) at the top. Enough will remain, however, to create a difference in level, h, between the fluid in the bowl and that in the tube.

The basic idea of the barometer is that this height difference is directly related to the atmospheric pressure that pushes down on the fluid in the bowl. To see how this works, first note that the pressure in the vacuum at the top of the tube is zero.

Hence, the pressure in the tube at a depth h below the vacuum is $0 + \rho g h = \rho g h$. Now, at the level of the fluid in the bowl we know that the pressure is one atmosphere, P_{at}. Therefore, it follows that

$$P_{at} = \rho g h$$

If these pressures were not the same, there would be a pressure difference between the fluid in the tube and that in the bowl, resulting in a net force and a flow of fluid. Thus, a measurement of h immediately gives atmospheric pressure.

A fluid that is often used in such a barometer is mercury (Hg), with a density of $\rho = 1.3595 \times 10^4\ \mathrm{kg/m^3}$. The corresponding height for a column of mercury is

$$h = \frac{P_{at}}{\rho g} = \frac{1.013 \times 10^5\ \mathrm{Pa}}{(1.3595 \times 10^4\ \mathrm{kg/m^3})(9.81\ \mathrm{m/s^2})} = 760\ \mathrm{mm}$$

In fact, atmospheric pressure is *defined* in terms of millimeters of mercury (mmHg):

$$1\ \text{atmosphere} = P_{at} = 760\ \mathrm{mmHg}$$

Table 15–2 summarizes the various expressions we have developed for atmospheric pressure.

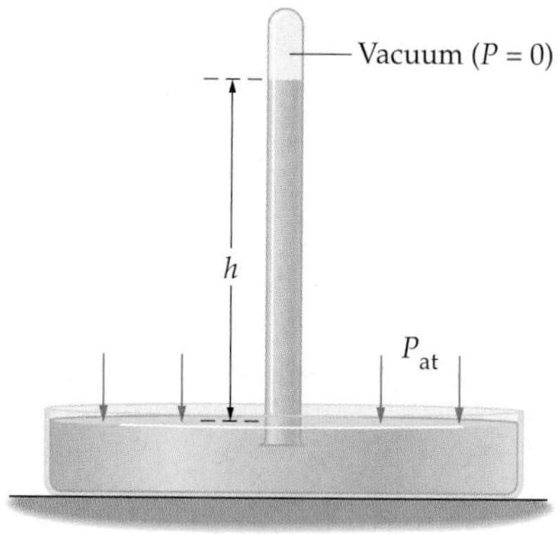

▲ **FIGURE 15–4 A simple barometer**
Atmospheric pressure, P_{at}, is related to the height of fluid in the tube by the relation $P_{at} = \rho g h$.

TABLE 15–2 Atmospheric Pressure

1 atmosphere	$= P_{at}$
	= 760 mmHg (definition)
	$= 14.7\ \mathrm{lb/in^2}$
	= 101 kPa
	$= 101\ \mathrm{kN/m^2}$
	~ 1 bar = 100 kPa

Water Seeks Its Own Level

We are all familiar with the aphorism that water seeks its own level. In order for this to hold true, however, it is necessary that the pressure at the surface of the water (or other fluid) be the same everywhere on the surface. This was not the case for the barometer just discussed, where the pressure was P_{at} on one portion of the surface and zero on another. Let's take a moment, then, to consider the level assumed by a fluid with constant pressure on its surface. In doing so, we shall apply considerations involving force, pressure, and energy.

First, the force–pressure point of view. In **Figure 15–5 (a)** we show a U-shaped tube containing a quantity of fluid of density ρ. The fluid rises to the same level in each arm of the U, where it is open to the atmosphere. Therefore, the pressure at the base of each arm is the same; $P_{at} + \rho g h$. Thus, the fluid in the horizontal section of the U is pushed with equal force from each side, giving zero net force. As a result, the fluid remains at rest.

On the other hand, in **Figure 15–5 (b)** the two arms of the U are filled to different levels. Therefore, the pressure at the base of the two arms is different, with the greater pressure at the base of the right arm. The fluid in the horizontal section, then, will experience a net force to the left, causing it to move in that direction. This will tend to equalize the fluid levels of the two arms.

We can arrive at the same conclusion on the basis of energy minimization. Consider a U tube that is initially filled to the same level in both arms, as in Figure 15–5 (a). Now, consider moving a small element of fluid from one arm to the other, to create different levels, as in **Figure 15–6**. In moving this fluid element to the other arm, it is necessary to lift it upward. This, in turn, causes its potential energy to increase. Since nothing else in the system has changed its position, the only change in potential energy is the increase experienced by the element. Thus, we conclude that the system has a minimum energy when the fluid levels are the same and a higher energy when the levels are different. Just as a ball rolls to the bottom of a hill, where its energy is minimized, the fluid seeks its own level and a minimum energy.

If two different liquids, with different densities, are combined in the same U tube, the levels in the arms are not the same. Still, the pressures at the base of each arm must be equal, as before. This is discussed in the next Example.

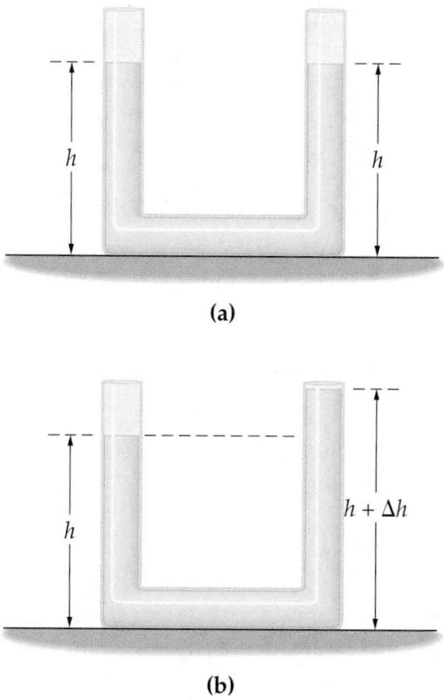

▶ **FIGURE 15–5 Fluids seek their own level**
(a) When the levels are equal, the pressure is the same at the base of each arm of the U tube. As a result, the fluid in the horizontal section of the U is in equilibrium. **(b)** With unequal heights, the pressures are different. In this case, the pressure is greater at the base of the right arm, hence fluid will flow toward the left and the levels will equalize.

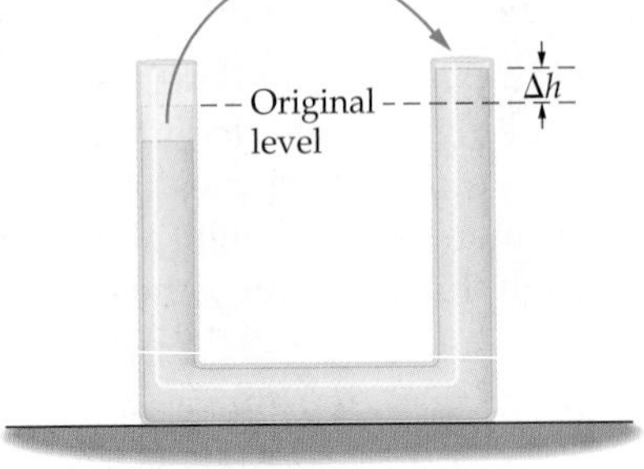

▲ **FIGURE 15–6 Gravitational potential energy of a fluid**

In order to create unequal levels in the two arms of the U tube, an element of fluid must be raised by the height Δh. This increases the gravitational potential energy of the system. The lowest potential energy corresponds to equal levels.

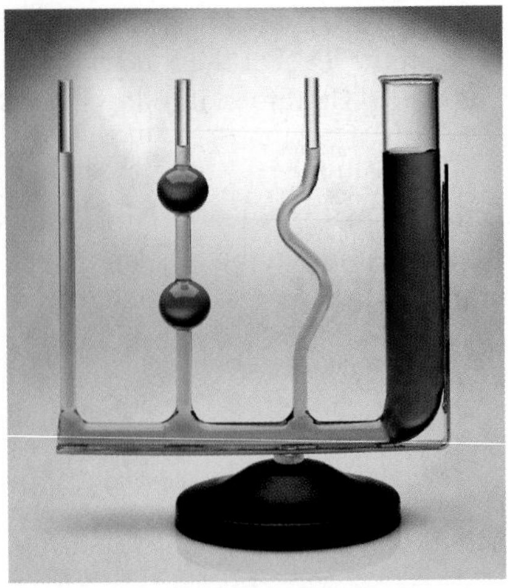

▲ The containers shown here are connected at the bottom by a hollow tube, which allows fluid to flow freely between them. As a result the fluid level is the same in each container regardless of its shape and size.

EXAMPLE 15–4 Oil and Water Don't Mix

A U-shaped tube is filled mostly with water, but a small amount of vegetable oil has been added to one side, as shown in the sketch. The density of the water is 1.00×10^3 kg/m^3, and the density of the vegetable oil is 9.20×10^2 kg/m^3. If the depth of the oil is 5.00 cm, what is the difference in level h between the top of the oil on one side of the U and the top of the water on the other side?

Picture the Problem

The U-shaped tube and the relevant dimensions are shown in our sketch. In particular, note that the depth of the oil is 5.00 cm, and that the oil rises to a greater height on its side of the U than the water does on its side. This is due to the oil having the lower density. Both sides of the U are open to the atmosphere, so the pressure at the top surfaces is P_{at}.

Strategy

For the system to be in equilibrium, it is necessary that the pressure be the same at the bottom of each side of the U; that is, at points C and D. If the pressure is the same at C and D, it will remain equal as one moves up through the water to the points A and B. Above this point the pressures will differ because of the presence of the oil.

Therefore, setting the pressure at point A equal to the pressure at point B determines the depth h_1 in terms of the known depth h_2. It follows that the difference in level between the two sides of the U is simply $h = h_2 - h_1$.

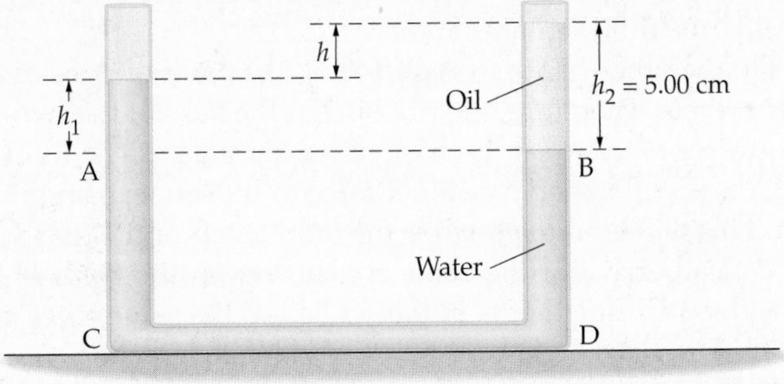

INTERACTIVE FIGURE

Solution

1. Find the pressure at point A, where the depth of the water is h_1: $P_A = P_{at} + \rho_{water} g h_1$

2. Find the pressure at point B, where the depth of the oil is $h_2 = 5.00$ cm: $P_B = P_{at} + \rho_{oil} g h_2$

3. Set P_A equal to P_B: $P_{at} + \rho_{water} g h_1 = P_{at} + \rho_{oil} g h_2$

4. Solve for the depth of the water, h_1, and substitute numerical values:

$$h_1 = h_2\left(\frac{\rho_{oil}}{\rho_{water}}\right)$$

$$= (5.00\ \text{cm})\left(\frac{9.20 \times 10^2\ \text{kg/m}^3}{1.00 \times 10^3\ \text{kg/m}^3}\right) = 4.60\ \text{cm}$$

5. Calculate the difference in levels between the water and oil sides of the U: $h = h_2 - h_1 = 5.00\ \text{cm} - 4.60\ \text{cm} = 0.40\ \text{cm}$

Insight

Note that the weight of the height h_1 of water is equal to the weight of the height h_2 of oil. This is a special case, however, and is due to the fact that our U has equal diameters on its two sides. If the oil side of the U had been wider, for example, the weight of the oil would have been greater, though the difference in height still would have been $h = 0.40$ cm. It is the pressure that matters in a system like this, not the weight. (For a similar situation, see the photo on page 484.)

Practice Problem

Find the pressure at points A and B. [**Answer:** $P_A = P_B = P_{at} + 451$ Pa]

Some related homework problems: Problem 16, Problem 22, Problem 23

Pascal's Principle

Recall from Equation 15–6 that if the surface of a fluid of density ρ is exposed to the atmosphere with a pressure P_{at}, the pressure at a depth h below the surface is

$$P = P_{at} + \rho g h$$

Suppose, now, that atmospheric pressure is increased from P_{at} to $P_{at} + \Delta P$. As a result, the pressure at the depth h is

$$P = P_{at} + \Delta P + \rho g h = (P_{at} + \rho g h) + \Delta P$$

Thus, by increasing the pressure at the top of the fluid by the amount ΔP, we have increased it by the same amount everywhere in the fluid. This is **Pascal's principle:**

> An external pressure applied to an enclosed fluid is transmitted unchanged to every point within the fluid.

A classic example of Pascal's principle at work is the **hydraulic lift,** which is shown schematically in **Figure 15–7**. Here we see two cylinders, one of cross-sectional area A_1, the other of cross-sectional area $A_2 > A_1$. The cylinders, each of which is fitted with a piston, are connected by a tube and filled with a fluid. Initially the pistons are at the same level and exposed to the atmosphere.

REAL-WORLD PHYSICS
The hydraulic lift

Now, suppose we push down on piston 1 with the force F_1. This increases the pressure in that cylinder by the amount

$$\Delta P = F_1/A_1$$

By Pascal's principle, the pressure in cylinder 2 increases by the *same* amount. Therefore, the increased upward force on piston 2 due to the fluid is

$$F_2 = (\Delta P)A_2$$

Substituting the increase in pressure, $\Delta P = F_1/A_1$, we find

$$F_2 = (F_1/A_1)A_2 = F_1\left(\frac{A_2}{A_1}\right) > F_1 \qquad \text{15–8}$$

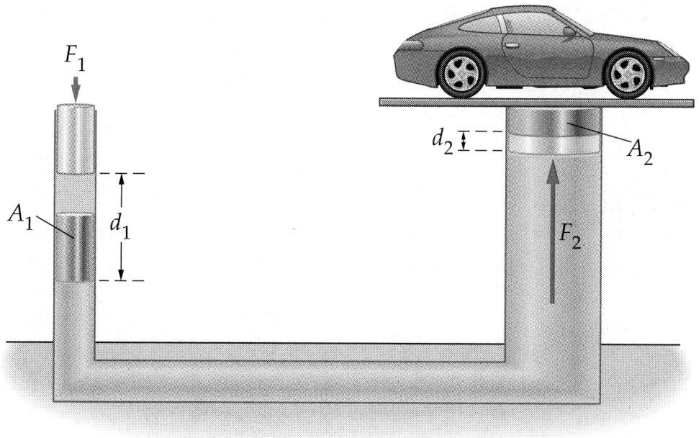

◀ **FIGURE 15–7 A hydraulic lift**
A force F_1 exerted on the small piston causes a much larger force, F_2, to act on the large piston.

To be specific, let's assume that A_2 is 100 times greater than A_1. Then, by pushing down on piston 1 with a force F_1, we push upward on piston 2 with a force of $100F_1$. Our force has been magnified 100 times!

If this sounds too good to be true, rest assured that we are not getting something for nothing. Just as with a lever, there is a tradeoff between the distance through which a force must be applied and the force magnification. This is illustrated in Figure 15–7, where we show piston 1 being pushed down through a distance d_1. This displaces a volume of fluid equal to A_1d_1. The same volume flows into cylinder 2, where it causes piston 2 to rise through a distance d_2. Equating the two volumes, we have

$$A_1d_1 = A_2d_2$$

or

$$d_2 = d_1\left(\frac{A_1}{A_2}\right)$$

Thus, in the example just given, if we move piston 1 down a distance d_1, piston 2 rises a distance $d_2 = d_1/100$. Our force at piston 2 has been magnified 100 times, but the distance it moves has been reduced 100 times.

EXERCISE 15–3

To inspect a 14,500-N car, it is raised with a hydraulic lift. If the radius of the small piston in Figure 15–7 is 4.0 cm, and the radius of the large piston is 17 cm, find the force that must be exerted on the small piston to lift the car.

Solution

Solving Equation 15–8 for F_1, and noting that the area is πr^2, we find

$$F_1 = F_2\left(\frac{A_1}{A_2}\right) = (14{,}500\text{ N})\left[\frac{\pi(0.040\text{ m})^2}{\pi(0.17\text{ m})^2}\right] = 800\text{ N}$$

15–4 Archimedes' Principle and Buoyancy

The fact that a fluid's pressure increases with depth leads to many interesting consequences. Among them is the fact that a fluid exerts a net upward force on any object it surrounds. This is referred to as a **buoyant force.**

To see the origin of buoyancy, consider a cubical block immersed in a fluid of density ρ, as in **Figure 15–8**. The surrounding fluid exerts normal forces on all of its faces. Clearly, the horizontal forces pushing to the right and to the left are equal, hence they cancel and have no effect on the block.

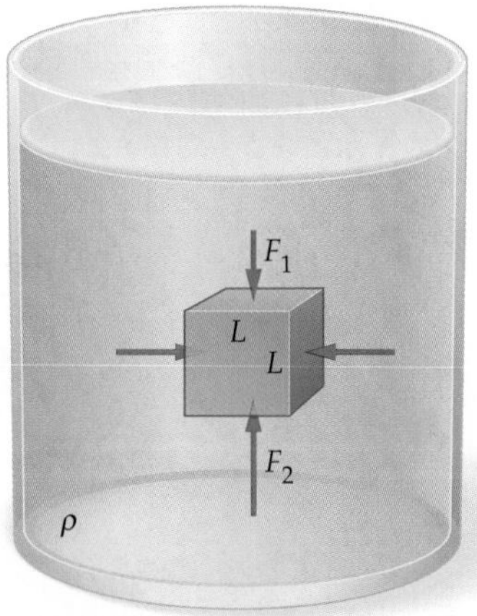

▲ FIGURE 15–8 Buoyant force due to a fluid

A fluid surrounding an object exerts a buoyant force in the upward direction. This is due to the fact that pressure increases with depth, and hence the upward force on the object, F_2, is greater than the downward force, F_1. Forces acting to the left and to the right cancel.

The situation is quite different for the vertical forces, however. Note, for example, that the downward force exerted on the top face is less than the upward force exerted on the lower face, since the pressure at the lower face is greater. This difference in forces gives rise to a net upward force—the buoyant force.

Let's calculate the buoyant force acting on the block. First, we assume that the cubical block is of length L on a side, and that the pressure on the top surface is P_1. The downward force on the block, then, is

$$F_1 = P_1A = P_1L^2$$

Note that we have used the fact that the area of a square face of side L is L^2. Next, we consider the bottom face. The pressure there is given by Equation 15–7, with a difference in depth of $h = L$:

$$P_2 = P_1 + \rho gL$$

Therefore, the upward force exerted on the bottom face of the cube is

$$F_2 = P_2A = (P_1 + \rho gL)L^2 = P_1L^2 + \rho gL^3 = F_1 + \rho gL^3$$

If we take upward as the positive direction, the net vertical force exerted by the fluid on the block—that is, the buoyant force, F_b—is

$$F_b = F_2 - F_1 = \rho g L^3$$

As expected, the block experiences a net upward force from the surrounding fluid.

The precise value of the buoyant force is of some significance, as we now show. First, note that the volume of the cube is L^3. It follows that $\rho g L^3$ is the weight of *fluid* that would occupy the same volume as the cube. Therefore, the buoyant force is equal to the weight of fluid that is displaced by the cube. This is a special case of **Archimedes' principle:**

> An object completely immersed in a fluid experiences an upward buoyant force equal in magnitude to the weight of fluid displaced by the object.

More generally, if a volume V of an object is immersed in a fluid of density ρ_{fluid}, the buoyant force can be expressed as follows:

Buoyant Force When a Volume V Is Submerged in a Fluid of Density ρ_{fluid}

$$F_b = \rho_{\text{fluid}} g V \qquad 15\text{–}9$$

SI unit: N

The volume V may be the total volume of the object, or any fraction of the total volume.

To see that Archimedes' principle is completely general, consider the submerged object shown in **Figure 15–9 (a)**. If we were to replace this object with an equivalent volume of fluid, as in **Figure 15–9 (b)**, the container would hold nothing but fluid and would be in static equilibrium. As a result, we conclude that the net buoyant force acting on this "fluid object" must be upward and equal in magnitude to its weight. Now here is the key idea: Since the original object and the fluid object occupy the same position, the forces acting on their surfaces are identical, and hence the net buoyant force is the same for both objects. Therefore, the original object experiences a buoyant force equal to the weight of fluid that it displaces—that is, equal to the weight of the fluid object.

PROBLEM-SOLVING NOTE

The Buoyant Force

Note that the buoyant force is equal to the weight of displaced fluid. It does not depend on the weight of the object that displaces the fluid.

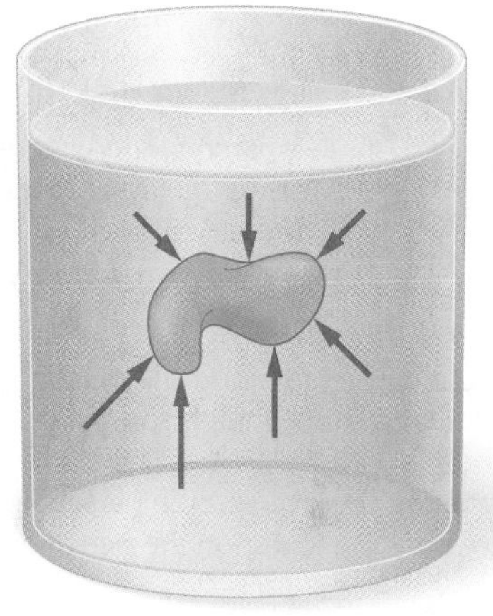

(a)

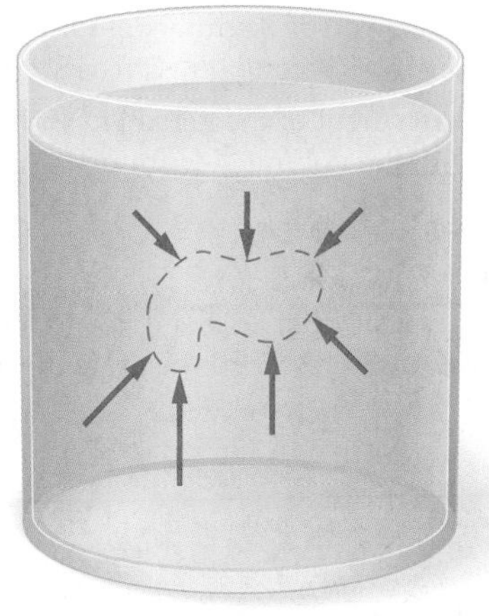

(b)

▶ FIGURE 15–9 Buoyant force equals the weight of displaced fluid
The buoyant force acting on the object in **(a)** is equal to the weight of the "fluid object" (with the same size and shape) in **(b)**. This is because the fluid in (b) is at rest; hence the buoyant force acting on the fluid object must cancel its weight. The same forces act on the original object in (a), however, and therefore the buoyant force it experiences is also equal to the weight of the fluid object.

CONCEPTUAL CHECKPOINT 15–3 How Is the Scale Reading Affected?

A flask of water rests on a scale. If you dip your finger into the water, without touching the flask, does the reading on the scale **(a)** increase, **(b)** decrease, or **(c)** stay the same?

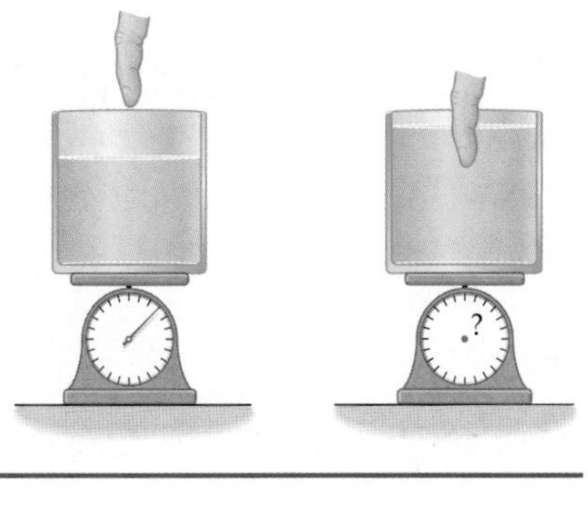

Reasoning and Discussion

Your finger experiences an upward buoyant force when it is dipped into the water. By Newton's third law, the water experiences an equal and opposite reaction force acting downward. This downward force is transmitted to the scale, which in turn gives a higher reading.

Another way to look at this result is to note that when you dip your finger into the water, its depth increases. This results in a greater pressure at the bottom of the flask, and hence a greater downward force on the flask. The scale reads this increased downward force.

Answer:

(a) The reading on the scale increases.

15–5 Applications of Archimedes' Principle

In this section we consider a variety of applications of Archimedes' principle. We begin with situations in which an object is fully immersed. Later we consider systems in which an object floats.

Complete Submersion

An interesting application of complete submersion can be found in an apparatus commonly used in determining a person's body-fat percentage. We consider the basic physics of the apparatus and the measurement procedure in the next Example. Following the Example, we derive the relation between overall body density and the body-fat percentage.

EXAMPLE 15–5 Measuring the Body's Density

REAL-WORLD PHYSICS: BIO

A person who weighs 720.0 N in air is lowered into a tank of water to about chin level. He sits in a harness of negligible mass suspended from a scale that reads his apparent weight. He now exhales as much air as possible and dunks his head underwater, submerging his entire body. If his apparent weight while submerged is 34.3 N, find **(a)** his volume and **(b)** his density.

Picture the Problem

The scale, the tank, and the person are shown in our sketch. We also show the free-body diagram for the person. Note that the weight of the person in air is designated by $W = mg = 720.0$ N, and the apparent weight in water, which is the upward force exerted by the scale on the person, is designated by $W_a = 34.3$ N. The buoyant force exerted by the water on the person is F_b.

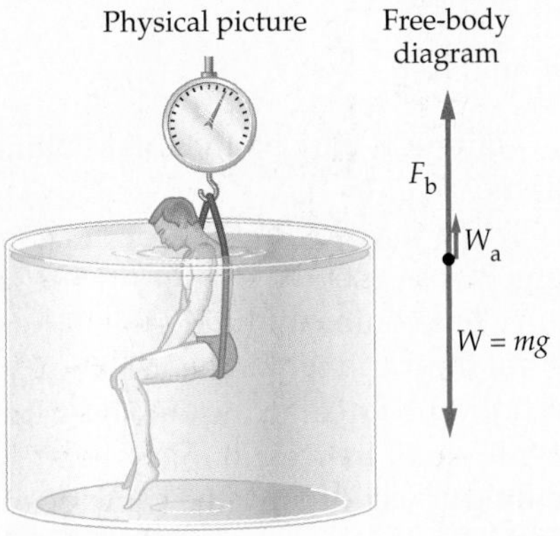

Strategy

To find the volume, V_p, and density, ρ_p, of the person, we must use two separate conditions. They are as follows:

(a) When the person is submerged, the surrounding water exerts an upward buoyant force given by Archimedes' principle: $F_b = \rho_{water} V_p g$. This relation, and Newton's second law, can be used to determine V_p.

(b) The weight of the person in air is $W = mg = \rho_p V_p g$. Combining this relation with the volume, V_p, found in part (a) allows us to determine the density, ρ_p.

Solution

Part (a)

1. Apply Newton's second law to the person. Note that the person remains at rest, and therefore the net force acting on him is zero:

$$W_a + F_b - W = 0$$

2. Substitute $F_b = \rho_{water} V_p g$ and solve for V_p:

$$W_a + \rho_{water} V_p g - W = 0$$

$$V_p = \frac{W - W_a}{\rho_{water} g} = \frac{720.0\ \text{N} - 34.3\ \text{N}}{(1.00 \times 10^3\ \text{kg/m}^3)(9.81\ \text{m/s}^2)}$$

$$= 6.99 \times 10^{-2}\ \text{m}^3$$

Part (b)

3. Use $W = \rho_p V_p g$ to solve for the density of the person, ρ_p:

$$W = \rho_p V_p g$$

$$\rho_p = \frac{W}{V_p g} = \frac{720.0\ \text{N}}{(6.99 \times 10^{-2}\ \text{m}^3)(9.81\ \text{m/s}^2)}$$

$$= 1050\ \text{kg/m}^3$$

Insight

As in Conceptual Checkpoint 15–3, the water exerts an upward buoyant force on the person, and an equal and opposite reaction force acts downward on the tank and water, making it press against the floor with a greater force.

In addition, notice that the density of the person (1050 kg/m^3) is only slightly greater than the density of seawater (1025 kg/m^3), as given in Table 15-1.

Practice Problem

The person can float in water if his lungs are partially filled with air, increasing the volume of his body. What volume must his body have to just float? [**Answer:** $V_p = 7.34 \times 10^{-2}\ \text{m}^3$]

Some related homework problems: Problem 33, Problem 34

Once the overall density of the body is determined, the percentage of body fat can be obtained by noting that body fat has a density of $\rho_f = 9.00 \times 10^2 \text{ kg/m}^3$, whereas the lean body mass (muscles and bone) has a density of $\rho_l = 1.10 \times 10^3 \text{ kg/m}^3$. Suppose, for example, that a fraction x_f of the total body mass M is fat mass, and a fraction $(1 - x_f)$ is lean mass; that is, the fat mass is $m_f = x_f M$ and the lean mass is $m_l = (1 - x_f)M$. The total volume of the body is $V = V_f + V_l$. Using the fact that $V = m/\rho$, we can write the total volume as $V = m_f/\rho_f + m_l/\rho_l = x_f M/\rho_f + (1 - x_f)M/\rho_l$. Combining these results, the overall density of a person's body, ρ_p, is

$$\rho_p = \frac{M}{V} = \frac{1}{\dfrac{x_f}{\rho_f} + \dfrac{(1 - x_f)}{\rho_l}}$$

Solving for the body-fat fraction, x_f, yields

$$x_f = \frac{1}{\rho_p}\left(\frac{\rho_l \rho_f}{\rho_l - \rho_f}\right) - \frac{\rho_f}{\rho_l - \rho_f}$$

Finally, substituting the values for ρ_f and ρ_l, we find

$$x_f = \frac{(4950 \text{ kg/m}^3)}{\rho_p} - 4.50$$

This result is known as *Siri's formula*. For example, if $\rho_p = 900 \text{ kg/m}^3$ (all fat), we find $x_f = 1$; if $\rho_p = 1100 \text{ kg/m}^3$ (no fat), we find $x_f = 0$. In the case of Example 15–5, where $\rho_p = 1050 \text{ kg/m}^3$, we find that this person's body-fat fraction is $x_f = 0.214$, for a percentage of 21.4%. This is a reasonable value for a healthy adult male.

A recent refinement to the measurement of body-fat percentage is the Bod Pod, an egg-shaped, air-tight chamber in which a person sits comfortably—high and dry, surrounded only by air. This device works on the same physical principle as submerging a person in water, only it uses air instead of water. Since air is about a thousand times less dense than water, the measurements of apparent weight must be roughly a thousand times more sensitive. Fortunately, this is possible with today's technology, allowing for a much more convenient means of measurement.

Next we consider a low-density object, such as a piece of wood, held down below the surface of a denser fluid.

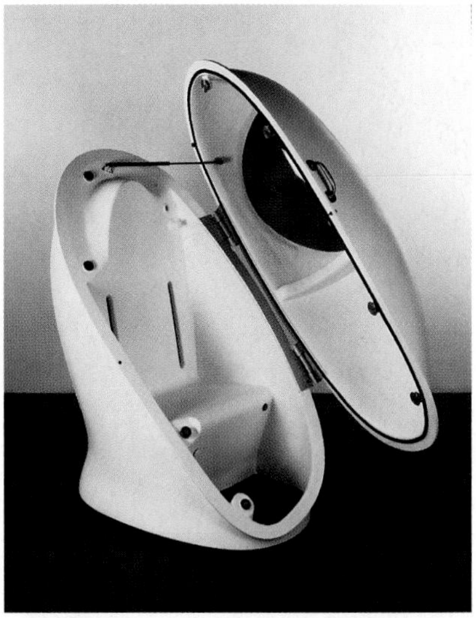

▲ This device, known as the Bod Pod, measures the body-fat percentage of a person inside it by varying the air pressure in the chamber and measuring the corresponding changes in the person's apparent weight. Archimedes' principle is at work here, just as it is in Example 15–5.

REAL-WORLD PHYSICS: BIO
Measuring body fat

ACTIVE EXAMPLE 15–1 Find the Tension in the String

A piece of wood with a density of 706 kg/m^3 is tied with a string to the bottom of a water-filled flask. The wood is completely immersed, and has a volume of $8.00 \times 10^{-6} \text{ m}^3$. What is the tension in the string?

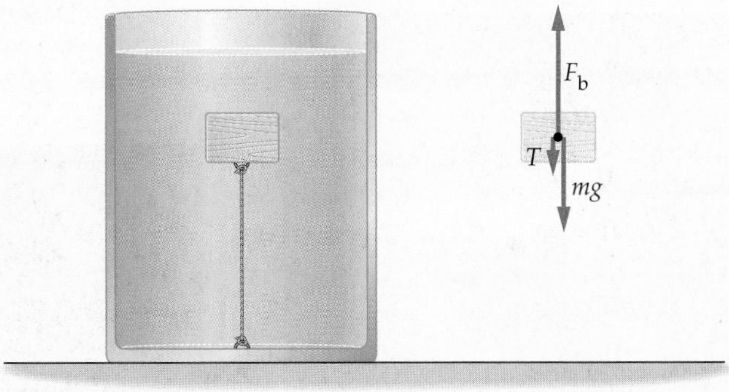

Solution *(Test your understanding by performing the calculations indicated in each step.)*

1. Apply Newton's second law to the wood:	$F_b - T - mg = 0$
2. Solve for the tension, T:	$T = F_b - mg$
3. Calculate the weight of the wood:	$mg = 0.0554 \text{ N}$
4. Calculate the buoyant force:	$F_b = 0.0785 \text{ N}$
5. Subtract to obtain the tension:	$T = 0.0231 \text{ N}$

Insight
Since the wood floats in water, its buoyant force when completely immersed is greater than its weight.

Your Turn
What is the tension in the string if the piece of wood has a density of 822 kg/m^3?

(Answers to **Your Turn** *problems can be found in the back of the book.)*

(a)

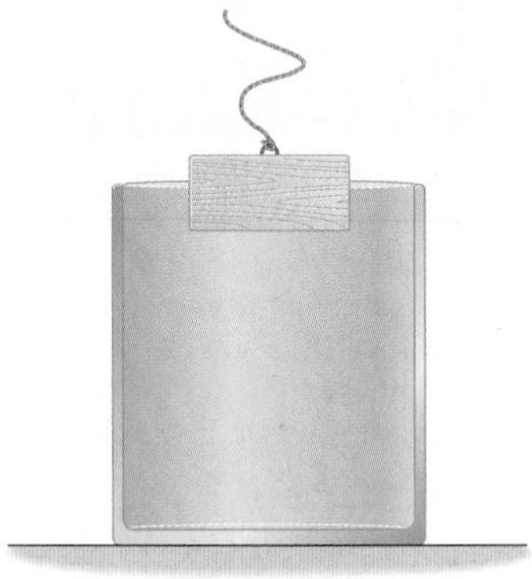

(b)

▲ **FIGURE 15–10 Floatation**

(a) The block of wood displaces some water, but not enough to equal its weight. Thus, the block would not float at this position. **(b)** The weight of displaced water equals the weight of the block in this case. The block floats now.

Floatation

When an object floats, the buoyant force acting on it equals its weight. For example, suppose we slowly lower a block of wood into a flask of water. At first, as in **Figure 15–10 (a)**, only a small amount of water is displaced and the buoyant force is a fraction of the block's weight. If we were to release the block now, it would drop farther into the water. As we continue to lower the block, more water is displaced, increasing the buoyant force.

Eventually, we reach the situation pictured in **Figure 15–10 (b)**, where the block begins to float. In this case, the buoyant force equals the weight of the wood. This, in turn, means that the weight of the displaced water is equal to the weight of the wood. In general,

> An object floats when it displaces an amount of fluid whose weight is equal to the weight of the object.

This is illustrated in **Figure 15–11 (a)**, where we show the volume of water equal to the weight of a block of wood. Similarly, in **Figure 15–11 (b)** we show the amount of water necessary to have the same weight as a block of metal. Clearly, if the metal is completely submerged, the buoyant force is only a fraction of its weight, and so it sinks. On the other hand, if the metal is formed into the shape of a bowl, as in **Figure 15–11 (c)**, it can displace a volume of water equal to its weight and float.

(a) (b) (c)

▲ **FIGURE 15–11 Floating an object that is more dense than water**

(a) A wood block and the volume of water that has the same weight. Because the wood has a larger volume than this, it floats. **(b)** A metal block and the volume of water that has the same weight. Since the metal displaces less water than this, it sinks. **(c)** If the metal in (b) is shaped like a bowl, it can displace more water than the volume of the metal itself. In fact, it can displace enough water to float.

▲ The water of the Dead Sea is unusually dense because of its great salt content. As a result, swimmers can float higher in the water than they are accustomed and engage in recreational activities that we don't ordinarily associate with a dip in the ocean.

▲ Although steel is denser than water, a ship's bowl-like hull (left) displaces enough water to allow it to float, so long as it is not too heavily loaded. (The boundary between the red and black areas of the hull is the Plimsoll line, which indicates where the ship should ride in the water when carrying its maximum safe load—see Conceptual Checkpoint 15–4.) For balloons (right), the key to buoyancy is the lower density and greater volume of hot air. As the air in the balloon is heated, it expands the bag, increasing the volume of air that the balloon displaces. At the same time, heated air spills out of the balloon, decreasing its weight. Eventually, the average density of the balloon plus the hot air it contains becomes lower than that of the surrounding air, and it starts to float.

Another way to change the buoyancy of an object is to alter its overall density. Consider, for example, the *Cartesian diver* shown in **Figure 15–12**. As illustrated, the diver is simply a small glass tube with an air bubble trapped inside. Initially, the overall density of the tube and the air bubble is less than the density of water, and the diver floats. When the bottle containing the diver is squeezed, however, the pressure in the water rises, and the air bubble is compressed to a smaller volume. Now, the overall density of the tube and air bubble is greater than that of water, and the diver descends. By adjusting the pressure on the bottle, the diver can be made to float at any depth in the bottle.

The same principle applies to the swim bladder of ray-finned bony fishes. The swim bladder is basically an air sac whose volume can be controlled by the fish. By adjusting the size of the swim bladder, the fish can give itself "neutral buoyancy"—that is, the fish can float without effort at a given depth, just like the Cartesian diver. Similar considerations apply to certain diving sea mammals, such as the bottlenose dolphin, Weddell seal, elephant seal, and blue whale. All of these animals are capable of diving to great depths—in fact, some of the seals have been observed at depths of nearly 400 m. In order to conserve energy on their long dives, they take advantage of the fact that the pressure of the surrounding water compresses their bodies and flattens the air sacs in their lungs. Just as with the Cartesian diver, this decreases their buoyancy to the point where they begin to sink. As a result, they can glide effortlessly to the desired depth, saving energy for the swim back to the surface.

REAL-WORLD PHYSICS: BIO
The swim bladder in fish

REAL-WORLD PHYSICS: BIO
Diving sea mammals

ACTIVE EXAMPLE 15–2 Floating a Block of Wood

How much water (density $1.00 \times 10^3\ \mathrm{kg/m^3}$) must be displaced to float a cubical block of wood (density $655\ \mathrm{kg/m^3}$) that is 15.0 cm on a side?

Solution *(Test your understanding by performing the calculations indicated in each step.)*

1. Calculate the volume of the wood: $V_{\text{wood}} = 3.38 \times 10^{-3}\ \mathrm{m^3}$
2. Find the weight of the wood: $\rho_{\text{wood}} V_{\text{wood}}\, g = 21.7\ \mathrm{N}$
3. Write an expression for the weight of a volume of water: $\rho_{\text{water}} V_{\text{water}}\, g$
4. Set the weight of water equal to the weight of the wood: $\rho_{\text{water}} V_{\text{water}}\, g = 21.7\ \mathrm{N}$
5. Solve for the volume of water: $V_{\text{water}} = 2.21 \times 10^{-3}\ \mathrm{m^3}$

Insight
As expected, only a fraction of the wood must be submerged in order for it to float.

Your Turn
What volume of water must be displaced if the density of the wood is $955\ \mathrm{kg/m^3}$? Compare this volume to the volume of the wood itself.

(Answers to **Your Turn** *problems can be found in the back of the book.)*

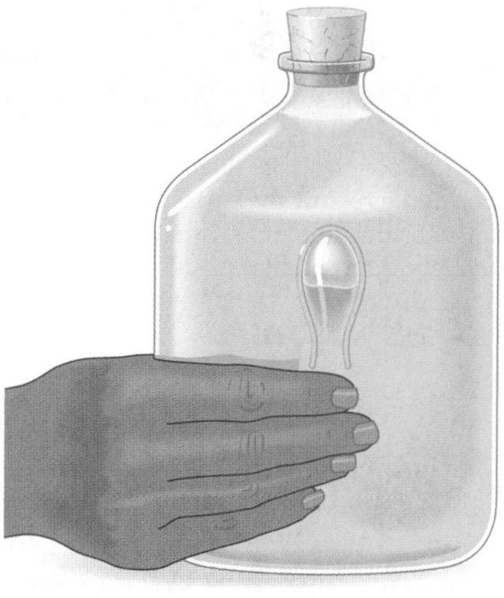

▲ **FIGURE 15–12 A Cartesian diver**
A Cartesian diver floats because of the bubble of air trapped within it. When the bottle is squeezed, increasing the pressure in the water, the bubble is reduced in size and the diver sinks.

CONCEPTUAL CHECKPOINT 15–4 The Plimsoll Mark

On the side of a cargo ship you may see a horizontal line indicating "maximum load." (It is sometimes known as the "Plimsoll mark," after the nineteenth-century British legislator who caused it to be adopted.) When a ship is loaded to capacity, the maximum load line is at water level. The ship shown here has two maximum load lines, one for fresh water and one for salt water. Which line should be marked "maximum load for salt water": **(a)** the top line or **(b)** the bottom line?

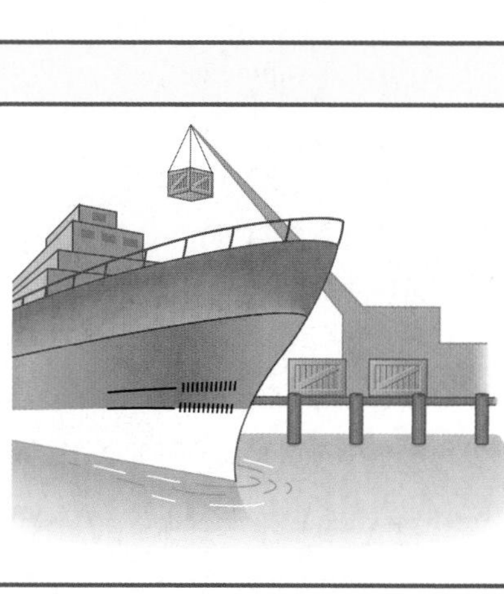

Reasoning and Discussion
If a ship sails from fresh water into salt water it floats higher, just as it is easier for you to float in an ocean than in a lake. The reason is that salt water is denser than fresh water; hence less of it needs to be displaced to provide a given buoyant force. Since the ship floats higher in salt water, the bottom line should be used to indicate maximum load.

Answer:
(b) The bottom line should be used in salt water.

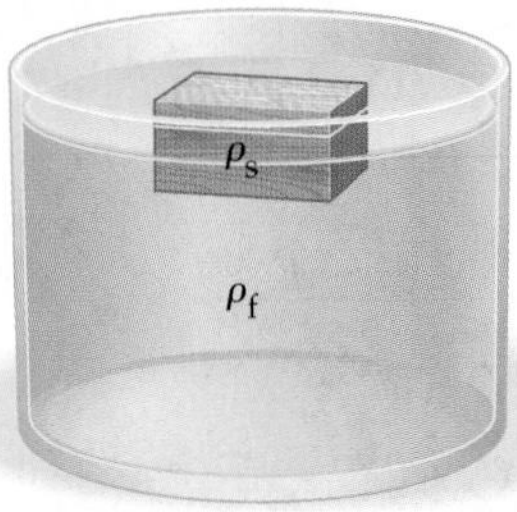

▲ **FIGURE 15–13 Submerged volume of a floating object**
A solid, of volume V_s and density ρ_s, floats in a fluid of density ρ_f. The volume of the solid that is submerged is $V_{sub} = V_s(\rho_s/\rho_f)$.

▲ Most people know that the bulk of an iceberg lies below the surface of the water. But as with ships and swimmers, the actual proportion that is submerged depends on whether the water is fresh or salt (see Example 15–6).

Tip of the Iceberg

As we have seen, an object floats when its weight is equal to the weight of the fluid it displaces. Let's use this condition to determine just how much of a floating object is submerged. We will then apply our result to the classic case of an iceberg.

Consider, then, a solid of density ρ_s floating in a fluid of density ρ_f, as in **Figure 15–13**. If the solid has a volume V_s, its total weight is

$$W_s = \rho_s V_s g$$

Similarly, the weight of a volume V_f of displaced fluid is

$$W_f = \rho_f V_f g$$

Equating these weights, we find the following:

$$W_s = W_f$$
$$\rho_s V_s g = \rho_f V_f g$$

Canceling g, and solving for the volume of displaced fluid, we have

$$V_f = V_s(\rho_s/\rho_f)$$

Since, by definition, the volume of displaced fluid, V_f, is the same as the volume of the solid that is submerged, V_{sub}, we find

Submerged Volume V_{sub} for a Solid of Volume V_s and Density ρ_s Floating in a Fluid of Density ρ_f

$$V_{sub} = V_s(\rho_s/\rho_f) \qquad \textbf{15–10}$$

SI unit: m^3

Note that this relation agrees with the results of Active Example 15–2.

We now apply this result to ice floating in water.

EXAMPLE 15–6 The Tip of the Iceberg

REAL-WORLD PHYSICS What percentage of a floating chunk of ice projects above the level of the water? Assume a density of $917\ kg/m^3$ for the ice and $1.00 \times 10^3\ kg/m^3$ for the water.

Picture the Problem

Our sketch shows a chunk of ice floating in a glass of water. In this case the solid object is ice, with a density $\rho_s = \rho_{ice} = 917\ kg/m^3$, and the fluid is water, with a density $\rho_f = \rho_{water} = 1.00 \times 10^3\ kg/m^3$.

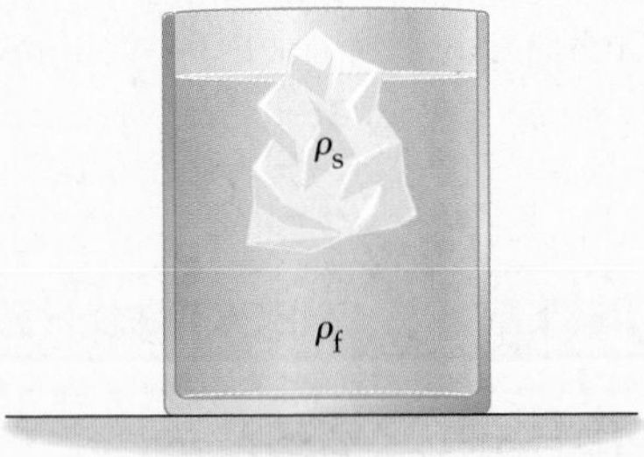

Strategy

We can apply Equation 15–10 to this system. First, the fraction of the total volume of the ice, V_s, that is submerged is $V_{sub}/V_s = \rho_s/\rho_f$. Hence the fraction that is above the water is $1 - V_{sub}/V_s = 1 - \rho_s/\rho_f$. Multiplying this fraction by 100 yields the percentage above water.

Solution

1. Calculate the fraction of the total volume of the ice that is submerged:

$$\frac{V_{sub}}{V_s} = \frac{\rho_s}{\rho_f} = \frac{917\ kg/m^3}{1.00 \times 10^3\ kg/m^3} = 0.917$$

2. Calculate the fraction of the ice that is above water:

$$1 - \frac{\rho_s}{\rho_f} = 1 - 0.917 = 0.083$$

3. Multiply by 100 to obtain a percentage:

$$100\left(1 - \frac{\rho_s}{\rho_f}\right) = 100(0.083) = 8.3\%$$

Insight

Because we seek a percentage, it is not necessary to know the total volume of the ice. Thus, our result that 8.3% of the ice is above the water applies whether we are talking about an ice cube in a drinking glass, or an iceberg floating in a freshwater lake. If an iceberg floats in the ocean, however, it will float higher due to the higher density of seawater. We consider this case in the following Practice Problem.

Practice Problem

What percentage of an ice chunk is above water level if it floats in seawater? (The density of seawater can be found in Table 15–1.) [**Answer:** 10.5%]

Some related homework problems: Problem 36, Problem 37

CONCEPTUAL CHECKPOINT 15–5 The New Water Level I

A cup is filled to the brim with water and a floating ice cube. When the ice melts, which of the following occurs? **(a)** Water overflows the cup, **(b)** the water level decreases, or **(c)** the water level remains the same.

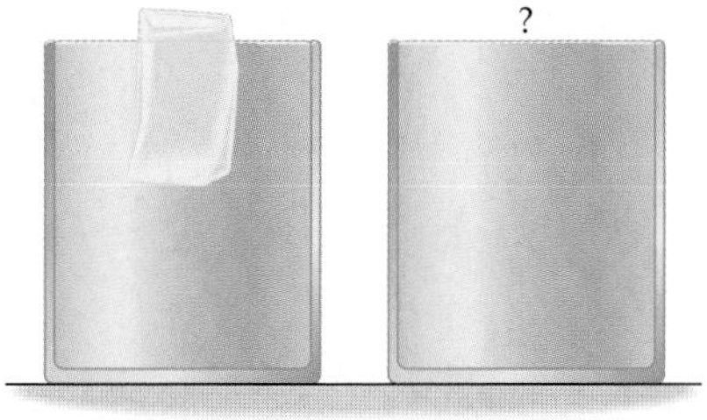

Reasoning and Discussion

Since the ice cube floats, it displaces a volume of water equal to its weight. But when it melts, it becomes water, and its weight is the same. Hence, the melted water fills exactly the same volume that the ice cube displaced when floating. As a result, the water level is unchanged.

Answer:

(c) The water level remains the same.

CONCEPTUAL CHECKPOINT 15–6 The New Water Level II

A cup is filled to the brim with water and a floating ice cube. Resting on top of the ice cube is a small pebble. When the ice melts, which of the following occurs? **(a)** Water overflows the cup, **(b)** the water level decreases, or **(c)** the water level remains the same.

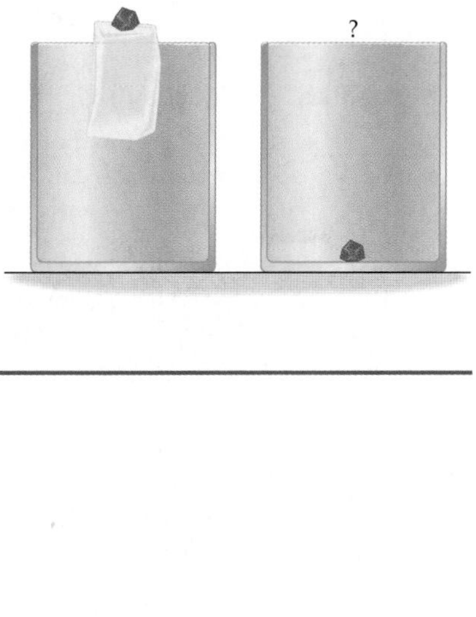

Reasoning and Discussion

We know from the previous Conceptual Checkpoint that the ice itself makes no difference to the water level. As for the pebble, when it floats on the ice it displaces an amount of water equal to its *weight*. When the ice melts, the pebble drops to the bottom of the cup, where it displaces a volume of water equal to its own *volume*. Since the volume of the pebble is less than the volume of water with the same weight, we conclude that less water is displaced after the ice melts. Hence, the water level decreases.

Answer:

(b) The water level decreases.

15–6 Fluid Flow and Continuity

Suppose you want to water the yard, but you don't have a spray nozzle for the end of the hose. Without a nozzle the water flows rather slowly from the hose and hits the ground within half a meter. But if you place your thumb over the end of the hose, narrowing the opening to a fraction of its original size, the water sprays out with a high speed and a large range. Why does decreasing the size of the opening have this effect?

To answer this question, we begin by considering a simple system that shows the same behavior. Imagine, then, that a fluid flows with a speed v_1 through a cylindrical pipe of cross-sectional area A_1, as in the left-hand portion of **Figure 15–14**. If the pipe narrows to a cross-sectional area A_2, as in the right-hand portion of Figure 15–14, the fluid will flow with a new speed, v_2.

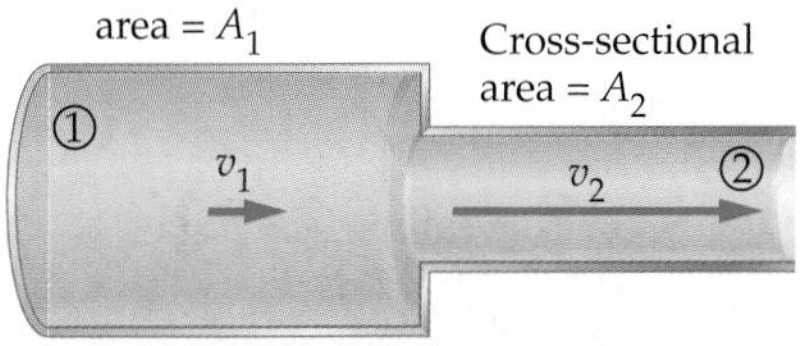

▲ **FIGURE 15–14 Fluid flow through a pipe of varying diameter**

As a fluid flows from a large pipe to a small pipe, the same mass of fluid passes a given point in a given amount of time. Thus, the speed in the small pipe is greater than it is in the large pipe.

We can find the speed in the narrow section of the pipe by assuming that any amount of fluid that passes point 1 in a given time, Δt, must also flow past the point 2 in the same time. If this were not the case, the system would be gaining or losing fluid. To find the mass of fluid passing point 1 in the time Δt, note that the fluid moves through a distance $v_1\,\Delta t$ in this time. As a result, the volume of fluid going past point 1 is

$$\Delta V_1 = A_1 v_1 \Delta t$$

▲ Narrowing the opening in a hose with a nozzle (or thumb) increases the velocity of flow, as one would expect from the equation of continuity.

Hence, the mass of fluid passing point 1 is

$$\Delta m_1 = \rho_1 \Delta V_1 = \rho_1 A_1 v_1 \Delta t$$

Similarly, the mass passing point 2 in the time Δt is

$$\Delta m_2 = \rho_2 \Delta V_2 = \rho_2 A_2 v_2 \Delta t$$

Note that we have allowed for the possibility of the fluid having different densities at points 1 and 2.

Finally, equating these two masses yields the relation between v_1 and v_2:

$$\Delta m_1 = \Delta m_2$$
$$\rho_1 A_1 v_1 \Delta t = \rho_2 A_2 v_2 \Delta t$$

Canceling Δt we find

Equation of Continuity

$$\rho_1 A_1 v_1 = \rho_2 A_2 v_2 \qquad 15\text{–}11$$

This relation is referred to as the **equation of continuity.**

Most gases are readily compressed, which means that their densities can change. In contrast, most liquids are practically incompressible, so their densities are essentially constant. Unless stated otherwise, we will assume all liquids discussed in this text to be perfectly incompressible. Thus, for liquids, ρ_1 and ρ_2 are the same in Equation 15–11, and the equation of continuity reduces to the following:

Equation of Continuity for an Incompressible Fluid

$$A_1 v_1 = A_2 v_2 \qquad 15\text{–}12$$

We next apply this relation to the case of water flowing through the nozzle of a fire hose.

PROBLEM-SOLVING NOTE

Continuity of Flow

The speed of an incompressible fluid is inversely proportional to the area through which it flows.

EXAMPLE 15–7 Spray I

Water travels through a 9.6-cm diameter fire hose with a speed of 1.3 m/s. At the end of the hose, the water flows out through a nozzle whose diameter is 2.5 cm. What is the speed of the water coming out of the nozzle?

Picture the Problem

In our sketch, we label the speed of the water in the hose with v_1 and the speed of the water coming out the nozzle with v_2. We are given that $v_1 = 1.3$ m/s. We also know that the diameter of the hose is $d_1 = 9.6$ cm and the diameter of the nozzle is $d_2 = 2.5$ cm.

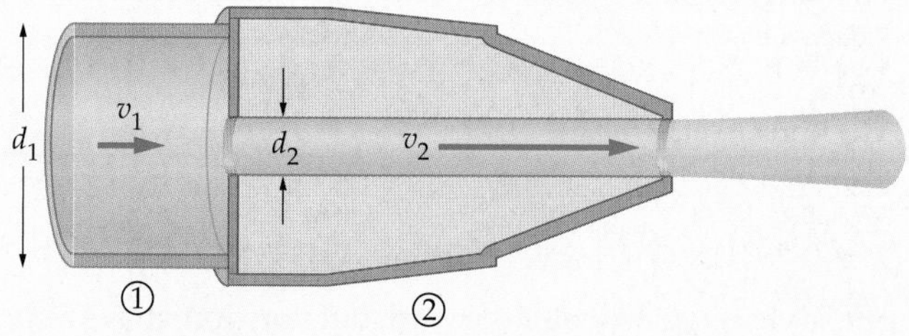

Strategy

We can find the water speed in the nozzle by applying $A_1 v_1 = A_2 v_2$. In addition, we assume that the hose and nozzle are circular in cross section; hence, their areas are given by $A = \pi d^2/4$, where d is the diameter.

Solution

1. Solve Equation 15–12 for v_2, the speed of the water in the nozzle: $v_2 = v_1(A_1/A_2)$

2. Replace the areas with $A = \pi d^2/4$: $v_2 = v_1\left(\frac{\pi d_1{}^2/4}{\pi d_2{}^2/4}\right) = v_1\left(\frac{d_1{}^2}{d_2{}^2}\right)$

3. Substitute numerical values: $v_2 = v_1\left(\frac{d_1{}^2}{d_2{}^2}\right) = (1.3\text{ m/s})\left(\frac{9.6\text{ cm}}{2.5\text{ cm}}\right)^2 = 19\text{ m/s}$

Insight

Note that a small-diameter nozzle can give very high speeds. In fact, the speed depends inversely on the diameter squared.

Practice Problem

What nozzle diameter would be required to give the water a speed of 21 m/s? [**Answer:** $d_2 = 2.4$ cm]

Some related homework problems: Problem 41, Problem 46, Problem 47

15–7 Bernoulli's Equation

In this section, we apply the work–energy theorem to fluids. The result is a relation between the pressure of a fluid, its speed, and its height. This relation is known as **Bernoulli's equation.**

Change in Speed

We begin by considering a system in which the speed of the fluid changes. To be specific, the system of interest is the same as that shown in Figure 15–14. We have already shown that the speed of the fluid increases as it flows from region 1 to region 2; we now investigate the corresponding change in pressure.

Our plan of attack is to first calculate the total work done on the fluid as it moves from one region to the next. This result will depend on the pressure in the fluid. Once the total work is obtained, the work-energy theorem allows us to equate it to the change in kinetic energy of the fluid. This will give the pressure-speed relationship we desire.

Consider an element of fluid of length Δx_1, as shown in **Figure 15–15**. This element is pushed in the direction of motion by the pressure P_1. Thus, the pressure does positive work, ΔW_1, on the fluid element. Noting that the force exerted on the element is $F_1 = P_1A_1$, and that work is force times distance, the work done on the element is

$$\Delta W_1 = F_1\Delta x_1 = P_1A_1\Delta x_1$$

The volume of the fluid element is $\Delta V_1 = A_1\Delta x_1$, so the work done by P_1 is

$$\Delta W_1 = P_1\Delta V_1$$

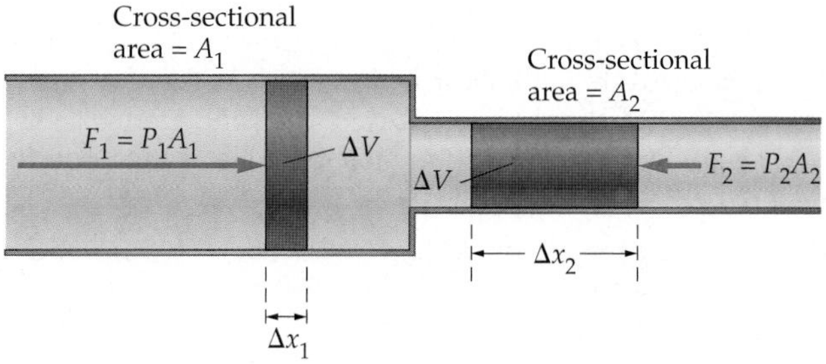

◀ **FIGURE 15–15 Work done on a fluid element**

As an incompressible fluid element of volume ΔV moves from pipe 1 to pipe 2, the pressure P_1 does a positive work $P_1\Delta V$ and the pressure P_2 does a negative work $P_2\Delta V$. Since P_1 is greater than P_2, the net result is that positive work is done, and the fluid element speeds up.

Next, when the fluid element emerges into region 2, it experiences a force opposite to its direction of motion due to the pressure P_2. Thus, P_2 does negative work on the element. Following the same steps given previously, we can write the work done by P_2 as

$$\Delta W_2 = -P_2\Delta V_2$$

Now, for an incompressible fluid, the volume of the element does not change as it goes from region 1 to region 2. Therefore,

$$\Delta V_1 = \Delta V_2 = \Delta V$$

Using this result, we can write the total work done on the fluid element as follows:

$$\Delta W_{\text{total}} = \Delta W_1 + \Delta W_2 = P_1\Delta V - P_2\Delta V = (P_1 - P_2)\Delta V$$

The final step is to equate the total work to the change in kinetic energy:

$$\Delta W_{\text{total}} = (P_1 - P_2)\Delta V = K_{\text{final}} - K_{\text{initial}} = K_2 - K_1 \qquad \textbf{15–13}$$

What is the kinetic energy of the fluid element? Well, the mass of the element is

$$\Delta m = \rho\Delta V$$

Thus, its kinetic energy is simply

$$K = \tfrac{1}{2}(\Delta m)v^2 = \tfrac{1}{2}(\rho\Delta V)v^2$$

Using this expression in Equation 15–13, we have

$$\Delta W_{\text{total}} = (P_1 - P_2)\Delta V = \left(\tfrac{1}{2}\rho v_2{}^2 - \tfrac{1}{2}\rho v_1{}^2\right)\Delta V$$

Canceling the common factor ΔV, and rearranging, we find

$$P_1 + \tfrac{1}{2}\rho v_1^2 = P_2 + \tfrac{1}{2}\rho v_2^2 \qquad \text{15–14}$$

Equation 15–14 is equivalent to saying that $P + \frac{1}{2}\rho v^2$ is constant. Thus, there is a tradeoff between the pressure in a fluid and its speed—as the fluid speeds up, its pressure decreases. If this seems odd, recall that P_1 acts to increase the speed of the fluid element and P_2 acts to decrease its speed. The element will speed up, then, only if P_2 is less than P_1.

EXAMPLE 15–8 Spray II

Referring to Example 15–7, suppose the pressure in the fire hose is 350 kPa. What is the pressure in the nozzle?

Picture the Problem

In our sketch, we use the same numbering system as in Example 15–7; that is, 1 refers to the hose, 2 to the nozzle. Therefore, $P_1 = 350$ kPa, and P_2 is to be determined.

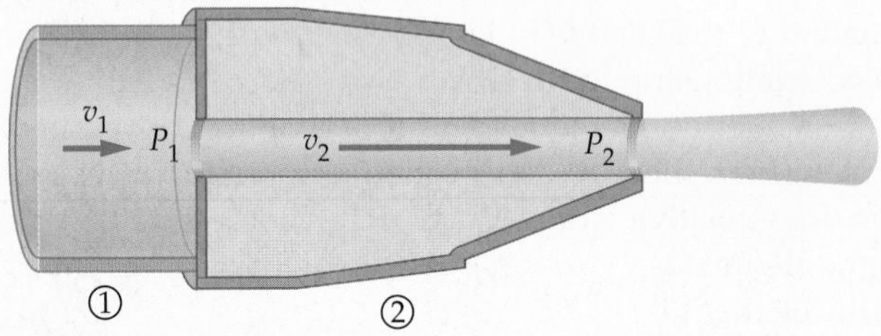

Strategy

We are given P_1, v_1, and the diameters of the hose and nozzle. From the equation of continuity, Equation 15–12, we find v_2 (as in Example 15–7). Now we use this result, plus Equation 15–14, to determine P_2.

Solution

1. Solve Equation 15–14 for the pressure in the nozzle, P_2: $P_2 = P_1 + \frac{1}{2}\rho(v_1^2 - v_2^2)$
2. Substitute numerical values, including v_2 from Example 15–7: $P_2 = 350\text{ kPa} + \frac{1}{2}(1.00 \times 10^3\text{ kg/m}^3)[(1.3\text{ m/s})^2 - (19\text{ m/s})^2]$ $= 170\text{ kPa}$

Insight

Note that the pressure in the nozzle is less than the pressure in the hose by roughly a factor of 2. What has happened is that part of the energy associated with the high pressure in the hose has been converted to kinetic energy as the water passes through the nozzle. The connection between pressure and energy will be explored in more detail later in this section.

Practice Problem

What nozzle speed would be required to give a nozzle pressure of 110 kPa? [**Answer:** $v_2 = 22$ m/s]

Some related homework problems: Problem 50, Problem 51

Change in Height

If a fluid flows through the pipe shown in **Figure 15–16,** its height increases from y_1 to y_2 as it goes from one region to the next. Since the cross-sectional area of the pipe is constant, however, the speed of the fluid is unchanged, according to Equation 15–12. Thus, the change in kinetic energy of the fluid element shown in Figure 15–16 is zero.

The total work done on the fluid element is the sum of the works done by the pressure in each of the two regions, plus the work done by gravity. As before, the work done by pressure is

$$\Delta W_{\text{pressure}} = \Delta W_1 + \Delta W_2 = (P_1 - P_2)\Delta V$$

As the fluid element rises, gravity does negative work on it. Recalling that the mass of the element is

$$\Delta m = \rho \Delta V$$

the work done by gravity is

$$\Delta W_{\text{gravity}} = -\Delta m g(y_2 - y_1) = -\rho \Delta V g(y_2 - y_1)$$

Setting the total work equal to zero (since $\Delta K = 0$), yields

$$\Delta W_{\text{total}} = \Delta W_{\text{pressure}} + \Delta W_{\text{gravity}} = (P_1 - P_2)\Delta V - \rho g(y_2 - y_1)\Delta V = 0$$

Canceling ΔV and rearranging gives

$$P_1 + \rho g y_1 = P_2 + \rho g y_2 \qquad \text{15–15}$$

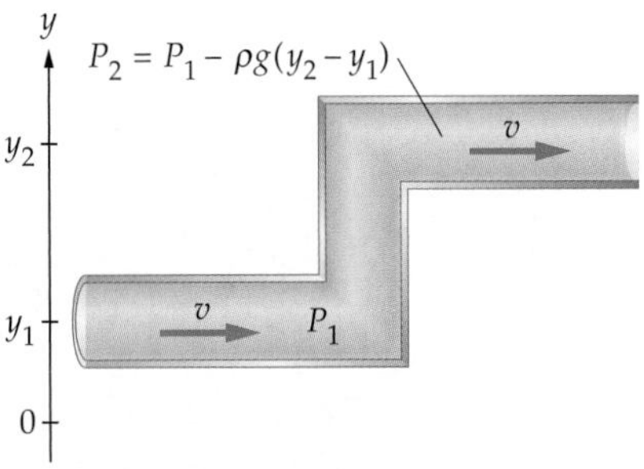

▲ **FIGURE 15–16 Fluid pressure in a pipe of varying elevation**

Fluid of density ρ flows in a pipe of uniform cross-sectional area from height y_1 to height y_2. As it does so, its pressure decreases by the amount $\rho g(y_2 - y_1)$.

In this case, it is $P + \rho g y$ that is constant—hence, pressure decreases as the height within a fluid increases. Note, in fact, that Equation 15–15 is precisely the same as Equation 15–7, which was obtained using force considerations. Here we obtained the result using the work-energy theorem.

EXERCISE 15–4

Water flows with constant speed through a garden hose that goes up a step 20.0-cm high. If the water pressure is 143 kPa at the bottom of the step, what is its pressure at the top of the step?

Solution

Apply Equation 15–15, letting subscript 1 refer to the bottom of the step and subscript 2 refer to the top of the step. Solve for P_2:

$$P_2 = P_1 + \rho g(y_1 - y_2)$$
$$= 143\text{ kPa} + (1.00 \times 10^3\text{ kg/m}^3)(9.81\text{ m/s}^2)(0 - 0.200\text{ m}) = 141\text{ kPa}$$

This is precisely the pressure difference that would be observed if the water had been at rest.

General Case

In a more general case, both the height of a fluid and its speed may change. Combining the results obtained in Equations 15–14 and 15–15 yields the full form of Bernoulli's equation:

Bernoulli's Equation

$$P_1 + \tfrac{1}{2}\rho v_1{}^2 + \rho g y_1 = P_2 + \tfrac{1}{2}\rho v_2{}^2 + \rho g y_2 \qquad \textbf{15–16}$$

Thus, in general, the quantity $P + \frac{1}{2}\rho v^2 + \rho g y$ is a constant within a fluid. This is basically a statement of energy conservation. For example, recalling the definition of density in Equation 15–1, we find that $\frac{1}{2}\rho v^2$ is $\frac{1}{2}(M/V)v^2 = \left(\frac{1}{2}Mv^2\right)/V$. Clearly, this term represents the kinetic energy per volume of the fluid. Similarly, the term $\rho g y$ can be written as $(M/V)gy = (Mgy)/V$, which is the gravitational potential energy per volume.

Finally, the first term in Bernoulli's equation—the pressure—can also be thought of as an energy per volume. Recall that $P = F/A$. If we multiply numerator and denominator by a distance, d, we have $P = Fd/Ad$. But Fd is the work done by the force F as it acts through the distance d, and Ad is the volume swept out by an area A moved through a distance d. Therefore, the pressure can be thought of as work (energy) per volume: $P = W/V$.

As a result, Bernoulli's equation is simply a restatement of the work–energy theorem in terms of quantities per volume. Of course, this relation holds only as long as we can ignore frictional losses, which would lead to heating. We will consider the energy aspects of heat in Chapter 16.

ACTIVE EXAMPLE 15–3 Find the Pressure

Repeat Exercise 15–4 with the following additional information: **(a)** the cross-sectional area of the hose on top of the step is half that at the bottom of the step, and **(b)** the speed of the water at the bottom of the step is 1.20 m/s.

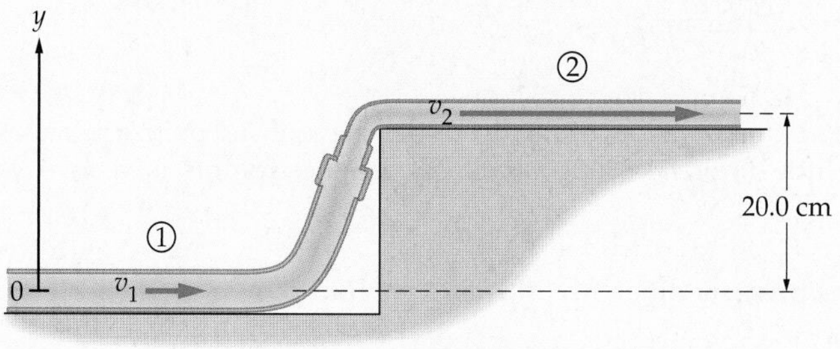

continued on next page

continued from previous page

Solution *(Test your understanding by performing the calculations indicated in each step.)*

1. Use the continuity equation to find the water's speed on top of the step: $v_2 = 2v_1 = 2.40 \text{ m/s}$
2. Solve Bernoulli's equation for P_2: $P_2 = P_1 - \rho g(y_2 - y_1) - \frac{1}{2}\rho(v_2^2 - v_1^2)$
3. Substitute numerical values: $P_2 = 139 \text{ kPa}$

Insight

The pressure on top of the step is less than in Exercise 15–4. This is to be expected because, in this case, the water speeds up as it rises over the step.

Your Turn

For what step height will the pressure at the top of the step be equal to atmospheric pressure?

(Answers to **Your Turn** *problems can be found in the back of the book.)*

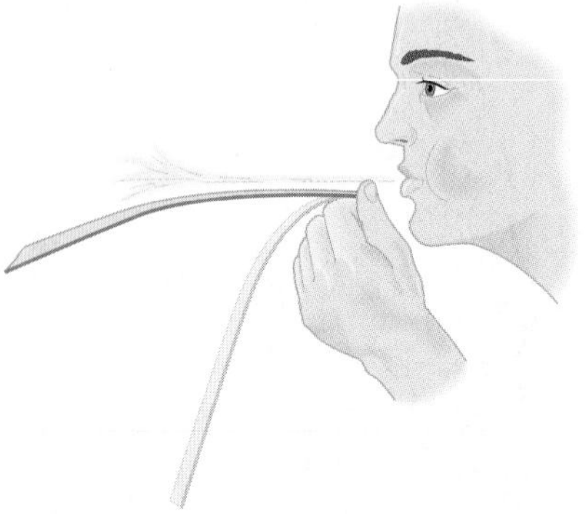

▲ **FIGURE 15–17 The Bernoulli effect on a sheet of paper**

If you hold a piece of paper by its end, it will bend downward. Blowing across the top of the paper reduces the pressure there, resulting in a net upward force which lifts the paper to a nearly horizontal position.

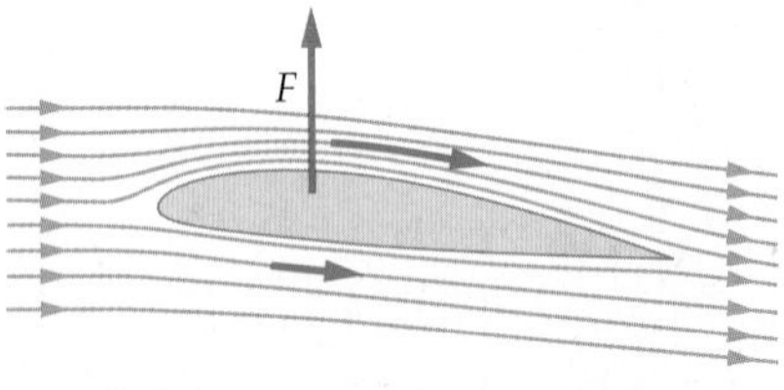

▲ **FIGURE 15–18 Airflow and lift in an airplane wing**

Cross section of an airplane wing with air flowing past it. The wing is shaped so that air flows more rapidly over the top of the wing than along the bottom. As a result, the pressure on top of the wing is reduced, and a net upward force (lift) is generated.

▲ **FIGURE 15–19 Force on a roof due to wind speed**

Wind blows across the roof of a house, but the air inside is at rest. The pressure over the roof is therefore less than the pressure inside, resulting in a net upward force on the roof.

15–8 Applications of Bernoulli's Equation

We now consider a variety of real-world examples that illustrate the application of Bernoulli's equation.

Pressure and Speed

As mentioned, it often seems counterintuitive that a fast-moving fluid should have less pressure than a slow-moving one. Remember, however, that pressure can be thought of as a form of energy. From this point of view, there is an energy tradeoff between pressure and kinetic energy.

Perhaps the easiest way to demonstrate the dependence of pressure on speed is to blow across the top of a piece of paper. If you hold the paper as shown in **Figure 15–17**, then blow over the top surface, the paper will lift upward. The reason is that there is a difference in air speed between the top and the bottom of the paper, with the higher speed on top. As a result, the pressure above the paper is lower. This pressure difference, in turn, results in a net upward force, referred to as **lift,** and the paper rises.

A similar example of pressure and speed is provided by the airplane wing. A cross section of a typical wing is shown in **Figure 15–18**. The shape of the wing is designed so that air flows more rapidly over the top surface than the lower surface. As a result, the pressure is less on top. As with the piece of paper, the pressure difference results in a net upward force (lift) on the wing.

Note that lift is a dynamic effect; it requires a flow of air. The greater the speed difference, the greater the upward force.

EXERCISE 15–5

During a windstorm, a 35.5 m/s wind blows across the flat roof of a small home, as in **Figure 15–19**. Find the difference in pressure between the air inside the home and the air just above the roof, assuming the doors and windows of the house are closed. (The density of air is 1.29 kg/m^3.)

Solution

Use Bernoulli's equation with point 1 just under the roof and point 2 just above the roof. Since there is little difference in elevation between these points, $y_1 = y_2 = y$. Thus,

$$P_1 + 0 + \rho g y = P_2 + \frac{1}{2}\rho v_2^2 + \rho g y$$

Solving for the pressure difference, $P_1 - P_2$, we find

$$P_1 - P_2 = \frac{1}{2}\rho v_2^2 = \frac{1}{2}(1.29 \text{ kg/m}^3)(35.5 \text{ m/s})^2 = 813 \text{ Pa}$$

A difference in pressure of 813 Pa might seem rather small, considering that atmospheric pressure is 101 kPa. However, it can still cause a significant force on a relatively large area, such as a roof. If a typical roof has an area of about 150 m^2, for example, a pressure difference of 813 Pa results in an upward force of over 27,000 pounds! This is why roofs are often torn from houses during severe windstorms.

On a lighter note, prairie dogs seem to *benefit* from the effects of Bernoulli's equation. A schematic prairie dog burrow is pictured in **Figure 15–20**. Note that one of the entrance/exit mounds is higher than the other. This is significant because the speed of air flow due to the incessant prairie winds varies with height; the speed goes to zero right at ground level, and increases to its maximum value within a few feet above the surface. As a result, the speed of air over the higher mound is greater than that over the lower mound. This causes the pressure to be less over the higher mound. With a pressure difference between the two mounds, air is drawn through the burrow, giving a form of natural air conditioning.

Similar effects are seen in an atomizer, which sprays perfume in a fine mist. As the bulb shoots a gust of air, as in **Figure 15–21**, it passes through a narrow orifice, which causes the air speed to increase. The pressure decreases as a result, and perfume is drawn up by the pressure difference into the stream of air.

▲ **FIGURE 15–20 Air circulation in a prairie dog burrow**

A prairie dog burrow typically has a high mound on one end and a low mound on the other. Since the wind speed increases with height above the ground, the pressure is smaller at the high-mound end of the burrow. The result is a very convenient circulation of fresh air through the burrow.

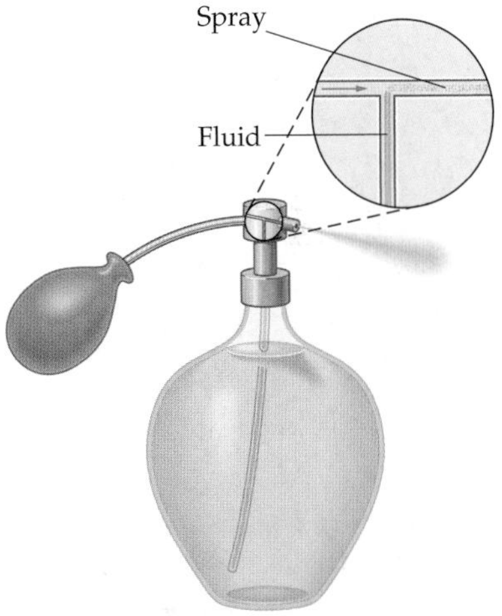

▲ **FIGURE 15–21 An atomizer**

The operation of an "atomizer" can be understood in terms of Bernoulli's equation. The high-speed jet of air created by squeezing the bulb creates a low pressure at the top of the vertical tube. This causes fluid to be drawn up the tube and expelled with the air jet as a fine spray.

CONCEPTUAL CHECKPOINT 15–7 A Ragtop Roof

A small ranger vehicle has a soft, ragtop roof. When the car is at rest, the roof is flat. When the car is cruising at highway speeds with its windows rolled up, does the roof **(a)** bow upward, **(b)** remain flat, or **(c)** bow downward?

Reasoning and Discussion

When the car is in motion, air flows over the top of the roof, while the air inside the car is at rest—since the windows are closed. Thus, there is less pressure over the roof than under it. As a result, the roof bows upward.

Answer:

(a) The roof bows upward.

▲ We often say that a hurricane or tornado "blew the roof off a house." However, the house at left lost its roof not because of the great pressure exerted by the wind, but rather the opposite. In accordance with the Bernoulli effect, the high speed of the wind passing over the roof created a region of reduced pressure. Normal atmospheric pressure inside the house then blew the roof off. The same phenomenon is exploited by prairie dogs to ventilate their burrows. One end of the burrow is always situated at a greater height than the other. Because the prairie wind blows much faster a few feet above ground level, the pressure at the elevated end of the burrow is reduced. The resulting pressure difference produces a flow of air through the burrow.

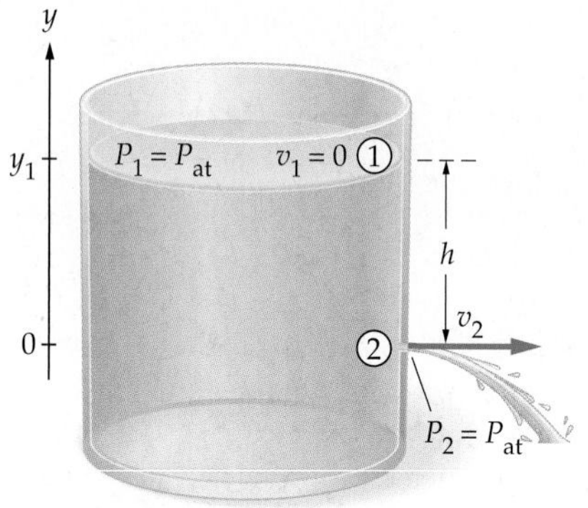

▲ FIGURE 15–22 Fluid emerging from a hole in a container

Since the fluid exiting the hole is in contact with the atmosphere, the pressure there is just as it is on the top surface of the fluid.

▲ FIGURE 15–23 Maximum height of a stream of water

If the fluid emerging from a hole in a container is directed upward, it has just enough speed to reach the surface level of the fluid. This is an example of energy conservation.

Torricelli's Law

Our final example of Bernoulli's equation deals with the speed of a fluid as it flows through a hole in a container. Consider, for example, the tank of water shown in **Figure 15–22**. If a hole is poked through the side of the tank at a depth h below the surface, what is the speed of the water as it emerges?

To answer this question, we apply Bernoulli's equation to the two points shown in the figure. First, at point 1 we note that the water is open to the atmosphere; thus $P_1 = P_{at}$. Next, with the origin at the level of the hole, the height of the water surface is $y_1 = h$. Finally, if the hole is relatively small and the tank is large, the top surface of the water will have essentially zero speed; thus we can set $v_1 = 0$. Collecting these results, we have the following for point 1:

$$P_1 + \tfrac{1}{2}\rho v_1^2 + \rho g y_1 = P_{at} + 0 + \rho g h$$

Now for point 2. At this point the height is $y_2 = 0$, by definition of the origin, and the speed of the escaping water is the unknown, v_2. Here is the key step: The pressure P_2 is *atmospheric pressure*, because the hole opens the water to the atmosphere. Thus, for point 2 we have the following:

$$P_2 + \tfrac{1}{2}\rho v_2^2 + \rho g y_2 = P_{at} + \tfrac{1}{2}\rho v_2^2 + 0$$

Equating these results yields

$$P_1 + \tfrac{1}{2}\rho v_1^2 + \rho g y_1 = P_2 + \tfrac{1}{2}\rho v_2^2 + \rho g y_2$$
$$P_{at} + \rho g h = P_{at} + \tfrac{1}{2}\rho v_2^2$$

Eliminating P_{at} and ρ we find

$$v_2 = \sqrt{2gh} \qquad \text{15–17}$$

This result is known as **Torricelli's law.**

This expression for v_2 should look familiar; it is the speed of an object that falls freely through a distance h. That is, the water emerges from the tank with the same speed as if it had fallen from the surface of the water to the hole. Similarly, if the emerging stream of water were to be directed upward, as in **Figure 15–23**, it would have just enough speed to rise through a height h—right back to the water's surface. This is precisely what one would expect on the basis of energy conservation.

EXAMPLE 15–9 A Water Fountain

In designing a backyard water fountain, a gardener wants a stream of water to exit from the bottom of one tub and land in a second one, as shown in the sketch. The top of the second tub is 0.500 m below the hole in the first tub, which has water in it to a depth of 0.150 m. How far to the right of the first tub must the second one be placed to catch the stream of water?

Picture the Problem

Our sketch labels the various pertinent quantities for this problem. We know that $h = 0.150$ m and $H = 0.500$ m. The distance D is to be found. We have also chosen an appropriate coordinate system, with the y direction vertical and the x direction horizontal.

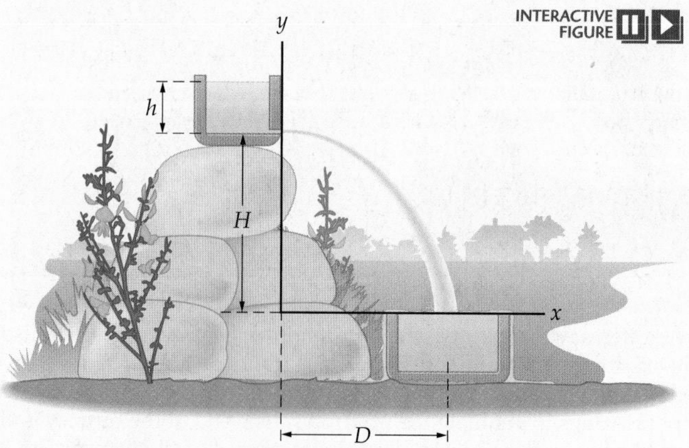

Strategy

This problem combines Torricelli's law and kinematics. First, we find the speed v of the stream of water as it leaves the first can, using Equation 15–17. Next, we find the time t required for the stream to fall freely through a distance H. Finally, since the stream moves with constant speed in the x direction, the distance D is given by $D = vt$.

Solution

1. Find the speed v of the stream when it leaves the first can: $v = \sqrt{2gh} = \sqrt{2(9.81\ \text{m/s}^2)(0.150\ \text{m})} = 1.72\ \text{m/s}$

2. Find the time t for free fall through a height H: $y = H - \frac{1}{2}gt^2 = 0$

$$t = \sqrt{\frac{2H}{g}} = \sqrt{\frac{2(0.500\ \text{m})}{9.81\ \text{m/s}^2}} = 0.319\ \text{s}$$

3. Multiply v times t to find the distance D: $x = vt = (1.72\ \text{m/s})(0.319\ \text{s}) = 0.549\ \text{m} = D$

Insight

Note that our solution for x can also be written as $x = vt = \sqrt{2gh}(\sqrt{2H/g}) = 2\sqrt{hH}$. Thus, if the values of h and H are interchanged, the distance D remains the same. Note also that x is independent of the acceleration of gravity; therefore, the fountain would work just as well on the Moon.

Practice Problem

Find the distance D for $h = 0.500$ m and $H = 0.150$ m. [**Answer:** $D = 2\sqrt{hH} = 0.548$ m, as expected.]

Some related homework problems: Problem 54, Problem 94

*15–9 Viscosity and Surface Tension

To this point we have considered only "ideal" fluids. In particular, we have assumed that fluids flow without frictional losses and that the molecules in a fluid have no interaction with one another. In this section, we consider the consequences that follow from relaxing these assumptions.

Viscosity

When a block slides across a rough floor, it experiences a frictional force opposing the motion. Similarly, a fluid flowing past a stationary surface experiences a force opposing the flow. This tendency to resist flow is referred to as the **viscosity** of a fluid. Fluids like air have low viscosities, thicker fluids like water are more viscous, and fluids like honey and motor oil are characterized by high viscosity.

To be specific, consider a situation of great practical importance—the flow of a fluid through a tube. Examples of this type of system include water flowing through a metal pipe in a house and blood flowing through an artery or a vein. If the fluid were ideal, with zero viscosity, it would flow through the tube with a speed that is the same throughout the fluid, as indicated in **Figure 15–24 (a)**. Real fluids with finite viscosity are found to have flow patterns like the one shown in **Figure 15–24 (b)**. In this case, the fluid is at rest next to the wall of the tube and flows with its greatest speed in the center of the tube. Because adjacent portions of the fluid flow past one another with different speeds, a force must be exerted on the fluid to maintain the flow, just as a force is required to keep a block sliding across a rough surface.

The force causing a viscous fluid to flow is provided by the pressure difference, $P_1 - P_2$, across a given length, L, of tube. Experiments show that the required pressure difference is proportional to the length of the tube and to the average speed, v, of the fluid. In addition, it is inversely proportional to the cross-sectional area, A, of the tube. Combining these observations, the pressure difference can be written in the following form:

$$P_1 - P_2 \propto \frac{vL}{A}$$

The constant of proportionality between the pressure difference and vL/A is related to the **coefficient of viscosity, η,** of a fluid. In fact, the viscosity is *defined* in such a way that the pressure difference is given by the following expression:

$$P_1 - P_2 = 8\pi\eta\frac{vL}{A} \qquad \text{15–18}$$

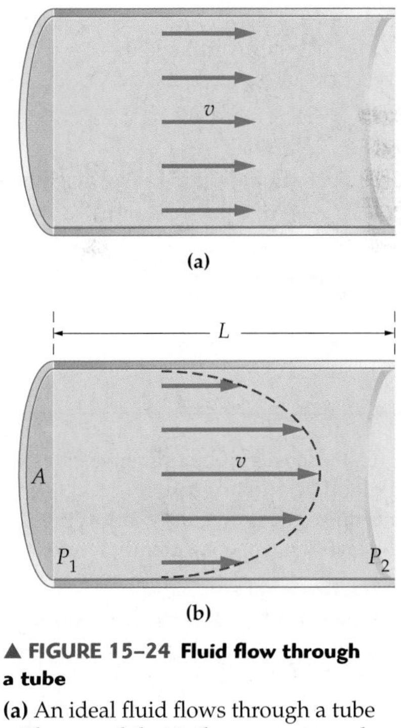

▲ **FIGURE 15–24 Fluid flow through a tube**

(a) An ideal fluid flows through a tube with a speed that is the same everywhere in the fluid. **(b)** In a fluid with finite viscosity, the speed of the fluid goes to zero on the walls of the tube and reaches its maximum value in the center of the tube. The average speed of the fluid depends on the pressure difference between the ends of the tube, $P_1 - P_2$, the length of the tube, L, the cross-sectional area of the tube, A, and the coefficient of viscosity of the liquid, η.

TABLE 15–3
Viscosities (η) of Various Fluids ($N \cdot s/m^2$)

Honey	10
Glycerine (20 °C)	1.50
10-wt motor oil (30 °C)	0.250
Whole blood (37 °C)	2.72×10^{-3}
Water (0 °C)	1.79×10^{-3}
Water (20 °C)	1.0055×10^{-3}
Water (100 °C)	2.82×10^{-4}
Air (20 °C)	1.82×10^{-5}

From this equation we can see that the dimensions of the coefficient of viscosity are $N \cdot s/m^2$. A common unit in the study of viscous fluids is the **poise,** named for the French physiologist Jean Louis Marie Poiseuille (1799–1869) and defined as:

$$1 \text{ poise} = 1 \text{ dyne} \cdot \text{s/cm}^2 = 0.1 \text{ N} \cdot \text{s/m}^2$$

For example, the viscosity of water at room temperature is $1.0055 \times 10^{-3}\ N \cdot s/m^2$ and the viscosity of blood at 37°C is $2.72 \times 10^{-3}\ N \cdot s/m^2$. A few additional viscosities are given in Table 15–3.

EXAMPLE 15–10 Blood Speed in the Pulmonary Artery

REAL-WORLD PHYSICS: BIO The pulmonary artery, which connects the heart to the lungs, is 8.5 cm long and has a pressure difference over this length of 450 Pa. If the inside radius of the artery is 2.4 mm, what is the average speed of blood in the pulmonary artery?

Picture the Problem

Our sketch shows a schematic representation of the pulmonary artery, not drawn to scale. The length (8.5 cm) and radius (2.4 mm) of the artery are indicated. In addition, we note that the pressure difference between the two ends of the artery is 450 Pa.

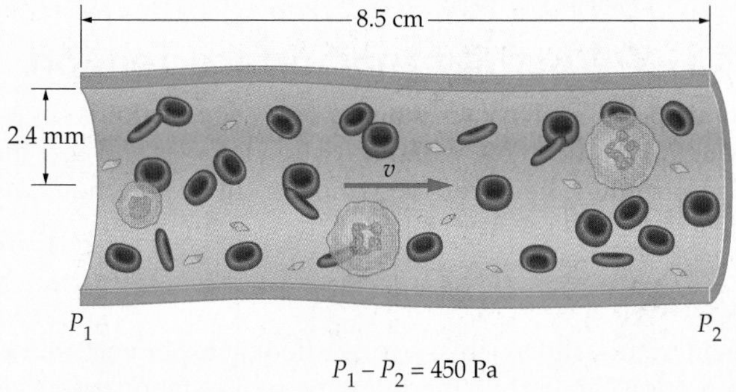

Strategy

The average speed of the blood can be found by using $P_1 - P_2 = 8\pi\eta(vL/A)$. Note that the pressure difference, $P_1 - P_2$, is given as 450 Pa = 450 N/m², and that the cross-sectional area of the blood vessel is $A = \pi r^2$.

Solution

1. Solve Equation 15–18 for the average speed, v: $$v = \frac{(P_1 - P_2)A}{8\pi\eta L}$$

2. Replace the cross-sectional area A with πr^2 and cancel π from the numerator and denominator: $$v = \frac{(P_1 - P_2)r^2}{8\eta L}$$

3. Substitute numerical values: $$v = \frac{(450 \text{ Pa})(0.0024 \text{ m})^2}{8(0.00272 \text{ N} \cdot \text{s/m}^2)(0.085 \text{ m})} = 1.4 \text{ m/s}$$

Insight

The viscosity of blood increases rapidly with its hematocrit value; that is, with the concentration of red blood cells in the whole blood (see Chapter 10 for more information on the hematocrit value of blood). Thus, thick blood, with a high hematocrit value, requires a significantly larger pressure difference for a given rate of blood flow. This higher pressure must be provided by the heart, which consequently works harder with each beat.

Practice Problem

What pressure difference is required to give the blood in this pulmonary artery an average speed of 1.5 m/s? [**Answer:** 480 Pa]

Some related homework problems: Problem 84, Problem 85

A convenient way to characterize the flow of a fluid is in terms of its volume flow rate—the volume of fluid that passes a given point in a given amount of time. Referring to Section 15–6 we see that the volume flow rate of a fluid is simply vA, where v is the average speed of the fluid and A is the cross-sectional area of the tube through which it flows. Solving Equation 15–18 for the average speed gives $v = (P_1 - P_2)A/8\pi\eta L$. Multiplying this result by the cross-sectional area of the tube yields the volume flow rate:

$$\text{volume flow rate} = \frac{\Delta V}{\Delta t} = vA = \frac{(P_1 - P_2)A^2}{8\pi\eta L}$$

Using the fact that the cross-sectional area of the tube is $A = \pi r^2$, where r is its radius, we obtain the result known as **Poiseuille's equation:**

$$\frac{\Delta V}{\Delta t} = \frac{(P_1 - P_2)\pi r^4}{8\eta L} \qquad 15\text{–}19$$

Note that the volume flow rate varies with the fourth power of the tube's radius, thus a small change in radius corresponds to a large change in volume flow rate.

To see the significance of the r^4 dependence, consider an artery that branches into an arteriole that has half the artery's radius. Letting r go to $r/2$ in Poiseuille's equation, and solving for the pressure difference, we find

$$P_1 - P_2 = \frac{8\eta L}{\pi(r/2)^4}\left(\frac{\Delta V}{\Delta t}\right) = 16\left[\frac{8\eta L}{\pi r^4}\left(\frac{\Delta V}{\Delta t}\right)\right]$$

Thus, the pressure difference across a given length of arteriole is 16 times what it is across the same length of artery. In fact, in the human body the pressure drop along an artery is small compared to the rather large pressure drop observed in the arterioles. This is a direct consequence of the increased viscous drag of the blood as it flows through narrower blood vessels.

Similarly, a narrowing, or *stenosis*, of an artery can produce significant increases in blood pressure. For example, a reduction in radius of only 20%, from r to $0.8r$, causes an increase in pressure by a factor of $(1/0.8)^4 \sim 2.4$. Thus, even a small narrowing of an artery can lead to an increased risk for heart disease and stroke.

▲ Surface tension causes the surface of a liquid to behave like an elastic skin or membrane. When a small force is applied to the liquid surface, it tends to stretch, resisting penetration. This phenomenon enables a fishing spider to launch itself vertically upward as much as 4 cm from the water surface, as if from a trampoline. Such an acrobatic maneuver can save the spider from the attack of a predatory fish.

Surface Tension

A small insect resting on the surface of a pond or a lake is a common sight in the summertime. If you look carefully, you can see that the insect creates tiny dimples in the water's surface, almost as if it were supported by a thin sheet of rubber. In fact, the surface of water and other fluids behaves in many respects as if it were an elastic membrane. This effect is known as **surface tension.**

To understand the origin of surface tension, we start by noting that the molecules in a fluid exert attractive forces on one another. Thus, a molecule deep within a fluid experiences forces in all directions, as indicated in **Figure 15–25**, due to the molecules that surround it on all sides. The net force on such a molecule is zero. As a molecule nears the surface, however, it experiences a net force away from the surface, since there are no fluid molecules on the other side of the surface to attract it in that direction. It follows that work must be done on a molecule to move it from within a fluid to the surface, and that the *energy* of a fluid is increased for every molecule on its surface.

In general, physical systems tend toward configurations of minimum energy. For example, a ball on a slope rolls downhill, lowering its gravitational potential energy. If the energy of a droplet of liquid is to be minimized, it must have the smallest surface area possible for a given volume; that is, it must have the shape of a sphere. This is the reason small drops of dew are always spherical. Larger drops of water may be distorted by the downward pull of Earth's gravity, but in orbit drops of all sizes are spherical.

Since energy is required to increase the area of a liquid surface, the situation is similar to the energy required to stretch a spring or to stretch a sheet of rubber. Thus, the surface of a liquid behaves as if it were elastic, resisting tendencies to increase its area. For example, if a drop of dew is distorted into an ellipsoid, it quickly returns to its original spherical shape. Similarly, when an insect alights on the surface of a pond, it creates dimples that increase the surface area. The water resists this distortion with a force sufficient to support the weight of the insect—if the insect is not too large. In fact, even a needle or a razor blade can be supported on the surface of water if they are put into place gently, even though they have densities significantly greater than the density of water.

Surface tension, which is important in many biological systems, plays a particularly important role in human breathing. For example, the crucial exchange of

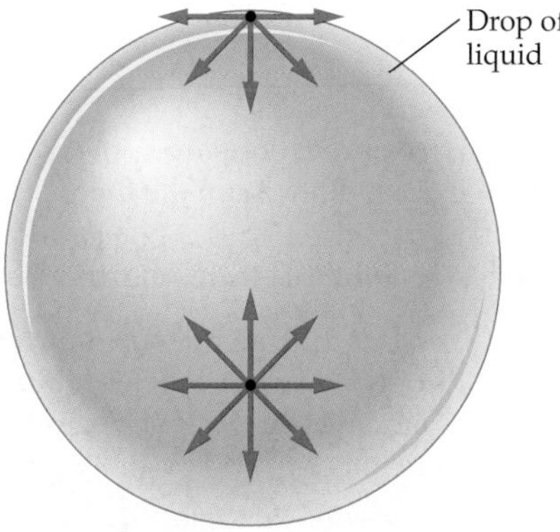

▲ **FIGURE 15–25 The origin of surface tension**

A molecule in the interior of a fluid experiences attractive forces of equal magnitude in all directions, giving a net force of zero. A molecule near the surface of the fluid experiences a net attractive force toward the interior of the fluid. This causes the surface to be pulled inward, resulting in a surface of minimum area.

REAL-WORLD PHYSICS: BIO

Breathing, alveoli, and premature birth

oxygen and carbon dioxide between inspired air and the blood occurs across the membranes of small balloonlike structures called *alveoli*. During inhalation, the alveoli expand from a radius of about 0.050 mm to 0.10 mm as they draw in fresh air. The walls of the alveoli are coated with a thin film of water, however, and in order to expand they must push outward against the water's surface tension—like trying to inflate a balloon against the surface tension of the rubber. To reduce the rather large surface tension of water, and make breathing easier, the lungs produce a substance called a *surfactant* that mixes with the water. This surfactant is produced rather late in the development of a fetus, however, and therefore premature infants may experience respiratory distress and even death as a result of too much surface tension in their alveoli.

Chapter Summary

15–1 Density

The density, ρ, of a material is its mass M per volume V:

$$\rho = M/V \qquad \text{15–1}$$

15–2 Pressure

Pressure, P, is force F per area A:

$$P = F/A \qquad \text{15–2}$$

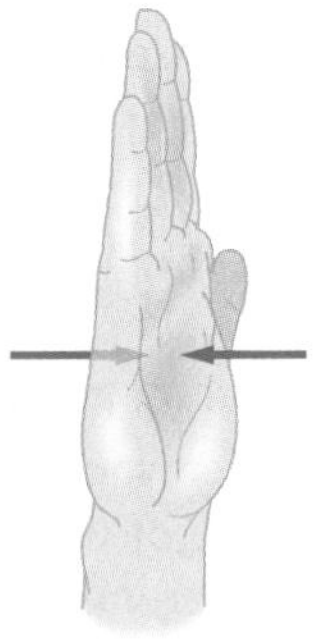

Atmospheric Pressure

The pressure exerted by the atmosphere is $P_{at} = 1.01 \times 10^5\ \text{N/m}^2 \approx 14.7\ \text{lb/in}^2$.

Pascals

Pressure is often given in terms of the pascal (Pa), where $1\ \text{Pa} = 1\ \text{N/m}^2$.

Gauge Pressure

Gauge pressure is the difference between the actual pressure and atmospheric pressure:

$$P_g = P - P_{at} \qquad \text{15–5}$$

15–3 Static Equilibrium in Fluids: Pressure and Depth

The pressure of a fluid in static equilibrium increases with depth.

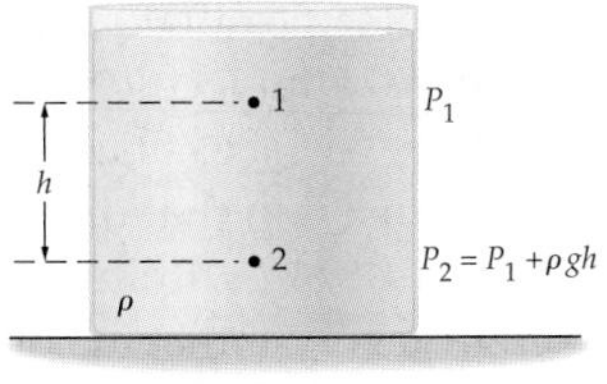

Pressure with Depth

If the pressure at one point in a fluid is P_1, the pressure at a depth h below that point is

$$P_2 = P_1 + \rho gh \qquad \text{15–7}$$

Pascal's Principle

An external pressure applied to an enclosed fluid is transmitted unchanged to every point within the fluid.

15–4 Archimedes' Principle and Buoyancy

The fact that the pressure in a fluid increases with depth leads to a net upward force on any object that is immersed in the fluid. This upward force is referred to as a buoyant force. The magnitude of the buoyant force is given by Archimedes' principle.

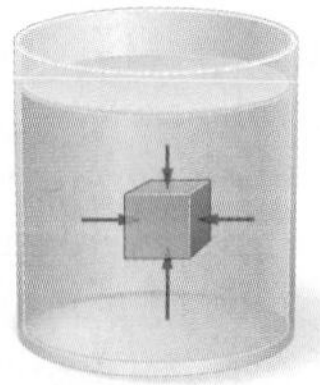

Archimedes' Principle

An object completely immersed in a fluid experiences an upward buoyant force equal in magnitude to the weight of fluid displaced by the object.

15–5 Applications of Archimedes' Principle

Archimedes' principle applies equally well to objects that are completely immersed, partially immersed, or floating.

Floatation

An object floats when it displaces an amount of fluid equal to its weight.

Submerged Volume
When a solid of volume V_s and density ρ_s floats in a fluid of density ρ_f, the volume of the solid that is submerged in the fluid, V_{sub}, is

$$V_{sub} = V_s(\rho_s/\rho_f) \qquad 15\text{–}10$$

15–6 Fluid Flow and Continuity

The speed of a fluid changes as the cross-sectional area of the pipe through which it flows changes.

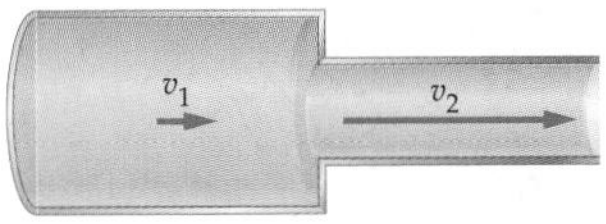

Equation of Continuity, Compressible Flow

$$\rho_1 A_1 v_1 = \rho_2 A_2 v_2 \qquad 15\text{–}11$$

Equation of Continuity, Incompressible Flow

$$A_1 v_1 = A_2 v_2 \qquad 15\text{–}12$$

15–7 Bernoulli's Equation

Bernoulli's equation can be thought of as energy conservation per volume for a fluid.

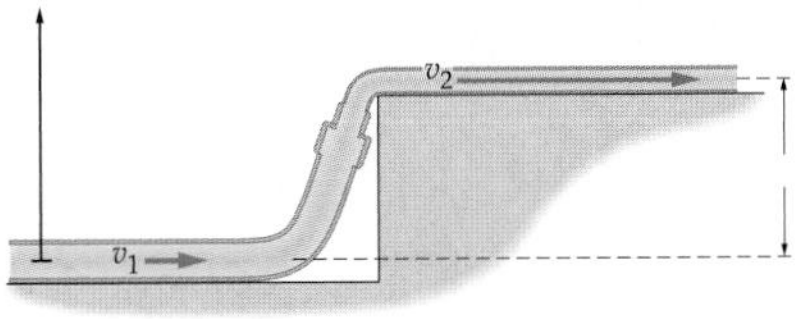

Change in Speed
If the speed of a fluid changes from v_1 to v_2 the corresponding pressures are related by:

$$P_1 + \tfrac{1}{2}\rho v_1^2 = P_2 + \tfrac{1}{2}\rho v_2^2 \qquad 15\text{–}14$$

Change in Height
If the height of a fluid changes from y_1 to y_2, the corresponding pressures are related by:

$$P_1 + \rho g y_1 = P_2 + \rho g y_2 \qquad 15\text{–}15$$

General Case
If both the height and the speed of a fluid change, the pressures are related by:

$$P_1 + \tfrac{1}{2}\rho v_1^2 + \rho g y_1 = P_2 + \tfrac{1}{2}\rho v_2^2 + \rho g y_2 \qquad 15\text{–}16$$

15–8 Applications of Bernoulli's Equation

Bernoulli's equation applies to a wide range of everyday situations, including airplane wings and prairie dog burrows.

Torricelli's Law
According to Torricelli's law, if a hole is poked in a container at a depth h below the surface, the fluid exits with the speed

$$v = \sqrt{2gh} \qquad 15\text{–}17$$

Note that this is the same speed as an object in free fall for a distance h.

15–9 Viscosity and Surface Tension

Viscosity and surface tension are two features of real fluids that are not found in an "ideal" fluid.

Viscosity
Viscosity in a fluid is similar to friction between two solid surfaces. A pressure difference, $P_1 - P_2$, is required to keep a viscous fluid flowing with a constant average speed, v. The relation between the volume flow rate of a fluid, $\Delta V/\Delta t$, and its coefficient of viscosity, η, is given by Poiseuille's equation:

$$\frac{\Delta V}{\Delta t} = \frac{(P_1 - P_2)\pi r^4}{8\eta L} \qquad 15\text{–}19$$

This expression applies to a tube of radius r and length L.

Surface Tension
A fluid tends to pull inward on its surface, resulting in a surface of minimum area. The surface of the fluid behaves much like an elastic membrane enclosing the fluid.

Problem-Solving Summary

Type of Problem	Relevant Physical Concepts	Related Examples
Relate pressure to depth in a static fluid.	The pressure in a static fluid increases uniformly with depth by the amount ρgh. Thus, if point 2 is a depth h below point 1, the pressure there is $P_2 = P_1 + \rho gh$.	Examples 15–3, 15–4
Calculate the buoyant force.	According to Archimedes' principle, the buoyant force is equal to the weight of displaced fluid.	Examples 15–5 Active Examples 15–1, 15–2
Calculate the speed of a fluid as the area through which it flows changes.	If a fluid is incompressible, its speed must vary inversely with the area through which it flows. Thus, if the area changes from A_1 to A_2 the speed of the fluid changes as follows: $A_1v_1 = A_2v_2$. This is the equation of continuity.	Example 15–7
Relate the pressure in a fluid to its height and speed.	Bernoulli's principle states that pressure, speed, and height in a fluid are related as follows: $P + \frac{1}{2}\rho v^2 + \rho gy = \text{constant}$. This is simply a statement of energy conservation per volume of the fluid.	Example 15–8 Active Example 15–3

Conceptual Questions

(Answers to odd-numbered Conceptual Questions can be found in the back of the book.)

1. Suppose you drink a liquid through a straw. Explain why the liquid moves upward, against gravity, into your mouth.
2. Considering your answer to the previous question, is it possible to sip liquid through a straw on the surface of the Moon? Explain.
3. Water towers on the roofs of buildings have metal bands wrapped around them for support. The spacing between bands is smaller near the base of a tower than near its top. Explain.
4. What holds a suction cup in place?
5. Suppose a force of 400 N is required to push the top off a wine barrel. In a famous experiment, Blaise Pascal attached a tall, thin tube to the top of a filled wine barrel, as shown in **Figure 15–26**. Water was slowly added to the tube until the barrel burst. The puzzling result found by Pascal was that the barrel broke when the weight of water in the tube was much less than 400 N. Explain Pascal's observation.

▲ **FIGURE 15–26** Conceptual Question 5 and Problem 17

6. Why is it more practical to use mercury in the barometer shown in Figure 15–3 than water?
7. An object's density can be determined by first weighing it in air, then in water (provided the density of the object is greater than the density of water, so that it is totally submerged when placed in water). Explain how these two measurements can give the desired result.
8. How does a balloonist control the vertical motion of a hot-air balloon?
9. Why is it possible for people to float without effort in Utah's Great Salt Lake?
10. **Physics in the Movies** In the movie *Voyage to the Bottom of the Sea,* the Earth is experiencing a rapid warming. In one scene, large icebergs break up into small, car-size chunks that drop downward through the water and bounce off the hull of the submarine *Seaview.* Is this an example of good, bad, or ugly physics? Explain.
11. One day, while snorkeling near the surface of a crystal-clear ocean, it occurs to you that you could go considerably deeper by simply lengthening the snorkel tube. Unfortunately, this does not work well at all. Why?
12. Since metal is more dense than water, how is it possible for a metal boat to float?
13. A sheet of water passing over a waterfall is thicker near the top than near the bottom. Similarly, a stream of water emerging from a water faucet becomes narrower as it falls. Explain.
14. It is a common observation that smoke rises more rapidly through a chimney when there is a wind blowing outside. Explain.
15. Is it best for an airplane to take off against the wind or with the wind? Explain.
16. If you have a hair dryer and a Ping Pong ball at home, try this demonstration. Direct the air from the dryer in a direction just above horizontal. Next, place the Ping Pong ball in the stream of air. If done just right, the ball will remain suspended in midair. Use the Bernoulli effect to explain this behavior.
17. Suppose a pitcher wants to throw a baseball so that it rises as it approaches the batter. How should the ball be spinning to accomplish this feat? Explain.

Conceptual Exercises

(Answers to odd-numbered Conceptual Exercises can be found in the back of the book.)

1. Two drinking glasses, A and B, are filled with water to the same depth. Glass A has twice the diameter of glass B. **(a)** Is the weight of the water in glass A greater than, less than, or equal to the weight of the water in glass B? Explain. **(b)** Is the pressure at the bottom of glass A greater than, less than, or equal to the pressure at the bottom of glass B? Explain. **(c)** Are your answers to parts (a) and (b) consistent? Explain.

2. **Figure 15–27** shows four containers, each filled with water to the same level. Rank the containers in order of increasing pressure at the depth h. Indicate ties where appropriate.

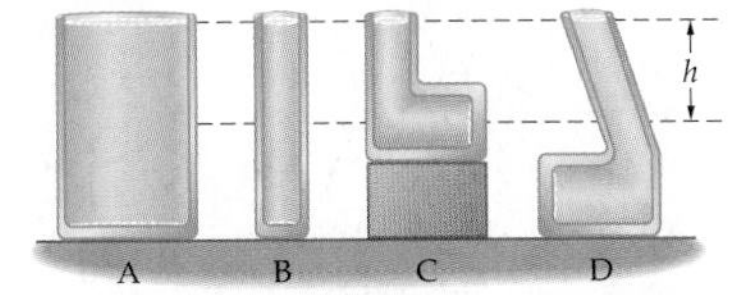

▲ **FIGURE 15–27** Conceptual Exercise 2

3. **Reading a Weather Glass** A weather glass, as shown in **Figure 15–28**, is used to give an indication of a change in the weather. Does the water level in the neck of the weather glass move up or move down when a low-pressure system approaches? Explain.

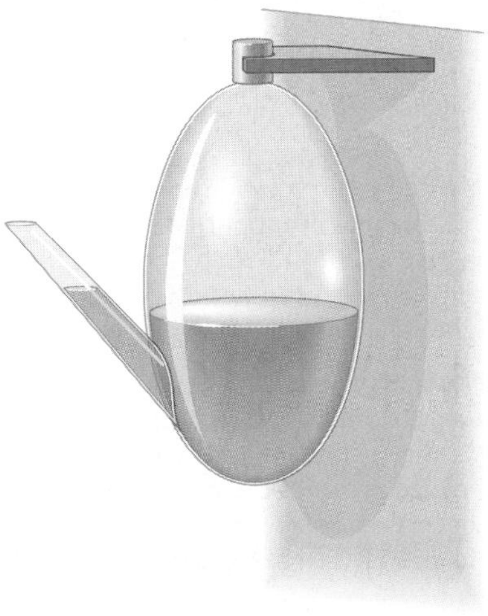

▲ **FIGURE 15–28** Conceptual Exercise 3

4. A helium-filled balloon for a birthday party is being brought home in a car. As the car accelerates away from a stop light, does the balloon remain vertical, tilt forward, or tilt backward? Explain.

5. A block of wood has a steel ball glued to one surface. The block can be floated with the ball "high and dry" on its top surface. When the block is inverted, and the ball immersed in water, does the volume of wood that is submerged increase, decrease, or stay the same? Explain.

6. In the preceding problem, suppose the block of wood with the ball "high and dry" is floating in a tank of water. When the block is inverted, does the water level in the tank increase, decrease, or stay the same? Explain.

7. **Measuring Density with a Hydrometer** A hydrometer, a device for measuring fluid density, is constructed as shown in **Figure 15–29**. If the hydrometer samples fluid A, the small float inside the tube is submerged to the level A. When fluid B is sampled, the float is submerged to level B. Is the density of fluid A greater than, less than, or equal to the density of fluid B? Explain. (This is how mechanics test your antifreeze level. Since antifreeze [ethylene glycol] is more dense than water, the higher the density of your coolant the more antifreeze protection you have.)

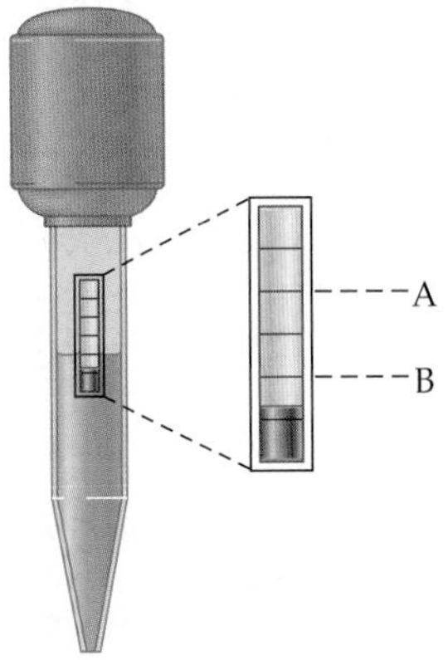

▲ **FIGURE 15–29** Conceptual Exercise 7

8. Referring to Active Example 15–1, suppose the flask with the wood tied to the bottom is placed on a scale. At some point the string breaks and the wood rises to the surface where it floats. When the wood is floating, is the reading on the scale greater than, less than, or equal to its previous reading? Explain.

9. A person floats in a boat in a small backyard swimming pool. Inside the boat with the person are some bricks. If the person drops the bricks overboard to the bottom of the pool, does the water level in the pool increase, decrease, or stay the same? Explain.

10. A person floats in a boat in a small backyard swimming pool. Inside the boat with the person are several blocks of wood. Suppose the person now throws the blocks of wood into the pool. **(a)** Does the boat float higher, lower, or at the same level relative to the water? Explain. **(b)** Does the water level in the pool increase, decrease, or stay the same? Explain.

11. The three identical containers in **Figure 15–30** are open to the air and filled with water to the same level. A block of wood floats in container A; an identical block of wood floats in container C, supporting a small lead weight; container B holds only water. **(a)** Rank the three containers in order of increasing weight of water they contain. Indicate ties where appropriate. **(b)** Rank the three containers in order of increasing weight of the container plus its contents. Indicate ties where appropriate.

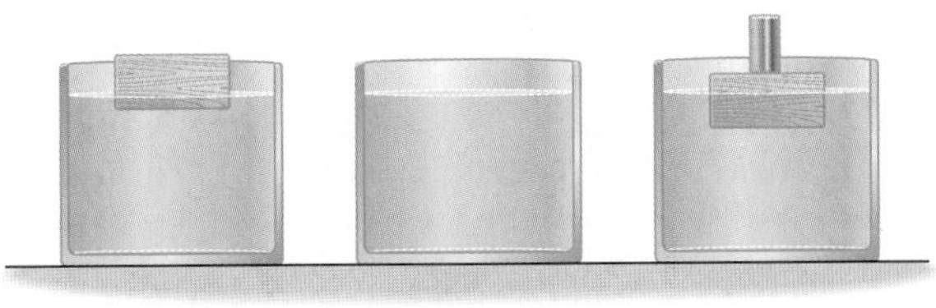

▲ **FIGURE 15–30** Conceptual Exercise 11

12. **Beebe and Barton** On Wednesday, August 15, 1934, William Beebe and Otis Barton made history by descending in the Bathysphere—basically a steel sphere 4.75 ft in diameter—3028 ft below the surface of the ocean, deeper than anyone had been before. As the Bathysphere was lowered, was the buoyant force

exerted on it at a depth of 10 ft greater than, less than, or equal to the buoyant force exerted on it at a depth of 50 ft? Explain.

13. Lead is more dense than aluminum. **(a)** Is the buoyant force on a solid lead sphere more than, less than, or the same as the buoyant force on a solid aluminum sphere of the same diameter? Explain. **(b)** Does your answer to part (a) depend on the fluid that is causing the buoyant force? Explain.

14. A fish carrying a pebble in its mouth swims with a small, constant velocity in a small bowl. When the fish drops the pebble to the bottom of the bowl, does the water level rise, fall, or stay the same? Explain.

15. On a planet in a different solar system, the acceleration of gravity is greater than it is on Earth. If you float in the water on this planet, is it easier to float, harder, or the same as on Earth? Explain.

16. A pan half-filled with water is placed near the rim of a rotating turntable. Is the surface of the water in the pan level, tilted inward toward the axis of rotation, or tilted outward away from the axis of rotation? Explain. (Refer to Conceptual Exercise 15 in Chapter 6 for a similar situation.)

17. In the system described in the previous problem, suppose the temperature is lowered below the freezing point of water. The turntable still rotates, but now the water is a solid block of ice. If a marble is placed on the surface of the ice and released from rest, will it move toward the axis of rotation, move away from the axis of rotation, or stay where it is released? Explain.

18. **BIO Sphygmomanometer** When a person's blood pressure is taken with a device known as a sphygmomanometer, it is measured on the arm, at approximately the same level as the heart. If the measurement were to be taken on the patient's leg instead, would the reading on the sphygmomanometer be greater than, less than, or the same as when the measurement is made on the arm? Explain.

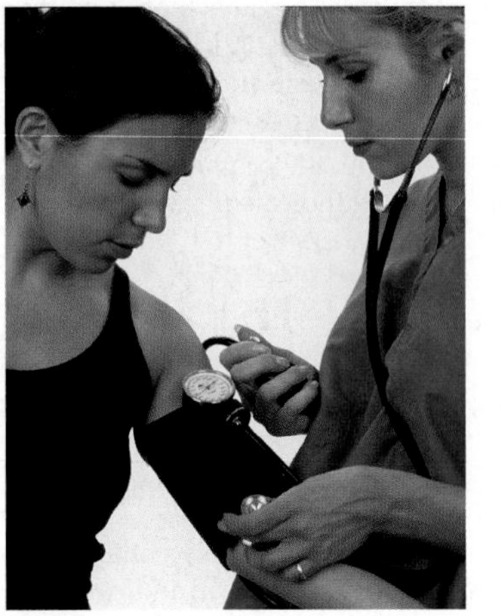

Sphygmomanometer (Conceptual Exercise 18)

Problems

Note: **IP** *denotes an integrated conceptual/quantitative problem.* **BIO** *identifies problems of biological or medical interest. Red bullets (•, ••, •••) are used to indicate the level of difficulty of each problem.*

Section 15–1 Density

1. • Estimate the weight of the air in your physics classroom.

2. • What weight of water is required to fill a 25-gallon aquarium?

3. • You buy a "gold" ring at a pawn shop. The ring has a mass of 0.014 g and a volume of 0.0022 cm^3. Is the ring solid gold?

4. • Estimate the weight of a treasure chest filled with gold doubloons.

5. • A cube of metal has a mass of 0.347 kg and measures 3.21 cm on a side. Calculate the density and identify the metal.

Section 15–2 Pressure

6. • What is the downward force exerted by the atmosphere on a football field, whose dimensions are 360 ft by 160 ft?

7. • **BIO Bioluminescence** Some species of dinoflagellate (a type of unicellular plankton) can produce light as the result of biochemical reactions within the cell. This light is an example of bioluminescence. It is found that bioluminescence in dinoflagellates can be triggered by deformation of the cell surface with a pressure as low as one dyne (10^{-5} N) per square centimeter. What is this pressure in **(a)** pascals and **(b)** atmospheres?

8. • A 72-kg person sits on a 3.8-kg chair. Each leg of the chair makes contact with the floor in a circle that is 1.3 cm in diameter. Find the pressure exerted on the floor by each leg of the chair, assuming the weight is evenly distributed.

9. • To prevent damage to floors (and to increase friction), a crutch will often have a rubber tip attached to its end. If the end of the crutch is a circle of radius 1.2 cm without the tip, and the tip is a circle of radius 2.5 cm, by what factor does the tip reduce the pressure exerted by the crutch?

10. • An inflated basketball has a gauge pressure of 9.9 lb/in^2. What is the actual pressure inside the ball?

11. •• Suppose that when you ride on your 7.70-kg bike the weight of you and the bike is supported equally by the two tires. If the gauge pressure in the tires is 70.5 lb/in^2 and the area of contact between each tire and the road is 7.13 cm^2. What is your weight?

12. •• **IP** The weight of your 1320-kg car is supported equally by its four tires, each inflated to a gauge pressure of 35.0 lb/in^2. **(a)** What is the area of contact each tire makes with the road? **(b)** If the gauge pressure is increased, does the area of contact increase, decrease, or stay the same? **(c)** What gauge pressure is required to give an area of contact of 116 cm^2 for each tire?

Section 15–3 Static Equilibrium in Fluids: Pressure and Depth

13. • Water in the lake behind Hoover Dam is 221 m deep. What is the water pressure at the base of the dam?

14. • In a classroom demonstration, the pressure inside a soft drink can is suddenly reduced to essentially zero. Assuming the can to be a cylinder with a height of 12 cm and a diameter of 6.5 cm, find the net inward force exerted on the vertical sides of the can due to atmospheric pressure.

15. •• As a storm front moves in, you notice that the column of mercury in a barometer rises to only 736 mm. **(a)** What is the air pressure? **(b)** If the mercury in this barometer is replaced with water, to what height does the column of water rise? Assume the same air pressure found in part (a).

16. •• In the hydraulic system shown in **Figure 15–31**, the piston on the left has a diameter of 4.5 cm and a mass of 1.7 kg. The piston on the right has a diameter of 12 cm and a mass of 3.2 kg. If the density of the fluid is 750 kg/m^3, what is the height difference h between the two pistons?

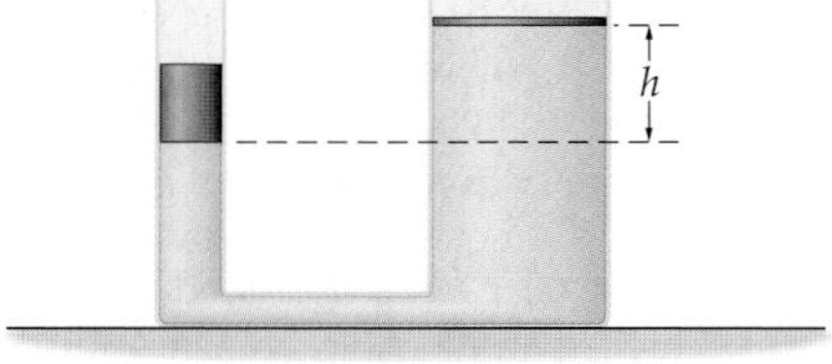

▲ **FIGURE 15–31** Problem 16

17. •• A circular wine barrel 75 cm in diameter will burst if the net upward force exerted on the top of the barrel is 643 N. A tube 1.0 cm in diameter extends into the barrel through a hole in the top, as indicated in Figure 15–26 (See Conceptual Question 5). Initially, the barrel is filled to the top and the tube is empty above that level. What weight of water must be poured into the tube in order to burst the barrel?

18. •• A cylindrical container with a cross-sectional area of 65.2 cm^2 holds a fluid of density 806 kg/m^3. At the bottom of the container the pressure is 116 kPa. **(a)** What is the depth of the fluid? **(b)** Find the pressure at the bottom of the container after an additional 2.05×10^{-3} m^3 of this fluid is added to the container. Assume that no fluid spills out of the container.

19. •• **IP Tourist Submarine** A submarine called the *Deep View 66* is currently being developed to take 66 tourists at a time on sightseeing trips to tropical coral reefs. According to guidelines of the American Society of Mechanical Engineers (ASME), to be safe for human occupancy the *Deep View 66* must be able to withstand a pressure of 10.0 N per square millimeter. **(a)** To what depth can the *Deep View 66* safely descend in seawater? **(b)** If the submarine is used in freshwater instead, is its maximum safe depth greater than, less than, or the same as in seawater? Explain.

20. •• **IP** A water storage tower, like the one shown in the accompanying photo, is filled with fresh water to a depth of 6.4 m. What is the pressure at **(a)** 4.5 m and **(b)** 5.5 m below the surface of the water? **(c)** Why are the metal bands on such towers more closely spaced near the base of the tower?

Roof-top water towers. (Problem 20)

21. •• **IP** You step onto an elevator holding a glass of water filled to a depth of 6.5 cm. After a moment, the elevator moves upward with constant acceleration, increasing its speed from zero to 2.2 m/s in 3.1 s. **(a)** During the period of acceleration, is the pressure exerted on the bottom of the glass greater than, less than, or the same as before the elevator began to move? Explain. **(b)** Find the change in the pressure exerted on the bottom of the glass as the elevator accelerates.

22. •• Suppose you pour water into a container until it reaches a depth of 12 cm. Next, you carefully pour in a 7.2-cm thickness of olive oil so that it floats on top of the water. What is the pressure at the bottom of the container?

23. •• Referring to Example 15–4, suppose that some vegetable oil has been added to both sides of the U-tube. On the right side of the tube, the depth of oil is 5.00 cm, as before. On the left side of the tube, the depth of the oil is 3.00 cm. Find the difference in fluid level between the two sides of the tube.

24. •• **IP** As a stunt, you want to sip some water through a very long, vertical straw. **(a)** First, explain why the liquid moves upward, against gravity, into your mouth when you sip. **(b)** What is the tallest straw that you could, in principle, drink from in this way?

25. •• **IP BIO** The patient in **Figure 15–32** is to receive an intravenous injection of medication. In order to work properly, the pressure of fluid containing the medication must be 109 kPa at the injection point. **(a)** If the fluid has a density of 1020 kg/m^3, find the height at which the bag of fluid must be suspended above the patient. Assume that the pressure inside the bag is one atmosphere. **(b)** If a less dense fluid is used instead, must the height of suspension be increased or decreased? Explain.

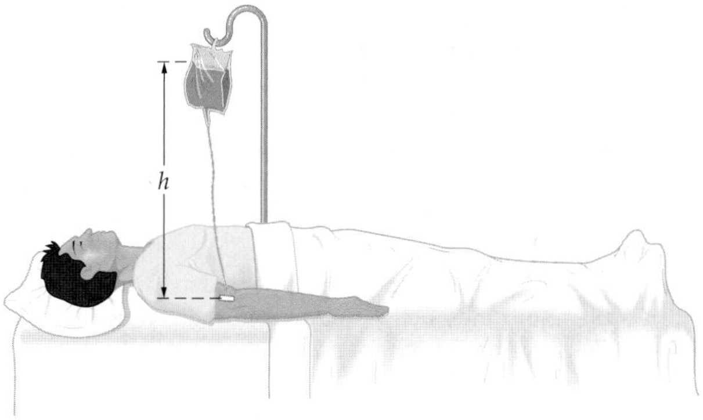

▲ **FIGURE 15–32** Problem 25

26. ••• A cylindrical container 1.0 m tall contains mercury to a certain depth, *d*. The rest of the cylinder is filled with water. If the pressure at the bottom of the cylinder is two atmospheres, what is the depth *d*?

Section 15–4 Archimedes' Principle and Buoyancy

27. • A raft is 4.2 m wide and 6.5 m long. When a horse is loaded onto the raft, it sinks 2.7 cm deeper into the water. What is the weight of the horse?

28. • To walk on water, all you need is a pair of water-walking boots shaped like boats. If each boot is 25-cm high and 35-cm wide, how long must they be to support a 75-kg person?

29. •• A 3.2-kg balloon is filled with helium (density = 0.179 kg/m^3). If the balloon is a sphere with a radius of 4.9 m, what is the maximum weight it can lift?

30. •• A hot-air balloon plus cargo has a mass of 1890 kg and a volume of 11,430 m^3. The balloon is floating at a constant height of 6.25 m above the ground. What is the density of the hot air in the balloon?

31. ••• In the lab you place a beaker that is half full of water (density ρ_w) on a scale. You now use a light string to suspend a piece of metal of volume *V* in the water. The metal is completely submerged, and none of the water spills out of the beaker. Give a symbolic expression for the change in reading of the scale.

Section 15–5 Applications of Archimedes' Principle

32. • An air mattress is 2.2 m long, 0.65 m wide, and 13 cm deep. If the air mattress itself has a mass of 0.22 kg, what is the maximum mass it can support in freshwater?

33. •• A solid block is attached to a spring scale. When the block is suspended in air, the scale reads 20.0 N; when it is completely

immersed in water, the scale reads 17.7 N. What is **(a)** the volume and **(b)** the density of the block?

34. •• As in the previous problem, a solid block is suspended from a spring scale. If the reading on the scale when the block is completely immersed in water is 25.0 N, and the reading when it is completely immersed in alcohol of density 806 kg/m^3 is 25.7 N, what is **(a)** the block's volume and **(b)** its density?

35. •• **BIO** A person weighs 756 N in air and has a body-fat percentage of 28.1%. **(a)** What is the overall density of this person's body? **(b)** What is the volume of this person's body? **(c)** Find the apparent weight of this person when completely submerged in water.

36. •• **IP** A log floats in a river with one-fourth of its volume above the water. **(a)** What is the density of the log? **(b)** If the river carries the log into the ocean, does the portion of the log above the water increase, decrease, or stay the same? Explain.

37. •• A person with a mass of 81 kg and a volume of 0.089 m^3 floats quietly in water. **(a)** What is the volume of the person that is above water? **(b)** If an upward force F is applied to the person by a friend, the volume of the person above water increases by 0.0018 m^3. Find the force F.

38. •• **IP** A block of wood floats on water. A layer of oil is now poured on top of the water to a depth that more than covers the block, as shown in **Figure 15–33**. **(a)** Is the volume of wood submerged in water greater than, less than, or the same as before? **(b)** If 90% of the wood is submerged in water before the oil is added, find the fraction submerged when oil with a density of 875 kg/m^3 covers the block.

▲ **FIGURE 15–33** Problem 38

39. •• A piece of lead has the shape of a hockey puck, with a diameter of 7.5 cm and a height of 2.5 cm. If the puck is placed in a mercury bath, it floats. How deep below the surface of the mercury is the bottom of the lead puck?

40. ••• **IP** A lead weight with a volume of 0.82×10^{-5} m^3 is lowered on a fishing line into a lake to a depth of 1.0 m. **(a)** What tension is required in the fishing line to give the weight an upward acceleration of 2.1 m/s^2? **(b)** If the initial depth of the weight is increased to 2.0 m, does the tension found in part (a) increase, decrease, or stay the same? Explain. **(c)** What acceleration will the weight have if the tension in the fishing line is 1.2 N? Give both direction and magnitude.

Section 15–6 Fluid Flow and Continuity

41. • To water the yard, you use a hose with a diameter of 3.4 cm. Water flows from the hose with a speed of 1.1 m/s. If you partially block the end of the hose so the effective diameter is now 0.57 cm, with what speed does water spray from the hose?

42. • Water flows through a pipe with a speed of 2.1 m/s. Find the flow rate in kg/s if the diameter of the pipe is 3.8 cm.

43. • To fill a child's inflatable wading pool, you use a garden hose with a diameter of 2.8 cm. Water flows from this hose with a speed of 1.1 m/s. How long will it take to fill the pool to a depth of 26 cm if the pool is circular and has a diameter of 2.0 m?

44. • **BIO Heart Pump Rate** When at rest, your heart pumps blood at the rate of 5.00 liters per minute (L/min). What is the volume and mass of blood pumped by your heart in one day?

45. •• **BIO Blood Speed in an Arteriole** A typical arteriole has a diameter of 0.030 mm and carries blood at the rate of 5.5×10^{-6} cm^3/s. **(a)** What is the speed of the blood in an arteriole? **(b)** Suppose an arteriole branches into 340 capillaries, each with a diameter of 4.0×10^{-6} m. What is the blood speed in the capillaries? (The low speed in capillaries is beneficial; it promotes the diffusion of materials to and from the blood.)

46. •• **IP** Water flows at the rate of 3.11 kg/s through a hose with a diameter of 3.22 cm. **(a)** What is the speed of water in this hose? **(b)** If the hose is attached to a nozzle with a diameter of 0.732 cm, what is the speed of water in the nozzle? **(c)** Is the number of kilograms per second flowing through the nozzle greater than, less than, or equal to 3.11 kg/s? Explain.

47. •• A river narrows at a rapids from a width of 12 m to a width of only 5.8 m. The depth of the river before the rapids is 2.7 m; the depth in the rapids is 0.85 m. Find the speed of water flowing in the rapids, given that its speed before the rapids is 1.2 m/s. Assume the river has a rectangular cross section.

48. •• **BIO How Many Capillaries?** The aorta has an inside diameter of approximately 2.1 cm, compared to that of a capillary, which is about 1.0×10^{-5} m (10 μm). In addition, the average speed of flow is approximately 1.0 m/s in the aorta and 1.0 cm/s in a capillary. Assuming that all the blood that flows through the aorta also flows through the capillaries, how many capillaries does the circulatory system have?

Section 15–7 Bernoulli's Equation

49. • **BIO Plaque in an Artery** The buildup of plaque on the walls of an artery may decrease its diameter from 1.1 cm to 0.75 cm. If the speed of blood flow was 15 cm/s before reaching the region of plaque buildup, find **(a)** the speed of blood flow and **(b)** the pressure drop within the plaque region.

50. • A horizontal pipe contains water at a pressure of 110 kPa flowing with a speed of 1.4 m/s. When the pipe narrows to one-half its original diameter, what is **(a)** the speed and **(b)** the pressure of the water?

51. •• **BIO A Blowhard** Tests of lung capacity show that adults are able to exhale 1.5 liters of air through their mouths in as little as 1.0 seconds. **(a)** If a person blows air at this rate through a drinking straw with a diameter of 0.60 cm, what is the speed of air in the straw? **(b)** If the air from the straw in part (a) is directed horizontally across the upper end of a second straw that is vertical, as shown in **Figure 15–34**, to what height does water rise in the vertical straw?

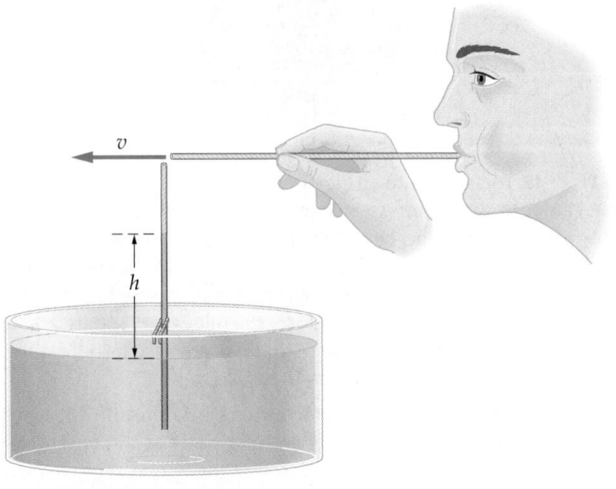

▲ **FIGURE 15–34** Problem 51

52. •• **IP** Water flows through a horizontal tube of diameter 2.8 cm that is joined to a second horizontal tube of diameter 1.6 cm. The pressure difference between the tubes is 7.5 kPa. **(a)** Which tube has the higher pressure? **(b)** Which tube has the higher speed of flow? **(c)** Find the speed of flow in the first tube.

53. •• A garden hose is attached to a water faucet on one end and a spray nozzle on the other end. The water faucet is turned on, but the nozzle is turned off so that no water flows through the hose. The hose lies horizontally on the ground, and a stream of water sprays vertically out of a small leak to a height of 0.68 m. What is the pressure inside the hose?

Section 15–8 Applications of Bernoulli's Equation

54. • A water tank springs a leak. Find the speed of water emerging from the hole if the leak is 2.5 m below the surface of the water, which is open to the atmosphere.

55. •• **(a)** Find the pressure difference on an airplane wing if air flows over the upper surface with a speed of 115 m/s, and along the bottom surface with a speed of 105 m/s. **(b)** If the area of the wing is 32 m^2, what is the net upward force exerted on the wing?

56. •• On a vacation flight, you look out the window of the jet and wonder about the forces exerted on the window. Suppose the air outside the window moves with a speed of approximately 170 m/s shortly after takeoff, and that the air inside the plane is at atmospheric pressure. **(a)** Find the pressure difference between the inside and outside of the window. **(b)** If the window is 25 cm by 42 cm, find the force exerted on the window by air pressure.

57. •• **IP** During a thunderstorm, winds with a speed of 47.7 m/s blow across a flat roof with an area of 668 m^2. **(a)** Find the magnitude of the force exerted on the roof as a result of this wind. **(b)** Is the force exerted on the roof in the upward or downward direction? Explain.

58. •• A garden hose with a diameter of 0.65 in. has water flowing in it with a speed of 0.55 m/s and a pressure of 1.2 atmospheres. At the end of the hose is a nozzle with a diameter of 0.25 in. Find **(a)** the speed of water in the nozzle and **(b)** the pressure in the nozzle.

59. ••• **IP** Water flows in a cylindrical, horizontal pipe. As the pipe narrows to half its initial diameter, the pressure in the pipe changes. **(a)** Is the pressure in the narrow region greater than, less than, or the same as the initial pressure? Explain. **(b)** Calculate the change in pressure between the wide and narrow regions of the pipe. Give your answer symbolically in terms of the density of the water, ρ, and its initial speed v.

*Section 15–9 Viscosity and Surface Tension

*60. • **BIO Vasodilation** When the body requires an increased blood flow rate in a particular organ or muscle, it can accomplish this by increasing the diameter of arterioles in that area. This is referred to as vasodilation. What percentage increase in the diameter of an arteriole is required to double the volume flow rate of blood, all other factors remaining the same?

*61. • **BIO** **(a)** Find the volume of blood that flows per second through the pulmonary artery described in Example 15–10. **(b)** If the radius of the artery is reduced by 16%, by what factor is the blood flow rate reduced? Assume that all other properties of the artery remain unchanged.

*62. • **BIO An Occlusion in an Artery** Suppose an occlusion in an artery reduces its diameter by 15%, but the volume flow rate of blood in the artery remains the same. By what factor has the pressure drop across the length of this artery increased?

*63. •• **IP** Water at 20 °C flows through a horizontal garden hose at the rate of 5.0×10^{-4} m^3/s. The diameter of the garden hose is 2.5 cm. **(a)** What is the water speed in the hose? **(b)** What is the pressure drop across a 15-m length of hose? Suppose the cross-sectional area of the hose is halved, but the length and pressure drop remain the same. **(c)** By what factor does the water speed change? **(d)** By what factor does the volume flow rate change? Explain.

General Problems

64. • At what depth below the ocean surface is the pressure equal to two atmospheres?

65. • **Supersonic Erosion** In waterjet cutting, a stream of supersonic water is used to slice through materials ranging from sheets of paper to solid steel plates. The water is held in a reservoir at 59,500 psi and allowed to exit through a small orifice at high speed. Find the exit speed of the water, and compare with the speed of sound.

66. • A water main broke on Lake Shore Drive in Chicago on November 8, 2002, shooting water straight upward to a height of 8.0 ft. What was the pressure in the pipe?

67. • **BIO Measuring Pain Threshold** A useful instrument for evaluating fibromyalgia and trigger-point tenderness is the Doloriometer or Algorimeter. This device consists of a force meter attached to a circular probe that is pressed against the skin until pain is experienced. If the reading of the force meter is 3.25 lb, and the diameter of the circular probe is 1.39 cm, what is the pressure applied to the skin? Give your answer in pascals.

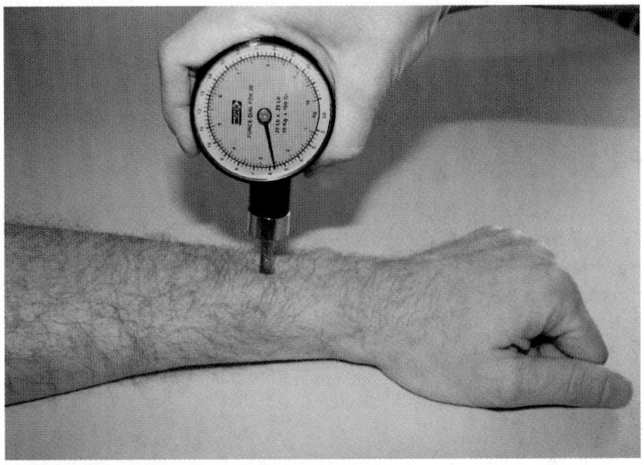

A Doloriometer/Algorimeter used to evaluate pain threshold. (Problem 67)

68. •• **BIO Power Output of the Heart** The power output of the heart is given by the product of the average blood pressure, 1.33 N/cm^2, and the flow rate, 105 cm^3/s. **(a)** Find the power of the heart. Give your answer in watts. **(b)** How much energy does the heart expend in a day? **(c)** Suppose the energy found in part (b) is used to lift a 72-kg person vertically to a height h. Find h, in meters.

69. •• An above-ground backyard swimming pool is shaped like a large hockey puck, with a circular bottom and a vertical wall forming its perimeter. The diameter of the pool is 4.9 m and its depth is 1.6 m. Find the total outward force exerted on the vertical wall of the pool by the water, assuming the pool is completely filled.

70. •• A solid block is suspended from a spring scale. When the block is in air, the scale reads 35.0 N, when immersed in water the scale reads 31.1 N, and when immersed in oil the scale reads 31.8 N. **(a)** What is the density of the block? **(b)** What is the density of the oil?

71. •• A wooden block with a density of 710 kg/m^3 and a volume of 0.012 m^3 is attached to the top of a vertical spring whose force constant is $k = 540$ N/m. Find the amount by which the spring is stretched or compressed if it and the wooden block are

(a) in air or (b) completely immersed in water. [The density of air may be neglected in part (a).]

72. •• IP **Floating a Ball and Block** A 1.25-kg wooden block has an iron ball of radius 1.22 cm glued to one side. **(a)** If the block floats in water with the iron ball "high and dry," what is the volume of wood that is submerged? **(b)** If the block is now inverted, so that the iron ball is completely immersed, does the volume of wood that is submerged in water increase, decrease, or remain the same? Explain. **(c)** Calculate the volume of wood that is submerged when the block is in the inverted position.

73. •• On a bet, you try to remove water from a glass by blowing across the top of a vertical straw immersed in the water. What is the minimum speed you must give the air at the top of the straw to draw water upward through a height of 1.8 cm?

74. •• **The Depth of the Atmosphere** Evangelista Torricelli (1608–1647) was the first to put forward the idea that we live at the bottom of an ocean of air. **(a)** Given the value of atmospheric pressure at the surface of the Earth, and the fact that there is zero pressure in the vacuum of space, determine the depth of the atmosphere, assuming that the density of air and the acceleration of gravity are constant. **(b)** According to this model, what is atmospheric pressure at the summit of Mt. Everest, 29,035 ft above sea level. (In fact, the density of air and the acceleration of gravity decrease with altitude, so the result obtained here is less than the actual depth of the atmosphere. Still this is a reasonable first estimate.)

75. •• **The Hydrostatic Paradox I** Consider the lightweight containers shown in **Figure 15–35**. Both containers have bases of area $A_{base} = 24\ \text{cm}^2$ and depths of water equal to 18 cm. As a result, the downward force on the base of container 1 is equal to the downward force on container 2, even though the containers clearly hold different weights of water. This is referred to as the hydrostatic paradox. **(a)** Given that container 2 has an annular (ring-shaped) region of area $A_{ring} = 72\ \text{cm}^2$, determine the *net* downward force exerted on the container by the water. **(b)** Show that your result from part (a) is equal to the weight of the water in container 2.

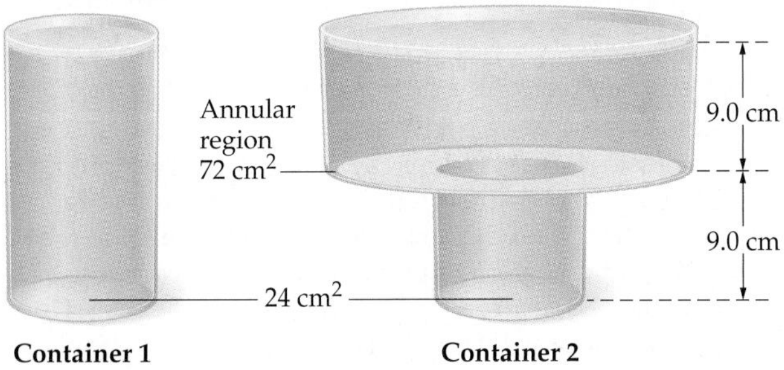

▲ **FIGURE 15–35** Problem 75

76. •• **The Hydrostatic Paradox II** Consider the two lightweight containers shown in **Figure 15–36**. As in the previous problem, these containers have equal forces on their bases but contain different weights of water. This is another version of the hydrostatic paradox. **(a)** Determine the *net* downward force exerted by the water on container 2. Note that the bases of the containers have an area $A_{base} = 24\ \text{cm}^2$, the annular region has an area $A_{ring} = 18\ \text{cm}^2$, and the depth of the water is 18 cm. **(b)** Show that your result from part (a) is equal to the weight of the water in container 2. **(c)** If a hole is poked in the annular region of container 2, how fast will water exit the hole? **(d)** How high above the hole will the stream of water rise?

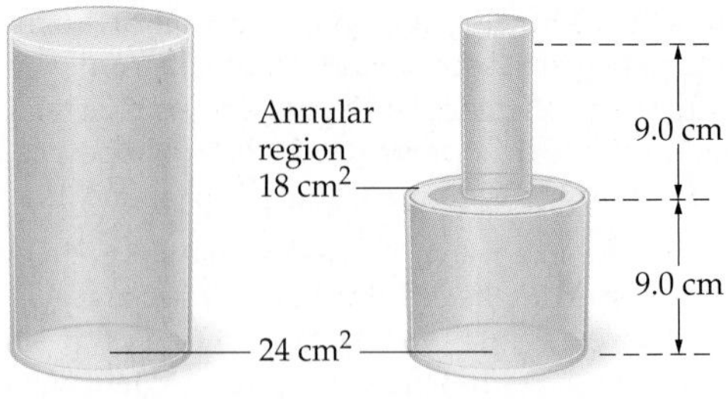

▲ **FIGURE 15–36** Problem 76

77. •• IP A backyard swimming pool is circular in shape and contains water to a uniform depth of 31 cm. It is 2.2 m in diameter and is not completely filled. **(a)** What is the pressure at the bottom of the pool? **(b)** If a person gets in the pool and floats peacefully, does the pressure at the bottom of the pool increase, decrease, or stay the same? **(c)** Calculate the pressure at the bottom of the pool if the floating person has a mass of 78 kg.

78. •• A prospector finds a solid rock composed of granite ($\rho = 2650\ \text{kg/m}^3$) and gold. If the volume of the rock is $3.55 \times 10^{-4}\ \text{m}^3$, and its mass is 3.81 kg, **(a)** what mass of gold is contained in the rock? What percentage of the rock is gold by **(b)** volume and **(c)** mass?

79. •• **The Maximum Depth of the Earth's Crust** Consider the crustal rocks of the Earth to be a fluid of density $3.0 \times 10^3\ \text{kg/m}^3$. Under this assumption, the pressure at a depth h within the crust is $P = P_{at} + \rho g h$. If the greatest pressure crustal rock can sustain before crumbling is 1.2×10^9 Pa, find the maximum depth of the Earth's crust. (Below this depth the crust changes from a solid to a plasticlike material.)

80. •• IP **(a)** If the tension in the string in Active Example 15–1 is 0.85 N, what is the volume of the wood? Assume that everything else remains the same. **(b)** If the string breaks and the wood floats on the surface, does the water level in the flask rise, drop, or stay the same? Explain. **(c)** Assuming the flask is cylindrical with a cross sectional area of 65 cm², find the change in water level after the string breaks.

81. •• IP **A Siphon for Irrigation** A siphon is a device that allows water to flow from one level to another. The siphon shown in **Figure 15–37** delivers water from an irrigation canal to a field of crops. To operate the siphon, water is first drawn through the length of the tube. After the flow is started in this way it continues on its own. **(a)** Using points 1 and 3 in Figure 15–34, find the speed v of the water leaving the siphon at its lower end. Give a symbolic answer. **(b)** Is the speed of the water at point 2 greater than, less than, or the same as its speed at point 3? Explain.

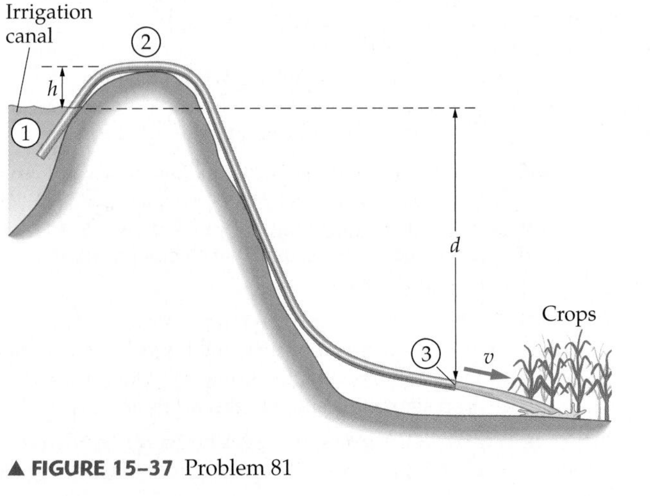

▲ **FIGURE 15–37** Problem 81

82. •• A tin can is filled with water to a depth of 37 cm. A hole 12 cm above the bottom of the can produces a stream of water that is directed at an angle of 34° above the horizontal. Find **(a)** the range and **(b)** the maximum height of this stream of water.

83. •• **BIO** A person weighs 682 N in air but only 498 N when standing in water up to the hips. Find **(a)** the volume of each of the person's legs and **(b)** the mass of each leg, assuming they have a density that is 1.05 times the density of water.

*84. •• A horizontal pipe carries oil whose coefficient of viscosity is $0.00012\ \mathrm{N \cdot s/m^2}$. The diameter of the pipe is 5.2 cm, and its length is 55 m. **(a)** What pressure difference is required between the ends of this pipe if the oil is to flow with an average speed of 1.2 m/s? **(b)** What is the volume flow rate in this case?

*85. •• **BIO** A patient is given an injection with a hypodermic needle 3.2 cm long and 0.28 mm in diameter. Assuming the solution being injected has the same density and viscosity as water at 20 °C, find the pressure difference needed to inject the solution at the rate of 1.5 g/s.

86. •• **Cooking Doughnuts** A popular brand of doughnut is made in an assembly-line fashion. First, the white, uncooked doughnut is dropped into hot vegetable oil ($\rho = 919\ \mathrm{kg/m^3}$) where it browns on one side as it cooks. It is then flipped over and browned on the other side. The final product still has a white stripe around the middle where the dough always stays out of the oil, as indicated in **Figure 15–38**. Assuming the doughnut has a cylindrical shape of height H and diameter D, and that the height of the white stripe is $0.22H$, find the density of the doughnut.

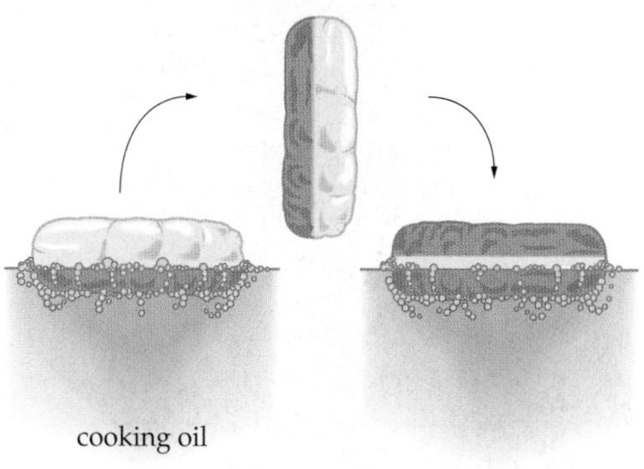

▲ **FIGURE 15–38** Problem 86

87. •• **An Airburst over Pennsylvania** On the evening of July 23, 2001, a meteor streaked across the skies of Pennsylvania, creating a spectacular fireball before exploding in the atmosphere with an energy release of 3 kilotons of TNT. The pressure wave from the airburst caused an increase in pressure of 0.50 kPa, enough to shatter some windows. Find the force that this "overpressure" would exert on a 34″ × 46″ window. Give your answer in newtons and pounds.

88. ••• **IP** A pan half-filled with water is placed in the back of an SUV. **(a)** When the SUV is driving on the freeway with a constant velocity, is the surface of the water in the pan level, tilted forward, or tilted backward? Explain. **(b)** Suppose the SUV accelerates in the forward direction with a constant acceleration a. Is the surface of the water tilted forward, or tilted backward? Explain. **(c)** Show that the angle of tilt, θ, in part (b) has a magnitude given by $\tan\theta = a/g$, where g is the acceleration of gravity.

89. ••• A wooden block of cross-sectional area A, height H, and density ρ_1 floats in a fluid of density ρ_2. If the block is displaced downward and then released, it will oscillate with simple harmonic motion. Find the period of its motion.

90. ••• A round wooden log with a diameter of 73 cm floats with one-half of its radius out of the water. What is the log's density?

91. ••• The hollow, spherical glass shell shown in **Figure 15–39** has an inner radius R and an outer radius $1.2\,R$. The density of the glass is ρ_g. What fraction of the shell is submerged when it floats in a liquid of density $\rho = 1.5\,\rho_g$? (Assume the interior of the shell is a vacuum.)

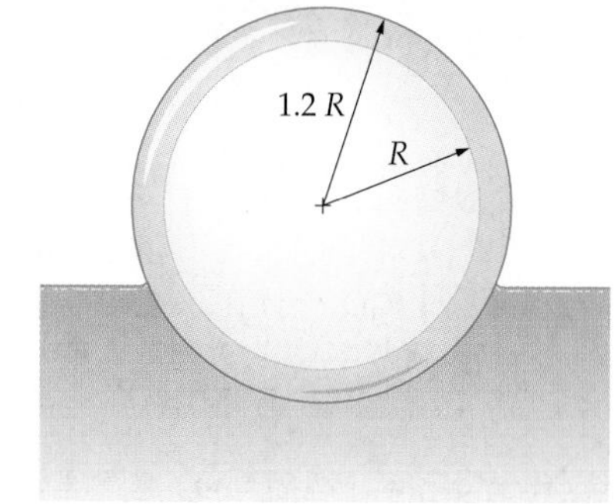

▲ **FIGURE 15–39** Problem 91

92. ••• A geode is a hollow rock with a solid shell and an air-filled interior. Suppose a particular geode weighs twice as much in air as it does when completely submerged in water. If the density of the solid part of the geode is $2500\ \mathrm{kg/m^3}$, what fraction of the geode's volume is hollow?

93. ••• A tank of water filled to a depth d has a hole in its side a height h above the table on which it rests. Show that water emerging from the hole hits the table at a horizontal distance of $2\sqrt{(d-h)h}$ from the base of the tank.

94. ••• The water tank in **Figure 15–40** is open to the atmosphere and has two holes in it, one 0.80 m and one 3.6 m above the floor on which the tank rests. If the two streams of water strike the floor in the same place, what is the depth of water in the tank?

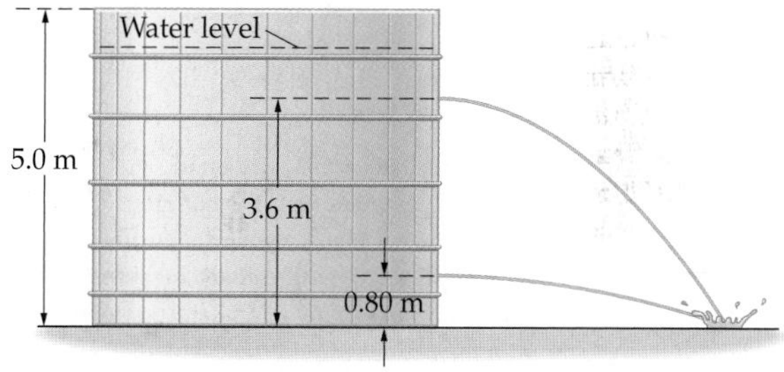

▲ **FIGURE 15–40** Problem 94

95. ••• A hollow cubical box, 0.29 m on a side, with walls of negligible thickness floats with 35% of its volume submerged. What mass of water can be added to the box before it sinks?

Interactive Problems

96. •• **IP** ***Referring to Example 15–4*** Suppose we use a different vegetable oil that has a higher density than the one in Example 15–4. **(a)** If everything else remains the same, will the height difference, h, increase, decrease. or remain the same? Explain. **(b)** Find the height difference for an oil that has a density of $9.60 \times 10^2\ \mathrm{kg/m^3}$.

97. •• ***Referring to Example 15–4*** Find the height difference, h, if the depth of the oil is increased to 7.50 cm. Assume everything else in the problem remains the same.

98. •• ***Referring to Example 15–9*** **(a)** Find the height H required to make $D = 0.655$ m. Assume everything else in the problem remains the same. **(b)** Find the depth h required to make $D = 0.455$ m. Assume everything else in the problem remains the same.

99. •• ***Referring to Example 15–9*** Suppose both h and H are increased by a factor of two. By what factor is the distance D increased?

Temperature and Heat

16

16–1 Temperature and the Zeroth Law of Thermodynamics 515
16–2 Temperature Scales 516
16–3 Thermal Expansion 520
16–4 Heat and Mechanical Work 526
16–5 Specific Heats 528
16–6 Conduction, Convection, and Radiation 531

In this computer-generated map displaying the variation in average global ocean temperatures, red represents the hottest and blue the coolest areas. The relatively high-temperature water in equatorial regions warms the air above it, greatly influencing worldwide climate and weather patterns. But what exactly is temperature, and how does thermal energy pass from the warm water to the cooler air? This chapter explores such questions and others related to the phenomenon known as heat.

To this point our study of physics has involved just three physical quantities: mass, length, and time. Every measurement, every calculation, has been in terms of M, L, and T, or some combination of the three. We now add a fourth physical quantity—*temperature*. With the introduction of temperature, we broaden the scope of physics, allowing the study of a wide variety of new physical situations that mechanics alone cannot address.

In this chapter we introduce the concept of temperature, and discuss its effects on macroscopic systems. We begin by showing how differences in temperature relate to a particular type of energy transfer we call *heat*. We also discuss the connection between changes in temperature and changes in other physical quantities, such as length, pressure, and volume. Finally, we consider various mechanisms by which thermal energy is exchanged. Later in the text, when we consider the microscopic aspects of temperature, we shall see that temperature is ultimately related to the rapidity of molecular motion.

16–1 Temperature and the Zeroth Law of Thermodynamics

Even as small children we learn to avoid objects that are "hot." We also discover early on that if we forget to wear our coats outside we can become "cold." Later, we associate high values of something called "temperature" with hot objects, and low values with cold objects. These basic notions about temperature carry over into physics, though with a bit more precision.

Similarly, when we put a cool pan of water on a hot stove burner we say that "heat" flows from the hot burner to the cool water. To be more precise, as required in physics, we will define **heat** as follows:

> Heat is the energy that is *transferred* between objects because of a temperature difference.

Therefore, when we say that there is a "transfer of heat" or a "heat flow" from object A to object B, we simply mean that the total energy of object A decreases and the total energy of object B increases. To summarize, an object does not "contain" heat—it has a certain energy content, and the energy it *exchanges* with other objects due to temperature differences is called heat.

Now, objects are said to be in **thermal contact** if heat can flow between them. In general, when a hot object is brought into thermal contact with a cold object heat will be exchanged. The result is that the hot object cools off—and its molecules move more slowly—while the cold object warms up—and its molecules move more rapidly. After some time in thermal contact, the transfer of heat ceases. At this point, we say that the objects are in **thermal equilibrium.** The study of physical processes involving the transfer of heat is the area of physics referred to as **thermodynamics.** Because thermodynamics deals with the flow of energy within and between objects, it has great practical importance in all areas of the life sciences, physical sciences, and engineering.

Note that thermal contact and physical contact are not necessarily the same. For example, thermal contact can occur even when there is no physical contact at all—as when you warm your hands near a fire. Various types of thermal contact will be discussed in detail in Section 16–6.

We now introduce perhaps the most fundamental law obeyed by thermodynamic systems—referred to, appropriately enough, as the zeroth law of thermodynamics. The zeroth law spells out the basic properties of temperature. Its name reflects not only its basic importance, but also its subtlety, in that it was not recognized as a separate law until after the other three laws of thermodynamics had been accepted. Later, in Chapter 18, we introduce the remaining three laws of thermodynamics. These laws enable us to analyze the behavior of engines and refrigerators, and to show, among other things, that perpetual motion machines are not possible.

Zeroth Law of Thermodynamics

The basic idea of the zeroth law of thermodynamics is that thermal equilibrium is determined by a single physical quantity—the **temperature.** For example, two objects in thermal contact are in equilibrium when they have the *same* temperature. If one or the other has a higher temperature, heat flows from that object to the other until their temperatures are equal.

This may seem almost too obvious to mention, at least until you give it more thought. Suppose, for example, that you have a piece of metal and a pool of water, and you want to know if heat will flow between them when you put the metal in the pool. You measure the temperature of each, and if they are the same, you can conclude that no heat will flow. If the temperatures are different, however, it follows that there will be a flow of heat. Nothing else matters—not the type of metal, its mass, its shape, the amount of water, whether the water is fresh or salt, and so on—all that matters is one number, the temperature.

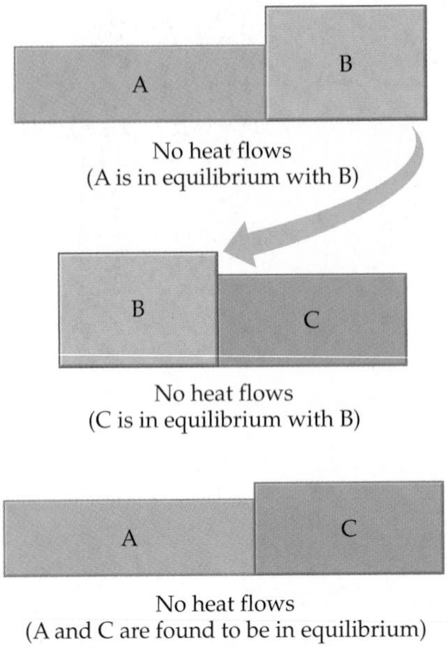

▲ **FIGURE 16–1 An illustration of the zeroth law of thermodynamics**
If A and C are each in thermal equilibrium with B, then A will be in thermal equilibrium with C when they are brought into thermal contact.

To summarize, the **zeroth law of thermodynamics** can be stated as follows:

> If object A is in thermal equilibrium with object B, and object C is also in thermal equilibrium with object B, then objects A and C will be in thermal equilibrium if brought into thermal contact.

This is illustrated in **Figure 16–1**. We begin with objects A and B in thermal contact and in equilibrium. Next, object B is separated from A, and placed in contact with C. Objects C and B are also found to be in equilibrium. Hence, by the zeroth law, we are assured that when A and C are placed in contact they also will be in equilibrium.

To apply this principle to our example, let object A be the piece of metal and object C be the pool of water. Object B, then, can be a thermometer, used to measure the temperature of the metal and the water. If A and C are each separately in equilibrium with B—which means that they have the same temperature—they will be in equilibrium with one another.

16–2 Temperature Scales

A variety of temperature scales is commonly used in both everyday situations and in physics. Some are related to familiar reference points, such as the temperature of boiling or freezing water. Others have more complex, historical rationale for their values. Here we consider three of the more frequently used temperature scales. We also examine the connections between them.

Later in this chapter we will discuss some of the physical phenomena—such as thermal expansion—that can be used to construct a thermometer. With a properly calibrated thermometer, we can determine the temperature on any of these scales. In the next chapter, we will explore more fully the question of just what temperature means on a conceptual and microscopic level.

The Celsius Scale

Perhaps the easiest temperature scale to remember is the Celsius scale, named in honor of the Swedish astronomer Anders Celsius (1701–1744). Originally, Celsius assigned zero degrees to boiling water and 100 degrees to freezing water. These values were later reversed by the biologist Carolus Linnaeus (1707–1778). Thus, today we say that water freezes at zero degrees Celsius, which we abbreviate as 0 °C, and boils at 100 °C.

Note that the choice of zero level for a temperature scale is quite arbitrary, as is the number of degrees between any two reference points. In the Celsius scale, as in others, there is no upper limit to the value a temperature may have. There is a lower limit, however. For the Celsius scale, the lowest possible temperature is −273.15 °C, as we shall see later in this section.

One bit of notation should be pointed out before we continue. When we write a Celsius temperature, we give a numerical value followed by the degree symbol, °, and the capital letter C. For example, 5 °C is the temperature five degrees Celsius. On the other hand, if the temperature of an object is *changed* by a given amount, we use the notation C°. Thus, if we increase the temperature by five degrees on the Celsius scale, we say that the change in temperature is 5 C°; that is, five Celsius degrees. This is summarized below:

A temperature T of five degrees is 5 °C
(five degrees Celsius)

A temperature change ΔT of five degrees is 5 C°
(five Celsius degrees)

The Fahrenheit Scale

The Fahrenheit scale was developed by Gabriel Fahrenheit (1686–1736), who chose zero to be the lowest temperature he was able to achieve in his laboratory. He also chose 96 degrees to be body temperature, though why he made this choice is not known. In the modern version of the Fahrenheit scale body temperature is 98.6 °F; in addition, water freezes at 32 °F and boils at 212 °F. Lastly, using the same convention as for °C and C°, we say that an increase of 180 F° is required to bring water from freezing to boiling.

Note that the Fahrenheit scale not only has a different zero than the Celsius scale, it also has a different "size" for its degree. As just noted, 180 Fahrenheit degrees are required for the same change in temperature as 100 Celsius degrees. Hence, the Fahrenheit degrees are smaller by a factor of 100/180 = 5/9.

To convert between a Fahrenheit temperature, T_F, and a Celsius temperature, T_C, we start by writing a linear relation between them. Thus, let

$$T_F = aT_C + b$$

We would like to determine the constants a and b. This requires two independent pieces of information, which we have in the freezing and boiling points of water. Using the freezing point, we find

$$32\ °F = a(0\ °C) + b = b$$

Thus, b is 32 °F. Next, the boiling point gives

$$212\ °F = a(100\ °C) + 32\ °F$$

Solving for the constant a we find

$$a = (212\ °F - 32\ °F)/(100\ °C) = \frac{180\ F°}{100\ C°} = \tfrac{9}{5}\,F°/C°$$

Combining our results gives the following conversion relationship:

Conversion Between Degrees Celsius and Degrees Fahrenheit

$$T_F = (\tfrac{9}{5}\,F°/C°)T_C + 32\ °F \qquad \textbf{16–1}$$

Similarly, this relation can be rearranged to convert from Fahrenheit to Celsius:

Conversion Between Degrees Fahrenheit and Degrees Celsius

$$T_C = (\tfrac{5}{9}\,C°/F°)(T_F - 32\ °F) \qquad \textbf{16–2}$$

Since conversion factors like $\tfrac{9}{5}\,F°/C°$ are a bit clumsy and tend to clutter up an equation, we will generally drop the degree symbols until the final result. For example, to convert 10 °C to degrees Fahrenheit, we write

$$T_F = \tfrac{9}{5}T_C + 32 = \tfrac{9}{5}(10) + 32 = 50\ °F$$

EXAMPLE 16–1 Temperature Conversions

(a) On a fine spring day you notice that the temperature is 75 °F. What is the corresponding temperature on the Celsius scale? **(b)** If the temperature on a brisk winter morning is −2.0 °C, what is the corresponding Fahrenheit temperature?

Picture the Problem

Our sketch shows a circular thermometer with both a Fahrenheit and a Celsius scale. The range of the scales extends well beyond the temperatures 75 °F and −2.0 °C needed for the problem.

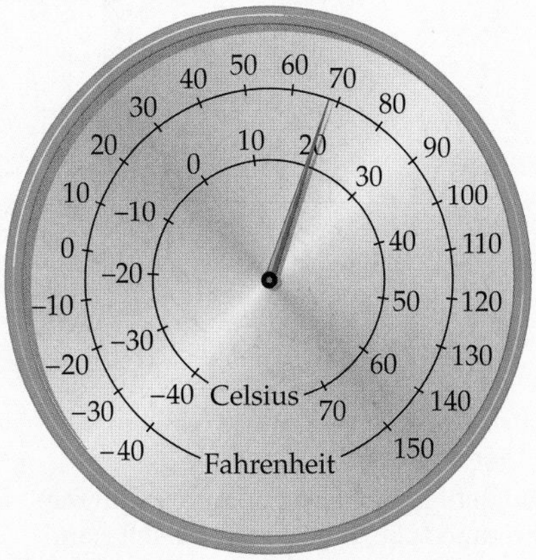

Strategy

The conversions asked for in this problem are straightforward applications of the relations between T_F and T_C. In particular, for **(a)** we use $T_C = (5/9)(T_F - 32)$, and for **(b)** we use $T_F = \tfrac{9}{5}T_C + 32$.

Solution

Part (a)

1. Substitute T_F = 75 °F into Equation 16–2: $T_C = \tfrac{5}{9}(75 - 32) = 24\ °C$

Part (b)

2. Substitute T_C = −2.0 °C in Equation 16–1: $T_F = \tfrac{9}{5}(-2.0) + 32 = 28\ °F$

continued on next page

continued from previous page

Insight
Note that the results given here agree with the scales shown in the drawing.

Practice Problem
Find the Celsius temperature that corresponds to 110 °F. [**Answer:** $T_C = 43$ °C]

Some related homework problems: Problem 1, Problem 2

ACTIVE EXAMPLE 16–1 Same Temperature

What temperature is the same on both the Celsius and Fahrenheit scales?

Solution *(Test your understanding by performing the calculations indicated in each step.)*

1. Set $T_F = T_C = t$ in Equation 16–1: $t = 9t/5 + 32$
2. Move all terms involving t to the left side of the equation: $-4t/5 = 32$
3. Solve for t: $t = -40$
4. As a check, substitute $T_F = -40$ °F in Equation 16–2: $T_C = (5/9)(-40 - 32) = -40$ °C

Insight
Thus, −40 °F is the same as −40 °C. This is consistent with the scale shown in Example 16–1.

Your Turn
Find the Fahrenheit temperature whose numerical value is three times greater than the corresponding Celsius temperature.

(Answers to **Your Turn** *problems can be found in the back of the book.)*

Absolute Zero

Experiments show conclusively that there is a lowest temperature below which it is impossible to cool an object. This is referred to as **absolute zero.** Though absolute zero can be approached from above arbitrarily closely, it can never be attained.

To give an idea of just where absolute zero is on the Celsius scale, we start with the following observation: If a given volume V of air—say the air in a balloon—is cooled from 100 °C to 0 °C, its volume decreases by roughly $V/4$. Imagine this trend continuing uninterrupted. In this case, cooling from 0 °C to −100 °C would reduce the volume by another $V/4$, from −100 °C to −200 °C by another $V/4$, and finally, from −200 °C to −300 °C by another $V/4$, which brings the volume down to zero. Clearly, it doesn't make sense for the volume to be less than zero, and hence absolute zero must be roughly −300 °C.

This result, though crude, is in the right ballpark. A precise determination of absolute zero can be made with a device known as a **constant-volume gas thermometer.** This instrument is shown in **Figure 16–2**. The basic idea is that by adjusting the level of mercury in the right-hand tube, the level of mercury in the left-hand tube can be set to a fixed reference level. With the mercury so adjusted, the gas occupies a constant volume and its pressure is simply $P_{gas} = P_{at} + \rho gh$ (Equation 15–7), where ρ is the density of mercury.

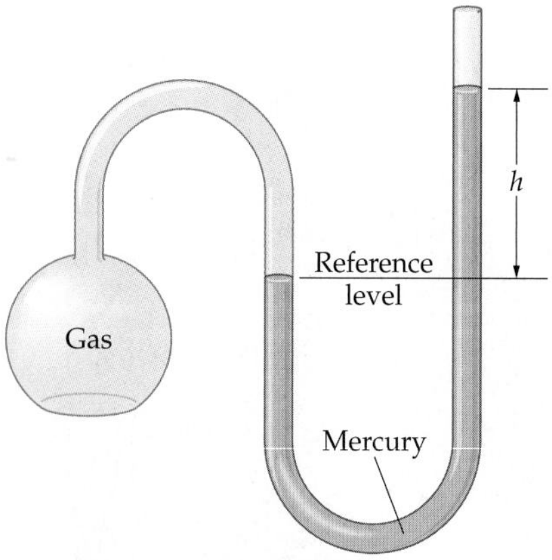

▲ **FIGURE 16–2 A constant-volume gas thermometer**

By adjusting the height of mercury in the right-hand tube, the level in the left-hand tube can be set at the reference level. This assures that the gas occupies a constant volume.

As the temperature of the gas is changed, the mercury level in the right-hand tube can be readjusted as described. The gas pressure can be determined again, and the process repeated. The results of a series of such measurements are shown in **Figure 16–3**.

Note that as a gas is cooled its pressure decreases, as one would expect. In fact, the decrease in pressure is approximately linear. At low enough temperatures the gas eventually liquefies, and its behavior changes, but if we extrapolate the

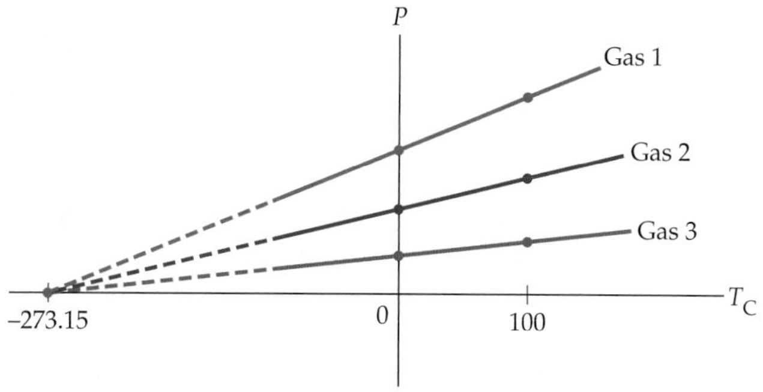

◀ FIGURE 16–3 Determining absolute zero

Different gases have different pressures at any given temperature. However, they all extend down to zero pressure at precisely the same temperature, −273.15 °C. This is the location of absolute zero.

straight line obtained before liquefaction, we see that it reaches zero pressure (the lowest pressure possible) at −273.15 °C.

What is remarkable about this result is that it is independent of the gas we use in the thermometer. For example, gas 2 and gas 3 in Figure 16–3 have pressures that are different from one another, and from gas 1. Yet all three gases extrapolate to zero pressure at precisely the same temperature. Thus, we conclude that there is indeed a *unique* value of absolute zero, below which further cooling is not possible.

EXAMPLE 16–2 It's a Gas

The gas in a constant-volume gas thermometer has a pressure of 80.0 kPa at 0.00 °C. Assuming ideal behavior, as in Figure 16–3, what is the pressure of this gas at 105 °C?

Picture the Problem

Our sketch plots the pressure of the gas as a function of temperature. Note that at $T_C = 0.00$ °C the pressure is 80.0 kPa, and that at $T_C = -273.15$ °C the pressure extrapolates linearly to zero. We also indicate the point corresponding to 105 °C on the graph.

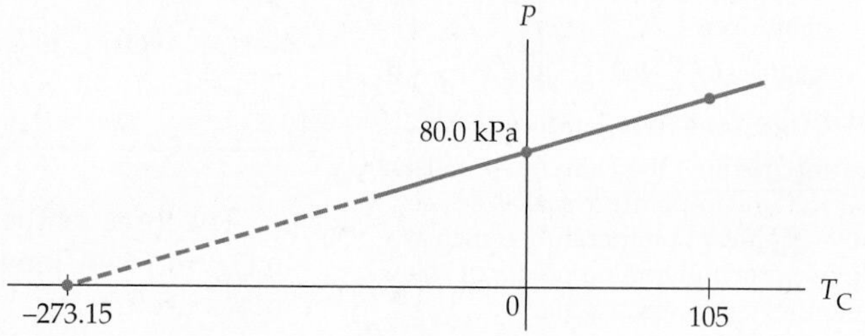

Strategy

We assume that the pressure lies on a straight line, as in Figure 16–3. To find the pressure at $T_C = 105$ °C we simply extend the straight line.

Thus, we start with the information that the pressure increases linearly from 0 to 80.0 kPa when the temperature increases from −273.15 °C to 0.00 °C. This rate of increase in pressure must also apply to an increase in temperature from −273.15 °C to 105 °C. Using this rate we can find the desired pressure.

Solution

1. Calculate the rate at which pressure increases for this gas: $\text{rate} = \dfrac{80.0\ \text{kPa}}{273.15\ \text{C}^\circ} = 0.293\ \text{kPa/C}^\circ$

2. Multiply this rate by the temperature change from −273.15 °C to 105 °C: $(0.293\ \text{kPa/C}^\circ)(378\ \text{C}^\circ) = 111\ \text{kPa}$

Insight

The pressure of this gas increases from slightly less than one atmosphere at 0.00 °C to slightly more than one atmosphere at 105 °C.

An alternative solution to this problem is to calculate the pressure difference from 0.00 °C to 105 °C. Specifically, we can say that $P = 80.0\ \text{kPa} + (0.293\ \text{kPa/C}^\circ)(105\ \text{C}^\circ) = 111\ \text{kPa}$.

Practice Problem

Find the temperature at which the pressure of the gas is 70.0 kPa. [**Answer:** $T_C = -34.1$ °C]

Some related homework problems: Problem 7, Problem 8

The Kelvin Scale

The Kelvin temperature scale, named for the Scottish physicist William Thomson, Lord Kelvin (1824–1907), is based on the existence of absolute zero. In fact, the zero of the Kelvin scale, abbreviated 0 K, is set exactly at absolute zero. Thus, in this scale there are no negative equilibrium temperatures. The Kelvin scale is also chosen to have the same degree size as the Celsius scale.

As mentioned, absolute zero occurs at $-273.15\ °C$, hence the conversion between a Kelvin-scale temperature, T, and a Celsius temperature, T_C, is as follows:

Conversion Between a Celsius Temperature and a Kelvin Temperature

$$T = T_C + 273.15 \qquad 16\text{–}3$$

Note that the difference between the Celsius and Kelvin scales is simply a difference in the zero level.

The notation for the Kelvin scale differs somewhat from that for the Celsius and Fahrenheit scales. In particular, by international agreement, the degree terminology and the degree symbol, °, are not used in the Kelvin scale. Instead, a temperature of 5 K is read simply as 5 kelvin. In addition, a change in temperature of 5 kelvin is written 5 K, the same as for a temperature of 5 kelvin.

Though the Celsius and Fahrenheit scales are the ones most commonly used in everyday situations, the Kelvin scale is used more than any other in physics. This stems from the fact that the Kelvin scale incorporates the significant concept of absolute zero. As a result, the thermal energy of a system depends in a very simple way on the Kelvin temperature. This will be discussed in detail in the next chapter.

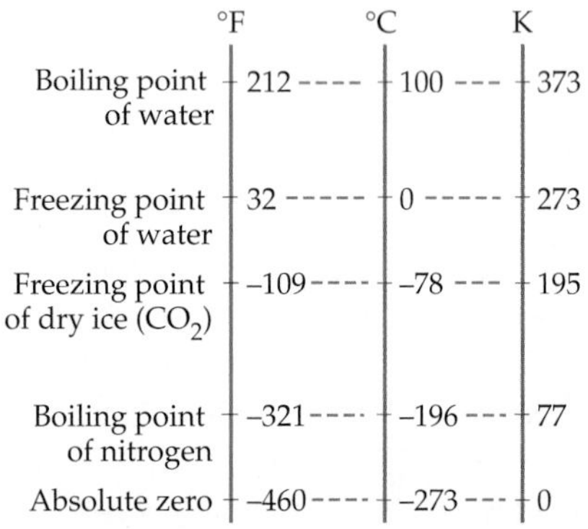

▲ **FIGURE 16–4 Temperature scales**
A comparison of the Fahrenheit, Celsius, and Kelvin temperature scales. Physically significant temperatures, such as the freezing and boiling points of water, are indicated for each scale.

EXERCISE 16–1

Convert 55 °F to the Kelvin temperature scale.

Solution

First, convert from °F to °C:

$$T_C = \tfrac{5}{9}(55 - 32) = 13\ °C$$

Next, convert °C to K:

$$T = 13 + 273.15 = 286\ K$$

The three temperature scales presented in this section are shown side by side in **Figure 16–4**. Temperatures of particular interest are indicated as well. This permits a useful visual comparison between the scales.

16–3 Thermal Expansion

Most substances expand when heated. For example, power lines on a hot summer day hang low compared to their appearance on a cold day in winter. In fact, thermal expansion is the basis for many thermometers, including the familiar thermometers used for measuring a fever. The expansion of a liquid, such as mercury or alcohol, results in a column of liquid of variable height within the glass neck of the thermometer. The height is read against markings on the glass, which gives the temperature.

You may wonder what bizarre substance could possibly be an exception to this common response to heating. The most important exception occurs in a substance you drink every day—water. This is just one of the many special properties that sets water apart from most other substances.

In this section we consider the physics of thermal expansion, including linear, area, and volume expansion. We also discuss briefly the unusual thermal behavior of water, and some of its more significant consequences.

Linear Expansion

Consider a rod whose length is L_0 at the temperature T_0. Experiments show that when the rod is heated or cooled, its length changes in direct proportion to the temperature change. Thus, if the change in temperature is ΔT, the change in length of the rod, ΔL, is

$$\Delta L = (\text{constant})\Delta T$$

The constant of proportionality depends, among other things, on the substance from which the rod is made.

CONCEPTUAL CHECKPOINT 16–1 Compare Expansions

When rod 1 is heated by an amount ΔT, its length increases by ΔL. If rod 2, which is twice as long as rod 1 and made of the same material, is heated by the same amount, does its length increase by **(a)** ΔL, **(b)** $2\Delta L$, or **(c)** $\Delta L/2$?

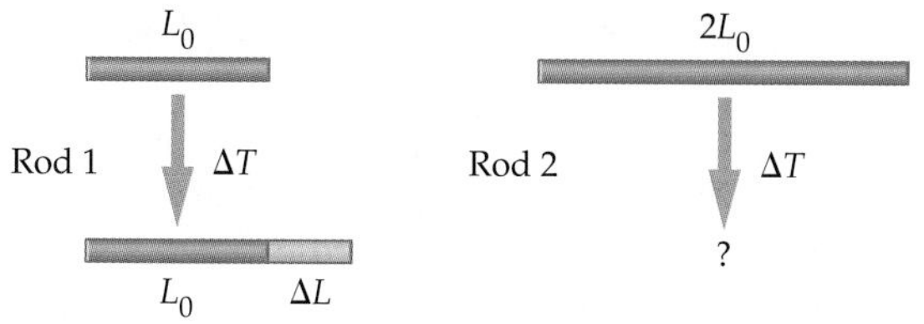

Reasoning and Discussion

We can imagine rod 2 to be composed of two copies of rod 1 placed end to end, as shown.

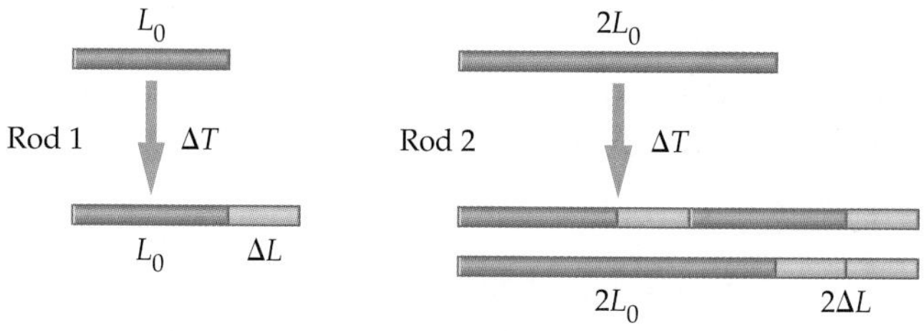

When the temperature is increased by ΔT, each copy of rod 1 expands by ΔL. Hence, the total expansion of the two copies is $2\Delta L$, as is the total expansion of rod 2.

Answer:

(b) The rod that is twice as long expands twice as much; $2\Delta L$.

We conclude, then, on the basis of the preceding Conceptual Checkpoint, that the change in length is proportional to *both* the initial length, L_0, and the temperature change, ΔT. The constant of proportionality is referred to as α, the **coefficient of linear expansion,** and is defined as follows:

Definition of Coefficient of Linear Expansion, α

$$\Delta L = \alpha L_0 \Delta T \qquad 16\text{–}4$$

SI unit for α: $K^{-1} = (C^\circ)^{-1}$

Table 16–1 gives values of α for a variety of substances.

TABLE 16–1 Coefficients of Thermal Expansion near 20 °C

Substance	Coefficient of linear expansion, $\alpha(K^{-1})$
Lead	29×10^{-6}
Aluminum	24×10^{-6}
Brass	19×10^{-6}
Copper	17×10^{-6}
Iron (steel)	12×10^{-6}
Concrete	12×10^{-6}
Window glass	11×10^{-6}
Pyrex glass	3.3×10^{-6}
Quartz	0.50×10^{-6}
Substance	**Coefficient of volume expansion, $\beta(K^{-1})$**
Ether	1.51×10^{-3}
Carbon tetrachloride	1.18×10^{-3}
Alcohol	1.01×10^{-3}
Gasoline	0.95×10^{-3}
Olive oil	0.68×10^{-3}
Water	0.21×10^{-3}
Mercury	0.18×10^{-3}

EXERCISE 16–2

The Eiffel Tower, constructed in 1889 by Alexandre Eiffel, is an impressive latticework structure made of iron. If the tower is 301 m high on a 22 °C day, how much does its height decrease when the temperature cools to 0.0 °C?

Solution

We can calculate the change in height with Equation 16–4. Note that the coefficient of linear expansion for iron, given in Table 16–1, is $12 \times 10^{-6}\ K^{-1}$ and the change in temperature is $\Delta T = -22\ C^\circ = -22\ K$:

$$\Delta L = \alpha L_0 \Delta T = (12 \times 10^{-6}\ K^{-1})(301\ m)(-22\ K) = -7.9\ cm$$

An interesting application of thermal expansion is in the behavior of a bimetallic strip. As the name suggests, a bimetallic strip consists of two metals bonded together to form a linear strip of metal. This is illustrated in **Figure 16–5**. Since two different metals will, in general, have different coefficients of linear expansion, α, the two sides of the strip will change lengths by different amounts when heated or cooled.

For example, suppose metal B in Figure 16–5 (a) has the larger coefficient of linear expansion. This means that its length will change by greater amounts than metal A for the same temperature change. Hence, if this strip is cooled, the B side will shrink more than the A side, resulting in the strip bending toward the B side,

▲ The Eiffel Tower in Paris gains about a thirteenth of an inch in height for each Fahrenheit degree that the temperature rises.

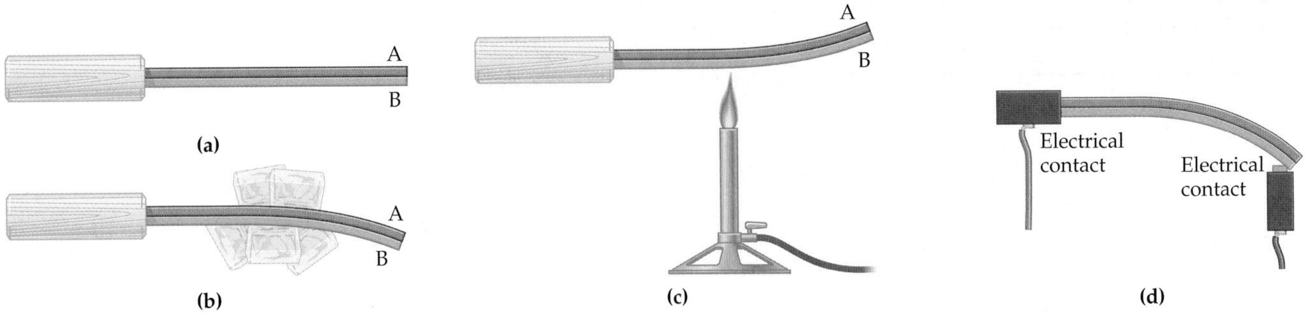

▲ FIGURE 16–5 A bimetallic strip

(a) A bimetallic strip composed of metals A and B. **(b)** If metal B has a larger coefficient of linear expansion than metal A, it will shrink more when cooled and **(c)** expand more when heated. **(d)** A bimetallic strip can be used to construct a thermostat. If the temperature falls, the strip bends downward and closes the electrical circuit, which then operates a heater. When the temperature rises, the strip deflects in the opposite direction, breaking the circuit and turning off the heater.

as in Figure 16–5 (b). On the other hand, if the strip is heated, the B side expands by a greater amount than the A side, and the strip curves toward the A side, as in Figure 16–5 (c). Thus, the shape of the bimetallic strip depends sensitively on temperature.

REAL-WORLD PHYSICS
Bimetallic strips

Because of this property, bimetallic strips are used in a variety of thermal applications. For example, a bimetallic strip can be used as a thermometer; as the strip changes its shape it can move a needle to indicate the temperature. Similarly, many thermostats have a bimetallic strip to turn on or shut off a heater. This is shown in Figure 16–5 (d). As the temperature of the room changes, the bimetallic strip deflects in one direction or the other, which either closes or breaks the electrical circuit connected to the heater.

REAL-WORLD PHYSICS
Antiscalding devices

Another common use of thermal expansion is the *antiscalding device* for water faucets. An antiscalding device is simply a valve inside a water faucet that is attached to a spring. When the water temperature is at a safe level, the valve permits water to flow freely through the faucet. If the temperature of the water reaches a dangerous level, however, the thermal expansion of the spring is enough to close the valve and stop the flow of water—thus preventing inadvertent scalding. When the water cools down again, the valve reopens and the flow of water resumes.

REAL-WORLD PHYSICS
Thermal expansion joints

Finally, thermal expansion can have undesirable effects in some cases. For example, you may have noticed that bridges often have gaps between different sections of the structure. When the air temperature rises in the summer, the sections of the bridge can expand freely into these gaps. If the gaps were not present, the

▲ Thermal expansion, though small, is far from negligible in many everyday situations. This is especially true when long objects such as railroad tracks, bridges, or pipelines are involved. Bridges and elevated highways (left), must include expansion joints to prevent the roadway from buckling when it expands in hot weather. Similarly, pipelines (right) typically include loops that allow for expansion and contraction when the temperature changes.

expansion of the different sections could cause the bridge to buckle and warp. Thus, these gaps, referred to as *expansion joints*, are a way of avoiding this type of heat-related damage. Expansion joints can also be found in railroad tracks and oil pipelines, to name just two other examples.

Area Expansion

Since the length of an object changes with temperature, it follows that its area changes as well. To see precisely how the area changes, consider a square piece of metal of length L on a side. The initial area of the square is $A = L^2$. If the temperature of the square is increased by ΔT the length of each side increases from L to $L + \Delta L = L + \alpha L \Delta T$. As a result, the square has an increased area, A', given by

$$A' = (L + \Delta L)^2 = (L + \alpha L \Delta T)^2$$
$$= L^2 + 2\alpha L^2 \Delta T + \alpha^2 L^2 \Delta T^2$$

Now, if $\alpha \Delta T$ is much less than one—which is certainly the case for typical changes in temperature, ΔT—then $\alpha^2 \Delta T^2$ is even smaller. Hence, if we ignore this small contribution, we find

$$A' \approx L^2 + 2\alpha L^2 \Delta T = A + 2\alpha A \Delta T$$

As a result, the change in area, ΔA, is

$$\Delta A = A' - A \approx 2\alpha A \Delta T \qquad 16\text{–}5$$

Note the similarity between this relation and Equation 16–4; the length L has been replaced with the area A, and the expansion coefficient has been doubled to 2α.

Though this calculation was done for the simple case of a square, the result applies to an area of any shape. For example, a circular disk of radius r and area $A = \pi r^2$ will increase its area by the amount $2\alpha A \Delta T$ with an increase in temperature of ΔT. What about a washer, however, which is a disk of metal with a circular hole cut out of its center? What happens to the area of the hole as the washer is heated? Does it expand along with everything else, or does the expanding washer "expand into the hole" to make it smaller? We consider this question in the next Conceptual Checkpoint.

CONCEPTUAL CHECKPOINT 16–2 Heating a Hole

A washer has a hole in the middle. As the washer is heated, does the hole (a) expand, (b) shrink, or (c) stay the same?

Reasoning and Discussion

To make a washer from a disk of metal, we can cut along a circular curve, as shown, and remove the inner disk. If we now heat the system, both the washer and the inner disk expand. On the other hand, if we had left the inner disk in place and heated the original disk, it would also expand. Removing the *heated* inner disk would create an expanded washer with an expanded hole in the middle. We obtain the same result whether we remove the inner disk and then heat, or heat first and then remove the inner disk.

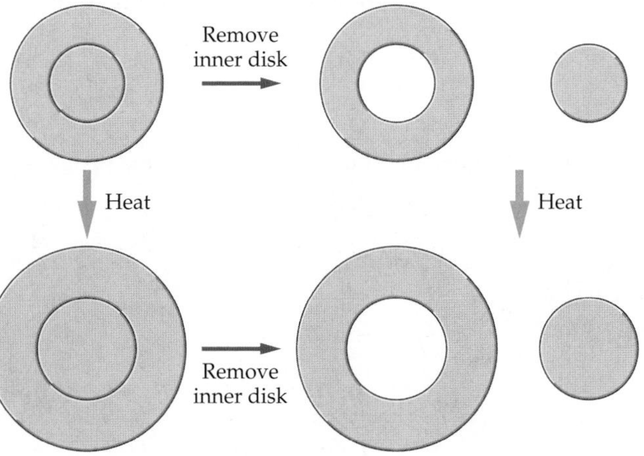

continued on next page

PROBLEM-SOLVING NOTE

Expansion of a Hole

A hole in a material expands the same as if it were made of the material itself. Thus, to find the expansion of a hole in a steel plate, we use the coefficient of linear expansion for steel.

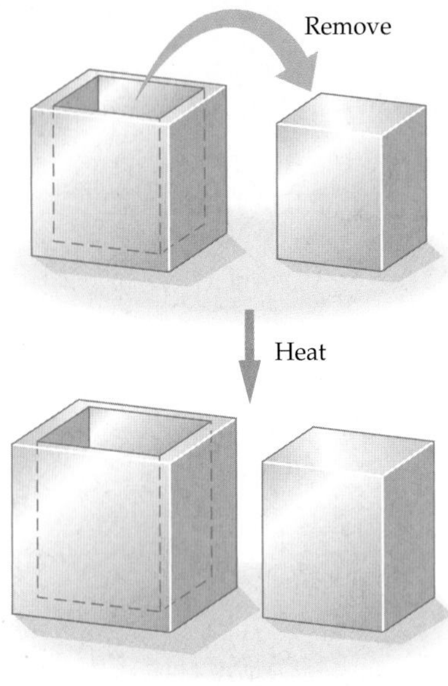

▲ **FIGURE 16–6 Volume expansion**
A portion of a cube is removed to create a container. When heated, the removed portion expands, just as the volume within the container expands.

PROBLEM-SOLVING NOTE
Expansion of a Volume

The empty volume inside a container expands the same as if it were made of the same material as the container. For example, to find the increase in volume of a steel tank, we use the coefficient of linear expansion for steel, just as if the tank were actually filled with steel.

PROBLEM-SOLVING NOTE
Temperature Change

Remember that a change in temperature of 1 C° is the same as a change in temperature of 1 K. Thus, when finding the thermal expansion of an object, the change in temperature ΔT can be expressed in terms of either the Celsius or the Kelvin temperature scale.

continued from previous page

Thus, heating the washer causes both it and its hole to expand, and they both expand with the same coefficient of linear expansion. Basically, the system behaves the same as if we had produced a photographic enlargement—everything expands.

Answer:
(a) The hole expands along with everything else.

Volume Expansion

Just as the hole in a washer increases in area with heating, so does the empty volume within a cup or other container. This is illustrated in **Figure 16–6**, where we show a block of material with a volume removed to convert it into a container. As the system is heated, there will be an expansion of the container, of the volume within it, and of the volume that was removed. As with the area of a hole, the volume within a container expands with the same coefficient of expansion as the container itself.

To calculate the change in volume, consider a cube of length L on a side. The initial volume of the cube is $V = L^3$. Increasing the temperature results in an increased volume given by

$$V' = (L + \Delta L)^3 = (L + \alpha L \Delta T)^3$$
$$= L^3 + 3\alpha L^3 \Delta T + 3\alpha^2 L^3 \Delta T^2 + \alpha^3 L^3 \Delta T^3$$

Neglecting the smaller contributions, as we did with the area, we find

$$V' \approx L^3 + 3\alpha L^3 \Delta T = V + 3\alpha V \Delta T$$

Therefore, the change in volume, ΔV, is

$$\Delta V = V' - V \approx 3\alpha V \Delta T$$

This expression, though calculated for a cube, applies to any volume.

In general, volume expansion is described in the same way as linear expansion, but with a **coefficient of volume expansion,** β, defined as follows:

Definition of Coefficient of Volume Expansion, β

$$\Delta V = \beta V \Delta T \qquad \text{16–6}$$

SI unit for β: $K^{-1} = (C°)^{-1}$

Typical values of β are given in Table 16–1 for a number of different liquids. If Table 16–1 lists a value of α for a given substance, but not for β, its change in volume is calculated as follows:

$$\Delta V = \beta V \Delta T \approx 3\alpha V \Delta T \qquad \text{16–7}$$

That is, we simply make the identification $\beta = 3\alpha$. For substances that have a specific value for β listed in Table 16–1, we simply use Equation 16–6. This is illustrated in the next Example.

EXAMPLE 16–3 Oil Spill

A copper flask with a volume of 150 cm^3 is filled to the brim with olive oil. If the temperature of the system is increased from 6.0 °C to 31 °C, how much oil spills from the flask?

Picture the Problem
Our sketch shows the flask filled to the top initially, then spilling over when heated by 25 C°. (*Note:* Since degrees have the same size on the Celsius and Kelvin scales, it follows that $\Delta T = 25\ C° = 25\ K$.)

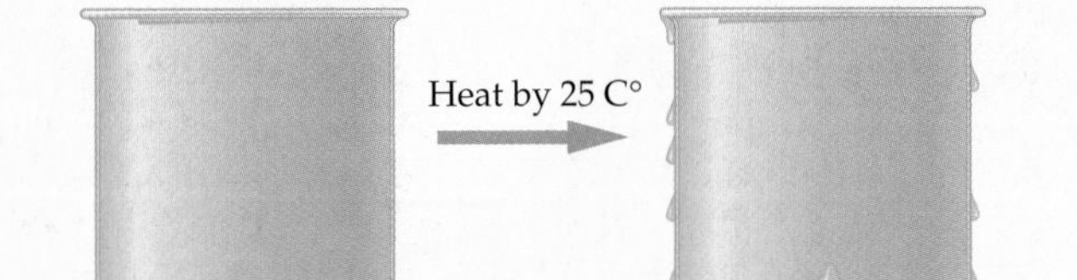

Strategy

As the system is heated, both the flask and the olive oil expand. Thus, we start by calculating the expansion of the oil and the flask separately. A quick glance at Table 16–1 shows that the olive oil will expand more, and the difference in expansion volumes is what spills out.

For the copper flask we find α in Table 16–1, then let $\beta = 3\alpha$.

Solution

1. Calculate the change in volume of the olive oil:

$$\Delta V_{\text{oil}} = \beta V \Delta T$$
$$= (0.68 \times 10^{-3}\,\text{K}^{-1})(150\,\text{cm}^3)(25\,\text{K}) = 2.6\,\text{cm}^3$$

2. Calculate the change in volume of the flask:

$$\Delta V_{\text{flask}} = 3\alpha V \Delta T$$
$$= 3(17 \times 10^{-6}\,\text{K}^{-1})(150\,\text{cm}^3)(25\,\text{K}) = 0.19\,\text{cm}^3$$

3. Find the difference in volume expansions. This is the volume of oil that spills out:

$$\Delta V_{\text{oil}} - \Delta V_{\text{flask}} = 2.6\,\text{cm}^3 - 0.19\,\text{cm}^3 = 2.4\,\text{cm}^3$$

Insight

If the system were cooled, the oil would lose volume more rapidly than the flask. This would result in a drop in oil level.

This Example illustrates why fuel tanks on cars are designed to stop filling before the gas reaches the top—otherwise, the tank could overflow on a hot day.

Practice Problem

Suppose the copper flask is initially filled to the brim with gasoline rather than olive oil. Referring to Table 16–1, do you expect the amount of liquid that spills with heating to be greater than, less than, or the same as in the case of olive oil? Calculate the volume of spilled gasoline for a temperature increase of 25 C°. [**Answer:** More liquid will spill. $\Delta V = 3.4\,\text{cm}^3$]

Some related homework problems: Problem 17, Problem 21

Special Properties of Water

As we have mentioned, water is a substance rich with unusual behavior. For example, in the last chapter we discussed the fact that the solid form of water (ice) is less dense than the liquid form. Hence, icebergs float. What is remarkable about icebergs floating is not that 90% is submerged, but that 10%, or any amount at all, is above the water. The solids of most substances are denser than their liquids; hence when they freeze, their solids immediately sink.

REAL-WORLD PHYSICS
Floating icebergs

Here we consider the unusual *thermal* behavior of water. **Figure 16–7** shows the density of water over a wide range of temperatures. Note that the density is a

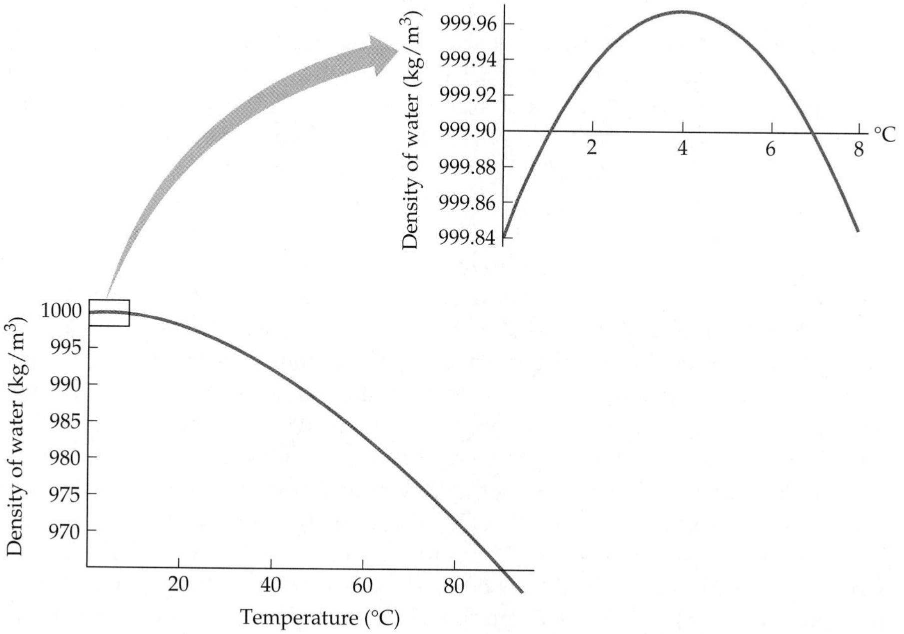

◀ FIGURE 16–7 The unusual behavior of water near 4 °C
The density of water actually *increases* as the water is heated between 0 °C and 4 °C. Maximum density occurs near 4 °C.

maximum at about 4 °C. Thus, when you *heat* water from 0 °C to 4 °C it actually *shrinks*, rather than expands, and becomes *more dense*. The reason is that water molecules that were once part of the rather open crystal structure of ice are now able to pack more closely together in the liquid.

REAL-WORLD PHYSICS: BIO
The ecology of lakes

This behavior has significant consequences for the ecology of lakes in northern latitudes. When temperatures drop in the winter, the surface waters of a lake cool first and sink, allowing warmer water to rise to the surface to be cooled in turn. Eventually, a lake can fill with water at 4 °C. Further drops in temperature result in cooler, less dense water near the surface, where it floats until it freezes. Thus, lakes freeze on the top surface first, with the bottom water staying relatively warm at about 4 °C. In addition, the ice and snow on top of a lake act as thermal insulation, slowing the continued growth of ice.

REAL-WORLD PHYSICS
Bursting water pipes

On the other hand, if water had more ordinary behavior—like shrinking when cooled, and a solid form that is more dense than the liquid—a lake would freeze from the bottom up. There would be no insulating layer of ice on top, and if the winter were long enough, and cold enough, the lake could freeze solid. This, of course, would be disastrous for fish and other creatures that live in the water.

Finally, the same physics that is responsible for floating icebergs and ice-capped lakes is to blame for water pipes that burst in the winter. Even a water pipe made of steel is not strong enough to keep from rupturing when the ice forming within it expands outward. Later, when the temperature rises above freezing again, the burst pipe will make itself known by springing a leak.

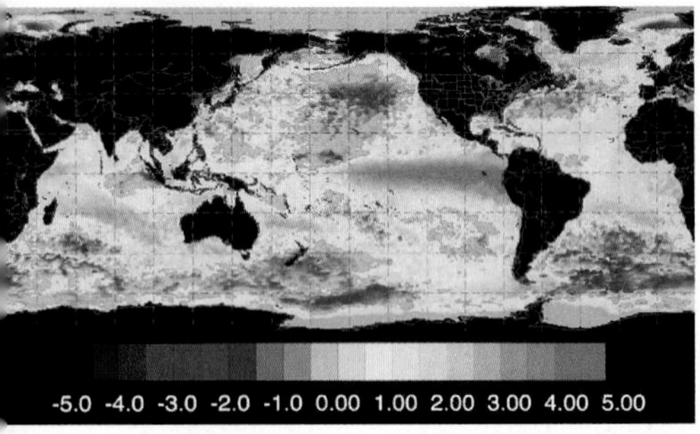

▲ Water exhibits unusual behavior around its freezing point, but above 4 °C it expands with increasing temperature like any normal liquid. This photo shows the expansion of water on a grand scale. It depicts an El Niño event: the appearance of a mass of unusually warm water in the equatorial region of the eastern Pacific Ocean every few years. An El Niño causes worldwide changes in climate and weather patterns. In this satellite image, the warmest water temperatures are represented by red, the coolest by violet. The El Niño is the red spike west of the South American coast. In this region, sea levels are as much as 20 cm higher than their average values, due to thermal expansion of the water.

16–4 Heat and Mechanical Work

In this section we consider the connection between heat and mechanical work. We also discuss the conservation of energy as it regards heat.

As mentioned previously, heat is the energy *transferred* from one object to another. At one time it was thought—erroneously—that an object contained a certain amount of "heat fluid," or caloric, that could flow from one place to another. This idea was overturned by the observations of Benjamin Thompson (1753–1814), also known as Count Rumford, the American-born physicist, spy, and social reformer who at one point in his eclectic career supervised the boring of cannon barrels by large drills. He observed that as long as mechanical work was done to turn the drill bits, they continued to produce heat in unlimited quantities. Clearly, the unlimited heat observed in boring the cannons was not present initially in the metal, but instead was produced by continually turning the drill bit.

With this observation, it became clear that heat was simply another form of energy that must be taken into account when applying conservation of energy. For example, if you rub sandpaper back and forth over a piece of wood, you do work. The energy associated with that work is not lost; instead, it produces an increase in temperature. Taking into account the energy associated with this temperature change, we find that energy is indeed conserved. In fact, no observation has ever indicated a situation in which energy is not conserved.

The equivalence between work and heat was first explored quantitatively by James Prescott Joule (1818–1889), the British physicist. In one of his experiments, Joule observed the increase in temperature in a device similar to that shown in **Figure 16–8**. Here, a total mass $2m$ falls through a certain distance h, during which gravity does the mechanical work, $2mgh$. As the mass falls, it turns the paddles in the water, which results in a slight warming of the water. By measuring the mechanical work, $2mgh$, and the increase in the water's temperature, ΔT, Joule was able to show that energy was indeed conserved—it had been converted from gravitational potential energy to an increased energy of the water, as indicated by a higher temperature. Joule's experiments established the precise amount of mechanical work that has the same effect as a given transfer of heat.

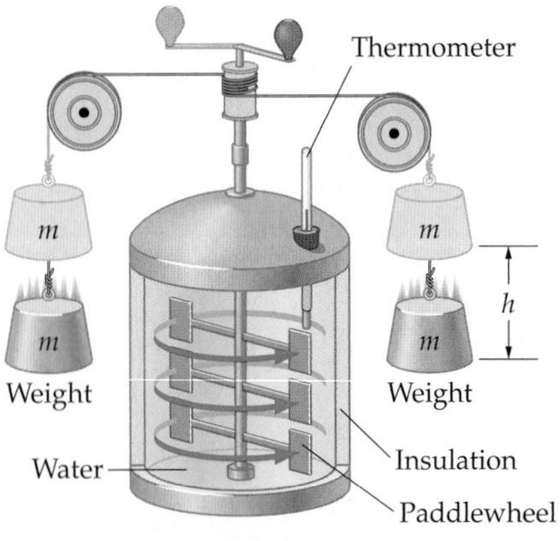

▲ **FIGURE 16–8 The mechanical equivalent of heat**

A device of this type was used by James Joule to measure the mechanical equivalent of heat.

Before Joule's work, heat was measured in a unit called the calorie (cal). In particular, one kilocalorie (kcal) was defined as the amount of heat needed to raise the temperature of 1 kg of water from 14.5 °C to 15.5 °C. With his experiments,

Joule was able to show that 1 kcal = 4186 J, or equivalently, that one calorie of heat transfer is the equivalent of 4.186 J of mechanical work. This is referred to as the **mechanical equivalent of heat:**

The Mechanical Equivalent of Heat

$$1 \text{ cal} = 4.186 \text{ J} \quad \text{16–8}$$

SI unit: J

In studies of nutrition a different calorie is used. It is the Calorie, with a capital C, and it is simply a kilocalorie; that is, 1 C = 1 kcal. Perhaps this helps people to feel a little better about their calorie intake. After all, a 250 C candy bar sounds a lot better than a 250,000 cal candy bar. They are equivalent, however.

Another common unit for measuring heat is the British thermal unit (Btu). By definition, a Btu is the energy required to heat 1 lb of water from 63 °F to 64 °F. In terms of calories and joules, a Btu is as follows:

$$1 \text{ Btu} = 0.252 \text{ kcal} = 1055 \text{ J} \quad \text{16–9}$$

Finally, we shall use the symbol Q to denote heat:

Heat, *Q*

$$Q = \text{heat} = \text{energy transferred due to temperature differences} \quad \text{16–10}$$

SI unit: J

Using the mechanical equivalent of heat as the conversion factor, we will typically give heat in either calories or joules.

EXAMPLE 16–4 Stair Master

A 74.0-kg person drinks a thick, rich, 305-C milkshake. How many stairs must this person climb to work off the shake? Let the height of a stair be 20.0 cm.

Picture the Problem

Our sketch shows the person climbing the stairs after drinking the milkshake. Note that each step has a height $h = 20.0$ cm. In addition, the total height to which the person climbs to work off the milkshake is designated by H. (As we noted in Conceptual Checkpoint 7–1, the horizontal distance is irrelevant here, as it does not affect the work done against gravity.)

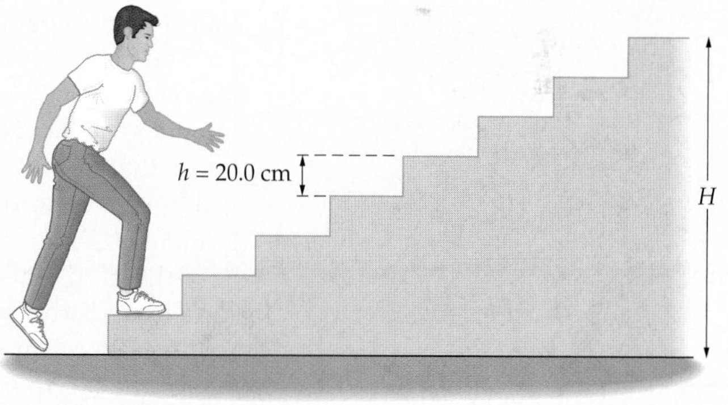

Strategy

We know that the energy intake by drinking the milkshake is the equivalent of the heat $Q = 305{,}000$ cal. This energy can be converted to joules by using the relation 1 cal = 4.186 J. Finally, we set the energy of the shake equal to the work done against gravity, mgH, in climbing to a height H.

Solution

1. Convert the energy of the milkshake to joules: $Q = 305{,}000 \text{ cal} = 305{,}000 \text{ cal}\left(\dfrac{4.186 \text{ J}}{1 \text{ cal}}\right) = 1.28 \times 10^6 \text{ J}$

2. Equate the energy of the milkshake with the work done against gravity: $Q = mgH$

   Solve for the height H: $H = Q/mg$

3. Substitute numerical values: $H = \dfrac{Q}{mg} = \dfrac{1.28 \times 10^6 \text{ J}}{(74.0 \text{ kg})(9.81 \text{ m/s}^2)} = 1760 \text{ m}$

4. Divide by the height of a stair to get the number of stairs: $\dfrac{1760 \text{ m}}{0.200 \text{ m/stair}} = 8800 \text{ stairs}$

continued on next page

continued from previous page

Insight

This is clearly a lot of stairs, and a significant height. In fact, 1760 m is more than a mile. Even assuming a metabolic efficiency of 25 percent, a height of about a quarter of a mile must be climbed to work off the shake. Put another way, the shake would be enough "fuel" for you to walk to the top of the Empire State Building with a little left over.

Practice Problem

If the person in the problem climbs 100 stairs, how many Calories have been burned? Assume 100 percent efficiency. [**Answer:** 3.47 C]

Some related homework problems: Problem 23, Problem 24

16–5 Specific Heats

In the previous section, we mentioned that it takes 4186 J of heat to raise the temperature of 1 kg of water by 1 C°. We have to be clear about the fact that we are heating water, however, because the heat required for a 1 C° increase varies considerably from one substance to another. For example, it takes only 128 J of heat to increase the temperature of 1 kg of lead by 1 C°. In general, the heat required for a given increase in temperature is given by the **heat capacity** of a substance.

Heat Capacity

Suppose we add the heat Q to a given object, and its temperature increases by the amount ΔT. The heat capacity, C, of this object is defined as follows:

Definition of Heat Capacity, C

$$C = \frac{Q}{\Delta T} \qquad \text{16–11}$$

SI unit: J/K = J/C°

Note that the units of heat capacity are joules per kelvin (J/K). Equivalently, since the degree size is the same for the Kelvin and Celsius scales, C can be expressed in units of joules per Celsius degree (J/C°).

The name "heat capacity" is perhaps a bit unfortunate. It derives from the mistaken idea of a "heat fluid," mentioned in the previous section. Objects were imagined to "contain" a certain amount of this nonexistent fluid. Today, we know that an object can readily gain or release heat when it is in thermal contact with other objects—objects cannot be thought of as holding a certain amount of heat.

Instead, the heat capacity should be viewed as the amount of heat necessary for a given temperature change. An object with a large heat capacity, like water, requires a large amount of heat for each increment in temperature. Just the opposite is true for an object with a small heat capacity, like a piece of lead.

To find the heat required for a given ΔT, we simply rearrange Equation 16–11 to solve for Q. This yields

$$Q = C\Delta T \qquad \text{16–12}$$

It should be noted that the *heat capacity is always positive*—just like a speed. Thus, Equation 16–12 shows that the heat Q and the temperature change ΔT have the same sign. This observation leads to the following sign conventions for Q:

Q is positive if ΔT is positive; that is, if heat is *added* to a system.

Q is negative if ΔT is negative; that is, if heat is *removed* from a system.

EXERCISE 16–3

The heat capacity of 1.00 kg of water is 4186 J/K. What is the temperature change of the water if **(a)** 505 J of heat is added to the system, or **(b)** 1010 J of heat is removed?

Solution

a. Calculate ΔT for $Q = 505$ J:

$$\Delta T = \frac{Q}{C} = \frac{505\text{ J}}{4186\text{ J/K}} = 0.121\text{ K}$$

b. Since heat is removed in this case, $Q = -1010$ J:

$$\Delta T = \frac{Q}{C} = \frac{-1010\text{ J}}{4186\text{ J/K}} = -0.241\text{ K}$$

Specific Heat

Since it takes 4186 J to increase the temperature of one kilogram of water by one degree Celsius, it takes twice that much to make the same temperature change in two kilograms of water, and so on. Thus, the heat capacity varies not only with the type of substance, but also with the mass of the substance.

We can therefore define a new quantity—the **specific heat,** c—that depends only on the substance, and not on its mass, as follows:

Definition of Specific Heat, *c*

$$c = \frac{Q}{m\Delta T} \qquad \text{16–13}$$

SI unit: $\text{J/(kg}\cdot\text{K)} = \text{J/(Kg}\cdot\text{C}^\circ)$

Thus, for example, the specific heat of water is

$$c_{\text{water}} = 4186\text{ J/(kg}\cdot\text{K)}$$

Specific heats for common substances are listed in Table 16–2. Note that the specific heat of water is by far the largest of any common material. This is just another of the many unusual properties of water. Having such a large specific heat means that water can give off or take in large quantities of heat with little change in temperature. It is for this reason that if you take a bite of a pie that is just out of the oven, you are much more likely to burn your tongue on the fruit filling (which has a high water content) than on the much drier crust.

Water's unusually large specific heat also accounts for the moderate climates experienced in regions near great bodies of water. In particular, the enormous volume and large heat capacity of an ocean serve to maintain a nearly constant temperature in the water, which in turn acts to even out the temperature of adjacent coastal areas. For example, the West Coast of the United States benefits from the moderating effect of the Pacific Ocean, aided by the prevailing breezes that come from the ocean onto the coastal regions. In the Midwest, on the other hand, temperature variations can be considerably greater as the land (with a relatively small specific heat) quickly heats up in the summer and cools off in the winter.

TABLE 16–2
Specific Heats at Atmospheric Pressure

Substance	Specific heat, c[J/(kg·K)]
Water	4186
Ice	2090
Steam	2010
Beryllium	1820
Air	1004
Aluminum	900
Glass	837
Silicon	703
Iron (steel)	448
Copper	387
Silver	234
Gold	129
Lead	128

REAL-WORLD PHYSICS
Water and the climate

Calorimetry

Let's use the specific heat to solve a practical problem. Suppose a block of mass m_b, specific heat c_b, and initial temperature T_b is dropped into a **calorimeter** (basically, a lightweight, insulated flask) containing water. If the water has a mass m_w, a specific heat c_w, and an initial temperature T_w, find the final temperature of the block and the water. Assume that the calorimeter is light enough that it can be ignored, and that no heat is transferred from the calorimeter to its surroundings.

There are two basic ideas involved in solving this problem: (a) the final temperatures of the block and water are equal, since the system will be in thermal equilibrium; and (b) the total energy of the system is conserved. In particular, the second condition means that the amount of heat lost by the block is equal to the heat gained by the water—or vice versa if the water's initial temperature is higher.

PROBLEM-SOLVING NOTE
Heat Flow and Thermal Equilibrium

To find the temperature of thermal equilibrium when two objects with different temperatures are brought into thermal contact, we simply use the idea that the heat that flows *out* of one of the objects flows *into* the other object.

Mathematically, we can write these conditions as follows:

$$Q_b + Q_w = 0 \tag{16–14}$$

This means that the heat flow from the block is equal and opposite to the heat flow from the water; in other words, energy is conserved. If we write the heats Q in terms of the specific heats and temperatures, letting the final temperature be T, we have

$$m_b c_b (T - T_b) + m_w c_w (T - T_w) = 0$$

Note that for each heat the change in temperature is T_{final} minus $T_{initial}$, as it should be. Solving for the final temperature, T, we find

$$T = \frac{m_b c_b T_b + m_w c_w T_w}{m_b c_b + m_w c_w} \tag{16–15}$$

This result need not be memorized, of course. Whenever solving a problem of this sort, one simply writes down energy conservation and solves for the desired unknown.

In some cases, we may wish to consider the influence of the container itself. This is illustrated in the following Active Example.

ACTIVE EXAMPLE 16–2 Find the Final Temperature

Suppose 550 g of water at 32 °C are poured into a 210-g aluminum can with an initial temperature of 15 °C. Find the final temperature of the system, assuming no heat is exchanged with the surroundings.

Solution *(Test your understanding by performing the calculations indicated in each step.)*

1. Write an expression for the heat flow out of the water: $Q_w = m_w c_w (T - T_w)$
2. Write an expression for the heat flow into the aluminum: $Q_a = m_a c_a (T - T_a)$
3. Apply energy conservation: $Q_w + Q_a = 0$
4. Solve for the final temperature: $T = 31\ °C$

Insight

As one might expect from water's large specific heat, the final common temperature (T) is much closer to the initial temperature of the water (T_w) than that of the aluminum (T_a).

Your Turn

Suppose the can is made from iron rather than aluminum. In this case, do you expect the final temperature to be greater than or less than the value for aluminum? Find the final temperature for the case of an iron can.

(Answers to **Your Turn** *problems can be found in the back of the book.)*

EXAMPLE 16–5 Cooling Off

A 0.50-kg block of metal with an initial temperature of 54.5 °C is dropped into a container holding 1.1 kg of water at 20.0 °C. If the final temperature of the block-water system is 21.4 °C, what is the specific heat of the metal? Assume the container can be ignored, and that no heat is exchanged with the surroundings.

Picture the Problem

Initially, when the block is first dropped into the water, the temperatures of the block and water are $T_b = 54.5\ °C$ and $T_w = 20.0\ °C$, respectively. When thermal equilibrium is established, both the block and the water have the same temperature, $T = 21.4\ °C$.

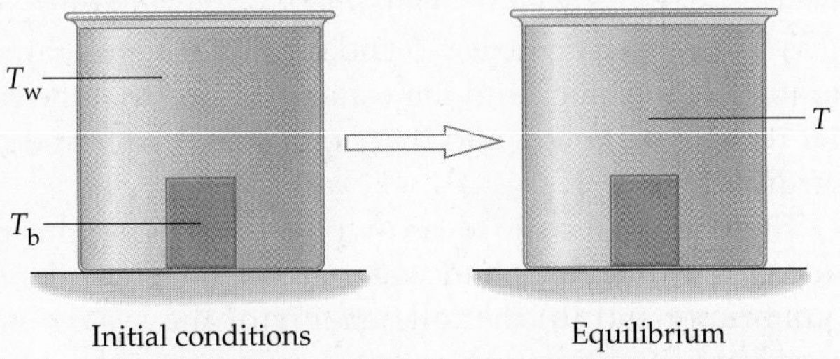

INTERACTIVE FIGURE

Strategy

Heat flows from the block to the water. Setting the heat flow *out of the block* plus the heat flow *into the water* equal to zero (conservation of energy) yields the block's specific heat.

Solution

1. Write an expression for the heat flow out of the block. Note that Q_{block} is negative, since T is less than T_b: $\quad Q_{\text{block}} = m_b c_b (T - T_b)$

2. Write an expression for the heat flow into the water. Note that Q_{water} is positive, since T is greater than T_w: $\quad Q_{\text{water}} = m_w c_w (T - T_w)$

3. Set the sum of the heats equal to zero: $\quad Q_{\text{block}} + Q_{\text{water}} = m_b c_b (T - T_b) + m_w c_w (T - T_w) = 0$

4. Solve for the specific heat of the block, c_b:
$$c_b = \frac{m_w c_w (T - T_w)}{m_b (T_b - T)}$$

5. Substitute numerical values:
$$c_b = \frac{(1.1\text{ kg})[4186\text{ J/(kg}\cdot\text{K)}](21.4\text{ °C} - 20.0\text{ °C})}{(0.50\text{ kg})(54.5\text{ °C} - 21.4\text{ °C})}$$
$$= 390\text{ J/(kg}\cdot\text{K)}$$

Insight

We note from Table 16–2 that the block is probably made of copper.

In addition, note that the final temperature is much closer to the initial temperature of the water than to the initial temperature of the block. This is due in part to the fact that the mass of the water is about twice that of the block; more important, however, is the fact that the water's specific heat is more than 10 times greater than that of the block. In the following Practice Problem we set the mass of the water equal to the mass of the block, so that we can see clearly the effect of the different specific heats.

Practice Problem

If the mass of the water is also 0.50 kg, what is the equilibrium temperature? [**Answer:** T = 23 °C. Still much closer to the water's initial temperature than to the block's.]

Some related homework problems: Problem 31, Problem 32

16–6 Conduction, Convection, and Radiation

Heat can be exchanged in a variety of ways. The Sun, for example, warms the Earth from across 93 million miles of empty space by a process known as radiation. As the sunlight strikes the ground and raises its temperature, the ground-level air gets warmer and begins to rise, producing a further exchange of heat by means of convection. Finally, if you walk across the ground in bare feet, you will feel the warming effect of heat entering your body by conduction. In this section we consider each of these three mechanisms of heat exchange in detail.

Conduction

Perhaps the most familiar form of heat exchange is **conduction,** which is the flow of heat directly through a physical material. For example, if you hold one end of a metal rod and put the other end in a fire, it doesn't take long before you begin to feel warmth on your end. The heat you feel is transported along the rod by conduction.

Let's consider this observation from a microscopic point of view. To begin, when you placed one end of the rod into the fire, the high temperature at that location caused the molecules to vibrate with an increased amplitude. These molecules in turn jostle their neighbors and cause them to vibrate with greater amplitude as well. Eventually, the effect travels from molecule to molecule across the length of the rod, resulting in the macroscopic phenomenon of conduction.

If you were to repeat the experiment, this time with a wooden rod, the hot end of the rod would heat up so much that it might even catch on fire, but your end would still be comfortably cool. Thus, conduction depends on the type of material involved. Some materials, called **conductors,** conduct heat very well, whereas other materials, called **insulators,** conduct heat poorly.

Just how much heat flows as a result of conduction? To answer this question we consider the simple system shown in **Figure 16–9**. Here we show a rod of

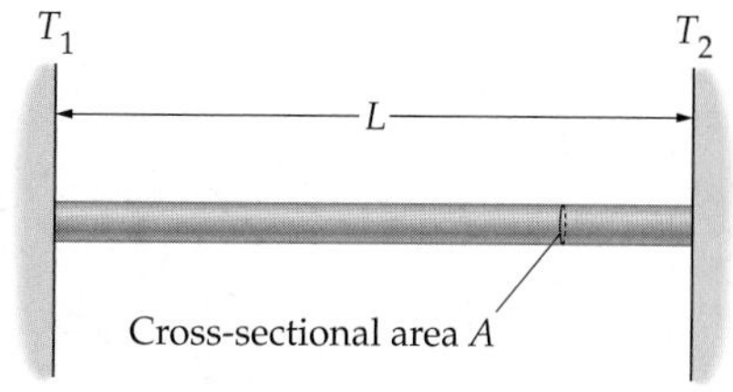

▲ **FIGURE 16–9 Heat conduction through a rod**
The amount of heat that flows through a rod of length L and cross-sectional area A per time is proportional to $A(T_2 - T_1)/L$.

▲ Maintaining proper body temperature in an environment that is often too hot or too cold is a problem for many animals. When the sand is blazing hot, this lizard (left) keeps its contact with the ground to a minimum. By standing on two legs instead of four, it reduces conduction of heat from the ground to its body. Polar bears (right) have the opposite problem. The loss of precious body heat to their surroundings is retarded by their thick fur, which is actually made up of hollow fibers. Air trapped within these fibers provides enhanced insulation, just as it does in our thermal blankets and double-paned windows.

length L and cross-sectional area A, with one end at the temperature T_1 and the other at the temperature $T_2 > T_1$. Experiments show that the amount of heat Q that flows through this rod:

- increases in proportion to the rod's cross-sectional area, A;
- increases in proportion to the temperature difference, $\Delta T = T_2 - T_1$;
- increases steadily with time, t;
- decreases with the length of the rod, L.

Combining these observations in a mathematical expression gives:

Heat Flow by Conduction

$$Q = kA\left(\frac{\Delta T}{L}\right)t \qquad \text{16–16}$$

The constant k is referred to as the **thermal conductivity** of the rod. It varies from material to material, as indicated in Table 16–3.

PROBLEM-SOLVING NOTE

Area and Length in Heat Conduction

When applying Equation 16–16, note that A is the area through which the heat flows. Thus, the plane of the area A is perpendicular to the direction of heat flow. The length L is the distance from the high-temperature side of an object to its low-temperature side.

TABLE 16–3 Thermal Conductivities

Substance	Thermal conductivity, $k[\mathrm{W/(m \cdot K)}]$
Silver	417
Copper	395
Gold	291
Aluminum	217
Steel, low carbon	66.9
Lead	34.3
Stainless steel—alloy 302	16.3
Ice	1.6
Concrete	1.3
Glass	0.84
Water	0.60
Asbestos	0.25
Wood	0.10
Wool	0.040
Air	0.0234

CONCEPTUAL CHECKPOINT 16–3 The Feel of Tile

You get up in the morning and walk barefoot from the bedroom to the bathroom. In the bedroom you walk on carpet, but in the bathroom the floor is tile. Does the tile feel **(a)** warmer, **(b)** cooler, or **(c)** the same temperature as the carpet?

Reasoning and Discussion

Everything in the house is at the same temperature, so it might seem that the carpet and the tile would feel the same. As you probably know from experience, however, the tile feels cooler. The reason is that tile has a much larger thermal conductivity than the carpet, which is actually a fairly good insulator. As a result, more heat flows from your skin to the tile than from your skin to the carpet. To your feet, then, it is as if the tile were much cooler than the carpet.

To get an idea of the thermal conductivities that would apply in this case, let's examine Table 16–3. For the tile, we might expect a thermal conductivity of roughly 0.84, the value appropriate for glass. For the carpet, the thermal conductivity might be as low as 0.04, the thermal conductivity of wool. Thus, the tile could have a thermal conductivity that is 20 times larger than that of the carpet.

Answer:

(b) The tile feels cooler.

Thermal conductivity is an important consideration when insulating a home. We consider some of these issues in the next Example.

EXAMPLE 16–6 What a Pane

One of the windows in a house has the shape of a square 1.0 m on a side. The glass in the window is 0.50 cm thick. How much heat is lost through this window in one day if the temperature in the house is 21 °C and the temperature outside is 0.0 °C?

Picture the Problem

The glass from the window is shown in our sketch, along with its relevant dimensions. Heat flows from the 21 °C side of the window to the 0.0 °C side.

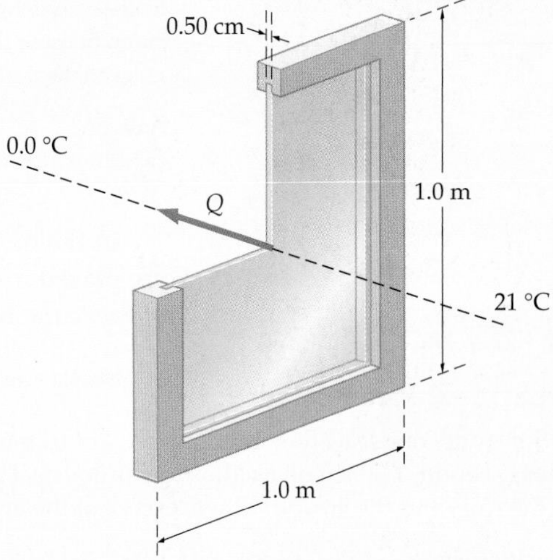

Strategy

The heat flow is given by Equation 16–16. Note that the area is $(1.0\text{ m})^2$ and that the length over which heat is conducted is, in this case, the thickness of the glass. Thus, $L = 0.0050$ m. The temperature difference is 21 C° = 21 K, and the thermal conductivity of glass (from Table 16–3) is 0.84 W/(m · K). (Recall from Section 7–4 that 1 W = 1 J/s.)

Solution

1. Calculate the heat flow for a given time, t:

$$Q = kA\left(\frac{\Delta T}{L}\right)t$$

$$= [0.84\text{ W}/(\text{m}\cdot\text{K})](1.0\text{ m})^2\left(\frac{21\text{ K}}{0.0050\text{ m}}\right)t = (3500\text{ W})t$$

2. Substitute the number of seconds in a day, 86,400 s, for the time t in the expression for Q:

$$Q = (3500\text{ W})t = (3500\text{ W})(86{,}400\text{ s}) = 3.0 \times 10^8\text{ J}$$

Insight

This is a sizable amount of heat, roughly equivalent to the energy released in burning a gallon of gasoline. A considerable reduction in heat loss can be obtained by using a double-pane window, which has an insulating layer of air (actually argon or krypton) sandwiched between the two panes of glass. This is discussed in more detail later in this section, and is explored in Problems 42 and 73.

Practice Problem

Suppose the window is replaced with a plate of solid silver. How thick must this plate be to have the same heat flow in a day as the glass? [**Answer:** The silver must have a thickness of $L = 2.5$ m.]

Some related homework problems: Problem 38, Problem 39

Now we consider the heat flow through a combination of two different materials with different thermal conductivities.

CONCEPTUAL CHECKPOINT 16–4 Compare the Heat Flow

Two metal rods are to be used to conduct heat from a region at 100 °C to a region at 0 °C. The rods can be placed in parallel, as shown on the left, or in series, as on the right. Is the heat conducted in the parallel arrangement **(a)** greater than, **(b)** less than, or **(c)** the same as the heat conducted with the rods in series?

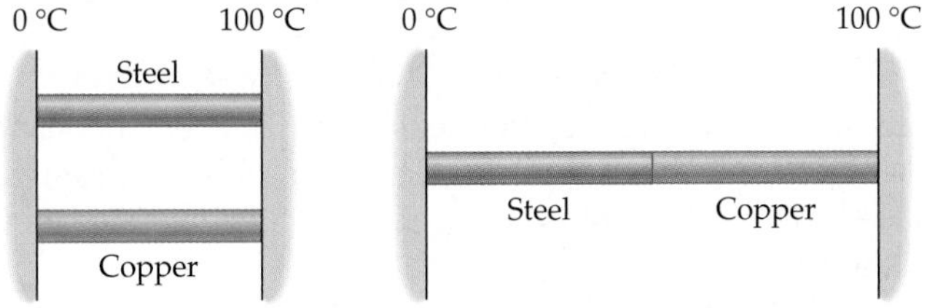

continued on next page

continued from previous page

Reasoning and Discussion

The parallel arrangement conducts more heat for two reasons. First, the cross-sectional area available for heat flow is twice as large for the parallel rods. A greater cross-sectional area gives a greater heat flow—everything else being equal. Second, more heat flows through each rod in the parallel configuration because they both have the full temperature difference of 100 C° between their ends. In the series configuration, each rod has a smaller temperature difference between its ends, so less heat flows.

Answer:

(a) More heat is conducted when the rods are in parallel.

In the next Example, we consider the case of heat conduction when two rods are placed in parallel. In the homework we shall consider the corresponding case of the same two rods conducting heat in series.

EXAMPLE 16–7 Parallel Rods

Two 0.525-m rods, one lead the other copper, are connected between metal plates held at 2.00 °C and 106 °C. The rods have a square cross section, 1.50 cm on a side. How much heat flows through the two rods in 1.00 s? Assume that no heat is exchanged between the rods and the surroundings, except at the ends.

Picture the Problem

The lead and the copper rods, each 0.525 m long, are shown in the sketch. Note that both rods have a temperature difference of 104 C° = 104 K between their ends.

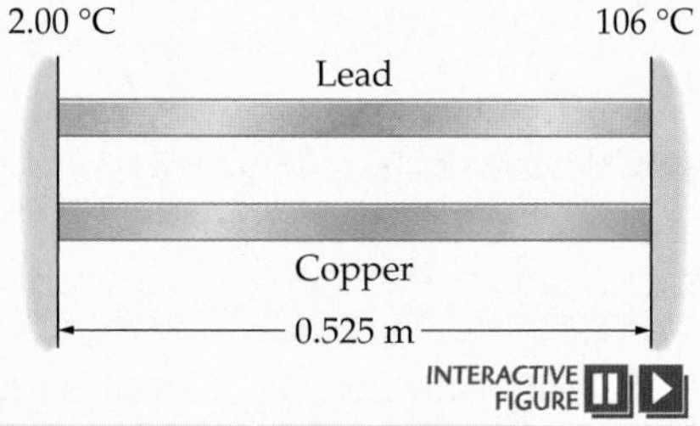

Strategy

The heat flowing through each rod can be calculated using $Q = kA(\Delta T/L)t$ and the value of k given in Table 16–3. The total heat flow is simply the sum of that calculated for each rod.

Solution

1. Calculate the heat flow in one second through the lead rod:

$$Q_l = k_l A\left(\frac{\Delta T}{L}\right)t$$

$$= [34.3\ \text{W}/(\text{m}\cdot\text{K})](0.0150\ \text{m})^2\left(\frac{104\ \text{K}}{0.525\ \text{m}}\right)(1.00\ \text{s})$$

$$= 1.53\ \text{J}$$

2. Calculate the heat flow in one second through the copper rod:

$$Q_c = k_c A\left(\frac{\Delta T}{L}\right)t$$

$$= [395\ \text{W}/(\text{m}\cdot\text{K})](0.0150\ \text{m})^2\left(\frac{104\ \text{K}}{0.525\ \text{m}}\right)(1.00\ \text{s})$$

$$= 17.6\ \text{J}$$

3. Sum the heats found in steps 1 and 2 to get the total heat:

$$Q_{total} = Q_l + Q_c = 1.53\ \text{J} + 17.6\ \text{J} = 19.1\ \text{J}$$

Insight

As our results show, the copper rod is by far the better conductor of heat. It is also a very good conductor of electricity, as we shall see in Chapter 21.

Practice Problem

What temperature difference would be required for the total heat flow in one second to be 15.0 J? [**Answer:** ΔT = 81.5 C° = 81.5 K]

Some related homework problems: Problem 43, Problem 44

As we shall see in Problem 45, the heat conducted by the same two rods connected in series is only 1.41 J, which is less than either of the two rods conduct when placed in parallel. This verifies the conclusion stated in Conceptual Checkpoint 16–4.

REAL-WORLD PHYSICS
Insulated windows

An application of thermal conductivity in series can be found in the *insulated window*. Most homes today have insulated windows as a means of increasing their energy efficiency. If you look closely at one of these windows, you will see that it is actually constructed from two panes of glass separated by an air-filled gap. Thus, heat flows through three different materials in series as it passes into or out of a home. The fact that the thermal conductivity of air is about 40 times smaller than that of glass means that the insulated window results in significantly less heat flow than would be experienced with a single pane of glass.

REAL-WORLD PHYSICS: BIO
Countercurrent exchange

As a final example of conduction, we note that many biological systems transfer heat by a mechanism known as *countercurrent exchange*. Consider, for example, an egret or other wading bird standing in cool water all day. As warm blood flows through the arteries on its way to the legs and feet of the bird, it passes through constricted regions where the legs join to the body. Here, where the arteries and veins are packed closely together, the body-temperature arterial blood flowing into the legs transfers heat to the much cooler venous blood returning to the body. Thus, the counter-flowing streams of blood serve to maintain the core body temperature of the bird, while at the same time keeping the legs and feet at much cooler temperatures. The feet still receive the oxygen and nutrients carried by the blood, but they stay at a relatively low temperature to reduce the amount of heat lost to the water.

REAL-WORLD PHYSICS: BIO
Cold hands and feet

Similar effects occur in humans. It is common, for example, to hear complaints that a person's hands or feet are cold. There is good reason for this, since they are in fact much cooler than the core body temperature. Just as with the wading birds, the warm arterial blood flowing to the hands and feet exchanges heat with the cool venous blood flowing in the opposite direction (**Figure 16–10**). This helps to reduce the heat loss to our surroundings, and to maintain the desired temperature in the core of the body.

▲ Many wading birds use countercurrent exchange in their circulatory systems. This mechanism allows them to keep the temperature of their legs well below that of their bodies. In this way they reduce the conductive loss of body heat to the water.

Convection

Suppose you want to heat a small room. To do so, you bring a portable electric heater into the room and turn it on. As the heating coils get red hot, they heat the air in their vicinity; as this air warms, it expands, becoming less dense. Because of its lower density, the warm air rises, to be replaced by cold dense air descending from overhead. This sets up a circulating flow of air that transports heat from the heating coils to the air throughout the room. Heat exchange of this type is referred to as **convection.**

In general, convection occurs when a fluid is unevenly heated. As with the room heater, the warm portions of the fluid rise because of their lower density and the cool portions sink because of their higher density. Thus, in convection, temperature differences result in a flow of fluid. It is this physical flow of matter that carries heat throughout the system.

Convection occurs on an enormous range of length scales. For example, the same type of uneven heating produced by an electric heater in a room can occur in the atmosphere of the Earth as well. The common seashore occurrence of sea breezes during the day and land breezes in the evening is one such example. (See **Figure 16–11** for an illustration of this effect.) On a larger scale, the Sun causes greater warming near the equator than near the poles; as a result, warm equatorial air rises, cool polar air descends, and global convection patterns are established. Similar convection patterns occur in ocean waters, and plate tectonics

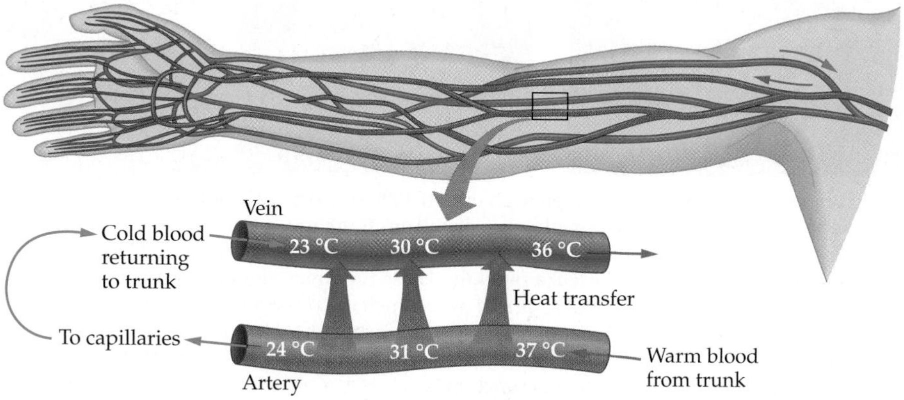

◀ **FIGURE 16–10 Countercurrent heat exchange in the human arm**

Arteries bringing warm blood to the limbs lie close to veins returning cooler blood to the body. This arrangement assures that a temperature difference (gradient) is maintained over the entire length that the vessels run parallel to one another, maximizing heat exchange between the warm arterial blood and the cooler venous blood.

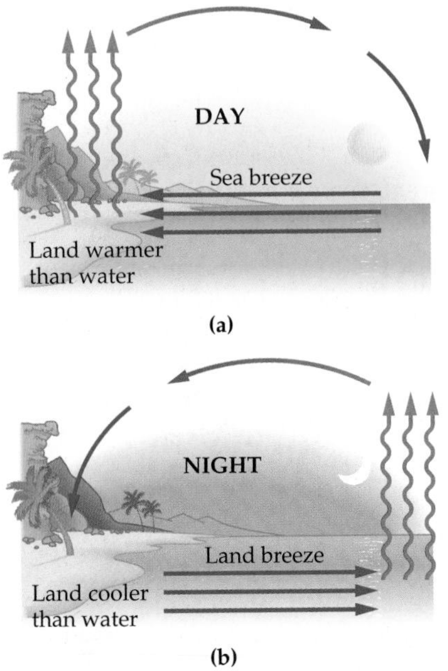

◀ **FIGURE 16–11 Alternating land and sea breezes**
(a) During the day, the sun warms the land more rapidly than the water. This is because the land, which is mostly rocks, has a lower specific heat than the water. The warm land heats the air above it, which becomes less dense and rises. Cooler air from over the water flows in to take its place, producing a "sea breeze." **(b)** At night, the land cools off more rapidly than the water—again because of its lower specific heat. Now it is the air above the relatively warm water that rises and is replaced by cooler air from over the land, producing a "land breeze."

is believed to be caused, at least in part, by convection currents in the Earth's mantle. The Sun also has convection currents, due to the intense heating that occurs in its interior, and disturbances in these currents are often visible as sunspots.

Radiation

Though convection and conduction occur primarily in specific situations, *all* objects give off energy as a result of **radiation.** The energy radiated by an object is in the form of electromagnetic waves (Chapter 25), which include visible light as well as infrared and ultraviolet radiation. Thus, unlike convection and conduction, radiation has no need for a physical material to mediate the energy transfer, since electromagnetic waves can propagate through empty space—that is, through a vacuum. Therefore, the heat you feel radiated from a hot furnace would reach you even if the air were suddenly removed—just as radiant energy from the Sun reaches the Earth across 150 million kilometers of vacuum.

REAL-WORLD PHYSICS
Using color to measure temperature

Since radiation can include visible light, it is often possible to "see" the temperature of an object. This is the physical basis of the **optical pyrometer,** invented by Josiah Wedgwood (1730–1795), the renowned English potter. When objects are about 800 °C they appear to be a dull "red hot." Examples include the heating coils in a range or oven. The filament in an incandescent lightbulb glows "yellow hot" at about 3000 °C, about the same temperature as the surface of the red supergiant star Betelgeuse in the constellation Orion. The surface of the Sun, in comparison, is about 6000 °C. Very hot stars, with surface temperatures in the vicinity of 10,000 to 30,000 °C, are "blue hot" and actually appear bluish in the night sky. Rigel, also in the constellation Orion, is an example of such a star. Look above and to the left of Orion's "belt" to see the red Betelgeuse, and below and to the right to see the blue Rigel.

REAL-WORLD PHYSICS
Temperatures of the stars

The energy radiated per time by an object—that is, the radiated power, P—is proportional to the surface area, A, over which the radiation occurs. It also depends on the temperature of the object. In fact, the dependence is on the fourth power of the temperature, T^4, where T is the Kelvin-scale temperature. Thus, for

▲ Stars have colors that are indicative of their surface temperatures. Thus, the blue supergiant Rigel (lower right) in the constellation Orion is much hotter than the red supergiant Betelgeuse (upper left).

▲ Red-hot volcanic lava is just hot enough (about 1000 °C) to radiate in the visible range. Even when it cools enough to stop glowing, it still emits energy, but most of it is in the form of invisible infrared radiation. Infrared radiation is also given off by the finned "radiators" attached to the supports of the Alaska pipeline. These fins are designed to prevent melting of the environmentally sensitive permafrost over which the pipeline runs. They function much like the radiator that cools your car engine, absorbing heat from the warm oil in the pipeline and dissipating it by radiation and convection into the atmosphere. (What design features can you see that facilitate this function?)

instance, if T is doubled, the radiated power increases by a factor of 16. All this behavior is contained in the **Stefan–Boltzmann law:**

Stefan–Boltzmann Law for Radiated Power, *P*

$$P = e\sigma AT^4 \qquad 16\text{–}17$$

SI unit: W

The constant σ in this expression is a fundamental physical constant, the **Stefan–Boltzmann constant:**

$$\sigma = 5.67 \times 10^{-8}\ \mathrm{W/(m^2 \cdot K^4)} \qquad 16\text{–}18$$

The other constant in the Stefan–Boltzmann law is the **emissivity,** e. The emissivity is a dimensionless number between 0 and 1 that indicates how effective the object is in radiating energy. A value of 1 means that the object is a perfect radiator. In general, a dark-colored object will have an emissivity near 1, and a light-colored object will have an emissivity closer to 0.

▲ Wear black in the desert? It actually helps, by radiating heat away from the wearer more efficiently.

EXERCISE 16–4

Calculate the radiated power from a sphere with a radius of 5.00 cm at the temperature 355 K. Assume the emissivity is unity.

Solution

Using $A = 4\pi r^2$ for a sphere, we have

$$P = e\sigma AT^4 = (1)[5.67 \times 10^{-8}\ \mathrm{W/(m^2 \cdot K^4)}]4\pi(0.0500\ \mathrm{m})^2(355\ \mathrm{K})^4 = 28.3\ \mathrm{W}$$

Experiments show that objects absorb radiation from their surroundings according to the same law, the Stefan–Boltzmann law, by which they emit radiation. Thus, if the temperature of an object is T, and its surroundings are at the temperature T_s, the *net* power radiated by the object is

Net Radiated Power, P_{net}

$$P_{net} = e\sigma A(T^4 - T_s{}^4) \qquad 16\text{–}19$$

SI unit: W

PROBLEM-SOLVING NOTE

Radiated Power

To correctly calculate the radiated power, Equations 16–17 and 16–19, the temperatures must be expressed in the Kelvin scale.

If the object's temperature is greater than its surroundings, it radiates more energy than it absorbs and P_{net} is positive. On the other hand, if its temperature is less than the surroundings, it absorbs more energy than it radiates and P_{net} is negative. When the object has the same temperature as its surroundings, it is in equilibrium and the net power is zero.

EXAMPLE 16–8 Human Polar Bears

On New Year's Day, several human "polar bears" prepare for their annual dip into the icy waters of Narragansett Bay. One of these hardy souls has a surface area of 1.15 m^2 and a surface temperature of 303 K (~30 °C). Find the net radiated power from this person **(a)** in a dressing room, where the temperature is 293 K (~20 °C), and **(b)** outside, where the temperature is 273 K (~0 °C). Assume an emissivity of 0.900 for the person's skin.

Picture the Problem

Our sketch shows the person radiating power in a room where the surroundings are at 293 K, and outside where the temperature is 273 K. The person also absorbs radiation from the surroundings; hence the net radiated power is greater when the surroundings are cooler.

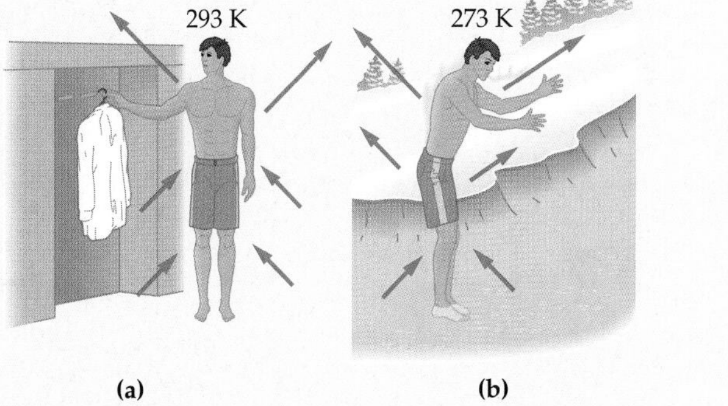

Strategy

A straightforward application of $P_{net} = e\sigma A(T^4 - T_s{}^4)$ can be used to find the net power for both parts (a) and (b). The only difference is the temperature of the surroundings, T_s.

continued on next page

continued from previous page

Solution

Part (a)

1. Calculate the net power using Equation 16–19 and $T_s = 293$ K:

$$P_{net} = e\sigma A(T^4 - T_s^4)$$
$$= (0.900)[5.67 \times 10^{-8}\ \mathrm{W/(m^2 \cdot K^4)}](1.15\ \mathrm{m^2}) \times [(303\ \mathrm{K})^4 - (293\ \mathrm{K})^4]$$
$$= 62.1\ \mathrm{W}$$

Part (b)

2. Calculate the net power using Equation 16–19 and $T_s = 273$ K:

$$P_{net} = e\sigma A(T^4 - T_s^4)$$
$$= (0.900)[5.67 \times 10^{-8}\ \mathrm{W/(m^2 \cdot K^4)}](1.15\ \mathrm{m^2}) \times [(303\ \mathrm{K})^4 - (273\ \mathrm{K})^4]$$
$$= 169\ \mathrm{W}$$

Insight

In the warm room the net radiated power is roughly that of a small lightbulb (about 60 W); outdoors, the net radiated power has more than doubled, and is comparable to that of a 150-W lightbulb.

Practice Problem

What temperature must the surroundings have if the net radiated power is to be 155 W? [**Answer:** $T = 276\ \mathrm{K} \approx 3\ °\mathrm{C}$]

Some related homework problems: Problem 40, Problem 48

REAL-WORLD PHYSICS
Thermos bottles and the Dewar

Note that the same emissivity e applies to both the emission and absorption of energy. Thus, a perfect emitter ($e = 1$) is also a perfect absorber. Such an object is referred to as a **blackbody.** As we shall see later, in Chapter 30, the study of blackbody radiation near the turn of the twentieth century ultimately led to one of the most fundamental revolutions in science—the introduction of quantum physics.

The opposite of a blackbody is an ideal reflector, which absorbs *no* radiation ($e = 0$). It follows that an ideal reflector also radiates no energy. This is why the inside of a Thermos bottle is highly reflective. As an almost ideal reflector, the inside of the bottle radiates very little of the energy contained in the hot liquid that it holds. In addition to its shiny interior, a Thermos bottle also has a vacuum between its inner and outer walls, as shown in **Figure 16–12**. This limits the flow of heat to radiation only, since convection and conduction cannot occur in a vacuum. This type of double-walled insulating container was invented by Sir James Dewar (1842–1923), a Scottish physicist and chemist.

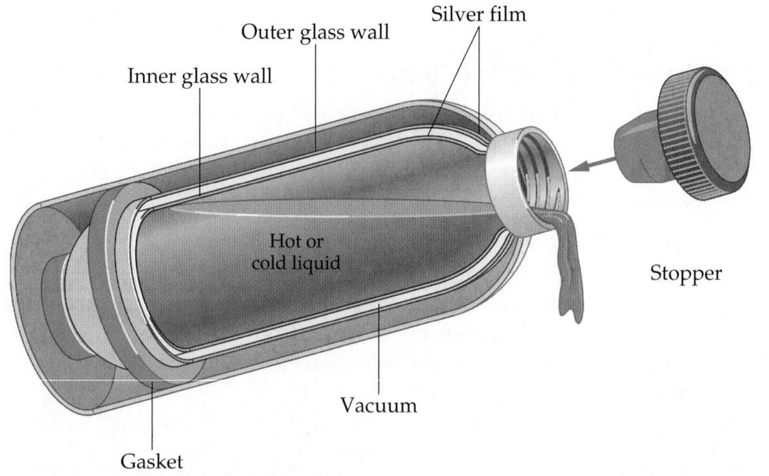

▲ **FIGURE 16–12 The Thermos bottle**

The hot or cold liquid stored in a Thermos bottle is separated from the outside world by a vacuum between the inner and outer glass walls. In addition, the inner glass wall has a reflective coating so that it is a good reflector and a poor radiator.

Chapter Summary

16–1 Temperature and the Zeroth Law of Thermodynamics

This section defines several new terms dealing with heat and temperature.

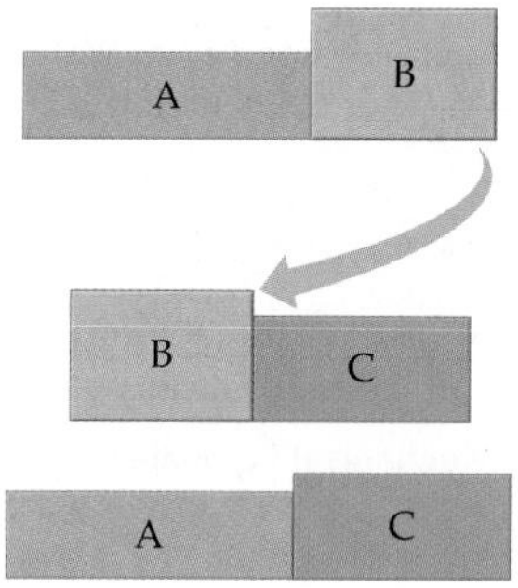

Heat
Heat is the energy transferred between objects because of a temperature difference.

Thermal Contact
Objects are in thermal contact if heat can flow between them.

Thermal Equilibrium
Objects that are in thermal contact, but have no heat exchange between them, are said to be in thermal equilibrium.

Thermodynamics
The study of physical processes involving the transfer of heat.

Zeroth Law of Thermodynamics
If objects A and C are in thermal equilibrium with object B, they are in thermal equilibrium with each other.

Temperature
Temperature is the quantity that determines whether or not two objects will be in thermal equilibrium.

16–2 Temperature Scales

Temperature is commonly measured in terms of several different scales.

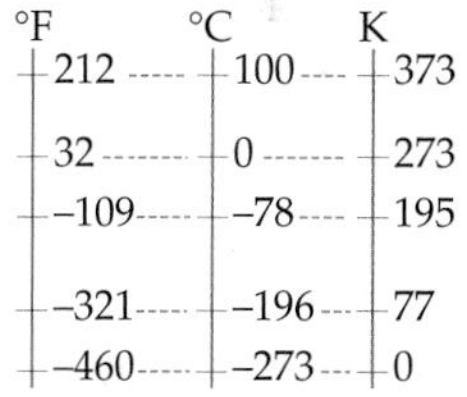

Celsius Scale
In the Celsius scale, water freezes at 0 °C and boils at 100 °C.

Fahrenheit Scale
In the Fahrenheit scale, water freezes at 32 °F and boils at 212 °F.

Absolute Zero
The lowest temperature attainable is referred to as absolute zero. It is impossible to cool an object to a temperature lower than absolute zero, which is −273.15 °C.

Kelvin Scale
In the Kelvin scale, absolute zero is 0 K. In addition, water freezes at 273.15 K and boils at 373.15 K. The degree size is the same for the Kelvin and Celsius scales.

Conversion Relations
The following relations convert between Celsius temperatures, T_C, Fahrenheit temperatures, T_F, and Kelvin temperatures, T:

$$T_F = \tfrac{9}{5}T_C + 32 \qquad \textbf{16–1}$$

$$T_C = \tfrac{5}{9}(T_F - 32) \qquad \textbf{16–2}$$

$$T = T_C + 273.15 \qquad \textbf{16–3}$$

16–3 Thermal Expansion

Most, though not all, substances expand when heated.

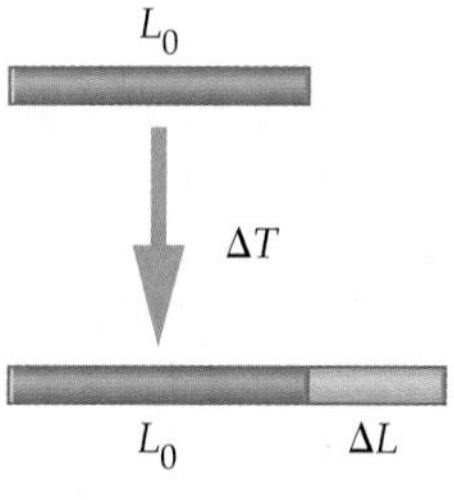

Linear Expansion
When an object of length L_0 is heated by the amount ΔT, its length increases by ΔL:

$$\Delta L = \alpha L_0 \Delta T \qquad \textbf{16–4}$$

The constant α is the coefficient of linear expansion (Table 16–1).

Volume Expansion
When an object of volume V is heated by the amount ΔT, its volume increases by ΔV:

$$\Delta V = \beta V \Delta T \qquad \textbf{16–6}$$

The constant β is the coefficient of volume expansion (Table 16–1).

Relation between α and β

If β is not listed for a particular substance, but α is listed, the volume expansion can be calculated using

$$\beta = 3\alpha$$

Special Properties of Water

Water is unusual in that it contracts when heated from 0 °C to 4 °C.

16–4 Heat and Mechanical Work

An important step forward in the understanding of heat was the recognition that it is a form of energy.

Mechanical Equivalent of Heat

$$1 \text{ cal} = 4.186 \text{ J} \qquad \text{16–8}$$

16–5 Specific Heats

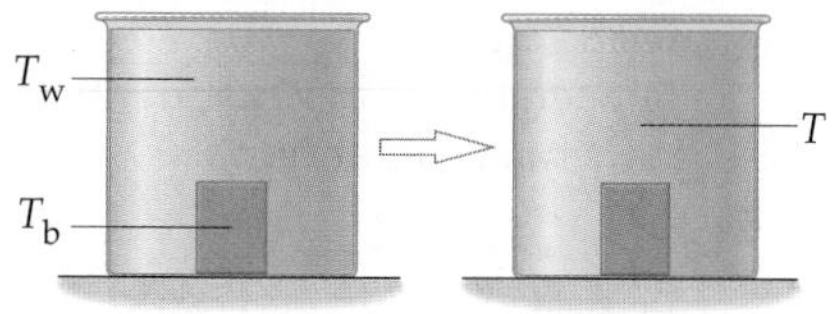

Heat Capacity

The heat capacity of an object is the heat Q divided by the associated temperature change, ΔT:

$$C = \frac{Q}{\Delta T} \qquad \text{16–11}$$

Specific Heat

The specific heat is the heat capacity per unit mass. Thus, the specific heat is independent of the quantity of a material in a given object:

$$c = \frac{Q}{m\Delta T} \qquad \text{16–13}$$

Conservation of Energy

In a system in which no heat is exchanged with the surroundings, the heat that flows out of one object equals the heat that flows into a second object. This is energy conservation.

16–6 Conduction, Convection, and Radiation

This section considers three common mechanisms of heat exchange.

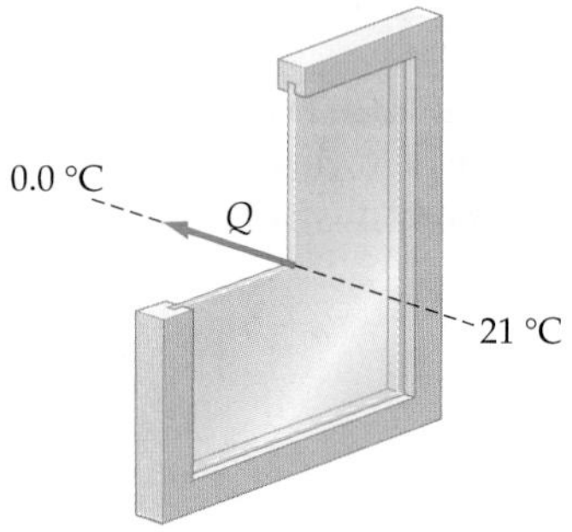

Conduction

In conduction, heat flows through a material with no bulk motion. The heat flows as a result of the interactions of individual atoms with their neighbors.

If the thermal conductivity of a material is k, its cross-sectional area is A, and its length L, the heat exchanged in the time t is

$$Q = kA\left(\frac{\Delta T}{L}\right)t \qquad \text{16–16}$$

Convection

Convection is heat exchange due to the bulk motion of an unevenly heated fluid.

Radiation

Radiation is the heat exchange due to electromagnetic radiation, such as infrared rays and light.

The energy per time, or power P, radiated by an object with a surface area A at the Kelvin temperature T is

$$P = e\sigma AT^4 \qquad \text{16–17}$$

where e is the emissivity (a constant between 0 and 1) and σ is the Stefan–Boltzmann constant, $\sigma = 5.67 \times 10^{-8}\ \text{W}/(\text{m}^2 \cdot \text{K}^4)$.

Since an object will also absorb radiation from its surroundings at the temperature T_s, the net radiated power is

$$P_{net} = e\sigma A(T^4 - T_s^{\,4}) \qquad \text{16–19}$$

Problem-Solving Summary

Type of Problem	Relevant Physical Concepts	Related Examples
Calculate the change in length of an object due to a change in temperature.	The change in an object's length is proportional to its initial length, L_0, and to the change in temperature, ΔT. In particular, $\Delta L = \alpha L_0 \Delta T$, where α is the coefficient of thermal expansion.	Exercise 16–2
Determine the change in volume of an object	If Table 16–1 gives a value of β for the substance in question, apply Equation 16–6. If a value of α is given instead, let $\beta = 3\alpha$ and use Equation 16–6.	Example 16–3
Relate the temperature change of an object to the amount of heat it absorbs or releases.	The heat capacity of an object determines the amount of heat, Q, it can absorb or give off for a given change in temperature, ΔT. The relationship is $Q = C\Delta T$.	Example 16–5, Active Example 16–2
Find the heat flow as a result of conduction.	Heat flow through a material is proportional to the temperature difference, the area through which the heat flows, and the time of flow; it is inversely proportional to the length over which the heat flows. Specifically, $Q = kA\Delta Tt/L$, where k is the thermal conductivity.	Examples 16–6, 16–7
Calculate the power given off by radiation.	The power, P, radiated by an object is proportional to its area, A, and to the fourth power of its temperature, T. The complete expression is $P = e\sigma AT^4$, where e is the emissivity and σ is the Stefan–Boltzmann constant.	Example 16–8

Conceptual Questions

(Answers to odd-numbered Conceptual Questions can be found in the back of the book.)

1. A cup of hot coffee is placed on the table. Is it in thermal equilibrium? What condition determines when the coffee is in equilibrium?
2. We know that -40 °F is the same as -40 °C. Is there a temperature for which the Kelvin and Celsius scales agree? Explain.
3. To find the temperature at the core of the Sun, you consult some Web sites on the Internet. One site says the temperature is about 15 million °C, another says it is 15 million kelvin. Is this a serious discrepancy? Explain.
4. Is it valid to say that a hot object contains more heat than a cold object?
5. If the glass in a glass thermometer had the same coefficient of volume expansion as mercury, the thermometer would not be very useful. Explain.
6. Suppose the glass in a glass thermometer expands more with temperature than the mercury it holds. What would happen to the mercury level as the temperature increased?
7. When a mercury-in-glass thermometer is inserted into a hot liquid the mercury column first drops and then rises. Explain this behavior.
8. Sometimes the metal lid on a glass jar has been screwed on so tightly that it is very difficult to open. Explain why holding the lid under hot running water often loosens it enough for easy opening.
9. Why do you hear creaking and groaning sounds in a house, particularly at night as the air temperature drops?
10. Two different objects receive different amounts of heat but experience the same increase in temperature. Give at least two possible reasons for this behavior.
11. Two different objects receive the same amount of heat. Give at least two reasons why their temperature changes may not be the same.
12. The specific heat of concrete is greater than that of soil. Given this fact, would you expect a major-league baseball field or the parking lot that surrounds it to cool off more in the evening following a sunny day?
13. Extending the result of the previous question to a larger scale, would you expect daytime winds to generally blow from a city to the surrounding suburbs or from the suburbs to the city? Explain.
14. When you touch a piece of metal and a piece of wood that are both at room temperature the metal feels cooler. Why?
15. After lighting a wooden match, you can hold onto the end of it for some time, until the flame almost reaches your fingers. Why aren't you burned as soon as the match is lit?
16. The rate of heat flow through a slab does *not* depend on which of the following? **(a)** The temperature difference between opposite faces of the slab. **(b)** The thermal conductivity of the slab. **(c)** The thickness of the slab. **(d)** The cross-sectional area of the slab. **(e)** The specific heat of the slab.
17. If a lighted match is held beneath a balloon inflated with air, the balloon quickly bursts. If, instead, the lighted match is held beneath a balloon filled with water, the balloon remains intact, even if the flame comes in contact with the balloon. Explain.
18. Updrafts of air allow hawks and eagles to glide effortlessly, all the while gaining altitude. What causes the updrafts?
19. When penguins huddle together during an Antarctic storm, they are warmer than if they are well separated. Explain.
20. **BIO** The fur of polar bears consists of hollow fibers. (Sometimes algae will grow in the hollow regions, giving the fur a green cast.) Explain why hollow hairs can be beneficial to the polar bears.
21. Object 2 has twice the emissivity of object 1, though they have the same size and shape. If the two objects radiate the same power, what is the ratio of their Kelvin temperatures?

Conceptual Exercises

(Answers to odd-numbered Conceptual Exercises can be found in the back of the book.)

1. Bimetallic strip A is made of copper and steel; bimetallic strip B is made of aluminum and steel. **(a)** Referring to Table 16–1, which strip bends more for a given change in temperature? Explain. **(b)** Which of the metals listed in Table 16–1 would give the greatest amount of bend when combined with steel in a bimetallic strip? Explain.
2. Referring to Table 16–1, which would be more accurate for all-season outdoor use, a tape measure made of steel or one made of aluminum? Explain.
3. A brass plate has a circular hole that is slightly smaller than the diameter of an aluminum ball. If the ball and the plate are always kept at the same temperature, should the temperature be increased or decreased in order for the ball to fit through the hole? Explain.
4. **Figure 16–13** shows five metal plates, all at the same temperature and all made from the same material. They are all placed in an oven and heated by the same amount. Rank the plates in order of increasing expansion in **(a)** the vertical and **(b)** the horizontal direction. Indicate ties where appropriate.

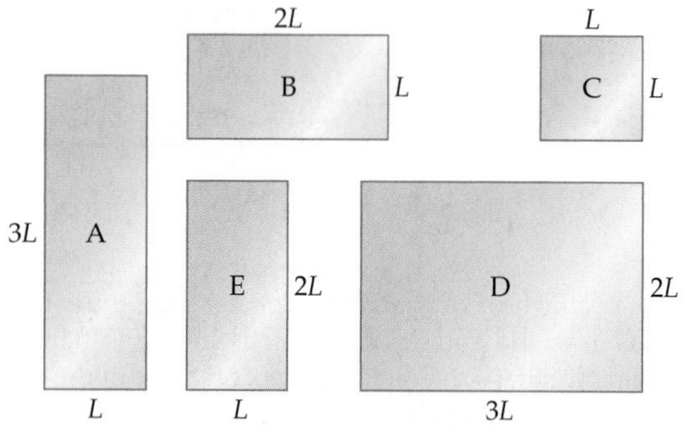

▲ **FIGURE 16–13** Conceptual Exercises 4 and 5

5. Referring to Exercise 4, rank the metal plates in order of increasing expansion in area. Indicate ties where appropriate.
6. A steel tape measure is marked in such a way that it gives accurate length measurements at a normal room temperature of 20 °C. If this tape measure is used outdoors on a cold day when the temperature is 0 °C, are its measurements too long, too short, or accurate? Explain.
7. A pendulum is made from an aluminum rod with a mass attached to its free end. If the pendulum is cooled, does the pendulum's period increase, decrease, or stay the same? Explain.
8. A copper ring stands on edge with a metal rod placed inside it, as shown in **Figure 16–14**. As this system is heated, will the rod ever touch the top of the ring? Answer yes or no for the case of a rod that is made of **(a)** copper, **(b)** aluminum, and **(c)** steel.

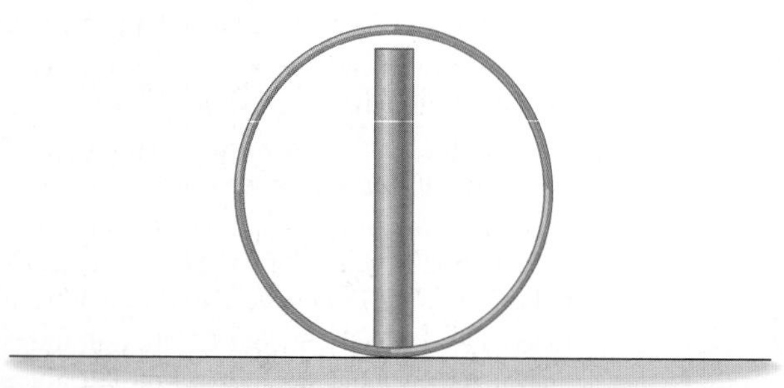

▲ **FIGURE 16–14** Conceptual Exercises 8 and 9

9. Referring to the copper ring in the previous problem, imagine that initially the ring is hotter than room temperature, and that an aluminum rod that is colder than room temperature fits snugly inside the ring. When this system reaches thermal equilibrium at room temperature, is the rod firmly wedged in the ring, or can it be removed easily? Explain.
10. Two objects are made of the same material but have different temperatures. Object 1 has a mass m, and object 2 has a mass $2m$. If the objects are brought into thermal contact, is the temperature change of object 1 greater than, less than, or equal to the temperature change of object 2? Explain.
11. A certain amount of heat is transferred to 2 kg of aluminum, and the same amount of heat is transferred to 1 kg of ice. Referring to Table 16–2, is the increase in temperature of the aluminum greater than, less than, or equal to the increase in temperature of the ice? Explain.
12. The specific heat of alcohol is about half that of water. Suppose you have 0.5 kg of alcohol at the temperature 20 °C in one container, and 0.5 kg of water at the temperature 30 °C in a second container. When these fluids are poured into the same container, and allowed to come to thermal equilibrium, is the final temperature greater than, less than, or equal to 25 °C? Explain.
13. The heat Q will warm 1 g of material A by 1 C°, the heat $2Q$ will warm 3 g of material B by 3 C°, the heat $3Q$ will warm 3 g of material C by 1 C°, and the heat $4Q$ will warm 4 g of material D by 2 C°. Rank these materials in order of increasing specific heat. Indicate ties where appropriate.
14. In a popular lecture demonstration, a sheet of paper is wrapped around a rod that is made from wood on one half and metal on the other half. If held over a flame, the paper on one half of the rod is burned while the paper on the other half is unaffected. Is the burned paper on the wooden half of the rod, or on the metal half? Explain.
15. **Figure 16–15** shows a composite slab of three different materials with equal thickness but different thermal conductivities. The opposite sides of the composite slab are held at the fixed temperatures T_1 and T_2. Given that $k_B > k_A > k_C$, rank the materials in order of the temperature difference across them, starting with the smallest. Indicate ties where appropriate.

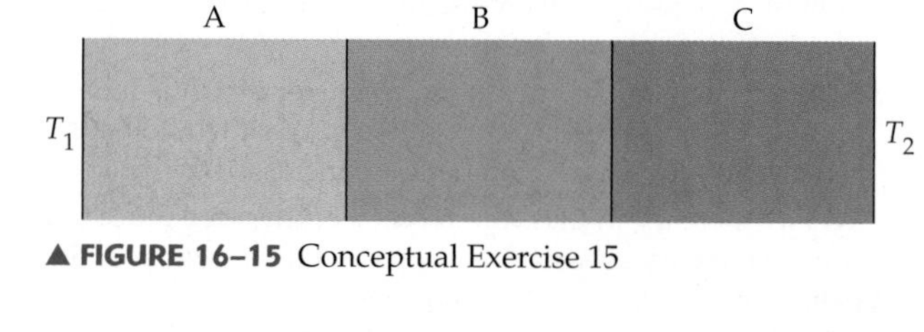

▲ **FIGURE 16–15** Conceptual Exercise 15

16. Heat is transferred from an area where the temperature is 20 °C to an area where the temperature is 0 °C through a composite slab consisting of four different materials, each with the same thickness. The temperatures at the interface between each of the materials is given in **Figure 16–16**. Rank the four

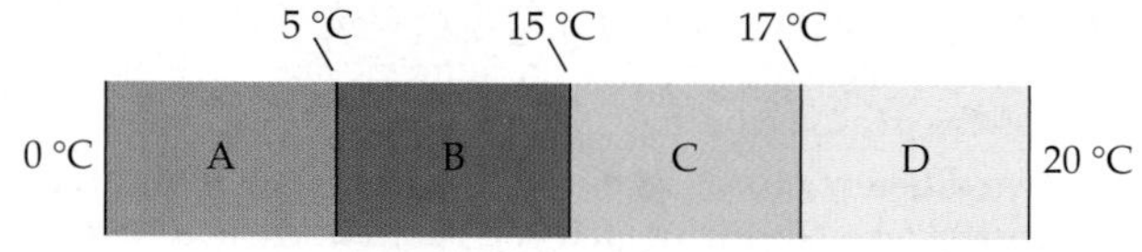

▲ **FIGURE 16–16** Conceptual Exercise 16

materials in order of increasing thermal conductivity. Indicate ties where appropriate.

17. On a sunny day identical twins wear different shirts. Twin 1 wears a dark shirt; twin 2 wears a light-colored shirt. Which twin has the warmer shirt? Explain.

18. Two bowls of soup with identical temperatures are placed on a table. Bowl 1 has a metal spoon in it; bowl 2 does not. After a few minutes, is the temperature of the soup in bowl 1 greater than, less than, or equal to the temperature of the soup in bowl 2? Explain.

19. Hot tea is poured from the same pot into two identical mugs. Mug 1 is filled to the brim; mug 2 is filled only halfway. Which mug cools off more rapidly?

20. **The "Blackbird"** The SR-71 reconnaissance aircraft is the fastest plane in the world. It is called "Blackbird" because of its distinctive black paint job (see Problem 12). When in flight, air resistance makes the skin of the airplane too hot to touch. If the SR-71 were painted white instead, would its in-flight temperature be greater than, less than, or the same as with black paint? Explain.

Problems

Note: **IP** *denotes an integrated conceptual/quantitative problem.* **BIO** *identifies problems of biological or medical interest. Red bullets (•, ••, •••) are used to indicate the level of difficulty of each problem.*

Section 16–2 Temperature Scales

1. • **Lowest Temperature on Earth** The official record for the lowest temperature ever recorded on Earth was set at Vostok, Antarctica, on July 21, 1983. The temperature on that day fell to −89.2 °C, well below the temperature of dry ice. What is this temperature in degrees Fahrenheit?

2. • More than likely, there is a glowing incandescent lightbulb in your room at this moment. The filament of that bulb, with a temperature of about 4500 °F, is almost half as hot as the surface of the Sun. What is this temperature in degrees Celsius?

3. • Normal body temperature for humans is 98.6 °F. What is the corresponding temperature in **(a)** degrees Celsius and **(b)** kelvins?

4. • What is the temperature 1.0 K on the Fahrenheit scale?

5. • The temperature at the surface of the Sun is about 6000 K. Convert this temperature to the **(a)** Celsius and **(b)** Fahrenheit scales.

6. • One day you notice that the outside temperature increased by 24 F° between your early morning jog and your lunch at noon. What is the corresponding change in temperature in the **(a)** Celsius and **(b)** Kelvin scales?

7. • The gas in a constant-volume gas thermometer has a pressure of 91.5 kPa at 101 °C. **(a)** What is the pressure of the gas at 50.0 °C? **(b)** At what temperature does the gas have a pressure of 115 kPa?

8. • **IP** A constant-volume gas thermometer has a pressure of 80.3 kPa at −10.0 °C and a pressure of 86.4 kPa at 10.0 °C. **(a)** At what temperature does the pressure of this system extrapolate to zero? **(b)** What are the pressures of the gas at the freezing and boiling points of water? **(c)** In general terms, how would your answers to parts (a) and (b) change if a different constant-volume gas thermometer is used? Explain.

9. •• **Greatest Change in Temperature** A world record for the greatest change in temperature was set in Spearfish, SD, on January 22, 1943. At 7:30 A.M. the temperature was −4.0 °F; two minutes later the temperature was 45 °F. Find the average rate of temperature change during those two minutes in kelvins per second.

10. •• We know that −40 °C corresponds to −40 °F. What temperature has the same value in both the Fahrenheit and Kelvin scales?

11. •• When the bulb of a constant-volume gas thermometer is placed in a beaker of boiling water at 100 °C, the pressure of the gas is 227 mmHg. When the bulb is moved to an ice-salt mixture, the pressure of the gas drops to 162 mmHg. Assuming ideal behavior, as in Figure 16–3, what is the Celsius temperature of the ice-salt mixture?

Section 16–3 Thermal Expansion

12. • **Faster than a Speeding Bullet** The SR-71 Blackbird, which is 107 feet 5 inches long, is the world's fastest airplane. It can fly at three times the speed of sound (Mach 3) at altitudes of 80,000 ft. When it lands after a long flight, it is too hot to be touched for about 30 minutes, and is 6.0 inches longer than at takeoff. How hot is the Blackbird when it lands, assuming its coefficient of linear expansion is $24 \times 10^{-6}\ K^{-1}$ and its temperature at takeoff is 23 °C?

The SR-71 Blackbird reconnaissance aircraft can reach speeds of 2350 mph. (Problem 12 and Conceptual Exercise 20)

13. • **Longest Suspension Bridge** The world's longest suspension bridge is the Akashi Kaikyo Bridge in Japan. The bridge is 3910 m long and is constructed of steel. How much longer is the bridge on a warm summer day (30.0 °C) than on a cold winter day (−5.00 °C)?

14. •• A hole in an aluminum plate has a diameter of 1.178 cm at 23.00 °C. **(a)** What is the diameter of the hole at 199.0 °C? **(b)** At what temperature is the diameter of the hole equal to 1.176 cm?

15. •• **IP** It is desired to slip an aluminum ring over a steel bar (**Figure 16–17**). At 10.00 °C the inside diameter of the ring is 4.000 cm and the diameter of the rod is 4.040 cm. **(a)** In order for the ring to slip over the bar, should the ring be heated or cooled? Explain. **(b)** Find the temperature of the ring at which it fits over the bar. The bar remains at 10.0 °C.

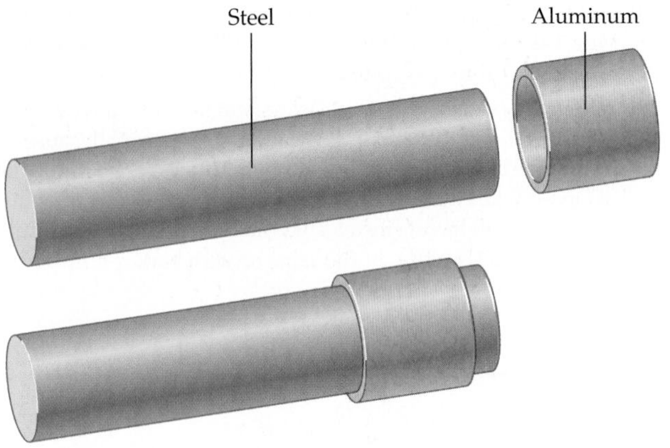

▲ **FIGURE 16–17** Problems 15 and 60

16. •• At 12.25 °C a brass sleeve has an inside diameter of 2.19625 cm and a steel shaft has a diameter of 2.19893 cm. It is desired to shrink-fit the sleeve over the steel shaft. **(a)** To what temperature must the sleeve be heated in order for it to slip over the shaft? **(b)** Alternatively, to what temperature must the shaft be cooled before it is able to slip through the sleeve?

17. •• Early in the morning, when the temperature is 5.0 °C, gasoline is pumped into a car's 51-L steel gas tank until it is filled to the top. Later in the day the temperature rises to 25 °C. Since the volume of gasoline increases more for a given temperature increase than the volume of the steel tank, gasoline will spill out of the tank. How much gasoline spills out in this case?

18. •• Some cookware has a stainless steel interior ($\alpha = 17.3 \times 10^{-6}\ K^{-1}$) and a copper bottom ($\alpha = 17.0 \times 10^{-6}\ K^{-1}$) for better heat distribution. Suppose an 8.0-in. pot of this construction is heated to 610 °C on the stove. If the initial temperature of the pot is 22 °C, what is the difference in diameter change for the copper and the steel?

19. •• **IP** You construct two wire-frame cubes, one using copper wire, the other using aluminum wire. At 21 °C the cubes enclose equal volumes of 0.016 m^3. **(a)** If the temperature of the cubes is increased, which cube encloses the greater volume? **(b)** Find the difference in volume between the cubes when their temperature is 97 °C.

20. •• A copper ball with a radius of 1.6 cm is heated until its diameter has increased by 0.18 mm. Assuming an initial temperature of 22 °C, find the final temperature of the ball.

21. •• **IP** An aluminum saucepan with a diameter of 23 cm and a height of 6.0 cm is filled to the brim with water. The initial temperature of the pan and water is 19 °C. The pan is now placed on a stove burner and heated to 88 °C. **(a)** Will water overflow from the pan, or will the water level in the pan decrease? Explain. **(b)** Calculate the volume of water that overflows or the drop in water level in the pan, whichever is appropriate.

Section 16–4 Heat and Mechanical Work

22. • **BIO Sleeping Metabolic Rate** When people sleep, their metabolic rate is about $2.6 \times 10^{-4}\ C/(s \cdot kg)$. How many Calories does a 75-kg person metabolize while getting a good night's sleep of 8.0 hr?

23. •• **BIO** An exercise machine indicates that you have worked off 2.5 Calories in a minute-and-a-half of running in place. What was your power output during this time? Give your answer in both watts and horsepower.

24. •• **BIO** During a workout, a person repeatedly lifts a 13-lb barbell through a distance of 1.4 ft. How many "reps" of this lift are required to burn off 150 C?

25. •• **IP** Consider the apparatus that Joule used in his experiments on the mechanical equivalent of heat, shown in Figure 16–8. Suppose both blocks have a mass of 0.95 kg and that they fall through a distance of 0.48 m. **(a)** Find the expected rise in temperature of the water, given that 6200 J are needed for every 1.0 C° increase. Give your answer in Celsius degrees. **(b)** Would the temperature rise in Fahrenheit degrees be greater than or less than the result in part (a)? Explain. **(c)** Find the rise in temperature in Fahrenheit degrees.

26. •• **BIO** It was shown in Example 16–8 that a typical person radiates about 62 W of power at room temperature. Given this result, how long does it take for a person to radiate away the energy acquired by consuming a 230-Calorie doughnut?

Section 16–5 Specific Heats

27. • Suppose 69.3 J of heat is added to a 121-g piece of aluminum at 22.5 °C. What is the final temperature of the aluminum?

28. • How much heat is required to raise the temperature of a 55-g glass ball by 15 C°?

29. • Estimate the heat required to heat a 0.13-kg apple from 15 °C to 35 °C. (Assume the apple is mostly water.)

30. • A 5.0-g lead bullet is fired into a fence post. The initial speed of the bullet is 250 m/s, and when it comes to rest, half its kinetic energy goes into heating the bullet. How much does the bullet's temperature increase?

31. •• **IP** Silver pellets with a mass of 1.0 g and a temperature of 85 °C are added to 220 g of water at 14 °C. **(a)** How many pellets must be added to increase the equilibrium temperature of the system to 25 °C? Assume no heat is exchanged with the surroundings. **(b)** If copper pellets are used instead, does the required number of pellets increase, decrease, or stay the same? Explain. **(c)** Find the number of copper pellets that are required.

32. •• A 235-g lead ball at a temperature of 84.2 °C is placed in a light calorimeter containing 177 g of water at 21.5 °C. Find the equilibrium temperature of the system.

33. •• If 2200 J of heat are added to a 190-g object, its temperature increases by 12 C°. **(a)** What is the heat capacity of this object? **(b)** What is the object's specific heat?

34. •• An 87.6-g lead ball is dropped from rest from a height of 5.43 m. The collision between the ball and the ground is totally inelastic. Assuming all the ball's kinetic energy goes into heating the ball, find its change in temperature.

35. •• To determine the specific heat of an object, a student heats it to 100 °C in boiling water. She then places the 38.0-g object in a 155-g aluminum calorimeter containing 103 g of water. The aluminum and water are initially at a temperature of 20.0 °C, and are thermally insulated from their surroundings. If the final temperature is 22.0 °C, what is the specific heat of the object? Referring to Table 16–2, identify the material in the object.

36. •• **IP** At the local county fair, you watch as a blacksmith drops a 0.50-kg iron horseshoe into a bucket containing 25 kg of water. **(a)** If the initial temperature of the horseshoe is 450 °C, and the initial temperature of the water is 23 °C, what is the equilibrium temperature of the system? Assume no heat is exchanged with the surroundings. **(b)** Suppose the 0.50-kg iron horseshoe had been a 1.0-kg lead horseshoe instead. Would the equilibrium temperature in this case be greater than, less than, or the same as in part (a)? Explain.

37. •• The ceramic coffee cup in **Figure 16–18**, with $m = 116$ g and $c = 1090\ J/(kg \cdot K)$, is initially at room temperature (24.0 °C). If 225 g of 80.3 °C coffee and 12.2 g of 5.00 °C cream are added to

▲ **FIGURE 16–18** Problem 37

the cup, what is the equilibrium temperature of the system? Assume that no heat is exchanged with the surroundings, and that the specific heat of coffee and cream are the same as the specific heat of water.

Section 16–6 Convection, Conduction, and Radiation

38. • A glass window 0.45 cm thick measures 88 cm by 32 cm. How much heat flows through this window per minute if the inside and outside temperatures differ by 15 C°?

39. • To compare the relative efficiency of air and glass as insulators, repeat the previous problem with a 0.45-cm thick layer of air instead of glass. By what factor is the rate of heat transfer reduced?

40. • **BIO** Assuming your skin temperature is 37.2 °C and the temperature of your surroundings is 21.8 °C, determine the length of time required for you to radiate away the energy gained by eating a 306-Calorie ice cream cone. Let the emissivity of your skin be 0.915 and its area be 1.22 m^2.

41. • Find the heat that flows in 1.0 s through a lead brick 15 cm long if the temperature difference between the ends of the brick is 8.5 C°. The cross-sectional area of the brick is 12 cm^2.

42. •• Consider a double-paned window consisting of two panes of glass, each with a thickness of 0.500 cm and an area of 0.725 m^2, separated by a layer of air with a thickness of 1.75 cm. The temperature on one side of the window is 0.00 °C; the temperature on the other side is 20.0 °C. In addition, note that the thermal conductivity of glass is roughly 36 times greater than that of air. **(a)** Approximate the heat transfer through this window by ignoring the glass. That is, calculate the heat flow per second through 1.75 cm of air with a temperature difference of 20.0 C°. (The exact result for the complete window is 19.1 J/s.) **(b)** Use the approximate heat flow found in part (a) to find an approximate temperature difference across each pane of glass. (The exact result is 0.157 C°.)

43. •• **IP** Two metal rods of equal length—one aluminum, the other stainless steel—are connected in parallel with a temperature of 20.0 °C at one end and 118 °C at the other end. Both rods have a circular cross section with a diameter of 3.50 cm. **(a)** Determine the length the rods must have if the combined rate of heat flow through them is to be 27.5 J per second. **(b)** If the length of the rods is doubled, by what factor does the rate of heat flow change?

44. •• Two cylindrical metal rods—one copper, the other lead—are connected in parallel with a temperature of 21.0 °C at one end and 112 °C at the other end. Both rods are 0.650 m in length, and the lead rod is 2.76 cm in diameter. If the combined rate of heat flow through the two rods is 33.2 J/s, what is the diameter of the copper rod?

45. •• **IP** Two metal rods—one lead, the other copper—are connected in series, as shown in **Figure 16–19**. These are the same two rods that were connected in parallel in Example 16–7. Note that each rod is 0.525 m in length and has a square cross section 1.50 cm on a side. The temperature at the lead end of the rods is 2.00 °C; the temperature at the copper end is 106 °C. **(a)** The average temperature of the two ends is 54.0 °C. Is the temperature in the middle, at the lead-copper interface, greater than, less than, or equal to 54.0 °C? Explain. **(b)** Given that the heat flow through each of these rods in 1.00 s is 1.41 J, find the temperature at the lead-copper interface.

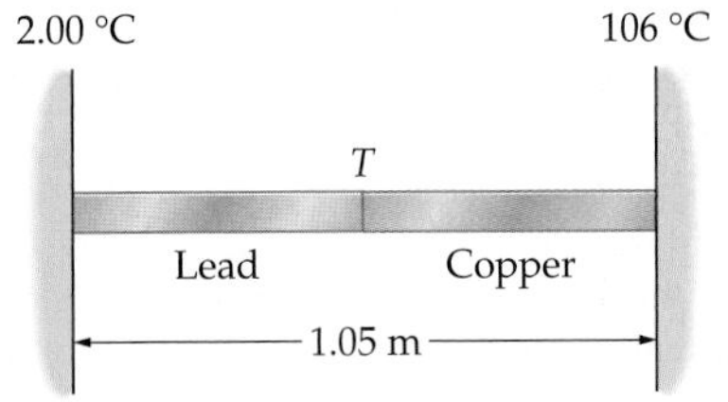

▲ **FIGURE 16–19** Problem 45

46. •• **IP** Consider two cylindrical metal rods with equal cross section—one lead, the other aluminum—connected in series. The temperature at the lead end of the rods is 20.0 °C; the temperature at the aluminum end is 80.0 °C. **(a)** Given that the temperature at the lead-aluminum interface is 50.0 °C, and that the lead rod is 14 cm long, what condition can you use to find the length of the aluminum rod? **(b)** Find the length of the aluminum rod.

47. •• A copper rod 85 cm long is used to poke a fire. The hot end of the rod is maintained at 105 °C and the cool end has a constant temperature of 21 °C. What is the temperature of the rod 23 cm from the cool end?

48. •• Two identical objects are placed in a room at 21 °C. Object 1 has a temperature of 95 °C and object 2 has a temperature of 25 °C. What is the ratio of the net power emitted by object 1 to that radiated by object 2?

49. ••• A block has the dimensions L, $2L$, and $3L$. When one of the $L \times 2L$ faces is maintained at the temperature T_1 and the other $L \times 2L$ face is held at the temperature T_2, the rate of heat conduction through the block is P. Answer the following questions in terms of P. **(a)** What is the rate of heat conduction in this block if one of the $L \times 3L$ faces is held at the temperature T_1 and the other $L \times 3L$ face is held at the temperature T_2? **(b)** What is the rate of heat conduction in this block if one of the $2L \times 3L$ faces is held at the temperature T_1 and the other $2L \times 3L$ face is held at the temperature T_2?

General Problems

50. • **Making Steel Sheets** In the continuous-caster process, steel sheets 25.4 cm thick, 2.03 m wide, and 10.0 m long are produced at a temperature of 872 °C. What are the dimensions of a steel sheet once it has cooled to 20.0 °C?

51. • **The Coldest Place in the Universe** The Boomerang nebula holds the distinction of having the lowest recorded temperature in the universe, a frigid −272 °C. What is this temperature in kelvins?

52. • When technicians work on a computer, they often ground themselves to prevent generating a spark. If an electrostatic discharge does occur, it can cause temperatures as high as 1500 °C in a localized area of a circuit. Temperatures this high can melt aluminum, copper, and silicon. What is this temperature in **(a)** degrees Fahrenheit and **(b)** kelvins?

53. • Two objects at the same initial temperature absorb equal amounts of heat. If the final temperature of the objects is different, it may be because they differ in which of the following properties: **(a)** mass, **(b)** coefficient of expansion, **(c)** thermal conductivity, or **(d)** specific heats. Explain.

54. • BIO **The Hottest Living Things** From the surreal realm of deep-sea hydrothermal vents 200 miles offshore from Puget Sound, comes a newly discovered hyperthermophilic—or extreme heat-loving—microbe that holds the record for the hottest existence known to science. This microbe is tentatively known as Strain 121 for the temperature at which is thrives: 121 °C. (At sea level, water at this temperature would boil vigorously, but the extreme pressures at the ocean floor prevent boiling from occurring.) What is this temperature in degrees Fahrenheit?

55. •• In many biological systems it is of more interest to know how much heat is required to raise the temperature of a given *volume* of material rather than a given *mass* of material. Calculate the heat needed to raise the temperature of one cubic meter of **(a)** air and **(b)** water by one degree Celsius. Compare with the corresponding specific heats (for a given mass) listed in Table 16–2.

56. •• BIO **Brain Power** As you read this problem, your brain is consuming about 22 W of power. **(a)** How many steps with a height of 21 cm must you climb to expend a mechanical energy equivalent to one hour of brain operation? **(b)** A typical human brain, which is 77% water, has a mass of 1.4 kg. Assuming that the 22 W of brain power is converted to heat, what temperature rise would you estimate for the brain in one hour of operation? Ignore the significant heat transfer that occurs between a human head and its surroundings, as well as the 23% of the brain that is not water.

57. •• BIO **The Cricket Thermometer** The rate of chirping of the snowy tree cricket (*Oecanthus fultoni* Walker) varies with temperature in a predictable way. A linear relationship provides a good match to the chirp rate, but an even more accurate relationship is the following:

$$N = (5.63 \times 10^{10})e^{-(6290\text{ K})/T}$$

In this expression, N is the number of chirps in 13.0 s and T is the temperature in kelvins. If a cricket is observed to chirp 185 times in 60.0 s, what is the temperature in degrees Fahrenheit?

58. •• If heat is transferred to 150 g of water at a constant rate for 2.5 min, its temperature increases by 13 C°. When heat is transferred at the same rate for the same amount of time to a 150-g object of unknown material, its temperature increases by 61 C°. **(a)** From what material is the object made? **(b)** What is the heating rate?

59. •• IP A pendulum consists of a large weight suspended by a steel wire that is 0.9500 m long. **(a)** If the temperature increases, does the period of the pendulum increase, decrease, or stay the same? Explain. **(b)** Calculate the change in length of the pendulum if the temperature increase is 150.0 C°. **(c)** Calculate the period of the pendulum before and after the temperature increase. (Assume that the coefficient of linear expansion for the wire is $12.00 \times 10^{-6}\text{ K}^{-1}$, and that $g = 9.810\text{ m/s}^2$ at the location of the pendulum.)

60. •• IP Once the aluminum ring in Problem 15 is slipped over the bar, the ring and bar are allowed to equilibrate at a temperature of 22 °C. The ring is now stuck on the bar. **(a)** If the temperature of both the ring and the bar are changed together, should the system be heated or cooled to remove the ring? **(b)** Find the temperature at which the ring can be removed.

61. •• A steel plate has a circular hole with a diameter of 1.000 cm. In order to drop a Pyrex glass marble 1.003 cm in diameter through the hole in the plate, how much must the temperature of the system be raised? (Assume the plate and the marble are always at the same temperature.)

62. •• A 206-kg rock sits in full sunlight on the edge of a cliff 5.00 m high. The temperature of the rock is 30.2 °C. If the rock falls from the cliff into a pool containing 6.00 m^3 of water at 15.5 °C, what is the final temperature of the rock-water system? Assume that the specific heat of the rock is 1010 J/(kg · K).

63. •• Water going over Iguaçu Falls on the border of Argentina and Brazil drops through a height of about 72 m. Suppose that all the gravitational potential energy of the water goes into raising its temperature. Find the increase in water temperature at the bottom of the falls as compared with the top.

64. •• IP A 0.22-kg steel pot on a stove contains 2.1 L of water at 22 °C. When the burner is turned on, the water begins to boil after 8.5 minutes. **(a)** At what rate is heat being transferred from the burner to the pot of water? **(b)** At this rate of heating, would it take more time or less time for the water to start to boil if the pot were made of gold rather than steel?

65. •• BIO Suppose you could convert the 525 Calories in the cheeseburger you ate for lunch into mechanical energy with 100% efficiency. **(a)** How high could you throw a 0.145-kg baseball with the energy contained in the cheeseburger? **(b)** How fast would the ball be moving at the moment of release?

66. •• You turn a crank on a device similar to that shown in Figure 16–8 and produce a power of 0.16 hp. If the paddles are immersed in 0.75 kg of water, for what length of time must you turn the crank to increase the temperature of the water by 5.0 C°?

67. •• IP BIO **Heat Transport in the Human Body** The core temperature of the human body is 37.0 °C, and the skin, with a surface area of 1.40 m^2, has a temperature of 34.0 °C. **(a)** Find the rate of heat transfer out of the body under the following assumptions: (i) The average thickness of tissue between the core and the skin is 1.20 cm; (ii) The thermal conductivity of the tissue is that of water. **(b)** Without repeating the calculation of part (a), what rate of heat transfer would you expect if the skin temperature were to fall to 31.0 °C? Explain.

68. •• **The Solar Constant** The surface of the Sun has a temperature of 5500 °C. **(a)** Treating the Sun as a perfect blackbody, with an emissivity of 1.0, find the power that it radiates into space. The radius of the Sun is 7.0×10^8 m, and the temperature of space can be taken to be 3.0 K. **(b)** The *solar constant* is the number of watts of sunlight power falling on a square meter of the Earth's upper atmosphere. Use your result from part (a) to calculate the solar constant, given that the distance from the Sun to the Earth is 1.5×10^{11} m.

69. ••• Bars of two different metals are bolted together, as shown in **Figure 16–20**. Show that the distance D does not change with temperature if the length of the two bars have the following ratio: $L_A/L_B = \alpha_B/\alpha_A$.

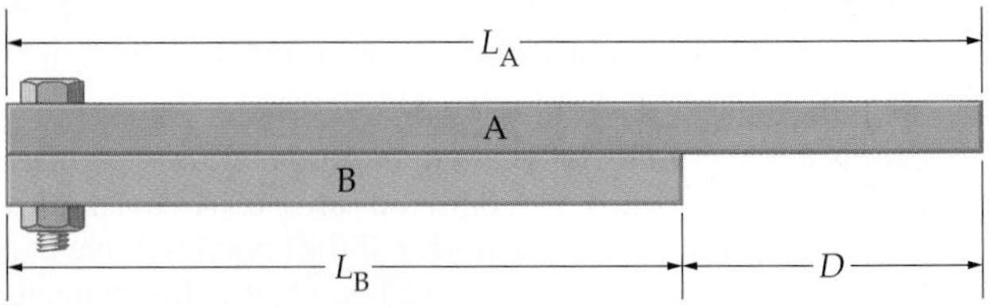

▲ **FIGURE 16–20** Problem 69

70. ••• A grandfather clock has a simple brass pendulum of length L. One night, the temperature in the house is 25.0 °C and the period of the pendulum is 1.00 s. The clock keeps correct time at this temperature. If the temperature in the house quickly drops to 18.1 °C just after 10 P.M., and stays at that value, what is the actual time when the clock indicates that it is 10 A.M. the next morning?

71. ••• IP A sheet of aluminum has a circular hole with a diameter of 10.0 cm. A 9.99 cm long steel rod is placed inside the hole, along a diameter of the circle, as shown in **Figure 16–21**. It is desired to change the temperature of this system until the steel rod just touches both sides of the circle. **(a)** Should the temperature of the system be increased or decreased? Explain. **(b)** By how much should the temperature be changed?

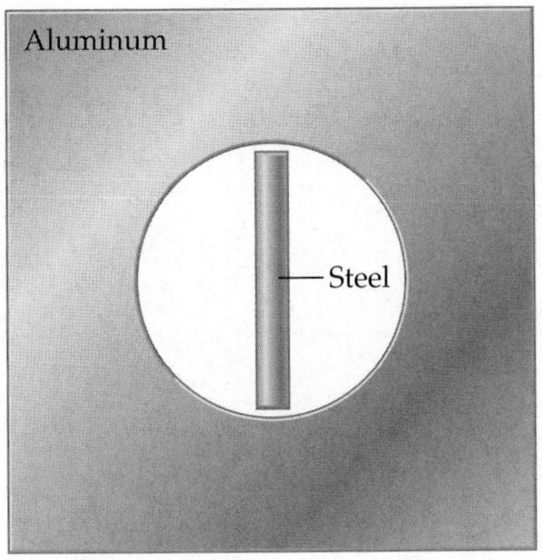

▲ **FIGURE 16–21** Problem 71

72. ••• A layer of ice has formed on a small pond. The air just above the ice is at −5.4 °C, the water-ice interface is at 0 °C, and the water at the bottom of the pond is at 4.0 °C. If the total depth from the top of the ice to the bottom of the pond is 1.4 m, how thick is the layer of ice? *Note*: The thermal conductivity of ice is 1.6 W/(m · C°) and that of water is 0.60 W/(m · C°).

73. ••• **A Double-Paned Window** An energy-efficient double-paned window consists of two panes of glass, each with thickness L_1 and thermal conductivity k_1, separated by a layer of air of thickness L_2 and thermal conductivity k_2. Show that the equilibrium rate of heat flow through this window per unit area, A, is

$$\frac{Q}{At} = \frac{(T_2 - T_1)}{2L_1/k_1 + L_2/k_2}$$

In this expression, T_1 and T_2 are the temperatures on either side of the window.

Interactive Problems

74. •• ***Referring to Example 16–5*** Suppose the mass of the block is to be increased enough to make the final temperature of the system equal to 22.5 °C. What is the required mass? Everything else in Example 16–5 remains the same.

75. •• ***Referring to Example 16–5*** Suppose the initial temperature of the block is to be increased enough to make the final temperature of the system equal to 22.5 °C. What is the required initial temperature? Everything else remains the same as in Example 16–5.

76. •• IP ***Referring to Example 16–7*** Suppose the lead rod is replaced with a second copper rod. **(a)** Will the heat that flows in 1.00 s increase, decrease, or stay the same? Explain. **(b)** Find the heat that flows in 1.00 s with two copper rods. Everything else remains the same as in Example 16–7.

77. •• IP ***Referring to Example 16–7*** Suppose the temperature of the hot plate is to be changed to give a total heat flow of 25.2 J in 1.00 s. **(a)** Should the new temperature of the hot plate be greater than or less than 106 °C? Explain. **(b)** Find the required temperature of the hot plate. Everything else is the same as in Example 16–7.

17 Phases and Phase Changes

17–1	Ideal Gases	549
17–2	Kinetic Theory	555
17–3	Solids and Elastic Deformation	560
17–4	Phase Equilibrium and Evaporation	565
17–5	Latent Heats	571
17–6	Phase Changes and Energy Conservation	574

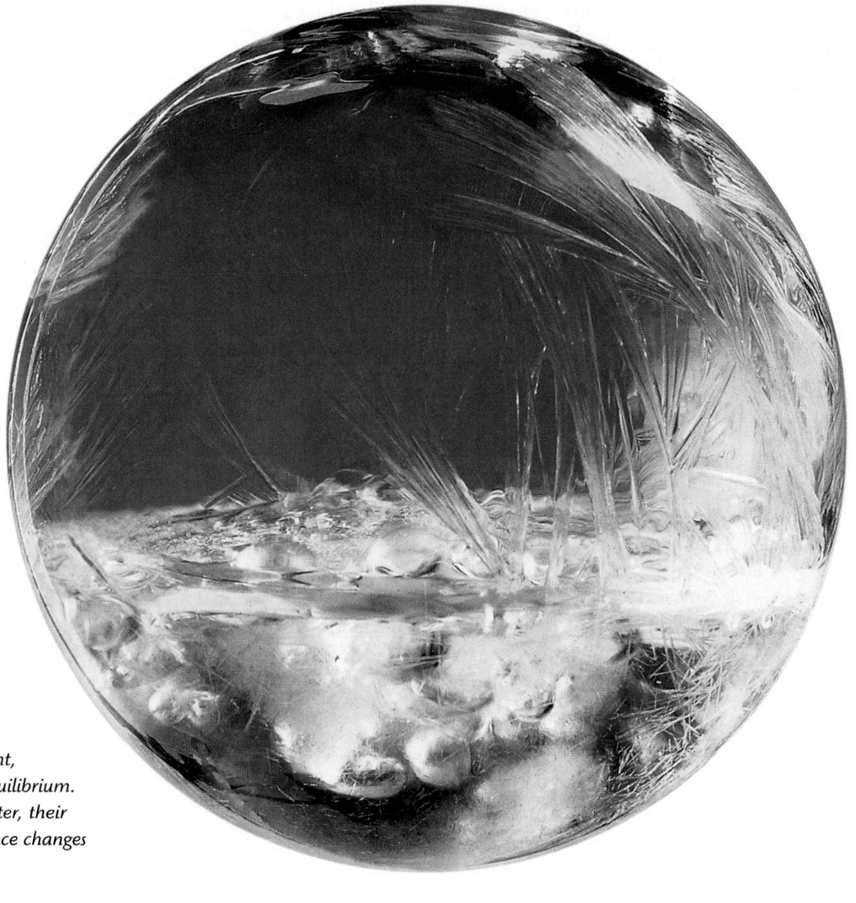

Though many people are surprised to learn that a liquid can boil and freeze at the same time, it is in fact true. At a unique combination of temperature and pressure called the triple point, all three phases of matter—solid, liquid, and gas—coexist in equilibrium. In this chapter we'll learn more about the three phases of matter, their differences and similarities, and what happens when a substance changes from one phase to another.

The matter we come into contact with every day is in one of three forms: solid, liquid, or gas. The air we breathe is a gas, the water we drink is a liquid, and the salt on our popcorn is a crystalline solid. In this chapter we consider these three *phases* of matter in detail.

We begin by considering the behavior of a gas, which is the easiest phase to describe physically. In particular, we show that an "ideal gas"—one that has no interactions between its molecules—is a good approximation to real gases. In addition, we show how the kinetic theory of gases allows us to relate the temperature of a substance to the kinetic energy of its molecules.

Next, we discuss some of the mechanical properties associated with the solid phase of matter. In particular, we explore the relationship between a force applied to a solid and the resulting deformation. For example, we shall determine the amount of stretch in your arm bones when you carry a heavy suitcase through an airport.

Finally, we turn our attention to the behavior of a substance when it changes from one phase to another—as when solid ice melts to form liquid water or water boils to produce gaseous clouds of steam. Remarkably, when a material is changing phase, the addition of heat does not result in a higher temperature. A significant amount of heat must first be absorbed by the material so that it can complete its phase change. Only after the change is accomplished can its temperature begin to rise again.

17–1 Ideal Gases

In Chapter 16 we discussed the thermal behavior of gases. In particular, we saw that the pressure of a constant volume of gas decreases linearly with decreasing temperature over a wide range of temperatures. Eventually, however, when the temperature is low enough, real gases change to liquid and then solid form, and their behavior changes. These changes are due to interactions between the molecules in a gas. Hence, the weaker these interactions, the wider the range of temperatures over which the simple, linear gas behavior persists. We now consider a simplified model of a gas, the **ideal gas,** in which intermolecular interactions are vanishingly small.

Though ideal gases do not actually exist in nature, the behavior of real gases is generally quite well approximated by that of an ideal gas, especially when the gas is dilute. By studying the simple "ideal" case, we can gain considerable insight into the workings of a real gas. This is similar to the kind of idealizations we made in mechanics. For example, we often considered a surface to be perfectly frictionless, when real surfaces have at least some friction, and we imagined springs to obey Hooke's law exactly, though real springs show some deviations. The ideal gas plays an analogous role in our study of thermodynamics.

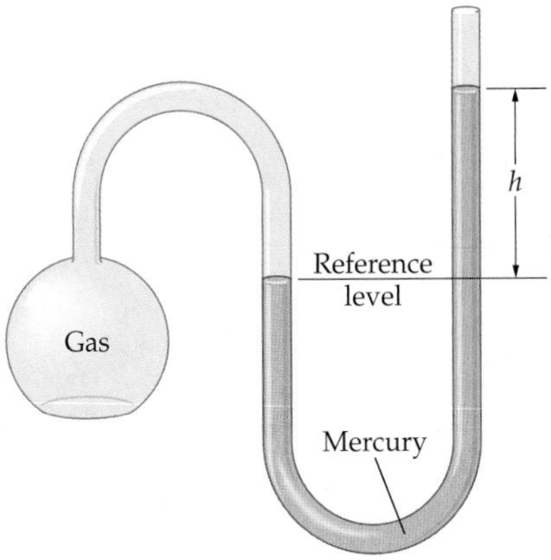

▲ **FIGURE 17–1 A constant-volume gas thermometer**

A device like this can be used as a thermometer. Note that both the volume of the gas and the number of molecules in the gas remain constant.

Equation of State

We can describe the way the pressure, P, of an ideal gas depends on temperature, T, number of molecules, N, and volume, V, from just a few simple observations. First, imagine holding the number of molecules and the volume of a gas constant, as in the constant-volume gas thermometer shown in **Figure 17–1**. As we have already noted, the pressure of a gas under these conditions varies linearly with temperature. Therefore, we conclude the following:

$$P = (\text{constant})T$$
$$(\textit{fixed volume, } V; \textit{ fixed number of molecules, } N)$$

In this expression, the constant multiplying the temperature depends on the number of molecules in the gas and its volume; the temperature T is measured on the Kelvin scale.

Second, imagine you have a basketball that is just slightly underinflated, as in **Figure 17–2**. The ball has the size and shape of a basketball, but is a bit too squishy. To increase the pressure you "pump" more molecules from the atmosphere into the ball. Thus, while the temperature and volume of the gas in the ball remain constant, its pressure increases as the number of molecules increases:

$$P = (\text{constant})N$$
$$(\textit{fixed volume, } V; \textit{ fixed temperature, } T)$$

▲ **FIGURE 17–2 Inflating a basketball**

A hand pump can be used to increase the number of molecules inside a basketball. This raises the pressure of the gas in the ball, causing it to inflate.

Our third and final observation concerns the volume dependence of pressure. Returning to the basketball, suppose that instead of pumping it up you sit on it, as pictured in **Figure 17–3**. This deforms it a bit, *reducing* its volume. At the same time, the pressure of the gas in the ball *increases*. Thus, when the number of molecules and temperature remain constant, the pressure varies *inversely* with volume:

$$P = \frac{(\text{constant})}{V}$$
$$(\textit{fixed number of molecules, } N; \textit{ fixed temperature, } T)$$

Combining these observations, we arrive at the following mathematical expression for the pressure of a gas:

$$P = k\frac{NT}{V}$$

▶ **FIGURE 17–3 Increasing pressure by decreasing volume**

Sitting on a basketball reduces its volume and increases the pressure of the gas it contains.

The constant k in this expression is a fundamental constant of nature, the **Boltzmann constant,** named for the Austrian physicist Ludwig Boltzmann (1844–1906):

Boltzmann Constant, *k*

$$k = 1.38 \times 10^{-23}\ \text{J/K} \qquad \textbf{17–1}$$

SI unit: J/K

In general, a relationship between the thermal properties of a substance is referred to as an **equation of state.** Rearranging slightly, the equation of state for an ideal gas can be written as

Equation of State for an Ideal Gas

$$PV = NkT \qquad \textbf{17–2}$$

We apply this result to the gas contained in a person's lungs in the next Example.

PROBLEM-SOLVING NOTE

Ideal Gas Law

When using the equation of state for an ideal gas, remember that the temperature must always be given in terms of the Kelvin scale.

EXAMPLE 17–1 Take a Deep Breath

REAL-WORLD PHYSICS: BIO A person's lungs might hold 6.0 L (1 L = 10^{-3} m^3) of air at body temperature (310 K) and atmospheric pressure (101 kPa). Given that the air is 21% oxygen, find the number of oxygen molecules in the lungs.

Picture the Problem

Our sketch shows a person's lungs, with their combined volume of $V = 6.0$ L. In addition, we indicate that the pressure in the lungs is $P = 101$ kPa and the temperature is $T = 310$ K.

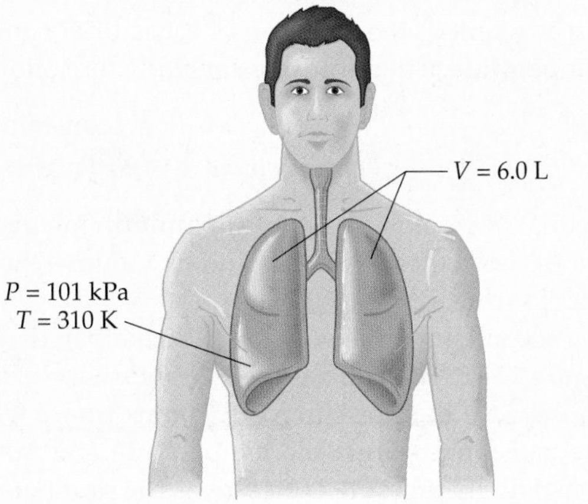

Strategy

We will treat the air in the lungs as an ideal gas. Given the volume, temperature, and pressure of the gas, we can use the equation of state, $PV = NkT$, to solve for the number, N.

Finally, only 21% of the molecules in the air are oxygen. Therefore, we multiply N by 0.21 to find the number of oxygen molecules.

Solution

1. Solve the equation of state for the number of molecules: $N = PV/kT$

2. Substitute numerical values to find the number of molecules in the lungs:
$$N = \frac{PV}{kT} = \frac{(1.01 \times 10^5\ \text{Pa})(6.0 \times 10^{-3}\ \text{m}^3)}{(1.38 \times 10^{-23}\ \text{J/K})(310\ \text{K})} = 1.4 \times 10^{23}$$

3. Multiply N by 0.21 to find the number of molecules that are oxygen, O_2:
$$0.21N = 0.21(1.4 \times 10^{23}) = 2.9 \times 10^{22}$$

Insight

As one might expect, the number of molecules in even an average-size pair of lungs is enormous. In fact, a number this large is difficult to compare to anything we are familiar with. For example, the number of stars in the Milky Way galaxy is estimated to be "only" about 10^{11}. Thus, if each star in the Milky Way were itself a galaxy of 10^{11} stars, then all of these stars combined would just begin to approach the number of oxygen molecules in your lungs at this very moment.

Practice Problem

If the person takes a particularly deep breath, so that the lungs hold a total of 1.5×10^{23} molecules, what is the new volume of the lungs? [**Answer:** $V = 6.4 \times 10^{-3}\ \text{m}^3 = 6.4$ L]

Some related homework problems: Problem 1, Problem 2, Problem 68

Another common way to write the ideal gas equation of state is in terms of the number of **moles** in a gas—as opposed to using the number of molecules, N, as in Equation 17–2. In the SI system of units, the mole (mol) is defined in terms of the most abundant isotope of carbon, which is referred to as carbon-12. The definition is as follows:

> A mole is the amount of a substance that contains as many elementary entities as there are atoms in 12 g of carbon-12.

The phrase "elementary entities" refers to molecules, which may contain more than one atom, as in water (H_2O), or just a single atom, as in carbon (C) and helium (He). Experiments show that the number of atoms in a mole of carbon-12 is 6.022×10^{23}. This number is known as **Avogadro's number,** N_A, named for the Italian physicist and chemist Amedeo Avogadro (1776–1856).

Avogadro's Number, N_A

$$N_A = 6.022 \times 10^{23}\ \text{molecules/mol} \qquad \text{17–3}$$

SI unit: mol^{-1}; the number of molecules is dimensionless

Examples of one mole of various substances are shown in **Figure 17–4**.

Now, if we let n denote the number of moles in a gas, the number of molecules it contains is

$$N = nN_A$$

Substituting this into the ideal-gas equation of state yields

$$PV = nN_AkT$$

The constants N_A and k are combined to form the **universal gas constant,** R, defined as follows:

Universal Gas Constant, R

$$R = N_Ak = (6.022 \times 10^{23}\ \text{molecules/mol})(1.38 \times 10^{-23}\ \text{J/K})$$
$$= 8.31\ \text{J/(mol} \cdot \text{K)} \qquad \text{17–4}$$

SI unit: $\text{J/(mol} \cdot \text{K)}$

Thus, an alternative form of the equation of state is

Equation of State for an Ideal Gas

$$PV = nRT \qquad \text{17–5}$$

This relation is used in the following Active Example.

▲ **FIGURE 17–4 Moles of various substances**

Counting atoms or molecules is obviously hard to do, but the mole concept provides a useful way of dealing with the difficulty. A mole of any substance contains the same number of elementary entities (atoms or molecules).

This number, referred to as Avogadro's number, has the value $N_A = 6.022 \times 10^{23}$. To count out N_A molecules of a gas at standard temperature and pressure, for example, all you need do is measure out a volume of gas equal to 22.4 liters. Similarly, you can effectively count one mole of any substance by measuring out a mass in grams that is equal to the atomic mass of that substance. Thus, the mole provides a convenient bridge between the realm of atoms and molecules and the macroscopic world of observable masses and volumes. This photo shows molar amounts of four different substances: hydrogen, copper, mercury, and sulfur.

ACTIVE EXAMPLE 17–1 The Amount of Air in a Basketball

How many moles of air are in an inflated basketball? Assume that the pressure in the ball is 171 kPa, the temperature is 293 K, and the diameter of the ball is 30.0 cm.

Solution *(Test your understanding by performing the calculations indicated in each step.)*

1. Solve $PV = nRT$ for the number of moles, n:	$n = PV/RT$
2. Calculate the volume of the ball:	$V = 4\pi r^3/3 = 0.0141\ \text{m}^3$
3. Substitute numerical values:	$n = 0.990\ \text{mol}$

Insight

Thus, an inflated basketball contains approximately one mole of air.

In this calculation we used $PV = nRT$ because we were interested in the number of *moles* (n) contained in the gas. In contrast, we used $PV = NkT$ in Example 17–1 because we wanted to know the number of *molecules* (N) in the gas.

Your Turn

Suppose we were to halve both the temperature and the diameter of the ball. By what factor would the number of moles contained within it change?

*(Answers to **Your Turn** problems are given in the back of the book.)*

As mentioned before, a mole of anything has precisely the same number (N_A) of particles. What differs from substance to substance is the mass of one mole. For example, one mole of helium atoms has a mass of 4.00260 g, and one mole of copper atoms has a mass of 63.546 g. In general, we define the **atomic or molecular mass,** M, of a substance to be the mass in grams of one mole of that substance. Thus, the atomic mass for helium is $M = 4.00260\ \text{g/mol}$ and for copper it is $M = 63.546\ \text{g/mol}$. The periodic table in Appendix E gives the atomic masses for all elements.

Note that the atomic mass provides a convenient bridge between the macroscopic world, where we measure the mass of a substance in grams, and the microscopic world, where the number of molecules is typically 10^{23} or greater. As we have seen, if you measure out a mass of copper equal to 63.546 g you have, in effect, counted out $N_A = 6.022 \times 10^{23}$ atoms of copper. It follows that the mass of an individual copper atom, m, is the atomic mass of copper divided by Avogadro's number; that is

$$m = \frac{M}{N_A} \qquad \text{17–6}$$

We use this relation to find the mass of a copper atom and an oxygen molecule in the following Exercise.

EXERCISE 17–1

Find the mass of **(a)** a copper atom and **(b)** a molecule of oxygen, O_2. Atomic masses are listed in Appendix E.

Solution

a. The atomic mass of copper is 63.546 g/mol. Thus, a copper atom has the mass

$$m_{Cu} = \frac{M}{N_A} = \frac{63.546\ \text{g/mol}}{6.022 \times 10^{23}\ \text{atoms/mol}} = 1.055 \times 10^{-22}\ \text{g/atom}$$

b. Since the atomic mass of oxygen is 16.00 g/mol (Appendix E), the molecular mass of O_2 is 32.00 g/mol. Therefore, the mass of O_2 is

$$m_{O_2} = \frac{M}{N_A} = \frac{32.00\ \text{g/mol}}{6.022 \times 10^{23}\ \text{molecules/mol}} = 5.314 \times 10^{-23}\ \text{g/molecule}$$

Finally, we consider a common situation to further illustrate the general features of the ideal-gas equation of state.

CONCEPTUAL CHECKPOINT 17–1 Air Pressure

Feeling a bit cool, you turn up the thermostat in your living room. A short time later the air is warmer. Assuming the room is well sealed, is the pressure of the air **(a)** greater than, **(b)** less than, or **(c)** the same as before you turned up the heat?

Reasoning and Discussion

We assume that the number of molecules and the volume occupied by the molecules is approximately constant. Thus, increasing T, while holding N and V fixed, leads to an increased pressure, P.

Answer:

(a) The air pressure increases.

Isotherms

Historically, the ideal-gas equation of state was arrived at piece by piece, as a result of the combined efforts of a number of researchers. For example, the English scientist Robert Boyle (1627–1691) established the fact that the pressure of a gas varies inversely with volume—as long as temperature and the number of molecules are held constant. This is known as **Boyle's law:**

$$P_i V_i = P_f V_f$$

$$(\textit{fixed number of molecules, N; fixed temperature, T}) \qquad \text{17–7}$$

To see that Boyle's law is consistent with $PV = NkT$, note that if N and T are constant, then so too is PV. Another way of saying that PV is a constant is to say that its initial value must be equal to its final value. This is Boyle's law.

When N and T are constant, the equation of state $PV = NkT$ implies that

$$P = \frac{NkT}{V} = \frac{\text{constant}}{V}$$

This result is plotted in **Figure 17–5**, where we show P as a function of V. Note that the larger the temperature, T, the larger the constant in the numerator. Therefore, the curves farther from the origin correspond to higher temperatures, as indicated.

Each of the curves in Figure 17–5 corresponds to a different, fixed temperature. As a result, these curves are known as **isotherms,** which means, literally, "constant temperature." We shall return to the topic of isotherms in the next chapter, when we consider various thermal processes.

PROBLEM-SOLVING NOTE

Pressure and Volume in an Isotherm

If the temperature of an ideal gas is constant, the pressure and volume change in such a way that their product has a constant value.

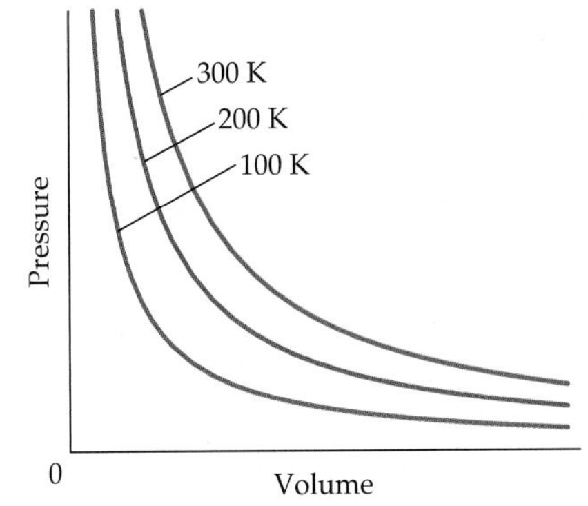

▶ **FIGURE 17–5 Ideal-gas isotherms**

Pressure versus volume isotherms for an ideal gas. Each isotherm is of the form $P = NkT/V = \text{constant}/V$. The three isotherms shown are for the temperatures 100 K, 200 K, and 300 K.

EXAMPLE 17–2 Under Pressure

A cylindrical flask of cross-sectional area A is fitted with an airtight piston that is free to slide up and down. Contained within the flask is an ideal gas. Initially the pressure applied by the piston is 130 kPa and the height of the piston above the base of the flask is 25 cm. When additional mass is added to the piston, the pressure increases to 170 kPa. Assuming the system is always at the temperature 290 K, find the new height of the piston.

Picture the Problem

The physical system is shown in our sketch. The initial pressure in the flask is $P_1 = 130$ kPa, and the initial height of the piston is $h_1 = 25$ cm. When additional mass is placed on the piston, the pressure increases to $P_2 = 170$ kPa and the height decreases to h_2. The temperature remains constant.

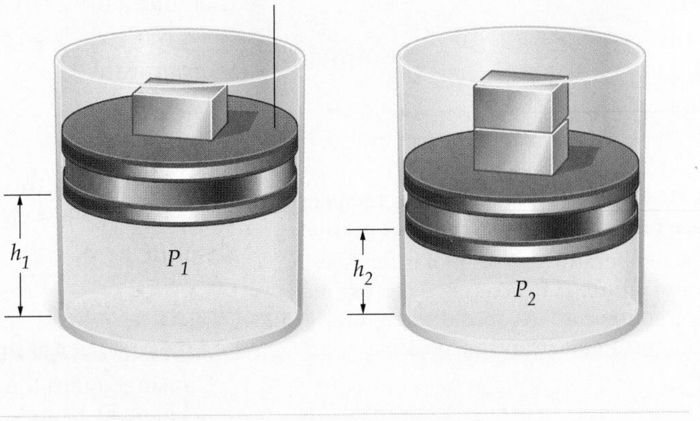

Strategy

Since the temperature is held constant in this system, it follows that $PV = NkT$ is also constant. Thus, as in Boyle's law, $P_1V_1 = P_2V_2$. This relation can be used to find V_2 (since we are given P_1, P_2, and V_1).

Next, the cylindrical volume is related to the height by $V = Ah$. Note that the area A is the same in both cases, hence it will cancel, allowing us to solve for the height, h_2.

Solution

1. Set the initial and final values of PV equal to one another, then solve for V_2:

$$P_1V_1 = P_2V_2$$
$$V_2 = V_1(P_1/P_2)$$

2. Substitute $V_1 = Ah_1$ and $V_2 = Ah_2$, and solve for h_2:

$$Ah_2 = Ah_1(P_1/P_2)$$
$$h_2 = h_1(P_1/P_2)$$

3. Substitute numerical values:

$$h_2 = h_1\frac{P_1}{P_2} = (25\text{ cm})\frac{130\text{ kPa}}{170\text{ kPa}} = 19\text{ cm}$$

Insight

In order to keep the temperature of this system constant, it is necessary to allow some heat to flow out of the cylindrical flask as the gas is compressed by the additional weight. Thus, the flask is not thermally insulated during this experiment.

Practice Problem

What pressure would be required to change the height of the piston to 29 cm? [**Answer:** $P = 110$ kPa]

Some related homework problems: Problem 13, Problem 14

▲ Charles's law states that the volume of a gas at constant pressure is proportional to its Kelvin temperature. For example, this balloon was fully inflated at room temperature (293 K) and atmospheric pressure. When cooled by vapors from liquid nitrogen (77 K), however, the volume of the air inside the balloon decreased markedly, causing it to shrivel.

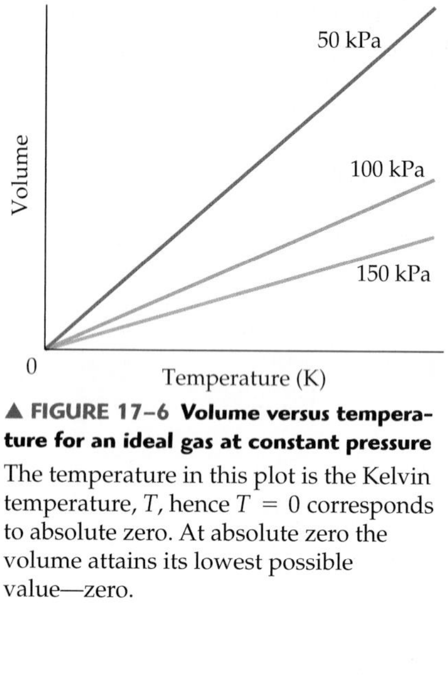

▲ **FIGURE 17–6 Volume versus temperature for an ideal gas at constant pressure**
The temperature in this plot is the Kelvin temperature, T, hence $T = 0$ corresponds to absolute zero. At absolute zero the volume attains its lowest possible value—zero.

Finally, recall that in Conceptual Checkpoint 15–2 we considered the behavior of an air bubble rising from a swimmer in shallow water. At the time, we said that the diameter of the bubble increases as it rises because the pressure of the surrounding water is decreasing. This is certainly the case, but now we can be more precise. If we assume that the water temperature is constant, and that no gas molecules are added to or removed from the bubble, then the volume of the bubble has the following dependence:

$$V = \frac{NkT}{P} = \frac{\text{constant}}{P}$$

This is just our isotherm, again, and we see that the volume indeed increases as pressure decreases.

Constant Pressure

Another aspect of ideal-gas behavior was discovered by the French scientist Jacques Charles (1746–1823) and later studied in greater detail by fellow Frenchman Joseph Gay-Lussac (1778–1850). Known today as **Charles's law,** their result is that the volume of a gas divided by its temperature is constant, as long as the pressure and number of molecules is constant:

$$\frac{V_i}{T_i} = \frac{V_f}{T_f}$$

$(fixed\ number\ of\ molecules,\ N;\ fixed\ pressure,\ P)$ **17–8**

As with Boyle's law, this result follows immediately from the ideal-gas equation of state. Solving Equation 17–2 for V/T, we have

$$\frac{V}{T} = \frac{Nk}{P}$$

If N and P are constant, then so is the quantity V/T.

Charles's law can be rewritten as a linear relation between volume and temperature:

$$V = (\text{constant})T$$

The constant in this expression is Nk/P. This result is illustrated in **Figure 17–6**, where we see the volume of an ideal gas vanishing as the temperature approaches absolute zero.

ACTIVE EXAMPLE 17–2 Find the Piston Height

Consider again the system described in Example 17–2. In this case the temperature is changed from an initial value of 290 K to a final value of 330 K. The pressure exerted on the gas remains constant at 130 kPa, and the initial height of the piston is 25 cm. Find the final height of the piston.

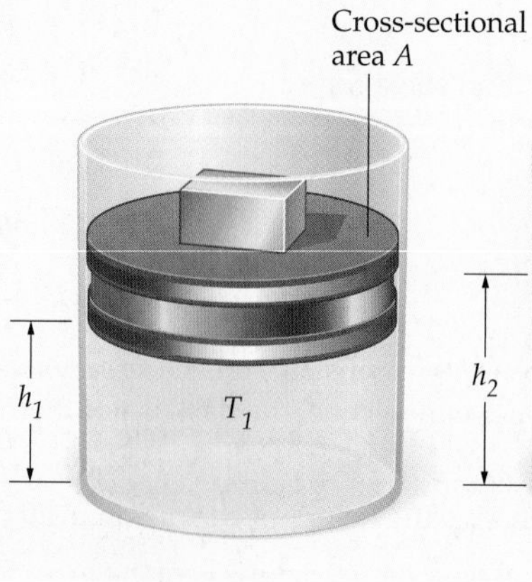

Solution *(Test your understanding by performing the calculations indicated in each step.)*

1. Set the initial value of V/T equal to the final value of V/T: $V_1/T_1 = V_2/T_2$
2. Solve for V_2: $V_2 = V_1(T_2/T_1)$
3. Use $V = Ah$ to solve for the height, h_2: $h_2 = h_1(T_2/T_1)$
4. Substitute numerical values: $h_2 = 28$ cm

Insight

The volume of the gas, and hence the height of the piston, increased in direct proportion to the increase in temperature. This is to be expected when the pressure is constant.

Your Turn

At what temperature is the piston height equal to 19 cm?

(*Answers to* **Your Turn** *problems are given in the back of the book.*)

PROBLEM-SOLVING NOTE

Volume and Temperature in an Ideal Gas

If the pressure of an ideal gas is constant, the ratio of the volume and temperature remains constant. This is true no matter what the value of the pressure.

17–2 Kinetic Theory

We can readily measure the pressure and temperature of a gas using a pressure gauge and a thermometer. These are *macroscopic* quantities that apply to the gas as a whole. It is not so easy, however, to measure *microscopic* quantities, such as the position or velocity of an individual molecule. Still, there must be some connection between what happens on the microscopic level and what we observe on the macroscopic level. This connection is described by the **kinetic theory of gases.**

In the kinetic theory, we imagine a gas to be made up of a collection of molecules moving about inside a container of volume V. In particular, we assume the following:

- The container holds a very large number N of identical molecules, each of mass m. The molecules themselves can be thought of as "point particles"—that is, the volume of the molecules is negligible compared to the volume of the container, and the size of the molecules is negligible compared to the distance between them. (This is why dilute real gases are the best approximation to the ideal case.)
- The molecules move about the container in a random manner. They obey Newton's laws of motion at all times.
- When molecules hit the walls of the container or collide with one another, they bounce elastically. Other than these collisions, the molecules have no interactions.

With these basic assumptions we can relate the pressure of a gas to the behavior of the molecules themselves.

The Origin of Pressure

As we shall see, the pressure exerted by a gas is due to the innumerable collisions between gas molecules and the walls of their container. Each collision results in a change of momentum for a given molecule, just like throwing a ball at a wall and having it bounce back. The total change in momentum of the molecules in a given time, divided by the time, is simply the force a wall must exert on the gas to contain it (see Section 9–2). The average of this force over time, and over the area of a wall, is the pressure of the gas.

To be specific, imagine a container that is a cube of length L on a side. Its volume, then, is $V = L^3$. In addition, consider a given molecule that happens to be moving in the negative x direction toward a wall, as in **Figure 17–7**. If its speed is v_x, its initial momentum is $p_{i,x} = -mv_x$. After bouncing from the wall, it moves in the positive

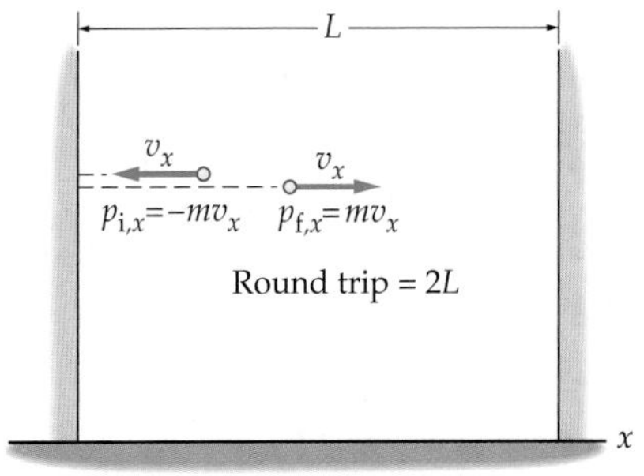

▲ **FIGURE 17–7 Force exerted by a molecule on the wall of a container**

A molecule bounces off a wall of a container, changing its momentum from $-mv_x$ to $+mv_x$; the change in momentum is $2mv_x$. A round trip will be completed in the time $\Delta t = 2L/v_x$, so the average force exerted on the molecule by the wall is $F = \Delta p/\Delta t = 2mv_x/(2L/v_x) = mv_x^2/L$.

x direction with the same speed (since the collision is elastic), and hence its final momentum is $p_{f,x} = +mv_x$. As a result, the molecule's change in momentum is

$$\Delta p_x = p_{f,x} - p_{i,x} = mv_x - (-mv_x) = 2mv_x$$

The wall exerts a force on the molecule to cause this momentum change.

After the bounce, the molecule travels to the other side of the container and back before bouncing off the same wall again. The time required for this round trip of length $2L$ is

$$\Delta t = 2L/v_x$$

Thus, by Newton's second law, the average force exerted by the wall on the molecule is

$$F = \frac{\Delta p_x}{\Delta t} = \frac{2mv_x}{2L/v_x} = \frac{mv_x^{\,2}}{L}$$

The average pressure exerted by this wall, then, is simply the force divided by the area. Since the area of the wall is $A = L^2$, we have

$$P = \frac{F}{A} = \frac{(mv_x^{\,2}/L)}{L^2} = \frac{mv_x^{\,2}}{L^3} = \frac{mv_x^{\,2}}{V} \qquad \text{17–9}$$

Note that we have used the fact that the volume of the container is L^3.

In this calculation we assumed that the molecule moves in the x direction. This was merely to simplify the derivation. If, instead, the molecule moves at some angle to the x-axis, the calculation applies to its x component of motion. The final conclusions are unchanged.

Speed Distribution of Molecules

In deriving Equation 17–9, we considered a single molecule with a particular speed. Other molecules, of course, will have different speeds. In addition, the speed of any given molecule changes with time as it collides with other molecules in the gas. What remains constant, however, is the overall **distribution of speeds.**

This is illustrated in **Figure 17–8**, where we present the results obtained by the Scottish physicist James Clerk Maxwell (1831–1879). This plot shows the relative probability that an O_2 molecule will have a given speed. For example, on the curve labeled 300 K, the most probable speed is about 390 m/s. When the temperature is increased to 1100 K, the most probable speed increases to roughly 750 m/s. Other speeds occur as well, from speeds near zero to those that are very large, but these have much lower probabilities.

Thus, in Equation 17–9, the term $v_x^{\,2}$ should be replaced with the average of $v_x^{\,2}$ over all the molecules in the gas. Writing this average as $(v_x^{\,2})_{av}$, we have

$$P = \frac{m(v_x^{\,2})_{av}}{V}$$

Since all N molecules in the gas follow the same distribution, the pressure exerted by the gas as a whole is N times this result:

$$P = N\frac{m(v_x^{\,2})_{av}}{V} = \left(\frac{N}{V}\right)m(v_x^{\,2})_{av} \qquad \text{17–10}$$

Of course, there is nothing special about the x direction—Equation 17–10 applies equally well with $(v_y^{\,2})_{av}$ or $(v_z^{\,2})_{av}$ in place of $(v_x^{\,2})_{av}$. Thus, it would be preferable to express the pressure of the gas in terms of the overall speed of the molecules, rather than in terms of a single component. The speed squared of a molecule is

$$v^2 = v_x^{\,2} + v_y^{\,2} + v_z^{\,2}$$

Hence, the average of v^2 is

$$(v^2)_{av} = (v_x^{\,2})_{av} + (v_y^{\,2})_{av} + (v_z^{\,2})_{av}$$

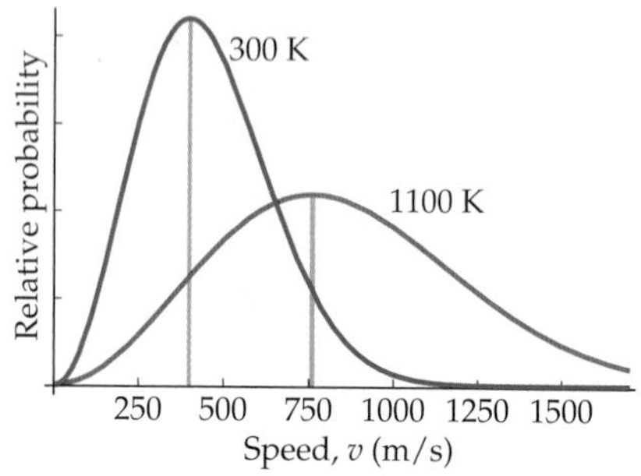

▲ **FIGURE 17–8 The Maxwell speed distribution**

The Maxwell distribution of molecular speeds for O_2 at the temperatures T = 300 K, and T = 1100 K. Note that the most probable speed increases with increasing temperature.

Since the x, y, and z directions are equivalent, it follows that

$$({v_x}^2)_{av} = ({v_y}^2)_{av} = ({v_z}^2)_{av}$$

As a result,

$$(v^2)_{av} = ({v_x}^2)_{av} + ({v_y}^2)_{av} + ({v_z}^2)_{av} = 3({v_x}^2)_{av}$$

or, equivalently

$$({v_x}^2)_{av} = \tfrac{1}{3}(v^2)_{av}$$

Using this replacement in Equation 17–10 yields

$$P = \tfrac{1}{3}\left(\frac{N}{V}\right)m(v^2)_{av}$$

The last part of this expression, $m(v^2)_{av}$, is simply twice the average kinetic energy of a molecule. Thus, using K for the kinetic energy, as in Chapters 7 and 8, we can write the pressure as follows:

Pressure in the Kinetic Theory of Gases

$$P = \tfrac{1}{3}\left(\frac{N}{V}\right)2K_{av} = \tfrac{2}{3}\left(\frac{N}{V}\right)\left(\tfrac{1}{2}mv^2\right)_{av} \qquad \text{17–11}$$

To summarize, using the kinetic theory, we have shown that the pressure of a gas is proportional to the number of molecules and inversely proportional to the volume. We discussed both of these dependences before when considering the ideal gas. In addition, we see that

> The pressure of a gas is directly proportional to the average kinetic energy of its molecules.

This is the key connection between microscopic behavior and macroscopic observables.

Kinetic Energy and Temperature

If we compare the ideal-gas equation of state, $PV = NkT$, with the result from kinetic theory, Equation 17–11, we find

$$PV = NkT = \tfrac{2}{3}N\left(\tfrac{1}{2}mv^2\right)_{av}$$

As a result,

$$\tfrac{2}{3}\left(\tfrac{1}{2}mv^2\right)_{av} = kT$$

Equivalently,

Kinetic Energy and Temperature

$$\left(\tfrac{1}{2}mv^2\right)_{av} = K_{av} = \tfrac{3}{2}kT \qquad \text{17–12}$$

This is one of the most important results of kinetic theory. It says that the average kinetic energy of a gas molecule is directly proportional to the Kelvin temperature, T. Thus, when we heat a gas, what happens on the microscopic level is that the molecules move with speeds that are, on average, greater. Similarly, cooling a gas causes the molecules to slow.

EXERCISE 17–2

Find the average kinetic energy of oxygen molecules in the air. Assume the air is at a temperature of 21 °C.

Solution

Using Equation 17–12, and the fact that 21 °C = 294 K, we find

$$K_{av} = \tfrac{3}{2}kT = \tfrac{3}{2}(1.38 \times 10^{-23}\ \text{J/K})(294\ \text{K}) = 6.09 \times 10^{-21}\ \text{J}$$

Note that this is also the average kinetic energy of nitrogen molecules in the air. In fact, the type of molecule is not important; all that matters is the temperature of the gas.

Returning to Equation 17–12, we find with a slight rearrangement that

$$(v^2)_{av} = 3kT/m$$

Now, the square root of $(v^2)_{av}$ is given a special name—it is called the **root mean square** (rms) speed. For gas molecules, then, the rms speed, v_{rms}, is the following:

RMS Speed of a Gas Molecule

$$v_{rms} = \sqrt{(v^2)_{av}} = \sqrt{\frac{3kT}{m}} \qquad \text{17–13}$$

SI unit: m/s

Rewriting this in terms of the molecular mass, M, we have

$$v_{rms} = \sqrt{\frac{3kT}{m}} = \sqrt{\frac{3kT}{(M/N_A)}} = \sqrt{\frac{3N_AkT}{M}}$$

Finally, using $N_Ak = R$ yields

$$v_{rms} = \sqrt{\frac{3RT}{M}} \qquad \text{17–14}$$

The rms speed is one of the characteristic speeds of the Maxwell speed distribution. As shown in **Figure 17–9**, v_{rms} is slightly greater than the most probable speed, v_{mp}, and the average speed, v_{av}. (Homework problem 73 illustrates the difference between v_{av} and v_{rms} for a simple system of molecules.)

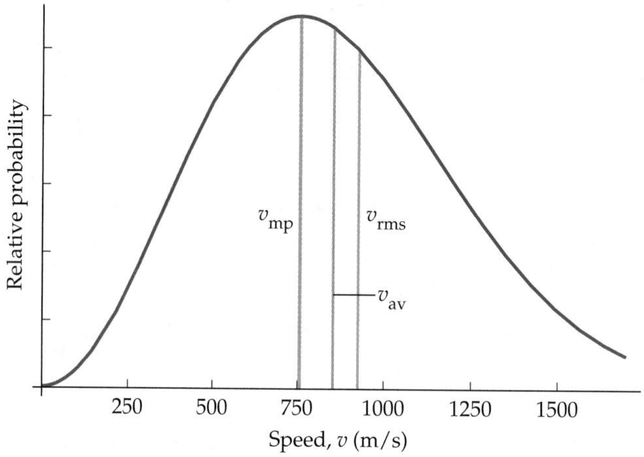

▶ **FIGURE 17–9 Most probable, average, and rms speeds**
Characteristic speeds for O_2 at the temperature $T = 1100$ K. From left to right, the indicated speeds are the most probable speed, v_{mp}, the average speed, v_{av}, and the rms speed, v_{rms}, respectively.

EXAMPLE 17–3 Fresh Air

The atmosphere is composed primarily of nitrogen N_2 (78%) and oxygen O_2 (21%). **(a)** Is the rms speed of N_2 (28.0 g/mol) greater than, less than, or the same as the rms speed of O_2 (32.0 g/mol)? **(b)** Find the rms speed of N_2 and O_2 at 293 K.

Picture the Problem

Our sketch shows molecules in the air bouncing off the walls of a room—they also bounce off a person in the room. Note that the molecules move in random directions with a variety of speeds. Most of the molecules are N_2, but about 21% are O_2.

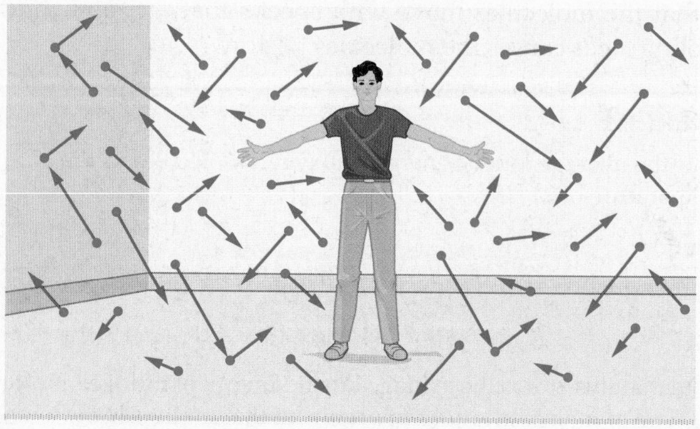

Strategy

The rms speeds are calculated by straightforward substitution into the relation $v_{rms} = \sqrt{3RT/M}$. The only point to be careful about is the molecular mass of each molecule. In particular, note that for nitrogen the molecular mass is 28.0 g/mol = 0.0280 kg/mol, and similarly for oxygen, the molecular mass is 32.0 g/mol = 0.0320 kg/mol.

Solution

Part (a)

Since both molecules are at the same temperature, they have the same kinetic energy. The nitrogen molecule has less mass, however; thus, if it is to have the same kinetic energy, it must have a higher speed. This is also what one would expect on the basis of Equation 17–14.

Part (b)

1. To find the rms speed of N_2, substitute $M = 0.0280\ \text{kg/mol}$ in Equation 17–14:

$$v_{rms} = \sqrt{\frac{3[8.31\ \text{J/(mol}\cdot\text{K)}](293\ \text{K})}{0.0280\ \text{kg/mol}}} = 511\ \text{m/s}$$

2. To find the rms speed of O_2, substitute $M = 0.0320\ \text{kg/mol}$ in Equation 17–14:

$$v_{rms} = \sqrt{\frac{3[8.31\ \text{J/(mol}\cdot\text{K)}](293\ \text{K})}{0.0320\ \text{kg/mol}}} = 478\ \text{m/s}$$

Insight

As expected, N_2 has the higher rms speed. For comparison, the speed of sound at 293 K is 343 m/s, which is just over 750 mi/h. Thus, the molecules in the air are bouncing off your skin with the speed of a supersonic jet.

Such high speeds can help to promote chemical reactions, which usually depend on collisions between molecules to take place. In general, higher temperatures mean higher molecular speeds and increased rates of chemical reactions.

Practice Problem

One of the substances found in air is monatomic, and has an rms speed of 428 m/s, at 293 K. What is this substance? [**Answer:** $M = 0.0399\ \text{kg/mol}$; thus the substance is argon, Ar, which comprises about 0.94% of the atmosphere.]

Some related homework problems: Problem 18, Problem 19

Chemical reactions usually depend on collisions between molecules to take place. Though the molecules in the air are not moving fast enough to trigger chemical reactions, high molecular speeds can help to promote chemical reactions in many circumstances. In general, higher temperatures mean higher molecular speeds and increased rates of chemical reactions. This is why the rate at which a cricket chirps increases with increasing temperature, making it a novel type of thermometer. (See Problem 76 in Chapter 13 and Problem 57 in Chapter 16.)

REAL-WORLD PHYSICS: BIO

Chemical reactions and temperature

CONCEPTUAL CHECKPOINT 17–2 **Compare Speeds**

Each of two containers of equal volume holds an ideal gas. Container A has twice as many molecules as container B. If the gas pressure is the same in the two containers, is the rms speed of the molecules in container A **(a)** greater than, **(b)** less than, or **(c)** the same as the rms speed of the molecules in container B?

Reasoning and Discussion

Since P and V are the same, you might think the rms speeds are the same as well. Recall, however, that *the pressure of a gas is caused by the collisions of gas molecules with the walls of the container.* If more molecules occupy a given volume, and they bounce off the walls with the same speed as in a container with fewer molecules, the pressure will be greater. Therefore, in order for the pressure in the two containers to be the same, it is necessary that the rms speed of the molecules in container A be less than the rms speed in container B.

Mathematically, we note that $P_A = P_B$ and $V_A = V_B$; hence $P_AV_A = P_BV_B$. Using the ideal-gas relation, $PV = NkT$, we can write this condition as $N_AkT_A = N_BkT_B$. Thus, if $N_A = 2N_B$, it follows that $T_A = T_B/2$. Since the temperature is lower in container A, its molecules have the smaller rms speed.

Answer:

(b) The rms speed is less in container A.

The Internal Energy of an Ideal Gas

The internal energy of a substance is the sum of all its potential and kinetic energies. In an ideal gas there are no interactions between molecules, other than perfectly elastic collisions; hence, there is no potential energy. As a result, the total energy of the system is the sum of the kinetic energy of each of its molecules. Thus, for an ideal gas of N pointlike molecules—that is, a monatomic gas—the internal energy is simply

Internal Energy of a Monatomic Ideal Gas

$$U = \tfrac{3}{2}NkT \qquad 17\text{–}15$$

SI unit: J

In terms of moles, this result is

$$U = \tfrac{3}{2}nRT \qquad 17\text{–}16$$

We shall return to this result in the next chapter.

EXERCISE 17–3

A basketball at 290 K holds 0.95 mol of air molecules. What is the internal energy of the air in the ball?

Solution

Applying Equation 17–16 we find

$$U = \tfrac{3}{2}nRT = \tfrac{3}{2}(0.95 \text{ mol})[8.31 \text{ J}/(\text{mol}\cdot\text{K})](290 \text{ K}) = 3400 \text{ J}$$

This is roughly the kinetic energy a basketball would have if you dropped it from a height of 700 m.

If an ideal gas is diatomic, there are additional contributions to the internal energy. For example, a diatomic molecule, which is shaped somewhat like a dumbbell, can have rotational kinetic energy. Diatomic molecules can also vibrate along the line joining the two atoms, which is yet another contribution to the total energy. Thus, the results in Equations 17–15 and 17–16 apply only to the simplest case, the ideal monatomic gas.

17–3 Solids and Elastic Deformation

The defining characteristic of a solid object is that it has a particular shape. For example, when Michelangelo carved the statue of David from a solid block of marble, he was confident it would retain its shape long after his work was done. Liquids and gases do not behave in this way. In contrast, they assume the shape of the container into which they are placed. On a molecular level, these differences arise from the fact that the intermolecular forces in a solid are strong enough to practically immobilize its molecules, whereas the intermolecular forces in liquids and gases are so weak the molecules can move about relatively freely.

Even so, the shape of a solid *can* be changed—though usually only slightly—if it is acted on by a force. In this section we consider various types of deformations that can occur in solids and the way these deformations are related to the forces that cause them.

Changing the Length of a Solid

A useful physical model for a solid is a lattice of small balls representing molecules connected to one another by springs representing the intermolecular forces. Pulling on a solid rod with a force F, for example, causes each "intermolecular spring" in the direction of the force to expand by an amount proportional to F. The net result is that the entire solid increases its length by an amount proportional to the force, $\Delta L \propto F$, as indicated in **Figure 17–10**.

The stretch ΔL is also proportional to the initial length of the rod, L_0. To see why, we first note that each intermolecular spring expands by the same amount, giving a total stretch that is proportional to the number of such springs. But the

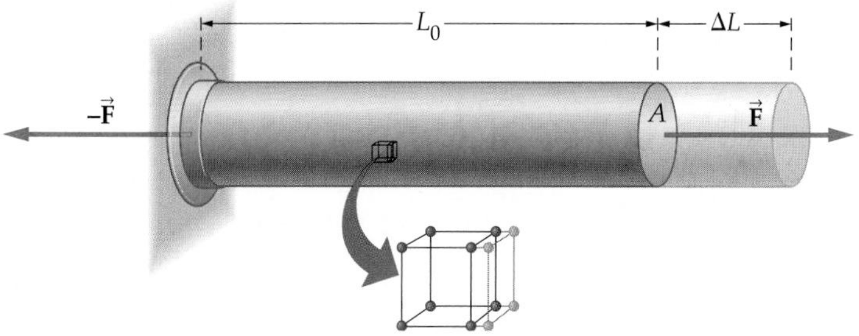

◀ **FIGURE 17–10 Stretching a rod**
Equal and opposite forces applied to the ends of a rod cause it to stretch. On the atomic level, the forces stretch the "intermolecular springs" in the solid, resulting in an overall increase in length. The stretch is proportional to the force F, the initial length L_0, and inversely proportional to the cross-sectional area A.

number of intermolecular springs is proportional to the total initial length of the rod. Thus, it follows that $\Delta L \propto FL_0$.

Finally, the amount of stretch for a given force F is inversely proportional to the cross-sectional area A of the rod. For example, a rod with a cross-sectional area $2A$ is like two rods of cross-sectional area A placed side by side. Thus, applying a force F to a rod of area $2A$ is equivalent to applying a force $F/2$ to two rods of area A. The result is half the stretch when the area is doubled; that is, $\Delta L \propto FL_0/A$. Solving this relation for the amount of force F required to produce a given stretch ΔL we find $F \propto (\Delta L/L_0)A$, or as an equality

$$F = Y\left(\frac{\Delta L}{L_0}\right)A \qquad \text{17–17}$$

The proportionality constant Y in this expression is **Young's modulus**, named for the English physicist Thomas Young (1773–1829). Comparing the two sides of Equation 17–17, we see that Young's modulus has the units of force per area (N/m^2).

Typical values for Young's modulus are given in Table 17–1. Notice that the values vary from material to material, but are all rather large. This means that a large force is required to cause even a small stretch in a solid. Of course, Equation 17–17 applies equally well to a compression or a stretch. However, some materials have a slightly different Young's modulus for compression and stretching. For example, human bones under tension (stretching) have a Young's modulus of $1.6 \times 10^{10}\ N/m^2$, while bones under compression have a slightly smaller Young's modulus of $9.4 \times 10^9\ N/m^2$.

TABLE 17–1
Young's Modulus for Various Materials

Material	Young's modulus, $Y(N/m^2)$
Tungsten	36×10^{10}
Steel	20×10^{10}
Copper	11×10^{10}
Brass	9.0×10^{10}
Aluminum	6.9×10^{10}
Pyrex glass	6.2×10^{10}
Lead	1.6×10^{10}
Bone	
Tension	1.6×10^{10}
Compression	0.94×10^{10}
Nylon	0.37×10^{10}

PROBLEM-SOLVING NOTE

Area and Young's Modulus

When using Young's modulus, remember that the area A is the area that is at right angles to the applied force.

EXAMPLE 17–4 Stretching a Bone

REAL-WORLD PHYSICS: BIO

At the local airport, a person carries a 21-kg duffel bag in one hand. Assuming the humerus (the upper arm bone) supports the entire weight of the bag, determine the amount by which it stretches. (The humerus may be assumed to be 33 cm in length and to have an effective cross-sectional area of $5.2 \times 10^{-4}\ m^2$.)

Picture the Problem

The humerus is oriented vertically, hence the weight of the duffel bag applies tension to the bone. The Young's modulus in this case is $1.6 \times 10^{10}\ N/m^2$. In addition, we note that the initial length of the bone is $L_0 = 0.33$ m and its cross-sectional area is $A = 5.2 \times 10^{-4}\ m^2$.

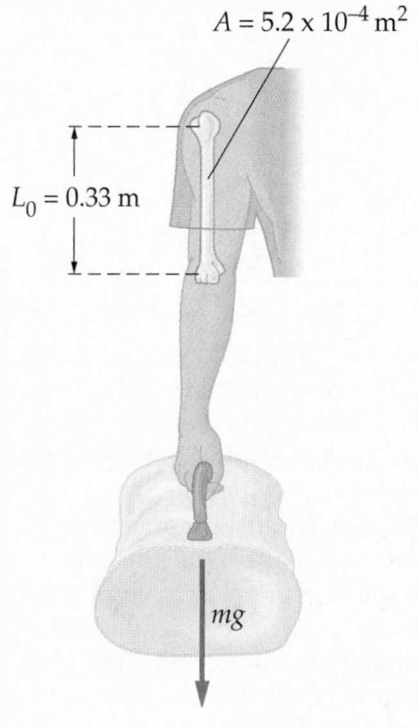

Strategy

We can find the amount of stretch by solving the relation $F = Y(\Delta L/L_0)A$ for the quantity ΔL. Note that the force applied to the bone is simply the weight of the suitcase, $F = mg$, with $m = 21$ kg. (We ignore the relatively small weight of the forearm and hand.)

continued on next page

continued from previous page

Solution

1. Solve Equation 17–17 for the amount of stretch, ΔL:

$$\Delta L = \frac{FL_0}{YA}$$

2. Calculate the force applied to the humerus:

$$F = mg = (21\text{ kg})(9.81\text{ m/s}^2) = 210\text{ N}$$

3. Substitute numerical values into the expression for ΔL:

$$\Delta L = \frac{FL_0}{YA} = \frac{(210\text{ N})(0.33\text{ m})}{(1.6 \times 10^{10}\text{ N/m}^2)(5.2 \times 10^{-4}\text{ m}^2)}$$
$$= 8.3 \times 10^{-6}\text{ m}$$

Insight

We find that the amount of stretch is imperceptibly small. The reason for this, of course, is that Young's modulus is such a large number. If the bone had been compressed rather than stretched, its change in length, though still minuscule, would have been greater by a factor of 16/9.4.

Practice Problem

Suppose the humerus is reduced uniformly in size by a factor of two. This means that both the length and diameter are halved. What is the amount of stretch in this case? [**Answer:** Since $L_0 \rightarrow L_0/2$ and $A \rightarrow A/4$, the stretch is doubled. Thus, $\Delta L \rightarrow 2\,\Delta L = 1.7 \times 10^{-5}$ m.]

Some related homework problems: Problem 27, Problem 28

There is a straightforward connection between Equation 17–17 and Hooke's law for a spring (Equation 6–4). To see the connection, we rewrite Equation 17–17 as follows:

$$F = \left(\frac{YA}{L_0}\right)\Delta L$$

Notice that the force required to cause a certain stretch is proportional to the stretch—just as in Hooke's law. In fact, if we identify ΔL with the displacement x of a spring from equilibrium and YA/L_0 with the force constant k, we have Hooke's law:

$$F = \left(\frac{YA}{L_0}\right)\Delta L = kx$$

Thus, we see that the force constant of a spring depends on the Young's modulus, Y, of the material from which it is made, the cross-sectional area A of the wire, and the length of the wire, L_0.

▶ The cables in this suspension bridge have significant forces pulling on them from either end. As a result, they are under tension, just like the rod in Figure 17–10. It follows that the length of each cable increases by an amount that is proportional to its initial length. The dam, on the other hand, experiences forces that tend to compress it. Still, the magnitude of its change in length can be described in terms of Equation 17–17, with the appropriate value of Young's modulus.

CONCEPTUAL CHECKPOINT 17–3 Compare Force Constants

Two identical springs are connected end to end. Is the force constant of the resulting compound spring (a) greater than, (b) less than, or (c) equal to that of a single spring?

Reasoning and Discussion

It might seem that the force constant would be the same, since we simply have twice the length of the same spring. On the other hand, it might seem that the force constant is greater, since two springs are exerting a force rather than just one. In fact, the force constant decreases.

The reason for the reduced force constant is that if we apply a force F to the compound spring, each individual spring stretches by a certain amount. The compound spring, then, stretches by twice this amount. By Hooke's law, $F = kx$, a spring that stretches twice as far for the same applied force has half the force constant.

We can also obtain this result by recalling that the force constant is $k = YA/L_0$. Thus, if the length of a spring is doubled—as in the compound spring—the force constant is halved.

Answer:

(b) The force constant of the compound spring is less, by a factor of 2.

Changing the Shape of a Solid

Another type of deformation, referred to as a **shear deformation,** changes the *shape* of a solid. Consider a book of thickness L_0 resting on a table, as shown in **Figure 17–11**. A force F is applied to the right on the top cover of the book, and static friction applies a force F to the left on the bottom cover of the book. The result is that the book remains at rest but becomes slanted by the amount Δx. The force required to cause a given amount of slant is proportional to Δx, inversely proportional to the thickness of the book L_0, and proportional to the surface area A of the book's cover; that is, $F \propto A\Delta x/L_0$. Writing this as an equality, we have

$$F = S\left(\frac{\Delta x}{L_0}\right)A \quad \text{17–18}$$

The constant of proportionality in this case is the **shear modulus,** S. Like Young's modulus, the shear modulus has the units N/m^2. Typical values of the shear modulus are collected in Table 17–2. As with Young's modulus, the shear modulus is large in magnitude, meaning that most solids require a large force to cause even a small amount of shear.

Equations 17–17 and 17–18 are similar in structure, but it is important to be aware of their differences as well. For example, the term L_0 in the Young's modulus equation refers to the length of a solid measured in the direction of the applied force. In contrast, L_0 in the shear modulus equation refers to the thickness of the solid as measured in a direction perpendicular to the applied force. Similarly, the area A in Equation 17–17 is the cross-sectional area of the solid, and this area is perpendicular to the applied force. On the other hand, the area A in Equation 17–18 is the area of the solid in the plane of the applied force.

TABLE 17–2
Shear Modulus for Various Materials

Material	Shear modulus, $S(N/m^2)$
Tungsten	15×10^{10}
Steel	8.1×10^{10}
Bone	8.0×10^{10}
Copper	4.2×10^{10}
Brass	3.5×10^{10}
Aluminum	2.4×10^{10}
Lead	0.54×10^{10}

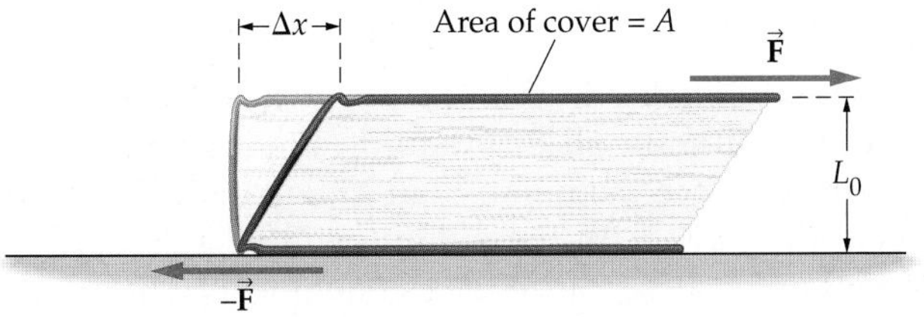

▲ **FIGURE 17–11 Shear deformation**

Equal and opposite forces applied to the top and bottom of a book result in a shear deformation. The amount of deformation is proportional to the force F, the thickness of the book L_0, and inversely proportional to the area A.

PROBLEM-SOLVING NOTE

Area and the Shear Modulus

When using the shear modulus, remember that the area A is the area that is parallel to the applied force.

ACTIVE EXAMPLE 17–3 Deforming a Stack of Pancakes: Find the Shear Modulus

A horizontal force of 1.2 N is applied to the top of a stack of pancakes 13 cm in diameter and 9.0 cm high. The result is a shear deformation of 2.5 cm. What is the shear modulus for these pancakes?

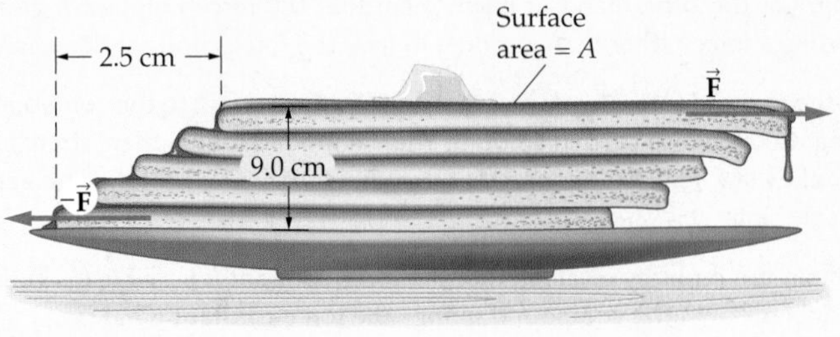

Solution *(Test your understanding by performing the calculations indicated in each step.)*

1. Solve Equation 17–18 for the shear modulus: $S = FL_0/A\,\Delta x$
2. Calculate the area of the pancakes: $A = \pi d^2/4 = 0.013\ \text{m}^2$
3. Substitute numerical values: $S = 330\ \text{N/m}^2$

Insight

Notice the small value of the pancakes' shear modulus, especially when compared to the shear modulus of a typical metal. This is a reflection of the fact that the pancake stack is easily deformed.

Your Turn

Suppose the stack of pancakes is doubled in height. By what factor does the shear deformation change?

(Answers to **Your Turn** *problems are given in the back of the book.)*

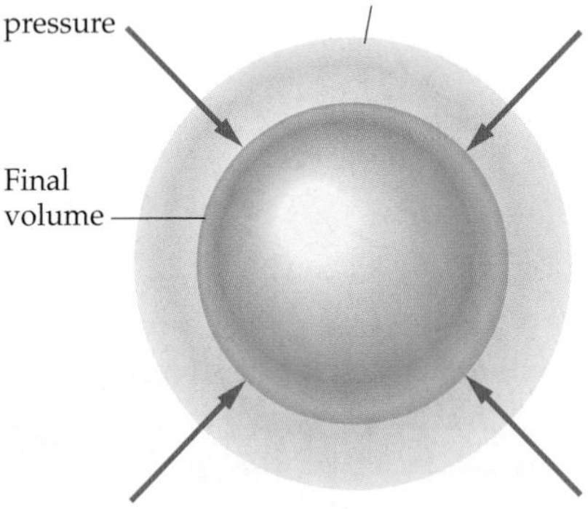

▲ **FIGURE 17–12 Changing the volume of a solid**

As the pressure surrounding an object increases, its volume decreases. The amount of volume change is proportional to the initial volume and to the change in pressure.

TABLE 17–3
Bulk Modulus for Various Materials

Material	Bulk modulus, $B(\text{N/m}^2)$
Gold	22×10^{10}
Tungsten	20×10^{10}
Steel	16×10^{10}
Copper	14×10^{10}
Aluminum	7.0×10^{10}
Brass	6.1×10^{10}
Ice	0.80×10^{10}
Water	0.22×10^{10}
Oil	0.17×10^{10}

Changing the Volume of a Solid

If a piece of Styrofoam is taken deep into the ocean, the tremendous pressure of the water causes it to shrink to a fraction of its original volume. This is an extreme example of the volume change that occurs in all solids when the pressure of their surroundings is changed. The general situation is illustrated in **Figure 17–12**, where we show a spherical solid whose volume decreases by the amount ΔV when the pressure acting on it increases by the amount ΔP. Experiments show that the pressure difference required to cause a given change in volume, ΔV, is proportional to ΔV and inversely proportional to the initial volume of the object, V_0. Therefore, we can write ΔP as follows:

$$\Delta P = -B\left(\frac{\Delta V}{V_0}\right) \qquad \text{17–19}$$

The constant of proportionality in this case, B, is called the **bulk modulus.** As with Young's modulus and the shear modulus, the bulk modulus is defined to be a positive quantity; hence the minus sign in Equation 17–19. For example, if the pressure increases ($\Delta P > 0$), the volume will decrease ($\Delta V < 0$) and the quantity $-\Delta V$ will be positive. Since ΔP is equal to $B(-\Delta V/V_0)$, and V_0 is always positive, it follows that B must be positive as well.

Table 17–3 gives a list of representative values of the bulk modulus. Note the large magnitudes of B, indicating that even small volume changes require large changes in pressure. Finally, the units of B are N/m^2, as is clear from Equation 17–19.

ACTIVE EXAMPLE 17–4 A Gold Doubloon: Find the Change in Volume

A gold doubloon 6.1 cm in diameter and 2.0 mm thick is dropped over the side of a pirate ship. When it comes to rest on the ocean floor at a depth of 770 m, how much has its volume changed?

Solution *(Test your understanding by performing the calculations indicated in each step.)*

1. Solve Equation 17–19 for the change in volume: $\Delta V = -V_0\,\Delta P/B$
2. Calculate the initial volume of the doubloon (a cylinder): $V_0 = 5.8 \times 10^{-6}\ \text{m}^3$
3. Find the change in pressure due to the depth of seawater. Use Equation 15–7: $\Delta P = \rho g h = 7.7 \times 10^{6}\ \text{N/m}^2$
4. Substitute numerical values: $\Delta V = -2.0 \times 10^{-10}\ \text{m}^3$

Insight

The change in volume is imperceptibly small. Clearly, enormous pressures must be applied to metals to cause a significant change in volume.

Your Turn

Suppose the "gold" doubloon is actually made of brass. Do you expect the change in volume in this case to be greater than or less than for gold? Verify your answer with a calculation.

(Answers to **Your Turn** *problems are given in the back of the book.)*

▲ Styrofoam has a very small bulk modulus, which means that even a relatively small increase in pressure can cause a large decrease in volume. The Styrofoam cup at left was immersed to a depth of 1955 m, where the water pressure is nearly 200 atm.

Stress and Strain

Equation 17–19 appears to differ from Equations 17–17 and 17–18 because of the absence of the area A on the right-hand side of the equation. When one recalls that pressure is a force per area, however, we can see that the area in Equation 17–19 is contained in the denominator of the left-hand side. Similarly, we can rewrite Equation 17–17 as $F/A = Y(\Delta L/L_0)$ and Equation 17–18 as $F/A = S(\Delta x/L_0)$. Written in this way, each of these equations states that a deformation of a particular type is proportional to a corresponding applied force per area.

In general, we refer to an applied force per area as a **stress** and the resulting deformation as a **strain.** If the stress applied to an object is not too large, the proportional relationship between the strain and stress is found to hold, and a plot of strain versus stress gives a straight line, as indicated in **Figure 17–13**. This straight-line relationship is simply a generalization of Hooke's law—extending the result for a spring to any solid object. As the stress becomes larger, the strain eventually begins to increase at a rate that is greater than the straight line.

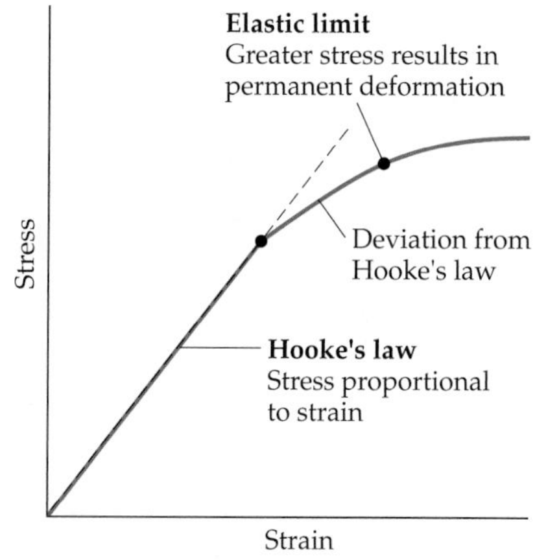

▲ **FIGURE 17–13 Stress versus strain**
When stress and strain are small they are proportional, as given by Hooke's law. Larger stresses can result in deviations from Hooke's law. Stresses greater than the elastic limit cause permanent deformation of the material.

A change in behavior occurs when the stress reaches the elastic limit. For stresses less than the elastic limit, the deformation of an object is reversible; that is, the deformation vanishes as the stress that caused it is reduced to zero. This is just like a spring returning to its equilibrium position when the applied force vanishes. When a deformation is reversible, we say that it is an **elastic deformation.** For stresses greater than the elastic limit, on the other hand, the object becomes permanently deformed, and it will not return to its original size and shape when the stress is removed. This is like a spring that has been stretched too far, or a car fender that has been dented. If the stress on an object is increased even further, the object eventually tears apart or fractures.

17–4 Phase Equilibrium and Evaporation

To this point, we have studied the behavior of substances when they are in a single phase of matter. For example, we considered the properties of liquids in Chapter 15, and in this chapter we have turned our attention to the characteristics of gases and solids. It is common, however, for a substance to exist in more than one phase at a

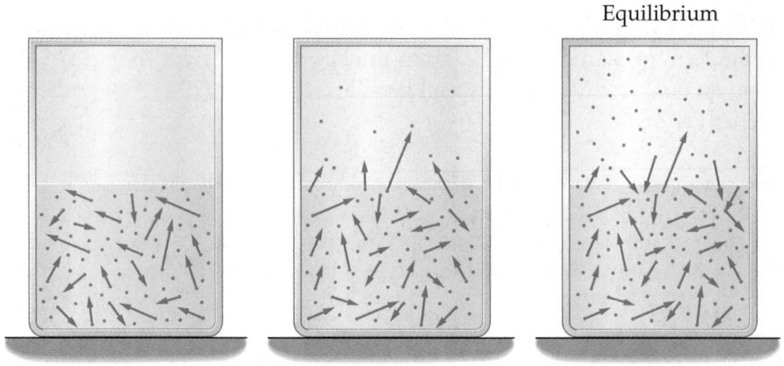

▶ FIGURE 17–14 A liquid in equilibrium with its vapor

Initially, a liquid is placed in a container, and the volume above it is a vacuum. High-speed molecules in the liquid are able to escape into the upper region of the container, forming a low-density gas. As the gas becomes more dense, the number of molecules leaving the liquid is balanced by the number returning to the liquid. At this point the system is in equilibrium.

time. In particular, if a substance has two or more phases that coexist in a steady, stable fashion, we say that the substance exhibits **phase equilibrium.**

To see just what this means in a specific case, consider a closed container that is partially filled with a liquid, as in **Figure 17–14**. The container is kept at the constant temperature T_0, and initially the volume above the liquid is empty of molecules—it is a vacuum. Soon, however, some of the faster molecules in the liquid begin to escape the relatively strong intermolecular forces of their neighbors in the liquid and start to form a low-density gas. Occasionally a gas molecule collides with the liquid and reenters it, but initially, more molecules are entering the gas than returning to the liquid.

This process continues until the gas is dense enough that the number of molecules returning to the liquid equals the number entering the gas. There is a constant "flow" of molecules in both directions, but when *phase equilibrium* is reached, these flows cancel, and the number of molecules in each phase remains constant. The pressure of the gas when equilibrium is established is referred to as the **equilibrium vapor pressure.** In **Figure 17–15** we plot the equilibrium vapor pressure of water.

What happens when we change the temperature? Well, if we increase the temperature, there will be more high-speed molecules in the liquid that can escape into the gas. Thus, to have an equal number of gas molecules returning to the liquid, it will be necessary for the pressure of the vapor to be greater. Thus, the equilibrium vapor pressure increases with temperature. This is also illustrated in Figure 17–15.

For each temperature there is just one equilibrium vapor pressure—that is, just one pressure where the precise balance between the phases is established. Thus, when we plot the equilibrium vapor pressure versus temperature, as in Figure 17–15, the result is a curve—the **vapor pressure curve.** The significance of this curve is that it determines the boiling point of a liquid. In particular:

> A liquid boils at the temperature at which its vapor pressure equals the external pressure.

Note in Figure 17–15 that a vapor pressure equal to atmospheric pressure, $P_{at} = 101$ kPa, occurs when the temperature of the water is 100 °C, as expected.

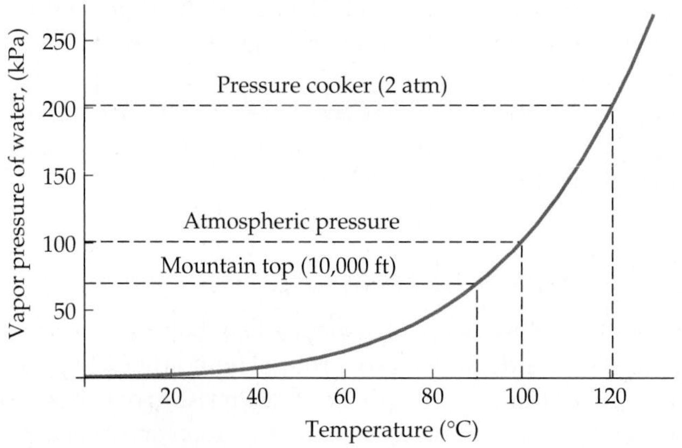

▶ FIGURE 17–15 The vapor-pressure curve for water

The vapor pressure of water increases with increasing temperature. In particular, at $T = 100$ °C the vapor pressure is one atmosphere, 101 kPa.

CONCEPTUAL CHECKPOINT 17–4 Boiling Temperature

When water boils at the top of a mountain, is its temperature **(a)** more than, **(b)** less than, or **(c)** equal to 100 °C?

Reasoning and Discussion

At the top of a mountain, air pressure is less than it is at sea level. Therefore, according to Figure 17–15, the boiling temperature will be less as well.

For example, the atmospheric pressure at the top of a 10,000-ft mountain is roughly three-quarters what it is at sea level. The corresponding boiling point, as shown in Figure 17–15, is about 90 °C. Thus, cooking food in boiling water may be very difficult at such altitudes. Similarly, by measuring the boiling temperature of a pot of water on the summit of an uncharted mountain, it is possible to gain a rough estimate of its altitude.

Answer:

(b) The temperature of the water is less than 100 °C.

REAL-WORLD PHYSICS: BIO
The autoclave

An *autoclave* (which is basically an elaborate pressure cooker) sterilizes surgical tools by using the same principle discussed in Conceptual Checkpoint 17–4, only in the opposite direction. If surgical tools were heated in boiling water open to the atmosphere, they would experience a temperature of 100 °C. In the autoclave, which is a sealed vessel, the pressure rises to values significantly greater than atmospheric pressure. As a result, the water has a much higher boiling temperature, and the sterilization is more effective. In the pressure cooker, this elevated temperature results in a reduced cooking time.

REAL-WORLD PHYSICS
The pressure cooker

Another way to increase the boiling temperature of water is to add a pinch of salt. The salt dissolves into sodium and chloride ions in the water. These ions have a strong interaction with the water molecules, making it harder for them to break free of the liquid and enter the vapor phase. As a result, the boiling temperature rises. This is why salt is often added to water when boiling eggs; the result is a higher temperature of boiling, which helps to solidify any material that might leak out of small cracks in the eggs.

REAL-WORLD PHYSICS
Adding salt to boiling water

The vapor-pressure curve separates areas of liquid and gas on a graph of pressure versus temperature, as shown in **Figure 17–16**. Notice that this curve comes to an end at a finite temperature and pressure. This end point is called the **critical point.** Beyond the critical point there is no longer a distinction between liquid and gas. They are one and the same, and are referred to, simply, as a fluid. Thus, as mentioned before, the liquid and gas phases are very similar—the liquid is just more dense than the gas and its molecules are somewhat less mobile.

A curve similar to the vapor-pressure curve indicates where the solid and liquid phases are in equilibrium. It is referred to as the **fusion curve.** Similarly, equilibrium between the solid and gas phases occurs along the **sublimation curve.** All

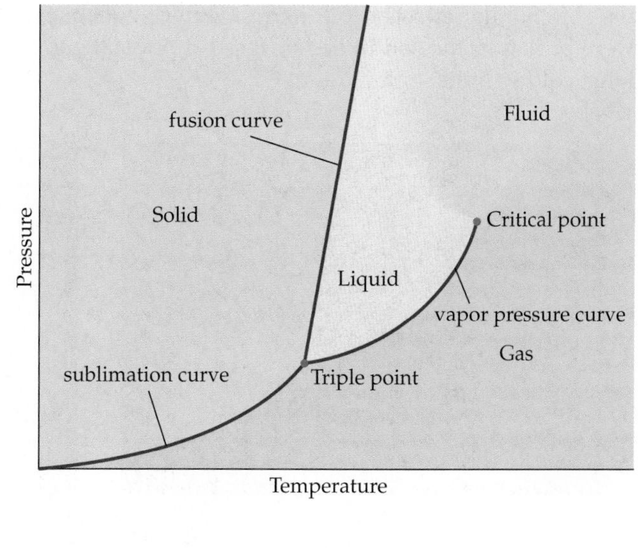

◀ FIGURE 17–16 A typical phase diagram
This phase diagram shows the regions corresponding to each of the three common phases of matter. Note that all three phases are in equilibrium at the triple point. In addition, the liquid and gas phases become indistinguishable beyond the critical point, where they are referred to as a fluid.

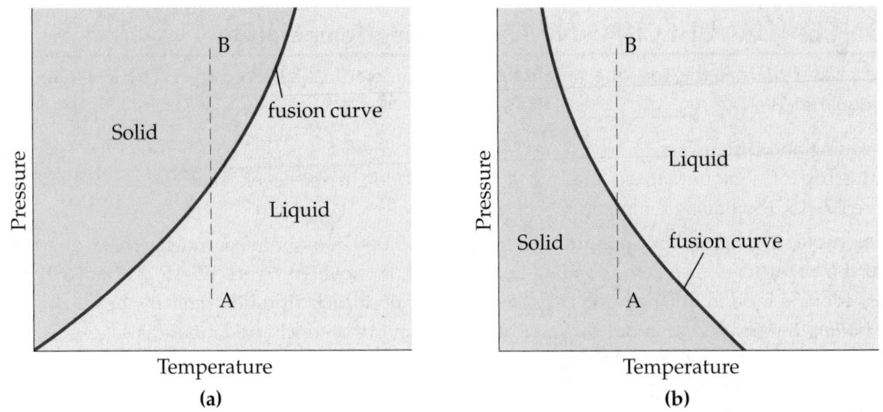

▶ **FIGURE 17–17 The solid–liquid phase boundary**

Fusion curves for a typical substance **(a)** and for water **(b)**. In (a), we note that an increase in pressure at constant temperature results in a liquid being converted to a solid. For the case of water, increasing the pressure exerted on ice at constant temperature can result in the ice melting.

three equilibrium curves are shown in Figure 17–16, on a plot that is referred to as a **phase diagram.**

Note that the phase diagram also indicates that there is one particular temperature and pressure where all three phases are in equilibrium. This point is called the **triple point.** In water, the triple point occurs at the temperature $T = 273.16\ \text{K}$ and the pressure $P = 611.2\ \text{Pa}$. At this temperature and pressure, ice, water, and steam are all in equilibrium with one another.

Finally, there is one feature of the fusion line that is of particular interest. In most substances, this line has a positive slope, as in Figure 17–16. This means that as the pressure is increased, the melting temperature of the substance also increases. This is sensible, because a solid is generally more dense than the corresponding liquid. Hence, if you apply pressure to a liquid—with the temperature held constant—the system will tend to become more dense and eventually solidify. This is indicated in **Figure 17–17 (a)**, where we see that an increase in pressure at constant temperature (as from A to B) results in crossing the fusion curve into the solid region.

REAL-WORLD PHYSICS
Ice melting under pressure

There are exceptions to this rule, however, and it will probably come as no surprise that water is one such exception. This is related to the fact that ice is less dense than water. As a result, the fusion curve for water has a negative slope coming out of the triple point. This is illustrated in **Figure 17–17 (b)**. What this means is that if the temperature is held constant, ice will melt when the pressure applied to it is increased. Thus, the pressure due to an ice skate's blade, for example, can cause the ice beneath it to melt.

REAL-WORLD PHYSICS
Frost wedging and frost heaving

The expansion of water as it freezes has important implications in geology as well. Suppose, for example, that water finds its way into cracks and fissures on a rocky cliff. If the temperature dips below freezing, the water in these cracks will begin to freeze and expand. This tends to split the rock farther apart in a process known as *frost wedging*. Over time, the repeated freezing and melting of water can break a "solid" rock cliff into a pile of debris, forming a talus slope at the base of the cliff. Similar effects occur in *frost heaving*, where expanding ice lifts rocks that were buried in the soil to the surface—a phenomenon well known in areas with long, cold winters.

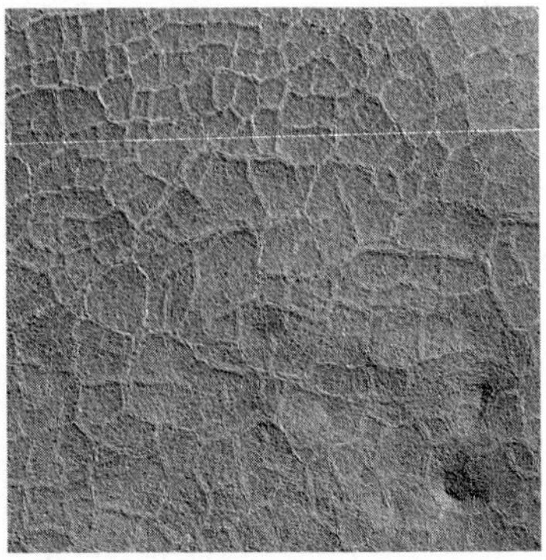

▶ The repeated expansion and contraction of water as it freezes and thaws over long periods of time can cause the ground to crack and buckle. Eventually, the resulting cracks may combine to form polygonal networks referred to as patterned ground (left), a common feature in arctic regions of the Earth. The same type of patterned ground has recently been observed on the surface of Mars (right), giving a strong indication that water may exist just below the surface in certain parts of the planet.

Finally, the expansion of ice can have devastating effects in living systems. If the blood cells of an organism were to freeze, for example, the resulting expansion could rupture the cells. The icefish, a transparent fish found in southern polar seas, has a high concentration of glycoproteins in its bloodstream that serve as a biological antifreeze, inhibiting the formation and growth of ice crystals.

REAL-WORLD PHYSICS: BIO
Biological antifreeze

Evaporation

If you think about it a moment, you realize that evaporation is a bit odd. After all, when you are hot and sweaty from physical exertion, steam rises from your skin. But how can this be, since water boils at 100 °C? How can a skin temperature of only 30 or 35 °C result in steam?

Recall that if a liquid is placed in a closed container with a vacuum above it, some of its fastest molecules will break loose and form a gas. When the gas becomes dense enough, its pressure rises to the equilibrium vapor pressure of the liquid, and the system attains equilibrium. But what if you open the container and let a breeze blow across it, removing much of the gas? In that case, the release of molecules from the liquid into the gas—that is, **evaporation**—continues without reaching equilibrium. As the molecules are continually removed, the liquid progressively evaporates until none is left. This is the basic mechanism of evaporation.

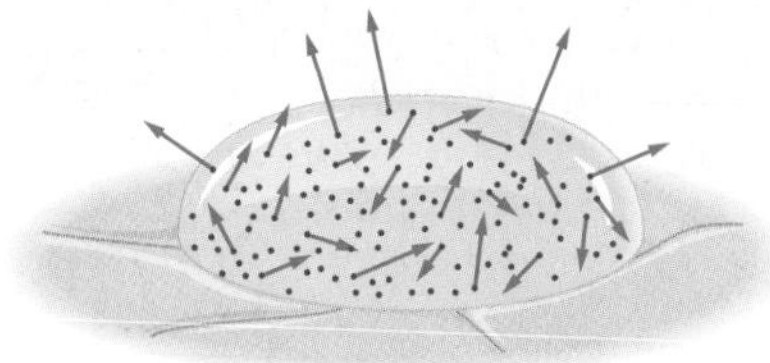

▲ **FIGURE 17–18 A droplet of sweat resting on the skin**
High-speed molecules in a droplet of sweat are able to escape the droplet and become part of the atmosphere. The average speed of the molecules that remain in the droplet is reduced, so the temperature of the droplet is reduced as well.

Let's investigate how evaporation helps to cool us when we exercise or work up a sweat. First, consider a droplet of sweat on the skin, as illustrated in **Figure 17–18**. As mentioned before, the high-speed molecules in the drop are the ones that will escape from the liquid into the surrounding air. The breeze takes these molecules away as soon as they escape; hence, the chance of their reentering the drop of sweat is very small.

REAL-WORLD PHYSICS: BIO
Cooling the body with evaporation

Now, what does this mean for the molecules that are left behind in the sweat droplet? Well, since the droplet is preferentially losing high-speed molecules, the average kinetic energy of the remaining molecules must decrease. As we know from kinetic theory, this means that the temperature of the droplet must also decrease. Since the droplet is now cooler than its surroundings, including the skin on which it rests, it draws in heat from the body. This warms the droplet, increasing the speed of its molecules, and continuing the evaporation process at more or less the same rate. Thus, sweat droplets are an effective means of drawing heat from the body, and releasing it into the surrounding air in the form of high-speed water molecules.

▲ Although the body temperature of these bison is well below the boiling point of water, water still evaporates rapidly from their skin. (Since water vapor is invisible, the "steam" we see in such photos, though it indicates the presence of evaporation, is not the water vapor itself. Rather, it is a cloud of tiny droplets that form when the water vapor loses heat to the cold air around it and condenses back to the liquid state.)

To look at this a bit more quantitatively, consider the Maxwell speed distribution for water molecules in a sweat droplet at 30 °C. This is shown in **Figure 17–19**. The rms speed at this temperature is 648 m/s. Also shown in Figure 17–19 is the speed distribution for water molecules at 100 °C. In this case the rms speed is 719 m/s, only slightly greater than that at 30 °C. Thus, it is clear that if water molecules at 100 °C have enough speed to escape into the gas phase, many molecules at 30 °C will be able to escape as well.

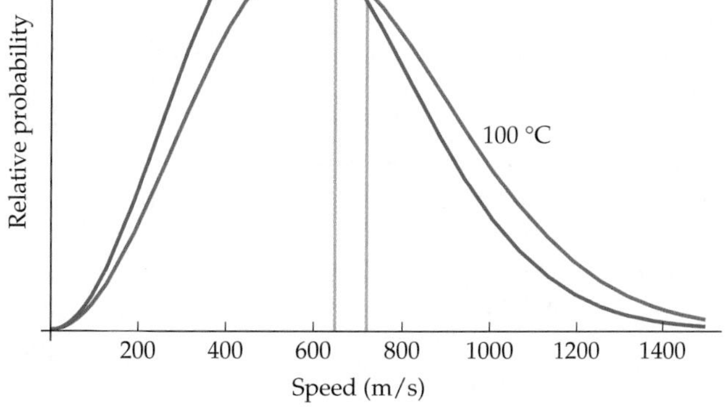

◀ **FIGURE 17–19 Speed distribution for water**
The blue curve shows the speed distribution for water at 30 °C, and the red curve corresponds to 100 °C. Note that the rms speed for 100 °C is only slightly greater than that for 30 °C.

▲ Human beings keep cool by sweating. Because the most energetic molecules are the ones most likely to escape by evaporation, significant quantities of heat are removed from the body when perspiration evaporates from the skin. Dogs, lacking sweat glands, nevertheless take advantage of the same mechanism to help regulate body temperature. In hot weather they pant to promote evaporation from the tongue. Of course, sitting in a cool pond can also help. (What mechanism is involved here?)

The Evaporating Atmosphere

REAL-WORLD PHYSICS
Stability of planetary atmospheres

The atmosphere of a planet or a moon can evaporate in much the same way a drop of sweat evaporates from your forehead. In the case of an atmosphere, however, it is the force of gravity that must be overcome. If an astronomical body has a weak gravitational field, some molecules in its atmosphere may be moving rapidly enough to escape; that is, some molecules may have speeds in excess of the escape speed for that body.

Consider the Earth, for example. If a rocket is to escape the Earth, it must have a speed of at least 11,200 m/s. Recall from Chapter 12, however, that the escape speed is independent of the mass of rocket. Thus, it applies equally well to molecules and rockets. As a result, molecules moving faster than 11,200 m/s may escape the Earth.

Now, let's compare the escape speed to the speeds of some of the molecules in the Earth's atmosphere. We have already seen in this chapter that the speed of nitrogen and oxygen are on the order of the speed of sound; that is, several hundred meters per second. This is much below the Earth's escape speed. Thus, the odds against a nitrogen or oxygen molecule having enough speed to escape the Earth are truly astronomical. Good thing, too, since this is why these molecules have persisted in our atmosphere for billions of years.

On the other hand, consider a lightweight molecule like hydrogen, H_2. The fact that its average kinetic energy is the same as that of all the other molecules in the air (see Equation 17–12) means that its speed will be much greater than the speed of, say, oxygen or nitrogen. This is illustrated in **Figure 17–20**, where we show the speed distribution for both O_2 and H_2. Clearly, it is very likely to find an H_2 molecule with a speed on the order of a couple thousand meters per second. Because of the higher speeds for H_2, the probability that an H_2 molecule will have enough speed to escape the Earth is about 300 orders of magnitude greater than the corresponding probability for O_2. It is no surprise, then, that Earth's atmosphere contains essentially no hydrogen.

On Jupiter, however, gravity is more intense than on Earth and the temperature is less. As a result, not even hydrogen moves quickly enough to escape. In fact, Jupiter's atmosphere is composed mostly of H_2 and He.

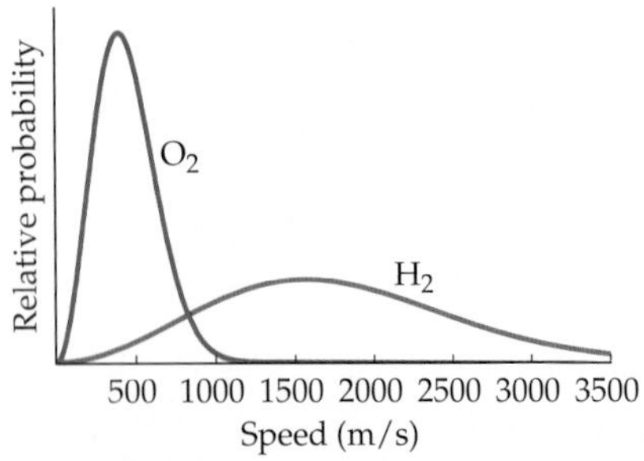

▲ FIGURE 17–20 Speed distribution for O_2 and H_2 at 20 °C

The typical speeds of an H_2 molecule (red curve) are much greater than those for an O_2 molecule (blue curve). In fact, some H_2 molecules move fast enough to escape from the Earth's atmosphere.

At the other extreme, the Moon has a rather weak gravitational field. In fact, it is unable to maintain any atmosphere at all. Whatever atmosphere it may have had early in its history has long since evaporated. You might say that the Moon's atmosphere is "lost in space."

17–5 Latent Heats

When two phases coexist something surprising happens—the temperature remains the same even when you add a small amount of heat. How can that be?

To understand this behavior, let's start by considering an ice cube initially at the temperature −10 °C. Adding an amount of heat Q results in a temperature increase ΔT given by the relation $Q = mc_{\text{ice}}\,\Delta T$, as discussed in Chapter 16. When the ice cube's temperature reaches 0 °C, however, adding more heat does not cause an additional increase in temperature. Instead, the heat goes into converting some of the ice into water. On a microscopic level, the added heat causes some of the molecules in the solid ice to vibrate with enough energy to break loose from neighboring molecules and become part of the liquid.

Thus, as long as any ice remains in a cup of water, *and* the water and ice are in equilibrium, you can be sure that both the ice and the water are at 0 °C. If heat is added to the system, the amount of ice decreases; if heat is removed, the amount of ice increases. The amount of heat required to completely convert 1 kg of ice to water is referred to as the **latent heat,** L. In general, the latent heat is defined as follows:

> The latent heat, L, is the heat that must be added to or removed from one kilogram of a substance to convert it from one phase to another.
>
> During the conversion process, the temperature of the system remains constant.

Just as with the specific heat, the latent heat is always a positive quantity.

In mathematical form, we can say that the heat Q required to convert a mass m from one phase to another is $Q = mL$. This gives the following relation:

Definition of Latent Heat, *L*

$$L = Q/m \qquad \text{17–20}$$

SI unit: J/kg

The value of the latent heat depends on which phases are involved. For example, the latent heat to melt (or fuse) a substance is referred to as the **latent heat of fusion,** L_f. Similarly, the latent heat required to convert a liquid to a gas is the **latent heat of vaporization,** L_v, and the latent heat needed to convert a solid directly to a gas is the **latent heat of sublimation,** L_s. Typical latent heats are given in Table 17–4.

The relationship between the temperature of a substance and the heat added to it is illustrated in **Figure 17–21**. Initially 1 kg of H_2O is in the form of ice at −20 °C.

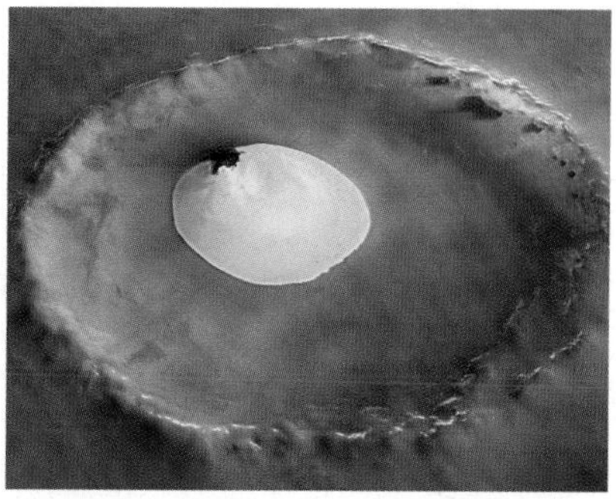

▲ This recently discovered ice lake on the surface of Mars lies on the floor of an impact crater. It grows or shrinks with the martian seasons. Because atmospheric pressure is so low on Mars, however, the ice does not melt during the martian summer—instead, it sublimates directly to the vapor phase. On Mars, water ice would have behavior similar to that of dry ice here on Earth.

TABLE 17–4 Latent Heats for Various Materials

Material	Latent heat of fusion, L_f(J/kg)	Latent heat of vaporization, L_v(J/kg)
Water	33.5×10^4	22.6×10^5
Ammonia	33.2×10^4	13.7×10^5
Copper	20.7×10^4	47.3×10^5
Benzene	12.6×10^4	3.94×10^5
Ethyl alcohol	10.8×10^4	8.55×10^5
Gold	6.28×10^4	17.2×10^5
Nitrogen	2.57×10^4	2.00×10^5
Lead	2.32×10^4	8.59×10^5
Oxygen	1.39×10^4	2.13×10^5

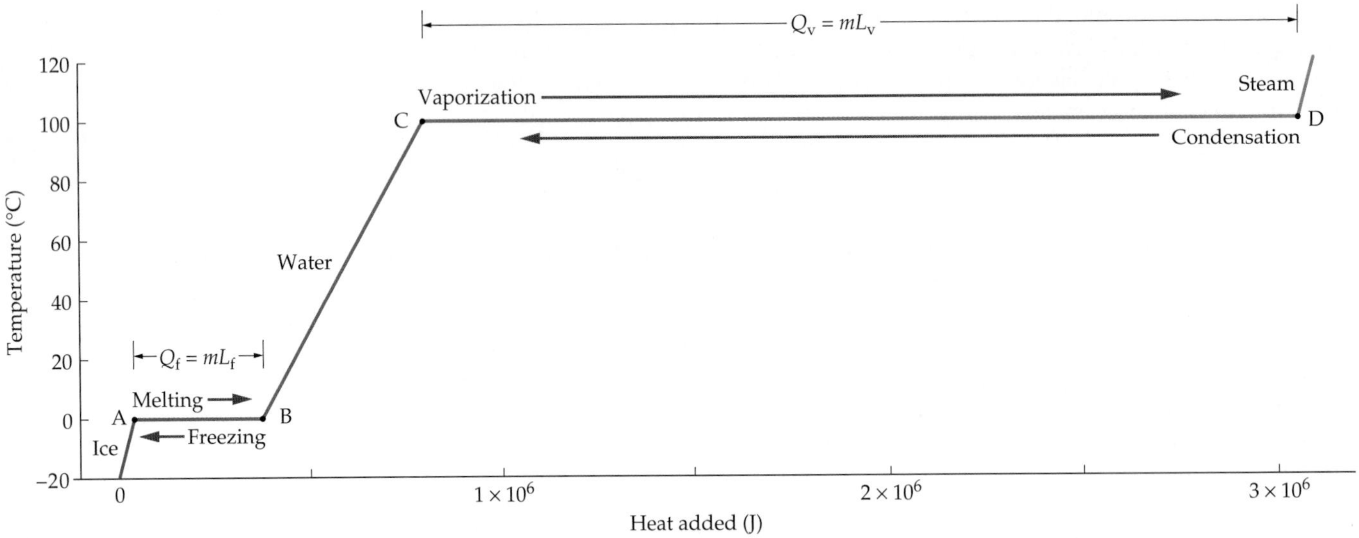

▲ FIGURE 17–21 Temperature versus heat added or removed
The temperature of $m = 1.000$ kg of water as heat is added to or removed from the system. Note that the temperature stays the same—even as heat is added—when the system is changing from one phase to another. The points A, B, C, and D are referred to in Problems 52 and 53.

As heat is added to the ice, its temperature rises until it begins to melt at 0 °C. The temperature then remains constant until the latent heat of fusion is supplied to the system. When all the ice has melted to water at 0 °C, continued heating results in a renewed increase in temperature. When the temperature of the water rises to 100 °C, boiling begins and the temperature again remains constant—this time until an amount of heat equal to the latent heat of vaporization is added to the system. Finally, with the entire system converted to steam, continued heating again produces an increasing temperature.

CONCEPTUAL CHECKPOINT 17–5 Severity of a Burn

Both water at 100 °C and steam at 100 °C can cause serious burns. Is a burn produced by steam likely to be **(a)** more severe, **(b)** less severe, or **(c)** the same as a burn produced by water?

Reasoning and Discussion

As the water or steam comes into contact with the skin, it cools from 100 °C to a skin temperature of something like 35 °C. For the case of water, this means that a certain amount of heat is transferred to the skin, which can cause a burn. The steam, on the other hand, must first give off the heat required for it to condense to water at 100 °C. After that, the condensed water cools to body temperature, as before. Thus, the heat transferred to the skin will be larger in the case of steam, resulting in a more serious burn.

Answer:

(a) The steam burn is worse.

As a numerical example of using latent heat, let's calculate the heat energy required to raise the temperature of 0.550 kg of ice from −20.0 °C to water at 20.0 °C; that is, from point A to point D in **Figure 17–22**. The way to approach a problem like this is to take it one step at a time—that is, each phase, and each conversion from one phase to another, should be treated separately.

Thus, the first step (A to B in Figure 17–22) is to find the heat necessary to warm the ice from −20.0 °C to 0 °C. Using the specific heat of ice, $c_{ice} = 2090\ \mathrm{J/(kg \cdot C°)}$, we find

$$Q_1 = mc_{ice}\,\Delta T = (0.550\ \mathrm{kg})[2090\ \mathrm{J/(kg \cdot C°)}](20.0\ \mathrm{C°}) = 23{,}000\ \mathrm{J}$$

The second step in the process (B to C) is to melt the ice at 0 °C. The heat required for this is found using the latent heat of fusion for water ($L_f = 33.5 \times 10^4\ \mathrm{J/kg}$):

$$Q_2 = mL_f = (0.550\ \mathrm{kg})(33.5 \times 10^4\ \mathrm{J/kg}) = 184{,}000\ \mathrm{J}$$

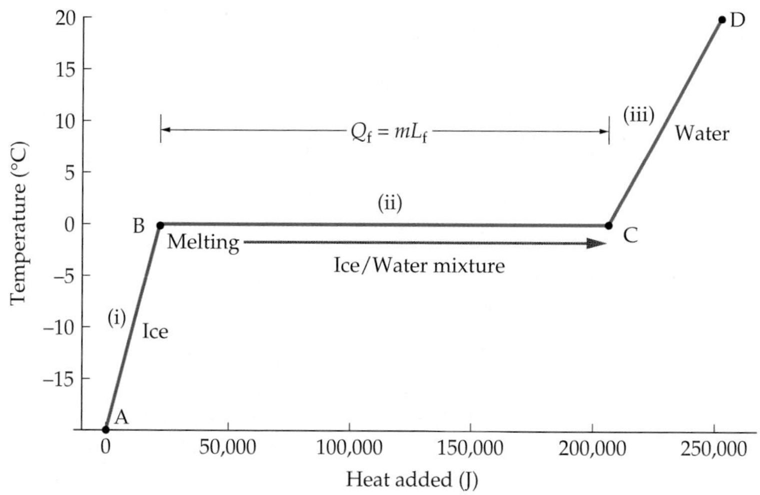

◀ **FIGURE 17–22 Heat required for a given change in temperature**
The amount of heat required to raise 0.550 kg of H_2O from ice at −20.0 °C to water at 20.0 °C is the heat difference between points A and D. To calculate this heat we sum the following three heats: (i) the heat to warm the ice from −20.0 °C to 0 °C; (ii) the heat to melt all the ice; (iii) the heat to warm the water from 0 °C to 20.0 °C.

Finally, the third step (C to D) is to heat the water at 0 °C to 20.0 °C. This time we use the specific heat for water, $c_{water} = 4186\ \text{J}/(\text{kg}\cdot\text{C}°)$:

$$Q_3 = mc_{water}\,\Delta T = (0.550\ \text{kg})[4186\ \text{J}/(\text{kg}\cdot\text{C}°)](20.0\ \text{C}°) = 46{,}000\ \text{J}$$

The total heat required for this process, then, is

$$Q_{total} = Q_1 + Q_2 + Q_3$$
$$= 23{,}000\ \text{J} + 184{,}000\ \text{J} + 46{,}000\ \text{J} = 253{,}000\ \text{J}$$

We consider a similar problem in the following Example.

PROBLEM-SOLVING NOTE

Specific Heats versus Latent Heats

In solving problems involving specific heats and latent heats, recall that specific heats give the heat related to a *change in temperature* in a given phase, and latent heats give the heat related to a *change in phase* at a given temperature.

EXAMPLE 17–5 Steam Heat

To make steam, you add 5.60×10^5 J of heat to 0.220 kg of water at an initial temperature of 50.0 °C. Find the final temperature of the steam.

Picture the Problem

Our sketch shows the temperature–versus–heat-added curve for water. The initial point for this system, placed at the origin, is at 50.0 °C. As we shall see, adding the given amount of heat, $Q = 5.60 \times 10^5$ J, raises the temperature to the point labeled "final" in the plot.

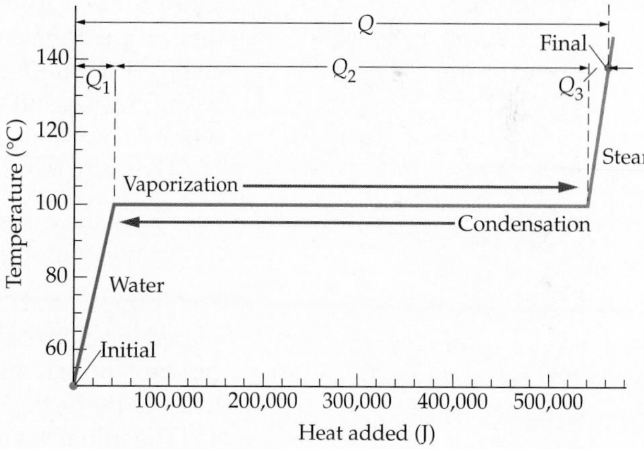

Strategy

To find the final temperature, we first calculate the amount of heat that must be added to heat the water to 100 °C. If this is less than the total heat added, we continue by calculating the amount of heat needed to vaporize all the water. If the sum of these two heats is still less than the total heat added to the water, we calculate the increase in temperature when the remaining heat is added to the steam.

Solution

1. Calculate the heat that must be added to the water to heat it to 100 °C. Call the result Q_1:

$$Q_1 = mc_{water}\,\Delta T$$
$$= (0.220\ \text{kg})[4186\ \text{J}/(\text{kg}\cdot\text{C}°)](50.0\ \text{C}°)$$
$$= 4.60 \times 10^4\ \text{J}$$

2. Next, calculate the heat that must be added to the water to convert it to steam. Let this result be Q_2:

$$Q_2 = mL_f = (0.220\ \text{kg})(22.6 \times 10^5\ \text{J/kg})$$
$$= 4.97 \times 10^5\ \text{J}$$

3. Determine the heat that is still to be added to the system. Let this remaining heat be Q_3:

$$Q_3 = 5.60 \times 10^5\ \text{J} - Q_1 - Q_2$$
$$= 5.60 \times 10^5\ \text{J} - 4.60 \times 10^4\ \text{J} - 4.97 \times 10^5\ \text{J}$$
$$= 17{,}000\ \text{J}$$

continued on next page

continued from previous page

4. Use Q_3 to find the increase in temperature of the steam:

$$Q_3 = mc_{\text{steam}}\,\Delta T$$

$$\Delta T = \frac{Q_3}{mc_{\text{steam}}} = \frac{17{,}000\text{ J}}{(0.220\text{ kg})[(2010\text{ J})/(\text{kg}\cdot\text{C}°)]} = 38\text{ C}°$$

Insight

Thus, the system ends up completely converted to steam at a temperature of 138 °C. If the amount of heat added to the system had been greater than Q_1, but less that $Q_1 + Q_2$, the final temperature of the system would have been 100 °C. In this case, the final state of the system is a mixture of liquid water and steam.

Practice Problem

Find the final temperature if the amount of heat added to the system is 3.40×10^5 J. [**Answer:** In this case, only 0.130 kg of water vaporizes into steam. Hence the final temperature is 100 °C, with water and steam in equilibrium.]

Some related homework problems: Problem 47, Problem 50

REAL-WORLD PHYSICS
Homemade ice cream

A pleasant application of latent heat is found in the making of homemade ice cream. As you may know, it is necessary to add salt to the ice–water mixture surrounding the container holding the ingredients for the ice cream. The dissolved salt molecules interact with water molecules in the liquid, impairing their ability to interact with one another and freeze. This means that ice and water are no longer in equilibrium at 0 °C; a lower temperature is required. The result is that ice begins to melt in the ice–water mixture, and in the process of melting it draws the required latent heat from its surroundings—which include the ice cream. Thus, the salt together with the ice produces a temperature lower than the melting temperature of ice alone.

17–6 Phase Changes and Energy Conservation

In the last section, we considered problems in which a given amount of heat is simply added to or removed from a system. We now turn to a more interesting type of problem involving energy conservation. In these problems, heat is exchanged within a system—that is, between its parts—but not with the external world. As a result, the total energy of the system is constant. Still, the heat flow within the system can cause changes in temperatures and phases.

The basic idea in solving energy conservation problems is the following:

> Set the magnitude of the heat *lost* by one part of the system equal to the magnitude of the heat *gained* by another.

For example, consider the following problem: A 0.0420-kg ice cube at −10.0 °C is placed in a Styrofoam cup containing 0.350 kg of water at 0 °C. Assuming the cup can be ignored, and that no heat is exchanged with the surroundings, find the mass of ice in the system when the equilibrium temperature of 0 °C is reached.

The initial setup for this problem is illustrated in **Figure 17–23 (a)**. Since the water is warmer than the ice, it follows that heat flows into the ice from the water,

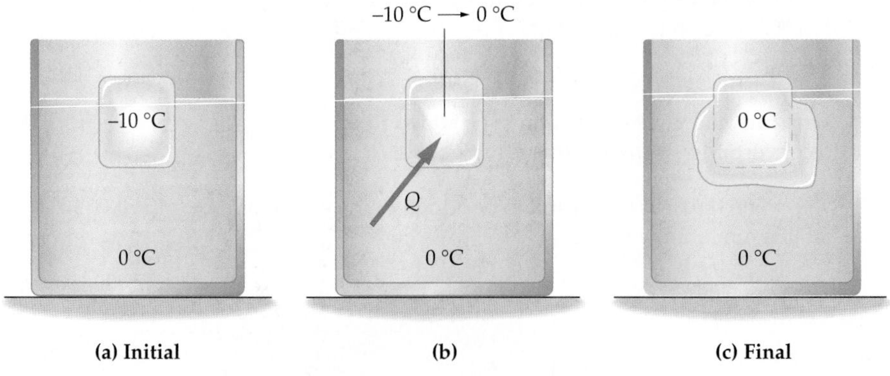

▶ FIGURE 17–23 Water freezing to form ice

An ice cube at −10.0 °C is placed in water at 0 °C. Since the ice is colder than the water, heat flows *from* the water *into* the ice. Heat that leaves the water results in the formation of additional ice in the system.

as indicated in **Figure 17–23 (b)**. The amount of heat will be just enough to raise the temperature of the ice from −10.0 °C to 0 °C. Thus,

$$Q = \text{heat gained by the ice}$$
$$= mc_{\text{ice}}\,\Delta T = (0.0420\text{ kg})[2090\text{ J}/(\text{kg}\cdot\text{C}°)](10.0\text{ C}°) = 878\text{ J}$$

By energy conservation, we can say that the heat lost by the water has the same magnitude as the heat gained by the ice. (In general, in these problems it is simplest to calculate the magnitude of each heat—so that all the heats are positive—and then set the "lost" and "gained" magnitudes equal.) In this case, then, we have

$$\text{heat lost by the water} = \text{heat gained by the ice} = 878\text{ J}$$

Thus, 878 J of heat are removed from the water.

Now, since the water is already at 0 °C, removing heat from it does not lower its temperature—instead, it merely converts some of the water to ice. How much is converted? The amount is determined by the latent heat of fusion. In particular, we have

$$Q = \text{heat lost by the water}$$
$$= mL_f = 878\text{ J}$$

In this expression, m is the mass of water that has been converted to ice. Solving for the mass yields

$$m = \frac{Q}{L_f} = \frac{878\text{ J}}{33.5 \times 10^4\text{ J/kg}} = 0.00262\text{ kg}$$

Thus, the final amount of ice in the system is 0.0420 kg + 0.00262 kg = 0.0446 kg. This is illustrated in **Figure 17–23 (c)**.

Finally, we consider a system in which a quantity of ice completely melts in warm water. The result is a container of liquid water at an intermediate temperature. Still, the basic idea is to apply energy conservation.

PROBLEM-SOLVING NOTE

Determining Equilibrium

In problems involving two different phases—like solid and liquid—it may not be clear in advance whether both phases or only one is present in the equilibrium state. It may be necessary to assume one case or the other and proceed with the calculation based on that assumption. If the result obtained in this way is not physical, the assumption must be changed.

EXAMPLE 17–6 Warm Punch

A large punch bowl holds 3.95 kg of lemonade (which is essentially just water) at 20.0 °C. A 0.0450-kg ice cube at 0 °C is placed in the lemonade. What is the final temperature of the system, and how much ice (if any) remains when the system reaches equilibrium? Ignore any heat exchange with the bowl or the surroundings.

Picture the Problem

Our sketch shows the various stages imagined for this problem, starting with the initial condition in which the ice is at 0 °C and the lemonade (water) is at 20.0 °C. As one might guess from the large amount of water and the small amount of ice, all of the ice will melt. Therefore, the final condition is a container of liquid water at a final temperature, T_f.

Strategy

To apply energy conservation to this problem, we first calculate the heat that would be lost by the water, Q_w, if we cooled it to 0 °C. We then imagine using part of this heat, Q_{ice}, to melt the ice.

As we shall see, a great deal of heat, $Q = Q_w - Q_{\text{ice}}$, is left after the ice is melted. We imagine adding this heat back into the system, which now contains 3.95 kg + 0.0450 kg of water at 0 °C. Calculating the increase in temperature caused by the heat Q gives us the final temperature of the system.

INTERACTIVE FIGURE

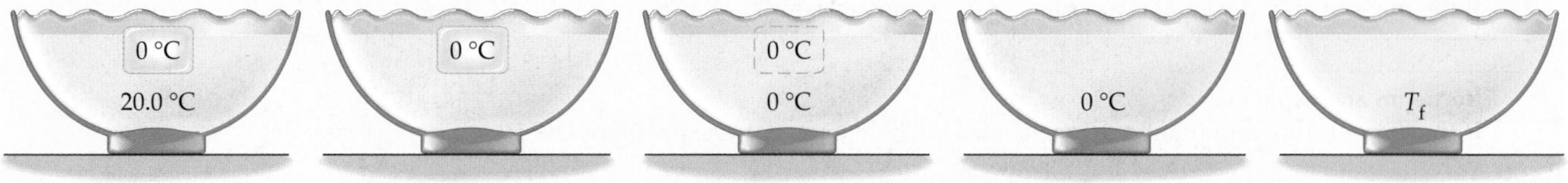

Initial condition | Calculate heat Q_w lost by water, 20 °C → 0 °C | Calculate heat Q_{ice} needed to melt ice cube | Calculate remaining heat $Q = Q_w - Q_{\text{ice}}$ | Calculate final temperature when heat Q is added to water and melted ice

continued on next page

continued from previous page

Solution

1. Find the heat lost by the water, Q_w, if it is cooled to 0 °C:

$$Q_w = m_{water} c_{water} \Delta T = (3.95 \text{ kg})[4186 \text{ J}/(\text{kg}\cdot\text{C}°)](20.0 \text{ C}°) = 3.31 \times 10^5 \text{ J}$$

2. Calculate the amount of heat, Q_{ice}, needed to melt all the ice:

$$Q_{ice} = m_{ice} L_f = (0.0450 \text{ kg})(33.5 \times 10^4 \text{ J/kg}) = 1.51 \times 10^4 \text{ J}$$

3. Determine the amount of heat, Q, that is left:

$$Q = Q_w - Q_{ice} = 3.31 \times 10^5 \text{ J} - 1.51 \times 10^4 \text{ J} = 3.16 \times 10^5 \text{ J}$$

4. Use the heat Q to warm the 3.95 kg + 0.0450 kg of water at 0 °C to its final temperature:

$$Q = (m_{water} + m_{ice}) c_{water} \Delta T$$

$$\Delta T = \frac{Q}{(m_{water} + m_{ice}) c_{water}} = \frac{3.16 \times 10^5 \text{ J}}{(3.95 \text{ kg} + 0.0450 \text{ kg})[4186 \text{ J}/(\text{kg}\cdot\text{C}°)]} = 18.9 \text{ C}°$$

Insight

Therefore, the final temperature of the system is 18.9 °C. As expected, the relatively small ice cube did not lower the temperature of the system very much.

Practice Problem

What would the final temperature of the system be if the ice cube's mass were 0.0750 kg? [**Answer:** $T_f = 18.2$ °C]

Some related homework problems: Problem 57, Problem 60

Chapter Summary

17–1 Ideal Gases

An ideal gas is a simplified model of a real gas in which interactions between molecules are ignored.

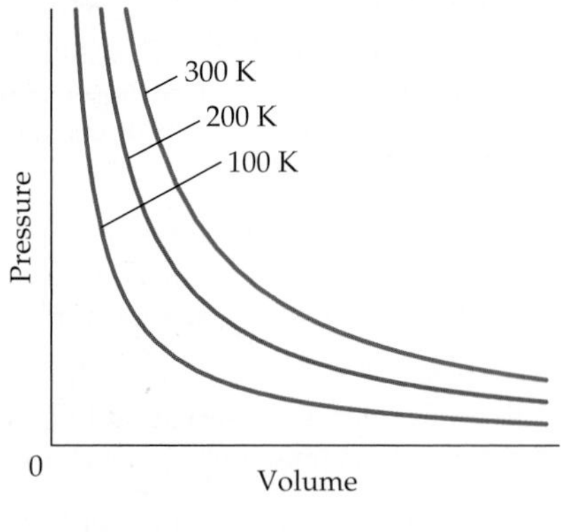

Equation of State

The equation of state for an ideal gas is

$$PV = NkT \quad \text{17–2}$$

In this expression, N is the number of molecules, T is the Kelvin temperature, and $k = 1.38 \times 10^{-23}$ J/K is Boltzmann's constant.

In terms of the universal gas constant, $R = 8.31$ J/(mol · K), and the number of moles in the gas, n, the ideal-gas equation of state is

$$PV = nRT \quad \text{17–5}$$

Avogadro's Number and Moles

The number of molecules in a mole (mol) is Avogadro's number, $N_A = 6.022 \times 10^{23}$.

Molecular Mass

If the mass of an individual molecule is m, its molecular mass, M, is

$$M = N_A m \quad \text{17–6}$$

Isotherms and Boyle's Law

If the temperature and number of molecules are held constant, the pressure and volume of an ideal gas satisfy Boyle's law:

$$PV = \text{constant} \quad \text{17–7}$$

Constant Pressure and Charles's Law

If the pressure and number of molecules are held constant, the temperature and volume of an ideal gas satisfy Charles's law:

$$\frac{V}{T} = \text{constant} \quad \text{17–8}$$

17–2 Kinetic Theory

In kinetic theory, a gas is imagined to be comprised of a large number of point-like molecules bouncing off the walls of a container.

The Origin of Pressure
The pressure exerted by a gas is a result of the momentum transfers that occur every time a molecule bounces off a wall of a container.

Speed Distribution of Molecules
The molecules in a gas have a range of speeds. The Maxwell distribution indicates which speeds are most likely to occur in a given gas.

Kinetic Energy and Temperature
Kinetic theory relates the average kinetic energy of the molecules in a gas to the Kelvin temperature of the gas, T:

$$\left(\tfrac{1}{2}mv^2\right)_{av} = K_{av} = \tfrac{3}{2}kT \quad \textbf{17–12}$$

RMS Speed
The rms (root mean square) speed of the molecules in a gas at the Kelvin temperature T is

$$v_{rms} = \sqrt{\frac{3kT}{m}} = \sqrt{\frac{3RT}{M}} \quad \textbf{17–13}$$

Internal Energy of an Ideal Gas
The internal energy of a monatomic ideal gas is

$$U = \tfrac{3}{2}NkT = \tfrac{3}{2}nRT \quad \textbf{17–15}$$

17–3 Solids and Elastic Deformation

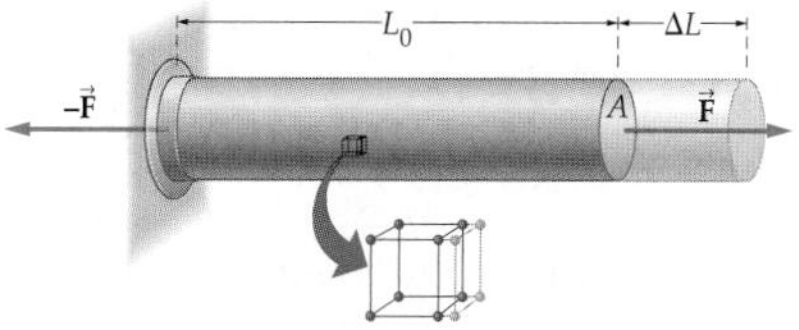

When a force is applied to a solid, its size and shape may change.

Changing the Length of a Solid
The force required to change the length of a solid by the amount ΔL is

$$F = Y\left(\frac{\Delta L}{L_0}\right)A \quad \textbf{17–17}$$

In this expression, Y is Young's modulus, L_0 is the initial length parallel to the applied force, and A is the cross-sectional area perpendicular to the applied force.

Shear Deformation
The force required to shear, or deform, a solid by the amount Δx is

$$F = S\left(\frac{\Delta x}{L_0}\right)A \quad \textbf{17–18}$$

In this expression, S is the shear modulus, L_0 is the initial length perpendicular to the applied force, and A is the cross-sectional area parallel to the applied force.

Changing the Volume of a Solid
The change in pressure required to change the volume of a solid by the amount ΔV is

$$\Delta P = -B\left(\frac{\Delta V}{V_0}\right) \quad \textbf{17–19}$$

In this expression, B is the bulk modulus and V_0 is the initial volume.

Stress and Strain
The applied force per area is the stress; the resulting deformation is the strain.

Elastic Deformation
An elastic deformation is one in which a solid returns to its original size and shape when the stress is removed.

17–4 Phase Equilibrium and Evaporation

The three most common phases of matter are the solid, liquid, and gas. Solids maintain a definite shape, whereas gases and liquids flow to take on the shape of their container.

Equilibrium Between Phases
When phases are in equilibrium, the number of molecules in each phase remains constant.

Evaporation
Evaporation occurs when some molecules in a liquid have speeds great enough to allow them to escape into the gas phase.

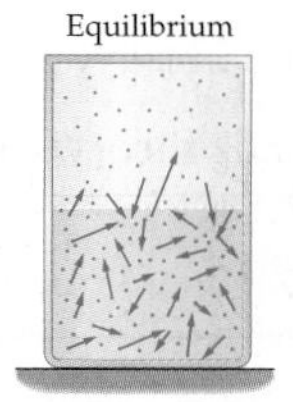

17–5 Latent Heats

The latent heat, L, is the amount of heat per unit mass that must be added to or removed from a substance to convert it from one phase to another.

Latent Heat of Fusion
The heat required for melting or freezing is called the latent heat of fusion, L_f.

Latent Heat of Vaporization
The heat required for vaporizing or condensing is the latent heat of vaporization, L_v.

Latent Heat of Sublimation
The heat required to sublime a solid directly to a gas, or to condense a gas to a solid, is the latent heat of sublimation, L_s.

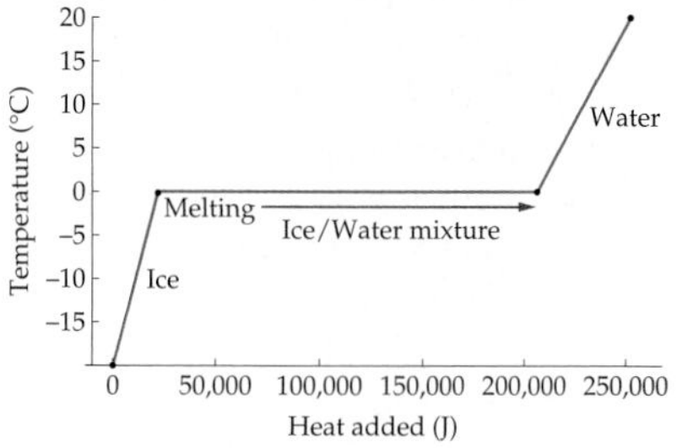

17–6 Phase Changes and Energy Conservation

When heat is exchanged within a system, with no exchanges with the surroundings, the energy of the system is conserved.

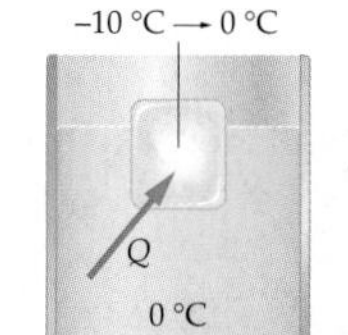

Problem-Solving Summary

Type of Problem	Relevant Physical Concepts	Related Examples
Find pressure, volume, or temperature in an ideal gas.	These basic quantities are related by the ideal gas equation of state, $PV = NkT = nRT$.	Examples 17–1, 17–2 Active Examples 17–1, 17–2
Relate the rms speed of a gas to the absolute temperature T.	Absolute temperature is directly related to the average kinetic energy, $3kT/2 = K_{av}$. From this connection we obtain the result $v_{rms} = \sqrt{3kT/m}$.	Example 17–3
Find the strain produced by a given stress.	Strain is proportional to stress, at least for small stress. The basic relations are $F/A = Y(\Delta L/L_0)$, $F/A = S(\Delta x/L_0)$, and $\Delta P = -B(\Delta V/V_0)$, where Y, S, and B are the Young's modulus, the shear modulus, and the bulk modulus, respectively.	Example 17–4, Active Examples 17–3, 17–4
Calculate the heat associated with a change in phase.	A certain amount of heat, called the latent heat, L, must be added to or taken away from a system to change it from one phase to another. The process occurs at constant temperature. The amount of heat involved in a change of phase is given by $Q = mL$, where m is the mass that changes phase.	Examples 17–5, 17–6

Conceptual Questions

(Answers to odd-numbered Conceptual Questions can be found in the back of the book.)

1. At the beginning of a typical airline flight you are instructed about the proper use of oxygen masks that will fall from the ceiling if the cabin pressure suddenly drops. You are advised that the oxygen masks are working properly, even if the bags do not fully inflate. In fact, the bags expand to their fullest if cabin pressure is lost at high altitude, but expand only partially if the plane is at low altitude. Explain.
2. How is the air pressure in a tightly sealed house affected by operating the furnace? Explain.
3. The average speed of air molecules in your room is on the order of the speed of sound. What is their average velocity?
4. Is it possible to change both the pressure and the volume of an ideal gas without changing the average kinetic energy of its molecules? If your answer is no, explain why not. If your answer is yes, give a specific example.
5. **An Airport at Great Elevation** One of the highest airports in the world is located in La Paz, Bolivia. Pilots prefer to take off

from this airport in the morning or the evening, when the air is quite cold. Explain.

6. A camping stove just barely boils water on a mountaintop. When the stove is used at sea level, will it be able to boil water? Explain your answer.

7. An autoclave is a device used to sterilize medical instruments. It is essentially a pressure cooker that heats the instruments in water under high pressure. This ensures that the sterilization process occurs at temperatures greater than the normal boiling point of water. Explain why the autoclave produces such high temperatures.

8. As the temperature of ice is increased, it changes first into a liquid and then into a vapor. On the other hand, dry ice, which is solid carbon dioxide, changes directly from a solid to a vapor as its temperature is increased. How might one produce liquid carbon dioxide?

9. **BIO** Isopropyl alcohol is sometimes rubbed onto a patient's arms and legs to lower their body temperature. Why is this effective?

10. If you toss an ice cube into a swimming pool, is the water in the pool now at 0 °C? Explain.

11. A drop of water on a kitchen counter evaporates in a matter of minutes. However, only a relatively small fraction of the molecules in the drop move rapidly enough to escape through the drop's surface. Why, then, does the entire drop evaporate rather than just a small fraction of it?

Conceptual Exercises

(Answers to odd-numbered Conceptual Exercises can be found in the back of the book.)

1. **(a)** Is the number of molecules in one mole of N_2 greater than, less than, or equal to the number of molecules in one mole of O_2? **(b)** Is the mass of one mole of N_2 greater than, less than, or equal to the mass of one mole of O_2? Explain.

2. Is the number of atoms in one mole of carbon greater than, less than, or the same as the number of atoms in one mole of oxygen gas (O_2)? Explain.

3. If you put a helium-filled balloon in the refrigerator, will its volume increase, decrease, or stay the same? Explain.

4. Plastic bubble wrap is used as a protective packing material. Is the bubble wrap more effective on a cold day or on a warm day? Explain.

5. Two containers hold ideal gases at the same temperature. Container B has twice the volume and half the number of molecules as container A. What is the ratio P_A/P_B, where P_A is the pressure in container A and P_B is the pressure in container B? Explain.

6. The air in your room is composed mostly of oxygen (O_2) and nitrogen (N_2) molecules. The oxygen molecules are more massive than the nitrogen molecules. Is the rms speed of the O_2 molecules greater than, less than, or equal to the rms speed of the N_2 molecules? Explain.

7. Two adjacent rooms in a hotel are equal in size and connected by an open door. Room 1 is warmer than room 2. Which room contains more air? Explain.

8. As you go up in altitude, do you expect the ratio of oxygen to nitrogen in the atmosphere to increase, decrease, or stay the same? Explain.

9. If the translational speed of molecules in an ideal gas is doubled, by what factor does the Kelvin temperature change? Explain.

10. Four ideal gases have the following pressures, P, volumes, V, and mole numbers, n: gas A, $P = 100$ kPa, $V = 1\ \text{m}^3$, $n = 10$ mol; gas B, $P = 200$ kPa, $V = 2\ \text{m}^3$, $n = 20$ mol; gas C, $P = 50$ kPa, $V = 1\ \text{m}^3$, $n = 50$ mol; gas D, $P = 50$ kPa, $V = 4\ \text{m}^3$, $n = 5$ mol. Rank these gases in order of increasing temperature. Indicate ties where appropriate.

11. Suppose the Celsius temperature of an ideal gas is doubled from 100 °C to 200 °C. Does the average kinetic energy of the molecules in this gas increase by a factor that is greater than, less than, or equal to 2? Explain.

12. Suppose the absolute temperature of an ideal gas is doubled from 100 K to 200 K. Does the average speed of the molecules in this gas increase by a factor that is greater than, less than, or equal to 2? Explain.

13. A piston held at the temperature T contains a gas mixture with molecules of three different types; A, B, and C. The corresponding molecular masses are $m_C > m_B > m_A$. Rank these molecular types in order of increasing **(a)** average kinetic energy and **(b)** rms speed. Indicate ties where appropriate.

14. Four cylindrical rods with various cross-sectional areas and initial lengths are stretched by an applied force, as in Figure 17–10. The resulting change in length of each rod is given in the following table. Rank these rods in order of increasing Young's modulus. Indicate ties where appropriate.

Rod	Applied Force	Cross-Sectional Area	Initial Length	Change in Length
1	F	A	L	ΔL
2	F	$2A$	$2L$	ΔL
3	$2F$	$2A$	L	$2\Delta L$
4	$3F$	A	$L/2$	ΔL

15. A brick has faces with the following dimensions: face 1 is 1 cm by 2 cm; face 2 is 2 cm by 3 cm; face 3 is 1 cm by 3 cm. On which face should the brick be placed if it is to have the smallest change in dimensions due to its own weight? Explain.

16. A hollow cylindrical rod (rod 1) and a solid cylindrical rod (rod 2) are made of the same material. The two rods have the same length and outer radius. If the same compressional force is applied to each rod, which one has the greater change in length? Explain.

Problems

Note: **IP** *denotes an integrated conceptual/quantitative problem.* **BIO** *identifies problems of biological or medical interest. Red bullets (•, ••, •••) are used to indicate the level of difficulty of each problem.*

Section 17–1 Ideal Gases

1. • Standard temperature and pressure (STP) is defined as a temperature of 0 °C and a pressure of 101.3 kPa. What is the volume occupied by one mole of an ideal gas at STP?

2. • **BIO** After emptying her lungs, a person inhales 4.2 L of air at 0.0 °C and holds her breath. How much does the volume of the air increase as it warms to her body temperature of 37 °C?

3. • In the morning, when the temperature is 286 K, a bicyclist finds that the absolute pressure in his tires is 501 kPa. That afternoon he finds that the pressure in the tire has increased to 554 kPa. Ignoring expansion of the tires, find the afternoon temperature.

4. • An automobile tire has a volume of 0.0185 m^3. At a temperature of 294 K the absolute pressure in the tire is 212 kPa. How many moles of air must be pumped into the tire to increase its pressure to 252 kPa, given that the temperature and volume of the tire remain constant?

5. • **Amount of Helium in a Blimp** The Goodyear blimp *Spirit of Akron* is 62.6 m long and contains 7023 m^3 of helium. When the temperature of the helium is 285 K, its absolute pressure is 112 kPa. Find the mass of the helium in the blimp.

6. • A compressed-air tank holds 0.500 m^3 of air at a temperature of 285 K and a pressure of 850 kPa. What volume would the air occupy if it were released into the atmosphere, where the pressure is 101 kPa and the temperature is 303 K?

7. • A typical region of interstellar space may contain 10^6 atoms per cubic meter (primarily hydrogen) at a temperature of 100 K. What is the pressure of this gas?

8. •• A balloon contains 3.7 liters of nitrogen gas at a temperature of 87 K and a pressure of 101 kPa. If the temperature of the gas is allowed to increase to 24 °C and the pressure remains constant, what volume will the gas occupy?

9. •• **IP** A balloon is filled with helium at a pressure of 2.4×10^5 Pa. The balloon is at a temperature of 18 °C and has a radius of 0.25 m. **(a)** How many helium atoms are contained in the balloon? **(b)** Suppose we double the number of helium atoms in the balloon, keeping the pressure and the temperature fixed. By what factor does the radius of the balloon increase? Explain.

10. •• **IP** A gas has a temperature of 310 K and a pressure of 101 kPa. **(a)** Find the volume occupied by 1.25 mol of this gas, assuming it is ideal. **(b)** Assuming the gas molecules can be approximated as small spheres of diameter 2.5×10^{-10} m, determine the fraction of the volume found in part (a) that is occupied by the molecules. **(c)** In determining the properties of an ideal gas, we assume that molecules are points of zero volume. Discuss the validity of this assumption for the case considered here.

11. •• A 515-cm^3 flask contains 0.460 g of a gas at a pressure of 153 kPa and a temperature of 322 K. What is the molecular mass of this gas?

12. •• **IP The Atmosphere of Mars** On Mars, the average temperature is −64 °F and the average atmospheric pressure is 0.92 kPa. **(a)** What is the number of molecules per volume in the Martian atmosphere? **(b)** Is the number of molecules per volume on the Earth greater than, less than, or equal to the number per volume on Mars? Explain your reasoning. **(c)** Estimate the number of molecules per volume in Earth's atmosphere.

13. •• The air inside a hot-air balloon has an average temperature of 77.2 °C. The outside air has a temperature of 20.3 °C. What is the ratio of the density of air in the balloon to the density of air in the surrounding atmosphere?

14. •• A cylindrical flask is fitted with an airtight piston that is free to slide up and down, as shown in **Figure 17–24**. **A mass rests on top of the piston.** The initial temperature of the system is 313 K and the pressure of the gas is held constant at 137 kPa. The temperature is now increased until the height of the piston rises from 23.4 cm to 26.0 cm. What is the final temperature of the gas?

15. •• Consider the system described in Problem 14. Contained within the flask is an ideal gas at a constant temperature of 313 K. Initially the pressure applied by the piston and the mass is 137 kPa and the height of the piston above the base of the flask is 23.4 cm. When additional mass is added to the piston, the height of the piston decreases to 20.0 cm. Find the new pressure applied by the piston.

▲ **FIGURE 17–24** Problems 14 and 15

16. ••• One mole of a monatomic ideal gas has an initial pressure of 210 kPa, an initial volume of 1.2×10^{-3} m^3, and an initial temperature of 350 K. The gas now undergoes three separate processes: (i) a constant-temperature expansion that triples its volume; (ii) a constant-pressure compression to its initial volume; and (iii) a constant-volume increase in pressure to its initial pressure. At the end of these three processes, the gas is back at its initial pressure, volume, and temperature. Plot these processes on a pressure-versus-volume graph, showing the values of P and V at the end points of each process.

Section 17–2 Kinetic Theory

17. • The molecules in a tank of hydrogen have the same rms speed as the molecules in a tank of oxygen. State whether each of the following statements is true, false, or unknowable with the given information: **(a)** the pressures are the same; **(b)** the hydrogen is at the higher temperature; **(c)** the hydrogen has the higher pressure; **(d)** the temperatures are the same; **(e)** the oxygen is at the higher temperature.

18. • At what temperature is the rms speed of H_2 equal to the rms speed that O_2 has at 323 K?

19. • Suppose a planet has an atmosphere of pure ammonia at 0.0 °C. What is the rms speed of the ammonia molecules?

20. •• **IP** Three moles of oxygen gas (that is, 3.0 mol of O_2) are placed in a portable container with a volume of 0.0035 m^3. If the temperature of the gas is 295 °C, find **(a)** the pressure of the gas and **(b)** the average kinetic energy of an oxygen molecule. **(c)** Suppose the volume of the gas is doubled, while the temperature and number of moles are held constant. By what factor do your answers to parts (a) and (b) change? Explain.

21. •• **IP** The rms speed of O_2 is 1550 m/s at a given temperature. **(a)** Is the rms speed of H_2O at this temperature greater than, less than, or equal to 1550 m/s? Explain. **(b)** Find the rms speed of H_2O at this temperature.

22. •• **IP** An ideal gas is kept in a container of constant volume. The pressure of the gas is also kept constant. **(a)** If the number of molecules in the gas is doubled, does the rms speed increase, decrease, or stay the same? Explain. **(b)** If the initial rms speed is 1200 m/s, what is the final rms speed?

23. •• What is the temperature of a gas of CO_2 molecules whose rms speed is 339 m/s?

24. •• The rms speed of a sample of gas is increased by one percent. **(a)** What is the percent change in the temperature of the gas? **(b)** What is the percent change in the pressure of the gas, assuming its volume is held constant?

25. •• **Enriching Uranium** In naturally occurring uranium atoms, 99.3% are ^{238}U (atomic mass = 238 u, where u = 1.6605×10^{-27} kg) and only 0.7% are ^{235}U (atomic mass = 235 u). Uranium-fueled reactors require an enhanced proportion of ^{235}U. Since both isotopes of uranium have identical chemical properties, they can be separated only by methods that depend on their differing masses. One such method is gaseous diffusion, in which uranium hexafluoride (UF_6) gas diffuses through a series of porous barriers. The lighter $^{235}UF_6$ molecules have a slightly higher rms speed at a given temperature than the heavier $^{238}UF_6$ molecules, and this allows the two isotopes to be separated. Find the ratio of the rms speeds of the two isotopes at 23.0 °C.

26. ••• A 350-mL spherical flask contains 0.075 mol of an ideal gas at a temperature of 293 K. What is the average force exerted on the walls of the flask by a single molecule?

Section 17–3 Solids and Elastic Deformation

27. • A rock climber hangs freely from a nylon rope that is 14 m long and has a diameter of 8.3 mm. If the rope stretches 4.9 cm, what is the mass of the climber?

28. • **BIO** To stretch a relaxed biceps muscle 2.5 cm requires a force of 25 N. Find the Young's modulus for the muscle tissue, assuming it to be a uniform cylinder of length 0.24 m and cross-sectional area 47 cm^2.

29. • A 22-kg chimpanzee hangs from the end of a horizontal, broken branch 1.1 m long, as shown in **Figure 17–25**. The branch is a uniform cylinder 4.6 cm in diameter, and the end of the branch supporting the chimp sags downward through a vertical distance of 13 cm. What is the shear modulus for this branch?

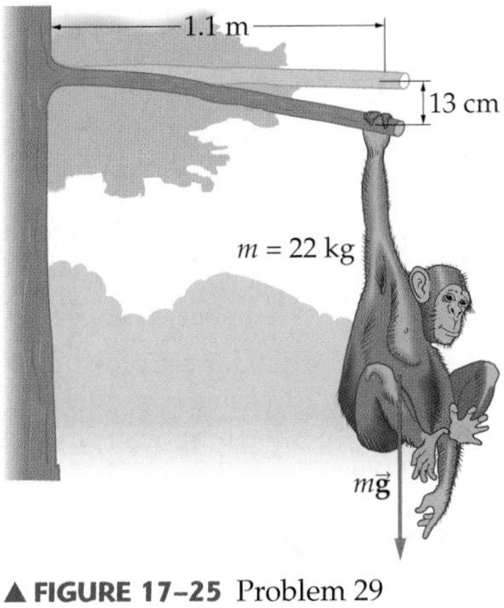

▲ **FIGURE 17–25** Problem 29

30. • **The Marianas Trench** The deepest place in all the oceans is the Marianas Trench, where the depth is 10.9 km and the pressure is 1.10×10^8 Pa. If a copper ball 15.0 cm in diameter is taken to the bottom of the trench, by how much does its volume decrease?

31. •• **IP** A steel wire 4.9 m long stretches 0.11 cm when it is given a tension of 360 N. **(a)** What is the diameter of the wire? **(b)** If it is desired that the stretch be less than 0.11 cm, should its diameter be increased or decreased? Explain.

32. •• **BIO Spiderweb** An orb weaver spider with a mass of 0.26 g hangs vertically by one of its threads. The thread has a Young's modulus of 4.7×10^9 N/m^2 and a radius of 9.8×10^{-6} m. **(a)** What is the fractional increase in the thread's length caused by the spider? **(b)** Suppose a 76-kg person hangs vertically from a nylon rope. What radius must the rope have if its fractional increase in length is to be the same as that of the spider's thread?

33. •• **IP** Two rods of equal length (0.45 m) and diameter (1.5 cm) are placed end to end. One rod is aluminum, the other is brass. If a compressive force of 8400 N is applied to the rods, **(a)** how much does their combined length decrease? **(b)** Which of the rods changes its length by the greatest amount? Explain.

34. •• A piano wire 0.82 m long and 0.93 mm in diameter is fixed on one end. The other end is wrapped around a tuning peg 3.5 mm in diameter. Initially the wire, whose Young's modulus is 2.4×10^{10} N/m^2, has a tension of 14 N. Find the tension in the wire after the tuning peg has been turned through one complete revolution.

35. •• **Diving in the Bathysphere** The American naturalist Charles William Beebe (1877–1962) set a world record in 1934 when he made a dive to a depth of 923 m below the surface of the ocean. The dive was made in a device known as the bathysphere, which was basically a steel sphere 4.75 ft in diameter. How much did the volume of the sphere change as it was lowered to its record depth?

Charles William Beebe (left) and Otis Barton with the bathysphere in 1934. (Problem 35)

36. ••• Referring to the previous problem, suppose the bathysphere and its occupants had a combined mass of 12,700 kg and that they were lowered into the ocean on a steel cable with a radius of 1.85 cm. How much did the cable stretch when the bathysphere was at the depth of 923 m? (Neglect the weight of the cable itself, but include the effects of buoyancy.)

Section 17–4 Phase Equilibrium and Evaporation

37. • The formation of ice from water is accompanied by which of the following: **(a)** an absorption of heat by the water; **(b)** an increase in temperature; **(c)** a decrease in volume; **(d)** a removal of heat from the water; **(e)** a decrease in temperature?

38. • **Vapor Pressure for Water** **Figure 17–26** shows a portion of the vapor pressure curve for water. Referring to the figure,

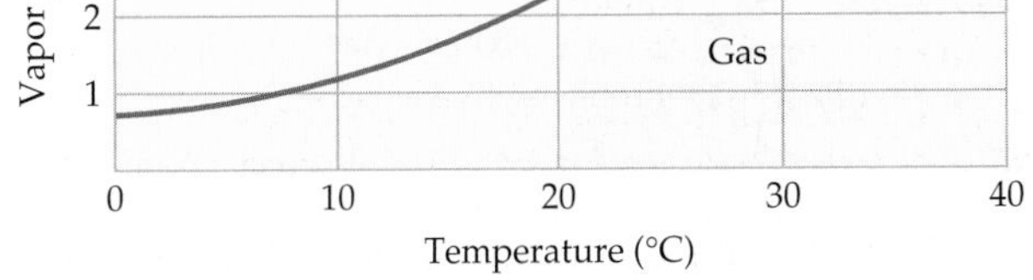

▲ **FIGURE 17–26** Problems 38 and 39

estimate the pressure that would be required for water to boil at 30 °C.

39. • Using the vapor pressure curve given in Figure 17–26, find the temperature at which water boils when the pressure is 1.5 kPa.

40. • **IP The Vapor Pressure of CO_2** A portion of the vapor pressure curve for carbon dioxide is given in **Figure 17–27**. **(a)** Estimate the pressure at which CO_2 boils at 0 °C. **(b)** If the temperature is increased, does the boiling pressure increase, decrease, or stay the same? Explain.

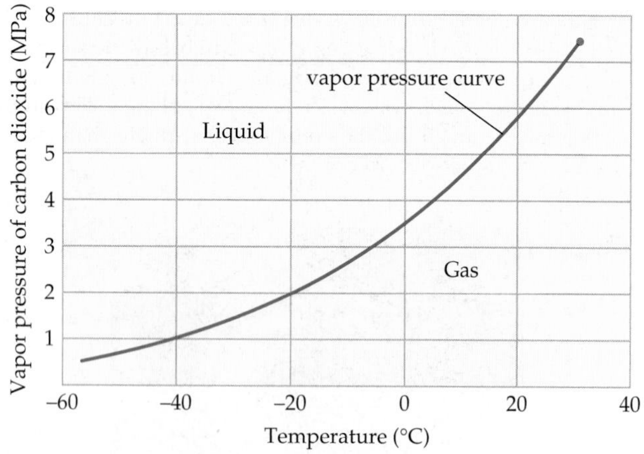

▲ **FIGURE 17–27** Problems 40 and 41

41. • **IP** Referring to the vapor pressure curve for carbon dioxide given in Figure 17–27, **(a)** estimate the temperature at which CO_2 boils when the pressure is 1.5×10^6 Pa. **(b)** If the pressure is increased, does the boiling temperature increase, decrease, or stay the same? Explain.

42. •• **Phase Diagram for Water** The phase diagram for water is shown in **Figure 17–28**. **(a)** What is the temperature T_1 on the phase diagram? **(b)** What is the temperature T_2 on the phase diagram? **(c)** What happens to the melting/freezing temperature of water if atmospheric pressure is *decreased*? Justify your answer by referring to the phase diagram. **(d)** What happens to the boiling/condensation temperature of water if atmospheric pressure is *increased*? Justify your answer by referring to the phase diagram.

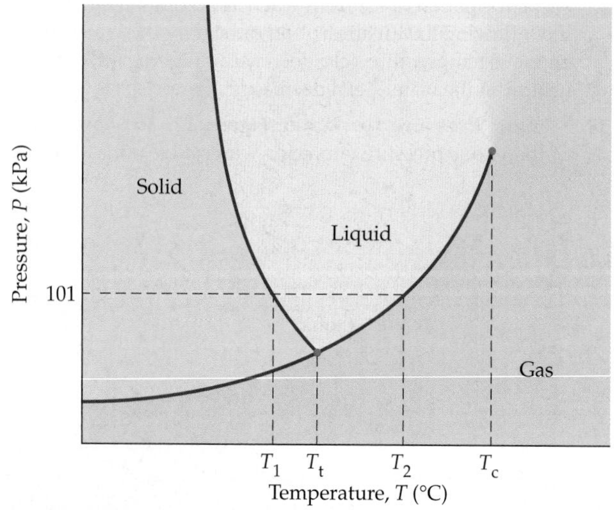

▲ **FIGURE 17–28** Problems 42, 44, and 45

43. •• **Phase Diagram for CO_2** The phase diagram for CO_2 is shown in **Figure 17–29**. **(a)** What is the phase of CO_2 at $T = 20$ °C and $P = 500$ kPa? **(b)** What is the phase of CO_2 at $T = -80$ °C and $P = 120$ kPa? **(c)** For reasons of economy and convenience, bulk CO_2 is often transported in liquid form in pressurized tanks. Using the phase diagram, determine the minimum pressure required to keep CO_2 in the liquid phase at 20 °C.

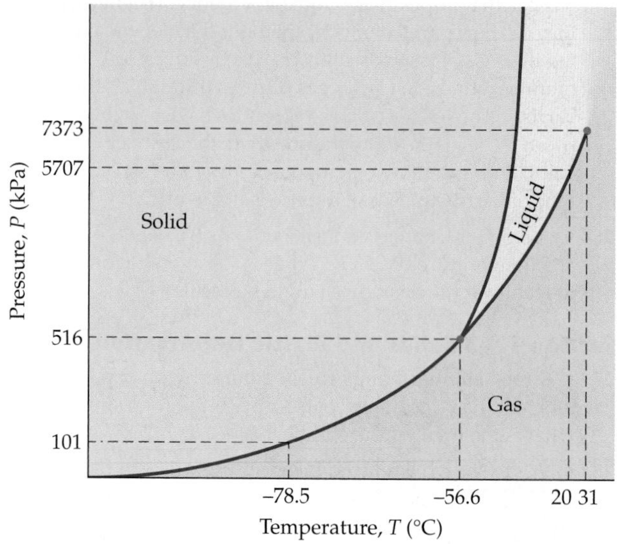

▲ **FIGURE 17–29** Problems 43 and 46

44. •• A sample of water ice at atmospheric pressure has a temperature just below the freezing point. Refer to Figure 17–28 to answer the following questions. **(a)** What phase changes occur if the temperature of the system is increased while the pressure is held constant? **(b)** Suppose, instead, that the temperature of the system is held constant just below the freezing point while the pressure is decreased. What phase changes occur now?

45. •• A sample of liquid water at atmospheric pressure has a temperature just above the freezing point. Refer to Figure 17–28 to answer the following questions. **(a)** What phase changes occur if the temperature of the system is increased while the pressure is held constant? **(b)** Suppose, instead, that the temperature of the system is held constant just above the freezing point while the pressure is decreased. What phase changes occur now?

46. •• A sample of solid carbon dioxide at a pressure of 5707 kPa has a temperature just below the freezing point. Refer to Figure 17–29 to answer the following questions. **(a)** What phase changes occur if the temperature of the system is increased while the pressure is held constant? **(b)** Suppose, instead, that the temperature of the system is held constant just below the freezing point while the pressure is decreased. What phase changes occur now?

Section 17–5 Latent Heats

47. • How much heat must be removed from 0.96 kg of water at 0 °C to make ice cubes at 0 °C?

48. • A heat transfer of 9.5×10^5 J is required to convert a block of ice at −15 °C to water at 15 °C. What was the mass of the block of ice?

49. • How much heat must be added to 1.5 kg of copper to change it from a solid at 1358 K to a liquid at 1358 K?

50. •• **IP** A 1.1-kg block of ice is initially at a temperature of −5.0 °C. **(a)** If 5.2×10^5 J of heat are added to the ice, what is the final temperature of the system? Find the amount of ice, if any, that remains. **(b)** Suppose the amount of heat added to the ice block is doubled. By what factor must the mass of the ice be increased if the system is to have the same final temperature? Explain.

51. •• **IP** Referring to the previous problem, suppose the amount of heat added to the block of ice is reduced by a factor of two to 2.6×10^5 J. Note that this amount of heat is still sufficient to melt at least some of the ice. **(a)** Do you expect the temperature

increase in this case to be one-half that found in the previous problem? Explain. **(b)** What is the final temperature of the system in this case? Find the amount of ice, if any, that remains.

52. •• Figure 17–21 shows a temperature-versus-heat plot for 1.000 kg of water. **(a)** Calculate the heat corresponding to the points A, B, C, and D. **(b)** Calculate the slope of the line from point B to point C. Show that this slope is equal to $1/c$, where c is the specific heat of liquid water.

53. •• **IP** Suppose the 1.000 kg of water in Figure 17–21 starts at point A at time zero. Heat is added to this system at the rate of 12,250 J/s. How long does it take for the system to reach **(a)** point B, **(b)** point C and **(c)** point D? **(d)** Describe the physical state of the system at time $t = 63.00$ s.

54. •• Figure 17–22 shows a temperature-versus-heat plot for 0.550 kg of water. **(a)** Calculate the slope of the line from point A to point B. Show that the slope is equal to $1/mc$, where c is the specific heat of ice. **(b)** Calculate the slope of the line from point C to point D. Show that the slope is equal to $1/mc$, where c is the specific heat of liquid water.

55. •• **BIO** In Conceptual Checkpoint 17–5 we pointed out that steam can cause more serious burns than water at the same temperature. Here we examine this effect quantitatively, noting that flesh becomes badly damaged when its temperature reaches 50.0 °C. **(a)** Calculate the heat released as 12.5 g of liquid water at 100 °C is cooled to 50.0 °C. **(b)** Calculate the heat released when 12.5 g of steam at 100 °C is condensed and cooled to 50.0 °C. **(c)** Find the mass of flesh that can be heated from 37.0 °C (normal body temperature) to 50.0 °C for the cases considered in parts **(a)** and **(b)**. (The average specific heat of flesh is 3500 J/kg · K.)

56. •• When you go out to your car one cold winter morning you discover a 0.57-cm thick layer of ice on the windshield, which has an area of 1.6 m^2. If the temperature of the ice is −2.0 °C, and its density is 917 kg/m^3, find the heat required to melt all the ice.

Section 17–6 Phase Changes and Energy Conservation

57. •• A large punch bowl holds 3.99 kg of lemonade (which is essentially water) at 20.5 °C. A 0.0550-kg ice cube at −10.2 °C is placed in the lemonade. What is the final temperature of the system, and the amount of ice (if any) remaining? Ignore any heat exchange with the bowl or the surroundings.

58. •• A 155-g aluminum cylinder is removed from a liquid nitrogen bath, where it has been cooled to −196 °C. The cylinder is immediately placed in an insulated cup containing 80.0 g of water at 15.0 °C. What is the equilibrium temperature of this system? If your answer is 0 °C, determine the amount of water that has frozen. The average specific heat of aluminum over this temperature range is 653 J/(kg · K).

59. •• An 825-g iron block is heated to 352 °C and placed in an insulated container (of negligible heat capacity) containing 40.0 g of water at 20.0 °C. What is the equilibrium temperature of this system? If your answer is 100 °C, determine the amount of water that has vaporized. The average specific heat of iron over this temperature range is 560 J/(kg · K).

60. •• **IP** A 35-g ice cube at 0.0 °C is added to 110 g of water in a 62-g aluminum cup. The cup and the water have an initial temperature of 23 °C. **(a)** Find the equilibrium temperature of the cup and its contents. **(b)** Suppose the aluminum cup is replaced with one of equal mass made from silver. Is the equilibrium temperature with the silver cup greater than, less than, or the same as with the aluminum cup? Explain.

61. •• A 48-g block of copper at −12 °C is added to 110 g of water in a 75-g aluminum cup. The cup and the water have an initial temperature of 4.1 °C. **(a)** Find the equilibrium temperature of the cup and its contents. **(b)** What mass of ice, if any, is present when the system reaches equilibrium?

62. •• A 0.075-kg ice cube at 0.0 °C is dropped into a Styrofoam cup holding 0.33 kg of water at 14 °C. **(a)** Find the final temperature of the system, and the amount of ice (if any) remaining. Assume the cup and the surroundings can be ignored. **(b)** Find the initial temperature of the water that would be enough to *just barely* melt all of the ice.

63. •• To help keep her barn warm on cold days, a farmer stores 845 kg of warm water in the barn. How many hours would a 2.00-kilowatt electric heater have to operate to provide the same amount of heat as is given off by the water as it cools from 20.0 °C to 0 °C and then freezes at 0 °C?

General Problems

64. • **Largest Raindrops** Atmospheric scientists studying clouds in the Marshall Islands have observed what they believe to be the world's largest raindrops, with a radius of 0.52 cm. How many molecules are in these monster drops?

65. • **Cooling Computers** Researchers are developing "heat exchangers" for laptop computers that take heat from the laptop—to keep it from being damaged by overheating—and use it to vaporize methanol. Given that 5100 J of heat is removed from the laptop when 4.6 g of methanol is vaporized, what is the latent heat of vaporization for methanol?

66. •• **Scuba Tanks** In scuba diving circles, "an 80" refers to a scuba tank that holds 80 cubic feet of air, a standard amount for recreational diving. Given that a scuba tank is a cylinder 2 feet long and half a foot in diameter, determine **(a)** the volume of a tank and **(b)** the pressure in a tank when 80 cubic feet of air is compressed into its relatively small volume. **(c)** What is the mass of air in a tank that holds 80 cubic feet of air. Assume the temperature is 21 °C and that the walls of the tank are of negligible thickness.

67. •• A reaction vessel contains 8.06 g of H_2 and 64.0 g of O_2 at a temperature of 125 °C and a pressure of 101 kPa. **(a)** What is the volume of the vessel? **(b)** The hydrogen and oxygen are now ignited by a spark, initiating the reaction $2\,H_2 + O_2 \rightarrow 2\,H_2O$. This reaction consumes all the hydrogen and oxygen in the vessel. What is the pressure of the resulting water vapor when it returns to its initial temperature of 125 °C?

68. •• A bicycle tire with a radius of 0.66 m has a gauge pressure of 41 lb/in^2. Treating the tire as a hollow hoop with a cross-sectional area of 0.0028 m^2, find the number of air molecules in the tire when its temperature is 23 °C.

69. •• Peter catches a 4.8-kg striped bass on a fishing line 0.54 mm in diameter and begins to reel it in. He fishes from a pier well above the water, and his fish hangs vertically from the line out of the water. The fishing line has a Young's modulus of $5.1 \times 10^9\ N/m^2$. **(a)** What is the fractional increase in length of the fishing line if the fish is at rest? **(b)** What is the fractional increase in the fishing line's length when the fish is pulled upward with a constant speed of 1.5 m/s? **(c)** What is the fractional increase in the fishing line's length when the fish is pulled upward with a constant acceleration of 1.5 m/s^2?

70. •• **IP** You use a steel socket wrench 28 cm long to loosen a rusty bolt, applying a force F at the end of the handle. The handle undergoes a shear deformation of 0.11 mm. **(a)** If the cross-sectional area of the handle is 2.3 cm^2, what is the magnitude of the applied force F? **(b)** If the cross-sectional area of the handle is doubled, by what factor does the shear deformation change? Explain.

71. •• A steel ball (density $= 7860\ \text{kg/m}^3$) with a diameter of 6.4 cm is tied to an aluminum wire 82 cm long and 2.5 mm in diameter. The ball is whirled about in a vertical circle with a tangential speed of 7.8 m/s at the top of the circle and 9.3 m/s at the bottom of the circle. Find the amount of stretch in the wire **(a)** at the top and **(b)** at the bottom of the circle.

72. •• A lead brick with the dimensions shown in **Figure 17–30** rests on a rough solid surface. A force of 2400 N is applied as indicated. Find **(a)** the change in height of the brick and **(b)** the amount of shear deformation.

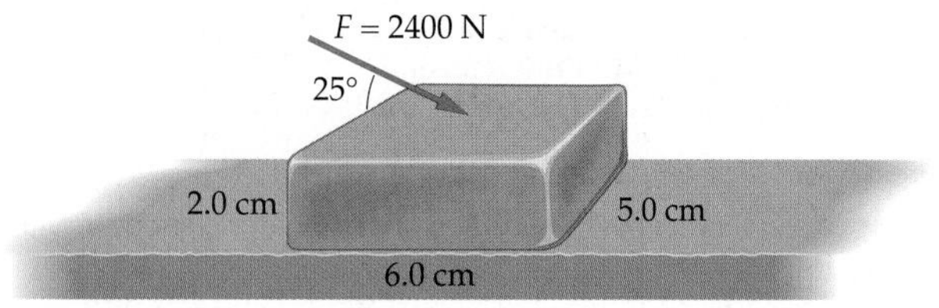

▲ **FIGURE 17–30** Problem 72

73. •• **IP** Five molecules have the following speeds: 221 m/s, 301 m/s, 412 m/s, 44.0 m/s, and 182 m/s. **(a)** Find v_{av} for these molecules. **(b)** Do you expect $(v^2)_{\text{av}}$ to be greater than, less than, or equal to $(v_{\text{av}})^2$? Explain. **(c)** Calculate $(v^2)_{\text{av}}$ and comment on your results. **(d)** Calculate v_{rms} and compare with v_{av}.

74. •• **(a)** Find the amount of heat that must be extracted from 1.0 kg of steam at 110 °C to convert it to ice at 0.0 °C. **(b)** What speed would this 1.0-kg block of ice have if its translational kinetic energy were equal to the thermal energy calculated in part (a)?

75. •• When water freezes into ice it expands in volume by 9.05 percent. Suppose a volume of water is in a household water pipe or a cavity in a rock. If the water freezes, what pressure must be exerted on it to keep its volume from expanding? (If the pipe or rock cannot supply this pressure, the pipe will burst and the rock will split.)

76. •• Suppose the 0.550 kg of ice in Figure 17–22 starts at point A. How much ice is left in the system after **(a)** 5.00×10^4 J, **(b)** 1.00×10^5 J, and **(c)** 1.50×10^5 J of heat are added to the system?

77. ••• Students on a spring break picnic bring a cooler that contains 5.1 kg of ice at 0.0 °C. The cooler has walls that are 3.8 cm thick and are made of Styrofoam, which has a thermal conductivity of 0.030 W/(m · C°). The surface area of the cooler is 1.5 m^2 and it rests in the shade where the air temperature is 21 °C. **(a)** Find the rate at which heat flows into the cooler. **(b)** How long does it take for the ice in the cooler to melt?

78. ••• A 5.0-kg block of ice at −1.5 °C slides on a horizontal surface with a coefficient of kinetic friction equal to 0.062. The initial speed of the block is 6.9 m/s and its final speed is 5.5 m/s. Assuming that all the energy dissipated by kinetic friction goes into melting a small mass m of the ice, and that the rest of the ice block remains at −1.5 °C, determine the value of m.

79. ••• A cylindrical copper rod 37 cm long and 7.5 cm in diameter is placed upright on a hot plate held at a constant temperature of 120 °C, as indicated in **Figure 17–31**. A small depression on top of the rod holds a 25-g ice cube at an initial temperature of 0.0 °C. How long does it take for the ice cube to melt? Assume there is no heat loss through the vertical surface of the rod, and that the thermal conductivity of copper is 390 W/(m · C°).

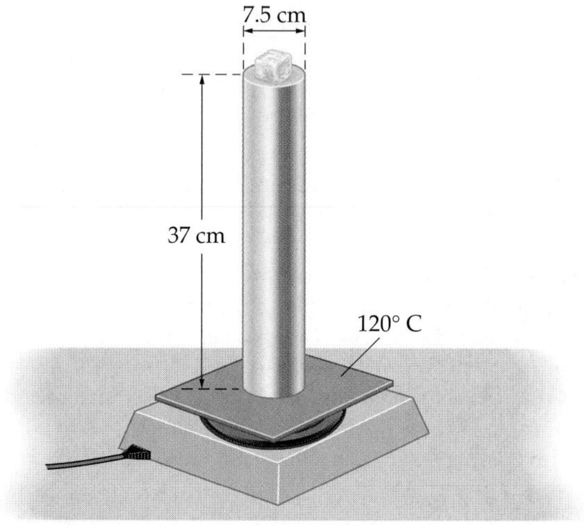

▲ **FIGURE 17–31** Problem 79

Interactive Problems

80. •• ***Referring to Example 17–6*** **(a)** Find the final temperature of the system if *two* 0.0450-kg ice cubes are added to the warm lemonade. The temperature of the ice is 0 °C; the temperature and mass of the warm lemonade are 20.0 °C and 3.95 kg, respectively. **(b)** How many 0.0450-kg ice cubes at 0 °C must be added to the original warm lemonade if the final temperature of the system is to be at least as cold as 15.0 °C?

81. •• ***Referring to Example 17–6*** **(a)** Find the final temperature of the system if a single 0.045-kg ice cube at 0 °C is added to 2.00 kg of lemonade at 1.00 °C. **(b)** What initial temperature of the lemonade will be just high enough to melt all of the ice in a single ice cube and result in an equilibrium temperature of 0 °C? The mass of the lemonade is 2.00 kg and the temperature of the ice cube is 0 °C.

The Laws of Thermodynamics

18

18–1 The Zeroth Law of Thermodynamics 586
18–2 The First Law of Thermodynamics 586
18–3 Thermal Processes 588
18–4 Specific Heats for an Ideal Gas: Constant Pressure, Constant Volume 596
18–5 The Second Law of Thermodynamics 600
18–6 Heat Engines and the Carnot Cycle 600
18–7 Refrigerators, Air Conditioners, and Heat Pumps 604
18–8 Entropy 608
18–9 Order, Disorder, and Entropy 612
18–10 The Third Law of Thermodynamics 614

Every day, green plants take small, simple molecules and use them to create large, complex molecules such as proteins and DNA. These in turn are assembled into even more highly ordered structures: cells and their components. Yet one of the most profound and far-reaching of physical laws holds that all processes decrease the amount of order in the universe. Are living things an exception to this principle? This chapter explores the laws of thermodynamics, and considers the question of whether they apply to biological organisms.

In this chapter we discuss the fundamental laws of nature that govern thermodynamic processes. One of these laws—the one dealing with temperature—was first introduced in Chapter 16. The others are presented here for the first time.

Of particular interest are the first and second laws of thermodynamics. The first law extends the basic principle of energy conservation to include the type of energy transfer we call heat. The real heart of thermodynamics, however, is embodied in the second law. This law introduces a fundamentally new concept to physics—the idea that there is a directionality to the behavior of nature. Melting ice, cooling lava, and the crumbling ruins of the Parthenon all illustrate the second law in action.

Finally, the third law of thermodynamics states that absolute zero is, in fact, the lowest temperature possible. Great efforts have been made over the years to reach lower and lower temperatures, and with remarkable success, but the third law sets the ultimate limit that we can only approach, but never attain.

18–1 The Zeroth Law of Thermodynamics

Though the zeroth law of thermodynamics has already been presented in Chapter 16, we repeat it here so that all the laws of thermodynamics can be collected together in one chapter. As you recall, the zeroth law states the conditions under which objects will be in thermal equilibrium with one another. To be precise:

Zeroth Law of Thermodynamics
If object A is in thermal equilibrium with object C, and object B is separately in thermal equilibrium with object C, then objects A and B will be in thermal equilibrium if they are placed in thermal contact.

The physical quantity that is equal when two objects are in thermal equilibrium is the temperature. In particular, if two objects have the same temperature, we can be assured that *no heat will flow* when they are placed in thermal contact. On the other hand, if heat does flow between two objects, it follows that they are not in thermal equilibrium, and they do not have the same temperature.

Any temperature scale can be used to determine whether objects will be in thermal equilibrium. As we saw in the previous chapter, however, the Kelvin scale is particularly significant in physics. For example, the average kinetic energy of a gas molecule is directly proportional to the Kelvin temperature, as is the volume of an ideal gas. In this chapter we present additional illustrations of the special significance of the Kelvin scale.

18–2 The First Law of Thermodynamics

The first law of thermodynamics is a statement of energy conservation that specifically includes heat. For example, consider the system shown in **Figure 18–1**. The internal energy of this system—that is, the sum of all its potential and kinetic energies—has the initial value U_i. If an amount of heat Q flows into the system, its internal energy increases to the final value $U_f = U_i + Q$. Thus,

$$\Delta U = U_f - U_i = Q \qquad \textbf{18–1}$$

Of course, if heat is removed from the system, its internal energy decreases. We can take this into account by giving Q a *positive* value when the system *gains* heat, and a *negative* value when it *loses* heat.

Similarly, suppose the system under consideration does a work W on the external world, as in **Figure 18–2**. If the system is insulated so that no heat can flow in or out, the energy to do the work must come from the internal energy of

▶ **FIGURE 18–1 The internal energy of a system**
A system initially has the internal energy U_i (left). After the heat Q is added, the system's new internal energy is $U_f = U_i + Q$ (right). The system has rigid walls; hence, it can do no work on the external world.

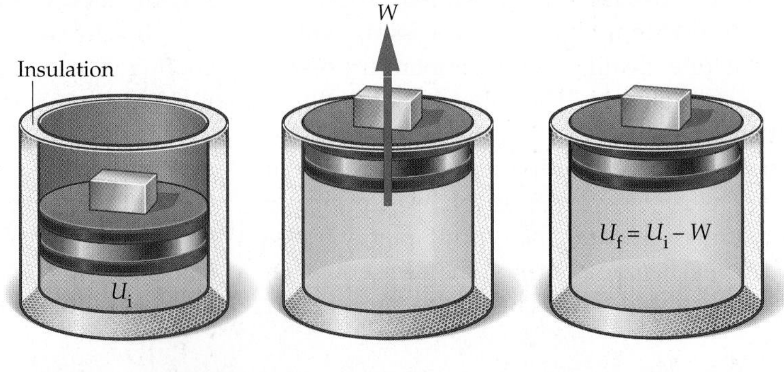

▶ **FIGURE 18–2 Work and internal energy**
A system initially has the internal energy U_i (left). After the system does the work W on the external world, its remaining internal energy is $U_f = U_i - W$ (right). Note that the insulation guarantees that no heat is gained or lost by the system.

the system. Thus, if the initial internal energy is U_i, the final internal energy is $U_f = U_i - W$. Therefore,

$$\Delta U = U_f - U_i = -W \qquad 18\text{–}2$$

On the other hand, if work is done *on* the system, its internal energy increases. Thus, we use the following sign convention for the work: W has a *positive* value when the system *does work on the external world*, and it has a *negative* value when *work is done on the system*. These sign conventions are summarized in Table 18–1.

Combining the results in Equations 18–1 and 18–2 yields the first law of thermodynamics:

First Law of Thermodynamics

The change in a system's internal energy, ΔU, is related to the heat Q and the work W as follows:

$$\Delta U = Q - W \qquad 18\text{–}3$$

If you apply the sign conventions given here, it is straightforward to verify that adding heat to a system, and/or doing work on it, increases the internal energy. On the other hand, if the system does work, and/or heat is removed, its internal energy decreases. Example 18–1 gives a specific numerical application of the first law.

TABLE 18–1 Signs of Q and W

Q positive	System *gains* heat
Q negative	System *loses* heat
W positive	Work done *by* system
W negative	Work done *on* system

PROBLEM-SOLVING NOTE

Proper Signs for Q and W

When applying the first law of thermodynamics, it is important to determine the proper signs for Q, W, and ΔU.

EXAMPLE 18–1 Heat, Work, and Internal Energy

(a) Jogging along the beach one day, you do 4.3×10^5 J of work and give off 3.8×10^5 J of heat. What is the change in your internal energy? **(b)** Switching over to walking, you give off 1.2×10^5 J of heat and your internal energy decreases by 2.6×10^5 J. How much work have you done while walking?

Picture the Problem

Our sketch shows a person jogging along the beach. The fact that the person does work on the external world means that W is positive. As for the heat, the fact that heat is given off by the person means that Q is negative.

Strategy

The signs of W and Q have been determined in our sketch, and the magnitudes are given in the problem statement. To find ΔU for part (a), we simply use the first law of thermodynamics, $\Delta U = Q - W$. To find the work W for part (b), we solve the first law for W, which yields $W = Q - \Delta U$.

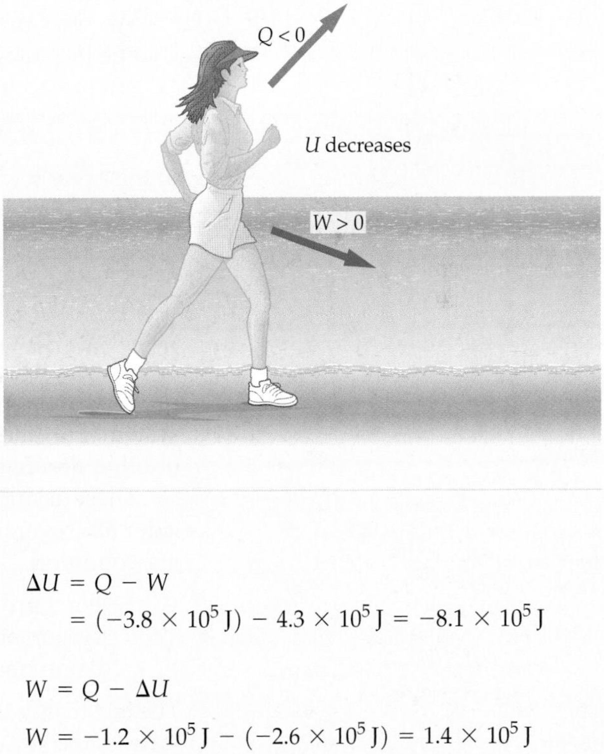

Solution

Part (a)

1. Calculate ΔU, using $Q = -3.8 \times 10^5$ J and $W = 4.3 \times 10^5$ J:

$$\Delta U = Q - W = (-3.8 \times 10^5\ \text{J}) - 4.3 \times 10^5\ \text{J} = -8.1 \times 10^5\ \text{J}$$

Part (b)

2. Solve $\Delta U = Q - W$ for W:

$$W = Q - \Delta U$$

3. Substitute $Q = -1.2 \times 10^5$ J and $\Delta U = -2.6 \times 10^5$ J:

$$W = -1.2 \times 10^5\ \text{J} - (-2.6 \times 10^5\ \text{J}) = 1.4 \times 10^5\ \text{J}$$

Insight

Note the importance of using the correct sign for Q, W, and ΔU. When the proper signs are used, the first law is simply a way of keeping track of all the energy exchanges—including mechanical work and heat—that can occur in a given system.

Practice Problem

After walking for a few minutes you begin to run, doing 5.1×10^5 J of work and decreasing your internal energy by 8.8×10^5 J. How much heat have you given off? [**Answer:** $Q = -3.7 \times 10^5$ J. The negative sign means you have given off heat, as expected.]

Some related homework problems: Problem 1, Problem 3

Just looking at the first law, $\Delta U = Q - W$, it is easy to get the false impression that U, Q, and W are basically the same type of physical quantity. Certainly, they are all measured in the same units (J). In other respects, however, they are quite different. For example, the heat Q represents energy that flows through thermal contact. In contrast, the work W indicates a transfer of energy by the action of a force through a distance.

The most important distinction between these quantities, however, is the way they depend on the **state of a system,** which is determined by its temperature, pressure, and volume. The internal energy, for example, depends only on the state of a system, and not on how the system is brought to that state. A simple example is the ideal gas, where U depends only on the temperature T, and not on any previous values T may have had. Since U depends only on the state of a system—whether the system is an ideal gas or something more complicated—it is referred to as a **state function.**

On the other hand, Q and W are *not* state functions; they depend on the precise way—that is, on the process—by which a system is changed from one state, A, to another state, B. For example, one process connecting the states A and B may result in a heat $Q_1 = 19\text{ J}$ and a work $W_1 = 32\text{ J}$. With a different process connecting the *same two states*, we may find a heat, $Q_2 = -24\text{ J} \neq Q_1$, and a work, $W_2 = -11\text{ J} \neq W_1$. Still, the difference in internal energy—which depends only on the initial and final states—must be the same for all processes connecting A and B. It follows that

$$\Delta U_{AB} = Q_1 - W_1 = Q_2 - W_2 = -13\text{ J}$$

Therefore, the energy of this system always decreases by 13 J when it changes from state A to state B.

We now turn our attention to specific types of thermal processes that can change the energy of a system.

18–3 Thermal Processes

In this section we consider a variety of thermodynamic processes that can be used to change the state of a system. Among these are ones that take place at constant pressure, constant volume, or constant temperature. We also consider processes in which no heat is allowed to flow into or out of the system.

All of the processes discussed in this section are assumed to be **quasi-static,** which is a fancy way of saying they occur so slowly that at any given time the system and its surroundings are essentially in equilibrium. Thus, in a quasi-static process, the pressure and temperature are always uniform throughout the system. Furthermore, we assume that the system in question is free from friction or other dissipative forces.

These assumptions can be summarized by saying that the processes we consider are **reversible.** To be precise, a reversible process is defined by the following condition:

> For a process to be reversible, it must be possible to return both the system and its surroundings to exactly the same states they were in before the process began.

The fact that both the system and its surroundings must be returned to their initial states is the key element of this definition. For example, if there is friction between a piston and the cylinder in which it slides, a reversible process is not possible. Even if the piston is returned to its original location, the heat generated by friction will warm the cylinder and eventually flow into the surrounding air. Thus, it is not possible to "undo" the effects of friction, and such a process is said to be **irreversible.** Even without friction, a process can be irreversible if it occurs rapidly enough to cause effects such as turbulence, or if the system is far from equilibrium.

In practice, then, all real processes are irreversible to some extent. It is still possible, however, to have a process that closely approximates a perfectly reversible process, just as we can have systems that are practically free of friction.

▲ **FIGURE 18–3 An idealized reversible process**

(a) A piston is slowly moved downward, compressing a gas. In order for its temperature to remain constant, a heat Q goes from the gas into the constant-temperature heat bath, which may be nothing more than a large volume of water. **(b)** As the piston is allowed to slowly rise back to its initial position, it draws the same amount of heat Q from the heat bath that it gave to the bath in (a). Hence, both the system (gas) and the surroundings (heat bath) return to their initial states.

For example, in **Figure 18–3 (a)**, we consider a "frictionless" piston that is slowly forced downward, while the gas in the cylinder is kept at constant temperature. An amount of heat, Q, goes from the gas to its surroundings in order to keep the temperature from rising. In **Figure 18–3 (b)** the piston is slowly moved back upward, drawing in the same heat Q from its surroundings to keep the temperature from dropping. In an "ideal" case like this, the process is reversible, since the system and its surroundings are left unchanged. We shall assume that all the processes described in this section are reversible.

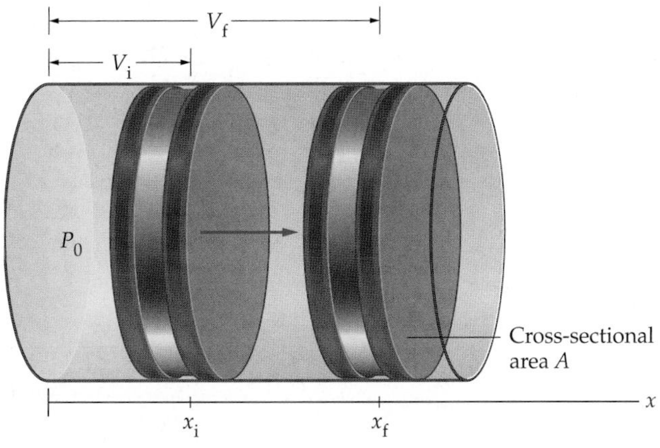

◀ **FIGURE 18–4 Work done by an expanding gas**

A gas in a cylinder of cross-sectional area A expands with a constant pressure of P_0 from an initial volume $V_i = Ax_i$ to a final volume $V_f = Ax_f$. As it expands, it does the work $W = P_0(V_f - V_i)$.

Constant Pressure/Constant Volume

We begin by considering a process that occurs at **constant pressure.** To be specific, suppose that a gas with the pressure P_0 is held in a cylinder of cross-sectional area A, as in **Figure 18–4**. If the piston moves outward, so that the volume of the gas increases from an initial value V_i to a final value V_f, the process can be represented graphically as shown in **Figure 18–5**. Here we plot pressure P versus volume V in a PV plot; the process just described is the horizontal line segment.

As the gas expands, it does work on the piston. First, the gas exerts a force on the piston equal to the pressure times the area:

$$F = P_0A$$

Second, the gas moves the piston from the position x_i to the position x_f, where $V_i = Ax_i$ and $V_f = Ax_f$, as indicated in Figure 18–4. Thus, the work done by the gas is the force times the distance through which the piston moves:

$$W = F(x_f - x_i) = P_0A(x_f - x_i) = P_0(Ax_f - Ax_i) = P_0(V_f - V_i)$$

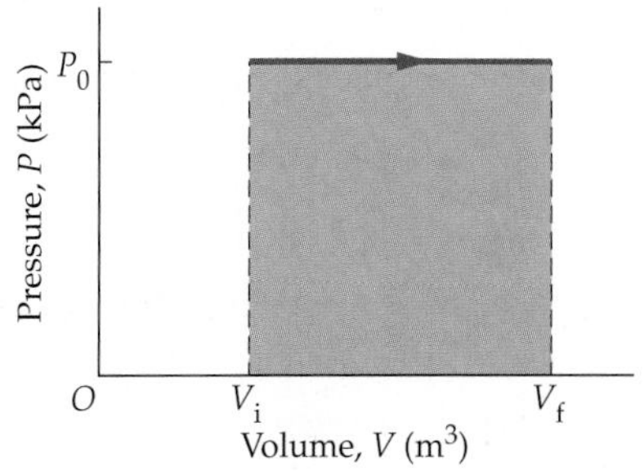

▲ **FIGURE 18–5 A constant-pressure process**

A PV plot representing the constant-pressure process shown in Figure 18–4. The area of the shaded region, $P_0(V_f - V_i)$, is equal to the work done by the expanding gas in Figure 18–4.

In general, if the pressure, P, is constant, and the volume changes by the amount ΔV, the work done by the gas is

$$W = P\Delta V \qquad \text{(constant pressure)} \qquad 18\text{–}4$$

EXERCISE 18–1

A gas with a constant pressure of 150 kPa expands from a volume of 0.76 m^3 to a volume of 0.92 m^3. How much work does the gas do?

Solution

Applying Equation 18–4 we find

$$W = P\,\Delta V = (150\text{ kPa})(0.92\text{ m}^3 - 0.76\text{ m}^3) = 24{,}000\text{ J}$$

This is the energy required to raise the temperature of 1.0 kg of water by 5.7 C°.

Looking closely at the PV plot in Figure 18–5, we see that the work done by the gas is equal to the area under the horizontal line representing the constant-pressure process. In particular, the shaded region is a rectangle of height P_0 and width $V_f - V_i$, and therefore its area is

$$\text{area} = P_0(V_f - V_i) = W$$

Though this result was obtained for the special case of constant pressure, it applies to any process; that is, *the work done by an expanding gas is equal to the area under the curve representing the process in a PV plot*. This result is applied in the next Example.

PROBLEM-SOLVING NOTE

Work and the *PV* Diagram

When finding the work done by calculating the area on a PV diagram, recall that a pressure of 1 Pa times a volume of 1 m^3 gives an energy equal to 1 J.

EXAMPLE 18–2 Work Area

A gas expands from an initial volume of 0.40 m^3 to a final volume of 0.62 m^3 as the pressure increases linearly from 110 kPa to 230 kPa. Find the work done by the gas.

Picture the Problem

The sketch shows the process for this problem. As the volume increases from $V_i = 0.40\text{ m}^3$ to $V_f = 0.62\text{ m}^3$, the pressure increases linearly from $P_i = 110$ kPa to $P_f = 230$ kPa.

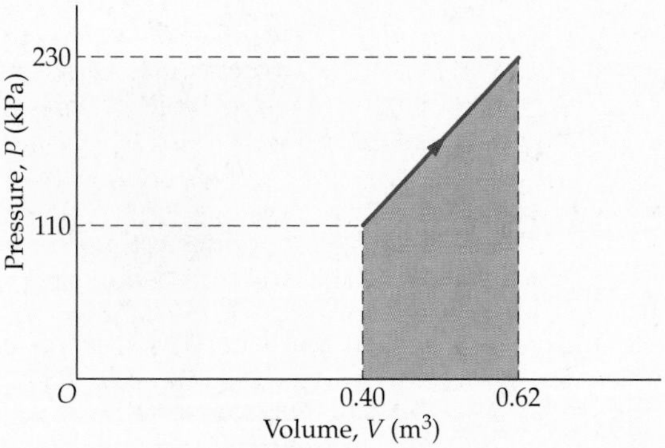

Strategy

The work done by this gas is equal to the shaded area in the sketch. We can calculate this area as the sum of the area of a rectangle plus the area of a triangle.

In particular, the rectangle has a height P_i and a width $(V_f - V_i)$. Similarly, the triangle has a height $(P_f - P_i)$ and a base of $(V_f - V_i)$.

Solution

1. Calculate the area of the rectangular portion of the total area:

$$A_{\text{rectangle}} = P_i(V_f - V_i) = (110\text{ kPa})(0.62\text{ m}^3 - 0.40\text{ m}^3) = 2.4 \times 10^4\text{ J}$$

2. Next, calculate the area of the triangular portion of the total area:

$$A_{\text{triangle}} = \tfrac{1}{2}(P_f - P_i)(V_f - V_i) = \tfrac{1}{2}(230\text{ kPa} - 110\text{ kPa})(0.62\text{ m}^3 - 0.40\text{ m}^3) = 1.3 \times 10^4\text{ J}$$

3. Sum these areas to find the work done by the gas:

$$W = A_{\text{rectangle}} + A_{\text{triangle}} = 2.4 \times 10^4\text{ J} + 1.3 \times 10^4\text{ J} = 3.7 \times 10^4\text{ J}$$

Insight

We could also have solved this problem by noting that since the pressure varies linearly, its average value is simply $P_{av} = \tfrac{1}{2}(P_f + P_i) = 170$ kPa. The work done by the gas, then, is $W = P_{av}\,\Delta V = (170\text{ kPa})(0.22\text{ m}^3) = 3.7 \times 10^4$ J, as before.

Practice Problem

Suppose the pressure varies linearly from 110 kPa to 260 kPa. How much work does the gas do in this case? [**Answer:** $W = 4.1 \times 10^4$ J]

Some related homework problems: Problem 17, Problem 18

◀ **FIGURE 18–6 Adding heat to a system of constant volume**

Heat is added to a system of constant volume, increasing its pressure from P_i to P_f. Since there is no displacement of the walls, there is no work done in this process. Therefore, $W = 0$ and the change in internal energy is simply equal to the heat added to or removed from the system, $\Delta U = Q$.

Next, we consider a **constant-volume** process. Suppose, for example, that heat is added to a gas in a container of fixed volume, as in **Figure 18–6**, causing the pressure to increase. Since there is no displacement of any of the walls, it follows that the force exerted by the gas does no work. Thus, for *any* constant-volume process, we have

$$W = 0 \qquad \textit{(constant volume)}$$

From the first law of thermodynamics, $\Delta U = Q - W$, it follows that $\Delta U = Q$; that is, the change in internal energy is equal to the amount of heat that is added to or removed from the system. Note that zero work in a constant-volume process is consistent with our earlier statement that the work is equal to the area under the curve representing the process. For example, in **Figure 18–7** we show a constant-volume process in which the pressure is increased from P_i to P_f. Since this line is vertical, the area under it is zero; that is, $W = 0$ as expected.

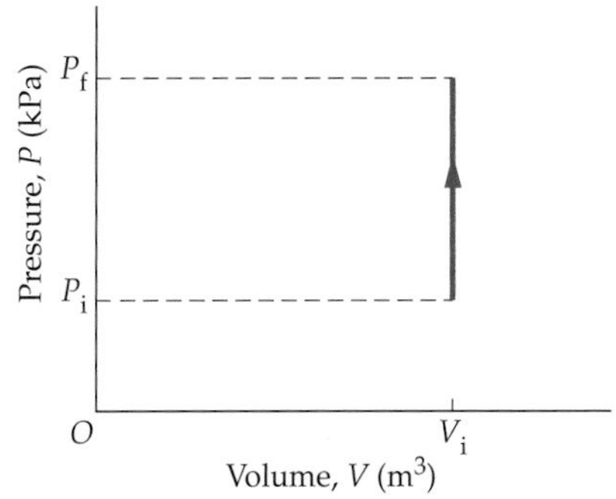

▲ **FIGURE 18–7 A constant-volume process**

The pressure increases from P_i to P_f, just as in Figure 18–6, while the volume remains constant at its initial value, V_i. The area under this process is zero, as is the work.

ACTIVE EXAMPLE 18–1 Find the Total Work

A gas undergoes the three-part process shown here, connecting the states A and B. Find the total work done by the gas during this process.

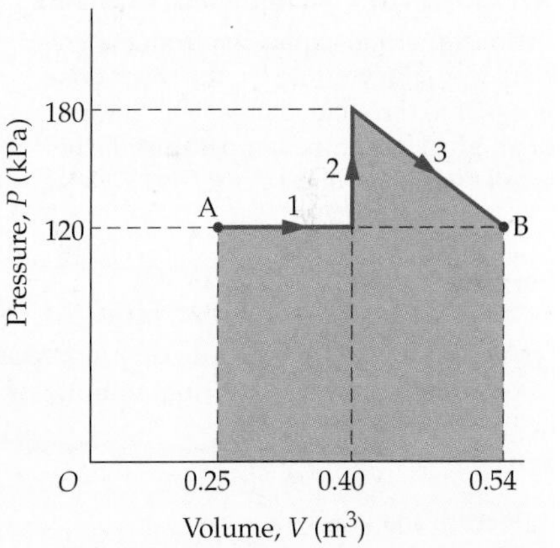

Solution *(Test your understanding by performing the calculations indicated in each step.)*

1. Calculate the work done during part 1 of the total process: 18,000 J
2. Calculate the work done during part 2 of the total process: 0 J
3. Calculate the work done during part 3 of the total process: 21,000 J
4. Sum the results to find the total work: 39,000 J

Insight

Note that the work done by the gas would be different if the process connecting A and B were a constant-pressure expansion with a pressure of 120 kPa. In that case, the total work done would be 35,000 J. Thus, the work W—which is not a state function—depends on the process, as expected.

Your Turn

Describe a process connecting the states A and B for which the work done by the gas is 58,000 J.

(Answers to **Your Turn** *problems are given in the back of the book.)*

Isothermal Processes

Another common process is one that takes place at constant temperature; that is, an **isothermal process.** For an ideal gas the isotherm has a relatively simple form. In particular, if T is constant, it follows that PV is a constant as well:

$$PV = NkT = \text{constant}$$

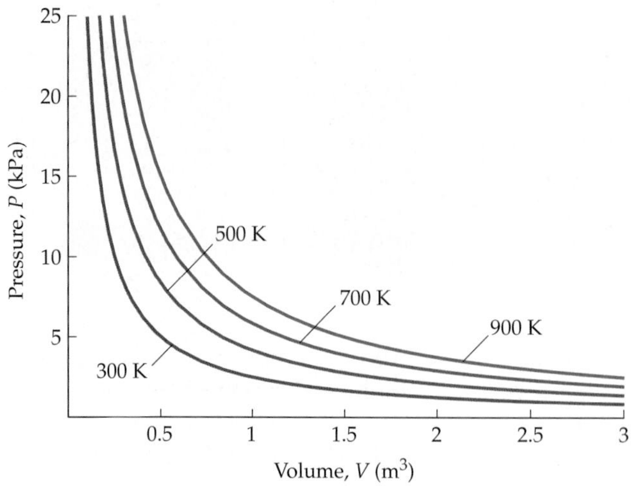

▶ **FIGURE 18–8 Isotherms on a *PV* plot**
These isotherms are for one mole of an ideal gas at the temperatures 300 K, 500 K, 700 K, and 900 K. Notice that each isotherm has the shape of a hyperbola. As the temperature is increased, however, the isotherms move farther from the origin. Thus, the pressure corresponding to a given volume increases with temperature, as one would expect.

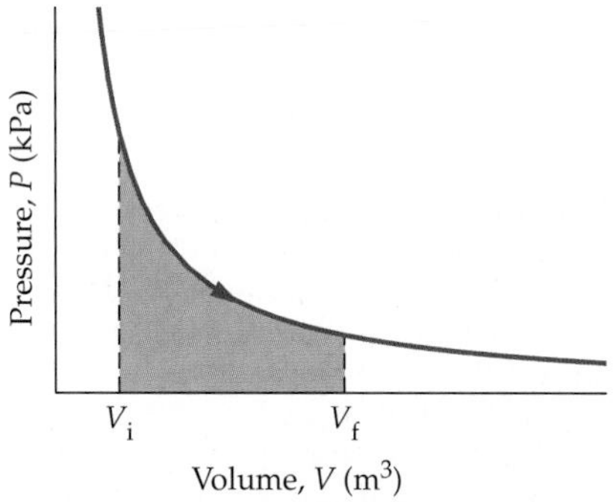

▲ **FIGURE 18–9 An isothermal expansion**
In an isothermal expansion from the volume V_i to the volume V_f, the work done is equal to the shaded area. For n moles of an ideal gas at the temperature T, the work by the gas is $W = nRT\ln(V_f/V_i)$.

Thus, ideal-gas isotherms have the following pressure-volume relationship:

$$P = \frac{NkT}{V} = \frac{\text{constant}}{V}$$

This is illustrated in **Figure 18–8**, where we show several isotherms corresponding to different temperatures. (Recall that we've seen isotherms before, in Figure 17–5.)

For any given isotherm, such as the one shown in **Figure 18–9**, the work done by an expanding gas is equal to the area under the curve, as usual. In particular, the work in expanding from V_i to V_f in Figure 18–9 is equal to the shaded area. This area may be derived by using the methods of calculus. The result is found to be

$$W = NkT\ln\left(\frac{V_f}{V_i}\right) = nRT\ln\left(\frac{V_f}{V_i}\right) \qquad \textit{(constant temperature)} \qquad \textbf{18–5}$$

Note that "ln" stands for the natural logarithm; that is, log to the base e. This result is utilized in the next Example.

EXAMPLE 18–3 Heat Flow

A cylinder holds 0.50 mol of a monatomic ideal gas at a temperature of 310 K. As the gas expands isothermally from an initial volume of 0.31 m^3 to a final volume of 0.45 m^3, determine the amount of heat that must be added to the gas to maintain a constant temperature.

Picture the Problem

The physical process is illustrated in the drawing to the left. Note that heat flows into the gas as it expands in order to keep its temperature constant at $T = 310$ K. The graph to the right is a PV plot showing the same process. The work done by the expanding gas is equal to the shaded area from $V_i = 0.31\ m^3$ to $V_f = 0.45\ m^3$.

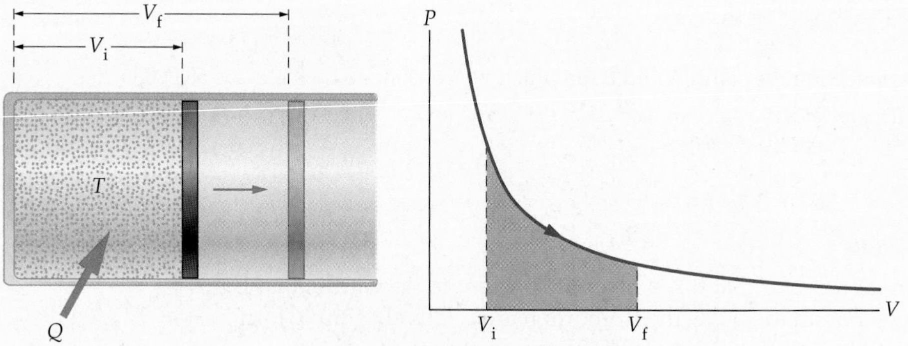

Strategy

We can use the first law, $\Delta U = Q - W$, to find the heat Q in terms of W and ΔU. We can find W and ΔU as follows:

First, the work W is found using $W = nRT \ln(V_f/V_i)$.

Next, recall that the internal energy of an ideal gas depends only on the temperature. For example, the internal energy of a monatomic ideal gas is $U = \frac{3}{2}nRT$. Since the temperature is constant in this process, there is no change in internal energy; that is, $\Delta U = 0$.

Solution

1. Solve the first law of thermodynamics for the heat, Q:

$$\Delta U = Q - W$$
$$Q = \Delta U + W$$

2. Calculate the work done by the expanding gas:

$$W = nRT \ln\left(\frac{V_f}{V_i}\right)$$
$$= (0.50\text{ mol})[8.31\text{ J}/(\text{mol}\cdot\text{K})](310\text{ K}) \ln\left(\frac{0.45\text{ m}^3}{0.31\text{ m}^3}\right)$$
$$= 480\text{ J}$$

3. Calculate the change in the gas's internal energy:

$$\Delta U = \tfrac{3}{2}nR(T_f - T_i) = 0$$

4. Substitute numerical values to find Q:

$$Q = \Delta U + W = 0 + 480\text{ J} = 480\text{ J}$$

Insight

We find, then, that when an ideal gas undergoes an isothermal expansion, the heat gained by the gas is equal to the work it does; that is, $Q = W$. This is a direct consequence of the first law of thermodynamics, $\Delta U = Q - W$, and the fact that $\Delta U = 0$ for an ideal gas at constant temperature.

Practice Problem

Find the final volume of this gas when it has expanded enough to do 590 J of work. [**Answer:** $V_f = 0.49\text{ m}^3$]

Some related homework problems: Problem 15, Problem 20

Note that if a gas is compressed at constant temperature, work is done *on* the gas, rather than *by* the gas. As a result, we expect the work to be negative. This is consistent with Equation 18–5, since in a compression the final volume is less than the initial volume. Thus, $V_f/V_i < 1$, and hence $W = nRT \ln(V_f/V_i)$ is negative.

Finally, we consider one last point regarding isothermal processes.

CONCEPTUAL CHECKPOINT 18–1 **Ideal or Not?**

The internal energy of a certain gas increases when it is compressed isothermally. Is the gas ideal?

Reasoning and Discussion

In an isothermal process, the internal energy of an ideal gas is unchanged, since the temperature is constant. Hence, if the internal energy of this gas changes during an isothermal compression, it must not be ideal.

Answer:

No, the gas is not ideal.

Adiabatic Processes

The final process we consider is one in which no heat flows into or out of the system. Such a process, in which $Q = 0$, is said to be **adiabatic.** One way to produce an adiabatic process is illustrated in **Figure 18–10 (a)**. Here we see a cylinder that is insulated well enough that no heat can pass through the insulation (adiabatic means, literally, "not passable"). When the piston is pushed downward in the cylinder—decreasing the volume—the gas heats up and its pressure increases. Similarly, in **Figure 18–10 (b)**, an adiabatic expansion causes the temperature of the gas to decrease, as does the pressure.

▲ **FIGURE 18–10 An adiabatic process**

In adiabatic processes, no heat flows into or out of the system. In the cases shown in this figure, heat flow is prevented by insulation. **(a)** An adiabatic compression increases both the pressure and the temperature. **(b)** An adiabatic expansion results in a decrease in pressure and temperature.

What does an adiabatic process look like on a *PV* plot? Certainly, its general shape must be similar to that of an isotherm; in particular, as the volume is decreased, the pressure increases. However, it can't be identical to an isotherm because, as we have pointed out, the temperature changes during an adiabatic process. The comparison between an adiabatic curve and an isothermal curve is the subject of the next Conceptual Checkpoint.

CONCEPTUAL CHECKPOINT 18–2 Pressure versus Volume

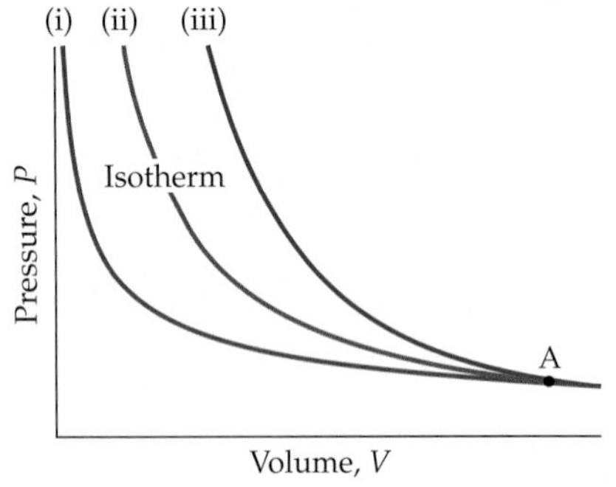

A certain gas has an initial volume and pressure given by point *A* in the *PV* plot. If the gas is compressed isothermally, its pressure rises as indicated by the curve labeled "isotherm." If, instead, the gas is compressed adiabatically from point A, does its pressure follow **(a)** curve i, **(b)** curve ii, or **(c)** curve iii?

Reasoning and Discussion

In an isothermal compression, some heat flows out of the system in order for its temperature to remain constant. No heat flows out of the system in the adiabatic process, however, and therefore its temperature rises. As a result, the pressure is greater for any given volume when the compression is adiabatic, as compared to isothermal. It follows, then, that the adiabatic curve, or adiabat, is represented by curve iii.

Answer:

(c) As volume is decreased, the pressure on an adiabat rises more rapidly than on an isotherm.

As we have seen, then, an adiabatic curve is similar to an isotherm, only steeper. The precise mathematical relationship describing an adiabat will be presented in the next section.

EXAMPLE 18–4 Work into Energy

When a certain gas is compressed adiabatically, the amount of work done on it is 640 J. Find the change in internal energy of the gas.

Picture the Problem

Our sketch shows a piston being pushed downward, compressing a gas in an insulated cylinder. The insulation ensures that no heat can flow—as required in an adiabatic process.

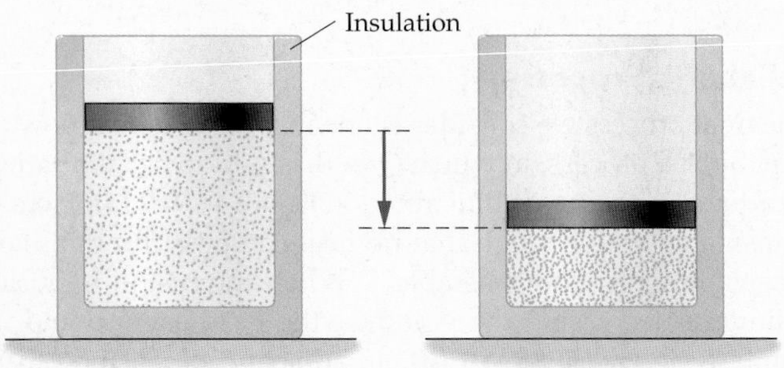

Strategy

We know that 640 J of work is done on the gas, and we know that no heat is exchanged ($Q = 0$) in an adiabatic process. Thus, we can find ΔU by substituting Q and W into the first law of thermodynamics. One note of caution: Be careful to use the correct sign for the work. In particular, recall that work done *on* a system is negative.

Solution

1. Identify the work and heat for this process: $W = -640\text{ J}$, $Q = 0$

2. Substitute Q and W into the first law of thermodynamics to find the change in internal energy, ΔU: $\Delta U = Q - W = 0 - (-640\text{ J}) = 640\text{ J}$

Insight

Because no energy can enter or leave the system in the form of heat, all the work done on the system goes into increasing its internal energy. This follows from the first law of thermodynamics, $\Delta U = Q - W$, which for $Q = 0$ reduces to $\Delta U = -W$. As a result, the temperature of the gas increases.

A familiar example of this type of effect is the heating that occurs when you pump air into a tire or a ball—the work done on the pump appears as an increased temperature. The effect occurs in the reverse direction as well. When air is let out of a tire, for example, it does work on the atmosphere as it expands, producing a cooling effect that can be quite noticeable. In extreme cases, the cooling can be great enough to create frost on the valve stem of the tire.

Practice Problem

If a system's internal energy decreases by 470 J in an adiabatic process, how much work was done by the system? [**Answer:** $W = +470\text{ J}$]

Some related homework problems: Problem 16, Problem 23

An adiabatic process can occur when the system is thermally insulated, as in Figure 18–10, or in a system where the change in volume occurs rapidly. For example, if an expansion or compression happens quickly enough, there is no time for heat flow to occur. As a result, the process is adiabatic, even if there is no insulation.

REAL-WORLD PHYSICS
Adiabatic heating and diesel engines

An example of a rapid process is shown in **Figure 18–11**. Here, a piston is fitted into a cylinder that contains a certain volume of gas and a small piece of tissue paper. If the piston is driven downward rapidly, by a sharp impulsive blow, for example, the gas is compressed before heat has a chance to flow. As a result, the temperature of the gas rises rapidly. In fact, the rise in temperature can be enough for the paper to burst into flames.

The same principle applies to the operation of a diesel engine. As you may know, a diesel differs from a standard internal combustion engine in that it has no spark plugs. It doesn't need them. Instead of using a spark to ignite the fuel in a cylinder, it uses adiabatic heating. Fuel and air are admitted into the cylinder, then the piston rapidly compresses the air-fuel mixture. Just as with the piece of paper in Figure 18–11, the rising temperature is sufficient to ignite the fuel and run the engine.

Adiabatic heating is one of the mechanisms being considered to explain the fascinating and enigmatic phenomenon known as *sonoluminescence*. Sonoluminescence occurs when an intense, high-frequency sound wave causes a small gas bubble in water to pulsate. When the sound wave collapses the bubble to its minimum size, which is about a thousandth of a millimeter, the bubble gives off an extremely short burst of light. The light is mostly in the ultraviolet (see Chapter 25), but enough is in the visible range of light to make the bubble appear blue to the eye. For an object to give off light in the ultraviolet it must be extremely hot (see Chapter 30). In fact, it is estimated that the temperature inside a collapsing bubble is at least 10,000 °F, about the same temperature as the surface of the Sun.

The characteristics of constant-pressure, constant-volume, isothermal, and adiabatic processes are summarized in Table 18–2.

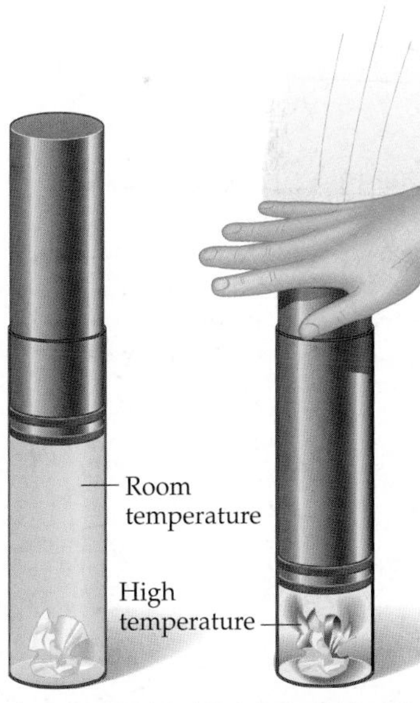

▲ **FIGURE 18–11 Adiabatic heating**
When a rod that fits snugly into a cylinder is pushed downward rapidly, the temperature of the gas increases before there is time for heat to flow out of the system. Thus, the process is essentially adiabatic. As a result, the temperature of the gas can increase enough to ignite bits of paper in the cylinder. In a diesel engine, the same principle is used to ignite an air-gasoline mixture.

TABLE 18–2 Thermodynamic Processes and Their Characteristics

Constant pressure	$W = P\Delta V$	$Q = \Delta U + P\Delta V$
Constant volume	$W = 0$	$Q = \Delta U$
Isothermal (constant temperature)	$W = Q$	$\Delta U = 0$
Adiabatic (no heat flow)	$W = -\Delta U$	$Q = 0$

18–4 Specific Heats for an Ideal Gas: Constant Pressure, Constant Volume

Recall that the specific heat of a substance is the amount of heat needed to raise the temperature of 1 kg of the substance by 1 Celsius degree. As we know, however, the amount of heat depends on the type of process used to raise the temperature. Thus, we should specify, for example, whether a specific heat applies to a process at constant pressure or constant volume.

If a substance is heated or cooled while open to the atmosphere, the process occurs at constant (atmospheric) pressure. This has been the case in all the specific heat discussions to this point; thus it was not necessary to make a distinction between different types of specific heats. We now wish to consider constant-volume processes as well, and the relationship between constant-volume and constant-pressure specific heats.

A constant-volume process is illustrated in **Figure 18–12**. Here we see an ideal gas of mass m in a container of fixed volume V. A heat Q flows into the container. As a result of this added heat, the temperature of the gas rises by the amount ΔT, and its pressure increases as well. Now, the specific heat at constant volume, c_{v}, is defined by the following relation:

$$Q_{\mathrm{v}} = mc_{\mathrm{v}}\Delta T$$

In what follows, it will be more convenient to use the **molar specific heat,** denoted by a capital letter C, which is defined in terms of the number of moles rather than the mass of the substance. Thus, if a gas contains n moles, its molar specific heat at constant volume is given by

$$Q_{\mathrm{v}} = nC_{\mathrm{v}}\Delta T$$

Similarly, a constant-pressure process is illustrated in **Figure 18–13**. In this case, the gas is held in a container with a moveable piston that applies a constant pressure P. As a heat Q is added to the gas, its temperature increases, which causes the piston to rise—after all, if the piston didn't rise, the pressure of the gas

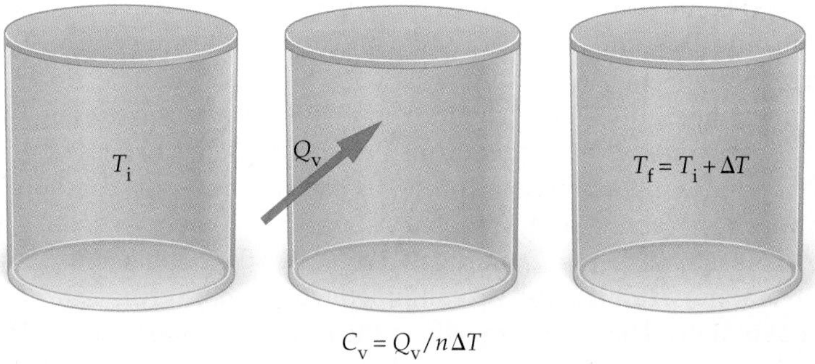

▶ **FIGURE 18–12 Heating at constant volume**

If the heat Q_{v} is added to n moles of gas, and the temperature rises by ΔT, the molar specific heat at constant volume is $C_{\mathrm{v}} = Q_{\mathrm{v}}/n\Delta T$. No mechanical work is done at constant volume.

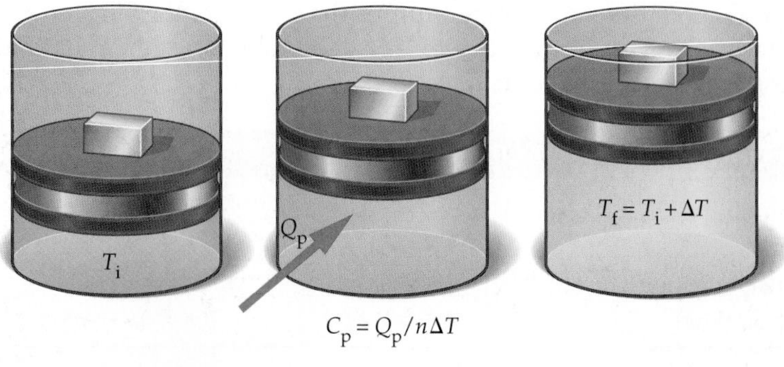

▶ **FIGURE 18–13 Heating at constant pressure**

If the heat Q_{p} is added to n moles of gas, and the temperature rises by ΔT, the molar specific heat at constant pressure is $C_{\mathrm{p}} = Q_{\mathrm{p}}/n\Delta T$. Note that the heat Q_{p} must increase the temperature *and* do mechanical work by lifting the piston.

would increase. If the temperature of the gas increases by the amount ΔT, the molar specific heat at constant pressure is given by

$$Q_p = nC_p\,\Delta T$$

We would now like to obtain a relation between C_p and C_v.

Before we carry out the mathematics, let's consider the qualitative relationship between these specific heats. This is addressed in the following Conceptual Checkpoint.

CONCEPTUAL CHECKPOINT 18–3 Comparing Specific Heats

How does the molar specific heat at constant pressure, C_p, compare with the molar specific heat at constant volume, C_v? (a) $C_p > C_v$; (b) $C_p = C_v$; (c) $C_p < C_v$.

Reasoning and Discussion

In a constant-volume process, as in Figure 18–12, the heat that is added to a system goes entirely into increasing the temperature, since no work is done. On the other hand, at constant pressure the heat added to a system increases the temperature *and* does mechanical work. This is illustrated in Figure 18–13 where we see that the heat must not only raise the temperature, but also supply enough energy to lift the piston. Thus, more heat is required in the constant-pressure process, and hence that specific heat is greater.

Answer:

(a) The specific heat at constant pressure is greater than the specific heat at constant volume.

We turn now to a detailed calculation of C_v for a monatomic ideal gas. To begin, rearrange the first law of thermodynamics, $\Delta U = Q - W$, to solve for the heat, Q:

$$Q = \Delta U + W$$

Recall from the previous section, however, that the work is zero, $W = 0$, for any constant volume process. Hence, for constant volume we have

$$Q_v = \Delta U$$

Finally, noting that $U = \frac{3}{2}NkT = \frac{3}{2}nRT$, yields

$$Q_v = \Delta U = \tfrac{3}{2}nR\Delta T$$

Comparing with the definition of the molar specific heat, we find

Molar Specific Heat for a Monatomic Ideal Gas at Constant Volume

$$C_v = \tfrac{3}{2}R \qquad \text{18–6}$$

Now we perform a similar calculation for constant pressure. In this case, referring again to the previous section, we find that $W = P\Delta V$. Since we are considering an ideal gas, in which $PV = nRT$, it follows that

$$W = P\Delta V = nR\Delta T$$

Combining this with the first law of thermodynamics yields

$$Q_p = \Delta U + W$$
$$= \tfrac{3}{2}nR\Delta T + nR\Delta T = \tfrac{5}{2}nR\Delta T$$

Applying the definition of molar specific heats yields

Molar Specific Heat for a Monatomic Ideal Gas at Constant Pressure

$$C_p = \tfrac{5}{2}R \qquad \text{18–7}$$

PROBLEM-SOLVING NOTE

Constant Volume versus Constant Pressure

The heat required to increase the temperature of an ideal gas depends on whether the process is at constant pressure or constant volume. More heat is required when the process occurs at constant pressure.

TABLE 18–3 $C_p - C_v$ for Various Gases

Helium	0.995 *R*
Nitrogen	1.00 *R*
Oxygen	1.00 *R*
Argon	1.01 *R*
Carbon dioxide	1.01 *R*
Methane	1.01 *R*

As expected, the specific heat at constant pressure is larger than the specific heat at constant volume, and the difference is precisely the extra contribution due to the work done in lifting the piston in the constant-pressure case. In particular, we see that

$$C_p - C_v = R \qquad \textbf{18–8}$$

Though this relation was derived for a monatomic ideal gas, it holds for all ideal gases, regardless of the structure of their molecules. It is also a good approximation for most real gases, as can be seen in Table18–3.

EXERCISE 18–2

Find the heat required to raise the temperature of 0.200 mol of a monatomic ideal gas by 5.00 C° at **(a)** constant volume and **(b)** constant pressure.

Solution

Applying Equations 18–6 and 18–7, we find

a. $Q_v = \frac{3}{2}nR\Delta T = \frac{3}{2}(0.200\text{ mol})[8.31\text{ J}/(\text{mol}\cdot\text{K})](5.00\text{ K}) = 12.5\text{ J}$

b. $Q_p = \frac{5}{2}nR\Delta T = \frac{5}{2}(0.200\text{ mol})[8.31\text{ J}/(\text{mol}\cdot\text{K})](5.00\text{ K}) = 20.8\text{ J}$

Adiabatic Processes

We return now briefly to a consideration of adiabatic processes. As we shall see, the relationship between C_p and C_v is important in determining the behavior of a system undergoing an adiabatic process.

Figure 18–14 shows an adiabatic curve and two isotherms. As mentioned before, the adiabatic curve is steeper and it cuts across the isotherms. For the isotherms, we recall that the curves are described by the equation

$$PV = \text{constant} \qquad (isothermal)$$

A similar equation applies to adiabats. In this case, using calculus, it can be shown that the appropriate equation is

$$PV^{\gamma} = \text{constant} \qquad (adiabatic) \quad \textbf{18–9}$$

In this expression, the constant γ is the ratio C_p/C_v:

$$\gamma = \frac{C_p}{C_v} = \frac{\frac{5}{2}R}{\frac{3}{2}R} = \frac{5}{3}$$

This value of γ applies to monatomic ideal gases—and is a good approximation for monatomic real gases as well. The value of γ is different, however, for gases that are diatomic, triatomic, and so on.

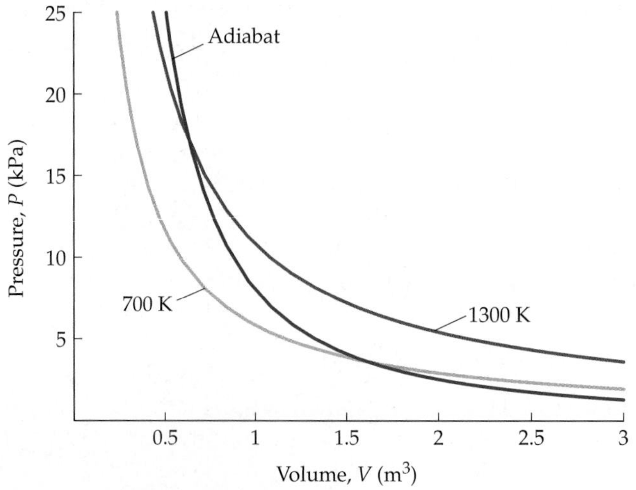

▶ **FIGURE 18–14 A comparison between isotherms and adiabats**

Two isotherms are shown, one for 700 K and one for 1300 K. An adiabat is also shown. Note that the adiabat is a steeper curve than the isotherms.

EXAMPLE 18–5 Hot Air

A container with an initial volume of $0.0625\ m^3$ holds 2.50 moles of a monatomic ideal gas at a temperature of 315 K. The gas is now compressed adiabatically to a volume of $0.0350\ m^3$. Find **(a)** the final pressure and **(b)** the final temperature of the gas.

Picture the Problem

Our sketch shows a gas being compressed from an initial volume of $0.0625\ m^3$ to a final volume of $0.0350\ m^3$. The gas starts with a temperature of 315 K, but because no heat can flow outward through the insulation (adiabatic process), the work done on the gas results in an increased temperature.

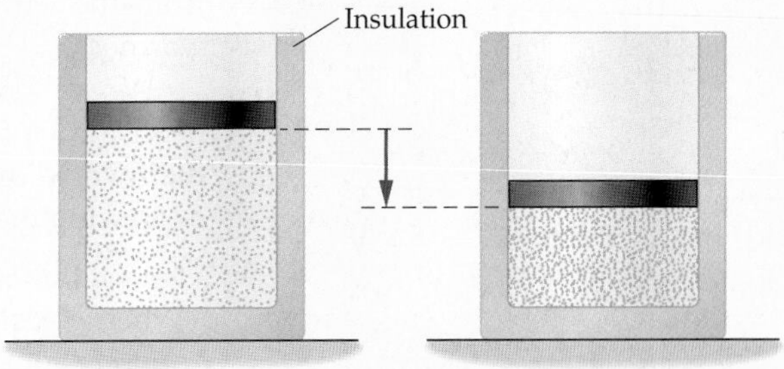

Strategy

a. We can find the final pressure as follows: First, find the initial pressure using the ideal-gas equation of state, $P_iV_i = nRT_i$.

Next, let $P_iV_i^{\gamma} = P_fV_f^{\gamma}$, since this is an adiabatic process. Solve this relation for the final pressure.

b. Use the final pressure and volume to find the final temperature, using the ideal-gas relation, $P_fV_f = nRT_f$.

Solution

Part (a)

1. Find the initial pressure, using $PV = nRT$:

$$P_i = \frac{nRT_i}{V_i} = \frac{(2.50\ \text{mol})[8.31\ \text{J}/(\text{mol}\cdot\text{K})](315\ \text{K})}{0.0625\ \text{m}^3} = 105\ \text{kPa}$$

2. Use $PV^{\gamma} = \text{constant}$ to find a relation for P_f:

$$P_iV_i^{\gamma} = P_fV_f^{\gamma}$$
$$P_f = P_i(V_i/V_f)^{\gamma}$$

3. Substitute numerical values:

$$P_f = (105\ \text{kPa})(0.0625\ \text{m}^3/0.0350\ \text{m}^3)^{5/3} = 276\ \text{kPa}$$

Part (b)

4. Use $PV = nRT$ to solve for the final temperature:

$$T_f = \frac{P_fV_f}{nR} = \frac{(276\ \text{kPa})(0.0350\ \text{m}^3)}{(2.50\ \text{mol})[8.31\ \text{J}/(\text{mol}\cdot\text{K})]} = 465\ \text{K}$$

Insight

Thus, decreasing the volume of the gas by a factor of roughly two has increased its pressure from 105 kPa to 276 kPa and increased its temperature from 315 K to 465 K. This is a specific example of adiabatic heating. Adiabatic cooling is the reverse effect, where the temperature of a gas decreases as its volume increases. For example, if the gas in this system is expanded back to its initial volume of $0.0625\ m^3$, its temperature will drop from 465 K to 315 K.

Practice Problem

To what volume must the gas be compressed to yield a final pressure of 425 kPa? [**Answer:** $V_f = 0.0270\ m^3$]

Some related homework problems: Problem 38, Problem 39

Adiabatic heating and cooling can have important effects on the climate of a given region. For example, moisture-laden winds blowing from the Pacific Ocean into western Oregon are deflected upward when they encounter the Cascade Mountains. As the air rises, the atmospheric pressure decreases (see Chapter 15), allowing the air to expand and undergo adiabatic cooling. The result is that the moisture in the air condenses to form clouds and precipitation on the west side of the mountains. (In some cases, where the air holds relatively little moisture, this mechanism may result in isolated, lens-shaped clouds just above the peak of a mountain, as shown in the accompanying photograph.) When the winds continue on to the east side of the mountains they have little moisture remaining; thus, eastern Oregon is in the *rain shadow* of the Cascade Mountains. In addition, as the air descends on the east side of the mountains, it undergoes adiabatic heating. These are the primary reasons why the summers in western Oregon are moist and mild, while the summers in eastern Oregon are hot and dry.

REAL-WORLD PHYSICS

Rain shadows

▲ A spectacular lenticular (lens-shaped) cloud floats above a mountain in Tierra del Fuego, at the southern tip of Chile. Lenticular clouds are often seen "parked" above and just downwind of high mountain peaks, even when there are no other clouds in the sky. The reason is that as moisture-laden winds are deflected upward by the mountain, the moisture they contain cools due to adiabatic expansion and condenses to form a cloud.

18–5 The Second Law of Thermodynamics

Have you ever warmed your hands by pressing them against a block of ice? Probably not. But if you think about it, you might wonder why it doesn't work. After all, the first law of thermodynamics would be satisfied if energy simply flowed from the ice to your hands. The ice would get colder while your hands got warmer, and the energy of the universe would remain the same.

As we know, however, this sort of thing just doesn't happen—the spontaneous flow of heat is *always* from warmer objects to cooler objects, and never in the reverse direction. This simple observation, in fact, is one of many ways of expressing the **second law of thermodynamics:**

Second Law of Thermodynamics: Heat Flow
When objects of different temperatures are brought into thermal contact, the spontaneous flow of heat that results is always from the high temperature object to the low temperature object. Spontaneous heat flow never proceeds in the reverse direction.

Thus, the second law of thermodynamics is more restrictive than the first law; it says that of all the processes that conserve energy, only those that proceed in a certain direction actually occur. In a sense, the second law implies a definite "directionality" to the behavior of nature. For this reason, the second law is sometimes referred to as the "arrow of time."

For example, suppose you saw a movie that showed a snowflake landing on a person's hand and melting to a small drop of water. Nothing would seem particularly noteworthy about the scene from a physics point of view. But if the movie showed a drop of water on a person's hand suddenly freeze into the shape of a snowflake, then lift off the person's hand into the air, it wouldn't take long to realize the film was running backward. It is clear in which direction time should "flow."

We shall study further consequences of the second law in the next few sections. As we do so, we shall find other more precise, but equivalent, ways of stating the second law.

18–6 Heat Engines and the Carnot Cycle

REAL-WORLD PHYSICS
The steam engine

A **heat engine,** simply put, is a device that converts heat into work. The classic example of this type of device is the steam engine, whose basic elements are illustrated in **Figure 18–15**. First, some form of fuel (oil, wood, coal, etc.) is used to vaporize water in the boiler. The resulting steam is then allowed to enter the engine itself, where it expands against a piston, doing mechanical work. As the piston moves, it causes gears or wheels to rotate, which delivers the mechanical

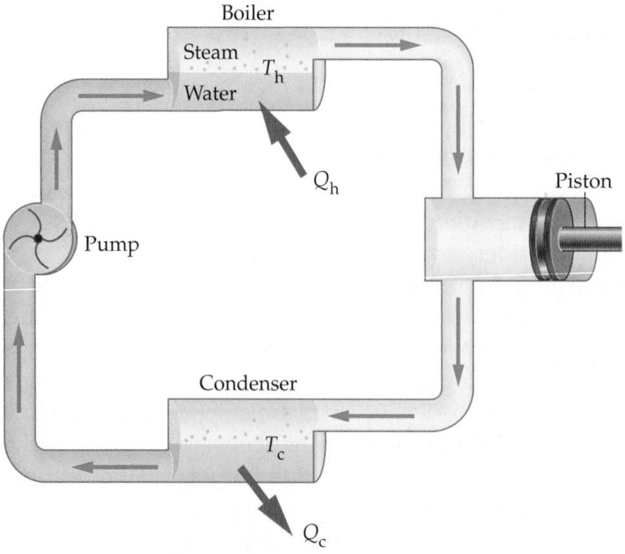

▶ **FIGURE 18–15 A schematic steam engine**
The basic elements of a steam engine are a boiler—where heat converts water to steam—and a piston that can be displaced by the expanding steam. In some engines, the steam is simply exhausted into the atmosphere after it has expanded against the piston. More sophisticated engines send the exhaust steam to a condenser, where it is cooled and condensed back to liquid water, then recycled to the boiler.

work to the external world. After leaving the engine, the steam proceeds to the condenser, where it gives off heat to the cool air in the atmosphere and condenses to liquid form.

What all heat engines have in common are: (i) A high temperature region, or reservoir, that supplies heat to the engine (the boiler in the steam engine); (ii) a low temperature reservoir where "waste" heat is released (the condenser in the steam engine); and (iii) an engine that operates in a cyclic fashion. These features are illustrated schematically in **Figure 18–16**. In addition, though not shown in the figure, heat engines have a working substance (steam in the steam engine) that causes the engine to operate.

To begin our analysis, we note that a certain amount of heat, Q_h, is supplied to the engine from the high temperature or "hot" reservoir during each cycle. Of this heat, a fraction appears as work, W, and the rest is given off as waste heat, Q_c, at a relatively low temperature to the "cold" reservoir. There is no change in energy for the engine, because it returns to its initial state at the completion of each cycle. Letting Q_h and Q_c denote magnitudes, so that both quantities are positive, energy conservation can be written as $Q_h = W + Q_c$, or

$$W = Q_h - Q_c \qquad \text{18–10}$$

As we shall see, the second law of thermodynamics requires that heat engines must *always* exhaust a finite amount of heat to a cold reservoir.

Of particular interest for any engine is its **efficiency,** e, which is simply the fraction of the heat supplied to the engine that appears as work. Thus, we define the efficiency to be

$$e = \frac{W}{Q_h} \qquad \text{18–11}$$

Using the energy-conservation result just derived in Equation 18–10, we find

Efficiency of a Heat Engine, *e*

$$e = \frac{W}{Q_h} = \frac{Q_h - Q_c}{Q_h} = 1 - \frac{Q_c}{Q_h} \qquad \text{18–12}$$

SI unit: dimensionless

For example, if $e = 0.20$, we say that the engine is 20% efficient. In this case, 20% of the input heat is converted to work, $W = 0.20Q_h$, and 80% goes to waste heat, $Q_c = 0.80Q_h$. Efficiency, then, can be thought of as the ratio of how much you receive (work) to how much you have to pay to run the engine (input heat).

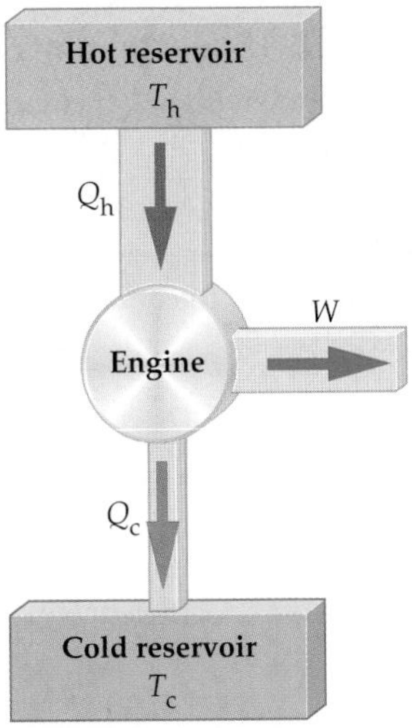

▲ **FIGURE 18–16 A schematic heat engine**

The engine absorbs a heat Q_h from the hot reservoir, performs the work W, and gives off the heat Q_c to the cold reservoir. Energy conservation gives $Q_h = W + Q_c$, where Q_h and Q_c are the magnitudes of the hot- and cold-temperature heats.

▲ At left, a modern version of Hero's engine, invented by the Greek mathematician and engineer Hero of Alexandria. In this simple heat engine, the steam that escapes from a heated vessel of water is directed tangentially, causing the vessel to rotate. This converts the thermal energy supplied to the water into mechanical energy, in the form of rotational motion. At right, a steam engine of slightly more recent design hauls passengers up and down Mt. Washington in New Hampshire. Note in the photo that the locomotive is belching two clouds, one black and one white. Can you explain their origin?

EXAMPLE 18–6 Heat into Work

A heat engine with an efficiency of 24.0% performs 1250 J of work. Find **(a)** the heat absorbed from the hot reservoir, and **(b)** the heat given off to the cold reservoir.

Picture the Problem

Our sketch shows a schematic of the heat engine. We know the amount of work that is done and the efficiency of the engine. We seek the heats Q_h and Q_c.

Note that an efficiency of 24.0% means that $e = 0.240$.

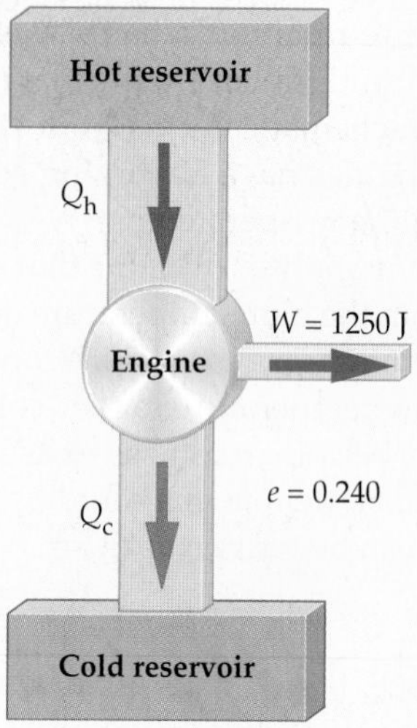

Strategy

a. We can find the heat absorbed from the hot reservoir directly from the definition of efficiency, $e = W/Q_h$.

b. We can find Q_c by using energy conservation, $W = Q_h - Q_c$, or by using the expression for efficiency in terms of the heats, $e = 1 - Q_c/Q_h$.

Solution

Part (a)

1. Use $e = W/Q_h$ to solve for the heat Q_h:

$$e = W/Q_h$$

$$Q_h = \frac{W}{e} = \frac{1250\ \text{J}}{0.240} = 5210\ \text{J}$$

Part (b)

2. Use energy conservation to solve for Q_c:

$$W = Q_h - Q_c$$

$$Q_c = Q_h - W = 5210\ \text{J} - 1250\ \text{J} = 3960\ \text{J}$$

3. Use the efficiency, expressed in terms of Q_h and Q_c, to find Q_c:

$$e = 1 - Q_c/Q_h$$

$$Q_c = (1 - e)Q_h = (1 - 0.240)(5210\ \text{J}) = 3960\ \text{J}$$

Insight

Note that when the efficiency of a heat engine is less than one-half (50%), as in this case, the amount of heat given off as waste to the cold reservoir is more than the amount of heat converted to work.

Practice Problem

What is the efficiency of a heat engine that does 1250 J of work and gives off 5250 J of heat to the cold reservoir? [**Answer:** $e = 0.192$]

Some related homework problems: Problem 43, Problem 44

A temperature difference is essential to the operation of a heat engine. As heat flows from the hot to the cold reservoir in Figure 18–16, for example, the heat engine is able to tap into that flow and convert part of it to work—the greater the efficiency of the engine, the more heat converted to work. The second law of thermodynamics imposes limits, however, on the maximum efficiency a heat engine can have. We explore these limits next.

The Carnot Cycle and Maximum Efficiency

In 1824, the French engineer Sadi Carnot (1796–1832) published a book entitled *Reflections on the Motive Power of Fire* in which he considered the following question: Under what conditions will a heat engine have maximum efficiency? To address this question, let's consider a heat engine that operates between a single hot reservoir at the fixed temperature T_h and a single cold reservoir at the fixed temperature T_c. Carnot's result, known today as **Carnot's theorem,** can be expressed as follows:

Carnot's Theorem

If an engine operating between two constant-temperature reservoirs is to have maximum efficiency, it must be an engine in which all processes are reversible.

In addition, all reversible engines operating between the same two temperatures, T_c and T_h, have the same efficiency.

We should point out that no real engine can ever be perfectly reversible, just as no surface can be perfectly frictionless. Nonetheless, the concept of a reversible engine is a useful idealization.

Carnot's theorem is remarkable for a number of reasons. First, consider what the theorem says: No engine, no matter how sophisticated or technologically advanced, can exceed the efficiency of a reversible engine. We can strive to improve the technology of heat engines, but there is an upper limit to the efficiency that can never be exceeded. Second, the theorem is just as remarkable for what it does not say. It says nothing, for example, about the working substance that is used in the engine—it is as valid for a liquid or solid working substance as for one that is gaseous. Furthermore, it says nothing about the type of reversible engine that is used, what the engine is made of, or how it is constructed. Diesel engine, jet engine, rocket engine—none of these things matter. In fact, all that *does* matter are the two temperatures, T_c and T_h.

Recall that the efficiency of a heat engine can be written as follows:

$$e = 1 - \frac{Q_c}{Q_h}$$

Since the efficiency e depends only on the temperatures T_c and T_h, according to Carnot's theorem, it follows that Q_c/Q_h must also depend only on T_c and T_h. In fact, Lord Kelvin used this observation to propose that, instead of using a thermometer to measure temperature, we measure the efficiency of a heat engine and from this determine the temperature. Thus, he suggested that we *define* the ratio of the temperatures of two reservoirs, T_c/T_h, to be equal to the ratio of the heats Q_c/Q_h:

$$\frac{Q_c}{Q_h} = \frac{T_c}{T_h}$$

If we choose the size of a degree in this temperature scale to be equal to 1 C°, then we have, in fact, defined the Kelvin temperature scale discussed in Chapter 16. Thus, if T_h and T_c are given in kelvins, the maximum efficiency of a heat engine is:

Maximum Efficiency of a Heat Engine

$$e_{max} = 1 - \frac{T_c}{T_h} \qquad \textbf{18–13}$$

Suppose for a moment that we *could* construct an ideal engine, perfectly reversible and free from all forms of friction. Would this ideal engine have 100% efficiency? No, it would not. From Equation 18–13 we can see that the only way the efficiency of a heat engine could be 100% (that is, $e_{max} = 1$) would be for T_c to be 0 K. As we shall see in the last section of this chapter, this is ruled out by the third law of thermodynamics. Hence, the maximum efficiency will always be less than 100%. No matter how perfect the engine, some of the input heat will always be wasted—given off as Q_c—rather than converted to work.

Since the efficiency is defined to be $e = W/Q_h$, it follows that the maximum work a heat engine can do with the input heat Q_h is

$$W_{max} = e_{max}Q_h = \left(1 - \frac{T_c}{T_h}\right)Q_h \qquad \textbf{18–14}$$

If the hot and cold reservoirs have the same temperature, so that $T_c = T_h$, it follows that the maximum efficiency is zero. As a result, the amount of work that such an engine can do is also zero. As mentioned before, a heat engine requires

PROBLEM-SOLVING NOTE

Maximum Efficiency

The maximum efficiency a heat engine can have is determined solely by the temperature of the hot (T_h) and cold (T_c) reservoirs. The numerical value of this efficiency is $e = 1 - T_c/T_h$. Remember, however, that the temperatures must be expressed in the Kelvin scale for this expression to be valid.

different temperatures in order to operate. For example, for a fixed T_c, the higher the temperature of T_h the greater the efficiency.

Finally, even though Carnot's theorem may seem quite different from the second law of thermodynamics, they are, in fact, equivalent. It can be shown, for example, that if Carnot's theorem were violated, it would be possible for heat to flow spontaneously from a cold object to a hot object.

CONCEPTUAL CHECKPOINT 18–4 Comparing Efficiencies

Suppose you have a heat engine that can operate in one of two different modes. In mode 1, the temperatures of the two reservoirs are $T_c = 200$ K and $T_h = 400$ K; in mode 2, the temperatures are $T_c = 400$ K and $T_h = 600$ K. Is the efficiency of mode 1 **(a)** greater than, **(b)** less than, or **(c)** equal to the efficiency of mode 2?

Reasoning and Discussion

At first, you might think that since the temperature difference is the same in the two modes, the efficiency is the same as well. This is not the case, however, since efficiency depends on the *ratio* of the two temperatures ($e = 1 - T_c/T_h$) rather than on their difference. In this case, the efficiency of mode 1 is $e_1 = 1 - 1/2 = 1/2$ and the efficiency of mode 2 is $e_2 = 1 - 2/3 = 1/3$. Thus, mode 1, even though it operates at the lower temperatures, is more efficient.

Answer:

(a) The efficiency of mode 1 is greater than the efficiency of mode 2.

ACTIVE EXAMPLE 18–2 Find the Temperature

If the heat engine in Example 18–6 is operating at maximum efficiency, and its cold reservoir is at a temperature of 295 K, what is the temperature of the hot reservoir?

Solution *(Test your understanding by performing the calculations indicated in each step.)*

1. Write the efficiency, e, in terms of the hot and cold temperatures: $e = 1 - T_c/T_h$
2. Solve for T_h: $T_h = T_c/(1 - e)$
3. Substitute numerical values for T_c and e to find T_h: $T_h = 388$ K

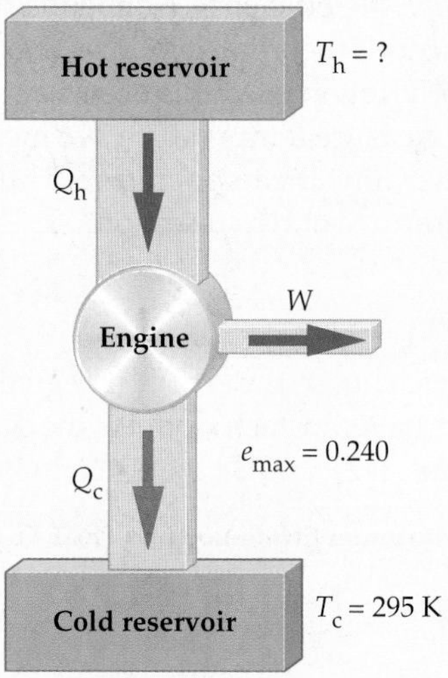

Insight

Though an efficiency of 24% may seem low, it is characteristic of many real engines.

Your Turn

What is the efficiency of this heat engine if T_h is increased by 20 K? What is the efficiency if, instead, T_c is reduced by 20 K?

(Answers to ***Your Turn*** *problems are given in the back of the book.)*

We can summarize the conclusions of this section as follows: The first law of thermodynamics states that you cannot get something for nothing. To be specific, you cannot get more work out of a heat engine than the amount of heat you put in. The best you can do is break even. The second law of thermodynamics is more restrictive than the first law; it says that you can't even break even—some of the input heat must be wasted. It's a law of nature.

18–7 Refrigerators, Air Conditioners, and Heat Pumps

When we stated the second law of thermodynamics in Section 18–5, we said that the spontaneous flow of heat is always from high temperature to low temperature. The key word here is "spontaneous." It *is* possible for heat to flow "uphill,"

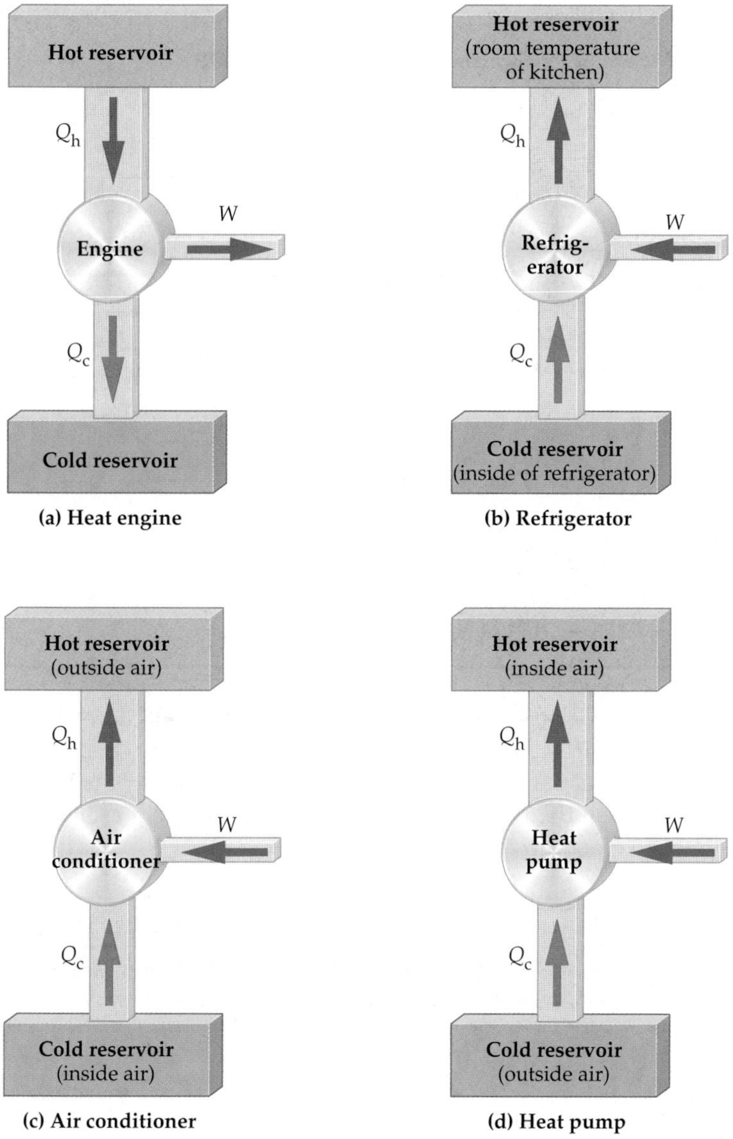

◀ **FIGURE 18–17 Cooling and heating devices**

Schematic comparison of a generalized heat engine **(a)** compared with a refrigerator, an air conditioner, and a heat pump. In the refrigerator **(b),** a work W is done to remove a heat Q_c from the cold reservoir (the inside of the refrigerator). By energy conservation, the heat given off to the hot reservoir (the kitchen) is $Q_h = W + Q_c$. An air conditioner **(c)** removes heat from the cool air inside a house and exhausts it into the hot air of the atmosphere. A heat pump **(d)** moves heat from a cold reservoir to a hot reservoir, just like an air conditioner. The difference is that the reservoirs are switched, so that heat is pumped into the house rather than out to the atmosphere.

from a cold object to a hot one, but it doesn't happen spontaneously—work must be done on the system to make it happen, just as work must be done to pump water from a well. Refrigerators, air conditioners, and heat pumps are devices that use work to transfer heat from a cold object to a hot object.

REAL-WORLD PHYSICS
Refrigerators

Let us first compare the operation of a heat engine, **Figure 18–17 (a)**, and a **refrigerator, Figure 18–17 (b)**. Note that all the arrows are reversed in the refrigerator; in effect, a refrigerator is a heat engine running backward. In particular, the refrigerator uses a work W to remove a certain amount of heat, Q_c, from the cold reservoir (the interior of the refrigerator). It then exhausts an even larger amount of heat, Q_h, to the hot reservoir (the air in the kitchen). By energy conservation, it follows that

$$Q_h = Q_c + W$$

Thus, as a refrigerator operates, it warms the kitchen at the same time that it cools the food stored within it.

To design an effective refrigerator, you would like it to remove as much heat from its interior as possible for the smallest amount of work. After all, the work is supplied by electrical energy that must be paid for each month. Thus, we define the **coefficient of performance, COP, for a refrigerator** as an indicator of its effectiveness:

Coefficient of Performance for a Refrigerator, COP

$$\text{COP} = \frac{Q_c}{W} \qquad \textbf{18–15}$$

SI unit: dimensionless

Typical values for the coefficient of performance are in the range 2 to 6.

EXERCISE 18–3

A refrigerator has a coefficient of performance of 2.50. How much work must be supplied to this refrigerator in order to remove 225 J of heat from its interior?

Solution

Solving Equation 18–15 for the work yields

$$W = \frac{Q_c}{\text{COP}} = \frac{225\ \text{J}}{2.50} = 90.0\ \text{J}$$

Thus, 90.0 J of work removes 225 J of heat from the refrigerator, and exhausts 90.0 J + 225 J = 315 J of heat into the kitchen.

REAL-WORLD PHYSICS
Air conditioners

An **air conditioner,** **Figure 18–17 (c)**, is basically a refrigerator in which the cold reservoir is the room that is being cooled. To be specific, the air conditioner uses electrical energy to "pump" heat from the cool room to the warmer air outside. As with the refrigerator, more heat is exhausted to the hot reservoir than is removed from the cold reservoir; that is, $Q_h = Q_c + W$, as before.

CONCEPTUAL CHECKPOINT 18–5 Room Temperature

You haven't had time to install your new air conditioner in the window yet, so as a short-term measure you decide to place it on the dining-room table and turn it on to cool things off a bit. As a result, does the air in the dining room **(a)** get warmer, **(b)** get cooler, or **(c)** stay at the same temperature?

Reasoning and Discussion

You might think the room stays at the same temperature, since the air conditioner draws heat from the room as usual, but then exhausts heat back into the room that would normally be sent outside. However, the motor of the air conditioner is doing work in order to draw heat from the room, and the heat that would normally be exhausted outdoors is equal to the heat drawn from the room *plus* the work done by the motor: $Q_h = Q_c + W$. Thus, the net effect is that the motor of the air conditioner is continually adding heat to the room, causing it to get warmer.

Answer:

(a) The air in the dining room gets warmer.

REAL-WORLD PHYSICS
Heat pumps

Finally, a **heat pump** can be thought of as an air conditioner with the reservoirs switched. As we see in **Figure 18–17 (d)**, a heat pump does a work W to remove an amount of heat Q_c from the cold reservoir of outdoor air, then exhausts a heat Q_h into the hot reservoir of air in the room. Just as with the refrigerator and the air conditioner, the heat going to the hot reservoir is $Q_h = Q_c + W$.

In an **ideal,** reversible heat pump with only two operating temperatures, T_c and T_h, the Carnot relationship $Q_c/Q_h = T_c/T_h$ holds, just as it does for a heat engine. Thus, if you want to add a heat Q_h to a room, the work that must be done to accomplish this is

$$W = Q_h - Q_c = Q_h\left(1 - \frac{Q_c}{Q_h}\right) = Q_h\left(1 - \frac{T_c}{T_h}\right) \qquad \textbf{18–16}$$

We use this result in the next Example.

EXAMPLE 18–7 Pumping Heat

An ideal heat pump, one that satisfies the Carnot relation given in Equation 18–16, is used to heat a room that is at 293 K. If the pump does 275 J of work, how much heat does it supply to the room if the outdoor temperature is **(a)** 273 K or **(b)** 263 K?

Picture the Problem

Our sketch shows the heat pump doing 275 J of mechanical work to transfer the heat Q_h to the hot reservoir at the temperature 293 K. The temperature of the cold reservoir is 273 K for part (a) and 263 K for part (b). We wish to determine Q_h for each of these cases.

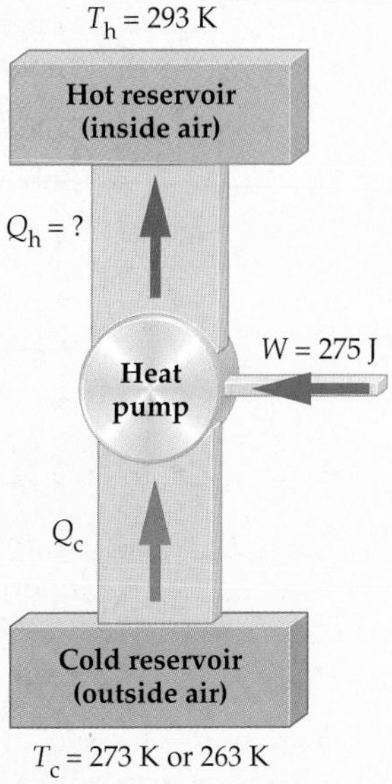

Strategy

For an ideal heat pump, we know that $W = Q_h(1 - T_c/T_h)$. Therefore, given the hot and cold temperatures, as well as the mechanical work, it is straightforward to determine the heat delivered to the hot reservoir, Q_h.

Solution

1. Solve Equation 18–16 for the heat Q_h:

$$Q_h = W/(1 - T_c/T_h)$$

Part (a)

2. Substitute $W = 275$ J, $T_c = 273$ K, and $T_h = 293$ K into the expression for Q_h:

$$Q_h = \frac{W}{1 - \frac{T_c}{T_h}} = \frac{275\text{ J}}{1 - \frac{273\text{ K}}{293\text{ K}}} = 4030\text{ J}$$

Part (b)

3. Substitute $W = 275$ J, $T_c = 263$ K, and $T_h = 293$ K into the expression for Q_h:

$$Q_h = \frac{W}{1 - \frac{T_c}{T_h}} = \frac{275\text{ J}}{1 - \frac{263\text{ K}}{293\text{ K}}} = 2690\text{ J}$$

Insight

As one might expect, the same amount of work provides less heat when the outside temperature is lower. That is, more work must be done on a colder day to provide the same heat to the inside air.

In addition, note that if 275 J of heat is supplied to an electric heater, then 275 J of heat is given to the air in the room. When that same energy is used to run a heat pump, a good deal more than 275 J of heat is added to the room.

Practice Problem

How much work must be done by this heat pump to supply 2550 J of heat on a day when the outside temperature is 253 K? [**Answer:** $W = 348$ J]

Some related homework problems: Problem 57, Problem 58

Since the purpose of a heat pump is to add heat to a room, and we want to add as much heat as possible for the least work, the **coefficient of performance, COP, for a heat pump** is defined as follows:

Coefficient of Performance for a Heat Pump, COP

$$\text{COP} = \frac{Q_h}{W} \qquad \text{18–17}$$

SI unit: dimensionless

The COP for a heat pump, which is usually in the range of 3 to 4, depends on the inside and outside temperatures. We use the COP in the next Exercise.

EXERCISE 18–4

A heat pump with a coefficient of performance equal to 3.5 supplies 2500 J of heat to a room. How much work is required?

Solution

Solving Equation 18–17 for the work, W, we find

$$W = \frac{Q_h}{\text{COP}} = \frac{2500\ \text{J}}{3.5} = 710\ \text{J}$$

18–8 Entropy

In this section we introduce a new quantity that is as fundamental to physics as energy or temperature. This quantity is referred to as the entropy, and it is related to the amount of disorder in a system. For example, a messy room has more entropy than a neat one, a pile of bricks has more entropy than a building constructed from the bricks, a freshly laid egg has more entropy than one that is just about to hatch, and a puddle of water has more entropy than the block of ice from which it melted. We begin by considering the connection between entropy and heat, and later develop more fully the connection between disorder and entropy.

When discussing heat engines, we saw that if an engine is reversible it satisfies the following relation:

$$\frac{Q_c}{Q_h} = \frac{T_c}{T_h}$$

Rearranging slightly, we can rewrite this as

$$\frac{Q_c}{T_c} = \frac{Q_h}{T_h}$$

Notice that the quantity Q/T is the same for both the hot and the cold reservoirs. This relationship prompted the German physicist Rudolf Clausius (1822–1888) to propose the following definition: The **entropy**, S, is a quantity whose change is given by the heat Q divided by the absolute temperature T:

PROBLEM-SOLVING NOTE

Calculating Entropy Change

When calculating the entropy change, $\Delta S = Q/T$, be sure to convert the temperature T to kelvins. It is *only* when this is done that the correct value for ΔS can be obtained.

Definition of Entropy Change, ΔS

$$\Delta S = \frac{Q}{T} \qquad \text{18–18}$$

SI unit: J/K

For this definition to be valid, the heat Q must be transferred reversibly at the fixed Kelvin temperature T. Note that if heat is added to a system ($Q > 0$), the entropy of the system increases; if heat is removed from a system ($Q < 0$), its entropy decreases.

Entropy is a state function, just like the internal energy, U. This means that the value of S depends only on the state of a system, and not on how the system gets to that state. It follows, then, that the *change* in entropy, ΔS, depends only on the initial and final states of a system. Thus, if a process is irreversible—so that Equation 18–18 *does not* hold—we can still calculate ΔS by using a reversible process to connect the same initial and final states.

EXAMPLE 18–8 Melts in Your Hand

Calculate the change in entropy when a 0.125-kg chunk of ice melts at 0 °C. Assume the melting occurs reversibly.

Picture the Problem

In our sketch we show a 0.125-kg chunk of ice at the temperature 0 °C. As the ice absorbs the heat Q from its surroundings, it melts to water at 0 °C. Because the system absorbs heat, its entropy increases.

Strategy

The entropy change is Q/T, where $T = 0\ °\text{C} = 273\ \text{K}$. To find the heat Q, we note that to melt the ice we must add to it the latent heat of fusion. Thus, the heat is $Q = mL_f$, where $L_f = 33.5 \times 10^4\ \text{J/kg}$.

Solution

1. Find the heat that must be absorbed by the ice for it to melt: $Q = mL_f = (0.125\ \text{kg})(33.5 \times 10^4\ \text{J/kg}) = 4.19 \times 10^4\ \text{J}$

2. Calculate the change in entropy: $\Delta S = \dfrac{Q}{T} = \dfrac{4.19 \times 10^4\ \text{J}}{273\ \text{K}} = 153\ \text{J/K}$

Insight

Note that we were careful to convert the temperature of the system from 0 °C to 273 K before we applied $\Delta S = Q/T$. This must always be done when calculating the entropy change. If we had neglected to do the conversion in this case, we would have found an infinite increase in entropy—which is clearly unphysical.

In addition, notice that the entropy change in this case is positive. This illustrates the general rule that entropy increases when heat is added to a system, and decreases when it is removed.

Practice Problem

Find the mass of ice that would be required to give an entropy change of 275 J/K. [**Answer:** $m = 0.224$ kg]

Some related homework problems: Problem 63, Problem 64

Let's apply the definition of entropy change to the case of a reversible heat engine. First, a heat Q_h leaves the hot reservoir at the temperature T_h. Thus, the entropy of this reservoir decreases by the amount Q_h/T_h:

$$\Delta S_h = -\frac{Q_h}{T_h}$$

Recall that Q_h is the magnitude of the heat leaving the hot reservoir; hence, the minus sign is used to indicate a decrease in entropy. Similarly, heat is added to the cold reservoir; hence, its entropy increases by the amount Q_c/T_c:

$$\Delta S_c = \frac{Q_c}{T_c}$$

The total entropy change for this system is

$$\Delta S_{\text{total}} = \Delta S_h + \Delta S_c = -\frac{Q_h}{T_h} + \frac{Q_c}{T_c}$$

Since we know that $Q_h/T_h = Q_c/T_c$ it follows that the total entropy change vanishes:

$$\Delta S_{\text{total}} = -\frac{Q_h}{T_h} + \frac{Q_c}{T_c} = 0$$

What is special about a reversible engine, then, is the fact that its entropy does not change.

On the other hand, a real engine will always have a lower efficiency than a reversible engine operating between the same temperatures. This means that in a real engine less of the heat from the hot reservoir is converted to work; hence, more heat is given off as waste heat to the cold reservoir. Thus, for a given value of Q_h, the heat Q_c is greater in an irreversible engine than in a reversible one. As a result, instead of $Q_c/T_c = Q_h/T_h$, we have

$$\frac{Q_c}{T_c} > \frac{Q_h}{T_h}$$

Therefore, if an engine is irreversible, the total entropy change is positive:

$$\Delta S_{total} = -\frac{Q_h}{T_h} + \frac{Q_c}{T_c} > 0$$

In general, *any irreversible process results in an increase of entropy.*

These results can be summarized in the following general statement:

Entropy in the Universe

The total entropy of the universe *increases* whenever an *irreversible* process occurs.

The total entropy of the universe is *unchanged* whenever a *reversible* process occurs.

PROBLEM-SOLVING NOTE

Entropy Change

Though it is tempting to treat entropy like energy, setting the final value equal to the initial value, this is not the case in general. Only in a reversible process is the entropy unchanged—otherwise it increases. Still, the entropy of part of a system can decrease, as long as the entropy of other parts increases by the same amount or more.

Since all *real* processes are irreversible (with reversible processes being a useful idealization), the total entropy of the universe continually increases. Thus, in terms of the entropy, the universe moves in only one direction—toward an ever-increasing entropy. This is quite different from the behavior with regard to energy, which remains constant no matter what type of process occurs.

In fact, this statement about entropy in the universe is yet another way of expressing the second law of thermodynamics. Recall, for example, that our original statement of the second law said that heat flows spontaneously from a hot object to a cold object. During this flow of heat the entropy of the universe increases, as we show in the next Example. Hence, the direction in which heat flows is seen to be the result of the general principle of entropy increase in the universe. Again, we see a directionality in nature; the ever-present "arrow of time."

EXAMPLE 18–9 Entropy Is Not Conserved!

A hot reservoir at the temperature 576 K transfers 1050 J of heat irreversibly to a cold reservoir at the temperature 305 K. Find the change in entropy of the universe.

Picture the Problem

The relevant physical situation is shown in our sketch. Note that the heat $Q = 1050$ J is transferred from the hot reservoir at the temperature $T_h = 576$ K directly to the cold reservoir at the temperature $T_c = 305$ K.

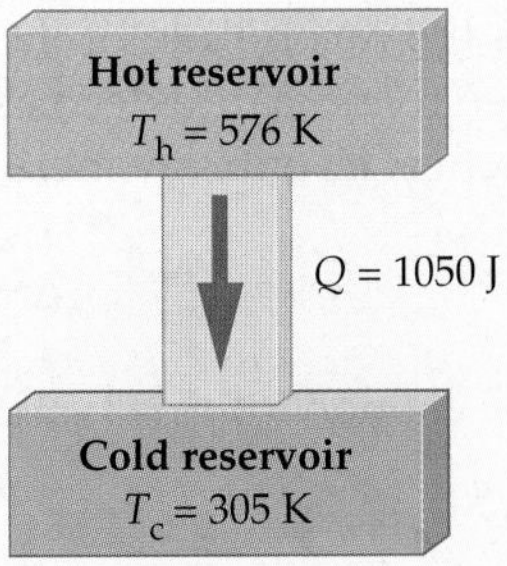

Strategy

As the heat Q leaves the hot reservoir, its entropy *decreases* by Q/T_h. When the same heat Q enters the cold reservoir, its entropy *increases* by the amount Q/T_c. Summing these two contributions gives the entropy change of the universe.

Solution

1. Calculate the entropy change of the hot reservoir:

$$\Delta S_h = -\frac{Q}{T_h} = -\frac{1050 \text{ J}}{576 \text{ K}} = -1.82 \text{ J/K}$$

2. Calculate the entropy change of the cold reservoir:

$$\Delta S_c = \frac{Q}{T_c} = \frac{1050 \text{ J}}{305 \text{ K}} = 3.44 \text{ J/K}$$

3. Sum these contributions to obtain the entropy change of the universe:

$$\Delta S_{universe} = \Delta S_h + \Delta S_c = -\frac{Q}{T_h} + \frac{Q}{T_c}$$

$$= -1.82 \text{ J/K} + 3.44 \text{ J/K} = 1.62 \text{ J/K}$$

Insight

Note that the decrease in entropy of the hot reservoir is more than made up for by the increase in entropy of the cold reservoir. This is a general result.

Practice Problem

What amount of heat must be transferred between these reservoirs for the entropy of the universe to increase by 1.50 J/K? [**Answer:** $Q = 972\ \text{J}$]

Some related homework problems: Problem 65, Problem 68

When certain processes occur, it sometimes appears as if the entropy of the universe has decreased. On closer examination, however, it can always be shown that there is a larger increase in entropy elsewhere that results in an overall increase. This issue is addressed in the next Conceptual Checkpoint.

CONCEPTUAL CHECKPOINT 18–6 Entropy Change

You put a tray of water in the kitchen freezer, and some time later it has turned to ice. Has the entropy of the universe **(a)** increased, **(b)** decreased, or **(c)** stayed the same?

Reasoning and Discussion

It might seem that the entropy of the universe has decreased. After all, heat is removed from the water to freeze it, and, as we know, removing heat from an object lowers its entropy. On the other hand, we also know that the freezer does work to draw heat from the water; hence, it exhausts more heat into the kitchen than it absorbs from the water. Detailed calculations always show that the entropy of the heated air in the kitchen increases by more than the entropy of the water decreases, thus the entropy of the universe increases—as it must for *any* real process.

Answer:

(a) The entropy of the universe has increased.

As the entropy in the universe increases, the amount of work that can be done is diminished. For example, the heat flow in Example 18–9 resulted in an increase in the entropy of the universe by the amount 1.62 J/K. If this same heat had been used in a reversible engine, however, it could have done work, and since the engine was reversible, the entropy of the universe would have stayed the same. In the next Active Example, we calculate the work that could be done with a reversible engine.

ACTIVE EXAMPLE 18–3 Find the Work

Suppose a reversible heat engine operates between the two heat reservoirs described in Example 18–9. Find the amount of work done by such an engine when 1050 J of heat is drawn from the hot reservoir.

Solution

(Test your understanding by performing the calculations indicated in each step.)

1. Calculate the efficiency of this engine: $e = 1 - T_c/T_h = 0.470$
2. Multiply the efficiency by Q_h to find the work done: $W = eQ_h = 494\ \text{J}$

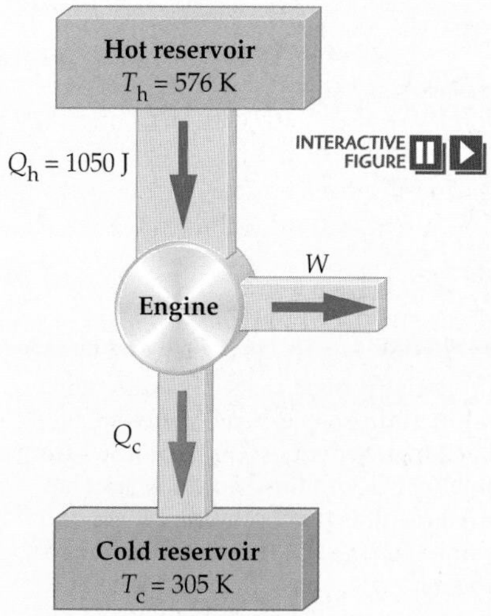

Insight

Since this engine is reversible, its total entropy change must be zero. The decrease in entropy of the hot reservoir is $-(1050\ \text{J})/576\ \text{K} = -1.82\ \text{J/K}$. It follows that the increase in entropy of the cold reservoir must have the same magnitude. The amount of heat that flows into the cold reservoir, Q_c, is $Q_h - W = 1050\ \text{J} - 494\ \text{J} = 556\ \text{J}$. This heat causes an entropy increase equal to $556\ \text{J}/305\ \text{K} = +1.82\ \text{J/K}$, as expected. Thus, the reason the engine exhausts the heat Q_c is to produce zero net change in entropy. If the engine were irreversible, it would exhaust a heat greater than 556 J, and this would create a net increase in entropy and a reduction in the amount of work done. Clearly, then, a reversible engine produces the maximum amount of work.

continued on following page

continued from previous page

Your Turn

Suppose this engine is not reversible, and that only 455 J of work is done when 1050 J of heat is drawn from the hot reservoir. What is the entropy increase of the universe in this case?

(Answers to **Your Turn** *problems are given in the back of the book.)*

Note that when 1050 J of heat is simply transferred from the hot reservoir to the cold reservoir, as in Example 18–9, the entropy of the universe increases by 1.62 J/K. When this same heat is transferred reversibly, with an ideal engine, the entropy of the universe stays the same, but 494 J of work is done. The connection between the entropy increase of the irreversible process and the work done by a reversible engine is very simple:

$$W = T_c \, \Delta S_{universe} = (305 \text{ K})(1.62 \text{ J/K}) = 494 \text{ J}$$

To see that this expression is valid in general, recall that in Example 18–9 the total change in entropy is $\Delta S_{universe} = Q/T_c - Q/T_h$. That is, the heat $Q_h = Q$ is withdrawn from the hot reservoir (lowering the entropy by the amount Q/T_h), and the same heat $Q_c = Q$ is added to the cold reservoir (increasing the entropy by the larger amount, Q/T_c). If we multiply this increase in entropy by the temperature of the cold reservoir, T_c, we have $T_c \, \Delta S_{universe} = Q - QT_c/T_h = Q(1 - T_c/T_h)$. Recalling that the efficiency of an ideal engine is $e = 1 - T_c/T_h$, we see that $T_c \, \Delta S_{universe} = Qe$. Finally, the work done by an ideal engine is $W = eQ_h$, or in this case, $W = eQ$, since $Q_h = Q$. Therefore, we see that $W = T_c \, \Delta S_{universe}$, as expected.

In general, a process in which the entropy of the universe increases is one in which less work is done than if the process had been reversible. Thus, we lose forever the ability for that work to be done, because to restore the universe to its former state would mean lowering its entropy, which cannot be done. Thus, with every increase in entropy, there is that much less work that can be done by the universe.

For this reason, entropy is sometimes referred to as a measure of the "quality" of energy. When an irreversible process occurs, and the entropy of the universe increases, we say that the energy of the universe has been "degraded" because less of it is available to do work. This process of increasing degradation of energy and increasing entropy in the universe is a continuing aspect of nature.

18–9 Order, Disorder, and Entropy

In the previous section we considered entropy from the point of view of thermodynamics. We saw that as heat flows from a hot object to a cold object the entropy of the universe increases. In this section we show that entropy can also be thought of as a measure of the amount of **disorder** in the universe.

We begin with the situation of heat flow from a hot to a cold object. In **Figure 18–18 (a)** we show two bricks, one hot and the other cold. As we know

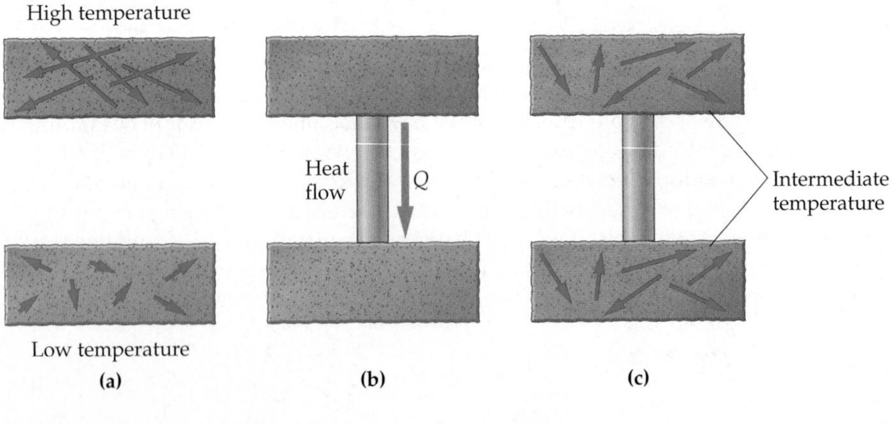

▶ **FIGURE 18–18 Heat flow and disorder**

(a) Initially, two bricks have different temperatures, and hence different average kinetic energies. **(b)** Heat flows from the hot brick to the cold brick. **(c)** The final result is that both bricks have the same intermediate temperature, and all the molecules have the same average kinetic energy. Thus, the initial orderly segregation of molecules by kinetic energy has been lost.

from kinetic theory, the molecules in the hot brick have more kinetic energy than the molecules in the cold brick. This means that the system is rather orderly, in that all the high kinetic energy molecules are grouped together in the hot brick, and all the low kinetic energy molecules are grouped together in the cold brick. There is a definite regularity, or order, to the distribution of the molecular speeds.

Now bring the bricks into thermal contact, as in **Figure 18–18 (b)**. The result is a flow of heat from the hot brick to the cold brick until the temperatures become equal. The final result is indicated in **Figure 18–18 (c)**. During the heat transfer the entropy of the universe increases, as we know, and the system loses the nice orderly distribution it had in the beginning. Now, all the molecules have the same average kinetic energy; hence, the system is randomized, or disordered. We are led to the following conclusion:

> As the entropy of a system increases, its disorder increases as well; that is, an *increase* in entropy is the same as a *decrease* in order.

Note that if heat had flowed in the opposite direction—from the cold brick to the hot brick—the ordered distribution of molecules would have been reinforced, rather than lost.

To take another example, consider the 0.12-kg cube of ice discussed in Example 18–8. As we saw there, the entropy of the universe increases as the ice melts. Now let's consider what happens on the molecular level. To begin, the molecules are well ordered in their crystalline positions. As heat is absorbed, however, the molecules begin to free themselves from the ice and move about randomly in the growing puddle of water. Thus, the regular order of the solid is lost. Again, we see that as entropy increases, so too does the disorder of the molecules.

Thus, the second law of thermodynamics can be stated as the principle that the disorder of the universe is continually increasing. Everything that happens in the universe is simply making it a more disorderly place. And there is nothing you can do to prevent it—nothing you can do will result in the universe being more ordered. Just as freezing a tray of water to make ice actually results in more entropy—and more disorder—in the universe, so does any action you take.

▲ All processes that occur spontaneously increase the entropy of the universe. In this case, the movements of the water molecules become more random and chaotic when they reach the tumultuous, swirling pool at the bottom of the falls. In addition, some of their kinetic energy is converted into thermal energy—the most disordered and degraded form of energy.

Heat Death

If one carries the previous discussion to its logical conclusion, it seems that the universe is "running down." That is, the disorder of the universe constantly increases, and as it does, the amount of energy available to do work decreases. If this process continues, might there come a day when no more work can be done? And if that day does come, what then?

This is one possible scenario for the fate of the universe, and it is referred to as the "heat death" of the universe. In this scenario, heat continues to flow from hotter regions in space (like stars) to cooler regions (like planets) until, after many billions of years, all objects in the universe have the same temperature. With no temperature differences left, there can be no work done, and the universe would cease to do anything of particular interest. Not a pretty picture, but certainly a possibility. The universe may simply continue with its present expansion until the stars burn out and the galaxies fade away like the dying embers of a scattered campfire.

Living Systems

So far we have focused on the rather gloomy prospect of the universe constantly evolving toward greater disorder. Is it possible, however, that life is an exception to this rule? After all, we know that an embryo utilizes simple raw materials to produce a complex, highly ordered living organism. Similarly, the well-known biological aphorism "ontogeny recapitulates phylogeny," while not strictly correct, expresses the fact that the development of an individual

REAL-WORLD PHYSICS: BIO

Entropy and life

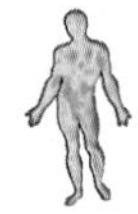

organism from embryo to adult often reflects certain aspects of the evolutionary development of the species as a whole. Thus, over time, species often evolve toward more complex forms. Finally, living systems are able to use disordered raw materials in the environment to produce orderly structures in which to live. It seems, then, that there are many ways in which living systems produce increasing order.

This conclusion is flawed, however, since it fails to take into account the entropy of the environment in which the organism lives. It is similar to the conclusion that a freezer violates the second law of thermodynamics because it reduces the entropy of water as it freezes it into ice. This analysis neglects the fact that the freezer exhausts heat into the room, increasing the entropy of the air by an amount that is greater than the entropy decrease of the water. In the same way, living organisms constantly give off heat to the atmosphere as a by-product of their metabolism, increasing its entropy. Thus, if we build a house from a pile of bricks—decreasing the entropy of the bricks—the heat we give off during our exertions increases the entropy of the atmosphere more than enough to give a net increase in entropy.

Finally, all living organisms can be thought of as heat engines, tapping into the flow of energy from one place to another to produce mechanical work. Plants, for example, tap into the flow of energy from the high temperature of the Sun to the cold of deep space and use a small fraction of this energy to sustain themselves and reproduce. Animals consume plants and generate heat within their bodies as they metabolize their food. A fraction of the energy released by the metabolism is in turn converted to mechanical work. Living systems, then, obey the same laws of physics as steam engines and refrigerators—they simply produce different results as they move the universe toward greater disorder.

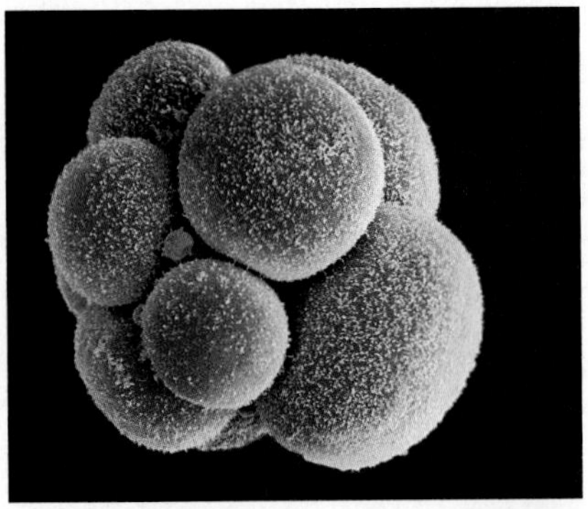

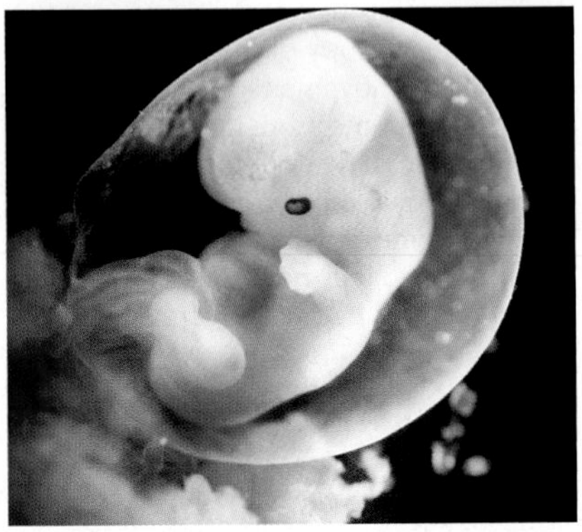

▲ Many species, including humans, develop from a single fertilized egg into a complex multicellular organism. In the process they create large, intricately ordered molecules such as proteins and DNA from smaller, simpler precursors. In the metabolic processes of living things, however, heat is produced, increasing the entropy of the universe as a whole. Thus, the second law is not violated.

18–10 The Third Law of Thermodynamics

Finally, we consider the **third law of thermodynamics,** which states that there is no temperature lower than absolute zero, and that absolute zero is unattainable. It is possible to cool an object to temperatures arbitrarily close to absolute zero—experiments have reached temperatures as low as 4.5×10^{-10} K—but no object can ever be cooled to precisely 0 K.

As an analogy to cooling toward absolute zero, imagine walking toward a wall, with each step half the distance between you and the wall. Even if you take an infinite number of steps, you will still not reach the wall. You can get arbitrarily close, of course, but you never get all the way there.

The same sort of thing happens when cooling. To cool an object, you can place it in thermal contact with an object that is colder. Heat transfer will occur, with your object ending up cooler and the other object ending up warmer. In particular, suppose you had a collection of objects at 0 K to use for cooling. You put your object in contact with one of the 0-K objects and your object cools, while the 0-K object warms slightly. You continue this process, each time throwing away the "warmed up" 0-K object and using a new one. Each time you cool your object it gets closer to 0 K, without ever actually getting there.

In light of this discussion, we can express the third law of thermodynamics as follows:

The Third Law of Thermodynamics
It is impossible to lower the temperature of an object to absolute zero in a finite number of steps.

As with the second law of thermodynamics, this law can be expressed in a number of different but equivalent ways. The essential idea, however, is always the same: Absolute zero is the limiting temperature and, though it can be approached arbitrarily closely, it can never be attained.

Chapter Summary

18–1 The Zeroth Law of Thermodynamics

When two objects have the same temperature, they are in thermal equilibrium.

18–2 The First Law of Thermodynamics

The first law of thermodynamics is a statement of energy conservation that includes heat.

If U is the internal energy of an object, Q is the heat added to it, and W is the work done by the object, the first law of thermodynamics can be written as follows:

$$\Delta U = Q - W \quad \textbf{18–3}$$

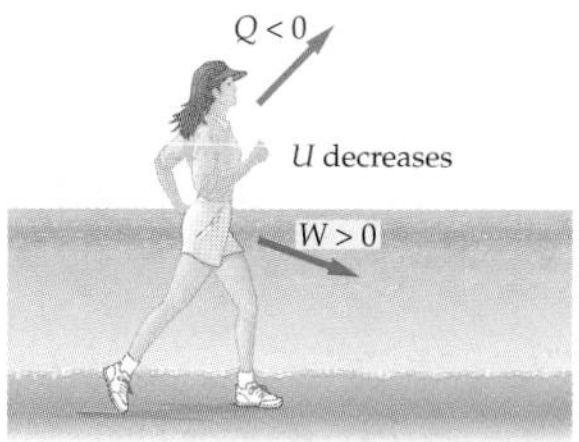

State Function
The internal energy U depends only on the state of a system; that is, on its temperature, pressure, and volume. The value of U is independent of how a system is brought to a certain state.

18–3 Thermal Processes

Quasi-Static
A quasi-static process is one in which a system is moved slowly from one state to another. The change of state is so slow that the system may be considered to be in equilibrium at any given time during the process.

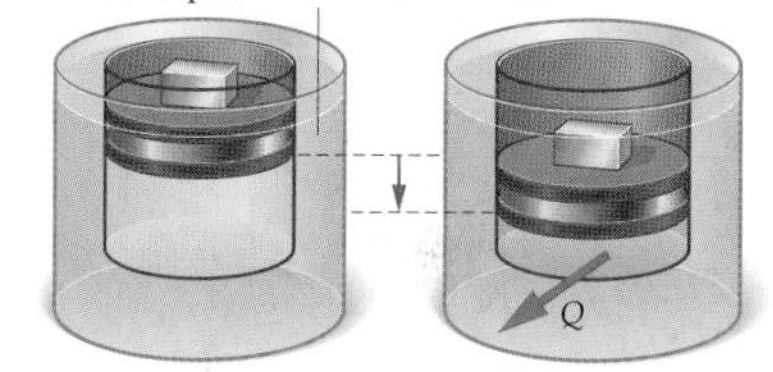

Reversible
In a reversible process it is possible to return the system and its surroundings to their initial states.

Irreversible
Irreversible processes cannot be "undone." When the system is returned to its initial state, the surroundings have been altered.

Work
In general, the work done during a process is equal to the area under the process curve in a PV plot.

Constant Pressure
In a PV plot, a constant-pressure process is represented by a horizontal line. The work done at constant pressure is $W = P\Delta V$.

Constant Volume
In a PV plot, a constant-volume process is represented by a vertical line. The work done at constant volume is zero; $W = 0$.

Isothermal Process
In a PV plot, an isothermal process is represented by PV = constant. The work done in an isothermal expansion from V_i to V_f is $W = nRT \ln(V_f/V_i)$.

Adiabatic Process
An adiabatic process occurs with no heat transfer; that is, $Q = 0$.

18–4 Specific Heats for an Ideal Gas: Constant Pressure, Constant Volume

Specific heats have different values depending on whether they apply to a process at constant pressure or a process at constant volume.

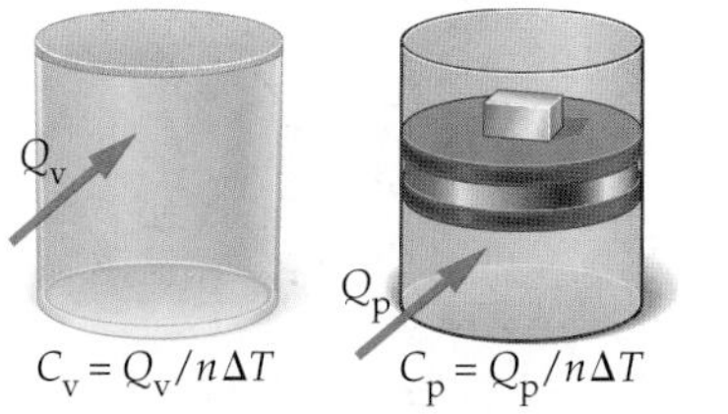

Molar Specific Heat
The molar specific heat, C, is defined by $Q = nC\Delta T$, where n is the number of moles.

Constant Volume
The molar specific heat for an ideal monatomic gas at constant volume is

$$C_v = \tfrac{3}{2}R \quad \textbf{18–6}$$

Constant Pressure
The molar specific heat for an ideal monatomic gas at constant pressure is

$$C_p = \tfrac{5}{2}R \quad \textbf{18–7}$$

Adiabatic Process
In a PV plot, an adiabatic process is represented by PV^γ = constant, where γ is the ratio C_p/C_v. For a monatomic, ideal gas, $\gamma = 5/3$.

18–5 The Second Law of Thermodynamics

When objects of different temperatures are brought into thermal contact, the spontaneous flow of heat that results is always from the high temperature object to the low temperature object. Spontaneous heat flow never proceeds in the reverse direction.

18–6 Heat Engines and the Carnot Cycle

A heat engine is a device that converts heat into work; for example, a steam engine.

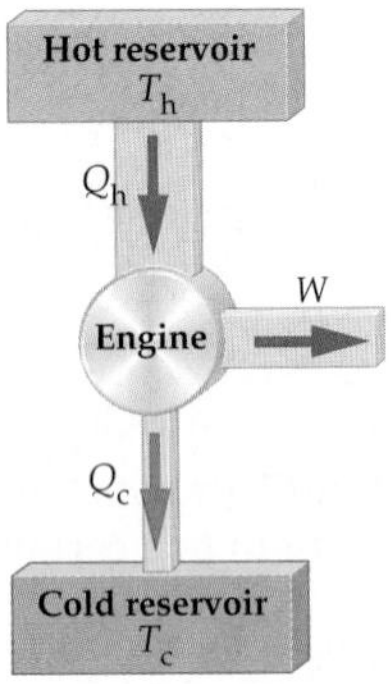

Efficiency
The efficiency e of a heat engine that takes in the heat Q_h from a hot reservoir, exhausts a heat Q_c to a cold reservoir, and does the work W is

$$e = \frac{W}{Q_h} = \frac{Q_h - Q_c}{Q_h} = 1 - \frac{Q_c}{Q_h} \quad \text{18–12}$$

Carnot's Theorem
If an engine operating between two constant-temperature reservoirs is to have maximum efficiency, it must be an engine in which all processes are reversible. In addition, all reversible engines operating between the same two temperatures have the same efficiency.

Maximum Efficiency
The maximum efficiency of a heat engine operating between the Kelvin temperatures T_h and T_c is

$$e_{max} = 1 - \frac{T_c}{T_h} \quad \text{18–13}$$

Maximum Work
If a heat engine takes in the heat Q_h from a hot reservoir, the maximum work it can do is

$$W_{max} = e_{max}Q_h = \left(1 - \frac{T_c}{T_h}\right)Q_h \quad \text{18–14}$$

18–7 Refrigerators, Air Conditioners, and Heat Pumps

Refrigerators, air conditioners, and heat pumps are devices that use work to make heat flow from a cold region to a hot region.

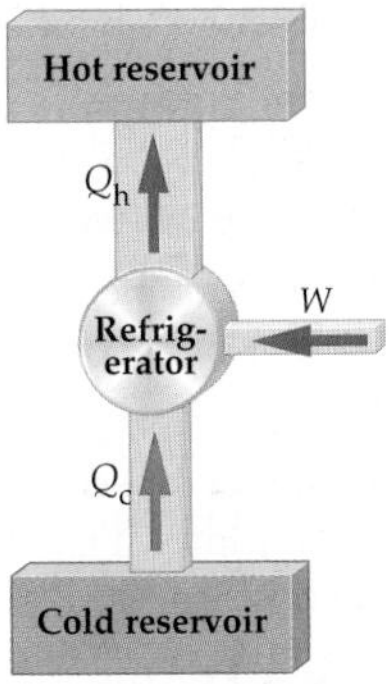

Coefficient of Performance
The coefficient of performance of a refrigerator or air conditioner doing the work W to remove a heat Q_c from a cold reservoir is

$$\text{COP} = \frac{Q_c}{W} \quad \text{18–15}$$

Heat Pump
In an ideal heat pump, the work W that must be done to deliver a heat Q_h to a hot reservoir at the temperature T_h by extracting a heat Q_c from a cold reservoir at the temperature T_c is

$$W = Q_h - Q_c = Q_h\left(1 - \frac{Q_c}{Q_h}\right) = Q_h\left(1 - \frac{T_c}{T_h}\right) \quad \text{18–16}$$

The coefficient of performance for a heat pump is

$$\text{COP} = \frac{Q_h}{W} \quad \text{18–17}$$

18–8 Entropy

Like the internal energy U, the entropy S is a state function.

Change in Entropy
The change in entropy during a reversible exchange of the heat Q at the Kelvin temperature T is

$$\Delta S = \frac{Q}{T} \quad \text{18–18}$$

Entropy in the Universe
The total entropy of the universe increases whenever an irreversible process occurs. *Note:* Entropy is not conserved.

In an idealized reversible process the entropy of the universe is unchanged.

18–9 Order, Disorder, and Entropy

Entropy is a measure of the disorder of a system. As entropy increases, a system becomes more disordered.

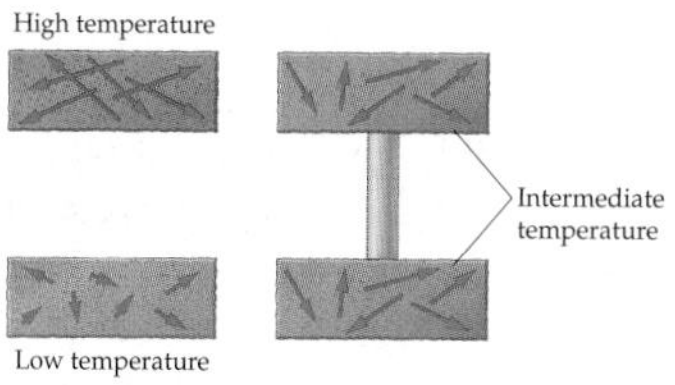

Heat Death

A possible fate of the universe is heat death, in which everything is at the same temperature and no more work can be done.

18–10 The Third Law of Thermodynamics

It is impossible to lower the temperature of an object to absolute zero in a finite number of steps.

Problem-Solving Summary

Type of Problem	Relevant Physical Concepts	Related Examples
Relate the heat and work exchanged by a system to its change in internal energy.	The heat, work, and internal energy of a system are related by the first law of thermodynamics; $\Delta U = Q - W$.	Example 18–1
Calculate the work done by a system during a given process in terms of the corresponding PV plot.	The work done by an expanding system is equal to the area under the curve that represents the process on a PV plot. If the system is compressed, the work is equal to the negative of the area, meaning that work is done on the system rather than by it.	Example 18–2 Active Example 18–1
Find the efficiency of a heat engine.	If a heat engine takes in a heat Q_h and does the work W, its efficiency, e, is given by $e = W/Q_h$. Since W is the difference between the heat going into the engine, Q_h, and the heat leaving the engine, Q_c, the efficiency can also be written as $e = (Q_h - Q_c)/Q_h = 1 - Q_c/Q_h$. Finally, the efficiency can also be related to the absolute temperature of the hot (T_h) and cold (T_c) reservoirs as follows: $e = 1 - T_c/T_h$.	Example 18–6 Active Example 18–2
Determine the change in entropy of a system.	If a system exchanges a heat Q at the absolute temperature T, its entropy S changes by the following amount: $\Delta S = Q/T$. If heat is added to the system, its entropy increases; if heat is removed from the system, its entropy decreases.	Examples 18–8, 18–9

Conceptual Questions

(Answers to odd-numbered Conceptual Questions can be found in the back of the book.)

1. If an engine has a reverse gear, does this make it reversible?
2. The temperature of a substance is held fixed. Is it possible for heat to flow **(a)** into or **(b)** out of this system? For each case, give an explanation if your answer is no. If your answer is yes, give a specific example.
3. A substance is thermally insulated, so that no heat can flow between it and its surroundings. Is it possible for the temperature of this substance to **(a)** increase or **(b)** decrease? For each case, give an explanation if your answer is no. If your answer is yes, give a specific example.
4. Heat is added to a substance. Is it safe to conclude that the temperature of the substance will rise? Give an explanation if your answer is no. If your answer is yes, give a specific example.
5. The temperature of a substance is increased. Is it safe to conclude that heat was added to the substance? Give an explanation if your answer is no. If your answer is yes, give a specific example.
6. Are there thermodynamic processes in which all the heat absorbed by an ideal gas goes completely into mechanical work? If so, give an example.
7. Is it possible to convert a given amount of mechanical work completely into heat? Explain.
8. An ideal gas is held in an insulated container at the temperature T. All the gas is initially in one-half of the container, with a partition separating the gas from the other half of the container, which is a vacuum. If the partition ruptures, and the gas expands to fill the entire container, what is its final temperature?
9. Which of the following processes are approximately reversible? **(a)** Lighting a match. **(b)** Pushing a block up a frictionless inclined plane. **(c)** Frying an egg. **(d)** Swimming from one end of a pool to the other. **(e)** Stretching a spring by a small amount. **(f)** Writing a report for class.
10. Which law of thermodynamics would be violated if heat were to spontaneously flow between two objects of equal temperature?
11. Heat engines always give off a certain amount of heat to a low-temperature reservoir. Would it be possible to use this "waste" heat as the heat input to a second heat engine, and

then use the "waste" heat of the second engine to run a third engine, and so on?

12. A heat pump uses 100 J of energy as it operates for a given time. Is it possible for the heat pump to deliver more than 100 J of heat to the inside of the house in this same time? Explain.

13. If you clean up a messy room, putting things back where they belong, you decrease the room's entropy. Does this violate the second law of thermodynamics? Explain.

14. Which law of thermodynamics is most pertinent to the statement that "all the king's horses and all the king's men couldn't put Humpty Dumpty back together again?"

15. Which has more entropy: **(a)** popcorn kernels, or the resulting popcorn; **(b)** two eggs in a carton, or an omelet made from the eggs; **(c)** a pile of bricks, or the resulting house; **(d)** a piece of paper, or the piece of paper after it has been burned?

Conceptual Exercises

(Answers to odd-numbered Conceptual Exercises can be found in the back of the book.)

1. Give the change in internal energy of a system if **(a)** $W = 50$ J, $Q = 50$ J, **(b)** $W = -50$ J, $Q = -50$ J, or **(c)** $W = 50$ J, $Q = -50$ J.

2. A gas expands, doing 100 J of work. How much heat must be added to this system for its internal energy to increase by 200 J?

3. You plan to add a certain amount of heat to a gas in order to raise its temperature. If you add the heat at constant volume, is the increase in temperature greater than, less than, or the same as if you add the heat at constant pressure? Explain.

4. An ideal gas is held in an insulated container at the temperature T. All the gas is initially in one-half of the container, with a partition separating the gas from the other half of the container, which is a vacuum. If the partition ruptures, and the gas expands to fill the entire container, is the final temperature greater than, less than, or equal to T? Explain.

5. **Figure 18–19** shows three different multi-step processes, labeled A, B, and C. Rank these processes in order of increasing work done by a gas that undergoes the process. Indicate ties where appropriate.

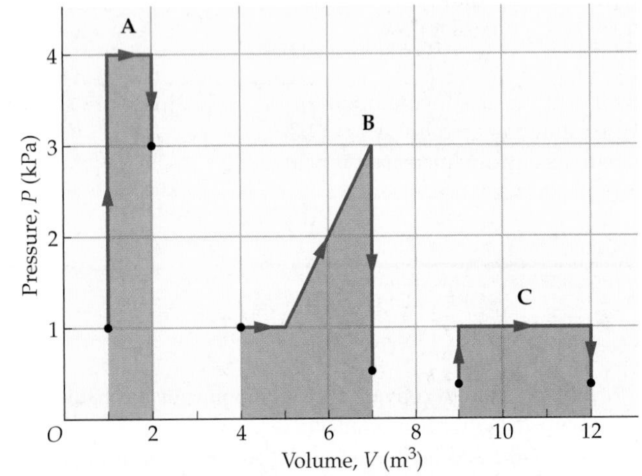

▲ **FIGURE 18–19** Conceptual Exercise 5 and Problem 69

6. If the temperature in the kitchen is decreased, is the cost to freeze a dozen ice cubes greater than, less than, or the same as it was before the kitchen was cooled? Explain.

7. A Carnot engine operates between a hot reservoir at the Kelvin temperature T_h and a cold reservoir at the Kelvin temperature T_c. **(a)** If both temperatures are doubled, does the efficiency of the engine increase, decrease, or stay the same? Explain. **(b)** If both temperatures are increased by 50 K, does the efficiency of the engine increase, decrease, or stay the same? Explain.

8. Consider the three-process cycle shown in **Figure 18–20.** For each process in the cycle, **(a)** A → B, **(b)** B → C, and **(c)** C → A, state whether the work done by the system is positive, negative, or zero.

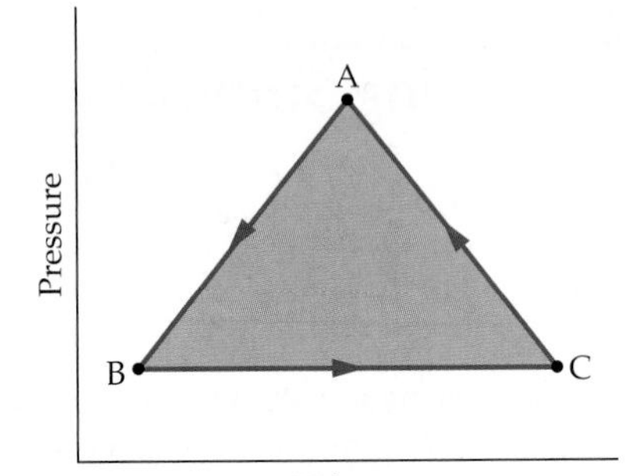

▲ **FIGURE 18–20** Conceptual Exercise 8

9. A Carnot engine can be operated with one of the following four sets of reservoir temperatures: **A**, 400 K and 800 K; **B**, 400 K and 600 K; **C**, 800 K and 1200 K; and **D**, 800 K and 1000 K Rank these reservoir temperatures in order of increasing efficiency of the Carnot engine. Indicate ties where appropriate.

10. You heat a pan of water on the stove. Rank the following temperature increases in order of increasing entropy change. Indicate ties where appropriate: **A**, 25 °C to 35 °C; **B**, 35 °C to 45 °C; **C**, 45 °C to 50 °C; and **D**, 50 °C to 55 °C.

11. An ideal gas has the pressure and volume indicated by point A in **Figure 18–21.** At this point its temperature is T_1. The temperature of the gas can be increased to T_2 by using the constant-volume process, A → B, or the constant-pressure process, A → C. Is the entropy change on the process A → B greater than, less than, or equal to the entropy change on the process A → C? Explain.

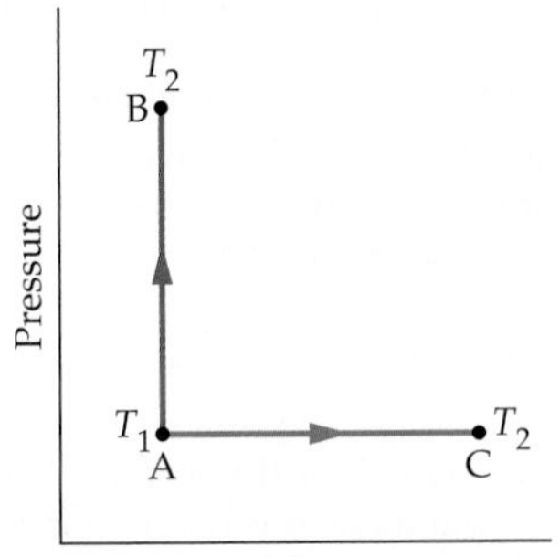

▲ **FIGURE 18–21** Conceptual Exercise 11

12. If you rub your hands together, does the entropy of the universe increase, decrease, or stay the same? Explain.

13. A gas is expanded isothermally. Does its entropy increase, decrease, or stay the same?

14. A gas is expanded adiabatically. Does its entropy increase, decrease, or stay the same?

Problems

Note: **IP** *denotes an integrated conceptual/quantitative problem.* **BIO** *identifies problems of biological or medical interest. Red bullets (•, ••, •••) are used to indicate the level of difficulty of each problem.*

Section 18–2 The First Law of Thermodynamics

1. • A swimmer does 6.7×10^5 J of work and gives off 4.1×10^5 J of heat during a workout. Determine ΔU, W, and Q for the swimmer.

2. • When 1010 J of heat are added to one mole of an ideal monatomic gas, its temperature increases from 272 K to 276 K. Find the work done by the gas during this process.

3. • Three different processes act on a system. **(a)** In process A, 42 J of work are done on the system and 77 J of heat are added to the system. Find the change in the system's internal energy. **(b)** In process B, the system does 42 J of work and 77 J of heat are added to the system. What is the change in the system's internal energy? **(c)** In process C, the system's internal energy decreases by 120 J while the system performs 120 J of work on its surroundings. How much heat was added to the system?

4. • An ideal gas is taken through the four processes shown in **Figure 18–22.** The changes in internal energy for three of these processes are as follows: $\Delta U_{AB} = +82$ J; $\Delta U_{BC} = +15$ J; $\Delta U_{DA} = -56$ J. Find the change in internal energy for the process from C to D.

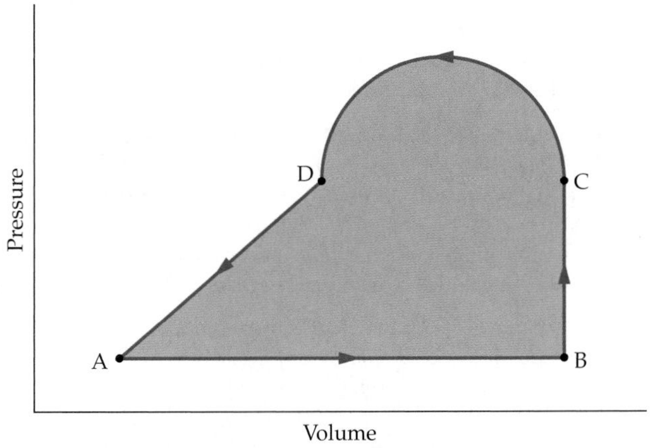

▲ **FIGURE 18–22** Problems 4 and 74

5. •• A basketball player does 2.13×10^5 J of work during her time in the game, and evaporates 0.110 kg of water. Assuming a latent heat of 2.26×10^6 J/kg for the perspiration (the same as for water) determine **(a)** the change in the player's internal energy and **(b)** the number of nutritional calories the player has converted to work and heat.

6. •• **IP** One mole of an ideal monatomic gas is initially at a temperature of 263 K. **(a)** Find the final temperature of the gas if 3280 J of heat are added to it and it does 722 J of work. **(b)** Suppose the amount of gas is doubled to two moles. Does the final temperature found in part (a) increase, decrease, or stay the same? Explain.

7. •• **IP Energy from Gasoline** Burning a gallon of gasoline releases 1.19×10^8 J of internal energy. If a certain car requires 5.20×10^5 J of work to drive one mile, **(a)** how much heat is given off to the atmosphere each mile, assuming the car gets 25.0 miles to the gallon? **(b)** If the miles per gallon of the car is increased, does the amount of heat released to the atmosphere increase, decrease, or stay the same? Explain.

8. •• A cylinder contains 4.0 moles of a monatomic gas at an initial temperature of 27 °C. The gas is compressed by doing 560 J of work on it, and its temperature increases by 130 C°. How much heat flows into or out of the gas?

9. •• An ideal gas is taken through the three processes shown in **Figure 18–23**. Fill in the missing entries in the following table:

	Q	W	ΔU
A → B	−53 J	(a)	(b)
B → C	−280 J	−130 J	(c)
C → A	(e)	150 J	(d)

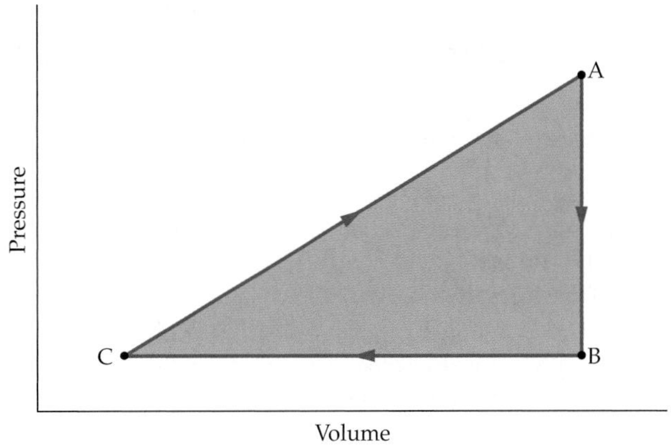

▲ **FIGURE 18–23** Problems 9 and 86

Section 18–3 Thermal Processes

10. • A system consisting of an ideal gas at the constant pressure of 110 kPa gains 920 J of heat. Find the change in volume of the system if the internal energy of the gas increases by **(a)** 920 J or **(b)** 360 J.

11. • An ideal gas is compressed at constant pressure to one-half its initial volume. If the pressure of the gas is 120 kPa, and 760 J of work is done on it, find the initial volume of the gas.

12. • As an ideal gas expands at constant pressure from a volume of 0.74 m^3 to a volume of 2.3 m^3 it does 93 J of work. What is the gas pressure during this process?

13. • The volume of a monatomic ideal gas doubles in an isothermal expansion. By what factor does its pressure change?

14. •• **IP** Consider a system in which 2.25 mol of an ideal monatomic gas is expanded at a constant pressure of 101 kPa from an initial volume of 2.15 L to a final volume of 3.30 L. **(a)** Is heat added to or removed from the system during this process? Explain. **(b)** Find the change in temperature for this process. **(c)** Determine the amount of heat added to or removed from the system during this process.

15. •• Three moles of an ideal monatomic gas expand isothermally at a temperature of 21°C. If the volume of the gas quadruples during this process, what is **(a)** the work done by the gas and **(b)** the heat flow into the gas?

16. •• **IP** **(a)** If the internal energy of a system increases as the result of an adiabatic process, is work done on the system or by the system? **(b)** Calculate the work done on or by the system in part (a) if its internal energy increases by 670 J.

17. •• **(a)** Find the work done by a monatomic ideal gas as it expands from point A to point C along the path shown in **Figure 18–24**. **(b)** If the temperature of the gas is 220 K at point A, what is its temperature at point C? **(c)** How much heat has been added to or removed from the gas during this process?

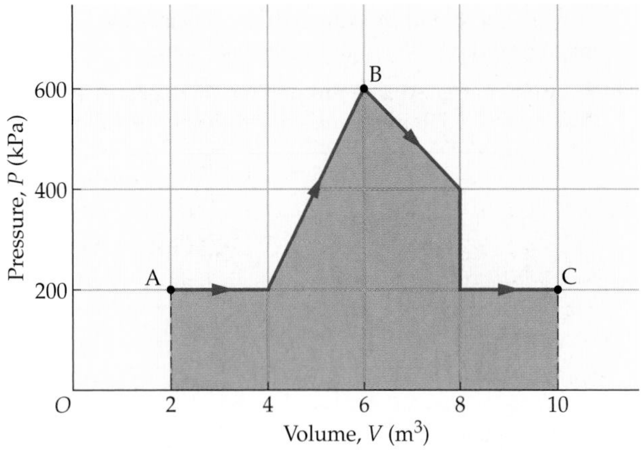

▲ **FIGURE 18–24** Problems 17 and 18

18. •• **IP** A fluid expands from point A to point B along the path shown in Figure 18–24. **(a)** How much work is done by the fluid during this expansion? **(b)** Does your answer to part (a) depend on whether the fluid is an ideal gas? Explain.

19. •• **IP** If 5.00 moles of a monatomic ideal gas at a temperature of 245 K are expanded isothermally from a volume of 1.12 L to a volume of 4.33 L, calculate **(a)** the work done and **(b)** the heat flow into or out of the gas. **(c)** If the number of moles is doubled, by what factors do your answers to parts (a) and (b) change? Explain.

20. •• Suppose 145 moles of a monatomic ideal gas undergo an isothermal expansion from 1.00 m^3 to 4.00 m^3, as shown in **Figure 18–25.** **(a)** What is the temperature at the beginning and at the end of this process? **(b)** How much work is done by the gas during this expansion?

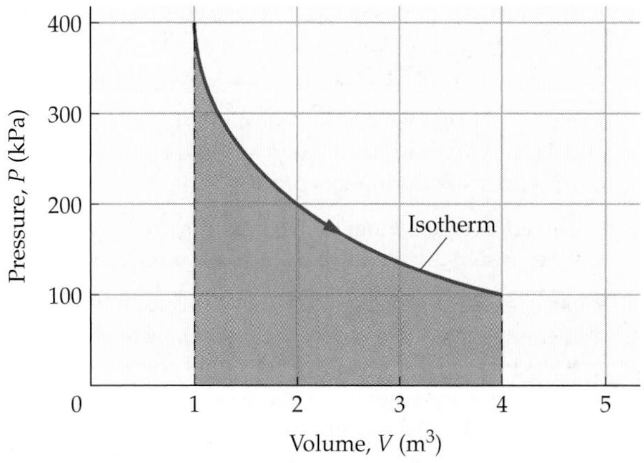

▲ **FIGURE 18–25** Problems 20, 21, and 73

21. •• **IP** A system consisting of 121 moles of a monatomic ideal gas undergoes the isothermal expansion shown in Figure 18–25. **(a)** During this process, does heat enter or leave the system? Explain. **(b)** Is the magnitude of the heat exchanged with the gas from 1.00 m^3 to 2.00 m^3 greater than, less than, or the same as it is from 3.00 m^3 to 4.00 m^3? Explain. Calculate the heat exchanged with the gas **(c)** from 1.00 m^3 to 2.00 m^3 and **(d)** from 3.00 m^3 to 4.00 m^3.

22. •• **IP** **(a)** A monatomic ideal gas expands at constant pressure. Is heat added to the system or taken from the system during this process? **(b)** Find the heat added to or taken from the gas in part (a) if it expands at a pressure of 130 kPa from a volume of 0.76 m^3 to a volume of 0.93 m^3.

23. •• During an adiabatic process, the temperature of 3.52 moles of a monatomic ideal gas drops from 485 °C to 205 °C. For this gas, find **(a)** the work it does, **(b)** the heat it exchanges with its surroundings, and **(c)** the change in its internal energy.

24. •• An ideal gas follows the three-part process shown in **Figure 18–26.** At the completion of one full cycle, find **(a)** the net work done by the system, **(b)** the net change in internal energy of the system, and **(c)** the net heat absorbed by the system.

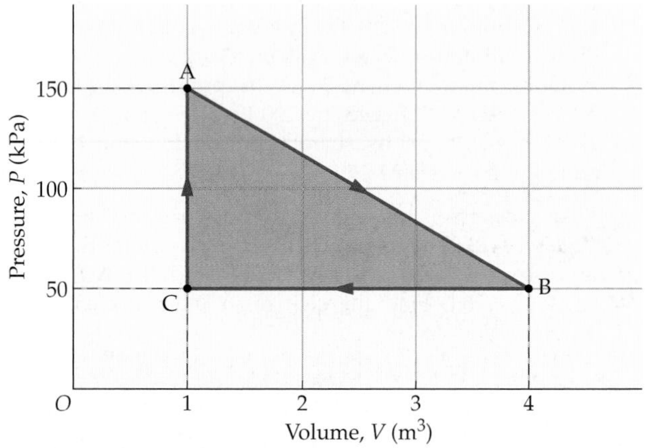

▲ **FIGURE 18–26** Problems 24 and 26

25. •• With the pressure held constant at 210 kPa, 49 mol of a monatomic ideal gas expands from an initial volume of 0.75 m^3 to a final volume of 1.9 m^3. **(a)** How much work was done by the gas during the expansion? **(b)** What were the initial and final temperatures of the gas? **(c)** What was the change in the internal energy of the gas? **(d)** How much heat was added to the gas?

26. •• **IP** Suppose 57.5 moles of an ideal monatomic gas undergo the series of processes shown in Figure 18–26. **(a)** Calculate the temperature at the points A, B, and C. **(b)** For each process, $A \rightarrow B$, $B \rightarrow C$, and $C \rightarrow A$, state whether heat enters or leaves the system. Explain in each case. **(c)** Calculate the heat exchanged with the gas during each of the three processes.

27. •• A gas is contained in a cylinder with a pressure of 120 kPa and an initial volume of 0.66 m^3. How much work is done by the gas as it **(a)** expands at constant pressure to twice its initial volume, or **(b)** is compressed to one-third its initial volume?

28. •• A system expands by 0.75 m^3 at a constant pressure of 125 kPa. Find the heat that flows into or out of the system if its internal energy **(a)** increases by 65 J or **(b)** decreases by 1850 J. In each case, give the direction of heat flow.

29. •• **IP** An ideal monatomic gas is held in a perfectly insulated cylinder fitted with a movable piston. The initial pressure of the gas is 110 kPa, and its initial temperature is 280 K. By pushing down on the piston, you are able to increase the pressure to 140 kPa. **(a)** During this process, did the temperature of the gas increase, decrease, or stay the same? Explain. **(b)** Find the final temperature of the gas.

30. ••• A certain amount of a monatomic ideal gas undergoes the process shown in **Figure 18–27**, in which its pressure doubles and its volume triples. In terms of the number of moles, n, the initial pressure, P_i, and the initial volume, V_i, determine **(a)** the

work done by the gas W, **(b)** the change in internal energy of the gas U, and **(c)** the heat added to the gas Q.

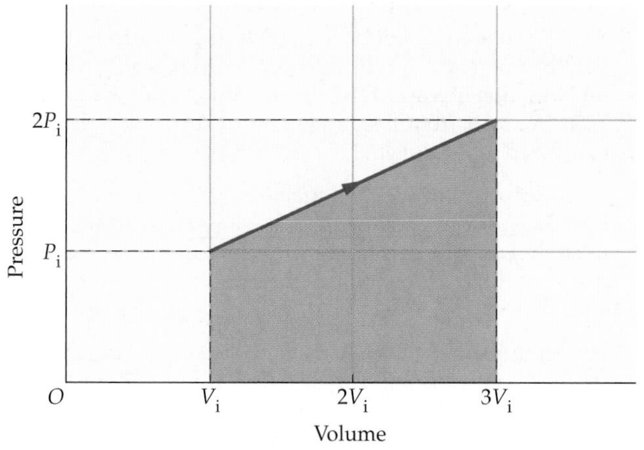

▲ **FIGURE 18–27** Problem 30

31. ••• An ideal gas doubles its volume in one of three different ways: (i) at constant pressure; (ii) at constant temperature; (iii) adiabatically. Explain your answers to each of the following questions: **(a)** In which expansion does the gas do the most work? **(b)** In which expansion does the gas do the least work? **(c)** Which expansion results in the highest final temperature? **(d)** Which expansion results in the lowest final temperature?

Section 18–4 Specific Heats for an Ideal Gas: Constant Pressure, Constant Volume

32. • Find the amount of heat needed to increase the temperature of 3.5 mol of an ideal monatomic gas by 23 K if **(a)** the pressure or **(b)** the volume is held constant.

33. • **(a)** If 535 J of heat are added to 45 moles of a monatomic gas at constant volume, how much does the temperature of the gas increase? **(b)** Repeat part (a), this time for a constant-pressure process.

34. • A system consists of 2.5 mol of an ideal monatomic gas at 325 K. How much heat must be added to the system to double its internal energy at **(a)** constant pressure or **(b)** constant volume.

35. • Find the change in temperature if 130 J of heat are added to 2.8 mol of an ideal monatomic gas at **(a)** constant pressure or **(b)** constant volume.

36. •• **IP** At constant pressure, it takes 110 J of heat to raise the temperature of a given ideal monatomic gas by 26 K. **(a)** Is the amount of heat required to produce the same increase in temperature at constant volume greater than, less than, or the same as at constant pressure? Explain. **(b)** Determine the amount of heat required for the same temperature increase at constant volume.

37. •• **IP** A cylinder contains 15 moles of a monatomic ideal gas at a constant pressure of 140 kPa. **(a)** How much work does the gas do as it expands 3200 cm^3, from 5400 cm^3 to 8600 cm^3? **(b)** If the gas expands by 3200 cm^3 again, this time from 2200 cm^3 to 5400 cm^3, is the work it does greater than, less than, or equal to the work found in part (a)? Explain. **(c)** Calculate the work done as the gas expands from 2200 cm^3 to 5400 cm^3.

38. •• **IP** The volume of a monatomic ideal gas doubles in an adiabatic expansion. By what factor do **(a)** the pressure and **(b)** the temperature of the gas change? **(c)** Verify your answers to parts (a) and (b) by considering 135 moles of gas with an initial pressure of 330 kPa and an initial volume of 1.2 m^3. Find the pressure and temperature of the gas after it expands adiabatically to a volume of 2.4 m^3.

39. •• A monatomic ideal gas is held in a thermally insulated container with a volume of 0.0750 m^3. The pressure of the gas is 105 kPa, and its temperature is 317 K. **(a)** To what volume must the gas be compressed to increase its pressure to 145 kPa? **(b)** At what volume will the gas have a temperature of 295 K?

40. ••• Consider the expansion of 60.0 moles of a monatomic ideal gas along processes 1 and 2 in **Figure 18–28.** On process 1 the gas is heated at constant volume from an initial pressure of 106 kPa to a final pressure of 212 kPa. On process 2 the gas expands at constant pressure from an initial volume of 1.00 m^3 to a final volume of 3.00 m^3. **(a)** How much heat is added to the gas during these two processes? **(b)** How much work does the gas do during this expansion? **(c)** What is the change in the internal energy of the gas?

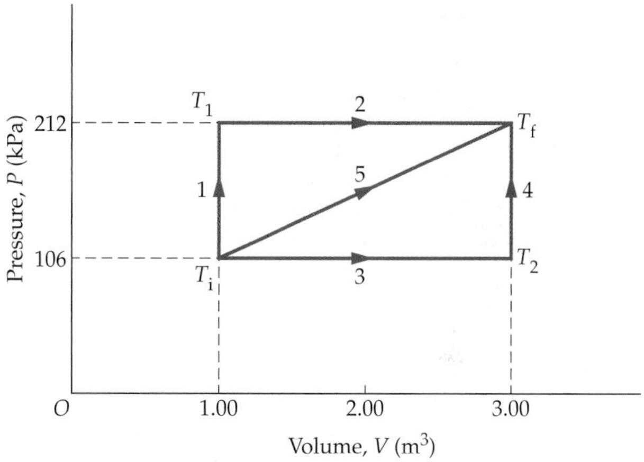

▲ **FIGURE 18–28** Problems 40, 41, and 75

41. ••• Referring to Problem 40, suppose the gas is expanded along processes 3 and 4 in Figure 18–28. On process 3 the gas expands at constant pressure from an initial volume of 1.00 m^3 to a final volume of 3.00 m^3. On process 4 the gas is heated at constant volume from an initial pressure of 106 kPa to a final pressure of 212 kPa. **(a)** How much heat is added to the gas during these two processes? **(b)** How much work does the gas do during this expansion? **(c)** What is the change in the internal energy of the gas?

Section 18–6 Heat Engines and the Carnot Cycle

42. • A heat engine takes in 1420 J of heat from the hot reservoir and exhausts 660 J of heat to the cold reservoir. How much work is done by the engine?

43. • What is the efficiency of an engine that exhausts 870 J of heat in the process of doing 340 J of work?

44. • An engine receives 690 J of heat from a hot reservoir and gives off 430 J of heat to a cold reservoir. What is **(a)** the work done and **(b)** the efficiency of this engine?

45. • A Carnot engine operates between the temperatures 410 K and 290 K. **(a)** How much heat must be given to the engine to produce 2500 J of work? **(b)** How much heat is discarded to the cold reservoir as this work is done?

46. • A nuclear power plant has a reactor that produces heat at the rate of 820 MW. This heat is used to produce 250 MW of mechanical power to drive an electrical generator. **(a)** At what rate is heat discarded to the environment by this power plant? **(b)** What is the thermal efficiency of the plant?

47. • At a coal-burning power plant a steam turbine is operated with a power output of 548 MW. The thermal efficiency of the power plant is 32.0%. **(a)** At what rate is heat discarded to the environment by this power plant? **(b)** At what rate must heat be supplied to the power plant by burning coal?

48. • **Energy from the Ocean** A research group plans to construct an engine that extracts energy from the ocean, using the fact that the water temperature near the surface is greater than the temperature in deep water. If the surface water temperature is 25.0 °C and the deep-water temperature is 12.0 °C, what is the maximum efficiency such an engine could have?

49. •• **IP** If a heat engine does 2700 J of work with an efficiency of 0.18, find **(a)** the heat taken in from the hot reservoir and **(b)** the heat given off to the cold reservoir. **(c)** If the efficiency of the engine is increased, do your answers to parts (a) and (b) increase, decrease, or stay the same? Explain.

50. •• The efficiency of a Carnot engine with a cold reservoir at a temperature of 295 K is 21.0%. Assuming the temperature of the hot reservoir remains the same, find the temperature the cold reservoir must have in order for the engine's efficiency to be 25.0%.

51. •• **IP** The efficiency of a particular Carnot engine is 0.300. **(a)** If the high-temperature reservoir is at a temperature of 545 K, what is the temperature of the low-temperature reservoir? **(b)** To increase the efficiency of this engine to 40.0%, must the temperature of the low-temperature reservoir be increased or decreased? Explain. **(c)** Find the temperature of the low-temperature reservoir that gives an efficiency of 0.400.

52. •• During each cycle a reversible engine absorbs 2500 J of heat from a high-temperature reservoir and performs 2200 J of work. **(a)** What is the efficiency of this engine? **(b)** How much heat is exhausted to the low-temperature reservoir during each cycle? **(c)** What is the ratio, T_h/T_c, of the two reservoir temperatures?

53. ••• The operating temperatures for a Carnot engine are T_c and $T_h = T_c + 55$ K. The efficiency of the engine is 11%. Find T_c and T_h.

54. ••• A certain Carnot engine takes in the heat Q_h and exhausts the heat $Q_c = 2Q_h/3$, as indicated in **Figure 18–29**. **(a)** What is the efficiency of this engine? **(b)** Using the Kelvin temperature scale, find the ratio T_c/T_h.

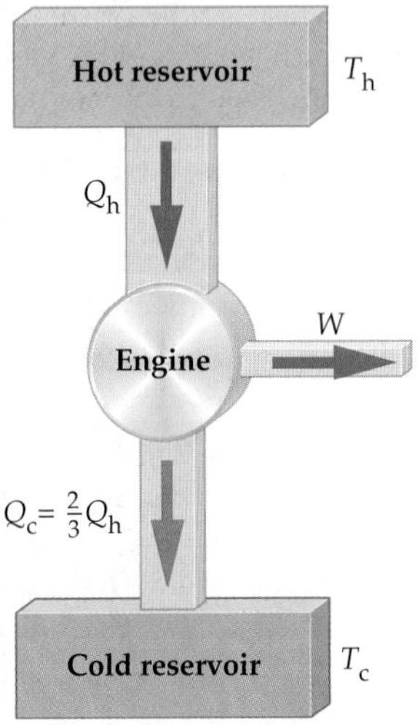

▲ **FIGURE 18–29** Problem 54

Section 18–7 Refrigerators, Air Conditioners, and Heat Pumps

55. • The refrigerator in your kitchen does 460 J of work to remove 110 J of heat from its interior. **(a)** How much heat does the refrigerator exhaust into the kitchen? **(b)** What is the refrigerator's coefficient of performance?

56. • A refrigerator with a coefficient of performance of 1.75 absorbs 3.45×10^4 J of heat from the low temperature reservoir during each cycle. **(a)** How much mechanical work is required to operate the refrigerator for a cycle? **(b)** How much heat does the refrigerator discard to the high-temperature reservoir during each cycle?

57. •• A Carnot air conditioner operates between an indoor temperature of 21.0 °C and an outdoor temperature of 32.0 °C. **(a)** How much work must the air conditioner do to remove 1550 J of heat from inside the house? **(b)** How much heat is exhausted to the outside?

58. •• To keep a room at a comfortable 21.0 °C, a Carnot heat pump does 345 J of work and supplies it with 3240 J of heat. **(a)** How much heat is removed from the outside air by the heat pump? **(b)** What is the temperature of the outside air?

59. •• An air conditioner is used to keep the interior of a house at a temperature of 21 °C while the outside temperature is 32 °C. If heat leaks into the house at the rate of 11 kW, and the air conditioner has the efficiency of a Carnot engine, what is the mechanical power required to keep the house cool?

60. •• A reversible refrigerator has a coefficient of performance equal to 10.0. What is its efficiency?

61. •• A freezer has a coefficient of performance equal to 4.0. How much electrical energy must this freezer use to produce 1.5 kg of ice at −5.0 °C from water at 15 °C?

62. •• If a Carnot engine has an efficiency of 0.23, what is its coefficient of performance if it is run backward as a heat pump?

Section 18–8 Entropy

63. • Find the change in entropy when 1.45 kg of water at 100 °C is boiled away to steam at 100 °C.

64. • Determine the change in entropy that occurs when 3.1 kg of water freezes at 0 °C.

65. •• On a cold winter's day heat leaks slowly out of a house at the rate of 20.0 kW. If the inside temperature is 22 °C, and the outside temperature is −14.5 °C, find the rate of entropy increase.

66. •• An 82-kg parachutist descends through a vertical height of 350 m with constant speed. Find the increase in entropy produced by the parachutist, assuming the air temperature is 21 °C.

67. •• **IP** Consider the air-conditioning system described in Problem 59. **(a)** Does the entropy of the universe increase, decrease, or stay the same as the air conditioner keeps the imperfectly insulated house cool? Explain. **(b)** What is the rate at which the entropy of the universe changes during this process?

68. •• A heat engine operates between a high-temperature reservoir at 610 K and a low-temperature reservoir at 320 K. In one cycle, the engine absorbs 6400 J of heat from the high-temperature reservoir and does 2200 J of work. What is the net change in entropy as a result of this cycle?

General Problems

69. • Find the work done by a monatomic ideal gas on each of the three multipart processes, **(a)**, **(b)**, and **(c)**, shown in Figure 18–19.

70. • Heat is added to a 0.16-kg block of ice at 0 °C, increasing its entropy by 98 J/K. How much ice melts?

71. • The heat that goes into a particular Carnot engine is 4.00 times greater than the work it performs. What is the engine's efficiency?

72. •• The heat that goes into a particular Carnot engine is 5.00 times greater than the work it performs. **(a)** If the hot reservoir has a temperature of 495 K, what is the temperature of the cold reservoir? **(b)** What fraction of the heat that goes into the engine is expelled to the cold reservoir?

73. •• IP Consider 132 moles of a monatomic gas undergoing the isothermal expansion shown in Figure 18–25. **(a)** What is the temperature T of this expansion? **(b)** Does the entropy of the gas increase, decrease, or stay the same during the process? Explain. **(c)** Calculate the change in entropy for the gas, ΔS, if it is nonzero. **(d)** Calculate the work done by the gas during this process, and compare to $T\Delta S$.

74. •• IP Consider a monatomic ideal gas that undergoes the four processes shown in Figure 18–22. Is the work done by the gas positive, negative, or zero on process **(a)** AB, **(b)** BC, **(c)** CD, and **(d)** DA? Explain in each case. **(e)** If the heat added to the gas on process AB is 27 J, how much work does the gas do during that process?

75. •• IP Referring to Figure 18–28, suppose 60.0 moles of a monatomic ideal gas are expanded along process 5. **(a)** How much work does the gas do during this expansion? **(b)** What is the change in the internal energy of the gas? **(c)** How much heat is added to the gas during this process?

76. •• IP Engine A has an efficiency of 66%. Engine B absorbs the same amount of heat from the hot reservoir and exhausts twice as much heat to the cold reservoir. **(a)** Which engine has the greater efficiency? Explain. **(b)** What is the efficiency of engine B?

77. •• A freezer with a coefficient of performance of 3.88 is used to convert 1.75 kg of water to ice in one hour. The water starts at a temperature of 20.0 °C and the ice that is produced is cooled to a temperature of −5.00 °C. **(a)** How much heat must be removed from the water for this process to occur? **(b)** How much electrical energy does the freezer use during this hour of operation? **(c)** How much heat is discarded into the room that houses the freezer?

78. •• Suppose 1600 J of heat are added to 3.7 mol of argon gas at a constant pressure of 120 kPa. Find the change in **(a)** internal energy and **(b)** temperature for this gas. **(c)** Calculate the change in volume of the gas. (Assume that the argon can be treated as an ideal monatomic gas.)

79. •• A system consists of 3.2 mol of the monatomic gas neon (which may be treated as an ideal gas). Initially the system has a temperature of 290 K and a volume of 1.2 m^3. Heat is added to this system at constant pressure until the volume triples, then more heat is added at constant volume until the pressure doubles. Find the total heat added to the system.

80. •• **Entropy and the Sun** The surface of the Sun has a temperature of 5500 °C and the temperature of deep space is 3.0 K. **(a)** Find the entropy increase produced by the Sun in one day, given that it radiates heat at the rate of 3.80×10^{26} W. **(b)** How much work could have been done if this heat had been used to run an ideal heat engine?

81. •• The following table lists results for various processes involving n moles of a monatomic ideal gas. Fill in the missing entries.

	Q	W	ΔU
Constant pressure	$\frac{5}{2}nR\Delta T$	(a)	$\frac{3}{2}nR\Delta T$
Adiabatic	(b)	$-\frac{3}{2}nR\Delta T$	(c)
Constant volume	$\frac{3}{2}nR\Delta T$	(d)	(e)
Isothermal	(f)	$nRT\ln(V_f/V_i)$	(g)

82. •• A cylinder with a moveable piston holds 2.50 mol of argon at a constant temperature of 295 K. As the gas is compressed isothermally, its pressure increases from 101 kPa to 121 kPa. Find **(a)** the final volume of the gas, **(b)** the work done by the gas, and **(c)** the heat added to the gas.

83. •• An inventor claims a new cyclic engine that uses organic grape juice as its working material. According to the claims, the engine absorbs 1250 J of heat from a 1010 K reservoir and performs 1120 J of work each cycle. The waste heat is exhausted to the atmosphere at a temperature of 302 K. **(a)** What is the efficiency that is implied by these claims? **(b)** What is the efficiency of a reversible engine operating between the same high and low temperatures used by this engine? (Should you invest in this invention?)

84. •• A nonreversible heat engine operates between a high-temperature reservoir at $T_h = 810$ K and a low-temperature reservoir at $T_c = 320$ K. During each cycle the engine absorbs 660 J of heat from the high-temperature reservoir and performs 250 J of work. **(a)** Calculate the total entropy change ΔS_{tot} for one cycle. **(b)** How much work would a reversible heat engine perform in one cycle if it operated between the same two temperatures and absorbed the same amount of heat? **(c)** Show that the difference in work between the nonreversible engine and the reversible engine is equal to $T_c\Delta S_{tot}$.

85. •• IP A small dish containing 510 g of water is placed outside for the birds. During the night the outside temperature drops to −5.0 °C and stays at that value for several hours. **(a)** When the water in the dish freezes, does its entropy increase, decrease, or stay the same? Explain. **(b)** Calculate the change in entropy that occurs as the water freezes. **(c)** When the water freezes, is there an entropy change anywhere else in the universe? If so, specify where the increase occurs.

86. •• IP An ideal gas is taken through the three processes shown in Figure 18–23. Fill in the missing entries in the following table:

	Q	W	ΔU
A → B	(a)	(b)	−38 J
B → C	(c)	−89 J	−82 J
C → A	332 J	(d)	(e)

87. ••• Which would make the greater change in the efficiency of a Carnot heat engine: **(a)** raising the temperature of the high-temperature reservoir by ΔT, or **(b)** lowering the temperature of the low-temperature reservoir by ΔT? Justify your answer by calculating the change in efficiency for each of these cases.

88. ••• One mole of an ideal monatomic gas follows the three-part cycle shown in **Figure 18–30**. **(a)** Fill in the following table:

	Q	W	ΔU
A → B			
B → C			
C → A			

(b) What is the efficiency of this cycle?

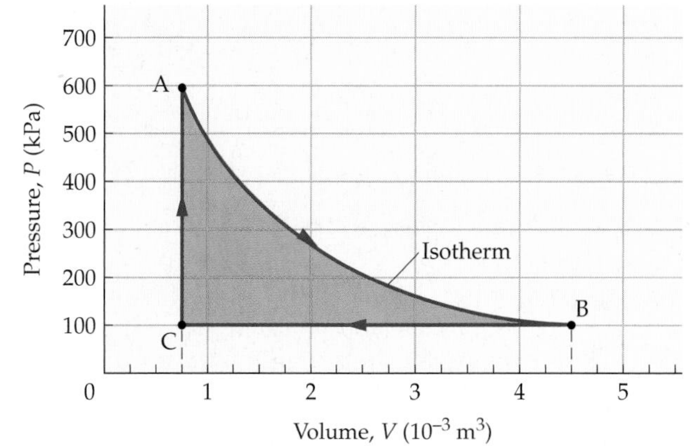

▲ **FIGURE 18–30** Problem 88

89. ••• When a heat Q is added to a monatomic ideal gas at constant pressure, the gas does a work W. Find the ratio, W/Q.

90. ••• A diesel engine uses adiabatic compression to heat the air in its cylinders to a temperature sufficiently high to cause combustion of the injected fuel. If the volume of air in a cylinder is reduced by a factor of 23, by what factor does its absolute temperature increase?

91. ••• **The Carnot Cycle Figure 18–31** shows an example of a Carnot cycle. The cycle consists of the following four processes: (1) an isothermal expansion from V_1 to V_2 at the temperature T_h; (2) an adiabatic expansion from V_2 to V_3 during which the temperature drops from T_h to T_c. (3) An isothermal compression from V_3 to V_4 at the temperature T_c; and (4) An adiabatic compression from V_4 to V_1 during which the temperature increases from T_c to T_h. Show that the efficiency of this cycle is $e = 1 - T_c/T_h$, as expected.

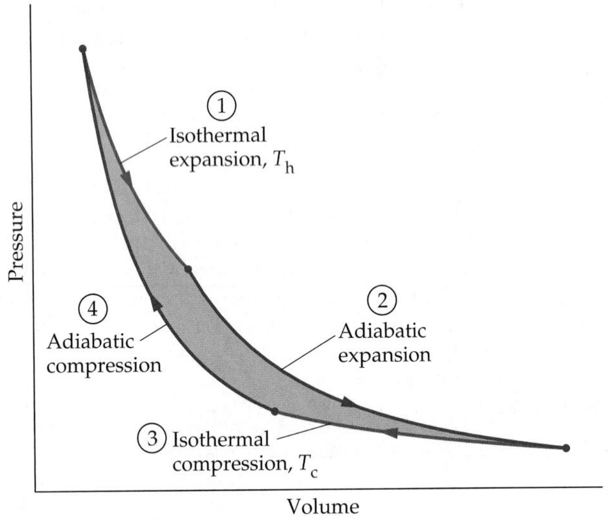

▲ **FIGURE 18–31** Problem 91

92. ••• A Carnot engine and a Carnot refrigerator operate between the same two temperatures. Show that the coefficient of performance, COP, for the refrigerator is related to the efficiency, e, of the engine by the following expression; COP $= (1 - e)/e$.

Interactive Problems

93. •• **IP** ***Referring to Active Example 18–3*** Suppose we lower the temperature of the cold reservoir to 295 K; the temperature of the hot reservoir is still 576 K. **(a)** Is the new efficiency of the engine greater than, less than, or equal to 0.470? Explain. **(b)** What is the new efficiency? **(c)** Find the work done by this engine when 1050 J of heat is drawn from the hot reservoir.

94. •• **IP** ***Referring to Active Example 18–3*** Suppose the temperature of the hot reservoir is increased by 16 K, from 576 K to 592 K, *and* that the temperature of the cold reservoir is also increased by 16 K, from 305 K to 321 K. **(a)** Is the new efficiency greater than, less than, or equal to 0.470? Explain. **(b)** What is the new efficiency? **(c)** What is the change in entropy of the hot reservoir when 1050 J of heat is drawn from it? **(d)** What is the change in entropy of the cold reservoir?

19 Electric Charges, Forces, and Fields

19–1	Electric Charge	626
19–2	Insulators and Conductors	629
19–3	Coulomb's Law	630
19–4	The Electric Field	637
19–5	Electric Field Lines	643
19–6	Shielding and Charging by Induction	646
19–7	Electric Flux and Gauss's Law	649

Amber, a form of fossilized tree resin long used to make beautiful beads and other ornaments, has also made contributions to two different sciences. Pieces of amber have preserved prehistoric insects and pollen grains for modern students of evolution. And over 2500 years ago, amber provided Greek scientists with their first opportunity to study electrical forces—the subject of this chapter.

We are all made up of electric charges. Every atom in every human body contains both positive and negative charges held together by an attractive force that is similar to gravity—only vastly stronger. Our atoms are bound together by electrical forces to form molecules; these molecules, in turn, interact with one another to produce solid bones and liquid blood. In a very real sense, we are walking, talking manifestations of electricity.

In this chapter, we discuss the basic properties of electric charge. Among these are that electric charge comes in discrete units (quantization) and that the total amount of charge in the universe remains constant (conservation). In addition, we present the force law that describes the interactions between electric charges. Finally, we introduce the idea of an electric *field*, and show how it is related to the distribution of charge.

19–1 Electric Charge

The effects of electric charge have been known since at least 600 B.C. About that time, the Greeks noticed that amber—a solid, translucent material formed from the fossilized resin of extinct coniferous trees—has a peculiar property. If a piece of amber is rubbed with animal fur, it attracts small, lightweight objects. This phenomenon is illustrated in **Figure 19–1**.

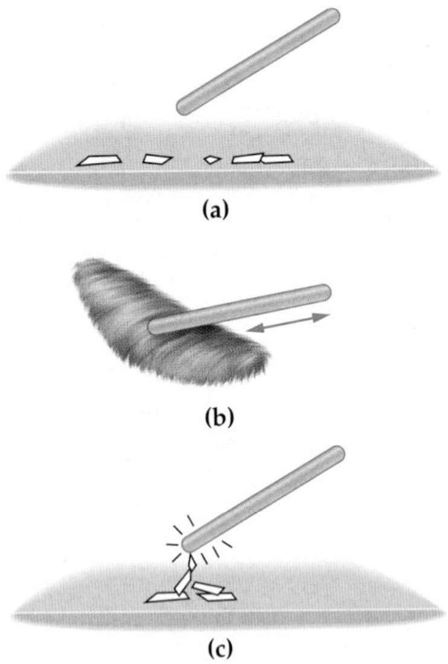

▲ **FIGURE 19–1 Charging an amber rod**
An uncharged amber rod **(a)** exerts no force on scraps of paper. When the rod is rubbed against a piece of fur **(b)**, it becomes charged and then attracts the paper **(c)**.

For some time, it was thought that amber was unique in its ability to become "charged." Much later, it was discovered that other materials can behave in this way as well. For example, if glass is rubbed with a piece of silk, it too can attract small objects. In this respect, glass and amber seem to be the same. It turns out, however, that these two materials have different types of charge.

To see this, imagine suspending a small, charged rod of amber from a thread, as in **Figure 19–2**. If a second piece of charged amber is brought near the rod, as shown in Figure 19–2 (a), the rod rotates away, indicating a repulsive force between the two pieces of amber. Thus, "like" charges repel. On the other hand, if a piece of charged glass is brought near the amber rod, the amber rotates toward the glass, indicating an attractive force. This is illustrated in Figure 19–2 (b). Clearly, then, the *different* charges on the glass and amber attract one another. We refer to different charges as being the "opposite" of one another, as in the familiar expression "opposites attract."

We know today that the two types of charge found on amber and glass are, in fact, the only types that exist, and we still use the purely arbitrary names—**positive** (+) charge and **negative** (−) charge—proposed by Benjamin Franklin (1706–1790) in 1747. In accordance with Franklin's original suggestion, the charge of amber is negative, and the charge of glass is positive (the opposite of negative). Calling the different charges + and − is actually quite useful mathematically; for example, an object that contains an equal amount of positive and negative charge has zero net charge. Objects with zero net charge are said to be electrically **neutral.**

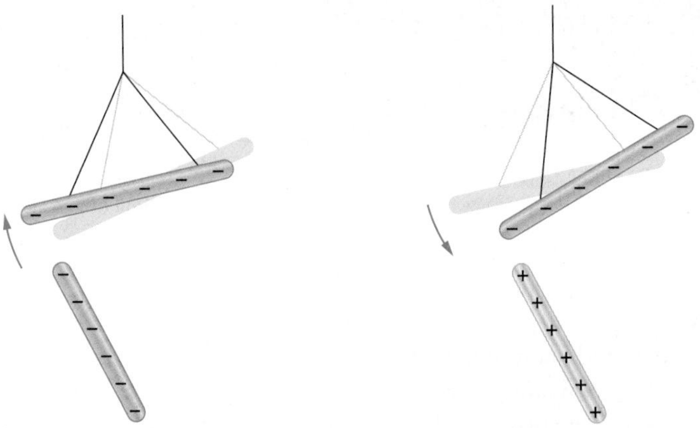

▶ **FIGURE 19–2 Likes repel, opposites attract**
A charged amber rod is suspended by a string. According to the convention introduced by Benjamin Franklin, the charge on the amber is designated as negative. **(a)** When another charged amber rod is brought near the suspended rod, it rotates away, indicating a repulsive force between like charges. **(b)** When a charged glass rod is brought close to the suspended amber rod, the amber rotates toward the glass, indicating an attractive force and the existence of a second type of charge, which we designate as positive.

A familiar example of an electrically neutral object is the atom. Atoms have a small, dense nucleus with a positive charge surrounded by a cloud of negatively charged electrons (from the Greek word for amber, *elektron*). A pictorial representation of an atom is shown in **Figure 19–3**.

All electrons have exactly the same electric charge. This charge is very small and is defined to have a magnitude, e, given by the following:

Magnitude of an Electron's Charge, *e*

$$e = 1.60 \times 10^{-19}\,\mathrm{C} \qquad \text{19–1}$$

SI unit: coulomb, C

In this expression, C is a unit of charge referred to as the **coulomb,** named for the French physicist Charles Augustin de Coulomb (1736–1806). (The precise definition of the coulomb is in terms of electric current, which we shall discuss in Chapter 21.) Clearly, the charge on an electron, which is negative, is $-e$. This is one of the defining, or *intrinsic*, properties of the electron. Another intrinsic property of the electron is its mass, m_e:

$$m_e = 9.11 \times 10^{-31}\text{ kg} \qquad \textbf{19–2}$$

In contrast, the charge on a proton—one of the main constituents of nuclei—is *exactly* $+e$. Therefore, the net charge on atoms, which have equal numbers of electrons and protons, is precisely zero. The mass of the proton is

$$m_p = 1.673 \times 10^{-27}\text{ kg} \qquad \textbf{19–3}$$

Note that this is about 2000 times larger than the mass of the electron. The other main constituent of the nucleus is the neutron, which, as its name implies, has zero charge. Its mass is slightly larger than that of the proton:

$$m_n = 1.675 \times 10^{-27}\text{ kg} \qquad \textbf{19–4}$$

Since the magnitude of the charge per electron is 1.60×10^{-19} C/electron, it follows that the number of electrons in 1 C of charge is enormous:

$$\frac{1\text{ C}}{1.60 \times 10^{-19}\text{ C/electron}} = 6.25 \times 10^{18}\text{ electrons}$$

As we shall see when we consider the force between charges, a coulomb is a significant amount of charge; even a powerful lightning bolt delivers only 20 to 30 C. A more common unit of charge is the microcoulomb, μC, where 1 μC $= 10^{-6}$ C. Still, the amount of charge contained in everyday objects is very large, even in units of the coulomb, as we show in the following Exercise.

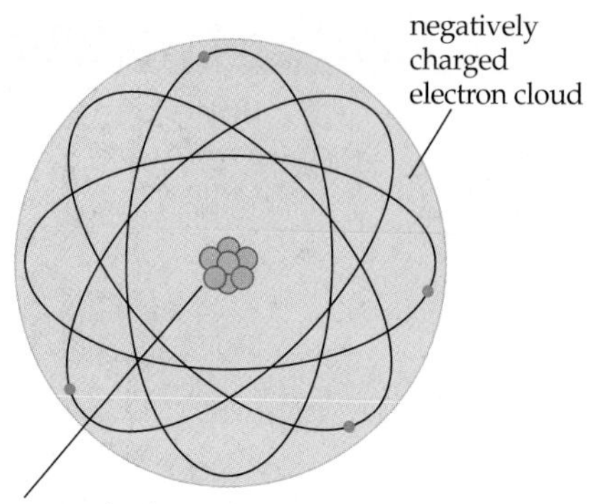

▲ **FIGURE 19–3 The structure of an atom**

A crude representation of an atom, showing the positively charged nucleus at its center and the negatively charged electrons orbiting about it. More accurately, the electrons should be thought of as forming a "cloud" of negative charge surrounding the nucleus.

EXERCISE 19–1

Find the amount of positive electric charge in one mole of helium atoms. (Note that the nucleus of a helium atom consists of two protons and two neutrons.)

Solution

Since each helium atom contains two positive charges of magnitude e, the total positive charge in a mole is

$$N_A(2e) = (6.02 \times 10^{23})(2)(1.60 \times 10^{-19}\text{ C}) = 1.93 \times 10^5\text{ C}$$

Thus, a mere 4 g of helium contains almost 200,000 C of positive charge, and the same amount of negative charge, as well.

Charge Separation

How is it that rubbing a piece of amber with fur gives the amber a charge? Originally, it was thought that the friction of rubbing *created* the observed charge. We now know, however, that rubbing the fur across the amber simply results in a *transfer* of charge from the fur to the amber—with the total amount of charge remaining unchanged. This is indicated in **Figure 19–4**. Before charging, the fur and the amber are both neutral. During the rubbing process some electrons are transferred from the fur to the amber, giving the amber a net negative charge, and leaving the fur with a net positive charge. At no time during this process is charge ever created or destroyed. This, in fact, is an example of one of the fundamental conservation laws of physics:

Conservation of Electric Charge

The total electric charge of the universe is constant. No physical process can result in an increase or decrease in the total amount of electric charge in the universe.

When charge is transferred from one object to another, it is generally due to the movement of electrons. In a typical solid, the nuclei of the atoms are fixed in position. The outer electrons of these atoms, however, are often weakly bound

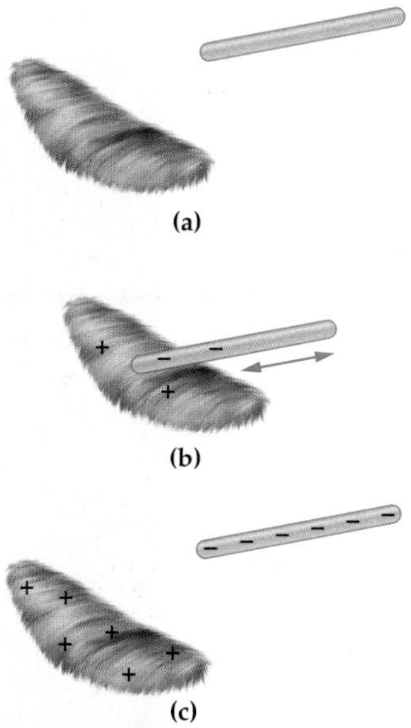

▲ **FIGURE 19–4 Charge transfer**

(a) Initially, an amber rod and a piece of fur are electrically neutral; that is, they each contain equal quantities of positive and negative charge. **(b)** As they are rubbed together, charge is transferred from one to the other. **(c)** In the end, the fur and the rod have charges of equal magnitude but opposite sign.

TABLE 19–1 Triboelectric charging

Material	Relative charging with rubbing
Rabbit fur	++++++
Glass	+++++
Human hair	++++
Nylon	+++
Silk	++
Paper	+
Cotton	−
Wood	−−
Amber	−−−
Rubber	−−−−
PVC	−−−−−
Teflon	−−−−−−

and fairly easily separated. As a piece of fur rubs across amber, for example, some of the electrons that were originally part of the fur are separated from their atoms and deposited onto the amber. The atom that loses an electron is now a **positive ion,** and the atom that receives an extra electron becomes a **negative ion.** This is charging by separation.

In general, when two materials are rubbed together, the magnitude *and* sign of the charge that each material acquires depends on how strongly it holds onto its electrons. For example, if silk is rubbed against glass, the glass acquires a positive charge, as was mentioned earlier in this section. It follows that electrons have moved from the glass to the silk, giving the silk a *negative* charge. If silk is rubbed against amber, however, the silk becomes *positively* charged, as electrons in this case pass from the silk to the amber.

These results can be understood by referring to Table 19–1, which presents the relative charging due to rubbing—also known as **triboelectric charging**—for a variety of materials. The more plus signs associated with a material, the more readily it gives up electrons and becomes positively charged. Similarly, the more minus signs for a material, the more readily it acquires electrons. For example, we know that amber becomes negatively charged when rubbed against fur, but a greater negative charge is obtained if rubber, PVC, or Teflon is rubbed with fur instead. In general, when two materials in Table 19–1 are rubbed together, the one higher in the list becomes positively charged, and the one lower in the list becomes negatively charged. The greater the separation on the list, the greater the magnitude of the charge.

Charge separation occurs not only when one object is rubbed against another, but also when objects collide. For example, colliding crystals of ice in a rain cloud can cause charge separation that may ultimately result in bolts of lightning to bring the charges together again. Similarly, particles in the rings of Saturn are constantly undergoing collisions and becoming charged as a result. In fact, when the *Voyager* spacecraft examined the rings of Saturn, it observed electrostatic discharges, similar to lightning bolts on Earth. In addition, ghostly radial "spokes" that extend across the rings of Saturn—which cannot be explained by gravitational forces alone—are also the result of electrostatic interactions.

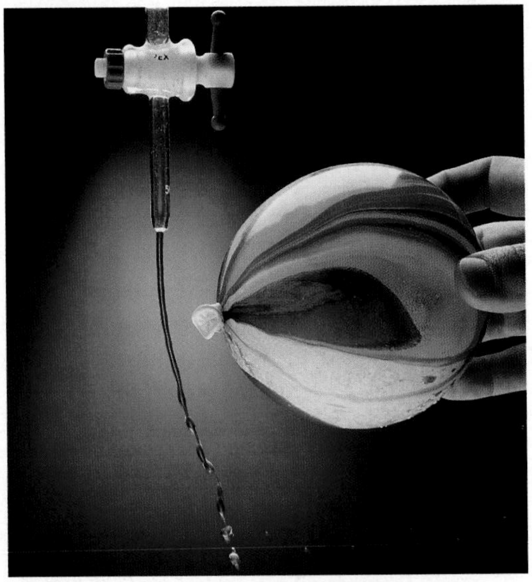

▲ The Van de Graaff generator (left) that these children are touching can produce very large charges of static electricity. Since they are clearly not frightened, why is their hair standing on end? On a smaller scale, if you rub a balloon against a cloth surface, the balloon acquires a negative electrical charge. The balloon can then attract a stream of water (right), even though water molecules themselves are electrically neutral. This phenomenon occurs because the water molecules, though they have no net charge, are polar: one end of the molecule has a slight positive charge and the other a slight negative charge. Under the influence of the balloon's negative charge, the water molecules orient themselves so that their positive ends point toward the balloon. This alignment ensures that the electrical attraction between the balloon and the positive part of each molecule exceeds the repulsion between the balloon and the negative part of each molecule.

CONCEPTUAL CHECKPOINT 19–1 Compare the Mass

Is the mass of an amber rod after charging with fur **(a)** more than, **(b)** less than, or **(c)** the same as its mass before charging?

Reasoning and Discussion

Since an amber rod becomes negatively charged, it has acquired electrons from the fur. Each electron has a small, but nonzero, mass. Therefore, the mass of the rod increases ever so slightly as it is charged.

Answer

(a) The mass of the amber rod is greater after charging.

Since electrons always have the charge $-e$, and protons always have the charge $+e$, it follows that all objects must have a net charge that is an integral multiple of e. This conclusion was confirmed early in the twentieth century by the American physicist Robert A. Millikan (1868–1953) in a classic series of experiments. He found that the charge on an object can be $\pm e$, $\pm 2e$, $\pm 3e$, and so on, but never $1.5e$ or $-9.3847e$, for example. We describe this restriction by saying that electric charge is **quantized.**

Polarization

We know that charges of opposite sign attract, but it is also possible for a charged rod to attract small objects that have zero net charge. The mechanism responsible for this attraction is called **polarization.**

To see how polarization works, consider **Figure 19–5**. Here we show a positively charged rod held close to an enlarged view of a neutral object. An atom near the surface of the neutral object will become elongated because the negative electrons in it are attracted to the rod while the positive protons are repelled. As a result, a net negative charge develops on the surface near the rod—the so-called polarization charge. The attractive force between the rod and this *induced* polarization charge leads to a net attraction between the rod and the entire neutral object.

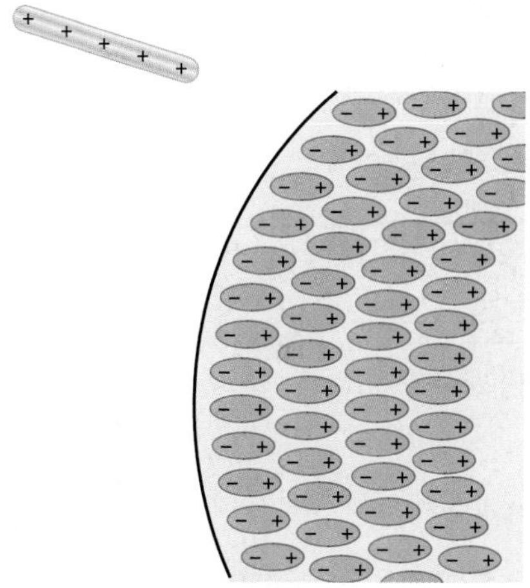

▲ **FIGURE 19–5 Electrical polarization**
When a charged rod is far from a neutral object, the atoms in the object are undistorted, as in Figure 19–3. As the rod is brought closer, however, the atoms distort, producing an excess of one type of charge on the surface of the object (in this case a negative charge). This induced charge is referred to as a polarization charge. Because the sign of the polarization charge is the opposite of the sign of the charge on the rod, there is an attractive force between the rod and the object.

Of course, the same conclusion is reached if we consider a negative rod held near a neutral object—except in this case the polarization charge is positive. Thus, the effect of polarization is to give rise to an attractive force regardless of the sign of the charged object. It is for this reason that both charged amber and charged glass attract neutral objects—even though their charges are opposite.

A potentially dangerous, and initially unsuspected, medical application of polarization occurs in endoscopic surgery. In these procedures, a tube carrying a small video camera is inserted into the body. The resulting video image is produced by electrons striking the inside surface of a computer monitor's screen, which is kept positively charged to attract the electrons. Minute airborne particles in the operating room—including dust, lint, and skin cells—are polarized by the positive charge on the screen, and are attracted to its exterior surface.

The problem comes when a surgeon touches the screen to point out an important feature to others in the medical staff. Even the slightest touch can transfer particles—many of which carry bacteria—from the screen to the surgeon's finger and from there to the patient. In fact, the surgeon's finger doesn't even have to touch the screen—as the finger approaches the screen, it too becomes polarized, and hence, it can attract particles from the screen, or directly from the air. Situations like these have resulted in infections, and surgeons are now cautioned not to bring their fingers near the video monitor.

REAL-WORLD PHYSICS: BIO
Bacterial infection from endoscopic surgery

19–2 Insulators and Conductors

Suppose you rub one end of an amber rod with fur, being careful not to touch the other end. The result is that the rubbed portion becomes charged, whereas the other end remains neutral. In particular, the negative charge transferred to the rubbed end stays put; it does not move about from one end of the rod to the other.

▲ People who work with electricity must be careful to use gloves made of non-conducting materials. Rubber, an excellent insulator, is often used for this purpose.

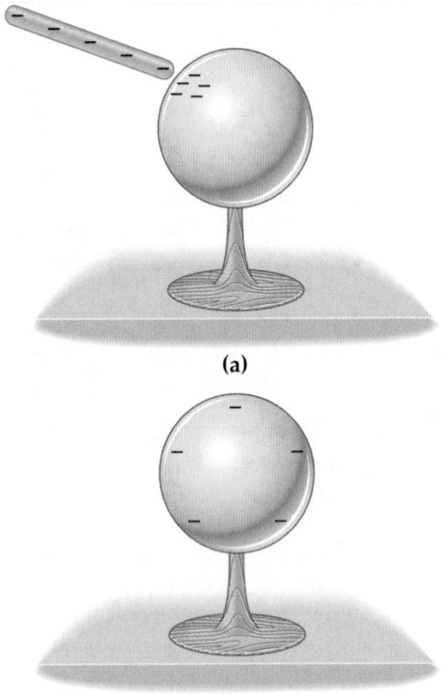

▲ **FIGURE 19–6 Charging a conductor**
(a) When an uncharged metal sphere is touched by a charged rod, some charge is transferred at the point of contact. **(b)** Because like charges repel, and charges move freely on a conductor, the transferred charge quickly spreads out and covers the entire surface of the sphere.

REAL-WORLD PHYSICS
Photocopiers and laser printers

Materials like amber, in which charges are not free to move, are referred to as **insulators.** Most insulators are nonmetallic substances, and most are also good thermal insulators.

In contrast, most metals are good **conductors** of electricity, in the sense that they allow charges to move about more or less freely. For example, suppose an uncharged metal sphere is placed on an insulating base. If a charged rod is brought into contact with the sphere, as in **Figure 19–6 (a)**, some charge will be transferred to the sphere at the point of contact. The charge does not stay put, however. Since the metal is a good conductor of electricity, the charges are free to move about the sphere, which they do because of their mutual repulsion. The result is a uniform distribution of charge over the surface of the sphere, as shown in **Figure 19–6 (b)**. Note that the insulating base prevents charge from flowing away from the sphere into the ground.

On a microscopic level, the difference between conductors and insulators is that the atoms in conductors allow one or more of their outermost electrons to become detached. These detached electrons, often referred to as "conduction electrons," can move freely throughout the conductor. In a sense, the conduction electrons behave almost like gas molecules moving about within a container. Insulators, in contrast, have very few, if any, free electrons; the electrons are bound to their atoms and cannot move from place to place within the material.

Some materials have properties that are intermediate between those of a good conductor and a good insulator. These materials, referred to as **semiconductors,** can be fine-tuned to display almost any desired degree of conductivity by controlling the concentration of the various components from which they are made. The great versatility of semiconductors is one reason they have found such wide areas of application in electronics and computers.

Exposure to light can sometimes determine whether a given material is an insulator or a conductor. An example of such a **photoconductive** material is selenium, which conducts electricity when light shines on it but is an insulator when in the dark. Because of this special property, selenium plays a key role in the production of photocopies. To see how, we first note that at the heart of every photocopier is a selenium-coated aluminum drum. Initially, the selenium is given a positive charge and kept in the dark—which causes it to retain its charge. When flash lamps illuminate a document to be copied, an image of the document falls on the drum. Where the document is light, the selenium is illuminated and becomes a conductor, and the positive charge flows away into the aluminum drum, leaving the selenium uncharged. Where the document is dark, the selenium is not illuminated, meaning that it is an insulator, and its charge remains in place. At this point, a negatively charged "toner" powder is wiped across the drum, where it sticks to those positively charged portions of the drum that were not illuminated. Next, the drum is brought into contact with paper, transferring the toner to it. Finally, the toner is fused into the paper fibers with heat, the drum is cleaned of excess toner, and the cycle repeats. Thus, a slight variation in electrical properties due to illumination is the basis of an entire technology.

The operation of a laser printer is basically the same as that of a photocopier, with the difference that in the laser printer the selenium-coated drum is illuminated with a computer-controlled laser beam. As the laser sweeps across the selenium, the computer turns the beam on and off to produce areas that will print light or dark, respectively.

19–3 Coulomb's Law

We have already discussed the fact that electric charges exert forces on one another. The precise law describing these forces was first determined by Coulomb in the late 1780s. His result is remarkably simple. Suppose, for example, that an idealized point charge q_1 is separated by a distance r from another point charge q_2. Both charges are at rest; that is, the system is **electrostatic.** According to Coulomb's law, the magnitude of the electrostatic force between these charges is proportional to the product of the magnitude of the charges, $|q_1||q_2|$, and inversely proportional to the square of the distance, r^2, between them:

Coulomb's Law for the Magnitude of the Electrostatic Force Between Point Charges

$$F = k\frac{|q_1||q_2|}{r^2} \qquad 19\text{–}5$$

SI unit: newton, N

In this expression, the proportionality constant k has the value

$$k = 8.99 \times 10^9 \text{ N} \cdot \text{m}^2/\text{C}^2 \qquad 19\text{–}6$$

Note that the units of k are simply those required for the force F to have the units of newtons.

The direction of the force in Coulomb's law is along the line connecting the two charges. In addition, we know from the observations described in Section 19–1 that like charges repel and opposite charges attract. These properties are illustrated in **Figure 19–7**, where force vectors are shown for charges of various signs. Thus, when applying Coulomb's law, we first calculate the magnitude of the force using Equation 19–5, and then determine its direction with the "likes repel, opposites attract" rule.

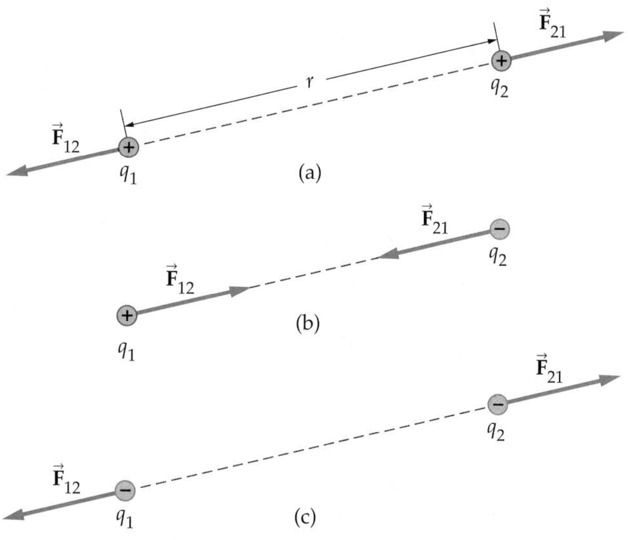

◀ **FIGURE 19–7 Forces between point charges**

The forces exerted by two point charges on one another are always along the line connecting the charges. If the charges have the same sign, as in **(a)** and **(c)**, the forces are repulsive; that is, each charge experiences a force that points away from the other charge. Charges of opposite sign, as in **(b)**, experience attractive forces. Notice that in all cases the forces exerted on the two charges form an action–reaction pair. That is, $\vec{F}_{21} = -\vec{F}_{12}$.

Finally, note how Newton's third law applies to each of the cases shown in Figure 19–7. For example, the force exerted on charge 1 by charge 2, $\vec{F}_{12}$, is always equal in magnitude and opposite in direction to the force exerted on charge 2 by charge 1, $\vec{F}_{21}$; that is, $\vec{F}_{21} = -\vec{F}_{12}$.

CONCEPTUAL CHECKPOINT 19–2 Where Do They Collide?

An electron and a proton, initially separated by a distance d, are released from rest simultaneously. The two particles are free to move. When they collide, are they **(a)** at the midpoint of their initial separation, **(b)** closer to the initial position of the proton, or **(c)** closer to the initial position of the electron?

Reasoning and Discussion

Because of Newton's third law, the forces exerted on the electron and proton are equal in magnitude and opposite in direction. For this reason, it might seem that the particles meet at the midpoint. The masses of the particles, however, are quite different. In fact, as mentioned in Section 19–1, the mass of the proton is about 2000 times greater than the mass of the electron; therefore, the proton's acceleration ($a = F/m$) is about 2000 times less than the electron's acceleration. As a result, the particles collide near the initial position of the proton. More specifically, they collide at the location of the center of mass of the system, which remains at rest throughout the process.

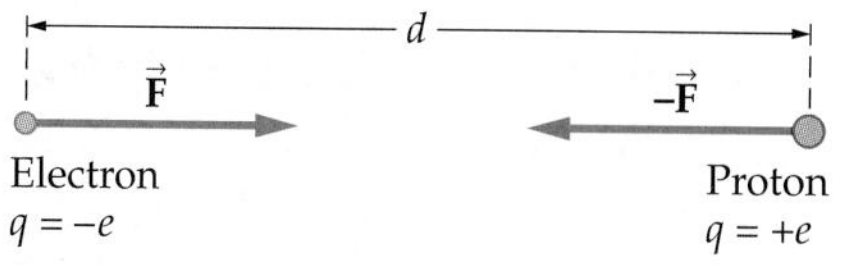

Answer

(b) The particles collide near the initial position of the proton.

It is interesting to note the similarities and differences between Coulomb's law, $F = k|q_1||q_2|/r^2$, and Newton's law of gravity, $F = Gm_1m_2/r^2$. In each case, the force decreases as the square of the distance between the two objects. In addition, both forces depend on a product of intrinsic quantities: in the case of the electric force the intrinsic quantity is the charge; in the case of gravity it is the mass.

Equally significant, however, are the differences. In particular, the force of gravity is always attractive, whereas the electric force can be attractive or repulsive. As a result, the net electric force between neutral objects, such as the Earth and the Moon, is essentially zero because attractive and repulsive forces cancel one another. Since gravity is always attractive, however, the net gravitational force between the Earth and the Moon is nonzero. Thus, in astronomy, gravity rules, and electric forces play hardly any role.

Just the opposite is true in atomic systems. To see this, let's compare the electric and gravitational forces between a proton and an electron in a hydrogen atom. Taking the distance between the two particles to be the radius of hydrogen, $r = 5.29 \times 10^{-11}$ m, we find that the gravitational force has a magnitude

$$\begin{aligned} F_g &= G\frac{m_e m_p}{r^2} \\ &= (6.67 \times 10^{-11}\ \text{N}\cdot\text{m}^2/\text{kg}^2)\frac{(9.11 \times 10^{-31}\ \text{kg})(1.673 \times 10^{-27}\ \text{kg})}{(5.29 \times 10^{-11}\ \text{m})^2} \\ &= 3.63 \times 10^{-47}\ \text{N} \end{aligned}$$

PROBLEM-SOLVING NOTE

Distance Dependence of the Coulomb Force

The Coulomb force has an inverse-square dependence on distance. Be sure to divide the product of the charges, $k|q_1||q_2|$, by r^2 when calculating the force.

Similarly, the magnitude of the electric force between the electron and the proton is

$$\begin{aligned} F_e &= k\frac{|q_1||q_2|}{r^2} \\ &= (8.99 \times 10^9\ \text{N}\cdot\text{m}^2/\text{C}^2)\frac{|-1.60 \times 10^{-19}\ \text{C}||1.60 \times 10^{-19}\ \text{C}|}{(5.29 \times 10^{-11}\ \text{m})^2} \\ &= 8.22 \times 10^{-8}\ \text{N} \end{aligned}$$

Taking the ratio, we find that the electric force is greater than the gravitational force by a factor of

$$\begin{aligned} \frac{F_e}{F_g} &= \frac{8.22 \times 10^{-8}\ \text{N}}{3.63 \times 10^{-47}\ \text{N}} = 2.26 \times 10^{39} \\ &= 2{,}260{,}000{,}000{,}000{,}000{,}000{,}000{,}000{,}000{,}000{,}000{,}000{,}000{,}000 \end{aligned}$$

This huge factor explains why a small piece of charged amber can lift bits of paper off the ground, even though the entire mass of the Earth is pulling downward on the paper.

Clearly, then, the force of gravity plays essentially no role in atomic systems. The reason gravity dominates in astronomy is that, even though the force is incredibly weak, it always attracts, giving a larger net force the larger the astronomical body. The electric force, on the other hand, is very strong but cancels for neutral objects.

Next, we use the electric force to get an idea of the speed of an electron in a hydrogen atom.

EXAMPLE 19–1 The Bohr Orbit

In an effort to better understand the behavior of atomic systems, the Danish physicist Niels Bohr (1885–1962) introduced a simple model for the hydrogen atom. In the Bohr model, as it is known today, the electron is imagined to move in a circular orbit about a stationary proton. The force responsible for the electron's circular motion is the electric force of attraction between the electron and the proton. Given that the radius of the electron's orbit is 5.29×10^{-11} m, and its mass is $m_e = 9.11 \times 10^{-31}$ kg, find the electron's speed.

Picture the Problem

Our sketch shows the electron moving with a speed v in its orbit of radius r. Because the proton is so much more massive than the electron, it is essentially stationary at the center of the orbit. Note that the electron has a charge $-e$ and the proton has a charge $+e$.

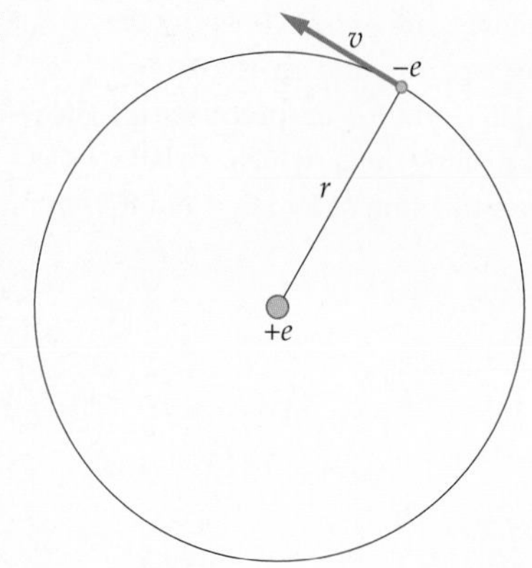

Strategy

The idea behind this model is that a force is required to make the electron move in a circular path, and this force is provided by the electric force of attraction between the electron and the proton. Thus, as with any circular motion, we set the force acting on the electron equal to its mass times its centripetal acceleration. This allows us to solve for the centripetal acceleration, $a_{cp} = v^2/r$ (Equation 6–14), which in turn gives us the speed v.

Solution

1. Set the Coulomb force between the electron and proton equal to the centripetal force required for the electron's circular orbit:

$$k\frac{|q_1||q_2|}{r^2} = m_e a_{cp}$$

$$k\frac{e^2}{r^2} = m_e\frac{v^2}{r}$$

2. Solve for the speed of the electron, v:

$$v = e\sqrt{\frac{k}{m_e r}}$$

3. Substitute numerical values:

$$v = (1.60 \times 10^{-19}\,\text{C})\sqrt{\frac{8.99 \times 10^9\,\text{N}\cdot\text{m}^2/\text{C}^2}{(9.11 \times 10^{-31}\,\text{kg})(5.29 \times 10^{-11}\,\text{m})}}$$
$$= 2.19 \times 10^6\,\text{m/s}$$

Insight

If you could travel around the world at this speed, your trip would take about 18 s. The acceleration of the electron in this orbit is about 10^{22} times greater than the acceleration of gravity on the surface of the Earth.

Practice Problem

The second Bohr orbit has a radius that is four times the radius of the first orbit. What is the speed of an electron in this orbit? [**Answer:** $v = 1.09 \times 10^6\,\text{m/s}$]

Some related homework problems: Problem 10, Problem 17, Problem 26

Another indication of the strength of the electric force is given in the following Exercise.

EXERCISE 19–2

Find the electric force between two 1.00-C charges separated by 1.00 m.

Solution

Substituting $q_1 = q_2 = 1.00\,\text{C}$ and $r = 1.00\,\text{m}$ in Coulomb's law, we find

$$F = k\frac{|q_1||q_2|}{r^2} = (8.99 \times 10^9\,\text{N}\cdot\text{m}^2/\text{C}^2)\frac{(1.00\,\text{C})(1.00\,\text{C})}{(1.00\,\text{m})^2} = 8.99 \times 10^9\,\text{N}$$

Exercise 19–2 shows that charges of one coulomb exert a force of about a million tons on one another when separated by a distance of a meter. If the charge in your body could be separated into a pile of positive charge on one side of the room and a pile of negative charge on the other side, the force needed to hold them apart would be roughly 10^{10} tons! Thus, everyday objects are never far from electrical neutrality, since disturbing neutrality requires such tremendous forces.

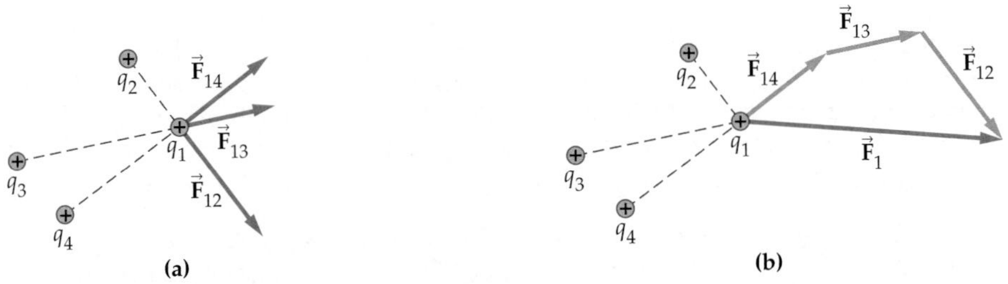

▶ **FIGURE 19–8 Superposition of forces**

(a) Forces are exerted on q_1 by the charges q_2, q_3, and q_4. These forces are $\vec{F}_{12}$, $\vec{F}_{13}$, and $\vec{F}_{14}$, respectively. **(b)** The net force acting on q_1, which we label $\vec{F}_1$, is the vector sum of $\vec{F}_{12}$, $\vec{F}_{13}$, and $\vec{F}_{14}$.

Superposition of Forces

The electric force, like all forces, is a vector quantity. Hence, when a charge experiences forces due to two or more other charges, the net force on it is simply the *vector* sum of the forces taken individually. For example, in **Figure 19–8**, the total force on charge 1, $\vec{F}_1$, is the vector sum of the forces due to charges 2, 3, and 4:

$$\vec{F}_1 = \vec{F}_{12} + \vec{F}_{13} + \vec{F}_{14}$$

This is referred to as the **superposition** of forces.

Notice that the total force acting on a given charge is the sum of interactions involving just *two* charges at a time, with the force between each pair of charges given by Coulomb's law. For example, the total force acting on charge 1 in Figure 19–8 is the sum of the forces between q_1 and q_2, q_1 and q_3, and q_1 and q_4. Therefore, superposition of forces can be thought of as the generalization of Coulomb's law to systems containing more than two charges. In our first numerical Example of superposition, we consider three charges in a line.

EXAMPLE 19–2 Net Force

A charge $q_1 = -5.4\ \mu C$ is at the origin, and a charge $q_2 = -2.2\ \mu C$ is on the x axis at $x = 1.00$ m. Find the net force acting on a charge $q_3 = +1.6\ \mu C$ located at $x = 0.75$ m.

Picture the Problem

The physical situation is shown in our sketch, with each charge at its appropriate location. Notice that the forces exerted on charge q_3 by the charges q_1 and q_2 are in opposite directions. We give the force on q_3 due to q_1 the label $\vec{F}_{31}$, and the force on q_3 due to q_2 the label $\vec{F}_{32}$.

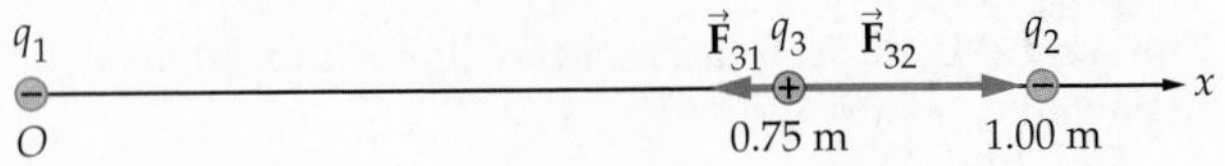

Strategy

The net force on q_3 is the vector sum of the forces due to q_1 and q_2. In particular, note that $\vec{F}_{31}$ points in the negative x direction $(-\hat{x})$, whereas $\vec{F}_{32}$ points in the positive x direction $(\hat{x})$. The magnitude of $\vec{F}_{31}$ is $k|q_1||q_3|/r^2$, with $r = 0.75$ m. Similarly, the magnitude of $\vec{F}_{32}$ is $k|q_2||q_3|/r^2$, with $r = 0.25$ m.

Solution

1. Find the force acting on q_3 due to q_1. Since this force is in the negative x direction, as indicated in the sketch, we give it a negative sign:

$$\vec{F}_{31} = -k\frac{|q_1||q_3|}{r^2}\hat{x}$$
$$= -(8.99 \times 10^9\ \mathrm{N \cdot m^2/C^2}) \times \frac{(5.4 \times 10^{-6}\ \mathrm{C})(1.6 \times 10^{-6}\ \mathrm{C})}{(0.75\ \mathrm{m})^2}\hat{x}$$
$$= -0.14\ \mathrm{N}\hat{x}$$

2. Find the force acting on q_3 due to q_2. Since this force is in the positive x direction, as indicated in the sketch, we give it a positive sign:

$$\vec{F}_{32} = k\frac{|q_2||q_3|}{r^2}\hat{x}$$
$$= (8.99 \times 10^9\ \mathrm{N \cdot m^2/C^2}) \times \frac{(2.2 \times 10^{-6}\ \mathrm{C})(1.6 \times 10^{-6}\ \mathrm{C})}{(0.25\ \mathrm{m})^2}\hat{x}$$
$$= 0.51\ \mathrm{N}\hat{x}$$

3. Superpose these forces to find the total force, $\vec{F}_3$, acting on q_3:

$$\vec{F}_3 = \vec{F}_{31} + \vec{F}_{32} = -0.14\ \mathrm{N}\hat{x} + 0.51\ \mathrm{N}\hat{x}$$
$$= 0.37\ \mathrm{N}\hat{x}$$

Insight
The net force acting on q_3 has a magnitude of 0.37 N, and it points in the positive x direction. As usual, notice that we use only magnitudes for the charges in the numerator of Coulomb's law.

Practice Problem
Find the net force on q_3 if it is at the location $x = 0.25$ m. [**Answer:** $\vec{\mathbf{F}}_3 = -1.2\ \mathrm{N}\hat{\mathbf{x}}$]

Some related homework problems: Problem 14, Problem 15, Problem 16

ACTIVE EXAMPLE 19–1 Find the Location of Zero Net Force

In Example 19–2, the net force acting on the charge q_3 is to the right. To what value of x should q_3 be moved for the net force on it to be zero?

Solution *(Test your understanding by performing the calculations indicated in each step.)*

1. Write the magnitude of the force due to q_1: $F_{31} = k|q_1||q_3|/x^2$
2. Write the magnitude of the force due to q_2: $F_{32} = k|q_2||q_3|/(1.00\ \mathrm{m} - x)^2$
3. Set these forces equal to one another, and cancel common terms: $|q_1|/x^2 = |q_2|/(1.00\ \mathrm{m} - x)^2$
4. Take the square root of both sides and solve for x: $x = 0.61$ m

Insight
Therefore, if q_3 is placed between $x = 0.61$ m and $x = 1.00$ m, the net force acting on it is to the right, in agreement with Example 19–2. On the other hand, if q_3 is placed between $x = 0$ and $x = 0.61$ m, the net force acting on it is to the left. This agrees with the result in the Practice Problem of Example 19–2.

Your Turn
If the magnitude of each charge in this system is doubled, does the point of zero net force move to the right, move to the left, or remain in the same place? Explain.

(Answers to **Your Turn** *problems are given in the back of the book.)*

Next we consider systems in which the individual forces are not along the same line. In such cases, it is often useful to resolve the individual force vectors into components and then perform the required vector sum component by component. This technique is illustrated in the following Example and Conceptual Checkpoint.

PROBLEM-SOLVING NOTE

Determining the Direction of the Electric Force

When determining the total force acting on a charge, begin by calculating the magnitude of each of the individual forces acting on it. Next, assign appropriate directions to the forces based on the principle that "opposites attract, likes repel" and perform a vector sum.

EXAMPLE 19–3 Superposition

Three charges, each equal to $+2.90\ \mu\mathrm{C}$, are placed at three corners of a square 0.500 m on a side, as shown in the diagram. Find the magnitude and direction of the net force on charge number 3.

Picture the Problem
The positions of the three charges are shown in the sketch. We also show the force produced by charge 1, $\vec{\mathbf{F}}_{31}$, and the force produced by charge 2, $\vec{\mathbf{F}}_{32}$. Note that $\vec{\mathbf{F}}_{31}$ is 45.0° above the x axis and that $\vec{\mathbf{F}}_{32}$ is in the positive x direction. Also, the distance from charge 2 to charge 3 is $r = 0.500$ m, and the distance from charge 1 to charge 3 is $\sqrt{2}r$.

Strategy
To find the net force, we first calculate the magnitudes of $\vec{\mathbf{F}}_{31}$ and $\vec{\mathbf{F}}_{32}$ and then their components. Summing these components yields the components of the net force, $\vec{\mathbf{F}}_3$. Once we know the components of $\vec{\mathbf{F}}_3$, we can calculate its magnitude and direction in the same way as for any other vector.

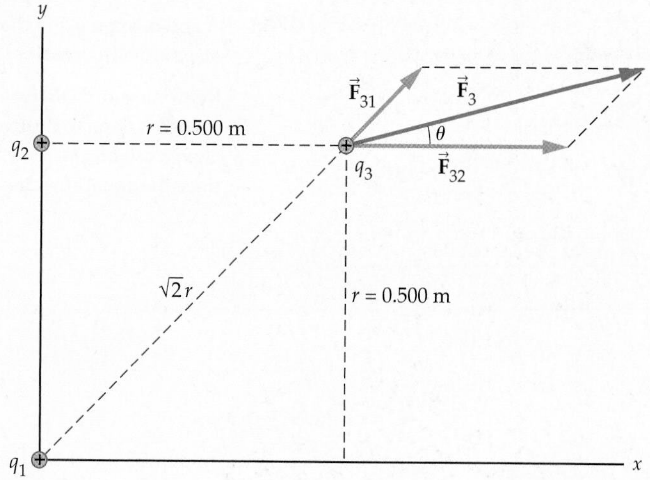

continued on next page

continued from previous page

Solution

1. Find the magnitude of $\vec{\mathbf{F}}_{31}$:

$$F_{31} = k\frac{|q_1||q_3|}{(\sqrt{2}r)^2}$$

$$= (8.99 \times 10^9\ \mathrm{N \cdot m^2/C^2})\frac{(2.90 \times 10^{-6}\ \mathrm{C})^2}{2(0.500\ \mathrm{m})^2}$$

$$= 0.151\ \mathrm{N}$$

2. Find the magnitude of $\vec{\mathbf{F}}_{32}$:

$$F_{32} = k\frac{|q_2||q_3|}{r^2}$$

$$= (8.99 \times 10^9\ \mathrm{N \cdot m^2/C^2})\frac{(2.90 \times 10^{-6}\ \mathrm{C})^2}{(0.500\ \mathrm{m})^2}$$

$$= 0.302\ \mathrm{N}$$

3. Calculate the components of $\vec{\mathbf{F}}_{31}$ and $\vec{\mathbf{F}}_{32}$:

$$F_{31,x} = F_{31}\cos 45.0° = (0.151\ \mathrm{N})(0.707) = 0.107\ \mathrm{N}$$

$$F_{31,y} = F_{31}\sin 45.0° = (0.151\ \mathrm{N})(0.707) = 0.107\ \mathrm{N}$$

$$F_{32,x} = F_{32}\cos 0° = (0.302\ \mathrm{N})(1) = 0.302\ \mathrm{N}$$

$$F_{32,y} = F_{32}\sin 0° = (0.302\ \mathrm{N})(0) = 0$$

4. Find the components of $\vec{\mathbf{F}}_3$:

$$F_{3,x} = F_{31,x} + F_{32,x} = 0.107\ \mathrm{N} + 0.302\ \mathrm{N} = 0.409\ \mathrm{N}$$

$$F_{3,y} = F_{31,y} + F_{32,y} = 0.107\ \mathrm{N} + 0 = 0.107\ \mathrm{N}$$

5. Find the magnitude of $\vec{\mathbf{F}}_3$:

$$F_3 = \sqrt{F_{3,x}{}^2 + F_{3,y}{}^2}$$

$$= \sqrt{(0.409\ \mathrm{N})^2 + (0.107\ \mathrm{N})^2} = 0.423\ \mathrm{N}$$

6. Find the direction of $\vec{\mathbf{F}}_3$:

$$\theta = \tan^{-1}\left(\frac{F_{3,y}}{F_{3,x}}\right) = \tan^{-1}\left(\frac{0.107\ \mathrm{N}}{0.409\ \mathrm{N}}\right) = 14.7°$$

Insight

Thus, the net force on charge 3 has a magnitude of 0.423 N and points in a direction 14.7° above the x axis. Note that charge 1, which is $\sqrt{2}$ times farther away from charge 3 than is charge 2, produces only half as much force as charge 2.

Practice Problem

Find the magnitude and direction of the net force on charge 3 if its magnitude is doubled to 5.80 μC. Assume that charge 1 and charge 2 are unchanged. [**Answer:** $F_3 = 2(0.423\ \mathrm{N}) = 0.846\ \mathrm{N}$, $\theta = 14.7°$. Note that the angle is unchanged.]

Some related homework problems: Problem 20, Problem 21

CONCEPTUAL CHECKPOINT 19–3 Compare the Force

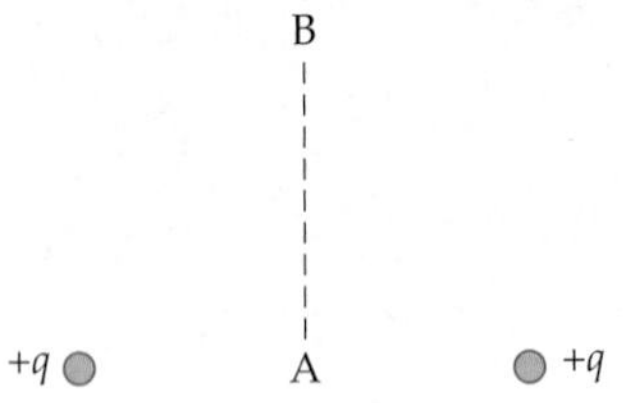

A charge $-q$ is to be placed at either point A or point B in the accompanying figure. Assume points A and B lie on a line that is midway between the two positive charges. Is the net force experienced at point A **(a)** greater than, **(b)** equal to, or **(c)** less than the net force experienced at point B?

Reasoning and Discussion

Point A is closer to the two positive charges than is point B. As a result, the force exerted by each positive charge will be greater when the charge $-q$ is placed at A. The *net* force, however, is zero at point A, since the equal attractive forces due to the two positive charges cancel, as shown in the diagram.

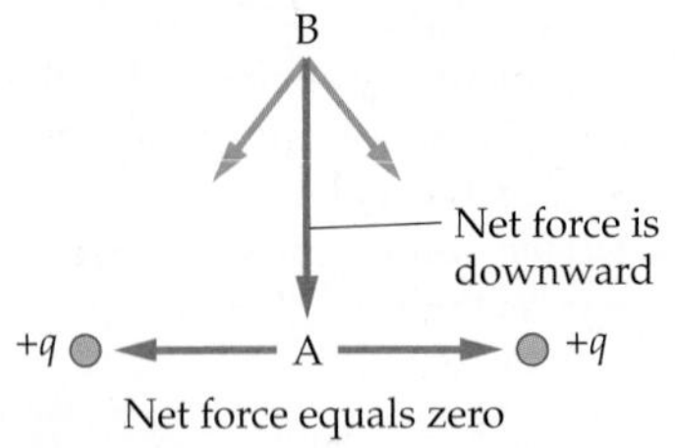

At point B, on the other hand, the attractive forces combine to give a net downward force. Hence, the charge $-q$ will experience a greater net force at point B.

Answer

(c) The net force at point A is less than the net force at point B.

Spherical Charge Distributions

Although Coulomb's law is stated in terms of point charges, it can be applied to any type of charge distribution by using the appropriate mathematics. For example, suppose a sphere has a charge Q distributed uniformly over its surface. If a point charge q is outside the sphere, a distance r from its center, the methods of calculus show that the magnitude of the force between the point charge and the sphere is simply

$$F = k\frac{|q||Q|}{r^2}$$

In situations like this, the spherical charge distribution behaves the same as if all its charge were concentrated in a point at its center. In general, the electrical behavior of spherical *charge* distributions is analogous to the gravitational behavior of spherical *mass* distributions.

In the next Active Example, we consider a system in which a charge Q is distributed uniformly over the surface of a sphere. In such a case it is often convenient to specify the amount of *charge per area* on the sphere. This is referred to as the **surface charge density,** σ. If a sphere has an area A and a surface charge density σ, its total charge is

$$Q = \sigma A \qquad \textbf{19–7}$$

Note that the SI unit of σ is C/m^2. If the radius of the sphere is R, then $A = 4\pi R^2$, and $Q = \sigma(4\pi R^2)$.

PROBLEM-SOLVING NOTE

Spherical Charge Distributions

Remember that a uniform spherical charge distribution can be replaced with a point charge only when considering points outside the charge distribution.

ACTIVE EXAMPLE 19–2 Find the Force Exerted by a Sphere

An insulating sphere of radius $R = 0.10$ m has a uniform surface charge density equal to 5.9 $\mu\text{C/m}^2$. A point charge of magnitude 0.71 μC is 0.45 m from the center of the sphere. Find the magnitude of the force exerted by the sphere on the point charge.

Solution *(Test your understanding by performing the calculations indicated in each step.)*

1. Find the area of the sphere: $A = 0.13\ \text{m}^2$
2. Calculate the total charge on the sphere: $Q = 0.77\ \mu\text{C}$
3. Use Coulomb's law to calculate the magnitude of the force between the sphere and the point charge: $F = 0.024$ N

Insight

As long as the point charge is outside the sphere, and the charge distribution remains spherically uniform, the sphere may be treated as a point charge.

Your Turn

Suppose the sphere in this problem is replaced by one with half the radius, but with the same surface charge density. Is the force exerted by this sphere greater than, less than, or the same as the force exerted by the original sphere? Explain.

(Answers to **Your Turn** *problems are given in the back of the book.)*

19–4 The Electric Field

You have probably encountered the notion of a "force field" in various science fiction novels and movies. A concrete example of a force field is provided by the force between electric charges. Consider, for example, a positive point charge q at the origin of a coordinate system, as in **Figure 19–9**. If a positive "test charge," q_0, is placed at point A, the force exerted on it by q is indicated by the vector $\vec{\mathbf{F}}_A$. On the other hand, if the test charge is placed at point B, the force it experiences there is $\vec{\mathbf{F}}_B$. At every point in space there is a corresponding force. In this sense, Figure 19–9 allows us to visualize the "force field" associated with the charge q.

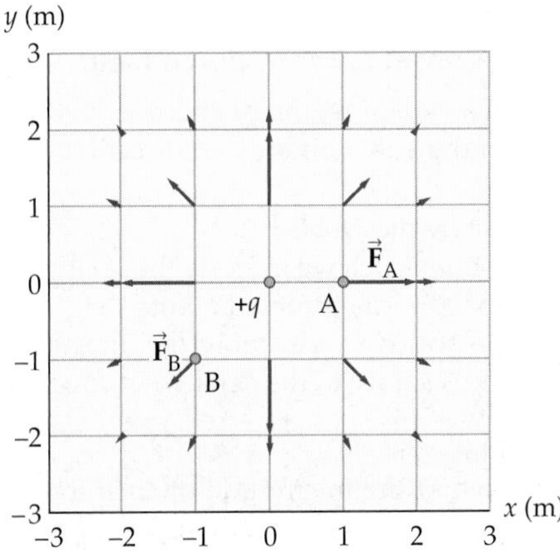

▲ **FIGURE 19–9 An electrostatic force field**

The positive charge $+q$ at the origin of this coordinate system exerts a different force on a given charge at every point in space. Here we show the force vectors associated with q for a grid of points.

Since the magnitude of the force at every point in Figure 19–9 is proportional to q_0 (due to Coulomb's law), it is convenient to divide by q_0 and define a *force per charge* at every point in space that is independent of q_0. We refer to the force per charge as the electric field, $\vec{E}$. Its precise definition is as follows:

Definition of the Electric Field, $\vec{E}$

If a test charge q_0 experiences a force $\vec{F}$ at a given location, the electric field $\vec{E}$ at that location is

$$\vec{E} = \frac{\vec{F}}{q_0} \qquad \text{19–8}$$

SI unit: N/C

It should be noted that this definition applies whether the force $\vec{F}$ is due to a single charge, as in Figure 19–9, or to a group of charges. In addition, it is assumed that the test charge is small enough that it does not disturb the position of any other charges in the system.

To summarize, *the electric field is the force per charge at a given location*. Therefore, if we know the electric field vector $\vec{E}$ at a given point, the force that a charge q experiences at that point is

$$\vec{F} = q\vec{E} \qquad \text{19–9}$$

Notice that the direction of the force depends on the sign of the charge. In particular,

- A positive charge experiences a force *in the direction* of $\vec{E}$.
- A negative charge experiences a force *in the opposite direction* of $\vec{E}$.

Finally, the magnitude of the force is the product of the magnitudes of q and $\vec{E}$:

- The magnitude of the force acting on a charge q is $F = |q|E$.

As we continue in this chapter, we will determine the electric field for a variety of different charge distributions. In some cases, $\vec{E}$ will decrease with distance as $1/r^2$ (a point charge), in other cases as $1/r$ (a line of charge), and in others $\vec{E}$ will be a constant (a charged plane). Before we calculate the electric field itself, however, we first consider the force exerted on charges by a constant electric field.

PROBLEM-SOLVING NOTE

The Force Exerted by an Electric Field

The force exerted on a charge by an electric field can point in only one of two directions—parallel or antiparallel to the direction of the field.

EXAMPLE 19–4 Force Field

In a certain region of space, a uniform electric field has a magnitude of 4.60×10^4 N/C and points in the positive x direction. Find the magnitude and direction of the force this field exerts on a charge of **(a)** $+2.80\ \mu C$ and **(b)** $-9.30\ \mu C$.

Picture the Problem

In our sketch we indicate the uniform electric field and the two charges mentioned in the problem. Note that the positive charge experiences a force in the positive x direction (the direction of $\vec{E}$), and the negative charge experiences a force in the negative x direction (opposite to $\vec{E}$).

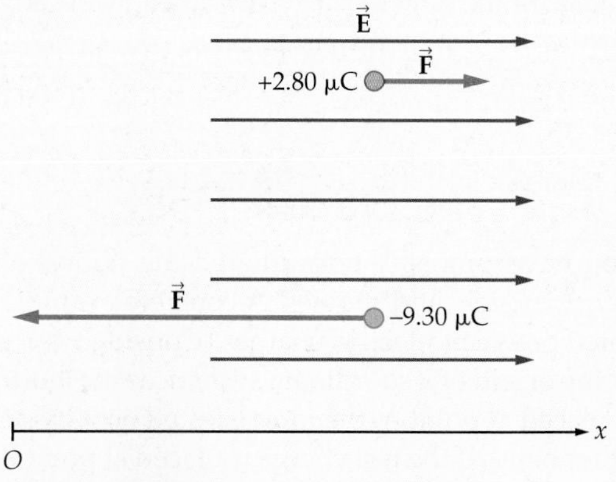

Strategy

To find the magnitude of each force, we use $F = |q|E$. The direction has already been indicated in our sketch.

Solution

Part (a)

1. Find the magnitude of the force on the $+2.80\text{-}\mu C$ charge: $F = |q|E = (2.80 \times 10^{-6}\,C)(4.60 \times 10^4\,N/C) = 0.129\,N$

Part (b)

2. Find the magnitude of the force on the $-9.30\text{-}\mu C$ charge: $F = |q|E = (9.30 \times 10^{-6}\,C)(4.60 \times 10^4\,N/C) = 0.428\,N$

Insight

To summarize, the force on the $+2.80\text{-}\mu C$ charge is of magnitude 0.129 N in the positive x direction; the force on the $-9.30\text{-}\mu C$ charge is of magnitude 0.428 N in the negative x direction.

Practice Problem

If the $+2.80\text{-}\mu C$ charge experiences a force of 0.25 N, what is the magnitude of the electric field? [**Answer:** $E = 8.9 \times 10^4\,N/C$]

Some related homework problems: Problem 33, Problem 35

REAL-WORLD PHYSICS
Electrodialysis for water purification

The fact that charges of opposite sign experience forces in opposite directions in an electric field is used to purify water in the process known as **electrodialysis.** This process depends on the fact that most minerals that dissolve in water dissociate into positive and negative ions. Probably the most common example is table salt (NaCl), which dissociates into positive sodium ions (Na^+) and negative chlorine ions (Cl^-). When brackish water is passed through a strong electric field in an electrodialysis machine, the mineral ions move in opposite directions and pass through two different types of semipermeable membrane—one that allows only positive ions to pass through, the other only negative ions. This process leaves water that is purified of dissolved minerals and suitable for drinking.

The Electric Field of a Point Charge

Perhaps the simplest example of an electric field is the field produced by an idealized point charge. To be specific, suppose a positive point charge q is at the origin in **Figure 19–10**. If a positive test charge q_0 is placed a distance r from the origin, the force it experiences is directed away from the origin and is of magnitude

$$F = k\frac{|q||q_0|}{r^2}$$

Applying our definition of the electric field in Equation 19–8, we find that the magnitude of the field is

$$E = \frac{F}{q_0} = \frac{\left(k\dfrac{|q||q_0|}{r^2}\right)}{q_0} = k\frac{|q|}{r^2}$$

Since a positive charge experiences a force that is radially outward, that too is the direction of $\vec{E}$.

In general, then, we can say that the electric field a distance r from a point charge q has the following magnitude:

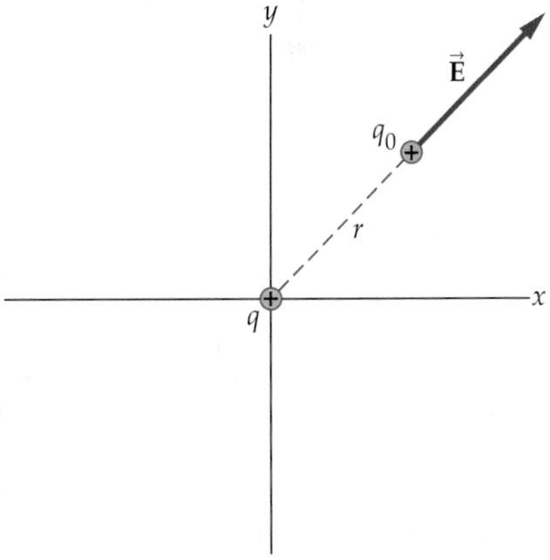

▲ **FIGURE 19–10 The electric field of a point charge**
The electric field $\vec{E}$ due to a positive charge q at the origin is radially outward. Its magnitude is $E = k|q|/r^2$.

Magnitude of the Electric Field Due to a Point Charge

$$E = k\frac{|q|}{r^2} \qquad 19\text{–}10$$

If the charge q is positive, the field points radially outward from the charge; if it is negative, the field is radially inward. This is illustrated in **Figure 19–11**. Thus, to determine the electric field due to a point charge, we first use Equation 19–10 to find its magnitude, and then use the rule illustrated in Figure 19–11 to find its direction.

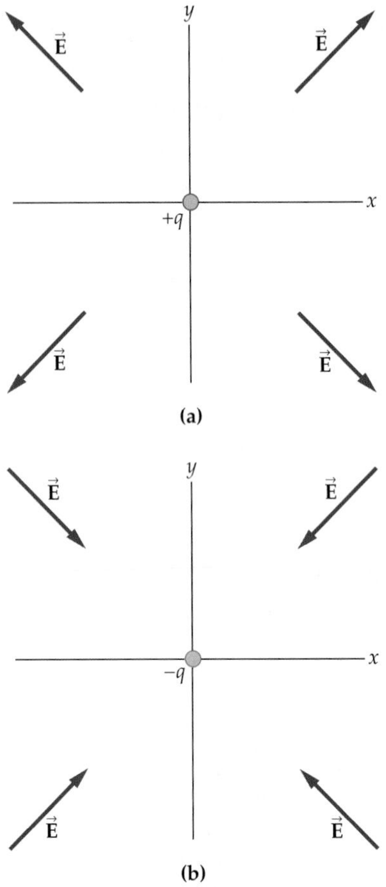

▲ **FIGURE 19–11 The direction of the electric field**

(a) The electric field due to a positive charge at the origin points radially outward. **(b)** If the charge at the origin is negative, the electric field is radially inward.

EXERCISE 19–3

Find the electric field produced by a 1.0-μC point charge at a distance of **(a)** 0.75 m and **(b)** 1.5 m.

Solution

a. Applying Equation 19–10 with $q = 1.0\ \mu\text{C}$ and $r = 0.75$ m yields

$$E = k\frac{|q|}{r^2} = (8.99 \times 10^9\ \text{N}\cdot\text{m}^2/\text{C}^2)\frac{(1.0 \times 10^{-6}\ \text{C})}{(0.75\ \text{m})^2} = 1.6 \times 10^4\ \text{N/C}$$

b. Noting that E depends on $1/r^2$, we see that doubling the distance from 0.75 m to 1.5 m results in a reduction in the electric field by a factor of 4:

$$E = \tfrac{1}{4}(1.6 \times 10^4\ \text{N/C}) = 0.40 \times 10^4\ \text{N/C}$$

Superposition of Fields

Many electrical systems consist of more than two charges. In such cases, the total electric field can be found by using superposition—just as when we find the total force due to a system of charges. In particular, the total electric field is found by calculating the vector sum—often using components—of the electric fields due to each charge separately.

For example, let's calculate the total electric field at point P in **Figure 19–12**. First we sketch the directions of the fields $\vec{\mathbf{E}}_1$ and $\vec{\mathbf{E}}_2$ due to the charges $q_1 = +q$ and $q_2 = +q$, respectively. In particular, if a positive test charge is at point P, the force due to q_1 is down and to the right, whereas the force due to q_2 is up and to the right. From the geometry of the figure we see that $\vec{\mathbf{E}}_1$ is at an angle θ below the x axis, and—by symmetry—$\vec{\mathbf{E}}_2$ is at the same angle θ above the axis. Since the two charges have the same magnitude, and the distances from P to the charges are the same, it follows that $\vec{\mathbf{E}}_1$ and $\vec{\mathbf{E}}_2$ have the same magnitude:

$$E_1 = E_2 = E = k|q|/d^2$$

To find the net electric field $\vec{\mathbf{E}}_{\text{net}}$, we use components. First, consider the y direction. In this case, we have $E_{1,y} = -E\sin\theta$ and $E_{2,y} = +E\sin\theta$. Hence, the y component of the net electric field is zero:

$$E_{\text{net},y} = E_{1,y} + E_{2,y} = -E\sin\theta + E\sin\theta = 0$$

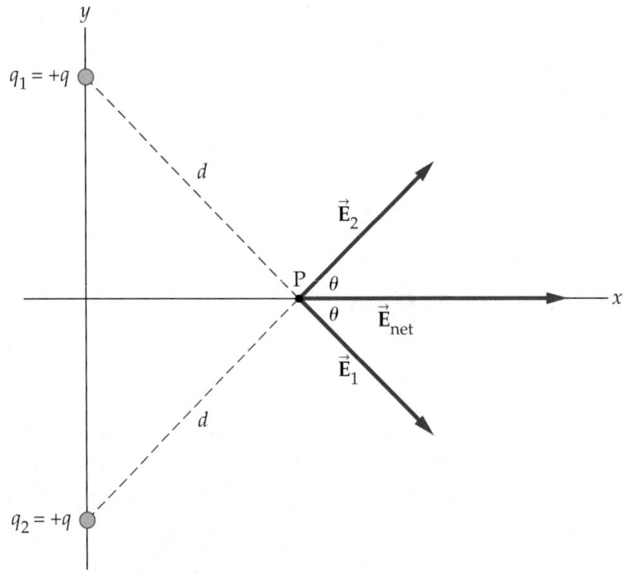

▶ **FIGURE 19–12 Superposition of the electric field**

The net electric field at the point P is the vector sum of the fields due to the charges q_1 and q_2. Note that $\vec{\mathbf{E}}_1$ and $\vec{\mathbf{E}}_2$ point away from the charges q_1 and q_2, respectively. This is as expected, since both of these charges are positive.

Referring again to Figure 19–12, it is apparent that this result could have been anticipated by symmetry considerations. Finally, we determine the x component of E_{net}.

$$E_{\text{net},x} = E_{1,x} + E_{2,x} = E\cos\theta + E\cos\theta = 2E\cos\theta$$

Thus, the net electric field at P is in the positive x direction, as shown in Figure 19–12, and has a magnitude equal to $2E\cos\theta$.

CONCEPTUAL CHECKPOINT 19–4 The Sign of the Charges

Two charges, q_1 and q_2, have equal magnitudes q and are placed as shown in the figure to the right. The net electric field at point P is vertically upward. We conclude that **(a)** q_1 is positive, q_2 is negative; **(b)** q_1 is negative, q_2 is positive; or **(c)** q_1 and q_2 have the same sign.

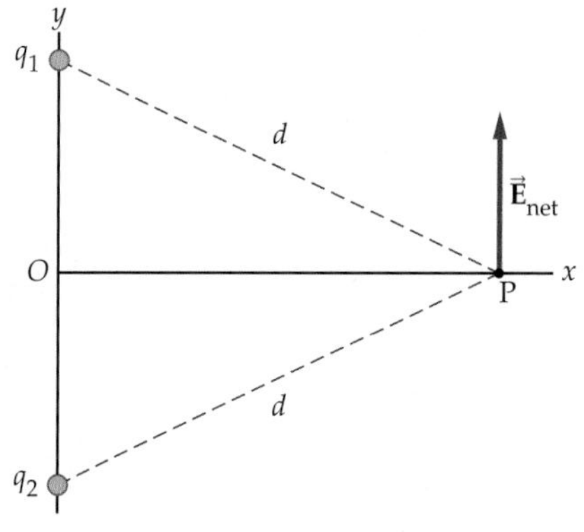

Reasoning and Discussion

If the net electric field at P is vertically upward, the x components of $\vec{E}_1$ and $\vec{E}_2$ must cancel, and the y components must both be in the positive y direction. The only way for this to happen is to have q_1 negative and q_2 positive, as shown in the following diagram.

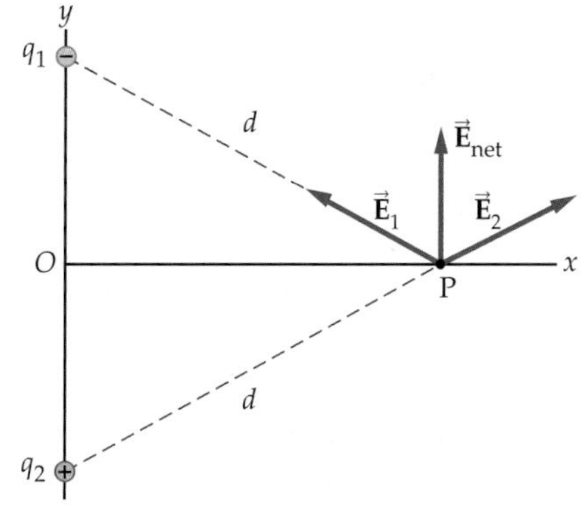

With this choice, a positive test charge at P is attracted to q_1 (so that $\vec{E}_1$ is up and to the left) and repelled from q_2 (so that $\vec{E}_2$ is up and to the right).

Answer

(b) q_1 is negative, q_2 is positive.

We conclude this section by considering the same physical system presented in Example 19–3, this time from the point of view of the electric field.

EXAMPLE 19–5 Superposition in the Field

Two charges, each equal to $+2.90\ \mu\text{C}$, are placed at two corners of a square 0.500 m on a side, as shown in the following sketch. Find the magnitude and direction of the net electric field at a third corner of the square, the point labeled 3 in the sketch.

Picture the Problem

The positions of the two charges are shown in the sketch. We also show the electric field produced by each charge. The key difference between this sketch and the one in Example 19–3 is that in this case there is no charge at point 3; the electric field still exists there, even though it has no charge on which to exert a force.

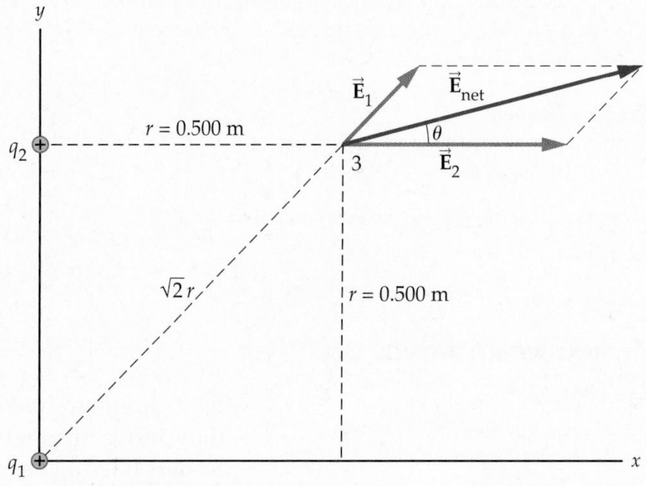

INTERACTIVE FIGURE

Strategy

In analogy with Example 19–3, we first calculate the magnitudes of $\vec{E}_1$ and $\vec{E}_2$ and then their components. Summing these components yields the components of the net electric field, $\vec{E}_{\text{net}}$. Once we know the components of $\vec{E}_{\text{net}}$, we find its magnitude and direction in the same way as for $\vec{F}_{\text{net}}$ in Example 19–3.

continued on next page

continued from previous page

Solution

1. Find the magnitude of $\vec{\mathbf{E}}_1$:

$$E_1 = k\frac{|q_1|}{(\sqrt{2}r)^2}$$
$$= (8.99 \times 10^9\ \mathrm{N\cdot m^2/C^2})\frac{(2.90 \times 10^{-6}\ \mathrm{C})}{2(0.500\ \mathrm{m})^2}$$
$$= 5.21 \times 10^4\ \mathrm{N/C}$$

2. Find the magnitude of $\vec{\mathbf{E}}_2$:

$$E_2 = k\frac{|q_2|}{r^2}$$
$$= (8.99 \times 10^9\ \mathrm{N\cdot m^2/C^2})\frac{(2.90 \times 10^{-6}\ \mathrm{C})}{(0.500\ \mathrm{m})^2}$$
$$= 1.04 \times 10^5\ \mathrm{N/C}$$

3. Calculate the components of $\vec{\mathbf{E}}_1$ and $\vec{\mathbf{E}}_2$:

$$E_{1,x} = E_1 \cos 45.0°$$
$$= (5.21 \times 10^4\ \mathrm{N/C})(0.707) = 3.68 \times 10^4\ \mathrm{N/C}$$
$$E_{1,y} = E_1 \sin 45.0°$$
$$= (5.21 \times 10^4\ \mathrm{N/C})(0.707) = 3.68 \times 10^4\ \mathrm{N/C}$$
$$E_{2,x} = E_2 \cos 0°$$
$$= (1.04 \times 10^5\ \mathrm{N/C})(1) = 1.04 \times 10^5\ \mathrm{N/C}$$
$$E_{2,y} = E_2 \sin 0° = (1.04 \times 10^5\ \mathrm{N/C})(0) = 0$$

4. Find the components of $\vec{\mathbf{E}}_{\mathrm{net}}$:

$$E_{\mathrm{net},x} = E_{1,x} + E_{2,x}$$
$$= 3.68 \times 10^4\ \mathrm{N/C} + 1.04 \times 10^5\ \mathrm{N/C}$$
$$= 1.41 \times 10^5\ \mathrm{N/C}$$
$$E_{\mathrm{net},y} = E_{1,y} + E_{2,y}$$
$$= 3.68 \times 10^4\ \mathrm{N/C} + 0 = 3.68 \times 10^4\ \mathrm{N/C}$$

5. Find the magnitude of $\vec{\mathbf{E}}_{\mathrm{net}}$:

$$E_{\mathrm{net}} = \sqrt{E_{\mathrm{net},x}{}^2 + E_{\mathrm{net},y}{}^2}$$
$$= \sqrt{(1.41 \times 10^5\ \mathrm{N/C})^2 + (3.68 \times 10^4\ \mathrm{N/C})^2}$$
$$= 1.46 \times 10^5\ \mathrm{N/C}$$

6. Find the direction of $\vec{\mathbf{E}}_{\mathrm{net}}$:

$$\theta = \tan^{-1}(E_{\mathrm{net},y}/E_{\mathrm{net},x})$$
$$= \tan^{-1}\left(\frac{3.68 \times 10^4\ \mathrm{N/C}}{1.41 \times 10^5\ \mathrm{N/C}}\right) = 14.6°$$

Insight

Note that, as one would expect, the direction of the net electric field is the same as the direction of the net force in Example 19–3 (except for a small discrepancy in the last decimal place due to rounding off in the calculations). In addition, the magnitude of the force exerted by the electric field on a charge of 2.90 μC is $F = qE_{\mathrm{net}} = (2.90\ \mu\mathrm{C})(1.46 \times 10^5\ \mathrm{N/C}) = 0.423\ \mathrm{N}$, the same as was found in Example 19–3.

Practice Problem

Find the magnitude and direction of the net electric field at the bottom right corner of the square. [**Answer:** $E_{\mathrm{net}} = 1.46 \times 10^5\ \mathrm{N/C}, \theta = -14.6°$]

Some related homework problems: Problem 38, Problem 39

REAL-WORLD PHYSICS: BIO
Electric fish

Many aquatic creatures are capable of producing electric fields. For example, African freshwater fishes in the family Mormyridae can generate weak electric fields from modified tail muscles and are able to detect variations in this field as they move through their environment. With this capability, these nocturnal feeders have an electrical guidance system that assists them in locating obstacles, enemies, and food. Much stronger electric fields are produced by electric eels and electric skates. In particular, the electric eel *Electrophorus electricus* generates electric fields great enough to kill small animals and to stun larger animals, including humans.

Sharks are well known for their sensitivity to weak electric fields in their surroundings. In fact, they possess specialized organs for this purpose, known as the ampullae of Lorenzini, which assist in the detection of prey. Recently, this sensitivity has been put to use as a method of repelling sharks in order to protect swimmers and divers. A device called the SharkPOD (Protective Oceanic Device) consists of two metal electrodes, one attached to a diver's air tank, the other to one of the diver's fins. These electrodes produce a strong electric field that completely surrounds the diver and causes sharks to turn away out to a distance of up to 7 m.

REAL-WORLD PHYSICS: BIO
Electrical shark repellent

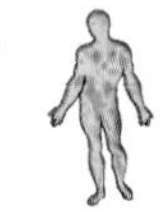

The SharkPOD was used in the 2000 Summer Olympic Games in Sydney to protect swimmers competing in the triathlon. The swimming part of the event was held in Sydney harbor, where great white sharks are a common sight. To protect the swimmers, divers wearing the SharkPOD swam along the course, a couple meters below the athletes. The race was completed without incident.

19–5 Electric Field Lines

When looking at plots like those in Figures 19–9 and 19–11, it is tempting to imagine a pictorial representation of the electric field. This thought is reinforced when one considers a photograph like **Figure 19–13**, which shows grass seeds suspended in oil. Because of polarization effects, the grass seeds tend to align in the direction of the electric field, much like the elongated atoms shown in Figure 19–5. In this case, the seeds are aligned radially, due to the electric field of the charged rod seen "end on" in the middle of the photograph. Clearly, a set of radial lines would seem to represent the electric field in this case.

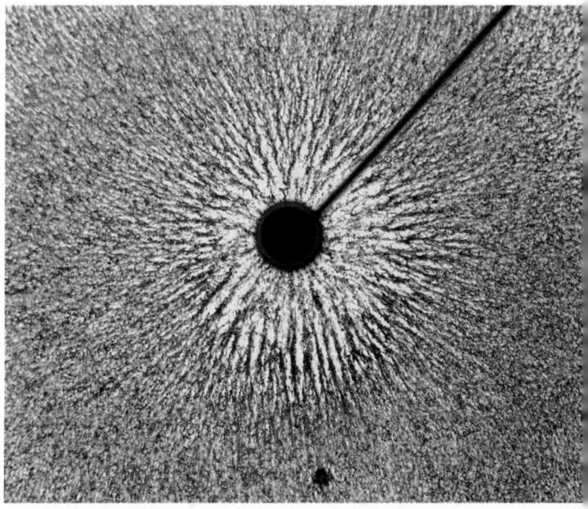

▲ **FIGURE 19–13 Grass seeds in an electric field**
Grass seeds aligning with electric field lines.

In fact, an entirely consistent method of drawing electric field lines is obtained by using the following set of rules:

Rules for Drawing Electric Field Lines
Electric field lines:

1. Point in the *direction* of the electric field vector $\vec{E}$ at every point;
2. *Start* at positive $(+)$ charges or at infinity;
3. *End* at negative $(-)$ charges or at infinity;
4. Are more *dense* where $\vec{E}$ has a greater magnitude. In particular, the number of lines entering or leaving a charge is proportional to the magnitude of the charge.

We now show how these rules are applied.

For example, the electric field lines for two different point charges are presented in **Figure 19–14**. First, we know that the electric field points directly away from the charge $(+q)$ in Figure 19–14 (a), hence from rule 1 the field lines are

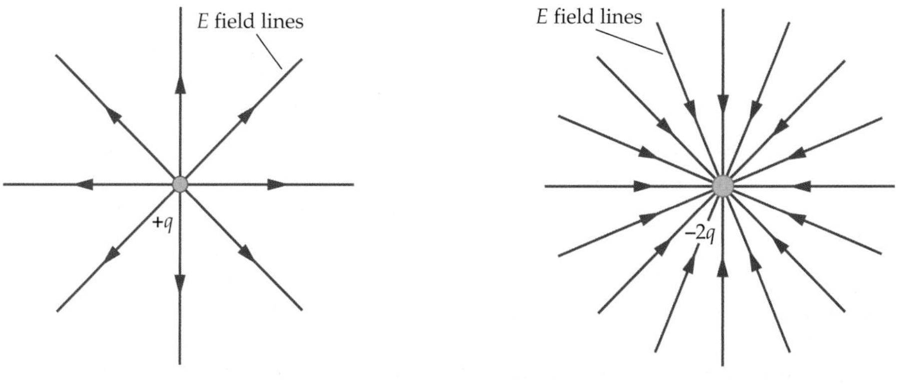

◀ **FIGURE 19–14 Electric field lines for a point charge**
(a) Near a positive charge the field lines point radially away from the charge. The lines start on the positive charge and end at infinity. **(b)** Near a negative charge the field lines point radially inward. They start at infinity and end on a negative charge and are more dense where the field is more intense. Notice that the number of lines drawn for part (b) is twice the number drawn for part (a), a reflection of the relative magnitudes of the charges.

radial. In agreement with rule 2 the field lines start on a + charge, and in agreement with rule 3 they end at infinity. Finally, as anticipated from rule 4, the field lines are closer together near the charge, where the field is more intense. Similar considerations apply to Figure 19–14 (b), where the charge is $-2q$. In this case, however, the direction of the field lines is reversed and the number of lines is doubled.

CONCEPTUAL CHECKPOINT 19–5 Intersect or Not?

Which of the following statements is correct: Electric field lines **(a)** can or **(b)** cannot intersect?

Reasoning and Discussion

By definition, electric field lines are always tangent to the electric field. Since the electric force, and hence the electric field, can point in only one direction at any given location, it follows that field lines cannot intersect. If they did, the field at the intersection point would have two conflicting directions.

Answer

(b) Electric field lines cannot intersect.

Figure 19–15 shows examples of electric field lines for various combinations of charges. In systems like these, we draw a set of curved field lines that are tangent to the electric field vector, $\vec{E}$, at every point. This is illustrated for a variety of points in Figure 19–15 (a), and similar considerations apply to all such field diagrams. In addition, note that the magnitude of $\vec{E}$ is greater in those regions of Figure 19–15 where the field lines are more closely packed together. Clearly, then, we expect an intense electric field between the charges in Figure 19–15 (b) and a vanishing field between the charges in Figure 19–15 (c).

Of particular interest is the $+q$ and $-q$ charge combination in Figure 19–15 (a). In general, a system of equal and opposite charges separated by a nonzero distance is known as an **electric dipole.** The total charge of the dipole is zero, but because the charges are separated, the electric field does not vanish. Instead, the field lines form "loops" that are characteristic of dipoles.

Many molecules are polar—water is a common example—which means they have an excess of positive charge near one end and a corresponding excess of

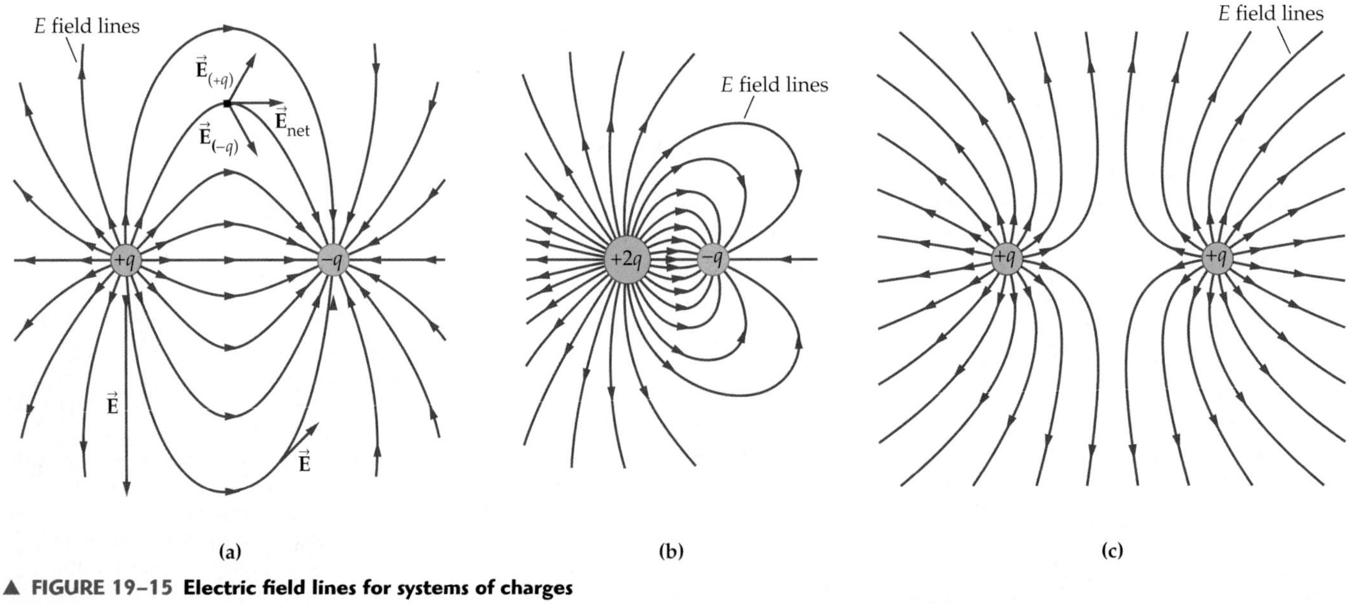

▲ **FIGURE 19–15 Electric field lines for systems of charges**

(a) The electric field lines for a dipole form closed loops that become more widely spaced with distance from the charges. Note that at each point in space, the electric field vector $\vec{E}$ is tangent to the field lines. **(b)** In a system with a net charge, some field lines extend to infinity. If the charges have opposite signs, some field lines start on one charge and terminate on the other charge. **(c)** All of the field lines in a system with charges of the same sign extend to infinity.

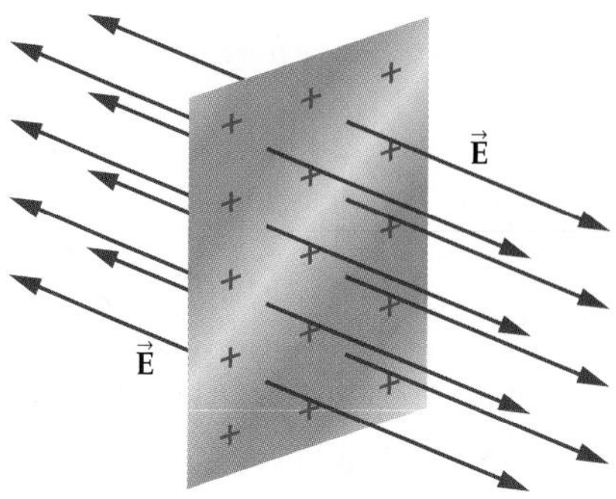

▶ **FIGURE 19–16 The electric field of a charged plate**
The electric field near a large charged plate is uniform in direction and magnitude.

negative charge near the other end. As a result, they produce an electric dipole field. Similarly, a typical bar magnet produces a *magnetic* dipole field, as we shall see in Chapter 22.

Finally, the electric field representations in Figures 19–14 and 19–15 are two-dimensional "slices" through the full field, which is three-dimensional. Therefore, one should imagine a similar set of field lines in these figures coming out of the page and going into the page.

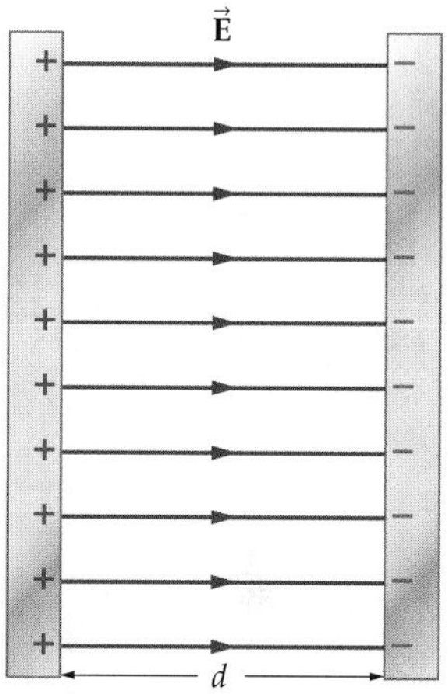

▲ **FIGURE 19–17 A parallel-plate capacitor**
In the ideal case, the electric field is uniform between the plates and zero outside.

Parallel-Plate Capacitor

A particularly simple and important field picture results when charge is spread uniformly over a large plate, as illustrated in **Figure 19–16**. At points that are not near the edge of the plate, the electric field is uniform in both direction and magnitude. That is, the field points in a single direction—perpendicular to the plate—and its magnitude is independent of the distance from the plate. This result can be proved using Gauss's law, as we show in Section 19–7.

If two such conducting plates with opposite charge are placed parallel to one another and separated by a distance d, as in **Figure 19–17**, the result is referred to as a **parallel-plate capacitor.** The field for such a system is uniform between the plates, and zero outside the plates. This is the ideal case, which is exactly true for an infinite plate and a good approximation for real plates. The field lines are illustrated in Figure 19–17. Parallel-plate capacitors are discussed further in the next chapter and will be of particular interest in Chapters 21 and 24, when we consider electric circuits.

EXAMPLE 19–6 Dangling by a Thread

The electric field between the plates of a parallel-plate capacitor is horizontal, uniform, and has a magnitude E. A small object of mass 0.0250 kg and charge $-3.10\ \mu C$ is suspended by a thread between the plates, as shown in the sketch. The thread makes an angle of 10.5° with the vertical. Find **(a)** the tension in the thread and **(b)** the magnitude of the electric field.

Picture the Problem

Our sketch shows the thread making an angle $\theta = 10.5°$ with the vertical. The inset to the right shows the free-body diagram for the suspended object, as well as our choice of positive x and y directions. Note that we label the charge of the object $-q$, where $q = 3.10\ \mu C$, in order to clearly indicate its sign.

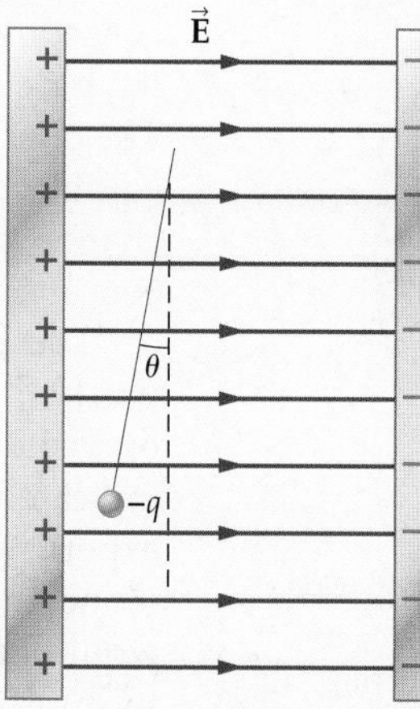

INTERACTIVE FIGURE

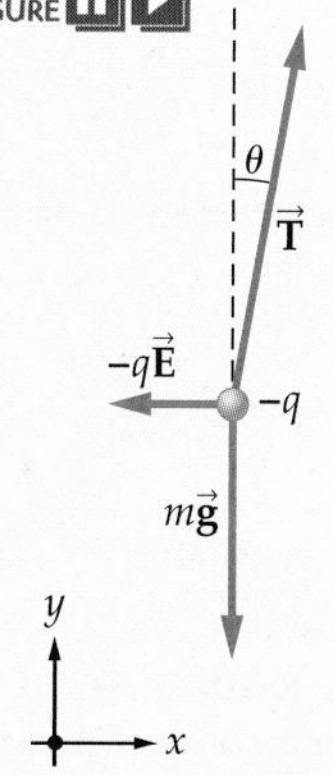

Strategy

The relevant physical principle in this problem is that because the object is at rest, the net force acting on it must vanish. Thus, setting the x and y components of the net force equal to zero yields two conditions, which can be used to solve for the two unknowns, T and E.

Solution

1. Set the net force in the x direction equal to zero: $-qE + T\sin\theta = 0$

2. Set the net force in the y direction equal to zero: $T\cos\theta - mg = 0$

Part (a)

3. Because we know all the quantities in the y force equation except for the tension, we use it to solve for T: $T = \dfrac{mg}{\cos\theta} = \dfrac{(0.0250\ \text{kg})(9.81\ \text{m/s}^2)}{\cos(10.5°)} = 0.249\ \text{N}$

continued on next page

continued from previous page

Part (b)

4. Now use the x force equation to find the magnitude of the electric field, E:

$$E = \frac{T \sin \theta}{q} = \frac{(0.249\text{ N}) \sin(10.5°)}{3.10 \times 10^{-6}\text{ C}} = 1.46 \times 10^{4}\text{ N/C}$$

Insight

As expected, the negatively charged object is attracted to the positively charged plate. This means that the electric force exerted on it is opposite in direction to the electric field.

Practice Problem

Suppose the electric field between the plates is 2.50×10^{4} N/C, but the charge of the object and the angle of the thread with the vertical are the same as before. Find the tension in the thread and the mass of the object. [**Answer:** $T = 0.425$ N, $m = 0.0426$ kg]

Some related homework problems: Problem 48, Problem 71

REAL-WORLD PHYSICS
Television screens and ink-jet printers

The fact that a charge experiences a force when it passes between two charged plates finds application in a wide variety of devices. For example, the image you see on many television screens is produced when a beam of electrons strikes the screen from behind and illuminates individual red, blue, or green *pixels*. Which pixels are illuminated and which remain dark is controlled by parallel charged plates that deflect the electron beam up or down and left or right. Thus, sending the appropriate electrical signals to the deflection plates makes the beam of electrons "paint" any desired picture on the screen.

Similar deflection plates are used in an ink-jet printer. In this case, the beam in question is not a beam of electrons but rather a beam of electrically charged ink droplets. The beam of droplets can be deflected as desired, so that individual letters can be constructed from a series of closely spaced dots on the page. A typical printer might produce as many as 600 dots per inch—that is, 600 dpi.

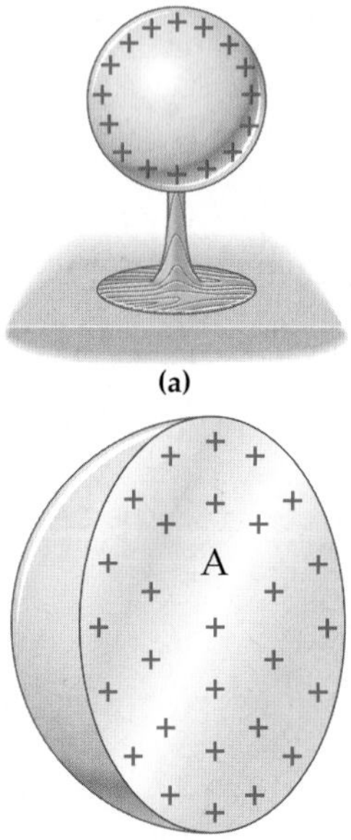

(a)

(b)

▲ **FIGURE 19–18 Charge distribution on a conducting sphere**

(a) A charge placed on a conducting sphere distributes itself uniformly on the surface of the sphere; none of the charge is within the volume of the sphere. **(b)** If the charge were distributed uniformly throughout the volume of a sphere, individual charges, like that at point A, would experience a force due to other charges in the volume. Since charges are free to move in a conductor, they will respond to these forces by moving as far from one another as possible; that is, to the surface of the conductor.

19–6 Shielding and Charging by Induction

In a perfect conductor there are enormous numbers of electrons completely free to move about within the conductor. This simple fact has some rather interesting consequences. Consider, for example, a solid metal sphere attached to an insulating base as in **Figure 19–18**. Suppose a positive charge Q is placed on the sphere. The question is: How does this charge distribute itself on the sphere when it is in equilibrium—that is, when all the charges are at rest? In particular, does the charge spread itself uniformly throughout the volume of the sphere, or does it concentrate on the surface?

The answer is that the charge concentrates on the surface, as shown in Figure 19–18 (a), but let's investigate why this should be the case. First, assume the opposite—that the charge is spread uniformly throughout the sphere's volume, as indicated in Figure 19–18 (b). If this were the case, a charge at location A would experience an outward force due to the spherical distribution of charge between it and the center of the sphere. Since charges are free to move, the charge at A would respond to this force by moving toward the surface. Clearly, then, a uniform distribution of charge within the sphere's volume is not in equilibrium. In fact, the argument that a charge at point A will move toward the surface can be applied to any charge within the sphere. Thus, the net result is that *all* the excess charge Q moves onto the surface of the sphere which, in turn, allows the individual charges to be spread as far from one another as possible.

The preceding result holds no matter what the shape of the conductor. In general,

Excess Charge on a Conductor

Excess charge placed on a conductor, whether positive or negative, moves to the exterior surface of the conductor.

We specify the exterior surface in this statement because a conductor may contain one or more cavities. When an excess charge is applied to such a conductor, all the charge ends up on the exterior surface, and none on the interior surfaces.

Electrostatic Shielding

The ability of electrons to move freely within a conductor has another important consequence; namely, the electric field within a conductor vanishes.

Zero Field within a Conductor

When electric charges are at rest, the electric field within a conductor is zero; $E = 0$.

By *within* a conductor, we mean a location in the actual material of the conductor, as opposed to a location in a cavity within the material.

The best way to see the validity of this statement is to again consider the opposite. If there were a nonzero field within a conductor, electrons would move in response to the field. They would continue to move until finally the field was reduced to zero, at which point the system would be in equilibrium and no more charges would move. Thus, equilibrium and $E = 0$ within a conductor go hand in hand.

A straightforward extension of this idea explains the phenomenon of **shielding,** in which a conductor "shields" its interior from external electric fields. For example, in **Figure 19–19 (a)** we show an uncharged, conducting metal sphere placed in an electric field. Because the positive ions in the metal do not move, the field tends to move negative charges to the left and leave excess positive charges on the right; hence, it causes the sphere to have an **induced** negative charge on its left half and an induced positive charge on its right half. The total charge on the sphere, of course, is still zero. Since field lines end on $(-)$ charges and begin on $(+)$ charges, the external electric field ends on the left half of the sphere and starts up again on the right half. In between, within the conductor, the field is zero, as expected. Thus, the conductor has shielded its interior from the applied electric field.

REAL-WORLD PHYSICS
Electrical shielding

Shielding occurs whether the conductor is solid, as in Figure 19–19, or hollow. In fact, even a thin sheet of metal foil formed into the shape of a box will shield its interior from external electric fields. This effect is put to use in numerous electrical devices, which often have a metal foil or wire mesh enclosure surrounding the sensitive electrical circuits. In this way, a given device can be isolated from the effects of other nearby devices that might otherwise interfere with its operation.

Notice also in Figure 19–19 that the field lines bend slightly near the surface of the sphere. In fact, on closer examination, as in **Figure 19–19 (b)**, we see that the field lines always contact the surface at right angles. This is true for any conductor:

Electric Fields at Conductor Surfaces

Electric field lines contact conductor surfaces at right angles.

If an electric field contacted a conducting surface at an angle other than 90°, the result would be a component of force parallel to the surface. This would result in a movement of electrons and, hence, would not correspond to equilibrium. Instead, electrons would move until the parallel component of the electric field was canceled.

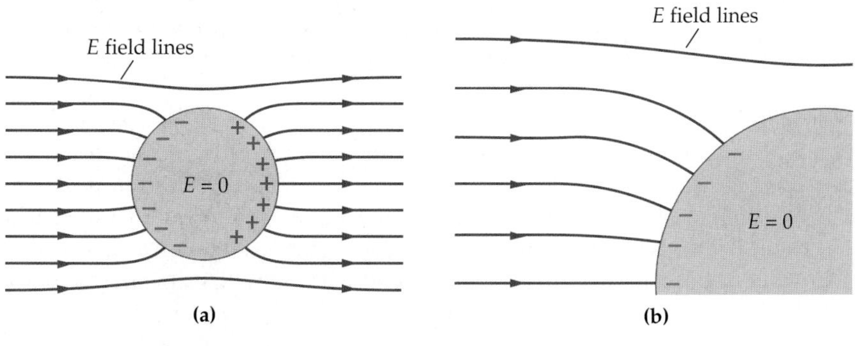

◀ **FIGURE 19–19 Electric field near a conducting surface**
(a) When an uncharged conductor is placed in an electric field, the field induces opposite charges on opposite sides of the conductor. The net charge on the conductor is still zero, however. The induced charges produce a field within the conductor that *exactly* cancels the external field, leading to $E = 0$ inside the conductor. This is an example of electrical shielding. **(b)** Electric field lines meet the surface of a conductor at right angles.

Further examples of field lines near conducting surfaces are shown in **Figure 19–20**. Notice that the field lines are more densely packed near a sharp point, indicating that the field is more intense in such regions. This effect illustrates the basic principle behind the operation of lightning rods. If you look closely, you will notice that all lightning rods have a sharply pointed tip. During an electrical storm, the electric field at the tip becomes so intense that electric charge is given off into the atmosphere. In this way, a lightning rod acts to discharge the area near a house—by giving off a steady stream of charge—thus preventing a strike by a bolt of lightning, which transfers charge in one sudden blast. Sharp points on the rigging of a ship at sea can also give off streams of charge during a storm, often producing glowing lights referred to as Saint Elmo's fire.

REAL-WORLD PHYSICS
Lightning rods and Saint Elmo's fire

The same principle is used to clean the air we breathe, in devices known as *electrostatic precipitators*. In an electrostatic precipitator, smoke and other airborne particles in a smokestack are given a charge as they pass by sharply pointed electrodes—like lightning rods—within the stack. Once the particles are charged, they are removed from the air by charged plates that exert electrostatic forces on them. The resultant emission from the smokestack contains drastically reduced amounts of potentially harmful particulates.

REAL-WORLD PHYSICS
Electrostatic precipitation

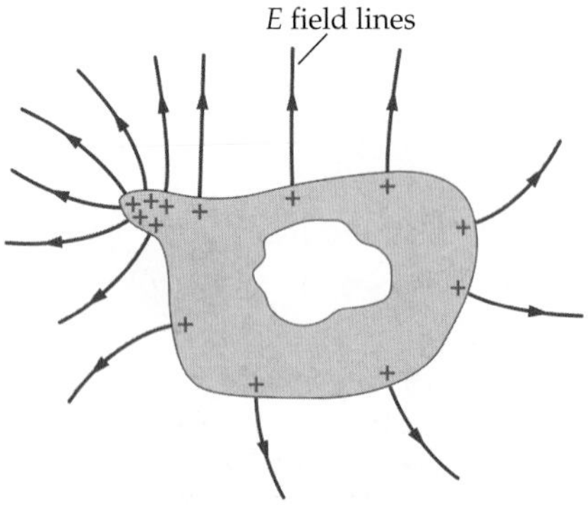

▲ **FIGURE 19–20 Intense electric field near a sharp point**

Electric charges and field lines are more densely packed near a sharp point. This means that the electric field is more intense in such regions as well. (Note that there are no electric charges on the interior surface surrounding the cavity.)

▲ In a dramatic science-museum demonstration of electrical shielding (left), the metal bars of a cage provide excellent protection from an artificially generated lightning bolt. A more practical safeguard is the lightning rod (right). Lightning rods always have sharp points, because that is where the electric field of a conductor is most intense. At the tip, the field can become so strong that charge leaks away into the atmosphere rather than building up to levels that will attract a lightning strike. If a strike does occur, it is conducted to the ground through the lightning rod, rather than through some part of the building itself.

One final note regarding shielding is that it works in one direction only: A conductor shields its interior from external fields, but it does not shield the external world from fields within it. This phenomenon is illustrated in **Figure 19–21** for the case of an uncharged conductor. First, the charge $+Q$ in the cavity induces a charge $-Q$ on the interior surface, in order for the field in the conductor to be zero. Since the conductor is uncharged, a charge $+Q$ will be induced on its exterior surface. As a result, the external world will experience a field due, ultimately, to the charge $+Q$ within the cavity of the conductor.

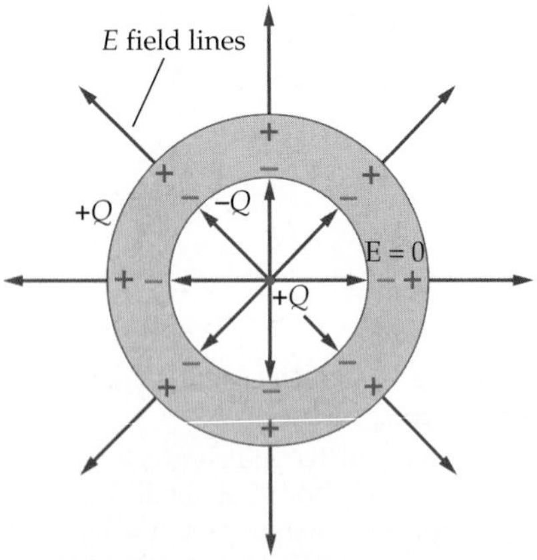

▲ **FIGURE 19–21 Shielding works in only one direction**

A conductor does not shield the external world from charges it encloses. Still, the electric field is zero within the conductor itself.

Charging by Induction

One way to charge an object is to touch it with a charged rod; but since electric forces can act at a distance, it is also possible to charge an object without making direct physical contact. This type of charging is referred to as **charging by induction.**

To see how this type of charging works, consider an uncharged metal sphere on an insulating base. If a negatively charged rod is brought close to the sphere without

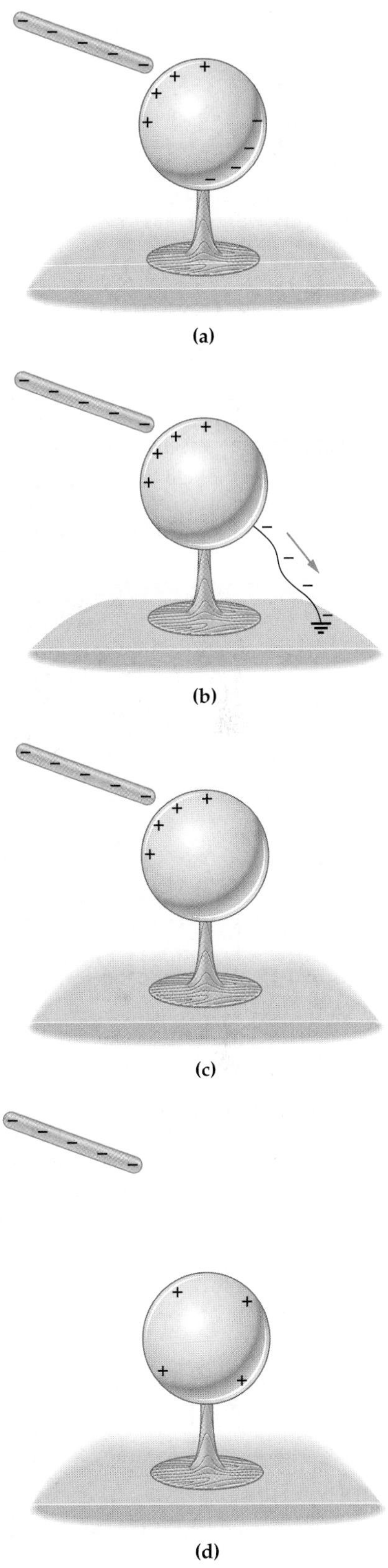

▶ **FIGURE 19–22 Charging by induction**
(a) A charged rod induces + and − charges on opposite sides of the conductor. **(b)** When the conductor is grounded, charges that are repelled by the rod enter the ground. There is now a net charge on the conductor. **(c)** Removing the grounding wire, with the charged rod still in place, traps the net charge on the conductor. **(d)** The charged rod can now be removed, and the conductor retains a charge that is opposite in sign to that on the charged rod.

touching it, as in **Figure 19–22 (a)**, electrons in the sphere are repelled. An induced positive charge is produced on the near side of the sphere, and an induced negative charge on the far side. At this point the sphere is still electrically neutral, however.

The key step in this charging process, shown in **Figure 19–22 (b)**, is to connect the sphere to the ground using a conducting wire. As one might expect, this is referred to as **grounding** the sphere, and is indicated by the symbol ⏚. (A table of electrical symbols can be found in Appendix D.) Since the ground is a fairly good conductor of electricity, and since the Earth can receive or give up practically unlimited numbers of electrons, the effect of grounding the sphere is that the electrons repelled by the charged rod enter the ground. Now the sphere has a net positive charge. With the rod kept in place, the grounding wire is removed, as in **Figure 19–22 (c)**, trapping the net positive charge on the sphere. The rod can now be pulled away, as in **Figure 19–22 (d)**.

Notice that the *induced* charge on the sphere is opposite in sign to the charge on the rod. In contrast, when the sphere is charged by *touch* as in Figure 19–6, it acquires a charge with the same sign as the charge on the rod.

19–7 Electric Flux and Gauss's Law

In this section, we introduce the idea of an electric flux and show that it can be used to calculate the electric field. The precise connection between electric flux and the charges that produce the electric field is provided by Gauss's law.

Electric Flux

Consider a uniform electric field $\vec{E}$, as in **Figure 19–23 (a)**, passing through an area A that is perpendicular to the field. Looking at the electric field lines with their arrows, we can easily imagine a "flow" of electric field through the area. Though there is no actual flow, of course, the analogy is a useful one. It is with this in mind that we define an **electric flux, Φ**, for this case as follows:

$$\Phi = EA$$

On the other hand, if the area A is parallel to the field lines, as in **Figure 19–23 (b)**, none of the $\vec{E}$ lines pierce the area, and hence there is no flux of electric field:

$$\Phi = 0$$

In an intermediate case, as shown in **Figure 19–23 (c)**, the $\vec{E}$ lines pierce the area A at an angle θ away from the perpendicular. As a result, the component of $\vec{E}$ perpendicular to the surface is $E\cos\theta$, and the component parallel to the surface is $E\sin\theta$. Since only the perpendicular component of $\vec{E}$ causes a flux (the parallel component does not pierce the area), the flux in the general case is the following:

Definition of Electric Flux, Φ

$$\Phi = EA\cos\theta \qquad 19\text{–}11$$

SI unit: $N \cdot m^2/C$

Finally, if the surface through which the flux is calculated is *closed*, the sign of the flux is as follows:

- The flux is *positive* for field lines that *leave* the enclosed volume of the surface.
- The flux is *negative* for field lines that *enter* the enclosed volume of the surface.

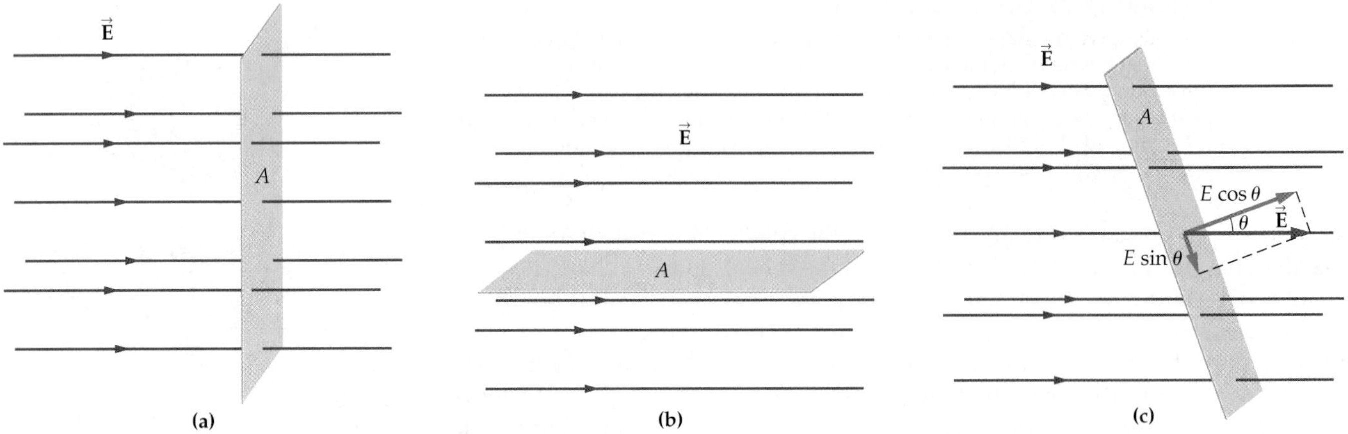

▲ **FIGURE 19–23 Electric flux**

(a) When an electric field $\vec{E}$ passes perpendicularly through the plane of an area A, the electric flux is $\Phi = EA$. **(b)** When the plane of an area is parallel to $\vec{E}$, so that no field lines "pierce" the area, the electric flux is zero, $\Phi = 0$. **(c)** When the normal to the plane of an area is tilted at an angle θ away from the electric field $\vec{E}$, only the perpendicular component of $\vec{E}$, $E \cos \theta$, contributes to the electric flux. Thus, the flux is $\Phi = (E \cos \theta)A$.

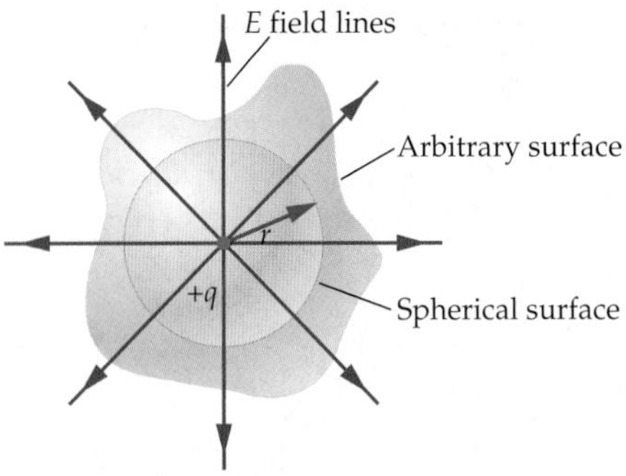

▲ **FIGURE 19–24 Electric flux for a point charge**

The electric flux through the spherical surface surrounding a positive point charge q is $\Phi = EA = (kq/r^2)(4\pi r^2) = q/\varepsilon_0$. The electric flux through an arbitrary surface is the same as for the sphere. The calculation of the flux, however, would be much more difficult for this surface.

Gauss's Law

As a simple example of electric flux, consider a positive point charge q and a spherical surface of radius r centered on the charge, as in **Figure 19–24**. The electric field on the surface of the sphere has the constant magnitude

$$E = k\frac{q}{r^2}$$

Since the electric field is everywhere perpendicular to the spherical surface, it follows that the electric flux is simply E times the area $A = 4\pi r^2$ of the sphere:

$$\Phi = EA = \left(k\frac{q}{r^2}\right)(4\pi r^2) = 4\pi kq$$

We will often find it convenient to express k in terms of another constant, ε_0, as follows: $k = 1/(4\pi\varepsilon_0)$. This new constant, which we call the **permittivity of free space,** is

$$\varepsilon_0 = \frac{1}{4\pi k} = 8.85 \times 10^{-12}\ \mathrm{C^2/N \cdot m^2} \qquad \textbf{19–12}$$

In terms of ε_0, the flux through the spherical surface reduces to

$$\Phi = 4\pi kq = \frac{q}{\varepsilon_0}$$

Thus, we find the very simple result that the electric flux through a sphere that encloses a charge q is the charge divided by the permittivity of free space, ε_0. This is a nice result, but what makes it truly remarkable is that it is equally true for *any* surface that encloses the charge q. For example, if one were to calculate the electric flux through the closed irregular surface also shown in Figure 19–24—which would be a difficult task—the result, nonetheless, would still be simply q/ε_0. This, in fact, is a special case of Gauss's law:

Gauss's Law

If a charge q is enclosed by an arbitrary surface, the total electric flux through the surface, Φ, is

$$\Phi = \frac{q}{\varepsilon_0} \qquad \textbf{19–13}$$

SI unit: $\mathrm{N \cdot m^2/C}$

Note that we use q rather than $|q|$ in Equation 19–13. This is because the electric flux can be positive or negative, depending on the sign of the charge. In particular, if the charge q is positive, the field lines leave the enclosed volume and the flux is positive; if the charge is negative, the field lines enter the enclosed volume and the flux is negative.

CONCEPTUAL CHECKPOINT 19–6 Sign of the Electric Flux

Consider the surface S shown in the sketch. Is the electric flux through this surface (a) negative, (b) positive, or (c) zero?

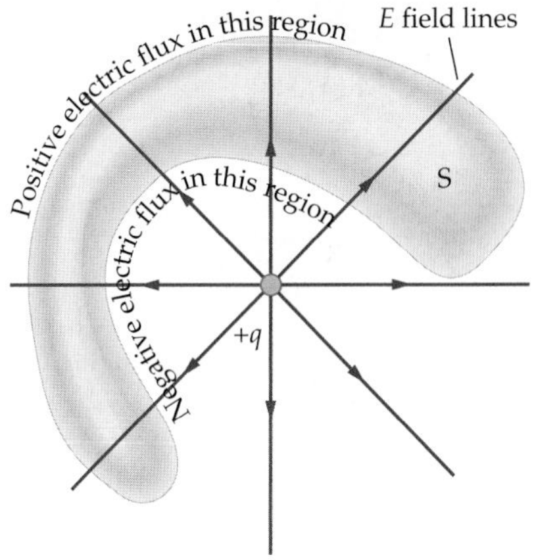

Reasoning and Discussion

Because the surface S encloses no charge, the net electric flux through it must be zero, by Gauss's law. The fact that a charge $+q$ is nearby is irrelevant, because it is outside the volume enclosed by the surface.

We can explain why the flux vanishes in another way. Notice that the flux on portions of S near the charge is negative, since field lines enter the enclosed volume there. On the other hand, the flux is positive on the outer portions of S where field lines exit the volume. The combination of these positive and negative contributions is a net flux of zero.

Answer

(c) The electric flux through the surface S is zero.

Although Gauss's law holds for an arbitrarily complex surface, its real utility is seen when the system in question has a simple symmetry. For example, consider a point charge $+Q$ in the center of a hollow, conducting spherical shell, as illustrated in **Figure 19–25**. The shell has an inside radius R_A and an outside radius R_B, and is uncharged. To calculate the field inside the cavity, where $r < R_A$, we consider an imaginary spherical surface—a so-called **Gaussian surface**—with radius r_1, and centered on the charge $+Q$, as indicated in Figure 19–25. The electric flux through this Gaussian surface is $\Phi = E(4\pi r_1^2) = Q/\varepsilon_0$. Therefore, the magnitude of the electric field, as expected, is

$$E = \frac{Q}{4\pi\varepsilon_0 r_1^2} = k\frac{Q}{r_1^2}$$

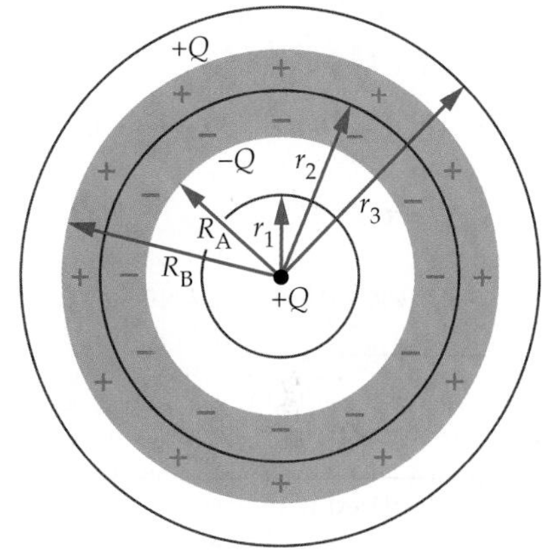

▲ **FIGURE 19–25 Gauss's law applied to a spherical shell**

A simple system with three different Gaussian surfaces.

Note, in particular, that the charges on the spherical shell do not affect the electric flux through this Gaussian surface, since they are not contained within the surface.

Next, consider a Gaussian surface within the shell, with $R_A < r_2 < R_B$, as in Figure 19–25. Since the field within a conductor is zero, $E = 0$, it follows that the electric flux for this surface is zero: $\Phi = EA = 0$. This means that the net charge within the Gaussian surface is also zero; that is, the induced charge on the inner surface of the shell is $-Q$.

Finally, consider a spherical Gaussian surface that encloses the entire spherical shell, with a radius $r_3 > R_B$. In this case, also shown in Figure 19–25, the flux is $\Phi = E(4\pi r_3^2) = (\text{enclosed charge})/\varepsilon_0$. What is the enclosed charge? Well, we know that the spherical shell is uncharged—the induced charges of $+Q$ and $-Q$ on its outer and inner surfaces sum to zero—hence the total enclosed charge is simply $+Q$ from the charge at the center of the shell. Therefore, $\Phi = E(4\pi r_3^2) = Q/\varepsilon_0$, and

$$E = \frac{Q}{4\pi\varepsilon_0 r_3^2} = k\frac{Q}{r_3^2}$$

Note that the field outside the shell is the same as if the shell were not present, showing that the conducting shell does not shield the external world from charges within it, in agreement with our conclusions in the previous section.

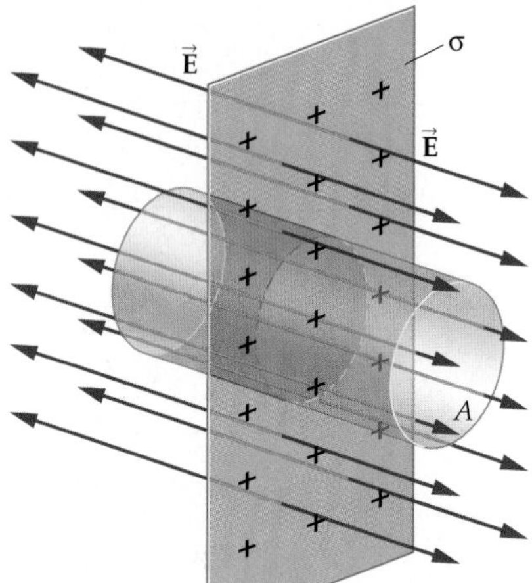

▲ **FIGURE 19–26 Gauss's law applied to a sheet of charge**

A charged sheet of infinite extent and the Gaussian surface used to calculate the electric field.

Gaussian surfaces do not need to be spherical, however. Consider, for example, a thin sheet of charge that extends to infinity, as in **Figure 19–26**. We expect the field to be at right angles to the sheet because, by symmetry, there is no reason for it to tilt in one direction more than in any other direction. Hence, we choose our Gaussian surface to be a cylinder, as in Figure 19–26. With this choice, no field lines pierce the curved surface of the cylinder. The electric flux through this Gaussian surface, then, is due solely to the contributions of the two end caps, each of area A. Hence, the flux is $\Phi = E(2A)$. If the charge per area on the sheet is σ, the enclosed charge is σA, and hence Gauss's law states that

$$\Phi = E(2A) = \frac{(\sigma A)}{\varepsilon_0}$$

Canceling the area, we find

$$E = \frac{\sigma}{2\varepsilon_0}$$

Note that E does not depend in any way on the distance from the sheet, as was mentioned in Section 19–5.

We conclude this chapter with an additional example of Gauss's law in action.

PROBLEM-SOLVING NOTE

Applying Gauss's Law

Gauss's law is useful only when the electric field is constant on a given surface. It is only in such cases that the electric flux can be calculated with ease.

ACTIVE EXAMPLE 19–3 Find the Electric Field

Use the cylindrical Gaussian surface shown in the diagram to calculate the electric field between the metal plates of a parallel-plate capacitor. Each plate has a charge per area of magnitude σ.

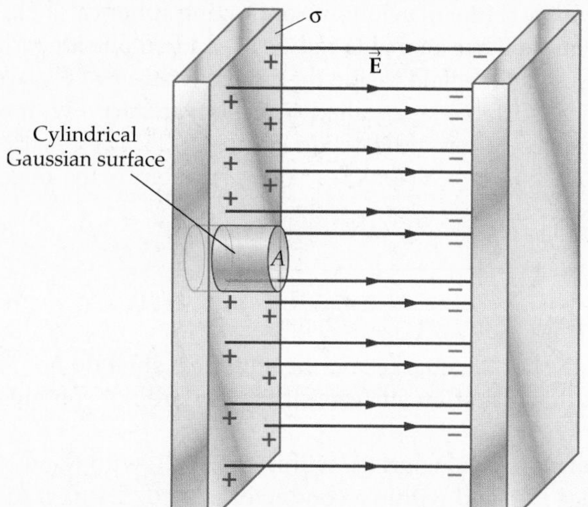

Solution *(Test your understanding by performing the calculations indicated in each step.)*

1. Calculate the electric flux through the curved surface of the cylinder: 0
2. Calculate the electric flux through the end caps of the cylinder: $0 + EA$
3. Determine the charge enclosed by the cylinder: σA
4. Apply Gauss's law to find the field, E: $E = \sigma/\varepsilon_0$

Insight

Note that the electric field is zero within the metal of the plates (because they are conductors). It is for this reason that the electric flux through the left end cap of the Gaussian surface is zero.

Your Turn

Suppose we extend the Gaussian surface so that its right end cap is within the metal of the right plate. The left end of the Gaussian surface remains in its original location. What is the electric flux through this new Gaussian surface? Explain.

(Answers to **Your Turn** *problems are given in the back of the book.)*

Chapter Summary

19–1 Electric Charge

Electric charge is one of the fundamental properties of matter. Electrons have a negative charge, $-e$, and protons have a positive charge, $+e$. An object with zero net charge, like a neutron, is said to be electrically neutral.

Magnitude of an Electron's Charge
The charge on an electron has the following magnitude:

$$e = 1.60 \times 10^{-19}\ \mathrm{C} \qquad \text{19–1}$$

The SI unit of charge is the coulomb, C.

Charge Conservation
The total charge in the universe is constant.

Charge Quantization
Charge comes in quantized amounts that are always integer multiples of e.

19–2 Insulators and Conductors

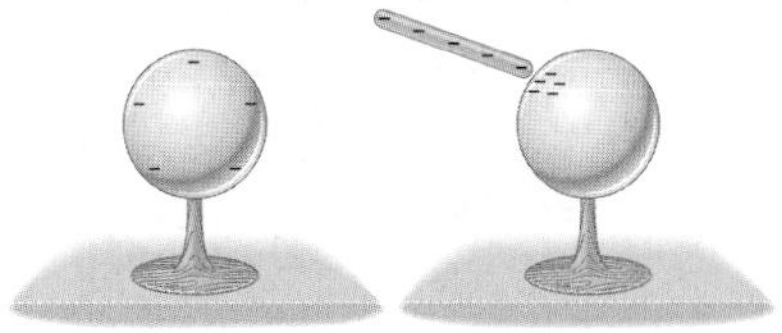

An insulator does not allow electrons within it to move from atom to atom. In conductors, each atom gives up one or more electrons that are then free to move throughout the material. Semiconductors have properties that are intermediate between those of insulators and conductors.

19–3 Coulomb's Law

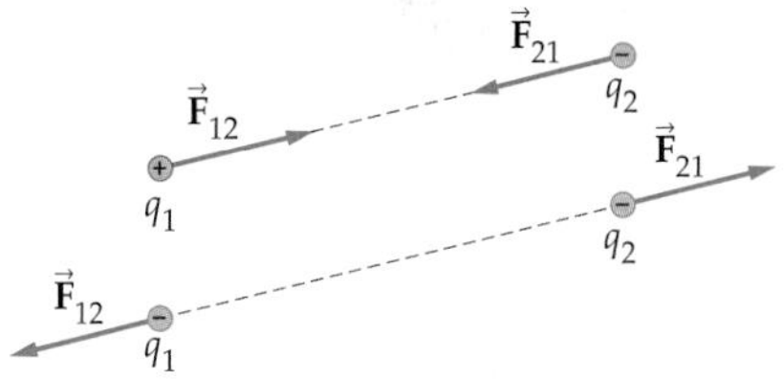

Electric charges exert forces on one another along the line connecting them: Like charges repel, opposite charges attract.

Coulomb's Law
The magnitude of the force between two point charges, q_1 and q_2, separated by a distance r is

$$F = k\frac{|q_1||q_2|}{r^2} \qquad \text{19–5}$$

The constant k in this expression is

$$k = 8.99 \times 10^9\ \mathrm{N \cdot m^2/C^2} \qquad \text{19–6}$$

Superposition
The electric force on one charge due to two or more other charges is the vector sum of each individual force.

Spherical Charge Distributions
A spherical distribution of charge, when viewed from outside, behaves the same as an equivalent point charge at the center of the sphere.

19–4 The Electric Field

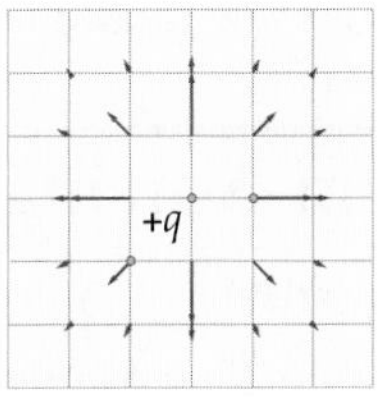

The electric field is the force per charge at a given location in space.

Direction of $\vec{E}$
$\vec{E}$ points in the direction of the force experienced by a *positive* test charge.

Point Charge
The electric field a distance r from a point charge q has a magnitude given by

$$E = k\frac{|q|}{r^2} \qquad \text{19–10}$$

Superposition
The total electric field due to two or more charges is given by the vector sum of the fields due to each charge individually.

19–5 Electric Field Lines

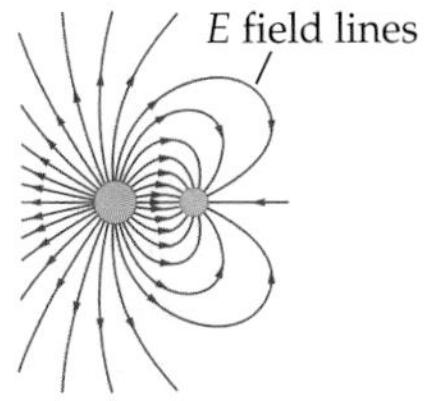

The electric field can be visualized by drawing lines according to a given set of rules.

Rules for Drawing Electric Field Lines
Electric field lines (i) point in the direction of the electric field vector $\vec{E}$ at all points; (ii) start at + charges or infinity; (iii) end at − charges or infinity; and (iv) are more dense the greater the magnitude of $\vec{E}$.

Parallel-Plate Capacitor
A parallel-plate capacitor consists of two oppositely charged, conducting parallel plates separated by a finite distance. The field between the plates is uniform in direction (perpendicular to the plates) and magnitude.

19–6 Shielding and Charging by Induction

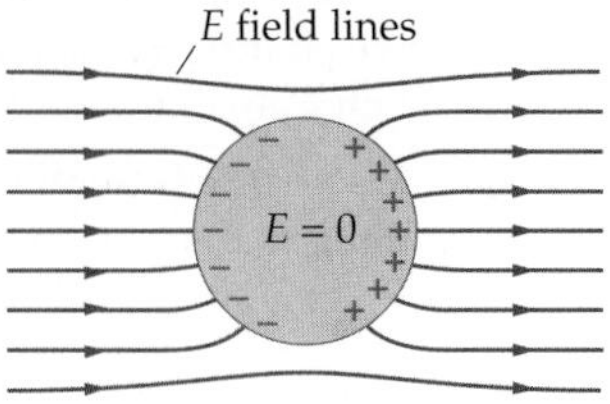

Ideal conductors have a range of interesting behaviors that arise because they have an enormous number of electrons that are free to move.

Excess Charge
Any excess charge placed on a conductor moves to its exterior surface.

Zero Field in a Conductor
The electric field within a conductor in equilibrium is zero.

Shielding
A conductor shields a cavity within it from external electric fields.

Electric Fields at Conductor Surfaces
Electric field lines contact conductor surfaces at right angles.

Charging by Induction
A conductor can be charged without direct physical contact with another charged object. This is charging by induction.

Grounding
Connecting a conductor to the ground is referred to as grounding. The ground itself is a good conductor, and it can give up or receive an unlimited number of electrons.

19–7 Electric Flux and Gauss's Law

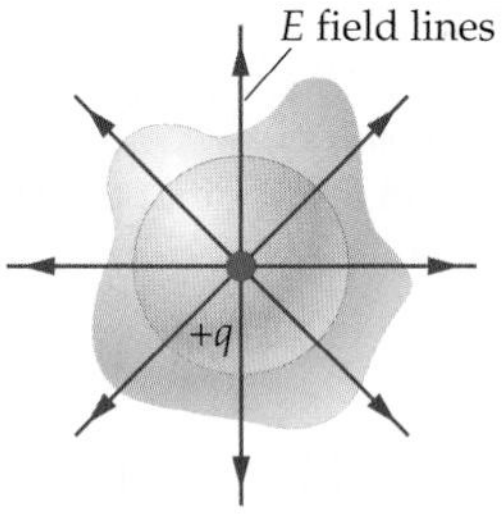

Gauss's law relates the charge enclosed by a given surface to the electric flux through the surface.

Electric Flux
If an area A is tilted at an angle θ to an electric field $\vec{\mathbf{E}}$, the electric flux through the area is

$$\Phi = EA\cos\theta \qquad \textbf{19–11}$$

Gauss's Law
Gauss's law states that if a charge q is enclosed by a surface, the electric flux through the surface is

$$\Phi = \frac{q}{\varepsilon_0} \qquad \textbf{19–13}$$

The constant appearing in this equation is the permittivity of free space, ε_0:

$$\varepsilon_0 = \frac{1}{4\pi k} = 8.85 \times 10^{-12}\ \mathrm{C^2/N \cdot m^2} \qquad \textbf{19–12}$$

Gauss's law is used to calculate the electric field in highly symmetric systems.

Problem-Solving Summary

Type of Problem	Relevant Physical Concepts	Related Examples
Find the electric force exerted by one or more point charges.	The magnitude of the electric force between point charges is $F = k\lvert q_1\rvert\lvert q_2\rvert/r^2$; the direction of the force is given by the expression "opposites attract, likes repel." When more than one charge exerts a force on a given charge, the net force is the vector sum of the individual forces.	Examples 19–1, 19–2, 19–3 Active Examples 19–1, 19–2
Find the electric force due to a spherical distribution of charge.	For points outside a spherical distribution of charge, the spherical distribution behaves the same as a point charge of the same amount at the center of the sphere.	Active Example 19–2
Calculate the force exerted by an electric field.	An electric field, $\vec{\mathbf{E}}$, produces a force, $\vec{\mathbf{F}} = q\vec{\mathbf{E}}$, on a point charge q. The direction of the force is in the direction of the field if the charge is positive and opposite to the field if the charge is negative.	Example 19–4

Type of Problem	Relevant Physical Concepts	Related Examples
Find the electric field due to one or more point charges.	The electric field due to a point charge q has a magnitude given by $E = k\|q\|/r^2$. The direction of the field is radially outward if q is positive, and radially inward if q is negative. When a group of point charges is being considered, the total electric field is the vector sum of the individual fields.	Example 19–5
Calculate the electric field using Gauss's law.	The electric field can be found by setting the electric flux through a given surface equal to the charge enclosed by the surface divided by ε_0.	Active 19–3

Conceptual Questions

(Answers to odd-numbered Conceptual Questions can be found in the back of the book.)

1. When an object that was neutral becomes charged, does the total charge of the universe change? Explain.
2. The fact that the electron has a negative charge and the proton has a positive charge is due to a convention established by Benjamin Franklin. Would there have been any significant consequences if Franklin had chosen the opposite convention? Is there any advantage to naming charges plus and minus as opposed to, say, A and B?
3. Explain why a comb that has been rubbed through your hair attracts small bits of paper, even though the paper is uncharged.
4. Small bits of paper are attracted to an electrically charged comb, but as soon as they touch the comb they are strongly repelled. Explain this behavior.
5. A charged rod is brought near a suspended object, which is repelled by the rod. Can we conclude that the suspended object is charged? Explain.
6. A charged rod is brought near a suspended object, which is attracted to the rod. Can we conclude that the suspended object is charged? Explain.
7. Describe some of the similarities and differences between Coulomb's law and Newton's law of gravity.
8. A point charge $+Q$ is fixed at a height H above the ground. Directly below this charge is a small ball with a charge $-q$ and a mass m. When the ball is at a height h above the ground, the net force (gravitational plus electrical) acting on it is zero. Is this a stable equilibrium for the object? Explain.
9. Four identical point charges are placed at the corners of a square. A fifth point charge placed at the center of the square experiences zero net force. Is this a stable equilibrium for the fifth charge? Explain.
10. A proton moves in a region of constant electric field. Does it follow that the proton's velocity is parallel to the electric field? Does it follow that the proton's acceleration is parallel to the electric field? Explain.
11. Describe some of the differences between charging by induction and charging by contact.
12. A system consists of two charges of equal magnitude and opposite sign separated by a distance d. Since the total electric charge of this system is zero, can we conclude that the electric field produced by the system is also zero? Does your answer depend on the separation d? Explain.
13. The force experienced by charge 1 at point A is different in direction and magnitude from the force experienced by charge 2 at point B. Can we conclude that the electric fields at points A and B are different? Explain.
14. Can an electric field exist in a vacuum? Explain.
15. Explain why electric field lines never cross.
16. Charge q_1 is inside a closed Gaussian surface; charge q_2 is just outside the surface. Does the electric flux through the surface depend on q_1? Does it depend on q_2? Explain.
17. In the previous question, does the electric field at a point on the Gaussian surface depend on q_1? Does it depend on q_2? Explain.
18. Gauss's law can tell us how much charge is contained within a Gaussian surface. Can it tell us where inside the surface it is located? Explain.
19. Explain why Gauss's law is not very useful in calculating the electric field of a charged disk.

Conceptual Exercises

(Answers to odd-numbered Conceptual Exercises can be found in the back of the book.)

1. Consider an electrically neutral object. If this object is given a positive charge, does its mass increase, decrease, or stay the same? Explain.
2. Consider an electrically neutral object. If this object is given a negative charge, does its mass increase, decrease, or stay the same? Explain.
3. **(a)** Based on the materials listed in Table 19–1, is the charge of the rubber balloon shown on page 628 more likely to be positive or negative? Explain. **(b)** If the charge on the balloon is reversed, will the stream of water deflect toward or away from the balloon? Explain.
4. **(a)** If rabbit fur is rubbed against glass, what is the sign of the charge each acquires? Explain. **(b)** Repeat part (a) for the case of glass and rubber. **(c)** Comparing the situations described in parts (a) and (b), in which case is the magnitude of the triboelectric charge the greatest? Explain. (Refer to Table 19–1.)
5. An electron and a proton are released from rest in space, far from any other objects. The particles move toward each other, due to their mutual electrical attraction. When they meet, is the kinetic energy of the electron greater than, less than, or the same as the kinetic energy of the proton? Explain.
6. Five point charges, $q_1 = +q, q_2 = +2q, q_3 = -3q, q_4 = -4q$, and $q_5 = -5q$, are placed in the vicinity of an insulating

spherical shell with a charge $+Q$, distributed uniformly over its surface, as indicated in **Figure 19–27**. Rank the point charges in order of increasing magnitude of the force exerted on them by the sphere. Indicate ties where appropriate.

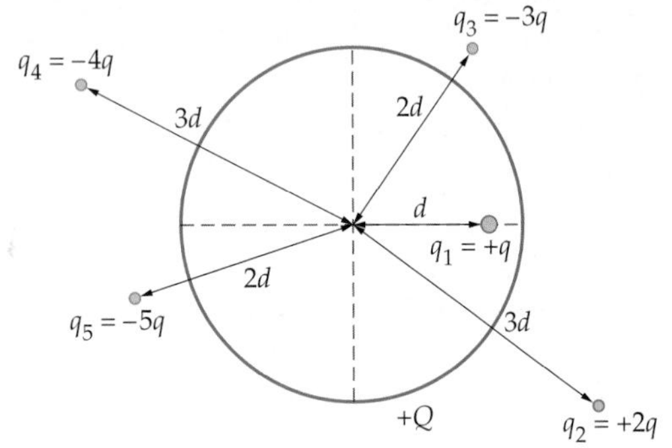

▲ **FIGURE 19–27** Conceptual Exercise 6

7. Three charges, $q_1 = +q$, $q_2 = -q$, and $q_3 = +q$, are at the vertices of an equilateral triangle, as shown in **Figure 19–28**. **(a)** Rank the three charges in order of increasing magnitude of the electric force they experience. Indicate ties where appropriate. **(b)** Give the direction angle, θ, of the net electric force experienced by charge 1. Note that θ is measured counterclockwise from the positive x axis. **(c)** Repeat part (b) for charge 2. **(d)** Repeat part (b) for charge 3.

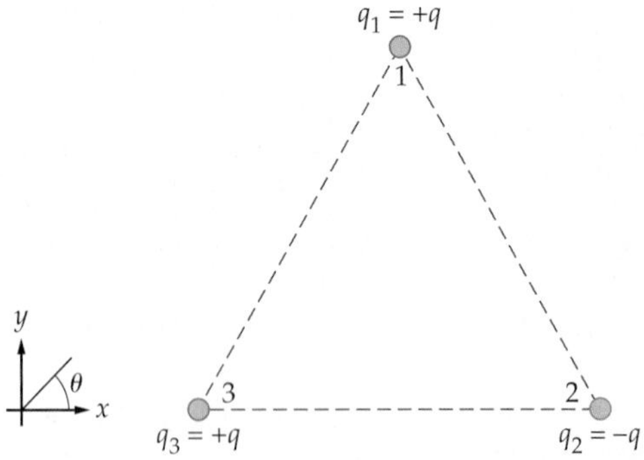

▲ **FIGURE 19–28** Conceptual Exercises 7 and 14

8. A charge $+q$ and a charge $-q$ are placed at opposite corners of a square. Will a third point charge experience a greater force if it is placed at one of the empty corners of the square, or at the center of the square? Explain.

9. Repeat the previous question, this time with charges $+q$ and $+q$ at opposite corners of a square.

10. Four lightweight, plastic spheres, labeled A, B, C, and D, are suspended from threads in various combinations, as illustrated in **Figure 19–29**. It is given that the net charge on sphere D is $+Q$, and that the other spheres have net charges of $+Q$, $-Q$, or 0. From the results of the four experiments shown in Figure 19-29, determine the net charge of **(a)** sphere A, **(b)** sphere B, and **(c)** sphere C.

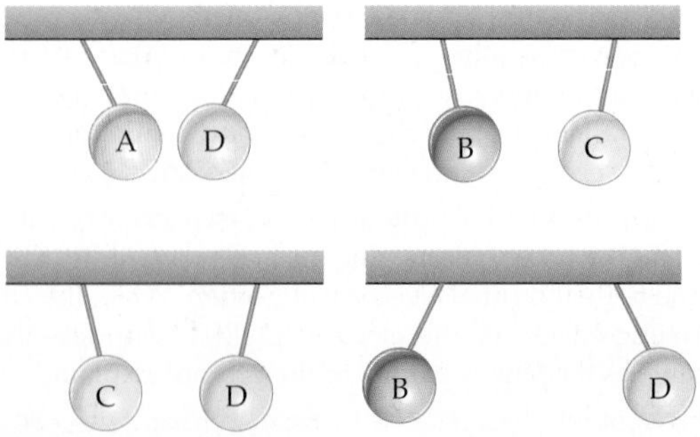

▲ **FIGURE 19–29** Conceptual Exercise 10

11. Consider the three electric charges shown in **Figure 19–30**. Rank the charges in order of increasing magnitude of the net force they experience. Indicate ties where appropriate.

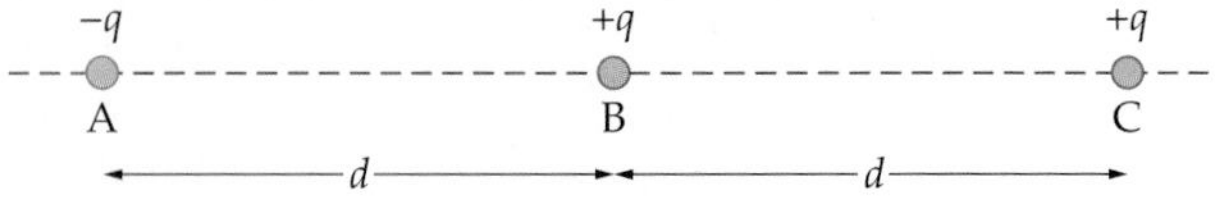

▲ **FIGURE 19–30** Conceptual Exercises 11 and 16

12. In Conceptual Checkpoint 19–3, suppose the charge that moves from point A to point B is $+q$ rather than $-q$. Is the net force experienced by this charge at point A greater than, less than, or the same as it is at point B? Explain.

13. An electron (charge $= -e$) orbits a helium nucleus (charge $= +2e$). Is the force exerted on the helium nucleus by the electron greater than, less than, or the same as the force exerted on the electron by the helium nucleus? Explain.

14. Find **(a)** the direction and **(b)** the magnitude of the net electric field at the center of the equilateral triangle in Figure 19–28. Give your answers in terms of the angle θ, as defined in Figure 19–28, and E, the magnitude of the electric field produced by any *one* of the charges at the center of the triangle.

15. Suppose the charged sphere in Active Example 19–2 is made from a conductor, rather than an insulator. Do you expect the magnitude of the force between the point charge and the sphere to be greater than, less than, or the same as in the case of an insulating sphere? Explain.

16. The electric field on the dashed line in Figure 19–30 vanishes at infinity, but also at two different points a finite distance from the charges. Identify the regions in which you can find $E = 0$ at a finite distance from the charges: **(a)** to the left of point A; **(b)** between points A and B; **(c)** between points B and C; **(d)** to the right of point C.

17. In the operating room, technicians and doctors must take care not to create an electric spark, since the presence of the oxygen gas used during an operation increases the risk of a deadly fire. Should the operating-room personnel wear shoes that are conducting or nonconducting? Explain.

18. Two electric charges are separated by a finite distance. Somewhere between the charges, on the line connecting them, the net electric field they produce is zero. **(a)** Do the charges have the same or opposite signs? Explain. **(b)** If the point of zero field is closer to charge 1, is the magnitude of charge 1 greater than or less than the magnitude of charge 2? Explain.

19. Under normal conditions, the electric field at the surface of the Earth points downward, into the ground. What is the sign of the electric charge on the ground?

20. Four pairs of conducting spheres, all with the same radius, are shown in **Figure 19–31**, along with the net charge placed on them initially. The spheres in each pair are now brought into contact, allowing charge to transfer between them. Rank the pairs of spheres in order of increasing magnitude of the charge transferred. Indicate ties where appropriate.

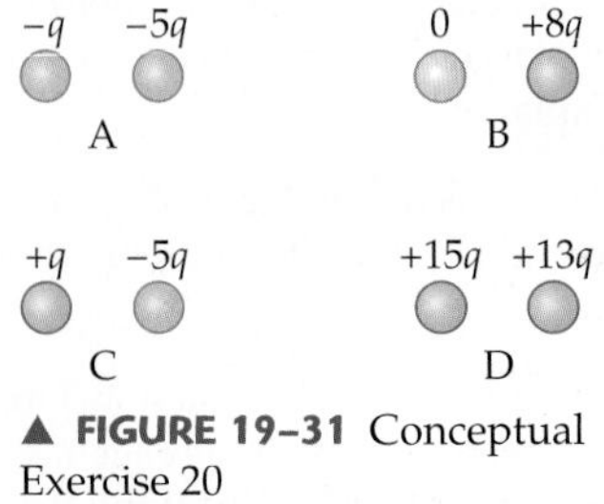

▲ **FIGURE 19–31** Conceptual Exercise 20

21. Two identical spheres are made of conducting material. Initially, sphere 1 has a net charge of $+35Q$ and sphere 2 has a net

charge of $-21Q$. If the spheres are now brought into contact, what is the final charge on sphere 1? Explain.

22. Gaussian surface 1 has twice the area of Gaussian surface 2. Both surfaces enclose the same charge Q. Is the electric flux through surface 1 greater than, less than, or the same as the electric flux through surface 2. Explain.

23. Suppose the conducting shell in Figure 19–25—which has a point charge $+Q$ at its center—has a nonzero net charge. How much charge is on the inner and outer surface of the shell when the net charge of the shell is **(a)** $-2Q$, **(b)** $-Q$, and **(c)** $+Q$?

24. Rank the Gaussian surfaces shown in **Figure 19–32** in order of increasing electric flux, starting with the most negative. Indicate ties where appropriate.

25. A Gaussian surface for the charges shown in Figure 19–32 has an electric flux equal to $+3q/\varepsilon_0$. Which charges are contained within this Gaussian surface?

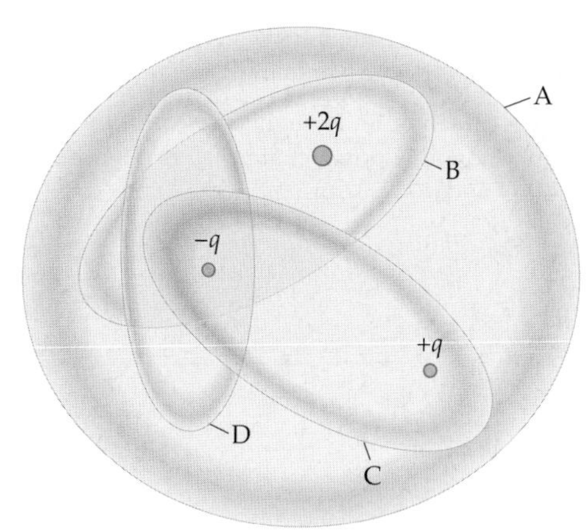

▲ **FIGURE 19–32** Conceptual Exercises 24 and 25

Problems

Note: **IP** *denotes an integrated conceptual/quantitative problem.* **BIO** *identifies problems of biological or medical interest. Red bullets (•, ••, •••) are used to indicate the level of difficulty of each problem.*

Section 19–1 Electric Charge

1. • Find the net charge of a system consisting of 3.9×10^7 electrons.

2. • Find the net charge of a system consisting of **(a)** 6.15×10^6 electrons and 7.44×10^6 protons or **(b)** 212 electrons and 165 protons.

3. • How much negative electric charge is contained in 2 moles of carbon?

4. • Find the total electric charge of 1.5 kg of **(a)** electrons and **(b)** protons.

5. • A container holds a gas consisting of 1.75 moles of oxygen molecules. One in a million of these molecules has lost a single electron. What is the net charge of the gas?

6. • **The Charge on Adhesive Tape** When adhesive tape is pulled from a dispenser, the detached tape acquires a positive charge and the remaining tape in the dispenser acquires a negative charge. If the tape pulled from the dispenser has 0.14 μC of charge per centimeter, what length of tape must be pulled to transfer 1.8×10^{13} electrons to the remaining tape?

7. •• A system of 1525 particles, each of which is either an electron or a proton, has a net charge of -5.456×10^{-17} C. **(a)** How many electrons are in this system? **(b)** What is the mass of this system?

Section 19–3 Coulomb's Law

8. • At what separation is the electrostatic force between a $+11.2$-μC point charge and a $+29.1$-μC point charge equal in magnitude to 1.77 N?

9. • The attractive electrostatic force between the point charges $+8.44 \times 10^{-6}$ C and Q has a magnitude of 0.975 N when the separation between the charges is 1.31 m. Find the sign and magnitude of the charge Q.

10. • If the speed of the electron in Example 19–1 were 7.3×10^5 m/s, what would be the corresponding orbital radius?

11. • **IP** Two point charges, the first with a charge of $+3.13 \times 10^{-6}$ C and the second with a charge of -4.47×10^{-6} C, are separated by 25.5 cm. **(a)** Find the magnitude of the electrostatic force experienced by the positive charge. **(b)** Is the magnitude of the force experienced by the negative charge greater than, less than, or the same as that experienced by the positive charge? Explain.

12. • When two identical ions are separated by a distance of 6.2×10^{-10} m, the electrostatic force each exerts on the other is 5.4×10^{-9} N. How many electrons are missing from each ion?

13. • A sphere of radius 4.22 cm and uniform surface charge density $+12.1\ \mu\text{C/m}^2$ exerts an electrostatic force of magnitude 46.9×10^{-3} N on a point charge $+1.75\ \mu$C. Find the separation between the point charge and the center of the sphere.

14. • Given that $q = +12\ \mu$C and $d = 16$ cm, find the direction and magnitude of the net electrostatic force exerted on the point charge q_1 in **Figure 19–33**.

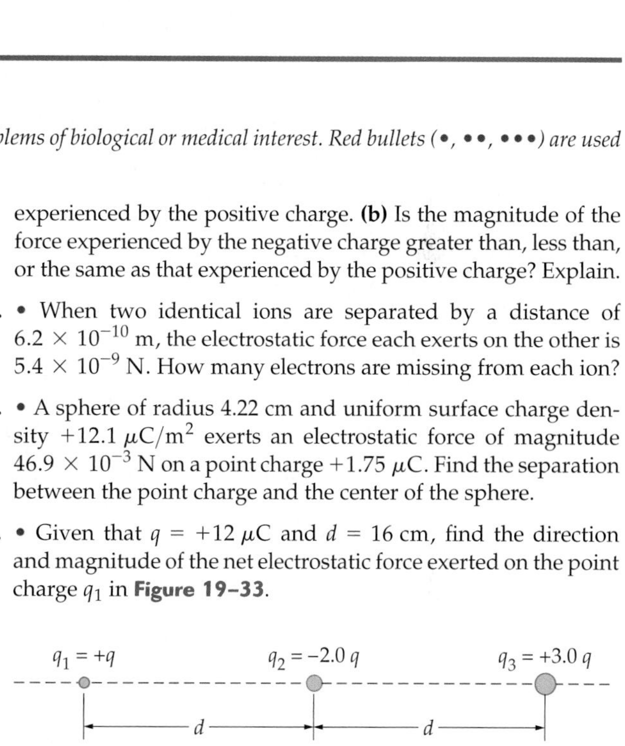

▲ **FIGURE 19–33** Problems 14, 15, 16, and 46

15. •• **IP** Given that $q = +12\ \mu$C and $d = 16$ cm, **(a)** find the direction and magnitude of the net electrostatic force exerted on the point charge q_2 in Figure 19–33. **(b)** How would your answers to part (a) change if the distance d were tripled?

16. •• Suppose the charge q_2 in Figure 19–33 can be moved left or right along the line connecting the charges q_1 and q_3. Given that $q = +12\ \mu$C, find the distance from q_1 where q_2 experiences a net electrostatic force of zero. (The charges q_1 and q_3 are separated by a fixed distance of 32 cm.)

17. •• Find the orbital radius for which the kinetic energy of the electron in Example 19–1 is 1.51 eV. (*Note:* 1 eV = 1 electron − volt = 1.6×10^{-19} J.)

18. •• A point charge $q = -0.35$ nC is fixed at the origin. Where must a proton be placed in order for the electric force acting on it to be exactly opposite to its weight? (Let the y axis be vertical and the x axis be horizontal.)

19. •• A point charge $q = -0.35$ nC is fixed at the origin. Where must an electron be placed in order for the electric force acting

on it to be exactly opposite to its weight? (Let the y axis be vertical and the x axis be horizontal.)

20. •• Find the direction and magnitude of the net electrostatic force exerted on the point charge q_2 in **Figure 19–34**. Let $q = +2.4\ \mu C$ and $d = 33$ cm.

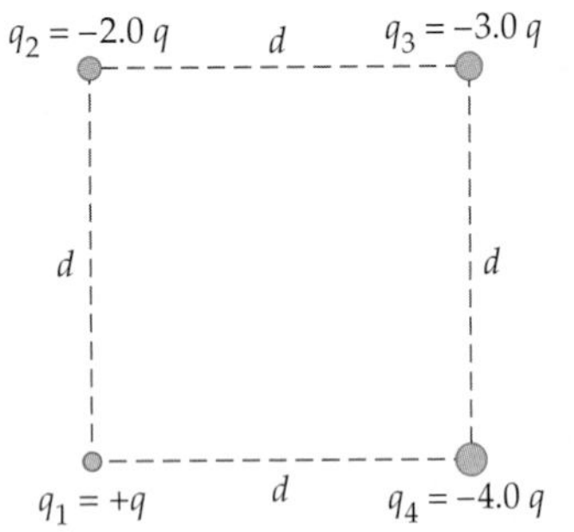

▲ **FIGURE 19–34** Problems 20 and 21

21. •• **IP (a)** Find the direction and magnitude of the net electrostatic force exerted on the point charge q_3 in Figure 19–34. Let $q = +2.4\ \mu C$ and $d = 33$ cm. **(b)** How would your answers to part (a) change if the distance d were doubled?

22. •• **IP** Two point charges lie on the x axis. A charge of $+9.9\ \mu C$ is at the origin, and a charge of $-5.1\ \mu C$ is at $x = 10.0$ cm. **(a)** At what position x would a third charge q_3 be in equilibrium? **(b)** Does your answer to part (a) depend on whether q_3 is positive or negative? Explain.

23. •• A system consists of two positive point charges, q_1 and $q_2 > q_1$. The total charge of the system is $+62.0\ \mu C$, and each charge experiences an electrostatic force of magnitude 75.0 N when the separation between them is 0.270 m. Find q_1 and q_2.

24. •• **IP** The point charges in **Figure 19–35** have the following values: $q_1 = +2.1\ \mu C$, $q_2 = +6.3\ \mu C$, $q_3 = -0.89\ \mu C$. **(a)** Given that the distance d in Figure 19–32 is 4.35 cm, find the direction and magnitude of the net electrostatic force exerted on the point charge q_1. **(b)** How would your answers to part (a) change if the distance d were doubled? Explain.

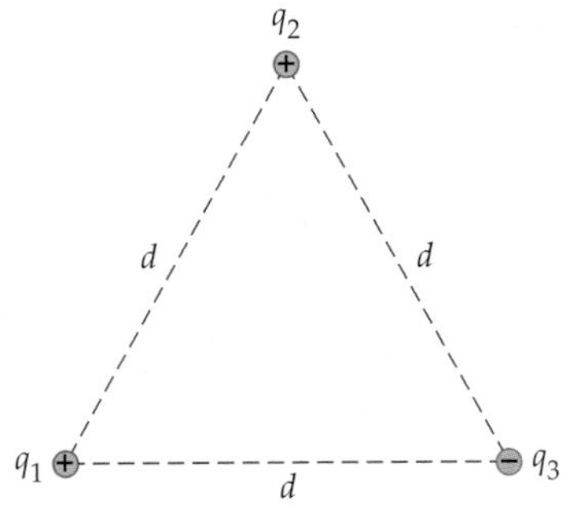

▲ **FIGURE 19–35** Problems 24, 25, 36, and 47

25. •• Referring to Problem 24, suppose that the magnitude of the net electrostatic force exerted on the point charge q_2 in Figure 19–35 is 0.65 N. **(a)** Find the distance d. **(b)** What is the direction of the net force exerted on q_2?

26. •• **IP (a)** If the nucleus in Example 19–1 had a charge of $+2e$ (as would be the case for a nucleus of helium), would the speed of the electron be greater than, less than, or the same as that found in the Example? Explain. (Assume the radius of the electron's orbit is the same.) **(b)** Find the speed of the electron for a nucleus of charge $+2e$.

27. •• Four point charges are located at the corners of a square with sides of length a. Two of the charges are $+q$, and two are $-q$. Find the magnitude and direction of the net electric force exerted on a charge $+Q$, located at the center of the square, for each of the following two arrangements of charge: **(a)** The charges alternate in sign ($+q, -q, +q, -q$) as you go around the square; **(b)** The two positive charges are on the top corners, and the two negative charges are on the bottom corners.

28. •• **BIO The Electric Honeybee** Measurements show that a honeybee in active flight can acquire an electrostatic charge as great as 93.0 pC. **(a)** How many electrons must be transferred to produce this charge? **(b)** Suppose two bees, each with a charge of 93.0 pC, are separated by a distance of 1.20 cm. Treating the bees as point charges, find the magnitude of the electrostatic force experienced by the bees. Compare this force to the weight of a 0.140-g bee.

29. •• **IP** Two identical point charges in free space are connected by a string 7.6 cm long. The tension in the string is 0.21 N. **(a)** Find the magnitude of the charge on each of the point charges. **(b)** Using the information given in the problem statement, is it possible to determine the sign of the charges? Explain. **(c)** Find the tension in the string if $+1.0\ \mu C$ of charge is transferred from one point charge to the other. Compare with your result from part (a).

30. ••• Two spheres with uniform surface charge density, one with a radius of 7.2 cm and the other with a radius of 4.7 cm, are separated by a center-to-center distance of 33 cm. The spheres have a combined charge of $+55\ \mu C$ and repel one another with a force of 0.75 N. What is the surface charge density on each sphere?

31. ••• Point charges, q_1 and q_2, are placed on the x axis, with q_1 at $x = 0$ and q_2 at $x = d$. A third point charge, $+Q$, is placed at $x = 3d/4$. If the net electrostatic force experienced by the charge $+Q$ is zero, how are q_1 and q_2 related?

Section 19–4 The Electric Field

32. • What is the magnitude of the electric field produced by a charge of magnitude $5.00\ \mu C$ at a distance of **(a)** 1.00 m and **(b)** 2.00 m?

33. • A $+5.0$-μC charge experiences a 0.44-N force in the positive y direction. If this charge is replaced with a -2.7-μC charge, what force will it experience?

34. • Two point charges lie on the x axis. A charge of $+6.2\ \mu C$ is at the origin, and a charge of $-9.5\ \mu C$ is at $x = 10.0$ cm. What is the net electric field at **(a)** $x = -4.0$ cm and at **(b)** $x = +4.0$ cm?

35. •• An object with a charge of $-3.6\ \mu C$ and a mass of 0.012 kg experiences an upward electric force, due to a uniform electric field, equal in magnitude to its weight. **(a)** Find the direction and magnitude of the electric field. **(b)** If the electric charge on the object is doubled while its mass remains the same, find the direction and magnitude of its acceleration.

36. •• **IP** Figure 19–35 shows a system consisting of three charges, $q_1 = +5.00\ \mu C$, $q_2 = +5.00\ \mu C$, and $q_3 = -5.00\ \mu C$, at the vertices of an equilateral triangle of side $d = 2.75$ cm. **(a)** Find the magnitude of the electric field at a point halfway between the charges q_1 and q_2. **(b)** Is the magnitude of the electric field halfway between the charges q_2 and q_3 greater than, less than, or the same as the electric field found in part (a)? Explain. **(c)** Find the magnitude of the electric field at the point specified in part (b).

37. •• Two point charges of equal magnitude are 8.0 cm apart. At the midpoint of the line connecting them, their combined electric field has a magnitude of 45 N/C. Find the magnitude of the charges.

38. •• **IP** A point charge $q = +4.7\ \mu C$ is placed at each corner of an equilateral triangle with sides 0.21 m in length. **(a)** What is the magnitude of the electric field at the midpoint of any of the three sides of the triangle? **(b)** Is the magnitude of the electric

field at the center of the triangle greater than, less than, or the same as the magnitude at the midpoint of a side? Explain.

39. ••• **IP** Four point charges, each of magnitude q, are located at the corners of a square with sides of length a. Two of the charges are $+q$, and two are $-q$. The charges are arranged in one of the following two ways: (1) The charges alternate in sign ($+q, -q, +q, -q$) as you go around the square; (2) the top two corners of the square have positive charges ($+q, +q$), and the bottom two corners have negative charges ($-q, -q$). **(a)** In which case will the electric field at the center of the square have the greatest magnitude? Explain. **(b)** Calculate the electric field at the center of the square for each of these two cases.

40. ••• The electric field at the point $x = 5.00$ cm and $y = 0$ points in the positive x direction with a magnitude of 10.0 N/C. At the point $x = 10.0$ cm and $y = 0$ the electric field points in the positive x direction with a magnitude of 15.0 N/C. Assuming this electric field is produced by a single point charge, find **(a)** its location and **(b)** the sign and magnitude of its charge.

Section 19–5 Electric Field Lines

41. • **IP** The electric field lines surrounding three charges are shown in **Figure 19–36**. The center charge is $q_2 = -10.0\ \mu\text{C}$. **(a)** What are the signs of q_1 and q_3? **(b)** Find q_1. **(c)** Find q_3.

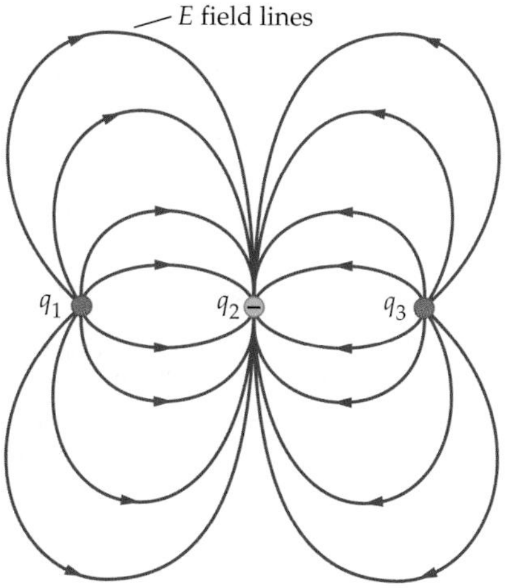

▲ **FIGURE 19–36** Problems 41 and 44

42. • Make a qualitative sketch of the electric field lines produced by two equal positive charges, $+q$, separated by a distance d.

43. • Make a qualitative sketch of the electric field lines produced by two charges, $+q$ and $-q$, separated by a distance d.

44. •• Referring to Figure 19–36, suppose q_2 is not known. Instead, it is given that $q_1 + q_2 = -2.5\ \mu\text{C}$. Find q_1, q_2, and q_3.

45. •• Make a qualitative sketch of the electric field lines produced by the four charges, $+q$, $-q$, $+q$, and $-q$, arranged clockwise on the four corners of a square with sides of length d.

46. •• Sketch the electric field lines for the system of charges shown in Figure 19–33.

47. •• Sketch the electric field lines for the system of charges shown in Figure 19–35.

48. •• Suppose the magnitude of the electric field between the plates in Example 19–6 is changed, and a new object with a charge of $-2.05\ \mu\text{C}$ is attached to the string. If the tension in the string is 0.350 N, and the angle it makes with the vertical is 12°, what are **(a)** the mass of the object and **(b)** the magnitude of the electric field?

Section 19–7 Electric Flux and Gauss's Law

49. • A uniform electric field of magnitude 25,000 N/C makes an angle of 27° with a plane surface of area 0.0133 m². What is the electric flux through this surface?

50. • A surface encloses the charges $q_1 = 3.2\ \mu\text{C}$, $q_2 = 6.9\ \mu\text{C}$, and $q_3 = -4.1\ \mu\text{C}$. Find the electric flux through this surface.

51. • **IP** A uniform electric field of magnitude 6.00×10^3 N/C points upward. An empty, closed shoe box has a top and bottom that are 35.0 cm by 25.0 cm, vertical ends that are 25.0 cm by 20.0 cm, and vertical sides that are 20.0 cm by 35.0 cm. **(a)** Which side of the box has the greatest positive electric flux? Which side has the greatest negative electric flux? Which sides have zero electric flux? **(b)** Calculate the electric flux through each of the six sides of the box.

52. • **BIO Nerve Cells** Nerve cells are long, thin cylinders along which electrical disturbances (nerve impulses) travel. The cell membrane of a typical nerve cell consists of an inner and an outer wall separated by a distance of 0.10 μm. The electric field within the cell membrane is 7.0×10^5 N/C. Approximating the cell membrane as a parallel-plate capacitor, determine the magnitude of the charge density on the inner and outer cell walls.

53. •• The electric flux through each of the six sides of a rectangular box are as follows:

$\Phi_1 = +150.0\ \text{N}\cdot\text{m}^2/\text{C}$; $\Phi_2 = +250.0\ \text{N}\cdot\text{m}^2/\text{C}$;
$\Phi_3 = -350.0\ \text{N}\cdot\text{m}^2/\text{C}$; $\Phi_4 = +175.0\ \text{N}\cdot\text{m}^2/\text{C}$;
$\Phi_5 = -100.0\ \text{N}\cdot\text{m}^2/\text{C}$; $\Phi_6 = +450.0\ \text{N}\cdot\text{m}^2/\text{C}$.

How much charge is in this box?

54. •• Consider a spherical Gaussian surface and three charges: $q_1 = 1.65\ \mu\text{C}$, $q_2 = -2.32\ \mu\text{C}$, and $q_3 = 3.71\ \mu\text{C}$. Find the electric flux through the Gaussian surface if it completely encloses **(a)** only charges q_1 and q_2, **(b)** only charges q_2 and q_3, and **(c)** all three charges. **(d)** Suppose a fourth charge, Q, is added to the situation described in part (c). Find the sign and magnitude of Q required to give zero electric flux through the surface.

55. ••• A thin wire of infinite extent has a charge per unit length of λ. Using the cylindrical Gaussian surface shown in **Figure 19–37**, show that the electric field produced by this wire at a radial distance r has a magnitude given by

$$E = \frac{\lambda}{2\pi\varepsilon_0 r}$$

Note that the direction of the electric field is always radially away from the wire.

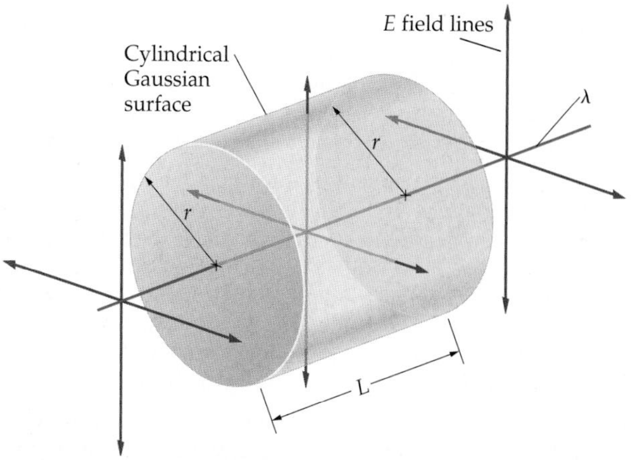

▲ **FIGURE 19–37** Problems 55 and 64

General Problems

56. • A proton is released from rest in a uniform electric field of magnitude 1.08×10^5 N/C. Find the speed of the proton after it has traveled **(a)** 1.00 cm and **(b)** 10.0 cm.

57. • **BIO Ventricular Fibrillation** If a charge of 0.30 C passes through a person's chest in 1.0 s, the heart can go into ventricular fibrillation—a nonrhythmic, "fluttering" of the ventricles that results in little or no blood being pumped to the body. If this rate of charge transfer persists for 4.5 s, how many electrons pass through the chest?

58. • A point charge at the origin of a coordinate system produces the electric field $\vec{\mathbf{E}} = (36{,}000 \text{ N/C})\hat{\mathbf{x}}$ on the x axis at the location $x = -0.50$ m. Determine the sign and magnitude of the charge.

59. • **BIO Bumblebees and Static Cling** Bumblebees actively foraging in the field have been found to be electrically charged, due largely to air resistance as they fly. The charge carried by a bee, which can be as great as 93 pC, is thought to play a significant role in pollination—the bee can attract grains of pollen from a distance, like a charged comb attracting bits of paper. Given that the force required to detach pollen from an avocado stigma is 4.0×10^{-10} N, find the maximum distance at which the electrostatic force between a bee and a grain of pollen is sufficient to detach the pollen. Treat the bee and pollen as point charges, and assume that the pollen has a charge opposite in sign and equal in magnitude to the bee.

A white-tailed bumblebee with static cling (Problem 59)

60. •• At the moment, the number of electrons in your body is essentially the same as the number of protons, giving you a net charge of zero. Suppose, however, that this balance of charges is off by one percent in both you and your friend, who is one meter away. Estimate the magnitude of the electrostatic force each of you experiences, and compare it with your weight.

61. •• A small object of mass 0.0150 kg and charge 3.1 μC hangs from the ceiling by a thread. A second small object, with a charge of 4.2 μC, is placed 1.2 m vertically below the first charge. Find **(a)** the electric field at the position of the upper charge due to the lower charge and **(b)** the tension in the thread.

62. •• **IP** Consider a system of three point charges on the x axis. Charge 1 is at $x = 0$, charge 2 is at $x = 0.20$ m, and charge 3 is at $x = 0.40$ m. In addition, the charges have the following values; $q_1 = -19\ \mu$C, $q_2 = q_3 = +19\ \mu$C. **(a)** The electric field vanishes at some point on the x axis between $x = 0.20$ m and $x = 0.40$ m. Is the point of zero field (i) at $x = 0.30$ m, (ii) to the left of $x = 0.30$ m or (iii) to the right of $x = 0.30$ m? Explain. **(b)** Find the point where $E = 0$ between $x = 0.20$ m and $x = 0.40$ m.

63. •• **IP** Consider the system of three point charges described in the previous problem. **(a)** The electric field vanishes at two different points on the x axis. One point is between $x = 0.20$ m and $x = 0.40$ m. Is the second point located to the left of charge 1 or to the right of charge 3? Explain. **(b)** Find the value of x at the second point where $E = 0$.

64. •• The electric field at a radial distance of 47.7 cm from the thin charged wire shown in Figure 19–37 has a magnitude of 25,400 N/C. **(a)** Using the result given in Problem 55, what is the magnitude of the charge per length on this wire? **(b)** At what distance from the wire is the magnitude of the electric field equal to $\frac{1}{2}(25{,}400 \text{ N/C})$?

65. •• A system consisting entirely of electrons and protons has a net charge of 1.84×10^{-15} C and a net mass of 4.56×10^{-23} kg. How many **(a)** electrons and **(b)** protons are in this system?

66. •• **IP** Three charges are placed at the vertices of an equilateral triangle of side $a = 0.63$ m, as shown in **Figure 19–38**. Charges 1 and 3 are $+7.3\ \mu$C; charge 2 is $-7.3\ \mu$C. **(a)** Find the magnitude and direction of the net force acting on charge 3. **(b)** If charge 3 is moved to the origin, will the net force acting on it there be greater than, less than, or equal to the net force found in part (a)? Explain. **(c)** Find the net force on charge 3 when it is at the origin.

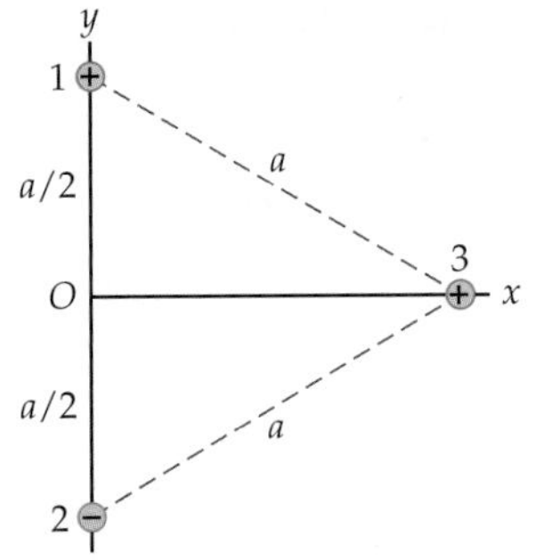

▲ **FIGURE 19–38** Problems 66 and 67

67. •• **IP** Consider the system of three charges described in the previous problem and shown in Figure 19–38. **(a)** Do you expect the net force acting on charge 1 to have a magnitude greater than, less than, or the same as the magnitude of the net force acting on charge 2? Explain. **(b)** Find the magnitude of the net force acting on charge 1. **(c)** Find the magnitude of the net force acting on charge 2.

68. •• **IP BIO Cell Membranes** The cell membrane in a nerve cell has a thickness of 0.12 μm. **(a)** Approximating the cell membrane as a parallel-plate capacitor with a surface charge density of 5.9×10^{-6} C/m^2, find the electric field within the membrane. **(b)** If the thickness of the membrane were doubled, would your answer to part (a) increase, decrease, or stay the same? Explain.

69. •• A square with sides of length L has a point charge at each of its four corners. Two corners that are diagonally opposite have charges equal to $+2.25\ \mu$C; the other two diagonal corners have charges Q. Find the magnitude and sign of the charges Q such that each of the $+2.25\ \mu$C charges experiences zero net force.

70. •• **IP** Suppose a charge $+Q$ is placed on the Earth, and another charge $+Q$ is placed on the Moon. **(a)** Find the value of Q needed to "balance" the gravitational attraction between the Earth and the Moon. **(b)** How would your answer to part (a) change if the distance between the Earth and the Moon were doubled? Explain.

71. •• Two small plastic balls hang from threads of negligible mass. Each ball has a mass of 0.14 g and a charge of magnitude q. The balls are attracted to each other, and the threads attached to the balls make an angle of 20.0° with the vertical,

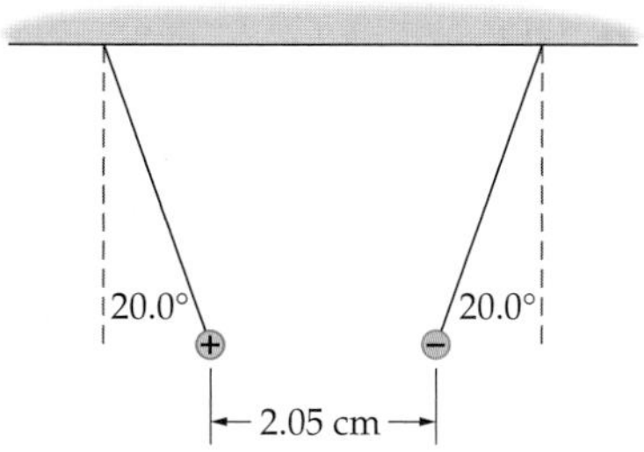

▲ **FIGURE 19–39** Problem 71

as shown in **Figure 19–39**. Find **(a)** the magnitude of the electric force acting on each ball, **(b)** the tension in each of the threads, and **(c)** the magnitude of the charge on the balls.

72. •• A small sphere with a charge of $+2.44\ \mu C$ is attached to a relaxed horizontal spring whose force constant is 89.2 N/m. The spring extends along the x axis, and the sphere rests on a frictionless surface with its center at the origin. A point charge $Q = -8.55\ \mu C$ is now moved slowly from infinity to a point $x = d > 0$ on the x axis. This causes the small sphere to move to the position $x = 0.124$ m. Find d.

73. •• Twelve identical point charges q are equally spaced around the circumference of a circle of radius R. The circle is centered at the origin. One of the twelve charges, which happens to be on the positive x axis, is now moved to the center of the circle. Find **(a)** the direction and **(b)** the magnitude of the net electric force exerted on this charge.

74. •• **BIO Nerve Impulses** When a nerve impulse propagates along a nerve cell, the electric field within the cell membrane changes from 7.0×10^5 N/C in one direction to 3.0×10^5 N/C in the other direction. Approximating the cell membrane as a parallel-plate capacitor, find the magnitude of the change in charge density on the walls on the cell membrane.

75. •• **IP The Electric Field of the Earth** The Earth produces an approximately uniform electric field at ground level. This electric field has a magnitude of 110 N/C and points radially inward, toward the center of the Earth. **(a)** Find the surface charge density (sign and magnitude) on the surface of the Earth. **(b)** Given that the radius of the Earth is 6.38×10^6 m, find the total electric charge on the Earth. **(c)** If the Moon had the same amount of electric charge distributed uniformly over its surface, would its electric field at the surface be greater than, less than, or equal to 110 N/C? Explain.

76. •• An object of mass $m = 3.7$ g and charge $Q = +44\ \mu C$ is attached to a string and placed in a uniform electric field that is inclined at an angle of 30.0° with the horizontal (**Figure 19–40**). The object is in static equilibrium when the string is horizontal. Find **(a)** the magnitude of the electric field and **(b)** the tension in the string.

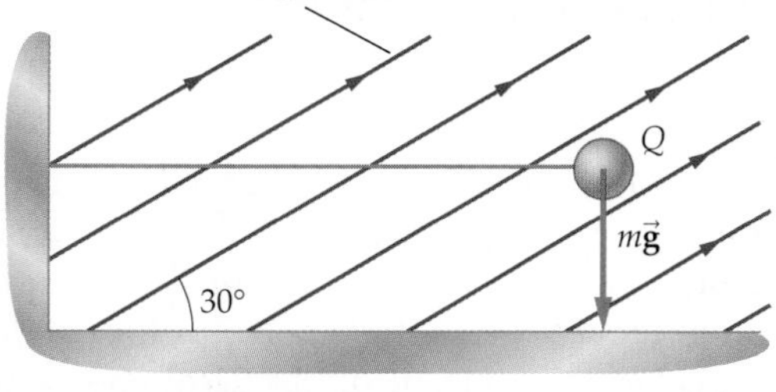

▲ **FIGURE 19–40** Problem 76

77. •• Four identical charges, $+Q$, occupy the corners of a square with sides of length a. A fifth charge, q, can be placed at any desired location. Find the location of the fifth charge, and the value of q, such that the net electric force acting on each of the original four charges, $+Q$, is zero.

78. •• **Figure 19–41** shows an electron entering a parallel-plate capacitor with a speed of 5.45×10^6 m/s. The electric field of the capacitor has deflected the electron downward by a distance of 0.618 cm at the point where the electron exits the capacitor. Find **(a)** the magnitude of the electric field in the capacitor and **(b)** the speed of the electron when it exits the capacitor.

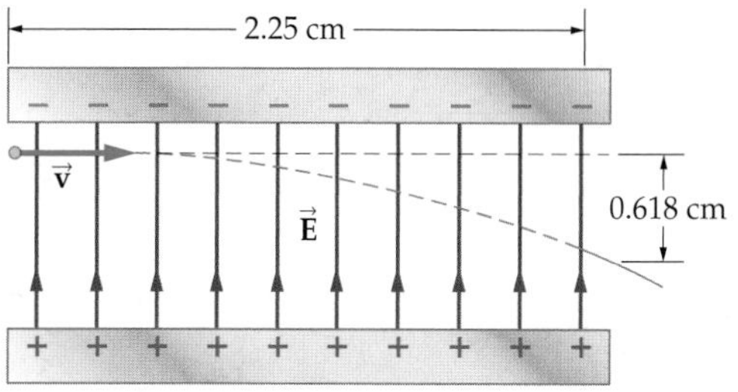

▲ **FIGURE 19–41** Problem 78

79. ••• Two identical conducting spheres are separated by a fixed center-to-center distance of 45 cm and have different charges. Initially, the spheres attract each other with a force of 0.095 N. The spheres are now connected by a thin conducting wire. After the wire is removed, the spheres are positively charged and repel one another with a force of 0.032 N. Find **(a)** the final and **(b)** the initial charges on the spheres.

Interactive Problems

80. •• **IP *Referring to Example 19–5*** Suppose $q_1 = +2.90\ \mu C$ is no longer at the origin, but is now on the y axis between $y = 0$ and $y = 0.500$ m. The charge $q_2 = +2.90\ \mu C$ is at $x = 0$ and $y = 0.500$ m, and the point 3 is at $x = y = 0.500$ m. **(a)** Is the magnitude of the net electric field at point 3, which we call E_{net}, greater than, less than, or equal to its previous value? Explain. **(b)** Is the angle θ that E_{net} makes with the x axis greater than, less than, or equal to its previous value? Explain. Find the new values of **(c)** E_{net} and **(d)** θ if q_1 is at $y = 0.250$ m.

81. •• **IP *Referring to Example 19–5*** In this system, the charge q_1 is at the origin, the charge q_2 is at $x = 0$ and $y = 0.500$ m, and the point 3 is at $x = y = 0.500$ m. Suppose that $q_1 = +2.90\ \mu C$, but that q_2 is increased to a value greater than $+\ 2.90\ \mu C$. As a result, do **(a)** E_{net} and **(b)** θ increase, decrease, or stay the same? Explain. If $E_{net} = 1.66 \times 10^5$ N/C, find **(c)** q_2 and **(d)** θ.

82. •• **IP *Referring to Example 19–6*** The magnitude of the charge is changed until the angle the thread makes with the vertical is $\theta = 15.0°$. The electric field is 1.46×10^4 N/C and the mass of the object is 0.0250 kg. **(a)** Is the new magnitude of q greater than or less than its previous value? Explain. **(b)** Find the new value of q.

83. •• ***Referring to Example 19–6*** Suppose the magnitude of the electric field is adjusted to give a tension of 0.253 N in the thread. This will also change the angle the thread makes with the vertical. **(a)** Find the new value of E. **(b)** Find the new angle between the thread and the vertical.

20 Electric Potential and Electric Potential Energy

20–1	Electric Potential Energy and the Electric Potential	663
20–2	Energy Conservation	666
20–3	The Electric Potential of Point Charges	669
20–4	Equipotential Surfaces and the Electric Field	673
20–5	Capacitors and Dielectrics	677
20–6	Electrical Energy Storage	683

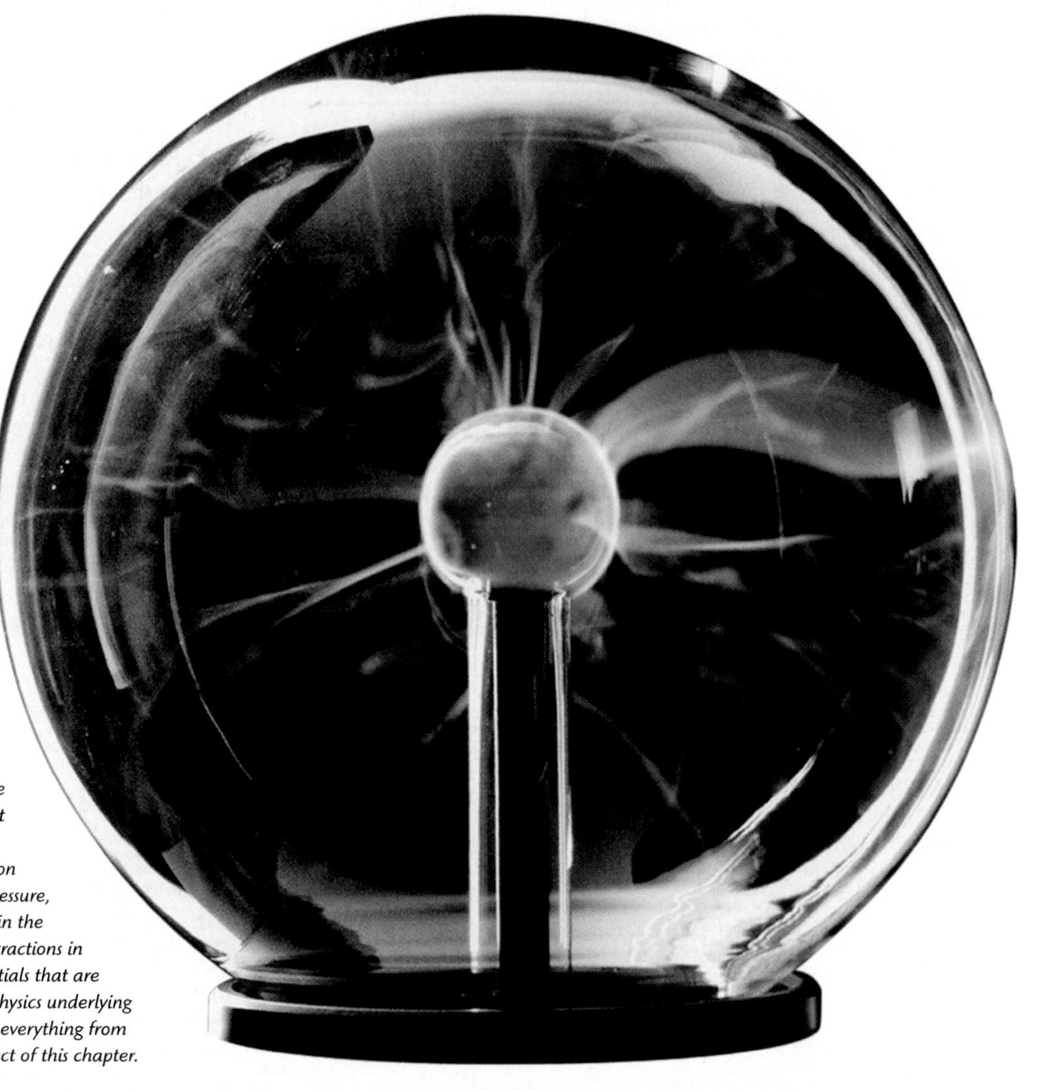

This familiar demonstration of "lightning in a jar" uses an electric potential of about 10,000 volts to create glowing filaments of low pressure gas that are eerily reminiscent of the elusive northern lights. If this same demonstration were to be carried out at atmospheric pressure, the required electric potential would be in the millions of volts. In contrast, muscle contractions in the human heart produce electric potentials that are only about a thousandth of a volt. The physics underlying electric potential—and its application to everything from lightning to auroras to EKGs—is the subject of this chapter.

When paramedics try to revive a heart-attack victim, they apply a jolt of electrical energy to the person's heart with a device known as a defibrillator. Before they can use the defibrillator, however, they must wait a few seconds for it to be "charged up." What exactly is happening as the defibrillator charges? The answer is that electric charge is building up on a capacitor, and in the process storing electrical energy. When the defibrillator is activated, the energy stored in the capacitor is released in a sudden surge of power that is often capable of saving a person's life. In this chapter we develop the concept of electrical energy and discuss how both energy and charge can be stored in a capacitor.

20–1 Electric Potential Energy and the Electric Potential

Electric and gravitational forces have many similarities. One of the most important of these is that both forces are conservative. As a result, there must be an **electric potential energy,** U, associated with the electric force, just as there is a gravitational potential energy, $U = mgy$, due to the force of gravity. (Recall that conservative forces and potential energies are discussed in Chapter 8.)

To illustrate the concept of electric potential energy, consider a uniform electric field, $\vec{\mathbf{E}}$, as shown in **Figure 20–1 (a)**. A positive test charge q_0 is placed in this field, where it experiences a downward electric force of magnitude $F = q_0E$. If the charge is moved upward through a distance d, the electric force and the displacement are in opposite directions; hence, the work done by the electric force is negative:

$$W = -q_0Ed$$

Using our definition of potential energy change given in Equation 8–1, $\Delta U = -W$, we find that the potential energy of the charge is changed by the amount

$$\Delta U = -W = q_0Ed \qquad \textbf{20–1}$$

Note that the electric potential energy increases, just as the gravitational potential energy of a ball increases when it is raised against the force of gravity to a greater altitude, as indicated in **Figure 20–1 (b)**.

On the other hand, if the charge q_0 is negative, the electric force acting on it will be upward. In this case, the electric force does *positive* work as the charge is raised through the distance d, and the change in potential energy is *negative*. Thus, the change in potential energy depends on the sign of a charge as well as on its magnitude.

The electric force depends on charge in the same way as does the change in electric potential energy. In the last chapter we found it convenient to define an electric field, $\vec{\mathbf{E}}$, as the force per charge:

$$\vec{\mathbf{E}} = \frac{\vec{\mathbf{F}}}{q_0}$$

Similarly, it is useful to define a quantity that is equal to the change in electric potential energy per charge, $\Delta U/q_0$. This quantity, which is independent of the test charge q_0, is referred to as the **electric potential,** V:

Definition of Electric Potential, *V*

$$\Delta V = \frac{\Delta U}{q_0} = \frac{-W}{q_0} \qquad \textbf{20–2}$$

SI unit: joule/coulomb = volt, V

In common usage, the electric potential is often referred to simply as the "potential."

PROBLEM-SOLVING NOTE

A "Potential" Cause of Confusion

When reading a problem statement, be sure to note whether it refers to the electric potential energy, U, the electric potential, V, or the potential—which is just a shorthand for the electric potential. The properties of these quantities are summarized below:

Quantity	Symbol	Units	Meaning
Electric potential energy	U	joules	energy associated with electric force
Electric potential	V	volts	electric potential energy per charge
Potential (shorthand for electric potential)	V	volts	same as electric potential

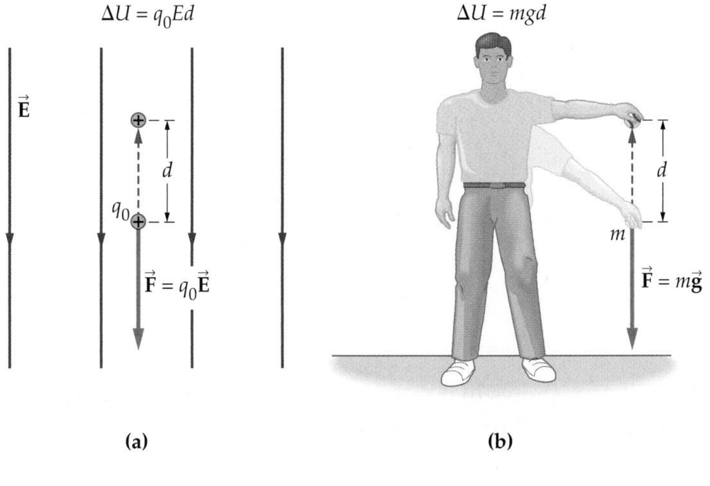

◀ FIGURE 20–1 Change in electric potential energy

(a) A positive test charge q_0 experiences a downward force due to the electric field $\vec{\mathbf{E}}$. If the charge is moved upward a distance d, the work done by the electric field is $-q_0Ed$. At the same time, the electric potential energy of the system increases by q_0Ed. The situation is analogous to that of an object in a gravitational field. **(b)** If a ball is lifted against the force exerted by gravity, the gravitational potential energy of the system increases.

PROBLEM-SOLVING NOTE

Electric Potential and Its Unit of Measurement

Be careful to distinguish between the electric potential V and the corresponding unit of measurement, the volt V, since both are designated by the same symbol. For example, in the expression $\Delta V = 12$ V, the first V refers to the electric potential, the second V refers to the units of the volt. Similarly, when someone says they have a "12-volt battery," what they really mean is that the battery produces a difference in electric potential, ΔV, of 12 volts.

Notice that our definition gives only the *change* in electric potential. As with gravitational potential energy, the electric potential can be set to zero at any desired location—only changes in electric potential are measurable. In addition, just as the potential energy is a scalar (that is, simply a number) so too is the potential V.

Potential energy is measured in joules, and charge is measured in coulombs; hence, the SI units of electric potential are joules per coulomb. This combination of units is referred to as the volt, in honor of Alessandro Volta (1745–1827), who invented a predecessor to the modern battery. To be specific, the volt (V) is defined as follows:

$$1 \text{ V} = 1 \text{ J/C} \qquad \textbf{20–3}$$

Rearranging slightly, we see that energy can be expressed as charge times voltage: $1 \text{ J} = (1 \text{ C})(1 \text{ V})$.

A convenient and commonly used unit of energy in atomic systems is the **electron volt** (eV), defined as the product of the electron charge and a potential difference of 1 volt:

$$1 \text{ eV} = (1.60 \times 10^{-19} \text{ C})(1 \text{ V}) = 1.60 \times 10^{-19} \text{ J}$$

It follows that the electron-volt is the energy change an electron experiences when it moves through a potential difference of 1 V. As we shall see in later chapters, typical atomic energies are in the eV range, whereas typical energies in nuclear systems are in the MeV (10^6 eV) range. Even the MeV is a minuscule energy, however, when compared to the joule.

EXERCISE 20–1

Find the change in electric potential energy, ΔU, as a charge of **(a)** 2.20×10^{-6} C or **(b)** -1.10×10^{-6} C moves from a point A to a point B, given that the change in electric potential between these points is $\Delta V = V_B - V_A = 24.0$ V.

Solution

a. Solving $\Delta V = \Delta U/q_0$ for ΔU, we find

$$\Delta U = q_0 \,\Delta V = (2.20 \times 10^{-6} \text{ C})(24.0 \text{ V}) = 5.28 \times 10^{-5} \text{ J}$$

b. Similarly, using $q_0 = -1.10 \times 10^{-6}$ C we obtain

$$\Delta U = q_0 \,\Delta V = (-1.10 \times 10^{-6} \text{ C})(24.0 \text{ V}) = -2.64 \times 10^{-5} \text{ J}$$

Electric Field and the Rate of Change of Electric Potential

There is a connection between the electric field and the electric potential that is both straightforward and useful. To obtain this relation, let's apply the definition $\Delta V = -W/q_0$ to the case of a test charge that moves through a distance Δs in the direction of the electric field, as in **Figure 20–2**. The work done by the electric field in this case is simply the magnitude of the electric force, $F = q_0 E$, times the distance, Δs:

$$W = q_0 E \Delta s$$

Therefore, the change in electric potential is

$$\Delta V = \frac{-W}{q_0} = \frac{-(q_0 E \Delta s)}{q_0} = -E\Delta s$$

Solving for the electric field, we find

Connection between the Electric Field and the Electric Potential

$$E = -\frac{\Delta V}{\Delta s} \qquad \textbf{20–4}$$

SI unit: volts/meter, V/m

This relation shows that the electric field, which can be expressed in units of N/C, also has the units of volts per meter. That is,

$$1 \text{ N/C} = 1 \text{ V/m} \qquad \textbf{20–5}$$

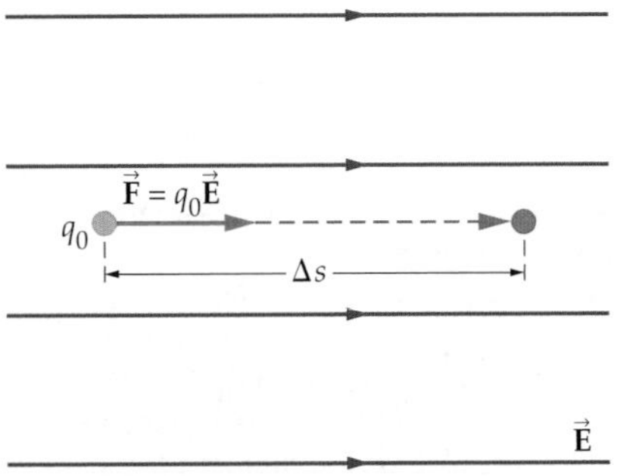

▲ **FIGURE 20–2 Electric field and electric potential**

As a charge q_0 moves in the direction of the electric field, $\vec{E}$, the electric potential, V, decreases. In particular, if the charge moves a distance Δs, the electric potential decreases by the amount $\Delta V = -E\Delta s$.

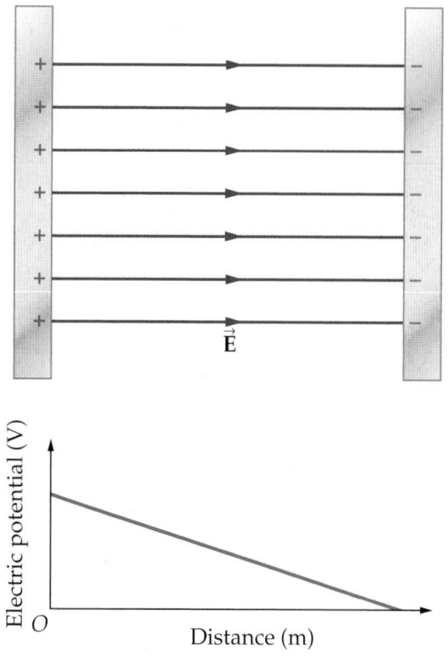

▶ **FIGURE 20–3 The electric potential for a constant electric field**
The electric potential, V, decreases as one moves in the direction of the electric field. In the case shown here, the electric field is constant; as a result, the electric potential decreases uniformly with distance. We have arbitrarily set the potential equal to zero at the right-hand plate.

To summarize, the electric field depends on *the rate of change of the electric potential with position.* In terms of our gravitational analogy, you can think of the potential V as the height of a hill and the electric field E as the slope of the hill.

In addition, it follows from Equation 20–4 that *the electric potential decreases as one moves in the direction of the electric field.* Specifically, notice that the change in potential, $\Delta V = -E\Delta s$, is negative when E and Δs are in the same direction—that is, when both E and Δs are positive or both are negative. For example, in cases where the electric field is *constant,* as between the plates of a parallel-plate capacitor (Section 19–5), the electric potential decreases *linearly* in the direction of the field. These observations are illustrated in **Figure 20–3**.

EXAMPLE 20–1 Plates at Different Potentials

A uniform electric field is established by connecting the plates of a parallel-plate capacitor to a 12-V battery. **(a)** If the plates are separated by 0.75 cm, what is the magnitude of the electric field in the capacitor? **(b)** A charge of $+6.24 \times 10^{-6}$ C moves from the positive plate to the negative plate. Find the change in electric potential energy for this charge. (In electrical systems we shall assume that gravity can be ignored, unless specifically instructed otherwise.)

Picture the Problem

Our sketch shows the parallel-plate capacitor connected to a 12-V battery. The battery guarantees that the potential difference between the plates is 12 V, with the positive plate at the higher potential. The separation of the plates is $d = 0.75$ cm $= 0.0075$ m, and the charge that moves from the positive to the negative plate is $q = +6.24 \times 10^{-6}$ C.

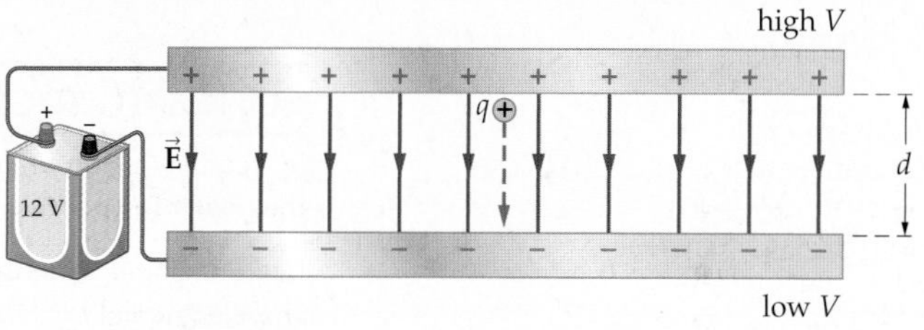

Strategy

a. The electric field can be calculated using $E = -\Delta V/\Delta s$. Note that if one moves in the direction of the field from the positive plate to the negative plate ($\Delta s = 0.75$ cm), the electric potential decreases by 12 V; that is, $\Delta V = -12$ V.

b. The change in electric potential energy is $\Delta U = q\,\Delta V$.

Solution

Part (a)

1. Substitute $\Delta s = 0.0075$ m and $\Delta V = -12$ V in $E = -\Delta V/\Delta s$:

$$E = -\frac{\Delta V}{\Delta s} = -\frac{(-12\text{ V})}{0.0075\text{ m}} = 1600\text{ V/m}$$

Part (b)

2. Evaluate $\Delta U = q\Delta V$:

$$\Delta U = q\Delta V = (6.24 \times 10^{-6}\text{ C})(-12\text{ V}) = -7.5 \times 10^{-5}\text{ J}$$

Insight

Note that the electric potential energy of the system decreases as the positive charge moves in the direction of the electric field, just as the gravitational potential energy of a ball decreases when it falls. In the next section, we show how this decrease in electric potential energy shows up as an increase in the charge's kinetic energy, just as the kinetic energy of a falling ball increases.

Practice Problem

Find the separation of the plates that results in an electric field of 2.0×10^3 V/m. [**Answer:** The separation should be 0.60 cm. Note that decreasing the separation *increases* the field.]

Some related homework problems: Problem 6, Problem 9

A similar problem is considered in the following Active Example.

ACTIVE EXAMPLE 20–1 Find the Electric Field and Potential Difference

The electric potential at point B in the parallel-plate capacitor shown below is less than the electric potential at point A by 4.50 V. The separation between points A and B is 0.120 cm, and the separation between the plates is 2.55 cm. Find **(a)** the electric field within the capacitor and **(b)** the potential difference between the plates.

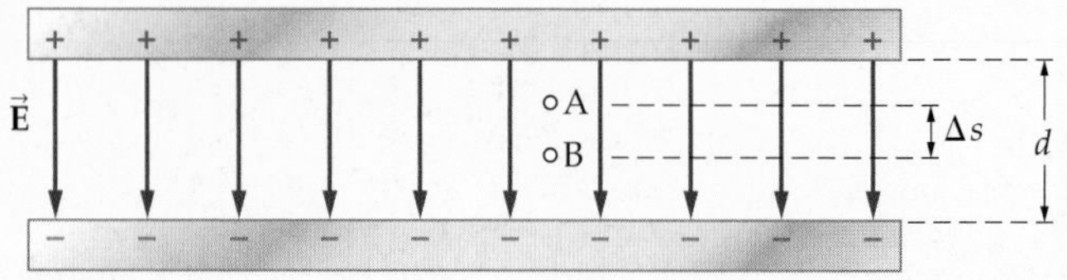

Solution *(Test your understanding by performing the calculations indicated in each step.)*

Part (a)

1. Calculate the electric field using $E = -\Delta V/\Delta s$, with $\Delta V = V_B - V_A = -4.50$ V and $\Delta s = 0.120$ cm: $E = 3750$ V/m

Part (b)

2. Calculate $\Delta V = V_{(-)} - V_{(+)}$ using $\Delta V = -E\,\Delta s$, with $E = 3750$ V/m and $\Delta s = 2.55$ cm: $\Delta V = -95.6$ V

Insight

Notice that in part (b) we use the same electric field found in part (a). This is valid because the field within an ideal parallel-plate capacitor is uniform; thus, the same value of E applies everywhere within the capacitor.

Finally, the negative value of ΔV in part (b) simply indicates that the electric potential of the negative plate is less than that of the positive plate by 95.6 V. Put another way, the electric potential increases as we move "against" the electric field, just as the gravitational potential energy increases when we move "against" gravity—that is, when we move uphill.

Your Turn

Is the electric potential energy of an electron at point A greater than or less than its electric potential energy at point B? Calculate the difference in electric potential energy as an electron moves from point A to point B.

(Answers to **Your Turn** *problems are given in the back of the book.)*

CONCEPTUAL CHECKPOINT 20–1 Constant Electric Potential

In a certain region of space the electric potential V is known to be constant. Is the electric field in this region **(a)** positive, **(b)** zero, or **(c)** negative?

Reasoning and Discussion

The electric field is related to the *rate of change* of the electric potential with position, not to the value of the potential. Since the rate of change of a constant potential is zero, so too is the electric field.

In particular, if one moves a distance Δs from one point to another in this region, the change in potential is zero; $\Delta V = 0$. Hence, the electric field vanishes; $E = -\Delta V/\Delta s = 0$.

Answer:

(b) The electric field is zero.

Finally, in our calculations of the potential difference $\Delta V = -E\Delta s$ in the examples to this point, we have always taken Δs to be a displacement in the direction of the electric field. More generally, Δs can be a displacement in any direction, as long as we replace E with the component of $\vec{E}$ in the direction of Δs. Thus, for example, the change in potential as we move in the x direction is $\Delta V = -E_x\,\Delta x$, and the corresponding change as we move in the y direction, is $\Delta V = -E_y\,\Delta y$. We shall point out the utility of these expressions in Section 20–4.

20–2 Energy Conservation

When a ball is dropped in a gravitational field, its gravitational potential energy decreases as it falls. At the same time, its kinetic energy increases. If nonconservative forces such as air resistance can be ignored, we know that the decrease in gravitational potential energy is equal to the increase in kinetic energy—in other words, the total energy of the ball is conserved.

Because the electric force is also conservative, the same considerations apply to a charged object in an electric field. Therefore, ignoring other forces, we can say that the total energy of an electric charge must be conserved. As a result, the sum of its kinetic and electric potential energies must be the same at any two points, say A and B:

$$K_A + U_A = K_B + U_B$$

Expressing the kinetic energy as $\frac{1}{2}mv^2$, we can write energy conservation as

$$\tfrac{1}{2}mv_A{}^2 + U_A = \tfrac{1}{2}mv_B{}^2 + U_B$$

This expression applies to any conservative force. In the case of a uniform gravitational field the potential energy is $U = mgy$; for an ideal spring it is $U = \frac{1}{2}kx^2$. When dealing with an electrical system, we can use Equation 20–2 to express the electric potential energy in terms of the electric potential as follows:

$$U = qV$$

To be specific, suppose a particle of mass $m = 1.75 \times 10^{-5}$ kg and charge $q = 5.20 \times 10^{-5}$ C is released from rest at a point A. As the particle moves to another point, B, the electric potential decreases by 60.0 V; that is, $V_A - V_B = 60.0$ V. The particle's speed at point B can be found using energy conservation. To do so, we first solve for the kinetic energy at point B:

$$\begin{aligned}\tfrac{1}{2}mv_B{}^2 &= \tfrac{1}{2}mv_A{}^2 + U_A - U_B \\ &= \tfrac{1}{2}mv_A{}^2 + q(V_A - V_B)\end{aligned}$$

Next we set $v_A = 0$, since the particle starts at rest, and solve for v_B:

$$\tfrac{1}{2}mv_B{}^2 = q(V_A - V_B)$$

$$v_B = \sqrt{\frac{2q(V_A - V_B)}{m}} = \sqrt{\frac{2(5.20 \times 10^{-5}\text{ C})(60.0\text{ V})}{1.75 \times 10^{-5}\text{ kg}}} = 18.9\text{ m/s} \qquad \textbf{20–6}$$

Thus, the decrease in electric potential energy appears as an increase in kinetic energy—and a corresponding increase in speed.

PROBLEM-SOLVING NOTE

Energy Conservation and Electric Potential Energy

Energy conservation applies to charges in an electric field. Therefore, to relate the speed or kinetic energy of a charge to its location simply requires setting the initial energy equal to the final energy. Note that the potential energy for a charge q is $U = qV$.

ACTIVE EXAMPLE 20–2 Find the Speed with a Running Start

Suppose the particle described in the preceding discussion has an initial speed of 5.00 m/s at point A. What is its speed at point B?

Solution *(Test your understanding by performing the calculations indicated in each step.)*

1. Write the equation for energy conservation:	$\frac{1}{2}mv_A{}^2 + qV_A = \frac{1}{2}mv_B{}^2 + qV_B$
2. Solve for the kinetic energy at point B:	$\frac{1}{2}mv_B{}^2 = \frac{1}{2}mv_A{}^2 + q(V_A - V_B)$
3. Solve for v_B:	$v_B = \sqrt{v_A{}^2 + 2q(V_A - V_B)/m}$
4. Substitute numerical values:	$v_B = 19.5$ m/s

Insight

The final speed is not 5.00 m/s greater than 18.9 m/s; in fact, it is only 0.6 m/s greater. As usual, this is due to the fact that the kinetic energy depends on v^2 rather than on v.

Your Turn

What initial speed is required to obtain a final speed of 18.9 m/s + 5.00 m/s = 23.9 m/s?

(Answers to **Your Turn** *problems are given in the back of the book.)*

Notice that in the preceding discussions a positive charge moves to a region where the electric potential is less, and its speed increases. As one might expect, the situation is just the opposite for a negative charge. In particular, a negative charge will move to a region of higher electric potential with an increase in speed.

For example, if the charge in Equation 20–6 is changed in sign to -5.20×10^{-5} C, and the potential difference is changed in sign to $V_A - V_B = -60.0$ V, the final speed remains the same. In general:

> *Positive* charges accelerate in the direction of *decreasing* electric potential;
>
> *Negative* charges accelerate in the direction of *increasing* electric potential.

In both cases, however, the charge moves to a region of lower electric potential energy.

EXAMPLE 20–2 From Plate to Plate

Suppose the charge in Example 20–1 is released from rest at the positive plate and that it reaches the negative plate with a speed of 3.4 m/s. What is **(a)** the mass of the charge and **(b)** its final kinetic energy?

Picture the Problem

The physical situation is the same as in Example 20–1. In this case, however, we know that the charge starts at rest at the positive plate, $v_A = 0$, and hits the negative plate with a speed of $v_B = 3.4$ m/s. Note that the electric potential of the positive plate is 12 V greater than that of the negative plate. Therefore $V_A - V_B = 12$ V.

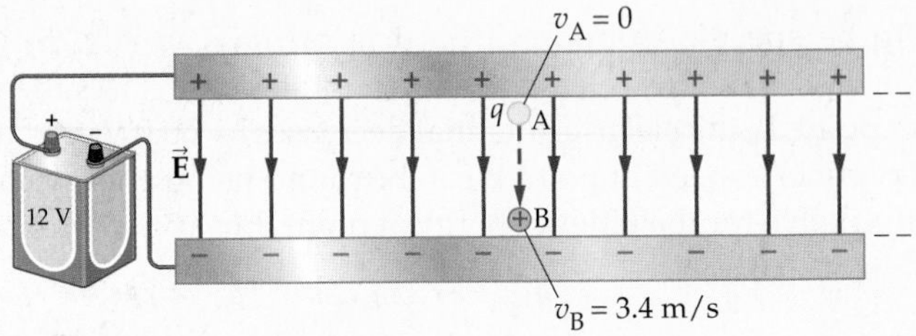

Strategy

a. The energy of the charge is conserved as it moves from one plate to the other. Setting the initial energy equal to the final energy gives us an equation in which there is only one unknown, the mass of the charge.

b. The final kinetic energy is simply $\frac{1}{2}mv_B^2$.

Solution

Part (a)

1. Apply energy conservation to this system:

$$\tfrac{1}{2}mv_A^2 + qV_A = \tfrac{1}{2}mv_B^2 + qV_B$$

2. Solve for the mass, m:

$$m = \frac{2q(V_A - V_B)}{v_B^2 - v_A^2}$$

3. Substitute numerical values:

$$m = \frac{2(6.24 \times 10^{-6}\ \text{C})(12\ \text{V})}{(3.4\ \text{m/s})^2 - 0} = 1.3 \times 10^{-5}\ \text{kg}$$

Part (b)

4. Calculate the final kinetic energy:

$$K_B = \tfrac{1}{2}mv_B^2$$
$$= \tfrac{1}{2}(1.3 \times 10^{-5}\ \text{kg})(3.4\ \text{m/s})^2 = 7.5 \times 10^{-5}\ \text{J}$$

Insight

We see that the final kinetic energy is precisely equal to the decrease in electric potential energy calculated in Example 20–1, as expected by energy conservation.

Practice Problem

If the mass of the charge had been 5.2×10^{-5} kg, what would be its **(a)** final speed and **(b)** final kinetic energy? [**Answer: (a)** $v_B = 1.7$ m/s, **(b)** $K_B = 7.5 \times 10^{-5}$ J. Note that the final kinetic energy is the same, regardless of the mass, because energy, not speed, is conserved.]

Some related homework problems: Problem 16, Problem 17

CONCEPTUAL CHECKPOINT 20–2 Final Speed

An electron, with a charge of -1.60×10^{-19} C, accelerates from rest through a potential difference V. A proton, with a charge of $+1.60 \times 10^{-19}$ C, accelerates from rest through a potential difference $-V$. Is the final speed of the electron **(a)** more than, **(b)** less than, or **(c)** the same as the final speed of the proton?

Reasoning and Discussion

The electron and proton have charges of equal magnitude, and therefore they have equal changes in electric potential energy. As a result, their final kinetic energies are equal. Since the electron has less mass than the proton, however, its speed must be greater.

Answer:

(a) The electron is moving faster than the proton.

20–3 The Electric Potential of Point Charges

Consider a point charge, $+q$, that is fixed at the origin of a coordinate system, as in **Figure 20–4**. Suppose, in addition, that a positive test charge, $+q_0$, is held at rest at point A, a distance r_A from the origin. At this location the test charge experiences a repulsive force with a magnitude given by Coulomb's law, $F = k|q_0||q|/r_A^2$.

If the test charge is now released, the repulsive force between it and the charge $+q$ will cause it to accelerate away from the origin. When it reaches an arbitrary point B, its kinetic energy will have increased by the same amount that its electric potential energy has decreased. Thus, we conclude that the electric potential energy is greater at point A than at point B. In fact, the methods of integral calculus can be used to show that the difference in electric potential energy between points A and B is

$$U_A - U_B = \frac{kq_0q}{r_A} - \frac{kq_0q}{r_B}$$

Note that the electric potential energy for point charges depends inversely on their separation, the same as the distance dependence of the gravitational potential energy for point masses (Equation 12–9).

The corresponding change in electric potential is found by dividing the electric potential energy by the test charge, q_0:

$$V_A - V_B = \frac{1}{q_0}(U_A - U_B) = \frac{kq}{r_A} - \frac{kq}{r_B}$$

If the test charge is moved infinitely far away from the origin, so that $r_B \rightarrow \infty$, the term kq/r_B vanishes, and the difference in electric potential becomes

$$V_A - V_B = \frac{kq}{r_A}$$

Since the potential can be set to zero at any convenient location, we choose to set V_B equal to zero; in other words, *we choose the electric potential to be zero infinitely far from a given charge*. With this choice, the potential of a point charge is $V_A = kq/r_A$. Dropping the subscript, we see that the electric potential at an arbitrary distance r is given by the following:

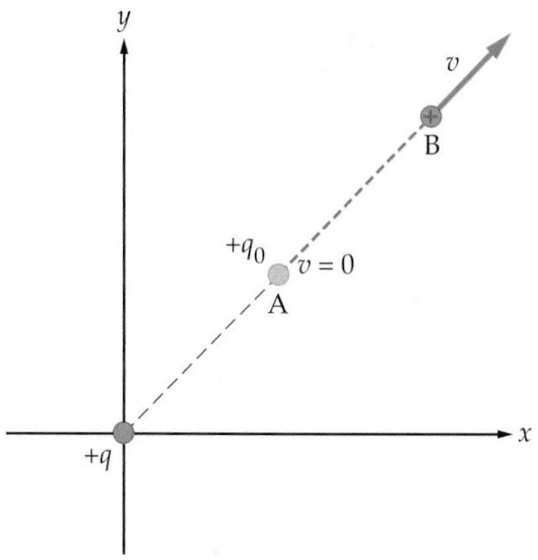

▲ **FIGURE 20–4 Energy conservation in an electrical system**

A test charge, $+q_0$, is released from rest at point A. When it reaches point B, its kinetic energy will have increased by the same amount that its electric potential energy has decreased.

Electric Potential for a Point Charge

$$V = \frac{kq}{r} \qquad 20\text{–}7$$

SI unit: volt, V

Recall that this expression for V actually represents a *change* in potential; in particular, V is the change in potential from a distance of infinity to a distance r. The corresponding difference in electric potential energy for the test charge q_0 is simply $U = q_0V$; that is,

Electric Potential Energy for Point Charges q and q_0 Separated by a Distance r

$$U = q_0V = \frac{kq_0q}{r} \qquad 20\text{–}8$$

SI unit: joule, J

Note that *the electric potential energy of two charges separated by an infinite distance is zero.*

One final point in regard to $V = kq/r$ and $U = kq_0q/r$ is that r is a *distance* and hence is always a positive quantity. For example, suppose a charge q is at the origin of a coordinate system. It follows that the potential at the point $x = 1$ m is the same as the potential at $x = -1$ m, since in both cases the distance is $r = 1$ m.

EXERCISE 20–2

Find the electric potential produced by a point charge of 6.80×10^{-7} C at a distance of 2.60 m.

Solution

Substituting $q = 6.80 \times 10^{-7}$ C and $r = 2.60$ m in $V = kq/r$ yields

$$V = \frac{kq}{r} = \frac{(8.99 \times 10^9\ \mathrm{N \cdot m^2/C^2})(6.80 \times 10^{-7}\ \mathrm{C})}{2.60\ \mathrm{m}} = 2350\ \mathrm{V}$$

Thus, the potential at $r = 2.60$ m due to this point charge is 2350 V greater than the potential at infinity.

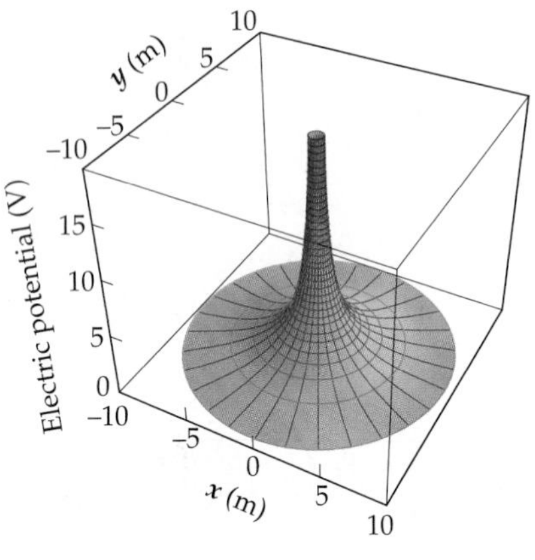

(a)

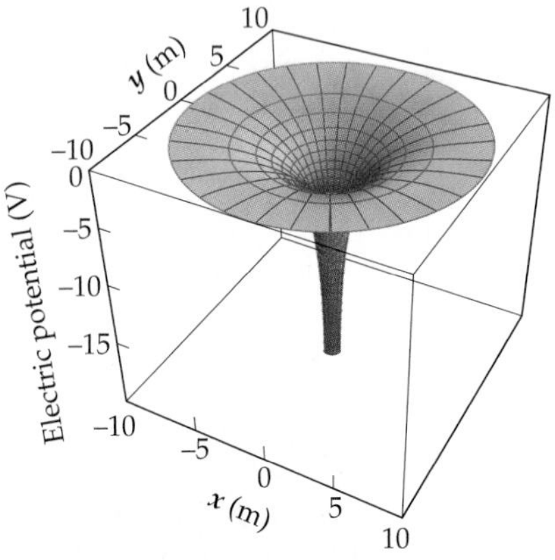

(b)

▲ **FIGURE 20–5 The electric potential of a point charge**

Electric potential near **(a)** a positive and **(b)** a negative charge at the origin. In the case of the positive charge, the electric potential forms a "potential hill" near the charge. Near the negative charge we observe a "potential well."

Note that V depends on the sign of the charge in question. This is shown in **Figure 20–5**, which represents the electric potential for a positive and a negative charge at the origin. The potential for the positive charge increases to positive infinity near the origin and decreases to zero far away, forming a "potential hill." On the other hand, the potential for the negative charge approaches negative infinity near the origin, forming a "potential well."

Thus, if the charge at the origin is positive, a positive test charge will move away from the origin, as if sliding "downhill" on the electric potential surface. Similarly, if the charge at the origin is negative, a positive test charge will move toward the origin, which again means that it slides downhill, this time into a potential well. Negative test charges, in contrast, always tend to slide "uphill" on electric potential curves, like bubbles rising in water.

Superposition of the Electric Potential

Like many physical quantities, the electric potential obeys a simple superposition principle. In particular:

> The total electric potential due to two or more charges is equal to the algebraic sum of the potentials due to each charge separately.

By *algebraic sum* we mean that the potential of a given charge may be positive or negative, and hence the *algebraic sign* of each potential must be taken into account when calculating the total potential. In particular, positive and negative contributions may cancel to give zero potential at a given location.

Finally, because the electric potential is a scalar, its superposition is as simple as adding numbers of various signs. This, in general, is easier than adding vectors, as is required for the superposition of electric fields.

EXAMPLE 20–3 Two Point Charges

A charge $q = 4.11 \times 10^{-9}$ C is placed at the origin, and a second charge equal to $-2q$ is placed on the x axis at the location $x = 1.00$ m. **(a)** Find the electric potential midway between the two charges. **(b)** The electric potential vanishes at some point between the charges; that is, for a value of x between 0 and 1.00 m. Find this value of x.

Picture the Problem

Our sketch shows the two charges placed on the x axis as described in the problem statement. Clearly, the electric potential is large and positive near the origin, and large and negative near $x = 1.00$ m. Thus, it follows that V must vanish at some point between $x = 0$ and $x = 1.00$ m.

A plot of V as a function of x is shown as an aid in visualizing the calculations given in this Example.

Strategy

a. As indicated in the sketch, an arbitrary point between $x = 0$ and $x = 1.00$ m is a distance x from the charge $+q$ and a distance $1.00 \text{ m} - x$ from the charge $-2q$. Thus, by superposition, the total electric potential at a point x is $V = kq/x + k(-2q)/(1.00 \text{ m} - x)$.

b. Setting $V = 0$ allows us to solve for the unknown, x.

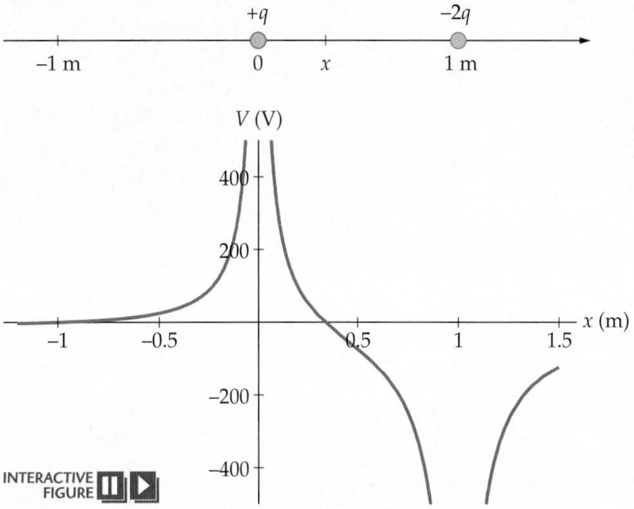

INTERACTIVE FIGURE

Solution

Part (a)

1. Use superposition to write an expression for V at an arbitrary point x between $x = 0$ and $x = 1.00$ m:

$$V = \frac{kq}{x} + \frac{k(-2q)}{1.00 \text{ m} - x}$$

2. Substitute numerical values into the expression for V. Note that the midway point between the charges is $x = 0.500$ m:

$$V = \frac{(8.99 \times 10^9 \text{ N}\cdot\text{m}^2/\text{C}^2)(4.11 \times 10^{-9} \text{ C})}{0.500 \text{ m}} + \frac{(8.99 \times 10^9 \text{ N}\cdot\text{m}^2/\text{C}^2)(-2)(4.11 \times 10^{-9} \text{ C})}{1.00 \text{ m} - 0.500 \text{ m}}$$

$$= -73.9 \text{ N}\cdot\text{m/C} = -73.9 \text{ V}$$

Part (b)

3. Set the expression for V in step 1 equal to zero, and simplify by canceling the common factor, kq:

$$V = \frac{kq}{x} + \frac{k(-2q)}{1.00 \text{ m} - x} = 0$$

$$\frac{1}{x} = \frac{2}{1.00 \text{ m} - x}$$

4. Solve for x:

$$1.00 \text{ m} - x = 2x$$

$$x = \tfrac{1}{3}(1.00 \text{ m}) = 0.333 \text{ m}$$

Insight

Suppose a small positive test charge is released from rest at $x = 0.500$ m. In which direction will it move? From the point of view of the Coulomb force, we know it will move to the right—repelled by the positive charge at the origin and attracted to the negative charge at $x = 1.00$ m. We come to the same conclusion when considering the electric potential, since we know that a positive test charge will "slide downhill" on the potential curve.

Practice Problem

The electric potential in this system also vanishes at a point on the negative x axis. Find this point. [**Answer:** $V = 0$ at $x = -1.00$ m. Note that the point $x = -1.00$ m is 1.00 m from the charge $+q$, and 2.00 m from the charge $-2q$. Hence, at $x = -1.00$ m we have $V = kq/(1.00 \text{ m}) + k(-2q)/(2.00 \text{ m}) = 0$.]

Some related homework problems: Problem 25, Problem 27

CONCEPTUAL CHECKPOINT 20–3 A Peak or a Valley?

Two point charges, each equal to $+q$, are placed on the x axis at $x = -1$ m and $x = +1$ m. As one moves along the x axis, does the potential look like a peak or a valley near the origin?

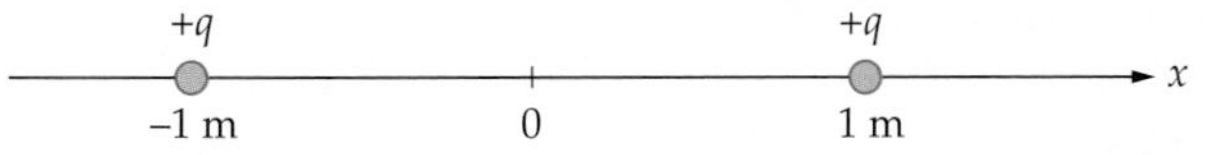

Reasoning and Discussion

We know that the potential is large and positive near each of the charges. As you move away from the charges, the potential tends to decrease. In particular, at very large positive or negative values of x the potential approaches zero.

Between $x = -1$ m and $x = +1$ m the potential has its lowest value when you are as far away from the two charges as possible. This occurs at the origin. Moving slightly to the left or the right simply brings you closer to one of the two charges, resulting in an increase in the potential; therefore, the potential has a minimum (bottom of a valley) at the origin. A plot of the potential for this case is shown below.

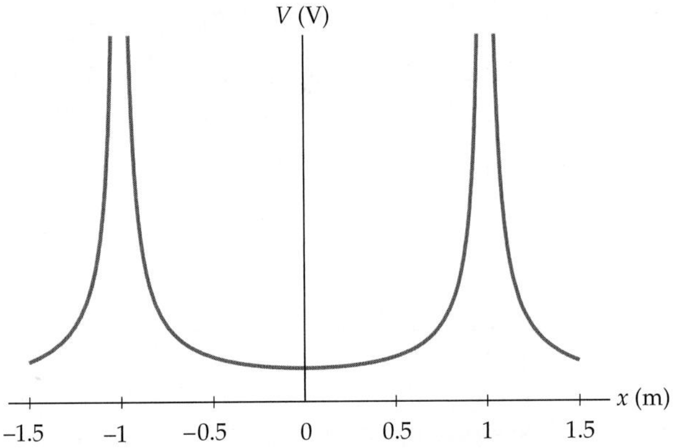

Finally, the electric field is associated with the slope of this curve, as mentioned in relation to Equation 20–4. Clearly, the field has a large magnitude near the charges and is zero at the origin.

Answer:

Near the origin, the potential looks like a valley.

PROBLEM-SOLVING NOTE

Superposition for the Electric Potential

Recall that to find the total electric potential due to a system of charges, you need only sum the potentials due to each charge separately—being careful to take into account the appropriate sign. Note that there are no components to deal with, as there are when we superpose electric fields.

Superposition applies equally well to the electric potential energy. The following Example illustrates superposition for both the electric potential and the electric potential energy.

EXAMPLE 20–4 Fly Away

Two charges, $+q$ and $+2q$, are held in place on the x axis at the locations $x = -d$ and $x = +d$, respectively. A third charge, $+3q$, is released from rest on the y axis at $y = d$. **(a)** Find the electric potential due to the first two charges at the initial location of the third charge. **(b)** Find the initial electric potential energy of the third charge. **(c)** What is the kinetic energy of the third charge when it has moved infinitely far away from the other two charges?

Picture the Problem

As indicated in the sketch, the third charge, $+3q$, is separated from the other two charges by the initial distance, $\sqrt{2}d$. When the third charge is released, the repulsive forces due to $+q$ and $+2q$ will cause it to move away to an infinite distance.

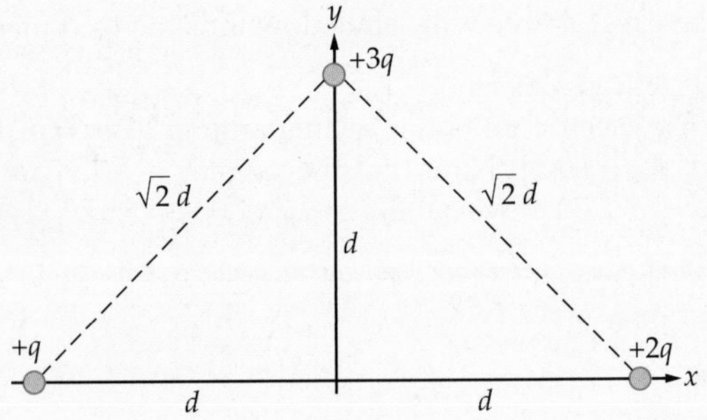

Strategy

a. The electric potential at the initial position of the third charge is the sum of the potentials due to $+q$ and $+2q$, each at a distance of $\sqrt{2}d$.

b. The initial electric potential energy of the third charge, U_i, is simply its charge, $+3q$, times the potential, V, found in part (a).

c. We can find the final kinetic energy using energy conservation; $U_i + K_i = U_f + K_f$. Because $K_i = 0$ (third charge starts at rest) and $U_f = 0$ (third charge infinitely far away), we find that $K_f = U_i$.

Solution

Part (a)

1. Calculate the net electric potential at the initial position of the third charge:

$$V_i = \frac{k(+q)}{\sqrt{2}d} + \frac{k(+2q)}{\sqrt{2}d} = \frac{3kq}{\sqrt{2}d}$$

Part (b)

2. Multiply V by $(+3q)$ to find the initial electric potential energy, U_i, of the third charge:

$$U_i = (+3q)V_i = (+3q)\frac{3kq}{\sqrt{2}d} = \frac{9kq^2}{\sqrt{2}d}$$

Part (c)

3. Use energy conservation to find the final (infinite separation) kinetic energy:

$$U_i + K_i = U_f + K_f$$

$$\frac{9kq^2}{\sqrt{2}d} + 0 = 0 + K_f$$

$$K_f = \frac{9kq^2}{\sqrt{2}d}$$

Insight

If the third charge had started closer to the other two charges, it would have been higher up on the "potential hill," and hence its kinetic energy at infinity would have been greater, as shown in the following Practice Problem.

Practice Problem

Suppose the third charge is released from rest just above the origin. What is its final kinetic energy in this case? [**Answer:** $K_f = U_i = 9kq^2/d$. As expected, this is greater than $9kq^2/\sqrt{2}d$.]

Some related homework problems: Problem 28, Problem 33

The electric potential energy for a pair of charges, q_1 and q_2, separated by a distance r is $U = kq_1q_2/r$. In systems that contain more than two charges, the total electric potential energy is the sum of terms like $U = kq_1q_2/r$ for each pair of charges in the system. This procedure is illustrated in the following Active Example.

ACTIVE EXAMPLE 20–3 Find the Electric Potential Energy

A system consists of the charges $-q$ at $(-d, 0)$, $+2q$ at $(d, 0)$, and $+3q$ at $(0, d)$. What is the total electric potential energy of the system?

Solution *(Test your understanding by performing the calculations indicated in each step.)*

1. Write the electric potential energy between $-q$ and $+2q$: $-kq(2q)/2d$

2. Write the electric potential energy between $-q$ and $+3q$: $-kq(3q)/\sqrt{2}d$

3. Write the electric potential energy between $+2q$ and $+3q$: $k(2q)(3q)/\sqrt{2}d$

4. Sum the contributions from each pair of charges to find the total electric potential energy: $U = 3kq^2/\sqrt{2}d - kq^2/d$

Your Turn

What is the total electric potential energy of this system if the charge $+3q$ is moved from $(0, d)$ to $(0, -d)$?

(Answers to **Your Turn** *problems are given in the back of the book.)*

20–4 Equipotential Surfaces and the Electric Field

A contour map is a useful tool for serious hikers and backpackers. The first thing you notice when looking at a map such as the one shown in Figure 8–13 is a series of closed curves—the contours—each denoting a different altitude. When the contours are closely spaced, the altitude changes rapidly, indicating steep terrain. Conversely, widely spaced contours indicate a fairly flat surface.

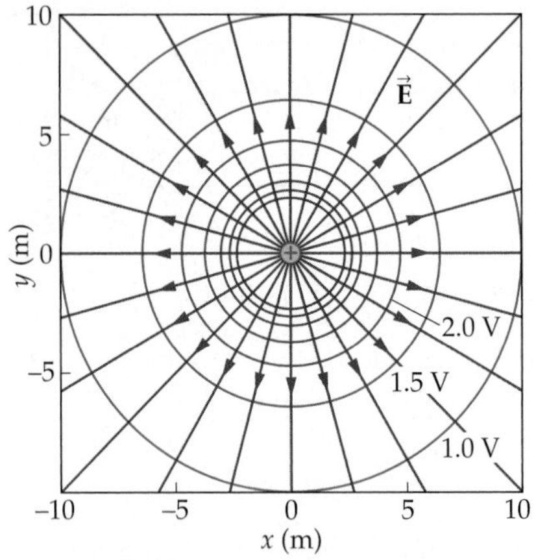

▲ **FIGURE 20–6 Equipotentials for a point charge**

Equipotential surfaces for a positive point charge located at the origin. Near the origin, where the equipotentials are closely spaced, the potential varies rapidly with distance and the electric field is large. This is a top view of Figure 20–5 (a).

A similar device can help visualize the electric potential due to one or more electric charges. Consider, for example, a single positive charge located at the origin. As we saw in the previous section, the electric potential due to this charge approaches zero far from the charge and rises to form an infinitely high "potential hill" near the charge. A three-dimensional representation of the potential is plotted in Figure 20–5. The same potential is shown as a contour map in **Figure 20–6**. In this case, the contours, rather than representing altitude, indicate the value of the potential. Since the value of the potential at any point on a given contour is equal to the value at any other point on the same contour, we refer to the contours as **equipotential surfaces,** or simply, **equipotentials.**

An equipotential plot also contains important information about the magnitude and direction of the electric field. For example, in Figure 20–6 we know that the electric field is more intense near the charge, where the equipotentials are closely spaced, than it is far from the charge, where the equipotentials are widely spaced. This simply illustrates that the electric field, $E = -\Delta V/\Delta s$, depends on the rate of change of the potential with position—the greater the change in potential, ΔV, over a given distance, Δs, the larger the magnitude of E.

The direction of the electric field is given by the minus sign in $E = -\Delta V/\Delta s$. For example, if the electric potential decreases over a distance Δs, it follows that ΔV is negative; $\Delta V < 0$. Thus, $-\Delta V$ is positive, and hence $E > 0$. To summarize:

> The electric field points in the direction of decreasing electric potential.

This is also illustrated in Figure 20–6, where we see that the electric field points away from the charge $+q$, in the direction of decreasing electric potential.

Not only does the field point in the direction of decreasing potential, it is, in fact, *perpendicular* to the equipotential surfaces:

> The electric field is always perpendicular to the equipotential surfaces.

To see why this is the case, we note that zero work is done when a charge is moved perpendicular to an electric field. That is, the work $W = Fd\cos\theta$ is zero when the angle θ is 90°. If zero work is done, it follows from $\Delta V = -W/q_0$ (Equation 20–2) that there is no change in potential. Therefore, the potential is constant (equipotential) in a direction perpendicular to the electric field.

CONCEPTUAL CHECKPOINT 20–4 Equipotential Surfaces

Is it possible for equipotential surfaces to intersect?

Reasoning and Discussion

To answer this question, it is useful to consider the analogous case of contours on a contour map (Figure 8–13). As we know, each contour corresponds to a different altitude. Because each point on the map has only a single value of altitude, it follows that it is impossible for contours to intersect.

Precisely the same reasoning applies to the electric potential, and hence equipotential surfaces never intersect.

Answer:

No. Equipotential surfaces cannot intersect.

▲ **FIGURE 20–7 Equipotential surfaces for a uniform electric field**

The electric field is perpendicular to the equipotentials and points in the direction of decreasing potential.

The graphical relationship between equipotentials and electric fields can be illustrated with a few examples. We begin with the simple case of a uniform electric field, as between the plates of a parallel-plate capacitor. In **Figure 20–7**, we plot both the electric field lines, which point in the positive x direction, and the corresponding equipotential surfaces. As expected, the field lines are perpendicular to the equipotentials, and $\vec{\mathbf{E}}$ points in the direction of decreasing V.

To see this last result more clearly, note that $\vec{\mathbf{E}} = E\hat{\mathbf{x}}$ in Figure 20–7, from which it follows that $E_x = E > 0$ and $E_y = 0$. Referring to Section 20–1, we see that if we move in the positive x direction, the change in electric potential is negative,

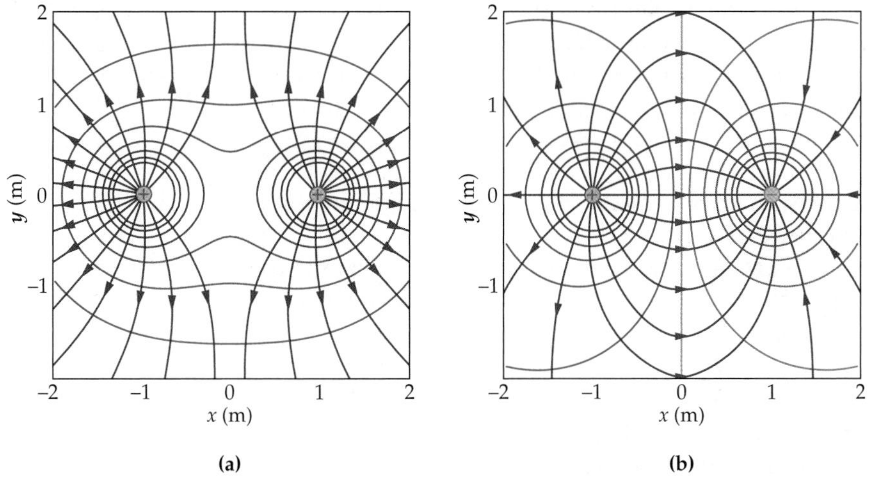

◀ **FIGURE 20–8 Equipotential surfaces for two point charges**
(a) In the case of two equal positive charges, the electric field between them is weak because the field produced by one charge effectively cancels the field produced by the other. As a result, the electric potential is practically constant between the charges. **(b)** For equal charges of opposite sign (a dipole), the electric field is strong between the charges, and the potential changes rapidly.

$\Delta V = -E_x \Delta x = -E\Delta x < 0$, as expected. On the other hand, if we move in the y direction, the change in potential is zero, $\Delta V = -E_y \Delta y = 0$. This is why the equipotentials, which are always perpendicular to $\vec{\mathbf{E}}$, are parallel to the y axis in Figure 20–7.

Figure 20–8 (a) shows a similar plot for the case of two positive charges of equal magnitude. Notice that the electric field lines always cross the equipotentials at right angles. In addition, the electric field is more intense where the equipotential surfaces are closely spaced. In the region midway between the two charges, where the electric field is essentially zero, the potential is practically constant.

Finally, in **Figure 20–8 (b)** we show the equipotentials for two charges of opposite sign, one $+q$ (at $x = -1$ m) and the other $-q$ (at $x = +1$ m), forming an electric dipole. In this case, the amber-color equipotentials denote negative values of V. Teal-color equipotentials correspond to positive V, and the tan-color equipotential has the value $V = 0$. The electric field is nonzero between the charges, even though the potential is zero there—recall that $E = -\Delta V/\Delta s$ is related to the *rate of change of V*, not to its value. The relatively large number of equipotential surfaces between the charges shows that V is indeed *changing rapidly* in that region.

Ideal Conductors

A charge placed on an ideal conductor is free to move. As a result, a charge can be moved from one location on a conductor to another with no work being done. Since the work is zero, the change in potential is also zero; therefore, every point on or within an ideal conductor is at the same potential:

> Ideal conductors are equipotential surfaces; every point on or within such a conductor is at the same potential.

For example, when a charge Q is placed on a conductor, as in **Figure 20–9**, it distributes itself over the surface in such a way that the potential is the same everywhere. If the conductor has the shape of a sphere, the charge spreads uniformly over its surface, as in Figure 20–9 (a). If, however, the conductor has a sharp end and a blunt end, as in Figure 20–9 (b), the charge is more concentrated near the sharp end, which results in a large electric field. We pointed out this effect earlier, in Section 19–6.

To see why this should be the case, consider a conducting sphere of radius R with a charge Q distributed uniformly over its surface, as in **Figure 20–10 (a)**. The charge density on this sphere is $\sigma = Q/4\pi R^2$, and the potential at its surface is the same as for a point charge Q at the center of the sphere; that is, $V = kQ/R$. Writing this in terms of the surface charge density, we have

$$V = \frac{kQ}{R} = \frac{k\sigma(4\pi R^2)}{R} = 4\pi k\sigma R$$

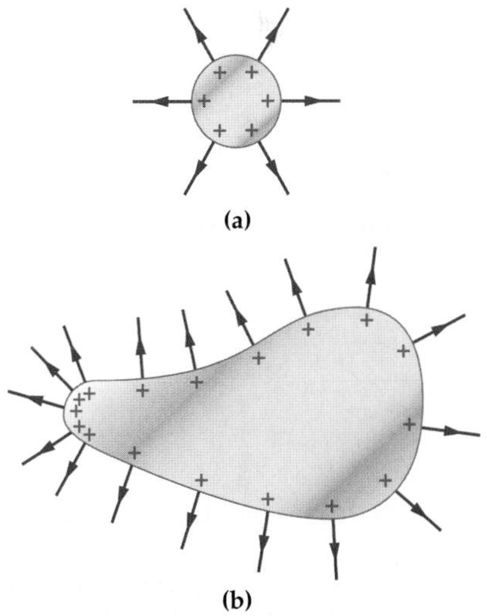

▲ **FIGURE 20–9 Electric charges on the surface of ideal conductors**
(a) On a spherical conductor, the charge is distributed uniformly over the surface. **(b)** On a conductor of arbitrary shape, the charge is more concentrated, and the electric field is more intense, where the conductor is more sharply curved. Note that in all cases the electric field is perpendicular to the surface of the conductor.

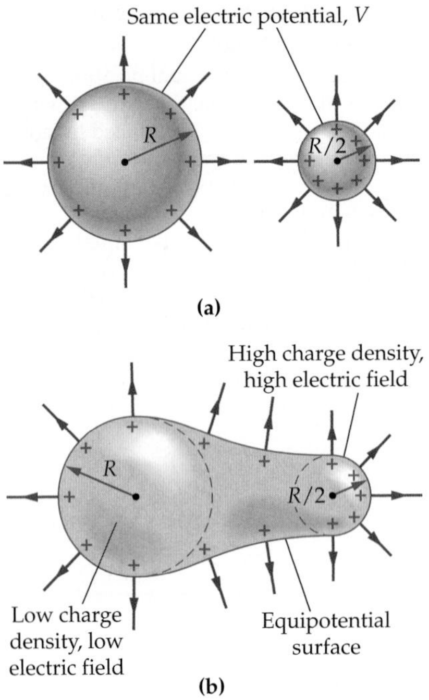

▲ **FIGURE 20–10 Charge concentration near points**

(a) If two spheres of different radii have the same electric potential at their surfaces, the sphere with the smaller radius of curvature has the greater charge density and the greater electric field. **(b)** An arbitrarily shaped conductor can be approximated by spheres with the same potential at the surface and varying radii of curvature. It follows that the more sharply curved end of a conductor has a greater charge density and a more intense field.

Now, consider a sphere of radius $R/2$. For this sphere to have the *same potential* as the large sphere, it must have twice the charge density, 2σ. In particular, letting σ go to 2σ and R go to $R/2$ in the expression $V = 4\pi k\sigma R$ yields the same result as for the large sphere:

$$V = 4\pi k(2\sigma)(R/2) = 4\pi k\sigma R$$

Clearly, then, the smaller the radius of a sphere with potential V, the larger its charge density.

The electric field is greater for the small sphere as well. At the surface of the large sphere in Figure 20–10 (a) the electric field has a magnitude given by

$$E = \frac{kQ}{R^2} = \frac{k\sigma(4\pi R^2)}{R^2} = 4\pi k\sigma$$

The small sphere in Figure 20–10 (a) has twice the charge density, hence it has twice the electric field at its surface.

The relevance of these results for a conductor of arbitrary shape can be seen in **Figure 20–10 (b)**. Here we show that even an arbitrarily shaped conductor has regions that approximate a spherical surface. In this case, the left end of the conductor is approximately a portion of a sphere of radius R, and the right end is approximately a sphere of radius $R/2$. Noting that every point on a conductor is at the *same potential*, we see that the situation in Figure 20–10 (b) is similar to that in Figure 20–10 (a). It follows that the sharper end of the conductor has the greater charge density and the greater electric field. If a conductor has a sharp point, as in a lightning rod, the corresponding radius of curvature is very small, which can result in an enormous electric field. We consider this possibility in more detail in the next section.

Finally, we note that because the surface of an ideal conductor is an equipotential, the electric field must meet the surface at right angles. This is also shown in Figures 20–9 and 20–10.

Electric Potential and the Human Body

The human body is a relatively good conductor of electricity, but it is not an ideal conductor. If it were, the entire body would be one large equipotential surface. Instead, muscle activity, the beating of the heart, and nerve impulses in the brain all lead to slight differences in electric potential from one point on the skin to another.

REAL-WORLD PHYSICS: BIO
Electrocardiograph

In the case of the heart, the powerful waves of electrical activity as the heart muscles contract result in potential differences that are typically in the range of 1 mV. These potential differences can be detected and displayed with an instrument known as an **electrocardiograph,** abbreviated ECG or EKG. (The K derives from the original Dutch name for the device.) A typical EKG signal is displayed in **Figure 20–11 (a)**.

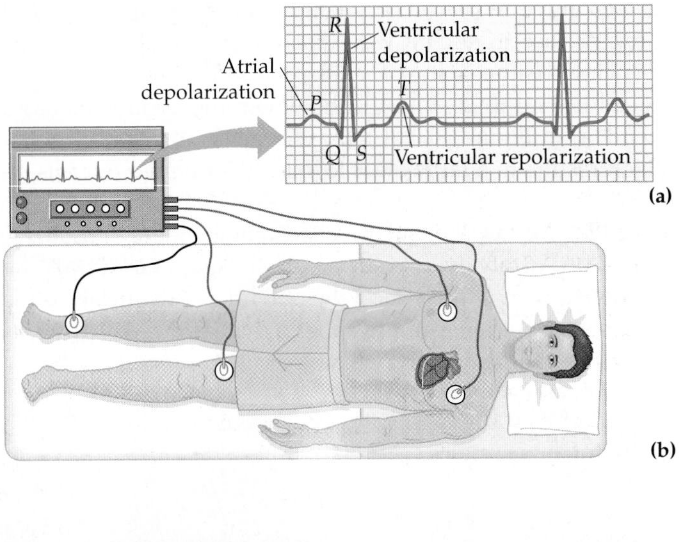

▶ **FIGURE 20–11 The electrocardiograph**

(a) A typical electrocardiograph tracing, with the major features labeled. The EKG records the electrical activity that accompanies the rhythmic contraction and relaxation of heart muscle tissue. The main pumping action of the heart is associated with the *QRS* complex: contraction of the ventricles begins just after the *R* peak. **(b)** The simplest arrangement of electrodes, or "leads," for an EKG. More precise information can be obtained by the use of additional leads—typically, a dozen in modern practice.

To understand the various features of an EKG signal, we note that a heartbeat begins when the heart's natural "pacemaker," the sinoatrial (SA) node in the right atrium, triggers a wave of muscular contraction across both atria that pumps blood into the ventricles. This activity gives rise to the pulse known as the *P wave* in Figure 20–11 (a). Following this contraction, the atrioventricular (AV) node, located between the ventricles, initiates a more powerful wave of contraction in the ventricles, sending oxygen-poor blood to the lungs and oxygenated blood to the rest of the body. These events are reflected in the series of pulses known as the *QRS complex*. Other features in the EKG signal can be interpreted as well. For example, the *T wave* is associated with repolarization of the ventricles in preparation for the next contraction cycle. Even small irregularities in the magnitude, shape, sequence, or timing of these features can provide an experienced physician with essential clues for the diagnosis of many cardiac abnormalities and pathologies. For example, an inverted *T* wave often indicates inadequate blood supply to the heart muscle (ischemia), possibly caused by a heart attack (myocardial infarction).

To record these signals, electrodes are attached to the body in a variety of locations. In the simplest case, illustrated in **Figure 20–11 (b)**, three electrodes are connected in an arrangement known as *Einthoven's triangle*, after the Dutch pioneer of these techniques. Two of these three electrodes are attached to the shoulders, the third to the left groin. For convenience, the two shoulder electrodes are often connected to the wrists, and the groin electrode is connected to the left ankle—the signal at these locations is practically the same as for the locations in Figure 20–11 (b) because of the high conductivity of the human limbs. By convention, the electrode on the right leg is used as the ground. In current applications of the EKG, it is typical to use 12 electrodes to gain more detailed information about the heart's activity.

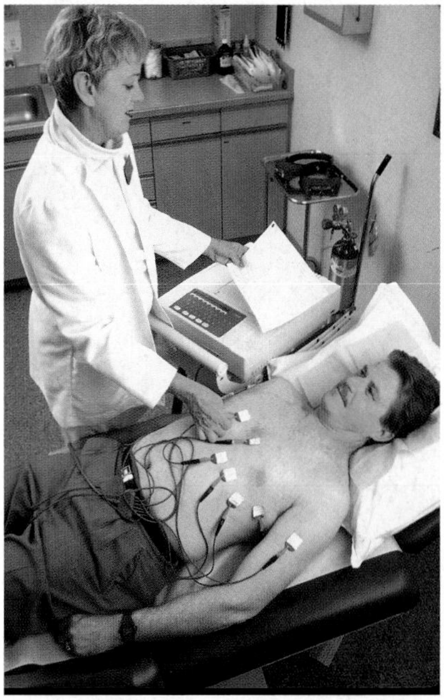

▲ An electrocardiograph can be made with as few as three electrodes and a ground. However, more detailed information about the heart's electrical activity can be obtained by using additional "leads" at precise anatomical locations. Modern EKGs often use a 12-lead array.

Electrical activity in the brain can be detected and displayed with an **electroencephalograph** or EEG. In a typical application of this technique, a regular array of 8 to 16 electrodes is placed around the head. The potential differences in this case—in the range of 0.1 mV—are much smaller than those produced by the heart, and much more complex to interpret. One of the key characteristics of an EEG signal, such as the one shown in **Figure 20–12**, is the frequency of the waves. For example, waves at 0.5 to 3.5 Hz, referred to as *D waves*, are common during sleep. A relaxed brain produces *a waves*, with frequencies in the range of 8 to 13 Hz, and an alert brain generates *b waves*, with frequencies greater than 13 Hz. Finally, *q waves*—with frequencies of 5 to 8 Hz—are common in newborns but indicate severe stress in adults.

REAL-WORLD PHYSICS: BIO
The electroencephalograph

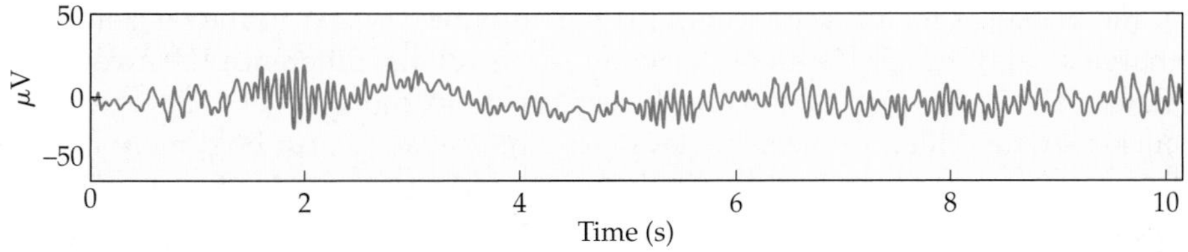

◀ **FIGURE 20–12 The electroencephalograph**
A typical EEG signal.

20–5 Capacitors and Dielectrics

A **capacitor** gets its name from the fact that it has a *capacity* to store both electric charge and energy. Capacitors are a common and important element in modern electronic devices. They can provide large bursts of energy to a circuit, or protect delicate circuitry from excess charge originating elsewhere.

In general, a capacitor is nothing more than two conductors, referred to as *plates*, separated by a finite distance. When the plates of a capacitor are connected to the terminals of a battery, they become charged—one plate acquires a charge $+Q$, the other an equal and opposite charge, $-Q$. The greater the charge Q for a given voltage V, the greater the **capacitance,** C, of the capacitor.

To be specific, suppose a certain battery produces a potential difference of V volts between its terminals. When this battery is connected to a capacitor, a charge of magnitude Q appears on each plate. If a different battery, with a voltage of $2V$, is connected to the same capacitor, the charge on the plates doubles in magnitude to $2Q$. Thus, the charge Q is proportional to the applied voltage V. We define the constant of proportionality to be the capacitance C:

$$Q = CV$$

Solving for the capacitance, we have

Definition of Capacitance, *C*

$$C = \frac{Q}{V} \qquad \textbf{20–9}$$

SI unit: coulomb/volt = farad, F

In this expression, Q is the magnitude of the charge on either plate, and V is the magnitude of the voltage difference between the plates. By definition, then, the capacitance is always a positive quantity.

As we can see from the relation $C = Q/V$, the units of capacitance are coulombs per volt. In the SI system this combination of units is referred to as the farad (F), in honor of the English physicist Michael Faraday (1791–1867), a pioneering researcher into the properties of electricity and magnetism. In particular,

$$1\ \text{F} = 1\ \text{C/V} \qquad \textbf{20–10}$$

Just as the coulomb is a rather large unit of charge, so too is the farad a rather large unit of capacitance. More typical values for capacitance are in the picofarad ($1\ \text{pF} = 10^{-12}\ \text{F}$) to microfarad ($1\ \mu\text{F} = 10^{-6}\ \text{F}$) range.

EXERCISE 20–3

A capacitor of 0.75 μF is charged to a voltage of 16 V. What is the magnitude of the charge on each plate of the capacitor?

Solution

Using $Q = CV$, we find

$$Q = CV = (0.75 \times 10^{-6}\ \text{F})(16\ \text{V}) = 1.2 \times 10^{-5}\ \text{C}$$

A bucket of water provides a useful analogy when thinking about capacitors. In this analogy we make the following identifications: (i) The cross-sectional area of the bucket is the capacitance, C; (ii) the amount of water in the bucket is the charge Q; and (iii) the depth of the water is the voltage difference V between the plates. Therefore, just as a wide bucket can hold more water than a narrow bucket—when filled to the same level—a large capacitor can hold more charge than a small capacitor when they both have the same potential difference. Charging a capacitor, then, is like pouring water into a bucket—if the capacitance is large, a large amount of charge can be placed on the plates with little potential difference between them.

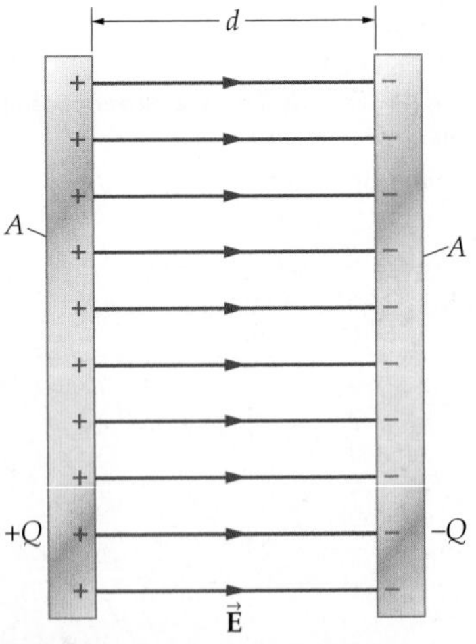

▲ **FIGURE 20–13 A parallel-plate capacitor**

A parallel-plate capacitor, with plates of area A, separation d, and charges of magnitude Q. The capacitance of such a capacitor is $C = \varepsilon_0 A/d$.

Parallel-Plate Capacitor

A particularly simple capacitor is the parallel-plate capacitor, first introduced in Section 19–5. In this device, two parallel plates of area A are separated by a distance d, as indicated in **Figure 20–13**. We would like to determine the capacitance of such a capacitor. As we shall see, the capacitance is related in a simple way to the parameters A and d.

To begin, we note that when a charge $+Q$ is placed on one plate, and a charge $-Q$ is placed on the other, a uniform electric field is produced between the plates. The magnitude of this field is given by Gauss's law, and was determined in Active Example 19–3. There we found that $E = \sigma/\varepsilon_0$. Noting that the charge per area is $\sigma = Q/A$, we have

$$E = \frac{Q}{\varepsilon_0 A} \qquad \textbf{20–11}$$

The corresponding potential difference between the plates is $\Delta V = -E\Delta s = -(Q/\varepsilon_0 A)d$. The magnitude of this potential difference is simply $V = (Q/\varepsilon_0 A)d$.

Now that we have determined the potential difference for a parallel-plate capacitor with a charge Q on its plates, we can apply $C = Q/V$ to find an expression for the capacitance:

$$C = \frac{Q}{V} = \frac{Q}{(Q/\varepsilon_0 A)d}$$

Canceling the charge Q and rearranging slightly gives the final result:

Capacitance of a Parallel-Plate Capacitor

$$C = \frac{\varepsilon_0 A}{d} \qquad \textbf{20–12}$$

As mentioned previously, the capacitance depends in a straightforward way on the area of the plates, A, and their separation, d. The capacitance does not depend separately on the amount of charge on the plates, Q, or the potential difference between the plates, V, but instead depends only on the *ratio* of these two quantities. We use this expression for capacitance in the next Example.

PROBLEM-SOLVING NOTE

Using Magnitudes with Capacitance

Note that the capacitance C is always positive. Therefore, when applying a relation like $C = Q/V$, we always use magnitudes for the charge and the electric potential.

EXAMPLE 20–5 All Charged Up

A parallel-plate capacitor is constructed with plates of area $0.0280\ \text{m}^2$ and separation 0.550 mm. Find the magnitude of the charge on each plate of this capacitor when the potential difference between the plates is 20.1 V.

Picture the Problem

The capacitor is shown in our sketch. Note that the plates of the capacitor have an area $A = 0.0280\ \text{m}^2$, a separation $d = 0.550$ mm, and a potential difference $V = 20.1$ V. The charge on each plate has a magnitude Q.

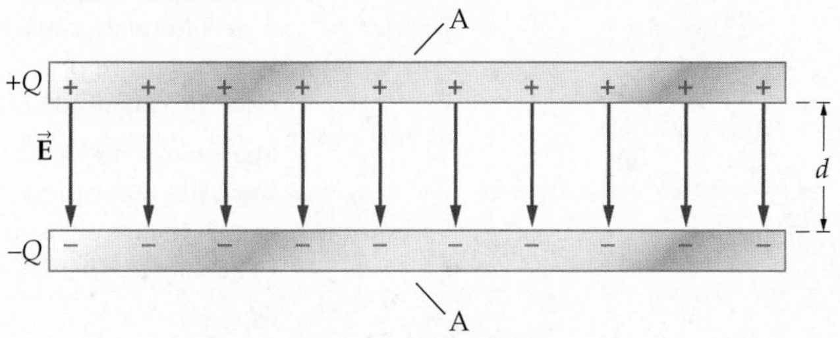

Strategy

The charge on the plates is given by $Q = CV$. We know V, but we must determine the capacitance. We can do this using the relation $C = \varepsilon_0 A/d$, along with the given information for A and d.

Solution

1. Calculate the capacitance of the capacitor:

$$C = \frac{\varepsilon_0 A}{d} = \frac{(8.85 \times 10^{-12}\ \text{C}^2/\text{N}\cdot\text{m}^2)(0.0280\ \text{m}^2)}{0.550 \times 10^{-3}\ \text{m}}$$
$$= 4.51 \times 10^{-10}\ \text{F}$$

2. Find the charge on the plates of the capacitor:

$$Q = CV = (4.51 \times 10^{-10}\ \text{F})(20.1\ \text{V}) = 9.06 \times 10^{-9}\ \text{C}$$

Insight

Notice the relatively small values for both the capacitance, C, and the magnitude of the charge on a plate, Q. The total amount of charge in the capacitor is zero, $+Q + (-Q) = 0$, but the fact that there is a charge separation—with one plate positive and the other negative—means that the capacitor stores energy, as we shall see in the next section.

If a capacitor is connected to a battery, the battery maintains a constant voltage between the plates of the capacitor. It follows from $Q = CV$ that any change in the capacitance—by changing the plate area, A, or separation, d, for example—results in a different amount of charge on the plates.

Practice Problem

What separation d is necessary to give a charge of magnitude 2.00×10^{-8} C on each plate of this capacitor? Assume all other quantities in the system remain the same. [**Answer:** $d = 0.249$ mm. Note that a *smaller* separation results in a *greater* amount of stored charge.]

Some related homework problems: Problem 40, Problem 41

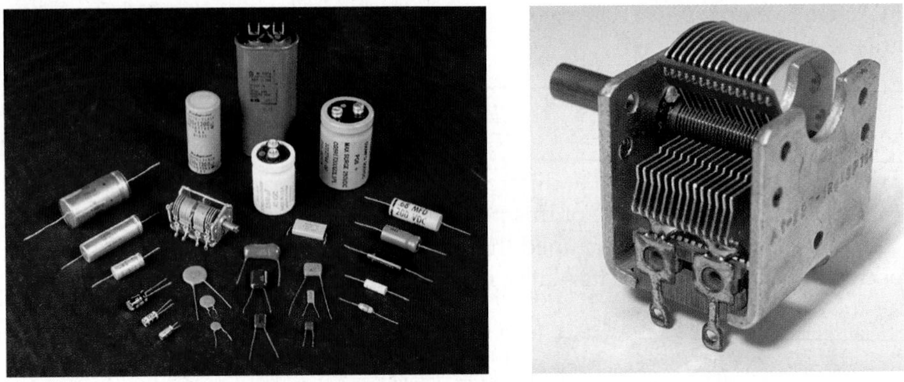

▶ **FIGURE 20–14 Capacitors**

Capacitors come in a variety of physical forms (left), with many different types of dielectric (insulating material) occupying the space between their plates. Capacitors are also rated for the maximum voltage that can be applied to them before the dielectric breaks down, allowing a spark to jump across the gap between the plates. A variable air capacitor (right), often used in early radios, has interleaved plates that can be rotated to vary their area of overlap. This changes the effective area of the capacitor, and hence its capacitance.

We see from Equation 20–12 that the capacitance of a parallel-plate capacitor increases with the area A of its plates—basically, the greater the area, the more room there is in the capacitor to hold charge. Again, this is like pouring water into a bucket with a large cross-sectional area. On the other hand, the capacitance decreases with increasing separation d. The reason for this dependence is that with the electric field between the plates constant (as we saw in Section 19–5), the potential difference between the plates is proportional to their separation: $V = Ed$. Thus, the capacitance, which is inversely proportional to the potential difference ($C = Q/V$), is also inversely proportional to the plate separation—the greater the separation, the greater the potential difference required to store a given amount of charge. All capacitors, regardless of their design, share these general features. Thus, to produce a large capacitance, one would like to have plates of large area close together. This is often accomplished by inserting a thin piece of paper between two large sheets of metal foil. The foil is then rolled up tightly to form a compact, large-capacity capacitor. Examples of common capacitors are shown in **Figure 20–14**.

CONCEPTUAL CHECKPOINT 20–5 Charge on the Plates

A parallel-plate capacitor is connected to a battery that maintains a constant potential difference V between the plates. If the plates are pulled away from each other, increasing their separation, does the magnitude of the charge on the plates **(a)** increase, **(b)** decrease, or **(c)** remain the same?

Reasoning and Discussion

Since the capacitance of a parallel-plate capacitor is $C = \varepsilon_0 A/d$, increasing the separation, d, decreases the capacitance. With a smaller value of C, and a constant value for V, the charge $Q = CV$ will decrease. The same general behavior can be expected with any capacitor.

Answer:

(b) The charge on the plates decreases.

Sometimes a capacitor is first connected to a battery to be charged and is then disconnected. In this case, the charge on the plates is "trapped"—it has no place to go—and hence Q must remain constant. If the capacitance is changed now, the result is a different potential difference, $V = Q/C$, between the plates.

Dielectrics

One way to increase the capacitance of a capacitor is to insert an insulating material, referred to as a **dielectric,** between its plates. With a dielectric in place, a capacitor can store more charge or energy in the same volume. Thus, dielectrics play an important role in miniaturizing electronic devices.

To see how this works, consider the parallel-plate capacitor shown in **Figure 20–15 (a)**. Initially the plates are separated by a vacuum and connected to a battery, giving the plates the charges $+Q$ and $-Q$. The battery is now removed, and the charge on the plates remains constant. The electric field between the plates is uniform and has a magnitude E_0. If the distance between the plates is d, the corresponding potential difference is $V_0 = E_0 d$, and the capacitance is

$$C_0 = \frac{Q}{V_0}$$

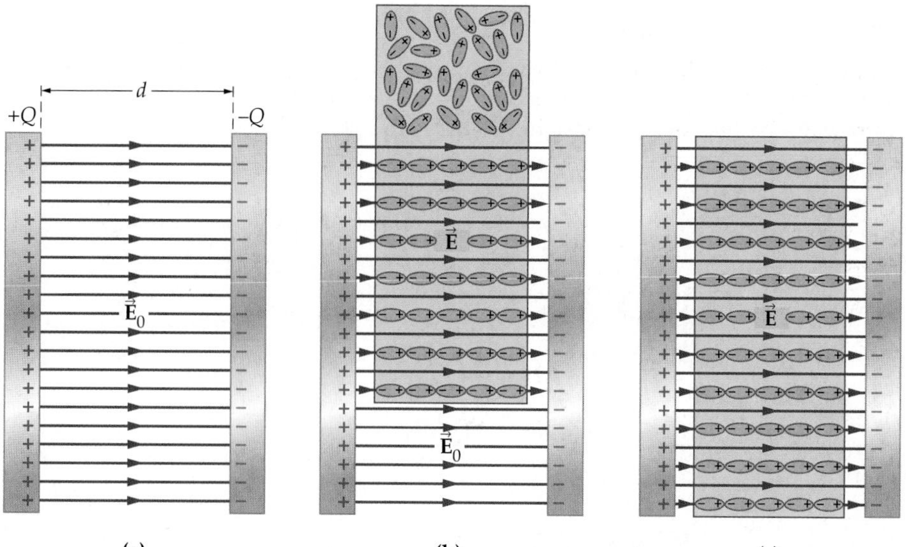

◀ **FIGURE 20–15 The effect of a dielectric on the electric field of a capacitor**
When a dielectric is placed in the electric field between the plates of a capacitor, the molecules of the dielectric tend to become oriented with their positive ends pointing toward the negatively charged plate and their negative ends pointing toward the positively charged plate. The result is a buildup of positive charge on one surface of the dielectric and of negative charge on the other. Since field lines start on positive charges and end on negative charges, we see that the number of field lines within the dielectric is reduced. Thus, within the dielectric the applied electric field $\vec{E}_0$, is partially canceled. Because the strength of the electric field is less, the voltage between the plates is less as well. Since V is smaller while Q remains the same, the capacitance, $C = Q/V$, is increased by the dielectric.

Now, insert a dielectric slab, as illustrated in **Figure 20–15 (b)** and **(c)**. If the molecules in the dielectric have a permanent dipole moment, they will align with the field, as shown in Figures 20–15 (b). Even without a permanent dipole moment, however, the molecules will become polarized by the field (see Section 19–1). This polarization leads to the same type of alignment, although the effect is weaker. The result of this alignment is a positive charge on the surface of the slab near the negative plate and a negative charge on the surface near the positive plate.

Recalling that electric field lines terminate on negative charges and start on positive charges, we can see from Figure 20–15 (c) that fewer field lines exist within the dielectric. Consequently, there is a reduced field, E, in a dielectric, which we characterize with a dimensionless **dielectric constant,** κ, as follows:

$$E = \frac{E_0}{\kappa} \qquad \text{20–13}$$

In the case of a vacuum, $\kappa = 1$, and $E = E_0$, as before. For an insulating material, however, the value of κ is greater than one. For example, paper has a dielectric constant of roughly 4, which means that the electric field within paper is about one-quarter what it would be in a vacuum. Typical values of κ are listed in Table 20–1.

Thus, a dielectric reduces the field between the plates of a capacitor by a factor of κ. This, in turn, decreases the *potential difference* between the plates by the same factor:

$$V = Ed = \left(\frac{E_0}{\kappa}\right)d = \frac{E_0 d}{\kappa} = \frac{V_0}{\kappa}$$

Finally, since the potential difference is smaller, the capacitance must be larger:

$$C = \frac{Q}{V} = \frac{Q}{(V_0/\kappa)} = \kappa\frac{Q}{V_0} = \kappa C_0 \qquad \text{20–14}$$

In effect, the dielectric partially shields one plate from the other, making it easier to build up a charge on the plates. If the space between the plates of a capacitor is filled with paper, for example, the capacitance will be about four times larger than if the space had been a vacuum.

The relation $C = \kappa C_0$ applies to any capacitor. For the special case of a parallel-plate capacitor filled with a dielectric, we have

Capacitance of a Parallel-Plate Capacitor Filled with a Dielectric

$$C = \frac{\kappa\varepsilon_0 A}{d} \qquad \text{20–15}$$

We apply this relation in the next Example.

TABLE 20–1 Dielectric constants

Substance	Dielectric constant, κ
Water	80.4
Neoprene rubber	6.7
Pyrex glass	5.6
Mica	5.4
Paper	3.7
Mylar	3.1
Teflon	2.1
Air	1.00059
Vacuum	1

PROBLEM-SOLVING NOTE

The Effects of a Dielectric

Dielectrics reduce the electric field in a capacitor, which results in a reduced potential difference between the plates. As a result, a dielectric always increases the capacitance.

EXAMPLE 20–6 Even More Charged-Up

A parallel-plate capacitor is constructed with plates of area 0.0280 m^2 and separation 0.550 mm. The space between the plates is filled with a dielectric with dielectric constant κ. When the capacitor is connected to a 12.0-V battery, each of the plates has a charge of magnitude 3.62×10^{-8} C. What is the value of the dielectric constant, κ?

Picture the Problem

The sketch shows the capacitor with a dielectric material inserted between the plates. In other respects, the capacitor is the same as the one considered in Example 20–5.

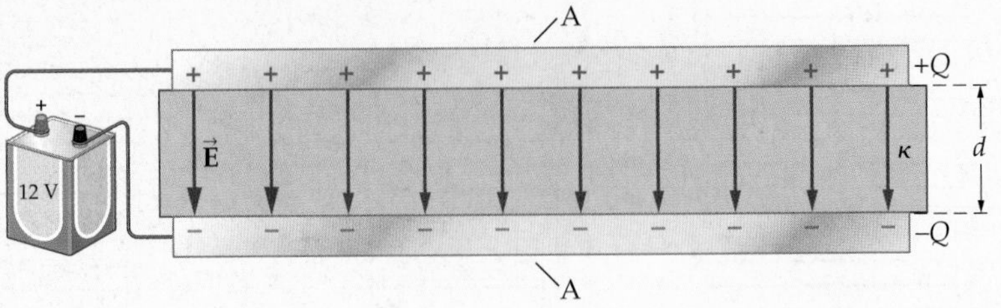

Strategy

Since we are given the potential difference V and the charge Q, we can find the capacitance using $C = Q/V$. Next, we relate the capacitance to the physical characteristics of the capacitor with $C = \kappa\varepsilon_0 A/d$. Using the given values for A and d, we solve for κ.

Solution

1. Determine the value of the capacitance:

$$C = \frac{Q}{V} = \frac{(3.62 \times 10^{-8}\ \text{C})}{12.0\ \text{V}} = 3.02 \times 10^{-9}\ \text{F}$$

2. Solve $C = \kappa\varepsilon_0 A/d$ for the dielectric constant, κ:

$$C = \kappa\varepsilon_0 A/d$$
$$\kappa = Cd/\varepsilon_0 A$$

3. Substitute numerical values to find κ:

$$\kappa = \frac{Cd}{\varepsilon_0 A}$$
$$= \frac{(3.02 \times 10^{-9}\ \text{F})(0.550 \times 10^{-3}\ \text{m})}{(8.85 \times 10^{-12}\ \text{C}^2/\text{N}\cdot\text{m}^2)(0.0280\ \text{m}^2)} = 6.70$$

Insight

Comparing our result with the dielectric constants given in Table 20–1, we see that the dielectric may be neoprene rubber.

Practice Problem

If a different dielectric with a smaller dielectric constant is inserted into the capacitor, does the charge on the plates increase, decrease, or remain the same? Find the charge on the plates for $\kappa = 3.5$. [**Answer:** The charge decreases to $Q = 1.89 \times 10^{-8}$ C.]

Some related homework problems: Problem 43, Problem 44

REAL-WORLD PHYSICS
Computer keyboards

The fact that the capacitance of a capacitor depends on the separation of its plates finds a number of interesting applications. For example, if you have ever typed on a computer keyboard, you have probably been utilizing the phenomenon of capacitance without realizing it. Many computer keyboards are designed in such a way that each key is connected to the upper plate of a parallel-plate capacitor, as illustrated in **Figure 20–16**. When you depress a given key, the separation between the plates of that capacitor decreases, and the corresponding capacitance increases. The circuitry of the computer can detect this change in capacitance, thereby determining which key you have pressed.

REAL-WORLD PHYSICS
The Theremin—a musical instrument you play without touching

Another, less well-known application of capacitance is the Theremin, a musical instrument that you play without touching! Two antennas on the theremin are used to control the sound it makes; one antenna adjusts the volume, the other adjusts the pitch. When a person places a hand near one of the antennas, the effect is similar to that of a parallel-plate capacitor, with the hand playing the role of one plate and the antenna playing the role of the other plate. Changing the separation between hand and antenna changes the capacitance, which the theremin's circuitry then converts into a corresponding change of volume or pitch. Theremins have been used to provide "ethereal" music for a number of science fiction films, and some popular bands use theremins in their musical arrangements.

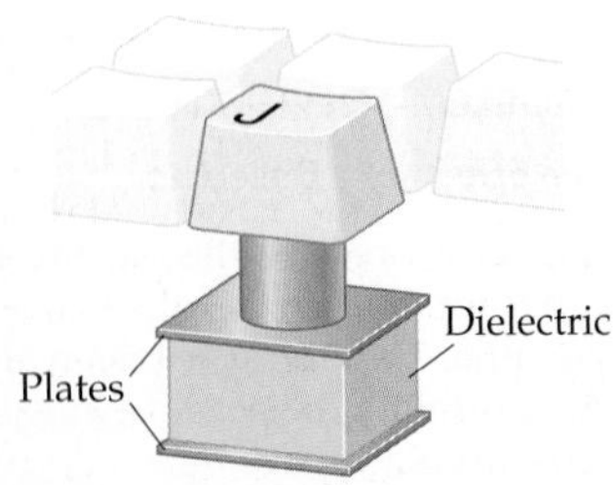

◀ **FIGURE 20–16 Capacitance and the computer keyboard**
The keys on many computer keyboards form part of a parallel-plate capacitor. Depressing the key changes the plate separation. The corresponding change in capacitance can be detected by the computer's circuitry.

◀ A musician plays a Theremin at an outdoor concert.

Dielectric Breakdown

If the electric field applied to a dielectric is large enough, it can literally tear the atoms apart, allowing the dielectric to conduct electricity. This condition is referred to as **dielectric breakdown.** The maximum field a dielectric can withstand before breakdown is called the **dielectric strength.** Typical values are given in Table 20–2.

For example, if the electric field in air exceeds about 3,000,000 V/m, dielectric breakdown will occur, leading to a spark on a small scale or a bolt of lightning on a larger scale. Next time you walk across a carpet and get a shock when reaching for the doorknob, think about the fact that you have just produced an electric field of roughly *3 million volts per meter!* The sharp tip of a lightning rod, which has a high electric field in its vicinity, helps initiate and guide lightning to the ground, or to dissipate charge harmlessly so that no lightning occurs at all. Saint Elmo's fire—the glow of light around the rigging of a ship in a storm—is another example of dielectric breakdown in air.

TABLE 20–2 Dielectric strengths

Substance	Dielectric strength (V/m)
Mica	100×10^6
Teflon	60×10^6
Paper	16×10^6
Pyrex glass	14×10^6
Neoprene rubber	12×10^6
Air	3.0×10^6

20–6 Electrical Energy Storage

As mentioned in the previous section, capacitors store more than just charge—they also store energy. To see how, consider a capacitor that has charges of magnitude Q on its plates, and a potential difference of V. Now, imagine transferring a small amount of charge, ΔQ, from one plate to the other, as in **Figure 20–17**. Since this charge must be moved across a potential difference of V, the change in electric potential energy is $\Delta U = (\Delta Q)V$. Thus, the potential energy of the capacitor increases by $(\Delta Q)V$ when the magnitude of the charge on its plates is increased from Q to $Q + \Delta Q$. As more charge is transferred from one plate to the other, more electrical potential energy is stored in the capacitor.

To find the total electric energy stored in a capacitor, we must take into account the fact that the potential difference between the plates increases as the charge on the plates increases. In fact, recalling that the potential difference is given by $V = Q/C$, it is clear that V increases linearly with the charge, as illustrated in **Figure 20–18**. In particular, if the final potential difference is V, the average potential during charging is $\frac{1}{2}V$. Therefore, the total energy U stored in a capacitor with charge Q and potential difference V can be written as follows:

$$U = QV_{av} = \tfrac{1}{2}QV \qquad \text{20–16}$$

Equivalently, since $Q = CV$, the energy stored in a capacitor of capacitance C and voltage V is

$$U = \tfrac{1}{2}CV^2 \qquad \text{20–17}$$

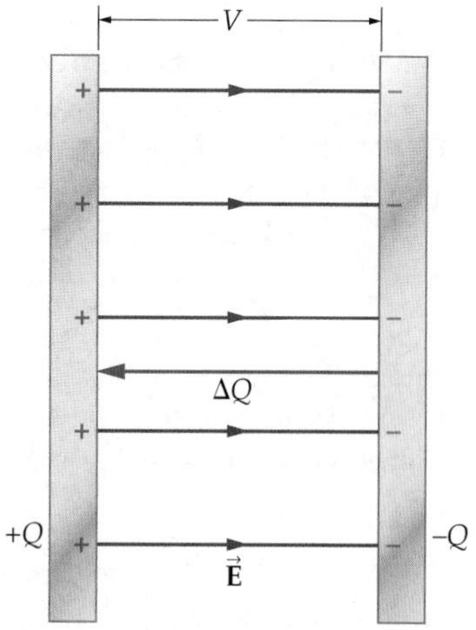

▲ **FIGURE 20–17 The energy required to charge a capacitor**

A capacitor has a charge of magnitude Q on its plates and a potential difference V between the plates. Transferring a small charge increment, $+\Delta Q$, from the negative plate to the positive plate increases the electric potential energy of the capacitor by the amount $\Delta U = (\Delta Q)V$.

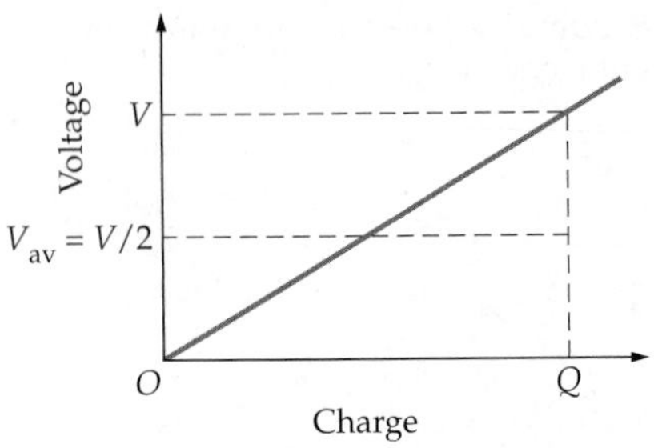

▲ **FIGURE 20–18 The voltage of a capacitor being charged**

The voltage V between the plates of a capacitor increases linearly with the charge Q on the plates, $V = Q/C$. Therefore, if a capacitor is charged to a final voltage of V, the average voltage during charging is $V_{av} = \frac{1}{2}V$.

Finally, using $V = Q/C$, we find that the energy stored in a capacitor of charge Q and capacitance C is

$$U = \frac{Q^2}{2C} \qquad 20\text{–}18$$

All these expressions are equivalent; they simply give the energy in terms of different variables.

REAL-WORLD PHYSICS
The electronic flash

The energy stored in a capacitor can be put to a number of practical uses. Any time you take a flash photograph, for example, you are triggering the rapid release of energy from a capacitor. The flash unit typically contains a capacitor with a capacitance of 100 to 400 μF. When fully charged to a voltage of about 300 V, the capacitor contains roughly 15 J of energy. Activating the flash causes the stored energy, which took several seconds to accumulate, to be released in less than a millisecond. Because of the rapid release of energy, the power output of a flash unit is impressively large—about 10 to 20 kW. This is far in excess of the power provided by the battery that operates the unit. Similar considerations apply to the defibrillator used in the treatment of heart attack victims, as we show in the next Example.

▶ An electronic flash unit like the one at left includes a capacitor (gray) that can store a large amount of charge. When the charge is released, the resulting flash can be as brief as a millisecond or less, allowing photographers to "freeze" motion, as in the photo at right. Even faster strobe units can be used to photograph explosions, shock waves, or speeding bullets.

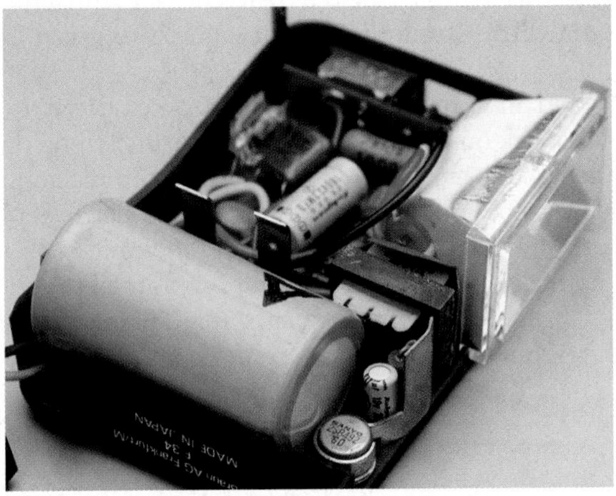

EXAMPLE 20–7 The Defibrillator: Delivering a Shock to the System

REAL WORLD PHYSICS: BIO

When a person's heart undergoes ventricular fibrillation—a rapid, uncoordinated twitching of the heart muscles—it often takes a strong jolt of electrical energy to restore the heart's regular beating and save the person's life. The device that delivers this jolt of energy is known as a defibrillator, and it uses a capacitor to store the necessary energy. In a typical defibrillator, a 175-μF capacitor is charged until the potential difference between the plates is 2240 V. **(a)** What is the magnitude of the charge on each plate of the fully charged capacitor? **(b)** Find the energy stored in the charged-up defibrillator.

Picture the Problem

Our sketch shows a simplified representation of a capacitor. The values of the capacitance and the potential difference are indicated.

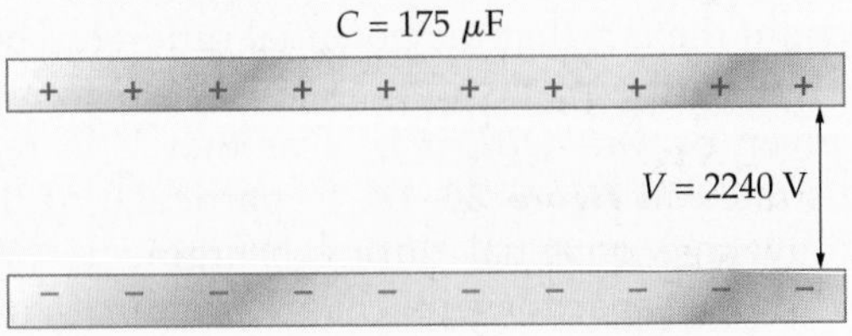

Strategy

a. We can find the charge stored on the capacitor plates using $Q = CV$.

b. The energy stored in the capacitor can be determined immediately using $U = \frac{1}{2}CV^2$. In addition, now that we know the charge on each plate of the capacitor, the energy can also be found with the relations $U = \frac{1}{2}QV$ and $U = Q^2/2C$.

Solution

Part (a)

1. Use $Q = CV$ to find the charge on the plates: $Q = CV = (175 \times 10^{-6}\ \text{F})(2240\ \text{V}) = 0.392\ \text{C}$

Part (b)

2. Find the stored energy using $U = \frac{1}{2}CV^2$: $U = \frac{1}{2}CV^2 = \frac{1}{2}(175 \times 10^{-6}\text{ F})(2240\text{ V})^2 = 439\text{ J}$

3. As a check, use $U = \frac{1}{2}QV$: $U = \frac{1}{2}QV = \frac{1}{2}(0.392\text{ C})(2240\text{ V}) = 439\text{ J}$

4. Finally, use the relation $U = Q^2/2C$: $U = \frac{Q^2}{2C} = \frac{(0.392\text{ C})^2}{2(175 \times 10^{-6}\text{ F})} = 439\text{ J}$

Insight

Of the 439 J stored in the defibrillator's capacitor, typically about 200 J will actually pass through the person's body in a pulse lasting about 2 ms. The power delivered by the pulse is approximately $P = U/t = (200\text{ J})/(0.002\text{ s}) = 100\text{ kW}$. This is significantly larger than the power delivered by the battery, which can take up to 30 s to fully charge the capacitor.

Practice Problem

Suppose the defibrillator is "fired" when the voltage is only half its maximum value of 2240 V. How much energy is stored in this case? [**Answer:** $E = (439\text{ J})/4 = 110\text{ J}$]

Some related homework problems: Problem 54, Problem 59

A defibrillator uses a capacitor to deliver a shock to a person's heart, restoring it to normal function. Capacitors can have the opposite effect as well, and it is for this reason that they can be quite dangerous, even in electrical devices that are turned off and unplugged from the wall. For example, a television set contains a number of capacitors, some of which store significant amounts of charge and energy. When a TV is unplugged, the capacitors retain their charge for long periods of time. Therefore, if you reach into the back of an unplugged television set there is a danger that you may come in contact with the terminals of a capacitor, which would then discharge its stored energy through your body. The resulting shock could be harmful or even fatal.

REAL-WORLD PHYSICS: BIO

Capacitor hazards

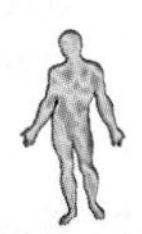

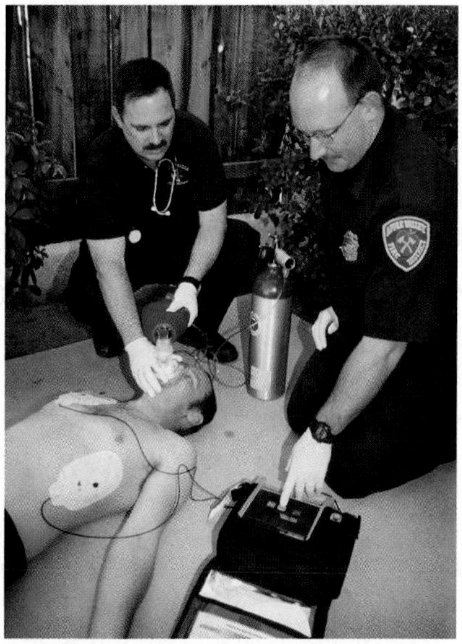

▲ A jolt of electric current from a defibrillator can restore normal heartbeat when the heart muscle has begun to twitch irregularly or has stopped beating altogether. A capacitor is used to store electricity, discharging it in a burst lasting only a couple of milliseconds.

Finally, we have discussed many examples of energy stored in a capacitor, but where exactly is the energy located? The answer is that the energy can be thought of as stored in the electric field, E, between the plates. To be specific, consider the relation

$$\text{energy} = \tfrac{1}{2}QV$$

In the case of a parallel-plate capacitor of area A and separation d, we know that $Q = \varepsilon_0 EA$ (Equation 20–11) and $V = Ed$. Thus, the energy stored in the capacitor can be written as

$$U = \text{energy} = \tfrac{1}{2}(\varepsilon_0 EA)(Ed) = \tfrac{1}{2}\varepsilon_0 E^2(Ad)$$

We have grouped A and d together because the product Ad is simply the total volume between the plates. Therefore, the **energy density** (energy per volume) is given by the following:

$$u_E = \text{electric energy density} = \frac{\text{electric energy}}{\text{volume}} = \tfrac{1}{2}\varepsilon_0 E^2 \qquad \textbf{20–19}$$

This result, though derived for a capacitor, is valid for any electric field, whether it occurs within a capacitor or anyplace else.

Chapter Summary

20–1 Electric Potential Energy and the Electric Potential

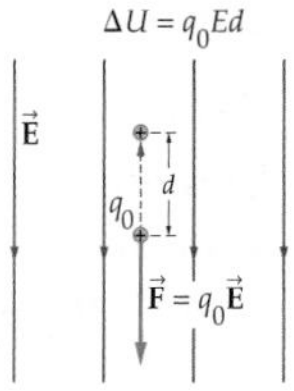

The electric force is conservative, just like the force of gravity. As a result, there is a potential energy U associated with the electric force.

Electric Potential Energy, U

The change in electric potential energy is defined by $\Delta U = -W$, where W is the work done by the electric field.

Electric Potential, V
The change in electric potential is defined to be $\Delta V = \Delta U/q_0$.

Relation between the Electric Field and the Electric Potential
The electric field is related to the rate of change of the electric potential. In particular, if the electric potential changes by the amount ΔV with a displacement Δs, the electric field in the direction of the displacement is

$$E = -\frac{\Delta V}{\Delta s} \qquad \textbf{20–4}$$

20–2 Energy Conservation

Another consequence of the fact that the electric force is conservative is that the total energy of an object is conserved—as long as nonconservative forces like friction can be ignored.

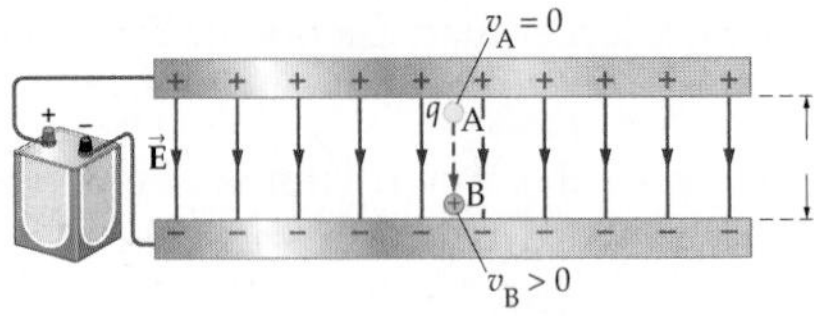

Energy Conservation
As usual, energy conservation can be expressed as follows:

$$\tfrac{1}{2}mv_A^2 + U_A = \tfrac{1}{2}mv_B^2 + U_B$$

In the case of the electric force, the potential energy is $U = q_0V$.

Direction of Acceleration
Positive charges accelerate in the direction of decreasing electric potential; negative charges accelerate in the direction of increasing electric potential.

20–3 The Electric Potential of Point Charges

If we define the electric potential of a point charge q to be zero at an infinite distance from the charge, the electric potential at a distance r is

$$V = \frac{kq}{r} \qquad \textbf{20–7}$$

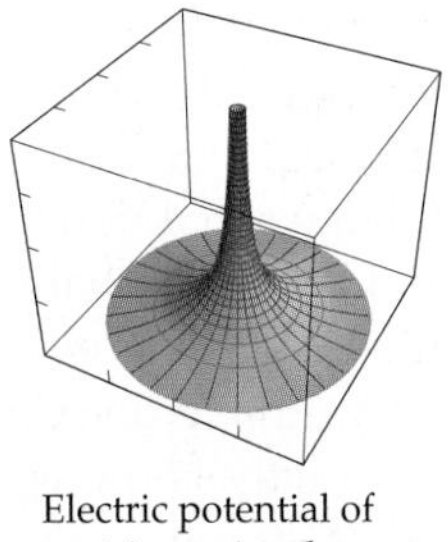
Electric potential of a positive point charge

Electric Potential Energy
We define the electric potential energy of two charges, q_0 and q, to be zero when the separation between them is infinite. When the charges are separated by a distance r, the potential energy of the system is

$$U = \frac{kq_0q}{r} \qquad \textbf{20–8}$$

Superposition
The electric potential of two or more point charges is simply the algebraic sum of the potentials due to each charge separately.

The total electric potential energy of two or more point charges is the sum of the potential energies due to each pair of charges.

20–4 Equipotential Surfaces and the Electric Field

Equipotential surfaces are defined as surfaces on which the electric potential is constant. Different equipotential surfaces correspond to different values of the potential.

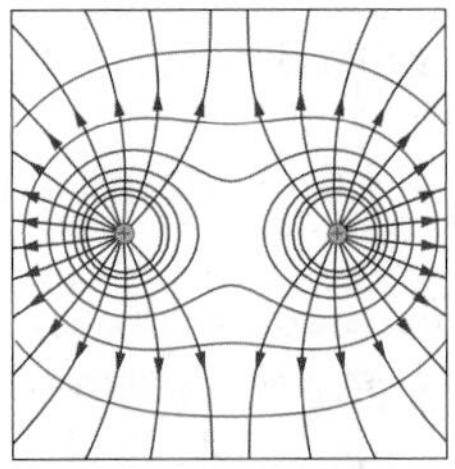
Equipotentials for two positive charges

Electric Field
The electric field is always perpendicular to the equipotential surfaces, and it points in the direction of decreasing electric potential.

Ideal Conductors
Ideal conductors are equipotential surfaces; every point on or within an ideal conductor is at the same potential. The electric field, therefore, is perpendicular to the surface of a conductor.

20–5 Capacitors and Dielectrics

A capacitor is a device that stores electric charge.

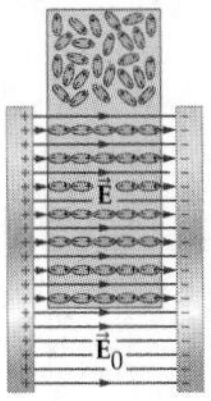

Capacitance
Capacitance is defined as the amount of charge Q stored in a capacitor per volt of potential difference V between the plates of the capacitor. Thus,

$$C = \frac{Q}{V} \qquad \textbf{20–9}$$

Parallel-Plate Capacitor
The capacitance of a parallel-plate capacitor, with plates of area A and separation d, is

$$C = \frac{\varepsilon_0 A}{d} \qquad 20\text{–}12$$

Dielectrics
A dielectric is an insulating material that increases the capacitance of a capacitor.

Dielectric Constant
A dielectric is characterized by the dimensionless dielectric constant, κ. In particular, the electric field in a dielectric is reduced by the factor κ, $E = E_0/\kappa$; the potential difference between capacitor plates is decreased by the factor κ, $V = V_0/\kappa$; and the capacitance is increased by the factor κ:

$$C = \kappa C_0 \qquad 20\text{–}14$$

Dielectric Breakdown/Dielectric Strength
A large electric field can cause a dielectric material to conduct electricity. This condition is referred to as dielectric breakdown. The strength of electric field required for dielectric breakdown is called the dielectric strength of the material.

20–6 Electrical Energy Storage

A capacitor, in addition to storing charge, also stores electrical energy.

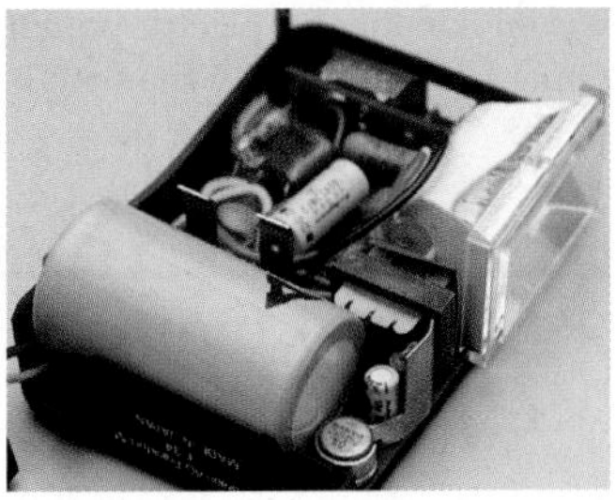

Energy Stored in a Capacitor
The electrical energy stored in a capacitor can be expressed as follows:

$$U = \tfrac{1}{2}QV = \tfrac{1}{2}CV^2 = Q^2/2C \qquad 20\text{–}16, 17, 18$$

Electrical Energy Density of an Electric Field
Electrical energy can be thought of as stored in the electric field. The electrical energy per volume, referred to as the electrical energy density, is given by the following relation:

$$u_E = \text{electrical energy density} = \tfrac{1}{2}\varepsilon_0 E^2 \qquad 20\text{–}19$$

Problem-Solving Summary

Type of Problem	Relevant Physical Concepts	Related Examples
Find the electric field corresponding to a change in electric potential.	The electric field is related to the change of electric potential with distance. The precise relation is $E = -\Delta V/\Delta s$.	Example 20–1 Active Example 20–1
Find the kinetic energy or speed of a particle moving in an electric field.	Apply energy conservation, including the electric potential energy, $U = qV$.	Examples 20–2, 20–4 Active Example 20–2
Calculate the electric potential due to a system of point charges.	The electric potential of a single point charge q at a distance r is $V = kq/r$. For a system of point charges, the total electric potential is the algebraic sum of the potentials calculated for each charge separately.	Examples 20–3, 20–4 Active Example 20–3
Determine the charge on the plates of a capacitor, or the potential difference between the plates.	The charge Q and potential difference V are related to the capacitance C by the expression $C = Q/V$.	Examples 20–5, 20–6
Determine the amount of energy stored in a capacitor.	The energy stored in a capacitor is given by three equivalent expressions: $U = \frac{1}{2}QV$; $U = \frac{1}{2}CV^2$; $U = Q^2/2C$.	Example 20–7

Conceptual Questions

(Answers to odd-numbered Conceptual Questions can be found in the back of the book.)

1. In one region of space the electric potential has a positive constant value. In another region of space the potential has a negative constant value. What can be said about the electric field within each of these two regions of space?

2. Two like charges a distance r apart have a positive electric potential energy. Conversely, two unlike charges a distance r apart have a negative electric potential energy. Explain the physical significance of these observations.

3. If the electric field is zero in some region of space is the electric potential zero there as well? Explain.
4. Sketch the equipotential surface that goes through point 1 in **Figure 20–19**. Repeat for point 2 and for point 3.

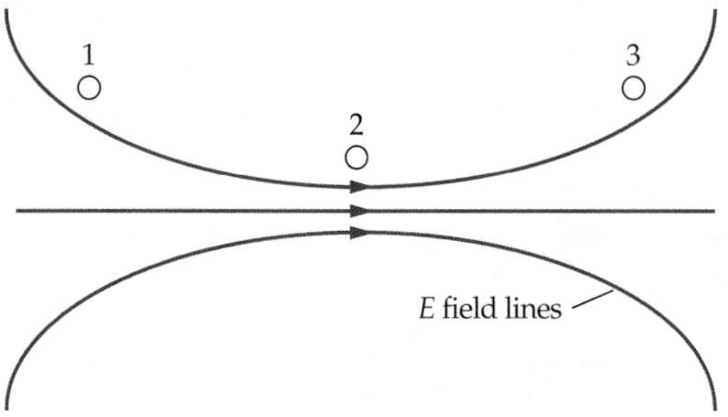

▲ **FIGURE 20–19** Conceptual Question 4 and Conceptual Exercises 12 and 13

5. How much work is required to move a charge from one location on an equipotential to another point on the same equipotential? Explain.
6. It is known that the electric potential is constant on a given two-dimensional surface. What can be said about the electric field on this surface?
7. Explain why equipotentials are always perpendicular to the electric field.
8. Two charges are at locations that have the same value of the electric potential. Is the electric potential energy the same for these charges? Explain.
9. A capacitor is connected to a battery and fully charged. What becomes of the charge on the capacitor when it is disconnected from the battery? What becomes of the charge when the two terminals of the capacitor are connected to one another?
10. It would be unwise to unplug a television set, take off the back, and reach inside. The reason for the danger is that if you happen to touch the terminals of a high-voltage capacitor you could receive a large electrical shock—even though the set is unplugged. Why?
11. On which of the following quantities does the capacitance of a capacitor depend: **(a)** the charge on the plates; **(b)** the separation of the plates; **(c)** the voltage difference between the plates; **(d)** the electric field between the plates; or **(e)** the area of the plates?
12. We say that a capacitor stores charge, yet the total charge in a capacitor is zero; that is, $Q + (-Q) = 0$. In what sense does a capacitor store charge if the net charge within it is zero?
13. The plates of a particular parallel-plate capacitor are uncharged. Is the capacitance of this capacitor zero? Explain.

Conceptual Exercises

(Answers to odd-numbered Conceptual Exercises can be found in the back of the book.)

1. An electron is released from rest in a region of space with a nonzero electric field. As the electron moves, does it experience an increasing or decreasing electric potential? Explain.
2. An electron is released from rest in a region of space with a nonzero electric field. As the electron moves, does the electric potential energy of the system increase, decrease, or stay the same? Explain.
3. A proton is released from rest in a region of space with a nonzero electric field. As the proton moves, does it experience an increasing or decreasing electric potential? Explain.
4. A proton is released from rest in a region of space with a nonzero electric field. As the proton moves, does the electric potential energy of the system increase, decrease, or stay the same? Explain.
5. In **Figure 20–20**, it is given that $q_1 = +Q$. **(a)** What value must q_2 have if the electric potential at point A is to be zero? **(b)** With the value for q_2 found in part (a), is the electric potential at point B positive, negative, or zero? Explain.

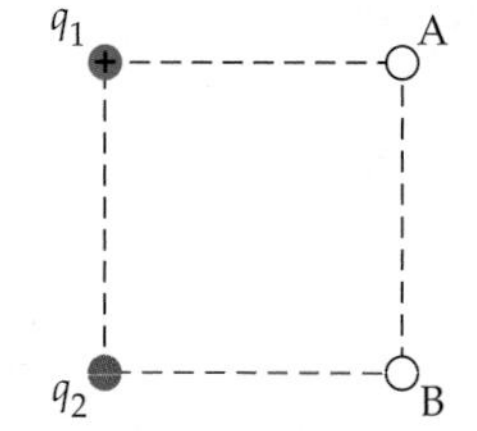

▲ **FIGURE 20–20** Conceptual Exercises 5, 6, and 7

6. The charge q_1 in Figure 20–20 has the value $+Q$. **(a)** What value must q_2 have if the electric potential at point B is to be zero? **(b)** With the value for q_2 found in part (a), is the electric potential at point A positive, negative, or zero? Explain.
7. It is given that the electric potential is zero at the center of the square in Figure 20–20. **(a)** If $q_1 = +Q$, what is the value of the charge q_2? **(b)** Is the electric potential at point A positive, negative, or zero? Explain. **(c)** Is the electric potential at point B positive, negative, or zero? Explain.
8. In the Bohr model of the hydrogen atom, a proton and an electron are separated by a constant distance r. **(a)** Would the electric potential energy of the system increase, decrease, or stay the same if the electron is replaced with a proton? Explain. **(b)** Suppose, instead, that the proton is replaced with an electron. Would the electric potential energy of the system increase, decrease, or stay the same? Explain.
9. Four different arrangements of point charges are shown in **Figure 20–21**. In each case the charges are the same distance from the origin. Rank the four arrangements in order of increasing electric potential at the origin, taking the potential at infinity to be zero. Indicate ties where appropriate.

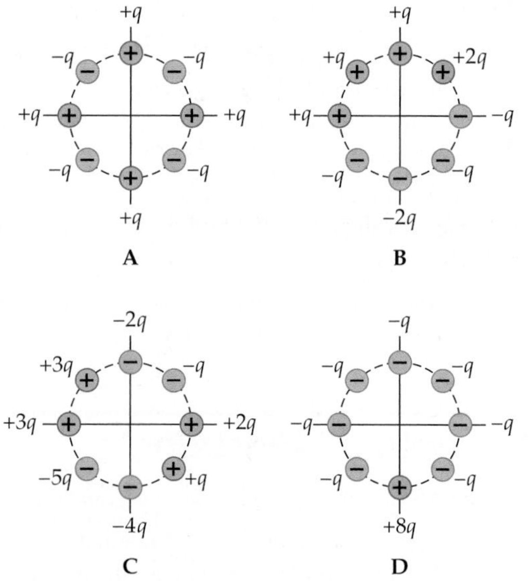

▲ **FIGURE 20–21** Conceptual Exercise 9

10. Two point charges are placed on the x axis. The charge $+2q$ is at $x = 1.5$ m, and the charge $-q$ is at $x = -1.5$ m. **(a)** There is a point on the x axis between the two charges where the electric potential is zero. Where is this point? **(b)** The electric potential also vanishes in one of the following regions: Region 1, $5.0\text{ m} > x > 1.5\text{ m}$; region 2, $-1.5\text{ m} > x > -3.0\text{ m}$; region 3, $-3.5\text{ m} > x > -5.0\text{ m}$. Identify the appropriate region. **(c)** Find the value of x referred to in part (b).
11. A charge is moved from one location on an equipotential to another point on the same equipotential. Is the work done on the charge positive, negative, or zero? Explain.
12. Is the electric potential at point 1 in Figure 20–19 greater than, less than, or the same as the electric potential at point 3? Explain.
13. Imagine sketching a large number of equipotential surfaces in Figure 20–19, with a constant difference in electric potential between adjacent surfaces. Would the spacing between the equipotentials at point 2 be greater than, less than, or the same as the spacing between equipotentials at point 1? Explain.
14. **Figure 20–22** shows a series of equipotentials in a particular region of space, and five different paths along which an electron is moved. **(a)** Does the electric field in this region point to the right, to the left, up, or down? Explain. **(b)** For each path, indicate whether the work done on the electron by the electric field is positive, negative, or zero. **(c)** Rank the paths in order of increasing amount of work done on the electron by the electric field. Indicate ties where appropriate. **(d)** Is the electric field near path A greater than, less than, or equal to the electric field near path E? Explain.

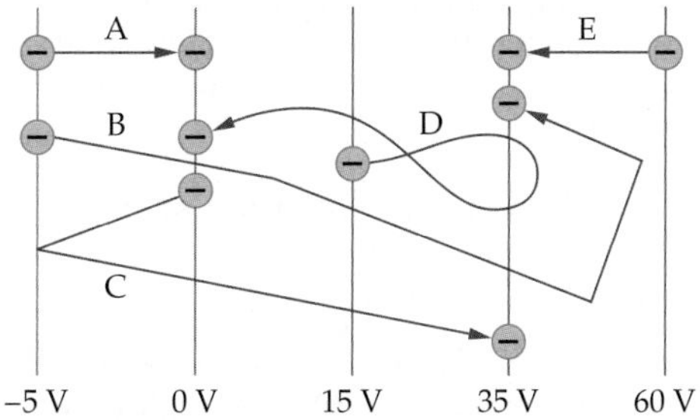

▲ **FIGURE 20–22** Conceptual Exercise 14

15. The plates of a parallel-plate capacitor have constant charges of $+Q$ and $-Q$. Do the following quantities increase, decrease, or remain the same as the separation of the plates is increased? **(a)** The electric field between the plates; **(b)** the potential difference between the plates; **(c)** the capacitance; **(d)** the energy stored in the capacitor.
16. A parallel-plate capacitor is connected to a battery that maintains a constant potential difference V between the plates. If the plates of the capacitor are pulled farther apart, do the following quantities increase, decrease, or remain the same? **(a)** The electric field between the plates; **(b)** the charge on the plates; **(c)** the capacitance; **(d)** the energy stored in the capacitor.
17. The plates of a parallel-plate capacitor have constant charges of $+Q$ and $-Q$. Do the following quantities increase, decrease, or remain the same as a dielectric is inserted between the plates? **(a)** The electric field between the plates; **(b)** the potential difference between the plates; **(c)** the capacitance; **(d)** the energy stored in the capacitor.
18. A parallel-plate capacitor is connected to a battery that maintains a constant potential difference V between the plates. If a dielectric is inserted between the plates of the capacitor, do the following quantities increase, decrease, or remain the same? **(a)** The electric field between the plates; **(b)** the charge on the plates; **(c)** the capacitance; **(d)** the energy stored in the capacitor.

Problems

Note: **IP** *denotes an integrated conceptual/quantitative problem.* **BIO** *identifies problems of biological or medical interest. Red bullets (•, ••, •••) are used to indicate the level of difficulty of each problem.*

Section 20–1 Electric Potential Energy and the Electric Potential

1. • A uniform electric field of magnitude 4.1×10^5 N/C points in the positive x direction. Find the change in electric potential energy of a 4.5-μC charge as it moves from the origin to the points **(a)** (0, 6.0 m); **(b)** (6.0 m, 0); and **(c)** (6.0 m, 6.0 m).
2. • A uniform electric field of magnitude 6.1×10^5 N/C points in the positive x direction. Find the change in electric potential between the origin and the points **(a)** (0, 6.0 m); **(b)** (6.0 m, 0); and **(c)** (6.0 m, 6.0 m).
3. • **BIO Electric Potential across a Cell Membrane** In a typical living cell, the electric potential inside the cell is 0.070 V lower than the electric potential outside the cell. The thickness of the cell membrane is 0.10 μm. What is the magnitude and direction of the electric field within the cell membrane?
4. • A computer monitor accelerates electrons and directs them to the screen in order to create an image. If the accelerating plates are 1.05 cm apart, and have a potential difference of 25,500 V, what is the magnitude of the uniform electric field between them?
5. • Find the change in electric potential energy for an electron that moves from one accelerating plate to the other in the computer monitor described in the previous problem.
6. • A parallel-plate capacitor has plates separated by 0.75 mm. If the electric field between the plates has a magnitude of **(a)** 1.2×10^5 V/m or **(b)** 2.4×10^4 N/C, what is the potential difference between the plates?
7. • When an ion accelerates through a potential difference of 2850 V, its electric potential energy decreases by 1.37×10^{-15} J. What is the charge on the ion?
8. • **The Electric Potential of the Earth** The Earth has a vertical electric field with a magnitude of approximately 100 V/m near its surface. What is the magnitude of the potential difference between a point on the ground and a point on the same level as the top of the Washington Monument (555 ft high)?
9. •• A uniform electric field with a magnitude of 6050 N/C points in the positive x direction. Find the change in electric potential energy when a $+12.5$-μC charge is moved 5.50 cm in **(a)** the positive x direction, **(b)** the negative x direction, and **(c)** in the positive y direction.
10. •• **IP** A spark plug in a car has electrodes separated by a gap of 0.025 in. To create a spark and ignite the air-fuel mixture in the engine, an electric field of 3.0×10^6 V/m is required in the gap. **(a)** What potential difference must be applied to the spark plug to initiate a spark? **(b)** If the separation between electrodes is increased, does the required potential difference increase,

decrease, or stay the same? Explain. **(c)** Find the potential difference for a separation of 0.050 in.

11. •• A uniform electric field with a magnitude of 1200 N/C points in the negative x direction, as shown in **Figure 20–23**. **(a)** What is the difference in electric potential, $\Delta V = V_B - V_A$, between points A and B? **(b)** What is the difference in electric potential, $\Delta V = V_B - V_C$, between points B and C? **(c)** What is the difference in electric potential, $\Delta V = V_C - V_A$, between points C and A? **(d)** From the information given in this problem, is it possible to determine the value of the electric potential at point A? If so, determine V_A; if not, explain why.

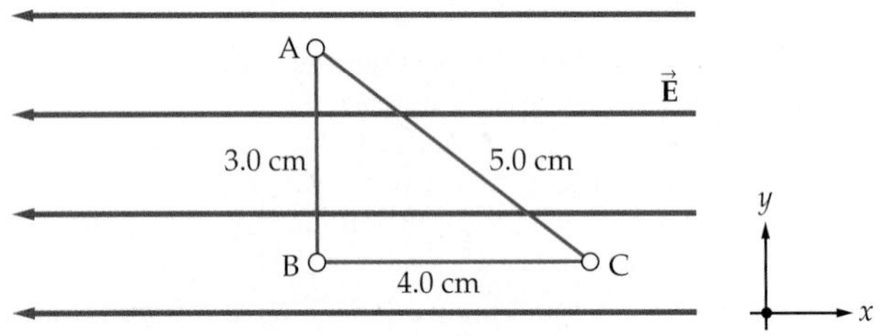

▲ **FIGURE 20–23** Problems 11 and 19

12. •• **A Charged Battery** A typical 12-V car battery can deliver 7.5×10^5 C of charge. If the energy supplied by the battery could be converted entirely to kinetic energy, what speed would it give to a 1200-kg car?

13. •• **IP BIO The Sodium Pump** Living cells actively "pump" positive sodium ions (Na^+) from inside the cell to outside the cell. This process is referred to as pumping because work must be done on the ions to move them from the negatively charged inner surface of the membrane to the positively charged outer surface. Given that the electric potential is 0.070 V higher outside the cell than inside the cell, and that the cell membrane is 0.10 μm thick, **(a)** calculate the work that must be done (in joules) to move one sodium ion from inside the cell to outside. **(b)** If the thickness of the cell membrane is increased, does your answer to part (a) increase, decrease, or stay the same. Explain. (It is estimated that as much as 20% of the energy we consume in a resting state is used in operating this "sodium pump.")

14. •• **IP** The electric potential of a system as a function of position along the x axis is given in **Figure 20–24**. **(a)** In which of the regions, 1, 2, 3, or 4, do you expect E_x to be greatest? In which region does E_x have its greatest magnitude? Explain. **(b)** Calculate the value of E_x in each of the regions, 1, 2, 3, and 4.

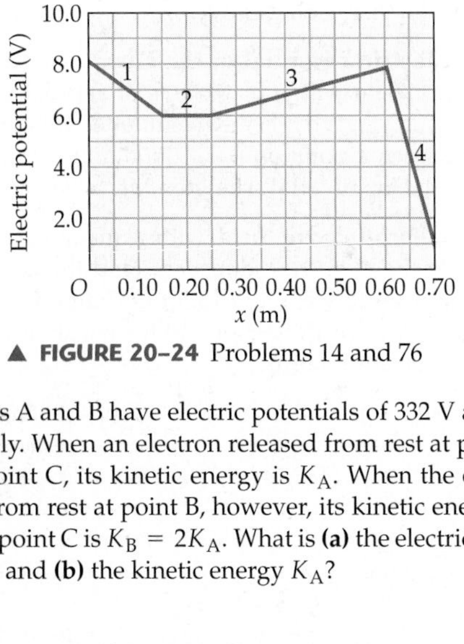

▲ **FIGURE 20–24** Problems 14 and 76

15. ••• Points A and B have electric potentials of 332 V and 149 V, respectively. When an electron released from rest at point A arrives at point C, its kinetic energy is K_A. When the electron is released from rest at point B, however, its kinetic energy when it reaches point C is $K_B = 2K_A$. What is **(a)** the electric potential at point C and **(b)** the kinetic energy K_A?

Section 20–2 Energy Conservation

16. • Calculate the speed of **(a)** a proton and **(b)** an electron after each particle accelerates from rest through a potential difference of 250 V.

17. • The electrons in a TV picture tube are accelerated from rest through a potential difference of 25 kV. What is the speed of the electrons after they have been accelerated by this potential difference?

18. • Find the potential difference required to accelerate protons from rest to 10% of the speed of light. (At this point, relativistic effects start to become significant.)

19. •• **IP** A particle with a mass of 3.5 g and a charge of +0.045 μC is released from rest at point A in Figure 20–23. **(a)** In which direction will this charge move? **(b)** What speed will it have after moving through a distance of 5.0 cm? **(c)** Suppose the particle continues moving for another 5.0 cm. Will its increase in speed for the second 5.0 cm be greater than, less than, or equal to its increase in speed in the first 5.0 cm? Explain.

20. •• A proton has an initial speed of 4.0×10^5 m/s. **(a)** What potential difference is required to bring the proton to rest? **(b)** What potential difference is required to reduce the initial speed of the proton by a factor of 2? **(c)** What potential difference is required to reduce the initial kinetic energy of the proton by a factor of 2?

Section 20–3 The Electric Potential of Point Charges

21. • The electric potential 1.8 m from a point charge q is 2.8×10^4 V. What is the value of q?

22. • A point charge of −7.2 μC is at the origin. What is the electric potential at **(a)** (3.0 m, 0); **(b)** (−3.0 m, 0); and **(c)** (3.0 m, −3.0 m)?

23. • **The Bohr Atom** The hydrogen atom consists of one electron and one proton. In the Bohr model of the hydrogen atom, the electron orbits the proton in a circular orbit of radius 0.529×10^{-10} m. What is the electric potential due to the proton at the electron's orbit?

24. • How far must the point charges $q_1 = +7.22$ μC and $q_2 = -26.1$ μC be separated for the electric potential energy of the system to be −126 J?

25. •• **IP** Point charges +4.1 μC and −2.2 μC are placed on the x axis at (11 m, 0) and (−11 m, 0), respectively. **(a)** Sketch the electric potential on the x axis for this system. **(b)** Your sketch should show one point on the x axis between the two charges where the potential vanishes. Is this point closer to the +4.1 μC charge or closer to the −2.2 μC charge? Explain. **(c)** Find the point referred to in part (b).

26. •• **IP** **(a)** In the previous problem, find the point to the left of the negative charge where the electric potential vanishes. **(b)** Is the electric field at the point found in part (a) positive, negative, or zero? Explain.

27. •• A dipole is formed by point charges +3.6 μC and −3.6 μC placed on the x axis at (0.20 m, 0) and (−0.20 m, 0), respectively. **(a)** Sketch the electric potential on the x axis for this system. **(b)** At what positions on the x axis does the potential have the value 7.5×10^5 V?

28. •• A charge of 3.05 μC is held fixed at the origin. A second charge of 3.05 μC is released from rest at the position (1.15 m, 0.550 m). **(a)** If the mass of the second charge is 2.66 g, what is its speed when it moves infinitely far from the origin? **(b)** At what distance from the origin does the 3.05-μC charge attain half the speed it will have at infinity?

29. •• **IP** A charge of 20.2 μC is held fixed at the origin. **(a)** If a -5.25 μC charge with a mass of 3.20 g is released from rest at the position (0.925 m, 1.17 m), what is its speed when it is halfway to the origin? **(b)** Suppose the -5.25 μC charge is released from rest at the point $x = \frac{1}{2}(0.925 \text{ m})$ and $y = \frac{1}{2}(1.17 \text{ m})$. When it is halfway to the origin, is its speed greater than, less than, or equal to the speed found in part (a)? Explain. **(c)** Find the speed of the charge for the situation described in part (b).

30. •• A charge of -2.205 μC is located at (3.055 m, 4.501 m), and a charge of 1.800 μC is located at $(-2.533 \text{ m}, 0)$. **(a)** Find the electric potential at the origin. **(b)** There is one point on the line connecting these two charges where the potential is zero. Find this point.

31. •• **IP Figure 20–25** shows three charges at the corners of a rectangle. **(a)** How much work must be done to move the $+2.7$-μC charge to infinity? **(b)** Suppose, instead, that we move the -6.1-μC charge to infinity. Is the work required in this case greater than, less than, or the same as when we moved the $+2.7$-μC charge to infinity? Explain. **(c)** Calculate the work needed to move the -6.1-μC charge to infinity.

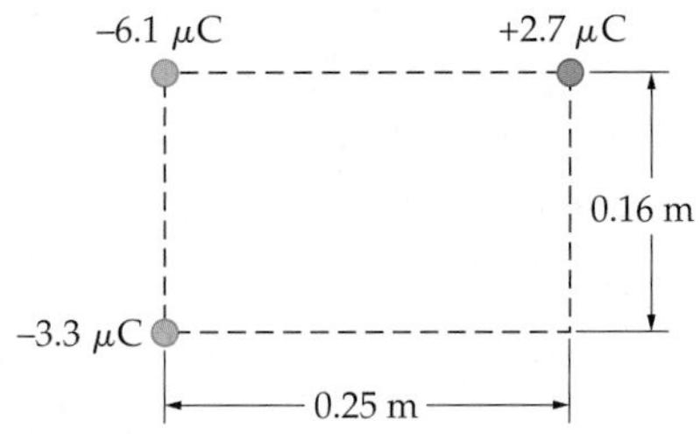

▲ **FIGURE 20–25** Problems 31 and 32

32. •• How much work must be done to move the three charges in Figure 20–25 infinitely far from one another?

33. •• **(a)** Find the electric potential at point P in **Figure 20–26**. **(b)** Suppose the three charges shown in Figure 20–26 are held in place. A fourth charge, with a charge of $+6.11$ μC and a mass of 4.71 g, is released from rest at point P. What is the speed of the fourth charge when it has moved infinitely far away from the other three charges?

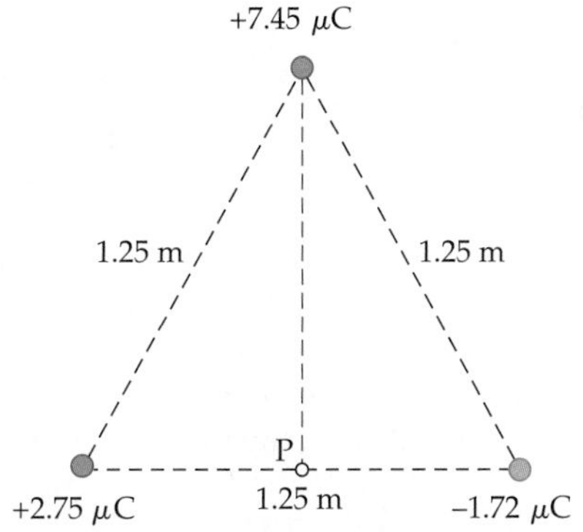

▲ **FIGURE 20–26** Problems 33 and 73

34. ••• A square of side a has a charge $+Q$ at each corner. What is the electric potential energy of this system of charges?

35. ••• A square of side a has charges $+Q$ and $-Q$ alternating from one corner to the next, as shown in **Figure 20–27**. Find the electric potential energy for this system of charges.

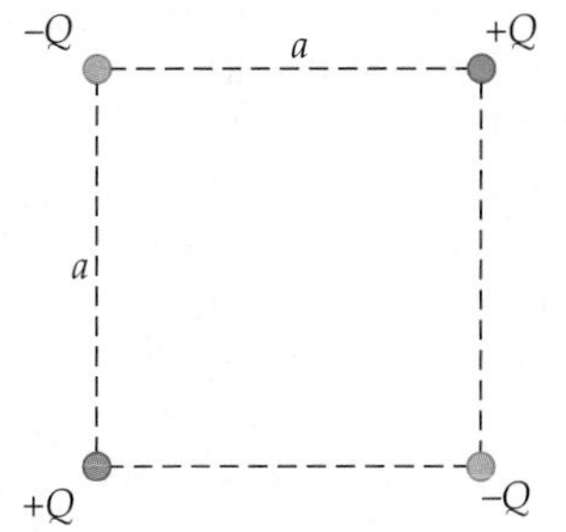

▲ **FIGURE 20–27** Problems 35 and 82

Section 20–4 Equipotential Surfaces and the Electric Field

36. • Two point charges are on the x axis. Charge 1 is $+q$ and is located at $x = -1.0$ m; charge 2 is $-2q$ and is located at $x = 1.0$ m. Make sketches of the equipotential surfaces for this system **(a)** out to a distance of about 2.0 m from the origin and **(b)** far from the origin. In each case, indicate the direction in which the potential increases.

37. • Two point charges are on the x axis. Charge 1 is $+q$ and is located at $x = -1.0$ m; charge 2 is $+2q$ and is located at $x = 1.0$ m. Make sketches of the equipotential surfaces for this system **(a)** out to a distance of about 2.0 m from the origin and **(b)** far from the origin. In each case, indicate the direction in which the potential increases.

38. •• **IP** Consider a region in space where a uniform electric field $E = 6500$ N/C points in the negative x direction. **(a)** What is the orientation of the equipotential surfaces? Explain. **(b)** If you move in the positive x direction, does the electric potential increase or decrease? Explain. **(c)** What is the distance between the $+14$-V and the $+16$-V equipotentials?

39. •• A given system has the equipotential surfaces shown in **Figure 20–28**. **(a)** What are the magnitude and direction of the electric field? **(b)** What is the shortest distance one can move to undergo a change in potential of 5.00 V?

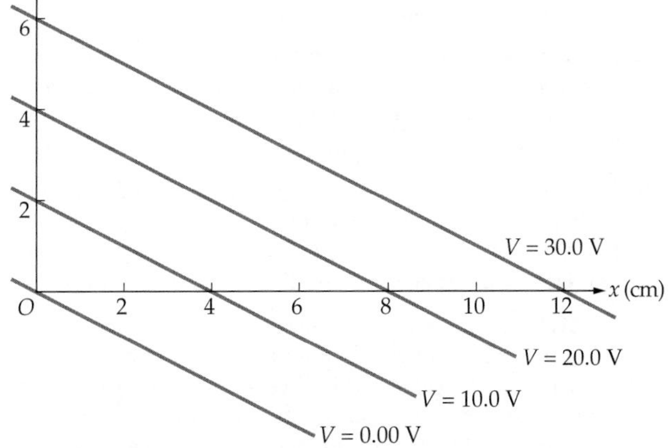

▲ **FIGURE 20–28** Problems 39 and 74

Section 20–5 Capacitors and Dielectrics

40. • A 0.40-μF capacitor is connected to a 9.0-V battery. How much charge is on each plate of the capacitor?

41. • It is desired that 5.8 μC of charge be stored on each plate of a 3.2-μF capacitor. What potential difference is required between the plates?

42. • To operate a given flash lamp requires a charge of 32 μC. What capacitance is needed to store this much charge in a capacitor with a potential difference between its plates of 9.0 V?

43. •• A parallel-plate capacitor is made from two aluminum-foil sheets, each 4.3 cm wide and 5.1 m long. Between the sheets is a Teflon strip of the same width and length that is 0.025 mm thick. What is the capacitance of this capacitor? (The dielectric constant of Teflon is 2.1.)

44. •• A parallel-plate capacitor is constructed with circular plates of radius 0.056 m. The plates are separated by 0.25 mm, and the space between the plates is filled with a dielectric with dielectric constant κ. When the charge on the capacitor is 1.2 μC the potential difference between the plates is 750 V. Find the value of the dielectric constant, κ.

45. •• **IP** A parallel-plate capacitor has plates with an area of 0.012 m^2 and a separation of 0.88 mm. The space between the plates is filled with a dielectric whose dielectric constant is 2.0. **(a)** What is the potential difference between the plates when the charge on the capacitor plates is 4.7 μC? **(b)** Will your answer to part (a) increase, decrease, or stay the same if the dielectric constant is increased? Explain. **(c)** Calculate the potential difference for the case where the dielectric constant is 4.0.

46. •• **IP** Consider a parallel-plate capacitor constructed from two circular metal plates of radius R. The plates are separated by a distance of 1.5 mm. **(a)** What radius must the plates have if the capacitance of this capacitor is to be 1.0 μF? **(b)** If the separation between the plates is increased, should the radius of the plates be increased or decreased to maintain a capacitance of 1.0 μF? Explain. **(c)** Find the radius of the plates that gives a capacitance of 1.0 μF for a plate separation of 3.0 mm.

47. •• A parallel-plate capacitor has plates of area 3.45×10^{-4} m^2. What plate separation is required if the capacitance is to be 1330 pF? Assume that the space between the plates is filled with **(a)** air or **(b)** paper.

48. •• **IP** A parallel-plate capacitor filled with air has plates of area 0.0066 m^2 and a separation of 0.45 mm. **(a)** Find the magnitude of the charge on each plate when the capacitor is connected to a 12-V battery. **(b)** Will your answer to part (a) increase, decrease, or stay the same if the separation between the plates is increased? Explain. **(c)** Calculate the magnitude of the charge on the plates if the separation is 0.90 mm.

49. •• Suppose that after walking across a carpeted floor you reach for a doorknob and just before you touch it a spark jumps 0.50 cm from your finger to the knob. Find the minimum voltage needed between your finger and the doorknob to generate this spark.

50. •• **(a)** What plate area is required if an air-filled, parallel-plate capacitor with a plate separation of 2.6 mm is to have a capacitance of 22 pF? **(b)** What is the maximum voltage that can be applied to this capacitor without causing dielectric breakdown?

51. •• **Lightning** As a crude model for lightning, consider the ground to be one plate of a parallel-plate capacitor and a cloud at an altitude of 550 m to be the other plate. Assume the surface area of the cloud to be the same as the area of a square that is 0.50 km on a side. **(a)** What is the capacitance of this capacitor? **(b)** How much charge can the cloud hold before the dielectric strength of the air is exceeded and a spark (lightning) results?

52. ••• A parallel-plate capacitor is made from two aluminum-foil sheets, each 3.00 cm wide and 10.0 m long. Between the sheets is a mica strip of the same width and length that is 0.0225 mm thick. What is the maximum charge that can be stored in this capacitor? (The dielectric constant of mica is 5.4, and its dielectric strength is 1.00×10^8 V/m.)

Section 20–6 Electrical Energy Storage

53. • Calculate the work done by a 3.0-V battery as it charges a 7.2-μF capacitor in the flash unit of a camera.

54. • **BIO Defibrillator** An automatic external defibrillator (AED) delivers 125 J of energy at a voltage of 1050 V. What is the capacitance of this device?

55. •• **IP BIO Cell Membranes** The membrane of a living cell can be approximated by a parallel-plate capacitor with plates of area 4.75×10^{-9} m^2, a plate separation of 8.5×10^{-9} m, and a dielectric with a dielectric constant of 4.5. **(a)** What is the energy stored in such a cell membrane if the potential difference across it is 0.0725 V? **(b)** Would your answer to part (a) increase, decrease, or stay the same if the thickness of the cell membrane is increased? Explain.

56. •• A 0.22-μF capacitor is charged by a 1.5-V battery. After being charged, the capacitor is connected to a small electric motor. Assuming 100% efficiency, **(a)** to what height can the motor lift a 5.0-g mass? **(b)** What initial voltage must the capacitor have if it is to lift a 5.0-g mass through a height of 1.0 cm?

57. •• Find the electric energy density between the plates of a 225-μF parallel-plate capacitor. The potential difference between the plates is 315 V, and the plate separation is 0.200 mm.

58. •• What electric field strength would store 15.0 J of energy in every 1.00 mm^3 of space?

59. •• An electronic flash unit for a camera contains a capacitor with a capacitance of 850 μF. When the unit is fully charged and ready for operation, the potential difference between the capacitor plates is 330 V. **(a)** What is the magnitude of the charge on each plate of the fully charged capacitor? **(b)** Find the energy stored in the "charged-up" flash unit.

60. ••• A parallel-plate capacitor has plates with an area of 405 cm^2 and an air-filled gap between the plates that is 2.25 mm thick. The capacitor is charged by a battery to 575 V and then is disconnected from the battery. **(a)** How much energy is stored in the capacitor? **(b)** The separation between the plates is now increased to 4.50 mm. How much energy is stored in the capacitor now? **(c)** How much work is required to increase the separation of the plates from 2.25 mm to 4.50 mm? Explain your reasoning.

General Problems

61. • Find the difference in electric potential, $\Delta V = V_B - V_A$, between the points A and B for the following cases: **(a)** The electric field does 0.052 J of work as you move a +5.7-μC charge from A to B. **(b)** The electric field does −0.052 J of work as you move a −5.7-μC charge from A to B. **(c)** You perform 0.052 J of work as you slowly move a +5.7-μC charge from A to B.

62. • The separation between the plates of a parallel-plate capacitor is doubled and the area of the plates is halved. How is the capacitance affected?

63. • A parallel-plate capacitor is connected to a battery that maintains a constant potential difference between the plates. If the spacing between the plates is doubled, how is the magnitude of charge on the plates affected?

64. •• A charge of 24.5 μC is located at (4.40 m, 6.02 m), and a charge of −11.2 μC is located at (−4.50 m, 6.75 m). What charge must be located at (2.23 m, −3.01 m) if the electric potential is to be zero at the origin?

65. •• **The Bohr Model** In the Bohr model of the hydrogen atom (see Problem 23) what is the smallest amount of work that must be done on the electron to move it from its circular orbit, with a

radius of 0.529×10^{-10} m, to an infinite distance from the proton? This value is referred to as the ionization energy of hydrogen.

66. •• **IP** A $+1.2$-μC charge and a -1.2-μC charge are placed at (0.50 m, 0) and (-0.50 m, 0), respectively. **(a)** In **Figure 20–29,** at which of the points A, B, C, or D is the electric potential smallest in value? At which of these points does it have its greatest value? Explain. **(b)** Calculate the electric potential at points A, B, C, and D.

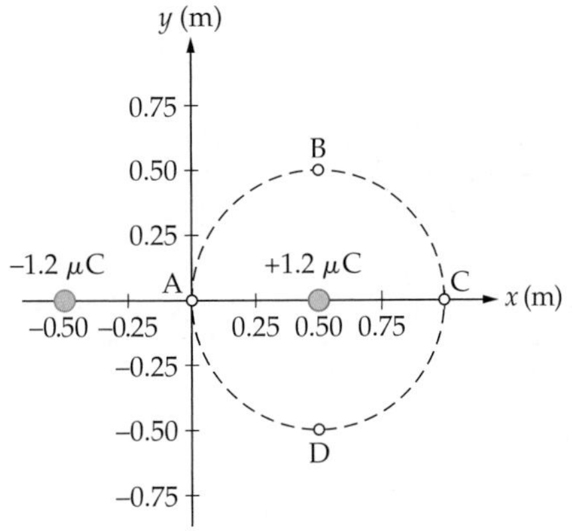

▲ **FIGURE 20–29** Problems 66, 67, and 85

67. •• Repeat Problem 66 for the case where both charges are $+1.2$ μC.

68. •• How much work is required to bring three protons, initially infinitely far apart, to a configuration where each proton is 1.5×10^{-15} m from the other two? (This is a typical separation for protons in a nucleus.)

69. •• A point charge $Q = +87.1$ μC is held fixed at the origin. A second point charge, with mass $m = 0.0526$ kg and charge $q = -2.37$ μC, is placed at the location (0.323 m, 0). **(a)** Find the electric potential energy of this system of charges. **(b)** If the second charge is released from rest, what is its speed when it reaches the point (0.121 m, 0)?

70. •• **Electron Escape Speed** An electron is at rest just above the surface of a sphere with a radius of 2.0 mm and a uniformly distributed positive charge of 1.8×10^{-15} C. Like a rocket blasting off from the Earth, the electron is given an initial speed v_e radially outward from the sphere. If the electron coasts to infinity, where its kinetic energy drops to zero, what is the escape speed, v_e?

71. •• **Quark Model of the Neutron** According to the quark model of fundamental particles, neutrons—the neutral particles in an atom's nucleus—are composed of three quarks. Two of these quarks are "down" quarks, each with a charge of $-e/3$; the third quark is an "up" quark, with a charge of $+2e/3$. This gives the neutron a net charge of zero. What is the electric potential energy of these three quarks, assuming they are equidistant from one another, with a separation distance of 1.3×10^{-15} m? (Quarks are discussed in Chapter 32.)

72. •• A parallel-plate capacitor is charged to an electric potential of 305 V by moving 3.75×10^{16} electrons from one plate to the other. How much work is done in charging the capacitor?

73. •• **IP** The three charges shown in Figure 20–26 are held in place as a fourth charge, q, is brought from infinity to the point P. The charge q starts at rest at infinity and is also at rest when it is placed at the point P. **(a)** If q is a positive charge, is the work required to bring it to the point P positive, negative, or zero? Explain. **(b)** Find the value of q if the work needed to bring it to point P is -1.3×10^{-11} J.

74. •• **(a)** In Figure 20–28 we see that the electric potential increases by 10.0 V as one moves 4.00 cm in the positive x direction. Use this information to calculate the x component of the electric field. (Ignore the y direction for the moment.) **(b)** Apply the same reasoning as in part (a) to calculate the y component of the electric field. **(c)** Combine the results from parts (a) and (b) to find the magnitude and direction of the electric field for this system.

75. •• **IP BIO Electric Fish** The electric catfish (*Malapterurus electricus*) is an aggressive fish, 1.0-m in length, found today in tropical Africa (and depicted in Egyptian hieroglyphics). The catfish is capable of generating jolts of electricity up to 350 V by producing a positively charged region of muscle near the head and a negatively charged region near the tail. **(a)** For the same amount of charge, can the catfish generate a higher voltage by separating the charge from one end of its body to the other, as it does, or from one side of the body to the other? Explain. **(b)** Estimate the charge generated at each end of a catfish as follows: Treat the catfish as a parallel-plate capacitor with plates of area 1.8×10^{-2} m^2, separation 1.0 m, and filled with a dielectric with a dielectric constant $\kappa = 95$.

76. •• As a $+6.2$-μC charge moves along the x axis from $x = 0$ to $x = 0.70$ m, the electric potential it experiences is shown in Figure 20–24. Find the approximate location(s) of the charge when its electric potential energy is **(a)** 2.6×10^{-5} J and **(b)** 4.3×10^{-5} J.

77. •• **IP** Many computer keyboards operate on the principle of capacitance. As shown in Figure 20–16, each key forms a small parallel-plate capacitor whose separation is reduced when the key is depressed. **(a)** Does depressing a key increase or decrease its capacitance? Explain. **(b)** Suppose the plates for each key have an area of 47.5 mm^2 and an initial separation of 0.550 mm. In addition, let the dielectric have a dielectric constant of 3.75. If the circuitry of the computer can detect a change in capacitance of 0.425 pF, what is the minimum distance a key must be depressed to be detected?

78. •• **IP** A point charge of mass 0.081 kg and charge $+6.77$ μC is suspended by a thread between the vertical parallel plates of a parallel-plate capacitor, as shown in **Figure 20–30.** **(a)** If the charge deflects to the right of vertical, as indicated in the figure, which of the two plates is at the higher electric potential? **(b)** If the angle of deflection is 22°, and the separation between the plates is 0.025 m, what is the potential difference between the plates?

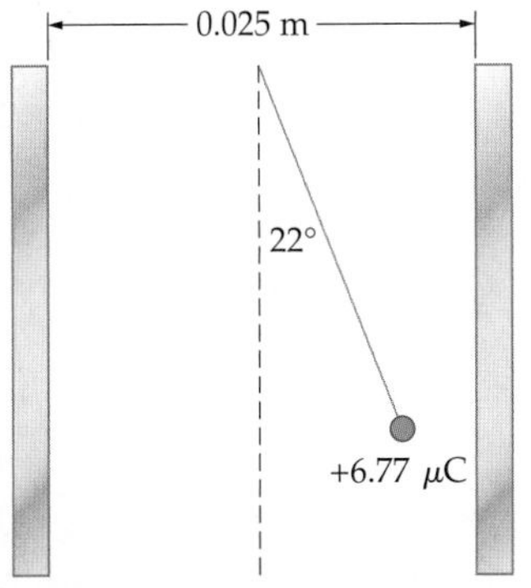

▲ **FIGURE 20–30** Problems 78 and 83

79. •• **BIO Cell Membranes and Dielectrics** Many cells in the body have a cell membrane whose inner and outer surfaces carry opposite charges, just like the plates of a parallel-plate capacitor. Suppose a typical cell membrane has a thickness of 7.5×10^{-9} m, and its inner and outer surfaces carry charge densities of -0.62×10^{-3} C/m^2 and $+0.62 \times 10^{-3}$ C/m^2,

respectively. In addition, assume that the material in the cell membrane has a dielectric constant of 5.5. **(a)** Find the direction and magnitude of the electric field within the cell membrane. **(b)** Calculate the potential difference between the inner and outer walls of the membrane, and indicate which wall of the membrane has the higher potential.

80. •• Long, long ago, on a planet far, far away, a physics experiment was carried out. First, a 0.250-kg ball with zero net charge was dropped from rest at a height of 1.00 m. The ball landed 0.552 s later. Next, the ball was given a net charge of 7.75 μC and dropped in the same way from the same height. This time the ball fell for 0.680 s before landing. What is the electric potential at a height of 1.00 m above the ground on this planet, given that the electric potential at ground level is zero? (Air resistance can be ignored.)

81. •• **Rutherford's Planetary Model of the Atom** In 1911, Ernest Rutherford developed a planetary model of the atom, in which a small positively charged nucleus is orbited by electrons. The model was motivated by an experiment carried out by Rutherford and his graduate students, Geiger and Marsden. In this experiment, they fired alpha particles with an initial speed of 1.75×10^7 m/s at a thin sheet of gold. (Alpha particles are obtained from certain radioactive decays. They have a charge of $+2e$ and a mass of 6.64×10^{-27} kg.) How close can the alpha particles get to a gold nucleus (charge $= +79e$), assuming the nucleus remains stationary? (This calculation sets an upper limit on the size of the gold nucleus. See Chapter 31 for further details.)

82. ••• **IP** **(a)** One of the $-Q$ charges in Figure 20–27 is given an outward "kick" that sends it off with an initial speed v_0 while the other three charges are held at rest. If the moving charge has a mass m, what is its speed when it is infinitely far from the other charges? **(b)** Suppose the remaining $-Q$ charge, which also has a mass m, is now given the same initial speed, v_0. When it is infinitely far away from the two $+Q$ charges, is its speed greater than, less than, or the same as the speed found in part (a)? Explain.

83. ••• Figure 20–30 shows a charge $q = +6.77\ \mu$C with a mass $m = 0.071$ kg suspended by a thread of length $L = 0.022$ m between the plates of a capacitor. **(a)** Plot the electric potential energy of the system as a function of the angle θ the thread makes with the vertical. (The electric field between the plates has a magnitude $E = 4.16 \times 10^4$ V/m.) **(b)** Repeat part (a) for the case of the gravitational potential energy of the system. **(c)** Show that the total potential energy of the system (electric plus gravitational) is a minimum when the angle θ satisfies the equilibrium condition for the charge, $\tan\theta = qE/mg$. This relation implies that $\theta = 22°$.

84. ••• The electric potential a distance r from a point charge q is 2.70×10^4 V. One meter farther away from the charge the potential is 6140 V. Find the charge q and the initial distance r.

85. ••• Referring to Problem 66, calculate and plot the electric potential on the circle centered at (0.50 m, 0). Give your results in terms of the angle θ, defined as follows: θ is the angle measured counterclockwise from a vertex at the center of the circle, with $\theta = 0$ at point C.

86. ••• When the potential difference between the plates of a capacitor is increased by 3.25 V, the magnitude of the charge on each plate increases by 13.5 μC. What is the capacitance of this capacitor?

87. ••• The electric potential a distance r from a point charge q is 155 V, and the magnitude of the electric field is 2240 N/C. Find the values of q and r.

Interactive Problems

88. •• **IP** ***Referring to Example 20–3*** Suppose the charge $-2q$ at $x = 1.00$ m is replaced with a charge $-3q$, where $q = 4.11 \times 10^{-9}$ C. The charge $+q$ is at the origin. **(a)** Is the electric potential positive, negative, or zero at the point $x = 0.333$ m? Explain. **(b)** Find the point between $x = 0$ and $x = 1.00$ m where the electric potential vanishes. **(c)** Is there a point in the region $x < 0$ where the electric potential passes through zero?

89. •• ***Referring to Example 20–3*** Suppose we can change the location of the charge $-2q$ on the x axis. The charge $+q$ (where $q = 4.11 \times 10^{-9}$ C) is still at the origin. **(a)** Where should the charge $-2q$ be placed to ensure that the electric potential vanishes at $x = 0.500$ m? **(b)** With the location of $-2q$ found in part (a), where does the electric potential pass through zero in the region $x < 0$?

90. •• **IP** ***Referring to Example 20–3*** Suppose the charge $+q$ at the origin is replaced with a charge $+5q$, where $q = 4.11 \times 10^{-9}$ C. The charge $-2q$ is still at $x = 1.00$ m. **(a)** Is there a point in the region $x < 0$ where the electric potential passes through zero? **(b)** Find the location between $x = 0$ and $x = 1.00$ m where the electric potential passes through zero. **(c)** Find the location in the region $x > 1.00$ m where the electric potential passes through zero.

21

Electric Current and Direct-Current Circuits

21–1	Electric Current	696
21–2	Resistance and Ohm's Law	699
21–3	Energy and Power in Electric Circuits	702
21–4	Resistors in Series and Parallel	705
21–5	Kirchhoff's Rules	711
21–6	Circuits Containing Capacitors	714
21–7	*RC* Circuits	717
*21–8	Ammeters and Voltmeters	720

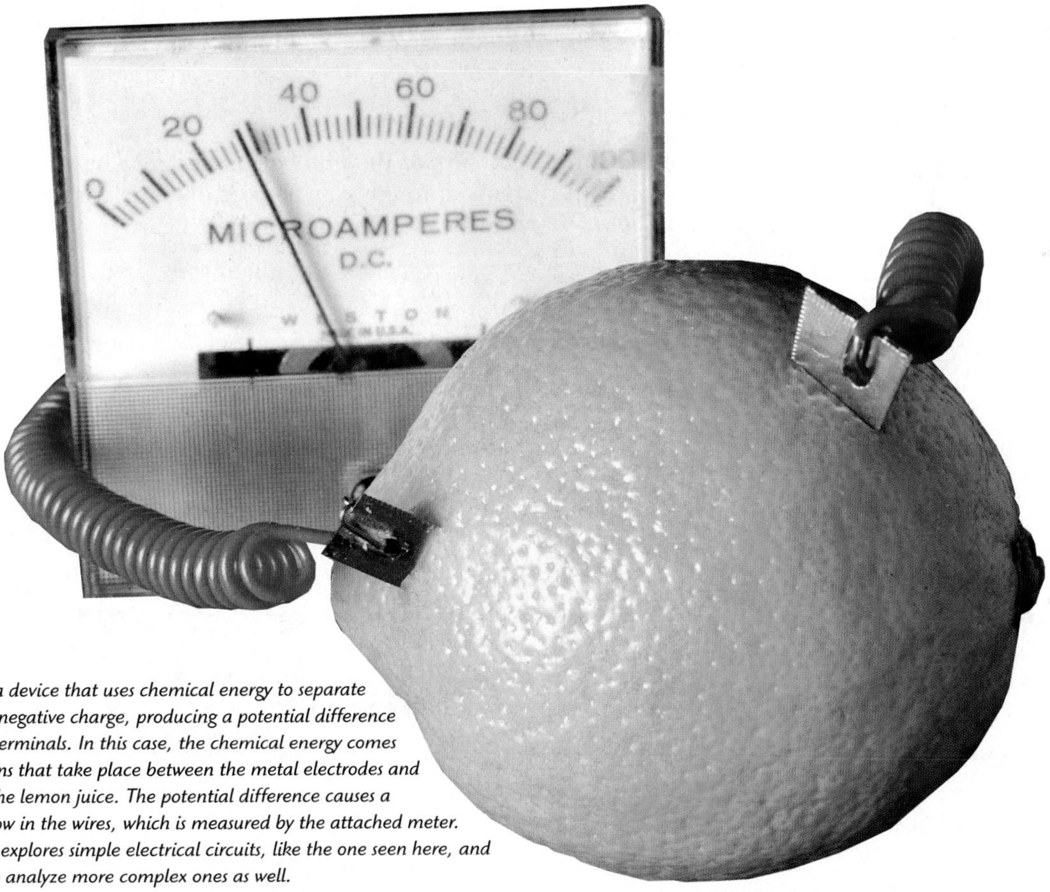

A battery is a device that uses chemical energy to separate positive and negative charge, producing a potential difference between its terminals. In this case, the chemical energy comes from reactions that take place between the metal electrodes and the acid in the lemon juice. The potential difference causes a current to flow in the wires, which is measured by the attached meter. This chapter explores simple electrical circuits, like the one seen here, and shows how to analyze more complex ones as well.

As you read this paragraph, your heart is pumping blood through the arteries and veins in your body. In a way, your heart is acting like a battery in an electrical circuit: A battery causes electric charge to flow through a closed circuit of wires; your heart causes blood to flow through your body. Just as the flow of blood is important to life, the flow of electric charge is of central importance to modern technology. In this chapter we consider some of the basic properties of moving electric charges, and we apply these results to simple electric circuits.

21–1 Electric Current

A flow of electric charge from one place to another is referred to as an **electric current.** Often, the charge is carried by electrons moving through a metal wire. Though the analogy should not be pushed too far, the electrons flowing through a wire are much like water molecules flowing through a garden hose or blood cells flowing through an artery.

To be specific, suppose a charge ΔQ flows past a given point in a wire in a time Δt. In such a case, we say that the electric current, I, in the wire is:

Definition of Electric Current, *I*

$$I = \frac{\Delta Q}{\Delta t} \qquad \text{21–1}$$

SI unit: coulomb per second, C/s = ampere, A

The unit of current, the ampere (A) or *amp* for short, is named for the French physicist André-Marie Ampère (1775–1836) and is defined simply as 1 coulomb per second:

$$1\ \text{A} = 1\ \text{C/s}$$

The following Example shows that the number of electrons involved in typical electric circuits, with currents of roughly an amp, is extremely large—not unlike the large number of water molecules flowing through a garden hose.

EXAMPLE 21–1 Mega Blaster

The disk drive in a portable CD player is connected to a battery that supplies it with a current of 0.22 A. How many electrons pass through the drive in 4.5 s?

Picture the Problem

Our sketch shows the CD drive with a current $I = 0.22$ A flowing through it. Also indicated is the time $\Delta t = 4.5$ s during which the current flows.

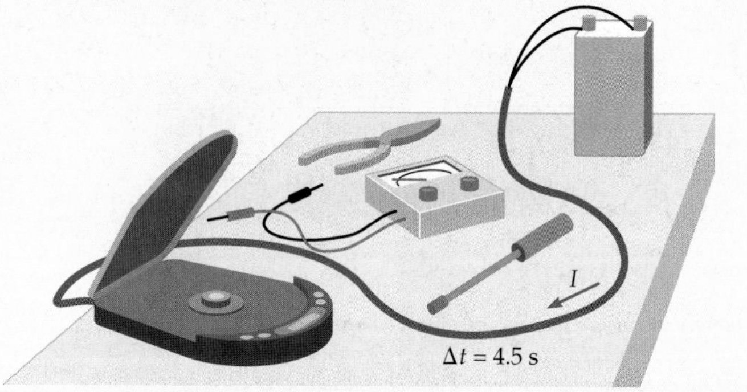

Strategy

Since we know both the current, I, and the length of time, Δt, we can use the definition of current, $I = \Delta Q/\Delta t$, to find the charge, ΔQ, that flows through the player. Once we know the charge, the number of electrons, N, is simply ΔQ divided by the magnitude of the electron's charge: $N = \Delta Q/e$.

Solution

1. Calculate the charge, ΔQ, that flows through the drive: $\Delta Q = I\,\Delta t = (0.22\ \text{A})(4.5\ \text{s}) = 0.99\ \text{C}$

2. Divide by the magnitude of the electron's charge, e, to find the number of electrons: $N = \frac{\Delta Q}{e} = \frac{0.99\ \text{C}}{1.60 \times 10^{-19}\ \text{C/electron}}$ $= 6.2 \times 10^{18}$ electrons

Insight

Thus, even a modest current flowing for a brief time corresponds to the transport of an extremely large number of electrons.

Practice Problem

How long must this current last if 7.5×10^{18} electrons are to flow through the disk drive? [**Answer:** 5.5 s]

Some related homework problems: Problem 1, Problem 2

When charge flows through a closed path and returns to its starting point, we refer to the closed path as an *electric circuit*. In this chapter we consider **direct-current circuits,** also known as DC circuits, in which the current always flows in the same direction. Circuits with currents that periodically reverse their direction

▲ Electric currents are not confined to the wires in our houses and machines, but occur in nature as well. A lightning bolt is simply an enormous, brief current. It flows when the difference in electric potential between cloud and ground (or cloud and cloud) becomes so great that it exceeds the breakdown strength of air. An enormous quantity of charge then leaps across the gap in a fraction of a second. Some organisms, such as this electric torpedo ray, have internal organic "batteries" that can produce significant electrical potentials. The resulting current is used to stun their prey.

are referred to as **alternating-current circuits.** These AC circuits are considered in detail in Chapter 24.

Batteries and Electromotive Force

Although electrons move rather freely in metal wires, they do not flow unless the wires are connected to a source of electrical energy. A close analogy is provided by water in a garden hose. Imagine that you and a friend each hold one end of a garden hose filled with water. If the two ends are held at the same level, as in **Figure 21–1 (a)**, the water does not flow. If, however, one end is raised above the other, as in **Figure 21–1 (b)**, water flows from the high end—where the gravitational potential energy is high—to the low end.

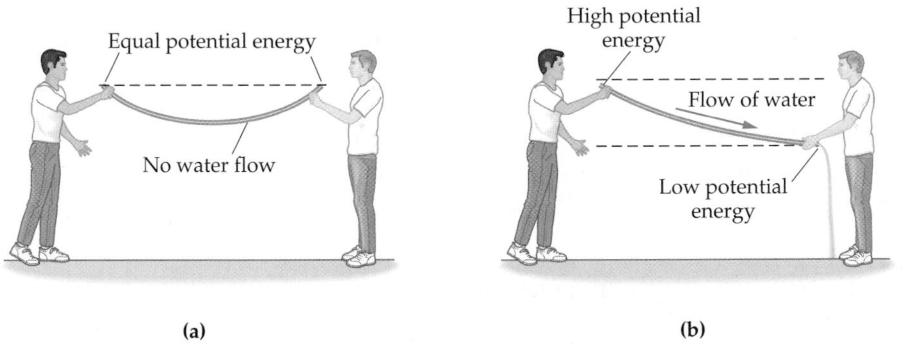

◀ **FIGURE 21–1 Water flow as an analogy for electric current**
Water can flow quite freely through a garden hose, but if both ends are at the same level **(a)**, there is no flow. If the ends are held at different levels **(b)**, the water flows from the region where the gravitational potential energy is high to the region where it is low.

A **battery** performs a similar function in an electric circuit. To put it simply, a battery uses chemical reactions to produce a difference in electric potential between its two ends, or **terminals.** The symbol for a battery is ⊣⊢. The terminal corresponding to a high electric potential is denoted by a +, and the terminal corresponding to a low electric potential is denoted by a −. When the battery is connected to a circuit, electrons move in a closed path from the negative terminal of the battery, through the circuit, and back to the positive terminal.

A simple example of an electrical system is shown in **Figure 21–2 (a)**, where we show a battery, a switch, and a lightbulb as they might be connected in a flashlight. In the schematic circuit shown in **Figure 21–2 (b)**, the switch is "open"—creating an **open circuit**—which means there is no closed path through which the electrons can flow. As a result, the light is off. When the switch is closed—which "closes" the circuit—charge flows around the circuit, causing the light to glow.

A mechanical analog to the flashlight circuit is shown in **Figure 21–3**. In this system, the person raising the water from a low to a high level is analogous to the battery, the paddle wheel is analogous to the lightbulb, and the water is analogous

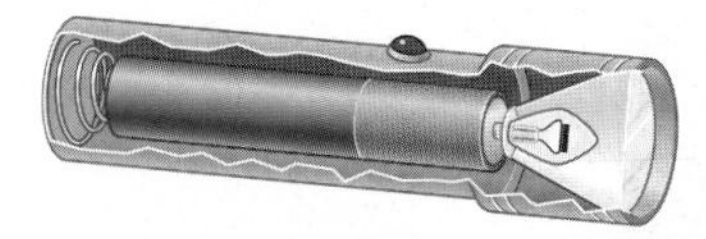

(a)

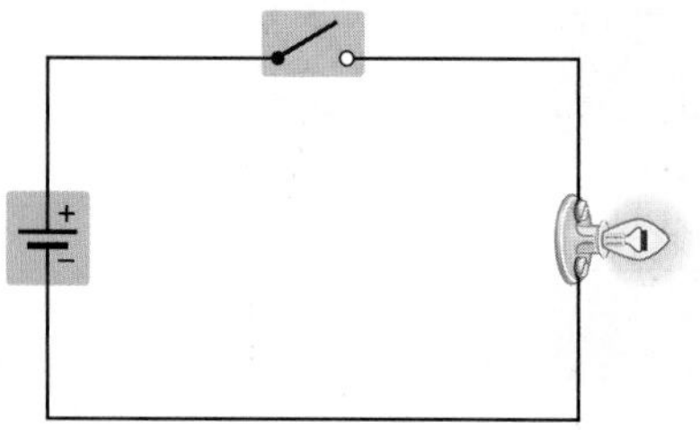

(b)

▲ **FIGURE 21–2 The flashlight: A simple electrical circuit**

(a) A simple flashlight, consisting of a battery, a switch, and a lightbulb.
(b) When the switch is in the open position, the circuit is "broken," and no charge can flow. When the switch is closed, electrons flow through the circuit and the light glows.

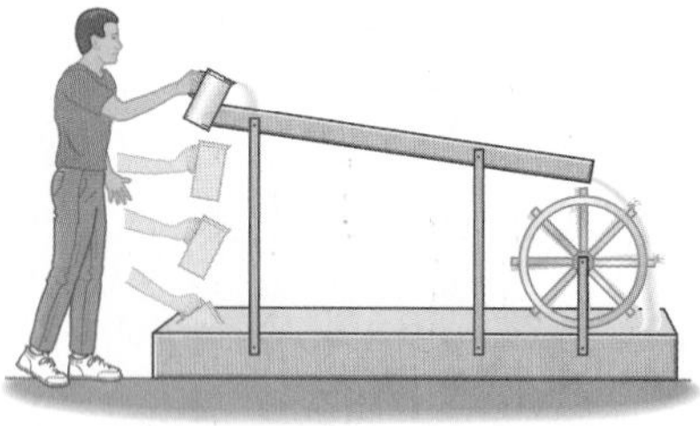

▲ **FIGURE 21–3 A mechanical analog to the flashlight circuit**

The person lifting the water corresponds to the battery in Figure 21–2, and the paddle wheel corresponds to the lightbulb.

to the electric charge. Notice that the person does work in raising the water; later, as the water falls to its original level, it does work on the external world by turning the paddle wheel.

When a battery is disconnected from a circuit and carries no current, the difference in electric potential between its terminals is referred to as its *electromotive force*, or *emf* ($\mathcal{E}$). It follows that the units of emf are the same as those of electric potential, namely, volts. Clearly, then, the electromotive force is not really a force at all. Instead, the emf determines the amount of work a battery does to move a certain amount of charge around a circuit (like the person lifting water in Figure 21–3). To be specific, the magnitude of the work done by a battery of emf $\mathcal{E}$ as a charge ΔQ moves from one of its terminals to the other is given by Equation 20–2:

$$W = \Delta Q\mathcal{E}$$

We apply this relation to a flashlight circuit in the following Active Example.

ACTIVE EXAMPLE 21–1 Operating a Flashlight: Find the Charge and the Work

A battery with an emf of 1.5 V delivers a current of 0.44 A to a flashlight bulb for 64 s (see Figure 21–2). Find **(a)** the charge that passes through the circuit and **(b)** the work done by the battery.

Solution *(Test your understanding by performing the calculations indicated in each step.)*

Part (a)

1. Use the definition of current, $I = \Delta Q/\Delta t$, to find the charge that flows through the circuit: $\Delta Q = 28\ \mathrm{C}$

Part (b)

2. Once we know ΔQ, we can use $W = \Delta Q\mathcal{E}$ to find the work: $W = 42\ \mathrm{J}$

Insight

Note that the more charge a battery moves through a circuit, the more work it does. Similarly, the greater the emf, the greater the work. We can see, then, that a car battery that operates at 12 volts and delivers several amps of current, does much more work than a flashlight battery—as expected.

Your Turn

How long must the flashlight battery operate to do 150 J of work?

(Answers to **Your Turn** *problems are given in the back of the book.)*

The emf of a battery is the potential difference it can produce between its terminals under ideal conditions. In real batteries, however, there is always some internal loss, leading to a potential difference that is less than the ideal value. In fact, the greater the current flowing through a battery the greater the reduction in potential difference between its terminals, as we shall see in Section 21–4. Only when the current is zero can a real battery produce its full emf. Because most batteries have relatively small internal losses, we shall treat batteries as ideal—always producing a potential difference precisely equal to $\mathcal{E}$—unless specifically stated otherwise.

When we draw an electric circuit, it will be useful to draw an arrow indicating the flow of current. By convention, the direction of the current arrow is given in terms of a positive test charge, in much the same way that the direction of the electric field is determined:

> The direction of the current in an electrical circuit is the direction in which a *positive* test charge would move.

Of course, in typical circuits the charges that flow are actually *negatively* charged electrons. As a result, the flow of electrons and the current arrow point in opposite directions, as indicated in **Figure 21–4**. Notice that a positive charge will flow from a region of high electric potential, near the positive terminal of the battery, to a region of low electric potential, near the negative terminal, as one would expect.

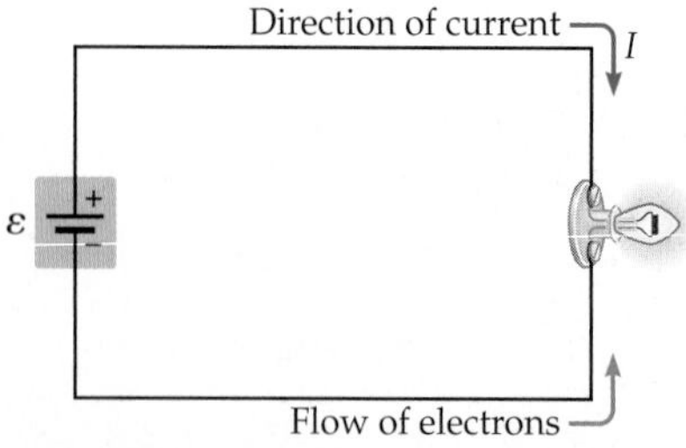

▲ **FIGURE 21–4 Direction of current and electron flow**

In the flashlight circuit, electrons flow from the negative terminal of the battery to the positive terminal. The direction of the current, I, is just the opposite: from the positive terminal to the negative terminal.

Finally, surprising as it may seem, electrons move rather slowly through a typical wire. They suffer numerous collisions with the atoms in the wire, and hence their path is rather tortuous and roundabout, as indicated in **Figure 21–5**. Like a car contending with a series of speed bumps, the electron's average speed, or **drift speed** as it is often called, is limited by the repeated collisions—in fact, their average speed is commonly about 10^{-4} m/s. Thus, if you switch on the headlights of a car, for example, an electron leaving the battery will take about an hour to reach the lightbulb, yet the lights seem to shine from the instant the switch is turned on. How is this possible?

▲ **FIGURE 21–5 Path of an electron in a wire**

Typical path of an electron as it bounces off atoms in a metal wire. Because of the tortuous path the electron follows, its average velocity is rather small.

The answer is that as an electron begins to move away from the battery, it exerts a force on its neighbors, causing them to move in the same general direction and, in turn, to exert a force on their neighbors, and so on. This process generates a propagating influence that travels through the wire at nearly the speed of light. The phenomenon is analogous to a bowling ball's hitting one end of a line of balls; the effect of the colliding ball travels through the line at roughly the speed of sound, although the individual balls have very little displacement. Similarly, the electrons in a wire move with a rather small average velocity as they collide with and bounce off the atoms making up the wire, whereas the influence they have on one another races ahead and causes the light to shine.

21–2 Resistance and Ohm's Law

Electrons flow through metal wires with relative ease. In the ideal case, nothing about the wire would prevent their free motion. Real wires, however, under normal conditions, always affect the electrons to some extent, creating a **resistance** to their motion in much the same way that friction slows a box sliding across the floor.

In order to cause electrons to move against the resistance of a wire, it is necessary to apply a potential difference between its ends. For a wire with constant resistance, R, the potential difference, V, necessary to create a current, I, is given by Ohm's law:

Ohm's Law

$$V = IR \qquad \text{21–2}$$

SI unit: volt, V

Ohm's law is named for the German physicist Georg Simon Ohm (1789–1854).

It should be noted at the outset that Ohm's law is not a law of nature but more on the order of a useful rule of thumb—like Hooke's law for springs or the ideal-gas laws that approximate the behavior of real gases. Materials that are well approximated by Ohm's law are said to be "ohmic" in their behavior; they show a simple linear relationship between the voltage applied to them and the current that results. In particular, if one plots current versus voltage for an ohmic material, the result is a straight line, with a constant slope equal to $1/R$. Nonohmic materials, on the other hand, have more complex relationships between voltage and current. A plot of current versus voltage for a nonohmic material is nonlinear; hence, the material does not have a constant resistance. (As an example, see Conceptual Exercises 5 and 8.) It is precisely these "nonlinearities," however, that can make such materials so useful in the construction of electronic devices, including the ubiquitous light-emitting diodes (LEDs).

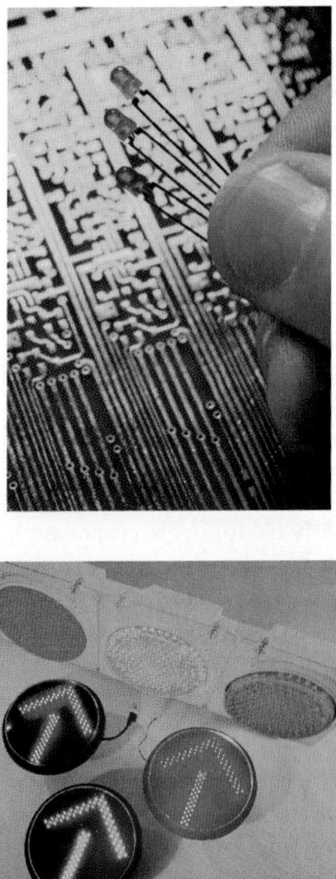

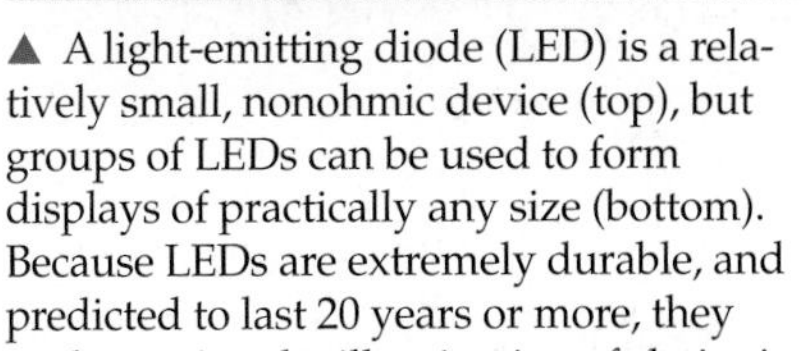

▲ A light-emitting diode (LED) is a relatively small, nonohmic device (top), but groups of LEDs can be used to form displays of practically any size (bottom). Because LEDs are extremely durable, and predicted to last 20 years or more, they are becoming the illumination of choice in high-reliability applications such as traffic lights, emergency exit signs, and brake lights. You'll probably see several on your way home today.

Solving Ohm's law for the resistance, we find

$$R = \frac{V}{I}$$

From this expression it is clear that the units of resistance are volts per amp. In particular, we define 1 volt per amp to be 1 **ohm.** Letting the Greek letter omega, Ω, designate the ohm, we have

$$1\ \Omega = 1\ \text{V/A}$$

A device for measuring resistance is called an ohmmeter. We describe the operation of an ohmmeter in Section 21–8.

EXERCISE 21–1

A potential difference of 24 V is applied to a 150-Ω resistor. How much current flows through the resistor?

Solution

Solving Ohm's law for the current, I, we find

$$I = \frac{V}{R} = \frac{24\ \text{V}}{150\ \Omega} = \frac{24\ \text{V}}{150\ \text{V/A}} = 0.16\ \text{A}$$

In an electrical circuit a resistor is signified by a zigzag line: . The straight lines in a circuit indicate ideal wires of zero resistance. To indicate the resistance of a real wire or device, we simply include a resistor of the appropriate value in the circuit.

TABLE 21–1 Resistivities

Substance	Resistivity, $\rho\,(\Omega \cdot \text{m})$
Insulators	
Quartz (fused)	7.5×10^{17}
Rubber	$1 \text{ to } 100 \times 10^{13}$
Glass	$1 \text{ to } 10{,}000 \times 10^{9}$
Semiconductors	
Silicon*	0.10 to 60
Germanium*	0.001 to 0.5
Conductors	
Lead	22×10^{-8}
Iron	9.71×10^{-8}
Tungsten	5.6×10^{-8}
Aluminum	2.65×10^{-8}
Gold	2.20×10^{-8}
Copper	1.68×10^{-8}
Silver	1.59×10^{-8}

*The resistivity of a semiconductor varies greatly with the type and amount of impurities it contains. This property makes them particularly useful in electronic applications.

Resistivity

Suppose you have a piece of wire of length L and cross-sectional area A. The resistance of this wire depends on the particular material from which it is constructed. If the wire is made of copper, for instance, its resistance will be less than if it is made from iron. The quantity that characterizes the resistance of a given material is its **resistivity,** ρ. For a wire of given dimensions, the greater the resistivity the greater the resistance.

The resistance of a wire also depends on its length and area. To understand the dependence on L and A, consider again the analogy of water flowing through a hose. If the hose is very long, the resistance it presents to the water will be correspondingly large, whereas a wider hose—one with a greater cross-sectional area—will offer less resistance to the water. After all, water flows more easily through a short fire hose than through a long soda straw; hence, the resistance of a hose—and similarly a piece of wire—should be proportional to L and inversely proportional to A; that is, proportional to (L/A).

Combining these observations, we can write the resistance of a wire of length L, area A, and resistivity ρ in the following way:

Definition of Resistivity, ρ

$$R = \rho\left(\frac{L}{A}\right) \qquad \text{21–3}$$

Since the units of L are m and the units of A are m^2, it follows that the units of resistivity are $(\Omega \cdot \text{m})$. Typical values for ρ are given in Table 21–1. Notice the enormous range in values of ρ, with the resistivity of an insulator like rubber about 10^{21} times greater than the resistivity of a good conductor like silver.

CONCEPTUAL CHECKPOINT 21–1 Compare the Resistance

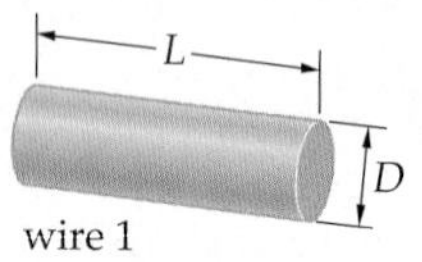

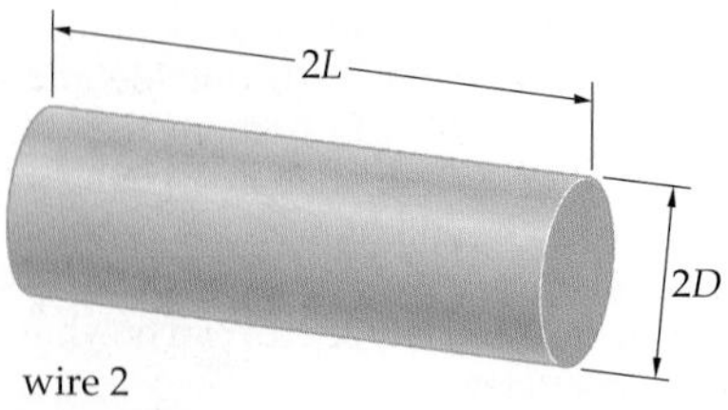

Wire 1 has a length L and a circular cross section of diameter D. Wire 2 is constructed from the same material as wire 1 and has the same shape, but its length is $2L$, and its diameter is $2D$. Is the resistance of wire 2 **(a)** the same as that of wire 1, **(b)** twice that of wire 1, or **(c)** half that of wire 1?

Reasoning and Discussion

First, the resistance of wire 1 is

$$R_1 = \rho\left(\frac{L}{A}\right) = \rho\frac{L}{(\pi D^2/4)}$$

Note that we have used the fact that the area of a circle of diameter D is $\pi D^2/4$. For wire 2 we replace L with $2L$ and D with $2D$:

$$R_2 = \rho\frac{2L}{[\pi(2D)^2/4]} = \left(\tfrac{1}{2}\right)\rho\frac{L}{(\pi D^2/4)} = \tfrac{1}{2}R_1$$

Thus, increasing the length by a factor of 2 increases the resistance by a factor of 2; on the other hand, increasing the diameter by a factor of 2 increases the area, and decreases the resistance, by a factor of 4. Overall, then, the resistance of wire 2 is half that of wire 1.

Answer:
(c) The resistance of wire 2 is half that of wire 1; $R_2 = R_1/2$.

EXAMPLE 21–2 A Current-Carrying Wire

A current of 1.82 A flows through a copper wire 1.75 m long and 1.10 mm in diameter. Find the potential difference between the ends of the wire. (The value of ρ for copper may be found in Table 21–1.)

Picture the Problem

The wire carries a current $I = 1.82$ A, and its total length L is 1.75 m. We assume that the wire has a circular cross section, with a diameter $D = 1.10$ mm.

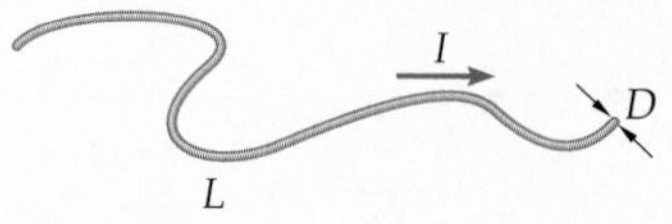

Strategy

We know from Ohm's law that the potential difference associated with a current I and a resistance R is $V = IR$. We are given the current in the wire, but not the resistance. The resistance is easily determined, however, using $R = \rho(L/A)$ with $A = \pi D^2/4$. Thus, we first calculate R and then substitute the result into $V = IR$ to obtain the potential difference.

Solution

1. Calculate the resistance of the wire:

$$R = \rho\left(\frac{L}{A}\right) = \rho\left(\frac{L}{\pi D^2/4}\right)$$

$$= (1.72 \times 10^{-8}\ \Omega \cdot \mathrm{m})\left[\frac{1.75\ \mathrm{m}}{\pi(0.00110\ \mathrm{m})^2/4}\right] = 0.0317\ \Omega$$

2. Multiply R by the current, I, to find the potential difference:

$$V = IR = (1.82\ \mathrm{A})(0.0317\ \Omega) = 0.0577\ \mathrm{V}$$

Insight

Copper is an excellent conductor; therefore, both the resistance and the potential difference are quite small.

Practice Problem

What diameter of copper wire is needed for there to be a potential difference of 0.100 V? Assume that all other quantities remain the same. [**Answer:** 0.835 mm]

Some related homework problems: Problem 12, Problem 13

Temperature Dependence and Superconductivity

We know from everyday experience that a wire carrying an electrical current can become warm—even quite hot, as in the case of a burner on a stove or the filament in an incandescent lightbulb. This follows from our earlier discussion of the fact that electrons collide with the atoms in a wire as they flow through an electrical circuit. These collisions cause the atoms to jiggle with greater kinetic energy about their equilibrium positions. As a result, the temperature of the wire increases (see Section 17–2, and Equation 17–21 in particular). For example, the wire filament in an incandescent lightbulb can reach temperatures of roughly 2800 °C (in comparison, the surface of the Sun has a temperature of about 5500 °C), and the heating coil on a stove has a temperature of about 750 °C.

As a wire is heated, its resistivity tends to increase. This is because atoms that are jiggling more rapidly are more likely to collide with electrons and slow their progress through the wire. In fact, many metals show an approximately linear increase of ρ over a wide range of temperature. Once the dependence of ρ on T is known for a given material, the change in resistivity can be used as a means of measuring temperature.

REAL-WORLD PHYSICS
The bolometer

The first practical application of this principle was in a device known as the **bolometer.** Invented in 1880, the bolometer is an extremely sensitive thermometer that uses the temperature variation in the resistivity of platinum, nickel, or bismuth as a means of detecting temperature changes as small as 0.0001 C°. Soon after its invention, a bolometer was used to detect infrared radiation from the stars.

REAL-WORLD PHYSICS
Thermistors and fever thermometers

Some materials, like semiconductors, actually show a drop in resistivity as temperature is increased. This is because the resistivity of a semiconductor is strongly dependent on the number of electrons that are free to move about and conduct a current. As the temperature is increased in a semiconductor, more electrons are able to break free from their atoms, leading to an increased current and a reduced resistivity. Electronic devices incorporating such temperature-dependent semiconductors are known as **thermistors.** The digital fever thermometer so common in today's hospitals uses a thermistor to provide accurate measurements of a patient's temperature.

Since resistivity typically increases with temperature, it follows that if a wire is cooled below room temperature, its resistivity will *decrease*. Quite surprising, however, was a discovery made in the laboratory of Heike Kamerlingh-Onnes in 1911. Measuring the resistance of a sample of mercury at temperatures just a few degrees above absolute zero, researchers found that at about 4.2 K the resistance of the mercury suddenly dropped to zero—not just to a very small value, but to *zero*. At this temperature, we say that the mercury becomes **superconducting,** a hitherto unknown phase of matter. Since that time many different superconducting materials have been discovered, with various different **critical temperatures,** T_c, at which superconductivity begins. Today we know that superconductivity is a result of quantum effects (Chapter 30).

REAL-WORLD PHYSICS
Superconductors and high-temperature superconductivity

When a material becomes superconducting, a current can flow through it with absolutely no resistance. In fact, if a current is initiated in a superconducting ring of material, it will flow undiminished for as long as the ring is kept cool enough. In some cases, circulating currents have been maintained for years, with absolutely no sign of diminishing.

In 1986 a new class of superconductors was discovered that has zero resistance at temperatures significantly greater than those of any previously known superconducting materials. At the moment, the highest temperature at which superconductivity has been observed is about 125 K. Since the discovery of these "high-T_c" superconductors, hopes have been raised that it may one day be possible to produce room-temperature superconductors. The practical benefits of such a breakthrough, including power transmission with no losses, improved MRI scanners, and magnetically-levitated trains, could be immense.

▲ When cooled below their critical temperature, superconductors not only lose their resistance to current flow but also exhibit new magnetic properties, such as repelling an external magnetic field. Here, a superconductor (bottom) levitates a small permanent magnet.

21–3 Energy and Power in Electric Circuits

When a charge ΔQ moves across a potential difference V, its electrical potential energy, U, changes by the amount

$$\Delta U = (\Delta Q)V$$

Recalling that power is the rate at which energy changes, $P = \Delta U/\Delta t$, we can write the electrical power as follows:

$$P = \frac{\Delta U}{\Delta t} = \frac{(\Delta Q)V}{\Delta t}$$

Since the electric current is given by $I = \Delta Q/\Delta t$, we have:

Electrical Power

$$P = IV \qquad 21\text{–}4$$

SI unit: watt, W

Thus, a current of 1 amp flowing across a potential difference of 1 V produces a power of 1 W.

EXERCISE 21–2

A handheld electric fan operates on a 3.00-V battery. If the power generated by the fan is 2.24 W, what is the current supplied by the battery?

Solution

Solving $P = IV$ for the current, we obtain

$$I = \frac{P}{V} = \frac{2.24\text{ W}}{3.00\text{ V}} = 0.747\text{ A}$$

▲ The heating element of an electric space heater is nothing more than a length of resistive wire coiled up for compactness. As electric current flows though the wire, the power it dissipates ($P = I^2R$) is converted to heat and light. The coils near the center are the hottest, and hence they glow with a higher-frequency, yellowish light.

The expression $P = IV$ applies to any electrical system. In the special case of a resistor, the electrical power is dissipated in the form of heat. Applying Ohm's law ($V = IR$) to this case, we can write the power dissipated in a resistor as

$$P = IV = I(IR) = I^2R \qquad 21\text{–}5$$

Similarly, using Ohm's law to solve for the current, $I = V/R$, we have

$$P = IV = \left(\frac{V}{R}\right)V = \frac{V^2}{R} \qquad 21\text{–}6$$

These relations also apply to incandescent lightbulbs, which are basically resistors that become hot enough to glow.

CONCEPTUAL CHECKPOINT 21–2 Compare Lightbulbs

A battery that produces a potential difference V is connected to a 5-W lightbulb. Later, the 5-W lightbulb is replaced with a 10-W lightbulb. **(a)** In which case does the battery supply the greatest current? **(b)** Which lightbulb has the greater resistance?

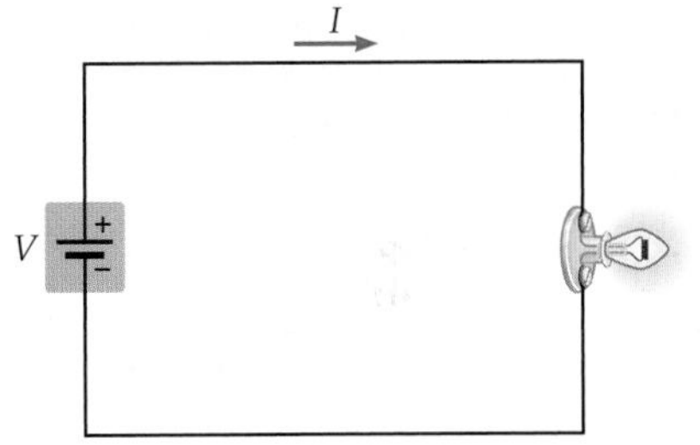

Reasoning and Discussion

a. To compare the currents, we need consider only the relation $P = IV$. Solving for the current yields $I = P/V$. When the voltage V is the same, it follows that the greater the power, the greater the current. In this case, then, the current in the 10-W bulb is twice the current in the 5-W bulb.

b. We now consider the relation $P = V^2/R$, which gives resistance in terms of voltage and power. In fact, $R = V^2/P$. Again, with V the same, it follows that the smaller the power the greater the resistance. Thus, the resistance of the 5-W bulb is twice that of the 10-W bulb.

Answer:

(a) When the battery is connected to the 10-W bulb, it delivers twice as much current as when it is connected to the 5-W bulb. **(b)** The 5-W bulb has twice as much resistance as the 10-W bulb.

On a microscopic level, the power dissipated by a resistor is the result of incessant collisions between electrons moving through the circuit and atoms making up the resistor. Specifically, the electric potential difference produced by the battery causes electrons to accelerate until they bounce off an atom of the resistor. At this point the electrons transfer energy to the atoms, causing them to jiggle more rapidly. The increased kinetic energy of the atoms is reflected in an increased temperature of the resistor (see Section 17–2). After each collision, the potential difference accelerates the electrons again and the process repeats—like a car bouncing through a series of speed bumps—resulting in a continuous transfer of energy from the electrons to the atoms.

EXAMPLE 21–3 Heated Resistance

A battery with an emf of 12 V is connected to a 545-Ω resistor. How much energy is dissipated in the resistor in 65 s?

Picture the Problem

The circuit, consisting of a battery and a resistor, is shown in our sketch. We show the current flowing from the positive terminal of the 12-V battery, through the 545-Ω resistor, and into the negative terminal of the battery.

continued on next page

continued from previous page

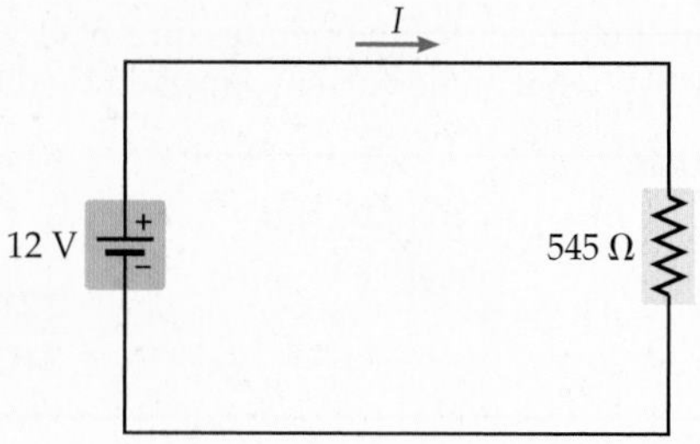

Strategy

We know that a current flowing through a resistor dissipates power (energy per time), which means that the energy it dissipates in a given time is simply the power multiplied by the time: $\Delta U = P\,\Delta t$. The time is given ($\Delta t = 65$ s), and the power can be found using $P = IV$, $P = I^2R$, or $P = V^2/R$. The last expression is most convenient in this case, because the problem statement gives us the voltage and resistance.

To summarize, we first calculate the power, then multiply by the time.

Solution

1. Calculate the power dissipated in the resistor: $P = V^2/R = (12\text{ V})^2/(545\ \Omega) = 0.26\text{ W}$
2. Multiply the power by the time to find the energy dissipated: $\Delta U = P\,\Delta t = (0.26\text{ W})(65\text{ s}) = 17\text{ J}$

Insight

The current in this circuit is $I = V/R = 0.022$ A. Using this result, we find that the power is $P = I^2R = IV = 0.26$ W, as expected.

Practice Problem

How much energy is dissipated in the resistor if the voltage is doubled to 24 V? [**Answer:** 4(17 J) = 68 J]

Some related homework problems: Problem 21, Problem 24

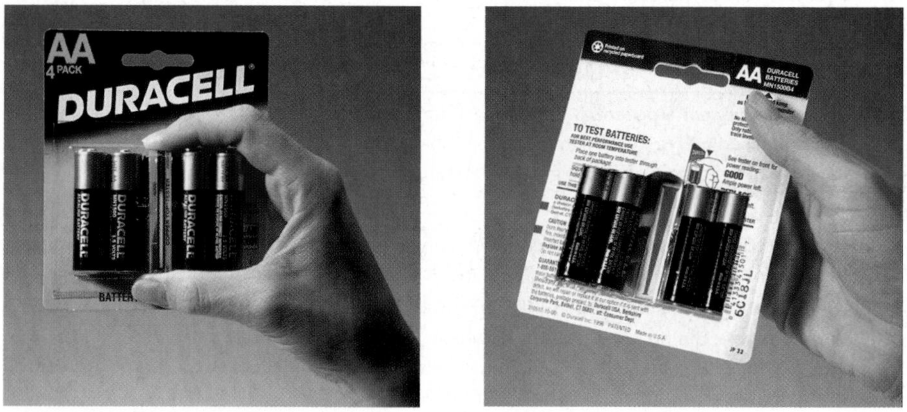

▶ The battery testers now often built into battery packages (left) employ a tapered graphite strip. The narrow end (at bottom in the right-hand photo) has the highest resistance, and thus produces the most heat when a current flows through the strip. The heat is used to produce the display on the front that indicates the strength of the battery—if the current is sufficient to warm even the top of the strip, where the resistance is lowest, the battery is fresh.

REAL-WORLD PHYSICS
"Battery check" meters

A commonly encountered application of resistance heating is found in the "battery check" meters often included with packs of batteries. To operate one of these meters, you simply press the contacts on either end of the meter against the corresponding terminals of the battery to be checked. This allows a current to flow through the main working element of the meter—a tapered strip of graphite.

The reason the strip is tapered is to provide a variation in resistance. According to the relation given in Equation 21–3, $R = \rho(L/A)$, the smaller the cross-sectional area A of the strip, the larger the resistance R. It follows that the narrow end has a higher resistance than the wide end. Because the same current I flows through all parts of the strip, the power dissipated is expressed most conveniently in the form $P = I^2R$. It follows that at the narrow end of the strip, where R is largest, the heating due to the current will be the greatest. Pressing the meter against the terminals of the battery, then, results in an overall warming of the graphite strip, with the narrow end warmer than the wide end.

The final element in the meter is a thin layer of liquid crystal (similar to the material used in LCD displays) that responds to small increases in temperature. In particular, this liquid crystal is black and opaque at room temperature but transparent when heated slightly. The liquid crystal is placed in front of a colored background, which can be seen in those regions where the graphite strip is warm enough to make the liquid crystal transparent. If the battery is weak, only the narrow portion of the strip becomes warm enough, and the meter shows only a small stripe of color. A strong battery, on the other hand, heats the entire strip enough to make the liquid crystal transparent, resulting in a colored stripe the full length of the meter.

Energy Usage

When you get a bill from the local electric company, you will find the number of kilowatt-hours of electricity that you have used. Notice that a kilowatt-hour (kWh) has the units of energy:

$$1 \text{ kilowatt-hour} = (1000 \text{ W})(3600 \text{ s}) = (1000 \text{ J/s})(3600 \text{ s})$$
$$= 3.6 \times 10^6 \text{ J}$$

Thus, the electric company is charging for the amount of energy you use—as one would expect—and not for the rate at which you use it. The following Example considers the energy and monetary cost for a typical everyday situation.

EXAMPLE 21–4 Your Goose Is Cooked

A holiday goose is cooked in the kitchen oven for 4.00 h. Assume that the stove draws a current of 20.0 A, operates at a voltage of 220.0 V, and uses electrical energy that costs $0.068 per kWh. How much does it cost to cook your goose?

Picture the Problem

We show a schematic representation of the stove cooking the goose in our sketch. The current in the circuit is 20.0 A, and the voltage difference across the heating coils is 220 V.

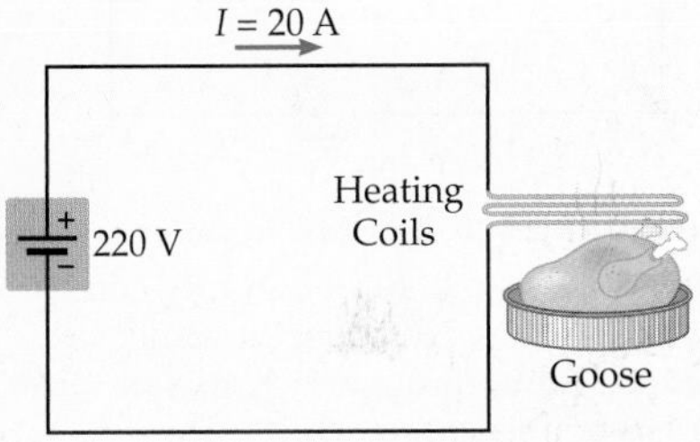

Strategy

The cost is simply the energy usage (in kWh) times the cost per kilowatt-hour ($0.068). To find the energy used, we note that energy is power multiplied by time. The time is given, and the power associated with a current I and a voltage V is $P = IV$.

Thus, we find the power, multiply by the time, and then multiply by $0.068 to find the cost.

Solution

1. Calculate the power delivered to the stove: $P = IV = (20.0 \text{ A})(220.0 \text{ V}) = 4.40 \text{ kW}$
2. Multiply power by time to determine the total energy supplied to the stove during the 4.00 h of cooking: $\Delta U = P\,\Delta t = (4.40 \text{ kW})(4.00 \text{ h}) = 17.6 \text{ kWh}$
3. Multiply by the cost per kilowatt-hour to find the total cost of cooking: $\text{cost} = (17.6 \text{ kWh})(\$0.068/\text{kWh}) = \$1.20$

Insight

Thus, your goose can be cooked for just over a dollar.

Practice Problem

If the voltage and current are reduced by a factor of two each, how long must the goose be cooked to use the same amount of energy? [**Answer:** 4(4.00 h) = 16.0 h. *Note*: You should be able to answer a question like this by referring to your previous solution, without repeating the calculation in detail.]

Some related homework problems: Problem 22, Problem 23

21–4 Resistors in Series and Parallel

Electrical circuits often contain a number of resistors connected in various ways. In this section we consider simple circuits containing only resistors and batteries. For each type of circuit considered, we calculate the **equivalent resistance** produced by a group of individual resistors.

Resistors in Series

When resistors are connected one after the other, end to end, we say that they are in *series*. **Figure 21–6 (a)** shows three resistors, R_1, R_2, and R_3, connected in series. The three resistors acting together have the same effect—that is, they draw the same current—as a single resistor, referred to as the equivalent resistor, R_{eq}. This equivalence is illustrated in **Figure 21–6 (b)**. We now calculate the value of the equivalent resistance.

The first thing to notice about the circuit in Figure 21–6 (a) is that the same current I must flow through each of the resistors—there is no other place for the

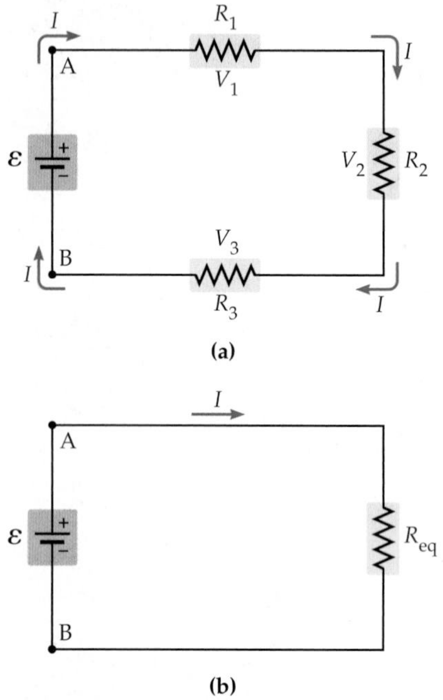

▲ **FIGURE 21–6 Resistors in series**
(a) Three resistors, R_1, R_2, and R_3, connected in series. Note that the same current I flows through each resistor. **(b)** The equivalent resistance, $R_{eq} = R_1 + R_2 + R_3$, has the same current I flowing through it as the current I in the original circuit.

current to go. As a result, the potential differences across the three resistors are $V_1 = IR_1$, $V_2 = IR_2$, and $V_3 = IR_3$, respectively. Since the total potential difference from point A to point B must be the emf of the battery, $\mathcal{E}$, it follows that

$$\mathcal{E} = V_1 + V_2 + V_3$$

Writing each of the potentials in terms of the current and resistance, we find

$$\mathcal{E} = IR_1 + IR_2 + IR_3 = I(R_1 + R_2 + R_3)$$

Now, let's compare this expression with the result we obtain for the equivalent circuit in Figure 21–6 (b). In this circuit, the potential difference across the battery is $V = IR_{eq}$. Since this potential must be the same as the emf of the battery, we have

$$\mathcal{E} = IR_{eq}$$

Comparing this expression with $\mathcal{E} = I(R_1 + R_2 + R_3)$, we see that the equivalent resistance is simply the sum of the individual resistances:

$$R_{eq} = R_1 + R_2 + R_3$$

In general, for any number of resistors in series, the equivalent resistance is

Equivalent Resistance for Resistors in Series

$$R_{eq} = R_1 + R_2 + R_3 + \cdots = \sum R \qquad \textbf{21–7}$$

SI unit: ohm, Ω

Note that the equivalent resistance is greater than the greatest resistance of any of the individual resistors. Connecting the resistors in series is like making a single resistor increasingly longer; as its length increases so does its resistance.

EXAMPLE 21–5 Three Resistors in Series

A circuit consists of three resistors connected in series to a 24.0-V battery. The current in the circuit is 0.0320 A. Given that $R_1 = 250.0\ \Omega$ and $R_2 = 150.0\ \Omega$, find **(a)** the value of R_3 and **(b)** the potential difference across each resistor.

Picture the Problem

The circuit is shown in our sketch. Note that the 24.0-V battery delivers the same current, $I = 0.0320$ A, to each of the three resistors. This is the key characteristic of a series circuit.

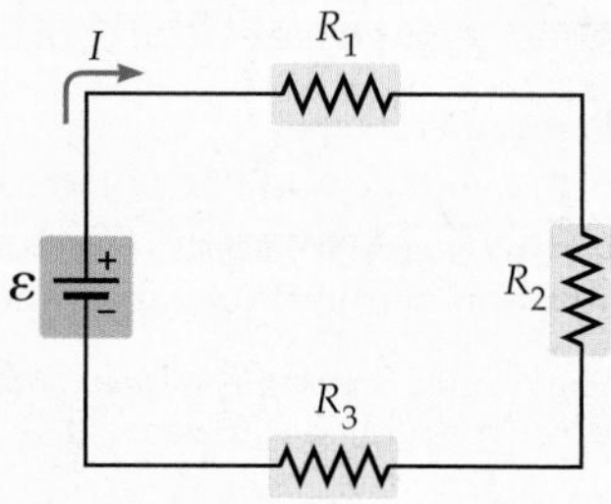

Strategy

a. First, we can obtain the equivalent resistance of the circuit using Ohm's law (as in Equation 21–2); $R_{eq} = \mathcal{E}/I$. Since the resistors are in series, we also know that $R_{eq} = R_1 + R_2 + R_3$. We can solve this relation for the only unknown, R_3.

b. We can then calculate the potential difference across each resistor using Ohm's law, $V = IR$.

Solution

Part (a)

1. Use Ohm's law to find the equivalent resistance of the circuit: $R_{eq} = \dfrac{\mathcal{E}}{I} = \dfrac{24.0\ \text{V}}{0.0320\ \text{A}} = 7.50 \times 10^2\ \Omega$

2. Set R_{eq} equal to the sum of the individual resistances, and solve for R_3:

$$R_{eq} = R_1 + R_2 + R_3$$
$$R_3 = R_{eq} - R_1 - R_2$$
$$= 7.50 \times 10^2\ \Omega - 250.0\ \Omega - 150.0\ \Omega = 3.50 \times 10^2\ \Omega$$

Part (b)

3. Use Ohm's law to determine the potential difference across R_1: $V_1 = IR_1 = (0.0320\ \text{A})(250.0\ \Omega) = 8.00\ \text{V}$

4. Find the potential difference across R_2: $V_2 = IR_2 = (0.0320\ \text{A})(150.0\ \Omega) = 4.80\ \text{V}$

5. Find the potential difference across R_3: $V_3 = IR_3 = (0.0320\ \text{A})(3.50 \times 10^2\ \Omega) = 11.2\ \text{V}$

Insight

Note that the greater the resistance, the greater the potential difference. In addition, the sum of the individual potential differences is 8.00 V + 4.80 V + 11.2 V = 24.0 V, as expected.

Practice Problem

Find the power dissipated in each resistor. [**Answer:** $P_1 = 0.256$ W, $P_2 = 0.154$ W, $P_3 = 0.358$ W]

Some related homework problems: Problem 31, Problem 32

An everyday example of resistors in series is the **internal resistance,** r, of a battery. As was mentioned in Section 21–1, real batteries have internal losses that cause the potential difference between their terminals to be less than $\mathcal{E}$ and to depend on the current in the battery. The simplest way to model a real battery is to imagine it to consist of an ideal battery of emf $\mathcal{E}$ in series with an internal resistance r, as shown in **Figure 21–7**. If this battery is then connected to an external resistance, R, the equivalent resistance of the circuit is $r + R$. As a result, the current flowing through the circuit is $I = \mathcal{E}/(r + R)$, and the potential difference between the terminals of the battery is $\mathcal{E} - Ir$. Thus, we see that the potential difference produced by the battery is less than $\mathcal{E}$ by an amount that is proportional to the current I. Only in the limit of zero current, or zero internal resistance, will the battery produce its full emf. (See Problems 39, 42, 85, and 91 for examples of batteries with internal resistance.)

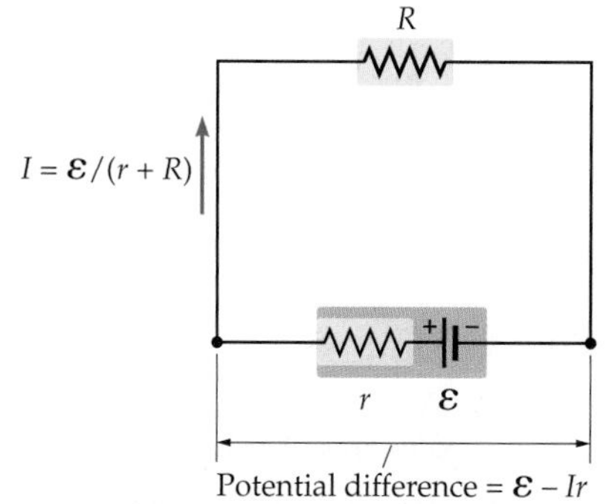

▲ **FIGURE 21–7 The internal resistance of a battery**

Real batteries always dissipate some energy in the form of heat. These losses can be modeled by a small "internal" resistance, r, within the battery. As a result, the potential difference between the terminals of a real battery is less than its ideal emf, $\mathcal{E}$. For example, if a battery produces a current I, the potential difference between its terminals is $\mathcal{E} - Ir$. In the case shown here, a battery is connected in series with the resistor R. Instead of producing the current $I = \mathcal{E}/R$, as in the ideal case, it produces the current $I = \mathcal{E}/(r + R)$.

Another application of resistors in series is the three-way lightbulb circuit shown in **Figure 21–8**. In this circuit, the two resistors represent two different filaments within a single bulb that are connected to a constant potential difference V. At the "high" setting, the lower-resistance filament, R_1, is connected to the electrical outlet via terminals A and B, and the brightest light is obtained ($P = V^2/R$). At the "middle" setting, the higher-resistance filament, R_2, is connected to the outlet via terminals B and C, resulting in a dimmer light. Finally, at the "low" setting, both filaments are connected in series via terminals A and C. This setting gives the greatest equivalent resistance, and thus the lowest light output.

An alternative method of producing a three-way lightbulb is to connect the resistors in parallel. This will be discussed in the next subsection.

REAL-WORLD PHYSICS

Three-way lightbulbs

Resistors in Parallel

Resistors are in *parallel* when they are connected across the same potential difference, as in **Figure 21–9 (a)**. In a case like this, the current has parallel paths through which it can flow. As a result, the total current in the circuit, I, is equal to the sum of the currents through each of the three resistors:

$$I = I_1 + I_2 + I_3$$

Since the potential difference is the same for each of the resistors, it follows that the currents flowing through them are as follows:

$$I_1 = \frac{\mathcal{E}}{R_1}, \quad I_2 = \frac{\mathcal{E}}{R_2}, \quad I_3 = \frac{\mathcal{E}}{R_3}$$

Summing these three currents, we find

$$I = \frac{\mathcal{E}}{R_1} + \frac{\mathcal{E}}{R_2} + \frac{\mathcal{E}}{R_3} = \mathcal{E}\left(\frac{1}{R_1} + \frac{1}{R_2} + \frac{1}{R_3}\right) \qquad \textbf{21–8}$$

Now, in the equivalent circuit shown in **Figure 21–9 (b)**, Ohm's law gives $\mathcal{E} = IR_{eq}$, or

$$I = \mathcal{E}\left(\frac{1}{R_{eq}}\right) \qquad \textbf{21–9}$$

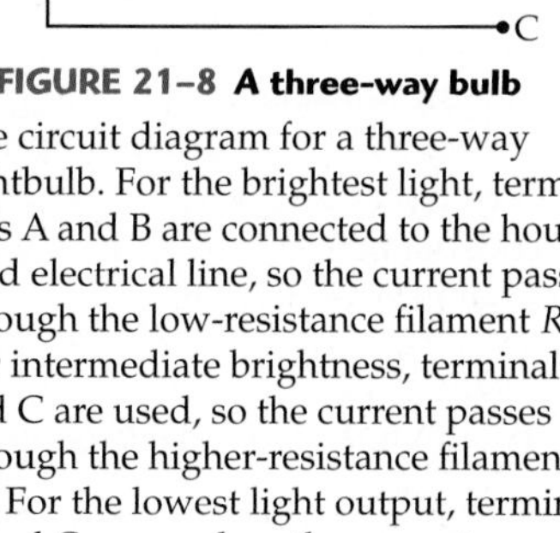

▲ **FIGURE 21–8 A three-way bulb**

The circuit diagram for a three-way lightbulb. For the brightest light, terminals A and B are connected to the household electrical line, so the current passes through the low-resistance filament R_1. For intermediate brightness, terminals B and C are used, so the current passes through the higher-resistance filament R_2. For the lowest light output, terminals A and C are used, so the current passes through both R_1 and R_2 in series.

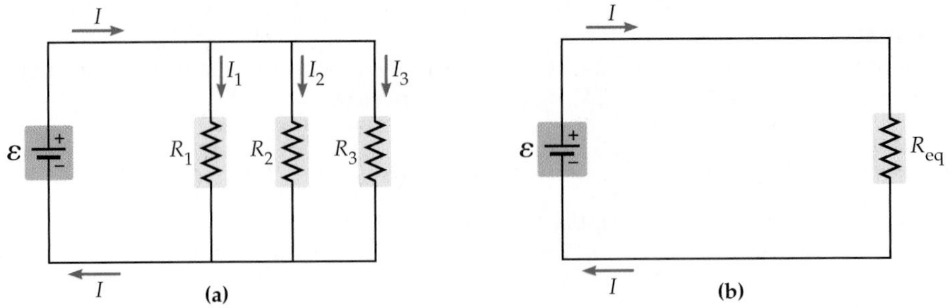

▶ **FIGURE 21–9 Resistors in parallel**
(a) Three resistors, R_1, R_2, and R_3, connected in parallel. Note that each resistor is connected across the same potential difference, $\mathcal{E}$. **(b)** The equivalent resistance, $1/R_{eq} = 1/R_1 + 1/R_2 + 1/R_3$, has the same current flowing through it as the total current I in the original circuit.

Comparing Equations 21–8 and 21–9, we find that the equivalent resistance for three resistors in parallel is

$$\frac{1}{R_{eq}} = \frac{1}{R_1} + \frac{1}{R_2} + \frac{1}{R_3}$$

In general, for any number of resistors in parallel, we have:

PROBLEM-SOLVING NOTE

The Equivalent Resistance of Resistors in Parallel

After summing the inverse of resistors in parallel, remember to take one more inverse at the end of your calculation to find the equivalent resistance.

Equivalent Resistance for Resistors in Parallel

$$\frac{1}{R_{eq}} = \frac{1}{R_1} + \frac{1}{R_2} + \frac{1}{R_3} + \cdots = \sum \frac{1}{R} \qquad 21\text{–}10$$

SI unit: ohm, Ω

As a simple example, consider a circuit with two identical resistors R connected in parallel. The equivalent resistance in this case is given by

$$\frac{1}{R_{eq}} = \frac{1}{R} + \frac{1}{R} = \frac{2}{R}$$

Solving for R_{eq}, we find $R_{eq} = \frac{1}{2}R$. If we connect three such resistors in parallel, the corresponding result is

$$\frac{1}{R_{eq}} = \frac{1}{R} + \frac{1}{R} + \frac{1}{R} = \frac{3}{R}$$

In this case, $R_{eq} = \frac{1}{3}R$. Thus, the more resistors we connect in parallel, the smaller the equivalent resistance. Each time we add a new resistor in parallel with the others, we give the battery a new path through which current can flow—analogous to opening an additional lane of traffic on a busy highway. Stated another way, giving the current multiple paths through which it can flow is equivalent to using a wire with a greater cross-sectional area. From either point of view, the fact that more current flows with the same potential difference means that the equivalent resistance has been reduced.

Finally, if any one of the resistors in a parallel connection is equal to zero, the equivalent resistance is also zero. This situation is referred to as a **short circuit,** and is illustrated in **Figure 21–10**. In this case, all of the current flows through the path of zero resistance.

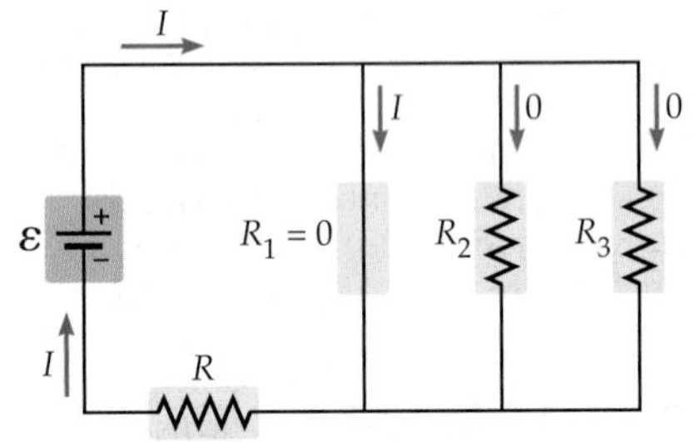

▲ **FIGURE 21–10 A short circuit**
If one of the resistors in parallel with others is equal to zero, all the current flows through that portion of the circuit, giving rise to a short circuit. In this case, resistors R_2 and R_3 are "shorted out," and the current in the circuit is $I = \mathcal{E}/R$.

EXAMPLE 21–6 Three Resistors in Parallel

Consider a circuit with three resistors, $R_1 = 250.0\ \Omega$, $R_2 = 150.0\ \Omega$, and $R_3 = 350.0\ \Omega$, connected in parallel with a 24.0-V battery. Find **(a)** the total current supplied by the battery and **(b)** the current through each resistor.

Picture the Problem

The accompanying sketch indicates the parallel connection of the resistors with the battery. Notice that each of the resistors experiences precisely the same potential difference; namely, the 24.0 V produced by the battery. This is the feature that characterizes parallel connections.

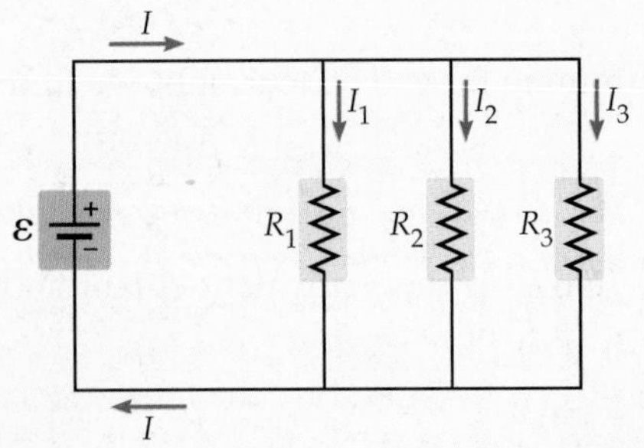

Strategy

a. We can find the total current from $I = \mathcal{E}/R_{eq}$, where $1/R_{eq} = 1/R_1 + 1/R_2 + 1/R_3$.

b. For each resistor, the current is given by Ohm's law, $I = \mathcal{E}/R$.

Solution

Part (a)

1. Find the equivalent resistance of the circuit:

$$\frac{1}{R_{eq}} = \frac{1}{R_1} + \frac{1}{R_2} + \frac{1}{R_3}$$

$$= \frac{1}{250.0\ \Omega} + \frac{1}{150.0\ \Omega} + \frac{1}{350.0\ \Omega} = 0.01352\ \Omega^{-1}$$

$$R_{eq} = (0.01352\ \Omega^{-1})^{-1} = 73.96\ \Omega$$

2. Use Ohm's law to find the total current:

$$I = \frac{\mathcal{E}}{R_{eq}} = \frac{24.0\ \text{V}}{73.96\ \Omega} = 0.325\ \text{A}$$

Part (b)

3. Calculate I_1 using $I_1 = \mathcal{E}/R_1$ with $\mathcal{E} = 24.0$ V:

$$I_1 = \frac{\mathcal{E}}{R_1} = \frac{24.0\ \text{V}}{250.0\ \Omega} = 0.0960\ \text{A}$$

4. Repeat the preceding calculation for resistors 2 and 3:

$$I_2 = \frac{\mathcal{E}}{R_2} = \frac{24.0\ \text{V}}{150.0\ \Omega} = 0.160\ \text{A}$$

$$I_3 = \frac{\mathcal{E}}{R_3} = \frac{24.0\ \text{V}}{350.0\ \Omega} = 0.0686\ \text{A}$$

Insight

As expected, the smallest resistor, R_2, carries the greatest current. The three currents combined yield the total current, as they must; that is, $I_1 + I_2 + I_3 = 0.0960\ \text{A} + 0.160\ \text{A} + 0.0686\ \text{A} = 0.325\ \text{A} = I$.

Practice Problem

Find the power dissipated in each resistor. [**Answer:** $P_1 = 2.30$ W, $P_2 = 3.84$ W, $P_3 = 1.65$ W]

Some related homework problems: Problem 33, Problem 34

In comparing Examples 21–5 and 21–6 note the differences in the power dissipated in each circuit. First, the total power dissipated in the parallel circuit is much greater than that dissipated in the series circuit. This is due to the fact that the equivalent resistance of the parallel circuit is smaller than the equivalent resistance of the series circuit, and the power delivered by a voltage V to a resistance R is inversely proportional to the resistance ($P = V^2/R$). In addition, note that the smallest resistor, R_2, has the smallest power in the series circuit but the largest power in the parallel circuit. These issues are explored further in the following Conceptual Checkpoint.

CONCEPTUAL CHECKPOINT 21–3 **Series versus Parallel**

Two identical lightbulbs are connected to a battery, either in series or in parallel. Are the bulbs in series **(a)** brighter, **(b)** dimmer, or **(c)** the same brightness as the bulbs in parallel?

Reasoning and Discussion

Both sets of lightbulbs are connected to the same potential difference, V; hence, the power delivered to the bulbs is V^2/R_{eq}, where R_{eq} is twice the resistance of a bulb in the series circuit and half the resistance of a bulb in the parallel circuit. As a result, more power is converted to light in the parallel circuit.

Answer:

(b) The bulbs connected in series are dimmer than the bulbs connected in parallel.

Finally, note that a three-way lightbulb can also be produced by simply wiring two filaments in parallel. For example, one filament might have a power of 50 W and the second filament a power of 100 W. One setting of the switch sends current through the 50-W filament, the next setting sends current through the 100-W filament, and the third setting connects the two filaments in parallel. With the third connection, each filament produces the same power as before—since each is connected to the same potential difference—giving a total power of 50 W + 100 W = 150 W.

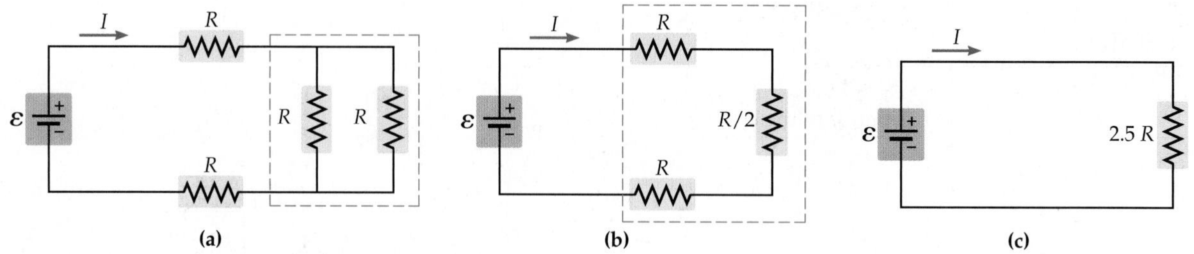

▲ **FIGURE 21–11 Analyzing a complex circuit of resistors**

(a) The two vertical resistors are in parallel with one another; hence, they can be replaced with their equivalent resistance, $R/2$. **(b)** Now the circuit consists of three resistors in series. The equivalent resistance of these three resistors is $2.5\ R$. **(c)** The original circuit reduced to a single equivalent resistance.

Combination Circuits

The rules we have developed for series and parallel resistors can be applied to more complex circuits as well. For example, consider the circuit shown in **Figure 21–11 (a)**, where four resistors, each equal to R, are connected in a way that combines series and parallel features. To analyze this circuit, we first note that the two vertically oriented resistors are connected in parallel with one another. Therefore, the equivalent resistance of this unit is given by $1/R_{eq} = 1/R + 1/R$, or $R_{eq} = R/2$. Replacing these two resistors with $R/2$ yields the circuit shown in **Figure 21–11 (b)**, which consists of three resistors in series. As a result, the equivalent resistance of the entire circuit is $R + R/2 + R = 2.5R$, as indicated in **Figure 21–11 (c)**. Similar methods can be applied to a wide variety of circuits.

PROBLEM-SOLVING NOTE

Analyzing a Complex Circuit

When considering an electric circuit with resistors in series and parallel, work from the smallest units of the circuit outward to ever larger units.

EXAMPLE 21–7 Combination Special

In the circuit shown in the diagram, the emf of the battery is 12.0 V, and each resistor has a resistance of 200.0 Ω. Find the current supplied by the battery to this circuit.

Picture the Problem

The circuit for this problem has three resistors connected to a battery. Note that the lower two resistors are in series with one another, and in parallel with the upper resistor. The battery has an emf of 12.0 V.

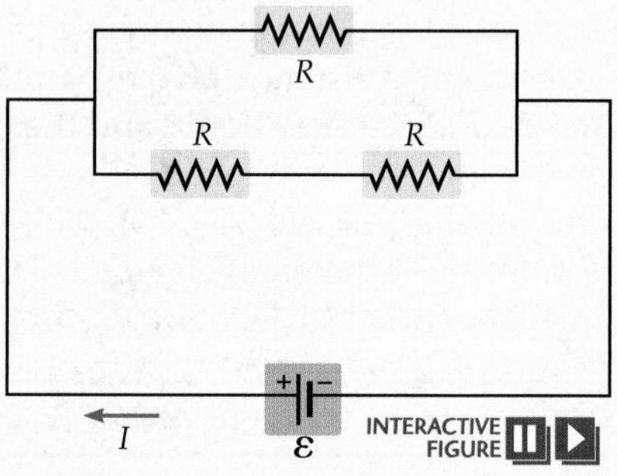

Strategy

The current I is given by Ohm's law, $I = \mathcal{E}/R_{eq}$, where R_{eq} is the equivalent resistance of the three resistors. To find R_{eq}, we first note that the lower two resistors are in series, giving a net resistance of $2R$. Next, the upper resistor, R, is in parallel with $2R$. Calculating this equivalent resistance yields the desired R_{eq}.

Solution

1. Calculate the equivalent resistance of the lower two resistors:

$$R_{eq,lower} = R + R = 2R$$

2. Calculate the equivalent resistance of R in parallel with $2R$:

$$\frac{1}{R_{eq}} = \frac{1}{R} + \frac{1}{2R} = \frac{3}{2R}$$

$$R_{eq} = \tfrac{2}{3}R = \tfrac{2}{3}(200.0\ \Omega) = 133.3\ \Omega$$

3. Find the current, I:

$$I = \frac{\mathcal{E}}{R_{eq}} = \frac{12.0\ \text{V}}{133.3\ \Omega} = 0.0900\ \text{A}$$

Insight

Note that the total resistance of the three 200.0-Ω resistors is only 133.3 Ω.

Practice Problem

Suppose the upper resistor is changed from R to $2R$. **(a)** Will the current in the circuit increase, decrease, or stay the same? **(b)** Find the new current in the circuit. [**Answer:** (a) The current will decrease, because there is greater resistance to its flow. (b) 0.0600 A]

Some related homework problems: Problem 36, Problem 37, Problem 39

21–5 Kirchhoff's Rules

To find the currents and voltages in a general electrical circuit, we use two rules first introduced by the German physicist Gustav Kirchhoff (1824–1887). The *Kirchhoff rules* are simply ways of expressing charge conservation (the junction rule) and energy conservation (the loop rule) in a closed circuit. Since these conservation laws are always obeyed in nature, the Kirchhoff rules are completely general.

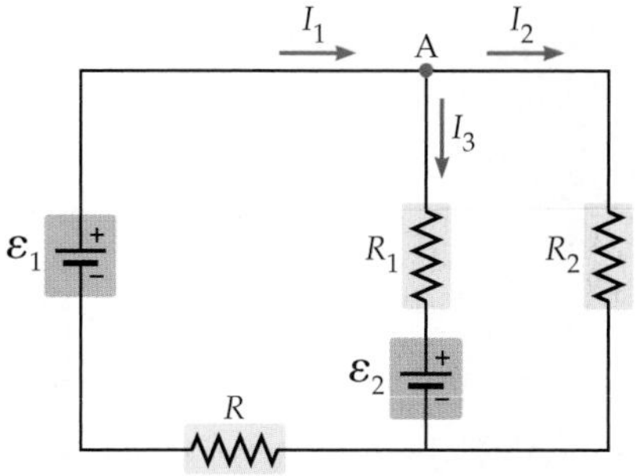

▲ **FIGURE 21–12 Kirchhoff's junction rule**

Kirchhoff's junction rule states that the sum of the currents entering a junction must equal the sum of the currents leaving the junction. In this case, for the junction labeled A, $I_1 = I_2 + I_3$, or $I_1 - I_2 - I_3 = 0$.

The Junction Rule

The junction rule follows from the observation that the current entering any point in a circuit must equal the current leaving that point. If this were not the case, charge would either build up or disappear from a circuit.

As an example, consider the circuit shown in **Figure 21–12**. At point A, three wires join to form a **junction.** (In general, a *junction* is any point in a circuit where three or more wires meet.) The current carried by each of the three wires is indicated in the figure. Notice that the current entering the junction is I_1; the current leaving the junction is $I_2 + I_3$. Setting the incoming and outgoing currents equal, we have $I_1 = I_2 + I_3$, or equivalently

$$I_1 - I_2 - I_3 = 0$$

This is Kirchhoff's junction rule applied to the junction at point A.

In general, if we associate a + sign with currents entering a junction and a − sign with currents leaving a junction, Kirchhoff's junction rule can be stated as follows:

> The algebraic sum of all currents meeting at any junction in a circuit must equal zero.

In the example just discussed, I_1 enters the junction (+), I_2 and I_3 leave the junction (−); hence, the algebraic sum of currents at the junction is $I_1 - I_2 - I_3$. Setting this sum equal to zero recovers our previous result.

In some cases we may not know the direction of all the currents meeting at a junction in advance. When this happens, we simply choose a direction for the unknown currents, apply the junction rule, and continue as usual. If the value we obtain for a given current is negative, it simply means that the direction we chose was wrong; the current actually flows in the opposite direction.

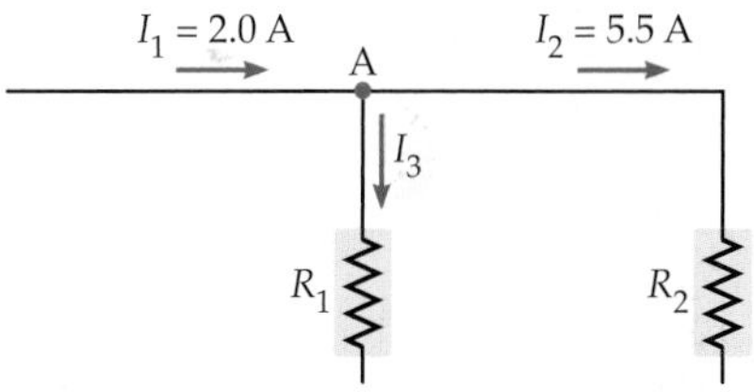

▲ **FIGURE 21–13 A specific application of Kirchhoff's junction rule**

Applying Kirchhoff's junction rule to the junction A, $I_1 - I_2 - I_3 = 0$, yields the result $I_3 = -3.5$ A. The minus sign indicates that I_3 flows opposite to the direction shown; that is, I_3 is upward.

For example, suppose we know both the direction and magnitude of the currents I_1 and I_2 in **Figure 21–13**. To find the third current, we apply the junction rule—but first we must choose a direction for I_3. If we choose I_3 to point downward, as shown in the figure, the junction rule gives

$$I_1 - I_2 - I_3 = 0$$

Solving for I_3, we have

$$I_3 = I_1 - I_2 = 2.0\text{ A} - 5.5\text{ A} = -3.5\text{ A}$$

Since I_3 is negative, we conclude that the actual direction of this current is upward; that is, the 2.0-A current and the 3.5-A current enter the junction and combine to yield the 5.5-A current that leaves the junction.

The Loop Rule

Imagine taking a day hike on a mountain path. First, you gain altitude to reach a scenic viewpoint; later you descend below your starting point into a valley; finally, you gain altitude again and return to the trailhead. During the hike you sometimes increase your gravitational potential energy, and sometimes you decrease it, but the net change at the end of the hike is zero—after all, you return to the same altitude from which you started. Kirchhoff's loop rule is an application of the same idea to an electrical circuit.

For example, consider the simple circuit shown in **Figure 21–14**. The electric potential increases by the amount $\mathcal{E}$ in going from point A to point B, since we

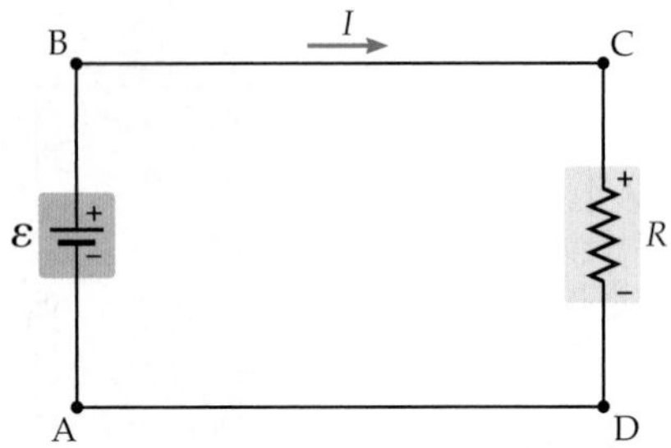

▲ **FIGURE 21–14 Kirchhoff's loop rule**

Kirchhoff's loop rule states that as one moves around a closed loop in a circuit, the algebraic sum of all potential differences must be zero. The electric potential increases as one moves from the − to the + plate of a battery; it decreases as one moves through a resistor in the direction of the current.

move from the low-potential (−) terminal of the battery to the high-potential (+) terminal. This is like gaining altitude in the hiking analogy. Next, there is no potential change as we go from point B to point C, since these points are connected by an ideal wire. As we move from point C to point D, however, the potential does change—recall that a potential difference is required to force a current through a resistor. We label the potential difference across the resistor ΔV_{CD}. Finally, there is no change in potential between points D and A, since they too are connected by an ideal wire.

We can now apply the idea that the net change in electric potential (the analog to gravitational potential energy in the hike) must be zero around any closed loop. In this case, we have

$$\mathcal{E} + \Delta V_{CD} = 0$$

Thus, we find that $\Delta V_{CD} = -\mathcal{E}$; that is, the electric potential *decreases* as one moves across the resistor *in the direction of the current*. To indicate this drop in potential, we label the side where the current enters the resistor with a + (indicating high potential) and the side where the current leaves the resistor with a − (indicating low potential). Finally, we can use Ohm's law to set the magnitude of the potential drop equal to IR and find the current in the circuit:

$$|\Delta V_{CD}| = \mathcal{E} = IR$$

$$I = \frac{\mathcal{E}}{R}$$

This, of course, is the expected result.

In general, Kirchhoff's loop rule can be stated as follows:

> The algebraic sum of all potential differences around any closed loop in a circuit is zero.

We now consider a variety of applications in which both the junction rule and the loop rule are used to find the various currents and potentials in a circuit.

PROBLEM-SOLVING NOTE

Applying Kirchhoff's Rules

When applying Kirchhoff's rules, be sure to use the appropriate sign for currents and potential differences.

Applications

We begin by considering the relatively simple circuit shown in **Figure 21–15**. The currents and voltages in this circuit can be found by considering various parallel and series combinations of the resistors, as we did in the previous section. Thus, Kirchhoff's rules are not strictly needed in this case. Still, applying the rules to this circuit illustrates many of the techniques that can be used when studying more complex circuits.

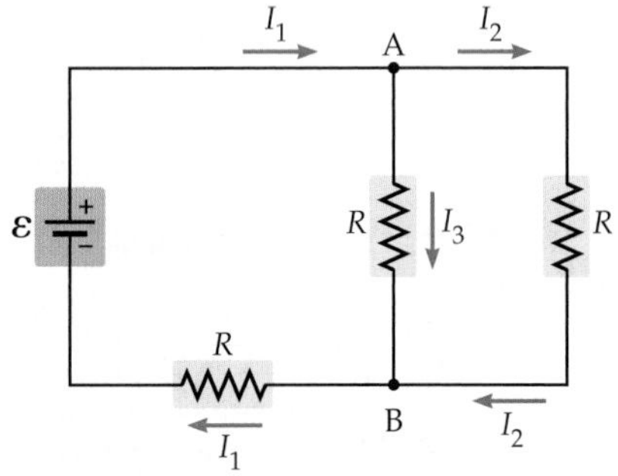

▲ **FIGURE 21–15 Analyzing a simple circuit**

A simple circuit that can be studied using either equivalent resistance or Kirchhoff's laws.

Let's suppose that all the resistors have the value $R = 100.0\ \Omega$, and that the emf of the battery is $\mathcal{E} = 15.0$ V. The equivalent resistance of the resistors can be obtained by noting that the vertical resistors are connected in parallel with one another and in series with the horizontal resistor. The vertical resistors combine to give a resistance of $R/2$, which, when added to the horizontal resistor, gives an equivalent resistance of $R_{eq} = 3R/2 = 150.0\ \Omega$. The current in the circuit, then, is $I = \mathcal{E}/R_{eq} = 15.0\ \text{V}/150.0\ \Omega = 0.100$ A.

Now we approach the same problem from the point of view of Kirchhoff's rules. First, we apply the junction rule to point A:

$$I_1 - I_2 - I_3 = 0 \qquad \text{(junction A)} \qquad \textbf{21–11}$$

Note that current I_1 splits at point A into currents I_2 and I_3, which combine again at point B to give I_1 flowing through the horizontal resistor. We can apply the junction rule to point B, which gives $-I_1 + I_2 + I_3 = 0$, but since this differs from Equation 21–11 by only a minus sign, no new information is gained.

Next, we apply the loop rule. Since there are three unknowns, I_1, I_2, and I_3, we need three independent equations for a full solution. One has already been given by the junction rule; thus, we expect that two loop equations will be required to complete the solution. To begin, we consider loop 1, which is shown in **Figure 21–16 (a)**. We choose to move around this loop in the clockwise direction. (If we were to choose the counterclockwise direction instead, the same information would be obtained.) For loop 1, then, we have an increase in potential as we move across the battery, a drop in potential across the vertical resistor of I_3R, and another drop in potential across the horizontal resistor, this time of magnitude I_1R. Applying the loop rule, we find the following:

$$\mathcal{E} - I_3R - I_1R = 0 \qquad \text{(loop 1)} \qquad \textbf{21–12}$$

Similarly, we can apply the loop rule to loop 2, shown in **Figure 21–16 (b)**. In this case we cross the right-hand vertical resistor in the direction of the current, implying a drop in potential, and we cross the left-hand vertical resistor against the current, implying an increase in potential. Therefore, the loop rule gives

$$I_3R - I_2R = 0 \qquad \text{(loop 2)} \qquad \textbf{21–13}$$

There is a third possible loop, shown in **Figure 21–16 (c)**, but the information it gives is not different from that already obtained. In fact, *any two of the three loops* complete our solution.

Note that R cancels in Equation 21–13; hence, we see that $I_3 - I_2 = 0$, or $I_3 = I_2$. Substituting this result into the junction rule (Equation 21–11), we obtain

$$\begin{aligned} I_1 - I_2 - I_3 &= I_1 - I_2 - I_2 \\ &= I_1 - 2I_2 = 0 \end{aligned}$$

Solving this equation for I_2 gives us $I_2 = I_1/2 = I_3$. Finally, using the first loop equation (Equation 21–12), we find

$$\mathcal{E} - (I_1/2)R - I_1R = \mathcal{E} - \tfrac{3}{2}I_1R = 0$$

Note that the only unknown in this equation is current I_1. Solving for this current, we find

$$I_1 = \frac{\mathcal{E}}{\tfrac{3}{2}R} = \frac{15.0\ \text{V}}{\tfrac{3}{2}(100.0\ \Omega)} = 0.100\ \text{A}$$

As expected, our result using Kirchhoff's rules agrees with the result obtained previously. Finally, the other two currents in the circuit are $I_2 = I_3 = I_1/2 = 0.0500$ A.

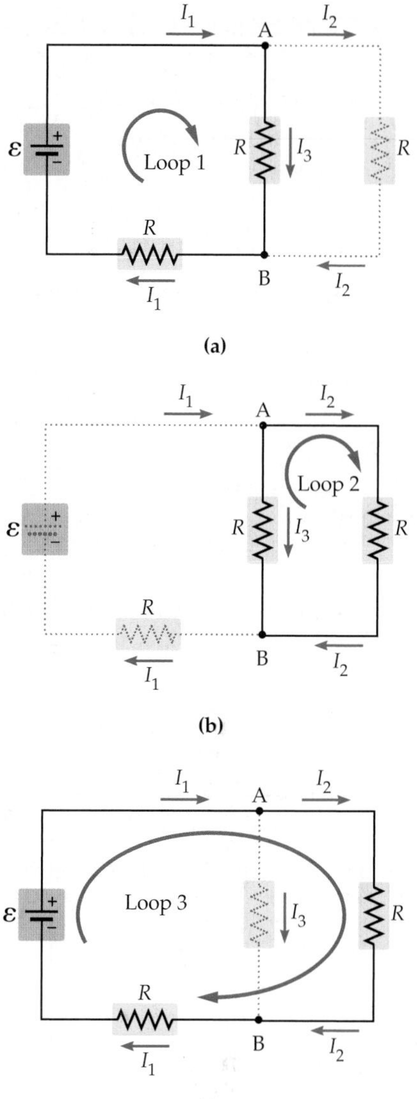

▲ **FIGURE 21–16 Using loops to analyze a circuit**

Three loops associated with the circuit in Figure 21–15.

EXERCISE 21–3

Write the loop equation for loop 3 in Figure 21–16 (c).

Solution

Proceeding in a clockwise direction, as indicated in the figure, we find

$$\mathcal{E} - I_2R - I_1R = 0$$

Since I_2 and I_3 are equal (loop 2), it follows that loop 1 ($\mathcal{E} - I_3R - I_1R = 0$) and loop 3 ($\mathcal{E} - I_2R - I_1R = 0$) give the same information. If we proceed in a counterclockwise direction around loop 3, we find

$$-\mathcal{E} + I_2R + I_1R = 0$$

Notice that this result is the same as the clockwise result except for an overall minus sign, and, therefore, it contains no new information. In general, it does not matter in which direction we choose to go around a loop.

Clearly, the Kirchhoff approach is more involved than the equivalent-resistance method. However, it is not possible to analyze all circuits in terms of equivalent resistances. In such cases, Kirchhoff's rules are the only option, as illustrated in the next Active Example.

ACTIVE EXAMPLE 21–2 Two Loops, Two Batteries: Find the Currents

Find the currents in the circuit shown.

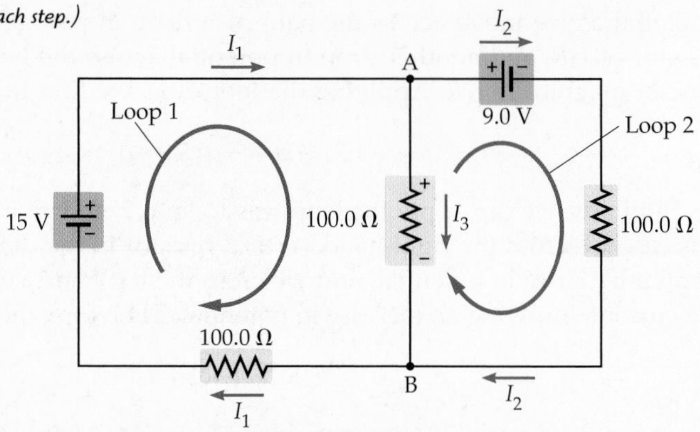

Solution *(Test your understanding by performing the calculations indicated in each step.)*

1. Apply the junction rule to point A: $I_1 - I_2 - I_3 = 0$
2. Apply the loop rule to loop 1 (let $R = 100.0\ \Omega$): $15\text{ V} - I_3R - I_1R = 0$
3. Apply the loop rule to loop 2 (let $R = 100.0\ \Omega$): $-9.0\text{ V} - I_2R + I_3R = 0$
4. Solve for I_1, I_2, and I_3: $I_1 = 0.070\text{ A}$, $I_2 = -0.010\text{ A}$, $I_3 = 0.080\text{ A}$

Insight

Note that I_2 is negative. This means that its direction is opposite to that shown in the circuit diagram.

Your Turn

Suppose the polarity of the 9.0-V battery is reversed. What are the currents in this case?

(Answers to **Your Turn** *problems are given in the back of the book.)*

21–6 Circuits Containing Capacitors

To this point we have considered only resistors and batteries in electrical circuits. Capacitors, which can also play an important role, are represented by a set of parallel lines (reminiscent of a parallel-plate capacitor): ⊣⊢ . We now investigate simple circuits involving batteries and capacitors, leaving for the next section circuits that combine all three circuit elements.

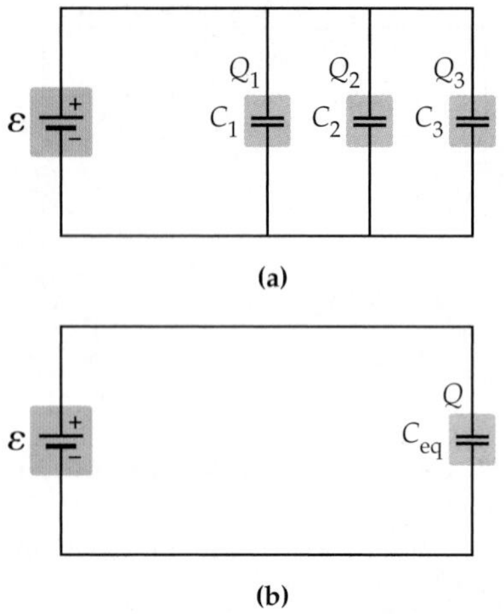

▲ **FIGURE 21–17 Capacitors in parallel**
(a) Three capacitors C_1, C_2, and C_3, connected in parallel. Note that each capacitor is connected across the same potential difference, $\mathcal{E}$. **(b)** The equivalent capacitance, $C_{eq} = C_1 + C_2 + C_3$, has the same charge on its plates as the total charge on the three original capacitors.

Capacitors in Parallel

The simplest way to combine capacitors, as we shall see, is by connecting them in parallel. For example, **Figure 21–17 (a)** shows three capacitors connected in parallel with a battery of emf $\mathcal{E}$. As a result, each capacitor has the same potential difference, $\mathcal{E}$, between its plates. The magnitudes of the charges on each capacitor are as follows:

$$Q_1 = C_1\mathcal{E}, \qquad Q_2 = C_2\mathcal{E}, \qquad Q_3 = C_3\mathcal{E}$$

As a result, the total charge on the three capacitors is

$$Q = Q_1 + Q_2 + Q_3 = \mathcal{E}C_1 + \mathcal{E}C_2 + \mathcal{E}C_3 = (C_1 + C_2 + C_3)\mathcal{E}$$

If an equivalent capacitor is used to replace the three in parallel, as in **Figure 21–17 (b)**, the charge on its plates must be the same as the total charge on the individual capacitors:

$$Q = C_{eq}\mathcal{E}$$

Comparing $Q = C_{eq}\mathcal{E}$ with $Q = (C_1 + C_2 + C_3)\mathcal{E}$, we see that the equivalent capacitance is simply

$$C_{eq} = C_1 + C_2 + C_3$$

In general, the equivalent capacitance of capacitors connected in parallel is the sum of the individual capacitances:

Equivalent Capacitance for Capacitors in Parallel

$$C_{eq} = C_1 + C_2 + C_3 + \cdots = \sum C \qquad \textbf{21–14}$$

SI unit: farad, F

Thus, connecting capacitors in parallel produces an equivalent capacitance greater than the greatest individual capacitance. It is as if the plates of the individual capacitors are connected together to give one large set of plates, with a correspondingly large capacitance.

EXAMPLE 21–8 Energy in Parallel

Two capacitors, one 12.0 μF and the other of unknown capacitance C, are connected in parallel across a battery with an emf of 9.00 V. The total energy stored in the two capacitors is 0.0115 J. What is the value of the capacitance C?

Picture the Problem

The circuit, consisting of one 9.00-V battery and two capacitors, is illustrated in the diagram. The total energy of 0.0115 J stored in the two capacitors is the same as the energy stored in the equivalent capacitance for this circuit.

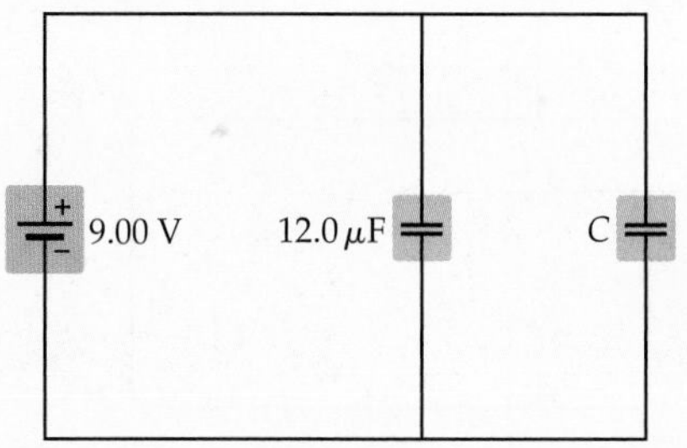

Strategy

Recall from Chapter 20 that the energy stored in a capacitor can be written as $U = \frac{1}{2}CV^2$. It follows, then, that for an equivalent capacitance, C_{eq}, the energy is $U = \frac{1}{2}C_{eq}V^2$. Since we know the energy and voltage, we can solve this relation for the equivalent capacitance. Finally, the equivalent capacitance is the sum of the individual capacitances, $C_{eq} = 12.0\ \mu\text{F} + C$. We use this relation to solve for C.

Solution

1. Solve $U = \frac{1}{2}C_{eq}V^2$ for the equivalent capacitance:

$$U = \frac{1}{2}C_{eq}V^2$$

$$C_{eq} = \frac{2U}{V^2}$$

2. Substitute numerical values to find C_{eq}:

$$C_{eq} = \frac{2U}{V^2} = \frac{2(0.0115\ \text{J})}{(9.00\ \text{V})^2} = 284\ \mu\text{F}$$

3. Solve for C in terms of the equivalent capacitance:

$$C_{eq} = 12.0\ \mu\text{F} + C$$

$$C = C_{eq} - 12.0\ \mu\text{F} = 284\ \mu\text{F} - 12.0\ \mu\text{F} = 272\ \mu\text{F}$$

Insight

The energy stored in the 12.0-μF capacitor is $U = \frac{1}{2}CV^2 = 0.000486$ J. In comparison, the 272-μF capacitor stores an energy equal to 0.0110 J. Thus, the larger capacitor stores the greater amount of energy. Though this may seem only natural, one needs to be careful. When we examine capacitors in *series* later in this section, we shall find exactly the opposite result.

Practice Problem

What is the total charge stored on the two capacitors? [**Answer:** $Q = C_{eq}\mathcal{E} = 2.56 \times 10^{-3}$ C]

Some related homework problems: Problem 55, Problem 56

REAL-WORLD PHYSICS
"Touch-sensitive" lamps

Although you probably haven't realized it, when you turn on a "touch sensitive" lamp, you are part of a circuit with capacitors in parallel. In fact, you are one of the capacitors! When you touch such a lamp, a small amount of charge moves onto your body—your body is like the plate of a capacitor. Because you have effectively increased the plate area—as always happens when capacitors are connected in parallel—the capacitance of the circuit increases. The electronic circuitry in the lamp senses this increase in capacitance and triggers the switch to turn the light on or off.

Capacitors in Series

You have probably noticed from Equation 21–14 that capacitors connected in *parallel* combine in the same way as resistors connected in *series*. Similarly, capacitors connected in *series* obey the same rules as resistors connected in *parallel*, as we now show.

Consider three capacitors—initially uncharged—connected in series with a battery, as in **Figure 21–18 (a)**. The battery causes the left plate of C_1 to acquire a positive charge, $+Q$. This charge, in turn, attracts a negative charge $-Q$ onto the right plate of

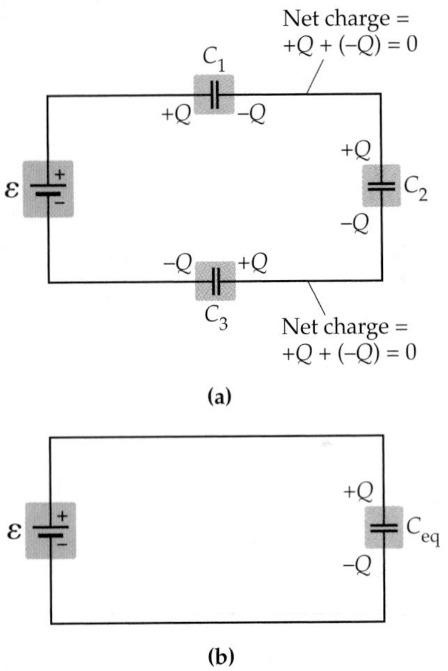

▲ **FIGURE 21–18 Capacitors in series**
(a) Three capacitors, C_1, C_2, and C_3, connected in series. Note that each capacitor has the same magnitude charge on its plates. **(b)** The equivalent capacitance, $1/C_{eq} = 1/C_1 + 1/C_2 + 1/C_3$, has the same charge as the original capacitors.

the capacitor. Because the capacitors start out uncharged, there is zero net charge between C_1 and C_2. As a result, the negative charge $-Q$ on the right plate of C_1 leaves a corresponding positive charge $+Q$ on the upper plate of C_2. The charge $+Q$ on the upper plate of C_2 attracts a negative charge $-Q$ onto its lower plate, leaving a corresponding positive charge $+Q$ on the right plate of C_3. Finally, the positive charge on the right plate of C_3 attracts a negative charge $-Q$ onto its left plate. The result is that all three capacitors have charge of the same magnitude on their plates.

With the same charge Q on all the capacitors, the potential difference for each is as follows:

$$V_1 = \frac{Q}{C_1}, \qquad V_2 = \frac{Q}{C_2}, \qquad V_3 = \frac{Q}{C_3}$$

Since the total potential difference across the three capacitors must equal the emf of the battery, we have

$$\mathcal{E} = V_1 + V_2 + V_3 = \frac{Q}{C_1} + \frac{Q}{C_2} + \frac{Q}{C_3} = Q\left(\frac{1}{C_1} + \frac{1}{C_2} + \frac{1}{C_3}\right) \qquad \textbf{21–15}$$

An equivalent capacitor connected to the same battery, as in **Figure 21–18 (b)**, will satisfy the relation $Q = C_{eq}\mathcal{E}$, or

$$\mathcal{E} = Q\left(\frac{1}{C_{eq}}\right) \qquad \textbf{21–16}$$

A comparison of Equations 21–15 and 21–16 yields the result

$$\frac{1}{C_{eq}} = \frac{1}{C_1} + \frac{1}{C_2} + \frac{1}{C_3}$$

Thus, in general, we have the following rule for combining capacitors in series:

Equivalent Capacitance for Capacitors in Series

$$\frac{1}{C_{eq}} = \frac{1}{C_1} + \frac{1}{C_2} + \frac{1}{C_3} + \cdots = \sum \frac{1}{C} \qquad \textbf{21–17}$$

SI unit: farad, F

It follows, then, that the equivalent capacitance of a group of capacitors connected in series is less than the smallest individual capacitance. In this case, it is as if the plate separations of the individual capacitors add to give a larger effective separation, and a correspondingly smaller capacitance.

More complex circuits, with some capacitors in series and others in parallel, can be handled in the same way as was done earlier with resistors. This is illustrated in the following Active Example.

PROBLEM-SOLVING NOTE

Finding the Equivalent Capacitance of a Circuit

When calculating the equivalent capacitance of capacitors in series, be sure to take one final inverse at the end of your calculation to find C_{eq}. Also, when considering circuits with capacitors in both series and parallel, start with the smallest units of the circuit and work your way out to the larger units.

ACTIVE EXAMPLE 21–3 Find the Equivalent Capacitance and the Stored Energy

Consider the electrical circuit shown below, consisting of a 12.0-V battery and three capacitors connected partly in series and partly in parallel. Find **(a)** the equivalent capacitance of this circuit and **(b)** the total energy stored in the capacitors.

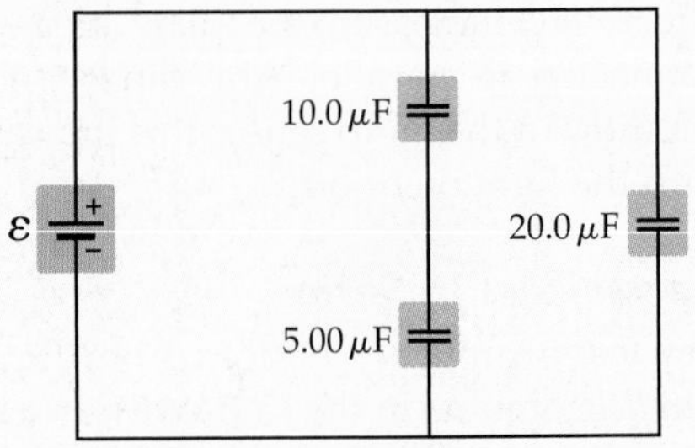

Solution *(Test your understanding by performing the calculations indicated in each step.)*

Part (a)

1. Find the equivalent capacitance of a 10.0-μF capacitor in series with a 5.00-μF capacitor: $\quad$ 3.33 μF

2. Find the equivalent capacitance of a 3.33-μF capacitor in parallel with a 20.0-μF capacitor: $\quad$ $C_{eq} = 23.3\ \mu F$

Part (b)

3. Calculate the stored energy using $U = \frac{1}{2}C_{eq}V^2$: $\quad$ $U = 1.68 \times 10^{-3}$ J

Insight

Notice that the 10.0-μF capacitor and the 5.00-μF capacitor are connected in series. As you might expect, one of these capacitors stores twice as much energy as the other. Which is it? Check the **Your Turn** question for the answer.

Your Turn

Is the energy stored in the 10.0-μF capacitor greater than or less than the energy stored in the 5.0-μF capacitor? Explain. Check your answer by calculating the energy stored in each of the capacitors.

(Answers to **Your Turn** *problems are given in the back of the book.)*

21–7 *RC* Circuits

When the switch is closed on a circuit containing only batteries and capacitors, the charge on the capacitor plates appears almost instantaneously—essentially at the speed of light. This is not the case, however, in circuits that also contain resistors. In these situations, the resistors limit the rate at which charge can flow, and an appreciable amount of time may be required before the capacitors acquire a significant charge. A useful analogy is the amount of time needed to fill a bucket with water. If you use a fire hose, which has little resistance to the flow of water, the bucket fills almost instantly. If you use a garden hose, which presents a much greater resistance to the water, filling the bucket may take a minute or more.

The simplest example of such a circuit, a so-called ***RC* circuit,** is shown in **Figure 21–19**. Initially (before $t = 0$) the switch is open, and there is no current in the resistor or charge on the capacitor. At $t = 0$ the switch is closed and current begins to flow. If the resistor was not present, the capacitor would immediately take on the charge $Q = C\mathcal{E}$. The effect of the resistor, however, is to slow the charging process—in fact, the larger the resistance, the longer it takes for the capacitor to charge. One way to think of this is to note that as long as a current flows in the circuit, as in Figure 21–19 (b), there is a potential drop across the resistor; hence, the potential difference between the plates of the capacitor is less than the emf of the battery. With less voltage across the capacitor there will be less charge on its plates compared with the charge that would result if the plates were connected directly to the battery.

The methods of calculus can be used to show that the charge on the capacitor in Figure 21–19 varies with time as follows:

$$q(t) = C\mathcal{E}(1 - e^{-t/\tau}) \tag{21–18}$$

In this expression, e is Euler's number ($e = 2.718\ldots$) or, more precisely, the base of natural logarithms (see Appendix A). The quantity τ is referred to as the **time constant** of the circuit. The time constant is related to the resistance and capacitance of a circuit by the following simple relation: $\tau = RC$. As we shall see, τ can be thought of as a characteristic time for the behavior of an *RC* circuit.

For example, at time $t = 0$ the exponential term is $e^{-0/\tau} = e^0 = 1$; therefore, the charge on the capacitor is zero at $t = 0$, as expected:

$$q(0) = C\mathcal{E}(1 - 1) = 0$$

In the opposite limit, $t \rightarrow \infty$, the exponential vanishes: $e^{-\infty/\tau} = 0$. Thus the charge in this limit is $C\mathcal{E}$:

$$q(t \rightarrow \infty) = C\mathcal{E}(1 - 0) = C\mathcal{E}$$

This is just the charge Q the capacitor would have had from $t = 0$ on if there had been no resistor in the circuit. Finally, at time $t = \tau$ the charge on the capacitor is $q = C\mathcal{E}(1 - e^{-1}) = C\mathcal{E}(1 - 0.368) = 0.632C\mathcal{E}$, which is 63.2% of its final charge. The charge on the capacitor as a function of time is plotted in **Figure 21–20**.

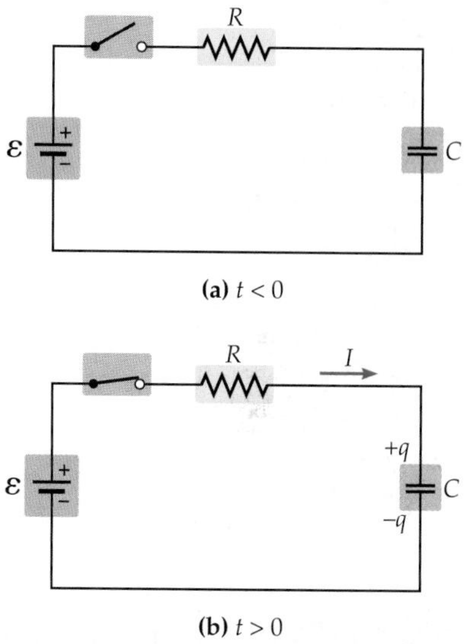

▲ **FIGURE 21–19 A typical *RC* circuit**

(a) Before the switch is closed ($t < 0$) there is no current in the circuit and no charge on the capacitor. **(b)** After the switch is closed ($t > 0$), current flows and the charge on the capacitor builds up over a finite time. As $t \rightarrow \infty$ the charge on the capacitor approaches $Q = C\mathcal{E}$.

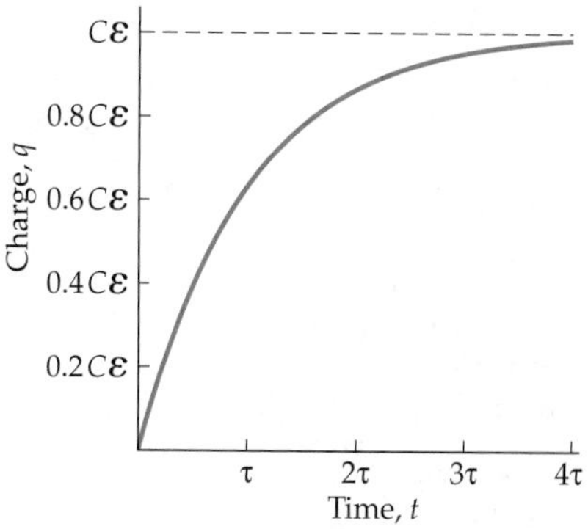

▲ **FIGURE 21–20 Charge versus time for an *RC* circuit**

Charge on the capacitor in the *RC* circuit of Figure 21–19 as a function of time. The horizontal axis shows time in units of the characteristic time, $\tau = RC$. The vertical axis shows the magnitude of the charge in units of $C\mathcal{E}$.

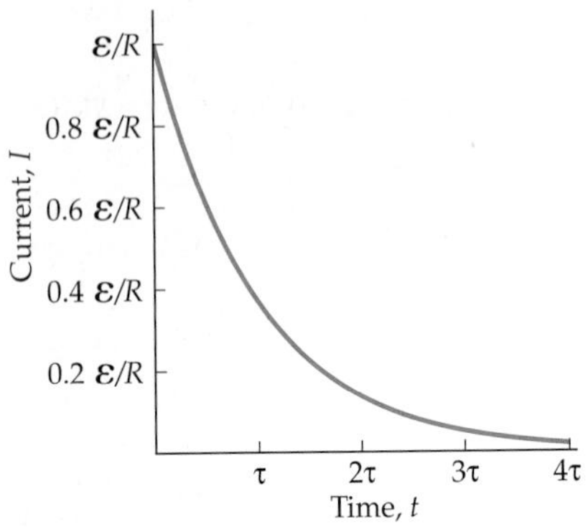

▲ **FIGURE 21–21 Current versus time in an *RC* circuit**

Current as a function of time in the *RC* circuit of Figure 21–19. Initially the current is $\mathcal{E}/R$, the same as if the capacitor were not present. The current approaches zero after a period equal to several time constants, $\tau = RC$.

PROBLEM-SOLVING NOTE

The Limiting Behavior of Capacitors

Capacitors in DC circuits act like short circuits at $t = 0$ and open circuits as $t \to \infty$.

Before we continue, let's check to see that the quantity $\tau = RC$ is in fact a time. Suppose, for example, that the resistor and capacitor in an RC circuit have the values $R = 120\ \Omega$ and $C = 3.5\ \mu\text{F}$, respectively. Multiplying R and C we find

$$\tau = RC = (120\text{ ohm})(3.5 \times 10^{-6}\text{ farad})$$
$$= \left(\frac{120\text{ volt}}{\text{coulomb/second}}\right)\left(\frac{3.5 \times 10^{-6}\text{ coulomb}}{\text{volt}}\right) = 4.2 \times 10^{-4}\text{ second}$$

The tick marks on the horizontal axis in Figure 21–20 indicate the times $\tau, 2\tau, 3\tau$, and 4τ. Notice that the capacitor is almost completely charged by the time $t = 4\tau$.

Figure 21–20 also shows that the charge on the capacitor increases rapidly initially, indicating a large current in the circuit. Eventually, the charging slows down, because the greater the charge on the capacitor, the harder it is to transfer additional charge against the electrical repulsive force. Later, the charge barely changes with time, which means that the current is essentially zero. In fact, the mathematical expression for the current—again derived from calculus—is the following:

$$I(t) = \left(\frac{\mathcal{E}}{R}\right)e^{-t/\tau} \qquad \text{21–19}$$

This expression is plotted in **Figure 21–21**, where we see that significant variation in the current occurs over times ranging from $t = 0$ to $t \sim 4\tau$. At time $t = 0$ the current is $I(0) = \mathcal{E}/R$, which is the value it would have if the capacitor were replaced by an ideal wire. As $t \to \infty$, the current approaches zero, as expected: $I(t \to \infty) \to 0$. In this limit, the capacitor is essentially fully charged, so that no more charge can flow onto its plates. Thus, in this limit, the capacitor behaves like an open switch.

EXAMPLE 21–9 Charging a Capacitor

A circuit consists of a 126-Ω resistor, a 275-Ω resistor, a 182-μF capacitor, a switch, and a 3.00-V battery all connected in series. Initially the capacitor is uncharged and the switch is open. At time $t = 0$ the switch is closed. **(a)** What charge will the capacitor have a long time after the switch is closed? **(b)** At what time will the charge on the capacitor be 80.0% of the value found in part (a)?

Picture the Problem

The circuit described in the problem statement is shown with the switch in the open position. Once the switch is closed at $t = 0$, current will flow in the circuit and charge will begin to accumulate on the capacitor plates.

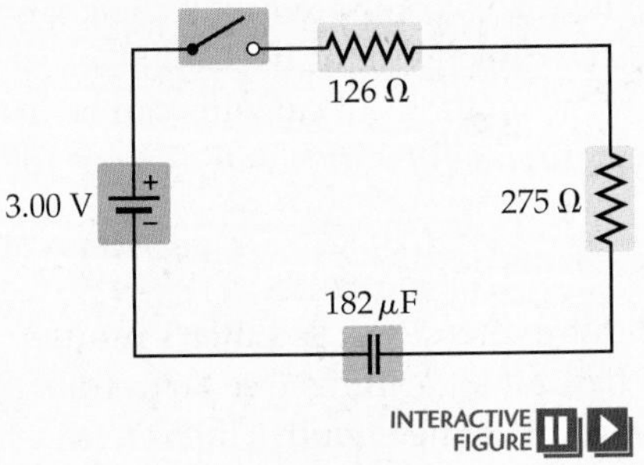

Strategy

a. A long time after the switch is closed, the current stops and the capacitor is fully charged. At this point, the voltage across the capacitor is equal to the emf of the battery. Therefore, the charge on the capacitor is $Q = C\mathcal{E}$.

b. To find the time when the charge will be 80.0% of the full charge, $Q = C\mathcal{E}$, we can set $q(t) = C\mathcal{E}(1 - e^{-t/\tau}) = 0.800C\mathcal{E}$ and solve for the desired time, t.

Solution

Part (a)

1. Evaluate $Q = C\mathcal{E}$ for this circuit: $Q = C\mathcal{E} = (182\ \mu\text{F})(3.00\text{ V}) = 546\ \mu\text{C}$

Part (b)

2. Set $q(t) = 0.800C\mathcal{E}$ in $q(t) = C\mathcal{E}(1 - e^{-t/\tau})$ and cancel $C\mathcal{E}$:
$$q(t) = 0.800C\mathcal{E} = C\mathcal{E}(1 - e^{-t/\tau})$$
$$0.800 = 1 - e^{-t/\tau}$$

3. Solve for t in terms of the time constant τ:
$$e^{-t/\tau} = 1 - 0.800 = 0.200$$
$$t = -\tau\ln(0.200)$$

4. Calculate τ and use the result to find the time t:
$$\tau = RC = (126\ \Omega + 275\ \Omega)(182\ \mu\text{F}) = 73.0\text{ ms}$$
$$t = -(73.0\text{ ms})\ln(0.200)$$
$$= -(73.0\text{ ms})(-1.61) = 118\text{ ms}$$

Insight

Note that the time required for the charge on a capacitor to reach 80.0% of its final value is 1.61 time constants. This result is independent of the values of R and C in an RC circuit.

Practice Problem

What is the current in this circuit at the time found in part (b)? [**Answer:** $I(t) = (\mathcal{E}/R)e^{-t/\tau} = [(3.00 \text{ V})/(126\ \Omega + 275\ \Omega)](0.200) = (7.48 \text{ mA})(0.200) = 1.50 \text{ mA}$]

Some related homework problems: Problem 62, Problem 64

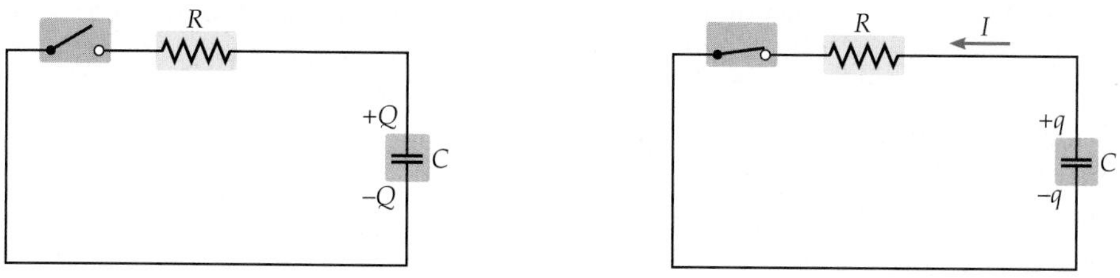

(a) $t < 0$ **(b)** $t > 0$

◀ **FIGURE 21–22 Discharging a capacitor**
(a) A charged capacitor is connected to a resistor. Initially the circuit is open, and no current can flow. **(b)** When the switch is closed, current flows from the + plate of the capacitor to the − plate. The charge remaining on the capacitor approaches zero after several time units, RC.

Similar behavior occurs when a charged capacitor is allowed to discharge, as in **Figure 21–22**. In this case, the initial charge on the capacitor is Q. If the switch is closed at $t = 0$, the charge for later times is

$$q(t) = Qe^{-t/\tau} \qquad \textbf{21–20}$$

Like charging, the discharging of a capacitor occurs with a characteristic time $\tau = RC$.

To summarize, circuits with resistors and capacitors have the following general characteristics:

- Charging and discharging occur over a finite, characteristic time given by the time constant, $\tau = RC$.
- At $t = 0$ current flows freely through a capacitor being charged; it behaves like a short circuit.
- As $t \to \infty$ the current flowing into a capacitor approaches zero. In this limit, a capacitor behaves like an open switch.

We explore these features further in the following Conceptual Checkpoint.

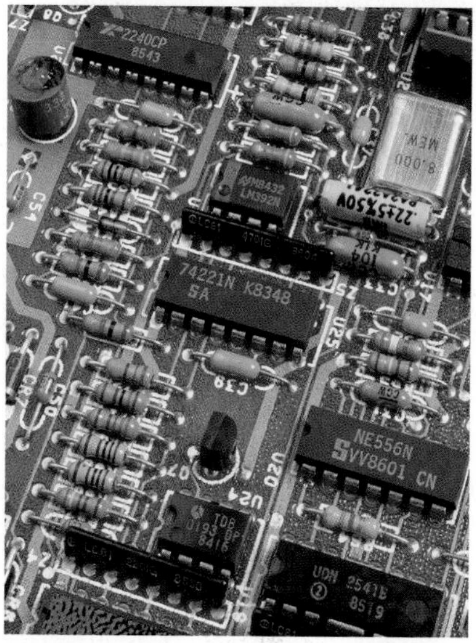

▲ A modern-day circuit board incorporates numerous resistors (cylinders with colored bands) and capacitors (yellow cylinders and metal container).

CONCEPTUAL CHECKPOINT 21–4 Current in an *RC* Circuit

What current flows through the battery in this circuit **(a)** immediately after the switch is closed and **(b)** a long time after the switch is closed?

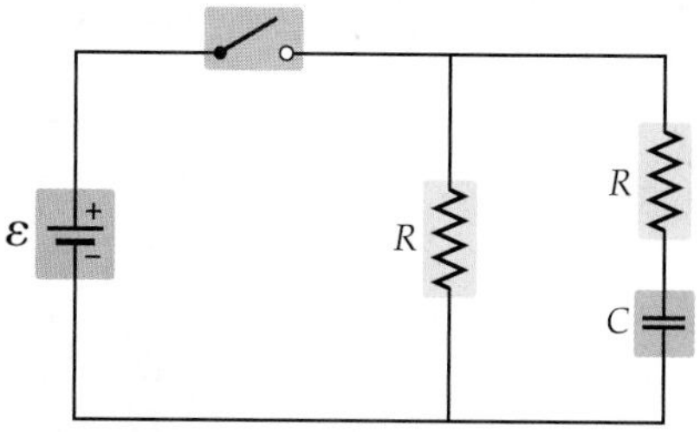

Reasoning and Discussion

a. Immediately after the switch is closed, the capacitor acts like a short circuit; that is, as if the battery were connected to two resistors R in parallel. The equivalent resistance in this case is $R/2$; therefore, the current is $I = \mathcal{E}/(R/2) = 2\mathcal{E}/R$.

b. After current has been flowing in the circuit for a long time, the capacitor acts like an open switch. Now current can flow only through the one resistor, R; hence, the current is $I = \mathcal{E}/R$, half of its initial value.

Answer:

(a) The current is $2\mathcal{E}/R$; **(b)** The current is $\mathcal{E}/R$.

REAL-WORLD PHYSICS
Delay circuits in windshield wipers and turn signals

The fact that RC circuits have a characteristic time makes them useful in a variety of different applications. On a rather mundane level, RC circuits are used to determine the time delay on windshield wipers. When you adjust the delay knob in your car, you change a resistance or a capacitance, which in turn changes the time constant of the circuit. This results in a greater or a smaller delay. The blinking rate of turn signals is also determined by the time constant of an RC circuit.

REAL-WORLD PHYSICS: BIO
Pacemakers

A more critical application of *RC* circuits is the heart pacemaker. In the simplest case, these devices use an *RC* circuit to deliver precisely timed pulses directly to the heart. The more sophisticated pacemakers available today can even "sense" when a patient's heart rate falls below a predetermined value. The pacemaker then begins sending appropriate pulses to the heart to increase its rate. Many pacemakers can even be reprogrammed after they are surgically implanted to respond to changes in a patient's condition.

Normally, the heart's rate of beating is determined by its own natural pacemaker, the sinoatrial or SA node, located in the upper right chamber of the heart. If the SA node is not functioning properly, it may cause the heart to beat slowly or irregularly. To correct the problem, a pacemaker is implanted just under the collarbone, and an electrode is introduced intravenously via the cephalic vein. The distal end of the electrode is positioned, with the aid of fluoroscopic guidance, in the right ventricular apex. From that point on, the operation of the pacemaker follows the basic principles of electric circuits, as described in this chapter.

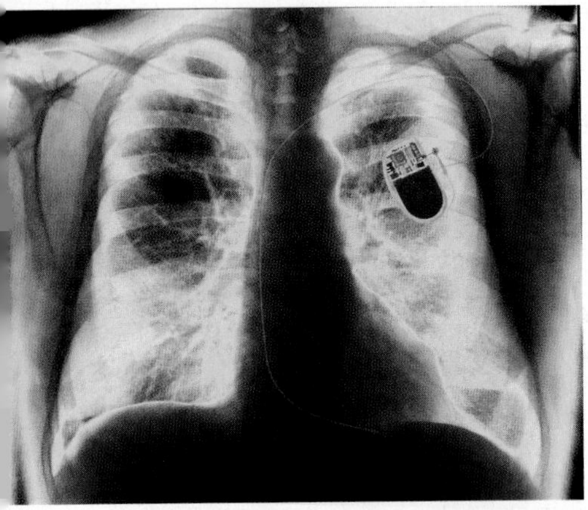

▲ An X-ray showing a pacemaker installed in a person's chest. The timing of the electrical pulses that keep the heart beating regularly is determined by an *RC* circuit powered by a small, long-lived battery.

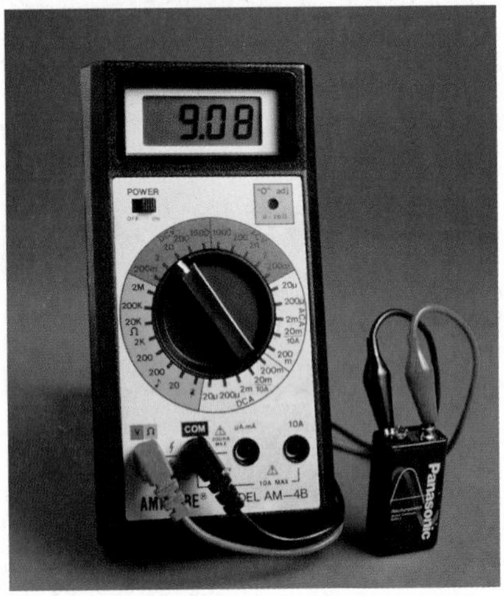

▲ A typical digital multimeter, which can measure resistance (teal settings), current (yellow settings), or voltage (red settings). This meter is measuring the voltage of a "9 volt" battery.

*21–8 Ammeters and Voltmeters

Devices for measuring currents and voltages in a circuit are referred to as **ammeters** and **voltmeters,** respectively. In each case, the ideal situation is for the meter to measure the desired quantity without altering the characteristics of the circuit being studied. This is accomplished in different ways for these two types of meters, as we shall see.

First, the ammeter is designed to measure the flow of current through a particular portion of a circuit. For example, we may want to know the current flowing between points A and B in the circuit shown in **Figure 21–23 (a)**. To measure this current, we insert the ammeter into the circuit in such a way that all the current flowing from A to B must also flow through the meter. This is done by connecting the meter "in series" with the other circuit elements between A and B, as indicated in **Figure 21–23 (b)**.

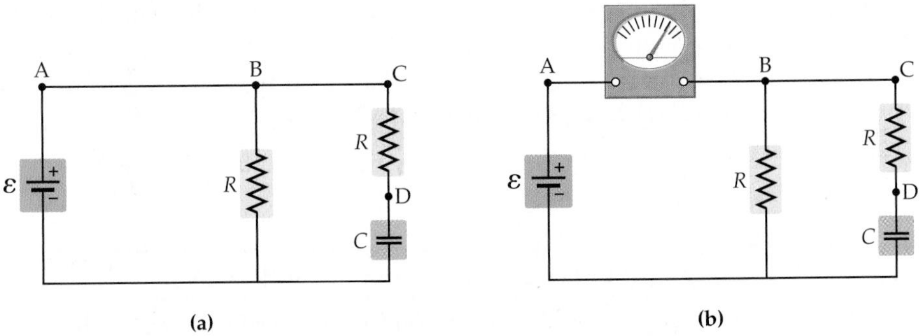

▲ **FIGURE 21–23 Measuring the current in a circuit**
To measure the current flowing between points A and B in **(a)**, an ammeter is inserted into the circuit, as shown in **(b)**. An ideal ammeter would have zero resistance.

If the ammeter has a finite resistance—which must be the case for real meters—the presence of the meter in the circuit will alter the current it is intended to measure. Thus, an *ideal* ammeter would be one with zero resistance. In practice, if the resistance of the ammeter is much less than the other resistances in the circuit, its reading will be reasonably accurate.

Second, a voltmeter measures the potential drop between any two points in a circuit. Referring again to the circuit in Figure 21–23 (a), we may wish to know the difference in potential between points C and D. To measure this voltage, we connect the voltmeter "in parallel" to the circuit at the appropriate points, as in **Figure 21–24**.

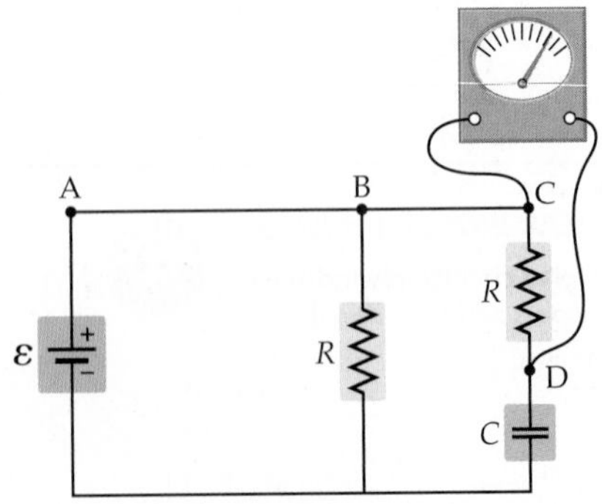

◀ **FIGURE 21–24 Measuring the voltage in a circuit**
The voltage difference between points C and D can be measured by connecting a voltmeter in parallel to the original circuit. An ideal voltmeter would have infinite resistance.

A real voltmeter always allows some current to flow through it, which means that the current flowing through the circuit is less than before the meter was connected. As a result, the measured voltage is altered from its ideal value. An *ideal* voltmeter, then, would be one in which the resistance is infinite, so that the current it draws from the circuit is negligible. In practical situations it is sufficient that the resistance of the meter be much greater than the resistances in the circuit.

Sometimes the functions of an ammeter, voltmeter, and ohmmeter are combined in a single device called a **multimeter.** Adjusting the settings on a multimeter allows a variety of circuit properties to be measured.

Chapter Summary

21–1 Electric Current

Electric current is the flow of electric charge.

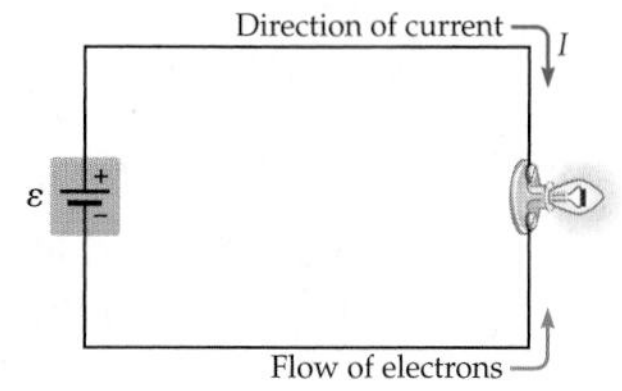

Definition

If a charge ΔQ passes a given point in the time Δt, the corresponding electric current is

$$I = \frac{\Delta Q}{\Delta t} \qquad \text{21–1}$$

Ampere

The unit of current is the ampere, or amp for short. By definition, 1 amp is one coulomb per second; 1 A = 1 C/s.

Battery

A battery is a device that uses chemical reactions to produce a potential difference between its two terminals.

Electromotive Force

The electromotive force, or emf, $\mathcal{E}$, is the potential difference between the terminals of a battery under ideal conditions.

Work Done by a Battery

As a battery moves a charge ΔQ around a circuit, it does the work $W = (\Delta Q)\mathcal{E}$.

Direction of Current

By definition, the direction of the current I in a circuit is the direction in which *positive* charges would move. The actual charge carriers, however, are generally electrons; hence, they move in the opposite direction to I.

21–2 Resistance and Ohm's Law

When electrons move through a wire, they encounter resistance to their motion. In order to move electrons against this resistance, it is necessary to apply a potential difference between the ends of the wire.

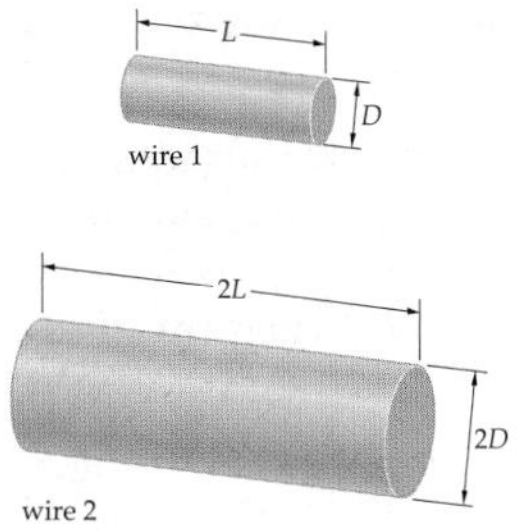

Ohm's Law

To produce a current I through a wire with resistance R the following potential difference, V, is required:

$$V = IR \qquad \text{21–2}$$

Resistivity

The resistivity ρ of a material determines how much resistance it gives to the flow of electric current.

Resistance of a Wire

The resistance of a wire of length L, cross-sectional area A, and resistivity ρ is

$$R = \rho\left(\frac{L}{A}\right) \qquad \text{21–3}$$

Temperature Dependence

The resistivity of most metals increases approximately linearly with temperature.

Superconductivity

Below a certain critical temperature, T_c, certain materials lose all electrical resistance. A current flowing in a superconductor can continue undiminished as long as its temperature is maintained below T_c.

21–3 Energy and Power in Electric Circuits

In general, energy is required to cause an electric current to flow through a circuit. The rate at which the energy must be supplied is the power.

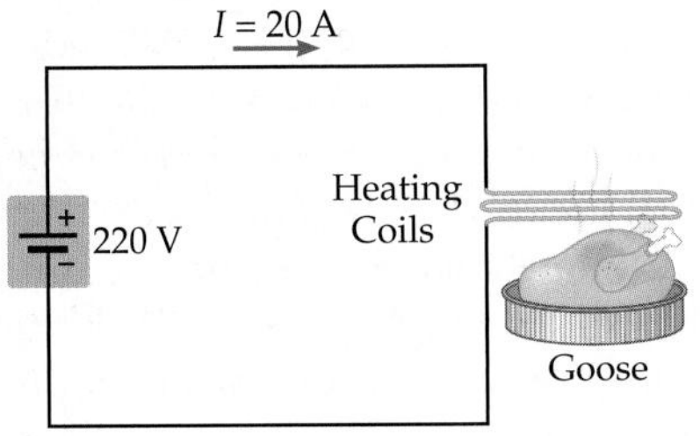

Electrical Power
If a current I flows across a potential difference V, the corresponding electrical power is

$$P = IV \qquad \text{21–4}$$

Power Dissipation in a Resistor
If a potential difference V produces a current I in a resistor R, the electrical power converted to heat is

$$P = I^2R = V^2/R \qquad \text{21–5, 21–6}$$

Energy Usage and the Kilowatt-Hour
The energy equivalent of one kilowatt-hour (kWh) is

$$1 \text{ kWh} = 3.6 \times 10^6 \text{ J}$$

21–4 Resistors in Series and Parallel

Resistors connected end to end—so that the same current flows through each one—are said to be in series. Resistors connected across the same potential difference—allowing parallel paths for the current to flow—are said to be connected in parallel.

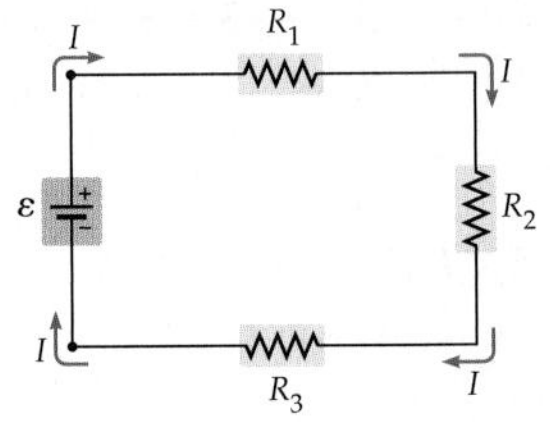

Series
The equivalent resistance, R_{eq}, of resistors connected in series is equal to the sum of the individual resistances:

$$R_{eq} = R_1 + R_2 + R_3 + \cdots = \sum R \qquad \text{21–7}$$

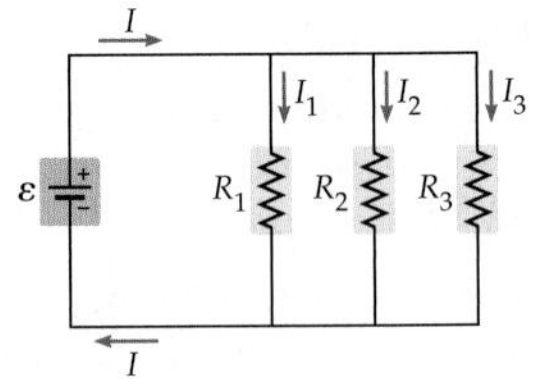

Parallel
The equivalent resistance, R_{eq}, of resistors connected in parallel is given by the following:

$$\frac{1}{R_{eq}} = \frac{1}{R_1} + \frac{1}{R_2} + \frac{1}{R_3} + \cdots = \sum \frac{1}{R} \qquad \text{21–10}$$

21–5 Kirchhoff's Rules

Kirchhoff's rules are statements of charge conservation and energy conservation as applied to closed electrical circuits.

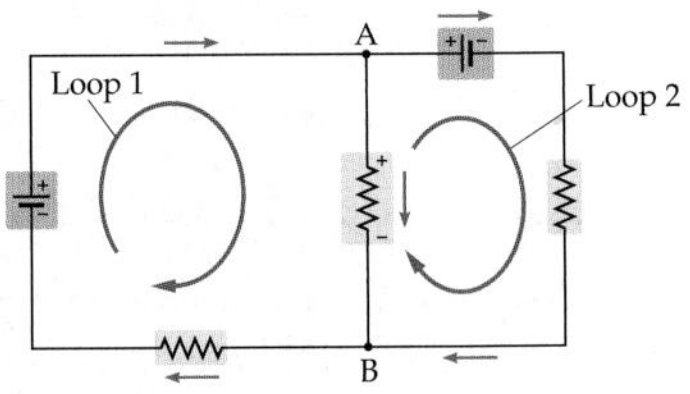

Junction Rule (Charge Conservation)
The algebraic sum of all currents meeting at a junction must equal zero. Currents entering the junction are taken to be positive; currents leaving are taken to be negative.

Loop Rule (Energy Conservation)
The algebraic sum of all potential differences around a closed loop is zero. The potential increases in going from the − to the + terminal of a battery and decreases when crossing a resistor in the direction of the current.

21–6 Circuits Containing Capacitors

Capacitors connected end to end—so that the same charge is on each one—are said to be in series. Capacitors connected across the same potential difference are said to be connected in parallel.

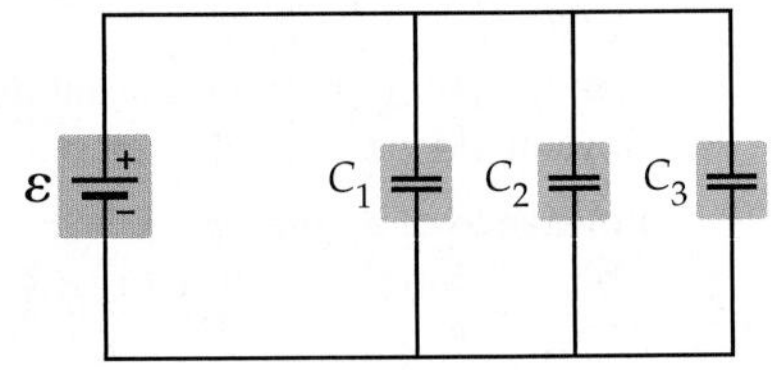

Parallel
The equivalent capacitance, C_{eq}, of capacitors connected in parallel is equal to the sum of the individual capacitances:

$$C_{eq} = C_1 + C_2 + C_3 + \cdots = \sum C \qquad \text{21–14}$$

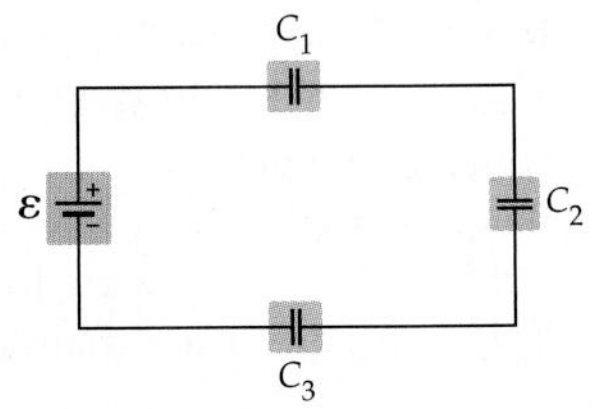

Series
The equivalent capacitance, C_{eq}, of capacitors connected in series is given by

$$\frac{1}{C_{eq}} = \frac{1}{C_1} + \frac{1}{C_2} + \frac{1}{C_3} + \cdots = \sum \frac{1}{C} \qquad \text{21–17}$$

21–7 *RC* Circuits

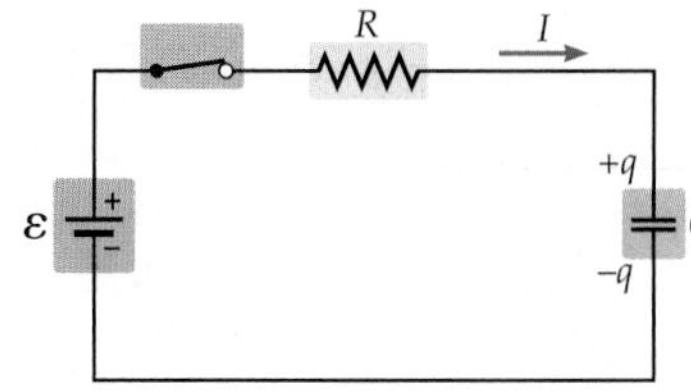

In circuits containing both resistors and capacitors, there is a characteristic time, $\tau = RC$, during which significant changes occur. This time is referred to as the time constant. The simplest such circuit, known as an *RC* circuit, consists of one resistor and one capacitor connected in series.

Charging a Capacitor
The charge on a capacitor in an *RC* circuit varies with time as follows:

$$q(t) = C\mathcal{E}(1 - e^{-t/\tau}) \qquad \textbf{21–18}$$

The corresponding current is given by

$$I(t) = \left(\frac{\mathcal{E}}{R}\right)e^{-t/\tau} \qquad \textbf{21–19}$$

Discharging a Capacitor
If a capacitor in an *RC* circuit starts with a charge Q at time $t = 0$, its charge at all later times is

$$q(t) = Qe^{-t/\tau} \qquad \textbf{21–20}$$

Behavior near $t = 0$
Just after the switch is closed in an *RC* circuit, capacitors behave like ideal wires—that is, they offer no resistance to the flow of current.

Behavior as $t \to \infty$
Long after the switch is closed in an *RC* circuit, capacitors behave like open circuits.

*21–8 Ammeters and Voltmeters

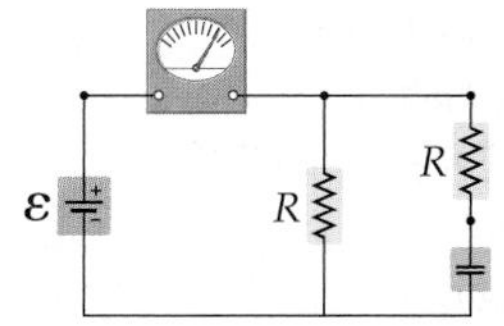

Ammeters and voltmeters are devices for measuring currents and voltages, respectively, in electrical circuits.

Ammeter
An ammeter is connected in series with the section of the circuit in which the current is to be measured. In the ideal case, an ammeter's resistance is zero.

Voltmeter
A voltmeter is connected in parallel with the portion of the circuit to be measured. In the ideal case, a voltmeter's resistance is infinite.

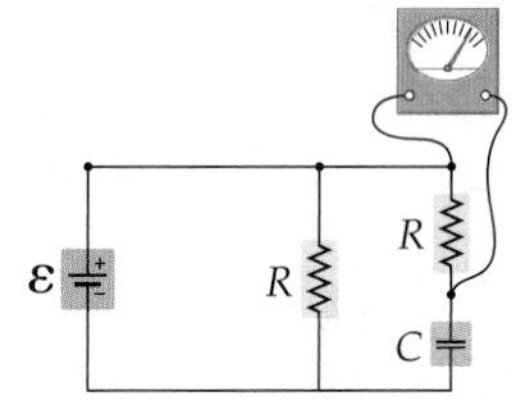

Problem-Solving Summary

Type of Problem	Relevant Physical Concepts	Related Examples
Find the work done by a battery.	The work done by a battery is the charge that passes through the battery times the emf of the battery: $W = \Delta Q\mathcal{E}$.	Active Example 21–1
Relate resistance to resistivity.	The resistance of a wire is its resistivity, ρ, times its length, divided by its cross-sectional area: $R = \rho(L/A)$.	Example 21–2
Relate the power in an electrical circuit to the current, voltage, and resistance.	The basic definition of electrical power is current times voltage: $P = IV$. Using Ohm's law when appropriate, the power can also be expressed as $P = I^2R$ and $P = V^2/R$.	Examples 21–3, 21–4
Determine the equivalent resistance of resistors in series and parallel.	Resistors in series simply add: $R_{eq} = R_1 + R_2 + \cdots$; resistors in parallel add in terms of inverses: $1/R_{eq} = 1/R_1 + 1/R_2 + \cdots$.	Examples 21–5, 21–6, 21–7
Find the current in a circuit containing resistors that are not simply in series or parallel.	Apply Kirchhoff's junction rule (the algebraic sum of currents at a junction must be zero) and loop rule (the algebraic sum of potential difference around a loop is zero).	Active Example 21–2
Determine the equivalent capacitance of capacitors in series and parallel.	Capacitors in parallel simply add: $C_{eq} = C_1 + C_2 + \cdots$; capacitors in series add in terms of inverses: $1/C_{eq} = 1/C_1 + 1/C_2 + \cdots$.	Example 21–8 Active Example 21–3
Find the charge and the current in an *RC* circuit as a function of time.	The charge and current in an *RC* circuit during charging vary exponentially with time as follows: $q(t) = C\mathcal{E}(1 - e^{-t/\tau})$; $I(t) = (\mathcal{E}/R)e^{-t/\tau}$. The characteristic time is $\tau = RC$.	Example 21–9

Conceptual Questions

(Answers to odd-numbered Conceptual Questions can be found in the back of the book.)

1. What is the direction of the electric current produced by an electron that falls toward the ground?
2. Your body is composed of electric charges. Does it follow, then, that you produce an electric current when you walk?
3. Suppose you charge a comb by rubbing it through your hair. Do you produce a current when you walk across the room carrying the comb?
4. Suppose you charge a comb by rubbing it through the fur on your dog's back. Do you produce a current when you walk across the room carrying the comb?
5. An electron moving through a wire has an average drift speed that is very small. Does this mean that its instantaneous velocity is also very small?
6. Are car headlights connected in series or parallel? Give an everyday observation that supports your answer.
7. Give an example of how four resistors of resistance R can be combined to produce an equivalent resistance of R.
8. Is it possible to connect a group of resistors of value R in such a way that the equivalent resistance is less than R? If so, give a specific example.
9. What physical quantity do resistors connected in series have in common?
10. What physical quantity do resistors connected in parallel have in common?
11. Explain how electrical devices can begin operating almost immediately after you throw a switch, even though individual electrons in the wire may take hours to reach the device.
12. Explain the difference between resistivity and resistance.
13. Explain why birds can roost on high-voltage wire without being electrocuted.
14. List two electrical applications that would benefit from room-temperature superconductors. List two applications for which room-temperature superconductivity would not be beneficial.
15. On what basic conservation laws are Kirchhoff's rules based?
16. What physical quantity do capacitors connected in series have in common?
17. What physical quantity do capacitors connected in parallel have in common?
18. Consider the circuit shown in **Figure 21–25**, in which a light of resistance R and a capacitor of capacitance C are connected in series. The capacitor has a large capacitance, and is initially uncharged. The battery provides enough power to light the bulb when connected to the battery directly. Describe the behavior of the light after the switch is closed.

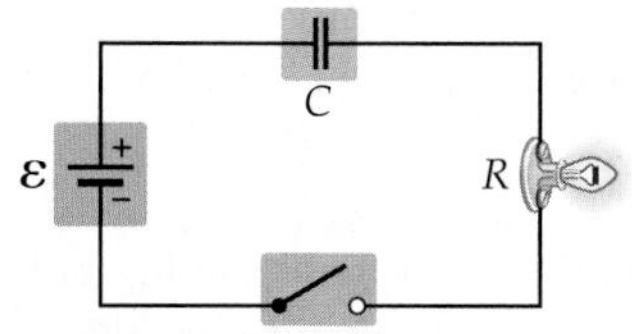

▲ **FIGURE 21–25** Conceptual Question 18

Conceptual Exercises

(Answers to odd-numbered Conceptual Exercises can be found in the back of the book.)

1. A given car battery is rated as 250 amp-hours. Does this rating indicate the amount of energy, power, charge, voltage, or current that can be drawn from the battery? Explain.
2. The four conducting cylinders shown in **Figure 21–26** are all made of the same material, though they differ in length and/or diameter. They are connected to four different batteries, which supply the necessary voltages to give the circuits the same current, I. Rank the four voltages, V_1, V_2, V_3, and V_4, in order of increasing value. Indicate ties where appropriate.

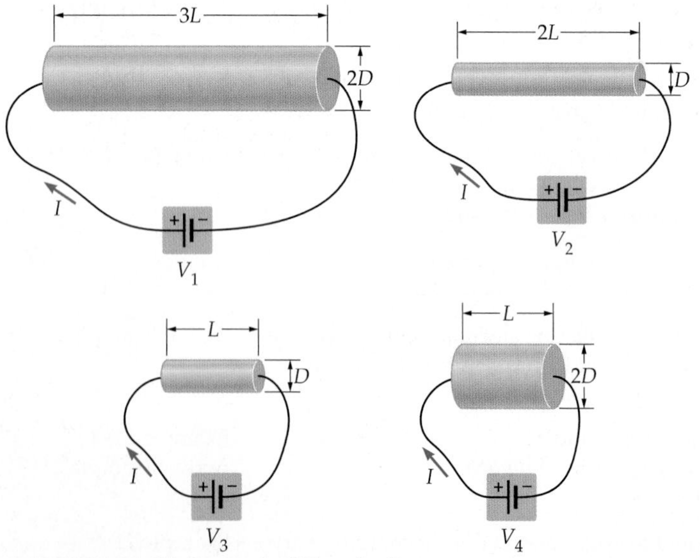

▲ **FIGURE 21–26** Conceptual Exercise 2

3. A conducting wire is quadrupled in length and tripled in diameter. **(a)** Does its resistance increase, decrease, or stay the same? Explain. **(b)** By what factor does its resistance change?
4. The resistivity of tungsten increases with temperature. When a light containing a tungsten filament heats up, does its power consumption increase, decrease, or stay the same? Explain.
5. **Figure 21–27** shows a plot of current versus voltage for two different materials, A and B. Which of these materials satisfies Ohm's law? Explain.

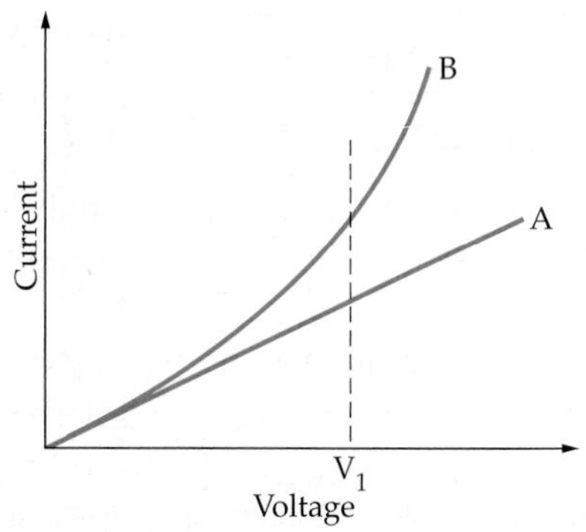

▲ **FIGURE 21–27** Conceptual Exercises 5 and 6

6. Current versus voltage plots for two materials, A and B, are shown in Figure 21–27. Which material has the greater resistance at the voltage V_1? Explain.

7. Two cylindrical wires are made of the same material and have the same length. If wire B is to have nine times the resistance of wire A, what must be the ratio of their radii, r_B/r_A?

8. A cylindrical wire is to be doubled in length, but it is desired that its resistance remain the same. **(a)** Must its radius be increased or decreased? Explain. **(b)** By what factor must the radius be changed?

9. An electric heater has a power rating of 500 W when connected to a given voltage V. If two of these heaters are connected in series to the same voltage, is the power consumed by the two heaters greater than, less than, or equal to 1000 W? Explain.

10. Light A has four times the power rating of light B when operated at the same voltage. **(a)** Is the resistance of light A greater than, less than, or equal to the resistance of light B? Explain. **(b)** What is the ratio of the resistance of light A to the resistance of light B?

11. Two lightbulbs operate on the same potential difference. Bulb A has four times the power output of bulb B. **(a)** Which bulb has the greater current passing through it? Explain. **(b)** What is the ratio of the current in bulb A to the current in bulb B?

12. Two lightbulbs operate on the same current. Bulb A has four times the power output of bulb B. **(a)** Is the potential difference across bulb A greater than or less than the potential difference across bulb B? Explain. **(b)** What is the ratio of the potential difference across bulb A to that across bulb B?

13. A dozen lights are to be connected to a given emf. Will the lights be brighter if connected in series or in parallel? Explain.

14. A *fuse* is a device to protect a circuit from the effects of a large current. The fuse is a small strip of metal that burns through when the current in it exceeds a certain value, thus producing an open circuit. Should a fuse be connected in series or in parallel with the circuit it is intended to protect? Explain.

15. A circuit consists of three resistors, $R_1 < R_2 < R_3$ connected in series to battery. Rank these resistors in order of increasing **(a)** current through them and **(b)** potential difference across them. Indicate ties where appropriate.

16. Two resistors are connected in parallel. If a third resistor is connected in parallel with the other two, does the equivalent resistance of the circuit increase, decrease, or remain the same? Explain.

17. Two resistors, $R_1 = R$ and $R_2 = 2R$, are connected to a battery. Which resistor dissipates more power when they are connected to the battery **(a)** in series or **(b)** in parallel? Explain.

18. Consider the circuit shown in **Figure 21–28**, in which three lights, each with a resistance R, are connected in series. The circuit also contains an open switch. **(a)** When the switch is closed, does the intensity of light 2 increase, decrease, or stay the same? Explain. **(b)** Do the intensities of lights 1 and 3 increase, decrease, or stay the same? Explain.

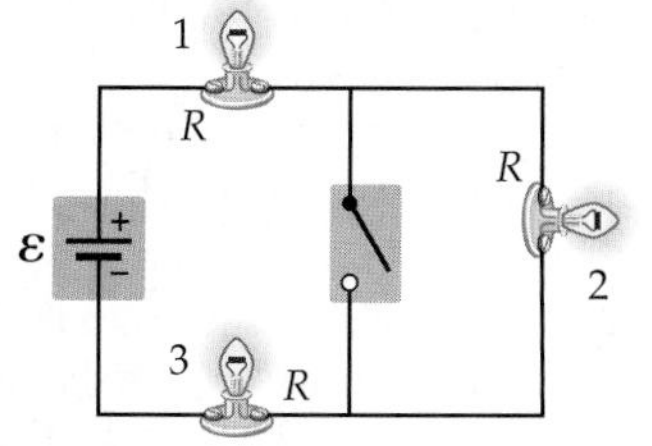

▲ **FIGURE 21–28** Conceptual Exercises 18, 19, and 20

19. Referring to Conceptual Exercise 18 and the circuit in Figure 21–28, does the current supplied by the battery increase, decrease, or remain the same when the switch is closed? Explain.

20. Referring to Conceptual Exercise 18 and the circuit in Figure 21–28, does the total power dissipated in the circuit increase, decrease, or remain the same when the switch is closed? Explain.

21. Consider the circuit shown in **Figure 21–29**, in which three lights, each with a resistance R, are connected in parallel. The circuit also contains an open switch. **(a)** When the switch is closed, does the intensity of light 3 increase, decrease, or stay the same? Explain. **(b)** Do the intensities of lights 1 and 2 increase, decrease, or stay the same? Explain.

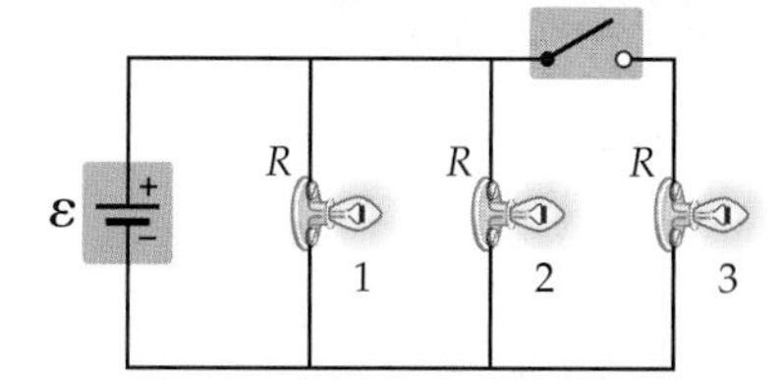

▲ **FIGURE 21–29** Conceptual Exercises 21, 22, and 23

22. When the switch is closed in the circuit shown in Figure 21–29, does the current supplied by the battery increase, decrease, or stay the same? Explain.

23. When the switch is closed in the circuit shown in Figure 21–29, does the total power dissipated in the circuit increase, decrease, or stay the same? Explain.

24. Two capacitors, $C_1 = C$ and $C_2 = 2C$, are connected to a battery. Which capacitor stores more energy when they are connected to the battery **(a)** in series or **(b)** in parallel? Explain.

25. Two capacitors are connected in series. If a third capacitor is connected in series with the other two, does the equivalent capacitance increase, decrease, or remain the same? Explain.

26. Two capacitors are connected in parallel. If a third capacitor is connected in parallel with the other two, does the equivalent capacitance increase, decrease, or remain the same? Explain.

27. You conduct a series of experiments in which you connect the capacitors C_1 and $C_2 > C_1$ to a battery in various ways. The experiments are as follows: A, C_1 alone connected to the battery; B, C_2 alone connected to the battery; C, C_1 and C_2 connected to the battery in series; D, C_1 and C_2 connected to the battery in parallel. Rank these four experiments in order of increasing equivalent capacitance. Indicate ties where appropriate.

28. Three different circuits, each containing a switch and two capacitors, are shown in **Figure 21–30**. Initially, the plates of the capacitors are charged as shown. The switches are then closed, allowing charge to move freely between the capacitors. Rank the circuits in order of increasing final charge on the left plate of **(a)** the upper capacitor and **(b)** the lower capacitor. Indicate ties where appropriate.

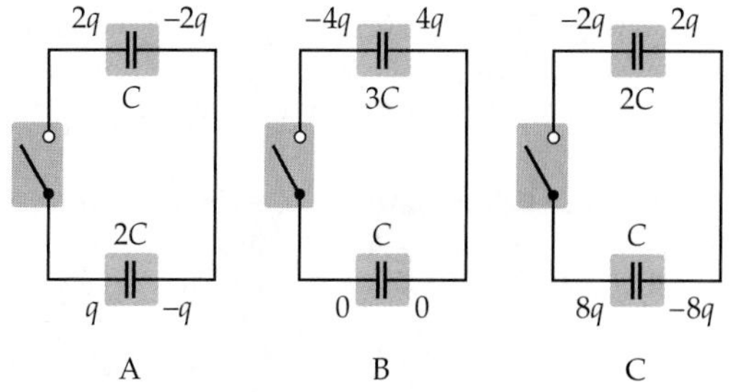

▲ **FIGURE 21–30** Conceptual Exercise 28

29. The circuit shown in **Figure 21–31** shows a resistor and two capacitors connected in series with a battery of voltage V. The circuit also has an ammeter and a switch. Initially, the switch is open and both capacitors are uncharged. The following questions refer to a time long after the switch is closed and current has ceased to flow. **(a)** In terms of V, what is the voltage

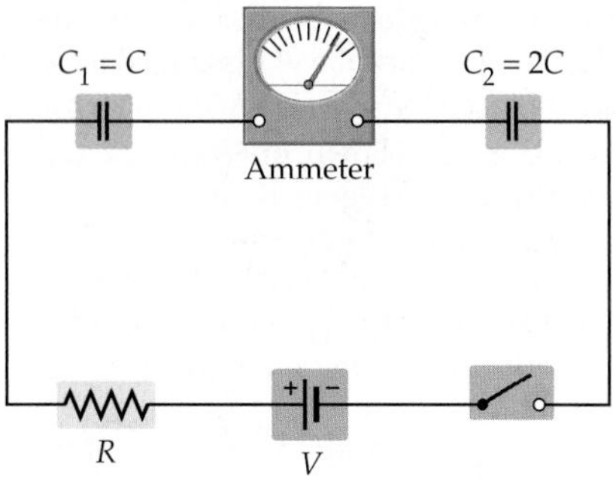

▲ **FIGURE 21–31** Conceptual Exercise 29

across the capacitor C_1? **(b)** In terms of CV, what is the charge on the right plate of C_2? **(c)** What is the net charge that flowed through the ammeter during charging? Give your answer in terms of CV.

30. Three RC circuits have the emf, resistance, and capacitance given in the accompanying table. Initially, the switch on the circuit is open and the capacitor is uncharged. Rank these circuits in order of increasing **(a)** initial current (immediately after the switch is closed) and **(b)** time for the capacitor to acquire half its final charge. Indicate ties where appropriate.

	$\mathcal{E}$(V)	$R(\Omega)$	$C(\mu F)$
Circuit A	12	4	3
Circuit B	9	3	1
Circuit C	9	9	2

31. The three circuits shown in **Figure 21–32** have identical batteries, resistors and capacitors. Initially, the switches are open and the capacitors are uncharged. Rank the circuits in order of increasing **(a)** final charge on the capacitor and **(b)** time for the current to drop to 90% of its final value. Indicate ties where appropriate.

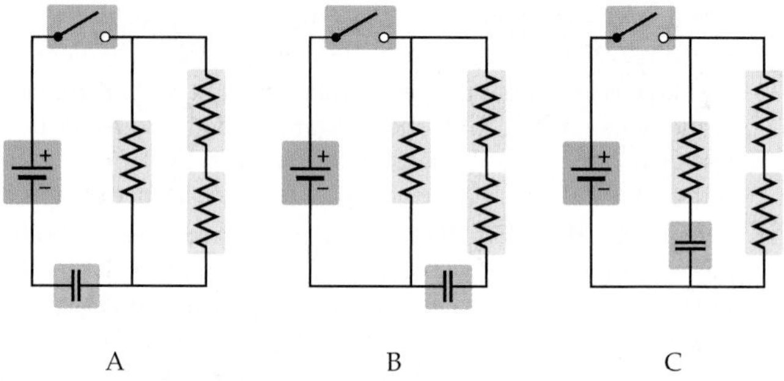

▲ **FIGURE 21–32** Conceptual Exercise 31

Problems

Note: **IP** *denotes an integrated conceptual/quantitative problem.* **BIO** *identifies problems of biological or medical interest. Red bullets (•, ••, •••) are used to indicate the level of difficulty of each problem.*

Section 21–1 Electric Current

1. • How many coulombs of charge are in one ampere-hour?

2. • A flashlight bulb carries a current of 0.18 A for 78 s. How much charge flows through the bulb in this time? How many electrons?

3. • The picture tube in a particular television draws a current of 15 A. How many electrons strike the viewing screen every second?

4. • **IP** A car battery does 260 J of work on the charge passing through it as it starts an engine. **(a)** If the emf of the battery is 12 V, how much charge passes through the battery during the start? **(b)** If the emf is doubled to 24 V, does the amount of charge passing through the battery increase or decrease? By what factor?

5. • Highly sensitive ammeters can measure currents as small as 10.0 fA. How many electrons per second flow through a wire with a 10.0-fA current?

6. •• A television set connected to a 120-V outlet consumes 85 W of power. **(a)** How much current flows through the television? **(b)** How long does it take for 10 million electrons to pass through the TV?

7. •• **BIO Pacemaker Batteries** Pacemakers designed for long-term use commonly employ a lithium–iodine battery capable of supplying 0.42 A·h of charge. **(a)** How many coulombs of charge can such a battery supply? **(b)** If the average current produced by the pacemaker is 5.6 μA, what is the expected lifetime of the device?

Section 21–2 Resistance and Ohm's Law

8. • A silver wire is 4.9 m long and 0.49 mm in diameter. What is its resistance?

9. • When a potential difference of 18 V is applied to a given wire, it conducts 0.35 A of current. What is the resistance of the wire?

10. • The tungsten filament of a lightbulb has a resistance of 0.07 Ω. If the filament is 27 cm long, what is its diameter?

11. • What is the resistance of 6.0 mi of copper wire with a diameter of 0.55 mm?

12. •• **IP** A bird lands on a bare copper wire carrying a current of 32 A. The wire is 8 gauge, which means that its cross-sectional area is 0.13 cm^2. **(a)** Find the difference in potential between the bird's feet, assuming they are separated by a distance of 6.0 cm. **(b)** Will your answer to part (a) increase or decrease if the separation between the bird's feet increases? Explain.

13. •• A current of 0.76 A flows through a copper wire 0.44 mm in diameter when it is connected to a potential difference of 15 V. How long is the wire?

14. •• **IP BIO Current through a Cell Membrane** A typical cell membrane is 8.0 nm thick and has an electrical resistivity of 1.3×10^7 Ω·m. **(a)** If the potential difference between the inner and outer surfaces of a cell membrane is 75 mV, how much current flows through a square area of membrane 1.0 μm on a side? **(b)** Suppose the thickness of the membrane is doubled, but the resistivity and potential difference remain the same. Does the current increase or decrease? By what factor?

15. •• When a potential difference of 12 V is applied to a wire 6.9 m long and 0.33 mm in diameter the result is an electric current of 2.1 A. What is the resistivity of the wire?

16. •• **IP (a)** What is the resistance per meter of an aluminum wire with a cross-sectional area of 2.4×10^{-7} m^2. **(b)** Would your answer to part (a) increase, decrease, or stay the same if the diameter of the wire were increased? Explain. **(c)** Repeat part (a) for a wire with a cross-sectional area of 3.6×10^{-7} m^2.

17. •• **BIO Resistance and Current in the Human Finger** The interior of the human body has an electrical resistivity of 0.15 Ω·m. **(a)** Estimate the resistance for current flowing the length of your index finger. (For this calculation, ignore the much higher resistivity of your skin.) **(b)** Your muscles will contract when they carry a current greater than 15 mA. What voltage is required to produce this current through your finger?

18. • • • Consider a rectangular block of metal of height A, width B, and length C, as shown in **Figure 21–33**. If a potential difference V is maintained between the two $A \times B$ faces of the block, a current I_{AB} is observed to flow. Find the current that flows if the same potential difference V is applied between the two $B \times C$ faces of the block. Give your answer in terms of I_{AB}.

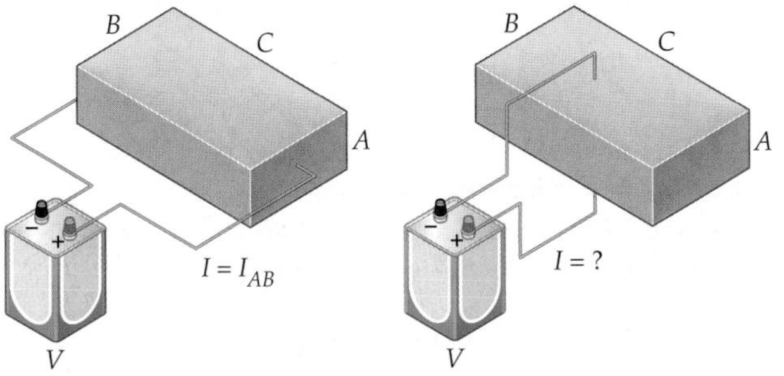

▲ **FIGURE 21–33** Problem 18

Section 21–3 Energy and Power in Electric Circuits

19. • A 65-V generator supplies 3.8 kW of power. How much current does the generator produce?

20. • A portable CD player operates with a current of 22 mA at a potential difference of 4.1 V. What is the power usage of the player?

21. • Find the power dissipated in a 25-Ω electric heater connected to a 120-V outlet.

22. • The current in a 120-V reading lamp is 2.3 A. If the cost of electrical energy is \$0.075 per kilowatt-hour, how much does it cost to operate the light for an hour?

23. • It costs 2.6 cents to charge a car battery at a voltage of 12 V and a current of 15 A for 120 minutes. What is the cost of electrical energy per kilowatt-hour at this location?

24. • • **IP** A 75-W lightbulb operates on a potential difference of 95 V. Find **(a)** the current in the bulb and **(b)** the resistance of the bulb. **(c)** If this bulb is replaced with one whose resistance is half the value found in part (b), is its power rating greater than or less than 75-W? By what factor?

25. • • **Rating Car Batteries** Car batteries are rated by the following two numbers: **(1)** cranking amps = current the battery can produce for 30.0 seconds while maintaining a terminal voltage of at least 7.2 V and **(2)** reserve capacity = number of minutes the battery can produce a 25-A current while maintaining a terminal voltage of at least 10.5 V. One particular battery is advertised as having 905 cranking amps and a 155-minute reserve capacity. Which of these two ratings represents the greater amount of energy delivered by the battery?

Section 21–4 Resistors in Series and Parallel

26. • Find the equivalent resistance between points A and B for the group of resistors shown in **Figure 21–34**.

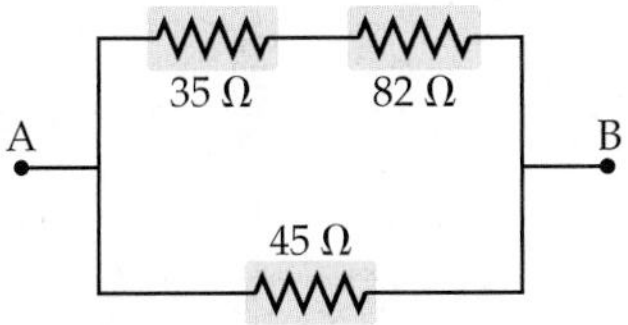

▲ **FIGURE 21–34** Problems 26 and 84

27. • What is the minimum number of 65-Ω resistors that must be connected in parallel to produce an equivalent resistance of 11 Ω or less?

28. • • Four lightbulbs (A, B, C, D) are connected together in a circuit of unknown arrangement. When each bulb is removed one at a time and replaced, the following behavior is observed:

	A	B	C	D
A removed	*	on	on	on
B removed	on	*	on	off
C removed	off	off	*	off
D removed	on	off	on	*

Draw a circuit diagram for these bulbs.

29. • • Your toaster has a power cord with a resistance of 0.020 Ω connected in series with a 9.6-Ω nichrome heating element. If the potential difference between the terminals of the toaster is 120 V, how much power is dissipated in **(a)** the power cord and **(b)** the heating element?

30. • • A hobbyist building a radio needs a 150-Ω resistor in her circuit, but has only a 220-Ω, a 79-Ω, and a 92-Ω resistor available. How can she connect these resistors to produce the desired resistance?

31. • • A circuit consists of a 9.0-V battery connected to three resistors (42 Ω, 17 Ω, and 110 Ω) in series. Find **(a)** the current that flows through the battery and **(b)** the potential difference across each resistor.

32. • • **IP** Three resistors, 11 Ω, 53 Ω, and R, are connected in series with a 24.0-V battery. The total current flowing through the battery is 0.16 A. **(a)** Find the value of resistance R. **(b)** Find the potential difference across each resistor. **(c)** If the voltage of the battery had been greater than 24.0 V, would your answer to part (a) have been larger or smaller? Explain.

33. • • A circuit consists of a battery connected to three resistors (65 Ω, 25 Ω, and 170 Ω) in parallel. The total current through the resistors is 1.8 A. Find **(a)** the emf of the battery and **(b)** the current through each resistor.

34. • • **IP** Three resistors, 22 Ω, 67 Ω, and R, are connected in parallel with a 12.0-V battery. The total current flowing through the battery is 0.88 A. **(a)** Find the value of resistance R. **(b)** Find the current through each resistor. **(c)** If the total current in the battery had been greater than 0.88 A, would your answer to part (a) have been larger or smaller? Explain.

35. • • An 84-Ω resistor has a current of 0.62 A and is connected in series with a 130-Ω resistor. What is the emf of the battery to which the resistors are connected?

36. • • The equivalent resistance between points A and B of the resistors shown in **Figure 21–35** is 26 Ω. Find the value of resistance R.

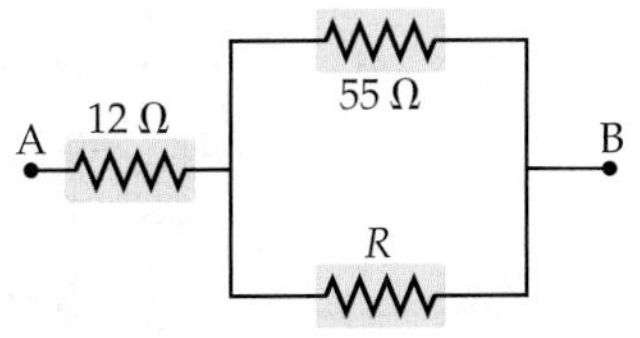

▲ **FIGURE 21–35** Problems 36, 40, and 69

37. • • Find the equivalent resistance between points A and B shown in **Figure 21–36**.

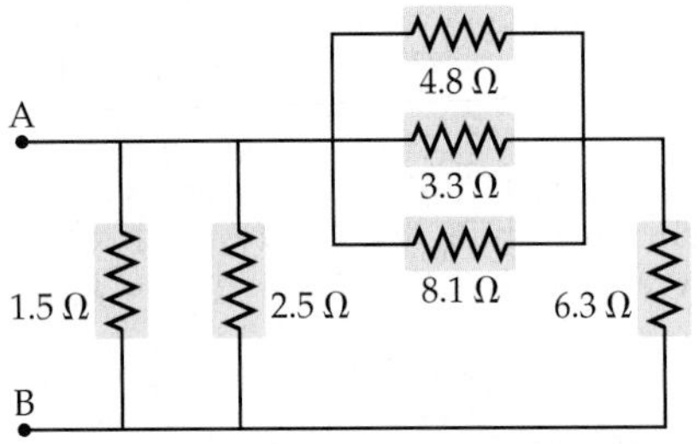

▲ **FIGURE 21–36** Problems 37 and 41

38. •• How many 65-W lightbulbs can be connected in parallel across a potential difference of 85 V before the total current in the circuit exceeds 2.1 A?

39. •• The circuit in **Figure 21–37** includes a battery with a finite internal resistance, $r = 0.50\ \Omega$. **(a)** Find the current flowing through the 7.1-Ω and the 3.2-Ω resistors. **(b)** How much current flows through the battery? **(c)** What is the potential difference between the terminals of the battery?

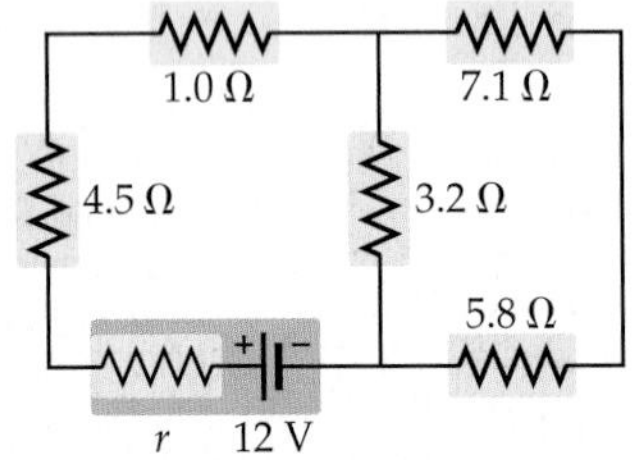

▲ **FIGURE 21–37** Problems 39 and 42

40. •• **IP** A 12-V battery is connected to terminals A and B in Figure 21–35. **(a)** Given that $R = 75\ \Omega$, find the current in each resistor. **(b)** Suppose the value of R is increased. For each resistor in turn, state whether the current flowing through it increases or decreases. Explain.

41. •• **IP** The terminals A and B in Figure 21–36 are connected to a 9.0-V battery. **(a)** Find the current flowing through each resistor. **(b)** Is the potential difference across the 6.3-Ω resistor greater than, less than, or the same as the potential difference across the 1.5-Ω resistor? Explain.

42. •• **IP** Suppose the battery in Figure 21–37 has an internal resistance $r = 0.25\ \Omega$. **(a)** How much current flows through the battery? **(b)** What is the potential difference between the terminals of the battery? **(c)** If the 3.2-Ω resistor is increased in value, will the current in the battery increase or decrease? Explain.

43. ••• **IP** The current flowing through the 8.45-Ω resistor in **Figure 21–38** is 1.52 A. **(a)** What is the voltage of the battery? **(b)** If the 17.2-Ω resistor is increased in value, will the current provided by the battery increase, decrease, or stay the same? Explain.

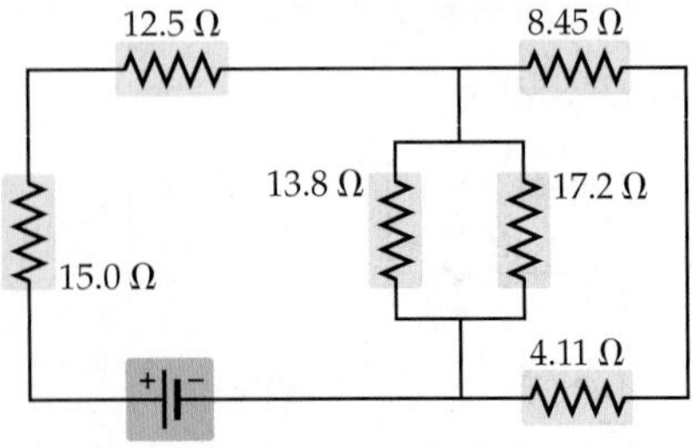

▲ **FIGURE 21–38** Problems 43 and 44

44. ••• The current in the 13.8-Ω resistor in Figure 21–38 is 0.775 A. Find the current in the other resistors in the circuit.

45. ••• **IP** Four identical resistors are connected to a battery as shown in **Figure 21–39**. When the switch is open, the current through the battery is I_0. **(a)** When the switch is closed, will the current through the battery increase, decrease, or stay the same? Explain. **(b)** Calculate the current that flows through the battery when the switch is closed. Give your answer in terms of I_0.

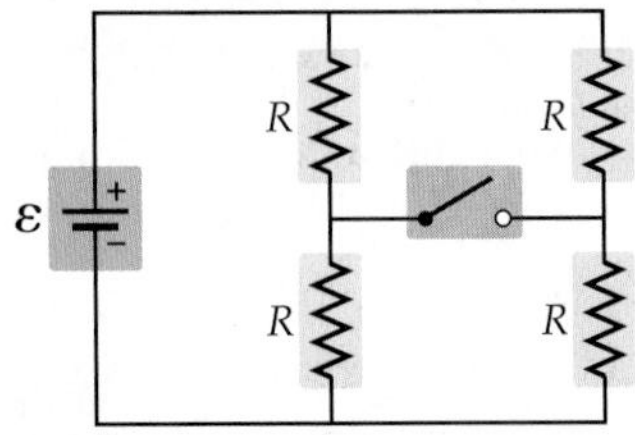

▲ **FIGURE 21–39** Problem 45

Section 21–5 Kirchhoff's Rules

46. • Find the magnitude and direction (clockwise or counterclockwise) of the current in **Figure 21–40**.

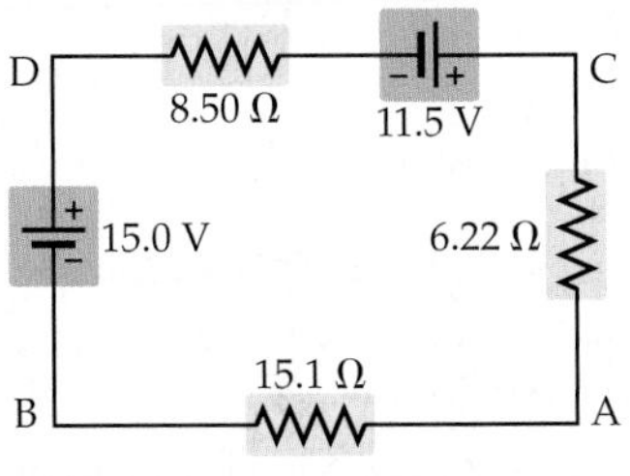

▲ **FIGURE 21–40** Problems 46, 47, and 48

47. • **IP** Suppose the polarity of the 11.5-V battery in Figure 21–40 is reversed. **(a)** Do you expect this to increase or decrease the amount of current flowing in the circuit? Explain. **(b)** Calculate the magnitude and direction (clockwise or counterclockwise) of the current in this case.

48. •• **IP** It is given that point A in Figure 21–40 is grounded ($V = 0$). **(a)** Is the potential at point B greater than or less than zero? Explain. **(b)** Is the potential at point C greater than or less than zero? Explain. **(c)** Calculate the potential at point D.

49. •• Consider the circuit shown in **Figure 21–41**. Find the current through each resistor using **(a)** the rules for series and parallel resistors and **(b)** Kirchhoff's rules.

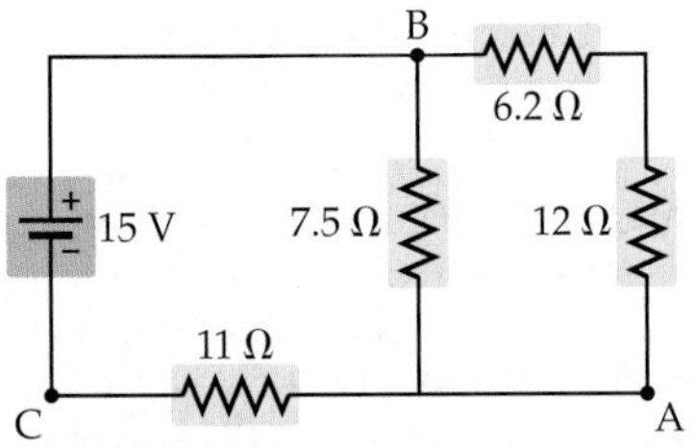

▲ **FIGURE 21–41** Problems 49 and 50

50. •• Suppose point A is grounded ($V = 0$) in Figure 21–41. Find the potential at points B and C.

51. •• **IP (a)** Find the current in each resistor in **Figure 21–42**. **(b)** Is the potential at point A greater than, less than, or equal to the

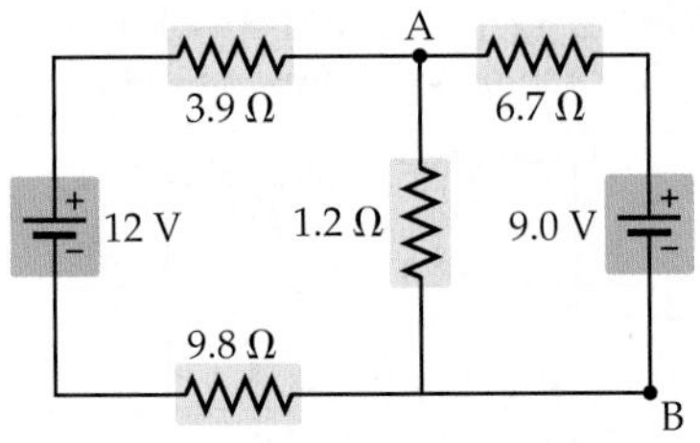

▲ **FIGURE 21–42** Problem 51

potential at point B? Explain. **(c)** Determine the potential difference between the points A and B.

52. ••• Two batteries and three resistors are connected as shown in **Figure 21–43**. How much current flows through each battery when the switch is **(a)** closed and **(b)** open?

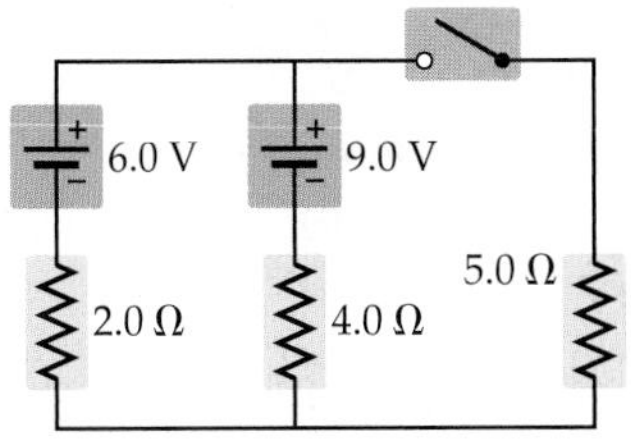

▲ **FIGURE 21–43** Problem 52

Section 21–6 Circuits Containing Capacitors

53. • Find the equivalent capacitance between points A and B for the group of capacitors shown in **Figure 21–44**.

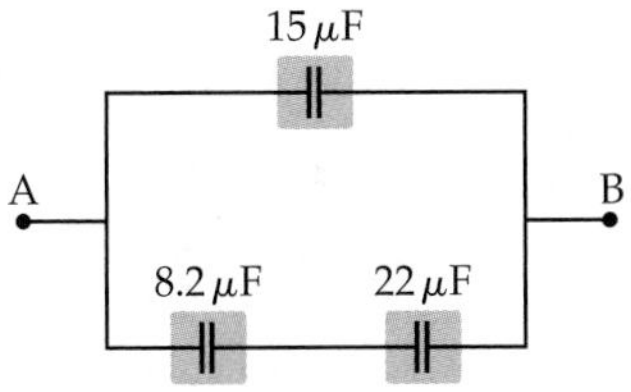

▲ **FIGURE 21–44** Problems 53 and 55

54. • An 18-V battery is connected to three capacitors in series. The capacitors have the following capacitances: 4.5 μF, 12 μF, and 32 μF. Find the voltage across the 32-μF capacitor.

55. •• Terminals A and B in Figure 21–44 are connected to a 9.0-V battery. Find the energy stored in each capacitor.

56. •• **IP** Two capacitors, one 7.5 μF the other 15 μF, are connected in parallel across a 12-V battery. **(a)** Find the equivalent capacitance of the two capacitors. **(b)** Which capacitor stores more charge? Explain. **(c)** Find the charge stored on each capacitor.

57. •• **IP** Two capacitors, one 7.5 μF the other 15 μF, are connected in series across a 12-V battery. **(a)** Find the equivalent capacitance of the two capacitors. **(b)** Which capacitor stores more charge? Explain. **(c)** Find the charge stored on each capacitor.

58. •• The equivalent capacitance of the capacitors shown in **Figure 21–45** is 9.22 μF. Find the value of capacitance C.

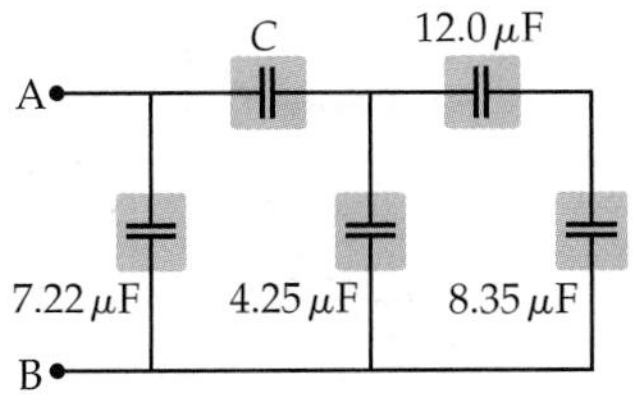

▲ **FIGURE 21–45** Problems 58 and 88

59. ••• Two capacitors, C_1 and C_2, are connected in series and charged by a battery. Show that the energy stored in C_1 plus the energy stored in C_2 is equal to the energy stored in the equivalent capacitor, C_{eq}, when it is connected to the same battery.

60. ••• With the switch in position A, the 11.2-μF capacitor in **Figure 21–46** is fully charged by the 12.0-V battery, and the 9.50-μF capacitor is uncharged. The switch is now moved to position B. As a result, charge flows between the capacitors until they have the same voltage across their plates. Find this voltage.

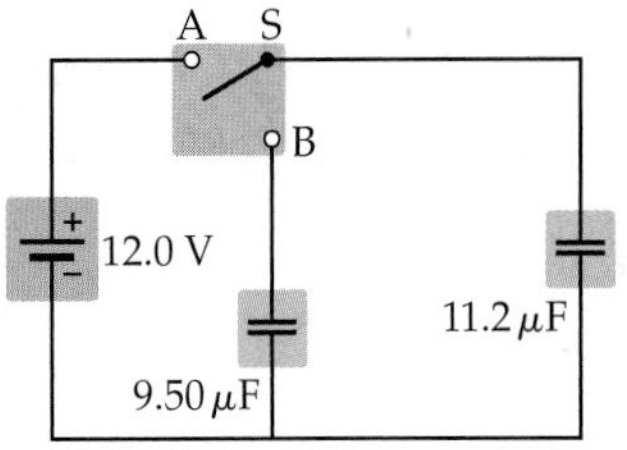

▲ **FIGURE 21–46** Problem 60

Section 21–7 *RC* Circuits

61. • The switch on an RC circuit is closed at $t = 0$. Given that $\mathcal{E} = 6.0$ V, $R = 150\ \Omega$, and $C = 23\ \mu$F, how much charge is on the capacitor at time $t = 4.2$ ms?

62. • The capacitor in an RC circuit ($R = 120\ \Omega$, $C = 45\ \mu$F) is initially uncharged. Find **(a)** the charge on the capacitor and **(b)** the current in the circuit one time constant ($\tau = RC$) after the circuit is connected to a 9.0-V battery.

63. •• Consider an RC circuit with $\mathcal{E} = 12.0$V, $R = 175\ \Omega$, and $C = 55.7\ \mu$F. Find **(a)** the time constant for the circuit, **(b)** the maximum charge on the capacitor, and **(c)** the initial current in the circuit.

64. •• The resistor in an RC circuit has a resistance of 145 Ω. **(a)** What capacitance must be used in this circuit if the time constant is to be 3.5 ms? **(b)** Using the capacitance determined in part (a), calculate the current in the circuit 7.0 ms after the switch is closed. Assume that the capacitor is uncharged initially and that the emf of the battery is 9.0 V.

65. •• A flash unit for a camera has a capacitance of 1200 μF. What resistance is needed in this RC circuit if the flash is to charge to 90% of its full charge in 21 s?

66. •• **Figure 21–47** shows a simplified circuit for a photographic flash unit. This circuit consists of a 9.0-V battery, a 50.0-kΩ resistor, a 140-μF capacitor, a flashbulb, and two switches. Initially, the capacitor is uncharged and the two switches are open. To charge the unit, switch S_1 is closed; to fire the flash, switch S_2 (which is connected to the camera's shutter) is closed. How long does it take to charge the capacitor to 5.0 V?

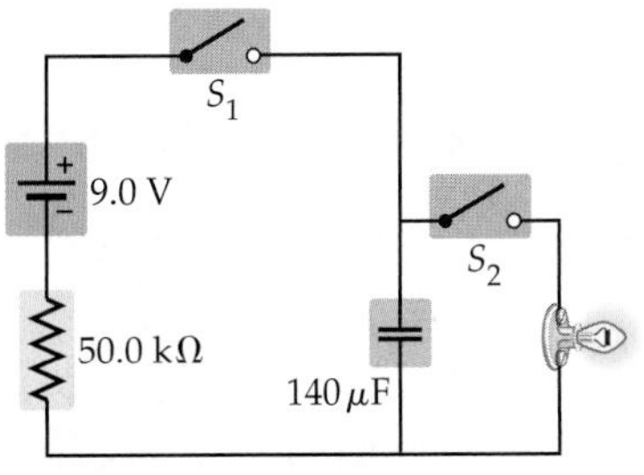

▲ **FIGURE 21–47** Problem 66

67. •• **IP** Consider the RC circuit shown in **Figure 21–48**. Find **(a)** the time constant and **(b)** the initial current for this circuit. **(c)** It is desired to increase the time constant of this circuit by adjusting

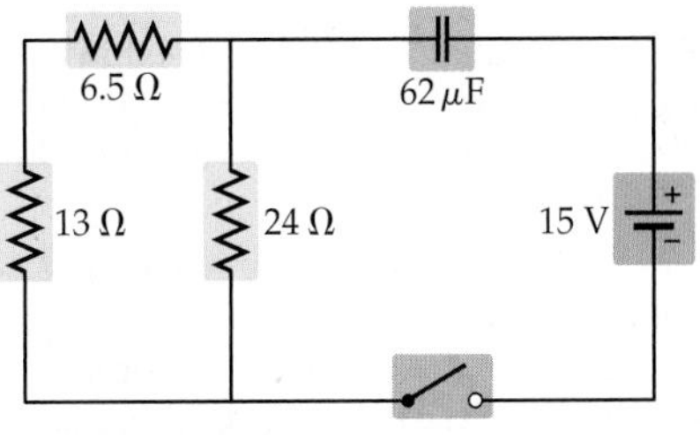

▲ **FIGURE 21–48** Problems 67 and 89

the value of the 6.5-Ω resistor. Should the resistance of this resistor be increased or decreased to have the desired effect? Explain.

68. ••• The capacitor in an *RC* circuit is initially uncharged. In terms of *R* and *C*, determine **(a)** the time required for the charge on the capacitor to rise to 50% of its final value and **(b)** the time required for the initial current to drop to 10% of its initial value.

General Problems

69. • Suppose that points A and B in Figure 21–35 are connected to a 12-V battery. Find the power dissipated in each of the resistors assuming that $R = 75\ \Omega$.

70. • You are given resistors of 413 Ω, 521 Ω, and 146 Ω. Describe how these resistors must be connected to produce an equivalent resistance of 255 Ω.

71. • You are given capacitors of 18 μF, 7.2 μF, and 9.0 μF. Describe how these capacitors must be connected to produce an equivalent capacitance of 22 μF.

72. • Suppose your car carries a charge of 65 μC. What current does it produce as it travels from Dallas to Fort Worth (35 mi) in 0.75 h?

73. •• It is desired to construct a 5.0-Ω resistor from a 1.7-m length of tungsten wire. What diameter is needed for this wire?

74. •• **Electrical Safety Codes** For safety reasons, electrical codes have been established that limit the amount of current a wire of a given size can carry. For example, an 18-gauge (cross-sectional area = 1.17 mm^2), rubber-insulated extension cord with copper wires can carry a maximum current of 5.0 A. Find the voltage drop in a 12-ft, 18-gauge extension cord carrying a current of 5.0 A. (*Note:* In an extension cord, the current must flow through two lengths—down and back.)

75. •• **A Three-Way Lightbulb** A three-way lightbulb has two filaments with resistances R_1 and R_2 connected in series. The resistors are connected to three terminals, as indicated in **Figure 21–49**, and the light switch determines which two of the three terminals are connected to a potential difference of 120 V at any given time. When terminals A and B are connected to 120 V the bulb uses 75.0 W of power. When terminals A and C are connected to 120 V the bulb uses 50.0 W of power. **(a)** What is the resistance R_1? **(b)** What is the resistance R_2? **(c)** How much power does the bulb use when 120 V is connected to terminals B and C?

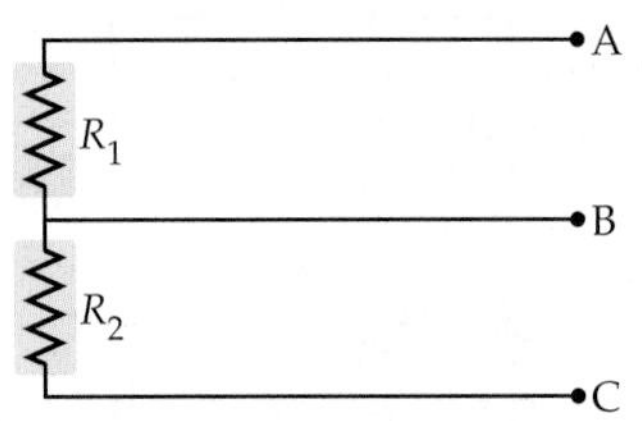

▲ **FIGURE 21–49** Problem 75

76. •• A portable CD player uses a current of 7.5 mA at a potential difference of 3.5 V. **(a)** How much energy does the player use in 35 s? **(b)** Suppose the player has a mass of 0.65 kg. For what length of time could the player operate on the energy required to lift it through a height of 1.0 m?

77. •• An electrical heating coil is immersed in 4.2 kg of water at 22 °C. The coil, which has a resistance of 250 Ω, warms the water to 32 °C in 3.5 min. What is the potential difference at which the coil operates?

78. •• **IP** Consider the circuit shown in **Figure 21–50**. **(a)** Is the current flowing through the battery immediately after the switch is closed greater than, less than, or the same as the current flowing through the battery long after the switch is closed? Explain. **(b)** Find the current flowing through the battery immediately after the switch is closed. **(c)** Find the current in the battery long after the switch is closed.

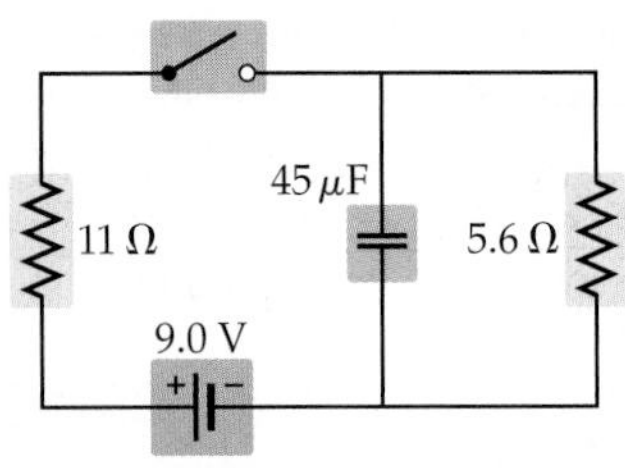

▲ **FIGURE 21–50** Problems 78 and 85

79. •• A silver wire and a copper wire have the same volume and the same resistance. Find the ratio of their radii, r_{silver}/r_{copper}.

80. •• Two resistors are connected in series to a battery with an emf of 12 V. The voltage across the first resistor is 2.7 V and the current through the second resistor is 0.15 A. Find the resistance of the two resistors.

81. •• **BIO Pacemaker Pulses** A pacemaker sends a pulse to a patient's heart every time the capacitor in the pacemaker charges to a voltage of 0.25 V. It is desired that the patient receive 75 pulses per minute. Given that the capacitance of the pacemaker is 110 μF and that the battery has a voltage of 9.0 V, what value should the resistance have?

82. •• A long, thin wire has a resistance *R*. The wire is now cut into three segments of equal length, which are connected in parallel. In terms of *R*, what is the equivalent resistance of the three wire segments?

83. •• Three resistors $\left(R, \frac{1}{2}R, 2R\right)$ are connected to a battery. **(a)** If the resistors are connected in series, which one has the greatest rate of energy dissipation? **(b)** Repeat part (a), this time assuming that the resistors are connected in parallel.

84. •• **IP** Suppose we connect a 9.0-V battery to terminals A and B in Figure 21–34. **(a)** Is the current in the 45-Ω resistor greater than, less than, or the same as the current in the 35-Ω resistor? Explain. **(b)** Calculate the current flowing through each of the three resistors in this circuit.

85. •• **IP** Suppose the battery in Figure 21–50 has an internal resistance of 0.73 Ω. **(a)** What is the potential difference across the terminals of the battery when the switch is open? **(b)** When the switch is closed, does the potential difference of the battery increase or decrease? Explain. **(c)** Find the potential difference across the battery after the switch has been closed a long time.

86. •• **Footwear Safety** The American National Standards Institute (ANSI) specifies safety standards for a number of potential workplace hazards. The regulation ANSI Z41-1999 states that the resistance of a person and their footwear be tested with the circuit shown in **Figure 21–51**, with $\mathcal{E} = 50.0$ V and

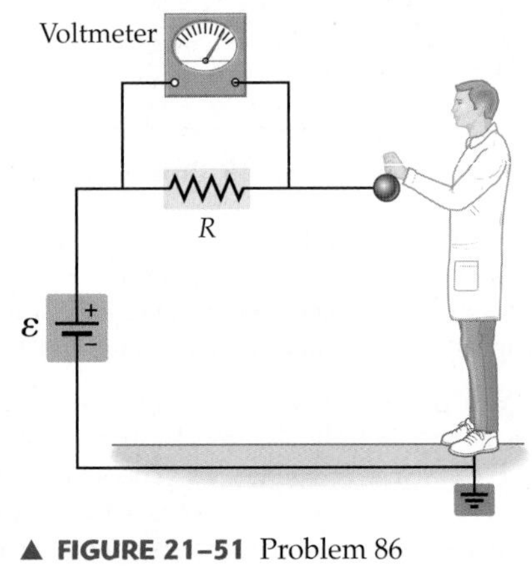

▲ **FIGURE 21–51** Problem 86

$R = 1.00\ \text{M}\Omega$. Suppose the voltmeter measures a potential difference of 3.70 V across the resistor. **(a)** What is the current that flows through the person's body? (To be safe, the current should not exceed 150 μA.) **(b)** What is the resistance of the person and the footwear? (Type II footwear, as defined by ANSI, should give a resistance of 10^6 to $10^9\ \Omega$.)

87. •• **National Electric Code** In the United States, the National Electric Code sets standards for maximum safe currents in insulated copper wires of various diameters. The accompanying table gives a portion of the code. Notice that wire diameters are identified by the *gauge* of the wire, and that 1 mil = 10^{-3} in. Find the maximum power dissipated per length in **(a)** an 8-gauge wire and **(b)** a 10-gauge wire.

Gauge	Diameter, mils	Safe current, A
8	129	35
10	102	25

88. ••• IP A 12.0-V battery is connected to terminals A and B in Figure 21–45. **(a)** Given that $C = 15.0\ \mu$F, find the charge on each of the capacitors. **(b)** Find the total energy stored in this system. **(c)** If the 7.22-μF capacitor is increased in value, will the total energy stored in the circuit increase or decrease? Explain.

89. ••• IP The switch in the *RC* circuit shown in Figure 21–48 is closed at $t = 0$. **(a)** How much power is dissipated in each resistor just after $t = 0$ and in the limit $t \rightarrow \infty$? **(b)** What is the charge on the capacitor at the time $t = 0.35$ ms? **(c)** How much energy is stored in the capacitor in the limit $t \rightarrow \infty$? **(d)** If the voltage of the battery is doubled, by what factor does your answer to part (c) change? Explain.

90. ••• Two resistors, R_1 and R_2, are connected in parallel and connected to a battery. Show that the power dissipated in R_1 plus the power dissipated in R_2 is equal to the power dissipated in the equivalent resistor, R_{eq}, when it is connected to the same battery.

91. ••• A battery has an emf $\mathcal{E}$ and an internal resistance r. When the battery is connected to a 25-Ω resistor, the current through the battery is 0.65 A. When the battery is connected to 55-Ω resistor, the current is 0.45 A. Find the battery's emf and internal resistance.

92. ••• When two resistors, R_1 and R_2, are connected in series across a 6.0-V battery, the potential difference across R_1 is 4.0 V. When R_1 and R_2 are connected in parallel to the same battery, the current through R_2 is 0.45 A. Find the values of R_1 and R_2.

93. ••• The circuit shown in **Figure 21–52** is known as a Wheatstone bridge. Find the value of the resistor R such that the current through the 85.0-Ω resistor is zero.

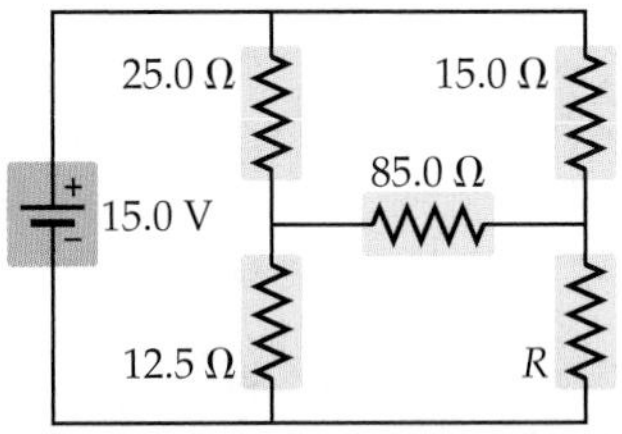

▲ **FIGURE 21–52** Problem 93

Interactive Problems

94. •• ***Referring to Example 21–7*** Suppose the three resistors in this circuit have the values $R_1 = 100.0\ \Omega$, $R_2 = 200.0\ \Omega$, and $R_3 = 300.0\ \Omega$, and that the emf of the battery is 12.0 V. **(a)** Find the potential difference across each resistor. **(b)** Find the current that flows through each resistor.

95. •• ***Referring to Example 21–7*** Suppose $R_1 = R_2 = 225\ \Omega$ and that $R_3 = R$. The emf of the battery is 12.0 V. **(a)** Find the value of R such that the current supplied by the battery is 0.0750 A. **(b)** Find the value of R that gives a potential difference of 2.65 V across resistor 2.

96. •• IP ***Referring to Example 21–9*** Suppose the resistance of the 126-Ω resistor is reduced by a factor of two. The other resistor is 275 Ω, the capacitor is 182 μF, and the battery has an emf of 3.00 V. **(a)** Does the final value of the charge on the capacitor increase, decrease, or stay the same? Explain. **(b)** Does the time for the capacitor to charge to 80.0% of its final value increase, decrease, or stay the same? Explain. **(c)** Find the time referred to in part (b).

97. •• IP ***Referring to Example 21–9*** Suppose the capacitance of the 182-μF capacitor is reduced by a factor of two. The two resistors are 126 Ω and 275 Ω, and the battery has an emf of 3.00 V. **(a)** Find the final value of the charge on the capacitor. **(b)** Does the time for the capacitor to charge to 80.0% of its final value increase, decrease, or stay the same? Explain. **(c)** Find the time referred to in part (b).

22 Magnetism

22–1	The Magnetic Field	733
22–2	The Magnetic Force on Moving Charges	735
22–3	The Motion of Charged Particles in a Magnetic Field	739
22–4	The Magnetic Force Exerted on a Current-Carrying Wire	744
22–5	Loops of Current and Magnetic Torque	746
22–6	Electric Currents, Magnetic Fields, and Ampère's Law	748
22–7	Current Loops and Solenoids	752
22–8	Magnetism in Matter	755

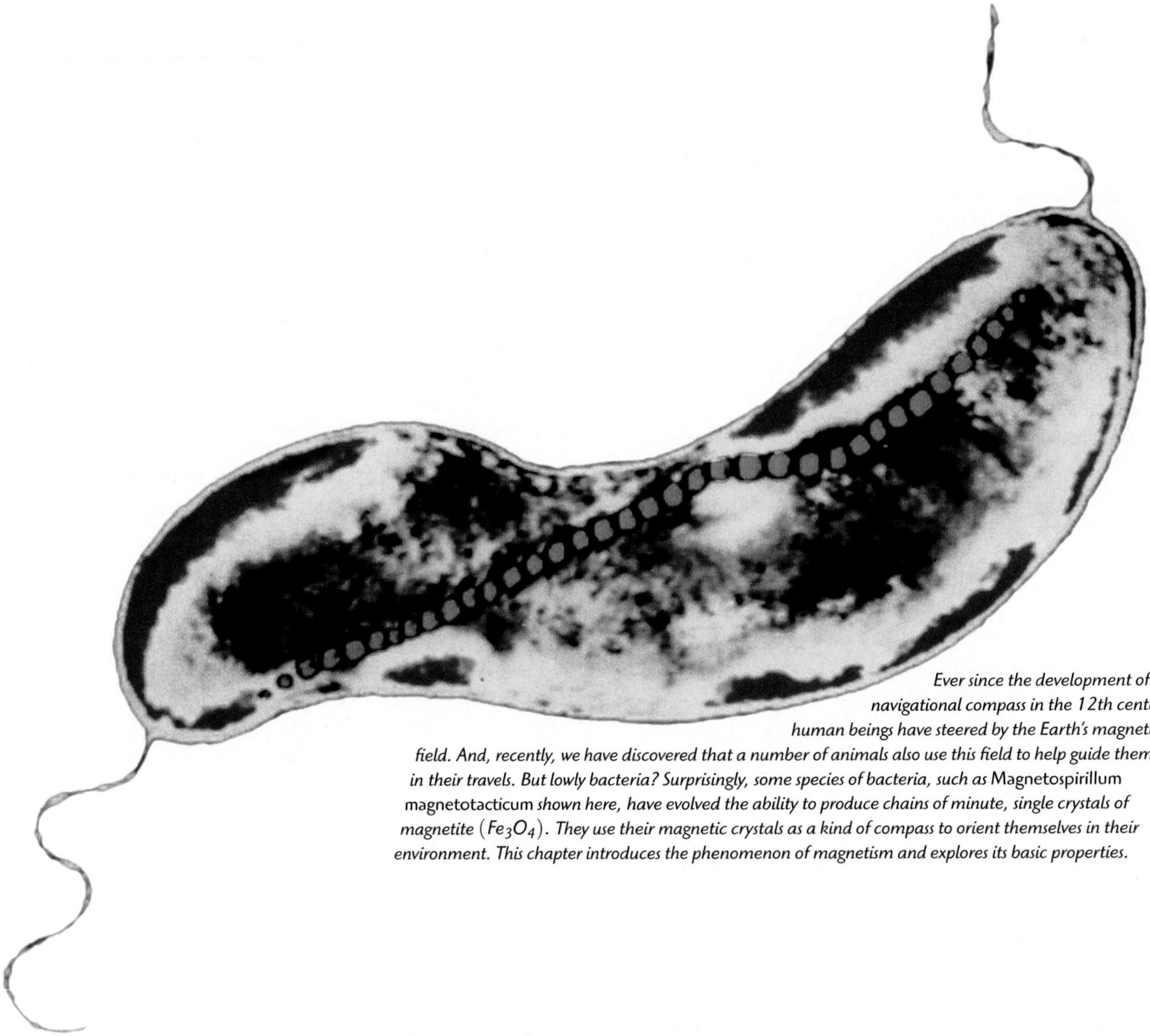

Ever since the development of the navigational compass in the 12th century, human beings have steered by the Earth's magnetic field. And, recently, we have discovered that a number of animals also use this field to help guide them in their travels. But lowly bacteria? Surprisingly, some species of bacteria, such as Magnetospirillum magnetotacticum *shown here, have evolved the ability to produce chains of minute, single crystals of magnetite* (Fe_3O_4)*. They use their magnetic crystals as a kind of compass to orient themselves in their environment. This chapter introduces the phenomenon of magnetism and explores its basic properties.*

The effects of magnetism have been known since antiquity. For example, a piece of lodestone, a naturally occurring iron oxide mineral, can behave just like a modern-day bar magnet. What has been discovered only much more recently, however, is the intimate connection that exists between electricity and magnetism. The first evidence of such a connection was obtained quite by accident, in 1820, as the Danish scientist Hans Christian Oersted (1777–1851) performed a demonstration during one of his popular public lectures. His observation, and the conclusions that followed from it, are the primary focus of this chapter.

22–1 The Magnetic Field

We begin our study of magnetism with a few general observations regarding magnets and the fields they produce. These observations apply over a wide range of scales, from the behavior of small, handheld bar magnets to the global effects associated with the magnetic field of the Earth. As we shall see later in this chapter, the ultimate source of any magnetic field is nothing more than the motion of electric charge.

▲ **FIGURE 22–1 The force between two bar magnets**
(a) Opposite poles attract each other. **(b)** The force between like poles is repulsive.

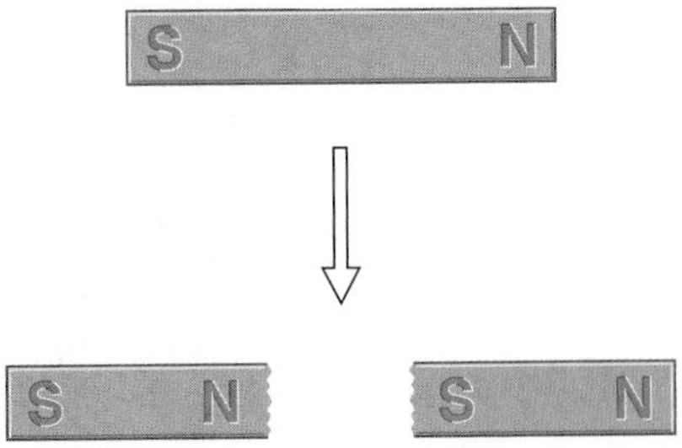

▲ **FIGURE 22–2 Magnets always have two poles**
When a bar magnet is broken in half, two new poles appear. Each half has both a north pole and a south pole, just like any other bar magnet.

Permanent Magnets

Our first direct experience with magnets is often a playful exploration of the way small permanent magnets, called bar magnets, affect one another. As we know, a bar magnet can attract another magnet or repel it, depending on which ends of the magnets are brought together. One end of a magnet is referred to as its *north pole;* the other end is its *south pole.* The poles are defined as follows: Imagine suspending a bar magnet from a string so that it is free to rotate in a horizontal plane—much like the needle of a compass. The end of the magnet that points toward the north geographic pole of the Earth we refer to as the "north-seeking pole," or simply the north pole. The opposite end of the magnet is the "south-seeking pole," or the south pole.

The rule for whether two bar magnets attract or repel each other is learned at an early age: opposites attract; likes repel. Thus, if two magnets are brought together so that opposite poles approach each other, as in **Figure 22–1 (a)**, the force they experience is attractive. Like poles brought close together, as in **Figure 22–1 (b)**, experience a repulsive force.

An interesting aspect of magnets is that they *always* have two poles. You might think that if you break a bar magnet in two, each of the halves will have just one pole. Instead, breaking a magnet in half results in the appearance of two new poles on either side of the break, as illustrated in **Figure 22–2**. This behavior is fundamentally different from that found in electricity, where the two types of charge can exist separately. Though physicists continue to look for the elusive "magnetic monopole," and speculate as to its possible properties, none has been found.

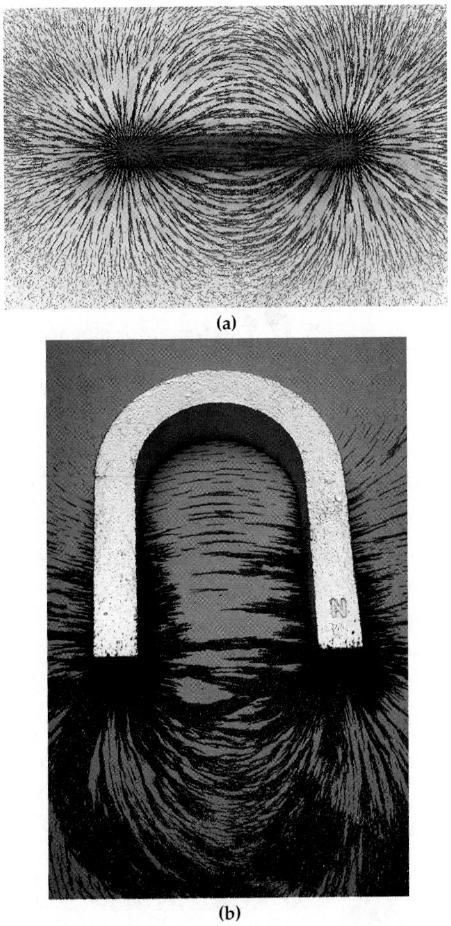

▲ **FIGURE 22–3 Magnetic field lines**
The field of a bar magnet **(a)** or horseshoe magnet **(b)** can be visualized using iron filings on a sheet of glass or paper. The filings orient themselves with the field lines, creating a "snapshot" of the magnetic field.

Magnetic Field Lines

Just as an electric charge creates an electric "field" in its vicinity, so too does a magnet create a magnetic field. As in the electric case, the magnetic field represents the effect a magnet has on its surroundings. For example, in Figure 19–15 we saw a visual indication of the electric field $\vec{E}$ of a point charge using grass seeds suspended in oil. Similarly, the magnetic field, which we represent with the symbol $\vec{B}$, can be visualized using small iron filings sprinkled onto a smooth surface. In **Figure 22–3 (a)**, for example, a sheet of glass or plastic is placed on top of a bar magnet. When iron filings are dropped onto the sheet they align with the magnetic field in their vicinity, giving a good representation of the overall field produced by the magnet. Similar effects are seen in **Figure 22–3 (b)**, with a magnet that has been bent to bring its poles close together. Because of its shape, this type of magnet is referred to as a horseshoe magnet.

In both cases shown in Figure 22–3, notice that the filings are bunched together near the poles of the magnets. This is where the magnetic field is most intense. We illustrate this by drawing field lines that are densely packed near the poles, as

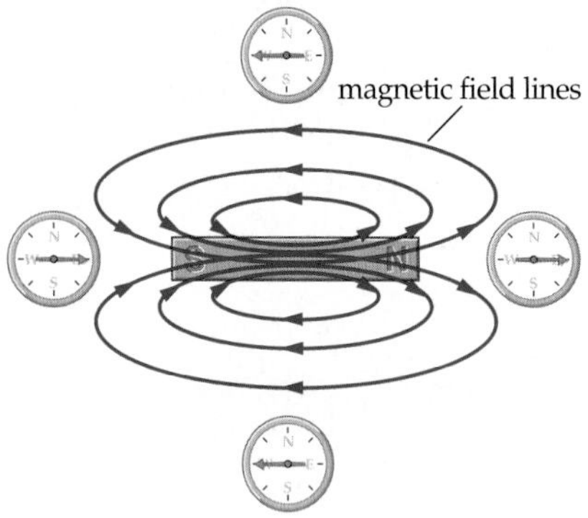

▲ **FIGURE 22–4 Magnetic field lines for a bar magnet**

The field lines are closely spaced near the poles, where the magnetic field $\vec{B}$ is most intense. In addition, the lines form closed loops that leave at the north pole of the magnet and enter at the south pole.

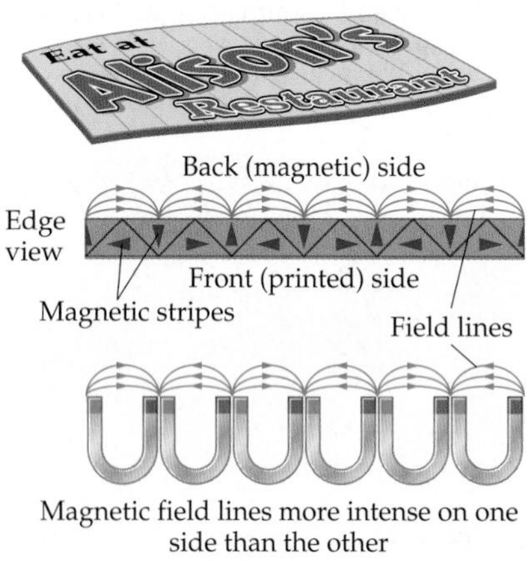

▲ **FIGURE 22–5 Refrigerator magnet**

A flexible refrigerator magnet is made from a large number of narrow magnetic stripes with magnetic fields in different directions. The net effect is a field similar to that of a series of parallel horseshoe magnets placed side by side.

REAL-WORLD PHYSICS
Refrigerator magnets

REAL-WORLD PHYSICS
The Earth's magnetic field

in **Figure 22–4**. As one moves away from a magnet in any direction the field weakens, as indicated by the increasingly wider separations between field lines. Thus, we indicate the magnitude of the vector $\vec{B}$ by the spacing of the field lines.

We can also assign a direction to the magnetic field. In particular, the direction of $\vec{B}$ is defined as follows:

> The direction of the magnetic field, $\vec{B}$, at a given location is the direction in which the north pole of a compass points when placed at that location.

We apply this definition to the bar magnet in Figure 22–4. Imagine, for example, placing a compass near its south pole. Because opposites attract, the north pole of the compass needle—the end with the arrowhead—will point toward the south pole of the magnet. Hence, according to our definition, the direction of the magnetic field at that location is toward the bar magnet's south pole. Similarly, one can see that the magnetic field must point away from the north pole of the bar magnet. In general,

> Magnetic field lines exit from the north pole of a magnet and enter at the south pole.

The field lines continue even within the body of a magnet. In fact, magnetic field lines always form closed loops; they never start or stop anywhere, in contrast to electric field lines. Again, this is related to the fact that there are no magnetic monopoles, where a magnetic field line could start or stop, whereas electric field lines start on positive charges and stop on negative charges.

CONCEPTUAL CHECKPOINT 22–1 **Magnetic Field Lines**

Can magnetic field lines cross one another?

Reasoning and Discussion

Recall that the direction in which a compass points at any given location is the direction of the magnetic field at that point. Since a compass can point in only one direction, there must be only one direction for the field $\vec{B}$. If field lines were to cross, however, there would be two directions for $\vec{B}$ at the crossing point, and this is not allowed.

Answer:

No. Magnetic field lines never cross.

An interesting example of magnetic field lines is provided by the humble refrigerator magnet. The flexible variety of these popular magnets has the unusual property that one side sticks quite strongly to the refrigerator, whereas the other side (the printed side) does not stick at all. Clearly, the magnetic field produced by such a magnet is not like that of a simple bar magnet, which generates a symmetrical field. Instead, these refrigerator magnets are composed of multiple magnetic stripes of opposite polarity, as indicated in **Figure 22–5**. The net effect is a magnetic field similar to the field that would be produced by a large number of tiny horseshoe magnets placed side by side, pointing in the same direction. In this way, the field is intense on the side containing the poles of the tiny magnets, as in Figure 22–3 (b), and very weak on the other side.

Geomagnetism

The Earth, like many planets, produces its own magnetic field. In many respects, the Earth's magnetic field is like that of a giant bar magnet, as illustrated in **Figure 22–6**, with a pole near each geographic pole of the Earth. The magnetic poles are not perfectly aligned with the rotational axis of the Earth, however, but are inclined at an angle that varies slowly with time. Presently, the magnetic poles are tilted away from the rotational axis by an angle of about 11.5°. The current location of the north magnetic pole is just west of Ellef Ringnes Island, one of the Queen Elizabeth Islands of extreme northern Canada.

Since the north pole of a compass needle points toward the north magnetic pole of the Earth, and since opposites attract, it follows that

> the *north* geographical pole of the Earth is actually near the *south* pole of the Earth's magnetic field.

This is indicated in Figure 22–6. The figure also shows that the field lines are essentially horizontal (parallel to the Earth's surface) near the equator but enter or leave the Earth vertically near the poles. Thus, for example, if you were to stand near the north geographic pole, your compass would try to point straight down.

Although the Earth's magnetic field is similar in many respects to that of a huge bar magnet, it is far more complex in both its shape and behavior—in fact, even the mechanism by which the field is produced is still not completely understood. It seems likely, however, that flowing currents of molten material in the Earth's core are the primary cause of the field, as expressed in the **dynamo theory** of magnetism.

One of the reasons for lingering uncertainty about the Earth's magnetic field is that its behavior over time is rather complicated. For example, we have already mentioned that the magnetic poles drift about slowly with time. This is just the beginning of the field's interesting behavior, however. We have learned in the last century that the Earth's field has actually *reversed* direction many times over the ages. In fact, the last reversal occurred about 780,000 years ago—from 980,000 years ago to 780,000 years ago your compass would have pointed opposite to its direction today.

These ancient field reversals have left a permanent record in the rocks of the ocean floors, among other places. By analyzing the evidence of these reversals, geologists have found strong support for the theories of continental drift and plate tectonics and have developed a whole new branch of geology referred to as **paleomagnetism.** We shall return to these topics again briefly near the end of this chapter.

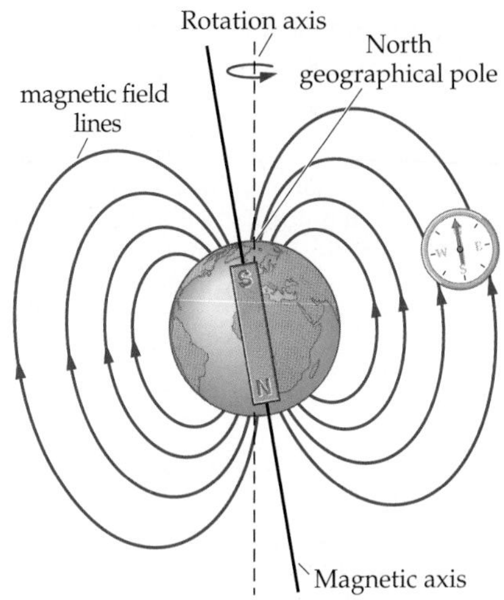

▲ **FIGURE 22–6 Magnetic field of the Earth**

The Earth's magnetic field is similar to that of a giant bar magnet tilted slightly from the rotational axis. Notice that the north geographic pole is actually near the south pole of the Earth's magnetic field.

22–2 The Magnetic Force on Moving Charges

We have discussed briefly the familiar forces that act between one magnet and another. In this section we consider the force a magnetic field exerts on a moving electric charge. As we shall see, both the magnitude and the direction of this force have rather unusual characteristics. We begin with the magnitude.

Magnitude of the Magnetic Force

Consider a magnetic field $\vec{\mathbf{B}}$ that points from left to right in the plane of the page, as indicated in **Figure 22–7**. A particle of charge q moves through this region with a velocity $\vec{\mathbf{v}}$. Note that the angle between $\vec{\mathbf{v}}$ and $\vec{\mathbf{B}}$ has a magnitude denoted by the symbol θ. Experiment shows that the magnitude of the force $\vec{\mathbf{F}}$ experienced by this particle is given by the following:

Magnitude of the Magnetic Force, *F*

$$F = |q|vB\sin\theta \tag{22–1}$$

SI unit: newton, N

Thus, the magnetic force depends on several factors. Two of these are the same as for the electric force:

(i) the charge of the particle, q;
(ii) the magnitude of the field, in this case the magnetic field, B;

However, the magnetic force also depends on two factors that do not affect the strength of the electric force:

(iii) the speed of the particle, v;
(iv) the magnitude of the angle between the velocity vector and the magnetic field vector, θ.

It follows that the behavior of particles in magnetic fields is significantly different from their behavior in electric fields.

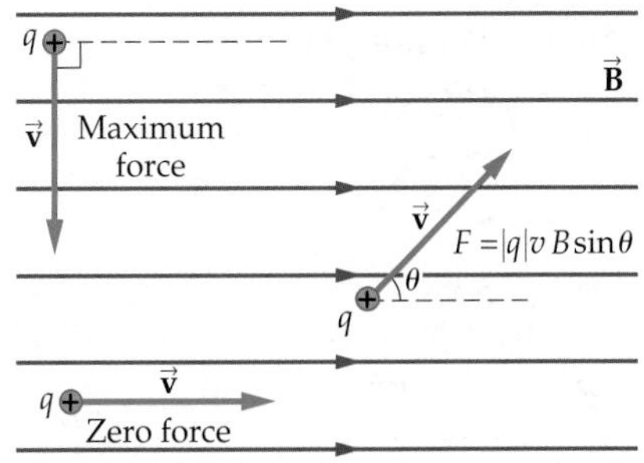

▲ **FIGURE 22–7 The magnetic force on a moving charged particle**

A particle of charge q moves through a region of magnetic field $\vec{\mathbf{B}}$ with a velocity $\vec{\mathbf{v}}$. The magnitude of the force experienced by the charge is $F = |q|vB\sin\theta$. Note that the force is a maximum when the velocity is perpendicular to the field and is zero when the velocity is parallel to the field.

PROBLEM-SOLVING NOTE

The Properties of Magnetic Forces

Remember that a magnetic force occurs only when a charged particle moves in a direction that is not along the line defined by the magnetic field. Stationary particles experience no magnetic force.

In particular, a particle must have a charge and must be moving if the magnetic field is to exert a force on it. Even then the force will vanish if the particle moves in the direction of the field (that is, $\theta = 0$), or in the direction opposite to the field ($\theta = 180°$). The maximum force is experienced when the particle moves at right angles to the field, so that $\theta = 90°$, and $\sin\theta = 1$.

Finally, the expression in Equation 22–1 gives only the magnitude of the force, and hence depends on the magnitude of the charge, $|q|$. As we shall see later in this section, the *sign* of q is important in determining the *direction* of the magnetic force. In addition, the angle θ in Equation 22–1 is the magnitude of the angle between $\vec{\mathbf{v}}$ and $\vec{\mathbf{B}}$, and always has a value in the range $0 \leq \theta \leq 180°$.

In practice, the magnitude—or strength—of a magnetic field is *defined* by the relation given in Equation 22–1. Thus, B is given by the following:

Definition of the Magnitude of the Magnetic Field, *B*

$$B = \frac{F}{|q|v\sin\theta} \qquad \text{22–2}$$

SI unit: 1 tesla = 1 T = 1 N/(A · m)

The units of B are those of force divided by the product of charge and speed; that is, N/[C · (m/s)]. Rearranging slightly, we can write these units as N/[(C/s) · m]. Finally, noting that 1 A = 1 C/s, we find that 1 N/[C · (m/s)] = 1 N/(A · m). The latter combination of units is named the **tesla,** in recognition of the pioneering electrical and magnetic studies of the Croatian-born American engineer Nikola Tesla (1856–1943). In particular,

$$1 \text{ tesla} = 1\text{ T} = 1\text{ N/(A}\cdot\text{m)}$$

We now give an example of the application of Equation 22–1.

EXAMPLE 22–1 A Tale of Two Charges

Particle 1, with a charge $q_1 = 3.60\ \mu\text{C}$ and a speed $v_1 = 862$ m/s, travels at right angles to a uniform magnetic field. The magnetic force it experiences is 4.25×10^{-3} N. Particle 2, with a charge $q_2 = 53.0\ \mu\text{C}$ and a speed $v_2 = 1.30 \times 10^3$ m/s, moves at an angle of 55.0° relative to the same magnetic field. Find **(a)** the strength of the magnetic field and **(b)** the magnitude of the magnetic force exerted on particle 2.

Picture the Problem

The two charged particles are shown in our sketch, along with the magnetic field lines. Notice that $q_1 = 3.60\ \mu\text{C}$ moves at right angles to $\vec{\mathbf{B}}$; the charge $q_2 = 53.0\ \mu\text{C}$ moves at an angle of 55.0° with respect to $\vec{\mathbf{B}}$. The magnetic force also depends on the speeds of the particles, $v_1 = 862$ m/s and $v_2 = 1.30 \times 10^3$ m/s.

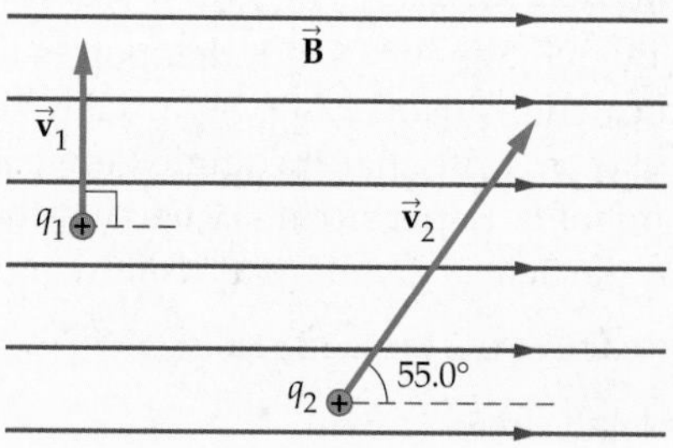

Strategy

a. We can find the strength of the magnetic field using the information given for particle 1. In particular, the particle moves at right angles to the magnetic field ($\theta_1 = 90°$), and, therefore, the magnetic force it experiences is $F_1 = |q_1|v_1 B\sin\theta_1 = q_1 v_1 B$. Given that we know F_1, q_1, and v_1, we can solve for B.

b. Once we have determined B in part (a), it is straightforward to calculate the magnetic force on particle 2 using $F_2 = |q_2|v_2 B\sin\theta_2$.

Solution

Part (a)

1. Set $\theta_1 = 90°$ in $F_1 = |q_1|v_1 B\sin\theta_1$, then solve for B. Use q_1 and v_1 for the magnitude of the charge and speed, respectively:

$$F_1 = |q_1|v_1 B\sin 90° = q_1 v_1 B$$

$$B = \frac{F_1}{q_1 v_1}$$

2. Substitute numerical values:

$$B = \frac{F_1}{q_1 v_1} = \frac{4.25 \times 10^{-3}\text{ N}}{(3.60 \times 10^{-6}\text{ C})(862\text{ m/s})} = 1.37\text{ T}$$

Part (b)

3. Use $B = 1.37\ \text{T}$ in $F_2 = \lvert q_2 \rvert v_2 B \sin\theta_2$ to find the magnetic force exerted on particle 2:	$F_2 = \lvert q_2 \rvert v_2 B \sin\theta_2$ $= (53.0 \times 10^{-6}\ \text{C})(1.30 \times 10^3\ \text{m/s})(1.37\ \text{T}) \sin 55.0°$ $= 0.0773\ \text{N}$

Insight

Notice that the charge and speed of a particle are not enough to determine the magnetic force acting on it—its direction of motion relative to the magnetic field is needed as well.

Practice Problem

At what angle relative to the magnetic field must particle 2 move if the magnetic force it experiences is to be 0.0500 N? [**Answer:** $\theta = 32.0°$]

Some related homework problems: Problem 2, Problem 5

The tesla is a fairly large unit of magnetic strength, especially when compared with the magnetic field at the surface of the Earth, which is roughly 5.0×10^{-5} T. Thus, another commonly used unit of magnetism is the gauss (G), defined as follows:

$$1\ \text{gauss} = 1\ \text{G} = 10^{-4}\ \text{T}$$

In terms of the gauss, the Earth's magnetic field on the surface of the Earth is approximately 0.5 G. It should be noted, however, that the gauss is not an SI unit of magnetic field. Even so, it finds wide usage because of its convenient magnitude. The magnitudes of some typical magnetic fields are given in Table 22–1.

TABLE 22–1 Typical Magnetic Fields

Physical system	Magnetic field (G)
Magnetar (a magnetic neutron star formed in a supernova explosion)	10^{15}
Strongest manmade magnetic field	6×10^5
High-field MRI	15,000
Low-field MRI	2000
Sunspots	1000
Bar magnet	100
Earth	0.50

Magnetic Force Right-Hand Rule (RHR)

We now consider the *direction* of the magnetic force, which is rather interesting and unexpected. Instead of pointing in the direction of the magnetic field, $\vec{\mathbf{B}}$, or the velocity, $\vec{\mathbf{v}}$, as one might expect, the following behavior is observed:

> The magnetic force $\vec{\mathbf{F}}$ points in a direction that is perpendicular to both $\vec{\mathbf{B}}$ and $\vec{\mathbf{v}}$.

As an example, consider the vectors $\vec{\mathbf{B}}$ and $\vec{\mathbf{v}}$ in **Figure 22–8 (a)**, which lie in the indicated plane. The force on a positive charge, $\vec{\mathbf{F}}$, as shown in the figure, is perpendicular to this plane and hence to both $\vec{\mathbf{B}}$ and $\vec{\mathbf{v}}$.

Notice that $\vec{\mathbf{F}}$ could equally well point downward in Figure 22–8 (a) and still be perpendicular to the plane. The way we determine the precise direction of $\vec{\mathbf{F}}$ is with a right-hand rule—similar to the right-hand rules used in calculating torques

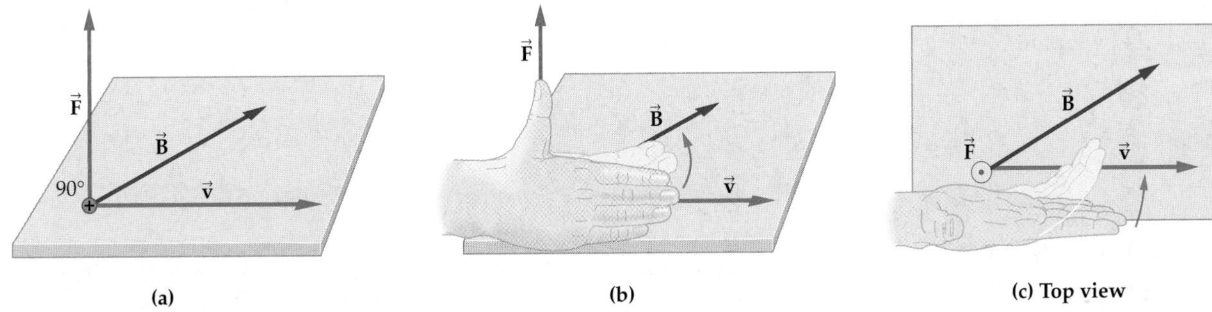

▲ **FIGURE 22–8 The magnetic force right-hand rule**
(a) The magnetic force, $\vec{\mathbf{F}}$, is perpendicular to both the velocity, $\vec{\mathbf{v}}$, and the magnetic field, $\vec{\mathbf{B}}$. (The force vectors shown in this figure are for the case of a positive charge. The force on a negative charge would be in the opposite direction.) **(b)** As the fingers of the right hand are curled from $\vec{\mathbf{v}}$ to $\vec{\mathbf{B}}$, the thumb points in the direction of $\vec{\mathbf{F}}$. **(c)** An overhead view, looking down on the plane defined by the vectors $\vec{\mathbf{v}}$ and $\vec{\mathbf{B}}$. In this two-dimensional representation, the force vector comes out of the page and is indicated by a circle with a dot inside. If the charge was negative, the force would point into the page, and the symbol indicating $\vec{\mathbf{F}}$ would be a circle with an X inside.

and angular momentum in Chapter 11. To be specific, the direction of $\vec{F}$ is found using the *magnetic force* right-hand rule:

Magnetic Force Right-Hand Rule

To find the direction of the magnetic force on a positive charge, start by pointing the fingers of your right hand in the direction of the velocity, $\vec{v}$. Now curl your fingers toward the direction of $\vec{B}$, as illustrated in **Figure 22–8 (b)**. Your thumb points in the direction of $\vec{F}$. If the charge is negative, the force points opposite to the direction of your thumb.

Applying this rule to Figure 22–8 (a), we see that $\vec{F}$ does indeed point upward, as indicated.

A mathematical way to write the magnetic force that includes both its magnitude and direction is in terms of the vector **cross product.** Details on the cross product are given in Appendix A, but basically we can write the magnetic force $\vec{F}$ as follows:

$$\vec{F} = q\vec{v} \times \vec{B}$$

In this expression, the term $\vec{v} \times \vec{B}$ is referred to as the cross product of $\vec{v}$ and $\vec{B}$. From the definition of the cross product, the magnitude of the force is $F = |q|vB \sin\theta$ where θ is the magnitude of the angle between $\vec{v}$ and $\vec{B}$, precisely as in Equation 22–1. In addition, the direction of $\vec{F}$ is given by the magnetic force RHR. Thus, Equation 22–1 plus the magnetic force RHR are all we need to calculate magnetic forces—the cross product simply provides a more compact way of saying the same thing.

Since three-dimensional plots like the one in Figure 22–8 (b) are somewhat difficult to sketch, we often use a shorthand for indicating vectors that point into or out of the page. (See Appendix A for a discussion and examples.) Imagining a vector as an arrow, with a pointed tip at one end and crossed feathers at the other end, we say that if a vector points out of the page—toward us—we see its tip. This situation is indicated by drawing a circle with a dot inside it on the page, as in **Figure 22–8 (c)**. If, on the other hand, the vector points into the page, we see the crossed feathers of its base; hence, we draw a circle with an X inside it to symbolize this situation.

As an example, **Figure 22–9** indicates a uniform magnetic field $\vec{B}$ that points into the page. A particle with a positive charge q moves to the right. Using the magnetic force RHR—extending our fingers to the right and then curling them into the page—we see that the force exerted on this particle is upward, as indicated. If the charge is negative, the direction of $\vec{F}$ is reversed, as is also illustrated in Figure 22–9.

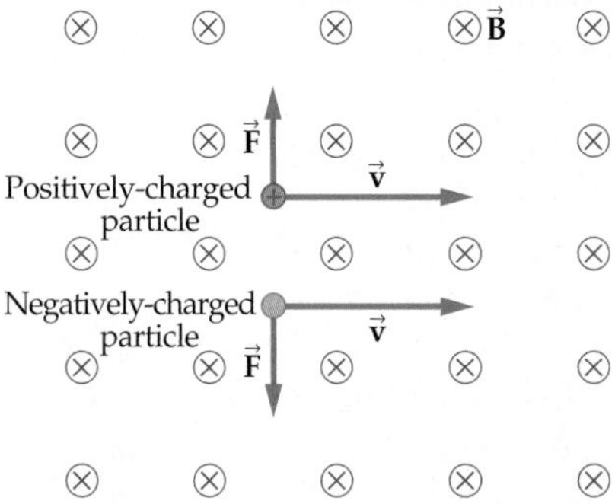

▲ **FIGURE 22–9 The magnetic force for positive and negative charges**

The direction of the magnetic force depends on the sign of the charge. Specifically, the force exerted on a negatively charged particle is opposite in direction to the force exerted on a positively charged particle.

CONCEPTUAL CHECKPOINT 22–2 Charge of a Particle

Three particles travel through a region of space where the magnetic field is out of the page, as shown below in the sketch to the left. For each of the three particles, state whether the particle's charge is positive, negative, or zero.

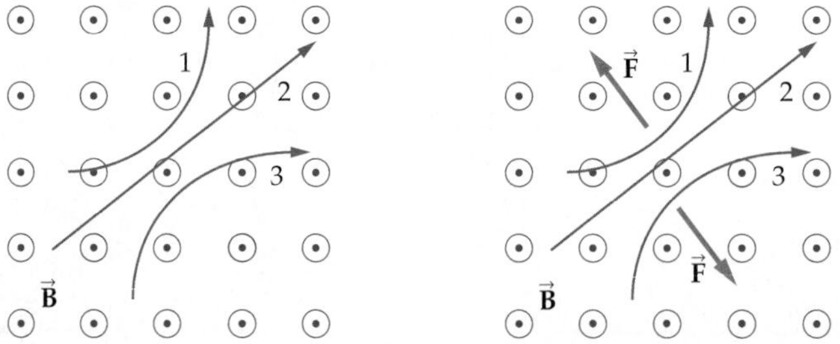

Reasoning and Discussion

In the second sketch, we indicate the general direction of the force required to cause the observed motion. The force indicated for particle 3 is in the direction given by the magnetic force RHR; hence, particle 3 must have a positive charge. The force acting on particle 1 is in the opposite direction; hence, that particle must be negatively charged. Finally, particle 2 is undeflected; hence, its charge, and the force acting on it, must be zero.

Answer:

Particle 1, negative; particle 2, zero; particle 3, positive.

PROBLEM-SOLVING NOTE

The Magnetic Force for Positive and Negative Charges

Note that the magnetic force right-hand rule is stated for the case of a positively charged particle. If the particle has a negative charge, the direction of the force is reversed.

We conclude this section with an Example illustrating the magnetic force RHR.

EXAMPLE 22–2 Electric and Magnetic Fields

A particle with a charge of 7.70 μC and a speed of 435 m/s is acted on by both an electric and a magnetic field. The particle moves along the x axis in the positive direction, the magnetic field has a strength of 3.20 T and points in the positive y direction, and the electric field points in the positive z direction with a magnitude of 8.10×10^3 N/C. Find the magnitude and direction of the net force acting on the particle.

Picture the Problem

The physical situation is shown in our sketch. In particular, we show a three-dimensional coordinate system with each of the three relevant vectors, $\vec{\mathbf{v}}$, $\vec{\mathbf{B}}$, and $\vec{\mathbf{E}}$, indicated. Note that the charge is positive, and therefore $|q| = q$.

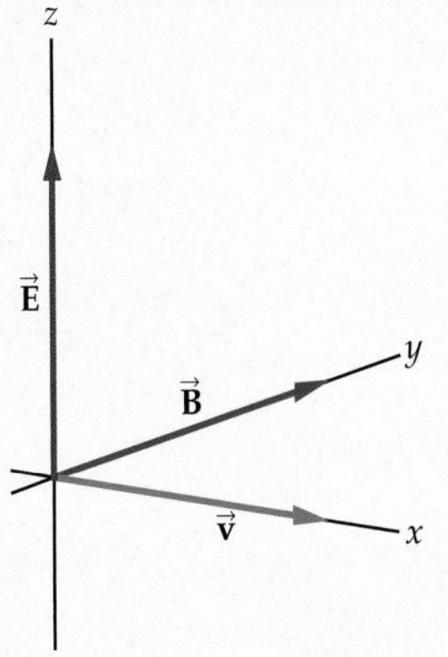

Strategy

First, the force due to the electric field is in the direction of $\vec{\mathbf{E}}$ (positive z direction), and it has a magnitude qE. Second, the direction of the magnetic force is given by the RHR. In this case curling the fingers of a right hand from $\vec{\mathbf{v}}$ to $\vec{\mathbf{B}}$ results in the thumb (the direction of $\vec{\mathbf{F}}$) pointing in the positive z direction. Thus, in this system the forces due to the electric and magnetic fields are in the same direction. Finally, because the angle between $\vec{\mathbf{v}}$ and $\vec{\mathbf{B}}$ is 90°, the magnitude of the magnetic force is qvB.

Solution

1. Calculate the magnitude of the electric force exerted on the particle:

$$F_E = qE = (7.70 \times 10^{-6}\ \text{C})(8.10 \times 10^3\ \text{N/C}) = 6.24 \times 10^{-2}\ \text{N}$$

2. Calculate the magnitude of the magnetic force exerted on the particle:

$$F_B = qvB = (7.70 \times 10^{-6}\ \text{C})(435\ \text{m/s})(3.20\ \text{T}) = 1.07 \times 10^{-2}\ \text{N}$$

3. Since both forces are in the positive z direction, simply add them to obtain the net force:

$$F_{\text{net}} = F_E + F_B = (6.24 \times 10^{-2}\ \text{N}) + (1.07 \times 10^{-2}\ \text{N}) = 7.31 \times 10^{-2}\ \text{N}$$

Insight

Note that the net force on the particle will be in the positive z direction as long as the velocity vector, $\vec{\mathbf{v}}$, points in *any* direction in the x-y plane. The magnitude of the net force, however, will depend on the direction of $\vec{\mathbf{v}}$.

In addition, the coordinate system shown in the Picture the Problem is a *right-handed* coordinate system. This means, for example, that the axes are set up so that $\hat{\mathbf{x}} \times \hat{\mathbf{y}} = \hat{\mathbf{z}}$, as you can verify with the right-hand rule. Further details are given in Appendix A. We will always use right-handed coordinate systems in this text.

Practice Problem

Find the net force on the particle if its velocity is reversed, and it moves in the negative x direction.
[**Answer:** $F_{\text{net}} = F_E + F_B = (6.24 \times 10^{-2}\ \text{N}) - (1.07 \times 10^{-2}\ \text{N}) = 5.17 \times 10^{-2}\ \text{N}$ in the positive z direction.]

Some related homework problems: Problem 8, Problem 10

22–3 The Motion of Charged Particles in a Magnetic Field

As we have seen, the magnetic force has characteristics that set it apart from the force exerted by electric or gravitational fields. In particular, the magnetic force depends not only on the speed of a particle but on its direction of motion as well. We now explore some of the consequences that follow from these characteristics and relate them to the type of motion that occurs in magnetic fields.

Electric versus Magnetic Forces

We begin by investigating the motion of a charged particle as it is projected into a region with either an electric or a magnetic field. For example, in **Figure 22–10 (a)** we consider a uniform electric field pointing downward. A positively charged particle moving horizontally into this region experiences a constant downward

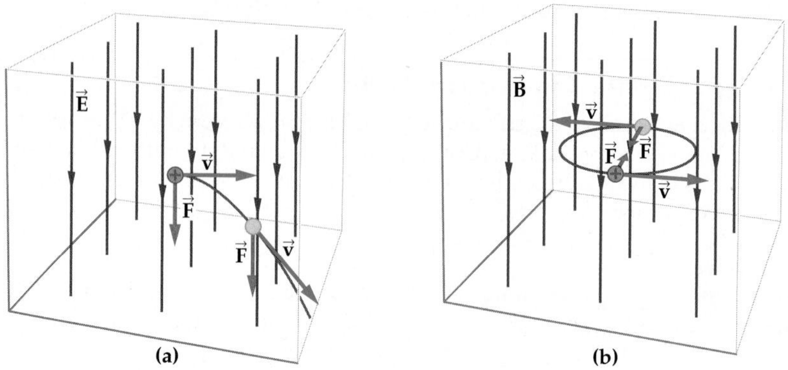

▶ **FIGURE 22–10 Differences between motion in electric and magnetic fields**
(a) A positively charged particle moving into a region with an electric field experiences a downward force that causes it to accelerate. **(b)** A positively charged particle entering a magnetic field experiences a horizontal force at right angles to its direction of motion. In this case, the speed of the particle remains constant.

force—much like a mass in a uniform gravitational field. As a result, the particle begins to accelerate downward and follow a parabolic path.

If the same particle, moving to the right, encounters a magnetic field instead, as in **Figure 22–10 (b)**, the resulting motion is quite different. We again assume that the field is uniform and pointing downward. In this case, however, the magnetic force RHR shows that $\vec{\mathbf{F}}$ points into the page. Our particle now begins to follow a horizontal path into the page. As we shall see, this path is circular.

Perhaps even more significant than these differences in motion is the fact that an electric field can do work on a charged particle, whereas a constant magnetic field cannot. In Figure 22–10 (a), for example, as soon as the particle begins to move downward, a component of its velocity is in the direction of the electric force. This means that the electric force does work on the particle, and its speed increases—again, just like a mass falling in a gravitational field. If the particle moves through a magnetic field, however, no work is done on it because the magnetic force is *always* at right angles to the direction of motion. Thus, the speed of the particle in the magnetic field remains constant.

CONCEPTUAL CHECKPOINT 22–3 Direction of the Magnetic Field

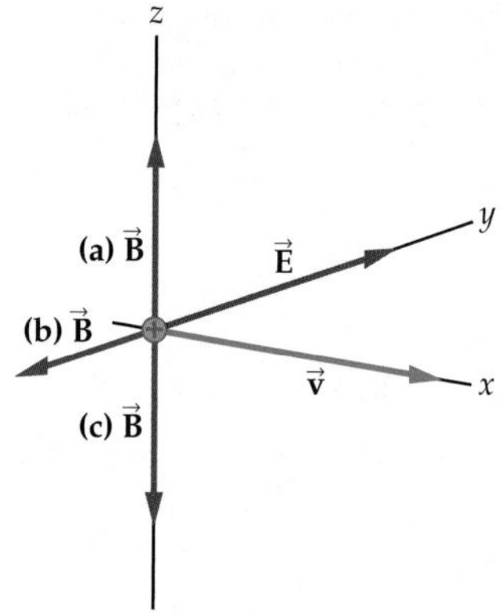

In a device called a **velocity selector**, charged particles move through a region of space with both an electric and a magnetic field. If the speed of the particle has a particular value, the net force acting on it is zero. Assume that a positively charged particle moves in the positive x direction, as shown in the sketch, and the electric field is in the positive y direction. Should the magnetic field be in **(a)** the positive z direction, **(b)** the negative y direction, or **(c)** the negative z direction in order to give zero net force?

Reasoning and Discussion

The force exerted by the electric field is in the positive y direction; hence, the magnetic force must be in the negative y direction if it is to cancel the electric force. If we simply try the three possible directions for $\vec{\mathbf{B}}$ one at a time, applying the magnetic force RHR in each case, we find that only a magnetic field along the positive z axis gives rise to a force in the negative y direction, as desired.

Answer:

(a) $\vec{\mathbf{B}}$ should point in the positive z direction.

To follow up on Conceptual Checkpoint 22–3, let's determine the speed for which the net force is zero. First, note that the electric force has a magnitude qE. Second, the magnitude of the magnetic force is qvB. With $\vec{\mathbf{E}}$ in the positive y direction and $\vec{\mathbf{B}}$ in the positive z direction, these forces are in opposite directions. Setting the magnitudes of the forces equal yields $qE = qvB$. Canceling the charge q and solving for v we find

$$v = \frac{E}{B} \qquad \text{(velocity selector)}$$

REAL-WORLD PHYSICS: BIO
The electromagnetic flowmeter

A particle with this speed, regardless of its charge, passes through the velocity selector with zero net force and hence no deflection.

The physical principle illustrated in the velocity selector can be used to measure the speed of blood with a device known as an *electromagnetic flowmeter.*

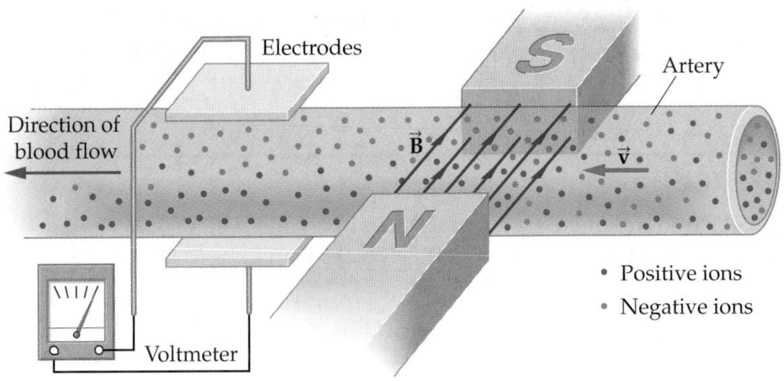

◀ **FIGURE 22–11 The electromagnetic flowmeter**

As blood flows through a magnetic field, the charged ions it contains are deflected. The deflection of charged particles results in an electric field opposing the deflection. If the speed of the blood is v, the electric field generated by ions moving through the magnetic field satisfies the relation $v = E/B$.

Suppose an artery passes between the poles of a magnet, as shown in **Figure 22–11**. Charged ions in the blood will be deflected at right angles to the artery by the magnetic field. The resulting charge separation produces an electric field that opposes the magnetic deflection. When the electric field is strong enough, the deflection ceases and the blood flows normally through the artery. If the electric field is measured, and the magnetic field is known, the speed of the blood flow is simply $v = E/B$, as in a standard velocity selector.

Constant-Velocity, Straight-Line Motion

Recall that the magnetic force is zero if a particle's velocity $\vec{v}$ is parallel (or antiparallel) to the magnetic field $\vec{B}$. In such a case the particle's acceleration is zero; therefore, its velocity remains constant. Thus, the simplest type of motion in a magnetic field is a constant-velocity, straight-line drift along the magnetic field lines.

Circular Motion

The next simplest case is a particle with a velocity that is perpendicular to the magnetic field. Consider, for example, the situation shown in **Figure 22–12**. Here a particle of mass m, charge $+q$, and speed v moves in a region with a constant magnetic field $\vec{B}$ pointing out of the page. Since $\vec{v}$ is at right angles to $\vec{B}$, the magnitude of the magnetic force is $F = |q|vB \sin 90° = |q|vB$.

Now we consider the particle's motion. At point 1 the particle is moving to the right; hence, the magnetic force is downward, causing the particle to accelerate in the downward direction. When the particle reaches point 2, it is moving downward, and now the magnetic force is to the left. This causes the particle to accelerate to the left. At point 3, the force exerted on it is upward, and so on, as the particle continues moving.

Thus, at every point on the particle's path the magnetic force is at right angles to the velocity, pointing toward a common center—but this is just the condition required for circular motion. As we saw in Section 6–5, in circular motion the acceleration is toward the center of the circle; for this reason, a centripetal force is required to cause the motion. In this case the centripetal force is supplied by the magnetic force—in the same way that a string exerts a centripetal force on a ball being whirled about in a circle.

Recall that the centripetal acceleration of a particle moving with a speed v in a circle of radius r is

$$a_{cp} = \frac{v^2}{r}$$

Therefore, setting ma_{cp} equal to the magnitude of the magnetic force, $|q|vB$, yields the following condition:

$$m\frac{v^2}{r} = |q|vB$$

Canceling one power of the speed, v, we find that the radius of the circular orbit is

$$r = \frac{mv}{|q|B} \qquad 22\text{–}3$$

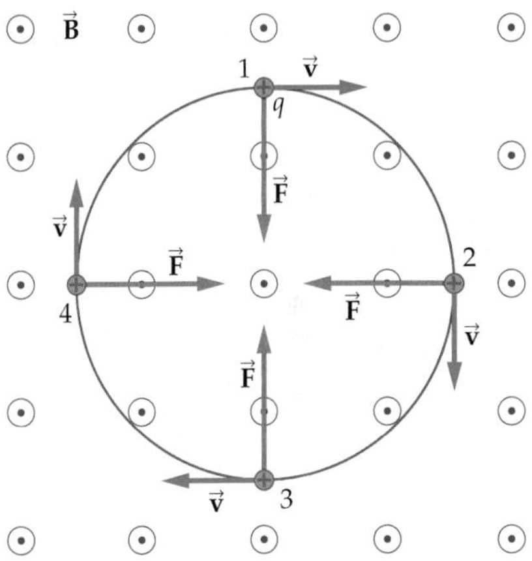

▲ **FIGURE 22–12 Circular motion in a magnetic field**

A charged particle moves in a direction perpendicular to the magnetic field. At each point on the particle's path the magnetic force is at right angles to the velocity and hence toward the center of a circle.

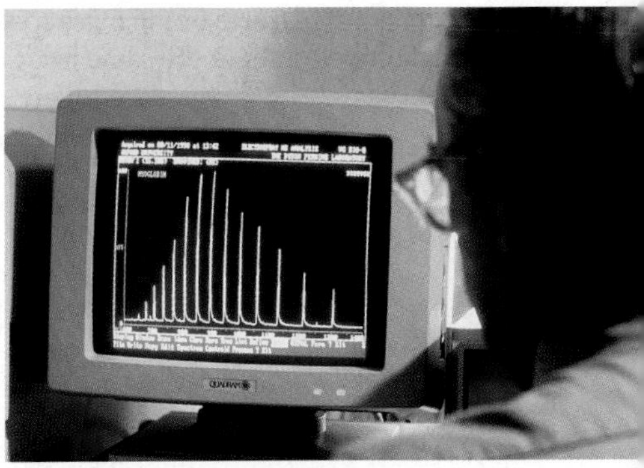

▲ In addition to separating and "weighing" isotopes, mass spectrometers can be used to study the chemical composition and structure of large, biologically important molecules. Here, a researcher studies the spectrum of myoglobin, an oxygen-carrying protein found in muscle tissue. Each peak represents a fragment of the molecule with its own characteristic combination of mass and charge.

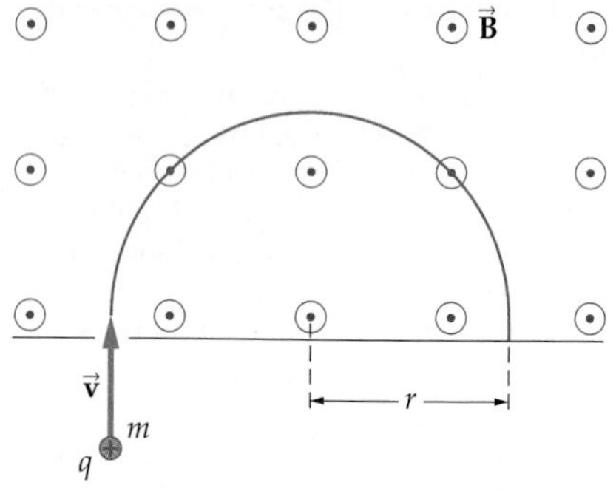

▲ **FIGURE 22–13 The operating principle of a mass spectrometer**

In a mass spectrometer, a beam of charged particles enters a region with a magnetic field perpendicular to the velocity. The particles then follow a circular orbit of radius $r = mv/|q|B$. Particles of different mass will follow different paths.

We can see, therefore, that the faster and more massive the particle, the larger the circle. Conversely, the stronger the magnetic field and the greater the charge, the smaller the circle.

EXERCISE 22–1

An electron moving perpendicular to a magnetic field of 4.60×10^{-3} T follows a circular path of radius 2.80 mm. What is the electron's speed?

Solution

Solving Equation 22–3 for v, we find

$$v = \frac{r|q|B}{m} = \frac{(2.80 \times 10^{-3}\ \text{m})(1.60 \times 10^{-19}\ \text{C})(4.60 \times 10^{-3}\ \text{T})}{9.11 \times 10^{-31}\ \text{kg}} = 2.26 \times 10^{6}\ \text{m/s}$$

Thus, the speed of this electron is about 1 percent of the speed of light.

REAL-WORLD PHYSICS
The mass spectrometer

One of the applications of circular motion in a magnetic field is in a device known as a **mass spectrometer.** A mass spectrometer can be used to separate isotopes (atoms of the same element with different masses) and to measure atomic masses. It finds many uses in medicine (anesthesiologists use it to measure respiratory gases), biology (to determine reaction mechanisms in photosynthesis), geology (to date fossils), space science (to determine the atmospheric composition of Mars), and a variety of other fields.

The basic principles of a mass spectrometer are illustrated in **Figure 22–13**. Here we see a beam of ions of mass m and charge $+q$ entering a region of constant magnetic field with a speed v. The field causes the ions to move along a circular path, with a radius that depends on the mass and charge of the ion, as described by Equation 22–3. Thus, different isotopes follow different paths and hence can be separated and identified. A specific example is considered next.

EXAMPLE 22–3 Uranium Separation

Two isotopes of uranium, ^{235}U ($m = 3.90 \times 10^{-25}$ kg) and ^{238}U ($m = 3.95 \times 10^{-25}$ kg), are sent into a mass spectrometer with a speed of 1.05×10^{5} m/s, as indicated in the accompanying sketch. Given that each isotope is singly ionized, and that the strength of the magnetic field is 0.750 T, what is the distance d between the two isotopes after they complete half a circular orbit?

Picture the Problem

The relevant features of the mass spectrometer are indicated in our sketch. Note that both isotopes enter at the same location with the same speed. Because they have different masses, however, they follow circular paths of different radii. This difference results in the separation d after half an orbit.

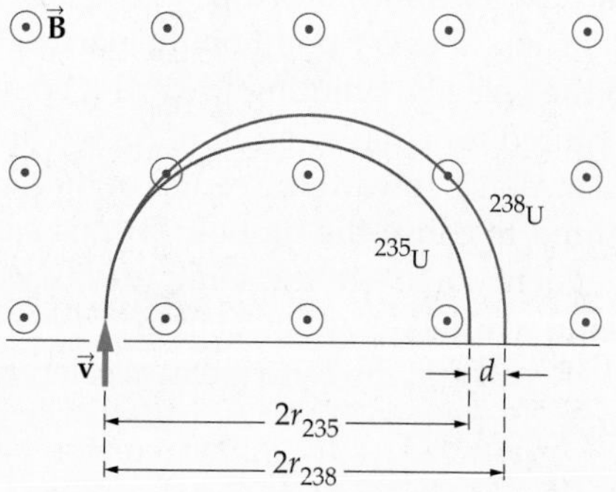

INTERACTIVE FIGURE

Strategy

We begin by calculating the radius of each isotope's circular path using $r = mv/|q|B$. The masses, speeds, and magnetic field are given. We can infer the charge from the fact that the isotopes are "singly ionized," which means that a single electron has been removed from each atom. Since the atoms were electrically neutral before the electron was removed, it follows that the charge of the isotopes is $q = e = 1.60 \times 10^{-19}$ C.

Finally, once we know the radii, the separation between the isotopes is given by $d = 2r_{238} - 2r_{235}$, as indicated in the sketch.

Solution

1. Determine the radius of the circular path of ^{235}U:

$$r_{235} = \frac{mv}{|q|B}$$

$$= \frac{(3.90 \times 10^{-25}\ \text{kg})(1.05 \times 10^{5}\ \text{m/s})}{(1.60 \times 10^{-19}\ \text{C})(0.750\ \text{T})} = 34.1\ \text{cm}$$

2. Determine the radius of the circular path of ^{238}U:

$$r_{238} = \frac{mv}{|q|B} = \frac{(3.95 \times 10^{-25}\ \text{kg})(1.05 \times 10^{5}\ \text{m/s})}{(1.60 \times 10^{-19}\ \text{C})(0.750\ \text{T})} = 34.6\ \text{cm}$$

3. Calculate the separation between the isotopes:

$$d = 2r_{238} - 2r_{235} = 2(34.6\ \text{cm} - 34.1\ \text{cm}) = 1\ \text{cm}$$

Insight

Notice that although the difference in masses is very small, the mass spectrometer converts this small difference into an easily measurable distance of separation.

Practice Problem

Does the separation d increase or decrease if the magnetic field is increased? Check your answer by calculating the separation for $B = 1.00$ T, everything else remaining the same. [**Answer:** A stronger field will decrease the radii and hence decrease the separation d. In particular, if B changes by a factor x, the separation changes by the factor $1/x$. In this case, the factor is $x = 4/3$. That is, $B = (4/3)(0.750\ \text{T}) = 1.00$ T; therefore, $d = (3/4)(1\ \text{cm}) = 3/4$ cm. Note that it is not necessary to repeat the entire calculation in detail.]

Some related homework problems: Problem 12, Problem 20

ACTIVE EXAMPLE 22–1 Find the Time for One Orbit

Calculate the time T required for a particle of mass m and charge q to complete a circular orbit in a magnetic field.

Solution *(Test your understanding by performing the calculations indicated in each step.)*

1. Write the speed, v, of a particle in terms of the time T to complete an orbit of radius r: $v = 2\pi r/T$

2. Substitute this expression for v into the condition for a circular orbit, $r = \dfrac{mv}{|q|B}$: $r = m(2\pi r/T)/|q|B$

3. Cancel r and solve for T: $T = \dfrac{2\pi m}{|q|B}$

Insight

Notice that the time required for an orbit is independent of its radius.

Your Turn

The circumference of an orbit depends on its radius, r. Why is it, then, that a particle completes an orbit in a time, T, that is independent of r?

(Answers to **Your Turn** *problems are given in the back of the book.)*

Helical Motion

The final type of motion we consider is a combination of the two motions discussed already. Suppose, for example, that a particle has an initial velocity at an angle to the magnetic field, as in **Figure 22–14 (a)**. In this case there is a component of velocity parallel to $\vec{\mathbf{B}}$ and a component perpendicular to $\vec{\mathbf{B}}$. The parallel component of the velocity remains constant with time (zero force in this direction), whereas the perpendicular component results in a circular motion, as just discussed. Combining the two motions, we can see that the particle follows a helical path, as shown in **Figure 22–14 (b)**.

If a magnetic field is curved, as in the case of a bar magnet or the magnetic field of the Earth, the helical motion of charged particles will be curved as well. Specifically, the axis of the helical motion will follow the direction of $\vec{\mathbf{B}}$. For example, electrons and protons emitted by the "solar wind" frequently encounter the Earth's magnetic field and begin to move in helical paths following the field lines. Near the poles, where the field lines are concentrated as they approach the Earth's surface, the circulating electrons begin to collide with atoms and molecules in the

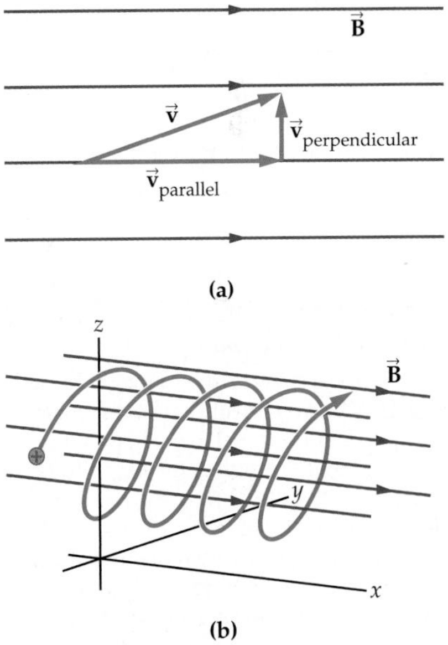

▲ **FIGURE 22–14 Helical motion in a magnetic field**
(a) A velocity at an angle to a magnetic field $\vec{\mathbf{B}}$ can be resolved into parallel and perpendicular components. The parallel component gives a constant-velocity drift in the direction of the field. The perpendicular component gives circular motion perpendicular to the field. **(b)** Helical motion is a combination of linear motion and circular motion.

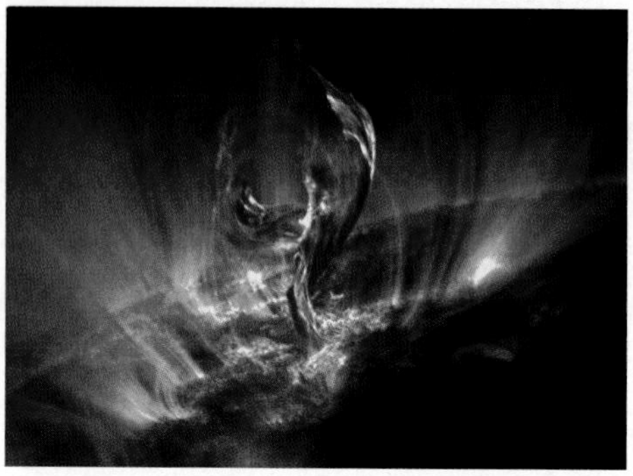
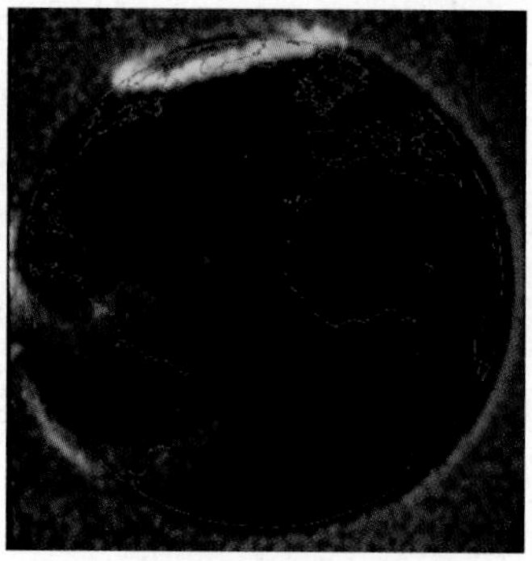

▲ (Left) An enormous eruption of matter from the surface of the Sun, large enough to encompass the entire Earth many times over. The loop structure is created by the Sun's own complex magnetic field. Charged particles from such eruptions often reach the Earth and become trapped in its magnetic field, producing auroras (see text). (Center) A glowing aurora borealis surrounds the Earth's north geographic pole in this photograph from space. (Right) An auroral display seen from Earth. The characteristic red and green colors are produced by ionized nitrogen and oxygen atoms, respectively.

REAL-WORLD PHYSICS

The aurora borealis and aurora australis

atmosphere. These collisions can excite and ionize the atmospheric atoms and molecules, resulting in the emission of light known as the **aurora borealis** (northern lights) in the northern hemisphere and the **aurora australis** (southern lights) in the southern hemisphere. The most common color of the auroras is a pale green, the color of light given off by excited oxygen atoms.

22–4 The Magnetic Force Exerted on a Current-Carrying Wire

A charged particle experiences a force when it moves across magnetic field lines. This is true whether it travels in a vacuum or in a current-carrying wire. Thus, a wire with a current will experience a force that is simply the resultant of all the forces experienced by the individual moving charges responsible for the current.

Specifically, consider a straight wire segment of length L with a current I flowing from left to right, as in **Figure 22–15 (a)**. Also present in this region of space is a magnetic field $\vec{\mathbf{B}}$ at an angle θ to the length of the wire. If the conducting charges move through the wire with an average drift speed v, the time required for them to move from one end of the wire segment to the other is $\Delta t = L/v$. The amount of charge that flows through the wire in this time is $q = I\,\Delta t = IL/v$. Therefore, the force exerted on the wire is

$$F = qvB\sin\theta = \left(\frac{IL}{v}\right)vB\sin\theta$$

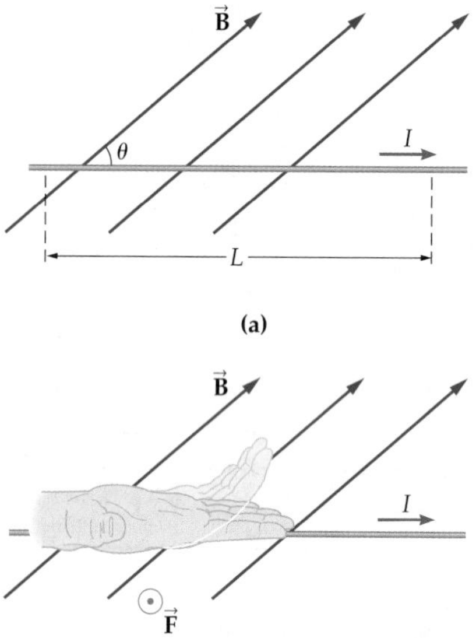

▲ **FIGURE 22–15 The magnetic force on a current-carrying wire**

A current-carrying wire in a magnetic field experiences a force, unless the current is parallel or antiparallel to the field. **(a)** For a wire segment of length L the magnitude of the force is $F = ILB\sin\theta$. **(b)** The direction of the force is given by the magnetic force RHR; the only difference is that you start by pointing the fingers of your right hand in the direction of the current I. In this case the force points out of the page.

Canceling v, we find that the force on a wire segment of length L with a current I at an angle θ to the magnetic field $\vec{\mathbf{B}}$ is

Magnetic Force on a Current-Carrying Wire

$$F = ILB\sin\theta \qquad 22\text{–}4$$

SI unit: newton, N

As with single charges, the maximum force occurs when the current is perpendicular to the magnetic field ($\theta = 90°$) and is zero if the current is in the same direction as $\vec{\mathbf{B}}(\theta = 0)$.

The direction of the magnetic force on a wire is given by the same right-hand rule used earlier for single charges. Thus, to find the direction of the force in **Figure 22–15 (b)**, start by pointing the fingers of your right hand in the direction of the current I. This assumes that positive charges are flowing in the direction of I, consistent with our convention from Chapter 21. Now, curl your fingers toward the direction of $\vec{\mathbf{B}}$. Your thumb, which points out of the page, indicates the direction of $\vec{\mathbf{F}}$.

PROBLEM-SOLVING NOTE

The Magnetic Force on a Current-Carrying Wire

Note that a current-carrying wire experiences no force when it is in the same direction as the magnetic field. The maximum force occurs when the wire is perpendicular to the magnetic field.

Of course, the current is actually caused by negatively charged electrons flowing in the opposite direction to the current. The magnetic force on these negatively charged particles moving in the opposite direction is the same as the force on positively charged particles moving in the direction of I. Thus, in all cases pertaining to current-carrying wires, we can simply think of the current as the direction in which positively charged particles move.

CONCEPTUAL CHECKPOINT 22–4 Magnetic Poles

When the switch is closed in the circuit shown in the sketch on the left, the wire between the poles of the horseshoe magnet deflects downward. Is the left end of the magnet **(a)** a north magnetic pole or **(b)** a south magnetic pole?

Reasoning and Discussion

Once the switch is closed, the current in the wire is into the page, as shown in the sketch below.

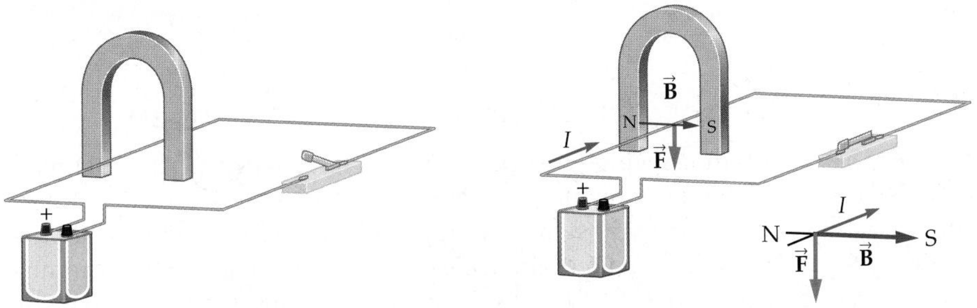

Applying the magnetic force RHR, we see that the magnetic field must point from left to right in order for the force to be downward. Since magnetic field lines leave from north poles and enter at south poles, it follows that the left end of the magnet must be a north magnetic pole.

Answer:

(a) The left end of the magnet is a north magnetic pole.

The magnetic force exerted on a current-carrying wire can be quite substantial. In the following Example, we consider the current necessary to levitate a copper rod.

EXAMPLE 22–4 Magnetic Levity

A copper rod 0.150 m long and with a mass of 0.0500 kg is suspended from two thin, flexible wires, as shown in the sketch. At right angles to the rod is a uniform magnetic field of 0.550 T pointing into the page. Find **(a)** the direction and **(b)** magnitude of the electric current needed to levitate the copper rod.

Picture the Problem

Our sketch shows the physical situation with relevant quantities labeled. The direction of the current has been indicated as well. Note that if the current is in this direction, the RHR gives an upward magnetic force, as required for the rod to be levitated.

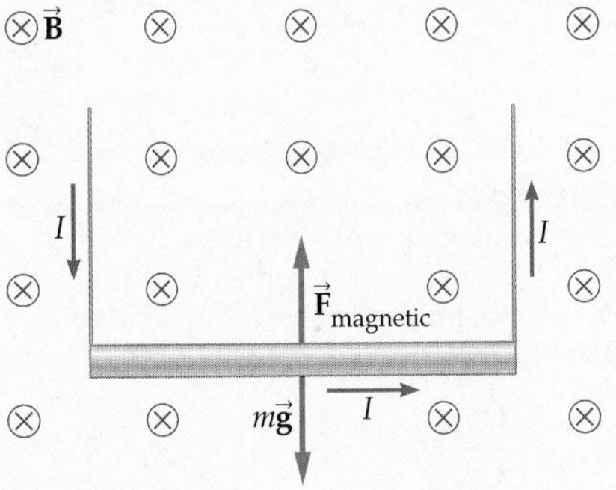

Strategy

To find the magnitude of the current, we set the magnetic force equal in magnitude to the force of gravity. For the magnetic force we have $F = ILB \sin \theta$. In this case the field is at right angles to the current; hence, $\theta = 90°$, and the force simplifies to $F = ILB$. Thus, setting ILB equal to mg determines the current.

Solution

Part (a)

1. Determine the direction of I:

The current, I, points to the right. To verify, point the fingers of your right hand to the right, curl into the page, and your thumb will point upward, as desired.

Part (b)

2. Set the magnitude of the magnetic force equal to the magnitude of the force of gravity:

$$ILB = mg$$

3. Solve for the current, I:

$$I = \frac{mg}{LB} = \frac{(0.0500\ \text{kg})(9.81\ \text{m/s}^2)}{(0.150\ \text{m})(0.550\ \text{T})} = 5.95\ \text{A}$$

continued on next page

continued from previous page

Insight

Magnetic forces, like electric forces, can easily exceed the force of gravity. In fact, when we consider atomic systems in Chapter 31 we shall see that gravity plays no role in the behavior of an atom—only electric and magnetic forces are important on the atomic level.

Practice Problem

Suppose the rod is doubled in length, which also doubles its mass. Does the current needed to levitate the rod increase, decrease, or stay the same? [**Answer:** Since the levitation current is given by $I = mg/LB$, it is clear that doubling both m and L has no effect on I; the current needed to levitate remains the same.]

Some related homework problems: Problem 23, Problem 27

22–5 Loops of Current and Magnetic Torque

The fact that a current-carrying wire experiences a force when placed in a magnetic field is one of the fundamental discoveries that makes modern applications of electric power possible. In most of these applications, including electric motors and generators, the wire is shaped into a current-carrying loop. We will examine some of these applications further in Chapter 23; in this section we lay the groundwork by considering what happens when a simple current loop is placed in a magnetic field.

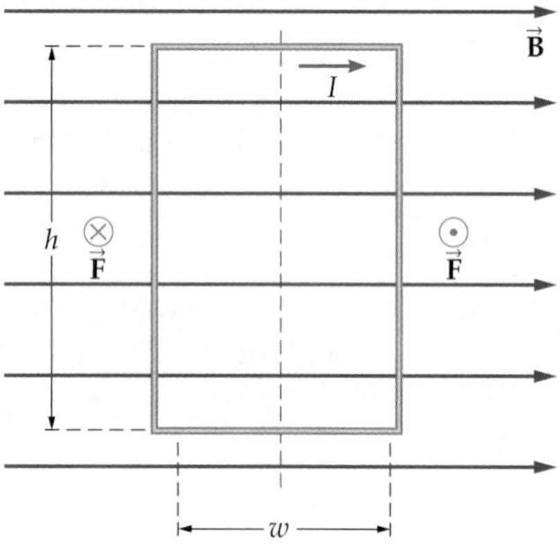

▲ **FIGURE 22–16 Magnetic forces on a current loop**

A rectangular current loop in a magnetic field. Only the vertical segments of the loop experience forces, and they tend to rotate the loop about a vertical axis.

Rectangular Current Loops

Consider a rectangular loop of height h and width w carrying a current I, as shown in **Figure 22–16**. The loop is placed in a region of space with a uniform magnetic field $\vec{B}$ that is parallel to the plane of the loop. From Figure 22–16, it is clear that the horizontal segments of the loop experience zero force, since they are parallel to the field. The vertical segments, on the other hand, are perpendicular to the field; hence, they experience forces of magnitude $F = IhB$. One of these forces is into the page (left side); the other is out of the page (right side).

Perhaps the best way to visualize the torque caused by these forces is to use a top view and look directly down on the loop, as in **Figure 22–17 (a)**. Here we can see more clearly that the forces on the vertical segments are equal in magnitude and opposite in direction. If we imagine an axis of rotation through the center of the loop, at the point O, it is clear that the forces exert a torque that tends to rotate the loop clockwise. The magnitude of this torque for each vertical segment is the force ($F = IhB$) times the moment arm ($w/2$). Noting that both vertical segments exert a torque in the same direction, we see that the total torque is simply the sum of the torque produced by each segment:

$$\tau = (IhB)\left(\frac{w}{2}\right) + (IhB)\left(\frac{w}{2}\right) = IB(hw)$$

Finally, observing that the area of the rectangular loop is $A = hw$, we can express the torque as follows:

$$\tau = IAB$$

As the loop begins to rotate, the situation will be like that shown in **Figure 22–17 (b)**. Here we see that the forces still have the same magnitude, IhB, but now the moment arms are $(w/2)\sin\theta$ rather than $w/2$. Thus, for a general angle, the torque must include the factor $\sin\theta$:

Torque Exerted on a Rectangular Loop of Area *A*

$$\tau = IAB\sin\theta \qquad 22\text{–}5$$

SI unit: N·m

Note that the angle θ is the angle between the plane of the loop and the magnetic force exerted on each side of the loop. Equivalently, θ is the angle between the normal and the magnetic field.

In the case shown in Figure 22–17, (a) the angle θ is 90°, and the normal to the plane of the loop is perpendicular to the magnetic field. In this orientation the torque attains its maximum value, $\tau = IBA$. When θ is zero, as in **Figure 22–17 (c)**, the torque vanishes because the moment arm of the magnetic forces is zero. Thus, there is no torque when the magnetic field is perpendicular to the plane of a loop.

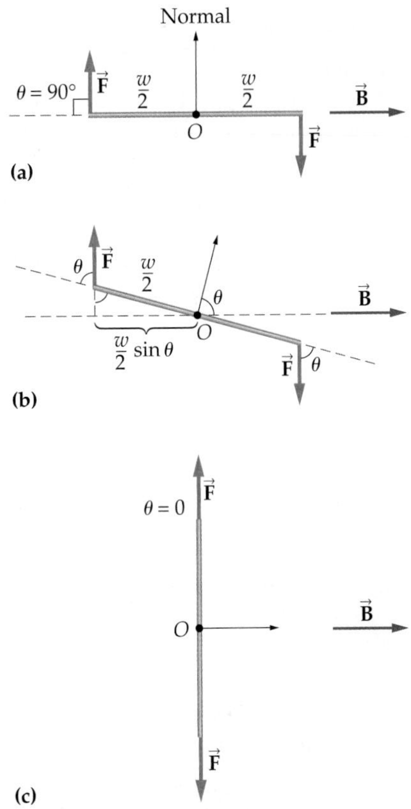

▲ **FIGURE 22–17 Magnetic torque on a current loop**

A current loop placed in a magnetic field produces a torque. **(a)** The torque is greatest when the plane of the loop is parallel to the magnetic field; that is, when the normal to the loop is perpendicular to the magnetic field. **(b)** As the loop rotates, the torque decreases by a factor of $\sin\theta$. **(c)** The torque vanishes when the plane of the loop is perpendicular to the magnetic field.

General Loops

We have shown that the torque exerted on a rectangular loop of area A is $\tau = IBA\sin\theta$. A more detailed derivation shows that the same relation applies to *any* planar loop, no matter what its shape. For example, a circular loop of radius r and area πr^2 experiences a torque given by the expression $\tau = IB\pi r^2 \sin\theta$.

In many applications it is desirable to produce as large a torque as possible. A simple way to increase the torque is to wrap a long wire around a loop N times, creating a coil of N "turns." Each of the N turns produces the same torque as a single loop; hence, the total torque is increased by a factor of N. In general, then, the torque produced by a loop with N turns is

Torque Exerted on a General Loop of Area *A* and *N* Turns

$$\tau = NIAB\sin\theta \qquad 22\text{–}6$$

SI unit: N·m

Notice that the torque depends on a number of factors in the system. First, it depends on the strength of the magnetic field, B, and on its orientation, θ, with respect to the normal of the loop. In addition, the torque depends on the current in the loop, I, the area of the loop, A, and the number of turns in the loop, N. The product of these "loop factors," NIA, is referred to as the **magnetic moment** of the loop. The magnetic moment, which has units of $A \cdot m^2$, is proportional to the amount of torque a given loop can exert.

EXAMPLE 22–5 Torque on a Coil

A rectangular coil with 200 turns is 5.0 cm high and 4.0 cm wide. When the coil is placed in a magnetic field of 0.35 T, its maximum torque is 0.22 N·m. What is the current in the coil?

Picture the Problem

The coil, along with its dimensions, is shown in the sketch. The sketch also reflects the fact that the maximum torque is produced when the magnetic field is in the plane of the coil.

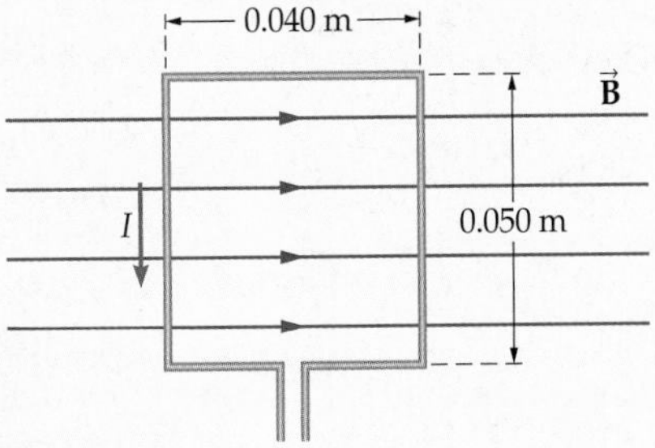

Strategy

The torque is given by the expression $\tau = NIAB\sin\theta$. Clearly, the maximum torque occurs when $\sin\theta = 1$; that is, $\tau_{max} = NIAB$. Solving this relation for I yields the current in the coil.

Solution

1. Write an expression for the maximum torque: $\tau_{max} = NIAB$

2. Solve for the current, I: $I = \dfrac{\tau_{max}}{NAB}$

3. Substitute numerical values: $I = \dfrac{0.22\ \text{N}\cdot\text{m}}{(200)(0.050\ \text{m})(0.040\ \text{m})(0.35\ \text{T})} = 1.6\ \text{A}$

continued on next page

continued from previous page

Insight

Note that this calculation gives the magnitude of the current and not its direction. The direction of the current will determine whether the torque on the coil is clockwise or counterclockwise.

Practice Problem

If the shape of this coil were changed to circular, keeping a constant perimeter, would the maximum torque increase, decrease, or stay the same? [**Answer:** In general, a circle has the greatest area for a given perimeter. Therefore, the maximum torque would increase if the coil were made circular, because its area would be larger.]

Some related homework problems: Problem 32, Problem 33

Applications of Torque

The torque exerted by a magnetic field finds a number of useful applications. For example, if a needle is attached to a coil, as in **Figure 22–18**, it can be used as part of a meter. When the coil is connected to an electrical circuit, it experiences a torque, and a corresponding deflection of the needle, that is proportional to the current. The result is that the current in a circuit can be indicated by the reading on the meter. A simple device of this type is referred to as a **galvanometer.**

REAL-WORLD PHYSICS
The galvanometer

Notice that the galvanometer can be a very sensitive instrument for two different reasons. First, the coil can have a large number of turns. Since the torque exerted by a coil is proportional to the number of turns, it follows that a coil with many turns produces a significant torque even when the current is small. Second, the needle amplifies the motion of the coil. Specifically, if the coil turns through an angle θ, and the needle has a length L, the tip of the needle moves through a distance $L\theta$. Thus, both the number of turns and the length of the needle increase the sensitivity of the meter.

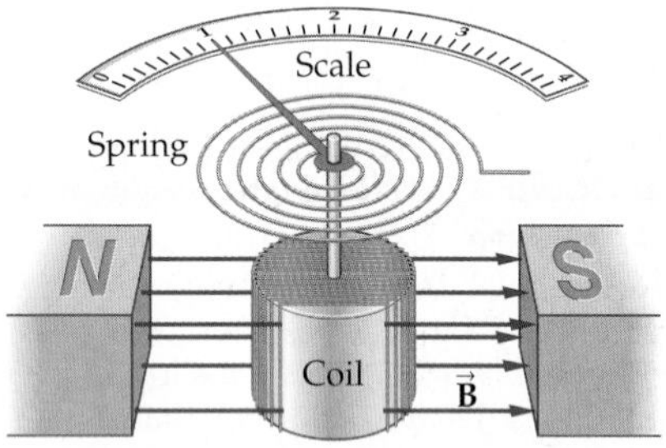

▲ **FIGURE 22–18 The galvanometer**
The basic elements of a galvanometer are a coil, a magnetic field, a spring, and a needle attached to the coil. As current passes through the coil, a torque acts on it, causing it to rotate. The spring ensures that the angle of rotation is proportional to the current in the coil.

Of even greater practical importance is the fact that magnetic torque can be used to power a motor. For example, an electric current passing through the coils of a motor causes a torque that rotates the axle of the motor. As the coils rotate, a device known as the commutator reverses the direction of the current as the orientation of the coil reverses, which ensures that the torque is always in the same direction and that the coils continue to turn in the same direction. Electric motors, which will be discussed in greater detail in Chapter 23, are used in everything from CD players to electric razors to electric cars.

22–6 Electric Currents, Magnetic Fields, and Ampère's Law

In the introduction to this chapter, we mentioned that a previously unexpected connection between electricity and magnetism was discovered accidentally by Hans Christian Oersted in 1820. Specifically, Oersted was giving a public lecture on various aspects of science when, at one point, he closed a switch and allowed a current to flow through a wire. What he noticed was that a nearby compass needle deflected from its usual orientation when the switch was closed—Oersted had just discovered that electric currents can create magnetic fields. In this section we focus on the connection between electric currents and magnetic fields. In so doing, our attention shifts from the effects of magnetic fields—the subject of the previous sections—to their production.

A Long, Straight Wire

We start with the simplest possible case—a straight, infinitely long wire that carries a current I. To visualize the magnetic field such a wire produces, we shake iron filing onto a sheet of paper that is pierced by the wire, as indicated in **Figure 22–19 (a)**. The result is that the filings form into circular patterns centered on the wire—evidently, the magnetic field "circulates" around the wire.

We can gain additional information about the field by placing a group of small compasses about the wire, as in **Figure 22–19 (b)**. In addition to confirming the circular shape of the field lines, the compass needles show the field's direction. To understand this direction, we must again utilize a right-hand rule—this time, we refer to the rule as the *magnetic field* right-hand rule:

Magnetic Field Right-Hand Rule

To find the direction of the magnetic field due to a current-carrying wire, point the thumb of your right hand along the wire in the direction of the current I. Your fingers are now curling around the wire in the direction of the magnetic field.

This rule is illustrated in **Figure 22–20**, where we see that it predicts the same direction as that indicated by the compass needles in Figure 22–19 (b).

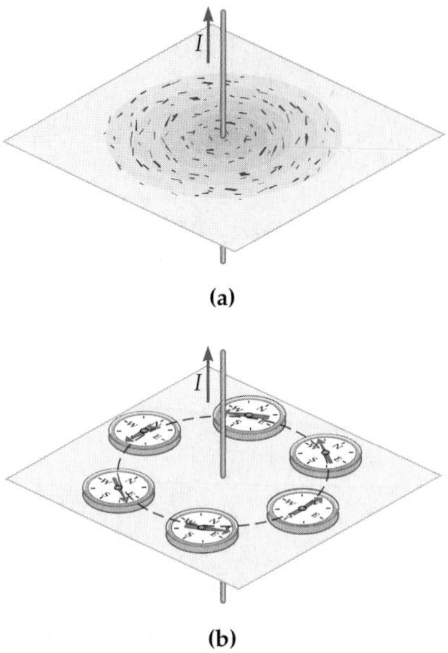

▲ **FIGURE 22–19 The magnetic field of a current-carrying wire**

(a) An electric current flowing through a wire produces a magnetic field. In the case of a long, straight wire, the field circulates around the wire. **(b)** Compass needles point along the circumference of a circle centered on the wire.

CONCEPTUAL CHECKPOINT 22–5 Direction of the Current

The magnetic field shown in the sketch is due to the horizontal, current-carrying wire. Does the current in the wire flow to the left or to the right?

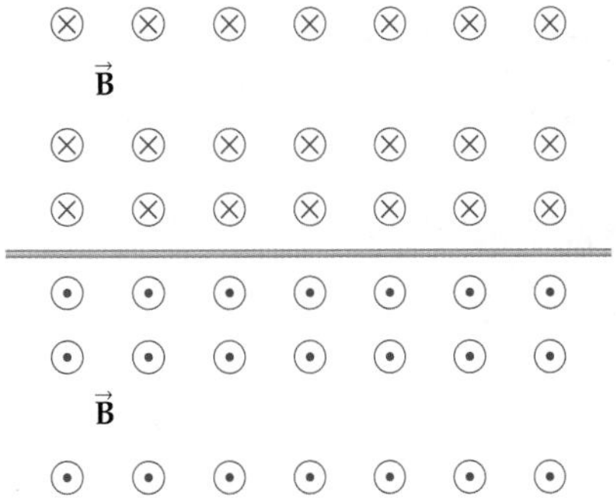

Reasoning and Discussion

If you point the thumb of your right hand along the wire to the left, your fingers curl into the page above the wire and out of the page below the wire, as shown in the figure. Thus, the current flows to the left.

Answer:

The current in the wire flows to the left.

Experiment shows that the field produced by a current-carrying wire doubles if the current I is doubled. In addition, the field decreases by a factor of 2 if the distance from the wire, r, is doubled. Hence, we conclude that the magnetic field B must be proportional to I/r; that is,

$$B = (\text{constant})\frac{I}{r}$$

The precise expression for B will now be derived using a law of nature known as Ampère's law.

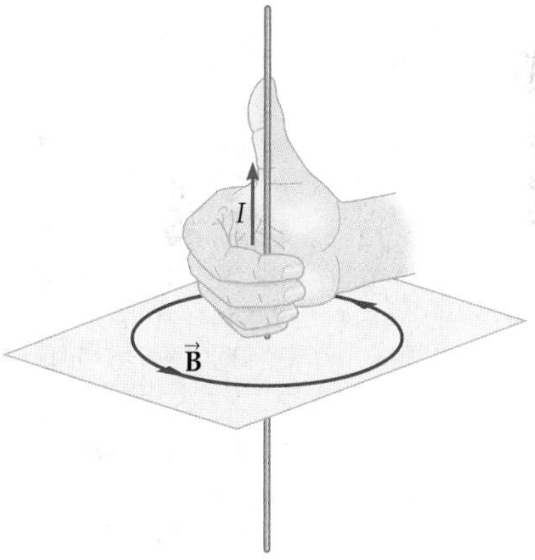

▲ **FIGURE 22–20 The magnetic-field right-hand rule**

The magnetic field right-hand rule determines the direction of the magnetic field produced by a current-carrying wire. With the thumb of the right hand pointing in the direction of the current, the fingers curl in the direction of the field.

Ampère's Law

Ampère's law relates the magnetic field along a closed path to the electric current enclosed by the path. Specifically, consider the current-carrying wires shown in **Figure 22–21**. These wires are enclosed by the closed path P, which can be divided into many small, straight-line segments of length ΔL. On each of these segments, the magnetic field $\vec{\mathbf{B}}$ can be resolved into a component parallel to the segment, $B_\parallel$, and a component perpendicular to the segment, $B_\perp$. Of particular interest is the product $B_\parallel \, \Delta L$. According to Ampère's law, the sum of $B_\parallel \, \Delta L$ over all segments of a closed path is equal to a constant times the current enclosed by the path. This relationship can be written symbolically as follows:

Ampère's Law

$$\sum B_\parallel \, \Delta L = \mu_0 I_{\text{enclosed}} \qquad \text{22–7}$$

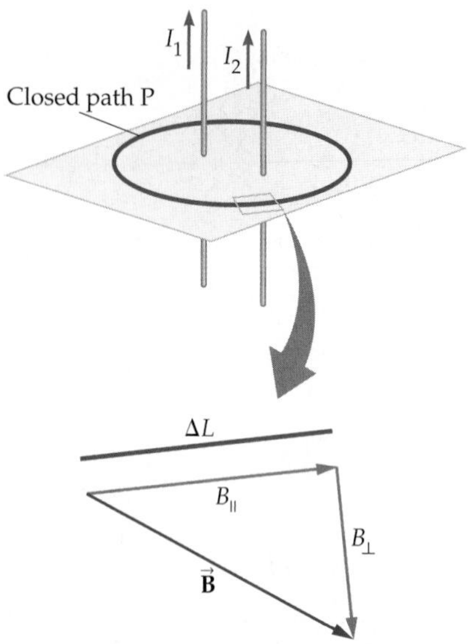

▲ **FIGURE 22–21 Illustrating Ampère's law**

A closed path P encloses the currents I_1 and I_2. According to Ampère's law, the sum of $B_{\|}\,\Delta L$ around the path P is equal to $\mu_0 I_{\text{enclosed}}$. In this case, $I_{\text{enclosed}} = I_1 + I_2$.

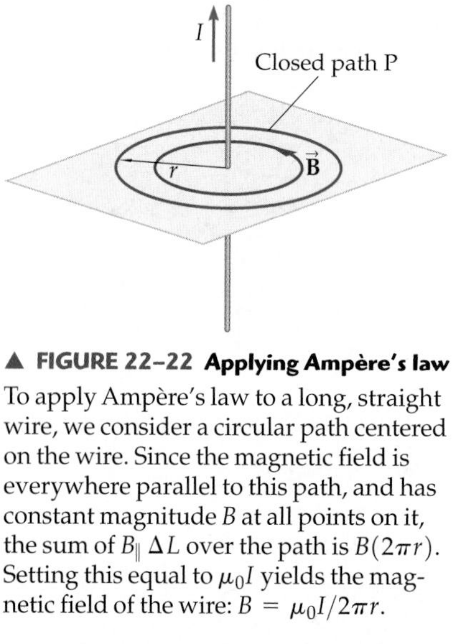

▲ **FIGURE 22–22 Applying Ampère's law**

To apply Ampère's law to a long, straight wire, we consider a circular path centered on the wire. Since the magnetic field is everywhere parallel to this path, and has constant magnitude B at all points on it, the sum of $B_{\|}\,\Delta L$ over the path is $B(2\pi r)$. Setting this equal to $\mu_0 I$ yields the magnetic field of the wire: $B = \mu_0 I/2\pi r$.

In this expression, μ_0 is a constant called the **permeability of free space.** Its value is

$$\mu_0 = 4\pi \times 10^{-7}\,\mathrm{T \cdot m/A} \qquad \text{22–8}$$

We should emphasize that Ampère's law is a law of nature—it is valid for all magnetic fields and currents that are constant in time.

Let's apply Ampère's law to the case of a long, straight wire carrying a current I. We already know that the field circulates around the wire, as illustrated in **Figure 22–22.** It is reasonable, then, to choose a circular path of radius r (and circumference $2\pi r$) to enclose the wire. Since the magnetic field is parallel to the circular path at every point, and all points on the path are the same distance from the wire, it follows that $B_{\|} = B =$ constant. Therefore, the sum of $B_{\|}\,\Delta L$ around the closed path gives

$$\sum B_{\|}\,\Delta L = B \sum \Delta L = B(2\pi r)$$

According to Ampère's law, this sum must equal $\mu_0 I_{\text{enclosed}} = \mu_0 I$. Therefore, we have

$$B(2\pi r) = \mu_0 I$$

Solving for B, we obtain

Magnetic Field for a Long, Straight Wire

$$B = \frac{\mu_0 I}{2\pi r} \qquad \text{22–9}$$

SI unit: tesla, T

As expected, the field is equal to a constant times I/r.

EXERCISE 22–2

Find the magnitude of the magnetic field 1 m from a long, straight wire carrying a current of 1 A.

Solution

Straightforward substitution in Equation 22–9 yields

$$B = \frac{\mu_0 I}{2\pi r} = \frac{(4\pi \times 10^{-7}\,\mathrm{T \cdot m/A})(1\,\mathrm{A})}{2\pi(1\,\mathrm{m})} = 2 \times 10^{-7}\,\mathrm{T}$$

This is a weak field, less than one hundredth the strength of the Earth's magnetic field.

EXAMPLE 22–6 An Attractive Wire

A 52-μC charged particle moves parallel to a long wire with a speed of 720 m/s. The separation between the particle and the wire is 13 cm, and the magnitude of the force exerted on the particle is 1.4×10^{-7} N. Find **(a)** the magnitude of the magnetic field at the location of the particle and **(b)** the current in the wire.

Picture the Problem

The physical situation is illustrated in our sketch. Note that the charged particle moves parallel to the current-carrying wire; hence, its velocity is at right angles to the magnetic field. This means that it will experience the maximum magnetic force. Finally, by pointing the thumb of your right hand in the direction of I, you can verify that $\vec{\mathbf{B}}$ points out of the page above the wire and into the page below it.

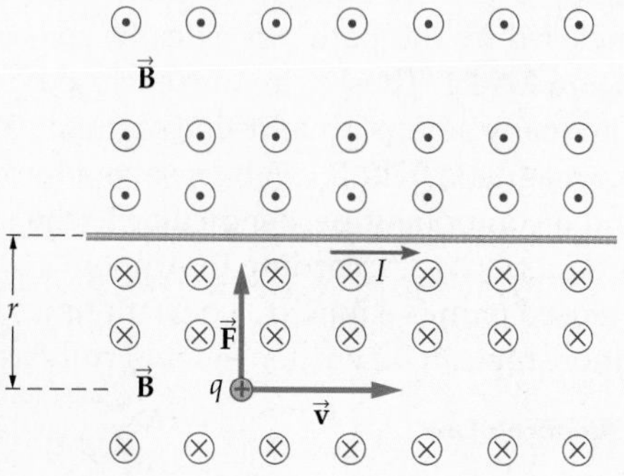

Strategy

a. The maximum magnetic force on the charged particle is $F = qvB$. This relation can be solved for B.

b. The magnetic field is produced by the current in the wire. Therefore, $B = \mu_0 I/2\pi r$. Using B from part (a), we can solve for the current I.

Solution

Part (a)

1. Use $F = qvB$ to solve for the magnetic field:

$$F = qvB$$

$$B = \frac{F}{qv} = \frac{1.4 \times 10^{-7}\ \text{N}}{(5.2 \times 10^{-5}\ \text{C})(720\ \text{m/s})} = 3.7 \times 10^{-6}\ \text{T}$$

Part (b)

2. Use the relation for the magnetic field of a wire to solve for the current:

$$B = \frac{\mu_0 I}{2\pi r}$$

$$I = \frac{2\pi r B}{\mu_0}$$

3. Substitute the value for B found in part (a) and evaluate I:

$$I = \frac{2\pi r B}{\mu_0} = \frac{2\pi(0.13\ \text{m})(3.7 \times 10^{-6}\ \text{T})}{4\pi \times 10^{-7}\ \text{T}\cdot\text{m/A}} = 2.4\ \text{A}$$

Insight

Using the magnetic force RHR, we can see that the direction of the force exerted on the charged particle is toward the wire, as illustrated in our sketch. If the particle had been negatively charged, the force would have been away from the wire.

Practice Problem

Suppose a particle with a charge of 52 μC is 13 cm above the wire and moving with a speed of 720 m/s to the left. Find the magnitude and direction of the force acting on this particle. [**Answer:** $F = 1.4 \times 10^{-7}$ N, away from wire]

Some related homework problems: Problem 39, Problem 45

ACTIVE EXAMPLE 22–2 Find the Magnetic Field

Two wires separated by a distance of 22 cm carry currents in the same direction. The current in one wire is 1.5 A, and the current in the other wire is 4.5 A. Find the magnitude of the magnetic field halfway between the wires.

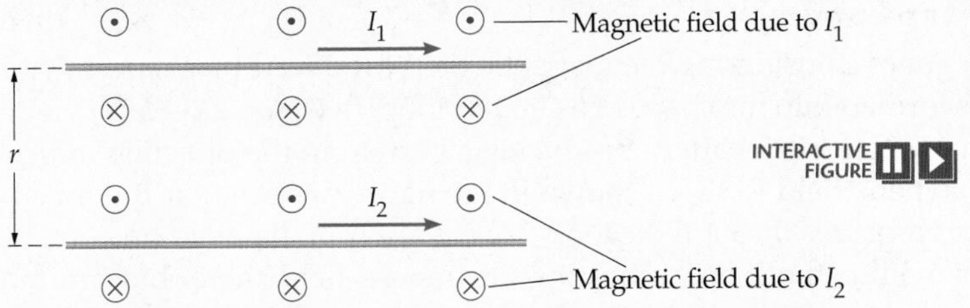

Solution *(Test your understanding by performing the calculations indicated in each step.)*

1. Find the magnitude and direction of the magnetic field produced by wire 1: $B_1 = 2.7 \times 10^{-6}$ T, into page

2. Find the magnitude and direction of the magnetic field produced by wire 2: $B_2 = 8.2 \times 10^{-6}$ T, out of page

3. Calculate the magnitude of the net field: $B = B_2 - B_1 = 5.5 \times 10^{-6}$ T

Insight

Since the field produced by I_2 has the greater magnitude, the net field is out of the page. If the currents were equal, the net field midway between the wires would be zero.

Your Turn

Find the magnitude and direction of the magnetic field 11 cm *below* wire 2.

(Answers to **Your Turn** *problems are given in the back of the book.)*

Forces between Current-Carrying Wires

We know that a current-carrying wire in a magnetic field experiences a force. We also know that a current-carrying wire produces a magnetic field. It follows, then, that one current-carrying wire will exert a force on another.

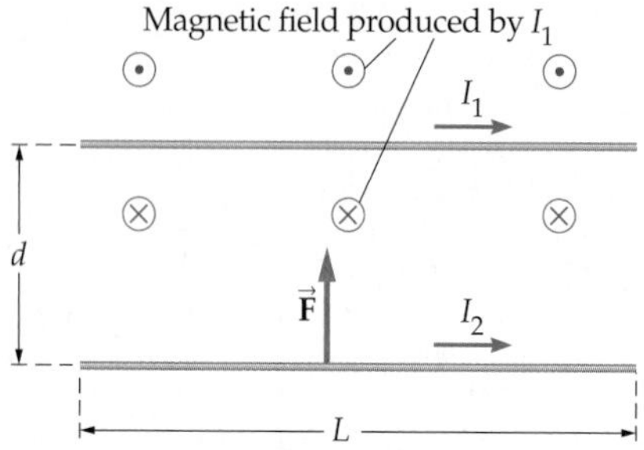

▲ **FIGURE 22–23 The magnetic force between current-carrying wires**

A current in wire 1 produces a magnetic field, $B_1 = \mu_0 I_1/2\pi d$, at the location of wire 2. The result is a force exerted on a length L of wire 2 of magnitude $F = \mu_0 I_1 I_2 L/2\pi d$.

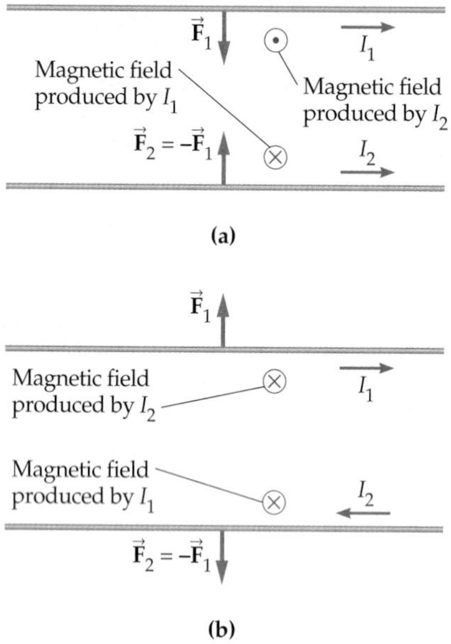

▲ **FIGURE 22–24 The direction of the magnetic force between current-carrying wires**

The forces between current-carrying wires depend on the relative direction of their currents. **(a)** If the currents are in the same direction, the force is attractive. **(b)** Wires with oppositely directed currents experience repulsive forces.

To work this relationship out in detail, consider the two wires with parallel currents and separation d shown in **Figure 22–23**. The magnetic field produced by wire 1 circulates around it, coming out of the page above the wire and entering the page below the wire. Thus, wire 2 experiences a magnetic field pointing into the page with a magnitude given by Equation 22–9; $B = \mu_0 I_1/2\pi d$. The force experienced by wire 2, therefore, has a magnitude given by Equation 22–4 with $\theta = 90°$:

$$F = I_2 LB = I_2 L\left(\frac{\mu_0 I_1}{2\pi d}\right) = \frac{\mu_0 I_1 I_2}{2\pi d}L \qquad \text{22–10}$$

The direction of the force acting on wire 2, as given by the magnetic force RHR, is upward, that is, toward wire 1. A similar calculation starting with the field produced by wire 2 gives a force of the same magnitude acting downward on wire 1. This is to be expected, because the forces acting on wires 1 and 2 form an action–reaction pair, as indicated in **Figure 22–24 (a)**. Hence, wires with parallel currents attract one another.

If the currents in wires 1 and 2 are in opposite directions, as in **Figure 22–24 (b)**, the situation is similar to that discussed in the preceding paragraph, except that the direction of the forces is reversed. Thus, wires with opposite currents repel one another.

22–7 Current Loops and Solenoids

We now consider the magnetic fields produced when a current-carrying wire has a circular or a helical geometry—as opposed to the straight wires considered in the previous section. As we shall see, there are many practical applications for such geometries.

Current Loop

We begin by considering the magnetic field produced by a current-carrying wire that is formed into the shape of a circular loop. In **Figure 22–25 (a)** we show a wire loop connected to a battery producing a current in the direction indicated. Using the magnetic field RHR, as shown in the figure, we see that $\vec{\mathbf{B}}$ points from left to right as it passes through the loop. Notice also that the field lines are bunched together within the loop, indicating an intense field there, but are more widely spaced outside the loop.

The most interesting aspect of the field produced by the loop is its close resemblance to the field of a bar magnet. This similarity is illustrated in **Figure 22–25 (b)**, where we see that one side of the loop behaves like a north magnetic pole and the other side like a south magnetic pole. Thus, if two loops with identical currents are

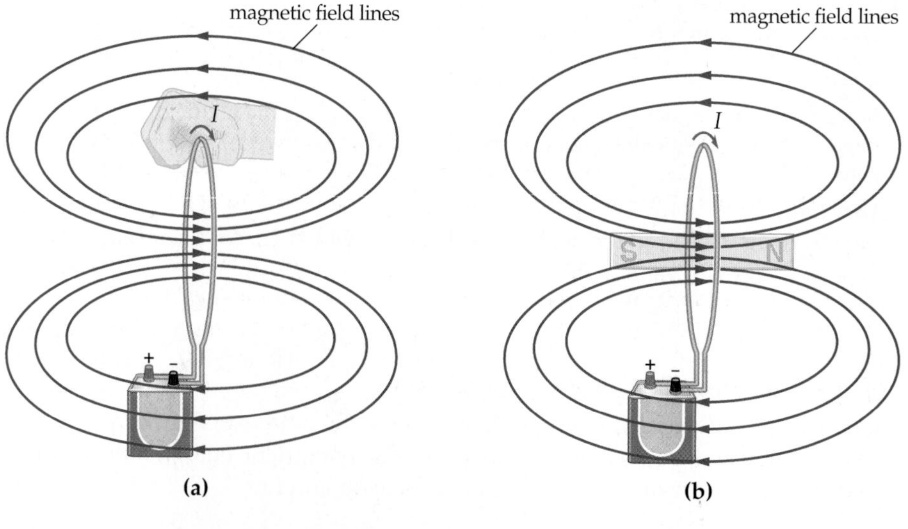

▶ **FIGURE 22–25 The magnetic field of a current loop**

(a) The magnetic field produced by a current loop is relatively intense within the loop and falls off rapidly outside the loop. **(b)** A permanent magnet produces a field that is very similar to the field of a current loop.

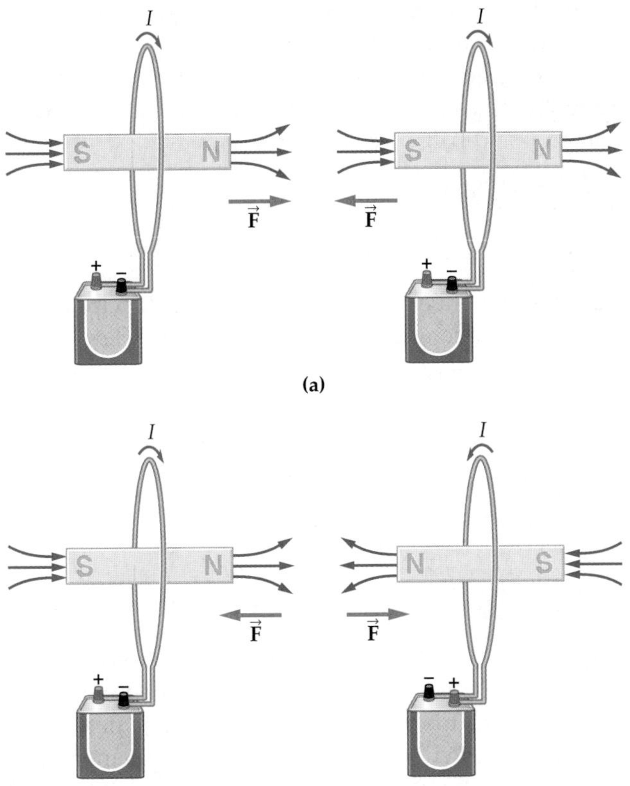

◀ **FIGURE 22–26 Magnetic forces between current loops**

To decide whether current loops will experience an attractive or repulsive force, it is useful to think in terms of the corresponding permanent magnets. **(a)** Current loops with currents in the same direction are like two bar magnets lined up in the same direction; they attract each other. **(b)** Current loops with opposite currents act like bar magnets with opposite orientations; they repel each other.

placed near each other, as in **Figure 22–26 (a)**, the force between them will be similar to the force between two bar magnets pointing in the same direction—that is, they will attract each other. If the loops have oppositely directed currents, as in **Figure 22–26 (b)**, the force between them is repulsive. Note the similarity of the results for straight wires, Figure 22–24, and for loops, Figure 22–26.

On a more fundamental level, the connection between a current loop and a permanent magnet is much more than accidental. In fact, the atomic origin of magnetic fields is due to circulating currents produced by electrons. We discuss this connection in greater detail in the next section.

The magnitude of the magnetic field produced by a circular loop of N turns, radius R, and current I varies from point to point; however, it can be shown that at the center of the loop, the field is given by the following simple expression:

$$B = \frac{N\mu_0 I}{2R} \quad \text{(center of circular loop of radius } R\text{)} \qquad 22\text{–}11$$

Notice that the field is proportional to the current in the loop and inversely proportional to its radius.

Solenoid

A **solenoid** is an electrical device in which a long wire has been wound into a succession of closely spaced loops with the geometry of a helix. Also referred to as an **electromagnet,** a solenoid carrying a current produces an intense, nearly uniform, magnetic field inside the loops, as indicated in **Figure 22–27**. Notice that each loop of the solenoid carries a current in the same direction; therefore, the magnetic force between loops is attractive and serves to hold the loops tightly together.

The magnetic field lines in Figure 22–27 are tightly packed inside the solenoid but are widely spaced outside. In the ideal case of a very long, tightly packed solenoid, the magnetic field outside is practically zero—especially when compared with the intense field inside the solenoid. We can use this idealization, in combination with Ampère's law, to calculate the magnitude of the field inside the solenoid.

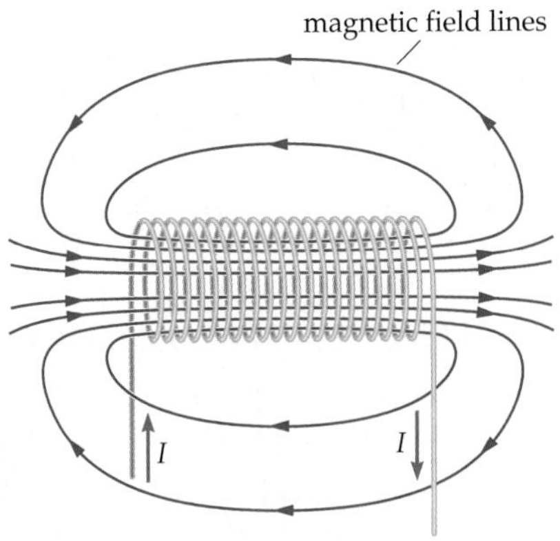

▲ **FIGURE 22–27 The solenoid**

A solenoid is formed from a long wire that is wound into a succession of current loops. The magnetic field inside the solenoid is relatively strong and uniform. Outside the solenoid the field is weak. In the ideal case, we consider the field outside the solenoid to be zero and that inside to be uniform and parallel to the solenoid's axis.

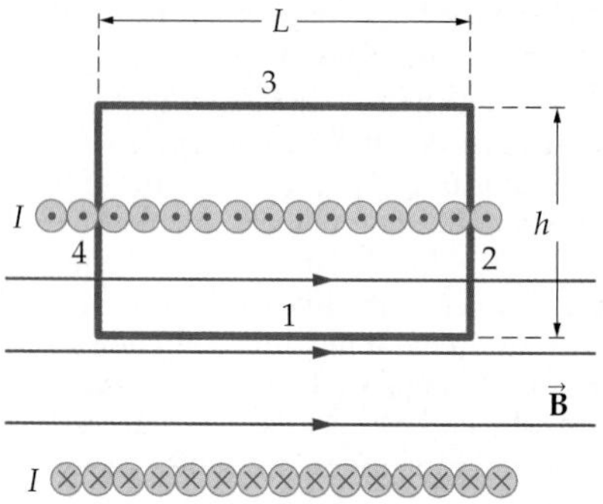

▲ **FIGURE 22–28 Ampère's law and the magnetic field in a solenoid**
To calculate the magnetic field inside a solenoid, we apply Ampère's law to the rectangular path shown here. The only side of the rectangle that has a nonzero, parallel component of $\vec{\mathbf{B}}$ is side 1.

To do so, consider the rectangular path of width L and height h shown in **Figure 22–28**. Notice that the parallel component of the field on side 1 is simply $\vec{\mathbf{B}}$. On sides 2 and 4 the parallel component is zero, since $\vec{\mathbf{B}}$ is perpendicular to those sides. Finally, on side 3 (which is outside the solenoid) the magnetic field is zero. Using these results, we obtain the sum of $B_{\parallel}\,\Delta L$ over the rectangular loops:

$$\sum B_{\parallel}\,\Delta L = \sum_{\text{side 1}} B_{\parallel}\,\Delta L + \sum_{\text{side 2}} B_{\parallel}\,\Delta L + \sum_{\text{side 3}} B_{\parallel}\,\Delta L + \sum_{\text{side 4}} B_{\parallel}\,\Delta L$$
$$= BL + 0 + 0 + 0 = BL$$

Next, the current enclosed by the rectangular circuit is NI, where N is the number of loops in the length L. Therefore, Ampère's law gives

$$BL = \mu_0 NI$$

Solving for B and letting the number of loops per length be $n = N/L$, we find

Magnetic Field of a Solenoid

$$B = \mu_0\left(\frac{N}{L}\right)I = \mu_0 nI \qquad \textbf{22–12}$$

SI unit: tesla, T

PROBLEM-SOLVING NOTE

The Magnetic Field inside a Solenoid

In an ideal solenoid, the magnetic field inside the solenoid points along the axis and is uniform. Outside the solenoid the field is zero. Therefore, when calculating the field produced by a solenoid, it is not necessary to specify a particular point inside the solenoid; all inside points have the same field.

Note that this result is independent of the cross-sectional area of the solenoid.

When used as an electromagnet, a solenoid has many useful properties. First and foremost, it produces a strong magnetic field that can be turned on or off at the flip of a switch—unlike the field of a permanent magnet. In addition, the magnetic field can be further intensified by filling the core of the solenoid with an iron bar. In such a case, the magnetic field of the solenoid magnetizes the iron bar, which then adds its field to the overall field of the system. These properties and others make solenoids useful devices in a variety of electrical circuits. Further examples are given in the next chapter.

CONCEPTUAL CHECKPOINT 22–6 Magnetic Field in a Solenoid

If you want to increase the strength of the magnetic field inside a solenoid, is it better to (a) double the number of loops, keeping the length the same, or (b) double the length, keeping the number of loops the same?

Reasoning and Discussion

Referring to the expression $B = \mu_0(N/L)I$, we see that doubling the number of loops $(N \rightarrow 2N)$ while keeping the length the same $(L \rightarrow L)$ results in a doubled magnetic field $(B \rightarrow 2B)$. On the other hand, doubling the length $(L \rightarrow 2L)$ while keeping the number of loops the same $(N \rightarrow N)$ reduces the magnetic field by a factor of $2 (B \rightarrow B/2)$. Hence, to increase the field one should pack more loops into the same length.

Answer:

(a) Double the number of loops with the same length.

EXAMPLE 22–7 Through the Core of a Solenoid

A solenoid is 20.0 cm long, has 200 loops, and carries a current of 3.25 A. Find the magnitude of the force exerted on a 15.0-μC charged particle moving at 1050 m/s through the interior of the solenoid, at an angle of 11.5° relative to the solenoid's axis.

Picture the Problem

Our sketch shows the solenoid, along with the uniform magnetic field it produces parallel to its axis. Inside the solenoid, a positively-charged particle moves at an angle of $\theta = 11.5°$ relative to the magnetic field.

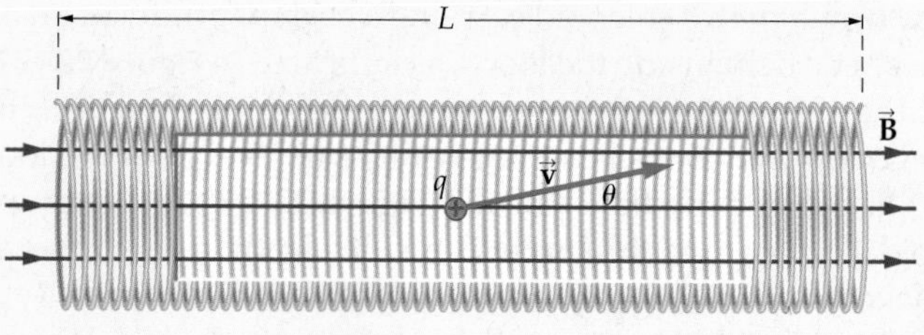

Strategy

The force exerted on the charged particle is magnetic; hence, we start by calculating the magnetic field produced by the solenoid, $B = \mu_0(N/L)I$. Next, we note that the magnetic field in a solenoid is parallel to its axis. It follows that the magnitude of the force exerted on the charge is given by $F = |q|vB\sin\theta$, with $\theta = 11.5°$.

Solution

1. Calculate the magnetic field inside the solenoid:

$$B = \mu_0\left(\frac{N}{L}\right)I$$
$$= (4\pi \times 10^{-7}\,\text{T}\cdot\text{m/A})\left(\frac{200}{0.200\,\text{m}}\right)(3.25\,\text{A}) = 4.08 \times 10^{-3}\,\text{T}$$

2. Use B to find the force exerted on the charged particle:

$$F = |q|vB\sin\theta$$
$$= (15.0 \times 10^{-6}\,\text{C})(1050\,\text{m/s})(4.08 \times 10^{-3}\,\text{T})\sin 11.5°$$
$$= 1.28 \times 10^{-5}\,\text{N}$$

Insight

Note that the magnetic field strength inside this modest solenoid is approximately 100 times greater than the magnetic field at the surface of the Earth.

Practice Problem

What current would be required to double the force acting on the particle to 2.56×10^{-5} N? [**Answer:** $I = 2(3.25\text{ A}) = 6.50\text{ A}$]

Some related homework problems: Problem 48, Problem 49

REAL-WORLD PHYSICS: BIO

MRI instruments

Magnetic resonance imaging (MRI) instruments utilize solenoids large enough to accommodate a person within their coils. Not only are these solenoids large in size, they are also capable of producing extremely powerful magnetic fields on the order of 1 or 2 tesla (see Table 22–1). So powerful are these fields, in fact, that metallic objects such as mop buckets and stretchers have been known to be pulled from across the room into the bore of the magnet. A metal oxygen bottle in the same room can be turned into a dangerous, high-speed projectile.

REAL-WORLD PHYSICS: BIO

Magnetic reed switches in pacemakers

In some cases, artificial pacemakers have been affected. Many types of pacemakers have what are known as *magnetic reed switches*. These switches allow a physician to change the operating mode of a pacemaker, without surgery, by simply placing a magnet at the appropriate location on a patient's chest. If a person with one of these pacemakers comes anywhere near an operating MRI instrument, the results can be serious.

22–8 Magnetism in Matter

Some materials have strong magnetic fields; others do not. To understand these differences, we must consider the behavior of matter on the atomic level.

To begin, recall that circulating electric currents can produce magnetic fields much like those produced by bar magnets. This is significant because atoms have electrons orbiting their nucleus, which means they have circulating electric currents. The fields produced by these orbiting electrons can be sizable—on the order of 10 T or so at the nucleus of the atom. In most atoms, though, the fields produced by the various individual electrons tend to cancel one another, resulting in a very small or zero net magnetic field.

Another type of circulating electric current present in atoms produces fields that occasionally do not cancel. These currents are associated with the *spin* that all electrons have. A simple model of the electron—which should *not* be taken literally—is of a spinning sphere of charge. The circulating charge associated with the spinning motion gives rise to a magnetic field. Electrons tend to "pair up" in atoms in such a way that their spins are opposite to one another, again resulting in zero net field. However, in some atoms—such as iron, nickel, and cobalt—the net field due to the spinning electrons is nonzero, and strong magnetic effects can occur as the magnetic field of one atom tends to align with the magnetic field of another. A full understanding of these types of effects requires the methods of quantum mechanics, which will be discussed in Chapters 30 and 31.

Ferromagnetism

If the tendency of magnetic atoms to self-align is strong enough, as it is in materials such as iron and nickel, the result can be an intense magnetic field—as in a bar

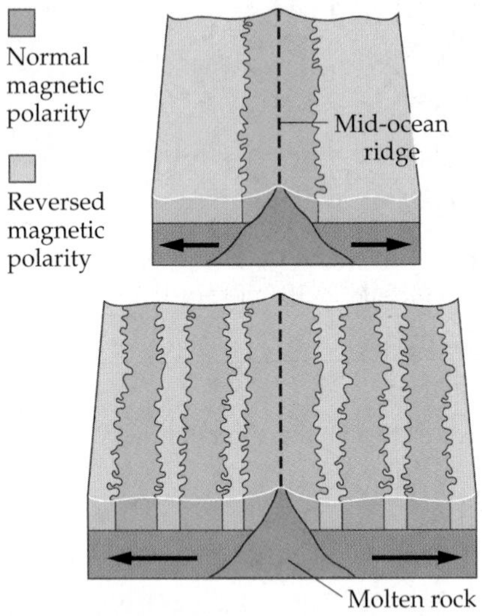

▲ **FIGURE 22–29 Mid-ocean ridge**

Molten rock extruded at a mid-ocean ridge magnetizes in the direction of the Earth's magnetic field when it cools to temperatures below about 770 °C. After cooling, the direction of the Earth's field remains "frozen" in the rocks. As these rocks move away from the ridge, due to seafloor spreading, newly extruded material near the ridge undergoes the same process. As a result, the magnetization of the rocks on either side of a mid-ocean ridge produces a geological record of the polarity of the Earth's magnetic field over time, as well as convincing confirmation of seafloor spreading.

magnet. Materials with this type of behavior are called **ferromagnets.** Counteracting the tendency of atoms to align, however, is the disorder caused by increasing temperature. In fact, all ferromagnets lose their magnetic field if the temperature is high enough to cause their atoms to orient in random directions. For example, the magnetic field of a bar magnet made of iron vanishes if its temperature exceeds 770 °C.

This type of temperature behavior shows that the simple picture of a large bar magnet within the Earth cannot be correct. As we know, temperature increases with depth below the Earth's surface. In fact, at depths of about 15 miles the temperature is already above 770 °C, so the magnetism of an iron magnet would be lost due to thermal effects. Of course, at even greater depths an iron magnet would melt to form a liquid. In fact, it appears that the magnetic field of the Earth is caused by circulating currents of molten iron, nickel, and other metals. These circulating currents create a magnetic field in much the same way as the circulating current in a solenoid.

Temperature also plays a key role in the magnetization that is observed in rocks on the ocean floor. As molten rock is extruded from mid-ocean ridges, it has no net magnetization because of its high temperature. When the rock cools, however, it becomes magnetized in the direction of the Earth's magnetic field. In effect, the Earth's magnetic field becomes "frozen" in the solidified rock. As the seafloor spreads, and more material is formed along the ridge, a continuous record of the Earth's magnetic field is formed. In particular, if the Earth's field reverses at some point in time, the field in the solidified rocks will record that fact. **Figure 22–29** shows a record of the magnetization of rock that has been formed by seafloor spreading, showing the regions of oppositely magnetized rocks that indicate the field reversals.

Another important feature of ferromagnets is that their magnetism is characterized by magnetic **domains** within the material, as illustrated in **Figure 22–30 (a)**. Each domain has a strong magnetic field, but different domains are oriented differently, so that the net effect may be small. The typical size of these domains is on the order of 10^{-4} cm to 10^{-1} cm. When an external field is applied to such a material, the magnetic domains that are pointing in the direction of the applied field often grow in size at the expense of domains with different orientations, as indicated in **Figure 22–30 (b)**. The result is that the applied external field produces a net magnetization in the material.

REAL-WORLD PHYSICS: BIO
Magnetite in living organisms

Many living organisms are known to incorporate small ferromagnetic crystals, consisting of *magnetite*, in their bodies. For example, some species of bacteria use magnetite crystals to help orient themselves with respect to the Earth's magnetic field. Magnetite has also been found in the brains of bees and pigeons, where it is suspected to play a role in navigation. It is even found in human brains, though its possible function there is unclear.

REAL-WORLD PHYSICS: BIO
Magnetism and the brain

Whatever the role of magnetite in people, observations show that a magnetic field can affect the way the human brain operates. In recent experiments involving people viewing optical illusions, it has been found that if the parietal lobe on one side of the brain is exposed to a highly focused magnetic field of about 1 T, a temporary interruption in much of the neural activity in that hemisphere results. Fortunately, fields of 1 T are generally not encountered in everyday life.

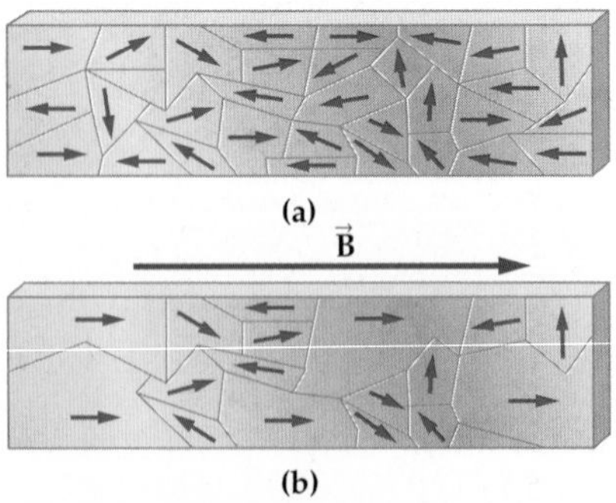

▲ **FIGURE 22–30 Magnetic domains**

(a) A ferromagnetic material tends to form into numerous domains with magnetization pointing in different directions. **(b)** When an external field is applied, the domains pointing in the direction of the field often grow in size at the expense of domains pointing in other directions.

Paramagnetism and Diamagnetism

Not all magnetic materials are ferromagnetic, however. In some cases a ferromagnet has zero magnetic field simply because it is at too high a temperature. In other cases, the tendency for the self-alignment of individual atoms in a given material is too weak to produce a net magnetic field, even at low temperatures. In either case, a strong external magnetic field applied to the material can cause alignment of the atoms and result in a magnetic field. Magnetic effects of this type are referred to as **paramagnetism.**

Finally, all materials display a further magnetic effect referred to as **diamagnetism.** In the diamagnetic effect, an applied magnetic field on a material produces an oppositely directed field in response. The resulting repulsive force is usually too weak to be noticed, except in superconductors. The basic mechanism responsible for diamagnetism will be discussed in greater detail in the next chapter.

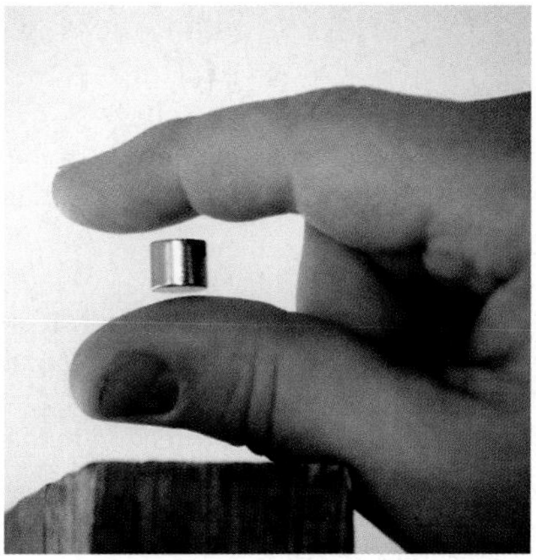

◀ (left) The repulsive diamagnetic forces produced by water molecules in the body of a frog are strong enough to levitate it in an intense magnetic field of 16 T. (Animals levitated in this way appear to suffer no harm or discomfort.) (right) This small magnet is suspended in the strong magnetic field produced by a larger magnet above it (outside the photo). Ordinarily, such an arrangement would be highly unstable. The addition of small, repulsive diamagnetic forces due to a person's fingers, however, is sufficient to convert it into a stable equilibrium.

If a magnetic field is strong enough, however, even the relatively weak repulsion of diamagnetism can lead to significant effects. For example, researchers at the Nijmegen High Field Magnet Laboratory have used a field of 16 T to levitate a strawberry, a cricket, and even a frog. The diamagnetic repulsion of the water in these organisms is great enough, in a field that strong, to counteract the gravitational force of the Earth. The researchers reported that the living frog showed no visible signs of discomfort during its levitation, and that it hopped away normally after the experiment.

REAL-WORLD PHYSICS
Magnetic levitation

A similar diamagnetic effect can be used to levitate a small magnet between a person's fingertips. In the case shown in the photograph, a powerful magnet about 8 ft above the person's hand applies a magnetic field strong enough to counteract the weight of the small magnet. By itself, this system of two magnets is not stable—if the small magnet is displaced upward slightly, the increased attractive force of the upper magnet raises it farther; if it is displaced downward, the force of gravity pulls it down even farther. With the diamagnetic effect of the fingers, however, the small magnet is stabilized. If it is displaced upward now, the repulsive diamagnetic effect of the finger pushes it back down; if it is displaced downward, the diamagnetic repulsion of the thumb pushes it back up. The result is a stable, magnetic levitation—with no smoke or mirrors.

Chapter Summary

22–1 The Magnetic Field

The magnetic field gives an indication of the effect a magnet will have in a given region.

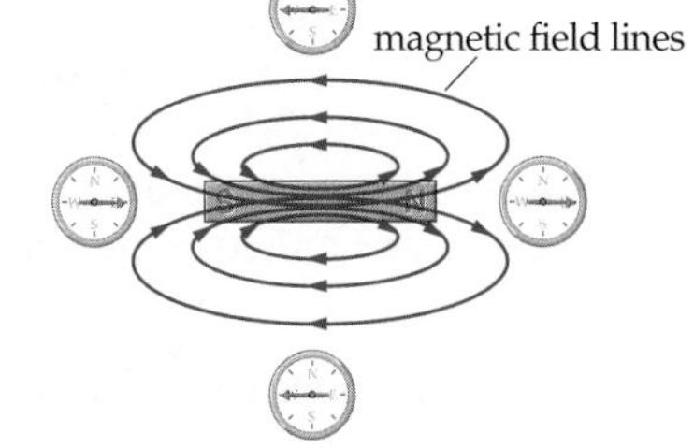

Magnetic Poles
A magnet is characterized by two poles, referred to as the north pole and the south pole. These poles cannot be separated—all magnets have both poles.

Magnetic Field Lines
Magnetic fields can be represented with lines in much the same way as electric fields can be portrayed. In particular, the more closely spaced the lines, the more intense the field. Magnetic field lines, which point away from north poles and toward south poles, always form closed loops.

Geomagnetism
The Earth produces its own magnetic field, which is inclined at an angle of about 11.5° with its rotational axis. The geographic north pole of the Earth is actually the south magnetic pole of the Earth's magnetic field.

22–2 The Magnetic Force on Moving Charges

In order for a magnetic field to exert a force on a particle, the particle must have charge and must be moving.

Magnitude of the Magnetic Force
The magnitude of the magnetic force is

$$F = |q|vB\sin\theta \qquad \textbf{22–1}$$

where q is the charge of the particle, v is its speed, B is the magnitude of the magnetic field, and θ is the angle between the velocity vector $\vec{\mathbf{v}}$ and the magnetic field vector $\vec{\mathbf{B}}$.

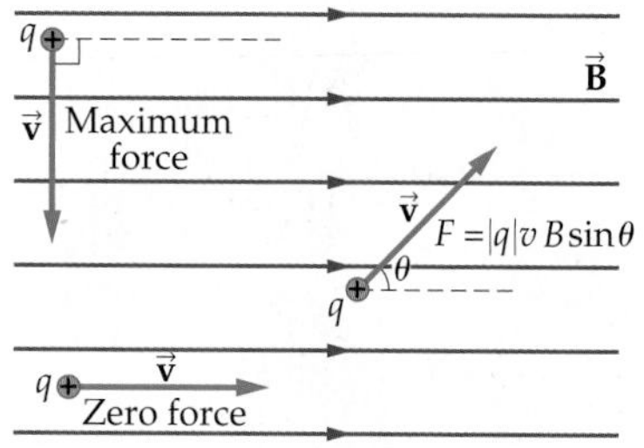

Magnetic Force Right-Hand Rule (RHR)
The magnetic force $\vec{\mathbf{F}}$ points in a direction that is perpendicular to both $\vec{\mathbf{B}}$ and $\vec{\mathbf{v}}$. For a positive charge, point the fingers of your right hand in the direction of $\vec{\mathbf{v}}$ and curl them toward the direction of $\vec{\mathbf{B}}$. Your thumb points in the direction of the force $\vec{\mathbf{F}}$. The force on a negative charge is in the opposite direction to that on a positive charge.

22–3 The Motion of Charged Particles in a Magnetic Field

The motion of a charged particle in a magnetic field is quite different from that in an electric field.

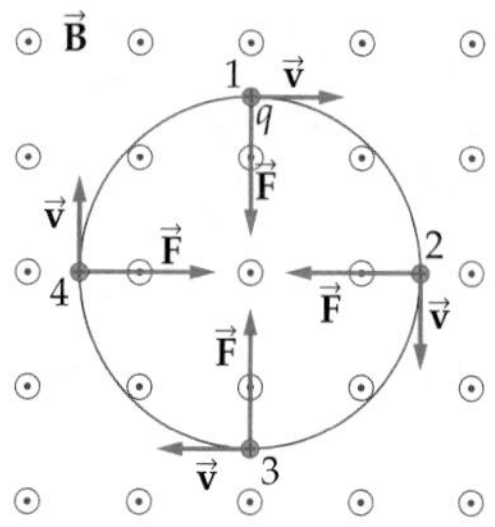

Electric versus Magnetic Forces
A charged particle in an electric field accelerates in the direction of the field; in a magnetic field the acceleration is perpendicular to the field and to the velocity. The electric field does work on a particle and changes its speed; a magnetic field does no work on a particle, and its speed remains constant.

Constant-Velocity Motion
If a charged particle moves parallel or antiparallel to a magnetic field, it experiences no force; hence, its velocity remains constant.

Circular Motion
If a charged particle moves perpendicular to a magnetic field, it will orbit with constant speed in a circle of radius $r = mv/|q|B$.

Helical Motion
When a particle's velocity has components both parallel and perpendicular to a magnetic field, it will follow a helical path.

22–4 The Magnetic Force Exerted on a Current-Carrying Wire

An electric current in a wire is caused by the movement of electric charges. Since moving electric charges experience magnetic forces, it follows that a current-carrying wire will as well.

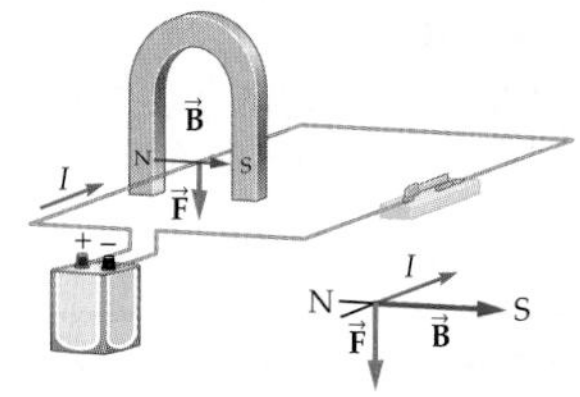

Force on a Current-Carrying Wire
A wire of length L carrying a current I at an angle θ to a magnetic field B experiences a force given by:

$$F = ILB\sin\theta \qquad \textbf{22–4}$$

22–5 Loops of Current and Magnetic Torque

A current loop placed in a magnetic field experiences a torque that depends on the relative orientation of the plane of the loop and the magnetic field.

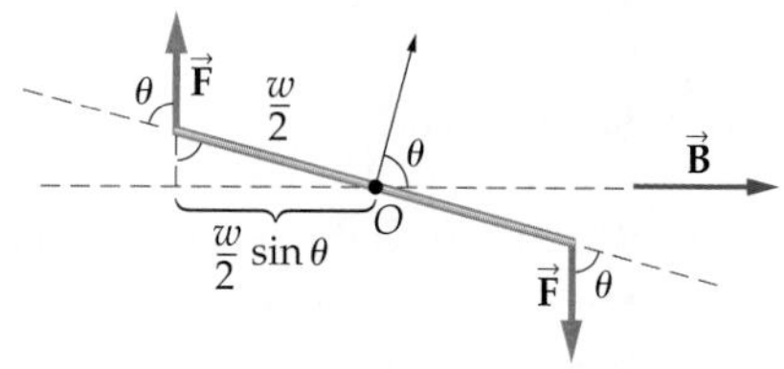

Torque on a General Loop
The magnetic torque exerted on a current loop is given by

$$\tau = NIAB\sin\theta \qquad \textbf{22–6}$$

where N is the number of turns around the loop, I is the current, A is the area of the loop, B is the strength of the magnetic field, and θ is the angle between the plane of the loop and the magnetic force.

22–6 Electric Currents, Magnetic Fields, and Ampère's Law

The key observation that serves to unify electricity and magnetism is that electric currents cause magnetic fields.

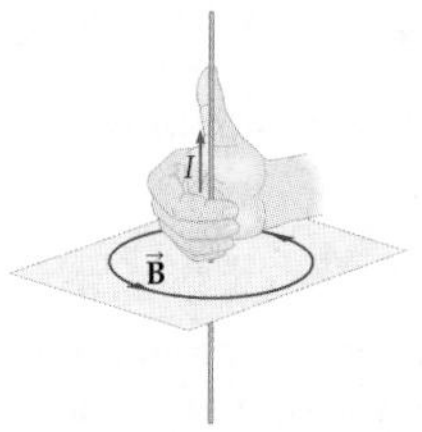

Magnetic Field Right-Hand Rule
The direction of the magnetic field produced by a current is found by pointing the thumb of the right hand in the direction of the current. The fingers of the right hand curl in the direction of the field.

Ampère's Law
Ampère's law can be expressed as follows:

$$\sum B_{\parallel}\,\Delta L = \mu_0 I_{\text{enclosed}} \qquad 22\text{–}7$$

where $B_{\parallel}$ is the component of the magnetic field parallel to a segment of a closed path of length ΔL, I is the current enclosed by the path, and $\mu_0 = 4\pi \times 10^{-7}\,\text{T}\cdot\text{m/A}$ is a constant called the permeability of free space.

Magnetic Field of a Long, Straight Wire
A long, straight wire carrying a current I produces a magnetic field of magnitude B given by

$$B = \frac{\mu_0 I}{2\pi r} \qquad 22\text{–}9$$

In this expression, r is the radial distance from the wire.

Forces between Current-Carrying Wires
Two wires, carrying the currents I_1 and I_2, exert forces on each other. If the wires are separated by a distance d, the force exerted on a length L is

$$F = \frac{\mu_0 I_1 I_2}{2\pi d} L \qquad 22\text{–}10$$

Wires that carry current in the same direction attract one another; wires with opposite-directed currents repel one another.

22–7 Current Loops and Solenoids

A single loop of current produces a magnetic field much like that of a permanent magnet. A succession of loops grouped together in a coil forms a solenoid.

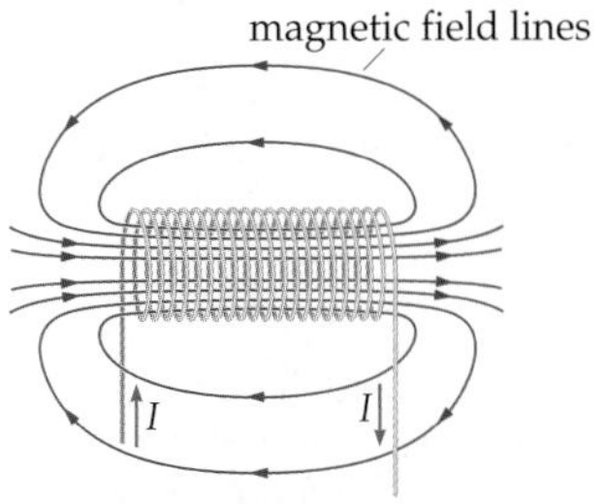

Current Loop
The magnetic field at the center of a current loop of N turns, radius R, and current I is

$$B = \frac{N\mu_0 I}{2R} \qquad 22\text{–}11$$

Magnetic Field of a Solenoid
The magnetic field inside a solenoid is nearly uniform and aligned along the solenoid's axis. If the solenoid has N loops in a length L and carries a current I, its magnetic field is

$$B = \mu_0\left(\frac{N}{L}\right)I = \mu_0 n I \qquad 22\text{–}12$$

In this expression, n is the number of loops per length; $n = N/L$. The magnetic field outside a solenoid is small, and in the ideal case can be considered to be zero.

22–8 Magnetism in Matter

The ultimate origin of the magnetic fields we observe around us is circulating electric currents on the atomic level.

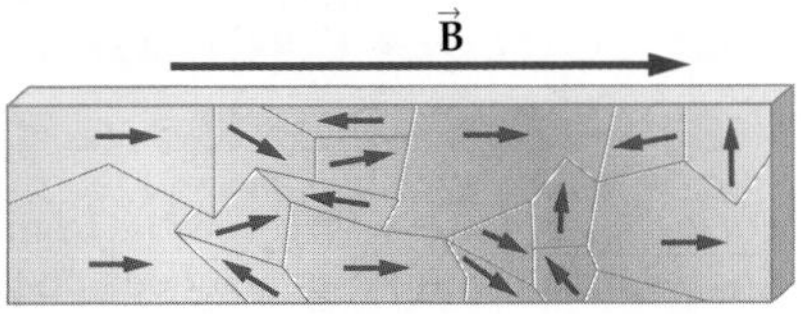

Paramagnetism
A paramagnetic material has no magnetic field unless an external magnetic field is applied to it. In this case, it develops a magnetization in the direction of the external field.

Ferromagnetism
A ferromagnetic material produces a magnetic field even in the absence of an external magnetic field. Permanent magnets are constructed of ferromagnetic materials.

Diamagnetism
Diamagnetism is the effect of the production by a material of a magnetic field in the opposite direction to an external magnetic field that is applied to it. All materials show at least a small diamagnetic effect.

Problem-Solving Summary

Type of Problem	Relevant Physical Concepts	Related Examples
Find the magnetic force exerted on a moving charge.	The magnitude of the magnetic force depends on the charge, the speed, the magnetic field, and the angle between the magnetic field and the velocity. The direction of the magnetic force is given by the right-hand rule.	Examples 22–1, 22–2
Determine the radius of the path followed by a charged particle in a magnetic field.	A charged particle moving at right angles to a uniform magnetic field moves on a circular path with a radius given by $r = mv/\|q\|B$.	Example 22–3
Calculate the magnetic force on a current-carrying wire.	If a wire segment of length L carries a current I at an angle θ to a magnetic field of strength B, the force it experiences is $F = ILB \sin \theta$.	Example 22–4
Find the torque exerted on a current loop.	If a current loop with current I, cross-sectional area A, and N turns is in a magnetic field B, the maximum torque is $\tau = NIAB$. The maximum torque occurs when the plane of the loop is parallel to the magnetic field ($\theta = 90°$). When the plane of the loop is at an angle θ to the magnetic force, the torque is $\tau = NIAB \sin \theta$.	Example 22–5
Determine the magnetic field produced by a long, straight wire carrying a current I.	The magnetic field produced by a long, straight wire with a current I circulates around the wire in a direction given by the right-hand rule. The magnitude of the magnetic field a radial distance r from the wire is $B = \mu_0 I/2\pi r$. This result follows from Ampère's law.	Example 22–6
Find the magnetic field inside a solenoid.	An ideal solenoid with n turns per length and current I produces a uniform magnetic field of magnitude $B = \mu_0 nI$. The field is parallel to the axis of the solenoid. In the ideal case, the magnetic field outside the solenoid is zero.	Example 22–7

Conceptual Questions

(Answers to odd-numbered Conceptual Questions can be found in the back of the book.)

1. Two charged particles move at right angles to a magnetic field and deflect in opposite directions. Can one conclude that the particles have opposite charges?
2. An electron moves with constant velocity through a region of space that is free of magnetic fields. Can one conclude that the electric field is zero in this region? Explain.
3. An electron moves with constant velocity through a region of space that is free of electric fields. Can one conclude that the magnetic field is zero in this region? Explain.
4. Describe how the motion of a charged particle can be used to distinguish between an electric and a magnetic field.
5. Explain how a charged particle moving in a circle of small radius can take the same amount of time to complete an orbit as an identical particle orbiting in a circle of large radius.
6. A current-carrying wire is placed in a region with a uniform magnetic field. The wire experiences zero magnetic force. Explain.

Conceptual Exercises

(Answers to odd-numbered Conceptual Exercises can be found in the back of the book.)

1. Proton 1 moves with a speed v from the east coast to the west coast in the continental United States; proton 2 moves with the same speed from the southern United States toward Canada. Is the magnitude of the magnetic force experienced by proton 2 greater than, less than, or equal to the force experienced by proton 1? Explain.
2. An electron moves west to east in the continental United States. Does the magnetic force experienced by the electron point in a direction that is generally north, south, east, west, upward, or downward? Explain.
3. Suppose particles A, B, and C in **Figure 22–31** have identical masses and charges of the same magnitude. Rank the particles in order of increasing speed. Indicate ties where appropriate.

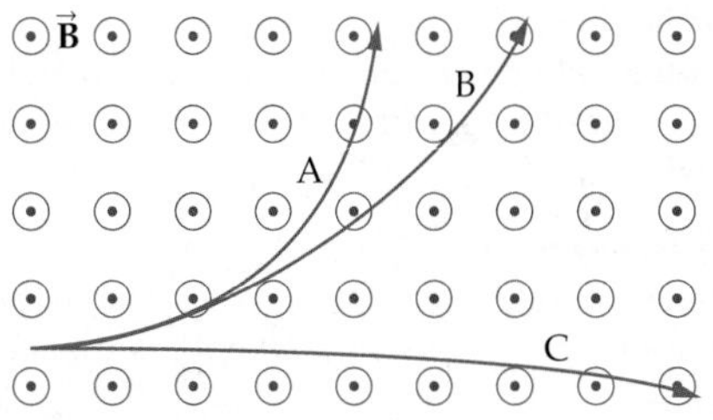

▲ **FIGURE 22–31** Conceptual Exercises 3, 6, and 7

4. At a point near the equator, the Earth's magnetic field is horizontal and points to the north. If an electron is moving vertically

upward at this point, does the magnetic force acting on it point north, south, east, west, upward, or downward? Explain.

5. A proton is to orbit the Earth at the equator using the Earth's magnetic field to supply part of the necessary centripetal force. Should the proton move eastward or westward? Explain.
6. Referring to Figure 22–31, what is the sign of the charge for each of the three particles? Explain.
7. Suppose the three particles in Figure 22–31 have the same mass and speed. Rank the particles in order of increasing magnitude of their charge. Indicate ties where appropriate.
8. A velocity selector is to be constructed using a magnetic field in the positive y direction. If positively charged particles move through the selector in the positive z direction, **(a)** what must be the direction of the electric field? **(b)** Repeat part (a) for the case of negatively charged particles.
9. An electron moving in the positive x direction, at right angles to a magnetic field, experiences a magnetic force in the positive y direction. What is the direction of the magnetic field?
10. A positively charged particle moves through a region with a uniform electric field pointing toward the top of the page and a uniform magnetic field pointing into the page. The particle can have one of the four velocities shown in **Figure 22–32**. **(a)** Rank the four possibilities in order of increasing magnitude of the net force (F_1, F_2, F_3, and F_4) the particle experiences. Indicate ties where appropriate. **(b)** Which of the four velocities could potentially result in zero net force?

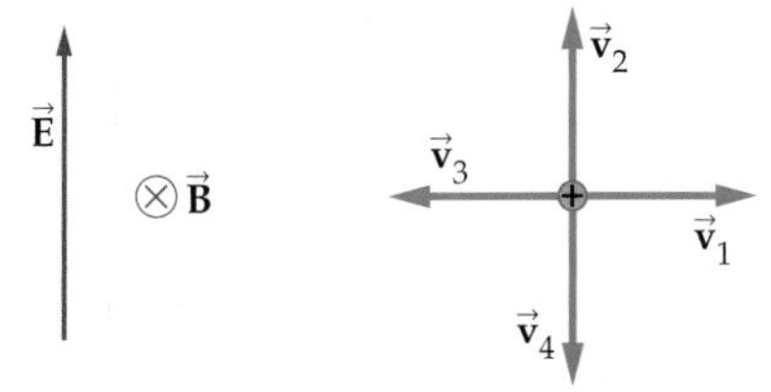

▲ **FIGURE 22–32** Conceptual Exercises 10 and 11

11. Suppose the fields in Figure 22–32 are interchanged, with the magnetic field pointing toward the top of the page and the electric field pointing into the page. **(a)** Rank the four possibilities in order of increasing magnitude of the net force (F_1, F_2, F_3, and F_4) the particle experiences. Indicate ties where appropriate. **(b)** Which of the four velocities could potentially result in zero net force?
12. The accompanying photograph shows an electron beam whose initial direction of motion is horizontal, from right to left. A magnetic field deflects the beam downward. What is the direction of the magnetic field?

A horizontal electron beam is deflected downward by a magnetic field. (Conceptual Exercise 12)

13. A proton follows the path shown in **Figure 22–33** as it moves through three regions with different uniform magnetic fields, B_1, B_2, and B_3. In each region the proton completes a half-circle, and the magnetic field is perpendicular to the page. **(a)** Rank the three fields in order of increasing magnitude. Indicate ties where appropriate. **(b)** Give the direction (into or out of the page) for each of the fields.

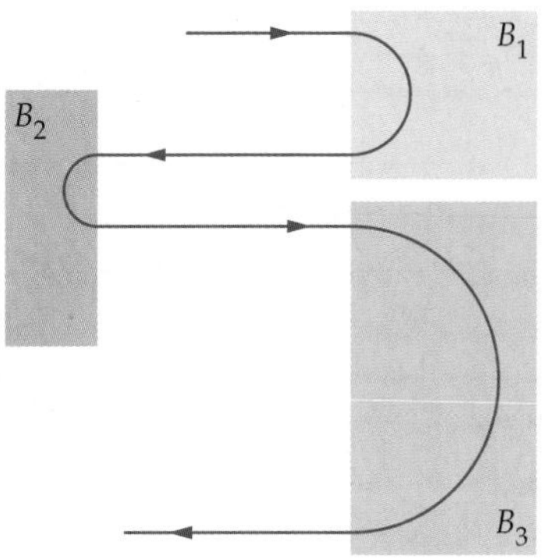

▲ **FIGURE 22–33** Conceptual Exercises 13 and 14

14. Suppose the speed of the proton in Figure 22–33 is increased. **(a)** Does the radius of each half-circular path segment increase, decrease, or stay the same? Explain. **(b)** Does the time spent in each of the field regions increase, decrease, or stay the same? Explain.
15. The three wires shown in **Figure 22–34** are long and straight, and they each carry a current of the same magnitude, I. The currents in wires 1 and 3 are out of the page; the current in wire 2 is into the page. What is the direction of the magnetic force experienced by wire 3?

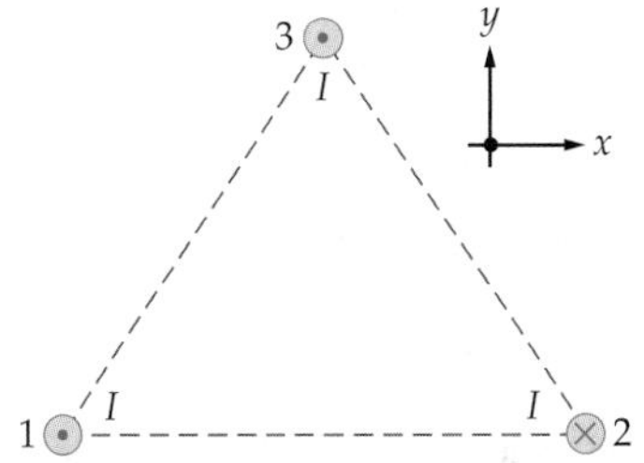

▲ **FIGURE 22–34** Conceptual Exercises 15 and 16

16. Each of the current-carrying wires in Figure 22–34 is long and straight, and carries the current I either into or out of the page, as shown. What is the direction of the net magnetic field produced by these three wires at the center of the triangle?
17. The four wires shown in **Figure 22–35** are long and straight, and they each carry a current of the same magnitude, I. The currents in wires 1, 2, and 3 are out of the page; the current in wire 4 is into the page. What is the direction of the magnetic force experienced by wire 2?

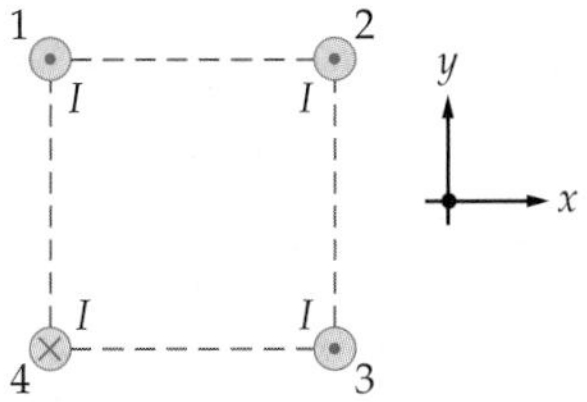

▲ **FIGURE 22–35** Conceptual Exercises 17 and 18

18. Each of the current-carrying wires in Figure 22–35 is long and straight, and carries the current I either into or out of the page, as shown. What is the direction of the net magnetic field produced by these four wires at the center of the square?
19. For each of the square current loops shown in **Figure 22–36**, indicate whether there will be a tendency for the loop to rotate clockwise, counterclockwise, or not at all, when viewed from above along the indicated axis.

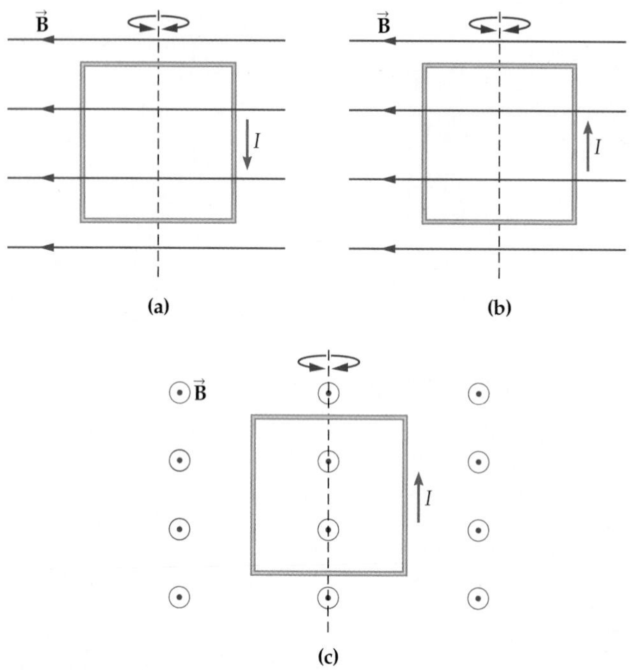

▲ **FIGURE 22–36** Conceptual Exercise 19 and Problems 26 and 35

20. A loop of wire is connected to the terminals of a battery, as indicated in **Figure 22–37**. If the loop is to attract the bar magnet, which of the terminals, A or B, should be the positive terminal of the battery? Explain.

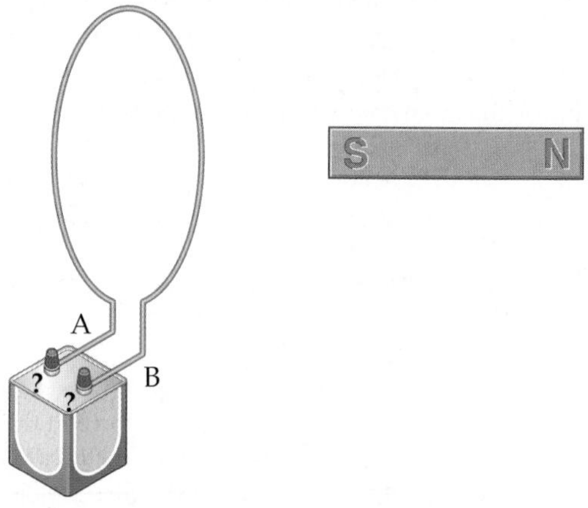

▲ **FIGURE 22–37** Conceptual Exercise 20

21. The number of turns in a solenoid is doubled, and at the same time its length is doubled. Does the magnetic field within the solenoid increase, decrease, or stay the same? Explain.

Problems

Note: **IP** *denotes an integrated conceptual/quantitative problem.* **BIO** *identifies problems of biological or medical interest. Red bullets (•, ••, •••) are used to indicate the level of difficulty of each problem.*

Section 22–2 The Magnetic Force on Moving Charges

1. • What is the acceleration of a proton moving with a speed of 9.5 m/s at right angles to a magnetic field of 1.6 T?

2. • An electron moves at right angles to a magnetic field of 0.12 T. What is its speed if the force exerted on it is 8.9×10^{-15} N?

3. • A negatively charged ion moves due north with a speed of 1.5×10^6 m/s at the Earth's equator. What is the magnetic force exerted on this ion?

4. • A proton high above the equator approaches the Earth moving straight downward with a speed of 355 m/s. Find the acceleration of the proton, given that the magnetic field at its altitude is 4.05×10^{-5} T.

5. •• A 0.32-μC particle moves with a speed of 16 m/s through a region where the magnetic field has a strength of 0.95 T. At what angle to the field is the particle moving if the force exerted on it is **(a)** 4.8×10^{-6} N, **(b)** 3.0×10^{-6} N, or **(c)** 1.0×10^{-7} N?

6. •• A particle with a charge of 14 μC experiences a force of 2.2×10^{-4} N when it moves at right angles to a magnetic field with a speed of 27 m/s. What force does this particle experience when it moves with a speed of 6.3 m/s at an angle of 25° relative to the magnetic field?

7. •• An ion experiences a magnetic force of 6.2×10^{-16} N when moving in the positive x direction but no magnetic force when moving in the positive y direction. What is the magnitude of the magnetic force exerted on the ion when it moves in the x-y plane along the line $x = y$? Assume that the ion's speed is the same in all cases.

8. •• An electron moving with a speed of 9.1×10^5 m/s in the positive x direction experiences zero magnetic force. When it moves in the positive y direction, it experiences a force of 2.0×10^{-13} N that points in the negative z direction. What is the direction and magnitude of the magnetic field?

9. •• **IP** Two charged particles with different speeds move one at a time through a region of uniform magnetic field. The particles move in the same direction and experience equal magnetic forces. **(a)** If particle 1 has four times the charge of particle 2, which particle has the greater speed? Explain. **(b)** Find the ratio of the speeds, v_1/v_2.

10. •• A 6.60-μC particle moves through a region of space where an electric field of magnitude 1250 N/C points in the positive x direction, and a magnetic field of magnitude 1.02 T points in the positive z direction. If the net force acting on the particle is 6.23×10^{-3} N in the positive x direction, find the magnitude and direction of the particle's velocity. Assume the particle's velocity is in the x-y plane.

11. ••• When at rest, a proton experiences a net electromagnetic force of magnitude 8.0×10^{-13} N pointing in the positive x direction. When the proton moves with a speed of 1.5×10^6 m/s in the positive y direction, the net electromagnetic force on it decreases in magnitude to 7.5×10^{-13} N, still pointing in the positive x direction. Find the magnitude and direction of **(a)** the electric field and **(b)** the magnetic field.

Section 22–3 The Motion of Charged Particles in a Magnetic Field

12. • Find the radius of an electron's orbit when it moves perpendicular to a magnetic field of 0.45 T with a speed of 6.27×10^5 m/s.

13. • Find the radius of a proton's orbit when it moves perpendicular to a magnetic field of 0.45 T with a speed of 6.27×10^5 m/s.

14. • Charged particles pass through a velocity selector with electric and magnetic fields at right angles to each other, as shown in

Figure 22–38. If the electric field has a magnitude of 450 N/C and the magnetic field has a magnitude of 0.18 T, what speed must the particles have to pass through the selector undeflected?

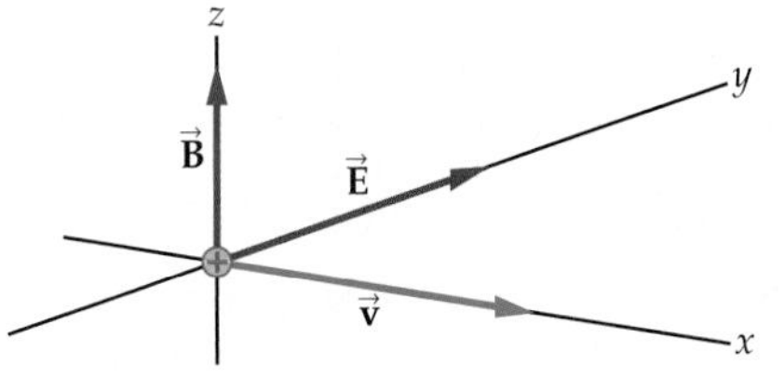

▲ **FIGURE 22–38** Problem 14

15. • The velocity selector in **Figure 22–39** is designed to allow charged particles with a speed of 4.5×10^3 m/s to pass through undeflected. Find the direction and magnitude of the required electric field, given that the magnetic field has a magnitude of 0.96 T.

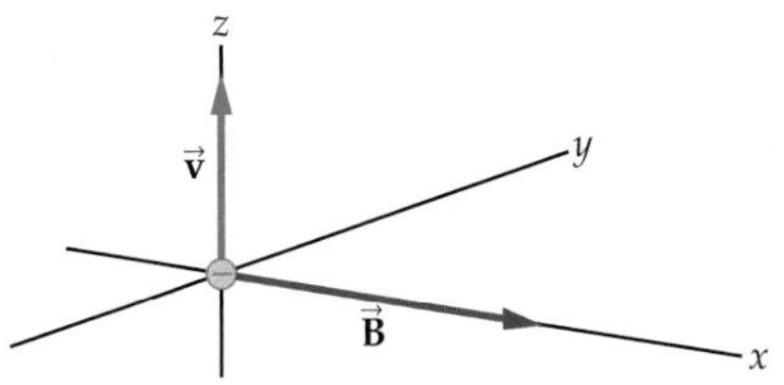

▲ **FIGURE 22–39** Problem 15

16. •• **IP BIO** The artery in Figure 22–11 has an inside diameter of 2.75 mm and passes through a region where the magnetic field is 0.065 T. **(a)** If the voltage difference between the electrodes is 195 μV, what is the speed of the blood? **(b)** Which electrode is at the higher potential? Does your answer depend on the sign of the ions in the blood? Explain.

17. •• An electron accelerated from rest through a voltage of 410 V enters a region of constant magnetic field. If the electron follows a circular path with a radius of 17 cm, what is the magnitude of the magnetic field?

18. •• A 12.5-μC particle with a mass of 2.80×10^{-5} kg moves perpendicular to a 1.01-T magnetic field in a circular path of radius 26.8 m. **(a)** How fast is the particle moving? **(b)** How long will it take the particle to complete one orbit?

19. •• **IP** When a charged particle enters a region of uniform magnetic field, it follows a circular path, as indicated in **Figure 22–40**. **(a)** Is this particle positively or negatively charged? Explain. **(b)** Suppose that the magnetic field has a magnitude of 0.180 T, the particle's speed is 6.0×10^6 m/s, and the radius of its path is 52.0 cm. Find the mass of the particle, given that its charge has a magnitude of 1.60×10^{-19} C. Give your result in atomic mass units, u, where 1 u = 1.67×10^{-27} kg.

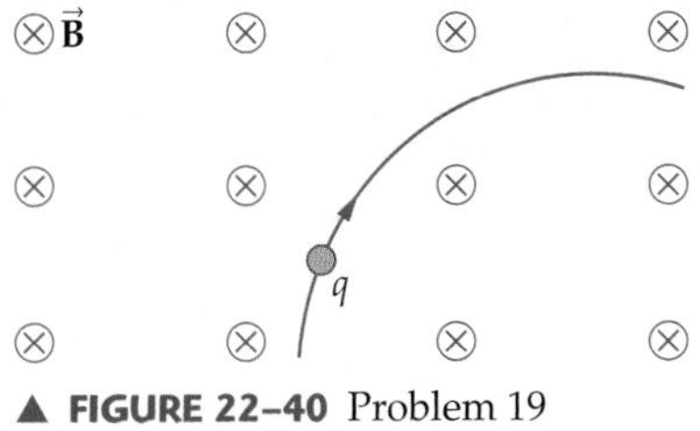

▲ **FIGURE 22–40** Problem 19

20. •• A proton with a kinetic energy of 4.9×10^{-16} J moves perpendicular to a magnetic field of 0.26 T. What is the radius of its circular path?

21. •• **IP** An alpha particle (the nucleus of a helium atom) consists of two protons and two neutrons, and has a mass of 6.64×10^{-27} kg. A horizontal beam of alpha particles is injected with a speed of 1.3×10^5 m/s into a region with a vertical magnetic field of magnitude 0.155 T. **(a)** How long does it take for an alpha particle to move halfway through a complete circle? **(b)** If the speed of the alpha particle is doubled, does the time found in part (a) increase, decrease, or stay the same? Explain. **(c)** Repeat part (a) for alpha particles with a speed of 2.6×10^5 m/s.

22. ••• An electron and a proton move in circular orbits in a plane perpendicular to a uniform magnetic field $\vec{B}$. Find the ratio of the radii of their circular orbits when the electron and the proton have **(a)** the same momentum and **(b)** the same kinetic energy.

Section 22–4 The Magnetic Force Exerted on a Current-Carrying Wire

23. • What is the magnetic force exerted on a 2.15-m length of wire carrying a current of 0.695 A perpendicular to a magnetic field of 0.720 T?

24. • A wire with a current of 2.8 A is at an angle of 36.0° relative to a magnetic field of 0.88 T. Find the force exerted on a 2.25-m length of the wire.

25. • The magnetic force exerted on a 1.2-m segment of straight wire is 1.6 N. The wire carries a current of 3.0 A in a region with a constant magnetic field of 0.50 T. What is the angle between the wire and the magnetic field?

26. • Consider a current loop immersed in a magnetic field, as in Figure 22–36 (a) (Conceptual Exercise 19). It is given that B = 0.34 T and I = 9.5 A. In addition, the loop is a square 0.46 m on a side. Find the magnitude of the magnetic force exerted on each side of the loop.

27. •• A 0.45-m copper rod with a mass of 0.17 kg carries a current of 11 A in the positive x direction. What are the magnitude and direction of the minimum magnetic field needed to levitate the rod?

28. •• The long, thin wire shown in **Figure 22–41** is in a region of constant magnetic field $\vec{B}$. The wire carries a current of 6.2 A and is oriented at an angle of 7.5° to the direction of the magnetic field. **(a)** If the magnetic force exerted on this wire per meter is 0.033 N, what is the magnitude of the magnetic field? **(b)** At what angle will the force exerted on the wire per meter be equal to 0.015 N?

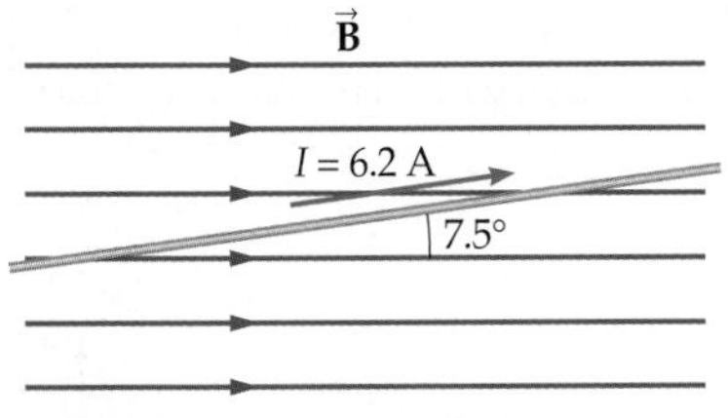

▲ **FIGURE 22–41** Problem 28

29. •• A wire with a length of 2.7 m and a mass of 0.75 kg is in a region of space with a magnetic field of 0.84 T. What is the minimum current needed to levitate the wire?

30. •• A high-voltage power line carries a current of 110 A at a location where the Earth's magnetic field has a magnitude of 0.59 G and points to the north, 72° below the horizontal. Find the direction and magnitude of the magnetic force exerted on a 250-m length of wire if the current in the wire flows **(a)** horizontally toward the east or **(b)** horizontally toward the south.

31. ••• A metal bar of mass m and length L is suspended from two conducting wires, as shown in **Figure 22–42**. A uniform magnetic field of magnitude B points vertically downward. Find the angle θ the suspending wires make with the vertical when the bar carries a current I.

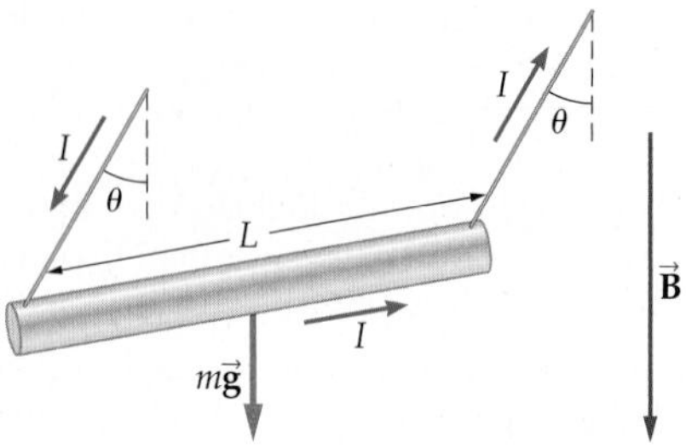

▲ **FIGURE 22–42** Problem 31

Section 22–5 Loops of Current and Magnetic Torque

32. • A rectangular loop of 260 turns is 22 cm wide and 16 cm high. What is the current in this loop if the maximum torque in a field of 0.48 T is 23 N · m?

33. • A single circular loop of radius 0.23 m carries a current of 2.6 A in a magnetic field of 0.95 T. What is the maximum torque exerted on this loop?

34. •• In the previous problem, find the angle the plane of the loop must make with the field if the torque is to be half its maximum value.

35. •• Consider a current loop in a region of uniform magnetic field, as shown in Figure 22–36 (a) (Conceptual Exercise 19). Find the magnitude of the torque exerted on the loop about the vertical axis of rotation, using the data given in Problem 26.

36. •• **IP** Two current loops, one square the other circular, have one turn made from wires of the same length. **(a)** If these loops carry the same current and are placed in magnetic fields of equal magnitude, is the maximum torque of the square loop greater than, less than, or the same as the maximum torque of the circular loop? Explain. **(b)** Calculate the ratio of the maximum torques, $\tau_{\text{square}}/\tau_{\text{circle}}$.

37. ••• **IP** Each of the 10 turns of wire in a vertical, rectangular loop carries a current of 0.22 A. The loop has a height of 8.0 cm and a width of 15 cm. A horizontal magnetic field of magnitude 0.050 T is oriented at an angle of $\theta = 65°$ relative to the normal to the plane of the loop, as indicated in **Figure 22–43**. Find **(a)** the magnetic force on each side of the loop, **(b)** the net magnetic force on the loop, and **(c)** the magnetic torque on the loop. **(d)** If the loop can rotate about a vertical axis with only a small amount of friction, will it end up with an orientation given by $\theta = 0$, $\theta = 90°$, or $\theta = 180°$? Explain.

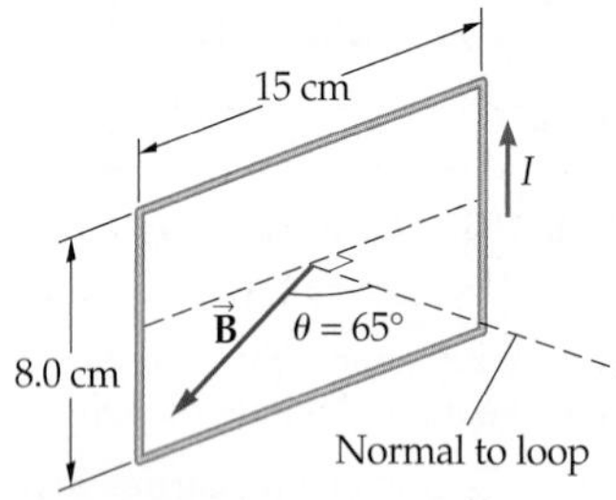

▲ **FIGURE 22–43** Problem 37

Section 22–6 Electric Currents, Magnetic Fields, and Ampère's Law

38. • Find the magnetic field 6.25 cm from a long, straight wire that carries a current of 5.81 A.

39. • A long, straight wire carries a current of 7.2 A. How far from this wire is the magnetic field it produces equal to the Earth's magnetic field, which is approximately 5.0×10^{-5} T?

40. • You travel to the north magnetic pole of the Earth, where the magnetic field points vertically downward. There, you draw a circle on the ground. Applying Ampère's law to this circle, show that zero current passes through its area.

41. • Two power lines, each 270 m in length, run parallel to each other with a separation of 25 cm. If the lines carry parallel currents of 110 A, what is the magnitude and direction of the magnetic force each exerts on the other?

42. • **BIO Pacemaker Switches** Some pacemakers employ magnetic reed switches to enable doctors to change their mode of operation without surgery. A typical reed switch can be switched from one position to another with a magnetic field of 5.0×10^{-4} T. What current must a wire carry if it is to produce a 5.0×10^{-4} T field at a distance of 0.50 m?

43. •• **IP** Consider the long, straight, current-carrying wires shown in **Figure 22–44**. One wire carries a current of 6.2 A in the positive y direction; the other wire carries a current of 4.5 A in the positive x direction. **(a)** At which of the two points, A or B, do you expect the magnitude of the net magnetic field to be greater? Explain. **(b)** Calculate the magnitude of the net magnetic field at points A and B.

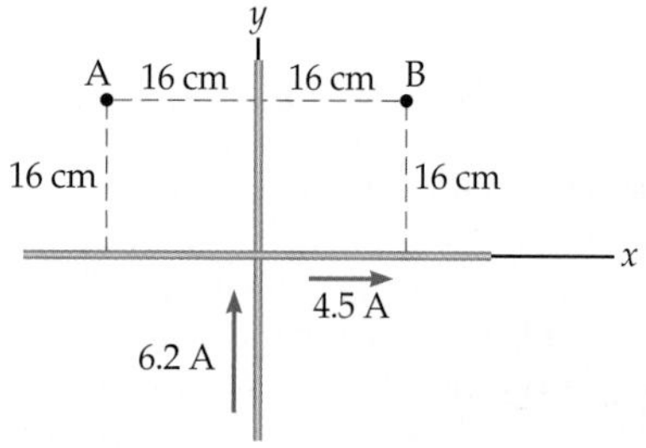

▲ **FIGURE 22–44** Problems 43 and 44

44. •• **IP** Repeat Problem 43 for the case where the 6.2-A current is reversed in direction.

45. •• In Oersted's experiment, suppose that the compass was 0.25 m from the current-carrying wire. If a magnetic field of half the Earth's magnetic field of 5.0×10^{-5} T was required to give a noticeable deflection of the compass needle, what current must the wire have carried?

46. •• **IP** Two long, straight wires are separated by a distance of 12.2 cm. One wire carries a current of 2.75 A, the other carries a current of 4.33 A. **(a)** Find the force per meter exerted on the 2.75-A wire. **(b)** Is the force per meter exerted on the 4.33-A wire greater than, less than, or the same as the force per meter exerted on the 2.75-A wire? Explain.

47. ••• Two long, straight wires are oriented perpendicular to the page, as shown in **Figure 22–45**. The current in one wire is $I_1 = 3.0$ A, pointing into the page, and the current in the other wire is $I_2 = 4.0$ A, pointing out of the page. Find the magnitude and direction of the net magnetic field at point P.

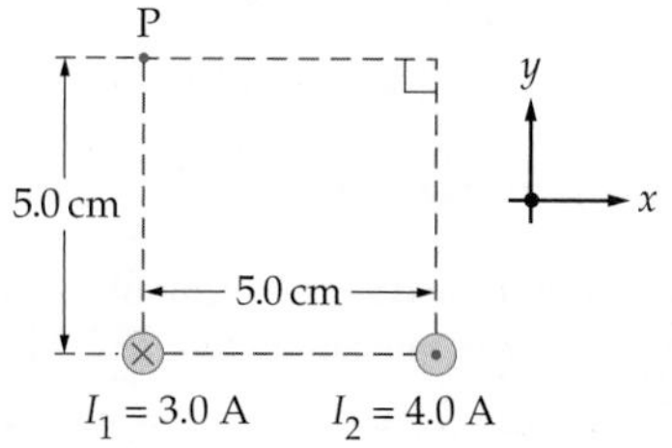

▲ **FIGURE 22–45** Problems 47 and 85

Section 22–7 Current Loops and Solenoids

48. • It is desired that a solenoid 38 cm long and with 430 turns produce a magnetic field within it equal to the Earth's magnetic field (5.0×10^{-5} T). What current is required?

49. • A solenoid that is 75 cm long produces a magnetic field of 1.3 T within its core when it carries a current of 8.4 A. How many turns of wire are contained in this solenoid?

50. • The maximum current in a superconducting solenoid can be as large as 3.75 kA. If the number of turns per meter in such a solenoid is 3250, what is the magnitude of the magnetic field it produces?

51. •• To construct a solenoid, you wrap insulated wire uniformly around a plastic tube 12 cm in diameter and 55 cm in length. You would like a 2.0-A current to produce a 2.5-kG magnetic field inside your solenoid. What is the total length of wire you will need to meet these specifications?

General Problems

52. • A stationary proton ($q = 1.60 \times 10^{-19}$ C) is located between the poles of a horseshoe magnet, where the magnetic field is 0.15 T. What is the magnitude of the magnetic force acting on the proton?

53. • **BIO Brain Function and Magnetic Fields** Experiments have shown that thought processes in the brain can be affected if the parietal lobe is exposed to a magnetic field with a strength of 1.0 T. How much current must a long, straight wire carry if it is to produce a 1.0-T magnetic field at a distance of 0.50 m? (For comparison, a typical lightning bolt carries a current of about 20,000 A, which would melt most wires.)

54. • **BIO Magnetoencephalography** Magnetoencephalography (MEG) is a noninvasive method for studying electrical activity in the brain. It is based on the principle that an electric current creates a magnetic field, whether the current is in a wire or in the neurons of a brain. Thus, your very thoughts at this moment are generating measurable magnetic fields outside your head. With today's technology, an MEG can detect magnetic fields as small as 1.0×10^{-15} T. Approximating a neuron by a straight wire, what electric current is needed to produce this magnetic field strength at a distance of 5.0 cm?

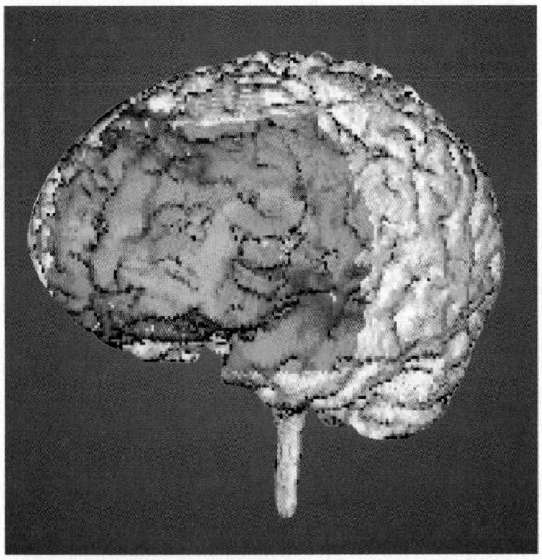

The magnetic fields produced by the human brain can be visualized in a procedure known as magnetoencephalography (MEG). (Problem 54)

55. • A mixture of two isotopes is injected into a mass spectrometer. One isotope follows a curved path of radius $R_1 = 48.9$ cm; the other follows a curved path of radius $R_2 = 51.7$ cm. Find the mass ratio, m_1/m_2, assuming that the two isotopes have the same charge and speed.

56. • High above the surface of the Earth, charged particles (such as electrons and protons) can become trapped in the Earth's magnetic field in regions known as Van Allen belts. A typical electron in a Van Allen belt has an energy of 45 keV and travels in a roughly circular orbit with an average radius of 220 m. What is the magnitude of the Earth's magnetic field where such an electron orbits?

57. • **Credit-Card Magnetic Strips** Experiments carried out on the television show *Mythbusters* determined that a magnetic field of 1000 gauss is needed to corrupt the information on a credit card's magnetic strip. (They also busted the myth that a credit card can be demagnetized by an electric eel or an eelskin wallet.) Suppose a long, straight wire carries a current of 2.5 A. How close can a credit card be held to this wire without damaging its magnetic strip?

58. • **Superconducting Solenoid** Cryomagnetics, Inc. advertises a high-field, superconducting solenoid that produces a magnetic field of 17 T with a current of 105 A. What is the number of turns per meter in this solenoid?

59. •• **BIO Magnetic Resonance Imaging** An MRI (magnetic resonance imaging) solenoid produces a magnetic field of 1.5 T. The solenoid is 2.5 m long, 1.0 m in diameter, and wound with insulated wires 2.2 mm in diameter. Find the current that flows in the solenoid. (Your answer should be rather large. A typical MRI solenoid uses niobium–titanium wire kept at liquid helium temperatures, where it is superconducting.)

60. •• **IP** A long, straight wire carries a current of 14 A. Next to the wire is a square loop with sides 1.0 m in length, as shown in **Figure 22–46**. The loop carries a current of 2.5 A in the direction indicated. **(a)** What is the direction of the net force exerted on the loop? Explain. **(b)** Calculate the magnitude of the net force acting on the loop.

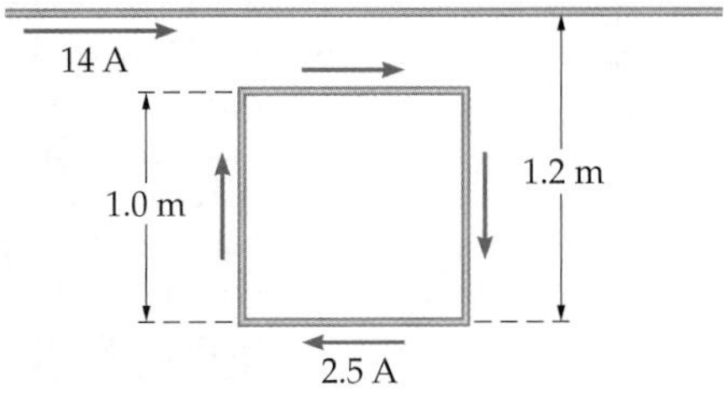

▲ **FIGURE 22–46** Problems 60 and 61

61. •• Suppose the 14-A current in the straight wire in Figure 22–46 is reversed in direction, but the current in the loop is unchanged. **(a)** Calculate the magnitude and direction of the net force acting on the loop. **(b)** If the loop is extended in the horizontal direction, so that it is 1.0 m high and 2.0 m wide, does the net force exerted on the loop increase or decrease? By what factor? Explain. **(c)** If, instead, the loop is extended in the vertical direction, so it is 2.0 m high and 1.0 m wide, does the net force exerted on the loop increase or decrease? Explain.

62. •• A charged particle moves through a region of space containing both electric and magnetic fields. The velocity of the particle is $\vec{\mathbf{v}} = (4.4 \times 10^3 \text{ m/s})\hat{\mathbf{x}} + (2.7 \times 10^3 \text{ m/s})\hat{\mathbf{y}}$ and the magnetic field is $\vec{\mathbf{B}} = (0.73 \text{ T})\hat{\mathbf{z}}$. Find the electric field vector $\vec{\mathbf{E}}$ necessary to yield zero net force on the particle. (*Note:* You may wish to use cross products in this problem. They are discussed in Appendix A.)

63. •• **IP Medical X-rays** An electron in a medical X-ray machine is accelerated from rest through a voltage of 10.0 kV. **(a)** Find the

maximum force a magnetic field of 0.957 T can exert on this electron. **(b)** If the voltage of the X-ray machine is increased, does the maximum force found in part (a) increase, decrease, or stay the same? Explain. **(c)** Repeat part (a) for an electron accelerated through a potential of 25.0 kV.

64. •• A particle with a charge of 34 μC moves with a speed of 62 m/s in the positive x direction. The magnetic field in this region of space has a component of 0.40 T in the positive y direction, and a component of 0.85 T in the positive z direction. What are the magnitude and direction of the magnetic force on the particle?

65. •• **IP** A beam of protons with various speeds is directed in the positive x direction. The beam enters a region with a uniform magnetic field of magnitude 0.52 T pointing in the negative z direction, as indicated in **Figure 22–47**. It is desired to use a uniform electric field (in addition to the magnetic field) to select from this beam only those protons with a speed of 1.42×10^5 m/s—that is, only these protons should be undeflected by the two fields. **(a)** Determine the magnitude and direction of the electric field that yields the desired result. **(b)** Suppose the electric field is to be produced by a parallel-plate capacitor with a plate separation of 2.5 cm. What potential difference is required between the plates? **(c)** Which plate in Figure 22–47 (top or bottom) should be positively charged? Explain.

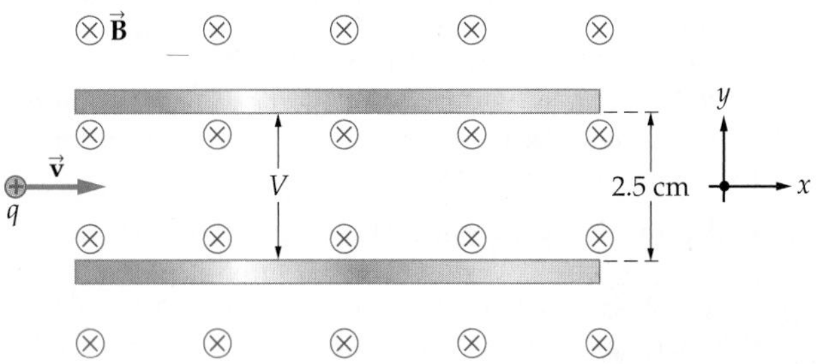

▲ **FIGURE 22–47** Problem 65

66. •• **IP** A charged particle moves in a horizontal plane with a speed of 8.70×10^6 m/s. When this particle encounters a uniform magnetic field in the vertical direction it begins to move on a circular path of radius 15.9 cm. **(a)** If the magnitude of the magnetic field is 1.21 T, what is the charge-to-mass ratio (q/m) of this particle? **(b)** If the radius of the circular path were greater than 15.9 cm, would the corresponding charge to mass ratio be greater than, less than, or the same as that found in part (a)? Explain. (Assume that the magnetic field remains the same.)

67. •• Two parallel wires, each carrying a current of 2.2 A in the same direction, are shown in **Figure 22–48**. Find the direction and magnitude of the net magnetic field at points A, B, and C.

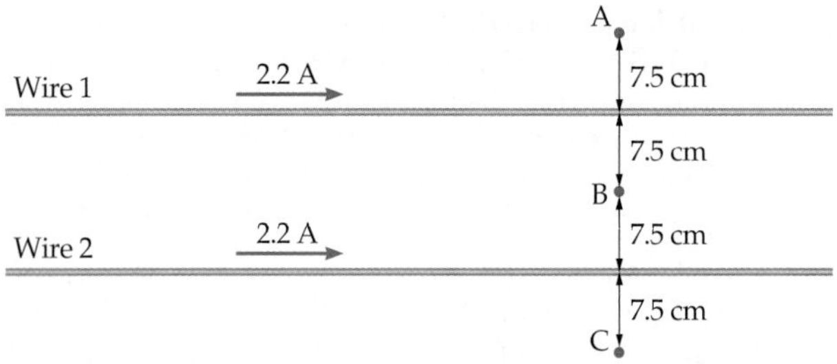

▲ **FIGURE 22–48** Problems 67 and 68

68. •• Repeat Problem 67 for the case where the current in wire 1 is reversed in direction.

69. •• **Lightning Bolts** A powerful bolt of lightning can carry a current of 225 kA. **(a)** Treating a lightning bolt as a long, thin wire, calculate the magnitude of the magnetic field produced by such a bolt of lightning at a distance of 35 m. **(b)** If two such bolts strike simultaneously at a distance of 35 m from each other, what is the magnetic force per meter exerted by one bolt on the other?

70. •• **IP** Consider the two current-carrying wires shown in **Figure 22–49**. The current in wire 1 is 3.7 A; the current in wire 2 is adjusted to make the net magnetic field at point A equal to zero. **(a)** Is the magnitude of the current in wire 2 greater than, less than, or the same as that in wire 1? Explain. **(b)** Find the magnitude and direction of the current in wire 2.

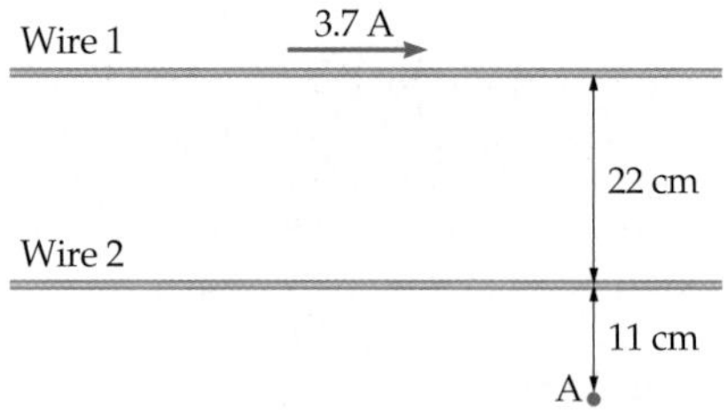

▲ **FIGURE 22–49** Problems 70 and 71

71. •• **IP** Consider the physical system shown in Figure 22–49, which consists of two current-carrying wires each with a length of 71 cm. **(a)** If the net magnetic field at the point A is out of the page, is the force between the wires attractive or repulsive? Explain. **(b)** Calculate the magnitude of the force exerted by each wire on the other wire, given that the magnetic field at point A is out of the page with a magnitude of 2.1×10^{-6} T.

72. •• **Magnetars** The astronomical object 4U0142 + 61 has the distinction of creating the most powerful magnetic field ever observed. This object is referred to as a "magnetar" (a subclass of pulsars), and its magnetic field is 1.3×10^{15} times greater than the Earth's magnetic field. **(a)** Suppose a 2.5-m straight wire carrying a current of 1.1 A is placed in this magnetic field at an angle of 65° to the field lines. What force does this wire experience? **(b)** A field this strong can significantly change the behavior of an atom. To see this, consider an electron moving with a speed of 2.2×10^6 m/s. Compare the maximum magnetic force exerted on the electron to the electric force a proton exerts on an electron in a hydrogen atom. The radius of the hydrogen atom is 5.29×10^{-11} m.

73. •• Consider a system consisting of two concentric solenoids, as illustrated in **Figure 22–50**. The current in the outer solenoid is $I_1 = 1.25$ A, and the current in the inner solenoid is $I_2 = 2.17$ A. Given that the number of turns per centimeter is 105 for the outer solenoid and 125 for the inner solenoid, find the magnitude and direction of magnetic field **(a)** between the solenoids and **(b)** inside the inner solenoid.

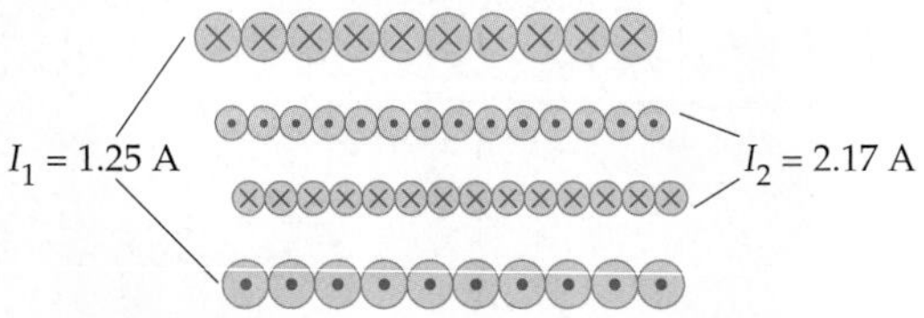

▲ **FIGURE 22–50** Problem 73

74. •• **IP** A long, straight wire on the x axis carries a current of 3.12 A in the positive x direction. The magnetic field produced by the wire combines with a uniform magnetic field of 1.45×10^{-6} T that points in the positive z direction. **(a)** Is the net magnetic field of this system equal to zero at a point on the positive y axis or at a point on the negative y axis? Explain. **(b)** Find the distance from the wire to the point where the field vanishes.

75. •• Find the angle between the plane of a loop and the magnetic field for which the magnetic torque acting on the loop is equal to x times its maximum value, where $0 \leq x \leq 1$.

76. •• Solenoids produce magnetic fields that are relatively intense for the amount of current they carry. To make a direct comparison, consider a solenoid with 55 turns per centimeter, a radius of 1.05 cm, and a current of 0.124 A. **(a)** Find the magnetic field at the center of the solenoid. **(b)** What current must a long, straight wire carry to have the same magnetic field as that found in part (a)? Let the distance from the wire be the same as the radius of the solenoid, 1.05 cm.

77. •• The current in a solenoid with 22 turns per centimeter is 0.50 A. The solenoid has a radius of 1.5 cm. A long, straight wire runs along the axis of the solenoid, carrying a current of 16 A. Find the magnitude of the net magnetic field a radial distance of 0.75 cm from the straight wire.

78. •• IP BIO **Transcranial Magnetic Stimulation** A recently developed method to study brain function is to produce a rapidly changing magnetic field within the brain. When this technique, known as transcranial magnetic stimulation (TMS), is applied to the prefrontal cortex, for example, it can reduce a person's ability to conjugate verbs, though other thought processes are unaffected. The rapidly varying magnetic field is produced with a circular coil of 21 turns and a radius of 6.0 cm placed directly on the head. The current in this loop increases at the rate of 1.2×10^7 A/s (by discharging a capacitor). **(a)** At what rate does the magnetic field at the center of the coil increase? **(b)** Suppose a second coil with half the area of the first coil is used instead. Would your answer to part (a) increase, decrease, or stay the same? By what factor?

79. ••• An electron with a velocity given by $\vec{v} = (1.5 \times 10^5 \text{ m/s})\hat{x} + (0.67 \times 10^4 \text{ m/s})\hat{y}$ moves through a region of space with a magnetic field $\vec{B} = (0.25 \text{ T})\hat{x} - (0.11 \text{ T})\hat{z}$ and an electric field $\vec{E} = (220 \text{ N/C})\hat{x}$. Using cross products, find the magnitude of the net force acting on the electron. (Cross products are discussed in Appendix A.)

80. ••• A thin ring of radius R and charge per length λ rotates with an angular speed ω about an axis perpendicular to its plane and passing through its center. Find the magnitude of the magnetic field at the center of the ring.

81. ••• A solenoid is made from a 25-m length of wire of resistivity 2.3×10^{-8} Ω · m. The wire, whose radius is 2.1 mm, is wrapped uniformly onto a plastic tube 4.5 cm in diameter and 1.65 m long. Find the emf to which the ends of the wire must be connected to produce a magnetic field of 0.015 T within the solenoid.

82. ••• IP A single current-carrying circular loop of radius R is placed next to a long, straight wire, as shown in **Figure 22–51**. The current in the wire points to the right and is of magnitude I. **(a)** In which direction must current flow in the loop to produce zero magnetic field at its center? Explain. **(b)** Calculate the magnitude of the current in part (a).

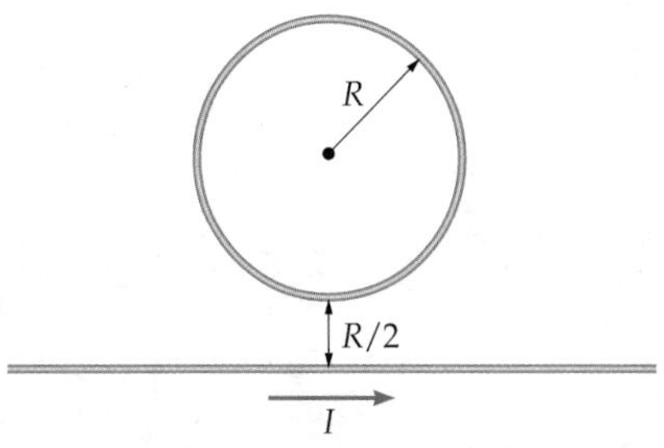

▲ **FIGURE 22–51** Problem 82

83. ••• **Magnetic Fields in the Bohr Model** In the Bohr model of the hydrogen atom, the electron moves in a circular orbit of radius 5.29×10^{-11} m about the nucleus. Given that the charge on the electron is -1.60×10^{-19} C, and that its speed is 2.2×10^6 m/s, find the magnitude of the magnetic field the electron produces at the nucleus of the atom.

84. ••• A single-turn square loop carries a current of 18 A. The loop is 15 cm on a side and has a mass of 0.035 kg. Initially the loop lies flat on a horizontal tabletop. When a horizontal magnetic field is turned on, it is found that only one side of the loop experiences an upward force. Find the minimum magnetic field, B_{min}, necessary to start tipping the loop up from the table.

85. ••• Consider the physical system shown in Figure 22–45. **(a)** Find the net magnetic field (direction and magnitude) at an arbitrary point on the bottom side of the square, a distance $0 < x < 5.0$ cm to the right of wire 1. **(b)** Find the magnitude of the net magnetic field at an arbitrary point on the left side of the square, a distance $0 < y < 5.0$ cm above wire 1.

Interactive Problems

86. •• IP ***Referring to Example 22–3*** Suppose the speed of the isotopes is doubled. **(a)** Does the separation distance, d, increase, decrease, or stay the same? Explain. **(b)** Find the separation distance for this case.

87. •• IP ***Referring to Example 22–3*** Suppose we change the initial speed of ^{238}U, leaving everything else the same. **(a)** If we want the separation distance to be zero, should the initial speed of ^{238}U be increased or decreased? Explain. **(b)** Find the required initial speed.

88. •• ***Referring to Active Example 22–2*** The current I_1 is adjusted until the magnetic field halfway between the wires has a magnitude of 2.5×10^{-6} T and points *into* the page. Everything else in the system remains the same as in Active Example 22–2. Find the magnitude and direction of I_1.

89. •• ***Referring to Active Example 22–2*** The current I_2 is adjusted until the magnetic field 5.5 cm below wire 2 has a magnitude of 2.5×10^{-6} T and points *out* of page. Everything else in the system remains the same as in Active Example 22–2. Find the magnitude and direction of I_2.

Magnetic Flux and Faraday's Law of Induction

23–1	Induced Electromotive Force	769
23–2	Magnetic Flux	770
23–3	Faraday's Law of Induction	772
23–4	Lenz's Law	775
23–5	Mechanical Work and Electrical Energy	778
23–6	Generators and Motors	781
23–7	Inductance	784
23–8	*RL* Circuits	787
23–9	Energy Stored in a Magnetic Field	788
23–10	Transformers	790

If you mention an electric guitar, everyone knows what you mean, whereas a reference to a "magnetic guitar" would probably evoke some very puzzled looks. Yet it could be argued that the second term is just as appropriate as the first. If the steel strings of the guitar didn't respond to the magnetic field of the pickups, there would be nothing to be "picked up"—no signal would go to the amp, and no sound would emerge from the speaker. This chapter focuses on the intimate connection between electricity and magnetism—more specifically, on the phenomenon of electromagnetic induction, which makes possible not only electric guitars but an enormous number of other devices that we use every day.

In the last chapter we saw a number of connections between electricity and magnetism. All the cases we considered, however, involved magnetic fields that were constant in time; that is, static fields. We now explore some of the interesting new phenomena that are produced when magnetic fields change with time.

Of particular interest is that changing magnetic fields can create electric fields, which, in turn, can cause stationary charges to move. When applied to an electrical circuit, a changing magnetic field can produce an electric current, just like a battery. Effects of this nature have led to a host of important applications in everyday life, from electric motors and generators to microphones, speakers, and computer disk drives.

Finally, a recurrent theme throughout this chapter is energy—its conservation and the various forms in which it can appear. As we shall see, mechanical energy can be converted to electrical energy—with the help of a magnetic field—and the voltage and current in one electric circuit can be converted, or "transformed," into different values in another circuit. Just as in earlier chapters, we again find that energy is one of the most fundamental and versatile of all physical concepts.

23–1 Induced Electromotive Force

When Oersted observed that an electric current produces a magnetic field, it was pure serendipity. In contrast, Michael Faraday (1791–1867), an English chemist and physicist who was aware of Oersted's results, purposefully set out to determine whether a similar effect operates in the reverse direction, that is, whether a magnetic field can produce an electric field. His careful experimentation showed that such a connection does indeed exist.

In **Figure 23–1** we show a simplified version of the type of experiment performed by Faraday. Two electrical circuits are involved. The first, called the **primary circuit,** consists of a battery, a switch, a resistor to control the current, and a coil of several turns around an iron bar. When the switch is closed on the primary circuit, a current flows through the coil, producing a magnetic field that is particularly intense within the iron bar.

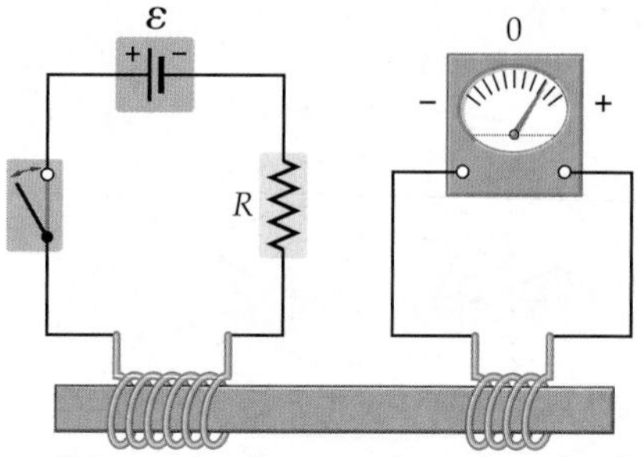

▲ **FIGURE 23–1 Magnetic induction**
Basic setup of Faraday's experiment on magnetic induction. When the position of the switch on the primary circuit is changed from open to closed or from closed to open, an electromotive force (emf) is induced in the secondary circuit. The induced emf causes a current in the secondary circuit, and the current is detected by the ammeter. If the current in the primary circuit does not change, no matter how large it may be, there is no induced current in the secondary circuit.

The **secondary circuit** also has a coil wrapped around the iron bar, and this coil is connected to an ammeter to detect any current in the circuit. Note, however, that there is no battery in the secondary circuit, and no direct physical contact between the two circuits. What does link the circuits, instead, is the magnetic field in the iron bar—it ensures that the field experienced by the secondary coil is approximately the same as that produced by the primary coil.

Now, let's look at the experimental results. When the switch is closed on the primary circuit, the magnetic field in the iron bar rises from zero to some finite amount, and the ammeter in the secondary coil deflects to one side briefly, and then returns to zero. As long as the current in the primary circuit is maintained at a constant value, the ammeter in the secondary circuit gives zero reading. If the switch on the primary circuit is now opened, so that the magnetic field decreases again to zero, the ammeter in the secondary circuit deflects briefly in the opposite direction, and then returns to zero. We can summarize these observations as follows:

- The current in the secondary circuit is zero as long as the current in the primary circuit is constant—which means, in turn, that the magnetic field in the iron bar is constant. It does not matter whether the constant value of the magnetic field is zero or nonzero.
- When the magnetic field passing through the secondary coil increases, a current is observed to flow in one direction in the secondary circuit; when the magnetic field decreases, a current is observed in the opposite direction.

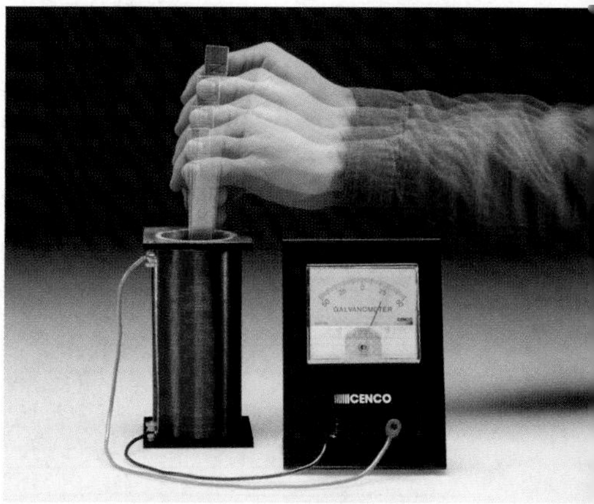

▲ When a magnet moves toward or away from a coil, the magnetic field within the coil changes. This, in turn, induces an electric current in the coil, which is detected by the meter.

Recall that the current in the secondary circuit appears without any direct contact between the primary and secondary circuits. For this reason we refer to the secondary current as an **induced current.** Since the induced current behaves the same as a current produced by a battery with an electromotive force (emf), we say that the changing magnetic field creates an **induced emf** in the secondary circuit. This leads to our final experimental observation:

- The magnitudes of the induced current and induced emf are found to be proportional to the rate of change of the magnetic field—the more rapidly the magnetic field changes, the greater the induced emf.

In Section 23–3 we return to these observations and state them in the form of a mathematical law.

Finally, in Faraday's experiment the changing magnetic field is caused by a changing current in the primary circuit. Any means of altering the magnetic field is just as effective, however. For example, in **Figure 23–2** we show a common classroom demonstration of induced emf. In this case, there is no primary circuit; instead, the magnetic field is changed by simply moving a permanent magnet toward or away from a coil connected to an ammeter. When the magnet is moved toward the coil, the meter deflects in one direction; when it is pulled away from the coil, the meter deflects in the opposite direction. There is no induced emf, however, when the magnet is held still.

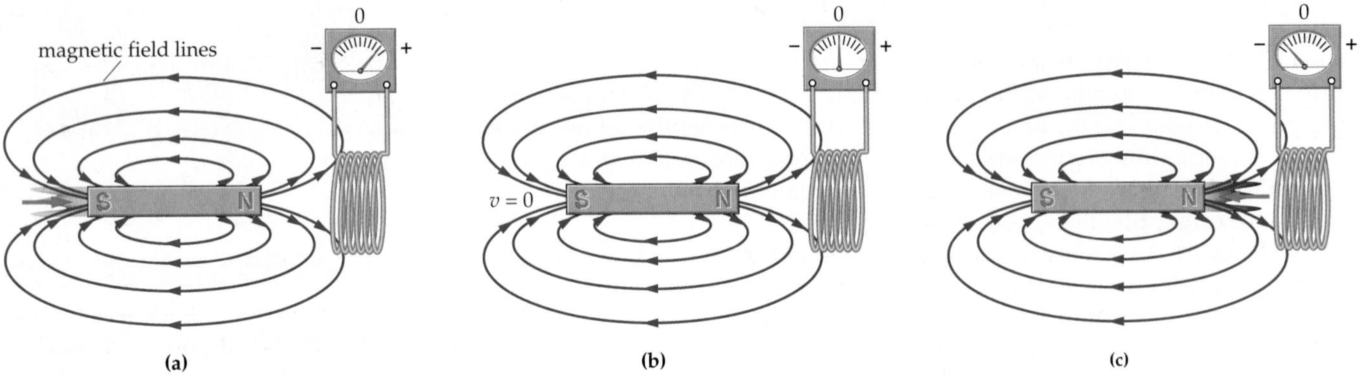

▲ **FIGURE 23–2 Induced current produced by a moving magnet**

A coil experiences an induced current when the magnetic field passing through it varies. **(a)** When the magnet moves toward the coil, the current is in one direction. **(b)** No current is induced while the magnet is held still. **(c)** When the magnet is pulled away from the coil, the current is in the other direction.

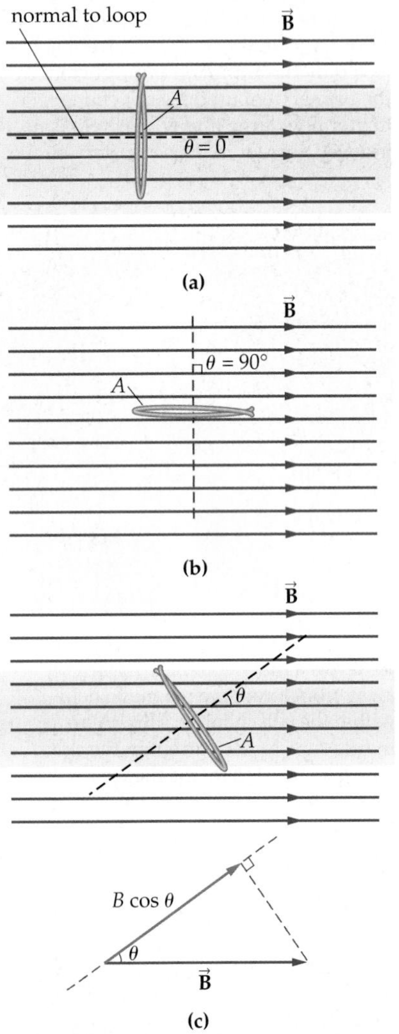

▲ **FIGURE 23–3 The magnetic flux through a loop**

The magnetic flux through a loop of area A is $\Phi = BA \cos \theta$, where θ is the angle between the normal to the loop and the magnetic field. **(a)** The loop is perpendicular to the field; hence, $\theta = 0$, and $\Phi = BA$. **(b)** The loop is parallel to the field; therefore, $\theta = 90°$ and $\Phi = 0$. **(c)** For a general angle θ, the component of the field that is perpendicular to the loop is $B \cos \theta$; hence, the flux is $\Phi = BA \cos \theta$.

23–2 Magnetic Flux

In the previous section, we saw that an emf can be induced in a coil of wire by simply changing the strength of the magnetic field that passes through the coil. This isn't the only way to induce an emf, however. More generally, we can change the direction of the magnetic field rather than its magnitude. We can even change the orientation of the coil in a constant magnetic field, or change the cross-sectional area of the coil. All of these situations can be described in terms of the change of a single physical quantity, the magnetic flux.

Basically, **magnetic flux** is a measure of the number of magnetic field lines that cross a given area, in complete analogy with the electric flux discussed in Section 19–7. Suppose, for example, that a magnetic field $\vec{B}$ crosses a surface area A at right angles, as in **Figure 23–3 (a)**. The magnetic flux, Φ, in this case is simply the magnitude of the magnetic field times the area:

$$\Phi = BA$$

If, on the other hand, the magnetic field is parallel to the surface, as in **Figure 23–3 (b)**, we see that *no field lines cross the surface*. In a case like this the magnetic flux is zero:

$$\Phi = 0$$

In general, only the component of $\vec{B}$ that is *perpendicular* to a surface contributes to the magnetic flux. The magnetic field in **Figure 23–3 (c)**, for example, crosses the surface at an angle θ relative to the normal, and hence its perpendicular component is $B \cos \theta$. The magnetic flux, then, is simply $B \cos \theta$ times the area A:

Definition of Magnetic Flux, Φ

$$\Phi = (B \cos \theta)A = BA \cos \theta \qquad \text{23–1}$$

SI unit: $1\ \text{T} \cdot \text{m}^2 = 1\ \text{weber} = 1\ \text{Wb}$

Note that $\Phi = BA \cos \theta$ gives $\Phi = BA$ when $\vec{B}$ is perpendicular to the surface ($\theta = 0$), and $\Phi = 0$ when $\vec{B}$ is parallel to the surface ($\theta = 90°$), as expected. Finally, the unit of magnetic flux is the **weber (Wb),** named after the German physicist Wilhelm Weber (1804–1891). It is defined as follows:

$$1\ \text{Wb} = 1\ \text{T} \cdot \text{m}^2 \qquad \text{23–2}$$

As we have seen, magnetic flux depends on the magnitude of the magnetic field, B, its orientation with respect to a surface, θ, and the area of the surface, A. A change in any of these variables results in a change in the flux. For example, in the case of the permanent magnet moved toward or away from the coil in the previous section, it is the change in *magnitude* of the field, B, that results in a change in flux. In the following Example, we consider the effect of changing the *orientation* of a wire loop in a region of constant magnetic field.

EXAMPLE 23–1 A System in Flux

Consider a circular loop with a 2.50-cm radius in a constant magnetic field of 0.625 T. Find the magnetic flux through this loop when its normal makes an angle of **(a)** 0°, **(b)** 30.0°, **(c)** 60.0°, and **(d)** 90.0° with the direction of the magnetic field $\vec{\mathbf{B}}$.

Picture the Problem

Our sketch shows four top views of the system, looking directly down on the top edge of the circular loop. The orientation of the loop and its normal is shown in each of the four panels. Note that the specified angles between the normal and the magnetic field direction are $\theta = 0°, 30.0°, 60.0°$, and 90.0°.

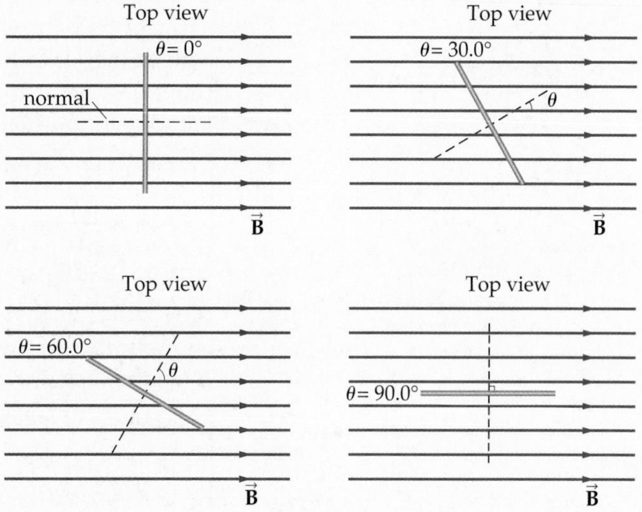

Strategy

To find the magnetic flux, we use the relation $\Phi = BA \cos \theta$. In this expression, B is the magnitude of the magnetic field and $A = \pi r^2$ is the area of a circular loop of radius r. The values of B, r, and θ are given in the problem statement.

Solution

Part (a)

1. Substitute $\theta = 0°$ in $\Phi = BA \cos \theta$:

$$\Phi = BA \cos \theta$$
$$= (0.625 \text{ T})\pi(0.0250 \text{ m})^2 \cos 0° = 1.23 \times 10^{-3} \text{ T} \cdot \text{m}^2$$

Part (b)

2. Substitute $\theta = 30.0°$ in $\Phi = BA \cos \theta$:

$$\Phi = BA \cos \theta$$
$$= (0.625 \text{ T})\pi(0.0250 \text{ m})^2 \cos 30.0° = 1.06 \times 10^{-3} \text{ T} \cdot \text{m}^2$$

Part (c)

3. Substitute $\theta = 60.0°$ in $\Phi = BA \cos \theta$:

$$\Phi = BA \cos \theta$$
$$= (0.625 \text{ T})\pi(0.0250 \text{ m})^2 \cos 60.0° = 6.14 \times 10^{-4} \text{ T} \cdot \text{m}^2$$

Part (d)

4. Substitute $\theta = 90.0°$ in $\Phi = BA \cos \theta$:

$$\Phi = BA \cos \theta$$
$$= (0.625 \text{ T})\pi(0.0250 \text{ m})^2 \cos 90.0° = 0$$

Insight

Thus, even if a magnetic field is uniform in space and constant in time, the magnetic flux through a given area will change if the orientation of the area changes. This is particularly relevant to the case of a coil that rotates in a field and constantly changes its orientation. As we shall see in the next section, a changing magnetic flux can create an electric current, and Section 23–6 shows how this applies to generators and motors.

Practice Problem

At what angle is the flux in this system equal to $1.00 \times 10^{-4} \text{ T} \cdot \text{m}^2$? [**Answer:** $\theta = 85.3°$]

Some related homework problems: Problem 1, Problem 3

CONCEPTUAL CHECKPOINT 23–1 Magnetic Flux

The three loops of wire shown in the sketch are all in a region of space with a uniform, constant magnetic field. Loop 1 swings back and forth as the bob on a pendulum; loop 2 rotates about a vertical axis; and loop 3 oscillates vertically on the end of a spring. Which loop or loops have a magnetic flux that changes with time?

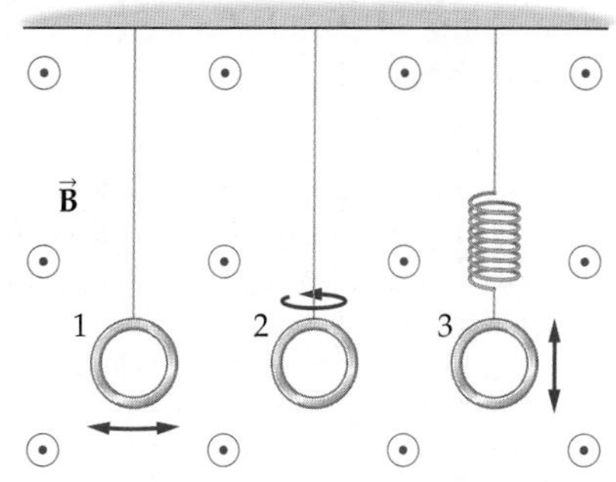

Reasoning and Discussion

Loop 1 moves back and forth, and loop 3 moves up and down, but since the magnetic field is uniform, the flux doesn't depend on the loop's position. Loop 2, on the other hand, changes its orientation relative to the field as it rotates; hence, its flux does change with time.

Answer:

Only loop 2 has a changing magnetic flux.

23–3 Faraday's Law of Induction

Now that the magnetic flux is defined, it is possible to be more precise about the experimental observations described in Section 23–1. In particular, Faraday found that the secondary coil in Figure 23–1 experiences an induced emf only when the *magnetic flux* through it changes with time. Furthermore, the induced emf for a given loop is found to be proportional to the *rate* at which the flux changes with time, $\Delta\Phi/\Delta t$. If there are N loops in a coil, each with the same magnetic flux, Faraday found that the induced emf is given by the following relation:

Faraday's Law of Induction

$$\mathcal{E} = -N\frac{\Delta\Phi}{\Delta t} = -N\frac{\Phi_{\text{final}} - \Phi_{\text{initial}}}{t_{\text{final}} - t_{\text{initial}}} \qquad 23\text{–}3$$

In this expression, known as **Faraday's law of induction,** the minus sign in front of N indicates that the induced emf opposes the change in magnetic flux, as we show in the next section. When we are concerned only with magnitudes, as is often the case, we use the form of Faraday's law given next:

Magnitude of Induced EMF

$$|\mathcal{E}| = N\left|\frac{\Delta\Phi}{\Delta t}\right| = N\left|\frac{\Phi_{\text{final}} - \Phi_{\text{initial}}}{t_{\text{final}} - t_{\text{initial}}}\right| \qquad 23\text{–}4$$

Notice that Faraday's law gives the *emf* that is induced in a circuit or a loop of wire. The current that is induced as a result of the emf depends on the characteristics of the circuit itself—for example, how much resistance it contains.

EXERCISE 23–1

The induced emf in a single loop of wire has a magnitude of 1.48 V when the magnetic flux is changed from $0.850\ \text{T}\cdot\text{m}^2$ to $0.110\ \text{T}\cdot\text{m}^2$. How much time is required for this change in flux?

Solution

Solving Equation 23–4 for the time, we obtain

$$|\Delta t| = N\frac{|\Delta\Phi|}{|\mathcal{E}|} = (1)\frac{|0.110\ \text{T}\cdot\text{m}^2 - 0.850\ \text{T}\cdot\text{m}^2|}{1.48\ \text{V}} = 0.500\ \text{s}$$

In the next Example we continue to deal with magnitudes, so we use Equation 23–4. We also consider a system in which the magnetic field varies in both magnitude and direction over the area of a coil. We can still calculate the magnetic flux in such a case if we simply replace the perpendicular component of the magnetic field ($B\cos\theta$) with its average value, B_{av}. With this substitution, the magnetic flux, $\Phi = BA\cos\theta$, becomes $\Phi = B_{\text{av}}A$.

EXAMPLE 23–2 Bar Magnet Induction

A bar magnet is moved rapidly toward a 40-turn circular coil of wire. As the magnet moves, the average value of $B\cos\theta$ over the area of the coil increases from 0.0125 T to 0.450 T in 0.250 s. If the radius of the coil is 3.05 cm, and the resistance of its wire is 3.55 Ω, find the magnitude of **(a)** the induced emf and **(b)** the induced current.

Picture the Problem

The motion of the bar magnet relative to the coil is shown in our sketch. As the magnet approaches the coil, the average value of the perpendicular component of the magnetic field increases. As a result, the magnetic flux through the coil increases with time.

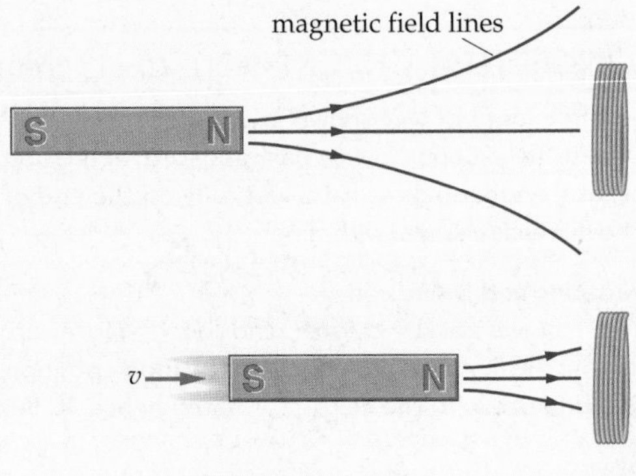

Strategy

To find the induced emf and the induced current we must first calculate the rate of change of the magnetic flux. To find the flux, we use $\Phi = B_{\text{av}}A$, where $A = \pi r^2$, and B_{av} is the average value of $B\cos\theta$. Once we have found the induced emf using Faraday's law, $|\mathcal{E}| = N|\Delta\Phi/\Delta t|$, we obtain the induced current using Ohm's law, $I = V/R$, with $V = |\mathcal{E}|$.

Solution

Part (a)

1. Calculate the initial and final values of the magnetic flux through the coil:

$$\Phi_i = B_i A$$
$$= (0.0125\ \text{T})\pi(0.0305\ \text{m})^2 = 3.65 \times 10^{-5}\ \text{T}\cdot\text{m}^2$$
$$\Phi_f = B_f A$$
$$= (0.450\ \text{T})\pi(0.0305\ \text{m})^2 = 1.32 \times 10^{-3}\ \text{T}\cdot\text{m}^2$$

2. Use Faraday's law to find the magnitude of the induced emf:

$$|\mathcal{E}| = N\left|\frac{\Phi_{\text{final}} - \Phi_{\text{initial}}}{t_{\text{final}} - t_{\text{initial}}}\right|$$
$$= (40)\left|\frac{1.32 \times 10^{-3}\ \text{T}\cdot\text{m}^2 - 3.65 \times 10^{-5}\ \text{T}\cdot\text{m}^2}{0.250\ \text{s}}\right|$$
$$= 0.205\ \text{V}$$

Part (b)

3. Use Ohm's law to calculate the induced current:

$$I = \frac{V}{R} = \frac{0.205\ \text{V}}{3.55\ \Omega} = 0.0577\ \text{A}$$

Insight

If the magnet is now pulled back to its original position in the same amount of time, the induced emf and current will have the same magnitudes; their directions will be reversed, however.

Practice Problem

How many turns of wire would be required in this coil to give an induced current of at least 0.100 A? [**Answer:** $N = 70$]

Some related homework problems: Problem 9, Problem 16, Problem 17

An everyday application of Faraday's law of induction can be found in a type of microphone known as a *dynamic microphone*. These devices employ a stationary permanent magnet and a wire coil attached to a movable diaphragm, as illustrated in **Figure 23–4**. When a sound wave strikes the microphone, it causes the diaphragm to oscillate, moving the coil alternately closer to and farther from the magnet. This movement changes the magnetic flux through the coil, which in turn produces an induced emf. Connecting the coil to an amplifier increases the magnitude of the induced emf enough that it can power a set of large speakers. The same principle is employed in the *seismograph*, except in this case the oscillations that produce the induced emf are generated by earthquakes and the vibrations they send through the ground.

REAL-WORLD PHYSICS
Dynamic microphones and seismographs

The American pop-jazz guitarist Les Paul (1915–) applied the same basic physics to musical instruments when he made the first solid-body *electric guitar* in 1941. The pickup in an electric guitar is simply a small permanent magnet with a coil wrapped around it, as shown in **Figure 23–5**. This magnet produces a field that is strong enough to produce a magnetization in the steel guitar string, which is the moving part in the system. When the string is plucked, the oscillating string

REAL-WORLD PHYSICS
Electric guitar pickups

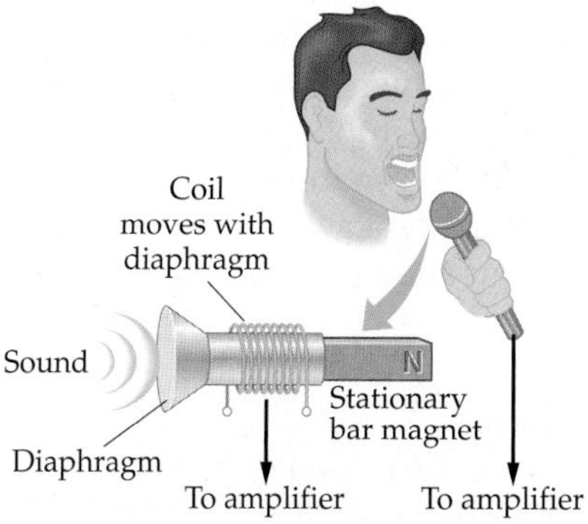

▲ **FIGURE 23–4 A dynamic microphone**

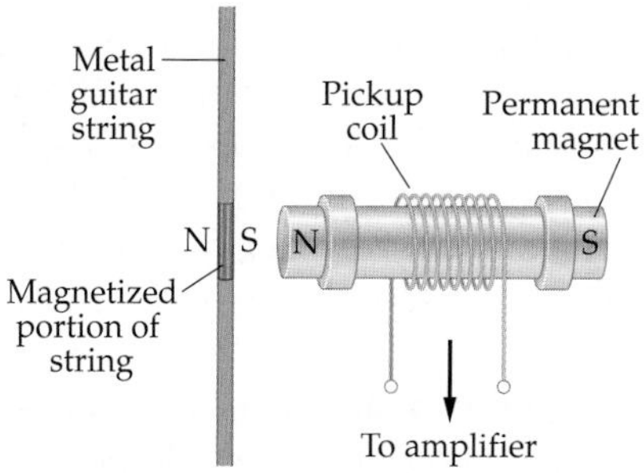

▲ **FIGURE 23–5 The pickup on an electric guitar**

▶ Many devices take advantage of Faraday's law to translate physical oscillations into electrical impulses. In all such devices, the oscillations are used to change the relative position of a coil and a magnet, thus causing the electric current in the coil to vary. For example, the vibrations produced by an earthquake or underground explosion produce an oscillating current in a seismograph (left) that can be amplified and used to drive a plotting pen. The magnetic pickups located under the strings of an electric guitar (right) function in much the same way. In this case, plucking a string (which is magnetized by the pickup) causes it to vibrate and generate an oscillating electrical signal in a pickup. This signal is amplified and fed to speakers, which produce most of the sound we hear.

changes the magnetic flux in the coil, inducing an emf that can be amplified. A typical electric guitar will have two or three sets of pickups, each positioned to amplify a different harmonic of the vibrating strings.

REAL-WORLD PHYSICS
Magnetic disk drives and credit card readers

Faraday's law of induction also applies to the operation of computer disk drives, credit card readers at grocery stores, audio tape players, and ATM machines. In these devices, information is recorded on a magnetic material by using an electromagnet to produce regions of opposite magnetic polarity, as indicated in **Figure 23–6**. When the information is to be "read out," the magnetic material is moved past the gap in an iron core that is wrapped with many turns of a wire coil. As magnetic regions of different polarity pass between the poles of the magnet, its field is alternately increased or decreased—resulting in a changing magnetic flux through the coil. The corresponding induced emf is sent to electronic circuitry to be translated into information a computer can process.

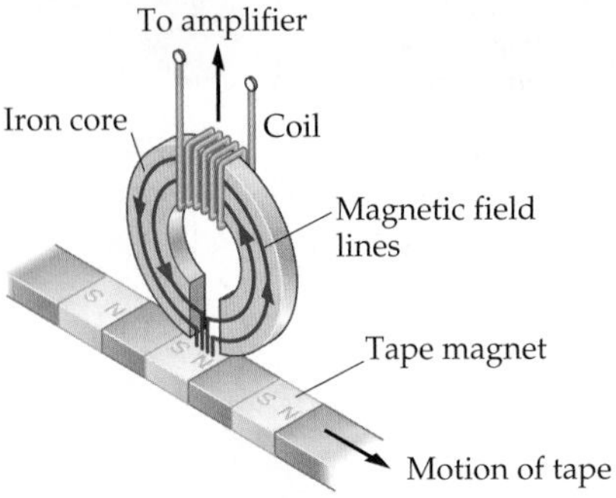

▲ **FIGURE 23–6 Magnetic tape recording**

REAL-WORLD PHYSICS
T coils and induction loops

Finally, induced emf can be of assistance to the hearing impaired. Many hearing aids have a "T switch" that activates a coil of wire in the device known as the telecoil, or **T coil** for short. This coil is designed to pick up the varying magnetic field produced by the speaker in a telephone and to convert the corresponding induced emf into sound for the user of the hearing aid. It should be emphasized that the varying magnetic field of the telephone speaker is simply a natural byproduct of the way it operates and is not produced specifically for the benefit of persons with hearing loss. Nonetheless, a person using the T coil on a hearing aid can usually hear more clearly than if the hearing aid merely amplifies the sound that comes from the speaker.

This principle is now being used on a larger scale in churches, auditoriums, and other public places. A single loop of wire, known as an **induction loop,** is placed on the floor of a room around its perimeter, as shown in **Figure 23–7**. When a person gives a talk in the room, the signal from the microphone is sent not only to speakers but to the induction loop as well. The loop produces a varying magnetic field throughout the room that hearing aids with the T switch can convert directly to sound. Ironically, it is often the hearing impaired in such a case who are able to hear the speech most clearly.

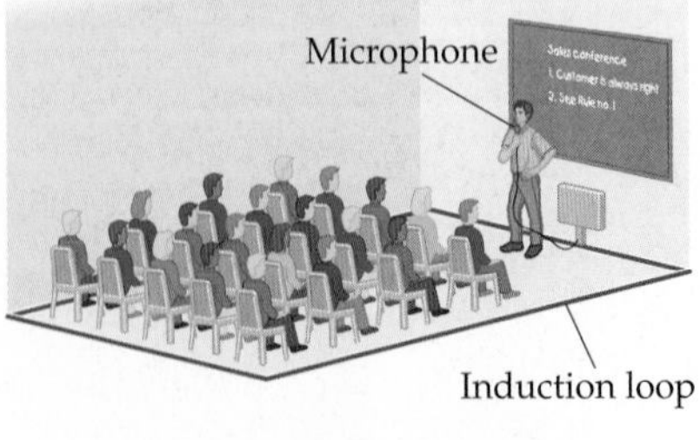

◀ **FIGURE 23–7 An induction loop for the hearing impaired**

A single loop of wire placed around the perimeter of a room can be used as an induction loop that creates a varying magnetic field. This field can be picked up by T coils in hearing aids.

23–4 Lenz's Law

In this section we discuss **Lenz's law,** a physical way of expressing the meaning of the minus sign in Faraday's law of induction. The basic idea of Lenz's law, first stated by the Estonian physicist Heinrich Lenz (1804–1865), is the following:

> **Lenz's Law**
> An induced current always flows in a direction that *opposes* the change that caused it.

This general physical principle is closely related to energy conservation, as we shall see at several points in our discussion. The remainder of this section, however, is devoted to specific applications of Lenz's law.

As a first example, consider a bar magnet that is moved toward a conducting ring, as in **Figure 23–8 (a)**. If the north pole of the magnet approaches the ring, a current is induced that tends to oppose the motion of the magnet. To be specific, the current in the ring creates a magnetic field that has a north pole (that is, diverging field lines) facing the north pole of the magnet, as indicated in the figure. There is now a repulsive force acting on the magnet, opposing its motion.

On the other hand, if the magnet is pulled away from the ring, as in **Figure 23–8 (b)**, the induced current creates a south pole facing the north pole of the magnet. The resulting attractive force again opposes the magnet's motion. We can obtain the same results using Faraday's law directly, but Lenz's law gives a simpler, more physical way to approach the problem.

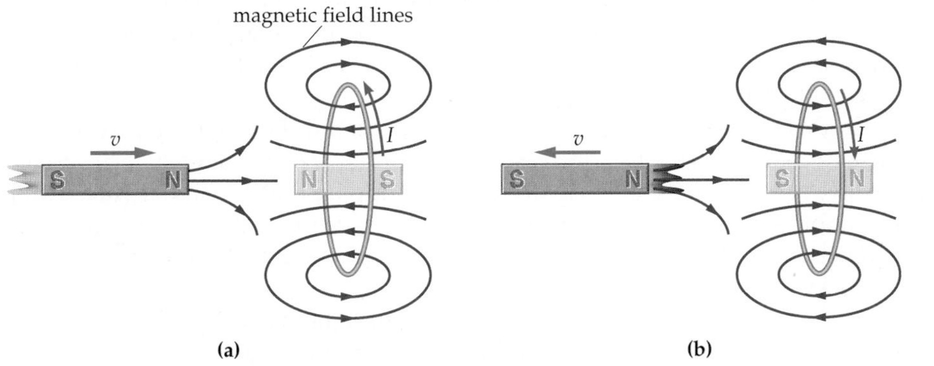

◀ **FIGURE 23–8 Applying Lenz's law to a magnet moving toward and away from a current loop**
(a) If the north pole of a magnet is moved toward a conducting loop, the induced current produces a north pole pointing toward the magnet's north pole. This creates a repulsive force opposing the change that caused the current. **(b)** If the north pole of a magnet is pulled away from a conducting loop, the induced current produces a south magnetic pole near the magnet's north pole. The result is an attractive force opposing the motion of the magnet.

CONCEPTUAL CHECKPOINT 23–2 Falling Magnets

The magnets shown in the sketch are dropped from rest through the middle of conducting rings. Notice that the ring on the right has a small break in it, whereas the ring on the left forms a closed loop. As the magnets drop toward the rings, does the magnet on the left have an acceleration that is **(a)** more than, **(b)** less than, or **(c)** the same as that of the magnet on the right?

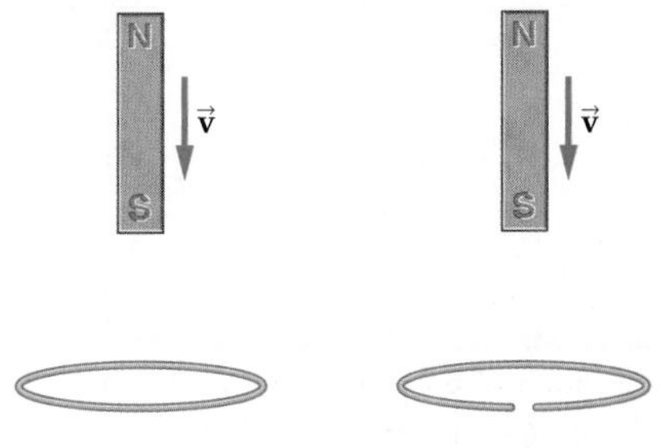

Reasoning and Discussion

As the magnet on the left approaches the ring, it induces a circulating current. According to Lenz's law, this current produces a magnetic field that exerts a repulsive force on the magnet—to oppose its motion. In contrast, the ring on the right has a break, so it cannot have a circulating current. As a result, it exerts no force on its magnet. Therefore, the magnet on the right falls with the acceleration of gravity; the magnet on the left falls with a smaller acceleration.

Answer:

(b) The magnet on the left has the smaller acceleration.

In the examples given so far in this section, the "change" involved the motion of a bar magnet, but Lenz's law applies no matter how the magnetic flux is changed. Suppose, for example, that a magnetic field is decreased with time, as

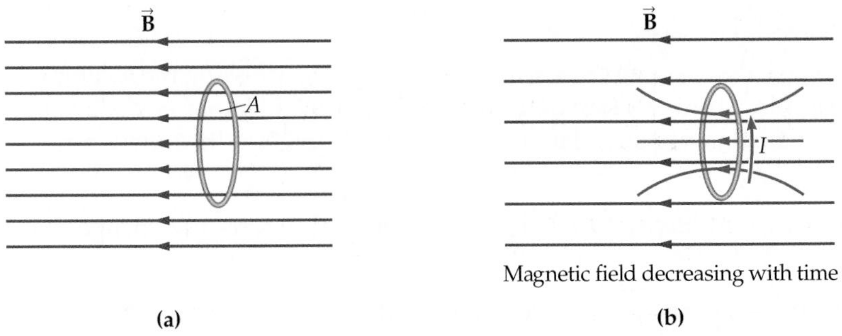

▶ **FIGURE 23–9 Lenz's law applied to a decreasing magnetic field**
As the magnetic field is decreased, the induced current produces a magnetic field that passes through the ring in the same direction as $\vec{B}$.

illustrated in **Figure 23–9**. In this case the change is a decrease of magnetic flux through the ring. The induced current flows in a direction that tends to oppose this change; that is, the induced current must produce a field within the ring that is in the same direction as the decreasing B field, as we see in Figure 23–9 (b).

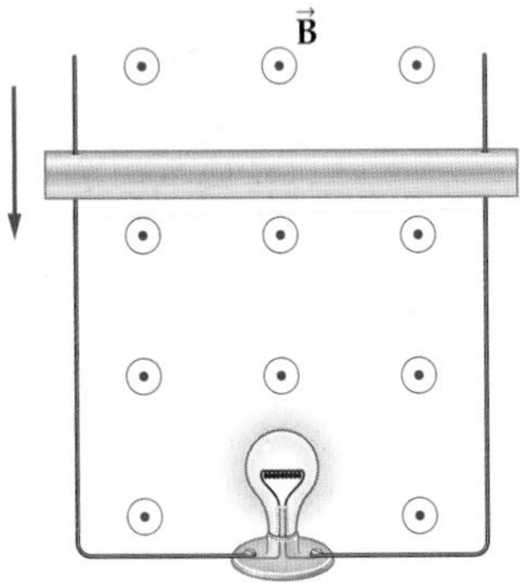

▲ **FIGURE 23–10 Motional emf**
Motional emf is created in this system as the rod falls. The result is an induced current, which causes the light to shine.

Motional EMF: Qualitative

Next, we consider the physical situation shown in **Figure 23–10**. Here we see a metal rod that is free to move in the vertical direction, and a constant magnetic field that points out of the page. The rod is in frictionless contact with two vertical wires; hence, a current can flow in a loop through the rod, the wires, and the lightbulb. Since the motion of the rod produces an emf in this system, it is referred to as **motional emf.** The question to be answered at this point is this: What is the direction of the current when the rod is released from rest and allowed to fall?

In this system the magnetic field is constant, but the magnetic flux through the current loop, $\Phi = BA$, decreases anyway. The reason is that as the rod falls, the area A enclosed by the loop decreases. To oppose this decrease in flux, Lenz's law implies that the induced current must flow in a direction that strengthens the magnetic field $\vec{B}$ within the loop, thus increasing the flux. This direction is counterclockwise, as indicated in **Figure 23–11**.

Notice that the current in the rod flows from right to left. This direction is also in agreement with Lenz's law, since it opposes the change that caused it. To be specific, the change for the rod is that it begins to fall downward. In order to oppose this change, the current through the rod must produce an upward magnetic force. As we see in Figure 23–11, using the right-hand rule, a current from right to left does just that.

When the rod is first released, there is no current and its downward acceleration is the acceleration of gravity—the rod speeds up as its gravitational potential energy is converted to kinetic energy. As the rod falls, the induced current begins to flow, however, and the rod is acted on by the upward magnetic force. This causes its acceleration to decrease. Ultimately, the magnetic force cancels the force of gravity, and the rod's acceleration vanishes. From this point on it falls with constant speed and hence constant kinetic energy. Now you may wonder what happens to the rod's gravitational potential energy. In fact, all the rod's gravitational potential energy is now being converted to heat and light in the lightbulb—none of it goes to increasing the kinetic energy of the rod.

Imagine for a moment the consequences that would follow if Lenz's law were not obeyed. If the current in the rod were from left to right, for example, the magnetic force would be downward, causing the rod to increase its downward acceleration. As the rod sped up, the induced current would become even larger and the downward force would increase as well. Thus, the rod would simply move faster and faster, without limit. Clearly, energy would not be conserved in such a case. The connections between mechanical and electrical energy are explored in greater detail in the next section.

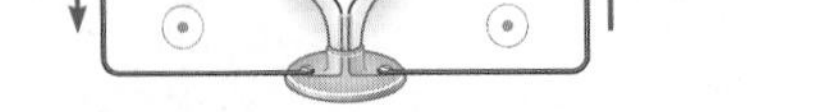

▲ **FIGURE 23–11 Determining the direction of an induced current**
The direction of the current induced by the rod's downward motion is counterclockwise, because this direction produces a magnetic field within the loop that points out of the page—in the same direction as the original field. Notice that a current flowing in this direction through the rod interacts with the original magnetic field to give an upward force, opposing the downward motion of the rod.

CONCEPTUAL CHECKPOINT 23–3 The Direction of Induced Current

Consider a system in which a metal ring is falling out of a region with a magnetic field and into a field-free region, as shown in our sketch to the right. According to Lenz's law, is the induced current in the ring (a) clockwise or (b) counterclockwise?

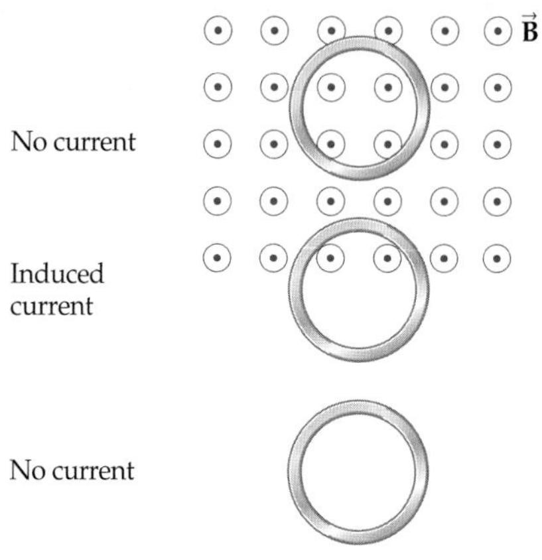

INTERACTIVE FIGURE

Reasoning and Discussion

The induced current must be in a direction that opposes the change in the system. In this case, the change is that fewer magnetic field lines are piercing the area of the loop and pointing out of the page. The induced current can oppose this change by generating more field lines out of the page *within the loop*. As shown to the left in the following figure, the induced current must be counterclockwise to accomplish this.

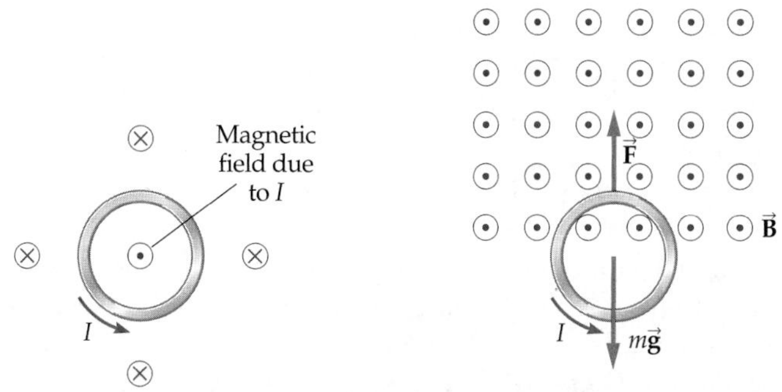

Finally, to the right in the preceding figure, note that the induced current generates an upward magnetic force at the top of the ring, but no magnetic force on the bottom, where the magnetic field is zero. Hence, the motion of the ring is retarded as it drops out of the field.

Answer:

(b) The induced current is counterclockwise.

The retarding effect on a ring leaving a magnetic field, as just described in the Conceptual Checkpoint, allows us to understand the behavior of **eddy currents.** Suppose, for example, that a sheet of metal falls from a region with a magnetic field to a region with no field, as in **Figure 23–12**. In the portion of the sheet that is just leaving the field, a circulating current—an eddy current—is induced in the metal, just like the circulating current set up in the ring in Conceptual Checkpoint 23–3. As in the case of the ring, the current will retard the motion of the sheet of metal, somewhat like a frictional force.

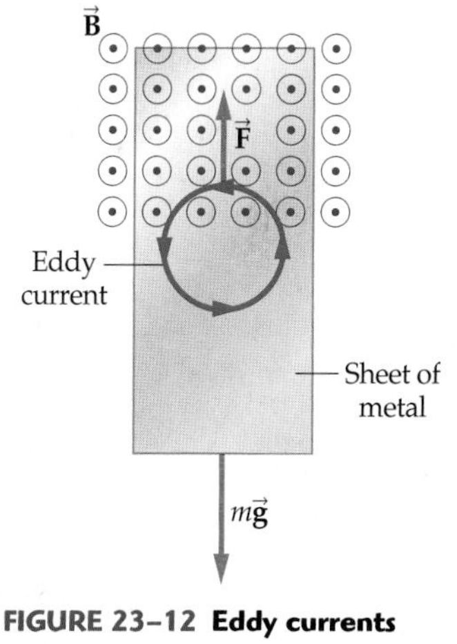

▲ **FIGURE 23–12 Eddy currents**
A circulating current is induced in a sheet of metal near where it leaves a region with a magnetic field. This "eddy current" is similar to the current induced in the ring described in Conceptual Checkpoint 23–3. In both cases, the induced current exerts a retarding force on the moving object, opposing its motion.

The frictionlike effect of eddy currents is the basis for the phenomenon known as *magnetic braking*. An important advantage of this type of braking is that no direct physical contact is needed, thus eliminating frictional wear. In addition, the force of magnetic braking is stronger the greater the speed of the metal with respect to the magnetic field. In contrast, kinetic friction is independent of the relative speed of the surfaces. Magnetic braking is used in everything from fishing reels to exercise bicycles to roller coasters.

Perhaps the most dramatic manifestation of magnetic braking is the slowing down of rotating magnetic stars. White dwarf stars, with intense magnetic fields, produce eddy currents in the clouds of ionized material that surround them. The result is a rapid slowing of their rotational speed, almost as if they were rotating in a viscous liquid. In some stars, magnetic braking has increased the rotational period from as little as a few hours to as much as 200 years.

REAL-WORLD PHYSICS
Magnetic braking and speedometers

On a more down-to-earth scale, magnetic braking is the operating principle behind the common analog *speedometer*. In these devices, a cable is connected at one end to a gear in the car's transmission and at the other end to a small permanent magnet. When the car is underway, the magnet rotates within a metal cup. The resulting "magnetic friction" between the rotating magnet and the metal cup—which is proportional to the speed of rotation—causes the cup to rotate in the same direction. A hairspring attached to the cup stops the rotation when the force of the spring equals the force of magnetic friction. Thus, the cup—and the needle attached to it—rotate through an angle that is proportional to the speed of the car.

REAL-WORLD PHYSICS
Induction stove

Finally, eddy currents are also used in the kitchen. In an *induction stove*, a metal coil is placed just beneath the cooking surface. When an alternating current is sent through the coil, it sets up an alternating magnetic field that, in turn, induces eddy currents in nearby metal objects. The finite resistance of a metal pan, for instance, will result in heating as the eddy current dissipates power according to the relation $P = I^2R$. Nonmetallic objects, like the surface of the stove and glassware, remain cool to the touch because they are poor electrical conductors and thus have negligible eddy currents.

23–5 Mechanical Work and Electrical Energy

To see how mechanical work can be converted to electrical energy, we turn to a detailed calculation of the forces and currents involved in a simple physical system. This analysis will form the basis for our understanding of electrical generators in the next section.

Motional EMF: Quantitative

We again consider motional emf, though this time in a somewhat simpler system where gravity plays no role. As shown in **Figure 23–13**, our setup consists of a rod that slides horizontally without friction on a U-shaped wire connected to a lightbulb of resistance R. A constant magnetic field points out of the page, and the rod is pushed by an external agent—your hand, for instance—so that it moves to the right with a constant speed v.

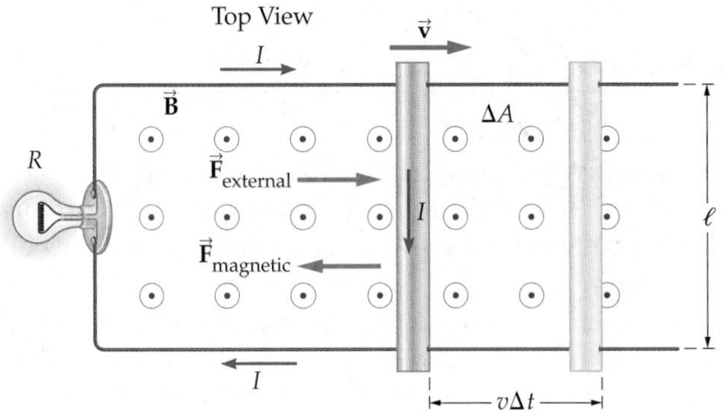

▶ **FIGURE 23–13 Force and induced current**

A conducting rod slides without friction on horizontal wires in a region where the magnetic field is uniform and pointing out of the page. The motion of the rod to the right induces a clockwise current and a corresponding magnetic force to the left. An external force of magnitude $F = B^2v\ell^2/R$ is required to offset the magnetic force and to keep the rod moving with a constant speed v.

Let's begin by calculating the rate at which the magnetic flux changes with time. Since $\vec{\mathbf{B}}$ is perpendicular to the area of the loop, it follows that $\Phi = BA$. Now the area increases with time because of the motion of the rod. In fact, the rod moves a horizontal distance $v\Delta t$ in the time Δt; therefore, the area of the loop increases by the amount $\Delta A = (v\Delta t)\ell$, as indicated in Figure 23–13. As a result, the increase in magnetic flux is

$$\Delta\Phi = B\,\Delta A = Bv\ell\,\Delta t$$

With this result in hand, we can use Faraday's law to calculate the magnitude of the induced emf. We find that for this single-turn loop

$$|\mathcal{E}| = N\left|\frac{\Delta\Phi}{\Delta t}\right| = (1)\frac{Bv\ell\,\Delta t}{\Delta t} = Bv\ell \qquad \textbf{23–5}$$

As one might expect, the induced emf is proportional to the strength of the magnetic field and to the speed of the rod.

Now let's calculate the electric field caused by the motion of the rod. We know that the emf from one end of the rod to the other is $\mathcal{E} = Bv\ell$, where $\mathcal{E}$ is an electric potential difference measured in volts. Recall that the difference in electric potential caused by an electric field E over a length ℓ is $V = E\ell$. Applying this to the emf in the rod, we have $E\ell = Bv\ell$, or

$$E = Bv \qquad \textbf{23–6}$$

Thus, a changing magnetic flux does indeed create an electric field, as Faraday suspected. In particular, moving through a magnetic field B at a speed v causes an electric field of magnitude $E = Bv$ in a direction that is at right angles to both the velocity and the magnetic field.

EXERCISE 23–2

As the rod in Figure 23–13 moves through a 0.445-T magnetic field, it experiences an induced electric field of 0.668 V/m. How fast is the rod moving?

Solution

Solving Equation 23–6 for v, we find

$$v = \frac{E}{B} = \frac{0.668\ \text{V/m}}{0.445\ \text{T}} = 1.50\ \text{m/s}$$

If we assume that the only resistance in the system is the lightbulb, R, it follows that the current in the circuit has the following magnitude:

$$I = \frac{|\mathcal{E}|}{R} = \frac{Bv\ell}{R} \qquad \text{23–7}$$

Using Lenz's law, we can determine that the direction of the current is clockwise, as indicated in Figure 23–13. Next, we show how both the magnitude and direction of the current play a key role in the energy aspects of this system.

Mechanical Work/Electrical Energy

We begin with a calculation of the force and power required to keep the rod moving with a constant speed v. First, note that the rod carries a current of magnitude $I = Bv\ell/R$ at right angles to a magnetic field of strength B. Since the length of the rod is ℓ, it follows that the magnetic force exerted on it has a magnitude

$$F = I\ell B = (Bv\ell/R)(\ell)B = \frac{B^2v\ell^2}{R} \qquad \text{23–8}$$

The direction of the current in the rod is downward on the page in Figure 23–13; hence, the magnetic-force *RHR* shows that $\vec{\mathbf{F}}_{\text{magnetic}}$ is directed to the left. As a result, an external force, $\vec{\mathbf{F}}_{\text{external}}$, must act to the right with equal magnitude to maintain the rod's constant speed. The mechanical power delivered by the external force is simply the force times the speed:

$$P_{\text{mechanical}} = Fv = \left(\frac{B^2v\ell^2}{R}\right)v = \frac{B^2v^2\ell^2}{R} \qquad \text{23–9}$$

Now let's compare this mechanical power with the electrical power converted to light and heat in the lightbulb. Recalling that the power dissipated in a resistor with a resistance R is $P = I^2R$, we have

$$P_{\text{electrical}} = I^2R = (Bv\ell/R)^2R = \frac{B^2v^2\ell^2}{R} \qquad \text{23–10}$$

Note that this is *exactly* the same as the mechanical power. Thus, with the aid of a magnetic field, we can convert *mechanical* power directly to *electrical* power. This simple example of motional emf illustrates the basic principle behind the generation of virtually all the world's electrical energy.

EXAMPLE 23–3 Light Power

The lightbulb in the circuit shown below has a resistance of 12 Ω and consumes 5.0 W of power; the rod is 1.25 m long and moves to the left with a constant speed of 3.1 m/s. **(a)** What is the strength of the magnetic field? **(b)** What external force is required to maintain the rod's constant speed?

continued on next page

continued from previous page

Picture the Problem

Our sketch shows the rod being pushed to the left by the force $\vec{\mathbf{F}}_{\text{external}}$. The motion of the rod, in turn, generates an electric current whose direction, according to Lenz's law, produces a magnetic force, $\vec{\mathbf{F}}_{\text{magnetic}}$, that opposes the motion of the rod. We know that the rod moves with constant speed, and therefore the applied force $\vec{\mathbf{F}}_{\text{external}}$ exactly cancels the magnetic force $\vec{\mathbf{F}}_{\text{magnetic}}$.

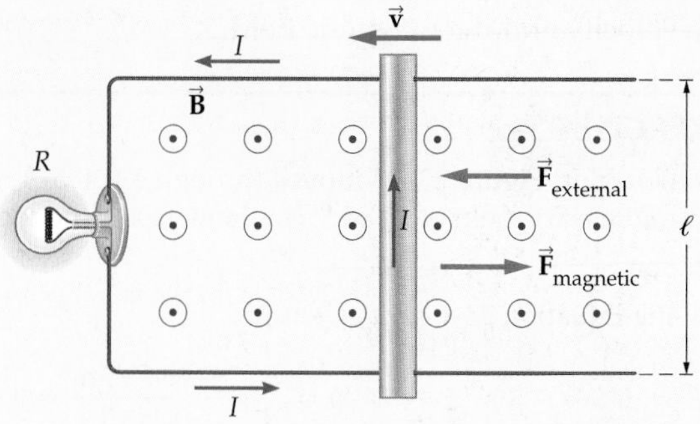

Strategy

a. The power consumed by the lightbulb is given by the expression $P = B^2v^2\ell^2/R$. This relation can be solved to give the strength of the magnetic field, B.

b. The magnitude of the external force is given by $F = B^2v\ell^2/R$, where B has the value obtained in part (a). Alternatively, we can solve for the force using the power relation, $P = Fv$.

Solution

Part (a)

1. Solve the power expression for the magnetic field, B:

$$P = \frac{B^2v^2\ell^2}{R}$$

$$B = \frac{\sqrt{PR}}{v\ell}$$

2. Substitute numerical values:

$$B = \frac{\sqrt{PR}}{v\ell} = \frac{\sqrt{(5.0\ \text{W})(12\ \Omega)}}{(3.1\ \text{m/s})(1.25\ \text{m})} = 2.0\ \text{T}$$

Part (b)

3. Calculate the magnitude of the external force using $F = B^2v\ell^2/R$:

$$F = \frac{B^2v\ell^2}{R}$$

$$= \frac{(2.0\ \text{T})^2(3.1\ \text{m/s})(1.25\ \text{m})^2}{12\ \Omega} = 1.6\ \text{N}$$

4. Find the magnitude of the external force using $P = Fv$:

$$F = \frac{P}{v} = \frac{5.0\ \text{W}}{3.1\ \text{m/s}} = 1.6\ \text{N}$$

Insight

Note that a constant external *force* is required to move the rod with constant *speed* and that the magnitude of the external force is proportional to the speed. This is similar to the behavior of a "drag" force like air resistance. Also, note that the *net* force acting on the rod is zero—as expected for an object that moves with constant speed.

Practice Problem

How fast will the rod move if the external force exerted on it is 1.5 N? [**Answer:** $v = 2.9\ \text{m/s}$]

Some related homework problems: Problem 31, Problem 32, Problem 33

ACTIVE EXAMPLE 23–1 Find the Work Converted to Light Energy

Referring to Example 23–3, find **(a)** the work done by the external force in 0.58 s and **(b)** the energy consumed by the lightbulb in the same time.

Solution *(Test your understanding by performing the calculations indicated in each step.)*

Part (a)

1. Find the distance covered by the rod in 0.58 s: 1.8 m
2. Find the work done using $W = Fd$: $(1.6\ \text{N})(1.8\ \text{m}) = 2.9\ \text{J}$

Part (b)

3. Calculate the energy used in the lightbulb by multiplying the power by the time: $(5.0\ \text{W})(0.58\ \text{s}) = 2.9\ \text{J}$

Insight
As expected, the energy used by the lightbulb is equal to the work done by the external force as it pushes the rod against the magnetic force.

Your Turn
What is the current flowing in this circuit? Obtain your answer in at least two different ways.

(Answers to **Your Turn** *problems are given in the back of the book.)*

Motional emfs are generated whenever a magnet moves relative to a conductor or when a conductor moves relative to a magnetic field. For example, *magnetic antitheft devices* in libraries and stores detect the motional emf produced by a small magnetic strip placed in a book or other item. If this strip is not demagnetized, it will trigger an alarm as it moves past a sensitive set of conductors designed to pick up the induced emf.

REAL-WORLD PHYSICS
Magnetic antitheft devices

In numerous biomedical applications of the same principle, it is the conductor that moves while the magnetic field remains stationary. For example, researchers have tracked the head and body movements of several flying insects, including blowflies, hover flies, and honeybees. They attach lightweight, flexible wires to a small metal coil on the insect's head, and another on its thorax, and then allow it to fly in a stationary magnetic field. As the coils move through the field, they experience induced emfs that can be analyzed by computer to determine the corresponding orientation of the head and thorax.

REAL-WORLD PHYSICS: BIO
Tracking the movement of insects

The same technique is used by researchers investigating the movements of the human eye. You may not be aware of it, but your eyes are jiggling back and forth rather rapidly at this very moment. This constant motion, referred to as **saccadic motion,** prevents an image from appearing at the same position on the retina for a period of time. Without saccadic motion, our brains would tend to ignore stationary images that we are focused on in favor of moving images of little interest. To study the dynamics of saccadic motion, a small coil of wire is attached to a modified contact lens. As the eye moves the coil through a stationary magnetic field, the induced motional emf gives the eye's orientation and rotational speed.

REAL-WORLD PHYSICS: BIO
Tracking the motion of the human eye

23–6 Generators and Motors

We have just seen that a changing magnetic flux can serve as a means of converting mechanical work to electrical energy. In this section we discuss this concept in greater detail as we explore the workings of an electrical generator. We also point out that the energy conversion can run in the other direction as well, with electrical energy being converted to mechanical work in an electric motor.

Electric Generators

An electric **generator** is a device designed to efficiently convert mechanical energy to electrical energy. The mechanical energy can be supplied by any of a number of sources, such as falling water in a hydroelectric dam, expanding steam in a coal-fired power plant, or the output of a small gasoline motor that powers a portable generator. In all cases the basic operating principle is the same—mechanical energy is used to move a conductor through a magnetic field to produce motional emf.

REAL-WORLD PHYSICS
Electric generators

The linear motion of a metal rod through a magnetic field, as in Figure 23–13, results in motional emf, but to continue producing an electric current in this way the rod must move through ever greater distances. A practical way to employ the same effect is to use a coil of wire that can be *rotated* in a magnetic field. By rotating the coil, which in turn changes the magnetic flux, we can continue the process indefinitely.

▲ Hydroelectric power provides an excellent example of energy transformations. First, the gravitational potential energy of water held behind a dam is converted to kinetic energy as the water descends through pipes in the generating station (left). Next, the kinetic energy of the falling water is used to spin turbines (center). Each turbine rotates the coil of a generator in a strong magnetic field, thus converting mechanical energy into an electric current in the coil (right). Eventually the electrical energy is delivered to homes, factories, and cities where it is converted to light (in incandescent bulbs and fluorescent tubes), heat (in devices such as toasters and broilers), and work (in various machines).

Imagine, then, a coil of area A located in the magnetic field between the poles of a magnet, as illustrated in **Figure 23–14**. Metal rings are attached to either end of the wire that makes up the coil, and carbon brushes are in contact with the rings to allow the induced emf to be communicated to the outside world. As the coil rotates with the angular speed ω, the emf produced in it is given by Faraday's law, $\mathcal{E} = -N(\Delta\Phi/\Delta t)$. Referring to the similar case of uniform circular motion in Chapter 13, and Equation 13–6 in particular, it can be shown that the explicit expression for the emf of a rotating coil is as follows:

$$\mathcal{E} = NBA\omega \sin \omega t \qquad \text{23–11}$$

This result is plotted in **Figure 23–15**. Notice that the induced emf in the coil alternates in sign, which means that the current in the coil alternates in direction. For this reason, this type of generator is referred to as an **alternating current generator** or, simply, an ac generator.

PROBLEM-SOLVING NOTE

Maximum and Minimum Values of Induced EMF

Note that the induced emf of a rotating coil has a maximum value of $NBA\omega$ and a minimum value of $-NBA\omega$. This follows from Equation 23–11 and the fact that the maximum and minimum values of $\sin \omega t$ are $+1$ and -1, respectively.

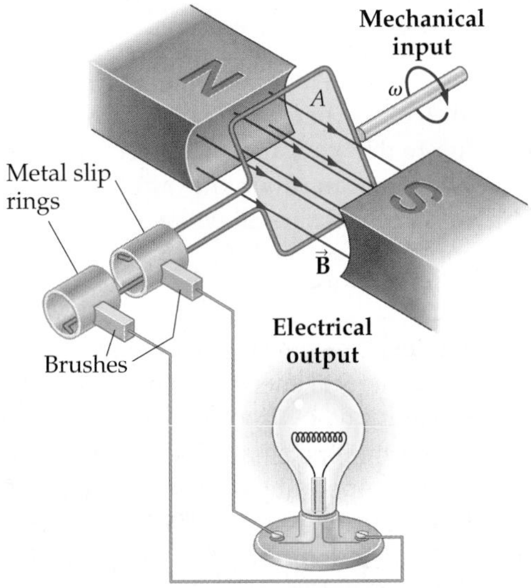

▲ **FIGURE 23–14 An electric generator**

The basic operating elements of an electric generator are shown in a schematic representation. As the coil is rotated by an external source of mechanical work, it produces an emf that can be used to power an electrical circuit.

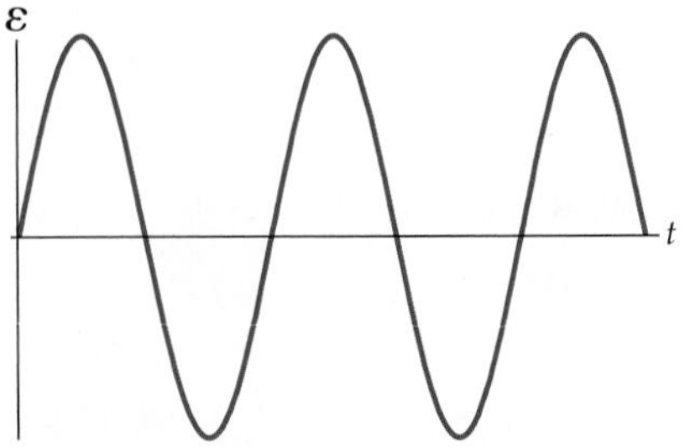

▲ **FIGURE 23–15 Induced emf of a rotating coil**

Induced emf produced by an electric generator like the one shown in Figure 23–14. The emf alternates in sign, so the current changes sign as well. We say that a generator of this type produces an "alternating current."

EXAMPLE 23–4 Generator Next

The coil in an electric generator has 100 turns and an area of $2.5 \times 10^{-3}\ m^2$. It is desired that the maximum emf of the coil be 120 V when it rotates at the rate of 60.0 cycles per second. Find the strength of the magnetic field B that is required for this generator.

Picture the Problem

In our sketch, a square conducting coil of area A and angular speed ω rotates between the poles of a magnet, where the magnetic field has a strength B. As the coil rotates, the magnetic flux through it changes continuously. The result is an induced emf that varies sinusoidally with time.

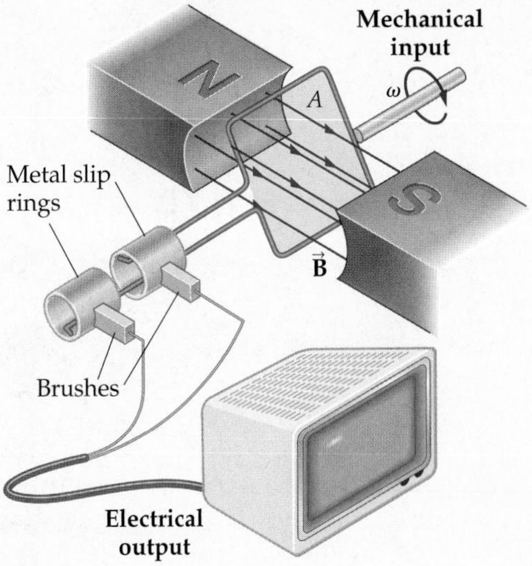

Strategy

First, the induced emf of the generator is given by $\mathcal{E} = NBA\omega \sin \omega t$. Therefore, the maximum emf occurs when $\sin \omega t$ has its maximum value of 1. It follows that $\mathcal{E}_{max} = NBA\omega$. This relation can be solved for the magnetic field, B.

Second, we need to find the angular speed ω, since it is not given directly. What we are given is that the frequency of the coil's rotation is $f = 60.0$ cycles per second $= 60.0$ Hz. The corresponding angular speed is $\omega = 2\pi f$.

Solution

1. Solve for the magnetic field in terms of the maximum emf:

$$\mathcal{E}_{max} = NBA\omega$$

$$B = \frac{\mathcal{E}_{max}}{NA\omega}$$

2. Calculate the angular speed of the coil:

$$\omega = 2\pi f = 2\pi(60.0\ \text{Hz}) = 377\ \text{rad/s}$$

3. Substitute numerical quantities into the expression for B:

$$B = \frac{\mathcal{E}_{max}}{NA\omega} = \frac{120\ \text{V}}{(100)(2.5 \times 10^{-3}\ \text{m}^2)(377\ \text{rad/s})} = 1.3\ \text{T}$$

Insight

Note that the emf of a generator coil depends on its area A but not on its shape. In addition, each turn of the coil generates the same induced emf; therefore, the total emf of the coil is directly proportional to the number of turns. Finally, the more rapidly the coil rotates, the more rapidly the magnetic flux through it changes. As a result, the induced emf is also proportional to the angular speed of the generator.

Practice Problem

What is the maximum emf of this generator if its rotation frequency is reduced to 50.0 Hz? [**Answer:** 100 V]

Some related homework problems: Problem 34, Problem 35

Electric Motors

REAL-WORLD PHYSICS
Electric motors

The basic physical principle behind the operation of electric motors has already been discussed. In particular, we saw in Section 22–5 that a current-carrying loop in a magnetic field experiences a torque that tends to make it rotate. If such a loop is mounted on an axle, as in **Figure 23–16**, it is possible for the magnetic torque to be applied to the external world.

To see just how this works in practice, notice that the torque exerted on the loop at the moment shown in Figure 23–16 causes it to rotate clockwise *toward* the vertical position. Once it reaches this orientation, and continues past it due to its angular momentum, the alternating current from the electrical input reverses direction. This reverses the torque on the loop and causes the loop to rotate *away* from the vertical—which means it is still rotating in the clockwise sense. The next time the loop becomes vertical, the current again reverses, causing the loop to continue rotating clockwise. The result is an axle continually turning in the same direction.

What we have just described is precisely the reverse of a generator—instead of doing work to turn a coil and produce an electric current, as in a generator, a motor uses an electric current to produce rotation in a coil, which then does work.

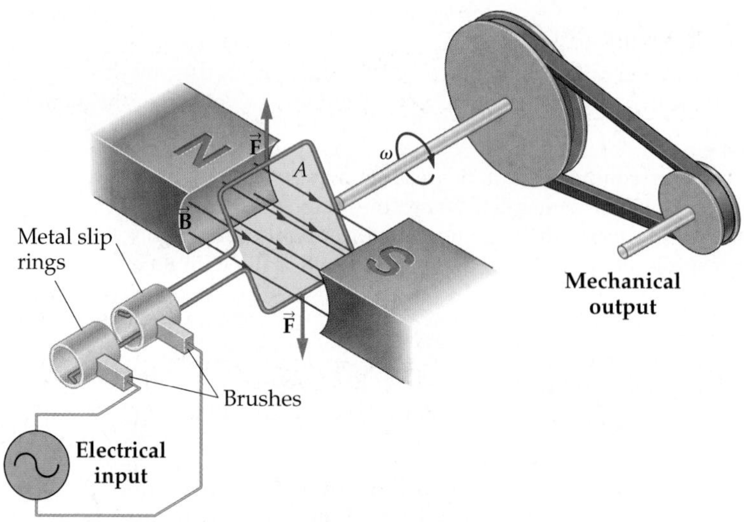

▶ **FIGURE 23–16 A simple electric motor**

An electric current causes the coil of a motor to rotate and deliver mechanical work to the outside world.

Thus, a motor, in some sense, is a generator run in reverse. In fact, it is possible to have one electric motor turn the axle of another identical motor and thereby produce electricity from the "motor"-turned generator.

REAL-WORLD PHYSICS
Energy recovery technology in cars

Similarly, if a car is powered by an electric motor, its motor can double as a generator when it helps to slow the car during braking. This results in a more efficient mode of transportation, because some of the car's kinetic energy that is normally converted to heat during braking can instead be used to recharge the batteries and extend the range of the car. Some of the newer "hybrid" cars, such as the Honda Insight and the Toyota Prius, make use of this "energy recovery" technology.

23–7 Inductance

We began this chapter with a description of Faraday's induction experiment, in which a changing current in one coil induces a current in another coil. This type of interaction between coils is referred to as **mutual inductance.** It is also possible, however, for a single, isolated coil to have a similar effect—that is, a coil with a changing current can induce an emf, and hence a current, in itself. This type of process is known as **self-inductance.**

To be more precise, consider a coil of wire in a region free of magnetic fields, as in **Figure 23–17 (a)**. Initially the current in the coil is zero, so the magnetic flux through the coil is zero as well. The current in the coil is now steadily increased from zero, as indicated in **Figure 23–17 (b)**. As a result, the magnetic flux through the coil increases with time, which, according to Faraday's law, induces an emf in the coil.

The direction of the induced emf is given by Lenz's law, which says that the induced current opposes the change that caused it. In this case, the change is the increasing current in the coil; hence, the induced emf is in a direction opposite to the current. This situation is illustrated in **Figure 23–18**, where we schematically show the effect of the induced emf (often referred to as a "back" emf) in the circuit. Note that in order to increase the current in the coil it is necessary to *force* the current against an opposing emf.

This is a rather interesting result. It says that even if the wires in a coil are ideal, with no resistance whatsoever, work must still be done to increase the current. Conversely, if a current already exists in a coil and it decreases with time, Lenz's law ensures that the self-induced current will oppose this change as well, in an attempt to keep the current constant. Hence, a coil tends to resist changes in its current, regardless of whether the change is an increase or a decrease.

To make a quantitative connection between a changing current and the associated self-induced emf, we start with Faraday's law in magnitude form:

$$|\mathcal{E}| = N\left|\frac{\Delta\Phi}{\Delta t}\right|$$

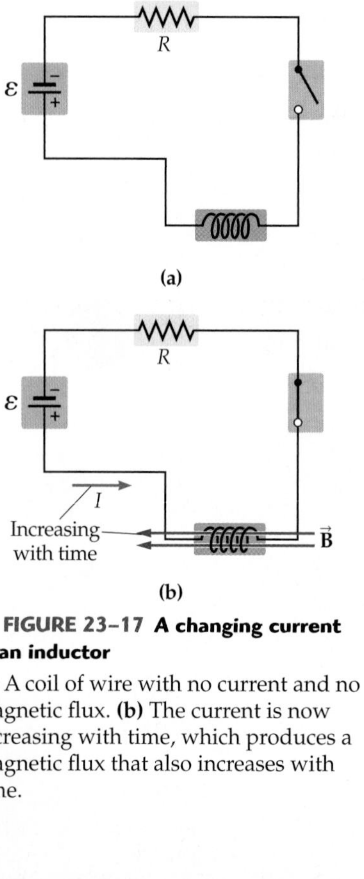

▲ **FIGURE 23–17 A changing current in an inductor**

(a) A coil of wire with no current and no magnetic flux. **(b)** The current is now increasing with time, which produces a magnetic flux that also increases with time.

Now, since the magnetic field is proportional to the current in a coil, it follows that the magnetic flux Φ is also proportional to I. As a result, the time rate of change of the magnetic flux, $\Delta\Phi/\Delta t$, must be proportional to the time rate of change of the current, $\Delta I/\Delta t$. Therefore, the induced emf can be written as

Definition of Inductance, *L*

$$|\mathcal{E}| = N\left|\frac{\Delta\Phi}{\Delta t}\right| = L\left|\frac{\Delta I}{\Delta t}\right| \qquad \text{23–12}$$

SI unit: $1\ \mathrm{V\cdot s/A} = 1\ \text{henry} = 1\ \mathrm{H}$

This expression defines the constant of proportionality, L, which is referred to as the **inductance** of a coil. Rearranging Equation 23–12, we see that

$$L = N\left|\frac{\Delta\Phi}{\Delta I}\right|$$

The SI unit of inductance is the **henry (H)**, named to honor the work of Joseph Henry (1797–1878), an American physicist. As can be seen from Equation 23–12, the henry is one volt-second per amp:

$$1\ \mathrm{H} = 1\ \mathrm{V\cdot s/A} \qquad \text{23–13}$$

Common inductances are generally in the millihenry (mH) range.

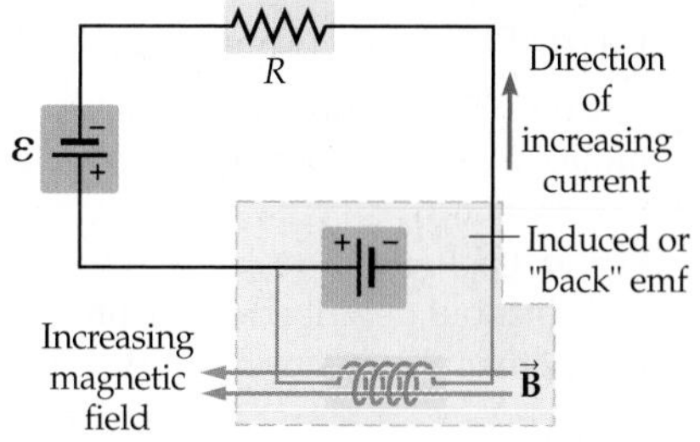

▲ **FIGURE 23–18 The back emf of an inductor**

The effect of an increasing current in a coil is an induced emf that opposes the increase. This is indicated schematically by replacing the coil with the opposing, or "back," emf.

EXERCISE 23–3

The coil in an electromagnet has an inductance of 2.9 mH and carries a constant direct current of 5.6 A. A switch is suddenly opened, allowing the current to drop to zero over a small interval of time, Δt. If the magnitude of the average induced emf during this time is 7.3 V, what is Δt?

Solution

Solving Equation 23–12 for Δt, and using only magnitudes, we find

$$\Delta t = \frac{L|\Delta I|}{|\mathcal{E}|} = \frac{(2.9\times 10^{-3}\ \mathrm{H})(5.6\ \mathrm{A})}{7.3\ \mathrm{V}} = 2.2\ \mathrm{ms}$$

Of particular interest is the inductance of an ideal solenoid with N turns and length ℓ. For example, suppose the initial current in a solenoid is zero and the final current is I. It follows that the initial magnetic flux is zero. The final flux is simply $\Phi_f = BA\cos\theta$, where, according to Equation 22–12, the magnetic field in the solenoid is $B = \mu_0(N/\ell)I$. Finally, we know that the field in a solenoid is perpendicular to its cross-sectional area, A; thus $\theta = 0$ and $\Phi_f = BA$. Combining these results, we obtain

$$L = \frac{N(BA - 0)}{(I - 0)} = \frac{N[\mu_0(N/\ell)I]A}{I} = \mu_0\left(\frac{N^2}{\ell}\right)A$$

Denoting the number of turns per length as $n = N/\ell$, we can express the inductance of a solenoid in the following forms:

Inductance of a Solenoid

$$L = \mu_0\left(\frac{N^2}{\ell}\right)A = \mu_0 n^2 A\ell \qquad \text{23–14}$$

Notice that doubling the number of turns per length quadruples the inductance. In addition, the inductance is proportional to the volume of a solenoid, $V = A\ell$.

In the following Examples we consider problems involving solenoids and their inductance.

EXAMPLE 23–5 Solenoid Self-Induction

A 500-turn solenoid is 8.0 cm long. When the current in this solenoid is increased from 0 to 2.5 A in 0.35 s, the magnitude of the induced emf is 0.012 V. Find **(a)** the inductance and **(b)** the cross-sectional area of the solenoid.

Picture the Problem

The solenoid described in the problem statement is shown in our sketch. Note that it is 8.0 cm in length and contains 500 turns.

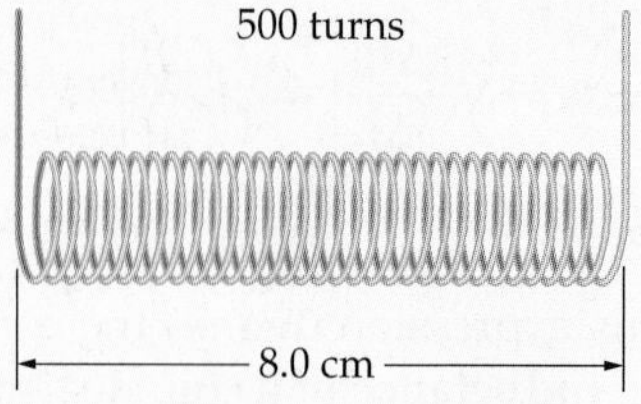

Strategy

a. We can use the definition of inductance, $|\mathcal{E}| = L|\Delta I/\Delta t|$, to solve for the inductance.

b. Once we know L from part (a), we can find the cross-sectional area from $L = \mu_0 n^2 A\ell$, where $n = N/\ell$.

Solution

Part (a)

1. Using magnitudes, solve $|\mathcal{E}| = L|\Delta I/\Delta t|$ for the inductance: $L = \dfrac{|\mathcal{E}|}{|\Delta I/\Delta t|} = \dfrac{0.012\text{ V}}{(2.5\text{ A}/0.35\text{ s})} = 1.7\text{ mH}$

Part (b)

2. Solve $L = \mu_0 n^2 A\ell$ for the cross-sectional area of the solenoid: $A = \dfrac{L}{\mu_0 n^2 \ell} = \dfrac{L\ell}{\mu_0 N^2}$

3. Substitute numerical values: $A = \dfrac{(1.7 \times 10^{-3}\text{ H})(0.080\text{ m})}{(4\pi \times 10^{-7}\text{ T}\cdot\text{m/A})(500)^2} = 4.3 \times 10^{-4}\text{ m}^2$

Insight

This cross-sectional area corresponds to a radius of about 1.2 cm. Therefore, it would take about 37 m of wire to construct this inductor.

Practice Problem

Find the inductance and induced emf for a solenoid with a cross-sectional area of $1.3 \times 10^{-3}\text{ m}^2$. The solenoid, which is 9.0 cm long, is made from a piece of wire that is 44 m in length. Assume that the rate of change of current remains the same. [**Answer:** $L = 2.2$ mH, $|\mathcal{E}| = 0.016$ V]

Some related homework problems: Problem 41, Problem 42

ACTIVE EXAMPLE 23–2 Find the Number of Turns

The inductance of a solenoid 5.00 cm long is 0.157 mH. If its cross-sectional area is $1.00 \times 10^{-4}\text{ m}^2$, how many turns does the solenoid have?

Solution *(Test your understanding by performing the calculations indicated in each step.)*

1. Solve $L = \mu_0(N^2/\ell)A$ for the number of turns, N: $N = (L\ell/\mu_0 A)^{1/2}$

2. Substitute numerical values: $N = 2.50 \times 10^2$

Your Turn

Which of the following changes would affect the value of the inductance more: **(a)** doubling the number of turns or **(b)** tripling the cross-sectional area? Verify your answer by giving the factor by which the inductance changes for each of these cases.

(Answers to **Your Turn** *problems are given in the back of the book.)*

A coil with a finite inductance—an **inductor** for short—resists changes in its electric current. Similarly, a particle of finite mass has the property of inertia, by which it resists changes in its velocity. These analogies, between inductance and mass on the one hand and current and velocity on the other, are useful ones, as we shall see in greater detail when we consider the behavior of ac circuits in the next chapter.

23–8 *RL* Circuits

The circuit shown in **Figure 23–19** is referred to as an ***RL* circuit**; it consists of a resistor, R, and an inductor, L, in series with a battery of emf $\mathcal{E}$. Note that the inductor is represented by a symbol reminiscent of a coil with several loops, as shown below:

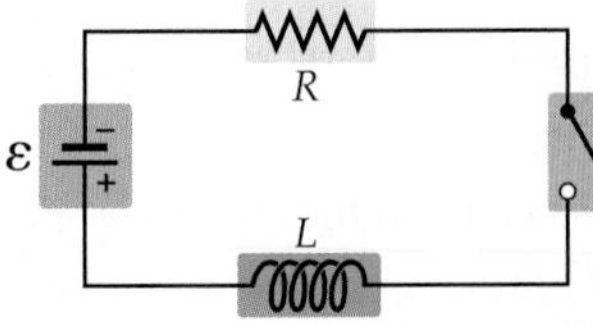

▲ **FIGURE 23–19 An *RL* circuit**

The characteristic time for an RL circuit is $\tau = L/R$. After the switch has been closed for a time much greater than τ, the current in the circuit is simply $I = \mathcal{E}/R$. That is, when the current is steady, the inductor behaves like a zero-resistance wire.

We assume the inductor to be ideal, which means that the wires forming its loops have no resistance—the only resistance in the circuit is the resistor R. It follows, then, that after the switch has been closed for a long time, the current in the circuit is simply $I = \mathcal{E}/R$.

You may have noticed that we said the current would be $\mathcal{E}/R$ after the switch had been closed "for a long time." Why did we make this qualification? The point is that because of the inductor, and its tendency to resist changes in the current, it is not possible for the current to rise immediately to its final value. The inductor slows the process, causing the current to rise over a finite period of time, just as the charge on a capacitor builds up over the characteristic time $\tau = RC$ in an RC circuit.

The corresponding characteristic time for an RL circuit is given by

$$\tau = \frac{L}{R} \qquad 23\text{–}15$$

As expected, the larger the inductance, the longer the time required for the current to build up. In addition, the current approaches its final value of $\mathcal{E}/R$ with an exponential time dependence, much like the case of an RC circuit. With the aid of calculus, it can be shown that the current as a function of time is given by

$$I = \frac{\mathcal{E}}{R}(1 - e^{-t/\tau}) = \frac{\mathcal{E}}{R}(1 - e^{-tR/L}) \qquad 23\text{–}16$$

Note the similarity between this expression, which is plotted in **Figure 23–20**, and the expression given in Equation 21–18 for an RC circuit.

PROBLEM-SOLVING NOTE

Opening or Closing a Switch on an *RL* Circuit

When the switch is first closed in an RL circuit, the inductor has a back emf equal to that of the battery. Long after the switch is closed current flows freely through the inductor, with no resistance at all.

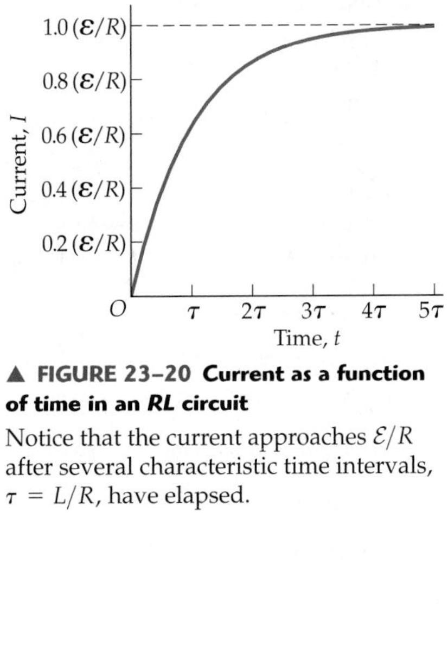

▲ **FIGURE 23–20 Current as a function of time in an *RL* circuit**

Notice that the current approaches $\mathcal{E}/R$ after several characteristic time intervals, $\tau = L/R$, have elapsed.

EXERCISE 23–4

Find the inductance in the RL circuit of Figure 23–19, given that the resistance is 35 Ω and the characteristic time is 7.5 ms.

Solution

Solving $\tau = L/R$ for the inductance, we get

$$L = \tau R = (7.5 \times 10^{-3}\ \text{s})(35\ \Omega) = 0.26\ \text{H}$$

In the next Example we consider a circuit with one inductor and two resistors. By replacing the resistors with their equivalent resistance, however, we can still treat the circuit as a simple RL circuit.

EXAMPLE 23–6 *RL* In Parallel

The circuit shown in the sketch consists of a 12-V battery, a 56-mH inductor, and two 150-Ω resistors in parallel. **(a)** Find the characteristic time for this circuit. **(b)** What is the current in this circuit one characteristic time interval after the switch is closed and **(c)** a long time after the switch is closed?

Picture the Problem

The circuit in question is shown in the first diagram. Noting that two resistors R connected in parallel have an equivalent resistance equal to $R/2$, we can replace the original circuit with the equivalent RL circuit shown in the second diagram.

continued on next page

continued from previous page

Strategy

a. We can find the characteristic time using $\tau = L/R_{eq}$, where $R_{eq} = R/2 = 75\ \Omega$.

b. To find the current after one characteristic time interval, we substitute $t = \tau$ in Equation 23–16.

c. A long time after the switch has been closed ($t \rightarrow \infty$), the exponential term in Equation 23–16 is essentially zero; hence, the current in this case is simply $I = \mathcal{E}/R_{eq}$.

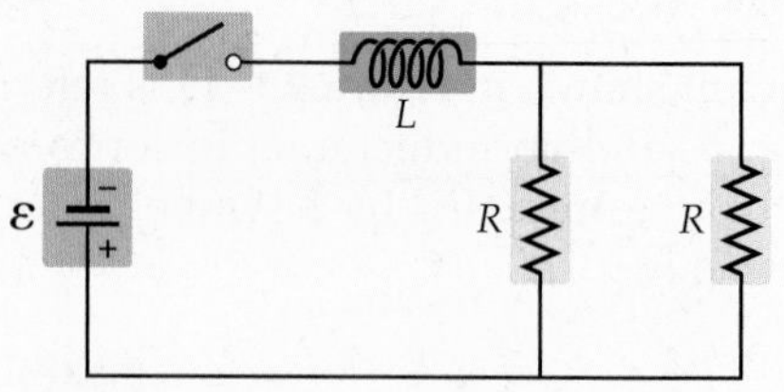

(a)

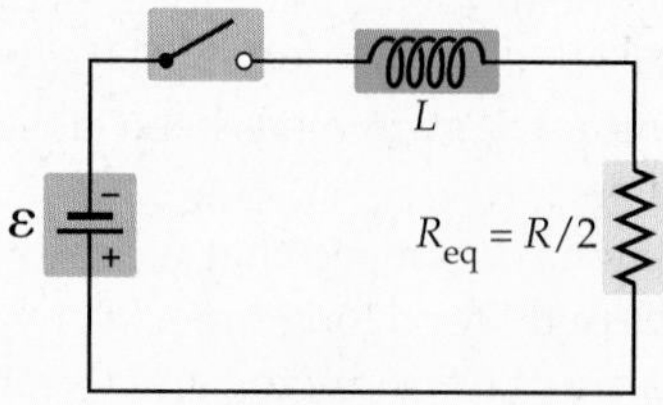

(b)

Solution

Part (a)

1. Calculate the characteristic time, $\tau = L/R_{eq}$, using 75 Ω for the resistance:

$$\tau = \frac{L}{R_{eq}} = \frac{56 \times 10^{-3}\ \text{H}}{75\ \Omega} = 7.5 \times 10^{-4}\ \text{s}$$

Part (b)

2. Substitute $t = \tau$ in Equation 23–16:

$$I = \frac{\mathcal{E}}{R_{eq}}(1 - e^{-t/\tau}) = \left(\frac{12\ \text{V}}{75\ \Omega}\right)(1 - e^{-1}) = 0.10\ \text{A}$$

Part (c)

3. Substitute $t \rightarrow \infty$ in Equation 23–16:

$$I = \frac{\mathcal{E}}{R_{eq}}(1 - e^{-t/\tau}) = \frac{\mathcal{E}}{R_{eq}}(1 - e^{-\infty}) = \frac{\mathcal{E}}{R_{eq}} = \frac{12\ \text{V}}{75\ \Omega} = 0.16\ \text{A}$$

Insight

Notice that the current rises to almost two-thirds of its final value after just one characteristic time interval. It takes only a few more time intervals for the current to essentially "saturate" at its final value, $\mathcal{E}/R_{eq}$.

Practice Problem

If L is doubled to 2(56 mH) = 112 mH, will the current after one characteristic time interval be greater than, less than, or the same as that found in part (b)? [**Answer:** The current is the same when $t = \tau$ because $\mathcal{E}$ and R are unchanged. Note that the value of τ is doubled, however, so it actually takes twice as long for the current to rise to 0.10 A.]

Some related homework problems: Problem 45, Problem 46

23–9 Energy Stored in a Magnetic Field

Work is required to establish a current in an inductor. For example, consider a circuit that consists of just two elements—a battery and an inductor. Even though there are no resistors in the circuit, the battery must still do electrical work to force charge to flow through the inductor, in opposition to its self-induced back emf. What happens to the energy expended to increase the current? It is not lost, nor is it dissipated, as there are no losses in this circuit. Instead, it is stored in the magnetic field, just as the energy in a capacitor is stored in its electric field.

To see how this energy storage works, imagine increasing the current in an inductor L from $I_i = 0$ to $I_f = I$ in the time $\Delta t = T$. The magnitude of the emf required to accomplish this increase in current is

$$\mathcal{E} = L\frac{\Delta I}{\Delta t} = L\frac{I - 0}{T} = \frac{LI}{T} \qquad 23\text{–}17$$

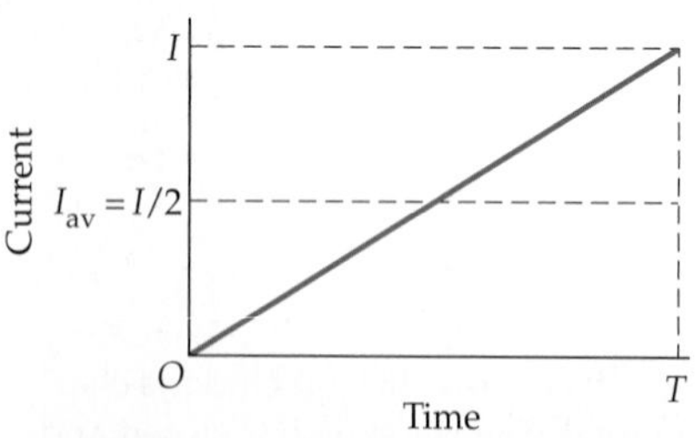

▲ **FIGURE 23–21 A circuit with only a battery and an inductor**

In a circuit with only a battery and an inductor, the current rises linearly with time, because the characteristic time ($\tau = L/R$) with $R = 0$ is infinite. If the current after a time T is I, the average current during this time is $I_{av} = I/2$.

During this time the average current is $I/2$, as indicated in **Figure 23–21**. Thus, the average power supplied by the emf, $P_{av} = I_{av}V$, is

$$P_{av} = \tfrac{1}{2}I\mathcal{E} = \tfrac{1}{2}LI^2/T \qquad 23\text{–}18$$

The total *energy* delivered by the emf is simply the average power times the total time; hence, the energy U required to increase the current in the inductor from 0 to I is

$$U = P_{av}T$$

Using the result given in Equation 23–18, we find

Energy Stored in an Inductor

$$U = \tfrac{1}{2}LI^2 \qquad \textbf{23–19}$$

SI unit: joule, J

Note the similarity of Equation 23–19 to $U = \frac{1}{2}CV^2$ for the energy stored in a capacitor.

EXERCISE 23–5

An inductor carrying a current of 2.5 A stores 0.10 J of energy. What is the inductance of this inductor?

Solution

Solving $U = \frac{1}{2}LI^2$ for the inductance, we find

$$L = \frac{2U}{I^2} = \frac{2(0.10\ \text{J})}{(2.5\ \text{A})^2} = 0.032\ \text{H}$$

As mentioned previously, the energy of an inductor is stored in its magnetic field. To be specific, consider a solenoid of length ℓ with n turns per unit length. The inductance of this solenoid, as determined in Section 23–7, is

$$L = \mu_0 n^2 A\ell$$

Therefore, the energy stored in the solenoid's magnetic field when it carries a current I is

$$U = \tfrac{1}{2}LI^2 = \tfrac{1}{2}(\mu_0 n^2 A\ell)I^2$$

Recalling that the magnetic field inside a solenoid has a magnitude given by $B = \mu_0 nI$, we can write the energy as follows:

$$U = \frac{1}{2\mu_0}B^2 A\ell$$

Finally, we note that the volume where the magnetic field exists—that is, the volume inside the solenoid—is area times length, $A\ell$. Therefore, the magnetic energy per volume is

$$u_B = \frac{\text{magnetic energy}}{\text{volume}} = \frac{B^2}{2\mu_0} \qquad \textbf{23–20}$$

Although derived for the case of an ideal solenoid, this expression applies to any magnetic field no matter how it is generated.

Finally, recall that in the case of an electric field we found that the energy density was $u_E = \varepsilon_0 E^2/2$. Note, in both cases, that the energy density depends on the magnitude of the field squared. We shall return to these expressions in the next chapter and again when we study electromagnetic waves, such as radio waves and light, in Chapter 25.

PROBLEM-SOLVING NOTE

Energy Stored in an Inductor

Remember that an inductor carrying a current always stores energy in its magnetic field. Thus, energy is required to establish a current in an inductor, and energy is released when the current in an inductor decreases.

EXAMPLE 23–7 An Unknown Resistance

After the switch in the circuit shown below has been closed a long time, the energy stored in the inductor is 3.11×10^{-3} J. What is the value of the resistance R?

Picture the Problem

Our sketch shows the circuit for this system, with a 36.0-V battery, a 92.5-Ω resistor, and a 75.0-mH inductor all connected in series with an unknown resistor, R. Initially, the switch is open, but after it has been closed a long time the current settles down to a constant value, as does the amount of energy stored in the inductor.

continued on next page

continued from previous page

Strategy

First, the energy stored in the inductor, $U = \frac{1}{2}LI^2$, gives us the current in the circuit.

Second, we note that after the switch has been closed a long time the current has a constant value. Thus, there is no opposing emf produced by the inductor. As a result, the final current is simply the emf divided by the equivalent resistance of two resistors in series; $I = \mathcal{E}/(92.5\ \Omega + R)$. This relation can be used to solve for R.

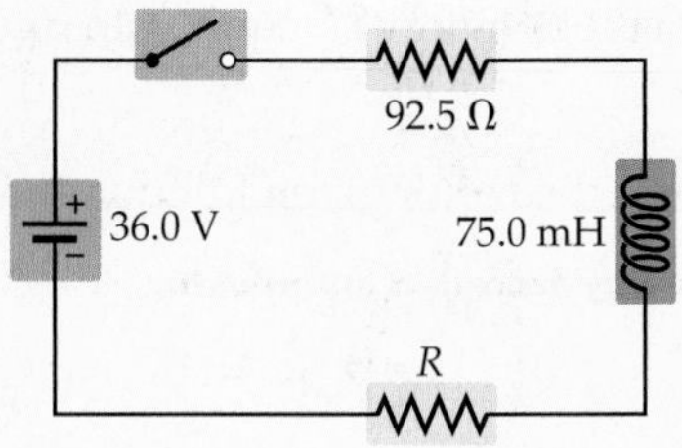

Solution

1. Use the energy stored in the inductor to solve for the current in the circuit:

$$U = \frac{1}{2}LI^2$$

$$I = \sqrt{\frac{2U}{L}}$$

2. Substitute numerical values:

$$I = \sqrt{\frac{2U}{L}} = \sqrt{\frac{2(3.11 \times 10^{-3}\ \text{J})}{75.0 \times 10^{-3}\ \text{H}}} = 0.288\ \text{A}$$

3. Use Ohm's law to solve for the resistance R:

$$I = \frac{\mathcal{E}}{92.5\ \Omega + R}$$

$$R = \frac{\mathcal{E}}{I} - 92.5\ \Omega$$

4. Substitute numerical values:

$$R = \frac{36.0\ \text{V}}{0.288\ \text{A}} - 92.5\ \Omega = 32.5\ \Omega$$

Insight

Note that as long as the switch is closed, energy is dissipated in the resistors at a constant rate. On the other hand, *no* energy is dissipated in the inductor—it simply stores a constant amount of energy, like a mass moving with a constant speed.

Practice Problem

If R is increased, does the energy stored in the inductor increase, decrease, or stay the same? Calculate the stored energy with $R = 106\ \Omega$ as a check on your answer. [**Answer:** With a larger R, the current is less; hence U is less. With $R = 106\ \Omega$ we find $U = 1.23 \times 10^{-3}$ J.]

Some related homework problems: Problem 49, Problem 52

23–10 Transformers

In electrical applications it is often useful to be able to change the voltage from one value to another. For example, high-voltage power lines may operate at voltages as high as 750,000 V, but before the electrical power can be used in the home it must be "stepped down" to 120 V. Similarly, the 120 V from a wall socket may be stepped down again to 9 V or 12 V to power a portable CD player, or stepped up to give the 15,000 V needed in a television tube. The electrical device that performs these voltage conversions is called a **transformer.**

The basic idea of a transformer has already been described in connection with Faraday's induction experiment in Section 23–1. In fact, the device shown in Figure 23–1 can be thought of as a crude transformer. Basically, the magnetic flux created by a primary coil induces a voltage in a secondary coil. If the number of turns is different in the two coils, however, the voltages will be different as well.

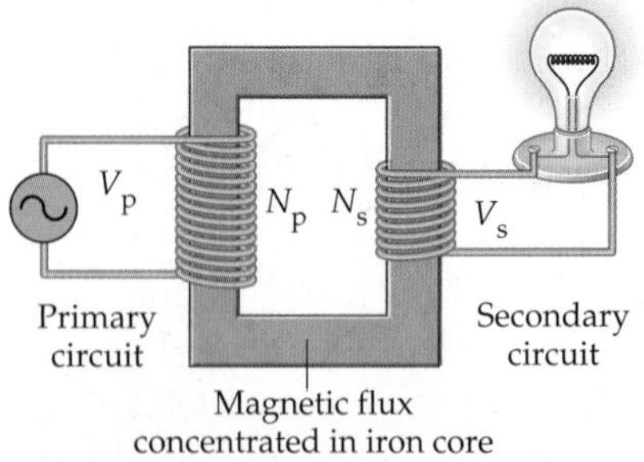

▲ **FIGURE 23–22 The basic elements of a transformer**

An alternating current in the primary circuit creates an alternating magnetic flux and, hence, an alternating induced emf in the secondary circuit. The ratio of emfs in the two circuits, V_s/V_p, is equal to the ratio of the number of turns in each circuit, N_s/N_p.

To see how this works, consider the system shown in **Figure 23–22.** Here an ac generator produces an alternating current in the primary circuit of voltage V_p. The primary circuit then loops around an iron core with N_p turns. The iron core intensifies and concentrates the magnetic flux and ensures, at least to a good approximation, that the secondary coil experiences the same magnetic flux as the primary coil. The secondary coil has N_s turns around its iron core and is connected to a secondary circuit that may operate a CD player, a lightbulb, or some other device. We would now like to determine the voltage, V_s, in the secondary circuit.

To find this voltage, we apply Faraday's law of induction to each of the coils. For the primary coil we have

$$\mathcal{E}_{\mathrm{p}} = -N_{\mathrm{p}}\frac{\Delta\Phi_{\mathrm{p}}}{\Delta t}$$

Similarly, the emf of the secondary coil is

$$\mathcal{E}_{\mathrm{s}} = -N_{\mathrm{s}}\frac{\Delta\Phi_{\mathrm{s}}}{\Delta t}$$

Now, recall that the magnetic fluxes are equal, $\Delta\Phi_{\mathrm{p}} = \Delta\Phi_{\mathrm{s}}$, because of the iron core. As a result, we can divide these two equations (which cancels the flux) to find

$$\frac{\mathcal{E}_{\mathrm{p}}}{\mathcal{E}_{\mathrm{s}}} = \frac{N_{\mathrm{p}}}{N_{\mathrm{s}}}$$

Assuming that the resistance in the coils is negligible, so that the emf in each is essentially the same as its voltage, we obtain the following result:

Transformer Equation

$$\frac{V_{\mathrm{p}}}{V_{\mathrm{s}}} = \frac{N_{\mathrm{p}}}{N_{\mathrm{s}}} \qquad 23\text{–}21$$

Equation 23–21 is known as the **transformer equation.**

It follows that the voltage in the secondary circuit is simply

$$V_{\mathrm{s}} = V_{\mathrm{p}}\left(\frac{N_{\mathrm{s}}}{N_{\mathrm{p}}}\right)$$

Therefore, if the number of turns in the secondary coil is less than the number of turns in the primary coil, the voltage is stepped down, and $V_{\mathrm{s}} < V_{\mathrm{p}}$. Conversely, if the number of turns in the secondary coil is higher, the voltage is stepped up, and $V_{\mathrm{s}} > V_{\mathrm{p}}$.

Now, before you begin to think that transformers give us something for nothing, we note that there is more to the story. Because energy must always be conserved, the average power in the primary circuit must be the same as the average power in the secondary circuit. Since power can be written as $P = IV$, it follows that

$$I_{\mathrm{p}}V_{\mathrm{p}} = I_{\mathrm{s}}V_{\mathrm{s}}$$

Therefore, the ratio of the current in the secondary coil to that in the primary coil is

Transformer Equation (Current and Voltage)

$$\frac{I_{\mathrm{s}}}{I_{\mathrm{p}}} = \frac{V_{\mathrm{p}}}{V_{\mathrm{s}}} = \frac{N_{\mathrm{p}}}{N_{\mathrm{s}}} \qquad 23\text{–}22$$

Thus, if the voltage is stepped up, the current is stepped down.

Suppose, for example, that the number of turns in the secondary coil is twice the number of turns in the primary coil. As a result, the transformer doubles the voltage in the secondary circuit, $V_{\mathrm{s}} = 2\,V_{\mathrm{p}}$, and at the same time halves the current, $I_{\mathrm{s}} = I_{\mathrm{p}}/2$. In general, there is always a tradeoff between voltage and current in a transformer:

> If a transformer increases the voltage by a given factor, it decreases the current by the same factor. Similarly, if it decreases the voltage, it increases the current.

This behavior is similar to that of a lever, in which there is a tradeoff between the force that can be exerted and the distance through which it is exerted. In the case of a lever, as in the case of a transformer, the tradeoff is due to energy conservation: The work done on one end of the lever, F_1d_1, must be equal to the work done on the other end of the lever, F_2d_2.

PROBLEM-SOLVING NOTE

Energy Conservation in Transformers

Energy conservation requires that the product of I and V be the same for each coil of a transformer. Thus, as the voltage of the secondary goes up with the number of turns, the current decreases by the same factor.

ACTIVE EXAMPLE 23–3 Find the Number of Turns on the Secondary

A common summertime sound in many backyards is the *zap* heard when an unfortunate insect flies into a high-voltage "bug zapper." Typically, such devices operate with voltages of about 4000 V obtained from a transformer plugged into a standard 120-V outlet. How many turns are on the secondary coil of the transformer if the primary coil has 27 turns?

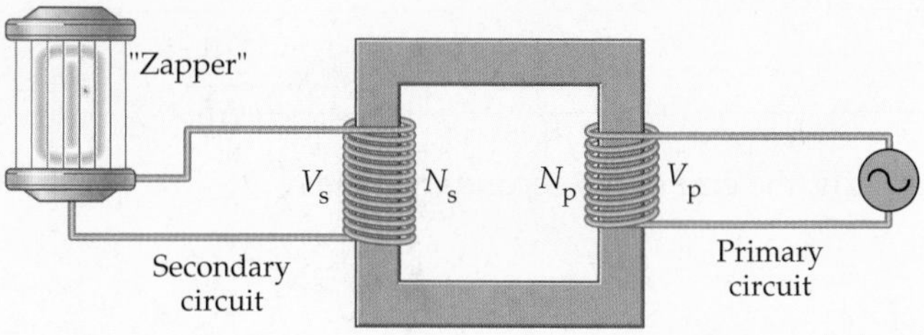

Solution *(Test your understanding by performing the calculations indicated in each step.)*

1. Solve the transformer equation for the number of turns in the secondary coil: $N_s = N_p(V_s/V_p)$
2. Substitute numerical values: $N_s = 900$

Insight

Of course, any transformer with a turns ratio of $N_s/N_p = 900/27$ will produce the same secondary voltage.

Your Turn

Suppose the primary voltage is doubled and the number of turns in the primary circuit is quadrupled. By what factor must the number of turns in the secondary circuit be changed if the secondary voltage is to remain the same?

(Answers to **Your Turn** *problems are given in the back of the book.)*

Notice that the operation of a transformer depends on a *changing* magnetic flux to create an induced emf in the secondary coil. If the current is constant in direction and magnitude, there is simply no induced emf. This is an important advantage that ac circuits have over dc circuits and is one reason that most electrical power systems today operate with alternating currents.

REAL-WORLD PHYSICS
High-voltage electric power transmission

Finally, transformers play an important role in the *transmission* of electrical energy from the power plants that produce it to the communities and businesses where it is used. As was pointed out in Equation 21–3, the resistance of a wire is directly proportional to its length. Therefore, when electrical energy is transmitted over a large distance, the finite resistivity of the wires that carry the current becomes significant. If the wire carries a current I and has a resistance R, the power dissipated as waste heat is $P = I^2R$. One way to reduce this energy loss in a given wire is to reduce the current, I. A transformer that steps up the voltage of a power plant by a factor of 15 (from 10,000 V to 150,000 V, for example) at the same time reduces the current by a factor of 15 and reduces the energy dissipation by a factor of $15^2 = 225$. When the electrical energy reaches the location where it is to be used, step-down transformers can convert the voltage to levels (such as 120 V or 240 V) typically used in the home or workplace.

▲ The transmission of electric power over long distances would not be feasible without transformers. A step-up transformer near the power plant (left) boosts the plant's output voltage from 12,000 V to the 240,000 V carried by high-voltage lines that criss-cross the country (center). A series of step-down transformers near the destination reduce the voltage, first to 2400 V at local substations for distribution to neighborhoods, and then to the 240 V supplied to most houses. The familiar gray cylinders commonly seen on utility poles (right) are the transformers responsible for this last voltage reduction.

Chapter Summary

23–1 Induced Electromotive Force

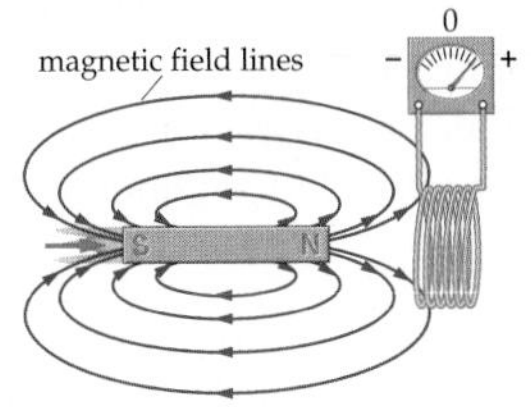

A changing magnetic field can induce a current in a circuit. The driving force behind the current is referred to as the induced electromotive force.

Basic Features of Magnetic Induction
An induced current occurs when there is a *change* in the magnetic field. The magnitude of the induced current depends on the *rate* of change of the magnetic field.

23–2 Magnetic Flux

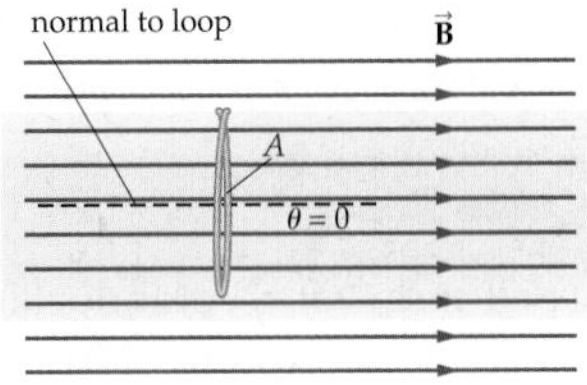

Magnetic flux is a measure of the number of magnetic field lines that cross a given area.

Magnetic Flux, Defined
If a magnetic field of strength B crosses an area A at an angle θ relative to the area's normal, the magnetic flux is

$$\Phi = (B\cos\theta)A = BA\cos\theta \qquad 23\text{–}1$$

Units of Magnetic Flux
Magnetic flux is measured in webers (Wb) defined as follows:

$$1\ \text{Wb} = 1\ \text{T}\cdot\text{m}^2 \qquad 23\text{–}2$$

23–3 Faraday's Law of Induction

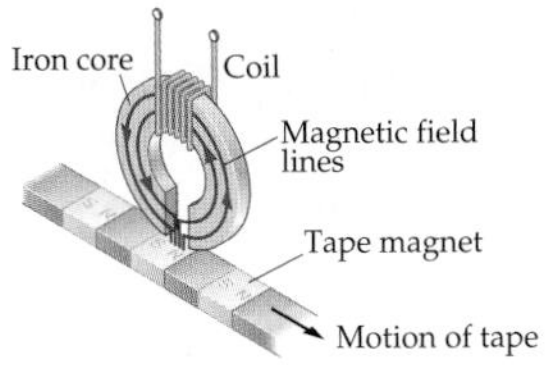

Faraday's law of induction gives precise mathematical form to the basic features of magnetic induction discussed in Section 23–1. Faraday's law is a law of nature, not an approximate rule of thumb.

Faraday's Law
If the magnetic flux in a coil of N turns changes by the amount $\Delta\Phi$ in the time Δt, the induced emf is

$$\mathcal{E} = -N\frac{\Delta\Phi}{\Delta t} = -N\frac{\Phi_{\text{final}} - \Phi_{\text{initial}}}{t_{\text{final}} - t_{\text{initial}}} \qquad 23\text{–}3$$

Magnitude of the Induced Emf
In many cases we are concerned only with the magnitude of the magnetic flux and the induced current. In such cases we use the following form of Faraday's law:

$$|\mathcal{E}| = N\left|\frac{\Delta\Phi}{\Delta t}\right| = N\left|\frac{\Phi_{\text{final}} - \Phi_{\text{initial}}}{t_{\text{final}} - t_{\text{initial}}}\right| \qquad 23\text{–}4$$

23–4 Lenz's Law

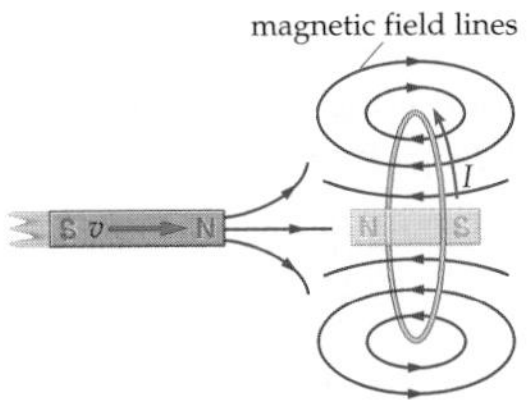

Lenz's law states that an induced current flows in the direction that opposes the change that caused the current.

Eddy Currents
Eddy currents are circulating electric currents formed in a conducting material that experiences a changing magnetic flux.

23–5 Mechanical Work and Electrical Energy

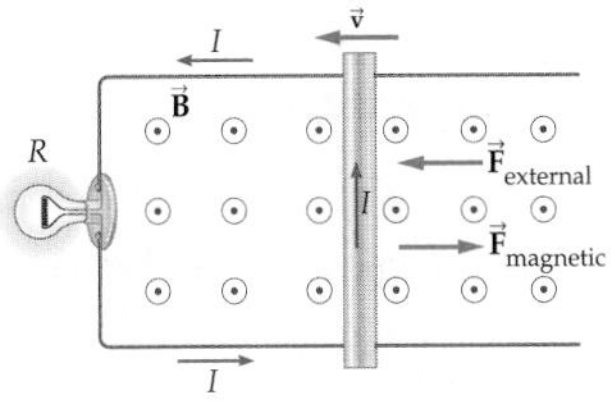

With the help of a magnetic field, mechanical work can be converted into electrical energy in the form of currents flowing through a circuit.

Motional Emf
If a conductor of length ℓ is moved at right angles to a magnetic field of strength B with a speed v, the magnitude of the induced emf in the conductor is

$$|\mathcal{E}| = Bv\ell \qquad 23\text{–}5$$

The corresponding electric field is

$$E = Bv \qquad 23\text{–}6$$

Mechanical Work/Electrical Energy
Assuming no frictional losses, the work done in moving a conductor through a magnetic field is equal to the electrical energy that can be delivered to a circuit.

23–6 Generators and Motors

Generators and motors are practical devices for converting between electrical and mechanical energy.

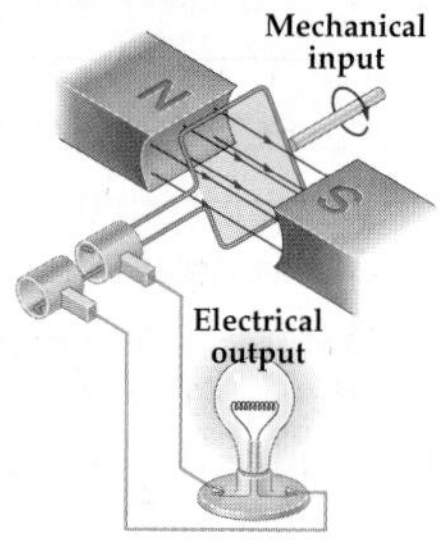

Electric Generators

Electric generators use mechanical work to produce electrical energy. The emf produced by a generator is

$$\mathcal{E} = NBA\omega \sin \omega t \qquad \text{23–11}$$

where N is the number of turns in the coil, B is the magnetic field, A is the area of the coil, and ω is the angular speed of the coil.

Electric Motors

An electric motor is basically an electric generator operated in reverse; it uses electrical energy to produce mechanical work.

23–7 Inductance

Self-inductance, or inductance for short, is the effect produced when a coil with a changing current induces an emf in itself.

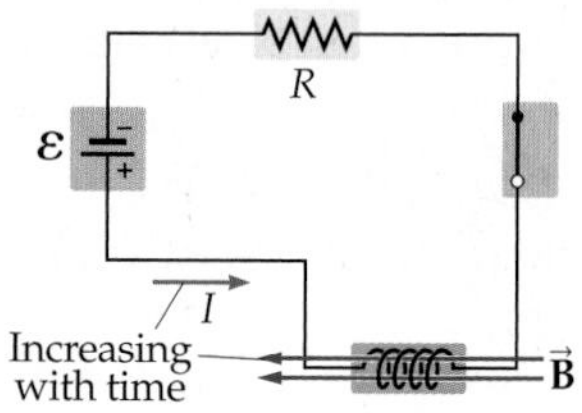

Inductance

Inductance, L, is defined as follows:

$$|\mathcal{E}| = L\left|\frac{\Delta I}{\Delta t}\right| \qquad \text{23–12}$$

The SI unit of inductance is the henry (H): $1\ \text{H} = 1\ \text{V}\cdot\text{s/A}$.

Inductance of a Solenoid

The inductance of a solenoid of length ℓ with N turns of wire and a cross-sectional area A, is

$$L = \mu_0 n^2 A\ell = \mu_0\left(\frac{N^2}{\ell}\right)A \qquad \text{23–14}$$

Note that the number of turns per length is $n = N/\ell$.

23–8 *RL* Circuits

A simple *RL* circuit consists of a battery, a switch, a resistor R, and an inductor L connected in series.

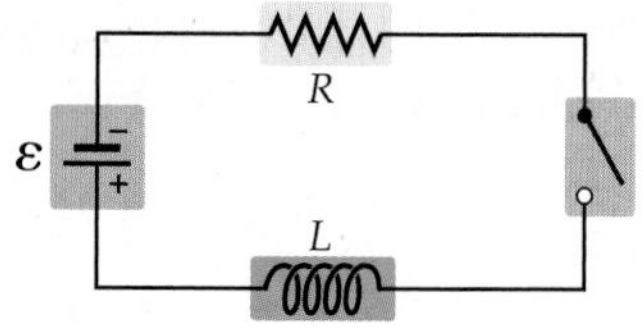

Time Constant

The characteristic time over which the current changes in an *RL* circuit is

$$\tau = \frac{L}{R} \qquad \text{23–15}$$

Increasing Current versus Time

When the switch is closed in an *RL* circuit at time $t = 0$, the current increases with time as follows:

$$I = \frac{\mathcal{E}}{R}(1 - e^{-t/\tau}) = \frac{\mathcal{E}}{R}(1 - e^{-tR/L}) \qquad \text{23–16}$$

In this expression $\mathcal{E}$ is the emf of the battery.

23–9 Energy Stored in a Magnetic Field

Energy can be stored in a magnetic field, just as it is in an electric field.

Energy Stored in an Inductor

An inductor of inductance L carrying a current I stores the energy

$$U = \tfrac{1}{2}LI^2 \qquad \text{23–19}$$

Energy Density

Whenever a magnetic field of strength B exists in a given region, there is a corresponding magnetic energy density given by

$$u_B = \frac{\text{magnetic energy}}{\text{volume}} = \frac{B^2}{2\,\mu_0} \qquad \text{23–20}$$

23–10 Transformers

A transformer is an electrical device used to convert the voltage and current in one circuit to a different voltage and current in another circuit using magnetic induction.

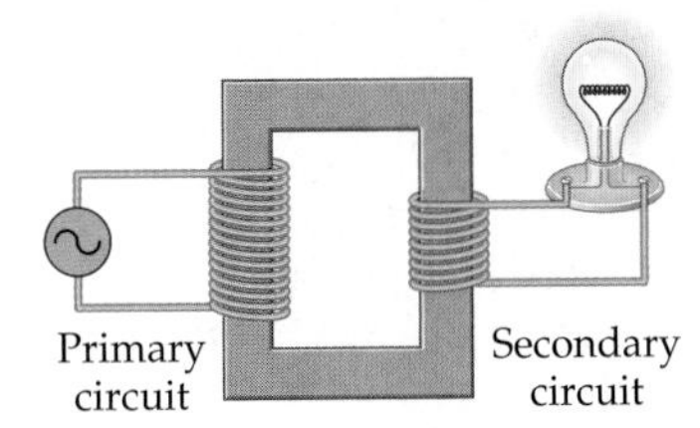

Transformer Equation
The equation relating the voltage V, current I, and number of turns N in the primary (p) and secondary (s) circuits of a transformer is

$$\frac{I_s}{I_p} = \frac{V_p}{V_s} = \frac{N_p}{N_s} \qquad 23\text{–}22$$

Problem-Solving Summary

Type of Problem	Relevant Physical Concepts	Related Examples
Calculate the magnetic flux.	The magnetic flux of a field B passing through an area A at an angle θ to the normal is $\Phi = BA\cos\theta$.	Example 23–1
Find the induced emf.	Faraday's law states that the induced emf is proportional to the rate of change of magnetic flux with time, $\mathcal{E} = -N\frac{\Delta\Phi}{\Delta t}$. In the special case of a rotating coil, the induced emf is $\mathcal{E} = NBA\omega\sin\omega t$.	Examples 23–2, 23–4
Calculate the inductance of a solenoid.	A solenoid with n turns per length, area A and length ℓ has an inductance given by $L = \mu_0 n^2 A\ell$.	Example 23–5 Active Example 23–2
Find the current as a function of time in an RL circuit.	The time constant of an RL circuit is $\tau = L/R$. When the current increases from zero, its time dependence is $I = \frac{\mathcal{E}}{R}(1 - e^{-t/\tau})$.	Example 23–6
Relate the energy stored in an inductor to its inductance and current.	The energy stored in an inductor L carrying a current I is $U = \frac{1}{2}LI^2$.	Example 23–7
Relate voltage and current to the number of turns in the primary and secondary coils of a transformer.	The voltage in the secondary coil is proportional to the number of turns in the secondary coil and inversely proportional to the number of turns in the primary coil. The current and voltage in the secondary coil are inversely proportional to each other.	Active Example 23–3

Conceptual Questions

(Answers to odd-numbered Conceptual Questions can be found in the back of the book.)

1. Explain the difference between a magnetic field and a magnetic flux.
2. A metal ring with a break in its perimeter is dropped from a field-free region of space into a region with a magnetic field. What effect does the magnetic field have on the ring?
3. In a common classroom demonstration, a magnet is dropped down a long, vertical copper tube. The magnet moves very slowly as it moves through the tube, taking several seconds to reach the bottom. Explain this behavior.
4. Many equal-arm balances have a small metal plate attached to one of the two arms. The plate passes between the poles of a magnet mounted in the base of the balance. Explain the purpose of this arrangement.
5. **Figure 23–23** shows a vertical iron rod with a wire coil of many turns wrapped around its base. A metal ring slides over the rod and rests on the wire coil. Initially the switch connecting the coil to a battery is open, but when it is closed, the ring flies into the air. Explain why this happens.

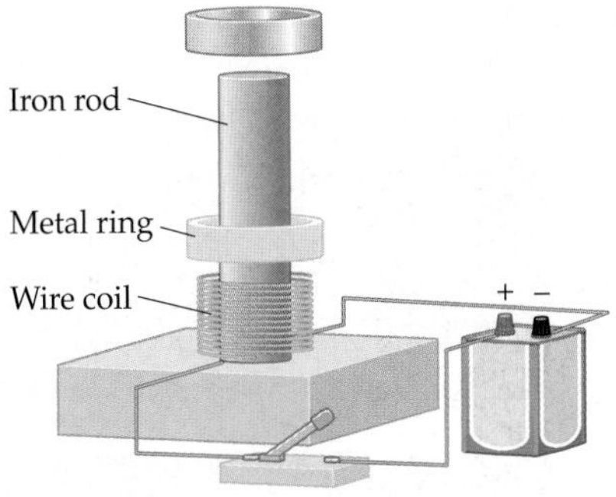

▲ **FIGURE 23–23** Conceptual Questions 5 and 6

6. Referring to Conceptual Question 5, suppose the metal ring has a break in its circumference. Describe what happens when the switch is closed in this case.
7. A metal rod of resistance R can slide without friction on two zero-resistance rails, as shown in **Figure 23–24**. The rod and the

rails are immersed in a region of constant magnetic field pointing out of the page. Describe the motion of the rod when the switch is closed. Your discussion should include the effects of motional emf.

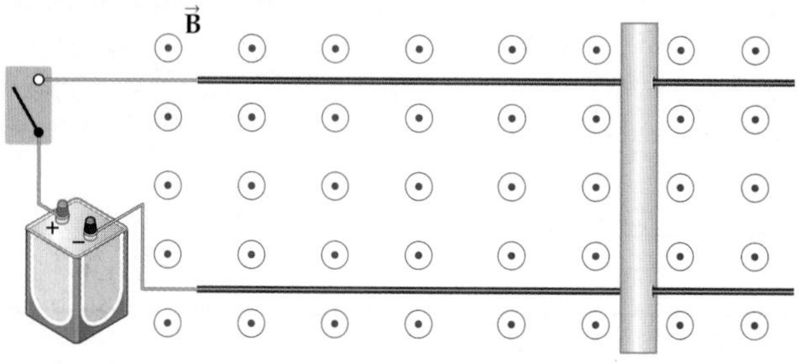

▲ **FIGURE 23–24** Conceptual Question 7

8. A penny is placed on edge in the powerful magnetic field of an MRI solenoid. If the penny is tipped over, it takes several seconds for it to land on one of its faces. Explain.
9. Recently, NASA tested a power generation system that involves connecting a small satellite to the space shuttle with a conducting wire several miles long. Explain how such a system can generate electrical power.
10. Explain what happens when the angular speed of the coil in an electric generator is increased.
11. The inductor in an *RL* circuit determines how long it takes for the current to reach a given value, but it has no effect on the final value of the current. Explain.
12. When the switch in a circuit containing an inductor is opened, it is common for a spark to jump across the contacts of the switch. Why?

Conceptual Exercises

(Answers to odd-numbered Conceptual Exercises can be found in the back of the book.)

1. An airplane flies level to the ground toward the north pole. Is the induced emf from wing tip to wing tip when the plane is at the equator greater than, less than, or equal to the wing-tip-to-wing-tip emf when it is at the latitude of New York? Explain.
2. A wire loop is placed in a magnetic field that is perpendicular to its plane. The field varies with time as shown in **Figure 23–25**. Rank the six regions of time in order of increasing magnitude of the induced emf. Indicate ties where appropriate.

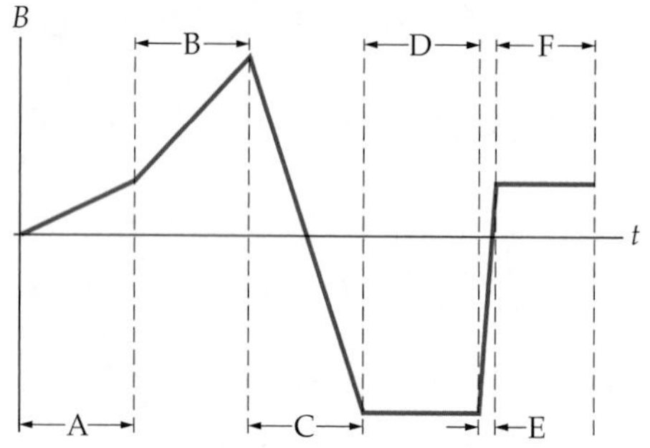

▲ **FIGURE 23–25** Conceptual Exercise 2

3. **Figure 23–26** shows four different situations in which a metal ring moves to the right with constant speed through a region with a varying magnetic field. The intensity of the color indicates the intensity of the field, and in each case the field either increases or decreases at a uniform rate from the left edge of the colored region to the right edge. The direction of the field in each region is indicated. For each of the four cases, state whether the induced emf is clockwise, counterclockwise, or zero.

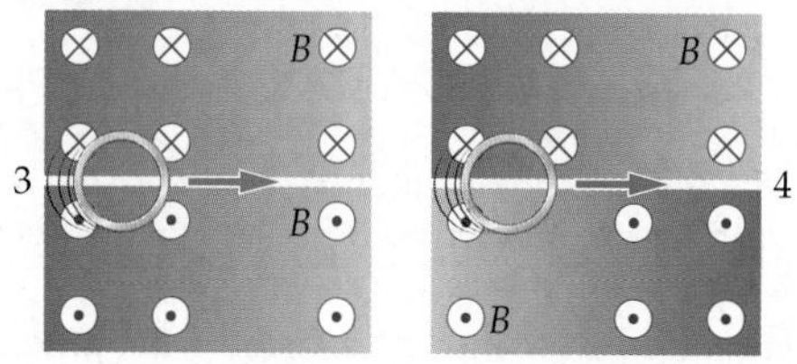

▲ **FIGURE 23–26** Conceptual Exercise 3

4. You hold a circular loop of wire at the north magnetic pole of the Earth. Consider the magnetic flux through this loop due to the Earth's magnetic field. Is the flux when the normal to the loop points horizontally greater than, less than, or equal to the flux when the normal points vertically downward? Explain.
5. You hold a circular loop of wire at the equator. Consider the magnetic flux through this loop due to the Earth's magnetic field. Is the flux when the normal to the loop points north greater than, less than, or equal to the flux when the normal points vertically upward? Explain.
6. A small metal ring is dropped into a region of constant magnetic field, as indicated in **Figure 23–27**. For each of the three indicated locations, **(a)** is the induced current clockwise, counterclockwise, or zero, and **(b)** is the magnetic force exerted on the ring upward, downward, or zero?

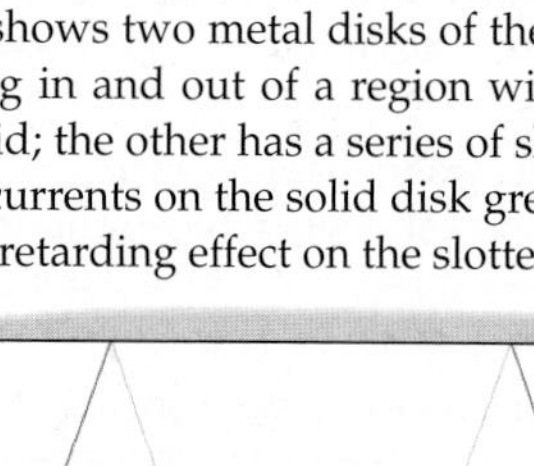

▲ **FIGURE 23–27** Conceptual Exercise 6

7. **Figure 23–28** shows two metal disks of the same size and material oscillating in and out of a region with a magnetic field. One disk is solid; the other has a series of slots. Is the retarding effect of eddy currents on the solid disk greater than, less than, or equal to the retarding effect on the slotted disk? Explain.

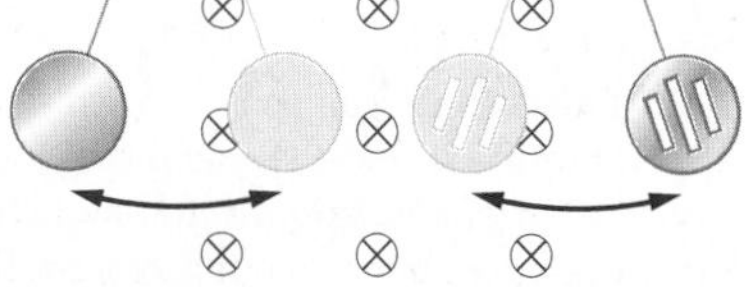

▲ **FIGURE 23–28** Conceptual Exercises 7, 8, and 9

8. Consider the solid disk in Figure 23–28. When this disk has swung to the right as far as it can go, is the induced current in it a maximum or a minimum? Explain.
9. As the solid metal disk in Figure 23–28 swings to the right, from the region with no field into the region with a finite magnetic field, does the induced current in the disk circulate clockwise or counterclockwise? Explain.
10. A conducting rod slides on two wires in a region with a magnetic field. The two wires are not connected. Is a force required to keep the rod moving with constant speed? Explain.
11. The number of turns per meter in a solenoid of fixed length is doubled. At the same time, the current in the solenoid is halved. Does the energy stored in the inductor increase, decrease, or stay the same? Explain.
12. The inductor shown in **Figure 23–29** is connected to an electrical circuit with a changing current. At the moment in question, the inductor has an induced emf with the indicated direction. Is the current in the circuit at this time **(a)** increasing and to the right, **(b)** increasing and to the left, **(c)** decreasing and to the right, **(d)** decreasing and to the left?

▲ **FIGURE 23–29** Conceptual Exercise 12

13. Transformer 1 has a primary voltage V_p and a secondary voltage V_s. Transformer 2 has twice the number of turns on both its primary and secondary coils compared with transformer 1. If the primary voltage on transformer 2 is $2V_p$, what is its secondary voltage? Explain.
14. Transformer 1 has a primary current I_p and a secondary current I_s. Transformer 2 has twice as many turns on its primary coil as transformer 1, and both transformers have the same number of turns on the secondary coil. If the primary current on transformer 2 is $3I_p$, what is its secondary current? Explain.
15. The four electrical circuits shown in **Figure 23–30** have identical batteries, resistors, and inductors. Rank the circuits in order of increasing current supplied by the battery long after the switch is closed. Indicate ties where appropriate.

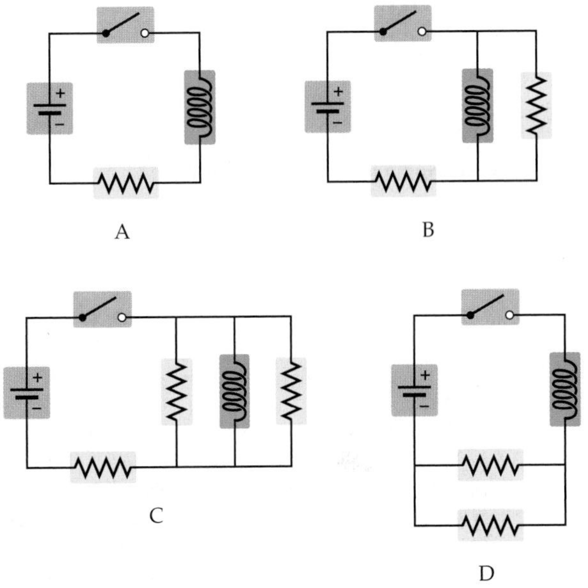

▲ **FIGURE 23–30** Conceptual Exercise 15

Problems

Note: **IP** *denotes an integrated conceptual/quantitative problem.* **BIO** *identifies problems of biological or medical interest. Red bullets (•, ••, •••) are used to indicate the level of difficulty of each problem.*

Section 23–2 Magnetic Flux

1. • A 0.055-T magnetic field passes through a circular ring of radius 3.1 cm at an angle of 16° with the normal. Find the magnitude of the magnetic flux through the ring.

2. • A uniform magnetic field of 0.0250 T points vertically upward. Find the magnitude of the magnetic flux through each of the five sides of the open-topped rectangular box shown in **Figure 23–31**, given that the dimensions of the box are $L = 32.5$ cm, $W = 12.0$ cm, and $H = 10.0$ cm.

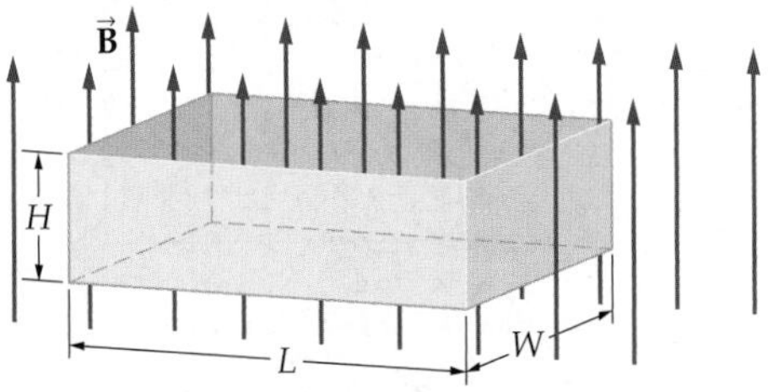

▲ **FIGURE 23–31** Problem 2

3. • A magnetic field is oriented at an angle of 42° to the normal of a rectangular area 5.5 cm by 6.8 cm. If the magnetic flux through this surface has a magnitude of $4.8 \times 10^{-5}\ \text{T} \cdot \text{m}^2$, what is the strength of the magnetic field?

4. • Find the magnitude of the magnetic flux through the floor of a house that measures 22 m by 18 m. Assume that the Earth's magnetic field at the location of the house has a horizontal component of 2.6×10^{-5} T pointing north, and a downward vertical component of 4.2×10^{-5} T.

5. • **MRI Solenoid** The magnetic field produced by an MRI solenoid 2.5 m long and 1.2 m in diameter is 1.7 T. Find the magnitude of the magnetic flux through the core of this solenoid.

6. •• At a certain location, the Earth's magnetic field has a magnitude of 5.9×10^{-5} T and points in a direction that is 72° below the horizontal. Find the magnitude of the magnetic flux through the top of a desk at this location that measures 130 cm by 82 cm.

7. •• **IP** A solenoid with 375 turns per meter and a diameter of 15.0 cm has a magnetic flux through its core of magnitude $1.28 \times 10^{-4}\ \text{T} \cdot \text{m}^2$. **(a)** Find the current in this solenoid. **(b)** How would your answer to part (a) change if the diameter of the solenoid were doubled? Explain.

8. ••• A single-turn square loop of side L is centered on the axis of a long solenoid. In addition, the plane of the square loop is perpendicular to the axis of the solenoid. The solenoid has 1250 turns per meter and a diameter of 6.00 cm, and carries a current of 2.50 A. Find the magnetic flux through the loop when **(a)** $L = 3.00$ cm, **(b)** $L = 6.00$ cm, and **(c)** $L = 12.0$ cm.

Section 23–3 Faraday's Law of Induction

9. • A 0.25-T magnetic field is perpendicular to a circular loop of wire with 53 turns and a radius of 15 cm. If the magnetic field is reduced to zero in 0.12 s, what is the magnitude of the induced emf?

10. • **Figure 23–32** shows the magnetic flux through a coil as a function of time. At what times shown in this plot do **(a)** the magnetic flux and **(b)** the induced emf have the greatest magnitude?

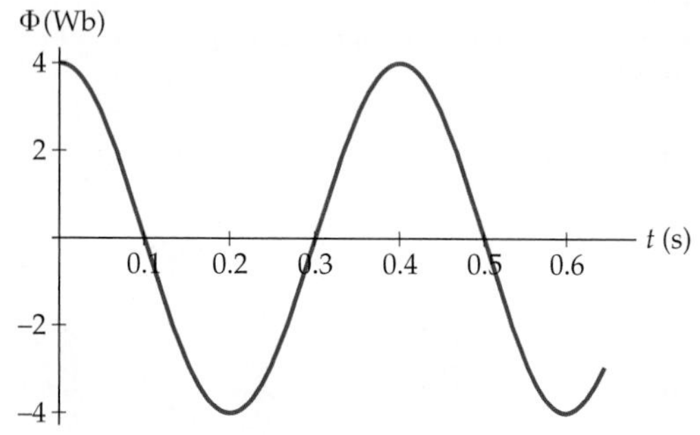

▲ **FIGURE 23–32** Problems 10 and 13

11. • **Figure 23–33** shows the magnetic flux through a single-loop coil as a function of time. What is the induced emf in the coil at **(a)** $t = 0.050$ s, **(b)** $t = 0.15$ s, and **(c)** $t = 0.50$ s?

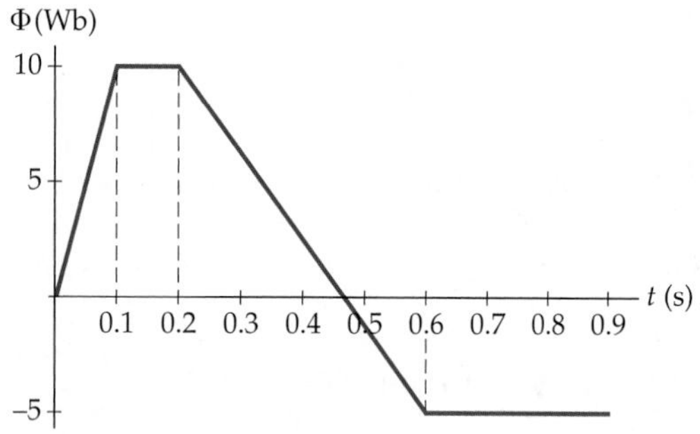

▲ **FIGURE 23–33** Problems 11 and 12

12. •• **IP** The magnetic flux through a single-loop coil is given by Figure 23–33. **(a)** Is the magnetic flux at $t = 0.25$ s greater than, less than, or the same as the magnetic flux at $t = 0.55$ s? Explain. **(b)** Is the induced emf at $t = 0.25$ s greater than, less than, or the same as the induced emf at $t = 0.55$ s? Explain. **(c)** Calculate the induced emf at the times $t = 0.25$ s and $t = 0.55$ s.

13. •• **IP** Consider a single-loop coil whose magnetic flux is given by Figure 23–32. **(a)** Is the magnitude of the induced emf in this coil greater near $t = 0.4$ s or near $t = 0.5$ s? Explain. **(b)** At what times in this plot do you expect the induced emf in the coil to have a maximum magnitude? Explain. **(c)** Estimate the induced emf in the coil at times near $t = 0.3$ s, $t = 0.4$ s and $t = 0.5$ s.

14. •• A single conducting loop of wire has an area of 7.4×10^{-2} m^2 and a resistance of 110 Ω. Perpendicular to the plane of the loop is a magnetic field of strength 0.28 T. At what rate (in T/s) must this field change if the induced current in the loop is to be 0.32 A?

15. •• The area of a 120-turn coil oriented with its plane perpendicular to a 0.20-T magnetic field is 0.050 m^2. Find the average induced emf in this coil if the magnetic field reverses its direction in 0.34 s.

16. •• An emf is induced in a conducting loop of wire 1.12 m long as its shape is changed from square to circular. Find the average magnitude of the induced emf if the change in shape occurs in 4.25 s and the local 0.105-T magnetic field is perpendicular to the plane of the loop.

17. •• A magnetic field increases from 0 to 0.20 T in 1.5 s. How many turns of wire are needed in a circular coil 12 cm in diameter to produce an induced emf of 6.0 V?

Section 23–4 Lenz's Law

18. • A bar magnet with its north pole pointing downward is falling toward the center of a horizontal conducting ring. As viewed from above, is the direction of the induced current in the ring clockwise or counterclockwise? Explain.

19. • **A Wire Loop and a Magnet** A loop of wire is dropped and allowed to fall between the poles of a horseshoe magnet, as shown in **Figure 23–34**. State whether the induced current in the loop is clockwise or counterclockwise when **(a)** the loop is above the magnet and **(b)** the loop is below the magnet.

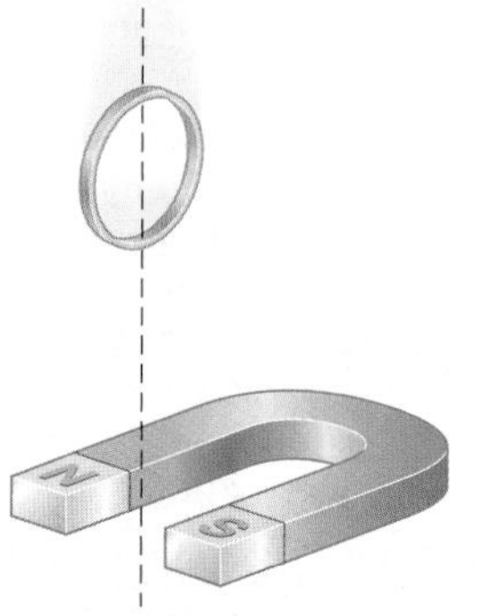

▲ **FIGURE 23–34** Problems 19, 20, and 21

20. •• Suppose we change the situation shown in Figure 23–34 as follows: Instead of allowing the loop to fall on its own, we attach a string to it and lower it with constant speed along the path indicated by the dashed line. Is the tension in the string greater than, less than, or equal to the weight of the loop? Give specific answers for times when **(a)** the loop is above the magnet and **(b)** the loop is below the magnet. Explain in each case.

21. •• Rather than letting the loop fall downward in Figure 23–34, suppose we attach a string to it and raise it upward with constant speed along the path indicated by the dashed line. Is the tension in the string greater than, less than, or equal to the weight of the loop? Give specific answers for times when **(a)** the loop is below the magnet and **(b)** the loop is above the magnet. Explain in each case.

22. •• **Figure 23–35** shows a current-carrying wire and a circuit containing a resistor R. **(a)** If the current in the wire is constant, is the induced current in the circuit clockwise, counterclockwise, or zero? Explain. **(b)** If the current in the wire increases, is the induced current in the circuit clockwise, counterclockwise, or zero? Explain.

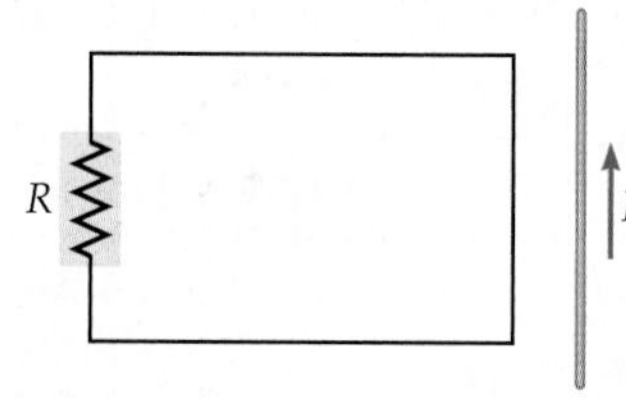

▲ **FIGURE 23–35** Problems 22 and 23

23. •• Consider the physical system shown in Figure 23–35. If the current in the wire changes direction, is the induced current in the circuit clockwise, counterclockwise, or zero? Explain.

24. •• A long, straight, current-carrying wire passes through the center of a circular coil. The wire is perpendicular to the plane of the coil. **(a)** If the current in the wire is constant, is the induced emf in the coil zero or nonzero? Explain. **(b)** If the current in the wire increases, is the induced emf in the coil zero or nonzero? Explain. **(c)** Does your answer to part (b) change if the

wire no longer passes through the center of the coil but is still perpendicular to its plane? Explain.

25. •• **Figure 23–36** shows a circuit containing a resistor and an uncharged capacitor. Pointing into the plane of the circuit is a uniform magnetic field $\vec{\mathbf{B}}$. If the magnetic field reverses direction in a short period of time, which plate of the capacitor (top or bottom) becomes positively charged? Explain.

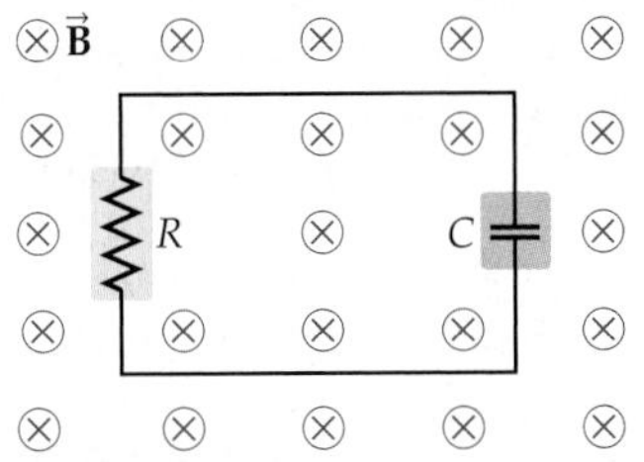

▲ **FIGURE 23–36** Problems 25 and 26

26. •• Referring to Problem 25, which plate of the capacitor (top or bottom) becomes positively charged if the magnetic field increases in magnitude with time? Explain.

27. •• A long, straight wire carries a current I, as indicated in **Figure 23–37**. Three small metal rings are placed near the current-carrying wire (A and C) or directly on top of it (B). If the current in the wire is increasing with time, indicate whether the induced emf in each of the rings is clockwise, counterclockwise, or zero. Explain your answer for each ring.

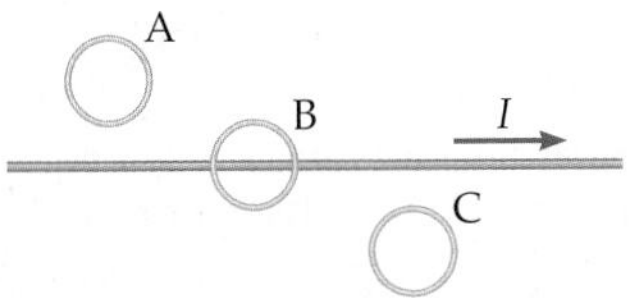

▲ **FIGURE 23–37** Problem 27

Section 23–5 Mechanical Work and Electrical Energy

28. • A metal rod 0.76 m long moves with a speed of 2.0 m/s perpendicular to a magnetic field. If the induced emf between the ends of the rod is 0.45 V, what is the strength of the magnetic field?

29. •• **Airplane EMF** A Boeing KC-135A airplane has a wingspan of 39.9 m and flies at constant altitude in a northerly direction with a speed of 850 km/h. If the vertical component of the Earth's magnetic field is 5.0×10^{-6} T, and its horizontal component is 1.4×10^{-6} T, what is the induced emf between the wing tips?

30. •• **IP Figure 23–38** shows a zero-resistance rod sliding to the right on two zero-resistance rails separated by the distance $L = 0.45$ m. The rails are connected by a 12.5-Ω resistor, and the entire system is in a uniform magnetic field with a magnitude of 0.750 T. **(a)** Find the speed at which the bar must be moved to produce a current of 0.125 A in the resistor. **(b)** Would your answer to part (a) change if the bar was moving to the left instead of to the right? Explain.

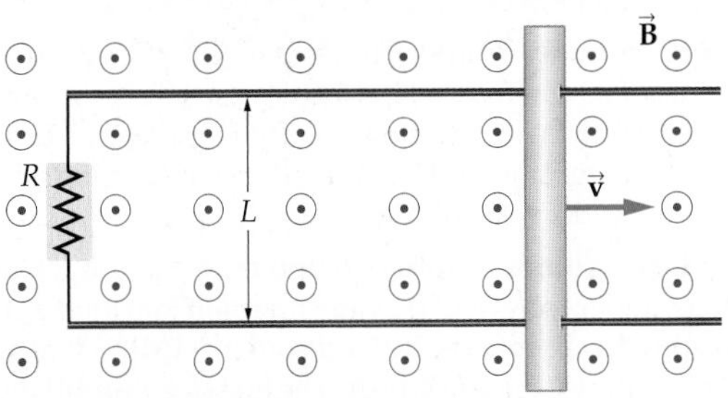

▲ **FIGURE 23–38** Problems 30 and 31

31. •• Referring to part (a) of Problem 30, **(a)** find the force that must be exerted on the rod to maintain a constant current of 0.125 A in the resistor. **(b)** What is the rate of energy dissipation in the resistor? **(c)** What is the mechanical power delivered to the rod?

32. •• **(a)** Find the current that flows in the circuit shown in Example 23–3. **(b)** What speed must the rod have if the current in the circuit is to be 1.0 A?

33. •• Suppose the mechanical power delivered to the rod in Example 23–3 is 7.9 W. Find **(a)** the current in the circuit and **(b)** the speed of the rod.

Section 23–6 Generators and Motors

34. • The maximum induced emf in a generator rotating at 210 rpm is 45 V. How fast must the rotor of the generator rotate if it is to generate a maximum induced emf of 55 V?

35. • A rectangular coil 25 cm by 35 cm has 120 turns. This coil produces a maximum emf of 65 V when it rotates with an angular speed of 190 rad/s in a magnetic field of strength B. Find the value of B.

36. • A 1.6-m wire is wound into a coil with a radius of 3.2 cm. If this coil is rotated at 95 rpm in a 0.070-T magnetic field, what is its maximum emf?

37. •• **IP** A circular coil with a diameter of 22.0 cm and 155 turns rotates about a vertical axis with an angular speed of 1250 rpm. The only magnetic field in this system is that of the Earth. At the location of the coil, the horizontal component of the magnetic field is 3.80×10^{-5} T, and the vertical component is 2.85×10^{-5} T. **(a)** Which component of the magnetic field is important when calculating the induced emf in this coil? Explain. **(b)** Find the maximum emf induced in the coil.

38. •• A generator is designed to produce a maximum emf of 170 V while rotating with an angular speed of 3600 rpm. Each coil of the generator has an area of 0.016 m^2. If the magnetic field used in the generator has a magnitude of 0.050 T, how many turns of wire are needed?

Section 23–7 Inductance

39. • Find the induced emf when the current in a 45.0-mH inductor increases from 0 to 515 mA in 16.5 ms.

40. • How many turns should a solenoid of cross-sectional area 0.035 m^2 and length 0.22 m have if its inductance is to be 45 mH?

41. •• The inductance of a solenoid with 450 turns and a length of 24 cm is 7.3 mH. **(a)** What is the cross-sectional area of the solenoid? **(b)** What is the induced emf in the solenoid if its current drops from 3.2 A to 0 in 55 ms?

42. •• Determine the inductance of a solenoid with 640 turns in a length of 25 cm. The circular cross section of the solenoid has a radius of 4.3 cm.

43. •• A solenoid with a cross-sectional area of 1.81×10^{-3} m^2 is 0.750 m long and has 455 turns per meter. Find the induced emf in this solenoid if the current in it is increased from 0 to 2.00 A in 45.5 ms.

44. ••• **IP** A solenoid has N turns of area A distributed uniformly along its length, ℓ. When the current in this solenoid increases at the rate of 2.0 A/s, an induced emf of 75 mV is observed. **(a)** What is the inductance of this solenoid? **(b)** Suppose the spacing between coils is doubled. The result is a solenoid that is twice as long but with the same area and number of turns. Will the induced emf in this new solenoid be greater than, less than, or equal to 75 mV when the current changes at the rate of 2.0 A/s? Explain. **(c)** Calculate the induced emf for part (b).

Section 23–8 *RL* Circuits

45. • How long does it take for the current in an *RL* circuit with $R = 130\ \Omega$ and $L = 63$ mH to reach half its final value?

46. •• The circuit shown in **Figure 23–39** consists of a 6.0-V battery, a 37-mH inductor, and four 55-Ω resistors. **(a)** Find the characteristic time for this circuit. **(b)** What is the current supplied by this battery two characteristic time intervals after closing the switch and **(c)** a long time after the switch is closed?

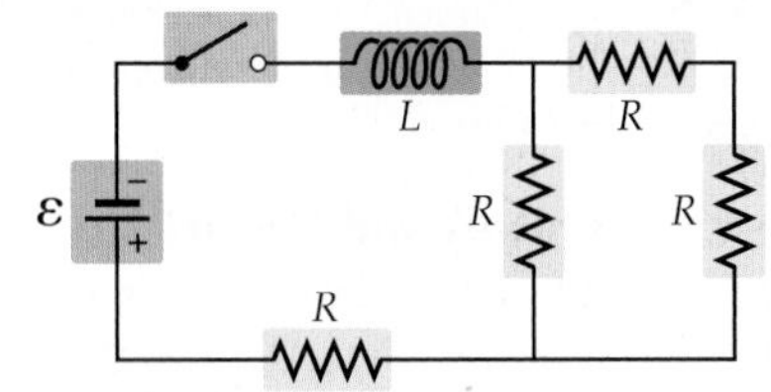

▲ **FIGURE 23–39** Problems 46 and 54

47. •• The current in an *RL* circuit increases to 95% of its final value 2.24 s after the switch is closed. **(a)** What is the time constant for this circuit? **(b)** If the inductance in the circuit is 0.275 H, what is the resistance?

48. ••• Consider the *RL* circuit shown in **Figure 23–40**. When the switch is closed, the current in the circuit is observed to increase from 0 to 0.32 A in 0.15 s. **(a)** What is the inductance *L*? **(b)** How long after the switch is closed does the current have the value 0.50 A? **(c)** What is the maximum current that flows in this circuit?

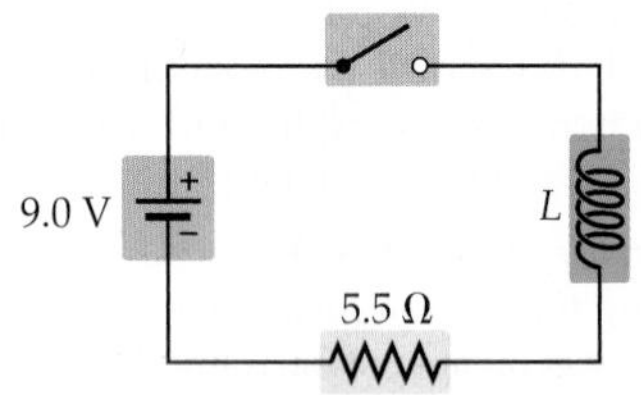

▲ **FIGURE 23–40** Problems 48 and 49

Section 23–9 Energy Stored in a Magnetic Field

49. • Consider the circuit shown in Figure 23–40. Assuming the inductor in this circuit has the value $L = 6.1$ mH, how much energy is stored in the inductor after the switch has been closed a long time?

50. • A solenoid is 1.5 m long and has 490 turns per meter. What is the cross-sectional area of this solenoid if it stores 0.31 J of energy when it carries a current of 12 A?

51. •• **Alcator Fusion Experiment** In the Alcator fusion experiment at MIT, a magnetic field of 50.0 T is produced. **(a)** What is the magnetic energy density in this field? **(b)** Find the magnitude of the electric field that would have the same energy density found in part (a).

52. •• **IP** After the switch in **Figure 23–41** has been closed for a long time, the energy stored in the inductor is 0.110 J. **(a)** What is the value of the resistance *R*? **(b)** If it is desired that more energy be stored in the inductor, should the resistance *R* be greater than or less than the value found in part (a)? Explain.

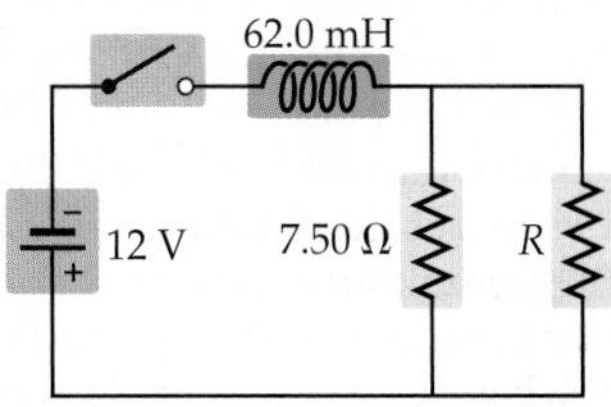

FIGURE 23–41 Problems 52 and 53

53. •• **IP** Suppose the resistor in Figure 23–41 has the value $R = 14\ \Omega$ and that the switch is closed at time $t = 0$. **(a)** How much energy is stored in the inductor at the time $t = \tau$? **(b)** How much energy is stored in the inductor at the time $t = 2\tau$? **(c)** If the value of *R* is increased, does the characteristic time, τ, increase or decrease? Explain.

54. •• **IP** Consider the circuit shown in Figure 23–39, which contains a 6.0-V battery, a 37-mH inductor, and four 55-Ω resistors. **(a)** Is more energy stored in the inductor just after the switch is closed, or long after the switch is closed? Explain. **(b)** Calculate the energy stored in the inductor one characteristic time interval after the switch is closed. **(c)** Calculate the energy stored in the inductor long after the switch is closed.

55. ••• You would like to store 8.9 J of energy in the magnetic field of a solenoid. The solenoid has 560 circular turns of diameter 7.2 cm distributed uniformly along its 28-cm length. **(a)** How much current is needed? **(b)** What is the magnitude of the magnetic field inside the solenoid? **(c)** What is the energy density (energy/volume) inside the solenoid?

Section 23–10 Transformers

56. • The electric motor in a toy train requires a voltage of 3.0 V. Find the ratio of turns on the primary coil to turns on the secondary coil in a transformer that will step the 110 V household voltage down to 3.0 V.

57. • **IP** A disk drive plugged into a 120-V outlet operates on a voltage of 9.0 V. The transformer that powers the disk drive has 125 turns on its primary coil. **(a)** Should the number of turns on the secondary coil be greater than or less than 125? Explain. **(b)** Find the number of turns on the secondary coil.

58. • A transformer with a turns ratio (secondary/primary) of 1:13 is used to step down the voltage from a 120-V wall socket to be used in a battery recharging unit. What is the voltage supplied to the recharger?

59. • A neon sign that requires a voltage of 11,000 V is plugged into a 120-V wall outlet. What turns ratio (secondary/primary) must a transformer have to power the sign?

60. •• A step-down transformer produces a voltage of 6.0 V across the secondary coil when the voltage across the primary coil is 120 V. What voltage appears across the primary coil of this transformer if 120 V is applied to the secondary coil?

61. •• A step-up transformer has 25 turns on the primary coil and 750 turns on the secondary coil. If this transformer is to produce an output of 4800 V with a 12-mA current, what input current and voltage are needed?

General Problems

62. • **Interstellar Magnetic Field** The *Voyager I* spacecraft moves through interstellar space with a speed of 8.0×10^3 m/s. The magnetic field in this region of space has a magnitude of 2.0×10^{-10} T. Assuming that the 5.0-m long antenna on the spacecraft is at right angles to the magnetic field, find the induced emf between its ends.

63. • **BIO Blowfly Flight** The coils used to measure the movements of a blowfly, as described in Section 23–5, have a diameter of 2.0 mm. In addition, the fly is immersed in a magnetic field of magnitude 0.15 mT. Find the maximum magnetic flux experienced by one of these coils.

64. • **Electrognathography** Computerized jaw tracking, or electrognathography (EGN), is an important tool for diagnosing and treating temporomandibular disorders (TMDs) that affect a person's ability to bite effectively. The first step in applying EGN is to attach a small permanent magnet to the patient's gum below the

lower incisors. Then, as the jaw undergoes a biting motion, the resulting change in magnetic flux is picked up by wire coils placed on either side of the mouth, as shown in **Figure 23–42**. Suppose this person's jaw moves to her right and that the north pole of the permanent magnet also points to her right. From her point of view, is the induced current in the coil to **(a)** her right and **(b)** her left clockwise or counterclockwise? Explain.

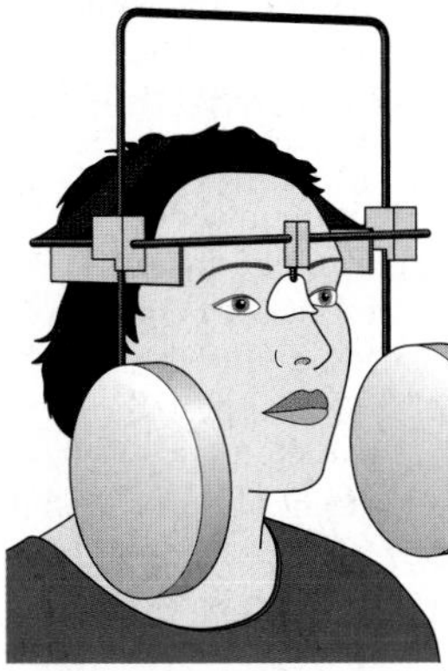

▲ **FIGURE 23–42** Problem 64

65\. •• A rectangular loop of wire 24 cm by 72 cm is bent into an L shape, as shown in **Figure 23–43**. The magnetic field in the vicinity of the loop has a magnitude of 0.035 T and points in a direction 25° below the y axis. The magnetic field has no x component. Find the magnitude of the magnetic flux through the loop.

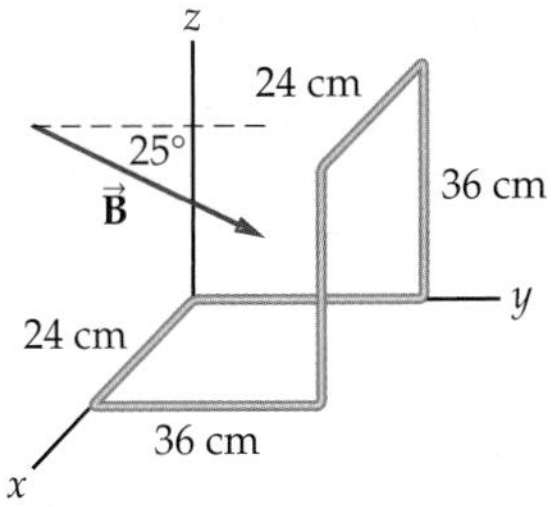

▲ **FIGURE 23–43** Problem 65

66\. •• **IP** A circular loop with a radius of 3.7 cm lies in the x-y plane. The magnetic field in this region of space is uniform and given by $\vec{\mathbf{B}} = (0.43\text{ T})\hat{\mathbf{x}} + (-0.11\text{ T})\hat{\mathbf{y}} + (0.52\text{ T})\hat{\mathbf{z}}$. **(a)** What is the magnitude of the magnetic flux through this loop? **(b)** Suppose we now increase the x component of $\vec{\mathbf{B}}$, leaving the other components unchanged. Does the magnitude of the magnetic flux increase, decrease, or stay the same? Explain. **(c)** Suppose, instead, that we increase the z component of $\vec{\mathbf{B}}$, leaving the other components unchanged. Does the magnitude of the magnetic flux increase, decrease, or stay the same? Explain.

67\. •• Consider a rectangular loop of wire 6.8 cm by 8.2 cm in a uniform magnetic field of magnitude 1.1 T. The loop is rotated from a position of zero magnetic flux to a position of maximum flux in 21 ms. What is the average induced emf in the loop?

68\. •• **IP** A car with a vertical radio antenna 85 cm long drives due east at 25 m/s. The Earth's magnetic field at this location has a magnitude of 5.9×10^{-5} T and points northward, 72° below the horizontal. **(a)** Is the top or the bottom of the antenna at the higher potential? Explain. **(b)** Find the induced emf between the ends of the antenna.

69\. •• The rectangular coils in a 305-turn generator are 14 cm by 17 cm. What is the maximum emf produced by this generator when it rotates with an angular speed of 525 rpm in a magnetic field of 0.45 T?

70\. •• A cubical box 22 cm on a side is placed in a uniform 0.35-T magnetic field. Find the net magnetic flux through the box.

71\. •• **BIO Transcranial Magnetic Stimulation** Transcranial magnetic stimulation (TMS) is a noninvasive method for studying brain function, and possibly for treatment as well. In this technique, a conducting loop is held near a person's head, as shown in **Figure 23–44**. When the current in the loop is changed rapidly, the magnetic field it creates can change at the rate of 3.00×10^4 T/s. This rapidly changing magnetic field induces an electric current in a restricted region of the brain that can cause a finger to twitch, bright spots to appear in the visual field (magnetophosphenes), or a feeling of complete happiness to overwhelm a person. If the magnetic field changes at the previously mentioned rate over an area of 1.13×10^{-2} m², what is the induced emf?

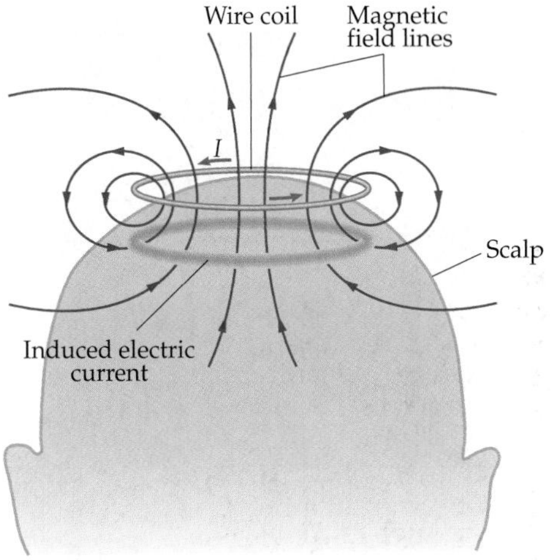

▲ **FIGURE 23–44** Transcranial magnetic stimulation (Problem 71)

72\. •• A magnetic field with the time dependence shown in **Figure 23–45** is at right angles to a 155-turn circular coil with a diameter of 3.75 cm. What is the induced emf in the coil at **(a)** $t = 2.50$ ms, **(b)** $t = 7.50$ ms, **(c)** $t = 15.0$ ms, and **(d)** $t = 25.0$ ms?

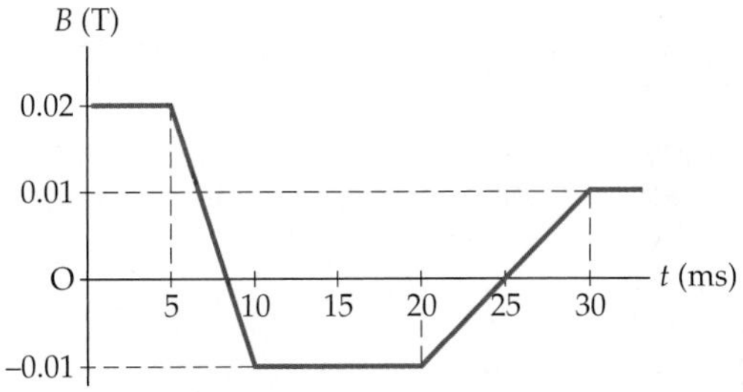

▲ **FIGURE 23–45** Problem 72

73\. •• You would like to construct a 50.0-mH inductor by wrapping insulated copper wire (diameter = 0.0332 cm) onto a tube with a circular cross section of radius 2.67 cm. What length of wire is required if it is wrapped onto the tube in a single, close-packed layer?

74\. •• The time constant of an RL circuit with $L = 25$ mH is twice the time constant of an RC circuit with $C = 45\ \mu$F. Both circuits have the same resistance R. Find **(a)** the value of R and **(b)** the time constant of the RL circuit.

75\. •• A 3.0-V battery is connected in series with a 25-mH inductor, a 110-Ω resistor, and an open switch. **(a)** How long after the switch is closed will the current in the circuit be equal to 12 mA? **(b)** How much energy is stored in the inductor when the current reaches its maximum value?

76. •• A 9.0-V battery is connected in series with a 38-mH inductor, a 170-Ω resistor, and an open switch. **(a)** What is the current in the circuit 0.120 ms after the switch is closed? **(b)** How much energy is stored in the inductor at this time?

77. •• IP **Loop Detectors on Roadways** "Smart" traffic lights are controlled by loops of wire embedded in the road (**Figure 23–46**). These "loop detectors" sense the change in magnetic field as a large metal object—such as a car or a truck—moves over the loop. Suppose a loop detector is circular, with a radius of 0.61 m, and that the downward vertical component of the magnetic field within the loop increases from 1.2×10^{-5} T to 2.6×10^{-5} T in 0.45 s as a car drives over a loop. **(a)** As viewed from above the road, what is the direction of the induced current in the loop? **(b)** What is the induced emf in a single turn of the loop?

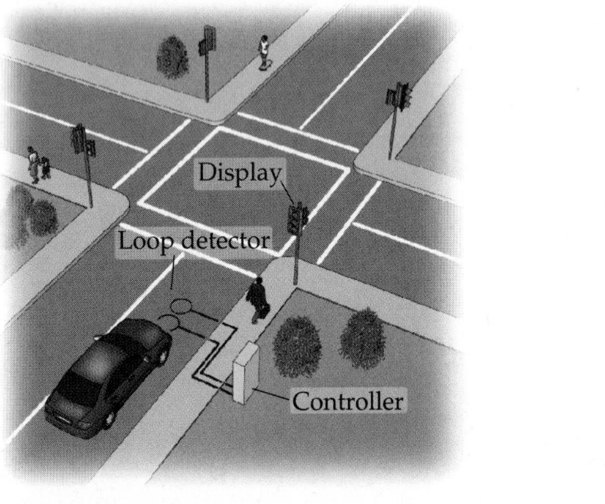

▲ **FIGURE 23–46** Problem 77

78. •• BIO **Blowfly Maneuvers** Suppose the fly described in Problem 63 turns through an angle of 90° in 32 ms. If the magnetic flux through one of the coils on the insect goes from a maximum to zero during this maneuver, and the coil has 95 turns of wire, find the magnitude of the induced emf.

79. ••• IP A conducting rod of mass m is in contact with two vertical conducting rails separated by a distance L, as shown in **Figure 23–47**. The entire system is immersed in a magnetic field of magnitude B pointing out of the page. Assuming the rod slides without friction, **(a)** describe the motion of the rod after it is released from rest. **(b)** What is the direction of the induced current (clockwise or counterclockwise) in the circuit? **(c)** Find the speed of the rod after it has fallen for a long time.

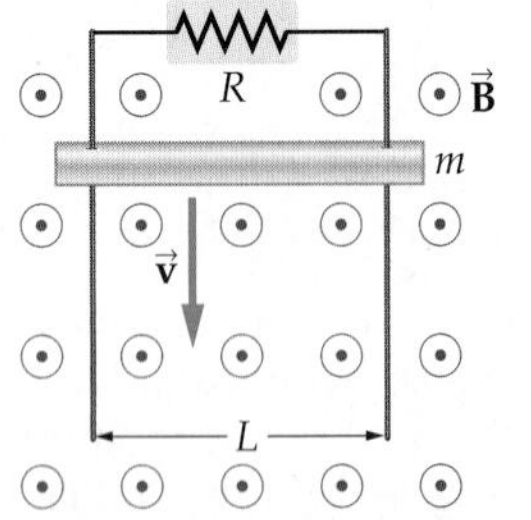

▲ **FIGURE 23–47** Problem 79

80. ••• IP A single-turn rectangular loop of width W and length L moves parallel to its length with a speed v. The loop moves from a region with a magnetic field $\vec{B}$ perpendicular to the plane of the loop to a region where the magnetic field is zero, as shown in **Figure 23–48**. Find the rate of change in the magnetic flux through the loop **(a)** before it enters the region of zero field, **(b)** just after it enters the region of zero field, and **(c)** once it is fully within the region of zero field. **(d)** For each of the cases considered in parts (a), (b), and (c), state whether the induced current in the loop is clockwise, counterclockwise, or zero. Explain in each case.

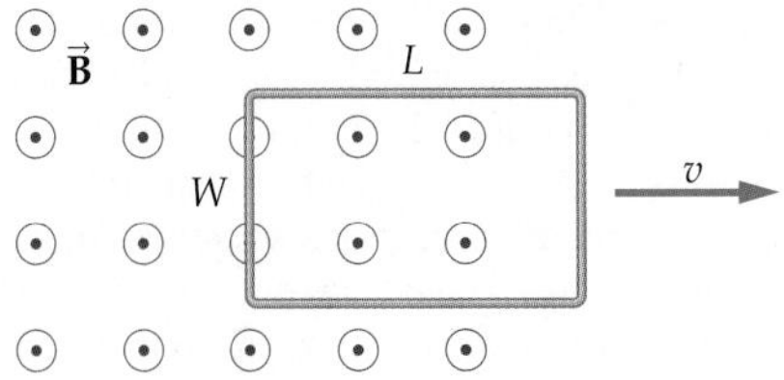

▲ **FIGURE 23–48** Problem 80

81. ••• IP The switch in the circuit shown in **Figure 23–49** is open initially. **(a)** Find the current in the circuit a long time after the switch is closed. **(b)** Describe the behavior of the lightbulb from the time the switch is closed until the current reaches the value found in part (a). **(c)** Now, suppose the switch is opened after having been closed for a long time. If the inductor is large, it is observed that the light flashes brightly and then burns out. Explain this behavior. **(d)** Find the voltage across the lightbulb just before and just after the switch is opened.

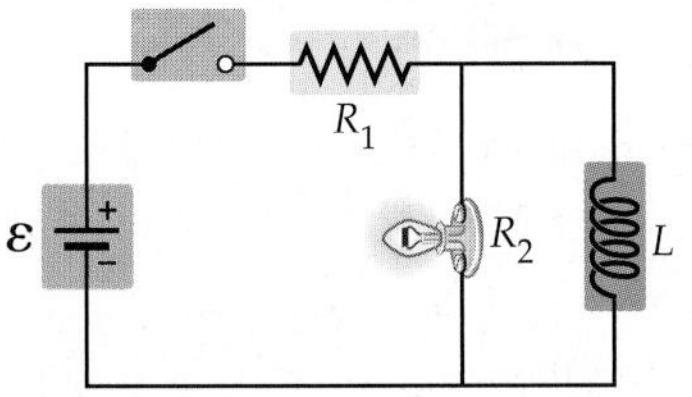

▲ **FIGURE 23–49** Problem 81

82. ••• **Energy Density in *E* and *B* Fields** An electric field E and a magnetic field B have the same energy density. **(a)** Express the ratio E/B in terms of the fundamental constants ε_0 and μ_0. **(b)** Evaluate E/B numerically, and compare your result with the speed of light.

Interactive Problems

83. •• ***Referring to Conceptual Checkpoint 23–3*** Suppose the ring is initially to the left of the field region, where there is no field, and is moving to the right. When the ring is partway into the field region, **(a)** is the induced current in the ring clockwise, counterclockwise, or zero, and **(b)** is the magnetic force exerted on the ring to the right, to the left, or zero? Explain.

84. •• ***Referring to Conceptual Checkpoint 23–3*** Suppose the ring is completely inside the field region initially and is moving to the right. **(a)** Is the induced current in the ring clockwise, counterclockwise, or zero and **(b)** is the magnetic force on the ring to the right, to the left, or zero? Explain. The ring now begins to emerge from the field region, still moving to the right. **(c)** Is the induced current in the ring clockwise, counterclockwise, or zero and **(d)** is the magnetic force on the ring to the right, to the left, or zero? Explain.

85. •• ***Referring to Example 23–3*** **(a)** What external force is required to give the rod a speed of 3.49 m/s, everything else remaining the same? **(b)** What is the current in the circuit in this case?

86. •• IP ***Referring to Example 23–3*** Suppose the direction of the magnetic field is reversed. Everything else in the system remains the same. **(a)** Is the magnetic force exerted on the rod to the right, to the left, or zero? Explain. **(b)** Is the direction of the induced current clockwise, counterclockwise, or zero? Explain. **(c)** Suppose we now adjust the strength of the magnetic field until the speed of the rod is 2.49 m/s, keeping the force equal to 1.60 N. What is the new magnitude of the magnetic field?

24 Alternating-Current Circuits

24–1	Alternating Voltages and Currents	804
24–2	Capacitors in AC Circuits	809
24–3	*RC* Circuits	812
24–4	Inductors in AC Circuits	817
24–5	*RLC* Circuits	820
24–6	Resonance in Electrical Circuits	824

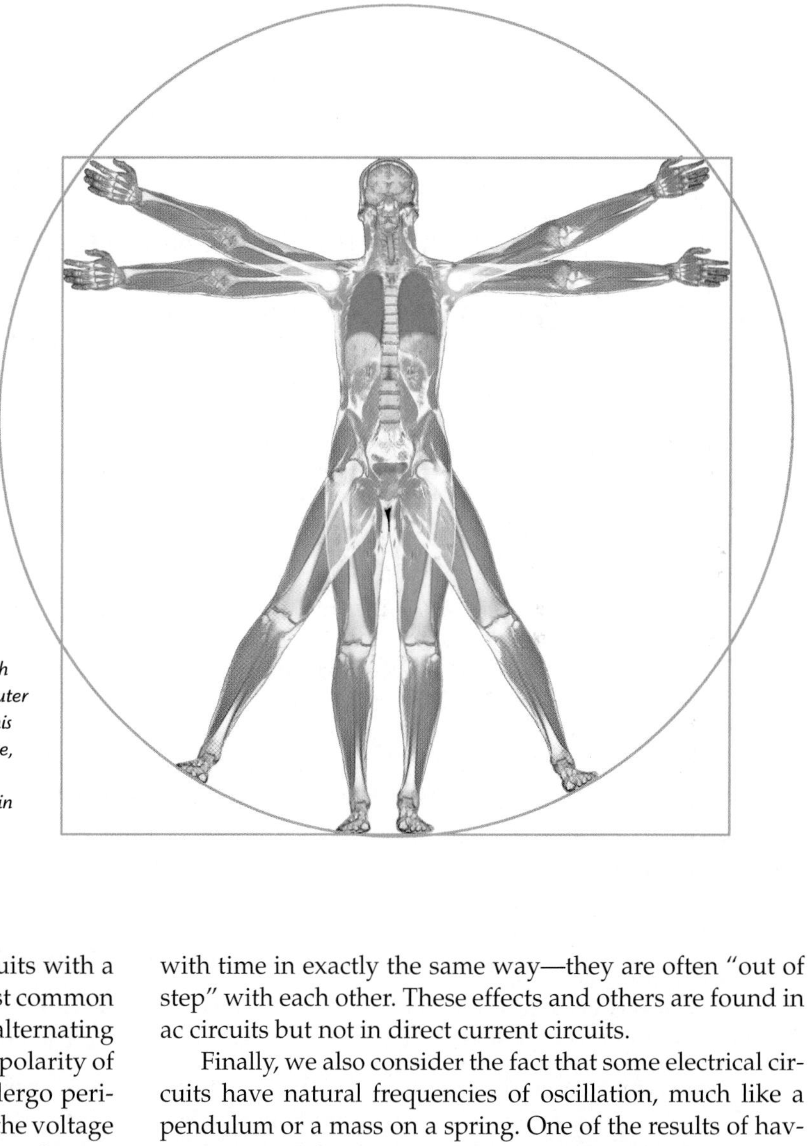

Magnetic resonance imaging (MRI), which produced the modern version of Leonardo da Vinci's Vitruvian Man *shown here, would not be possible without alternating-current (ac) electrical circuits. In fact, the key component of all MRI machines is an ac circuit that oscillates with a frequency in the range of 15 to 80 MHz. At these frequencies, the circuit resonates with hydrogen atoms within a person's body. By detecting how much each portion of the body responds to the resonance effect, a computer can construct detailed images of the body's internal structure. In this chapter, we explore many aspects of ac circuits, including resonance, and point out the close connections between inductors, capacitors, and resistors in electrical circuits, and masses, springs, and friction in mechanical systems.*

We conclude our study of electrical circuits with a consideration of the type of circuit most common in everyday usage—a circuit with an alternating current. In an alternating-current (ac) circuit, the polarity of the voltage and the direction of the current undergo periodic reversals. Typically, the time dependence of the voltage and current is sinusoidal, with a frequency of 60 cycles per second being the standard in the United States.

As we shall see, ac circuits require that we generalize the notion of resistance to include the time-dependent effects associated with capacitors and inductors. In particular, the voltage and current in an ac circuit may not vary with time in exactly the same way—they are often "out of step" with each other. These effects and others are found in ac circuits but not in direct current circuits.

Finally, we also consider the fact that some electrical circuits have natural frequencies of oscillation, much like a pendulum or a mass on a spring. One of the results of having a natural frequency is that resonance effects are to be expected. Indeed, electrical circuits do show resonance—in fact, this is how you are able to tune your radio or television to the desired station. Clearly, then, ac circuits have a rich variety of behaviors that make them not only very useful but also most interesting from a physics point of view.

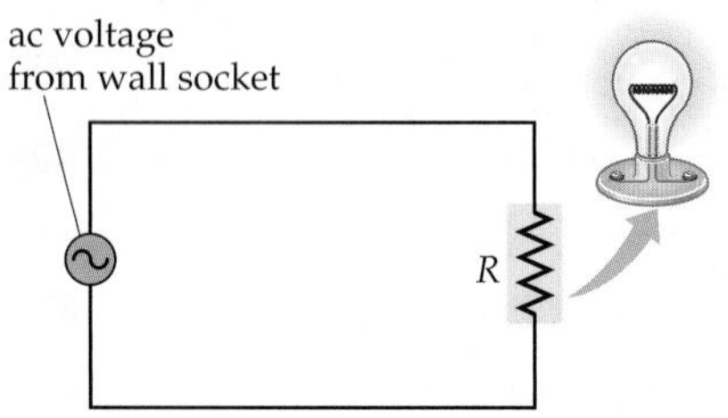

▲ **FIGURE 24–1 An ac generator connected to a lamp**
Simplified alternating-current circuit diagram for a lamp plugged into a wall socket. The lightbulb is replaced in the circuit with its equivalent resistance, R.

24–1 Alternating Voltages and Currents

When you plug a lamp into a wall socket, the voltage and current supplied to the lightbulb vary sinusoidally with a steady frequency, completing 60 cycles each second. Because the current periodically reverses direction, we say that the wall socket provides an **alternating current.** A simplified alternating-current (ac) circuit diagram for the lamp is shown in **Figure 24–1**. In this circuit, we indicate the bulb by its equivalent resistance, R, and the wall socket by an **ac generator,** represented as a circle enclosing one cycle of a sine wave.

The voltage delivered by an ac generator, which is plotted in **Figure 24–2 (a)**, can be represented mathematically as follows:

$$V = V_{\text{max}} \sin \omega t \qquad \text{24–1}$$

In this expression, V_{max} is the largest value attained by the voltage during a cycle, and the angular frequency is $\omega = 2\pi f$, where $f = 60$ Hz. (Note the similarity between the sinusoidal time dependence in an ac circuit and the time dependence given in Section 13–2 for simple harmonic motion.) From Ohm's law, $I = V/R$, it follows that the current in the lightbulb is

$$I = \frac{V}{R} = \left(\frac{V_{\text{max}}}{R}\right) \sin \omega t = I_{\text{max}} \sin \omega t \qquad \text{24–2}$$

This result is plotted in **Figure 24–2 (b)**.

Notice that the voltage and current plots have the same time variation. In particular, the voltage reaches its maximum value at precisely the same time as the current. We express this relationship between the voltage and current in a resistor by saying that they are *in phase* with one another. As we shall see later in this chapter, other circuit elements, like capacitors and inductors, have different phase relationships between the current and voltage. For these elements, the current and voltage reach maximum values at different times.

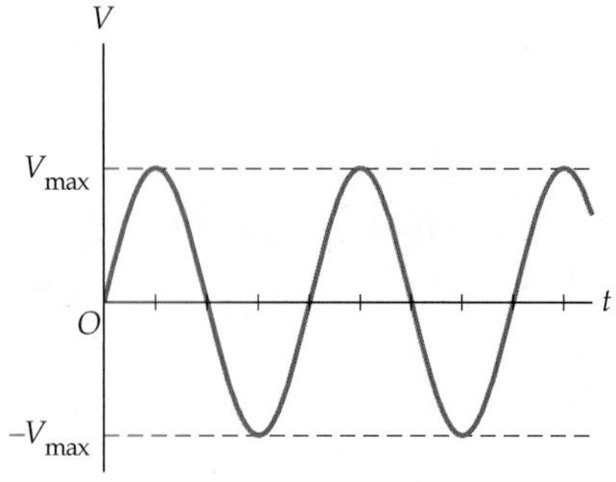

(a)

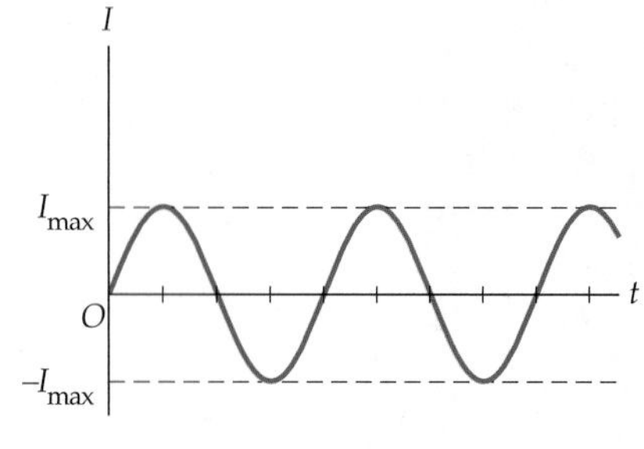

(b)

▲ **FIGURE 24–2 AC voltage and current for a resistor circuit**
(a) An ac voltage described by $V = V_{\text{max}} \sin \omega t$. **(b)** The alternating current, $I = I_{\text{max}} \sin \omega t = (V_{\text{max}}/R) \sin \omega t$, corresponding to the ac voltage in (a). Note that the voltage of the generator and the current in the resistor are in phase with each other; that is, their maxima and minima occur at precisely the same times.

Phasors

A convenient way to represent an alternating voltage and the corresponding current is with counterclockwise rotating vectors referred to as **phasors.** Phasors allow us to take advantage of the connection between uniform circular motion and linear, sinusoidal motion—just as we did when we studied oscillations in Chapter 13. In that case, we related the motion of a peg on a rotating turntable to the oscillating motion of a mass on a spring by projecting the position of the peg onto a screen. We use a similar projection with phasors.

For example, **Figure 24–3** shows a vector of magnitude V_{max} rotating about the origin with an angular speed ω. This rotating vector is the voltage phasor. If the voltage phasor makes an angle $\theta = \omega t$ with the x axis, it follows that its y component is $V_{\text{max}} \sin \omega t$; that is, if we project the voltage phasor onto the y axis, the projection gives the instantaneous value of the voltage, $V = V_{\text{max}} \sin \omega t$, in agreement with Equation 24–1. In general,

> the instantaneous value of a quantity represented by a phasor is the projection of the phasor onto the y axis.

Figure 24–3 also shows the current phasor, represented by a rotating vector of magnitude $I_{\text{max}} = V_{\text{max}}/R$. The current phasor points in the same direction as the voltage phasor; hence, the instantaneous current is $I = I_{\text{max}} \sin \omega t$, as in Equation 24–2.

The fact that the voltage and current phasors always point in the same direction for a resistor is an equivalent way of saying that the voltage and current are in phase:

> The voltage phasor for a resistor points in the same direction as the current phasor.

In circuits that also contain capacitors or inductors, the current and voltage phasors will usually point in different directions.

Root Mean Square (rms) Values

Notice that both the voltage and the current in Figure 24–2 have average values that are zero. Thus, V_{av} and I_{av} give very little information about the actual behavior of V and I. A more useful type of average, or mean, is the **root mean square,** or **rms** for short.

To see the significance of a *root mean square*, consider the current as a function of time, as given in Equation 24–2. First, we *square* the current to obtain

$$I^2 = I_{max}^2 \sin^2 \omega t$$

Clearly, I^2 is always positive; hence, its average value will not vanish. Next, we calculate the *mean* value of I^2. This can be done by inspecting **Figure 24–4**, where we plot I^2 as a function of time. As we see, I^2 varies *symmetrically* between 0 and I_{max}^2; that is, it spends equal amounts of time above and below the value $\frac{1}{2}I_{max}^2$. Hence, the mean value of I^2 is half of I_{max}^2:

$$(I^2)_{av} = \tfrac{1}{2}I_{max}^2$$

Finally, we take the square *root* of this average so that our final result is a current rather than a current squared. This calculation yields the rms value of the current:

$$I_{rms} = \sqrt{(I^2)_{av}} = \frac{1}{\sqrt{2}}I_{max} \qquad \text{24–3}$$

In general, any quantity x that varies with time as $x = x_{max} \sin \omega t$ or $x = x_{max} \cos \omega t$ obeys the same relationships among its average, maximum, and rms values:

RMS Value of a Quantity with Sinusoidal Time Dependence

$$(x^2)_{av} = \tfrac{1}{2}x_{max}^2$$

$$x_{rms} = \frac{1}{\sqrt{2}}x_{max} \qquad \text{24–4}$$

As an example, the rms value of the voltage in an ac circuit is

$$V_{rms} = \frac{1}{\sqrt{2}}V_{max} \qquad \text{24–5}$$

This result is applied to standard household voltages in the following Exercise.

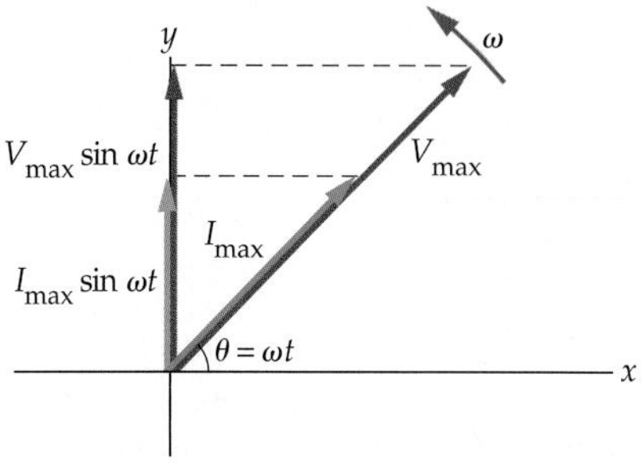

▲ **FIGURE 24–3 Phasor diagram for an ac resistor circuit**

Since the current and voltage are in phase in a resistor, the corresponding phasors point in the same direction at all times. Both phasors rotate about the origin with an angular speed ω, and the vertical component of each is the instantaneous value of that quantity.

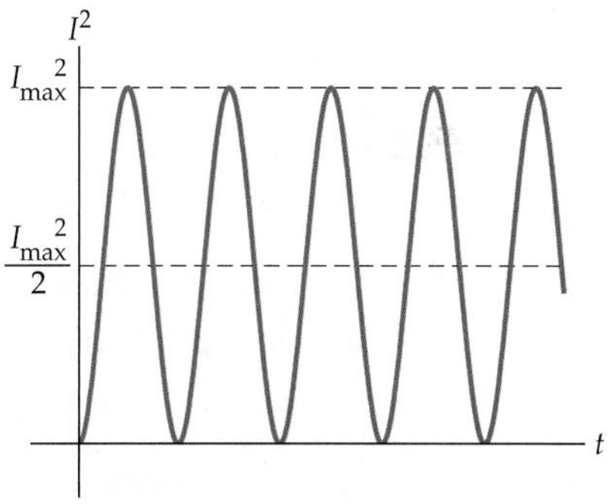

▲ **FIGURE 24–4 The square of a sinusoidally varying current**

Note that I^2 varies symmetrically about the value $\frac{1}{2}I_{max}^2$. The average of I^2 over time, then, is $\frac{1}{2}I_{max}^2$.

EXERCISE 24–1

Typical household circuits operate with an rms voltage of 120 V. What is the maximum, or peak, value of the voltage in these circuits?

Solution

Solving Equation 24–5 for the maximum voltage, V_{max}, we find

$$V_{max} = \sqrt{2}V_{rms} = \sqrt{2}(120 \text{ V}) = 170 \text{ V}$$

Whenever we refer to an ac generator in this chapter, we shall assume that the time variation is sinusoidal, so that the relations given in Equation 24–4 hold. If a different form of time variation is to be considered, it will be specified explicitly. Because the rms and maximum values of a sinusoidally varying quantity are proportional, it follows that *any* relation between rms values, like $I_{rms} = V_{rms}/R$, is equally valid as a relation between maximum values, $I_{max} = V_{max}/R$.

We now show how rms values are related to the average power consumed by a circuit. Referring again to the lamp circuit shown in Figure 24–1, we see that the instantaneous power dissipated in the resistor is $P = I^2R$. Using the time dependence $I = I_{max} \sin \omega t$, we find

$$P = I^2R = I_{max}^2 R \sin^2 \omega t$$

Note that P is always positive, as one might expect; after all, a current always dissipates energy as it passes through a resistor, regardless of its direction. To find the average power dissipated in the resistor, we note again that the average of $\sin^2 \omega t$ is $\frac{1}{2}$, thus

$$P_{av} = I_{max}^2 R(\sin^2 \omega t)_{av} = \tfrac{1}{2} I_{max}^2 R$$

In terms of the rms current, $I_{rms} = I_{max}/\sqrt{2}$, it follows that

$$P_{av} = I_{rms}^2 R$$

Therefore, we arrive at the following conclusion:

> $P = I^2R$ gives the instantaneous power consumption in both ac and dc circuits. To find the *average* power in an ac circuit, we simply replace I with I_{rms}.

Similar conclusions apply to many other dc and instantaneous formulas as well. For example, the power can also be written as $P = V^2/R$. Using the time-dependent voltage of an ac circuit, we have

$$P = \frac{V^2}{R} = \left(\frac{V_{max}^2}{R}\right) \sin^2 \omega t$$

Clearly, the average power is

$$P_{av} = \left(\frac{V_{max}^2}{R}\right)\left(\frac{1}{2}\right) = \frac{V_{rms}^2}{R} \qquad \text{24–6}$$

PROBLEM-SOLVING NOTE

Maximum versus RMS Values

When reading a problem statement, be sure to determine whether a given voltage or current is a maximum value or an rms value. If an rms current is given, for example, it follows that the maximum current is $I_{max} = \sqrt{2} I_{rms}$.

As before, we convert the dc power, $P = V^2/R$, to an average ac power by using the rms value of the voltage.

The close similarity between dc expressions and the corresponding ac expressions with rms values is one of the advantages of working with rms values. Another is that electrical meters, such as ammeters and voltmeters, generally give readings of rms values, rather than peak values, when used in ac circuits. In the remainder of this chapter, we shall make extensive use of rms quantities.

EXAMPLE 24–1 A Resistor Circuit

An ac generator with a maximum voltage of 24.0 V and a frequency of 60.0 Hz is connected to a resistor with a resistance $R = 265\ \Omega$. Find **(a)** the rms voltage and **(b)** the rms current in the circuit. In addition, determine **(c)** the average and **(d)** the maximum power dissipated in the resistor.

Picture the Problem

The circuit in this case consists of a 60.0-Hz ac generator connected directly to a 265-Ω resistor. The maximum voltage of the generator is 24.0 V.

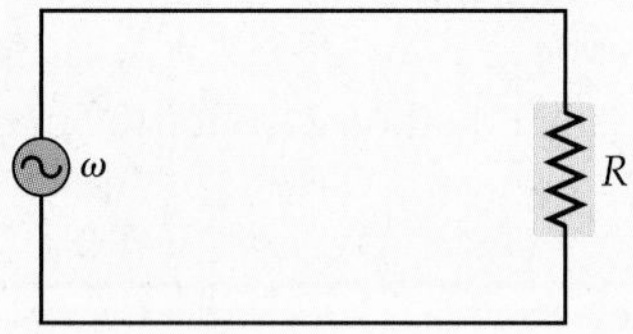

Strategy

a. The rms voltage is simply $V_{rms} = V_{max}/\sqrt{2}$.

b. Ohm's law gives the rms current, $I_{rms} = V_{rms}/R$.

c. The average power can be found using $P_{av} = I_{rms}^2 R$ or $P_{av} = V_{rms}^2/R$.

d. The maximum power, $P_{max} = V_{max}^2/R$, is twice the average power, from Equation 24–6.

Solution

Part (a)

1. Use $V_{rms} = V_{max}/\sqrt{2}$ to find the rms voltage:

$$V_{rms} = \frac{V_{max}}{\sqrt{2}} = \frac{24.0\ \text{V}}{\sqrt{2}} = 17.0\ \text{V}$$

Part (b)

2. Divide the rms voltage by the resistance, R, to find the rms current:

$$I_{rms} = \frac{V_{rms}}{R} = \frac{17.0\ \text{V}}{265\ \Omega} = 0.0642\ \text{A}$$

Part (c)

3. Use $I_{rms}^2 R$ to find the average power: $P_{av} = I_{rms}^2 R = (0.0642\ \text{A})^2(265\ \Omega) = 1.09\ \text{W}$

4. Use V_{rms}^2/R to find the average power: $P_{av} = \frac{V_{rms}^2}{R} = \frac{(17.0\ \text{V})^2}{265\ \Omega} = 1.09\ \text{W}$

Part (d)

5. Use $P_{max} = V_{max}^2/R$ to find the maximum power: $P_{max} = \frac{V_{max}^2}{R} = 2\left(\frac{V_{rms}^2}{R}\right) = 2(1.09\ \text{W}) = 2.18\ \text{W}$

Insight

Note that the instantaneous power in the resistor, $P = (V_{max}^2/R)\sin^2 \omega t$, oscillates symmetrically between 0 and twice the average power. This is another example of the type of averaging illustrated in Figure 24–4.

Practice Problem

Suppose we would like the average power dissipated in the resistor to be 5.00 W. Should the resistance be increased or decreased, assuming the same ac generator? Find the required value of R. [**Answer:** The resistance should be decreased. The required resistance is $R = 57.8\ \Omega$.]

Some related homework problems: Problem 3, Problem 4

CONCEPTUAL CHECKPOINT 24–1 Compare Average Power

If the frequency of the ac generator in Example 24–1 is increased, does the average power dissipated in the resistor **(a)** increase, **(b)** decrease, or **(c)** stay the same?

Reasoning and Discussion

None of the results in Example 24–1 depend on the frequency of the generator. For example, the relation $V_{rms} = V_{max}/\sqrt{2}$ depends only on the fact that the voltage varies sinusoidally with time and not at all on the frequency of the oscillations. The same frequency independence applies to the rms current and the average power.

These results are due to the fact that resistance is independent of frequency. In contrast, we shall see later in this chapter that the behavior of capacitors and inductors does indeed depend on frequency.

Answer:

(c) The average power remains the same.

Safety Features in Household Electrical Circuits

In today's technological world, with electrical devices as common as a horse and buggy in an earlier era, we sometimes forget that household electrical circuits pose potential dangers to homes and their occupants. For example, if several electrical devices are plugged into a single outlet, the current in the wires connected to that outlet may become quite large. The corresponding power dissipation in the wires ($P = I^2R$) can turn them red hot and lead to a fire.

To protect against this type of danger, household circuits use fuses and circuit breakers. In the case of a fuse, the current in a circuit must flow through a thin metal strip enclosed within the fuse. If the current exceeds a predetermined amount (typically 15 A) the metal strip becomes so hot that it melts and breaks the circuit. Thus, when a fuse "burns out," it is an indication that too many devices are operating on that circuit.

Circuit breakers provide similar protection with a switch that incorporates a bimetallic strip (Figure 16–5). When the bimetallic strip is cool, it closes the switch, allowing current to flow. When the strip is heated by a large current, however, it bends enough to open the switch and stop the current. Unlike the fuse, which cannot be used after it burns out, the circuit breaker can be reset when the bimetallic strip cools and returns to its original shape.

Household circuits also pose a threat to the occupants of a home, and at much lower current levels than the 15 A it takes to trigger a fuse or a circuit breaker. For example, it takes a current of only about 0.001 A to give a person a mild tingling sensation. Currents in the range of 0.01 A to 0.02 A produce muscle spasms that

REAL-WORLD PHYSICS: BIO
Electric shock hazard

make it difficult to let go of the wire delivering the current, or may even result in respiratory arrest. When currents reach 0.1 A to 0.2 A, the heartbeat is interrupted by an uncoordinated twitching referred to as ventricular fibrillation. As a result, currents in this range can prove to be fatal in a matter of seconds.

REAL-WORLD PHYSICS
Polarized plugs and grounded plugs

Several strategies are employed to reduce the danger of electrical shock. The first line of defense is the *polarized plug,* in which one prong of the plug is wider than the other prong. The corresponding wall socket will accept the plug in only one orientation, with the wide prong in the wide receptacle. The narrow receptacle of the outlet is wired to the high-potential side of the circuit; the wide receptacle is connected to the low-potential side, which is essentially at "ground" potential. A polarized plug provides protection by ensuring that the case of an electrical appliance, which manufacturers design to be connected to the wide prong, is at low potential. Furthermore, when an electrical device with a polarized plug is turned off, the high potential extends only from the wall outlet to the switch, leaving the rest of the device at zero potential.

The next line of defense against accidental shock is the three-prong grounded plug. In this plug, the rounded third prong is connected directly to ground when plugged into a three-prong receptacle. In addition, the third prong is wired to the case of an electrical appliance. If something goes wrong within the appliance, and a high-potential wire comes into contact with the case, the resulting current flows through the third prong, rather than through the body of a person who happens to touch the case.

REAL-WORLD PHYSICS
Ground fault circuit interrupter

An even greater level of protection is provided by a device known as a *ground fault circuit interrupter* (GFCI). The basic operating principle of an interrupter is illustrated in **Figure 24–5**. Note that the wires carrying an ac current to the protected appliance pass through a small iron ring. When the appliance operates normally, the two wires carry equal currents in opposite directions—in one wire the current goes to the appliance, in the other the same current returns from the appliance. Each of these wires produces a magnetic field (Equation 22–9), but because their currents are in opposite directions the magnetic fields are in opposite directions as well. As a result, the magnetic fields of the two wires cancel. If a malfunction occurs in the appliance—say, a wire frays and contacts the case of the appliance—current that would ordinarily return through the power cord may pass through the user's body instead and into the ground. In such a situation, the wire carrying current to the appliance now produces a net magnetic field within the iron ring that varies with the frequency of the ac generator. The changing magnetic field in the ring induces a current in the sensing coil wrapped around the ring, and the induced current triggers a circuit breaker in the interrupter. This cuts the flow of current to the appliance within a millisecond, protecting the user. In newer homes, interrupters are built directly into the wall sockets. The same protection can be obtained, however, by plugging a ground fault interrupter into an unprotected wall socket and then plugging an appliance into the interrupter.

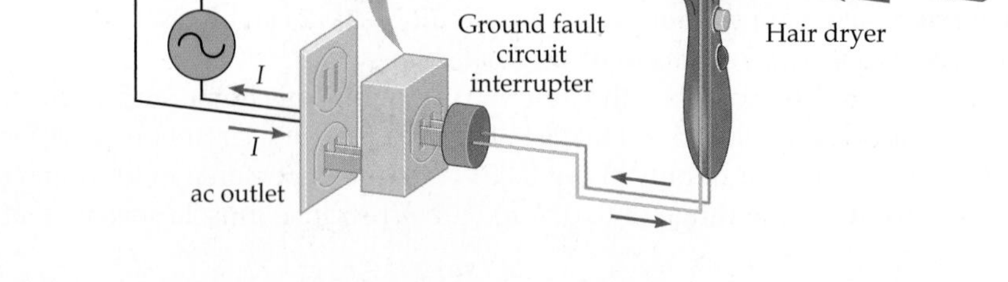

▶ **FIGURE 24–5 The ground fault circuit interrupter**

A short circuit in an appliance will ordinarily blow the fuse or trip the circuit breaker in the electrical line supplying power to the circuit. However, although fuses and circuit breakers help prevent electrical fires, these devices are not very good at preventing electric shock. This is because they are activated by the heat produced when the current becomes abnormally high—for example, when a "hot" wire touches a conductor that is in contact with the ground. If you happen to be that conductor, the fuse or breaker may act too slowly to prevent serious injury or even death. A ground fault circuit interrupter, by contrast, can cut off the current in a shorted circuit in less than a millisecond, before any harm can be done.

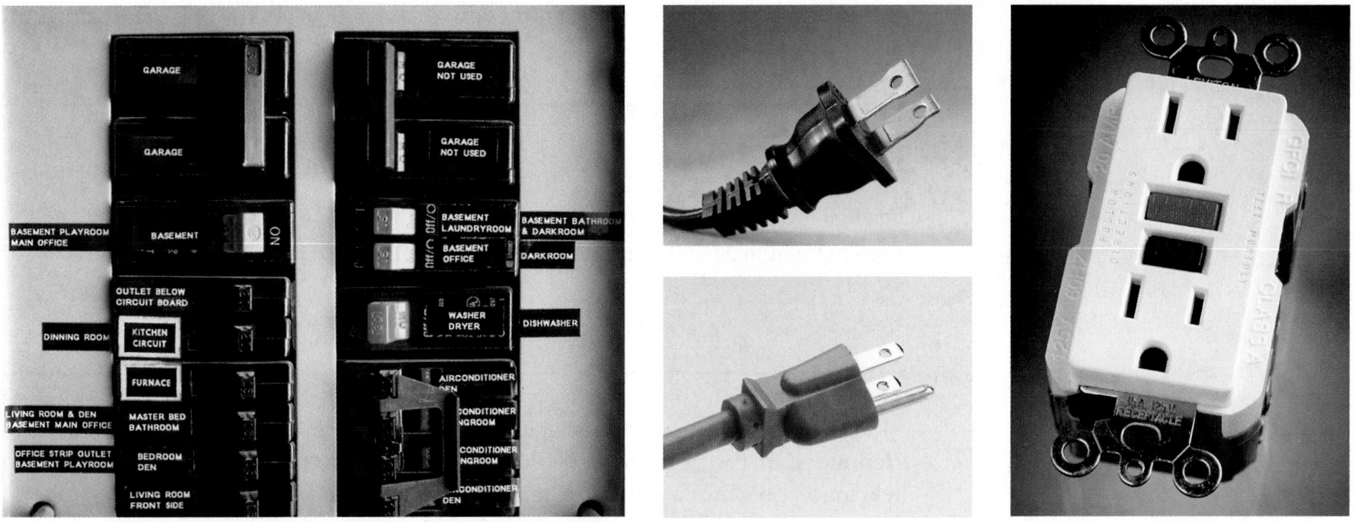

▲ Electricity is everywhere in the modern house, and electricity can be dangerous. Many common safety devices help us to minimize the risks associated with this very convenient form of energy. Whereas older homes typically have fuse boxes, most new houses are protected by circuit breaker panels (left). Both function in a similar way—if too much current is being drawn (perhaps as the result of a short circuit), the heat produced "blows" the fuse (by melting a metal strip) or "trips" the circuit breaker (by bending a bimetallic strip). Either way, the circuit is interrupted and current is cut off before the wires can become hot enough to start a fire.

The danger of electric shock can be reduced by the use of polarized plugs (center, top) or three-prong, grounded plugs (center, bottom) on appliances. Each of these provides a low-resistance path to ground that can be wired to the case of an appliance. In the event that a "hot" wire touches the case, most of the current will flow through the grounded wire rather than through the user. Even more protection is afforded by a ground fault circuit interrupter, or GFCI (right). This device, which is much faster and more sensitive than an ordinary circuit breaker, utilizes magnetic induction to interrupt the current in a circuit that has developed a short.

24–2 Capacitors in AC Circuits

In an ac circuit containing only a generator and a capacitor, the relationship between current and voltage is different in many important respects from the behavior seen with a resistor. In this section we explore these differences and show that the average power consumed by a capacitor is zero.

Capacitive Reactance

Consider a simple circuit consisting of an ac generator and a capacitor, as in **Figure 24–6**. The generator supplies an rms voltage, V_{rms}, to the capacitor. We would like to answer the following question: How is the rms current in this circuit related to the capacitance of the capacitor, C, and the frequency of the generator, ω?

To answer this question requires the methods of calculus, but the final result is quite straightforward. In fact, the rms current is simply

$$I_{rms} = \omega C V_{rms} \qquad \text{24–7}$$

In analogy with the expression $I_{rms} = V_{rms}/R$ for a resistor, we will find it convenient to rewrite this result in the following form:

$$I_{rms} = \frac{V_{rms}}{X_C} \qquad \text{24–8}$$

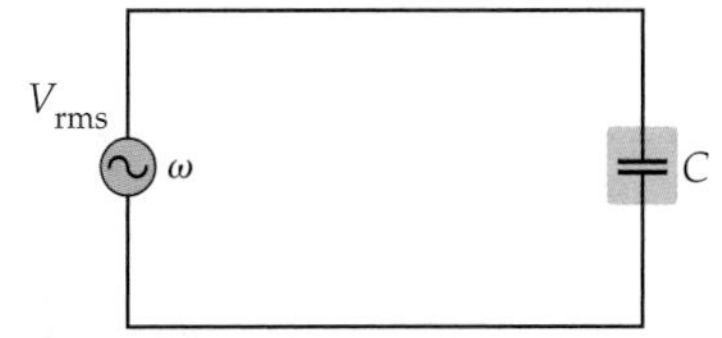

▲ **FIGURE 24–6 An ac generator connected to a capacitor**

The rms current in an ac capacitor circuit is $I_{rms} = V_{rms}/X_C$, where the capacitive reactance is $X_C = 1/\omega C$. Therefore, the rms current in this circuit is proportional to the frequency of the generator.

In this expression, X_C is referred to as the **capacitive reactance;** it plays the same role for a capacitor that resistance does for a resistor. In particular, to find the rms current in a resistor we divide V_{rms} by R; to find the rms current that flows into one side of a capacitor and out the other side, we divide V_{rms} by X_C. Comparing Equations 24–7 and 24–8, we see that the capacitive reactance can be written as follows:

Capacitive Reactance, X_C

$$X_C = \frac{1}{\omega C} \qquad \text{24–9}$$

SI unit: ohm, Ω

It is straightforward to show that the unit of $X_C = 1/\omega C$ is the ohm, the same unit as for resistance.

EXERCISE 24–2

Find the capacitive reactance of a 22-μF capacitor in a 60.0-Hz circuit.

Solution

Using Equation 24–9, and recalling that $\omega = 2\pi f$, we find

$$X_C = \frac{1}{\omega C} = \frac{1}{2\pi(60.0\ \text{s}^{-1})(22 \times 10^{-6}\ \text{F})} = 120\ \Omega$$

Unlike resistance, the capacitive reactance of a capacitor depends on the frequency of the ac generator. For example, at low frequencies the capacitive reactance becomes very large, and hence the rms current, $I_{rms} = V_{rms}/X_C$, is very small. This is to be expected. In the limit of low frequency, the current in the circuit becomes constant; that is, the ac generator becomes a dc battery. In this case we know that the capacitor becomes fully charged, and the current ceases to flow.

In the limit of large frequency, the capacitive reactance is small and the current is large. The reason is that at high frequency the current changes direction so rapidly that there is never enough time to fully charge the capacitor. As a result, the charge on the capacitor is never very large, and therefore it offers essentially no resistance to the flow of charge.

The behavior of a capacitor as a function of frequency is considered in the next Exercise.

EXERCISE 24–3

Suppose the capacitance in Figure 24–6 is 4.5 μF and the rms voltage of the generator is 120 V. Find the rms current in the circuit when the frequency of the generator is **(a)** 60.0 Hz and **(b)** 6.00 Hz.

Solution

Applying Equations 24–8 and 24–9, we find for part (a)

$$I_{rms} = \frac{V_{rms}}{X_C} = \omega C V_{rms} = (2\pi)(60.0\ \text{s}^{-1})(4.5 \times 10^{-6}\ \text{F})(120\ \text{V}) = 0.20\ \text{A}$$

Similarly, for part (b)

$$I_{rms} = \frac{V_{rms}}{X_C} = \omega C V_{rms} = (2\pi)(6.00\ \text{s}^{-1})(4.5 \times 10^{-6}\ \text{F})(120\ \text{V}) = 0.020\ \text{A}$$

When the frequency is reduced, the reactance becomes larger and the current decreases, as expected.

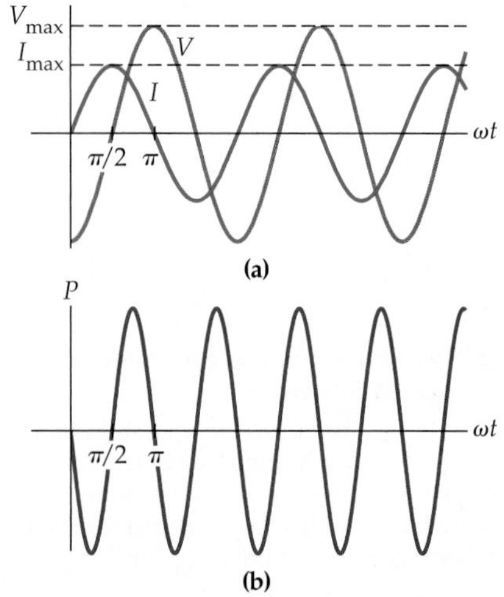

▲ FIGURE 24–7 Time variation of voltage, current, and power for a capacitor in an ac circuit

(a) Note that the time dependences are different for the voltage and the current. In particular, the voltage reaches its maximum value $\pi/2$ rad, or 90°, *after* the current. Thus we say that the voltage *lags* the current by 90°. **(b)** The power consumed by a capacitor in an ac circuit, $P = IV$, has an average value of zero.

Notice that the rms current in a capacitor is proportional to its capacitance, C, for all frequencies. A capacitor with a large capacitance can store and release large amounts of charge, which results in a large current. Finally, recall that rms expressions like $I_{rms} = V_{rms}/X_C$ (Equation 24–8) are equally valid in terms of maximum quantities; that is, $I_{max} = V_{max}/X_C$.

Phasor Diagrams: Capacitor Circuits

Figure 24–7 (a) shows the time dependence of the voltage and current in an ac capacitor circuit. Notice that the voltage and current are not in phase. For example, the current has its maximum value at the time $\omega t = \pi/2 = 90°$, whereas the

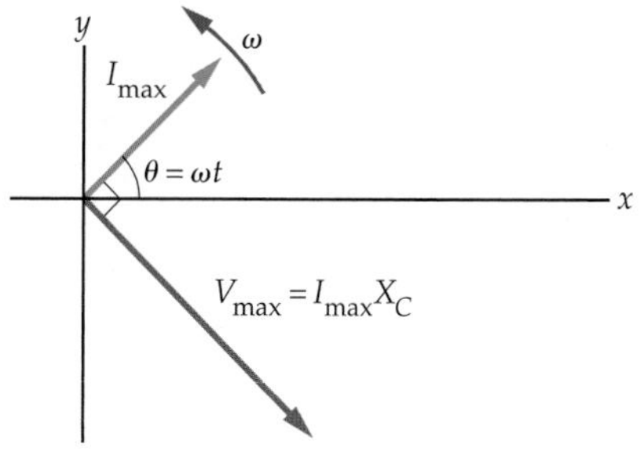

▶ **FIGURE 24–8 Current and voltage phasors for a capacitor**
Both the current and voltage phasors rotate counterclockwise about the origin. The voltage phasor lags the current phasor by 90°. This means that the voltage points in a direction that is 90° clockwise from the direction of the current.

voltage does not reach its maximum value until a later time, when $\omega t = \pi = 180°$. In general,

the voltage across a capacitor *lags* the current by 90°.

This 90° difference between current and voltage can probably best be seen in a phasor diagram. For example, in **Figure 24–8** we show both the current and the voltage phasors for a capacitor. The current phasor, with a magnitude I_{max}, is shown at the angle $\theta = \omega t$; it follows that the instantaneous current is $I = I_{max} \sin \omega t$. (As a matter of consistency, all phasor diagrams in this chapter will show the current phasor at the angle $\theta = \omega t$.) The voltage phasor, with a magnitude $V_{max} = I_{max}X_C$, is at right angles to the current phasor, pointing in the direction $\theta = \omega t - 90°$. The instantaneous value of the voltage is $V = V_{max} \sin(\omega t - 90°)$. Because the phasors rotate counterclockwise, we see that the voltage phasor lags the current phasor.

PROBLEM-SOLVING NOTE

Drawing a Capacitor Phasor

When drawing a voltage phasor for a capacitor, always draw it at right angles to the current phasor. In addition, be sure the capacitor phasor is 90° *clockwise* from the current phasor.

EXAMPLE 24–2 Instantaneous Voltage

Consider a capacitor circuit, as in Figure 24–6, where the capacitance is $C = 50.0\ \mu F$, the maximum current is 2.10 A, and the frequency of the generator is 60.0 Hz. At a given time, the current in the capacitor is 0.500 A and increasing. What is the voltage across the capacitor at this time?

Picture the Problem

The appropriate phasor diagram for this system is shown in our sketch. To see how this diagram was drawn, we first recall that the current has a positive value and is increasing. The fact that it is positive means it must point in a direction θ between 0 and 180°. The additional fact that it is increasing means θ must be between 0 and 90°; if it were between 90° and 180°, it would be decreasing with time. Finally, the voltage phasor lags the current phasor by 90° in a capacitor.

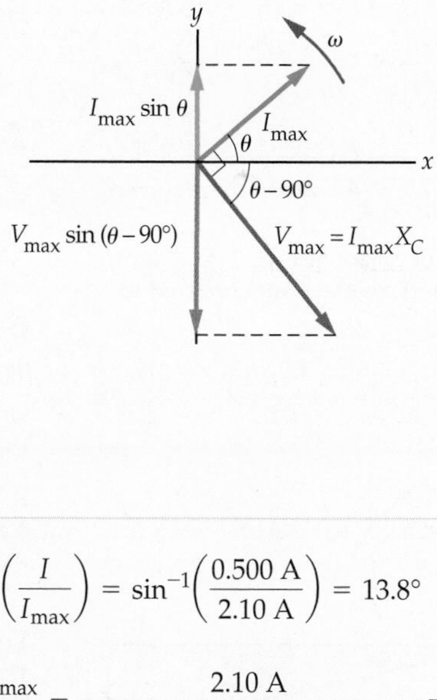

Strategy

There are several pieces that we must put together to complete this solution. First, we know the maximum current, I_{max}, and the instantaneous current, $I = I_{max} \sin \omega t = I_{max} \sin \theta$. We can solve this relation for the angle θ.

Second, the maximum voltage is given by $V_{max} = I_{max}X_C = I_{max}/\omega C$.

Finally, the instantaneous voltage is $V = V_{max} \sin(\theta - 90°)$, as indicated in the sketch.

Solution

1. Use $I = I_{max} \sin \theta$ to find the angle θ of the current phasor above the x axis:

$$\theta = \sin^{-1}\left(\frac{I}{I_{max}}\right) = \sin^{-1}\left(\frac{0.500\text{ A}}{2.10\text{ A}}\right) = 13.8°$$

2. Find the maximum voltage in the circuit using $V_{max} = I_{max}/\omega C$:

$$V_{max} = \frac{I_{max}}{\omega C} = \frac{2.10\text{ A}}{2\pi(60.0\text{ s}^{-1})(50.0 \times 10^{-6}\text{ F})} = 111\text{ V}$$

3. Calculate the instantaneous voltage across the capacitor with $V = V_{max} \sin(\theta - 90°)$:

$$V = V_{max} \sin(\theta - 90°)$$
$$= (111\text{ V}) \sin(13.8° - 90°) = -108\text{ V}$$

Insight

We can see from the phasor diagram that the current is positive and increasing in magnitude, whereas the voltage is negative and decreasing in magnitude. It follows that both the current and the voltage are becoming more positive with time.

Practice Problem

What is the voltage across the capacitor when the current in the circuit is 2.10 A? Is the voltage increasing or decreasing at this time? [**Answer:** $V = 0$, increasing]

Some related homework problems: Problem 12, Problem 13

To gain a qualitative understanding of the phase relation between the voltage and current in a capacitor, note that at the time when $\omega t = \pi/2$ the voltage across the capacitor in Figure 24–7 (a) is zero. Since the capacitor offers no resistance to the flow of current at this time, the current in the circuit is now a maximum. As the current continues to flow, charge builds up on the capacitor, and its voltage increases. This causes the current to decrease. At the time when $\omega t = \pi$ the capacitor voltage reaches a maximum and the current vanishes. As the current begins to flow in the opposite direction, charge flows out of the capacitor and its voltage decreases. When the voltage goes to zero, the current is once again a maximum, though this time in the opposite direction. It follows, then, that the variation of current and voltage are 90° out of phase—that is, when one is a maximum or a minimum, the other is zero—just like position and velocity in simple harmonic motion.

Power

As a final observation on the behavior of a capacitor in an ac circuit, we consider the power it consumes. Recall that the instantaneous power for any circuit can be written as $P = IV$. In **Figure 24–7 (b)** we plot the power $P = IV$ corresponding to the current and voltage shown in Figure 24–7 (a). The result is a power that changes sign with time.

In particular, note that the power is negative when the current and voltage have opposite signs, as between $\omega t = 0$ and $\omega t = \pi/2$, but is positive when they have the same sign, as between $\omega t = \pi/2$ and $\omega t = \pi$. This means that between $\omega t = 0$ and $\omega t = \pi/2$ the capacitor draws energy from the generator, but between $\omega t = \pi/2$ and $\omega t = \pi$ it delivers energy back to the generator. As a result, the average power as a function of time is zero, as can be seen by the symmetry about zero power in Figure 24–7 (b). Thus, *a capacitor in an ac circuit consumes zero net energy.*

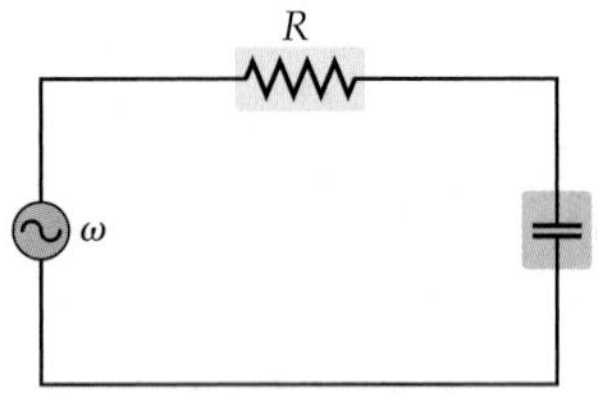

▲ **FIGURE 24–9 An alternating-current *RC* circuit**

An ac *RC* circuit consists of a generator, a resistor, and a capacitor connected in series.

24–3 *RC* Circuits

We now consider a resistor and a capacitor in series in the same ac circuit. This leads to a useful generalization of resistance known as the impedance.

Impedance

The circuit shown in **Figure 24–9** consists of an ac generator, a resistor, R, and a capacitor, C, connected in series. It is assumed that the values of R and C are known, as well as the maximum voltage, V_{max}, and angular frequency, ω, of the generator. In terms of these quantities we would like to determine the maximum current in the circuit and the maximum voltages across both the resistor and the capacitor.

To begin, we note that the magnitudes of the voltages are readily determined. For instance, the magnitude of the maximum voltage across the resistor is $V_{max,R} = I_{max}R$, and for the capacitor it is $V_{max,C} = I_{max}X_C = I_{max}/\omega C$. The total voltage in this circuit is *not* the sum of these two voltages, however, because they are not in phase—they do not attain their maximum values at the same time. To take these phase differences into account we turn to a phasor diagram.

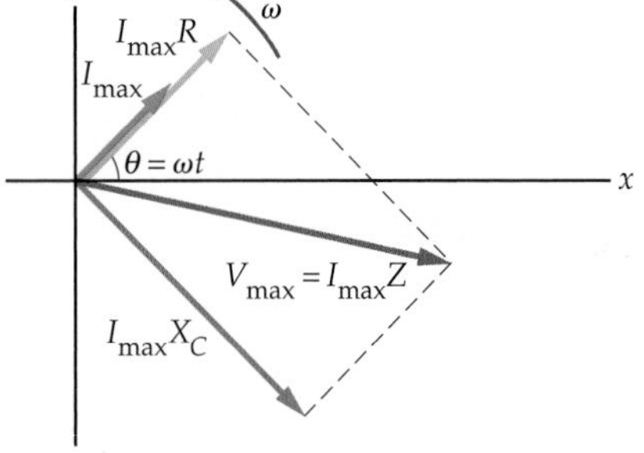

▲ **FIGURE 24–10 Phasor diagram for an *RC* circuit**

The maximum current in an *RC* circuit is $I_{max} = V_{max}/Z$, where V_{max} is the maximum voltage of the ac generator and $Z = \sqrt{R^2 + X_C{}^2}$ is the impedance of the circuit. The maximum voltage across the resistor is $V_{max,R} = I_{max}R$ and the maximum voltage across the capacitor is $V_{max,C} = I_{max}X_C$.

Figure 24–10 shows the phasor diagram for a simple *RC* circuit. To construct this diagram, we start by drawing the current phasor with a length I_{max} at the angle $\theta = \omega t$, as indicated in the figure. Next, we draw the voltage phasor associated with the resistor. This has a magnitude of $I_{max}R$ and points in the same direction as the current phasor. Finally, we draw the voltage phasor for the capacitor. This phasor has a magnitude of $I_{max}X_C$ and points in a direction that is rotated 90° clockwise from the current phasor.

Now, to obtain the total voltage in the circuit in terms of the individual voltages, we perform a vector sum of the resistor-voltage phasor and the capacitor-voltage phasor, as indicated in Figure 24–10. The total voltage, then, is the hypotenuse of the right triangle formed by these two phasors. Its magnitude, using the Pythagorean theorem, is simply

$$V_{max} = \sqrt{V_{max,R}^2 + V_{max,C}^2} \qquad 24\text{–}10$$

Substituting the preceding expressions for the voltages across R and C, we find

$$V_{max} = \sqrt{(I_{max}R)^2 + (I_{max}X_C)^2} = I_{max}\sqrt{R^2 + X_C{}^2}$$

The last quantity in the preceding expression is given a special name; it is called the **impedance,** Z:

Impedance in an *RC* Circuit

$$Z = \sqrt{R^2 + X_C{}^2} = \sqrt{R^2 + \left(\frac{1}{\omega C}\right)^2} \qquad 24\text{–}11$$

SI unit: ohm, Ω

Clearly, the impedance has units of ohms, the same as those of resistance and reactance.

EXERCISE 24–4

A given RC circuit has $R = 135\ \Omega$, $C = 28.0\ \mu F$, and $f = 60.0$ Hz. What is the impedance of the circuit?

Solution

Applying Equation 24–11 with $\omega = 2\pi f$, we obtain

$$Z = \sqrt{R^2 + X_C{}^2} = \sqrt{R^2 + \left(\frac{1}{\omega C}\right)^2}$$

$$= \sqrt{(135\ \Omega)^2 + \left(\frac{1}{2\pi(60.0\ s^{-1})(28.0 \times 10^{-6}\ F)}\right)^2} = 165\ \Omega$$

We are now in a position to calculate the maximum current in an RC circuit. First, given R, C, and $\omega = 2\pi f$, we can determine the value of Z using Equation 24–11. Next, we solve for the maximum current using the relation

$$V_{max} = I_{max}\sqrt{R^2 + X_C{}^2} = I_{max}Z$$

which yields

$$I_{max} = \frac{V_{max}}{Z}$$

In a circuit containing only a resistor, the maximum current is $I_{max} = V_{max}/R$; thus, we see that the impedance Z is indeed a generalization of resistance that can be applied to more complex circuits.

To check some limits of Z, note that in the capacitor circuit discussed in the previous section (Figure 24–6), the resistance is zero. Hence, in that case the impedance is

$$Z = \sqrt{0 + X_C{}^2} = X_C$$

The maximum voltage across the capacitor, then, is $V_{max} = I_{max}/Z = V_{max}/X_C$, as expected. Similarly, in a resistor circuit with no capacitor, as in Figure 24–1, the capacitive reactance is zero; hence,

$$Z = \sqrt{R^2 + 0} = R$$

Thus, Z includes X_C and R as special cases.

EXAMPLE 24–3 Find the Frequency

An ac generator with an rms voltage of 110 V is connected in series with a 35-Ω resistor and a 11-μF capacitor. At what frequency, f, must the generator operate if the rms current in the circuit is to be 1.2 A?

Picture the Problem

The appropriate circuit is shown in our sketch. We are given the rms voltage of the generator, $V_{rms} = 110$ V; the resistance, $R = 35\ \Omega$; and the capacitance, $C = 11\ \mu F$. The only remaining variable that affects the current is the frequency of the generator, f.

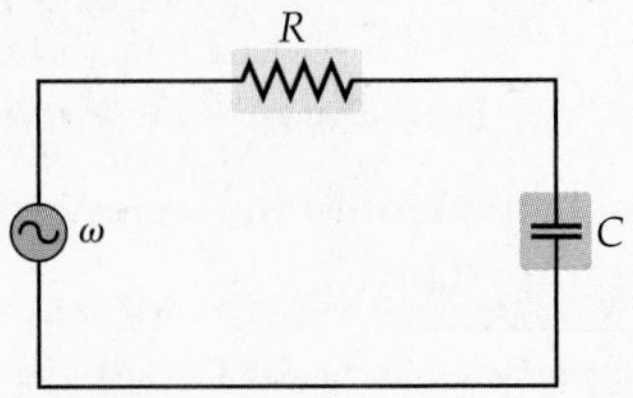

Strategy

Note that the frequency appears in the capacitive reactance, $X_C = 1/\omega C = 1/2\pi fC$. Thus, if we can determine X_C, we can find the frequency. To calculate X_C, we first use $V_{rms} = I_{rms}Z$ to find Z, then we solve $Z = \sqrt{R^2 + X_C{}^2}$ for X_C.

Solution

1. Use $V_{rms} = I_{rms}Z$ to find the impedance, Z:

$$Z = \frac{V_{rms}}{I_{rms}} = \frac{110\text{ V}}{1.2\text{ A}} = 92\ \Omega$$

2. Solve for X_C in terms of Z and R:

$$Z = \sqrt{R^2 + X_C{}^2}$$
$$X_C = \sqrt{Z^2 - R^2} = \sqrt{(92\ \Omega)^2 - (35\ \Omega)^2} = 85\ \Omega$$

3. Use this value of X_C to find the frequency, f:

$$X_C = \frac{1}{\omega C} = \frac{1}{2\pi fC}$$
$$f = \frac{1}{2\pi X_C C} = \frac{1}{2\pi(85\ \Omega)(11 \times 10^{-6}\text{ F})} = 170\text{ Hz}$$

Insight

Note that the rms voltage across the resistor is $V_{rms,R} = I_{rms}R = 42$ V, and the rms voltage across the capacitor is $V_{rms,C} = I_{rms}X_C = 102$ V. As expected, these voltages *do not* add up to the generator voltage of 110 V—in fact, they add up to a considerably larger voltage. The point is, however, that this sum is physically meaningless, since the voltages are 90° out of phase. If the voltages are combined as in Equation 24–10, which takes into account the 90° phase difference, we find $V_{rms} = \sqrt{(V_{rms,R}^2 + V_{rms,C}^2)} = \sqrt{(42\text{ V})^2 + (102\text{ V})^2} = 110$ V, as expected.

Practice Problem

If the frequency in this circuit is increased, do you expect the rms current to increase, decrease, or stay the same? What is the rms current if the frequency is increased to 250 Hz? [**Answer:** The rms current will increase with frequency. At 250 Hz the current is 1.6 A.]

Some related homework problems: Problem 20, Problem 21

ACTIVE EXAMPLE 24–1 Find the Resistance

An ac generator with an rms voltage of 110 V and a frequency of 60.0 Hz is connected in series with a 270-μF capacitor and a resistor of resistance R. What value must R have if the rms current in this circuit is to be 1.7 A?

Solution *(Test your understanding by performing the calculations indicated in each step.)*

1. Solve for the resistance: $R = \sqrt{Z^2 - X_C{}^2}$
2. Find the impedance of the circuit: $Z = 65\ \Omega$
3. Find the capacitive reactance: $X_C = 9.8\ \Omega$
4. Substitute numerical values: $R = 64\ \Omega$

Insight

At this frequency the capacitor has relatively little effect in the circuit, as can be seen by comparing X_C, R, and Z.

Your Turn

At what frequency is the capacitive reactance equal to 64 Ω? What is the current in the circuit at this frequency, assuming all other variables remain the same?

(Answers to **Your Turn** *problems are given in the back of the book.)*

We can also gain considerable insight about an ac circuit with qualitative reasoning—that is, without going through detailed calculations like those just given. This is illustrated in the following Conceptual Checkpoint.

CONCEPTUAL CHECKPOINT 24–2 Compare Brightness: Capacitor Circuit

Shown in the sketch are two circuits with identical ac generators and lightbulbs. Circuit 2 differs from circuit 1 by the addition of a capacitor in series with the light. Does the lightbulb in circuit 2 shine **(a)** more brightly, **(b)** less brightly, or **(c)** with the same intensity as that in circuit 1?

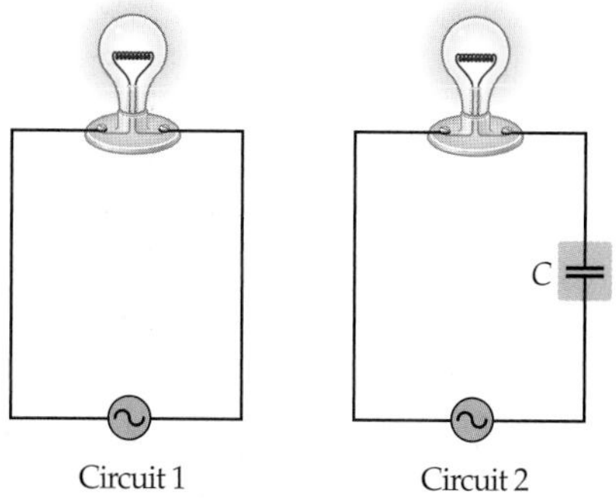

Reasoning and Discussion

Since circuit 2 has both a resistance (in the lightbulb) and a capacitive reactance, its impedance, $Z = \sqrt{R^2 + X_C^2}$, is greater than that of circuit 1, which has only the resistance R. Therefore, the current in circuit 2, $I_{rms} = V_{rms}/Z$, is less than the current in circuit 1. As a result, the average power dissipated in the lightbulb, $P_{av} = I_{rms}^2 R$, is less in circuit 2, so its bulb shines less brightly.

Note that the bulb dims, even though no power is consumed by the capacitor.

Answer:

(b) The light bulb in circuit 2 shines less brightly.

Phase Angle and Power Factor

We have seen how to calculate the current in an *RC* circuit and how to find the voltages across each element. Next we consider the phase relation between the total voltage in the circuit and the current. As we shall see, there is a direct connection between this phase relation and the power consumed by a circuit.

To do this, consider the phasor diagram shown in **Figure 24–11 (a)**. The **phase angle,** ϕ, between the voltage and the current can be read off the diagram as indicated in **Figure 24–11 (b)**. Clearly, the cosine of ϕ is given by the following:

$$\cos\phi = \frac{I_{max}R}{I_{max}Z} = \frac{R}{Z} \qquad \text{24–12}$$

As we shall see, both the magnitude of the voltage and its phase angle relative to the current play important roles in the behavior of a circuit.

Consider, for example, the power consumed by the circuit shown in Figure 24–9. One way to obtain this result is to recall that no power is consumed by the capacitor at all, as was shown in the previous section. Thus the total power of the circuit is simply the power dissipated by the resistor. This power can be written as

$$P_{av} = I_{rms}^2 R$$

An equivalent expression for the power is obtained by replacing one power of I_{rms} with the expression $I_{rms} = V_{rms}/Z$. This replacement yields

$$P_{av} = I_{rms}^2 R = I_{rms} I_{rms} R = I_{rms}\left(\frac{V_{rms}}{Z}\right) R$$

Finally, recalling that $R/Z = \cos\phi$, we can write the power as follows:

$$P_{av} = I_{rms} V_{rms} \cos\phi \qquad \text{24–13}$$

Thus, a knowledge of the current and voltage in a circuit, along with the value of $\cos\phi$, gives the power. The multiplicative factor $\cos\phi$ is referred to as the **power factor.**

Writing the power in terms of ϕ allows one to get a feel for the power in a circuit simply by inspecting the phasor diagram. For example, in the case of a circuit with only a capacitor, the phasor diagram (Figure 24–8) shows that the angle between the current and voltage has a magnitude of 90°, so the power factor is zero; $\cos\phi = \cos 90° = 0$. Thus no power is consumed in this circuit, as expected. In contrast, in a purely resistive circuit the phasor diagram shows that $\phi = 0$, as in Figure 24–3, giving a power factor of 1. In this case, the power is simply $P_{av} = I_{rms}V_{rms}$. Therefore, the angle between the current and the voltage in a phasor diagram gives an indication of the power being used by the circuit. We explore this feature in greater detail in the next Example and in **Figure 24–12**.

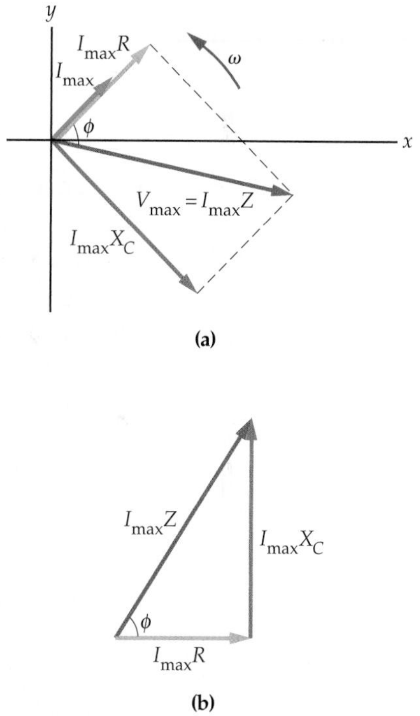

▲ **FIGURE 24–11 Phase angle for an *RC* circuit**

(a) The phase angle, ϕ, is the angle between the voltage phasor, $V_{max} = I_{max}Z$, and the current phasor, I_{max}. Because I_{max} and $I_{max}R$ are in the same direction, we can say that ϕ is the angle between $I_{max}Z$ and $I_{max}R$. **(b)** From this triangle we can see that $\cos\phi = R/Z$, a result that is valid for any circuit.

PROBLEM-SOLVING NOTE

The Angle in the Power Factor

The angle in the power factor, $\cos\phi$, is the angle between the voltage phasor and the current phasor. Be careful not to identify ϕ with the angle between the voltage phasor and the x or y axis.

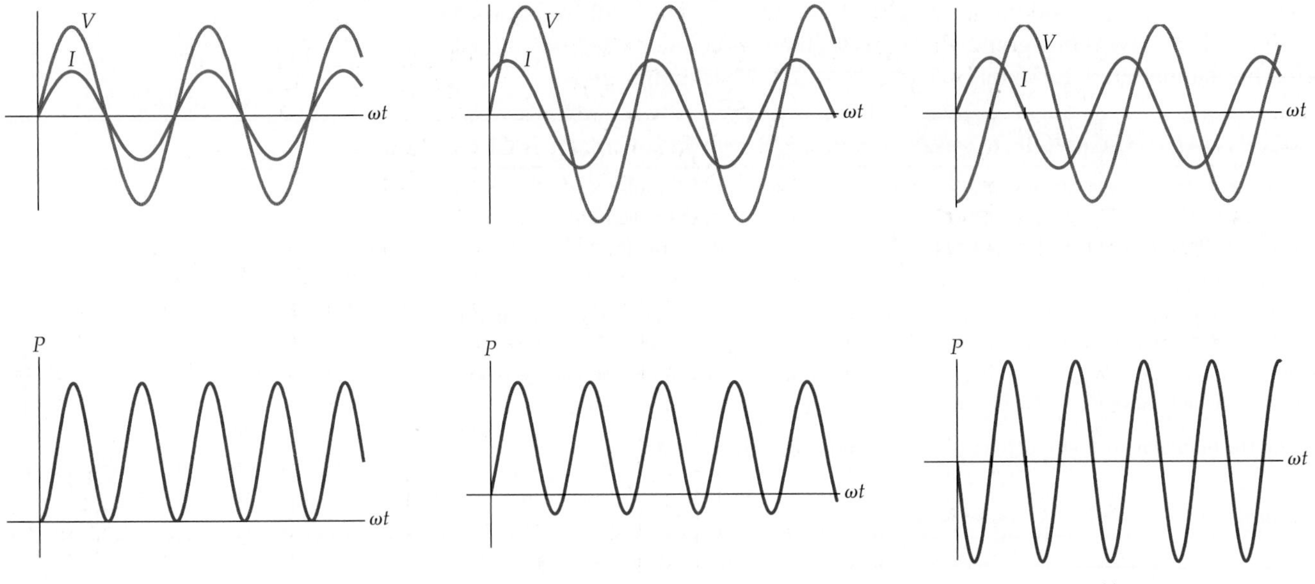

▲ **FIGURE 24–12 Voltage, current, and power for various phase angles, ϕ**

(a) $\phi = 0$. In this case the power is always positive. **(b)** $\phi = 45°$. The power oscillates between positive and negative values, with an average that is positive. **(c)** $\phi = 90°$. The power oscillates symmetrically about its average value of zero.

EXAMPLE 24–4 Power and Resistance

A certain *RC* circuit has an ac generator with an rms voltage of 240 V. The rms current in the circuit is 2.5 A, and it leads the voltage by 56°. Find **(a)** the value of the resistance, *R*, and **(b)** the average power consumed by the circuit.

Picture the Problem

The phasor diagram appropriate to this circuit is shown in the sketch. Note that the voltage lags the current by the angle $\phi = 56°$.

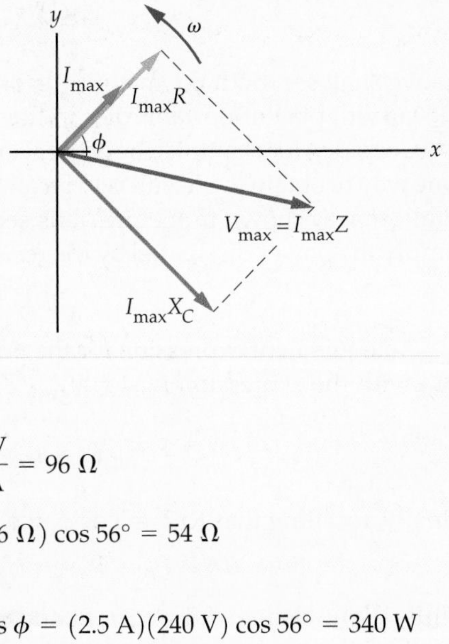

Strategy

a. To find the resistance in the circuit, recall that $\cos\phi = R/Z$; thus, $R = Z\cos\phi$. We can find the impedance, *Z*, from $V_{rms} = I_{rms}Z$, since we know both V_{rms} and I_{rms}.

b. The average power consumed by the circuit is simply $P_{av} = I_{rms}V_{rms}\cos\phi$.

Solution

Part (a)

1. Use $V_{rms} = I_{rms}Z$ to find the impedance of the circuit:

$$Z = \frac{V_{rms}}{I_{rms}} = \frac{240\text{ V}}{2.5\text{ A}} = 96\ \Omega$$

2. Now use $\cos\phi = R/Z$ to find the resistance:

$$R = Z\cos\phi = (96\ \Omega)\cos 56° = 54\ \Omega$$

Part (b)

3. Calculate the average power with the expression $P_{av} = I_{rms}V_{rms}\cos\phi$:

$$P_{av} = I_{rms}V_{rms}\cos\phi = (2.5\text{ A})(240\text{ V})\cos 56° = 340\text{ W}$$

Insight

Of course, the average power consumed by this circuit is simply the average power dissipated in the resistor: $P_{av} = I_{rms}^2R = (2.5\text{ A})^2(54\ \Omega) = 340$ W. Note that care must be taken if the average power dissipated in the resistor is calculated using $P_{av} = V_{rms}^2/R$. The potential pitfall is that one might use 240 V for the rms voltage; but this is the rms voltage of the generator, *not* the rms voltage across the resistor. This problem didn't arise with $P_{av} = I_{rms}^2R$ because the same current flows through both the generator and the resistor.

Practice Problem

What is the capacitive reactance in this circuit? [**Answer:** $X_C = 79\ \Omega$]

Some related homework problems: Problem 22, Problem 23, Problem 24

24–4 Inductors in AC Circuits

We turn now to the case of ac circuits containing inductors. As we shall see, the behavior of inductors is, in many respects, just the opposite of that of capacitors.

Inductive Reactance

The voltage across a capacitor is given by $V = IX_C$, where $X_C = 1/\omega C$ is the capacitive reactance. Similarly, the voltage across an inductor connected to an ac generator, as in **Figure 24–13**, can be written as $V = IX_L$, which defines the **inductive reactance,** X_L. The precise expression for X_L, in terms of frequency and inductance, can be derived using the methods of calculus. The result of such a calculation is the following:

Inductive Reactance, X_L

$$X_L = \omega L \qquad \text{24–14}$$

SI unit: ohm, Ω

A plot of X_L versus frequency is given in **Figure 24–14**, where it is compared with X_C and R. The rms current in an ac inductor circuit is $I_{rms} = V_{rms}/X_L = V_{rms}/\omega L$.

Note that X_L increases with frequency, in contrast with the behavior of X_C. This is easily understood when one recalls that the voltage across an inductor has a magnitude given by $\mathcal{E} = L\,\Delta I/\Delta t$. Thus the higher the frequency, the more rapidly the current changes with time, and hence the greater the voltage across the inductor.

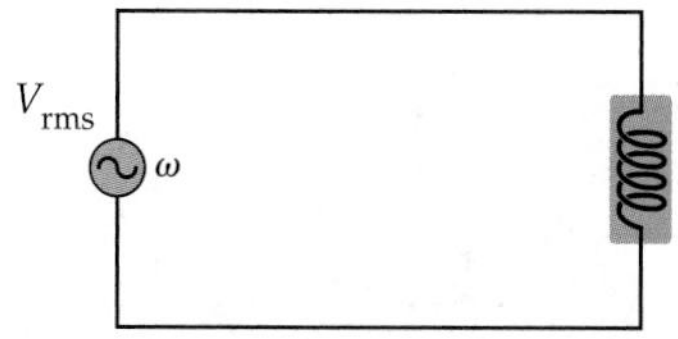

▲ **FIGURE 24–13 An ac generator connected to an inductor**

The rms current in an ac inductor circuit is $I_{rms} = V_{rms}/X_L$, where the inductive reactance is $X_L = \omega L$.

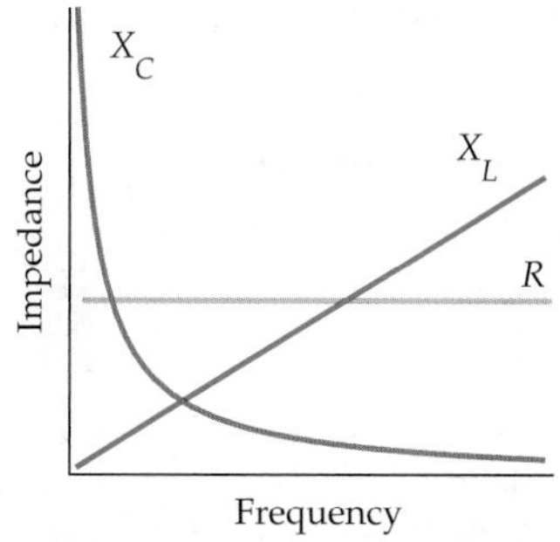

▲ **FIGURE 24–14 Frequency variation of the inductive reactance, X_L, the capacitive reactance, X_C, and the resistance, R**

The resistance is independent of frequency. In contrast, the capacitive reactance becomes large with decreasing frequency, and the inductive reactance becomes large with increasing frequency.

EXERCISE 24–5

A 21-mH inductor is connected to an ac generator with an rms voltage of 24 V and a frequency of 60.0 Hz. What is the rms current in the inductor?

Solution

The rms current is $I_{rms} = V_{rms}/X_L = V_{rms}/\omega L$. Substituting numerical values, we find

$$I_{rms} = \frac{V_{rms}}{\omega L} = \frac{24\ \text{V}}{2\pi(60.0\ \text{s}^{-1})(21 \times 10^{-3}\ \text{H})} = 3.0\ \text{A}$$

Phasor Diagrams: Inductor Circuits

Now that we can calculate the magnitude of the voltage across an inductor, we turn to the question of the phase of the inductor's voltage relative to the current in the circuit. With both phase and magnitude determined, we can construct the appropriate phasor diagram for an inductor circuit.

Suppose the current in an inductor circuit is as shown in **Figure 24–15 (a)**. At time zero the current is zero, but increasing at its maximum rate. Since the voltage across an inductor depends on the rate of change of current, it follows that the inductor's voltage is a maximum at $t = 0$. When the current reaches a maximum value, at the time $\omega t = \pi/2$, its rate of change becomes zero; hence, the voltage across the inductor falls to zero at that point, as is also indicated in Figure 24–15 (a).

PROBLEM-SOLVING NOTE

Drawing an Inductor Phasor

When drawing a voltage phasor for an inductor, always draw it at right angles to the current phasor. In addition, be sure the inductor phasor is 90° *counterclockwise* from the current phasor.

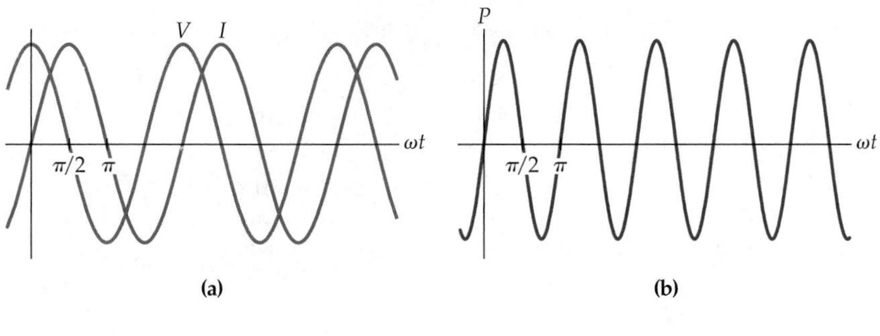

◀ **FIGURE 24–15 Voltage, current, and power in an ac inductor circuit**

(a) Voltage V and current I in an inductor. Note that V reaches a maximum $\pi/2$ rad $= 90°$ *before* the current. Thus, the voltage *leads* the current by 90°.
(b) Power consumed by an inductor. Note that the average power is zero.

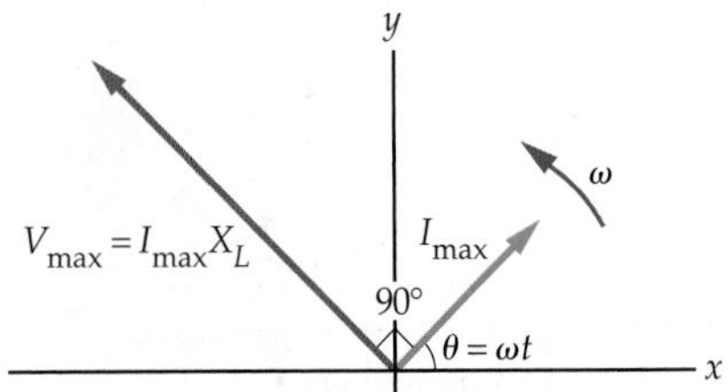

▲ **FIGURE 24–16 Phasor diagram for an ac inductor circuit**

The maximum value of the inductor's voltage is $I_{max}X_L$, and its angle is 90° ahead (counterclockwise) of the current.

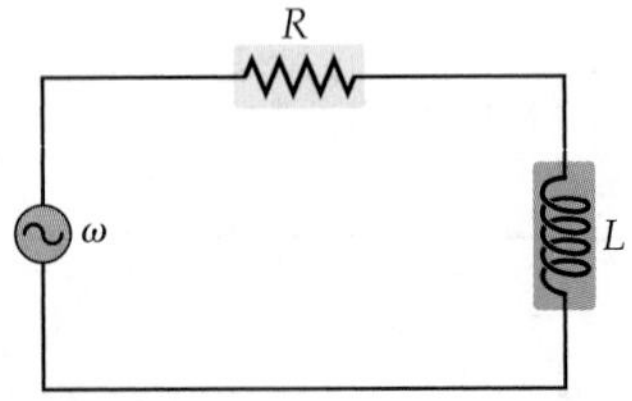

▲ **FIGURE 24–17 An alternating-current *RL* circuit**

An ac *RL* circuit consists of a generator, a resistor, and an inductor connected in series.

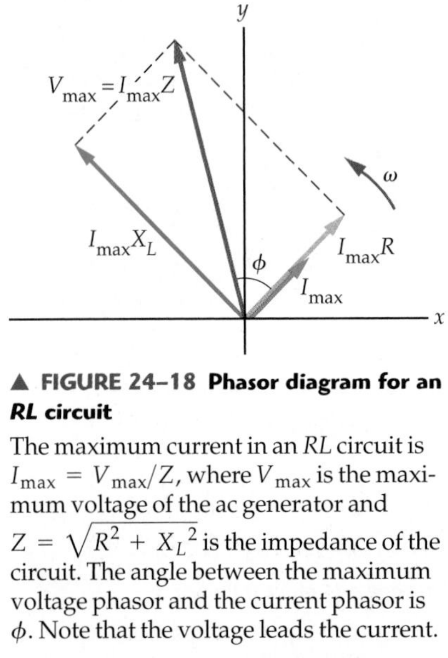

▲ **FIGURE 24–18 Phasor diagram for an *RL* circuit**

The maximum current in an *RL* circuit is $I_{max} = V_{max}/Z$, where V_{max} is the maximum voltage of the ac generator and $Z = \sqrt{R^2 + X_L^2}$ is the impedance of the circuit. The angle between the maximum voltage phasor and the current phasor is ϕ. Note that the voltage leads the current.

Thus we see that the current and the inductor's voltage are a quarter of a cycle (90°) out of phase. More specifically, since the voltage reaches its maximum *before* the current reaches its maximum, we say that the voltage *leads* the current:

The voltage across an inductor *leads* the current by 90°.

Note that this, again, is just the opposite of the behavior in a capacitor.

The phase relation between current and voltage in an inductor is shown with a phasor diagram in **Figure 24–16**. Here we see the current, I_{max}, and the corresponding inductor voltage, $V_{max} = I_{max}X_L$. Note that the voltage is rotated counterclockwise (that is, ahead) of the current by 90°.

Because of the 90° angle between the current and voltage, the power factor for an inductor is zero; cos 90° = 0. Thus, an ideal inductor—like an ideal capacitor—consumes zero average power. That is,

$$P_{av} = I_{rms}V_{rms}\cos\phi = I_{rms}V_{rms}\cos 90° = 0$$

The instantaneous power in an inductor alternates in sign, as shown in **Figure 24–15 (b)**. Thus energy enters the inductor at one time, only to be given up at a later time, for a net gain on average of zero energy.

RL Circuits

Next, we consider an ac circuit containing both a resistor and an inductor, as shown in **Figure 24–17**. The corresponding phasor diagram is drawn in **Figure 24–18**, where we see the resistor voltage in phase with the current, and the inductor voltage 90° ahead. The total voltage, of course, is given by the vector sum of these two phasors. Therefore, the magnitude of the total voltage is

$$V_{max} = \sqrt{(I_{max}R)^2 + (I_{max}X_L)^2} = I_{max}\sqrt{R^2 + X_L{}^2} = I_{max}Z$$

This expression defines the impedance, just as with an *RC* circuit. In this case, the impedance is

Impedance in an *RL* Circuit

$$Z = \sqrt{R^2 + X_L{}^2} = \sqrt{R^2 + (\omega L)^2} \qquad \textbf{24–15}$$

SI unit: ohm, Ω

Note that the impedance for an *RL* circuit has the same form as for an *RC* circuit, except that X_L replaces X_C. Similarly, the power factor for an *RL* circuit can be written as follows:

$$\cos\phi = \frac{R}{Z} = \frac{R}{\sqrt{R^2 + (\omega L)^2}}$$

We consider an *RL* circuit in the following Example.

EXAMPLE 24–5 *RL* Circuit

A 0.380-H inductor and a 225-Ω resistor are connected in series to an ac generator with an rms voltage of 30.0 V and a frequency of 60.0 Hz. Find the rms values of **(a)** the current in the circuit, **(b)** the voltage across the resistor, and **(c)** the voltage across the inductor.

Picture the Problem

Our sketch shows a 60.0-Hz generator connected in series with a 225-Ω resistor and a 0.380-H inductor. Because of the series connection, the same current flows through each of the circuit elements.

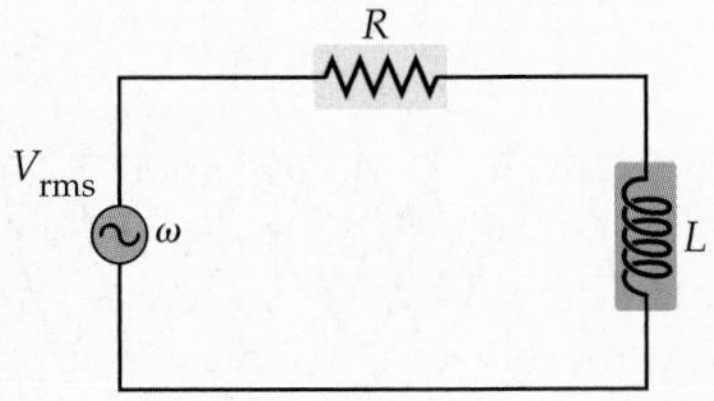

Strategy

a. The rms current in the circuit is $I_{rms} = V_{rms}/Z$, where the impedance is $Z = \sqrt{R^2 + (\omega L)^2}$.

b. The rms voltage across the resistor is $V_{rms,R} = I_{rms}R$.

c. The rms voltage across the inductor is $V_{rms,L} = I_{rms}X_L = I_{rms}\,\omega L$.

Solution

Part (a)

1. Calculate the impedance of the circuit:

$$Z = \sqrt{R^2 + X_L{}^2} = \sqrt{R^2 + (\omega L)^2}$$
$$= \sqrt{(225\ \Omega)^2 + [2\pi(60.0\ \text{s}^{-1})(0.380\ \text{H})]^2} = 267\ \Omega$$

2. Use Z to find the rms current:

$$I_{\text{rms}} = \frac{V_{\text{rms}}}{Z} = \frac{30.0\ \text{V}}{267\ \Omega} = 0.112\ \text{A}$$

Part (b)

3. Multiply I_{rms} by the resistance, R, to find the rms voltage across the resistor:

$$V_{\text{rms},R} = I_{\text{rms}}R = (0.112\ \text{A})(225\ \Omega) = 25.2\ \text{V}$$

Part (c)

4. Multiply I_{rms} by the inductive reactance, X_L, to find the rms voltage across the inductor:

$$V_{\text{rms},L} = I_{\text{rms}}X_L = I_{\text{rms}}\omega L$$
$$= (0.112\ \text{A})2\pi(60.0\ \text{s}^{-1})(0.380\ \text{H}) = 16.0\ \text{V}$$

Insight

As with the RC circuit, the individual voltages *do not* add up to the generator voltage. However, the generator rms voltage *is* equal to $\sqrt{V_{\text{rms},R}^2 + V_{\text{rms},L}^2}$.

Practice Problem

If the frequency in this circuit is increased, do you expect the rms current to increase, decrease, or stay the same? What is the current if the frequency is increased to 125 Hz? [**Answer:** The current will decrease with frequency. At 125 Hz the current is 0.0803 A.]

Some related homework problems: Problem 31, Problem 32

The preceding Practice Problem shows that the current in an RL circuit decreases with increasing frequency. Conversely, as we saw in Example 24–3, an RC circuit has the opposite behavior. These results are summarized and extended in **Figure 24–19**, where we show the current in RL and RC circuits as a function of frequency. It is assumed in this plot that both circuits have the same resistance, R. The horizontal line at the top of the plot indicates the current that would flow if the circuits contained *only* the resistor.

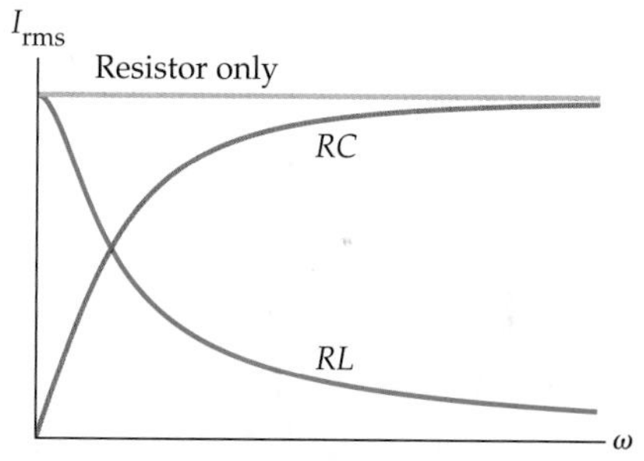

▲ **FIGURE 24–19 RMS currents in *RL* and *RC* circuits as a function of frequency**

A circuit with only a resistor has the same rms current regardless of frequency. In an RC circuit the current is low where the capacitive reactance is high (low frequency), and in the RL circuit the current is low where the inductive reactance is high (high frequency).

CONCEPTUAL CHECKPOINT 24–3 Compare Brightness: Inductor Circuit

Shown in the sketch below are two circuits with identical ac generators and lightbulbs. Circuit 2 differs from circuit 1 by the addition of an inductor in series with the light. Does the lightbulb in circuit 2 shine **(a)** more brightly, **(b)** less brightly, or **(c)** with the same intensity as that in circuit 1?

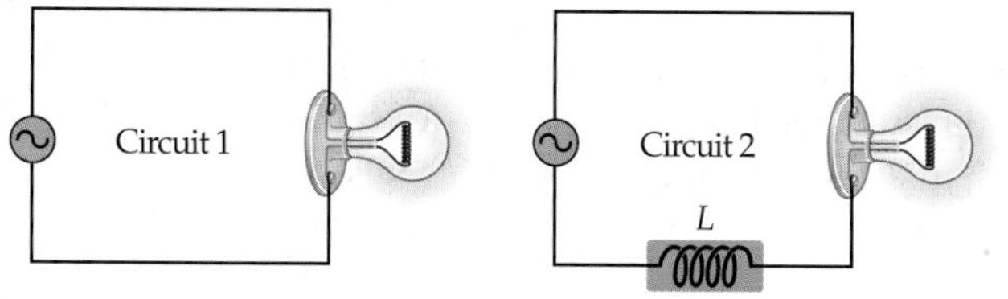

Reasoning and Discussion

Since circuit 2 has both a resistance and an inductive reactance, its impedance, $Z = \sqrt{R^2 + X_L{}^2}$, is greater than that of circuit 1. Therefore, the current in circuit 2, $I_{\text{rms}} = V_{\text{rms}}/Z$, is less than the current in circuit 1, and hence the average power dissipated in lightbulb 2 is less.

Answer:

(b) The lightbulb in circuit 2 shines less brightly.

REAL-WORLD PHYSICS
Light dimmers

The basic principle illustrated in Conceptual Checkpoint 24–3 is used commercially in light dimmers. When you rotate the knob on a light dimmer in one direction or the other, adjusting the light's intensity to the desired level, what you are actually doing is moving an iron rod into or out of the coils of an inductor. This changes both the inductance of the inductor and the intensity of the light.

For example, if the inductance is increased—by moving the iron rod deeper within the inductor's coil—the current in the circuit is decreased and the light dims. The advantage of dimming a light in this way is that no energy is dissipated by the inductor. In contrast, if one were to dim a light by placing a resistor in the circuit, the reduction in light would occur at the expense of wasted energy in the form of heat in the resistor. This needless dissipation of energy is avoided with the inductive light dimmer.

24–5 *RLC* Circuits

After having considered *RC* and *RL* circuits separately, we now consider circuits with all three elements, *R*, *L*, and *C*. In particular, if *R*, *L*, and *C* are connected in series, as in **Figure 24–20**, the resulting circuit is referred to as an ***RLC* circuit.** We now consider the behavior of such a circuit, using our previous results as a guide.

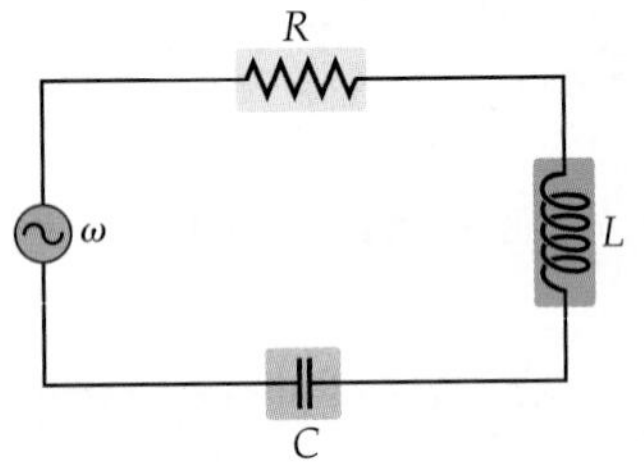

▲ **FIGURE 24–20 An alternating-current *RLC* circuit**

An ac *RLC* circuit consists of a generator, a resistor, an inductor, and a capacitor connected in series. The voltage may lead or lag the current, depending on the frequency of the generator and the values of *L* and *C*.

Phasor Diagram

As one might expect, a useful way to analyze the behavior of an *RLC* circuit is with the assistance of a phasor diagram. Thus, the phasor diagram for the circuit in Figure 24–20 is shown in **Figure 24–21**. Note that in addition to the current phasor, we show three separate voltage phasors corresponding to the resistor, inductor, and capacitor. The phasor diagram also shows that the voltage of the resistor is in phase with the current, the voltage of the inductor is 90° ahead of the current, and the voltage of the capacitor is 90° behind the current.

To find the total voltage in the system we must, as usual, perform a vector sum of the three voltage phasors. This process can be simplified if we first sum the inductor and capacitor phasors, which point in opposite directions along the same line. In the case shown in Figure 24–21 we assume for specificity that X_L is greater than X_C, so that the sum of these two voltage phasors is $I_{max}X_L - I_{max}X_C$. Combining this result with the phasor for the resistor voltage, and applying the Pythagorean theorem, we obtain the total maximum voltage:

$$V_{max} = \sqrt{(I_{max}R)^2 + (I_{max}X_L - I_{max}X_C)^2} = I_{max}\sqrt{R^2 + (X_L - X_C)^2} = I_{max}Z$$

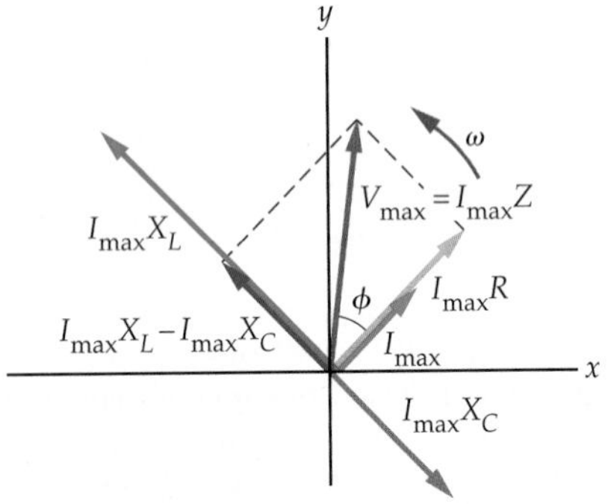

▲ **FIGURE 24–21 Phasor diagram for a typical *RLC* circuit**

In the case shown here, we assume that X_L is greater than X_C. All the results are the same for the opposite case, X_C greater than X_L, except that the phase angle ϕ changes sign, as can be seen in Equation 24–17.

The impedance of the circuit is thus defined as follows:

Impedance of an *RLC* Circuit

$$Z = \sqrt{R^2 + (X_L - X_C)^2} = \sqrt{R^2 + \left(\omega L - \frac{1}{\omega C}\right)^2} \qquad \textbf{24–16}$$

SI unit: ohm, Ω

Note that this result for the impedance contains the expressions given for the *RC* and *RL* circuits as special cases. For example, in the *RC* circuit the inductance is zero; hence, $X_L = 0$, and the impedance becomes

$$Z = \sqrt{R^2 + X_C{}^2}$$

This relation is identical with the result given for the *RC* circuit in Equation 24–11.

In addition to the magnitude of the total voltage, we are also interested in the phase angle ϕ between it and the current. From the phasor diagram it is clear that this angle is given by

$$\tan\phi = \frac{I_{max}(X_L - X_C)}{I_{max}R} = \frac{X_L - X_C}{R} \qquad \textbf{24–17}$$

In particular, if X_L is greater than X_C, as in Figure 24–21, then ϕ is positive and the voltage leads the current. On the other hand, if X_C is greater than X_L it follows that ϕ is negative, so the voltage lags the current. In the special case $X_C = X_L$ the phase angle is zero, and the current and voltage are in phase. There is special significance to this last case, as we shall see in greater detail in the next section.

Finally, the power factor, $\cos\phi$, can also be obtained from the phasor diagram. Referring again to Figure 24–21, we see that

$$\cos\phi = \frac{R}{Z}$$

Note that this is precisely the result given earlier for *RC* and *RL* circuits.

EXAMPLE 24–6 Draw Your Phasors!

An ac generator with a frequency of 60.0 Hz and an rms voltage of 120.0 V is connected in series with a 175-Ω resistor, a 90.0-mH inductor, and a 15.0-μF capacitor. Draw the appropriate phasor diagram for this system and calculate the phase angle, ϕ.

Picture the Problem

Our sketch shows the *RLC* circuit described in the problem statement. In particular, the 60.0-Hz generator is connected in series with a 175-Ω resistor, a 90.0-mH inductor, and a 15.0-μF capacitor.

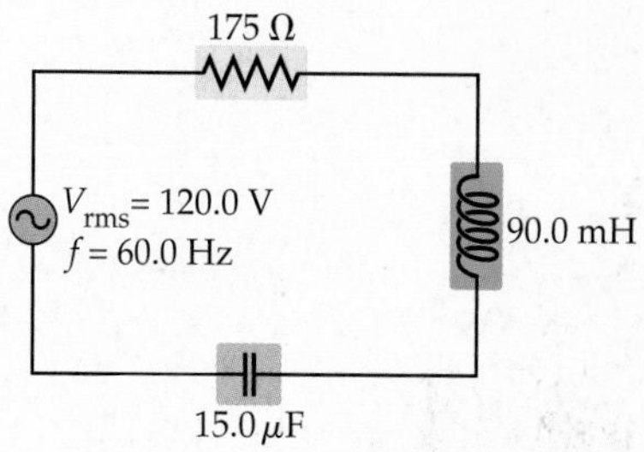

Strategy

To draw an appropriate phasor diagram we must first determine the values of X_C and X_L. These values determine immediately whether the voltage leads the current or lags it.

The precise value of the phase angle between the current and voltage is given by $\tan\phi = (X_L - X_C)/R$.

Solution

1. Calculate the capacitive and inductive reactances:

$$X_C = \frac{1}{\omega C} = \frac{1}{2\pi(60.0\ \text{s}^{-1})(15.0\times10^{-6}\ \text{F})} = 177\ \Omega$$

$$X_L = \omega L = 2\pi(60.0\ \text{s}^{-1})(90.0\times10^{-3}\ \text{H}) = 33.9\ \Omega$$

2. Use $\tan\phi = (X_L - X_C)/R$ to find the phase angle, ϕ:

$$\phi = \tan^{-1}\left(\frac{X_L - X_C}{R}\right)$$

$$= \tan^{-1}\left(\frac{33.9\ \Omega - 177\ \Omega}{175\ \Omega}\right) = -39.3°$$

Insight

We can now draw the phasor diagram for this circuit. First, the fact that X_C is greater than X_L means that the voltage of the circuit lags the current and that the phase angle is negative. In fact, we know that the phase angle has a magnitude of 39.3°. It follows that the phasor diagram for this circuit is as shown in the diagram.

Note that the lengths of the phasors $I_{max}R$, $I_{max}X_C$, and $I_{max}X_L$ in this diagram are drawn in proportion to the values of R, X_C, and X_L, respectively.

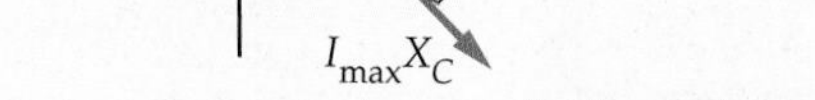

Practice Problem

Find the impedance and rms current for this circuit. [**Answer:** $Z = 226\ \Omega$, $I_{rms} = V_{rms}/Z = 0.531$ A]

Some related homework problems: Problem 46, Problem 47

The phase angle can also be obtained from the power factor: $\cos\phi = R/Z = (175\ \Omega)/(226\ \Omega) = 0.774$. Notice, however, that $\cos^{-1}(0.774) = \pm39.3°$. Thus, because of the symmetry of cosine, the power factor determines only the magnitude of ϕ, not its sign. The sign can be obtained from the phasor diagram, of course, but using $\phi = \tan^{-1}(X_L - X_C)/R$ yields both the magnitude and sign of ϕ.

ACTIVE EXAMPLE 24–2 Find the Inductor Voltage

The circuit elements shown in the sketch are connected to an ac generator at points A and B. If the rms voltage of the generator is 41 V and its frequency is 75 Hz, what is the rms voltage across the inductor?

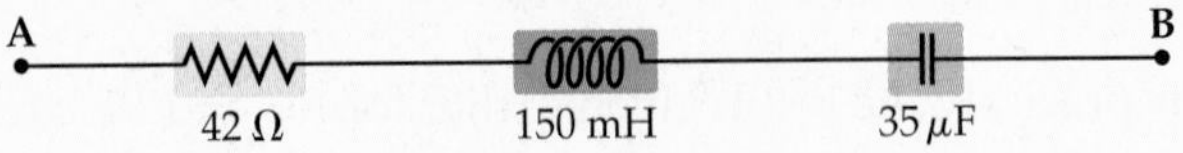

Solution *(Test your understanding by performing the calculations indicated in each step.)*

1. Calculate the capacitive reactance: $X_C = 61\ \Omega$
2. Calculate the inductive reactance: $X_L = 71\ \Omega$
3. Use Equation 24–16 to determine the impedance of the circuit: $Z = 43\ \Omega$
4. Divide V_{rms} by Z to find the rms current: $I_{rms} = 0.95\ \text{A}$
5. Multiply I_{rms} by X_L to find the rms voltage of the inductor: $V_{rms,L} = 67\ \text{V}$

Insight

Note that the voltage across individual components in a circuit can be larger than the applied voltage. In this case the applied voltage is 41 V, whereas the voltage across the inductor is 67 V.

Your Turn

Find the rms voltage across **(a)** the resistor and **(b)** the capacitor.

(Answers to **Your Turn** *problems are given in the back of the book.)*

Table 24–1 summarizes the various results for ac circuit elements (R, L, and C) and their combinations (RC, RL, RLC).

TABLE 24–1 Properties of AC Circuit Elements and Their Combinations

Circuit element	Impedance, Z	Average power, P_{av}	Phase angle, ϕ	Phasor diagram
	$Z = R$	$P_{av} = I_{rms}^2 R = V_{rms}^2/R$	$\phi = 0°$	Figure 24–3
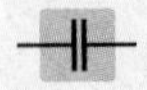	$Z = X_C = \dfrac{1}{\omega C}$	$P_{av} = 0$	$\phi = -90°$	Figure 24–8
	$Z = X_L = \omega L$	$P_{av} = 0$	$\phi = +90°$	Figure 24–16
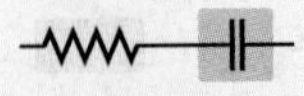	$Z = \sqrt{R^2 + X_C{}^2} = \sqrt{R^2 + \left(\dfrac{1}{\omega C}\right)^2}$	$P_{av} = I_{rms}V_{rms}\cos\phi$	$-90° < \phi < 0°$	Figure 24–10
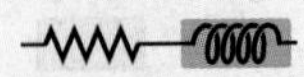	$Z = \sqrt{R^2 + X_L{}^2} = \sqrt{R^2 + (\omega L)^2}$	$P_{av} = I_{rms}V_{rms}\cos\phi$	$0° < \phi < 90°$	Figure 24–18
	$Z = \sqrt{R^2 + (X_L - X_C)^2}$ $= \sqrt{R^2 + \left(\omega L - \dfrac{1}{\omega C}\right)^2}$	$P_{av} = I_{rms}V_{rms}\cos\phi$	$-90° < \phi < 0°$ $(X_C > X_L)$ $0° < \phi < 90°$ $(X_C < X_L)$	Figure 24–21

Large and Small Frequencies

In the limit of very large or small frequencies, the behavior of capacitors and inductors is quite simple, allowing us to investigate more complex circuits containing R, L, and C. For example, in the limit of large frequency the reactance of an inductor becomes very large, whereas that of a capacitor becomes vanishingly small. This means that a capacitor acts like a segment of ideal wire, with no resistance, and an inductor behaves like a very large resistor with practically no current—essentially, an open circuit.

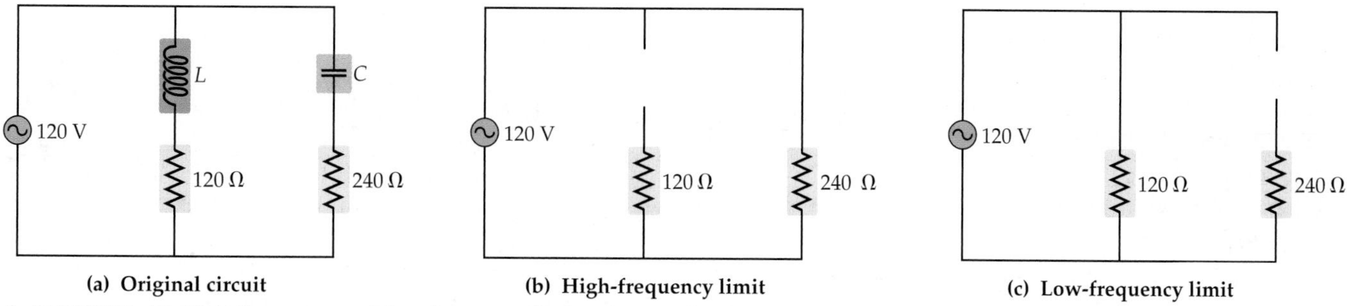

▲ **FIGURE 24–22 High-frequency and low-frequency limits of an ac circuit**
(a) A complex circuit containing resistance, inductance, and capacitance. Although this is not a simple *RLC* circuit, we can still obtain useful results about the circuit in the limits of high and low frequencies. **(b)** The high-frequency limit, in which the inductor is essentially an open circuit, and the capacitor behaves like an ideal wire. **(c)** The low-frequency limit. In this case, the inductor is like an ideal wire, and the capacitor acts like an open circuit.

By applying these observations to the circuit in **Figure 24–22 (a)**, for example, we can predict the current supplied by the generator at high frequencies. First, we replace the capacitor with an ideal wire and the inductor with an open circuit. This results in the simplified circuit shown in **Figure 24–22 (b)**. Clearly, the current in this circuit is 0.50 A; hence, we expect the current in the original circuit to approach this value as the frequency is increased.

In the opposite extreme of very small frequency we obtain the behavior that would be expected if the ac generator were replaced with a battery. Specifically, the reactance of an inductor vanishes as the frequency goes to zero, whereas that of a capacitor becomes extremely large. Thus, the roles of the inductor and capacitor are now reversed; it is the inductor that acts like an ideal wire, and the capacitor that behaves like an open circuit. For small frequencies, then, the circuit in Figure 24–22 (a) will behave the same as the circuit shown in **Figure 24–22 (c)**. The current in this circuit, and in the original circuit at low frequency, is 1.0 A.

EXAMPLE 24–7 Find *R*

The circuit shown in the sketch is connected to an ac generator with an rms voltage of 120 V. What value must *R* have if the rms current in this circuit is to approach 1.0 A at high frequency?

Picture the Problem

The top diagram shows the original circuit with its various elements. The high-frequency behavior of this circuit is indicated in the bottom diagram, where the inductor has been replaced with an open circuit, and the capacitors have been replaced with ideal wires.

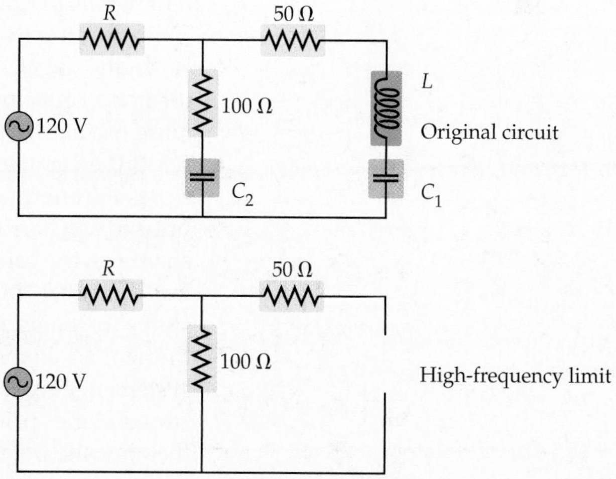

Strategy

The high-frequency circuit has only one remaining path through which current can flow. On this path the resistors with resistance R and 100 Ω are in series. Therefore, the total resistance of the circuit is $R_{\text{total}} = R + 100\ \Omega$. Finally, the rms current is $I_{\text{rms}} = V_{\text{rms}}/R_{\text{total}}$. Setting I_{rms} equal to 1.0 A gives us the value of R.

Solution

1. Calculate the total resistance of the high-frequency circuit:

$$R_{\text{total}} = R + 100\ \Omega$$

2. Write an expression for the rms current in the circuit:

$$I_{\text{rms}} = \frac{V_{\text{rms}}}{R_{\text{total}}} = \frac{V_{\text{rms}}}{R + 100\ \Omega}$$

3. Solve for the resistance R:

$$R = \frac{V_{\text{rms}}}{I_{\text{rms}}} - 100\ \Omega = \frac{120\ \text{V}}{1.0\ \text{A}} - 100\ \Omega = 20\ \Omega$$

continued on next page

continued from previous page

Insight

Note that no values are given for the capacitances and the inductance. At high enough frequencies the precise values of these quantities are unimportant.

Practice Problem

What is the rms current in this circuit in the limit of small frequency? [**Answer:** In the limit of small frequency, the current approaches zero.]

Some related homework problems: Problem 44, Problem 45

24–6 Resonance in Electrical Circuits

Many physical systems have natural frequencies of oscillation. For example, we saw in Chapter 13 that a child on a swing oscillates about the vertical with a definite, characteristic frequency determined by the length of the swing and the acceleration of gravity. Similarly, an object attached to a spring oscillates about its equilibrium position with a frequency determined by the "stiffness" of the spring and the mass of the object. Certain electrical circuits have analogous behavior—their electric currents oscillate with certain characteristic frequencies. In this section we consider some examples of "oscillating" electrical circuits.

LC Circuits

Perhaps the simplest circuit that displays an oscillating electric current is an ***LC* circuit** with no generator; that is, one that consists of nothing more than an inductor and a capacitor. Suppose, for example, that at $t = 0$ a charged capacitor has just been connected to an inductor and that there is no current in the circuit, as shown in **Figure 24–23 (a)**. Since the capacitor is charged, it has a voltage, $V = Q/C$, which causes a current to begin flowing through the inductor, as in **Figure 24–23 (b)**. Soon all the charge drains from the capacitor and its voltage drops to zero, but the current continues to flow because an inductor resists changes in its current. In fact, the current continues flowing until the capacitor becomes charged enough in the *opposite* direction to stop the current, as in **Figure 24–23 (c)**. At this point, the current begins to flow back the way it came, and the same sequence of events occurs again, leading to a *series of oscillations* in the current.

In the ideal case, the oscillations can continue forever, since neither an inductor nor a capacitor dissipates energy. The situation is completely analogous to a mass oscillating on a spring with no friction, as we indicate in Figure 24–23. At $t = 0$ the capacitor has a charge of magnitude Q on its plates, which means it stores the energy $U_C = Q^2/2C$. This situation is analogous to a spring being compressed a distance x and storing the potential energy $U = \frac{1}{2}kx^2$. At a later time the charge on the capacitor is zero, so it no longer stores any energy. The energy is not lost, however. Instead, it is now in the inductor, which carries a current I and stores the energy $U_L = \frac{1}{2}LI^2 = U_C$. In the mass–spring system, this corresponds to the mass being at the equilibrium position of the spring. At this time all the system's energy is the kinetic energy of the mass, $K = \frac{1}{2}mv^2 = U$, and none is stored in the spring. As the current continues to flow, it charges the capacitor with the opposite polarity until it reaches the magnitude Q and stores the same energy, U_C, as at $t = 0$. In the mass–spring system, this corresponds to the spring being stretched by the same distance x that it was originally compressed, which again stores all the initial energy as potential energy.

Thus, we see that a close analogy exists between a capacitor and a spring, and between an inductor and a mass. In addition, the charge on the capacitor is analogous to the displacement of the spring, and the current in the inductor is analogous to the speed of the mass. Thus, for example, the energy stored in the inductor, $\frac{1}{2}LI^2$, corresponds precisely to the kinetic energy of the mass, $\frac{1}{2}mv^2$. Comparing the potential energy of a spring, $\frac{1}{2}kx^2$, and the energy stored in a capacitor, $Q^2/2C$, we see that the stiffness of a spring is analogous to $1/C$. This

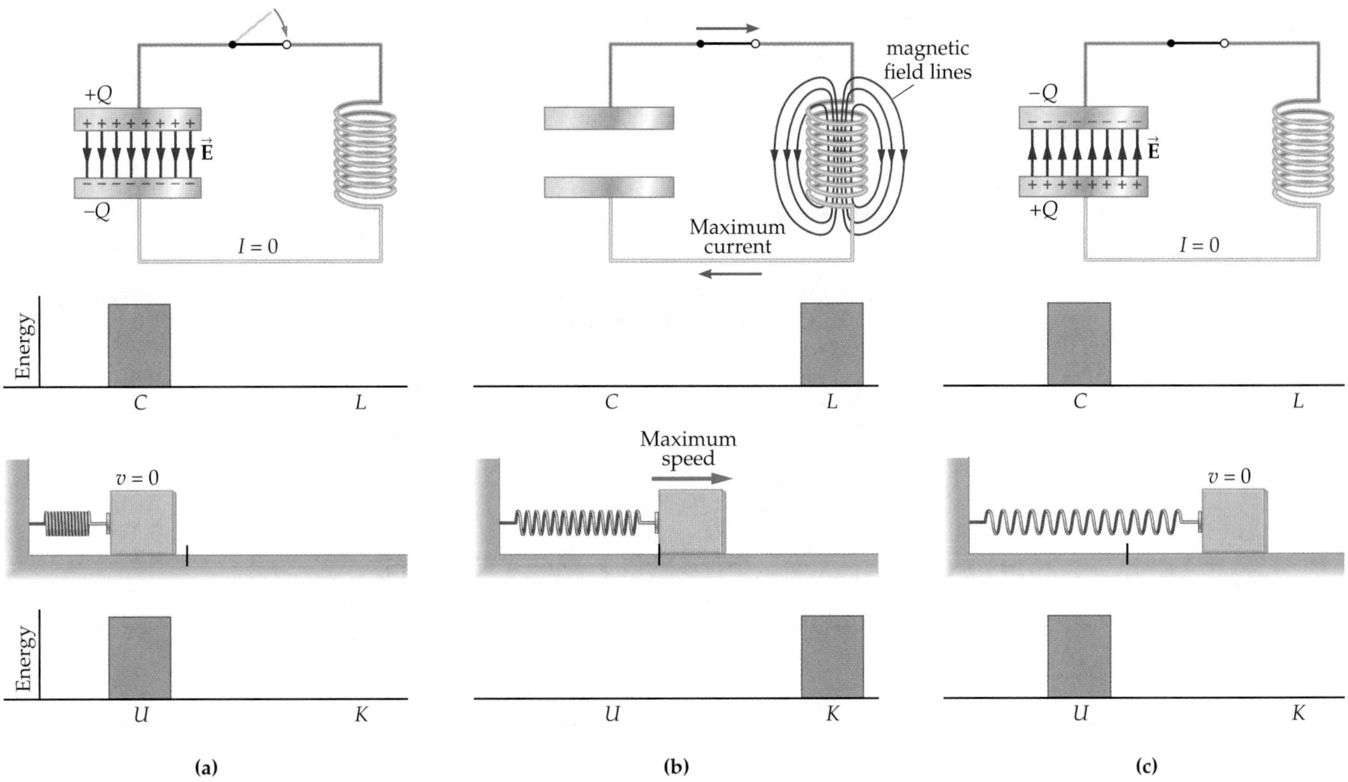

▲ **FIGURE 24–23 Oscillations in an *LC* circuit with no generator**

The current oscillations in an *LC* circuit are analogous to the oscillations of a mass on a spring. **(a)** At the moment the switch is closed, all the energy in the circuit is stored in the electric field of the charged capacitor. This is analogous to a mass at rest against a compressed spring, where all the energy of the system is stored in the spring. **(b)** A quarter of the way through a cycle, the capacitor is uncharged and the current in the inductor is a maximum. At this time, all the energy in the circuit is stored in the magnetic field of the inductor. In the mass-spring analog, the spring is at equilibrium and the mass has its maximum speed. All the system's energy is now in the form of kinetic energy. **(c)** After half a cycle, the capacitor is fully charged in the opposite direction and holds all the system's energy. This corresponds to a fully extended spring (with all the system's energy) and the mass at rest.

makes sense, because a capacitor with a large capacitance, C, can store large quantities of charge with ease, just as a spring with a small force constant (if C is large, then $k = 1/C$ is small) can be stretched quite easily.

In the mass-spring system the natural angular frequency of oscillation is determined by the characteristics of the system. In particular, recall from Section 13–8 that

$$f_0 = \frac{1}{2\pi}\sqrt{\frac{k}{m}} \qquad \text{or} \qquad \omega_0 = 2\pi f_0 = \sqrt{\frac{k}{m}}$$

The natural frequency of the *LC* circuit can be determined by noting that the rms voltage across the capacitor C in Figure 24–23 must be the same as the rms voltage across the inductor L. This condition can be written as follows:

$$V_{\text{rms},C} = V_{\text{rms},L}$$
$$I_{\text{rms}}X_C = I_{\text{rms}}X_L$$
$$I_{\text{rms}}\left(\frac{1}{\omega_0 C}\right) = I_{\text{rms}}(\omega_0 L)$$

Solving for ω_0, we find

Natural Frequency of an *LC* Circuit

$$\omega_0 = \frac{1}{\sqrt{LC}} = 2\pi f \qquad \text{24–18}$$

SI unit: s^{-1}

Note again the analogy between this result and that for a mass on a spring: if we make the following replacements, $m \rightarrow L$ and $k \rightarrow 1/C$, in the mass-spring result we find the expected LC result:

$$\omega_0 = \sqrt{\frac{k}{m}} = \sqrt{\frac{1}{LC}}$$

We summarize the mass-spring/LC circuit analogies in Table 24–2.

TABLE 24–2 Analogies between a mass on a spring and an *LC* circuit

Mass-spring system		*LC* circuit	
Position	x	Charge	q
Velocity	$v = \Delta x/\Delta t$	Current	$I = \Delta q/\Delta t$
Mass	m	Inductance	L
Force constant	k	Inverse capacitance	$1/C$
Natural frequency	$\omega_0 = \sqrt{k/m}$	Natural frequency	$\omega_0 = \sqrt{1/LC}$

EXERCISE 24–6

It is desired to tune the natural frequency of an LC circuit to match the 88.5-MHz broadcast signal of an FM radio station. If a 1.50-μH inductor is to be used in the circuit, what capacitance is required?

Solution

Solving $\omega_0 = 1/\sqrt{LC}$ for the capacitance, we find

$$C = \frac{1}{\omega_0^2 L} = \frac{1}{[2\pi(88.5 \times 10^6\ \text{s}^{-1})]^2(1.50 \times 10^{-6}\ \text{H})} = 2.16 \times 10^{-12}\ \text{F}$$

Resonance

Whenever a physical system has a natural frequency, we can expect to find resonance when it is driven near that frequency. In a mass-spring system, for example, if we move the top end of the spring up and down with a frequency near the natural frequency, the displacement of the mass can become quite large. Similarly, if we push a person on a swing at the right frequency, the amplitude of motion will increase. These are examples of resonance in mechanical systems.

To drive an electrical circuit, we can connect it to an ac generator. As we adjust the frequency of the generator, the current in the circuit will be a maximum at the natural frequency of the circuit. For example, consider the circuit shown in Figure 24–20. Here an ac generator drives a circuit containing an inductor, a capacitor, and a resistor. As we have already seen, the inductor and capacitor together establish a natural frequency $\omega_0 = 1/\sqrt{LC}$, and the resistor provides for energy dissipation.

The phasor diagram for this circuit is like the one shown in Figure 24–21. Recall that the maximum current in this circuit is

$$I_{\text{max}} = \frac{V_{\text{max}}}{Z}$$

Thus, the smaller the impedance, the larger the current. Hence, to obtain the largest possible current, we must have the smallest Z. Recall that the impedance is given by

$$Z = \sqrt{R^2 + (X_L - X_C)^2}$$

Writing Z in terms of frequency, we have

$$Z = \sqrt{R^2 + \left(\omega L - \frac{1}{\omega C}\right)^2} \qquad \text{24–19}$$

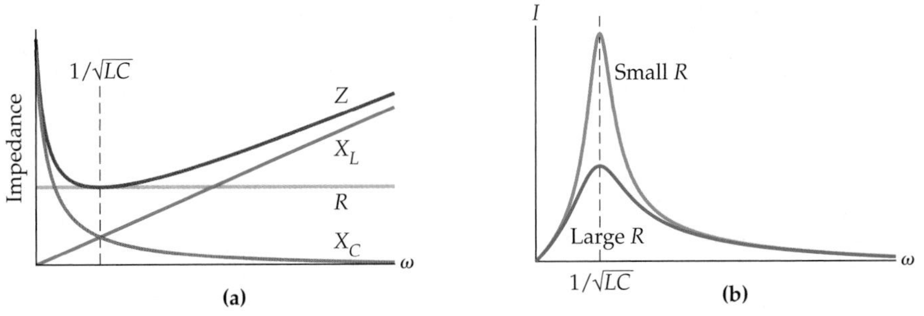

◀ **FIGURE 24–24 Resonance in an *RLC* circuit**

(a) The impedance, Z, of an RLC circuit varies as a function of frequency. The minimum value of Z—which corresponds to the largest current—occurs at the resonance frequency $\omega_0 = 1/\sqrt{LC}$, where $X_C = X_L$. At this frequency $Z = R$. **(b)** Typical resonance peaks for an RLC circuit. The location of the peak is independent of resistance, but the resonance effect becomes smaller and more spread out with increasing R.

This expression for Z is plotted in **Figure 24–24 (a)**. Notice that the smallest value of the impedance is $Z = R$, and that this value is attained precisely at the frequency where $X_L = X_C$. We can see this mathematically by setting $X_L = X_C$ in the expression $Z = \sqrt{R^2 + (X_L - X_C)^2}$, which yields $Z = \sqrt{R^2 + 0} = R$, as expected. The frequency at which $X_L = X_C$ is the frequency for which $\omega L = 1/\omega C$. This frequency is $\omega = 1/\sqrt{LC} = \omega_0$, the *natural frequency* found in Equation 24–18 for LC circuits.

PROBLEM-SOLVING NOTE

A Circuit at Resonance

The resonance frequency of an RLC circuit depends only on the inductance, L, and the capacitance, C; that is, $\omega_0 = 1/\sqrt{LC}$. On the other hand, the impedance at resonance depends only on the resistance, R; that is, $Z = R$.

Figure 24–24 (b) shows typical plots of the current in an RLC circuit. Note that the current peaks at the resonance frequency. Note also that increasing the resistance, although it reduces the maximum current in the circuit, does not change this frequency. It does, however, make the resonance peak flatter and broader. As a result, the resonance effect occurs over a wide range of frequencies and gives only a small increase in current. If the resistance is small, however, the peak is high and sharp. In this case, the resonance effect yields a large current that is restricted to a very small range of frequencies.

REAL-WORLD PHYSICS

Tuning a radio or television

Radio and television tuners use low-resistance RLC circuits so they can pick up one station at a frequency f_1 without also picking up a second station at a nearby frequency f_2. In a typical radio tuner, for example, turning the knob of the tuner rotates one set of capacitor plates between a second set of plates, effectively changing both the plate separation and the plate area. This changes the capacitance in the circuit and the frequency at which it resonates. If the resonance peak is high and sharp—as occurs with low resistance—only the station broadcasting at the resonance frequency will be picked up and amplified. Other stations at nearby frequencies will produce such small currents in the tuning circuit that they will be undetectable.

REAL-WORLD PHYSICS

Metal detectors

Metal detectors also use resonance in RLC circuits, although in this case it is the inductance that changes rather than the capacitance. For example, when you walk through a metal detector at an airport, you are actually walking through a large coil of wire—an inductor. Metal objects on your person cause the inductance of the coil to increase slightly. This increase in inductance results in a small decrease in the resonance frequency of the RLC circuit to which the coil is connected. If the resonance peak is sharp and high, even a slight change in frequency results in a large change in current. It is this change in current that sets off the detector, indicating the presence of metal.

CONCEPTUAL CHECKPOINT 24–4 Phase of the Voltage

An RLC circuit is driven at its resonance frequency. Is its voltage (a) ahead of, (b) behind, or (c) in phase with the current?

Reasoning and Discussion

At resonance the capacitive and inductive reactances are equal, which means that the voltage across the capacitor is equal in magnitude and opposite in direction to the voltage across the inductor. As a result, the net voltage in the system is simply the voltage across the resistor, which is in phase with the current.

Answer:

(c) The voltage and current are in phase.

EXAMPLE 24–8 A Circuit in Resonance

An ac generator with an rms voltage of 25 V is connected in series to a 10.0-Ω resistor, a 53-mH inductor, and a 65-μF capacitor. Find **(a)** the resonance frequency of the circuit and **(b)** the rms current at resonance. In addition, sketch the phasor diagram at resonance.

Picture the Problem

As mentioned in the previous Conceptual Checkpoint, at resonance the magnitude of the voltage across the inductor is equal to the magnitude of the voltage across the capacitor. Since the phasors corresponding to these voltages point in opposite directions, however, they cancel. This leads to a net voltage phasor that is simply $I_{max}R$ in phase with the current. Finally, note that the lengths of the phasors $I_{max}R$, $I_{max}X_C$, and $I_{max}X_L$ are drawn in proportion to the values of R, X_C, and X_L, respectively.

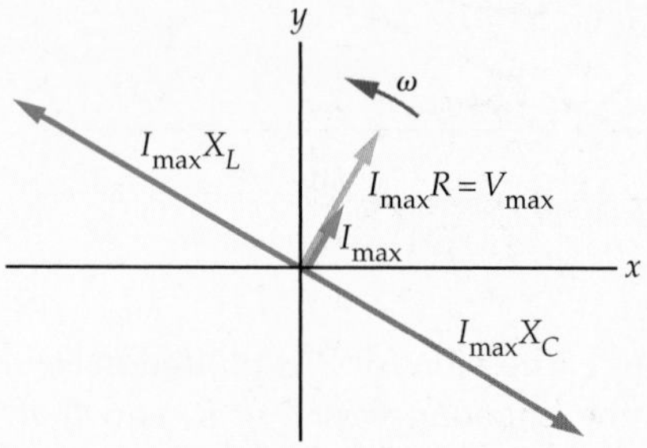

Strategy

a. We find the resonance frequency by substituting numerical values into $\omega_0 = 1/\sqrt{LC} = 2\pi f_0$.

b. At resonance the impedance of the circuit is simply the resistance; that is, $Z = R$. Thus, the rms current in the circuit is $I_{rms} = V_{rms}/Z = V_{rms}/R$.

Solution

Part (a)

1. Calculate the resonance frequency:

$$\omega_0 = \frac{1}{\sqrt{LC}} = 2\pi f_0$$

$$f_0 = \frac{1}{2\pi\sqrt{LC}}$$

$$= \frac{1}{2\pi\sqrt{(53 \times 10^{-3}\,\text{H})(65 \times 10^{-6}\,\text{F})}} = 86\ \text{Hz}$$

Part (b)

2. Determine the impedance at resonance:

$$Z = R = 10.0\ \Omega$$

3. Divide the rms voltage by the impedance to find the rms current:

$$I_{rms} = \frac{V_{rms}}{Z} = \frac{25\ \text{V}}{10.0\ \Omega} = 2.5\ \text{A}$$

Insight

If the frequency of this generator is increased above resonance, the inductive reactance, X_L, will be larger than the capacitive reactance, X_C, and hence the voltage will lead the current. If the frequency is lowered below resonance the voltage will lag the current.

Practice Problem

What is the magnitude of the rms voltage across the capacitor? [**Answer:** 71 V. Note again that the voltage across a given circuit element can be much larger than the applied voltage.]

Some related homework problems: Problem 56, Problem 58

ACTIVE EXAMPLE 24–3 Find the Off-Resonance Current

Referring to the system in the previous Example, find the rms current when the generator operates at a frequency that is twice the resonance frequency.

Solution *(Test your understanding by performing the calculations indicated in each step.)*

1. Calculate the inductive reactance at the frequency $f = 2(86\ \text{Hz})$: $X_L = 57\ \Omega$

2. Calculate the capacitive reactance at the frequency $f = 2(86\ \text{Hz})$: $X_C = 14\ \Omega$

3. Use these results plus the resistance R to find the impedance, Z: $Z = 44\ \Omega$

4. Divide the rms voltage by the impedance to find the rms current: $I_{rms} = 0.57\ \text{A}$

Insight

As expected, the inductive reactance is greater than the capacitive reactance at this frequency. In addition, note that the rms current has been reduced by roughly a factor of 5 compared with its value in Example 24–8.

Your Turn

At what two frequencies is the current in this circuit equal to 1.5 A?

(Answers to **Your Turn** *problems are given in the back of the book.)*

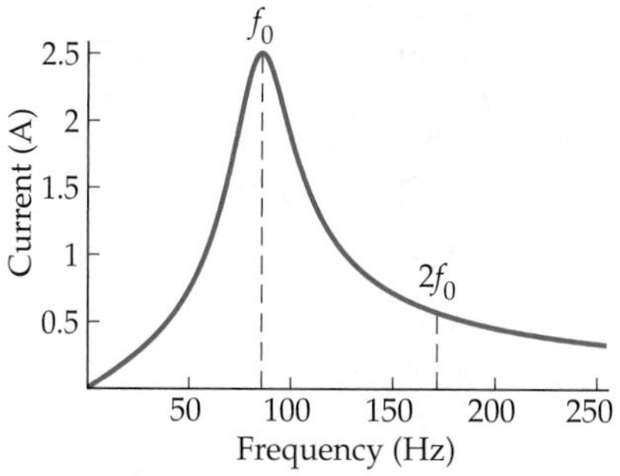

▲ **FIGURE 24–25 RMS current in an *RLC* circuit**

This plot shows the rms current versus frequency for the circuit considered in Example 24–8 and Active Example 24–3. The currents determined in these Examples occur at the frequencies indicated by the vertical dashed lines.

The rms current as a function of frequency for the system considered in the previous Example and Active Example is shown in **Figure 24–25**. Notice that the two currents just calculated, at $f_0 = 86$ Hz and $f = 2f_0 = 2(86\text{ Hz}) = 172$ Hz, are indicated on the graph.

Chapter Summary

24–1 Alternating Voltages and Currents

An ac generator produces a voltage that varies with time as

$$V = V_{\text{max}} \sin \omega t \qquad \text{24–1}$$

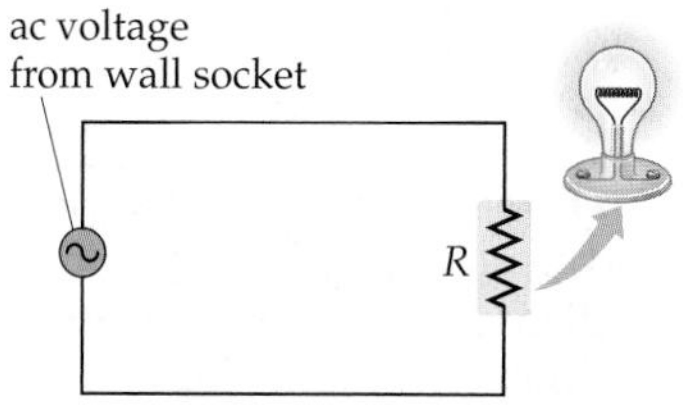

Phasors

A phasor is a rotating vector representing a voltage or a current in an ac circuit. Phasors rotate counterclockwise about the origin in the x-y plane with an angular speed ω. The y component of a phasor gives the instantaneous value of that quantity.

rms Values

The rms, or root mean square, of a quantity x is the square root of the average value of x^2. For any quantity x that varies with time as a sine or a cosine, the rms value is

$$x_{\text{rms}} = \frac{1}{\sqrt{2}} x_{\text{max}} \qquad \text{24–4}$$

Standard dc formulas like $P = I^2R$ can be converted to ac average formulas by using rms values. For example, $P_{\text{av}} = I_{\text{rms}}^2 R$.

24–2 Capacitors in AC Circuits

A capacitor in an ac circuit has a current that depends on the frequency and is out of phase with the voltage.

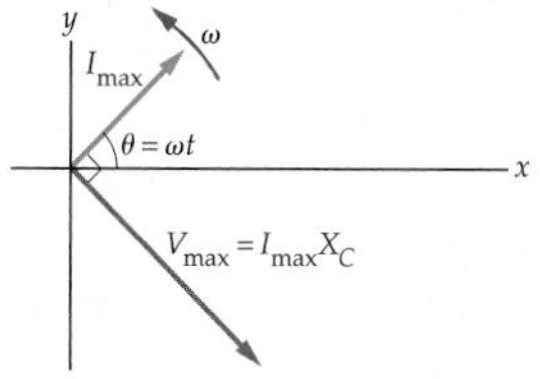

Capacitive Reactance

The rms current in a capacitor is

$$I_{\text{rms}} = \frac{V_{\text{rms}}}{X_C} \qquad \text{24–8}$$

where X_C is the capacitive reactance. It is defined as follows:

$$X_C = \frac{1}{\omega C} \qquad \text{24–9}$$

Phase Relation between Voltage and Current in a Capacitor

The voltage across a capacitor lags the current by 90°.

Phasor Diagram for Capacitors

In a phasor diagram, the voltage of a capacitor is drawn at an angle that is 90° clockwise from the direction of the current phasor.

24–3 *RC* Circuits

To analyze an *RC* circuit, the 90° phase difference between the resistor and capacitor voltages must be taken into account.

Impedance
The impedance, Z, of an RC circuit is analogous to the resistance in a simple resistor circuit and takes into account both the resistance, R, and the capacitive reactance, X_C:

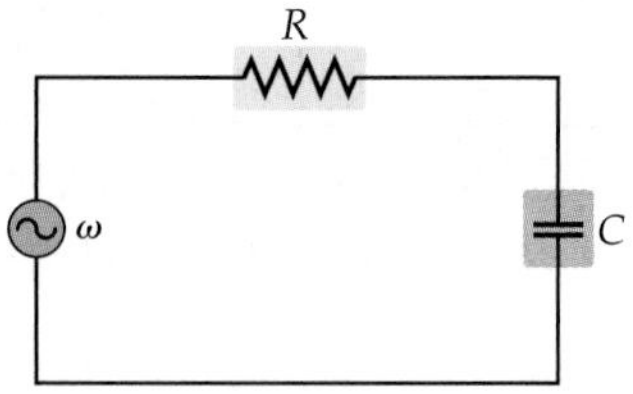

$$Z = \sqrt{R^2 + X_C{}^2} = \sqrt{R^2 + \left(\frac{1}{\omega C}\right)^2} \quad \text{24–11}$$

Voltage and Current
The rms voltage and current in an RC circuit are related by:

$$V_{rms} = I_{rms}\sqrt{R^2 + X_C{}^2} = I_{rms}Z$$

Power Factor
If the phase angle between the current and voltage in an RC circuit is ϕ, the average power consumed by the circuit is

$$P_{av} = I_{rms}V_{rms}\cos\phi \quad \text{24–13}$$

where $\cos\phi$ is referred to as the power factor.

24–4 Inductors in AC Circuits

Inductors in ac circuits have frequency-dependent currents, and voltages that are out of phase with the current.

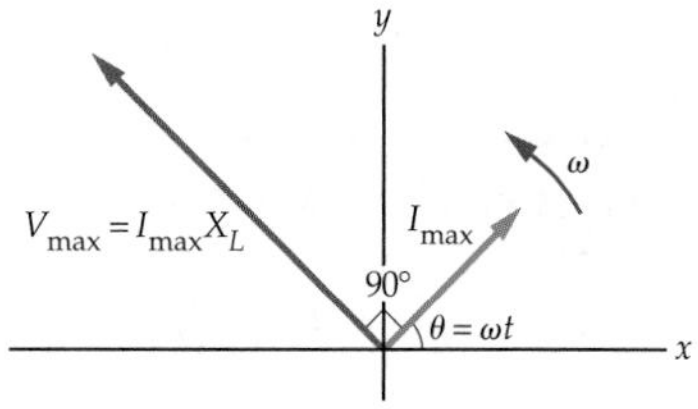

Inductive Reactance
The inductive reactance is

$$X_L = \omega L \quad \text{24–14}$$

The rms current in an inductor is

$$I_{rms} = \frac{V_{rms}}{X_L}$$

Phasor Diagram for Inductors
The voltage across an inductor leads the current by 90°. Thus, in a phasor diagram, the voltage phasor for an inductor is rotated 90° counterclockwise from the current phasor.

***RL* Circuits**
The impedance of an RL circuit is

$$Z = \sqrt{R^2 + X_L{}^2} = \sqrt{R^2 + (\omega L)^2} \quad \text{24–15}$$

24–5 *RLC* Circuits

An RLC circuit consists of a resistor, R, an inductor, L, and a capacitor, C, connected in series to an ac generator.

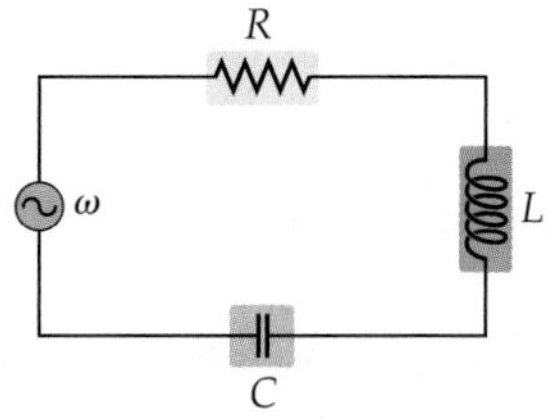

Impedance
The impedance of an RLC circuit is

$$Z = \sqrt{R^2 + (X_L - X_C)^2} = \sqrt{R^2 + \left(\omega L - \frac{1}{\omega C}\right)^2} \quad \text{24–16}$$

Large and Small Frequencies
In the limit of large frequencies, inductors behave like open circuits and capacitors are like ideal wires. For very small frequencies, the behaviors are reversed; inductors are like ideal wires, and capacitors act like open circuits.

24–6 Resonance in Electrical Circuits

Electrical circuits can have natural frequencies of oscillation, just like a pendulum or a mass on a spring.

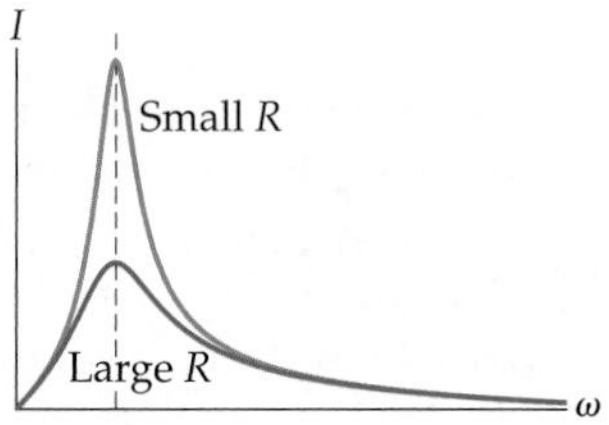

***LC* Circuits**
Circuits containing only an inductor and a capacitor have a natural frequency given by

$$\omega_0 = \frac{1}{\sqrt{LC}} = 2\pi f_0 \quad \text{24–18}$$

Resonance
An RLC circuit connected to an ac generator has maximum current at the frequency $\omega = 1/\sqrt{LC}$. This effect is referred to as resonance.

Problem-Solving Summary

Type of Problem	Relevant Physical Concepts	Related Examples
Find the average power dissipated in a resistor.	To find the average power dissipated in a resistor, simply replace I with I_{rms} in $P = I^2R$, or replace V with V_{rms} in $P = V^2/R$.	Example 24–1
Calculate the rms voltage across a capacitor in an ac circuit.	In analogy with $V = IR$ for a resistor, the rms voltage across a capacitor is $V_{rms} = I_{rms}X_C$, where $X_C = 1/\omega C$.	Example 24–2
Find the impedance and rms voltage or rms current in an RC circuit.	The impedance in an RC circuit is $Z = \sqrt{R^2 + X_C{}^2} = \sqrt{R^2 + (1/\omega C)^2}$. The voltage and current in an RC circuit are related by the relation $V_{rms} = I_{rms}Z$.	Example 24–3 Active Example 24–1
Find the average power consumed by an ac circuit.	In an ac circuit, the average power consumed is $P_{av} = I_{rms}V_{rms}\cos\phi$, where the power factor $(\cos\phi)$ is defined as follows: $\cos\phi = R/Z$.	Example 24–4
Find the impedance and rms voltage or rms current in an RL circuit.	The impedance in an RL circuit is $Z = \sqrt{R^2 + X_L{}^2} = \sqrt{R^2 + (\omega L)^2}$. The voltage and current in an RL circuit are related by $V_{rms} = I_{rms}Z$.	Example 24–5
Find the impedance and rms voltage or rms current in an RLC circuit.	The impedance in an RLC circuit is $Z = \sqrt{R^2 + (X_L - X_C)^2} = \sqrt{R^2 + \left(\omega L - \frac{1}{\omega C}\right)^2}$. The voltage and current in an RLC circuit are related by the relation $V_{rms} = I_{rms}Z$.	Example 24–6 Active Example 24–2
Determine the resonance frequency and impedance of an RLC circuit.	In a circuit with a capacitance C and an inductance L, the resonance frequency is $\omega_0 = 1/\sqrt{LC}$. At resonance, the impedance of an RLC circuit is simply equal to its resistance; that is, $Z = R$.	Example 24–8 Active Example 24–3

Conceptual Questions

(Answers to odd-numbered Conceptual Questions can be found in the back of the book.)

1. How can the rms voltage of an ac circuit be nonzero when its average value is zero? Explain.
2. Why is the current in an ac circuit not always in phase with its voltage?
3. Does an LC circuit consume any power? Explain.
4. An LC circuit is driven at a frequency higher than its resonance frequency. What can be said about the phase angle, ϕ, for this circuit?
5. An LC circuit is driven at a frequency lower than its resonance frequency. What can be said about the phase angle, ϕ, for this circuit?
6. In Conceptual Checkpoint 24–3 we considered an ac circuit consisting of a lightbulb in series with an inductor. The effect of the inductor was to cause the bulb to shine less brightly. Would the same be true in a direct-current (dc) circuit? Explain.
7. How do the resistance, capacitive reactance, and inductive reactance change when the frequency in a circuit is increased?
8. In the analogy between an RLC circuit and a mass on a spring, what is the analog of the current in the circuit? Explain.
9. In the analogy between an RLC circuit and a mass on a spring, the mass is analogous to the inductance, and the spring constant is analogous to the inverse of the capacitance. Explain.
10. Two RLC circuits have different values of L and C. Is it possible for these two circuits to have the same resonance frequency? Explain.
11. Can an RLC circuit have the same impedance at two different frequencies? Explain.

Conceptual Exercises

(Answers to odd-numbered Conceptual Exercises can be found in the back of the book.)

1. Consider a circuit consisting of a lightbulb and a capacitor, as shown in Conceptual Checkpoint 24–2. If the frequency of the generator is increased, does the intensity of the lightbulb increase, decrease, or stay the same? Explain.
2. **BIO Persistence of Vision** Although an incandescent lightbulb appears to shine with constant intensity, this is an artifact of the eye's persistence of vision. In fact, the intensity of a bulb's light rises and falls with time due to the alternating current used in household circuits. If you could perceive these oscillations, would you see the light attain maximum brightness 60 or 120 times per second? Explain.
3. When a long copper wire of finite resistance is connected to an ac generator, as shown in **Figure 24–26 (a)**, a certain amount of current flows through the wire. The wire is now wound into a coil of many loops and reconnected to the generator, as indicated in **Figure 24–26 (b)**. Is the current supplied to the coil more than, less than, or the same as the current supplied to the uncoiled wire? Explain.

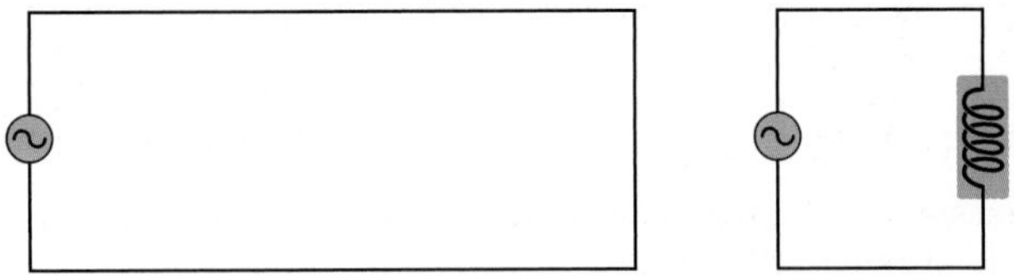

(a) (b)

▲ **FIGURE 24–26** Conceptual Exercise 3

4. Consider a circuit consisting of a lightbulb and an inductor, as shown in Conceptual Checkpoint 24–3. If the frequency of the generator is increased, does the intensity of the lightbulb increase, decrease, or stay the same? Explain.
5. An inductor and a capacitor are to be connected to a generator. Will the generator supply more current at *high* frequency if the inductor and capacitor are connected in series or in parallel? Explain.
6. An inductor and a capacitor are to be connected to a generator. Will the generator supply more current at *low* frequency if the inductor and capacitor are connected in series or in parallel? Explain.
7. An inductor in an *LC* circuit has a maximum current of 2.4 A and a maximum energy of 32 mJ. When the current in the inductor is 1.2 A, what is the energy stored in the capacitor?
8. A capacitor and an inductor connected in series have a period of oscillation given by T. At the time $t = 0$ the capacitor has its maximum charge. In terms of T, what is the first time after $t = 0$ that **(a)** the current in the circuit has its maximum value and **(b)** the energy stored in the electric field is a maximum?
9. An *RLC* circuit is driven at its resonance frequency. Is its impedance greater than, less than, or equal to R? Explain.
10. When the ac generator in **Figure 24–27** operates at high frequency, is the rms current in the circuit more than, less than, or the same as when the generator operates at low frequency? Explain.

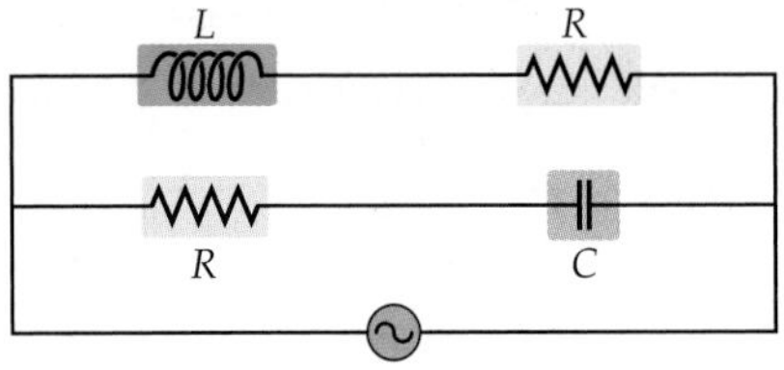

▲ **FIGURE 24–27** Conceptual Exercises 10 and 13

11. When the ac generator in **Figure 24–28** operates at high frequency, is the rms current in the circuit more than, less than, or the same as when the generator operates at low frequency? Explain.

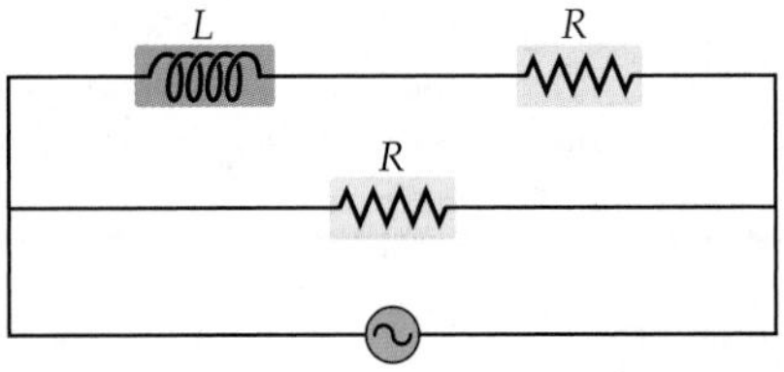

▲ **FIGURE 24–28** Conceptual Exercises 11 and 14, Problem 44

12. When the ac generator in **Figure 24–29** operates at high frequency, is the rms current in the circuit more than, less than, or the same as when the generator operates at low frequency? Explain.

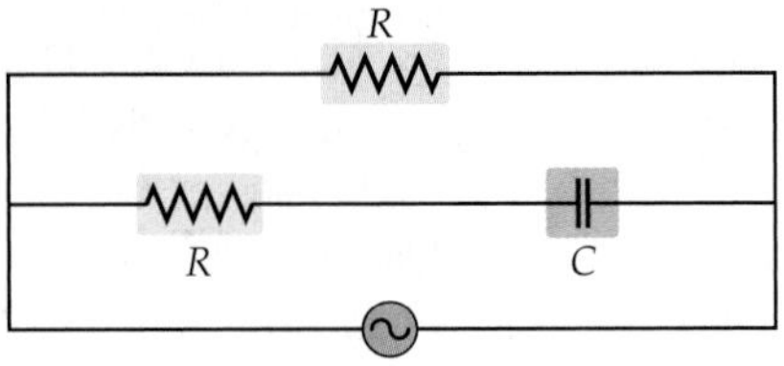

▲ **FIGURE 24–29** Conceptual Exercises 12, 13, and 14, Problem 45

13. Suppose the circuits shown in Figures 24–29 and 24–27 are connected to identical batteries, rather than to ac generators. Assuming the values of *R*, *L*, and *C* are the same in the two circuits, is the current in Figure 24–29 more than, less than, or the same as the current in Figure 24–27? Explain.
14. Suppose the circuits shown in Figures 24–29 and 24–28 are connected to identical batteries, rather than to ac generators. Assuming the values of *R*, *L*, and *C* are the same in the two circuits, is the current in Figure 24–29 more than, less than, or the same as the current in Figure 24–28? Explain.
15. In an *RLC* circuit a second capacitor is added in *series* to the capacitor already present. Does the resonance frequency increase, decrease, or stay the same? Explain.
16. In an *RLC* circuit a second capacitor is added in *parallel* to the capacitor already present. Does the resonance frequency increase, decrease, or stay the same? Explain.
17. A generator drives an *RLC* circuit with the voltage V shown in **Figure 24–30**. The corresponding current I is also shown in the figure. **(a)** Is the inductive reactance of this circuit greater than, less than, or equal to its capacitive reactance? Explain. **(b)** Is the frequency of this generator greater than, less than, or equal to the resonance frequency of the circuit? Explain.

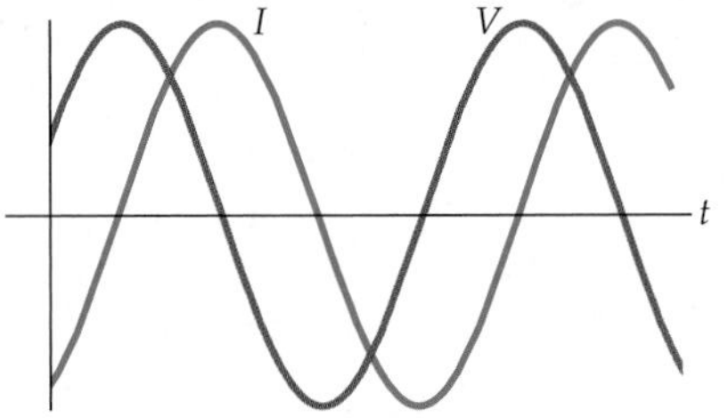

▲ **FIGURE 24–30** Conceptual Exercise 17

18. The resistance in an *RLC* circuit is doubled. **(a)** Does the resonance frequency increase, decrease, or stay the same? Explain. **(b)** Does the maximum current in the circuit increase, decrease, or stay the same? Explain.
19. The voltage in a sinusoidally driven *RLC* circuit leads the current. **(a)** If we want to bring this circuit into resonance by changing the frequency of the generator, should the frequency be increased or decreased? Explain. **(b)** If we want to bring this circuit into resonance by changing the inductance instead, should the inductance be increased or decreased? Explain.

Problems

Note: **IP** *denotes an integrated conceptual/quantitative problem.* **BIO** *identifies problems of biological or medical interest. Red bullets (•, ••, •••) are used to indicate the level of difficulty of each problem.*

Section 24–1 Alternating Voltages and Currents

1. • An ac generator produces a peak voltage of 55 V. What is the rms voltage of this generator?
2. • In many European homes the rms voltage available from a wall socket is 240 V. What is the maximum voltage in this case?

3. • An rms voltage of 120 V produces a maximum current of 2.8 A in a certain resistor. Find the resistance of this resistor.

4. • The rms current in an ac circuit with a resistance of 150 Ω is 0.85 A. What are the **(a)** average and **(b)** maximum power consumed by this circuit?

5. • A 3.33-kΩ resistor is connected to a generator with a maximum voltage of 121 V. Find **(a)** the average and **(b)** the maximum power delivered to this circuit.

6. •• A "75-watt" lightbulb uses an average power of 75 W when connected to an rms voltage of 120 V. **(a)** What is the resistance of the lightbulb? **(b)** What is the maximum current in the bulb? **(c)** What is the maximum power used by the bulb at any given instant of time?

7. ••• The relationship $V_{rms} = V_{max}/\sqrt{2}$ is valid only for voltages that vary sinusoidally. Find the relationship between V_{rms} and V_{max} for the "square wave" voltage shown in **Figure 24–31**.

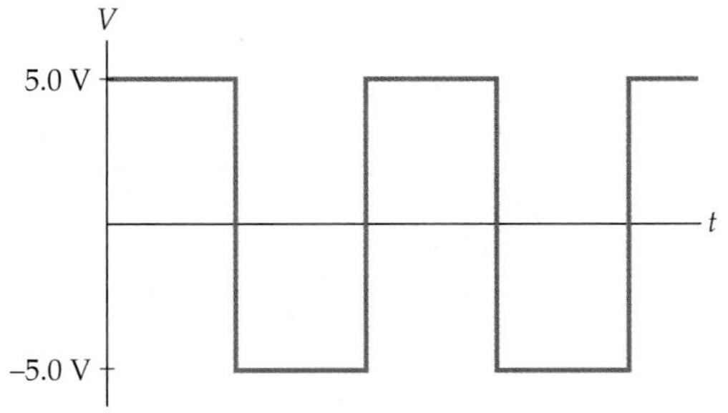

▲ **FIGURE 24–31** Problem 7

Section 24–2 Capacitors in AC Circuits

8. • The reactance of a capacitor is 65 Ω at a frequency of 57 Hz. What is its capacitance?

9. • The capacitive reactance of a capacitor at 60.0 Hz is 105 Ω. At what frequency is its capacitive reactance 72.5 Ω?

10. • A 105-μF capacitor is connected to an ac generator with an rms voltage of 20.0 V and a frequency of 100.0 Hz. What is the rms current in this circuit?

11. • The rms voltage across a 0.010-μF capacitor is 1.4 V at a frequency of 52 Hz. What are **(a)** the rms and **(b)** the maximum current through the capacitor?

12. •• An ac generator with a frequency of 30.0 Hz and an rms voltage of 12.0 V is connected to a 45.5-μF capacitor. **(a)** What is the maximum current in this circuit? **(b)** What is the current in the circuit when the voltage across the capacitor is 5.25 V and increasing? **(c)** What is the current in the circuit when the voltage across the capacitor is 5.25 V and decreasing?

13. •• The maximum current in a 22-μF capacitor connected to an ac generator with a frequency of 120 Hz is 0.15 A. **(a)** What is the maximum voltage of the generator? **(b)** What is the voltage across the capacitor when the current in the circuit is 0.10 A and increasing? **(c)** What is the voltage across the capacitor when the current in the circuit is 0.10 A and decreasing?

14. •• **IP** An rms voltage of 10.0 V with a frequency of 1.00 kHz is applied to a 0.395-μF capacitor. **(a)** What is the rms current in this circuit? **(b)** By what factor does the current change if the frequency of the voltage is doubled? **(c)** Calculate the current for a frequency of 2.00 kHz.

15. •• A circuit consists of a 1.00-kHz generator and a capacitor. When the rms voltage of the generator is 0.500 V, the rms current in the circuit is 0.430 mA. **(a)** What is the reactance of the capacitor at 1.00 kHz? **(b)** What is the capacitance of the capacitor? **(c)** If the rms voltage is maintained at 0.500 V, what is the rms current at 2.00 kHz? At 10.0 kHz?

16. •• **IP** A capacitor has an rms current of 21 mA at a frequency of 60.0 Hz when the rms voltage across it is 14 V. **(a)** What is the capacitance of this capacitor? **(b)** If the frequency is increased, will the current in the capacitor increase, decrease, or stay the same? Explain. **(c)** Find the rms current in this capacitor at a frequency of 410 Hz.

17. •• A 0.22-μF capacitor is connected to an ac generator with an rms voltage of 12 V. For what range of frequencies will the rms current in the circuit be less than 1.0 mA?

18. •• At what frequency will a generator with an rms voltage of 504 V produce an rms current of 7.50 mA in a 0.0150-μF capacitor?

Section 24–3 *RC* Circuits

19. • Find the impedance of a 60.0-Hz circuit with a 25.5-Ω resistor connected in series with a 95.0-μF capacitor.

20. • An ac generator with a frequency of 105 Hz and an rms voltage of 22.5 V is connected in series with a 10.0-kΩ resistor and a 0.250-μF capacitor. What is the rms current in this circuit?

21. • The rms current in an *RC* circuit is 0.72 A. The capacitor in this circuit has a capacitance of 13 μF and the ac generator has a frequency of 150 Hz and an rms voltage of 95 V. What is the resistance in this circuit?

22. •• A 65.0-Hz generator with an rms voltage of 115 V is connected in series to a 3.35-kΩ resistor and a 1.50-μF capacitor. Find **(a)** the rms current in the circuit and **(b)** the phase angle, ϕ, between the current and the voltage.

23. •• **(a)** At what frequency must the circuit in Problem 22 be operated for the current to lead the voltage by 23.0°? **(b)** Using the frequency found in part (a), find the average power consumed by this circuit.

24. •• **(a)** Sketch the phasor diagram for an ac circuit with a 105-Ω resistor in series with a 32.2-μF capacitor. The frequency of the generator is 60.0 Hz. **(b)** If the rms voltage of the generator is 120 V, what is the average power consumed by the circuit?

25. •• Find the power factor for an *RC* circuit connected to a 70.0-Hz generator with an rms voltage of 125 V. The values of *R* and *C* in this circuit are 105 Ω and 82.4 μF, respectively.

26. •• **IP (a)** Determine the power factor for an *RC* circuit with $R = 4.0$ kΩ and $C = 0.35$ μF that is connected to an ac generator with an rms voltage of 24 V and a frequency of 150 Hz. **(b)** Will the power factor for this circuit increase, decrease, or stay the same if the frequency of the generator is increased? Explain.

27. ••• The "square wave" voltage shown in **Figure 24–32** is applied to an *RC* circuit. Sketch the shape of the instantaneous voltage across the capacitor, assuming the time constant of the circuit is equal to the period of the applied voltage.

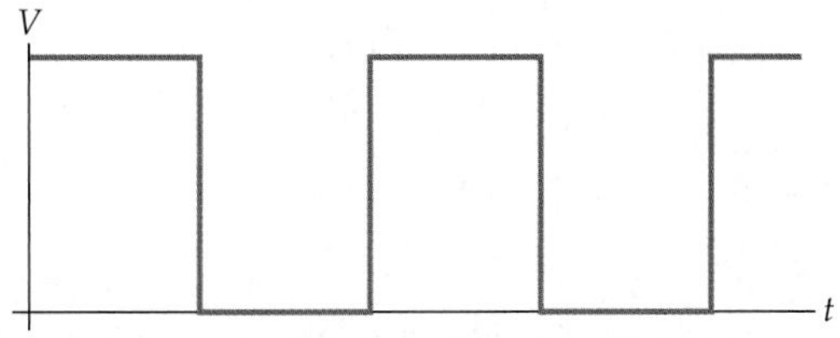

▲ **FIGURE 24–32** Problems 27, 41, and 79

Section 24–4 Inductors in AC Circuits

28. • An inductor has a reactance of 56.5 Ω at 75.0 Hz. What is its reactance at 60.0 Hz?

29. • What is the rms current in a 77.5-mH inductor when it is connected to a 60.0-Hz generator with an rms voltage of 115 V?

30. • What rms voltage is required to produce an rms current of 2.1 A in a 66-mH inductor at a frequency of 25 Hz?

31. •• A 525-Ω resistor and a 255-mH inductor are connected in series with an ac generator with an rms voltage of 20.0 V and a frequency of 60.0 Hz. What is the rms current in this circuit?

32. •• The rms current in an *RL* circuit is 0.26 A when it is connected to an ac generator with a frequency of 60.0 Hz and an rms voltage of 25 V. **(a)** Given that the inductor has an inductance of 145 mH, what is the resistance of the resistor? **(b)** Find the rms voltage across the resistor. **(c)** Find the rms voltage across the inductor. **(d)** Use your results from parts (b) and (c) to show that $\sqrt{V_{\text{rms},R}^2 + V_{\text{rms},L}^2}$ is equal to 25 V.

33. •• An ac generator with a frequency of 1.34 kHz and an rms voltage of 24.2 V is connected in series with a 2.00-kΩ resistor and a 315-mH inductor. **(a)** What is the power factor for this circuit? **(b)** What is the average power consumed by this circuit?

34. •• **IP** An rms voltage of 12.2 V with a frequency of 1.00 kHz is applied to a 0.290-mH inductor. **(a)** What is the rms current in this circuit? **(b)** By what factor does the current change if the frequency of the voltage is doubled? **(c)** Calculate the current for a frequency of 2.00 kHz.

35. •• A 0.22-μH inductor is connected to an ac generator with an rms voltage of 12 V. For what range of frequencies will the rms current in the circuit be less than 1.0 mA?

36. •• The phase angle in a certain *RL* circuit is 76° at a frequency of 60.0 Hz. If $R = 2.7\ \Omega$ for this circuit, what is the value of the inductance, *L*?

37. •• An *RL* circuit consists of a resistor $R = 68\ \Omega$, an inductor, $L = 31$ mH, and an ac generator with an rms voltage of 120 V. **(a)** At what frequency will the rms current in this circuit be 1.5 A? For this frequency, what is **(b)** the rms voltage across the resistor, $V_{\text{rms},R}$, and **(c)** the rms voltage across the inductor, $V_{\text{rms},L}$? **(d)** Show that $V_{\text{rms},R} + V_{\text{rms},L} > 120$ V, but that $\sqrt{V_{\text{rms},R}^2 + V_{\text{rms},L}^2} = 120$ V.

38. •• **(a)** Sketch the phasor diagram for an ac circuit with a 105-Ω resistor in series with a 22.5-mH inductor. The frequency of the generator is 60.0 Hz. **(b)** If the rms voltage of the generator is 120 V, what is the average power consumed by the circuit?

39. •• **IP** In Problem 36, does the phase angle increase, decrease, or stay the same when the frequency is increased? Verify your answer by calculating the phase angle at 70.0 Hz.

40. •• A large air conditioner has a resistance of 7.0 Ω and an inductive reactance of 15 Ω. If the air conditioner is powered by a 60.0-Hz generator with an rms voltage of 240 V, find **(a)** the impedance of the air conditioner, **(b)** its rms current, and **(c)** the average power consumed by the air conditioner.

41. ••• The "square wave" voltage shown in Figure 24–32 is applied to an *RL* circuit. Sketch the shape of the instantaneous voltage across the inductor, assuming the time constant of the circuit is much less than the period of the applied voltage.

Section 24–5 *RLC* Circuits

42. • Find the rms voltage across each element in an *RLC* circuit with $R = 9.9\ \text{k}\Omega$, $C = 0.15\ \mu\text{F}$, and $L = 25$ mH. The generator supplies an rms voltage of 115 V at a frequency of 60.0 Hz.

43. • What is the impedance of a 1.50-kΩ resistor, a 155-mH inductor, and a 12.5-μF capacitor connected in series with a 60.0-Hz ac generator?

44. • Consider the circuit shown in Figure 24–28 (Conceptual Exercise 11). The ac generator in this circuit has an rms voltage of 65 V. Given that $R = 15\ \Omega$ and $L = 0.22$ mH, find the rms current in this circuit in the limit of **(a)** high frequency and **(b)** low frequency.

45. • Consider the circuit shown in Figure 24–29 (Conceptual Exercise 12). The ac generator in this circuit has an rms voltage of 65 V. Given that $R = 15\ \Omega$ and $C = 41\ \mu\text{F}$, find the rms current in this circuit in the limit of **(a)** high frequency and **(b)** low frequency.

46. •• What is the phase angle in an *RLC* circuit with $R = 9.9\ \text{k}\Omega$, $C = 1.5\ \mu\text{F}$, and $L = 250$ mH? The generator supplies an rms voltage of 115 V at a frequency of 60.0 Hz.

47. •• **IP** An *RLC* circuit has a resistance of 105 Ω, an inductance of 85.0 mH, and a capacitance of 13.2 μF. **(a)** What is the power factor for this circuit when it is connected to a 125-Hz ac generator? **(b)** Will the power factor increase, decrease, or stay the same if the resistance is increased? Explain. **(c)** Calculate the power factor for a resistance of 525 Ω.

48. •• An ac voltmeter, which displays the rms voltage between the two points touched by its leads, is used to measure voltages in the circuit shown in **Figure 24–33**. In this circuit, the ac generator has an rms voltage of 6.00 V and a frequency of 30.0 kHz. The inductance in the circuit is 0.300 mH, the capacitance is 0.100 μF, and the resistance is 2.50 Ω. What is the reading on a voltmeter when it is connected to points **(a)** A and B, **(b)** B and C, **(c)** A and C, and **(d)** A and D?

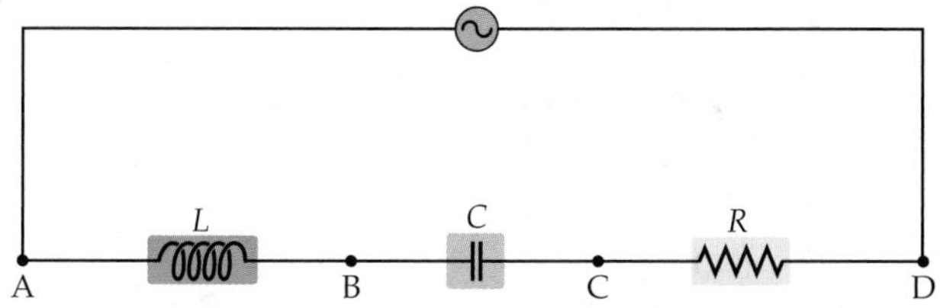

▲ **FIGURE 24–33** Problems 48 and 49

49. •• **IP** Consider the ac circuit shown in Figure 24–33, where we assume that the values of *R*, *L*, and *C* are the same as in the previous problem, and that the rms voltage of the generator is still 6.00 V. The frequency of the generator, however, is doubled to 60.0 kHz. Calculate the rms voltage across **(a)** the resistor, *R*, **(b)** the inductor, *L*, and **(c)** the capacitor, *C*. **(d)** Do you expect the sum of the rms voltages in parts (a), (b), and (c) to be greater than, less than, or equal to 6.00 V? Explain.

50. •• **(a)** Sketch the phasor diagram for an ac circuit with a 105-Ω resistor in series with a 22.5-mH inductor and a 32.2-μF capacitor. The frequency of the generator is 60.0 Hz. **(b)** If the rms voltage of the generator is 120 V, what is the average power consumed by the circuit?

51. •• A generator connected to an *RLC* circuit has an rms voltage of 120 V and an rms current of 34 mA. If the resistance in the circuit is 3.3 kΩ and the capacitive reactance is 6.6 kΩ, what is the inductive reactance of the circuit?

52. ••• A manufacturing plant uses 2.22 kW of electric power provided by a 60.0-Hz ac generator with an rms voltage of 485 V. The plant uses this power to run a number of high-inductance electric motors. The plant's total resistance is $R = 25.0\ \Omega$ and its inductive reactance is $X_L = 45.0\ \Omega$. **(a)** What is the total impedance of the plant? **(b)** What is the plant's power factor? **(c)** What is the rms current used by the plant? **(d)** What capacitance,

connected in series with the power line, will increase the plant's power factor to unity? **(e)** If the power factor is unity, how much current is needed to provide the 2.22 kW of power needed by the plant? Compare your answer with the current found in part (c). (Because power-line losses are proportional to the square of the current, a utility company will charge an industrial user with a low power factor a higher rate per kWh than a company with a power factor close to unity.)

Section 24–6 Resonance in Electrical Circuits

53. • An *RLC* circuit has a resonance frequency of 2.4 kHz. If the capacitance is 47 μF, what is the inductance?

54. • At resonance, the rms current in an *RLC* circuit is 1.8 A. If the rms voltage of the generator is 120 V, what is the resistance, *R*?

55. •• A 115-Ω resistor, a 57.6-mH inductor, and a 179-μF capacitor are connected in series to an ac generator. **(a)** At what frequency will the current in the circuit be a maximum? **(b)** At what frequency will the impedance of the circuit be a minimum?

56. •• **IP** An ac generator of variable frequency is connected to an *RLC* circuit with $R = 12\ \Omega$, $L = 0.15$ mH, and $C = 0.20$ mF. At a frequency of 1.0 kHz, the rms current in the circuit is larger than desired. Should the frequency of the generator be increased or decreased to reduce the current? Explain.

57. •• **(a)** Find the frequency at which a 33-μF capacitor has the same reactance as a 33-mH inductor. **(b)** What is the resonance frequency of an *LC* circuit made with this inductor and capacitor?

58. •• Consider an *RLC* circuit with $R = 105\ \Omega$, $L = 518$ mH, and $C = 0.200\ \mu$F. **(a)** At what frequency is this circuit in resonance? **(b)** Find the impedance of this circuit if the frequency has the value found in part (a), but the capacitance is increased to 0.220 μF. **(c)** What is the power factor for the situation described in part (b)?

59. •• **IP** An *RLC* circuit has a resonance frequency of 125 Hz. **(a)** If both *L* and *C* are doubled, does the resonance frequency increase, decrease, or stay the same? Explain. **(b)** Find the resonance frequency when *L* and *C* are doubled.

60. •• An *RLC* circuit has a capacitance of 0.29 μF. **(a)** What inductance will produce a resonance frequency of 95 MHz? **(b)** It is desired that the impedance at resonance be one-fifth the impedance at 11 kHz. What value of *R* should be used to obtain this result?

General Problems

61. • A 4.40-μF and an 8.80-μF capacitor are connected in *parallel* to a 60.0-Hz generator operating with an rms voltage of 115 V. What is the rms current supplied by the generator?

62. • A 4.40-μF and an 8.80-μF capacitor are connected in *series* to a 60.0-Hz generator operating with an rms voltage of 115 V. What is the rms current supplied by the generator?

63. • A 10.0-μF capacitor and a 20.0-μF capacitor are connected in parallel to an ac generator with a frequency of 60.0 Hz. What is the capacitive reactance of this pair of capacitors?

64. •• **IP** Consider the *RLC* circuit shown in Example 24–6, and the corresponding phasor diagram given in the Insight. **(a)** On the basis of the phasor diagram, can you conclude that the resonance frequency of this circuit is greater than, less than, or equal to 60.0 Hz? Explain. **(b)** Calculate the resonance frequency for this circuit. **(c)** The impedance of this circuit at 60.0 Hz is 226 Ω. What is the impedance at resonance?

65. •• **IP** When a certain resistor is connected to an ac generator with a maximum voltage of 15 V, the average power dissipated in the resistor is 22 W. **(a)** What is the resistance of the resistor? **(b)** What is the rms current in the circuit? **(c)** We know that $P_{av} = I_{rms}^2 R$, and hence it seems that reducing the resistance should reduce the average power. On the other hand, we also know that $P_{av} = V_{rms}^2/R$, which suggests that reducing *R* increases P_{av}. Which conclusion is correct? Explain.

66. •• A 7.5-Hz generator is connected to a capacitor. If the current in the generator has its maximum value at $t = 0$, what is the earliest possible time that the voltage across the capacitor is a maximum?

67. •• The voltage across an inductor reaches its maximum value 22 ms before the current supplied by the generator reaches *its* maximum value. What is the lowest possible frequency at which the generator operates?

68. •• Find the average power consumed by an *RC* circuit connected to a 60.0-Hz generator with an rms voltage of 112 V. The values of *R* and *C* in this circuit are 3.30 kΩ and 2.75 μF, respectively.

69. •• A 1.15-kΩ resistor and a 505-mH inductor are connected in series to a 1250-Hz generator with an rms voltage of 14.2 V. **(a)** What is the rms current in the circuit? **(b)** What capacitance must be inserted in series with the resistor and inductor to reduce the rms current to half the value found in part (a)?

70. •• **IP** The phasor diagram for an *RLC* circuit is shown in **Figure 24–34**. **(a)** If the resistance in this circuit is 525 Ω, what is the impedance? **(b)** If the frequency in this circuit is increased, will the impedance increase, decrease, or stay the same? Explain.

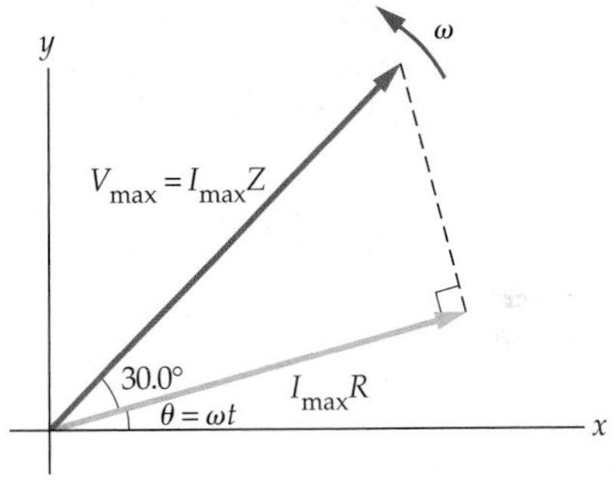

▲ **FIGURE 24–34** Problems 70, 71, and 72

71. •• **IP** Figure 24–34 shows the phasor diagram for an *RLC* circuit in which the impedance is 337 Ω. **(a)** What is the resistance, *R*, in this circuit? **(b)** Is this circuit driven at a frequency that is greater than, less than, or equal to the resonance frequency of the circuit? Explain.

72. •• **IP** An *RLC* circuit has a resistance $R = 25\ \Omega$ and an inductance $L = 160$ mH, and is connected to an ac generator with a frequency of 55 Hz. The phasor diagram for this circuit is shown in Figure 24–34. Find **(a)** the impedance, *Z*, and **(b)** the capacitance, *C*, for this circuit. **(c)** If the value of *C* is decreased, will the impedance of the circuit increase, decrease, or stay the same? Explain.

73. •• **IP** You are given a sealed box with two electrical terminals. The box contains a 5.00-Ω resistor in series with either an inductor or a capacitor. When you attach an ac generator with an rms voltage of 0.750 V to the terminals of the box, you find that the current increases with increasing frequency. **(a)** Does the box contain an inductor or a capacitor? Explain. **(b)** When the frequency of the generator is 25.0 kHz, the rms current is 87.2 mA. What is the capacitance or inductance of the unknown component in the box?

74. •• IP A circuit is constructed by connecting a 1.00-kΩ resistor, a 252-μF capacitor, and a 515-mH inductor in series. **(a)** At what frequency will the impedance of this circuit be equal to 2.00 kΩ? **(b)** If it is desired to reduce the impedance of this circuit, should the frequency be increased or decreased? Explain.

75. •• An *RLC* circuit with $R = 25.0\ \Omega$, $L = 325$ mH, and $C = 45.2\ \mu$F is connected to an ac generator with an rms voltage of 24 V. Determine the average power delivered to this circuit when the frequency of the generator is **(a)** equal to the resonance frequency, **(b)** twice the resonance frequency, and **(c)** half the resonance frequency.

76. •• The intensity of a lightbulb with a resistance of 120 Ω is controlled by connecting it in series with an inductor whose inductance can be varied from $L = 0$ to $L = L_{max}$. This "light dimmer" circuit is connected to an ac generator with a frequency of 60.0 Hz and an rms voltage of 110 V. **(a)** What is the average power dissipated in the lightbulb when $L = 0$? **(b)** The inductor is now adjusted so that $L = L_{max}$. In this case, the average power dissipated in the lightbulb is one-fourth the value found in part (a). What is the value of L_{max}?

77. ••• An electric motor with a resistance of 15 Ω and an inductance of 53 mH is connected to a 60.0-Hz ac generator. **(a)** What is the power factor for this circuit? **(b)** In order to increase the power factor of this circuit to 0.80, a capacitor is connected in series with the motor and inductor. Find the required value of the capacitance.

78. ••• IP A radio tuning circuit contains an *RLC* circuit with $R = 5.0\ \Omega$ and $L = 2.8\ \mu$H. **(a)** What capacitance is needed to produce a resonance frequency of 85 MHz? **(b)** If the capacitance is increased above the value found in part (a), will the impedance increase, decrease, or stay the same? Explain. **(c)** Find the impedance of the circuit at resonance. **(d)** Find the impedance of the circuit when the capacitance is 1 percent higher than the value found in part (a).

79. ••• If the maximum voltage in the square wave shown in Figure 24–32 is V_{max}, what are **(a)** the average voltage, V_{av}, and **(b)** the rms voltage, V_{rms}?

80. ••• An ac generator supplies an rms voltage of 5.00 V to an *RC* circuit. At a frequency of 20.0 kHz the rms current in the circuit is 45.0 mA; at a frequency of 25.0 kHz the rms current is 50.0 mA. What are the values of *R* and *C* in this circuit?

81. ••• An ac generator supplies an rms voltage of 5.00 V to an *RL* circuit. At a frequency of 20.0 kHz the rms current in the circuit is 45.0 mA; at a frequency of 25.0 kHz the rms current is 40.0 mA. What are the values of *R* and *L* in this circuit?

82. ••• An *RC* circuit consists of a resistor $R = 32\ \Omega$, a capacitor $C = 25\ \mu$F, and an ac generator with an rms voltage of 120 V. **(a)** At what frequency will the rms current in this circuit be 2.9 A? For this frequency, what is **(b)** the rms voltage across the resistor, $V_{rms,R}$, and **(c)** the rms voltage across the capacitor, $V_{rms,C}$? **(d)** Show that $V_{rms,R} + V_{rms,C} > 120$ V, but that $\sqrt{V_{rms,R}^2 + V_{rms,C}^2} = 120$ V.

Interactive Problems

83. •• IP ***Referring to Example 24–6*** Suppose we would like to change the phase angle for this circuit to $\phi = -25.0°$, and that we would like to accomplish this by changing the resistor to a value other than 175 Ω. The inductor is still 90.0 mH, the capacitor is 15.0 μF, the rms voltage is 120.0 V, and the ac frequency is 60.0 Hz. **(a)** Should the resistance be increased or decreased? Explain. **(b)** Find the resistance that gives the desired phase angle. **(c)** What is the rms current in the circuit with the resistance found in part (b)?

84. •• IP ***Referring to Example 24–6*** You plan to change the frequency of the generator in this circuit to produce a phase angle of smaller magnitude. The resistor is still 175 Ω, the inductor is 90.0 mH, the capacitor is 15.0 μF, and the rms voltage is 120.0 V. **(a)** Should you increase or decrease the frequency? Explain. **(b)** Find the frequency that gives a phase angle of −22.5°. **(c)** What is the rms current in the circuit at the frequency found in part (b)?

25 Electromagnetic Waves

25–1 The Production of Electromagnetic Waves 838

25–2 The Propagation of Electromagnetic Waves 841

25–3 The Electromagnetic Spectrum 845

25–4 Energy and Momentum in Electromagnetic Waves 849

25–5 Polarization 853

Most people think they know exactly what the world looks like—all you have to do, after all, is open your eyes and look. There's more to it than that, however. We all know, for example, that the simple yellow flower shown here (left) is a daisy. But is that really what a daisy looks like? To a bee—with its ability to see ultraviolet light invisible to us—the very same daisy is a flower with three concentric circles forming a "bull's-eye" centered on the nectar (right). Such "nectar guides" are a common feature of flowers as viewed by bees, but are generally invisible to humans. This chapter explores the nature and properties of electromagnetic radiation—the kind we know as visible light, and many other kinds as well.

Electricity and magnetism can seem very different in many ways, but they are actually intimately related—in fact, electric and magnetic fields can be considered as different aspects of the same thing, like the two sides of a coin. For example, we have seen that an electric current produces a magnetic field, and a changing magnetic field produces an electric field. Because of fundamental connections like these, we refer to the phenomena of electricity and magnetism together as **electromagnetism.**

In this chapter we consider one of the most significant manifestations of electromagnetism; namely, that electric and magnetic fields can work together to create traveling waves called **electromagnetic waves.** As we shall see, these waves are responsible for everything from radio and TV signals, to the visible light we see all around us, to the X-rays that reveal our internal structure, and much more. In addition, the prediction, discovery, and technological development of electromagnetic waves is a fascinating success story in the history of science.

25–1 The Production of Electromagnetic Waves

Electromagnetic waves were predicted, and their properties were studied theoretically, decades before they were first produced with electrical circuits in the lab. The prediction came from Scottish physicist James Clerk Maxwell (1831–1879), who, in 1864, hypothesized that since a changing magnetic field produces an electric field (Faraday's law) a changing electric field should similarly produce a magnetic field. In effect, Maxwell suggested a sort of "symmetry" between electric and magnetic fields.

Maxwell followed up on his suggestion by working out its mathematical consequences. Among these was that electric and magnetic fields, acting together, could produce an *electromagnetic wave* that travels with the speed of light. As a result, he proposed that visible light—which had previously been thought of as a completely separate phenomenon from electricity and magnetism—was, in fact, an electromagnetic wave. His theory also implied that electromagnetic waves would not be limited to visible light and that it should be possible to produce them with oscillating electric circuits similar to those studied in the previous chapter.

The first production and observation of electromagnetic waves in the lab was carried out by the German physicist Heinrich Hertz (1857–1894) in 1887. Hertz used what was basically an *LC* circuit to generate an alternating current and found that energy could be transferred from this circuit to a similar circuit several meters away. He was able to show, in addition, that the energy transfer exhibited such standard wave phenomena as reflection, refraction, interference, diffraction, and polarization. There could be no doubt that waves were produced by the first circuit and that they propagated across the room to the second circuit. Even more significantly, he was able to show that the speed of the waves was roughly the speed of light, as predicted by Maxwell.

It took only a few years for Hertz's experimental apparatus to be refined and improved to the point where it could be used in practical applications. The first to do so was Guglielmo Marconi (1874–1937), who immediately recognized the implications of the electromagnetic-wave experiments—namely, that waves could be used for communications, eliminating the wires necessary for telegraphy. He patented his first system in 1896 and gained worldwide attention when, in 1901, he received a radio signal in St. John's, Newfoundland, that had been sent from Cornwall, England. When Maxwell died, electromagnetic waves were still just a theory; twenty years later, they were revolutionizing communications.

▲ Electromagnetic waves are produced by (and detected as) oscillating electric currents in a wire or similar conducting element. The actual antenna is often much smaller than is commonly imagined—the bowl-shaped structures that we tend to think of as antennas, such as these microwave relay dishes, serve to focus the transmitted beam in a particular direction or concentrate the received signal on the actual detector.

To gain an understanding of electromagnetic waves, consider the simple electrical circuit shown in **Figure 25–1**. Here we show an ac generator of period T connected to the center of an antenna, which is basically a long, straight wire with a break in the middle. Suppose at time $t = 0$ the generator gives the upper segment of the antenna a maximum positive charge and the lower segment a maximum negative charge, as shown in Figure 25–1 (a). A positive test charge placed on the x axis at point P will experience a downward force; hence, the electric field there is downward. A short time later, when the charge on the antenna is reduced in magnitude, the electric field at P also has a smaller magnitude. We show this result in Figure 25–1 (b).

More importantly, Figure 25–1 (b) also shows that the electric field produced at time $t = 0$ has not vanished, nor has it simply been replaced with the new, reduced-magnitude field. Instead, the original field has *moved farther away from the antenna,* to point Q. The reason that the reduction in charge on the antenna is felt at point P *before* it is felt at point Q is simply that it takes a finite time for this change in charge to be felt at a distance. This is analogous to the fact that a person near a lightning strike hears the thunder before a person who is half a mile away, or that a wave pulse on a string takes a finite time to move from one end of the string to the other.

After the generator has completed one-quarter of a cycle, at time $t = \frac{1}{4}T$, the antenna is uncharged and the field vanishes, as in Figure 25–1 (c). Still later the charges on the antenna segments change sign, giving rise to an electric field that

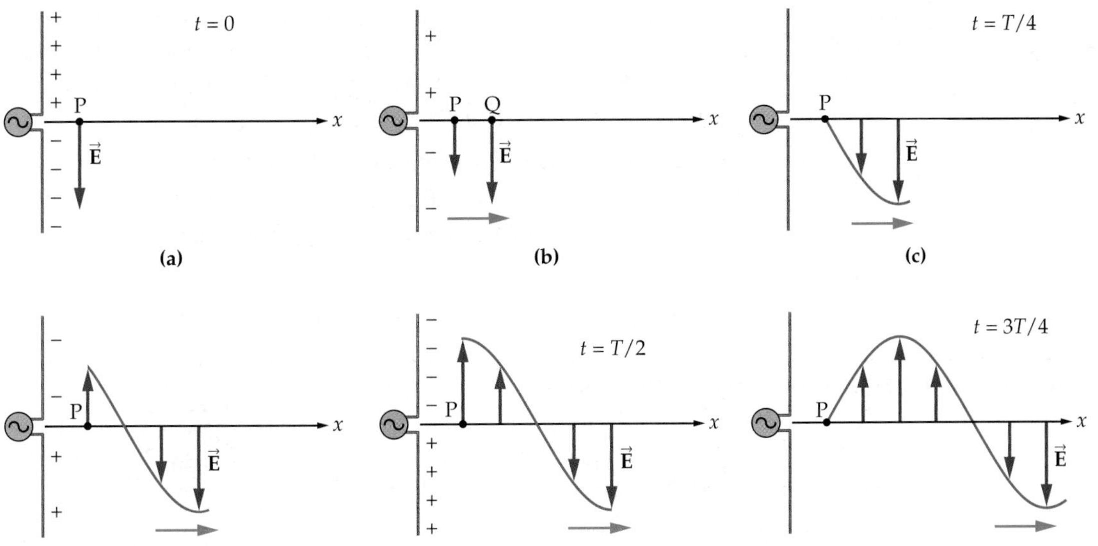

▲ **FIGURE 25–1 Producing an electromagnetic wave**

A traveling electromagnetic wave produced by an ac generator attached to an antenna. **(a)** At $t = 0$ the electric field at point P is downward. **(b)** A short time later, the electric field at P is still downward, but now with a reduced magnitude. Note that the field created at $t = 0$ has moved to point Q. **(c)** After one quarter of a cycle, at $t = \frac{1}{4}T$, the electric field at P vanishes. **(d)** The charge on the antenna has reversed polarity now, and the electric field at P points upward. **(e)** When the oscillator has completed half a cycle, $t = \frac{1}{2}T$, the field at point P is upward and of maximum magnitude. **(f)** At $t = \frac{3}{4}T$ the field at P vanishes again. The fields produced at earlier times continue to move away from the antenna.

points upward, as we see in Figures 25–1 (d) and (e). The field vanishes again after three-quarters of a cycle, at $t = \frac{3}{4}T$, as shown in Figure 25–1 (f). Immediately after this time, the electric field begins to point downward once more. The net result is a wavelike electric field moving steadily away from the antenna. To summarize,

> The electric field produced by an antenna connected to an ac generator propagates away from the antenna, analogous to a wave on a string moving away from your hand as you wiggle it up and down.

This is really only half of the electromagnetic wave, however; the other half is a similar wave in the magnetic field. To see this, consider **Figure 25–2**, where we show the current in the antenna flowing upward at a time when the upper segment is positive. Pointing the thumb of the right hand in the direction of the current, and curling the fingers around the wire, as specified in the magnetic field RHR, we see that $\vec{\mathbf{B}}$ points into the page at the same time that $\vec{\mathbf{E}}$ points downward. It follows, then, that $\vec{\mathbf{E}}$ and $\vec{\mathbf{B}}$ are at right angles to each other. A more detailed analysis shows that $\vec{\mathbf{E}}$ and $\vec{\mathbf{B}}$ are perpendicular to each other at all times, and that they are also in phase; that is, when the magnitude of $\vec{\mathbf{E}}$ is at its maximum, so is the magnitude of $\vec{\mathbf{B}}$.

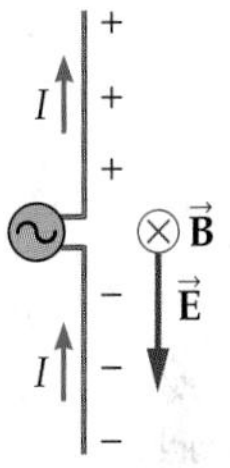

▲ **FIGURE 25–2 Field directions in an electromagnetic wave**

At a time when the electric field produced by the antenna points downward, the magnetic field points into the page. In general, the electric and magnetic fields in an electromagnetic wave are always at right angles to each other.

Combining the preceding results, we can represent the electric and magnetic fields in an electromagnetic wave as shown in **Figure 25–3**. Notice that not only

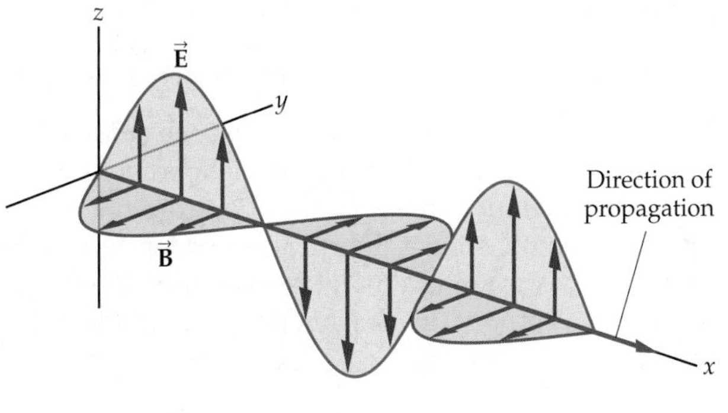

▶ **FIGURE 25–3 The right-hand rule applied to an electromagnetic wave**

An electromagnetic wave propagating in the positive x direction. Note that $\vec{\mathbf{E}}$ and $\vec{\mathbf{B}}$ are perpendicular to each other and in phase. The direction of propagation is given by the thumb of the right hand, after pointing the fingers in the direction of $\vec{\mathbf{E}}$ and curling them toward $\vec{\mathbf{B}}$.

are $\vec{E}$ and $\vec{B}$ perpendicular to each other, they are also perpendicular to the direction of propagation; hence, electromagnetic waves are **transverse** waves. (See Section 14–1 for a comparison of various types of waves.) Finally, the direction of propagation is given by another right-hand rule:

Direction of Propagation for Electromagnetic Waves

Point the fingers of your right hand in the direction of $\vec{E}$, curl your fingers toward $\vec{B}$, and your thumb will point in the direction of propagation.

This rule is consistent with the direction of propagation shown in Figure 25–3.

CONCEPTUAL CHECKPOINT 25–1 Direction of Magnetic Field

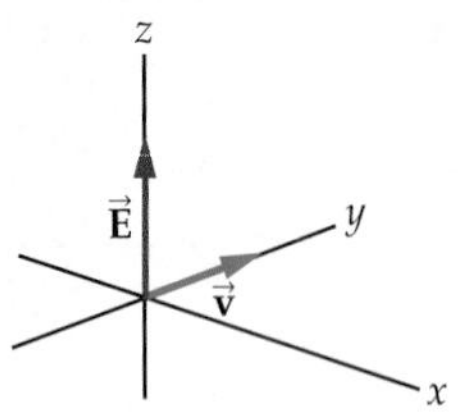

An electromagnetic wave propagates in the positive *y* direction, as shown in the sketch. If the electric field at the origin is in the positive *z* direction, is the magnetic field at the origin in **(a)** the positive *x* direction, **(b)** the negative *x* direction, or **(c)** the negative *y* direction?

Reasoning and Discussion

Pointing the fingers of the right hand in the positive *z* direction (the direction of $\vec{E}$), we see that in order for the thumb to point in the direction of propagation (the positive *y* direction) the fingers must be curled toward the positive *x* direction. Therefore, $\vec{B}$ points in the positive *x* direction.

Answer:

(a) $\vec{B}$ is in the positive *x* direction.

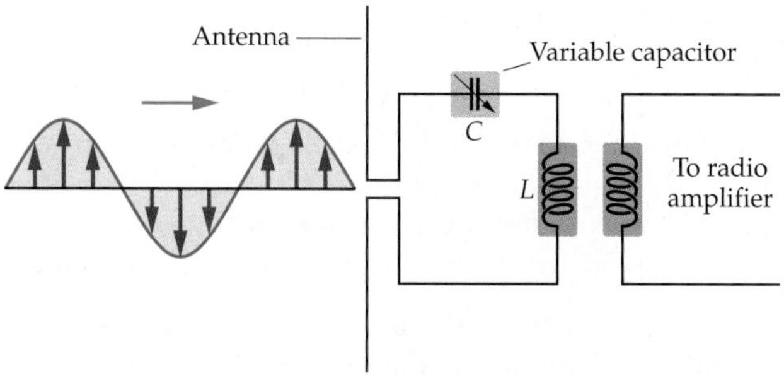

▶ **FIGURE 25–4 Receiving radio waves**
Basic elements of a tuning circuit used to receive radio waves. First, an incoming wave sets up an alternating current in the antenna. Next, the resonance frequency of the *LC* circuit is adjusted to match the frequency of the radio wave, resulting in a relatively large current in the circuit. This current is then fed into an amplifier to further increase the signal.

Electromagnetic waves can be detected in much the same way they are generated. Suppose, for instance, that an electromagnetic wave moves to the right, as in **Figure 25–4**. As the wave continues to move, its electric field exerts a force on electrons in the antenna that is alternately up and down, resulting in an alternating current. Thus the electromagnetic field makes the antenna behave much like an ac generator. If the antenna is connected to an *LC* circuit, as indicated in the figure, the resulting current can be relatively large if the resonant frequency of the circuit matches the frequency of the wave. This is the basic principle behind radio and television tuners. In fact, when you turn the tuning knob on a radio, you are actually changing the capacitance or the inductance in an *LC* circuit and, therefore, changing the resonance frequency.

REAL-WORLD PHYSICS
Radio and television communications

Finally, though we have discussed the production of electromagnetic waves by means of an electric circuit and an antenna, this is certainly not the only way such waves can be generated. In fact, any time an electric charge is accelerated, it will radiate:

Accelerated charges radiate electromagnetic waves.

This condition applies no matter what the cause of the acceleration. In addition, the intensity of radiated electromagnetic waves depends on the orientation of the acceleration relative to the viewer. For example, viewing the antenna perpendicular to its length, so that the charges accelerate at right angles to the line of sight, results in maximum intensity, as illustrated in **Figure 25–5**. Conversely, viewing the antenna straight down from above, in the same direction as the acceleration, results in zero intensity.

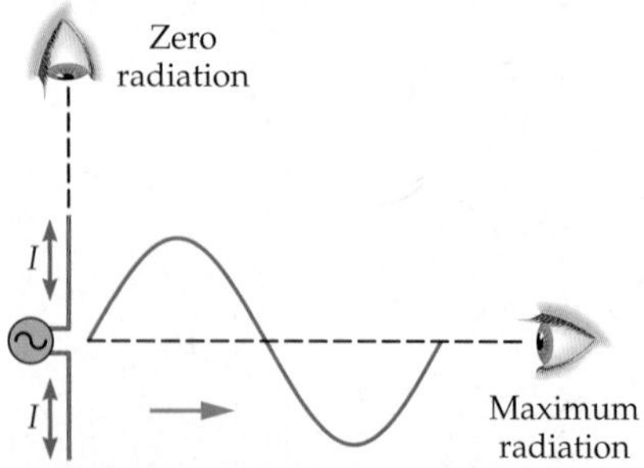

▲ **FIGURE 25–5 Electromagnetic waves and the line of sight**
Electromagnetic radiation is greatest when charges accelerate at right angles to the line of sight. Zero radiation is observed when the charges accelerate along the line of sight. These observations apply to electromagnetic waves of all frequencies.

25–2 The Propagation of Electromagnetic Waves

A sound wave or a wave on a string requires a medium through which it can propagate. For example, when the air is pumped out of a jar containing a ringing bell, its sound can no longer be heard. In contrast, we can still *see* that the bell is ringing. Thus, light can propagate through a vacuum, as can all other types of electromagnetic waves, such as radio waves and microwaves. In fact, electromagnetic waves travel through a vacuum with the maximum speed that *any* form of energy can have, as we discuss in detail in Chapter 29.

The Speed of Light

All electromagnetic waves travel through a vacuum with precisely the same speed, c. Since light is the form of electromagnetic wave most familiar to us, we refer to c as the *speed of light in a vacuum*. The approximate value of this speed is as follows:

Speed of Light in a Vacuum

$$c = 3.00 \times 10^8 \text{ m/s} \qquad \textbf{25–1}$$

This is a large speed, corresponding to about 186,000 mi/s. Put another way, a beam of light could travel around the world about seven times in a single second. In air the speed of light is slightly less than it is in a vacuum, and in denser materials, such as glass or water, the speed of light is reduced to about two-thirds of its vacuum value.

EXERCISE 25–1

The distance between Earth and the Sun is 1.50×10^{11} m. How long does it take for light to cover this distance?

Solution

Recalling that speed is distance divided by time, it follows that the time t to cover a distance d is $t = d/v$. Using $v = c$, we find

$$t = \frac{d}{c} = \frac{1.50 \times 10^{11}\ \text{m}}{3.00 \times 10^8\ \text{m/s}} = 500\ \text{s}$$

Noting that 500 s is $8\frac{1}{3}$ min, we say that Earth is about 8 light-minutes from the Sun.

Because the speed of light is so large, its value is somewhat difficult to determine. The first scientific attempt to measure the speed of light was made by Galileo (1564–1642), who used two lanterns for the experiment. Galileo opened the shutters of one lantern, and an assistant a considerable distance away was instructed to open the shutter on the second lantern as soon as he observed the light from Galileo's lantern. Galileo then attempted to measure the time that elapsed before he saw the light from his assistant's lantern. Since there was no perceptible time lag, beyond the normal human reaction time, Galileo could conclude only that the speed of light must be very great indeed.

The first to give a finite, numerical value to the speed of light was the Danish astronomer Ole Romer (1644–1710), though he did not set out to measure the speed of light at all. Romer was measuring the times at which the moons of Jupiter disappeared behind the planet, and he noticed that these eclipses occurred earlier when Earth was closer to Jupiter and later when Earth was farther away from Jupiter. This difference is illustrated in **Figure 25–6**. From the results of Exercise 25–1, we know that light requires about 16 minutes to travel from one side of Earth's orbit to the other, and this is roughly the discrepancy in eclipse times observed by Romer. In 1676 he announced a value for the speed of light of 2.25×10^8 m/s.

The first laboratory measurement of the speed of light was performed by the French scientist Armand Fizeau (1819–1896). The basic elements of his experiment, shown in **Figure 25–7**, are a mirror and a rotating, notched wheel. Light passing through one notch travels to a mirror a considerable distance away, is reflected

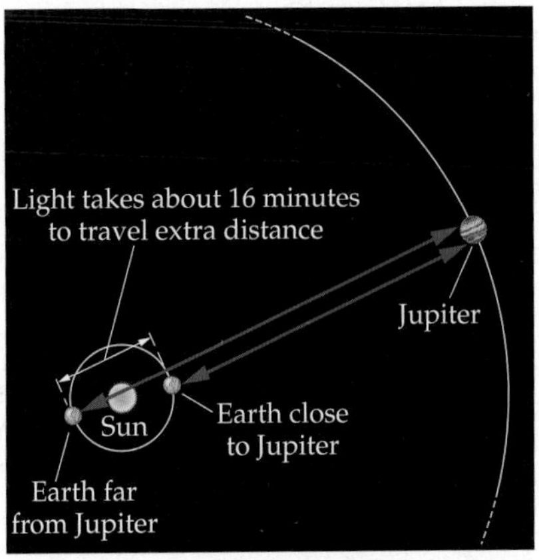

▲ **FIGURE 25–6 Using Jupiter to determine the speed of light**
When the Earth is at its greatest distance from Jupiter, light takes about 16 minutes longer to travel between them. This time lag allowed Ole Romer to estimate the speed of light.

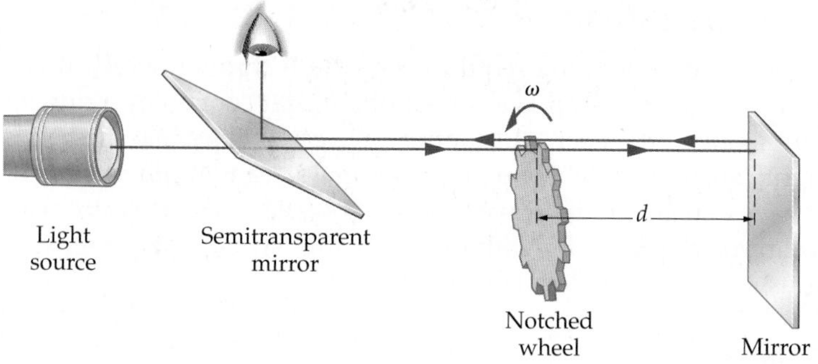

▶ **FIGURE 25–7 Fizeau's experiment to measure the speed of light**
If the time required for light to travel to the far mirror and back is equal to the time it takes the wheel to rotate from one notch to the next, light will pass through the wheel and on to the observer.

back, and then, if the rotational speed of the wheel is adjusted properly, passes through the *next notch* in the wheel. By measuring the rotational speed of the wheel and the distance from the wheel to the mirror, Fizeau was able to obtain a value of 3.13×10^8 m/s for the speed of light.

Today, experiments to measure the speed of light have been refined to such a degree that we now use it to *define* the meter, as was mentioned in Chapter 1. Thus, by definition, the speed of light in a vacuum is

$$c = 299\,792\,458 \text{ m/s}$$

For most routine calculations, however, the value $c = 3.00 \times 10^8$ m/s is adequate.

Maxwell's theoretical description of electromagnetic waves allowed him to obtain a simple expression for c in terms of previously known physical quantities. In particular, he found that c could be written as follows:

$$c = \frac{1}{\sqrt{\varepsilon_0 \mu_0}} \qquad \text{25–2}$$

Recall that $\varepsilon_0 = 8.85 \times 10^{-12}\ \text{C}^2/(\text{N}\cdot\text{m}^2)$ occurs in the expression for the electric field due to a point charge; in fact, ε_0 determines the strength of the electric field. The constant $\mu_0 = 4\pi \times 10^{-7}\ \text{T}\cdot\text{m/A}$ plays an equivalent role for the magnetic field. Thus, Maxwell was able to show that these two constants, which were determined by electrostatic and magnetostatic measurements, also combine to yield the speed of light—again demonstrating the symmetrical role that electric and magnetic fields play in electromagnetic waves. Substituting the values for ε_0 and μ_0 we find

$$c = \frac{1}{\sqrt{(8.85 \times 10^{-12}\ \text{C}^2/(\text{N}\cdot\text{m}^2))(4\pi \times 10^{-7}\ \text{T}\cdot\text{m/A})}} = 3.00 \times 10^8 \text{ m/s}$$

Clearly, Maxwell's theoretical expression agrees with experiment.

EXAMPLE 25–1 Fizeau's Results

Consider a Fizeau experiment in which the wheel has 450 notches and rotates with a speed of 35 rev/s. Light passing through one notch travels to the mirror and back just in time to pass through the next notch. If the distance from the wheel to the mirror is 9500 m, what is the speed of light obtained by this measurement?

Picture the Problem

Our sketch shows an experimental setup similar to Fizeau's. The notched wheel is 9500 m from a mirror and spins with an angular speed of 35 rev/s. We show a few notches in our sketch, which represent the 450 notches on the actual wheel.

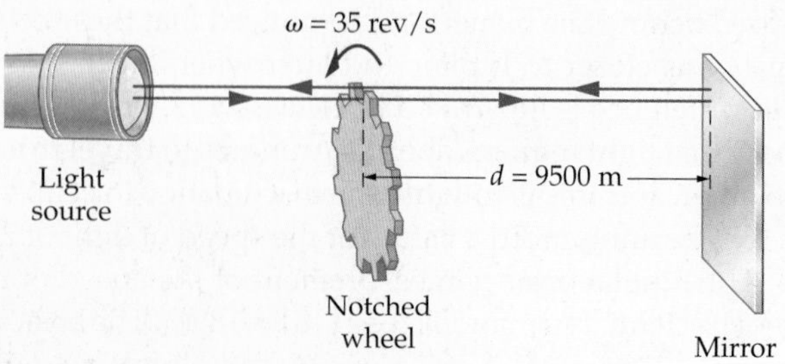

Strategy

The speed of light is the distance traveled, $2d$, divided by the time required, Δt. To find the time, we note that the wheel rotates from one notch to the next during this time; that is, it rotates through an angle $\Delta\theta = (1/450)$ rev. Knowing the rotational speed, ω, of the wheel, we can find the time using the relation $\Delta\theta = \omega\,\Delta t$ (Section 10–1).

Solution

1. Find the time required for the wheel to rotate from one notch to the next:
$$\Delta t = \frac{\Delta\theta}{\omega} = \frac{(1/450)\text{ rev}}{35\text{ rev/s}} = 6.3 \times 10^{-5}\text{ s}$$

2. Divide the time into the distance to find the speed of light:
$$c = \frac{2d}{\Delta t} = \frac{2(9500\text{ m})}{6.3 \times 10^{-5}\text{ s}} = 3.0 \times 10^{8}\text{ m/s}$$

Insight

Note that even with a rather large distance for the round trip, the travel time of the light is small, only 0.063 milliseconds. This illustrates the great difficulty experimentalists faced in attempting to make an accurate measurement of c.

Practice Problem

If the wheel has 430 notches, what rotational speed is required for the return beam of light to pass through the next notch? [**Answer:** $\omega = 37$ rev/s]

Some related homework problems: Problem 14, Problem 15

Although the speed of light is enormous by earthly standards, it is useful to look at it from an astronomical perspective. Imagine, for example, that you could shrink the solar system to fit onto a football field, with the Sun at one end zone and Pluto at the other. On this scale, Earth would be a grain of sand located at the 2.5-yard line from the Sun, and light would take 8 min to cover that distance. To travel to Pluto, at the other end of the field, light would require about 5.5 hr. Thus, on this scale, the speed of light is like the crawl of a small caterpillar. When one recalls that the solar system is but a speck on the outskirts of the Milky Way galaxy, and that the nearest major galaxy to our own—the Andromeda galaxy—is about 2.2 million light-years away, the speed of light doesn't appear so great after all.

The Doppler Effect

In Section 14–5 we discussed the Doppler effect for sound waves—the familiar increase or decrease in frequency as a source of sound approaches or recedes. A similar Doppler effect applies to electromagnetic waves. There are two fundamental differences, however. First, sound waves require a medium through which to travel, whereas light can propagate across a vacuum. Second, the speed of sound

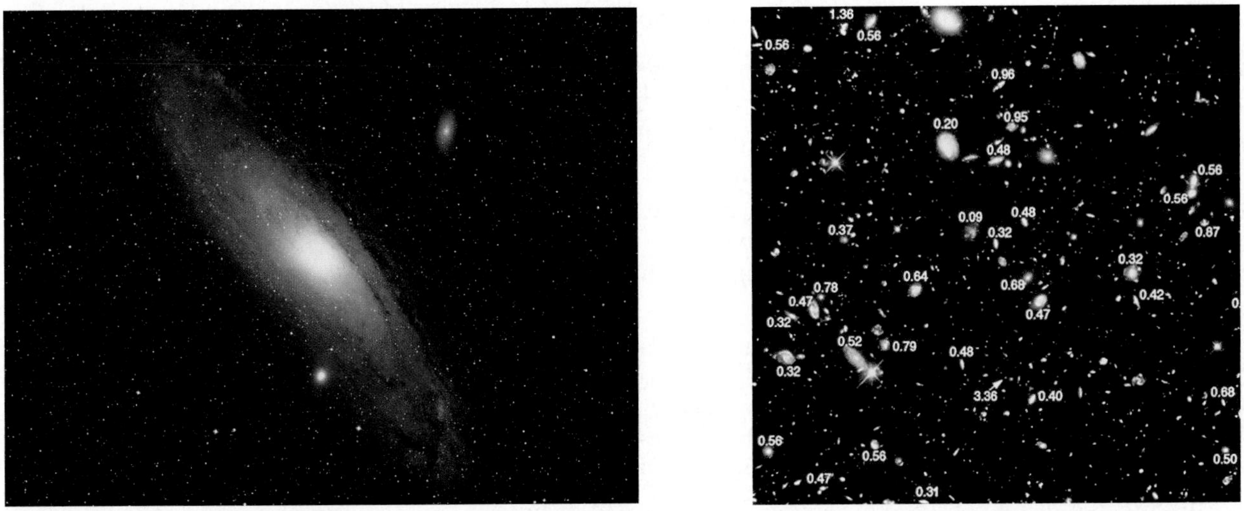

▲ Even traveling at 300 million meters per second, light from the Andromeda galaxy (left) takes over 2 million years to reach us. Yet Andromeda is one of our nearest cosmic neighbors. A sense of the true vastness of the universe is provided by the image known as the Hubble Deep Field (right). This long-exposure photograph, taken from orbit by the Hubble Space Telescope, shows over 1600 galaxies when examined closely. Most of them exhibit a Doppler red-shift—that is, their light is shifted to lower frequencies by the Doppler effect, indicating that they are receding from Earth as the universe expands. The red shifts marked on the photo correspond to distances ranging from about 1.3 billion light years to over 13 billion light years (nearly 10^{23} miles).

can be different for different observers. For example, an observer approaching a source of sound measures an increased speed of sound, whereas an observer detecting sound from a moving source measures the usual speed of sound. For this reason, the Doppler effect with sound is different for a moving observer than it is for a moving source (see Figure 14–18 for a direct comparison). In contrast, the speed of electromagnetic waves is *independent* of the motion of the source and observer, as we shall see in Chapter 29. Therefore, there is just one Doppler effect for electromagnetic waves, and it depends only on the *relative speed* between the observer and source.

For source speeds u that are small compared with the speed of light, the observed frequency f' from a source with frequency f is

$$f' = f\left(1 \pm \frac{u}{c}\right) \qquad 25\text{–}3$$

PROBLEM-SOLVING NOTE

Evaluating the Doppler Shift

Since everyday objects generally move with speeds much less than the speed of light, the Doppler-shifted frequency differs little from the original frequency. To see the Doppler effect more clearly, it is often useful to calculate the difference in frequency, $f' - f$, rather than the Doppler-shifted frequency itself.

Note that u in this expression is a speed and hence is always positive. The appropriate sign in front of the term u/c is chosen for a given situation—the plus sign applies to a source that is approaching the observer, the minus sign to a receding source. In addition, u is a *relative* speed between the source and the observer, both of which may be moving. For example, if an observer is moving in the positive x direction with a speed of 5 m/s, and a source ahead of the observer is moving in the positive x direction with a speed of 12 m/s, the relative speed is $u = 12\text{ m/s} - 5\text{ m/s} = 7\text{ m/s}$. Since the distance between the observer and source is increasing with time in this case, we would choose the minus sign in Equation 25–3.

EXERCISE 25–2

An FM radio station broadcasts at a frequency of 88.5 MHz. If you drive your car toward the station at 32.0 m/s, what change in frequency do you observe?

Solution

We can find the change in frequency, $f' - f$, using Equation 25–3:

$$f' - f = f\frac{u}{c} = (88.5 \times 10^6\text{ Hz})\frac{32.0\text{ m/s}}{3.00 \times 10^8\text{ m/s}} = 9.44\text{ Hz}$$

Thus, the frequency changes by only 9.44 Hz = 0.00000944 MHz.

REAL-WORLD PHYSICS

Doppler radar

Common applications of the Doppler effect include the radar units used to measure the speed of automobiles, and the Doppler radar that is used to monitor the weather. In Doppler radar, electromagnetic waves are sent out into the atmosphere and are reflected back to the receiver. The change in frequency of the reflected beam relative to the outgoing beam provides a way of measuring the speed of the clouds and precipitation that reflected the beam. Thus, Doppler radar gives more information than just where a rainstorm is located; it also tells how it is moving. Measurements of this type are particularly important for airports, where information regarding areas of possible wind shear can be crucial for safety.

EXAMPLE 25–2 Nexrad

REAL-WORLD PHYSICS

The Doppler weather radar used by the National Weather Service is referred to as Nexrad, which stands for next generation radar. Nexrad commonly operates at a frequency of 2.7 GHz. If a Nexrad wave reflects from an approaching weather system moving with a speed of 28 m/s, find the difference in frequency between the outgoing and returning waves.

Picture the Problem

Our sketch shows the outgoing radar waves, the incoming weather system, and the returning radar waves. The speed of the weather system relative to the radar station is $u = 28$ m/s, and the frequency of the outgoing waves is $f = 2.7$ GHz.

Strategy

Two Doppler effects are involved in this system. First, the outgoing wave is seen to have a frequency $f' = f(1 + u/c)$ by the weather system, since it is moving toward the source. The waves reflected by the weather system, then, have the frequency f'. Since the weather system acts like a moving source of radar with frequency f', an observer at the radar facility detects a frequency $f'' = f'(1 + u/c)$. Thus, given u and f, we can calculate the difference, $f'' - f$.

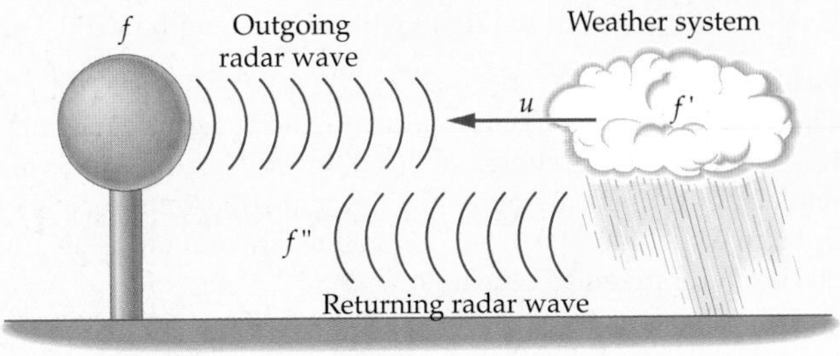

Solution

1. Use $f' = f(1 + u/c)$ to calculate the difference in frequency, $f' - f$:

$$f' - f = fu/c = \frac{(2.7 \times 10^9\ \text{Hz})(28\ \text{m/s})}{3.00 \times 10^8\ \text{m/s}} = 250\ \text{Hz}$$

2. Now, use $f'' = f'(1 + u/c)$ to find the difference between f'' and f':

$$f'' - f' = f'u/c$$

3. Use the results of step 1 to replace f' with $f + 250$ Hz:

$$f'' - (f + 250\ \text{Hz}) = (f + 250\ \text{Hz})u/c$$

4. Solve for the frequency difference, $f'' - f$:

$$f'' - f = (f + 250\ \text{Hz})u/c + 250\ \text{Hz}$$
$$= \frac{(2.7 \times 10^9\ \text{Hz} + 250\ \text{Hz})(28\ \text{m/s})}{3.00 \times 10^8\ \text{m/s}} + 250\ \text{Hz}$$
$$= 500\ \text{Hz}$$

Insight

Notice that we focus on the *difference* in frequency between the very large numbers $f = 2{,}700{,}000{,}000$ Hz and $f'' = 2{,}700{,}000{,}500$ Hz. Clearly, it is more convenient to simply write $f'' - f = 500$ Hz.

In addition, note that the two Doppler shifts in this problem are analogous to the two Doppler shifts we found for the case of a train approaching a tunnel in Example 14–6.

Practice Problem

Find the difference in frequency if the weather system is receding with a speed of 21 m/s. [**Answer:** $f'' - f = -380$ Hz]

Some related homework problems: Problem 21, Problem 23

25–3 The Electromagnetic Spectrum

When white light passes through a prism it spreads out into a rainbow of colors, with red on one end and violet on the other. All these various colors of light are electromagnetic waves, of course; they differ only in their frequency and, hence, their wavelength. The relationship between frequency and wavelength for any wave with a speed v is simply $v = f\lambda$, as was shown in Section 14–11. Because all electromagnetic waves in a vacuum have the same speed, c, it follows that f and λ are related as follows:

$$c = f\lambda \qquad \textbf{25–4}$$

Thus, as the frequency of an electromagnetic wave increases, its wavelength decreases.

In the following Example we calculate the frequency of red and violet light, given the corresponding wavelengths. Notice that the wavelengths are given in units of nanometers (nm), where 1 nm $= 10^{-9}$ m. Occasionally the wavelength of light is given in terms of a non-SI unit referred to as the *angstrom* (Å), defined as follows: 1Å $= 10^{-10}$ m.

EXAMPLE 25–3 Roses are Red, Violets are Violet

Find the frequency of red light, with a wavelength of 700.0 nm, and violet light, with a wavelength of 400.0 nm.

Picture the Problem

The visible electromagnetic spectrum, along with representative wavelengths, is shown in our diagram. In addition to the wavelengths of 700.0 nm for red light and 400.0 nm for violet light, we include 600.0 nm for yellowish orange light and 500.0 nm for greenish blue light.

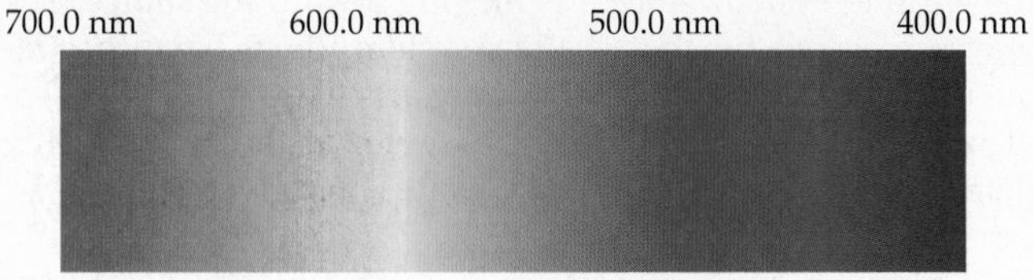

Strategy

We obtain the frequency by rearranging $c = f\lambda$ to yield $f = c/\lambda$.

Solution

1. Substitute $\lambda = 700.0$ nm for red light: $$f = \frac{c}{\lambda} = \frac{3.00 \times 10^8 \text{ m/s}}{700.0 \times 10^{-9} \text{ m}} = 4.29 \times 10^{14} \text{ Hz}$$

2. Substitute $\lambda = 400.0$ nm for violet light: $$f = \frac{c}{\lambda} = \frac{3.00 \times 10^8 \text{ m/s}}{400.0 \times 10^{-9} \text{ m}} = 7.50 \times 10^{14} \text{ Hz}$$

Insight

The frequency of visible light is extremely large. In fact, even for the relatively low frequency of red light, it takes only 2.33×10^{-15} s to complete one cycle. The *range* of visible frequencies is relatively small, however, when compared with other portions of the electromagnetic spectrum.

Practice Problem

What is the wavelength of light with a frequency of 5.25×10^{14} Hz? [**Answer:** 571 nm]

Some related homework problems: Problem 25, Problem 26

In principle, the frequency of an electromagnetic wave can have any positive value, and this full range of frequencies is known as the **electromagnetic spectrum.** Certain bands of the spectrum are given special names, as indicated in **Figure 25–8**. For example, we have just seen that visible light occupies a relatively narrow band of frequencies from 4.29×10^{14} Hz to 7.50×10^{14} Hz. In what follows, we discuss the various regions of the electromagnetic spectrum of most relevance to humans and our technology, in order of increasing frequency.

▲ Since the development of the first radiotelescopes in the 1950s, the radio portion of the electromagnetic spectrum has provided astronomers with a valuable new window on the universe. These antennas are part of the Very Large Array (VLA), located in San Augustin, New Mexico.

Radio Waves

($f \sim 10^6$ Hz to 10^9 Hz, $\lambda \sim 300$ m to 0.3 m) The lowest-frequency electromagnetic waves of practical importance are *radio* and *television waves* in the frequency range of roughly 10^6 Hz to 10^9 Hz. Waves in this frequency range are produced in a variety of ways. For example, molecules and accelerated electrons in space give off radio waves, which radio astronomers detect with large dish receivers. Radio waves are also produced as a piece of adhesive tape is slowly peeled from a surface, as you can confirm by holding a transistor radio near the tape and listening for pops and snaps coming from the speaker. Most commonly, the radio waves we pick up with our radios and televisions are produced by alternating currents in metal antennas.

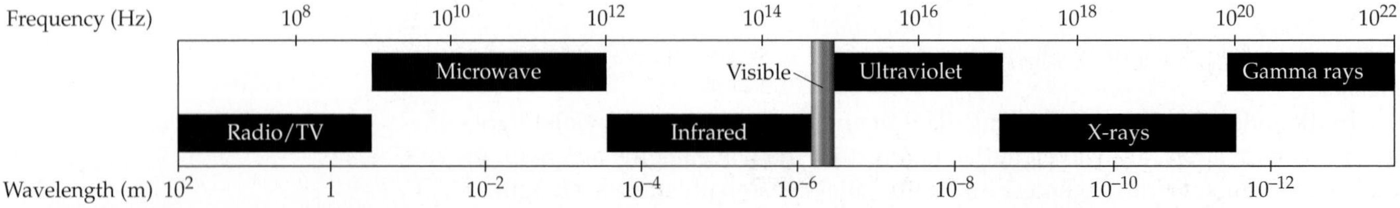

▲ **FIGURE 25–8 The electromagnetic spectrum**

Note that the visible portion of the spectrum is relatively narrow. The boundaries between various bands of the spectrum are not sharp but, instead, are somewhat arbitrary.

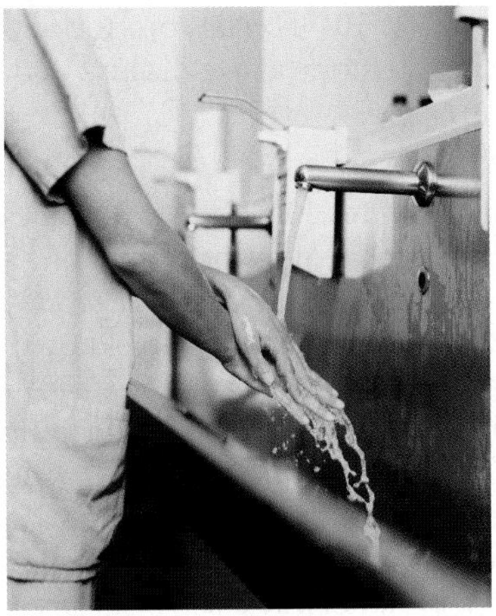

◀ We use infrared rays all the time, even though they are invisible to us. If you change the channel with a remote control, your signal is sent by an infrared ray; if you move your hand in front of a no-touch water faucet, an infrared ray detects the motion. In contrast, snakes called pit vipers can actually "see" infrared rays with the "pit" organs located just in front of their eyes. What must a remote control look like to one of these creatures?

Microwaves

($f \sim 10^9$ Hz to 10^{12} Hz, $\lambda \sim$ 300 mm to 0.3 mm) Electromagnetic radiation with frequencies from 10^9 Hz to about 10^{12} Hz are referred to as *microwaves*. Waves in this frequency range are used to carry long-distance telephone conversations, as well as to cook our food. Microwaves, with wavelengths of about 1 mm to 30 cm, are the highest-frequency electromagnetic waves that can be produced by electronic circuitry.

Infrared Waves

($f \sim 10^{12}$ Hz to 4.3×10^{14} Hz, $\lambda \sim$ 0.3 mm to 700 nm) Electromagnetic waves with frequencies just below that of red light—roughly 10^{12} Hz to 4.3×10^{14} Hz—are known as *infrared* rays. These waves can be felt as heat on our skin but cannot be seen with our eyes. Many creatures, including various types of pit vipers, have specialized infrared receptors that allow them to "see" the infrared rays given off by a warm-blooded prey animal, even in total darkness. Infrared rays are often generated by the rotations and vibrations of molecules. In turn, when infrared rays are absorbed by an object, its molecules rotate and vibrate more vigorously, resulting in an increase in the object's temperature. Finally, many remote controls—for items ranging from TVs to DVD players to gas fireplaces—operate on a beam of infrared light with a wavelength of about 1000 nm. This infrared light is so close to the visible spectrum and so low in intensity that it cannot be felt as heat.

REAL-WORLD PHYSICS: BIO
Infrared receptors in pit vipers

Visible Light

($f \sim 4.3 \times 10^{14}$ Hz to 7.5×10^{14} Hz, $\lambda \sim$ 700 nm to 400 nm) The portion of the electromagnetic spectrum most familiar to us is the spectrum of visible light, represented by the full range of colors seen in a rainbow. Each of the different colors, as perceived by our eyes and nervous system, is nothing more than an electromagnetic wave with a different frequency. Waves in this frequency range

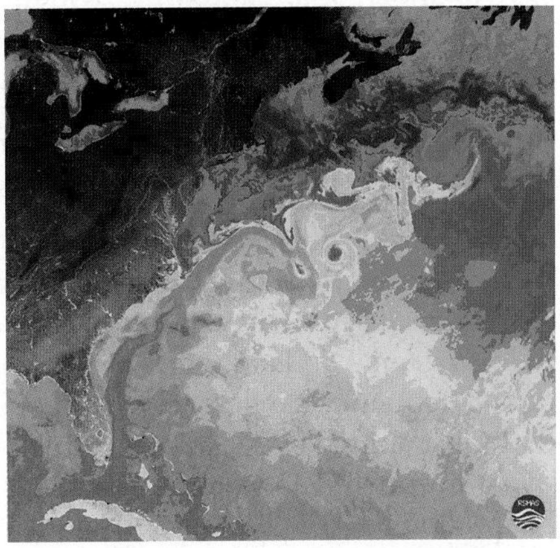

◀ Photographs made with infrared radiation are often called thermograms, since most infrared wavelengths can be felt as heat by the human skin. Thermograms provide a useful remote sensing technique for measuring temperature. In the photo on the left, the areas of the cat's head that are warmest (pink) and coolest (blue) can be clearly identified. In the photo on the right, an infrared satellite image of the Atlantic Ocean off the coast of North America, warmer colors are used to indicate higher sea surface temperatures. The swirling red streak running from lower left toward the upper right is the Gulf Stream.

(4.3×10^{14} to 7.5×10^{14} Hz) are produced primarily by electrons changing their positions within an atom, as we discuss in detail in Chapter 31.

REAL WORLD-PHYSICS: BIO
Biological effects of ultraviolet light

Ultraviolet Light

($f \sim 7.5 \times 10^{14}$ Hz to 10^{17} Hz, $\lambda \sim 400$ nm to 3 nm) When electromagnetic waves have frequencies just above that of violet light—from about 7.5×10^{14} Hz to 10^{17} Hz—they are called *ultraviolet* or *UV rays*. Although these rays are invisible, they often make their presence known by causing suntans with moderate exposure. More prolonged or intense exposure to UV rays can have harmful consequences, including an increased probability of developing a skin cancer. Fortunately, most of the UV radiation that reaches Earth from the Sun is absorbed in the upper atmosphere by ozone (O_3) and other molecules. A significant reduction in the ozone concentration in the stratosphere could result in an unwelcome increase of UV radiation on Earth's surface.

▲ Most ultraviolet radiation cannot penetrate Earth's atmosphere, but ultraviolet astronomical photographs, such as the one shown above, have been taken by cameras in orbit. The bright spots in this image of a spiral galaxy are areas of intense star formation, populated by hot young stars that radiate heavily in the ultraviolet.

X-Rays

($f \sim 10^{17}$ Hz to 10^{20} Hz, $\lambda \sim 3$ nm to 0.003 nm) As the frequency of electromagnetic waves is raised even higher, into the range between about 10^{17} Hz to 10^{20} Hz, we reach the part of the spectrum known as *X-rays*. Typically, the X-rays used in medicine are generated by the rapid deceleration of high-speed electrons projected against a metal target, as we show in Section 31–17. These energetic rays, which are only weakly absorbed by the skin and soft tissues, pass through our bodies rather freely, except when they encounter bones, teeth, or other relatively dense material. This property makes X-rays most valuable for medical diagnosis, research, and treatment. Still, X-rays can cause damage to human tissue, and it is desirable to reduce unnecessary exposure to these rays as much as possible.

Gamma Rays

($f \sim 10^{20}$ Hz and higher, $\lambda \sim 0.003$ nm and smaller) Finally, electromagnetic waves with frequencies above about 10^{20} Hz are generally referred to as *gamma* (γ) *rays*. These rays, which are even more energetic than X-rays, are often produced as neutrons and protons rearrange themselves within a nucleus, or when a particle collides with its antiparticle, and the two annihilate each other. These processes are discussed in detail in Chapter 32. Gamma rays are also highly penetrating and destructive to living cells. It is for this reason that they are used to kill cancer cells and, more recently, microorganisms in food. Irradiated food, however, is a concept that has yet to become popular with the general public, even though NASA has irradiated astronauts' food since the 1960s. If you happen to see irradiated food in the grocery store, you will know that it has been exposed to γ rays from cobalt-60 for 20 to 30 minutes.

REAL-WORLD PHYSICS: BIO
Irradiated food

Notice that the visible part of the electromagnetic spectrum, so important to life on Earth, is actually the smallest of the frequency bands we have named. This accounts for the fact that a rainbow produces only a narrow band of color in the sky—if the visible band were wider, the rainbow would be wider as well. It should be remembered, however, that there is nothing particularly special about the visible band; in fact, it is even species dependent. For example, some bees and butterflies can see ultraviolet light, and, as mentioned previously, certain snakes can form images from infrared radiation.

▲ The use of radiation to preserve food can be quite effective, but the technique is still controversial. Both boxes of strawberries shown here were stored for about 2 weeks at refrigerator temperature. Before storage, the box at right was irradiated to kill microorganisms and mold spores.

One of the main factors in determining the visible range of frequencies is Earth's atmosphere. For example, if one examines the transparency of the atmosphere as a function of frequency, it is found that there is a relatively narrow range of frequencies for which the atmosphere is highly transparent. As eyes evolved in living systems on Earth, they could have evolved to be sensitive to various different frequency ranges. It so happens, however, that the range of frequencies that most animal eyes can detect match nicely with the range of frequencies that the atmosphere allows to reach Earth's surface. This is a nice example of natural adaptation.

25–4 Energy and Momentum in Electromagnetic Waves

All waves transmit energy, and electromagnetic waves are no exception. When you walk outside on a sunny day, for example, you feel warmth where the sunlight strikes your skin. The energy creating this warm sensation originated in the Sun, and it has just completed a 93-million-mile trip when it reaches your body. In fact, the energy necessary for most of the life on Earth is transported here across the vacuum of space by electromagnetic waves traveling at the speed of light.

That electromagnetic waves carry energy with them is no surprise when you recall that they are composed of electric and magnetic fields, each of which has an associated energy density. For example, in Section 20–6 we showed that the energy density of an electric field of magnitude E is

$$u_E = \tfrac{1}{2}\varepsilon_0 E^2$$

Similarly, we showed in Section 23–9 that the magnetic energy density of a magnetic field of magnitude B is

$$u_B = \frac{1}{2\mu_0}B^2$$

It follows that the total energy density, u, of an electromagnetic wave is simply

$$u = u_E + u_B = \tfrac{1}{2}\varepsilon_0 E^2 + \frac{1}{2\mu_0}B^2 \qquad \text{25–5}$$

As expected, both E and B contribute to the total energy carried by a wave. Not only that, but it can be shown that the electric and magnetic energy densities in an electromagnetic wave are, in fact, equal to each other—again demonstrating the symmetrical role played by the electric and magnetic fields. Thus, the total energy density of an electromagnetic field can be written in the following equivalent forms:

$$u = \tfrac{1}{2}\varepsilon_0 E^2 + \frac{1}{2\mu_0}B^2 = \varepsilon_0 E^2 = \frac{1}{\mu_0}B^2 \qquad \text{25–6}$$

Since E and B vary sinusoidally with time, as indicated in Figure 25–3, it follows that their average values are zero—just like the current or voltage in an ac circuit. Therefore, to find the average energy density of an electromagnetic wave, we must use the rms values of E and B:

$$u_{av} = \tfrac{1}{2}\varepsilon_0 E_{rms}^2 + \frac{1}{2\mu_0}B_{rms}^2 = \varepsilon_0 E_{rms}^2 = \frac{1}{\mu_0}B_{rms}^2 \qquad \text{25–7}$$

Recall that the rms value of a sinusoidally varying quantity x is related to its maximum value as $x_{rms} = x_{max}/\sqrt{2}$. Thus, for the electric and magnetic fields in an electromagnetic wave we have

$$E_{rms} = \frac{E_{max}}{\sqrt{2}}$$

$$B_{rms} = \frac{B_{max}}{\sqrt{2}} \qquad \text{25–8}$$

The fact that the electric and magnetic energy densities are equal in an electromagnetic wave has a further interesting consequence. Setting u_E equal to u_B we obtain

$$\tfrac{1}{2}\varepsilon_0 E^2 = \frac{1}{2\mu_0}B^2$$

Rearranging slightly, and taking the square root, we find

$$E = \frac{1}{\sqrt{\varepsilon_0\mu_0}}B$$

Finally, from the fact that the speed of light is given by the relation $c = 1/\sqrt{\varepsilon_0\mu_0}$, it follows that

$$E = cB \qquad \textbf{25–9}$$

Thus, not only does an electromagnetic wave have both an electric and a magnetic field, the fields must also have the specific ratio $E/B = c$.

EXERCISE 25–3

At a given instant of time the electric field in a beam of sunlight has a magnitude of 510 N/C. What is the magnitude of the magnetic field at this instant?

Solution

Using $E = cB$, or $B = E/c$, we find

$$B = \frac{E}{c} = \frac{510\ \text{N/C}}{3.00 \times 10^8\ \text{m/s}} = 1.7 \times 10^{-6}\ \text{T}$$

Note that the units are consistent, since $1\ \text{T} = 1\ \text{N}/(\text{C}\cdot\text{m/s})$.

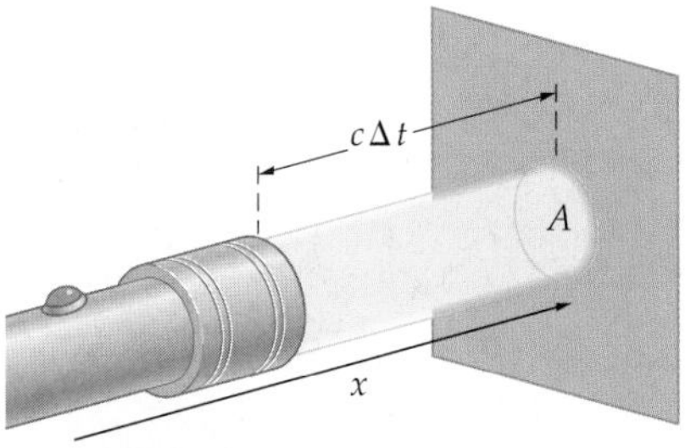

▲ **FIGURE 25–9 The energy in a beam of light**

A beam of light of cross-sectional area A shines on a surface. All the light energy contained in the volume $\Delta V = A(c\Delta t)$ strikes the surface in the time Δt.

The amount of energy a wave delivers to a unit area in a unit time is referred to as its **intensity,** I. (Equivalently, since power is energy per time, the intensity of a wave is the power per unit area.) Imagine, for example, an electromagnetic wave of area A moving in the positive x direction, as in **Figure 25–9**. In the time Δt the wave moves through a distance $c\Delta t$; hence, all the energy in the volume $\Delta V = A(c\Delta t)$ is deposited on the area A in this time. Because energy is equal to the energy density times the volume, it follows that the energy in the volume ΔV is $\Delta U = u\Delta V$. Therefore, the intensity of the wave (energy per area per time) is

$$I = \frac{\Delta U}{A\Delta t} = \frac{u(Ac\Delta t)}{A\Delta t} = uc$$

Averaged over time, the intensity is $I_{av} = u_{av}c$. In terms of the electric and magnetic fields, we have

$$I = uc = \tfrac{1}{2}c\varepsilon_0E^2 + \frac{1}{2\mu_0}cB^2 = c\varepsilon_0E^2 = \frac{c}{\mu_0}B^2 \qquad \textbf{25–10}$$

As before, to calculate an average intensity, we must replace E and B with their rms values.

Notice that the intensity is proportional to the square of the fields. This is analogous to the case of simple harmonic motion, where the energy of oscillation is proportional to the square of the amplitude, as we found in Section 14–15.

EXAMPLE 25–4 Lightbulb Fields

A garage is illuminated by a single incandescent lightbulb dangling from a wire. If the bulb radiates light uniformly in all directions, and consumes an average electrical power of 50.0 W, what are **(a)** the average intensity of the light and **(b)** the rms values of E and B at a distance of 1.00 m from the bulb? (Assume that 5.00% of the electrical power consumed by the bulb is converted to light.)

Picture the Problem

The physical situation is pictured in the sketch. We assume that all the power radiated by the bulb passes uniformly through the area of a sphere of radius $r = 1.00$ m centered on the bulb.

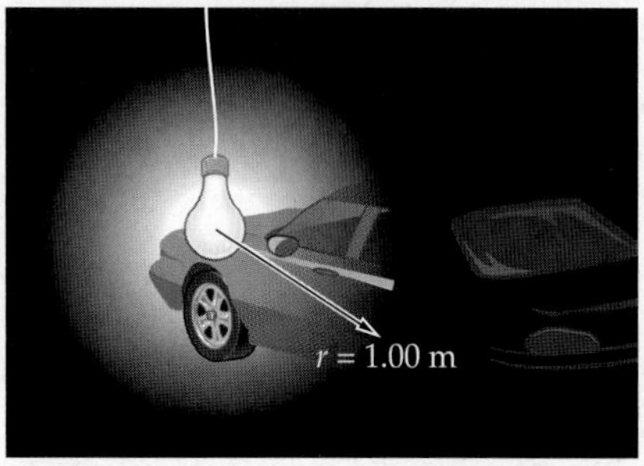

Strategy

a. Recall that intensity is power per unit area; therefore, $I_{av} = P_{av}/A$. In this case the area is $A = 4\pi r^2$, the surface area of a sphere of radius r. The average power of the light is 5.00% of 50.0 W.

b. Once we know the intensity, I_{av}, we obtain the fields using Equation 25–10. Since the intensity is an average value, the corresponding fields are rms values.

Solution

Part (a)

1. Calculate the average intensity at the surface of the sphere of radius $r = 1.00$ m:

$$I_{av} = \frac{P_{av}}{A} = \frac{(50.0\text{ W})(0.0500)}{4\pi(1.00\text{ m})^2} = 0.199\text{ W/m}^2$$

Part (b)

2. Use $I_{av} = c\varepsilon_0 E_{rms}^2$ to find E_{rms}:

$$I_{av} = c\varepsilon_0 E_{rms}^2$$

$$E_{rms} = \sqrt{\frac{I_{av}}{c\varepsilon_0}} = \sqrt{\frac{0.199\text{ W/m}^2}{(3.00 \times 10^8\text{ m/s})(8.85 \times 10^{-12}\text{ C}^2/\text{N}\cdot\text{m}^2)}}$$

$$= 8.66\text{ N/C}$$

2. Use $I_{av} = cB_{rms}^2/\mu_0$ to find B_{rms}:

$$I_{av} = cB_{rms}^2/\mu_0$$

$$B_{rms} = \sqrt{\frac{\mu_0 I_{av}}{c}} = \sqrt{\frac{(4\pi \times 10^{-7}\text{ N}\cdot\text{s}^2/\text{C}^2)(0.199\text{ W/m}^2)}{(3.00 \times 10^8\text{ m/s})}}$$

$$= 2.89 \times 10^{-8}\text{ T}$$

3. As a check, use the relation $E = cB$ and the results from steps 2 and 3 to calculate the speed of light, c:

$$c = \frac{E_{rms}}{B_{rms}} = \frac{8.66\text{ N/C}}{2.89 \times 10^{-8}\text{ T}} = 3.00 \times 10^8\text{ m/s}$$

Insight

Notice that the electric field in the light from the bulb has a magnitude (8.66 N/C = 8.66 V/m) that is relatively easy to measure. The magnetic field, however, is so small that a direct measurement would be difficult. This is common in electromagnetic waves because $B = E/c$, and c is such a large number.

Practice Problem

If the distance from the bulb is doubled, does E_{rms} increase, decrease, or stay the same? Check your answer by calculating E_{rms} at $r = 2.00$ m. [**Answer:** E_{rms} decreases with increasing distance from the bulb. For $r = 2.00$ m, we find $E_{rms} = \frac{1}{2}(8.66\text{ N/C}) = 4.33\text{ N/C}$.]

Some related homework problems: Problem 47, Problem 58

ACTIVE EXAMPLE 25–1 Find *E* and *B* in a Beam of Light

A small laser emits a cylindrical beam of light 1.00 mm in diameter with an average power of 5.00 mW. Find the maximum values of E and B in this beam.

Solution *(Test your understanding by performing the calculations indicated in each step.)*

1. Divide the average power of the beam by its area to find the intensity: $I_{av} = 6370\text{ W/m}^2$
2. Use $I_{av} = c\varepsilon_0 E_{rms}^2$ to find the rms electric field: $E_{rms} = 1550\text{ N/C}$
3. Multiply E_{rms} by $\sqrt{2}$ to find E_{max}: $E_{max} = 2190\text{ N/C}$
4. Divide E_{max} by c to find B_{max}: $B_{max} = 7.30 \times 10^{-6}\text{ T}$

Insight

To find the cross-sectional area of the cylindrical beam in step 1 we used $A = \pi d^2/4$.

Your Turn

Suppose the beam spreads out to twice its initial diameter. By what factor do E_{max} and B_{max} change?

(Answers to **Your Turn** *problems are given in the back of the book.)*

Finally, an electromagnetic wave also carries momentum. In fact, it can be shown that if a total energy U is absorbed by a given area, the momentum, p, that it receives is

$$p = \frac{U}{c} \qquad 25\text{–}11$$

For an electromagnetic wave absorbed by an area A, the total energy received in the time Δt is $U = u_{av}Ac\Delta t$; hence, the momentum, Δp, received in this time is

$$\Delta p = \frac{u_{av}Ac\Delta t}{c} = \frac{I_{av}A\Delta t}{c}$$

Since the average force is $F_{av} = \Delta p/\Delta t = u_{av}A = I_{av}A/c$, it follows that the average pressure (force per area) is simply

$$\text{pressure}_{av} = \frac{I_{av}}{c} \qquad \text{25–12}$$

The pressure exerted by light is commonly referred to as **radiation pressure.**

CONCEPTUAL CHECKPOINT 25–2 Momentum Transfer

When an electromagnetic wave carrying an energy U is absorbed by an object, the momentum the object receives is $p = U/c$. If, instead, the object reflects the wave, is the momentum the object receives **(a)** more than, **(b)** less than, or **(c)** the same as when it absorbs the wave?

Reasoning and Discussion

When the object reflects the wave, it must supply not only enough momentum to stop the wave, $p = U/c$, but an equivalent amount of momentum to send the wave back in the opposite direction. Thus, the momentum the object receives is $p = 2U/c$. This situation is completely analogous to the momentum transfer of a ball that either sticks to a wall or bounces back the way it came.

Answer:

(a) The object receives twice as much momentum.

Radiation pressure is a very real effect, though in everyday situations it is too small to be noticed. In principle, turning on a flashlight should give the user a "kick," like the recoil from firing a gun. In practice, the effect is much too small to be felt. In the next Exercise, we calculate the radiation pressure due to sunlight to get a feel for the magnitudes involved.

EXERCISE 25–4

On a sunny day, the average intensity of sunlight on Earth's surface is about $1.00 \times 10^3\ \text{W/m}^2$. Find **(a)** the average radiation pressure due to the sunlight and **(b)** the average force exerted by the light on a 1.00-m × 2.50-m beach towel. For part (b), assume that the towel absorbs all the light that falls on it.

Solution

a. Divide the average intensity by the speed of light to obtain the average pressure:

$$\text{pressure}_{av} = \frac{I_{av}}{c} = \frac{1.00 \times 10^3\ \text{W/m}^2}{3.00 \times 10^8\ \text{m/s}} = 3.33 \times 10^{-6}\ \text{N/m}^2$$

b. Multiply pressure times area to find the force exerted on the towel:

$$F_{av} = \text{pressure}_{av}A = (3.33 \times 10^{-6}\ \text{N/m}^2)(1.00\ \text{m} \times 2.50\ \text{m}) = 8.33 \times 10^{-6}\ \text{N}$$

Since a newton is about a quarter of a pound, it follows that the force exerted on the beach towel by sunlight is incredibly small.

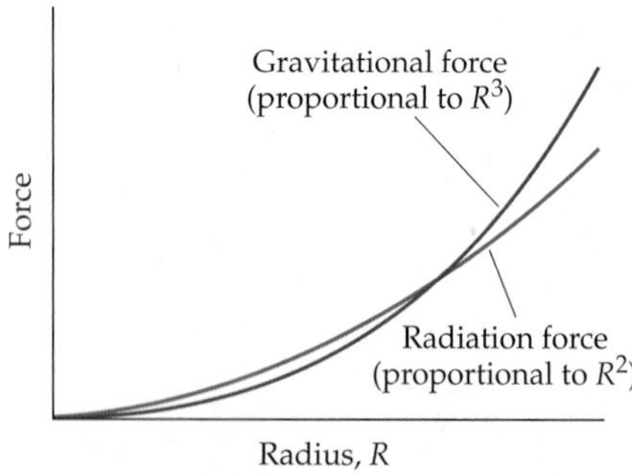

▲ **FIGURE 25–10 Radiation versus gravitational forces**

The forces exerted on a small particle of radius R by gravity and radiation pressure. For large R the gravitational force dominates; for small R the radiation force is larger.

Even though the radiation force on a beach towel is negligible, the effects of radiation pressure can be significant on particles that are very small. Consider a small object of radius R drifting through space somewhere in our solar system. The gravitational attraction it feels from the Sun depends on its mass. Since mass is proportional to volume, and volume is proportional to the cube of the particle's radius, the gravitational force on the particle varies as R^3. On the other hand, the radiation pressure is exerted over the area of the object, and the area is proportional to R^2. As R becomes smaller, the radiation pressure, with its R^2 dependence, decreases less rapidly than the gravitational force, with its R^3 dependence, as illustrated in **Figure 25–10**. Thus, for small enough particles, the radiation pressure from sunlight is actually more important than the gravitational force. It is for this reason that dust particles given off by a comet are "blown" away by sunlight, giving the comet a long tail that streams away from the sun. Some imaginative thinkers have envisaged a "sailing" ship designed to travel through the cosmos using this light-pressure "wind."

Finally, it has been discovered that the energy carried by light comes in discrete units, as if there were "particles" of energy in a light beam. So, in many ways, light behaves just like any other wave, showing the effects of diffraction and interference, whereas in other ways it acts like a particle. Therefore, light, and all electromagnetic waves, have a "dual" nature, exhibiting both wave and particle properties. As we shall see when we consider quantum physics in Chapter 30, this *wave–particle duality* plays a fundamental role in our understanding of modern physics.

▲ Like many comets, Comet Hale-Bopp, which passed relatively close to Earth in 1997, developed two tails as it approached the Sun. The straighter, blue tail in this photograph is gas boiling off the comet's head as volatile material is vaporized by the Sun's heat. The curved, whiter tail consists mostly of dust particles, blown away by the pressure of sunlight.

25–5 Polarization

When looking into the blue sky of a crystal-clear day, we see light that is fairly uniform—as long as we refrain from looking too close to the Sun. However, for some animals, like the common honeybee, the light in the sky is far from uniform. The reason is that honeybees are sensitive to the **polarization** of light, an ability that aids in their navigation from hive to flower and back.

To understand what is meant by the polarization of light, or any other electromagnetic wave, consider the electromagnetic waves pictured in **Figure 25–11**. Each of these waves has an electric field that points along a single line. For example, the electric field in Figure 25–11 (a) points in either the positive or negative z direction. We say, then, that this wave is **linearly polarized** in the z direction. A wave of this sort might be produced by a straight-wire antenna oriented along the z axis. Similarly, the direction of polarization for the wave in Figure 25–11 (b) is in the y-z plane at an angle of 60° relative to the y direction. In general, *the polarization of an electromagnetic wave refers to the direction of its electric field.*

A convenient way to represent the polarization of a beam of light is shown in **Figure 25–12**. In part (a) of this figure, we indicate light that is polarized in the vertical direction. Not all light is polarized, however. In part (b) of Figure 25–12 we show light that is a combination of many waves with polarizations in different, random directions. Such light is said to be **unpolarized.** A common incandescent lightbulb produces unpolarized light because each atom in the heated filament sends out light of random polarization. Similarly, the light from the Sun is unpolarized.

Passing Light through Polarizers

A beam of unpolarized light can be polarized in a number of ways, including by passing it through a **polarizer.** To be specific, a polarizer is a material that is composed of long, thin, electrically conductive molecules oriented in a specific

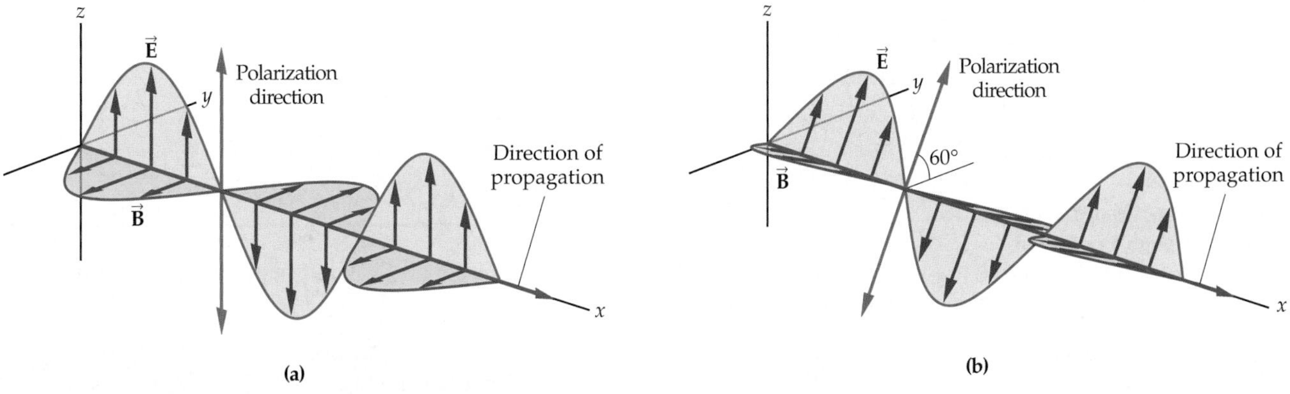

▲ **FIGURE 25–11 Polarization of electromagnetic waves**

The polarization of an electromagnetic wave is the direction along which its electric field vector, $\vec{E}$, points. The cases shown illustrate **(a)** polarization in the z direction and **(b)** polarization in the y-z plane, at an angle of 60° with respect to the y axis.

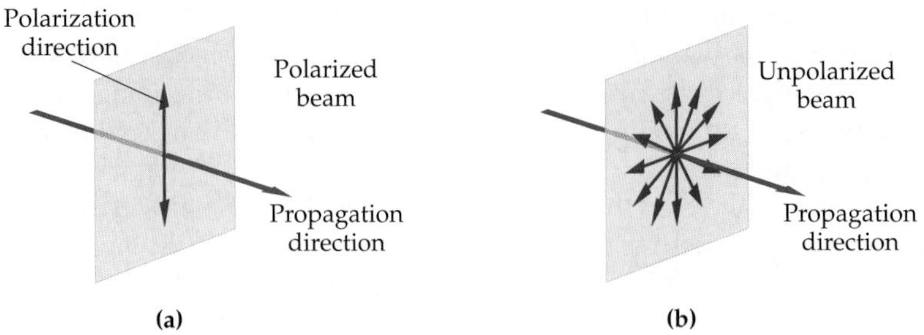

▶ **FIGURE 25–12 Polarized versus unpolarized light**

A beam of light that is **(a)** polarized in the vertical direction and **(b)** unpolarized.

direction. When a beam of light strikes a polarizer, it is readily absorbed if its electric field is parallel to the molecules; light whose electric field is perpendicular to the molecules passes through the material with little absorption. As a result, the light that passes through a polarizer is preferentially polarized along a specific direction. Common examples of a polarizer are the well-known Polaroid sheets used to make Polaroid sunglasses.

A simple mechanical analog of a polarizer is shown in **Figure 25–13**. Here we see a wave that displaces a string in the vertical direction as it propagates toward a slit cut into a block of wood. If the slit is oriented vertically, as in Figure 25–13 (a), the wave passes through unhindered. Conversely, when the slit is oriented horizontally it stops the wave, as indicated in Figure 25–13 (b). A polarizer performs a similar function on a beam of light.

We now consider what happens when a beam of light polarized in one direction encounters a polarizer oriented in a different direction. The situation is illustrated in **Figure 25–14**, where we see light with a vertical polarization and intensity I_0 passing through a polarizer with its preferred direction—its **transmission axis**—at an angle θ to the vertical. As shown in the figure, the component of $\vec{\mathbf{E}}$ along the transmission axis is $E\cos\theta$. Recalling that the intensity of light is proportional to the electric field squared, we see that the intensity, I, of the transmitted beam is reduced by the factor $\cos^2\theta$. Therefore,

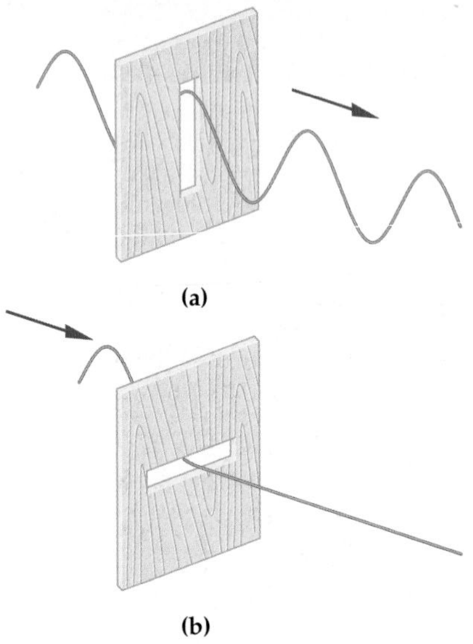

▲ **FIGURE 25–13 A mechanical analog of a polarizer**

(a) The polarization of the wave is in the same direction as the polarizer; hence, the wave passes through unaffected. **(b)** The polarization of the wave is at right angles to the direction of the polarizer. In this case the wave is absorbed.

Law of Malus

$$I = I_0\cos^2\theta \qquad \textbf{25–13}$$

This result is known as the **law of Malus,** after the French engineer Etienne-Louis Malus (1775–1812). Notice that the intensity is unchanged if $\theta = 0$, and is zero if $\theta = 90°$.

EXERCISE 25–5

Vertically polarized light with an intensity of 515 W/m^2 passes through a polarizer oriented at an angle θ to the vertical. Find the transmitted intensity of the light for **(a)** $\theta = 10.0°$, **(b)** $\theta = 45.0°$, and **(c)** $\theta = 90.0°$.

Solution

Applying $I = I_0\cos^2\theta$ we obtain

a. $I = (515\ \text{W/m}^2)\cos^2 10.0° = 499\ \text{W/m}^2$
b. $I = (515\ \text{W/m}^2)\cos^2 45.0° = 258\ \text{W/m}^2$
c. $I = (515\ \text{W/m}^2)\cos^2 90.0° = 0$

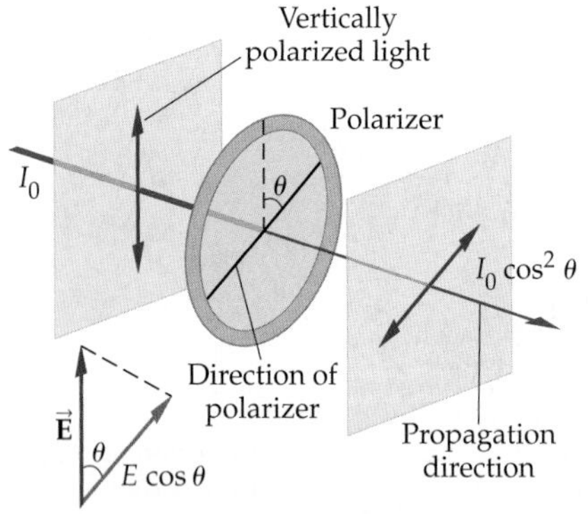

▲ **FIGURE 25–14 Transmission of polarized light through a polarizer**

A polarized beam of light, with intensity I_0, encounters a polarizer oriented at an angle θ relative to the polarization direction. The intensity of light transmitted through the polarizer is $I = I_0\cos^2\theta$. After passing through the polarizer, the light is polarized in the same direction as the polarizer.

Just as important as the change in intensity is what happens to the *polarization* of the transmitted light:

> The transmitted beam of light is no longer polarized in its original direction; it is now polarized in the direction of the polarizer.

This effect also is illustrated in Figure 25–14. Thus, a polarizer changes both the intensity *and* the polarization of a beam of light.

Figure 25–15 shows an unpolarized beam of light with an intensity I_0 encountering a polarizer. In this case, there is no single angle θ; instead, to obtain the transmitted intensity, we must average $\cos^2\theta$ over all angles. This has already been done in Section 24–1, where we considered rms values in ac circuits. As was shown there, the average of $\cos^2\theta$ is one-half; thus, the transmitted intensity for an unpolarized beam with an intensity of I_0 is

Transmitted Intensity for an Unpolarized Beam

$$I = \tfrac{1}{2}I_0 \qquad 25\text{–}14$$

A common type of polarization experiment is shown in **Figure 25–16**. An unpolarized beam is first passed through a polarizer to give the light a specified polarization. The light next passes through a second polarizer, referred to as the **analyzer**, whose transmission axis is at an angle θ relative to the polarizer. The orientation of the analyzer can be adjusted to give a beam of light of variable intensity and polarization. We consider a situation of this type in the next Example.

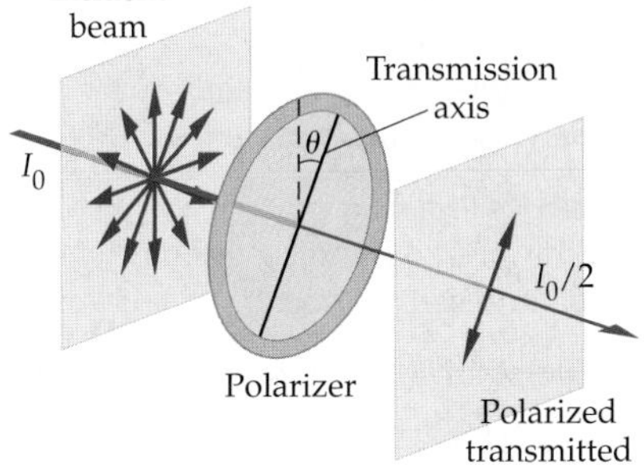

▲ **FIGURE 25–15 Transmission of unpolarized light through a polarizer**
When an unpolarized beam of intensity I_0 passes through a polarizer, the transmitted beam has an intensity of $\frac{1}{2}I_0$ and is polarized in the direction of the polarizer.

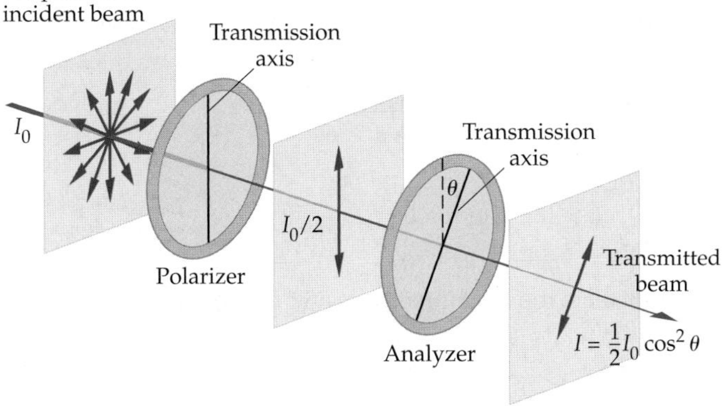

◀ **FIGURE 25–16 A polarizer and an analyzer**
An unpolarized beam of intensity I_0 is polarized in the vertical direction by a polarizer with a vertical transmission axis. Next, it passes through another polarizer, the analyzer, whose transmission axis is at an angle θ relative to the transmission axis of the polarizer. The final intensity of the beam is $I = \frac{1}{2}I_0\cos^2\theta$.

EXAMPLE 25–5 Analyze This

In the polarization experiment shown in our sketch, the final intensity of the beam is $0.200\,I_0$. What is the angle θ between the transmission axes of the analyzer and polarizer?

Picture the Problem

The experimental setup is shown in the sketch. As indicated, the intensity of the unpolarized incident beam, I_0, is reduced to $\frac{1}{2}I_0$ after passing through the first polarizer. The second polarizer reduces the intensity further, to $0.200\,I_0$.

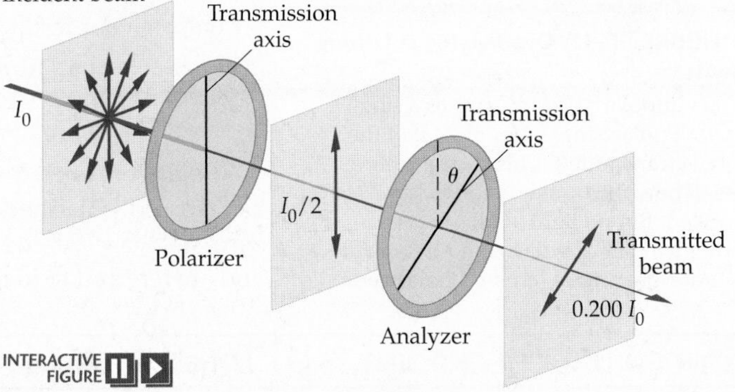

Strategy

It is clear from the sketch that the analyzer reduces the intensity of the light by a factor of $1/2.50$; that is, the intensity is reduced from $I_0/2$ to $(1/2.50)(I_0/2) = I_0/5.00 = 0.200\,I_0$. Thus we must find the angle θ that gives this reduction. Recalling that an analyzer reduces intensity according to the relation $I = I_0\cos^2\theta$, we set $\cos^2\theta = 1/2.50$ and solve for θ.

Solution

1. Set $\cos^2\theta$ equal to $1/2.50$ and solve for $\cos\theta$:

$$\cos^2\theta = \frac{1}{2.50}$$

$$\cos\theta = \frac{1}{\sqrt{2.50}}$$

2. Solve for the angle θ:

$$\theta = \cos^{-1}\left(\frac{1}{\sqrt{2.50}}\right) = 50.8°$$

continued on next page

continued from previous page

Insight

Since the analyzer absorbs part of the light as the beam passes through, it also absorbs energy. Therefore, in principle, the analyzer would experience a slight heating. As always, energy must be conserved.

Practice Problem

If the angle θ is increased slightly, does the final intensity increase, decrease, or stay the same? Check your answer by finding the final intensity for 60.0°. [**Answer:** The final intensity decreases. For $\theta = 60.0°$, the final intensity is $I = 0.125\,I_0$.]

Some related homework problems: Problem 66, Problem 68, Problem 72

PROBLEM-SOLVING NOTE

Transmission through Polarizing Filters

The intensity of light transmitted through a pair of polarizing filters depends only on the relative angle θ between the filters; it is independent of whether the filters are rotated clockwise or counterclockwise relative to each other.

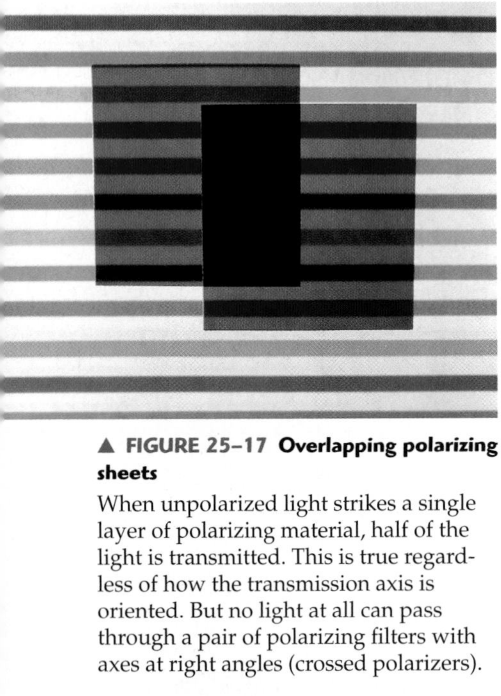

▲ **FIGURE 25–17 Overlapping polarizing sheets**

When unpolarized light strikes a single layer of polarizing material, half of the light is transmitted. This is true regardless of how the transmission axis is oriented. But no light at all can pass through a pair of polarizing filters with axes at right angles (crossed polarizers).

ACTIVE EXAMPLE 25–2 Find the Transmitted Intensity

Calculate the transmitted intensity for the following two cases: **(a)** A vertically polarized beam of intensity I_0 passes through a polarizer with its transmission axis at 60° to the vertical. **(b)** A vertically polarized beam of intensity I_0 passes through two polarizers, the first with its transmission axis at 30° to the vertical, and the second with its transmission axis rotated an additional 30° to the vertical. (*Note:* In both cases, the final beam is polarized at 60° to the vertical.)

Solution *(Test your understanding by performing the calculations indicated in each step.)*

Part (a)

1. Calculate the final intensity using $I = I_0 \cos^2\theta$, with $\theta = 60°$: $I = \frac{1}{4}I_0$

Part (b)

2. Calculate the intermediate intensity using $I = I_0 \cos^2\theta$, with $\theta = 30°$: $I = \frac{3}{4}I_0$
3. Calculate the final intensity using $I = I_0 \cos^2\theta$, with $\theta = 30°$ again: $I = \frac{9}{16}I_0$

Insight

Even though the direction of polarization is rotated by a total of 60° in each case, the final intensity is more than twice as great when two polarizers are used instead of just one. In general, the more polarizers that are used—and, hence, the more smoothly the direction of polarization changes—the greater the final intensity.

Your Turn

Calculate the transmitted intensity when three polarizers are used, the first at 20.0° to the vertical, the second at 40.0° to the vertical, and the third at 60.0° to the vertical. Compare your result with $\frac{9}{16}I_0 = 0.563\,I_0$ found with two polarizers.

(Answers to **Your Turn** *problems are given in the back of the book.)*

Polarizers with transmission axes at right angles to one another are referred to as "crossed polarizers." The transmission through a pair of crossed polarizers is zero, since $\theta = 90°$ in Malus's law. Crossed polarizers are illustrated in **Figure 25–17** and are referred to in the following Conceptual Checkpoint.

CONCEPTUAL CHECKPOINT 25–3 Light Transmission

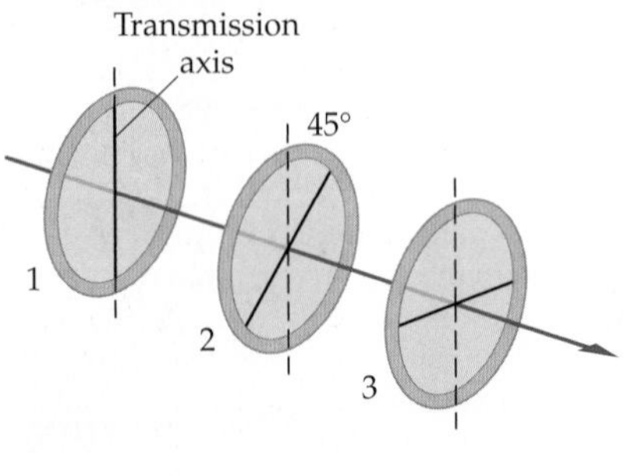

Consider a set of three polarizers. Polarizer 1 has a vertical transmission axis, and polarizer 3 has a horizontal transmission axis. Taken together, polarizers 1 and 3 are a pair of crossed polarizers. Polarizer 2, with a transmission angle at 45° to the vertical, is placed between polarizers 1 and 3, as shown in the sketch. A beam of unpolarized light shines on polarizer 1 from the left. Is the transmission of light through the three polarizers (a) zero or (b) nonzero?

Reasoning and Discussion

Since polarizers 1 and 3 are still crossed, it might seem that no light can be transmitted. When one recalls, however, that a polarizer causes a beam to have a polarization in the direction of its transmission axis, it becomes clear that transmission is indeed possible.

To be specific, some of the light that passes through polarizer 1 will also pass through polarizer 2, since the angle between their transmission axes is less than 90°. After passing through polarizer 2, the light is polarized at 45° to the vertical. As a result, some of this light can pass through polarizer 3 because, again, the angle between the polarization direction and the transmission axis is less than 90°.

In fact, the intensity of the incident light is reduced by a factor of 2 when it passes through polarizer 1, by a factor of 2 when it passes through polarizer 2 (since $\cos^2 45° = \frac{1}{2}$), and again by a factor of 2 when it passes through polarizer 3. The final intensity, then, is one-eighth the incident intensity.

Answer:
(b) The transmission is nonzero.

REAL-WORLD PHYSICS
Photoelastic stress analysis

There are many practical uses for crossed polarizers. For example, engineers often construct a plastic replica of a building, bridge, or similar structure to study the relative amounts of stress in its various parts in a technique known as *photoelastic stress analysis*. Dentists use the same technique to study stresses in teeth, and doctors use stress analysis when they design prosthetic joints. In this application, the plastic replica plays the role of polarizer 2 in Conceptual Checkpoint 25–3. In particular, in those regions of the structure where the stress is high, the plastic acts to rotate the plane of polarization and—like polarizer 2 in the Conceptual Checkpoint—allows light to pass through the system. By examining such models with crossed polarizers, engineers can gain valuable insight into the safety of the structures they plan to build, dentists can determine where a tooth is likely to break, and doctors can see where an artificial hip joint needs to be strengthened.

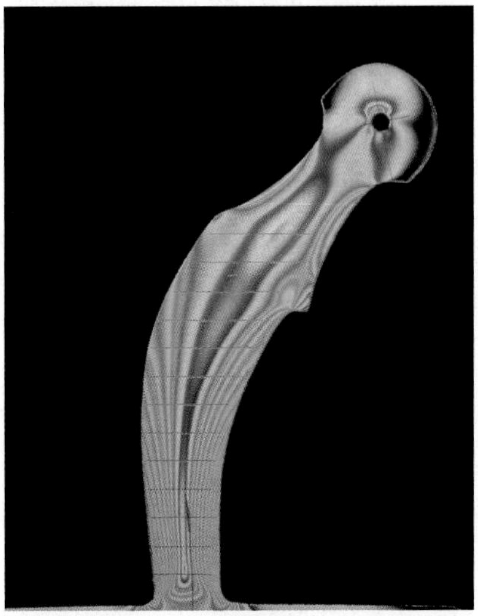

▲ In photoelastic stress analysis, a plastic model of an object being studied is placed between crossed polarizers. In this case, the object is a prosthetic hip joint. If the polarization of the light is unchanged by the plastic, it will not pass through the second polarizer. In areas where the plastic is stressed, however, it rotates the plane of polarization, allowing some of the light to pass through.

Another use of crossed polarizers is in the operation of a liquid-crystal display (LCD). There are basically three essential elements to each active area of an LCD—two crossed polarizers and a thin cell that holds a fluid composed of long, thin molecules known as a **liquid crystal.** The liquid crystal is selected for its ability to rotate the direction of polarization, and the thickness of the liquid-crystal cell is adjusted to give a rotation of 90°. Thus, in its "off" state, the liquid crystal rotates the direction of polarization and light passes through the crossed polarizers, as illustrated in **Figure 25–18 (a)**. In this state, the LCD is transparent—it

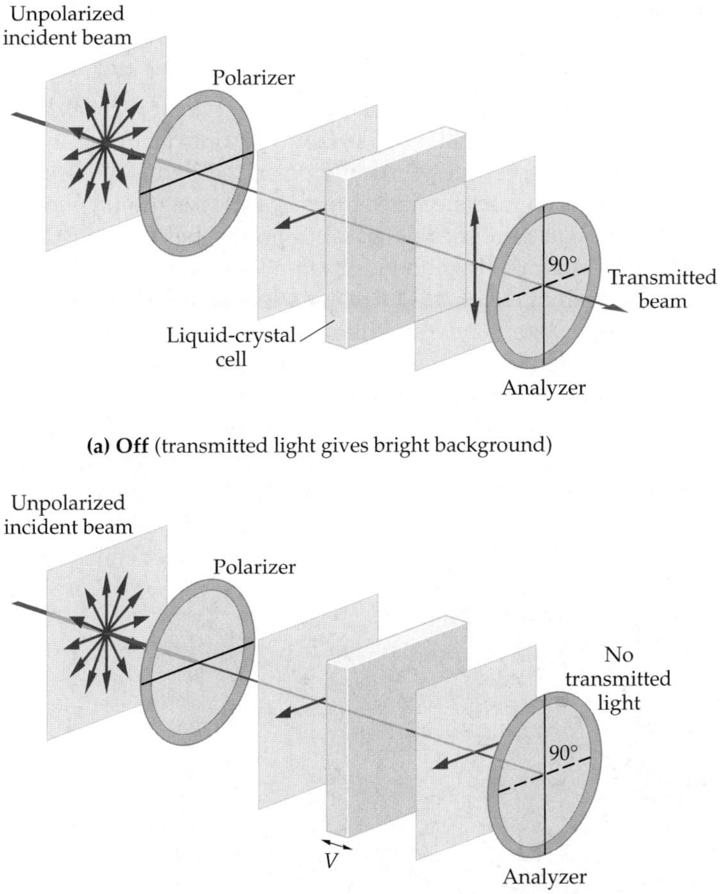

◀ **FIGURE 25–18 Basic operation of a liquid-crystal display (LCD)**
(a) In the "off" state, the liquid crystal in the cell rotates the polarization of the light by 90°, allowing light to pass through the display. This produces the light background of the display. **(b)** In the "on" state, a voltage V is applied to the liquid-crystal cell, which means that the polarization of the light is no longer rotated. As a result, no light is transmitted and this element of the display appears black. This is how the dark characters are created on the display.

REAL-WORLD PHYSICS
Liquid crystal displays (LCDs)

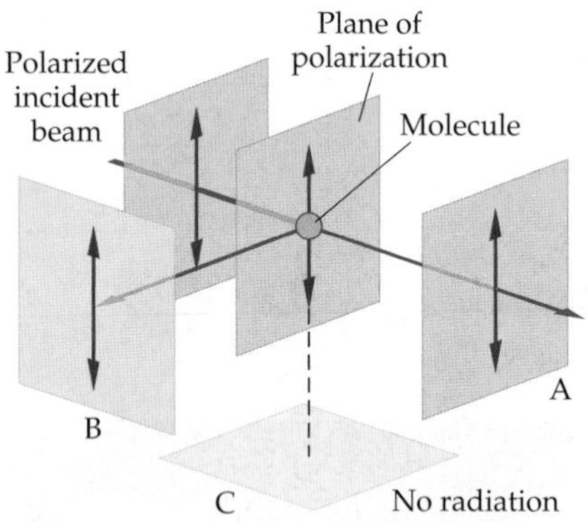

▲ **FIGURE 25–19 Vertically polarized light scattering from a molecule**

The molecule radiates scattered light, much like a small antenna. At points A and B strong scattered rays are observed; at point C no radiation is seen.

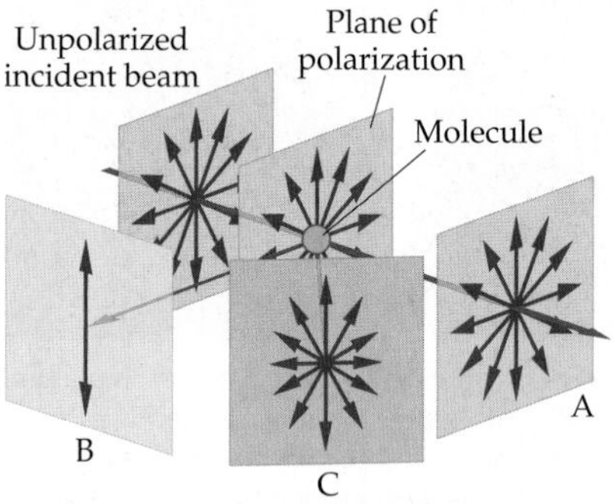

▲ **FIGURE 25–20 Unpolarized light scattering from a molecule**

In the forward direction, A, the scattered light is unpolarized. At right angles to the initial beam of light, B, the scattered light is polarized. Along other directions the light is only partially polarized.

REAL-WORLD PHYSICS: BIO
Navigating using polarized light from the sky

allows ambient light to enter the display, reflect off the back, and exit the display, giving it the characteristic light background. When a voltage is applied to the liquid crystal, it no longer rotates the direction of polarization, and light is no longer transmitted through that area of the display. Thus, in the "on" state, shown in **Figure 25–18 (b)**, an area of the LCD appears black, which is how the black characters are formed on the light background. Since very little energy is required to give the voltage necessary to turn a liquid crystal cell "on," the LCD is very energy efficient. In addition, the LCD uses light already present in the environment; it does not need to produce its own light, as do some displays.

Finally, there are many organic compounds that are capable of rotating the polarization direction of a beam of light. Such compounds, which include sugar, turpentine, and tartaric acid, are said to be **optically active.** When a solution containing optically active molecules is placed between crossed polarizers, the amount of light that passes through the system gives a direct measure of the concentration of the active molecules in the solution.

Polarization by Scattering and Reflection

When unpolarized light is scattered from atoms or molecules, as in the atmosphere, or reflected by a solid or liquid surface, the light can acquire some degree of polarization. The basic reason for this is illustrated in **Figure 25–19**. In this case, we consider a vertically polarized beam of light that encounters a molecule. The light causes electrons in the molecule to oscillate in the vertical direction, that is, in the direction of the light's electric field. As we know, an accelerating charge radiates—thus, the molecule radiates as if it were a small vertical antenna, and this radiation is what we observe as scattered light. We also know, however, that an antenna gives off maximum radiation in the direction perpendicular to its length, and no radiation at all along its axis. Therefore, an observer at point A or B in Figure 25–19 sees maximum scattered radiation, whereas an observer at point C sees no radiation at all.

If we apply this same idea to an initially unpolarized beam of light scattering from a molecule, we find the situation pictured in **Figure 25–20**. In this case, electrons in the molecule oscillate in all directions within the plane of polarization; hence, an observer in the forward direction, at point A, sees scattered light of all polarizations—that is, unpolarized light. An observer at point B, however, sees no radiation from electrons oscillating horizontally, only from those oscillating vertically. Hence this observer sees vertically polarized light. An observer at an intermediate angle sees an intermediate amount of polarization.

This mechanism produces polarization in the light coming from the sky. In particular, maximum polarization is observed in a direction at right angles to the Sun, as can be seen in **Figure 25–21**. Thus, to a creature that is sensitive to the polarization of light, like a bee or certain species of birds, the light from the sky varies with the direction relative to the Sun. Experiments show that such creatures can use polarized light as an aid in navigation.

Another point of interest follows from the general observation that the amount of light scattered from a molecule is greatest when the wavelength of the light is comparable to the size of the molecule. The molecules in the atmosphere are generally much smaller than the wavelength of visible light, but blue light, with its relatively short wavelength, scatters more effectively than red light, with

▶ **Figure 25–21 The effects of scattering on sunlight**

The scattering of sunlight by the atmosphere produces polarized light for an observer looking at right angles to the direction of the Sun. This observer also sees more blue light than red. An observer looking toward the Sun sees unpolarized light that contains more red light than blue.

its longer wavelength. Similarly, microscopic particles of dust in the upper atmosphere scatter the short-wavelength blue light most effectively. This basic mechanism is the answer to the age-old question, Why is the sky blue? Similarly, a sunset appears red because you are viewing the Sun through the atmosphere, and most of the Sun's blue light has been scattered off in other directions. In fact, the blue light that is missing from your sunset—giving it a red color—is the blue light of someone else's blue sky, as indicated in Figure 25–21.

REAL-WORLD PHYSICS
Why the sky is blue

Polarization also occurs when light reflects from a smooth surface, like the top of a table or the surface of a calm lake. A typical reflection situation, with unpolarized light from the Sun reflecting from the surface of a lake, is shown in **Figure 25–22**. When the light encounters molecules in the water, their electrons oscillate in the plane of polarization. For an observer at point A, however, only oscillations at right angles to the line of sight give rise to radiation. As a result, the reflected light from the lake is polarized horizontally. Polaroid sunglasses take advantage of this effect by using sheets of Polaroid material

REAL WORLD PHYSICS
How Polaroid sunglasses cut reflected glare

◀ As it is scattered by air molecules and dust particles in the atmosphere, sunlight becomes polarized. The shorter wavelengths at the blue end of the spectrum are scattered most effectively, creating our familiar blue skies. Sunsets are red (top) because much of the blue light has been scattered as it passes through the atmosphere on the way to our eyes. The polarization of light from the sky can be demonstrated with a pair of photographs like those at left—one taken with a polarizing filter (left bottom), the other without (left center). A smooth surface such as a lake (right) can also act as a polarizing filter. That is why the sky is darker and the clouds more easily seen in the reflected image than in the direct view.

▶ **FIGURE 25–22 Polarization by reflection**

Because the observer sees no radiation from electrons moving along the line of sight, the radiation that is observed is polarized horizontally. Polaroid sunglasses with a vertical transmission axis reduce this kind of reflected glare.

with a vertical transmission axis. With this orientation, the horizontally polarized reflected light—the glare—is not transmitted.

This presents a potential problem for the makers of digital watches and electronic calculators with LCD displays. Suppose the person using such a device is wearing Polaroid sunglasses—not an unusual circumstance. As we saw in Figure 25–18, the light emerging from an LCD display is linearly polarized. If the polarization direction is vertical, the light will pass through the sunglasses and the display can be read as usual. On the other hand, if the polarization direction of the display is horizontal, it will appear completely black through a pair of Polaroid sunglasses—no light will pass through at all. The same effect can be seen by observing an LCD display through a pair of Polaroid sunglasses and slowly rotating the glasses or the display through 90°. Try this experiment sometime with a calculator or a digital watch, and you will see the result shown in the accompanying photographs. Clearly, it would not be wise to make an LCD display with a horizontal polarization direction—a person wearing Polaroid sunglasses would think the display was broken.

▶ To eliminate the glare from reflected light, which is horizontally polarized, the lenses of Polaroid sunglasses have a vertical transmission axis. LCD displays are usually constructed so that the polarized light that comes from them is also vertically polarized. Thus a person wearing Polaroid glasses can view an LCD display (left). If the pair of glasses or the LCD device is rotated 90°, the display becomes invisible (right).

Chapter Summary

25–1 The Production of Electromagnetic Waves

Electromagnetic waves are traveling waves of oscillating electric and magnetic fields.

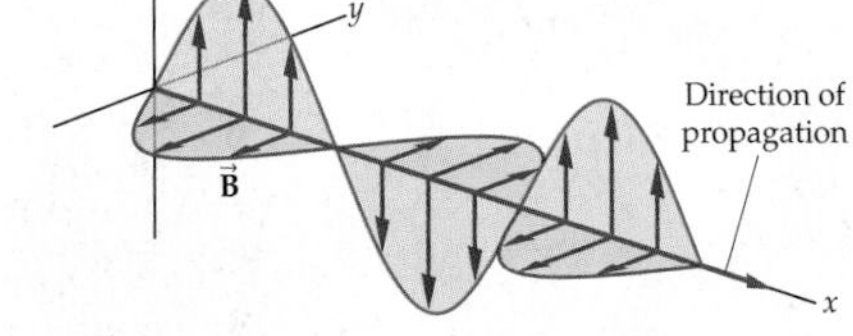

$\vec{E}$ and $\vec{B}$

The electric and magnetic fields in an electromagnetic wave are perpendicular to each other and to the direction of propagation. They are also in phase.

Direction of Propagation

To find the direction of propagation of an electromagnetic wave, point the fingers of your right hand in the direction of $\vec{E}$, then curl them toward $\vec{B}$. Your thumb will be pointing in the direction of propagation.

Generation of Electromagnetic Waves
Any accelerated charge will radiate energy in the form of electromagnetic waves.

25–2 The Propagation of Electromagnetic Waves

Electromagnetic waves can travel through a vacuum, and all electromagnetic waves in a vacuum have precisely the same speed, $c = 3.00 \times 10^8$ m/s. This is referred to as the *speed of light in a vacuum*.

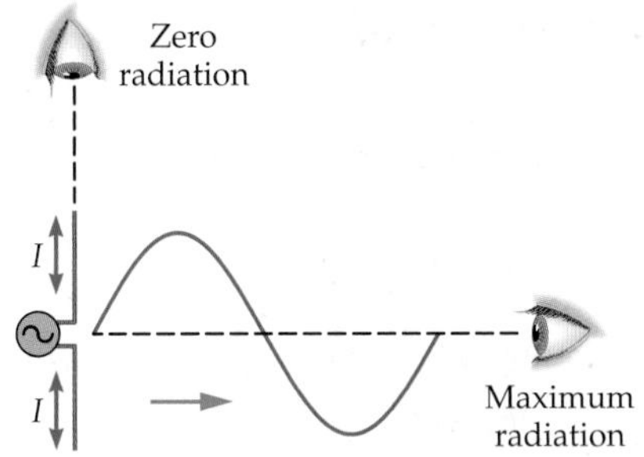

Doppler Effect
Electromagnetic radiation experiences a Doppler effect that is analogous to that observed in sound waves. For relative speeds, u, that are small compared with the speed of light, c, the Doppler effect shifts the frequency of an electromagnetic wave as follows:

$$f' = f\left(1 \pm \frac{u}{c}\right) \qquad \text{25–3}$$

where the plus sign is used when source and receiver are approaching, the minus sign when they are receding.

25–3 The Electromagnetic Spectrum

Electromagnetic waves can have any frequency from zero to infinity. The entire range of waves with different frequencies is referred to as the electromagnetic spectrum. Several "bands" of frequency are given special names.

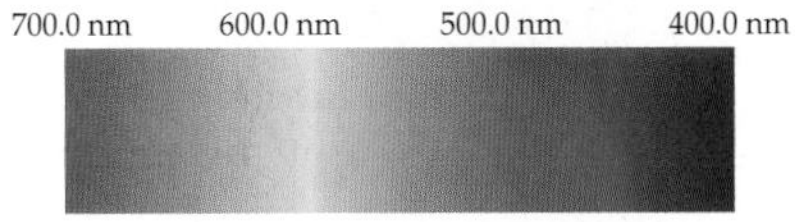

Frequency Bands in the Electromagnetic Spectrum
Radio/TV, 10^6 Hz to 10^9 Hz; microwaves, 10^9 Hz to 10^{12} Hz; infrared, 10^{12} Hz to 10^{14} Hz; visible, 4.29×10^{14} Hz to 7.50×10^{14} Hz; ultraviolet, 10^{15} Hz to 10^{17} Hz; X-rays, 10^{17} Hz to 10^{20} Hz; γ rays, above 10^{20} Hz.

Frequency–Wavelength Relationship
The frequency, f, and wavelength, λ, of all electromagnetic waves in a vacuum are related as follows:

$$c = f\lambda \qquad \text{25–4}$$

25–4 Energy and Momentum in Electromagnetic Waves

Electromagnetic waves carry both energy and momentum, shared equally between the electric and magnetic fields.

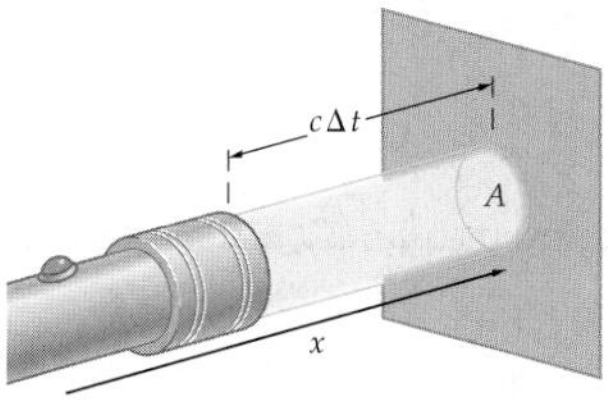

Energy Density
The energy density, u, of an electromagnetic wave can be written in several equivalent forms:

$$u = \tfrac{1}{2}\varepsilon_0 E^2 + \frac{1}{2\mu_0}B^2 = \varepsilon_0 E^2 = \frac{1}{\mu_0}B^2 \qquad \text{25–6}$$

Ratio of Electric and Magnetic Fields
In an electromagnetic field, the magnitudes E and B are related as follows:

$$E = cB \qquad \text{25–9}$$

Intensity
The intensity, I, of an electromagnetic field is the power per area. It can be expressed in the following forms:

$$I = uc = \tfrac{1}{2}c\varepsilon_0 E^2 + \frac{1}{2\mu_0}cB^2 = c\varepsilon_0 E^2 = \frac{c}{\mu_0}B^2 \qquad \text{25–10}$$

Momentum
An electromagnetic wave that delivers an energy U to an object transfers the momentum

$$p = \frac{U}{c} \qquad \text{25–11}$$

Radiation Pressure

If an electromagnetic wave shines on an object with an average intensity I_{av}, the average pressure exerted on the object by the radiation is

$$\text{pressure}_{av} = \frac{I_{av}}{c} \qquad 25\text{–}12$$

25–5 Polarization

The polarization of a beam of light is the direction along which its electric field points. An unpolarized beam has components with polarization in random directions.

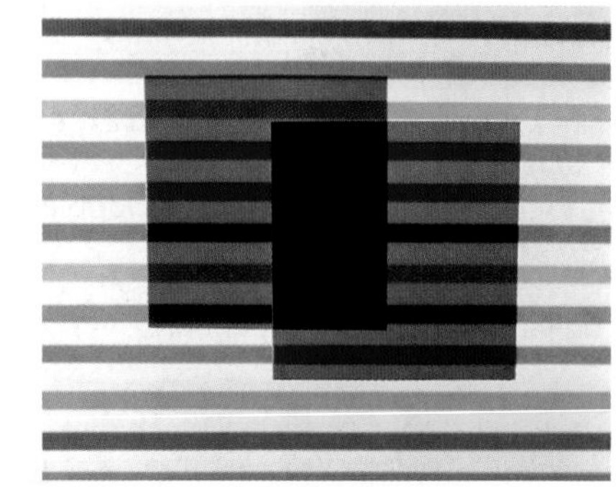

Polarizer

A polarizer transmits only light whose electric field has a component in the direction of the polarizer's transmission axis.

Law of Malus

If light with intensity I_0 encounters a polarizer with a transmission axis at a direction θ relative to its polarization, the transmitted intensity, I, is

$$I = I_0 \cos^2 \theta \qquad 25\text{–}13$$

Unpolarized Light Passing Through a Polarizer

When an unpolarized beam of light of intensity I_0 passes through a polarizer, the transmitted intensity, I, is reduced by half:

$$I = \tfrac{1}{2} I_0 \qquad 25\text{–}14$$

Polarization by Scattering

Light scattered by the atmosphere is polarized when viewed at right angles to the Sun.

Polarization by Reflection

When light reflects from a horizontal surface, like a tabletop or the surface of a lake, it is partially polarized in the horizontal direction.

Problem-Solving Summary

Type of Problem	Relevant Physical Concepts	Related Examples
Find the difference in frequency produced by the Doppler effect.	If an object moves with speed u and emits electromagnetic waves of speed c and frequency f, the Doppler-shifted frequency is $f' = f(1 \pm u/c)$. In this expression, the plus sign corresponds to a source approaching the observer; the minus sign corresponds to the source receding from the observer.	Example 25–2
Relate the frequency, f, wavelength, λ, and speed of an electromagnetic wave.	Electromagnetic waves in a vacuum always travel at the speed of light, c, regardless of their frequency. If either the frequency or wavelength of a wave is known, the other quantity can be obtained from the general relation $c = \lambda f$.	Example 25–3
Relate the intensity of an electromagnetic wave to its electric and magnetic fields.	An electromagnetic wave always consists of both electric and magnetic fields. The intensity of a wave, I, (energy per area per time) can be related to either field as follows: $I = c\varepsilon_0 E^2 = cB^2/\mu_0$.	Example 25–4 Active Example 25–1
Find the average radiation pressure exerted by an electromagnetic wave.	When electromagnetic waves fall on an object, they exert a pressure known as the radiation pressure. The average radiation pressure, P_{av}, is directly proportional to the average intensity of the wave; that is, $P_{av} = I_{av}/c$.	Exercise 25–4
Find the transmitted intensity of unpolarized light passing through a polarizing filter.	When unpolarized light of intensity I_0 passes through a polarizing filter, its intensity is cut in half; that is, $I = \frac{1}{2} I_0$.	Example 25–5
Find the transmitted intensity of polarized light passing through a polarizing filter.	When polarized light of intensity I_0 passes through a polarizing filter, its transmitted intensity is $I = I_0 \cos^2 \theta$, where θ is the angle between the polarization direction of the light and the polarization direction of the filter.	Example 25–5 Active Example 25–2

Conceptual Questions

(Answers to odd-numbered Conceptual Questions can be found in the back of the book.)

1. Explain why the "invisible man" would be unable to see.
2. The magnitude of the Doppler effect tells how rapidly a weather system is moving. What determines whether the system is approaching or receding?
3. Explain why radiation pressure is more significant on a grain of dust in interplanetary space when the grain is very small.
4. While wearing your Polaroid sunglasses at the beach, you notice that they reduce the glare from the water better when you are sitting upright than when you are lying on your side. Explain.
5. You want to check the time while wearing your Polaroid sunglasses. If you hold your forearm horizontally, you can read the time easily. If you hold your forearm vertically, however, so that you are looking at your watch sideways, you notice that the display is black. Explain.
6. **BIO Polarization and the Ground Spider** The ground spider *Drassodes cupreus*, like many spiders, has several pairs of eyes. It has been discovered that one of these pairs of eyes acts as a set of polarization filters, with one eye's polarization direction oriented at 90° to the other eye's polarization direction. In addition, experiments show that the spider uses these eyes to aid in navigating to and from its burrow. Explain how such eyes might aid navigation.
7. The electromagnetic wave we pick up on our radios are typically polarized. In contrast, the indoor light we see everyday is typically unpolarized. Explain.
8. You are given a sheet of Polaroid material. Describe how to determine the direction of its transmission axis if none is indicated on the sheet.
9. Can sound waves be polarized? Explain.
10. At a garage sale you find a pair of "Polaroid" sunglasses priced to sell. You are not sure, however, if the glasses are truly Polaroid, or if they simply have tinted lenses. How can you tell which is the case? Explain.
11. **3-D Movies** Modern-day 3-D movies are produced by projecting two different images onto the screen, with polarization directions that are at 90° relative to one another. Viewers must wear headsets with polarizing filters to experience the 3-D effect. Explain how this works.

Conceptual Exercises

(Answers to odd-numbered Conceptual Exercises can be found in the back of the book.)

1. If the electric field in an electromagnetic wave is increasing in magnitude at a particular time, is the magnitude of the magnetic field at the same time increasing or decreasing? Explain.
2. Suppose the magnitude of the electric field in an electromagnetic wave is doubled. **(a)** By what factor does the magnitude of the magnetic field change? **(b)** By what factor does the maximum intensity of the wave change?
3. If the rms value of the electric field in an electromagnetic wave is doubled, **(a)** by what factor does the rms value of the magnetic field change? **(b)** By what factor does the average intensity of the wave change?
4. Three electromagnetic waves have electric and magnetic fields pointing in the directions shown in **Figure 25–23**. For each of the three cases, state whether the wave propagates in the $+x$, $-x$, $+y$, $-y$, $+z$, or $-z$ direction.

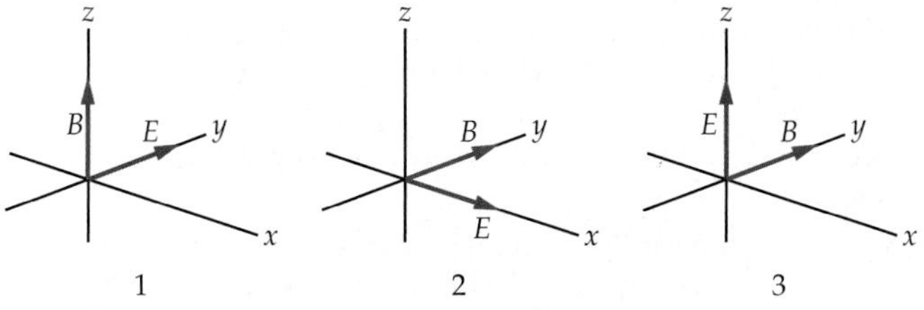

▲ **FIGURE 25–23** Conceptual Exercise 4

5. The radiation pressure exerted by beam of light 1 is half the radiation pressure of beam of light 2. If the rms electric field of beam 1 has the value E_0, what is the rms electric field in beam 2?
6. If "sailors" of the future use radiation pressure to propel their ships, should the surfaces of their sails be absorbing or reflecting? Explain.
7. Consider the two polarization experiments shown in **Figure 25–24**. If the incident light is unpolarized, is the transmitted intensity in case **(a)** greater than, less than, or the same as the transmitted intensity in case **(b)**? Explain.

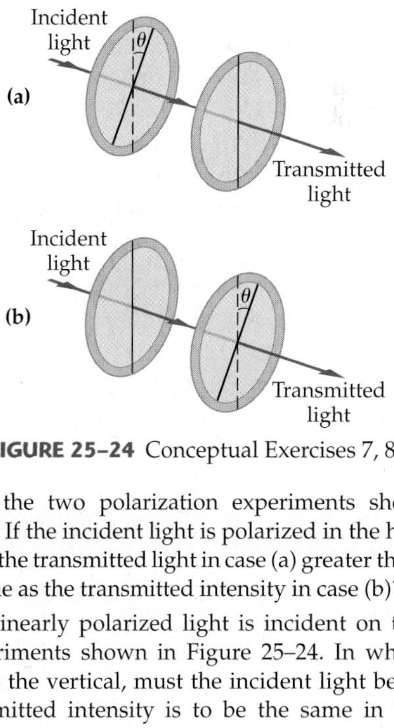

▲ **FIGURE 25–24** Conceptual Exercises 7, 8, and 9

8. Consider the two polarization experiments shown in Figure 25–24. If the incident light is polarized in the horizontal direction, is the transmitted light in case (a) greater than, less than, or the same as the transmitted intensity in case (b)? Explain.
9. Suppose linearly polarized light is incident on the polarization experiments shown in Figure 25–24. In what direction, relative to the vertical, must the incident light be polarized if the transmitted intensity is to be the same in both experiments? Explain.
10. **Figure 25–25** shows four polarization experiments in which unpolarized incident light passes through two polarizing filters with different orientations. Rank the four cases in order of increasing amount of transmitted light. Indicate ties where appropriate.

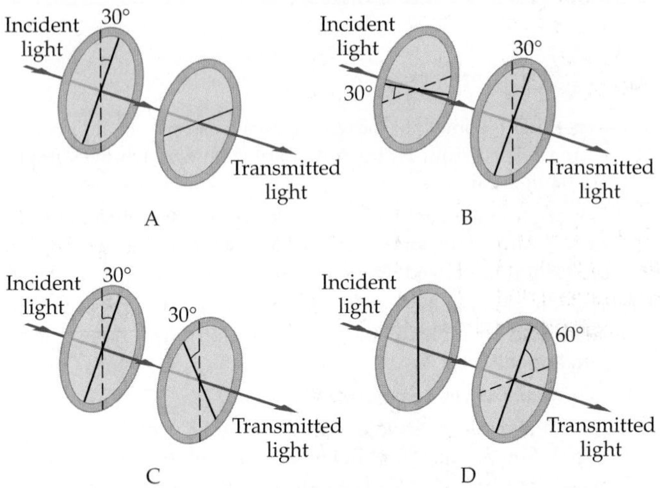

▲ **FIGURE 25–25** Conceptual Exercise 10

11. An incident beam of light with an intensity I_0 passes through a polarizing filter whose transmission axis is at an angle θ to the vertical. As the angle is changed from $\theta = 0$ to $\theta = 90°$, the intensity as a function of angle is given by one of the curves in **Figure 25–26**. Give the color of the curve corresponding to an incident beam that is **(a)** unpolarized, **(b)** vertically polarized, and **(c)** horizontally polarized.

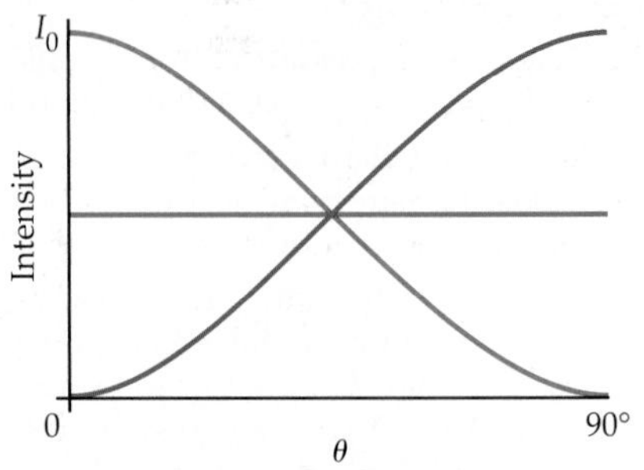

▲ **FIGURE 25–26** Conceptual Exercise 11

Problems

Note: **IP** *denotes an integrated conceptual/quantitative problem.* **BIO** *identifies problems of biological or medical interest. Red bullets (•, ••, •••) are used to indicate the level of difficulty of each problem.*

Section 25–1 The Production of Electromagnetic Waves

1. • The electric field of an electromagnetic wave points in the positive y direction. At the same time, the magnetic field of this wave points in the positive z direction. In what direction is the wave traveling?

2. • An electric charge on the x axis oscillates sinusoidally about the origin. A distant observer is located at a point on the $+y$ axis. **(a)** In what direction will the electric field oscillate at the observer's location? **(b)** In what direction will the magnetic field oscillate at the observer's location? **(c)** In what direction will the electromagnetic wave propagate at the observer's location?

3. • An electric charge on the z axis oscillates sinusoidally about the origin. A distant observer is located at a point on the $+y$ axis. **(a)** In what direction will the electric field oscillate at the observer's location? **(b)** In what direction will the magnetic field oscillate at the observer's location? **(c)** In what direction will the electromagnetic wave propagate at the observer's location?

4. •• Give the direction (N, S, E, W, up, or down) of the missing quantity for each of the four electromagnetic waves listed in Table 25–1.

TABLE 25–1 Problem 4

Direction of propagation	Direction of electric field	Direction of magnetic field
N	W	**(a)**
N	**(b)**	W
up	S	**(c)**
(d)	down	S

5. •• Give the direction ($\pm x$, $\pm y$, $\pm z$) of the missing quantity for each of the four electromagnetic waves listed in Table 25–2.

TABLE 25–2 Problem 5

Direction of propagation	Direction of electric field	Direction of magnetic field
$+x$	$+y$	**(a)**
$+x$	**(b)**	$+y$
$-y$	$+z$	**(c)**
(d)	$+z$	$+y$

6. •• **IP** At a particular instant of time, a light beam traveling in the positive z direction has an electric field given by $\vec{\mathbf{E}} = (6.22\text{ N/C})\hat{\mathbf{x}} + (2.87\text{ N/C})\hat{\mathbf{y}}$. The magnetic field in the beam has a magnitude of 2.28×10^{-8} T at the same time. **(a)** Does the magnetic field at this time have a z component that is positive, negative, or zero? Explain. **(b)** Write $\vec{\mathbf{B}}$ in terms of unit vectors.

7. •• **IP** A light beam traveling in the negative z direction has a magnetic field $\vec{\mathbf{B}} = (2.32 \times 10^{-9}\text{ T})\hat{\mathbf{x}} + (-4.06 \times 10^{-9}\text{ T})\hat{\mathbf{y}}$ at a given instant of time. The electric field in the beam has a magnitude of 1.40 N/C at the same time. **(a)** Does the electric field at this time have a z component that is positive, negative, or zero? Explain. **(b)** Write $\vec{\mathbf{E}}$ in terms of unit vectors.

Section 25–2 The Propagation of Electromagnetic Waves

8. • The light-year (ly) is a unit of distance commonly used in astronomy. It is defined as the distance traveled by light in a vacuum in one year. **(a)** Express 1 ly in km. **(b)** Express the speed of light, c, in units of ly per year. **(c)** Express the speed of light in feet per nanosecond.

9. • Alpha Centauri, the closest star to the sun, is 4.3 ly away. How far is this in meters?

10. • **Mars Rover** When the Mars rover was deployed on the surface of Mars in July 1997, radio signals took about 12 min to travel from Earth to the rover. How far was Mars from Earth at that time?

11. •• A distant star is traveling directly away from Earth with a speed of 36,500 km/s. By what factor are the wavelengths in this star's spectrum changed?

12. •• IP A distant star is traveling directly toward Earth with a speed of 36,500 km/s. **(a)** When the wavelengths in this star's spectrum are measured on Earth, are they greater or less than the wavelengths we would find if the star were at rest relative to us? Explain. **(b)** By what fraction are the wavelengths in this star's spectrum shifted?

13. •• IP The frequency of light reaching Earth from a particular galaxy is 15% lower than the frequency the light had when it was emitted. **(a)** Is this galaxy moving toward or away from Earth? Explain. **(b)** What is the speed of this galaxy relative to the Earth? Give your answer as a fraction of the speed of light.

14. •• **Measuring the Speed of Light** Galileo attempted to measure the speed of light by measuring the time elapsed between his opening a lantern and his seeing the light return from his assistant's lantern. The experiment is illustrated in **Figure 25–27**. What distance, d, must separate Galileo and his assistant in order for the human reaction time, $\Delta t = 0.2$ s, to introduce no more than a 15% error in the speed of light?

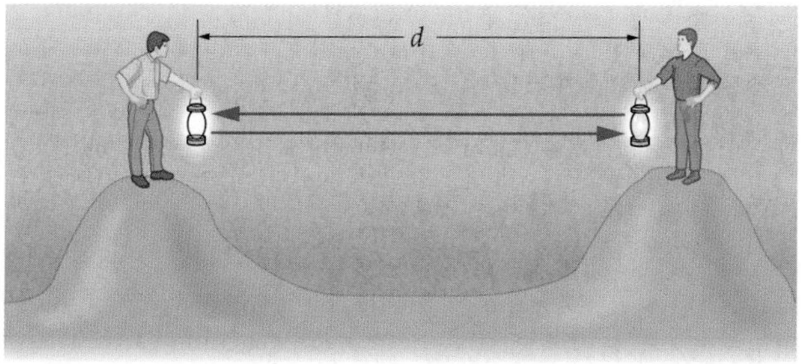

▲ **FIGURE 25–27** Problem 14

15. •• **Measuring the Speed of Light: Michelson** In 1926, Albert Michelson measured the speed of light with a technique similar to that used by Fizeau. Michelson used an eight-sided mirror rotating at 528 rev/s in place of the toothed wheel, as illustrated in **Figure 25–28**. The distance from the rotating mirror to a distant reflector was 35.5 km. If the light completed the 71.0-km round trip in the time it took the mirror to complete one-eighth of a revolution, what is the speed of light?

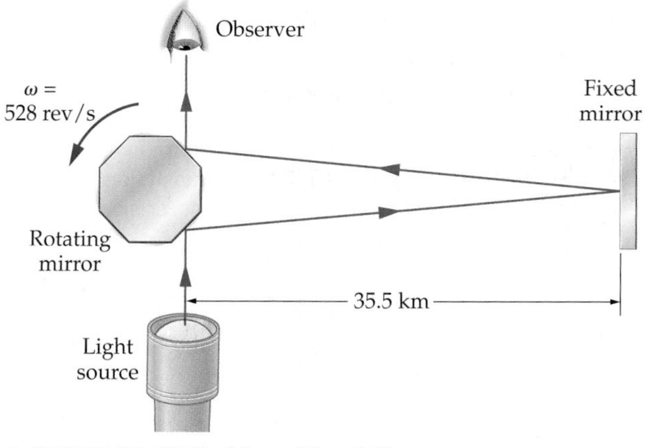

▲ **FIGURE 25–28** Problems 15 and 81

16. •• **Communicating with the *Voyager* Spacecraft** When the *Voyager I* and *Voyager II* spacecraft were exploring the outer planets, NASA flight controllers had to plan the crafts' moves well in advance. How many seconds elapse between the time a command is sent from Earth and the time the command is received by *Voyager* at Neptune? Assume the distance from Earth to Neptune is 4.5×10^{12} m.

17. •• A father and his daughter are interested in the same baseball game. The father sits next to his radio at home and listens to the game; his daughter attends the game and sits in the outfield bleachers. In the bottom of the ninth inning a home run is hit. If the father's radio is 132 km from the radio station, and the daughter is 115 m from home plate, who hears the home run first? (Assume that there is no time delay between the baseball being hit and its sound being broadcast by the radio station. In addition, let the speed of sound in the stadium be 343 m/s.)

18. •• IP **(a)** How fast would a motorist have to be traveling for a yellow ($\lambda = 590$ nm) traffic light to appear green ($\lambda = 550$ nm) because of the Doppler shift? **(b)** Should the motorist be traveling toward or away from the traffic light to see this effect? Explain.

19. •• Most of the galaxies in the universe are observed to be moving away from Earth. Suppose a particular galaxy emits orange light with a frequency of 5.000×10^{14} Hz. If the galaxy is receding from Earth with a speed of 3325 km/s, what is the frequency of the light when it reaches Earth?

20. •• Two starships, the *Enterprise* and the *Constitution*, are approaching each other head-on from a great distance. The separation between them is decreasing at a rate of 722.5 km/s. The *Enterprise* sends a laser signal toward the *Constitution*. If the *Constitution* observes a wavelength $\lambda = 670.3$ nm, what wavelength was emitted by the *Enterprise?*

21. •• Baseball scouts often use a radar gun to measure the speed of a pitch. One particular model of radar gun emits a microwave signal at a frequency of 10.525 GHz. What will be the increase in frequency if these waves are reflected from a 90.0 mi/h fastball headed straight toward the gun? (*Note*: 1 mi/h = 0.447 m/s)

22. •• A state highway patrol car radar unit uses a frequency of 8.00×10^9 Hz. What frequency difference will the unit detect from a car receding at a speed of 44.5 m/s from a stationary patrol car?

23. •• Consider a spiral galaxy that is moving directly away from Earth with a speed $V = 3.600 \times 10^5$ m/s at its center, as shown in **Figure 25–29**. The galaxy is also rotating about its center, so that points in its spiral arms are moving with a speed $v = 6.400 \times 10^5$ m/s relative to the center. If light with a frequency of 8.230×10^{14} Hz is emitted in both arms of the galaxy, what frequency is detected by astronomers observing the arm that is moving **(a)** toward and **(b)** away from Earth? (Measurements of this type are used to map out the speed of various regions in distant, rotating galaxies.)

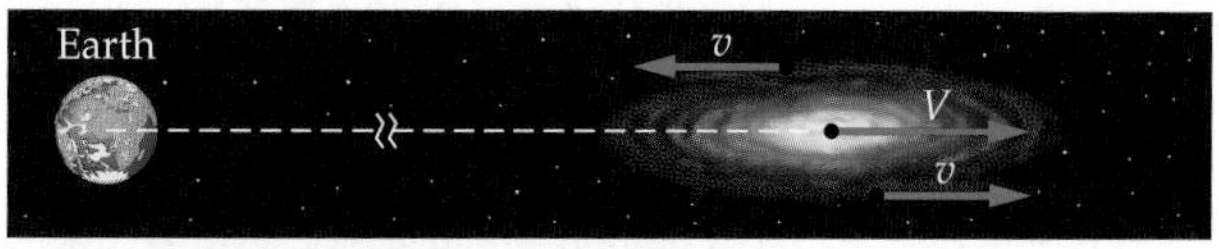

▲ **FIGURE 25–29** Problems 23 and 82

24. ••• IP A highway patrolman sends a 24.150-GHz radar beam toward a speeding car. The reflected wave is lower in frequency by 4.04 kHz. **(a)** Is the car moving toward or away from the radar gun? Explain. **(b)** What is the speed of the car? [*Hint*: For small values of x, the following approximation may be used: $(1 + x)^2 \approx 1 + 2x$.]

Section 25–3 The Electromagnetic Spectrum

25. • BIO **Dental X-rays** X-rays produced in the dentist's office typically have a wavelength of 0.30 nm. What is the frequency of these rays?

26. • Find the frequency of blue light with a wavelength of 460 nm.

27. • Yellow light has a wavelength $\lambda = 590$ nm. How many of these waves would span the 1.0-mm thickness of a dime?

28. • How many red wavelengths ($\lambda = 705$ nm) tall are you?

29. • A cell phone transmits at a frequency of 1.25×10^8 Hz. What is the wavelength of the electromagnetic wave used by this phone?

30. • BIO **Human Radiation** Under normal conditions, humans radiate electromagnetic waves with a wavelength of about 9.0 microns. **(a)** What is the frequency of these waves? **(b)** To what portion of the electromagnetic spectrum do these waves belong?

31. • BIO **UV Radiation** Ultraviolet light is typically divided into three categories. UV-A, with wavelengths between 400 nm and 320 nm, has been linked with malignant melanomas. UV-B radiation, which is the primary cause of sunburn and other skin cancers, has wavelengths between 320 nm and 280 nm. Finally, the region known as UV-C extends to wavelengths of 100 nm. **(a)** Find the range of frequencies for UV-B radiation. **(b)** In which of these three categories does radiation with a frequency of 7.9×10^{14} Hz belong?

32. • **Communicating with a Submarine** Normal radio-frequency waves cannot penetrate more than a few meters below the surface of the ocean. One method of communicating with submerged submarines uses very low frequency (VLF) radio waves. What is the wavelength (in air) of a 10.0-kHz VLF radio wave?

33. •• IP When an electromagnetic wave travels from one medium to another with a different speed of propagation, the frequency of the wave remains the same. Its wavelength, however, changes. **(a)** If the wave speed decreases, does the wavelength increase or decrease? Explain. **(b)** Consider a case where the wave speed decreases from c to $\frac{3}{4}c$. By what factor does the wavelength change?

34. •• IP **(a)** Which color of light has the higher frequency, red or violet? **(b)** Calculate the frequency of blue light with a wavelength of 470 nm, and red light with a wavelength of 680 nm.

35. •• ULF (ultra low frequency) electromagnetic waves, produced in the depths of outer space, have been observed with wavelengths in excess of 29 million kilometers. What is the period of such a wave?

36. •• A television is tuned to a station broadcasting at a frequency of 6.60×10^7 Hz. For best reception, the rabbit-ear antenna used by the TV should be adjusted to have a tip-to-tip length equal to half a wavelength of the broadcast signal. Find the optimum length of the antenna.

37. •• An AM radio station's antenna is constructed to be $\lambda/4$ tall, where λ is the wavelength of the radio waves. How tall should the antenna be for a station broadcasting at a frequency of 880 kHz?

38. •• As you drive by an AM radio station, you notice a sign saying that its antenna is 112 m high. If this height represents one quarter-wavelength of its signal, what is the frequency of the station?

39. •• Find the difference in wavelength ($\lambda_1 - \lambda_2$) for each of the following pairs of radio waves: **(a)** $f_1 = 50$ kHz and $f_2 = 52$ kHz. **(b)** $f_1 = 500$ kHz and $f_2 = 502$ kHz.

40. •• Find the difference in frequency ($f_1 - f_2$) for each of the following pairs of radio waves: **(a)** $\lambda_1 = 300.0$ m and $\lambda_2 = 300.5$ m. **(b)** $\lambda_1 = 30.0$ m and $\lambda_2 = 30.5$ m.

Section 25–4 Energy and Momentum in Electromagnetic Waves

41. • The maximum magnitude of the electric field in an electromagnetic wave is 0.0400 V/m. What is the maximum magnitude of the magnetic field in this wave?

42. • What is the rms value of the electric field in a sinusoidal electromagnetic wave that has a maximum electric field of 65 V/m?

43. •• The magnetic field in an electromagnetic wave has a peak value given by $B = 2.7\ \mu$T. For this wave, find **(a)** the peak electric field strength, **(b)** the peak intensity, and **(c)** the average intensity.

44. •• What is the maximum value of the electric field in an electromagnetic wave whose maximum intensity is 5.00 W/m^2?

45. •• What is the maximum value of the electric field in an electromagnetic wave whose average intensity is 5.00 W/m^2?

46. •• IP Electromagnetic wave 1 has a maximum electric field of $E_0 = 52$ V/m, and electromagnetic wave 2 has a maximum magnetic field of $B_0 = 1.5\ \mu$T. **(a)** Which wave has the greater intensity? **(b)** Calculate the intensity of each wave.

47. •• A 55-kW radio station broadcasts its signal uniformly in all directions. **(a)** What is the average intensity of its signal at a distance of 250 m from the antenna? **(b)** What is the average intensity of its signal at a distance of 2500 m from the antenna?

48. •• At what distance will a 75-W lightbulb have the same apparent brightness as a 120-W bulb viewed from a distance of 25 m? (Assume that both bulbs convert electrical power to light with the same efficiency, and radiate light uniformly in all directions.)

49. •• What is the ratio of the sunlight intensity reaching Pluto compared with the sunlight intensity reaching Earth? (On average, Pluto is 39 times as far from the Sun as is Earth.)

50. •• IP In the following, assume that lightbulbs radiate uniformly in all directions and that 5.0% of their power is converted to light. **(a)** Find the average intensity of light at a point 2.0 m from a 120-W red lightbulb ($\lambda = 710$ nm). **(b)** Is the average intensity 2.0 m from a 120-W blue lightbulb ($\lambda = 480$ nm) greater than, less than, or the same as the intensity found in part (a)? Explain. **(c)** Calculate the average intensity for part (b).

51. •• A 5.0-mW laser produces a narrow beam of light. How much energy is contained in a 1.0 m length of its beam?

52. •• What length of a 5.0-mW laser's beam will contain 8.5 mJ of energy?

53. •• **Sunlight Intensity** After filtering through the atmosphere, the Sun's radiation illuminates Earth's surface with an average intensity of 1.0 kW/m^2. Assuming this radiation strikes the 15-m × 45-m black, flat roof of a building at normal incidence, calculate the average force the radiation exerts on the roof.

54. •• IP **(a)** Find the electric and magnetic field amplitudes in an electromagnetic wave that has an average energy density of 1.0 J/m^3. **(b)** By what factor must the field amplitudes be increased if the average energy density is to be doubled to 2.0 J/m^3?

55. •• **Lasers for Fusion** Two of the most powerful lasers in the world are used in nuclear fusion experiments. The NOVA laser produces 40.0 kJ of energy in a pulse that lasts 2.50 ns, and the NIF laser (under construction) will produce a 10.0-ns pulse with 3.00 MJ of energy. **(a)** Which laser produces more energy in each pulse? **(b)** Which laser produces the greater average power during each pulse? **(c)** If the beam diameters are the same, which laser produces the greater average intensity?

56. •• **BIO** You are standing 2.5 m from a 150-W lightbulb. **(a)** If the pupil of your eye is a circle 5.0 mm in diameter, how much energy enters your eye per second? (Assume that 5.0% of the lightbulb's power is converted to light.) **(b)** Repeat part (a) for the case of a 1.0-mm diameter laser beam with a power of 0.50-mW.

57. •• **BIO Laser Safety** A 0.75-mW laser emits a narrow beam of light that enters the eye, as shown in **Figure 25–30**. **(a)** How much energy is absorbed by the eye in 0.2 s? **(b)** The eye focuses this beam to a tiny spot on the retina, perhaps 5.0 μm in diameter. What is the average intensity of light (in W/cm^2) at this spot? **(c)** Damage to the retina can occur if the average intensity of light exceeds $1.0 \times 10^{-2}\ W/cm^2$. By what factor has the intensity of this laser beam exceeded the safe value?

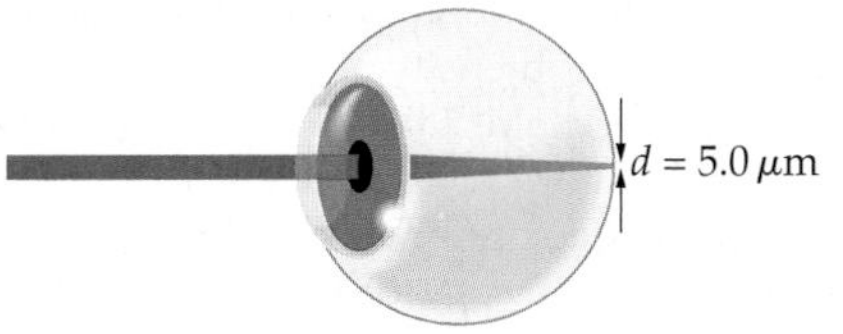

▲ **FIGURE 25–30** Problems 57 and 83

58. •• Find the rms electric and magnetic fields at a point 3.50 m from a lightbulb that radiates 75.0 W of light uniformly in all directions.

59. •• A 0.50-mW laser produces a beam of light with a diameter of 1.5 mm. **(a)** What is the average intensity of this beam? **(b)** At what distance does a 150-W lightbulb have the same average intensity as that found for the laser beam in part (a)? (Assume that 5.0% of the bulb's power is converted to light.)

60. •• A laser emits a cylindrical beam of light 2.4 mm in diameter. If the average power of the laser is 2.8 mW, what is the rms value of the electric field in the laser beam?

61. •• **(a)** If the laser in Problem 60 shines its light on a perfectly absorbing surface, how much energy does the surface receive in 12 s? **(b)** What is the radiation pressure exerted by the beam?

62. ••• **BIO Laser Surgery** Each pulse produced by an argon–fluoride excimer laser used in PRK and LASIK ophthalmic surgery lasts only 10.0 ns but delivers an energy of 2.50 mJ. **(a)** What is the power produced during each pulse? **(b)** If the beam has a diameter of 0.850 mm, what is the average intensity of the beam during each pulse? **(c)** If the laser emits 55 pulses per second, what is the average power it generates?

63. ••• A pulsed laser produces brief bursts of light. One such laser emits pulses that carry 0.350 J of energy but last only 225 fs. **(a)** What is the average power during one of these pulses? **(b)** Assuming the energy is emitted in a cylindrical beam of light 2.00 mm in diameter, calculate the average intensity of this laser beam. **(c)** What is the rms electric field in this wave?

Section 25–5 Polarization

64. • Vertically polarized light with an intensity of $0.55\ W/m^2$ passes through a polarizer whose transmission axis is at an angle of 35.0° with the vertical. What is the intensity of the transmitted light?

65. • A person riding in a boat observes that the sunlight reflected by the water is polarized parallel to the surface of the water. The person is wearing polarized sunglasses with the polarization axis vertical. If the wearer leans at an angle of 27.5° to the vertical, what fraction of the reflected light intensity will pass through the sunglasses?

66. •• Unpolarized light passes through two polarizers whose transmission axes are at an angle of 30.0° with respect to each other. What fraction of the incident intensity is transmitted through the polarizers?

67. •• In Problem 66, what should be the angle between the transmission axes of the polarizers if it is desired that one-tenth of the incident intensity be transmitted?

68. •• Unpolarized light with intensity I_0 falls on a polarizing filter whose transmission axis is vertical. The axis of a second polarizing filter makes an angle of θ with the vertical. Plot a graph that shows the intensity of light transmitted by the second filter (expressed as a fraction of I_0) as a function of θ. Your graph should cover the range $\theta = 0°$ to $\theta = 360°$.

69. •• **IP** A beam of vertically polarized light encounters two polarizing filters, as shown in **Figure 25–31**. **(a)** Rank the three cases, A, B, and C, in order of increasing transmitted intensity. Indicate ties where appropriate. **(b)** Calculate the transmitted intensity for each of the cases in Figure 25–31, assuming that the incident intensity is $37.0\ W/m^2$. Verify that your numerical results agree with the rankings in part (a).

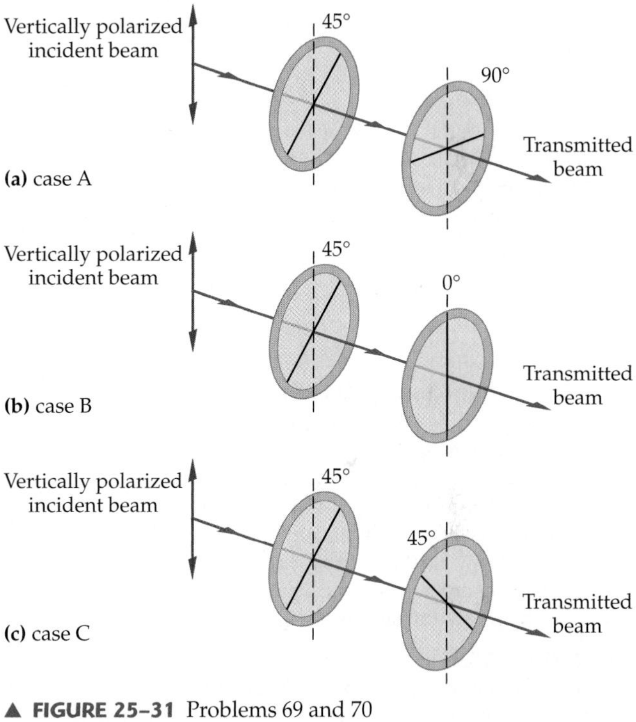

▲ **FIGURE 25–31** Problems 69 and 70

70. •• **IP** Repeat Problem 69, this time assuming that the polarizers to the left in Figure 25–31 are at an angle of 22.5° to the vertical rather than 45°. The incident intensity is again $37.0\ W/m^2$.

71. •• **IP BIO Optical Activity** Optically active molecules have the property of rotating the direction of polarization of linearly polarized light. Many biologically important molecules have this property, some causing a counterclockwise rotation (negative rotation angle), others causing a clockwise rotation (positive rotation angle). For example, a 5.00 gram per 100 mL solution of *l*-leucine causes a rotation of −0.550°; the same concentration of *d*-glutamic acid causes a rotation of 0.620°. **(a)** If placed between crossed polarizers, which of these solutions transmits the greater intensity? Explain. **(b)** Find the transmitted intensity for each of these solutions when placed between crossed polarizers. The incident beam is unpolarized and has an intensity of $12.5\ W/m^2$.

72. •• A helium–neon laser emits a beam of unpolarized light that passes through three Polaroid filters, as shown in **Figure 25–32**. The intensity of the laser beam is I_0. **(a)** What is the intensity of

the beam at point A? **(b)** What is the intensity of the beam at point B? **(c)** What is the intensity of the beam at point C? **(d)** If filter 2 is removed, what is the intensity of the beam at point C?

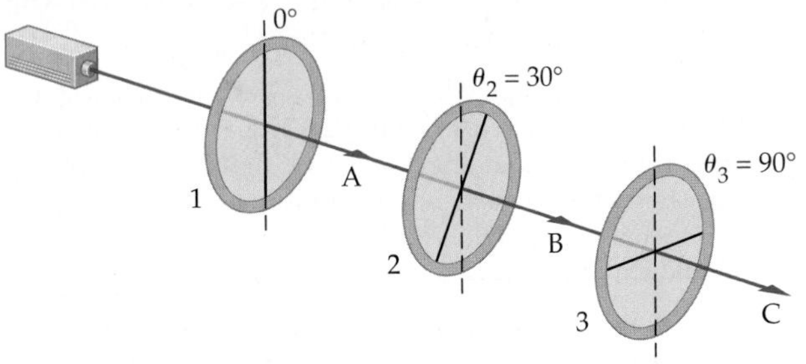

▲ **FIGURE 25–32** Problems 72, 73, 89, and 90

73. ••• Referring to Figure 25–32, suppose that filter 3 is at a general angle θ with the vertical, rather than the angle 90°. **(a)** Find an expression for the transmitted intensity as a function of θ. **(b)** Plot your result from part (a), and determine the maximum transmitted intensity. **(c)** At what angle θ does maximum transmission occur?

General Problems

74. • Sunlight at the surface of Earth has an average intensity of about 1.00×10^3 W/m^2. Find the rms values of the electric and magnetic fields in the sunlight.

75. • **BIO** A typical medical X-ray has a frequency of 1.50×10^{19} Hz. What is the wavelength of such an X-ray?

76. • How many hydrogen atoms, 0.10 nm in diameter, must be placed end to end to fit into one wavelength of 410-nm violet light?

77. • **BIO Radiofrequency Ablation** In radiofrequency (RF) ablation, a small needle is inserted into a cancerous tumor. When radiofrequency oscillating currents are sent into the needle, ions in the neighboring tissue respond by vibrating rapidly, causing local heating to temperatures as high as 100 °C. This kills the cancerous cells and, because of the small size of the needle, relatively few of the surrounding healthy cells. A typical RF ablation treatment uses a frequency of 750 kHz. What is the wavelength that such radio waves would have in a vacuum?

78. •• **IP (a)** What minimum intensity must a laser beam have if it is to levitate a tiny black (100% absorbing) sphere of radius $r = 0.5\ \mu$m and mass $= 1.6 \times 10^{-15}$ kg? Comment on the feasibility of such levitation. **(b)** If the radius of the sphere is doubled but its mass remains the same, will the minimum intensity be greater than, less than, or equal to the value found in part (a)? Explain. **(c)** Find the minimum intensity for the situation described in part (b).

79. •• **The *Apollo 11* reflector** One of the experiments placed on the Moon's surface by *Apollo 11* astronauts was a reflector that is used to measure the Earth–Moon distance with high accuracy. A laser beam on Earth is bounced off the reflector, and its round-trip travel time is recorded. If the travel time can be measured to within an accuracy of 0.030 ns, what is the uncertainty in the Earth–Moon distance?

80. •• The H_β line of the hydrogen atom's spectrum has a normal wavelength $\lambda_\beta = 486$ nm. This same line is observed in the spectrum of a distant quasar, but lengthened by 20.0 nm. What is the speed of the quasar relative to Earth, assuming it is moving along our line of sight?

81. •• **IP** Suppose the distance to the fixed mirror in Figure 25–28 is decreased to 20.5 km. **(a)** Should the angular speed of the rotating mirror be increased or decreased to ensure that the experiment works as described in Problem 15? **(b)** Find the required angular speed, assuming the speed of light is 3.00×10^8 m/s.

82. •• **IP** Suppose the speed of the galaxy in Problem 23 is increased by a factor of ten; that is, $V = 3.600 \times 10^6$ m/s. The speed of the arms, v, and the frequency of the light remain the same. **(a)** Does the arm near the top of Figure 25–29 show a red shift (toward lower frequency) or a blue shift (toward higher frequency)? Does the lower arm show a red or a blue shift? Explain. What frequency is detected by astronomers observing **(b)** the upper arm and **(c)** the lower arm?

83. •• **IP BIO** Consider the physical situation illustrated in Figure 25–30. **(a)** Is E_{rms} in the incident laser beam greater than, less than, or the same as E_{rms} where the beam hits the retina? Explain. **(b)** If the intensity of the beam at the retina is equal to the damage threshold, 1.0×10^{-2} W/cm^2, what is the value of E_{rms} at that location? **(c)** If the diameter of the spot on the retina is reduced by a factor of two, by what factor does the intensity increase? By what factor does E_{rms} increase?

84. •• **BIO Polaroid Vision in a Spider** Experiments show that the ground spider *Drassodes cupreus* uses one of its several pairs of eyes as a polarization detector. In fact, the two eyes in this pair have polarization directions that are at right angles to one another. Suppose linearly polarized light with an intensity of 825 W/m^2 shines from the sky onto the spider, and that the intensity transmitted by one of the polarizing eyes is 212 W/m^2. **(a)** For this eye, what is the angle between the polarization direction of the eye and the polarization direction of the incident light? **(b)** What is the intensity transmitted by the other polarizing eye?

85. •• A state highway patrol car radar unit uses a frequency of 9.00×10^9 Hz. What frequency difference will the unit detect from a car approaching a parked patrol car with a speed of 35.0 m/s?

86. •• What is the ratio of the sunlight intensity reaching Mercury compared with the sunlight intensity reaching Earth? (On average, Mercury's distance from the Sun is 0.39 that of Earth's).

87. •• What area is needed for a solar collector to absorb 25.0 kW of power from the Sun's radiation if the collector is 80.0% efficient? (At the surface of Earth, sunlight has an average intensity of 1.00×10^3 W/m^2.)

88. •• **BIO Near-Infrared Brain Scans** Light in the near-infrared (close to visible red) can penetrate surprisingly far through human tissue, a fact that is being used to "illuminate" the interior of the brain in a noninvasive technique known as near-infrared spectroscopy (NIRS). In this procedure, illustrated in **Figure 25–33**, an optical fiber carrying a beam of infrared laser light with a power of 1.5 mW and a cross-sectional diameter of 1.2 mm is placed against the skull. Some of the light enters

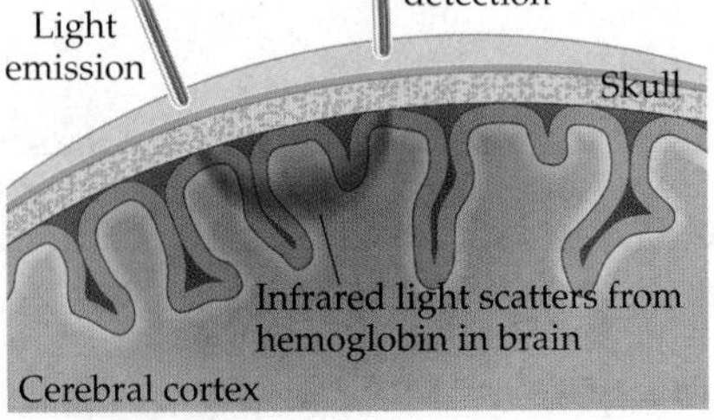

▲ **FIGURE 25–33** The basic elements of a near-infrared brain scan. (Problem 88)

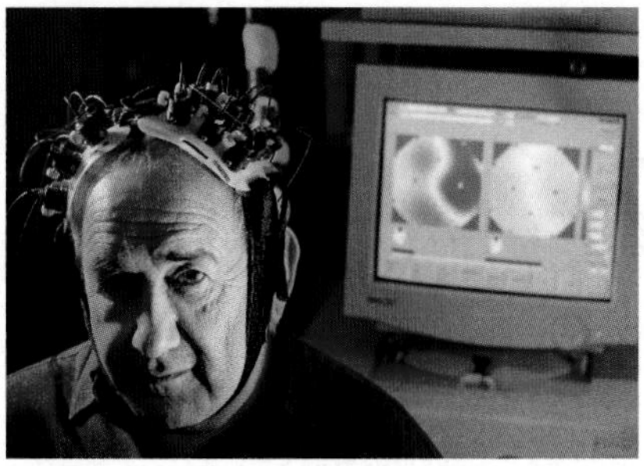

A patient undergoing a near-infrared brain scan. (Problem 88)

the brain, where it scatters from hemoglobin in the blood. The scattered light is picked up by a detector and analyzed by a computer. **(a)** According to the **Beer-Lambert law,** the intensity of light, I, decreases with penetration distance, d, as $I = I_0 e^{-\mu d}$, where I_0 is the initial intensity of the beam and $\mu = 4.7\ \text{cm}^{-1}$ for a typical case. Find the intensity of the laser beam after it penetrates through 3.0 cm of tissue. **(b)** Find the electric field of the initial light beam.

89. •• Three polarizers are arranged as shown in Figure 25–32. If the incident beam of light is unpolarized, and has an intensity of $1.60\ \text{W/m}^2$, find the transmitted intensity **(a)** when $\theta_2 = 25.0°$ and $\theta_3 = 50.0°$, and **(b)** when $\theta_2 = 50.0°$ and $\theta_3 = 25.0°$.

90. •• Repeat Problem 89, this time assuming an incident beam that is vertically polarized. The intensity of the incident beam is $1.60\ \text{W/m}^2$.

91. •• A lightbulb emits light uniformly in all directions. If the rms electric field of this light is 18.0 N/C at a distance of 1.35 m from the bulb, what is the average total power radiated by the bulb?

92. •• **IP** A beam of light is a mixture of unpolarized light with intensity I_u and linearly polarized light with intensity I_p. The polarization direction for the polarized light is vertical. When this mixed beam of light is passed through a polarizer that is vertical, the transmitted intensity is $16.8\ \text{W/m}^2$; when the polarizer is at an angle of 55.0° with the vertical, the transmitted intensity is $8.68\ \text{W/m}^2$. **(a)** Is I_u greater than, less than, or equal to I_p? Explain. **(b)** Calculate I_u and I_p.

93. ••• **BIO** As mentioned in Problem 84, one pair of eyes in a particular species of ground spider has polarization directions that are at right angles to one another. Suppose that linearly polarized light is incident on such a spider. **(a)** Prove that the transmitted intensity of one eye plus the transmitted intensity from the other eye is equal to the incident intensity. **(b)** If the transmitted intensities for the two eyes are $183\ \text{W/m}^2$ and $669\ \text{W/m}^2$, through what angle must the spider rotate to make the transmitted intensities equal to one another?

94. ••• A typical home may require a total of 2.00×10^3 kWh of energy per month. Suppose you would like to obtain this energy from sunlight, which has an average daylight intensity of $1.00 \times 10^3\ \text{W/m}^2$. Assuming that sunlight is available 8.0 h per day, 25 d per month (accounting for cloudy days), and that you have a way to store energy from your collector when the Sun isn't shining, determine the smallest collector size that will provide the needed energy, given a conversion efficiency of 25%.

95. ••• At the top of Earth's atmosphere, sunlight has an average intensity of $1360\ \text{W/m}^2$. If the average distance from Earth to the Sun is 1.50×10^{11} m, at what rate does the Sun radiate energy?

96. ••• **IP** A typical laser used in introductory physics laboratories produces a continuous beam of light about 1.0 mm in diameter. The average power of such a laser is 0.75 mW. What is **(a)** the average intensity, **(b)** the peak intensity, and **(c)** the average energy density of this beam? If the beam is reflected from a mirror, what is **(d)** the maximum force the laser beam can exert on it? **(e)** Describe the orientation of the laser beam relative to the mirror for the case of maximum force.

97. ••• Four polarizers are set up so that the transmission axis of each successive polarizer is rotated clockwise by an angle θ relative to the previous polarizer. Find the angle θ for which unpolarized light is transmitted through these four polarizers with its intensity reduced by a factor of 25.

98. ••• **BIO Optical Activity of Sugar** The sugar concentration in a solution (e.g., in a urine specimen) can be measured conveniently by using the *optical activity* of sugar and other asymmetric molecules. In general, an optically active molecule, like sugar, will rotate the plane of polarization through an angle that is proportional to the thickness of the sample and to the concentration of the molecule. To measure the concentration of a given solution, a sample of known thickness is placed between two polarizing filters that are at right angles to each other, as shown in **Figure 25–34**. The intensity of light transmitted through the two filters can be compared with a calibration chart to determine the concentration. **(a)** What percentage of the incident (unpolarized) light will pass through the first filter? **(b)** If no sample is present, what percentage of the initial light will pass through the second filter? **(c)** When a particular sample is placed between the two filters, the intensity of light emerging from the second filter is 40.0% of the incident intensity. Through what angle did the sample rotate the plane of polarization? **(d)** A second sample has half the sugar concentration of the first sample. Find the intensity of light emerging from the second filter in this case.

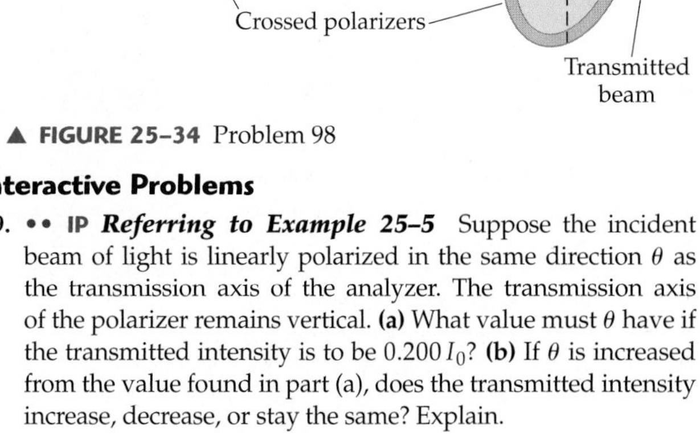

▲ **FIGURE 25–34** Problem 98

Interactive Problems

99. •• **IP** ***Referring to Example 25–5*** Suppose the incident beam of light is linearly polarized in the same direction θ as the transmission axis of the analyzer. The transmission axis of the polarizer remains vertical. **(a)** What value must θ have if the transmitted intensity is to be $0.200\,I_0$? **(b)** If θ is increased from the value found in part (a), does the transmitted intensity increase, decrease, or stay the same? Explain.

100. •• ***Referring to Example 25–5*** Suppose the incident beam of light is linearly polarized in the vertical direction. In addition, the transmission axis of the analyzer is at an angle of 80.0° to the vertical. What angle should the transmission axis of the polarizer make with the vertical if the transmitted intensity is to be a maximum?

26 Geometrical Optics

26–1 The Reflection of Light 871
26–2 Forming Images with a Plane Mirror 872
26–3 Spherical Mirrors 875
26–4 Ray Tracing and the Mirror Equation 877
26–5 The Refraction of Light 884
26–6 Ray Tracing for Lenses 891
26–7 The Thin-Lens Equation 894
26–8 Dispersion and the Rainbow 896

We all learn early in life that light travels in straight lines—but we also learn, when we first look into a mirror or through a glass of water, that these lines can take some odd bends and turns. The mirrors shown here, for example, present a distorted view of their surroundings, simply because light striking their surface is reflected onto new directions. As we shall see in this chapter, optical devices that change the direction of light—such as mirrors and lenses—follow simple geometrical laws that account for their ability to form images. By analyzing the direction of light rays before and after they encounter a mirror or lens, we can determine the precise location, size, and orientation of the image it produces.

When you look into a mirror, or through a pair of binoculars, you see images of various objects in your surroundings. These images are formed by mirrors or lenses that redirect light coming from the objects. In the case of mirrors, light is *reflected* onto a new path. In lenses the speed of light is reduced, resulting in a change of direction referred to as *refraction*. By changing the direction in which light moves, both mirrors and lenses can be used to create images of various sizes and orientations.

26–1 The Reflection of Light

Perhaps the simplest way to change the direction of light is by reflection from a shiny surface. To understand this process in detail, it is convenient to describe light in terms of "wave fronts" and "rays." As we shall see later in this chapter, these concepts are equally useful in understanding the behavior of lenses.

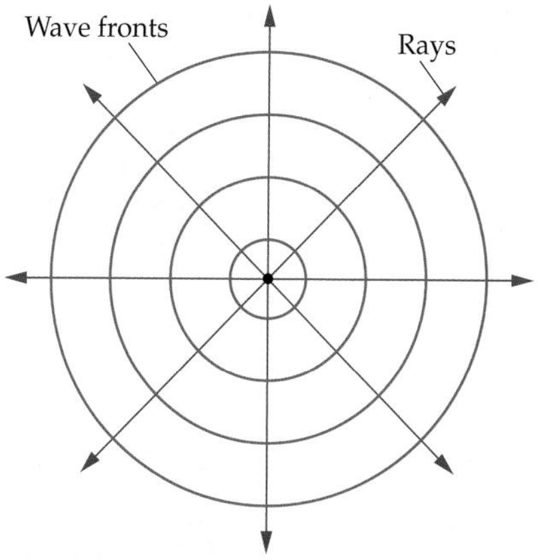

▲ **FIGURE 26–1 Wave fronts and rays**
In this case, the wave fronts indicate the crests of water waves. The rays indicate the local direction of motion.

Wave Fronts and Rays

Consider the waves created by a rock dropped into a still pool of water. As we know, these waves form concentric outward-moving circles. A simplified representation of this system is given in **Figure 26–1**, where the circles indicate the crests of the waves. We refer to these circles as **wave fronts.** In addition, the radial motion of the waves is indicated by the outward-pointing arrows, referred to as **rays.** Notice that the rays are always at right angles to the wave fronts.

A similar situation applies to electromagnetic waves radiated by a small source, as illustrated in **Figure 26–2 (a)**. In this case, however, the waves move outward in three dimensions, giving rise to spherical wave fronts. As expected, **spherical waves** such as these have rays that point radially outward.

In **Figure 26–2 (b)** we show that as one moves farther from the source of spherical waves the wave fronts become increasingly flat and the rays more nearly parallel. In the limit of increasing distance, the wave fronts approach perfectly flat planes. Such **plane waves,** with their planar wave fronts and parallel rays, are useful idealizations for investigating the properties of mirrors and lenses.

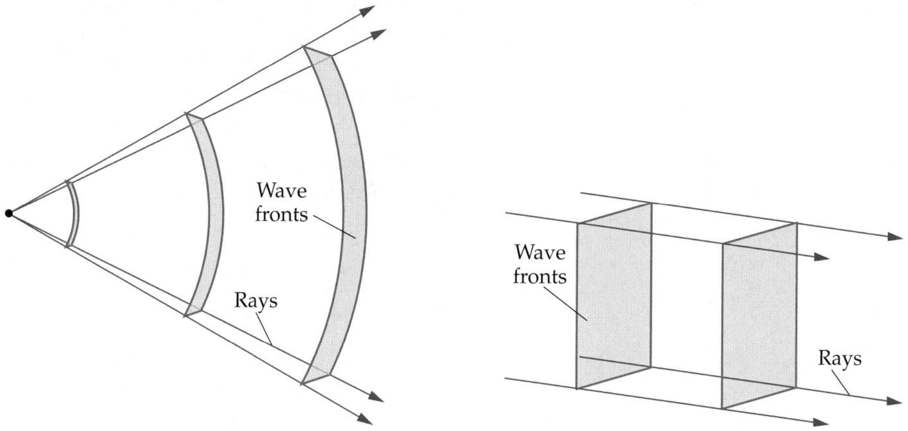

◀ **FIGURE 26–2 Spherical and planar wave fronts**
(a) As spherical waves move farther from their source, the wave fronts become increasingly flat. **(b)** In the limit of large distances the wave fronts are planes, and the corresponding rays are parallel to one another.

Finally, we usually simplify our representation of light beams even further by omitting the wave fronts and plotting only one, or a few, rays. For example, in **Figure 26–3** both the incident plane wave and the reflected plane wave are shown as single rays pointing in the direction of propagation. The direction of these rays is considered next.

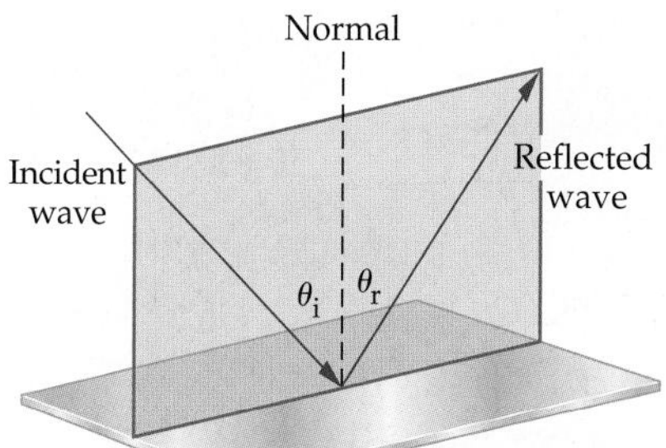

▲ **FIGURE 26–3 Reflection from a smooth surface**
In this simplified representation, the incident and reflected waves are indicated by single rays pointing in the direction of propagation. Notice that the angle of reflection, θ_r, is equal to the angle of incidence, θ_i. In addition, the incident ray, reflected ray, and the normal all lie in the same plane.

The Law of Reflection

To characterize the behavior of light as it reflects from a mirror or other shiny object, we begin by drawing the normal, which is simply a line perpendicular to the reflecting surface at the point of incidence. Relative to the normal, the incident ray strikes the surface at the angle θ_i, the *angle of incidence,* as shown in Figure 26–3. Similarly, the *angle of reflection,* θ_r, is the angle the reflected ray makes with the normal. The relationship between these two angles is very simple—they are equal:

Law of Reflection

$$\theta_r = \theta_i \qquad 26\text{–}1$$

Note, in addition, that the incident ray, the normal, and the reflected ray all lie in the same plane, as is also clear from Figure 26–3.

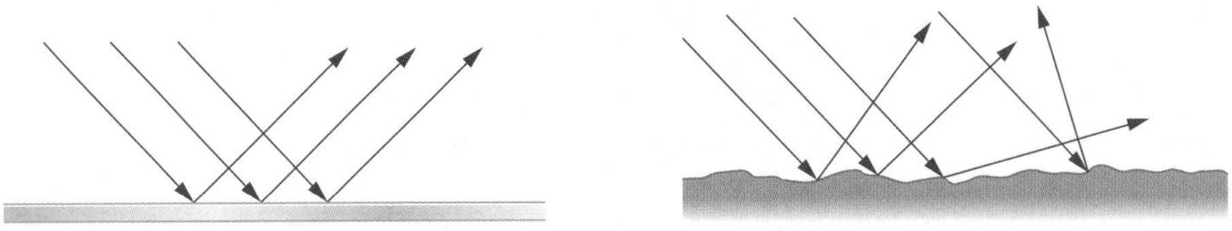

▶ **FIGURE 26–4 Reflection from smooth and rough surfaces**
(a) A smooth surface gives specular reflection, in which all the reflected light propagates in a single direction. **(b)** A rough surface gives rise to reflected waves propagating over a range of directions. This process is referred to as diffuse reflection.

The reflection of light from a smooth shiny surface, as in **Figure 26–4 (a)**, is referred to as **specular reflection.** Notice that all the reflected light moves in the same direction. In contrast, if a surface is rough, as in **Figure 26–4 (b)**, the reflected light is sent out in a variety of directions, giving rise to **diffuse reflection.** For example, when the surface of a road is wet, the water creates a smooth surface, and headlights reflecting from the road undergo specular reflection. As a result, the reflected light goes in a single direction, giving rise to an intense glare. When the same road is dry its surface is microscopically rough; hence, light is reflected in many directions and glare is not observed. The law of reflection is obeyed in either case, of course; it is the surface that is different, not the underlying physics.

▲ When rays of light are reflected from a smooth, flat surface, both the incident and reflected rays make identical angles with the normal. Consequently, parallel rays are reflected in the same direction, as shown in Figure 26–4. The result is specular reflection, which produces a "mirror image" of the source of the rays.

A novel variation on specular versus diffuse reflection occurs in a new type of electronic chip known as a Digital Micromirror Device (DMD). These small devices consist of as many as 1.3 million microscopic plane mirrors, each of which, though smaller than the diameter of a human hair, can be oriented independently in response to electrical signals. The reflection from each micromirror is specular, and if all 1.3 million mirrors are oriented in the same direction, the DMD acts like a small plane mirror. Conversely, if the mirrors are oriented randomly, the reflection from the DMD is diffuse. When a DMD is used to project a movie, as will soon be the case in certain theaters, each micromirror will play the role of a single pixel in the projected image. In such a system, the light directed onto the DMD cycles rapidly from red to green to blue, and each of the mirrors reflects only the appropriate colors for that pixel onto the screen (**Figure 26–5**). The result is a projected image of great brilliance and vividness that eliminates the need for film.

REAL-WORLD PHYSICS
Micromirror devices and digital movie projection

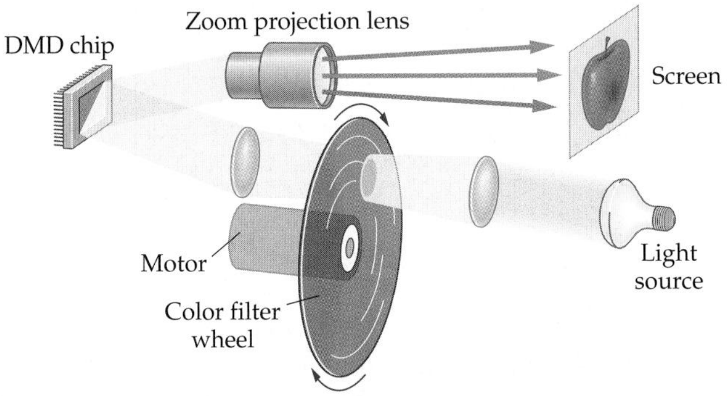

▲ **FIGURE 26–5 A Digital Micromirror Projection System**
A digital projection system based on micromirrors reflects incoming light onto a distant screen. Each micromirror produces one pixel of the final image. The color and intensity of a given pixel is determined by the amount of red, green, and blue light the corresponding micromirror reflects to the screen.

▲ An ant's leg provides a sense of scale in this photo of an array of Digital Micromirror Device (DMD) mirrors. Each mirror has an area of 16 μm^2 and can pivot 10° in either direction about a diagonal axis.

26–2 Forming Images with a Plane Mirror

We are all familiar with looking at ourselves in a mirror. If the mirror is perfectly flat—that is, a **plane mirror**—we see an upright image of ourselves as far behind the mirror as we are in front. In addition, the image is reversed right to left; for example, if we raise our right hand, the mirror image raises its left hand. Lettering read in a mirror is reversed right to left as well. This is the reason ambulances and other emergency vehicles use mirror-image writing on the front of their vehicles—when viewed in the rear-view mirror of a car, the writing looks normal. In this section we use the law of reflection to derive these and other results.

Before we delve into the details of mirror images, however, let's pause for a moment to consider the process by which physical objects produce images in our eyes. First, any object that is near you at this moment is bathed in light coming at it from all directions. As this object reflects the incoming light back into the room, every point on it acts like an omnidirectional source of light. When we view the object, the light coming from any given point enters our eyes and is focused to a point on our retina. This is the case *for every point* that we can see on the object—each point on the object is detected by a different point on the retina. This results in a one-to-one mapping between the physical object and the image on the retina.

The formation of a mirror image occurs in a similar manner, except that the light from an object reflects from a mirror before it enters our eyes. This is illustrated in **Figure 26–6 (a)**, where we show an object—a small flower—placed in front of a plane mirror. Rays of light leaving the top of the flower at point P are shown reflecting from the mirror and entering the eye of an observer. To the observer, it appears that the rays are emanating from point P′ *behind the mirror*.

We can show that the image is the same distance behind the mirror as the object is in front. Consider **Figure 26–6 (b)**, where we indicate the distance of the object from the mirror by d_o, and the distance from the image to the mirror by d_i. One ray from the top of the flower is shown reflecting from the mirror and entering the observer's eye. We also show the extension of the reflected ray back to the image. By the law of reflection, if the angle of incidence at point A is θ, the angle of reflection is also θ. Therefore, the straight line from the observer to the image cuts across the normal line with an angle θ on either side, as shown. Clearly, then, the angles indicated by ϕ must also be equal to one another. Combining these results, we see that triangle PAQ shares a side and two adjacent angles with triangle P′AQ; hence, the two triangles are equal. It follows that the distance d_o is equal to the distance d_i, as expected.

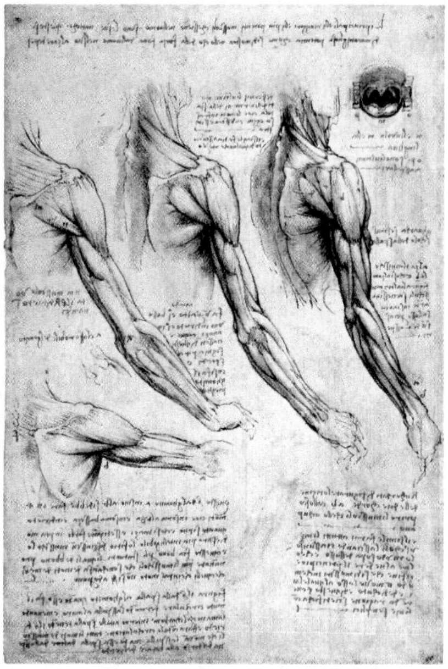

▲ Leonardo da Vinci (1452–1519), the quintessential Renaissance man, tried to keep the contents of his notebooks private by setting down his observations in "mirror writing."

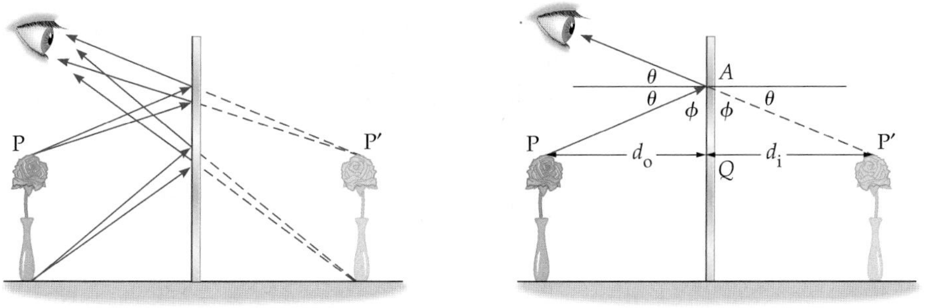

(a) (b)

◀ **FIGURE 26–6 Locating a mirror image**
(a) Rays of light from point P at the top of the flower appear to originate from point P′ behind the mirror. **(b)** Construction showing that the distance from the object to the mirror is the same as the distance from the image to the mirror.

In the following Example, we once again apply the law of reflection to the flower and its mirror image.

EXAMPLE 26–1 Reflecting on a Flower

An observer is at table level, a distance d to the left of a flower of height h. The flower itself is a distance d to the left of a mirror, as shown in the sketch. Note that a ray of light propagating from the top of the flower to the observer's eye reflects from the mirror at a height y above the table. Find y in terms of the height of the flower, h.

Picture the Problem

The physical situation is shown in the diagram, along with a ray from the top of the flower to the eye of the observer. The point where the ray hits the mirror is a height y above the table. The flower is a distance d to the left of the mirror, and its image is a distance d to the right of the mirror. Finally, the observer's eye is a distance $2d$ to the left of the mirror.

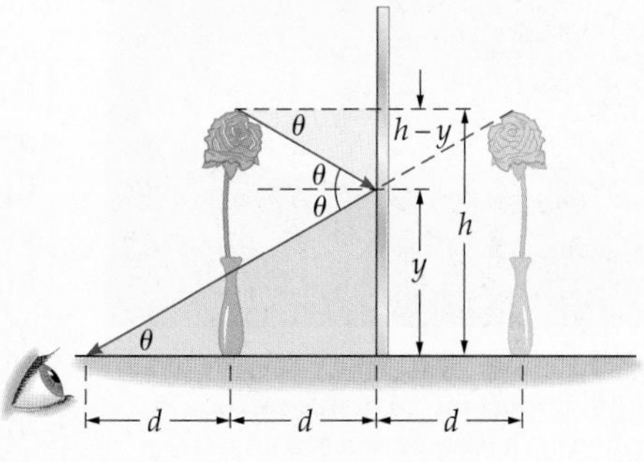

Strategy

To find y, we must use the fact that the angle of reflection is equal to the angle of incidence. In the diagram we indicate these two angles by the symbol θ. It follows, then, that $\tan\theta$ obtained from the small yellow triangle must be equal to $\tan\theta$ obtained from the larger green triangle. Setting these expressions for $\tan\theta$ equal to one another yields a relation that can be solved for y.

continued on next page

continued from previous page

Solution

1. Write an expression for $\tan\theta$ using the yellow triangle: $\tan\theta = \dfrac{h-y}{d}$

2. Now, write a similar expression for $\tan\theta$ using the green triangle: $\tan\theta = \dfrac{y}{2d}$

3. Set the two expressions for $\tan\theta$ equal to one another: $\dfrac{h-y}{d} = \dfrac{y}{2d}$

4. Rearrange the preceding equation and solve for y: $h - y = \dfrac{y}{2}$

$$y = \frac{2}{3}h$$

Insight

Note that the observer will see the entire image of the flower in a section of mirror that is only two-thirds the height of the flower.

Practice Problem

If the observer moves farther from the base of the flower, does the point of reflection move upward, downward, or stay at the same location? As a check on your answer, calculate the point of reflection for the case where the distance between the observer and the base of the flower is $2d$. [**Answer:** The point of reflection moves upward. In this case, we find $y = \frac{3}{4}h$.]

Some related homework problems: Problem 5, Problem 7

To summarize, the basic features of reflection by a plane mirror are the following:

Properties of Mirror Images Produced by Plane Mirrors

- A mirror image is upright, but appears reversed right to left.
- A mirror image appears to be the same distance behind the mirror that the object is in front of the mirror.
- A mirror image is the same size as the object.

CONCEPTUAL CHECKPOINT 26–1 Height of Mirror

To save expenses, you would like to buy the shortest mirror that will allow you to see your entire body. Should the mirror be **(a)** half your height, **(b)** two-thirds your height, or **(c)** equal to your height?

Reasoning and Discussion

First, to see your feet, the mirror must extend from your eyes downward to a point halfway between your eyes and feet, as shown in the sketch.

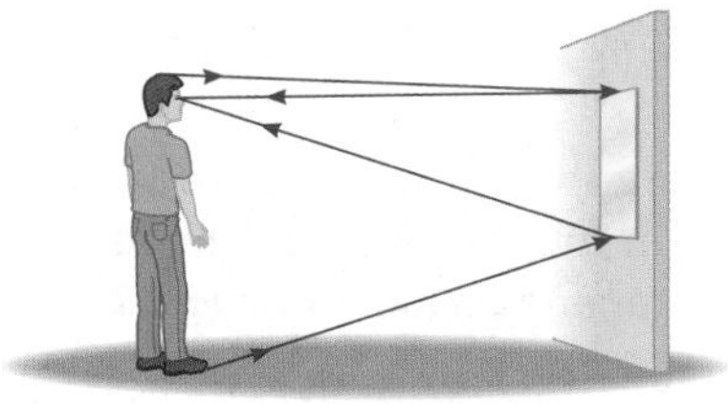

Similarly, the mirror must extend upward from your eyes half the distance to the top of your head. Altogether, then, the mirror must have a height equal to half the distance from your eyes to your feet plus half the distance from your eyes to the top of your head—that is, half your total height.

Answer:

(a) The mirror needs to be only half your height.

▲ In this heads-up display in an airplane cockpit, important flight information is reflected on a transparent screen in the windshield, enabling the pilot to view the data without diverting attention from the scene ahead.

An application of mirror images that has been used in military aircraft for many years, and is now beginning to appear in commercial automobiles, is the *heads-up display*. In the case of a car, a small illuminated display screen is recessed in the dashboard, out of direct sight of the driver. The screen shows important information, like the speed of the car, in mirror image. The driver sees the information not by looking directly at the screen—which is hidden from view—but by looking at its reflection in the windshield. Thus, while still looking straight ahead (heads up), the driver can see both the road and the reading of the speedometer.

A similar device is used in theaters to provide subtitles for the hearing impaired. In this case, a transparent plastic screen is mounted on the arm of a person's chair. This screen is adjusted in such a way that the person can look through it to see the movie, and at the same time see the reflection of a screen in the back of the theater that gives the subtitles in mirror-image form.

EXAMPLE 26–2 Two-Dimensional Corner Reflector

Two mirrors are placed at right angles, as shown in the diagram. An incident ray of light makes an angle of 30° with the x axis and reflects from the lower mirror. Find the angle the outgoing ray makes with the y axis after it reflects once from each mirror.

Picture the Problem

The physical system is shown in the diagram. Notice that the normal for the first reflection is vertical, whereas it is horizontal for the second reflection. It follows that the angle of incidence for the first reflection is $\theta = 90° - 30° = 60°$. Similarly, the angle of incidence for the second reflection is $90° - \theta = 30°$.

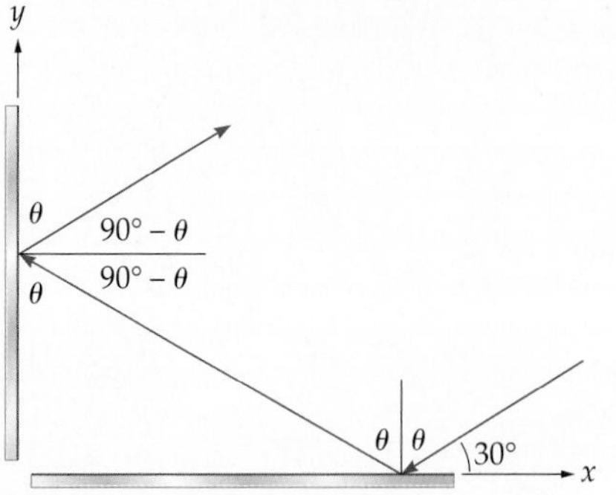

Solution

We apply the law of reflection to each of the two reflections. For the first reflection, the angle of incidence is $\theta = 60°$; hence, the reflected ray also makes a 60° angle with the vertical.

At the second reflection, the normal is horizontal; hence, the angle of incidence is $90° - \theta = 30°$. After the second reflection, the outgoing ray is 30° above the horizontal and hence it makes an angle $\theta = 60°$ with respect to the y axis, as shown.

Insight

Note that the outgoing ray travels parallel to the incoming ray, but in exactly the opposite direction. This is true regardless of the value of the initial angle θ.

Practice Problem

If the incoming ray hits the horizontal mirror farther to the right, is the angle the outgoing ray makes with the vertical greater than θ, equal to θ, or less than θ? [**Answer:** The angle of the outgoing ray is still equal to θ. The distance between the incoming and outgoing rays would be increased, however.]

Some related homework problems: Problem 8, Problem 15

If three plane mirrors are joined at right angles, as in **Figure 26–7**, the result is a **corner reflector.** A corner reflector behaves in three dimensions the same as two mirrors at right angles in two dimensions; namely, a ray incident on the corner reflector is sent back in the same direction from which it came. This type of behavior has led to many useful applications for corner reflectors.

One of the more interesting applications involves the only Apollo experiments still returning data from the Moon. On *Apollos 11, 14*, and *15* the astronauts placed retroreflector arrays consisting of 100 corner reflectors on the lunar surface. Scientists at observatories on Earth can send a laser beam to the appropriate location on the Moon, where the retroreflector sends it back to its source. By measuring the round-trip travel time of the light, it is possible to determine the Earth–Moon distance to an accuracy of about 3 cm out of a total distance of roughly 385,000 km! Over the years, these measurements have revealed that the Moon is moving away from Earth at the rate of 3.8 cm per year.

REAL-WORLD PHYSICS
Corner reflectors and the Earth–Moon distance

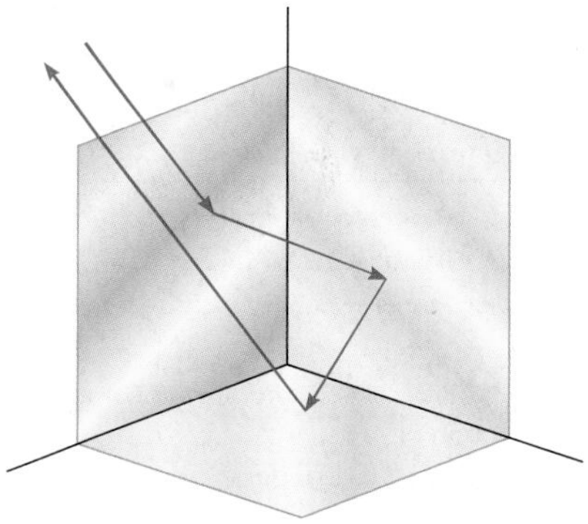

▲ **FIGURE 26–7 A corner reflector**
A three-dimensional corner reflector constructed from three plane mirrors at right angles to one another. A ray entering the reflector is sent back in the direction from which it came.

26–3 Spherical Mirrors

We now consider another type of mirror that is encountered frequently in everyday life—the spherical mirror. These mirrors get their name from the fact that they have the same shape as a section of a sphere, as is shown in **Figure 26–8**, where a portion of a spherical shell of radius R is cut away from the rest of the shell. If the outside of this spherical section is a reflecting surface, the result is a **convex** mirror; if the inside surface is reflecting, we have a **concave** mirror.

Convex and concave spherical mirrors are illustrated in **Figure 26–9**, where we also indicate the **center of curvature,** C, and the **principal axis.** The center of curvature is the center of the sphere with radius R of which the mirror is a section, and the principal axis is a straight line drawn through the center of curvature and

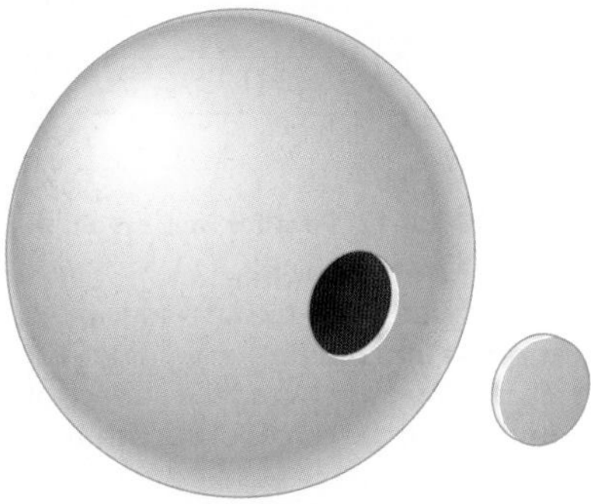

▶ **FIGURE 26–8 Spherical mirrors**
A spherical mirror has the same shape as a section of a sphere. If the outside surface of this section is reflecting, the mirror is convex. If the inside surface reflects, the mirror is concave.

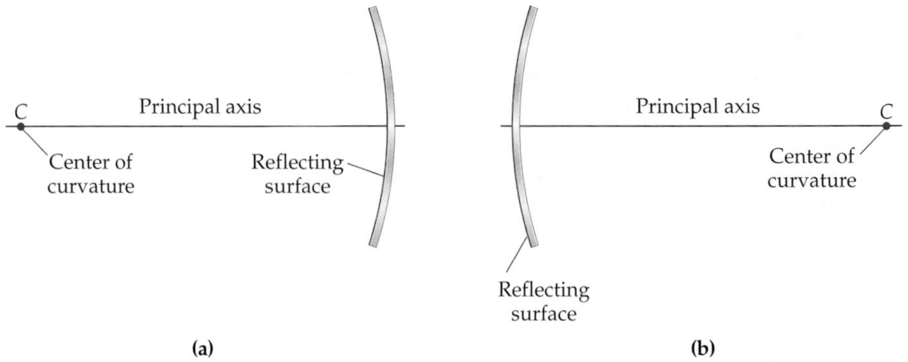

▶ **FIGURE 26–9 Concave and convex mirrors**

(a) A concave mirror and its center of curvature, *C*, which is on the same side as the reflecting surface. **(b)** A convex mirror and its center of curvature, *C*. In this case *C* is on the opposite side of the mirror from its reflecting surface.

the midpoint of the mirror. Note that the principal axis intersects the mirror at right angles.

To investigate the behavior of spherical mirrors, suppose that a beam of light is directed toward either a convex or a concave mirror along its principal axis. For example, several parallel rays of light are approaching a convex mirror in **Figure 26–10**. After reflecting from the mirror, the rays diverge as if they originated from a single point behind the mirror called the **focal point,** *F*. (This is strictly true only for light rays close to the principal axis, as we point out in greater detail later in this section. All of our results for spherical mirrors assume rays close to the axis.)

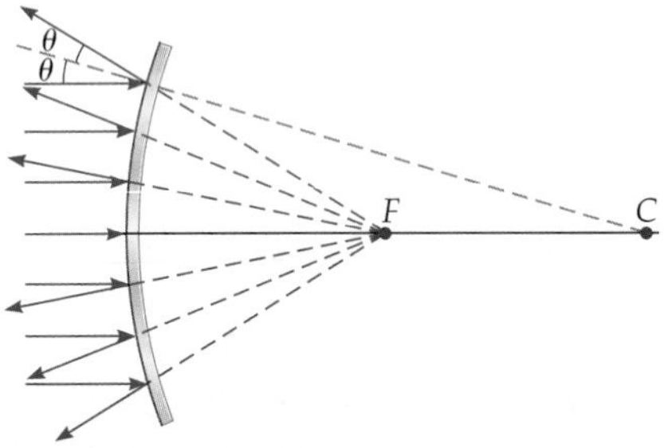

▲ **FIGURE 26–10 Parallel rays on a convex mirror**

When parallel rays of light reflect from a convex mirror, they diverge as if originating from a focal point halfway between the surface of the mirror and its center of curvature.

Now, let's find the **focal length;** that is, the distance from the surface of the mirror to the focal point. This can be done with the aid of **Figure 26–11**, which shows a single ray reflecting from the mirror. The first thing to notice about this diagram is that a straight line drawn through the center of curvature always intersects the mirror at right angles; hence, the line through *C* is the normal to the surface at the point of incidence, *A*. Since the incoming ray is parallel to the principal axis, it follows that the angle of incidence, θ, is equal to the angle *FCA*. Next, the law of reflection states that the angle of reflection must equal θ, which means that the angle *CAF* is also θ. We see, then, that *CAF* is an isosceles triangle, with the sides *CF* and *FA* of equal length. Finally, for small angles θ, the length *CF* is approximately equal to half the length $CA = R$; that is $CF \sim \frac{1}{2}R$. Therefore, to this same approximation, the distance *FB* is also $\frac{1}{2}R$.

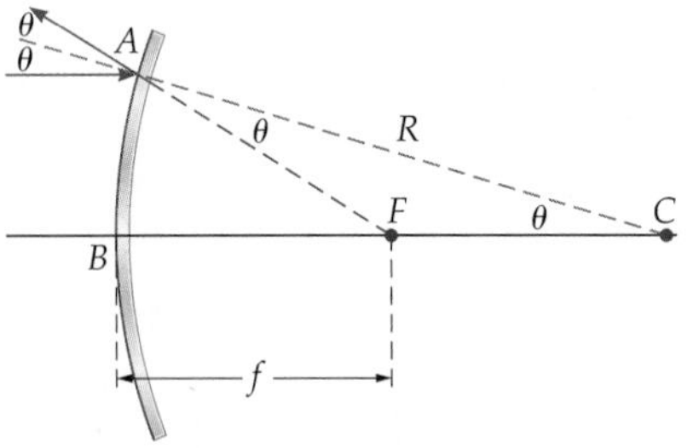

▲ **FIGURE 26–11 Ray diagram for a convex mirror**

Ray diagram used to locate the focal point, *F*.

Thus, when considering a convex mirror of radius *R*, we will always use the following result for the focal length:

Focal Length for a Convex Mirror of Radius *R*

$$f = -\tfrac{1}{2}R \qquad \text{26–2}$$

SI unit: m

The minus sign in this expression is used to indicate that the focal point lies behind the mirror. This is part of a general sign convention for mirrors that will be discussed in detail in the next section.

The situation is similar for a concave mirror. First, rays parallel to the principal axis are reflected by the mirror and brought together at a focal point, *F*, as shown in **Figure 26–12**. The same type of analysis used for the convex mirror can be applied to the concave mirror as well, with the result that the focal point is a distance $\frac{1}{2}R$ in front of the mirror. Thus, for a concave mirror,

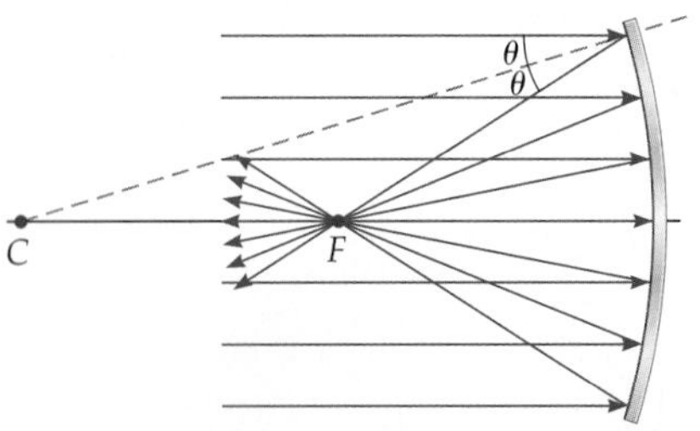

▲ **FIGURE 26–12 Parallel rays on a concave mirror**

Parallel rays reflecting from a concave mirror pass through the focal point *F*, which is halfway between the surface of the mirror and the center of curvature.

Focal Length for a Concave Mirror of Radius *R*

$$f = \tfrac{1}{2}R \qquad \text{26–3}$$

SI unit: m

This time *f* is positive, since the focal point is in front of the mirror. Note that in this case the rays of light actually pass through the focal point, in contrast to the behavior of the convex mirror.

CONCEPTUAL CHECKPOINT 26–2 Starting a Fire

Suppose you would like to use the Sun to start a fire in the wilderness. Which type of mirror, concave or convex, would work best?

Reasoning and Discussion

First, note that rays from the Sun are essentially parallel, since it is at such a great distance. Therefore, if a convex mirror is used, the situation is like that shown in Figure 26–10, in which the rays are spread out after reflection. Conversely, a concave mirror will bring the rays together at a point, as in Figure 26–12. Clearly, by focusing the sunlight at one point, the concave mirror will stand a better chance of starting a fire.

Answer:

The concave mirror is the one to use.

A final point regards the approximation made earlier in this section; namely, that the angle θ is small. This is equivalent to saying that the distance between the principal axis of the mirror and the incoming rays is much less than the radius of curvature of the mirror, R. When rays are displaced from the axis by distances comparable to the radius R, as in **Figure 26–13 (a)**, the result is that they do not all pass through the focal point—the farther a ray is from the axis, the more it misses the focal point. In a case like this, the mirror will produce a blurred image, an effect known as **spherical aberration.** This effect can be reduced to undetectable levels by restricting the incoming rays to only those near the axis. These axis-hugging rays are referred to as **paraxial rays.**

REAL-WORLD PHYSICS
Parabolic mirrors

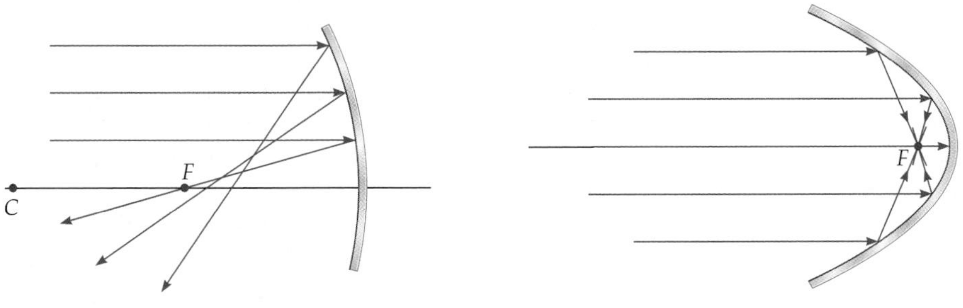

(a) (b)

◀ FIGURE 26–13 Spherical aberration and the parabolic mirror

(a) Rays far from the axis of a spherical mirror are not reflected through its focal point. The result is a blurred image referred to as spherical aberration. **(b)** A mirror with a parabolic cross section brings all parallel rays to a focus at one point, regardless of how far they are from the axis of the mirror.

Another way to eliminate spherical aberration is to construct a mirror with a *parabolic* cross section, as shown in **Figure 26–13 (b)**. One of the key properties of a parabola is that rays parallel to its axis are reflected through the same point F regardless of their distance from the axis. Thus a parabolic mirror produces a sharp image from all the rays coming into it. For this reason, astronomical mirrors, like that of the Hale telescope on Mount Palomar, California, are polished to a parabolic shape to give the greatest possible light-gathering ability and the sharpest possible images.

The same principle works in reverse as well. For example, if a source of light is placed at the focal point of a parabolic mirror, as at point F in Figure 26–13 (b), the mirror will redirect the light into an intense, unidirectional beam that can be aimed in a precise direction. Applications of this effect include flashlights, car headlights, and the giant arc lights that sweep across the sky to announce a grand opening.

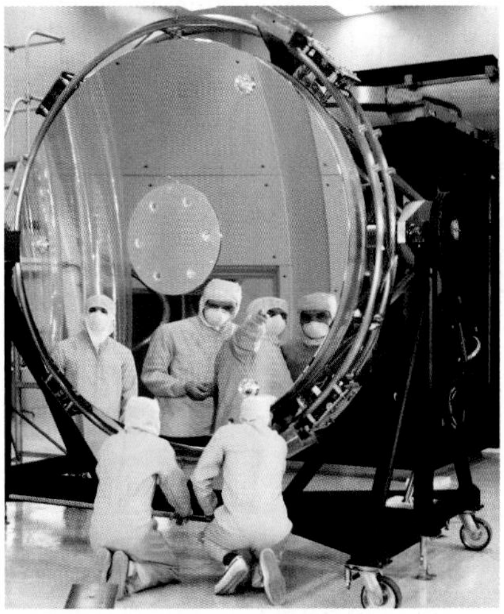

▲ The mirror of the Hubble Space Telescope (HST) was to have a perfectly parabolic shape. Unfortunately, owing to a mistake in the grinding of the mirror, the images produced by the HST were marred by spherical aberration. This necessitated a repair mission to the orbiting telescope, during which astronauts installed corrective optics to compensate for the defect.

26–4 Ray Tracing and the Mirror Equation

The images formed by a spherical mirror can be more varied than those produced by a plane mirror. In a plane mirror the image is always upright, the same size as the object, and the same distance from the mirror as the object. In the case of spherical mirrors, the image can be either upright or inverted, larger or smaller than the object, and closer or farther from the mirror than the object.

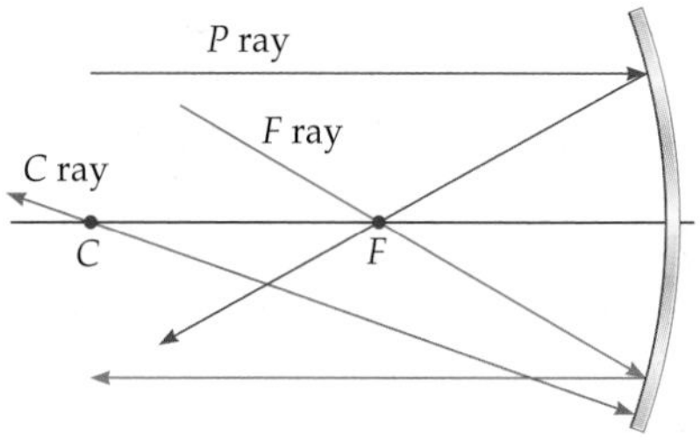

▲ **FIGURE 26–14 Principal rays used in ray tracing for a concave mirror**

The parallel ray (*P* ray) reflects through the focal point. The focal ray (*F* ray) reflects parallel to the axis, and the center-of-curvature ray (*C* ray) reflects back along its incoming path.

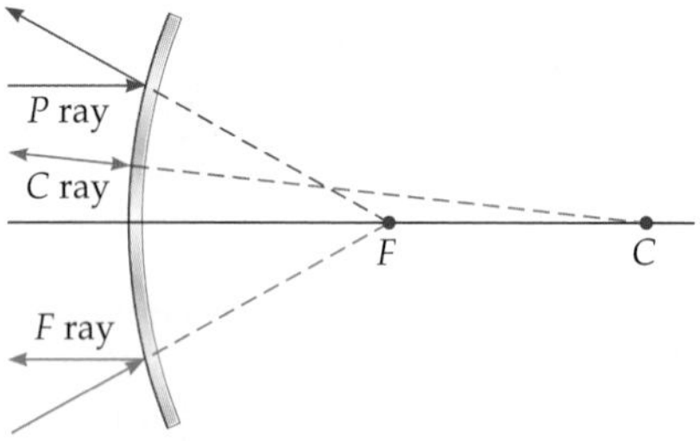

▲ **FIGURE 26–15 Principal rays used in ray tracing for a convex mirror**

The parallel ray (*P* ray) reflects along a direction that extends back to the focal point. Similarly, a focal ray (*F* ray) moves toward the focal point until it reflects from the mirror, after which it moves parallel to the axis. A center-of-curvature ray (*C* ray) is directed toward the center of curvature. It reflects from the mirror back along its incoming path.

To find the orientation, size, and location of an image in a spherical mirror, we use two techniques. The first, referred to as **ray tracing,** gives the orientation of the image as well as qualitative information on its location and size. If drawn carefully to scale, a ray diagram can also give quantitative results. The second method, using a relation referred to as the **mirror equation,** provides precise quantitative information without the need for accurate scale drawings. Both methods are presented in this section.

Ray Tracing

The basic idea behind ray tracing is to follow the path of representative rays of light as they reflect from a mirror and form an image. This was done for the plane mirror in Section 26–2. Ray tracing for spherical mirrors is a straightforward extension of the same basic techniques.

There are three rays with simple behavior that are used most often in ray tracing with spherical mirrors. These rays are illustrated in **Figure 26–14** for a concave mirror, and in **Figure 26–15** for a convex mirror. We start with the parallel ray (*P* ray), which, as its name implies, is parallel to the principal axis of the mirror. As we know from the previous section, a parallel ray is reflected through the focal point of a concave mirror, as shown by the purple ray in Figure 26–14. Similarly, a *P* ray reflects from a convex mirror along a line that *extends back* through the focal point, as with the purple ray in Figure 26–15.

Next, a ray that passes through the focal point of a concave mirror is reflected parallel to the axis, as indicated by the green ray in Figure 26–14. Thus, in a sense, a focal-point ray (*F* ray) is the reverse of a *P* ray. In general, any ray is equally valid in either the forward or reverse direction. The corresponding *F* ray for a convex mirror is shown in green in Figure 26–15.

Finally, note that any straight line extending from the center of curvature intersects the mirror at right angles. Thus, a ray moving along such a path is reflected back along the same path. Center-of-curvature rays (*C* rays) are illustrated in red in Figures 26–14 and 26–15.

To see how these rays can be used to obtain an image, consider the convex mirror shown in **Figure 26–16**. In front of the mirror is an object, represented symbolically by the red arrow. Also indicated in the figure are the three rays described above. Note that these rays diverge from the mirror as if they had originated from the tip of the orange arrow behind the mirror. This arrow is the image of the object; in fact, since no light passes through the image, we call it a **virtual image.** As we can see from the diagram, the virtual image is upright, smaller than the object, and located between the mirror and the focal point *F*.

▶ **FIGURE 26–16 Image formation with a convex mirror**

Ray diagram showing an image formed by a convex mirror. The three outgoing rays (*P*, *F*, and *C*) extend back to a single point at the top of the image.

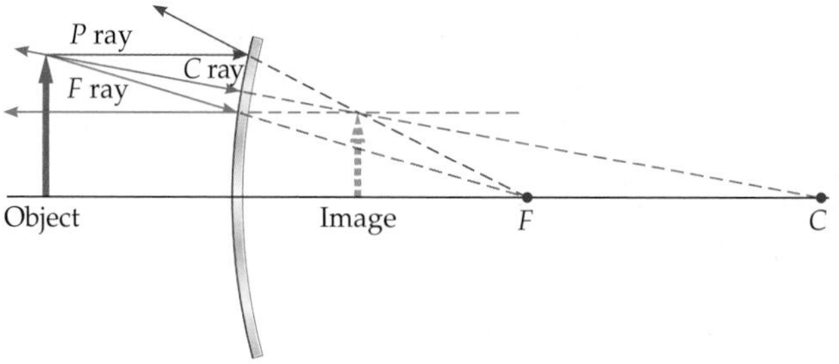

Even though we drew three rays in Figure 26–16, any two would have given the intersection point at the tip of the virtual image. This is commonly the case with ray diagrams. When possible, it is useful to draw all three rays as a check on your results.

Finally, in the limit that the object is very close to a convex mirror, the mirror is essentially flat and behaves like a plane mirror. Thus the virtual image will be about the same distance behind the mirror that the image is in front, and about the same size as the object. Conversely, if the object is far from the mirror, the image is very small and practically at the focal point. These limits are illustrated in **Figure 26–17**.

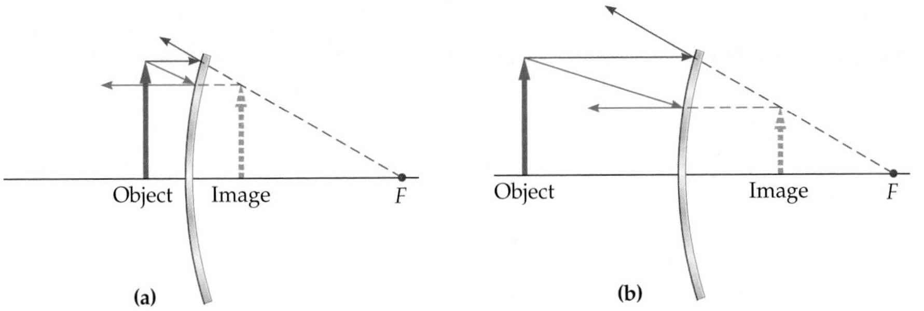

◀ **FIGURE 26–17 Image size and location in a convex mirror**

(a) When an object is close to a convex mirror, the image is practically the same size and distance from the mirror. **(b)** In the limit that the object is very far from the mirror, the image is small and close to the focal point.

Next we consider the image formed by a concave mirror. In particular, we examine the following three cases: (i) The object is farther from the mirror than the center of curvature; (ii) the object is between the center of curvature and the focal point; and (iii) the object is between the mirror and the focal point. The *F* and *P* rays for case (i) are drawn in **Figure 26–18 (a)**, showing that the image is inverted, closer to the mirror, and smaller than the object. In addition, because light rays pass through the image, we call it a **real image.** The corresponding ray diagram for case (ii) is shown in **Figure 26–18 (b)**. Note that the image is again real and inverted, but it is now farther from the mirror and larger than the object—as the object moves closer to the focal point, an observer will see a large inverted image of the object far in front of the mirror. The *C* ray was not useful in these cases and therefore was omitted from the diagrams.

PROBLEM-SOLVING NOTE

Using Rays to Locate the Image of a Spherical Mirror

To find the top of an image, draw the three rays, *P*, *F*, and *C*, from the top of the object. The rays will either intersect at the top of the image (real image) or extend backward to the top of the image (virtual image). When drawing the rays, remember that a *P* ray is *p*arallel to the axis of the mirror, the *C* ray goes either through or toward the mirror's *c*enter of curvature, and the *F* ray goes either through or toward the *f*ocal point of the mirror.

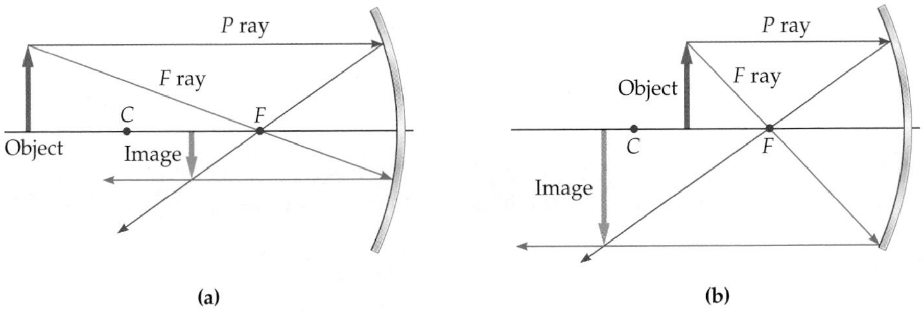

◀ **FIGURE 26–18 Image formation with a concave mirror**

Ray diagrams showing the image formed by a concave mirror when the object is **(a)** beyond the center of curvature and **(b)** between the center of curvature and the focal point.

The final case, in which the object is between the mirror and the focal point, is considered in the next Example.

EXAMPLE 26–3 Image Formation

Use a ray diagram to find the location, size, and orientation of the image formed by a concave mirror when the object is between the mirror and the focal point.

Picture the Problem

The physical system is shown in the diagram, along with the three principal rays. Notice that after reflection these three rays diverge from one another, just as if they had originated at a point behind the mirror.

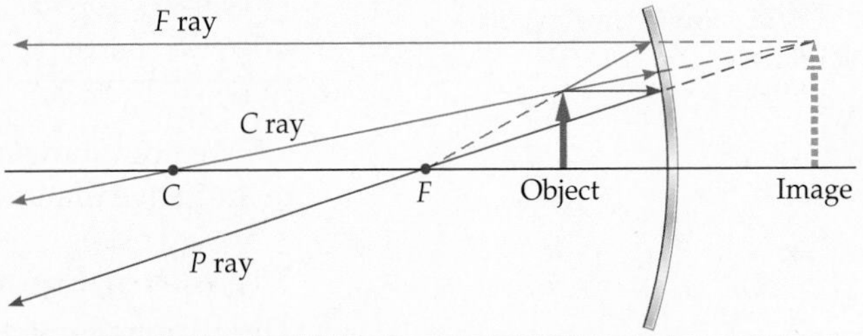

INTERACTIVE FIGURE

Solution

The three outgoing rays (*P*, *F*, and *C*) extend back to form an image behind the mirror that is upright (same orientation as the object) and enlarged. We now discuss these three rays one at a time:

***P* ray:** The *P* ray is the most straightforward of the three. It starts parallel to the axis, then reflects through the focal point.

***F* ray:** The *F* ray does not go through the focal point, as is usually the case. Instead, it starts on a line that *extends back* to the focal point, then reflects parallel to the axis.

continued on next page

continued from previous page

***C* ray:** The *C* ray starts at the top of the object, contacts the mirror at right angles, then reflects back along its initial path and through the center of curvature.

Insight

Makeup mirrors are concave mirrors with fairly large focal lengths. The person applying makeup is between the mirror and its focal point, as is the object in this Example, and therefore sees an upright and enlarged image, as desired.

Practice Problem

If the object in the diagram is moved closer to the mirror, does the image increase or decrease in size? [**Answer:** As the object moves closer, the mirror behaves more like a plane mirror. Thus, the image becomes smaller, so that it is closer in size to the object.]

Some related homework problems: Problem 20, Problem 23

CONCEPTUAL CHECKPOINT 26–3 Rear-View Mirrors

The passenger-side rearview mirrors in newer cars often have warning labels that read, OBJECTS IN MIRROR ARE CLOSER THAN THEY APPEAR. Are these rearview mirrors concave or convex?

Reasoning and Discussion

Objects in the mirror are closer than they appear because the mirror produces an image that is reduced in size, which makes the object look as if it is farther away. In addition, we know that the rearview mirror always gives an upright image, no matter how close or far away the object. The mirror that always produces upright and reduced images is the convex mirror.

Answer:

The mirrors are convex.

The imaging characteristics of convex and concave mirrors are summarized in Table 26–1.

TABLE 26–1 Imaging Characteristics of Convex and Concave Spherical Mirrors

CONVEX MIRROR			
Object location	**Image orientation**	**Image size**	**Image type**
Arbitrary	Upright	Reduced	Virtual
CONCAVE MIRROR			
Object location	**Image orientation**	**Image size**	**Image type**
Beyond *C*	Inverted	Reduced	Real
C	Inverted	Same as object	Real
Between *F* and *C*	Inverted	Enlarged	Real
Just beyond *F*	Inverted	Approaching infinity	Real
Just inside *F*	Upright	Approaching infinity	Virtual
Between mirror and *F*	Upright	Enlarged	Virtual

We now show how to obtain results about the images formed by mirrors in a quantitative manner.

The Mirror Equation

The mirror equation is a precise mathematical relationship between the object distance and the image distance for a given mirror. To obtain this relation, we use the ray diagrams shown in **Figure 26–19**. Note that the distance from the mirror to the object is d_o, the distance from the mirror to the image is d_i, and the distance from the mirror to the center of curvature is R. In addition, the height of the object is h_o, and the height of the image is h_i. Since the image is inverted, its height is negative; thus $-h_i$ is positive.

The ray in Figure 26–19 (a) hits the mirror at its midpoint, where the principal axis is the normal to the mirror. As a result, the ray reflects at an angle θ below

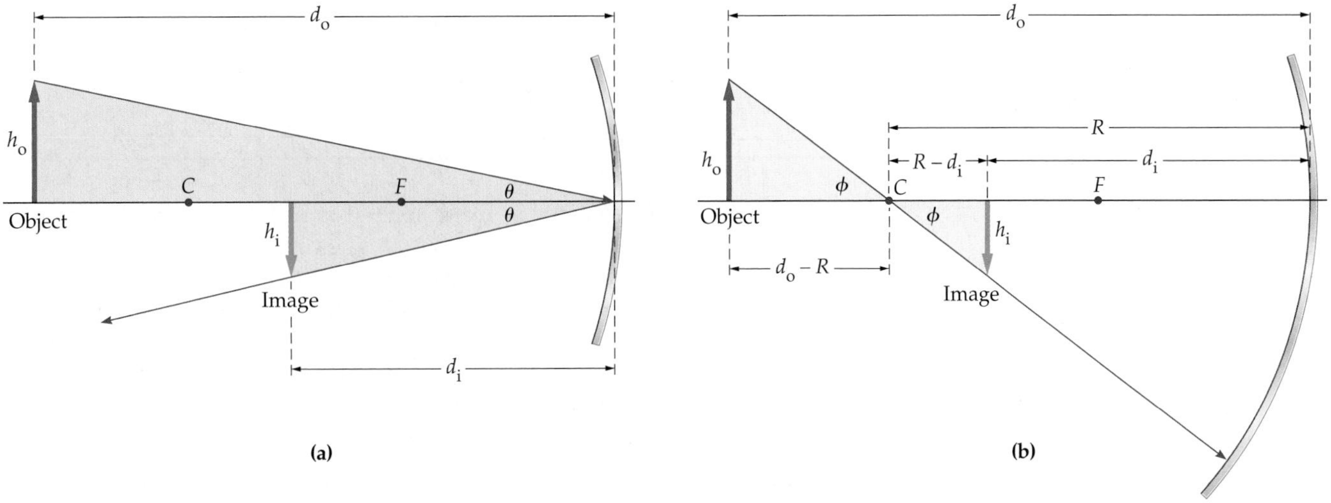

▲ **FIGURE 26–19 Ray diagrams used to derive the mirror equation**
(a) The two similar triangles in this case are used to obtain Equation 26–4. **(b)** These similar triangles yield Equation 26–5.

the principal axis that is equal to its incident angle θ above the axis. Therefore, the two green triangles in this diagram are similar, from which it follows that $h_o/d_o = (-h_i)/d_i$, or

$$\frac{h_o}{-h_i} = \frac{d_o}{d_i} \qquad \textbf{26–4}$$

Next, Figure 26–19 (b) shows the C ray for this mirror. From the figure it is clear that the two yellow triangles in this diagram are also similar, since they are both right triangles and share the common angle ϕ. Thus, $h_o/(d_o - R) = (-h_i)/(R - d_i)$, or

$$\frac{h_o}{-h_i} = \frac{d_o - R}{R - d_i} \qquad \textbf{26–5}$$

Setting these two expressions for $h_o/(-h_i)$ equal gives us

$$\frac{d_o}{d_i} = \frac{d_o - R}{R - d_i} \quad \text{or} \quad 1 = \frac{1 - \dfrac{R}{d_o}}{\dfrac{R}{d_i} - 1}$$

Rearranging, we find $R/d_o + R/d_i = 2$. Finally, dividing by R and recalling that $f = \frac{1}{2}R$ for a concave mirror, we get the **mirror equation:**

The Mirror Equation

$$\frac{1}{d_o} + \frac{1}{d_i} = \frac{1}{f} \qquad \textbf{26–6}$$

We apply the mirror equation to a simple system in the following Exercise.

EXERCISE 26–1

The concave side of a spoon has a focal length of 5.00 cm. Find the image distance for this "mirror" when the object distance is **(a)** 25.0 cm, **(b)** 9.00 cm, and **(c)** 2.00 cm. These three cases correspond to Figure 26–18 (a), Figure 26–18 (b), and Example 26–3, respectively.

Solution

Solving Equation 26–6 for d_i, we obtain the following results:

a. $d_i = \dfrac{d_o f}{d_o - f} = \dfrac{(25.0\text{ cm})(5.00\text{ cm})}{25.0\text{ cm} - 5.00\text{ cm}} = 6.25\text{ cm}$

b. $d_i = \dfrac{d_o f}{d_o - f} = \dfrac{(9.00\text{ cm})(5.00\text{ cm})}{9.00\text{ cm} - 5.00\text{ cm}} = 11.3\text{ cm}$

c. $d_i = \dfrac{d_o f}{d_o - f} = \dfrac{(2.00\text{ cm})(5.00\text{ cm})}{2.00\text{ cm} - 5.00\text{ cm}} = -3.33\text{ cm}$

Note that the image distance in Exercise 26–1 is negative when the object is closer to the mirror than the focal point [$d_o < f$, as in part (c)]. We know from Example 26–3 that this is also the case where the image is *behind* the mirror. Thus, the *sign* of the image distance indicates the *side* of the mirror on which the image is located. As long as we are discussing signs, it should be noted that the mirror equation applies equally well to a *convex* mirror, as long as we recall that the focal length in this case is *negative*. The following Exercise calculates the image distance for the convex mirror shown in Figure 26–16.

EXERCISE 26–2

The convex mirror in Figure 26–16 has a 20.0-cm radius of curvature. Find the image distance for this mirror when the object distance is 6.33 cm, as it is in Figure 26–16.

Solution

Recalling that $f = -\frac{1}{2}R$ for a convex mirror, we find $f = -10.0$ cm and

$$d_i = \frac{d_o f}{d_o - f} = \frac{(6.33\text{ cm})(-10.0\text{ cm})}{6.33\text{ cm} - (-10.0\text{ cm})} = -3.88\text{ cm}$$

This result agrees with the image shown in Figure 26–16.

Next, we consider the height of an image, which is given by the relation in Equation 26–4. Solving this equation for h_i we find

$$h_i = -\left(\frac{d_i}{d_o}\right)h_o \qquad \textbf{26–7}$$

The ratio of the height of the image to the height of the object is defined as the **magnification,** m; that is, $m = h_i/h_o$. From Equation 26–7, we see that

Magnification, *m*

$$m = \frac{h_i}{h_o} = -\frac{d_i}{d_o} \qquad \textbf{26–8}$$

The sign of the magnification gives the orientation of the image. For example, if both d_o and d_i are positive, as in Figure 26–18, the magnification is negative and the image is inverted. Conversely, if the image is behind the mirror, so that d_i is negative, the magnification is positive, and the image is upright. An example of this case is shown in Example 26–3 and Figure 26–16. Finally, the magnitude of the magnification gives the factor by which the size of the image is increased or decreased compared with the object. In the special case of an image with the same size and orientation as the object, as in a plane mirror, the magnification is one.

The sign conventions for mirrors are summarized below:

Focal Length
f is positive for concave mirrors.
f is negative for convex mirrors.

Magnification
m is positive for upright images.
m is negative for inverted images.

Image Distance
d_i is positive for images in front of a mirror (real images).
d_i is negative for images behind a mirror (virtual images).

Object Distance

d_o is positive for objects in front of a mirror (real objects).

d_o is negative for objects behind a mirror (virtual objects).

The case of a negative object distance—that is, a virtual object—can occur when the image from one mirror serves as the object for another mirror or lens. For example, if mirror 1 produces an image that is behind mirror 2, we say that the image of mirror 1 is a virtual object for mirror 2. Such situations will be considered in the next chapter.

We now apply the mirror and magnification equations to specific Examples.

PROBLEM-SOLVING NOTE

Applying the Mirror Equation

To use the mirror equation correctly, be careful to use the appropriate signs for all the known quantities. The final answer will also have a sign, which gives additional information about the system.

EXAMPLE 26–4 Checking It Twice

After leaving some presents under the tree, Santa notices his image in a shiny, spherical Christmas ornament. The ornament is 8.50 cm in diameter and 1.10 m away from Santa. Curious to know the location and size of his image, Santa consults a book on physics. Knowing that Santa likes to "check it twice," what results should he obtain, assuming his height is 1.75 m?

Picture the Problem

The physical situation is illustrated in the sketch, along with the image Santa sees in the ornament. Because the spherical ornament is a convex mirror, the image it forms is upright and reduced.

Strategy

The ornament is a convex mirror of radius $R = \frac{1}{2}(8.50\text{ cm}) = 4.25\text{ cm}$. We can find the location of the image, then, by using the mirror equation with the focal length given by $f = -\frac{1}{2}R$.

Once we have determined the location of the image, we can find its size using $h_i = -(d_i/d_o)h_o = mh_o$, where $h_o = 1.75\text{ m}$.

Solution

1. Calculate the focal length for the ornament: $f = -\frac{1}{2}R = -\frac{1}{2}(4.25\text{ cm}) = -0.0213\text{ m}$

2. Use the mirror equation to find the image distance, d_i:

$$\frac{1}{d_i} = \frac{1}{f} - \frac{1}{d_o} = \frac{1}{(-0.0213\text{ m})} - \frac{1}{1.10\text{ m}} = -47.9\text{ m}^{-1}$$

$$d_i = \frac{1}{(-47.9\text{ m}^{-1})} = -0.0209\text{ m}$$

3. Determine the magnification of the image using $m = -(d_i/d_o)$:

$$m = -\left(\frac{d_i}{d_o}\right) = -\left(\frac{-0.0209\text{ m}}{1.10\text{ m}}\right) = 0.0190$$

4. Find the image height with $h_i = mh_o$: $h_i = mh_o = (0.0190)(1.75\text{ m}) = 0.0333\text{ m}$

Insight

Thus, Santa's image is 2.09 cm behind the surface of the ornament—about halfway between the surface and the center—and 3.33 cm high. Note that his image fits on the surface of the ornament with room to spare, and therefore Santa can see the reflection of his entire body.

Practice Problem

How far from the ornament must Santa stand if his image is to be 1.75 cm behind its surface? [**Answer:** 9.81 cm]

Some related homework problems: Problem 24, Problem 27, Problem 32

ACTIVE EXAMPLE 26–1 Say Ahhh: Find the Magnification of the Tooth

A dentist uses a small mirror attached to a thin rod to examine one of your teeth. When the tooth is 1.20 cm in front of the mirror, the image it forms is 9.25 cm behind the mirror. Find **(a)** the focal length of the mirror and **(b)** the magnification of the image.

Solution *(Test your understanding by performing the calculations indicated in each step.)*

1. Use the information given to identify the object and image distances: $d_o = 1.20\text{ cm}, d_i = -9.25\text{ cm}$

continued on next page

continued from previous page

Part (a)

2. Use the mirror equation to calculate the focal length: $f = 1.38$ cm

Part (b)

3. Use $m = -(d_i/d_o)$ to find the magnification: $m = 7.71$

Insight

Because the focal length is positive, and the magnification is greater than 1, we conclude that the dentist's mirror is concave. Note that it will make the tooth look 7.71 times larger than life. A convex mirror, in contrast, always produces a magnification less than one.

Your Turn

If the mirror is moved closer to the tooth, will the magnification of the tooth increase or decrease? Find the magnification if the distance from the tooth to the mirror is reduced to 1.00 cm.

(Answers to **Your Turn** *problems are given in the back of the book.)*

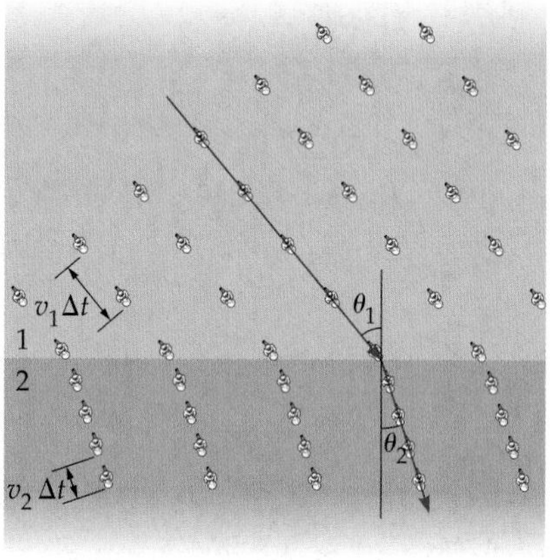

▲ **FIGURE 26–20 An analogy for refraction**

As a marching band moves from an area where the ground is solid to one where it is soft and muddy, the direction of motion changes.

Finally, recall that a spherical mirror behaves like a plane mirror when the object is close to the mirror. But just what exactly do we mean by *close*? Well, in this case, *close* means that the object distance should be small in comparison with the radius of curvature. For example, if the radius of curvature were to go to infinity, the mirror would behave like a plane mirror for all object distances. Simply put, a sphere with $R \to \infty$ has a surface that is essentially as flat as a plane. Letting $f = \frac{1}{2}R$ go to infinity in the mirror equation yields

$$\frac{1}{d_o} + \frac{1}{d_i} = \frac{1}{f} \to 0$$

Therefore, in this limit $d_i = -d_o$, as expected for a plane mirror.

26–5 The Refraction of Light

When a wave propagates from a medium in which its speed is v_1 to another in which its speed is $v_2 \neq v_1$, it will, in general, change its direction of motion. This phenomenon is called **refraction.** To understand the cause of refraction, consider the behavior of a marching band as it moves from a solid section of ground to an area where the ground is muddy, as indicated in **Figure 26–20**. In this analogy, the rows of the marching band correspond to the wave fronts of a traveling wave, and the solid and muddy sections of the ground represent media in which the wave speed is different.

If the speed of the marchers on solid ground is v_1, then after a time Δt they have advanced a distance $v_1\Delta t$. On the other hand, if the marchers in the mud move at the reduced speed v_2, they advance only a distance $v_2\Delta t$ in the same time. This causes a bend in the line of marchers, as can be seen in Figure 26–20, and therefore a change in the direction of motion.

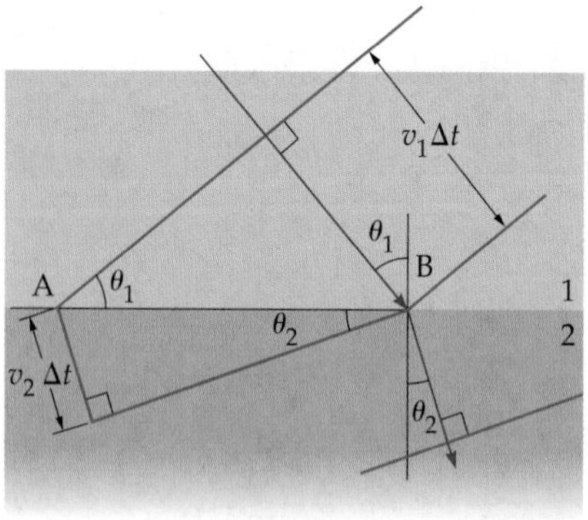

▲ **FIGURE 26–21 The basic mechanism of refraction**

Refraction is the bending of wave fronts and a change in direction of propagation due to a change in speed.

Figure 26–21 shows a simplified version of Figure 26–20, with three "wave fronts" taking the place of the marchers. Also shown is a ray drawn at right angles to the wave fronts, and a normal to the interface between the two types of ground. The angle of incidence is θ_1, and from the figure we can see that this is also the angle between the incoming wave front and the interface. The outgoing wave front and its ray are characterized by a different angle, θ_2. From the geometry of the figure, we see that the green and brown triangles share a common side, AB. Therefore,

$$\sin\theta_1 = \frac{v_1\Delta t}{AB} \quad \text{and} \quad \sin\theta_2 = \frac{v_2\Delta t}{AB}$$

Eliminating the common factor $\Delta t/AB$ yields

$$\frac{\sin\theta_1}{v_1} = \frac{\sin\theta_2}{v_2} \qquad \textbf{26–9}$$

Thus, we see that the direction of propagation is directly related to the speed of propagation.

The speed of light, in turn, depends on the medium through which it travels. For example, we know that in a vacuum the speed of light is $c = 3.00 \times 10^8$ m/s. When light propagates through water, however, its speed is reduced by a factor of 1.33. In general, the speed of light in a given medium, v, is determined by the medium's **index of refraction,** n, defined as follows:

Definition of the Index of Refraction, *n*

$$v = \frac{c}{n} \qquad \textbf{26–10}$$

Representative values of the index of refraction for a variety of media are given in Table 26–2.

TABLE 26–2 Index of Refraction for Common Substances

Substance	Index of refraction, n
SOLIDS	
Diamond	2.42
Flint glass	1.66
Crown glass	1.52
Fused quartz (glass)	1.46
Ice	1.31
LIQUIDS	
Benzene	1.50
Ethyl alcohol	1.36
Water	1.33
GASES	
Carbon dioxide	1.00045
Air	1.000293

EXERCISE 26–3

How long does it take for light to travel 2.50 m in water?

Solution

The speed of light in water is c/n, where $n = 1.33$. Therefore, the time required to cover 2.50 m is

$$t = \frac{d}{v} = \frac{d}{(c/n)} = \frac{2.50\text{ m}}{\left(\dfrac{3.00 \times 10^8\text{ m/s}}{1.33}\right)} = 1.11 \times 10^{-8}\text{ s}$$

▲ The water in this tank contains a fluorescent dye, making it easier to see the refraction of the beam as it passes from the air into the water.

Returning to the direction of propagation, let's suppose light has the speed $v_1 = c/n_1$ in one medium and $v_2 = c/n_2$ in a second medium. The direction of propagation in these two media is related by Equation 26–9:

$$\frac{\sin\theta_1}{(c/n_1)} = \frac{\sin\theta_2}{(c/n_2)}$$

Elimination of the common factor c yields the following relation, known as Snell's law:

Snell's Law

$$n_1 \sin\theta_1 = n_2 \sin\theta_2 \qquad \textbf{26–11}$$

A typical application of Snell's law is given in the following Exercise.

EXERCISE 26–4

A beam of light in air enters **(a)** water ($n = 1.33$) or **(b)** diamond ($n = 2.42$) at an angle of 60.0° relative to the normal. Find the angle of refraction for each case.

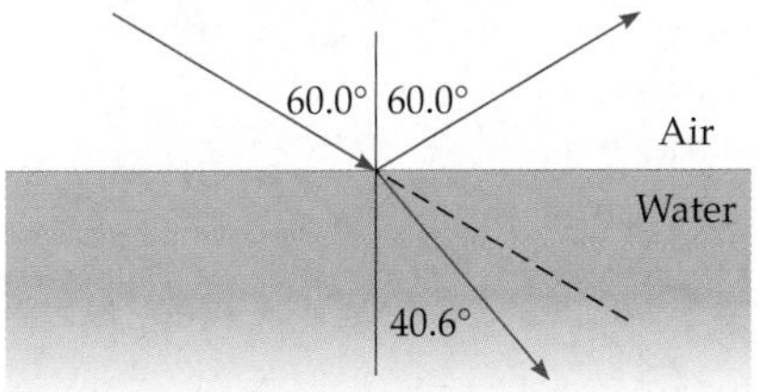

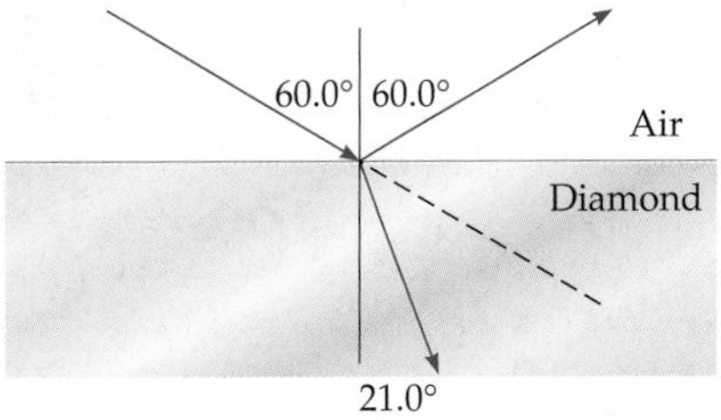

Solution

Since the beam starts in air, we refer to Table 26–2 and set $n_1 = 1.000293$, or simply $n_1 = 1.00$ to three significant figures.

a. With $n_2 = 1.33$ we find

$$\theta_2 = \sin^{-1}\left(\frac{1.00}{n_2}\sin\theta_1\right) = \sin^{-1}\left(\frac{1.00}{1.33}\sin 60.0^\circ\right) = 40.6^\circ$$

b. Setting $n_2 = 2.42$, we get

$$\theta_2 = \sin^{-1}\left(\frac{1.00}{n_2}\sin\theta_1\right) = \sin^{-1}\left(\frac{1.00}{2.42}\sin 60.0^\circ\right) = 21.0^\circ$$

From the preceding Exercise we can see that the greater the difference in index of refraction between two different materials, the greater the difference in direction of propagation. In addition, light is bent closer to the normal in the medium where its speed is less. Of course, the opposite is true when light passes into a medium in which its speed is greater, as can be seen by reversing the incident and refracted rays. The qualitative features of refraction are as follows:

- When a ray of light enters a medium where the index of refraction is increased, and hence the speed of the light is *decreased*, the ray is bent *toward* the normal.
- When a ray of light enters a medium where the index of refraction is decreased, and hence the speed of the light is *increased*, the ray is bent *away* from the normal.
- There is no change in direction of propagation if there is no change in index of refraction. The greater the change in index of refraction, the greater the change in propagation direction.
- If a ray of light goes from one medium to another along the normal, it is undeflected, regardless of the index of refraction.

The last property listed follows directly from Snell's law: If θ_1 is zero, then $0 = n_2 \sin\theta_2$, which means that $\theta_2 = 0$. Refraction is explored further in the following Example.

PROBLEM-SOLVING NOTE

Applying Snell's Law

To apply Snell's law correctly, recall that the two angles in $n_1 \sin\theta_1 = n_2 \sin\theta_2$ are always measured relative to the normal at the interface. Also, note that each angle is associated with its corresponding index of refraction. For example, the angle θ_1 is the angle to the normal in the substance with an index of refraction equal to n_1.

EXAMPLE 26–5 Sitting on a Dock of the Bay

One night, while on vacation in the Caribbean, you walk to the end of a dock and, for no particular reason, shine your laser pointer into the water. When you shine the beam of light on the water a horizontal distance of 2.4 m from the dock, you see a glint of light from a shiny object on the sandy bottom—perhaps a gold doubloon. If the pointer is 1.8 m above the surface of the water, and the water is 5.5 m deep, what is the horizontal distance from the end of the dock to the shiny object?

Picture the Problem

The person standing at the end of the dock and the shiny object on the bottom are shown in the sketch. Note also that all the known distances are indicated, along with the angle of incidence, θ_1, and the angle of refraction, θ_2. Finally, the appropriate indices of refraction from Table 26–2 are given as well.

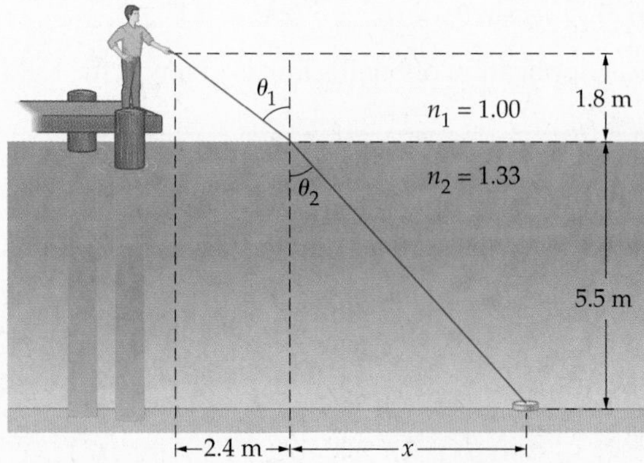

Strategy

We can use Snell's law and basic trigonometry to find the horizontal distance to the shiny object.

First, the information given in the problem determines the angle of incidence, θ_1. In particular, we can see from the sketch that $\tan\theta_1 = (2.4\text{ m})/(1.8\text{ m})$.

Second, Snell's law, $(n_1 \sin\theta_1 = n_2 \sin\theta_2)$, gives the angle of refraction, θ_2.

Finally, the sketch shows that the horizontal distance to the shiny object is $2.4\text{ m} + x$. We can find the distance x from θ_2, since $\tan\theta_2 = x/(5.5\text{ m})$.

Solution

1. Find the angle of incidence from the information given in the problem: $\theta_1 = \tan^{-1}\left(\frac{2.4\text{ m}}{1.8\text{ m}}\right) = 53°$

2. Use Snell's law to calculate the angle of refraction: $\theta_2 = \sin^{-1}\left(\frac{n_1}{n_2}\sin\theta_1\right)$

$$= \sin^{-1}\left[\left(\frac{1.00}{1.33}\right)\sin 53°\right] = 37°$$

3. Calculate x using $\tan\theta_2 = x/(5.5\text{ m})$: $\tan\theta_2 = \frac{x}{5.5\text{ m}}$

$$x = (5.5\text{ m})\tan\theta_2 = (5.5\text{ m})\tan 37° = 4.1\text{ m}$$

4. Add 2.4 m to x to find the total horizontal distance to the shiny object: $\text{distance} = 2.4\text{ m} + x = 2.4\text{ m} + 4.1\text{ m} = 6.5\text{ m}$

Insight

Notice that if the water were to be removed, the incident beam of light would continue along its original path, characterized by the angle θ_1, and overshoot the hoped-for doubloon by a significant distance. Similarly, if you were to simply stand at the end of the dock and look out into the water at a glint of gold on the bottom, you would be looking in a direction (again characterized by θ_1) that is too high—the gold would be below your line of sight and closer to you than you think.

Practice Problem

If the index of refraction for water were 1.35 instead of 1.33, would the doubloon be farther from the dock, nearer the dock, or the same distance calculated above? Check your answer by calculating the distance with $n_2 = 1.35$. [**Answer:** The doubloon would be closer to the dock. The horizontal distance in this case would be 6.4 m.]

Some related homework problems: Problem 39, Problem 43, Problem 47

REAL-WORLD PHYSICS
Apparent depth

Refraction is responsible for a number of common "optical illusions." For example, we all know that a pencil placed in a glass of water appears to be bent, though it is still perfectly straight. The cause of this illusion is shown in **Figure 26–22**, where we see that rays leaving the water bend away from the normal and make the pencil appear to be above its actual position. This is an example of what is known as *apparent depth*, in which an object appears to be closer to the water's surface than it really is. The relation between the true depth and apparent depth is considered in Problem 48.

REAL-WORLD PHYSICS
Mirages

Similarly, refraction can cause a mirage, which makes hot, dry ground in the distance appear to be covered with water. Basically, hot air near the surface is less dense—and hence has a smaller index of refraction—than the cooler air higher

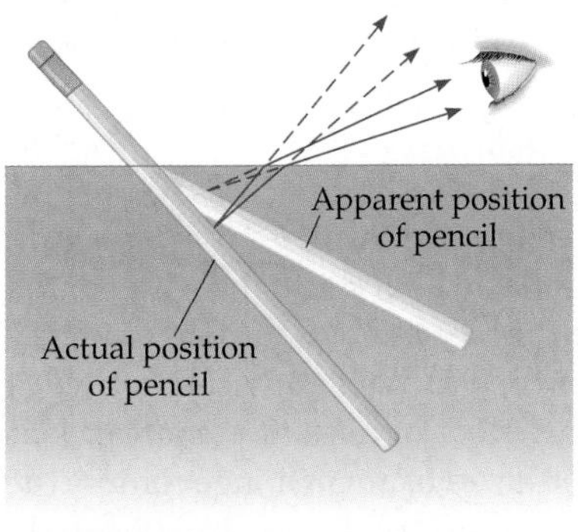

▲ **FIGURE 26–22 Refraction and the "bent" pencil**

Refraction causes a pencil to appear bent when placed in water. Note that rays leaving the water are bent away from the normal and hence extend back to a point that is higher than the actual position of the pencil.

▲ One of the most common mirages, often seen in hot weather, makes a stretch of road look like the surface of a lake. The blue color that so resembles water to our eyes is actually an image of the sky, refracted by the hot, low-density air above the road.

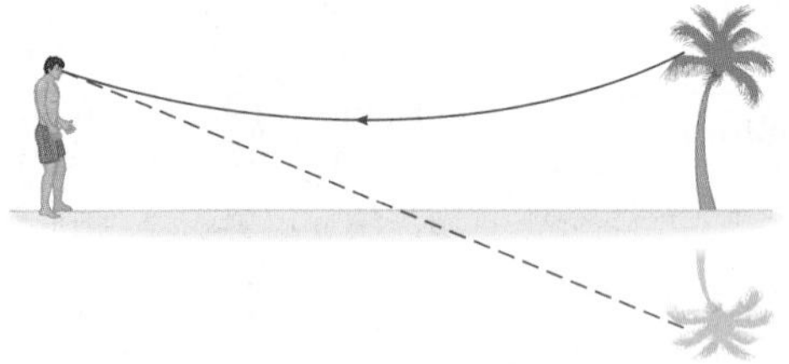

▶ **FIGURE 26–23 A mirage**

A mirage is produced when light bends upward due to the low index of refraction of heated air near the ground.

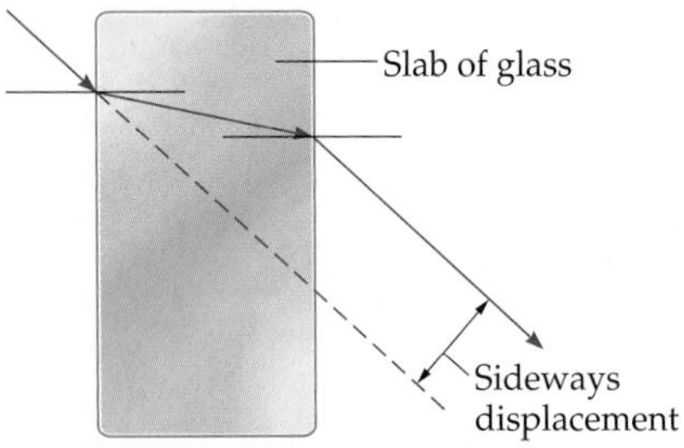

▲ **FIGURE 26–24 Light propagating through a glass slab**

When a ray of light passes through a glass slab, it first refracts toward the normal, then away from the normal. The net result is that the ray continues in its original direction but is displaced sideways by a finite distance.

up. Thus, as light propagates toward the ground, it bends away from the normal until, eventually, it travels upward and enters the eye of an observer, as indicated in **Figure 26–23**. What appears to be a reflecting pool of water in the distance, then, is actually an image of the sky.

Finally, if light passes through a refracting slab, like the sheet of glass in **Figure 26–24**, it undergoes two refractions—one at each surface of the slab. The first refraction bends the light rays closer to the normal, and the second refraction bends the rays away from the normal. As can be seen in the figure, the two changes in direction cancel, so that the final direction of the light is the same as its original direction. The light has been displaced slightly, however, by an amount proportional to the thickness of the slab. The displacement distance is calculated in Problem 98.

CONCEPTUAL CHECKPOINT 26–4 Refraction in a Prism

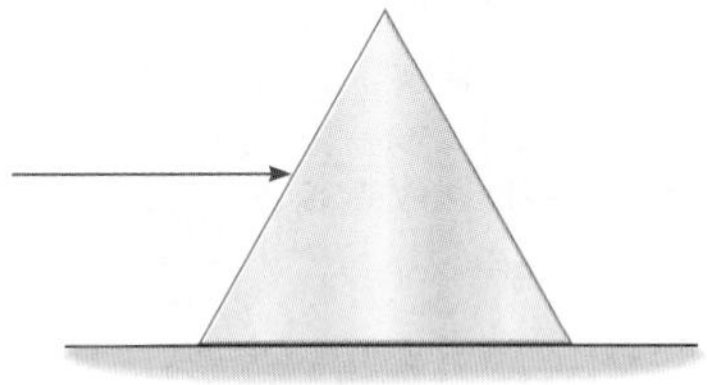

A horizontal ray of light encounters a prism, as shown in the first diagram. After passing through the prism, is the ray **(a)** deflected upward, **(b)** still horizontal, or **(c)** deflected downward?

Reasoning and Discussion

When the ray enters the prism, it is bent toward the normal, which deflects it *downward*, as shown in the second diagram. When it leaves through the opposite side of the prism, it is bent away from the normal. Because the sides of a prism are angled in opposite directions, however, bending away from the normal in the second refraction also causes a *downward* deflection.

The net result, then, is a downward deflection of the ray.

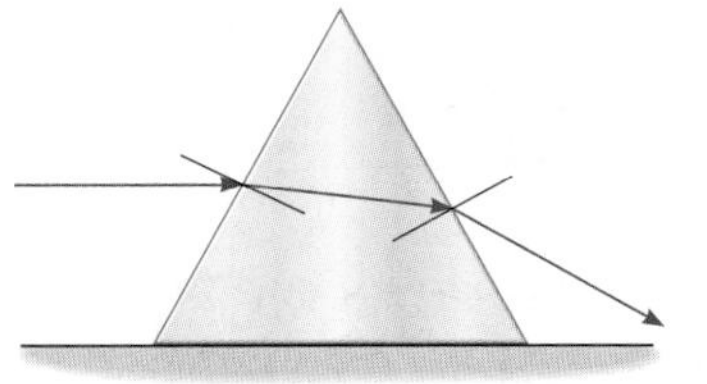

Answer:

(c) The ray deflects downward.

We will use the results of this Conceptual Checkpoint in the next section when we investigate the behavior of lenses. For now, we turn to two additional phenomena associated with refraction.

Total Internal Reflection

Figure 26–25 (a) shows a ray of light in water encountering a water–air interface. In such a case, it is observed that part of the light is reflected back into the water at the interface—as from the surface of a mirror—while the rest emerges into the air along a direction that is bent away from the normal according to Snell's law. If the angle of incidence is increased, as in **Figure 26–25 (b)**, the angle of refraction increases as well. At some critical angle of incidence, θ_c, the refracted beam no longer enters the air but instead is directed parallel to the water–air interface (**Figure 26–25 (c)**). In this case, the angle of refraction is 90°. For angles of incidence greater than the critical angle, as in **Figure 26–25 (d)**, it is observed that all the light is reflected back into the water. This phenomenon is referred to as **total internal reflection.**

We can find the critical angle for total internal reflection by setting $\theta_2 = 90°$ and applying Snell's law:

$$n_1 \sin \theta_c = n_2 \sin 90° = n_2$$

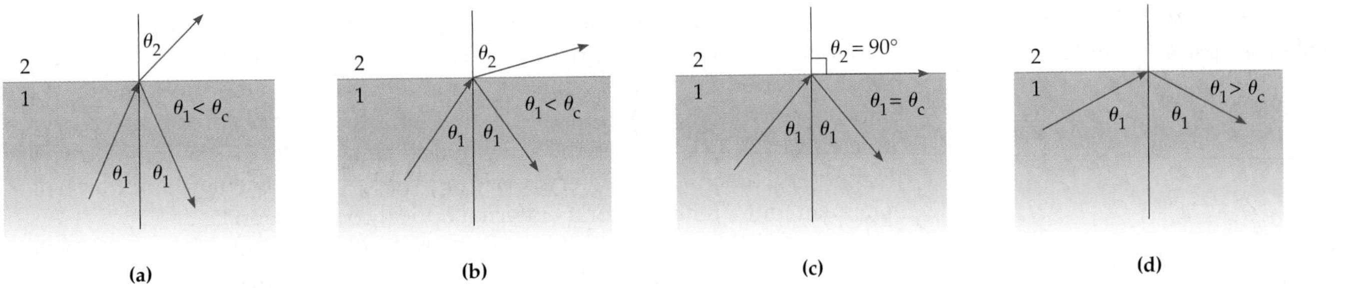

▲ **FIGURE 26–25 Total internal reflection**

Total internal reflection can occur when light propagates from a region with a high index of refraction to one with a lower index of refraction. Part **(a)** shows an incident ray in medium 1 encountering the interface with medium 2, which has a lower index of refraction. A portion of the ray is transmitted to region 2 at the angle θ_2, given by Snell's law, and a portion of the ray is reflected back into medium 1 at the angle of incidence, θ_1. The sum of the intensities of the refracted and reflected rays equals the intensity of the incident ray. In part **(b)** the angle of incidence has been increased, and the refracted beam makes a smaller angle with the interface. When the angle of incidence equals the critical angle, θ_c, as in part **(c)**, the refracted ray propagates parallel to the interface. For incident angles greater than θ_c there is no refracted ray at all, as shown in part **(d)**, and all of the incident intensity goes into the reflected ray.

Therefore, the critical angle is given by the following relation:

Critical Angle for Total Internal Reflection, θ_c

$$\sin\theta_c = \frac{n_2}{n_1} \qquad \textbf{26–12}$$

Since $\sin\theta$ is always less than or equal to 1, the index of refraction, n_1, must be larger than the index of refraction n_2 if the Equation 26–12 is to give a physical solution. Thus, total internal reflection can occur only when light in one medium encounters an interface with another medium in which the speed of light is greater. For example, light moving from water to air can undergo total internal reflection, as shown in Figure 26–25, but light moving from air to water cannot.

PROBLEM-SOLVING NOTE

Total Internal Reflection

Remember that total internal reflection can occur only on the side of an interface between two different substances that has the greater index of refraction.

EXAMPLE 26–6 Light Totally Reflected

Find the critical angle for light traveling from glass ($n = 1.50$) to **(a)** air ($n = 1.00$) and **(b)** water ($n = 1.33$).

Picture the Problem

Our sketch shows the two cases considered in the problem. Note that in each case the incident medium is glass; therefore $n_1 = 1.50$. For part (a) it follows that $n_2 = 1.00$, and for part (b) $n_2 = 1.33$.

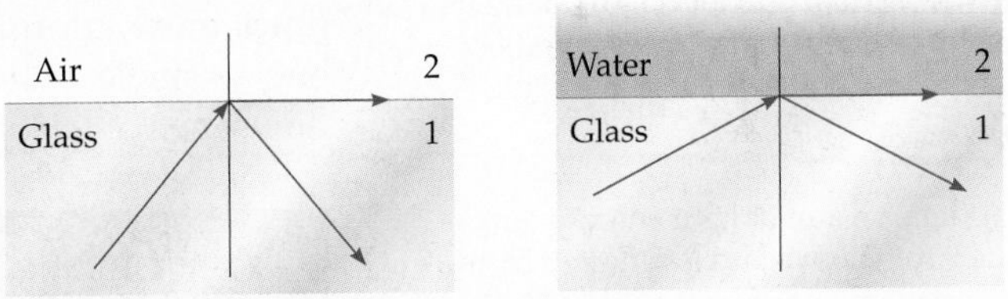

Strategy

The critical angle, θ_c, is defined in Equation 26–12. It is straightforward to obtain θ_c by simply substituting the appropriate indices of refraction for each case in the relation $\theta_c = \sin^{-1}(n_2/n_1)$.

Solution

Part (a)

1. Solve Equation 26–12 for θ_c, using $n_1 = 1.50$ and $n_2 = 1.00$:

$$\theta_c = \sin^{-1}\left(\frac{n_2}{n_1}\right) = \sin^{-1}\left(\frac{1.00}{1.50}\right) = 41.8°$$

Part (b)

2. Solve Equation 26–12 for θ_c, using $n_1 = 1.50$ and $n_2 = 1.33$:

$$\theta_c = \sin^{-1}\left(\frac{n_2}{n_1}\right) = \sin^{-1}\left(\frac{1.33}{1.50}\right) = 62.5°$$

Insight

Note that in the case of water the two indices of refraction are closer in value; hence, light escapes from glass to water over a wider range of incident angles (0° to 62.5°) than from glass to air (0° to 41.8°). In general, if the indices of refraction of two media are close in value, only light rays with large angles of incidence will undergo total internal reflection.

Practice Problem

Suppose the incident ray is in a different type of glass, with a glass–air critical angle of 40.0°. Is the index of refraction of this glass more than or less than 1.50? Verify your answer with a calculation. [**Answer:** The index of refraction is greater than 1.50. The calculated value is 1.56.]

Some related homework problems: Problem 49, Problem 50, Problem 51

REAL-WORLD PHYSICS
Porro prisms in binoculars

Total internal reflection is frequently put to practical use. For example, many binoculars contain a set of prisms—referred to as *Porro prisms*—that use total internal reflection to "fold" a relatively long light path into the short length of the binoculars, as shown in **Figure 26–26**. This allows the user to hold with ease a relatively short device that has the same optical behavior as a set of long, unwieldy telescopes. Thus, the characteristic zigzag shape of a binocular is not a fashion statement, but a reflection of its internal optical construction.

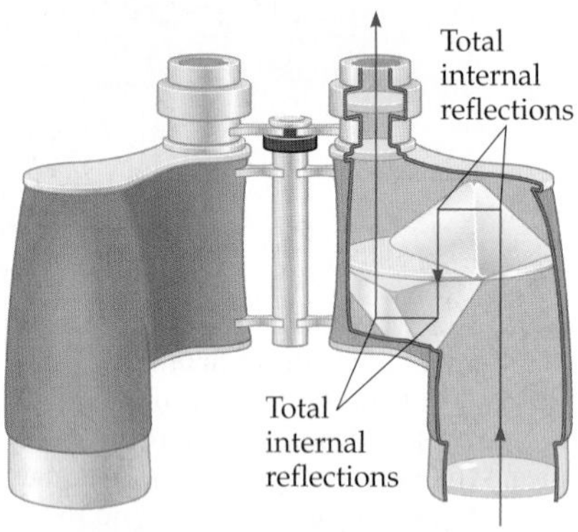

▲ **FIGURE 26–26 Porro prisms**
Prisms are used to "fold" the light path within a pair of binoculars. This makes the binoculars easier to handle.

REAL-WORLD PHYSICS: BIO
Optical fibers and endoscopes

Optical fibers are another important application of total internal reflection. These thin fibers are generally composed of a glass or plastic core with a high index of refraction surrounded by an outer coating, or cladding, with a low index of refraction. Light is introduced into the core of the fiber at one end. It then propagates along the fiber in a zigzag path, undergoing one total internal reflection after another, as indicated in **Figure 26–27**. The core is so transparent that even after light propagates through a 1-km length of fiber, the amount of absorption is roughly the same as if the light had simply passed through a glass window. In addition, the total internal reflections allow the fiber to go around corners, and even to be tied into knots, and still deliver the light to the other end.

The ability of optical fibers to convey light along curved paths has been put to good use in various fields of medicine. In particular, devices known as *endoscopes* allow physicians to examine the interior of the body by snaking a flexible tube containing optical fibers into the part of the body to be examined. For example, a type of endoscope called the bronchoscope can be inserted into the nose or throat, threaded through the bronchial tubes, and eventually placed in the lungs. There, the bronchoscope delivers light through one set of fibers and returns light to the physician through another set of fibers. In some cases, the bronchoscope can even be used to retrieve small samples from the lung for further analysis. Similarly, the colonoscope can be used to examine the colon, making it one of the most important weapons in the fight against colon cancer.

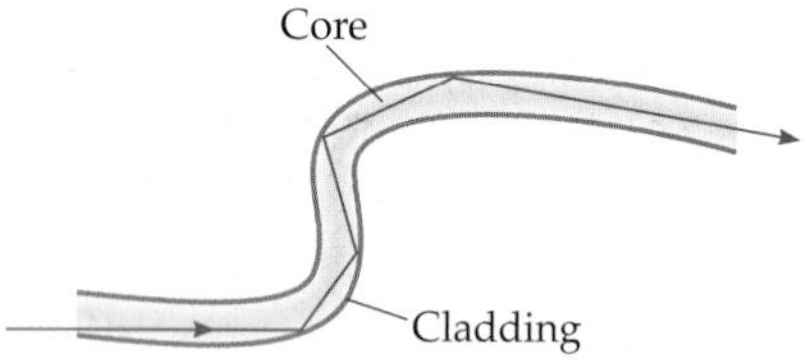

▲ **FIGURE 26–27 An optical fiber**
An optical fiber channels light along its core by a series of total internal reflections between the core and the cladding.

Finally, optical fibers are important components of telecommunication systems. Not only are they small, light, and flexible, but they are also immune to the type of electrical interference that can degrade information carried on copper wires. Even more important, however, is that optical fibers can carry thousands of times more information than an electric current in a wire. For example, it takes only a single optical fiber to transmit several television programs and tens of thousands of telephone conversations, all at the same time.

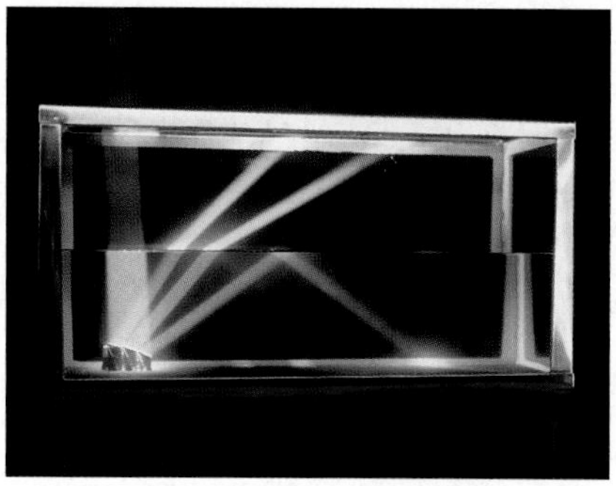

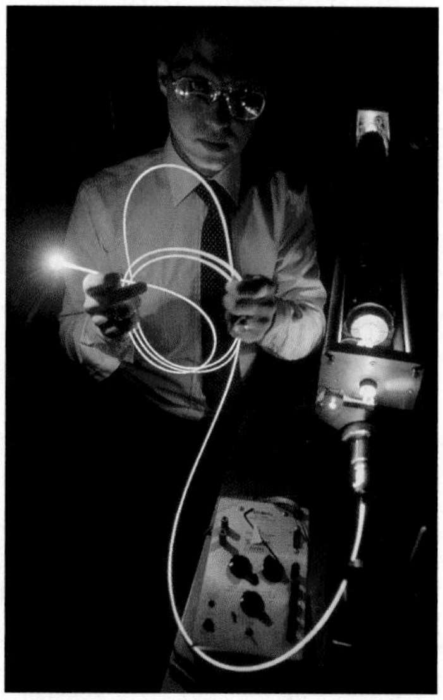

▶ At left, a beam of light enters a tank of water from above and is reflected by mirrors oriented at different angles. Most of the light in the first two beams passes through the water-air interface, undergoing refraction as it leaves the water. Only a small portion of the light is reflected back down into the tank. (The weak beams are hard to see, but the spots they make on the bottom of the tank are clearly visible.) The third beam, however, strikes the interface at an angle of incidence greater than the critical angle. As a result, all of the beam is reflected, as if the surface of the water were a mirror. This phenomenon of total internal reflection makes it possible to "pipe" light through tiny optical fibers such as the one shown at right.

Total Polarization

As explained in Section 25–5, light reflected from a nonmetallic surface is generally polarized to some degree. For example, the light reflected from the surface of a lake is preferentially polarized in the horizontal direction. The polarization of reflected light is complete for one special angle of incidence, **Brewster's angle,** θ_B, defined as follows:

> Reflected light is completely polarized when the reflected and refracted beams are at right angles to one another. The direction of polarization is parallel to the reflecting surface.

This situation is illustrated in **Figure 26–28**. Brewster's angle is named for its discoverer, and the inventor of the kaleidoscope, the Scottish physicist Sir David Brewster (1781–1868).

To calculate Brewster's angle, we begin by applying Snell's law with an incident angle equal to θ_B:

$$n_1 \sin \theta_B = n_2 \sin \theta_2$$

Next, we note from Figure 26–28 that $\theta_B + 90° + \theta_2 = 180°$; that is, $\theta_B + \theta_2 = 90°$, or $\theta_2 = 90° - \theta_B$. Therefore, a standard trigonometric identity (Appendix A) leads to the following relation: $\sin \theta_2 = \sin(90° - \theta_B) = \cos \theta_B$. Combining this result with the preceding equation, we obtain

Brewster's Angle, θ_B

$$\tan \theta_B = \frac{n_2}{n_1} \qquad \text{26–13}$$

We calculate Brewster's angle for a typical situation in the next Exercise.

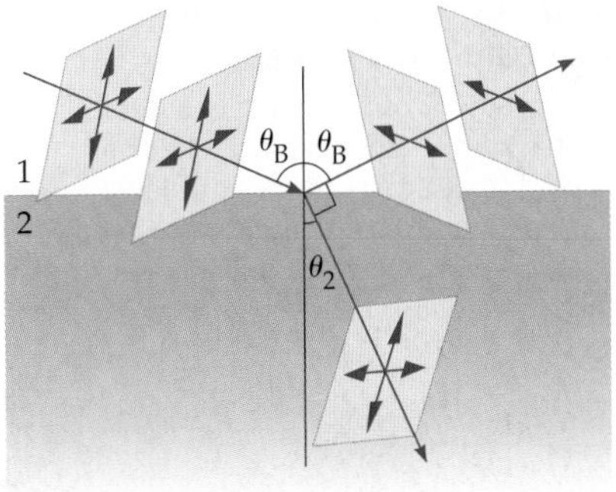

▲ **FIGURE 26–28 Brewster's angle**
When light is incident at Brewster's angle, θ_B, the reflected and refracted rays are perpendicular to each other. In addition, the reflected light is completely polarized parallel to the reflecting surface.

EXERCISE 26–5

Find Brewster's angle for light reflected from the top of a glass ($n = 1.50$) coffee table.

Solution

Letting $n_1 = 1.00$ and $n_2 = 1.50$, we find

$$\theta_B = \tan^{-1}\left(\frac{n_2}{n_1}\right) = 56.3°$$

26–6 Ray Tracing for Lenses

As we have seen, a ray of light can be redirected as it passes from one medium to another. A device that takes advantage of this effect, and uses it to focus light and form images, is a **lens.** Typically, a lens is a thin piece of glass or other transparent substance that can be characterized by the effect it has on light. In particular, converging lenses take parallel rays of light and bring them together at a focus; diverging lenses cause parallel rays to spread out as if diverging from a point source. A variety of converging and diverging lenses are illustrated in **Figure 26–29**, though we consider only the most basic types here—namely, the double concave (or simply concave) and the double convex (or simply convex).

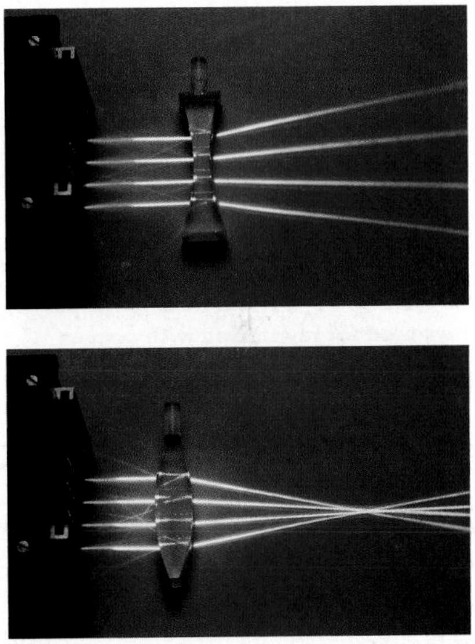

▲ The paths of light rays through a concave (diverging) lens and a convex (converging) lens.

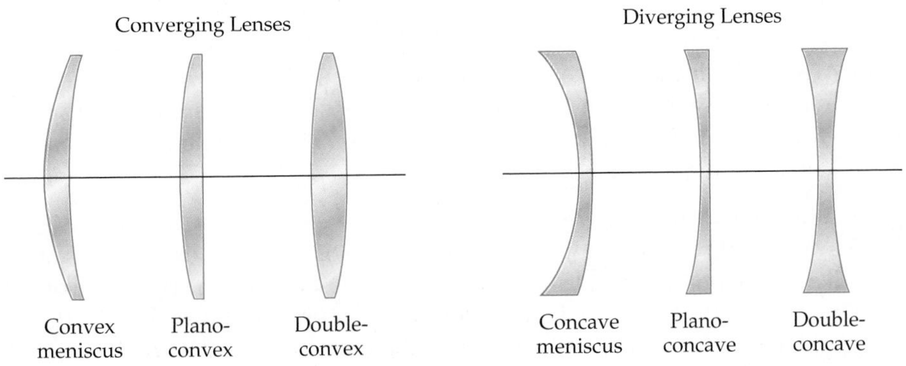

◀ **FIGURE 26–29 A variety of converging and diverging lenses**
Converging and diverging lenses come in a variety of shapes. Generally speaking, converging lenses are thicker in the middle than at the edges, and diverging lenses are thinner in the middle. We will use double convex and double concave lenses when we present examples of converging and diverging lenses, respectively.

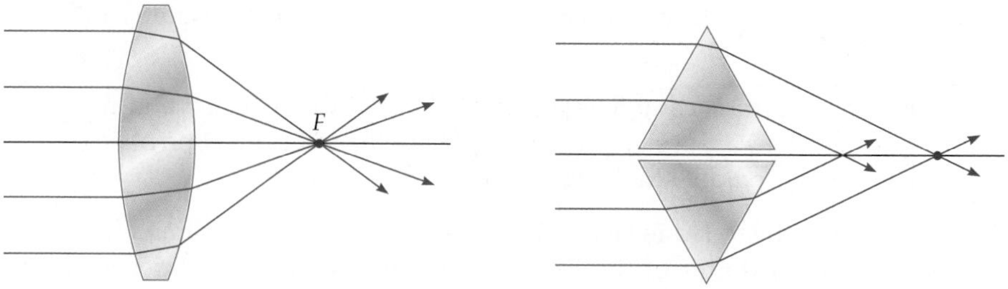

▶ **FIGURE 26–30 A convex lens compared with a pair of prisms**

The behavior of a convex lens is similar to that of two prisms placed back to back. In both cases, light parallel to the axis is made to converge. Note that the lens, because of its curved shape, brings light to a focus at the focal point, *F*.

▲ Drops of dew can serve as double-convex lenses, producing tiny, inverted images of objects beyond their focal points.

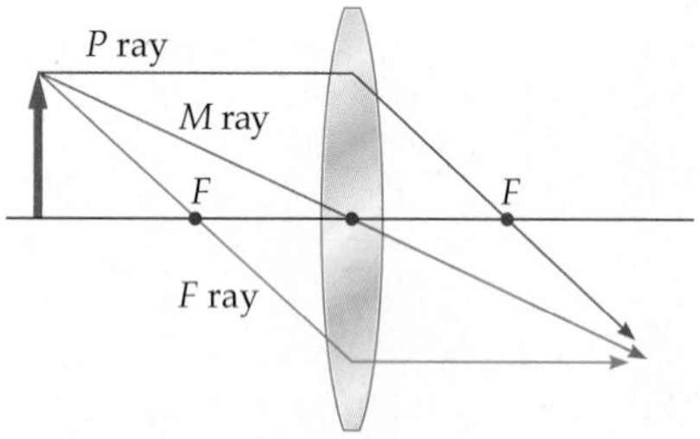

▲ **FIGURE 26–32 The three principal rays used for ray tracing with convex lenses**

The image formed by a convex lens can be found by using the rays shown here. The *P* ray propagates parallel to the principal axis until it encounters the lens, where it is refracted to pass through the focal point on the far side of the lens. The *F* ray passes through the focal point on the near side of the lens, then leaves the lens parallel to the principal axis. The *M* ray passes through the middle of the lens with no deflection. Note that in each case we consider the lens to be of negligible thickness (thin-lens approximation). Therefore, the *P* and *F* rays are depicted as undergoing refraction at the center of the lens, rather than at its front and back surfaces, and the *M* ray has zero displacement. This convention is adopted in all subsequent figures.

First, consider a convex lens, as shown in **Figure 26–30**. To see qualitatively why such a lens is converging, note that it is similar to two prisms placed back-to-back. Recalling the bending of light described for a prism in Conceptual Checkpoint 26–4, we expect parallel rays of light to be brought together. In fact, lenses are shaped so that they bring parallel light to a focus at a **focal point,** *F*, along the center line, or axis, of the lens, as indicated in the figure.

Similarly, a concave lens is qualitatively the same as two prisms brought together point to point, as we see in **Figure 26–31**. In this case, parallel rays are bent away from the axis of the lens. When the diverging rays from a lens are extended back, they appear to originate at a focal point *F* on the axis of the lens.

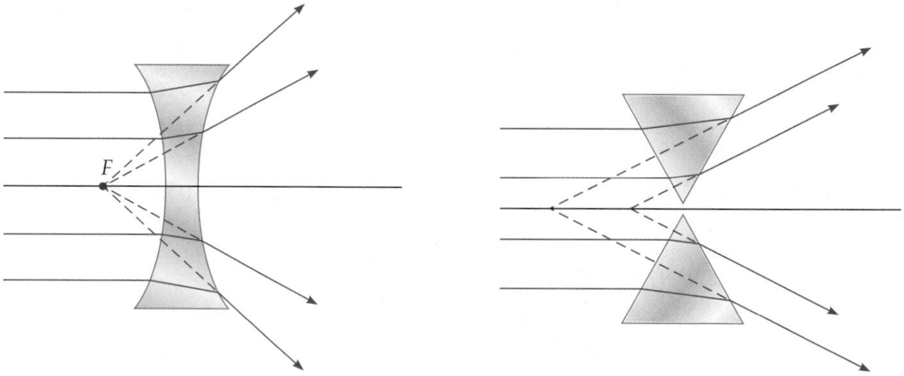

▲ **FIGURE 26–31 A concave lens compared with a pair of prisms**

A concave lens and two prisms placed point to point have similar behavior. In both cases, parallel light is made to diverge.

To determine the type of image formed by a convex or concave lens, we can use ray tracing, as we did with mirrors. The three principal rays for lenses are shown in **Figures 26–32** and **26–33**. Their properties are as follows:

- The *P* ray—or parallel ray—approaches the lens parallel to its axis. The *P* ray is bent so that it passes through the focal point of a convex lens or extrapolates back to the focal point on the same side of a concave lens.
- The *F* ray (focal-point ray) on a convex lens is drawn through the focal point and on to the lens, as pictured in Figure 26–32. The lens bends the ray parallel to the axis—basically the reverse of a *P* ray. For a concave lens, the *F* ray is drawn toward the focal point *on the other side* of the lens, as in Figure 26–33. Before it gets there, however, it passes through the lens and is bent parallel to the axis.
- The midpoint ray (*M* ray) goes through the middle of the lens, which is basically like a thin slab of glass. For ideal lenses, which are infinitely thin, the *M* ray continues in its original direction with negligible displacement after passing through the lens.

To illustrate ray tracing, we start with the concave lens shown in **Figure 26–34**. Notice that the three rays originating from the top of the object extend backward to a single point on the left side of the lens—to an observer on the right side of the lens this point is the top of the image. Our ray diagram also shows that the image is upright and reduced in size. In addition, the image is virtual, since it is on the

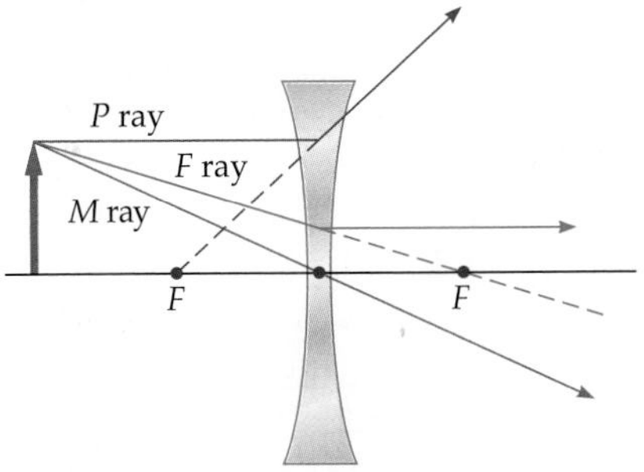

▲ **FIGURE 26–33 The three principal rays used for ray tracing with concave lenses**

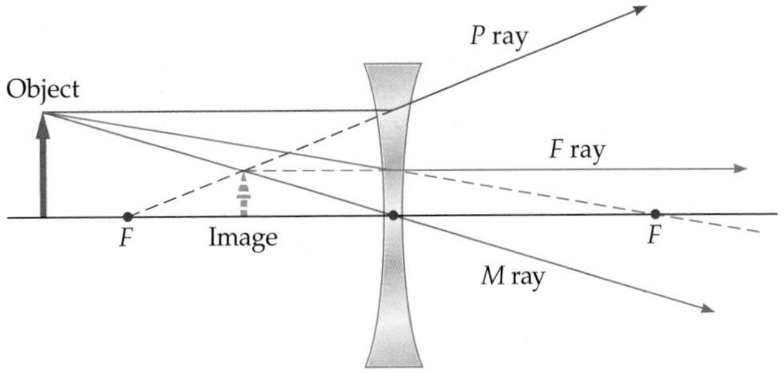

▲ **FIGURE 26–34 The image formed by a concave lens**
Ray tracing can be used to find the image produced by a concave lens. Note that the *P*, *F*, and *M* rays all extend back to the top of the virtual image, which is upright and reduced in size.

same side of the lens as the object. These are general features of the image formed by a concave lens.

The behavior of a convex lens is more interesting in that the type of image it forms depends on the location of the object. For example, if the object is placed beyond the focal point, as in **Figure 26–35 (a)**, the image is on the opposite side of the lens and light passes through it—it is a real image. Notice as well that the image is inverted. If the object is placed between the lens and the focal point, the result, shown in **Figure 26–35 (b)**, is an image that is virtual (on the same side as the object) and upright.

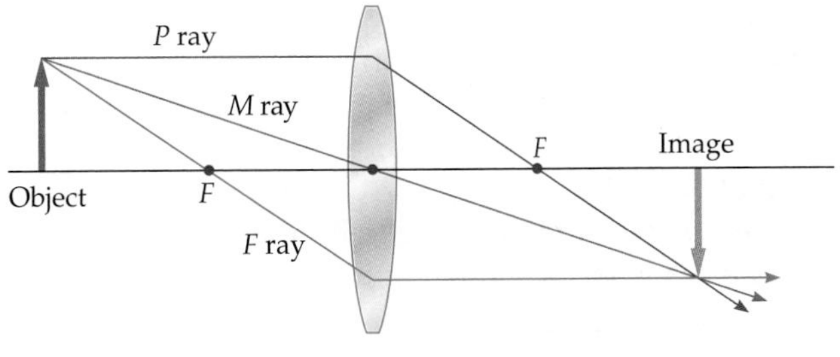

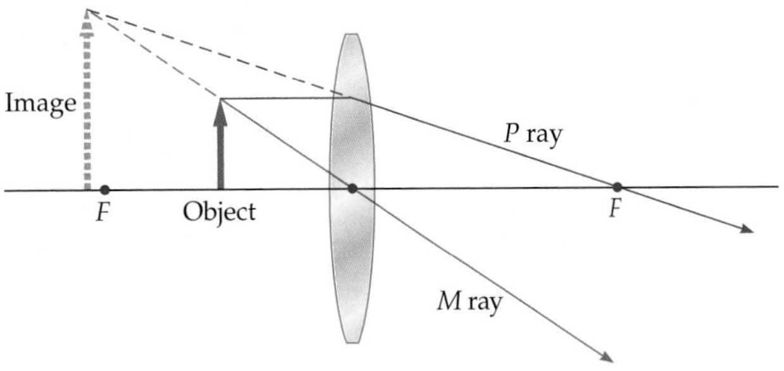

▲ **FIGURE 26–35 Ray tracing for a convex lens**
(a) The object is beyond the focal point. The image in this case is real and inverted. **(b)** The object is between the lens and the focal point. In this case the image is virtual, upright, and enlarged.

The imaging characteristics of concave and convex lenses are summarized in Table 26–3. Compare with Table 26–1, which presents the same information for mirrors.

Finally, the location of the focal point depends on the index of refraction of the lens as well as that of the surrounding medium. This effect is considered in the next Conceptual Checkpoint.

TABLE 26–3 Imaging Characteristics of Concave and Convex Lenses

CONCAVE LENS			
Object location	**Image orientation**	**Image size**	**Image type**
Arbitrary	Upright	Reduced	Virtual
CONVEX LENS			
Object location	**Image orientation**	**Image size**	**Image type**
Beyond *F*	Inverted	Reduced or enlarged	Real
Just beyond *F*	Inverted	Approaching infinity	Real
Just inside *F*	Upright	Approaching infinity	Virtual
Between lens and *F*	Upright	Enlarged	Virtual

CONCEPTUAL CHECKPOINT 26–5 A Lens in Water

The lens shown in the diagram below (left) is generally used in air. If it is placed in water instead, does its focal length (a) increase, (b) decrease, or (c) stay the same?

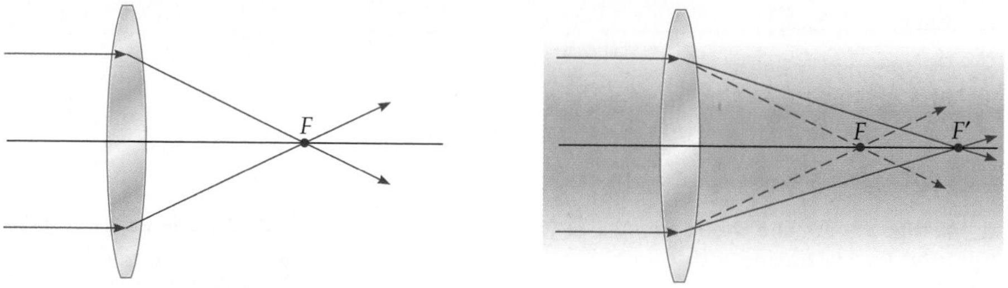

REAL-WORLD PHYSICS: BIO
Underwater vision

Reasoning and Discussion

In water, the difference in index of refraction between the lens and its surroundings is less than when it is in air. Therefore, recalling the discussion following Exercise 26–4, we conclude that light is bent less by the lens when it is in water, as illustrated in the second diagram (right).

As a result, the focal length of the lens is increased.

This explains why our vision is so affected by immersing our eyes in water—the focusing ability of the eye is greatly altered by the water, as we can see from the diagram above. On the other hand, if we wear goggles, so that our eyes are still in contact with air, our vision is normal.

Answer:

(a) The focal length increases.

26–7 The Thin-Lens Equation

To calculate the precise location and size of the image formed by a lens, we use an equation that is analogous to the mirror equation. This equation can be derived by referring to **Figure 26–36**, which shows the image produced by a convex lens, along with the P and M rays that locate the image.

First, note that the P ray creates two similar triangles on the right side of the lens in Figure 26–36(a). Since the triangles are similar, it follows that

$$\frac{h_o}{f} = \frac{-h_i}{d_i - f} \qquad 26\text{–}14$$

In this expression, f is the **focal length**—that is, the distance from the lens to the focal point, F,—and we use $-h_i$ on the right side of the equation, since h_i is negative for an inverted image. Next, the M ray forms another pair of similar triangles, shown in Figure 26–36(b), from which we obtain the following:

$$\frac{h_o}{d_o} = \frac{-h_i}{d_i} \qquad 26\text{–}15$$

Combining these two relations, we obtain the thin-lens equation:

Thin-Lens Equation

$$\frac{1}{d_o} + \frac{1}{d_i} = \frac{1}{f} \qquad 26\text{–}16$$

Finally, the magnification, m, of the image is defined in the same way as for a mirror:

$$h_i = mh_o \qquad 26\text{–}17$$

Rearranging Equation 26–15, we find that $h_i = -(d_i/d_o)h_o$. Therefore, the magnification for a lens, just as for a mirror, is

Magnification, *m*

$$m = -\frac{d_i}{d_o} \qquad 26\text{–}18$$

As before, the sign of the magnification indicates the orientation of the image, and the magnitude gives the amount by which its size is enlarged or reduced compared with the object.

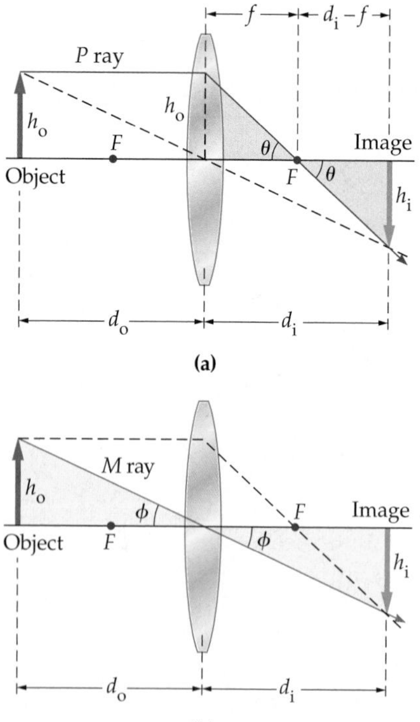

▲ **FIGURE 26–36 Ray diagrams used to derive the thin-lens equation**

(a) The two similar triangles in this case are used to obtain Equation 26–14.
(b) These similar triangles yield Equation 26–15.

A summary of the sign conventions for lenses is as follows:

Focal Length
f is positive for converging (convex) lenses.
f is negative for diverging (concave) lenses.

Magnification
m is positive for upright images (same orientation as object).
m is negative for inverted images (opposite orientation of object).

Image Distance
d_i is positive for real images (images on the opposite side of the lens from the object).
d_i is negative for virtual images (images on the same side of the lens as the object).

Object Distance
d_o is positive for real objects (from which light diverges).
d_o is negative for virtual objects (toward which light converges).

Examples of virtual objects will be presented in Chapter 27.

We now apply the thin-lens equation and the definition of magnification to typical lens systems.

PROBLEM-SOLVING NOTE

Applying the Thin-Lens Equation

To use the thin-lens equation correctly, be careful to use the appropriate signs for all the known quantities. The final answer will also have a sign, which gives additional information about the system.

EXAMPLE 26–7 Object Distance and Focal Length

A lens produces a real image that is twice as large as the object and is located 15 cm from the lens. Find the focal length of the lens and the object distance.

Picture the Problem

Because the image is *real*, the lens must be convex, and the object must be outside the focal point, as indicated in our sketch. [Compare with Figure 26–35 (a).] Note that the image is inverted, which means that the magnification is −2. In addition, the distance to the real image is $d_i = 15$ cm.

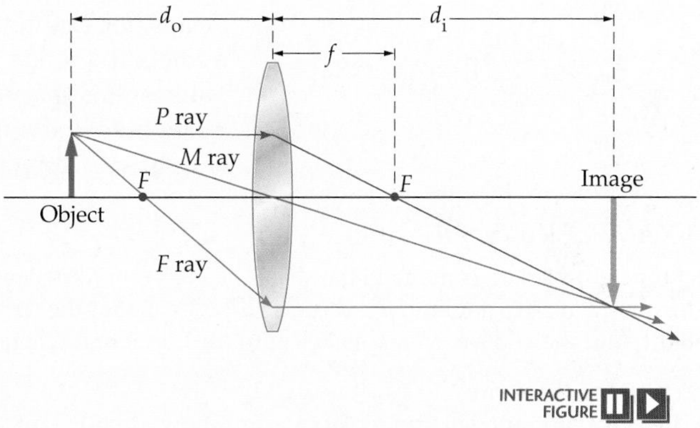

Strategy

To find both f and d_o requires two independent relations. One is provided by the magnification, the other by the thin-lens equation.

Specifically, we can use the magnification, $m = -d_i/d_o$, to find the object distance, d_o. As noted before, $m = -2$ in this case.

Next, we can use the values of d_i and d_o in the thin-lens equation, $1/d_o + 1/d_i = 1/f$, to find the focal length, f.

Solution

1. Use $m = -d_i/d_o$ to find the object distance, d_o:

$$m = -\frac{d_i}{d_o} = -2 \qquad d_o = -\frac{d_i}{m} = -\left(\frac{15\text{ cm}}{-2}\right) = 7.5\text{ cm}$$

2. Use the thin-lens equation to find $1/f$:

$$\frac{1}{f} = \frac{1}{d_o} + \frac{1}{d_i} = \frac{1}{7.5\text{ cm}} + \frac{1}{15\text{ cm}} = \frac{1}{5.0\text{ cm}}$$

3. Invert $1/f$ to find the focal length, f:

$$f = \left(\frac{1}{5.0\text{ cm}}\right)^{-1} = 5.0\text{ cm}$$

Insight

As expected for a convex lens, the focal length is positive. In addition, note that the object distance is greater than the focal length, in agreement with both Figure 26–35 (a) and our sketch for this Example. Finally, the magnification produced by this lens is not always −2. In fact, it depends on the precise location of the object, as we see in the following Practice Problem.

Practice Problem

Suppose we would like to have a magnification of −3 using the same lens. Should the object be moved closer to the lens or farther from it? Calculate the object and image distances for this case. [**Answer:** The object should be moved closer to the lens. This moves the image farther from the lens, and makes it larger in size. We find $d_o = 6.67$ cm and $d_i = 3d_o = 20.0$ cm.]

Some related homework problems: Problem 65, Problem 68, Problem 92

ACTIVE EXAMPLE 26–2 Find the Magnification

An object is placed 12 cm in front of a diverging lens with a focal length of −7.9 cm. Find **(a)** the image distance and **(b)** the magnification.

Solution *(Test your understanding by performing the calculations indicated in each step.)*

1. Use the thin-lens equation to find the image distance, d_i: $d_i = -4.8$ cm
2. Use $m = -d_i/d_o$ to find the magnification: $m = 0.40$

Insight

Since the image distance is negative, it follows that the image is virtual and, hence, on the same side of the lens as the object, as expected for a concave (diverging) lens. In addition, the fact that the magnification is positive means that the image is upright. These numerical values correspond to the system illustrated in Figure 26–34.

Your Turn

If we would like a larger magnification, should the object be moved closer to the lens or farther from it? Calculate the object and image distances that give a magnification of 0.75.

(Answers to **Your Turn** *problems are given in the back of the book.)*

26–8 Dispersion and the Rainbow

As discussed earlier in this chapter, different materials, such as air, water, and glass, have different indices of refraction. There is more to the story, however. The index of refraction for a given material also depends on the frequency of the light—in general, the higher the frequency, the higher the index of refraction. This means, for example, that violet light—with its high frequency and large index of refraction—bends more when refracted by a given material than does red light. The result is that white light, with its mixture of frequencies, is spread out by refraction, so that different colors travel in different directions. This "spreading out" of light according to color is known as **dispersion.**

EXAMPLE 26–8 Prismatics

A flint-glass prism is made in the shape of a 30°-60°-90° triangle, as shown in the diagram. Red and violet light are incident on the prism at right angles to its vertical side. Given that the index of refraction of flint glass is 1.66 for red light and 1.70 for violet light, find the angle each ray makes with the horizontal when it emerges from the prism.

Picture the Problem

The prism and the red and violet rays are shown in our sketch. Note that the angle of incidence on the vertical side of the prism is 0°; hence, the angle of refraction is also 0° for both rays. On the slanted side of the prism, the rays have an angle of incidence equal to 30.0°. Their angles of refraction are different, however.

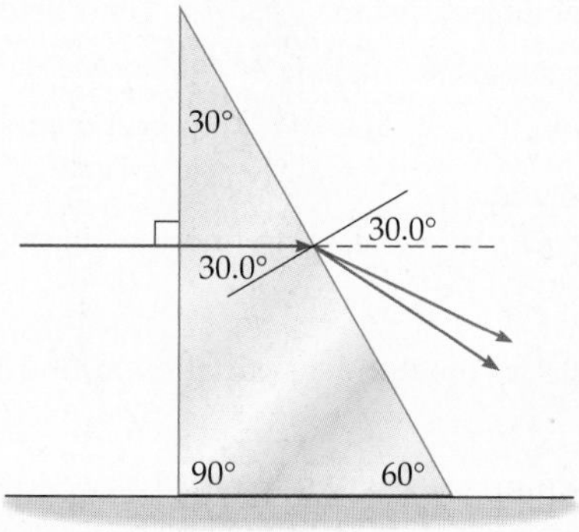

Strategy

To find the final angle for each ray, we apply Snell's law with the appropriate index of refraction. Note, however, that the angle of refraction is measured relative to the normal, which itself is 30.0° above the horizontal. Therefore, the angle each ray makes with the horizontal is the angle of refraction minus 30.0°.

Solution

1. Apply Snell's law with $n_1 = 1.66$, $\theta_1 = 30.0°$, and $n_2 = 1.00$ to find the angle of refraction, θ_2, for red light:

$$n_1 \sin\theta_1 = n_2 \sin\theta_2$$

$$\theta_2 = \sin^{-1}\left(\frac{n_1}{n_2}\sin\theta_1\right) = \sin^{-1}\left(\frac{1.66}{1.00}\sin 30.0°\right) = 56.1°$$

2. Subtract 30.0° to find the angle relative to the horizontal: $56.1° - 30.0° = 26.1°$

3. Next, apply Snell's law with $n_1 = 1.70$, $\theta_1 = 30.0°$, and $n_2 = 1.00$ to find the angle of refraction, θ_2, for violet light:

$$\theta_2 = \sin^{-1}\left(\frac{n_1}{n_2}\sin\theta_1\right) = \sin^{-1}\left(\frac{1.70}{1.00}\sin 30.0°\right) = 58.2°$$

4. Subtract 30.0° to find the angle relative to the horizontal: $58.2° - 30.0° = 28.2°$

Insight

This is the reason for the familiar "rainbow" of colors seen with a prism. It is also the cause of a common defect of simple lenses—referred to as chromatic aberration—in which different colors focus at different points. In the next chapter, we show how two or more lenses in combination can be used to correct this problem.

Practice Problem

If green light emerges from the prism at an angle of 27.0° below the horizontal, what is the index of refraction for this color of light? [**Answer:** $n = 1.68$]

Some related homework problems: Problem 77, Problem 78

REAL-WORLD PHYSICS
The rainbow

Perhaps the most famous and striking example of dispersion is provided by the rainbow, which is caused by the dispersion of light in droplets of rain. The physical situation is illustrated in **Figure 26–37**, which shows a single drop of rain and an incident beam of sunlight. When sunlight enters the drop, it is separated into its red and violet components by dispersion, as shown. The light then reflects from the back of the drop and finally refracts and undergoes additional dispersion as it leaves the drop.

Note that the final direction of the light is almost opposite to its incident direction, falling short by only 40° to 42°, depending on the color of the light. To be specific, violet light—which is bent the most by refraction—changes its direction of propagation by 320°, so that it is 40° away from moving in the direction of the Sun. Red light, which is not bent as much as violet light, changes its direction by only 318° and hence moves in a direction that is 42° away from the Sun.

To see how the rainbow is formed in the sky, imagine standing with your back to the setting Sun, looking toward an area where rain is falling. Consider a single drop as it falls toward the ground. This drop, like all the other drops in the area, is sending out light of all colors in different directions. When the drop is at an angle of 42° above the horizontal, we see the red light coming from it, as indicated in **Figure 26–38**. As the drop continues to fall, its angle above the horizontal decreases. Eventually it reaches a height where its angle with the horizontal is 40°, at which point the violet light from the drop reaches our eye. In between, the drop has sent all the other colors of the rainbow to us for our enjoyment.

▲ A rainbow over Isaac Newton's childhood home in the manor house of Woolsthorpe, near Grantham, Lincolnshire, England. (Note the apple tree near the right side of the house.)

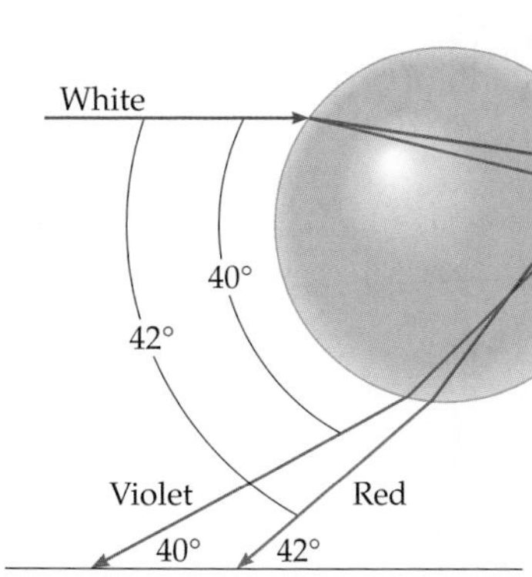

▲ **FIGURE 26–37 Dispersion in a raindrop**
White light entering a raindrop is spread out by dispersion into its various color components—like red and violet. (The angles shown in this figure are exaggerated for clarity.) This is the basic mechanism responsible for the formation of a rainbow. Rays of light that reflect twice inside the raindrop before exiting produce the "secondary" rainbow, which is above the normal, or "primary" rainbow. The sequence of colors in a secondary rainbow is reversed from that in the primary rainbow. A faint secondary bow can be seen in the photograph above in the margin.

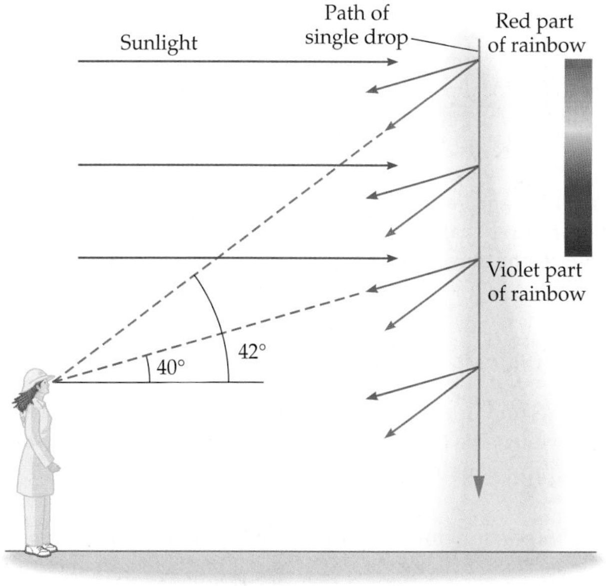

▲ **FIGURE 26–38 How rainbows are produced**
As a single drop of rain falls toward the ground, it sends all the colors of the rainbow to an observer. Note that the top of the rainbow is red; the bottom is violet. (The angles in this figure have been exaggerated for clarity.)

Chapter Summary

26–1 The Reflection of Light

The direction of light can be changed by reflecting it from a shiny surface.

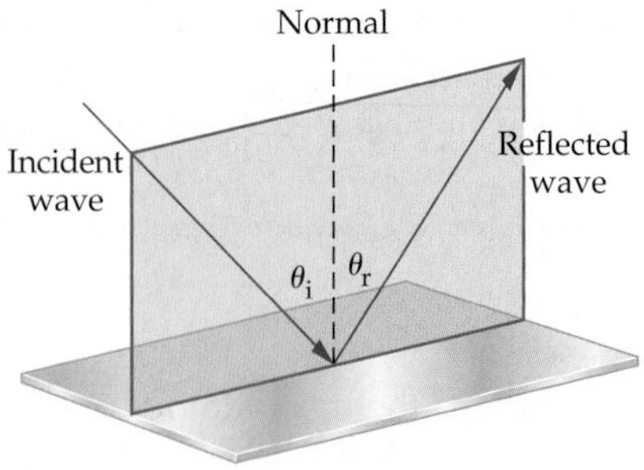

Wave Fronts
A wave front is a surface on which the phase of a wave is constant.

Rays
The direction of wave propagation is indicated by rays, which are always at right angles to wave fronts.

Law of Reflection
The law of reflection states that the angle of reflection, θ_r, is equal to the angle of incidence, θ_i; that is, $\theta_r = \theta_i$.

Specular/Diffuse Reflection
A smooth surface reflects light in a single direction (specular reflection). A rough surface reflects light in many directions (diffuse reflection).

26–2 Forming Images with a Plane Mirror

The image formed by a plane mirror has the following characteristics:

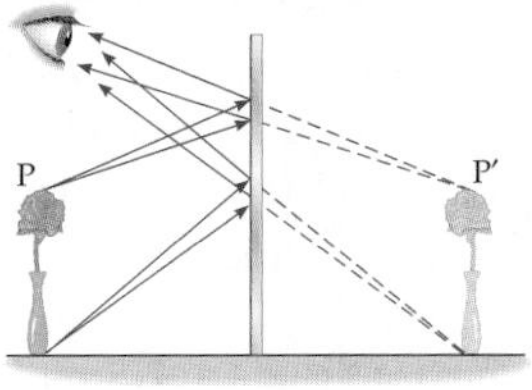

- The image is upright, but appears reversed right to left.
- The image appears to be the same distance behind the mirror that the object is in front of the mirror.
- The image is the same size as the object.

26–3 Spherical Mirrors

A spherical mirror has a spherical reflecting surface. A convex spherical mirror has a reflecting surface that bulges outward. A concave spherical mirror has a hollowed reflecting surface.

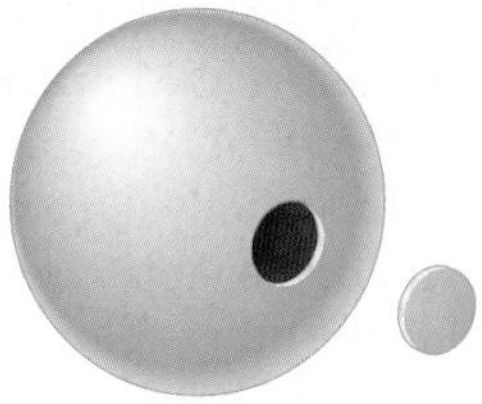

Focal Point and Focal Length for a Convex Mirror
A convex mirror reflects rays that are parallel to its principal axis so that they diverge, as if they had originated from a focal point, F, behind the mirror. The focal length of a convex mirror is

$$f = -\tfrac{1}{2}R \qquad 26\text{–}2$$

Focal Point and Focal Length for a Concave Mirror
A concave mirror reflects rays that are parallel to its principal axis so that they pass through a point known as the focal point, F. The distance from the surface of the mirror to the focal point is the focal length, f. The focal length for a mirror with a radius of curvature R is

$$f = \tfrac{1}{2}R \qquad 26\text{–}3$$

Spherical Aberration and Paraxial Rays
Paraxial rays are rays that are close to the principal axis of a mirror. Rays that are farther from the axis produce a blurred effect known as spherical aberration.

26–4 Ray Tracing and the Mirror Equation

The location, size, and orientation of an image produced by a mirror can be found qualitatively using a ray diagram or quantitatively using a relation known as the mirror equation.

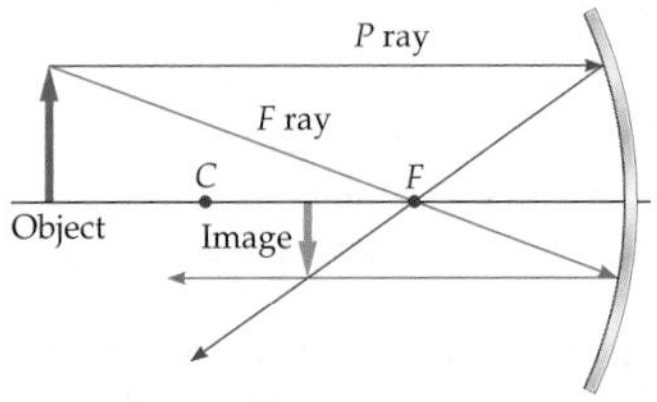

Ray Tracing
Ray tracing involves drawing two or three of the rays that have particularly simple behavior. These rays originate at a point on the object and intersect at the corresponding point on the image.

Real/Virtual Images
An image is said to be real if light passes through the apparent position of the image itself; it is virtual if light does not pass through the image.

Mirror Equation
The mirror equation relates the object distance, d_o, image distance, d_i, and focal length, f:

$$\frac{1}{d_o} + \frac{1}{d_i} = \frac{1}{f} \qquad \text{26–6}$$

The focal length is positive for a concave mirror, negative for a convex mirror. Similarly, the image distance is positive for an image in front of the mirror and negative for an image behind the mirror.

Magnification
The magnification of an image is

$$m = -\frac{d_i}{d_o} \qquad \text{26–8}$$

26–5 The Refraction of Light

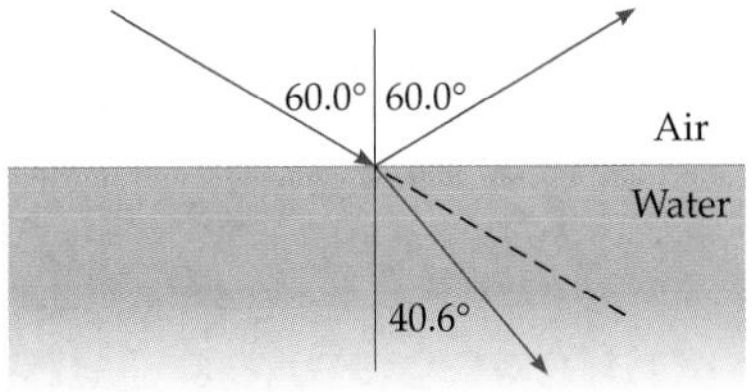

Refraction is the change in direction of light due to a change in its speed.

Index of Refraction
The index of refraction, n, quantifies how much a medium slows the speed of light. In particular, the speed of light in a medium is

$$v = \frac{c}{n} \qquad \text{26–10}$$

Snell's Law
Snell's law relates the index of refraction and angle of incidence in one medium (n_1, θ_1) to the index of refraction and angle of refraction in another medium (n_2, θ_2):

$$n_1 \sin\theta_1 = n_2 \sin\theta_2 \qquad \text{26–11}$$

Qualitative Properties of Refraction
Refracted light is bent closer to the normal in a medium where its speed is reduced and away from the normal in a medium where its speed is increased.

Total Internal Reflection
When light in a medium in which its speed is relatively low encounters a medium in which its speed is greater, the light will be totally reflected back into its original medium if its angle of incidence exceeds the critical angle, θ_c, given by

$$\sin\theta_c = \frac{n_2}{n_1} \qquad \text{26–12}$$

Total Polarization
Reflected light is totally polarized parallel to the surface when the reflected and refracted rays are at right angles. This condition occurs at Brewster's angle, θ_B, given by

$$\tan\theta_B = \frac{n_2}{n_1} \qquad \text{26–13}$$

26–6 Ray Tracing for Lenses

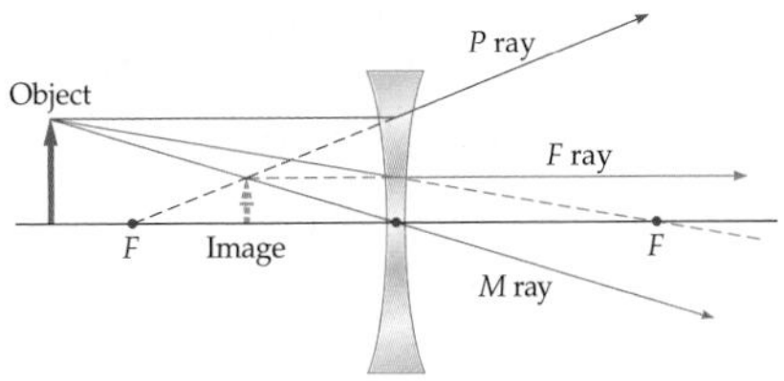

As with mirrors, ray tracing is a convenient way to determine the qualitative features of an image formed by a lens.

Lens
A lens is an object that uses refraction to bend light and form images.

26–7 The Thin-Lens Equation

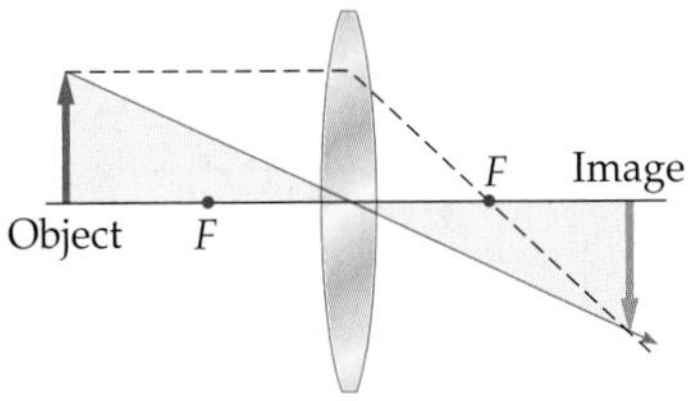

The precise location of an image formed by a lens can be obtained using the thin-lens equation, which has the same form as the mirror equation.

Thin-Lens Equation
The thin-lens equation relates the object distance, d_o, the image distance, d_i, and the focal length, f, for a lens:

$$\frac{1}{d_o} + \frac{1}{d_i} = \frac{1}{f} \qquad \text{26–16}$$

Magnification

The magnification, m, of an image formed by a lens is given by the same expression used for the images produced by mirrors:

$$m = -\frac{d_i}{d_o} \qquad 26\text{–}18$$

Sign Conventions

The focal length is positive for a converging lens, negative for a diverging lens. The magnification is positive for an upright image, negative for an inverted image. The image distance is positive when the image is on the opposite side of a lens from the object, negative when it is on the same side of the lens.

26–8 Dispersion and the Rainbow

The index of refraction depends on frequency—generally, the higher the frequency, the higher the index of refraction. This difference in index of refraction causes light of different colors to be refracted in different directions (dispersion). Rainbows are caused by dispersion within raindrops.

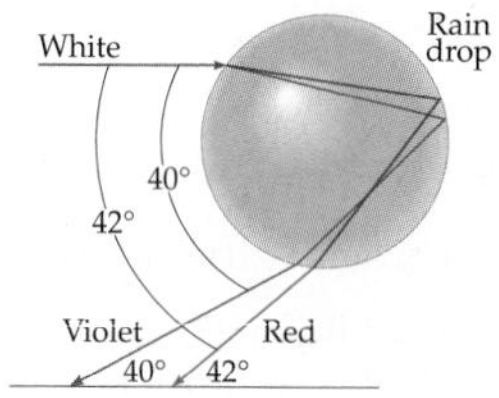

Problem-Solving Summary

Type of Problem	Relevant Physical Concepts	Related Examples
Find the approximate location and size of the image produced by a spherical mirror.	Begin by drawing the three principal rays, P, F, and C from the top of the object. The intersection of these rays gives the location of the top of the image.	Example 26–3
Determine precise values for the location and size of the image produced by a spherical mirror.	The mirror equation, $1/d_o + 1/d_i = 1/f$, relates the object distance, d_o, the image distance, d_i, and the focal length, f. The magnification of the image is given by $m = -(d_i/d_o)$.	Example 26–4 Active Example 26–1
Relate the angles of incidence and refraction.	The angles of incidence and refraction as a beam of light passes from one medium to another are related by Snell's law: $n_1 \sin\theta_1 = n_2 \sin\theta_2$.	Example 26–5
Find the critical angle for total internal reflection.	A light beam in a medium with an index of refraction n_1 may undergo total internal reflection when it encounters a second medium with an index of refraction $n_2 < n_1$. The critical angle at which total internal reflection begins is given by the relation $\sin\theta_c = n_2/n_1$.	Example 26–6
Determine precise values for the location and size of the image produced by a convex or concave lens.	The thin-lens equation, $1/d_o + 1/d_i = 1/f$, relates the object distance, d_o, the image distance, d_i, and the focal length, f. The magnification of the image is given by $m = -(d_i/d_o)$.	Example 26–7 Active Example 26–2

Conceptual Questions

(Answers to odd-numbered Conceptual Questions can be found in the back of the book.)

1. Two plane mirrors meet at right angles at the origin, as indicated in **Figure 26–39**. Suppose an L-shaped object has the position and orientation labeled A. Draw the location and orientation of *all* the images of object A formed by the two mirrors.

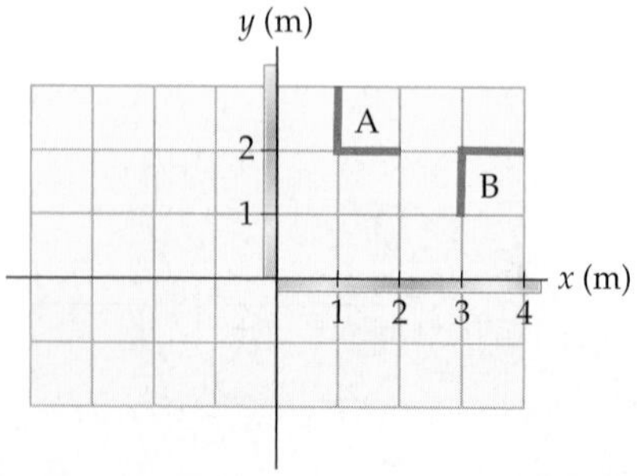

▲ **FIGURE 26–39** Conceptual Questions 1 and 2

2. Two plane mirrors meet at right angles at the origin, as indicated in Figure 26–39. Suppose an L-shaped object has the position and orientation labeled B. Draw the location and orientation of *all* the images of object B formed by the two mirrors.
3. What is the radius of curvature of a plane mirror? What is its focal length? Explain.
4. Dish receivers for satellite TV always use the concave side of the dish, never the convex side. Explain.
5. Suppose you would like to start a fire by focusing sunlight onto a piece of paper. In Conceptual Checkpoint 26–2 we saw that a

concave mirror would be better than a convex mirror for this purpose. At what distance from the mirror should the paper be held for best results?

6. When light propagates from one medium to another, does it always bend toward the normal? Explain.
7. A swimmer at point B in **Figure 26–40** needs help. Two lifeguards depart simultaneously from their tower at point A, but they follow different paths. Although both lifeguards run with equal speed on the sand and swim with equal speed in the water, the lifeguard who follows the longer path, ACB, arrives at point B before the lifeguard who follows the shorter, straight-line path from A to B. Explain.

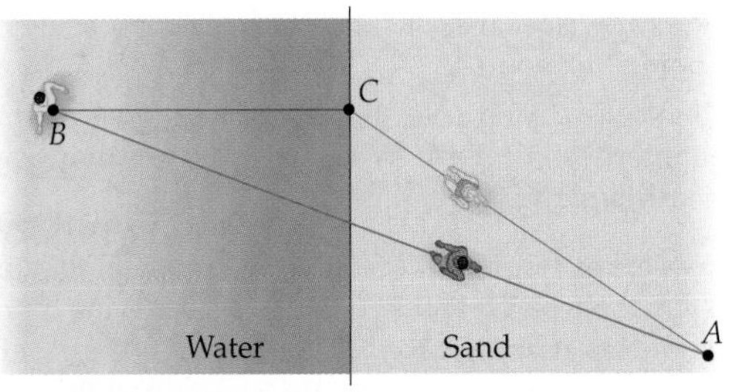

▲ **FIGURE 26–40** Conceptual Question 7

8. When you observe a mirage on a hot day, what are you actually seeing when you gaze at the "pool of water" in the distance?
9. Explain the difference between a virtual and a real image.
10. Sitting on a deserted beach one evening, you watch as the last bit of the Sun approaches the horizon. Just before the Sun disappears from sight, is the top of the Sun actually above or below the horizon? That is, if Earth's atmosphere could be instantly removed just before the Sun disappeared, would the Sun still be visible, or would it be below the horizon? Explain.
11. A large, empty coffee mug sits on a table. From your vantage point the bottom of the mug is not visible. When the mug is filled with water, however, you *can* see the bottom of the mug. Explain.
12. **The Disappearing Eyedropper** The accompanying photograph shows eyedroppers partially immersed in oil (left) and water (right). Explain why the dropper is invisible in the oil.

What happened to the dropper? (Conceptual Questions 12 and 13)

13. **The Invisible Man** In the H. G. Wells novel *The Invisible Man*, a person becomes invisible by altering his index of refraction to match that of air. This is the idea behind the disappearing eyedropper in Conceptual Question 12. If the invisible man could actually do this, would he be able to see? Explain.
14. **What's the Secret?** The top of **Figure 26–41** shows the words SECRET CODE written in different colors. If you place a cylindrical rod of glass or plastic just above the words, you find that SECRET appears inverted, but CODE does not. Explain.

▲ **FIGURE 26–41** Conceptual Question 14

Conceptual Exercises

(Answers to odd-numbered Conceptual Exercises can be found in the back of the book.)

1. If you view a clock in a mirror, do the hands rotate clockwise or counterclockwise?

Which way do the hands go? (Conceptual Exercise 1)

2. Astronomers often use large mirrors in their telescopes to gather as much light as possible from faint distant objects. Should the mirror in their telescopes be concave or convex? Explain.
3. You hold a shiny tablespoon at arm's length and look at the back side of the spoon. **(a)** Is the image you see of yourself upright or inverted? **(b)** Is the image enlarged or reduced? **(c)** Is the image real or virtual?
4. You hold a shiny tablespoon at arm's length and look at the front side of the spoon. **(a)** Is the image you see of yourself upright or inverted? **(b)** Is the image enlarged or reduced? **(c)** Is the image real or virtual?
5. An object is placed in front of a convex mirror whose radius of curvature is R. What is the greatest distance behind the mirror that the image can be formed?
6. An object is placed to the left of a concave mirror, beyond its focal point. In which direction will the image move when the object is moved farther to the left?

7. An object is placed to the left of a convex mirror. In which direction will the image move when the object is moved farther to the left?
8. **Jurassic Park** A *T. rex* chases the heroes of Steven Spielberg's *Jurassic Park* as they desperately try to escape in their Jeep. The *T. rex* is closing in fast, as they can see in the outside rearview mirror. Near the bottom of the mirror they also see the following helpful message: OBJECTS IN THE MIRROR ARE CLOSER THAN THEY APPEAR. Is this mirror concave or convex? Explain.
9. The receiver for a dish antenna is placed in front of the concave surface of the dish. If the radius of curvature of the dish is R, how far in front of the dish should the receiver be placed? Explain.
10. Is the image you see in a three-dimensional corner reflector upright or inverted?
11. If a lens is immersed in water, its focal length changes, as discussed in Conceptual Checkpoint 26–5. If a spherical mirror is immersed in water, does its focal length increase, decrease, or stay the same? Explain.
12. When a ray of light enters a glass lens, it slows down. When it leaves the glass, does its speed increase, decrease, or stay the same? Explain.
13. A glass slab surrounded by air causes a sideways displacement in a beam of light. If the slab is now placed in water, does the displacement it causes increase, decrease, or stay the same? Explain.
14. **Inverse Lenses** Suppose we mold a hollow piece of plastic into the shape of a double concave lens. The "lens" is watertight, and its interior is filled with air. We now place this lens in water and shine a beam of light on it. **(a)** Does the lens converge or diverge the beam of light? Explain. **(b)** If our hollow lens is double convex, instead, does it converge or diverge a beam of light when immersed in water? Explain.
15. Referring to Conceptual Question 12, suppose the glass from an eyedropper is used to make a convex lens with a focal length f. If this lens is immersed in the oil of the bottle on the left, is its focal length $0, f/2, 2f$, or ∞? (*Hint:* See Conceptual Checkpoint 26–5.)
16. **Samurai Fishing** A humorous scene in Akira Kurosawa's classic film *The Seven Samurai* shows the young samurai Kikuchiyo wading into a small stream and plucking a fish from it for his dinner. **(a)** As Kikuchiyo looks through the water to the fish, does he see it in the general vicinity of point 1 or point 2 in **Figure 26–42**? **(b)** If the fish looks up at Kikuchiyo, does it see Kikuchiyo's head in the general vicinity of point 3 or point 4?

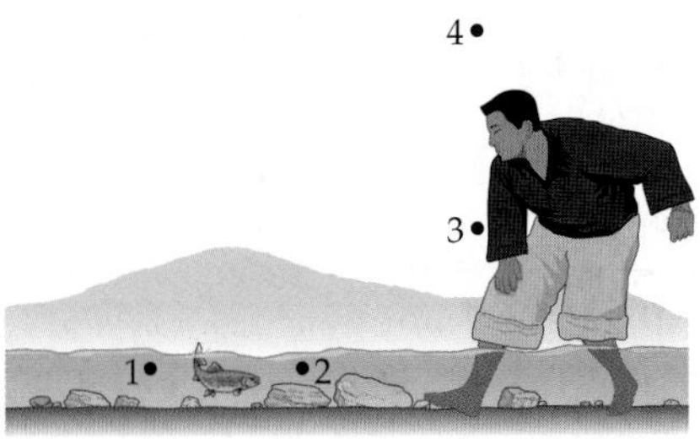

▲ **FIGURE 26–42** Conceptual Exercise 16

17. When color A and color B are sent through a prism, color A is bent more than color B. Which color travels more rapidly in the prism? Explain.
18. **Day versus Night** **(a)** Imagine for a moment that the Earth has no atmosphere. Over the period of a year, is the number of daylight hours at your home greater than, less than, or equal to the number of nighttime hours? **(b)** Repeat part (a), only this time take into account the Earth's atmosphere.
19. A kitchen has twin side-by-side sinks. The left sink is filled with water, and the right sink is empty. Does the left sink appear to be deeper, shallower, or the same depth as the sink on the right? Explain.
20. Two identical containers are filled with different transparent liquids. The container with liquid A appears to have a greater depth than the container with liquid B. Which liquid has the greater index of refraction? Explain.
21. A light beam undergoes total internal reflection at the interface between medium A, in which it propagates, and medium B, on the other side of the interface. Which medium has the greater index of refraction? Explain.
22. You take a picture of a rainbow with an infrared camera, and your friend takes a picture at the same time with visible light. Is the height of the rainbow in the infrared picture greater than, less than, or the same as the height of the rainbow in the visible-light picture? Explain.

Problems

Note: **IP** *denotes an integrated conceptual/quantitative problem.* **BIO** *identifies problems of biological or medical interest. Red bullets (•, ••, •••) are used to indicate the level of difficulty of each problem.*

The outside medium is assumed to be air, with an index of refraction of 1.00, *unless specifically stated otherwise.*

Section 26–1 The Reflection of Light

1. • A laser beam is reflected by a plane mirror. It is observed that the angle between the incident and reflected beams is 28°. If the mirror is now rotated so that the angle of incidence increases by 5.0°, what is the new angle between the incident and reflected beams?
2. •• The reflecting surfaces of two mirrors form a vertex with an angle of 120°. If a ray of light strikes mirror 1 with an angle of incidence of 55°, find the angle of reflection of the ray when it leaves mirror 2.
3. •• A ray of light reflects from a plane mirror with an angle of incidence of 37°. If the mirror is rotated by an angle θ, through what angle is the reflected ray rotated?
4. •• **IP** A small vertical mirror hangs on the wall, 1.40 m above the floor. Sunlight strikes the mirror, and the reflected beam forms a spot on the floor 2.50 m from the wall. Later in the day, you notice that the spot has moved to a point 3.75 m from the wall. **(a)** Were your two observations made in the morning or in the afternoon? Explain. **(b)** What was the change in the Sun's angle of elevation between your two observations?
5. •• Sunlight enters a room at an angle of 32° above the horizontal and reflects from a small mirror lying flat on the floor. The reflected light forms a spot on a wall that is 2.0 m behind the mirror, as shown in **Figure 26–43**. If you now place a pencil under the edge of the mirror nearer the wall, tilting it upward by 5.0°, how much higher on the wall (Δy) is the spot?

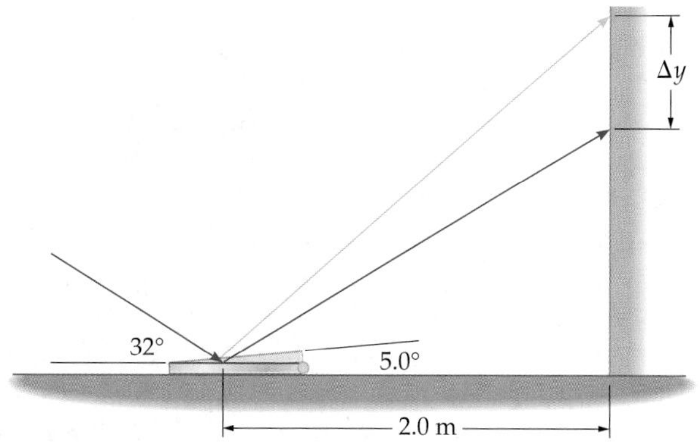

▲ **FIGURE 26–43** Problem 5

6\. •• You stand 1.50 m in front of a wall and gaze downward at a small vertical mirror mounted on it. In this mirror you can see the reflection of your shoes. If your eyes are 1.75 m above your feet, through what angle should the mirror be tilted for you to see your eyes reflected in the mirror? (The location of the mirror remains the same, only its angle to the vertical is changed.)

7\. •• **IP** Standing 2.0 m in front of a small vertical mirror, you see the reflection of your belt buckle, which is 0.70 m below your eyes. **(a)** What is the vertical location of the mirror relative to the level of your eyes? **(b)** What angle do your eyes make with the horizontal when you look at the buckle? **(c)** If you now move backward until you are 6.0 m from the mirror, will you still see the buckle, or will you see a point on your body that is above or below the buckle? Explain.

8\. •• How many times does the light beam shown in **Figure 26–44** reflect from **(a)** the top and **(b)** the bottom mirror?

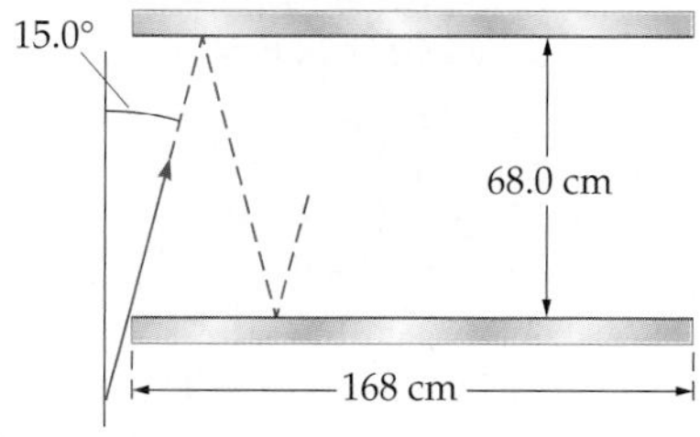

▲ **FIGURE 26–44** Problems 8 and 81

Section 26–2 Forming Images with a Plane Mirror

9\. • A 12.0-foot-long, nearsighted python is stretched out perpendicular to a plane mirror, admiring its reflected image. If the greatest distance to which the snake can see clearly is 27.0 ft, how close must its head be to the mirror for it to see a clear image of its tail?

10\. •• **(a)** How rapidly does the distance between you and your mirror image decrease if you walk directly toward a mirror with a speed of 2.6 m/s? **(b)** Repeat part (a) for the case in which you walk toward a mirror but at an angle of 38° to its normal.

11\. •• You are 1.8 m tall and stand 3.0 m from a plane mirror that extends vertically upward from the floor. On the floor 1.5 m in front of the mirror is a small table, 0.80 m high. What is the minimum height the mirror must have for you to be able to see the top of the table in the mirror?

12\. •• The rear window in a car is approximately a rectangle, 1.3 m wide and 0.30 m high. The inside rearview mirror is 0.50 m from the driver's eyes, and 1.50 m from the rear window. What are the minimum dimensions for the rearview mirror if the driver is to be able to see the entire width and height of the rear window in the mirror without moving her head?

13\. •• **IP** You hold a small plane mirror 0.50 m in front of your eyes, as shown in **Figure 26–45** (not to scale). The mirror is 0.32 cm high, and in it you see the image of a tall building behind you. **(a)** If the building is 95 m behind you, what vertical height of the building, H, can be seen in the mirror at any one time? **(b)** If you move the mirror closer to your eyes, does your answer to part (a) increase, decrease, or stay the same? Explain.

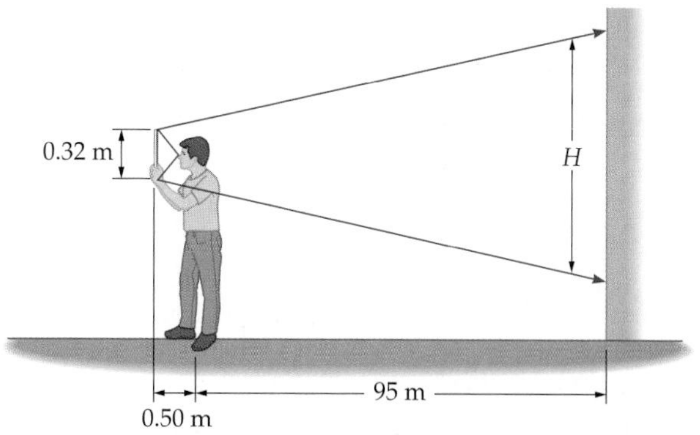

▲ **FIGURE 26–45** Problems 13 and 92

14\. •• Two rays of light converge toward each other, as shown in **Figure 26–46**, forming an angle of 27°. Before they intersect, however, they are reflected from a circular plane mirror with a diameter of 11 cm. If the mirror can be moved horizontally to the left or right, what is the greatest possible distance d from the mirror to the point where the reflected rays meet?

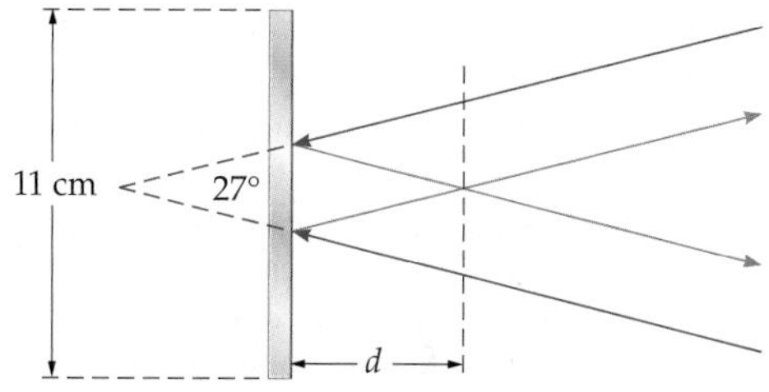

▲ **FIGURE 26–46** Problem 14

15\. •• For a corner reflector to be effective, its surfaces must be precisely perpendicular. Suppose the surfaces of a corner reflector left on the Moon's surface by the Apollo astronauts formed a 90.001° angle with each other. If a laser beam is bounced back to Earth from this reflector, how far (in kilometers) from its starting point will the reflected beam strike Earth? For simplicity, assume the beam reflects from only two sides of the reflector, and that it strikes the first surface at precisely 45°.

Section 26–3 Spherical Mirrors

16\. • A section of a sphere has a radius of curvature of 0.86 m. If this section is painted with a reflective coating on both sides, what is the focal length of **(a)** the convex side and **(b)** the concave side?

17\. • A mirrored-glass gazing globe in a garden is 31.2 cm in diameter. What is the focal length of the globe?

18\. • Sunlight reflects from a concave piece of broken glass, converging to a point 15 cm from the glass. What is the radius of curvature of the glass?

Section 26–4 Ray Tracing and the Mirror Equation

19\. • A small object is located 30.0 cm in front of a concave mirror with a radius of curvature of 40.0 cm. Where will the image be formed?

20\. • Use ray diagrams to show whether the image formed by a convex mirror increases or decreases in size as an object is brought closer to the mirror's surface.

21. • An object with a height of 42 cm is placed 2.0 m in front of a concave mirror with a focal length of 0.50 m. **(a)** Determine the approximate location and size of the image using a ray diagram. **(b)** Is the image upright or inverted?

22. • Find the location and magnification of the image produced by the mirror in Problem 21 using the mirror and magnification equations.

23. • An object with a height of 42 cm is placed 2.0 m in front of a convex mirror with a focal length of −0.50 m. **(a)** Determine the approximate location and size of the image using a ray diagram. **(b)** Is the image upright or inverted?

24. • Find the location and magnification of the image produced by the mirror in Problem 23 using the mirror and magnification equations.

25. •• During a daytime football game you notice that a player's reflective helmet forms an image of the Sun 4.8 cm behind the surface of the helmet. What is the radius of curvature of the helmet, assuming it to be roughly spherical?

26. •• **IP** A magician wishes to create the illusion of a 2.74-m-tall elephant. He plans to do this by forming a virtual image of a 50.0-cm-tall model elephant with the help of a spherical mirror. **(a)** Should the mirror be concave or convex? **(b)** If the model must be placed 3.00 m from the mirror, what radius of curvature is needed? **(c)** How far from the mirror will the image be formed?

27. •• A person 1.6 m tall stands 0.60 m from a reflecting globe in a garden. **(a)** If the diameter of the globe is 18 cm, where is the image of the person, relative to the surface of the globe? **(b)** How large is the person's image?

28. •• Shaving/makeup mirrors typically have one flat and one concave (magnifying) surface. You find that you can project a magnified image of a lightbulb onto the wall of your bathroom if you hold the mirror 1.8 m from the bulb and 3.5 m from the wall. **(a)** What is the magnification of the image? **(b)** Is the image erect or inverted? **(c)** What is the focal length of the mirror?

29. •• **The Hale Telescope** The 200-inch-diameter concave mirror of the Hale telescope on Mount Palomar has a focal length of 16.9 m. An astronomer stands 20.0 m in front of this mirror. **(a)** Where is her image located? Is it in front of or behind the mirror? **(b)** Is her image real or virtual? How do you know? **(c)** What is the magnification of her image?

30. •• A concave mirror produces a virtual image that is three times as tall as the object. **(a)** If the object is 22 cm in front of the mirror, what is the image distance? **(b)** What is the focal length of this mirror?

31. •• A concave mirror produces a real image that is three times as large as the object. **(a)** If the object is 22 cm in front of the mirror, what is the image distance? **(b)** What is the focal length of this mirror?

32. •• The virtual image produced by a convex mirror is one-quarter the size of the object. **(a)** If the object is 32 cm in front of the mirror, what is the image distance? **(b)** What is the focal length of this mirror?

33. •• **IP** A 5.7-ft-tall shopper in a department store is 17 ft from a convex security mirror. The shopper notices that his image in the mirror appears to be only 6.4 in. tall. **(a)** Is the shopper's image upright or inverted? Explain. **(b)** What is the mirror's radius of curvature?

34. •• You view a nearby tree in a concave mirror. The inverted image of the tree is 3.5 cm high and is located 7.0 cm in front of the mirror. If the tree is 21 m from the mirror, what is its height?

35. ••• A shaving/makeup mirror produces an erect image that is magnified by a factor of 2.0 when your face is 25 cm from the mirror. What is the mirror's radius of curvature?

36. ••• A concave mirror with a focal length of 36 cm produces an image whose distance from the mirror is one-third the object distance. Find the object and image distances.

Section 26–5 The Refraction of Light

37. • Light travels a distance of 0.960 m in 4.00 ns in a given substance. What is the index of refraction of this substance?

38. • Find the ratio of the speed of light in water to the speed of light in diamond.

39. • **Ptolemy's Optics** One of the many works published by the Greek astronomer Ptolemy (A.D. ca. 100–170) was *Optics*. In this book Ptolemy reports the results of refraction experiments he conducted by observing light passing from air into water. His results are as follows: angle of incidence = 10.0°, angle of refraction = 8.00°; angle of incidence = 20.0°, angle of refraction = 15.5°. Find the percentage error in the calculated index of refraction of water for each of Ptolemy's measurements.

40. • Light enters a container of benzene at an angle of 43° to the normal; the refracted beam makes an angle of 27° with the normal. Calculate the index of refraction of benzene.

41. • The angle of refraction of a ray of light traveling through an ice cube is 31°. Find the angle of incidence.

42. • **IP** **(a)** Referring to Problem 41, suppose the ice melts, but the angle of refraction remains the same. Is the corresponding angle of incidence greater than, less than, or the same as it was for ice? Explain. **(b)** Calculate the angle of incidence for part (a).

43. •• A submerged scuba diver looks up toward the calm surface of a freshwater lake and notes that the Sun appears to be 35° from the vertical. The diver's friend is standing on the shore of the lake. At what angle above the horizon does the friend see the sun?

44. •• A pond with a total depth (ice + water) of 3.00 m is covered by a transparent layer of ice, with a thickness of 0.32 m. Find the time required for light to travel vertically from the surface of the ice to the bottom of the pond.

45. •• Light is refracted as it travels from a point A in medium 1 to a point B in medium 2. If the index of refraction is 1.33 in medium 1 and 1.51 in medium 2, how long does it take light to go from A to B, assuming it travels 331 cm in medium 1 and 151 cm in the medium 2?

46. •• You have a semicircular disk of glass with an index of refraction of $n = 1.52$. Find the incident angle θ for which the beam of light in **Figure 26–47** will hit the indicated point on the screen.

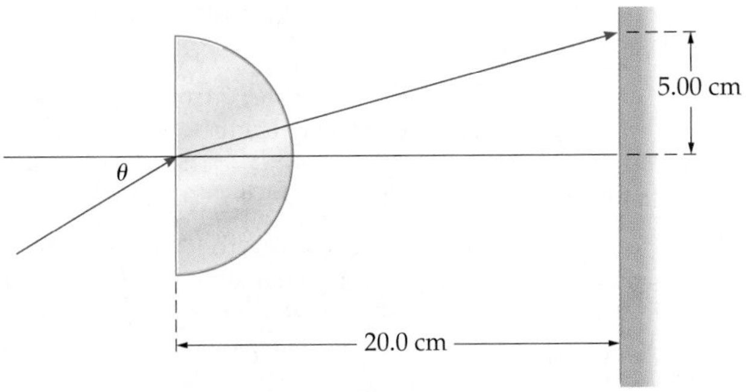

▲ **FIGURE 26–47** Problems 46 and 53

47. •• The observer in **Figure 26–48** is positioned so that the far edge of the bottom of the empty glass (not to scale) is just visible. When the glass is filled to the top with water, the center of the bottom of the glass is just visible to the observer. Find the height, H, of the glass, given that its width is $W = 6.2$ cm.

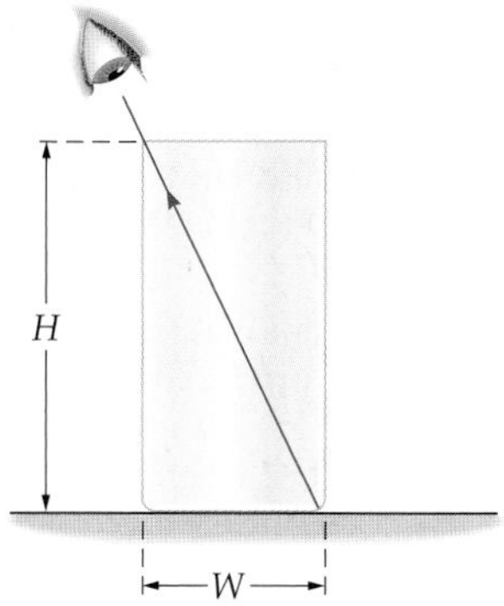

▲ **FIGURE 26–48** Problem 47

48. •• A coin is lying at the bottom of a pool of water that is 6.5 feet deep. Viewed from directly above the coin, how far below the surface of the water does the coin appear to be? (The coin is assumed to be small in diameter; therefore, we can use the small angle approximations $\sin\theta \cong \tan\theta \cong \theta$.)

49. •• A ray of light enters the long side of a 45°-90°-45° prism and undergoes two total internal reflections, as indicated in **Figure 26–49**. The result is a reversal in the ray's direction of propagation. Find the minimum value of the prism's index of refraction, n, for these internal reflections to be total.

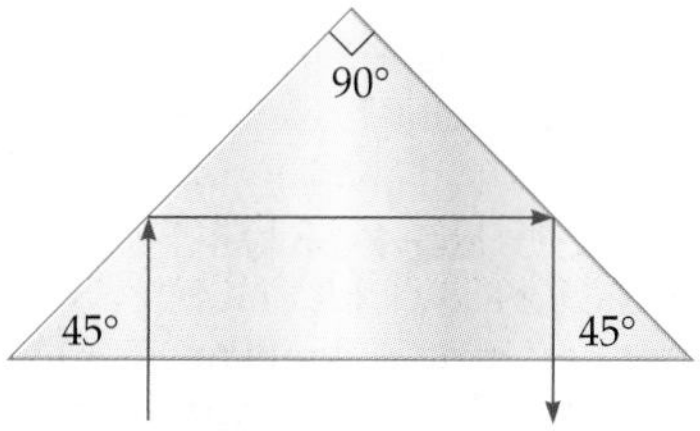

▲ **FIGURE 26–49** Problems 49 and 50

50. •• When the prism in Problem 49 is immersed in a fluid with an index of refraction of 1.21, the internal reflections shown in Figure 26–48 are still total. The reflections are no longer total, however, when the prism is immersed in a fluid with $n = 1.43$. Use this information to set upper and lower limits on the possible values of the prism's index of refraction.

51. •• **IP** A glass paperweight with an index of refraction n rests on a desk, as shown in **Figure 26–50**. An incident ray of light enters the horizontal top surface of the paperweight at an angle $\theta = 77°$ to the vertical. **(a)** Find the minimum value of n for which the internal reflection on the vertical surface of the paperweight is total. **(b)** If θ is decreased, is the minimum value of n increased or decreased? Explain.

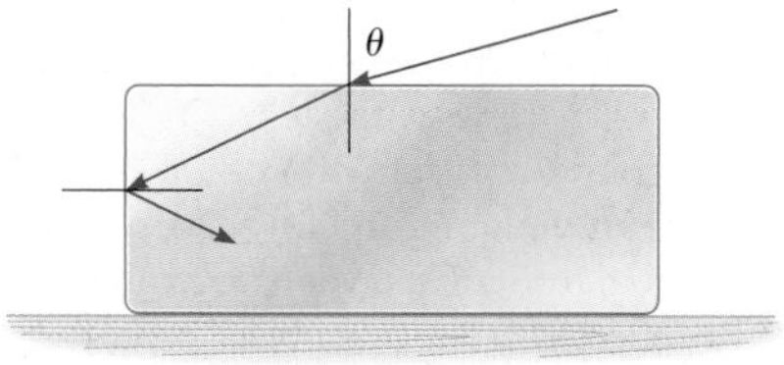

▲ **FIGURE 26–50** Problems 51 and 52

52. •• **IP** Suppose the glass paperweight in Figure 26–50 has an index of refraction $n = 1.38$. **(a)** Find the value of θ for which the reflection on the vertical surface of the paperweight exactly satisfies the condition for total internal reflection. **(b)** If θ is increased, is the reflection at the vertical surface still total? Explain.

53. •• **IP BIO** Consider the physical system shown in Figure 26–47 and described in Problem 46. **(a)** If the index of refraction of the glass is increased, will the desired value of θ increase or decrease? Explain. **(b)** Find the value of θ for the case of flint glass $(n = 1.66)$.

54. •• While studying physics at the library late one night, you notice the image of the desk lamp reflected from the varnished tabletop. When you turn your Polaroid sunglasses sideways, the reflected image disappears. If this occurs when the angle between the incident and reflected rays is 110°, what is the index of refraction of the varnish?

55. ••• A horizontal beam of light enters a 45°-90°-45° prism at the center of its long side, as shown in **Figure 26–51**. The emerging ray moves in a direction that is 34° below the horizontal. What is the index of refraction of this prism?

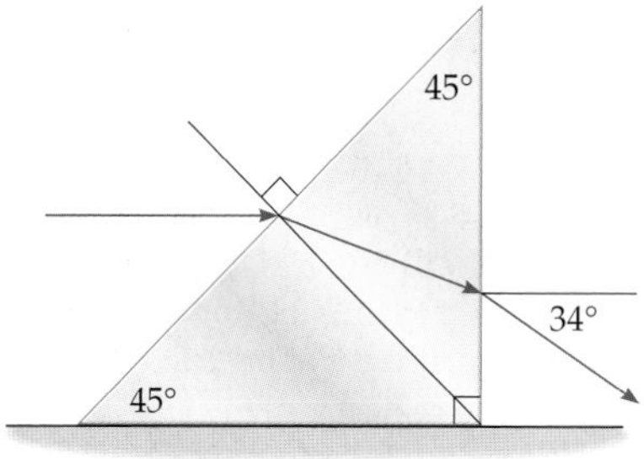

▲ **FIGURE 26–51** Problems 55 and 102

56. ••• A laser beam enters one of the sloping faces of the equilateral glass prism $(n = 1.42)$ in **Figure 26–52** and refracts through the prism. Within the prism the light travels horizontally. What is the angle θ between the direction of the incident ray and the direction of the outgoing ray?

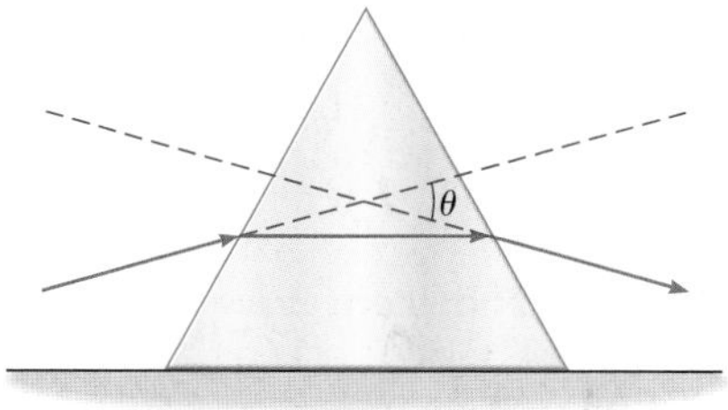

▲ **FIGURE 26–52** Problems 56 and 78

Section 26–6 Ray Tracing for Lenses

57. • **(a)** Use a ray diagram to determine the approximate location of the image produced by a concave lens when the object is at a distance $\frac{1}{2}|f|$ from the lens. **(b)** Is the image upright or inverted? **(c)** Is the image real or virtual? Explain.

58. • **(a)** Use a ray diagram to determine the approximate location of the image produced by a concave lens when the object is at a distance $2|f|$ from the lens. **(b)** Is the image upright or inverted? **(c)** Is the image real or virtual? Explain.

59. • An object is a distance $f/2$ from a convex lens. **(a)** Use a ray diagram to find the approximate location of the image. **(b)** Is the image upright or inverted? **(c)** Is the image real or virtual? Explain.

60. • An object is a distance $2f$ from a convex lens. **(a)** Use a ray diagram to find the approximate location of the image. **(b)** Is the image upright or inverted? **(c)** Is the image real or virtual? Explain.

61. •• Two lenses that are 35 cm apart are used to form an image, as shown in **Figure 26–53**. Lens 1 is converging and has a focal length $f_1 = 14$ cm; lens 2 is diverging and has a focal length $f_2 = -7.0$ cm. The object is placed 24 cm to the left of lens 1. **(a)** Use a ray diagram to find the approximate location of the image. **(b)** Is the image upright or inverted? **(c)** Is the image real or virtual? Explain.

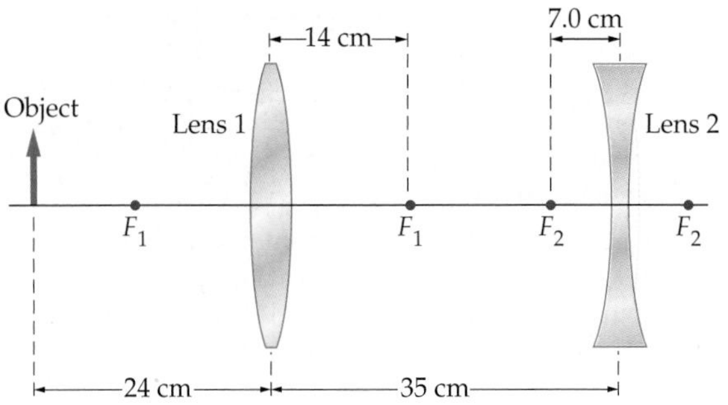

▲ **FIGURE 26–53** Problems 61 and 70

62. •• Two lenses that are 35 cm apart are used to form an image, as shown in **Figure 26–54**. Lens 1 is diverging and has a focal length $f_1 = -7.0$ cm; lens 2 is converging and has a focal length $f_2 = 14$ cm. The object is placed 24 cm to the left of lens 1. **(a)** Use a ray diagram to find the approximate location of the image. **(b)** Is the image upright or inverted? **(c)** Is the image real or virtual? Explain.

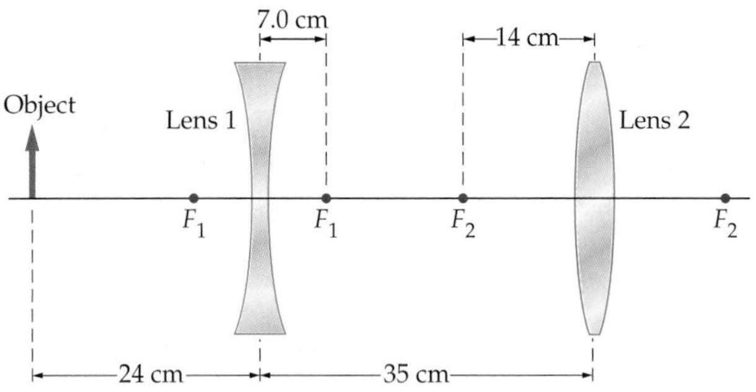

▲ **FIGURE 26–54** Problems 62 and 71

Section 26–7 The Thin-Lens Equation

63. • A convex lens is held over a piece of paper outdoors on a sunny day. When the paper is held 26 cm below the lens, the sunlight is focused on the paper and the paper ignites. What is the focal length of the lens?

64. • A concave lens has a focal length of -32 cm. Find the image distance and magnification that results when an object is placed 23 cm in front of the lens.

65. • When an object is located 42 cm to the left of a lens, the image is formed 17 cm to the right of the lens. What is the focal length of the lens?

66. • An object with a height of 2.54 cm is placed 36.3 mm to the left of a lens with a focal length of 35.0 mm. **(a)** Where is the image located? **(b)** What is the height of the image?

67. •• A lens for a 35-mm camera has a focal length given by $f = 45$ mm. **(a)** How close to the film should the lens be placed to form a sharp image of an object that is 5.0 m away? **(b)** What is the magnification of the image on the film?

68. •• **IP** An object is located to the left of a convex lens whose focal length is 36 cm. The magnification produced by the lens is $m = 3.0$. **(a)** To increase the magnification to 4.0, should the object be moved closer to the lens or farther away? Explain. **(b)** Calculate the distance through which the object should be moved.

69. •• **IP** You have two lenses at your disposal, one with a focal length $f_1 = +40.0$ cm, the other with a focal length $f_2 = -40.0$ cm. **(a)** Which of these two lenses would you use to project an image of a lightbulb onto a wall that is far away? **(b)** If you want to produce an image of the bulb that is enlarged by a factor of 2.00, how far from the wall should the lens be placed?

70. •• **(a)** Determine the distance from lens 1 to the final image for the system shown in Figure 26–53. **(b)** What is the magnification of this image?

71. •• **(a)** Determine the distance from lens 1 to the final image for the system shown in Figure 26–54. **(b)** What is the magnification of this image?

72. •• **IP** An object is located to the left of a concave lens whose focal length is -36 cm. The magnification produced by the lens is $m = \frac{1}{3}$. **(a)** To decrease the magnification to $m = \frac{1}{4}$, should the object be moved closer to the lens or farther away? **(b)** Calculate the distance through which the object should be moved.

73. •• **IP BIO** Albert is nearsighted, and without his eyeglasses he can focus only on objects less than 2.2 m away. **(a)** Are Albert's eyeglasses concave or convex? Explain. **(b)** To correct Albert's nearsightedness, his eyeglasses must produce a virtual, upright image at a distance of 2.2 m when viewing an infinitely distant object. What is the focal length of Albert's eyeglasses?

74. •• A small insect viewed through a convex lens is 1.2 cm from the lens and appears twice its actual size. What is the focal length of the lens?

75. ••• **IP** A friend tells you that when he takes off his eyeglasses and holds them 21 cm above a printed page the image of the print is erect but reduced to 0.67 of its actual size. **(a)** Is the image real or virtual? How do you know? **(b)** What is the focal length of your friend's glasses? **(c)** Are the lenses in the glasses concave or convex? Explain.

76. ••• **IP** A friend tells you that when she takes off her eyeglasses and holds them 21 cm above a printed page the image of the print is erect but enlarged to 1.5 times its actual size. **(a)** Is the image real or virtual? How do you know? **(b)** What is the focal length of your friend's glasses? **(c)** Are the lenses in the glasses concave or convex? Explain.

Section 26–8 Dispersion and the Rainbow

77. •• The index of refraction for red light in a certain liquid is 1.320; the index of refraction for violet light in the same liquid is 1.332. Find the dispersion $(\theta_v - \theta_r)$ for red and violet light when both are incident on the flat surface of the liquid at an angle of 45.00° to the normal.

78. •• A horizontal incident beam consisting of white light passes through an equilateral prism, like the one shown in Figure 26–52. What is the dispersion $(\theta_v - \theta_r)$ of the outgoing beam if the prism's index of refraction is $n_v = 1.505$ for violet light and $n_r = 1.421$ for red light?

79. •• The focal length of a lens is inversely proportional to the quantity $(n - 1)$, where n is the index of refraction of the lens material. The value of n, however, depends on the wavelength of the light that passes through the lens. For example, one type of flint glass has an index of refraction of $n_r = 1.572$ for red light and $n_v = 1.605$ in violet light. Now, suppose a white object is placed 24.00 cm in front of a lens made from this type of glass. If the red light reflected from this object produces a sharp image 55.00 cm from the lens, where will the violet image be found?

General Problems

80. •• **Focal Length** A lens is constructed from material with an index of refraction n_{lens}. When used in air ($n \cong 1$) the lens has a focal length f_{air}. It can be shown that if this lens is immersed in fluid with an index of refraction n_{fluid}, its focal length is changed to f_{fluid} given by the following relation:

$$f_{fluid} = \left[\frac{(n_{lens} - 1)n_{fluid}}{n_{lens} - n_{fluid}}\right] f_{air}$$

(a) Use this relation to show that if a converging lens is immersed in a fluid with $n_{fluid} > n_{lens}$, it will become a diverging lens.

Similarly, show that a diverging lens in a fluid with $n_{fluid} > n_{lens}$ will become a converging lens. **(b)** Calculate the focal length of lens in a fluid with $n_{fluid} = 1.33$, given that the index of refraction of the lens is $n_{lens} = 1.52$ and its focal length in air is 25.0 cm.

81. •• **IP** Suppose the separation between the two mirrors in Figure 26–44 is increased by moving the top mirror upward. **(a)** Will this affect the number of reflections made by the beam of light? If so, how? **(b)** What is the total number of reflections made by the beam of light when the separation between the mirrors is 125 cm?

82. •• Standing 2.0 m in front of a small vertical mirror you see the reflection of your belt buckle, which is 0.70 m below your eyes. If you remain 2.0 m from the mirror but climb onto a stool, how high must the stool be to allow you to see your knees in the mirror? Assume that your knees are 1.2 m below your eyes.

83. •• **IP Size of Floats in a Termometro Lentos** The Galileo thermometer, or Termometro Lentos (slow thermometer in Italian), consists of a vertical, cylindrical flask containing a fluid and several glass floats. The floats all have the same dimensions, but they appear to differ in size depending on their location within the cylinder. **(a)** Does a float near the front surface of the cylinder (closest to you) appear to be larger or smaller than a float near the back surface? **(b) Figure 26–55** shows a ray diagram for a float near the front surface of the cylinder. Draw a ray diagram for the float at the center of the cylinder, and show that the change in apparent size agrees with your answer to part (a).

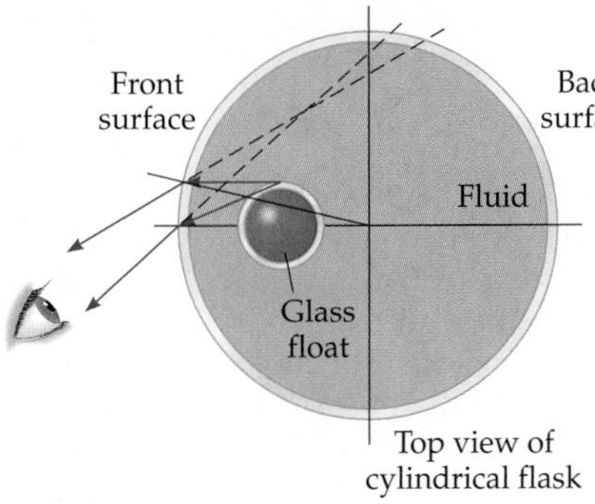

▲ **FIGURE 26–55** Problem 83

84. •• **(a)** Find the two locations where an object can be placed in front of a concave mirror with a radius of curvature of 39 cm such that its image is twice its size. **(b)** In each of these cases, state whether the image is real or virtual, upright or inverted.

85. •• A convex mirror with a focal length of −75 cm is used to give a truck driver a view behind the vehicle. **(a)** If a person who is 1.7 m tall stands 2.2 m from the mirror, where is the person's image located? **(b)** Is the image upright or inverted? **(c)** What is the size of the image?

86. •• **IP** The three laser beams shown in **Figure 26–56** meet at a point at the back of a solid, transparent sphere. **(a)** What is the index of refraction of the sphere? **(b)** Is there a finite index of refraction that will make the three beams come to a focus at the center of the sphere? If your answer is yes, give the required index of refraction; if your answer is no, explain why not.

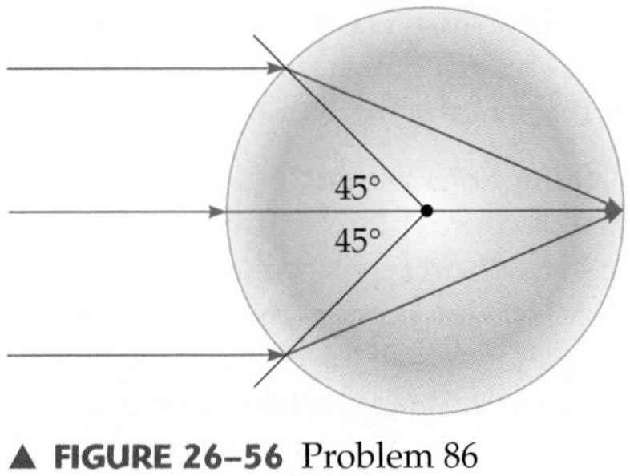

▲ **FIGURE 26–56** Problem 86

87. •• The speed of light in substance A is x times greater than the speed of light in substance B. Find the ratio n_A/n_B in terms of x.

88. •• **IP** A film of oil, with an index of refraction of 1.48 and a thickness of 1.50 cm, floats on a pool of water, as shown in **Figure 26–57**. A beam of light is incident on the oil at an angle of 60.0° to the vertical. **(a)** Find the angle θ the light beam makes with the vertical as it travels through the water. **(b)** How does your answer to part (a) depend on the thickness of the oil film? Explain.

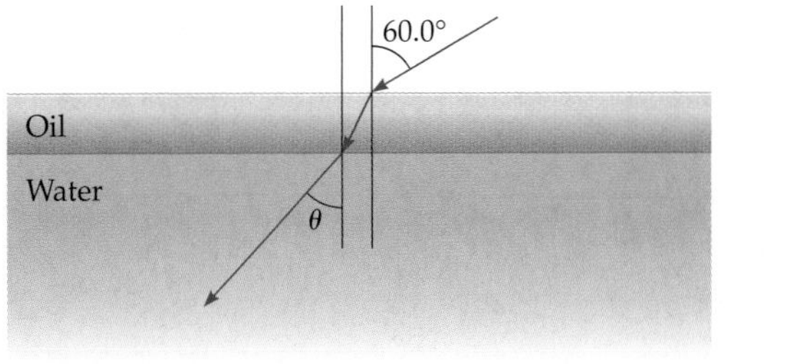

▲ **FIGURE 26–57** Problems 88, 89, and 90

89. •• **IP** Consider the physical system shown in Figure 26–57. For this problem we assume that the angle of incidence at the air-oil interface can be varied from 0° to 90°. **(a)** What is the maximum possible value for θ, the angle of refraction in the water? **(b)** If an oil with a larger index of refraction is used, does your answer to part (a) increase or decrease? Explain.

90. •• **IP** Consider the physical system shown in Figure 26–57, only this time let the direction of the light rays be reversed. **(a)** Find the angle of incidence θ at the water-oil interface such that the condition for total internal reflection at the oil-air surface is exactly satisfied. **(b)** If θ is decreased, is the reflection at the oil-air interface still total? Explain.

91. •• **Figure 26–58** shows a ray of light entering one end of an optical fiber at an angle of incidence $\theta_i = 50.0°$. The index of refraction of the fiber is 1.62. **(a)** Find the angle θ the ray makes with the normal when it reaches the curved surface of the fiber. **(b)** Show that the internal reflection from the curved surface is total.

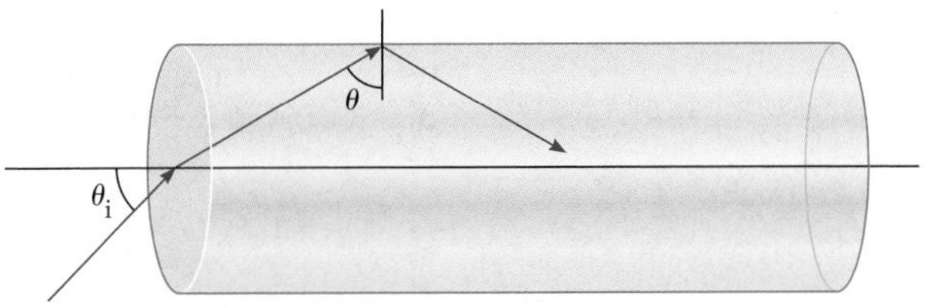

▲ **FIGURE 26–58** Problems 91 and 99

92. •• Suppose the person's eyes in Figure 26–45 are 1.6 m above the ground and that the small plane mirror can be moved up or down. **(a)** Find the height of the bottom of the mirror such that the lowest point the person can see on the building is 19.6 m above the ground. **(b)** With the mirror held at the height found in part (a), what is the highest point on the building the person can see?

93. •• An arrow 2.00 cm long is located 75.0 cm from a lens that has a focal length $f = 30.0$ cm. **(a)** If the arrow is perpendicular to the principal axis of the lens, as in **Figure 26–59 (a)**, what is its lateral magnification, defined as h_i/h_o? **(b)** Suppose, instead, that the arrow lies *along* the principal axis, extending from 74.0 cm to 76.0 cm from the lens, as indicated in **Figure 26–59 (b)**. What is the longitudinal magnification of the arrow, defined as L_i/L_o? (*Hint*: Use the thin-lens equation to locate the image of each end of the arrow.)

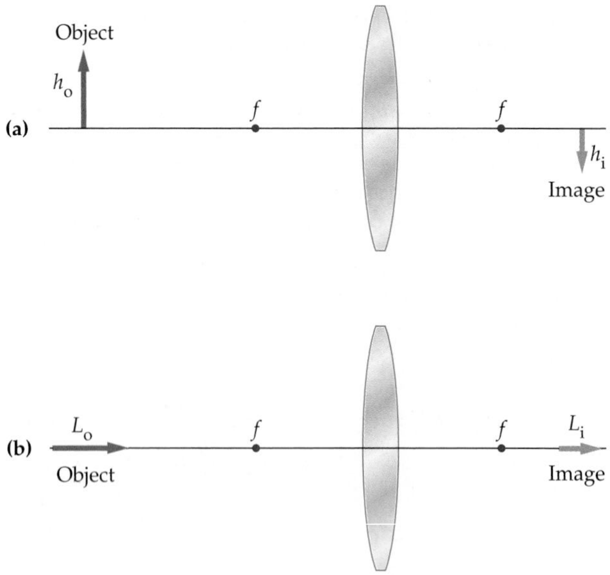

▲ **FIGURE 26–59** Problems 93 and 94

94. •• Repeat Problem 93, this time for a diverging lens with a focal length $f = -30.0$ cm.

95. ••• A convex lens with $f_1 = 20.0$ cm is mounted 40.0 cm to the left of a concave lens. When an object is placed 30.0 cm to the left of the convex lens, a real image is formed 60.0 cm to the right of the concave lens. What is the focal length f_2 of the concave lens?

96. ••• Two thin lenses, with focal lengths f_1 and f_2, are placed in contact. What is the effective focal length of the double lens?

97. ••• When an object is placed a distance d_o in front of a curved mirror, the resulting image has a magnification m. Find an expression for the focal length of the mirror, f, in terms of d_o and m.

98. ••• **A Slab of Glass** Give a symbolic expression for the sideways displacement d of a light ray passing through the slab of glass shown in **Figure 26–60**. The thickness of the glass is t, its index of refraction is n, and the angle of incidence is θ.

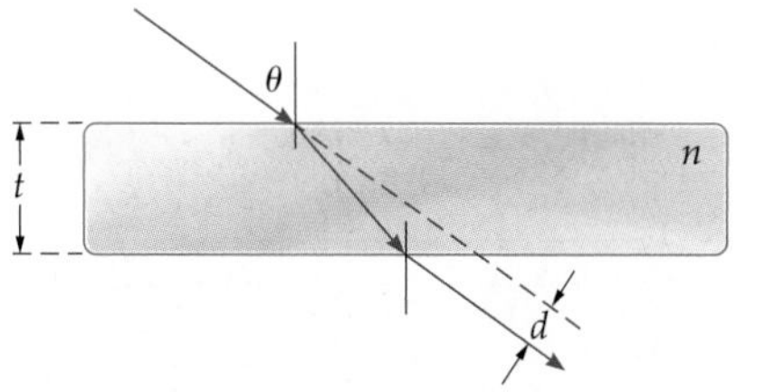

▲ **FIGURE 26–60** Problem 98

99. ••• Referring to Figure 26–58, show that the internal reflection from the curved surface of the fiber is always total for any incident angle θ_i, provided the index of refraction of the fiber exceeds $\sqrt{2}$.

100. ••• **Least Time** A beam of light propagates from point A in medium 1 to point B in medium 2, as shown in **Figure 26–61**. The index of refraction is different in these two media; therefore, the light follows a refracted path that obeys Snell's law. **(a)** Calculate the time required for light to travel from A to B along the refracted path. **(b)** Compare the time found in part (a) with the time it takes for light to travel from A to B along a straight-line path. (Note that the time on the straight-line path is longer than the time on the refracted path. In general, the shortest time between two points in different media is along the path given by Snell's law.)

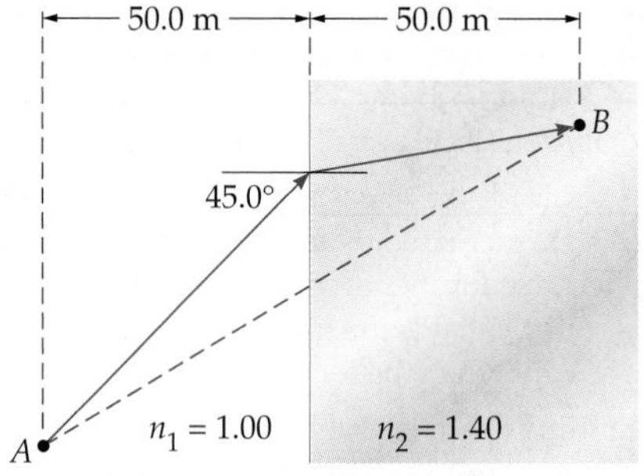

▲ **FIGURE 26–61** Problem 100

101. ••• The ray of light shown in **Figure 26–62** passes from medium 1 to medium 2 to medium 3. The index of refraction in medium 1 is n_1, in medium 2 it is $n_2 > n_1$, and in medium 3 it is $n_3 > n_2$. Show that medium 2 can be ignored when calculating the angle of refraction in medium 3; that is, show that $n_1 \sin\theta_1 = n_3 \sin\theta_3$.

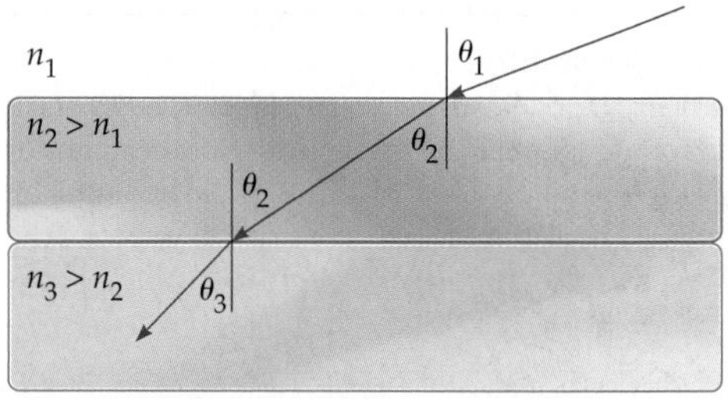

▲ **FIGURE 26–62** Problem 101

102. ••• **IP** A beam of light enters the sloping side of a 45°-90°-45° glass prism with an index of refraction $n = 1.66$. The situation is similar to that shown in Figure 26–51, except that the angle of incidence of the incoming beam can be varied. **(a)** Find the angle of incidence for which the reflection on the vertical side of the prism exactly satisfies the condition for total internal reflection. **(b)** If the angle of incidence is increased, is the reflection at the vertical surface still total? Explain. **(c)** What is the minimum value of n such that a horizontal beam like that in Figure 26–51 undergoes total internal reflection at the vertical side of the prism?

Interactive Problems

103. •• ***Referring to Example 26–3*** Suppose the radius of curvature of the mirror is 5.0 cm. **(a)** Find the object distance that gives an upright image with a magnification of 1.5. **(b)** Find the object distance that gives an inverted image with a magnification of −1.5.

104. •• **IP** ***Referring to Example 26–3*** An object is 4.5 cm in front of the mirror. **(a)** What radius of curvature must the mirror have if the image is to be 2.2 cm in front of the mirror? **(b)** What is the magnification of the image? **(c)** If the object is moved closer to the mirror, does the magnification of the image increase in magnitude, decrease in magnitude, or stay the same?

105. •• ***Referring to Example 26–7*** **(a)** What object distance is required to give an image with a magnification of +2.0? Assume that the focal length of the lens is +5.0 cm. **(b)** What is the location of the image in this case?

106. •• **IP** ***Referring to Example 26–7*** Suppose the convex lens is replaced with a concave lens with a focal length of −5.0 cm. **(a)** Where must the object be placed to form an image with a magnification of 0.50? **(b)** What is the location of the image in this case? **(c)** If we now move the object closer to the lens, does the magnification of the image increase, decrease, or stay the same?

27 Optical Instruments

27–1 The Human Eye and the Camera 910
27–2 Lenses in Combination and Corrective Optics 913
27–3 The Magnifying Glass 919
27–4 The Compound Microscope 923
27–5 Telescopes 924
27–6 Lens Aberrations 927

Large modern astronomical telescopes are seldom used for visual observations. More often, the light they capture is directed to a scientific instrument, such as a camera, spectrograph, or semiconducting device. When a human observer is involved, it's often within the telescope rather than behind it. The astronomer in this photo is actually perched above the 200-inch mirror of the great Hale reflecting telescope on Mount Palomar, at the "prime focus" of the telescope. This chapter explores how mirrors and lenses are used to create a variety of optical devices, from telescopes that let us scan the cosmos, to microscopes that give us access to the world of the very small—and even to eyeglasses that help us read a newspaper.

Human vision can be aided in a number of ways by the practical application of optics. For example, a pair of glasses or contact lenses can correct a person's faulty eyesight to produce normal vision. Optics can also extend the abilities of human sight by magnifying very small objects to a visible size or by making distant objects seem near at hand. In this chapter, after discussing the optics of a normal eye and the closely related behavior of a camera, we consider a variety of optical instruments designed to either correct or extend our abilities to see the natural world.

27–1 The Human Eye and the Camera

The eye is a marvelously sensitive and versatile optical instrument, allowing us to observe objects as distant as the stars or as close as the book in our hands. Perhaps more amazing is the fact that the eye can accomplish all this even though its basic structure is that of a spherical bag of water 2.5 cm in diameter. The slight differences that set the eye apart from a bag of water make all the difference in terms of optical performance, however.

The fundamental elements of an eye are illustrated in **Figure 27–1**. Basically, light enters the eye through the transparent outer coating of the eye, the *cornea*, and then passes through the *aqueous humor*, the adjustable *lens*, and the jellylike *vitreous humor* before reaching the light-sensitive *retina* at the back of the eye, as illustrated in **Figure 27–2**. The retina is covered with millions of small structures known as *rods* and *cones*, which, when stimulated by light, send electrical impulses along the *optic nerve* to the brain. How the nerve impulses are processed, so that we interpret the upside-down image on the retina as a right-side-up object, is another story altogether; here we concentrate on the optical properties of the eye.

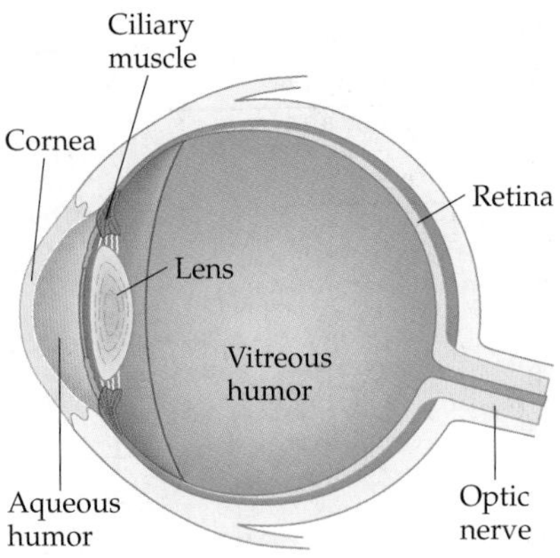

▲ **FIGURE 27–1 Basic elements of the human eye**

Light enters the eye through the cornea and the lens. It is focused onto the retina by the ciliary muscles, which change the shape of the lens.

REAL-WORLD PHYSICS: BIO

Optical properties of the human eye

To begin, we note that most of the refraction needed to produce an image occurs at the cornea, as light first enters the eye. The reason is that the difference in index of refraction is greater at the air–cornea interface than at any interface within the eye. Specifically, the index of refraction of air is $n = 1.00$, whereas that of the cornea is about $n = 1.38$, just slightly greater than the index of refraction of water ($n = 1.33$). When light passes from the cornea to the aqueous humor, the index of refraction changes from $n = 1.38$ to $n = 1.33$. Next, light encounters the lens ($n = 1.40$) and then the vitreous humor ($n = 1.34$) before arriving at the retina.

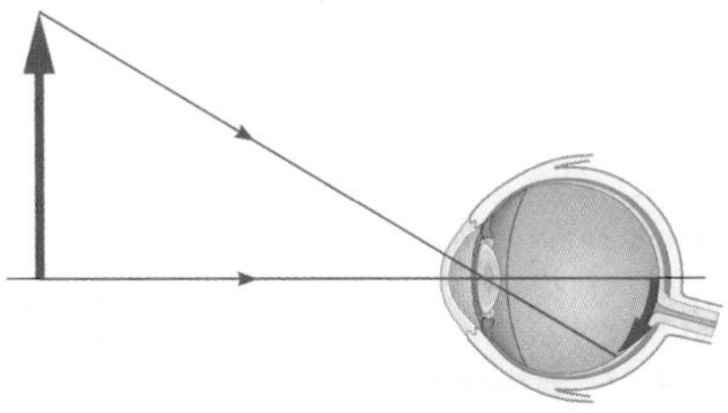

▲ **FIGURE 27–2 Image production in the eye**

The eye produces a real, inverted image on the retina.

In the end, the lens accounts for only about a quarter of the total refraction produced by the eye—but it is a crucial contribution nonetheless. By altering the shape of the lens with the *ciliary muscles*, we are able to change the precise amount of refraction the lens produces, which, in turn, changes its focal length. Specifically, when we view a distant object, our ciliary muscles are *relaxed*, as shown in **Figure 27–3 (a)**, allowing the lens to be relatively flat. As a result, it causes little refraction and its focal length is at its greatest. When we view a nearby object, the lens must shorten its focal length and cause more refraction, as shown in **Figure 27–3 (b)**. Thus, the ciliary muscles *tense* to give the lens a greater curvature. The process of changing the shape of the lens, and hence adjusting its focal length, is referred to as **accommodation.** Producing the proper accommodation is no easy feat for a newborn but is automatic for an adult.

The fact that the ciliary muscles must be tensed to focus on nearby objects means that our eyes can "tire" from muscular strain. That is why it is beneficial

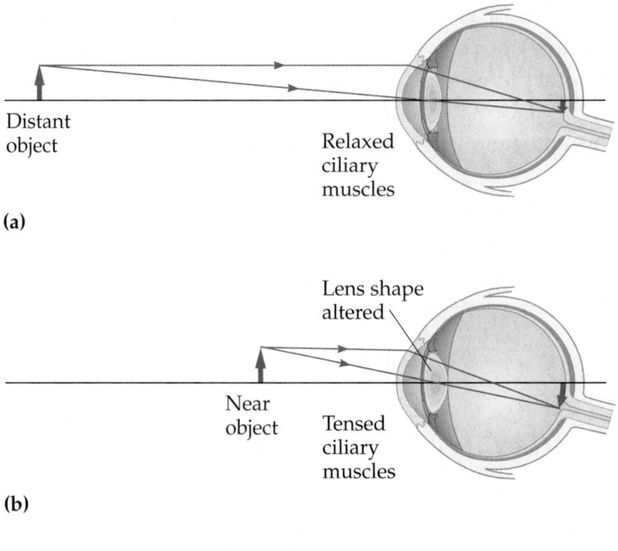

▶ **FIGURE 27–3 Accommodation in the human eye**

(a) When the eye is viewing a distant object, the ciliary muscles are relaxed and the focal length of the lens is at its greatest. **(b)** When the eye is focusing on a near object, the ciliary muscles are tensed, changing the shape and reducing the focal length of the lens.

to pause occasionally from reading and to look off into the distance. Viewing distant objects allows the ciliary muscles to relax, thus reducing the strain on our eyes to a minimum.

The lens can be distorted only so much, however; hence, there is a limit to how close the eye can focus. The shortest distance at which a sharp focus can be obtained is the **near point**—anything closer will appear fuzzy no matter how hard we try to focus on it. For young people the near point distance, N, is typically about 25 cm, but it increases with age. Persons 40 years of age may experience a near point that is 40 cm from the eye, and in later years the near point may move to 500 cm or more. The extension of the near point to greater distances with age is referred to as *presbyopia* (or "short arm" syndrome) and is due to the lens becoming less flexible. Thus, as one ages it is not uncommon to have to move a piece of paper away from the eyes in order to focus, and eventually reading glasses may be necessary.

At the other end of the scale, the **far point** is the greatest distance an object can be from the eye and still be in focus. Since we can focus on the Moon and stars, it is clear that the normal far point is essentially infinity.

Finally, the amount of light that reaches the retina is controlled by a colored diaphragm called the *iris*. As the iris expands or contracts it adjusts the size of the *pupil*, the opening through which light enters the eye. In bright light the pupil closes down to about 1 mm in diameter. On the darkest nights, the dark-adapted pupil can open up to a diameter of about 7.0 mm.

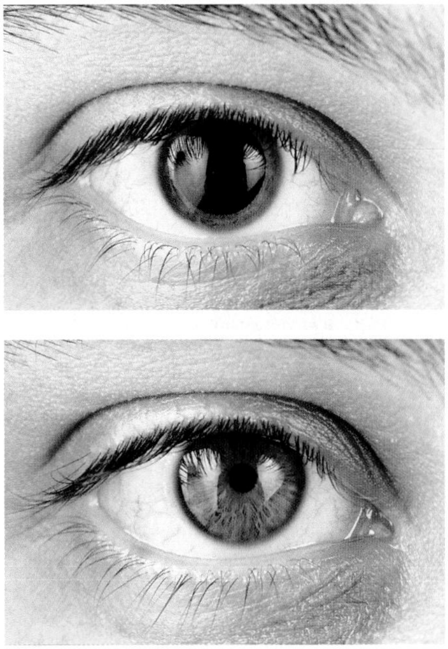

▲ The pigmented iris of the human eye responds automatically to changing levels of illumination, dilating the pupil in dim light and contracting it in bright light.

EXAMPLE 27–1 Journey to Near Point

The near-point distance of a given eye is $N = 25$ cm. Treating the eye as if it were a single thin lens a distance 2.5 cm from the retina, find the focal length of the lens when it is focused on an object **(a)** at the near point and **(b)** at infinity. (Typical values for the effective lens–retina distance range from 1.7 cm to 2.5 cm.)

Picture the Problem

The eye, and the simplified thin-lens equivalent, are shown in the sketch. Note that the horizontal axis is broken in order to bring the object and eye into the same sketch. The object and image distances are indicated as well.

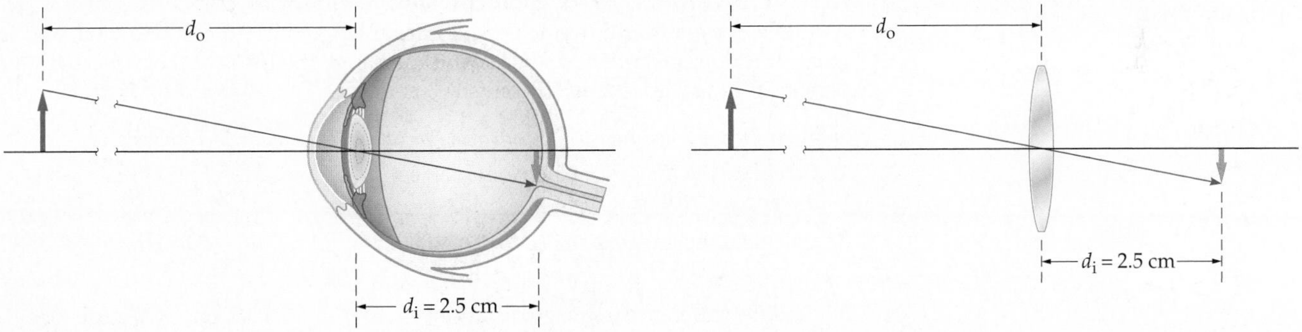

Strategy

The focal length can be found using the thin-lens equation, $1/d_o + 1/d_i = 1/f$. The image distance is $d_i = 2.5$ cm for both (a) and (b). For part (a) the object distance is $d_o = 25$ cm; for part (b) the object distance is $d_o = \infty$.

Solution

Part (a)

1. Substitute $d_o = 25$ cm and $d_i = 2.5$ cm into the thin-lens equation and solve for f:

$$\frac{1}{f} = \frac{1}{d_o} + \frac{1}{d_i} = \frac{1}{25\text{ cm}} + \frac{1}{2.5\text{ cm}} = 0.44\text{ cm}^{-1}$$

$$f = \frac{1}{0.44\text{ cm}^{-1}} = 2.3\text{ cm}$$

Part (b)

2. Substitute $d_o = \infty$ and $d_i = 2.5$ cm into the thin-lens equation and solve for f:

$$\frac{1}{f} = \frac{1}{d_o} + \frac{1}{d_i} = \frac{1}{\infty} + \frac{1}{2.5\text{ cm}} = \frac{1}{2.5\text{ cm}}$$

$$f = 2.5\text{ cm}$$

continued on next page

continued from previous page

Insight

Note that the effective focal length of the eye changes by only about 2 mm in changing focus from the near point to infinity. Thus, if the shape of the eye is changed even slightly—so that the distance from the lens to the retina is increased or decreased by a couple millimeters—the eye will no longer be able to function properly. We return to this point in the next section when we discuss near- and farsightedness.

Practice Problem

At what object distance is the effective focal length of the eye equal to 2.4 cm? [**Answer:** $d_o = 60$ cm]

Some related homework problems: Problem 3, Problem 4

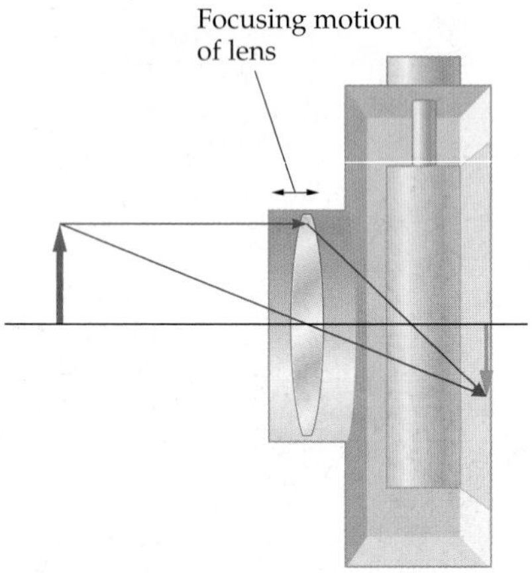

▲ **FIGURE 27–4 Basic elements of a camera**

A camera forms a real, inverted image on photographic film. The camera focuses by moving the lens back and forth relative to the film. Unlike the adjustable shape of the eye, the shape of the camera lens remains fixed.

Perhaps you have noticed that when you are looking at a light-colored background, or a clear sky, one or more "spots" may float across your field of vision. Many people have these "floaters." In fact, some people are occasionally fooled by one of these spots into thinking a fly is buzzing about their head. For this reason, these floaters are called *muscae volitantes*, which means, literally, "flying flies." Muscae volitantes are caused by cells, cell fragments, or other small impurities suspended in either the vitreous humor of the eye or in the lens itself. These are usually harmless but can be symptoms of a detached retina.

The Camera

A simple camera, such as the one illustrated in **Figure 27–4**, operates in much the same way as the eye. In particular, the lens of the camera forms a real, inverted image on a light-sensitive material—which in this case is either photographic film or, more commonly these days, the charge-coupled device (CCD) in a digital camera. The focusing mechanism is different, however. To focus a digital camera at different distances the lens is moved either toward or away from the CCD. Thus, the eye focuses by changing the shape of a stationary lens; the camera focuses by moving a lens of fixed shape.

ACTIVE EXAMPLE 27–1 Find the Displacement of the Lens

A simple camera uses a thin lens with a focal length of 50.0 mm. How far, and in what direction, must the lens be moved to change the focus of the camera from a person 20.0 m away to a person only 3.00 m away?

Solution *(Test your understanding by performing the calculations indicated in each step.)*

1. Calculate the image distance for an object distance of 20.0 m: $d_{i1} = 5.01$ cm
2. Calculate the image distance for an object distance of 3.00 m: $d_{i2} = 5.08$ cm
3. Find the difference in image distance: 0.07 cm = 0.7 mm

Insight

Since the image distance is greater for the person at 3.00 m, it follows that the lens must be moved *away* from the film by 0.7 mm to change the focus the desired amount. Note how little displacement is required.

Your Turn

Suppose the lens is moved an additional 0.7 mm away from the film. At what distance is the camera focused now?

(Answers to **Your Turn** *problems are given in the back of the book.)*

REAL-WORLD PHYSICS

Speed and aperture settings on a camera

The aperture of a camera is analogous to the pupil of an eye, and like the pupil, its size can be adjusted—the greater the size the more light that is available to make an image. Photographers often characterize the size of the aperture with a dimensionless quantity called the ***f*-number,** which is defined as follows:

$$f\text{-number} = \frac{\text{focal length}}{\text{diameter of aperture}} = \frac{f}{D} \qquad \text{27–1}$$

Notice that the larger the diameter of the aperture, the smaller the f-number; in fact, $D = f/(f\text{-number})$.

Professional-grade cameras have aperture settings indicated by a sequence of f-numbers such as the following:

$$2, \quad 2.8, \quad 4, \quad 5.6, \quad 8, \quad 11, \quad 16$$

For example, a camera lens with a focal length of 50.0 mm and an aperture setting of 4 has an aperture diameter of $D = (50.0\text{ mm})/4 = 12.5\text{ mm}$. This is often referred to by photographers as an $f/4$ setting for the aperture. Similarly, turning the aperture ring to the $f/2$ setting opens the aperture to a diameter of $(50.0\text{ mm})/2 = 25.0\text{ mm}$. Since the aperture is a circular opening, the area through which light enters the camera, $A = \pi D^2/4$, varies as the square of the aperture diameter D and thus inversely as the square of the f-number.

The amount of light that falls on the photographic film or CCD in a camera is also controlled by the shutter speed. A shutter speed of 1/500, for example, means that the shutter is open for only 1/500 of a second—very effective for "freezing" the motion of a high-speed object. Changing the shutter speed to 1/250 doubles the time the shutter is open and, hence, doubles the light received by the film. Typical camera shutter speeds are 1/1000, 1/500, 1/250, 1/125, and so on. (On professional-grade cameras, these speeds are indicated on a dial as 1000, 500, 250, 125 to save space.)

Suppose, for example, that a photograph receives the proper amount of light when the shutter speed is 1/500 and the aperture f-number is 4. If the photographer decides to take a second shot with a shutter speed of 1/250, what f-number is required to maintain the correct exposure? Since the slower shutter speed doubles the light entering the camera, the area of the aperture must be halved to compensate. To halve the area, the diameter must be *reduced* by a factor of $\sqrt{2} \approx 1.4$, which means the f-number must be *increased* by a factor of 1.4. Hence, the photographer must change the aperture setting to $4 \times (1.4) = 5.6$. Note that each of the f-numbers listed earlier is larger than the previous one by a factor of roughly 1.4, leading to a factor of 2 difference in the light received by the film.

27–2 Lenses in Combination and Corrective Optics

Whereas a normal eye can provide sharp images for objects over a wide range of distances, some eyes have difficulty focusing on distant objects, and others are unable to focus as close as the normal near point. Problems such as these can be corrected with glasses or contact lenses used in combination with the eye's lens. In general, a combination of lenses can have beneficial properties not possible with a single lens, as we shall see throughout the rest of this chapter. First, in this section, we discuss how to analyze a system with more than one lens; we then apply these results to correcting near- and farsightedness.

The basic operating principle for a system consisting of more than one lens is the following:

> The *image* produced by one lens acts as the *object* for the next lens. This is true regardless of whether the image produced by the first lens is real or virtual, or whether it is in front of or behind the second lens.

This principle finds many applications, since most optical instruments—like cameras, microscopes, and telescopes—use a number of lenses to produce the desired results.

As an example of a lens system, consider the two lenses shown in **Figure 27–5**. An object is 20.0 cm to the left of the convex lens, which is 50.0 cm from a concave lens to its right. Given that the focal lengths of the convex and concave lenses are 10.0 cm and −12.5 cm, respectively, we would like to find the location and orientation of the image produced by the two lenses acting together.

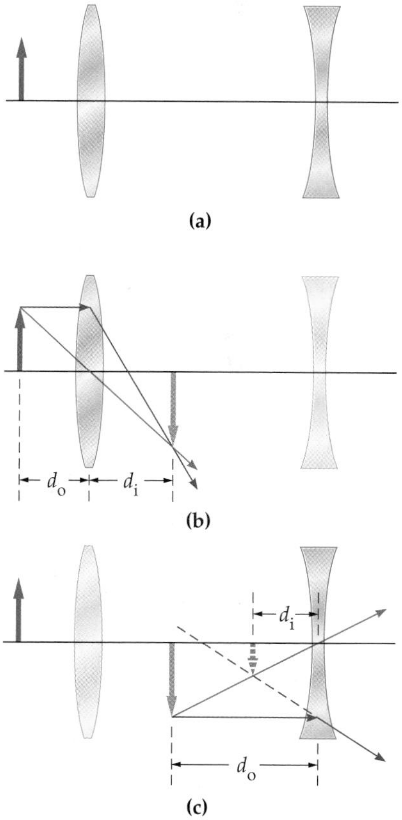

▲ **FIGURE 27–5 A two-lens system**
In this system, a convex and a concave lens are separated by 50.0 cm. **(a)** An object is placed 20.0 cm to the left of the convex lens, whose focal length is 10.0 cm. **(b)** The image formed by the convex lens is 20.0 cm to its right. This image is the object for the concave lens. **(c)** The object for this lens is 30.0 cm to its left. Because the focal length of this lens is −12.5 cm, it forms an image 8.82 cm to its left. This is the final image of the system.

The first step is illustrated in Figure 27–5 (b), where we determine the image formed by the convex lens using a ray diagram. The precise location of "image 1" is given by the thin-lens equation:

$$\frac{1}{d_o} + \frac{1}{d_i} = \frac{1}{f} \quad \text{or} \quad \frac{1}{20.0\text{ cm}} + \frac{1}{d_i} = \frac{1}{10.0\text{ cm}}$$

$$d_i = 20.0\text{ cm}$$

Note that image 1 is inverted and 20.0 cm to the right of the convex lens.

The next step is to note that image 1 is 50.0 cm − 20.0 cm = 30.0 cm to the left of the concave lens. Considering image 1 to be the object for the concave lens, we obtain the ray diagram shown in Figure 27–5 (c). The thin-lens equation, with $d_o = 30.0$ cm and $f = -12.5$ cm, yields the following image distance:

$$\frac{1}{d_o} + \frac{1}{d_i} = \frac{1}{f} \quad \text{or} \quad \frac{1}{30.0\text{ cm}} + \frac{1}{d_i} = \frac{1}{-12.5\text{ cm}}$$

$$d_i = -8.82\text{ cm}$$

Thus the final image of the two-lens system is 8.82 cm to the left of the concave lens.

The orientation and size of the final image can be found as follows:

> The total magnification produced by a lens system is equal to the *product* of the magnifications produced by each lens individually.

Using Equation 26–18, $m = -d_i/d_o$, we find that the first lens produces magnification $m_1 = -(20.0\text{ cm})/(20.0\text{ cm}) = -1$. Similarly, the second lens causes a magnification of $m_2 = -(-8.82\text{ cm})/(30.0\text{ cm}) = 0.294$. The total magnification of the system, then, is $m = m_1 m_2 = -0.294$, showing that the final image is inverted and reduced in size by a factor of 0.294 compared with the original object. This value is in agreement with the results shown in the ray diagrams of Figure 27–5.

The following Active Example considers the effect of reversing the order of the two lenses.

PROBLEM-SOLVING NOTE

Finding the Image for a Multilens System

To find the final image of a multilens system, consider the lenses one at a time. In particular, find the image for the first lens, then use that image as the object for the next lens, and so on.

ACTIVE EXAMPLE 27–2 Find the Final Image

Find the location and orientation of the final image produced by the system shown in Figure 27–5 if the positions of the two lenses are reversed.

Solution *(Test your understanding by performing the calculations indicated in each step.)*

1. Use the thin-lens equation to find the image distance for the concave lens, with $d_{o1} = 20.0$ cm: $d_{i1} = -7.69$ cm
2. Find the object distance for the convex lens: $d_{o2} = 57.7$ cm
3. Use this object distance and the thin-lens equation to find the location of the final image: $d_{i2} = 12.1$ cm

Insight

The final image is 12.1 cm to the right of the convex lens. The magnification produced by the first lens is $m_1 = -d_{i1}/d_{o1} = 0.385$, and that produced by the second lens is $m_2 = -d_{i2}/d_{o2} = -0.210$. Hence, the total magnification is $m = m_1 m_2 = -0.0809$, showing that the final image is reduced in size by a factor of 0.0809 and inverted. Notice how different the results are when we simply reverse the order of the lenses—just like looking through the wrong end of a pair of binoculars.

Your Turn

Consider again the system shown in Figure 27–5. If we replace the concave lens with a concave mirror whose focal length is $f = 12.5$ cm, what is the location and magnification of the final image?

(Answers to **Your Turn** *problems are given in the back of the book.)*

We now show how a two-lens system (the eye plus an external lens) can correct abnormal vision.

Nearsightedness

When a person with normal vision relaxes the ciliary muscles of the eye, an object at infinity is in focus. In a nearsighted (myopic) person, however, a totally relaxed eye focuses only out to a finite distance from the eye—the far point. Thus a person with this condition is said to be nearsighted because objects near the eye can be focused, whereas objects beyond the far point are fuzzy.

The problem in this situation is that the eye converges the light coming into it in too short a distance—in other words, the focal length of the eye is less than the distance from the lens to the retina. This condition is illustrated in **Figure 27–6**, where we see that an object at infinity forms an image in front of the retina, because of the elongation of the eye. The effect need not be large; as we saw in the previous section, an elongation of only a millimeter or two is enough to cause a problem.

To correct this condition, we need to "undo" some of the excess convergence produced by the eye, so that images again fall on the retina. This can be done by placing a *diverging* lens in front of the eye. Specifically, consider an object at infinity—which would ordinarily appear blurry to a nearsighted person. If a concave lens with the proper focal length produces a virtual image of this object at the nearsighted person's far point, as in **Figure 27–7**, the person's relaxed eye can now focus on the object. We consider this situation in the next Example.

PROBLEM-SOLVING NOTE

Correcting Nearsightedness

To correct for nearsightedness, one must use a lens that produces an image at the person's far-point distance when the object is at infinity. Note that the far point is closer to the lens in a pair of glasses than it is to the eye, since the glasses are a finite distance in front of the eyes.

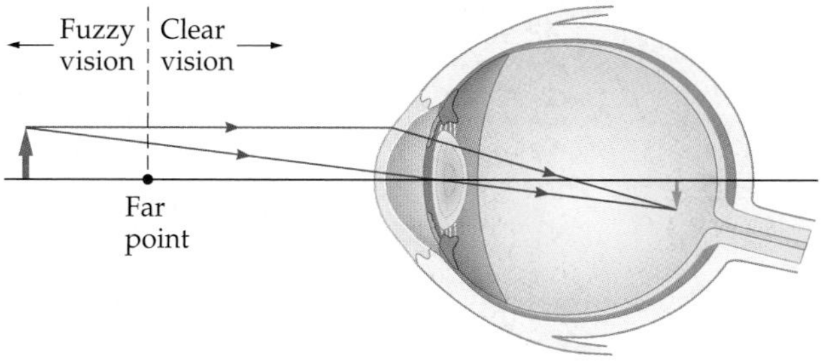

▲ **FIGURE 27–6 Eye shape and nearsightedness**
An eye that is elongated can cause nearsightedness. In this case, an object at infinity comes to a focus in front of the retina.

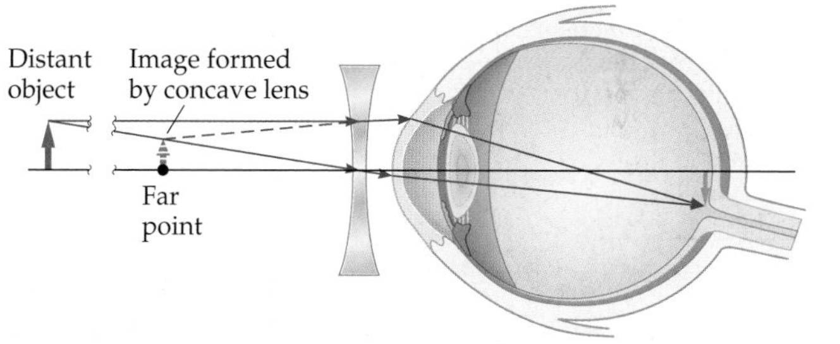

▲ **FIGURE 27–7 Correcting nearsightedness**
A diverging lens in front of the eye can correct for nearsightedness. The concave lens focuses light from an object beyond the far point to produce an image that is at the far point. The eye can now focus on the image of the object.

EXAMPLE 27–2 Extended Vision

REAL-WORLD PHYSICS: BIO

A nearsighted person has a far point that is 323 cm from her eye. If the lens in a pair of glasses is 2.00 cm in front of this person's eye, what focal length must it have to allow her to focus on distant objects?

Picture the Problem

In our sketch, we show an object at infinity, and the corresponding image produced by the concave lens. As is usual with a concave lens, the image is upright, reduced, and on the same side of the lens as the object. In addition, the image is placed at the person's far point, where the eye can focus on it.

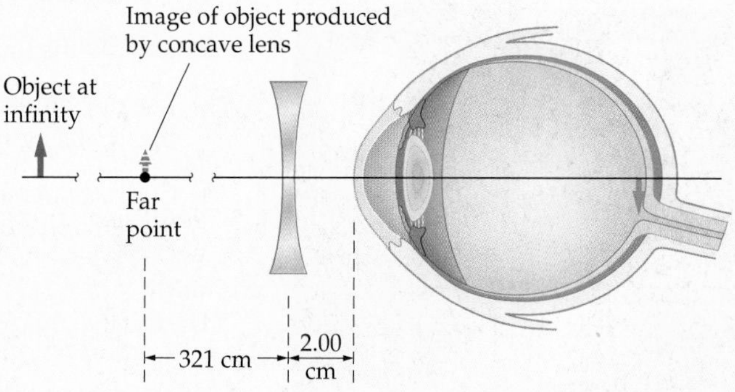

INTERACTIVE FIGURE

Strategy

We can find the focal length of the lens using the thin-lens equation, $1/d_o + 1/d_i = 1/f$.

In this case, $d_o = \infty$, since the object is infinitely far away.

As for the image, it must be 323 cm from the eye, which is 323 cm − 2.00 cm = 321 cm in front of the lens. Thus, the image distance is $d_i = -321$ cm, where the minus sign is required because the image is on the same side of the lens as the object.

With these values for d_o and d_i, it is straightforward to find the focal length, f.

continued on next page

continued from previous page

Solution

1. Substitute $d_o = \infty$ and $d_i = -321$ cm in the thin-lens equation: $\frac{1}{d_o} + \frac{1}{d_i} = \frac{1}{f} = \frac{1}{\infty} + \frac{1}{-321 \text{ cm}}$
2. Solve for the focal length, f: $f = -321$ cm

Insight

With a lens of this focal length, the person can focus on distant objects with a relaxed eye. In addition, note that the focal length is equal to the image distance, $f = d_i$. This is always the case when the object distance is infinite.

Practice Problem

If these glasses are used to view an object 525 cm from the eye, how far from the eye is the image produced by the concave lens? [**Answer:** The image is 201 cm in front of the eye.]

Some related homework problems: Problem 27, Problem 29, Problem 30

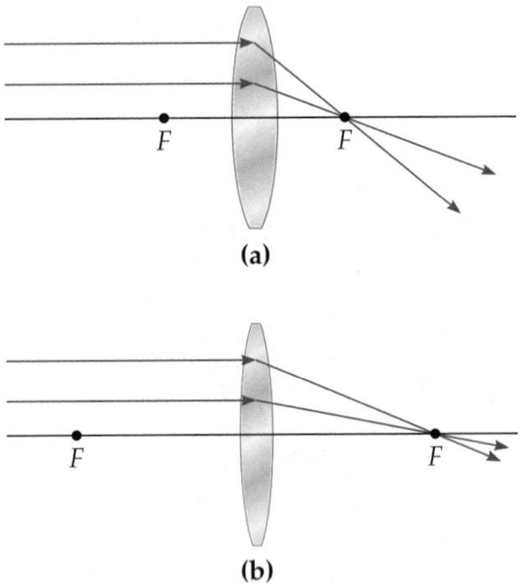

▲ **FIGURE 27–8 Refractive power**
Light is bent (refracted) more by a lens with a short focal length than one with a long focal length. Therefore, lens **(a)** has a greater refractive power than lens **(b)**.

The ability of a lens to refract light—its **refractive power**—is related to its focal length. For example, the shorter the focal length, the more strongly a lens refracts light, as indicated in **Figure 27–8**. Thus refractive power depends inversely on the focal length. By definition, then, we say that the refractive power of a lens is $1/f$, where f is measured in meters:

Refractive Power

$$\text{refractive power} = \frac{1}{f} \qquad \text{27–2}$$

SI unit: diopter $= \text{m}^{-1}$

Lenses are typically characterized by optometrists in terms of diopters rather than in terms of focal length.

As an example of the meaning of diopters, a lens with a refractive power of 10.0 diopters has a focal length of $1/(10.0 \text{ m}^{-1}) = 10.0$ cm (a converging lens), and a lens with a refractive power of -10.0 diopters has a focal length of -10.0 cm (a diverging lens). In Example 27–2, the lens required to correct nearsightedness had a refractive power of $1/(-3.21 \text{ m}) = -0.312$ diopter.

ACTIVE EXAMPLE 27–3 Find the Refractive Power

A person has a far point that is 5.50 m from his eyes. If this person is to wear glasses that are 2.00 cm from his eyes, what refractive power, in diopters, must his lenses have?

Solution *(Test your understanding by performing the calculations indicated in each step.)*

1. Identify the object distance: $d_o = \infty$
2. Identify the image distance: $d_i = -548$ cm
3. Use the thin-lens equation to calculate the focal length of the lenses: $f = -548$ cm
4. Convert f to meters and invert to find the refractive power: refractive power $= -0.182$ diopter

Insight

The fact that this person's far point is farther away (closer to infinity) than the far point in Example 27–2 means that the lenses do not have to be as "strong" to correct the vision. As a result, the magnitude of the refractive power in this case (0.182 diopter) is less than the corresponding magnitude (0.312 diopter) in Example 27–2.

Your Turn

A person wears glasses with a refractive power of -0.500 diopter. If the glasses are 2.00 cm in front of the eyes, what is the distance from the eyes to the far point?

*(Answers to **Your Turn** problems are given in the back of the book.)*

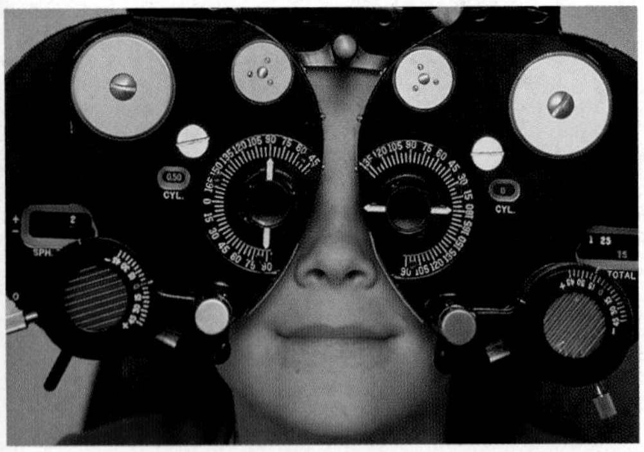

▲ In order to prescribe the right corrective lenses, it is necessary to measure the refractive properties of the patient's eyes. This device, known as a *phoropter*, allows the optometrist to see how lenses with various optical characteristics affect the patient's vision.

Today, in addition to glasses and contact lenses, a number of medical procedures are available to correct nearsightedness. Some of these procedures involve using laser beams to reshape the cornea of the eye. These techniques are discussed in Chapter 31, where we consider lasers in detail. Here we present two alternative procedures that change the shape of a cornea by mechanical means.

REAL-WORLD PHYSICS: BIO
Intracorneal rings and Intacts

Perhaps the simplest such technique is the implantation of either an *intracorneal ring* or an *Intact* in the cornea of an eye. An Intact, in particular, consists of two small, clear crescents made of the same material used in contact lenses. These crescents are slipped into "tunnels" that are cut into the cornea. When the crescents are in place they tend to stretch the cornea outward and flatten its surface. This flattening increases the focal length of the eye and corrects for nearsightedness by -1.0 to -3.0 diopters. If necessary, the Intacts can be removed and replaced with others that cause more or less flattening of the cornea. Intracorneal rings are similar, except that they consist of a single ring rather than the two crescents used in Intacts.

REAL-WORLD PHYSICS: BIO
Radial keratotomy

A more complex method involves making radial incisions in the cornea with a highly precise diamond blade that cuts to a specified depth. This method is referred to as *radial keratotomy* or RK (**Figure 27–9**). The cuts allow the peripheral parts of the cornea to bulge outward, which, in turn, causes the central portion to flatten. As with an Intact, the flattening of the cornea corrects for nearsightedness.

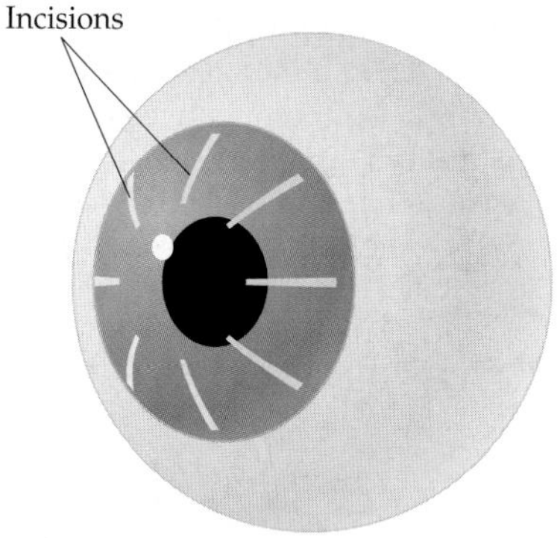

▲ **FIGURE 27–9 Radial keratotomy**
In radial keratotomy, a series of radial incisions is made around the periphery of the cornea. This allows the cornea to expand outward, resulting in a flattening of the cornea's central region. The reduced curvature of the cornea increases the focal length of the eye, allowing it to focus on distant objects.

Farsightedness

A person who is farsighted (hyperopic) can see clearly beyond a certain distance—the near point—but cannot focus on closer objects. Basically, the vision of a farsighted person differs from that of a person with normal vision by having a near point that is much farther from the eye than the usual 25 cm. As a result, a farsighted person is typically unable to read clearly, since a book would have to be held at such a great distance to come into focus.

Farsightedness can be caused by an eyeball that is shorter than normal, as illustrated in **Figure 27–10**, or by a lens that becomes sufficiently stiff with age that it can no longer take on the shape required to focus on nearby objects. In such cases, rays from an object inside the near point are brought to a focus behind the retina. Thus the focal length of the farsighted eye is too large—stated another way, the farsighted eye does not converge the incoming light strongly enough to focus on the retina.

This problem can be corrected by "preconverging" the light—that is, by using a converging lens in front of the eye to add to its insufficient convergence. For example, suppose an object is inside a person's near point, as in **Figure 27–11**. If a

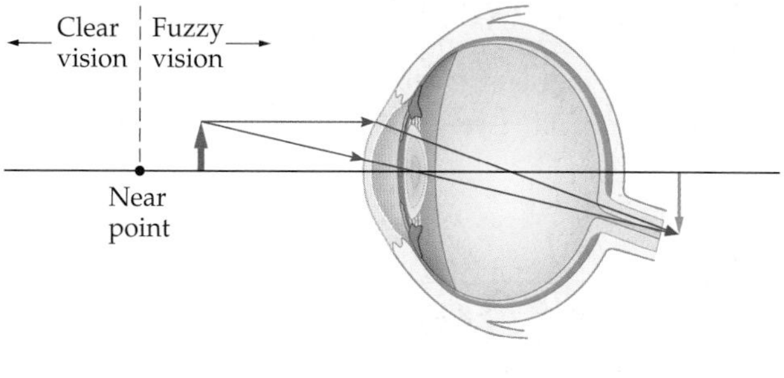

◀ **FIGURE 27–10 Eye shape and farsightedness**
An eye that is shorter than normal can cause farsightedness. Note that an object inside the near point comes to a focus behind the retina.

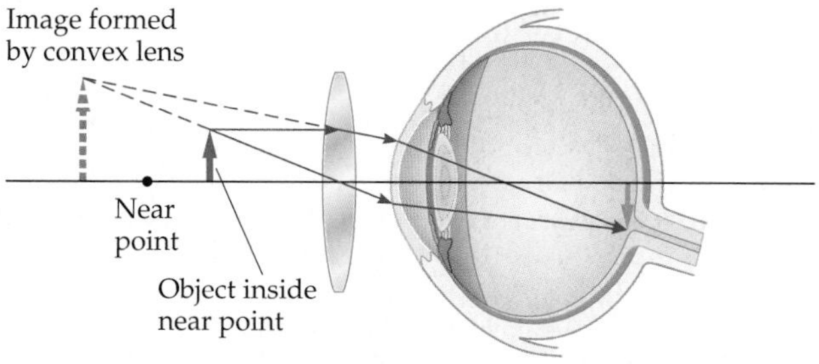

◀ **FIGURE 27–11 Correcting Farsightedness**
A converging lens in front of the eye can correct for farsightedness. The convex lens focuses light from an object inside the near point to produce an image that is beyond the near point. The eye can now focus on the image of the object.

converging lens placed in front of the eye can produce an image that is far away—that is, beyond the near point—the farsighted person can view the object with ease. Such a system is considered in the next Example.

EXAMPLE 27–3 His Vision Is a Far Sight Better

REAL-WORLD PHYSICS: BIO A farsighted person wears glasses that enable him to read a book held at a distance of 25.0 cm from his eyes, even though his near-point distance is 57.0 cm. If his glasses are at a distance of 2.00 cm from his eyes, find the focal length and refractive power required of his lenses to place the image of the book at the near point.

Picture the Problem

The physical situation is shown in our sketch. Notice that we use a convex lens, and that the image produced by the lens is upright and farther from the eye than the book. These features are in agreement with the qualitative characteristics shown in Figure 27–11. In our case, however, the image of the book is exactly at the near point.

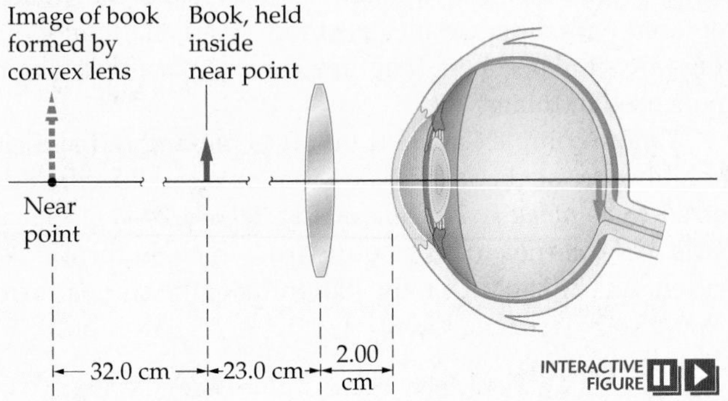

INTERACTIVE FIGURE

Strategy

The focal length of the lens can be found using the thin-lens equation. First, the object distance is $d_o = 23.0$ cm, taking into account the 2.00 cm between the glasses and the eye. Similarly, the desired image distance is $d_i = -55.0$ cm. Note that the minus sign is required on the image distance, since the image is on the same side of the lens as the object.

Solution

1. Use the thin-lens equation to find the focal length, f:

$$\frac{1}{f} = \frac{1}{d_o} + \frac{1}{d_i} = \frac{1}{23.0 \text{ cm}} + \frac{1}{-55.0 \text{ cm}} = 0.0253 \text{ cm}^{-1}$$

$$f = \frac{1}{0.0253 \text{ cm}^{-1}} = 39.5 \text{ cm}$$

2. The refractive power is $1/f$, with f measured in meters:

$$\text{refractive power} = \frac{1}{f} = \frac{1}{0.395 \text{ m}} = 2.53 \text{ diopters}$$

Insight

Notice that the book is between the lens and its focal point. As a result, the image is upright [see Figure 26–35 (b)], as desired for reading glasses. In addition, the image is virtual, as was also the case for the concave lens in Example 27–2. Thus, even though a "virtual image" can sound like one that isn't real or of practical importance, you are viewing a virtual image every time you look through a pair of glasses.

Practice Problem

A second person has a near-point distance that is greater than 57.0 cm. Is the refractive power of the lenses required for this person greater than or less than 2.53 diopters? To check your answer, calculate the refractive strength for a near-point distance of 67.0 cm. [**Answer:** The refractive power must be greater than 2.53 diopters. We find 2.81 diopters for a near-point distance of 67.0 cm.]

Some related homework problems: Problem 26, Problem 28

PROBLEM-SOLVING NOTE

Correcting Farsightedness

To correct for farsightedness, one must use a lens that produces an image at the person's near-point distance when the object is closer than the near point. Note that the distance from the object to the lens in a pair of glasses is less than the distance from the object to the eye.

CONCEPTUAL CHECKPOINT 27–1 Eyeglasses to Start a Fire

Bill and Ted are on an excellent camping trip when they decide to start a fire by focusing sunlight with a pair of eyeglasses. If Bill is nearsighted and Ted is farsighted, should they use (a) Bill's glasses or (b) Ted's glasses?

Reasoning and Discussion

To focus the parallel rays of light from the Sun to a point requires a converging lens. As we have seen, nearsightedness is corrected with a diverging lens; farsightedness is corrected with a converging lens. Ted's eyeglasses are converging; therefore, they should be the ones used to start a fire.

Answer:

(b) Ted's eyeglasses would be the more excellent choice.

To consider the effect of a pair of contact lenses, we simply take into account that contacts are placed directly against the eye. This means that the eye–object distance is the same as the lens–object distance. We make use of this fact in the following Active Example.

PROBLEM-SOLVING NOTE

Finding the Focal Length of Contact Lenses

Contact lenses are so named because they are in direct contact with the eye. Therefore, the object distance for a contact lens is the same as the distance from the eye to the object. Similarly, the near point is the same distance from both the contact lens and the eye.

ACTIVE EXAMPLE 27–4 Contact: Find the Focal Length of Contact Lenses

Find the focal length of a pair of contact lenses that will allow a person with a near-point distance of 145 cm to read a newspaper held 25.1 cm from the eyes.

Solution *(Test your understanding by performing the calculations indicated in each step.)*

1. Identify the object distance: $d_o = 25.1\text{ cm}$
2. Identify the image distance: $d_i = -145\text{ cm}$
3. Use the thin-lens equation to calculate the focal length: $f = 30.4\text{ cm}$

Insight

Note that the image distance is negative, since the image is on the same side of the lens as the object.

Your Turn

Suppose a second person has a near-point distance of 205 cm. Is the focal length of the second person's contacts greater than or less than the focal length of the first person's contacts? Find the focal length of the second person's contacts.

(Answers to **Your Turn** *problems are given in the back of the book.)*

An important step in assuring that contact lenses fit a patient's eye properly is to measure the radius of curvature of the cornea. This is accomplished with a device known as a *keratometer*. To begin the procedure, a brightly lit object is brought near the eye. The light from the object reflects from the front surface of the cornea, just as light reflects from a convex, spherical mirror. In the next step of the procedure, the keratometer measures the magnification of the mirror image produced by the cornea. Finally, a straightforward application of the mirror equation determines the radius of curvature. A specific example of a keratometer in use is given in Problem 78.

REAL-WORLD PHYSICS: BIO

Keratometers

Another eye condition that requires corrective optics is **astigmatism.** In most cases, astigmatism is due to an irregular curvature of the cornea, with a greater curvature in one direction than in another. For example, the curvature in the vertical plane may be greater than the curvature in the horizontal plane. As a result, if the eye is focused for light coming into it from one direction, it will be out of focus for light arriving from a different direction. Almost everyone has some degree of astigmatism, usually quite mild, but in serious cases it can cause distorted or blurry vision at all distances.

27–3 The Magnifying Glass

A **magnifying glass** is nothing more than a simple convex lens. Working together with the eye as part of a two-lens system, a *magnifier* can make objects appear to be many times larger than their actual size. As we shall see, the magnifying glass works by moving the near point closer to the eye—much as a converging lens corrects farsightedness. Basically, the magnifier allows an object to be viewed from a reduced distance, and this is what makes it appear larger.

To be more specific about the apparent size of an object, we consider the angle it subtends on the retina—after all, the more area an image takes up on the retina, the larger it will seem to us. For example, suppose an object of height h_o is a distance d_o from the eye, as in **Figure 27–12 (a)**. The image formed by this object subtends an angle θ on the retina of the eye, as shown in the figure. If the angle is small, as is often

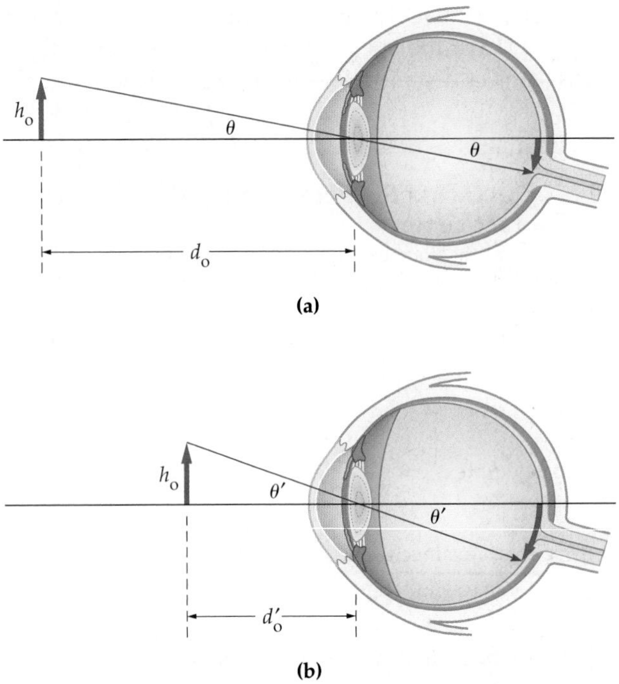

▶ **FIGURE 27–12 Angular size and distance**

The angular size of an object depends on its distance from the eye, even though its height, h_o, remains the same.

the case, it can be approximated by the tangent of the angle, since $\tan\theta \approx \theta$ for small angles. Referring to the figure, we see that the angle θ is approximately

$$\theta \approx \frac{h_o}{d_o}$$

If the object is now moved closer to the eye, to a distance d'_o the angle subtended by the image is greater. Using the small-angle approximation again, and referring to **Figure 27–12 (b)**, we find that the new angle is

$$\theta' \approx \frac{h_o}{d'_o} > \theta$$

Thus, moving an object closer to the eye increases its apparent size, since its image covers a larger portion of the retina. There is a limit, however, to how close an object can come to the unaided eye and still be in focus—the near point. This is where a magnifier comes into play.

Suppose, then, that you would like to see as much detail as possible on a small object—a feather, perhaps, or a flower petal. With the unaided eye you can bring the object to the near point, a distance N from the eye, as in **Figure 27–13 (a)**. If the height of the object is h_o, the angular size of the object on the retina is approximately

$$\theta = \frac{h_o}{N}$$

Now, consider placing a convex lens of focal length $f < N$ just in front of the eye, as shown in **Figure 27–13 (b)**. If the object is brought to the focal point of this lens, its image will be infinitely far from the eye, where it can be viewed in focus with ease. As we can see from the figure, the angular size of the image is approximately

$$\theta' = \frac{h_o}{f}$$

Note that θ' is greater than θ—since f is less than N—and hence the object appears larger. The factor by which the object is enlarged, referred to as the **angular magnification,** M, is defined as follows:

Angular Magnification, M

$$M = \frac{\theta'}{\theta} \qquad \textbf{27–3}$$

SI unit: dimensionless

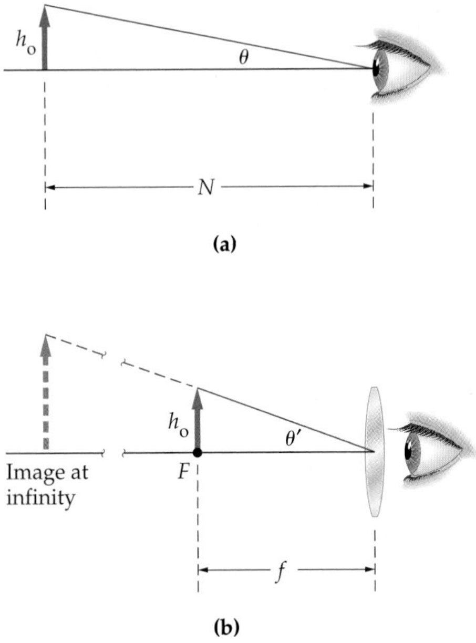

▲ **FIGURE 27–13 How a simple magnifier works**

(a) An object viewed with the unaided eye at the near point of the eye subtends an angle $\theta \approx h_o/N$. **(b)** With a magnifier, the object can be viewed from the distance f, which is less than N. As a result, the angular size of the object is $\theta' \approx h_o/f$, which is greater than θ.

Using the angles θ and θ' obtained previously, we find

$$M = \frac{h_o/f}{h_o/N} = \frac{N}{f} \qquad \text{27–4}$$

A typical situation is considered in the following Exercise.

EXERCISE 27–1

A person with a near-point distance of 30 cm examines a stamp with a magnifying glass. If the magnifying glass produces an angular magnification of 6, what is its focal length?

Solution

Solving $M = N/f$ for the focal length we find

$$f = \frac{N}{M} = \frac{30 \text{ cm}}{6} = 5 \text{ cm}$$

Note that an object examined with a magnifier appears larger precisely because it is closer to the eye. In fact, the angular size of the object in Figure 27–13 (b) is h_o/f regardless of whether the magnifier is present or not. If the magnifier is not present, however, the object is out of focus, and then its increased angular size is of no practical value. The magnifier is beneficial in that it brings the object into focus at this close distance.

In addition, since the image produced by the magnifier is at infinity, the rays entering the eye are parallel. This means that a person can view the image with a completely relaxed eye. If the same object is viewed with the unaided eye at the near point, the ciliary muscles are fully tensed, causing eye strain. Thus, not only does the magnifier enlarge the object, it also makes it more comfortable to view.

Now, of course, it is also possible to produce an enlarged image with a convex lens without holding the lens close to the eye. In fact, if the lens is held at some distance from the eye, and just under a focal length from the object to be viewed, we see an enlarged image. There are two distinct disadvantages to holding a magnifier far from the eye, however. First, it is difficult to hold the lens motionless at a distance, as opposed to bracing it against the face. Second, when the lens is held at a distance, only a small portion of the object can be viewed at any given time. When the lens is held close up, however, the entire object can be viewed at once.

As we shall see in the next two sections, the magnifier plays an important role in both the microscope and the telescope.

▲ A magnifying glass produces an enlarged image when held near an object. The image is upright and virtual.

CONCEPTUAL CHECKPOINT 27–2 Using a Magnifier

Person 1, with a near-point distance of 25 cm, and person 2, with a near-point distance of 50 cm, both use the same magnifying glass. Does **(a)** person 1 or **(b)** person 2 benefit more from using the magnifier?

Reasoning and Discussion

The person with the greater near-point distance cannot see an object as closely with the unaided eye as can the person with the smaller near-point distance. With the magnifier, however, both people can view an object from the same close-in distance. Hence, the person with the larger near-point distance benefits more from the magnifier.

Answer:

(b) Person 2, with the 50-cm near-point distance, benefits more.

It is possible to obtain a magnification that is slightly greater than the value N/f given in Equation 27–4. After all, this magnification is for an image formed at infinity; if the image were closer to the eye, it would appear larger. The closest the image can be—and still be in focus—is the near point. In the next Example we compare the magnifications that result when the image is at infinity and at the near point.

EXAMPLE 27–4 Comparing Magnifications

A biologist with a near-point distance of $N = 26$ cm examines an insect wing through a magnifying glass whose focal length is 4.3 cm. Find the angular magnification when the image produced by the magnifier is **(a)** at infinity and **(b)** at the near point.

Picture the Problem

Our sketch shows the situation in which the image formed by the magnifying glass is at the near point. To produce an image at this point, the object must be closer to the magnifier than its focal point. Therefore, the object appears larger in this case than it does in the case where the object is at the focal length and the image is at infinity.

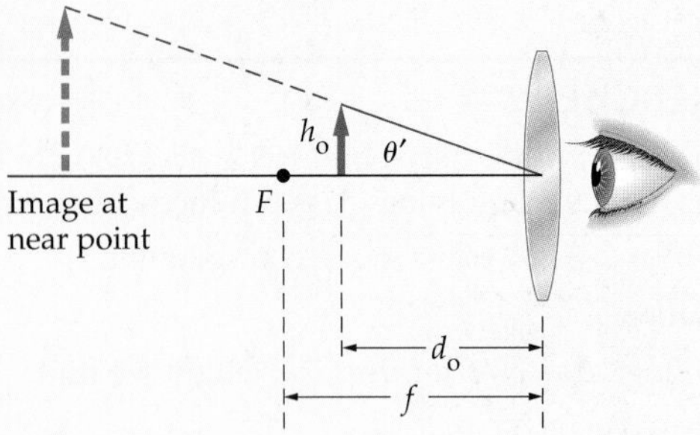

Strategy

a. The magnification when the image is at infinity is $M = N/f$, as given in Equation 27–4.

b. The angular size of the object in this case is $\theta' \approx h_o/d_o$, where d_o is the object distance that places the image at the near point. Thus the magnification is $M = \theta'/\theta = (h_o/d_o)/(h_o/N) = N/d_o$. Note that this result differs from the magnification in part (a) only in that f has been replaced with d_o. All that remains is to calculate d_o using the thin-lens equation.

Solution

Part (a)

1. Substitute $N = 26$ cm and $f = 4.3$ cm into $M = N/f$:

$$M = \frac{N}{f} = \frac{26 \text{ cm}}{4.3 \text{ cm}} = 6.0$$

Part (b)

2. Use the thin-lens equation to find d_o, given $f = 4.3$ cm and $d_i = -26$ cm:

$$\frac{1}{d_o} = \frac{1}{f} - \frac{1}{d_i} = \frac{1}{4.3 \text{ cm}} - \frac{1}{-26 \text{ cm}} = 0.27 \text{ cm}^{-1}$$

$$d_o = \frac{1}{0.27 \text{ cm}^{-1}} = 3.7 \text{ cm}$$

3. Calculate the magnification using $M = N/d_o$:

$$M = \frac{N}{d_o} = \frac{26 \text{ cm}}{3.7 \text{ cm}} = 7.0$$

Insight

We see that moving the object closer to the lens, which brings the image in from infinity to the near point, results in an increase in magnification of 1.0—from 6.0 to 7.0.

Practice Problem

What is the magnification if the object is placed 4.0 cm in front of the magnifying glass? [**Answer:** $M = N/d_o = 6.5$]

Some related homework problems: Problem 46, Problem 48

Thus we see from Example 27–4 that the magnification of a magnifying glass with the object at a distance d_o is $M = N/d_o$, where N is the near-point distance of the observer. In the special case of an image at infinity, the object distance is the focal length, and $M = N/f$. It can be shown (see Problem 95) that the greatest magnification—which occurs when the image is at the near point—is $M = 1 + N/f$. The magnification results are summarized here:

$$M = \frac{N}{f} \quad \text{(image at infinity)}$$

$$M = 1 + \frac{N}{f} \quad \text{(image at near point)} \qquad \textbf{27–5}$$

Note that Example 27–4 is simply a special case of these general expressions.

The early microscopes produced by Antoni van Leeuwenhoek (1632–1723) were, in fact, simply powerful magnifying glasses using a single lens mounted in a hole in a flat metal plate. The object to be examined with the microscope was placed on the head of a small, movable pin placed just below the lens. The observer would then hold the top surface of the lens close to the eye for the most stable, wide-angle view. Leeuwenhoek's instruments were capable of magnifying objects by as much as 275 times, enough that he could make the first detailed microscopic descriptions of single-celled animals, red blood cells, plant cells, and much more. Microscopes in common use today have more than one lens, as we describe in the next section.

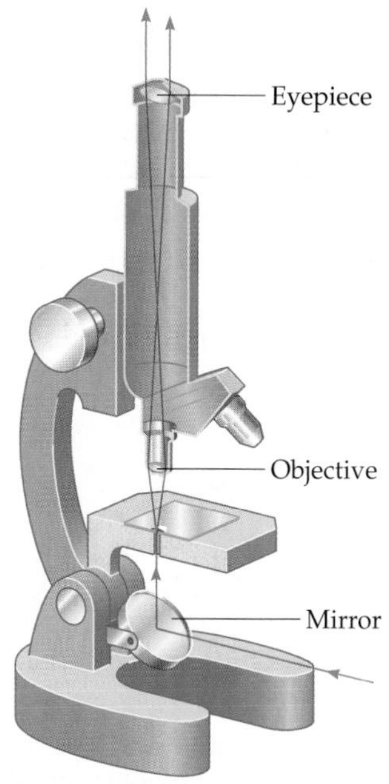

▲ **FIGURE 27–14 Basic elements of a compound microscope**

A compound microscope consists of two lenses—an objective and an eyepiece—fixed at either end of a movable tube.

27–4 The Compound Microscope

Although a magnifying glass is a useful device, higher magnifications and improved optical quality can be obtained with a **microscope.** The simplest microscope consists of two converging lenses fixed at either end of a tube. Such an instrument, illustrated in **Figure 27–14**, is sometimes referred to as a *compound microscope.*

The basic optical elements of a microscope are the **objective** and the **eyepiece.** The objective is a converging lens with a relatively short focal length that is placed near the object to be viewed. It forms a real, inverted, and enlarged image, as shown in **Figure 27–15**. The precise location of the image is adjusted when the microscope is focused by moving the objective up or down. This image serves as the object for the second lens in the microscope—the eyepiece. In fact, the eyepiece is simply a magnifier that views the image of the objective, giving it an additional enlargement.

The final magnification of the microscope, then, is simply the product of the magnification of the objective and the magnification of the eyepiece. For example, a microscope might have a 10× eyepiece (meaning it magnifies 10 times) and a 50× objective. When these two lenses are used together, the magnification of the microscope is 500×.

In a typical situation, the object to be examined is placed only a small distance beyond the focal point of the objective, which means that $d_o \approx f_{\text{objective}}$. The magnification produced by the objective is given by Equation 26–18:

$$m_{\text{objective}} = -\frac{d_i}{d_o} \approx -\frac{d_i}{f_{\text{objective}}}$$

Next, the image formed by the objective is essentially at the focal point of the eyepiece. This means that the eyepiece forms a virtual image at infinity that the observer can view with a relaxed eye. The angular magnification of the eyepiece is given by Equation 27–4:

$$M_{\text{eyepiece}} = \frac{N}{f_{\text{eyepiece}}}$$

Multiplying these magnifications, we find the total magnification of the microscope:

$$M_{\text{total}} = m_{\text{objective}} M_{\text{eyepiece}} = \left(-\frac{d_i}{f_{\text{objective}}}\right)\left(\frac{N}{f_{\text{eyepiece}}}\right)$$

$$= -\frac{d_i N}{f_{\text{objective}} f_{\text{eyepiece}}} \qquad \text{27–6}$$

The minus sign indicates that the image is inverted.

Typical numerical values are considered in the following Example.

PROBLEM-SOLVING NOTE

Lens Placement in a Microscope

In a working microscope, the distance between the lenses is greater than the sum of their focal lengths.

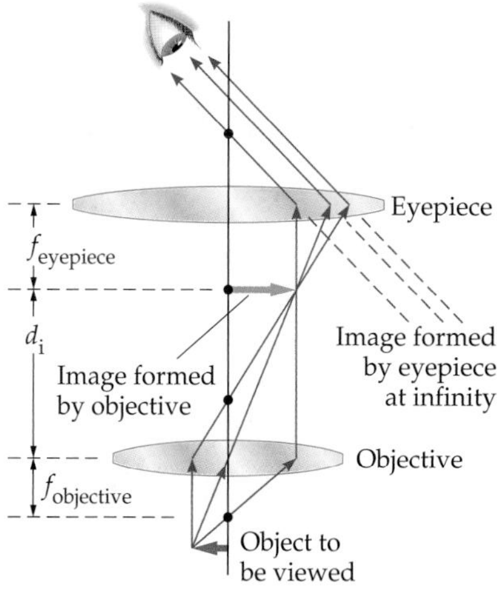

▶ **FIGURE 27–15 The operation of a compound microscope**

In a compound microscope the object is placed just outside the focal point of the objective. The resulting enlarged image is then enlarged further by the eyepiece, which is basically a magnifying glass.

EXAMPLE 27–5 A Microscopic View

In biology class, a student with a near-point distance of $N = 25$ cm uses a microscope to view an amoeba. If the objective has a focal length of 1.0 cm, the eyepiece has a focal length of 2.5 cm, and the amoeba is 1.1 cm from the objective, what is the magnification produced by the microscope?

continued on next page

continued from previous page

Picture the Problem

The lenses of the microscope are shown in the sketch. Note that the distance from the objective to the object is d_o and the distance from the objective to its image is d_i.

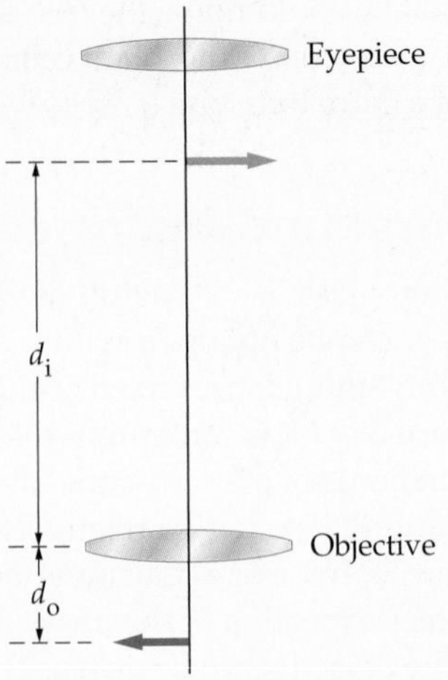

Strategy

The magnification of the microscope is given by Equation 27–6. The only unknown in this expression is the image distance, d_i. We can find this by using the thin-lens equation.

Solution

1. Use the thin-lens equation to find the image distance, d_i:

$$\frac{1}{d_i} = \frac{1}{f} - \frac{1}{d_o} = \frac{1}{1.0 \text{ cm}} - \frac{1}{1.1 \text{ cm}} = 0.091 \text{ cm}^{-1}$$

$$d_i = \frac{1}{0.091 \text{ cm}^{-1}} = 11 \text{ cm}$$

2. Use Equation 27–6 to find the magnification of the microscope:

$$M_{\text{total}} = -\frac{d_i N}{f_{\text{objective}} f_{\text{eyepiece}}} = -\frac{(11 \text{ cm})(25 \text{ cm})}{(1.0 \text{ cm})(2.5 \text{ cm})} = -110$$

Insight

Thus the amoeba appears 110 times larger and is inverted. If the amoeba is to be viewed with a relaxed eye, the image formed by the objective should be at the focal point of the eyepiece, which will then form an image at infinity. Therefore, the length of the tube containing the objective and eyepiece is 11 cm + 2.5 cm = 13.5 cm in this case.

Practice Problem

If the focal length of the eyepiece is increased, does the magnitude of the magnification increase or decrease? Check your response by calculating the magnification when the focal length of the eyepiece is 3.5 cm. [**Answer:** The magnification is reduced in magnitude. Its new value is −79.]

Some related homework problems: Problem 53, Problem 56

27–5 Telescopes

A telescope is similar in many respects to a microscope—both use two converging lenses to give a magnified image of an object with small angular size. In the case of a microscope the object itself is small and close at hand; in the case of the telescope the object may be as large as a galaxy, but its angular size can be very small because of its great distance. The major difference between these instruments is that the telescope must deal with an object that is essentially infinitely far away.

For this reason, it is clear that the light entering the objective of a telescope from a distant object is focused at the focal point of the objective, as shown in **Figure 27–16**. If the image formed by the objective has a height h_i, the angular size of the object is approximately

$$\theta = -\frac{h_i}{f_{\text{objective}}}$$

Note that the minus sign is included because h_i is negative for the inverted image formed by the objective.

As in the microscope, the image of the objective is the object for the eyepiece, which is basically a magnifier. Thus, if the image of the objective is placed

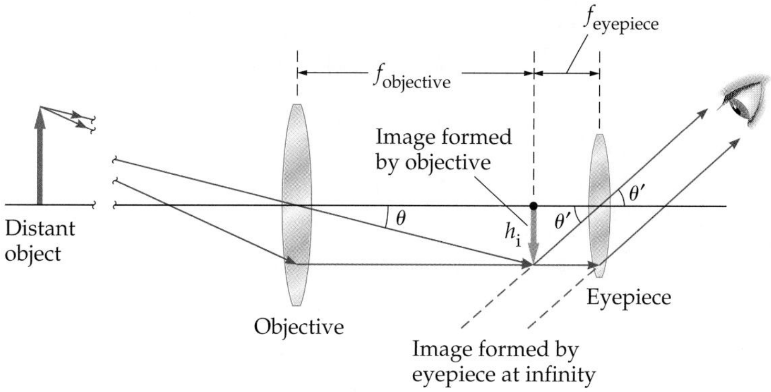

◀ **FIGURE 27–16 Basic elements of a telescope**

A telescope focuses light from a distant object at its focal point. The image of the objective is placed at the focal point of the eyepiece to produce an enlarged image that can be viewed with a relaxed eye.

at the focal point of the eyepiece, it will form an image that is at infinity, as indicated in Figure 27–16. In this configuration, the observer can view the final image of the telescope with a completely relaxed eye. The angular size of the image formed by the eyepiece, θ', is shown in Figure 27–16. Clearly, this angle is approximately

$$\theta' = -\frac{h_{\text{i}}}{f_{\text{eyepiece}}}$$

To find the total angular magnification of the telescope, we take the ratio of θ' to θ:

$$M_{\text{total}} = \frac{\theta'}{\theta} = \frac{f_{\text{objective}}}{f_{\text{eyepiece}}} \qquad \text{27–7}$$

Thus, for example, a telescope with an objective whose focal length is 1500 mm and an eyepiece whose focal length is 10.0 mm produces an angular magnification of 150.

PROBLEM-SOLVING NOTE

Lens Placement in a Telescope

In a working telescope, the distance between the lenses is approximately equal to the sum of their focal lengths. In addition, the focal length of the eyepiece is significantly less than that of the objective.

CONCEPTUAL CHECKPOINT 27–3 Comparing Telescopes

Two telescopes have identical eyepieces, but telescope A is twice as long as telescope B. Is the magnification of telescope A **(a)** greater than the magnification of telescope B, **(b)** less than the magnification of telescope B, or **(c)** is there no way to tell?

Reasoning and Discussion

Note in Figure 27–16 that the total length of the telescope is $f_{\text{objective}}$ plus f_{eyepiece}. Thus, because the scopes have identical eyepieces, it follows that telescope A must have a greater objective focal length than telescope B. We know, then, from $M_{\text{total}} = f_{\text{objective}}/f_{\text{eyepiece}}$, that the magnification of telescope A is greater than the magnification of telescope B.

Answer:

(a) Telescope A has a greater magnification than telescope B.

Telescopes using two or more lenses, as in Figure 27–16, are referred to as *refractors*. In fact, the first telescopes constructed for astronomical purposes, made by Galileo starting in 1609, were refractors. Galileo's telescopes differed from the telescopes in common use today, however, in that they used a diverging lens for the eyepiece. We consider this type of telescope in Problems 68, 74, and 83. By the end of 1609, Galileo had produced a telescope whose angular magnification was 20. This was more than enough to enable him to see—for the first time in human history—mountains on the Moon, stars in the Milky Way, the phases of Venus, and moons orbiting Jupiter. As a result of his telescopic observations, Galileo became a firm believer in the Copernican model of the solar system.

In the next Example we consider a standard refractor with two converging lenses. In particular, we show how the length of such a telescope is related to the focal lengths of its objective and eyepiece.

EXAMPLE 27–6 Considering the Telescope at Length

A telescope has a magnification of 40.0 and a length of 1230 mm. What are the focal lengths of the objective and eyepiece?

Picture the Problem

Our sketch shows that the overall length of the telescope, L, is equal to the sum of the focal lengths of the objective and the eyepiece. In addition, note that the objective forms an inverted image of the distant object, which the eyepiece enlarges. The final image seen by the observer, then, is inverted.

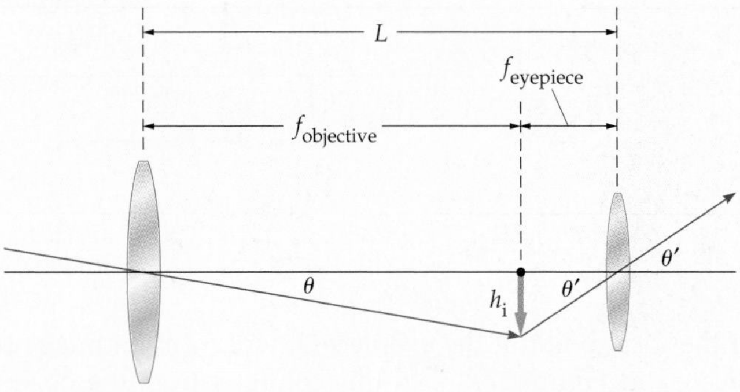

Strategy

We can determine the two unknowns, $f_{\text{objective}}$ and f_{eyepiece}, from the two independent bits of information given in the problem statement:

a. The magnification, $M_{\text{total}} = f_{\text{objective}}/f_{\text{eyepiece}}$, is equal to 40.0.

b. The total length, $L = f_{\text{objective}} + f_{\text{eyepiece}}$, is equal to 1230 mm.

Combining this information gives us the two focal lengths.

Solution

1. Use the magnification equation to write $f_{\text{objective}}$ in terms of f_{eyepiece}:

$$M_{\text{total}} = \frac{f_{\text{objective}}}{f_{\text{eyepiece}}} = 40.0$$

$$f_{\text{objective}} = 40.0 f_{\text{eyepiece}}$$

2. Substitute $f_{\text{objective}} = 40.0 f_{\text{eyepiece}}$ into the expression for the total length of the telescope:

$$L = f_{\text{objective}} + f_{\text{eyepiece}}$$

$$= 40.0 f_{\text{eyepiece}} + f_{\text{eyepiece}} = 41.0 f_{\text{eyepiece}} = 1230 \text{ mm}$$

3. Divide 1230 mm by 41.0 to find the focal length of the eyepiece:

$$f_{\text{eyepiece}} = \frac{1230 \text{ mm}}{41.0} = 30.0 \text{ mm}$$

4. Multiply f_{eyepiece} by 40.0 to find the focal length of the objective:

$$f_{\text{objective}} = 40.0 f_{\text{eyepiece}} = 40.0(30.0 \text{ mm}) = 1200 \text{ mm}$$

Insight

The focal lengths found in this example are typical of those used in many popular amateur telescopes. A telescope with a magnification of 40.0 can easily show the moons of Jupiter and the rings of Saturn.

Practice Problem

A telescope 1820 mm long has an objective with a focal length of 1780 mm. Find the focal length of the eyepiece and magnification of the telescope. [**Answer:** $f_{\text{eyepiece}} = 40.0$ mm, $M_{\text{total}} = 44.5$]

Some related homework problems: Problem 61, Problem 62, Problem 71

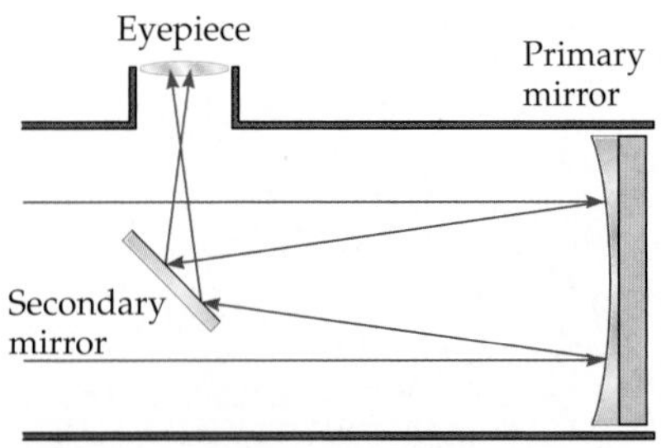

▲ **FIGURE 27–17 A Newtonian reflecting telescope**

The large primary mirror collects incoming light and reflects it off a small secondary mirror into the eyepiece.

Because telescopes are typically used to view objects that are very dim, it is desirable to have an objective with as large a diameter, or aperture, as possible. If the diameter is doubled, for example, the light gathered by a telescope increases by the same factor as the area of its objective: $2^2 = 4$. In addition, a larger aperture results in a higher-resolution image, as we shall see in the next chapter.

For a refractor, a large aperture means a very large and heavy piece of glass. In fact, the world's largest refractor, the Yerkes refractor at Williams Bay, Wisconsin, has an objective that is only 1 m across. If refractors were made much larger, the objective lens would sag and distort under its own weight.

The telescopes with the largest apertures are reflectors, one example of which is illustrated in **Figure 27–17**. Invented by Isaac Newton in 1671, and referred to as a Newtonian reflector, this type of telescope uses a mirror in place of an objective lens. As with the refractor, the mirror forms an image which the eyepiece then magnifies. Since a mirror can be much thinner and lighter than a lens, and can be supported all over its back surface instead of just around the edges as with a lens, it has many advantages for a large scope. In fact, the largest telescopes in the

▲ At left, the 1-m objective lens of the Yerkes refractor—the largest refracting telescope ever constructed—being dusted to remove spiders. If lenses were made much larger than this, they would sag and distort under their own weight. At right, one of the twin 10-m Keck reflecting telescopes atop Mauna Kea on the Big Island of Hawaii, where the air is cold, thin, and dry—ideal conditions for astronomical observation. Clearly, the use of interchangeable objectives, commonly found in microscopes, is not feasible for large telescopes. Instead, magnification is varied by using eyepieces with different focal lengths. It is rare for a person to actually look through one of these telescopes, however—instead, they are usually fitted with sophisticated cameras, spectrographs, or other instruments.

world are the twin Keck reflecting telescopes atop Hawaii's Mauna Kea. These telescopes have hexagonal objective mirrors that are 10 m across, which means that each gathers 100 times as much light as the Yerkes refractor. When used together, the Keck telescopes constitute the world's largest pair of binoculars!

27–6 Lens Aberrations

An ideal lens brings all parallel rays of light that strike it together at a single focal point. Real lenses, however, never quite live up to the ideal. Instead, a real lens blurs the focal point into a small but finite region of space. This, in turn, blurs the image it forms. The deviation of a lens from ideal behavior is referred to as an **aberration.**

One example is **spherical aberration,** which is related to the shape of a lens. **Figure 27–18** shows parallel rays of light passing through a lens with spherical aberration; note that the rays do not meet at a single focal point. This is completely analogous to the spherical aberration present in mirrors with a spherical cross section, as shown in Figure 26–13. To prevent spherical aberration, a lens must be ground and polished to a very precise nonspherical shape.

Another common aberration in lenses is **chromatic aberration,** which is due to the unavoidable dispersion present in any refracting material. Just as a prism splits white light into a spectrum of colors, each color bent by a different amount, so too does a lens bend light of different colors by different amounts as shown in **Figure 27–19**. The result is that white light passing through a lens does not focus

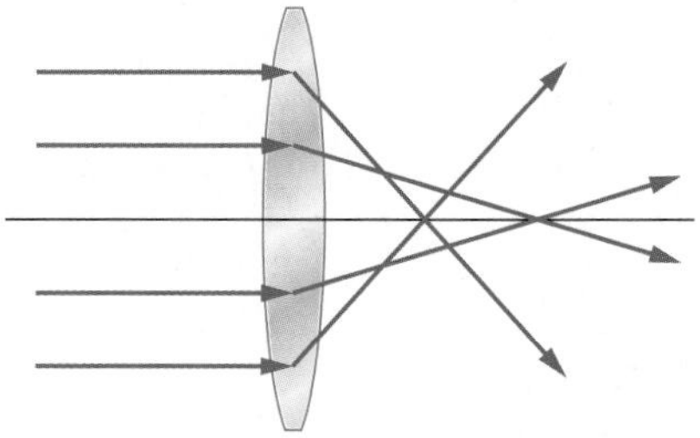

▲ **FIGURE 27–18 Spherical aberration**
In a lens with spherical aberration, light striking the lens at different locations comes together at different focal points.

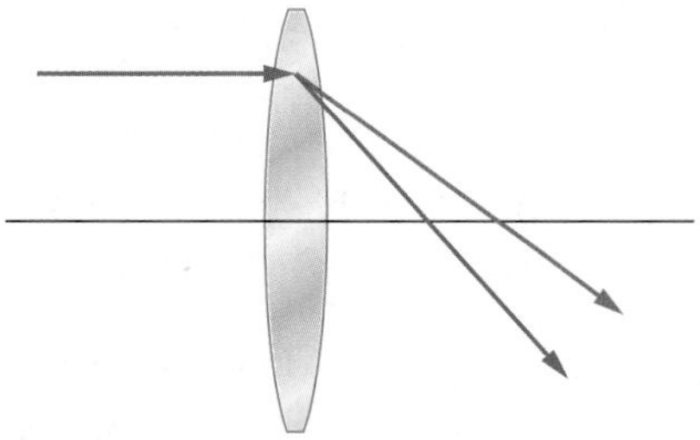

▲ **FIGURE 27–19 Chromatic aberration in a converging lens**
Chromatic aberration is caused by the fact that blue light bends more than red light as it passes through a lens. As a result, different colors have different focal points.

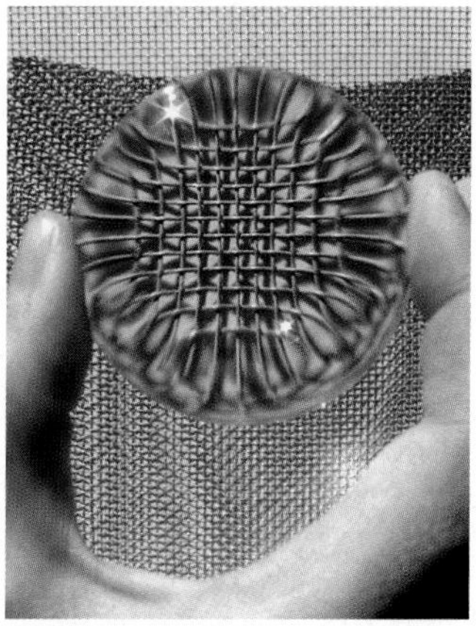

▲ Chromatic aberration typical of a simple lens.

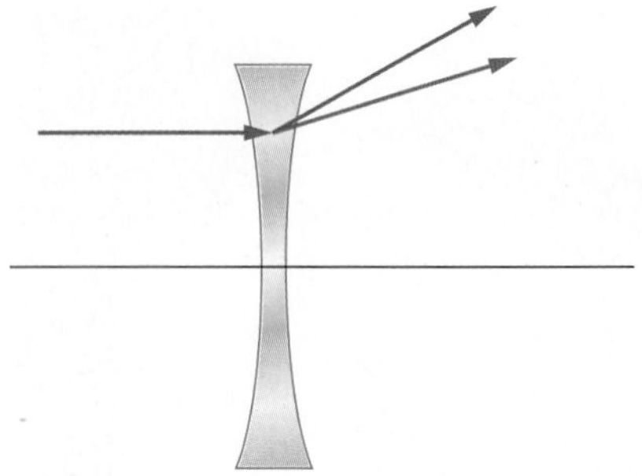

▲ **FIGURE 27–20 Chromatic aberration in a diverging lens**
A diverging lens also bends blue light more than red light, though the rays are bent in the opposite direction compared with the bending by a converging lens.

to a single point. This is why you sometimes see a fringe of color around an image seen through a simple lens.

Chromatic aberration can be corrected by combining two or more lenses to form a compound lens. For example, suppose a convex lens made from glass A bends red and violet light as shown in Figure 27–19. Note that the violet light is bent more, as expected. A second lens, concave this time, is made of glass B. This lens also bends violet light more than red light, but since it is a diverging lens, it bends both the red and violet light in the opposite direction compared with the convex lens, as we see in **Figure 27–20**. If the amount of bending of violet light compared with red light is different for glasses A and B, it is possible to construct a convex and a concave lens in such a way that the opposite directions in which they bend red and violet light can be made to cancel while still causing the light to come to a focus. This combination is illustrated in **Figure 27–21**. Such a lens is said to be **achromatic.** Three lenses connected similarly can produce even better correction for chromatic aberration. Lenses of this type are referred to as **apochromatic.**

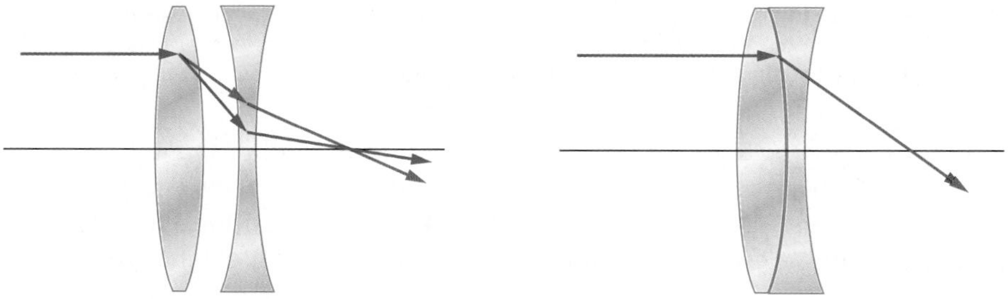

▶ **FIGURE 27–21 Correcting for chromatic aberration**
An achromatic lens is a combination of a converging and a diverging lens made of different types of glass. In the converging lens the blue light is bent toward the axis of the lens more than the red light. In the diverging lens the opposite is the case. The net result is that the red and blue light pass through the same focal point.

Many lenses in optical instruments are not single lenses at all but, instead, are compound achromatic lenses. For example, the "lens" in a 35-mm camera may actually contain five or more individual lenses—some of them converging, some diverging, some made of one type of glass, some made of another. A zoom lens for a camera not only must correct for chromatic aberration, but must also be able to change its focal length. Thus, it is not uncommon for such a lens to contain as many as twelve individual lenses.

Chapter Summary

27–1 The Human Eye and the Camera

The human eye forms a real, but inverted, image on the retina; a camera forms a real, but inverted, image on photographic film.

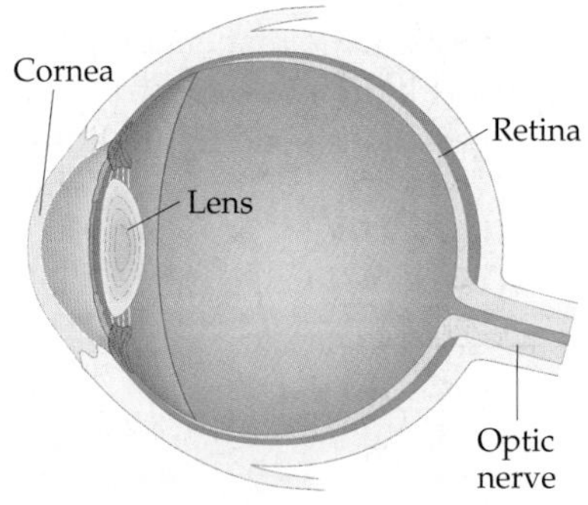

Focusing the Eye
The eye is focused by the ciliary muscles, which change the shape of the lens. This process is referred to as accommodation.

Focusing a Camera
A camera is focused by moving the lens closer to or farther away from the photographic film. The shape of the lens is unchanged.

Near Point
The near point is the closest distance to which the eye can focus. A typical value for the near-point distance, N, is 25 cm.

Far Point
The far point is the greatest distance at which the eye can focus. In a normal eye the far point is infinity.

***f*-Number**
The f-number of a lens relates the diameter of the aperture, D, to the focal length, f, as follows:

$$f\text{-number} = \frac{\text{focal length}}{\text{diameter of aperture}} = \frac{f}{D} \qquad \textbf{27–1}$$

27–2 Lenses in Combination and Corrective Optics

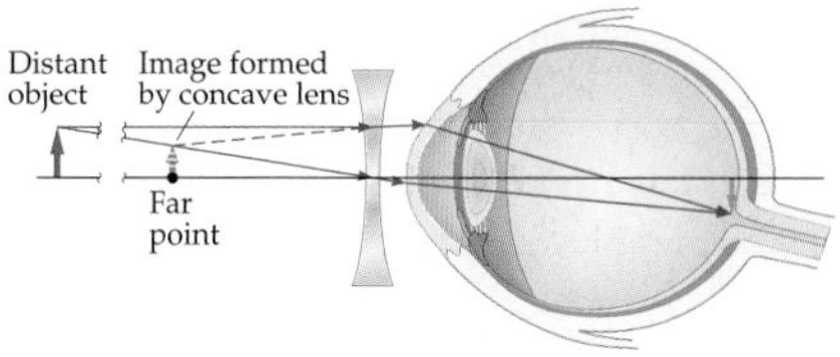

The basic idea used in analyzing systems with more than one lens is the following: The *image* produced by one lens acts as the *object* for the next lens.

In addition, the total magnification of a system of lenses is given by the product of the magnifications produced by each lens individually.

Nearsightedness
Nearsightedness is a condition in which clear vision is restricted to a region relatively close to the eye—in other words, the far point is not at infinity but instead is a finite distance from the eye. This condition can be corrected with diverging lenses placed in front of the eyes.

Farsightedness
A person who is farsighted can see clearly only at relatively large distances from the eye—that is, the person's near point is much farther from the eye than the more typical values of 25 to 40 cm. Farsightedness can be corrected by placing converging lenses in front of the eyes.

Refractive Power and Diopters
The refractive power of a lens refers to its ability to bend light and is measured in diopters. It is defined as follows:

$$\text{refractive power} = \frac{1}{f} \qquad \textbf{27–2}$$

In this expression, the focal length, f, must be measured in meters.

The greater the magnitude of the refractive power, the more strongly the lens bends light. A positive refractive power indicates a converging lens; a negative refractive power indicates a diverging lens.

27–3 The Magnifying Glass

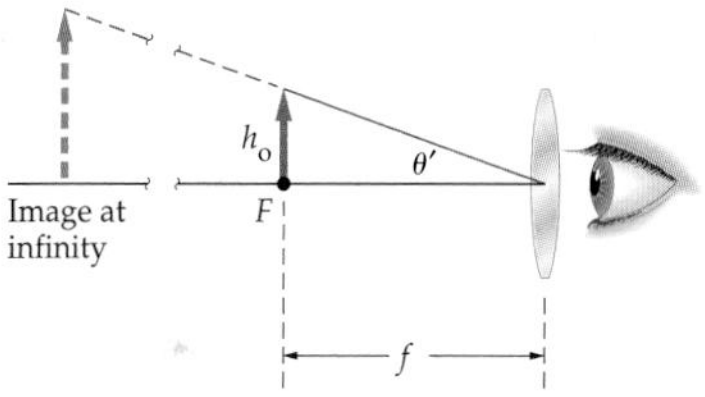

A magnifying glass is simply a converging lens. It works by allowing an object to be viewed at a distance less than the near-point distance. Since the distance is reduced, the angular size is increased.

Magnification
A person with a near-point distance N using a magnifying glass with a focal length f experiences the following magnifications, M:

$$M = \frac{N}{f} \quad \text{(image at infinity)}$$

$$M = 1 + \frac{N}{f} \quad \text{(image at near point),} \qquad \textbf{27–5}$$

27–4 The Compound Microscope

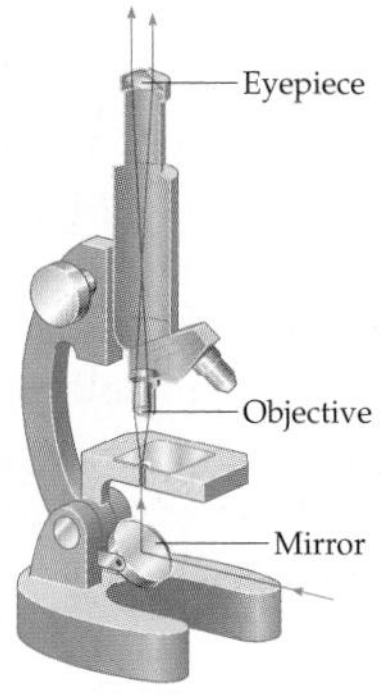

A compound microscope uses two lenses in combination—an objective and an eyepiece—to produce a magnified image.

The object to be viewed is placed just outside the focal length of the objective. The image formed by the objective is then viewed by the eyepiece, giving additional magnification.

Magnification
The magnification produced by a compound microscope is

$$M_{\text{total}} = -\frac{d_i N}{f_{\text{objective}} f_{\text{eyepiece}}} \qquad \textbf{27–6}$$

where N is the near-point distance and d_i is the image distance for the objective. The magnifications quoted for microscopes assume a near-point distance of 25 cm.

27–5 Telescopes

A telescope provides magnified views of distant objects using two lenses. The objective lens focuses the incoming light at its focal point; the eyepiece magnifies the image formed by the objective.

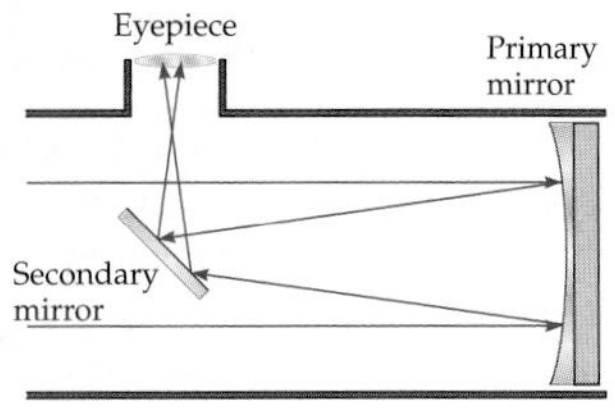

Magnification

The total magnification produced by a telescope is

$$M_{total} = \frac{f_{objective}}{f_{eyepiece}} \quad 27\text{–}7$$

Length of a Telescope

The length, L, of a telescope is the sum of the focal lengths of its two lenses:

$$L = f_{objective} + f_{eyepiece}$$

Reflecting Telescopes

A reflecting telescope uses a mirror in place of an objective lens. The largest telescopes are reflectors, as are many amateur telescopes.

27–6 Lens Aberrations

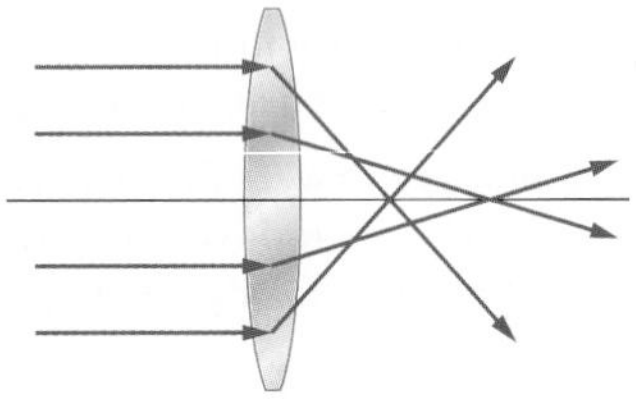

Any deviation of a lens from ideal behavior is referred to as an aberration.

Spherical Aberration

In spherical aberration, parallel rays of light passing through a lens fail to go through a single focal point. Spherical aberration is related to the shape of a lens.

Chromatic Aberration

Chromatic aberration results from dispersion within a refracting material. It causes different colors to focus at different points. Achromatic lenses correct for chromatic aberration by combining two lenses with different refractive properties.

Problem-Solving Summary

Type of Problem	Relevant Physical Concepts	Related Examples
Find the effective focal length of an eye when focused on a given object.	Use the thin-lens equation with an image distance of 1.7 cm to 2.5 cm to take into account the distance from the lens to the retina.	Example 27–1
Determine the optical behavior of a multilens system.	Apply the thin-lens equation to each lens one at a time. Let the image of one lens be the object for the next lens.	Active Example 27–2
Find the focal length of corrective glasses.	In the case of a nearsighted person, make sure the image is no farther away from the eye than the far point. For a farsighted person, make sure the image is no closer to the eye than the near point.	Examples 27–2, 27–3 Active Examples 27–3, 27–4

Conceptual Questions

(Answers to odd-numbered Conceptual Questions can be found in the back of the book.)

1. Why is it restful to your eyes to gaze off into the distance?
2. If a lens is cut in half through a plane perpendicular to its surface, does it show only half an image?
3. If your near-point distance is N, how close can you stand to a mirror and still be able to focus on your image?
4. When you open your eyes underwater, everything looks blurry. Can this be thought of as an extreme case of nearsightedness or farsightedness? Explain.
5. Would you benefit more from a magnifying glass if your near-point distance is 25 cm or if it is 15 cm? Explain.
6. When you use a simple magnifying glass, does it matter whether you hold the object to be examined closer to the lens than its focal length or farther away? Explain.
7. Is the final image produced by a telescope real or virtual? Explain.
8. Does chromatic aberration occur in mirrors? Explain.

Conceptual Exercises

(Answers to odd-numbered Conceptual Exercises can be found in the back of the book.)

1. Two people are stranded on a deserted island. Both people wear glasses, though one is nearsighted and the other is farsighted. Which person's glasses should be used to focus the rays of the Sun and start a fire? Explain.
2. A clerk at the local grocery store wears glasses that make her eyes look larger than they actually are. Is the clerk nearsighted or farsighted? Explain.

3. The umpire at a baseball game wears glasses that make his eyes look smaller than they actually are. Is the umpire nearsighted or farsighted? Explain.

4. **BIO Intracorneal Ring** An intracorneal ring is a small plastic device implanted in a person's cornea to change its curvature. By changing the shape of the cornea, the intracorneal ring can correct a person's vision. If a person is nearsighted, should the ring increase or decrease the cornea's curvature? Explain.

5. **BIO** The lens in a normal human eye, with aqueous humor on one side and vitreous humor on the other side, has a refractive power of 15 diopters. Suppose a lens is removed from an eye and surrounded by air. In this case, is its refractive power greater than, less than, or equal to 15 diopters? Explain.

6. You have two lenses: lens 1 with a focal length of 0.45 cm and lens 2 with a focal length of 1.9 cm. If you construct a microscope with these lenses, which one should you use as the objective? Explain.

7. An optical system consists of two lenses, one with a focal length of 0.50 cm and the other with a focal length of 2.3 cm. If the separation between the lenses is 12 cm, is the instrument a microscope or a telescope? Explain.

8. An optical system consists of two lenses, one with a focal length of 50 cm and the other with a focal length of 2.5 cm. If the separation between the lenses is 52.5 cm, is the instrument a microscope or a telescope? Explain.

9. Two telescopes of different length produce the same angular magnification. Is the focal length of the long telescope's eyepiece greater than or less than the focal length of the short telescope's eyepiece? Explain.

10. To construct a telescope, you are given a lens with a focal length of 32 mm and a lens with a focal length of 1600 mm. **(a)** On the basis of focal length alone, which lens should be the objective and which the eyepiece? Explain. **(b)** What magnification would this telescope produce?

11. **BIO Treating Cataracts** When the lens in a person's eye becomes clouded by a cataract, the lens can be removed with a process called phacoemulsification and replaced with a man-made intraocular lens. The intraocular lens restores clear vision, but its focal length cannot be changed to allow the user to focus at different distances. In most cases, the intraocular lens is adjusted for viewing of distant objects, and corrective glasses are worn when viewing nearby objects. Should the refractive power of the corrective glasses be positive or negative? Explain.

12. **BIO Octopus Eyes** To focus its eyes, an octopus does not change the shape of its lens, as is the case in humans. Instead, the octopus moves its rigid lens back and forth. This changes the distance from the lens to the retina and brings an object into focus. If an object moves closer to an octopus, must the octopus move its lens closer to or farther from its retina to keep the object in focus? Explain.

Problems

Note: **IP** *denotes an integrated conceptual/quantitative problem.* **BIO** *identifies problems of biological or medical interest. Red bullets (•, ••, •••) are used to indicate the level of difficulty of each problem.*

The outside medium is assumed to be air, with an index of refraction of 1.00*, unless specifically stated otherwise.*

Section 27–1 The Human Eye and the Camera

1. • Your friend is 1.8 m tall. **(a)** When she stands 3.5 m from you, what is the height of her image formed on the retina of your eye? (Consider the eye to consist of a thin lens 2.5 cm from the retina.) **(b)** What is the height of her image when she is 4.5 m from you?

2. • Which forms the larger image on the retina of your eye: a 43-ft tree seen from a distance of 210 ft, or a 12-in. flower viewed from a distance of 2.0 ft?

3. • Approximating the eye as a single thin lens 2.60 cm from the retina, find the eye's near-point distance if the smallest focal length the eye can produce is 2.20 cm.

4. • Referring to Problem 3, what is the focal length of the eye when it is focused on an object at a distance of **(a)** 250 cm and **(b)** 25 cm?

5. • Four camera lenses have the following focal lengths and *f*-numbers:

Lens	Focal length (mm)	*f*-number
A	150	$f/1.2$
B	150	$f/5.6$
C	35	$f/1.2$
D	35	$f/5.6$

Rank these lenses in order of increasing aperture diameter. Indicate ties where appropriate.

6. • **BIO** The focal length of the human eye is approximately 1.7 cm. **(a)** What is the *f*-number for the human eye in bright light, when the pupil diameter is 2.0 mm? **(b)** What is the *f*-number in dim light, when the pupil diameter has expanded to 7.0 mm?

7. • **IP** A camera with a 55-mm focal length lens has aperture settings of 2.8, 4, 8, 11, and 16. **(a)** Which setting has the largest aperture diameter? **(b)** Calculate the five possible aperture diameters for this camera.

8. • The actual frame size of "35-mm" film is 24 mm × 36 mm. You want to take a photograph of your friend, who is 1.9 m tall. Your camera has a 55-mm focal length lens. How far from the camera should your friend stand in order to produce a 36-mm-tall image on the film?

9. • To completely fill a frame of "35-mm" film, the image produced by a camera must be 36 mm high. If a camera has a focal length of 150 mm, how far away must a 2.0-m-tall person stand to produce an image that fills the frame?

10. •• You are taking a photograph of a poster on the wall of your dorm room, so you can't back away any farther than 3.0 m to take the shot. The poster is 0.80 m wide and 1.2 m tall, and you want the image to fit in the 24 × 36-mm frame of the film in your camera. What is the longest focal length lens that will work?

11. •• A photograph is properly exposed when the aperture is set to $f/8$ and the shutter speed is 125. Find the approximate shutter speed needed to give the same exposure if the aperture is changed to $f/2.4$.

12. •• You are taking pictures of the beach at sunset. Just before the Sun sets, a shutter speed of 1/100 s at $f/11$ produces a properly exposed picture. Shortly after the Sun sets, however, your light

meter indicates that the scene is only one quarter as bright as before. **(a)** If you don't change the aperture, what approximate shutter speed is needed for your second shot? **(b)** If, instead, you keep the shutter speed at 1/100 s, what approximate *f*-stop will be needed for the second shot?

13. •• IP You are taking a photograph of a horse race. A shutter speed of 125 at *f*/5.6 produces a properly exposed image, but the running horses give a blurred image. Your camera has *f*-stops of 2, 2.8, 4, 5.6, 8, 11, and 16. **(a)** To use the shortest possible exposure time (i.e., highest shutter speed), which *f*-stop should you use? **(b)** What is the shortest exposure time you can use and *still* get a properly exposed image?

14. •• **The Hale Telescope** The 200-in. (5.08-m) diameter mirror of the Hale telescope on Mount Palomar has a focal length $f = 16.9$ m. **(a)** When the detector is placed at the focal point of the mirror (the "prime focus"), what is the *f*-ratio for this telescope? **(b)** The coudé focus arrangement uses additional mirrors to bend the light path and increase the effective focal length to 155.4 m. What is the *f*-ratio of the telescope when the coudé focus is being used? (*Coudé* is French for elbow, since the light path is "bent like an elbow." This arrangement is useful when the light needs to be focused onto a distant instrument.)

Section 27–2 Lenses in Combination and Corrective Optics

15. • Construct a ray diagram for Active Example 27–2.

16. • The cornea of a normal human eye has an optical power of +43.0 diopters. What is its focal length?

17. • A myopic student is shaving without his glasses. If his eyes have a far point of 1.6 m, what is the greatest distance he can stand from the mirror and still see his image clearly?

18. • An eyeglass prescription calls for a lens with an optical power of +2.3 diopters. What is the focal length of this lens?

19. •• Two thin lenses, with $f_1 = +25.0$ cm and $f_2 = -42.5$ cm, are placed in contact. What is the focal length of this combination?

20. •• Two thin lenses have refractive powers of +4.00 diopters and −2.35 diopters, respectively. What is the refractive power of the two if they are placed in contact? (Note that these are the same two lenses described in the previous problem.)

21. •• Two concave lenses, each with $f = -12$ cm, are separated by 6.0 cm. An object is placed 24 cm in front of one of the lenses. Find **(a)** the location and **(b)** the magnification of the final image produced by this lens combination.

22. •• IP BIO The focal length of a relaxed human eye is approximately 1.7 cm. When we focus our eyes on a close-up object, we can change the refractive power of the eye by about 16 diopters. **(a)** Does the refractive power of our eyes increase or decrease by 16 diopters when we focus closely? Explain. **(b)** Calculate the focal length of the eye when we focus closely.

23. •• IP BIO **Diopter Change in Diving Cormorants** Double-crested cormorants (*Phalacrocorax auritus*) are extraordinary birds—they can focus on objects in the air, just like we can, but they can also focus underwater as they pursue their prey. To do so, they have one of the largest *accommodation ranges* in nature—that is, they can change the focal length of their eyes by amounts that are greater than is possible in other animals. When a cormorant plunges into the ocean to catch a fish, it can change the refractive power of its eyes by about 45 diopters, as compared to only 16 diopters of change possible in the human eye. **(a)** Should this change of 45 diopters be an increase or a decrease? Explain. **(b)** If the focal length of the cormorant's eyes is 4.2 mm before it enters the water, what is the focal length after the refractive power changes by 45 diopters?

The double-crested cormorant. (Problem 23)

24. •• A converging lens of focal length 8.000 cm is 20.0 cm to the left of a diverging lens of focal length −6.00 cm. A coin is placed 12.0 cm to the left of the converging lens. Find **(a)** the location and **(b)** the magnification of the coin's final image.

25. •• Repeat Problem 24, this time with the coin placed 18.0 cm to the right of the diverging lens.

26. •• Find the focal length of contact lenses that would allow a farsighted person with a near-point distance of 176 cm to read a book at a distance of 10.1 cm.

27. •• Find the focal length of contact lenses that would allow a nearsighted person with a 135-cm far point to focus on the stars at night.

28. •• What focal length should a pair of contact lenses have if they are to correct the vision of a person with a near point of 56 cm?

29. •• A nearsighted person wears contacts with a focal length of −8.5 cm. If this person's far-point distance with her contacts is 8.5 m, what is her uncorrected far-point distance?

30. •• Without his glasses, Isaac can see objects clearly only if they are less than 4.5 m from his eyes. What focal length glasses worn 2.1 cm from his eyes will allow Isaac to see distant objects clearly?

31. •• A person whose near-point distance is 39 cm wears a pair of glasses that are 2.0 cm from her eyes. With the aid of these glasses, she can now focus on objects 25 cm away from her eyes. Find the focal length and refractive power of her glasses.

32. •• A pair of eyeglasses is designed to allow a person with a far-point distance of 2.50 m to read a road sign at a distance of 25.0 m. Find the focal length required of these glasses if they are to be worn **(a)** 2.00 cm or **(b)** 1.00 cm from the eyes.

33. •• IP Your favorite aunt can read a newspaper only if it is within 12.0 cm of her eyes. **(a)** Is your aunt nearsighted or farsighted? Explain. **(b)** Should your aunt wear glasses that are converging or diverging to improve her vision? Explain. **(c)** How many diopters of refractive power must her glasses have if they are worn 2.00 cm from the eyes and allow her to read a newspaper at a distance of 25.0 cm?

34. •• IP The relaxed eyes of a patient have a refractive power of 48.5 diopters. **(a)** Is this patient nearsighted or farsighted? Explain. **(b)** If this patient is nearsighted, find the far point. If this person is farsighted, find the near point. (For the purposes of this problem, treat the eye as a single-lens system, with the retina 2.40 cm from the lens.)

35. •• IP You are comfortably reading a book at a distance of 24 cm. **(a)** What is the refractive power of your eyes? **(b)** Does the refractive power of your eyes increase or decrease when you

move the book farther away? Explain. (For the purposes of this problem, treat the eye as a single-lens system, with the retina 2.40 cm from the lens.)

36. •• Without glasses, your Uncle Albert can see things clearly only if they are between 25 cm and 150 cm from his eyes. **(a)** What power eyeglass lens will correct your uncle's myopia? Assume the lenses will sit 2.0 cm from his eyes. **(b)** What is your uncle's near point when wearing these glasses?

37. •• A 2.05-cm-tall object is placed 30.0 cm to the left of a converging lens with a focal length $f_1 = 20.5$ cm. A diverging lens, with a focal length $f_2 = -42.5$ cm, is placed 30.0 cm to the right of the first lens. How tall is the final image of the object?

38. •• A simple camera telephoto lens consists of two lenses. The objective lens has a focal length $f_1 = +39.0$ cm. Precisely 36.0 cm behind this lens is a concave lens with a focal length $f_2 = -10.0$ cm. The object to be photographed is 4.00 m in front of the objective lens. **(a)** How far behind the concave lens should the film be placed? **(b)** What is the linear magnification of this lens combination?

39. •• **IP** With unaided vision, a librarian can focus only on objects that lie at distances between 5.0 m and 0.50 m. **(a)** Which type of lens (converging or diverging) is needed to correct his nearsightedness? Explain. **(b)** Which type of lens will correct his farsightedness? Explain. **(c)** Find the refractive power needed for each part of the bifocal eyeglass lenses that will give the librarian normal visual acuity from 25 cm out to infinity. (Assume the lenses rest 2.0 cm from his eyes.)

40. •• **IP** With unaided vision, a physician can focus only on objects that lie at distances between 5.0 m and 0.50 m. **(a)** Which type of lens (converging or diverging) is needed to correct her nearsightedness? Explain. **(b)** Which type of lens will correct her farsightedness? Explain. **(c)** Find the refractive power needed for each part of the bifocal contact lenses that will give the physician normal visual acuity from 25 cm out to infinity.

41. •• A person's prescription for her new bifocal glasses calls for a refractive power of -0.425 diopters in the distance-vision part, and a power of $+1.55$ diopters in the close-vision part. What are the near and far points of this person's uncorrected vision? Assume the glasses are 2.00 cm from the person's eyes, and that the person's near-point distance is 25.0 cm when wearing the glasses.

42. •• A person's prescription for his new bifocal eyeglasses calls for a refractive power of -0.0625 diopters in the distance-vision part and a power of $+1.05$ diopters in the close-vision part. Assuming the glasses rest 2.00 cm from his eyes and that the corrected near-point distance is 25.0 cm, determine the near and far points of this person's uncorrected vision.

43. ••• Two lenses, with $f_1 = +20.0$ cm and $f_2 = +30.0$ cm, are placed on the x axis, as shown in **Figure 27–22**. An object is fixed 50.0 cm to the left of lens 1, and lens 2 is a variable distance x to the right of lens 1. Find the lateral magnification and location of the final image relative to lens 2 for the following cases:

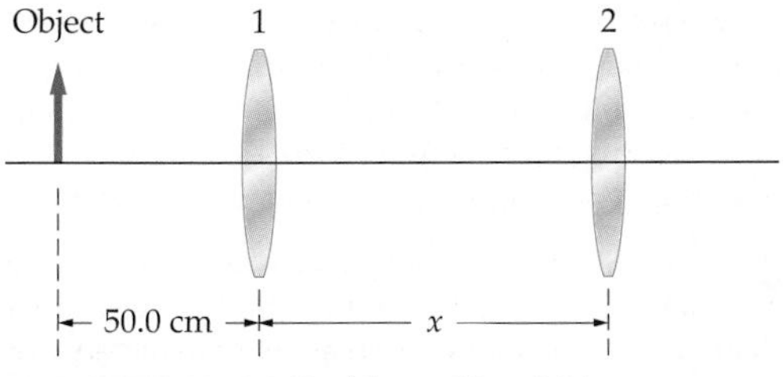

▲ **FIGURE 27–22** Problems 43 and 91

(a) $x = 115$ cm; **(b)** $x = 30.0$ cm; **(c)** $x = 0$. **(d)** Show that your result for part (c) agrees with the relation for two lenses in contact, $1/f_{\text{eff}} = 1/f_1 + 1/f_2$.

44. ••• A converging lens with a focal length of 4.0 cm is to the left of a second identical lens. When a feather is placed 12 cm to the left of the first lens, the final image is the same size and orientation as the feather itself. What is the separation between the lenses?

Section 27–3 The Magnifying Glass

45. • The Moon is 3476 km in diameter and orbits the Earth at an average distance of 384,400 km. **(a)** What is the angular size of the Moon as seen from Earth? **(b)** A penny is 19 mm in diameter. How far from your eye should the penny be held to produce the same angular diameter as the Moon?

46. •• A magnifying glass is a single convex lens with a focal length of $f = +12.0$ cm. **(a)** What is the angular magnification when this lens forms a (virtual) image at $-\infty$? How far from the object should the lens be held? **(b)** What is the angular magnification when this lens forms a (virtual) image at the person's near point (assumed to be 25 cm)? How far from the object should the lens be held in this case?

47. •• **IP** A student has two lenses, one of focal length $f_1 = 5.0$ cm and the other with focal length $f_2 = 13$ cm. **(a)** When used as a simple magnifier, which of these lenses can produce the greater magnification? Explain. **(b)** Find the maximum magnification produced by each of these lenses.

48. •• A beetle 4.13 mm long is examined with a simple magnifier of focal length $f = 10.4$ cm. If the observer's eye is relaxed while using the magnifier, and has a near-point distance of 25 cm, what is the apparent length of the beetle?

49. •• To engrave wishes of good luck on a watch, an engraver uses a magnifier whose focal length is 8.65 cm. If the image formed by the magnifier is at the engraver's near point of 25.6 cm, find **(a)** the distance between the watch and the magnifier and **(b)** the angular magnification of the engraving. Assume the magnifying glass is directly in front of the engraver's eyes.

50. •• A jeweler examines a diamond with a magnifying glass. If the near-point distance of the jeweler is 20.8 cm, and the focal length of the magnifying glass is 7.50 cm, find the angular magnification when the diamond is held at the focal point of the magnifier. Assume the magnifying glass is directly in front of the jeweler's eyes.

51. •• In Problem 50, find the angular magnification when the diamond is held 5.59 cm from the magnifying glass.

52. ••• A person with a near-point distance of 25 cm finds that a magnifying glass gives an angular magnification that is 1.5 times larger when the image of the magnifier is at the near point than when the image is at infinity. What is the focal length of the magnifying glass?

Section 27–4 The Compound Microscope

53. • A compound microscope has an objective lens with a focal length of 2.2 cm and an eyepiece with a focal length of 5.4 cm. If the image produced by the objective is 12 cm from the objective, what magnification does this microscope produce?

54. • **BIO** A typical red blood cell subtends an angle of only 1.9×10^{-5} rad when viewed at a person's near-point distance of 25 cm. Suppose a red blood cell is examined with a compound microscope in which the objective and eyepiece are separated by a distance of 12.0 cm. Given that the focal length of the eyepiece

is 2.7 cm, and the focal length of the objective is 0.49 cm, find the magnitude of the angle subtended by the red blood cell when viewed through this microscope.

55. •• The medium-power objective lens in a laboratory microscope has a focal length $f_{\text{objective}} = 4.00$ mm. **(a)** If this lens produces a lateral magnification of -40.0, what is its "working distance"; that is, what is the distance from the object to the objective lens? **(b)** What is the focal length of an eyepiece lens that will provide an overall magnification of 125?

56. •• A compound microscope has the objective and eyepiece mounted in a tube that is 18.0 cm long. The focal length of the eyepiece is 2.62 cm, and the near-point distance of the person using the microscope is 25.0 cm. If the person can view the image produced by the microscope with a completely relaxed eye, and the magnification is -4525, what is the focal length of the objective?

57. •• In Problem 56, what is the distance between the objective lens and the object to be examined?

58. •• The barrel of a compound microscope is 15 cm in length. The specimen will be mounted 1.0 cm from the objective, and the eyepiece has a 5.0-cm focal length. Determine the focal length of the objective lens.

59. •• A compound microscope uses a 75.0-mm lens as the objective and a 2.0-cm lens as the eyepiece. The specimen will be mounted 122 mm from the objective. Determine **(a)** the barrel length and **(b)** the total magnification produced by the microscope.

60. ••• The "tube length" of a microscope is defined to be the difference between the (objective) image distance and objective focal length: $L = d_i - f_{\text{objective}}$. Many microscopes are standardized to a tube length of $L = 160$ mm. Consider such a microscope whose objective lens has a focal length $f_{\text{objective}} = 7.50$ mm. **(a)** How far from the object should this lens be placed? **(b)** What focal length eyepiece would give an overall magnification of -55? **(c)** What focal length eyepiece would give an overall magnification of -110?

Section 27–5 Telescopes

61. • A grade school student plans to build a 35-power telescope as a science fair project. She starts with a magnifying glass with a focal length of 5.0 cm as the eyepiece. What focal length is needed for her objective lens?

62. • A 55-power refracting telescope has an eyepiece with a focal length of 5.0 cm. How long is the telescope?

63. • An amateur astronomer wants to build a small refracting telescope. The only lenses available to him have focal lengths of 5.00 cm, 10.0 cm, 20.0 cm, and 30.0 cm. **(a)** What is the greatest magnification that can be obtained using two of these lenses? **(b)** How long is the telescope with the greatest magnification?

64. • A pirate sights a distant ship with a spyglass that gives an angular magnification of 33. If the focal length of the eyepiece is 11 mm, what is the focal length of the objective?

65. •• A telescope has lenses with focal lengths $f_1 = +30.0$ cm and $f_2 = +5.0$ cm. **(a)** What distance between the two lenses will allow the telescope to focus on an infinitely distant object and produce an infinitely distant image? **(b)** What distance between the lenses will allow the telescope to focus on an object that is 5.0 m away and to produce an infinitely distant image?

66. •• Jason has a 25-power telescope whose objective lens has a focal length of 120 cm. To make his sister appear smaller than normal, he turns the telescope around and looks through the objective lens. What is the angular magnification of his sister when viewed through the "wrong" end of the telescope?

67. •• **Roughing It with Science** A professor shipwrecked on Hooligan's Island decides to build a telescope from his eyeglasses and some coconut shells. Fortunately, the professor's eyes require different prescriptions, with the left lens having a power of +5.0 diopters and the right lens having a power of +2.0 diopters. **(a)** Which lens should he use as the objective? **(b)** What is the angular magnification of the professor's telescope?

68. •• **Galileo's Telescope** Galileo's first telescope used a convex objective lens with a focal length $f = 1.7$ m and a concave eyepiece, as shown in **Figure 27–23**. When this telescope is focused on an infinitely distant object, and produces an infinitely distant image, its angular magnification is $+3.0$. **(a)** What is the focal length of the eyepiece? **(b)** How far apart are the two lenses?

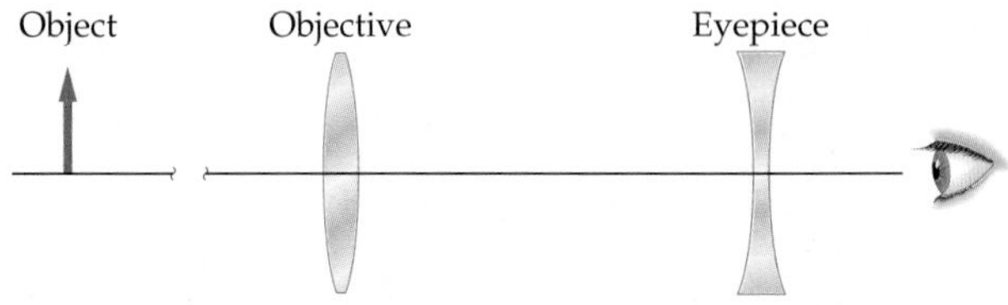

▲ **FIGURE 27–23** Problems 68, 74, and 83

69. •• The Moon has an angular size of 0.50° when viewed with unaided vision from Earth. Suppose the Moon is viewed through a telescope with an objective whose focal length is 53 cm and an eyepiece whose focal length is 25 mm. What is the angular size of the Moon as seen through this telescope?

70. •• In Problem 69, an eyepiece is selected to give the Moon an angular size of 15°. What is the focal length of this eyepiece?

71. •• A telescope is 275 mm long and has an objective lens with a focal length of 257 mm. **(a)** What is the focal length of the eyepiece? **(b)** What is the magnification of this telescope?

General Problems

72. •• **IP** The greatest refractive power a patient's eyes can produce is 43.1 diopters. **(a)** Is this patient nearsighted or farsighted? Explain. **(b)** If this patient is nearsighted, find the far point. If this person is farsighted, find the near point. (For the purposes of this problem, treat the eye as a single-lens system, with the retina 2.40 cm from the lens.)

73. •• **IP** You are observing a rare species of bird in a distant tree with your unaided eyes. **(a)** What is the refractive power of your eyes? **(b)** Does the refractive power of your eyes increase or decrease when you shift your view to the guidebook in your hands? Explain. (For the purposes of this problem, treat the eye as a single-lens system, with the retina 2.40 cm from the lens.)

74. •• Galileo's original telescope (Figure 27–23) used a convex objective and a concave eyepiece. Use a ray diagram to show that this telescope produces an upright image when a distant object is being viewed. Assume that the eyepiece is to the right of the object and that the right-hand focal point of the eyepiece is just to the left of the objective's right-hand focal point. In addition, assume that the focal length of the eyepiece has a magnitude that is about one-quarter the focal length of the objective.

75. •• **IP** For each of the following cases, use a ray diagram to show that the angular sizes of the image and the object are identical if both angles are measured *from the center of the lens.* **(a)** A convex lens with the object outside the focal length. **(b)** A convex lens with the object inside the focal length. **(c)** A concave lens with the

object outside the focal length. **(d)** Given that the angular size does not change, how does a simple magnifier work? Explain.

76. •• **IP** You have two lenses, with focal lengths $f_1 = +2.60$ cm and $f_2 = +20.4$ cm. **(a)** How would you arrange these lenses to form a magnified image of the Moon? **(b)** What is the maximum angular magnification these lenses could produce? **(c)** How would you arrange the same two lenses to form a magnified image of an insect? **(d)** If you use the magnifier of part (c) to view an insect, what is the angular magnification when the insect is held 2.35 cm from the objective lens?

77. •• **BIO** The eye is actually a multiple-lens system, but we can approximate it with a single-lens system for most of our purposes. When the eye is focused on a distant object, the optical power of the equivalent single lens is +41.4 diopters. **(a)** What is the effective focal length of the eye? **(b)** How far in front of the retina is this "equivalent lens" located?

78. •• **BIO Fitting Contact Lenses with a Keratometer** When a patient is being fitted with contact lenses, the curvature of the patient's cornea is measured with an instrument known as a keratometer. A lighted object is held near the eye, and the keratometer measures the magnification of the image formed by reflection from the front of the cornea. If an object is held 10.0 cm in front of a patient's eye, and the reflected image is magnified by a factor of 0.035, what is the radius of curvature of the patient's cornea?

79. •• **Pricey Stamp** A rare 1918 "Jenny" stamp, depicting a misprinted, upside-down Curtiss JN-4 "Jenny" airplane, sold at auction for $525,000. A collector uses a simple magnifying glass to examine the "Jenny," obtaining a linear magnification of 2.5 when the stamp is held 2.76 cm from the lens. What is the focal length of the magnifying glass?

80. •• **IP** A person needs glasses with a refractive power of −1.15 diopters to be able to focus on distant objects. **(a)** Is this person nearsighted or farsighted? Explain. **(b)** What is this person's (unaided) far point?

81. •• **IP BIO Intraocular Lens** A person's vision can be severely degraded if a cataract forms in the lens of an eye. One form of treatment is to remove the affected lens with a technique referred to as phacoemulsification and then to replace it with a man-made intraocular lens. The replacement lens cannot be focused, so it is usually chosen to have a focal length that allows the eye to see clearly at a distance. Corrective glasses or contacts must be worn for close vision. **(a)** If corrective contacts are used, should their refractive power be positive or negative? Explain. **(b)** Find the refractive power of contacts that will allow a person to focus on a book at a distance of 23 cm.

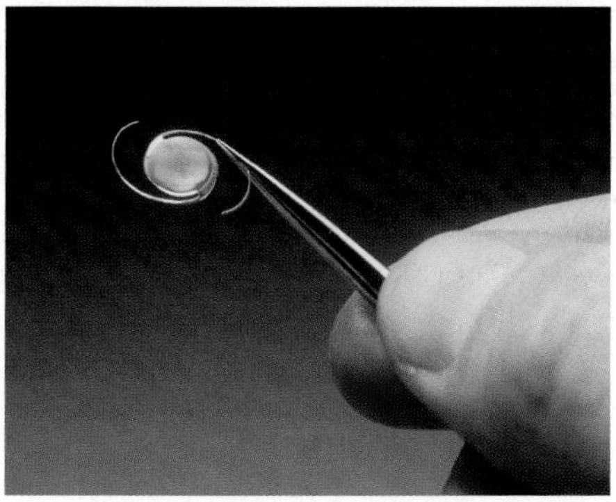

An intraocular lens. Spring-loaded lenses like this are implanted in an eye whose natural lens has been removed due to a severe cataract. (Problem 81)

82. •• **IP BIO A Big Eye** The largest eye ever to exist on Earth belonged to an extinct species of ichthyosaur, *Temnodontosaurus platyodon*. This creature had an eye that was 26.4 cm in diameter. It is estimated that this ichthyosaur also had a relatively large pupil, giving it an effective aperture setting of about $f/1.1$. **(a)** Assuming its pupil was one-third the diameter of the eye, what was the approximate focal length of the ichthyosaur's eye? **(b)** When the ichthyosaur narrowed its pupil in bright light, did its *f*-number increase or decrease? Explain.

83. •• Consider a Galilean telescope, as illustrated in Figure 27–23, constructed from two lenses with focal lengths of 75.6 cm and −18.0 mm. **(a)** What is the distance between these lenses if an infinitely distant object is to produce an infinitely distant image? **(b)** What is the angular magnification when the lenses are separated by the distance calculated in part (a)?

84. •• A virtual object is 12 cm from a 3.60-diopter lens. **(a)** Where is the image? **(b)** Is the image real or virtual? Explain.

85. •• A farsighted person uses glasses with a refractive power of 3.4 diopters. The glasses are worn 2.5 cm from his eyes. What is this person's near point when not wearing glasses?

86. ••• **Landing on an Aircraft Carrier** The Long-Range Lineup System (LRLS) used to ensure safe landings on aircraft carriers consists of a series of Fresnel lenses of different colors. Each lens focuses light in a different, specific direction, and hence which light a pilot sees on approach determines whether the plane is above, below, or on the proper landing path. The basic idea behind a Fresnel lens, which has the same optical properties as an ordinary lens, is shown in **Figure 27–24**, along with a photo of the LRLS. Suppose an object (a light bulb in this case) is 17.1 cm behind a Fresnel lens, and that the corresponding image is a

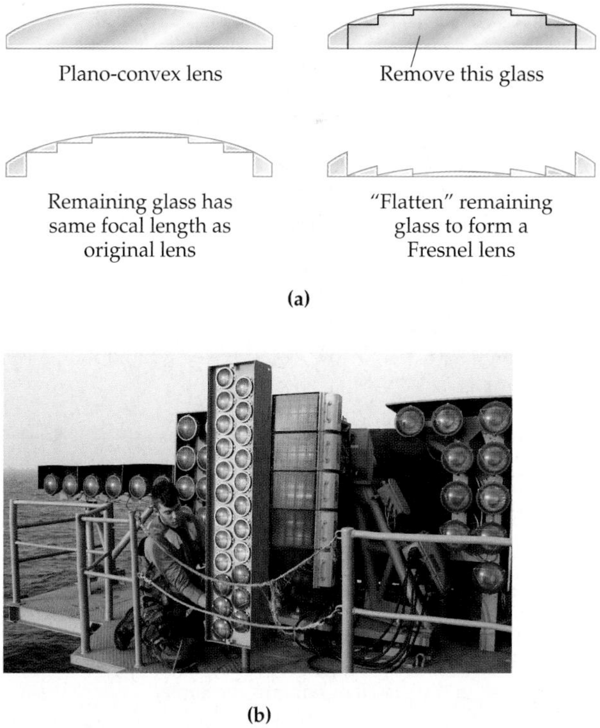

▲ **FIGURE 27–24 Fresnel lenses and the Long-Range Lineup System**

(a) A lens causes light to refract at its surface; therefore, the interior glass can be removed without changing its optical properties. This produces a Fresnel lens, which is much lighter than the original lens. **(b)** If an airplane is on the correct approach path, the pilot will see an amber light, called the "meatball," in line with the row of green lights.

distance $d_i = d$ in front of the lens. If the object is moved to a distance of 12.0 cm behind the lens, the image distance doubles to $d_i = 2d$. In the LRLS, it is desired to have the image of the light bulb at infinity. What object distance will give this result for this particular lens?

87. ••• When using a telescope to photograph a faint astronomical object, you need to maximize the amount of light energy that falls on each square millimeter of the image on the film. For a given telescope and object, the total light that falls on the film is proportional to the length of the exposure, so a long exposure will reveal fainter objects than a short exposure. Show that for a given length of exposure, the brightness of the image is inversely proportional to the *square* of the *f*-number of the telescope system.

88. ••• A Cassegrain astronomical telescope uses two mirrors to form the image. The larger (concave) objective mirror has a focal length $f_1 = +50.0$ cm. A small convex secondary mirror is mounted 43.0 cm in front of the primary. As shown in **Figure 27–25**, light is reflected from the secondary through a hole in the center of the primary, thereby forming a real image 8.00 cm behind the primary mirror. What is the radius of curvature of the secondary mirror?

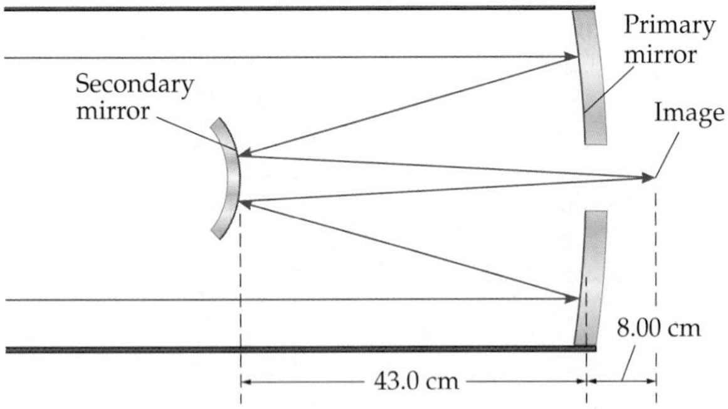

▲ **FIGURE 27–25** Problem 88

89. ••• **IP** A convex lens ($f = 20.0$ cm) is placed 10.0 cm in front of a plane mirror. A matchstick is placed 25.0 cm in front of the lens, as shown in **Figure 27–26**. **(a)** If you look through the lens toward the mirror, where will you see the image of the matchstick? **(b)** Is the image real or virtual? Explain. **(c)** What is the magnification of the image? **(d)** Is the image upright or inverted?

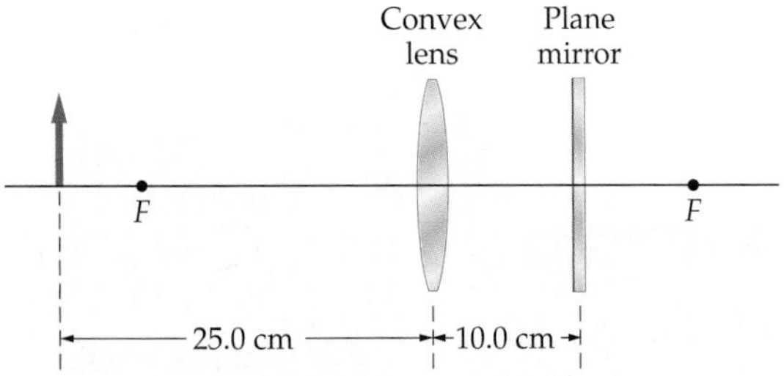

▲ **FIGURE 27–26** Problems 89 and 90

90. ••• Repeat Problem 89 for the case where the converging lens is replaced with a diverging lens with $f = -20.0$ cm. Everything else in the problem remains the same.

91. ••• Repeat Problem 43 for the case where lens 1 is replaced with a diverging lens with $f_1 = -20.0$ cm. Everything else in the problem remains the same.

92. ••• The diameter of a collimated laser beam can be expanded or reduced by using two converging lenses, with focal lengths f_1 and f_2, mounted a distance $f_1 + f_2$ from each other, as shown in **Figure 27–27**. What is the ratio of the two beam diameters, (d_1/d_2), expressed in terms of the focal lengths?

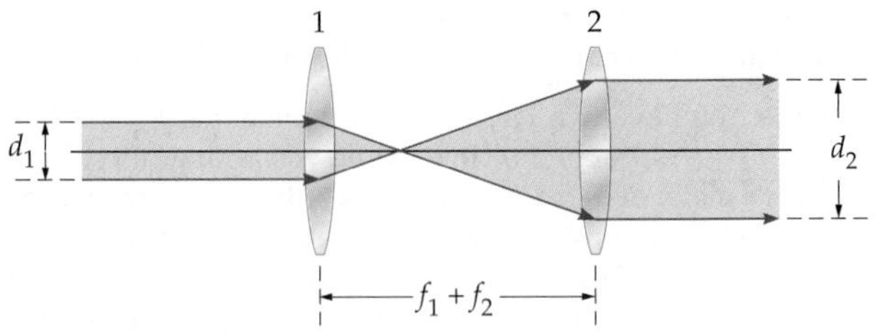

▲ **FIGURE 27–27** Problem 92

93. ••• Consider three lenses with focal lengths of 25.0 cm, −15.0 cm, and 11.0 cm positioned on the x axis at $x = 0$, $x = 0.400$ m, and $x = 0.500$ m, respectively. An object is at $x = -122$ cm. Find **(a)** the location and **(b)** the orientation and magnification of the final image produced by this lens system.

94. ••• Because a concave lens cannot form a real image of a real object, it is difficult to measure its focal length precisely. One method uses a second, convex, lens to produce a virtual object for the concave lens. Under the proper conditions, the concave lens will form a real image of the virtual object! A student conducting a laboratory project on concave lenses makes the following observations: When a lamp is placed 42.0 cm to the left of a particular convex lens, a real (inverted) image is formed 37.5 cm to the right of the lens. The lamp and convex lens are kept in place while a concave lens is mounted 15.0 cm to the right of the convex lens. A real image of the lamp is now formed 35.0 cm to the right of the concave lens. What is the focal length of each lens?

95. ••• A person with a near-point distance N uses a magnifying glass with a focal length f. Show that the greatest magnification that can be achieved with this magnifier is $M = 1 + N/f$.

Interactive Problems

96. •• **IP** ***Referring to Example 27–2*** Suppose a person's eyeglasses have a focal length of −301 cm, are 2.00 cm in front of the eyes, and allow the person to focus on distant objects. **(a)** Is this person's far point greater than or less than 323 cm, which is the far point for glasses the same distance from the eyes and with a focal length of −321 cm? Explain. **(b)** Find the far point for this person.

97. •• **IP** ***Referring to Example 27–2*** In Example 27–2, a person has a far-point distance of 323 cm. If this person wears glasses 2.00 cm in front of the eyes with a focal length of −321 cm, distant objects can be brought into focus. Suppose a second person's far point is 353 cm. **(a)** Is the magnitude of the focal length of the eyeglasses that allow this person to focus on distant objects greater than or less than 321 cm? Assume the glasses are 2.00 cm in front of the eyes. **(b)** Find the required focal length for the second person's eyeglasses.

98. •• **IP** ***Referring to Example 27–3*** Suppose a person's eyeglasses have a refractive power of 2.75 diopters and that they allow the person to focus on an object that is just 25.0 cm from the eye. The glasses are 2.00 cm in front of the eyes. **(a)** Is this person's near point greater than or less than 57.0 cm, which is the near-point distance when the glasses have a refractive power of 2.53 diopters? Explain. **(b)** Find the near point for this person.

99. •• **IP** ***Referring to Example 27–3*** Suppose a person's near-point distance is 67.0 cm. **(a)** Is the refractive power of the eyeglasses that allow this person to focus on an object just 25.0 cm from the eye greater than or less than 2.53 diopters, which is the refractive power when the near-point distance is 57.0 cm? The glasses are worn 2.00 cm in front of the eyes. **(b)** Find the required refractive power for this person's eyeglasses.

28 Physical Optics: Interference and Diffraction

28–1	Superposition and Interference	938
28–2	Young's Two-Slit Experiment	940
28–3	Interference in Reflected Waves	944
28–4	Diffraction	950
28–5	Resolution	954
28–6	Diffraction Gratings	957

Most of the colors that we see every day can be understood in terms of absorption and reflection—an old-fashioned fire engine absorbs green light and reflects red light, a grassy lawn absorbs red light and reflects green light, and so on. Yet the distinctive iridescent hues of this blue morpho butterfly are created by different physical mechanisms. These mechanisms, known as interference and diffraction, cannot be understood solely in terms of geometrical optics. Instead, they have their origin in the wave nature of light, the subject of this chapter.

In our discussion of optics to this point we have treated light in terms of "rays" that propagate in straight lines. This description is valid in a wide variety of circumstances, as we have seen, though it ignores the fact that light is an electromagnetic *wave*. There are situations, however, in which the ray model fails and the wave nature of light is of central importance. For example, any time we deal with objects or sources that have characteristic dimensions comparable to the wavelength of light, we shall find new effects not predicted by ray optics. In this chapter we show that the wave properties of light are responsible for such phenomena as the operation of a CD, the appearance of images on a television screen, and the brilliant iridescent colors of a butterfly's wing.

28–1 Superposition and Interference

One of the fundamental aspects of wave behavior is **superposition,** in which the net displacement caused by a combination of waves is the algebraic sum of the displacements caused by each wave individually. If waves add to cause a larger displacement, we say that they interfere **constructively;** if the net displacement is reduced, the interference is **destructive.** Examples of wave superposition on a string are given in Figures 14–19 and 14–20. Because light also exhibits wave behavior, with propagating electric and magnetic fields, it too can show interference effects and a resulting increase or decrease in brightness.

Interference is noticeable, however, only if certain conditions are met. In the case of light, for example, the light should be **monochromatic;** that is, it should have a single color and hence a single frequency. In addition, if two or more sources of light are to show interference, they must maintain a constant phase relationship with one another—that is, they must be **coherent.** Sources whose relative phases vary randomly with time show no discernible interference patterns and are said to be **incoherent.** Incoherent light sources include incandescent and fluorescent lights. In contrast, lasers emit light that is both monochromatic and coherent.

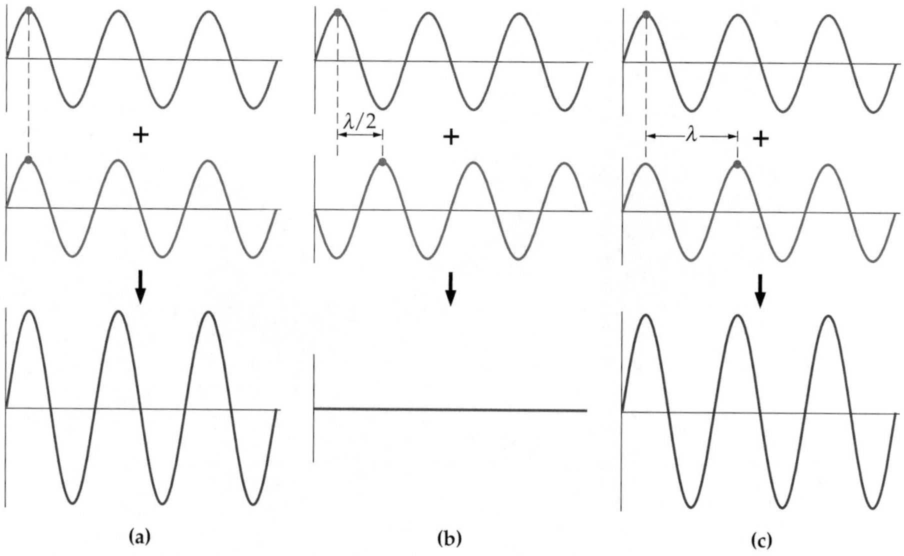

▶ **FIGURE 28–1 Constructive and destructive interference**

(a) Waves that are in phase add to give a larger amplitude. This is constructive interference. **(b)** Waves that are half a wavelength out of phase interfere destructively. If the individual waves have equal amplitudes, as here, their sum will have zero amplitude. **(c)** When waves are one wavelength out of phase, the result is again constructive interference, exactly the same as when the waves are in phase.

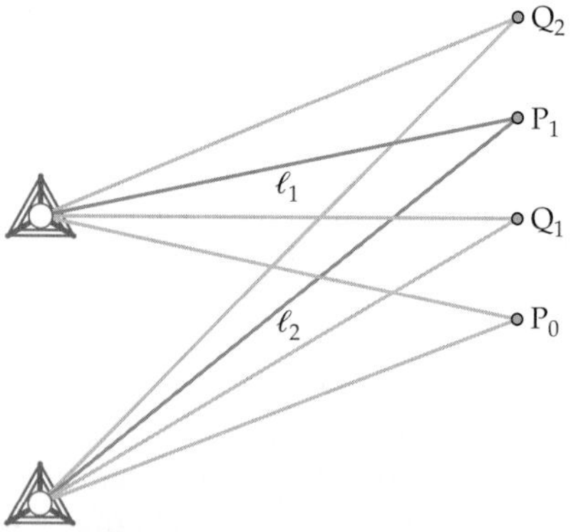

▲ **FIGURE 28–2 Two radio antennas transmitting the same signal**

At point P_0, midway between the antennas, the waves travel the same distance, and hence they interfere constructively. At point P_1 the distance ℓ_2 is greater than the distance ℓ_1 by one wavelength; thus, P_1 is also a point of constructive interference. At Q_1 the distance ℓ_2 is greater than the distance ℓ_1 by half a wavelength, and the waves interfere destructively at that point.

The basic principle that determines whether waves will interfere constructively or destructively is their phase relative to one another. For example, if two waves have zero phase difference—that is, the waves are in phase—they add constructively, and the net result is an increased amplitude, as **Figure 28–1 (a)** indicates. If waves of equal amplitude are 180° out of phase, however, the net result is zero amplitude and destructive interference, as indicated in **Figure 28–1 (b).** Notice that a *180° difference in phase* corresponds to waves being out of step by *half a wavelength*. Similarly, if the phase difference between two waves is 360°—which corresponds to one full wavelength—the interference is again constructive, as in **Figure 28–1 (c).**

Finally, Figure 28–1 shows the blue wave shifted to the right, meaning it is *ahead* of the red wave by half a wavelength in part (b), and *ahead* by a full wavelength in part (c). The same results are obtained, however, if the blue wave is *behind* the red wave. Thus, for example, a phase difference of −180° produces destructive interference, the same as in Figure 28–1 (b), and a phase difference of −360° produces constructive interference, just as in Figure 28–1 (c).

Let's apply the preceding observations to a system consisting of two radio antennas radiating electromagnetic waves of frequency f and wavelength λ, as in **Figure 28–2.** If the antennas are connected to the same transmitter, they emit waves that are in phase and coherent. When waves from each antenna reach point P_0, they have traveled the same distance—that is, the same number of wavelengths—and hence they are still in phase. As a result, point P_0 experiences constructive interference and the radio signal is strong. This location corresponds to the situation shown in Figure 28–1 (a).

To reach point P_1, the waves from the two antennas must travel different distances, ℓ_1 and ℓ_2, as indicated in Figure 28–2. If the difference in these distances is one wavelength, $\ell_2 - \ell_1 = \lambda$, the waves are 360° out of phase, and point P_1 also has constructive interference. This location corresponds to Figure 28–1 (c). Similarly, if the difference in path length to a point P_m is an integer $m = 0, 1, 2, \ldots$ times the wavelength, $\ell_2 - \ell_1 = m\lambda$, that point will also be a location of constructive interference.

Finally, the difference in path length to point Q_1 is half a wavelength, $\ell_2 - \ell_1 = \lambda/2$, and hence the waves cancel there, just as in Figure 28–1 (b). Destructive interference also occurs at Q_2, where the difference in path length is one and a half wavelengths, $\ell_2 - \ell_1 = 3\lambda/2$. In general, for $m = 1, 2, 3, \ldots$, destructive interference occurs at points Q_m where the difference in path lengths is $\ell_2 - \ell_1 = \left(m - \frac{1}{2}\right)\lambda$.

To summarize, *constructive interference* satisfies the conditions

$$\ell_2 - \ell_1 = m\lambda \qquad m = 0, 1, 2, \ldots \qquad \textit{(constructive interference)}$$

Similarly, *destructive interference* satisfies the conditions

$$\ell_2 - \ell_1 = \left(m - \tfrac{1}{2}\right)\lambda \quad m = 1, 2, 3, \ldots \qquad \textit{(destructive interference)}$$

These conditions are applied to a specific physical system in the following Example.

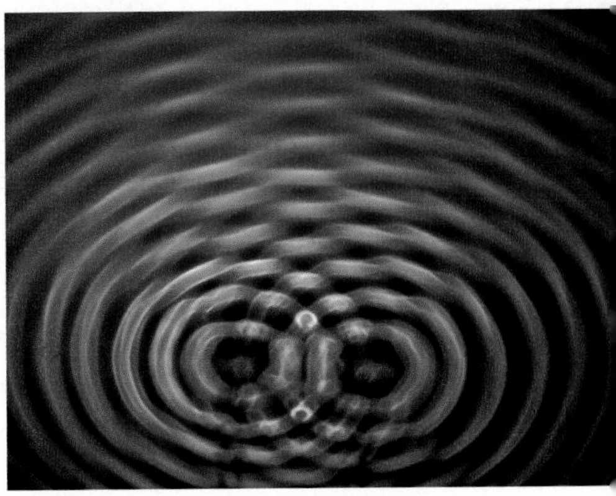

▲ Two sets of water waves, radiating outward in circular patterns from point sources, create an interference pattern where they overlap.

EXAMPLE 28–1 Two May Not Be Better than One

Two friends tune their radios to the same frequency and pick up a signal transmitted simultaneously by a pair of antennas. The friend who is equidistant from the antennas, at P_0, receives a strong signal. The friend at point Q_1 receives a very weak signal. Find the wavelength of the radio waves if $d = 7.50$ km, $L = 14.0$ km, and $y = 1.88$ km. Assume that Q_1 is the first point of minimum signal as one moves away from P_0 in the y direction.

Picture the Problem

Our sketch shows the radio antennas and the two locations mentioned in the problem statement. Notice that the radio antennas are separated by a distance $d = 7.50$ km in the y direction, and that the points P_0 and Q_1 have a y-direction separation of $y = 1.88$ km. The distance to P_0 and Q_1 in the x direction is $L = 14.0$ km.

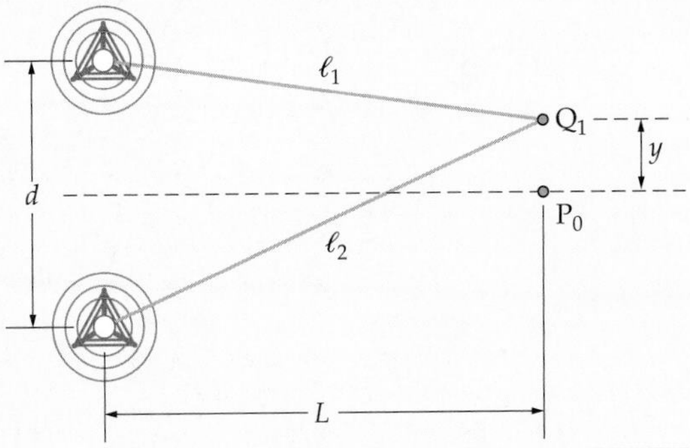

Strategy

Since point Q_1 is the first *minimum* in the y direction from the *maximum* at point P_0, we know that the path difference, $\ell_2 - \ell_1$, is half a wavelength. Thus we can determine λ by calculating the lengths ℓ_2 and ℓ_1 and setting their difference equal to $\lambda/2$.

Solution

1. Calculate the path length ℓ_1:

$$\ell_1 = \sqrt{L^2 + \left(\frac{d}{2} - y\right)^2}$$

$$= \sqrt{(14.0\text{ km})^2 + \left(\frac{7.50\text{ km}}{2} - 1.88\text{ km}\right)^2} = 14.1\text{ km}$$

2. Calculate the path length ℓ_2:

$$\ell_2 = \sqrt{L^2 + \left(\frac{d}{2} + y\right)^2}$$

$$= \sqrt{(14.0\text{ km})^2 + \left(\frac{7.50\text{ km}}{2} + 1.88\text{ km}\right)^2} = 15.1\text{ km}$$

3. Set $\ell_2 - \ell_1$ equal to $\lambda/2$ and solve for the wavelength:

$$\ell_2 - \ell_1 = \tfrac{1}{2}\lambda$$

$$\lambda = 2(\ell_2 - \ell_1) = 2(15.1\text{ km} - 14.1\text{ km}) = 2.0\text{ km}$$

continued on next page

continued from previous page

Insight

We see that radio waves are rather large. In fact, the distance from one crest to the next (the wavelength) for these waves is about 1.2 miles. Recalling that radio waves travel at the speed of light, the corresponding frequency is $f = c/\lambda = 150$ kHz.

Practice Problem

Suppose the wavelength broadcast by these antennas is changed, and that the y distance between P_0 and Q_1 (first minimum) increases as a result. Is the new wavelength greater than or less than 2.0 km? Find the wavelength for the case $y = 2.91$ km. [**Answer:** The wavelength is greater than 2.0 km. For $y = 2.91$ km we find $\lambda = 3.0$ km.]

Some related homework problems: Problem 1, Problem 3, Problem 4

28–2 Young's Two-Slit Experiment

We now consider a classic physics experiment that not only demonstrates the wave nature of light, but also allows one to determine the wavelength of a beam of light, just as we did for the radio waves in Example 28–1. The experiment was first performed in 1801 by the English physician and physicist Thomas Young (1773–1829), whose medical background contributed to his studies of vision, and whose love of languages made him a key figure in deciphering the Rosetta Stone. The experiment, in simplest form, consists of a beam of monochromatic light that passes through a small slit in a screen and then illuminates two slits, S_1 and S_2, in a second screen. After the light passes through the two slits, it shines on a distant screen, as **Figure 28–3** shows, where an interference pattern of bright and dark "fringes" is observed.

In this "two-slit experiment" the slit in the first screen serves only to produce a small source of light that prevents the interference pattern on the distant screen from becoming smeared out. The key elements in the experiment are the two slits

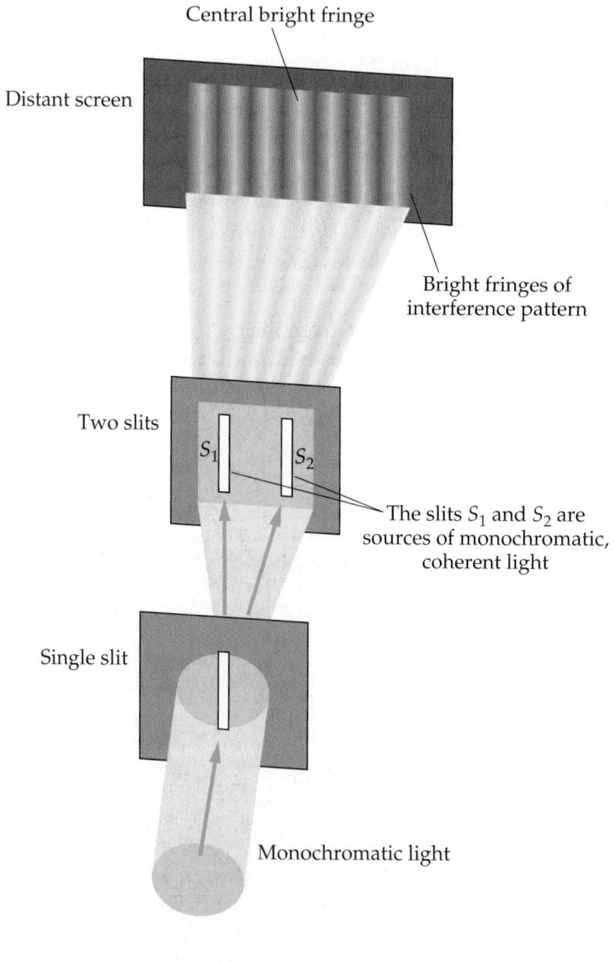

▶ **FIGURE 28–3 Young's two-slit experiment**

The first screen produces a small source of light that illuminates the two slits, S_1 and S_2. After passing through these slits, the light spreads out into an interference pattern of alternating bright and dark fringes on a distant screen.

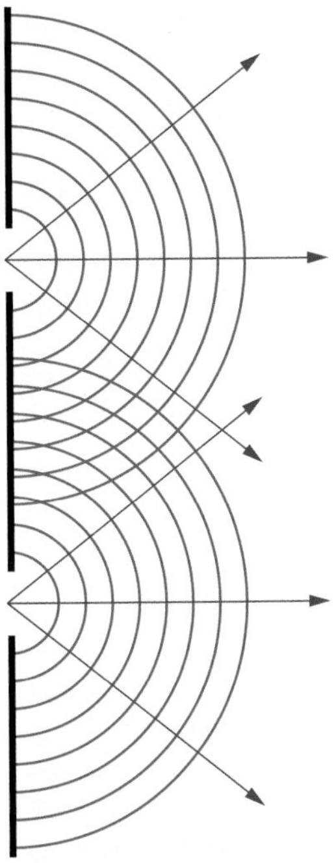

▲ **FIGURE 28–4 Huygens's principle**
According to Huygens's principle, each of the two slits in Young's experiment acts as a source of light waves propagating outward in all forward directions. It follows that light from the two sources can overlap, resulting in an interference pattern.

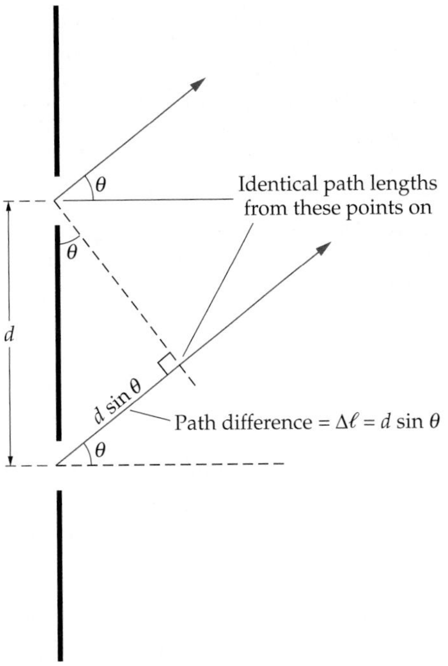

▲ **FIGURE 28–5 Path difference in the two-slit experiment**
Light propagating from two slits to a distant screen along parallel paths; note that the paths make an angle θ relative to the normal to the slits. The difference in path length is $\Delta\ell = d \sin\theta$, where d is the slit separation.

in the second screen. Since they are equidistant from the single slit, as shown in Figure 28–3, the light passing through them has the same phase. Thus the two slits act as monochromatic, coherent sources of light—analogous to the two radio antennas in Example 28–1—as needed to produce interference.

If light were composed of small particles, or "corpuscles" as Newton referred to them, they would simply pass straight through the two slits and illuminate the distant screen directly behind each slit. If light is a wave, on the other hand, each slit acts as the source of new waves, analogous to water waves passing through two small openings. This is referred to as **Huygens's principle** and is illustrated in **Figure 28–4**. Notice that light radiates away from the slits in all forward directions—not just in the direction of the incoming light. The result is that light is spread out over a large area on the distant screen; it is not localized in small regions directly behind the slits. Thus, an experiment like this can readily distinguish between the two models of light.

Of key importance is the fact that the illumination of the distant screen is not only spread out, but also consists of alternating bright and dark fringes, as indicated in Figure 28–3. These fringes are the direct result of constructive and destructive interference. For example, the central bright fringe is midway between the two slits; hence, the path lengths for light from the slits are equal. Because light coming from the slits is in phase, it follows that the light is also in phase at the midpoint, giving rise to constructive interference—just like at point P_0 in Figure 28–2.

The next bright fringe occurs when the difference in path length from the two slits is equal to one wavelength of light, as with point P_1 in Figure 28–2. In most experimental situations the distance to the screen is much greater than the separation d of the slits; hence, the light travels to a point on the screen along approximately parallel paths, as indicated in **Figure 28–5**. Therefore, the path difference, $\Delta\ell$, is

$$\Delta\ell = d \sin\theta$$

As a result, the bright fringe closest to the midpoint occurs at the angle θ given by the condition $d \sin\theta = \lambda$, or $\sin\theta = \lambda/d$. In general, a bright fringe occurs whenever

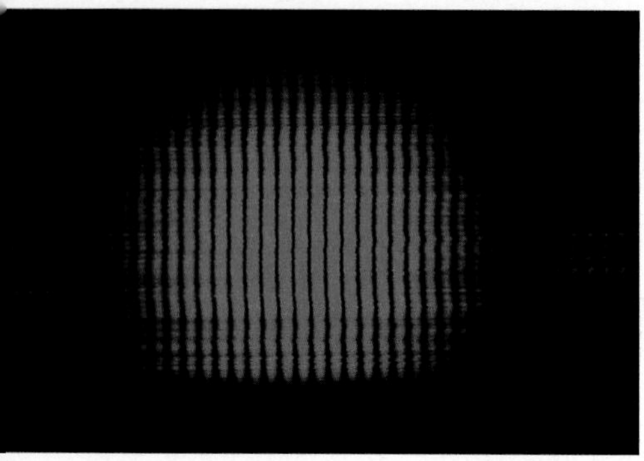

▲ An interference pattern created by monochromatic laser light passing through two slits.

the path difference, $\Delta\ell = d \sin\theta$, is equal to $m\lambda$, where $m = 0, \pm1, \pm2, \ldots$. Therefore, we find that bright fringes satisfy the following conditions:

Conditions for Bright Fringes (Constructive Interference) in a Two-Slit Experiment

$$d \sin\theta = m\lambda \qquad m = 0, \pm1, \pm2, \ldots \qquad \textbf{28–1}$$

Note that the $m = 0$ fringe occurs at $\theta = 0$, which corresponds to the central fringe. In addition, positive values of m indicate fringes above the central bright fringe; negative values indicate fringes below the central bright fringe.

Between the bright fringes we find dark fringes, where destructive interference occurs. The condition for a dark fringe is that the difference in path lengths be $\pm\lambda/2, \pm3\lambda/2, \pm5\lambda/2, \ldots$, and so on. Notice that the dark fringes are analogous to the points $Q_1, Q_2, \ldots$ in Figure 28–2. It follows that the angles corresponding to dark fringes are given by the following conditions:

Conditions for Dark Fringes (Destructive Interference) in a Two-Slit Experiment

$$d \sin\theta = \left(m - \tfrac{1}{2}\right)\lambda \quad m = 1, 2, 3, \ldots \quad \text{(above central bright fringe)}$$
$$d \sin\theta = \left(m + \tfrac{1}{2}\right)\lambda \quad m = -1, -2, -3, \ldots \quad \text{(below central bright fringe)} \qquad \textbf{28–2}$$

Clearly, $m = +1$ corresponds to a path difference of $\Delta\ell = \lambda/2$, the value $m = +2$ corresponds to a path difference of $\Delta\ell = 3\lambda/2$, and so on. These dark fringes are above the central bright fringe. Similarly, $m = -1$ corresponds to a path difference of $-\lambda/2$, and $m = -2$ corresponds to a path difference of $-3\lambda/2$. These dark fringes are below the central bright fringe. **Figure 28–6** shows the numbering systems for both bright and dark fringes. (All problems in this text refer to fringes above the central bright fringe, so only the first set of conditions in Equation 28–2 will be needed.)

Since $\sin\theta$ is less than or equal to 1, it follows from Equation 28–1 that d must be greater than or equal to λ to show the $m = \pm1$ bright fringes. In a typical situation d may be 100 times larger than λ, which means that the angle to the first dark or bright fringe will be roughly half a degree. If d is too much larger than λ, however, the angle between successive minima and maxima is so small that they tend to merge together, making the interference pattern difficult to discern.

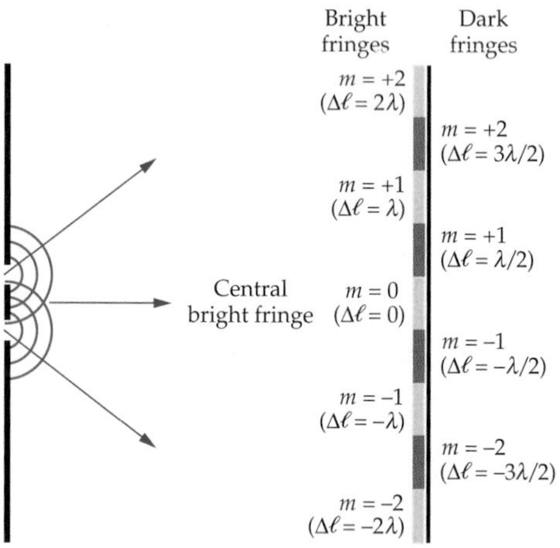

▲ **FIGURE 28–6 The two-slit pattern**
Numbering systems for bright and dark fringes.

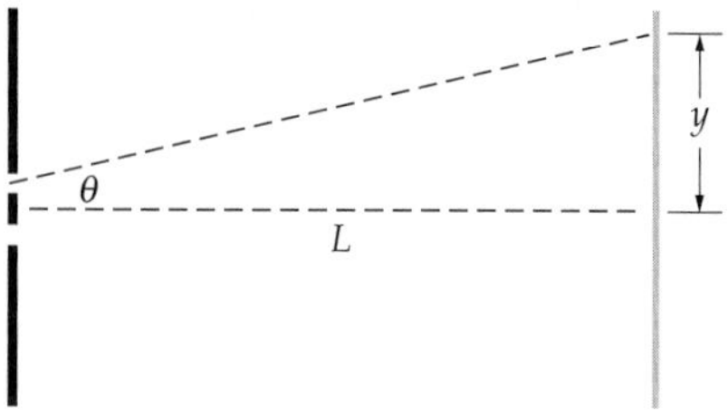

▲ **FIGURE 28–7 Linear distance in an interference pattern**
If light propagates at an angle θ relative to the normal to the slits, it is displaced a linear distance $y = L \tan\theta$ on the distant screen.

EXERCISE 28–1

Red light ($\lambda = 752$ nm) passes through a pair of slits with a separation of 6.20×10^{-5} m. Find the angles corresponding to **(a)** the first bright fringe and **(b)** the second dark fringe above the central bright fringe.

Solution

a. Referring to Figure 28–6 we find that $m = +1$ corresponds to the first bright fringe above the central bright fringe; hence,

$$\theta = \sin^{-1}\left(m\frac{\lambda}{d}\right) = \sin^{-1}\left[(1)\frac{7.52 \times 10^{-7}\ \text{m}}{6.20 \times 10^{-5}\ \text{m}}\right] = 0.695^\circ$$

b. In this case $m = +2$, therefore

$$\theta = \sin^{-1}\left[\left(m - \tfrac{1}{2}\right)\frac{\lambda}{d}\right] = \sin^{-1}\left[\left(2 - \tfrac{1}{2}\right)\frac{7.52 \times 10^{-7}\ \text{m}}{6.20 \times 10^{-5}\ \text{m}}\right] = 1.04^\circ$$

A convenient way to characterize the location of interference fringes is in terms of their linear distance from the central fringe, as indicated in **Figure 28–7**. If the distance to the screen is L—and L is much greater than the slit separation d—it follows that the linear distance y is given by the following expression:

Linear Distance from Central Fringe

$$y = L \tan\theta \qquad \textbf{28–3}$$

In the next Example we show how a measurement of the linear distance between fringes can determine the wavelength of light.

PROBLEM-SOLVING NOTE

Angular versus Linear Position of Fringes

The angle at which a bright or dark fringe occurs is determined by the wavelength of the light and the separation of the slits. The linear position of a fringe on a screen is determined by the distance from the slits to the screen.

EXAMPLE 28–2 Blue Light Special

Two slits with a separation of 8.5×10^{-5} m create an interference pattern on a screen 2.3 m away. If the tenth bright fringe above the central fringe is a linear distance of 12 cm from it, what is the wavelength of light used in the experiment?

Picture the Problem

Our sketch shows that the first bright fringe above the central fringe corresponds to $m = +1$ in Equation 28–1, the second bright fringe corresponds to $m = +2$, and so on. Therefore, $m = +10$ gives the position of the tenth bright fringe. In addition, we note that the separation of the slits is $d = 8.5 \times 10^{-5}$ m, the distance to the screen is $L = 2.3$ m, and the vertical distance to the $m = +10$ bright fringe is $y = 12$ cm $= 0.12$ m.

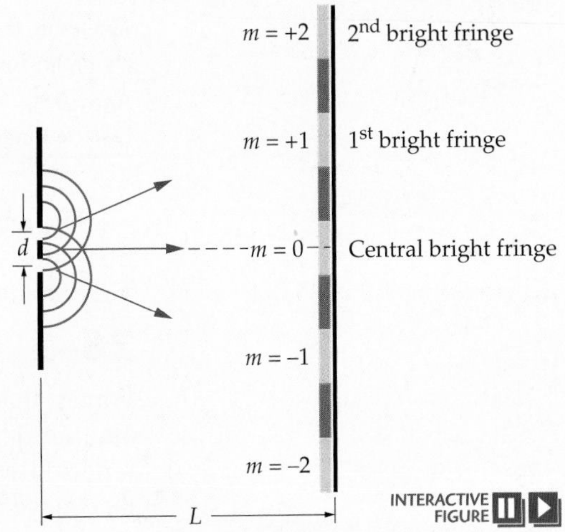

Strategy

To find the wavelength, we first determine the angle to the tenth fringe using $y = L \tan\theta$ (Equation 28–3). Once we know θ, we use the condition for bright fringes to determine the wavelength. That is, we set $m = +10$ in Equation 28–1 ($d \sin\theta = m\lambda$) and solve for λ.

Solution

1. Calculate the angle to the tenth bright fringe:

$$y = L \tan\theta$$

$$\theta = \tan^{-1}\left(\frac{y}{L}\right) = \tan^{-1}\left(\frac{0.12\text{ m}}{2.3\text{ m}}\right) = 3.0^\circ$$

2. Use $\sin\theta = m\lambda/d$ to find the wavelength:

$$\lambda = \frac{d}{m}\sin\theta = \left(\frac{8.5 \times 10^{-5}\text{ m}}{10}\right)\sin(3.0^\circ)$$

$$= 4.4 \times 10^{-7}\text{ m} = 440\text{ nm}$$

Insight

Note that we have expressed the wavelength of the light in nanometers, a common unit of measure for light waves. Referring to the electromagnetic spectrum shown in Example 25–3, we see that light with a wavelength of 440 nm is dark blue in color.

Practice Problem

If the wavelength of light used in this experiment is increased, does the linear distance to the tenth bright fringe above the central fringe increase, decrease, or stay the same? Check your reasoning by calculating the linear distance to the tenth bright fringe for a wavelength of 550 nm. [**Answer:** The linear distance will increase. For $\lambda = 550$ nm we find $y = 0.15\text{ m} > 0.12\text{ m}$.]

Some related homework problems: Problem 13, Problem 19, Problem 21

Finally, we consider the effect of changing the medium through which the light propagates in a two-slit experiment.

CONCEPTUAL CHECKPOINT 28–1 Fringe Spacing

A two-slit experiment is performed in the air. Later, the same apparatus is immersed in water and the experiment is repeated. When the apparatus is in water, are the interference fringes **(a)** more closely spaced, **(b)** more widely spaced, or **(c)** spaced the same as when the apparatus was in air?

Reasoning and Discussion

The angles corresponding to bright fringes are related to the wavelength by the equation $d \sin\theta = m\lambda$. From this relation it is clear that if λ is increased, the angle θ (and hence the spacing between fringes) also increases; if λ is decreased, the angle θ decreases. Thus the behavior of the two-slit experiment in water depends on how the wavelength of light changes in water.

Recall that when light goes from air ($n = 1.00$) to a medium in which the index of refraction is $n > 1$, the speed of propagation decreases by the factor n:

$$v = \frac{c}{n}$$

The frequency of light, f, is unchanged throughout as it goes from one medium to another. Therefore, the fact that the speed $v = \lambda f$ decreases by a factor n means that the wavelength λ decreases by the same factor. Hence, if the wavelength of light is λ when $n = 1$, its wavelength in a medium with an index of refraction $n > 1$ is

$$\lambda_n = \frac{\lambda}{n} \qquad 28\text{–}4$$

As a result, the wavelength of light is less in water than in air, and therefore, the interference fringes are more closely spaced when the experiment is performed in water.

Answer:
(a) The fringes are more closely spaced.

The relation $\lambda_n = \lambda/n$ plays an important role in the interference observed in thin films, as we shall see in the next section.

28–3 Interference in Reflected Waves

Waves that reflect from objects at different locations can interfere with one another, just like the light from two different slits. In fact, interference due to reflected waves is observed in many everyday circumstances, as we show next. Before we can understand the physics behind this type of interference, however, we must note that reflected waves change their phase in two completely different ways. First, the phase changes in proportion to the distance the waves travel—just as with light in the two-slit experiment. For example, the phase of a wave that travels half a wavelength changes by 180°, and the phase of a wave that travels one wavelength changes by 360°. Second, the phase of a reflected wave can change as a result of the reflection process itself. We begin by considering the latter type of phase change.

Phase Changes due to Reflection

Phase changes due to reflection have been discussed before in connection with waves on a string in Chapter 14. In particular, we observed at that time that a wave on a string reflects differently depending on whether the end of the string is tied to a solid support, as in Figure 14–7, or is free to move up and down, as in Figure 14–8. Specifically, the wave with a loose end is reflected back exactly as it approached the end; that is, there is no phase change. Conversely, a wave on a string that is tied down is inverted when reflected. This is equivalent to changing the phase of the wave by 180°, or half a wavelength.

Since light is a wave, it undergoes an analogous phase change on reflection. As indicated in **Figure 28–8 (a)**, a light wave that encounters a region with a lower index of refraction is reflected with no phase change, like a wave on a string whose end is free to move. In contrast, when light encounters a region with a larger index of refraction, as in **Figure 28–8 (b)**, it is reflected with a phase change of half a wavelength, like a wave on a string whose end is fixed. This half-wavelength phase change also applies to light reflected from a solid surface, such as a mirror.

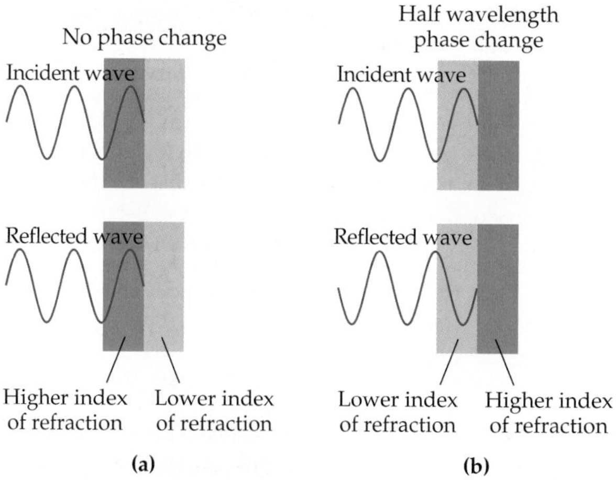

▶ FIGURE 28–8 Phase change with reflection

(a) An electromagnetic wave reflects with no phase change when it encounters a medium with a lower index of refraction. **(b)** An electromagnetic wave reflects with a half-wavelength (180°) phase change when it encounters a medium with a larger index of refraction.

To summarize:

> There is no phase change when light reflects from a region with a lower index of refraction.
>
> There is a half-wavelength phase change when light reflects from a region with a higher index of refraction, or from a solid surface.

We now apply these observations to the case of an air wedge.

Air Wedge

An interesting example of reflection interference is provided by two plates of glass that touch at one end and have a small separation at the other, as shown in **Figure 28–9**. Note that the air between the plates occupies a thin, wedge-shaped region; hence, this type of arrangement is referred to as an **air wedge.**

The predominant interference effect in this system is between light reflected from the bottom surface of the top glass plate and light reflected from the upper surface of the lower plate, since these surfaces are physically so close together. Consider, for example, the two rays illustrated in Figure 28–9. Ray 1 reflects at the glass-to-air interface; it experiences no phase change. Ray 2 travels a distance d through the air ($n = 1.00$), reflects from the air-to-glass interface, then travels essentially the same distance d in the opposite direction before rejoining ray 1 (the rays in Figure 28–9 are shown widely separated for clarity). Since the reflection from air to glass results in a 180° phase change—the same phase change as if the wave had traveled half a wavelength—the *effective* path length of ray 2 is

$$\Delta\ell_{\text{eff}} = d + \tfrac{1}{2}\lambda + d = \tfrac{1}{2}\lambda + 2d$$

If the effective path length is an integer number of wavelengths, $\lambda/2 + 2d = m\lambda$, rays 1 and 2 will interfere constructively. Dividing by the wavelength, we have the following condition for **constructive interference:**

$$\frac{1}{2} + \frac{2d}{\lambda} = m \qquad m = 1, 2, 3, \ldots \qquad \textbf{28–5}$$

Similarly, if the effective path length of ray 2 is an odd half integer there will be **destructive interference:**

$$\frac{1}{2} + \frac{2d}{\lambda} = m + \frac{1}{2} \qquad m = 0, 1, 2, \ldots \qquad \textbf{28–6}$$

Since the distance between the plates, d, increases linearly with the distance from the point where the glass plates touch, it follows that the dark and bright interference fringes are evenly spaced, as shown in **Figure 28–10**.

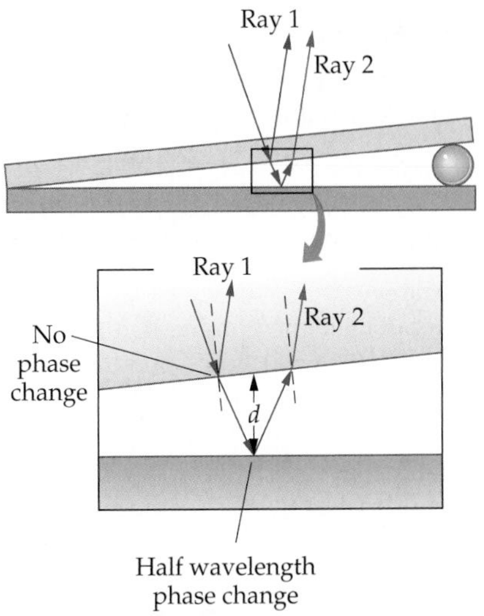

▲ **FIGURE 28–9 An air wedge**

In an air wedge, interference occurs between the light reflected from the bottom surface of the top plate of glass (ray 1) and light reflected from the top surface of the bottom plate of glass (ray 2). (The two rays are shown widely separated for clarity; in reality, they should be almost on top of one another.)

PROBLEM-SOLVING NOTE

Interference of Reflected Light

When determining whether reflected light interferes constructively or destructively, it is essential to take into account the phase changes that can occur under reflection.

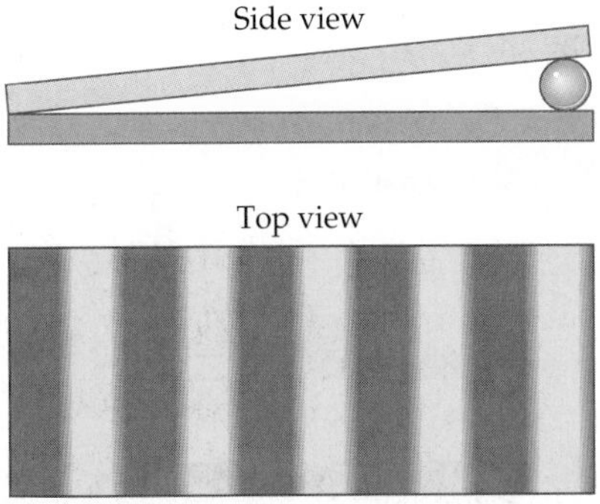

▲ **FIGURE 28–10 Interference fringes in an air wedge**

The interference fringes in an air wedge are regularly spaced, as shown in the top view of the wedge.

CONCEPTUAL CHECKPOINT 28–2 Dark or Bright Fringe?

Is the point where the glass plates touch in an air wedge **(a)** a dark fringe or **(b)** a bright fringe?

Reasoning and Discussion

At the point where the glass plates touch, the separation d is zero. Setting d equal to zero in Equation 28–5 gives $\frac{1}{2} = m$, which can never be satisfied with an integer value of m. As a result, we conclude that the point of contact is not a bright fringe.

Considering Equation 28–6 in the limit of $d = 0$ yields $\frac{1}{2} = m + \frac{1}{2}$, which is satisfied by $m = 0$. Therefore, the point of contact of the glass plates is the first dark fringe in the system.

We can understand the origin of the dark fringe at the point of contact by recalling that ray 1 undergoes no phase change on reflection, whereas ray 2 experiences a 180° phase change due to reflection. It follows, then, that when the path difference, $2d$, approaches zero, rays 1 and 2 will cancel with destructive interference.

Answer:

(a) A dark fringe occurs where the two glass plates touch.

The next Example uses the number of fringes observed in an air wedge to calculate the thickness of a human hair.

EXAMPLE 28–3 Splitting Hairs

An air wedge is formed by placing a human hair between two glass plates on one end, and allowing them to touch on the other end. When this wedge is illuminated with red light ($\lambda = 771$ nm), it is observed to have 179 bright fringes. How thick is the hair?

Picture the Problem

The air wedge used in this experiment is shown in the sketch. Note that the separation of the plates on the end with the hair is equal to the thickness of the hair, t. It is at this point that the 179th bright fringe is observed.

Strategy

Recall that the condition for bright fringes is $\frac{1}{2} + 2d/\lambda = m$, and that $m = 1$ corresponds to the first bright fringe, $m = 2$ corresponds to the second bright fringe, and so on. Clearly, then, the 179th bright fringe is given by $m = 179$. Substituting this value for m in $\frac{1}{2} + 2d/\lambda = m$, and setting the plate separation, d, equal to the thickness of the hair, t, allows us to solve for t.

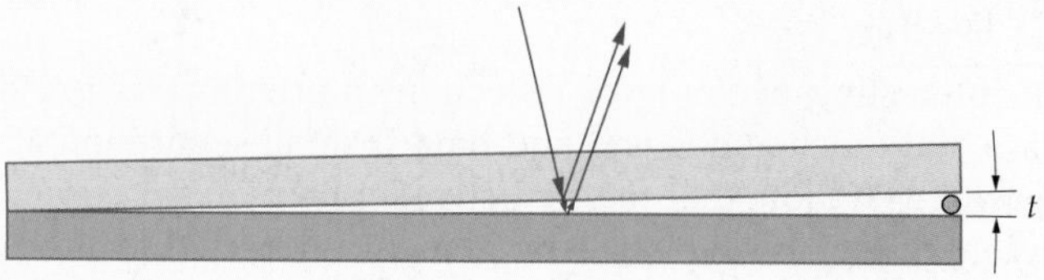

Solution

1. Solve the bright-fringe condition, $\frac{1}{2} + 2d/\lambda = m$, for the plate separation, d:

$$\frac{1}{2} + \frac{2d}{\lambda} = m$$

$$d = \frac{\lambda}{2}\left(m - \frac{1}{2}\right)$$

2. Set $m = 179$, and solve for the hair thickness, $t = d$:

$$t = \frac{\lambda}{2}\left(m - \frac{1}{2}\right)$$

$$= \frac{(771 \times 10^{-9}\ \text{m})}{2}\left(179 - \frac{1}{2}\right)$$

$$= 6.88 \times 10^{-5}\ \text{m} = 68.8\ \mu\text{m}$$

Insight

Thus, a human hair has a diameter of roughly 70 micrometers. Note that to measure the thickness of a hair we have used a "ruler" with units that are comparable to the distance to be measured. In this case, the hair has a thickness about 100 times larger than the wavelength of the light that we used.

Practice Problem

If a thicker hair is used in this experiment, will the number of bright fringes increase, decrease, or stay the same? How many bright fringes will be observed if the hair has a thickness of 80.0 μm? **[Answer:** The number of fringes increases with thickness. In this case there will be 208 fringes.]

Some related homework problems: Problem 31, Problem 36

REAL-WORLD PHYSICS
Newton's rings

Clearly, the location of interference fringes is very sensitive to even extremely small changes in plate separation. This can be illustrated dramatically by simply pressing down lightly on the upper plate with a finger. Even though the finger causes no visible change in the glass plate, the fringes are observed to move. By measuring the displacement of the fringes it is possible to calculate the tiny deflection, or bend, the finger caused in the plate. Devices using this type of mechanism are frequently used to show small displacements that would be completely invisible to the naked eye.

Newton's Rings

A system similar to an air wedge, but with a slightly different geometry, is shown in **Figure 28–11 (a)**. In this case, the upper glass plate is replaced by a curved piece of glass with a spherical cross section. Still, the mechanism producing interference is the same as before. The result is a series of circular interference fringes, as shown in **Figure 28–11 (b)**, referred to as **Newton's rings.**

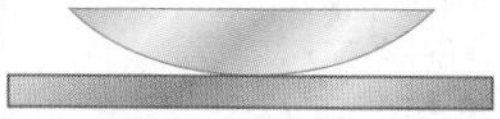

(a)

(b)

◀ **FIGURE 28–11 A system for generating Newton's rings**
(a) A variation on the air wedge is produced by placing a piece of glass with a spherical cross section on top of a plate of glass. **(b)** A top view of the system shown in part (a). The circular interference fringes are referred to as Newton's rings. Here again, the radiation used is monochromatic light—in this case from sodium atoms.

Notice that the fringes become more closely spaced as one moves farther from the center of the pattern. The reason is that the curved surface of the upper piece of glass moves away from the lower plate at a progressively faster rate. As a result, the horizontal distance required to go from one fringe to the next becomes less as one moves away from the center.

Newton's rings can be used to test the shape of a lens. Imperfections in the ring pattern indicate slight distortions in the lens. As in the case of an air wedge, even a very small change in shape can cause a significant displacement of the interference fringes.

Thin Films

Perhaps the most commonly observed examples of interference are provided by thin films, such as those found in soap bubbles and oil slicks. We are all familiar, for example, with the swirling patterns of color that are seen on the surface of a bubble. These colors are the result of the constructive and destructive interference that can occur when white light reflects from a thin film. In particular, some colors undergo destructive interference and are *eliminated* from the incident light, while others colors are *enhanced* by constructive interference.

To determine the conditions for constructive and destructive interference in a thin film, consider **Figure 28–12**. Here we show a thin film of thickness t and index of refraction $n > 1$, with air ($n = 1.00$) on either side. To analyze this system, we proceed in much the same way as we did for the air wedge earlier in this section. Specifically, we focus on the phase change of rays 1 and 2, taking into account phase changes due to both reflection and path-length difference.

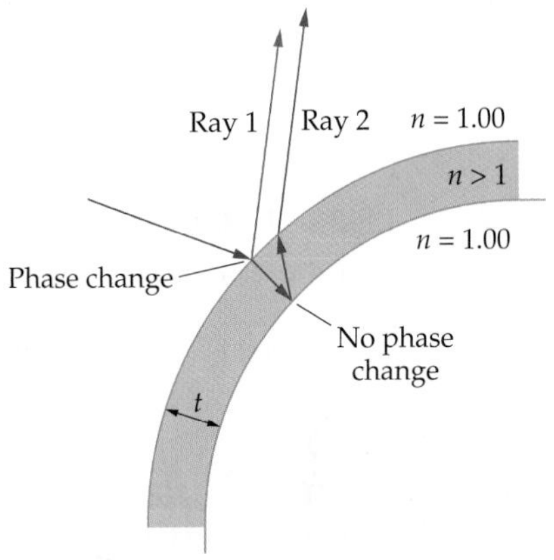

▲ **FIGURE 28–12 Interference in thin films**

The phase of ray 1 changes by half a wavelength due to reflection; the phase of ray 2 changes by $2t/\lambda_n$, where λ_n is the wavelength of light within the thin film.

First, note that ray 1 reflects from the air-to-film interface; hence, its phase changes by half a wavelength. The effective path length for ray 1, then, is

$$\ell_{\text{eff},1} = \tfrac{1}{2}\lambda$$

Dividing by the wavelength, λ, gives the phase change of ray 1 in terms of the wavelength:

$$\frac{\ell_{\text{eff},1}}{\lambda} = \frac{1}{2} \qquad \text{28–7}$$

Recall that if two rays of light differ in phase by half a wavelength, the result is destructive interference.

Next, ray 2 reflects from the film-to-air interface; hence, it experiences no phase change due to reflection. It does, however, have a phase change as a result of traveling an extra distance $2t$ through the film. Thus the effective path length for ray 2 is

$$\ell_{\text{eff},2} = 2t$$

To put this path length in terms of the wavelength, we must recall that if the wavelength of light in a vacuum is λ_{vacuum}, its wavelength in a medium with an index of refraction n is

$$\lambda_n = \frac{\lambda_{\text{vacuum}}}{n}$$

Dividing the path length by λ_n gives us the phase change of ray 2 in terms of the wavelength within the film:

$$\frac{\ell_{\text{eff},2}}{\lambda_n} = \frac{2t}{\lambda_n} = \frac{2nt}{\lambda_{\text{vacuum}}} \qquad \text{28–8}$$

Finally, we can calculate the difference in phase changes for rays 1 and 2 using the preceding results:

$$\text{difference in phase changes} = \frac{2nt}{\lambda_{\text{vacuum}}} - \frac{1}{2} \qquad \text{28–9}$$

Note that in the limit of zero film thickness, $t = 0$, the phase difference is $-\tfrac{1}{2}$. This corresponds to destructive interference, since moving a wave back half a

▲ The swirling colors typical of soap bubbles are created by interference, both destructive and constructive, which eliminates certain wavelengths from the reflected light while enhancing others. Which colors are removed or enhanced at a given point depends on the precise thickness of the film in that region.

wavelength is equivalent to moving it ahead half a wavelength. In general, then, destructive interference occurs if any of the following conditions are satisfied:

$$\frac{2nt}{\lambda_{\text{vacuum}}} - \frac{1}{2} = -\frac{1}{2}, \frac{1}{2}, \frac{3}{2}, \ldots$$

Adding $\frac{1}{2}$ to each side yields our final result for **destructive interference:**

$$\frac{2nt}{\lambda_{\text{vacuum}}} = m \qquad m = 0, 1, 2, \ldots \tag{28–10}$$

Similarly, if the phase difference between the rays is equal to an integer, m, the result is **constructive interference:**

$$\frac{2nt}{\lambda_{\text{vacuum}}} - \frac{1}{2} = m \qquad m = 0, 1, 2, \ldots \tag{28–11}$$

These conditions are applied in the next Example.

EXAMPLE 28–4 Red Light Special

A beam consisting of red light ($\lambda_{\text{vacuum}} = 662$ nm) and blue light ($\lambda_{\text{vacuum}} = 465$ nm) is directed at right angles onto a thin soap film. If the film has an index of refraction $n = 1.33$ and is suspended in air ($n = 1.00$), find the smallest nonzero thickness for which it appears red in reflected light.

Picture the Problem

Our sketch shows a soap film of thickness t and index of refraction $n = 1.33$ suspended in air ($n = 1.00$). We consider each of the colors in the incoming beam of light separately, with one blue ray and one red ray. If the film is to look red in reflected light, it follows that the reflected blue light must cancel due to destructive interference, as indicated.

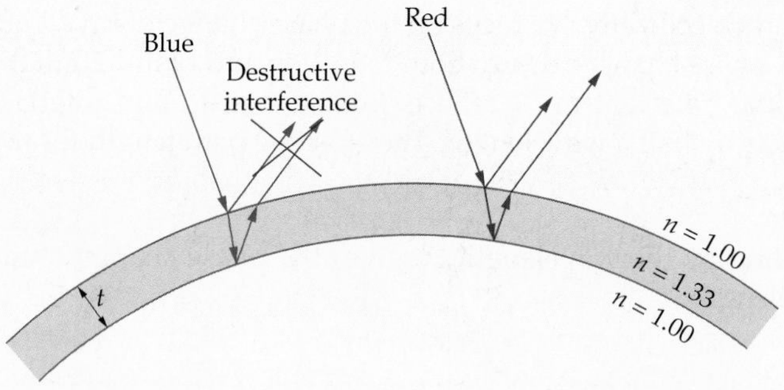

Strategy

As mentioned above, the desired thickness of the film is such that blue light satisfies one of the conditions for destructive interference given in Equation 28–10. These conditions are $2nt/\lambda_{\text{vacuum}} = m$, where m is equal to 0, 1, 2, and so on. Because the case $m = 0$ corresponds to zero thickness, $t = 0$, it follows that the smallest nonzero thickness is given by $m = 1$.

Solution

1. Solve $2nt/\lambda_{\text{vacuum}} = m$ for the thickness t:

$$\frac{2nt}{\lambda_{\text{vacuum}}} = m \quad \text{or} \quad t = \frac{m\lambda_{\text{vacuum}}}{2n}$$

2. Calculate the thickness using $m = 1$:

$$t = \frac{m\lambda_{\text{vacuum}}}{2n} = \frac{(1)(465 \text{ nm})}{2(1.33)} = 175 \text{ nm}$$

Insight

Although blue light is canceled at this thickness, red light is not. In fact, repeating the preceding calculation shows that red light is not canceled until the thickness of the film is 249 nm.

Practice Problem

What is the smallest thickness of film that gives *constructive* interference for the blue light in this system? [**Answer:** $t = 87.4$ nm]

Some related homework problems: Problem 26, Problem 32

The connection between the thickness of a film and the color it shows in reflected light is illustrated in **Figure 28–13**. Notice that in thicker regions of the film, the longer wavelengths of light interfere destructively, resulting in reflected light that is bluish. In the limit of zero thickness the condition for destructive interference, $2nt/\lambda_{\text{vacuum}} = m$, is satisfied for all wavelengths with m set equal to zero. Hence, an extremely thin film appears dark in reflected light—it is essentially one large dark fringe.

Another example of a thin-film interference is provided by a film floating on a liquid or coating a solid surface. For example, **Figure 28–14** shows a thin film floating on water. If the index of refraction of the film is greater than that of water,

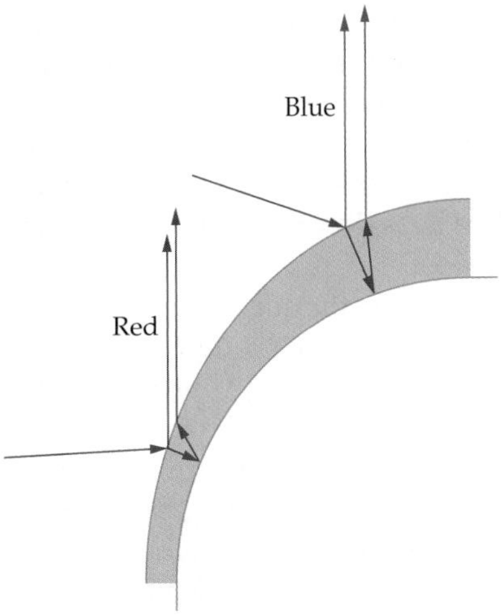

▲ **FIGURE 28–13 Thickness and color in a thin film**

Thicker portions of thin film appear blue, since the long-wavelength red light experiences destructive interference. Thinner regions appear red because the short-wavelength blue light interferes destructively.

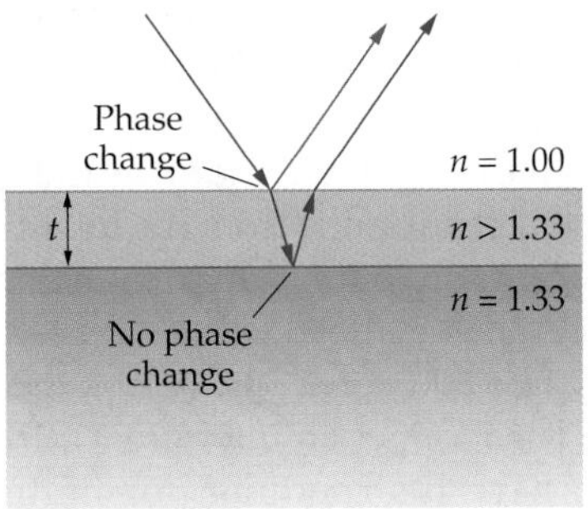

▲ **FIGURE 28–14 A thin film with one phase change**

If the index of refraction of the film is greater than that of the water, the situation in terms of phase changes is essentially the same as for a thin film suspended in air.

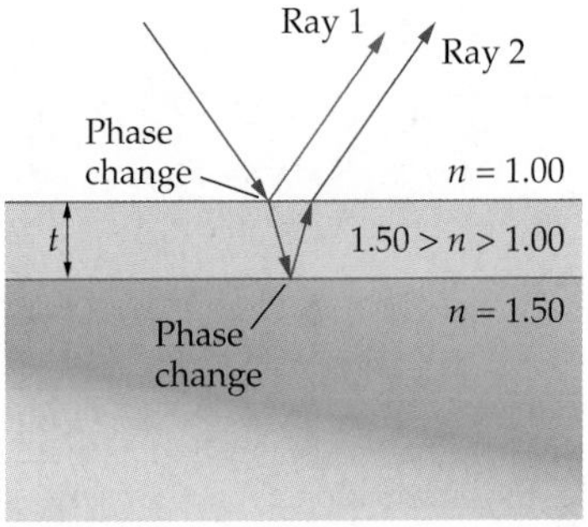

▲ **FIGURE 28–15 A thin film with two phase changes**

A thin film is applied to a material with a relatively large index of refraction. If the index of refraction of the film is less than that of the material that supports it, there will be a phase change for reflections from both surfaces of the film. Films of this type are often used in nonreflective coatings.

the situation in terms of interference is essentially the same as for a thin film suspended in air. In particular, the ray reflected from the top surface of the film undergoes a phase change of half a wavelength; the ray reflected from the bottom of the film has no phase change due to reflection. Thus the interference conditions given in Equations 28–10 and 28–11 still apply.

On the other hand, suppose a thin film has an index of refraction that is greater than 1.00 but less than the index of refraction of the material on which it is supported, as in **Figure 28–15**. In a case like this, there is a reflection phase change at both the upper and the lower surface of the film. The condition for destructive interference for this type of system is discussed in the following Active Example.

▲ The lenses of binoculars and cameras often have a blue, purple, or amber tint, the product of their antireflection coating. The coating is a thin film that reduces reflection from the lens surfaces by destructive interference.

REAL-WORLD PHYSICS
Nonreflective coating

ACTIVE EXAMPLE 28–1 Nonreflective Coating: Find the Thickness

Camera lenses ($n = 1.52$) are often coated with a thin film of magnesium fluoride ($n = 1.38$). These "nonreflective coatings" use destructive interference to reduce unwanted reflections. Find the condition for destructive interference in this case, and calculate the minimum thickness required to give destructive interference for light in the middle of the visible spectrum (yellow-green light, $\lambda_{\text{vacuum}} = 565$ nm).

Solution *(Test your understanding by performing the calculations indicated in each step.)*

1. Give the phase change (in units of the wavelength) for ray 1 in Figure 28–15: $\frac{1}{2}$
2. Give the phase change (in units of the wavelength) for ray 2 in Figure 28–15: $\frac{1}{2} + 2t/\lambda_n$
3. Set the difference in phase changes equal to one-half: $2t/\lambda_n = \frac{1}{2}$
4. Solve for the thickness: $t = \lambda_n/4$
5. Substitute numerical values: $t = 102$ nm

Insight

Because the thickness of the film should be a quarter of a wavelength, these nonreflective films are often referred to as quarter-wave coatings.

Your Turn

Suppose a different coating material with a larger index of refraction is used on this lens. Is the desired minimum thickness greater than or less than it was with magnesium fluoride? Calculate the new minimum thickness, assuming the index of refraction for the coating material is $n = 1.45$.

(Answers to **Your Turn** *problems are given in the back of the book.)*

Interference in CDs

Destructive interference plays a crucial role in the operation of a CD. The basic idea behind these devices is that information is encoded in the form of a series of "bumps" on an otherwise smooth reflecting surface. A laser beam directed onto the surface is reflected back to a detector, and as the intensity of the reflected beam varies due to the bumps, the information on the CD is decoded—similar to using dots and dashes to send information in Morse code.

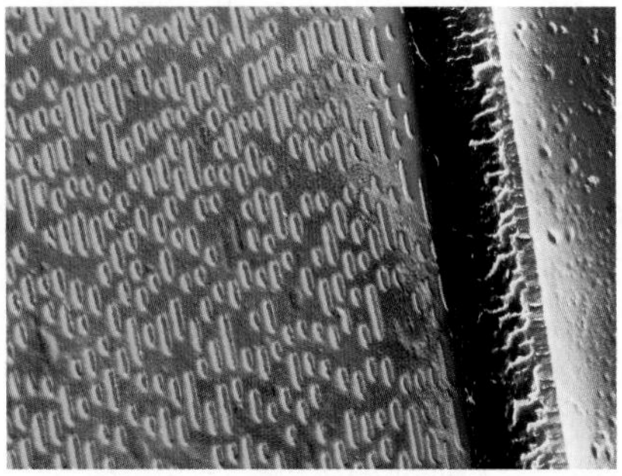

▲ This microscopic view of the surface of a CD reveals the pattern of tiny bumps that encodes the information. From the back, these areas take the form of indentations—hence they are commonly known as "pits," even though to the laser beam scanning the CD they appear as raised areas.

Imagine the laser beam illuminating a small area on the surface of a CD. As a bump moves into the beam, as in **Figure 28–16**, there are two components to the reflected beam—one from the top of the bump, the other from the bottom. If these two beams are out of phase by half a wavelength, there will be destructive interference and the detector will receive a weak signal. When the beam is reflected solely from the top of the bump, the detector again receives a strong signal, since there is no longer an interference effect. As the bump moves out of the beam we again have the condition for destructive interference, and the detector signal again falls. Thus the bumps give the detector a series of "on" and "off" signals that can be converted to sound, pictures, or other types of information.

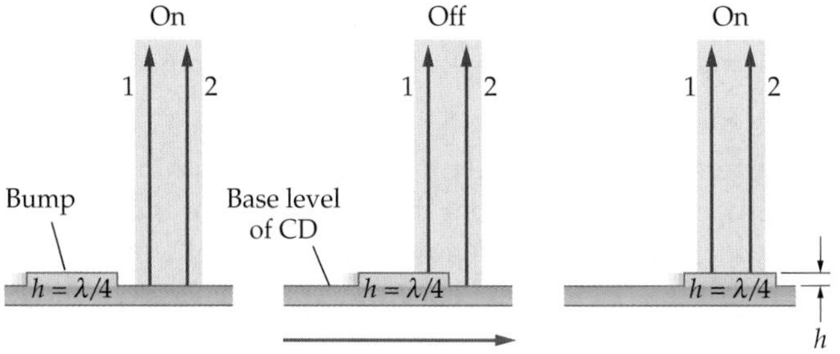

▶ **FIGURE 28–16 Reading information on a CD**

As a "bump" on a CD moves through the laser beam, the detector receives a weak "off" signal when the bump enters and leaves the beam. When the beam reflects entirely from the top of a bump, or from the base level of the CD, the detector receives a strong "on" signal.

REAL-WORLD PHYSICS
Reading the information on a CD

Let's find the necessary height h of a bump if the red light of a ruby laser, with a wavelength of 694 nm, is to have destructive interference. Figure 28–16 shows the situation, in which ray 1 reflects from the top of a bump, and ray 2 reflects from the base level of the CD. The path difference between the rays is $2h$; thus, to have destructive interference, $2h$ must be equal to half a wavelength, $2h = \lambda/2$, or

$$h = \frac{\lambda}{4}$$

In the case of a ruby laser, the required height of a bump is about 174 nm.

28–4 Diffraction

If light is indeed a wave, it must exhibit behavior similar to that displayed by the water waves in **Figure 28–17**. Note that the waves are initially traveling directly to the right. After passing through the gap in the barrier, however, they spread out and travel in all possible forward directions, in accordance with Huygens's principle. Thus, an observer at point P detects waves even though she is not on a direct line with the initial waves and the gap. In general, waves always bend, or **diffract,** when they pass by a barrier or through an opening.

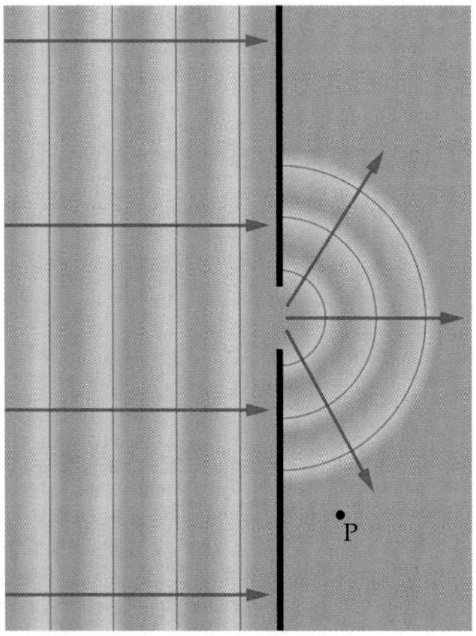

▲ **FIGURE 28–17 Diffraction of water waves**

As water waves pass through an opening, they diffract, or change direction. Thus an observer at point P detects waves even though this point is not on a line with the original direction of the waves and the opening. All waves exhibit similar diffraction behavior.

A familiar example of diffraction is the observation that you can hear a person talking even when that person is out of sight around a corner. The sound waves from the person bend around the corner, just like the water waves in Figure 28–17. It might seem, then, that light cannot be a wave, since it does not bend around a corner along with the sound. There is a significant difference between sound and light waves, however; namely, their wavelengths differ by many orders of magnitude—from a meter or so for sound to about 10^{-7} m for light. As we shall see, the angle through which a wave bends is greater the larger the wavelength of the wave; hence, diffraction effects in light are typically small compared with those in sound waves and water waves.

To investigate the diffraction of light we start by considering the behavior of a beam of light as it passes through a single, narrow slit in a screen.

Single-Slit Diffraction

Consider monochromatic light of wavelength λ passing through a narrow slit of width W, as shown in **Figure 28–18**. After passing through the slit the light shines on a distant screen, which we assume to be much farther from the slit than the width W. According to Huygens's principle, each point within the slit can be considered as a source of new waves that radiate toward the screen. The interference of these sources with one another generates a diffraction pattern.

We can understand the origin of a single-slit diffraction pattern by referring to **Figure 28–19**, where we show light propagating to the screen from various points in the slit. For example, consider waves from points 1 and 1′ propagating to the screen at an angle θ relative to the initial direction of the light. Since the screen is distant, the waves from 1 and 1′ travel on approximately parallel paths to reach the screen. From the figure, then, it is clear that the path difference for these waves is $(W/2)\sin\theta$. Similarly, the same path difference applies to the "wave pairs" from points 2 and 2′, the wave pairs from points 3 and 3′, and so on through all points in the slit.

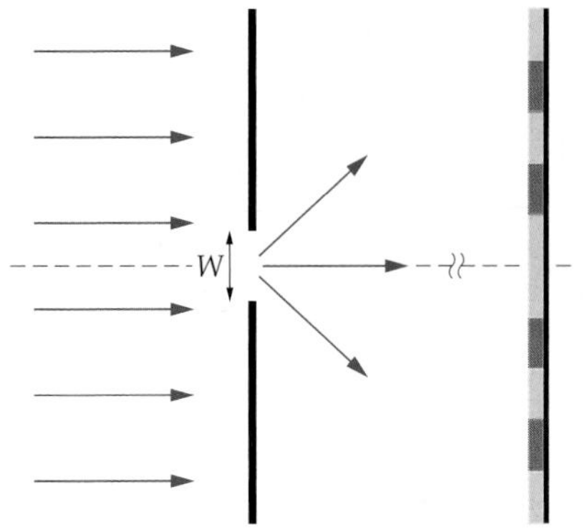

▲ **FIGURE 28–18 Single-slit diffraction**
When light of wavelength λ passes through a slit of width W a "diffraction pattern" of bright and dark fringes is formed.

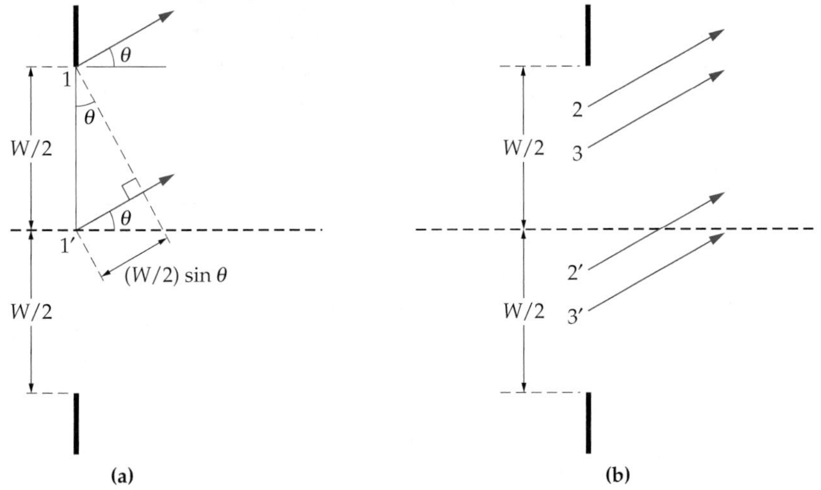

◀ **FIGURE 28–19 Locating the first dark fringe in single-slit diffraction**
The location of the first dark fringe in single-slit diffraction can be determined by considering pairs of waves radiating from the top and bottom half of a slit. **(a)** A wave pair originating at points 1 and 1′ has a path difference of $(W/2)\sin\theta$. These waves interfere destructively if the path difference is equal to half a wavelength. **(b)** The rest of the light coming from the slit can be considered to consist of additional wave pairs, like 2 and 2′, 3 and 3′, and so on. Each wave pair has the same path difference.

In the forward direction, $\theta = 0°$, the path difference is zero; $(W/2)\sin 0° = 0$. As a result, all wave pairs interfere constructively, giving maximum intensity at the center of the diffraction pattern. However, if θ is increased until the path difference is half a wavelength, $(W/2)\sin\theta = \lambda/2$, each wave pair experiences destructive interference. Thus, the *first minimum*, or dark fringe, in the diffraction pattern occurs at the angle given by

$$W\sin\theta = \lambda$$

To find the second dark fringe, imagine dividing the slit into four regions, as illustrated in **Figure 28–20**. Within the upper two regions, we perform the same wave-pair construction described above; the same is done with the lower two regions. In this case, the path difference between wave pairs 1 and 1′ is $(W/4)\sin\theta$. As before, destructive interference first occurs when the path difference is $\lambda/2$. Solving for $\sin\theta$ in this case gives us the condition for the *second dark fringe:*

$$W\sin\theta = 2\lambda$$

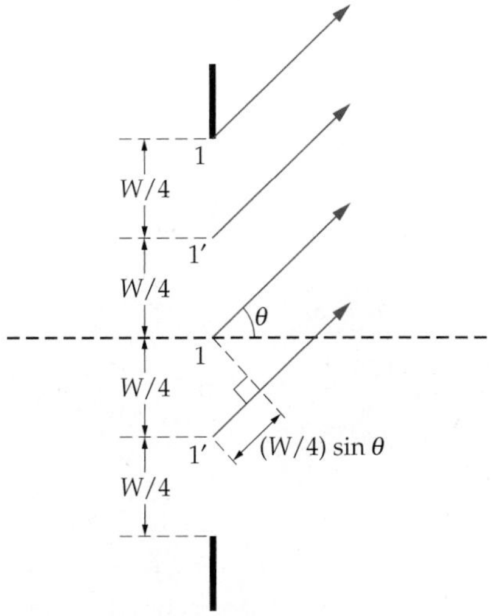

▲ **FIGURE 28–20 Locating the second dark fringe in single-slit diffraction**
To find the second dark fringe in a single-slit diffraction pattern, we divide the slit into four regions and consider wave pairs that originate from points separated by a distance $W/4$. Destructive interference occurs when the path difference, $(W/4)\sin\theta$, is half a wavelength.

The next dark fringe can be found by dividing the slit into six regions, with a path difference between wave pairs of $(W/6)\sin\theta$. In this case, the condition for destructive interference is $W\sin\theta = 3\lambda$. In general, then, dark fringes satisfy the following conditions:

Conditions for Dark Fringes in Single-Slit Interference

$$W\sin\theta = m\lambda \qquad m = \pm 1, \pm 2, \pm 3 \ldots \qquad \textbf{28–12}$$

Note that by including both positive and negative values for m, we have taken into account the symmetry of the diffraction pattern about its midpoint.

EXERCISE 28–2

Monochromatic light passes through a slit of width 1.2×10^{-5} m. If the first dark fringe of the resulting diffraction pattern is at angle $\theta = 3.25°$, what is the wavelength of the light?

Solution

Solving Equation 28–12 for the wavelength gives us

$$\lambda = \frac{W \sin\theta}{m} = \frac{(1.2 \times 10^{-5}\ \text{m})\sin(3.25°)}{(1)} = 680\ \text{nm}$$

Notice that we use $m = 1$, since this is the *first* dark fringe.

The bright fringes in a diffraction pattern consist of the central fringe plus additional bright fringes approximately halfway between successive dark fringes. Note, in addition, that the central fringe is about twice as wide as the other bright fringes. In the small-angle approximation ($\sin\theta \sim \theta$), the central fringe extends from $\theta = \lambda/W$ to $\theta = -\lambda/W$, and hence its width is given by the following:

$$\text{approximate angular width of central fringe} = 2\frac{\lambda}{W} \qquad \textbf{28–13}$$

Thus we see that the wavelength λ plays a crucial role in diffraction patterns and that λ/W gives the characteristic angle of "bending" produced by diffraction. In the following Conceptual Checkpoint we consider the role played by the width of the slit.

CONCEPTUAL CHECKPOINT 28–3 Width of Central Bright Fringe

If the width of the slit through which light passes is reduced, does the central bright fringe **(a)** become wider, **(b)** become narrower, or **(c)** remain the same size?

Reasoning and Discussion

It might seem that making the slit narrower will cause the diffraction pattern to be narrower as well. Recall, however, that the diffraction pattern is produced by waves propagating from all parts of the slit. If the slit is wide, the incoming wave passes through with little deflection. If it is small, on the other hand, it acts like a point source, and light is radiated over a broad range of angles. Therefore, the smaller slit produces a wider central fringe.

This result is also confirmed by considering Equation 28–13, where we see that a smaller value of W results in a wider central fringe.

Answer:

(a) The central bright fringe is wider.

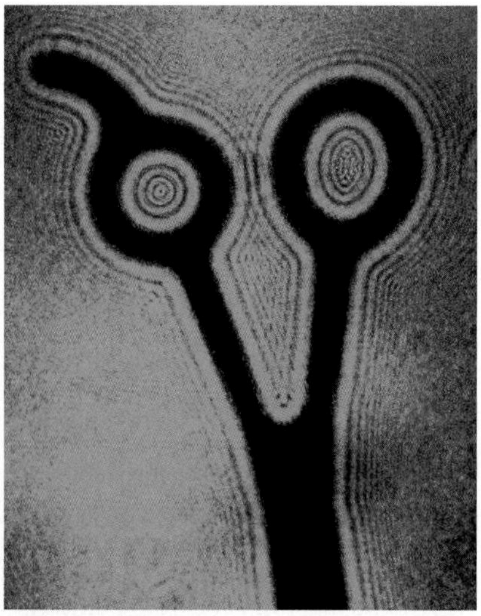

▲ Upon close inspection, the shadow of a sharp edge is seen to consist of numerous fringes produced by diffraction.

As mentioned earlier, waves diffract whenever they encounter some sort of barrier or opening. It follows, then, that the shadow cast by an object is really not as sharp as it may seem. On closer examination, as shown in the accompanying photo, the shadow of an object such as a pair of scissors actually consists of a tightly spaced series of diffraction fringes. Thus, shadows are not the sharp boundaries implied by geometrical optics but, instead, are smeared out on a small length scale. Under ordinary conditions the diffraction pattern in a shadow is not readily visible. However, a similar diffraction pattern can be observed by simply holding two fingers close together before your eyes—try it.

EXAMPLE 28–5 Exploring the Dark Side

Light with a wavelength of 511 nm forms a diffraction pattern after passing through a single slit of width 2.20×10^{-6} m. Find the angle associated with **(a)** the first and **(b)** the second dark fringe above the central bright fringe.

Picture the Problem

In our sketch we identify the first and second dark fringes above the central bright fringe. Note that the first dark fringe corresponds to $m = 1$, and the second corresponds to $m = 2$. Finally, the width of the slit is $W = 2.20 \times 10^{-6}$ m.

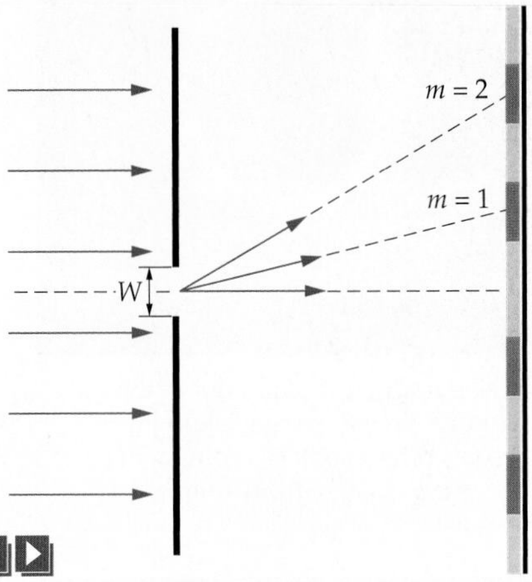

Strategy

We can find the desired angles by using the condition for dark fringes, $W \sin\theta = m\lambda$ (Equation 28–12). As mentioned above, we use $m = 1$ for part **(a)**, and $m = 2$ for part **(b)**. The values of λ and W are given in the problem statement.

Solution

Part (a)

1. Solve for θ using $m = 1$:

$$\theta = \sin^{-1}\left(\frac{m\lambda}{W}\right) = \sin^{-1}\left[\frac{(1)(511 \times 10^{-9}\ \text{m})}{2.20 \times 10^{-6}\ \text{m}}\right] = 13.4^\circ$$

Part (b)

2. Solve for θ using $m = 2$:

$$\theta = \sin^{-1}\left(\frac{m\lambda}{W}\right) = \sin^{-1}\left[\frac{(2)(511 \times 10^{-9}\ \text{m})}{2.20 \times 10^{-6}\ \text{m}}\right] = 27.7^\circ$$

Insight

Notice that the angle to the second dark fringe is *not* simply twice the angle to the first dark fringe. This is because the angle θ depends on the sine function, which is not linear. If you look at a plot of the sine function, as in Figure 13–15, you will see that $\sin\theta$ is slightly less than θ for the angles considered here. Therefore, to double the value of $\sin\theta$ you must increase θ by slightly more than a factor of two.

Practice Problem

Suppose the wavelength of light in this experiment is changed to give the first dark fringe at an angle greater than 13.4°. Is the required wavelength greater than or less than 511 nm? Check your answer by calculating the wavelength required to give the first dark fringe at $\theta = 15.0^\circ$. [**Answer:** The wavelength must be larger. For $\theta = 15.0^\circ$ we find $\lambda = 569$ nm.]

Some related homework problems: Problem 37, Problem 38, Problem 43

ACTIVE EXAMPLE 28–2 Find the Linear Distance

In a single-slit experiment, light passes through the slit and forms a diffraction pattern on a screen 2.31 m away. If the wavelength of light is 632 nm, and the width of the slit is 4.20×10^{-5} m, find the linear distance on the screen from the center of the diffraction pattern to the first dark fringe.

Solution *(Test your understanding by performing the calculations indicated in each step.)*

1. Find the angle to the first dark fringe: $\theta = 0.862^\circ$

2. Use $y = L\tan\theta$ to find the linear distance: $y = 3.48$ cm

Insight

Note that the linear distance is found using Equation 28–3, just as for the two-slit experiment.

Your Turn

Find the linear distance from the center of the pattern to the second dark fringe.

(Answers to **Your Turn** *problems are given in the back of the book.)*

▲ The paradoxical bright dot at the center of a circular shadow, known as Poisson's bright spot, is a convincing proof of the wave nature of light.

The photo at the top of this page shows a particularly interesting diffraction phenomenon: the shadow produced by a penny. As expected, the edges of the shadow show diffraction fringes, but of even greater interest is the bright point of light seen in the center of the shadow. This is referred to as "Poisson's bright spot," after the French scientist Siméon D. Poisson (1781–1840) who predicted its existence. It should be noted that Poisson did not believe that light is a wave. In fact, he used his prediction of a bright spot to show that the wave model of light was absurd and must be rejected—after all, how could a bright spot occur in the darkest part of a shadow? When experiments soon after his prediction showed conclusively that the bright spot does exist, however, the wave model of light gained almost universal acceptance. We know today that light has both wave and particle properties, a fact referred to as the *wave–particle duality*. This will be discussed in detail in Chapter 30.

28–5 Resolution

Diffraction affects, and ultimately limits, the way we see the world. There is a difference, for example, in how sharply we can see a scene depending on the size of our pupil—in general, the larger the pupil, the sharper the vision. For example, an eagle, which has pupils that are even larger than ours, can see a small creature on the ground with greater acuity than is possible for a person. Similarly, a camera or a telescope with a large aperture can "see" with greater detail than the human eye. The sharpness of vision—in particular, the ability to visually separate closely spaced objects—is referred to as **resolution.**

The way in which diffraction affects resolution can be seen by considering the diffraction pattern created by a circular aperture—such as the pupil of an eye. Just as with the diffraction pattern of a slit, a circular aperture creates a pattern of alternating bright and dark regions. The difference, as one might expect, is that the diffraction pattern of a circular opening is circular in shape, as we see in **Figure 28–21**.

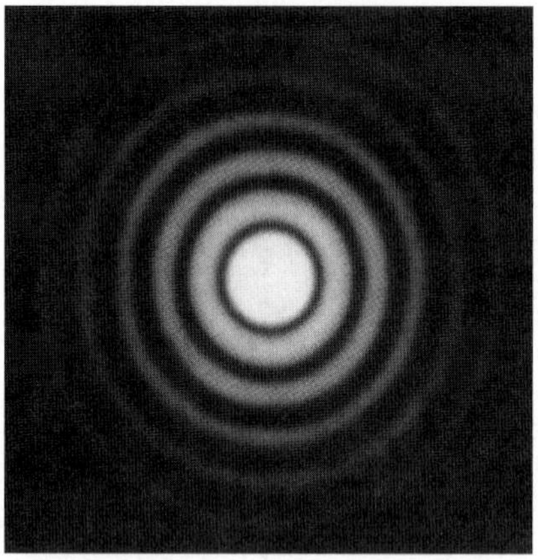

▲ **FIGURE 28–21 Diffraction from a circular opening**

Light passing through a circular aperture creates a circular diffraction pattern of alternating bright and dark regions.

In a slit pattern, the first dark fringe occurs at the angle given by $\sin\theta = \lambda/W$, where W is the width of the slit. A circular aperture of diameter D produces a central bright spot and a dark fringe at an angle θ from the center line given by the following expression:

First Dark Fringe for the Diffraction Pattern of a Circular Opening

$$\sin\theta = 1.22\frac{\lambda}{D} \qquad \text{28–14}$$

So we see that the change in geometry from a slit of width W to a circular opening of diameter D is reflected in the replacement of $(1)\lambda/W$ with $(1.22)\lambda/D$.

The physical significance of this result is that even if you focus perfectly on a point source of light, it will form a circular image of finite size on the retina. This blurs the image, replacing a point with a small circle. Thus, the diffraction of light through the pupil limits the resolution of your eye. In addition, it should be noted that the wavelength in Equation 28–14 refers to the wavelength in the medium in which the diffraction pattern is observed. For example, the wavelength that should be used in considering the eye is $\lambda_n = \lambda/n$, where n is the eye's average index of refraction (approximately $n = 1.36$).

Diffraction-induced smearing also makes it difficult to visually separate objects that are close to one another. In particular, if two closely spaced sources of light are smeared by diffraction, the circles they produce may overlap, making it difficult to tell if there are two sources or only one. The condition that is used to determine whether two sources can be visually separated is called **Rayleigh's criterion:**

> If the first dark fringe of one circular diffraction pattern passes through the center of a second diffraction pattern, the two sources responsible for the patterns will appear to be a single source.

This condition is illustrated in **Figure 28–22**.

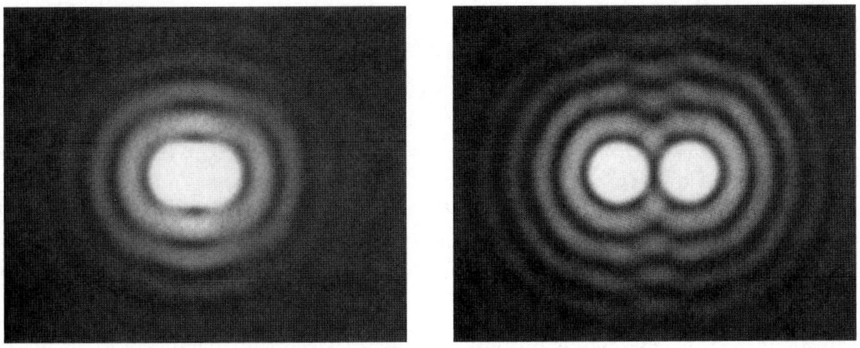

◀ **FIGURE 28–22 Resolving two point sources: Rayleigh's criterion**

If the angular separation between two sources is not great enough (left), their diffraction patterns overlap to the point where they appear as a single elongated source. With greater angular separation (right), the individual sources can be distinguished as separate.

To put Rayleigh's criterion in quantitative terms, note that for small angles (as is usually the case) the location of the first dark fringe is given by $\theta = 1.22\lambda/D$. Therefore, *two objects can be seen as separate only if their angular separation is greater than the following minimum:*

Rayleigh's Criterion

$$\theta_{\text{min}} = 1.22\frac{\lambda}{D} \qquad \textbf{28–15}$$

Note that the larger the diameter D of the aperture, the smaller the angular separation that can be resolved and, hence, the greater the resolution.

EXERCISE 28–3

Find θ_{min} for yellow light (551 nm) and an aperture diameter of 5.00 mm.

Solution

Substituting into Equation 28–15 we find

$$\theta_{\text{min}} = 1.22\frac{\lambda}{D} = 1.22\left(\frac{551 \times 10^{-9}\text{ m}}{5.00 \times 10^{-3}\text{ m}}\right) = 0.000134\text{ rad} = 0.00770°$$

Because our pupils have diameters of about 5.00 mm, the small value of θ_{min} indicates that human vision has the potential for relatively high resolution.

PROBLEM-SOLVING NOTE

Angular Resolution and the Index of Refraction

When applying the condition $\theta_{\text{min}} = 1.22\lambda/D$, it is important to remember that λ refers to the wavelength in the region between the aperture and the screen (retina, film, etc.) on which the diffraction pattern is observed. If the index of refraction in this region is n, the wavelength is reduced from λ to λ/n. (See Conceptual Checkpoint 28–1.)

An example of diffraction-limited resolution is illustrated in the accompanying photos. In the first photo we see a brilliant light in the distance that may be the single headlight of an approaching motorcycle or the unresolved image of two headlights on a car. If we are seeing the headlights of a car, the angular separation between them will increase as the car approaches. When the angular separation exceeds $1.22\ \lambda/D$, as in the second photo, we are able to distinguish the two headlights as separate sources of light. As the car continues to approach, its individual headlights become increasingly distinct, as shown in the third photo.

▲ Resolving the headlights of an approaching car. If the headlights were true point sources and the atmosphere perfectly transparent (or absent), the individual headlights could be distinguished at a much greater distance, as Example 28–6 shows.

The distance at which the two headlights can be resolved increases as the size of the aperture increases. This dependence is considered in detail in the following Example.

EXAMPLE 28–6 Motorcycle or Car?

The linear distance separating the headlights of a car is 1.1 m. Assuming light of 460 nm, a pupil diameter of 5.0 mm, and an average index of refraction for the eye of 1.36, find the maximum distance at which the headlights can be distinguished as two separate sources of light.

Picture the Problem

Our sketch shows the car a distance L from the observer, with the headlights separated by a linear distance $y = 1.1$ m. The condition for the headlights to be resolved is that their angular separation be at least θ_{min}.

Strategy

To find the maximum distance, we must first determine the minimum angular separation, which is given by $\theta_{min} = 1.22\lambda_n/D$. In this expression, $\lambda_n = \lambda/1.36$.

Once we know θ_{min}, we can find the distance L using the trigonometric relation $\tan\theta_{min} = y/L$, which follows directly from our sketch.

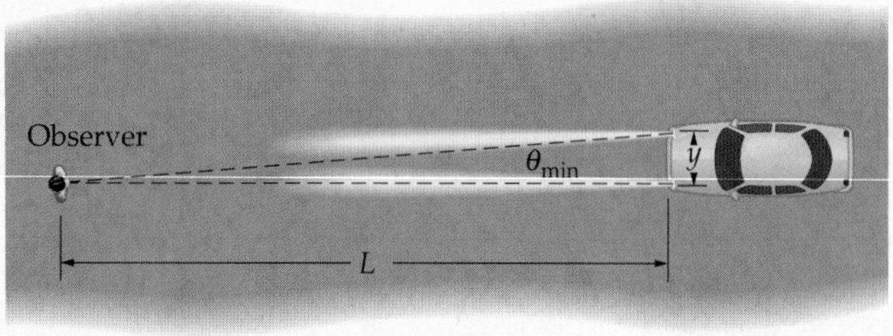

Solution

1. Find the minimum angular separation for the headlights to be resolved:

$$\theta_{min} = 1.22\frac{\lambda/n}{D}$$

$$= 1.22\left[\frac{(460 \times 10^{-9}\ \text{m})/1.36}{5.0 \times 10^{-3}\ \text{m}}\right] = 8.3 \times 10^{-5}\ \text{rad}$$

2. Solve the relation $\tan\theta_{min} = y/L$ for the distance L:

$$L = \frac{y}{\tan\theta_{min}}$$

3. Substitute numerical values to find L:

$$L = \frac{y}{\tan\theta_{min}} = \frac{1.1\ \text{m}}{\tan(8.3 \times 10^{-5}\ \text{rad})} = 13{,}000\ \text{m}$$

Insight

Thus, the car must be about 8 mi away before the headlights appear to merge. This, of course, is the ideal case. In the real world, the finite size of the headlights and the blurring effects of the atmosphere greatly reduce the maximum distance.

Practice Problem

If the pupil diameter for an eagle is 6.2 mm, from what distance can it resolve the car's headlights under ideal conditions? [**Answer:** About 9.9 mi.]

Some related homework problems: Problem 46, Problem 53

ACTIVE EXAMPLE 28–3 Resolving Ida and Dactyl

The asteroid Ida is orbited by its own small "moon" called Dactyl. If the separation between these two asteroids is 2.5 km, what is the maximum distance at which the Hubble Space Telescope (aperture diameter = 2.4 m) can still resolve them with 550-nm light?

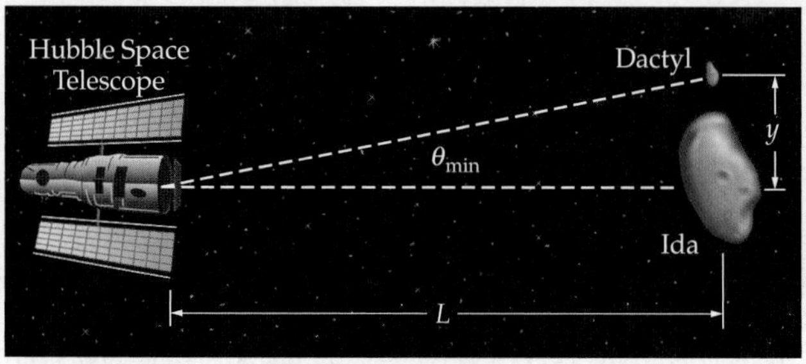

Solution *(Test your understanding by performing the calculations indicated in each step.)*

1. Calculate the minimum angular separation for the asteroids to be resolved: $\theta_{min} = 2.8 \times 10^{-7}$ rad
2. Express L in terms of y and θ_{min}: $L = y/\tan(\theta_{min})$
3. Substitute numerical values to find L: $L = 8.9 \times 10^9$ m $= 5.5 \times 10^6$ mi

Insight

If the asteroids are farther away than this distance, the Hubble Space Telescope will not be able to image them as two separate objects. In fact, the asteroids are close enough to be viewed separately, as we see in the photo on page 370.

Your Turn

If the aperture diameter of the telescope is increased, does the maximum resolution distance increase or decrease? Calculate the maximum resolution distance for an aperture diameter of 3.0 m.

(Answers to **Your Turn** *problems are given in the back of the book.)*

▲ **FIGURE 28–23 Pointillism**

Paul Signac's *The Mills at Owerschie* (1905), an example of pointillism.

We conclude by considering two interesting examples of diffraction-limited resolution. First, in the artistic style known as *pointillism*, the artist applies paint to a canvas in the form of small dots of color. When viewed from a distance the individual dots cannot be resolved and the painting appears to be painted with continuous colors. An example is given in **Figure 28–23**. If this painting is viewed from a distance of more than a few meters, the color dots blend into one another.

REAL-WORLD PHYSICS: BIO
Pointillism and painting

The second example is the formation of a picture on the screen of a television. Although a television can show all the colors of the rainbow, it in fact produces only three colors—red, green, and blue, the so-called additive primaries. These three colors are grouped together tightly to form the "pixels" on the screen, as **Figure 28–24** shows. From a distance the three individual color spots can no longer be distinguished, and the eye sees the net effect of the three colors combined. Since any color can be created with the proper amounts of the three primary colors—red, green, and blue—the television screen can reproduce any picture.

REAL-WORLD PHYSICS: BIO
Color television image

▲ **FIGURE 28–24 Pixels on a television screen**

A typical pixel on the screen of a color television consists of three closely spaced color spots: one red, one blue, and one green. These are the only colors the television actually produces.

To see this effect in action, try the following: look for a region on a television screen or a computer monitor where the picture is yellow. Since yellow is created by mixing red and green light equally, you will see on close examination (perhaps with the aid of a magnifying glass) that pixels in the yellow region of the screen have both the red and green dots illuminated, but the blue dots are dark. As you slowly move away from the screen, note that the red and green dots merge, leaving the brain with the sensation of a yellow light, even though there are no yellow color dots on the screen.

28–6 Diffraction Gratings

As we saw earlier in this chapter, a screen with one or two slits can produce striking patterns of interference fringes. It is natural to wonder what interference effects may be produced if the number of slits is increased significantly. In general, we refer to a system with a large number of slits as a **diffraction grating.** There are many ways of producing a grating; for example, one might use a diamond stylus to cut thousands of slits in the aluminum coating on a sheet of glass. Alternatively, one might photoreduce an image of parallel lines onto a strip of film. In some cases it is possible to produce gratings with as many as 40,000 slits—or "lines," as they are often called—per centimeter.

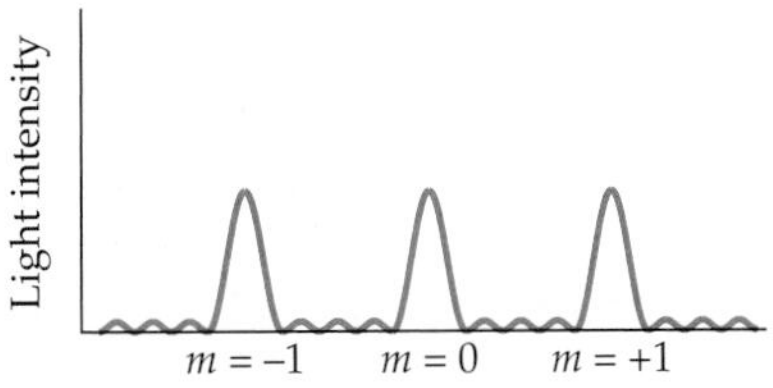

▲ **FIGURE 28–25 Diffraction pattern for five slits**

The interference pattern produced by a diffraction grating with five slits. The large "principal" maxima are sharper than the maxima in the two-slit apparatus. The small "secondary" maxima are negligible compared with the principal maxima.

The interference pattern formed by a diffraction grating consists of a series of sharp, widely spaced bright fringes—called *principal maxima*—separated by relatively dark regions with a number of weak secondary maxima, as indicated in **Figure 28–25** for the case of five slits. In the limit of a large number of slits, the principal maxima become more sharply peaked, and the secondary maxima

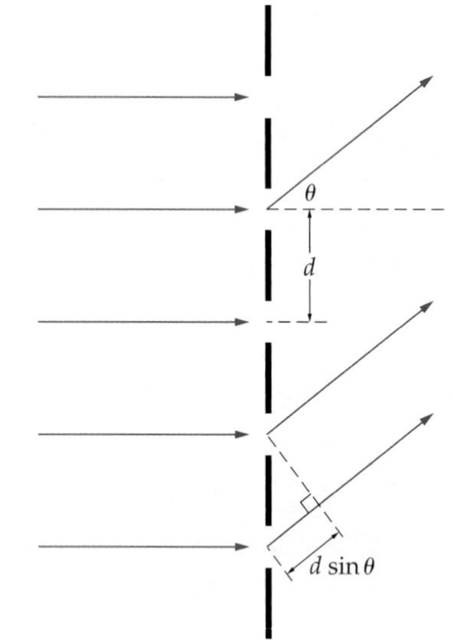

▲ **FIGURE 28–26 Path-length difference in a diffraction grating**
A simple diffraction grating consists of a number of slits with a spacing d. The difference in path length for rays from neighboring slits is $d \sin \theta$.

become insignificant. As one might expect, the angle at which a principal maximum occurs depends on the wavelength of light that passes through the grating. In this way, a grating acts much like a prism—sending the various colors of white light off in different directions. A grating, however, can spread the light out over a wider range of angles than a prism.

To determine the angles at which principal maxima are found, consider a grating with a large number of slits, each separated from the next by a distance d, as shown in **Figure 28–26**. A beam of light with wavelength λ is incident on the grating from the left and is diffracted onto a distant screen. At an angle θ to the incident direction the path difference between successive slits is $d \sin \theta$, as Figure 28–26 shows. Therefore, constructive interference, and a principal maximum, occurs when the path difference is an integral number of wavelengths, λ:

Constructive Interference in a Diffraction Grating

$$d \sin \theta = m\lambda \qquad m = 0, \pm 1, \pm 2, \ldots \qquad \textbf{28–16}$$

Notice that the angle θ becomes larger as d is made smaller. In particular, if a grating has more lines per centimeter (smaller d), light passing through it will be spread out over a larger range of angles.

EXERCISE 28–4

Find the slit spacing necessary for 450-nm light to have a first-order ($m = 1$) principal maximum at 15°.

Solution

Solving Equation 28–16 for d we obtain

$$d = \frac{m\lambda}{\sin \theta} = \frac{(1)(450 \times 10^{-9}\ \text{m})}{\sin 15°} = 1.7 \times 10^{-6}\ \text{m}$$

A grating is often characterized in terms of its number of lines per unit length, N. For example, a particular grating might have 2250 lines per centimeter. The corresponding slit separation, d, is simply the inverse of the number of lines per length. In this case the slit separation is $d = 1/N = 1/(2250\ \text{cm}^{-1}) = 4.44 \times 10^{-4}\ \text{cm} = 4.44 \times 10^{-6}\ \text{m}$.

EXAMPLE 28–7 A Second-Order Maximum

When 546-nm light passes through a particular diffraction grating, a second-order principal maximum is observed at an angle of 16.0°. How many lines per centimeter does this grating have?

Picture the Problem

Our sketch shows the first few principal maxima on either side of the center of the diffraction pattern. The second-order maximum ($m = 2$) is the second maximum above the central peak, and it occurs at an angle of 16.0°.

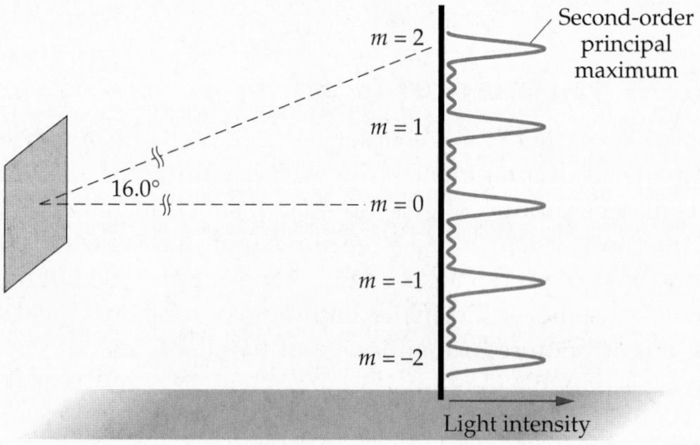

Strategy

First, we can use $\sin \theta = m\lambda/d$ to find the necessary spacing d, given that m, λ, and θ are specified in the problem statement. Next, the number of lines per centimeter, N, is simply the inverse of the spacing d; that is, $N = 1/d$.

Solution

1. Calculate the distance d between slits:

$$d = \frac{m\lambda}{\sin \theta} = \frac{(2)(546 \times 10^{-9}\ \text{m})}{\sin 16.0°} = 3.96 \times 10^{-6}\ \text{m}$$

2. Take the inverse of d to find the number of lines per meter: $N = \frac{1}{d} = \frac{1}{3.96 \times 10^{-6}\ \mathrm{m}} = 2.53 \times 10^{5}\ \mathrm{m}^{-1}$

3. Convert to lines per centimeter: $N = 2.53 \times 10^{5}\ \mathrm{m}^{-1}\left(\frac{1\ \mathrm{m}}{100\ \mathrm{cm}}\right) = 2530\ \mathrm{cm}^{-1}$

Insight

Thus, this diffraction grating must have 2530 lines per centimeter. Though this is a lot of lines to pack into a distance of one centimeter, it is common to find this many lines or more in a typical diffraction grating.

Practice Problem

If the grating is ruled with more lines per centimeter, does the angle to the second-order maximum increase, decrease, or stay the same? Check your answer by calculating the angle for a grating with 3530 lines per centimeter. [**Answer:** The angle will increase. For 3530 lines per centimeter we find $\theta = 22.7°$.]

Some related homework problems: Problem 54, Problem 55, Problem 65

Of the many ways to produce a diffraction grating, one of the more novel is by *acousto-optic modulation* (AOM). In this technique, light is diffracted not by a series of slits but by a series of high-density wave fronts produced by a sound wave propagating through a solid or a liquid. For example, in the AOM device shown in **Figure 28–27**, sound waves propagate through a quartz crystal, producing a series of closely spaced, parallel wave fronts. An incoming beam of light diffracts from these wave fronts, giving rise to an intense outgoing beam. If the sound is turned off, however, the incoming light passes through the crystal without being deflected. Thus, simply turning the sound on or off causes the diffracted beam to be switched on or off, whereas changing the frequency of the sound can change the angle of the diffracted beam. Many laser printers use AOMs to control the laser beam responsible for "drawing" the desired image on a light-sensitive surface.

REAL-WORLD PHYSICS
Acousto-optic modulation

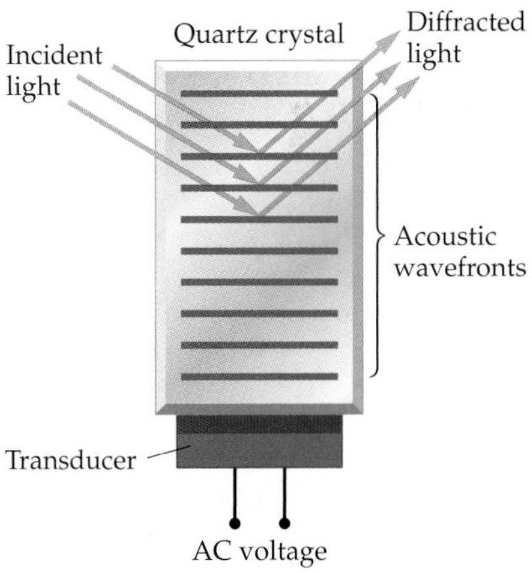

▲ **FIGURE 28–27 An acousto-optical modulator**
Acousto-optical modulators use sound waves, and the density variations they produce, to diffract light. The angle at which the light diffracts can be controlled by the frequency of the sound.

X-ray Diffraction

There is another type of diffraction grating that is not made by clever applications of technology but occurs naturally—the crystal. The key characteristic of a crystal is that it has a regular, repeating structure. In particular, crystals generally consist of regularly spaced planes of atoms or ions; these planes, just like the wave fronts in an AOM, can diffract an incoming beam of electromagnetic radiation.

For a diffraction grating to be effective, however, the wavelength of the radiation, λ, must be comparable to the spacing, d, in Equation 28–16. In a typical crystal, the spacing between atomic planes is roughly an angstrom; that is, $d \sim 10^{-10}\ \mathrm{m} \sim 0.1\ \mathrm{nm}$. Notice that this distance is much less than the wavelength of visible light, which is roughly 400 to 700 nm. Therefore, visible light will not produce useful diffraction effects from a crystal. However, if we consider the full electromagnetic spectrum, as presented in Figure 25–8, we see that wavelengths of 0.1 nm fall within the X-ray portion of the spectrum. Indeed, X-rays produce vivid diffraction patterns when sent through crystals (**Figure 28–28**).

Today, *X-ray diffraction* is a valuable scientific tool. First, the angle at which principal maxima occur in an X-ray diffraction pattern can determine the precise distance between various planes of atoms in a particular crystal. Second, the symmetry of the pattern determines the type of crystal structure. More sophisticated analysis of X-ray diffraction patterns—in particular, the angles and intensities of diffraction maxima—can be used to help determine the structures of even large organic molecules. In fact, it was in part through examination of X-ray diffraction patterns that J. D. Watson and F. H. C. Crick were able to deduce the double-helix structure of DNA in 1953.

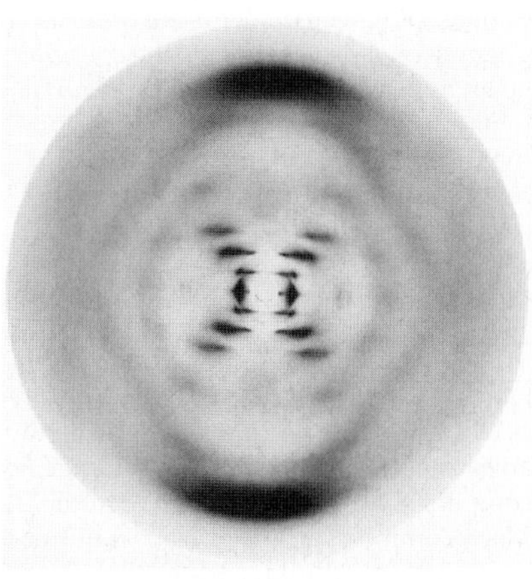

▲ **FIGURE 28–28 X-ray diffraction**
X-ray diffraction pattern produced by DNA. A photo like this one helped Watson and Crick deduce the double-helix structure of the DNA molecule in 1953.

Grating Spectroscopes

As noted earlier in this section, a grating can produce a wide separation in the various colors contained in a beam of light. This phenomenon is used as a means of

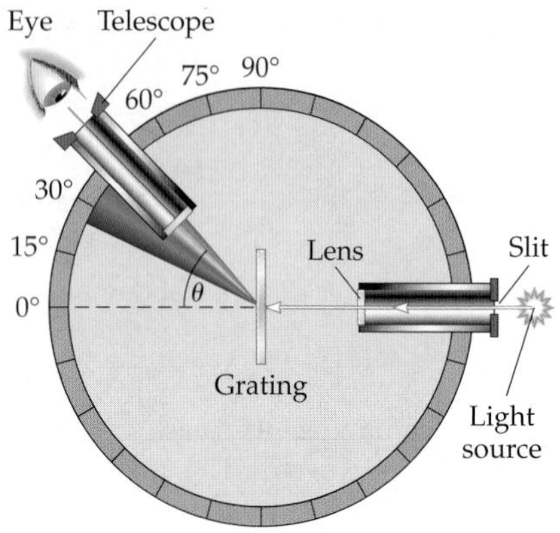

▲ **FIGURE 28–29 A grating spectroscope**
Because gratings can spread light out over a wider angle than prisms, they are used in most modern spectroscopes.

measuring the corresponding wavelengths with an instrument known as a *grating spectroscope*. As we see in **Figure 28–29**, light entering a grating spectroscope is diffracted as it passes through a grating. Next, the angle of diffraction of a given color is determined by a small telescope mounted on a rotating base. Finally, application of Equation 28–16 allows one to determine the wavelength of the light to great precision. Devices of this type have played a key role in elucidating the expanding nature of the universe, as we show in the following Active Example.

REAL-WORLD PHYSICS
Measuring the red shift of a quasar

ACTIVE EXAMPLE 28–4 Find the Wavelengths

A grating spectroscope with a line separation of 4.600×10^{-6} m is used to analyze light from the distant quasar (quasi-stellar object) designated 3C 273. With this instrument it is determined that light from the quasar exhibits a Doppler "red shift" (Section 25–2), indicating motion away from Earth. For example, some of the light given off by hydrogen atoms in the lab has a principal maximum at an angle of 6.067°. When this same hydrogen light is analyzed from the quasar, it is found to have a principal maximum at an angle of 7.030°. Find the wavelength in the lab, and the red-shifted wavelength from the quasar.

Solution *(Test your understanding by performing the calculations indicated in each step.)*

1. Use Equation 28–16 to solve for λ: $\lambda = d \sin\theta$
2. Substitute $\theta = 6.067°$: $\lambda = 486.2$ nm
3. Substitute $\theta = 7.030°$: $\lambda = 563.0$ nm

Insight

Thus we find that the wavelength from the quasar is 15.8% longer than the wavelength in the lab. If we combine this information with the Doppler effect for light (Section 25–2), we can determine the speed at which the quasar is receding from Earth.

Your Turn

Find the recession speed of the quasar.

(Answers to **Your Turn** *problems are given in the back of the book.)*

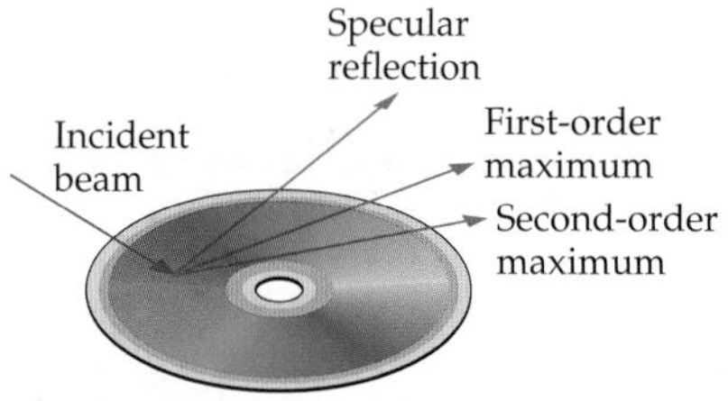

▲ **FIGURE 28–30 Reflection from a CD**
When a laser beam shines on a CD, a number of reflected beams are observed. The most intense of these is the specular beam, whose angle of reflection is equal to its angle of incidence—the same as if the CD were a plane mirror. Additional reflected beams are observed at angles corresponding to the principal maxima of the grating.

Reflection Gratings

Yet another way to produce a diffraction grating is to inscribe lines on a reflecting surface, with the regions between the lines acting as coherent sources of light. An everyday example of this type of *reflection grating* is a CD. As we saw earlier in this chapter, the information on a CD is encoded in the form of a series of bumps, and these bumps spiral around the CD, creating a tightly spaced set of lines. When a beam of monochromatic light is incident on a CD, as in **Figure 28–30**, a series of reflected beams is created. One beam reflects at an angle equal to the incident angle—this is referred to as the *specular beam*, since it is the beam that would be expected from a smooth plane mirror. In addition, a number of other reflected beams are observed, each corresponding to a different principal maximum of the grating.

If white light is reflected from a CD, different colors in the incident light are reflected at different angles. This is why light reflected from a CD shows the colors of the rainbow. A similar effect occurs in light reflected from feathers, or from the

▶ Diffraction can occur when light falls on any surface having grooves with a spacing comparable to the wavelength of the light. Many common surfaces can act as diffraction gratings, including artificially created ones, such as the CDs at left, and natural ones, such as the fly's eye at right. If the incident light is white, comprising a range of different wavelengths, diffraction gives rise to a rainbow effect.

wing of a butterfly like the blue morpho shown on page 937. For example, a microscopic examination of a butterfly wing shows that it consists of a multitude of plates, much like shingles on a roof. On each of these plates is a series of closely spaced ridges. These ridges act like the grooves on a CD, producing reflected light of different color in different directions. This type of coloration is referred to as **iridescence.** The next time you are able to examine an iridescent object, notice how the color changes as you change the angle from which it is viewed.

REAL-WORLD PHYSICS: BIO
Iridescence in nature

Chapter Summary

28–1 Superposition and Interference

The simple addition of two or more waves to give a resultant wave is referred to as superposition. When waves are superposed, the result may be a wave of greater amplitude (constructive interference) or of reduced amplitude (destructive interference).

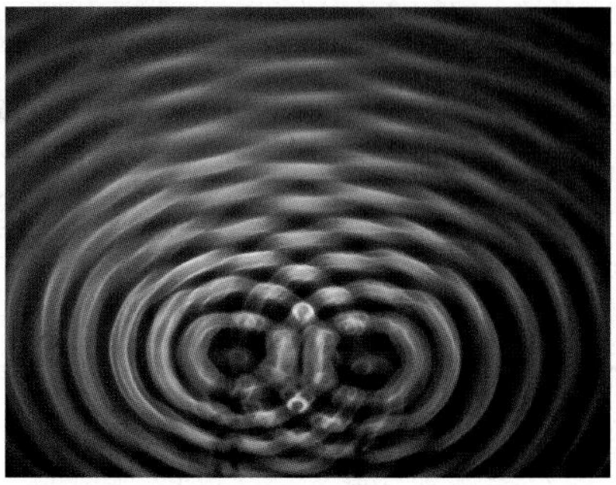

Monochromatic Light
Monochromatic light consists of waves with a single frequency and, hence, a single color.

Coherent/Incoherent Light
Light waves that maintain a constant phase relationship with one another are referred to as coherent. Light waves in which the relative phases vary randomly with time are said to be incoherent.

28–2 Young's Two-Slit Experiment

Interference effects in light are shown clearly in Young's two-slit experiment, in which light passing through two slits forms bright and dark interference "fringes."

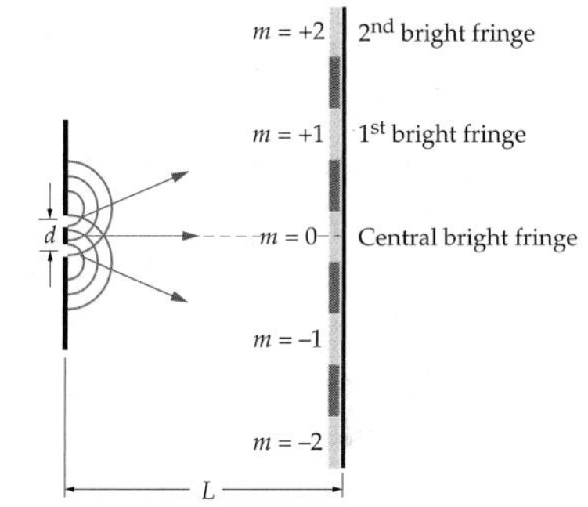

Conditions for Bright Fringes
Bright fringes in a two-slit experiment occur at angles θ given by the following relation:

$$d \sin\theta = m\lambda \qquad m = 0, \pm1, \pm2, \ldots \qquad \textbf{28–1}$$

In this expression, λ is the wavelength of the light, and d is the separation of the slits. The various values of the integer m correspond to different bright fringes.

Conditions for Dark Fringes
The locations of dark fringes in a two-slit experiment are given by the following:

$$d \sin\theta = \left(m - \tfrac{1}{2}\right)\lambda \quad m = 1, 2, 3, \ldots \quad \text{(above central bright fringe)}$$
$$d \sin\theta = \left(m + \tfrac{1}{2}\right)\lambda \quad m = -1, -2, -3, \ldots \quad \text{(below central bright fringe)} \qquad \textbf{28–2}$$

Linear Distance
If the screen on which the interference pattern is projected in a two-slit experiment is a distance L from the slits, the linear distance to a given bright or dark fringe is $y = L \tan\theta$.

28–3 Interference in Reflected Waves

Light waves reflected from different locations can interfere, just like light from the slits in a two-slit experiment.

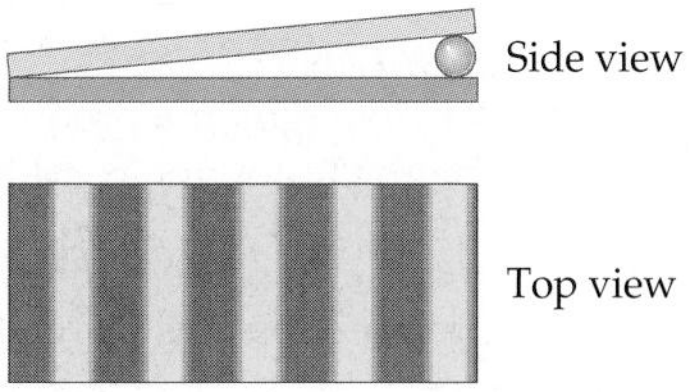

Phase Changes Due to Reflection
No phase change occurs when light is reflected from a region with a lower index of refraction, whereas a 180° (half-wavelength) phase change occurs when light reflects from a region with a higher index of refraction, or from a solid surface.

Air Wedge
Two plates of glass that touch on one end and have a small separation on the other end form an air wedge. When light of wavelength λ shines on an air wedge, bright fringes occur when the separation between the plates, d, is such that

$$\frac{1}{2} + \frac{2d}{\lambda} = m \qquad m = 1, 2, 3, \ldots \qquad \textbf{28–5}$$

Similarly, dark fringes occur when the following conditions are satisfied:

$$\frac{1}{2} + \frac{2d}{\lambda} = m + \frac{1}{2} \qquad m = 0, 1, 2, \ldots \qquad \textbf{28–6}$$

Newton's Rings
When a piece of glass with a spherical cross section is placed on a flat sheet of glass, the resulting interference fringes form a set of concentric circles known as Newton's rings.

Thin Films
Thin films, like those in a soap bubble, can produce colors in reflected light by eliminating other colors with destructive interference.

28–4 Diffraction

When a wave encounters an obstacle, or passes through an opening, it changes direction. This phenomenon is referred to as diffraction.

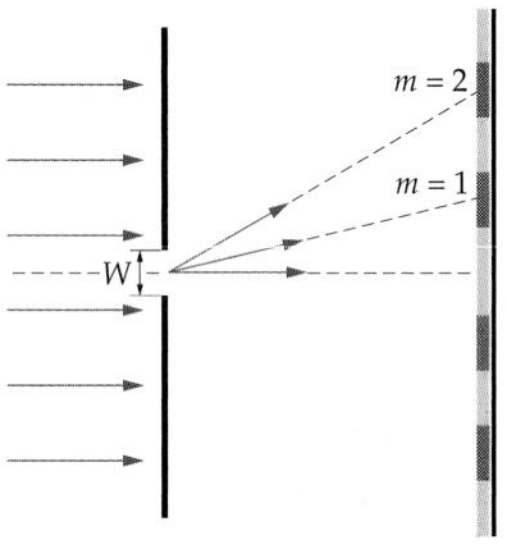

Single-Slit Diffraction
When monochromatic light of wavelength λ passes through a single slit of width W, it forms a diffraction pattern of alternating bright and dark fringes.

Condition for Dark Fringes
The condition that determines the location of dark fringes in single-slit diffraction is

$$W \sin\theta = m\lambda \qquad m = \pm 1, \pm 2, \pm 3, \ldots \qquad \textbf{28–12}$$

Bright Fringes
Bright fringes are located approximately halfway between successive dark fringes. In addition, the central bright fringe is approximately twice as wide as the other bright fringes.

28–5 Resolution

Resolution refers to the ability of a visual system, like the eye or a camera, to distinguish closely spaced objects.

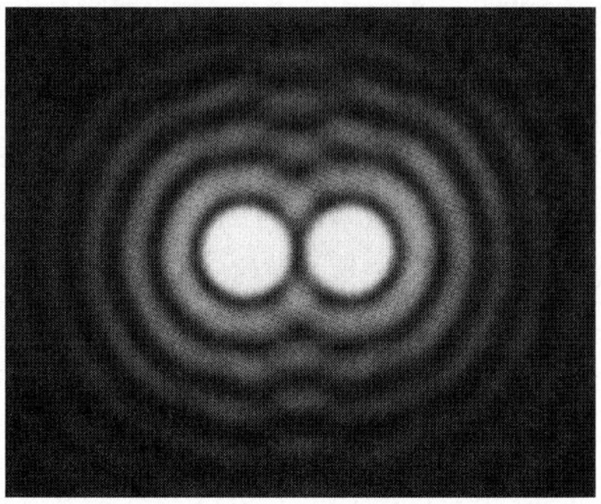

First Dark Fringe
A circular aperture of diameter D produces a circular diffraction pattern in which the first dark fringe occurs at the angle θ given by the following condition:

$$\sin\theta = 1.22\frac{\lambda}{D} \qquad \textbf{28–14}$$

Rayleigh's Criterion: Qualitative Statement
Rayleigh's criterion states that two objects become blurred together when the first dark fringe of one object's diffraction pattern passes through the center of the other object's diffraction pattern.

Rayleigh's Criterion: Quantitative Statement
In quantitative terms, Rayleigh's criterion states that if the angular separation between two objects is less than a certain minimum, $\theta_{min} = 1.22\lambda/D$, they will appear to be a single object.

28–6 Diffraction Gratings

A diffraction grating is a large number of slits through which a beam of light can pass.

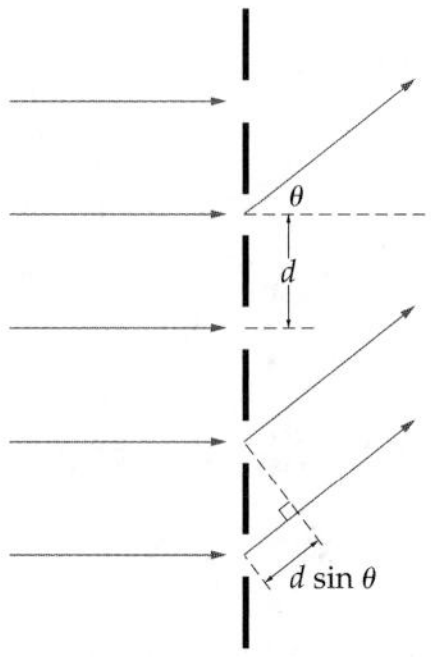

Principal Maxima
The principal maxima produced by a diffraction grating occur at the angles given by the following conditions:

$$d \sin\theta = m\lambda \qquad m = 0, \pm 1, \pm 2, \ldots \qquad \textbf{28–16}$$

In this expression, d is the distance between successive slits and λ is the wavelength of light.

Number of Lines per Centimeter
Diffraction gratings are often characterized by the number of lines, or slits, they have per centimeter. If the number of lines per centimeter is N, the spacing between slits is $d = 1/N$, where d is measured in centimeters.

Reflection Gratings
Diffraction gratings can also be constructed from a reflecting surface with a large number of reflecting lines, like a CD or a butterfly wing.

Iridescence
When white light shines on a reflecting grating, different colors in the light are reflected at different angles. The color effects produced in this way are referred to as iridescence.

Problem-Solving Summary

Type of Problem	Relevant Physical Concepts	Related Examples
Determine the angular and linear positions of fringes in a two-slit experiment.	The angular positions are determined by constructive and destructive interference, with the results given in Equations 28–1 and 28–2. The linear distance is given by simple geometry, as in Equation 28–3.	Example 28–2
Determine whether reflection results in constructive or destructive interference	When reflection is involved, the condition for constructive versus destructive interference involves both the difference in path length and the phase change that may result from reflection.	Examples 28–3, 28–4 Active Example 28–1
Find the angle corresponding to dark fringes in a diffraction pattern.	A single slit produces a dark fringe when a ray from the top of the slit follows a path that is an integer number of half wavelengths longer than the path followed by a ray starting at the center of the slit. The result is the condition given in Equation 28–12.	Example 28–5 Active Example 28–2
Determine whether two nearby objects can be resolved.	To be resolved, nearby objects must have an angular separation that is at least $1.22\lambda/D$, where D is the diameter of the aperture, and λ is the wavelength of the light.	Example 28–6 Active Example 28–3
Find the location of the principal maxima produced by a diffraction grating.	The principal maxima in a diffracting grating pattern are at the same angular positions as the bright fringes in a two-slit experiment, as given in Equation 28–1.	Example 28–7 Active Example 28–4

Conceptual Questions

(Answers to odd-numbered Conceptual Questions can be found in the back of the book.)

1. When two light waves interfere destructively, what happens to their energy?
2. What happens to the two-slit interference pattern if the separation between the slits is less than the wavelength of light?
3. If a radio station broadcasts its signal through two different antennas simultaneously, does this guarantee that the signal you receive will be stronger than from a single antenna? Explain.
4. How would you expect the interference pattern of a two-slit experiment to change if white light is used instead of monochromatic light?
5. Suppose a sheet of glass is placed in front of one of the slits in a two-slit experiment. If the thickness of the glass is such that the light reaching the two slits is 180° out of phase, how does this affect the interference pattern?
6. Describe the changes that would be observed in the two-slit interference pattern if the entire experiment were to be submerged in water.
7. Explain why the central spot in Newton's rings is dark.
8. Two identical sheets of glass are coated with films of different materials but equal thickness. The colors seen in reflected light from the two films are different. Give a reason that can account for this observation.
9. Spy cameras use lenses with very large apertures. Why are large apertures advantageous in such applications?
10. A cat's eye has a pupil that is elongated in the vertical direction. How does the resolution of a cat's eye differ in the horizontal and vertical directions?
11. Which portion of the soap film in the accompanying photograph is thinnest? Explain.

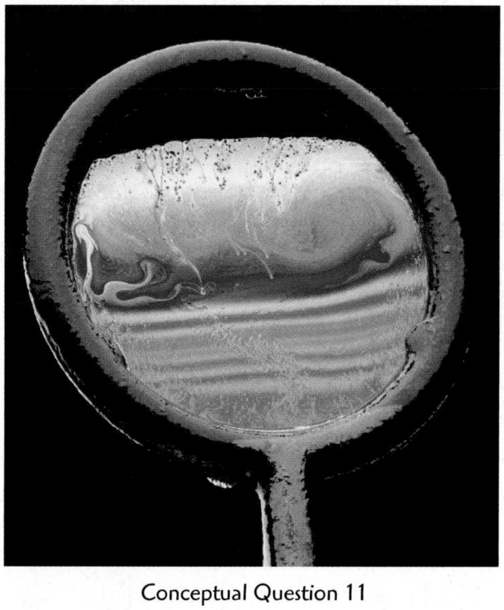

Conceptual Question 11

12. The color of an iridescent object, like a butterfly wing or a feather, appears to be different when viewed from different directions. The color of a painted surface appears the same from all viewing angles. Explain the difference.

Conceptual Exercises

(Answers to odd-numbered Conceptual Exercises can be found in the back of the book.)

1. Consider a two-slit interference pattern, with monochromatic light of wavelength λ. What is the path difference $\Delta\ell$ for **(a)** the fourth bright fringe and **(b)** the third dark fringe above the central bright fringe? Give your answer in terms of the wavelength of the light.
2. **(a)** Does the path length difference $\Delta\ell$ increase or decrease as you move from one bright fringe of a two-slit experiment to the next bright fringe farther out? **(b)** What is $\Delta\ell$ in terms of the wavelength λ of the light?
3. A two-slit experiment with red light produces a set of bright fringes. Will the spacing between the fringes increase, decrease, or stay the same if **(a)** the color of the light is changed to blue, or **(b)** the separation of the slits is decreased, or **(c)** the experiment is immersed in water?
4. **Figure 28–31** shows four different cases where light of wavelength λ reflects from both the top and the bottom of a thin film of thickness d. The indices of refraction of the film and the media above and below it are indicated in the figure. For which of the cases will the two reflected rays undergo constructive interference if **(a)** $d = \lambda/4$ or **(b)** $d = \lambda/2$?

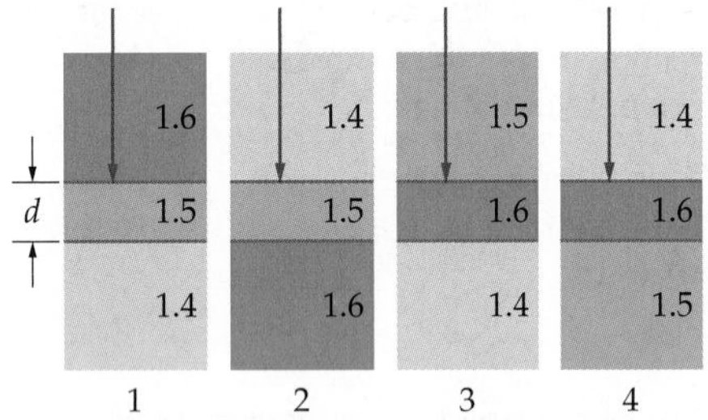

▲ **FIGURE 28–31** Conceptual Exercise 4

5. The oil film floating on water in the accompanying photo appears dark near the edges, where it is thinnest. Is the index of refraction of the oil greater than or less than that of the water? Explain.

Conceptual Exercise 5

6. If a thin film on water has an index of refraction less than that of water, will the film appear bright or dark in reflected light as its thickness goes to zero? Explain.
7. A single-slit diffraction pattern is formed on a distant screen. Assuming the angles involved are small, by what factor will the width of the central bright spot on the screen change if **(a)** the wavelength is doubled, **(b)** the slit width is doubled, or **(c)** the distance from the slit to the screen is doubled?
8. Monochromatic light with a wavelength λ passes through a single slit of width W and forms a diffraction pattern of alternating bright and dark fringes. **(a)** If the width of the slit is decreased, do the dark fringes move outward or inward? Explain. **(b)** What width is necessary for the first dark fringe to move outward to infinity? Give your answer in terms of λ.
9. In principle, do your eyes have greater resolution on a dark cloudy day or on a bright sunny day? Explain.
10. If the index of refraction of an eye could be magically reduced, would the eye's resolution increase or decrease? Explain.
11. In order to increase the resolution of a camera, should its f-number be increased or decreased? Explain.
12. Is resolution greater with blue light or red light, all other factors being equal? Explain.

Problems

Note: **IP** *denotes an integrated conceptual/quantitative problem.* **BIO** *identifies problems of biological or medical interest. Red bullets (•, ••, •••) are used to indicate the level of difficulty of each problem.*

(In all problems involving sound waves, take the speed of sound to be 343 m/s.)

Section 28–1 Superposition and Interference

1. • Two sources emit waves that are coherent, in phase, and have wavelengths of 26.0 m. Do the waves interfere constructively or destructively at an observation point 78.0 m from one source and 143 m from the other source?
2. • Repeat Problem 1 for observation points that are **(a)** 91.0 m and 221 m and **(b)** 44.0 m and 135 m from the two sources.
3. •• Two sources emit waves that are in phase with each other. What is the largest wavelength that will give constructive interference at an observation point 161 m from one source and 295 m from the other source?
4. •• A person driving at 17 m/s crosses the line connecting two radio transmitters at right angles, as shown in **Figure 28–32**. The transmitters emit identical signals in phase with each other, which the driver receives on the car radio. When the car is at point A, the radio picks up a maximum net signal. **(a)** What is the longest possible wavelength of the radio waves? **(b)** How long after the car passes point A does the radio experience a minimum in the net signal? Assume that the wavelength has the value found in part (a).

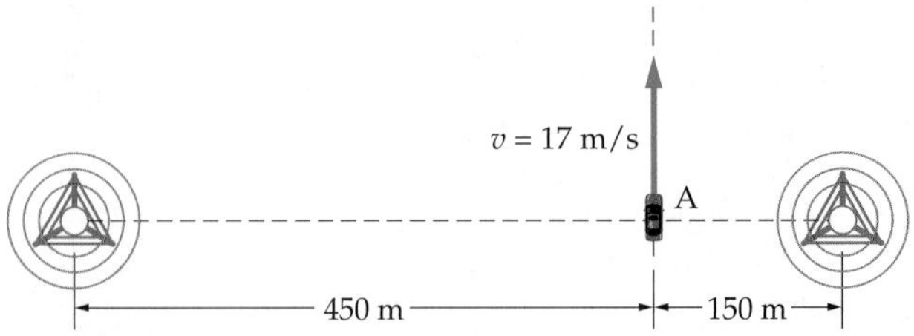

▲ **FIGURE 28–32** Problems 4 and 11

5. •• Two students in a dorm room listen to a pure tone produced by two loudspeakers that are in phase. Students A and B in **Figure 28–33** hear a maximum sound. What is the lowest possible frequency of the loudspeakers?

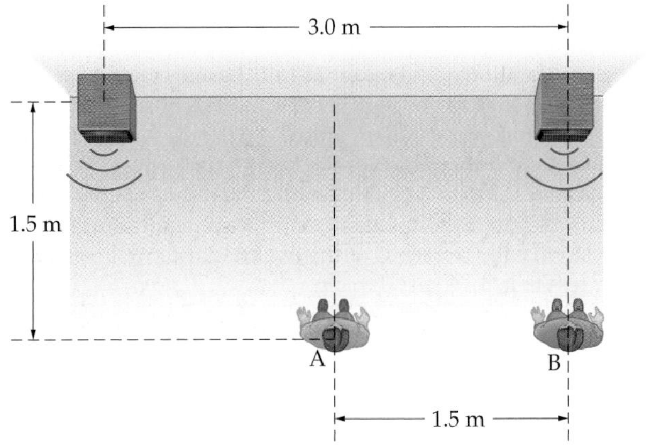

▲ **FIGURE 28–33** Problems 5 and 6

6. •• If the loudspeakers in Problem 5 are 180° out of phase, determine whether a 185-Hz tone heard at location B is a maximum or a minimum.

7. •• A microphone is located on the line connecting two speakers that are 0.750 m apart and oscillating in phase. The microphone is 2.25 m from the midpoint of the two speakers. What are the lowest two frequencies that produce an interference maximum at the microphone's location?

8. •• A microphone is located on the line connecting two speakers that are 0.750 m apart and oscillating 180° out of phase. The microphone is 2.00 m from the midpoint of the two speakers. What are the lowest two frequencies that produce an interference maximum at the microphone's location?

9. •• Moe, Larry, and Curly stand in a line with a spacing of 1.00 m. Larry is 3.00 m in front of a pair of stereo speakers 0.800 m apart, as shown in **Figure 28–34**. The speakers produce a single-frequency tone, vibrating in phase with each other. What are the two lowest frequencies that allow Larry to hear a loud tone while Moe and Curly hear very little?

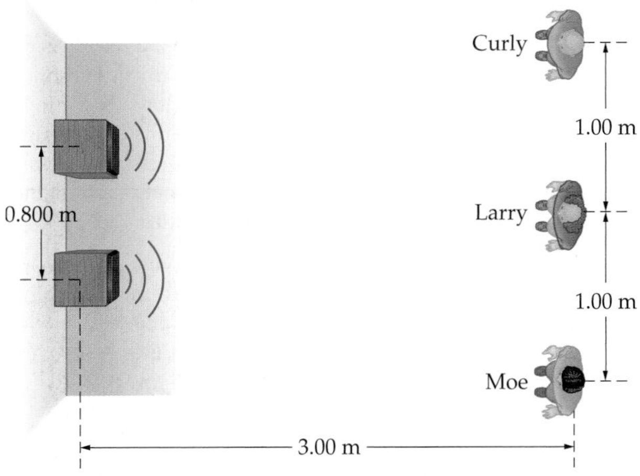

▲ **FIGURE 28–34** Problems 9 and 10

10. •• **IP** In Figure 28–34 the two speakers emit sound that is 180° out of phase and of a single frequency, f. **(a)** Does Larry hear a sound intensity that is a maximum or a minimum? Does your answer depend on the frequency of the sound? Explain. **(b)** Find the lowest two frequencies that produce a maximum sound intensity at the positions of Moe and Curly.

11. •• **IP** Suppose the car radio in Problem 4 picks up a minimum net signal at point A. **(a)** What is the largest possible value for the wavelength of the radio waves? **(b)** If the radio transmitters use a wavelength that is half the value found in part (a), will the car radio pick up a net signal at point A that is a maximum or a minimum? Explain. **(c)** What is the second largest wavelength that will result in a minimum signal at point A?

Section 28–2 Young's Two-Slit Experiment

12. • Laser light with a wavelength $\lambda = 670$ nm illuminates a pair of slits at normal incidence. What slit separation will produce first-order maxima at angles of $\pm 35°$ from the incident direction?

13. • Monochromatic light passes through two slits separated by a distance of 0.0334 mm. If the angle to the third maximum above the central fringe is 3.21°, what is the wavelength of the light?

14. • In Young's two-slit experiment, the first dark fringe above the central bright fringe occurs at an angle of 0.29°. What is the ratio of the slit separation, d, to the wavelength of the light, λ?

15. •• **IP** A two-slit experiment with slits separated by 48.0×10^{-5} m produces a second-order maximum at an angle of 0.0990°. **(a)** Find the wavelength of the light used in this experiment. **(b)** If the slit separation is increased but the second-order maximum stays at the same angle, does the wavelength increase, decrease, or stay the same? Explain. **(c)** Calculate the wavelength for a slit separation of 68.0×10^{-5} m.

16. •• A two-slit pattern is viewed on a screen 1.00 m from the slits. If the two third-order minima are 25.0 cm apart, what is the width (in cm) of the central bright fringe?

17. •• Light from a He–Ne laser ($\lambda = 632.8$ nm) strikes a pair of slits at normal incidence, forming a double-slit interference pattern on a screen located 1.40 m from the slits. **Figure 28–35** shows the interference pattern observed on the screen. What is the slit separation?

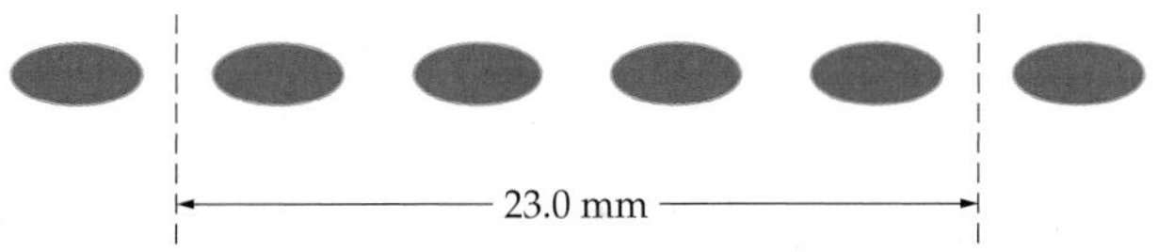

▲ **FIGURE 28–35** Problems 17 and 20

18. •• Light with a wavelength of 546 nm passes through two slits and forms an interference pattern on a screen 8.75 m away. If the linear distance on the screen from the central fringe to the first bright fringe above it is 5.16 cm, what is the separation of the slits?

19. •• A set of parallel slits for optical interference can be made by holding two razor blades together (carefully!) and scratching a pair of lines on a glass microscope slide that has been painted black. When monochromatic light strikes these slits at normal incidence, an interference pattern is formed on a distant screen. The thickness of each razor blade used to make the slits is 0.230 mm, and the screen is 2.50 m from the slits. If the center-to-center separation of the fringes is 7.15 mm, what is the wavelength of the light?

20. •• **IP** Suppose the interference pattern shown in Figure 28–35 is produced by monochromatic light passing through two slits, with a separation of 135 μm, and onto a screen 1.20 m away. **(a)** What is the wavelength of the light? **(b)** If the frequency of this light is increased, will the bright spots of the pattern move closer together or farther apart? Explain.

21. •• A physics instructor wants to produce a double-slit interference pattern large enough for her class to see. For the size of the room, she decides that the distance between successive bright fringes on the screen should be at least 2.50 cm. If the slits have a separation $d = 0.0220$ mm, what is the minimum distance from the slits to the screen when 632.8-nm light from a He–Ne laser is used?

22. •• **IP** When green light ($\lambda = 505$ nm) passes through a pair of double slits, the interference pattern shown in **Figure 28–36 (a)** is observed. When light of a different color passes through the same pair of slits, the pattern shown in **Figure 28–36 (b)** is observed. **(a)** Is the wavelength of the second color greater than or less than 505 nm? Explain. **(b)** Find the wavelength of the second color. (Assume that the angles involved are small enough to set $\sin\theta \approx \tan\theta$.)

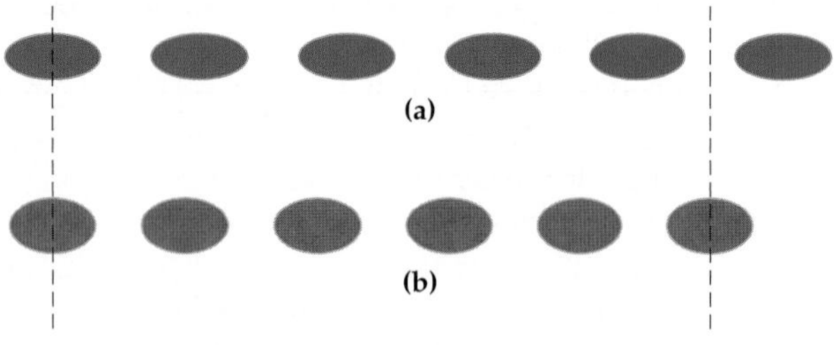

▲ **FIGURE 28–36** Problems 22 and 23

23. •• **IP** The interference pattern shown in Figure 28–36 (a) is produced by green light with a wavelength of $\lambda = 505$ nm passing through two slits with a separation of 127 μm. After passing through the slits, the light forms a pattern of bright and dark spots on a screen located 1.25 m from the slits. **(a)** What is the distance between the two vertical, dashed lines in Figure 28–36 (a)? **(b)** If it is desired to produce a more tightly packed interference pattern, like the one shown in Figure 28–36 (b), should the frequency of the light be increased or decreased? Explain.

Section 28–3 Interference in Reflected Waves

24. • A soap bubble with walls 401 nm thick floats in air. If this bubble is illuminated perpendicularly with sunlight, what wavelength (and color) will be absent in the reflected light? Assume that the index of refraction of the soap film is 1.33. (Refer to Example 25–3 for the connection between wavelength and color.)

25. • A soap film ($n = 1.33$) is 815 nm thick. White light strikes the film at normal incidence. What visible wavelengths will be constructively reflected if the film is surrounded by air on both sides? (Refer to Example 25–3 for the range of visible wavelengths.)

26. • White light is incident on a soap film ($n = 1.30$) in air. The reflected light looks bluish because the red light ($\lambda = 670$ nm) is absent in the reflection. What is the minimum thickness of the soap film?

27. • A 772-nm-thick soap film ($n_{film} = 1.33$) rests on a glass plate ($n_{glass} = 1.52$). White light strikes the film at normal incidence. What visible wavelengths will be constructively reflected from the film? (Refer to Example 25–3 for the range of visible wavelengths.)

28. • An oil film ($n = 1.38$) floats on a water puddle. You notice that green light ($\lambda = 521$ nm) is absent in the reflection. What is the minimum thickness of the oil film?

29. •• A radio broadcast antenna is 36.00 km from your house. Suppose an airplane is flying 2.230 km above the line connecting the broadcast antenna and your radio, and that waves reflected from the airplane travel 88.00 wavelengths farther than waves that travel directly from the antenna to your house. **(a)** Do you observe constructive or destructive interference between the direct and reflected waves? (*Hint:* Does a phase change occur when the waves are reflected?) **(b)** The situation just described occurs when the plane is above a point on the ground that is two-thirds of the way from the antenna to your house. What is the wavelength of the radio waves?

30. •• **IP Newton's Rings** Monochromatic light with $\lambda = 648$ nm shines down on a planoconvex lens lying on a piece of plate glass, as shown in **Figure 28–37**. When viewed from above, one sees a set of concentric dark and bright fringes, referred to as Newton's rings (See Figure 28–11 for a photo of Newton's rings.). **(a)** If the radius of the twelfth dark ring from the center is measured to be 1.56 cm, what is the radius of curvature, R, of the lens? **(b)** If light with a greater wavelength is used with this system, will the radius of the twelfth dark ring be greater than or less than 1.56 cm? Explain.

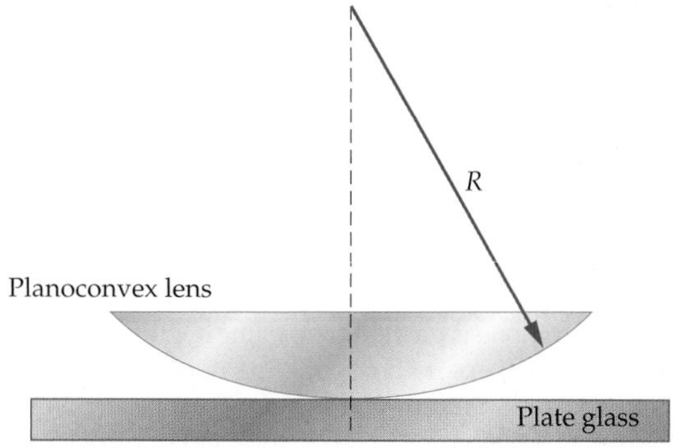

▲ **FIGURE 28–37** Problems 30 and 78

31. •• Light is incident from above on two plates of glass, separated on both ends by small wires of diameter $d = 0.600$ μm. Considering only interference between light reflected from the bottom surface of the upper plate and light reflected from the upper surface of the lower plate, state whether the following wavelengths give constructive or destructive interference: **(a)** $\lambda = 600.0$ nm; **(b)** $\lambda = 800.0$ nm; **(c)** $\lambda = 343.0$ nm.

32. •• **(a)** What is the minimum soap-film thickness ($n = 1.33$) in air that will produce constructive interference in reflection for red ($\lambda = 652$ nm) light? **(b)** Which visible wavelengths will destructively interfere when reflected from this film? (Refer to Example 25–3 for the range of visible wavelengths.)

33. •• **IP** A thin layer of magnesium fluoride ($n = 1.38$) is used to coat a flint-glass lens ($n = 1.61$). **(a)** What thickness should the magnesium fluoride film have if the reflection of 565-nm light is to be suppressed? Assume that the light is incident at right angles to the film. **(b)** If it is desired to suppress the reflection of light with a higher frequency, should the coating of magnesium fluoride be made thinner or thicker? Explain.

34. ••• White light is incident normally on a thin soap film ($n = 1.33$) suspended in air. **(a)** What are the two minimum thicknesses that will constructively reflect yellow ($\lambda = 590$ nm) light? **(b)** What are the two minimum thicknesses that will *destructively* reflect yellow ($\lambda = 590$ nm) light?

35. ••• A thin coating ($t = 340.0$ nm, $n = 1.480$) is placed on a glass lens. Which visible ($400 \text{ nm} < \lambda < 700 \text{ nm}$) wavelength(s) will be absent in the reflected beam if **(a)** the glass has an index of refraction $n = 1.350$, and **(b)** the glass has an index of refraction $n = 1.675$?

36. ••• Two glass plates are separated by fine wires with diameters $d_1 = 0.0500$ mm and $d_2 = 0.0520$ mm, as indicated in **Figure 28–38**. The wires are parallel and separated by a distance of 7.00 cm. If monochromatic light with $\lambda = 589$ nm is incident from above, what is the distance (in cm) between adjacent dark bands in the reflected light? (Consider interference only between light reflected from the bottom surface of the upper plate and light reflected from the upper surface of the lower plate.)

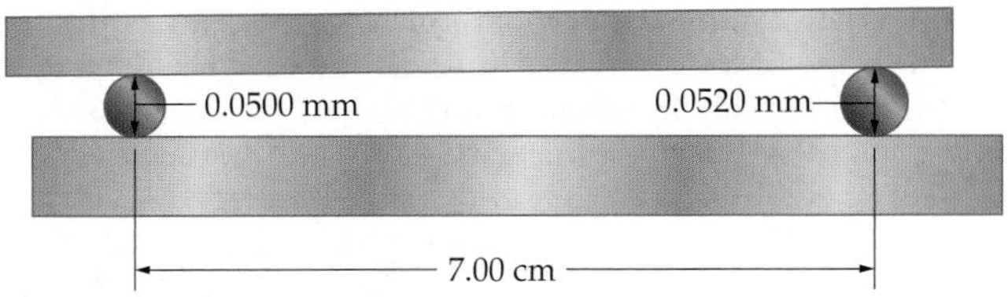

▲ **FIGURE 28–38** Problem 36

Section 28–4 Diffraction

37. • What width single slit will produce first-order diffraction minima at angles of $\pm 23°$ from the central maximum with 690-nm light?

38. • Diffraction also occurs with sound waves. Consider 1300-Hz sound waves diffracted by a door that is 84 cm wide. What is the angle between the two first-order diffraction minima?

39. • Green light ($\lambda = 546$ nm) strikes a single slit at normal incidence. What width slit will produce a central maximum that is 2.50 cm wide on a screen 1.60 m from the slit?

40. • Light with a wavelength of 626 nm passes through a slit 7.64 μm wide and falls on a screen 1.85 m away. Find the linear distance on the screen from the central bright fringe to the first bright fringe above it.

41. • Repeat Problem 40, only this time find the distance on the screen from the central bright fringe to the third dark fringe above it.

42. •• **IP** A single slit is illuminated with 610-nm light, and the resulting diffraction pattern is viewed on a screen 2.3 m away. **(a)** If the linear distance between the first and second dark fringes of the pattern is 12 cm, what is the width of the slit? **(b)** If the slit is made wider, will the distance between the first and second dark fringes increase or decrease? Explain.

43. •• How many dark fringes will be produced on either side of the central maximum if green light ($\lambda = 553$ nm) is incident on a slit that is 8.00 μm wide?

44. •• **IP** The diffraction pattern shown in **Figure 28–39** is produced by passing He–Ne laser light ($\lambda = 632.8$ nm) through a single slit and viewing the pattern on a screen 1.50 m behind the slit. **(a)** What is the width of the slit? **(b)** If monochromatic yellow light with a wavelength of 591 nm is used with this slit instead, will the distance indicated in Figure 28–39 be greater than or less than 15.2 cm? Explain.

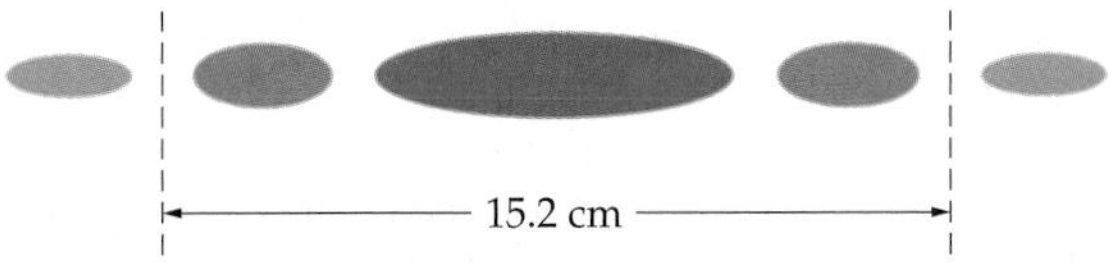

▲ **FIGURE 28–39** Problems 44 and 79

45. •• A screen is placed 1.00 m behind a single slit. The central maximum in the resulting diffraction pattern on the screen is 1.60 cm wide—that is, the two first-order diffraction minima are separated by 1.60 cm. What is the distance between the two second-order minima?

Section 28–5 Resolution

46. • Two point sources of light are separated by 5.0 cm. As viewed through a 12-μm diameter pinhole, what is the maximum distance from which they can be resolved **(a)** if red light ($\lambda = 690$ nm) is used, or **(b)** if violet light ($\lambda = 420$ nm) is used?

47. • A spy camera is said to be able to read the numbers on a car's license plate. If the numbers on the plate are 5.0 cm apart, and the spy satellite is at an altitude of 160 km, what must be the diameter of the camera's aperture? (Assume light with a wavelength of 550 nm.)

48. • **Splitting Binary Stars** As seen from Earth, the red dwarfs Krüger 60A and Krüger 60B form a binary star system with an angular separation of 2.5 arc seconds. What is the smallest diameter telescope that could theoretically resolve these stars using 550-nm light? (*Note:* 1 arc sec $= 1/3600°$)

49. • Find the minimum aperture diameter of a camera that can resolve detail on the ground the size of a person (2.0 m) from an SR-71 Blackbird airplane flying at an altitude of 27 km. (Assume light with a wavelength of 450 nm.)

50. • **The Resolution of Hubble** The Hubble Space Telescope (HST) orbits Earth at an altitude of 613 km. It has an objective mirror that is 2.4 m in diameter. If the HST were to look down on Earth's surface (rather than up at the stars), what is the minimum separation of two objects that could be resolved using 550-nm light? [*Note:* The HST is used only for astronomical work, but a (classified) number of similar telescopes are in orbit for spy purposes.]

51. •• A lens that is "optically perfect" is still limited by diffraction effects. Suppose a lens has a diameter of 130 mm and a focal length of 640 mm. **(a)** Find the angular width (that is, the angle from the bottom to the top) of the central maximum in the diffraction pattern formed by this lens when illuminated with 540-nm light. **(b)** What is the linear width (diameter) of the central maximum at the focal distance of the lens?

52. •• The resolution of a telescope is ultimately limited by the diameter of its objective lens or mirror. A typical amateur astronomer's telescope may have a 6.0-in.-diameter mirror. **(a)** What is the minimum angular separation (in arc seconds) of two stars that can be resolved with a 6.0-in. scope? (Take λ to be at the center of the visible spectrum, about 550 nm, and see Problem 48 for the definition of an arc second.) **(b)** What is the minimum distance (in km) between two points on the Moon's surface that can be resolved by a 6.0-in. scope? (*Note:* The average distance from Earth to the Moon is 384,400 km.)

53. •• Early cameras were little more than a box with a pinhole on the side opposite the film. **(a)** What angular resolution would you expect from a pinhole with a 0.50-mm diameter? **(b)** What is the greatest distance from the camera at which two point objects 15 cm apart can be resolved? (Assume light with a wavelength of 520 nm.)

Section 28–6 Diffraction Gratings

54. • A grating has 787 lines per centimeter. Find the angles of the first three principal maxima above the central fringe when this grating is illuminated with 655-nm light.

55. • Suppose you want to produce a diffraction pattern with X-rays whose wavelength is 0.030 nm. If you use a diffraction grating, what separation between lines is needed to generate a pattern with the first maximum at an angle of 14°? (For comparison, a typical atom is a few tenths of a nanometer in diameter.)

56. • A diffraction grating has 2200 lines/cm. What is the angle between the first-order maxima for red light ($\lambda = 680$ nm) and blue light ($\lambda = 410$ nm)?

57. • A diffraction grating with 355 lines/mm is 1.00 m in front of a screen. What is the wavelength of light whose first-order maxima will be 16.4 cm from the central maximum on the screen?

58. • The yellow light from a helium discharge tube has a wavelength of 587.5 nm. When this light illuminates a certain diffraction grating it produces a first-order principal maximum at

an angle of 1.250°. Calculate the number of lines per centimeter on the grating.

59. •• IP The second-order maximum produced by a diffraction grating with 560 lines per centimeter is at an angle of 3.1°. **(a)** What is the wavelength of the light that illuminates the grating? **(b)** If a grating with a larger number of lines per centimeter is used with this light, is the angle of the second-order maximum greater than or less than 3.1°? Explain.

60. •• White light strikes a grating with 7400 lines/cm at normal incidence. How many complete visible spectra will be formed on either side of the central maximum? (Refer to Example 25–3 for the range of visible wavelengths.)

61. •• White light strikes a diffraction grating (890 lines/mm) at normal incidence. What is the highest-order visible maximum that is formed? (Refer to Example 25–3 for the range of visible wavelengths.)

62. •• White light strikes a diffraction grating (760 lines/mm) at normal incidence. What is the longest wavelength that forms a second-order maximum?

63. •• A light source emits two distinct wavelengths [$\lambda_1 = 430$ nm (violet); $\lambda_2 = 630$ nm (orange)]. The light strikes a diffraction grating with 450 lines/mm at normal incidence. Identify the colors of the first eight interference maxima on either side of the central maximum.

64. •• A laser emits two wavelengths ($\lambda_1 = 420$ nm; $\lambda_2 = 630$ nm). When these two wavelengths strike a grating with 450 lines/mm, they produce maxima (in different orders) that coincide. **(a)** What is the order (m) of each of the two overlapping lines? **(b)** At what angle does this overlap occur?

65. •• IP When blue light with a wavelength of 465 nm illuminates a diffraction grating, it produces a first-order principal maximum but no second-order maximum. **(a)** Explain the absence of higher-order principal maxima. **(b)** What is the maximum spacing between lines on this grating?

66. •• Monochromatic light strikes a diffraction grating at normal incidence before illuminating a screen 2.00 m away. If the first-order maxima are separated by 1.53 m on the screen, what is the distance between the two second-order maxima?

67. ••• A diffraction grating with a slit separation d is illuminated by a beam of monochromatic light of wavelength λ. The diffracted beam is observed at an angle ϕ relative to the incident direction. If the plane of the grating bisects the angle between the incident and diffracted beams, show that the mth maximum will be observed at an angle that satisfies the relation $m\lambda = 2d\sin(\phi/2)$, with $m = 0, \pm1, \pm2, \ldots$.

General Problems

68. • Diffraction effects often involve small angles, and we usually make the approximation $\sin\theta \cong \tan\theta$. To see how accurate this approximation is, complete the following table.

θ (deg)	θ (rad)	$\sin\theta$	$\tan\theta$	$\sin\theta/\tan\theta$
0.0100°				
1.00°				
5.00°				
10.0°				
20.0°				
30.0°				
40.0°				

69. •• When reading the printout from a laser printer, you are actually looking at an array of tiny dots. If the pupil of your eye is 4.2 mm in diameter when reading a page held 25 cm from your eye, what is the minimum separation of adjacent dots that can be resolved? (Assume light with a wavelength of 540 nm.)

70. •• The headlights of a pickup truck are 1.32 m apart. What is the greatest distance at which these headlights can be resolved as separate points of light on a photograph taken with a camera whose aperture has a diameter of 12.5 mm? (Take $\lambda = 555$ nm.)

71. •• **Antireflection Coating** A glass lens ($n_{glass} = 1.52$) has an antireflection coating of MgF_2 ($n = 1.38$). **(a)** For 517-nm light, what minimum thickness of MgF_2 will cause the reflected rays R_2 and R_4 in **Figure 28–40** to interfere destructively, assuming normal incidence? **(b)** Interference will also occur between the forward-moving rays, R_1 and R_3 in Figure 28–40. What minimum thickness of MgF_2 will cause these two rays to interfere constructively?

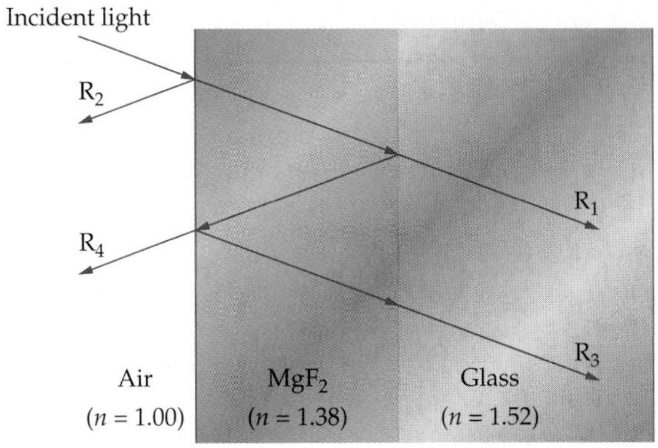

▲ **FIGURE 28–40** Problem 71

72. •• IP White light reflected at normal incidence from a soap bubble ($n = 1.33$) in air produces an interference maximum at $\lambda = 575$ nm but no interference minima in the visible spectrum. **(a)** Explain the absence of interference minima in the visible. **(b)** What are the possible thicknesses of the soap film? (Refer to Example 25–3 for the range of visible wavelengths.)

73. •• A thin film of oil ($n = 1.30$) floats on water ($n = 1.33$). When sunlight is incident at right angles to this film, the only colors that are enhanced by reflection are blue (458 nm) and red (687 nm). Estimate the thickness of the oil film.

74. •• **Resolving Lines on HDTV** The American Television Systems Committee (ATSC) sets the standards for high-definition television (HDTV). One of the approved HDTV formats is 720p, which means 720 horizontal lines scanned progressively—one line after another in sequence. Suppose the 720 lines are displayed on a television with a screen that is 24 inches high, and that the light coming from the screen has a wavelength of 570 nm. If the pupils of your eyes have a diameter of 5.5 mm, what is the closest you can be to the TV before seeing the individual horizontal lines? This is the minimum acceptable viewing distance.

75. •• The yellow light of sodium, with wavelengths of 588.99 nm and 589.59 nm, is normally incident on a grating with 474 lines/cm. Find the linear distance between the first-order maxima for these two wavelengths on a screen 2.85 m from the grating.

76. •• IP A thin soap film ($n = 1.33$) suspended in air has a uniform thickness. When white light strikes the film at normal incidence, violet light ($\lambda_V = 420$ nm) is constructively reflected. **(a)** If we would like green light ($\lambda_G = 560$ nm) to be constructively reflected, instead, should the film's thickness be increased or decreased? **(b)** Find the new thickness of the film.

(Assume the film has the minimum thickness that can produce these reflections.)

77. •• IP A thin film of oil ($n = 1.40$) floats on water ($n = 1.33$). When sunlight is incident at right angles to this film, the only colors that are absent from the reflected light are blue (458 nm) and red (687 nm). Estimate the thickness of the oil film.

78. •• IP Sodium light, with a wavelength of $\lambda = 589$ nm, shines downward onto the system shown in Figure 28–37. When viewed from above, you see a series of concentric circles known as Newton's rings. **(a)** Do you expect a bright or a dark spot at the center of the pattern? Explain. **(b)** If the radius of curvature of the planoconvex lens is $R = 26.1$ m, what is the radius of the tenth largest dark ring? (Only rings of nonzero radius will be counted as "rings.")

79. •• IP Figure 28–39 shows a single-slit diffraction pattern formed by light passing through a slit of width $W = 11.2\ \mu\text{m}$ and illuminating a screen 0.895 m behind the slit. **(a)** What is the wavelength of the light? **(b)** If the width of the slit is decreased, will the distance indicated in Figure 28–39 be greater than or less than 15.2 cm? Explain.

80. •• BIO **Entoptic Halos** Images produced by structures within the eye (like lens fibers or cell fragments) are referred to as entoptic images. These images can sometimes take the form of "halos" around a bright light seen against a dark background. The halo in such a case is actually the bright outer rings of a circular diffraction pattern, like Figure 28–21, with the central bright spot not visible because it overlaps the direct image of the light. Find the diameter of the eye structure that causes a circular diffraction pattern with the first dark ring at an angle of 3.7° when viewed with monochromatic light of wavelength 630 nm. (Typical eye structures of this type have diameters on the order of 10 μm. Also, the index of refraction of the vitreous humor is 1.336.)

81. ••• White light is incident on a soap film ($n = 1.33$, thickness $= 800.0$ nm) suspended in air. If the incident light makes a 45° angle with the normal to the film, what visible wavelength(s) will be constructively reflected? (Refer to Example 25–3 for the range of visible wavelengths.)

82. ••• IP A system like that shown in Figure 28–26 consists of N slits, each transmitting light of intensity I_0. The light from each slit has the same phase and the same wavelength. The net intensity I observed at an angle θ due to all N slits is

$$I = I_0\left[\frac{\sin(N\phi/2)}{\sin(\phi/2)}\right]^2$$

In this expression, $\phi = (2\pi d/\lambda)\sin\theta$, where λ is the wavelength of the light. **(a)** Show that the intensity in the limit $\theta \rightarrow 0$ is $I = N^2 I_0$. This is the maximum intensity of the interference pattern. **(b)** Show that the first points of zero intensity on either side of $\theta = 0$ occur at $\phi = 2\pi/N$ and $\phi = -2\pi/N$. **(c)** Does the central maximum ($\theta = 0$) of this pattern become narrower or broader as the number of slits is increased? Explain.

83. ••• Two plates of glass are separated on both ends by small wires of diameter d. Derive an expression for the condition for constructive interference when light of wavelength λ is incident normally on the plates. Consider only interference between waves reflected from the bottom of the top plate and the top of the bottom plate.

84. ••• A curved piece of glass with a radius of curvature R rests on a flat plate of glass. Light of wavelength λ is incident normally on this system. Considering only interference between waves reflected from the curved (lower) surface of glass and the top surface of the plate, show that the radius of the nth dark ring is

$$r_n = \sqrt{n\lambda R - n^2\lambda^2/4}$$

85. ••• BIO **The Resolution of the Eye** The resolution of the eye is ultimately limited by the pupil diameter. What is the smallest diameter spot the eye can produce on the retina if the pupil diameter is 4.25 mm? Assume light with a wavelength of $\lambda = 550$ nm. (*Note*: The distance from the pupil to the retina is 25.4 mm. In addition, the space between the pupil and the retina is filled with a fluid whose index of refraction is $n = 1.336$.)

Interactive Problems

86. •• IP ***Referring to Example 28–2*** Suppose we change the slit separation to a value other than 8.5×10^{-5} m, with the result that the linear distance to the tenth bright fringe above the central bright fringe increases from 12 cm to 18 cm. The screen is still 2.3 m from the slits, and the wavelength of the light is 440 nm. **(a)** Did we increase or decrease the slit separation? Explain. **(b)** Find the new slit separation.

87. •• IP ***Referring to Example 28–2*** The wavelength of the light is changed to a value other than 440 nm, with the result that the linear distance to the *seventh* bright fringe above the central bright fringe is 12 cm. The screen is still 2.3 m from the slits, and the slit separation is 8.5×10^{-5} m. **(a)** Is the new wavelength greater than or less than 440 nm? Explain. **(b)** Find the new wavelength.

88. •• IP ***Referring to Example 28–5*** The light used in this experiment has a wavelength of 511 nm. **(a)** If the width of the slit is decreased, will the angle to the first dark fringe above the central bright fringe increase or decrease? Explain. **(b)** Find the angle to the first dark fringe if the reduced slit width is 1.50×10^{-6} m.

89. •• IP ***Referring to Example 28–5*** The width of the slit in this experiment is 2.20×10^{-6} m. **(a)** If the frequency of the light is decreased, will the angle to the first dark fringe above the central bright fringe increase or decrease? Explain. **(b)** Find the angle to the first dark fringe if the reduced frequency is 5.22×10^{14} Hz.

29 Relativity

29–1	The Postulates of Special Relativity	971
29–2	The Relativity of Time and Time Dilation	972
29–3	The Relativity of Length and Length Contraction	978
29–4	The Relativistic Addition of Velocities	981
29–5	Relativistic Momentum	983
29–6	Relativistic Energy and $E = mc^2$	986
29–7	The Relativistic Universe	991
29–8	General Relativity	991

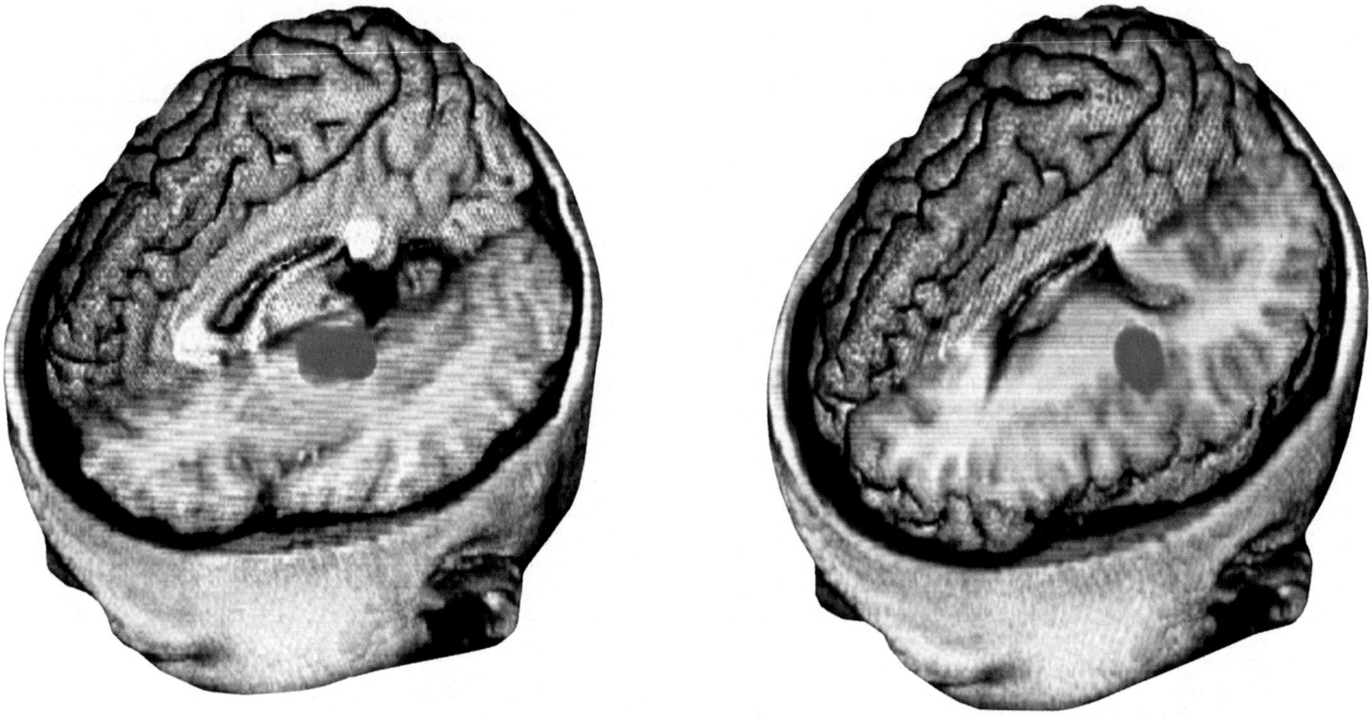

If someone reads this caption to you, a specific region of your brain is activated as you process the information. On the other hand, you may try to recall the words that were read to you after a period of time has elapsed, in which case a different region of your brain is activated as you retrieve the words from your memory. How do we know? The images shown here are the result of positron emission tomography (PET) scans of the brain, in which a radioactive tracer is injected into the bloodstream. In the image on the left, a person is listening to words being read, and the hippocampus region is activated, causing a higher concentration of radioactive blood flow in that area; in the image on the right, the person tries to recall the words, and the temporoparietal region is activated. What is really amazing about these images, however, is that they show areas where matter (electrons) and antimatter (positrons emitted by the radioactive tracer) are annihilating inside the brain and sending out bursts of energy. The energy released in these annihilations is given by the familiar equation ($E = mc^2$), developed by a 26-year-old patent clerk who would soon become the most famous scientist of our time. This chapter explores some of the major contributions of Albert Einstein, including his special and general theories of relativity, and shows how they have radically altered our views of space, time, matter, and energy.

What we refer to today as "modern physics" can be thought of as having started around the beginning of the twentieth century, when two fundamentally new ways of looking at nature were introduced. One of these developments was the introduction of the quantum hypothesis by Max Planck. This will be considered in detail in the next chapter. The other revolutionary development was Albert Einstein's theory of relativity.

The one thing these new theories had in common was that they showed Newton's laws to be incomplete. As it turns out, Newton's laws apply only to objects of macroscopic size and relatively small speeds. In this chapter we explore the surprising behavior associated with speeds approaching the speed of light. We find, in fact, that clocks run slow, metersticks become shorter, and objects become more massive as their speed increases. Although this may seem like science fiction rather than science fact, experiment shows that the predictions of Einstein's theory are indeed correct.

29–1 The Postulates of Special Relativity

At a time when some scientists thought physics was almost completely understood, with only minor details to be straightened out, physics was changed forever with the introduction of the **special theory of relativity.** Published in 1905 by Albert Einstein (1879–1955), a 26-year-old patent clerk (third class) in Berne, Switzerland, relativity fundamentally altered our understanding of such basic physical concepts as time, length, mass, and energy. It may come as a surprise, then, that Einstein's theory of relativity is based on just two simply stated postulates, and that algebra is all the mathematics required to work out its main results.

The postulates of special relativity put forward by Einstein can be stated as follows:

Equivalence of Physical Laws
The laws of physics are the same in all inertial frames of reference.

Constancy of the Speed of Light
The speed of light in a vacuum, $c = 3.00 \times 10^8$ m/s, is the same in all inertial frames of reference, independent of the motion of the source or the receiver.

All the consequences of relativity explored in this chapter follow as a direct result of these postulates.

The first postulate is certainly reasonable. Recall from Section 5–2 that an inertial frame of reference is one in which Newton's laws of motion are obeyed. Specifically, an object with no force acting on it has zero acceleration in all inertial frames. Einstein's first postulate simply extends this notion of an inertial frame to cover *all* the known laws of physics, including those dealing with thermodynamics, electricity, magnetism, and electromagnetic waves. For example, a mechanics experiment performed on the surface of the Earth (which is approximately an inertial frame) gives the same results as when the same experiment is carried out in an airplane moving with constant velocity. In addition, the behavior of heat, magnets, and electrical circuits is the same in the airplane as on the ground, as indicated in **Figure 29–1**.

All inertial frames of reference move with constant velocity (that is, zero acceleration) relative to one another. Hence, the special theory of relativity is "special" in the sense that it restricts our considerations to frames with no acceleration. The more general case, in which accelerated motion is considered, is the subject of the *general* theory of relativity, which we discuss later in this chapter. In the case of Earth, the accelerations associated with its orbital and rotational motions are small enough to be ignored in most experiments. Thus, unless otherwise stated, we shall consider the Earth and objects moving with constant velocity relative to it to be inertial frames of reference.

The second postulate of relativity is less intuitive than the first. Specifically, it states that light travels with the same speed, c, regardless of whether the source or observer is in motion. To understand the implications of this assertion, consider for a moment the case of waves on water. In **Figure 29–2 (a)** we see an observer at

▲ Albert Einstein in his twenties.

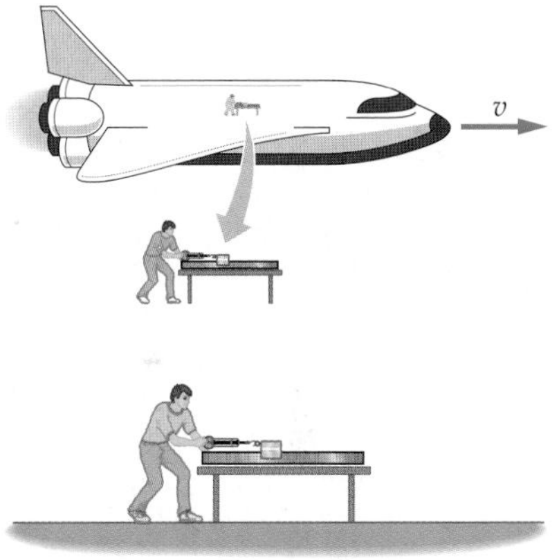

▲ **FIGURE 29–1 Inertial frames of reference**

The two observers shown are in different inertial frames of reference. According to the first postulate of relativity, physical experiments will give identical laws of nature in the two frames.

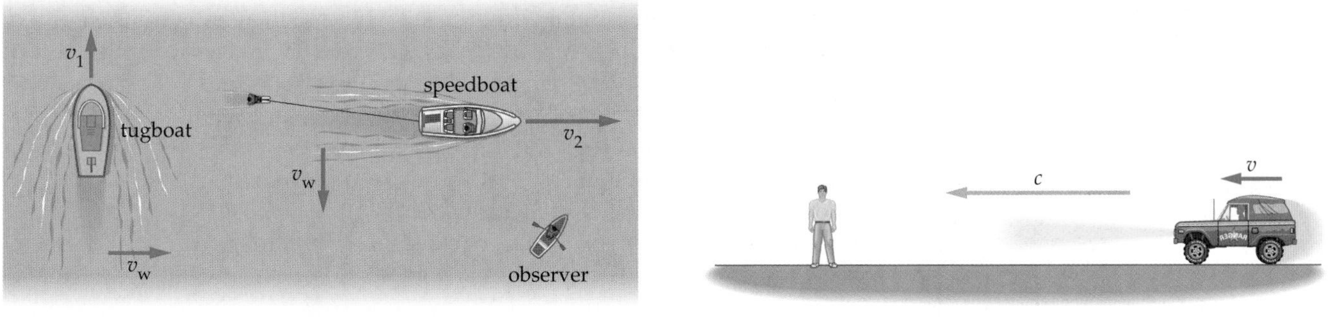

▲ **FIGURE 29–2 Wave speed versus source speed**

The speed of waves is independent of the speed of the source that generates it. **(a)** Water waves produced by a slow-moving tugboat have the same speed as those produced by a high-powered speedboat. **(b)** The speed of a beam of light, c, is independent of the speed of its source.

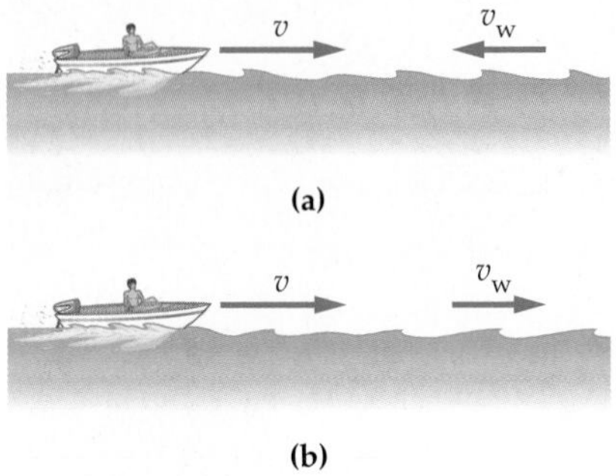

▲ **FIGURE 29–3 Wave speed versus observer speed**

The speed of a wave depends on the speed of the observer relative to the medium through which the wave propagates. **(a)** The water waves move relative to the observer with speed $v + v_w$. **(b)** In this case, the waves move relative to the observer with speed $v - v_w$.

rest relative to the water, and two moving sources generating waves. The waves produced by both the speedboat and the tugboat travel at the characteristic speed of water waves, v_w, once they are generated. Thus the observer sees a wave speed that is independent of the speed of the source—just as postulated for light, and shown in **Figure 29–2 (b)**.

On the other hand, suppose the observer is in motion with a speed v with respect to the water. If the observer is moving to the right, and water waves are moving to the left with a speed v_w, as in **Figure 29–3 (a)**, the waves move past the observer with a speed $v + v_w$. Similarly, if the water waves are moving to the right, as in **Figure 29–3 (b)**, the observer finds them to have a speed $v - v_w$. Clearly, the fact that the observer is in motion with respect to the medium through which the waves are traveling (water in this case) means that the speed of the water waves depends on the speed of the observer.

Before Einstein's theory of relativity, it was generally accepted that a similar situation would apply to light waves. In particular, light was thought to propagate through a hypothetical medium, referred to as the *luminiferous ether*, or the ether for short, that permeates all space. Since the Earth rotates about its axis with a speed of roughly 1000 mi/h at the equator, and orbits the Sun with a speed of about 67,000 mi/h, it follows that it must move relative to the ether. If this is the case, it should be possible to detect this motion by measuring differences in the speed of light propagating in different directions—just as in the case of water waves. Extremely precise experiments were carried out to this end by the American physicists A. A. Michelson (1852–1931) and E. W. Morley (1838–1923) from 1883 to 1887. They were unable to detect *any* difference in the speed of light. More recent and accurate experiments have come to precisely the same conclusion; namely, the second postulate of relativity is an accurate description of the way light behaves.

To see how counterintuitive the second postulate can be, consider the situation illustrated in **Figure 29–4**. In this case a ray of light is propagating to the right with a speed c relative to observer 1. A second observer is moving to the right as well, with a speed of $0.9c$. Although it seems natural to think that observer 2 should see the ray of light passing with a speed of only $0.1c$, this is not the case. Observer 2, like observers in all inertial frames of reference, sees the ray go by with the speed of light, c.

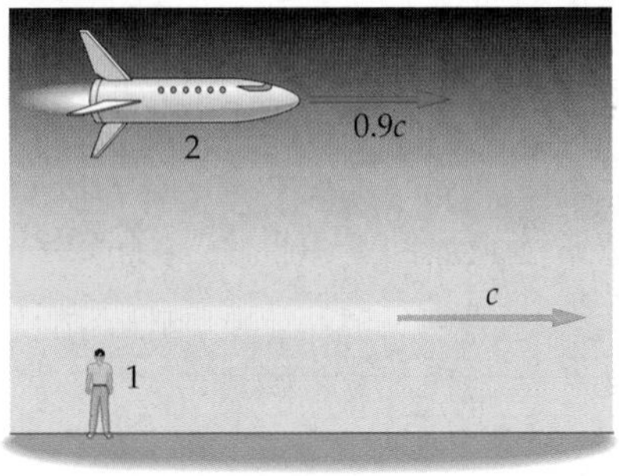

▲ **FIGURE 29–4 The speed of light for different observers**

A beam of light is moving to the right with a speed c relative to observer 1. Observer 2 is moving to the right with a speed of $0.9c$. Still, from the point of view of observer 2, the beam of light is moving to the right with a speed of c, in agreement with the second postulate of relativity.

For the observations given in Figure 29–4 to be valid—that is, for both observers to measure the same speed of light—the behavior of space and time must differ from our everyday experience when speeds approach the speed of light. This is indeed the case, as we shall see in considerable detail in the next few sections. In everyday circumstances, however, the physics described by Newton's laws are perfectly adequate. In fact, Newton's laws are valid in the limit of very small speeds, whereas Einstein's theory of relativity gives correct results for all speeds from zero up to the speed of light.

Since all inertial observers measure the same speed for light, they are all equally correct in claiming that they are at rest. For example, observer 1 in Figure 29–4 may say that he is at rest and that observer 2 is moving to the right with a speed of $0.9c$. Observer 2, however, is equally justified in saying that she is at rest and that observer 1 is in motion with a speed of $0.9c$ to the left. From the point of view of relativity, both observers are equally correct. There is no absolute rest or absolute motion, only motion relative to something else.

Finally, note that it would not make sense for observer 2 in Figure 29–4 to have a speed greater than that of light. If this were the case, it would not be possible for the light ray to pass the observer, much less to pass with the speed c. Thus we conclude that *the ultimate speed in the universe is the speed of light in a vacuum*. In the next several sections of this chapter, we shall see several more ways of arriving at precisely the same conclusion.

29–2 The Relativity of Time and Time Dilation

We generally think of time as moving forward at a constant rate, as suggested by our everyday experience. This is simply not the case, however, when dealing with speeds approaching the speed of light. If you were to observe a spaceship moving past you with a speed of $0.5c$, for example, you would notice that the clocks on the ship run slow compared with your clocks—even if they were identical in all other respects.

To calculate the difference between the rate of a moving clock and one at rest, consider the "light clock" shown in **Figure 29–5**. In this clock, a cycle begins when a burst of light is emitted from the light source S. The light then travels a distance d to a mirror, where it is reflected. It travels back a distance d to the detector D, and triggers the next burst of light. Each round trip of light can be thought of as one "tick" of the clock.

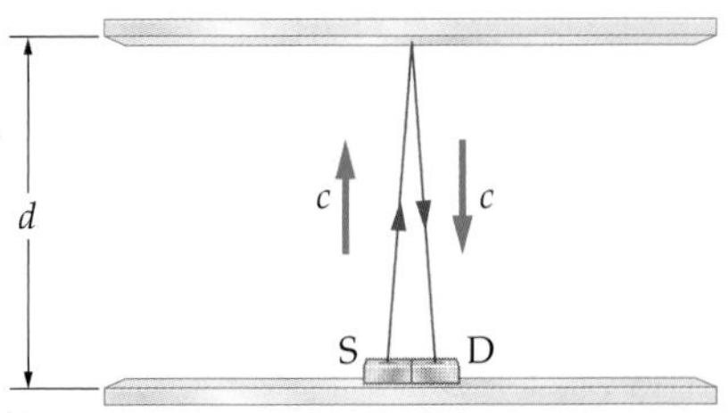

▲ **FIGURE 29–5 A stationary light clock**
Light emitted by the source S travels to a mirror a distance d away and is reflected back into the detector D. The time between emission and detection is one cycle, or one "tick," of the clock.

We begin by calculating the time interval between the ticks of this clock when it is at rest; that is, when its speed relative to the observer making the measurement is zero. Since the light covers a total distance $2d$ with a constant speed c, the time between ticks is simply

$$\Delta t_0 = \frac{2d}{c} \qquad \text{29–1}$$

The subscript 0 indicates the clock is at rest ($v = 0$) when the measurement is made.

In contrast, consider the same light clock moving with a finite speed v, as in **Figure 29–6**. Notice that the light must now follow a zigzag path in order to complete a tick of the clock. Since this path is clearly longer than $2d$, and the speed of the light is still the same—according to the second postulate of relativity—the time between ticks must be greater than Δt_0. With more time elapsing between ticks, the clock runs slow. We refer to this phenomenon as **time dilation,** because the time interval for one tick has been increased—or dilated—from Δt_0 for a clock at rest relative to an observer to $\Delta t > \Delta t_0$ for a clock in motion relative to an observer.

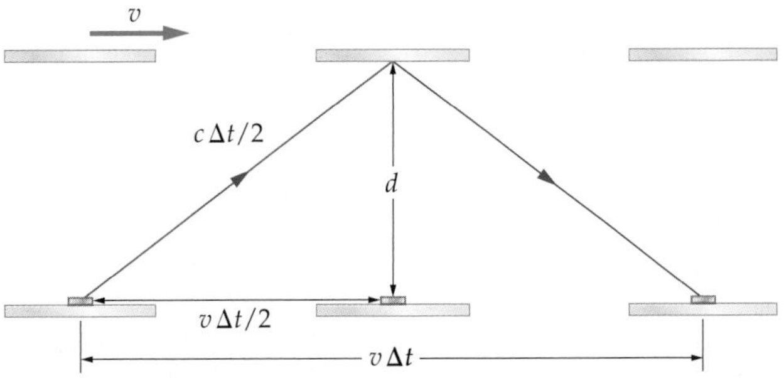

◀ **FIGURE 29–6 A moving light clock**
A moving light clock requires a time Δt to complete one cycle. Note that the light follows a zigzag path that is longer than $2d$; hence, the time between ticks is greater for the moving clock than it is for the clock at rest.

To calculate the dilated time, Δt, notice that in the time $\Delta t/2$ the clock moves a horizontal distance $v\Delta t/2$, which is halfway to its position at the end of the tick. The distance traveled by the light in this time is $c\Delta t/2$, which is the hypotenuse of the right triangle shown in Figure 29–6. Applying the Pythagorean theorem to this triangle, we find the following relation:

$$\left(\frac{v\Delta t}{2}\right)^2 + d^2 = \left(\frac{c\Delta t}{2}\right)^2$$

Solving for the time Δt, we find

$$\Delta t = \frac{2d}{\sqrt{c^2 - v^2}} = \frac{2d}{c\sqrt{1 - v^2/c^2}}$$

Recalling that $\Delta t_0 = 2d/c$, we can relate the two time intervals as follows:

Time Dilation

$$\Delta t = \frac{\Delta t_0}{\sqrt{1 - v^2/c^2}} \qquad \text{29–2}$$

SI unit: s

Notice that $\Delta t = \Delta t_0$ for $v = 0$, as expected. For speeds v that are greater than zero but less than c, the denominator in Equation 29–2 is less than 1. As a result, it

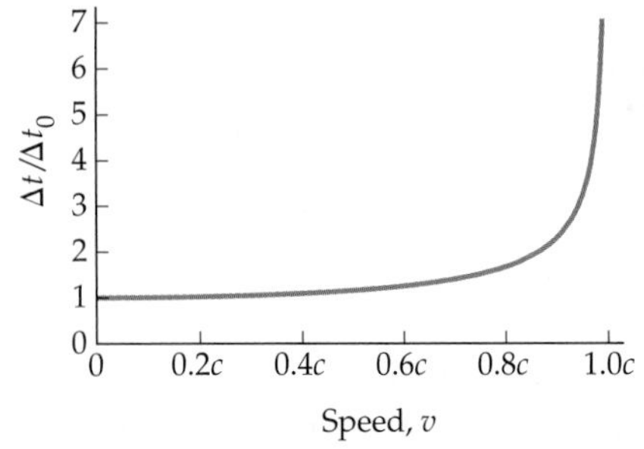

▲ **FIGURE 29–7 Time dilation as a function of speed**

As the speed of a clock increases, the time required for it to advance by 1 s increases slowly at first and then rapidly near the speed of light.

follows that Δt is greater than Δt_0. Finally, in the limit that the speed v approaches the speed of light, c, we observe that the denominator of Equation 29–2 vanishes, and the time interval Δt goes to infinity. This behavior is illustrated in **Figure 29–7**, where we show the ratio $\Delta t/\Delta t_0$ as a function of the speed v. The fact that Δt goes to infinity means that it takes an infinite amount of time for one tick—in other words, as v approaches the speed of light a clock slows to the point of stopping. Clearly, then, the speed of light provides a natural upper limit to the possible speed of an object.

EXERCISE 29–1

A spaceship carrying a light clock moves with a speed of $0.500c$ relative to an observer on Earth. According to this observer, how long does it take for the spaceship's clock to advance 1.00 s?

Solution

Substituting $\Delta t_0 = 1.00$ s and $v = 0.500c$ in Equation 29–2, we obtain

$$\Delta t = \frac{\Delta t_0}{\sqrt{1 - v^2/c^2}} = \frac{1.00\text{ s}}{\sqrt{1 - (0.500c)^2/c^2}} = \frac{1.00\text{ s}}{\sqrt{1 - 0.25}} = 1.15\text{ s}$$

Even at this high speed, the relativistic effect is relatively small—only about 15 percent.

Equation 29–2 applies to any type of clock, not just the light clock. If this were not the case—if different clocks ran at different rates when in motion with constant velocity—the first postulate of relativity would be violated.

CONCEPTUAL CHECKPOINT 29–1 The Rate of Time

A clock moving with a finite speed v is observed to run slow. If the speed of light were twice as large as it actually is, would the factor by which the clock runs slow be **(a)** increased, **(b)** decreased, or **(c)** unchanged?

Reasoning and Discussion

As Figure 29–7 shows, the factor by which time is dilated increases as the speed of a clock approaches the speed of light. If the speed of light were twice as large, the speed of a moving clock would be a smaller fraction of the speed of light; hence, the time dilation factor would be less.

Extending the preceding argument, it follows that in the limit of an infinite speed of light there would be no time dilation at all. This conclusion is verified by noting that in the limit $c \to \infty$ the denominator in Equation 29–2 is equal to 1, so that Δt and Δt_0 are equal. Since the speed of light is practically infinite in everyday terms, the relativistic time dilation effect is negligible for everyday objects.

Answer:

(b) The factor by which the clock runs slow would be decreased.

Before we proceed, it is useful to introduce some of the terms commonly employed in relativity. First, an **event** is a physical occurrence that happens at a *specified location* at a *specified time*. In three dimensions, for example, we specify an event by giving the values of the coordinates x, y, and z as well as the time t. If two events occur at the same location but at different times, the time between the events is referred to as the **proper time:**

> The *proper time* is the amount of time separating two events that occur at the *same location*.

As an example, the proper time between ticks in a light clock is the time between the emission of light (event 1) and its detection (event 2) when the clock is at rest relative to the observer. Thus, Δt_0 in Equation 29–2 is the proper time, and Δt is the corresponding time when the clock moves relative to the observer with a speed v.

In Exercise 29–1 we calculated the relativistic effect of time dilation for the case of a clock moving with a speed equal to half the speed of light. In everyday circumstances, however, speeds are never so large. In fact, the greatest speed a

PROBLEM-SOLVING NOTE

Identifying the Proper Time

The key to solving problems involving time dilation is to correctly identify the proper time, Δt_0. Simply put, the proper time is the time between events that occur at the same location.

human might reasonably attain today is the speed of the space shuttle in orbit. As we saw in Chapter 12, this speed is only about 7700 m/s, or 17,000 mi/h. Although this is a rather large speed, it is still only 1/39,000th the speed of light.

To find the time dilation in a case like this we cannot simply substitute $v = 7700$ m/s into Equation 29–2, since a typical calculator does not have enough decimal places to give the correct answer. You might want to try it yourself; your calculator will probably give the incorrect result that Δt is equal to Δt_0. To find the correct answer, we must use the binomial expansion (Appendix A) to reexpress Equation 29–2 as follows:

$$\Delta t = \frac{\Delta t_0}{\sqrt{1 - v^2/c^2}} \approx \Delta t_0\left(1 + \frac{1}{2}\frac{v^2}{c^2} + \cdots\right)$$

Substituting the speed of the space shuttle and the speed of light, we find

$$\Delta t \approx \Delta t_0\left[1 + \frac{1}{2}\left(\frac{7700 \text{ m/s}}{3.00 \times 10^8 \text{ m/s}}\right)^2\right] = \Delta t_0(1 + 3.3 \times 10^{-10})$$

Thus, a clock on board the space shuttle runs slow by a factor of 1.00000000033. At this rate, it would take almost 100 years for the clock on the shuttle to lose 1 s compared with a clock on Earth.

Clearly, such small differences in time cannot be measured with an ordinary clock. In recent years, however, atomic clocks have been constructed that have an accuracy sufficient to put relativity to a direct test. The physicists J. C. Hafele and R. E. Keating conducted such a test in 1971 by placing one atomic clock on board a jet airplane and leaving an identical clock at rest in the laboratory. After flying the moving clock at high speed for many hours, the experimenters found it to have run slower than the clock left in the lab. The discrepancy in times agreed with the predictions of relativity. Today, if an atomic clock is transported from one location to another, the relativistic effects of time dilation must be taken into account; otherwise the clock will give a time that is behind the correct value.

Another aspect of time dilation is the fact that different observers disagree on *simultaneity*. For example, suppose observer 1 notes that two events at different locations occur at the same time. For observer 2, who is moving with a speed v relative to observer 1, these same two events are not simultaneous. Thus, relativity not only changes the rate at which time progresses for different observers, it also changes the amount of time that separates events.

Space Travel and Biological Aging

To this point, our discussion of time dilation has been applied solely to clocks. Clocks are not the only objects that show time dilation, however. In fact,

> relativistic time dilation applies equally to *all physical processes*, including chemical reactions and biological functions.

Thus an astronaut in a moving spaceship *ages more slowly* than one who remains on Earth and by precisely the same factor that a clock on the spaceship runs more slowly than one at rest. To the astronaut, however, time seems to progress as usual. The implications of time dilation for a high-speed trip to a nearby star are considered in the next Example.

EXAMPLE 29–1 Benny and Jenny—Separated at Launch

Astronaut Benny travels to Vega, the fifth brightest star in the night sky, leaving his 35.0-year-old twin sister Jenny behind on Earth. Benny travels with a speed of 0.990c, and Vega is 25.3 light-years from Earth. **(a)** How long does the trip take from the point of view of Jenny? **(b)** How much has Benny aged when he arrives at Vega?

Picture the Problem

Our sketch shows the spacecraft as it travels in a straight line from Earth to Vega. The speed of the spacecraft relative to Earth is 99% of the speed of light, $v = 0.990c$. In addition, the distance to Vega is 25.3 light-years; that is, $d = 25.3$ ly. [*Note:* One light-year is the distance light travels with a speed c in one year. Specifically, 1 light-year = 1 ly = $c(1 \text{ y})$.]

continued on next page

continued from previous page

Strategy

The two events of interest in this problem are (1) leaving Earth and (2) arriving at Vega. For Jenny, these two events clearly occur at different locations. It follows that the time interval for her is Δt, and not the proper time, Δt_0.

For Benny, however, the two events occur at the same location—namely, just outside the spacecraft door. (In fact, from Benny's point of view the spacecraft is at rest, and the stars are in motion.) Therefore, the time interval measured by Benny is the proper time, Δt_0.

Finally, we note that Δt and Δt_0 are related by Equation 29–2; that is, $\Delta t = \Delta t_0/\sqrt{1 - v^2/c^2}$.

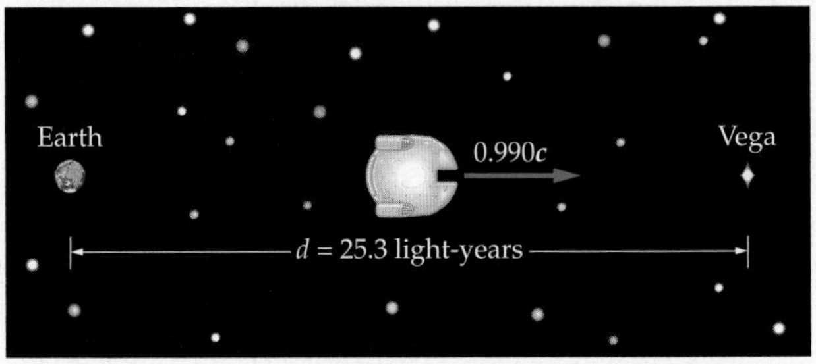

Solution

Part (a)

1. The spacecraft covers a distance d in a time Δt with a speed $v = 0.990c$. Use $v = d/\Delta t$ to solve for the time, Δt:

$$v = \frac{d}{\Delta t}$$

$$\Delta t = \frac{d}{v} = \frac{25.3 \text{ ly}}{0.990c} = \frac{25.3c(1\text{y})}{0.990c} = 25.6 \text{ y}$$

Part (b)

2. Rearrange $\Delta t = \Delta t_0/\sqrt{1 - v^2/c^2}$ (Equation 29–2) to solve for the proper time, Δt_0:

$$\Delta t = \frac{\Delta t_0}{\sqrt{1 - v^2/c^2}}$$

$$\Delta t_0 = \Delta t\sqrt{1 - v^2/c^2}$$

3. Substitute $v = 0.990c$ and $\Delta t = 25.6$ y to find Δt_0:

$$\Delta t_0 = \Delta t\sqrt{1 - v^2/c^2}$$

$$= (25.6 \text{ y})\sqrt{1 - \frac{(0.990c)^2}{c^2}} = 3.61 \text{ y}$$

Insight

Thus when Benny reaches Vega, he is only 38.6 y old, whereas Jenny, who stayed behind on Earth, is 60.6 y old.

From the point of view of Benny, the trip took 3.61 y at a speed of 0.990c. As a result, he would say that the distance covered in traveling to Vega was only $(3.61 \text{ y})(0.990c) = 3.57$ ly. We consider this result in greater detail in the next section.

Practice Problem

How fast must Benny travel if he wants to age only 2.00 y during his trip to Vega? [**Answer:** $v = 0.997c$]

Some related homework problems: Problem 4, Problem 10

In the following Active Example, we determine the speed of an astronaut by considering the change in her heart rate.

ACTIVE EXAMPLE 29–1 Heartthrob: Find the Astronaut's Speed

An astronaut traveling with a speed v relative to Earth takes her pulse and finds that her heart beats once every 0.850 s. Mission control on Earth, which monitors her heart activity, observes one heartbeat every 1.40 s. What is the astronaut's speed relative to Earth?

Solution *(Test your understanding by performing the calculations indicated in each step.)*

1. Identify the proper time, Δt_0, and the dilated time, Δt: $\Delta t_0 = 0.850$ s, $\Delta t = 1.40$ s

2. Solve the time-dilation expression, $\Delta t = \Delta t_0/\sqrt{1 - v^2/c^2}$, for the speed, v: $v = c\sqrt{1 - \Delta t_0^2/\Delta t^2}$

3. Substitute numerical values: $v = 0.795c$

Your Turn

What speed is required for mission control to measure a time between heartbeats that is $2(1.40 \text{ s}) = 2.80 \text{ s}$?

(Answers to **Your Turn** *problems are given in the back of the book.)*

The Decay of the Muon

A particularly interesting example of time dilation involves subatomic particles called *muons* that are created by cosmic radiation high in Earth's atmosphere. A muon is an unstable particle; in fact, a muon at rest exists for only about 2.20×10^{-6} s, on average, before it decays. Suppose, for example, that a muon is created at an altitude of 5.00 km above the surface of the Earth. If this muon travels toward the ground with a speed of $0.995c$ for 2.20×10^{-6} s, it will cover a distance of only 657 m before decaying. Thus, without time dilation, one would conclude that the muons produced at high altitude should not reach the surface of the Earth. It is found, however, that large numbers of muons do in fact reach the ground. The reason is that they age more slowly due to their motion—just like the astronaut traveling to Vega considered in Example 29–1. The next Example examines this time dilation effect.

EXAMPLE 29–2 The Life and Times of a Muon

Consider muons traveling toward Earth with a speed of $0.995c$ from their point of creation at a height of 5.00 km. **(a)** Find the average lifetime of these muons, assuming that a muon at rest has an average lifetime of 2.20×10^{-6} s. **(b)** Calculate the average distance these muons can cover before decaying.

Picture the Problem

Our sketch shows a muon that has just been created at an altitude of 5.00 km. The muon is heading straight for the ground with a speed $v = 0.995c$. As a result of its high speed, the muon's "internal clock" runs slow, allowing it to live longer as seen by an observer on Earth.

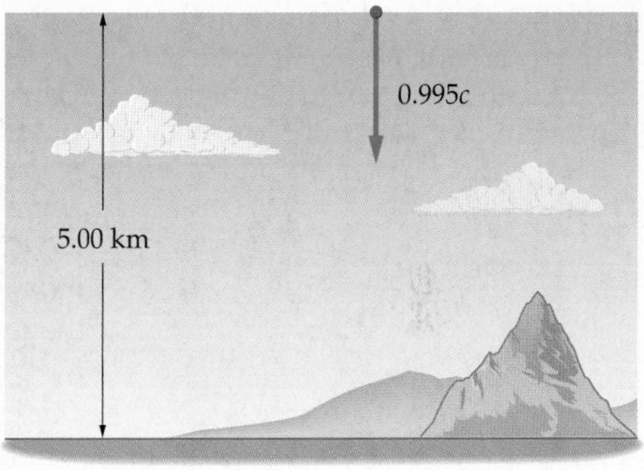

Strategy

The two events to be considered in this case are (1) the creation of a muon and (2) the decay of a muon. From the point of view of an observer on Earth, these two events occur at different locations. It follows that the corresponding time is Δt.

From the muon's point of view, it is at rest and the Earth is moving toward it at $0.995c$. Hence, for the muon, the two events occur at the same location, and the time between them is $\Delta t_0 = 2.20 \times 10^{-6}$ s. We can find the time Δt by using time dilation, $\Delta t = \Delta t_0/\sqrt{1 - v^2/c^2}$.

Solution

Part (a)

1. Substitute $v = 0.995c$ and $\Delta t_0 = 2.20 \times 10^{-6}$ s into Equation 29–2:

$$\Delta t = \frac{\Delta t_0}{\sqrt{1 - v^2/c^2}} = \frac{2.20 \times 10^{-6}\ \text{s}}{\sqrt{1 - (0.995c)^2/c^2}} = 22.0 \times 10^{-6}\ \text{s}$$

Part (b)

2. Multiply $v = 0.995c$ times the time 22.0×10^{-6} s to find the average distance covered:

$$d_{av} = (0.995c)(22.0 \times 10^{-6}\ \text{s}) = 6570\ \text{m}$$

Insight

Thus relativistic time dilation allows the muons to travel about 10 times farther (6570 m instead of 657 m) than would have been expected from nonrelativistic physics. As a result, muons are detected on Earth's surface.

Practice Problem

Muons produced at the laboratory of the European Council for Nuclear Research (CERN) in Geneva are accelerated to high speeds. At these speeds, the muons are observed to have a lifetime 30.00 times greater than the lifetime of muons at rest. What is the speed of the muons at CERN? **[Answer:** $v = 0.9994c$]

Some related homework problems: Problem 7, Problem 9, Problem 13

29–3 The Relativity of Length and Length Contraction

Just as time is altered for an observer moving with a speed close to the speed of light, so too is distance. For example, a meterstick moving with a speed of $0.5c$ would appear noticeably shorter than a meterstick at rest. As the speed of the meterstick approaches c, its length diminishes toward zero.

To see why lengths contract, and to calculate the amount of contraction, recall the example of the twins Benny and Jenny and the trip to Vega. From Jenny's point of view on Earth, Benny's trip took 25.6 y and covered a distance of $(0.990c)(25.6\text{ y}) = 25.3\text{ ly}$. From Benny's point of view, however, the trip took only 3.61 y. Since both twins agree on their relative velocity, it follows that as far as Benny is concerned, his trip covered a distance of only $(0.990c)(3.61\text{ y}) = 3.57\text{ ly}$. Thus, from the point of view of the astronaut, Earth and Vega move by at a speed of $0.990c$, and the distance between them is not 25.3 ly, but only 3.57 ly. This is an example of length contraction, as illustrated in **Figure 29–8**.

▲ **FIGURE 29–8 A relativistic trip to Vega**
(a) From the Earth observer's point of view the spaceship is traveling with a speed of $0.990c$, covering a distance of 25.3 ly in a time of 25.6 y. **(b)** From the astronaut's point of view the spaceship is at rest, and Earth and Vega are moving with a speed of $0.990c$. For the astronaut the trip takes only 3.61 y and covers a contracted distance of only 3.57 ly.

In general, we would like to determine the contracted length L of an object moving with a speed v. When the object is at rest ($v = 0$), we say that its length is the **proper length,** L_0:

> The proper length is the distance between two points as measured by an observer who is at rest with respect to them.

In the Benny and Jenny Example, Jenny is at rest with respect to Earth and Vega. As a result, the distance between them as measured by Jenny is the proper length; that is, $L_0 = 25.3\text{ ly}$. The contracted length, $L = 3.57\text{ ly}$, is measured by Benny.

As for the times measured for the trip, from Jenny's point of view the two events (event 1, departing Earth; event 2, arriving at Vega) occur at different locations. As a result, she measures the *dilated time*, $\Delta t = 25.6\text{ y}$, even though she also measures the *proper length*, L_0. In contrast, Benny measures the *proper time*, $\Delta t_0 = 3.61\text{ y}$, and the *contracted length*, L. Note that one must be careful to determine from the definitions given previously which observer measures the proper time and which observer measures the proper length—*it should never be assumed*, for example, that just because one observer measures the proper time that that observer also measures the proper length.

We now use these observations to obtain a general expression relating L and L_0. To begin, note that both observers measure the same relative speed, v. For Jenny, the speed is $v = L_0/\Delta t$, and for Benny it is $v = L/\Delta t_0$. Setting these speeds equal we obtain

$$v = \frac{L_0}{\Delta t} = \frac{L}{\Delta t_0}$$

PROBLEM-SOLVING NOTE
Identifying the Proper Length

The key to solving problems involving length contraction is to correctly identify the proper length, L_0. Specifically, the proper length is the distance between two points that are at rest relative to the observer.

PROBLEM-SOLVING NOTE
Measuring Proper Length and Proper Time

Keep in mind that the observer who measures the proper length is *not* necessarily the same observer who measures the proper time.

Solving for L in terms of L_0, we find $L = L_0(\Delta t_0/\Delta t)$. Finally, using Equation 29–2 to express Δt in terms of Δt_0 we obtain the following:

Length Contraction

$$L = L_0\sqrt{1 - \frac{v^2}{c^2}} \qquad 29\text{–}3$$

SI unit: m

Substituting the numerical values from the Benny and Jenny Example gives us $L = (25.3\text{ ly})\sqrt{1 - (0.990c)^2/c^2} = 3.57$ ly, as expected.

Note that if $v = 0$ in Equation 29–3, we find that $L = L_0$. As v approaches the speed of light, the contracted length L approaches zero. In general, the length of a moving object is always less than its proper length. **Figure 29–9** shows L as a function of speed v for a meterstick, where we see again that the speed of light is the ultimate speed possible.

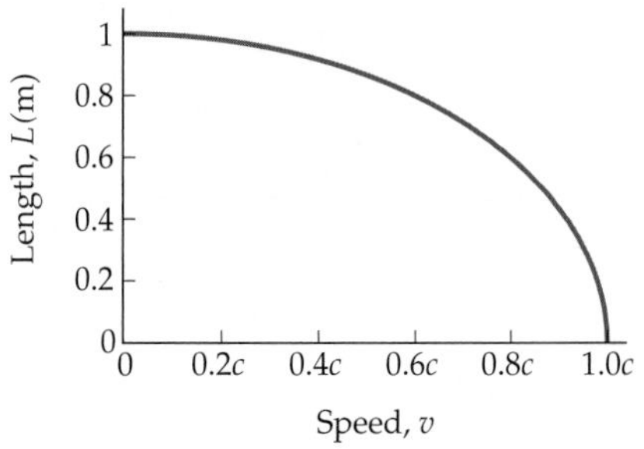

▲ **FIGURE 29–9 Length contraction**
The length of a meterstick as a function of its speed. Note that the length shrinks to zero in the limit that its speed approaches the speed of light.

EXAMPLE 29–3 Half a Meter

Find the speed for which the length of a meterstick is 0.500 m.

Picture the Problem
Our sketch shows a moving meterstick with a contracted length $L = 0.500$ m. The meterstick at rest has its proper length, $L_0 = 1.00$ m.

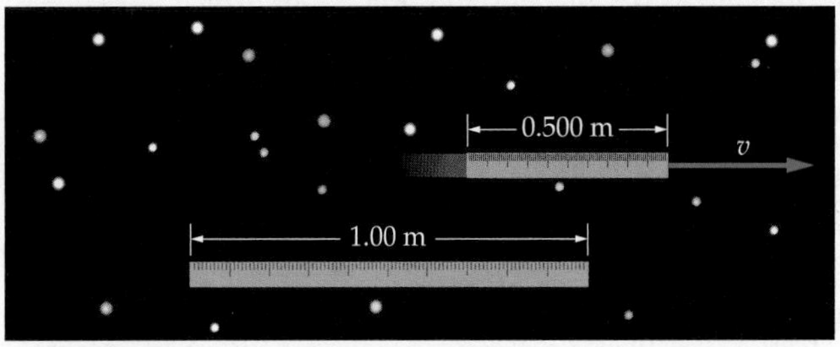

Strategy
We can find the desired speed by applying length contraction (Equation 29–3) to the moving meterstick. In particular, given L and L_0 we can solve $L = L_0\sqrt{1 - v^2/c^2}$ for the speed, v.

Solution

1. Solve Equation 29–3 for the speed v:
$$L = L_0\sqrt{1 - \frac{v^2}{c^2}}$$
$$v = c\sqrt{1 - \frac{L^2}{L_0^{\,2}}}$$

2. Substitute numerical values:
$$v = c\sqrt{1 - \frac{L^2}{L_0^{\,2}}} = c\sqrt{1 - \frac{(0.500\text{ m})^2}{(1.00\text{ m})^2}} = 0.866c$$

Insight
Note that a person traveling along with the moving meterstick would see it as having its proper length of 1.00 m. From this person's point of view, it is the *other* meterstick that is only 0.500 m long. Thus, lengths—like times—are relative, and depend on the observer making the measurement.

Practice Problem
Find the length of a meterstick that is moving with half the speed of light. [**Answer:** $L = 0.866$ m]

Some related homework problems: Problem 19, Problem 24

In **Figure 29–10** we illustrate the effects of length contraction for the case of a muon traveling toward Earth's surface. From the point of view of the Earth, Figure 29–10 (a), the muon travels a distance of 5.00 km at a speed of 0.995c. To cover this distance, the muon must live about 10 times longer than it would at rest. From the point of view of the muon, Figure 29–10 (b), the Earth is moving upward at a speed of 0.995c, and the distance the Earth must travel to reach the muon is only $L = (5.00\text{ km})\sqrt{1 - (0.995c)^2/c^2} = 499$ m. The Earth can easily cover this distance during the muon's resting lifetime of 2.20 μs.

One final comment: The length contraction calculated in Equation 29–3 *pertains only to lengths in the direction of relative motion*. Lengths at right angles to the relative motion are unaffected.

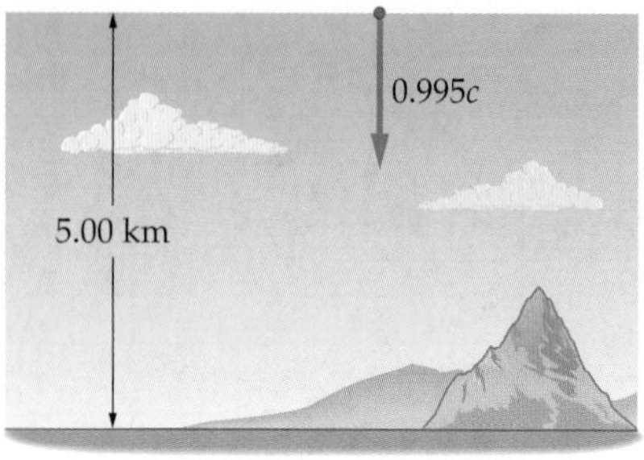

(a) Earth observer's point of view

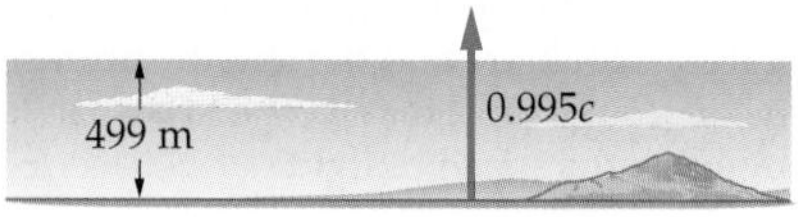

(b) Muon's point of view

▲ **FIGURE 29–10 A muon reaches Earth's surface**

(a) From Earth's point of view a muon travels downward with a speed of 0.995c for a distance of 5.00 km. The muon can reach Earth's surface only if it ages slowly due to its motion. **(b)** From the muon's point of view, Earth is moving upward with a speed of 0.995c, and the distance to Earth's surface is only 499 m. From this point of view the muon reaches Earth's surface because the distance is contracted, not because the muon lives longer.

CONCEPTUAL CHECKPOINT 29–2 Angle of Repose

An astronaut is resting on a bed inclined at an angle θ above the floor of a spaceship, as shown in the first sketch. From the point of view of an observer who sees the spaceship moving to the right with a speed approaching c, is the angle the bed makes with the floor **(a)** greater than, **(b)** less than, or **(c)** equal to the angle θ observed by the astronaut?

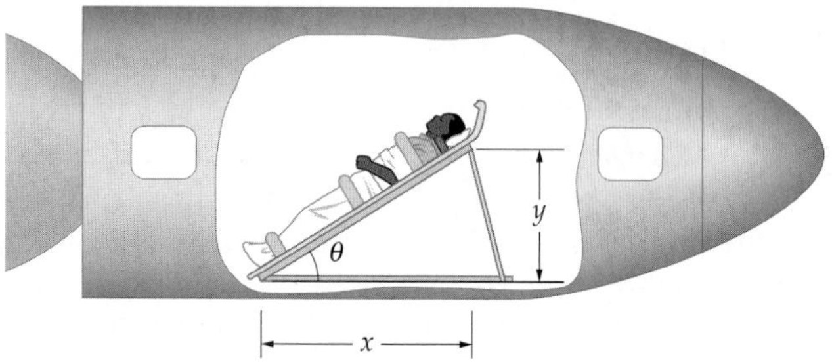

Reasoning and Discussion

A person observing the spaceship moving with a speed v notices a contracted length, x', in the direction of motion, but an unchanged length, y, perpendicular to the direction of motion, as shown in the second sketch.

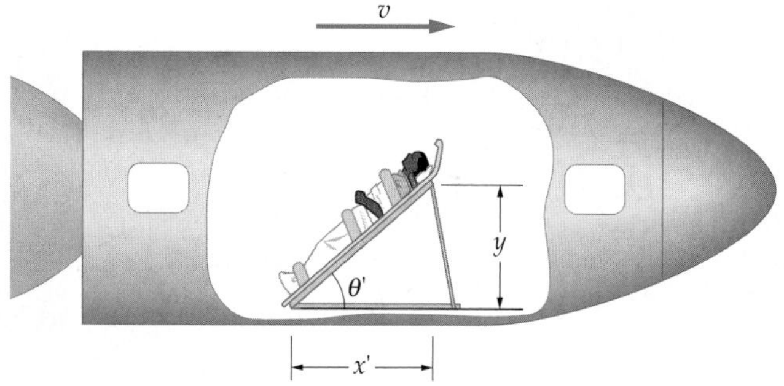

As a result of the contraction in just one direction, the bed is inclined at an angle greater than the angle θ measured by the astronaut.

Answer:

(a) The angle will be greater than θ.

The fact that lengths in different directions contract differently has interesting effects on the way a rapidly moving object appears to the eye. Additional effects are related to the finite time it takes for light to propagate to the eye from different parts of an object. An example of the way an object would actually look if it moved past us at nearly the speed of light is shown in **Figure 29–11.**

▲ **FIGURE 29–11 Relativistic distortions**

If the streetcar depicted at rest in part **(a)** moved to the left at nearly the speed of light, it would look like the computer-simulated image in part **(b)**. The streetcar would appear compressed in the direction of its motion. Because the streetcar would move significantly in the time it took for light from it to reach you, however, you would also observe other odd effects—for example, you would be able to see its back surface even from a position directly alongside it.

29–4 The Relativistic Addition of Velocities

Suppose you are piloting a spaceship in deep space, moving toward an asteroid with a speed of 25 mi/h. To signal a colleague on the asteroid, you activate a beam of light on the nose of the ship and point it in the direction of motion. Since light in a vacuum travels with the same speed c relative to all inertial observers, the speed of the light beam relative to the asteroid is simply c, *not* $c + 25$ mi/h. Clearly, then, simple addition of velocities, which seems to work just fine for everyday speeds, no longer applies.

The correct way to add velocities, valid for all speeds from zero to the speed of light, was obtained by Einstein. Consider, for example, a one-dimensional system consisting of a spaceship (1), a probe (2), and an asteroid (3), as indicated in **Figure 29–12**. Suppose the spaceship moves toward the asteroid with the velocity v_{13}, where the subscripts mean "the velocity of object 1 relative to object 3." Since the spaceship moves to the right in Figure 29–12 (a), its velocity is positive ($v_{13} > 0$); if it moved to the left, its velocity would be negative ($v_{13} < 0$). Now imagine that the ship sends out a probe, whose velocity relative to the ship is v_{21}, as shown in Figure 29–12 (b). The question is, how do we add velocities correctly to get the velocity of the probe relative to the asteroid, v_{23}?

The "classical" answer is to simply add the velocities, as we did in Chapter 3; that is, $v_{23} = v_{21} + v_{13}$. Einstein showed that the correct result, taking into account relativity, contains an additional factor:

Relativistic Addition of Velocities

$$v_{23} = \frac{v_{21} + v_{13}}{1 + \dfrac{v_{21}v_{13}}{c^2}} \qquad \text{29–4}$$

SI unit: m/s

Notice that if the speed of light were infinite, $c \rightarrow \infty$, the denominator would be $1 + v_{13}v_{21}/(\infty)^2 = 1$ and we would recover classical velocity addition. Thus, it is the *finite* speed of light that is responsible for relativistic effects.

To return to the example at the beginning of this section, what if the probe is actually a beam of light, with the velocity $v_{21} = c$? In this case, relativistic velocity addition gives the speed of the probe (light) relative to the asteroid as

$$v_{23} = \frac{v_{21} + v_{13}}{1 + \dfrac{v_{13}v_{21}}{c^2}} = \frac{c + v_{13}}{1 + \dfrac{v_{13}c}{c^2}} = \frac{c\left(\dfrac{v_{13}}{c} + 1\right)}{1 + \dfrac{v_{13}}{c}} = c$$

Thus, as expected, both the observer in the spaceship (1) and the observer on the asteroid (3) measure the same velocity for the light beam, $v_{21} = v_{23} = c$, regardless of the velocity of the ship relative to the asteroid, v_{13}.

PROBLEM-SOLVING NOTE

Adding Velocities

To add velocities correctly, it is important to identify each velocity in terms of the object or observer relative to which it is measured, just as for velocity addition in nonrelativistic physics (Section 3–6).

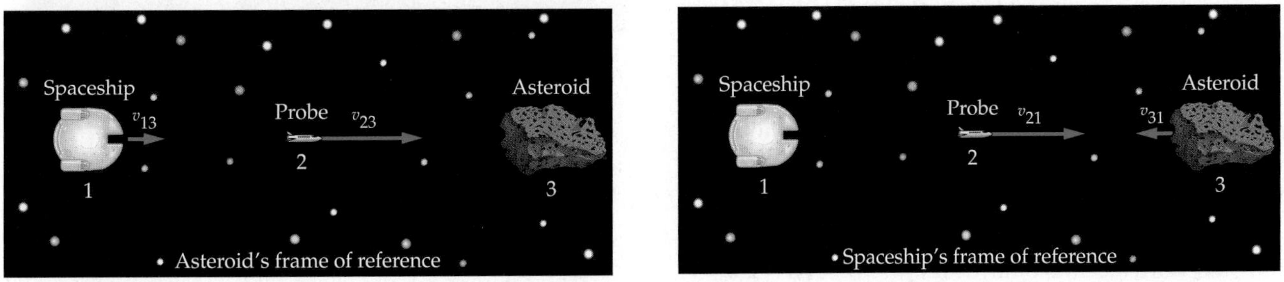

(a) (b)

▲ **FIGURE 29–12 The velocity of a probe: Two frames of reference**

A spaceship approaches an asteroid and sends a probe to investigate it. **(a)** In the asteroid's frame of reference, where the asteroid is at rest, the spaceship approaches with the velocity v_{13} and the probe approaches with the velocity v_{23}. **(b)** In the spaceship's frame of reference, the probe is launched with the velocity v_{21}, and the asteroid approaches with the velocity $v_{31} = -v_{13}$. The relativistic sum of the velocities v_{13} and v_{21}, using Equation 29–4, yields the probe's velocity relative to the asteroid, v_{23}.

We've seen that Equation 29–4 gives the correct result for a beam of light, but how does it work when applied to speeds much smaller than the speed of light? As an example, suppose a person on a spaceship moving at 25 mi/h throws a ball with a speed of 15 mi/h in the direction of the asteroid. According to the classical addition of velocities, the velocity of the ball relative to the asteroid is 15 mi/h + 25 mi/h = 40 mi/h. Application of Equation 29–4 gives the following result:

$$v = 39.99999999999997 \text{ mi/h}$$

Thus, in any practical measurement, the velocity of the ball relative to the asteroid will be 40 mi/h. We conclude that the classical result, $v_{23} = v_{21} + v_{13}$, although not strictly correct, is appropriate for small speeds.

EXERCISE 29–2

Suppose the spaceship described previously is approaching an asteroid with a speed of $0.750c$. If the spaceship launches a probe toward the asteroid with a speed of $0.800c$ relative to the ship, what is the speed of the probe relative to the asteroid?

Solution

Substituting $v_{13} = 0.750c$ and $v_{21} = 0.800c$ in Equation 29–4 we obtain

$$v_{23} = \frac{v_{21} + v_{13}}{1 + \dfrac{v_{13}v_{21}}{c^2}} = \frac{0.800c + 0.750c}{1 + \dfrac{(0.750c)(0.800c)}{c^2}} = 0.969c$$

As expected, the speed relative to the asteroid is less than c. Note, however, that classical velocity addition gives a strikingly different prediction for the speed relative to the asteroid; namely, $v = 0.800c + 0.750c = 1.550c$. Thus, velocity addition is one more way that experiments can verify the predictions of relativity.

In the following Example and Active Example, we apply relativistic velocity addition to a variety of physical systems.

EXAMPLE 29–4 Generation Next

At starbase Faraway Point, you observe two spacecraft approaching from the same direction. The *La Forge* is approaching with a speed of $0.906c$, and the *Picard* is approaching with a speed of $0.806c$. Find the velocity of the *La Forge* relative to the *Picard*.

Picture the Problem

Our sketch shows the two spacecraft approaching the starbase along the same line, moving in the positive direction. The speed of the *La Forge* is $0.906c$, and the speed of the *Picard* is $0.806c$. In addition, note that we have numbered the objects in this system as follows: *Picard* (1); *La Forge* (2); Faraway Point (3). Finally, we show the *La Forge* ahead of the *Picard* in our sketch, but the final result is the same even if their positions are reversed.

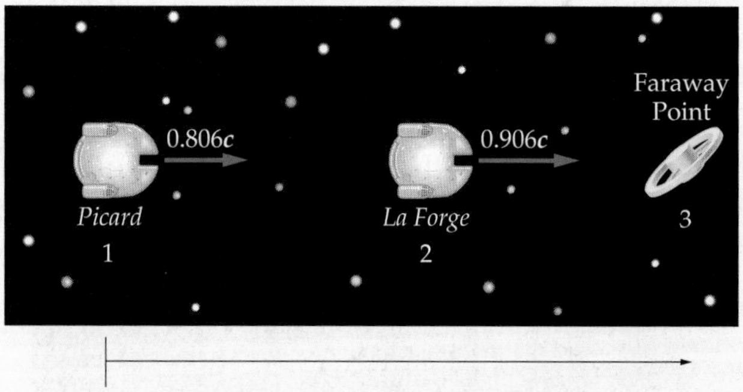

INTERACTIVE FIGURE

Strategy

The key to solving a problem like this is to choose the velocities v_{13}, v_{21}, and v_{23} in a way that is consistent with our numbering scheme. In this case, we know that $v_{13} = 0.806c$ and $v_{23} = 0.906c$. The velocity we want to find, *La Forge* relative to *Picard*, is v_{21}. This can be determined by substituting the known quantities into Equation 29–4.

Solution

1. Use relativistic velocity addition (Equation 29–4) and straightforward algebra to solve for v_{21}:

$$v_{23} = \frac{v_{21} + v_{13}}{1 + \dfrac{v_{13}v_{21}}{c^2}} \quad \text{or} \quad v_{21} = \frac{v_{23} - v_{13}}{1 - \dfrac{v_{13}v_{23}}{c^2}}$$

2. Substitute $v_{13} = 0.806c$ and $v_{23} = 0.906c$ to find v_{21}, the velocity of the *La Forge* relative to the *Picard*:

$$v_{21} = \frac{v_{23} - v_{13}}{1 - \dfrac{v_{13}v_{23}}{c^2}} = \frac{0.906c - 0.806c}{1 - \dfrac{(0.806c)(0.906c)}{c^2}} = 0.371c$$

Insight

A nonrelativistic calculation of the velocity of the *La Forge* relative to the *Picard* gives $0.100c$, considerably less than the correct value of $0.371c$ obtained from relativistic velocity addition.

Also, when we set up Equation 29–4, we arbitrarily chose the *Picard*, *La Forge*, and Faraway Point to be 1, 2, and 3, respectively. It is important to note, however, that any assignment of 1, 2, and 3 is equally valid and will yield precisely the same results.

Practice Problem

Find the velocity of the *La Forge* relative to the *Picard* if the *Picard*'s direction of motion is reversed. [**Answer:** In this case $v_{13} = -0.806c$; hence, $v_{21} = 0.989c$.]

Some related homework problems: Problem 35, Problem 36, Problem 37

ACTIVE EXAMPLE 29–2 Find the Relativistic Length

As a spaceship approaches a distant planet with a speed of $0.445c$, it launches a probe toward the planet. The proper length of the probe is 10.0 m, and its length as measured by an observer on the spaceship is 7.50 m. What is the length of the probe as measured by an observer on the planet?

Solution *(Test your understanding by performing the calculations indicated in each step.)*

1. Use length contraction, Equation 29–3, to find the velocity of the probe relative to the spaceship: $v_{21} = 0.661c$
2. Use relativistic velocity addition, Equation 29–4, to find the velocity of the probe relative to the planet: $v_{13} = 0.445c$, $v_{21} = 0.661c$, $v_{23} = 0.855c$
3. Use the velocity of the probe relative to the planet to calculate its contracted length: $L = 5.19$ m

Insight

Therefore, the probe has a length of 10.0 m when it is at rest relative to an observer. After it is launched from the spaceship, it has a length of 7.50 m relative to the ship, and a length of 5.19 m relative to the planet the ship is approaching.

Your Turn

Find the length of the probe relative to the planet if the spaceship moves *away* from the planet with a speed of $0.445c$ when it launches the probe.

(Answers to **Your Turn** *problems are given in the back of the book.)*

To get a better feeling for relativistic velocity addition, consider a spacecraft, initially at rest, that increases its speed by $0.1c$ when it fires its rockets. At first, the speed of the spacecraft increases linearly with the number of times the rockets are fired, as indicated by the line labeled "Classical velocity addition" in **Figure 29–13**. As the spacecraft's speed approaches the speed of light, however, further rocket firings have less and less effect, as seen in the curve labeled "Relativistic velocity addition." In the limit of infinite time—and an infinite number of rocket firings—the speed of the spacecraft approaches c, without ever attaining that value.

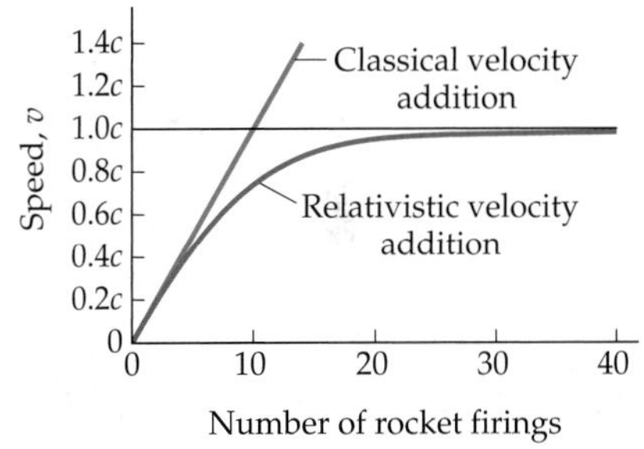

▲ **FIGURE 29–13 Relativistic velocity addition**

A comparison of classical velocity addition and relativistic velocity addition. Classically, a spacecraft that continues to fire its rockets attains a greater and greater speed. The correct, relativistic, result is that the spacecraft approaches the speed of light in the limit of an infinite number of rocket firings.

29–5 Relativistic Momentum

The first postulate of relativity states that the laws of physics are the same for all observers in all inertial frames of reference. Among the most fundamental of these laws are the conservation of momentum and the conservation of energy for an isolated system. In this section we consider the relativistic expression for momentum, leaving energy conservation for the next section.

When one considers the unusual way that relativistic velocities add, as given in Equation 29–4, it comes as no surprise that the classical expression for momentum, $p = mv$, is not valid for all speeds. As an example, we saw in Chapter 9 that

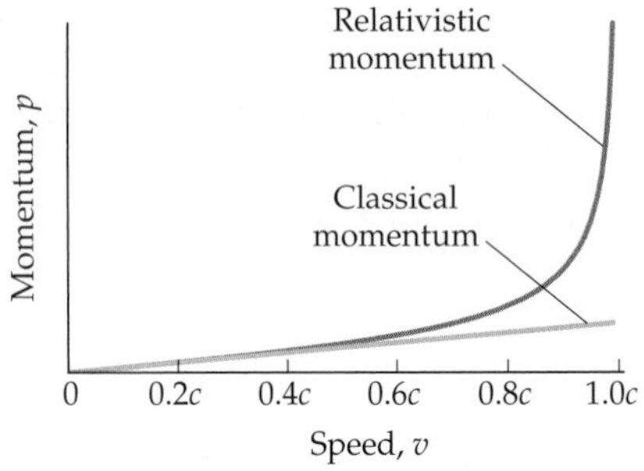

▲ **FIGURE 29–14 Relativistic momentum**

The magnitude of momentum as a function of speed. Classically, the momentum increases linearly with speed, as shown by the straight line. The correct relativistic momentum increases to infinity as the speed of light is approached.

if a large mass with a speed v collides elastically with a small mass at rest, the small mass is given a speed $2v$. Clearly, this cannot happen if the speed of the large mass is greater than $0.5c$, since the small mass cannot have a speed greater than the speed of light. Thus the nonrelativistic relation $p = mv$ must be modified for speeds comparable to c.

A detailed analysis shows that the correct relativistic expression for the magnitude of momentum is the following:

Relativistic Momentum

$$p = \frac{mv}{\sqrt{1 - \frac{v^2}{c^2}}} \qquad 29\text{–}5$$

SI unit: kg · m/s

As v approaches the speed of light, the relativistic momentum becomes significantly larger than the classical momentum, eventually diverging to infinity as $v \rightarrow c$, as shown in **Figure 29–14**. For low speeds the classical and relativistic results agree.

EXERCISE 29–3

Find **(a)** the classical and **(b)** the relativistic momentum of a 2.4-kg mass moving with a speed of $0.81c$.

Solution

a. Evaluate $p = mv$:

$$p = mv = (2.4\text{ kg})(0.81)(3.00 \times 10^8\text{ m/s}) = 5.8 \times 10^8\text{ kg}\cdot\text{m/s}$$

b. Evaluate $p = mv/\sqrt{1 - v^2/c^2}$:

$$p = \frac{mv}{\sqrt{1 - \frac{v^2}{c^2}}} = \frac{(2.4\text{ kg})(0.81)(3.00 \times 10^8\text{ m/s})}{\sqrt{1 - \frac{(0.81c)^2}{c^2}}} = 9.9 \times 10^8\text{ kg}\cdot\text{m/s}$$

As expected, the relativistic momentum is larger in magnitude than the classical momentum.

Next we consider a system in which the relativistic momentum is conserved.

EXAMPLE 29–5 The Unknown Mass

A satellite, initially at rest in deep space, explodes into two pieces. One piece has a mass of 150 kg and moves away from the explosion with a speed of $0.76c$. The other piece moves away from the explosion in the opposite direction with a speed of $0.88c$. Find the mass of the second piece of the satellite.

Picture the Problem

The two pieces of the satellite and their speeds ($0.76c$ and $0.88c$) are indicated in the sketch. Note that the two pieces move in opposite directions, which means that they move away from one another along the same line.

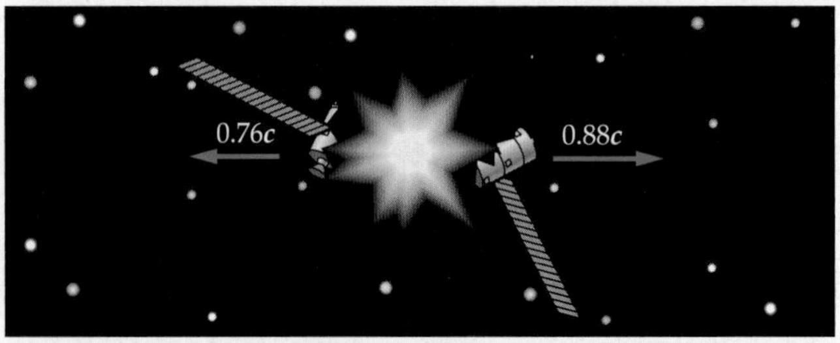

Strategy

The basic idea in this system is that because no external forces act on the satellite, its total momentum must be conserved. The initial momentum is zero; hence, the final momentum must be zero as well. This means that the pieces will move in opposite directions, as mentioned, and with momenta of equal magnitude.

Thus, we begin by calculating the magnitude of the momentum for the first piece of the satellite. Next, we set the momentum of the second piece equal to the same magnitude and solve for the mass.

Solution

1. Calculate the magnitude of the momentum for the piece of the satellite with a mass $m_1 = 150$ kg and a speed $v_1 = 0.76c$:

$$p_1 = \frac{m_1 v_1}{\sqrt{1 - \frac{v_1^2}{c^2}}} = \frac{(150\text{ kg})(0.76)(3.00 \times 10^8\text{ m/s})}{\sqrt{1 - \frac{(0.76c)^2}{c^2}}} = 5.3 \times 10^{10}\text{ kg}\cdot\text{m/s}$$

2. Set the momentum of the second piece of the satellite equal to the momentum of the first piece:

$$p_2 = \frac{m_2 v_2}{\sqrt{1 - \frac{v_2^2}{c^2}}} = p_1$$

3. Solve the above relation for m_2, the mass of the second piece of the satellite, and substitute the numerical values $v_2 = 0.88c$ and $p_1 = 5.3 \times 10^{10}$ kg·m/s:

$$m_2 = \left(\frac{p_1}{v_2}\right)\sqrt{1 - \frac{v_2^2}{c^2}} = \left[\frac{5.3 \times 10^{10}\text{ kg}\cdot\text{m/s}}{(0.88)(3.00 \times 10^8\text{ m/s})}\right]\sqrt{1 - \frac{(0.88c)^2}{c^2}} = 95\text{ kg}$$

Insight

In comparison, a classical calculation, with $p = mv$, gives the erroneous result that the mass of the second piece of the satellite is 130 kg. This overestimate of the mass is due to the omission of the factor $1/\sqrt{1 - v^2/c^2}$ in the classical momentum.

Practice Problem

If the mass of the second piece of the satellite had been 210 kg, what would its speed have been? **[Answer:** $v_2 = 0.64c$.]

Some related homework problems: Problem 44, Problem 45

Equation 29–5 is sometimes thought of in terms of a mass that increases with speed. Suppose, for example, that an object has the mass m_0 when it is at rest; that is, its **rest mass** is m_0. When the speed of the object is v, its momentum is

$$p = \frac{m_0 v}{\sqrt{1 - \frac{v^2}{c^2}}} = \left(\frac{m_0}{\sqrt{1 - \frac{v^2}{c^2}}}\right)v$$

If we now make the identification $m = m_0/\sqrt{1 - v^2/c^2}$ we can write the momentum as follows:

$$p = mv$$

Thus the classical expression for momentum can be used for all speeds if we simply interpret the mass as increasing with speed according to the expression

$$m = \frac{m_0}{\sqrt{1 - \frac{v^2}{c^2}}} \qquad \textbf{29–6}$$

Note that m approaches infinity as $v \to c$. Hence a constant force acting on an object generates less and less acceleration, $a = F/m$, as the speed of light is approached. This gives one further way of seeing that the speed of light cannot be exceeded.

Though Equation 29–6 helps to show why an object of finite mass can never be accelerated to speeds exceeding the speed of light, it must be noted that the concept of a relativistic mass has its limitations. For example, the relativistic kinetic energy of an object is *not* obtained by simply replacing m in $\frac{1}{2}mv^2$ with the expression given in Equation 29–6, as we shall see in detail in the next section.

29–6 Relativistic Energy and $E = mc^2$

We have just seen that, from the point of view of momentum, an object's mass increases as its speed increases. Therefore, when work is done on an object, part of the work goes into increasing its speed, and part goes into increasing its mass. It follows, then, that *mass is another form of energy*. This result, like time dilation, was completely unanticipated before the introduction of the theory of relativity.

Consider, for example, an object whose mass while at rest is m_0. Einstein was able to show that when the object moves with a speed v, its **total energy,** E, is given by the following expression:

Relativistic Energy

$$E = \frac{m_0 c^2}{\sqrt{1 - \frac{v^2}{c^2}}} = mc^2 \qquad \text{29–7}$$

SI unit: J

This is Einstein's most famous result from relativity; that is, $E = mc^2$, where m is the relativistic mass given in Equation 29–6.

Note that the total energy, E, does not vanish when the speed goes to zero, as does the classical kinetic energy. Instead, the energy of an object at rest—its **rest energy,** E_0—is

Rest Energy

$$E_0 = m_0 c^2 \qquad \text{29–8}$$

SI unit: J

Because the speed of light is so large, it follows that the mass of an object times the *speed of light squared* is an enormous amount of energy, as illustrated in the following Exercise.

EXERCISE 29–4

Find the rest energy of a 0.12-kg apple.

Solution

Substituting $m_0 = 0.12$ kg and $c = 3.00 \times 10^8$ m/s in Equation 29–8 we find

$$E_0 = m_0 c^2 = (0.12\ \text{kg})(3.00 \times 10^8\ \text{m/s})^2 = 1.1 \times 10^{16}\ \text{J}$$

To put the result of Exercise 29–4 in context, let's compare it with the total yearly energy usage in the United States, which is about 10^{20} J. This means that if the rest energy of the apple could be converted entirely to usable forms of energy, it could supply the energy needs of the entire United States for about an hour. Put another way, if the rest energy of the apple could be used to light a 100-W light bulb, it would stay lit for about 10 million years.

REAL-WORLD PHYSICS

Nuclear power—converting mass to energy

This example illustrates the basic principle behind the operation of nuclear power plants, in which small decreases in mass—due to various nuclear reactions—are used to generate electrical energy. The type of reactions that power such plants are referred to as **fission reactions,** in which a large nucleus splits into smaller nuclei and neutrons, as described in Section 32–5. For example, the nucleus of a uranium-235 atom may decay into two smaller nuclei and a number of neutrons. Since the mass of the uranium nucleus is greater than the sum of the masses of the fragments of the decay, the reaction releases an enormous amount of energy. In fact, 1 lb of uranium can produce about 3×10^6 kWh of electrical energy, compared with the 1 kWh that can be produced by the combustion of 1 lb of coal.

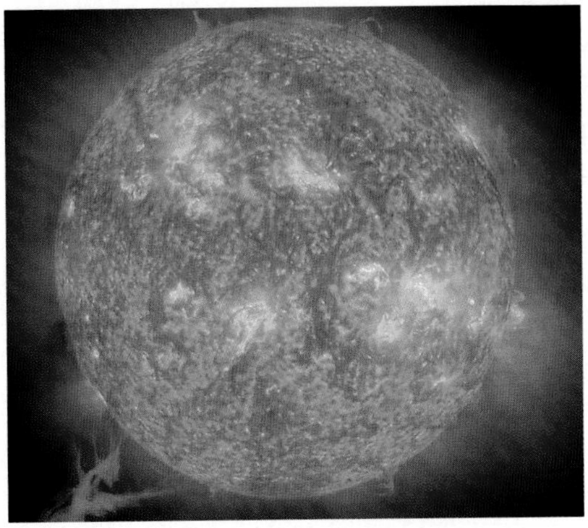

▲ A nuclear reactor (left) converts mass to energy by means of fission reactions, in which large nuclei (such as those of uranium or plutonium) are split into smaller fragments. (The photo shows the Three Mile Island nuclear plant in Pennsylvania; a touch-screen control is in the foreground.) The Sun (right) and other stars are powered by fusion reactions, in which small nuclei (such as those of hydrogen) combine to form heavier ones (such as those of helium). Although this process is utilized in the hydrogen bomb, it has not yet been successfully harnessed as a practical source of power on Earth.

The Sun is also powered by the conversion of mass to energy. In this case, however, the energy is released by **fusion reactions,** in which two very small nuclei combine to form a larger nucleus. The detailed reactions are presented in Section 32–6. In the following Example, we determine the amount of mass lost by the Sun per second.

REAL-WORLD PHYSICS

Converting mass to energy to power the Sun

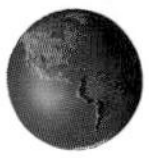

EXAMPLE 29–6 The Prodigal Sun

Energy is radiated by the Sun at the rate of about 3.92×10^{26} W. Find the corresponding decrease in the Sun's mass for every second that it radiates.

Picture the Problem

Our sketch indicates energy radiated continuously by the Sun at the rate $P = 3.92 \times 10^{26}\ \text{W} = 3.92 \times 10^{26}\ \text{J/s}$. We label the energy given off in the time interval $\Delta t = 1.00$ s with ΔE. The corresponding decrease in mass is Δm.

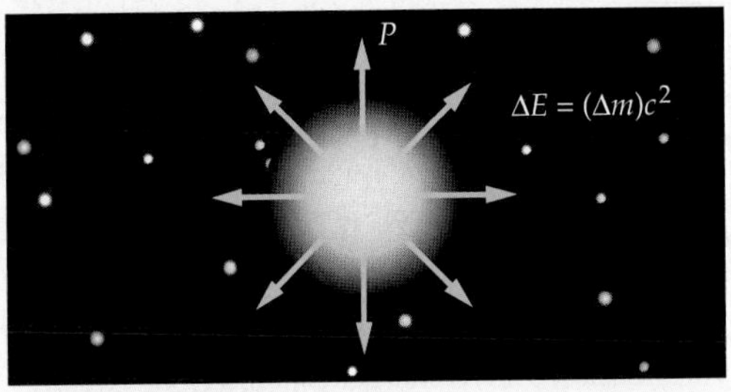

Strategy

If the energy radiated by the Sun in 1.00 s is ΔE, the corresponding decrease in mass, according to the relation $E = mc^2$, is given by $\Delta m = \Delta E/c^2$.

To find ΔE, we simply recall (Chapter 7) that power is energy per time, $P = \Delta E/\Delta t$. Thus the energy radiated by the Sun in 1.00 s is $\Delta E = P\Delta t$, with $\Delta t = 1.00$ s.

Solution

1. Calculate the energy radiated by the Sun in 1.00 s:

$$\Delta E = P\Delta t = (3.92 \times 10^{26}\ \text{J/s})(1.00\ \text{s}) = 3.92 \times 10^{26}\ \text{J}$$

2. Divide ΔE by the speed of light squared, c^2, to find the decrease in mass:

$$\Delta m = \frac{\Delta E}{c^2} = \frac{3.92 \times 10^{26}\ \text{J}}{(3.00 \times 10^8\ \text{m/s})^2} = 4.36 \times 10^9\ \text{kg}$$

Insight

Thus the Sun loses a rather large amount of mass each second—in fact, roughly the equivalent of 2000 space shuttles. Since the Sun has a mass of 1.99×10^{30} kg, however, the mass it loses in 1500 y is only 10^{-10} of its total mass. Even after 1.5 billion years of radiating at its present rate, the Sun will lose a mere 0.01% of its mass. Clearly, the Sun will not evaporate into space anytime soon.

Practice Problem

Find the power radiated by a star whose mass decreases by the mass of the Moon (7.36×10^{22} kg) in half a million years (5.00×10^5 y). [**Answer:** $P = 4.21 \times 10^{26}$ W, slightly more than the power of the Sun.]

Some related homework problems: Problem 52, Problem 57, Problem 63

CONCEPTUAL CHECKPOINT 29–3 Compare the Mass

When you compress a spring between your fingers, does its mass (a) increase, (b) decrease, or (c) stay the same?

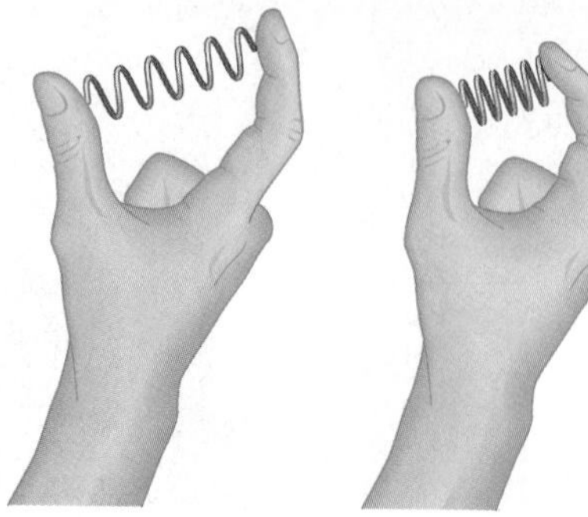

Reasoning and Discussion

When the spring is compressed by an amount x, its energy is increased by the amount $\Delta E = \frac{1}{2}kx^2$, as we saw in Chapter 8. Since the energy of the spring has increased, its mass increases as well, by the amount $\Delta m = \Delta E/c^2$.

Answer:

(a) The mass of the spring increases.

As one might expect, the increase in the mass of a compressed spring is generally too small to be measured. For example, if the energy of a spring increases by 1.00 J, its mass increases by only $\Delta m = (1.00\ \text{J})/c^2 = 1.11 \times 10^{-17}$ kg.

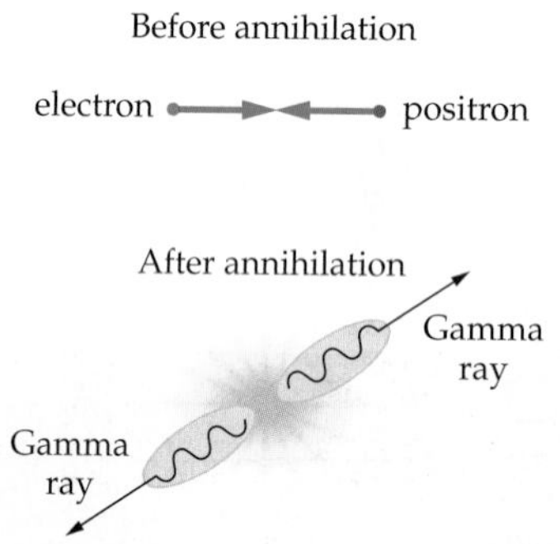

▲ **FIGURE 29–15 Electron-positron annihilation**

An electron and a positron annihilate when they come into contact. The result is the emission of two energetic gamma rays with no mass. The mass of the original particles has been converted into the energy of the gamma rays.

Matter and Antimatter

A particularly interesting aspect of the equivalence of mass and energy is the existence of **antimatter.** For every elementary particle known to exist, there is a corresponding antimatter particle that has precisely the same mass but exactly the opposite charge. For example, an *electron* has a mass $m_e = 9.11 \times 10^{-31}$ kg and a charge $-e = -1.60 \times 10^{-19}$C; an *antielectron* has a mass of 9.11×10^{-31} kg and a charge equal to $+1.60 \times 10^{-19}$ C. Since an antielectron has a positive charge, it is generally referred to as a **positron.**

Antimatter is frequently created in accelerators, where particles collide at speeds approaching the speed of light. In fact, it is possible to create antiatoms in the lab made entirely of antimatter. An intriguing possibility is that the universe may actually contain entire antigalaxies of antimatter.

If this is indeed the case, one would have to be a bit careful about visiting such a galaxy, because particles of matter and antimatter have a rather interesting behavior when they meet—they **annihilate** one another. This situation is illustrated in **Figure 29–15**, where we show an electron and a positron coming into contact. The result is that the particles cease to exist, which satisfies charge conservation, since the net charge of the system is zero before and after the annihilation. As for energy conservation, the mass of the two particles is converted into two gamma rays, which are similar to X-rays only more energetic. Each of the gamma rays must have an energy that is at least $E = m_e c^2$. Thus, in matter-antimatter annihilation the particles vanish in a burst of radiation.

REAL-WORLD PHYSICS: BIO
Positron-emission tomography

Electron-positron annihilation is the basis for the diagnostic imaging technique called positron-emission tomography (PET), which is often used to examine biological processes within the brain, heart, and other organs. In a typical PET brain scan, for example, a patient is injected with glucose (the primary energy source for brain activity) that has been "tagged" with radioactive tracers. These tracers emit positrons in a nuclear reaction described in Section 32–7, and the resulting positrons, in turn, encounter electrons in the brain and undergo annihilation. The resulting gamma rays exit through the patient's skull and are monitored by the PET scanner (**Figure 29–16**), which converts them into false-color images showing the glucose metabolism levels within the brain. Thus, surprising as it may seem, this powerful diagnostic tool actually relies on the annihilation of matter and antimatter inside a person's brain.

The conversion between mass and energy can go the other way, as well. That is, an energetic gamma ray, which has zero mass, can be converted into a particle-antiparticle pair. For example, a gamma ray with an energy of at least $(2m_e)c^2$ can be converted into an electron and a positron; that is, the energy of the gamma ray can be converted into the rest energy of the two particles. Thus we can see that the equivalence of mass and energy, as given by the relation $E = mc^2$, has far-reaching implications.

▲ **FIGURE 29–16 A PET scanner**

In a PET scan, positrons emitted by a radioactively labeled metabolite are annihilated when they collide with electrons. The mass of the two particles is converted to energy in the form of a pair of gamma rays, which are always emitted in diametrically opposite directions. A ring of detectors surrounding the patient records the radiation and uses it to construct an image. In this way it is possible to map the location of particular metabolic activities in the body.

Relativistic Kinetic Energy

When work is done on an isolated object, accelerating it from rest to a finite speed v, its total energy increases, as given by Equation 29–7. We refer to the *increase* in the object's energy as its *kinetic energy*. Thus the total energy, E, of an object is the sum of its rest energy, m_0c^2, and its kinetic energy, K. In particular,

$$E = \frac{m_0c^2}{\sqrt{1 - \frac{v^2}{c^2}}} = m_0c^2 + K$$

Solving for the kinetic energy, we find

Relativistic Kinetic Energy

$$K = \frac{m_0c^2}{\sqrt{1 - \frac{v^2}{c^2}}} - m_0c^2 \qquad \textbf{29–9}$$

SI unit: J

As a check, note that the kinetic energy is zero when the speed is zero, as expected. A comparison between the relativistic and classical kinetic energies is presented in **Figure 29–17**.

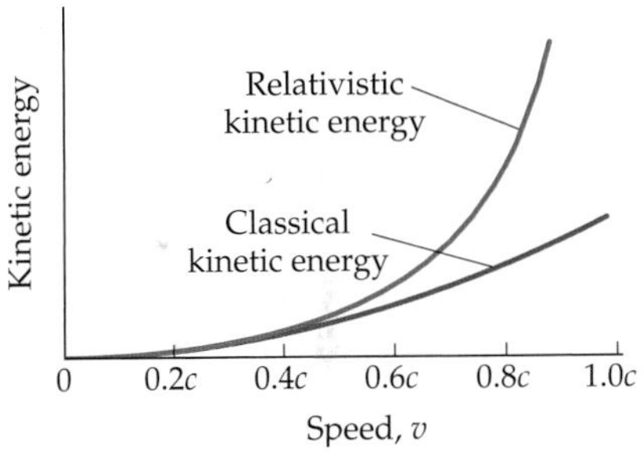

▲ **FIGURE 29–17 Relativistic and classical kinetic energies**

The relativistic kinetic energy (upper curve) goes to infinity as the speed of light is approached. The classical kinetic energy (lower curve) agrees with the relativistic result when the speed is small compared with the speed of light.

Although the expression in Equation 29–9 looks nothing like the familiar classical kinetic energy, $\frac{1}{2}mv^2$, it does approach this value in the limit of small speeds. To see this, we can expand Equation 29–9 for small v using the binomial expansion given in Appendix A. The result is as follows:

$$K = \frac{m_0c^2}{\sqrt{1 - \frac{v^2}{c^2}}} - m_0c^2 = m_0c^2\left[\frac{1}{\sqrt{1 - \frac{v^2}{c^2}}}\right] - m_0c^2$$

$$= m_0c^2\left[1 + \frac{1}{2}\left(\frac{v^2}{c^2}\right) + \frac{3}{8}\left(\frac{v^2}{c^2}\right)^2 + \cdots\right] - m_0c^2$$

The second term in the square brackets is much smaller than the first term for everyday speeds. For example, if the speed v is equal to $0.00001c$ (a rather large everyday speed of roughly 6000 mi/h), the second term is only one ten-millionth of a percent of the first term. Later terms in the expansion are smaller still. Hence, for all practical purposes, the kinetic energy for low speeds is

$$K = m_0c^2\left[1 + \frac{1}{2}\left(\frac{v^2}{c^2}\right)\right] - m_0c^2 = m_0c^2 + \tfrac{1}{2}m_0v^2 - m_0c^2$$

Canceling the rest energy, m_0c^2, we find

$$K = \tfrac{1}{2}m_0v^2$$

Note that the subscript 0 in this expression simply emphasizes the fact that the mass to be used is the rest mass.

We apply the relativistic kinetic energy in the next Example.

PROBLEM-SOLVING NOTE

Rest Mass

To correctly evaluate the relativistic kinetic energy, it is necessary to use the rest mass in Equation 29–9.

EXAMPLE 29–7 Relativistic Kinetic Energy

An observer watching a high-speed spaceship pass by notices that a clock on board runs slow by a factor of 1.50. **(a)** Find the speed of the clock relative to the observer. **(b)** If the rest mass of the clock is 0.320 kg, what is its kinetic energy?

Picture the Problem

Our sketch shows the high-speed spaceship moving by an observer on a distant asteroid. To the observer, the clock in the ship is running slow. In fact, 1.50 s elapse on the observer's clock in the time it takes the spaceship's clock to advance only 1.00 s.

Strategy

a. To begin, we can find the speed of the spaceship, v, by using time dilation, as given in Equation 29–2. In particular, we set $\Delta t = \Delta t_0/\sqrt{1 - v^2/c^2} = 1.50\,\Delta t_0$ and solve for v.

b. Now that we know both the speed and the rest mass ($m_0 = 0.320$ kg) of the clock, we can find its kinetic energy by applying Equation 29–9; $K = m_0c^2/\sqrt{1 - v^2/c^2} - m_0c^2$.

Solution

Part (a)

1. Use time dilation to solve for the speed, v:

$$\Delta t = \frac{\Delta t_0}{\sqrt{1 - \dfrac{v^2}{c^2}}}$$

$$v = c\sqrt{1 - \left(\frac{\Delta t_0}{\Delta t}\right)^2}$$

2. The dilated time, Δt, is greater than the proper time, Δt_0, by a factor of 1.50. Therefore, we substitute $\Delta t_0/\Delta t = 1/1.50$ in the expression for v from step 1:

$$v = c\sqrt{1 - \left(\frac{1}{1.50}\right)^2} = 0.745c$$

Part (b)

3. Substitute $v = 0.745c$ and $m_0 = 0.320$ kg into the expression for the relativistic kinetic energy (Equation 29–9):

$$K = \frac{m_0c^2}{\sqrt{1 - \dfrac{v^2}{c^2}}} - m_0c^2 = m_0c^2\left(\frac{1}{\sqrt{1 - \dfrac{v^2}{c^2}}} - 1\right)$$

$$= m_0c^2\left(\frac{1}{\sqrt{1 - \dfrac{(0.745c)^2}{c^2}}} - 1\right) = m_0c^2(1.50 - 1)$$

$$= (0.320\text{ kg})(3.00 \times 10^8\text{ m/s})^2(1.50 - 1) = 1.44 \times 10^{16}\text{ J}$$

Insight

In comparison, the classical kinetic energy at this speed is only $\frac{1}{2}m_0v^2 = \frac{1}{2}m_0(0.745c)^2 = m_0c^2(0.278) = 7.99 \times 10^{15}\text{ J} < m_0c^2(1.50 - 1) = 1.44 \times 10^{16}$ J. In fact, the classical kinetic energy is *always* less than the relativistic kinetic energy at any given speed, as we can see in Figure 29–17. This is because more work must be done to accelerate a particle that is becoming more massive with speed, as is the case with relativity.

Practice Problem

How fast must the clock be moving if its relativistic kinetic energy is to be 5.00×10^{16} J? **[Answer:** $v = 0.931c$]

Some related homework problems: Problem 49, Problem 59, Problem 60

Once again, we see that the speed of light is the ultimate speed possible for an object of finite rest mass. As is clear from Figure 29–17, the kinetic energy of an object goes to infinity as its speed approaches c. Thus to accelerate an object to the speed of light would require an infinite amount of energy. Any finite amount of work will increase the speed only to a speed less than c.

29–7 The Relativistic Universe

It is a profound understatement to say that relativity has revolutionized our understanding of the universe. If we think back over the results presented in the last several sections—time dilation, length contraction, increasing mass, mass-energy equivalence—it is clear that relativity reveals to us a universe that is far richer and more varied in its behavior than was ever imagined before. In fact, it is often said that the universe is not only stranger than we imagine, but stranger than we *can* imagine.

By way of analogy, it is almost as if we had spent our lives on a small island on the equator. We would have no knowledge of snow or deserts or mountain ranges. Although our knowledge of Earth would be valid for our small island, we would have an incomplete picture of the world. Our situation with respect to relativity is similar. Before relativity, our knowledge of the physical universe seemed complete, with only minor details to be worked out. After all, Newton's laws and other fundamental principles of physics gave correct predictions for virtually everything we experienced. What Einstein revealed with his theory of relativity, however, was that we were seeing only a small part of the whole and that the behavior we experience at low speeds cannot be extended to high speeds. We weren't missing small details—like the mid-ocean islanders, we were missing most of the picture.

Now, it might seem that relativity plays no significant role in our daily lives, since we do not move at speeds approaching the speed of light. In many respects this observation is correct—relativity is not used to design better cars or airplanes, nor is it used to calculate the orbits needed to send astronauts to the Moon or to Mars. On the other hand, every major hospital has a particle accelerator in its basement to produce radioactive elements for various types of treatments. The accelerator brings particles to speeds very close to the speed of light; hence, relativistic effects cannot be ignored. In fact, for an accelerator to work properly, it must be constructed with relativistic effects taken into account. Similarly, GPS technology can provide accurate positional information only by taking into account relativistic effects—both those having to do with speed, as we have seen in the previous sections, as well as relativistic effects due to gravity, to be discussed in the next section. Thus, as these examples show, we now live in a world where relativity is truly an important part of our everyday lives.

▲ **FIGURE 29–18 A frame of reference in a gravitational field**
An observer in an elevator at rest on Earth's surface. If the observer drops or throws a ball, it falls with a downward acceleration of g.

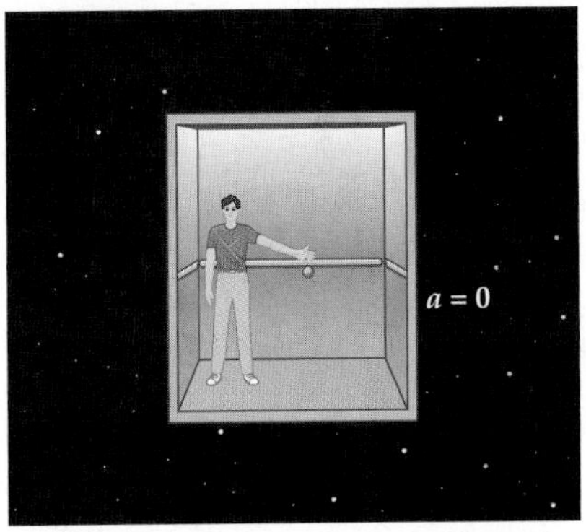

▲ **FIGURE 29–19 An inertial frame of reference with no gravitational field**
An observer in an elevator in deep space experiences weightlessness. If the observer releases an object, it remains at rest.

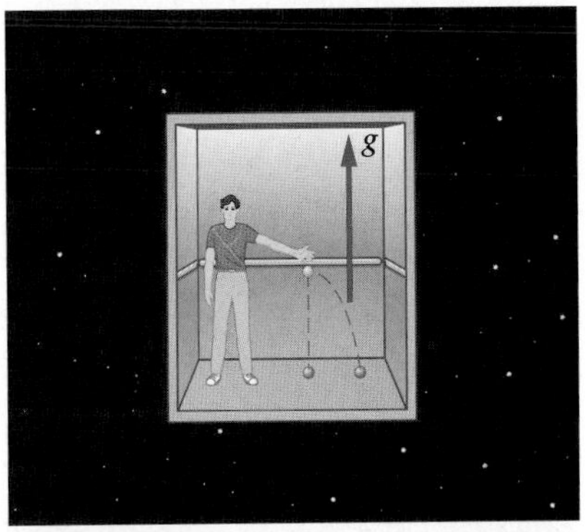

▲ **FIGURE 29–20 An accelerated frame of reference**
If the elevator in Figure 29–19 is given an upward acceleration of magnitude g, the observer in the elevator will note that objects that are dropped or thrown fall toward the floor of the elevator with an acceleration g, just as for the observer in Figure 29–18.

29–8 General Relativity

Einstein's **general theory of relativity** applies to accelerated frames of reference and to gravitation. In fact, the theory provides a link between these two types of physical processes that leads to a new interpretation of gravity.

Consider two observers, both standing within closed elevators. Observer 1 is in an elevator that is at rest on the surface of the Earth, as **Figure 29–18** shows. If this observer drops or throws an object, it falls toward the floor of the elevator with an acceleration equal to the acceleration of gravity.

Observer 2 stands in an identical elevator located in deep space. If this elevator is at rest, or moving with a constant velocity, the observer experiences weightlessness within the elevator, as shown in **Figure 29–19**. If an object is released, it remains in place. Now suppose the elevator is given an upward acceleration equal to the acceleration of gravity, g, as indicated in **Figure 29–20**. An object that is released now remains at rest relative to the background stars while the floor of the elevator accelerates upward toward it with the acceleration g. Similarly, if observer 2 throws the ball horizontally it will follow a parabolic path to the floor, just as for observer 1. In addition, the floor of the elevator exerts a force mg on the feet of observer 1 to give that observer (whose mass is m) an upward acceleration g.

We conclude, then, that when observer 2 conducts an experiment in his accelerating elevator, the results are the same as those obtained by observer 1 in her

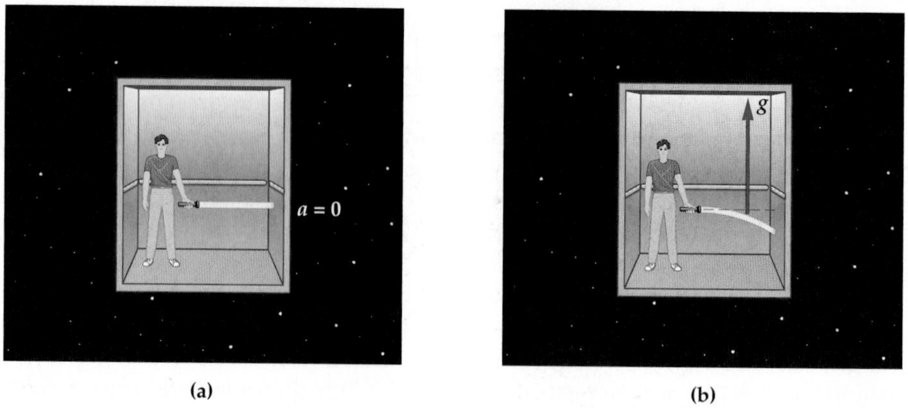

▶ **FIGURE 29–21 A light experiment in two different frames of reference**

(a) In a nonaccelerating elevator, a beam of light travels on a straight line as it crosses the elevator. **(b)** In an accelerated elevator, the elevator moves upward as the light crosses the elevator; hence, the light strikes the opposite wall at a lower level. The path of the light in this case appears parabolic to the observer riding in the elevator.

▲ **FIGURE 29–22 The principle of equivalence**

By the principle of equivalence, a beam of light in a gravitational field should follow a parabolic path, just as in the accelerated elevator in Figure 29–21. The amount of bending of the light's path has been exaggerated here for clarity.

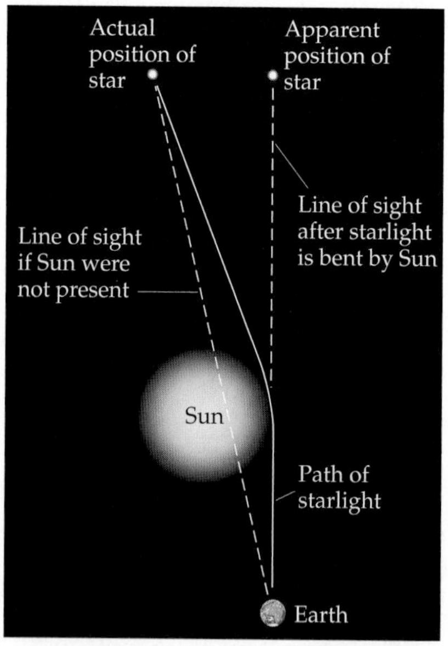

▲ **FIGURE 29–23 Gravitational bending of light**

As light from a distant star passes close to the Sun, it is bent. The result is that an observer on Earth sees the star along a line of sight that is displaced away from the center of the Sun.

elevator at rest on Earth. Einstein extended these observations to a general principle, the **principle of equivalence:**

Principle of Equivalence

All physical experiments conducted in a uniform gravitational field and in an accelerated frame of reference give identical results.

Thus the two observers cannot tell, without looking outside the elevator, whether they are at rest in a gravitational field or in deep space in an accelerating elevator.

Now let's apply the principle of equivalence to a simple experiment involving light. If the observer in **Figure 29–21 (a)** shines a flashlight toward the opposite wall of the elevator (which has zero acceleration), the light strikes the wall at its initial height. If the same experiment is conducted in an accelerated elevator, as in **Figure 29–21 (b)**, the elevator is accelerating upward during the time the light travels across the elevator. Thus, by the time the light reaches the far wall, it strikes it at a lower level. In fact, the light has followed a parabolic path, just as one would expect for a ball that was projected horizontally.

Applying the principle of equivalence, Einstein concluded that a beam of light in a gravitational field must also bend downward, just as it does in an accelerated elevator; that is, *gravity bends light*. This phenomenon is illustrated in **Figure 29–22**, where the amount of bending has been exaggerated for clarity.

In order to put Einstein's prediction to the test, it is necessary to increase the amount of bending as much as possible to make it large enough to be measured. Thus, we need to use the strongest gravitational field available. In our solar system, the strongest gravitational field is provided by the Sun; hence, experiments were planned to look for the bending of light produced by the Sun.

To see what effect the Sun's gravitational field might have on light, consider the Sun and a ray of light from a distant star, as shown in **Figure 29–23**. As the light passes the Sun, it is bent, as indicated. So an observer on Earth must look in a direction that is farther from the Sun than the actual direction of the distant star—the Sun's gravitational field displaces the distant stars to apparent positions farther from the Sun. If we imagine the Sun moving in front of a background field of stars, as in **Figure 29–24**, the stars near the Sun are displaced outward. It is almost as if the Sun were a lens, slightly distorting the scene behind it.

Because the Sun is so bright, it is possible to carry out an experiment like that shown in Figures 29–23 and 29–24 only during a total eclipse of the Sun, when the Sun's light is blocked by the Moon. During the eclipse, photographs can be taken to show the positions of the background stars. Later, these photographs can be compared with photographs of the same star field taken 6 months later, when the Sun is on the other side of Earth. Comparing the photographs allows one to measure the displacement of the stars. This experiment was carried out during an expedition to Africa in 1919 by Sir Arthur Eddington. His results confirmed the predictions of the general theory of relativity and made Einstein a household name.

▲ **FIGURE 29–24 Bending of light near the Sun**
As the Sun moves across a starry background, the stars near it appear to be displaced outward, away from the center of the Sun. This is the effect that was used in the first experimental confirmation of general relativity.

Since gravity can bend light, the more powerful the gravitational force, the more the bending, and the more dramatic the results. In Figure 12–18 we saw what can happen when a large galaxy or a cluster of galaxies, with its immense gravitational field, lies between us and a more distant galaxy. The intermediate galaxy can produce significant bending of light, resulting in multiple images of the distant galaxy that can form arcs or crosses. Some examples of such *gravitational lensing* are shown in **Figure 29–25**.

REAL-WORLD PHYSICS
Gravitational lensing

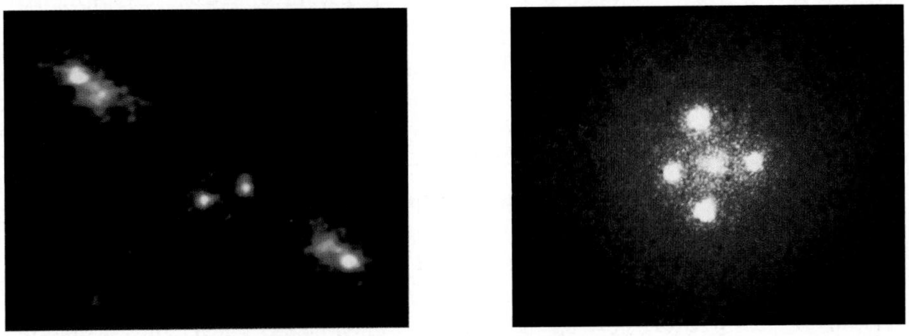

▲ **FIGURE 29–25 Gravitational lensing**
The images created by gravitational lenses take a number of forms. Sometimes the light from a distant object is stretched out into an arc or even a complete ring. In other cases, such as that of the distant quasar at left, we may see a pair of images. (The quasar appears at the upper left and lower right. The lensing galaxy is not visible in the photo—the small dots at center are unrelated objects.) In still other instances, four images may be produced, as in the famous "Einstein cross" shown at right. The lensing galaxy at the center is some 400 million light-years from us, while the quasar whose multiple images surround it is about 20 times farther away.

An intense gravitational field can also be produced when a star burns up its nuclear fuel and collapses to a very small size. In such a case, the gravitational field can become strong enough to actually trap light; that is, to bend it around to the point that it cannot escape the star. Since such a star cannot emit light of its own, it is referred to as a **black hole.** Black holes, by definition, cannot be directly observed; however, their presence can be inferred by their gravitational effects on other bodies. It is also possible to detect the intense radiation emitted by ionized matter as it falls into a black hole. (Recall from Section 25–1 that accelerated charges produce electromagnetic radiation.) By these and other means, the existence of black holes in the centers of many galaxies has been firmly established. In fact, it is now thought that black holes may be relatively common in the universe.

REAL-WORLD PHYSICS
Black holes

Recall that in Chapter 12 we calculated the escape speed for Earth. The result was

$$v_e = \sqrt{\frac{2GM_E}{R_E}}$$

▲ Because they swallow up all light, black holes are themselves invisible. However, they can be detected indirectly in several ways. If one member of a binary star collapses to become a black hole, an accretion disk may form around it, as shown in the artist's conception at left. The accretion disk is a ring of matter wrenched from its companion, whirling around at ever greater speeds as it spirals into the black hole. The radiation emitted by this matter as it falls into the abyss is the signature of a black hole. The same mechanism, on a much vaster scale, probably accounts for the enormous radiation from active galaxies, such as M87 (right), a giant elliptical galaxy with an enormous jet of matter emanating from its nucleus. Black holes millions of times more massive than the Sun are thought to be present in the cores of such galaxies—and possibly in all galaxies.

Replacing M_E with M and R_E with R gives the escape speed for any spherical body of mass M and radius R. Now, if we set the escape speed equal to the speed of light, $v_e = c$, we find the following:

$$v_e = \sqrt{\frac{2GM}{R}} = c$$

Solving for R, we find that for an astronomical body to be a black hole, its radius must be no greater than

$$R = \frac{2GM}{c^2} \qquad \textbf{29–10}$$

Although this calculation is too simplistic and not entirely correct, the final result given in Equation 29–10, known as the **Schwarzschild radius,** does indeed agree with the results of general relativity. In the following Exercise, we calculate the radius of a black hole for a specific case.

EXERCISE 29–5

To what radius must Earth be condensed for it to become a black hole?

Solution

Substituting G, M_E, and c in Equation 29–10 we get

$$R = \frac{2GM_E}{c^2} = \frac{2(6.67 \times 10^{-11}\,\text{N}\cdot\text{m}^2/\text{kg}^2)(5.98 \times 10^{24}\,\text{kg})}{(3.00 \times 10^8\,\text{m/s})^2} = 8.86\ \text{mm}$$

Thus Earth would have to be reduced to roughly the size of a walnut for it to become a black hole.

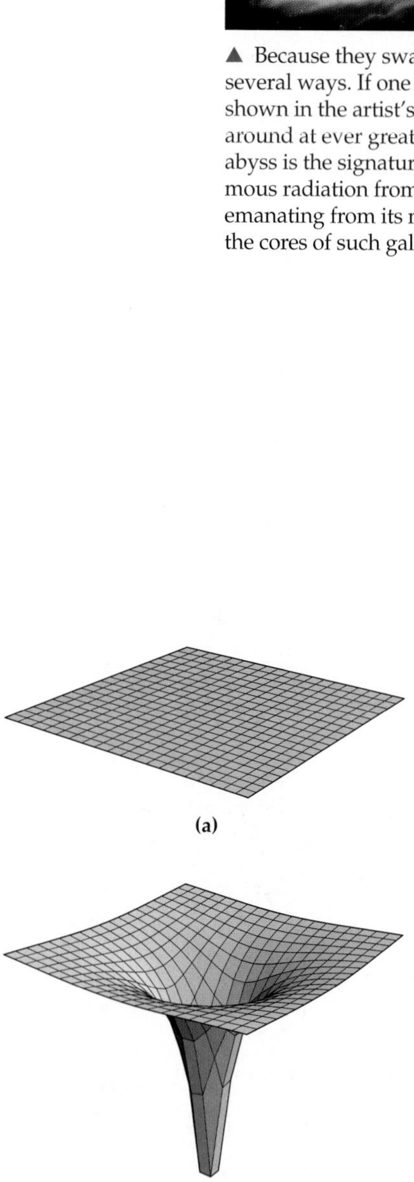

(b)

▲ **FIGURE 29–26 Warped space and black holes**

(a) Regions of space that are far from any large masses can be thought of as flat. In these regions, light propagates in straight lines. **(b)** Near a large concentrated mass, such as a black hole, space can be thought of as "warped." In these regions, the paths of light rays are bent.

A convenient way to visualize the effects of intense gravitational fields is to think of space as a sheet of rubber with a square array of grid lines, as shown in **Figure 29–26**. In the absence of mass, the sheet is flat, and the grid lines are straight, as in Figure 29–26 (a). In this case, a beam of light follows a straight-line path, parallel to a grid line. If a large mass is present, however, the sheet of rubber is deformed, as in Figure 29–26 (b), and light rays follow the curved paths of the grid lines. In cases where one large mass orbits another, as in **Figure 29–27**, the result is a series of ripples moving outward through the rubber sheet. These ripples represent **gravity waves,** one of the many intriguing predictions of general relativity.

When a gravity wave passes through a given region of space, it causes a local deformation of space, as Figure 29–27 suggests. Early attempts to detect gravity waves were based on measuring the distortion a gravity wave would produce in a large metal bar. Unfortunately, the sensitivity of these devices was too low to detect the weak waves that are thought to pass through the Earth all the time. Ironically, these detectors *were* sensitive enough to detect the relatively strong gravity waves that must have accompanied the 1987 supernova explosion in the nearby Large Magellanic Cloud (a small satellite galaxy of our Milky Way). As luck would have it, however, none of the detectors were operating at the time.

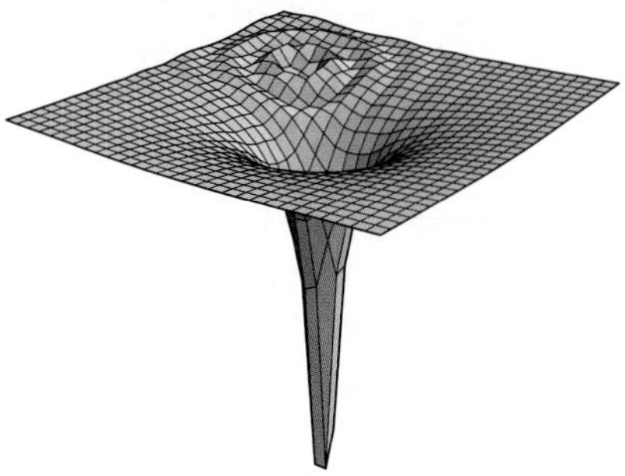

▲ **FIGURE 29–27 Gravity waves**
Gravity waves can be thought of as "ripples" in the warped space described in Figure 29–26. In the case illustrated here, a neutron star orbits a black hole. As a result of its acceleration, the neutron star emits gravity waves.

The next generation of gravity wave detectors is now nearing completion. These detectors, which go by the name of Laser Interferometer Gravitational Wave Observatory, or LIGO for short, should be sensitive enough to detect several gravitational wave events per year. One type of event that LIGO will be looking for is the final death spiral of a neutron star as it plunges into a black hole. The neutron star might orbit the black hole for 150,000 years, but only in the last 15 minutes of its life does it find fame, because during those few minutes its acceleration is great enough to produce gravity waves detectable by LIGO.

The basic operating principle of LIGO is to send a laser beam in two different directions along 4-km-long vacuum tubes. At the far end of each tube the beam is reflected back to its starting point. If the two beams travel equal distances, they will interfere constructively when reunited, just like the two rays that meet at the center of a two-slit interference pattern, as discussed in Section 28–2. If the path lengths along the two tunnels are slightly different, however, the beams will have at least partial destructive interference when they combine. The resulting difference in intensity is what LIGO will measure.

The connection with gravity waves is indicated in **Figure 29–28**. Here we see that as a gravity wave passes a LIGO facility, it will cause one tube to increase in length and the other tube to decrease in length. Hence, if the observatory can detect a small enough change in length, it will "see" the gravity wave. Just how much change in length is a gravity wave expected to produce? Incredibly, a typical strong gravity wave will change the length of a tube by less than the diameter of an atomic nucleus. Thus, the task for LIGO is to measure this small change in length in a tube that is 4 km long—no easy task, but the LIGO scientists are confident it can be done.

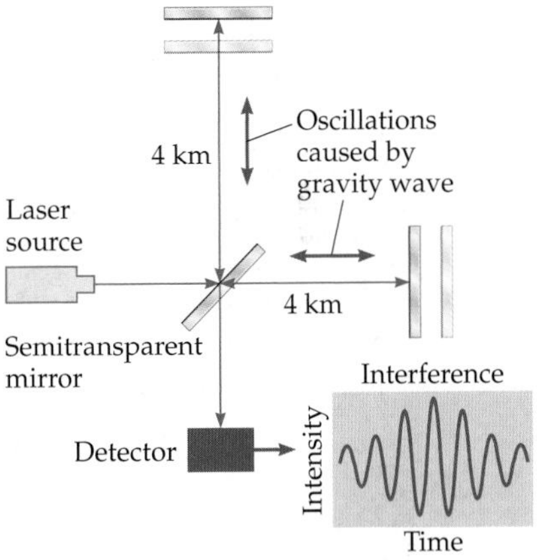

▲ **FIGURE 29–28 How LIGO detects a gravity wave**
When a gravity wave passes through the Earth, it "distorts" everything in its path. The idea behind LIGO is to detect these distortions using interference effects similar to those observed in Young's two-slit experiment (Chapter 28). In particular, LIGO sends laser light along two 4-km vacuum tubes oriented at right angles to one another. When a gravity wave passes, one arm will be lengthened and the other arm will be shortened. The resulting difference in path length changes the interference pattern, allowing us to "see" the gravity wave.

If gravity waves are indeed observed in the early years of the twenty-first century, as expected, it will open an entirely new window on the universe. So far, information about the universe comes to our telescopes by way of electromagnetic waves. We have certainly learned a lot about the universe in this way; in fact, whenever we observe in a different part of the electromagnetic spectrum, we find surprising new phenomena. One can only imagine what additional wonders will be observed when an entirely different spectrum—the gravity wave spectrum—is explored.

Chapter Summary

29–1 The Postulates of Special Relativity

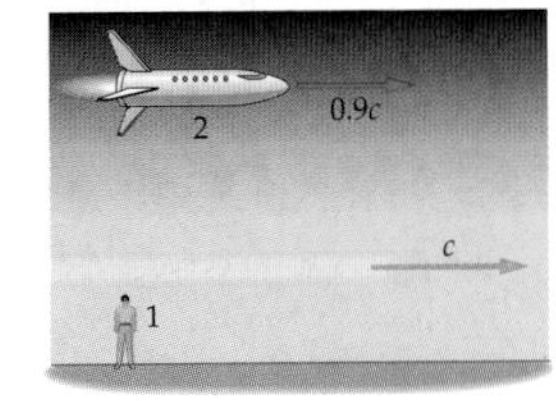

Einstein's theory of relativity is based on just two postulates.

Equivalence of Physical Laws
The laws of physics are the same in all inertial frames of reference.

Constancy of the Speed of Light
The speed of light in a vacuum, $c = 3.00 \times 10^8$ m/s, is the same in all inertial frames of reference, independent of the motion of the source or the receiver.

29–2 The Relativity of Time and Time Dilation

Clocks that move relative to one another keep time at different rates. In particular, a moving clock runs slower than one that is at rest relative to a given observer.

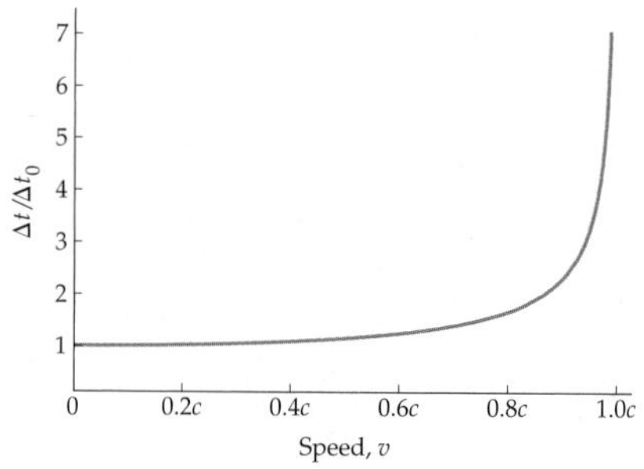

Proper Time
The proper time, Δt_0, is the amount of time separating two events that occur at the same location.

Time Dilation
If two events separated by a proper time Δt_0 occur in a frame of reference moving with a speed v relative to an observer, the dilated time measured by the observer, Δt, is given by

$$\Delta t = \frac{\Delta t_0}{\sqrt{1 - v^2/c^2}} \quad \text{29–2}$$

Space Travel and Biological Aging
Time dilation applies equally to all physical processes, including chemical reactions and biological functions.

29–3 The Relativity of Length and Length Contraction

The length of an object depends on its speed relative to a given observer.

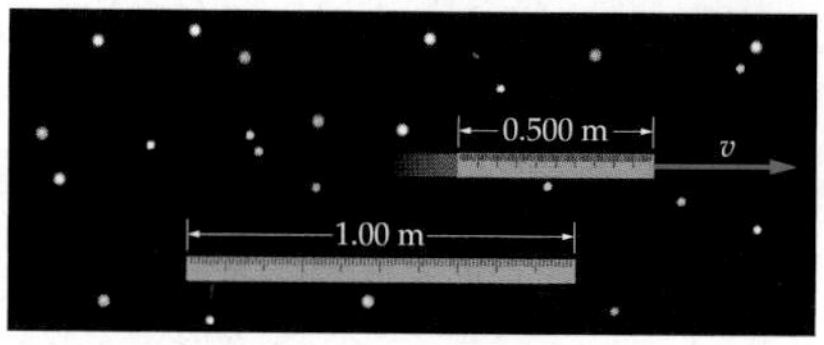

Proper Length
The proper length, L_0, is the distance between two points as measured by an observer who is at rest with respect to them.

Contracted Length
An object with a proper length L_0 moving with a speed v relative to an observer has a contracted length L given by:

$$L = L_0\sqrt{1 - \frac{v^2}{c^2}} \quad \text{29–3}$$

Direction of Contraction
Lengths contract only in the direction of motion.

29–4 The Relativistic Addition of Velocities

Simple velocity addition, $v = v_1 + v_2$, is valid only in the limit of very small speeds.

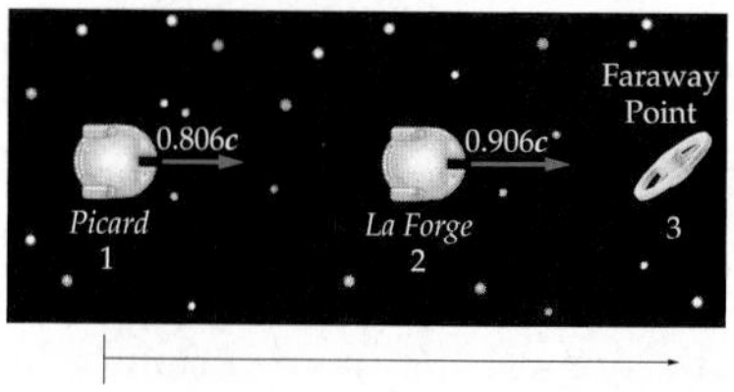

Relativistic Addition of Velocities
Suppose object 1 moves with a velocity v_{13} relative to object 3. If object 2 moves along the same straight line with a velocity v_{21} relative to object 1, the velocity of object 2 relative to object 3, v_{23}, is

$$v_{23} = \frac{v_{21} + v_{13}}{1 + \dfrac{v_{21}v_{13}}{c^2}} \quad \text{29–4}$$

Ultimate Speed
If two velocities, v_1 and v_2, are less than the speed of light, c, then their relativistic sum, v, is also less than c. Thus it is not possible to increase the speed of an object from a value less than c to a value greater than c.

29–5 Relativisitic Momentum

The momentum of an object of mass m and speed v is

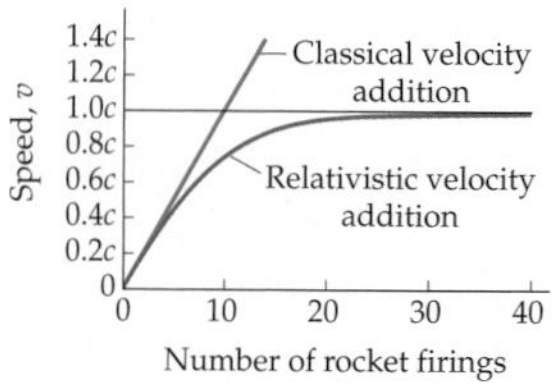

$$p = \frac{mv}{\sqrt{1 - \dfrac{v^2}{c^2}}} \quad \text{29–5}$$

This expression is valid for all speeds between zero and the speed of light and reduces to $p = mv$ in the limit of small speeds.

29–6 Relativisitic Energy and $E = mc^2$

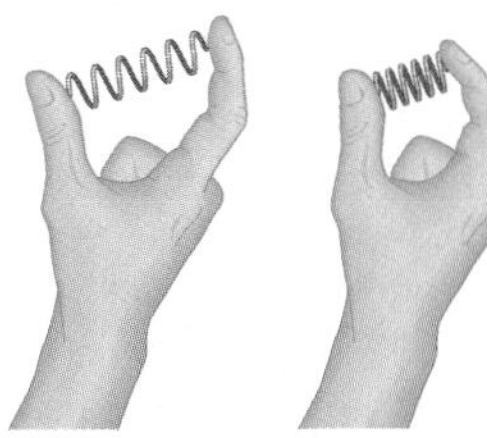

One of the most important results of relativity is that mass is another form of energy. To put it another way, mass and energy are two different aspects of the same quantity.

Relativistic Energy
The total energy, E, of an object with rest mass m_0 and speed v is

$$E = \frac{m_0c^2}{\sqrt{1 - \frac{v^2}{c^2}}} = mc^2 \qquad \text{29–7}$$

Rest Energy
When an object is at rest, its energy E_0 is

$$E_0 = m_0c^2 \qquad \text{29–8}$$

Relativistic Kinetic Energy
The relativistic kinetic energy, K, of an object of rest mass m_0 moving with a speed v is its total energy, E, minus its rest energy, E_0. In particular,

$$K = \frac{m_0c^2}{\sqrt{1 - \frac{v^2}{c^2}}} - m_0c^2 \qquad \text{29–9}$$

29–8 General Relativity

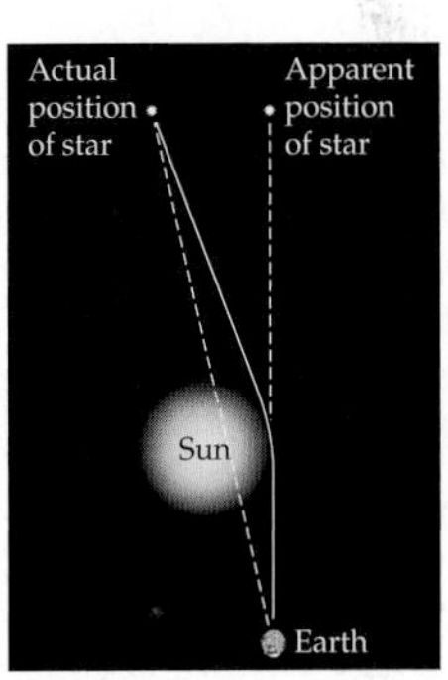

General relativity deals with accelerated frames of reference and with gravity.

Principle of Equivalence
One of the basic principles on which general relativity is founded is the following: All physical experiments conducted in a gravitational field and in an accelerated frame of reference give identical results.

Radius of a Black Hole
For an astronomical body of mass M and radius R to be a black hole, its radius must be less than or equal to the following value, known as the Schwarzschild radius:

$$R = \frac{2GM}{c^2} \qquad \text{29–10}$$

Problem-Solving Summary

Type of Problem	Relevant Physical Concepts	Related Examples
Find the time between two events as measured by different observers.	The time between events that occur at the same location is the proper time, Δt_0. The time measured by an observer with a relative speed v is $\Delta t = \Delta t_0/\sqrt{1 - v^2/c^2}$.	Examples 29–1, 29–2 Active Example 29–1
Find the distance between two points as measured by different observers.	The distance between points that are at rest relative to the observer is the proper length, L_0. The distance measured by an observer with a relative speed v is $L = L_0\sqrt{1 - v^2/c^2}$.	Example 29–3
Add velocities relativistically.	The correct rule for adding two velocities, v_{21} and v_{13}, to obtain the total velocity v_{23} is $v_{23} = (v_{21} + v_{13})/(1 + v_{21}v_{13}/c^2)$. In the limit of small velocities, this equation reduces to simple addition, as one would expect. For velocities comparable to the speed of light, simple addition fails, because the final velocity can never be greater than c.	Example 29–4 Active Example 29–2
Find the relativistic momentum.	The momentum of an object with a rest mass m_0 and speed v is $p = m_0v/\sqrt{1 - v^2/c^2}$.	Example 29–5
Find the energy equivalence of a given mass.	The rest energy of an object with a rest mass m_0 is $E_0 = m_0c^2$.	Example 29–6
Find the relativistic kinetic energy.	The kinetic energy of an object with a rest mass m_0 and speed v is $K = m_0c^2/\sqrt{1 - v^2/c^2} - m_0c^2$.	Example 29–7

Conceptual Questions

(Answers to odd-numbered Conceptual Questions can be found in the back of the book.)

1. Some distant galaxies are moving away from us at speeds greater than $0.5c$. What is the speed of the light received on Earth from these galaxies? Explain.
2. The speed of light in glass is less than c. Why is this not a violation of the second postulate of relativity?
3. How would velocities add if the speed of light were infinitely large? Justify your answer by considering Equation 29–4.
4. Describe some of the everyday consequences that would follow if the speed of light were 35 mi/h.
5. When we view a distant galaxy, we notice that the light coming from it has a longer wavelength (it is "red shifted") than the corresponding light here on Earth. Is this consistent with the postulate that all observers measure the same speed of light? Explain.
6. According to the theory of relativity, the maximum speed for any particle is the speed of light. Is there a similar restriction on the maximum energy of a particle? Is there a maximum momentum? Explain.
7. Give an argument that shows that an object of finite mass cannot be accelerated from rest to a speed greater than the speed of light in a vacuum.

Conceptual Exercises

(Answers to odd-numbered Conceptual Exercises can be found in the back of the book.)

1. You are in a spaceship, traveling directly away from the Moon with a speed of $0.9c$. A light signal is sent in your direction from the surface of the Moon. As the signal passes your ship, do you measure its speed to be greater than, less than, or equal to $0.1c$? Explain.
2. A street performer tosses a ball straight up into the air (event 1) and then catches it in his mouth (event 2). For each of the following observers, state whether the time they measure between these two events is the proper time or the dilated time: **(a)** the street performer; **(b)** a stationary observer on the other side of the street; **(c)** a person sitting at home watching the performance on TV; **(d)** a person observing the performance from a moving car.
3. Two observers are moving relative to one another. Which of the following quantities will they always measure to have the same value: **(a)** their relative speed; **(b)** the time between two events; **(c)** the length of an object; **(d)** the speed of light in a vacuum; **(e)** the speed of a third observer?
4. A clock in a moving rocket runs slow. If the rocket reverses direction, does the clock run slow or fast? Explain.
5. You are standing next to a runway as an airplane lands. **(a)** If you and the pilot observe a clock in the cockpit, which of you measures the proper time? **(b)** If you and the pilot observe a large clock on the control tower, which of you measures the proper time? **(c)** Which of you measures the proper length of the airplane? **(d)** Which of you measures the proper length of the runway?
6. Suppose you are a traveling salesman for the Spacely Sprockets Company. You travel on a spaceship that reaches speeds near the speed of light, and you are paid by the hour. When you return to Earth after a sales trip, would you prefer to be paid according to clock 1, at Spacely Sprockets headquarters on Earth, or according to clock 2, on the spaceship in which you travel? Explain.
7. Which clock runs slower relative to a clock on the North Pole, clock 1 on an airplane flying from New York to Los Angeles, or clock 2 on an airplane flying from Los Angeles to New York? Explain.
8. If the speed of light were larger than 3.00×10^8 m/s, would the effects of length contraction be greater or less than they are now? Explain.
9. Particles A through D have the following rest energies and total energies:

Particle	Rest Energy	Total Energy
A	$6E$	$6E$
B	$2E$	$4E$
C	$4E$	$6E$
D	$3E$	$4E$

Rank these particles in order of increasing **(a)** rest mass, **(b)** kinetic energy, and **(c)** speed. Indicate ties where appropriate.

10. An apple drops from the bough of a tree to the ground. Is the mass of the apple near the top of its fall greater than, less than, or the same as its mass after it has landed? Explain.
11. Consider two apple pies that are identical in every respect, except that pie 1 is piping hot and pie 2 is at room temperature. If identical forces are applied to the two pies, is the acceleration of pie 1 greater than, less than, or equal to the acceleration of pie 2? Explain.
12. Is the mass of a warm cup of tea greater than, less than, or the same as the mass of the same cup of tea when it has cooled? Explain.
13. An uncharged capacitor is charged by moving some electrons from one plate of the capacitor to the other plate. Is the mass of the charged capacitor greater than, less than, or the same as the mass of the uncharged capacitor? Explain.

Problems

Note: **IP** *denotes an integrated conceptual/quantitative problem.* **BIO** *identifies problems of biological or medical interest. Red bullets (•, ••, •••) are used to indicate the level of difficulty of each problem.*

(For speeds, v, that are much less than the speed of light, c, the following expansion may be used: $\dfrac{1}{\sqrt{1 - v^2/c^2}} \approx 1 + \frac{1}{2}\dfrac{v^2}{c^2}$.*)*

Section 29–1 The Postulates of Special Relativity

1. • Albert is piloting his spaceship, heading east with a speed of $0.90c$. Albert's ship sends a light beam in the forward (eastward) direction, which travels away from his ship at a speed c. Meanwhile, Isaac is piloting his ship in the westward direction, also at $0.90c$, toward Albert's ship. With what speed does Isaac see Albert's light beam pass his ship?

Section 29–2 The Relativity of Time and Time Dilation

2. • A neon sign in front of a café flashes on and off once every 4.5 s, as measured by the head cook. How much time elapses between flashes of the sign as measured by an astronaut in a spaceship moving toward Earth at a speed of $0.84c$?

3. • A lighthouse sweeps its beam of light around in a circle once every 7.5 s. To an observer in a spaceship moving away from Earth, the beam of light completes one full circle every 12 s. What is the speed of the spaceship relative to Earth?

4. • Refer to Example 29–1. How much does Benny age if he travels to Vega with a speed of $0.9995c$?

5. • As a spaceship flies past at a speed v, you observe that 1.0000 s elapses on the ship's clock in the same time that 1.0000 min elapses on Earth. How fast is the ship traveling, relative to the Earth? (Express your answer as a fraction of the speed of light.)

6. • Donovan Bailey set a world record for the 100-m dash on July 27, 1996. If observers on a spaceship moving with a speed of $0.7705c$ relative to Earth saw Donovan Bailey's run and measured his time to be 15.44 s, find the time that was recorded on Earth.

7. • Find the average distance (in the Earth's frame of reference) covered by the muons in Example 29–2 if their speed relative to Earth is $0.750c$.

8. •• An elementary particle called a pi meson (or pion for short) has an average lifetime of 2.6×10^{-8} s when at rest. If a pion moves with a speed of $0.99c$ relative to Earth, find **(a)** the average lifetime of the pion as measured by an observer on Earth and **(b)** the average distance traveled by the pion as measured by the same observer. **(c)** How far would the pion have traveled relative to Earth if relativistic time dilation did not occur?

9. •• The Σ^- is an exotic particle that has a lifetime (when at rest) of 0.15 ns. How fast would it have to travel in order for its lifetime, as measured by laboratory clocks, to be 0.25 ns?

10. •• **IP** **(a)** Is it possible for you to travel far enough and fast enough so that when you return from a trip, you are younger than your stay-at-home sister, who was born 5.0 y after you? **(b)** Suppose you fly on a rocket with a speed $v = 0.99c$ for 1 y, according to the ship's clocks and calendars. How much time elapses on Earth during your 1-y trip? **(c)** If you were 22 y old when you left home and your sister was 17, what are your ages when you return?

11. •• The radar antenna on a navy ship rotates with an angular speed of 0.29 rad/s. What is the angular speed of the antenna as measured by an observer moving away from the antenna with a speed of $0.72c$?

12. •• An observer moving toward Earth with a speed of $0.95c$ notices that it takes 5.0 min for a person to fill her car with gas. Suppose, instead, that the observer had been moving away from Earth with a speed of $0.80c$. How much time would the observer have measured for the car to be filled in this case?

13. •• **IP** An astronaut moving with a speed of $0.65c$ relative to Earth measures her heart rate to be 72 beats per minute. **(a)** When an Earth-based observer measures the astronaut's heart rate, is the result greater than, less than, or equal to 72 beats per minute? Explain. **(b)** Calculate the astronaut's heart rate as measured on Earth.

14. •• **BIO** Newly sprouted sunflowers can grow at the rate of 0.30 in. per day. One such sunflower is left on Earth, and an identical one is placed on a spacecraft that is traveling away from Earth with a speed of $0.94c$. How tall is the sunflower on the spacecraft when the one on Earth is 2.0 in. high?

15. •• An astronaut travels to Mars with a speed of 8350 m/s. After 1 mo (30.0 d) of travel, as measured by clocks on Earth, how much difference is there between the Earth clock and the spaceship clock? Give your answer in seconds.

16. •• As measured in Earth's frame of reference, two planets are 424,000 km apart. A spaceship flies from one planet to the other with a constant velocity, and the clocks on the ship show that the trip lasts only 1.00 s. How fast is the ship traveling?

17. •• Captain Jean-Luc is piloting the USS *Enterprise XXIII* at a constant speed $v = 0.825c$. As the *Enterprise* passes the planet Vulcan, he notices that his watch and the Vulcan clocks both read 1:00 P.M. At 3:00 P.M., according to his watch, the *Enterprise* passes the planet Endor. If the Vulcan and Endor clocks are synchronized with each other, what time do the Endor clocks read when the *Enterprise* passes by?

18. •• **IP** A plane flies with a constant velocity of 212 m/s. The clocks on the plane show that it takes exactly 2.00 h to travel a certain distance. **(a)** According to ground-based clocks, will the flight take slightly more or slightly less than 2.00 h? **(b)** Calculate how much longer or shorter than 2.00 h this flight will last, according to clocks on the ground.

Section 29–3 The Relativity of Length and Length Contraction

19. • How fast does a 250-m spaceship move relative to an observer who measures the ship's length to be 150 m?

20. • Suppose the speed of light in a vacuum were only 25.0 mi/h. Find the length of a bicycle being ridden at a speed of 20.0 mi/h as measured by an observer sitting on a park bench, given that its proper length is 1.89 m.

21. • A rectangular painting is 124 cm wide and 80.5 cm high, as indicated in **Figure 29–29**. At what speed, v, must the painting move parallel to its width if it is to appear to be square?

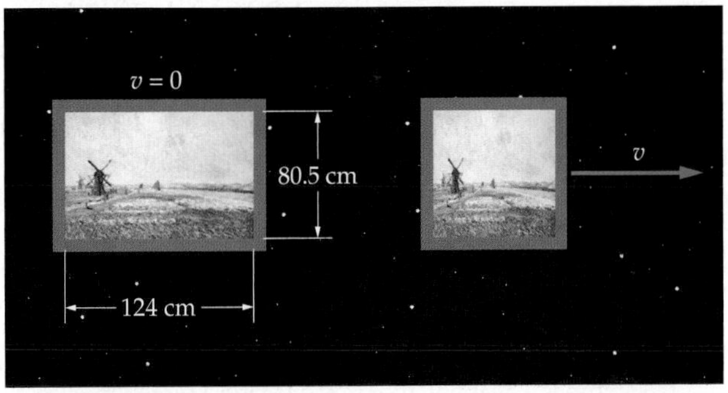

▲ **FIGURE 29–29** Problem 21

22. • The Linac portion of the Fermilab Tevatron contains a high-vacuum tube that is 64 m long, through which protons travel with an average speed $v = 0.65c$. How long is the Linac tube, as measured in the proton's frame of reference?

23. •• A cubical box is 0.50 m on a side. **(a)** What are the dimensions of the box as measured by an observer moving with a speed of $0.88c$ parallel to one of the edges of the box? **(b)** What is the volume of the box, as measured by this observer?

24. •• When parked, your car is 5.0 m long. Unfortunately, your garage is only 4.0 m long. **(a)** How fast would your car have to be moving for an observer on the ground to find your car shorter than your garage? **(b)** When you are driving at this speed, how long is your garage, as measured in the car's frame of reference?

25. •• An astronaut travels to a distant star with a speed of $0.55c$ relative to Earth. From the astronaut's point of view, the star is 7.5 ly from Earth. On the return trip, the astronaut travels with a speed of $0.89c$ relative to Earth. What is the distance covered

on the return trip, as measured by the astronaut? Give your answer in light-years.

26. •• **IP** Laboratory measurements show that an electron traveled 3.50 cm in a time of 0.200 ns. **(a)** In the rest frame of the electron, did the lab travel a distance greater than or less than 3.50 cm? Explain. **(b)** What is the electron's speed? **(c)** In the electron's frame of reference, how far did the laboratory travel?

27. •• You and a friend travel through space in identical spaceships. Your friend informs you that he has made some length measurements and that his ship is 150 m long but that yours is only 120 m long. From your point of view, **(a)** how long is your friend's ship, **(b)** how long is your ship, and **(c)** what is the speed of your friend's ship relative to yours?

28. •• A ladder 5.0 m long leans against a wall inside a spaceship. From the point of view of a person on the ship, the base of the ladder is 3.0 m from the wall, and the top of the ladder is 4.0 m above the floor. The spaceship moves past the Earth with a speed of $0.90c$ in a direction parallel to the floor of the ship. Find the angle the ladder makes with the floor as seen by an observer on Earth.

29. ••• When traveling past an observer with a relative speed v, a rocket is measured to be 8.00 m long. When the rocket moves with a relative speed $2v$, its length is measured to be 5.00 m. **(a)** What is the speed v? **(b)** What is the proper length of the rocket?

30. ••• **IP** The starships *Picard* and *La Forge* are traveling in the same direction toward the Andromeda galaxy. The *Picard* moves with a speed of $0.90c$ relative to the *La Forge*. A person on the *La Forge* measures the length of the two ships and finds the same value. **(a)** If a person on the *Picard* also measures the lengths of the two ships, which of the following is observed: (i) the *Picard* is longer; (ii) the *La Forge* is longer; or (iii) both ships have the same length? Explain. **(b)** Calculate the ratio of the proper length of the *Picard* to the proper length of the *La Forge*.

Section 29–4 The Relativistic Addition of Velocities

31. • A spaceship moving toward Earth with a speed of $0.90c$ launches a probe in the forward direction with a speed of $0.10c$ relative to the ship. Find the speed of the probe relative to Earth.

32. • Suppose the probe in Problem 31 is launched in the opposite direction to the motion of the spaceship. Find the speed of the probe relative to Earth in this case.

33. • A spaceship moving relative to an observer with a speed of $0.60c$ shines a beam of light in the forward direction, directly toward the observer. Use Equation 29–4 to calculate the speed of the beam of light relative to the observer.

34. • Suppose the speed of light is 35 mi/h. A paper girl riding a bicycle at 22 mi/h throws a rolled-up newspaper in the forward direction, as shown in **Figure 29–30**. If the paper is thrown with a speed of 19 mi/h relative to the bike, what is its speed, v, with respect to the ground?

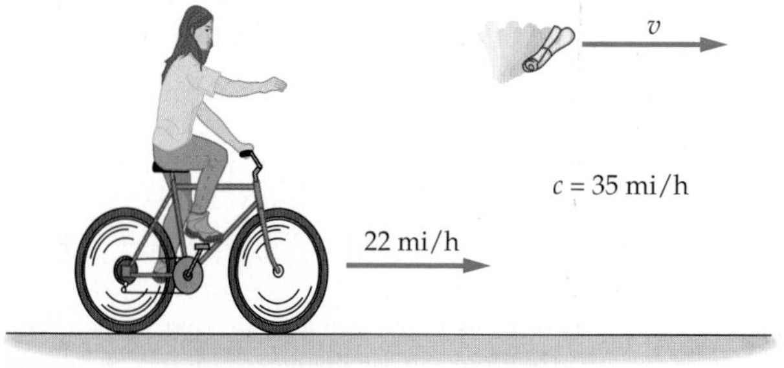

▲ **FIGURE 29–30** Problem 34

35. •• Two asteroids head straight for Earth from the same direction. Their speeds relative to Earth are $0.80c$ for asteroid 1 and $0.60c$ for asteroid 2. Find the speed of asteroid 1 relative to asteroid 2.

36. •• Two rocket ships approach Earth from opposite directions, each with a speed of $0.6c$ relative to Earth. What is the speed of one ship relative to the other?

37. •• A spaceship and an asteroid are moving in the same direction away from Earth with speeds of $0.77c$ and $0.41c$, respectively. What is the relative speed between the spaceship and the asteroid?

38. •• An electron moves to the right in a laboratory accelerator with a speed of $0.84c$. A second electron in a different accelerator moves to the left with a speed of $0.43c$ relative to the first electron. Find the speed of the second electron relative to the lab.

39. •• **IP** Two rocket ships are racing toward Earth, as shown in **Figure 29–31**. Ship A is in the lead, approaching the Earth at $0.80c$ and separating from ship B with a relative speed of $0.50c$. **(a)** As seen from Earth, what is the speed, v, of ship B? **(b)** If ship A increases its speed by $0.10c$ relative to the Earth, does the relative speed between ship A and ship B increase by $0.10c$, by more than $0.10c$, or by less than $0.10c$? Explain. **(c)** Find the relative speed between ships A and B for the situation described in part (b).

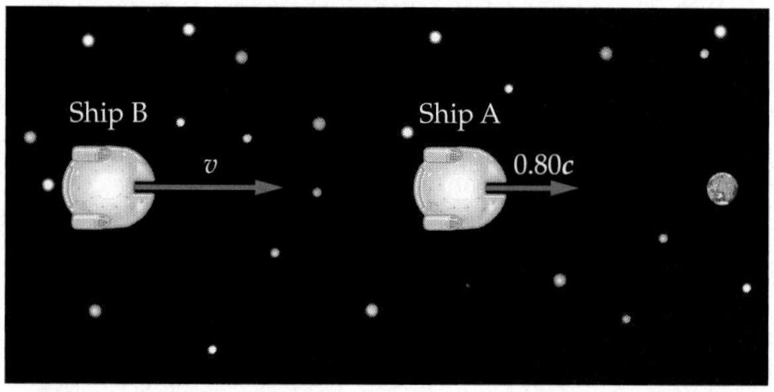

▲ **FIGURE 29–31** Problem 39

40. •• **IP** An inventor has proposed a device that will accelerate objects to speeds greater than c. He proposes to place the object to be accelerated on a conveyor belt whose speed is $0.80c$. Next, the entire system is to be placed on a second conveyor belt that also has a speed of $0.80c$, thus producing a final speed of $1.6c$. **(a)** Construction details aside, should you invest in this scheme? **(b)** What is the actual speed of the object relative to the ground?

Section 29–5 Relativistic Momentum

41. • A 4.5×10^6-kg spaceship moves away from Earth with a speed of $0.65c$. What is the magnitude of the ship's **(a)** classical and **(b)** relativistic momentum?

42. • An asteroid with a mass of 8.2×10^{11} kg is observed to have a relativistic momentum of magnitude 7.74×10^{20} kg·m/s. What is the speed of the asteroid relative to the observer?

43. •• An object has a relativistic momentum that is 5.0 times greater than its classical momentum. What is its speed?

44. •• A football player with a mass of 88 kg and a speed of 2.0 m/s collides head-on with a player from the opposing team whose mass is 120 kg. The players stick together and are at rest after the collision. Find the speed of the second player, assuming the speed of light is 3.0 m/s.

45. •• In the previous problem, suppose the speed of the second player is 1.2 m/s. What is the speed of the players after the collision?

46. •• A space probe with a rest mass of 8.2×10^7 kg and a speed of $0.50c$ smashes into an asteroid at rest and becomes embedded within it. If the speed of the probe-asteroid system is $0.26c$ after the collision, what is the rest mass of the asteroid?

47. •• At what speed does the classical momentum, $p = mv$, give an error, when compared with the relativistic momentum, of **(a)** 1.00% and **(b)** 5.00%?

48. •• A proton has 1836 times the rest mass of an electron. At what speed will an electron have the same momentum as a proton moving at $0.0100c$?

Section 29–6 Relativistic Energy and $E = mc^2$

49. • Find the work that must be done on a proton to accelerate it from rest to a speed of $0.90c$.

50. • If a neutron moves with a speed of $0.99c$, what is its **(a)** total energy, **(b)** rest energy, and **(c)** kinetic energy?

51. • A spring with a force constant of 544 N/m is compressed a distance of 38 cm. Find the resulting increase in the spring's mass.

52. • When a certain capacitor is charged, its mass increases by 8.3×10^{-16} kg. How much energy is stored in the capacitor?

53. • What minimum energy must a gamma ray have to create an electron-antielectron pair?

54. • When a proton encounters an antiproton, the two particles annihilate each other, producing two gamma rays. Assuming the particles were at rest when they annihilated, find the energy of each of the two gamma rays produced. (*Note*: The rest energies of an antiproton and a proton are identical.)

55. • A rocket with a mass of 2.7×10^6 kg has a relativistic kinetic energy of 2.7×10^{23} J. How fast is the rocket moving?

56. •• An object has a total energy that is 4.5 times its rest energy. What is its speed?

57. •• A nuclear power plant produces an average of 1.0×10^3 MW of power during a year of operation. Find the corresponding change in mass of reactor fuel, assuming all of the energy released by the fuel can be converted directly to electrical energy. (In a practical reactor, only a relatively small fraction of the energy can be converted to electricity.)

58. •• A helium atom has a rest mass of $m_{He} = 4.002603$ u. When disassembled into its constituent particles (2 protons, 2 neutrons, 2 electrons), the well-separated individual particles have the following masses: $m_p = 1.007276$ u, $m_n = 1.008665$ u, $m_e = 0.000549$ u. How much work is required to completely disassemble a helium atom? (*Note:* 1 u of mass has a rest energy of 931.49 MeV.)

59. •• What is the percent difference between the classical kinetic energy, $K_{cl} = \frac{1}{2}m_0v^2$, and the correct relativistic kinetic energy, $K = m_0c^2/\sqrt{1 - v^2/c^2} - m_0c^2$, at a speed of **(a)** $0.10c$ and **(b)** $0.90c$?

60. •• A proton has 1836 times the rest mass of an electron. At what speed will an electron have the same kinetic energy as a proton moving at $0.0250c$?

61. •• **IP** Consider a baseball with a rest mass of 0.145 kg. **(a)** How much work is required to increase the speed of the baseball from 25.0 m/s to 35.0 m/s? **(b)** Is the work required to increase the speed of the baseball from 200,000,025 m/s to 200,000,035 m/s greater than, less than, or the same as the amount found in part (a)? Explain. **(c)** Calculate the work required for the increase in speed indicated in part (b).

62. •• **IP** A particle has a kinetic energy equal to its rest energy. **(a)** What is the speed of this particle? **(b)** If the kinetic energy of this particle is doubled, does its speed increase by a more than, less than, or exactly a factor of two? Explain. **(c)** Calculate the speed of a particle whose kinetic energy is twice its rest energy.

63. ••• A lump of putty with a mass of 0.240 kg and a speed of $0.980c$ collides head-on and sticks to an identical lump of putty moving with the same speed. After the collision the system is at rest. What is the mass of the system after the collision?

Section 29–8 General Relativity

64. • Find the radius to which the Sun must be compressed for it to become a black hole.

65. • There is strong evidence that the center of the Milky Way galaxy contains a giant black hole with a mass of perhaps one million Suns. What is the maximum radius of a million-solar-mass black hole?

66. •• Recent measurements show that the black hole at the center of the Milky Way galaxy, which is believed to coincide with the powerful radio source Sagittarius A^*, is 2.6 million times more massive than the Sun; that is, $M = 5.2 \times 10^{36}$ kg. **(a)** Find the acceleration of gravity at the Schwarzschild radius of this black hole, using the expression for R given in Equation 29–10. **(b)** How does your answer to part (a) change if the mass of the black hole is doubled? Explain.

General Problems

67. •• **Cosmic Rays** Protons in cosmic rays have been observed with kinetic energies as large as 1.0×10^{20} eV. **(a)** How fast are these protons moving? Give your answer as a fraction of the speed of light. **(b)** Show that the kinetic energy of a single one of these protons is much greater than the kinetic energy of a 15-mg ant walking with a speed of 8.8 mm/s.

68. •• An apple falls from a tree, landing on the ground 3.7 m below. How long is the apple in the air, as measured by an observer moving toward Earth with a speed of $0.89c$?

69. •• What is the momentum of a proton with 1.50×10^3 MeV of kinetic energy? (*Note*: The rest energy of a proton is 938 MeV.)

70. •• **IP** A container holding 2.00 moles of an ideal monatomic gas is heated at constant volume until the temperature of the gas increases by 112 F°. **(a)** Does the mass of the gas increase, decrease, or stay the same? Explain. **(b)** Calculate the change in mass of the gas, if any.

71. •• A ^{14}C nucleus, initially at rest, emits a beta particle. The beta particle is an electron with 156 keV of kinetic energy. **(a)** What is the speed of the beta particle? **(b)** What is the momentum of the beta particle? **(c)** What is the momentum of the nucleus after it emits the beta particle? **(d)** What is the speed of the nucleus after it emits the beta particle?

72. •• A clock at rest has a rectangular shape, with a width of 24 cm and a height of 12 cm. When this clock moves parallel to its width with a certain speed v its width and height are the same. Relative to a clock at rest, how long does it take for the moving clock to advance by 1.0 s?

73. •• **IP TV Accelerator** Electrons in a television set are accelerated through a potential difference of 25.0 kV. **(a)** Find the speed of the electrons when they reach the screen. Ignore any relativistic effects. **(b)** When relativistic effects are included, do you expect the speed of the electrons to be greater than, less than, or the same as the result found in part (a)? Explain. **(c)** Find the speed of the electrons when they reach the screen, using a correct relativistic calculation.

74. •• A starship moving toward Earth with a speed of $0.65c$ launches a shuttle craft in the forward direction. The shuttle, which has a proper length of 12.5 m, is only 6.25 m long as viewed from Earth. What is the speed of the shuttle relative to the starship?

75. •• When a particle of charge q and momentum p enters a uniform magnetic field at right angles it follows a circular path of radius $R = p/qB$, as shown in **Figure 29–32**. What radius does this expression predict for a proton traveling with a speed $v = 0.99c$ through a magnetic field $B = 0.20$ T if you use **(a)** the nonrelativistic momentum ($p = mv$) or **(b)** the relativistic momentum ($p = mv/\sqrt{1 - v^2/c^2}$)?

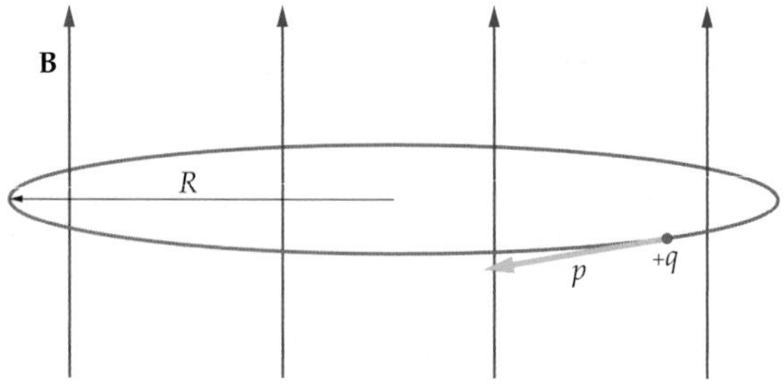

▲ **FIGURE 29–32** Problem 75

76. •• **IP** A starship moving away from Earth with a speed of $0.75c$ launches a shuttle craft in the reverse direction, that is, toward Earth. **(a)** If the speed of the shuttle relative to the starship is $0.40c$, and its proper length is 13 m, how long is the shuttle as measured by an observer on Earth? **(b)** If the shuttle had been launched in the forward direction instead, would its length as measured by an observer on Earth be greater than, less than, or the same as the length found in part (a)? Explain. **(c)** Calculate the length for the case described in part (b).

77. •• A 2.2-m titanium rod in a moving spacecraft is at an angle of 45° with respect to the direction of motion. The craft moves directly toward Earth at $0.97c$. As viewed from Earth, **(a)** how long is the rod and **(b)** what angle does the rod make with the direction of motion?

78. •• Electrons are accelerated from rest through a potential difference of 276,000 V. What is the final speed predicted **(a)** classically and **(b)** relativistically?

79. •• The rest energy, m_0c^2, of a particle with a kinetic energy K and a momentum p can be determined as follows:

$$m_0c^2 = \frac{(pc)^2 - K^2}{2K}$$

Suppose a pion (a subatomic particle) is observed to have a kinetic energy $K = 35.0$ MeV and a momentum $p = 5.61 \times 10^{-20}$ kg·m/s $= 105$ MeV/c. What is the rest energy of the pion? Give your answer in MeV.

80. •• A small star of mass m orbits a supermassive black hole of mass M. **(a)** Find the orbital speed of the star if its orbital radius is $2R$, where R is the Schwarzschild radius (Equation 29–10.) **(b)** Repeat part (a) for an orbital radius equal to R.

81. •• **IP** Consider a "relativistic air track" on which two identical air carts undergo a completely inelastic collision. One cart is initially at rest; the other has an initial speed of $0.650c$. **(a)** In classical physics, the speed of the carts after the collision would be $0.325c$. Do you expect the final speed in this relativistic collision to be greater than or less than $0.325c$? Explain. **(b)** Use relativistic momentum conservation to find the speed of the carts after they collide and stick together.

82. ••• **IP** In Conceptual Checkpoint 29–2 we considered an astronaut at rest on an inclined bed inside a moving spaceship. From the point of view of observer 1, on board the ship, the astronaut has a length L_0 and is inclined at an angle θ_0 above the floor. Observer 2 sees the spaceship moving to the right with a speed v. **(a)** Show that the length of the astronaut as measured by observer 2 is

$$L = L_0\sqrt{1 - \left(\frac{v^2}{c^2}\right)\cos^2\theta_0}$$

(b) Show that the angle θ the astronaut makes with the floor of the ship, as measured by observer 2, is given by

$$\tan\theta = \frac{\tan\theta_0}{\sqrt{1 - v^2/c^2}}$$

83. ••• A pulsar is a collapsed, rotating star that sends out a narrow beam of radiation, like the light from a lighthouse. With each revolution, we see a brief, intense pulse of radiation from the pulsar. Suppose a pulsar is receding directly away from Earth with a speed of $0.800c$, and the starship *Endeavor* is sent out toward the pulsar with a speed of $0.950c$ relative to Earth. If an observer on Earth finds that 153 pulses are emitted by the pulsar every second, at what rate does an observer on the *Endeavor* see pulses emitted?

84. ••• Show that the total energy of an object is related to its momentum by the relation $E^2 = p^2c^2 + (m_0c^2)^2$.

85. ••• Show that if $0 < v_1 < c$ and $0 < v_2 < c$ are two velocities pointing in the same direction, the relativistic sum of these velocities, v, is greater than v_1 and greater than v_2 but less than c. In particular, show that this is true even if v_1 and v_2 are greater than $0.5c$.

86. ••• Show that an object with momentum p and rest mass m_0 has a speed given by

$$v = \frac{c}{\sqrt{1 + (m_0c/p)^2}}$$

87. ••• When at rest, the Σ^- particle has a lifetime of 0.15 ns before it decays into a neutron and a pion. One particular Σ^- particle is observed to travel 3.0 cm in the lab before decaying. What was its speed? (*Hint:* Its speed was not $\frac{2}{3}c$.)

Interactive Problems

88. •• ***Referring to Example 29–4*** The *Picard* approaches starbase Faraway Point with a speed of $0.806c$ and the *La Forge* approaches the starbase with a speed of $0.906c$. Suppose the *Picard* now launches a probe toward the starbase. **(a)** What velocity must the probe have relative to the *Picard* if it is to be at rest relative to the *La Forge*? **(b)** What velocity must the probe have relative to the *Picard* if its velocity relative to the *La Forge* is to be $0.100c$? **(c)** For the situation described in part (b), what is the velocity of the probe relative to the Faraway Point starbase?

89. •• ***Referring to Example 29–4*** Faraway Point starbase launches a probe toward the approaching starships. The probe has a velocity relative to the *Picard* of $-0.906c$. The *Picard* approaches starbase Faraway Point with a speed of $0.806c$ and the *La Forge* approaches the starbase with a speed of $0.906c$. **(a)** What is the velocity of the probe relative to the *La Forge*? **(b)** What is the velocity of the probe relative to Faraway Point starbase?

30 Quantum Physics

30–1	Blackbody Radiation and Planck's Hypothesis of Quantized Energy	1004
30–2	Photons and the Photoelectric Effect	1007
30–3	The Mass and Momentum of a Photon	1013
30–4	Photon Scattering and the Compton Effect	1014
30–5	The de Broglie Hypothesis and Wave-Particle Duality	1017
30–6	The Heisenberg Uncertainty Principle	1021
30–7	Quantum Tunneling	1025

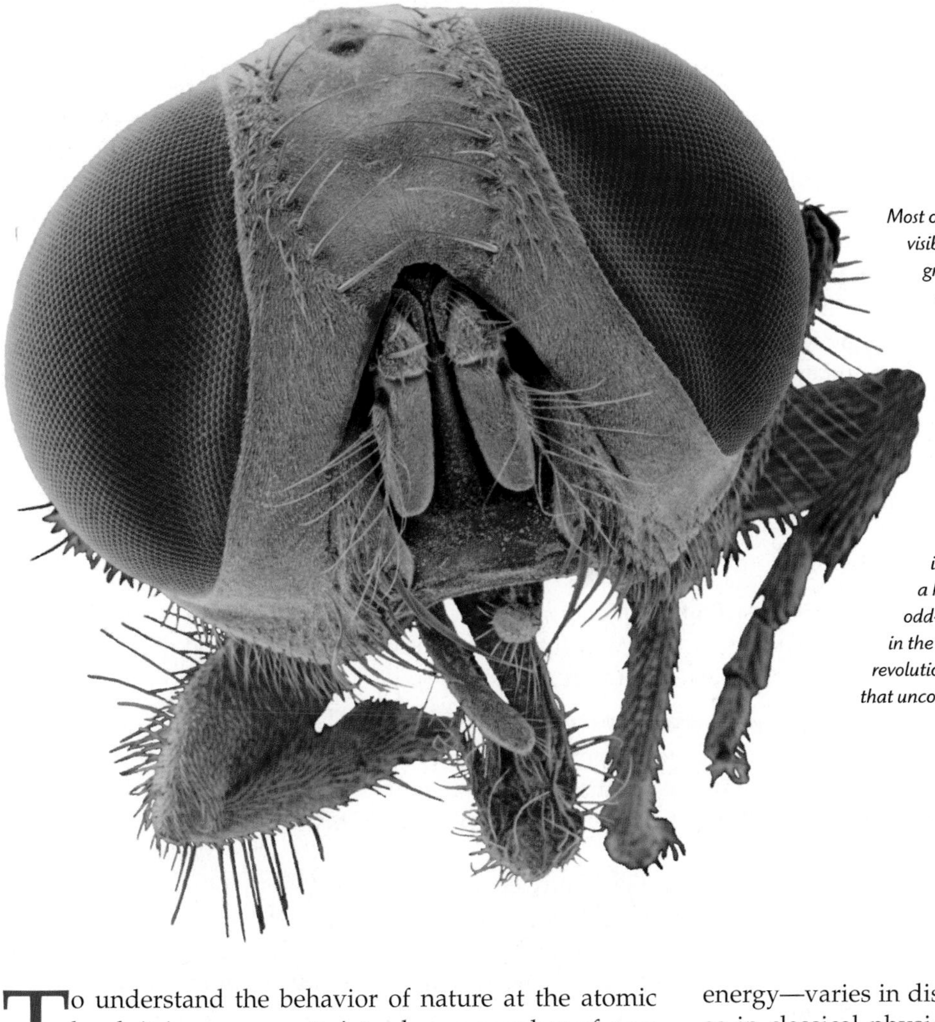

Most of the images that we encounter are made with visible light. Even those that are not, such as thermograms and X-rays, employ other kinds of electromagnetic radiation. Until about 80 years ago, the idea of making a picture by means of particles rather than radiation would have seemed absurd—like trying to create a portrait by bouncing paintballs off the subject and onto a canvas. Yet by the 1920s, physicists discovered that light waves often behave like particles and, conversely, that particles, such as electrons, often behave like waves. Indeed, the wave properties of electrons can be exploited to create remarkably detailed images, such as this scanning electron micrograph of a housefly. This chapter explores the sometimes odd-seeming laws that describe the behavior of nature in the atomic and subatomic realms, and the series of revolutions in physics between the 1890s and the 1930s that uncovered them.

To understand the behavior of nature at the atomic level, it is necessary to introduce a number of new concepts to physics and to modify many others. In this chapter we consider the basic ideas of quantum physics and show that they lead to a deeper understanding of microscopic systems—in much the same way that relativity extends physics into the realm of high speeds. Relativity and quantum physics, taken together, provide the basis for what we refer to today as modern physics.

We begin this chapter by introducing the concept of *quantization*, in which a physical quantity—such as energy—varies in discrete steps rather than continuously, as in classical physics. This concept leads to the idea of the *photon*, which can be thought of as a "particle" of light. Next, we find that just as light can behave like a particle, particles—such as electrons, protons, and neutrons—can behave like waves. Finally, the wave nature of matter introduces a fundamental uncertainty to our knowledge of physical quantities and allows for such classically "forbidden" behavior as *quantum tunneling*.

30–1 Blackbody Radiation and Planck's Hypothesis of Quantized Energy

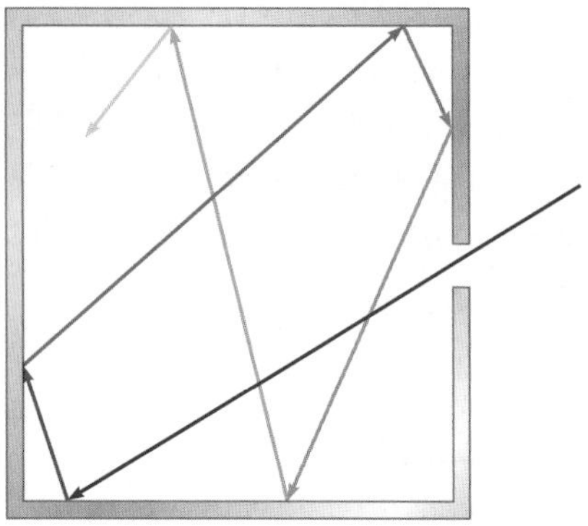

▲ **FIGURE 30–1 An ideal blackbody** In an ideal blackbody, incident light is completely absorbed. In the case shown here, the absorption occurs as the result of multiple reflections within a cavity. The blackbody, and the electromagnetic radiation it contains, are in thermal equilibrium at a temperature T.

If you have ever looked through a small opening into a hot furnace, you have seen the glow of light associated with its high temperature. As unlikely as it may seem, this light played a central role in the revolution of physics that occurred in the early 1900s. It was through the study of such systems that the idea of *energy quantization*—energy taking on only discrete values—was first introduced to physics.

More precisely, physicists in the late 1800s were actively studying the electromagnetic radiation given off by a physical system known as a **blackbody.** An example of a blackbody is illustrated in **Figure 30–1**. Note that this blackbody has a cavity with a small opening to the outside world—much like a furnace. Light that enters the cavity through the opening is reflected multiple times from the interior walls until it is completely absorbed. It is for this reason that the system is referred to as "black," even though the material from which it is made need not be black at all.

An ideal blackbody absorbs all the light that is incident on it.

Objects that absorb much of the incident light—though not all of it—are reasonable approximations to a blackbody; objects that are highly reflective and shiny are poor representations of a blackbody.

As we saw in Section 16–6, objects that are effective at absorbing radiation are also effective at giving off radiation. Thus an ideal blackbody is also an ideal radiator. In fact, the basic experiment performed with a blackbody is the following: Heat the blackbody to a fixed temperature, T, and measure the amount of electromagnetic radiation it gives off at a given frequency, f. Repeat this measurement for a number of different frequencies, then plot the intensity of radiation versus frequency. The results of a typical blackbody experiment are shown in **Figure 30–2**, for a variety of different temperatures. Note that there is little radiation at small frequencies, a peak in the radiation at intermediate frequencies, and finally a fall-off to little radiation again at high frequencies.

Now, what is truly remarkable about the blackbody experiment is the following:

> The distribution of energy in blackbody radiation is *independent* of the material from which the blackbody is constructed—it depends only on the temperature, T.

Therefore, a blackbody of steel and one of wood give precisely the same results when held at the same temperature. When physicists observe a phenomenon that is independent of the details of the system, it is a clear signal that they are observing something of fundamental significance. This was certainly the case with blackbody radiation.

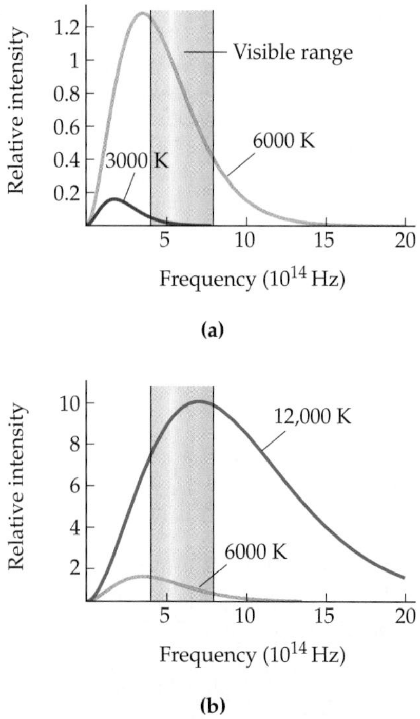

▲ **FIGURE 30–2 Blackbody radiation** Blackbody radiation as a function of frequency for various temperatures. **(a)** 3000 K and 6000 K. **(b)** 6000 K and 12,000 K. Note that as the temperature is increased, the peak in the radiation shifts toward higher frequency.

Two aspects of the blackbody curves in Figure 30–2 are of particular importance. First, note that as the temperature is increased, the area under the curve increases. Since the total area under the curve is a measure of the total energy emitted by the blackbody, it follows that an object radiates more energy as it becomes hotter.

Second, note that the peak in the blackbody curve moves to higher frequency as the absolute temperature T is increased. This movement, or displacement, of the peak with temperature is described by **Wien's displacement law:**

Wien's Displacement Law

$$f_{\text{peak}} = (5.88 \times 10^{10}\ \text{s}^{-1} \cdot \text{K}^{-1})T \qquad 30\text{–}1$$

SI unit: $\text{Hz} = \text{s}^{-1}$

Thus, there is a direct connection between the temperature of an object and the frequency of radiation it emits most strongly.

CONCEPTUAL CHECKPOINT 30–1 Compare Temperatures

Betelgeuse is a red-giant star in the constellation Orion; Rigel is a bluish white star in the same constellation. Is the surface temperature of Betelgeuse **(a)** greater than, **(b)** less than, or **(c)** the same as the surface temperature of Rigel?

Reasoning and Discussion

Recall that red light has a lower frequency than blue light, as can be seen in Figure 25–8. It follows, from Wien's displacement law, that a red star has a lower temperature than a blue star. Therefore, Betelgeuse has the lower surface temperature.

Answer:

(b) The surface temperature of Betelgeuse is less than that of Rigel.

▲ All objects emit electromagnetic radiation over a range of frequencies. The frequency that is radiated most intensely depends on the object's temperature, as specified by Wien's law. The glowing bolt in this picture radiates primarily in the infrared part of the spectrum, but it is hot enough (a few thousand kelvin) so that a significant portion of its radiation falls within the red end of the visible region. The other bolts are too cool to radiate any detectable amount of visible light.

To be more specific about the conclusion given in Conceptual Checkpoint 30–1, let's consider Figure 30–2 in greater detail. At the lowest temperature shown, 3000 K, the radiation is more intense at the red end of the visible spectrum than at the blue end. An object at this temperature—like the heating coil on a stove, for example—would appear "red hot" to the eye. Even so, most of the radiation at this temperature is in the infrared, and thus is not visible to the eye at all. A blackbody at 6000 K, like the surface of the Sun, gives out strong radiation throughout the visible spectrum, though there is still more radiation at the red end than at the blue end. As a result, the light of the Sun appears somewhat yellowish. Finally, at 12,000 K a blackbody appears bluish white, and most of its radiation is in the ultraviolet. The temperature of the star Rigel is determined from the location of its radiation peak in the following Exercise.

EXERCISE 30–1

Find the surface temperature of Rigel, given that its radiation peak occurs at a frequency of 1.17×10^{15} Hz.

REAL-WORLD PHYSICS

Measuring the temperature of a star

Solution

Solving Equation 30–1 for T, we find

$$T = \frac{f_{\text{peak}}}{5.88 \times 10^{10}\,\text{s}^{-1}\cdot\text{K}^{-1}} = \frac{1.17 \times 10^{15}\,\text{Hz}}{5.88 \times 10^{10}\,\text{s}^{-1}\cdot\text{K}^{-1}} = 19{,}900\,\text{K}$$

This is a little more than three times the surface temperature of the Sun. Thus blackbody radiation allows us to determine the temperature of a distant star that we may never visit.

Planck's Quantum Hypothesis

Although experimental understanding of blackbody radiation was quite extensive in the late 1800s, there was a problem. Attempts to explain the blackbody curves of Figure 30–2 theoretically, using classical physics, failed—and failed miserably. To see the problem, consider the curves shown in **Figure 30–3**. The green curve is the experimental result for a blackbody at a given temperature. In contrast, the blue curve shows the prediction of classical physics. Clearly, the classical result cannot be valid, since its curve diverges to infinity at high frequency, which in turn implies that the blackbody radiates an infinite amount of energy. This unphysical divergence at high frequencies is referred to as the *ultraviolet catastrophe*.

The German physicist Max Planck (1858–1947) worked long and hard on this problem. Eventually, he was able to construct a mathematical formula that agreed with experiment for all frequencies. His next problem was to "derive" the equation. The only way he could do this, it turned out, was to make the following bold and unprecedented assumption: The radiation energy in a blackbody at the frequency f must be an integral multiple of a constant (h) times the frequency; that is, energy is *quantized*:

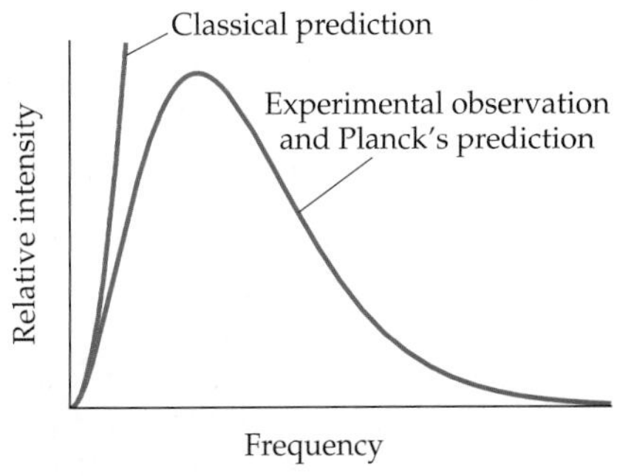

▲ **FIGURE 30–3 The ultraviolet catastrophe**

Classical physics predicts a blackbody radiation curve that rises without limit as the frequency increases. This outcome is referred to as the ultraviolet catastrophe. By assuming energy quantization, Planck was able to derive a curve in agreement with experimental results.

Quantized Energy

$$E_n = nhf \qquad n = 0, 1, 2, 3, \ldots \qquad \text{30–2}$$

The constant, h, in this expression is known as **Planck's constant,** and it has the following value:

Planck's Constant, h

$$h = 6.63 \times 10^{-34}\ \text{J} \cdot \text{s} \qquad \textbf{30–3}$$

SI unit: J · s

This constant is recognized today as one of the fundamental constants of nature, on an equal footing with other such constants as the speed of light in a vacuum and the rest mass of an electron.

The assumption of energy quantization is quite a departure from classical physics, in which energy can take on any value at all and is related to the amplitude of a wave rather than its frequency. In Planck's calculation, the energy can have only the discrete values hf, $2hf$, $3hf$, and so on. Because of this quantization, it follows that the energy can change only in *quantum jumps* of energy no smaller than hf as the system goes from one quantum state to another. The fundamental increment, or *quantum*, of energy, hf, is incredibly small, as can be seen from the small magnitude of Planck's constant. The next Example explores the size of the quantum and the value of the *quantum number, n*, for a typical macroscopic system.

EXAMPLE 30–1 Quantum Numbers

Suppose the maximum speed of a 1.2-kg mass attached to a spring with a force constant of 35 N/m is 0.95 m/s. **(a)** Find the frequency of oscillation and total energy of this mass–spring system. **(b)** Determine the size of one quantum of energy in this system. **(c)** Assuming the energy of this system satisfies $E_n = nhf$, find the quantum number, n.

Picture the Problem

Our sketch shows a 1.2-kg mass oscillating on a spring with a force constant of 35 N/m. The mass has its maximum speed of $v_{max} = 0.95$ m/s when it passes through the equilibrium position. At this moment, the total energy of the system is simply the kinetic energy of the mass.

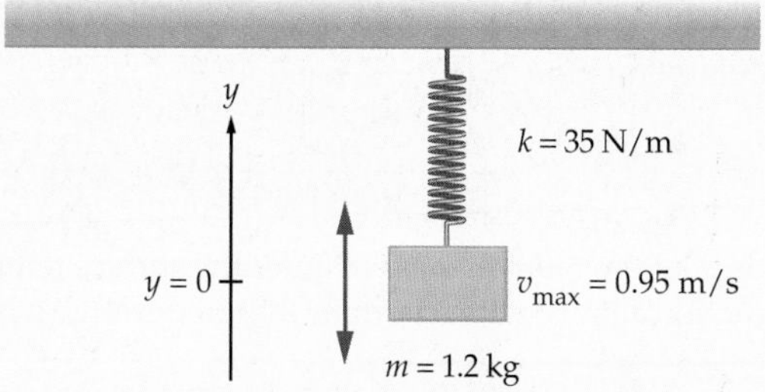

Strategy

a. We can find the frequency of oscillation using $\omega = \sqrt{k/m}$ (Equation 13–10) and $\omega = 2\pi f$ (Equation 13–15). The total energy is simply the kinetic energy as the mass passes through equilibrium, $E = K_{max} = \frac{1}{2}mv_{max}^2$.

b. The energy of one quantum is hf, where f is the frequency found in part (a).

c. We determine the quantum number by solving $E_n = nhf$ for n.

Solution

Part (a)

1. Calculate the frequency of oscillation using $\omega = \sqrt{k/m} = 2\pi f$:

$$\omega = \sqrt{\frac{k}{m}} = 2\pi f$$

$$f = \frac{1}{2\pi}\sqrt{\frac{k}{m}} = \frac{1}{2\pi}\sqrt{\frac{35\ \text{N/m}}{1.2\ \text{kg}}} = 0.86\ \text{Hz}$$

2. Calculate the maximum kinetic energy of the mass $\left(\frac{1}{2}mv_{max}^2\right)$ to find the total energy, E, of the system:

$$E = \tfrac{1}{2}mv_{max}^2 = \tfrac{1}{2}(1.2\ \text{kg})(0.95\ \text{m/s})^2 = 0.54\ \text{J}$$

Part (b)

3. The energy of one quantum is hf, where $f = 0.86$ Hz:

$$hf = (6.63 \times 10^{-34}\ \text{J} \cdot \text{s})(0.86\ \text{Hz}) = 5.7 \times 10^{-34}\ \text{J}$$

Part (c)

4. Set $E_n = nhf$ equal to the total energy of the system and solve for n:

$$E_n = nhf$$

$$n = \frac{E_n}{hf} = \frac{0.54\ \text{J}}{5.7 \times 10^{-34}\ \text{J}} = 9.5 \times 10^{32}$$

Insight

The numbers found in parts (b) and (c) are incredible for their size. For example, the quantum is on the order of 10^{-34} J, as compared with the energy required to break a bond in a DNA molecule, which is on the order of 10^{-20} J. Thus the quantum for a macroscopic system is about 10^{14} times smaller than the energy needed to affect a molecule. Similarly, the number of quanta in the system, roughly 10^{33}, is comparable to the number of atoms in four Olympic-size swimming pools.

Practice Problem

If the quantum of energy for a 1.5-kg mass on a spring is 0.80×10^{-33} J, what is the force constant of the spring? [**Answer:** $k =$ 86 N/m]

Some related homework problems: Problem 9, Problem 72

Clearly, then, the quantum numbers in typical macroscopic systems are incredibly large. As a result, a change of one in the quantum number is completely insignificant and undetectable. Similarly, the change in energy from one quantum state to the next is so small that it cannot be measured in a typical experiment; hence, for all practical purposes, the energy of a macroscopic system seems to change continuously, even though it actually changes by small increments. In contrast, in an atomic system, the energy jumps are of great importance, as we shall see in the next section.

Returning to the ultraviolet catastrophe for a moment, we can now see how Planck's hypothesis removes the unphysical divergence at high frequency predicted by classical physics. In Planck's theory, the higher the frequency f, the greater the quantum of energy, hf. Therefore, as the frequency is increased, the amount of energy required for even the smallest quantum jump increases as well. Since a blackbody has only a finite amount of energy, however, it simply cannot supply the large amount of energy required to produce an extremely high-frequency quantum jump. As a result, the amount of radiation at high frequency drops off toward zero.

Planck's theory of energy quantization leads to an adequate description of the experimental results for blackbody radiation. Still, the theory was troubling and somewhat unsatisfying to Planck and to many other physicists as well. Although the idea of energy quantization worked, at least in this case, it seemed ad hoc and more of a mathematical trick than a true representation of nature. With the work of Einstein, however, which we present in the next section, the well-founded misgivings about quantum theory began to fade away.

30–2 Photons and the Photoelectric Effect

From Max Planck's point of view, energy quantization in a blackbody was probably related to quantized vibrations of atoms in the walls of the blackbody. We are familiar, for example, with the fact that a string tied at both ends can produce standing waves at only certain discrete frequencies (Chapter 14), so perhaps atoms vibrating in a blackbody behave in a similar way, vibrating only with certain discrete energies. Certainly, Planck did not think the light in a blackbody had a quantized energy, since most physicists thought of light as being a wave, which can have any energy.

A brash young physicist named Albert Einstein, however, took the idea of quantized energy seriously and applied it to the radiation in the blackbody. Einstein proposed that light comes in bundles of energy, called **photons,** that obey Planck's hypothesis of energy quantization; that is, light of frequency f consists of photons with an energy given by the following relation:

Energy of a Photon of Frequency f

$$E = hf \qquad 30\text{–}4$$

SI unit: J

Thus the energy in a beam of light of frequency f can have only the values hf, $2hf$, $3hf$, and so on. Planck's initial reaction to Einstein's suggestion was that he had

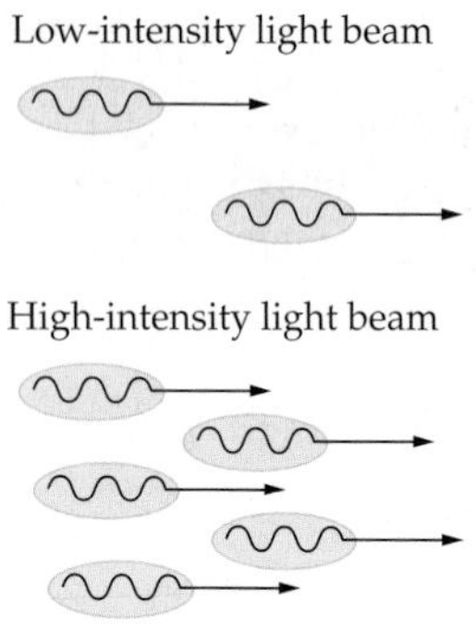

▲ **FIGURE 30–4 The photon model of light**
In the photon model of light, a beam of light consists of many photons, each with an energy hf. The more intense the beam, the more tightly packed the photons.

gone too far with the idea of quantization. As it turns out, nothing could have been further from the truth.

In Einstein's photon model, a beam of light can be thought of as a beam of particles, each carrying the energy hf, as indicated in **Figure 30–4**. If the beam of light is made more intense while keeping the frequency the same, the result is that the photons in the beam are more tightly packed, so that more photons pass a given point in a given time. In this way, more photons shine on a given surface in a given time, increasing the energy delivered to the surface per time. Even so, each photon in the more intense beam has exactly the same amount of energy as those in the less intense beam. The energy of a typical photon of visible light is calculated in the next Exercise.

EXERCISE 30–2

Calculate the energy of a photon of yellow light with a frequency of 5.25×10^{14} Hz. Give the energy in both joules and electron volts.

Solution

Applying Equation 30–4, we find

$$E = hf = (6.63 \times 10^{-34}\,\text{J}\cdot\text{s})(5.25 \times 10^{14}\,\text{s}^{-1}) = 3.48 \times 10^{-19}\,\text{J}$$

$$= 3.48 \times 10^{-19}\,\text{J}\left(\frac{1\,\text{eV}}{1.60 \times 10^{-19}\,\text{J}}\right) = 2.18\,\text{eV}$$

Note that the energy of a visible photon is on the order of an electron volt (eV). This is also the typical energy scale for atomic and molecular systems, as we show in detail in the following Example.

EXAMPLE 30–2 When Oxygens Split

Molecular oxygen (O_2) is a diatomic molecule. The energy required to dissociate 1 mol of O_2 to form 2 mol of atomic oxygen is 118 kcal. **(a)** Find the energy (in joules and electron volts) required to dissociate one O_2 molecule. **(b)** Assuming the dissociation energy for one molecule is supplied by a single photon, find the frequency of the photon.

Picture the Problem

In our sketch we show a single photon dissociating an O_2 molecule to form two O atoms. The energy of the photon is $E = hf$.

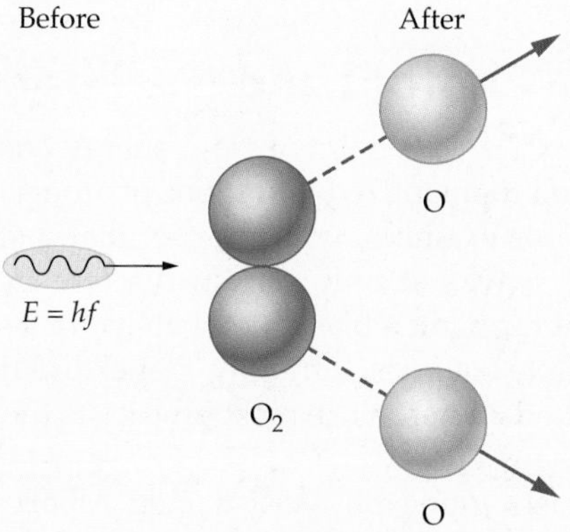

Strategy

a. This part of the problem is simply a matter of converting from kcal per mole to joules per molecule. This can be accomplished by using the fact that 1 kcal = 4186 J and that Avogadro's number (Section 17–1) is 6.02×10^{23} molecules/mol. We can then convert to electron volts using $1\,\text{eV} = 1.60 \times 10^{-19}\,\text{J}$.

b. We can find the frequency of the photon by setting hf equal to the energy E found in part (a). Since Planck's constant is given in units of J·s, we must use the energy expressed in joules from part (a).

Solution

Part (a)

1. Convert 118 kcal/mol to J/molecule:

$$\left(118\frac{\text{kcal}}{\text{mol}}\right)\left(\frac{4186\,\text{J}}{1\,\text{kcal}}\right)\left(\frac{1}{6.02 \times 10^{23}\,\text{molecules/mol}}\right)$$
$$= 8.21 \times 10^{-19}\,\text{J/molecule}$$

2. Convert the preceding result to eV/molecule:

$$8.21 \times 10^{-19}\frac{\text{J}}{\text{molecule}}\left(\frac{1\,\text{eV}}{1.60 \times 10^{-19}\,\text{J}}\right)$$
$$= 5.13\,\text{eV/molecule}$$

Part (b)

3. Use $E = 8.21 \times 10^{-19}\ \text{J} = hf$ to solve for the frequency, f: $f = \dfrac{E}{h} = \dfrac{8.21 \times 10^{-19}\ \text{J}}{6.63 \times 10^{-34}\ \text{J}\cdot\text{s}} = 1.24 \times 10^{15}\ \text{Hz}$

Insight

This frequency is in the ultraviolet. In fact, ultraviolet rays in Earth's upper atmosphere cause O_2 molecules to dissociate, freeing up atomic oxygen which can then combine with O_2 to form ozone, O_3.

Practice Problem

An infrared photon has a frequency of 1.00×10^{13} Hz. How much energy is carried by one mole of these photons? [**Answer:** 3990 J = 0.953 kcal]

Some related homework problems: Problem 19, Problem 20

Since photons typically have rather small amounts of energy on a macroscopic scale, it follows that enormous numbers of photons must be involved in everyday situations, as demonstrated in the following Active Example.

REAL-WORLD PHYSICS
Dark-adapted vision

ACTIVE EXAMPLE 30–1 Dark Vision: Find the Number of Photons

Dark-adapted (scotopic) vision is possible in humans with as little as $4.00 \times 10^{-11}\ \text{W/m}^2$ of 505-nm light entering the eye. If light of this intensity and wavelength enters the eye through a pupil that is 6.00 mm in diameter, how many photons enter the eye per second?

Solution *(Test your understanding by performing the calculations indicated in each step.)*

1. Calculate the area of the pupil: $2.83 \times 10^{-5}\ \text{m}^2$
2. Multiply the intensity by the area of the pupil to find the energy entering the eye per second: $1.13 \times 10^{-15}\ \text{J/s}$
3. Calculate the energy of a photon: $3.94 \times 10^{-19}\ \text{J}$
4. Divide the energy of a photon into the energy per second to find the number of photons per second: 2870 photons/s

Insight

Thus, even though a typical lightbulb gives off roughly 10^{18} photons per second, we need only about 10^3 photons per second to see. Our eyes are extraordinary instruments, sensitive to an incredibly wide range of intensities.

Finally, suppose an astronomer views a dim, distant galaxy with scotopic vision. The 2870 photons that enter the astronomer's eye each second are separated from one another by about 65 miles. It follows that only one photon at a time from the distant galaxy traverses the astronomer's telescope.

Your Turn

Suppose we consider light with a wavelength greater than 505 nm by a factor of 1.25. Will more or fewer photons be required per second with this new wavelength? By what factor will the required number of photons per second change?

*(Answers to **Your Turn** problems are given in the back of the book.)*

The Photoelectric Effect

Einstein applied his photon model of light to the **photoelectric effect,** in which a beam of light (photo-) hits the surface of a metal and ejects an electron (-electric). The effect can be measured using a device like that pictured in **Figure 30–5**. Note that incoming light ejects an electron—referred to as a photoelectron—from a metal plate called the emitter (E); the electron is then attracted to a collector plate (C), which is at a positive potential relative to the emitter. The result is an electric current that can be measured with an ammeter.

The minimum amount of energy necessary to eject an electron from a particular metal is referred to as the **work function,** W_0, for that metal. Work functions

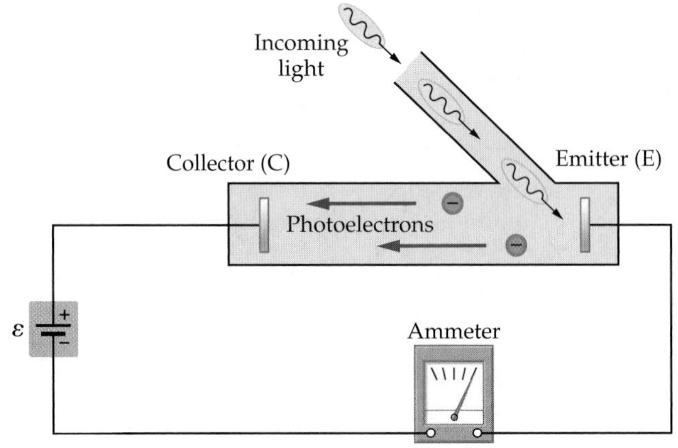

▶ **FIGURE 30–5 The photoelectric effect**
The photoelectric effect can be studied with a device like that shown. Light shines on a metal plate, ejecting electrons, which are then attracted to a positively charged "collector" plate. The result is an electric current that can be measured with an ammeter.

vary from metal to metal but are typically on the order of a few electron volts. If an electron is given an energy E by the beam of light that is greater than W_0, the excess energy goes into kinetic energy of the ejected electron. The maximum kinetic energy (K) a photoelectron can have, then, is

$$K_{max} = E - W_0 \qquad \textbf{30–5}$$

Just as with blackbody radiation, the photoelectric effect exhibits behavior that is at odds with classical physics. Two of the main areas of disagreement are the following:

- Classical physics predicts that a beam of light of *any* color (frequency) can eject electrons, as long as the beam has sufficient intensity. That is, if a beam is intense enough, the energy it delivers to an electron will exceed the work function and cause it to be ejected.
- Classical physics also predicts that the maximum kinetic energy of an ejected electron should increase as the intensity of the light beam is increased. In particular, the more energy the beam delivers to the metal, the more energy that any given electron can have as it is ejected.

Although both of these predictions are reasonable—necessary, in fact, from the classical physics point of view—they simply do not agree with experiments on the photoelectric effect. In fact, experiments show the following behavior:

- To eject electrons, the incident light beam must have a frequency greater than a certain minimum value, referred to as the **cutoff frequency,** f_0. If the frequency of the light is less than f_0, it will not eject electrons, no matter how intense the beam.
- If the frequency of light is greater than the cutoff frequency, f_0, the effect of increasing the intensity is to increase the *number* of electrons that are emitted per second. The maximum kinetic energy of the electrons does not increase with the intensity of the light; the kinetic energy depends only on the frequency of the light.

As we shall see, these observations are explained quite naturally with the photon model.

First, in Einstein's model each photon has an energy determined solely by its frequency. Therefore, making a beam of a given frequency more intense simply means increasing the number of photons hitting the metal in a given time—not increasing the energy carried by a photon. An electron, then, is ejected only if an incoming photon has an energy that is at least equal to the work function: $E = hf_0 = W_0$. The *cutoff frequency* is thus defined as follows:

Cutoff Frequency, f_0

$$f_0 = \frac{W_0}{h} \qquad \textbf{30–6}$$

SI unit: $\mathrm{Hz} = \mathrm{s}^{-1}$

If the frequency of the light is greater than f_0, the electron can leave the metal with a finite kinetic energy; if the frequency is less than f_0, no electrons are ejected, no matter how intense the beam. We determine a typical cutoff frequency in the next Exercise.

EXERCISE 30–3

The work function for a gold surface is 4.58 eV. Find the cutoff frequency, f_0, for a gold surface.

Solution

Substitution in Equation 30–6 yields

$$f_0 = \frac{W_0}{h} = \frac{(4.58 \text{ eV})\left(\dfrac{1.60 \times 10^{-19}\text{ J}}{1 \text{ eV}}\right)}{6.63 \times 10^{-34}\text{ J}\cdot\text{s}} = 1.11 \times 10^{15}\text{ Hz}$$

This frequency is in the near ultraviolet.

Second, the fact that a more intense beam of monochromatic light delivers more photons per time to the metal just means that more electrons are ejected per time. Since each electron receives precisely the same amount of energy, however, the maximum kinetic energy is the same regardless of the intensity. In fact, if we return to Equation 30–5 and replace the energy, E, with the energy of a photon, hf, we find

$$K_{max} = hf - W_0 \qquad 30\text{–}7$$

Note that K_{max} depends linearly on the frequency but is independent of the intensity. A plot of K_{max} for sodium (Na) and gold (Au) is given in **Figure 30–6.** Clearly, both lines have the same slope, h, as expected from Equation 30–7, but have different cutoff frequencies. Therefore, with the result given in Equation 30–7, Einstein was able to show that Planck's constant, h, appears in a natural way in the photoelectric effect and is not limited in applicability to the blackbody.

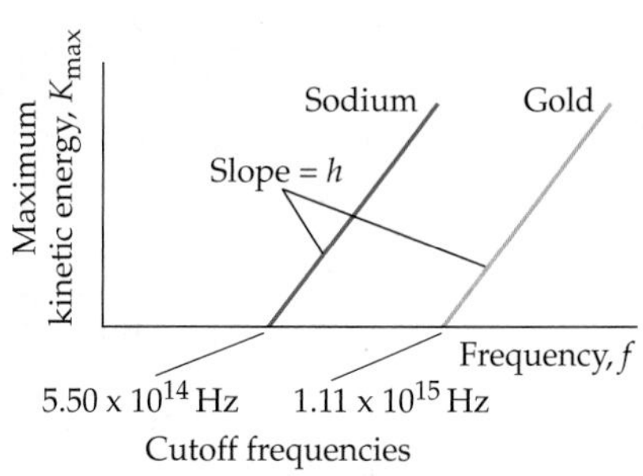

▲ **FIGURE 30–6 The kinetic energy of photoelectrons**
The maximum kinetic energy of photoelectrons as a function of the frequency of light. Note that sodium and gold have different cutoff frequencies, as one might expect for different materials. On the other hand, the slope of the two lines is the same, h, as predicted by Einstein's photon model of light.

EXAMPLE 30–3 White Light on Sodium

A beam of white light containing frequencies between 4.00×10^{14} Hz and 7.90×10^{14} Hz is incident on a sodium surface, which has a work function of 2.28 eV. **(a)** What is the range of frequencies in this beam of light for which electrons are ejected from the sodium surface? **(b)** Find the maximum kinetic energy of the "photoelectrons" that are ejected from this surface.

Picture the Problem

Our sketch shows a beam of white light, represented by photons with different frequencies, incident on a sodium surface. Photoelectrons are ejected from this surface with a kinetic energy that depends on the frequency of the photon that was absorbed.

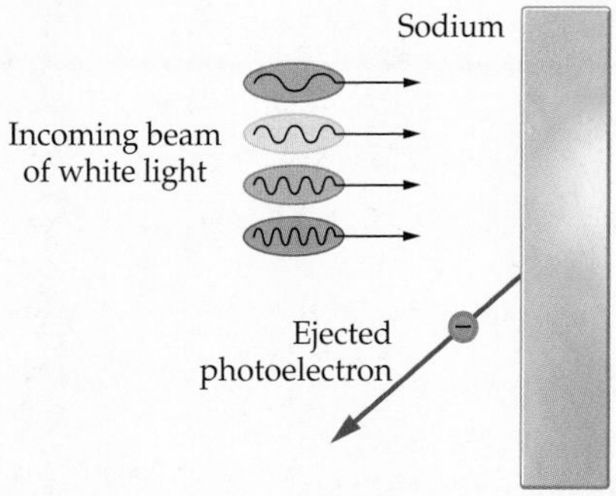

Strategy

a. We can find the cutoff frequency, f_0, for sodium using $f_0 = W_0/h$, with $W_0 = 2.28$ eV. Frequencies between the cutoff frequency and the maximum frequency in the beam of light, 7.90×10^{14} Hz, will eject electrons.

b. We can obtain the maximum kinetic energy for a given frequency, f, from Equation 30–7: $K_{max} = hf - W_0$. Clearly, the higher the frequency, the greater the maximum kinetic energy. It follows, then, that the greatest possible maximum kinetic energy corresponds to the highest frequency in the beam, 7.90×10^{14} Hz.

Solution

Part (a)

1. Use $f_0 = W_0/h$ to calculate the cutoff frequency for sodium:

$$f_0 = \frac{W_0}{h} = \frac{(2.28 \text{ eV})\left(\dfrac{1.60 \times 10^{-19}\text{ J}}{1 \text{ eV}}\right)}{6.63 \times 10^{-34}\text{ J}\cdot\text{s}} = 5.50 \times 10^{14}\text{ Hz}$$

2. The frequencies that eject electrons are those between the cutoff frequency and the highest frequency in the beam:

frequencies in this beam that eject electrons: 5.50×10^{14} Hz to 7.90×10^{14} Hz

continued on next page

continued from previous page

Part (b)

3. Using $K_{max} = hf - W_0$, calculate K_{max} for the maximum frequency in the beam, $f = 7.90 \times 10^{14}$ Hz:

$$K_{max} = hf - W_0$$
$$= (6.63 \times 10^{-34}\ \text{J}\cdot\text{s})(7.90 \times 10^{14}\ \text{Hz}) - (2.28\ \text{eV})\left(\frac{1.60 \times 10^{-19}\ \text{J}}{1\ \text{eV}}\right) = 1.59 \times 10^{-19}\ \text{J}$$

Insight

Note that most of the photons in a beam of white light will eject electrons from sodium, and that the maximum kinetic energy of one of these photoelectrons is about 1 eV.

Practice Problem

What frequency of light would be necessary to give a maximum kinetic energy of 2.00 eV to the photoelectrons from this surface? [**Answer:** 1.03×10^{15} Hz]

Some related homework problems: Problem 25, Problem 26, Problem 27

CONCEPTUAL CHECKPOINT 30–2 Ejected Electrons

Consider a photoelectric experiment such as the one illustrated in Figure 30–5. A beam of light with a frequency greater than the cutoff frequency shines on the emitter. If the frequency of this beam is increased while the intensity is held constant, does the number of electrons ejected per second from the metal surface **(a)** increase, **(b)** decrease, or **(c)** stay the same?

Reasoning and Discussion

Increasing the frequency of the beam means that each photon carries more energy; however, we know that the intensity of the beam remains constant. It follows, then, that fewer photons hit the surface per time—otherwise the intensity would increase. Since fewer photons hit the surface per time, fewer electrons are ejected per time.

Answer:

(b) The number of electrons ejected per second decreases.

REAL-WORLD PHYSICS
Photocells

Applications of the photoelectric effect are in common use all around us. For example, if you have ever dashed into an elevator as its doors were closing, you were probably saved from being crushed by the photoelectric effect. Many elevators and garage-door systems use a beam of light and a photoelectric device known as a *photocell* as a safety feature. As long as the beam of light strikes the photocell, the photoelectric effect generates enough ejected electrons to produce a detectable electric current. When the light beam is blocked—by a late arrival at the elevator, for example—the electric current produced by the photocell is interrupted and the doors are signaled to open. Similar photocells automatically turn on streetlights at dusk and measure the amount of light entering a camera.

▶ The photoelectric effect is the basic mechanism used by photovoltaic cells, which are now used to power both terrestrial devices such as pay phones (left) and the solar panels that supply electricity to the Hubble Space Telescope (right).

Photocells are also the basic unit in the *solar energy panels* that convert some of the energy in sunlight into electrical energy. A small version of a solar energy panel can be found on many pocket calculators. These panels are efficient enough to operate their calculators with nothing more than dim indoor lighting. Larger outdoor panels can operate billboards and safety lights in remote areas far from commercial power lines. Truly large solar panels, 240 ft in length, power the International Space Station and are so large that they make the station visible to the naked eye from Earth's surface. These applications of solar panels may only hint at the potential for solar energy in the future, however, especially when one considers that sunlight delivers about 200,000 times more energy to Earth each day than all the world's electrical energy production combined.

REAL-WORLD PHYSICS
Solar energy panels

Finally, it is interesting to note that though Einstein is best known for his development of the theory of relativity, he was awarded the Nobel Prize in physics not for relativity but for the photoelectric effect.

30–3 The Mass and Momentum of a Photon

When we say that a photon is like a "particle" of light, we put the word particle in quotes because a photon is quite different from everyday particles in many important respects. For example, a typical particle can be held at rest in the hand. It can also be placed on a scale to have its mass determined. These operations are not possible with a photon.

First, photons travel with the speed of light, which means that all observers see them as having the same speed. It is not possible to stop a photon and hold it in your hand. In contrast, particles with a finite rest mass can never attain the speed of light. It follows, then, that photons must have *zero rest mass*. This condition can be seen mathematically by rewriting Equation 29–7 for the total energy, E, as follows:

$$E\sqrt{1-\frac{v^2}{c^2}} = m_0c^2 \qquad \textbf{30–8}$$

Since photons travel at the speed of light, $v = c$, the left side of the equation is zero. As a result, the right side of the equation must also be zero, which can happen only if the rest mass is zero:

Rest Mass of a Photon

$$m_0 = 0 \qquad \textbf{30–9}$$

Second, photons differ from everyday particles in that they have a finite momentum even though they have no mass. To see how this can be, note that the relativistic equation for momentum, Equation 29–5, can be rewritten as follows:

$$p\sqrt{1-\frac{v^2}{c^2}} = m_0v \qquad \textbf{30–10}$$

Dividing Equation 30–10 by Equation 30–8 we get

$$\frac{p}{E} = \frac{v}{c^2}$$

Once again setting $v = c$ for a photon, we find that the momentum of a photon is related to its total energy as follows:

$$p = \frac{E}{c}$$

Finally, recalling that $E = hf$, and that $f = c/\lambda$, we obtain the following result:

Momentum of a Photon

$$p = \frac{hf}{c} = \frac{h}{\lambda} \qquad \textbf{30–11}$$

PROBLEM-SOLVING NOTE

The Momentum of a Photon

The relation $p = mv$ is not valid for photons. In fact, photons have a finite momentum even though their rest mass is zero.

Note that a photon's momentum increases with its frequency, and thus with its energy.

As one might expect, the momentum of a typical photon of visible light is quite small, as is illustrated in the next Exercise.

EXERCISE 30–4

Calculate the momentum of a photon of yellow light with a frequency of 5.25×10^{14} Hz.

Solution

Substituting $f = 5.25 \times 10^{14}$ Hz in Equation 30–11, we obtain

$$p = \frac{hf}{c} = \frac{(6.63 \times 10^{-34}\,\mathrm{J \cdot s})(5.25 \times 10^{14}\,\mathrm{Hz})}{3.00 \times 10^{8}\,\mathrm{m/s}} = 1.16 \times 10^{-27}\,\mathrm{kg \cdot m/s}$$

REAL-WORLD PHYSICS
Sailing on a beam of light

As small as the momentum of a photon is, it can still have a significant impact if the number of photons is large. For example, NASA is studying the feasibility of constructing spaceships that would be powered by huge "light sails" that transfer the momentum of photons from the Sun to the ship. A sail would reflect photons, creating a change in momentum, and hence a reaction force on the sail due to the **radiation pressure** of the photons. With a large enough sail, it may one day be possible to cruise among the stars with space-faring sailboats. In fact, the designs under consideration by NASA would result in the largest and fastest spacecraft ever constructed—10 times faster than the space shuttle.

REAL-WORLD PHYSICS
Optical tweezers

A more "down to Earth," though no less exotic, application of radiation pressure is the *optical tweezers*. Basically, an optical tweezers is a laser beam that is made more intense in the middle than near the edges. When a small, translucent object is placed in this beam, the recoil produced by photons passing through the object exerts a small force on it directed toward the center of the beam. The situation is similar to a ball suspended in the "beam" of air produced by a hair dryer. Although the force exerted by such tweezers is typically on the order of only 10^{-12} N, it is still large enough to manipulate cells, DNA, and other subcellular particles. In fact, optical tweezers can even capture, lift, and separate a single bacterium from a bacterial culture for further study.

30–4 Photon Scattering and the Compton Effect

Einstein's photon model of the photoelectric effect, published in 1905, focused on the energy of a photon, $E = hf$. The momentum of a photon, $p = hf/c = h/\lambda$, plays a key role in a different type of experiment, in which an X-ray photon undergoes a collision with an electron initially at rest. The result of this collision is that the photon is scattered, which changes its direction and energy. This type of process, referred to as the **Compton effect,** was explained in terms of the photon model of light by the American physicist Arthur Holly Compton (1892–1962) in 1923. The success of the photon model in explaining both the Compton effect and the photoelectric effect gained it widespread acceptance.

Figure 30–7 shows an X-ray photon striking an electron at rest and scattering at an angle θ with respect to the incident direction. To understand the behavior of a system like this, we treat the photon as a particle, with a certain energy and momentum, that collides with another particle (the electron) of mass m_e and initial speed zero. If we assume the electron is free to move, this collision conserves both energy and momentum.

First, to conserve energy the following relation must be satisfied:

energy of incident photon = energy of scattered photon + final kinetic energy of electron

$$hf = hf' + K \qquad \text{30–12}$$

Note that, in general, the frequency of the scattered photon, f', is less than the frequency of the incident photon, f, since part of the energy of the incident photon

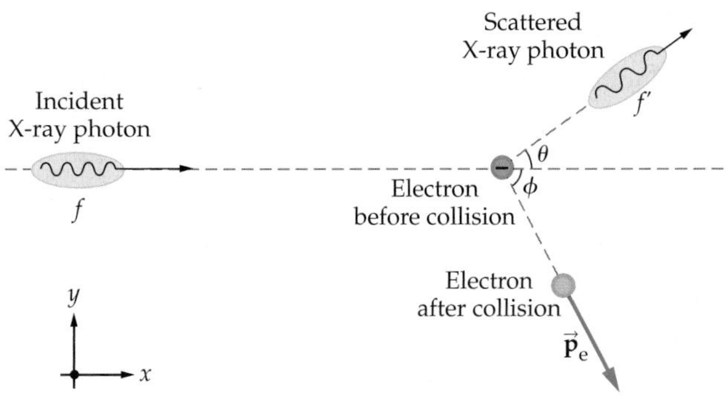

◀ **FIGURE 30–7 The Compton effect**
An X-ray photon scattering from an electron at rest can be thought of as a collision between two particles. The result is a change of wavelength for the scattered photon. This is referred to as the Compton effect.

has gone into kinetic energy of the electron. Because the frequency of the scattered photon is reduced, its wavelength, $\lambda' = c/f'$, is increased.

Next, we conserve both the x and y components of momentum, using $p = h/\lambda$ as the magnitude of the incident photon's momentum, and $p' = h/\lambda'$ as the magnitude of the scattered photon's momentum. For the x component of momentum we have the following relation:

$$\frac{h}{\lambda} = \frac{h}{\lambda'}\cos\theta + p_e \cos\phi \qquad \textbf{30–13}$$

Note that we assume the electron has a momentum p_e at an angle ϕ to the incident direction. Conserving the y component of momentum, which initially is zero, we have

$$0 = \frac{h}{\lambda'}\sin\theta - p_e \sin\phi \qquad \textbf{30–14}$$

Given the initial wavelength and frequency, $\lambda = c/f$, and the scattering angle θ, we can use the three relations, Equations 30–12, 30–13, and 30–14, to solve for the three unknowns; λ', ϕ, and K.

Of particular interest is the change in wavelength produced by the scattering. We find the following result:

Compton Shift Formula

$$\Delta\lambda = \lambda' - \lambda = \frac{h}{m_e c}(1 - \cos\theta) \qquad \textbf{30–15}$$

SI unit: m

In this expression, the quantity $h/m_e c = 2.43 \times 10^{-12}$ m is referred to as the *Compton wavelength of an electron*. The maximum change in the photon's wavelength occurs when it scatters in the reverse direction ($\theta = 180°$), in which case the change is twice the Compton wavelength, $\Delta\lambda = 2(h/m_e c)$. On the other hand, if the scattering angle is zero ($\theta = 0$)—in which case there really is no scattering at all—the change in wavelength is zero; $\Delta\lambda = 0$.

PROBLEM-SOLVING NOTE

Scattering Angle

The angle θ in the Compton shift formula is the angle between the incident direction of an X-ray and its direction of propagation after scattering.

CONCEPTUAL CHECKPOINT 30–3 Change in Wavelength

If X-rays are scattered from protons instead of from electrons, is the change in their wavelength for a given angle of scattering **(a)** increased, **(b)** decreased, or **(c)** unchanged?

Reasoning and Discussion

As can be seen from Equation 30–15, the change in wavelength for a given angle is proportional to $h/m_e c$. If a proton is substituted for the electron, the change in wavelength will be proportional to $h/m_p c$ instead. Since protons have about 2000 times the mass of electrons, the change in wavelength will be reduced by a factor of about 2000.

Answer:

(b) The change in wavelength is decreased when X-rays scatter from protons rather than from electrons.

The next Example gives a detailed analysis of a photon-electron collision.

EXAMPLE 30–4 Scattering X-rays

An X-ray photon with a wavelength of 0.650 nm scatters from a free electron at rest. After scattering, the photon moves at an angle of 152° relative to the incident direction. Find **(a)** the wavelength and **(b)** the energy of the scattered photon. **(c)** Determine the kinetic energy of the recoiling electron.

Picture the Problem

As shown in the sketch, the incoming photon is scattered at an angle of 152° relative to its initial direction; thus, it almost heads back the way it came. After the collision, the electron moves with a speed v at an angle ϕ relative to the forward direction.

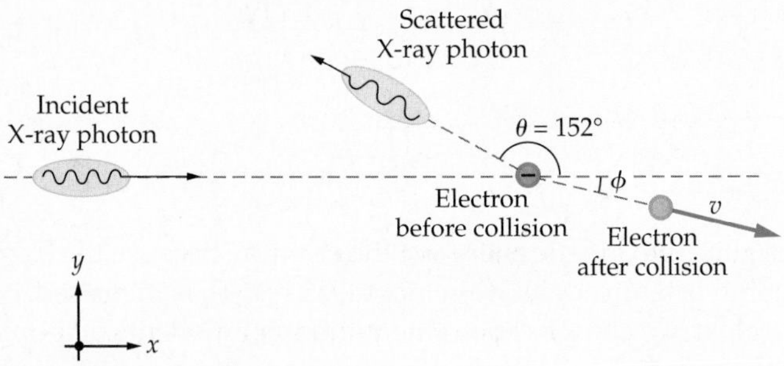

INTERACTIVE FIGURE

Strategy

a. We can find the wavelength after scattering, λ', by using the Compton shift formula, Equation 30–15. Note that we are given the initial wavelength, λ, as well as the scattering angle, θ.

b. The energy of the scattered photon is $E' = hf' = hc/\lambda'$.

c. Since energy is conserved, the kinetic energy of the electron is given by $hf = hf' + K$. We can solve this relation to yield the kinetic energy, K.

Solution

Part (a)

1. Use Equation 30–15 to find the wavelength of the scattered photon:

$$\lambda' = \lambda + \frac{h}{m_e c}(1 - \cos\theta) = 0.650 \times 10^{-9}\ \text{m} + \left(\frac{6.63 \times 10^{-34}\ \text{J}\cdot\text{s}}{(9.11 \times 10^{-31}\ \text{kg})(3.00 \times 10^{8}\ \text{m/s})}\right)(1 - \cos 152°)$$
$$= 6.55 \times 10^{-10}\ \text{m} = 0.655\ \text{nm}$$

Part (b)

2. The energy of the scattered photon is given by $E' = hf' = hc/\lambda'$:

$$E' = hf' = \frac{hc}{\lambda'} = \frac{(6.63 \times 10^{-34}\ \text{J}\cdot\text{s})(3.00 \times 10^{8}\ \text{m/s})}{6.55 \times 10^{-10}\ \text{m}} = 3.04 \times 10^{-16}\ \text{J}$$

Part (c)

3. Find the initial energy of the photon using $E = hf = hc/\lambda$:

$$E = hf = \frac{hc}{\lambda} = \frac{(6.63 \times 10^{-34}\ \text{J}\cdot\text{s})(3.00 \times 10^{8}\ \text{m/s})}{6.50 \times 10^{-10}\ \text{m}} = 3.06 \times 10^{-16}\ \text{J}$$

4. Subtract the final energy of the photon from its initial energy to find the kinetic energy of the electron:

$$K = hf - hf' = 3.06 \times 10^{-16}\ \text{J} - 3.04 \times 10^{-16}\ \text{J} = 2 \times 10^{-18}\ \text{J} = 10\ \text{eV}$$

Insight

Notice that the energy of the X-ray photon is roughly 1900 eV = 1.9 keV both before and after scattering. Only a very small fraction of its energy is delivered to the electron. Still, the electron acquires about 10 eV of energy—enough to ionize an atom.

Practice Problem

At what scattering angle will the photon have a wavelength of 0.652 nm? **[Answer:** $\theta = 79.9°$**]**

Some related homework problems: Problem 46, Problem 47

As we have seen in both the photoelectric effect and the Compton effect, a photon can behave as a particle with a well-defined energy and momentum. We also know, however, that light exhibits wavelike behavior, as when it produces interference fringes in Young's two-slit experiment (Chapter 28). That light can have such seemingly opposite attributes is one of the deepest mysteries of quantum physics, and at the same time, one of its most significant insights. In the next section we expand on this insight and extend it to the behavior of matter.

30–5 The de Broglie Hypothesis and Wave-Particle Duality

As we have seen, the Compton effect was explained in terms of photons in 1923. Another significant advance in quantum physics occurred that same year, when a French graduate student, Louis de Broglie (1892–1987), put forward a most remarkable hypothesis that would later win him the Nobel Prize in physics. His suggestion was basically the following:

> Since light, which we usually think of as a wave, can exhibit particle-like behavior, perhaps a particle of matter, like an electron, can exhibit wavelike behavior.

In particular, de Broglie proposed that the *same* relation between wavelength and momentum that Compton applied to the photon, $p = h/\lambda$, should apply to particles as well. Thus, if the momentum of a particle is p, its de Broglie wavelength is

de Broglie Wavelength

$$\lambda = \frac{h}{p} \qquad \text{30–16}$$

SI unit: m

Note that the greater a particle's momentum, the smaller its de Broglie wavelength.

How can the idea of a wavelength for matter make sense, however, when we know that objects like baseballs and cars behave like particles, not like waves? To see how this is possible, we calculate the de Broglie wavelength of a typical macroscopic object: a 0.13-kg apple moving with a speed of 5.0 m/s. Substituting these values into $p = mv$, and using $\lambda = h/p$, we find the following wavelength: $\lambda = 1.0 \times 10^{-33}$ m. Clearly, this wavelength, which is smaller than the diameter of an atom by a factor of 10^{23}, is much too small to be observed in any macroscopic experiment. Thus, an apple could have a wavelength as given by the de Broglie relation, and we would never notice it.

In contrast, consider an electron with a kinetic energy of 10.0 eV, a typical atomic energy. Using $K = \frac{1}{2}mv^2 = p^2/2m$ to solve for the momentum p, we find that the de Broglie wavelength in this case is $\lambda = 3.88 \times 10^{-10}\text{ m} = 3.88\text{ Å}$. Now this wavelength, which is comparable to the size of an atom or molecule, would clearly be significant in such systems. Therefore, the de Broglie wavelength may be unobservable in macroscopic systems but all-important in atomic systems.

ACTIVE EXAMPLE 30–2 The Speed and Wavelength of an Electron

How fast is an electron moving if its de Broglie wavelength is 3.50×10^{-7} m?

Solution *(Test your understanding by performing the calculations indicated in each step.)*

1. Write Equation 30–16 in terms of the electron's mass and speed: $\lambda = h/mv$
2. Rearrange to solve for the speed, v: $v = h/m\lambda$
3. Substitute numerical values: $v = 2080\text{ m/s}$

Insight

The relatively small value obtained for the electron's speed justifies using the non-relativistic expression for momentum, $p = mv$.

Your Turn

If the electron's speed is doubled, does its de Broglie wavelength increase or decrease? By what factor?

(Answers to **Your Turn** *problems are given in the back of the book.)*

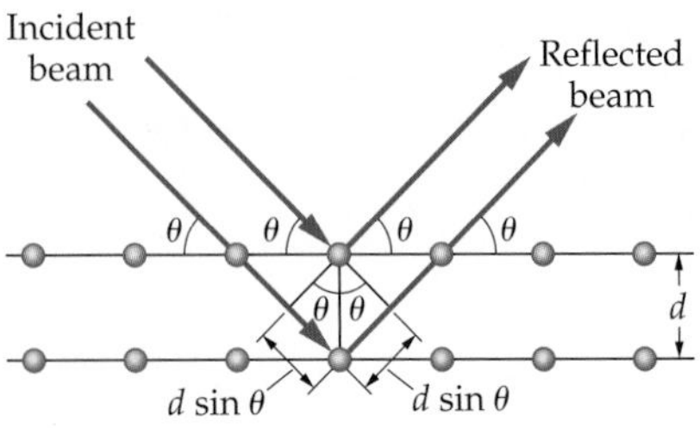

▲ **FIGURE 30–8 Scattering from a crystal**
Scattering of X-rays or particles from a crystal. Note that waves reflecting from the lower plane of atoms have a path length that is longer than the path of the upper waves by the amount $2d \sin \theta$.

In order for the de Broglie wavelength to be taken seriously, however, it must be observed experimentally. We next consider ways in which this can be done.

Diffraction of X-rays and Particles by Crystals

An especially powerful way to investigate wave properties is through the study of interference patterns. Consider, for example, directing a coherent beam of X-rays onto a crystalline substance composed of regularly spaced planes of atoms, as indicated in **Figure 30–8**. Notice that the reflected beam combines rays that have followed different paths, with different path lengths. If the difference in path lengths is half a wavelength, destructive interference occurs; if the difference in path lengths is one wavelength, constructive interference results, and so on. From Figure 30–8 it is clear that the difference in path lengths for rays reflecting from adjacent planes that have a spacing d is $2d \sin \theta$. Thus, constructive interference occurs when the following conditions are met:

Constructive Interference when Scattering from a Crystal

$$2d \sin \theta = m\lambda \quad m = 1, 2, 3, \ldots \tag{30–17}$$

This is very similar to the condition for constructive interference in a diffraction grating, as derived in Equation 28–16. The resulting pattern of interference maxima—referred to as a **diffraction pattern**—can be projected onto photographic film, as indicated in **Figure 30–9**, and used to determine the geometric properties of a crystal. Examples of diffraction patterns are shown in **Figure 30–10**.

Now, if particles—like electrons, for example—have a wavelength comparable to atomic distances, it should be possible to produce diffraction patterns with them in much the same way as with X-rays. This is indeed the case, as was shown by the American physicists C. J. Davisson and L. H. Germer, who, in 1928, produced diffraction patterns by scattering low-energy electrons (about 54 eV) from crystals of nickel. The spacing between spots in the electron diffraction pattern allowed the researchers to determine the electron's wavelength, which verified

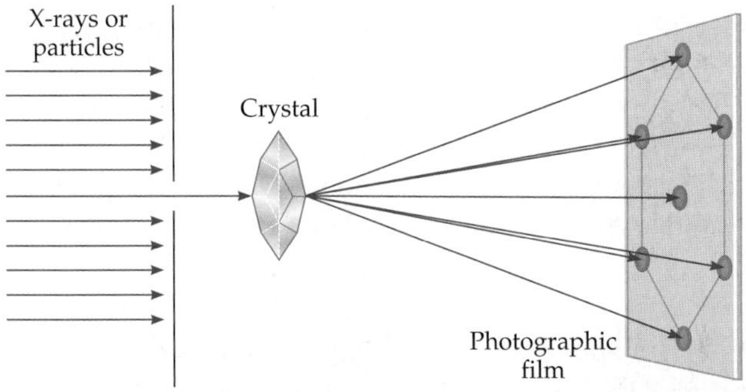

▶ **FIGURE 30–9 Diffraction patterns**
Diffraction patterns can be observed by passing a beam of X-rays or particles through a crystal. The beams emerging from the crystal are at specific angles, due to constructive interference, and can be recorded on photographic film.

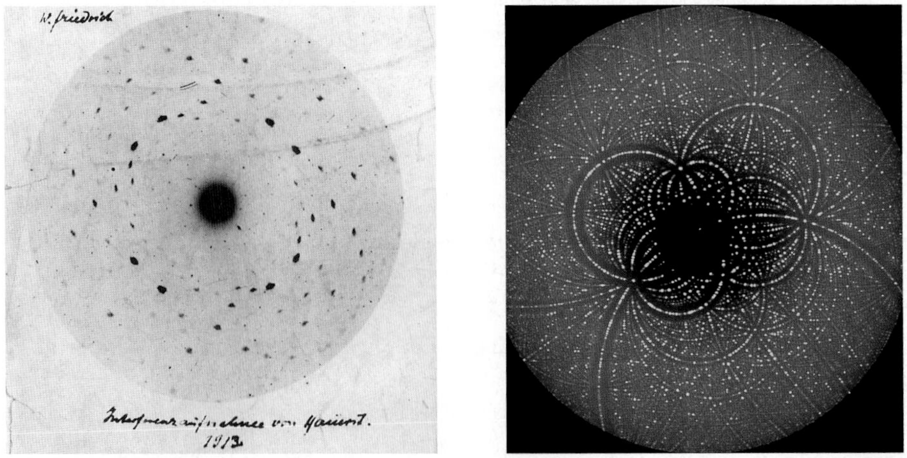

▶ **FIGURE 30–10 X-ray diffraction**
Because the spacing of atoms in a crystal is of the same order of magnitude as the wavelengths of X-rays, the planes of a crystal can serve as a diffraction grating for X-rays. When properly interpreted, the resulting diffraction patterns can provide remarkably detailed information about the structure of the crystal. The historic photograph at left, made by two of the pioneers of X-ray crystallography in 1912, records the diffraction pattern from a simple inorganic salt. The photograph at right, made nearly a century later, shows the much more intricate pattern produced by a large protein molecule.

◀ **FIGURE 30–11 Neutron diffraction**
The neutron diffraction pattern of the protein lysozyme, a digestive enzyme.

in detail the de Broglie relation, $\lambda = h/p$. Similar diffraction patterns have since been observed with neutrons, hydrogen atoms, and helium atoms. An example of a neutron diffraction pattern is shown in **Figure 30–11**.

EXAMPLE 30–5 Neutron Diffraction

A beam of neutrons moving with a speed of 1450 m/s is diffracted from a crystal of table salt, (sodium chloride) which has an interplanar spacing of $d = 0.282$ nm. **(a)** Find the de Broglie wavelength of the neutrons. **(b)** Find the angle of the first interference maximum.

Picture the Problem

Our sketch shows a beam of neutrons reflecting from two adjacent atomic planes of table salt. The distance d between the planes is 0.282 nm.

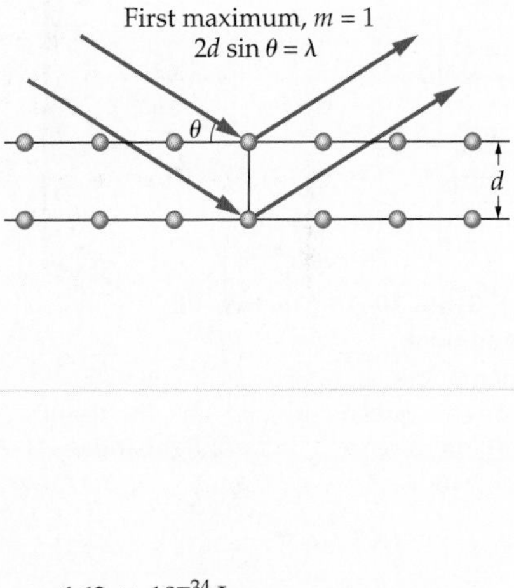

Strategy

a. We can find the de Broglie wavelength using Equation 30–16; $\lambda = h/p$. Since the speed of the neutrons is much less than the speed of light, we can write the momentum as $p = m_n v$, where m_n is the mass of a neutron.

b. Referring to Equation 30–17, we see that the first interference maximum occurs when $m = 1$ in the relation $2d \sin\theta = m\lambda$. Thus we set $2d \sin\theta$ equal to λ and solve for θ.

Solution

Part (a)

1. Calculate the de Broglie wavelength using $\lambda = h/p = h/m_n v$:

$$\lambda = \frac{h}{p}$$

$$= \frac{h}{m_n v} = \frac{6.63 \times 10^{-34}\,\mathrm{J \cdot s}}{(1.67 \times 10^{-27}\,\mathrm{kg})(1450\,\mathrm{m/s})} = 0.274\,\mathrm{nm}$$

Part (b)

2. Set $m = 1$ in $2d \sin\theta = m\lambda$:

$$2d \sin\theta = \lambda$$

3. Solve for the angle θ:

$$\theta = \sin^{-1}\left(\frac{\lambda}{2d}\right) = \sin^{-1}\left(\frac{0.274\,\mathrm{nm}}{2(0.282\,\mathrm{nm})}\right) = 29.1^\circ$$

Insight

Note that if we change the speed of the neutrons, we will change their wavelength. This, in turn, changes the angle of the interference maxima. This connection between the neutron speed and the interference maxima provides a detailed and precise way of verifying the de Broglie relation.

Practice Problem

Find the angle of the first interference maximum if the neutron speed is increased to 2250 m/s. [**Answer:** 18.2°]

Some related homework problems: Problem 52, Problem 55

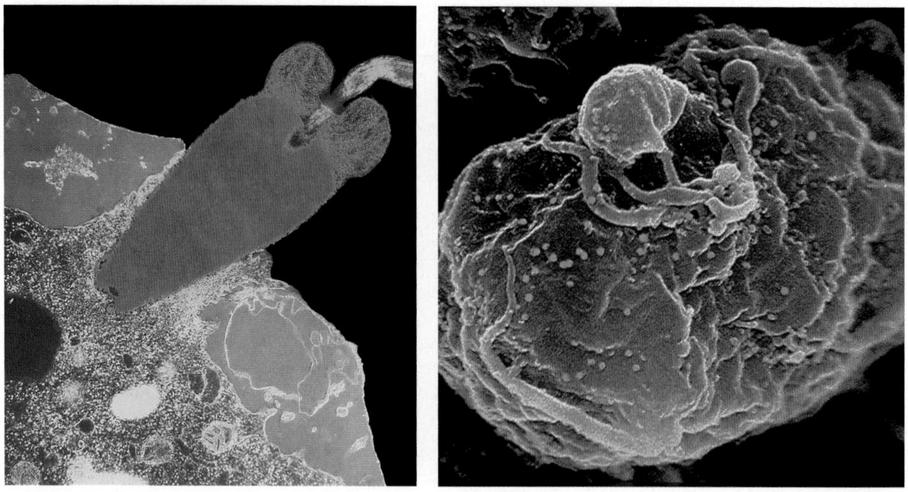

▶ **FIGURE 30–12 Electron micrography**

Electron micrographs can be produced in several different ways. In a transmission electron micrograph (TEM), such as the image at left, the beam of electrons passes through the specimen, as in an ordinary light microscope. This colorized photo shows a sea urchin sperm cell penetrating the membrane of an egg. In a scanning electron micrograph (SEM), like the image at right, the electrons are reflected from the surface of the specimen (which is usually coated with a thin layer of metal atoms first) and used to produce a startlingly detailed three-dimensional image. This photo, also colorized for greater clarity, shows particles of the AIDS virus HIV (purple) emerging from an infected human cell.

REAL-WORLD PHYSICS
Electron microscopes

In *electron microscopes*, a beam of electrons is used to form an image of an object in much the same way that a beam of light is used in a light microscope. One of the differences, however, is that the wavelength of the electrons can be much shorter than the wavelength of visible light. For example, the shortest visible wavelength is that of blue light, which is about 380 nm. In comparison, we have seen that the wavelength of an electron with an energy of only 10.0 eV is 0.388 nm—about 1000 times smaller than the wavelength of blue light. Since the ability to resolve small objects depends on using a wavelength that is smaller than the object to be imaged, an electron microscope can see much finer detail than a light microscope, as **Figure 30–12** indicates.

Wave-Particle Duality

Notice that we have now come full circle in our study of waves and particles, from considering light as a wave, and then noting that it has particle-like properties, to investigating particles of matter like electrons, and finding that they have wave-like properties. This type of behavior is referred to as the **wave-particle duality.** Thus, as strange as it may seem, light is a wave, but it also comes in discrete units called photons. Electrons come in discrete units of well-defined mass and charge, but they also have wave properties.

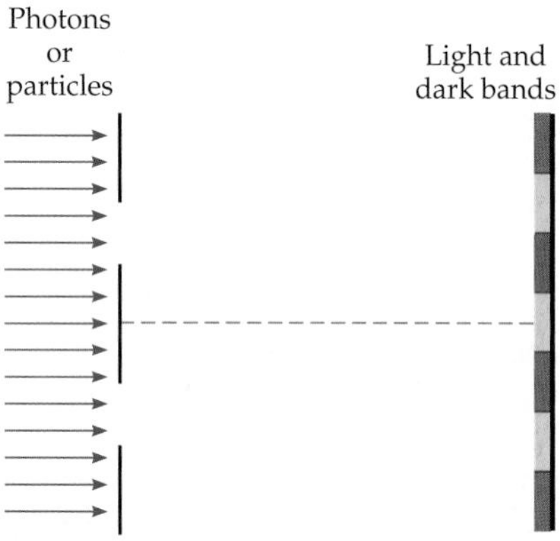

▲ **FIGURE 30–13 The two-slit experiment**

When photons or particles are passed through a screen with two slits, the result is an interference pattern of light and dark bands.

To illustrate the wave-particle duality, we consider a two-slit experiment, as shown in **Figure 30–13**. If light is passed through these slits it forms interference patterns of dark and light fringes, as we saw in Chapter 28. If the intensity of the light is reduced to a very low level, it is possible to have only a single photon at a time passing through the apparatus. This photon lands on the distant screen. Eventually, as more and more photons land on the screen, the interference pattern emerges.

Similar behavior can be observed with electrons passing through a pair of slits. The results, as shown in **Figure 30–14**, are the direct analog of the results obtained with light. Notice that the dark portions of the interference pattern are those places where electron waves passing through the two slits have combined

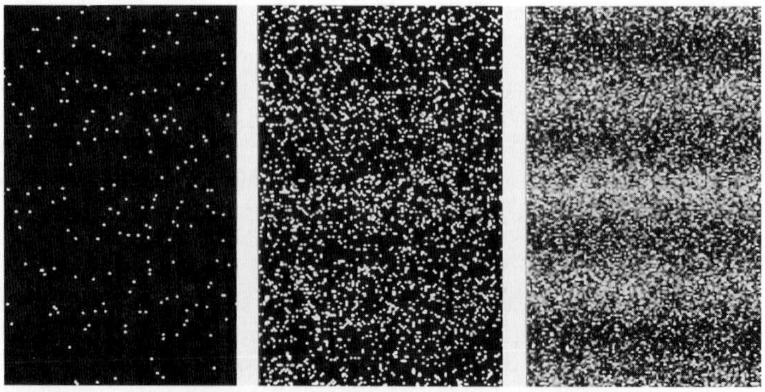

▶ **FIGURE 30–14 Creation of an interference pattern by electrons passing through two slits**

At first, the electrons seem to arrive at the screen in random locations. As the number of electrons increases, however, the interference pattern becomes more evident. (Compare Figures 28–3 and 28–6.)

to produce destructive interference—that is, at these special points an electron is essentially able to "cancel itself out" to produce a dark fringe.

Clearly, then, subatomic particles and light are quite different from objects and waves that we observe on a macroscopic level. In fact, one of the most profound and unexpected insights of quantum physics is simply that even though baseballs, apples, and people are composed of electrons, protons, and other particles, the behavior of these subatomic particles is nothing like the behavior of baseballs, apples, and people. In short, an electron is not like a BB reduced in size. An electron has properties that are different from those of any object we experience on the macroscopic level. To try to force light and electrons into categories like waves and particles is to miss the essence of their existence—they are neither one nor the other, though they have characteristics of both. Again, one is reminded of the saying "The universe is not only stranger than we imagine, it is stranger than we can imagine."

30–6 The Heisenberg Uncertainty Principle

One of the more interesting aspects of Figure 30–14 is that the dots, corresponding to observations of individual electrons, appear on the screen in random order. To be specific, each time this experiment is run, the dots appear in different locations and in different sequence—all that remains the same is the final pattern that emerges after a large number of electrons are observed. The point is that as any given electron passes through the two-slit apparatus, *it is not possible to predict exactly where that one electron will land on the screen*. We can give the probability that it will land at different locations, but the fate of each individual electron is uncertain.

This kind of "uncertainty" is inherent in quantum physics and is due to the fact that matter has wavelike properties. As a simple example, consider a beam of electrons moving in the x direction and passing through a single slit, as in **Figure 30–15**. The result of this experiment is a diffraction pattern similar to that found for light in Chapter 28 (see Figure 28–18). In particular, if the beam passes through a slit of width W, it produces a large central maximum—where the probability of detecting an electron is high—with a dark fringe on either side at an angle θ given by Equation 28–12:

$$\sin\theta = \frac{\lambda}{W} \qquad \textbf{30–18}$$

If the incoming electrons have a momentum p_x, the wavelength in Equation 30–18 is given by the de Broglie relation: $\lambda = h/p_x$.

We interpret this experiment as follows: By passing the beam of electrons through the slit, we have determined their location in the y direction to within an uncertainty of roughly $\Delta y \sim W$. After passing through the slit, however, the beam spreads out to form a diffraction pattern—with some electrons acquiring a y component of momentum. Thus there is now an uncertainty in the y component of momentum, Δp_y.

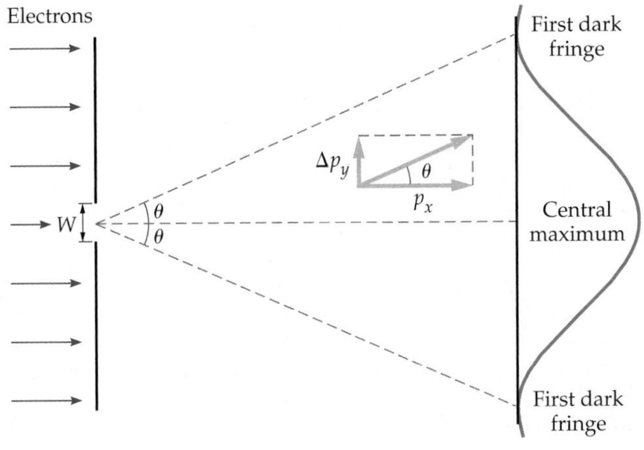

◀ **FIGURE 30–15 Diffraction pattern of electrons**

Central region of the diffraction pattern formed by electrons passing through a single slit in a screen. The curve to the right is a measure of the number of electrons detected at any given location. Note that no electrons are detected at the dark fringes. This diffraction pattern is identical in form with that produced by light passing through a single slit.

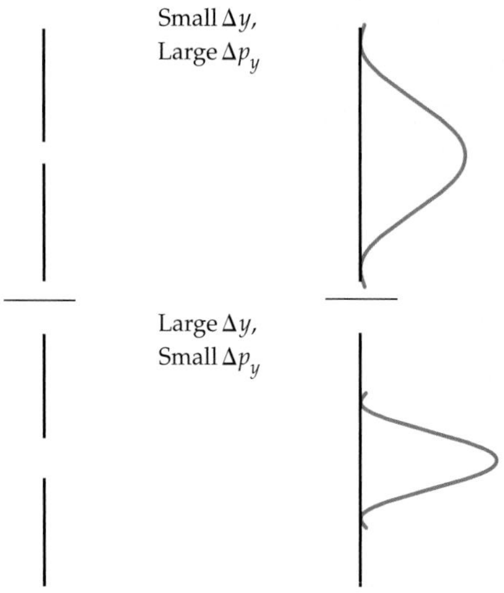

▲ **FIGURE 30–16 Uncertainty in position and momentum**

Reciprocal relationship between the uncertainty in position (Δy) and the uncertainty in momentum (Δp_y). As in Figure 29–15, the curves at the right indicate the number of electrons detected at any given location.

There is a reciprocal relationship between Δy and Δp_y, as we show in **Figure 30–16**. As the width W of the slit is made smaller, to decrease Δy, the angle in Equation 30–18 increases, and the diffraction pattern spreads out, increasing Δp_y. Conversely, if the width of the slit is increased, the diffraction pattern becomes narrower; that is, increasing Δy results in a decrease in Δp_y. To summarize;

> If we know the position of a particle with greater precision, its momentum is more uncertain; if we know the momentum of a particle with greater precision, its position is more uncertain.

We can give this conclusion in mathematical form by returning to Equation 30–18 and Figure 30–15. First, assuming small angles, θ, we can use the approximation that $\sin\theta \sim \theta$ to write Equation 30–18 as $\theta \sim \lambda/W$. Similarly, from Figure 30–15 we see that $\tan\theta = \Delta p_y/p_x$. Since $\tan\theta \sim \theta$ for small angles, it follows that $\theta \sim \Delta p_y/p_x$. Equating these two expressions for θ we get

$$\theta \sim \frac{\lambda}{W} \sim \frac{\Delta p_y}{p_x}$$

Setting $p_x = h/\lambda$ and $W \sim \Delta y$, we find

$$\frac{\lambda}{\Delta y} \sim \frac{\Delta p_y}{h/\lambda}$$

Finally, canceling λ and rearranging, we obtain

$$\Delta p_y\,\Delta y \sim h$$

Thus, the product of uncertainties in position and momentum cannot be less than a certain finite amount that is approximately equal to Planck's constant.

A thorough treatment of this system yields the more precise relation first given by the German physicist Werner Heisenberg (1901–1976) in 1927, known as the **Heisenberg uncertainty principle:**

The Heisenberg Uncertainty Principle: Momentum and Position

$$\Delta p_y\,\Delta y \geq \frac{h}{2\pi} \qquad \text{30–19}$$

In fact, Heisenberg showed that this relation is a general principle and not restricted in any way to the single-slit system considered here. There is simply an unremovable, intrinsic uncertainty in nature that is the result of the wave behavior of matter. Since the wavelength of matter, $\lambda = h/p$, depends directly on the magnitude of h, so too does the uncertainty.

One way of stating the uncertainty principle is that it is impossible to know both the position and momentum of a particle with arbitrary precision at any given time. For example, if the position is known precisely, so that Δy approaches zero, it follows that Δp_y must approach infinity, as shown in Figure 30–16. This result implies that the y component of momentum is completely uncertain. Similarly, complete knowledge of $p_y(\Delta p_y \rightarrow 0)$ implies that the position y is completely uncertain.

As one might expect, the uncertainty restrictions given in Equation 30–19 have negligible impact on macroscopic systems, but are all-important in atomic and nuclear systems. This is illustrated in the following Example.

EXAMPLE 30–6 What Is Your Position?

If the speed of an object is 35.0 m/s, with an uncertainty of 5.00%, what is the minimum uncertainty in the object's position if it is **(a)** an electron or **(b)** a volleyball ($m = 0.350$ kg)?

Picture the Problem

As shown in the sketch, the electron and the volleyball have the same speed. The fact that the electron has the smaller mass means that the values of its momentum, and of its uncertainty in momentum, are less than the corresponding values for the volleyball. As a result, the uncertainty in position will be larger for the electron.

Strategy

The 5.00% uncertainty in speed means that the magnitude of the momentum (mass times speed) of the electron and the volleyball are also uncertain by 5.00%. The minimum uncertainty in position, then, is given by the Heisenberg uncertainty principle; $\Delta y = h/(2\pi\ \Delta p_y)$.

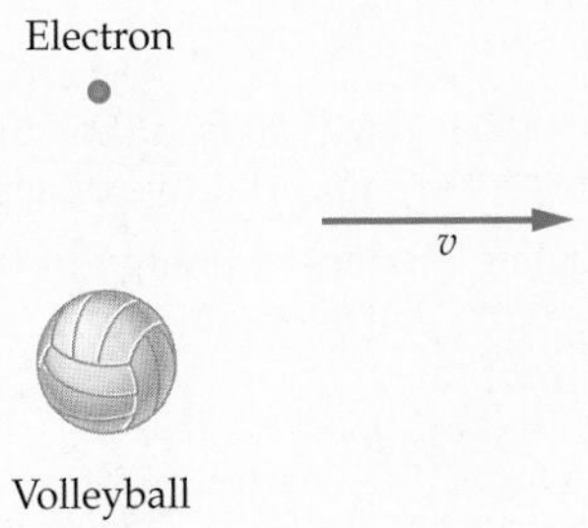

Solution

Part (a)

1. Calculate the uncertainty in the electron's momentum:

$$\Delta p_y = 0.0500 m_e v = 0.0500(9.11 \times 10^{-31}\ \text{kg})(35.0\ \text{m/s}) = 1.59 \times 10^{-30}\ \text{kg} \cdot \text{m/s}$$

2. Use the uncertainty principle to find the minimum uncertainty in position:

$$\Delta y = \frac{h}{2\pi\ \Delta p_y} = \frac{6.63 \times 10^{-34}\ \text{J} \cdot \text{s}}{2\pi(1.59 \times 10^{-30}\ \text{kg} \cdot \text{m/s})} = 6.64 \times 10^{-5}\ \text{m}$$

Part (b)

3. Calculate the uncertainty in the volleyball's momentum:

$$\Delta p_y = 0.0500 m_{\text{volleyball}} v = 0.0500(0.350\ \text{kg})(35.0\ \text{m/s}) = 0.613\ \text{kg} \cdot \text{m/s}$$

4. As before, use the uncertainty principle to find the minimum uncertainty in position:

$$\Delta y = \frac{h}{2\pi\ \Delta p_y} = \frac{6.63 \times 10^{-34}\ \text{J} \cdot \text{s}}{2\pi(0.613\ \text{kg} \cdot \text{m/s})} = 1.72 \times 10^{-34}\ \text{m}$$

Insight

To put these results in perspective, the minimum uncertainty in the position of the electron is roughly 100,000 times the size of an atom. Clearly, this is a significant uncertainty on the atomic level. On the other hand, the minimum uncertainty of the volleyball is smaller than an atom by a factor of about 10^{24}. It follows, then, that the uncertainty principle does not have measurable consequences on typical macroscopic objects.

Practice Problem

Repeat this problem, assuming the uncertainty in speed is reduced by a factor of 2 to 2.50%. [**Answer:** The uncertainties in position increase by a factor of 2. Electron, $\Delta y = 1.33 \times 10^{-4}$ m; volleyball, $\Delta y = 3.44 \times 10^{-34}$ m.]

Some related homework problems: Problem 60, Problem 62

The Heisenberg uncertainty principle also sets the typical energy scales in atomic and nuclear systems. For example, if an electron is known to be confined to an atom, the uncertainty in its position Δy, will be roughly 1 Å. This implies a finite uncertainty for Δp_y, which in turn implies a finite kinetic energy—even though the average value of p_y is zero.

EXAMPLE 30–7 An Electron in a Box

Suppose an electron is confined to a box that is about 0.50 Å on a side. If this distance is taken as the uncertainty in position of the electron, **(a)** calculate the corresponding minimum uncertainty in momentum. **(b)** Because the electron is confined to a stationary box, its average momentum is zero. The magnitude of the electron's momentum is nonzero, however. Assuming the magnitude of the electron's momentum is the same as its uncertainty in momentum, calculate the corresponding kinetic energy.

Picture the Problem

In our sketch we show an electron bouncing around inside a box that is 0.50 Å on a side. The rapid motion of the electron is a quantum effect, due to the uncertainty principle.

continued on next page

continued from previous page

Strategy

a. Letting $\Delta y = 0.50$ Å, we can find the minimum uncertainty in momentum by using the Heisenberg uncertainty principle: $\Delta p_y = h/(2\pi\,\Delta y)$.

b. We can calculate the kinetic energy of the electron from its momentum by using $K = p^2/2m$. For the momentum, p, we will use the value of Δp_y obtained in part (a).

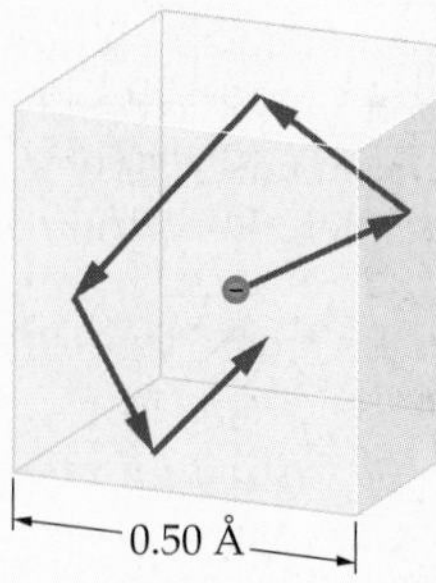

Solution

Part (a)

1. Use $\Delta y = 0.50$ Å and the uncertainty principle to find Δp_y:

$$\Delta p_y = \frac{h}{2\pi\,\Delta y} = \frac{6.63 \times 10^{-34}\,\text{J}\cdot\text{s}}{2\pi(0.50 \times 10^{-10}\,\text{m})} = 2.1 \times 10^{-24}\,\text{kg}\cdot\text{m/s}$$

Part (b)

2. Set $p = \Delta p_y$, and use $K = p^2/2m$ to find the kinetic energy:

$$K = \frac{p^2}{2m} = \frac{(2.1 \times 10^{-24}\,\text{kg}\cdot\text{m/s})^2}{2(9.11 \times 10^{-31}\,\text{kg})} = 2.4 \times 10^{-18}\,\text{J} = 15\,\text{eV}$$

Insight

The important point of this rough estimate of an electron's kinetic energy is that it is of the order of typical atomic energies. Thus, the fact that an electron is confined to an object the size of an atom means that its energy must be on the order of 10 eV. This is why the electron in our sketch is bouncing around inside the box rather than resting on its floor—as would be the case for a tennis ball inside a cardboard box. The uncertainty principle implies that the electron must have about 10 eV of kinetic energy when it is confined in this way; hence, it must be moving quite rapidly. We consider the case of a tennis ball in a cardboard box in the following Practice Problem.

Practice Problem

A tennis ball ($m = 0.06$ kg) is confined within a cardboard box 0.5 m on a side. Estimate the kinetic energy of this ball and its corresponding speed. [**Answer:** $\Delta K \sim 4 \times 10^{-67}$ J, $\Delta v \sim 4 \times 10^{-33}$ m/s. For all practical purposes, the tennis ball will appear to be at rest in the box.]

Some related homework problems: Problem 61, Problem 67, Problem 68

Finally, the uncertainty relation $\Delta y \Delta p_y \geq h/2\pi$ is just one of many forms the uncertainty principle takes. There are a number of quantities like y and p_y that satisfy the same type of relation between their uncertainties. Perhaps the most important of these uncertainty principles is the one relating the uncertainty in energy to the uncertainty in time:

The Heisenberg Uncertainty Principle: Energy and Time

$$\Delta E \Delta t \geq \frac{h}{2\pi} \qquad \textbf{30–20}$$

For example, the shorter the half-life of an unstable particle, the greater the uncertainty in its energy.

CONCEPTUAL CHECKPOINT 30–4 Uncertainties

If Planck's constant were magically reduced to zero, would the uncertainties in position/momentum and energy/time be **(a)** increased or **(b)** decreased?

Reasoning and Discussion

If h were zero, the product of uncertainties—$\Delta y \Delta p_y$ and $\Delta E \Delta t$—would be reduced to zero as well. The result is that position and momentum, for example, could be determined simultaneously with arbitrary accuracy; that is, with zero uncertainty. In this limit, particles would behave as predicted in classical physics, with well-defined positions and momenta at all times.

Answer:

(b) The uncertainties would decrease to zero.

Thus, if h were zero, the classical description of particles moving along well-defined trajectories, and having no wavelike properties, would be valid. Notice, however, that Planck's constant has an extremely small magnitude, on the order of 10^{-34} in SI units. It is this small difference from zero that is responsible for the quantum behavior seen on the atomic scale.

If h were relatively large, however, the wavelike properties of matter would be apparent even on the macroscopic level. For example, consider pitching a baseball toward the catcher's glove in a universe where h is rather large. If the ball is thrown with a well-defined momentum toward the glove, its position will be uncertain, and the ball may end up anywhere. Similarly, if the catcher gives the glove a relatively small uncertainty in position, its uncertainty in momentum is large, meaning that the glove is moving rapidly about its average position. So, even if the pitcher could aim the ball to go where desired—which is not possible—there is no way to know where to aim it. If h were significantly larger than it actually is, our experience of the natural world would be very different indeed.

30–7 Quantum Tunneling

Because particles have wavelike properties, any behavior seen in waves can be found in particles as well, under the right conditions. An example is the phenomenon known as **tunneling,** in which a wave, or a quantum particle, "tunnels" through a region of space that would be forbidden to it if it were simply a small piece of matter like a classical particle.

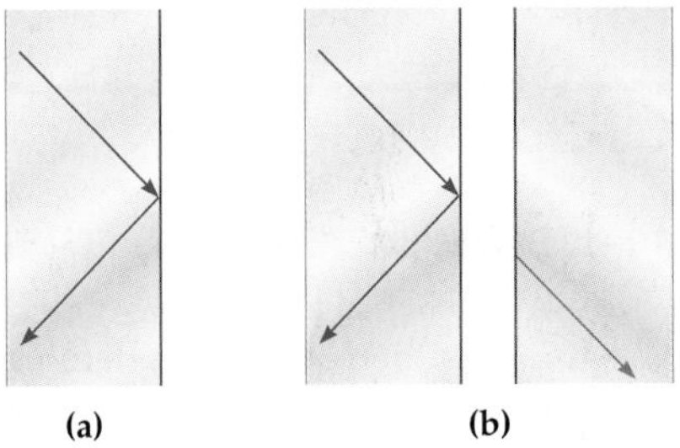

▲ **FIGURE 30–17 Optical tunneling**

(a) An incident beam of light undergoes total internal reflection from the glass-air interface. **(b)** If a second piece of glass is brought near the first piece, a weak beam of light is observed continuing in the same direction as the incident beam. We say that the light has "tunneled" across the gap.

Figure 30–17 shows a case of tunneling by light. In Figure 30–17 (a) we see a beam of light propagating within glass and undergoing total internal reflection (Chapter 26) when it encounters the glass–air interface. If light were composed of classical particles this would be the end of the story; the particles of light would simply change direction at the interface and continue propagating within the glass.

What can actually happen in such a system is more interesting, however. In Figure 30–17 (b) we show a second piece of glass brought near the first piece, but with a small vacuum gap between them. If the gap is small, but still finite, a weak beam of light is observed in the second piece of glass, propagating in the same direction as the original beam. We say that the light waves have "tunneled" across the gap. The strength of the beam of light in the second piece of glass depends very sensitively on the size of the gap through which it tunnels, decreasing exponentially as the width of the gap is increased.

Just as with light, electrons and other particles can tunnel across gaps that would be forbidden to them if they were classical particles. An electron, for instance, is not simply a point of mass; instead, it has wave properties that extend outward from it like ripples on a pond. These waves, like those of light, can "feel" the surroundings of an electron, allowing it to "tunnel" through a barrier to a region on the other side where it can propagate.

REAL-WORLD PHYSICS
Scanning tunneling microscopy

An example of electron tunneling is shown in **Figure 30–18**, where we illustrate the operation of a scanning tunneling microscope, or STM. In the lower part of the figure we see the material, or specimen, to be investigated with the microscope. The upper portion of the figure shows the key element of the microscope—a small, pointed tip of metal that can be moved up and down with piezoelectric supports. This tip is brought very close to the specimen being observed, leaving a small vacuum gap between them. Classically, electrons in the specimen are not able to move across the gap to the tip, but in reality they can tunnel to the tip and create a small electric current. The number of electrons that tunnel, and thus the magnitude of the tunneling current, depends on the width of the gap.

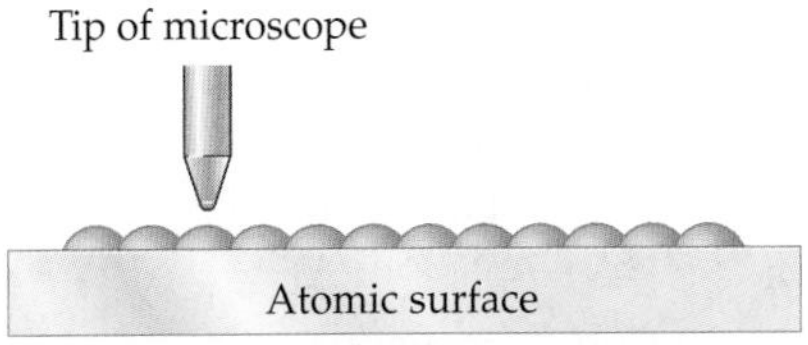

▲ **FIGURE 30–18 Operation of a scanning tunneling microscope**

Schematic operation of a scanning tunneling microscope.

In one version of the STM, the tunneling current between the specimen and the tip is held constant. Imagine, for example, that the tip in Figure 30–18 is scanned horizontally from left to right. Since the tunneling current is very sensitive to the size of the gap between the tip and the specimen, the tip must be moved up and down with the contours of the specimen in order to maintain a gap of constant width. The tip is moved by sending electrical voltage to the piezoelectric supports; thus, the voltage

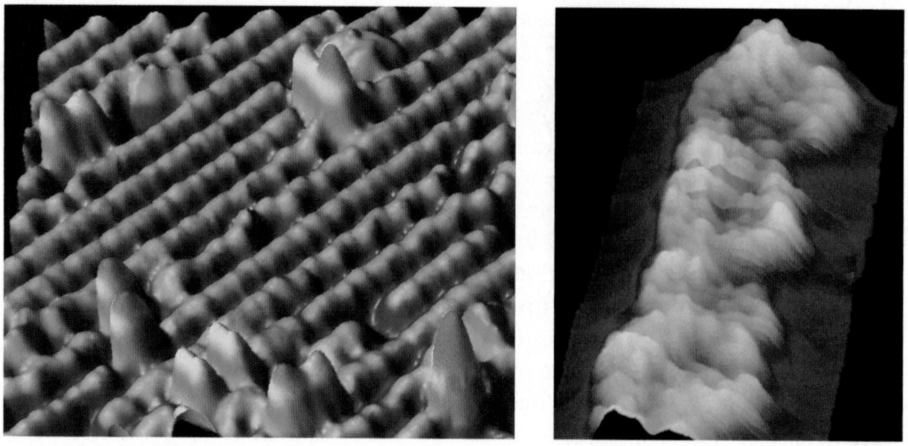

▶ **FIGURE 30–19 Scanning tunneling microscopy**

STM images are particularly good at recording the "terrain" of surfaces at the atomic level. The image at left shows clusters of antimony atoms on a silicon surface. At right, a DNA molecule sits on a substrate of graphite. Three turns of the double helix are visible, magnified about 2,000,000 times.

going to the supports is a measure of the height of the surface being scanned and can be converted to a visual image of the surface, as in **Figure 30–19**. The resolution of these microscopes is on the atomic level, showing the hills and valleys created by atoms on the surface of the material being examined.

Chapter Summary

30–1 Blackbody Radiation and Planck's Hypothesis of Quantized Energy

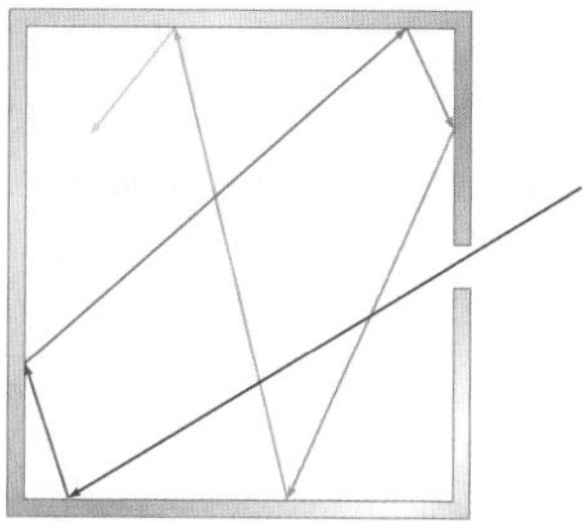

An ideal blackbody is an object that absorbs all the light incident on it. The distribution of energy as a function of frequency within a blackbody is independent of the material from which the blackbody is made and depends only on the temperature, T.

Wien's Displacement Law

The frequency at which the radiation from a blackbody is maximum is given by the following relation:

$$f_{\text{peak}} = (5.88 \times 10^{10}\ \text{s}^{-1} \cdot \text{K}^{-1})T \qquad \text{30–1}$$

Planck's Quantum Hypothesis

Planck hypothesized that the energy in a blackbody at a frequency f must be an integer multiple of the constant $h = 6.63 \times 10^{-34}\ \text{J} \cdot \text{s}$; that is,

$$E_n = nhf \quad n = 0, 1, 2, 3, \ldots \qquad \text{30–2}$$

The constant h is known as Planck's constant.

30–2 Photons and the Photoelectric Effect

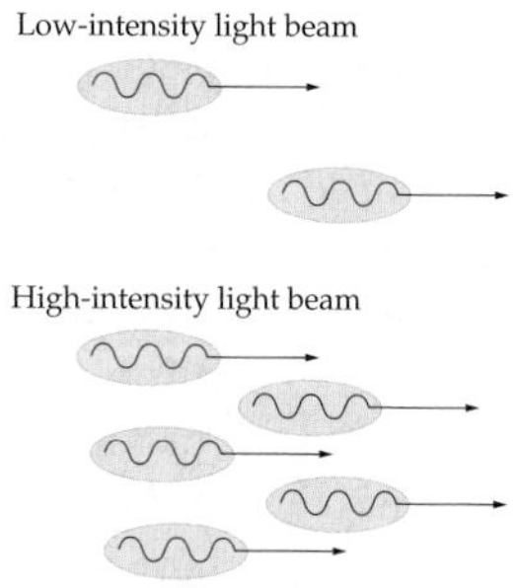

Light is composed of particle-like photons, which carry energy in discrete amounts.

Energy of a Photon

The energy of a photon depends on its frequency. A photon with the frequency f has the energy

$$E = hf \qquad \text{30–4}$$

Noting the relation $\lambda f = c$, we can also express the energy of a photon in terms of its wavelength, $E = hc/\lambda$.

The Photoelectric Effect

The photoelectric effect occurs when photons of light eject electrons from the surface of a metal.

Work Function

The minimum energy required to eject an electron from a particular metal is the work function, W_0.

Cutoff Frequency

To eject an electron, a photon must have an energy at least as great as W_0, and thus the minimum, or cutoff, frequency to eject an electron is

$$f_0 = \frac{W_0}{h} \qquad \text{30–6}$$

If the frequency of the photon is greater than f_0, the ejected electron has a finite kinetic energy.

Maximum Kinetic Energy
The maximum kinetic energy of an electron ejected from a metal by a photon of frequency $f > f_0$ is

$$K_{max} = hf - W_0 \qquad 30\text{–}7$$

30–3 The Mass and Momentum of a Photon

A photon is like a "typical" particle in some respects but different in others. In particular, a photon has zero rest mass, yet it still has a nonzero momentum.

Rest Mass of a Photon
Photons, which travel through a vacuum at the speed of light, c, have zero rest mass:

$$m_0 = 0 \qquad 30\text{–}9$$

Only objects with zero rest mass can propagate at the speed of light.

Momentum of a Photon
The momentum, p, of a photon of frequency f and wavelength $\lambda = c/f$ is given by

$$p = \frac{hf}{c} = \frac{h}{\lambda} \qquad 30\text{–}11$$

30–4 Photon Scattering and the Compton Effect

Photons can undergo collisions with particles, in much the same way that particles can collide with other particles. In order to conserve the total energy and momentum of a system during such a collision, the frequency and wavelength of a photon will change.

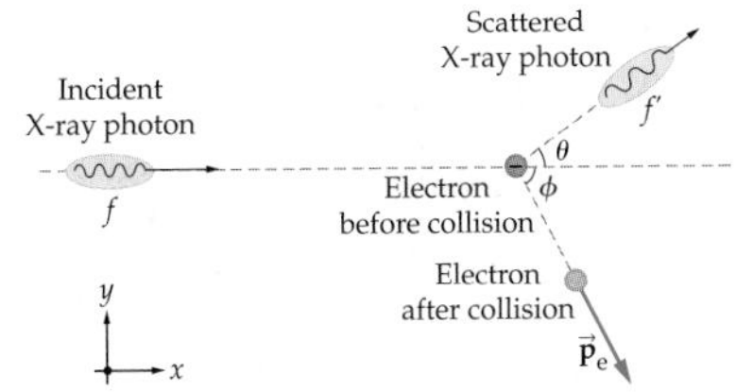

The Compton Effect
If a photon of wavelength λ undergoes a collision with an electron (mass $= m_e$) and scatters into a new direction at an angle θ from its incident direction, its new wavelength, λ', is given by the Compton shift formula:

$$\Delta\lambda = \lambda' - \lambda = \frac{h}{m_e c}(1 - \cos\theta) \qquad 30\text{–}15$$

The Compton shift formula can be applied to scattering from particles other than the electron by changing the mass in Equation 30–15.

30–5 The de Broglie Hypothesis and Wave-Particle Duality

The de Broglie hypothesis is basically the following: Since light displays particle-like behavior, perhaps particles display wavelike behavior. In particular, de Broglie hypothesized that particles have wavelengths.

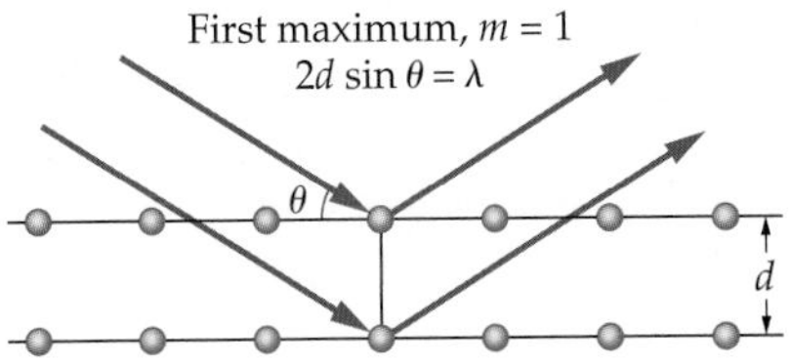

de Broglie Wavelength
According to de Broglie, the relationship between momentum and wavelength should be the same for both light and particles. Thus, the de Broglie wavelength of a particle of momentum p is

$$\lambda = \frac{h}{p} \qquad 30\text{–}16$$

Diffraction of X-Rays and Particles by Crystals
X-rays and particles can show interference effects when reflected from the atomic layers of a crystal. If the wavelength of the X-rays or particles is λ, and the spacing between atomic layers is d, the angles at which constructive interference occurs are given by

$$2d\sin\theta = m\lambda \quad m = 1, 2, 3, \ldots \qquad 30\text{–}17$$

The resulting patterns produced by the constructive interference are referred to as a diffraction pattern.

Wave-Particle Duality
Light and matter display both wavelike and particle-like properties.

30–6 The Heisenberg Uncertainty Principle

Because particles have wavelengths and can behave as waves, their position and momentum cannot be determined simultaneously with arbitrary precision.

The Heisenberg Uncertainty Principle: Momentum and Position

In terms of momentum and position, the Heisenberg uncertainty principle states the following:

$$\Delta p_y \Delta y \geq \frac{h}{2\pi} \qquad \textbf{30–19}$$

Thus, as momentum is determined more precisely, the position becomes more uncertain, and vice versa.

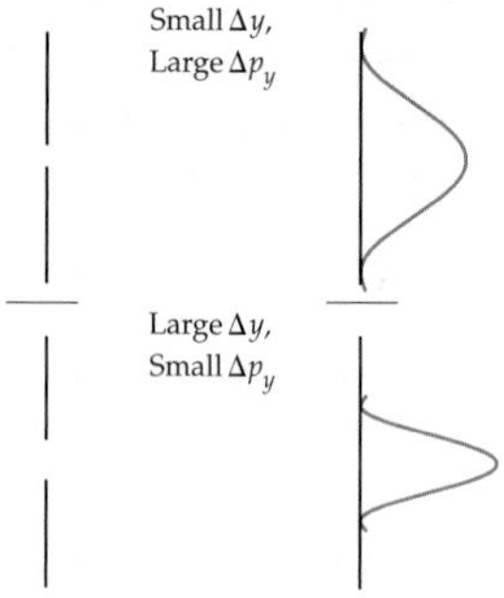

The Heisenberg Uncertainty Principle: Energy and Time

In terms of energy and time, the Heisenberg uncertainty principle states the following:

$$\Delta E \Delta t \geq \frac{h}{2\pi} \qquad \textbf{30–20}$$

30–7 Quantum Tunneling

Particles, because of their wavelike behavior, can pass through regions of space that would be forbidden to a classical particle. This phenomenon is referred to as tunneling.

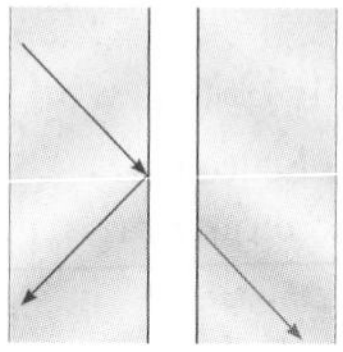

Problem-Solving Summary

Type of Problem	Relevant Physical Concepts	Related Examples
Relate the peak frequency of blackbody radiation to the temperature of the blackbody.	A blackbody at the absolute temperature T has a peak in its radiation spectrum at the frequency $f_{peak} = (5.88 \times 10^{10}\ \mathrm{s^{-1} \cdot K^{-1}})T$ This is true regardless of the blackbody's composition.	Exercise 30–1
Find the quantum number of a macroscopic system.	If a macroscopic system has the energy E and oscillates with frequency f, the corresponding quantum number is $n = E/hf$.	Example 30–1
Relate the energy and frequency of a photon.	A photon of frequency f has an energy hf, where h is Planck's constant.	Example 30–2 Active Example 30–1
Find the maximum kinetic energy of electrons given off in the photoelectric effect.	When light of frequency f illuminates a metal surface with a work function W_0, the maximum kinetic energy of ejected electrons is $K_{max} = hf - W_0$.	Example 30–3
Relate the change in wavelength of an X-ray to the angle through which it scatters from an electron.	The change in wavelength is related to the scattering angle θ by the following relation: $\Delta\lambda = \lambda' - \lambda = \frac{h}{m_e c}(1 - \cos\theta)$.	Example 30–4
Determine the de Broglie wavelength of a particle.	The de Broglie wavelength of a particle with momentum p is $\lambda = h/p$.	Example 30–5 Active Example 30–2
Relate the uncertainty in position to the uncertainty in momentum.	The Heisenberg uncertainty principle states that the uncertainty in momentum, Δp_y, and the uncertainty in position, Δy, are related as follows: $\Delta p_y \Delta y \geq h/2\pi$. Similarly, the uncertainties in time and energy obey $\Delta E\ \Delta t \geq h/2\pi$.	Examples 30–6, 30–7

Conceptual Questions

(Answers to odd-numbered Conceptual Questions can be found in the back of the book.)

1. Give a brief description of the "ultraviolet catastrophe."
2. How does Planck's hypothesis of energy quantization resolve the "ultraviolet catastrophe?"
3. Is there a lowest temperature below which blackbody radiation is no longer given off by an object? Explain.

4. How can an understanding of blackbody radiation allow us to determine the temperature of distant stars?
5. **Differential Fading** Many vehicles in the United States have a small American flag decal in one of their windows. If the decal has been in place for a long time, the colors will show some fading from exposure to the Sun. In fact, the red stripes are generally more faded than the blue background for the stars, as shown in the accompanying photo. Photographs and posters react in the same way, with red colors showing the most fading. Explain this effect in terms of the photon model of light.

Differential fading. (Conceptual Question 5)

6. A source of light is monochromatic. What can you say about the photons emitted by this source?
7. The relative intensity of radiation given off by a blackbody is shown in Figure 30–2. Notice that curves corresponding to different temperatures never cross one another. If two such curves did intersect, however, it would be possible to violate the second law of thermodynamics. Explain.
8. **(a)** Is it possible for a photon from a green source of light to have more energy than a photon from a blue source of light? Explain. **(b)** Is it possible for a photon from a green source of light to have more energy than a photon from a red source of light? Explain.
9. Light of a given wavelength ejects electrons from the surface of one metal but not from the surface of another metal. Give a possible explanation for this observation.
10. Why does the existence of a cutoff frequency in the photoelectric effect argue in favor of the photon model of light?
11. Why can an electron microscope resolve smaller objects than a light microscope?
12. A proton is about 2000 times more massive than an electron. Is it possible for an electron to have the same de Broglie wavelength as a proton? Explain.

Conceptual Exercises

(Answers to odd-numbered Conceptual Exercises can be found in the back of the book.)

1. The blackbody spectrum from blackbody A peaks at a longer wavelength than that of blackbody B. Is the temperature of blackbody A greater than, less than, or equal to the temperature of blackbody B? Explain.
2. A source of red light, a source of green light, and a source of blue light each produce beams of light with the same power. Rank these sources in order of increasing **(a)** wavelength of light, **(b)** frequency of light, and **(c)** number of photons emitted per second. Indicate ties where appropriate.
3. A source of red light has a higher wattage than a source of green light. **(a)** Is the energy of photons emitted by the red source greater than, less than, or equal to the energy of photons emitted by the green source? Explain. **(b)** Is the number of photons emitted per second by the red source greater than, less than, or equal to the number of photons emitted per second by the green source? Explain.
4. Light of a particular wavelength does not eject electrons from the surface of a given metal. Should the wavelength of the light be increased or decreased in order to cause electrons to be ejected? Explain.
5. Light of a particular wavelength and intensity does not eject electrons from the surface of a given metal. Can electrons be ejected from the metal by increasing the intensity of the light? Explain.
6. Suppose you perform an experiment on the photoelectric effect using light with a frequency high enough to eject electrons. If the intensity of the light is increased while the frequency is held constant, describe whether the following quantities increase, decrease, or stay the same: **(a)** The maximum kinetic energy of an ejected electron; **(b)** the minimum de Broglie wavelength of an electron; **(c)** the number of electrons ejected per second; **(d)** the electric current in the phototube.
7. Suppose you perform an experiment on the photoelectric effect using light with a frequency high enough to eject electrons. If the frequency of the light is increased while the intensity is held constant, describe whether the following quantities increase, decrease, or stay the same: **(a)** The maximum kinetic energy of an ejected electron; **(b)** the minimum de Broglie wavelength of an electron; **(c)** the number of electrons ejected per second; **(d)** the electric current in the phototube.
8. In a Compton scattering experiment, the scattered electron is observed to move in the same direction as the incident X-ray photon. What is the scattering angle of the photon? Explain.
9. If the momentum of a particle with finite mass is doubled, its kinetic energy increases by a factor of 4. If the momentum of a photon is doubled, by what factor does its energy increase?
10. As you accelerate your car away from a stoplight, does the de Broglie wavelength of the car increase, decrease, or stay the same? Explain.
11. By what factor does the de Broglie wavelength of a particle change if **(a)** its momentum is doubled or **(b)** its kinetic energy is doubled? Assume the particle is nonrelativistic.
12. An electron that is accelerated from rest through a potential difference V_0 has a de Broglie wavelength λ_0. What potential difference will double the electron's wavelength? (Express your answer in terms of V_0.)
13. A beam of particles diffracts from a crystal, producing an interference maximum at the angle θ. **(a)** If the mass of the particles is increased, with everything else remaining the same, does the angle of the interference maximum increase, decrease, or stay the same? Explain **(b)**. If the energy of the particles is increased, with everything else remaining the same, does the angle of the interference maximum increase, decrease, or stay the same? Explain.

Problems

Note: **IP** *denotes an integrated conceptual/quantitative problem.* **BIO** *identifies problems of biological or medical interest. Red bullets (•, ••, •••) are used to indicate the level of difficulty of each problem.*

Section 30–1 Blackbody Radiation and Planck's Hypothesis of Quantized Energy

1. • **The Surface Temperature of Betelgeuse** Betelgeuse, a red-giant star in the constellation Orion, has a peak in its radiation at a frequency of 3.09×10^{14} Hz. What is the surface temperature of Betelgeuse?

2. • What is the frequency of the most intense radiation emitted by your body? Assume a skin temperature of 95 °F. What is the wavelength of this radiation?

3. • **The Cosmic Background Radiation** Outer space is filled with a sea of photons, created in the early moments of the universe. The frequency distribution of this "cosmic background radiation" matches that of a blackbody at a temperature near 2.7 K. **(a)** What is the peak frequency of this radiation? **(b)** What is the wavelength that corresponds to the peak frequency?

4. • The Sun has a surface temperature of about 5800 K. At what frequency does the Sun emit the most radiation?

5. •• **(a)** By what factor does the peak frequency change if the Kelvin temperature of an object is doubled from 10.0 K to 20.0 K? **(b)** By what factor does the peak frequency change if the Celsius temperature of an object is doubled from 10.0 °C to 20.0 °C?

6. •• **IP A Famous Double Star** Albireo in the constellation Cygnus, which appears as a single star to the naked eye, is actually a beautiful double-star system. The brighter of the two stars is referred to as A (or Beta-01 Cygni), with a surface temperature of $T_A = 4700$ K; its companion is B (or Beta-02 Cygni), with a surface temperature of $T_B = 13{,}000$ K. **(a)** When viewed through a telescope, one star is a brilliant blue color, and the other has a warm golden color, as shown in the accompanying photo. Is the blue star A or B? Explain. **(b)** What is the ratio of the peak frequencies emitted by the two stars, (f_A/f_B)?

The double star Albireo in the constellation Cygnus. (Problem 6)

7. •• **IP Halogen Lightbulbs** Modern halogen lightbulbs allow their filaments to operate at a higher temperature than the filaments in standard incandescent bulbs. For comparison, the filament in a standard lightbulb operates at about 2900 K, whereas the filament in a halogen bulb may operate at 3400 K. **(a)** Which bulb has the higher peak frequency? **(b)** Calculate the ratio of peak frequencies (f_{hal}/f_{std}). **(c)** The human eye is most sensitive to a frequency around 5.5×10^{14} Hz. Which bulb produces a peak frequency closer to this value?

8. •• **IP** A typical lightbulb contains a tungsten filament that reaches a temperature of about 2900 K, roughly half the surface temperature of the Sun. **(a)** Treating the filament as a blackbody, determine the frequency for which its radiation is a maximum. **(b)** Do you expect the lightbulb to radiate more energy in the visible or in the infrared part of the spectrum? Explain.

9. •• **Exciting an Oxygen Molecule** An oxygen molecule (O_2) vibrates with an energy identical to that of a single particle of mass $m = 1.340 \times 10^{-26}$ kg attached to a spring with a force constant of $k = 1215$ N/m. The energy levels of the system are uniformly spaced, as indicated in **Figure 30–20**, with a separation given by hf. **(a)** What is the vibration frequency of this molecule? **(b)** How much energy must be added to the molecule to excite it from one energy level to the next higher level?

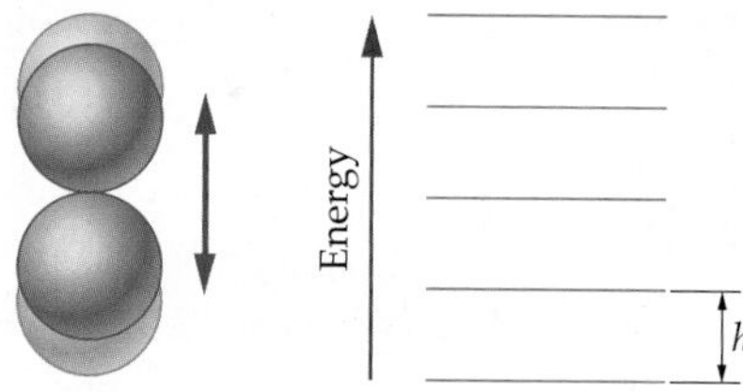

▲ **FIGURE 30–20** Problem 9

Section 30–2 Photons and the Photoelectric Effect

10. • When a person visits the local tanning salon, they absorb photons of ultraviolet (UV) light to get the desired tan. What is the frequency and wavelength of a UV photon whose energy is 6.5×10^{-19} J?

11. • An AM radio station operating at a frequency of 950 kHz radiates 270 kW of power from its antenna. How many photons are emitted by the antenna every second?

12. • A photon with a wavelength of less than 50.4 nm can ionize a helium atom. What is the ionization potential of helium?

13. • A flashlight emits 1.5 W of light energy. Assuming a frequency of 5.2×10^{14} Hz for the light, determine the number of photons given off by the flashlight per second.

14. • Light of frequency 9.95×10^{14} Hz ejects electrons from the surface of silver. If the maximum kinetic energy of the ejected electrons is 0.180×10^{-19} J, what is the work function of silver?

15. • The work function of gold is 4.58 eV. What frequency of light must be used to eject electrons from a gold surface with a maximum kinetic energy of 6.48×10^{-19} J?

16. • **(a)** How many 350-nm (UV) photons are needed to provide a total energy of 2.5 J? **(b)** How many 750-nm (red) photons are needed to provide the same energy?

17. •• **(a)** How many photons per second are emitted by a monochromatic lightbulb ($\lambda = 650$ nm) that emits 75 W of power? **(b)** If you stand 15 m from this bulb, how many photons enter each of your eyes per second? Assume your pupil is 5.0 mm in diameter and that the bulb radiates uniformly in all directions.

18. •• **IP** Two 57.5-kW radio stations broadcast at different frequencies. Station A broadcasts at a frequency of 892 kHz, and station B broadcasts at a frequency of 1410 kHz. **(a)** Which station emits more photons per second? Explain. **(b)** Which station emits photons of higher energy?

19. •• **Dissociating the Hydrogen Molecule** The energy required to separate a hydrogen molecule into its individual atoms is 104.2 kcal per mole of H_2. **(a)** If the dissociation energy for a single H_2 molecule is provided by one photon, determine

its frequency and wavelength. **(b)** In what region of the electromagnetic spectrum does the photon found in part (a) lie? (Refer to the spectrum shown in Figure 25–8.)

20. •• **(a)** How many photons are emitted per second by a He-Ne laser that emits 1.0 mW of power at a wavelength $\lambda = 632.8$ nm? **(b)** What is the frequency of the electromagnetic waves emitted by a He-Ne laser?

21. •• **IP** You have two lightbulbs of different power and color, as indicated in **Figure 30–21**. One is a 150-W red bulb, and the other is a 25-W blue bulb. **(a)** Which bulb emits more photons per second? **(b)** Which bulb emits photons of higher energy? **(c)** Calculate the number of photons emitted per second by each bulb. Take $\lambda_{red} = 650$ nm and $\lambda_{blue} = 460$ nm. (Most of the electromagnetic radiation given off by incandescent lightbulbs is in the infrared portion of the spectrum. For the purposes of this problem, however, assume that all of the radiated power is at the wavelengths indicated.)

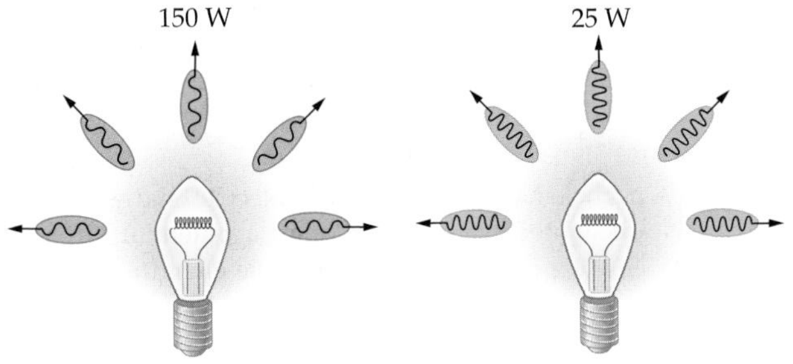

▲ **FIGURE 30–21** Problem 21

22. •• The maximum wavelength an electromagnetic wave can have and still eject an electron from a copper surface is 264 nm. What is the work function of a copper surface?

23. •• **IP** Aluminum and calcium have photoelectric work functions of $W_{Al} = 4.28$ eV and $W_{Ca} = 2.87$ eV, respectively. **(a)** Which metal requires higher-frequency light to produce photoelectrons? Explain. **(b)** Calculate the minimum frequency that will produce photoelectrons from each surface.

24. •• **IP** Two beams of light with different wavelengths $(\lambda_A > \lambda_B)$ are used to produce photoelectrons from a given metal surface. **(a)** Which beam produces photoelectrons with greater kinetic energy? Explain. **(b)** Find K_{max} for cesium $(W_0 = 1.9 \text{ eV})$ if $\lambda_A = 620$ nm, and $\lambda_B = 410$ nm.

25. •• **IP** Zinc and cadmium have photoelectric work functions given by $W_{Zn} = 4.33$ eV and $W_{Cd} = 4.22$ eV, respectively. **(a)** If both metals are illuminated by UV radiation of the same wavelength, which one gives off photoelectrons with the greater maximum kinetic energy? Explain. **(b)** Calculate the maximum kinetic energy of photoelectrons from each surface if $\lambda = 275$ nm.

26. •• White light, with frequencies ranging from 4.00×10^{14} Hz to 7.90×10^{14} Hz, is incident on a potassium surface. Given that the work function of potassium is 2.24 eV, find **(a)** the maximum kinetic energy of electrons ejected from this surface and **(b)** the range of frequencies for which no electrons are ejected.

27. •• Electromagnetic waves, with frequencies ranging from 4.00×10^{14} Hz to 9.00×10^{16} Hz, are incident on an aluminum surface. Given that the work function of aluminum is 4.28 eV, find **(a)** the maximum kinetic energy of electrons ejected from this surface and **(b)** the range of frequencies for which no electrons are ejected.

28. •• **IP** Platinum has a work function of 6.35 eV, and iron has a work function of 4.50 eV. Light of frequency 1.88×10^{15} Hz ejects electrons from both of these surfaces. **(a)** From which surface will the ejected electrons have a greater maximum kinetic energy? Explain. **(b)** Calculate the maximum kinetic energy of ejected electrons for each surface.

29. •• When light with a frequency $f_1 = 547.5$ THz illuminates a metal surface, the most energetic photoelectrons have 1.260×10^{-19} J of kinetic energy. When light with a frequency $f_2 = 738.8$ THz is used instead, the most energetic photoelectrons have 2.480×10^{-19} J of kinetic energy. Using these experimental results, determine the approximate value of Planck's constant.

30. •• **BIO Owl Vision** Owls have large, sensitive eyes for good night vision. Typically, the pupil of an owl's eye can have a diameter of 8.5 mm (as compared with a maximum diameter of about 7.0 mm for humans). In addition, an owl's eye is about 100 times more sensitive to light of low intensity than a human eye, allowing owls to detect light with an intensity as small as 5.0×10^{-13} W/m^2. Find the minimum number of photons per second an owl can detect, assuming a frequency of 7.0×10^{14} Hz for the light.

An apt pupil. (Problem 30)

Section 30–3 The Mass and Momentum of a Photon

31. • The photons used in microwave ovens have a momentum of 5.1×10^{-33} kg·m/s. **(a)** What is their wavelength? **(b)** How does the wavelength of the microwaves compare with the size of the holes in the metal screen on the door of the oven?

32. • What speed must an electron have if its momentum is to be the same as that of an X-ray photon with a wavelength of 0.29 nm?

33. • What is the wavelength of a photon that has the same momentum as an electron moving with a speed of 1400 m/s?

34. • What is the frequency of a photon that has the same momentum as a neutron moving with a speed of 1500 m/s?

35. •• A hydrogen atom, initially at rest, emits an ultraviolet photon with a wavelength of $\lambda = 122$ nm. What is the recoil speed of the atom after emitting the photon?

36. •• A blue-green photon $(\lambda = 486 \text{ nm})$ is absorbed by a free hydrogen atom, initially at rest. What is the recoil speed of the hydrogen atom after absorbing the photon?

37. •• **IP (a)** Which has the greater momentum, a photon of red light or a photon of blue light? Explain. **(b)** Calculate the momentum of a photon of red light $(f = 4.0 \times 10^{14} \text{ Hz})$ and a photon of blue light $(f = 7.9 \times 10^{14} \text{ Hz})$.

38. •• **IP** Photon A has twice the momentum of photon B. **(a)** Which photon has the greater wavelength? Explain. **(b)** If the wavelength of photon A is 333 nm, what is the wavelength of photon B?

39. ••• A laser produces a 5.00-mW beam of light, consisting of photons with a wavelength of 632.8 nm. **(a)** How many photons

are emitted by the laser each second? **(b)** The laser beam strikes a black surface and is absorbed. What is the change in the momentum of each photon that is absorbed? **(c)** What force does the laser beam exert on the black surface?

40. ••• A laser produces a 7.50-mW beam of light, consisting of photons with a wavelength of 632.8 nm. **(a)** How many photons are emitted by the laser each second? **(b)** The laser beam strikes a mirror at normal incidence and is reflected. What is the change in momentum of each reflected photon? Give the magnitude only. **(c)** What force does the laser beam exert on the mirror?

Section 30–4 Photon Scattering and the Compton Effect

41. • An X-ray photon has 36 keV of energy before it scatters from a free electron, and 22 keV after it scatters. What is the kinetic energy of the recoiling electron?

42. • In the Compton effect, an X-ray photon scatters from a free electron. Find the *change* in the photon's wavelength if it scatters at an angle of **(a)** $\theta = 30.0°$, **(b)** $\theta = 90.0°$, and **(c)** $\theta = 180.0°$ relative to the incident direction.

43. • An X-ray scattering from a free electron is observed to change its wavelength by 3.33 pm. At what angle to the incident direction does the scattered X-ray move?

44. •• The maximum Compton shift in wavelength occurs when a photon is scattered through 180°. What scattering angle will produce a wavelength shift of one-fourth the maximum?

45. •• **IP** Consider two different photons that scatter through an angle of 180° from a free electron. One is a visible-light photon with $\lambda = 520$ nm, the other is an X-ray photon with $\lambda = 0.030$ nm. **(a)** Which (if either) photon experiences the greater change in wavelength as a result of the scattering? Explain. **(b)** Which photon experiences the greater percentage change in wavelength? Explain. **(c)** Calculate the percentage change in wavelength of each photon.

46. •• An X-ray photon with a wavelength of 0.240 nm scatters from a free electron at rest. The scattered photon moves at an angle of 105° relative to its incident direction. Find **(a)** the initial momentum and **(b)** the final momentum of the photon.

47. •• An X-ray photon scatters from a free electron at rest at an angle of 175° relative to the incident direction. **(a)** If the scattered photon has a wavelength of 0.320 nm, what is the wavelength of the incident photon? **(b)** Determine the energy of the incident and scattered photons. **(c)** Find the kinetic energy of the recoil electron.

48. •• **IP** An X-ray photon scatters through 180° from (i) an electron or (ii) a helium atom. **(a)** In which case is the change in wavelength of the X-ray greater? Explain. **(b)** Calculate the change in wavelength for each of these two cases.

49. ••• A photon has an energy E and wavelength λ before scattering from a free electron. After scattering through a 135° angle, the photon's wavelength has increased by 10.0%. Find the initial wavelength and energy of the photon.

50. ••• Find the direction of propagation of the scattered electron in Problem 43, given that the incident X-ray has a wavelength of 0.525 nm and propagates in the positive x direction.

Section 30–5 The de Broglie Hypothesis and Wave-Particle Duality

51. • A particle with a mass of 6.69×10^{-27} kg has a de Broglie wavelength of 7.62 pm. What is the particle's speed?

52. • What speed must a neutron have if its de Broglie wavelength is to be equal to the interionic spacing of table salt (0.282 nm)?

53. • A 77-kg jogger runs with a speed of 4.1 m/s. What is the jogger's de Broglie wavelength?

54. • Find the kinetic energy of an electron whose de Broglie wavelength is 1.5 Å.

55. •• A beam of neutrons with a de Broglie wavelength of 0.250 nm diffracts from a crystal of table salt, which has an interionic spacing of 0.282 nm. **(a)** What is the speed of the neutrons? **(b)** What is the angle of the second interference maximum?

56. •• **IP** An electron and a proton have the same speed. **(a)** Which has the longer de Broglie wavelength? Explain. **(b)** Calculate the ratio (λ_e/λ_p).

57. •• **IP** An electron and a proton have the same de Broglie wavelength. **(a)** Which has the greater kinetic energy? Explain. **(b)** Calculate the ratio of the electron's kinetic energy to the kinetic energy of the proton.

58. •• Diffraction effects become significant when the width of an aperture is comparable to the wavelength of the waves being diffracted. **(a)** At what speed will the de Broglie wavelength of a 65-kg student be equal to the 0.76-m width of a doorway? **(b)** At this speed, how long will it take the student to travel a distance of 1.0 mm? (For comparison, the age of the universe is approximately 4×10^{17} s.)

59. ••• A particle has a mass m and an electric charge q. The particle is accelerated from rest through a potential difference V. What is the particle's de Broglie wavelength, expressed in terms of m, q, and V?

Section 30–6 The Heisenberg Uncertainty Principle

60. • A baseball (0.15 kg) and an electron both have a speed of 41 m/s. Find the uncertainty in position of each of these objects, given that the uncertainty in their speed is 5.0%.

61. • The uncertainty in position of a proton confined to the nucleus of an atom is roughly the diameter of the nucleus. If this diameter is 7.5×10^{-15} m, what is the uncertainty in the proton's momentum?

62. • The position of a 0.26-kg air-track cart is determined to within an uncertainty of 2.2 mm. What speed must the cart acquire as a result of the position measurement?

63. • The measurement of an electron's energy requires a time interval of 1.0×10^{-8} s. What is the smallest possible uncertainty in the electron's energy?

64. • A particle's energy is measured with an uncertainty of 0.0010 eV. What is the smallest possible uncertainty in our knowledge of when the particle had this energy?

65. • An excited state of a particular atom has a mean lifetime of 0.60×10^{-9} s, which we may take as the uncertainty Δt. What is the minimum uncertainty in any measurement of the energy of this state?

66. • The Σ^+ is an unstable particle, with a mean lifetime of 2.5×10^{-10} s. Its lifetime defines the uncertainty Δt for this particle. What is the minimum uncertainty in this particle's energy?

67. •• The uncertainty in an electron's position is 0.15 nm. **(a)** What is the minimum uncertainty Δp in its momentum? **(b)** What is the kinetic energy of an electron whose momentum is equal to this uncertainty $(\Delta p = p)$?

68. •• The uncertainty in a proton's position is 0.15 nm. **(a)** What is the minimum uncertainty Δp in its momentum? **(b)** What is the kinetic energy of a proton whose momentum is equal to this uncertainty $(\Delta p = p)$?

69. •• An electron has a momentum $p \approx 1.7 \times 10^{-25}$ kg·m/s. What is the minimum uncertainty in its position that will keep the relative uncertainty in its momentum $(\Delta p/p)$ below 1.0%?

General Problems

70. • You want to construct a photocell that works with visible light. Three materials are readily available: aluminum ($W_0 = 4.28$ eV), lead ($W_0 = 4.25$ eV), and cesium ($W_0 = 2.14$ eV). Which material(s) would be suitable?

71. • **BIO Human Vision** Studies have shown that some people can detect 545-nm light with as few as 100 photons entering the eye per second. What is the power delivered by such a beam of light?

72. •• A pendulum consisting of a 0.15-kg mass attached to a 0.88-m string undergoes simple harmonic motion. **(a)** What is the frequency of oscillation for this pendulum? **(b)** Assuming the energy of this system satisfies $E_n = nhf$, find the maximum speed of the 0.15-kg mass when the quantum number is 1.0×10^{33}.

73. •• To listen to a radio station, a certain home receiver must pick up a signal of at least 1.0×10^{-10} W. **(a)** If the radio waves have a frequency of 96 MHz, how many photons must the receiver absorb per second to get the station? **(b)** How much force is exerted on the receiving antenna for the case considered in part (a)?

74. •• The latent heat for converting ice at 0 °C to water at 0 °C is 80.0 kcal/kg (Chapter 17). **(a)** How many photons of frequency 6.0×10^{14} Hz must be absorbed by a 1.0-kg block of ice at 0 °C to melt it to water at 0 °C? **(b)** How many molecules of H_2O can one photon convert from ice to water?

75. •• How many 520-nm photons would have to be absorbed to raise the temperature of 1 g of water by 1.0 C°?

76. •• A microwave oven can heat 205 mL of water from 20.0 °C to 90.0 °C in 2.00 min. If the wavelength of the microwaves is $\lambda = 12.2$ cm, how many photons were absorbed by the water? (Assume no loss of heat by the water.)

77. •• Light with a frequency of 2.11×10^{15} Hz ejects electrons from the surface of lead, which has a work function of 4.25 eV. What is the minimum de Broglie wavelength of the ejected electrons?

78. •• An electron moving with a speed of 2.7×10^6 m/s has the same momentum as a photon. Find **(a)** the de Broglie wavelength of the electron and **(b)** the wavelength of the photon.

79. •• **BIO The Cold Light Of Fireflies** Fireflies are often said to give off "cold light." Given that the peak in a firefly's radiation occurs at about 5.4×10^{14} Hz, determine the temperature of a blackbody that would have the same peak frequency. From your result, would you say that firefly radiation is well approximated by blackbody radiation? Explain.

How cool is that? (Problem 79)

80. •• **IP** When light with a wavelength of 545 nm shines on a metal surface, electrons are ejected with speeds of 3.10×10^5 m/s or less. **(a)** Give a strategy that allows you to use the preceding information to calculate the work function and cutoff frequency for this surface. **(b)** Carry out your strategy and determine the work function and cutoff frequency.

81. •• **IP** A hydrogen atom absorbs a 486.2-nm photon. A short time later, the same atom emits a photon with a wavelength of 97.23 nm. **(a)** Has the net energy of the atom increased or decreased? Explain. **(b)** Calculate the change in energy of the hydrogen atom.

82. •• The American physicist Robert A. Millikan (1868–1953) won the 1923 Nobel Prize in physics, in part for his work on the photoelectric effect. In experiments on lithium, Millikan observed a maximum kinetic energy of 0.550 eV when electrons were ejected with 433.9-nm light. When light of 253.5 nm was used, he observed a maximum kinetic energy of 2.57 eV. Using these results, find **(a)** the work function, W_0, for lithium and **(b)** the predicted value of Planck's constant. **(c)** What maximum kinetic energy do you predict Millikan found when he used light with a wavelength of 365.0 nm?

83. •• When a beam of atoms emerges from an oven at the absolute temperature T, the most probable de Broglie wavelength for a given atom is

$$\lambda_{mp} = \frac{h}{\sqrt{5mkT}}$$

In this expression, m is the mass of an atom, and k is Boltzmann's constant (Chapter 17). What is the most probable speed of a hydrogen atom emerging from an oven at 450 K?

84. •• **IP (a)** Does the de Broglie wavelength of a particle increase or decrease as its kinetic energy increases? Explain. **(b)** Show that the de Broglie wavelength of an electron in nanometers can be written as $\lambda = (1.23 \text{ nm})/\sqrt{K}$, where K is the kinetic energy of the electron in eV. Use classical expressions for momentum and kinetic energy.

85. ••• A jar is filled with monatomic helium gas at a temperature of 25 °C. The pressure inside the jar is one atmosphere; that is, 101 kPa. **(a)** Find the average de Broglie wavelength of the helium atoms. **(b)** Calculate the average separation between helium atoms in the jar. (*Note*: The fact that the spacing between atoms is much greater than the de Broglie wavelength means quantum effects are negligible, and the atoms can be treated as particles.)

86. ••• **The Compton Wavelength** The *Compton wavelength*, λ_C, of a particle of mass m is defined as follows: $\lambda_C = h/mc$. **(a)** Calculate the Compton wavelength of a proton. **(b)** Calculate the energy of a photon that has the same wavelength as found in part (a). **(c)** Show, in general, that a photon with a wavelength equal to the Compton wavelength of a particle has an energy that is equal to the rest energy of the particle.

87. ••• **IP** Light of frequency 8.22×10^{14} Hz ejects electrons from surface A with a maximum kinetic energy that is 2.00×10^{-19} J greater than the maximum kinetic energy of electrons ejected from surface B. **(a)** If the frequency of the light is increased, does the difference in maximum kinetic energy observed from the two surfaces increase, decrease, or stay the same? Explain. **(b)** Calculate the difference in work function for these two surfaces.

Interactive Problems

88. •• **IP** ***Referring to Example 30–4*** An X-ray photon with $\lambda = 0.6500$ nm scatters from an electron, giving the electron a kinetic energy of 7.750 eV. **(a)** Is the scattering angle of the photon greater than, less than, or equal to 152°? **(b)** Find the scattering angle.

89. •• **IP** ***Referring to Example 30–4*** An X-ray photon with $\lambda = 0.6500$ nm scatters from an electron. The wavelength of the scattered photon is 0.6510 nm. **(a)** Is the scattering angle in this case greater than, less than, or equal to 152°? **(b)** Find the scattering angle.

31 Atomic Physics

31–1	Early Models of the Atom	1035
31–2	The Spectrum of Atomic Hydrogen	1036
31–3	Bohr's Model of the Hydrogen Atom	1039
31–4	de Broglie Waves and the Bohr Model	1046
31–5	The Quantum Mechanical Hydrogen Atom	1047
31–6	Multielectron Atoms and the Periodic Table	1050
31–7	Atomic Radiation	1055

A number of animals, including many species of coral, sponges, and even scorpions, emit beautiful and sometimes eerie colors when illuminated by ultraviolet light. We do not ordinarily expect rabbits to fall into this category. However, scientists recently inserted a jellyfish gene that directs the production of green fluorescent protein (GFP) into the fertilized egg of a rabbit. The result was Alba, the "GFP bunny"—the world's first fluorescent rabbit. This chapter explores our modern understanding of the structure of the atom, which makes it possible to explain not only the phenomenon of fluorescence but many others as well, including the properties of the chemical elements and the generation of laser light.

In today's world it is taken for granted that we, along with everything else on Earth, are made of atoms. Although it may seem surprising at first, this belief in atoms has not always been universal. As recently as the first part of the twentieth century there was still serious debate about the microscopic nature of matter. With the advent of quantum physics, however, the debate quickly faded away as atomic structure came to be understood in ever greater detail.

In this chapter we begin by developing the quantum model of the simplest of all atoms—the hydrogen atom. We then show that the basic features of hydrogen apply to more complex atoms as well. As a result, we are able to understand—in detail—the arrangement of elements in the periodic table. That quantum physics can describe the structure of an atom, and show why the various elements have their characteristic properties, is one of the greatest successes of modern science.

31–1 Early Models of the Atom

Speculations about the microscopic structure of matter have intrigued humankind for thousands of years. Ancient Greek philosophers, including Leucippus and Democritus, considered the question of what would happen if you took a small object, like a block of copper, and cut it in half, then cut it in half again, and again, for many subsequent divisions. They reasoned that eventually you would reduce the block to a single speck of copper that could not be divided further. This smallest piece of an element was called the **atom** (a + tom), which means, literally, "without division."

It was not until the late nineteenth century, however, that the question of atoms began to yield to direct scientific investigation. We now consider some of the more important early developments in atomic models that helped lead to our current understanding.

The Thomson Model: Plum Pudding

In 1897 the English physicist J. J. Thomson (1856–1940) discovered a "particle" that is smaller in size and thousands of times less massive than even the smallest atom. The **electron,** as this particle was named, was also found to have a negative electric charge—in contrast with atoms, which are electrically neutral. Thomson proposed, therefore, that atoms have an internal structure that includes both electrons and a quantity of positively charged matter. The latter would account for most of the mass of an atom, and would have a charge equal in magnitude to the charge on the electrons.

The picture of an atom that Thomson settled on is one he referred to as the "plum-pudding model." In this model, electrons are embedded in a more or less uniform distribution of positive charge—like raisins spread throughout a pudding. This model is illustrated in **Figure 31–1**. Although the plum-pudding model was in agreement with everything Thomson knew about atoms at the time, new experiments were soon to rule out this model and replace it with one that was more like the solar system than a pudding.

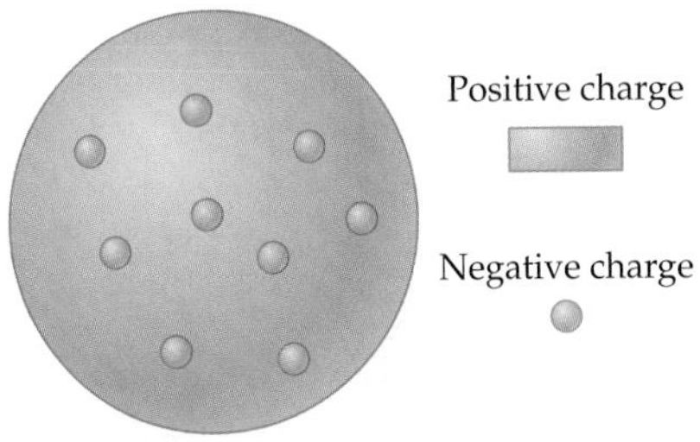

▲ **FIGURE 31–1 The plum-pudding model of an atom**

The model of an atom proposed by J. J. Thomson consists of a uniform positive charge, which accounts for most of the mass of an atom, with small negatively charged electrons scattered throughout, like raisins in a pudding.

The Rutherford Model: A Miniature Solar System

Inspired by the findings and speculations of Thomson, other physicists began to investigate atomic structure. In 1909, Ernest Rutherford (1871–1937) and his coworkers Hans Geiger (1882–1945) and Ernest Marsden (1889–1970) (at that time a twenty-year-old undergraduate) decided to test Thomson's model by directing a beam of positively charged particles, known as **alpha particles,** at a thin gold foil. Since alpha particles—which were later found to be the nuclei of helium atoms—carry a positive charge, they should be deflected as they pass through the positively charged "pudding" in the gold foil. The deflection should have the following properties: (i) it should be relatively small, since the alpha particles have a substantial mass and the positive charge in the atom is spread out; and (ii) all the alpha particles should be deflected in roughly the same way, since the positive pudding fills virtually all space.

When Geiger and Marsden performed the experiment, their results were not in agreement with these predictions. In fact, most of the alpha particles passed right through the foil as if it were not there—as if the atoms in the foil were mostly empty space. Because the results were rather surprising, Rutherford suggested that the experiment be modified to look not only for alpha particles with small angles of deflection—as originally expected—but for ones with large deflections as well.

This suggestion turned out to be an inspired hunch. Not only were large-angle deflections observed, but some of the alpha particles, in fact, were found to have practically reversed their direction of motion. Rutherford was stunned. In his own words, "It was almost as incredible as if you fired a fifteen-inch shell at a piece of tissue paper and it came back and hit you."

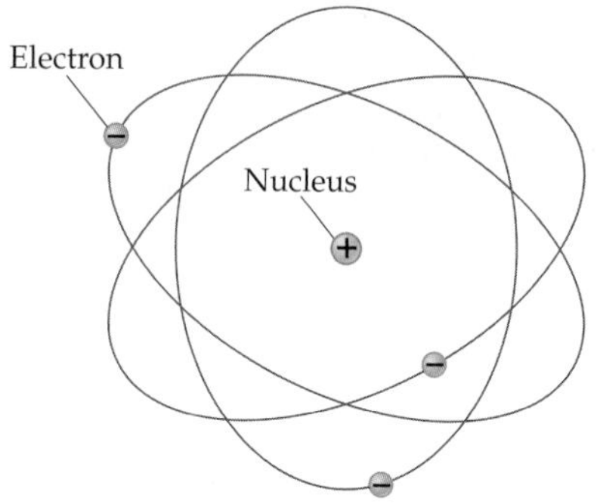

▲ **FIGURE 31–2 The solar system model of an atom**

Ernest Rutherford proposed that an atom is like a miniature solar system, with a massive positively charged nucleus orbited by lightweight negatively charged electrons.

To account for the results of these experiments, Rutherford proposed that an atom has a structure similar to that of the solar system, as illustrated in **Figure 31–2**. In particular, he imagined that the lightweight, negatively charged electrons orbit a small, positively charged **nucleus** containing almost all the atom's mass. In this nuclear model of the atom, most of the atom is indeed empty space, allowing the majority of the alpha particles to pass right through. Furthermore, the positive charge of the atom is now highly concentrated in a small nucleus, rather than spread throughout the atom. This means that an alpha particle that happens to make a head-on collision with the nucleus can actually be turned around, as observed in the experiments.

To see just how small the nucleus must be in his model, Rutherford combined the experimental data with detailed theoretical calculations. His result was that the radius of a nucleus must be smaller than the diameter of the atom by a factor of about 10,000. To put this value into perspective, imagine an atom magnified in size until its nucleus is as large as the Sun. At what distance would an electron orbit in this "atomic" solar system? Using the factor given by Rutherford, we find that the orbit of the electron would have a radius essentially the same as the orbit of Pluto—inside this radius would be only empty space and the nucleus. Thus an atom must have an even larger fraction of empty space than the solar system!

Although Rutherford's nuclear model of the atom seems reasonable, it contains fatal flaws. First, an orbiting electron undergoes a centripetal acceleration toward the nucleus (Chapter 6). As we know from Section 25–1, however, any electric charge that accelerates gives off energy in the form of electromagnetic radiation. Thus, an electron continually radiating energy as it orbits is similar to a satellite losing energy to air resistance when it orbits too close to the Earth's atmosphere. Just as in the case of a satellite, an electron would spiral inward and eventually plunge into the nucleus. Since the entire process of collapse would occur in a fraction of a second (about 10^{-9} s in fact), the atoms in Rutherford's model would simply not be stable—in contrast with the observed stability of atoms in nature.

Even if we ignore the stability problem for a moment, there is another serious discrepancy between Rutherford's model and experiment. Maxwell's equations state that the frequency of radiation from an orbiting electron should be the same as the frequency of its orbit. In the case of an electron spiraling inward the frequency would increase continuously. Thus if we look at light coming from an atom, the Rutherford model indicates that we should see a continuous range of frequencies. This prediction is in striking contrast with experiments, which show that light coming from an atom has only certain discrete frequencies and wavelengths, as we discuss in the next section.

▲ Emission nebulas, like the Lagoon Nebula in Sagittarius shown here, are masses of glowing interstellar gas. The gas is excited by high-energy radiation from nearby stars and emits light at wavelengths characteristic of the atoms present, chiefly hydrogen. Much of the visible light from such nebulas is contributed by the red Balmer line of hydrogen with a wavelength of 656.3 nm, known as H-alpha.

31–2 The Spectrum of Atomic Hydrogen

A red-hot piece of metal glows with a ruddy light that represents only a small fraction of its total radiation output. As we saw in Chapter 30, the metal gives off blackbody radiation that extends in a continuous distribution over all possible frequencies. This blackbody distribution, or spectrum, of radiation is characteristic of the entire collection of atoms that make up the metal—it is not characteristic of the spectrum of light that would be given off by a single, isolated metal atom.

To see the light produced by an isolated atom, we turn our attention from a solid—where the atoms are close together and strongly interacting—to a low-pressure gas—where the atoms are far apart and have little interaction with one another. Consider, then, an experiment in which we seal a low-pressure gas in a tube. If we apply a large voltage between the ends of the tube, the gas will emit electromagnetic radiation characteristic of the individual gas atoms. When this radiation is passed through a diffraction grating (Chapter 28), it is separated into

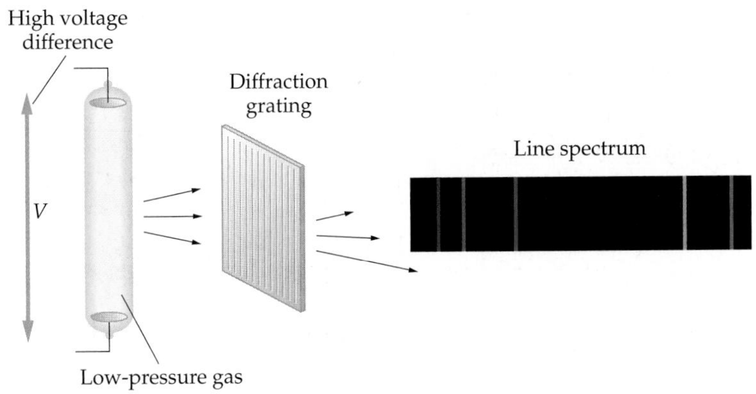

◀ **FIGURE 31–3 The line spectrum of an atom**

The light given off by individual atoms, as in a low-pressure gas, consists of a series of discrete wavelengths corresponding to different colors.

its various wavelengths, as indicated in **Figure 31–3**. The result of such an experiment is that a series of bright "lines" is observed, reminiscent of the bar codes used in supermarkets. The precise wavelength associated with each of these lines provides a sort of "fingerprint" identifying a particular type of atom, just as each product in a supermarket has its own unique bar code.

This type of spectrum, with its bright lines in different colors, is referred to as a **line spectrum.** As an example, we show the visible part of the line spectrum of atomic hydrogen in **Figure 31–4 (a)**. Hydrogen produces additional lines in the infrared and ultraviolet parts of the electromagnetic spectrum.

The line spectrum shown in Figure 31–4 (a) is an *emission spectrum,* since it shows light that is emitted by the hydrogen atoms. Similarly, if light of all colors is passed through a tube of hydrogen gas, some wavelengths will be absorbed by the atoms, giving rise to an *absorption spectrum,* which consists of dark lines (where the atoms absorb the radiation) against an otherwise bright background. The absorption lines occur at precisely the same wavelengths as the emission lines. **Figure 31–4 (b)** shows the absorption spectrum of hydrogen.

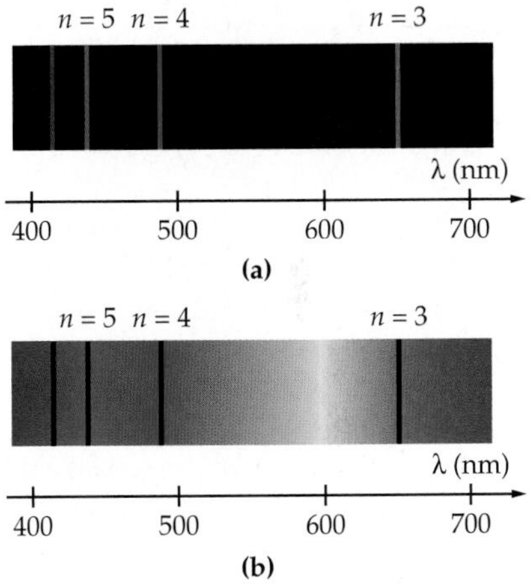

▲ **FIGURE 31–4 The line spectrum of hydrogen**

The emission **(a)** and absorption **(b)** spectra of hydrogen. Note that the wavelengths absorbed by hydrogen (dark lines) are the same as those emitted by hydrogen (colored lines). The location of these lines is predicted by the Balmer series (Equation 31–1) with the appropriate values of n.

The first step in developing a quantitative understanding of the hydrogen spectrum occurred in 1885, when Johann Jakob Balmer (1825–1898), a Swiss schoolteacher, used trial-and-error methods to discover the following simple formula that gives the wavelength of the visible lines in the spectrum:

$$\frac{1}{\lambda} = R\left(\frac{1}{2^2} - \frac{1}{n^2}\right) \qquad n = 3, 4, 5, \ldots \text{(Balmer series)} \qquad \textbf{31–1}$$

The constant, R, in this expression is known as the *Rydberg constant.* Its value is

$$R = 1.097 \times 10^7 \text{ m}^{-1}$$

Each integer value of n (3, 4, 5, . . .) in Balmer's formula corresponds to the wavelength, λ, of a different spectral line. For example, if we set $n = 5$ in Equation 31–1 we find

$$\frac{1}{\lambda} = (1.097 \times 10^7 \text{ m}^{-1})\left(\frac{1}{2^2} - \frac{1}{5^2}\right)$$

Solving for the wavelength, we have

$$\lambda = 4.341 \times 10^{-7} \text{ m} = 434.1 \text{ nm}$$

This is the bluish line, second from the left in Figure 31–4 (a).

The collection of all lines predicted by the Balmer formula is referred to as the **Balmer series.** We consider the Balmer series in detail in the following Example.

PROBLEM-SOLVING NOTE

Calculating Wavelengths for the Balmer Series

Note that the formula for the Balmer series gives the inverse of the wavelength, rather than the wavelength itself.

EXAMPLE 31–1 The Balmer Series

Find the longest and shortest wavelengths in the Balmer series of spectral lines.

Picture the Problem

In our sketch we indicate the first several lines in the Balmer series, along with their corresponding colors, using the results given in Figure 31–4 as a guide. There are an infinite number of lines in the Balmer series, as indicated by the ellipsis (three dots) to the right of the $n \rightarrow \infty$ line.

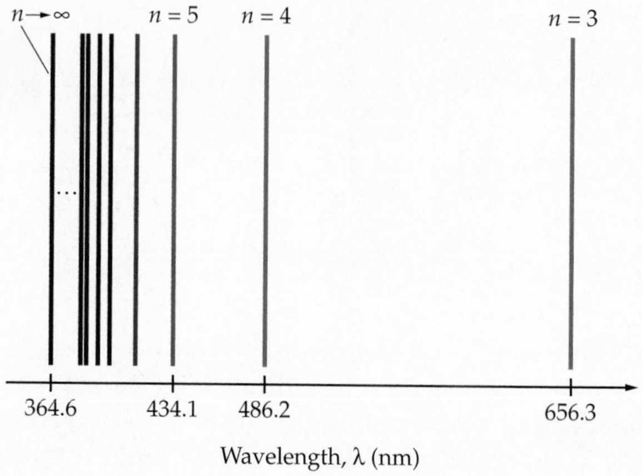

Strategy

By substituting the values $n = 3, n = 4$, and $n = 5$ in the Balmer series (Equation 31–1), we find that the wavelength decreases with increasing n. Hence, the longest wavelength corresponds to $n = 3$, and the shortest wavelength corresponds to $n \rightarrow \infty$.

Solution

1. To find the longest wavelength in the Balmer series, substitute $n = 3$ in Equation 31–1:

$$\frac{1}{\lambda} = R\left(\frac{1}{2^2} - \frac{1}{3^2}\right) = (1.097 \times 10^7 \text{ m}^{-1})\left(\frac{5}{36}\right)$$

$$\lambda = \frac{36}{5(1.097 \times 10^7 \text{ m}^{-1})} = 656.3 \text{ nm}$$

2. The shortest wavelength is found in the limit $n \rightarrow \infty$ or, equivalently, $(1/n^2) \rightarrow 0$. Make this substitution in Equation 31–1:

$$\frac{1}{\lambda} = R\left(\frac{1}{2^2} - 0\right) = (1.097 \times 10^7 \text{ m}^{-1})\left(\frac{1}{4}\right)$$

$$\lambda = \frac{4}{(1.097 \times 10^7 \text{ m}^{-1})} = 364.6 \text{ nm}$$

Insight

The longest wavelength corresponds to visible light with a reddish hue, whereas the shortest wavelength is well within the ultraviolet portion of the electromagnetic spectrum—it is invisible to our eyes.

Practice Problem

What value of n corresponds to a wavelength of 377.1 nm in the Balmer series? [**Answer:** $n = 11$]

Some related homework problems: Problem 5, Problem 6

TABLE 31–1
Common Spectral Series of Hydrogen

n'	Series name
1	Lyman
2	Balmer
3	Paschen
4	Brackett
5	Pfund

PROBLEM-SOLVING NOTE

Correctly Applying Equation 31–2

Notice that n and n' are integers in Equation 31–2 and that the integer n must always be larger than n'.

Figure 31–5 shows that the Balmer series is not the only series of lines produced by atomic hydrogen. The series with the shortest wavelengths is the **Lyman series**—all its lines are in the ultraviolet. Similarly, the series with wavelengths just greater than those in the Balmer series is the **Paschen series.** The lines in this series are all in the infrared. The formula that gives the wavelength in all the series of hydrogen is

$$\frac{1}{\lambda} = R\left(\frac{1}{n'^2} - \frac{1}{n^2}\right) \quad n' = 1, 2, 3, \ldots$$
$$n = n' + 1, n' + 2, n' + 3, \ldots \qquad \textbf{31–2}$$

Referring to Equation 31–1, we see that the Balmer series corresponds to the choice $n' = 2$. Similarly, the Lyman series is given by Equation 31–2 with $n' = 1$, and the Paschen series corresponds to $n' = 3$. As we shall see later in this chapter, there is an infinite number of series in hydrogen, each corresponding to a different choice for the integer n'. The names of the most common spectral series of hydrogen are summarized in Table 31–1.

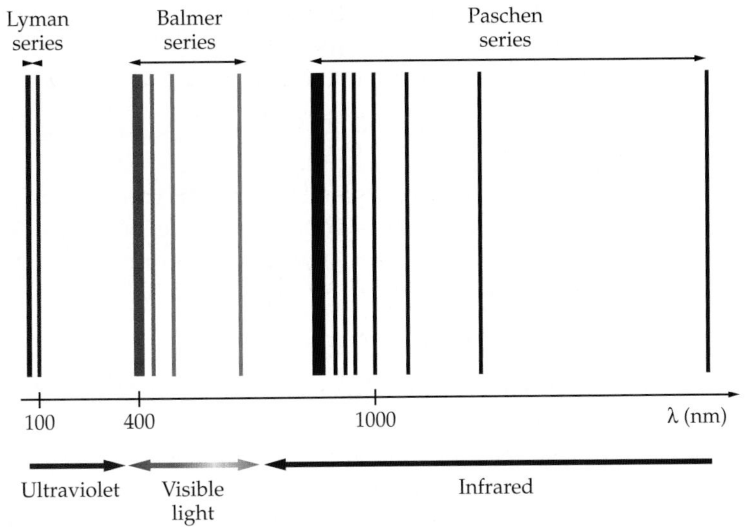

◀ **FIGURE 31–5 The Lyman, Balmer, and Paschen series of spectral lines**

The first three series of spectral lines in the spectrum of hydrogen. The shortest wavelengths appear in the Lyman series. There is no upper limit to the number of series in hydrogen or to the wavelengths that can be emitted.

EXERCISE 31–1

Find **(a)** the shortest wavelength in the Lyman series and **(b)** the longest wavelength in the Paschen series.

Solution

a. Substitute $n' = 1$ and $n \to \infty$ in Equation 31–2:

$$\frac{1}{\lambda} = R\left(\frac{1}{1^2} - 0\right) = (1.097 \times 10^7\ \text{m}^{-1})$$

$$\lambda = \frac{1}{(1.097 \times 10^7\ \text{m}^{-1})} = 91.16\ \text{nm}$$

b. Substitute $n' = 3$ and $n = 4$ in Equation 31–2:

$$\frac{1}{\lambda} = R\left(\frac{1}{3^2} - \frac{1}{4^2}\right) = (1.097 \times 10^7\ \text{m}^{-1})\left(\frac{7}{144}\right)$$

$$\lambda = \frac{144}{7(1.097 \times 10^7\ \text{m}^{-1})} = 1875\ \text{nm}$$

As successful as Equation 31–2 is in giving the various wavelengths of radiation produced by hydrogen, it is still just an empirical formula. It gives no insight as to *why* these particular wavelengths, and no others, are produced. The goal of atomic physicists in the early part of the twentieth century was to *derive* Equation 31–2 from basic physical principles. The first significant step in that direction is the topic of the next section.

31–3 Bohr's Model of the Hydrogen Atom

Our scientific understanding of the hydrogen atom took a giant leap forward in 1913, when Niels Bohr (1885–1962), a Danish physicist who had just earned his doctorate in physics in 1911, introduced a model that allowed him to derive Equation 31–2. Bohr's model combined elements of classical physics with the ideas of quantum physics introduced by Planck and Einstein about ten years earlier. As such, his model is a hybrid that spanned the gap between the classical physics of Newton and Maxwell and the newly emerging quantum physics.

▲ Niels Bohr, applying the principles of classical mechanics, with some members of his family.

Assumptions of the Bohr Model

Bohr's model of the hydrogen atom is based on four assumptions. Two are specific to his model and do not apply to the full quantum mechanical picture of hydrogen that will be introduced in Section 31–5. The remaining two assumptions are quite general—they apply not only to hydrogen but to all atoms.

The two specific assumptions of the Bohr model are as follows:

- The electron in a hydrogen atom moves in a circular orbit about the nucleus.
- Only certain circular orbits are allowed. In these orbits the angular momentum of the electron is equal to an integer times Planck's constant divided by 2π. That is, the angular momentum of an electron in the nth allowed orbit is $L_n = nh/2\pi$, where $n = 1, 2, 3, \ldots$.

The next two assumptions are more general:

- Electrons do not give off electromagnetic radiation when they are in an allowed orbit. Thus, the orbits are stable.
- Electromagnetic radiation is given off or absorbed only when an electron changes from one allowed orbit to another. If the energy difference between two allowed orbits is ΔE, the frequency, f, of the photon that is emitted or absorbed is given by the relation $|\Delta E| = hf$.

Notice that Bohr's model retains the classical picture of an electron orbiting a nucleus, as in Rutherford's model. It adds the stipulations, however, that only certain orbits are allowed and that no radiation is given off from these orbits. Radiation is given off *only* when an electron shifts from one orbit to another, and then the radiation is in the form of a photon that obeys Einstein's quantum relation $E = hf$. Thus, as mentioned before, the Bohr model is a hybrid that includes ideas from both classical and quantum physics. We now use this model to determine the behavior of hydrogen.

Bohr Orbits

We begin by determining the radii of the allowed orbits in the Bohr model and the speed of the electrons in these orbits. There are two conditions that we must apply. First, for the electron to move in a circular orbit of radius r with a speed v, as depicted in **Figure 31–6**, the electrostatic force of attraction between the electron and the nucleus must be equal in magnitude to the mass of the electron times its centripetal acceleration, mv^2/r. Recalling Coulomb's law (Equation 19–5), we see that the electrostatic force between the electron (with charge $-e$) and the nucleus (with charge $+e$) has a magnitude given by ke^2/r^2. It follows that $mv^2/r = ke^2/r^2$, or, canceling one power of r,

$$mv^2 = k\frac{e^2}{r} \qquad \text{31–3}$$

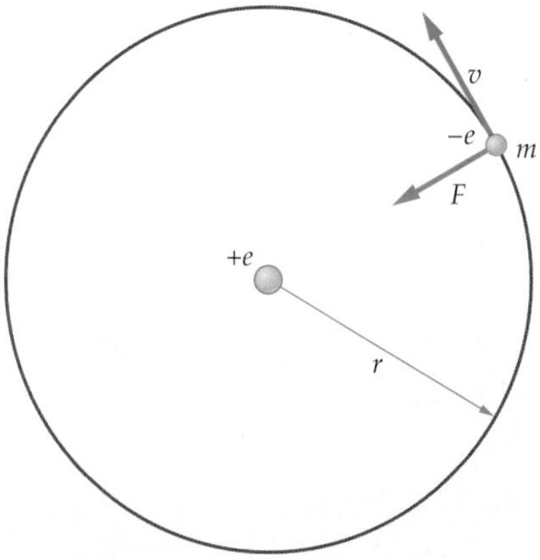

▲ **FIGURE 31–6 A Bohr orbit**
In the Bohr model of hydrogen, electrons orbit the nucleus in circular orbits. The centripetal acceleration of the electron, v^2/r, is produced by the Coulomb force of attraction between the electron and the nucleus.

Note that this relation is completely analogous to the one that was used to derive Kepler's third law (Chapter 12), except in that case the force of attraction was provided by gravity.

The second condition for an allowed orbit is that the angular momentum of the electron must be a nonzero integer n times $h/2\pi$. Since the electron moves with a speed v in a circular path of radius r, its angular momentum is $L = rmv$ (Equation 11–12). Hence, the condition for the nth allowed orbit is $L_n = r_n m v_n = nh/2\pi$, or, solving for v_n, we have

$$v_n = \frac{nh}{2\pi m r_n} \qquad n = 1, 2, 3, \ldots \qquad \text{31–4}$$

Combining these two conditions allows us to solve for the two unknowns, r_n and v_n.

For example, if we substitute v_n from Equation 31–4 into Equation 31–3, we can solve for r_n. Specifically, we find the following:

$$m\left(\frac{nh}{2\pi m r_n}\right)^2 = k\frac{e^2}{r_n}$$

Rearranging and solving for r_n, we get

$$r_n = \left(\frac{h^2}{4\pi^2 mke^2}\right)n^2 \qquad n = 1, 2, 3, \ldots \qquad \textbf{31–5}$$

The quantity in parenthesis is the radius for the smallest ($n = 1$) orbit. Substitution of numerical values for h, m, e, and k gives the following value for r_1:

$$r_1 = \frac{h^2}{4\pi^2 mke^2}$$

$$= \frac{(6.626 \times 10^{-34}\ \text{J}\cdot\text{s})^2}{4\pi^2(9.109 \times 10^{-31}\ \text{kg})(8.988 \times 10^9\ \text{N}\cdot\text{m}^2/\text{C}^2)(1.602 \times 10^{-19}\ \text{C})^2}$$

$$= 5.29 \times 10^{-11}\ \text{m}$$

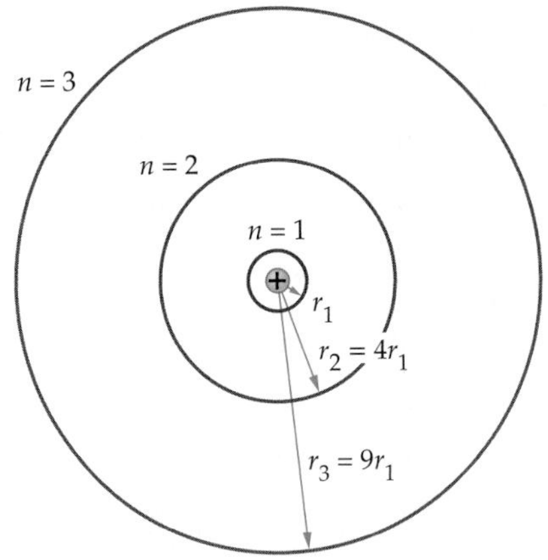

▲ **FIGURE 31–7 The first three Bohr orbits**

The first Bohr orbit has a radius $r_1 = 5.29 \times 10^{-11}$ m. The second and third Bohr orbits have radii $r_2 = 2^2 r_1 = 4r_1$, and $r_3 = 3^2 r_1 = 9r_1$, respectively. (*Note*: For clarity, the nucleus is drawn larger than its true scale relative to the size of the atom.)

This radius, which is about half an angstrom and is referred to as the **Bohr radius,** is in agreement with the observed size of hydrogen atoms. Note the n^2 dependence in the radii of allowed orbits: $r_n = r_1 n^2 = (5.29 \times 10^{-11}\ \text{m})n^2$. This dependence is illustrated in **Figure 31–7**.

To complete the solution, we can substitute our result for r_n (Equation 31–5) into the expression for v_n (Equation 31–4). This yields

$$v_n = \frac{nh}{2\pi m}\left(\frac{4\pi^2 mke^2}{n^2h^2}\right) = \frac{2\pi ke^2}{nh} \qquad n = 1, 2, 3, \ldots \qquad \textbf{31–6}$$

Note that the speed of the electron is smaller in orbits farther from the nucleus.

EXAMPLE 31–2 First and Second Bohr Orbits

Find the speed and kinetic energy of the electron in **(a)** the first Bohr orbit ($n = 1$) and **(b)** the second Bohr orbit ($n = 2$).

Picture the Problem

The first two orbits of the Bohr model are shown in the sketch. Note that the second orbit has a radius four times greater than the radius of the first orbit. In addition, the speed of the electron in the second orbit is half its value in the first orbit.

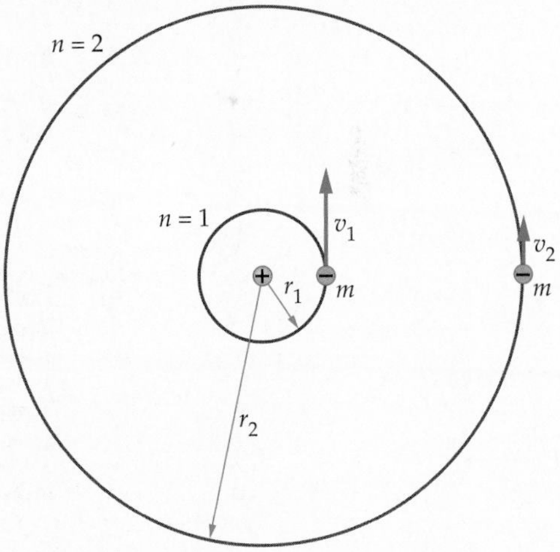

Strategy

The speed of the electron can be determined by direct substitution in $v_n = 2\pi ke^2/nh$ (Equation 31–6). Once v is determined, the kinetic energy is simply $K = \frac{1}{2}mv^2$.

Note that because the speed in the second orbit is half the speed in the first orbit, the kinetic energy is smaller by a factor of 4.

Solution

Part (a)

1. Substitute $n = 1$ in Equation 31–6:

$$v_1 = \frac{2\pi ke^2}{h}$$

$$= \frac{2\pi(8.99 \times 10^9\ \text{N}\cdot\text{m}^2/\text{C}^2)(1.60 \times 10^{-19}\ \text{C})^2}{(6.63 \times 10^{-34}\ \text{J}\cdot\text{s})}$$

$$= 2.18 \times 10^6\ \text{m/s}$$

2. The corresponding kinetic energy is $K_1 = \frac{1}{2}mv_1^{\,2}$:

$$K_1 = \tfrac{1}{2}mv_1^{\,2} = \tfrac{1}{2}(9.11 \times 10^{-31}\ \text{kg})(2.18 \times 10^6\ \text{m/s})^2$$

$$= 2.16 \times 10^{-18}\ \text{J}$$

continued on next page

continued from previous page

Part (b)

3. Divide the speed found in part (a) by 2: $v_2 = \frac{1}{2}v_1 = \frac{1}{2}(2.18 \times 10^6 \text{ m/s}) = 1.09 \times 10^6 \text{ m/s}$

4. Similarly, divide the kinetic energy found in part (a) by 4: $K_2 = \frac{1}{2}mv_2{}^2 = \frac{1}{2}m(v_1/2)^2$
$= \frac{1}{4}K_1 = \frac{1}{4}(2.16 \times 10^{-18} \text{ J}) = 5.40 \times 10^{-19} \text{ J}$

Insight

The speed of the electron in the first Bohr orbit is smaller than the speed of light by a factor of about 137. In higher orbits, the electron's speed is even less. It follows that relativistic effects are small for the hydrogen atom.

In addition, note that the kinetic energy of the electron in the first Bohr orbit is approximately 13.6 eV. We shall encounter this particular energy again later in the section.

Practice Problem

An electron in a Bohr orbit has a kinetic energy of 8.64×10^{-20} J. Find the radius of this orbit. [**Answer:** $r = 1.32 \times 10^{-9}$ m, corresponding to $n = 5$.]

Some related homework problems: Problem 11, Problem 15, Problem 16

Bohr's model applies equally well to singly ionized helium, doubly ionized lithium, and other ions with only a single electron. In the case of singly ionized helium (one electron removed) the charge on the nucleus is $+2e$, for doubly ionized lithium (two electrons removed) it is $+3e$, and so on. In the general case, we may consider a nucleus that contains Z protons and has a charge of $+Ze$, where Z is the atomic number associated with that nucleus. Hydrogen, which has only a single proton in its nucleus, corresponds to $Z = 1$.

To be more explicit about the Z dependence, notice that the electrostatic force between an electron and a nucleus with Z protons has a magnitude of $k(e)(Ze)/r^2 = kZe^2/r^2$. Thus the results derived earlier in this section can be applied to the more general case if we simply replace e^2 with Ze^2. For example, Equation 31–3 becomes $mv^2 = kZe^2/r$. Similarly, the radius of an allowed orbit is

$$r_n = \left(\frac{h^2}{4\pi^2 mkZe^2}\right)n^2 \qquad n = 1, 2, 3, \ldots \qquad \textbf{31–7}$$

For example, the radius of the $n = 1$ orbit of singly ionized helium is half the radius of the $n = 1$ orbit of hydrogen.

The Energy of a Bohr Orbit

To find the energy of an electron in a Bohr orbit, we simply note that its total mechanical energy, E, is the sum of its kinetic and potential energies:

$$E = K + U = \tfrac{1}{2}mv^2 + U$$

Using the fact that $mv^2 = kZe^2/r$ for a hydrogen-like atom of atomic number Z (Equation 31–3), and the fact that the electrostatic potential energy of a charge $-e$ and a charge $+Ze$ a distance r apart is $U = -kZe^2/r$ (Equation 20–8), we find that the total mechanical energy is

$$E = \frac{1}{2}\left(\frac{kZe^2}{r}\right) - \frac{kZe^2}{r} = -\frac{kZe^2}{2r}$$

Finally, substituting the radius of a Bohr orbit, as given in Equation 31–7, we obtain the corresponding energy for the nth orbit:

$$E_n = -\frac{kZe^2}{2r_n} = -\left(\frac{kZe^2}{2}\right)\left(\frac{4\pi^2 mkZe^2}{h^2}\right)\frac{1}{n^2}$$
$$= -\left(\frac{2\pi^2 mk^2e^4}{h^2}\right)\frac{Z^2}{n^2} \qquad n = 1, 2, 3, \ldots \qquad \textbf{31–8}$$

Using the numerical values for m, k, e, and h, as we did in calculating the Bohr radius, we have

$$E_n = -(13.6\ \text{eV})\frac{Z^2}{n^2} \qquad n = 1, 2, 3, \ldots \qquad \textbf{31–9}$$

Let's first consider the specific case of hydrogen. With $Z = 1$ we find that the energy of the orbits in hydrogen are given by the relation $E_n = -(13.6\ \text{eV})/n^2$. We plot these energies in **Figure 31–8** for various values of n. This type of plot is referred to as an **energy-level diagram.** Notice that the **ground state** ($n = 1$) corresponds to the lowest possible energy of the system. The higher energy levels are referred to as **excited states.** As the integer n tends to infinity, the energy of the excited states approaches zero—the energy the electron and proton would have if they were at rest and separated by an infinite distance. Thus to **ionize** hydrogen—that is, to remove the electron from the atom—requires a minimum energy of 13.6 eV. This value, which is a specific prediction of the Bohr model, is in complete agreement with experiment.

PROBLEM-SOLVING NOTE

The Z Dependence of Hydrogen-like Ions

Note that the energy of a hydrogen-like ion depends on Z^2, the square of the atomic number.

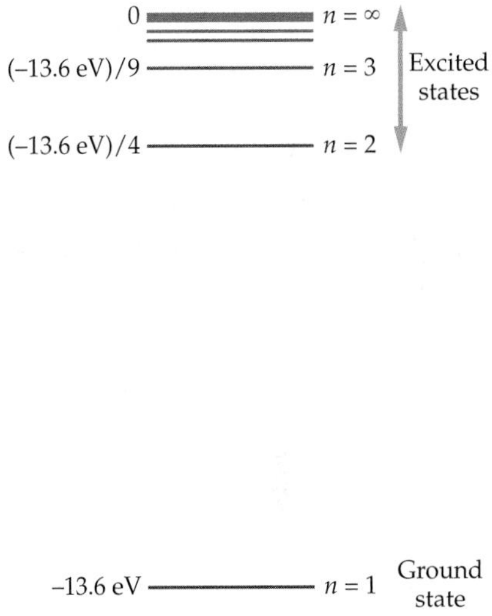

▲ **FIGURE 31–8 Energy-level diagram for the Bohr model of hydrogen**

The energy of the ground state of hydrogen is -13.6 eV. Excited states of hydrogen approach zero energy. Note that the difference in energy from the ground state to the first excited state is $\frac{3}{4}(13.6\ \text{eV})$, and the energy difference from the first excited state to the zero level is only $\frac{1}{4}(13.6\ \text{eV})$.

EXERCISE 31–2

In doubly ionized lithium, a single electron orbits a lithium nucleus. Calculate the minimum energy required to remove this electron.

Solution

The nucleus of lithium has a charge of $+3e$. Substitution of $Z = 3$ and $n = 1$ in Equation 31–9 yields

$$E_1 = -(13.6\ \text{eV})\left(\frac{3^2}{1^2}\right) = -122\ \text{eV}$$

Therefore, 122 eV must be added to remove the electron.

The electron in doubly ionized lithium experiences a stronger attractive force than the single electron in hydrogen. In fact, the force is greater by a factor of 3. In addition, the $n = 1$ orbit of doubly ionized lithium has one-third the radius of the $n = 1$ orbit in hydrogen. As a result, nine times as much energy is required to remove the electron from the lithium ion.

At room temperature, most hydrogen atoms are in the ground state. This is because the typical thermal energy of such atoms is too small to cause even the lowest-energy excitation from the ground state. Specifically, a typical thermal energy, k_BT, corresponding to room temperature ($T \sim 300$ K) is only about $\frac{1}{40}$ eV. In comparison, the energy required to excite an electron from the ground state of hydrogen to the first excited state is roughly 10 eV; that is, $\Delta E = E_2 - E_1 = (-3.40\ \text{eV}) - (-13.6\ \text{eV}) = 10.2$ eV. Excitations to higher excited states require even more energy. As a result, typical intermolecular collisions are simply not energetic enough to produce an excited state in hydrogen.

The Spectrum of Hydrogen

To find a formula describing the spectrum of hydrogen, we use Bohr's assumption that the frequency of emitted radiation for a change in energy equal to ΔE is given by $|\Delta E| = hf$. Since $\lambda f = c$ for electromagnetic radiation (Equation 25–4), we can rewrite this relation in terms of the wavelength as $|\Delta E| = hc/\lambda$. To find $|\Delta E|$, we recall that the energy for hydrogen is given by Equation 31–9 with $Z = 1$. Therefore, the *change* in energy as the electron moves from an excited outer orbit with $n = n_i$ to a lower orbit with $n = n_f < n_i$ has the following magnitude:

$$|\Delta E| = \left(\frac{2\pi^2 mk^2e^4}{h^2}\right)\left(\frac{1}{n_f^{\,2}} - \frac{1}{n_i^{\,2}}\right)$$

Using $|\Delta E| = hc/\lambda$ we can now solve for $1/\lambda$:

$$\frac{1}{\lambda} = \left(\frac{2\pi^2 mk^2e^4}{h^3c}\right)\left(\frac{1}{n_f^{\,2}} - \frac{1}{n_i^{\,2}}\right) \qquad \textbf{31–10}$$

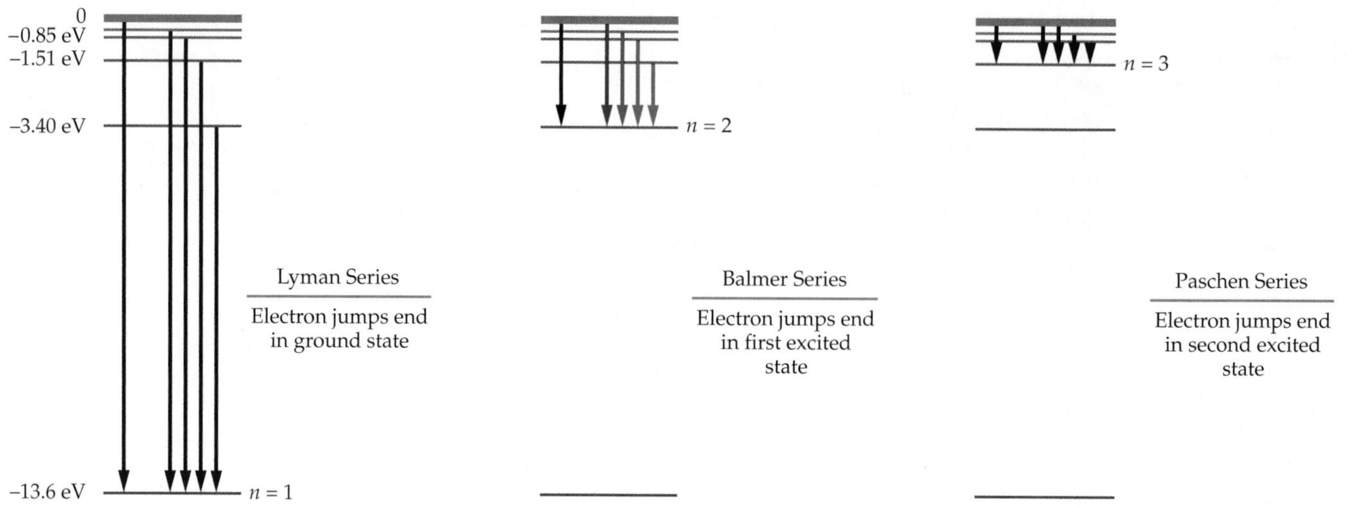

▲ **FIGURE 31–9 The origin of spectral series in hydrogen**

Each series of spectral lines in hydrogen is the result of electrons jumping from an excited state to a particular lower level. For the Lyman series the lower level is the ground state. The lower level for the Balmer series is the first excited state ($n = 2$), and the lower level for the Paschen series is the second excited state ($n = 3$).

Comparing Equation 31–10 with Equation 31–2, we see that the expressions have precisely the same form, provided we identify n_f with n', and n_i with n. In addition, we see that the Rydberg constant in Equation 31–2, $R = 1.097 \times 10^7\ \text{m}^{-1}$, has been replaced by the rather unusual constant $2\pi^2 mk^2e^4/h^3c$ in Equation 31–10. It is remarkable that when the known values of the fundamental constants m, k, e, h, and c are substituted into $2\pi^2 mk^2e^4/h^3c$, the resulting value is precisely $1.097 \times 10^7\ \text{m}^{-1}$. This completes the derivation of Equation 31–2, one of the most significant accomplishments of the Bohr model.

The origin of the line spectrum of hydrogen can be visualized in **Figure 31–9**. Notice that transitions involving an electron jumping from an excited state ($n_i = n > 1$) to the ground state ($n_f = n' = 1$) result in the Lyman series of lines in the ultraviolet. Jumps ending in the $n = 2$ level give rise to the Balmer series, and jumps ending in the $n = 3$ level give the Paschen series. The largest energy jump in each series occurs when an electron falls from $n = \infty$ to the final level. Thus each series of spectral lines has a well-defined shortest wavelength.

CONCEPTUAL CHECKPOINT 31–1 Compare Wavelengths

The wavelength of the photon emitted when an electron in hydrogen jumps from the $n_i = 100$ state to the $n_f = 2$ state is **(a)** greater than, **(b)** less than, or **(c)** equal to the wavelength of the photon when the electron jumps from the $n_i = 2$ state to the $n_f = 1$ state.

Reasoning and Discussion

We begin by noting that wavelength is inversely proportional to the energy difference between levels ($|\Delta E| = hc/\lambda$), thus the smaller the energy difference, the greater the wavelength.

Next, we note that the full range of energies in the hydrogen atom extends from −13.6 eV to 0, and that the difference in energy between $n_i = 2$ and $n_f = 1$ is three quarters of this range. It follows that the energy difference between $n_i = 2$ and $n_f = 1$ is greater than the energy difference between *any* state with $n_i > 2$ and $n_f = 2$. In fact, the maximum energy difference ending in the state 2 is $\frac{1}{4}(-13.6\text{ eV})$, corresponding to $n_i = \infty$ and $n_f = 2$.

Since the energy difference for $n_i = 100$ and $n_f = 2$ is less than the energy difference for $n_i = 2$ and $n_f = 1$, the wavelength for $n_i = 100$ and $n_f = 2$ is the greater of the two.

Answer:

(a) The wavelength for $n_i = 100$, $n_f = 2$ is greater than the wavelength for $n_i = 2$, $n_f = 1$.

We calculate a variety of specific wavelengths in the following Active Example.

ACTIVE EXAMPLE 31–1 Find the Wavelengths

An electron in a hydrogen atom is in the initial state $n_i = 4$. Calculate the wavelength of the photon emitted by this electron if it jumps to the final state **(a)** $n_f = 3$, **(b)** $n_f = 2$, or **(c)** $n_f = 1$. (*Note*: To simplify the calculations, replace the constant $2\pi^2 mk^2e^4/h^3c$ in Equation 31–10 with its numerical equivalent, the Rydberg constant, $R = 1.097 \times 10^7\ \text{m}^{-1}$.)

Solution *(Test your understanding by performing the calculations indicated in each step.)*

Part (a)

1. Substitute $n_i = 4$ and $n_f = 3$ in Equation 31–10: $\lambda = 1875\ \text{nm}$

Part (b)

2. Repeat with $n_f = 2$: $\lambda = 486.2\ \text{nm}$

Part (c)

3. Finally, use $n_f = 1$: $\lambda = 97.23\ \text{nm}$

Insight

Other wavelengths are possible when an electron in the $n_i = 4$ state jumps to a lower state. For example, after dropping from the $n = 4$ state to the $n = 3$ state, the electron might then jump from the $n = 3$ state to the $n = 2$ state, and finally from the $n = 2$ state to the $n = 1$ state. Alternatively, the electron might first jump from the $n = 4$ state to the $n = 2$ state, and then from the $n = 2$ state to the $n = 1$ state. Thus an electron in an excited state may result in the emission of a variety of different wavelengths.

Your Turn

A hydrogen atom with its electron in the initial state $n_i = 5$ emits a photon with a wavelength of 434 nm. To which state did the electron jump?

(Answers to **Your Turn** *problems are given in the back of the book.)*

PROBLEM-SOLVING NOTE

The Frequency and Wavelength of a Photon

Recall that the frequency of a photon is given by $hf = |\Delta E|$, and its wavelength is given by $hc/\lambda = |\Delta E|$.

Just as an electron can *emit* a photon when it jumps to a lower level, it can also *absorb* a photon and jump to a higher level. This process occurs, however, only if the photon has the proper energy. In particular, the photon must have an energy that precisely matches the energy difference between the lower level of the electron and the higher level to which it is raised. This situation is explored in more detail in the next Active Example.

ACTIVE EXAMPLE 31–2 Absorbing a Photon: What Is the Frequency?

Find the frequency a photon must have if it is to raise an electron in a hydrogen atom from the $n = 3$ state to the $n = 5$ state.

Solution *(Test your understanding by performing the calculations indicated in each step.)*

1. Calculate the energy of the $n = 5$ state in joules: $-8.70 \times 10^{-20}\ \text{J}$
2. Calculate the energy of the $n = 3$ state: $-2.42 \times 10^{-19}\ \text{J}$
3. Calculate the difference in energy between these states: $1.55 \times 10^{-19}\ \text{J}$
4. Set the energy of a photon equal to this energy difference: $hf = 1.55 \times 10^{-19}\ \text{J}$
5. Solve for the frequency of the photon: $f = 2.34 \times 10^{14}\ \text{Hz}$

Insight

Of course, this frequency corresponds to one of the lines in the absorption spectrum of hydrogen. In this case, the line is in the infrared.

Your Turn

Suppose an electron jumps from the $n = 3$ state to the $n = 5$ state in singly ionized helium. Does the required frequency of the absorbed photon increase or decrease? By what factor?

(Answers to **Your Turn** *problems are given in the back of the book.)*

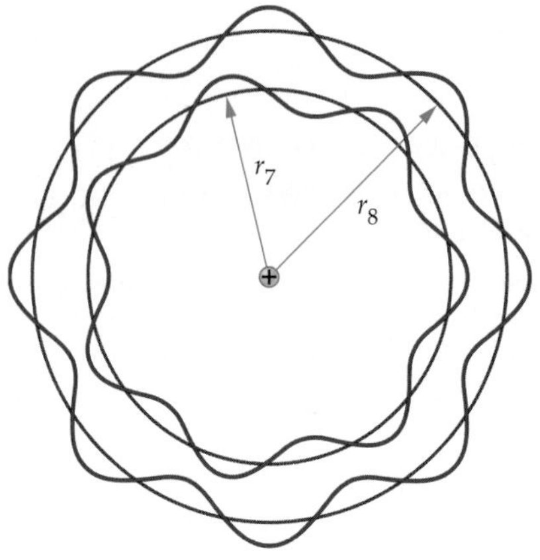

(a)

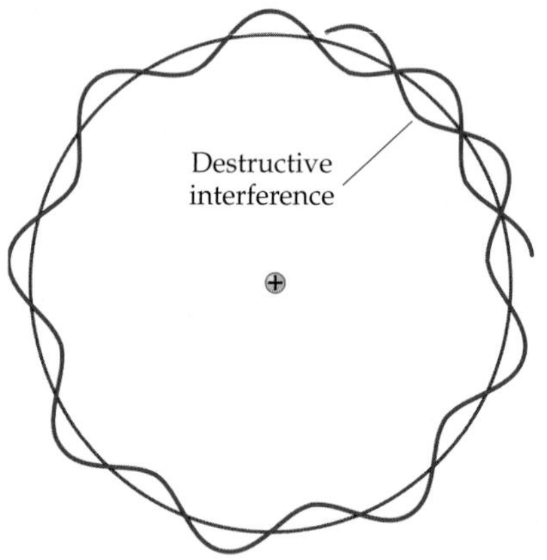

(b)

▲ **FIGURE 31–10 de Broglie wavelengths and Bohr orbits**

Bohr's condition that the angular momentum of an allowed orbit must be an integer n times $h/2\pi$ is equivalent to the condition that n de Broglie wavelengths must fit into the circumference of an orbit. **(a)** De Broglie waves for the $n = 7$ and $n = 8$ orbits. **(b)** If an integral number of wavelengths do not fit the circumference of an orbit, the result is destructive interference.

31–4 de Broglie Waves and the Bohr Model

The fact that hydrogen emits radiation only at certain well-defined wavelengths is reminiscent of the harmonics of standing waves on a string. Recall that a vibrating string tied down at both ends produces a standing wave only if an integral number of half-wavelengths fit within its length (Section 14–8). Perhaps the behavior of hydrogen can be understood in similar terms.

In 1923 de Broglie used his idea of matter waves (Section 30–5) to show that one of Bohr's assumptions could indeed be thought of as a condition for standing waves. As we saw earlier in this chapter, Bohr assumed that the angular momentum of an electron in an allowed orbit must be a nonzero integer times $h/2\pi$. Specifically,

$$rmv = n\frac{h}{2\pi} \qquad n = 1, 2, 3, \ldots$$

In Bohr's model there is no particular reason for this condition other than that it produces results in agreement with experiment.

Now, de Broglie imagined his matter waves as analogous to a wave on a string—except that in this case the "string" is not tied down at both ends. Instead, it is formed into a circle of radius r representing an electron's orbit about the nucleus, as illustrated in **Figure 31–10**. The condition for a standing wave in this case is that an integral number of wavelengths fit into the circumference of the orbit. Stated mathematically, the condition is $n\lambda = 2\pi r$, as shown in Figure 31–10 (a) for the cases $n = 7$ and $n = 8$. Other wavelengths would result in destructive interference, as Figure 31–10 (b) shows.

Finally, de Broglie combined the standing wave condition with his matter-wave relationship $p = h/\lambda$ (Equation 30–16). The result is as follows:

$$p = mv = \frac{h}{\lambda} = \frac{h}{(2\pi r/n)} = \frac{nh}{2\pi r} \qquad n = 1, 2, 3, \ldots$$

Multiplying both sides of this equation by r to obtain the angular momentum, we find

$$L = rmv = \frac{nh}{2\pi} \qquad n = 1, 2, 3, \ldots$$

This is precisely the Bohr orbital condition, now understood as a reflection of the wave nature of matter.

EXAMPLE 31–3 The Wavelength of an Electron

Find the wavelength associated with an electron in the $n = 4$ state of hydrogen.

Picture the Problem

Our sketch shows that four wavelengths fit around the circumference of the $n = 4$ orbit. Recall that the radius of this orbit is $4^2 = 16$ times the radius of the ground-state orbit.

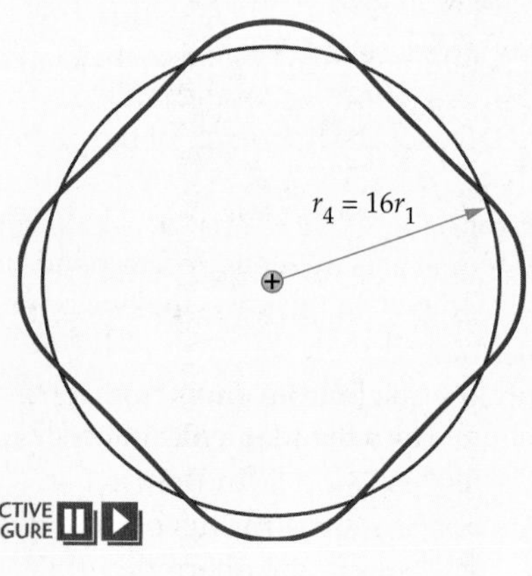

INTERACTIVE FIGURE

Strategy

To find the wavelength of this matter wave we simply calculate the circumference of the $n = 4$ orbit, then divide by 4.

Solution

1. Calculate the radius of the $n = 4$ orbit: $r_4 = 4^2 r_1 = 16(5.29 \times 10^{-11}\text{ m}) = 8.46 \times 10^{-10}\text{ m}$

2. Use this result to find the circumference of the orbit: $2\pi r_4 = 2\pi(8.46 \times 10^{-10}\text{ m}) = 5.32 \times 10^{-9}\text{ m}$

3. Divide the circumference by four to find the wavelength: $\lambda_4 = \frac{1}{4}(2\pi r_4) = \frac{1}{4}(5.32 \times 10^{-9}\text{ m}) = 1.33 \times 10^{-9}\text{ m}$

Insight

An equivalent way of determining the wavelength is to use the de Broglie relation, $p = mv = h/\lambda$. Solving for the wavelength yields $\lambda_4 = h/mv_4$, and substituting v_4 from Equation 31–6 yields the same result given in step 3.

Practice Problem

In which state of hydrogen does the electron have a wavelength of 2.66×10^{-9} m? **[Answer:** $n = 8$**]**

Some related homework problems: Problem 29, Problem 30

The striking success of de Broglie's matter waves in deriving Bohr's angular momentum condition encouraged physicists to give the idea of matter waves serious consideration. If we accept matter waves as being real, however, a large number of new questions must be addressed. For example, if particles like electrons can be described by matter waves, how do the matter waves behave? What determines the value of a matter wave at a particular location? What is the physical significance of a matter wave having a large value at one location and a small value at another location?

These questions were answered by Erwin Schrödinger (1887–1961), Max Born (1882–1970), and others. In particular, Schrödinger introduced an equation—similar in many respects to the equation that describes sound waves—to describe the behavior of matter waves. Today, this equation, known as **Schrödinger's equation,** forms the basis for quantum mechanics, which is the quantum physics version of classical mechanics. In fact, Schrödinger's equation plays the same role in quantum mechanics that Newton's laws play in classical mechanics and Maxwell's equations play in electromagnetism.

After the introduction of the Schrödinger equation, Born developed an interpretation of the matter waves that was quite different from that for mechanical waves. For example, in the case of a wave on a string, the amplitude of a wave simply represents the displacement of the string from its equilibrium position. For a matter wave, on the other hand, the amplitude is related to the *probability* of finding a particle in a particular location. Thus, matter waves do not tell us precisely where a particle is located; rather, they give the probability of finding the particle at a given place, as we shall see in detail in the next section.

31–5 The Quantum Mechanical Hydrogen Atom

Although Schrödinger's equation and its solution for the hydrogen atom are beyond the scope of this text, we present here some of the main features obtained by this analysis. Other than relativistic effects, the Schrödinger equation presents our most complete understanding of the hydrogen atom and of behavior at the atomic level in general. As we shall see, many aspects of the Bohr model survive in this analysis, though there are also significant differences.

To begin, we note that whereas the Bohr model was characterized by a single quantum number, n, the quantum mechanical description of the hydrogen atom requires four quantum numbers. They are as follows:

- **The principal quantum number, n:** The quantum number $n = 1, 2, 3, \ldots$ plays a similar role in the quantum mechanical hydrogen atom and in Bohr's model. In Bohr's model, n is the only quantum number, and it determines the radius of an orbit, its angular momentum, and its energy. In particular, the energy in the Bohr model is $E = (-13.6\text{ eV})/n^2$. The energy given by Schrödinger's equation is precisely the same, if we neglect small relativistic effects and small magnetic interactions within the atom.

- **The orbital angular momentum quantum number, ℓ:** In the Bohr model an electron's orbital angular momentum is determined by the quantum number n. In particular, $L_n = nh/2\pi$, where $n = 1, 2, 3, \ldots$. In the quantum mechanical solution, there is a separate quantum number, ℓ, for the orbital angular momentum. This quantum number can take on the following values for any given value of the principal quantum number, n:

$$\ell = 0, 1, 2, \ldots, (n - 1) \qquad \text{31–11}$$

The magnitude of the angular momentum for any given value of ℓ is given by the following relation:

$$L = \sqrt{\ell(\ell + 1)}\frac{h}{2\pi} \qquad \text{31–12}$$

Note that the angular momentum of the electron can have a range of values for a given n, in contrast with the Bohr model, where the angular momentum has just a single value. In particular, the electron in a hydrogen atom can have zero angular momentum; an orbiting electron in the Bohr model always has a nonzero angular momentum.

Finally, although the energy of the hydrogen atom does not depend on ℓ, the energy of more complex atoms does have an ℓ dependence, as we shall see in the next section.

- **The magnetic quantum number, m_ℓ:** If a hydrogen atom is placed in an external magnetic field, its energy is found to depend not only on n but on an additional quantum number as well. This quantum number, m_ℓ, is referred to as the magnetic quantum number. The allowed values of m_ℓ are as follows:

$$m_\ell = -\ell, -\ell + 1, -\ell + 2, \ldots, -1, 0, 1, \ldots, \ell - 2, \ell - 1, \ell \qquad \text{31–13}$$

This quantum number gives the component of orbital angular momentum vector along a specified direction, usually chosen to be the z axis. With this choice, L_z has the following values:

$$L_z = m_\ell \frac{h}{2\pi}$$

Only a single component of the orbital angular momentum can be known precisely; it is not possible to know all three components of the angular momentum simultaneously, due to the Heisenberg uncertainty principle.

- **The electron spin quantum number, m_s:** The final quantum number needed to describe the hydrogen atom is related to the angular momentum of the electron itself. Just as Earth spins on its axis at the same time that it orbits the Sun, the electron can be thought of as having both an orbital and a "spin" angular momentum. The spin quantum number for an electron takes on just two values:

$$m_s = -\tfrac{1}{2}, \tfrac{1}{2}$$

These two values correspond to the electron's spin being "up" $\left(m_s = \frac{1}{2}\right)$ or "down" $\left(m_s = -\frac{1}{2}\right)$ with respect to the z axis.

Spin angular momentum is an *intrinsic* property of an electron, like its mass and its charge—*all* electrons have exactly the *same* mass, the *same* charge, and the *same* spin angular momentum. Thus, we do not imagine the electron to be a small spinning sphere, like a microscopic planet. You can't speed up or slow down the spin of an electron. Instead, spin is simply one of the properties that defines an electron.

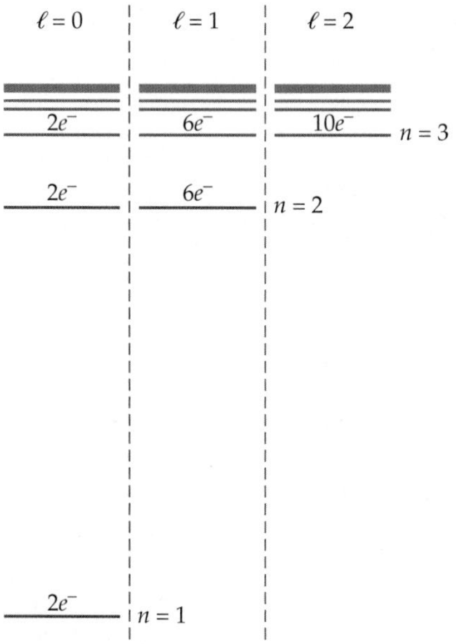

▲ **FIGURE 31–11 Energy-level structure of hydrogen**

The values of the quantum mechanical energy levels for hydrogen are in complete agreement with the Bohr model. In the quantum model, however, each energy level has associated with it a specific number of "quantum states" determined by the values of all four quantum numbers, as specified in Table 31–2. In multielectron atoms, these states lead to the formation of the periodic table of elements, as we show in Section 31–6.

TABLE 31–2 States of Hydrogen for $n = 1$ and $n = 2$

$n = 1, \ell = 0$			**Two states**
$n = 1$	$\ell = 0$	$m_\ell = 0$	$m_s = \frac{1}{2}$
$n = 1$	$\ell = 0$	$m_\ell = 0$	$m_s = -\frac{1}{2}$
$n = 2, \ell = 0$			**Two states**
$n = 2$	$\ell = 0$	$m_\ell = 0$	$m_s = \frac{1}{2}$
$n = 2$	$\ell = 0$	$m_\ell = 0$	$m_s = -\frac{1}{2}$
$n = 2, \ell = 1$			**Six states**
$n = 2$	$\ell = 1$	$m_\ell = 1$	$m_s = \frac{1}{2}$
$n = 2$	$\ell = 1$	$m_\ell = 1$	$m_s = -\frac{1}{2}$
$n = 2$	$\ell = 1$	$m_\ell = 0$	$m_s = \frac{1}{2}$
$n = 2$	$\ell = 1$	$m_\ell = 0$	$m_s = -\frac{1}{2}$
$n = 2$	$\ell = 1$	$m_\ell = -1$	$m_s = \frac{1}{2}$
$n = 2$	$\ell = 1$	$m_\ell = -1$	$m_s = -\frac{1}{2}$

The energy-level structure of hydrogen in zero magnetic field is shown in **Figure 31–11**, along with the corresponding quantum numbers. Since the energies are the same as in the Bohr model, it follows that the spectrum will be the same as the Bohr model, and experiment.

We define a **state** of hydrogen to be a specific assignment of values for each of the four quantum numbers. For example, there are two states that correspond to the lowest possible energy level of hydrogen. These are $n = 1, \ell = 0, m_\ell = 0, m_s = \frac{1}{2}$; and $n = 1, \ell = 0, m_\ell = 0, m_s = -\frac{1}{2}$. Similarly, there are two states corresponding to the $n = 2, \ell = 0$ energy level, and six states corresponding to the $n = 2, \ell = 1$ level. These states are listed in Table 31–2, and the corresponding numbers are shown in Figure 31–11. When we consider multielectron atoms in the next section, we shall see that the number of states associated with a given energy level determines the number of electrons (e^-) it can accommodate. Once an energy level is "filled," additional electrons must occupy higher levels. This progressive filling of energy levels ultimately leads to the periodic table of elements.

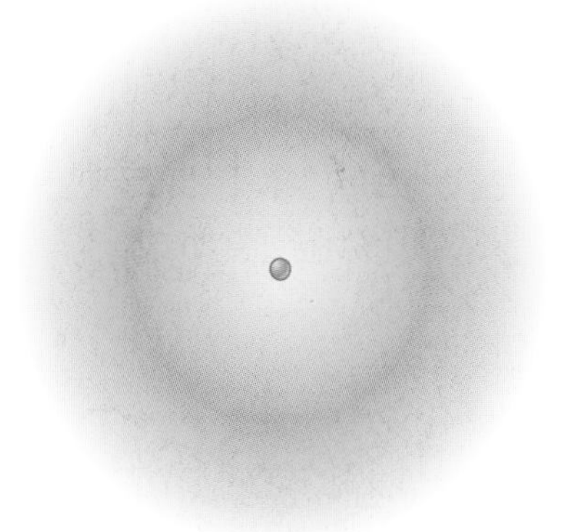

▲ **FIGURE 31–12 Probability cloud for the ground state of hydrogen**

In the quantum mechanical model of hydrogen, the electron can be found at any distance from the nucleus. The probability of finding the electron at a given location is proportional to the density of the "probability cloud."

Electron Probability Clouds: Three-Dimensional Standing Waves

As mentioned in the previous section, the solution to Schrödinger's equation gives a matter wave, or **wave function,** as it is known, corresponding to a particular physical system. The wave function for hydrogen gives the probability of finding the electron at a particular location. The best way to visualize this probability distribution is in terms of a "probability cloud," as shown in **Figure 31–12**. The probability of finding the electron is greatest where the cloud is densest. In the case shown in Figure 31–12, corresponding to the ground state, $n = 1, \ell = 0, m_\ell = 0$, the probability of finding the electron is distributed with spherical symmetry. Note that the probability decreases rapidly far from the nucleus, as one would expect.

Figure 31–13 gives a different way of looking at the probability distribution for the ground state. Here we plot the probability versus distance from the nucleus. It is interesting to note that the maximum probability occurs at a distance from the nucleus equal to the Bohr radius. Thus certain aspects of the Bohr model find their way into the final solution for hydrogen. The difference, however, is that in the Bohr model the electron is always at a particular distance from the nucleus as it moves in its circular orbit. In the quantum solution to hydrogen, the electron can be found at virtually any distance from the nucleus, not just one distance.

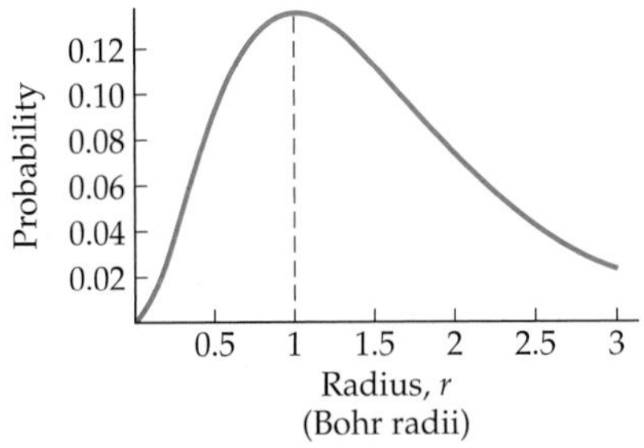

▲ **FIGURE 31–13 Probability as a function of distance**

This plot shows the probability of finding an electron in the ground state of hydrogen at a given distance from the nucleus. Note that the probability is greatest at a distance equal to the Bohr radius, r_1.

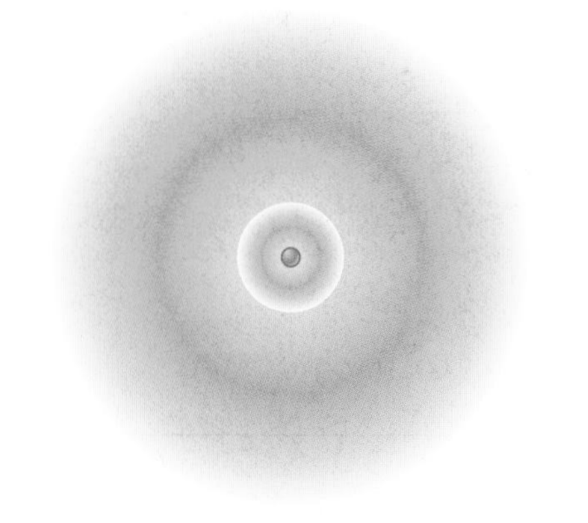

(a)

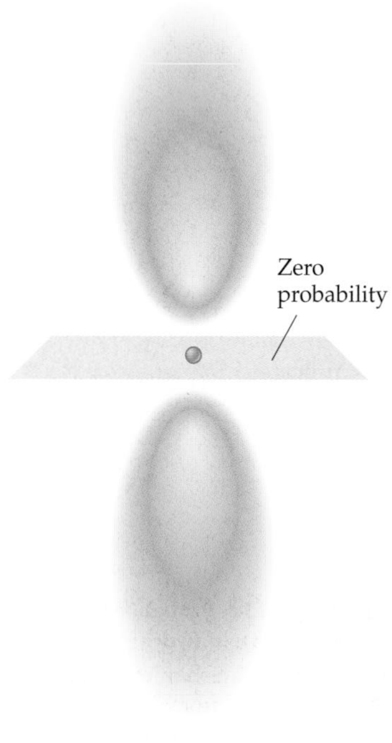

(b)

▲ **FIGURE 31–14 Probability clouds for excited states of hydrogen**

The "probability cloud" for an electron in hydrogen increases in complexity as the quantum numbers increase. **(a)** $n = 2$, $\ell = 0$; **(b)** $n = 2$, $\ell = 1$.

States of higher quantum number have increasingly complex probability distributions, as indicated in **Figure 31–14**. As the quantum number n is increased, for example, the basic shape of the distribution remains the same, but additional nodes appear. This is illustrated in Figure 31–14 (a) for the case $n = 2$, $\ell = 0$. Note that the distribution is spherically symmetric, as in the state $n = 1$, $\ell = 0$, but now there is a node—where the probability is zero—separating an inner and an outer portion of the distribution. As the quantum number ℓ is increased, the distributions become more complex in their shape. An example is given in Figure 31–14 (b) for the case $n = 2$, $\ell = 1$, and $m_\ell = 0$.

CONCEPTUAL CHECKPOINT 31–2 Finding the Electron

The probability cloud for the $n = 2$, $\ell = 1$ state of hydrogen is shown in Figure 31–14 (b). Notice that this cloud consists of two lobes of high probability separated by a plane of zero probability. Given that an electron in this state can never be halfway between the two lobes, how is it possible that the electron is as likely to be found in the upper lobe as in the lower lobe?

Reasoning and Discussion

The probability lobes of an electron are the result of a standing wave pattern, analogous to the standing waves found on a string tied down at both ends. Both the node on a string and the region of zero probability of an electron are the result of destructive interference. For example, the displacement of a string may be of equal amplitude on either side of a node—where the displacement is always zero—just as the electron probability may be high on either side of a probability node and zero in the middle. In summary, an electron does not simply move from place to place in an atom, like a small ball of charge; instead, it forms a standing wave pattern with nodes in certain locations.

31–6 Multielectron Atoms and the Periodic Table

In this section we extend our considerations of atomic physics to atoms with more than one electron. As we shall see, certain regularities arise in the properties of multielectron atoms, and these regularities are intimately related to the quantum numbers described in the previous section.

Multielectron Atoms

One of the great simplicities of the hydrogen atom is that the only electrostatic force in the atom is the attractive force between the electron and the proton. In multielectron atoms the situation is more complex. Specifically, the electrons in such atoms experience repulsive electrostatic interactions with one another, in addition to their interaction with the nucleus. Thus the simple expression for the energy of a single-electron atom given in Equation 31–9 cannot be applied to atoms with multiple electrons, since it does not include energy contributions due to the forces between electrons.

Although no simple formula analogous to Equation 31–9 exists for multielectron atoms, the energy levels of these atoms can be understood in terms of the four quantum numbers (n, ℓ, m_ℓ, m_s) used to describe hydrogen. In fact, by applying Schrödinger's equation to such atoms, we have discovered that the energy levels of multielectron atoms depend on the principal quantum number, n, and on the orbital quantum number, ℓ. For example, increasing n for a fixed value of ℓ results in an increase in energy. This relationship is illustrated in **Figure 31–15**, where we see that the energy for $n = 2$ and $\ell = 0$ is greater than the energy for $n = 1$ and $\ell = 0$. Similarly, the energy for $n = 3$ and $\ell = 1$ is greater than the energy for $n = 2$ and $\ell = 1$. It is also found that the energy increases with increasing ℓ for fixed n. Thus the energy for $n = 2$ and $\ell = 1$ is greater than the energy for $n = 2$ and $\ell = 0$.

Figure 31–15 also shows that in some cases the energy levels corresponding to different values of n can cross. For example, the energy for $n = 3$ and $\ell = 2$ is greater than the energy for $n = 4$ and $\ell = 0$. A similar crossing occurs with the $n = 4$, $\ell = 2$ state and the $n = 5$, $\ell = 0$ state. Note in all cases, however, that the energy still increases with n for fixed ℓ, and increases with ℓ for fixed n. We shall

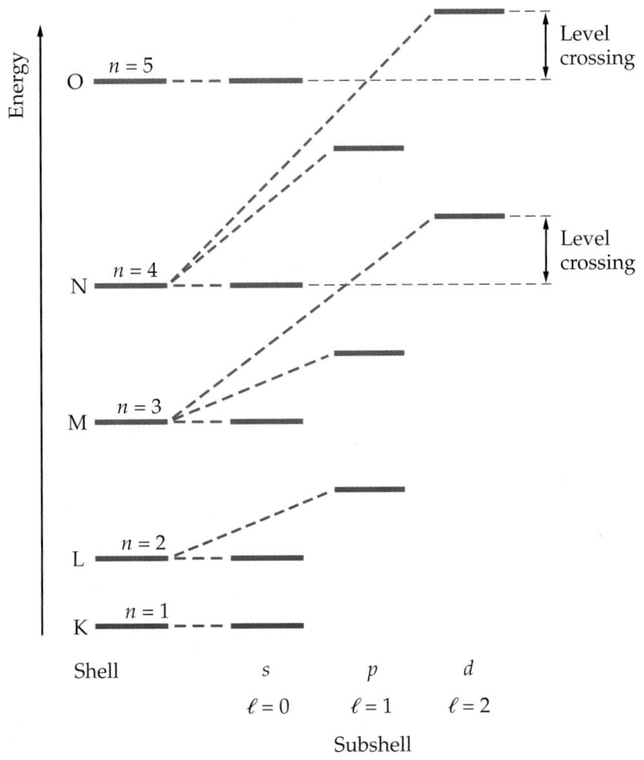

FIGURE 31–15 Energy levels in multielectron atoms

In multielectron atoms the energy increases with n for fixed ℓ and increases with ℓ for fixed n. Note the possibility of energy level crossing. For example, the $n = 3$, $\ell = 2$ energy level is higher than the $n = 4$, $\ell = 0$ energy level.

see the effect of these energy-level crossings in terms of the periodic table later in this section.

Because the energy levels of multielectron atoms depend on n and ℓ, we give specific names and designations to the various values of these quantum numbers. For example, all electrons that have the same value of n are said to be in the same **shell.** Specifically, electrons with $n = 1$ are said to be in the K shell, those with $n = 2$ are in the L shell, those with $n = 3$ are in the M shell, and so on. These designations are summarized in Table 31–3 and displayed in Figure 31–15.

Similarly, electrons in a given shell with the same value of ℓ are said to be in the same **subshell,** and different values of ℓ have different alphabetical designations. For example, electrons with $\ell = 0$ are said to be in the s subshell, those with $\ell = 1$ are in the p subshell, and those with $\ell = 2$ are in the d subshell. These names, though not particularly logical, are used for historical reasons. After the f subshell ($\ell = 3$), the names of subsequent subshells continue in alphabetical order, as indicated in Table 31–3.

TABLE 31–3
Shell and Subshell Designations

n	Shell
1	K
2	L
3	M
4	N
...	...
ℓ	**Subshell**
0	s
1	p
2	d
3	f
4	g
...	...

The Pauli Exclusion Principle

As mentioned in Section 31–3, most hydrogen atoms are in their ground state at room temperature, since typical thermal energies are not great enough to excite the electron to higher energy levels. The same is true of multielectron atoms—they too are generally found in their ground state at room temperature. The question is this: What is the ground state of a multielectron atom?

The answer to this question involves an entirely new fundamental principle of physics put forward by the Austrian physicist Wolfgang Pauli (1900–1958) in 1925. Pauli's "exclusion principle" states that no two electrons in an atom can be in the same state at the same time. That is, once an electron occupies a given state, as defined by the values of its quantum numbers, other electrons are *excluded* from occupying the same state:

The Pauli Exclusion Principle

Only one electron at a time may have a particular set of quantum numbers, n, ℓ, m_ℓ, and m_s. Once a particular state is occupied, other electrons are excluded from that state.

Because of the exclusion principle, the ground state of a multielectron atom is *not* obtained by placing all the electrons in the lowest possible energy state, as one might at first suppose. Once the lowest-energy states are occupied, additional electrons in the atom must occupy levels of higher energy. As more electrons are added to an atom, they fill up one subshell after another until all the electrons are accommodated. The situation is analogous to placing marbles into a jar, as illustrated in **Figure 31–16**. The first few marbles occupy the lowest level of the jar, but as more marbles are added they must occupy levels of higher gravitational potential energy, simply because the lower levels are already occupied.

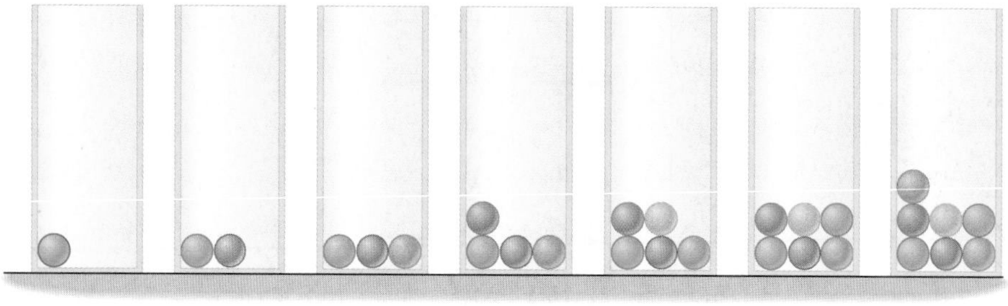

▶ FIGURE 31–16 Filling a jar with marbles
As marbles are added to a jar, they fill in one level, then another, and another. Once a given level is filled with marbles, additional marbles are excluded from that level, analogous to the filling of energy levels in multielectron atoms.

In the case of atoms, the lowest energy level corresponds to $n = 1$ and $\ell = 0$; that is, to the s subshell of the K shell. To completely define a state, however, we must specify all four quantum numbers, n, ℓ, m_ℓ, and m_s. First, recall that $m_\ell = 0, \pm1, \pm2, \ldots, \pm\ell$ for general ℓ. It follows that in the $\ell = 0$ state the only possible value of m_ℓ is 0. The quantum number m_s, however, can always take on two values, $m_s = +\frac{1}{2}$ and $m_s = -\frac{1}{2}$. Therefore, two electrons can occupy the $n = 1, \ell = 0$ energy level, since two different states—$n = 1, \ell = 0, m_\ell = 0, m_s = -\frac{1}{2}$; and $n = 1, \ell = 0, m_\ell = 0, m_s = +\frac{1}{2}$—correspond to that level, as indicated in **Figure 31–17**.

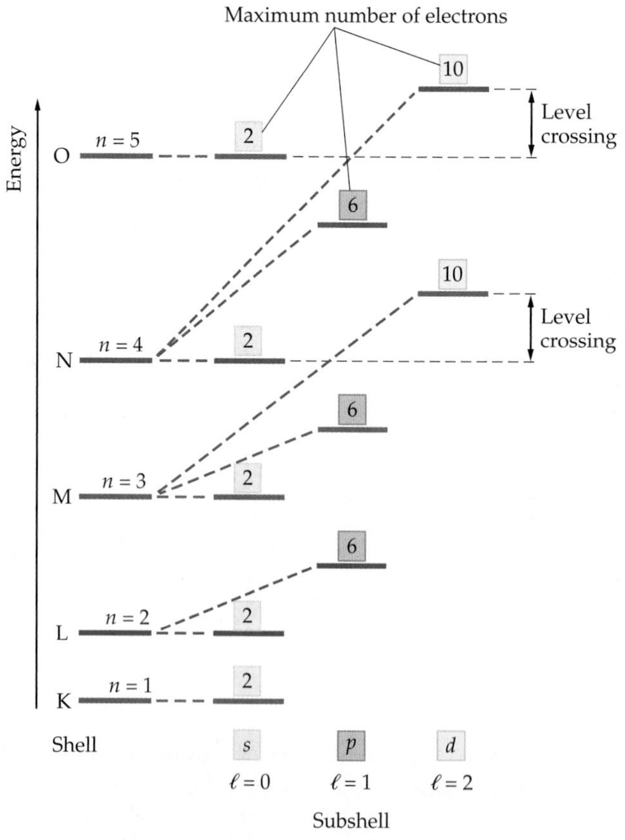

▶ FIGURE 31–17 Maximum number of electrons in an energy level
The maximum number of electrons that can occupy a given energy level in a multielectron atom is $2(2\ell + 1)$. Occupancy by more than this number of electrons would violate the Pauli exclusion principle.

The next higher energy level corresponds to $n = 2$ and $\ell = 0$. Again, two states correspond to this level, allowing it to hold two electrons. Lest we conclude that all levels can hold two electrons, consider the next level up: $n = 2$ and $\ell = 1$. In this case, m_ℓ can take on three values: $m_\ell = 0, \pm 1$. For each of these three values, m_s can take on two values. Therefore, the $n = 2$, $\ell = 1$ energy level can accommodate 6 electrons, as shown in Figure 31–17. For general ℓ, the number of possible values of $m_\ell(0, \pm 1, \pm 2, \ldots, \pm \ell)$ is $2\ell + 1$. When we multiply by 2 (for the number of values of m_s), we find a total number of states equal to $2(2\ell + 1)$. For example, note in Figure 31–17 that the $n = 3$, $\ell = 2$ energy level can hold $2(2 \cdot 2 + 1) = 10$ electrons.

Figure 31–18 presents the ground-state electron arrangements for the following elements: hydrogen (1 electron); helium (2 electrons); lithium (3 electrons); beryllium (4 electrons); boron (5 electrons); carbon (6 electrons); nitrogen (7 electrons); and oxygen (8 electrons). Notice that the energy levels are filled from the bottom upward, like marbles in a jar, with each level having a predetermined maximum number of electrons.

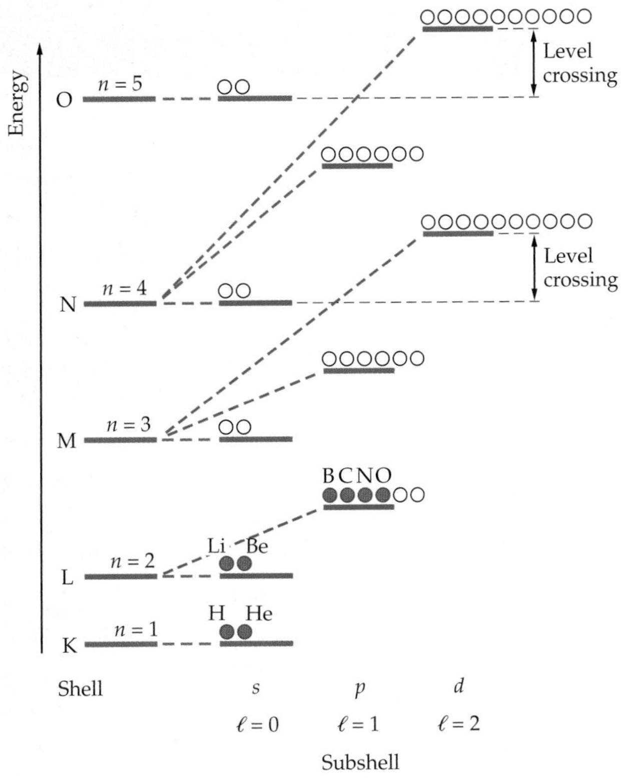

◀ **FIGURE 31–18 First eight elements of the periodic table**
The elements hydrogen and helium fill the $n = 1$, $\ell = 0$ level; the elements lithium and beryllium fill the $n = 2$, $\ell = 0$ level; the elements boron, carbon, nitrogen, and oxygen fill four of the six available states in the $n = 2$, $\ell = 1$ level.

Electronic Configurations

Indicating the arrangements of electrons as in Figure 31–18 is instructive, but also somewhat cumbersome. This is especially so when we consider elements with a large number of electrons. To streamline the process, we introduce a shorthand notation that can be applied to all elements.

As an example of this notation, consider the element lithium, which has two electrons in the $n = 1$, $\ell = 0$ state and one electron in the $n = 2$, $\ell = 0$ state. This arrangement of electrons—referred to as an **electronic configuration**—will be abbreviated as follows:

$$1s^2 2s^1$$

In this expression, the $1s^2$ part indicates $n = 1$ ($1s^2$), $\ell = 0$ ($1s^2$), and an occupancy of two electrons ($1s^2$). Similarly, the $2s^1$ part indicates one electron ($2s^1$) in the $n = 2$ ($2s^1$) and $\ell = 0$ ($2s^1$) state. The general labeling is indicated in **Figure 31–19**.

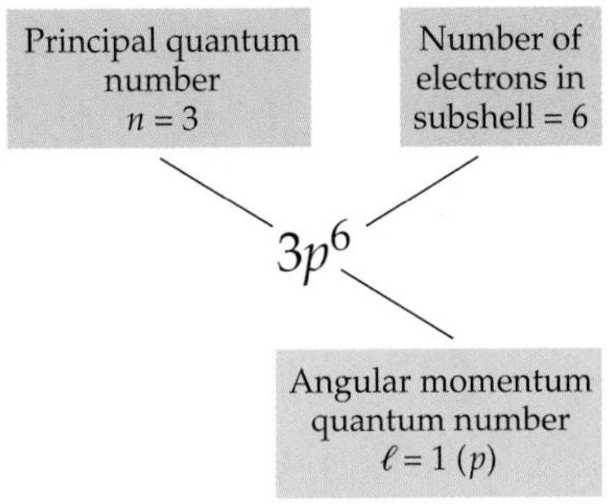

▲ **FIGURE 31–19 Designation of electronic configuration**
The example shown here indicates six electrons in the $n = 3$, $\ell = 1(p)$ energy level.

EXERCISE 31–3

Use the shorthand notation just introduced to write the electronic configuration of **(a)** nitrogen and **(b)** sodium.

Solution

a. The 1*s* and 2*s* subshells of nitrogen ($Z = 7$) are filled; three electrons are in the 2*p* subshell:

$$1s^2\,2s^2\,2p^3$$

b. In sodium ($Z = 11$), one electron is in the 3*s* subshell; all the inner subshells are fully occupied:

$$1s^2\,2s^2\,2p^6\,3s^1$$

Table 31–4 presents a list of electronic configurations for the elements hydrogen (H) through potassium (K). Note that the 3*d* levels in potassium have no electrons, in agreement with the level crossing shown in Figure 31–18.

TABLE 31–4 Electronic Configurations of the Elements Hydrogen through Potassium

Atomic number	Element	Electronic configuration
1	Hydrogen (H)	$1s^1$
2	Helium (He)	$1s^2$
3	Lithium (Li)	$1s^2\,2s^1$
4	Beryllium (Be)	$1s^2\,2s^2$
5	Boron (B)	$1s^2\,2s^2\,2p^1$
6	Carbon (C)	$1s^2\,2s^2\,2p^2$
7	Nitrogen (N)	$1s^2\,2s^2\,2p^3$
8	Oxygen (O)	$1s^2\,2s^2\,2p^4$
9	Fluorine (F)	$1s^2\,2s^2\,2p^5$
10	Neon (Ne)	$1s^2\,2s^2\,2p^6$
11	Sodium (Na)	$1s^2\,2s^2\,2p^6\,3s^1$
12	Magnesium (Mg)	$1s^2\,2s^2\,2p^6\,3s^2$
13	Aluminum (Al)	$1s^2\,2s^2\,2p^6\,3s^2\,3p^1$
14	Silicon (Si)	$1s^2\,2s^2\,2p^6\,3s^2\,3p^2$
15	Phosphorus (P)	$1s^2\,2s^2\,2p^6\,3s^2\,3p^3$
16	Sulfur (S)	$1s^2\,2s^2\,2p^6\,3s^2\,3p^4$
17	Chlorine (Cl)	$1s^2\,2s^2\,2p^6\,3s^2\,3p^5$
18	Argon (Ar)	$1s^2\,2s^2\,2p^6\,3s^2\,3p^6$
19	Potassium (K)	$1s^2\,2s^2\,2p^6\,3s^2\,3p^6\,4s^1$

The Periodic Table

Referring to the elements listed in Table 31–4, we observe a number of interesting patterns. For example, notice that the elements hydrogen, lithium, sodium, and potassium all have the same type of electronic configuration for the final (and outermost) electron in the atom. In particular, hydrogen's outermost (and only) electron is the $1s^1$ electron. In the case of lithium, we see that the outermost electron is $2s^1$, the outermost electron of sodium is $3s^1$, and for potassium it is $4s^1$. Continuing through the list of elements, we note that rubidium has a $5s^1$ outer electron, cesium has a $6s^1$ outer electron, and francium has a $7s^1$ outer electron. In each case the outermost electron is a single electron in an otherwise empty *s* subshell.

What makes this similarity in *electronic configuration* of particular interest is that these elements have similar *chemical properties* as well. In particular, the metallic members of this group of elements (lithium, sodium, potassium, rubidium, cesium, and francium) are referred to as the *alkali metals*. In each of these metals, the outermost electron is easily removed from the atom, leading to a stable, positively charged ion, as in the familiar case of the sodium ion, Na^+. Thus the regular pattern in the filling of shells leads to a regular pattern in the properties of the elements.

Grouping various elements with similar chemical properties was the motivation behind the development of the **periodic table** of elements by the Russian chemist Dmitri Mendeleev (1834–1907). The periodic table is presented in Appendix E. Notice that the elements just mentioned form Group I in the leftmost column of the table. Although Mendeleev grouped these elements strictly on the basis of their chemical properties, we can now see that the grouping also corresponds to the filling of shells, in accordance with the Pauli exclusion principle.

Many groups of elements appear in the periodic table. As another example, Group VII consists of the *halogens:* fluorine, chlorine, bromine, and iodine. Note that these elements also have similar configurations of their outer electrons. In this case, the outer electrons are $2p^5$, $3p^5$, and so on. Thus we see that these elements are just one electron short of filling one of the p subshells. As a result, halogens are highly reactive—they can readily acquire a single electron from another element to form a stable negative ion, as in the case of chlorine, which forms the chloride ion, Cl^-.

Group VIII consists of the *noble gases*. These elements all have completely filled subshells. Thus they do not readily gain or lose an electron. It is for this reason that the noble gases are relatively inert.

Finally, the *transition elements* represent those cases where a crossing of energy levels occurs. For example, the 4s subshell is filled at the element calcium ($Z = 20$). The next electron, rather than going into the 4p subshell, goes into the 3d subshell. In fact, the 10 elements from scandium ($Z = 21$) to zinc ($Z = 30$) correspond to filling of the 3d subshell. After this subshell is filled, additional electrons go into the 4p subshell in the elements gallium ($Z = 31$) to krypton ($Z = 36$).

Figure 31–20 shows the meaning of the various symbols used in the periodic table. Note that each box in the table gives the symbol for the element, its atomic mass, and its atomic number. Each box also includes the configuration of the outermost electrons in the element. In the case of iron, shown in Figure 31–20, this configuration is $3d^6 4s^2$.

▲ Neon (atomic number 10), one of the "noble gases," is most widely known for its role in neon tubes. Electrons in neon atoms are excited by electrical discharge through the tube. When they return to the ground state, they emit electromagnetic radiation, much of it in the red part of the visible spectrum. The other colors familiarly found in "neon signs" or "neon lights" are produced by adding other elements of the same chemical family: argon, krypton, or xenon.

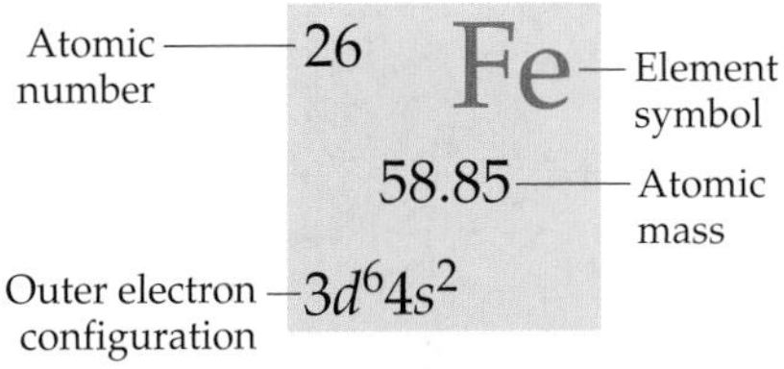

▲ **FIGURE 31–20 Designation of elements in the periodic table**

An explanation of the various entries to be found for each element in the periodic table of Appendix E. The example shown here is for the element iron (Fe).

CONCEPTUAL CHECKPOINT 31–3 Compare the Energy

The energy required to remove the outermost electron from sodium is 5.1 eV. Is the energy required to remove the outermost electron from potassium **(a)** greater than, **(b)** less than, or **(c)** equal to 5.1 eV?

Reasoning and Discussion

Referring to the periodic table, we see that potassium has one electron in its outermost shell, just like sodium. In this respect the two elements are alike. The difference between the elements is that the outermost electron in potassium is in a higher energy state; that is, its electron is in an $n = 4$ state, as opposed to the $n = 3$ state in sodium. Less energy is required to remove an electron from a state of higher energy; hence, the outermost electron of potassium can be removed with less than 5.1 eV. In fact, the required energy is only 4.3 eV.

Answer:

(b) Less energy is required to remove the outermost electron from potassium.

31–7 Atomic Radiation

We conclude this chapter with a brief investigation of various types of radiation associated with multielectron atoms. Examples range from X-rays that are energetic enough to pass through a human body, to the soft white light of a fluorescent lightbulb.

X-rays

X-rays were discovered quite by accident by the German physicist Wilhelm Roentgen (1845–1923) on November 8, 1895. Within months of their discovery

REAL-WORLD PHYSICS: BIO
Medical X-ray tubes

they were being used in medical applications, and they have played an important role in medicine ever since. Today, the X-rays used to give diagnostic images in hospitals and dentist offices are produced by an X-ray tube similar to the one shown in **Figure 31–21**. The basic operating principle of this device is that an energetic beam of electrons is generated and directed at a metal target—when the electrons collide with the target, X-rays are emitted. **Figure 31–22** shows a typical plot of X-ray intensity per wavelength versus the wavelength for such a device.

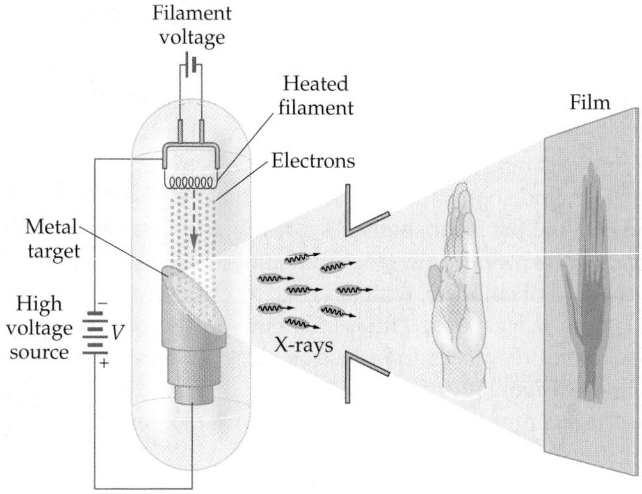

▶ **FIGURE 31–21 X-ray tube**
An X-ray tube accelerates electrons in a vacuum through a potential difference and then directs the electrons onto a metal target. X-rays are produced when the electrons decelerate in the target, and as a result of the excitations they cause when they collide with the metal atoms of the target.

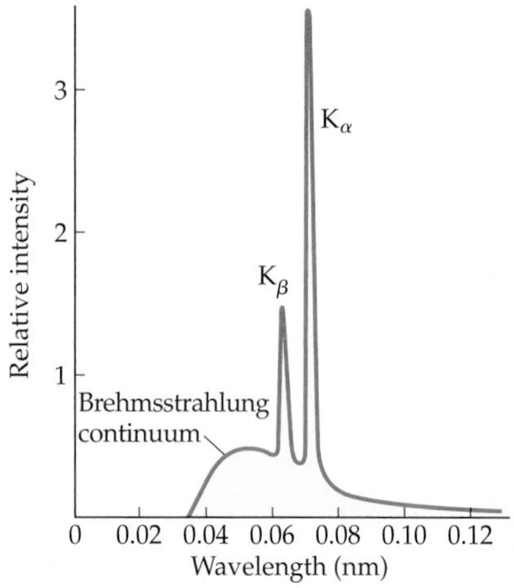

▲ **FIGURE 31–22 X-ray spectrum**
The spectrum produced by an X-ray tube in which electrons are accelerated from rest through a potential difference of 35,000 V and directed against a molybdenum target.

The radiation produced by an X-ray tube is created by two completely different physical mechanisms. The first mechanism is referred to as **bremsstrahlung,** which is German for "braking radiation." What is meant by this expression is that as the energetic electrons impact the target, they undergo a rapid deceleration. This is the "braking." As we know from Chapter 25, an accelerated charge gives off electromagnetic radiation; hence, as the electrons suddenly come to rest in the target they give off high-energy radiation in the form of X-rays. These X-rays cover a wide range of wavelengths, giving rise to the continuous part of the spectrum shown in Figure 31–22.

The sharp peaks in Figure 31–22 are produced by the second physical mechanism. To understand their origin, imagine what happens if one of the electrons in the incident beam is energetic enough to knock an electron out of a target atom. In addition, suppose the ejected electron comes from the lowest energy level of the atom, that is, from the K shell. This "vacancy" is filled almost immediately when an electron from an outer shell drops to the K shell, with the energy difference given off as a photon. In an atom of large atomic number [molybdenum ($Z = 42$) is often used for a target, as is tungsten ($Z = 74$)], the photon that emerges is an X-ray. If an electron drops from the $n = 2$ level to the K shell, the sharp peak of radiation that results is called the K_α line. Similarly, if an electron drops from the $n = 3$ level to the K shell, the resulting peak is called the K_β line. These lines are shown in Figure 31–22, and the corresponding electron jumps are indicated in **Figure 31–23**. Because the wavelengths of these lines vary from element to element—that is, they are characteristic of a certain element—they are referred to as **characteristic X-rays.** Similar characteristic X-rays can be emitted when electrons drop in energy to fill a vacancy in the L shell, the M shell, and so on.

The energy an incoming electron must have to dislodge a K-shell electron from an atom can be estimated using results from the Bohr model. The basic idea is that the energy of a K-shell electron in an atom of atomic number Z is given approximately by Equation 31–9 [$E_n = -(13.6\text{ eV})Z^2/n^2$], with one minor modification: since there are two electrons in the K shell, each electron shields the other from the nucleus. That is, the negative charge on one electron, $-e$, partially cancels the positive charge of the nucleus, $+Ze$, giving an effective charge experienced by the second electron of $+(Z - 1)e$. Thus, replacing Z with $Z - 1$ in Equation 31–9, and setting $n = 1$, we obtain a reasonable estimate for the energy of a K-shell electron:

$$E_K = -(13.6\text{ eV})\frac{(Z-1)^2}{1^2} \qquad \textbf{31–14}$$

We apply this result in the following Active Example.

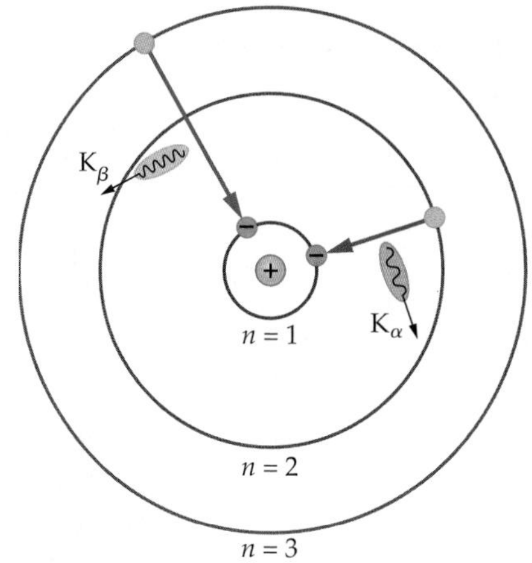

▲ **FIGURE 31–23 Production of characteristic X-rays**

When an electron strikes a metal atom in the target of an X-ray tube, it may knock one of the two K-shell ($n = 1$) electrons out of the atom. The resulting vacancy in the K shell will be filled by an electron dropping from a higher shell. If an electron drops from the $n = 2$ shell to the $n = 1$ shell, we say that the resulting photon is a K_α X-ray. Similarly, if an electron in the $n = 3$ shell drops to the $n = 1$ shell, the result is a K_β X-ray. Clearly, the K_β X-ray has the greater energy and the shorter wavelength, as we see in Figure 31–22.

ACTIVE EXAMPLE 31–3 Find the Voltage of an X-ray Tube

Estimate the minimum energy an incoming electron must have to knock a K-shell electron out of a tungsten atom ($Z = 74$).

Solution *(Test your understanding by performing the calculations indicated in each step.)*

1. Calculate $(Z-1)^2$:	5329
2. Multiply −13.6 eV by $(Z-1)^2$:	−72,500 eV

Insight

An electron at rest an infinite distance from a tungsten atom has an energy of zero. Therefore, a minimum energy of 72,500 eV must be supplied to the electron to remove it from the K shell of tungsten.

Recall that an electron gains an energy of 1 eV when it accelerates through a potential difference of 1 V. Thus a potential difference of at least 72,500 V = 72.5 kV is required to produce characteristic X-rays with a tungsten target. Typical X-ray tubes, like those used in dental offices, operate in the 100-kV range.

Your Turn

Suppose an X-ray tube has a voltage of only 35 kV. What is the largest value of Z for which this tube can knock out a K-shell electron?

(Answers to **Your Turn** *problems are given in the back of the book.)*

Equation 31–14 can also be used to obtain an estimate of the K_α wavelength for a given element. We show how this can be done for the case of molybdenum in the next Example.

EXAMPLE 31–4 K_α for Molybdenum

Estimate the K_α wavelength for molybdenum ($Z = 42$).

Picture the Problem

Our sketch indicates the electron jump that is responsible for the K_α X-ray; that is, from $n = 2$ to $n = 1$. Note that the net charge from the K shell outward is $+(Z-1)e$.

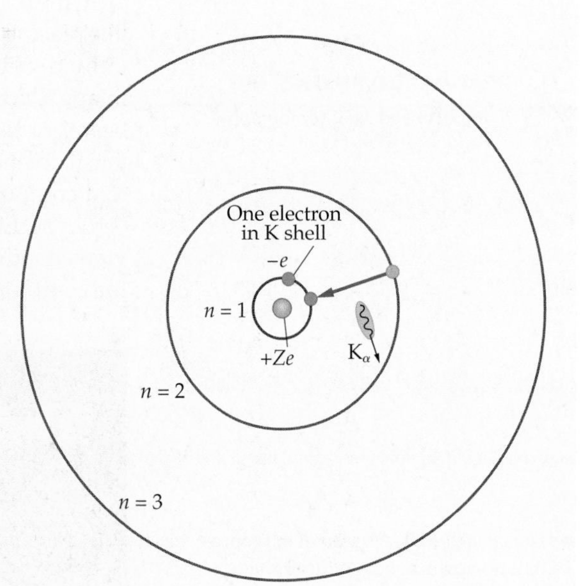

Strategy

To find the wavelength of a K_α X-ray, we start with the relationship between the change in energy of an electron and the wavelength of the corresponding photon, $|\Delta E| = hf = hc/\lambda$. Once we have calculated ΔE, we can find the wavelength using $\lambda = hc/|\Delta E|$.

To find ΔE, we first calculate the energy of an electron in the K shell of molybdenum, using $E_K = -(13.6\text{ eV})(Z-1)^2/1^2$ with $Z = 42$.

Next, we calculate the energy of an electron in the L shell ($n = 2$) of molybdenum, since it is an L-shell electron that fills the vacancy in the K shell and emits a K_α X-ray. Note that an L-shell electron sees a nucleus with an effective charge of $+(Z-1)e$. This follows because there is one electron in the K shell, and this electron partially screens the nucleus. As a result, the energy of an L-shell electron is given by the following expression: $E_L = -(13.6\text{ eV})(Z-1)^2/2^2$.

Finally, with these two energies determined, the change in energy is simply $\Delta E = E_K - E_L$.

continued on next page

continued from previous page

Solution

1. Calculate the energy of a K-shell electron, E_K:

$$E_K = -(13.6\text{ eV})\frac{(Z-1)^2}{1^2} = -(13.6\text{ eV})\frac{(42-1)^2}{1^2}$$
$$= -22{,}900\text{ eV}$$

2. Calculate the energy of an L-shell electron, E_L:

$$E_L = -(13.6\text{ eV})\frac{(Z-1)^2}{2^2} = -(13.6\text{ eV})\frac{(42-1)^2}{2^2}$$
$$= -5720\text{ eV}$$

3. Determine the change in energy, ΔE, of an electron that jumps from the L shell to the K shell:

$$\Delta E = E_K - E_L = -22{,}900\text{ eV} - (-5720\text{ eV}) = -17{,}200\text{ eV}$$

4. Calculate the wavelength corresponding to ΔE:

$$\lambda = \frac{hc}{|\Delta E|} = \frac{(6.63\times10^{-34}\text{ J}\cdot\text{s})(3.00\times10^{8}\text{ m/s})}{(17{,}200\text{ eV})(1.60\times10^{-19}\text{ J/eV})}$$
$$= 7.23\times10^{-11}\text{ m} = 0.0723\text{ nm}$$

Insight

Comparing our result with Figure 31–22, which shows the X-ray spectrum for molybdenum, we see that our approximate wavelength of 0.0723 nm is in good agreement with experiment.

Practice Problem

Which element has a K_α peak at a wavelength of approximately 0.155 nm? [**Answer:** Copper, $Z = 29$]

Some related homework problems: Problem 50, Problem 51, Problem 66

As mentioned earlier in this section, X-rays were put to medical use as soon as people found out about their properties and how to produce them. In fact, Roentgen himself, whose first article about X-rays was published in late December 1895, produced an X-ray image of his wife's left hand, clearly showing the bones in her fingers and her wedding ring. Less than two months later, in February 1896, American physicians were starting to test them on patients. One of the earliest patients was a young boy named Eddie McCarthy, who had his broken forearm X-rayed. A New Yorker by the name of Tolson Cunningham had a bullet removed from his leg after its position was determined by a 45-minute X-ray exposure.

REAL-WORLD PHYSICS: BIO
Computerized axial tomography

Standard X-rays can be difficult to interpret, however, since they cast shadows of all the body materials they pass through onto a single sheet of film. It is somewhat like placing several transparencies on top of one another and trying to decipher their individual contents. With the advent of high-speed computers, a new type of X-ray image is now possible. In a **computerized axial tomography scan** (CAT scan), thin beams of X-rays are directed through the body from a variety of directions. The intensity of the transmitted beam is detected for each direction, and the results are sent to a computer for processing. The result is an image that shows the physician a "cross-sectional slice" through the body. In this way, each part of the body can be viewed individually and with clarity. If a series of such slices are stacked together in the computer, they can give a three-dimensional view of the body's interior.

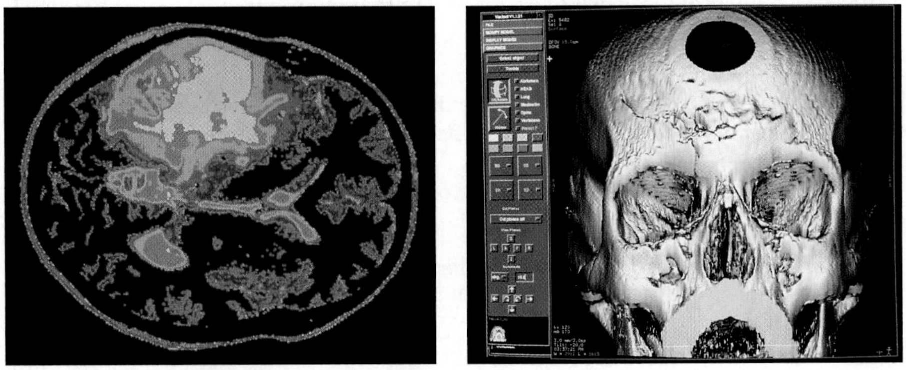

▶ The false-color CAT scan at left represents a horizontal section through the brain, revealing a large benign tumor (the white and orange area at top). A series of CAT scans can also be combined to create remarkable three-dimensional images such as the one at right.

Lasers

The production of light by humans advanced significantly when flames were replaced by the lightbulb. An even greater advancement occurred in 1960, however, when the first laser was developed. Lasers produce light that is intense, highly collimated (uni-directional), and pure in its color. Because of these properties, lasers are used in a multitude of technological applications, ranging from supermarket scanners to CDs, from laser pointers to eye surgery. In fact, lasers are now almost as common in everyday life as the lightbulb.

To understand just what a laser is and what makes it so special, we start with its name. The word **laser** is an acronym for **l**ight **a**mplification by the **s**timulated **e**mission of **r**adiation. As we shall see, the properties of stimulated emission lead directly to the amplification of light.

To begin, consider two energy levels in an atom, and suppose an electron occupies the higher of the two levels. If the electron is left alone, it will eventually drop to the lower level in a time that is typically about 10^{-8} s, giving off a photon in the process referred to as **spontaneous emission.** The photon given off in this process can propagate in any direction.

In contrast, suppose the electron in the excited state just described is not left alone. For example, a photon with an energy equal to the energy difference between the two energy levels might pass near the electron, which *enhances* the probability that the electron will drop to the lower level. That is, the incident photon can *stimulate* the emission of a second photon by the electron. The photon given off in this process of **stimulated emission** has the same energy as the incident photon, the same phase, and propagates in the same direction, as indicated in **Figure 31–24**. This accounts for the fact that laser light is highly focused and of a single color.

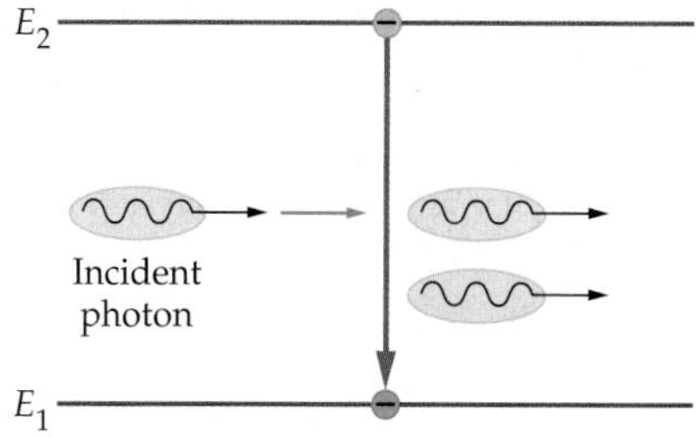

▲ **FIGURE 31–24 Stimulated emission**
A photon with an energy equal to the difference $E_2 - E_1$ can enhance the probability that an electron in the state E_2 will drop to the state E_1 and emit a photon. When a photon stimulates the emission of a second photon in this way, the new photon has the same frequency, direction, and phase as the incident photon. If each of the two photons resulting from this process in turn produces two photons, the total number of photons can increase exponentially, resulting in an intense beam of light.

As for the amplification of light, notice that a single photon entering an excited atom can cause two identical photons to exit the atom. If each of these two photons encounters another excited atom and undergoes the same process, the number of photons increases to four. Continuing in this manner, the photons undergo a sort of "chain reaction" that doubles the number of photons with each generation. It is this property of stimulated emission that results in **light amplification.**

In order for the light amplification process to work, it is necessary that photons continue to encounter atoms with electrons in excited states. Under ordinary conditions this will not be the case, since most electrons are in the lowest possible energy levels. For laser action to occur, atoms must first be prepared in an excited state. Then, before the electrons have a chance to drop to a lower level by way of spontaneous emission, the process of stimulated emission can proceed. This requires what is known as a **population inversion,** in which more electrons are found in the excited state than in a lower state. In addition, the excited state must be one that lasts for a relatively long time so that photons will continue to encounter excited atoms. A long-lived excited state is referred to as a **metastable state.** So to produce a laser, one needs a metastable excited state with a population inversion.

REAL-WORLD PHYSICS
Helium-neon laser

A specific example of a laser is the **helium-neon laser,** shown schematically in **Figure 31–25 (a)**, in which the neon atoms produce the laser light. The appropriate energy level diagrams for neon are shown in **Figure 31–25 (b)**. The excited state E_3 is metastable—electrons promoted to that level stay in the level for a relatively long time. Electrons are excited to this level by an electrical power supply connected to the tube containing the helium-neon mixture. The power supply causes electrons to move through the tube, colliding with neon atoms and exciting them. Similarly, excited helium atoms colliding with the neon atoms also cause excitations to the E_3 level. Since this level is metastable, the excitation processes can cause a population inversion, setting the stage for laser action. Stimulated emission then occurs, allowing the electrons to drop to the lower level, E_2. The electrons subsequently proceed through various intermediate steps to the ground state. It is the emission of light between the levels E_3 and E_2 that results in laser light. Since the difference in energy between these levels is 1.96 eV, the light coming out of the laser is red, with a wavelength of 633 nm.

▶ FIGURE 31–25 The helium-neon laser
(a) A schematic representation of the basic features of a helium-neon laser.
(b) The relevant energy levels in helium and neon that result in the laser action.

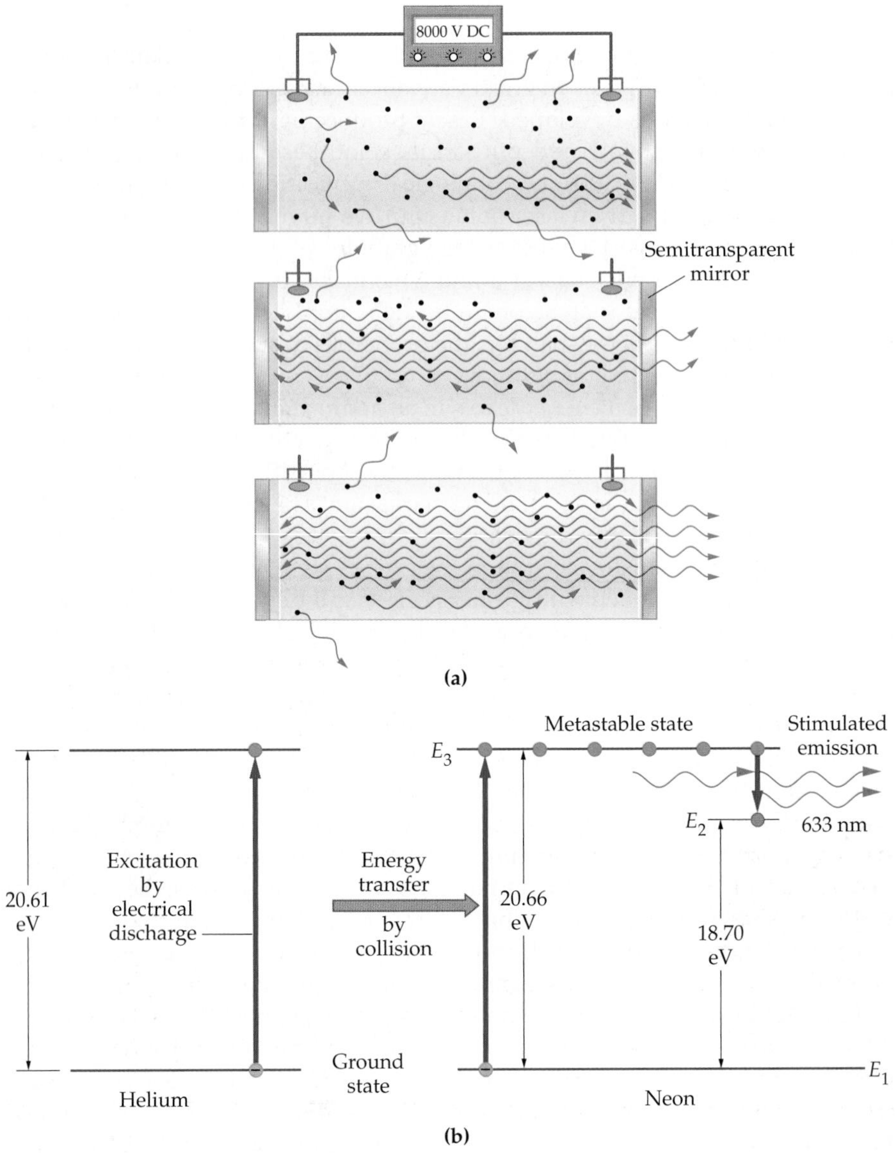

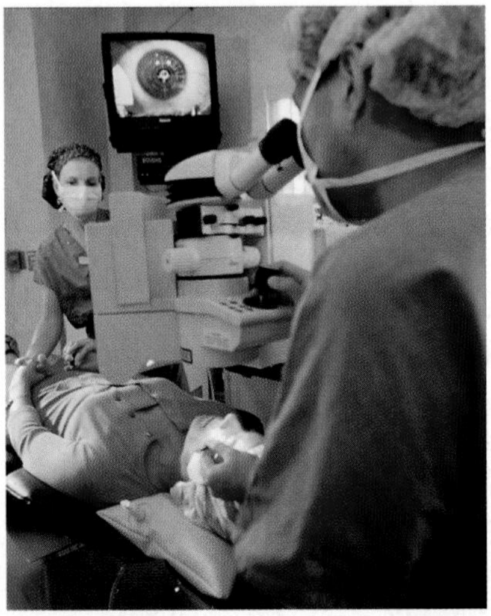

▲ A doctor uses a microscope and video monitor to ensure precise cutting of the corneal flap in the LASIK procedure. A UV laser will then remove some of the material under the flap, flattening the cornea and correcting the patient's nearsightedness.

REAL-WORLD PHYSICS: BIO
Laser eye surgery

To enhance the output of a laser, the light is reflected back and forth between mirrors. In fact, this reflection produces a resonant condition in the light, much like standing waves in an organ pipe. A schematic of a laser in operation is shown in Figure 31–25 (a).

A recent medical application of lasers involves several types of *laser eye surgery*. In these techniques, a laser that emits high-energy photons in the UV range (typically at wavelengths of 193 nm) is used to reshape the cornea and correct nearsightedness. For example, in LASIK (**la**ser in **si**tu **k**eratomileusis) eye surgery the procedure begins with a small mechanical shaver known as a microkeratome cutting a flap in the cornea, leaving a portion of the cornea uncut to serve as a hinge. After the mechanical cut is made, the corneal flap is folded back, exposing the middle portion of the cornea as shown in **Figure 31–26 (a)**. Next, an *excimer laser* sends pulses of UV light onto the cornea, each pulse vaporizing a small layer of corneal material (0.1 to 0.5 μm in thickness) with no heating. This process continues until the cornea is flattened just enough to correct the nearsightedness, after which the corneal flap is put back into place.

Photorefractive keratectomy (PRK) is similar to LASIK eye surgery, except that material is removed directly from the surface of the cornea, without the use of a corneal flap, as shown in **Figure 31–26 (b)**. To correct nearsightedness, the laser beam is directed onto the central portion of the cornea (left), resulting in a flattening of the cornea. To correct farsightedness, it is necessary to increase the curvature

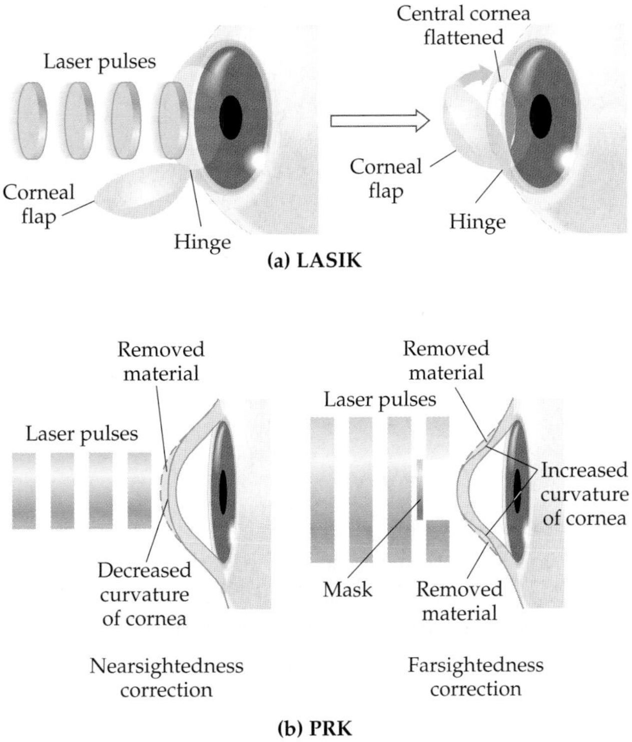

◀ **FIGURE 31–26 Laser vision correction**
(a) In LASIK eye surgery, a flap of the cornea is cut and folded back. Next, an ultraviolet excimer laser is used to vaporize some of the underlying corneal material. When the flap is replaced, the cornea is flatter than it was, correcting the patient's nearsightedness. **(b)** In the PRK procedure, the laser removes material directly from the corneal surface. If the cornea is too curved, producing nearsightedness, the laser beam is directed at its center and the cornea is flattened. If the cornea is too flat, producing farsightedness, the central region is masked and material is removed from the periphery, increasing the curvature.

of the cornea. This is accomplished by masking the central portion of the cornea so that the laser removes only peripheral portions of the cornea (right). In both cases, it is necessary to keep the beam focused at the desired location on the eye. This is difficult, because the eye routinely moves by small amounts roughly every 15 ms. In the most sophisticated application of PRK, these eye movements can be tracked and the aiming of the laser beam corrected accordingly.

REAL-WORLD PHYSICS: BIO
Photodynamic therapy

Another medical application of lasers is known as *photodynamic therapy,* or PDT. In this type of therapy, light-sensitive chemicals (such as porphyrins) are injected into the bloodstream and are taken up by cells throughout the body. These chemicals are found to remain in cancerous cells for greater periods of time than in normal cells. Thus, after an appropriate time interval, the light-sensitive chemicals are preferentially concentrated in cancerous cells. If a laser beam with the precise wavelength absorbed by the light-sensitive chemicals illuminates the cancer cells, the resulting chemical reactions kill the cancer cells without damaging the adjacent normal cells. The laser beam can be directed to the desired location using a flexible fiber-optic cable in conjunction with a bronchoscope to treat lung cancer or with an endoscope to treat esophageal cancer.

PDT is also used to treat certain types of age-related macular degeneration (AMD). For most people, macular degeneration results when abnormal blood vessels behind the retina leak fluid and blood into the central region of the retina, or macula, causing it to degenerate. In this case, light-sensitive chemicals are preferentially taken up by the abnormal blood vessels; hence, a laser beam of the proper wavelength can destroy these blood vessels without damaging the normal structures in the retina.

▲ A hologram creates a three-dimensional image in empty space. The image can be viewed from different angles to reveal different parts of the original subject.

REAL-WORLD PHYSICS
Holography

Finally, lasers can also be used to take three-dimensional photographs known as **holograms.** A typical setup for taking a hologram is shown in **Figure 31–27**. Notice that the hologram is produced with no focusing lenses, in contrast with normal photography. The basic procedure in holography begins with the splitting of a laser beam into two separate beams. One beam, the reference beam, is directed onto the photographic film. The second beam, called the object beam, is directed onto the object to be recorded in the hologram. The object beam reflects from various parts of the object and then combines with the reference beam on the film. Because the laser light is coherent, and because the object and reference beams

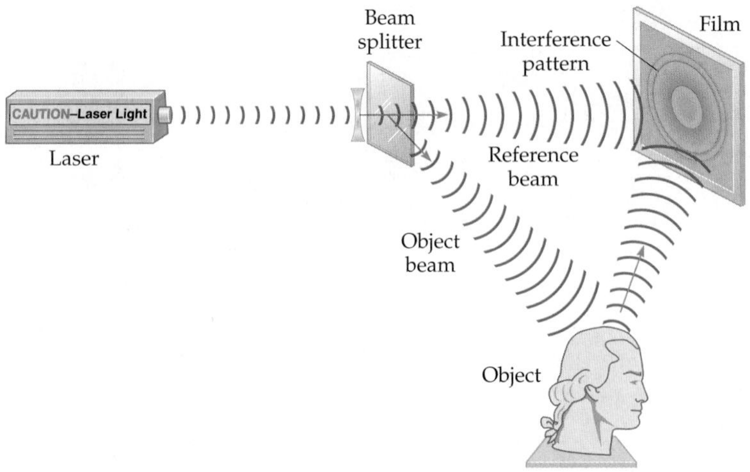

▶ **FIGURE 31–27 Holography**

To create a hologram, laser light is split into two beams. One, the reference beam, is directed onto the photographic film. The other is reflected from the surface of an object onto the film, where it combines with the reference beam to create an interference pattern. When this pattern is illuminated with laser light of the same wavelength, a three-dimensional image of the original object is produced.

travel different distances, the combined light results in an interference pattern. In fact, if you look at a hologram in normal light it simply looks like a confusing mass of swirls and lines. When a laser beam illuminates the hologram, however, the interference pattern causes the laser light to propagate away from the hologram in exactly the same way as the light that originally produced the interference pattern. Thus, a person viewing the hologram sees precisely the same patterns of light that would have been observed when the hologram was recorded.

Holograms give a true three-dimensional image. When you view a hologram, you can move your vantage point to see different parts of the scene that is recorded. In particular, viewing the hologram from one angle may obscure an object in the background, but by moving your head, you can look around the foreground objects to get a clear view of the obscured object. In addition, you have to adjust your focus as you shift your gaze from foreground to background objects, just as in the real world. Finally, if you cut a hologram into pieces, each piece still shows the entire scene! This is analogous to your ability to see everything in your front yard through a small window just as you can through a large window—both show the entire scene.

The holograms on your credit cards are referred to as *rainbow holograms*, because they are designed to be viewed with white light containing all the colors of the rainbow. Although these holograms give an impression of three-dimensionality, they do not compare in quality to a hologram viewed with a laser.

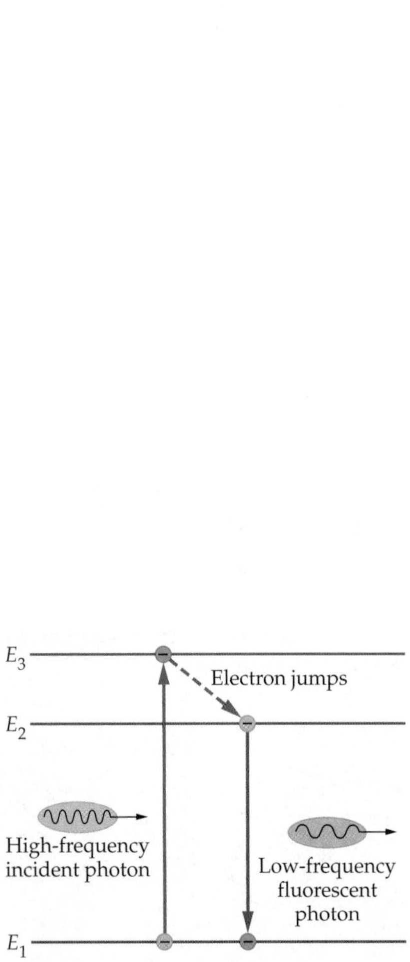

▲ **FIGURE 31–28 The mechanism of fluorescence**

In fluorescence, a high-frequency photon raises an electron to an excited state. When the electron drops back to the ground state, it may do so by way of various intermediate states. The jumps between intermediate states produce photons of lower frequency, which are observed as the phenomenon of fluorescence.

Fluorescence and Phosphorescence

In the Insight to Active Example 31–1 it was noted that an electron in an excited state can emit photons of various energies as it falls to the ground state. This type of behavior is at the heart of fluorescence and phosphorescence.

Consider the energy levels shown in **Figure 31–28**. If an atom with these energy levels absorbs a photon of energy $E_3 - E_1$, it can excite an electron from state E_1 to state E_3. In some atoms, the most likely way for the electron to return to the ground state is by first jumping to level E_2 and then jumping to level E_1. The photons emitted in these jumps have less energy than the photon that caused the excitation in the first place. Hence, in a system like this, an atom is illuminated with a photon of one frequency, and it subsequently emits photons of lower energy and lower frequency. The emission of light of lower frequency after illumination by a higher frequency is referred to as **fluorescence.** In essence, fluorescence can be thought of as a conversion process, in which photons of high frequency are converted to photons of lower frequency.

REAL-WORLD PHYSICS
Fluorescent lightbulbs

Perhaps the most common example of fluorescence is the *fluorescent lightbulb*. This device uses fluorescence to convert high-frequency ultraviolet light to lower-frequency visible light. In particular, the tube of a fluorescent light contains mercury vapor. When electricity is applied to one of these tubes, a filament is

▲ The fingerprints on this mug become clearly visible when treated with fluorescent dye and illuminated by ultraviolet light.

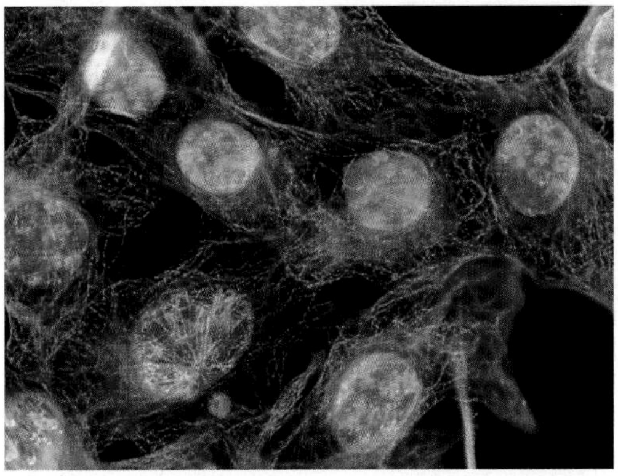

▲ In the technique known as immunofluorescence, an antibody molecule is linked to a fluorescent dye molecule, making it possible to visualize cellular structures and components that are otherwise largely invisible. This photograph utilized antibodies that bind to a protein found in intermediate filaments. Such filaments (seen here as lacy fibers resembling a spiderweb) are a component of the cytoskeleton, the cellular scaffolding that enables cells to maintain or change their shape and move materials about.

heated, producing electrons. These electrons are accelerated by an applied voltage. The electrons strike mercury atoms in the tube, exciting them, and they give off ultraviolet light as they decay to their ground state. This process is not particularly useful in itself, since the ultraviolet light is invisible to us. However, the inside of the tube is coated with a phosphor that absorbs the ultraviolet light and then emits a lower-frequency light that is visible. Thus, several different physical processes must take place before a fluorescent lightbulb produces visible light.

Fluorescence finds many other less familiar applications as well. In forensics, the analysis of a crime scene is enhanced by the fact that human bones and teeth are fluorescent. Thus, illuminating a crime scene with ultraviolet light can make items of interest stand out for easy identification. In addition, the use of a fluorescent dye can make fingerprints visible with great clarity.

REAL-WORLD PHYSICS: BIO
Applications of fluorescence in forensics

Many creatures produce fluorescence in their bodies, as well. For example, several types of coral glow brightly when illuminated with ultraviolet light. It is also well known that scorpions are strongly fluorescent, giving off a distinctive green light. In fact, it is often possible to discern a greenish cast when viewing a scorpion in sunlight. At night in the desert, scorpions stand out with a bright green glow when a person illuminates the area with a portable ultraviolet light. This aids researchers who would like to find certain scorpions for study, and campers who are just as interested in avoiding scorpions altogether.

REAL-WORLD PHYSICS: BIO
Detecting scorpions at night

The green fluorescence produced by the jellyfish *Aequorea victoria* finds many uses in biological experiments. The gene that produces the green fluorescent protein (GFP) can serve as a marker to identify whether an organism has incorporated a new segment of DNA into its genome. For example, bacterial colonies that incorporate the GFP gene can be screened by eye simply by viewing the

◀ A variety of creatures, including the scorpion at left, are naturally fluorescent when illuminated with ultraviolet light. So too are many minerals, such as the one at right.

REAL-WORLD PHYSICS: BIO
The GFP Bunny

colony under an ultraviolet light. Recently, GFP has been inserted into the genome of a white rabbit, giving rise to the "GFP bunny." The bunny appears normal in white light, but when viewed under light with a wavelength of 392 nm, it glows with a bright green light at 509 nm (see page 1034). The fluorescence spectrum of GFP is shown in **Figure 31–29**.

Phosphorescence is similar to fluorescence, except that phosphorescent materials continue to give off a secondary glow long after the initial illumination that excited the atoms. In fact, phosphorescence may persist for periods of time ranging from a few seconds to several hours, as on the hands of a watch that glow in the dark.

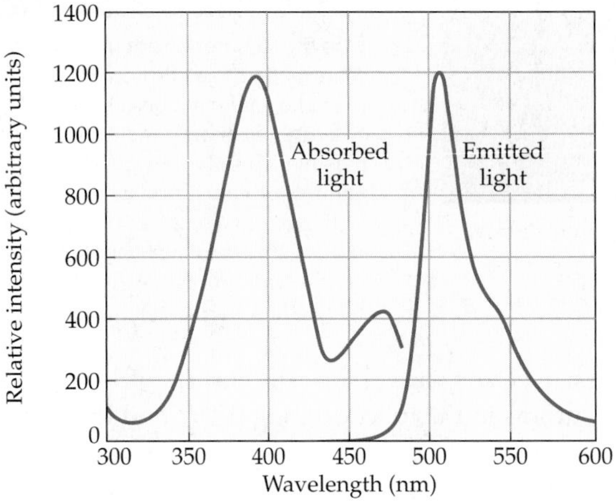

▶ **FIGURE 31–29 The fluorescence spectrum of GFP**
The green fluorescent protein (GFP) strongly absorbs light with a wavelength of about 400 nm (violet). It reemits green light with a wavelength of 509 nm.

Chapter Summary

31–1 Early Models of the Atom

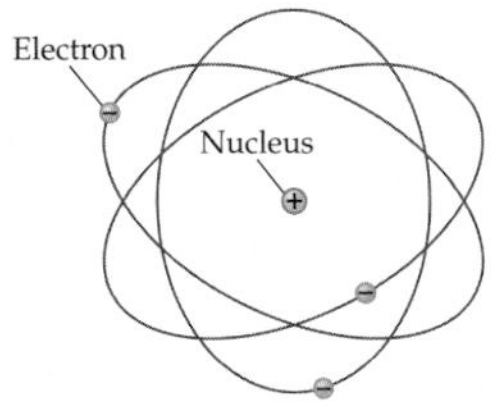

Atoms are the smallest unit of a given element. If an atom is broken down into smaller pieces, it loses the properties that characterized the element.

The Thomson Model: Plum Pudding
In Thomson's model, an atom is imagined to be like a positively charged pudding with negatively charged electrons scattered throughout.

The Rutherford Model: A Miniature Solar System
Rutherford discovered that an atom is somewhat like an atomic-scale solar system: mostly empty space, with most of its mass concentrated in the nucleus. The electrons were thought to orbit the nucleus.

31–2 The Spectrum of Atomic Hydrogen

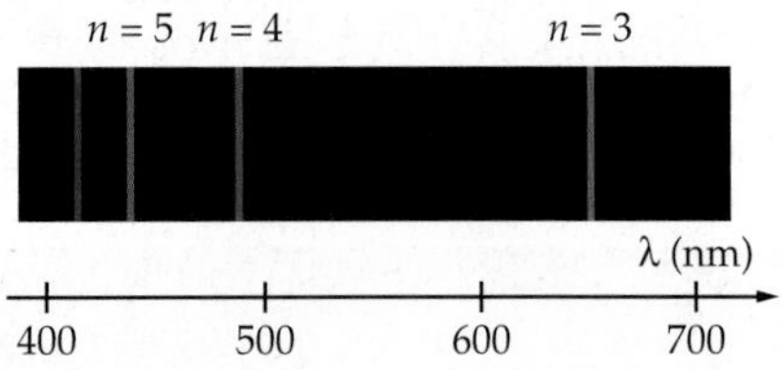

Excited atoms of hydrogen in a low-pressure gas give off light of specific wavelengths. This is referred to as the spectrum of hydrogen.

Line Spectra
The spectrum of hydrogen is a series of bright lines of well-defined wavelengths.

Series
Hydrogen's line spectrum is formed by a series of lines that are grouped together. The wavelengths of these series are given by the expression

$$\frac{1}{\lambda} = R\left(\frac{1}{n'^2} - \frac{1}{n^2}\right) \qquad n' = 1, 2, 3, \ldots$$

$$n = n' + 1, n' + 2, n' + 3, \ldots \qquad \textbf{31–2}$$

Each line in a given series corresponds to a different value of n. The different series correspond to different values of n'. For example, $n' = 1$ is the Lyman series, $n' = 2$ is the Balmer series, and $n' = 3$ is the Paschen series.

31–3 Bohr's Model of the Hydrogen Atom

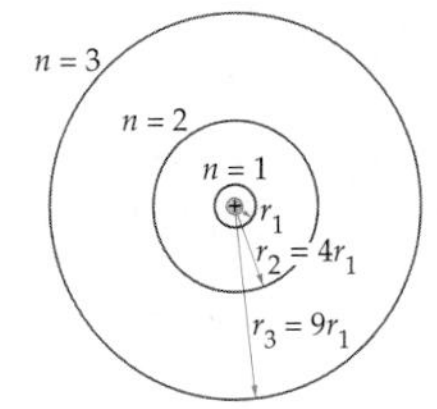

Bohr's model of hydrogen is basically a solar-system model, with the electron orbiting the nucleus. In Bohr's model, however, only certain orbits are allowed.

Assumptions of the Bohr Model

The Bohr model assumes the following: (i) Electrons move in circular orbits about the nucleus; (ii) allowed orbits must have an angular momentum equal to $L_n = nh/2\pi$, where $n = 1, 2, 3, \ldots$; (iii) electrons in allowed orbits do not give off electromagnetic radiation; and (iv) radiation is emitted only when electrons jump from one orbit to another.

Bohr Orbits

The radii of allowed orbits in the Bohr model are given by

$$r_n = \left(\frac{h^2}{4\pi^2 mkZe^2}\right)n^2 = (5.29 \times 10^{-11}\text{ m})n^2 \qquad n = 1, 2, 3, \ldots \qquad \textbf{31–7}$$

The Energy of a Bohr Orbit

The energy of an allowed Bohr orbit is

$$E_n = -(13.6\text{ eV})\frac{Z^2}{n^2} \qquad n = 1, 2, 3, \ldots \qquad \textbf{31–9}$$

These expressions correspond to hydrogen when $Z = 1$.

31–4 de Broglie Waves and the Bohr Model

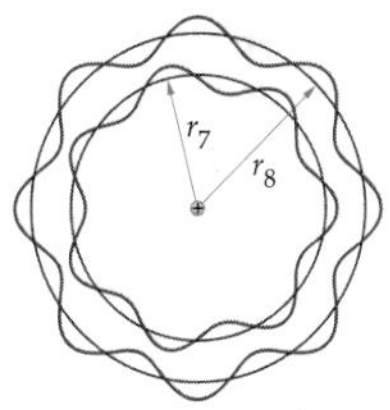

De Broglie was able to show that the allowed orbits of the Bohr model correspond to standing matter waves of the electrons. In particular, an allowed orbit in Bohr's model has a circumference equal to an integer times the wavelength of the electron in that orbit.

31–5 The Quantum Mechanical Hydrogen Atom

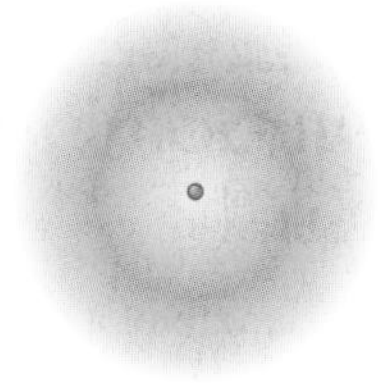

The correct description of the hydrogen atom is derived from Schrödinger's equation. It agrees in many key ways with the Bohr model but has significant differences as well.

Quantum Numbers

The quantum mechanical hydrogen atom is described by four quantum numbers. They are as follows: (i) the principal quantum number, n, which is analogous to n in the Bohr model; (ii) the orbital angular momentum quantum number, ℓ, which takes on the values $\ell = 0, 1, 2, \ldots, (n-1)$; the orbital angular momentum has a magnitude given by $L = \sqrt{\ell(\ell+1)}(h/2\pi)$; (iii) the magnetic quantum number, m_ℓ, the allowed values of m_ℓ are $m_\ell = -\ell, -\ell+1, -\ell+2, \ldots, -1, 0, 1, \ldots, \ell-2, \ell-1, \ell$, and the z component of the orbital angular momentum is $L_z = m_\ell(h/2\pi)$; and (iv) the electron spin quantum number, m_s, which can have the values $m_s = -\frac{1}{2}, \frac{1}{2}$.

Electron Probability Clouds: Three-Dimensional Standing Waves

In the quantum mechanical hydrogen atom, the electron does not orbit at a precise distance from the nucleus. Instead, the electron distribution is represented by a probability cloud, where the densest regions of the cloud correspond to regions of highest probability.

31–6 Multielectron Atoms and the Periodic Table

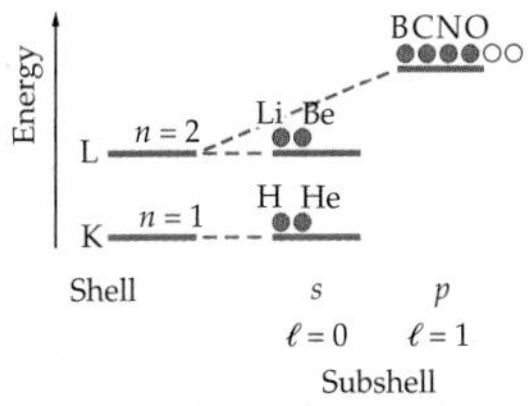

As electrons are added to atoms, the properties of the atoms change in a regular and predictable way.

Multielectron Atoms

Energy levels in a multielectron atom depend on n and ℓ. The energy increases with increasing n for fixed ℓ, and with increasing ℓ for fixed n.

Shells and Subshells

Electrons with the same value of n are said to be in the same shell. Electrons in a given shell with the same value of ℓ are said to be in the same subshell.

The Pauli Exclusion Principle
The Pauli exclusion principle states that only a single electron may have a particular set of quantum numbers. This means that it is not possible for all the electrons in a multielectron atom to occupy the lowest energy level.

Electronic Configurations
The arrangement of electrons is indicated by the electronic configuration. For example, the configuration $1s^2$ indicates that 2 electrons ($1s^2$) are in the $n = 1\ (1s^2)$, $\ell = 0\ (1s^2)$ state.

The Periodic Table
As electrons fill subshells of progressively higher energy, they produce the elements of the periodic table. Atoms with the same configuration of outermost electrons generally have similar chemical properties.

31–7 Atomic Radiation

Atoms can give off radiation ranging from X-rays to visible light to infrared rays.

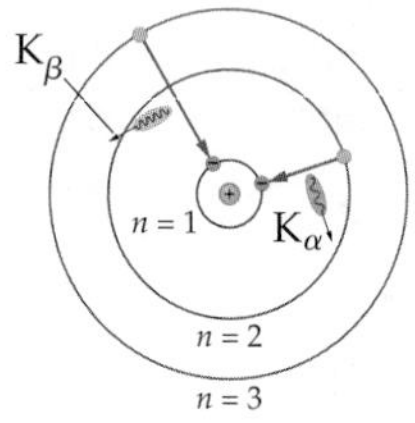

X-rays
X-rays characteristic of a particular element are given off when an electron in an inner shell is knocked out of the atom, and an electron from an outer shell drops down to take its place.

Lasers
A laser is a device that produces light amplification by the stimulated emission of radiation.

Fluorescence and Phosphorescence
When an electron in an atom is excited to a high energy level, it may return to the ground state through a series of lower-energy jumps. These jumps give off radiation of longer wavelength than the radiation that caused the original excitation.

Problem-Solving Summary

Type of Problem	Relevant Physical Concepts	Related Examples
Find the wavelength of a spectral line in hydrogen.	All the spectral lines of hydrogen are described by Equation 31–2. Note that n must be greater than n', and that $n' = 1$ gives the Lyman series, $n' = 2$ gives the Balmer series, $n' = 3$ gives the Paschen series, and so on.	Example 31–1, Exercise 31–1
Find the radius and energy of a Bohr orbit in hydrogen. Also determine the speed of the electron.	The radius of a Bohr orbit varies with n^2 as follows: $r_n = r_1 n^2 = (5.29 \times 10^{-11}\ \text{m})n^2$. The speed of the electron varies as $1/n$, as shown in Equation 31–6. The energy associated with an orbit is the kinetic energy of the electron plus the potential energy of the system, $U = -ke^2/r$.	Example 31–2 Active Example 31–2
Determine the wavelength (or frequency) of a photon that is emitted or absorbed when an electron jumps from one Bohr orbit to another.	When an electron jumps from an initial state n_i to a final state n_f, the wavelength of the associated photon is given by $\frac{1}{\lambda} = \left(\frac{2\pi^2 mk^2e^4}{h^3c}\right)\left(\frac{1}{n_f^2} - \frac{1}{n_i^2}\right) = (1.097 \times 10^7\ \text{m}^{-1})\left(\frac{1}{n_f^2} - \frac{1}{n_i^2}\right).$ If $n_i > n_f$, the photon is emitted; if $n_i < n_f$, the photon is absorbed. The photon's wavelength and frequency are related by the expression $\lambda f = c$.	Active Examples 31–1, 31–2
Calculate the wavelength of an electron in a Bohr orbit.	An integral number of wavelengths fits around the circumference of a Bohr orbit, starting with one wavelength for the ground state.	Example 31–3
Estimate the energy of a K-shell electron and the wavelength of a K_α X-ray for an element with atomic number Z.	The nuclear charge of $+Ze$ is partially screened by a K-shell electron, giving an effective nuclear charge of $+(Z-1)e$. Therefore, we use $Z - 1$ in Equation 31–9, which results in Equation 31–14.	Example 31–4 Active Example 31–3

Conceptual Questions

(Answers to odd-numbered Conceptual Questions can be found in the back of the book.)

1. Give a reason why the Thomson plum-pudding model does not agree with experimental observations.
2. Give a reason why the Rutherford solar-system model does not agree with experimental observations.
3. Cite one example of how the Bohr model disagrees with the quantum mechanical model of the hydrogen atom.
4. What observation led Rutherford to propose that atoms have a small nucleus containing most of the atom's mass?
5. Do you expect the light given off by **(a)** a neon sign or **(b)** an incandescent lightbulb to be continuous in distribution or in the form of a line spectrum? Explain.
6. In principle, how many spectral lines are there in any given series of hydrogen? Explain.
7. Is there an upper limit to the radius of an allowed Bohr orbit? Explain.
8. **(a)** Is there an upper limit to the wavelength of lines in the spectrum of hydrogen? Explain. **(b)** Is there a lower limit? Explain.
9. The principal quantum number, n, can increase without limit in the hydrogen atom. Does this mean that the energy of the hydrogen atom also can increase without limit? Explain.
10. For each of the following configurations of outermost electrons, state whether the configuration is allowed by the rules of quantum mechanics. If the configuration is not allowed, give the rule or rules that are violated. **(a)** $2d^1$, **(b)** $1p^7$, **(c)** $3p^5$, **(d)** $4g^6$.
11. **(a)** In the quantum mechanical model of the hydrogen atom, there is one value of n for which the angular momentum of the electron must be zero. What is this value of n? **(b)** Can the angular momentum of the electron be zero in states with other values of n? Explain.
12. Would you expect characteristic X-rays to be emitted by **(a)** helium atoms or **(b)** lithium atoms in their ground state? Explain.
13. The elements fluorine, chlorine, and bromine are found to exhibit similar chemical properties. Explain.

Conceptual Exercises

(Answers to odd-numbered Conceptual Exercises can be found in the back of the book.)

1. Consider the following three transitions in a hydrogen atom: (A) $n_i = 5, n_f = 2$; (B) $n_i = 7, n_f = 2$; (C) $n_i = 7, n_f = 6$. Rank the transitions in order of increasing **(a)** wavelength and **(b)** frequency of the emitted photon. Indicate ties where appropriate.
2. If the mass of the electron were to be doubled, would the ionization energy of hydrogen increase, decrease, or stay the same? Explain.
3. Consider the Bohr model as applied to the following three atoms: (A) neutral hydrogen in the state $n = 2$; (B) singly ionized helium in the state $n = 1$; (C) doubly ionized lithium in the state $n = 3$. Rank these three atoms in order of increasing Bohr radius. Indicate ties where appropriate.
4. Consider the Bohr model as applied to the following three atoms: (A) neutral hydrogen in the state $n = 3$; (B) singly ionized helium in the state $n = 2$; (C) doubly ionized lithium in the state $n = 1$. Rank these three atoms in order of increasing energy. Indicate ties where appropriate.
5. Is the de Broglie wavelength of an electron in the $n = 2$ Bohr orbit of hydrogen greater than, less than, or equal to the de Broglie wavelength in the $n = 1$ Bohr orbit? Explain.
6. In the Bohr model, the potential energy of a hydrogen atom in the nth orbit has a value we will call U_n. What is the potential energy of a hydrogen atom when the electron is in the $(n + 1)$th Bohr orbit? Give your answer in terms of U_n and n.
7. An electron in the $n = 1$ Bohr orbit has the kinetic energy K_1. In terms of K_1, what is the kinetic energy of an electron in the $n = 2$ Bohr orbit?
8. Suppose an electron is in the ground state of hydrogen. **(a)** What is the highest-energy photon this system can absorb without dissociating the electron from the proton? Explain. **(b)** What is the lowest-energy photon this system can absorb? Explain.
9. How many electrons can occupy **(a)** the $2p$ subshell and **(b)** the $3p$ subshell?
10. **(a)** How many electrons can occupy the $3d$ subshell? **(b)** How many electrons can occupy the $n = 2$ shell?
11. The electronic configuration of a given atom is $1s^2 2s^2 2p^6 3s^2 3p^1$. How many electrons are in this atom?
12. The electronic configuration of a particular carbon atom is $1s^2 2s^2 2p^1 3s^1$. Is this atom in its ground state or in an excited state? Explain.
13. The electronic configuration of a particular potassium atom is $1s^2 2s^2 2p^6 3s^2 3p^6 3d^1$. Is this atom in its ground state or in an excited state? Explain.
14. In an X-ray tube, do you expect the wavelength of the characteristic X-rays to increase, decrease, or stay the same if the energy of the electrons striking the target is increased? Explain.
15. Do you expect the ionization energy of sodium (Na) to be greater than, less than, or equal to the ionization energy of lithium (Li)? Explain.
16. Is the wavelength of the radiation that excites a fluorescent material greater than, less than, or equal to the wavelength of the radiation the material emits? Explain.

Problems

Note: **IP** *denotes an integrated conceptual/quantitative problem.* **BIO** *identifies problems of biological or medical interest. Red bullets (•, ••, •••) are used to indicate the level of difficulty of each problem.*

Section 31–1 Early Models of the Atom

1. • The electron in a hydrogen atom is typically found at a distance of about 5.3×10^{-11} m from the nucleus, which has a diameter of about 1.0×10^{-15} m. If you assume the hydrogen atom to be a sphere of radius 5.3×10^{-11} m, what fraction of its volume is occupied by the nucleus?
2. • Referring to Problem 1, suppose the nucleus of the hydrogen atom were enlarged to the size of a baseball (diameter = 7.3 cm).

At what typical distance from the center of the baseball would you expect to find the electron?

3. •• Copper atoms have 29 protons in their nuclei. If the copper nucleus is a sphere with a diameter of 4.8×10^{-15} m, find the work required to bring an alpha particle (charge $= +2e$) from rest at infinity to the "surface" of the nucleus.

4. •• In Rutherford's scattering experiments, alpha particles (charge $= +2e$) were fired at a gold foil. Consider an alpha particle with an initial kinetic energy K heading directly for the nucleus of a gold atom (charge $= +79e$). The alpha particle will come to rest when all its initial kinetic energy has been converted to electrical potential energy. Find the distance of closest approach between the alpha particle and the gold nucleus for the case $K = 3.0$ MeV.

Section 31–2 The Spectrum of Atomic Hydrogen

5. • Find the wavelength of the Balmer series spectral line corresponding to $n = 15$.

6. • What is the smallest value of n for which the wavelength of a Balmer series line is less than 400 nm?

7. • Find the wavelength of the three longest-wavelength lines of the Lyman series.

8. • Find the wavelength of the three longest-wavelength lines of the Paschen series.

9. • Find **(a)** the longest wavelength in the Lyman series and **(b)** the shortest wavelength in the Paschen series.

10. •• In Table 31–1 we see that the Paschen series corresponds to $n' = 3$ in Equation 31–2, and that the Brackett series corresponds to $n' = 4$. **(a)** Show that the ranges of wavelengths of these two series overlap. **(b)** Is there a similar overlap between the Balmer series and the Paschen series? Verify your answer.

Section 31–3 Bohr's Model of the Hydrogen Atom

11. • Find the ratio v/c for an electron in the first excited state $(n = 2)$ of hydrogen.

12. • Find the magnitude of the force exerted on an electron in the ground-state orbit of the Bohr model.

13. • How much energy is required to ionize hydrogen when it is in the $n = 4$ state?

14. • Find the energy of the photon required to excite a hydrogen atom from the $n = 2$ state to the $n = 5$ state.

15. •• A hydrogen atom is in its second excited state, $n = 3$. Using the Bohr model of hydrogen, find **(a)** the linear momentum and **(b)** the angular momentum of the electron in this atom.

16. •• Referring to Problem 15, find **(a)** the kinetic energy of the electron, **(b)** the potential energy of the atom, and **(c)** the total energy of the atom. Give your results in eV.

17. •• Initially, an electron is in the $n = 3$ state of hydrogen. If this electron acquires an additional 1.23 eV of energy, what is the value of n in the final state of the electron?

18. •• Identify the initial and final states if an electron in hydrogen emits a photon with a wavelength of 656 nm.

19. •• **IP** An electron in hydrogen absorbs a photon and jumps to a higher orbit. **(a)** Find the energy the photon must have if the initial state is $n = 3$ and the final state is $n = 5$. **(b)** If the initial state was $n = 5$ and the final state $n = 7$, would the energy of the photon be greater than, less than, or the same as that found in part (a)? Explain. **(c)** Calculate the photon energy for part (b).

20. •• **IP** Consider the following four transitions in a hydrogen atom:

(i) $n_i = 2, n_f = 6$ (ii) $n_i = 2, n_f = 8$
(iii) $n_i = 7, n_f = 8$ (iv) $n_i = 6, n_f = 2$

Find **(a)** the longest- and **(b)** the shortest-wavelength photon that can be emitted or absorbed by these transitions. Give the value of the wavelength in each case. **(c)** For which of these transitions does the atom lose energy? Explain.

21. •• **IP Muonium** Muonium is a hydrogen-like atom in which the electron is replaced with a muon, a fundamental particle with a charge of $-e$ and a mass equal to $207m_e$. (The muon is sometimes referred to loosely as a "heavy electron.") **(a)** What is the Bohr radius of muonium? **(b)** Will the wavelengths in the Balmer series of muonium be greater than, less than, or the same as the wavelengths in the Balmer series of hydrogen? Explain. **(c)** Calculate the longest wavelength of the Balmer series in muonium.

22. •• **IP (a)** Find the radius of the $n = 4$ Bohr orbit of a doubly ionized lithium atom (Li^{2+}, $Z = 3$). **(b)** Is the energy required to raise an electron from the $n = 4$ state to the $n = 5$ state in Li^{2+} greater than, less than, or equal to the energy required to raise an electron in hydrogen from the $n = 4$ state to the $n = 5$ state? Explain. **(c)** Verify your answer to part (b) by calculating the relevant energies.

23. •• Applying the Bohr model to a triply ionized beryllium atom (Be^{3+}, $Z = 4$), find **(a)** the shortest wavelength of the Lyman series for Be^{3+} and **(b)** the ionization energy required to remove the final electron in Be^{3+}.

24. •• **(a)** Calculate the time required for an electron in the $n = 2$ state of hydrogen to complete one orbit about the nucleus. **(b)** The typical "lifetime" of an electron in the $n = 2$ state is roughly 10^{-8} s—after this time the electron is likely to have dropped back to the $n = 1$ state. Estimate the number of orbits an electron completes in the $n = 2$ state before dropping to the ground state.

25. •• **IP** The kinetic energy of an electron in a particular Bohr orbit of hydrogen is 1.35×10^{-19} J. **(a)** Which Bohr orbit does the electron occupy? **(b)** Suppose the electron moves away from the nucleus to the next higher Bohr orbit. Does the kinetic energy of the electron increase, decrease, or stay the same? Explain. **(c)** Calculate the kinetic energy of the electron in the orbit referred to in part (b).

26. •• **IP** The potential energy of a hydrogen atom in a particular Bohr orbit is -1.20×10^{-19} J. **(a)** Which Bohr orbit does the electron occupy in this atom? **(b)** Suppose the electron moves away from the nucleus to the next higher Bohr orbit. Does the potential energy of the atom increase, decrease, or stay the same? Explain. **(c)** Calculate the potential energy of the atom for the orbit referred to in part (b).

27. ••• Consider a head-on collision between two hydrogen atoms, both initially in their ground state and moving with the same speed. Find the minimum speed necessary to leave both atoms in their $n = 2$ state after the collision.

28. ••• A hydrogen atom is in the initial state $n_i = n$, where $n > 1$. **(a)** Find the frequency of the photon that is emitted when the electron jumps to state $n_f = n - 1$. **(b)** Find the frequency of the electron's orbital motion in the state n. **(c)** Compare your results for parts (a) and (b) in the limit of large n.

Section 31–4 de Broglie Waves and the Bohr Model

29. • Find the de Broglie wavelength of an electron in the ground state of the hydrogen atom.

30. •• Find an expression for the de Broglie wavelength of an electron in the nth state of the hydrogen atom.

31. •• What is the radius of the hydrogen-atom Bohr orbit shown in **Figure 31–30**?

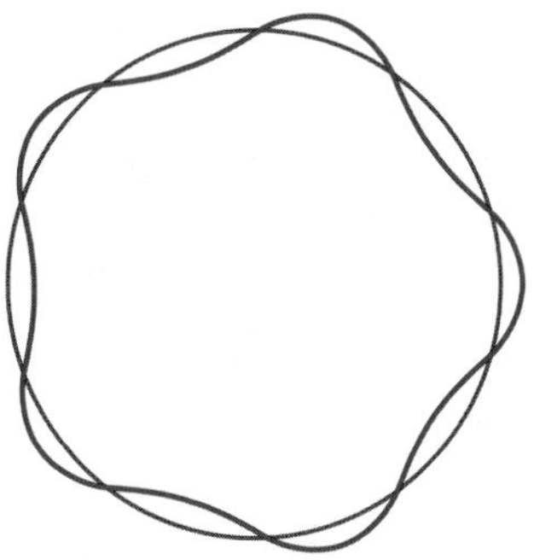

▲ FIGURE 31–30 Problem 31

32. •• (a) Find the kinetic energy (in eV) of an electron whose de Broglie wavelength is equal to 0.5 A, a typical atomic size. (b) Repeat part (a) for an electron with a wavelength equal to 10^{-15} m, a typical nuclear size.

Section 31–5 The Quantum Mechanical Hydrogen Atom

33. • What are the allowed values of ℓ when the principal quantum number is $n = 5$?

34. • How many different values of m_ℓ are possible when the principal quantum number is $n = 4$?

35. • Give the value of the quantum number ℓ, if one exists, for a hydrogen atom whose orbital angular momentum has a magnitude of (a) $\sqrt{6}(h/2\pi)$, (b) $\sqrt{15}(h/2\pi)$, (c) $\sqrt{30}(h/2\pi)$, or (d) $\sqrt{36}(h/2\pi)$.

36. •• IP Hydrogen atom number 1 is known to be in the $4f$ state. (a) What is the energy of this atom? (b) What is the magnitude of this atom's orbital angular momentum? (c) Hydrogen atom number 2 is in the $5d$ state. Is this atom's energy greater than, less than, or the same as that of atom 1? Explain. (d) Is the magnitude of the orbital angular momentum of atom 1 greater than, less than, or the same as that of atom 2? Explain.

37. •• IP A hydrogen atom has an orbital angular momentum with a magnitude of $10\sqrt{57}(h/2\pi)$. (a) Determine the value of the quantum number ℓ for this atom. (b) What is the minimum possible value of this atom's principal quantum number, n? Explain. (c) If $10\sqrt{57}(h/2\pi)$ is the *maximum* orbital angular momentum this atom can have, what is its energy?

38. •• IP The electron in a hydrogen atom with an energy of -0.544 eV is in a subshell with 18 states. (a) What is the principal quantum number, n, for this atom? (b) What is the maximum possible orbital angular momentum this atom can have? (c) Is the number of states in the subshell with the next lowest value of ℓ equal to 16, 14, or 12? Explain.

39. •• IP Consider two different states of a hydrogen atom. In state I the maximum value of the magnetic quantum number is $m_\ell = 3$; in state II the corresponding maximum value is $m_\ell = 2$. Let L_{I} and L_{II} represent the magnitudes of the orbital angular momentum of an electron in states I and II, respectively. (a) Is L_{I} greater than, less than, or equal to L_{II}? Explain. (b) Calculate the ratio $L_{\mathrm{I}}/L_{\mathrm{II}}$.

Section 31–6 Multielectron Atoms and the Periodic Table

40. • Give the electronic configuration for the ground state of carbon.

41. • List the values of the four quantum numbers (n, ℓ, m_ℓ, m_s) for each of the electrons in the ground state of neon.

42. • Give the electronic configuration for the ground state of nitrogen.

43. • Give a list of all possible sets of the four quantum numbers (n, ℓ, m_ℓ, m_s) for electrons in the $3s$ subshell.

44. • Give a list of all possible sets of the four quantum numbers (n, ℓ, m_ℓ, m_s) for electrons in the $3p$ subshell.

45. •• List the values of the four quantum numbers (n, ℓ, m_ℓ, m_s) for each of the electrons in the ground state of magnesium.

46. •• The configuration of the outer electrons in Ni is $3d^8 4s^2$. Write out the complete electronic configuration for Ni.

47. •• Determine the number of different sets of quantum numbers possible for each of the following shells; (a) $n = 2$, (b) $n = 3$, (c) $n = 4$.

48. ••• Generalize the results of Problem 47 and show that the number of different sets of quantum numbers for the nth shell is $2n^2$.

49. •• Suppose that the $5d$ subshell is filled in a certain atom. Write out the 10 sets of four quantum numbers (n, ℓ, m_ℓ, m_s) for the electrons in this subshell.

Section 31–7 Atomic Radiation

50. • Using the Bohr model, estimate the wavelength of the K_α X-ray in nickel $(Z = 28)$.

51. • Using the Bohr model, estimate the energy of a K_α X-ray emitted by lead $(Z = 82)$.

52. •• The K-shell ionization energy of iron is 8500 eV, and its L-shell ionization energy is 2125 eV. What is the wavelength of K_α X-rays emitted by iron?

53. •• An electron drops from the L shell to the K shell and gives off an X-ray with a wavelength of 0.0205 nm. What is the atomic number of this atom?

54. •• Consider an X-ray tube that uses platinum $(Z = 78)$ as its target. (a) Use the Bohr model to estimate the minimum kinetic energy electrons must have in order for K_α X-rays to just appear in the X-ray spectrum of the tube. (b) Assuming the electrons are accelerated from rest through a voltage V, estimate the minimum voltage necessary to produce the K_α X-rays.

55. •• BIO **Photorefractive Keratectomy** A person's vision may be improved significantly by having the cornea reshaped with a laser beam, in a procedure known as photorefractive keratectomy. The excimer laser used in these treatments produces ultraviolet light with a wavelength of 193 nm. (a) What is the difference in energy between the two levels that participate in stimulated emission in the excimer laser? (b) How many photons from this laser are required to deliver an energy of 1.58×10^{-13} J to the cornea?

56. •• IP BIO **Welding a Detached Retina** A detached retina can be "welded" back into place using a laser beam that emits photons with a wavelength of 514 nm. (a) If such a laser emits 1.36×10^{19} photons per second, what is its power output in watts? (b) Suppose a second laser emits the same number of photons per second. If the wavelength of photons from the second laser is 414 nm, is its power output greater than, less than, or the same as the power found in part (a)? Explain. (c) Calculate the power output of the second laser.

General Problems

57. • Find the minimum frequency a photon must have if it is to ionize the ground state of the hydrogen atom.

58. •• It was pointed out in Section 31–3 that intermolecular collisions at room temperature do not have enough energy to cause an excitation in hydrogen from the $n = 1$ state to the $n = 2$ state. Given that the average kinetic energy of a hydrogen atom in a high-temperature gas is $\frac{3}{2}kT$ (where k is Boltzmann's constant), find the minimum temperature required for atoms to have enough thermal energy to excite electrons from the ground state to the $n = 2$ state.

59. •• IP Referring to Problem 60, find (a) the energy of the emitted photon and (b) the kinetic energy of the hydrogen atom after the photon is emitted. (c) Do you expect the sum of the energies in parts (a) and (b) to be greater than, less than, or the same as

the difference in energy between the $n = 4$ and $n = 2$ states of hydrogen? Explain.

60. •• The electron in a hydrogen atom makes a transition from the $n = 4$ state to the $n = 2$ state, as indicated in **Figure 31–31**. **(a)** Determine the linear momentum of the photon emitted as a result of this transition. **(b)** Using your result to part (a), find the recoil speed of the hydrogen atom, assuming it was at rest before the photon was emitted.

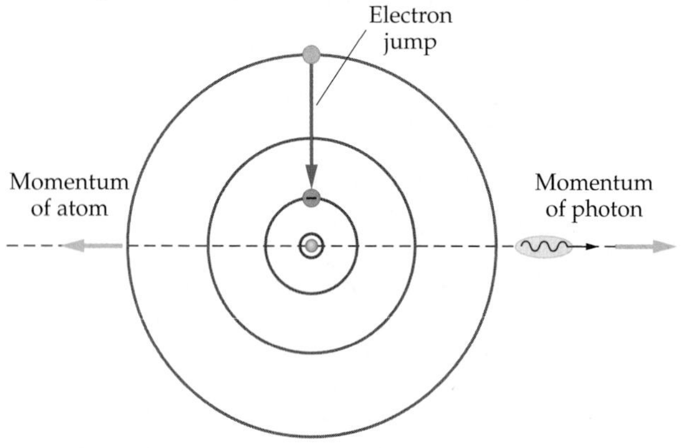

▲ **FIGURE 31–31** Problems 59 and 60

61. •• **BIO Laser Eye Surgery** In laser eye surgery, the laser emits a 1.25-ns pulse focused on a spot that is 32.0 μm in diameter. **(a)** If the energy contained in the pulse is 2.75 mJ, what is the power per square meter (the irradiance) associated with this beam? **(b)** Suppose a molecule with a diameter of 0.650 nm is irradiated by the laser beam. How much energy does the molecule receive in one pulse from the laser? (This much energy is likely to dissociate the molecule.)

62. •• Consider an electron in the ground-state orbit of the Bohr model of hydrogen. **(a)** Find the time required for the electron to complete one orbit about the nucleus. **(b)** Calculate the current (in amperes) corresponding to the electron's motion.

63. •• A particular Bohr orbit in a hydrogen atom has a total energy of -0.85 eV. What is **(a)** the kinetic energy of the electron in this orbit and **(b)** the electric potential energy of the system?

64. •• The element helium is named for the Sun because that is where it was first observed. **(a)** What is the shortest wavelength that one would expect to observe from a singly ionized helium atom in the atmosphere of the Sun? **(b)** Suppose light with a wavelength of 388.9 nm is observed from singly ionized helium. What are the initial and final values of the quantum number n corresponding to this wavelength?

65. •• An ionized atom has only a single electron. The $n = 6$ Bohr orbit of this electron has a radius of 2.72×10^{-10} m. Find **(a)** the atomic number Z of this atom and **(b)** the total energy E of its $n = 3$ Bohr orbit.

66. •• Find the approximate wavelength of K_β X-rays emitted by molybdenum ($Z = 42$), and compare your result with Figure 31–22. (*Hint*: An electron in the M shell is shielded from the nucleus by the single electron in the K shell, plus all the electrons in the L shell.)

67. •• Referring to the hint given in Problem 66, estimate the wavelength of L_α X-rays in molybdenum.

68. •• **IP The Pickering Series** In 1896, the American astronomer Edward C. Pickering (1846–1919) discovered an unusual series of spectral lines in light from the hot star Zeta Puppis. After some time, it was determined that these lines are produced by singly ionized helium. In fact, the "Pickering series" is produced when electrons drop from higher levels to the $n = 4$ level of He^+. Spectral lines in the Pickering series have wavelengths given by

$$\frac{1}{\lambda} = C\left(\frac{1}{16} - \frac{1}{n^2}\right)$$

In this expression, $n = 5, 6, 7, \ldots$. **(a)** Do you expect the constant C to be greater than, less than, or equal to the Rydberg constant R? Explain. **(b)** Find the numerical value of C. **(c)** Pickering lines with $n = 6, 8, 10, \ldots$ correspond to Balmer lines in hydrogen with $n = 3, 4, 5, \ldots$. Verify this assertion for the $n = 6$ Pickering line.

69. •• **IP Rydberg Atoms** There is no limit to the size a hydrogen atom can attain, provided it is free from disruptive outside influences. In fact, radio astronomers have detected radiation from large, so-called "Rydberg atoms" in the diffuse hydrogen gas of interstellar space. **(a)** Find the smallest value of n such that the Bohr radius of a single hydrogen atom is greater than 8.0 microns, the size of a typical single-celled organism. **(b)** Find the wavelength of radiation this atom emits when its electron drops from level n to level $n - 1$. **(c)** If the electron drops one more level, from $n - 1$ to $n - 2$, is the emitted wavelength greater than or less than the value found in part (b)? Explain.

70. ••• Consider a particle of mass m, charge q, and constant speed v moving perpendicular to a uniform magnetic field of magnitude B, as shown in **Figure 31–32**. The particle follows a circular path. Suppose the angular momentum of the particle about the center of its circular motion is quantized in the following way: $mvr = n\hbar$, where $n = 1, 2, 3, \ldots$, and $\hbar = h/2\pi$.

a. Show that the radii of its allowed orbits have the following values:

$$r_n = \sqrt{\frac{n\hbar}{qB}}$$

b. Find the speed of the particle in each allowed orbit.

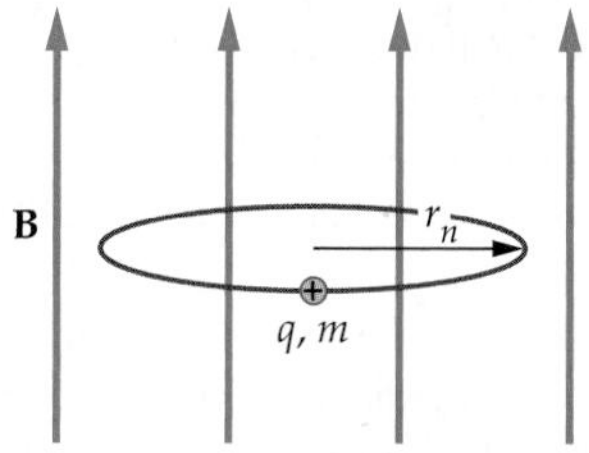

▲ **FIGURE 31–32** Problem 70

71. ••• Consider a particle of mass m confined in a one-dimensional box of length L. In addition, suppose the matter wave associated with this particle is analogous to a wave on a string of length L that is fixed at both ends. Using the de Broglie relationship, show that **(a)** the quantized values of the linear momentum of the particle are

$$p_n = \frac{nh}{2L} \qquad n = 1, 2, 3, \ldots$$

and **(b)** the allowed energies of the particle are

$$E_n = n^2\left(\frac{h^2}{8mL^2}\right) \qquad n = 1, 2, 3, \ldots$$

72. ••• Show that the time required for an electron in the nth Bohr orbit of hydrogen to circle the nucleus once is given by

$$T = T_1 n^3 \qquad n = 1, 2, 3, \ldots$$

where $T_1 = h^3/4\pi^2 mk^2e^4$.

Interactive Problems

73. •• **IP *Referring to Example 31–3*** Suppose the electron is in a state whose standing wave consisting of two wavelengths. **(a)** Is the wavelength of this standing wave greater than or less than 1.33×10^{-9} m? **(b)** Find the wavelength of this standing wave.

74. •• ***Referring to Example 31–3*** **(a)** Which state has a de Broglie wavelength of 3.99×10^{-9} m? **(b)** What is the Bohr radius of this state?

32 Nuclear Physics and Nuclear Radiation

32–1 The Constituents and Structure of Nuclei 1072
32–2 Radioactivity 1076
32–3 Half-Life and Radioactive Dating 1083
32–4 Nuclear Binding Energy 1089
32–5 Nuclear Fission 1090
32–6 Nuclear Fusion 1093
32–7 Practical Applications of Nuclear Physics 1095
32–8 Elementary Particles 1099
32–9 Unified Forces and Cosmology 1102

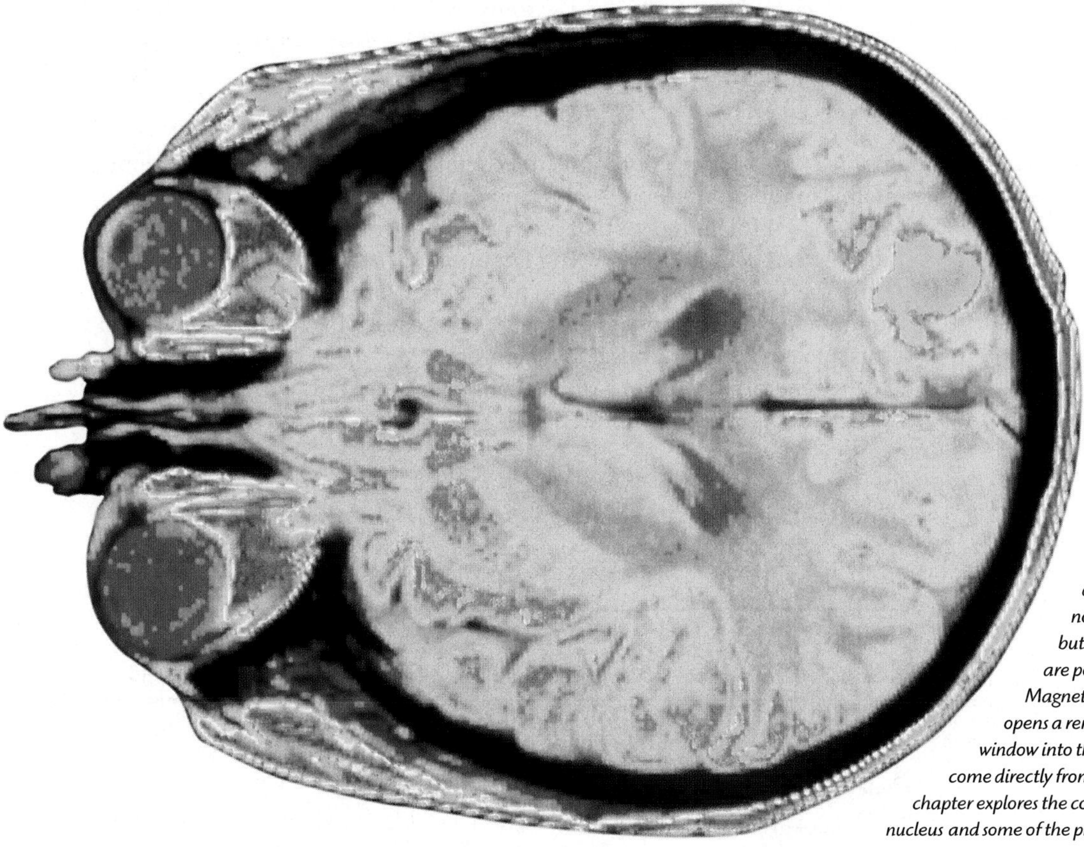

Until about a century ago, the only way for physicians to explore the inside of the human body was to cut it open. The discovery of X-rays gave us our first noninvasive imaging technique, but at a price—high-energy X-rays are potentially harmful to living tissues. Magnetic resonance imaging (MRI) opens a remarkable (and very safe) new window into the body by utilizing signals that come directly from the nuclei of its atoms. This chapter explores the composition of the atomic nucleus and some of the phenomena associated with it.

In the previous chapter our focus was almost entirely on the electrons in an atom. We studied their orbits and energies, their jumps from orbit to orbit, and the photons they emitted or absorbed. The nucleus played little role in these considerations. It was treated as a point object at the center of the atom, providing the electrostatic force necessary to hold the atom together.

The nucleus is much more than a point, however. Most nuclei contain a number of strongly interacting particles packed closely together in a more or less spherical assembly. The energies associated with changes inside a nucleus are orders of magnitude greater than those involved in chemical reactions, which involve only the electrons. It is for this reason that the Sun, which is powered by nuclear reactions, can burn for many billions of years—if the Sun were powered by chemical reactions it would have burned out after giving off light for only a few million years. This chapter considers the physics at play in the nucleus, discusses the nuclear reactions that occur in the Sun and in nuclear power plants, and describes a number of biomedical applications related to the nucleus.

32–1 The Constituents and Structure of Nuclei

The simplest nucleus is that of the hydrogen atom. This nucleus consists of a single **proton,** whose mass is about 1836 times greater than the mass of an electron and whose electric charge is $+e$. All other nuclei contain neutrons in addition to protons. The **neutron** is an electrically neutral particle (its electric charge is zero) with a mass just slightly greater than that of the proton. No other particles are found in nuclei. Collectively, protons and neutrons are referred to as **nucleons.**

Nuclei are characterized by the number and type of nucleons they contain. First, the **atomic number,** Z, is defined as the number of protons in a nucleus. In an electrically neutral atom, the number of electrons will also be equal to Z. Next, the number of neutrons in a nucleus is designated by the **neutron number,** N. Finally, the total number of nucleons in a nucleus is the **mass number,** A. These definitions are summarized in Table 32–1. Clearly, the mass number is the sum of the atomic number and the neutron number:

TABLE 32–1
Numbers That Characterize a Nucleus

Z	Atomic number = number of protons in nucleus
N	Neutron number = number of neutrons in nucleus
A	Mass number = number of nucleons in nucleus

$$A = Z + N \qquad 32\text{–}1$$

A special notation is used to indicate the composition of a nucleus. Consider, for example, an unstable but very useful form of carbon known as carbon-14. The nucleus of carbon-14 is written as follows:

$$^{14}_{6}\text{C}$$

In this expression, C represents the chemical element carbon. The number 6 is the atomic number of carbon, $Z = 6$, and the number 14 is the mass number of this nucleus, $A = 14$. This means that carbon-14 has 14 nucleons in its nucleus. The neutron number can be found by solving Equation 32–1 for N: $N = A - Z = 14 - 6 = 8$. Thus the nucleus of carbon-14 consists of 6 protons and 8 neutrons. The most common form of carbon is carbon-12, whose nucleus is designated as follows: $^{12}_{6}\text{C}$. This nucleus has 6 protons and 6 neutrons.

In general, the nucleus of an arbitrary element, X, with atomic number Z and mass number A, is represented as

$$^{A}_{Z}\text{X}$$

Note that once a given element is specified, the value of Z is known. As a result, the subscript Z is sometimes omitted.

EXERCISE 32–1

a. Give the symbol for a nucleus of aluminum that contains 14 neutrons.

b. Tritium is a type of "heavy hydrogen." The nucleus of tritium can be written as $^{3}_{1}\text{H}$. What is the number of protons and neutrons in a tritium nucleus?

Solution

a. Looking up aluminum in the periodic table in Appendix E, we find that $Z = 13$. In addition, we are given that $N = 14$. Therefore, $A = Z + N = 27$, and hence the symbol for this nucleus is $^{27}_{13}\text{Al}$.

b. We obtain the number of protons from the subscript; therefore, $Z = 1$. The number of neutrons, from Equation 32–1, is $N = A - Z$, where A is the superscript. Therefore, the number of neutrons is $N = 3 - 1 = 2$.

All nuclei of a given element have the same number of protons, Z. They may have different numbers of neutrons, N, however. Nuclei with the same value of Z but different values of N are referred to as **isotopes.** For example, $^{12}_{6}\text{C}$ and $^{13}_{6}\text{C}$ are two isotopes of carbon, with $^{12}_{6}\text{C}$ being the most common one, constituting about 98.89% of naturally occurring carbon. About 1.11% of natural carbon is $^{13}_{6}\text{C}$. Values for the percentage abundance of various isotopes can be found in Appendix F.

Also given in Appendix F are the atomic masses of many common isotopes. These masses are given in terms of the **atomic mass unit,** u, defined so that the mass of one atom of $^{12}_{6}C$ is exactly 12 u. The value of u is as follows:

Definition of Atomic Mass Unit, u

$$1\ \text{u} = 1.660540 \times 10^{-27}\ \text{kg} \qquad \textbf{32–2}$$

SI unit: kg

Protons have a mass just slightly greater than 1 u, and the neutron is slightly more massive than the proton. The precise masses of the proton and neutron are given in Table 32–2, along with the mass of the electron.

TABLE 32–2 Mass and Charge of Particles in the Atom

Particle	Mass (kg)	Mass (MeV/c^2)	Mass (u)	Charge (C)
Proton	1.672623×10^{-27}	938.28	1.007276	$+1.6022 \times 10^{-19}$
Neutron	1.674929×10^{-27}	939.57	1.008665	0
Electron	9.109390×10^{-31}	0.511	0.0005485799	-1.6022×10^{-19}

When we consider nuclear reactions later in this chapter, an important consideration will be the energy equivalent of a given mass, as given by Einstein's famous relation, $E = mc^2$ (Equation 30–7). The energy equivalent of one atomic mass unit is

$$\begin{aligned} E = mc^2 &= (1\ \text{u})c^2 \\ &= (1.660540 \times 10^{-27}\ \text{kg})(2.998 \times 10^8\ \text{m/s})^2\left(\frac{1\ \text{eV}}{1.6022 \times 10^{-19}\ \text{J}}\right) \\ &= 931.5\ \text{MeV} \end{aligned}$$

where 1 MeV $= 10^6$ eV. When we consider that the ionization energy of hydrogen is only 13.6 eV, it is clear that the energy equivalent of nucleons is enormous compared with typical atomic energies. In general, energies involving the nucleus are on the order of MeV, and energies associated with the electrons in an atom are on the order of eV. Finally, because mass and energy can be converted from one form to the other, it is common to express the atomic mass unit in terms of energy as follows:

$$1\ \text{u} = 931.5\ \text{MeV}/c^2 \qquad \textbf{32–3}$$

The masses of the proton, neutron, and electron are also given in units of MeV/c^2 in Table 32–2.

Nuclear Size and Density

To obtain an estimate for the size of a nucleus, Rutherford did a simple calculation using energy conservation. He considered the case of a particle of charge $+q$ and mass m approaching a nucleus of charge $+Ze$ with a speed v. He further assumed that the approach was head-on. At some distance, d, from the center of the nucleus, the incoming particle comes to rest instantaneously, before turning around. It follows that the radius of the nucleus is less than d. In Example 32–1 we obtain a symbolic expression for the distance d.

EXAMPLE 32–1 Setting a Limit on the Radius of a Nucleus

A particle of mass m, charge $+q$, and speed v heads directly toward a distant, stationary nucleus of charge $+Ze$. Find the distance of closest approach between the incoming particle and the center of the nucleus.

Picture the Problem

The incoming particle moves on a line that passes through the center of the nucleus. Far from the nucleus, the particle's speed is v. The particle turns around (comes to rest instantaneously) a distance d from the center of the nucleus.

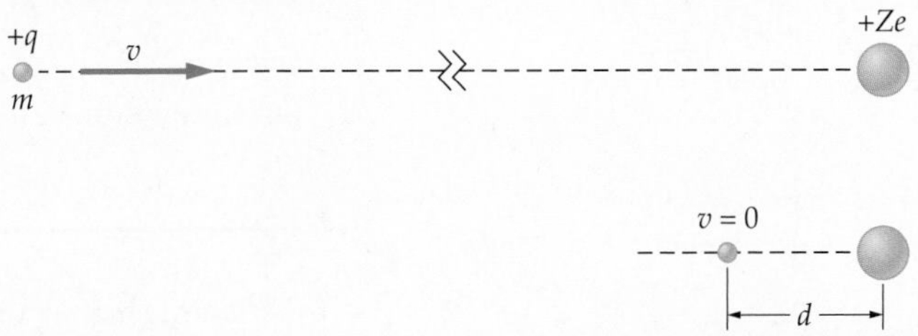

continued on next page

continued from previous page

Strategy

We can find the distance d by applying energy conservation. In particular, the initial energy of the system is the kinetic energy of the particle, $\frac{1}{2}mv^2$, assuming the particle approaches the nucleus from infinity. The final energy is the electric potential energy, $U = kq_1q_2/r = k(Ze)q/d$. Setting these energies equal to each other allows us to solve for d.

Solution

1. Write an expression for the initial energy of the system: $E_i = \frac{1}{2}mv^2$

2. Write an expression for the final energy of the system: $E_f = \dfrac{k(Ze)q}{d}$

3. Set the final energy equal to the initial energy and solve for d: $\frac{1}{2}mv^2 = \dfrac{k(Ze)q}{d}$

$$d = \frac{kZeq}{\left(\frac{1}{2}mv^2\right)}$$

Insight

Notice that the distance of closest approach is inversely proportional to the initial kinetic energy of the incoming particle and directly proportional to the charge on the nucleus.

Practice Problem

Find the distance at which the speed of the incoming particle is equal to $\frac{1}{2}v$. [**Answer:** $\dfrac{kZeq}{\frac{3}{4}\left(\frac{1}{2}mv^2\right)}$]

Some related homework problems: Problem 6, Problem 7

Using the result obtained in Example 32–1, Rutherford found that for an alpha particle approaching a gold nucleus in one of his experiments, the distance of closest approach was 3.2×10^{-14} m. A similar calculation for alpha particles fired at silver atoms gives a closest approach distance of 2.0×10^{-14} m. This suggests that the size of the nucleus varies from element to element; in particular, the nucleus of silver is smaller than that of gold. In fact, more careful measurements since Rutherford's time have established that the average radius of a nucleus of mass number A is given approximately by the following expression:

$$r = (1.2 \times 10^{-15}\ \text{m})A^{1/3} \qquad 32\text{–}4$$

Notice that the length scale of the nucleus is on the order of 10^{-15} m, as opposed to the length scale of an atom, which is on the order of 10^{-10} m. Recall that 10^{-15} m is referred to as a *femtometer* (fm). To honor the pioneering work of Enrico Fermi (1901–1954) in the field of nuclear physics, the femtometer is often referred to as the **fermi:**

Definition of the Fermi, fm

$$1\ \text{fermi} = 1\ \text{fm} = 10^{-15}\ \text{m}$$

SI unit: m

We now use Equation 32–4 to find the radius of a particular nucleus.

EXERCISE 32–2

Find the radius of a ${}^{14}_{6}C$ nucleus.

Solution

Substitute $A = 14$ in Equation 32–4:

$$r = (1.2 \times 10^{-15}\ \text{m})(14)^{1/3} = 2.9 \times 10^{-15}\ \text{m} = 2.9\ \text{fm}$$

The fact that the radius of a nucleus depends on $A^{1/3}$ has interesting consequences for the density of the nucleus, and these are explored in the next Example.

EXAMPLE 32–2 Nuclear Density

Using the expression $r = (1.2 \times 10^{-15}\ \text{m})A^{1/3}$, calculate the density of a nucleus with mass number A.

Picture the Problem

Our sketch shows a collection of neutrons and protons in a densely packed nucleus of radius r.

Strategy

To find the density of a nucleus, we must divide its mass, M, by its volume, V. Ignoring the small difference in mass between a neutron and a proton, we can express the mass of a nucleus as $M = Am$, where $m = 1.67 \times 10^{-27}$ kg. The volume of a nucleus is simply the volume of a sphere of radius r; $V = 4\pi r^3/3$.

Solution

1. Write an expression for the mass of a nucleus: $M = Am = A(1.67 \times 10^{-27}\ \text{kg})$

2. Write an expression for the volume of a nucleus:
$$V = \tfrac{4}{3}\pi r^3 = \tfrac{4}{3}\pi[(1.2 \times 10^{-15}\ \text{m})A^{1/3}]^3 = \tfrac{4}{3}\pi(1.7 \times 10^{-45}\ \text{m}^3)A$$

3. Divide the mass by the volume to find the density:
$$\rho = \frac{M}{V} = \frac{Am}{\tfrac{4}{3}\pi(1.7 \times 10^{-45}\ \text{m}^3)A} = \frac{(1.67 \times 10^{-27}\ \text{kg})}{\tfrac{4}{3}\pi(1.7 \times 10^{-45}\ \text{m}^3)} = 2.3 \times 10^{17}\ \text{kg/m}^3$$

Insight

Note that the density of a nucleus is found to be *independent* of the mass number, A. This means that a nucleus can be thought of as a collection of closely packed nucleons, much like a group of marbles in a bag. The neutrons in a nucleus serve to separate the protons, thereby reducing their mutual electrostatic repulsion.

The density of a nucleus is incredibly large. For example, a single teaspoon of nuclear matter would weigh about a trillion tons.

Practice Problem

Find the surface area of a nucleus in terms of the mass number, A, assuming it to be a sphere.
[**Answer:** area $= (1.8 \times 10^{-29}\ \text{m}^2)A^{2/3}$]

Some related homework problems: Problem 8, Problem 9

Nuclear Stability

We know that like charges repel one another, and that the force of repulsion increases rapidly with decreasing distance. It follows that protons in a nucleus, with a separation of only about a fermi, must exert relatively large forces on one another. Applying Coulomb's law (Equation 19–5), we find the following force for two protons (charge $+e$) separated by a distance of 10^{-15} m:

$$F = \frac{ke^2}{r^2} = 230\ \text{N}$$

The acceleration such a force would give to a proton is $a = F/m = (230\ \text{N})/(1.67 \times 10^{-27}\ \text{kg}) = 1.4 \times 10^{29}\ \text{m/s}^2$, which is about 10^{28} times greater than the acceleration of gravity! Thus, if protons in the nucleus experienced only the electrostatic force, the nucleus would fly apart in an instant. It follows that large attractive forces must also act within the nucleus.

The attractive force that holds a nucleus together is called the **strong nuclear force.** This force has the following properties:

- The strong force is short range, acting only to distances of a couple fermis.
- The strong force is attractive and acts with nearly equal strength between protons and protons, protons and neutrons, and neutrons and neutrons.

In addition, the strong nuclear force does not act on electrons. As a result, it has no effect on the chemical properties of an atom.

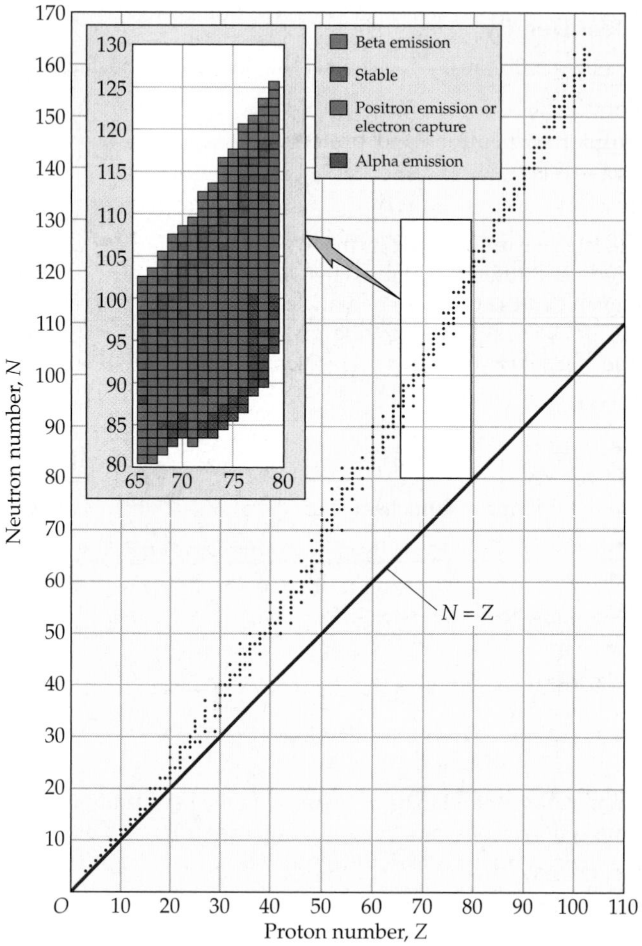

▶ **FIGURE 32–1** ***N* and *Z* for stable and unstable nuclei**

Stable nuclei with proton numbers less than 104 are indicated by small red dots. Notice that large nuclei have significantly more neutrons, N, than protons, Z. The inset shows unstable nuclei and their decay modes for proton numbers between 65 and 80.

It is the competition between the repulsive electrostatic forces and the attractive strong nuclear forces that determines whether a given nucleus is stable. **Figure 32–1** shows the neutron number, N, and atomic number (proton number), Z, for nuclei that are stable. Note that nuclei of relatively small atomic number are most stable when the numbers of protons and neutrons in the nucleus are approximately equal, $N = Z$. For example, ${}^{12}_{6}C$ and ${}^{13}_{6}C$ are both stable. As the atomic number increases, however, we see that the points corresponding to stable nuclei deviate from the line $N = Z$. In fact, we see that large stable nuclei tend to contain significantly more neutrons than protons, as in the case of ${}^{185}_{75}Re$. Since all nucleons experience the strong nuclear force, but only the protons experience the electrostatic force, the neutrons effectively "dilute" the nuclear charge density, reducing the effect of the repulsive forces that otherwise would make the nucleus disintegrate.

As the number of protons in a nucleus increases, however, a point is reached at which the strong nuclear forces are no longer able to compensate for the repulsive forces between protons. In fact, the largest number of protons in a stable nucleus is $Z = 83$, corresponding to the element bismuth. Nuclei with more than 83 protons are simply not stable, as can be seen by noting that all elements with $Z > 83$ in Appendix F decay in a finite time—that is, they have a finite half-life. (We shall discuss the half-life of unstable nuclei in detail in Section 32–3.) The nuclei of many well-known elements, such as radon and uranium, disintegrate—decay—in a finite time. We turn now to a discussion of the various ways in which an unstable nucleus decays.

32–2 Radioactivity

An unstable nucleus does not last forever—sooner or later it changes its composition by emitting a particle of one type or another. Alternatively, a nucleus in an excited state may rearrange its nucleons into a lower-energy state and emit a high-energy photon. We refer to such processes as the **decay** of a nucleus, and the various emissions that result are known collectively as **radioactivity.**

When a nucleus undergoes radioactive decay, the mass of the system decreases. That is, the mass of the initial nucleus before decay is greater than the mass of the resulting nucleus plus the mass of the emitted particle. The difference in mass, $\Delta m < 0$, appears as a release of energy, according to the relation $E = |\Delta m|c^2$. The mass difference for any given decay can be determined by referring to Appendix F. Note that the atomic masses listed in Appendix F are the masses of neutral atoms; that is, the values given in the table include the mass of the electrons in an atom. This factor must be considered whenever the mass difference of a reaction is calculated, as we shall see later in this section.

Three types of particles with mass are given off during the various processes of radioactive decay. They are as follows:

- Alpha (α) particles, which are the nuclei of ${}^{4}_{2}He$. Note that an alpha particle consists of two protons and two neutrons. When a nucleus decays by giving off alpha particles, we say that it emits α rays.
- Electrons, also referred to as beta (β) particles. The electrons given off by a nucleus are called β rays, or β^- rays to be more precise. (The minus sign is a reminder that the charge of an electron is $-e$.)
- **Positrons,** which have the same mass as an electron but a charge of $+e$. If a nucleus gives off positrons, we say that it emits β^+ rays. (A positron, which is short for "positive electron," is the **antiparticle** of the ordinary electron. Positrons will be considered in greater detail in Section 32–7.)

Finally, radioactivity may take the form of a photon rather than a particle with nonzero mass:

- A nucleus in an excited state may emit a high-energy photon, a gamma (γ) ray, and drop to a lower-energy state.

The following Conceptual Checkpoint examines the behavior of radioactivity in a magnetic field.

CONCEPTUAL CHECKPOINT 32–1 Identify the Radiation

A sample of radioactive material is placed at the bottom of a small hole drilled into a piece of lead. The sample emits α rays, β^- rays, and γ rays into a region of constant magnetic field. It is observed that the radiation follows three distinct paths, 1, 2, and 3, as shown in the sketch. Identify each path with the corresponding type of radiation: **(a)** path 1, α rays; path 2, β^- rays; path 3, γ rays; **(b)** path 1, β^- rays; path 2, γ rays; path 3, α rays; or **(c)** path 1, α rays; path 2, γ rays; path 3, β^- rays.

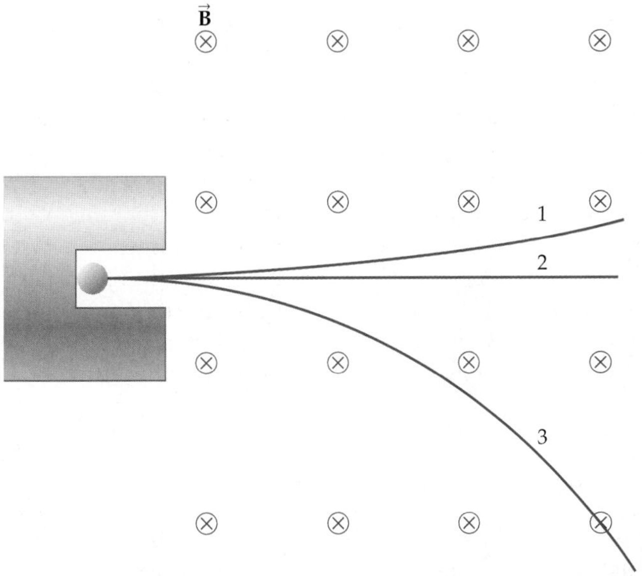

continued on next page

continued from previous page

Reasoning and Discussion

First, because γ rays are uncharged, they are not deflected by the magnetic field. It follows that path 2 corresponds to γ rays.

Next, the right-hand rule for the magnetic force (Section 22–2) indicates that positively charged particles will be deflected upward, and negatively charged particles will be deflected downward. As a result, path 1 corresponds to α rays, and path 3 corresponds to β^- rays.

Answer:

(c) Path 1, α rays; path 2, γ rays; path 3, β^- rays.

Radioactivity was discovered by the French physicist Antoine Henri Becquerel (1852–1908) in 1896 when he observed that uranium was able to expose photographic emulsion, even when the emulsion was covered. Thus radioactivity has the ability to penetrate various materials. In fact, the various types of radioactivity were initially named according to their ability to penetrate, starting with α rays, which are the least penetrating. Typical penetrating abilities for α, β, and γ rays are as follows:

- α rays can barely penetrate a sheet of paper.
- β rays (both β^- and β^+) can penetrate a few millimeters of aluminum.
- γ rays can penetrate several centimeters of lead.

We turn now to a detailed examination of each of these types of decay.

PROBLEM-SOLVING NOTE

The Effects of Alpha Decay

In alpha decay, the total number of protons is the same before and after the reaction. The same is true of the total number of neutrons. On the other hand, the mass number of the daughter nucleus is 4 less than the mass number of the parent nucleus. Similarly, the atomic number of the daughter nucleus is 2 less than the atomic number of the parent nucleus.

Alpha Decay

When a nucleus decays by giving off an α particle (^4_2He), it loses two protons and two neutrons. As a result, its atomic number, Z, decreases by 2, and its mass number decreases by 4. Symbolically, we can write this process as follows:

$$^A_Z X \longrightarrow {}^{A-4}_{Z-2}Y + {}^4_2\text{He}$$

where X is referred to as the **parent nucleus,** and Y is the **daughter nucleus.** Notice that the sum of the atomic numbers on the right side of this process is equal to the atomic number on the left side; similar remarks apply to the mass numbers.

The next Example considers the alpha decay of uranium-238. We first use conservation of atomic number and mass number to determine the identity of the daughter nucleus. Next, we use the mass difference to calculate the amount of energy released by the decay.

PROBLEM-SOLVING NOTE

Atomic Masses Include Electrons

When calculating the mass difference in a nuclear reaction, be sure to note that the atomic masses in Appendix F include the electrons that would be present in a neutral atom. The only item in Appendix F to which this does *not* apply is the neutron.

EXAMPLE 32–3 Uranium Decay

Determine **(a)** the daughter nucleus and **(b)** the energy released when $^{238}_{92}\text{U}$ undergoes alpha decay.

Picture the Problem

Our sketch shows the specified decay of $^{238}_{92}\text{U}$ into a daughter nucleus plus an α particle. Note that the number of neutrons and protons is indicated for $^{238}_{92}\text{U}$. The α particle consists of two neutrons and two protons.

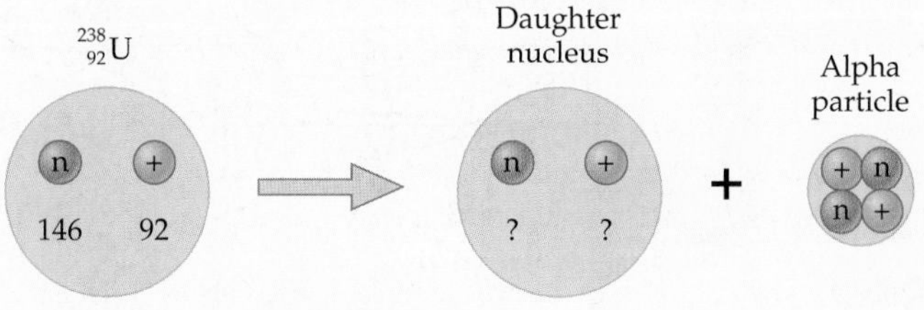

Strategy

a. We can identify the daughter nucleus by requiring that the total number of neutrons and protons be the same before and after the decay.

b. To find the energy, we first calculate the mass before and after the decay. The magnitude of the difference in mass, $|\Delta m|$, times the speed of light squared, c^2, gives the amount of energy released.

Solution

Part (a)

1. Determine the number of neutrons and protons in the daughter nucleus. Add these numbers together to obtain the mass number of the daughter nucleus:

$N = 146 - 2 = 144$

$Z = 92 - 2 = 90$

$A = N + Z = 144 + 90 = 234$

2. Referring to Appendix F, we see that the daughter nucleus is thorium-234:

${}^{234}_{90}\mathrm{Th}$

Part (b)

3. Use Appendix F to find the initial mass of the system; that is, the mass of a ${}^{238}_{92}\mathrm{U}$ atom:

$m_i = 238.050786 \text{ u}$

4. Use Appendix F to find the final mass of the system; that is, the mass of a ${}^{234}_{90}\mathrm{Th}$ atom plus the mass of a ${}^{4}_{2}\mathrm{He}$ atom:

$m_f = 234.043596 \text{ u} + 4.002603 \text{ u} = 238.046199 \text{ u}$

5. Calculate the mass difference and the corresponding energy release (recall that $1 \text{ u} = 931.5 \text{ MeV}/c^2$):

$$\Delta m = m_f - m_i$$
$$= 238.046199 \text{ u} - 238.050786 \text{ u} = -0.004587 \text{ u}$$
$$E = |\Delta m| c^2 = (0.004587 \text{ u})\left(\frac{931.5 \text{ MeV}/c^2}{1 \text{ u}}\right)c^2$$
$$= 4.273 \text{ MeV}$$

Insight

Each of the masses used in this decay includes the electrons of the corresponding neutral atom. Since the number of electrons initially (92) is the same as the number of electrons in a thorium atom (90) plus the number of electrons in a helium atom (2), the electrons make no contribution to the total mass difference, Δm.

Practice Problem

Find the daughter nucleus and energy released when ${}^{226}_{88}\mathrm{Ra}$ undergoes alpha decay. [**Answer:** ${}^{222}_{86}\mathrm{Rn}$, 4.871 MeV]

Some related homework problems: Problem 16, Problem 17

A considerable amount of energy is released in the alpha decay of uranium-238. As indicated in **Figure 32–2**, this energy appears as kinetic energy of the daughter nucleus and the α particle as they move off in opposite directions. The following Conceptual Checkpoint compares the kinetic energy of the daughter nucleus with that of the α particle.

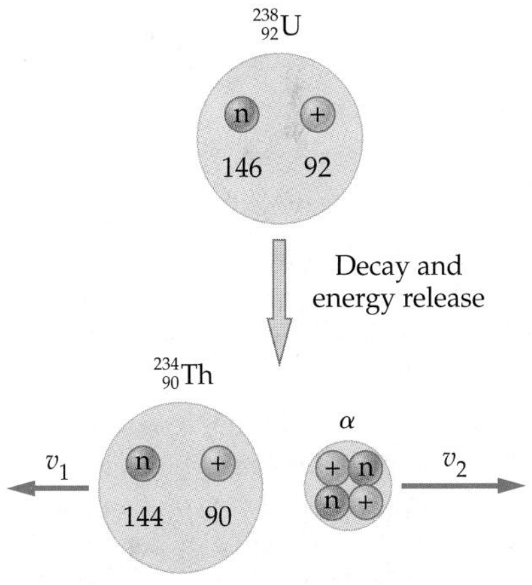

▲ **FIGURE 32–2 Alpha decay of uranium-238**

When ${}^{238}_{92}\mathrm{U}$ decays into ${}^{234}_{90}\mathrm{Th}$ and an alpha particle, the mass of the system decreases. The "lost" mass is actually converted into energy; it appears as the kinetic energy of the ${}^{234}_{90}\mathrm{Th}$ nucleus and the alpha particle.

CONCEPTUAL CHECKPOINT 32–2 **Compare Kinetic Energies**

When a stationary ${}^{238}_{92}\mathrm{U}$ nucleus decays into a ${}^{234}_{90}\mathrm{Th}$ nucleus and an α particle, is the kinetic energy of the α particle **(a)** greater than, **(b)** less than, or **(c)** the same as the kinetic energy of the ${}^{234}_{90}\mathrm{Th}$ nucleus?

Reasoning and Discussion

Because no external forces are involved in the decay process, it follows that the momentum of the system is conserved. Letting subscript 1 refer to the ${}^{234}_{90}\mathrm{Th}$ nucleus, and subscript 2 to the α particle, the condition for momentum conservation is $m_1v_1 = m_2v_2$. Solving for the speed of the α particle, we have $v_2 = (m_1/m_2)v_1$; that is, the α particle has the greater speed, since m_1 is greater than m_2.

The fact that the α particle has the greater speed does not, in itself, ensure that its kinetic energy is the larger of the two. After all, the α particle also has the smaller mass. To compare the kinetic energies, note that the kinetic energy of the ${}^{234}_{90}\mathrm{Th}$ nucleus is $\frac{1}{2}m_1v_1^2$. The kinetic energy of the α particle is $\frac{1}{2}m_2v_2^2 = \frac{1}{2}m_2[(m_1/m_2)v_1]^2 = \frac{1}{2}m_1v_1^2(m_1/m_2) > \frac{1}{2}m_1v_1^2$. Therefore, the α particle carries away the majority of the kinetic energy released in the decay. In this particular case, the α particle has a kinetic energy that is $m_1/m_2 = 234/4 = 58.5$ times greater than the kinetic energy of the ${}^{234}_{90}\mathrm{Th}$ nucleus.

Answer:

(a) The α particle has the greater kinetic energy.

REAL-WORLD PHYSICS

Smoke detector

Although you may not be aware of it, many homes are protected from the hazards of fire by a small device—a smoke detector—that uses the alpha decay

▲ Smoke detectors like this one make use of a synthetic radioactive isotope, americium-241. The alpha particles emitted when this isotope decays ionize air molecules, making them able to conduct a small current. Smoke particles neutralize the ions, interrupting the current and setting off an alarm.

of a manmade radioactive isotope, ${}^{241}_{95}\text{Am}$. In this type of smoke detector, a minute quantity of ${}^{241}_{95}\text{Am}$ is placed between two metal plates connected to a battery or other source of emf. The α particles emitted by the radioactive source ionize the air, allowing a measurable electric current to flow between the plates. As long as this current flows, the smoke detector remains silent. When smoke enters the detector, however, the ionized air molecules tend to stick to the smoke particles and become neutralized. This reduces the current and triggers the alarm. These "ionization" smoke detectors are more sensitive than the "photoelectric" detectors that rely on the thickness of smoke to dim a beam of light.

Beta Decay

The basic process that occurs in beta decay is the conversion of a neutron to a proton and an electron:

$$ {}^{1}_{0}\text{n} \longrightarrow {}^{1}_{1}\text{p} + \text{e}^{-} $$

Thus when a nucleus decays by giving off an electron, its mass number is unchanged (since protons and neutrons count equally in determining A), but its atomic number increases by 1. This process can be represented as follows:

$$ {}^{A}_{Z}X \longrightarrow {}_{Z+1}^{\;\;A}Y + \text{e}^{-} $$

Similarly, if a nucleus undergoes a different type of decay in which it gives off a positron, the process can be written

$$ {}^{A}_{Z}X \longrightarrow {}_{Z-1}^{\;\;A}Y + \text{e}^{+} $$

In the next Example we determine the energy that is released as carbon-14 undergoes beta decay.

EXAMPLE 32–4 Beta Decay of Carbon-14

Find **(a)** the daughter nucleus and **(b)** the energy released when ${}^{14}_{6}\text{C}$ undergoes β^{-} decay.

Picture the Problem

In our sketch we show ${}^{14}_{6}\text{C}$ giving off a β^{-} particle and converting into a daughter nucleus. The number of neutrons and protons in the ${}^{14}_{6}\text{C}$ nucleus is indicated. Note that the β^{-} particle is not a nucleon.

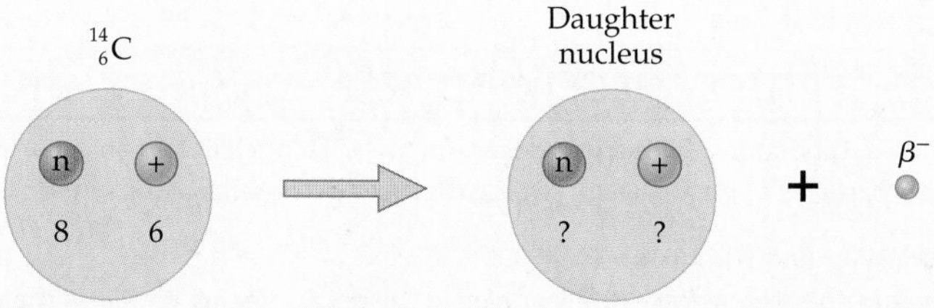

Strategy

a. We can identify the daughter nucleus by requiring that the total number of nucleons be the same before and after the decay. The number of neutrons will be decreased by one, and the number of protons will be increased by one.

b. To find the energy, we begin by calculating the mass before and after the decay. The magnitude of the difference in mass, $|\Delta m|$, times the speed of light squared, c^2, gives the amount of energy released.

Solution

Part (a)

1. Determine the number of neutrons and protons in the daughter nucleus. Add these numbers together to obtain the mass number of the daughter nucleus:

$N = 8 - 1 = 7$

$Z = 6 + 1 = 7$

$A = N + Z = 7 + 7 = 14$

2. Referring to Appendix F, we see that the daughter nucleus is nitrogen-14:

${}^{14}_{7}\text{N}$

Part (b)

3. Use Appendix F to find the initial mass of the system; that is, the mass of a $^{14}_{6}C$ atom:

$$m_i = 14.003242 \text{ u}$$

4. Use Appendix F to find the final mass of the system, which is simply the mass of a $^{14}_{7}N$ atom (the mass of the β^- particle is included in the mass of $^{14}_{7}N$, as we point out in the Insight):

$$m_f = 14.003074 \text{ u}$$

$$\Delta m = m_f - m_i$$

$$= 14.003074 \text{ u} - 14.003242 \text{ u} = -0.000168 \text{ u}$$

5. Calculate the mass difference and the corresponding energy release (recall that $1 \text{ u} = 931.5 \text{ MeV}/c^2$):

$$E = |\Delta m|c^2 = (0.000168 \text{ u})\left(\frac{931.5 \text{ MeV}/c^2}{1 \text{ u}}\right)c^2$$

$$= 0.156 \text{ MeV}$$

Insight

With regard to the masses used in this calculation, note that the mass of $^{14}_{6}C$ includes the mass of its 6 electrons. Similarly, the mass of $^{14}_{7}N$ includes the mass of 7 electrons in the neutral $^{14}_{7}N$ atom. However, when the $^{14}_{6}C$ nucleus converts to a $^{14}_{7}N$ nucleus, the number of electrons orbiting the nucleus is still 6. In effect, the newly created $^{14}_{7}N$ atom is missing one electron; that is, the mass of the $^{14}_{7}N$ atom includes the mass of one too many electrons. Therefore, it is not necessary to add the mass of an electron (representing the β^- particle) to the final mass of the system, because this extra electron mass is already included in the mass of the $^{14}_{7}N$ atom.

Practice Problem

Find the daughter nucleus and the energy released when $^{234}_{90}Th$ undergoes β^- decay. [**Answer:** $^{234}_{91}Pa$, 0.274 MeV]

Some related homework problems: Problem 18, Problem 20

Referring to Conceptual Checkpoint 32–2, we would expect the kinetic energy of an electron emitted during beta decay to account for most of the energy released by the decay process. In fact, energy conservation allows us to predict the precise amount of kinetic energy the electron should have. It turns out, however, that when the kinetic energy of emitted electrons is measured, a range of values is obtained, as indicated in **Figure 32–3**. Specifically, we find that all electrons given off in beta decay have energies that are less than would be predicted by energy conservation. On closer examination it is found that beta decay seems to violate conservation of linear and angular momentum as well! For these reasons, beta decay was an interesting and intriguing puzzle for physicists.

The resolution of this puzzle was given by Pauli in 1930, when he proposed that the "missing" energy and momentum were actually carried off by a particle that was not observed in the experiments. For this particle to have been unobserved, it must have zero charge and little or no mass. Fermi dubbed Pauli's hypothetical particle the **neutrino,** meaning, literally, "little neutral one." We now know that neutrinos do in fact exist and that they account exactly for the missing energy and momentum. They interact so weakly with matter, however, that it wasn't until 1950 that they were observed experimentally. Recent experiments on neutrinos given off by the Sun provide the best evidence yet that the mass of a neutrino is in fact finite—though extremely small. In fact, the best estimate of the neutrino mass at this time is that it is less than about 7 eV/c^2. For comparison, the mass of the electron is 511,000 eV/c^2.

To give an indication of just how weakly neutrinos interact with matter, only one in every 200 million neutrinos that pass through the Earth interacts with it in any way. As far as the neutrinos are concerned, it is almost as if the Earth did not exist. Right now, in fact, billions of neutrinos are passing through your body every second without the slightest effect.

We can now write the correct expression for the decay of a neutron. Indicating the electron neutrino with the symbol ν_e, we have the following:

$$^{1}_{0}n \longrightarrow {}^{1}_{1}p + e^- + \overline{\nu}_e \qquad 32\text{–}5$$

The bar over the neutrino symbol indicates that the neutrino given off in β^- decay is actually an **antineutrino,** the antiparticle counterpart of the neutrino

PROBLEM-SOLVING NOTE

The Effects of Beta Decay

In β^- decay, the number of neutrons decreases by 1, and the number of protons increases by 1. The mass number is unchanged.

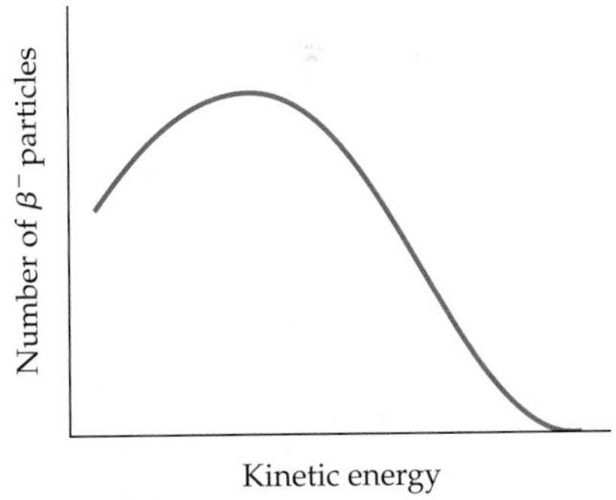

▲ **FIGURE 32–3 Energy of electrons emitted in β^- decay**

When electrons are emitted during β^- decay, they come off with a range of energies. This indicates that another particle (the neutrino) must also be taking away some of the energy.

(just as the positron is the antiparticle of the electron). The neutrino itself is given off in β^+ decay.

Gamma Decay

An atom in an excited state can emit a photon when one of its electrons drops to a lower-energy level. Similarly, a nucleus in an excited state can emit a photon as it decays to a state of lower energy. Since nuclear energies are so much greater than typical atomic energies, the photons given off by a nucleus are highly energetic. In fact, these photons have energies that place them well beyond X-rays in the electromagnetic spectrum. We refer to such high-energy photons as **gamma (γ) rays.**

As an example of a situation in which a γ ray can be given off, consider the following beta decay:

$$^{14}_{6}\mathrm{C} \longrightarrow {}^{14}_{7}\mathrm{N}^* + \mathrm{e}^- + \bar{\nu}_\mathrm{e}$$

The asterisk on the nitrogen symbol indicates that the nitrogen nucleus has been left in an excited state as a result of the beta decay. Subsequently, the nitrogen nucleus may decay to its ground state with the emission of a γ ray:

$$^{14}_{7}\mathrm{N}^* \longrightarrow {}^{14}_{7}\mathrm{N} + \gamma$$

Notice that neither the atomic number nor the mass number is changed by the emission of a γ ray.

ACTIVE EXAMPLE 32–1 Gamma-Ray Emission: Find the Change in Mass

A $^{226}_{88}\mathrm{Ra}$ nucleus in an excited state emits a γ ray with a wavelength of 6.67×10^{-12} m. Find the decrease in mass of the $^{226}_{88}\mathrm{Ra}$ nucleus as a result of this process.

Solution *(Test your understanding by performing the calculations indicated in each step.)*

1. Find the frequency of the γ ray: $f = c/\lambda = 4.50 \times 10^{19}$ Hz
2. Calculate the energy of the γ ray photon: $E = hf = 0.186$ MeV
3. Determine the mass difference corresponding to the energy of the photon: $|\Delta m| = E/c^2 = 0.000200$ u

Insight

As a result of emitting this γ ray, the mass of the $^{226}_{88}\mathrm{Ra}$ nucleus decreases by an amount that is about one-third the mass of the electron.

Your Turn

If the wavelength of the emitted gamma ray is doubled, by what factor does the mass difference change?

(Answers to **Your Turn** *problems are given in the back of the book.)*

Radioactive Decay Series

Consider an unstable nucleus that decays and produces a daughter nucleus. If the daughter nucleus is also unstable, it will eventually decay and produce its own daughter nucleus, which may in turn be unstable. In such cases, an original parent nucleus can produce a series of related nuclei referred to as a **radioactive decay series.** An example of a radioactive decay series is shown in **Figure 32–4**. In this case, the parent nucleus is $^{235}_{92}\mathrm{U}$, and the final nucleus of the series is $^{207}_{82}\mathrm{Pb}$, which is stable.

Notice that several of the intermediate nuclei in this series can decay in two different ways—either by alpha decay or by beta decay. Thus there is a variety of "paths" a $^{235}_{92}\mathrm{U}$ nucleus can follow as it transforms into a $^{207}_{82}\mathrm{Pb}$ nucleus. In addition, the intermediate nuclei in this series decay fairly rapidly, at least on a geological time scale. For example, any actinium-227 that was present when the Earth formed would have decayed away long ago. The fact that actinium-227 is still found on the Earth today in natural uranium deposits is due to its continual production in this and other decay series.

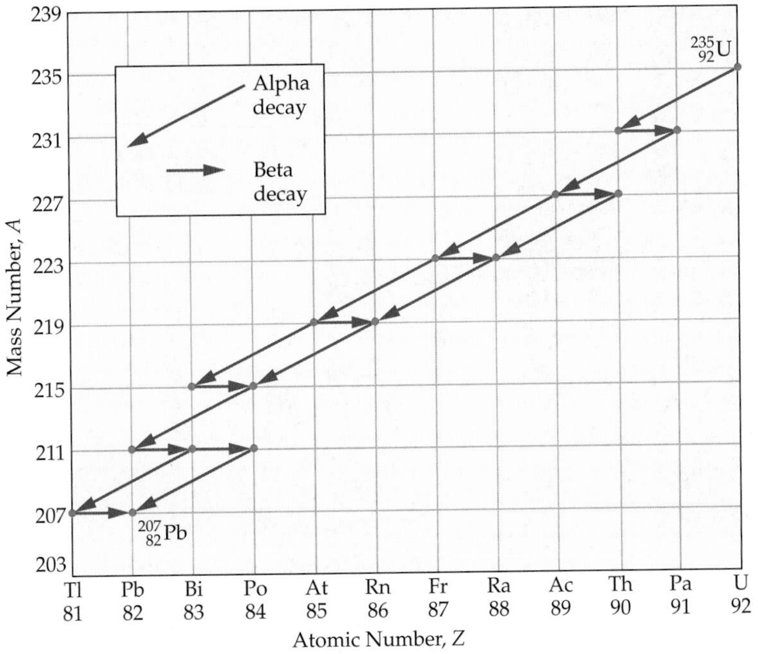

◀ **FIGURE 32–4 Radioactive decay series of $^{235}_{92}U$**

When $^{235}_{92}U$ decays, it passes through a number of intermediate nuclei before reaching the stable end of the series, $^{207}_{82}Pb$. Note that some intermediary nuclei can decay in only one way, whereas others have two decay possibilities.

Activity

The rate at which nuclear decay occurs—that is, the number of decays per second—is referred to as the **activity.** A highly active material has many nuclear decays occurring every second. For example, a typical sample of radium (usually a fraction of a gram) might have 10^5 to 10^{10} decays per second.

The unit we use to measure activity is the curie, named in honor of Pierre (1859–1906) and Marie (1867–1934) Curie, pioneers in the study of radioactivity. The **curie (Ci)** is defined as follows:

$$1 \text{ curie} = 1 \text{ Ci} = 3.7 \times 10^{10} \text{ decays/s} \qquad \textbf{32–6}$$

The reason for this choice is that 1 Ci is roughly the activity of 1 g of radium. In SI units, we measure activity in terms of the **becquerel** (Bq):

$$1 \text{ becquerel} = 1 \text{ Bq} = 1 \text{ decay/s} \qquad \textbf{32–7}$$

The units of activity most often encountered in practical applications are the millicurie ($1 \text{ mCi} = 10^{-3} \text{ Ci}$) and the microcurie ($1 \, \mu\text{Ci} = 10^{-6} \text{ Ci}$).

EXERCISE 32–3

A sample of radium has an activity of 15 μCi. How many decays per second occur in this sample?

Solution

Using the definition given in Equation 32–6, we find

$$15 \, \mu\text{Ci} = 15 \times 10^{-6} \text{ Ci}$$

$$= (15 \times 10^{-6} \text{ Ci})\left(\frac{3.7 \times 10^{10} \text{ decays/s}}{1 \text{ Ci}}\right) = 5.6 \times 10^5 \text{ decays/s}$$

32–3 Half-Life and Radioactive Dating

The phenomenon of radioactive decay, though fundamentally random in its behavior, has certain properties that make it useful as a type of "nuclear clock." In fact, it has been discovered that radioactive decay can be used to date numerous items of interest from the recent—and not so recent—past. In this section we

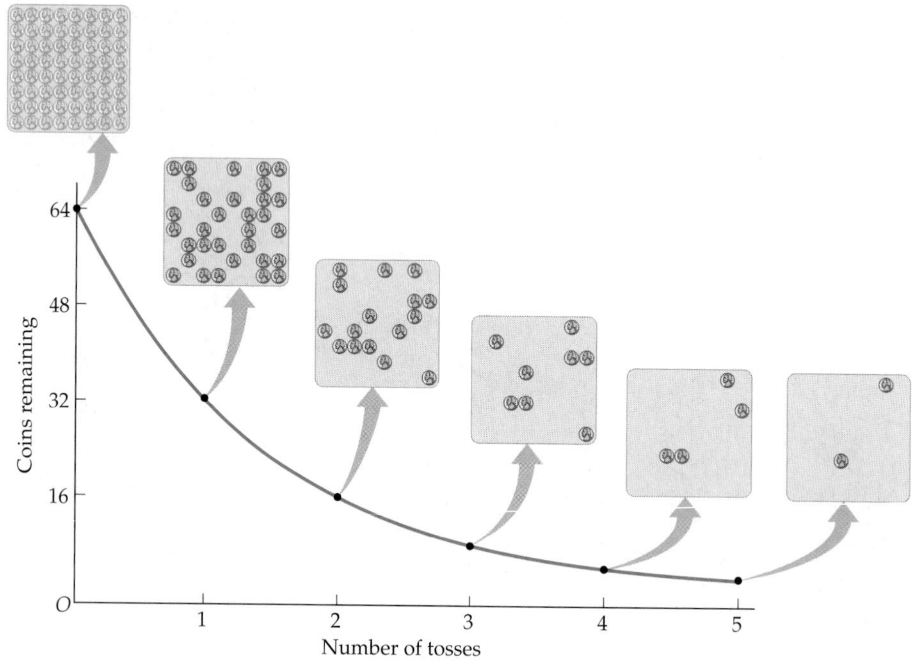

▶ **FIGURE 32–5 Tossing coins as an analogy for nuclear decay**

The points on this graph show the number of coins remaining (on average) if one starts with 64 coins and removes half with each round of tosses. The curve is a plot of the mathematical function $64e^{-(0.693)t}$, where $t = 1$ means one round of tosses, $t = 2$ means two rounds, and so on.

consider the behavior of radioactive decay as a function of time, introduce the concept of a half-life, and show explicitly how these concepts can be applied to dating.

To begin, consider an analogy in which coins represent nuclei, and the side that comes up when a given coin is tossed determines whether the corresponding nucleus decays. Suppose, for example, that we toss a group of 64 coins and remove any coin that comes up tails. We expect that—on average—32 coins will be removed after the first round of tosses. Which coins will be removed—and the precise number that will be removed—cannot be known, because the flip of a coin, like the decay of a nucleus, is a random process. When we toss the remaining 32 coins, we expect an average of 16 more to be removed, and so on, with each round of tosses decreasing the number of coins by a factor of 2. The results after the first few rounds are shown by the points in **Figure 32–5**.

Also shown in Figure 32–5 is a smooth curve representing the following mathematical function:

$$N = (64)e^{-(\ln 2)t} = (64)e^{-(0.693)t} \qquad \textbf{32–8}$$

where N represents the number of coins, and the time variable, t, represents the number of rounds of tosses. For example, if we set $t = 1$ in Equation 32–8, we find $N = 32$, and if we set $t = 2$, we find $N = 16$. This type of "exponential dependence" is a general feature whenever the number of some quantity increases or decreases by a constant factor with each constant interval of time. Examples include the balance in a bank account with compounding interest and the population of the Earth as a function of time.

When nuclei decay, their behavior is much like that of the coins in our analogy. Which nucleus will decay in a given interval of time, and the precise number that will decay, are controlled by a random process that causes the decay—on average—of a given fraction of the original number of nuclei. Thus the number of nuclei, N, remaining at time t is given by an expression analogous to Equation 32–8:

$$N = N_0 e^{-\lambda t} \qquad \textbf{32–9}$$

where N_0 is the number of nuclei present at time $t = 0$, and the constant λ is referred to as the **decay constant.** In the analogy of the coins, $N_0 = 64$ and $\lambda = 0.693\ \text{s}^{-1}$. Note that the larger the value of the decay constant, the more rapidly the number of nuclei decreases with time. **Figure 32–6** shows the dependence on λ graphically.

PROBLEM-SOLVING NOTE

Consistent Units and the Decay Constant

When calculating the number of nuclei present at a given time, be sure to use consistent units for the time, t, and the decay constant, λ. For example, if you express λ in units of y^{-1}, measure the time in units of y.

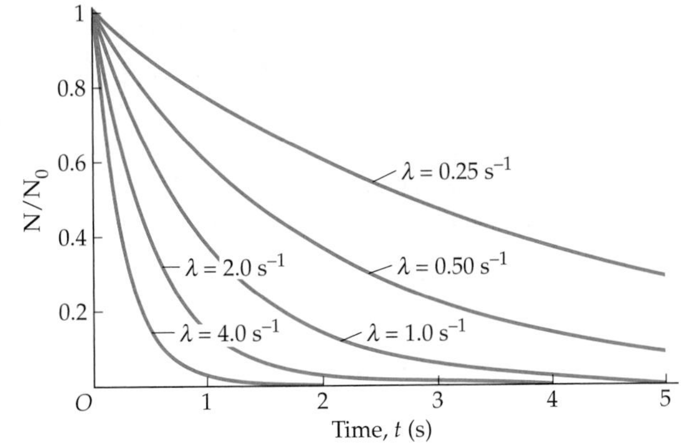

FIGURE 32–6 Dependence on the decay constant
The larger the decay constant, λ, the more rapidly the population of a group of nuclei decreases. In this plot, the value of λ doubles as we move downward from one curve to the next.

ACTIVE EXAMPLE 32–2 Find the Radon Level

Radon can pose a health risk when high levels become trapped in the basement of a house. Suppose 4.75×10^7 radon atoms are in a basement at a time when it is sealed to prevent any additional radon from entering. Given that the decay constant of radon is $0.181\ \text{d}^{-1}$, how many radon atoms remain in the basement after **(a)** 7 d and **(b)** 14 d?

Solution *(Test your understanding by performing the calculations indicated in each step.)*

Part (a)

1. Evaluate Equation 32–9 with $t = 7$ d: $N = 1.34 \times 10^7$

Part (b)

2. Evaluate Equation 32–9 with $t = 14$ d: $N = 3.77 \times 10^6$

Insight

Notice that after 7 d the number of radon atoms has decreased by a factor of about 3.55. After 14 d, the number has decreased by a factor of about $3.55^2 = 12.6$. It follows that every 7 d the number of radon atoms decreases by another factor of 3.55.

Your Turn

What period of time is required for the number of radon atoms to decrease by a factor of 2.00?

(Answers to **Your Turn** *problems are given in the back of the book.)*

A useful way to characterize the rate at which a given type of nucleus decays is in terms of its half-life, which is defined as follows:

> The **half-life** of a given type of radioactive nucleus is the time required for the number of such nuclei to decrease by a factor of 2; that is, for the number to decrease from N_0 to $\frac{1}{2}N_0$, from $\frac{1}{2}N_0$ to $\frac{1}{4}N_0$, and so on.

We can solve for this time, call it $T_{1/2}$, by setting $N = \frac{1}{2}N_0$ in Equation 32–9:

$$\tfrac{1}{2}N_0 = N_0 e^{-\lambda T_{1/2}}$$

Canceling N_0 and taking the natural logarithm of both sides of the equation, we find

$$T_{1/2} = \frac{\ln 2}{\lambda} = \frac{0.693}{\lambda} \qquad \text{32–10}$$

Notice that a large decay constant corresponds to a short half-life, in agreement with the plots shown in Figure 32–6.

▲ Radon-222 ($^{222}_{86}\text{Rn}$) is an isotope produced in a radioactive decay series that includes uranium-238 and radium-226. Since uranium is naturally present in certain kinds of rocks and the soils derived from them, radon, which is a gas, can accumulate in basements and similar enclosed underground spaces that lack adequate ventilation. Radon-222 is itself radioactive, undergoing alpha decay with a half-life of about 4 days, and although its concentration is generally small, it may produce radiation levels great enough to be a health hazard if exposure is prolonged. Homeowners in many parts of the country use test kits to monitor the radiation levels produced by radon.

CONCEPTUAL CHECKPOINT 32–3 How Many Nuclei?

A system consists of N_0 radioactive nuclei at time $t = 0$. The number of nuclei remaining after *half* a half-life (that is, at time $t = \frac{1}{2}T_{1/2}$) is **(a)** $\frac{1}{4}N_0$, **(b)** $\frac{3}{4}N_0$, or **(c)** $\frac{1}{\sqrt{2}}N_0$?

Reasoning and Discussion

Referring to the Insight following Active Example 32–2, we note that if the number of nuclei decreases by a factor f in the time $\frac{1}{2}T_{1/2}$, it will decrease by the factor f^2 in the time $2\left(\frac{1}{2}T_{1/2}\right) = T_{1/2}$. We know, however, that the number of nuclei remaining at the time $T_{1/2}$ is $\frac{1}{2}N_0$. It follows that $f^2 = \frac{1}{2}$, or that $f = \frac{1}{\sqrt{2}}$. Therefore, the number of nuclei remaining at the time $\frac{1}{2}T_{1/2}$ is $\frac{1}{\sqrt{2}}N_0$.

Answer:

(c) At half a half-life, the number of nuclei remaining is $\frac{1}{\sqrt{2}}N_0$.

Now, the property that makes radioactivity so useful as a clock is that its **decay rate, *R*,** or **activity,** depends on time in a straightforward way. To see this, think back to the analogy of the coins. The number of coins that decay (are removed) on the first round of tosses is $32 = \frac{1}{2}(64)$, the number that decay on the second round is $16 = \frac{1}{2}(32)$, and so on. That is, the number that decay in any given interval of time is proportional to the number present at the beginning of the interval.

The same type of analysis applies to nuclei. Therefore, the number of nuclei that decay, ΔN, in a given time interval, Δt, is proportional to the number, N, that is present at time t:

$$R = \left|\frac{\Delta N}{\Delta t}\right| = \lambda N \qquad \text{32–11}$$

There are two points to note regarding this equation. First, observe that the number of nuclei is decreasing, $\Delta N < 0$. It is for this reason that we take the absolute value of the quantity $\Delta N/\Delta t$. Second, notice that the proportionality constant is simply λ, the decay constant. That λ is the correct constant of proportionality can be shown using calculus.

Combining Equation 32–11 with Equation 32–9, we obtain the time dependence of the activity, R:

$$R = \lambda N_0 e^{-\lambda t} = R_0 e^{-\lambda t} \qquad \text{32–12}$$

Note that $R_0 = \lambda N_0$ is the initial value of the activity. We apply this relation in the following Active Example.

ACTIVE EXAMPLE 32–3 Find the Activity of Radon

Referring to Active Example 32–2, calculate how many radon atoms disintegrate per second **(a)** initially and **(b)** after 7 d.

Solution *(Test your understanding by performing the calculations indicated in each step.)*

Part (a)

1. Calculate the activity for $N_0 = 4.75 \times 10^7$. Be sure to convert the decay constant to the unit s^{-1}: $R = \lambda N_0 = 99.5$ decays/s

Part (b)

2. Repeat the calculation for $N = 1.34 \times 10^7$: $R = \lambda N = 28.1$ decays/s

Insight

We see that the initial activity (number of decays per time) is 99.5 Bq. In terms of the curie, the initial activity is 0.00269 μCi.

Your Turn

How long does it take after the basement is sealed for the activity of the radon to decrease to 10.0 Bq?

(Answers to **Your Turn** *problems are given in the back of the book.)*

Referring to Equation 32–12, we see that the basic idea of radioactive dating is simply this: If we know the initial activity of a sample, R_0, and we also know the sample's activity now, R, we can find the corresponding time, t, as follows:

$$\frac{R}{R_0} = e^{-\lambda t}$$

$$t = -\frac{1}{\lambda}\ln\frac{R}{R_0} = \frac{1}{\lambda}\ln\frac{R_0}{R} \qquad \text{32–13}$$

The current activity, R, can be measured in the lab, but how can we know the initial activity, R_0? We address the question next for the specific case of carbon-14.

Carbon-14 Dating

To determine the initial activity of carbon-14 requires a basic knowledge of the role it plays in Earth's biosphere. First, we note that carbon-14 is unstable, with a half-life of 5730 y. It follows that the carbon-14 initially present in any *closed* system will decay away to practically nothing in a time of several half-lives, yet the ratio of carbon-14 to carbon-12 in Earth's atmosphere remains approximately constant at the value 1.20×10^{-12}. Evidently, Earth's atmosphere is not a closed system, at least as far as carbon-14 is concerned.

This is indeed the case. Cosmic rays, which are high-energy particles from outer space, are continuously entering Earth's upper atmosphere and initiating nuclear reactions in nitrogen-14 (a stable isotope). These reactions result in a steady production of carbon-14. Thus, the steady level of carbon-14 in the atmosphere is a result of the balance between the *production rate* due to cosmic rays and the *decay rate* due to the properties of the carbon-14 nucleus.

We note that living organisms have the same ratio of carbon-14 to carbon-12 as the atmosphere, since they continuously exchange carbon with their surroundings. When an organism dies, however, the exchange of carbon ceases and the carbon-14 in the organism (wood, bone, shell, etc.) begins to decay. This process is illustrated in **Figure 32–7**, where we see that the carbon-14 activity of an organism is constant until it dies, at which point it decreases exponentially with a half-life of 5730 y.

All that remains to implement carbon-14 dating is to determine the initial activity, R_0. For convenience, we calculate R_0 for a 1-g sample of carbon. As we know from Chapter 31, 12 g of carbon-12 consists of Avogadro's number of atoms, 6.02×10^{23}. Therefore, 1 g consists of $(6.02 \times 10^{23})/12 = 5.02 \times 10^{22}$ carbon-12 atoms. Multiplying this number of atoms by the ratio of carbon-14 to carbon-12, 1.20×10^{-12}, shows that the number of carbon-14 atoms in the 1-g sample is 6.02×10^{10}. Next, we need the decay constant, which we obtain by rearranging Equation 32–10:

$$\lambda = \frac{\ln 2}{T_{1/2}} = \frac{0.693}{T_{1/2}}$$

▲ The Iceman, found in the Italian Alps in 1991. His age was established by means of radiocarbon dating.

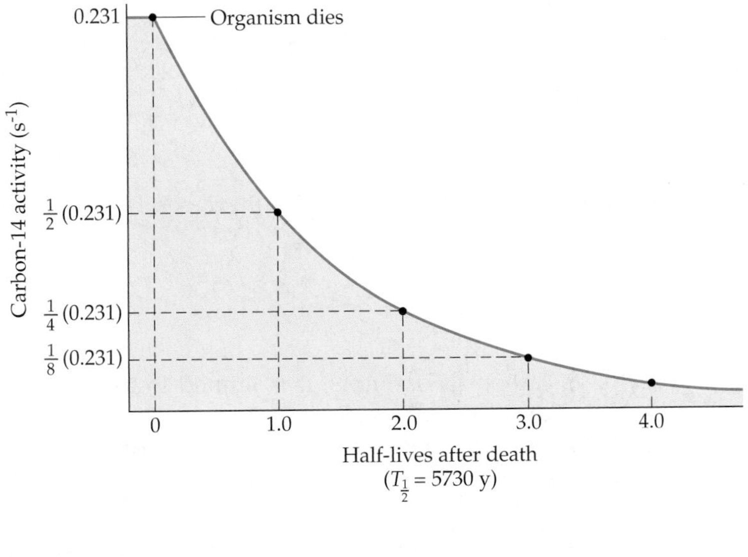

◀ **FIGURE 32–7 Activity of carbon-14**
While an organism is living and exchanging carbon with the atmosphere, its carbon-14 activity remains constant. When the organism dies, the carbon-14 activity decays exponentially with a half-life of 5730 years.

Using $T_{1/2} = 5730\text{ y} = 1.81 \times 10^{11}\text{ s}$, we find $\lambda = 3.83 \times 10^{-12}\text{ s}^{-1}$. Finally, the initial activity of a 1-g sample of carbon is

$$R_0 = \lambda N_0 = (3.83 \times 10^{-12}\text{ s}^{-1})(6.02 \times 10^{10}) = 0.231\text{ Bq} \qquad \textbf{32–14}$$

This is the initial activity used in Figure 32–7. It follows that 5730 y after an organism dies, its carbon-14 activity per gram of carbon will have decreased to about $\frac{1}{2}(0.231\text{ Bq}) = 0.116\text{ Bq}$.

REAL-WORLD PHYSICS
Dating the Iceman

The next Example applies this basic idea to a real-world case of some interest—the Iceman of the Alps.

EXAMPLE 32–5 Age of the Iceman: You Don't Look a Day over 5000

Early in the afternoon of September 19, 1991, a German couple hiking in the Italian Alps noticed something brown sticking out of the ice 8 to 10 m ahead of them. At first they took the object to be a doll or some rubbish. As they got closer, however, it became apparent that the object they had discovered was the body of a person trapped in the ice, with only the top part of the body exposed. Subsequent investigation revealed the remarkably well-preserved body to be that of a Stone Age man who had died in the mountains and become entombed in the ice. When the carbon-14 dating method was applied to the remains of the Iceman and some of the materials he had carried with him, it was found that the carbon-14 activity was about 0.121 Bq per gram of carbon. Using this information, date the remains of the Iceman.

Picture the Problem

Our sketch shows the decay of the carbon-14 activity of a gram of carbon as a function of time. The initial activity of carbon-14 in such a sample is 0.231 Bq.

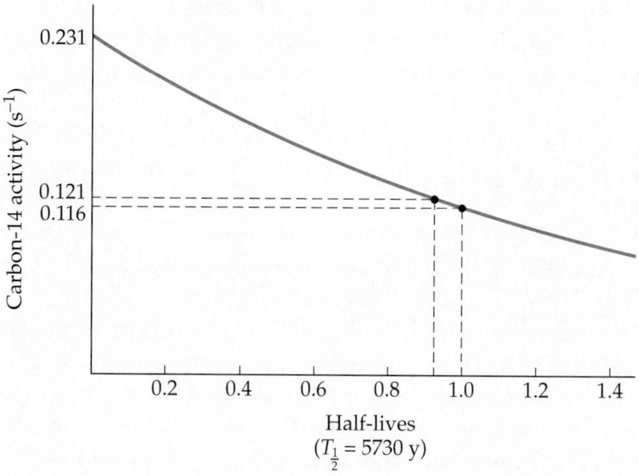

Strategy

We can obtain the age of the remains directly from Equation 32–13: $t = (1/\lambda)\ln(R_0/R)$. In this case, $R_0 = 0.231\text{ Bq}$, and $R = 0.121\text{ Bq}$. Since an answer in years would be most useful, we express the decay constant as $\lambda = 0.693/T_{1/2}$, with $T_{1/2} = 5730\text{ y}$.

Note that the observed activity of 0.121 Bq is slightly greater than $\frac{1}{2}(0.231\text{ Bq}) = 0.116\text{ Bq}$; hence, we expect the age of the remains to be slightly less than the half-life of 5730 y.

Solution

1. Determine the value of the decay constant, λ, in units of y^{-1}:

$$\lambda = \frac{0.693}{5730\text{ y}} = 1.21 \times 10^{-4}\text{ y}^{-1}$$

2. Substitute λ, R_0, and R into Equation 32–13:

$$t = \frac{1}{\lambda}\ln\left(\frac{R_0}{R}\right)$$

$$= \frac{1}{(1.21 \times 10^{-4}\text{ y}^{-1})}\ln\left(\frac{0.231\text{ Bq}}{0.121\text{ Bq}}\right) = 5340\text{ y}$$

Insight

We conclude that the Iceman, who has been dubbed Ötzi, died in the mountains during the Stone Age, some 5340 y ago. Detailed examination of Ötzi's body and possessions indicate he was probably an itinerant sheepherder and/or hunter. He met his end in a violent fashion, however. Recent CT scans confirm that Ötzi was killed by an arrow that entered through his shoulder blade and lodged less than an inch from his left lung.

Practice Problem

If the remains of another iceman of the same age are found in the year 2991, what will be the carbon-14 activity of 1 g of carbon? [**Answer:** 0.107 Bq]

Some related homework problems: Problem 29, Problem 30

As useful as carbon-14 dating is, it is limited to time spans of only a few half-lives, say, 10,000 to 15,000 y. Beyond that range, the current activity will be so small that accurate measurements will be difficult. To measure dates on different time scales, different radioactive isotopes must be used. Other frequently used isotopes and their half-lives are ${}^{210}_{82}\text{Pb}(22.3\text{ y})$, ${}^{40}_{19}\text{K}(1.28 \times 10^9\text{ y})$, and ${}^{238}_{92}\text{U}(4.468 \times 10^9\text{ y})$.

32–4 Nuclear Binding Energy

An α particle consists of two protons and two neutrons. Does it follow that the mass of an α particle is twice the mass of a proton plus twice the mass of a neutron? One would certainly think so, but in fact this is not the case. Alpha particles, and all other stable nuclei containing more than one nucleon, have a mass that is *less* than the mass of the individual nucleons added together.

As strange as this result may seem, it is just one more manifestation of Einstein's theory of relativity. In particular, the reduction in mass of a nucleus, compared with the mass of its constituents, corresponds to a reduction in its energy, according to the relation $E = mc^2$. This reduction in energy is referred to as the **binding energy** of the nucleus. To separate a nucleus into its individual nucleons requires an energy at least as great as the binding energy; therefore, the binding energy indicates how firmly a given nucleus is held together. We illustrate this concept in **Figure 32–8**.

The next Example uses $E = (\Delta m)c^2$ and the atomic masses given in Appendix F to calculate the binding energy of tritium.

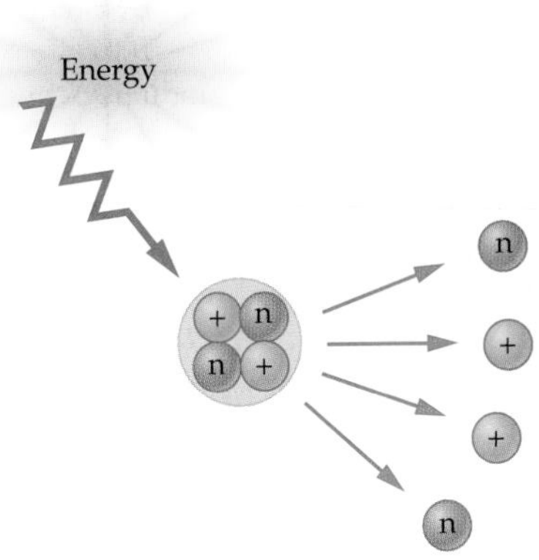

▲ **FIGURE 32–8 The concept of binding energy**
The minimum energy that must be supplied to a stable nucleus to break it into its constituent nucleons is referred to as the binding energy. Because the stable nucleus is the lower-energy state, it has less mass than the sum of the masses of its individual constituents.

EXAMPLE 32–6 The Binding Energy of Tritium

Using the information given in Appendix F, calculate the binding energy of tritium, ${}^3_1\text{H}$.

Picture the Problem
In our sketch we show energy being added to the nucleus of tritium, which consists of one proton and two neutrons. The result is one hydrogen atom and two neutrons.

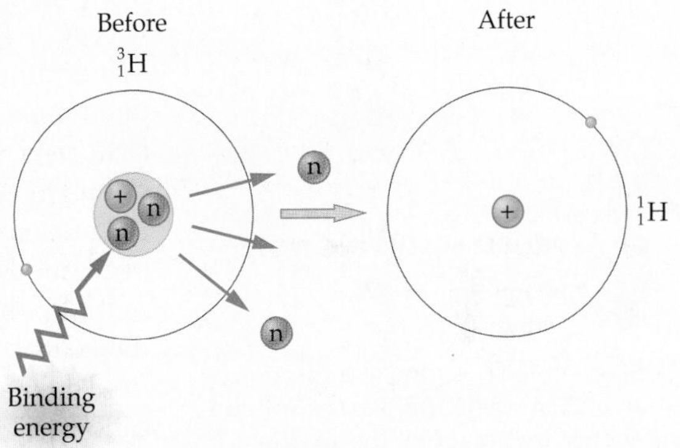

Strategy
To find the binding energy, we simply calculate Δm and multiply by c^2. The only point to be careful about is that, with the exception of the neutron, all the masses given in Appendix F are for neutral atoms. This means that the mass of tritium includes the mass of one electron. The same is true of the hydrogen atom, however. It follows that the electron mass will cancel when we calculate Δm.

Solution

1. Let the initial mass be the mass of tritium:
$$m_i = 3.016049\text{ u}$$

2. The final mass is the mass of hydrogen plus two neutrons:
$$m_f = 1.007825\text{ u} + 2(1.008665\text{ u}) = 3.025155\text{ u}$$

3. Calculate Δm:
$$\Delta m = m_f - m_i$$
$$= 3.025155\text{ u} - 3.016049\text{ u} = 0.009106\text{ u}$$

4. Find the corresponding binding energy (Recall that $1\text{ u} = 931.5\text{ MeV}/c^2$.):
$$E = (\Delta m)c^2 = (0.009106\text{ u})\left(\frac{931.5\text{ MeV}/c^2}{1\text{ u}}\right)c^2$$
$$= 8.482\text{ MeV}$$

Insight
We see that it takes about 8.5 MeV to separate the nucleons of tritium. Compare this with the fact that only 13.6 eV is required to remove the electron from tritium. As we have noted before, nuclear processes typically involve energies in the MeV range, whereas atomic processes require energies in the eV range.

Practice Problem
Calculate the binding energy of an α particle. [**Answer:** 28.296 MeV]

Some related homework problems: Problem 35, Problem 36

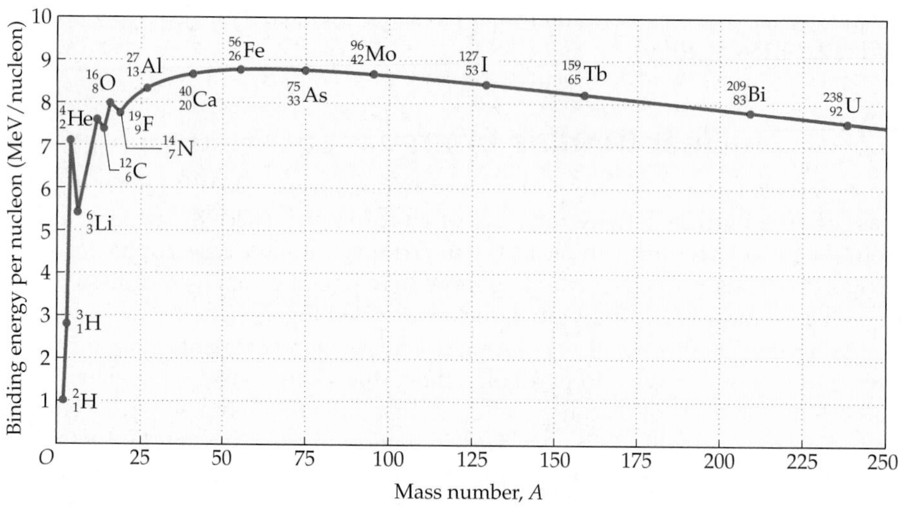

▶ FIGURE 32–9 The curve of binding energy

This graph shows the binding energy per nucleon for a variety of nuclei. Notice that the binding energy per nucleon for tritium is approximately 2.8 MeV, in agreement with Example 32–6.

In addition to knowing the binding energy of a given nucleus, it is also of interest to know the binding energy per nucleon. In the case of tritium, there are three nucleons; hence, the binding energy per nucleon is $\frac{1}{3}(8.482 \text{ MeV}) = 2.827 \text{ MeV}$. **Figure 32–9** presents the binding energy per nucleon for various stable nuclei as a function of the mass number, A. Notice that the binding-energy curve rises rapidly to a maximum near $A = 60$ and then decreases slowly to a value near 7.4 MeV per nucleon for larger A.

It follows that nuclei with A in the range $A = 50$ to $A = 75$ are the most stable nuclei in nature. In addition, the fact that the binding energy per nucleon changes very little for large A means that the strong nuclear force "saturates," in the sense that adding more nucleons does not add to the binding energy per nucleon. This phenomenon can be understood as a consequence of the short range of the nuclear force. Because of this short range, each nucleon interacts with only a few nucleons that are close neighbors to it in the nucleus. Nucleons on the other side of the nucleus are too far away to interact. Thus, from the point of view of a given nucleon, the attractive energy it feels due to other nucleons in the nucleus is essentially the same whether the nucleus has 150 nucleons or 200 nucleons—only the nearby nucleons interact.

The fact that the binding-energy curve has a maximum has important implications, as will be shown in the next two sections. For example, energy can be released when large nuclei split into smaller nuclei (fission), or when small nuclei combine to form a larger nucleus (fusion).

PROBLEM-SOLVING NOTE

Interpreting the Mass Difference

If the mass difference due to a reaction is negative, $\Delta m = m_f - m_i < 0$, it follows that energy is *released* by the reaction. If Δm is positive, energy must be *supplied* to the system to make the reaction occur.

32–5 Nuclear Fission

A new type of physical phenomenon was discovered in 1939 when Otto Hahn and Fritz Strassman found that, under certain conditions, a uranium nucleus can split apart into two smaller nuclei. This process is called **nuclear fission.** As we shall see, nuclear fission releases an amount of energy that is many orders of magnitude greater than the energy released in chemical reactions. This fact has had a profound impact on the course of human events over the last several decades.

To obtain a rough estimate of just how much energy is released as a result of a fission reaction, consider the binding-energy curve in Figure 32–9. For a large nucleus like uranium-235, $^{235}_{92}$U, we see that the binding energy per nucleon is approximately 7.5 MeV. If this nucleus splits into two nuclei with roughly half the mass number ($A \sim 235/2 \sim 115$), we see that the binding energy per nucleon increases to roughly 8.3 MeV. If all 235 nucleons in the original uranium nucleus release $(8.3 \text{ MeV} - 7.5 \text{ MeV}) = 0.8 \text{ MeV}$ of energy, the total energy release is

$$(235 \text{ nucleons})(0.8 \text{ MeV/nucleon}) = 200 \text{ MeV}$$

Note that this is the energy released by a *single nucleus* undergoing fission. In contrast, the combustion of a *single molecule* of gasoline releases an energy of only about 2 eV. Thus the nuclear reaction gives off about one hundred million times more energy than the chemical reaction!

The first step in a typical fission reaction occurs when a slow neutron is absorbed by a uranium-235 nucleus. This step increases the mass number of the nucleus by one and leaves it in an excited state:

$$ {}^{1}_{0}\text{n} + {}^{235}_{92}\text{U} \longrightarrow {}^{236}_{92}\text{U}^{*} $$

The excited nucleus oscillates wildly and becomes highly distorted, as depicted in **Figure 32–10**. In many respects, the nucleus behaves like a spinning drop of water. Like a drop of water, the nucleus can distort only so much before it breaks apart into smaller pieces; that is, before it undergoes fission.

There are about 90 different ways in which the uranium-235 nucleus can undergo fission. Typically, 2 or 3 neutrons (2.47 on average) are released during the fission process, in addition to the two smaller nuclei that are formed. The reason that neutrons are released can be seen by examining Figure 32–1, where we show N and Z for various nuclei. A large nucleus, like uranium-235, contains a higher percentage of neutrons than a smaller nucleus—that is, the larger nuclei deviate more from the $N = Z$ line. Thus, if ${}^{235}_{92}\text{U}$ were to simply break in two—keeping the same percentage of neutrons and protons in each piece—the smaller nuclei would have too many neutrons to be stable. As a result, neutrons are typically given off in a fission reaction.

One of the possible fission reactions for ${}^{235}_{92}\text{U}$ is considered in the following Example.

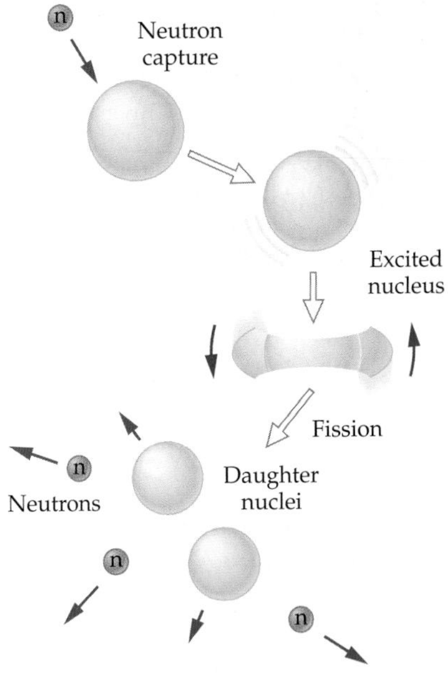

▲ **FIGURE 32–10 Nuclear fission**
When a large nucleus captures a neutron, it may become excited and ultimately split into two smaller nuclei. This is the process of fission.

EXAMPLE 32–7 A Fission Reaction of Uranium-235

When uranium-235 captures a neutron, it may undergo the following fission reaction: ${}^{1}_{0}\text{n} + {}^{235}_{92}\text{U} \longrightarrow {}^{236}_{92}\text{U}^{*} \longrightarrow {}^{141}_{56}\text{Ba} + ? + 3{}^{1}_{0}\text{n}$. **(a)** Complete the reaction and **(b)** determine the energy it releases.

Picture the Problem

The indicated fission reaction is shown in our sketch, starting with the ${}^{236}_{92}\text{U}^{*}$ nucleus. The known numbers of protons and neutrons are indicated for both ${}^{236}_{92}\text{U}^{*}$ and ${}^{141}_{56}\text{Ba}$. The corresponding numbers for the unidentified nucleus are to be determined.

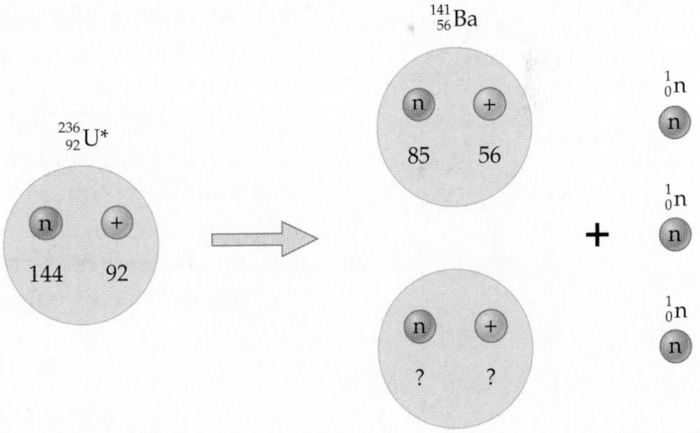

Strategy

a. The total number of protons must be the same before and after the reaction, as must the number of neutrons. By conserving protons and neutrons, we can determine the missing numbers on the unidentified nucleus.

b. As in other reactions, we calculate the difference in mass, $|\Delta m|$ and multiply by c^2. The same number of electrons appears on both sides of the reaction; hence, the electron mass will cancel when we calculate $|\Delta m|$.

Solution

Part (a)

1. Determine the number of protons, Z, neutrons, N, and nucleons, A, in the unidentified nucleus. When calculating the number of neutrons, be sure to include the three individual neutrons given off in the reaction:

$$Z = 92 - 56 = 36$$
$$N = 144 - 85 - 3 = 56$$
$$A = Z + N = 36 + 56 = 92$$

2. Use Z and A to specify the unidentified nucleus: ${}^{92}_{36}\text{Kr}$

Part (b)

3. Use Appendix F to find the initial mass of the system; that is, the mass of a ${}^{235}_{62}\text{U}$ atom plus the mass of a neutron:

$$m_{\text{i}} = 235.043925\ \text{u} + 1.008665\ \text{u} = 236.052590\ \text{u}$$

continued on next page

continued from previous page

4. Use Appendix F to find the final mass of the system, including both nuclei and the three neutrons:

$$m_f = 140.914406\text{ u} + 91.926111\text{ u} + 3(1.008665\text{ u}) = 235.866512\text{ u}$$

5. Calculate the mass difference and the corresponding energy release (recall that $1\text{ u} = 931.5\text{ MeV}/c^2$):

$$\Delta m = m_f - m_i = 235.866512\text{ u} - 236.052590\text{ u} = -0.186078\text{ u}$$

$$E = |\Delta m|c^2 = (0.186078\text{ u})\left(\frac{931.5\text{ MeV}/c^2}{1\text{ u}}\right)c^2 = 173.3\text{ MeV}$$

Insight

Thus, as suggested by our crude calculation earlier in this section, the energy given off by a typical fission reaction is on the order of 200 MeV.

Practice Problem

Complete the following reaction, and determine the energy it releases: ${}^1_0\text{n} + {}^{235}_{92}\text{U} \longrightarrow {}^{236}_{92}\text{U}^* \longrightarrow {}^{140}_{54}\text{Xe} + ? + 2{}^1_0\text{n}$. [**Answer:** ${}^{94}_{38}\text{Sr}$, 184.7 MeV]

Some related homework problems: Problem 42, Problem 43

Chain Reactions

The fact that fission reactions of ${}^{235}_{92}\text{U}$ give off more than one neutron on average has significant implications. To see why, recall that the fission of ${}^{235}_{92}\text{U}$ is initiated by the absorption of a neutron in the first place. So the neutrons given off by one ${}^{235}_{92}\text{U}$ fission reaction may cause additional fission reactions in other ${}^{235}_{92}\text{U}$ nuclei. A reaction that proceeds from one nucleus to another in this fashion is referred to as a **chain reaction.**

Suppose, for the sake of discussion, that a fission reaction gives off two neutrons and that both neutrons induce additional fissions in other nuclei. These nuclei in turn give off two neutrons. Starting with one nucleus to begin the chain reaction, we have two nuclei in the second generation of the chain, four nuclei in the third generation, and so on, as indicated in **Figure 32–11**. After only 100 generations, the number of nuclei undergoing fission is 1.3×10^{30}. If each of these reactions gives off 200 MeV of energy, the total energy release after just 100 generations is 4.1×10^{19} J. To put this into everyday terms, this is enough energy to supply the needs of the entire United States for half a year. Clearly, a rapidly developing **runaway** chain reaction, like the one just described, would result in the explosive release of an enormous amount of energy.

A great deal of effort has gone into controlling the chain reaction of ${}^{235}_{92}\text{U}$. If the release of energy can be kept to a manageable and usable level—avoiding

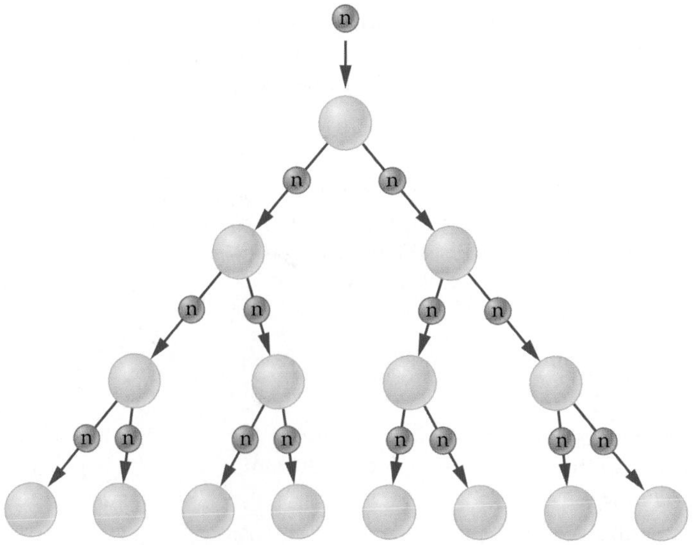

▶ FIGURE 32–11 A chain reaction
A chain reaction occurs when a neutron emitted by one fission reaction induces a fission reaction in another nucleus. In a runaway chain reaction, the number of neutrons given off by one reaction that cause an additional reaction is greater than one. In the case shown here, two neutrons from one reaction induce additional reactions. In a controlled chain reaction, like those used in nuclear power plants, only one neutron, on average, induces additional reactions.

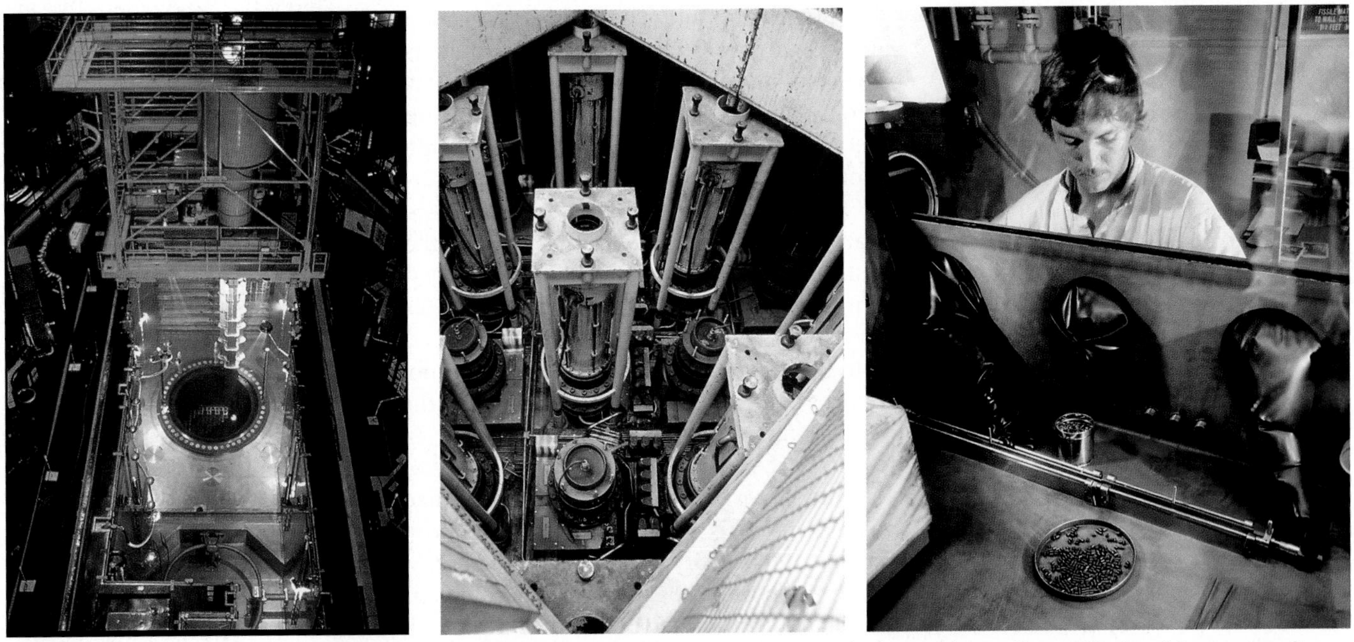

▲ (Left) The core of a nuclear reactor at a power plant. The core sits in a pool of water, which provides cooling while absorbing stray radiation. The blue glow suffusing the pool is radiation from electrons traveling through the water at relativistic velocities. Above the core is a crane used to replace the fuel rods when their radioactive material becomes depleted. (Center) A closer view of a reactor core. The structures projecting above the core house the mechanism that lowers and raises the control rods, thereby regulating the rate of fission in the reactor. The tubes that are nearly flush with the top of the core house the fuel rods. (Right) A technician loads pellets of fissionable material into fuel rods.

explosions or meltdowns—it follows that a powerful source of energy is at our disposal.

The first controlled nuclear chain reaction was achieved by Fermi in 1942, using a racquetball court at the University of Chicago for his improvised laboratory. His reactor consisted of blocks of uranium (the fuel) stacked together with blocks of graphite (the moderator) to form a large "pile." In such a **nuclear reactor,** the **moderator** slows the neutrons given off during fission, making it more likely that they will be captured by other uranium nuclei and cause additional fissions. The reaction rate is adjusted with **control rods** made of a material (such as cadmium) that is very efficient at absorbing neutrons. With the control rods fully inserted into the pile, any reaction that begins quickly dies out because neutrons are absorbed rather than allowed to cause additional fissions. As the control rods are pulled partway out of the pile, more neutrons become available to induce reactions. When, on average, one neutron given off by any fission reaction produces an additional reaction, the reactor is said to be **critical**. If the control rods are pulled out even farther from the pile, the number of neutrons causing fission reactions in the next generation becomes greater than one, and a runaway reaction begins to occur. Nuclear reactors that produce power for practical applications are operated near their critical condition by continuous adjustment of the placement of the control rods.

REAL-WORLD PHYSICS
Nuclear reactors

Fermi's original reactor, which was designed to test the basic scientific principles underlying fission reactions, produced a power of only about 0.5 W when operating near its critical condition. This is the power necessary to operate a flashlight bulb. In comparison, modern-day nuclear power reactors typically produce 1000 MW of power, enough to power an entire city.

32–6 Nuclear Fusion

When two light nuclei combine to form a more massive nucleus, the reaction is referred to as **nuclear fusion.** The binding-energy curve shown in Figure 32–9 shows that when light nuclei undergo fusion, the resulting nucleus will have a greater binding energy per nucleon than the original nuclei. This means that *the larger nucleus formed by fusion has less mass than the sum of the masses of the original light nuclei.* The mass difference appears as an energy ($E = mc^2$) given off by the reaction.

It is not easy to initiate a fusion reaction. The reason is that to combine two small nuclei, say two protons, into a single nucleus requires that the initial nuclei have a kinetic energy great enough to overcome the Coulomb repulsion the protons exert on each other. The temperature required to give protons the needed kinetic energy is about 10^7 K. When the temperature is high enough to initiate fusion, we say that the resulting process is a **thermonuclear fusion reaction.**

One place where such high temperatures may be encountered is at the core of a star. In fact, all stars generate energy by the process of thermonuclear reactions. Most stars (including the Sun) fuse hydrogen to produce helium. At this very moment, the Sun is converting roughly 600,000,000 tons of hydrogen into helium every second. Some stars also fuse helium or other heavier elements.

In its early stages a star begins as an enormous cloud of gas and dust. As the cloud begins to fall inward under the influence of its own gravity, it converts gravitational potential energy into kinetic energy. This means that the gas heats up. When the temperature becomes high enough to begin the fusion process, the resulting release of energy tends to stabilize the star, preventing its further collapse. A star like the Sun can burn its hydrogen for about 10 billion years, producing a remarkably stable output of energy so important to life on Earth. When its hydrogen fuel is depleted, the Sun will enter a new phase of its life, becoming a red giant and converting its fusion process to one that fuses helium. When the Sun enters its red giant phase, it will expand greatly in size, eventually engulfing the Earth and vaporizing it.

REAL-WORLD PHYSICS

Powering the Sun: the proton-proton cycle

The Sun is powered by the **proton-proton cycle** of fusion reactions, as first described by Hans Bethe (1906–2005). This cycle consists of three steps. The first two steps are as follows:

$$ {}^{1}_{1}\text{H} + {}^{1}_{1}\text{H} \longrightarrow {}^{2}_{1}\text{H} + e^{+} + \nu_e $$

$$ {}^{1}_{1}\text{H} + {}^{2}_{1}\text{H} \longrightarrow {}^{3}_{2}\text{He} + \gamma $$

A number of reactions are possible for the third step, but the dominant one is

$$ {}^{3}_{2}\text{He} + {}^{3}_{2}\text{He} \longrightarrow {}^{4}_{2}\text{He} + {}^{1}_{1}\text{H} + {}^{1}_{1}\text{H} $$

The overall energy production during this cycle is about 27 MeV.

In the next Active Example we show how the mass difference in a deuterium-deuterium reaction can be used to calculate the amount of energy released.

ACTIVE EXAMPLE 32–4 The Deuterium-Deuterium Reaction

Find the energy released in the deuterium-deuterium reaction, ${}^{2}_{1}\text{H} + {}^{2}_{1}\text{H} \longrightarrow {}^{3}_{1}\text{H} + {}^{1}_{1}\text{H}$.

Solution *(Test your understanding by performing the calculations indicated in each step.)*

1. Calculate the initial mass:	$m_i = 4.028204$ u
2. Calculate the final mass:	$m_f = 4.023874$ u
3. Find the difference in mass:	$\Delta m = -0.004330$ u
4. Convert the mass difference to an energy release:	$E = \lvert\Delta m\rvert c^2 = 4.033$ MeV

Insight

The tritium produced in this reaction has a half-life of 12.33 y. Deuterium is stable.

Your Turn

Find the energy released in the reaction ${}^{3}_{2}\text{He} + {}^{3}_{2}\text{He} \longrightarrow {}^{4}_{2}\text{He} + {}^{1}_{1}\text{H} + {}^{1}_{1}\text{H}$.

(Answers to **Your Turn** *problems are given in the back of the book.)*

Just as nuclear fission reactions can be controlled, allowing for the generation of usable power, it would be desirable to control fusion as well. In fact, there are many potential advantages of fusion over fission. For example, fusion reactions yield more energy from a given mass of fuel than fission reactions. In addition, one type of fuel for a fission reactor, ${}^{2}_{1}H$, is readily obtained from seawater. To date, however, more energy is required to produce sustained fusion reactions in the lab than is released by the reactions. Still, researchers are close to the break-even point, and many are confident that in time controlled fusion reactions that give off usable amounts of energy will be possible.

REAL-WORLD PHYSICS
Manmade fusion

Most attempts at controlled nuclear fusion employ one of two basic methods. In both methods the basic idea is to overcome the repulsive electrical forces that keep nuclei apart by giving them sufficiently large kinetic energies, which can be accomplished by heating the fuel to temperatures on the order of 10^7 K.

At temperatures like these, all atoms are completely ionized, forming a gas of electrons and nuclei known as a *plasma*. To initiate fusion in the plasma, one must maintain the plasma long enough for collisions to occur between the appropriate nuclei. This can be accomplished using a technique known as **magnetic confinement,** in which powerful magnetic fields confine a plasma within a "magnetic bottle." The trapped plasma is kept away from the walls of the container—preventing their melting—and is heated until fusion begins.

Another approach to reaching the high temperatures and pressures required for fusion is by way of **inertial confinement.** In this technique, one begins by dropping a small solid pellet of fuel into a vacuum chamber. When the pellet reaches the center of the chamber it is bombarded from all sides by high-power laser beams. These beams heat and vaporize the surface of the pellet in an almost instantaneous event. The heating that causes fusion is still to come, however. As the vaporized exterior of the pellet expands rapidly outward, it exerts an inward "thrust" that causes the remainder of the pellet to implode. This implosion is so violent that the pellet's temperature and pressure rise to levels sufficient to ignite the desired fusion reactions.

▲ Nuclear fusion, which powers the stars, may eventually turn out to be the clean, inexpensive, and renewable energy source that our society needs. But fusion requires enormous temperatures and pressures, and so far the problems involved in creating a practical fusion technology have not been overcome. Several different approaches have been explored in the effort to produce sustained nuclear fusion. One of them, embodied in the Z machine at Sandia National Laboratories in New Mexico, shown here, involves the compression of a plasma by an intense burst of X-rays. So far, temperatures of nearly 2 million degrees have been attained, but only for very brief periods of time.

32–7 Practical Applications of Nuclear Physics

Nuclear physics, though it involves objects none of us will ever touch or see, nevertheless has significant impact on our everyday life. In this section we consider a number of ways in which nuclear physics affects us all, either directly or indirectly.

Biological Effects of Radiation

Although nuclear radiation can be beneficial when used to image our bodies, or to treat a variety of diseases, it can also be harmful to living tissues. For example, the high-energy photons, electrons, and α particles given off by radioactive decay can ionize a neutral atom by literally "knocking" one or more of its electrons out of the atom. In fact, since typical nuclear decays give off energies in the MeV range, and typical ionization energies are in the 10 eV range, it follows that a single α, β, or γ particle can ionize thousands of atoms or molecules.

REAL-WORLD PHYSICS: BIO
Radiation and cells

Such ionization can be harmful to a living cell by altering the structure of some of its molecules. The result can be a cell that no longer functions or behaves normally, or even a cell that will soon die. This effect is the basis for radiation treatments of cancer, which seek to concentrate high doses of radiation on a cancerous tumor in order to kill the malignant cells. But just what is a "high dose," and how do we quantify radiation dosage?

The first radiation unit to be defined, called the **roentgen** (R), is directly related to the amount of ionization caused by X-rays or γ rays. Suppose such radiation is sent through a mass, m, of dry air at standard temperature and pressure (STP). If the radiation creates ions in the air with a total charge of q, we say that the dose of radiation delivered to the air is proportional to q/m. Specifically, we say that the dosage of X-rays or γ rays is 1 R if an ionization charge of 2.58×10^{-4} C is produced in 1 kg of dry air.

Definition of the Roentgen, R

$$1\ \text{R} = 2.58 \times 10^{-4}\ \text{C/kg} \qquad \text{(X-rays or } \gamma \text{ rays in dry air at STP)} \qquad 32\text{–}15$$

SI unit: C/kg

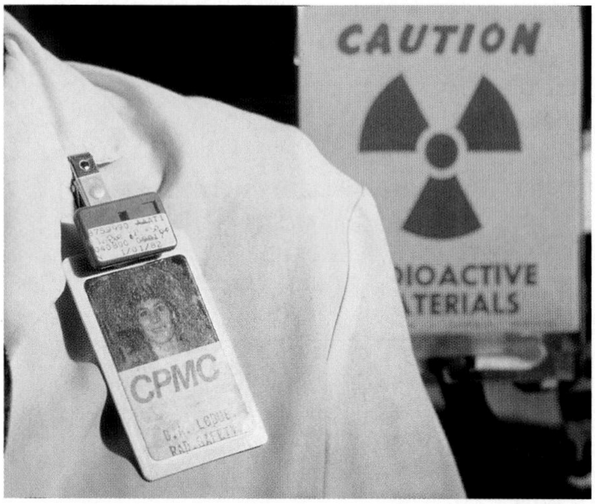

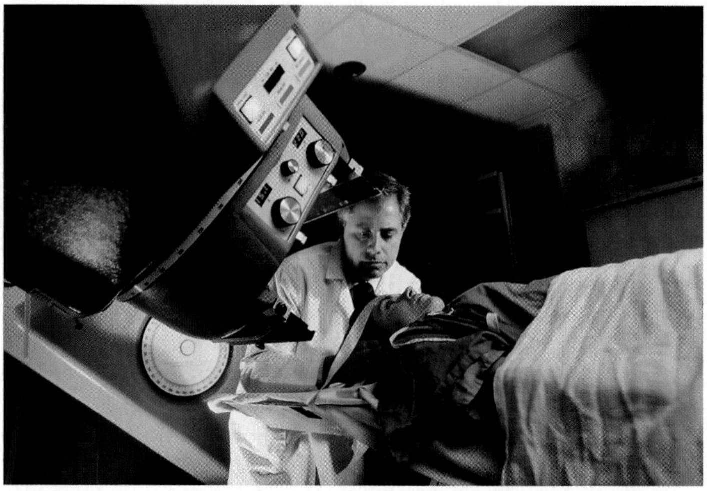

▲ Because radiation can kill or damage living cells, people who work with radioactive materials or other sources of ionizing radiation (such as X-ray tubes) must be careful to monitor their cumulative exposure. One way to do this is with a simple film badge (left). When the photographic film is developed, the degree of darkening records the amount of radiation it has received. Ironically, the fact that radiation can be lethal to cells also makes it a valuable therapeutic tool in the fight against cancer. At right, radiation from cobalt-60 is being used to attack a malignant tumor. Today, many hospitals have their own particle accelerators to manufacture short-lived radioactive isotopes for the treatment of different kinds of cancer or to use as tracers.

The roentgen is one key measure of radiation dosage, but there are others. The focus of the **rad,** an acronym for **r**adiation **a**bsorbed **d**ose, is the amount of energy that is absorbed by the irradiated material, regardless of the type of radiation that delivers the energy. For example, if a 1-kg sample of any material absorbs 0.01 J of energy, we say that it has received a dose of 1 rad:

Definition of the Rad

$$1 \text{ rad} = 0.01 \text{ J/kg} \qquad \text{(any type of radiation)} \tag{32–16}$$

SI unit: J/kg

Because the rad depends only on the amount of energy absorbed, and not on the type of radiation, more information is needed if we are to have an indication of the biological effect a certain dosage will produce. For example, a 1-rad dose of X-rays is far less likely to cause a cataract in the cornea of the eye than a 1-rad dose of neutrons. To take into account such differences, we introduce a quantity called the **r**elative **b**iological **e**ffectiveness, or **RBE.** The standard used for comparison is the biological effect produced by a 1-rad dose of 200-keV X-rays. Thus, the RBE is defined as follows:

Definition of Relative Biological Effectiveness, RBE

$$\text{RBE} = \frac{\text{the dose of 200-keV X-rays necessary to produce a given biological effect}}{\text{the dose of a particular type of radiation necessary to produce the same biological effect}} \tag{32–17}$$

SI unit: dimensionless

Representative RBE values are given in Table 32–3. Note that the larger the RBE for a certain type of radiation, the greater the biological damage caused by a given dosage of that radiation.

Combining the dosage in rad and the RBE value for a given radiation yields a new unit of radiation referred to as the **biologically equivalent dose.** This unit is measured in **rem,** which stands for **r**oentgen **e**quivalent in **m**an. To be specific, dosage in rem is defined as follows:

Definition of Roentgen Equivalent in Man, rem

$$\text{dose in rem} = \text{dose in rad} \times \text{RBE} \tag{32–18}$$

SI unit: J/kg

TABLE 32–3
Relative Biological Effectiveness, RBE, for Different Types of Radiation

Type of radiation	RBE
Heavy ions	20
α rays	10–20
Protons	10
Fast neutrons	10
Slow neutrons	4–5
β rays	1.0–1.7
γ rays	1
200-keV X-rays	1

Defined in this way, 1 rem of radiation produces the same amount of biological damage, no matter which type of radiation is involved. In addition, the larger the dosage in rem, the greater the biological damage. Referring to Table 32–3, we see that 1 rad of 200-keV X-rays produces a radiation dose of 1 rem, whereas 1 rad of protons produces a dose of 10 rem. The dosage of radiation we receive per year from cosmic rays is about 28 mrem = 0.028 rem. Other typical values of radiation dosages are presented in Table 32–4. (For comparison, a dose of 50 to 100 rem damages blood-forming tissues, whereas a dose of 500 rem usually results in death.)

TABLE 32–4 Typical Radiation Dosages

Source of radiation	Radiation dose (rem/y)
Inhaled radon	~0.200
Medical/dental examinations	0.040
Cosmic rays	0.028
Natural radioactivity in the Earth and atmosphere	0.028
Nuclear medicine	0.015

EXERCISE 32–4

A biological sample receives a dose of 456 rad from neutrons with an RBE of 6.20. **(a)** Find the dosage in rem. **(b)** If this same dosage in rem is to be delivered by α rays with an RBE of 13.0, what dosage in rad is required?

Solution

a. The dose in rem is the product of the dose in rad and the RBE:

$$(456 \text{ rad})(6.20) = 2830 \text{ rem}$$

b. The dose in rad is the dose in rem divided by the RBE:

$$\frac{2830 \text{ rem}}{13.0} = 218 \text{ rad}$$

Radioactive Tracers

The phenomenon of radioactivity has found applications of many types in medicine and biology. One of these applications employs a radioactive isotope as a sort of "identification tag" or "tracer" to determine the location and quantity of a substance of interest.

REAL-WORLD PHYSICS: BIO
Radioactive tracers

For example, an artificially produced radioactive isotope of iodine, $^{131}_{53}\text{I}$, is used to determine the condition of a person's thyroid gland. A healthy thyroid plays an important role in the distribution of iodine—a necessary nutrient—throughout the body. To see if the thyroid is functioning as it should, a patient drinks a small quantity of radioactive sodium iodide, which incorporates the isotope $^{131}_{53}\text{I}$. A couple of hours later, the radiation given off by the patient's thyroid gland is measured, giving a direct indication of the amount of sodium iodide the gland has processed.

Because radioactive tracers differ from their nonradioactive counterparts only in the composition of their nuclei, they undergo the same chemical and metabolic reactions. This makes them useful in diagnosing and treating a variety of different conditions. For example, chromium-51 is used to label red blood cells and to quantify gastrointestinal protein loss, copper-64 is used to study genetic diseases affecting copper metabolism, and yttrium-90 is used for liver cancer therapy.

PET Scans

REAL-WORLD PHYSICS: BIO
Positron-emission tomography

Perhaps the most amazing medical use of radioactive decay is in the diagnostic technique known as **positron-emission tomography** (PET). Like the images made using radioactive tracers, PET scans are produced with radiation that *emerges from within* the body, as opposed to radiation that is generated externally and is then passed through the body. Remarkably, the radiation in a PET scan is produced by the *annihilation of matter and antimatter* within the patient's body! It sounds like science fiction, but it is, in fact, a valuable medical tool.

To produce a PET scan, a patient is given a radiopharmaceutical, like fluorodeoxyglucose (FDG), which contains an atom that decays by giving off a positron (the antiparticle to the electron). In the case of FDG, a fluorine-18 atom is attached to a molecule of glucose, the basic energy fuel of cells. The fluorine-18 atom undergoes the following decay with a half-life of 110 minutes:

$$^{18}_{9}\text{F} \longrightarrow {}^{18}_{8}\text{O} + \text{e}^{+} + \nu_{\text{e}}$$

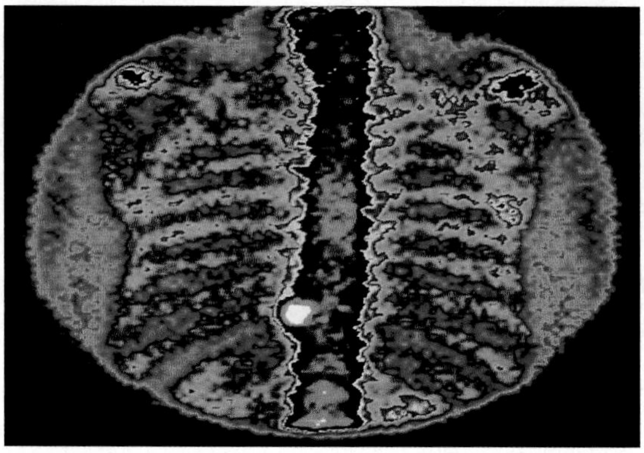

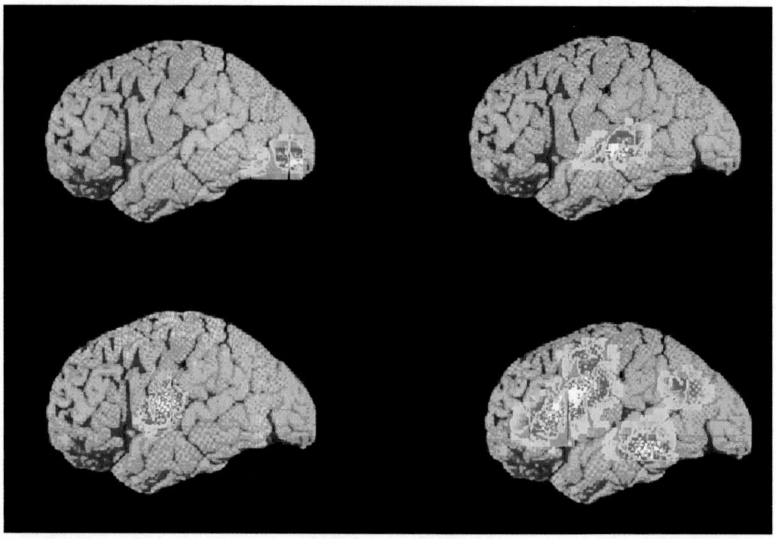

▲ Whereas X-ray images are created by passing a beam of radiation through the body, other imaging techniques make use of radiation produced within the body. One method involves the use of radioactive isotopes that tend to become concentrated in particular tissues or structures, such as the thyroid gland or the bones. Radiation from that component can then be measured, or even used to create an image, as in the bone scan at left, where areas of abnormally high radiation (white) indicate the presence of malignancy. A sophisticated variant of this approach is used to produce PET scans (see text), which are most commonly employed to visualize areas of the body where cellular energy production is most intense. The series of images at right records brain activity associated with seeing words (upper left), hearing words (upper right), speaking words (lower left), and thinking about words while uttering them (lower right). (The image of the brain, superimposed for reference, is not part of the PET scan.)

Almost immediately after the positron is emitted in this decay, it encounters an electron, and the two particles annihilate each other in a burst of energy. In fact, the annihilation process generates two powerful γ rays moving in opposite directions. As these γ rays emerge from the body, specialized detectors observe them and determine their point of origin. The resulting computerized image shows the areas in the body where glucose metabolism is most intense.

PET scans are particularly useful in examining the brain. For example, a PET scan using FDG can show which regions of the brain are most active when a person is performing a specific mental task, like counting, speaking, or translating a foreign language. The scans can also show abnormality in brain function, as when one side of a brain becomes more active than the other during an epileptic seizure. Finally, PET scans can be used to locate tumors in the brain and other parts of the body, and to monitor the progress of their treatment.

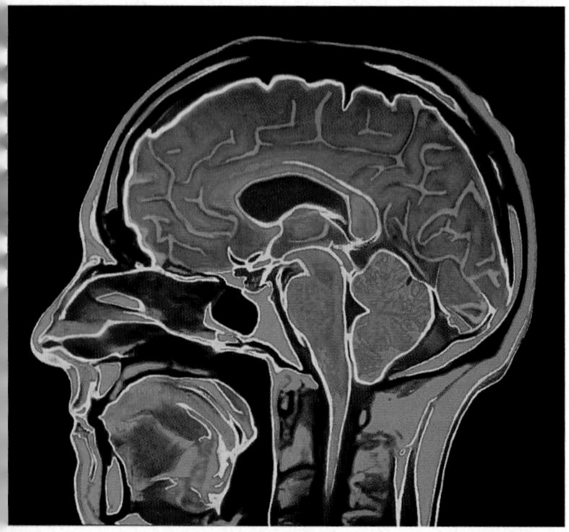

▲ Magnetic resonance imaging (MRI) is a safe, noninvasive technique for visualizing internal body structures. No ionizing radiation is involved, so the risk of tissue damage is minimal. MRI images generally show soft tissue with greater clarity than do ordinary X-rays.

Magnetic Resonance Imaging

A diagnostic technique that gives images similar to those obtained with computerized tomography (Section 31–7) is **magnetic resonance imaging,** or MRI. The basic physical mechanism used in MRI is the interaction of a nucleus with an externally applied magnetic field.

Consider the simplest nucleus—a single proton. Protons have a magnetic moment, like a small bar magnet, and when they are placed in a constant magnetic field they precess about the field with a frequency proportional to the field strength. In addition, protons have a spin of one-half, like electrons; hence, their spin can be either "up" or "down" relative to the direction of the constant magnetic field. When an oscillating magnetic field is applied perpendicular to the constant magnetic field, it can cause protons to "flip" from one spin state to the other if the frequency of oscillation is in resonance with the proton's precessional frequency. These spin flips result in the absorption or release of energy in the radio portion of the electromagnetic spectrum, which can be detected electronically.

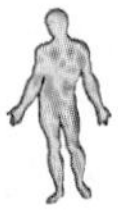

REAL-WORLD PHYSICS: BIO
Magnetic resonance imaging (MRI)

By varying the strength of the constant magnetic field as a function of position, spin flips can be detected in various parts of the body being examined. Using a computer to combine signals from various positions allows for the generation of detailed cross-sectional images, as in CAT scans.

One of the advantages of MRI, however, is that the photons associated with the magnetic fields used in the imaging process are very low in energy. In fact, typical energies are in the range of 10^{-7} eV, much less than typical ionization energies; hence, MRI causes very little cellular damage. In contrast, photons used in CAT scans have energies that can range from 10^4 to 10^6 eV, more than enough to produce cellular damage.

32–8 Elementary Particles

Scientists have long sought to identify the fundamental building blocks of all matter, the **elementary particles.** At one point, it was thought that atoms were elementary particles—one for each element. As we saw in the previous chapter, however, this idea was put to rest when atoms were discovered to be made up of electrons, protons, and neutrons. Of these three particles, only the electron is presently considered to be elementary—protons and neutrons are now known to be composed of still smaller particles. In addition, approximately 300 new particles were discovered in the last half of the twentieth century, most of which are unstable and have lifetimes of only 10^{-6} to 10^{-23} s.

Although a complete accounting of the current theories of elementary particles is beyond the scope of this text, we shall outline the basic insights that have been derived from these theories and from related experiments. We begin by describing the four fundamental forces of nature, since particles can be categorized according to which of these forces they experience.

The Fundamental Forces of Nature

Although nature presents us with a myriad of different physical phenomena—from tornadoes and volcanoes to sunspots and comets to galaxies and black holes—all are the result of just *four* fundamental forces. This is one example of the simplicity that physicists see in nature. These forces, in order of diminishing strength, are the strong nuclear force, the electromagnetic force, the weak nuclear force, and gravity. If we assign a strength of 1 to the strong nuclear force, for purposes of comparison, the strength of the electromagnetic force is 10^{-2}, the strength of the weak nuclear force is 10^{-6}, and the strength of the gravitational force is an incredibly tiny 10^{-43}. These results are summarized in Table 32–5.

TABLE 32–5 The Fundamental Forces

Force	Relative Strength	Range
Strong	1	≈ 1 fm
Electromagnetic	10^{-2}	Infinite ($\propto 1/r^2$)
Weak	10^{-6}	$\approx 10^{-3}$ fm
Gravitational	10^{-43}	Infinite ($\propto 1/r^2$)

All objects of finite mass experience gravitational forces. This is one reason why gravity is such an important force in the universe, even though it is spectacularly weak. Similarly, objects with a finite charge experience electromagnetic forces. As for the weak and strong nuclear forces, some particles experience only the weak force, whereas others experience both the weak and the strong force. We turn now to a discussion of particles that fall into these latter two categories.

Leptons

Particles that are acted on by the weak nuclear force but not by the strong nuclear force are referred to as **leptons.** There are only six leptons known to exist, all of which are listed in Table 32–6. The most familiar of these are the electron and its corresponding neutrino—both of which are stable. No internal structure has ever been detected in any of these particles. As a result, all six leptons have the status of true elementary particles.

TABLE 32–6 Leptons

Particle	Particle Symbol	Antiparticle Symbol	Rest Energy (MeV)	Lifetime (s)
Electron	e^- or β^-	e^+ or β^+	0.511	Stable
Muon	μ^-	μ^+	105.7	2.2×10^{-6}
Tau	τ^-	τ^+	1784	10^{-13}
Electron neutrino	v_e	$\overline{v}_e$	≈ 0	Stable
Muon neutrino	v_μ	$\overline{v}_\mu$	≈ 0	Stable
Tau neutrino	v_τ	$\overline{v}_\tau$	≈ 0	Stable

The weak nuclear force is responsible for most radioactive decay processes, such as beta decay. It is also a force of extremely short range. In fact, the weak force can be felt only by particles that are separated by roughly one thousandth the diameter of a nucleus. Beyond that range, the weak force has practically no effect at all.

Hadrons

Hadrons are particles that experience both the weak and the strong nuclear force. They are also acted on by gravity, since all hadrons have finite mass. The two most familiar hadrons, of course, are the proton and the neutron. A partial list of the hundreds of hadrons known to exist is given in Table 32–7. Notice that the proton is the only stable hadron (though some theories suggest that even it may decay with the incredibly long half-life of 10^{35} y).

TABLE 32–7 Hadrons

	Particle	Particle Symbol	Antiparticle Symbol	Rest Energy (MeV)	Lifetime (s)
MESONS					
	Pion	π^+	π^-	139.6	2.6×10^{-8}
		π^0	π^0	135.0	0.8×10^{-16}
	Kaon	K^+	K^-	493.7	1.2×10^{-8}
		$K_S{}^0$	$\overline{K}_S{}^0$	497.7	0.9×10^{-10}
		$K_L{}^0$	$\overline{K}_L{}^0$	497.7	5.2×10^{-8}
	Eta	η^0	η^0	548.8	$<10^{-18}$
BARYONS					
	Proton	p	$\overline{p}$	938.3	Stable
	Neutron	n	$\overline{n}$	939.6	900
	Sigma	Σ^+	$\overline{\Sigma}^-$	1189	0.8×10^{-10}
		Σ^0	$\overline{\Sigma}^0$	1192	6×10^{-20}
		Σ^-	$\overline{\Sigma}^+$	1197	1.6×10^{-10}
	Omega	Ω^-	Ω^+	1672	0.8×10^{-10}

The strong nuclear force is the only force powerful enough to hold a nucleus together. It is a short-range force, extending only to distances comparable to the diameter of a nucleus, but within that range it is strong enough to counteract the intense electromagnetic repulsion between positively charged protons. Outside the nucleus, however, the strong nuclear force is of negligible strength.

In striking contrast with leptons, none of the hadrons are elementary particles. In fact, all hadrons are composed of either two or three smaller particles called **quarks.** Hadrons formed from two quarks are referred to as **mesons;** those formed from three quarks are **baryons.** The properties of quarks will be considered next.

Quarks

To account for the internal structure observed in hadrons, Murray Gell-Mann (1929–) and George Zweig (1937–) independently proposed in 1963 that all

TABLE 32–8 Quarks and Antiquarks

		Quarks		Antiquarks	
Name	Rest Energy (MeV)	Symbol	Charge	Symbol	Charge
Up	360	u	$+\frac{2}{3}e$	$\overline{\text{u}}$	$-\frac{2}{3}e$
Down	360	d	$-\frac{1}{3}e$	$\overline{\text{d}}$	$+\frac{1}{3}e$
Charmed	1500	c	$+\frac{2}{3}e$	$\overline{\text{c}}$	$-\frac{2}{3}e$
Strange	540	s	$-\frac{1}{3}e$	$\overline{\text{s}}$	$+\frac{1}{3}e$
Top	173,000	t	$+\frac{2}{3}e$	$\overline{\text{t}}$	$-\frac{2}{3}e$
Bottom	5000	b	$-\frac{1}{3}e$	$\overline{\text{b}}$	$+\frac{1}{3}e$

hadrons are composed of a number of truly elementary particles that Gell-Mann dubbed quarks. Originally, it was proposed that there are three types of quarks, arbitrarily named up (u), down (d), and strange (s). Discoveries of new and more massive hadrons, such as the J/ψ particle discovered in 1974, have necessitated the addition of three more quarks. The equally whimsical names for these new quarks are charmed (c), top or truth (t), and bottom or beauty (b). Table 32–8 lists the six quarks, along with some of their more important properties.

The antiparticles to the quarks are also given in Table 32–8. Notice that antiquarks are indicated with a bar over the symbol for the corresponding quark. For example, the symbol for the up quark is u; the symbol for the corresponding antiquark is $\overline{\text{u}}$.

Quarks are unique among the elementary particles in a number of ways. For example, they all have charges that are fractions of the charge of the electron. As can be seen in Table 32–8, some quarks have a charge of $+\left(\frac{2}{3}\right)e$ or $-\left(\frac{2}{3}\right)e$; others have a charge of $+\left(\frac{1}{3}\right)e$ or $-\left(\frac{1}{3}\right)e$. No other particles are known to have charges that differ from integer multiples of the electron's charge.

Now it might seem that the fractional charge of a quark would make it easy to identify experimentally. In fact, a number of experiments have searched for quarks in just that way, by looking for particles with fractional charge. No such particle has ever been observed, however. It is now believed that a free, independent quark cannot exist; quarks must always be bound with other quarks. This concept is referred to as **quark confinement.** The physical reason behind confinement is that the force between two quarks increases with separation—like two particles connected by a spring. Hence, an infinite amount of energy is required to increase the separation between two quarks to infinity.

The smallest system of bound quarks that can be observed as an independent particle is a pair of quarks. In fact, mesons consist of bound pairs of quarks and antiquarks, as illustrated schematically in **Figure 32–12**. For example, the π^+ meson is composed of a u$\overline{\text{d}}$ pair of quarks. Note that this combination of quarks gives the π^+ meson a net charge of $+e$. The π^- meson, the antiparticle to the π^+ meson, consists of a $\overline{\text{u}}$d pair with a charge of $-e$. Quarks are always bound in configurations that result in integer charges.

Baryons are bound systems consisting of three quarks, as shown in Figure 32–12. The proton, for example, has the composition uud, with a net charge of $+e$. The neutron, on the other hand, is formed from the combination udd, with a net charge of 0. A variety of hadrons and their corresponding quark compositions is given in Table 32–9.

Finally, not long after the quark model of elementary particles was introduced, it was found that some quark compositions implied a violation of the Pauli exclusion principle. To resolve these discrepancies, it was suggested that quarks must come in three different varieties, which were given the completely arbitrary but colorful names red, green, and blue. Though these quark "colors" have nothing to do with visible colors in the electromagnetic spectrum, they bring quarks into agreement with the exclusion principle and explain other experimental observations that were difficult to understand before the introduction of this new

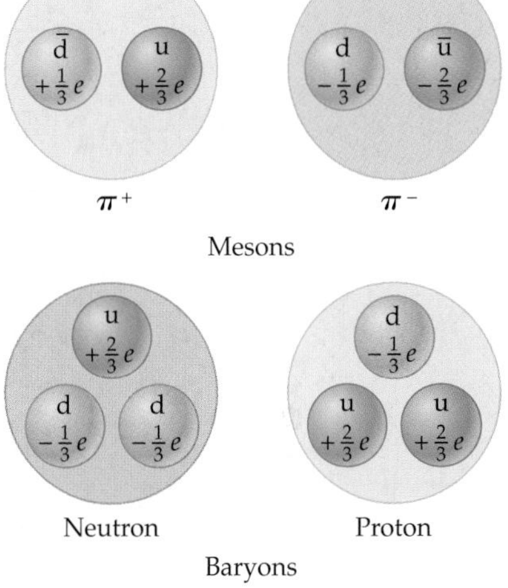

▲ **FIGURE 32–12 The quark composition of mesons and baryons**

Meson and baryons are composed of various quark combinations—mesons always have a quark and an antiquark, baryons always have three quarks. Note that even though quarks have fractional charges (in units of the electron charge, e), the resulting mesons and baryons always have integer charges.

TABLE 32–9 Quark Composition of Some Hadrons

Particle	Quark Composition
MESONS	
π^+	u$\overline{\text{d}}$
π^-	$\overline{\text{u}}$d
K^+	u$\overline{\text{s}}$
K^-	$\overline{\text{u}}$s
K^0	d$\overline{\text{s}}$
BARYONS	
p	uud
n	udd
Σ^+	uus
Σ^0	uds
Σ^-	dds
Ξ^0	uss
Ξ^-	dss
Ω^-	sss

property. The theory of how colored quarks interact with one another is called **quantum chromodynamics,** or QCD, in analogy with the theory describing interactions between charged particles, which is known as **quantum electrodynamics,** or QED.

32–9 Unified Forces and Cosmology

As discussed earlier, the universe as we see it today has four fundamental forces through which various particles interact with one another. This has not always been the case, however. Shortly after the Big Bang these four forces were combined into a single force sometimes referred to as the **unified force.** This situation lasted for only a brief interval of time. As the early universe expanded and cooled, it eventually underwent a type of "phase transition" in which the gravitational force took on a separate identity. This transition occurred at a time of approximately 10^{-43} s after the Big Bang, when the temperature of the universe was about 10^{32} K.

The phase transition just described was the first of three such transitions to occur in the early universe, as we see in **Figure 32–13**. At 10^{-35} s, when the temperature was 10^{28} K, the strong nuclear force became a separate force. Similarly, the weak nuclear force became a separate force at 10^{-10} s, when the temperature was 10^{15} K. From 10^{-10} s until the present, the situation has remained the same, even as the temperature of the universe has dipped to a chilly 2.7 K.

Let's look at these forces and the transitions between them more carefully. First, the electromagnetic force combines the forces associated with both electricity and magnetism. Although electricity and magnetism were originally thought to be separate forces, the work of Maxwell and others showed that these forces are simply different aspects of the same underlying force. For example, changing electric fields generate magnetic fields, and changing magnetic fields generate electric fields. In fact, the theory of electromagnetism can be thought of as the first *unified field theory*, in which seemingly different forces are combined into one all-encompassing theory.

At times earlier than 10^{-10} s the weak nuclear force was indistinguishable from the electromagnetic force. Thus, even though these forces seem very different today, we can recognize them as different aspects of the same underlying force—much like the two faces of a coin look very different but are part of the same physical object. The theory that encompasses the weak nuclear force and the electromagnetic force is called the **electroweak theory.** It was developed by Sheldon Glashow, Abdus Salam, and Steven Weinberg.

Going further back in time, the strong nuclear force was indistinguishable from the electroweak force before 10^{-35} s. Although no one has yet succeeded in producing a theory combining the electroweak force and the strong nuclear force, most physicists feel confident that such a theory exists. This hypothetical theory is referred to as the **grand unified theory,** or GUT.

Finally, a theory that encompasses gravity along with the other forces of nature is one of the ultimate goals of physics. Many physicists, including Einstein in his later years, have worked long and hard toward such an end, but so far with little success.

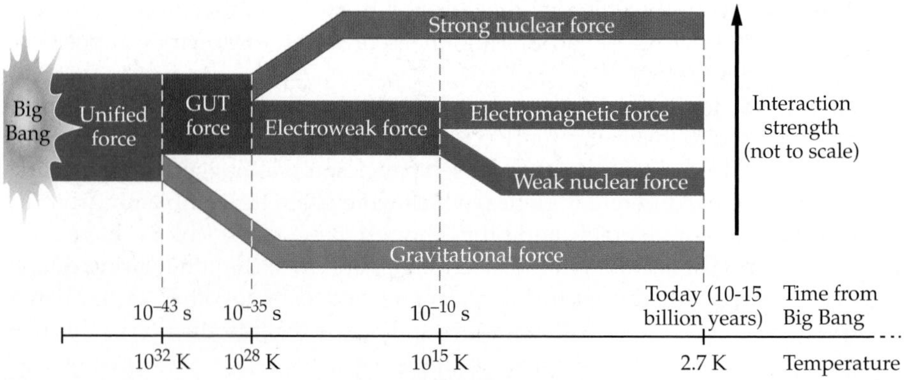

▶ **FIGURE 32–13 The evolution of the four fundamental forces**
The four forces we observe in today's universe began as a single unified force at the time of the Big Bang. As the universe cooled, the unified force evolved through a series of "transitions" in which the various forces took on different characteristics.

Chapter Summary

32–1 The Constituents and Structure of Nuclei

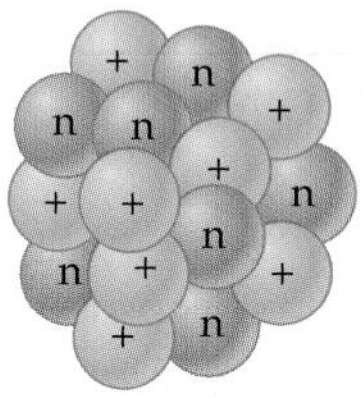

Nuclei are composed of just two types of particles: protons and neutrons. These particles are referred to collectively as nucleons.

Atomic Number, *Z*; Neutron Number, *N*; Mass Number, *A*

The atomic number, Z, is equal to the number of protons in a nucleus. The neutron number, N, is the number of neutrons in a nucleus. The mass number of a nucleus, A, is the total number of nucleons it contains. Thus, $A = N + Z$.

Designation of Nuclei

A nucleus with atomic number Z and mass number A is designated as follows:

$$^{A}_{Z}\text{X}$$

Isotopes

Isotopes are nuclei with the same atomic number but different neutron number.

Atomic Mass Unit, u

A convenient mass unit for nucleons and nuclei is the atomic mass unit, u, which is defined so that the mass of $^{12}_{6}\text{C}$ is exactly 12 u. The value of u is as follows:

$$1\text{ u} = 1.660540 \times 10^{-27}\text{ kg} \qquad \textbf{32–2}$$

Neutrons and protons have masses slightly greater than 1 u, with the neutron slightly more massive than the proton.

Energy/Mass Equivalence

The atomic mass unit can be expressed in terms of energy (MeV) as follows:

$$1\text{ u} = 931.5\text{ MeV}/c^2 \qquad \textbf{32–3}$$

Nuclear Size and Density

The approximate radius of a nucleus of mass number A is given by

$$r = (1.2 \times 10^{-15}\text{ m})A^{1/3} \qquad \textbf{32–4}$$

All nuclei have roughly the same density, regardless of their mass number.

Nuclear Forces and Stability

Nuclei are held together by the strong nuclear force. This force is attractive between all nucleons and has a range of only a few fermis.

32–2 Radioactivity

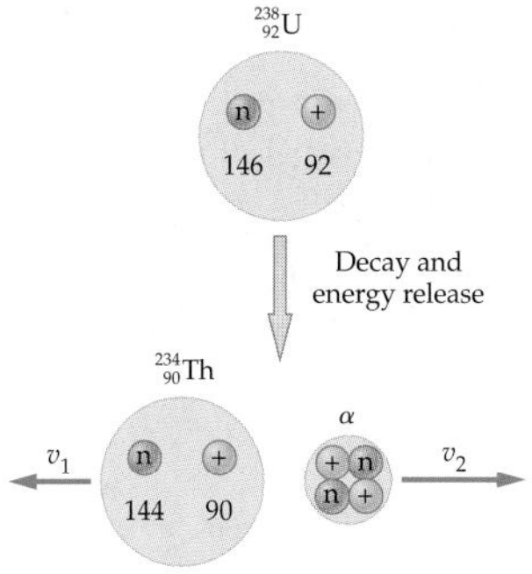

Radioactivity refers to the emissions observed when an unstable nucleus changes its composition or when an excited nucleus decays to a lower-energy state.

Alpha Decay

An α particle (the nucleus of a helium atom) consists of two protons and two neutrons. A nucleus that emits an α particle decreases its mass number by 4 and its atomic number by 2:

$$^{A}_{Z}X \longrightarrow {}^{A-4}_{Z-2}Y + {}^{4}_{2}\text{He}$$

where X is the parent nucleus and Y is the daughter nucleus.

Beta Decay

Beta decay refers to the emission of an electron, as when a neutron decays into a proton, an electron, and an antineutrino:

$$^{1}_{0}\text{n} \longrightarrow {}^{1}_{1}\text{p} + \text{e}^- + \overline{\nu}$$

This type of decay, which increases the atomic number by 1 but leaves the mass number unchanged, is referred to as β^- decay. If a positron and a neutrino are given off instead, we refer to the process as β^+ decay.

Gamma Decay

Gamma decay occurs when an excited nucleus drops to a lower-energy state and emits a photon. In this case, neither the mass number nor the atomic number is changed.

Activity

The activity of a radioactive sample is equal to the number of decays per second. The units of activity are the curie (Ci) and the becquerel (Bq):

$$1\text{ curie} = 1\text{ Ci} = 3.7 \times 10^{10}\text{ decays/s} \qquad \textbf{32–6}$$

$$1\text{ becquerel} = 1\text{ Bq} = 1\text{ decay/s} \qquad \textbf{32–7}$$

32–3 Half-Life and Radioactive Dating

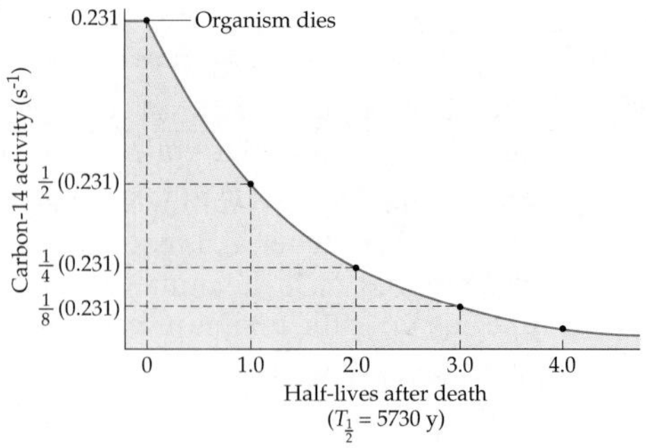

Radioactive nuclei decay with time in a well-defined way. As a result, many radioactive nuclei can be used as a type of "nuclear clock."

Nuclei as a Function of Time and the Decay Constant, λ
If the number of radioactive nuclei in a sample at time $t = 0$ is N_0, the number, N, at a later time is

$$N = N_0 e^{-\lambda t} \qquad \textbf{32–9}$$

The constant, λ, in this expression is referred to as the decay constant.

Half-life, $T_{1/2}$
The half-life of a radioactive material is the time required for half of its nuclei to decay. In terms of the decay constant, the half-life is

$$T_{1/2} = \frac{\ln 2}{\lambda} = \frac{0.693}{\lambda} \qquad \textbf{32–10}$$

Decay Rate, or Activity, R
The rate at which radioactive nuclei decay is proportional to the number of nuclei present at any given time, and to the decay constant:

$$R = \left|\frac{\Delta N}{\Delta t}\right| = \lambda N = \lambda N_0 e^{-\lambda t} = R_0 e^{-\lambda t} \qquad \textbf{32–11, 32–12}$$

Carbon-14 Dating
Carbon-14 can be used to date organic materials with ages up to about 15,000 y. The age can be found using

$$t = -\frac{1}{\lambda}\ln\frac{R}{R_0} = \frac{1}{\lambda}\ln\frac{R_0}{R} \qquad \textbf{32–13}$$

where $R_0 = 0.231$ is the initial activity, R is the present activity, and $\lambda = 1.21 \times 10^{-4}\ \text{y}^{-1}$.

32–4 Nuclear Binding Energy

The binding energy of a nucleus is the energy that must be supplied to separate it into its component nucleons.

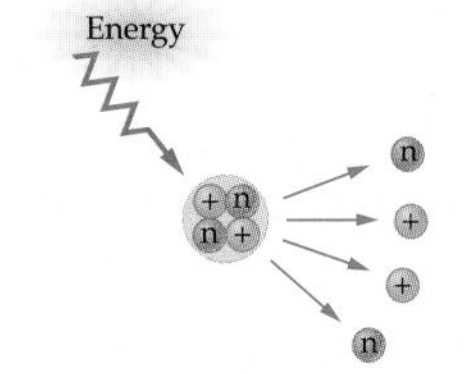

32–5 Nuclear Fission

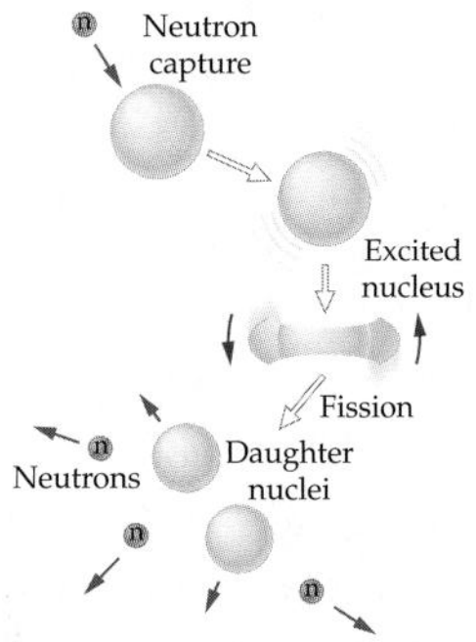

Nuclear fission is the process in which a large nucleus captures a neutron and then divides into two smaller "daughter" nuclei. Typical fission reactions emit two or three neutrons along with the daughter nuclei.

Chain Reactions
When neutrons given off by one fission reaction initiate additional fission reactions, we refer to the process as a chain reaction.

32–6 Nuclear Fusion

Nuclear fusion occurs when two small nuclei merge to form a larger nucleus. To initiate a fusion reaction, it is necessary to give the small nuclei enough energy to overcome their mutual Coulomb repulsion.

32–7 Practical Applications of Nuclear Physics

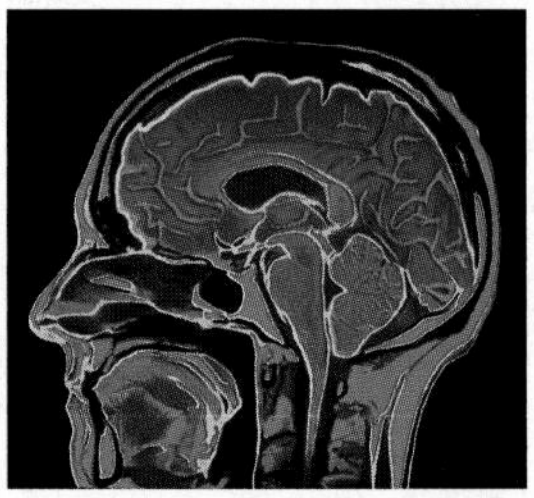

Nuclear radiation can have both harmful and useful effects. An important way to characterize exposure to radiation is in terms of dosage, which can be defined in a number of ways.

Roentgen, R
The first unit of radiation, the roentgen, is related to the amount of ionization charge produced by 200-keV X-rays in 1 kg of dry air at STP:

$$1\ \text{R} = 2.58 \times 10^{-4}\ \text{C/kg} \quad \text{(X-rays or } \gamma \text{ rays in dry air at STP)} \qquad \textbf{32–15}$$

Radiation Absorbed Dose (rad)

The rad is a measure of the amount of energy absorbed by an irradiated material, regardless of the type of radiation:

$$1 \text{ rad} = 0.01 \text{ J/kg} \quad \text{(any type of radiation)} \qquad \textbf{32–16}$$

Relative Biological Effectiveness (RBE)

The RBE takes into account that different types of radiation produce different amounts of biological damage. It is defined as follows:

$$\text{RBE} = \frac{\text{the dose of 200-keV X-rays necessary to produce a given biological effect}}{\text{the dose of a particular type of radiation necessary to produce the same biological effect}} \qquad \textbf{32–17}$$

Roentgen Equivalent in Man (rem)

Combining the rad and the RBE yields the rem:

$$\text{dose in rem} = \text{dose in rad} \times \text{RBE} \qquad \textbf{32–18}$$

A dose of 1 rem of any type of radiation causes the same amount of biological damage.

32–8 Elementary Particles

Elementary particles are the fundamental building blocks of all matter.

The Fundamental Forces of Nature

There are just four fundamental forces in nature. In order of decreasing strength, they are the strong nuclear force, the electromagnetic force, the weak nuclear force, and the gravitational force.

Leptons

Leptons are elementary particles that experience the weak nuclear force but not the strong nuclear force.

Hadrons

Hadrons are composite particles that experience both the weak and the strong nuclear force.

Quarks

Quarks are elementary particles that combine to form hadrons. Mesons are formed from quark-antiquark pairs; baryons are formed from combinations of three quarks.

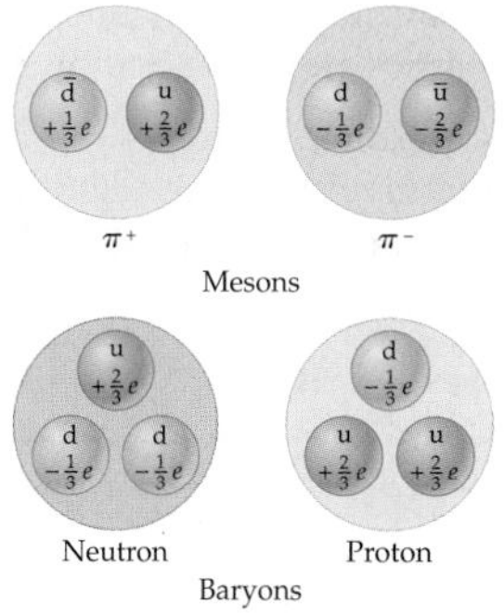

32–9 Unified Forces and Cosmology

The four fundamental forces observed in the universe today began as a single force at the time of the Big Bang. As the universe expanded and cooled, the single force split into four different forces with different characteristics.

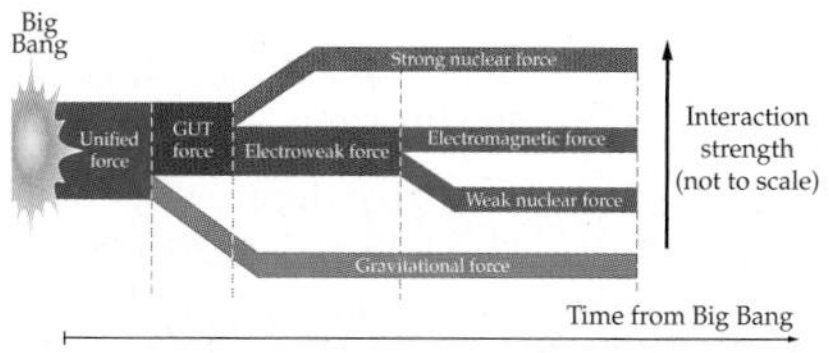

Problem-Solving Summary

Type of Problem	Relevant Physical Concepts	Related Examples
Identify a missing term in a nuclear reaction.	The number of protons and neutrons must be the same before and after an alpha decay. In a β^- decay, the number of neutrons after the decay is decreased by 1, and the number of protons is increased by 1.	Examples 32–3, 32–4, 32–7
Find the energy released in a nuclear reaction.	Calculate the difference in mass before and after the reaction. Multiply the difference in mass by the speed of light squared.	Examples 32–3, 32–4, 32–6, 32–7 Active Example 32–4
Determine the number of radioactive nuclei as a function of time.	The number of nuclei decreases exponentially with time according to the relation $N = N_0 e^{-\lambda t}$.	Active Example 32–2
Determine the activity of radioactive nuclei as a function of time.	Activity, which is the rate of decay, is the decay constant times the number of nuclei present at a given time.	Active Example 32–3
Find the age of an organic sample using carbon-14 decay.	The age can be found using $t = \frac{1}{\lambda}\ln\frac{R_0}{R}$, where the initial activity of a 1-g sample of carbon is 0.231 Bq, and $\lambda = 1.21 \times 10^{-4}\ \text{y}^{-1}$.	Example 32–5

Conceptual Questions

(Answers to odd-numbered Conceptual Questions can be found in the back of the book.)

1. Nucleus A and nucleus B have different numbers of protons and different numbers of neutrons. Explain how it is still possible for these nuclei to have equal radii.
2. When α particles are emitted in a nuclear decay, they have well-defined energies. In contrast, β particles are found to be emitted with a range of energies. Explain this difference.
3. Is it possible for a form of heavy hydrogen to decay by emitting an α particle? Explain.
4. Which is more likely to expose film kept in a cardboard box, α particles or β particles? Explain.
5. It is not possible for a stable nucleus to contain more than one proton without also having at least one neutron. Explain why neutrons are necessary in a stable, multiparticle nucleus.
6. Different isotopes of a given element have different masses, but they have the same chemical properties. Explain why chemical properties are unaffected by a change of isotope.
7. **(a)** Give three examples of objects for which carbon-14 dating would give useful results. **(b)** Give three examples of objects for which carbon-14 dating would not be useful.
8. Explain why the large, stable nuclei in Figure 32–1 are found to lie above the $N = Z$ line, rather than below the line.
9. Suppose each of the following items is about 10,000 years old: a feather, a tooth, an obsidian arrowhead, a deer hide moccasin. Which of these items cannot be dated with carbon-14? Explain.
10. Can carbon-14 dating give the age of fossil dinosaur skeletons? Explain.
11. Two different samples contain the same radioactive isotope. Is it possible for these samples to have different activities? Explain.
12. Two samples contain different radioactive isotopes. Is it possible for these samples to have the same activity? Explain.
13. Two different types of radiation deliver the same amount of energy to a sample of tissue. Does it follow that each of these types of radiation has the same RBE? Explain.

Conceptual Exercises

(Answers to odd-numbered Conceptual Exercises can be found in the back of the book.)

1. Consider a nucleus that can undergo either **(a)** α decay or **(b)** β decay. In each case, state whether the radius of the resulting daughter nucleus is greater than, less than, or the same as that of the original nucleus. Explain.
2. An α particle (charge $+2e$) and a β particle (charge $-e$) deflect in opposite directions when they pass through a magnetic field. Which particle deflects by a greater amount, given that both particles have the same speed? Explain.
3. The two radioactive decay series that begin ${}^{232}_{90}\text{Th}$ and end with ${}^{208}_{82}\text{Pb}$ are shown in **Figure 32–14**. Identify the 10 intermediary nuclei that appear in these series.

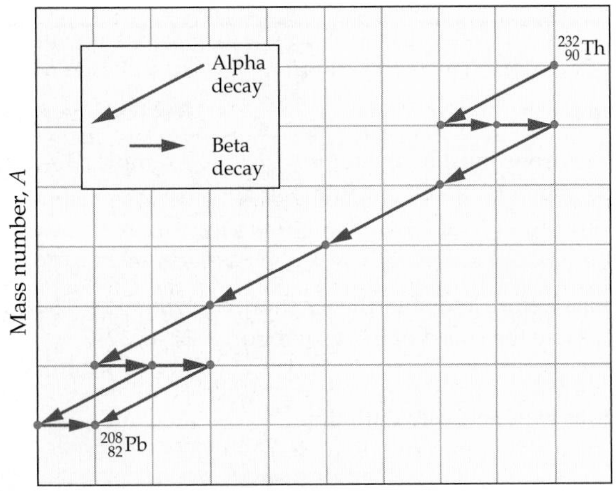

▲ **FIGURE 32–14** Conceptual Exercise 3

4. One possible decay series for ${}^{238}_{92}\text{U}$ is ${}^{234}_{90}\text{Th}$, ${}^{234}_{91}\text{Pa}$, ${}^{234}_{92}\text{U}$, ${}^{230}_{90}\text{Th}$, ${}^{226}_{88}\text{Ra}$, ${}^{222}_{86}\text{Rn}$, ${}^{218}_{84}\text{Po}$, ${}^{218}_{85}\text{At}$, ${}^{218}_{86}\text{Rn}$, ${}^{214}_{84}\text{Po}$, ${}^{210}_{82}\text{Pb}$, ${}^{210}_{83}\text{Bi}$, ${}^{206}_{81}\text{Tl}$, and ${}^{206}_{82}\text{Pb}$. Identify, in the order given, each of the 14 decays that occur in this series.
5. Which of the three decay processes (α, β, or γ) results in a new element? Explain.
6. The half-life of carbon-14 is 5730 y. **(a)** Is it possible for a particular nucleus in a sample of carbon-14 to decay after only 1 s has passed? Explain. **(b)** Is it possible for a particular nucleus to decay after 10,000 y? Explain.
7. Suppose we were to discover that the ratio of carbon-14 to carbon-12 in the atmosphere was significantly smaller 10,000 years ago than it is today. How would this affect the ages we have assigned to objects on the basis of carbon-14 dating? In particular, would the true age of an object be greater than or less than the age we had previously assigned to it? Explain.
8. A radioactive sample is placed in a closed container. Two days later only one-quarter of the sample is still radioactive. What is the half-life of this sample?
9. Radioactive samples A and B have equal half-lives. The initial activity of sample A is twice that of sample B. What is the ratio of the activity of sample A to that of sample B after two half-lives have elapsed?
10. The initial activity of sample A is twice that of sample B. After two half-lives of sample A have elapsed, the two samples have the same activity. What is the ratio of the half-life of B to the half-life of A?
11. To produce a given amount of electrical energy, is the amount of coal burned in a coal-burning power plant greater than, less than, or the same as the amount of ${}^{235}_{92}\text{U}$ consumed in a nuclear power plant? Explain.

Problems

Note: **IP** *denotes an integrated conceptual/quantitative problem.* **BIO** *identifies problems of biological or medical interest. Red bullets (•, ••, •••) are used to indicate the level of difficulty of each problem. Refer to Appendix F for the masses and half-lives of relevant isotopes. Remember that the masses in Appendix F include the mass of the electrons associated with the neutral atoms.*

Section 32–1 The Constituents and Structure of Nuclei

1. • Identify Z, N, and A for the following isotopes: **(a)** $^{238}_{92}U$, **(b)** $^{239}_{94}Pu$, **(c)** $^{144}_{60}Nd$.

2. • Identify Z, N, and A for the following isotopes: **(a)** $^{202}_{80}Hg$, **(b)** $^{220}_{86}Rn$, **(c)** $^{93}_{41}Nb$.

3. • What are the nuclear radii of **(a)** $^{197}_{79}Au$ and **(b)** $^{60}_{27}Co$?

4. • A certain chlorine nucleus has a radius of approximately 4.0×10^{-15} m. How many neutrons are in this nucleus?

5. •• **IP (a)** What is the nuclear density of $^{228}_{90}Th$? **(b)** Do you expect the nuclear density of an alpha particle to be greater than, less than, or the same as that of $^{228}_{90}Th$? Explain. **(c)** Calculate the nuclear density of an alpha particle.

6. •• **IP (a)** What initial kinetic energy must an alpha particle have if it is to approach a stationary gold nucleus to within a distance of 25 fm? **(b)** If the initial speed of the alpha particle is reduced by a factor of two, by what factor is the distance of closest approach changed? Explain.

7. •• **IP** An α particle with a kinetic energy of 0.75 MeV approaches a stationary gold nucleus. **(a)** What is the speed of the α particle? **(b)** What is the distance of closest approach between the α particle and the gold nucleus? **(c)** If this same α particle were fired at a copper nucleus instead, would its distance of closest approach be greater than, less than, or the same as that found in part (b)? Explain. (To obtain the mass of an alpha particle, refer to Appendix F and subtract the mass of two electrons from the mass of $^{4}_{2}He$.)

8. •• Suppose a marble with a radius of 1.2 cm has the density of a nucleus, as given in Example 32–2. **(a)** What is the mass of this marble? **(b)** How many of these marbles would be required to have a mass equal to the mass of Earth?

9. •• **IP (a)** Find the nuclear radius of $^{30}_{15}P$. **(b)** What mass number would be required for a nucleus to have twice the radius found in part (a)? **(c)** Verify your answer to part (b) with an explicit calculation.

10. •• **IP** An alpha particle is the nucleus of a $^{4}_{2}He$ atom. **(a)** How many nucleons are in a nucleus with twice the radius of an alpha particle? Explain. **(b)** Write the symbol for a phosphorus nucleus that has twice the radius of an alpha particle.

11. •• **IP** Suppose a uranium-236 nucleus undergoes fission by splitting into two smaller nuclei of equal size. **(a)** Is the radius of each of the smaller nuclei one half, more than one-half, or less than one-half the radius of the uranium-236 nucleus? Explain. **(b)** Calculate the radius of the uranium-236 nucleus. **(c)** Calculate the radii of the two smaller nuclei.

12. •• A hypothetical nucleus weighs 1 lb. **(a)** How many nucleons are in this nucleus? **(b)** What is the radius of this nucleus?

Section 32–2 Radioactivity

13. • Complete the following nuclear reaction:

$$^{7}_{3}Li + ^{1}_{1}H \longrightarrow ^{4}_{2}He + ?$$

14. • Complete the following nuclear reaction:

$$^{234}_{90}Th \longrightarrow ^{230}_{88}Ra + ?$$

15. • Complete the following nuclear reaction:

$$? \longrightarrow ^{14}_{7}N + e^{-} + \overline{\nu}$$

16. •• Complete the following nuclear reaction and determine the amount of energy it releases:

$$^{3}_{1}H \longrightarrow ^{3}_{2}He + ? + ?$$

Be sure to take into account the mass of the electrons associated with the neutral atoms.

17. •• The following nuclei are observed to decay by emitting an α particle: **(a)** $^{212}_{84}Po$; **(b)** $^{239}_{94}Pu$. Write out the decay process for each of these nuclei, and determine the energy released in each reaction. Be sure to take into account the mass of the electrons associated with the neutral atoms.

18. •• The following nuclei are observed to decay by emitting a β^- particle: **(a)** $^{35}_{16}S$; **(b)** $^{212}_{82}Pb$. Write out the decay process for each of these nuclei, and determine the energy released in each reaction. Be sure to take into account the mass of the electrons associated with the neutral atoms.

19. •• The following nuclei are observed to decay by emitting a β^+ particle: **(a)** $^{18}_{9}F$; **(b)** $^{22}_{11}Na$. Write out the decay process for each of these nuclei, and determine the energy released in each reaction. Be sure to take into account the mass of the electrons associated with the neutral atoms.

20. •• Find the energy released when $^{211}_{82}Pb$ undergoes β^- decay to become $^{211}_{83}Bi$. Be sure to take into account the mass of the electrons associated with the neutral atoms.

21. ••• It is observed that $^{66}_{28}Ni$, with an atomic mass of 65.9291 u, decays by β^- emission. **(a)** Identify the nucleus that results from this decay. **(b)** If the nucleus found in part (a) has an atomic mass of 65.9289 u, what is the maximum kinetic energy of the emitted electron?

Section 32–3 Half-Life and Radioactive Dating

22. • Radon gas has a half-life of 3.82 d. What is the decay constant for radon?

23. • A radioactive substance has a decay constant equal to $8.9 \times 10^{-3}\ s^{-1}$. What is the half-life of this substance?

24. • The number of radioactive nuclei in a particular sample decreases over a period of 18 d to one-sixteenth the original number. What is the half-life of these nuclei?

25. • The half-life of $^{15}_{8}O$ is 122 s. How long does it take for the number of $^{15}_{8}O$ nuclei in a given sample to decrease by a factor of 10^{-4}?

26. •• **IP BIO Treating a Thyroid** The half-life of iodine-131, an isotope used in the treatment of thyroid disorders, is 8.04 d. **(a)** If a sample of iodine-131 contains 4.5×10^{16} nuclei, what is the activity of the sample? Express your answer in curies. **(b)** If the half-life of iodine-131 were only half of its actual value, would the activity of this sample be increased or decreased? Explain. **(c)** Calculate the factor by which the activity of this sample would change under the assumptions stated in part (b).

27. •• **BIO A Radioactive Tag** A drug prepared for a patient is tagged with $^{99}_{43}Tc$, which has a half-life of 6.05 h. **(a)** What is the decay constant of this isotope? **(b)** How many $^{99}_{43}Tc$ nuclei are required to give an activity of 1.50 μCi?

28. •• **BIO** Referring to Problem 27, suppose the drug containing $^{99}_{43}Tc$ with an activity of 1.50 μCi is injected into the patient 2.05 h after it is prepared. What is its activity at the time it is injected?

29. •• An archeologist on a dig finds a fragment of an ancient basket woven from grass. Later, it is determined that the carbon-14

content of the grass in the basket is 9.25% that of an equal carbon sample from present-day grass. What is the age of the basket?

30. •• The bones of a saber-toothed tiger are found to have an activity per gram of carbon that is 15.0% of what would be found in a similar live animal. How old are these bones?

31. •• Charcoal from an ancient fire pit is found to have a carbon-14 content that is only 17.5% that of an equivalent sample of carbon from a living tree. What is the age of the fire pit?

32. •• One of the many isotopes used in cancer treatment is $^{198}_{79}\text{Au}$, with a half-life of 2.70 d. Determine the mass of this isotope that is required to give an activity of 225 Ci.

33. •• **Smoke Detectors** The radioactive isotope $^{241}_{95}\text{Am}$, with a half-life of 432 y, is the active element in many smoke detectors. Suppose such a detector will no longer function if the activity of the $^{241}_{95}\text{Am}$ it contains drops below $\frac{1}{525}$ of its initial activity. How long will this smoke detector work?

34. •• **BIO Radioactivity in the Bones** Because of its chemical similarity to calcium, $^{90}_{38}\text{Sr}$ can collect in the bones and present a health risk. What percentage of $^{90}_{38}\text{Sr}$ present initially still exists after a period of **(a)** 50.0 y, **(b)** 60.0 y, and **(c)** 70.0 y?

Section 32–4 Nuclear Binding Energy

35. • The atomic mass of gold-197 is 196.96654 u. How much energy is required to completely separate the nucleons in a gold-197 nucleus?

36. • The atomic mass of lithium-7 is 7.016003 u. How much energy is required to completely separate the nucleons in a lithium-7 nucleus?

37. •• Calculate the average binding energy per nucleon of **(a)** $^{56}_{26}\text{Fe}$ and **(b)** $^{238}_{92}\text{U}$.

38. •• Calculate the average binding energy per nucleon of **(a)** $^{4}_{2}\text{He}$ and **(b)** $^{64}_{30}\text{Zn}$.

39. •• Find the energy required to remove one neutron from $^{16}_{8}\text{O}$.

40. ••• **IP** **(a)** Consider the following nuclear process, in which a proton is removed from an oxygen nucleus:

$$^{16}_{8}\text{O} + \text{energy} \longrightarrow {}^{15}_{7}\text{N} + {}^{1}_{1}\text{H}$$

Find the energy required for this process to occur. **(b)** Now consider a process in which a neutron is removed from an oxygen nucleus:

$$^{16}_{8}\text{O} + \text{energy} \longrightarrow {}^{15}_{8}\text{O} + {}^{1}_{0}\text{n}$$

Find the energy required for this process to occur. **(c)** Which particle, the proton or the neutron, do you expect to be more tightly bound in the oxygen nucleus? Verify your answer.

Section 32–5 Nuclear Fission

41. • Find the number of neutrons released by the following fission reaction:

$$^{1}_{0}\text{n} + {}^{235}_{92}\text{U} \longrightarrow {}^{132}_{50}\text{Sn} + {}^{101}_{42}\text{Mo} + (?)\text{neutrons}$$

42. •• Complete the following fission reaction and determine the amount of energy it releases:

$$^{1}_{0}\text{n} + {}^{235}_{92}\text{U} \longrightarrow {}^{133}_{51}\text{Sb} + ? + 5{}^{1}_{0}\text{n}$$

43. •• Complete the following fission reaction and determine the amount of energy it releases:

$$^{1}_{0}\text{n} + {}^{235}_{92}\text{U} \longrightarrow {}^{88}_{38}\text{Sr} + {}^{136}_{54}\text{Xe} + (?)\text{neutrons}$$

44. •• A gallon of gasoline releases about 2.0×10^8 J of energy when it is burned. How many gallons of gas must be burned to release the same amount of energy as is released when 1.0 lb of $^{235}_{92}\text{U}$ undergoes fission. (Assume that each fission reaction in $^{235}_{92}\text{U}$ releases 173 MeV.)

45. •• Assuming a release of 173 MeV per fission reaction, determine the minimum mass of $^{235}_{92}\text{U}$ that must undergo fission to supply the annual energy needs of the United States. (The amount of energy consumed in the United States each year is 8.4×10^{19} J.)

46. •• Assuming a release of 173 MeV per fission reaction, calculate how many reactions must occur per second to produce a power output of 150 MW.

Section 32–6 Nuclear Fusion

47. • Consider a fusion reaction in which two deuterium nuclei fuse to form a tritium nucleus and a proton. How much energy is released in this reaction?

48. • Consider a fusion reaction in which a proton fuses with a neutron to form a deuterium nucleus. How much energy is released in this reaction?

49. • Find the energy released in the following fusion reaction:

$$^{1}_{1}\text{H} + {}^{2}_{1}\text{H} \longrightarrow {}^{3}_{2}\text{He} + \gamma$$

50. •• **(a)** Complete the following fusion reaction and determine the energy it releases:

$$^{2}_{1}\text{H} + {}^{3}_{1}\text{H} \longrightarrow ? + {}^{1}_{0}\text{n}$$

(b) How many of these reactions must occur per second to produce a power output of 25 MW?

51. ••• **The Evaporating Sun** The Sun radiates energy at the prodigious rate of 3.90×10^{26} W. **(a)** At what rate, in kilograms per second, does the Sun convert mass into energy? **(b)** Assuming that the Sun has radiated at this same rate for its entire lifetime of 4.50×10^9 y, and that the current mass of the Sun is 2.00×10^{30} kg, what percentage of its original mass has been converted to energy?

Section 32–7 Practical Applications of Nuclear Physics

52. • **BIO Radiation Damage** A sample of tissue absorbs a 52-rad dose of α particles (RBE = 20). How many rad of protons (RBE = 10) cause the same amount of damage to the tissue?

53. • **BIO X-ray Damage** How many rad of 200-keV X-rays cause the same amount of biological damage as 50 rad of heavy ions?

54. •• **IP BIO** **(a)** Find the energy absorbed by a 78-kg person who is exposed to 52 mrem of α particles with an RBE of 15. **(b)** If the RBE of the α particles is increased, does the energy absorbed increase, decrease, or stay the same? Explain.

55. •• **BIO** A patient undergoing radiation therapy for cancer receives a 225-rad dose of radiation. **(a)** Assuming the cancerous growth has a mass of 0.17 kg, calculate how much energy it absorbs. **(b)** Assuming the growth to have the specific heat of water, determine its increase in temperature.

56. •• **BIO** Alpha particles with an RBE of 13 deliver a 32-mrad whole-body radiation dose to a 69-kg patient. **(a)** What dosage, in rem, does the patient receive? **(b)** How much energy is absorbed by the patient?

57. ••• **BIO A Radioactive Pharmaceutical** As part of a treatment program, a patient ingests a radioactive pharmaceutical containing $^{32}_{15}\text{P}$, which emits β rays with an RBE of 1.50. The half-life of $^{32}_{15}\text{P}$ is 14.28 d, and the initial activity of the medication is 1.34 MBq. **(a)** How many electrons are emitted over the period of 7.00 d? **(b)** If the β rays have an energy of 705 keV, what is the

total amount of energy absorbed by the patient's body in 7.00 d? **(c)** Find the absorbed dosage in rem, assuming the radiation is absorbed by 125 g of tissue.

General Problems

58. • Determine the number of neutrons and protons in **(a)** $^{232}_{90}\text{Th}$, **(b)** $^{211}_{82}\text{Pb}$, and **(c)** $^{60}_{27}\text{Co}$.

59. • Identify the daughter nucleus that results when **(a)** $^{210}_{82}\text{Pb}$, undergoes α decay, **(b)** $^{239}_{92}\text{U}$ undergoes β^- decay, and **(c)** $^{11}_{6}\text{C}$ undergoes β^+ decay.

60. • Suppose it is desired to give a cancerous tumor a dose of 3800 rem. How many rads are needed if the tumor is exposed to alpha radiation?

61. • A patient is exposed to 250 rad of gamma rays. What is the dose the patient receives in rem?

62. •• In one of the rocks brought back from the Moon, it is found that 80.5% of the initial potassium-40 in the rock has decayed to argon-40. **(a)** If the half-life for this decay is 1.20×10^9 years, how old is the rock? **(b)** How much longer will it take before only 10.0% of the original potassium-40 is still present in the rock?

63. •• IP **Mantles in Gas Lanterns** Gas lanterns used on camping trips have mantles (small lacy bags that give off light) made from a rayon mesh impregnated with thorium and other materials. Thorium is used, even though it is radioactive, because it forms an oxide that can withstand being incandescent for long periods of time. Almost all natural thorium is $^{232}_{90}\text{Th}$, which has a half-life of 1.405×10^{10} y. **(a)** A typical mantle contains 325 mg of thorium. What is the activity of the mantle? **(b)** If the half-life of thorium had been double its actual value, by what factor would the activity of a mantle be changed? Explain.

The glowing mantle of a gas lantern. (Problem 63)

64. •• Identify the nucleus whose β^- decay produces the same nucleus as that produced by the α decay of $^{214}_{84}\text{Po}$.

65. •• An α particle fired head-on at a stationary nickel nucleus approaches to a radius of 15 fm before being turned around. **(a)** What is the maximum Coulomb force exerted on the α particle? **(b)** What is the electric potential energy of the α particle at its point of closest approach? **(c)** Find the initial kinetic energy of the α particle.

66. •• Calculate the number of disintegrations per second that one would expect from a 1.5-g sample of $^{226}_{88}\text{Ra}$. What is the activity of this sample in curies?

67. •• IP Initially, a sample of radioactive nuclei of type A contains four times as many nuclei as a sample of radioactive nuclei of type B. Two days later (2.00 d) the two samples contain the same number of nuclei. **(a)** Which type of nucleus has the longer half-life? Explain. **(b)** Determine the half-life of type B nuclei if the half-life of type A nuclei is known to be 0.500 d.

68. •• Stable nuclei have mass numbers that range from a minimum of 1 to a maximum of 209. **(a)** Find the corresponding range in nuclear radii. **(b)** Assuming all nuclei to be spherical, determine the ratio of the surface area of the largest stable nucleus to the surface area of the smallest nucleus. **(c)** Repeat part (b), only this time find the ratio of the volumes.

69. •• **Radius of a Neutron Star** Neutron stars are so named because they are composed of neutrons and have a density the same as that of a nucleus. Referring to Example 32–2 for the nuclear density, find the radius of a neutron star whose mass is 0.50 that of the Sun.

70. •• A specimen taken from the wrappings of a mummy contains 7.82 g of carbon and has an activity of 1.38 Bq. How old is the mummy? (Refer to pages 1087 and 1088 for relevant information regarding the isotopes of carbon.)

71. •• **(a)** How many fission reactions are required to light a 100-W lightbulb for 2.5 d? Assume an energy release of 212 MeV per fission reaction and a 32% conversion efficiency. **(b)** What mass of $^{235}_{92}\text{U}$ corresponds to the number of fission reactions found in part (a)?

72. •• IP Energy is released when three α particles fuse to form carbon-12. **(a)** Is the mass of carbon-12 greater than, less than, or the same as the mass of three α particles? Explain. **(b)** Calculate the energy given off in this fusion reaction.

73. •• Find the dose of γ rays that must be absorbed by a block of ice at 0 °C to convert it to water at 0 °C. Give the dosage in rad.

74. •• IP **(a)** What dosage (in rad) must a 1.0-kg sample of water absorb to increase its temperature by 1.0 C°? **(b)** If the mass of the water sample is increased, does the dosage found in part (a) increase, decrease, or stay the same? Explain.

75. •• BIO **Chest X-rays** A typical chest X-ray uses X-rays with an RBE of 0.85. If the radiation dosage is 35 mrem, find the energy absorbed by a 72-kg patient, assuming one-quarter of the patient's body is exposed to the X-rays.

76. ••• A γ ray photon emitted by $^{226}_{88}\text{Ra}$ has an energy of 0.186 MeV. Use conservation of linear momentum to calculate the recoil speed of a $^{226}_{88}\text{Ra}$ nucleus after such a γ ray is emitted. Assume that the nucleus is at rest initially, and that relativistic effects can be ignored.

77. ••• The energy released by α decay in a 50.0-g sample of $^{239}_{94}\text{Pu}$ is to be used to heat 4.50 kg of water. Assuming all the energy released by the radioactive decay goes into heating the water, find how much the temperature of the water increases in 1 h.

78. ••• Consider a solid sphere of $^{235}_{92}\text{U}$ with a radius of 2.00 cm in a room with a temperature of 293 K. Assume that all the energy released by α decay goes into heating the sphere, and that the sphere radiates heat to its surroundings as a blackbody. What is the change in temperature of the sphere as a result of the α decays? (*Note*: The density of uranium is 18.95 g/cm^3.)

Appendix A Basic mathematical tools

This text is designed for students with a working knowledge of basic algebra and trigonometry. Even so, it is useful to review some of the mathematical tools that are of particular importance in the study of physics. In this Appendix we cover a number of topics related to mathematical notation, trigonometry, algebra, mathematical expansions, and vector multiplication.

Mathematical Notation

Common mathematical symbols

In Table A–1 we present some of the more common mathematical symbols, along with a translation into English. Though these symbols are probably completely familiar, it is worthwhile to be sure we all interpret them in the same way.

TABLE A–1 Mathematical Symbols

$=$	is equal to
$\neq$	is not equal to
$\approx$	is approximately equal to
$\propto$	is proportional to
$>$	is greater than
$\geq$	is greater than or equal to
$\gg$	is much greater than
$<$	is less than
$\leq$	is less than or equal to
$\ll$	is much less than
$\pm$	plus or minus
$\mp$	minus or plus
x_{av} or $\overline{x}$	average value of x
Δx	change in $x(x_f - x_i)$
$\lvert x \rvert$	absolute value of x
Σ	sum of
$\rightarrow 0$	approaches 0
∞	infinity

A couple of the symbols in Table A–1 warrant further discussion. First, Δx, which means "change in x," is used frequently, and in many different contexts. Pronounced "delta x," it is defined as the final value of x, x_f, minus the initial value of x, x_i:

$$\Delta x = x_f - x_i \qquad \text{A–1}$$

Thus, Δx is not Δ times x; it is a shorthand way of writing $x_f - x_i$. The same delta notation can be applied to any quantity—it does not have to be x. In general, we can say that

$$\Delta(anything) = (anything)_f - (anything)_i$$

For example, $\Delta t = t_f - t_i$ is the change in time, $\Delta\vec{\mathbf{v}} = \vec{\mathbf{v}}_f - \vec{\mathbf{v}}_i$ is the change in velocity, and so on. Throughout this text, we use the delta notation whenever we want to indicate the change in a given quantity.

Second, the Greek letter Σ (capital sigma) is also encountered frequently. In general, Σ is shorthand for "sum." For example, suppose we have a system comprised of nine masses, m_1 through m_9. The total mass of the system, M, is simply

$$M = m_1 + m_2 + m_3 + m_4 + m_5 + m_6 + m_7 + m_8 + m_9$$

This is a rather tedious way to write M, however, and would be even more so if the number of masses were larger. To simplify our equation, we use the Σ notation:

$$M = \sum_{i=1}^{9} m_i \qquad \text{A–2}$$

With this notation we could sum over any number of masses, simply by changing the upper limit of the sum.

In addition, Σ is often used to designate a general summation, where the number of terms in the sum may not be known, or may vary from one system to another. In a case like this we would simply write Σ without specific upper and lower limits. Thus, a general way of writing the total mass of a system is as follows:

$$M = \sum m \qquad \text{A–3}$$

Vector notation

When we draw a vector to represent a physical quantity, we typically use an arrow whose length is proportional to the magnitude of the quantity, and whose direction is the direction of the quantity. (This and other aspects of vector notation are discussed in Chapter 3.) A slight problem arises, however, when a physical quantity points into or out of the page. In such a case, we use the conventions illustrated in **Figure A–1**.

Figure A–1 (a) shows a vector pointing out of the page. Note that we see only the tip. Below, we show the corresponding convention, which is a dot set off by a circle. The dot represents the point of the vector's arrow coming out of the page toward you.

A similar convention is employed in Figure A–1 (b) for a vector pointing into the page. In this case, the arrow moves directly away from you, giving a view of its "tail feathers." The feathers are placed in an X-shaped pattern, so we represent the vector as an X set off by a circle.

These conventions are used in Chapter 22 to represent the magnetic field vector, $\vec{\mathbf{B}}$, and in other locations in the text as well.

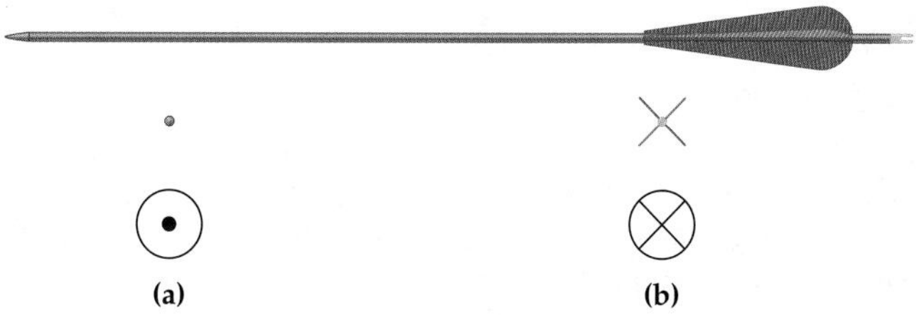

▲ **FIGURE A–1 Vectors pointing out of and into the page**

(a) A vector pointing out of the page is represented by a dot in a circle. The dot indicates the tip of the vector's arrow. **(b)** A vector pointing into the page is represented by an X in a circle. The X indicates the "tail feathers" of the vector's arrow.

Scientific notation

In physics, the numerical value of a physical quantity can cover an enormous range, from the astronomically large to the microscopically small. For example, the mass of the Earth is roughly

$$M_E = 5970000000000000000000000 \text{ kg}$$

In contrast, the mass of a hydrogen atom is approximately

$$M_{\text{hydrogen}} = 0.00000000000000000000000000167 \text{ kg}$$

Clearly, representing such large and small numbers with a long string of zeroes is clumsy and prone to error.

The preferred method for handling such numbers is to replace the zeros with the appropriate power of ten. For example, the mass of the Earth can be written as follows:

$$M_E = 5.97 \times 10^{24} \text{ kg}$$

The factor of 10^{24} simply means that the decimal point for the mass of the Earth is 24 places to the right of its location in 5.97. Similarly, the mass of a hydrogen atom is

$$M_{\text{hydrogen}} = 1.67 \times 10^{-27} \text{ kg}$$

In this case, the correct location of the decimal point is 27 places to the left of its location in 1.67. This type of representation, using powers of ten, is referred to as **scientific notation.**

Scientific notation also simplifies various mathematical operations, such as multiplication and division. For example, the product of the mass of the Earth and the mass of a hydrogen atom is

$$\begin{aligned} M_E M_{\text{hydrogen}} &= (5.97 \times 10^{24} \text{ kg})(1.67 \times 10^{-27} \text{ kg}) \\ &= (5.97 \times 1.67)(10^{24} \times 10^{-27}) \text{ kg}^2 \\ &= 9.99 \times 10^{24-27} \text{ kg}^2 \\ &= 9.99 \times 10^{-3} \text{ kg}^2 \end{aligned}$$

Similarly, the mass of a hydrogen atom divided by the mass of the Earth is

$$\begin{aligned} \frac{M_{\text{hydrogen}}}{M_E} &= \frac{1.67 \times 10^{-27} \text{ kg}}{5.97 \times 10^{24} \text{ kg}} = \frac{1.67}{5.97} \times \frac{10^{-27}}{10^{24}} \\ &= 0.280 \times 10^{-27-24} \\ &= 0.280 \times 10^{-51} = 2.80 \times 10^{-52} \end{aligned}$$

Note the change in location of the decimal point in the last two expressions, and the corresponding change in the power of ten.

Exponents and their manipulation are discussed in greater detail later in this Appendix.

Trigonometry

Degrees and radians

We all know the definition of a degree; there are 360 degrees in a circle. The definition of a radian is somewhat less well known; there are 2π radians in a circle. An equivalent definition of the radian is the following:

> A radian is the angle for which the corresponding arc length is equal to the radius.

To visualize this definition, consider a pie with a piece cut out, as shown in **Figure A–2 (a)**. Note that a piece of pie has three sides—two radial lines from the center, and an arc of crust. If a piece of pie is cut with an angle of one radian, all three sides are equal in length, as shown in **Figure A–2 (b)**. Since a radian is about 57.3°, this amounts to a fairly good-sized piece of pie. Thus, if you want a healthy helping of pie, just tell the server, "One radian, please."

(a) (b)

▲ **FIGURE A–2 The definition of a radian**

(a) This piece of pie is cut with an angle less than a radian. Thus, the two radial sides (coming out from the center) are longer than the arc of crust. **(b)** The angle for this piece of pie is equal to one radian (about 57.3°). Thus, all three sides of the piece are of equal length.

Now, radians are particularly convenient when we are interested in the length of an arc. In **Figure A–3** we show a circular arc corresponding to the radius r and the angle θ. *If the angle θ is measured in radians*, the length of the arc, s, is given by

$$s = r\theta \qquad \text{A–4}$$

Note that this simple relation is *not valid* when θ is measured in degrees. For a full circle, in which case $\theta = 2\pi$, the length of the arc (which is the circumference of the circle) is $2\pi r$, as expected.

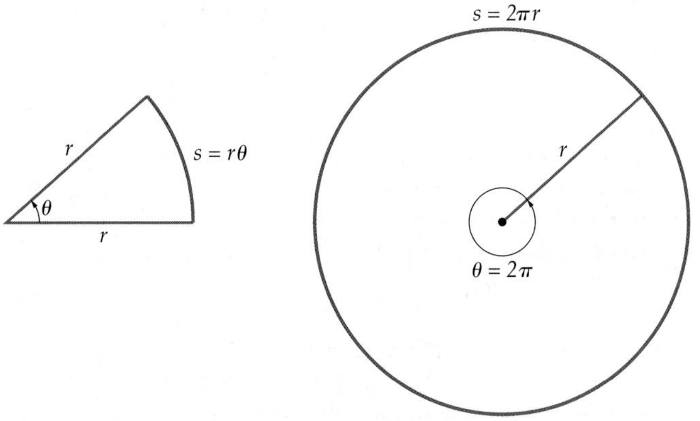

▲ **FIGURE A–3 The Length of an arc**

The arc created by a radius r rotated through an angle θ has a length $s = r\theta$. If the radius is rotated through a full circle, the angle is $\theta = 2\pi$, and the arc length is the circumference of a circle, $s = 2\pi r$.

Trigonometric functions and the Pythagorean theorem

Next, we consider some of the more important and frequently used results from trigonometry. We start with the right triangle, shown in **Figure A–4**, and the basic **trigonometric functions,** sin θ (sine theta), cos θ (cosine theta), and tan θ (tangent theta). The cosine of an angle θ is defined to be the side adjacent to the angle divided by the hypotenuse; $\cos\theta = x/r$. Similarly, the sine is defined to be the opposite side divided by the hypotenuse, $\sin\theta = y/r$, and the tangent is the opposite side divided by the adjacent side, $\tan\theta = y/x$. These relations are summarized in the following equations:

$$\cos\theta = \frac{x}{r}$$

$$\sin\theta = \frac{y}{r} \qquad \text{A–5}$$

$$\tan\theta = \frac{y}{x} = \frac{\sin\theta}{\cos\theta}$$

Note that each of the trigonometric functions is the ratio of two lengths, and hence is dimensionless.

According to the **Pythagorean theorem,** the sides of the right triangle in Figure A–4 are related as follows:

$$x^2 + y^2 = r^2 \qquad \text{A–6}$$

Dividing by r^2 yields

$$\frac{x^2}{r^2} + \frac{y^2}{r^2} = 1$$

This can be re-written in terms of sine and cosine to give

$$\sin^2\theta + \cos^2\theta = 1$$

Figure A–4 also shows how sine and cosine are used in a typical calculation. In many cases, the hypotenuse of a triangle, r, and one of its angles, θ, are given. To find the short sides of the triangle we rearrange the relations given in Equation A–5. For example, in Figure A–4 we see that $x = r\cos\theta$ is the length of the short side adjacent to the angle, θ, and $y = r\sin\theta$ is the length of the short side opposite the angle. The following Example applies this type of calculation to the case of an inclined roadway.

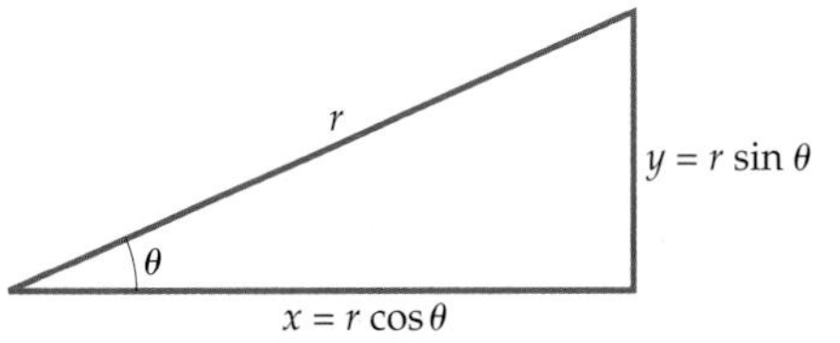

▲ **FIGURE A–4 Relating the short sides of a right triangle to its hypotenuse**

The trigonometric functions sin θ and cos θ and the Pythagorean theorem are useful in relating the lengths of the short sides of a right triangle to the length of its hypotenuse.

EXAMPLE A–1 Highway to Heaven

You are driving on a long straight road that slopes uphill at an angle of 6.4° above the horizontal. At one point you notice a sign that reads, "Elevation 1500 feet." What is your elevation after you have driven another 1.0 mi?

Picture the Problem

From our sketch, we see that the car is moving along the hypotenuse of a right triangle. The length of the hypotenuse is one mile.

Strategy

The elevation gain is the vertical side of the triangle, y. We find y by multiplying the hypotenuse, r, by the sine of theta. That is, since $\sin\theta = y/r$ it follows that $y = r\sin\theta$.

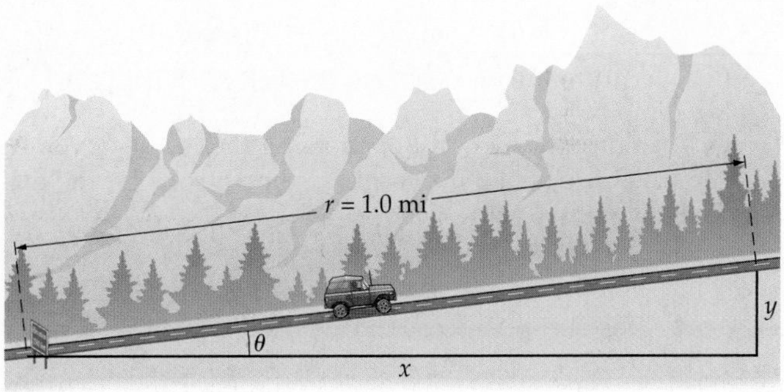

Solution

1. Calculate the elevation gain, y:

$$y = r\sin\theta = (1.0\text{ mi})\sin 6.4° = (1.0\text{ mi})(0.11) = 0.11\text{ mi}$$

2. Convert y from miles to feet:

$$y = (0.11\text{ mi})\left(\frac{5280\text{ ft}}{1\text{ mi}}\right) = 580\text{ ft}$$

3. Add the elevation gain to the original elevation to obtain the new elevation:

$$\text{elevation} = 1500\text{ ft} + 580\text{ ft} = 2100\text{ ft}$$

Insight

As surprising as it may seem, the horizontal distance covered by the car is $r\cos\theta = (5280\text{ ft})\cos 6.4° = 5200$ ft, only about 80 ft less than the total distance driven by the car. At the same time, the car rises a distance of 580 ft.

Practice Problem

How far up the road from the first sign should the road crew put another sign reading "Elevation 3500 ft"?
[**Answer:** 18,000 ft = 3.4 mi]

In some problems, the sides of a triangle (x and y) are given and it is desired to find the corresponding hypotenuse, r, and angle, θ. For example, suppose that $x = 5.0$ m and $y = 2.0$ m. Using the Pythagorean theorem, we find $r = \sqrt{x^2 + y^2} = \sqrt{(5.0\text{ m})^2 + (2.0\text{ m})^2} = 5.4$ m. Similarly, to find the angle we use the definition of tangent: $\tan\theta = y/x$. The inverse of this relation is $\theta = \tan^{-1}(y/x) = \tan^{-1}(2.0\text{ m}/5.0\text{ m}) = \tan^{-1}(0.40)$. Note that the expression $\tan^{-1}$ is the *inverse tangent function*—it does not mean 1 divided by tangent, but rather "the angle whose tangent is——." Your calculator should have a button on it labeled $\tan^{-1}$. If you enter 0.40 and then press $\tan^{-1}$, you should get 22° (to two significant figures), which means that $\tan 22° = 0.40$. Inverse sine and cosine functions work in the same way.

Trigonometric identities

In addition to the basic definitions of sine, cosine, and tangent just given, there are a number of useful relationships involving these functions referred to as **trigonometric identities.** First, consider changing the sign of an angle. This corresponds to flipping the triangle in Figure A–4 upside-down, which changes the sign of y but leaves x unaffected. The result is that sine changes its sign, but cosine does not. Specifically, for a general angle A we find the following:

$$\sin(-A) = -\sin A$$
$$\cos(-A) = \cos A \qquad \textbf{A–7}$$

Next, we consider trigonometric identities relating to the sum or difference of two angles. For example, consider two general angles A and B. The sine and cosine of the sum of these angles, $A + B$, are given below:

$$\sin(A + B) = \sin A \cos B + \sin B \cos A$$
$$\cos(A + B) = \cos A \cos B - \sin A \sin B \qquad \textbf{A–8}$$

By changing the sign of B, and using the results given in Equation A–7, we obtain the corresponding results for the difference between two angles:

$$\sin(A - B) = \sin A \cos B - \sin B \cos A$$
$$\cos(A - B) = \cos A \cos B + \sin A \sin B \qquad \textbf{A–9}$$

Applications of these relations can be found in Chapters 4, 14, 23, and 24.

To see how one might use a relation like $\sin(A + B) = \sin A \cos B + \sin B \cos A$, consider the case where $A = B = \theta$. With this substitution we find

$$\sin(\theta + \theta) = \sin\theta\cos\theta + \sin\theta\cos\theta$$

Simplifying somewhat yields the commonly used double-angle formula

$$\sin 2\theta = 2\sin\theta\cos\theta \qquad \textbf{A–10}$$

This expression is used in deriving Equation 4–16.

As a final example of using trigonometric identities, let $A = 90°$ and $B = \theta$. Making these substitutions in Equations A–9 yields

$$\sin(90° - \theta) = \sin 90° \cos\theta - \sin\theta\cos 90° = \cos\theta$$
$$\cos(90° - \theta) = \cos 90° \cos\theta + \sin 90° \sin\theta = \sin\theta \qquad \textbf{A–11}$$

Algebra

The quadratic equation

A well-known result that finds many uses in physics is the solution to the **quadratic equation**

$$\mathrm{a}x^2 + \mathrm{b}x + \mathrm{c} = 0 \qquad \textbf{A–12}$$

In this equation, a, b, and c are constants and x is a variable. When we refer to the solution of the quadratic equation, we mean the values of x that satisfy Equation A–12. These values are given by the following expression:

Solutions to the Quadratic Equation

$$x = \frac{-\mathrm{b} \pm \sqrt{\mathrm{b}^2 - 4\mathrm{ac}}}{2\mathrm{a}} \qquad \textbf{A–13}$$

Note that there are two solutions to the quadratic equation, in general, corresponding to the plus and minus sign in front of the square root. In the special case that the quantity under the square root vanishes, there will be only a single solution. If the quantity under the square root is negative the result for x is not physical, which means a mistake has probably been made in the calculation.

To illustrate the use of the quadratic equation and its solution, we consider a standard one-dimensional kinematics problem, such as one might encounter in Chapter 2:

> A ball is thrown straight upward with an initial speed of 11 m/s. How long does it take for the ball to first reach a height of 4.5 m above its launch point?

The first step in solving this problem is to write the equation giving the height of the ball, y, as a function of time. Referring to Equation 2–11, we have

$$y = y_0 + v_0 t - \tfrac{1}{2}gt^2$$

To make this look more like a quadratic equation, we move all the terms onto the left-hand side, which yields

$$\tfrac{1}{2}gt^2 - v_0 t + y - y_0 = 0$$

This is the same as Equation A–12 if we make the following identifications: $x = t$; $\mathrm{a} = \tfrac{1}{2}g$; $\mathrm{b} = -v_0$; $\mathrm{c} = y - y_0$. The desired solution, then, is given by making these substitutions in Equation A–13:

$$t = \frac{v_0 \pm \sqrt{v_0^{\,2} - 2g(y - y_0)}}{g}$$

The final step is to use the appropriate numerical values; $g = 9.81\text{ m/s}^2$, $v_0 = 11$ m/s, $y - y_0 = 4.5$ m. Straightforward calculation gives $t = 0.54$ s and $t = 1.7$ s. Therefore, the time it takes to first reach a height of 4.5 m is 0.54 s; the second solution is the time when the ball is again at a height of 4.5 m, this time on its way down.

Two equations in two unknowns

In some problems, two unknown quantities are determined by two interlinked equations. In such cases it often seems at first that you have not been given enough information to obtain a solution. By patiently writing out what is known, however, you can generally use straightforward algebra to solve the problem.

As an example, consider the following problem: A father and daughter share the same birthday. On one birthday the father announces to his daughter, "Today I am four times older than you, but in 5 years I will be only three times older." How old are the father and daughter now?

You might be able to solve this problem by guessing, but here's how to approach it systematically. First, write what is given in the form of equations. Letting F be the father's age in years, and D the daughter's age in years, we know that on this birthday

$$F = 4D \qquad \text{A–14}$$

In 5 years, the father's age will be $F + 5$, the daughter's age will be $D + 5$, and the following will be true:

$$F + 5 = 3(D + 5)$$

Multiplying through the parenthesis gives

$$F + 5 = 3D + 15 \qquad \text{A–15}$$

Now if we subtract Equation A–15 from Equation A–14 we can eliminate one of the unknowns, F:

$$\begin{array}{r} F = 4D \\ -F + 5 = 3D + 15 \\ \hline -5 = D - 15 \end{array}$$

The solution to this new equation is clearly $D = 10$, and thus the father's age is $F = 4D = 40$.

The following Example investigates a similar problem. In this case, we use the fact that if you drive with a speed v for a time t the distance covered is $d = vt$.

EXAMPLE A–2 **Hit the Road**

It takes 1.50 h to drive with a speed v from home to a nearby town, a distance d away. Later, on the way back, the traffic is lighter, and you are able to increase your speed by 15 mi/h. With this higher speed, you get home in just 1.00 h. Find your initial speed v, and the distance to the town, d.

Picture the Problem

Our sketch shows home and the town, separated by a distance d. Going to town the speed is v, returning home the speed is $v + 15$ mi/h.

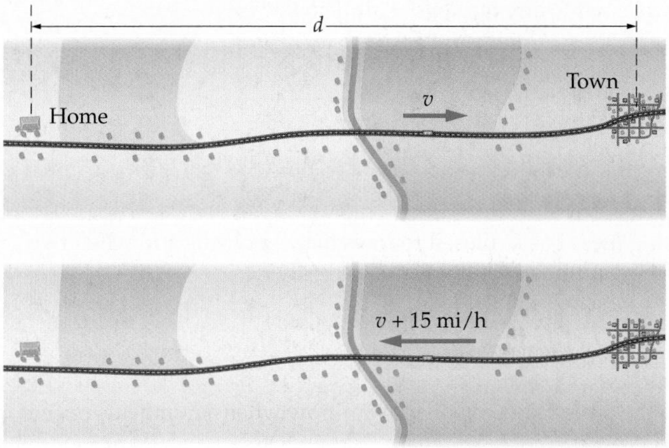

Strategy

To determine the two unknowns, v and d, we need two separate equations. One equation corresponds to what we know about the trip to the town, the second equation corresponds to what we know about the return trip.

Solution

1. Write an equation for the trip to the town. Recall that this trip takes one and a half hours:

$$d = vt = v(1.50\text{ h})$$

2. Write an equation for the trip home. This trip takes one hour, and covers the same distance d:

$$d = (v + 15\text{ mi/h})t = (v + 15\text{ mi/h})(1.00\text{ h})$$

3. Subtract these two equations to eliminate d:

$$\begin{array}{rl} & d = v(1.50\text{ h}) \\ - & d = (v + 15\text{ mi/h})(1.00\text{ h}) \\ \hline & 0 = v(1.50\text{ h}) - v(1.00\text{ h}) - (15\text{ mi/h})(1.00\text{ h}) \end{array}$$

4. Solve this new equation for v:

$$0 = v(1.50\text{ h}) - v(1.00\text{ h}) - (15\text{ mi/h})(1.00\text{ h})$$
$$0 = v(0.50\text{ h}) - (15\text{ mi/h})(1.00\text{ h})$$
$$v = \frac{(15\text{ mi/h})(1.00\text{ h})}{(0.50\text{ h})} = 30\text{ mi/h}$$

5. Use the first equation to solve for d:

$$d = vt = (30\text{ mi/h})(1.50\text{ h}) = 45\text{ mi}$$

continued on next page

continued from previous page

Insight

We could also use the second equation to solve for d. The algebra is a bit messier, but it provides a good double check on our results: $d = (30\text{ mi/h} + 15\text{ mi/h})(1.00\text{ h}) = (45\text{ mi/h})(1.00\text{ h}) = 45\text{ mi}$.

Practice Problem

Suppose the speed going home is 12 mi/h faster than the speed on the way to town, but the times are the same as above. What are v and d in this case? [**Answer:** $v = 24\text{ mi/h}$, $d = 36\text{ mi}$]

Exponents and logarithms

An **exponent** is the power to which a number is raised. For example, in the expression 10^3, we say that the exponent of 10 is 3. To evaluate 10^3 we simply multiply 10 by itself three times:

$$10^3 = 10 \times 10 \times 10 = 1000$$

Similarly, a negative exponent implies an inverse, as in the relation $10^{-1} = 1/10$. Thus, to evaluate a number like 10^{-4}, for example, we multiply $1/10$ by itself four times:

$$10^{-4} = \frac{1}{10} \times \frac{1}{10} \times \frac{1}{10} \times \frac{1}{10} = \frac{1}{10{,}000} = 0.0001$$

The relations just given apply not just to powers of 10, of course, but to any number at all. Thus, x^4 is

$$x^4 = x \times x \times x \times x$$

and x^{-3} is

$$x^{-3} = \frac{1}{x} \times \frac{1}{x} \times \frac{1}{x} = \frac{1}{x^3}$$

Using these basic rules, it follows that exponents add when two or more numbers are multiplied together:

$$\begin{aligned} x^2x^3 &= (x \times x)(x \times x \times x) \\ &= x \times x \times x \times x \times x = x^5 = x^{2+3} \end{aligned}$$

On the other hand, exponents multiply when a number is raised to a power:

$$\begin{aligned} (x^2)^3 &= (x \times x) \times (x \times x) \times (x \times x) \\ &= x \times x \times x \times x \times x \times x = x^6 = x^{2\times 3} \end{aligned}$$

In general, the rules obeyed by exponents can be summarized as follows:

$$\begin{aligned} x^n x^m &= x^{n+m} \\ x^{-n} &= \frac{1}{x^n} \\ \frac{x^n}{x^m} &= x^{n-m} \qquad \text{A–16} \\ (xy)^n &= x^n y^n \\ (x^n)^m &= x^{nm} \end{aligned}$$

Fractional exponents, such as $1/n$, indicate the nth root of a number. Specifically, the square root of x is written as

$$\sqrt{x} = x^{1/2}$$

For n greater than 2 we write the nth root in the following form:

$$\sqrt[n]{x} = x^{1/n} \qquad \text{A–17}$$

Thus, the nth root of a number, x, is the value that gives x when multiplied by itself n times: $(x^{1/n})^n = x^{n/n} = x^1 = x$.

A general method for calculating the exponent of a number is provided by the **logarithm.** For example, suppose x is equal to 10 raised to the power n:

$$x = 10^n$$

In this expression, 10 is referred to as the *base*. The exponent, n, is equal to the logarithm (log) of x:

$$n = \log x$$

The notation "log" is known as the *common logarithm*, and it refers specifically to base 10.

As an example, suppose that $x = 1000 = 10^n$. Clearly, we can write x as 10^3, which means that the exponent of x is 3:

$$\log x = \log 1000 = \log 10^3 = 3$$

When dealing with a number this simple, the exponent can be determined without a calculator. Suppose, however, that $x = 1205 = 10^n$. To find the exponent for this value of x we use the "log" button on a calculator. The result is

$$n = \log 1205 = 3.081$$

Thus, 10 raised to the 3.081 power gives 1205.

Another base that is frequently used for calculating exponents is $e = 2.718\ldots$. To represent $x = 1205$ in this base we write

$$x = 1205 = e^m$$

The logarithm to base e is known as the *natural logarithm*, and it is represented by the notation "ln." Using the "ln" button on a calculator, we find

$$m = \ln 1205 = 7.094$$

Thus, e raised to the 7.094 power gives 1205. The connection between the common and natural logarithms is as follows:

$$\ln x = 2.3026 \log x \qquad \text{A–18}$$

In the example just given, we have $\ln 1205 = 7.094 = 2.3026 \log 1205 = 2.3026(3.081)$.

The basic rules obeyed by logarithms follow directly from the rules given for exponents in Equation A–16. In particular,

$$\begin{aligned} \ln(xy) &= \ln x + \ln y \\ \ln\left(\frac{x}{y}\right) &= \ln x - \ln y \qquad \text{A–19} \\ \ln x^n &= n \ln x \end{aligned}$$

Though these rules are stated in terms of natural logarithms, they are satisfied by logarithms with any base.

Mathematical Expansions

We conclude with a brief consideration of small quantities in mathematics. Consider the following equation:

$$(1 + x)^3 = 1 + 3x + 3x^2 + x^3$$

This expression is valid for all values of x. However, if x is much smaller than one, $x \ll 1$, we can say to a good approximation that

$$(1 + x)^3 \approx 1 + 3x$$

Now, just how good is this approximation? After all, it ignores two terms that would need to be included to produce an equality. In the case $x = 0.001$, for example, the two terms that are neglected, $3x^2$ and x^3, have a combined contribution of only about 3 ten-thousandths of a percent! Clearly, then, little error is made in the approximation $(1 + 0.001)^3 \sim 1 + 3(0.001) = 1.003$. This can be seen visually in **Figure A–5 (a)**, where we plot $(1 + x)^3$ and $1 + 3x$ for x ranging from 0 to 1. Note that there is little difference in the two expressions for x less than about 0.1.

This is just one example of a general result in mathematics that can be derived from the **binomial expansion.** In general, we can say that the following approximation is valid for $x \ll 1$:

$$(1 + x)^n \approx 1 + nx \qquad \textbf{A–20}$$

This result holds for arbitrary n, not just for the case of $n = 3$. For example, if $n = -1$ we have

$$(1 + x)^{-1} = \frac{1}{1 + x} \approx 1 - x$$

We plot $(1 + x)^{-1}$ and $1 - x$ in **Figure A–5 (b)**, and again we see that the results are in good agreement for x less than about 0.1.

An example of an expansion that arises in the study of relativity concerns the following quotient:

$$\frac{1}{\sqrt{1 - \dfrac{v^2}{c^2}}}$$

In this expression v is the speed of an object and c is the speed of light. Since objects we encounter generally have speeds much less than the speed of light, the ratio v/c is much less than one, and v^2/c^2 is even smaller than v/c. Thus, if we let $x = v^2/c^2$ we have

$$\frac{1}{\sqrt{1 - x}}$$

We can apply the binomial expansion to this result if we replace n with $-1/2$ and x with $-x$ in Equation A–20. This yields

$$\frac{1}{\sqrt{1 - \dfrac{v^2}{c^2}}} \approx 1 + \frac{1}{2}\frac{v^2}{c^2}$$

The two sides of this approximate equality are plotted in **Figure A–5 (c)**, showing the accuracy of the approximation for small v/c.

Another type of mathematical expansion leads to the following useful results:

$$\sin\theta \approx \theta$$
$$\cos\theta \approx 1 - \tfrac{1}{2}\theta^2 \qquad \textbf{A–21}$$

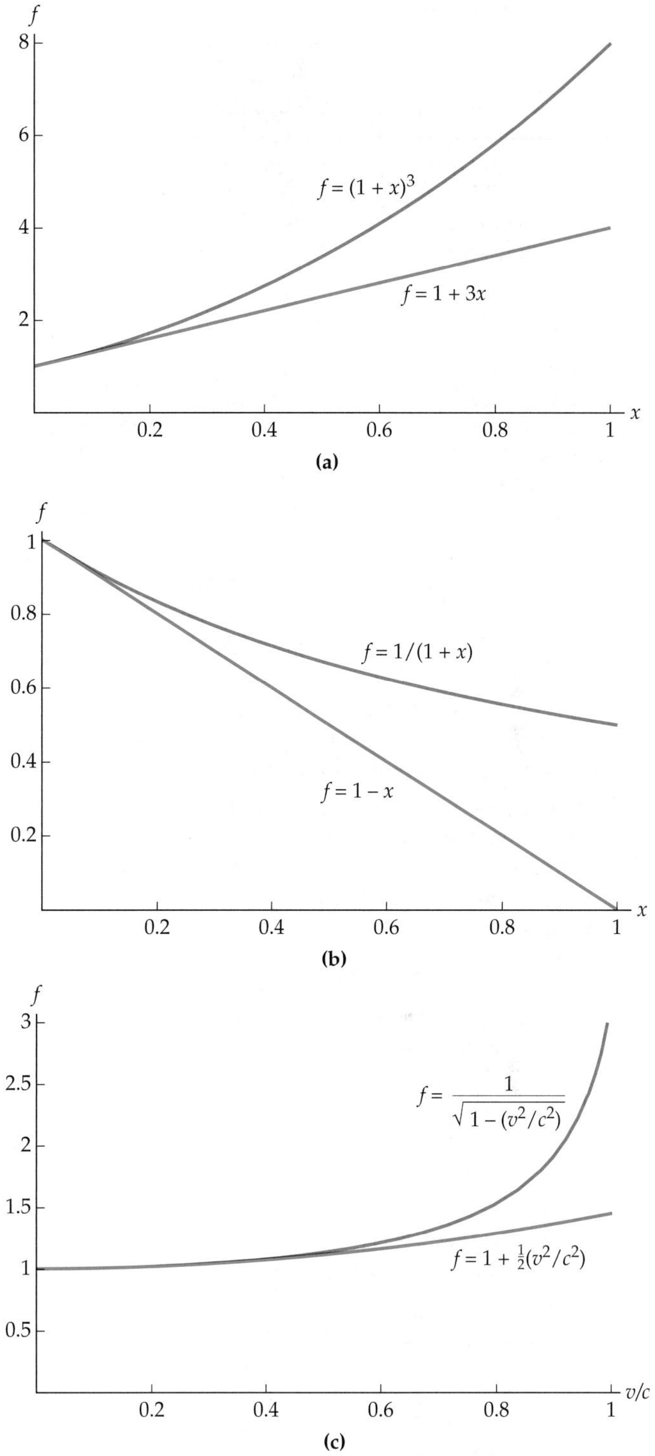

▲ **FIGURE A–5 Examples of mathematical expansions**

(a) A comparison between $(1 + x)^3$ and the result obtained from the binomial expansion, $1 + 3x$. **(b)** A comparison between $1/(1 + x)$ and the result obtained from the binomial expansion, $1 - x$. **(c)** A comparison between $1/\sqrt{1 - (v^2/c^2)}$ and the result obtained from the binomial expansion, $1 + \frac{1}{2}(v^2/c^2)$.

These expansions are valid for small angles θ measured in radians. Note that the result $\sin\theta \approx \theta$ is used to derive Equations 6–13 and 13–19. (See Table 6–2, p. 162, and Figure 13–15, p. 414, for more details on this expansion.)

Vector Multiplication

There are two distinct ways to multiply vectors, referred to as the **dot product** and the **cross product.** The difference between these two types of multiplication is that the dot product yields a scalar (a number) as its result, whereas the cross product results in a vector. Both types of product have important applications in physics. In what follows, we present the basic techniques associated with dot and cross products, and point out places in the text where they are used.

The dot product

Consider two vectors, $\vec{\mathbf{A}}$ and $\vec{\mathbf{B}}$, as shown in **Figure A–6 (a)**. The magnitudes of these vectors are A and B, respectively, and the angle between them is θ. We define the dot product of $\vec{\mathbf{A}}$ and $\vec{\mathbf{B}}$ as follows:

$$\vec{\mathbf{A}} \cdot \vec{\mathbf{B}} = AB\cos\theta \qquad \text{A–22}$$

In words, the dot product of two vectors is a scalar equal to the magnitude of one vector times the magnitude of the second vector times the cosine of the angle between them.

A geometric interpretation of the dot product is presented in **Figure A–6 (b)**. We begin by projecting the vector $\vec{\mathbf{A}}$ onto the direction of vector $\vec{\mathbf{B}}$. This is done by dropping a perpendicular from the tip of $\vec{\mathbf{A}}$ onto the line that passes through $\vec{\mathbf{B}}$, as shown in **Figure A–6 (b)**. Note that the projection of $\vec{\mathbf{A}}$ on the direction of $\vec{\mathbf{B}}$ has a length given by $A\cos\theta$. It follows that the dot product is simply the projection of $\vec{\mathbf{A}}$ onto $\vec{\mathbf{B}}$ times the magnitude of $\vec{\mathbf{B}}$; that is, $(A\cos\theta)B = AB\cos\theta = \vec{\mathbf{A}} \cdot \vec{\mathbf{B}}$. Equivalently, the dot product can be thought of as the projection of $\vec{\mathbf{B}}$ onto $\vec{\mathbf{A}}$ times the magnitude of $\vec{\mathbf{A}}$.

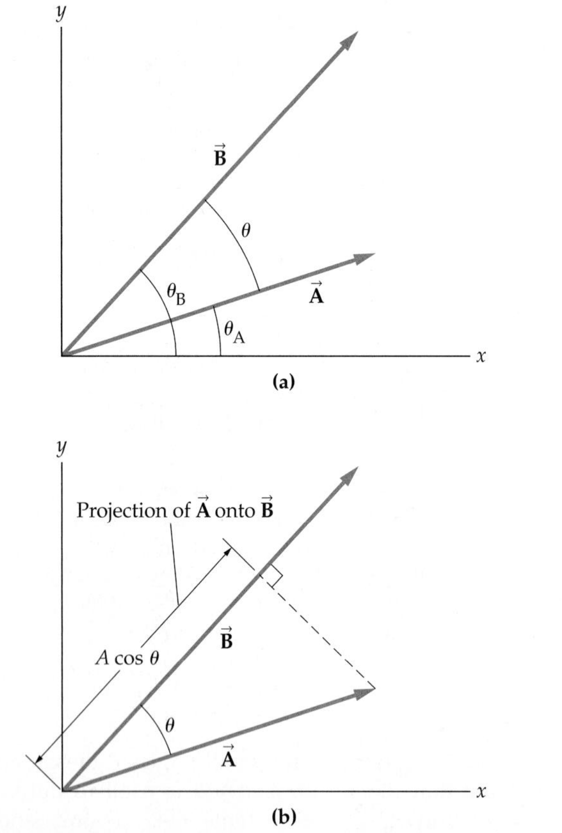

▲ **FIGURE A–6 The dot product between vectors $\vec{\mathbf{A}}$ and $\vec{\mathbf{B}}$.**

A few special cases will help to clarify the dot product. Suppose, for example, that $\vec{\mathbf{A}}$ and $\vec{\mathbf{B}}$ are parallel. In this case, $\theta = 0$ and $\vec{\mathbf{A}} \cdot \vec{\mathbf{B}} = AB$. Therefore, when vectors are parallel, the dot product is simply the product of their magnitudes. On the other hand, suppose $\vec{\mathbf{A}}$ and $\vec{\mathbf{B}}$ point in opposite directions. Now we have $\theta = 180$, and therefore $\vec{\mathbf{A}} \cdot \vec{\mathbf{B}} = -AB$. In general, the sign of $\vec{\mathbf{A}} \cdot \vec{\mathbf{B}}$ is positive if the angle between $\vec{\mathbf{A}}$ and $\vec{\mathbf{B}}$ is less than 90°, and is negative if the angle between them is greater than 90°. Finally, if $\vec{\mathbf{A}}$ and $\vec{\mathbf{B}}$ are perpendicular to one another—that is, if $\theta = 90°$—we see that $\vec{\mathbf{A}} \cdot \vec{\mathbf{B}} = AB\cos 90° = 0$. In this case, neither vector has a nonzero projection onto the other vector.

Dot products have a particularly simple form when applied to unit vectors. Recall, for example, that $\hat{\mathbf{x}}$ and $\hat{\mathbf{y}}$ have unit magnitude and are perpendicular to one another. It follows that

$$\hat{\mathbf{x}} \cdot \hat{\mathbf{x}} = 1, \quad \hat{\mathbf{x}} \cdot \hat{\mathbf{y}} = \hat{\mathbf{y}} \cdot \hat{\mathbf{x}} = 0, \quad \hat{\mathbf{y}} \cdot \hat{\mathbf{y}} = 1 \qquad \text{A–23}$$

These results can be applied to the general two-dimensional vectors $\vec{\mathbf{A}} = A_x\hat{\mathbf{x}} + A_y\hat{\mathbf{y}}$ and $\vec{\mathbf{B}} = B_x\hat{\mathbf{x}} + B_y\hat{\mathbf{y}}$ to give

$$\begin{aligned} \vec{\mathbf{A}} \cdot \vec{\mathbf{B}} &= (A_x\hat{\mathbf{x}} + A_y\hat{\mathbf{y}}) \cdot (B_x\hat{\mathbf{x}} + B_y\hat{\mathbf{y}}) \\ &= A_xB_x\hat{\mathbf{x}} \cdot \hat{\mathbf{x}} + A_xB_y\hat{\mathbf{x}} \cdot \hat{\mathbf{y}} + A_yB_x\hat{\mathbf{y}} \cdot \hat{\mathbf{x}} + A_yB_y\hat{\mathbf{y}} \cdot \hat{\mathbf{y}} \qquad \text{A–24} \\ &= A_xB_x + A_yB_y \end{aligned}$$

Thus, the dot product of two-dimensional vectors is simply the product of their x components plus the product of their y components.

At first glance the result $\vec{\mathbf{A}} \cdot \vec{\mathbf{B}} = AB\cos\theta$ looks quite different from the result $\vec{\mathbf{A}} \cdot \vec{\mathbf{B}} = A_xB_x + A_yB_y$. They are identical, however, as we now show. Suppose that $\vec{\mathbf{A}}$ is at an angle θ_A to the positive x axis, and that $\vec{\mathbf{B}}$ is at an angle $\theta_B > \theta_A$ to the x axis, from which it follows that the angle between $\vec{\mathbf{A}}$ and $\vec{\mathbf{B}}$ is $\theta = \theta_B - \theta_A$. Noting that $A_x = A\cos\theta_A$ and $A_y = A\sin\theta_A$, and similarly for B_x and B_y, we have

$$\vec{\mathbf{A}} \cdot \vec{\mathbf{B}} = AB(\cos\theta_A\cos\theta_B + \sin\theta_A\sin\theta_B)$$

The second trigonometric identity in Equation A–9 can be applied to the quantity in brackets, with the result that $\vec{\mathbf{A}} \cdot \vec{\mathbf{B}} = A_xB_x + A_yB_y = AB\cos(\theta_B - \theta_A) = AB\cos\theta$, as desired.

The most prominent application of dot products in this text is in Chapter 7, where in Equation 7–3 we define the work to be $W = Fd\cos\theta$, with θ the angle between $\vec{\mathbf{F}}$ and $\vec{\mathbf{d}}$. Clearly, this is simply a statement that work is the dot product of force and displacement:

$$W = \vec{\mathbf{F}} \cdot \vec{\mathbf{d}} = Fd\cos\theta$$

Later in the text, in Equation 19–11, we define the electric flux to be $\Phi = EA\cos\theta$. If we let $\vec{\mathbf{A}}$ represent a vector that has a magnitude equal to the area, A, and points in the direction of the normal to the area, we can write the electric flux as a dot product:

$$\Phi = \vec{\mathbf{E}} \cdot \vec{\mathbf{A}} = EA\cos\theta$$

Similar remarks apply to the magnetic flux, defined in Equation 23–1.

The cross product

When two vectors are multiplied with the cross product, the result is a third vector that is perpendicular to both original vectors. An example is shown in **Figure A–7**, where we see a vector $\vec{\mathbf{A}}$, a vector $\vec{\mathbf{B}}$, and their cross product, $\vec{\mathbf{C}}$:

$$\vec{\mathbf{C}} = \vec{\mathbf{A}} \times \vec{\mathbf{B}} \qquad \text{A–25}$$

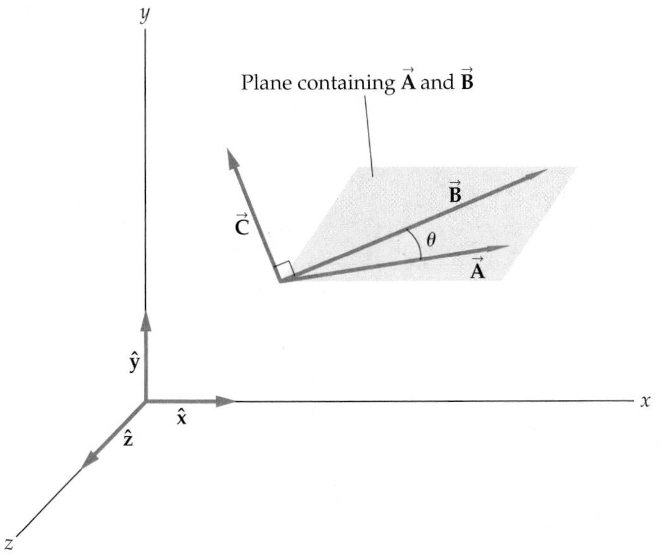

▲ **FIGURE A–7 The vector cross product of $\vec{A}$ and $\vec{B}$.**

Notice that $\vec{\mathbf{C}}$ is perpendicular to the plane formed by the vectors $\vec{\mathbf{A}}$ and $\vec{\mathbf{B}}$. In addition, the direction of $\vec{\mathbf{C}}$ is given by the following right-hand rule:

> To find the direction of $\vec{\mathbf{C}} = \vec{\mathbf{A}} \times \vec{\mathbf{B}}$, point the fingers of your right hand in the direction of $\vec{\mathbf{A}}$ and curl them toward $\vec{\mathbf{B}}$. Your thumb is now pointing in the direction of $\vec{\mathbf{C}}$.

It is clear from this rule that if $\vec{\mathbf{A}} \times \vec{\mathbf{B}} = \vec{\mathbf{C}}$, then $\vec{\mathbf{B}} \times \vec{\mathbf{A}} = -\vec{\mathbf{C}}$.

The magnitude of $\vec{\mathbf{C}} = \vec{\mathbf{A}} \times \vec{\mathbf{B}}$ depends on the magnitudes of the vectors $\vec{\mathbf{A}}$ and $\vec{\mathbf{B}}$, and on the angle θ between them. In particular,

$$C = AB\sin\theta \qquad \text{A–26}$$

Comparing with Equation A–22, we see that the cross product involves a $\sin\theta$, whereas the dot product depends on $\cos\theta$. As a result, it follows that the cross product has zero magnitude when $\vec{\mathbf{A}}$ and $\vec{\mathbf{B}}$ point in the same direction ($\theta = 0°$) or in opposite directions ($\theta = 180°$). On the other hand, the cross product has its greatest magnitude, $C = AB$, when $\vec{\mathbf{A}}$ and $\vec{\mathbf{B}}$ are perpendicular to one another ($\theta = 90°$).

When we apply these rules to unit vectors, which are at right angles to one another and of unit magnitude, the results are particularly simple. For example, consider the cross product $\hat{\mathbf{x}} \times \hat{\mathbf{y}}$. Referring to Figure A–7, we see that this cross product points in the positive z direction. In addition, the magnitude of $\hat{\mathbf{x}} \times \hat{\mathbf{y}}$ is $(1)(1)\sin 90° = 1$. It follows, therefore, that $\hat{\mathbf{x}} \times \hat{\mathbf{y}} = \hat{\mathbf{z}}$. On the other hand, $\hat{\mathbf{x}} \times \hat{\mathbf{x}} = \hat{\mathbf{y}} \times \hat{\mathbf{y}} = \hat{\mathbf{z}} \times \hat{\mathbf{z}} = 0$ because $\theta = 0°$ in each of these cases. To summarize:

$$\begin{aligned}
&\hat{\mathbf{x}} \times \hat{\mathbf{y}} = \hat{\mathbf{z}}, \quad &&\hat{\mathbf{y}} \times \hat{\mathbf{z}} = \hat{\mathbf{x}}, \quad &&\hat{\mathbf{z}} \times \hat{\mathbf{x}} = \hat{\mathbf{y}} \\
&\hat{\mathbf{y}} \times \hat{\mathbf{x}} = -\hat{\mathbf{z}}, \quad &&\hat{\mathbf{z}} \times \hat{\mathbf{y}} = -\hat{\mathbf{x}}, \quad &&\hat{\mathbf{x}} \times \hat{\mathbf{z}} = -\hat{\mathbf{y}} \qquad \text{A–27} \\
&\hat{\mathbf{x}} \times \hat{\mathbf{x}} = 0, \quad &&\hat{\mathbf{y}} \times \hat{\mathbf{y}} = 0, \quad &&\hat{\mathbf{z}} \times \hat{\mathbf{z}} = 0
\end{aligned}$$

As an example of how to use these unit-vector results, consider the cross product of the two-dimensional vectors shown in Figure A–6, $\vec{\mathbf{A}} = A_x\hat{\mathbf{x}} + A_y\hat{\mathbf{y}}$ and $\vec{\mathbf{B}} = B_x\hat{\mathbf{x}} + B_y\hat{\mathbf{y}}$. Straightforward application of Equation A–27 yields

$$\begin{aligned}
\vec{\mathbf{C}} = \vec{\mathbf{A}} \times \vec{\mathbf{B}} &= (A_x\hat{\mathbf{x}} + A_y\hat{\mathbf{y}}) \times (B_x\hat{\mathbf{x}} + B_y\hat{\mathbf{y}}) \\
&= A_xB_x(\hat{\mathbf{x}} \times \hat{\mathbf{x}}) + A_xB_y(\hat{\mathbf{x}} \times \hat{\mathbf{y}}) \\
&\quad + A_yB_x(\hat{\mathbf{y}} \times \hat{\mathbf{x}}) + A_yB_y(\hat{\mathbf{y}} \times \hat{\mathbf{y}}) \\
&= (A_xB_y - A_yB_x)\hat{\mathbf{z}} \qquad \text{A–28}
\end{aligned}$$

Notice that $\vec{\mathbf{C}}$ is perpendicular to both $\vec{\mathbf{A}}$ and $\vec{\mathbf{B}}$, as required for a cross product. In addition, the magnitude of $\vec{\mathbf{C}}$ is

$$A_xB_y - A_yB_x = AB(\cos\theta_A \sin\theta_B - \sin\theta_A\cos\theta_B)$$

If we now apply the first trigonometric identity in Equation A–9, we recover the result given in Equation A–26:

$$\begin{aligned}
C &= AB(\cos\theta_A \sin\theta_B - \sin\theta_a \cos\theta_B) \\
&= AB\sin(\theta_B - \theta_A) = AB\sin\theta
\end{aligned}$$

The first application of cross products in this text is torque, which is discussed in Chapter 11. In fact, Equation 11–2 defines the magnitude of the torque, τ, as follows: $\tau = r(F\sin\theta)$. As one might expect by referring to Equations A–25 and A–26, the torque vector, $\vec{\boldsymbol{\tau}}$, can be written as the following cross product:

$$\vec{\boldsymbol{\tau}} = \vec{\mathbf{r}} \times \vec{\mathbf{F}}$$

Similarly, the angular momentum vector, $\vec{\mathbf{L}}$, whose magnitude is given in Equation 11–13, is simply

$$\vec{\mathbf{L}} = \vec{\mathbf{r}} \times \vec{\mathbf{p}}$$

Finally, the cross product appears again in magnetism. In fact, the magnitude of the magnetic force on a charge q with a velocity $\vec{\mathbf{v}}$ in a magnetic field $\vec{\mathbf{B}}$ is $F = qvB\sin\theta$, as given in Equation 22–1. As one might guess, the vector form of this force is

$$\vec{\mathbf{F}} = q\vec{\mathbf{v}} \times \vec{\mathbf{B}}$$

The advantage of a cross product expression like this is that it contains both the direction and magnitude of a vector in one compact equation. In fact, we can now see the origin of the right-hand rule for magnetic forces given in Section 22–2.

Appendix B Typical values

Mass

Sun	2.00×10^{30} kg
Earth	5.97×10^{24} kg
Moon	7.35×10^{22} kg
747 airliner (maximum takeoff weight)	3.5×10^{5} kg
blue whale	178,000 kg = 197 tons
elephant	5400 kg
mountain gorilla	180 kg
human	70 kg
bowling ball	7 kg
half gallon of milk	1.81 kg = 4 lbs
baseball	0.141–0.148 kg
golf ball	0.045 kg
female calliope hummingbird (smallest bird in North America)	3.5×10^{-3} kg $= \frac{1}{8}$ oz
raindrop	3×10^{-5} kg
antibody molecule (IgG)	2.5×10^{-22} kg
hydrogen atom	1.67×10^{-27} kg

Length

orbital radius of Earth (around Sun)	1.5×10^{8} km
orbital radius of Moon (around Earth)	3.8×10^{5} km
altitude of geosynchronous satellite	35,800 km = 22,300 mi
radius of Earth	6370 km
altitude of Earth's ozone layer	50 km
height of Mt. Everest	8848 m
height of Washington Monument	169 m = 555 ft
pitcher's mound to home plate	18.44 m
baseball bat	1.067 m
CD (diameter)	120 mm
aorta (diameter)	18 mm
period in sentence (diameter)	0.5 mm
red blood cell	$7.8\ \mu m = \frac{1}{3300}$ in.
typical bacterium (*E. coli*)	$2\ \mu m$
wavelength of green light	550 nm
virus	20–300 nm
large protein molecule	25 nm
diameter of DNA molecule	2.0 nm
radius of hydrogen atom	5.29×10^{-11} m

Time

estimated age of Earth	approx. 4.6 billion y $\approx 10^{17}$ s
estimated age of human species	approx. 150,000 y $\approx 5 \times 10^{12}$ s
half life of carbon-14	5730 y $= 1.81 \times 10^{11}$ s
period of Halley's comet	76 y $= 2.40 \times 10^{9}$ s
half life of technetium-99	6 h $= 2.16 \times 10^{4}$ s
time for driver of car to apply brakes	0.46 s
human reaction time	60–180 ms
air bag deployment time	10 ms

period of middle C sound wave	3.9 ms
collision time for batted ball	2 ms
decay of excited atomic state	10^{-8} s
period of green light wave	1.8×10^{-15} s

Speed

light	3×10^{8} m/s
meteor	35–95 km/s
space shuttle (orbital velocity)	8.5 km/s = 19,000 mi/h
rifle bullet	700–750 m/s
sound in air (STP)	340 m/s
fastest human nerve impulses	140 m/s
747 at takeoff	80.5 m/s
kangaroo	18.1 m/s = 40.5 mi/h
200-m dash (Olympic record)	10.1 m/s
butterfly	1 m/s
blood speed in aorta	0.35 m/s
giant tortoise	0.076 m/s = 0.170 mi/h
Mer de Glace glacier (French Alps)	4×10^{-6} m/s

Acceleration

protons in particle accelerator	9×10^{13} m/s^2
ultracentrifuge	3×10^{6} m/s^2
meteor impact	10^{5} m/s^2
baseball struck by bat	3×10^{4} m/s^2
loss of consciousness	$7.14g$ = 70 m/s^2
acceleration of gravity on Earth (g)	9.81 m/s^2
braking auto	8 m/s^2
acceleration of gravity on the moon	1.62 m/s^2
rotation of Earth at equator	3.4×10^{-2} m/s^2

Appendix C Planetary data

Name	Equatorial Radius (km)	Mass (Relative to Earth's)*	Mean Density (km/m^3)	Surface Gravity (Relative to Earth's)	Orbital Semimajor Axis $\times 10^6$ km	Orbital Semimajor Axis A. U.	Escape Speed (km/s)	Orbital Period (Years)	Orbital Eccentricity
Mercury	2440	0.0553	5430	0.38	57.9	0.387	4.2	0.240	0.206
Venus	6052	0.816	5240	0.91	108.2	0.723	10.4	0.615	0.007
Earth	6370	1	5510	1	149.6	1	11.2	1.000	0.017
Mars	3394	0.108	3930	0.38	227.9	1.523	5.0	1.881	0.093
Jupiter	71,492	318	1360	2.53	778.4	5.203	60	11.86	0.048
Saturn	60,268	95.1	690	1.07	1427.0	9.539	36	29.42	0.054
Uranus	25,559	14.5	1270	0.91	2871.0	19.19	21	83.75	0.047
Neptune	24,776	17.1	1640	1.14	4497.1	30.06	24	163.7	0.009
Pluto	1137	0.0021	2060	0.07	5906	39.84	1.2	248.0	0.249

*Mass of Earth = 5.97×10^{24} kg

Appendix D Elements of electrical circuits

Circuit Element	Symbol	Physical Characteristics
resistor		Resists the flow of electric current. Converts electrical energy to thermal energy.
capacitor		Stores electrical energy in the form of an electric field.
inductor		Stores electrical energy in the form of a magnetic field.
incandescent lightbulb		A device containing a resistor that gets hot enough to give off visible light.
battery		A device that produces a constant difference in electrical potential between its terminals.
ac generator		A device that produces a potential difference between its terminals that oscillates with time.
switches (open and closed)	 	Devices to control whether electric current is allowed to flow through a portion of a circuit.
ground		Sets the electric potential at a point in a circuit equal to a constant value usually taken to be $V = 0$.

Appendix E Periodic table of the elements

Key: Atomic number — 26 **Fe** — Symbol; 58.85 — Atomic mass; $3d^64s^2$ — Outer electron configuration

PERIODS	s: GROUP I	s: GROUP II	d: Transition elements										p: GROUP III	p: GROUP IV	p: GROUP V	p: GROUP VI	p: GROUP VII	p: GROUP VIII
1	1 **H** 1.01 $1s^1$																	2 **He** 4.00 $1s^2$
2	3 **Li** 6.94 $2s^1$	4 **Be** 9.01 $2s^2$											5 **B** 10.81 $2p^1$	6 **C** 12.01 $2p^2$	7 **N** 14.01 $2p^3$	8 **O** 16.00 $2p^4$	9 **F** 19.00 $2p^5$	10 **Ne** 20.18 $2p^6$
3	11 **Na** 22.99 $3s^1$	12 **Mg** 24.31 $3s^2$											13 **Al** 26.98 $3p^1$	14 **Si** 28.09 $3p^2$	15 **P** 30.97 $3p^3$	16 **S** 32.07 $3p^4$	17 **Cl** 35.45 $3p^5$	18 **Ar** 39.95 $3p^6$
4	19 **K** 39.10 $4s^1$	20 **Ca** 40.08 $4s^2$	21 **Sc** 44.96 $3d^14s^2$	22 **Ti** 47.88 $3d^24s^2$	23 **V** 50.94 $3d^34s^2$	24 **Cr** 52.00 $3d^54s^1$	25 **Mn** 54.94 $3d^54s^2$	26 **Fe** 55.85 $3d^64s^2$	27 **Co** 58.93 $3d^74s^2$	28 **Ni** 58.69 $3d^84s^2$	29 **Cu** 63.55 $3d^{10}4s^1$	30 **Zn** 65.39 $3d^{10}4s^2$	31 **Ga** 69.72 $4p^1$	32 **Ge** 72.61 $4p^2$	33 **As** 74.92 $4p^3$	34 **Se** 78.96 $4p^4$	35 **Br** 79.90 $4p^5$	36 **Kr** 83.80 $4p^6$
5	37 **Rb** 85.47 $5s^1$	38 **Sr** 87.62 $5s^2$	39 **Y** 88.96 $4d^15s^2$	40 **Zr** 91.22 $4d^25s^2$	41 **Nb** 92.91 $4d^45s^1$	42 **Mo** 95.94 $4d^55s^1$	43 **Tc** (98) $4d^55s^2$	44 **Ru** 101.07 $4d^75s^1$	45 **Rh** 102.91 $4d^85s^1$	46 **Pd** 106.42 $4d^{10}5s^6$	47 **Ag** 107.87 $4d^{10}5s^1$	48 **Cd** 112.41 $4d^{10}5s^2$	49 **In** 114.82 $5p^1$	50 **Sn** 118.71 $5p^2$	51 **Sb** 121.76 $5p^3$	52 **Te** 127.60 $5p^4$	53 **I** 126.90 $5p^5$	54 **Xe** 131.29 $5p^6$
6	55 **Cs** 132.91 $6s^1$	56 **Ba** 137.33 $6s^2$	57 **La** 138.91 $5d^16s^2$ *	72 **Hf** 178.49 $5d^26s^2$	73 **Ta** 180.95 $5d^36s^2$	74 **W** 183.85 $5d^46s^2$	75 **Re** 186.21 $5d^56s^2$	76 **Os** 190.2 $5d^66s^2$	77 **Ir** 192.22 $5d^76s^2$	78 **Pt** 195.08 $5d^96s^1$	79 **Au** 196.97 $5d^{10}6s^1$	80 **Hg** 200.59 $5d^{10}6s^2$	81 **Tl** 204.36 $6p^1$	82 **Pb** 207.2 $6p^2$	83 **Bi** 208.98 $6p^3$	84 **Po** (209) $6p^4$	85 **At** (210) $6p^5$	86 **Rn** (222) $6p^6$
7	87 **Fr** (223) $7s^1$	88 **Ra** 226.03 $7s^2$	89 **Ac** 227.03 $6d^17s^2$ †	104 **Rf** (261) $6d^27s^2$	105 **Db** (262) $6d^37s^2$	106 **Sg** (266) $6d^47s^2$	107 **Bh** (264) $6d^57s^2$	108 **Hs** (269) $6d^67s^2$	109 **Mt** (268) $6d^77s^2$	110 (271)	111 (272)	112 (277)		114 (289)		116 (289)		118 (293)

f

*	58 **Ce** 140.12 $5d^14f^16s^2$	59 **Pr** 140.91 $4f^36s^2$	60 **Nd** 144.24 $4f^46s^2$	61 **Pm** (145) $4f^56s^2$	62 **Sm** 150.36 $4f^66s^2$	63 **Eu** 151.96 $4f^76s^2$	64 **Gd** 157.25 $5d^14f^76s^2$	65 **Tb** 158.93 $5d^14f^86s$	66 **Dy** 162.50 $4f^{10}6s^2$	67 **Ho** 164.93 $4f^{11}6s^2$	68 **Er** 167.26 $4f^{12}6s^2$	69 **Tm** 168.93 $4f^{13}6s^2$	70 **Yb** 173.04 $4f^{14}6s^2$	71 **Lu** 174.97 $5d^14f^{14}6s^2$	Lanthanides
†	90 **Th** 232.04 $6d^27s^2$	91 **Pa** 231.04 $5f^26d^17s^2$	92 **U** 238.03 $5f^36d^17s^2$	93 **Np** 237.05 $5f^46d^17s^2$	94 **Pu** (244) $5f^66d^07s^2$	95 **Am** (243) $5f^76d^07s^2$	96 **Cm** (247) $5f^76d^17s^2$	97 **Bk** (247) $5f^86d^17s^2$	98 **Cf** (251) $5f^{10}6d^07s^2$	99 **Es** (252) $5s^{11}6d^07s^2$	100 **Fm** (257) $5f^{12}6d^07s^2$	101 **Md** (258) $5f^{13}6d^07s^2$	102 **No** (259) $5f^{14}6d^07s^2$	103 **Lr** (262) $5f^{14}6d^17s^2$	Actinides

Appendix F Properties of selected isotopes

Atomic Number (Z)	Element	Symbol	Mass Number (A)	Atomic Mass*	Abundance (%) or Decay Mode† (if radioactive)	Half-Life (if radioactive)
0	(Neutron)	n	1	1.008665	β^-	10.6 min
1	Hydrogen	H	1	1.007825	99.985	
	Deuterium	D	2	2.014102	0.015	
	Tritium	T	3	3.016049	β^-	12.33 y
2	Helium	He	3	3.016029	0.00014	
			4	4.002603	≈100	
3	Lithium	Li	6	6.015123	7.5	
			7	7.016003	92.5	
4	Beryllium	Be	7	7.016930	EC, γ	53.3 d
			8	8.005305	2α	6.7×10^{-17} s
			9	9.012183	100	
5	Boron	B	10	10.012938	19.9	
			11	11.009305	80.1	
			12	12.014353	β^-	20.2 ms
6	Carbon	C	11	11.011433	β^+, EC	20.3 min
			12	12.000000	98.89	
			13	13.003355	1.11	
			14	14.003242	β^-	5730 y
7	Nitrogen	N	13	13.005739	β^-	9.96 min
			14	14.003074	99.63	
			15	15.000109	0.37	
8	Oxygen	O	15	15.003065	β^+, EC	122 s
			16	15.994915	99.76	
			18	17.999159	0.204	
9	Fluorine	F	19	18.998403	100	
			18	18.000938	EC	109.77 min
10	Neon	Ne	20	19.992439	90.51	
			22	21.991384	9.22	
11	Sodium	Na	22	21.994435	β^+, EC, γ	2.602 y
			23	22.989770	100	
			24	23.990964	β^-, γ	15.0 h
12	Magnesium	Mg	24	23.985045	78.99	
13	Aluminum	Al	27	26.981541	100	
14	Silicon	Si	28	27.976928	92.23	
			31	30.975364	β^-, γ	2.62 h
15	Phosphorus	P	31	30.973763	100	
			32	31.973908	β^-	14.28 d
16	Sulfur	S	32	31.972072	95.0	
			35	34.969033	β^-	87.4 d
17	Chlorine	Cl	35	34.968853	75.77	
			37	36.965903	24.23	
18	Argon	Ar	40	39.962383	99.60	
19	Potassium	K	39	38.963708	93.26	
			40	39.964000	β^-, EC, γ, β^+	1.28×10^9 y

Atomic Number (Z)	Element	Symbol	Mass Number (A)	Atomic Mass*	Abundance (%) or Decay Mode† (if radioactive)	Half-Life (if radioactive)
20	Calcium	Ca	30	39.962591	96.94	
24	Chromium	Cr	52	51.940510	83.79	
25	Manganese	Mn	55	54.938046	100	
26	Iron	Fe	56	55.934939	91.8	
27	Cobalt	Co	59	58.933198	100	
			60	59.933820	β^-, γ	5.271 y
28	Nickel	Ni	58	57.935347	68.3	
			60	59.930789	26.1	
			64	63.927968	0.91	
29	Copper	Cu	63	62.929599	69.2	
			64	63.929766	β^-, β^+	12.7 h
			65	64.927792	30.8	
30	Zinc	Zn	64	63.929145	48.6	
			66	65.926035	27.9	
33	Arsenic	As	75	74.921596	100	
35	Bromine	Br	79	78.918336	50.69	
36	Krypton	Kr	84	83.911506	57.0	
			89	88.917563	β^-	3.2 min
			92	91.926153	β^-	1.84 s
38	Strontium	Sr	86	85.909273	9.8	
			88	87.905625	82.6	
			90	89.907746	β^-	28.8 y
39	Yttrium	Y	89	89.905856	100	
41	Niobium	Nb	98	97.910331	β^-	2.86 s
43	Technetium	Tc	98	97.907210	β^-, γ	4.2×10^6 y
47	Silver	Ag	107	106.905095	51.83	
			109	108.904754	48.17	
48	Cadmium	Cd	114	113.903361	28.7	
49	Indium	In	115	114.90388	95.7; β^-	5.1×10^{14} y
50	Tin	Sn	120	119.902199	32.4	
51	Antimony	Sb	133	132.915237	β^-	2.5 min
53	Iodine	I	127	126.904477	100	
			131	130.906118	β^-, γ	8.04 d
54	Xenon	Xe	132	131.90415	26.9	
			136	135.90722	8.9	
55	Cesium	Cs	133	132.90543	100	
56	Barium	Ba	137	136.90582	11.2	
			138	137.90524	71.7	
			141	140.914406	β^-	18.27 min
			144	143.92273	β^-	11.9 s
61	Promethium	Pm	145	144.91275	EC, α, γ	17.7 y
74	Tungsten (Wolfram)	W	184	183.95095	30.7	
76	Osmium	Os	191	190.96094	β^-, γ	15.4 d
			192	191.96149	41.0	
78	Platinum	Pt	195	194.96479	33.8	
79	Gold	Au	197	196.96656	100	

Atomic Number (Z)	Element	Symbol	Mass Number (A)	Atomic Mass*	Abundance (%) or Decay Mode† (if radioactive)	Half-Life (if radioactive)
81	Thallium	Tl	205	204.97441	70.5	
			210	209.990069	β^-	1.3 min
82	Lead	Pb	204	203.973044	β^-, 1.48	1.4×10^{17} y
			206	205.97446	24.1	
			207	206.97589	22.1	
			208	207.97664	52.3	
			210	209.98418	α, β^-, γ	22.3 y
			211	210.98874	β^-, γ	36.1 min
			212	211.99188	β^-, γ	10.64 h
			214	213.99980	β^-, γ	26.8 min
83	Bismuth	Bi	209	208.98039	100	
			211	210.98726	α, β^-, γ	2.15 min
			212	211.991272	α	60.55 min
84	Polonium	Po	210	209.98286	α, γ	138.38 d
			212	211.988852	α	0.299 μs
			214	213.99519	α, γ	164 μs
86	Radon	Rn	222	222.017574	α, β	3.8235 d
87	Francium	Fr	223	223.019734	α, β^-, γ	21.8 min
88	Radium	Ra	226	226.025406	α, γ	1.60×10^3 y
			228	228.031069	β^-	5.76 y
89	Actinium	Ac	227	227.027751	α, β^-, γ	21.773 y
90	Thorium	Th	228	228.02873	α, γ	1.9131 y
			231	231.036297	α, β^-	25.52 h
			232	232.038054	100; α, γ	1.41×10^{10} y
			234	234.043596	β^-	24.10 d
91	Protactium	Pa	234	234.043302	β^-	6.70 h
92	Uranium	U	232	232.03714	α, γ	72 y
			233	233.039629	α, γ	1.592×10^5 y
			235	235.043925	0.72; α, γ	7.038×10^8 y
			236	236.045563	α, γ	2.342×10^7 y
			238	238.050786	99.275; α, γ	4.468×10^9 y
			239	239.054291	β^-, γ	23.5 min
93	Neptunium	Np	239	239.052932	β^-, γ	2.35 d
94	Plutonium	Pu	239	239.052158	α, γ	2.41×10^4 y
95	Americium	Am	243	243.061374	α, γ	7.37×10^3 y
96	Curium	Cm	245	245.065487	α, γ	8.5×10^3 y
97	Berkelium	Bk	247	247.07003	α, γ	1.4×10^3 y
98	Californium	Cf	249	249.074849	α, γ	351 y
99	Einsteinium	Es	254	254.08802	α, γ, β^-	276 d
100	Fermium	Fm	253	253.08518	EC, α, γ	3.0 d
101	Mendelevium	Md	255	255.0911	EC, α	27 min
102	Nobelium	No	255	255.0933	EC, α	3.1 min
103	Lawrencium	Lr	257	257.0998	α	≈35 s

*The masses given throughout this table are those for the neutral atom, including the Z electrons.

†EC stands for electron capture.

Answers to Your Turn Problems

Chapter 1

Active Example 1–1 1.26 km/h

Chapter 2

Active Example 2–1 **(a)** 10.7 mi. **(b)** $\Delta\vec{\mathbf{x}} = (2.1 \text{ mi})\hat{\mathbf{x}}$. Note that these results are independent of the location of the origin.

Chapter 3

Active Example 3–1 D will decrease by a factor of two, but θ is unchanged. Numerical calculation gives $D = 2.33$ m and $\theta = 20.5°$.

Active Example 3–2 Reducing the time interval by a factor of two increases the magnitude of the acceleration by a factor of two, but does not change the direction. We find $a_{av} = 2.68 \text{ m/s}^2$, and $\theta = 53.6°$ north of east.

Chapter 4

Active Example 4–1 4.89 s. As expected, more time is required in this case.

Active Example 4–2 With $v_0 = 22$ m/s we find $R = 49$ m, for an increase of 20 percent (to two significant figures). In general, the fact that R depends on the *second* power of v_0 means that an increase of p percent in v_0 will result in roughly a $2p$ percent increase in R.

Chapter 5

Active Example 5–1 Doubling the force doubles the acceleration. This, in turn, reduces the stopping distance by a factor of two. We find $\Delta x = 61.0$ m.

Active Example 5–2 **(a)** In this case, the horizontal component of force will be greater than before. Therefore, the final speed will be greater as well. **(b)** $v_x = 1.40$ m/s.

Chapter 6

Active Example 6–1 Referring to Active Example 6–1, we see that the angle must be less than 20.0°. Numerical calculation yields $\theta = 14.0°$.

Active Example 6–2 For a tension greater than that found in Active Example 6–2, the sag angle must be less than 3.50°. We find $\theta = 2.96°$.

Active Example 6–3 Referring to Active Example 6–3, we see that the speed must be greater than 17.0 m/s. In fact, we find $v = 19.4$ m/s.

Chapter 7

Active Example 7–1 Doubling the mass doubles the initial kinetic energy of the block. Because the potential energy of the spring depends on the compression squared, the compression will be increased by a factor of the square root of 2. Therefore, the new compression is 0.17 m.

Active Example 7–2 Recalling that $P = Fv$, it follows that increasing the speed from 29.1 m/s to 32.0 m/s requires an increase in power by the factor (32.0/29.1). Thus, the needed power is 4.10×10^4 W.

Chapter 8

Active Example 8–1 If we set the energy at the halfway position equal to $-mg(d/2) + \frac{1}{2}k(d/2)^2 + \frac{1}{2}mv^2$, we find $v = 0.664$ m/s. Notice that this is *greater* than the speed when the block reaches the position $y = 0$. The reason is that when the block is between $y = -mg/k$ and $y = 0$ the net force acting on it is downward, and hence it decelerates before it reaches $y = 0$.

Active Example 8–2 Using $d = 3.50$ m in step 3, we find $W_{nc} = -6060$ J.

Active Example 8–3 In this case, we find $h = 14.8$ m. Note that the height is less for an increased mass.

Chapter 9

Active Example 9–1 This time of contact would result in a final speed of 74.4 m/s.

Active Example 9–2 Doubling the mass of the stick means that the bee's velocity will be doubled in magnitude. Therefore, $\vec{\mathbf{v}}_b = \vec{\mathbf{p}}_b/m_b = (7.60 \text{ cm/s})\hat{\mathbf{x}}$.

Active Example 9–3 No. The stick will move in the opposite direction with a greater speed, but the center of mass will remain at rest.

Chapter 10

Active Example 10–1 The time required is one-half the time for the pulley to come to rest. During this time, the pulley rotates through three-quarters of its total angular displacement; that is, $\theta - \theta_0 = \frac{3}{4}(6.94 \text{ rad}) = 5.21$ rad.

Active Example 10–2 Notice that both a_{cp} and a_t depend linearly on the radius, r. Therefore, at half the radius to the bottom of the tubes, the total acceleration has half the magnitude given in Active Example 10–2. The direction of the acceleration is unchanged, however. Thus, $a = 5.20 \text{ m/s}^2$ and $\phi = 33.9°$.

Active Example 10–3 The final speed decreases if the moment of inertia is increased. This is because with a larger moment of inertia, there is more kinetic energy of rotation for given speed. In this case, we find $v = 0.74$ m/s.

Chapter 11

Active Example 11–1 The child moved to the left (away from the father). The distance from the child to the father is now 0.40 L; therefore, the child moved to the left a distance of 0.15 L.

Active Example 11–2 Zero force condition: $F_2 - Mg - mg = 0$. Zero torque condition: $-Mg(1.56 \text{ m}) + F_2(2.06 \text{ m}) - mg(3.56 \text{ m}) = 0$.

Active Example 11–3 Zero force condition, horizontal: $f_2 - f_3 = 0$. Zero force condition, vertical: $f_1 - mg = 0$. Zero torque condition, with L = distance from base of ladder to wall: $mg(L - b) + f_2(a) - f_1(L) = 0$.

Active Example 11–4 This period corresponds to an angular speed of 420 rad/s; therefore, the radius of the star must be less than 20.0 km. Straightforward substitution yields $R_f = 17$ km.

Active Example 11–5 If it takes 22.5 s to complete one revolution, the angular speed of the merry-go-round is 0.279 rad/s. Noting that the angular speed of the merry-go-round is proportional to the initial speed of the child, we find that $v = 2.46$ m/s. As expected, this result is less than 2.80 m/s.

Chapter 12

Active Example 12–1 8.67×10^7 m

Active Example 12–2 7.63 km/s

Chapter 13

Active Example 13–1 A direct application of $K_{max} = \frac{1}{2}mv_{max}{}^2$ yields $m = 0.275$ kg.

Active Example 13–2 We know that $v_{max} = A\omega$ and $\omega = 2\pi/T$, with $T = 0.700$ s. Therefore, $A = 0.0877$ m.

Active Example 13–3 The amplitude is directly proportional to the initial speed; therefore, doubling the initial speed doubles the amplitude. The period, however, is independent of the amplitude. As a result, the time required to come to rest does not change.

Active Example 13–4 The moment of inertia of a solid sphere is less than that for a hollow sphere. Therefore, the solid sphere has the smaller period. Substituting $I = \frac{7}{5}MR^2$ in Equation 13–21 yields $T = 2\pi\sqrt{\frac{R}{g}}\left(\sqrt{\frac{7}{5}}\right)$.

Chapter 14

Active Example 14–1 Comparing to a whisper at 1 m, we find $r = 8.63$ km. Of course, other intervening sounds over such a large distance would surely drown out the "whisper."

Active Example 14–2 The next higher frequency resulting in constructive interference corresponds to a path length difference of 1.5λ. This means that $\lambda = 1.10$ m, and hence $f = 312$ Hz.

Chapter 15

Active Example 15–1 0.0140 N

Active Example 15–2 The volume of the water must be 0.955 times the volume of the wood, or $V_{water} = 3.22 \times 10^{-3} \text{ m}^3$. The volume of the wood is $(0.150 \text{ m})^3$; that is, $V_{wood} = 3.38 \times 10^{-3} \text{ m}^3$.

Active Example 15–3 4.06 m

Chapter 16

Active Example 16–1 Referring to the figure in Example 16–1, we see that the correct temperature should be somewhere between 70 °F and 90 °F. In fact, if we solve the equation $3t = 9t/5 + 32$, we find $t = T_C = (80/3)$ °C and $3t = T_F = 80$ °F.

Active Example 16–2 The iron has a lower specific heat than aluminum; therefore less heat is required for a given change in temperature. This implies that the final temperature of the system with an iron can will be greater than with an aluminum can. Numerical calculation yields $T = 31$ °C to two significant figures. If three significant figures are used, we find $T = 30.7$ °C for the aluminum can and $T = 31.3$ °C for the iron can.

Chapter 17

Active Example 17–1 Recall that $n = PV/RT$. Halving the temperature, by itself, doubles the number of moles. On the other hand, halving the diameter reduces the volume—and the number of moles—by a factor of eight. Combining the two effects, we find that the number of moles is reduced by a factor of 4.

Active Example 17–2 The height is proportional to the temperature. Therefore, changing the final height by a factor of (19/28) changes the final temperature by the same factor. The final temperature, then, is 330 K (19/28) = 220 K. Note that this is less than the initial temperature, as expected.

Active Example 17–3 The shear deformation, Δx, is proportional to the height of the pancakes, L_0. Therefore, the shear deformation doubles to 5.0 cm.

Active Example 17–4 Brass has a smaller bulk modulus than gold, and hence its volume changes more for a given change in pressure. The change in volume of a brass doubloon is $\Delta V = -7.2 \times 10^{-10}\ \text{m}^3$.

Chapter 18

Active Example 18–1 There are an infinite number of ways in which this can be done. Perhaps the simplest is to move straight up from point A to a pressure of 200 kPa, then expand at constant pressure from 0.25 m^3 to 0.54 m^3. Finally, drop straight down in pressure back to the value of 120 kPa and point B.

Active Example 18–2 Increasing T_h by 20 K gives $e = 0.277$; decreasing T_c by 20 K gives $e = 0.291$. Clearly, it is more effective to decrease the temperature of the cold reservoir.

Active Example 18–3 In this case, we find that the heat released to the cold reservoir is 595 J. This increases the entropy of the cold reservoir by 1.95 J/K. Therefore, the total entropy increase of the universe is 1.95 J/K − 1.82 J/K = 0.13 J/K. Note that this is a positive value, as expected.

Chapter 19

Active Example 19–1 The point of zero force remains in the same place. This can be seen most clearly in step 3, where we see that doubling each charge simply yields an additional factor of two on each side of the equation. Since the factor of two appears on both sides of the equation, it cancels.

Active Example 19–2 The new sphere exerts less force than the original sphere. Specifically, the new sphere has half the radius of the original sphere, and one quarter its surface area. Therefore, it has one quarter the total charge, and exerts one quarter as much force.

Active Example 19–3 The new Gaussian surface has zero electric flux—there is no flux through either end cap, nor through the curved sides of the cylindrical surface. This means that the net charge contained within the Gaussian surface is zero, which is evident when we note that the two plates of the capacitor have opposite charge densities of equal magnitude.

Chapter 20

Active Example 20–1 We are given that the electric potential at point A is *higher* than at point B; therefore, the electric potential energy of an electron (with its negative charge) is *less* at point A than at point B. As an electron moves from point A to point B its electric potential energy increases by 7.2×10^{-19} J.

Active Example 20–2 The initial speed must be considerably greater than 5.00 m/s. In fact, it must be 14.6 m/s.

Active Example 20–3 There is no change in the electric potential energy. The reason is that moving the charge as described does not change the *separation* between any pair of charges in the system. Therefore, the total electric potential energy is as given in step 4.

Chapter 21

Active Example 21–1 The required time is 230 s.

Active Example 21–2 In this case, we find $I_1 = 0.13$ A, $I_2 = 0.11$ A, and $I_3 = 0.020$ A. Notice that each current flows in the direction indicated in the sketch of the circuit.

Active Example 21–3 The 5.00-μF capacitor stores more energy than the 10.0-μF capacitor because more work is required to force a given amount of charge onto its plates. In fact, we find that the 5.00-μF capacitor stores 1.60×10^{-4} J of energy, twice as much (to three significant figures) as the 7.98×10^{-5} J stored in the 10.0-μF capacitor.

Chapter 22

Active Example 22–1 The reason is that the orbital speed is directly proportional to the radius, as we see in Equation 22–3. Therefore, reducing the radius (or circumference) by a given factor reduces the speed by the same factor. It follows that the time required to travel a distance equal to one circumference is independent of the radius.

Active Example 22–2 In this case, the fields produced by the two wires are in the same direction; namely, into the page. The net field is 9.1×10^{-6} T, into the page.

Chapter 23

Active Example 23–1 One way to calculate the current is to note that the light bulb consumes a power of 5.0 W and has a resistance of 12 Ω. We can solve $P = I^2R$ to find $I = 0.65$ A. A second way is to note from Example 23–3 that the external force acting on the rod has a magnitude of 1.6 N. The rod moves with constant speed, however, and hence the magnetic force acting on it, $F = ILB$, must have the same magnitude. If we equate these magnitudes, we find $I = 0.64$ A. The slight discrepancy between these answers is due to round-off error, as can be verified by repeating the calculations with more significant figures.

Active Example 23–2 The value of the inductance changes by a greater factor if we double the number of turns. This is because the inductance depends on the square of the number of turns, but depends only linearly on the cross-sectional area. Specifically, we find the following: (a) doubling N quadruples the inductance; (b) tripling A triples the inductance.

Active Example 23–3 From Equation 23–22 we see that the voltage in the secondary circuit is $V_s = V_p(N_s/N_p)$. Therefore, if $V_p \rightarrow 2V_p$ and $N_p \rightarrow 4N_p$, it is clear that we must double N_s to keep V_s the same.

Chapter 24

Active Example 24–1 The capacitive reactance is equal to 64 Ω when the frequency is reduced to 9.2 Hz. At this frequency, the current is 1.2 A.

Active Example 24–2 **(a)** 40 V. **(b)** 58 V.

Active Example 24–3 Setting the impedance of the circuit equal to 2.50 V/1.50 A = 1.67 Ω, we find $f = 68.0$ Hz and $f = 108$ Hz, to three significant figures.

Chapter 25

Active Example 25–1 If the beam spreads out to twice its initial diameter, its area quadruples. This means, in turn, that the intensity of the beam decreases by a factor of four. The intensity, however, depends on the fields squared. Therefore, it follows that both E_{max} and B_{max} decrease by a factor of two.

Active Example 25–2 With three equally rotated polarizers, we find a transmitted intensity of $0.689I_0$. This is considerably greater than the intensity found with two polarizers. In general, the more smoothly and continuously the plane of polarization is rotated, the greater the transmitted intensity.

Chapter 26

Active Example 26–1 The magnification of the tooth will decrease. After all, as the object moves closer to the mirror, the mirror behaves more and more like a plane mirror, in which case the magnification is 1. With $f = 1.38$ cm and $d_o = 1.00$ cm we find $d_i = -3.63$ cm and $m = 3.63$.

Active Example 26–2 Referring to Figure 26–34, it is clear that to obtain a larger magnification we must move the object closer to the lens. To obtain $m = 0.75$, we find that the object distance must be reduced from 12 cm to $d_o = 2.633$ cm, to four significant figures. The corresponding image distance is -1.975 cm.

Chapter 27

Active Example 27–1 The camera is now focussed at a distance of 1.72 m.

Active Example 27–2 The only change from the analysis given in the text for Figure 27–5, is that the concave mirror has a positive focal length (+12.5 cm) rather than a negative focal length. Therefore, the image distance for the mirror is 21.4 cm. This image is then an object for the lens, producing the final image of the system 15.4 cm to the left of the lens. The final magnification is -0.384, indicating an inverted image 38.4% of its original height.

Active Example 27–3 In this case, the far point is 202 cm from the person's eyes. Note that this far point is closer to the eye than the far point in Example 27–2; therefore, the required refractive power has a magnitude that is greater than 0.312.

Active Example 27–4 The second person's vision needs more correction, since the near point is farther from the eyes. Therefore, the refractive power of the second person's contacts must be greater, which, in turn, means that the focal length must be smaller. In fact, we find $f = 28.6$ cm.

Chapter 28

Active Example 28–1 The desired minimum thickness is one-quarter the wavelength in the material. Recall, however, that the wavelength in a material with an index of refraction n is $\lambda_n = \lambda/n$ (Equation 28–4). Therefore, if the index of refraction is increased, the minimum thickness will be decreased. In this case, we find a minimum thickness of 97.4 nm.

Active Example 28–2 The second dark fringe corresponds to $m = 2$ in Equation 28–12. With this substitution, we find that the linear distance is $y = 6.96$ cm.

Active Example 28–3 As the aperture of a telescope increases, the minimum angular separation that can be resolved decreases, as can be seen from Equation 28–15. If a telescope can resolve smaller angular separations, it follows that its maximum resolution distance is greater. For the case $D = 3.0$ m, we find $L = 1.1 \times 10^{10}$ m.

Active Example 28–4 First, convert the wavelength 486.2 nm to the frequency $f = 6.170 \times 10^{14}$ Hz and the wavelength 563.0 nm to the frequency $f' = 5.329 \times 10^{14}$ Hz. With these results, we can now apply Equation 25–3 to find $u = 4.089 \times 10^7$ m/s. Since this is only 13.6% of the speed of light, the approximations used to derive Equation 25–3 should be valid.

Chapter 29

Active Example 29–1 The speed in this case is $v = 0.953c$.

Active Example 29–2 In this case, the velocity of the probe relative to the planet is $v = 0.306c$. The corresponding length is $L = 9.52$ m.

Chapter 30

Active Example 30–1 Increasing the wavelength by a factor of 1.25 results in a reduction in the frequency by a factor of 1.25. Similarly, the energy of a photon (which is proportional to frequency) is reduced by a factor of 1.25. Since each photon carries less energy, it follows that more photons will be required per second at this new wavelength. In fact, the minimum number of photons per second will be increased by the factor 1.25.

Active Example 30–2 The de Broglie wavelength is inversely proportional to speed; therefore, doubling the speed results in a wavelength that is reduced by a factor of two.

Chapter 31

Active Example 31–1 In this case, the final state is $n_f = 2$.

Active Example 31–2 In singly ionized helium, the charge of the nucleus is $+Ze = +2e$; therefore, $Z = 2$. Referring to Equation 31–9, we see that the energy of any given energy level depends on Z^2. It follows that the energy—and frequency—of the absorbed photon increases by a factor of $2^2 = 4$.

Active Example 31–3 Direct substitution in Equation 31–14 shows that $Z = 51$ is the largest value of Z that requires an acceleration voltage less than 35 kV.

Chapter 32

Active Example 32–1 Doubling the wavelength reduces the frequency of the gamma ray (and its energy) by a factor of two. Since the mass difference is proportional to the energy of the gamma ray, it too will be reduced by a factor of two.

Active Example 32–2 Direct substitution in Equation 32–9 shows that the number of radon atoms has decreased by a factor of two when $t = 3.83$ d. Note that this result is in agreement with Equation 32–10.

Active Example 32–3 Using Equation 32–12, we see that the activity of the radon decreases to 10.0 Bq after 12.7 d.

Active Example 32–4 This reaction releases 12.9 MeV.

Answers to Odd-Numbered Conceptual Questions

Chapter 1

1. No. The factor of 2 is dimensionless.
3. **(a)** Not possible, since units have dimensions. For example, seconds can only have the dimension of time. **(b)** Possible, since different units can be used to measure the same dimensions. For example, time can be measured in seconds, minutes, or hours.
5. To the nearest power of ten: **(a)** 1 m; **(b)** 10^{-2} m; **(c)** 10 m; **(d)** 100 m; **(e)** 10^7 m.

Chapter 2

1. The displacement is the same for you and your dog; the distance covered by the dog is greater.
3. **(a)** Yes. If you drive in a complete circle your distance is the circumference of the circle, but your displacement is zero. **(b)** Yes. The distance and the magnitude of the displacement are equal if you drive in a straight line. **(c)** No. Any deviation from a straight line results in a distance that is greater than the magnitude of the displacement.
5. Their velocities are different because they travel in different directions.
7. Since the car circles the track its direction of motion must be changing. Therefore, its velocity changes as well. Its speed, however, can be constant.
9. Constant-velocity motion; that is, straight-line motion with constant speed.
11. **(a)** The time required to stop is doubled. **(b)** The distance required to stop increases by a factor of four.
13. Yes, if it moves with constant velocity.
15. **(a)** No. If air resistance can be ignored, the acceleration of the ball is the same at each point on its flight. **(b)** Same answer as part (a).
17. Ignoring air resistance, the two gloves have the same acceleration.

Chapter 3

1. **(a)** scalar; **(b)** vector; **(c)** vector; **(d)** scalar.
3. **(a)** $\vec{\mathbf{A}}$ and $\vec{\mathbf{B}}$ have the same magnitude. **(b)** $\vec{\mathbf{A}}$ and $\vec{\mathbf{B}}$ have opposite directions.
5. Yes, if they have the same magnitude and point in opposite directions.
7. Note that the magnitudes A, B, and C satisfy the Pythagorean theorem. It follows that $\vec{\mathbf{A}}$, $\vec{\mathbf{B}}$, and $\vec{\mathbf{C}}$ form a right triangle, with $\vec{\mathbf{C}}$ as the hypotenuse. Thus, $\vec{\mathbf{A}}$ and $\vec{\mathbf{B}}$ are perpendicular to one another.
9. $\vec{\mathbf{A}}$ and $\vec{\mathbf{B}}$ must be collinear and point in opposite directions. In addition, the magnitude of $\vec{\mathbf{A}}$ must be greater than the magnitude of $\vec{\mathbf{B}}$; that is, $|A| > |B|$.
11. The vector $\vec{\mathbf{A}}$ points in one of two directions: (1) 135° counterclockwise from the x axis; (2) 45° clockwise from the x axis. Case (1) corresponds to $A_x < 0$ and case (2) corresponds to $A_x > 0$.
13. Tilt the umbrella forward so that it points in the opposite direction of the rain's velocity relative to you.

Chapter 4

1. Ignoring air resistance, the acceleration of a projectile is vertically downward at all times.
3. The projectile was launched at an angle of 30° above the positive x axis; when it landed its direction of motion was 30° below the positive x axis. Hence, its change in direction was 60° clockwise.
5. At its highest point, the projectile is moving horizontally. This means that gravity has reduced its y component of velocity to zero. The x component of velocity is unchanged by gravity, however. Therefore, the projectile has a velocity equal to $\vec{\mathbf{v}} = (4\text{ m/s})\hat{\mathbf{x}}$ at the highest point in its trajectory.
7. Less than 45°. See Figure 4–9.
9. **(a)** From the child's point of view the scoop of ice cream falls straight downward. **(b)** From the point of view of the parents the scoop of ice cream follows a parabolic trajectory.
11. **(a)** At its highest point the projectile moves horizontally. Therefore, its launch angle was 50°. **(b)** In this case, the launch angle was 150°, or, equivalently, 30° above the negative x axis.

Chapter 5

1. The force exerted on the car by the brakes causes it to slow down, but your body continues to move forward with the same velocity (due to inertia) until the seat belt exerts a force on it to decrease its speed.
3. No. You are at rest relative to your immediate surroundings, but you are in motion relative to other objects in the universe.
5. When the magnitude of the force exerted on the girl by the rope equals the magnitude of her weight, the net force acting on her is zero. As a result, she moves with constant velocity.
7. **(a)** The upper string breaks, because the tension in it is equal to the applied force on the lower string plus the weight of the block. When the tension in the upper string reaches the breaking point, the tension in the lower string is below this value. **(b)** The lower string breaks in this case, because of the inertia of the block. As you move your hand downward rapidly the lower string stretches and breaks before the block can move a significant distance and stretch the upper string.
9. Each time the astronauts throw or catch the ball they exert a force on it, and it exerts an equal and opposite force on them. This causes the astronauts to move farther apart from one another with increasing speed as the game progresses.
11. Mr. Ed's reasoning is incorrect because he is adding two action-reaction forces that act on *different* objects. Wilbur should point out that the net force exerted on the cart is simply the force exerted on it by Mr. Ed. Thus the cart will accelerate. The equal and opposite reaction force acts on Mr. Ed, and does not cancel the force acting on the cart.
13. The whole brick has twice the force acting on it, but it also has twice the mass. Since the acceleration of an object is proportional to the force exerted on it and inversely proportional to its mass, the whole brick has the same acceleration as the half brick.
15. Yes. An object with zero net force acting on it has a constant velocity. This velocity may or may not have zero magnitude.
17. The acceleration of an object is inversely proportional to its mass. The diver and the Earth experience the same force, but the Earth—with its much larger mass—has a much smaller acceleration.
19. An astronaut can tell which of two objects is more massive by pushing on both with the same force. Since acceleration is inversely proportional to mass, the more massive object can be recognized by its smaller acceleration.
21. On solid ground you come to rest in a much smaller distance than when you plunge into water. The smaller the distance over which you come to rest the greater the acceleration, and the greater the acceleration the greater the force exerted on you. Thus, when you land on solid ground the large force it exerts on you may be enough to cause injury.
23. Yes, in fact it happens all the time. Whenever you throw a ball upward it moves in the opposite direction to the net force acting on it, until it reaches the top of its trajectory.
25. This is an example of bad physics. Though the rocket-powered backpack may well produce enough force to give the truck a large acceleration, that force must be transmitted to the truck through the boy's arms. A force large enough to rapidly accelerate a truck would crush a person's arms.

Chapter 6

1. The clothesline has a finite mass, and so the tension in the line must have an upward component to oppose the downward force of gravity. Thus, the line sags much the same as if a weight were hanging from it.
3. The braking distance of a skidding car depends on its initial speed and the coefficient of kinetic friction. Thus, if the coefficient of friction is known reasonably well, the initial speed can be determined from the length of the skid marks.
5. The force that ultimately is responsible for stopping a train is the frictional force between its metal wheels and the metal track. These are fairly smooth surfaces. In contrast, the frictional force that stops a car is between the rubber tires and the concrete roadway. These are rougher surfaces with a greater coefficient of friction.
7. As you brake harder your car has a greater acceleration. The greater the acceleration of the car, the greater the force required to give the flat of strawberries the same acceleration. When the required force exceeds the maximum force of static friction the strawberries begin to slide.
9. For a drop of water to stay on a rotating wheel an inward force is required to give the drop the necessary centripetal acceleration. Since the force between a drop of water and the wheel is small, the drop will separate from the wheel rather than follow its circular path.
11. A centripetal force is required to make the motorcycle follow a circular path, and this force increases rapidly with the speed of the cycle. If the necessary centripetal force exceeds the weight of the motorcycle, because its speed is high enough, then the track must exert a downward force on the cycle at the top of the circle. This keeps the cycle in firm contact with the track.
13. Since the passengers are moving in a circular path a centripetal force must be exerted on them. This force, which is radially inward, is supplied by the wall of the cylinder.
15. This helps because the students sitting on the trunk increase the normal force between your tires and the road. Since the force of friction is proportional to the normal force, this increases the frictional force enough (one hopes) to allow your car to move.
17. The normal force exerted on a gecko by a vertical wall is zero. If the gecko is to stay in place, however, the force of static friction must exert an upward force equal to the gecko's weight. For this to happen when the normal force is zero would require an infinite coefficient of static friction. Thus, we conclude that the physics of gecko feet is more complex than our simple models of friction.
19. Yes. The steering wheel can accelerate a car—even if its speed remains the same—by changing its direction of motion.
21. When a bicycle rider leans inward on a turn, the force applied to the wheels of the bicycle by the ground is both upward and inward. It is this inward force that produces the centripetal acceleration of the rider.

Chapter 7

1. No. Work requires that a force acts through a distance.
3. True. To do work on an object a force must have a nonzero component along its direction of motion.
5. Yes, you must do work against the force of gravity to raise your body upward out of bed.
7. The frictional force between your shoes and the ground does positive work on you whenever you begin to walk.
9. Gravity exerts an equal and opposite force on the package, and hence the net work done on it is zero. The result is no change in kinetic energy.
11. No, we must also know how much time it takes for engine 1 to do the work. For example, if engine 1 takes twice as much time to do twice the work of engine 2, it has the same power output as engine 2. If it takes more than twice as much time, then engine 1 actually produces less power than engine 2.

Chapter 8

1. The kinetic energy cannot be negative, since m and v^2 are always positive or zero. The gravitational potential energy can be negative since any level can be chosen to be zero.
3. Since both distance and average force are doubled, the work done in stretching the spring is quadrupled. Equivalently, the potential energy of the spring is proportional to $(\Delta x)^2$, and hence doubling the stretch distance quadruples the spring's potential energy.
5. If the spring is permanently deformed, it will not return to its original length. As a result, the work that was done to stretch the spring is not fully recovered—some of it goes into the energy of deformation. For this reason, the spring force is not conservative during the deformation. If the spring is now stretched or compressed by a small amount about its new equilibrium position, its force is again conservative—though the force constant will be different.
7. When the term "energy conservation" is used in everyday language, it doesn't refer to the total amount of energy in the universe. Instead, it refers to using energy wisely, especially when a particular source of energy—like oil or natural gas—is finite and nonrenewable.
9. A variety of conservative and nonconservative forces are involved in the situation shown in the photo. First, the engine of the earth mover does positive nonconservative work as it digs out and lifts a load of rocks. At the same time, gravity does negative conservative work on the rocks as the gravitational potential energy of the system increases. Next, the earth mover does positive nonconservative work to transport the rocks to the dump truck. When the rocks are released, gravity does positive conservative work as the gravitational potential energy of the system is converted to kinetic energy. Nonconservative frictional forces do negative work to convert the kinetic energy of the rocks into sound and heat when they land in the truck. Finally, the increased load in the truck does conservative work as it compresses the springs, storing part of the system's energy in the form of spring potential energy.
11. Zero force implies that the rate of change in the potential energy with distance is zero—that is, the potential energy is constant—but the value of the potential energy can be anything at all. Similarly, if the potential energy is zero, it does not mean that the force is zero. Again, what matters is the rate of change of the potential energy with distance.
13. The dive begins with the diver climbing the ladder to the diving board, which converts chemical energy in the muscles into an increased gravitational potential energy. Next, by jumping on the board the diver causes the board to flex and to store potential energy. As the board rebounds, the diver springs into the air, using the kinetic energy derived from the leg muscles and the potential energy released by the board. The diver's kinetic energy is then converted into an increased gravitational potential energy until the highest point of the dive is reached. After that, gravitational potential energy is converted back to kinetic energy as the diver moves downward. Finally, the kinetic energy of the diver is converted into heat, sound, and flowing water as splashdown occurs.

Chapter 9

1. The momentum of the keys increases as they fall because a net force acts on them. The momentum of the universe is unchanged because an equal and opposite force acts on the Earth.
3. If the kinetic energy is zero the speeds must be zero as well. This means that the momentum is zero.
5. Yes, in much the same way that a propeller in water can power a speedboat.

7. When a heavy object and a light object collide they exert equal and opposite forces on one another. Since the light object has less mass, its acceleration is greater. This can result in more severe injuries for the light vehicle.
9. No. The fact that the initial momentum of the system is nonzero means that the final momentum must also be nonzero. Thus, it is not possible for both objects to be at rest after the collision.
11. **(a)** Yes. Suppose two objects have momenta of equal magnitude. If these objects collide in a head-on, completely inelastic collision, they will be at rest after the collision. In this case, all of the initial kinetic energy is converted to other forms of energy. **(b)** No. In order for its momentum to change, an external force must act on the system. We are given, however, that the system is isolated. Therefore, the only forces acting on it are internal forces.
13. The kinetic energy of the bullet is much greater than that of the gun. Thus, less energy is dissipated in stopping the gun.
15. The plane weighs the same whether the fly lands on the dashboard or flies about the cockpit. The reason is that the fly must exert a downward force on the air in the cockpit equal in magnitude to its own weight in order to stay aloft. This force ultimately acts downward on the plane, just as if the fly had landed.
17. Your center of mass is somewhere directly above the area of contact between your foot and the ground.

Chapter 10

1. All points on the rigid object have the same angular speed. Not all points have the same linear speed, however. The farther a given point is from the axis of rotation the greater its linear speed.
3. No. As long as you are driving in a circular path you will have a nonzero centripetal acceleration.
5. **(a)** No, because your direction of motion is constantly changing. **(b)** Yes, your linear speed is simply the radius of the wheel times your angular speed. **(c)** Yes. It is equal to your linear speed squared divided by the radius of the wheel. **(d)** No. Your centripetal acceleration is always toward the center of the wheel, but this is in a different direction as you rotate to different locations.
7. **(a)** A basketball thrown with no spin. **(b)** A spinning airplane propeller on a plane that is at rest. **(c)** A bicycle wheel on a moving bicycle.
9. When the chunky stew is rolled down the aisle, all of the contents of the can roll together with approximately the same angular speed. This is because the chunky stew is thick and almost solid. The beef broth, however, is little more than water. Therefore, when the broth is rolled down the aisle, almost all that is actually rolling is the metal can itself. It follows that the stew has the greater initial kinetic energy, and hence it rolls a greater distance.

Chapter 11

1. No. Torque depends both on the magnitude of the force and on the distance from the axis of rotation, or moment arm, at which it is applied. A small force can produce the same torque as a large force if it is applied farther from the axis of rotation.
3. The long pole has a large moment of inertia, which means that for a given applied torque the walker and pole have a small angular acceleration. This allows more time for the walker to "correct" his balance.
5. A force applied radially to a wheel produces zero torque, though the net force is nonzero.
7. No. In most cars the massive engine is located in the front, thus the car's center of mass is not in the middle of the car, but is closer to the front end. This means that the force exerted on the front tires is greater than the force exerted on the rear tires. (This situation is analogous to Active Example 11–1.)
9. You are in static equilibrium as you sit in your chair; so is the building where you have your physics class.
11. The angular speed of the dust cloud increases, just like a skater pulling in her arms, due to conservation of angular momentum.
13. Yes. Imagine turning on a ceiling fan. This increases the fan's angular momentum, without changing its linear momentum.

Chapter 12

1. No. The force of Earth's gravity is practically as strong in orbit as it is on the surface of the Earth. The astronauts experience weightlessness because they are in constant free fall.
3. If the gravitational force depended on the sum of the two masses, it would predict a nonzero force even when one of the masses is zero. That is, there would be a gravitational force between a mass and a point in empty space, which is certainly not what is observed.
5. No. The amount of area swept out per time varies from planet to planet; what is constant is the amount of area swept out by a *given* planet per time.
7. Once the period of Charon is determined, the mass of the body it orbits (Pluto) can be calculated using Equation 12–7.
9. On the Moon, where there is no atmosphere, a rock can orbit at any altitude where it clears the mountains—as long as it has sufficient speed. Thus, if you could give the rock enough speed, it would orbit the Moon and come up to you from behind.
11. No. In the weightless environment of the Shuttle there would be no convection, which is needed to bring fresh oxygen to the flame. Without convection a flame usually goes out very quickly. In carefully controlled experiments on the Shuttle, however, small flames have been maintained for considerable times. These "weightless" flames are spherical in shape, as opposed to the tear-shaped flames here on Earth.
13. The net force acting on the Moon is always directed toward the Sun, never away from the Sun. Therefore, the Moon's orbit must always curve toward the Sun. The path shown in the upper part of Figure 12–20, though it seems "intuitive," sometimes curves toward the Sun, sometimes away from the Sun. The correct path, shown in the lower part of Figure 12–20, curves sharply toward the Sun when both the Sun and the Earth pull inward on the Moon, and curves only slightly toward the Sun when the Moon is pulled in opposite directions by the Sun and the Earth.

Chapter 13

1. The motion is periodic. It is not simple harmonic, however, because the position and velocity of the ball do not vary sinusoidally with time.
3. The motion of the air cart is periodic; it repeats after a fixed length of time. It is not simple harmonic motion, however, because the position and velocity of the cart do not vary sinusoidally with time.
5. The period remains the same because, even though the distance traveled by the object is doubled, its speed at any given time is also doubled.
7. Referring to Equation 13–6, we see that the constant C represents the maximum speed of the object, $C = v_{max} = A\omega$; similarly, the constant D is the angular frequency, $D = \omega = 2\pi/T$. It follows that the amplitude is $A = C/D$ and the period is $T = 2\pi/D$.
9. If soldiers march in synchrony, the bridge will oscillate with the frequency of their step. If this frequency is near a resonance frequency of the bridge, the amplitude of oscillation could increase to dangerous levels.

Chapter 14

1. To generate a longitudinal wave, hit the nail on the head in a direction parallel to its length. To generate a transverse wave, hit the nail in a direction that is perpendicular to its length.
3. Typical waves at stadiums are transverse, since people move vertically up and down while the wave moves horizontally. To produce a longitudinal wave people could move back-and-forth to their left or right.
5. If the speed of sound depended on frequency, the sound in the first

row—where the travel time is small—would not be affected significantly. Farther back from the stage, however, sounds with different frequencies would arrive at different times—the bass would be "out of sync" with the treble.

7. The part of the nail that vibrates most freely is the portion not yet in the wood. As you drive the nail farther into the wood, therefore, the part that vibrates becomes shorter and shorter. The vibrating portion of the nail is similar to the vibrating air column in an organ pipe, and hence the frequency goes up as the length decreases.
9. When you tune a violin you change the tension in the string. This causes the speed of waves in the string to change. The wavelength of a given mode of oscillation is unchanged, however, due to the fixed length of string. Thus, changing the speed while keeping the wavelength constant results in a change in frequency, according to the relation $v = f\lambda$.
11. No, the energy of oscillation is the same at all times. When the string is flat, the energy of oscillation is purely kinetic.
13. You hear no beats because the difference in frequency between these notes is too great to produce detectable beats.

Chapter 15

1. To draw a liquid up a straw you expand your lungs, which reduces the air pressure inside your mouth to less than atmospheric pressure. The resulting difference in pressure produces a net upward force on the liquid in the straw.
3. The pressure in a tank of water increases with depth, hence the pressure is greatest near the bottom. To provide sufficient support there, the metal bands must be spaced more closely together.
5. This experiment shows that a certain pressure is needed at the bottom of the water column and not just a certain weight of water. To blow the top off the barrel it is necessary to increase the pressure in the barrel enough so that the increase in pressure times the surface area of the top exceeds 400 N. Thus, the required height of water is the height that gives the necessary increase in pressure. But the increase in pressure, $\rho g h$, depends only on the height of the water in the tube, not on its weight.
7. Two quantities are unknown; the object's density and its volume. The two weight measurements provide two independent conditions that can be solved for the two unknowns.
9. The Great Salt Lake has water with a higher salinity, and hence a higher density, than ocean water. In fact, the density of its water is somewhat greater than the density of a typical human body. This means that a person can float in the Great Salt Lake much like a block of wood floats in fresh water.
11. The problem is that as you go deeper into the water, the pressure pushing against your chest and lungs increases rapidly. Even if you had a long tube on your snorkel, you would find it difficult to expand your lungs to take a breath. The air coming through the snorkel is at atmospheric pressure, but the water pushing against your chest might have twice that pressure or more. Thus, scuba gear not only holds air for you to breathe in a tank, it also feeds this air to you under pressure.
13. As the water falls it speeds up. Still, the amount of water that passes a given point in a given time is the same at any height. If the thickness of the water stayed the same, and its speed increased, the amount of water per time would increase. Hence, the thickness of the water must decrease to offset the increase in speed.
15. If you takeoff into the wind the air speed over the wings is greater than if you takeoff with the wind. This means that more lift is produced when taking off into the wind, which is clearly the preferable situation.
17. The ball should spin so that the side facing the batter is moving upward and the side facing the pitcher is moving downward.

Chapter 16

1. The coffee is not in equilibrium because its temperature is different from that of its surroundings. Over time the temperature of the coffee will decrease, until finally it is the same as room temperature. At this point it will be in equilibrium—as long as the room stays at the same temperature.
3. The discrepancy is not serious. After all, a temperature of 15,000,000 °C is equivalent to a temperature of 15,000,273.15 K. A difference of 273.15 out of 15 million is generally insignificant.
5. If the glass and the mercury had the same coefficient of volume expansion, the level of mercury in the glass would not change with temperature. This is because the volume of the cavity in the glass would expand by the same amount as the volume of mercury.
7. The mercury level drops at the beginning because the glass is the first to increase its temperature when it comes into contact with the hot liquid. Therefore, the glass expands before the mercury, leading to a drop in level. As the mercury attains the same temperature few moments later, its level will increase.
9. As the temperature of the house decreases, the length of the various pieces of wood from which it is constructed will decrease as well. As the house adjusts to these changing lengths, it will often creak or groan.
11. If the objects have different masses, the less massive object will have a greater temperature change, since it has the smaller heat capacity. On the other hand, the objects may have the same mass but differ in the material from which they are made. In this case, the object with the smaller specific heat will have the greater temperature change.
13. As the ground warms up on a sunny day, the ground of the surrounding suburbs warms up faster, since it has a smaller specific heat. This would lead to a wind blowing from the city to the suburbs.
15. Even though the flame at the far end of the match is very hot, the wood from which it is made is a poor conductor of heat. The air between the flame and your finger is an even poorer conductor of heat.
17. Two important factors work in favor of the water-filled balloon. First, the water has a large heat capacity, hence it can take on a large amount of heat with little change in temperature. Second, water is a better conductor of heat than air, hence the heat from the flame is conducted into a large volume of water—which gives it a larger effective heat capacity.
19. When penguins (or people) huddle together their rate of heat production is the same, but the surface area over which this heat is radiated to the surroundings is decreased. This results in the penguins being warmer.
21. Object 1 must have the higher temperature, to compensate for object 2's greater emissivity. Since radiated power depends on temperature raised to the fourth power, the temperature of object 1 must be greater by a factor of 2 to the one fourth power.

Chapter 17

1. The volume of an ideal gas is given by $V = nRT/P$. Thus, we expect the volume of the oxygen bags to vary inversely with pressure—the lower the pressure in the cabin, the larger the bags.
3. Recall that velocity takes into account both the speed of an object and its direction of motion. Therefore, the average *velocity* of air molecules in a room is zero, since they move randomly in all directions.
5. Airplanes can have a difficult time taking off from high-altitude airports because the air is thin and provides less lift than air at sea level. When the air is cool, however, its density is greater than when it is warm. Therefore, taking off in the morning or evening will provide the airplane with more lift—which can be a very important advantage at high altitude.

7. The boiling temperature of water depends on the pressure at which the boiling occurs—the higher the pressure, the higher the boiling temperature. Thus, in the autoclave, where the pressure is greater than atmospheric pressure, the temperature of boiling water is greater than 100 °C.
9. The alcohol, besides having antiseptic qualities, evaporates readily. As it evaporates, it draws heat from the body.
11. As high-speed molecules leave the drop, it draws heat from its surroundings to keep its temperature constant. As long as the temperature of the drop is constant, it will continue to have the same fraction of molecules moving quickly enough to escape.

Chapter 18

1. No. If the engine has friction it is not reversible.
3. The answer, in both cases, is yes. **(a)** Compress a gas in a thermally insulated cylinder. This will cause its temperature to rise. **(b)** If you expand a gas in a thermally insulated cylinder, its temperature will decrease.
5. One cannot conclude that heat was added to the system. For example, if a gas in a thermally insulated cylinder is compressed, its temperature will rise with no heat transfer.
7. Yes. You can convert mechanical work completely into heat by rubbing your hands together.
9. **(a)** No; **(b)** yes; **(c)** no; **(d)** no; **(e)** yes; **(f)** no.
11. Yes, this is possible. The problem is that you would need low-temperature reservoirs of ever lower temperature to keep the process going.
13. No. As you do work to put things in order you give off heat to the atmosphere. This increases the entropy of the air by more than the decrease in entropy of the room, for a net increase in entropy.
15. **(a)** Popped popcorn; **(b)** an omelet; **(c)** a pile of bricks; **(d)** a burned piece of paper.

Chapter 19

1. No. When an object becomes charged it is because of a transfer of charge between it and another object.
3. The charged comb causes the paper to become polarized, with the side nearest the comb acquiring a charge opposite to the charge of the comb. The result is an attractive interaction between the comb and the paper.
5. Yes. If the suspended object were neutral, it would be attracted to the charged rod by polarization effects. The fact that the suspended object is repelled indicates it has a charge of the same sign as that of the rod.
7. Both force laws depend on the product of specific properties of the objects involved; in the case of gravity it is the mass that is relevant, in the case of electrostatics it is the electric charge. In addition, both forces decrease with increasing distance as $1/r^2$. The extremely important difference between the forces, however, is that gravity is always attractive, whereas electrostatic forces can be attractive or repulsive.
9. No. Only for very special displacements will the electrostatic force act in a direction that points back toward the equilibrium point. For a general displacement the electrostatic force does not point toward the equilibrium point, and the fifth charge would move farther from equilibrium, making the system unstable.
11. One difference is that when an object is charged by induction, there is no physical contact between the object being charged and the object used to do the charging. In contrast, charging by contact—as the name implies—involves direct physical contact to transfer charge from one object to another. The main difference is that when an object is charged by induction, the sign of the charge the object acquires is opposite to that of the object used to do the charging. Charging by contact gives the object being charged the same sign of charge as the original charged object.
13. No. The direction of the forces might be different simply because the sign of the charges are different. The magnitude of the forces might be different simply because the magnitudes of the charges are different.
15. By definition, electric field lines point in the direction of the electric force on a positive charge at any given location is space. This force can point in only one direction at any one location, however. Therefore, electric field lines cannot cross, because if they did, it would imply two different directions for the electric force at the same location.
17. The electric field depends on both charges. The total electric field at any point is simply the superposition of the electric field produced by each charge separately.
19. Gauss's law is useful as a calculational tool only in cases of high symmetry, where one can produce a gaussian surface on which the electric field is either constant or has no perpendicular component. It is not possible to do this in any simple way for the case of a charged disk. Gauss's law still applies—it's just not particularly useful.

Chapter 20

1. The electric field is a measure of how much the electric potential changes from one position to another. Therefore, the electric field in each of these regions is zero.
3. Not necessarily. The electric field is related to the rate of change of electric potential, not to the value of the electric potential. Therefore, if the electric field is zero in some region of space, it follows that the electric potential is constant in that region. The constant value of the electric potential may be zero, but it may also be positive or negative.
5. Zero. The electric field is perpendicular to an equipotential, therefore the work done in moving along an equipotential is zero.
7. If the electric field is not perpendicular to an equipotential, the field would do work on a charge that moves along the equipotential. In this case, the potential energy of the charge would change, and the surface would not in fact be an equipotential.
9. When the capacitor is disconnected from the battery, the charge on the capacitor plates simply remains where it is—there is no way for it to go anywhere else. When the terminals are connected to one another the charges flow from plate to plate until both plates have zero charge.
11. The capacitance of a capacitor depends on **(b)** the separation of the plates and **(e)** the area of the plates. The capacitance does not depend on **(a)** the charge on the plates, **(c)** the voltage difference between the plates, or **(d)** the electric field between the plates.
13. No. As an example, note that the volume of a milk container is not zero just because the container happens to be empty of milk. The same can be said about the capacitance of a capacitor that happens to be uncharged.

Chapter 21

1. Electric current is in the opposite direction to the motion of negative charge, therefore the electric current of the falling electron is upward.
3. No. By rubbing the comb through your hair you have transferred charge from your hair to the comb, but the net charge of you and the comb together is still zero. Therefore, no current is produced when you walk.
5. No. An electron may have a fairly large velocity at any given time, but because its direction of motion keeps changing—due to its collisions with atoms in the wire—its average velocity is almost zero.
7. Connect the four resistors in a parallel arrangement with two branches, each branch containing two resistors connected in series. In this way, the equivalent resistance of each branch is $2R$, and the equivalent resistance of two resistors of $2R$ in parallel is simply R, as desired.
9. Resistors connected in series have the same current flowing through them.

11. Each electron in the wire affects its neighbors by exerting a force on them, causing them to move. Thus, when electrons begin to move out of a battery their motion sets up a propagating influence that moves through the wire at nearly the speed of light, causing electrons everywhere in the wire to begin moving.
13. A number of factors come into play here. First, the bottom of a bird's foot is tough, and definitely not a good conductor of electricity. Second, and more important, is the fact that a potential difference is required for there to be a flow of current. Just being in contact with a high-voltage wire isn't enough to cause a problem; somewhere else there must be contact with a lower voltage. But the bird is in contact with essentially the same high voltage in two different places (where its feet touch the wire), which doesn't lead to a potential difference. The only potential difference the bird experiences is due to the very small voltage drop along the segment of wire between the bird's two feet.
15. The junction rule is based on conservation of electric charge; the loop rule is based on the conservation of energy.
17. Capacitors in parallel have the same potential difference between their plates.

Chapter 22

1. No. The particles may have charge of the same sign but move in opposite directions along the same line. In this way, they would both move perpendicular to the field, but would deflect in opposite directions.
3. No. If the electron moves in the same direction as the magnetic field, or opposite to the direction of the field, the magnetic force exerted on it will be zero. As a result, its velocity will remain constant.
5. The radius of curvature is proportional to the speed of the particle. It follows that the particle moving in a circle of large radius (and large circumference) has a proportionally larger speed than the particle moving in a circle of small radius (and small circumference). Therefore, the time required for an orbit ($t = d/v$) is the same for both particles.

Chapter 23

1. The magnetic field indicates the strength and direction of the magnetic force that a charged particle moving with a certain velocity would experience at a given point in space. The magnetic flux, on the other hand, can be thought of as a measure of the "amount" of magnetic field that passes through a given area.
3. As the magnet falls, it induces eddy currents in the copper tube. These induced currents produce a magnetic field opposite in direction to that of the magnet. This results in a magnetic repulsion that slows the fall of the magnet. In fact, the motion of the magnet is much the same as if it had been dropped into a tube filled with honey.
5. When the switch is closed, a magnetic field is produced in the wire coil and in the iron rod. This results in an increasing magnetic flux through the metal ring, and a corresponding induced emf. The current produced by the induced emf generates a magnetic field opposite in direction to the field in the iron rod. The resulting magnetic repulsion propels the ring into the air.
7. Initially, the rod accelerates to the left, due to the downward current it carries. As it speeds up, however, the motional emf it generates will begin to counteract the emf of the battery. Eventually the two emfs balance one another, and current stops flowing in the rod. From this point on, the rod continues to move with constant speed.
9. As the shuttle orbits, it moves through the Earth's magnetic field at high speed. A long conducting wire moving through the field can generate an induced emf. In fact, the emf is given by the product of the length of the wire, the speed of the shuttle, and the perpendicular component of the magnetic field. With such large values for the speed and length, the induced emf can be great enough to provide substantial electrical power.
11. The final current in an *RL* circuit is determined only by the resistor *R* and the emf of the battery. The reason is that when the current stops changing, the back emf in an inductor vanishes. Thus, the inductor behaves like an ideal wire of zero resistance when the current reaches its final value.

Chapter 24

1. The average voltage in an ac circuit is zero because it oscillates symmetrically between positive and negative values. To calculate the rms voltage, however, one first squares the voltage. This gives values that are always greater than or equal to zero. Therefore, the rms voltage will be nonzero unless the voltage in the circuit is zero at all times.
3. An *LC* circuit consumes zero power because ideal inductors and capacitors have no resistance. From a different point of view, the phase angle in an *LC* circuit is either $\phi = +90°$ or $\phi = -90°$, depending on whether the frequency is less than or greater than the resonance frequency, respectively. In either case, the power factor, $\cos\phi$, is zero; hence the power consumed by the circuit is zero.
5. In the phasor diagram for an *LC* circuit, the impedance is always perpendicular to the current—the only question, therefore, is whether ϕ is $+90°$ or $-90°$. At a frequency less than the resonance frequency the capacitive reactance ($X_C = 1/\omega C$) is greater than the inductive reactance ($X_L = \omega L$). Therefore, the situation is similar to that of a circuit with only a capacitor, in which case the phase angle is $\phi = -90°$.
7. As frequency is increased there is no change in resistance, *R*. On the other hand, the capacitive reactance ($X_C = 1/\omega C$) decreases and the inductive reactance ($X_L = \omega L$) increases.
9. Mass resists changes in its motion due to its inertia. Similarly, an inductor resists changes in the current flowing through it due to its inductance. Therefore, mass and inductance are analogous. As for the spring constant, a stiff spring (large spring constant) gives little stretch for a given force. Similarly, a capacitor with a small capacitance stores little charge for a given voltage. Since charge is what moves in a circuit, it follows that displacement and charge are analogous. Therefore, the spring constant and the inverse of the capacitance are analogous.
11. Yes. Recall that the impedance of an *RLC* circuit is $Z = \sqrt{R^2 + (X_L - X_C)^2}$. Because the quantity in parenthesis is squared, we get the same impedance when X_L is greater than X_C by a certain amount as when X_C is greater than X_L by the same amount.

Chapter 25

1. Presumably, an "invisible man" would be invisible because light passes through his body unimpeded, just as if it were passing through thin air. If some light were deflected or absorbed, we would see this effect and the person would no longer be invisible. For a person to see, however, some light must be absorbed by the retina. This absorption would cause the invisible man to be visible.
3. As a grain of dust becomes smaller, its volume—and therefore its mass—decreases more rapidly than does its area. It follows that radiation pressure, which acts on the surface of the grain, becomes increasingly important as the size of the grain is decreased. Gravity, which acts on the mass of the grain, becomes less important in this limit.
5. Your watch must have an LCD display. In such watches, the light coming from the display is linearly polarized. If the polarization direction of the display and the sunglasses align, you can read the time. If these directions are at 90° to one another no light will pass through the sunglasses, and the display will appear black.
7. Radio stations generate their electromagnetic waves with large vertical antennas. As a result, their waves are polarized in the vertical direction, as in Figure 25–1. On the other hand, the light

we see from the sun and from light bulbs is unpolarized because the atoms emitting the light can have any orientation relative to one another. Hence, even if individual atoms emit polarized light, the net result from a group of atoms is light with no preferred direction of polarization.

9. Sound waves cannot be polarized. This is because sound is a longitudinal wave, and the molecules can move in only a single direction—back and forth along the direction of propagation. In contrast, the electric field in an electromagnetic wave, which is transverse, can point in any direction within the plane that is transverse to the direction of propagation. This means that the electromagnetic wave can have different polarizations.

11. The two projected images give a view of a scene from slightly different angles, just as your eyes view a three-dimensional object from different angles. Without the headsets, the screen is a confusing superposition of the two images; with the headsets, your right eye sees one view of the scene and your left eye sees the other view. As these views are combined in your brain, you experience a realistic three-dimensional effect.

Chapter 26

1. Three images are formed of object A. One extends from $(-2\text{ m}, 2\text{ m})$ to $(-1\text{ m}, 2\text{ m})$ to $(-1\text{ m}, 3\text{ m})$. Another image forms an "L" from $(1\text{ m}, -3\text{ m})$ to $(1\text{ m}, -2\text{ m})$ to $(2\text{ m}, -2\text{ m})$. Finally, the third image extends from $(-1\text{ m}, -3\text{ m})$ to $(-1\text{ m}, -2\text{ m})$ to $(-2\text{ m}, -2\text{ m})$.

3. A plane mirror is flat, which means that its radius of curvature is infinite. This means that the focal length of the mirror is also infinite. Whether you consider the focal length to be positive infinity (the limit of a concave mirror) or negative infinity (the limit of a convex mirror) doesn't matter, because in either case the term $1/f$ in the mirror equation will be zero.

5. We can consider the sun to be infinitely far from the mirror. As a result, its parallel rays will be focused at the focal point of the mirror. Therefore, the distance from the mirror to the paper should be $f = \frac{1}{2}R$.

7. The key to this system is the fact that the lifeguards can run much faster on the sand than they can swim in the water. Therefore, the path ACB—even though it is longer—has a shorter travel time because more time is spent on the sand and less time is spent in the water.

9. In a real image, light passes through the location of the image before reaching the eye. In a virtual image, light propagates *as if* it were coming from the image—though the reflected or refracted light never actually passes through the image location.

11. When the mug is filled with water, light coming upward from the bottom of the mug is bent toward the horizontal when it passes from water to air. If the light had not been bent, it would have passed over your head—placing the bottom of the mug out of sight. With the bending, however, the light can now propagate to your eyes—making the bottom of the mug visible.

13. No. In order to see, a person's eyes must first bring light to a focus, and then must absorb the light to convert it to nervous impulses that can travel to the brain. Both the bending of the light and its absorption would give away the presence of the invisible man. Therefore, if the man were truly invisible, he would be unable to see.

Chapter 27

1. When you focus on an object at infinity your ciliary muscles are relaxed. It takes muscular effort to focus your eyes on nearby objects.

3. If you are a distance D in front of a mirror your image is a distance D behind the mirror. Therefore, you can see your image clearly if the distance from you to your image, $2D$, is equal to N. In other words, the minimum distance to the mirror is $D = N/2$.

5. The person with the smaller near-point distance can examine an object at closer range than the person with the larger near-point distance. Therefore, the person with the larger near-point distance benefits more from the magnifier.

7. The image you see when looking through a telescope is virtual. First, the objective forms a real image of a distant object, as shown in Figure 27–16. Next, the eyepiece forms an upright and enlarged image of the objective's image. The situation with the eyepiece is essentially the same as that shown in Figure 26–35 (b). Therefore, it is clear that the final image is virtual in this case.

Chapter 28

1. At a point where destructive interference is complete, there is no energy. The total energy in the system is unchanged, however, because those regions with constructive interference have increased amounts of energy. One can simply think of the energy as being redistributed.

3. No. The net signal could be near zero if the waves from the two antennas interfere destructively.

5. In this case, the interference pattern would be reversed. That is, bright fringes in the ordinary pattern would now be the location of dark fringes, and dark fringes would be replaced with bright fringes. For example, in the usual case there is a bright fringe halfway between the slits because the path lengths are equal. Changing the phase of the light from one slit by 180°, however, results in destructive interference (a dark fringe) at the center of the pattern.

7. A ray of light reflected from the lower surface of the curved piece of glass has no phase change. On the other hand, a ray of light reflected from the top surface of the flat piece of glass undergoes a phase change of half a wavelength. Near the center of the pattern the path difference for these two rays goes to zero. As a result, they are half a wavelength out of phase there, and undergo destructive interference.

9. In general, the larger the aperture in an optical instrument, the greater the resolution. This follows directly from Equation 28–14, where we see that a large aperture diameter D implies a small angle θ. The angular separation that can be resolved is decreased—by making θ smaller—the resolution is increased.

11. The soap film in the photograph is thinnest near the top (as one might expect) because in that region the film appears black. Specifically, the light reflected from the front surface of the film has its phase changed by 180°; light that reflects from the back surface of the film has no change in phase. Therefore, light from the front and back surfaces of the film will undergo destructive interference as the path length between the surfaces goes to zero. This is why the top of the film, where the film is thinnest, appears black in the photo.

Chapter 29

1. The light received from these galaxies moves with the speed c, as is true for all light in a vacuum.

3. Velocities add by simple addition in the limit $c \to \infty$. This is evident from Equation 29–4 when one notices that the second term in the denominator vanishes in this limit, making the denominator equal to 1.

5. Yes. All light, regardless of its wavelength, has the same speed in a vacuum. The *frequency* of this "red shifted" light will be affected, however. Recalling that $v = \lambda f$, we see that a longer wavelength also implies a smaller frequency.

7. Since the total energy of an object of finite mass goes to infinity in the limit that $v \to c$, an infinite amount of energy is required to accelerate an object to the speed c.

Chapter 30

1. The "ultraviolet catastrophe" refers to the classical prediction that the intensity of light emitted by a blackbody increases without limit as the frequency is increased.

3. No, all objects with finite temperature give off blackbody radiation. Only an object at absolute zero—which is unattainable—gives no blackbody radiation.

5. If you look at a painting, photo, decal, or similar object, you are viewing light that it reflects to you. Therefore, when an area appears red, it is because the pigments there absorb blue photons and reflect red photons. Similarly, a blue area absorbs red photons. Blue photons carry considerably more energy than red photons, however, and hence blue photons are more likely to cause damage to the pigment molecules, or to alter their structure. It follows that red pigments (which absorb blue photons) are more likely to become faded when exposed to intense light.
7. If two blackbody curves intersected, there would be a range of frequencies where the low-temperature blackbody gives off more energy than the high-temperature blackbody. In this frequency range, then, it would be possible for energy to be spontaneously transferred from the low-temperature body to the high-temperature body, in violation of the second law of thermodynamics.
9. The likely explanation is that the second metal has a greater work function than the first metal. In this case, a shorter-wavelength photon—that is, a photon with higher frequency and higher energy—would be required to supply the additional energy needed to eject an electron.
11. The resolution of a microscope is determined by the wavelength of the imaging radiation—the smaller the wavelength the greater the resolution. Since the typical wavelength of an electron in an electron microscope is much smaller than the wavelength of visible light, the electron microscope has the greater resolution.

Chapter 31

1. Rutherford's alpha particle scattering experiments indicate that the positive charge in an atom is concentrated in a small volume, rather than spread throughout the atom, as in Thomson's model.
3. In the Bohr model, the electron orbits at a well-defined radius; in the quantum mechanical model, the electron can be found at virtually any distance from the nucleus.
5. **(a)** The glass tube of a neon sign contains a low-pressure gas. Therefore, we expect the light from the sign to be in the form of a line spectrum. **(b)** The light from an incandescent lightbulb is basically blackbody radiation from a hot object; therefore, its radiation is distributed continuously as a function of frequency.
7. No, there is no upper limit to the radius of a Bohr orbit. In fact, the radius increases as n^2 for $n = 1, 2, 3, \ldots$.
9. No, the energy does not increase without limit. The energy of a given level in hydrogen ranges from a low of -13.6 eV to a maximum of 0.
11. **(a)** The angular momentum in the quantum mechanical model of the hydrogen atom is zero if the quantum number ℓ is zero. In the $n = 1$ state, the only allowed value for ℓ is 0, and hence the angular momentum must be zero for $n = 1$. **(b)** Yes. For $n > 1$, there are n allowed values for ℓ. One of these values is always zero, therefore the angular momentum can be zero for any value of n.
13. These elements all have similar configurations of their outermost electrons. In fact, the outermost electrons in fluorine, chlorine, and bromine are $2p^5$, $3p^5$, and $4p^5$, respectively. Therefore, each of these atoms is one electron shy of completing the p subshell. This accounts for their similar chemical behavior.

Chapter 32

1. The radius of a nucleus is given by the following expression: $r = (1.2 \times 10^{-15}\ \text{m})A^{1/3}$. Therefore, the radius depends only on the total number of nucleons in the nucleus, A, and not on the number of protons and neutrons separately. If the number of protons plus the number of neutrons is the same for two nuclei, their radii will be equal as well.
3. No. An alpha particle contains two protons, whereas any form of hydrogen contains only a single proton. Therefore, hydrogen cannot give off an alpha particle.
5. A nucleus that contained more than one proton and no neutrons would be unstable because the electrostatic repulsion between the protons would blow the nucleus apart. Neutrons tend to push the protons farther apart, reducing their mutual repulsion, and at the same time add more to the attractive strong nuclear force that holds nuclei together.
7. Carbon-14 dating is useful for objects that are of biological origin and—at most—are on the order of thousands of years old. **(a)** It is useful for dating such things as human or animal remains, plant tissue, clothing, and charcoal from a fire. **(b)** It is not useful for dating inorganic materials, like rocks and minerals, or biological materials that are millions of years old, like dinosaur fossils.
9. The obsidian arrowhead cannot be dated with carbon-14, because it is not of biological origin.
11. Yes. The two samples may contain different quantities of the radioactive isotope, and hence their activities may be different.
13. No. The RBE is related to the amount of biological effect produced by a given type of radiation, not to the amount of energy it delivers.

Answers to Odd-Numbered Conceptual Exercises

Chapter 1

1. Each of the equations is dimensionally consistent.
3. (a), (b), (d)
5. (b), (c), (d)

Chapter 2

1. The velocity is the slope of the position-versus-time curve, and it can be positive, negative or zero. Therefore, we have C < A = B = D = F < E.
3. Your average speed is 20 m/s since you spent equal amounts of time at 15 m/s and 25 m/s.
5. **(a)** 3 and 4. **(b)** 2. **(c)** 2 and 3.
7. The areas are equal. This is because (1) the areas give the distance covered by the hammer and (2) the distances are the same because the windows are of equal height.
9. The answer is (a). Specifically, ball A has the greater increase in speed because it takes more time to land. As a result, it is accelerated for the greater period of time.
11. (b) The initial speed of the first ball (at the height h) is zero; the initial speed of the second ball is v_0. Therefore, the second ball covers more distance in a given amount of time, and the balls meet at a height greater than $h/2$. In fact, they meet at the height $3h/4$.

Chapter 3

1. **(a)** The magnitude of the vector doubles; that is, it changes by a multiplicative factor of 2. **(b)** The direction of the vector is unchanged. Put another way, the direction angle changes by a multiplicative factor of 1.
3. $D_x \leq C_x < B_x < A_x$
5. **(a)** $\vec{\mathbf{A}}$ has the larger x component, simply because its x component is positive and the x component of $\vec{\mathbf{B}}$ is negative. **(b)** The magnitude of $\vec{\mathbf{B}}$ is greater than the magnitude of $\vec{\mathbf{A}}$, because 3.4 is greater than 1.2.
7. **(a)** $\vec{\mathbf{E}}$, since $\vec{\mathbf{A}} + \vec{\mathbf{B}}$ points up and to the right; **(b)** $\vec{\mathbf{F}}$, since $\vec{\mathbf{A}} - \vec{\mathbf{B}}$ points up and to the left; **(c)** $\vec{\mathbf{D}}$, since $\vec{\mathbf{B}} - \vec{\mathbf{A}}$ points down and to the right.
9. (c) The direction angle for $\vec{\mathbf{A}}$ is between 180° and 270°.
11. **(a)** 125 m/s due east. **(b)** zero.

Chapter 4

1. You and the ball have the same horizontal component of velocity. Therefore, the ball will come down in your hand if you (a) throw it straight upward relative to yourself.
3. The divers are in free fall for the same amount of time. It follows that diver 2 covers (b) twice the distance.
5. **(a)** Projectile A, projectile B, projectile C. **(b)** Projectile C, projectile B, projectile A.
7. Since the penguin lands at an elevation that is above the water, its speed will be (b) less than when it left the water.
9. In this case, the coin's time of flight would be (b) less than when the person was at rest. The reason is that the coin now has a greater acceleration relative to the floor of the elevator, and hence it will reach the floor in less time.
11. Ignoring air resistance, the acceleration of snowball 2 is (b) equal to the acceleration of snowball 1, regardless of the direction in which they are thrown.

Chapter 5

1. **(a)** These forces form an action-reaction pair, and hence the force experienced by the car has the same magnitude as the force experienced by the truck. The forces point in opposite directions, however. **(b)** The car, with its smaller mass, experiences the larger acceleration.
3. The net force acting on object A has a direction angle of 90°; that is, it points in the positive y direction. The net force on object B has a direction angle of 45°. Finally, the direction angle for the net force on object C is 135°.
5. No adjustment is necessary because you are moving with constant velocity. If you throw an object in the elevator it falls with the same downward acceleration relative to you as it would if you were to throw it while standing on the surface of the Earth.
7. **(a)** northward; **(b)** northward; **(c)** zero; **(d)** zero; **(e)** downward.
9. To cause an acceleration, the upward force exerted on you by the parachute must be greater in magnitude than your weight. In this case, the net force acting on you is upward, as is your acceleration. Since your acceleration and velocity are in different directions, you will decelerate. See Figure 2–11 for further details.
11. The acceleration a force produces is independent of the speed of the object. Therefore, A = B = C.

Chapter 6

1. Graph B, because T_2 must always be greater than T_1. Also, at 90° all of the weight is supported by wire 2, therefore $T_2 = mg$ and $T_1 = 0$.
3. **(a)** Multiplying both m_1 and m_2 by the same factor increases the total mass of the system, $m_1 + m_2$, by that factor. It also increases the net force acting on the system, $(m_2 - m_1)g$, by the same factor. Therefore, the acceleration of the blocks stays the same. **(b)** Adding a certain amount to each mass increases the total mass, but does not change the net force acting on the system. As a result, the acceleration of the blocks will decrease.
5. The box is at rest in each case. Therefore, the force of static friction must balance the horizontal component of the net applied force. This gives box 3 < box 2 < box 1.
7. **(a)** The shape of the object's path is circular. **(b)** Since the object is accelerating, its velocity changes. **(c)** The object's speed remains constant since there is no component of acceleration in the direction of motion.
9. The stopping distance of driver 1 is greater than that of driver 2. The reason is that for driver 2 the force stopping the car is static friction which, in general, is greater than the force of kinetic (skidding) friction.
11. **(a)** The car's velocity is not constant since its direction of motion changes. **(b)** The magnitude of the car's acceleration is v^2/r. Since the speed and radius are constant, so is the magnitude of the acceleration. **(c)** The direction of the car's acceleration is toward the center of the circular path. Thus the direction of the acceleration changes as the car moves around the circle.
13. Noting that the centripetal acceleration is v^2/r, we have A < C < B.
15. Growing plants tilt inward toward the axis of rotation. The reason is that they grow in a direction *opposite* to the effective direction of gravity at their location. To see this, imagine a pendulum suspended at the rim of the turntable. If the turntable is at rest the pendulum hangs straight down and plants grow straight up. When the turntable rotates the pendulum slants outward, and plants slant inward.

Chapter 7

1. The work done on the Space Station by gravity in one complete orbit is zero.
3. **(a)** In going from point B to point C the pendulum bob moves upward. Since this is opposite to the direction of the force of gravity, it follows that gravity does negative work. **(b)** The tension in the string is always at right angles to the

direction of motion, and hence the tension does zero work on the bob.

5. If the work to go from 0 to 50 km/h is W_0, the work to go from 0 to 150 km/h is $(3)^2W_0 = 9W_0$. It follows that the work to go from 50 km/h to 150 km/h is $8W_0$.
7. The forces and displacements all have equal magnitude; therefore the deciding factor is the component of the force in the direction of the displacement. This gives C < A < B.
9. Since the kinetic energy of the youngster increased, the work done was positive.
11. The statement of the Exercise tells us that $\frac{1}{2}(2m)v^2 = \frac{1}{2}mv_2{}^2$. It follows that $v_2 = 2v^2$, or $v_2 = \sqrt{2}v$.
13. Halving the mass of the block and doubling its speed means that its kinetic energy changes by a multiplicative factor of $(1/2)(2)^2 = 2$. This means, in turn, that the work done by the spring to stop the block also doubles. Therefore, noting that the work done by the spring is proportional to $(\Delta x)^2$, the new compression is $\sqrt{2}\,\Delta x$, since $\left(\sqrt{2}\,\Delta x\right)^2 = 2(\Delta x)^2$.
15. Noting that power is work per time, we have $F_4 < F_3 < F_1 < F_2$.

Chapter 8

1. The work done in going from A to B must be the same, regardless of the path. From the middle path, we see that the work is 15 J. Similarly, the work to go from B to A must be −15 J. **(a)** For the total work along this path to be 15 J, the work on Path 1 must be −3 J. **(b)** The work done on Path 2 is −4 J.
3. The change in gravitational potential energy for ball 1 is equal to the change for ball 2. In general, the change in gravitational potential energy depends only on the change in elevation, not on how the change in elevation comes about.
5. **(a)** The potential energy of the spring increases. **(b)** The gravitational potential energy of the mass–Earth system decreases.
7. **(a)** With a smaller acceleration of gravity, it would be necessary for the skier to rise to a greater height in order to gain the same amount of gravitational potential energy. Thus, a higher hill would be needed. **(b)** The final speed would be the same because the values of the initial kinetic energy and the final gravitational potential energy are the same, assuming we are using the higher hill discussed for part (a).
9. The correct statement is (3). First, the mass is immaterial, since both the kinetic energy and the gravitational potential energy depend on mass in the same way. Therefore, the speed of the balls depends only on their initial speed and their height. Since these are equal for all three balls, the speeds are equal as well.
11. **(a)** No nonconservative work is done on your bicycle. Therefore, your mechanical energy stays the same, even though your speed decreases. **(b)** Your friend does positive nonconservative work to maintain a constant speed while going uphill. Hence, your friend's mechanical energy increases.
13. As the leaf falls, air resistance exerts a force on it opposite to its direction of motion. We know this because the leaf's speed remains constant, which means that the net force acting on it is zero. Hence, a negative nonconservative work (due to air resistance) is done on the leaf, and therefore its initial mechanical energy $(U_i + K_i)$ is greater than its final mechanical energy $(U_f + K_f)$.
15. **(a)** The mechanical energy of block 2 decreases as it drops through the distance d. This follows because when the block hits the ground, its speed is less than it would have been if it had dropped in free fall, in which case its mechanical energy would have been conserved. **(b)** The tension in the rope acts in a direction that is opposite to the direction of motion of block 2. Therefore, the tension does a negative nonconservative work on block 2. This, of course, accounts for the decrease in the mechanical energy of block 2, as discussed in part (a).
17. The speed of the object depends on the height of the potential energy curve, with larger speeds at lower points on the curve. Therefore, F = D = B < E < C.

Chapter 9

1. Recall that kinetic energy can be written as $K = p^2/2m$. Since these objects have the same momentum, it follows that their kinetic energies are inversely proportional to their masses. Hence, we have B < A < C.
3. The rubber bullet is more likely to knock the block over. The reason is that the change in momentum is twice as great when an object rebounds as it is when the object is simply brought to rest.
5. The boulder and the pebble have the same rate of momentum change, since the same force acts on both objects. Force, in fact, is the rate of change of momentum.
7. The two balls have the same momentum. However, the first ball has a kinetic energy equal to $\frac{1}{2}mv^2$ and the second ball has half that much kinetic energy; $\frac{1}{2}(2m)(v/2)^2 = \frac{1}{2}\left(\frac{1}{2}mv^2\right)$. Thus, less energy is dissipated in stopping the second ball, and so it has less "sting."
9. Impulse is force times time. Therefore, D < C < B < A.
11. **(a)** If we take the direction of fall to be positive, the horizontal component of the contact force is in the positive direction. This is because as the pencil falls its center of mass accelerates both downward and in the positive direction. **(b)** Since the center of mass of the pencil has a nonzero, downward acceleration, it follows that the net vertical force acting on it is downward. Thus, the upward force exerted by the table is less than the weight of the pencil.
13. The net force acting on the system of the two connected blocks—including gravity, the normal force, and the force exerted by the pulley—is constant in both magnitude and direction. Therefore, since the center of mass starts at rest, it will accelerate in a straight line in the direction of the net force. Only the green curve satisfies this condition.

Chapter 10

1. In the time $t/2$ the object rotates through the angle $\theta/4$. This follows because of the t^2 time dependence associated with constant angular acceleration.
3. Jason and Betsy both complete one revolution in the same amount of time. Therefore, their periods of rotation are the same.
5. Your angular speed due to the Earth's rotation is the same regardless of your elevation.
7. The reading on the speedometer gives the speed of the axles of your car. This is the same as the speed of the occupants inside the car. If the top of the tires have a speed v, the axles have a speed $v/2$. (See Figure 10–11.)
9. In principle, the Earth's moment of inertia increases because both its mass and radius increase.
11. The hard-boiled egg spins longer, because it is a rigid body. When you give a hard-boiled egg a spin the entire egg rotates with the same angular speed. When you spin a raw egg, however, the outside rotates more rapidly than the liquid interior. The drag between the interior and the exterior results in the raw egg slowing down more rapidly than the solid hard-boiled egg.
13. The moment of inertia of an object is smaller when it is rotated about its center of mass than when rotated about a parallel axis through any other point. Therefore, the moment of inertia is greater about the rim of the wheel than about its axle.
15. Its height will be less than the original height. In this case, all of the original gravitational potential energy is converted to linear kinetic energy when the ball reaches the bottom of the track. It then begins to spin as it skids on the no-slip surface until it rolls without slipping. This process dissipates some of the mechanical energy. The remaining mechanical energy is converted to gravitational potential energy, but since mechanical energy has been lost the height attained will be less.
17. The moment of inertia is the sum of mr^2 for each of the masses, where r is the distance from the specified axis of rotation. Therefore, case 3 < case 2 < case 1.

Chapter 11

1. **(a)** Since the mass accelerates downward, we conclude that the upward acting tension in the string is less than the weight of the mass. **(b)** The mass in the left-hand system drops with a smaller acceleration than the mass in the right-hand system. Therefore, the tension in the left-hand string is greater than the tension in the right-hand string. This produces a smaller net force on the left-hand mass and a correspondingly smaller acceleration.
3. The angular acceleration is inversely proportional to the moment of inertia. Hence, the greatest angular acceleration occurs when the object is rotated about the x axis (case A), and the smallest angular acceleration occurs when the object is rotated about the z axis (case C). The ranking, then, is as follows: case C < case B < case A.
5. **(a)** The disk rotates counterclockwise, as one would expect from a force exerted to the right on the bottom of the disk. **(b)** The center of the disk moves to the right, since that is the direction of the net force exerted on the disk.
7. The block moves to the left. The reason is that the hoop, with the larger moment of inertia, has the smaller angular acceleration. Since less string unwinds from the hoop, the block moves in that direction.
9. His angular speed is unaffected. The angular momentum of the student–weights system decreases when the floor exerts a torque on the weights to bring them to rest.
11. **(a)** Since the string is always at right angles to the motion of the puck, the tension does no work on it. Therefore, the linear speed of the puck, v, remains constant. As the string becomes shorter the angular speed, $\omega = v/r$, increases. **(b)** The angular momentum of the puck is $L = I\omega = (mr^2)(v/r) = rmv$. As r decreases, with v remaining constant, the angular momentum of the puck decreases as well.
13. Since there are no external torques acting on the system, its angular momentum must remain constant. Thus, as the beetle walks toward the axis of rotation, which reduces the moment of inertia of the system, the angular speed of the turntable increases.
15. The hollow sphere (sphere 2) is harder to stop because it—with its greater moment of inertia—has more kinetic energy for a given speed. The more the kinetic energy, the more work that must be done to bring it to rest. Therefore, the work required to stop sphere 1 is less than the work required to stop sphere 2.

Chapter 12

1. The gravitational force is proportional to m_1m_2/r^2. Therefore, B < A < C < D.
3. The gravitational force is proportional to m_1m_2/r^2, and is always attractive. Therefore, A < B < C.
5. The net force acting on the Moon is directed toward the Sun. This is due to the fact that the Sun exerts more force on the Moon than does the Earth.
7. **(a)** The apogee distance increases. **(b)** The perigee distance stays the same. (See Figure 12–13.)
9. We know from Kepler's third law that the period of an orbit increases with its radius. Therefore, the length of the month would increase—as, in fact, it is.
11. **(a)** In one complete orbit, a satellite returns to its starting point, resulting in zero net displacement. Since gravity is a conservative force, it does zero net work when the displacement is zero. **(b)** No. The answer to part (a) is independent of the shape of the orbit—all that mattered was that the orbit was closed.
13. We can see from Equation 12–13 that if the radius of the Earth is decreased, with its mass remaining the same, the escape speed increases. The reason is that in this case the rocket starts closer to the center of the Earth, and therefore experiences a greater attractive force. It follows that a greater speed is required to overcome the increased force.

Chapter 13

1. **(a)** In the time T a mass that starts at equilibrium travels a distance A to its maximum displacement in one direction, a distance A back to equilibrium, a distance A to the maximum displacement in the other direction, and finally a distance A back to equilibrium again. The total distance is $4A$. The distance is the same regardless of where the mass starts in its cycle. **(b)** In the time $5T/2$, the mass travels a distance of $10A$.
3. **(a)** Speed is given by the magnitude of the slope. Thus, C = F < B = E < A = D. **(b)** Velocity is the slope itself, which can be positive, negative or zero. Hence, D < E < C = F < B < A. **(c)** Acceleration is proportional to the curvature of the x-versus-t plot, and it can be positive, negative, or zero. Therefore, C < B < A = D < E < F.
5. **(a)** zero. **(b)** zero. **(c)** positive. **(d)** negative.
7. **(a)** No change in angular frequency; **(b)** no change in frequency; **(c)** no change in period; **(d)** maximum speed doubles; **(e)** maximum acceleration doubles; **(f)** total mechanical energy quadruples.
9. The longer spring has a smaller force constant; hence, its period of oscillation is greater than T.
11. **(a)** When the elevator moves upward with constant *speed* there is no change in the behavior of the mass and the spring. Therefore, the period remains the same. **(b)** When the elevator moves upward with constant *acceleration* the system behaves as if the acceleration of gravity were larger. Therefore, the spring is stretched by a greater amount, but from Equation 13–11 we see that its period of oscillation remains the same.
13. The period of block 1 is the same as the period of block 2 because both blocks experience exactly the same restoring force for a given displacement from equilibrium.
15. Since gravity is weaker on top of a mountain, the period of the pendulum would increase. As a result, the clock would run slow.
17. **(a)** When the elevator moves upward with constant *speed* there is no change in the behavior of the pendulum. Therefore, its period remains the same. **(b)** When the elevator moves upward with constant *acceleration* the system behaves as if the acceleration of gravity were larger. Therefore, we can see from Equation 13–20 that the period of the pendulum will decrease.

Chapter 14

1. **(a)** The speed of a wave on the string does not change as long as the tension remains constant. Therefore, the speed changes by a factor of 1; that is, it is unchanged. **(b)** Doubling the frequency reduces the wavelength by a factor of two.
3. The tension in the strings is the same, and therefore the deciding factor in determining the wave speed is the mass per length. Recalling that a larger mass per length results in a smaller speed, it follows that waves on string 1 (the thick string) have a lower speed than waves on string 2 (the thin string).
5. **(a)** The kinetic energy of the string is a minimum (zero) at a point where the displacement of the string is a maximum, because the string is instantaneously at rest at this point. **(b)** The kinetic energy of the string is a maximum where the displacement of the string is zero. This follows because the string has its greatest speed at this point.
7. With equal tensions and equal mass per length, waves on the three strings have the same speed, $v = \lambda f$. **(a)** When speed is constant, the frequency is inversely proportional to the wavelength. This gives B < C < A. **(b)** Wavelength is simply the distance from crest to crest. Therefore, A < C < B. **(c)** As stated previously, the speeds are the same; A = B = C.
9. **(a)** Setting $Bx - Ct$ = constant yields $x = (C/B)t$ + constant. Recalling that B and C are positive, it follows that the wave moves in the positive x direction. **(b)** Referring to Equation 14–4, we note that $B = 2\pi/\lambda$. It follows that $\lambda = 2\pi/B$. **(c)** Equation 14–4 shows that $C = 2\pi/T = 2\pi f$. Therefore, $f = C/2\pi$. **(d)** At $t = 0$ we have $y(x, 0) = A\sin(Bx)$.

It follows that $y = 0$ when $x = 0$, $x = \pm\pi/B = \pm\lambda/2$, $x = \pm 2\pi/B = \pm\lambda$, $x = \pm 3\pi/B = \pm 3\lambda/2 \ldots$. The smallest positive value is $x = \pi/B$.

11. The intensity decreases with the square of the distance. Hence, doubling the distance reduces the intensity by a factor of $2^2 = 4$. Therefore, the multiplicative factor is $1/4$.
13. The lowest frequency is the frequency of the horn itself. The person on the island hears a higher frequency due to the motion of the boat toward the island. This same frequency goes out as the echo. The person in the boat hears the echo, but Doppler shifted up to yet a higher frequency. Noting that the speed of the sound is constant, it follows that the wavelength is inversely proportional to the frequency. Hence, the ranking is $\lambda_3 < \lambda_2 < \lambda_1$.
15. **(a)** The fundamental frequency of this string is 50 Hz. **(b)** The frequency of the third harmonic is 150 Hz.
17. The beat frequencies are the same. All that matters as far as the beat frequency is concerned is the absolute value of the difference between the two frequencies.

Chapter 15

1. **(a)** Glass A holds four times as much water as glass B. **(b)** The pressure at the bottom of the two glasses is the same. **(c)** Recall that pressure is force per area. Glass A exerts four times as much force on its base, because it holds four times as much water, but it also has a base that is four times greater in area. These two factors of four cancel in determining the pressure.
3. Different water levels in the tube and the main body of the weather glass indicate different pressures in the atmosphere and the interior of the weather glass. As a low-pressure system approaches, atmospheric pressure drops. This allows the water level in the tube to rise.
5. The volume of wood that is submerged decreases when the steel ball is immersed in the water. In the original situation, the block had to be submerged far enough to provide enough buoyant force to support both the block and the ball. In the inverted situation, the ball experiences a buoyant force that partially supports its weight. Thus, the buoyant force that must be provided by the block is less, and so is its submerged volume.
7. In both cases, the weight of displaced fluid is equal to the weight of the small float. Since less of fluid B is displaced, it must be more dense.
9. When a brick is in the boat it displaces a volume of water equal to its weight. When the brick is at the bottom of the pool it displaces a volume of water equal to its own volume. Thus, the brick displaces a greater volume of water when it is in the boat. As a result, the water level decreases when the brick is thrown overboard.
11. **(a)** The containers are filled to the same level, but less water is needed for containers A and C, since the floating objects displace a certain volume of water. This gives C < A < B. **(b)** Each of the containers has the same pressure on its bottom surface, meaning that the downward force exerted by each container (which is its weight) is the same; A = B = C.
13. **(a)** The buoyant force, which is equal to the weight of displaced fluid, is the same for both the lead and aluminum balls since they have equal diameters. **(b)** The result of part (a) is true for any fluid (as long as both objects sink in the fluid), though the magnitude of the buoyant force varies from fluid to fluid.
15. As far as floating goes, this planet and the Earth are the same. The increased acceleration of gravity increases your weight and the weight of water by the same factor.
17. The marble will stay where it is released. The reason is that the surface of the water is perpendicular to the local "effective" gravity of the rotating turntable. Even the water itself does not flow inward or outward when it is tilted at this angle. If the water is frozen, and a marble is placed on its surface, the marble will stay put just as the liquid water did before it was frozen.

Chapter 16

1. **(a)** The amount of bend in a bimetallic strip depends on the difference in the coefficients of thermal expansion for the two metals—the greater the difference in thermal expansion, the greater the bend. For this reason, strip 2 (aluminum-steel) bends more than strip 1 (copper-steel). **(b)** A steel-lead bimetallic strip would give the greatest amount of bend for the metals in Table 16–1.
3. Aluminum has the greater coefficient of thermal expansion, and hence its diameter changes more with temperature than does the diameter of the hole in the brass plate. It follows that the temperature should be lowered, thus shrinking the aluminum ball enough to fit through the hole.
5. The expansion in area is proportional to the initial area. Hence, C < B = E < A < D.
7. The period of the pendulum decreases, since its length is reduced as the temperature is lowered.
9. The rod will be firmly wedged in the ring. Recall that the cool rod starts out fitting snugly inside the hot ring. As the ring and the rod come to room temperature, the ring shrinks and the rod expands. The result is a very tight fit.
11. From Table 16–2 we see that 900 J of heat will raise the temperature of 1 kg of aluminum by 1 K; therefore, 1800 J are required for the same temperature change in 2 kg of aluminum. On the other hand, 2090 J of heat are needed to raise the temperature of 1 kg of ice by 1 K. Clearly, less heat is needed for a given temperature change with the 2 kg of aluminum; hence, a given amount of heat will increase the temperature of the aluminum by a greater amount than the ice.
13. Specific heat is $c = Q/(m\,\Delta T)$. It follows that B < D < C = A.
15. The smallest temperature difference occurs in the material with the greatest thermal conductivity. Hence, in order of increasing temperature difference, we have the following: B < A < C.
17. The darker shirt is a more effective absorber of radiation, and hence it will be warmer.
19. The tea in mug 2 (the one that is filled only halfway) cools off more rapidly. The reason is that its surface area is greater in proportion to its mass than the tea in the full mug, and therefore it loses heat more rapidly. (In general, small objects have more surface area per mass than larger objects of the same basic shape.)

Chapter 17

1. **(a)** By the definition of a mole, the number of molecules in one mole of N_2 is the same as the number of molecules in one mole of O_2. **(b)** The mole of O_2 has the greater mass, however, since each of its molecules is more massive than an N_2 molecule.
3. Treating the helium as an ideal gas, it satisfies the relation $PV = nRT$. When we put the balloon in the refrigerator, the pressure and the number of moles remain constant. Under these conditions, a decrease in temperature implies a decrease in volume.
5. The pressure of an ideal gas is given by $P = nRT/V$. If $P_A = n_A RT/V_A$, it follows that $P_B = (n_A/2)RT/(2V_A) = P_A/4$.
7. Assuming the air in the rooms is an ideal gas, we have that the number of moles is given by $n = PV/RT$. The two rooms have the same pressure (atmospheric) and the same volume. Hence, room 2 (the one with the lower temperature) is the one with the greater number of molecules.
9. The Kelvin temperature is proportional to the average kinetic energy of the molecules, which in turn is proportional to the speed of the molecules squared. Hence, doubling the speed implies an increase in temperature by a factor of $2^2 = 4$.
11. Though the Celsius temperature is doubled, it is the Kelvin temperature that is directly related to the average kinetic energy. Increasing the temperature from

100 °C to 200 °C corresponds to an increase from 373.15 K to 473.15 K. Since 473.15/373.15 is clearly less than 2, it follows that the average kinetic energy increases by a factor that is less than 2.

13. **(a)** Kinetic energy depends only on the absolute temperature of the gas; A = B = C. **(b)** Noting that $K = \frac{1}{2}mv^2$, we see that speed is inversely proportional to mass for a given kinetic energy. Hence, C < B < A.

15. For a given force, the change in length is proportional to the initial length, L_0, and inversely proportional to the cross-sectional area, A. Hence, the change in dimensions of the brick will be least when it is placed on the side that is 2 cm by 3 cm. This gives the smallest height ($L_0 = 1$ cm) and the largest cross-sectional area ($A = 2 \text{ cm} \times 3 \text{ cm} = 6 \text{ cm}^2$).

Chapter 18

1. Recall that $\Delta U = Q - W$. Therefore, we have the following results: **(a)** 0; **(b)** 0; **(c)** −100 J.

3. You get a greater increase in temperature when you add the heat at constant volume. At constant pressure, the gas expands and does work as the heat is added. Hence, only part of the heat goes into increasing the internal energy. When heat is added at constant volume no work is done. In this case, all the heat goes into increasing the internal energy, and hence the temperature.

5. The work done by the gas is equal to the area under the process. Hence, the process with the least work is C, with 3 J, followed by A, with 4 J, and finally B, with 4.5 J. Writing this out as a ranking, we have C < A < B.

7. **(a)** The efficiency of a heat engine ($e = 1 - T_c/T_h$) depends on the ratio of T_c to T_h. Doubling both temperatures leaves this ratio, and the efficiency, unchanged. **(b)** Adding the same temperature to both T_c and T_h means that the ratio T_c/T_h will have a value that is closer to 1. Therefore, the efficiency of the engine will decrease.

9. Recall that the efficiency of a Carnot engine is $e = 1 - T_c/T_h$. This gives D < B = C < A.

11. More heat is required to increase the temperature on a constant pressure process; therefore, the entropy change on the process A → B (constant volume) is less than the entropy change on the process A → C (constant pressure).

13. The entropy of the gas increases because heat must be added to it to keep its temperature constant.

Chapter 19

1. Its mass decreases. This is because a positive charge is generally due to a loss of electrons, each of which has a small mass.

3. **(a)** The balloon is made from rubber, and the cloth that is rubbed against it is probably made of cotton or a similar material. It follows from Table 19–1 that the rubber balloon will have a negative charge. In fact, the only two materials more "negative" than rubber in Table 19–1 are materials one would not associate with "cloth". **(b)** The stream of water will deflect toward the balloon regardless of whether its charge is positive or negative, just as bits of paper are attracted equally well to a charged amber (−) rod or a charged glass (+) rod.

5. The electron has the greater kinetic energy (by about a factor of 2000.) This is due to the fact that the momentum of the system is conserved; hence $m_e v_e = m_p v_p$ when the particles meet. It follows that $\frac{1}{2}m_e v_e^2 = (m_p/m_e)(\frac{1}{2}m_p v_p^2)$, where $m_p/m_e \sim 2000$.

7. The net force on the positive charge in the lower left corner is directed to the right and 60° below the horizontal; the net force on the upper positive charge is horizontal and to the right; the force on the negative charge is directed toward the center of the triangle. The two positive charges experience forces that are smaller by a factor of $\sqrt{3}$ than that experienced by the negative charge. **(a)** 1 = 3 < 2. **(b)** 0. **(c)** 150°. **(d)** −60° or 300°.

9. The third charge experiences a greater force at one of the empty corners of the square. At the center of the square the forces exerted by the two $+q$ charges cancel, giving a net force of zero.

11. The charge in the middle experiences the greatest force. Next in magnitude is the net force exerted on the charge on the left end, and the smallest net force is exerted on the charge to the right. Thus, the ranking is C < A < B.

13. By Newton's third law, the forces exerted on the electron and the nucleus are equal in magnitude and opposite in direction.

15. The magnitude of the force will be less in the case of a conducting sphere, because its charge is free to move. To see this, first note that both the sphere and the point charge are positive. Therefore, the positive charge on the sphere will move as far away from the point charge as possible (as opposed to the uniform charge on an insulator, which cannot move). This will result in a greater charge separation, and hence a reduced force.

17. Their shoes should be conducting, so that any charge they transfer during walking can flow away into the ground.

19. Electric field lines point in the direction of negative charges. We conclude, therefore, that the charge on the Earth is negative.

21. The net charge of the two-sphere system is $+35Q - 21Q = +14Q$. When the spheres are brought into contact, this charge will be shared equally between the two spheres, giving each a charge of $7Q$.

23. **(a)** Outer surface, $-Q$; inner surface, $-Q$. **(b)** Outer surface, 0; inner surface, $-Q$. **(c)** Outer surface, $+2Q$; inner surface, $-Q$.

25. The electric flux through each surface is proportional to the total charge it encloses. The charge enclosed by each surface is: A, $+2q$; B, $+q$; C, 0; D, $-q$. Therefore, the ranking of electric flux is as follows: D < C < B < A.

Chapter 20

1. The electron moves in a direction opposite to the electric field. Because the electric field points in the direction of decreasing electric potential, it follows that the electron experiences an increasing potential.

3. As the proton accelerates from rest its kinetic energy increases. It follows that its electric potential energy decreases by the same amount.

5. **(a)** The distance from q_2 to the point A is $\sqrt{2}$ times greater than the distance from q_1 to the point A. Therefore, for the electric potential to be zero at point A, it is necessary that q_2 have the charge $-\sqrt{2}Q$. **(b)** With this value for q_2, the electric potential at point B will be negative.

7. **(a)** The center of the square is equidistant from q_1 and q_2; therefore, if the electric potential is zero at that location it follows that $q_2 = -Q$. **(b)** The electric potential at point A, which is closer to q_1 than to q_2, is positive. **(c)** Similarly, the electric potential at point B is negative.

9. The electric potential at the origin is proportional to the sum of the charges, since they are all the same distance from the origin. The sum of the charges in the four cases are: A, 0; B, 0; C, $-3q$; D, $+q$. Therefore, the ranking of electric potential at the origin is as follows: C < A = B < D.

11. Zero. The electric field is perpendicular to an equipotential, therefore the work done in moving along an equipotential is zero.

13. The electric field is more intense near point 2 than near point 1, because the field lines are packed more closely together near point 2. This means, in turn, that the electric potential changes more rapidly with position at point 2. As a result, the spacing between equipotential surfaces at point 2 is less than it is at point 1.

15. **(a)** The electric field remains the same. **(b)** The potential difference increases linearly with separation. **(c)** The capacitance decreases inversely with the separation. **(d)** The energy stored in the capacitor increases linearly with separation due to the work required to move the plates farther apart.

17. **(a)** The electric field decreases; **(b)** the potential difference decreases; **(c)** the capacitance increases; **(d)** the energy stored decreases.

Chapter 21

1. Recalling that 1 A = 1 C/s, it follows that an amp-hour has the dimensions of charge, and hence 250 amp-hours is a measure of the amount of charge that can be drawn from the battery.
3. **(a)** The length L of the wire increases by a factor of 4, and its cross-sectional area A increases by a factor of $3^2 = 9$. The resistance, which is proportional to L/A, decreases because the larger area more than makes up for the greater length. **(b)** The factor by which the resistance changes is 4/9, so the new resistance is a little less than half the original resistance.
5. Material A satisfies Ohm's law because the relationship between current and voltage for this material is linear; that is, current is proportional to voltage.
7. First, if wire B is to have the greater resistance with the same length, its cross-sectional area must be smaller than that of wire A. This is because resistance depends on L/A. In fact, the ratio of the areas must be 1/9; that is, $A_B/A_A = 1/9$. Finally, the cross-sectional area depends on the radius squared, and hence $r_B/r_A = 1/\sqrt{9} = 1/3$.
9. If two heaters, each with resistance R, are connected in series, the equivalent resistance is now $2R$. The heaters are connected to the same potential difference, however, and therefore they draw half the original current. From $P = I^2R$ we can see that each heater now dissipates one-quarter the power of the single heater, for a total power consumption of one-half the original value.
11. **(a)** In terms of potential difference V and current I, the power output of the light bulbs can be written as $P = IV$. It is given that V is the same for both bulbs. It follows, then, that the current in bulb A is greater than the current in bulb B. **(b)** Since power is proportional to current, the current in bulb A is four times greater than the current in bulb B.
13. The lights are brighter when connected in parallel. We can see this by writing the power in the form $P = V^2/R$, which takes advantage of the fact that the potential difference is the same in either case. It follows that parallel lights, which have a smaller equivalent resistance than series lights, have the greater power output.
17. **(a)** When connected in series, the two resistors carry the same current. Therefore, from $P = I^2R$ we see that the larger the resistance, the greater the power dissipated. In our case, the resistor $2R$ dissipates more power. **(b)** When connected in parallel, the two resistors have the same voltage across their terminals. Therefore, from $P = V^2/R$ we see that the smaller the resistance, the greater the power dissipated. In our case, the resistor R dissipates more power. Note that this is just opposite to our conclusion for the case of the resistors connected in series.
19. When the switch is closed, light 2 is shorted out and the equivalent resistance of the circuit drops from $3R$ to $2R$. It follows that the current in the circuit increases.
21. **(a)** Originally, light 3 is dark; after the switch is closed, light 3 is illuminated. Therefore, its intensity increases. **(b)** The intensity of lights 1 and 2 stays the same when the switch is closed. After all, they still have the same potential difference and the same resistance. Therefore, it is clear from $P = V^2/R$ that they will continue to dissipate the same amount of power.
23. The total power dissipated in the circuit increases when the switch is closed. Before the switch was closed, light 1 dissipated the power $P = \mathcal{E}^2/R$, as did light 2. Now all three lights dissipate the power $P = \mathcal{E}^2/R$.
25. Because adding another capacitor in series simply increases the effective separation between the plates, we expect the equivalent capacitance to decrease.
27. The ranking is C < A < B < D. This follows because two capacitors in series have an equivalent capacitance less than the smallest of the capacitors, and two capacitors connected in parallel have an equivalent capacitance greater than the largest of the capacitors.
29. **(a)** The voltage across C_1 is $2V/3$. The two capacitors have the same charge, and hence the smaller capacitor will have the larger voltage. **(b)** The equivalent capacitance of C_1 and C_2 in series is $2C/3$. Therefore, the charge on the equivalent capacitance is $Q = C_{eq}V = 2CV/3$, and the charge on the right plate of C_2 is $-2CV/3$. **(c)** The magnitude of the net charge that flowed through the ammeter is simply the magnitude of the charge on each plate of the capacitors, $2CV/3$. More specifically, in terms of direction and sign, the charge $-2CV/3$ flowed to the left through the ammeter.
31. **(a)** A = B = C. In each of the three circuits, the capacitor has the same potential difference as the battery—in order to stop the flow of current. Therefore, they each have the same charge. **(b)** A < C < B. The time required is proportional to the time constant, RC; therefore, the ranking follows the equivalent resistance in the branch of the circuit where the current will come to a stop.

Chapter 22

1. The force experienced by proton 2 is less than that experienced by proton 1. The reason is that proton 2 moves in a direction that is nearly the same as the magnetic field, which leads to a weak force. On the other hand, proton 1 moves in a direction that is essentially perpendicular to the magnetic field. This results in a force that is close to the maximum possible for a proton of speed v moving in any direction.
3. Since the radius of curvature ($r = mv/qB$) is proportional to the speed, it follows that the particle with the larger radius has the greater speed. In this case, the ranking is A < B < C.
5. We want the magnetic force on the proton to be toward the center of the Earth, so that it provides some of the necessary centripetal force. It follows that the proton must move in a westward direction.
7. The radius of curvature is $r = mv/qB$, and hence a large charge implies a small radius. It follows that the ranking in terms of the magnitude of charge is C < B < A.
9. The magnetic field is in the positive z direction.
11. **(a)** $F_1 < F_2 = F_4 < F_3$. **(b)** Velocity 1.
13. **(a)** $B_3 < B_1 < B_2$. **(b)** B_1, out of the page; B_2, into the page; B_3, out of the page.
15. Wire 3 experiences an attractive force from wire 1 and a repulsive force of equal magnitude from wire 2. The result is a net force on wire 3 pointing horizontally to the left.
17. The force between wires carrying currents in the same direction is attractive, and inversely proportional to the distance between the wires. Similarly, the force between wires with oppositely directed currents is repulsive. It follows from geometrical considerations, then, that the net force acting on wire 2 is directed toward wire 4.
19. **(a)** The force on the left vertical wire is out of the page; the force on the right vertical wire is into the page; the horizontal wires experience zero force. As a result, the loop tends to rotate counterclockwise, as viewed from above. **(b)** This loop tends to rotate clockwise. **(c)** This loop will not tend to rotate at all.
21. The magnetic field depends on the number of turns per length, n. Doubling both the number of turns and the length leaves n unchanged. It follows that the magnetic field within the solenoid remains the same.

Chapter 23

1. The rate of change of magnetic flux for the airplane's wings is the product of the following three quantities: (1) the downward component of the magnetic field, (2) the wingspan of the plane, and (3) the speed of the plane. At the equator

the field is horizontal; hence, the first of the three quantities is zero. At the latitude of New York all three quantities are nonzero. Therefore, the induced emf is less at the equator than it is at the latitude of New York .

3. A, counterclockwise; B, zero; C, zero; D, clockwise.

5. The magnetic flux through a loop is greatest when the normal to the loop points in the direction of the field. At the equator the field points to the north. Therefore, the flux when the normal points north is greater than when the normal points vertically upward.

7. The retarding effect is greatest on the solid disk. The reason is that the holes in the slotted disk interrupt the flow of eddy currents, and it is the eddy currents that produce the opposing magnetic field responsible for the retarding effect.

9. As the disk enters the field, which points into the page, a current will be induced that produces a field pointing out of the page. Therefore, the induced current circulates in the counterclockwise direction.

11. The energy stored in the inductor remains the same. Doubling the number of turns per length quadruples the inductance of the solenoid, as we can see from Equation 23–14. The energy stored in an inductor, however, depends on both the inductance of the inductor and the current it carries, as we see in Equation 23–19. In fact, the energy stored in an inductor depends on the square of the current. Therefore, halving the current reduces the stored energy by a factor of four. The net effect, then, is no change in the energy.

13. The two transformers have the same ratio of the number of turns (N_p/N_s) in their primary and secondary circuits. Therefore, the ratio of their primary and secondary voltages must be the same as well. It follows that the secondary voltage in transformer 2 is $2V_s$; that is, $V_p/V_s = 2V_p/2V_s$.

15. D < A = B = C. Long after the switch is closed the inductors act like ideal wires. Thus, in circuits B and C the resistors in parallel with the inductor are shorted out, and make no contribution to the current in the circuit. The equivalent resistance in these circuits, then, becomes R. Circuit A has an equivalent resistance R, and circuit D has an equivalent resistance $R/2$.

Chapter 24

1. As the frequency is increased, the capacitor behaves more and more like a short circuit. Therefore, the intensity of the bulb will increase.

3. When the wire is coiled, it becomes an inductor—in addition to having the same resistance it had before being coiled. Therefore, the coiled wire has a greater impedance than the uncoiled wire. Since current is inversely proportional to impedance, it follows that less current is supplied to the coiled wire.

5. More current is supplied when the inductor and capacitor are connected in parallel. At high frequency, the inductor behaves the same as an open circuit. If the components are connected in series the flow of current will cease in the high-frequency limit. If the components are connected in parallel, current can still flow through the branch of the circuit containing the capacitor.

7. Reducing the current in the inductor by a factor of two reduces the energy it stores by a factor of four (see Equation 23–19). Therefore, when the current in this inductor is 1.2 A, the energy it stores is (32 mJ)/4 = 8 mJ.

9. When an *RLC* circuit is driven at resonance its impedance is equal to its resistance; that is, $Z = R$. This is a direct consequence of the fact that at resonance the inductive and capacitive reactances are equal, $X_C = X_L$.

11. The current at high frequency is less than the current at low frequency. The reason is that at high frequency the inductor behaves like an open circuit, cutting off the upper branch of the circuit. At low frequency current flows through both branches, giving a larger total current.

13. When connected to a battery, an inductor behaves the same as a zero-resistance wire, and a capacitor behaves the same as an open circuit. Therefore, the circuit in Figure 24–27 has a equivalent resistance R, due to current flowing through the upper branch of the circuit. The same is true of the circuit in Figure 24–29. Therefore, the two circuits draw the same current from their batteries.

15. Adding a second capacitor in series with the first capacitor results in a total capacitance that is less than the original capacitance. As a result, the resonance frequency increases (see Equation 24–18).

17. **(a)** Notice that the voltage leads the current, which means that the inductance (whose voltage leads the current) dominates in this circuit. More specifically, this means that the inductive reactance is greater than the capacitive reactance. **(b)** The result of part (a) means that the generator frequency is greater than the resonance frequency of the circuit. In fact, the voltage leads by close to 90°, meaning the frequency is much greater than the resonance frequency.

19. **(a)** Since the voltage leads the current, it follows that the inductive reactance is greater than the capacitive reactance. This means, in turn, that the frequency should be decreased if we want to bring the circuit into resonance. **(b)** Again, we would like to decrease the inductive reactance. This can be accomplished by decreasing the inductance.

Chapter 25

1. The electric and magnetic fields in an electromagnetic wave vary in phase with one another. Therefore, when the electric field is increasing, the magnetic field is increasing as well.

3. **(a)** The electric and magnetic fields in an electromagnetic field are proportional to one another; in fact, $E = cB$. Therefore, doubling the rms value of the electric field doubles the rms value of the magnetic field. **(b)** The intensity of an electromagnetic wave, on the other hand, depends on the square of the electric and magnetic fields. As a result, doubling the fields quadruples the intensity of the wave.

5. We note from Equation 25–12 that the radiation pressure depends on the intensity of a wave. The intensity, in turn, depends on the square of the fields, as Equation 25–10 shows. Therefore, the electric field in beam 2 has a magnitude of $\sqrt{2}E_0$.

7. If the incident light is unpolarized, the transmitted intensity in case (a) is the same as in case (b). Specifically, the first filter in both cases reduces the intensity of the unpolarized light by a factor of 2. It also leaves the transmitted light polarized in the direction of its transmission axis, which is at an angle θ relative to the transmission axis of the second filter. The second filter then reduces the transmitted intensity by a further factor of $(\cos\theta)^2$.

9. The transmitted intensities will be the same if the incident light is polarized in a direction that makes an angle of $\frac{1}{2}\theta$ with the vertical. In this case, both experiments give a transmitted intensity equal to $I_0(\cos\frac{1}{2}\theta)^2(\cos\theta)^2$, where I_0 is the incident intensity.

11. **(a)** Green. Unpolarized light will be reduced in intensity by a factor of 2 regardless of the angle of the transmission axis. **(b)** Red. At $\theta = 0$ we have complete transmission, meaning that the light is vertically polarized. **(c)** Blue. At $\theta = 0$ there is no transmission, which means the incident light is horizontally polarized.

Chapter 26

1. As can be seen in the photo accompanying this question, the hands on the mirror-image clock rotate counterclockwise.

3. The back of the spoon behaves like a convex mirror. Therefore, your image is **(a)** upright and **(b)** reduced in size. **(c)** The fact that no light passes through the image (which is behind the spoon) means that the image is virtual.

5. The greatest image distance occurs when the object is infinitely far from the mirror. In this case, the image is at the focal point. Therefore, the greatest distance the image can be behind the mirror is $f = \frac{1}{2}R$.
7. Referring to Figure 26–17, we see that the image produced by a convex mirror is always behind the mirror, and between its surface (when the object approaches the surface) and the focal point (when the object moves to infinity). Therefore, as the object is moved farther to the left, the image will move to the right, toward the focal point.
9. The dish antenna receives signals from essentially an infinite distance (typically a satellite in geosynchronous orbit), and therefore the signals are focused at the focal point of the dish. The focal point is a distance $R/2$ in front of the mirror, and this is where the receiver should be placed.
11. The mirror's focal length remains the same. The difference is that the focal length of a lens depends on how much it bends light, and the amount of bending, in turn, depends on the difference in index of refraction between the lens and its surroundings. For the mirror, the focal length is determined solely by its radius of curvature, and the fact that the angle of incidence and the angle of reflection are the same—which is true regardless of the medium in which the mirror is immersed.
13. The index of refraction of water is closer to the index of refraction of glass than is the index of refraction of air. It follows that light passing from water to glass or from glass to water undergoes less of a change in direction than is the case for air and glass. Therefore, the displacement will decrease when the glass is placed in water. In fact, if the slab is placed in a fluid with the same index of refraction as the glass there is no displacement at all.
15. The answer is $f \to \infty$. We can see from Conceptual Question 12 that the index of refraction of the eye dropper glass perfectly matches the index of refraction of the oil. That is, light is not deflected as it passes from the glass to the oil or from the oil to the glass. If a lens doesn't bend light, it follows that its focal point is at infinity.
17. Color B travels more rapidly. The reason is that color B, which is "bent" less than color A, must have the smaller index of refraction. Since the speed of light in a material is c/n, the color with the smaller index of refraction has the greater speed.
19. Water causes the apparent depth to decrease. Therefore, the left sink (with water) appears to be shallower than the sink on the right (without water).
21. Medium A has the greater index of refraction. In general, total internal reflection can occur only when light is moving from a region with a larger index of refraction to a region with a smaller index of refraction.

Chapter 27

1. The farsighted person's glasses are converging, and therefore they are the ones that should be used to start a fire.
3. To produce a reduced image of his eyes, the umpire's glasses must be diverging. Therefore, the umpire is nearsighted.
5. When the lens is in air the difference in index of refraction between the lens and its surroundings is greater than when it is surrounded by the aqueous and vitreous humors. It follows that the lens bends rays of light more in air; hence, its refractive power will be greater than 15 diopters when surrounded by air.
7. Since the separation between the lenses is greater than the sum of the focal lengths, as would be the case for a telescope (see Figure 27–16), we conclude that the instrument is a microscope.
9. The length of a telescope is equal to the sum of the focal lengths of its objective and eyepiece. On the other hand, the angular magnification is the ratio of the focal length of the objective to that of the eyepiece. Now, if the angular magnification is the same, and the overall length is greater, it follows that both the objective and eyepiece have greater focal lengths in the long telescope.
11. To view close objects, the eye must refract light by a greater amount than when viewing distant objects. Therefore, the corrective glasses must produce additional convergence of the incoming light; that is, the corrective glasses must have a positive refractive power. The situation is similar to correcting the vision of a farsighted person, as illustrated in Figure 27–11.

Chapter 28

1. **(a)** The path difference for bright fringes is a multiple of the wavelength. Referring to Figure 28–6, or Example 28–2, we see that the fourth bright fringe above the central bright fringe corresponds to $m = +4$, which means the path difference is four wavelengths; that is $\Delta\ell = 4\lambda$. **(b)** Dark fringes correspond to path differences that are an odd multiple of half a wavelength. For the case of the third dark fringe, the path difference is $\Delta\ell = \left(3 - \frac{1}{2}\right)\lambda = \frac{5}{2}\lambda$.
3. **(a)** Blue light has a smaller wavelength; therefore, Equations 28–1 and 28–2 show that the angles to the fringes will be smaller. As a result, the spacing between the fringes decreases. **(b)** If the separation d between the slits is increased, we see from Equations 28–1 and 28–2 that the angles to fringes will decrease. Therefore, the spacing between fringes will decrease. **(c)** In water, the wavelength of the light is less than it is in air. As in parts (a) and (b), this will result in a decreased spacing between the fringes.
5. Light reflecting from the top surface of the oil has its phase changed by half a wavelength, since the oil has a greater index of refraction than air. At the boundary between oil and water the reflected wave experiences no phase change if the index of refraction of oil is greater than that of water. This must be the case, because where the oil is thinnest the phase difference between light reflecting from the top and the bottom of the oil film is just the phase difference due to reflection—the phase difference due to path length difference goes to zero. With a half wavelength phase change between light reflected from the top and bottom of the oil, light will undergo destructive interference and the film will appear dark. Therefore, a dark appearance near the edges implies oil has the greater index of refraction.
7. **(a)** The width is doubled. **(b)** The width is halved. **(c)** The width is doubled.
9. The resolution of your eyes is greater on a cloudy day than on a sunny day. The reason is that resolution goes up as the aperture of an optical system increases. On a cloudy day your pupils will be open wider, due to the lower level of light, and this, in turn, will result in a higher resolution.
11. To increase resolution, one must increase the diameter of the aperture. As we saw in Chapter 27, this corresponds to a smaller *f*-number.

Chapter 29

1. A light signal will always pass your ship with a speed of c, regardless of the speed of the ship. Therefore, you measure the speed to be greater than $0.1c$.
3. The two observers will always measure **(a)** the relative speed and **(d)** the speed of light in a vacuum to have the same value. The time between events, the length of an object, and the speed of a third observer will, in general, have different value for different observers.
5. **(a)** The pilot measures the proper time of the cockpit clock, since the pilot is at rest relative to it. **(b)** You measure the proper time of the control tower clock, since you are at rest relative to it. **(c)** The pilot measures the proper length of the airplane. **(d)** You measure the proper length of the runway.
7. The clock on the airplane traveling from Los Angeles to New York runs slower because its speed relative to the axis of the spinning Earth is greater than the speed of the other airplane.
9. **(a)** Rest energy is proportional to the mass, and therefore, $B < D < C < A$. **(b)** The kinetic energy is the total

energy minus the rest energy, and therefore A < D < B = C.
(c) The total energy divided by the rest energy is $1/\sqrt{1 - v^2/c^2}$, which increases with increasing speed. Therefore, A < D < C < B.

11. The extra energy content of pie 1 (piping hot) means that its mass is greater than that of pie 2, as a result of mass-energy equivalence. Therefore, the acceleration of pie 1 is less than the acceleration of pie 2.

13. The charged capacitor stores energy, as we saw in Chapter 21. Therefore, by the mass-energy equivalence, it follows that the charged capacitor is more massive than the uncharged capacitor.

Chapter 30

1. Because blackbody A has the longer wavelength it also has the lower frequency. We know from Equation 30–1, however, that the peak frequency of a blackbody spectrum is proportional to the absolute temperature of the blackbody. Therefore, the temperature of blackbody A is lower than the temperature of blackbody B.

3. **(a)** The energy of a photon of green light is *always* greater than the energy of a photon of red light, regardless of the wattage of the source. This conclusion follows from the relation $E = hf$ and the fact that green light has a higher frequency than red light. **(b)** The number of photons emitted per time by the red source is greater than the number of photons emitted per second by the green source. There are two reasons for this. First, its photons have less energy than the photons from the green source, as pointed out in part (a). Therefore, even if the two sources had the same power, the red source would emit more photons per second. Second, the red source has the greater power, requiring even more photons per second.

5. No. If the intensity of the light is increased there will be more photons striking the surface per second. Each photon, however, will still have too little energy to eject an electron.

7. Increasing the frequency of the light increases the energy carried by each photon. Therefore, **(a)** the maximum kinetic energy of an ejected electron increases and **(b)** its minimum de Broglie wavelength decreases. Since the intensity is unchanged, however, there must be fewer of these higher-energy photons reaching the surface each second. As a result, **(c)** the number of electrons ejected per second decreases and **(d)** the electric current in the phototube decreases.

9. Both the energy and the momentum of a photon are proportional to its frequency. Therefore, doubling the momentum of a photon also doubles its energy.

11. **(a)** Since the de Broglie wavelength is $\lambda = h/p$, it follows that doubling the momentum halves the wavelength. Therefore, the wavelength changes by a factor of $1/2$. **(b)** Doubling the kinetic energy means that the momentum increases by a factor of the square root of two. As a result, the de Broglie wavelength changes by a factor of $1/\sqrt{2}$.

13. **(a)** Increasing the mass of the particles implies that their momentum increases as well. It follows from Equation 30–16 that their de Broglie wavelength decreases. Therefore, from $2d \sin \theta = m\lambda$ we see that the angle of the interference maximum decreases. **(b)** Increasing the energy of the particles also implies that their momentum increases. Therefore, we again expect that the angle of the interference maximum decreases.

Chapter 31

1. **(a)** The shortest-wavelength photon corresponds to the greatest energy difference. Therefore, B < A < C.
(b) All photons travel with the same speed, $c = \lambda f$, and hence the frequency f is inversely proportional to the wavelength λ. As a result, C < A < B.

3. Noting that the radius of Bohr orbits is proportional to n^2/Z, we find B < C < A.

5. The de Broglie wavelength in the $n = 2$ Bohr orbit is greater than the wavelength in the $n = 1$ orbit. First, the $n = 2$ Bohr orbit has a circumference that is 4 times greater than that of the $n = 1$ orbit. Second, two de Broglie wavelengths fit into the $n = 2$ orbit, compared to one in the $n = 1$ orbit. Combining these results, we see that the de Broglie wavelength in the $n = 2$ orbit is twice as long as the de Broglie wavelength in the $n = 1$ orbit.

7. We can see from the derivation leading up to Equation 31–8 that the kinetic energy of a Bohr orbit is equal to minus one times the total energy of the orbit; that is, $K = -E$. In addition, recall that E is proportional to $1/n^2$. Therefore, the kinetic energy of an electron in the $n = 2$ Bohr orbit is $\frac{1}{4}K_1$.

9. The maximum number of electrons allowed in any given subshell depends only on the quantum number $\ell(s, p, d, \ldots)$, and not on the quantum number $n(\mathrm{K}, \mathrm{L}, \mathrm{M}, \ldots)$. Thus, **(a)** 6 electrons can occupy the $2p$ subshell, and **(b)** 6 electrons can occupy the $3p$ subshell.

11. Adding the numbers in the exponents, we find that this atom has 2 + 2 + 6 + 2 + 1 = 13 electrons.

13. The atom is in an excited state. In the ground state of potassium, the outermost electron is in the $4s^1$ state. For the atom in question, however, the outermost electron is in the $3d^1$ state. It is clear from Figure 31–18 that the energy of the $3d$ level is higher than the energy of the $4s$ level; therefore, the potassium atom in question is in an excited state.

15. The outermost electron in sodium is in a higher energy level than the outermost electron of lithium. Therefore, the ionization energy of sodium is less than that of lithium.

Chapter 32

1. **(a)** In α decay, the radius of the daughter nucleus is less than that of the original nucleus, because the daughter nucleus contains four fewer nucleons. **(b)** In β decay the radius of the daughter nucleus is the same as that of the original nucleus, because the daughter nucleus contains the same number of nucleons.

3. The nine intermediary nuclei on the upper series are as follows: $^{228}_{88}\mathrm{Ra}$, $^{228}_{89}\mathrm{Ac}$, $^{228}_{90}\mathrm{Th}$, $^{224}_{88}\mathrm{Ra}$, $^{220}_{86}\mathrm{Rn}$, $^{216}_{84}\mathrm{Po}$, $^{212}_{82}\mathrm{Pb}$, $^{212}_{83}\mathrm{Bi}$, $^{212}_{84}\mathrm{Po}$. The tenth intermediary, on the lower series, is $^{208}_{81}\mathrm{Tl}$.

5. Both alpha (α) and beta decay (β) result in a new element, since the atomic number Z changes. For example, in alpha decay the atomic number of a nucleus changes from Z to $Z - 2$. In beta decay, the atomic number changes from Z to $Z + 1$. In contrast, gamma decay (γ) is simply a release of energy with no change in atomic number.

7. When we assign ages based on carbon-14 dating, we assume that the ratio of carbon-14 to carbon-12 in the atmosphere has remained constant. If the ratio had been smaller 10,000 years ago, the initial decay rate of carbon-14 would have been smaller as well. As a result, less time would be required for the decay rate to decrease to its current value. Therefore, the true age of an object would be less than the age we have assigned to it.

9. The two samples have equal half lives, therefore their numbers decrease with time by the same factor, $e^{-\lambda t}$. It follows that the activity of sample A will always be twice that of sample B.

11. The amount of coal burned in a conventional power plant is much greater than the amount of uranium consumed in a nuclear power plant for a given amount of energy production. The reason is that in a coal-powered plant, energy is released as a result of chemical reactions. In a nuclear-powered plant the reactions occur within the nucleus, and therefore they release millions of times more energy than comparable chemical reactions. As a result, the amount of coal that is burned is much greater than the amount of uranium that is consumed.

Answers to Odd-Numbered Problems

Note: In cases where an ambiguity might arise, numbers in this text are assumed to have the smallest possible number of significant figures. For example, a number like 150 is assumed to have just two significant figures—the zero simply indicates the location of the decimal point. To represent this number with three significant figures, we would use the form 1.50×10^2.

Chapter 1

1. **(a)** 0.114 gigadollar
(b) 1.14×10^{-4} teradollar
3. 3×10^8 m/s
5. $p = 1$
9. $\frac{[M]}{[T]^2}$
11. 3.00×10^8 m/s
13. 28.7 lb
15. **(a)** 648.7 m^2 **(b)** 45 m^2
17. **(a)** 4.83×10^8 ft^3 **(b)** 1.37×10^7 m^3
19. 108.7827757 μs
21. 1×10^5 times
23. 0.23 lb
25. 6.71×10^8 mi/h
27. **(a)** 3.3357×10^{-11} s
(b) 1.7987×10^{12} jiffy
29. 2.0×10^6 m^2
31. **(a)** 65.6 ft/s **(b)** 44.7 mi/h
33. 10^5 seats
35. **(a)** 10^3 mi/h **(b)** 10^4 mi **(c)** 10^3 mi
37. **(a)** 6.75×10^{-4} mm **(b)** 2.66×10^{-5} mm
39. **(a)** 3.63×10^4 beats/min
(b) 1.52×10^7 cycles/beat
41. **(a)** 39 ft/s^2 **(b)** 1.6×10^5 km/h^2
43. **(a)** 2.3 g **(b)** 2.4 days
45. **(a)** 51 revolutions **(b)** 8.7 ft
47. $q = -\frac{1}{2}; p = \frac{1}{2}$

Chapter 2

1. **(a)** 1.95 mi **(b)** 0.75 mi
3. **(a)** 15.0 m **(b)** 10.0 m
5. **(a)** 130 m; 100 m **(b)** 260 m; 0
7. 10.13 m/s; 22.66 mi/h
9. **(a)** 3.5 km **(b)** 14 s
11. 2.57 s
13. 10 ms
15. 6.0 m/s
17. 11 m
19. **(a)**

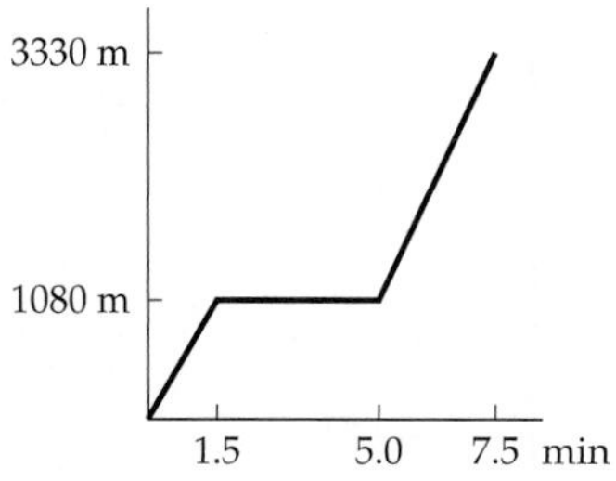

(b) 7.4 m/s
21.

	A	B	C	D
(a)–(d)	positive	zero	positive	negative
(e)–(h)	2.0 m/s	0 m/s	1.0 m/s	−1.5 m/s

23. **(a)**

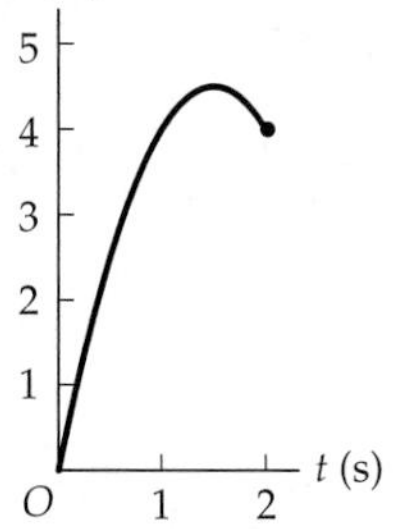

(b) 4 m/s **(c)** 4 m/s
25. 35 mi/h
27. **(a)**

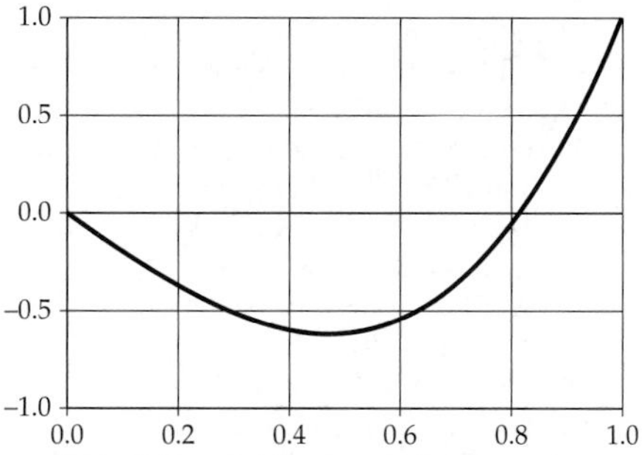

(b) −1.63 m/s **(c)** −1.64 m/s
(d) Closer to −1.64 m/s
29. **(a)** 3.8 m/s **(b)** 9.9 m/s
31. **(a)** 27.9 m/s north **(b)** 9.48 m/s north
33. **(a)** 10 m **(b)** 20 m **(c)** 40 m
35. **(a)** a factor of two **(b)** 3.8 s **(c)** 7.6 s
37. 10.3 m/s
39. 3.53 m/s^2 due north
41. 5.8 s
43. **(a)** 77.8 m **(b)** 12.5 m/s **(c)** 6.25 m/s, 18.8 m/s, **(d)** 19.4 m, 58.5 m
45. 9.1 m/s^2
47. 2.8×10^5 m/s^2
49. **(a)** $x_1 = (20.0\text{ m/s})t + (1.25\text{ m/s}^2)t^2$;
$x_2 = (1000\text{ m}) + (-30.0\text{ m/s})t + (1.60\text{ m/s}^2)t^2$
(b) 24.0 s
51. **(a)** 17 m/s^2 upward **(b)** 51 m/s
53. **(a)** 5.0 s **(b)** 8.0 m/s **(c)** 0.94 s
55. **(a)** 2.1 m/s^2 due east **(b)** 0.97 s
(c) less than 3.4 m/s
57. **(a)** 5.7 s **(b)** −0.18 m/s^2
(c) more than 2.1 m/s
59. **(a)** 5.3 m/s^2 toward third base **(b)** 3.2 m
61. **(a)** 5.2 m **(b)** 2.6 m **(c)** 2.6 m/s
63. **(a)**

angle	10.0°	20.0°	30.0°
acceleration, m/s^2	1.71	3.37	4.88
$g \sin \theta$, m/s^2	1.70	3.35	4.91

65. The statement is accurate.
67. 4.9 m/s
69. **(a)** 8.4 m/s, +8.4 m/s
(b) 1.4 m/s, −1.4 m/s
71. 0.10 s
73. $x_{\text{Bill}} = (4.9\text{ m/s}^2)t^2$
$x_{\text{Ted}} = (2.0\text{ m}) + (-4.2\text{ m/s})t + (4.9\text{ m/s}^2)t^2$
75. **(a)** 58 m/s **(b)** 5.9 s
77. **(a)** 1.2 s **(b)** 0.61 s
79. 24.8 m/s
81. **(a)** 2.8 s **(b)** −30 m/s
83. **(a)** 58 m **(b)** 34 m/s **(c)** 8.1 s
85. 11 m/s
87. 2.01 m/s
89. 5.9 km
91. **(a)** 6.8 ft/s **(b)** 1.2 s
93. **(a)** 2 **(b)** 4
(c) 0.41 s and 0.82 s; 0.20 m and 0.82 m
95. **(a)** 2.64 s **(b)** more than
(c) 1.87 s > 0.77 s
97. **(a)** 9.81 m/s^2 downward
(b) 13.9 m
(c) 2.21 s **(d)** 16.5 m/s
99. 3.0 m
101. **(a)** 13.4 m **(b)** 16.2 m/s
103. 6.67 cm
105. **(a)** 3.0 m; 4.4 m/s **(b)** 130 drops/min
107. 5.5 m; 11 m/s
109. **(a)** 10.0 ms
(b) 4.50 m/s **(c)** 10.0 cm
113. 4.3 m/s^2
115. **(a)** 2.3 sec
(b) 20 m/s

Chapter 3

1. 142 ft
3. 3°
5. **(a)** $(90\text{ ft})\hat{\mathbf{x}} + (90\text{ ft})\hat{\mathbf{y}}$ **(b)** $(90\text{ ft})\hat{\mathbf{y}}$
(c) $(0\text{ ft})\hat{\mathbf{x}} + (0\text{ ft})\hat{\mathbf{y}}$
7. 1.5 Å
9. **(a)** 3.5 cm
(b) less than
(c) 1.7 cm
11. **(a)** $\vec{A}$ **(b)** $\vec{B}$
13. **(a)** 51 m deep
(b) 140 m
15. **(a)**

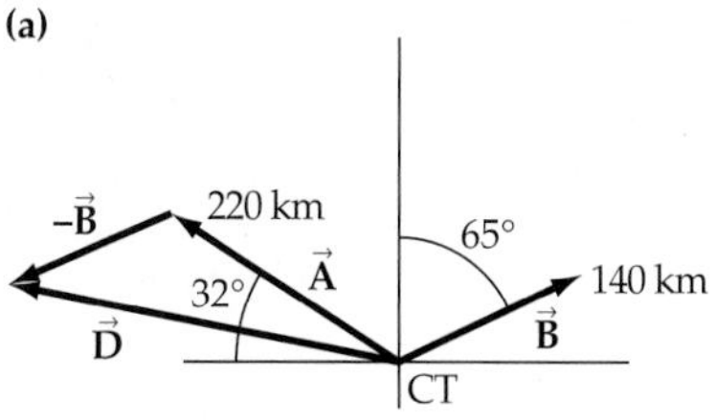

(b) 320 km, 10° north of west

17. **(a)**

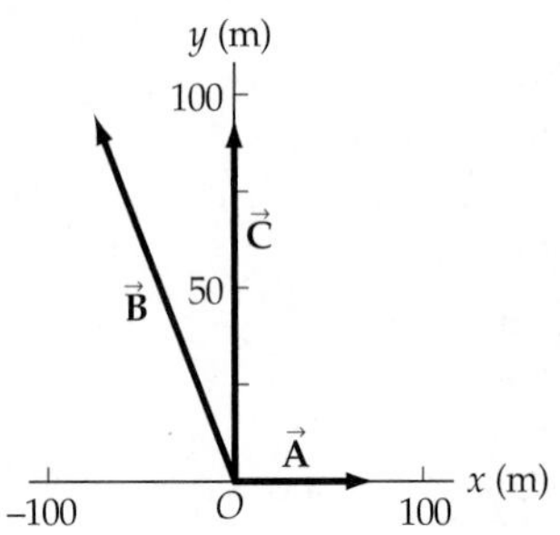

(b) about 120 m; about 130°
(c) 121 m; 128°

19. **(a)** −27°, 11 units **(b)** 207°, 11 units **(c)** 27°, 11 units

21. $(40\ \text{m})\hat{\mathbf{x}} - (36\ \text{m})\hat{\mathbf{y}}$

23. **(a)** 23 m **(b)** 23 m

25. **(a)** −22°, 5.4 m **(b)** 110°, 5.4 m **(c)** 45°, 4.2 m

27. **(a)** $(23\ \text{m})\hat{\mathbf{x}} - (27\ \text{m})\hat{\mathbf{y}}$ **(b)** $-(23\ \text{m})\hat{\mathbf{x}} + (27\ \text{m})\hat{\mathbf{y}}$

29. $(2.1\ \text{m})\hat{\mathbf{x}} + (0.74\ \text{m})\hat{\mathbf{y}}$

31. **(a)** 140 m, 59° south of east; **(b)** no change

33. 14° south of west, 3.6 km

35. **(a)** 2.81 m/s, 1.63 m/s **(b)** The components will be halved.

39. **(a)** 852 m/s, 135° **(b)** greater

41. 15.3 m/s

43. 25 s

45. **(a)** 11° west of north
(b)

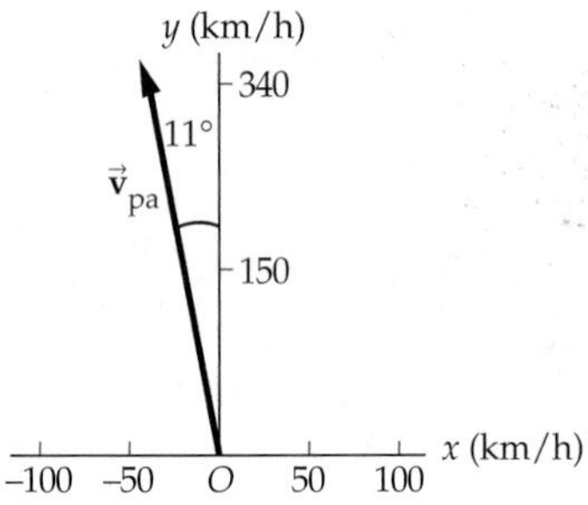

(c) increased

47. 11 m/s

49. **(a)** Jet ski A **(b)** $\dfrac{t_A}{t_B} = 0.82$

51. −53.1°; 40.3 m

53. **(a)** toward the rear **(b)** 14.3 m/s **(c)** 11.7 m/s

55. $A_x = 44$ m, $A_y = 31$ m, $A_z = 37$ m

57. **(a)** $-9.81\ \text{m/s}^2$ **(b)** $-9.81\ \text{m/s}^2$ **(c)** $-9.81\ \text{m/s}^2$

59. **(a)** ≈38 ft **(b)** 39 ft

61. **(a)** 3.7 **(b)** 67 m/s

63. **(a)** 10° **(b)** $(7.1\ \text{m/s})\hat{\mathbf{x}}$ **(c)** decrease

65. 28 m; 19 m

67. **(a)** 12° upstream **(b)** decrease

Chapter 4

1. **(a)** 3.6 km **(b)** 3.2 km

3. **(a)** 140 m **(b)** 14 m

5. **(a)** 2.95×10^{-9} s **(b)** 2.31 cm

7. 46 m/s

9. 7.87 m/s

11. **(a)** 1.6 m **(b)** decreases **(c)** decreases

13. **(a)** 2.70 m/s **(b)** −20.6 m/s **(c)** The speed of the clam in the x-direction would increase with the speed of the crow, but the speed in the y-direction would stay the same. The speed of the crow determines v_x and gravity determines v_y.

15. **(a)** 0.378 m **(b)** decreases

17. **(a)** −66°, 8.1 m/s **(b)** −76°, 14 m/s

19. **(a)** 1.06 m **(b)** the same time

21. 1.20 m

23. **(a)** 13.9 m/s **(b)** 1.99 s

25. 1.27 m/s

27. 1.3 m

29. **(a)** the same **(b)** 18 m/s, 18 m/s

31. **(a)** 91.7 m **(b)** 21.2 m/s

33. 24.8 m/s

35. **(a)** 12 m/s, 22° **(b)** No; the y-component of the velocity is still positive.

37. 1.07 m

39. 111 m/s

41. **(a)** 32° **(b)** 0.76 s

43. **(a)** 2.98 m/s **(b)** 4.29 s

45. **(a)** 849 m/s **(b)** less than

47. 46.2°

49. **(a)** 0.405 s **(b)** 1.02 s **(c)** increase

51. 6.1 m/s

53. **(a)** $(5.00\ \text{m/s})\hat{\mathbf{x}} + (2.00\ \text{m/s})\hat{\mathbf{y}}$ **(b)** 5.39 m/s at 21.8° above horizontal **(c)** 6.20 m **(d)** 1.33 s

55. **(a)** Use the vector magnitude formula and $v_y = v_{0y} - gt$ to determine v_{0y}. Then use the relationship of v_0 and v_{0y} to determine θ. **(b)** 29.2° above horizontal

57. **(a)** 0.616 s **(b)** 0.794 m

59. **(a)** 11.3 m **(b)** 13.3 m **(c)** 14.2 m

61. **(a)** 12.3 m **(b)** 13.2 m

63. **(a)** 12.1 m/s, top **(b)** 12.9 m/s, 19.9° **(c)** 0.978 m **(d)** 0.978 m

65. 14.5 m/s

67. **(a)** To find the initial speed of the puck, eliminate t from the equations $x = (v_0 \cos\theta)t$ and $y = (v_0 \sin\theta)t - (1/2)gt^2$, then solve for v_0. **(b)** 25.1 m/s

69. **(a)** 64.9° **(b)** 21 m/s **(c)** 1.8 s

71. **(a)** 2.7 cm **(b)** 0.021 s

73. 76.0°

75. $v_{landing} = \sqrt{v_0^{\,2} + 2gh}$ independent of angle θ

79. **(a)** top, 9.9 m/s **(b)** 11 m/s, 24° **(c)** 0.98 m **(d)** 0.98 m

81. **(a)** 60.5° **(b)** 15.9 m **(c)** 41.6° **(d)** 18.4 m

83. **(a)** 42.3° **(b)** 3.12 m **(c)** 10.4 m/s

Chapter 5

1. $1.24\ \text{m/s}^2$ upward

3. 37 kg

5. 0.26 kN

7. **(a)** 85 N **(b)** increased

9. **(a)** 5.1 kN opposite to the direction of motion **(b)** 15.3 m

11. **(a)** 20.8 kN opposite to the direction of motion **(b)** Determine the acceleration from the speeds and displacement and the force from the acceleration and mass.

13. **(a)** two forces **(b)** gravity and your hand **(c)** No **(d)** No

15. **(a)** same **(b)** more **(c)** $0.70\ \text{m/s}^2$

17. **(a)** 1.04 N **(b)** 3.59 N

19. 11 kN

21. 28 N

23. $1.5\ \text{m/s}^2$

25. **(a)** 0.24 kN, downhill, parallel to the slope **(b)** increases

27. **(a)** 4.7 s **(b)** 7.4 m

29. **(a)** 10.9 kg **(b)** 107 N

31. 3.5 kN

33. **(a)** 0.0119 N **(b)** same

35. **(a)** upward **(b)** $1.9\ \text{m/s}^2$ **(c)** elevator is speeding up if it is traveling upwards, slowing down if it is traveling downwards

37. 0.11 kN

39. **(a)** 91 N

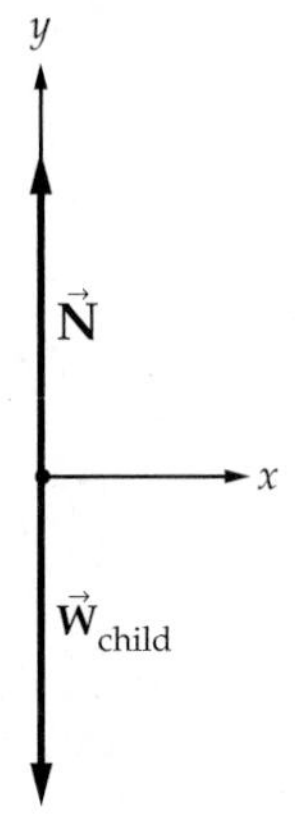

(b) 0.13 kN

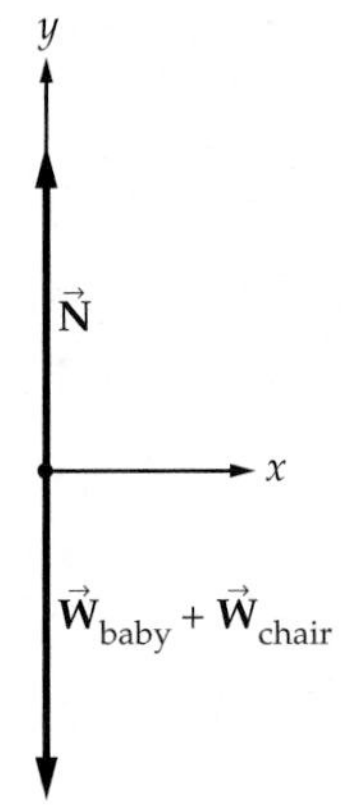

41. The free-body diagram does not change for the moving cart.

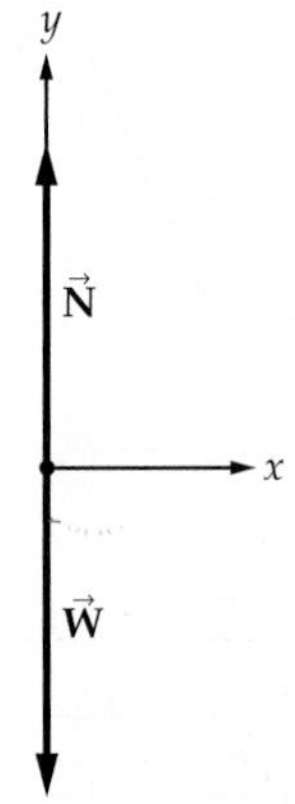

43. **(a)** 0.29 kN **(b)** increases

45. 54 N

47. **(a)** $5.5 \times 10^{-3}\ \text{m/s}^2$ **(b)** 8.3×10^{9} m **(c)** 1/3

49. 9.1 kg

51. **(a)** 99 m/s^2 **(b)** 0.22 N **(c)** stay the same **(d)** increase
53. **(a)** 0.42 kN opposite the direction of motion **(b)** 2
55. **(a)** 2.2 N **(b)** 1.4 s
57. 290 kg
59. **(a)** 2.8×10^5 N
(b) Determine the acceleration during takeoff using the given data (x, t). Then calculate the force using $F = ma$.
61. **(a)** 2.06 kg **(b)** 5.00 kg
63. 74 kg
65. $F_1 = \frac{1}{2}m(a_1 + a_2)$
$F_2 = \frac{1}{2}m(a_1 - a_2)$
67. **(a)** 11 m/s^2 to the left **(b)** 6.0 m/s^2 to the right **(c)** 0.078 s **(d)** 0.47 m/s **(e)** 0.18 kg·m/s, 0.18 kg·m/s
69. **(a)** 13.0 kg **(b)** 3.71 kg
71. **(a)** increased **(b)** 0.081 m/s^2

Chapter 6

1. 1.8 m
3. 1.2
5. 0.75 N opposite the direction of the push
7. 0.25 kN
9. **(a)** 0.42 **(b)** 23°
11. **(a)** 2.4 m/s^2 **(b)** 6.4 s
13. **(a)** 2.55 m/s^2 opposite the direction of motion **(b)** 4.26 m/s
15. 49.3 N
17. 8.3 kN/m
19. **(a)** 3.0 N **(b)** No
21. **(a)** 0.26 kN **(b)** 0.51 kN **(c)** 0.51 kN
23. **(a)** 4.8 cm **(b)** Yes; the spring displacement would double.
25. **(a)** 0.124 N **(b)** 60°
27. **(a)** 45 N **(b)** No
29. 35.9 lb
31. 42.5 N
33. 0.85 N
35. **(a)** 60° **(b)** 106 N
37. 0.076 m/s^2 upward
39. 4.9 N; 15 N
41. **(a)** 2.9 m/s^2 **(b)** 2.4 N **(c)** decrease
43. 4.7 kN
45. 36 m/s
47. 0.65 kN
49. **(b)** 0.52 kN at top; 0.56 kN at bottom
51. 19 m/s
53. 1.9 kg
55. 0.140
57. 5.62 kg
59. 14°
61. **(a)** 3.81 N **(b)** 195 N
63. **(a)** 0.025 N down the incline **(b)** 0.025 N up the incline
65. $F_1 = mg \sin 20°$; $F_2 = mg \cos 20°$
67. **(a)** 22 N **(b)** stays the same
69. **(a)** greater **(b)** 3.4 N **(c)** 0.41 kg
71. **(a)** 19° **(b)** 0.78 N
73. **(a)** 32 N **(b)** increase
75. **(a)** greater **(b)** 1.88 kN
77. **(a)** static **(b)** 0.50
79. 26 m/s^2
81. 7.6°; The mass of the dice drops out of the equations.
83. 28°
85. $T = mg \sin \theta$
87. $\mu_s g(2m + M)$
89. $\dfrac{\mu_s}{1 + \mu_s}$
91. **(a)** $\dfrac{\mu_s mg}{\cos \theta - \mu_s \sin \theta}$
93. **(a)** 0.163 s **(b)** 0.102 m
95. **(a)** $m(v^2/r + g)$ **(b)** mv^2/r **(c)** $m(v^2/r - g)$
97. **(a)** 0.315 **(b)** 0.127
99. **(a)** 15 m/s **(b)** equal to

Chapter 7

1. 0.29 kJ
3. **(a)** 38 J **(b)** 0
5. **(a)** 60 J **(b)** 0 **(c)** −60 J
7. **(a)** negative **(b)** −7.8 kJ
9. 0.48 kJ
11. 57.4°
13. **(a)** 0.55 J **(b)** 0.28 J **(c)** −1.4 J
15. 1.9 kJ
17. **(a)** 8.03 kJ **(b)** 2.01 kJ **(c)** 32.1 kJ
19. **(a)** −10 J **(b)** 0.63 N upward
21. **(a)** −587 J **(b)** 0.284
23. **(a)** by less **(b)** −1.4 m/s
25. **(a)** 44 J **(b)** 44 J
27. 91 N/m
29. **(a)** 0.90 m **(b)** 0.13 m
31. **(a)** 16 kN/m **(b)** more than 160 J; 480 J
33. **(a)** 0.82 kJ **(b)** 1.1 kJ
35. 3.6 MJ
37. 5.11 min
39. 0.143 hp
41. **(a)** 2.3 MJ **(b)** 2 Snickers
43. 1.4 hp
45. 6.0×10^{-20} J
47. 6.3 km/s
49. **(a)** 0.18 kW **(b)** 6.3 kJ
51. 21.6 mJ
53. $K/2$
55. **(a)** 13.1 J **(b)** 13.1 J **(c)** stay the same
57. **(a)** $mg(h - h_{max})$ **(b)** mgh **(c)** $\frac{1}{2}m{v_0}^2 + mgh$
59. 0.37 mW
61. **(a)** 5.5 W **(b)** 0.50 kJ
63. **(a)** 2.9×10^4 kg **(b)** 38 MW
65. **(a)** 1.8 m/s^2 **(b)** decrease **(c)** 0.89 m/s^2
67. **(a)** kg/m **(b)** $\sqrt[3]{\dfrac{P}{b}}$ **(c)** 1.26
69. $\dfrac{1}{2}(k_1 + k_2)x^2$
71. **(a)** 0.93 kN/m **(b)** 1.7 cm
73. **(a)** 7.8 s **(b)** less than **(c)** 3.9 s

Chapter 8

1. −63 J; −63 J; −63 J
3. **(a)** −0.11 J; −0.11 J **(b)** The results have no dependence on the mass of the block.
5. **(a)** −0.096 J; −0.25 J **(b)** −0.096 J; −0.085 J
7. 7.2 MJ
9. **(a)** 0.63 N/m **(b)** 7.2 μJ **(c)** 4.2 mN
11. **(a)** 200 N/m **(b)** 0.92 cm
13. 37.3 J
15. 7.16 m/s
17. **(a)** 0.942 m **(b)** Doubling the ball's mass would cause no change to (a).
19.

y (m)	4.0	3.0	2.0	1.0	0
U (J)	8.2	6.2	4.1	2.1	0
K (J)	0	2.1	4.1	6.2	8.2
E (J)	8.2	8.2	8.2	8.2	8.2

21. **(a)** 15 m/s **(b)** 43 m
23. **(a)** 0; 98.1 J; 98.1 J **(b)** 98.1 J; 196 J; 98.1 J
25. **(a)** 0.95 J **(b)** 0.95 J **(c)** 41°
27. 4.0 cm
29. 1.8 m/s
31. −57 MJ
33. 1.8 m/s
35.

	W_{nc}(J)	U(J)	K(J)	E(J)
(a)	0	34	0	34
(b)	−2.3	24	7.0	31
(c)	−4.6	15	14	29

37. **(a)** uphill **(b)** 3.65 m
39. **(a)** −29.4 MJ **(b)** 5.51 MJ **(c)** No
41. At point A, the object is at rest. As the object travels from point A to point B, some of its potential energy is converted into kinetic energy and the object's speed increases. As the object travels from point B to point C, some of its kinetic energy is converted back into potential energy and its speed decreases. From point C to point D, the speed increases again, and from point D to point E, the speed decreases.
43. At approximately x = 0.5 m and x = 4.7 m
45. ±12°
47. **(a)**

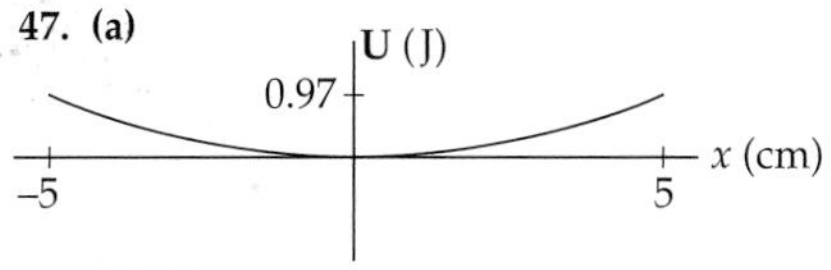

(b) ±4.6 cm
49. $L \pm v_{max}\sqrt{\dfrac{2m}{k}}$
51. 2.87 m
53. 190 N/m
55. **(a)** 1.5 mm **(b)** 0.46 J
57. **(a)** −0.86 J **(b)** greater than **(c)** −1.2 J
59. **(a)** 53.0 MJ **(b)** 218 m/s
61. 1.04 m
63. **(a)** 634 N **(b)** decrease
65. 7.20 m/s
67. 0.14 m
69. 1.1 kN
71. **(a)** ±0.11 m **(b)** $\dfrac{1}{\sqrt{2}}$
73. 0.85 m
75. 32 cm
77. 5.3 cm
79. **(b)** Tension depends upon a_{cp}, which is directly proportional to v^2 and inversely proportional to the radius r. Since both v^2 and r are directly proportional to l, l cancels out.
81. 48.2°
83. **(a)** 0.79 m/s **(b)** positive **(c)** 0.36 J
85. 0.688 m
87. **(a)** negative **(b)** −46 J **(c)** 38 N
89. **(a)** 0.99 m/s **(b)** 0.48 s
91. **(a)** decrease **(b)** −7.31 J **(c)** −7.31 J **(d)** 14.6 N **(e)** 7.31 J

Chapter 9

1. 2.49×10^5 mi/h
3. 73.3°, 0.746 m/s
5. 1.38 m
7. 2.51 kg, 5.21 kg
9. 3.0 kN
11. 0.33 kg
13. 2.7 kg·m/s
15. **(a)** 121° from x axis **(b)** 2.1 kg·m/s
17. 31 kg
19. mass ratio = 2. The piece with the smaller kinetic energy has the larger mass.
21. **(a)** 2.2 m/s **(b)** greater than **(c)** 2.3 m/s
23. $\frac{1}{4}mv^2$
25. 2.82 m/s
27. **(a)** 1.0×10^2 m/s **(b)** less than **(c)** 850 J; 47 J
29. 3.71 cm
31. $v_{truck} = 5.56$ m/s; $v_{car} = 20.1$ m/s
33. 1.31 m/s; 107° counterclockwise from the $+x$ axis
35. **(a)** 16.8 m/s **(b)** Kinetic energy has been transferred from the elephant to the ball.
37. **(a)** $\frac{v_0}{2}$ **(b)** $\frac{v_0}{\sqrt{2}}$
39. 0.60 m from the left end
41. $L/10$ units below the center of the box
43. $(0, 3.6 \times 10^{-11}\text{ m})$
45. $y_{cm} = (0.164\text{ m/s}^2)t^2$ $v_{cm} = (0.328\text{ m/s}^2)t$
47. **(a)** 13.2 N **(b)** 13.2 N
49. 840 kg/s
51. 0.503 m/s
53. **(a)** 12.8 N **(b)** 12.7 N
55. 6.73×10^{-21} ft
57. 2.36 m
59. 0.24 N
61. 0.354 mm
63. **(a)** greater than **(b)** 3.6 m
65. **(a)** 13.2 N **(b)** 12.9 N
67. $2m$
69. $(6.5 \times 10^{-12}\text{ m}, 0)$
71. **(a)** 27.8 m/s **(b)** $(19.7\text{ m/s})\hat{\mathbf{y}}$ both before and after **(c)** $(-9.81\text{ m/s}^2)\hat{\mathbf{y}}$ both before and after
73. $\frac{25}{24}L$
75. **(a)** 0.23 m/s **(b)** 0.064 J **(c)** 0.45 m/s, −0.23 m/s
79. **(a)** $v + 2u$
83. **(a)** $a_{cm} = \frac{v^2}{L}$ upward **(b)** upward
85. 1.6 kg
87. **(a)** $\frac{3}{4}v_0$ **(b)** 10% lost **(c)** $v_{1,f} = v_0/2$; $v_{2,f} = v_0$

Chapter 10

1. $\frac{\pi}{6}, \frac{\pi}{4}, \frac{\pi}{2}, \pi$
3. **(a)** 1 rev/h **(b)** 2 rev/day
5. tire, propeller, drill
7. 190 rad/s
9. 1.90×10^{-6} rev/min
11. **(a)** 1.3×10^2 rad/s **(b)** 2.1×10^2 rad/s **(c)** 3.0×10^2 rad/s **(d)** positive **(e)** 85 rad/s²; 85 rad/s²
13. 4.41 rad
15. 48 rad
17. **(a)** 59 rev **(b)** 45 rev
19. 3:08:11
21. -6.14×10^{-22} rad/s²
23. **(a)** −40.3 rev/s² **(b)** 211 ft **(c)** 10.0 in.
25. 12 μm/s
27. 23 rad/s
29. **(a)** 1.18 rad/s **(b)** 10.0 m/s² **(c)** the vine
31. **(a)** 0.503 m/s² **(b)** 0.503 m/s²
33. **(a)** 0.70 m/s **(b)** It would also double.
35. 2.1 m/s²; 170° below direction of motion
37. **(a)** 68.8 m/s **(b)** 7.31×10^{-5} s
39. **(a)** 4.12×10^{16} rad/s **(b)** 6.56×10^{15} orbits **(c)** 8.98×10^{22} m/s²
41. 39 rad/s
43. 28.5 rad/s
45. **(a)** 2.03 rad/s² **(b)** greater
47. 4.81 J
49. **(a)** 1.19 J **(b)** 1.19 J **(c)** 2.39 J
51. $K_t = 160$ J; $K_r = 0.069$ J
53. -3.59×10^{13}
55. 0.17 kg·m²
57. **(a)** 3.3 m/s **(b)** 2.8 m/s
59. **(a)** 1.2×10^2 rad/s **(b)** 0.54 m
61. **(a)** 3.0 m/s **(b)** decrease
63. **(a)** 15 J **(b)** 4.9 J **(c)** 9.8 J
65. 1 hr 24 min
67. 7.5 rad/s
69. 874 m
71. **(a)** 22 m/s **(b)** 190 m/s **(c)** 60 cm **(d)** 7.0 N
73. **(a)** decrease **(b)** 50.0 rad/s **(c)** 20.8 rad/s **(d)** -7.31×10^{-3} rad/s²
75. **(a)** 0.30 rev **(b)** It does not depend on her initial speed.
77. **(a)** 3.6×10^2 rad/s² **(b)** 5.3×10^2 rad/s²
79. **(a)** 91 s **(b)** 220° cw from North **(c)** 15 m
81. **(a)** 6.9 rad/s² **(b)** It doubles.
83. **(a)** 0.334 m **(b)** 1.01 kN
85. **(a)** 15 rad/s² **(b)** 4.0×10^3 rad
87. **(a)** 8.3×10^{-16} rad/s **(b)** 1.7×10^{17} mi
89. **(a)** $0.95g$ **(b)** 8.9 m/s
91. **(a)** $\sqrt{3g/L}$ **(b)** $\sqrt{3gL}$
93. **(a)** 0.50 m **(b)** 2.4 rev
95. **(a)** 1.5 m **(b)** 2.7 rev **(c)** increase
97. **(a)** solid sphere wins, then disk, then hollow sphere **(b)** all kinetic energies are equal
99. **(a)** greater **(b)** 5.3×10^{-5} kg·m²

Chapter 11

1. 60 N
3. **(a)** 8.96 N·m **(b)** 8.28 N·m
5. **(a)** −2.14 N·m **(b)** clockwise **(c)** increase
7. 11 rad/s²
9. 0.0018 N·m
11. **(a)** −0.25 N·m **(b)** decrease
13. **(a)** greatest about the x axis, least about the z axis **(b)** 7.4 rad/s² **(c)** 5.1 rad/s² **(d)** 3.0 rad/s²
15. **(a)** 54 rad/s² **(b)** 0.093 m
17. 57 N
19. **(a)** more **(b)** 90.3 N
21. 1.1 m
23. 9.14 cm
25. 3.2 kN; −2.2 kN
27. **(a)** 32 N **(b)** 32 N **(c)** 30 N
29. $f_1 = 900$ N; $f_2 = f_3 = 170$ N
31. $F_1 = 1.5$ N; $F_2 = 3.3$ N
33. **(a)** 0.983 N **(b)** 0.492 N **(c)** 1.25 N
35. **(a)** 1.06 m **(b)** 90.7 N
37. 0.25 m
39. $(11/12)L$
41. **(a)** 8.68 m/s² **(b)** 71.7 rad/s² **(c)** 9.77 m
43. 213 kg
45. 14 m/s²
47. $\left(\frac{m_2 - m_1}{m_1 + m_2 + \frac{1}{2}M}\right)g$
49. 5.9×10^{-4} kg·m²/s
51. **(a)** 219 kg·m²/s **(b)** 1.09×10^3 kg·m²/s
53. **(a)** greater for point B **(b)** same **(c)** 0; 2.55×10^3 kg·m²/s; 2.55×10^3 kg·m²/s
55. 210 N·m
57. 0.560
59. 2
61. 2.84 rad/s
63. 0.37 rad/s
65. **(a)** faster **(b)** 4.0 rad/s.
67. **(a)** $\frac{Iv}{I + mR^2}$ **(b)** As $I \to 0$, $v_g \to 0$; as $I \to \infty$, $v_g \to v$.
69. 0.076 N·m
71. 24 J
73. **(a)** 25 J **(b)** 28 J **(c)** 52 J
75. **(a)** 15.8 N·m **(b)** greater **(c)** 2560 rpm
77. 3.15 m
79. **(a)** 31.1 N·m **(b)** 286 N **(c)** 199 N **(d)** −123 N
81. **(a)** 5250 ft·lb·rev/min/hp **(b)** 259 ft·lb
83. **(a)** the force from the index finger **(b)** finger 0.49 N; thumb 0.25 N
85. **(a)** 229 N **(b)** $F_x = 216$ N; $F_y = 78.5$ N
87. $\vec{\mathbf{F}}_1 = [-(0.59\text{ kN/m})x + 0.33\text{ kN}]\hat{\mathbf{y}}$; $\vec{\mathbf{F}}_2 = [(0.59\text{ kN/m})x + 1.4\text{ kN}]\hat{\mathbf{y}}$
89. 0.16 kN
91. 962 N
93. **(a)** increase **(b)** 261 s
95. $T_1 = 1.25$ N; $T_2 = 0.224$ N; $m = 22.8$ g
97. **(a)** 7.5 N
101. $(2/3)L$
103. **(a)** increase **(b)** 1.0 N
105. 121°

Chapter 12

1. **(a)** 7.2×10^{-11} N **(b)** 1.8×10^{-11} N
3. **(a)** 4.4 kN **(b)** 0.15 kN
5. **(a)** 2.8×10^{-12} N **(b)** 2.8×10^{-12} N
7. **(a)** 3.56×10^{22} N toward the Sun **(b)** 2.40×10^{20} N toward the Sun **(c)** 3.58×10^{22} N toward the Earth-Moon system
9. 3.58×10^{22} N, 0.145° toward the Earth off the ray from the Sun to the Moon
11. **(a)** 4.7×10^{-8} N, 74° below horizontal, down and to the left **(b)** All forces are reduced by a factor of $2^2 = 4$; the directions of the forces are unchanged.
13. **(a)** 3.70 m/s² **(b)** 8.88 m/s²
15. 6.2×10^{-8} N
17. 1.36 m/s²
19. $(1/96)M_E$
21. **(a)** 3.46×10^8 m **(b)** The astronauts are accelerated at same rate as the spaceship and so they will appear to float, not walk.

23. 1.98 h
25. 1.4×10^{11} m
27. 1.90×10^{27} kg
29. **(a)** 12 h **(b)** 3.9 km/s
31. **(a)** 3.98 h **(b)** 7.31 h **(c)** The periods do not depend on the mass of the satellite. **(d)** The periods depend inversely on the square root of the mass of the Earth.
33. 1.91×10^{30} kg
35. -4.94×10^{8} J
37. **(a)** -4.81×10^{8} J **(b)** -4.56×10^{8} J **(c)** 2.5×10^{7} J; $mgh = 2.6 \times 10^{7}$ J; 4% difference
39. **(a)** 1.1×10^{11} J **(b)** 2.4×10^{12} J
41. 5.04 km/s
43. 7.91 km/s
45. **(a)** 4.25 km/s **(b)** 10.4 km/s
47. 1.73 km/s
49. It is 10 times that of the Earth.
51. 2.96 km
53. $\sqrt{(1.00 \times 10^{7}\text{ m})h}$
57. -6.34×10^{-10} J
59. **(a)** 1.98×10^{20} N **(b)** 4.36×10^{20} N **(c)** It makes more sense to think of the Moon as orbiting the Sun, with a small effect due to the Earth.
61. 7.27×10^{-5} m/s
63. **(a)** less **(b)** 4.91 m/s^2
67. $\sqrt{\dfrac{72\pi^2 r_1{}^3}{Gm_1}}$
69. 1.71×10^{7} m
71. 0.967 m/s
75. **(a)** No; m drops out of the expression for v. **(b)** 7.8 km/s **(c)** 1.5 h
77. **(a)** $\sqrt{\dfrac{GM_E}{r}}$ **(c)** Yes
79. $2\pi\sqrt{\dfrac{\sqrt{3}R^3}{GM}}$
81. $-\dfrac{(1.17 \times 10^{9}\text{ N}\cdot\text{m}^2)x}{[x^2 + (1500\text{ m})^2]^{3/2}}$
83. 8.3×10^{10} m
85. **(a)** increase **(b)** 15.8 km/s **(c)** stay the same

Chapter 13

1. 13 s; 0.075 Hz
3. 0.45 s
5. 0.81 s; 1.2 Hz
7. **(a)** 0.024 s; 42 Hz **(b)** 1800 rpm
9. **(a)** 1.5 Hz **(b)** 0.34 s
11. **(a)** $x = (3.50\text{ nm})\cos(\omega t)$ where $\omega = 4.00\pi \times 10^{14}$ rad/s **(b)** sine
13. **(a)** 0.88 s **(b)** -1.4 cm
15. T/3
17. **(a)** $v_{max}{}^2/a_{max}$ **(b)** $2\pi v_{max}/a_{max}$
19. **(a)** 28 m **(b)** 42 s
21. **(a)** 0.0495 g **(b)** 0.128 m/s **(c)** minimum
23. $12g$
25. **(a)** 1.1 km/s^2 **(b)** 6.2 m/s
27. **(a)** The rider must begin hanging on when a_{max} (at the top of the cycle) equals g. **(b)** 0.14 m
29. 29 N/m
31. 0.47 kg
33. **(a)** 1.06×10^{3} kg **(b)** 932 kg
35. 7.68 cm
37. **(a)** more **(b)** $\sqrt{2}T$
39. 1.37 m/s
41. 1.2 J
43. **(a)** 0.26 m/s **(b)** 2.8 cm
45. 0.32 m/s
47. **(a)** 897 m/s **(b)** 0.0687 s
49. 9.6 m/s^2
51. 24.8 cm
53. $\pi\sqrt{\dfrac{8R}{g}}$
55. **(a)** greater **(b)** 2/3 m
57. 1.2 s; 0.91 m/s
59. 0.53 m/s
61. 4×10^{14} m/s^2; $(4 \times 10^{13})g$
63. 5.52 N/m
65. **(a)** 0.384 Hz **(b)** 0.125 m/s **(c)** 5.20 m
67. **(a)** less **(b)** 0.240 km/s **(c)** 0.433 s
69. $T/6$
71. 3
73. 1.1×10^{3} kg/s
75. **(a)** 1.0 s, 3.0 s, 5.0 s **(b)** 4.7 N **(c)** 0, 2.0 s, 4.0 s, 6.0 s **(d)** -3.3 N
77. **(a)** 15 N/m **(b)** 4.8 kg
79. **(a)** less than **(b)** $\pi\sqrt{\ell/g} + \pi\sqrt{L/g}$ **(c)** 1.5 s
83. **(a)** the pencil begines to rattle when the maximum acceleration of the speaker exceeds the acceleration due to gravity **(b)** $\dfrac{1}{2\pi}\sqrt{\dfrac{g}{A}}$
85. **(a)** $1/\sqrt{2}$ **(b)** 5.90 cm **(c)** 0.0702 s

Chapter 14

1. **(a)** 52 cm **(b)** 5.5 cm
3. 1.2 m
5. **(a)** 61 cm **(b)** 1.1 m **(c)** The answer in part (a) is unchanged; the answer in part (b) is halved.
7. 0.51 m/s; 67 Hz
9. 0.19 s
11. **(a)** 0.70 N **(b)** larger **(c)** 0.78 N
13. $v \propto \sqrt{\dfrac{T}{R^2\rho}}$
15. $y = (0.12\text{ m}) \times \cos\left(\dfrac{2\pi}{2.8\text{ m}}x + \dfrac{2\pi}{0.21\text{ s}}t + \phi\right)$
17. **(a)**

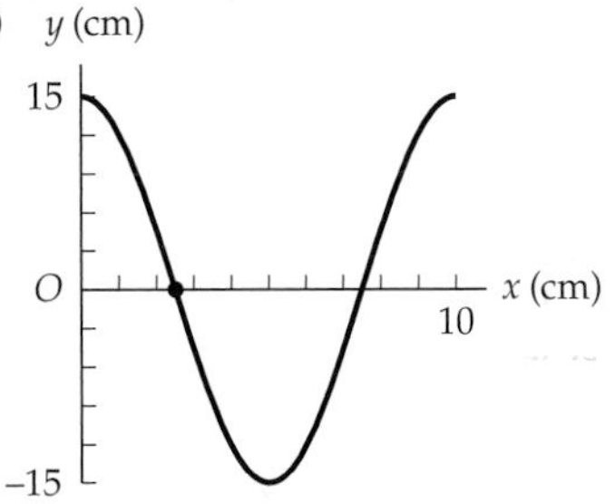

(b)

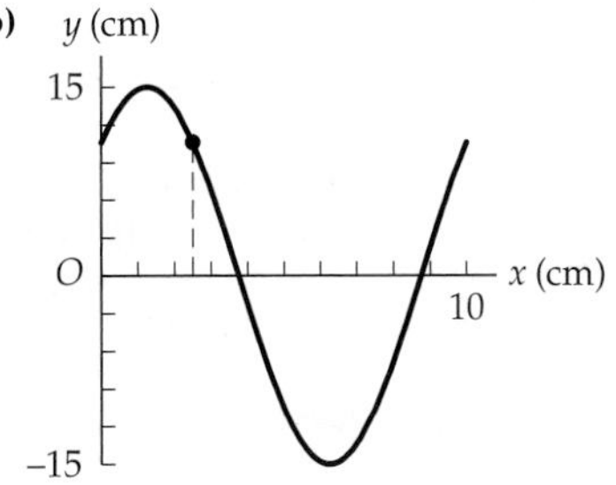

(c)

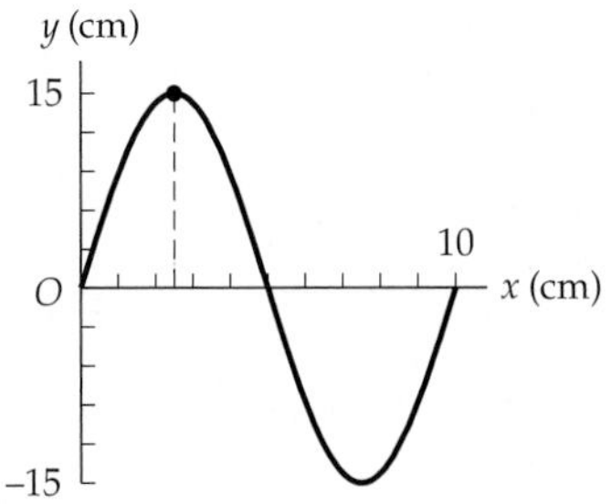

(d) 6.0 s
19. 317 m
21. 27.50 Hz, 12.5 m; 4.187 kHz, 8.19 cm
23. **(a)** 11 m **(b)** less than
25. 1.6 mW/m^2
27. 67.6 W/m^2
29. **(a)** 104 dB **(b)** 99.6 dB **(c)** 2.0×10^{6} m
31. **(a)** 69.5 dB **(b)** less than
33. 21.7 m
35. 1.50×10^{2} Hz
37. **(a)** 35.3 kHz **(b)** higher **(c)** 35.4 kHz
39. 535 Hz
41. 51 m/s
43. **(a)** 0.33 kHz **(b)** Bicyclist A speeding up.
45. 11.2 m/s
47.

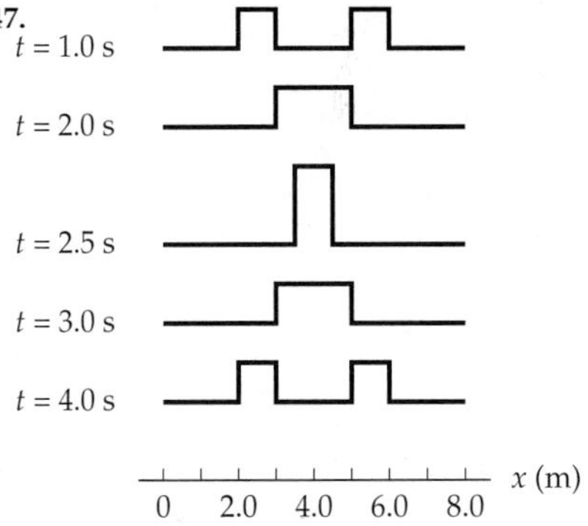

49.

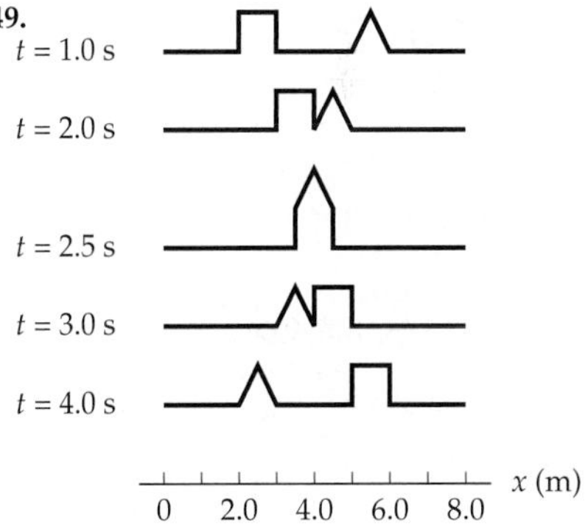

51. 1.2 kHz
53. 0.335 m
55. 0.31 kHz
57. 49 Hz
59. **(a)** 3.6 kHz, 9.6 cm **(b)** 11 kHz, 3.2 cm **(c)** greater
61. **(a)** 7.60 Hz **(b)** 15.2 Hz **(c)** increase.
63. **(a)** 93.5 Hz **(b)** 31.2 Hz
65. **(a)** 88 Hz **(b)** 1.9 m
67. 264 Hz or 258 Hz
69. **(a)** 13.8 Hz **(b)** increase **(c)** 18.3 Hz
71. 1.21 m
73. 2.9 km
75. 6.72×10^{6} m
77. 4
79. 9.0 km
81. **(a)** increase **(b)** 618 Hz

83. 32 m/s
85. 6.73 m/s if approaching, 7.00 m/s if receding
87. 0.13 km
89. **(a)** 0.41 N/m **(b)** stay the same **(c)** 0.75 s
91. **(a)** The length determines its resonant frequency. **(b)** 0.21 kHz **(c)** 126 dB
93. 6.28 Hz
95. **(a)** 14.75 m **(b)** 3.30 m
97. $v = \sqrt{gd}$
99. **(a)** 257 Hz, 255 Hz **(b)** 2 Hz **(c)** 0.670 m **(d)** 2 times per second
101. **(a)** 617 Hz **(b)** 583 Hz
103. **(a)** increase **(b)** 446 Hz

Chapter 15

1. 10^3 N
3. The ring is not solid gold.
5. 1.05×10^4 kg/m^3; silver
7. **(a)** 10^{-1} Pa **(b)** 10^{-6} atm
9. 4.3
11. 615 N
13. 2.17×10^6 Pa
15. **(a)** 97.8 kPa **(b)** 9.97 m
17. 1.1 N
19. **(a)** 995 m **(b)** greater
21. **(a)** greater **(b)** 46 Pa
23. 0.16 cm
25. **(a)** 0.770 m **(b)** Increased; the gauge pressure in a fluid is proportional to its density.
27. 7.2 kN
29. 5.3 kN
31. $\rho_w V_{metal}\, g$
33. **(a)** 2.34×10^{-4} m^3 **(b)** 8.70×10^3 kg/m^3
35. **(a)** 1.04×10^3 kg/m^3 **(b)** 0.0745 m^3 **(c)** 26 N
37. **(a)** 0.008 m^3 **(b)** 18 N
39. 2.1 cm
41. 39 m/s
43. 20 min
45. **(a)** 0.78 cm/s **(b)** 0.13 cm/s
47. 7.9 m/s
49. **(a)** 32 cm/s **(b)** 43 Pa
51. **(a)** 53 m/s **(b)** 19 cm
53. 1.08×10^5 Pa
55. **(a)** 1.42 kPa **(b)** 45 kN
57. **(a)** 980 kN **(b)** upward
59. **(a)** less **(b)** $\Delta P = -\frac{15}{2}\rho v^2$
61. **(a)** 25 cm^3 **(b)** reduced by a factor of 2
63. **(a)** 1.0 m/s **(b)** 0.79 kPa **(c)** reduced by a factor of 2 **(d)** reduced by a factor of 4
65. 904 m/s; $v/v_s = 2.64$
67. 95.3 kPa
69. 1.9×10^5 N
71. **(a)** compressed 0.15 m **(b)** stretched 0.063 m
73. 17 m/s
75. **(a)** 5.3 N
77. **(a)** 1.04×10^5 Pa **(b)** increases **(c)** 1.05×10^5 Pa
79. 41 km
81. **(a)** $v = \sqrt{2gd}$ **(b)** the same
83. **(a)** 9.38×10^{-3} m^3 **(b)** 9.85 kg
85. 3.2×10^5 Pa
87. 0.50 kN; 1.1×10^2 lb
89. $2\pi\sqrt{\frac{\rho_1 H}{\rho_2 g}}$
91. 0.28
95. 16 kg
97. 0.60 cm
99. increased by a factor of 2

Chapter 16

1. −128.6 °F
3. **(a)** 37.0 °C **(b)** 310.2 K
5. **(a)** 5.7×10^3 °C **(b)** 1.0×10^4 °F
7. **(a)** 79.0 kPa **(b)** 197 °C
9. 0.23 K/s
11. −6.85 °C
13. 1.6 m
15. **(a)** heated **(b)** 430 °C
17. 0.93 L
19. **(a)** aluminum **(b)** $0.0016V_0$
21. **(a)** The water will overflow. **(b)** 24 cm^3
23. 0.12 kW or 0.16 hp
25. **(a)** 1.4×10^{-3} C° **(b)** greater **(c)** 2.6×10^{-3} F°
27. 23.1 °C
29. 11 kJ
31. **(a)** 7.2×10^2 pellets **(b)** decrease **(c)** 4.4×10^2 pellets
33. **(a)** 0.18 kJ/K **(b)** 0.96 kJ/(kg·K)
35. 385 J/(kg·K), copper
37. 70.5 °C
39. 1.3 kJ/min
41. 2.3 J
43. **(a)** 0.800 m **(b)** decreases by a factor of 2
45. **(a)** greater than **(b)** 98 °C
47. 44 °C
49. **(a)** $(9/4)P$ **(b)** $9P$
51. 1 K
53. mass and/or specific heat
55. **(a)** 1.30×10^3 J, 1.29 times bigger than the specific heat **(b)** 4.19×10^6 J, 1000 times bigger than the specific heat
57. 77.9 °F
59. **(a)** increase **(b)** 1.710 mm **(c)** 1.955 s before; 1.957 s after
61. 3.5×10^2 C°
63. 0.17 C°
65. **(a)** 1.54×10^6 m **(b)** 5.51 km/s
67. **(a)** 0.21 kW **(b)** 0.42 kW
71. **(a)** decreased **(b)** −83 C°
75. 82 °C
77. **(a)** greater **(b)** 139 °C

Chapter 17

1. 0.0224 m^3
3. 316 K
5. 1.33×10^3 kg
7. 10^{-15} Pa
9. **(a)** 3.9×10^{24} atoms **(b)** 0.26 or 26%
11. 15.6 g/mol
13. 0.838
15. 1.60×10^5 Pa
17. **(a)** unknowable **(b)** false **(c)** unknowable **(d)** false **(e)** true
19. 632 m/s
21. **(a)** greater **(b)** 2.07 km/s
23. 203 K
25. $v_{238}/v_{235} = 0.996$
27. 71 kg
29. 1.1×10^6 N/m^2
31. **(a)** 0.27 cm **(b)** increased
33. **(a)** 0.55 mm **(b)** aluminum
35. -9.2×10^{-5} m^3
37. (d)
39. about 13 °C
41. **(a)** about −28 °C **(b)** increase
43. **(a)** gas **(b)** solid **(c)** 5707 kPa
45. **(a)** The liquid turns to gas. **(b)** First the liquid turns to solid, then the solid sublimates to gas.
47. 3.2×10^5 J
49. 3.1×10^5 J
51. **(a)** No. **(b)** 0 °C; 0.4 kg
53. **(a)** 27.3 s **(b)** 61.5 s **(c)** 246 s **(d)** boiling water
55. **(a)** 2.62 kJ **(b)** 30.9 kJ **(c)** 0.058 kg; 0.68 kg
57. 19.1 °C
59. 123 °C; All of the water has vaporized.
61. **(a)** 3.6 °C **(b)** no ice is present.
63. 49.1 h
65. 1.1×10^6 J/kg
67. **(a)** 0.196 m^3 **(b)** 67.5 kPa
69. **(a)** 0.040 **(b)** 0.040 **(c)** 0.046
71. **(a)** 0.16 mm **(b)** 0.29 mm
73. **(a)** 232 m/s **(b)** greater **(c)** 6.88×10^4 m^2/s^2; **(d)** 262 m/s
75. 6.6×10^8 Pa
77. **(a)** 25 W **(b)** 19 h
79. 15 s
81. **(a)** 0.0 °C **(b)** 1.8 °C

Chapter 18

1. $\Delta U = -10.8 \times 10^5$ J; $W = 6.7 \times 10^5$ J; $Q = -4.1 \times 10^5$ J
3. **(a)** 119 J **(b)** 35 J **(c)** 0
5. **(a)** −462 kJ **(b)** 110 C
7. **(a)** 4.24 MJ/mi **(b)** decrease
9. **(a)** 0 **(b)** −53 J **(c)** −150 J **(d)** 200 J **(e)** 350 J
11. 1.3×10^{-2} m^3
13. $P_f/P_i = \frac{1}{2}$
15. **(a)** 10 kJ **(b)** 10 kJ
17. **(a)** 2.6 MJ **(b)** 1100 K **(c)** 5.0 MJ
19. **(a)** 13.8 kJ **(b)** 13.8 kJ **(c)** both answers double
21. **(a)** enters **(b)** greater **(c)** 277 kJ **(d)** 115 kJ
23. **(a)** 12.3 kJ **(b)** 0 **(c)** −12.3 kJ
25. **(a)** 2.4×10^5 J **(b)** 3.9×10^2 K, 9.8×10^2 K **(c)** 3.6×10^5 J **(d)** 6.0×10^5 J
27. **(a)** 79 kJ **(b)** −53 kJ
29. **(a)** increase **(b)** 310 K
31. **(a)** constant-pressure **(b)** adiabatic **(c)** constant-pressure **(d)** adiabatic
33. **(a)** 0.95 K **(b)** 0.57 K
35. **(a)** 2.2 K **(b)** 3.7 K
37. **(a)** 0.45 kJ **(b)** equal to **(c)** 0.45 kJ
39. **(a)** 0.0618 m^3 **(b)** 0.0835 m^3
41. **(a)** 530 kJ, 477 kJ **(b)** 212 kJ **(c)** 795 kJ
43. 0.28
45. **(a)** 8.5 kJ **(b)** 6.0 kJ
47. **(a)** 1.16 GW **(b)** 1.71 GW
49. **(a)** 15 kJ **(b)** 12 kJ **(c)** decrease
51. **(a)** 382 K **(b)** decreased **(c)** 327 K
53. 4.5×10^2 K, 5.0×10^2 K

55. **(a)** 0.57 kJ **(b)** 0.24
57. **(a)** 58.0 J **(b)** 1.61 kJ
59. 0.41 kW
61. 1.5×10^5 J
63. 8.78 kJ/K
65. 9.6 W/K
67. **(a)** increasing
(b) 37 W/K
69. **(a)** 4.0 kJ
(b) 4.5 kJ
(c) 3.0 kJ
71. 0.250
73. **(a)** 365 K
(b) increase
(c) 1.52 kJ/K
(d) 555 kJ, 555 kJ
75. **(a)** 318 kJ **(b)** 795 kJ **(c)** 1113 kJ
77. **(a)** 751 kJ **(b)** 194 kJ **(c)** 945 kJ
79. 73 kJ
81. **(a)** $nR\Delta T$ **(b)** 0
(c) $\frac{3}{2}nR\Delta T$ **(d)** 0
(e) $\frac{3}{2}nR\Delta T$ **(f)** $nRT\ln\left(\frac{V_f}{V_i}\right)$ **(g)** 0
83. **(a)** 0.896
(b) 0.701 (Don't invest.)
85. **(a)** decrease
(b) −0.63 kJ/K
(c) Wherever the heat flows there will be an increase in entropy.
87. (ii) lowering the temperature of the low temperature reservoir
89. 2/5
93. **(a)** greater **(b)** 0.488 **(c)** 512 J

Chapter 19

1. -6.2×10^{-12} C
3. -1×10^6 C
5. 0.169 C
7. **(a)** 933 **(b)** 9.91×10^{-25} kg
9. -2.21×10^{-5} C
11. **(a)** 1.93 N **(b)** the same as
13. 30.1 cm
15. **(a)** 200 N towards q_3
(b) Part (a) would be cut to a ninth.
17. 4.76×10^{-10} m
19. 2.4×10^5 m above q
21. **(a)** 66° (q_1 at origin, q_4 on positive x-axis); 5.7 N
(b) The magnitude would be cut to one-fourth and the direction would be unchanged.
23. $q_1 = 12.2\ \mu$C, $q_2 = 49.8\ \mu$C
25. **(a)** 40 cm
(b) 35° (q_1 at origin, q_3 on positive x-axis)
27. **(a)** $\vec{\mathbf{F}}_{\text{net}} = 0$ **(b)** $-\frac{4\sqrt{2}kqQ}{a^2}\hat{\mathbf{y}}$
29. **(a)** 3.7×10^{-7} C **(b)** No
(c) The tension will be zero.
31. $q_1 = 9q_2$
33. $(-0.24\ \text{N})\hat{\mathbf{y}}$
35. **(a)** $(-3.3 \times 10^4\ \text{N/C})\hat{\mathbf{y}}$
(b) $(9.81\ \text{m/s}^2)\hat{\mathbf{y}}$
37. 4.0×10^{-12} C
39. **(a)** (2) **(b)** (1) $\vec{\mathbf{E}} = 0$; (2) $\vec{\mathbf{E}} = -\frac{4\sqrt{2}kq}{a^2}\hat{\mathbf{y}}$
41. **(a)** positive
(b) 5.00 μC **(c)** 5.00 μC

43.
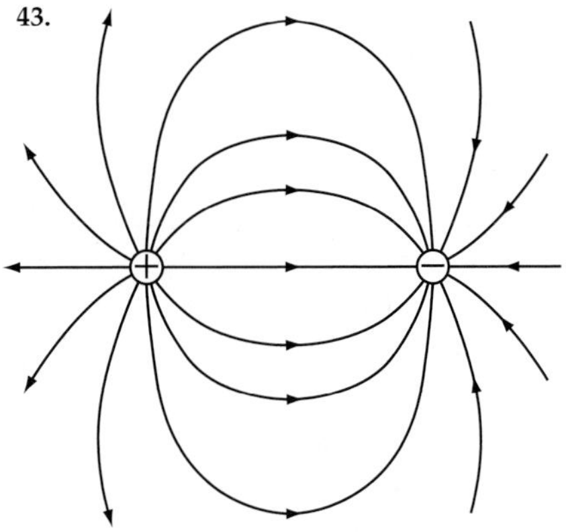

45.
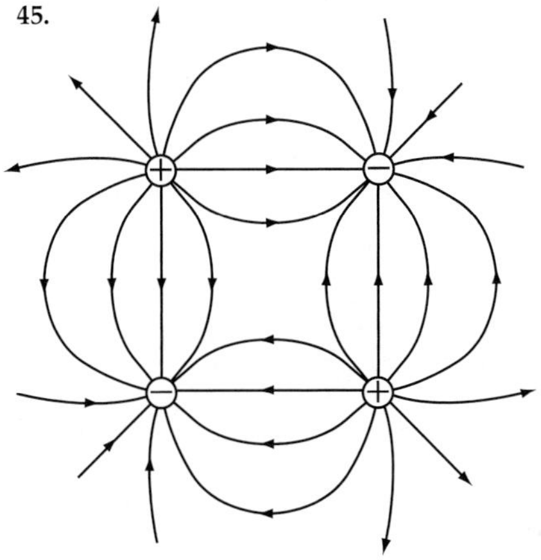

47.
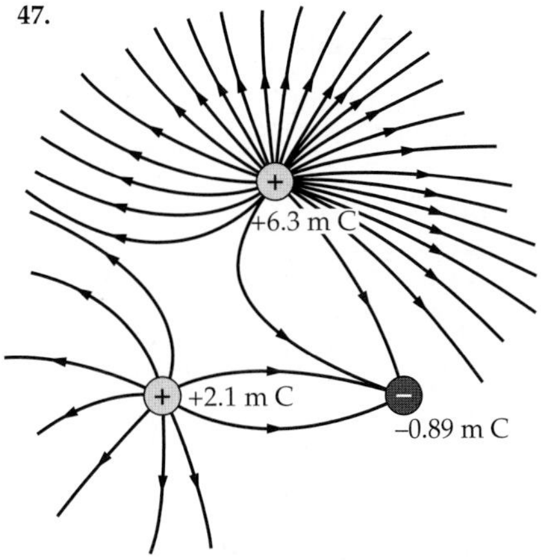

49. $0.15\ \text{kN}\cdot\text{m}^2/\text{C}$
51. **(a)** positive: the top, negative: the bottom, zero: the sides
(b) top: $525\ \text{N}\cdot\text{m}^2/\text{C}$, bottom: $-525\ \text{N}\cdot\text{m}^2/\text{C}$, ends: 0, sides: 0
53. 5.09×10^{-9} C
57. 8.4×10^{18} electrons
59. 0.44 m
61. **(a)** 2.6×10^4 N/C **(b)** 0.066 N
63. **(a)** to the left of charge 1. **(b)** −0.688 m
65. **(a)** 1.57×10^4 electrons
(b) 2.73×10^4 protons
67. **(a)** less than **(b)** 1.2 N **(c)** 2.1 N
69. -7.95×10^{-7} C
71. **(a)** 5.0×10^{-4} N **(b)** 1.5×10^{-3} N
(c) 4.8×10^{-9} C
73. **(a)** along the positive x-axis **(b)** $\frac{kq^2}{R^2}$
75. **(a)** $-9.7 \times 10^{-10}\ \text{C/m}^2$
(b) -5.0×10^5 C
(c) greater than
77. The fifth charge is located at the center of the square and has a charge of $-\frac{2\sqrt{2}+1}{4}Q$.
79. **(a)** 8.5×10^{-7} C on each
(b) 2.5×10^{-6} C, -8.4×10^{-7} C
81. **(a)** increase **(b)** decrease **(c)** 3.48 μC
(d) 12.8°
83. **(a)** 2.00×10^4 N/C **(b)** 14.2°

Chapter 20

1. **(a)** 0 **(b)** −11 J **(c)** −11 J
3. 7.0×10^5 V/m directed inward
5. −25,500 eV
7. $-3.00e$
9. **(a)** −4.16 mJ **(b)** 4.16 mJ **(c)** 0
11. **(a)** 0 **(b)** −48 V **(c)** 48 V **(d)** No
13. **(a)** 1.1×10^{-20} J **(b)** stays the same
15. **(a)** 515 V **(b)** 183 eV
17. 9.4×10^7 m/s
19. **(a)** negative x-direction
(b) 3.9 cm/s **(c)** less
21. 5.6 μC
23. 27.2 V
25. **(a)**
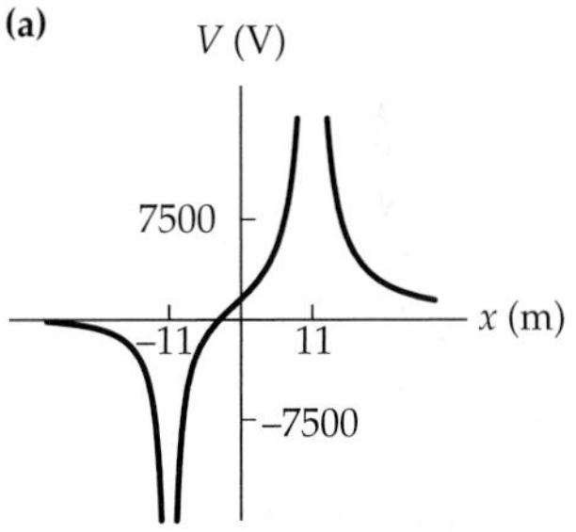

(b) closer to the negative charge
(c) $x = -3.3$ m
27. **(a)**
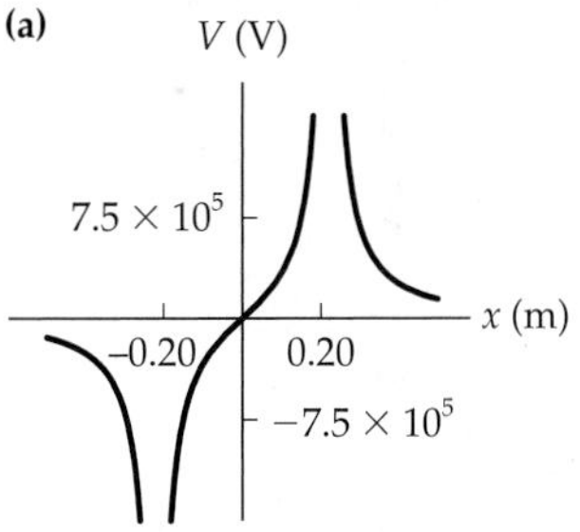

(b) 0.24 m, 0.16 m
29. **(a)** 20.0 m/s **(b)** greater **(c)** 28.3 m/s
31. **(a)** 0.86 J **(b)** less **(c)** −0.54 J
33. **(a)** 76.7 kV **(b)** 14.1 m/s
35. $-\left(4 - \sqrt{2}\right)\frac{kQ^2}{a}$
37. **(a)**
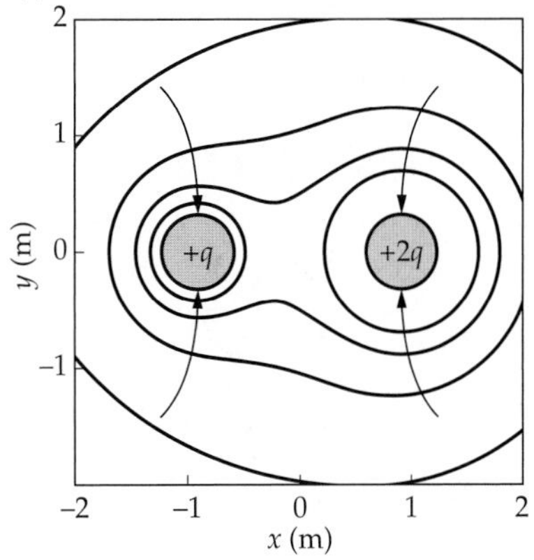

(b)

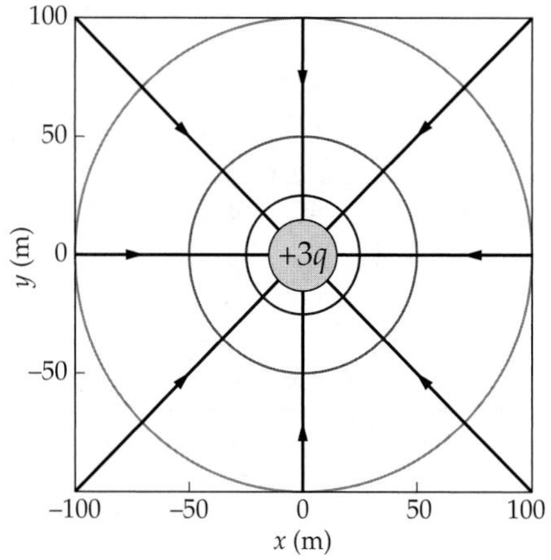

39. **(a)** 559 V/m; 243°
(b) 8.94 mm
41. 1.8 V
43. 0.16 μF
45. **(a)** 19 kV
(b) decrease
(c) 9.7 kV
47. **(a)** 2.30×10^{-6} m
(b) 8.5×10^{-6} m
49. 15 kV
51. **(a)** 4.0 nF **(b)** 6.6 C
53. 32 μJ
55. **(a)** 5.8×10^{-14} J
(b) decrease
57. 11.0 J/m^3
59. **(a)** 0.28 C **(b)** 46 J
61. **(a)** −9.1 kV
(b) −9.1 kV
(c) 9.1 kV
63. The magnitude of the charge is cut in half.
65. 2.18×10^{-18} J
67. **(a)** smallest at C and greatest at A
(b)

Location	A	B	C	D
Potential (kV)	43	31	29	31

69. **(a)** −5.75 J **(b)** 19.1 m/s
71. -5.9×10^{-14} J
73. **(a)** positive
(b) -1.7×10^{-16} C
75. **(a)** end-to-end **(b)** 5.3×10^{-9} C
77. **(a)** increases **(b)** 0.071 mm
79. **(a)** 1.3×10^7 N/C into the cell
(b) 96 mV; the outer wall
81. 3.58×10^{-14} m
83. **(a)**

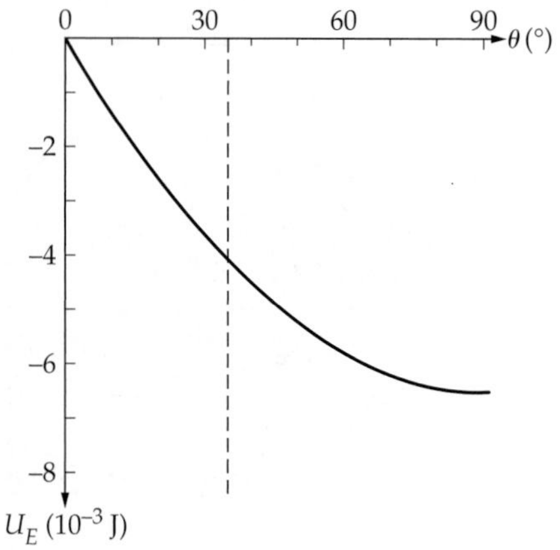

(b)

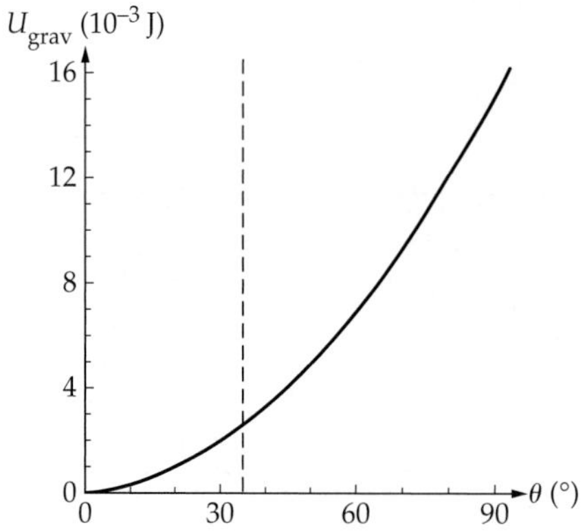

(c)

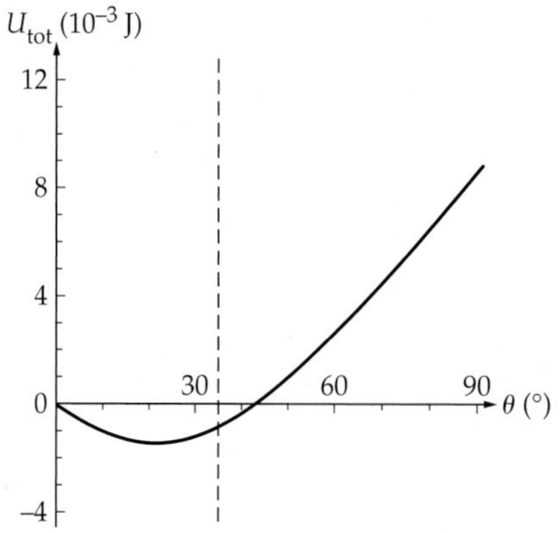

85. $V(\theta) = (22\text{ kV})\left(1 - \dfrac{1}{\sqrt{5 + 4\cos\theta}}\right)$
87. 6.92 cm; 1.19 nC
89. **(a)** 1.50 m **(b)** −1.50 m

Chapter 21

1. 3600 C
3. 9.4×10^{19} electrons/s
5. 62,500 electrons/s
7. **(a)** 1.5 kC **(b)** 8.6 y
9. 51 Ω
11. 0.68 kΩ
13. 0.18 km
15. 7.1×10^{-8} Ω·m
17. **(a)** 10^2 Ω **(b)** 1 V
19. 58 A
21. 0.58 kW
23. 7.2 cents/kWh
25. The 155-minute reserve capacity
27. 6 resistors
29. **(a)** 3.1 W **(b)** 1.5 kW
31. **(a)** 53 mA
(b) $V_{42} = 2.2$ V, $V_{17} = 0.91$ V, $V_{110} = 5.9$ V
33. **(a)** 29 V
(b) $I_{65} = 0.45$ A, $I_{25} = 1.2$ A, $I_{170} = 0.17$ A
35. 0.13 kV
37. 0.838 Ω
39. **(a)** $I_{3.2} = 1.1$ A, $I_{7.1} = 0.3$ A **(b)** 1.4 A
(c) 11.3 V
41. **(a)** $I_{1.5} = 6.0$ A, $I_{2.5} = 3.6$ A, $I_{6.3} = 1.1$ A, $I_{4.8} = 0.38$ A, $I_{3.3} = 0.55$ A, $I_{8.1} = 0.22$ A
(b) less
43. **(a)** 129 V **(b)** decrease
45. **(a)** stay the same **(b)** I_0
47. **(a)** decrease **(b)** 0.12 A, clockwise
49. **(a)** 11 Ω: 0.92 A; 7.5 Ω: 0.65 A; 6.2 Ω, 12 Ω: 0.27 A
(b) 11 Ω: 0.92 A; 7.5 Ω: 0.65 A; 6.2 Ω, 12 Ω: 0.27 A
51. **(a)** 3.9 Ω, 9.8 Ω: 0.72 A; 1.2 Ω: 2 A; 6.7 Ω: 1.0 A
(b) greater **(c)** 2 V
53. 21 μF
55. 8.2 μF: 0.18 mJ; 15 μF: 0.61 mJ; 22 μF: 66 μJ
57. **(a)** 5.0 μF **(b)** Their charges are the same. **(c)** 60 μC
61. 97 μC
63. **(a)** 9.75 ms **(b)** 6.68×10^{-4} C **(c)** 68.6 mA
65. 7.6 kΩ
67. **(a)** 6.7×10^{-4} s **(b)** 1.4 A **(c)** increased
69. 12 Ω: 0.90 W; 75 Ω: 1.0 W; 55 Ω: 1.4 W
71. Connect the 7.2 μF and 9.0 μF capacitors in series. Then, connect this pair in parallel with the 18 μF capacitor.
73. 0.16 mm
75. **(a)** 190 Ω **(b)** 96 Ω **(c)** 150 W
77. 0.46 kV
79. 0.986
81. 0.26 MΩ
83. **(a)** $2R$ **(b)** $R/2$
85. **(a)** 9.0 V **(b)** decreases **(c)** 8.6 V
87. **(a)** 2.4 W/m **(b)** 2.0 W/m
89. **(a)** Initially—24 Ω: 9.4 W; 13 Ω: 7.7 W; 6.5 Ω: 3.8 W. Finally—0 in all three
(b) 0.38 mC **(c)** 7.0 mJ
(d) It quadruples.
91. 44 V; 43 Ω
93. 7.50 Ω
95. **(a)** 329 Ω **(b)** 794 Ω
97. **(a)** 273 μC **(b)** decrease **(c)** 58.7 ms

Chapter 22

1. 1.5×10^9 m/s^2
3. 0
5. **(a)** 81° **(b)** 38° **(c)** 1.2°
7. 4.4×10^{-16} N
9. **(a)** particle 2 **(b)** $\dfrac{1}{4}$
11. **(a)** $(5.0 \times 10^6\text{ N/C})\hat{\mathbf{x}}$ **(b)** $(-0.2\text{ T})\hat{\mathbf{z}}$
13. 1.5 cm
15. $(-4300\text{ N/C})\hat{\mathbf{y}}$
17. 0.40 mT
19. **(a)** negatively charged **(b)** 1.5 u
21. **(a)** 4.21×10^{-7} s
(b) stay the same
(c) 4.21×10^{-7} s
23. 1.08 N
25. 63°
27. If up is in the positive y-direction, $\vec{\mathbf{B}} = (-0.34\text{ T})\hat{\mathbf{z}}$.
29. 3.2 A
31. $\tan^{-1}\left(\dfrac{ILB}{mg}\right)$
33. 0.41 N·m
35. 0.68 N·m
37. **(a)** $\vec{\mathbf{F}}_{top} = (7.0 \times 10^{-3}\text{ N})\hat{\mathbf{z}}$, $\vec{\mathbf{F}}_{bottom} = (-7.0 \times 10^{-3}\text{ N})\hat{\mathbf{z}}$, $\vec{\mathbf{F}}_{left} = (-8.8 \times 10^{-3}\text{ N})\hat{\mathbf{y}}$, $\vec{\mathbf{F}}_{right} = (8.8 \times 10^{-3}\text{ N})\hat{\mathbf{y}}$
(b) 0 **(c)** 1.2×10^{-3} N·m **(d)** $\theta = 0$

39. 2.9 cm
41. 2.6 N towards each other
43. **(a)** A **(b)** A: 13 μT, B: 21 μT
45. 31 A
47. 8.9 μT, 63° below the dashed line to the right of P
49. 92,000
51. 21 km
53. 2.5×10^6 A
55. 0.946
57. 4.5 μm
59. 2.6 kA
61. **(a)** 3×10^{-5} N, away from the wire **(b)** increases by a factor of two **(c)** increases
63. **(a)** 9.08×10^{-12} N **(b)** increase **(c)** 1.43×10^{-11} N
65. **(a)** $(-74 \text{ kV/m})\hat{\mathbf{y}}$ **(b)** 1.8 kV **(c)** top
67. $\vec{\mathbf{B}}_A = 7.8\ \mu$T out of the page; $\vec{\mathbf{B}}_B = 0$; $\vec{\mathbf{B}}_C = 7.8\ \mu$T into the page
69. **(a)** 1.3 mT **(b)** 0.29 kN/m
71. **(a)** repulsive **(b)** 5.7 μN
73. **(a)** 16.5 mT to the left **(b)** 17.6 mT to the right
75. $\cos^{-1} x$
77. 2.7×10^{-15} N
79. 1.4 mT
81. 4.6 V
83. 13 T
85. **(a)** down, $(2.0 \times 10^{-5}\ \text{T}\cdot\text{cm/A}) \times \left[\dfrac{3.0\ \text{A}}{x} + \dfrac{4.0\ \text{A}}{5.0\ \text{cm} - x}\right]$
(b) $\left(2 \times 10^{-5}\dfrac{\text{T}\cdot\text{cm}}{\text{A}}\right) \times$
$$\sqrt{\left(\frac{3.0\ \text{A}}{y} - \frac{(4.0\ \text{A})y}{(5.0\ \text{cm})^2 + y^2}\right)^2 + \left(\frac{20\ \text{A}\cdot\text{cm}}{(5.0\ \text{cm})^2 + y^2}\right)^2}$$
87. **(a)** decreased **(b)** 104 km/s
89. 0.99 A to the left

Chapter 23

1. 1.6×10^{-4} Wb
3. 17 mT
5. 1.9 Wb
7. **(a)** 15.4 A **(b)** The current would be cut to a fourth.
9. 7.8 V
11. **(a)** -0.10 kV **(b)** 0 **(c)** 0.040 kV
13. **(a)** near $t = 0.5$ s **(b)** $t = 0.1$ s, 0.3 s, 0.5 s, ... **(c)** -0.06 kV; 0; 0.06 kV
15. 7.1 V
17. 4.0×10^3
19. **(a)** clockwise **(b)** counterclockwise
21. **(a)** greater **(b)** greater
23. counterclockwise
25. top
27. A: clockwise; B: zero; C: counterclockwise
29. 47 mV
31. **(a)** 42 mN **(b)** 195 mW **(c)** 195 mW
33. **(a)** 0.81 A **(b)** 3.9 m/s
35. 33 mT
37. **(a)** horizontal **(b)** 29.3 mV
39. -1.40 V
41. **(a)** $6.9 \times 10^{-3}\ \text{m}^2$ **(b)** 0.42 V
43. -15.5 mV
45. 3.4×10^{-4} s
47. **(a)** 0.75 s **(b)** 0.37 Ω
49. 8.2 mJ
51. **(a)** $9.95 \times 10^8\ \text{J/m}^3$ **(b)** 1.50×10^{10} V/m
53. **(a)** 0.075 J **(b)** 0.14 J **(c)** decrease
55. **(a)** 56 A **(b)** 0.14 T **(c)** 7.8 kJ/m^3
57. **(a)** less than **(b)** 9.4 turns
59. 92
61. 0.36 A; 0.16 kV
63. 4.7×10^{-10} Wb
65. 4.0 mWb
67. 0.29 V
69. 0.18 kV
71. 339 V
73. 0.990 km
75. **(a)** 0.13 ms **(b)** 9.3 μJ
77. **(a)** counterclockwise **(b)** 36 μV
79. **(a)** Initially, the rod falls with the acceleration of gravity. As the rod falls, the induced current begins to flow, and the rod is acted upon by an upward magnetic force. This causes the acceleration to decrease. When the acceleration reaches zero, the rod falls with constant speed.
(b) clockwise **(c)** $\dfrac{mgR}{L^2B^2}$
81. **(a)** $\dfrac{\mathcal{E}}{R_1}$
(b) Just after the switch is closed, the current begins to increase. Back emf from the inductor prevents current from flowing in the inductor, so most flows through the light bulb and it glows brightly. As time goes by, the back emf reduces and most of the current flows through the near-zero resistance of the inductor.
(c) After the switch is opened, the current continues to flow for a finite time due to the back emf of the inductor. The current flows through the light bulb, causing it to flash and then go out.
(d) before: 0; after: $\mathcal{E}\dfrac{R_2}{R_1}$
83. **(a)** clockwise **(b)** to the left
85. **(a)** 1.8 N **(b)** 0.73 A

Chapter 24

1. 39 V
3. 61 Ω
5. **(a)** 2.20 W **(b)** 4.40 W
7. $V_{rms} = V_{max}$
9. 86.9 Hz
11. **(a)** 4.6 μA **(b)** 6.5 μA
13. **(a)** 9.0 V **(b)** -6.7 V **(c)** 6.7 V
15. **(a)** 1.16 kΩ **(b)** 0.137 μF **(c)** For $f = 2.00$ kHz, $I_{rms} = 0.860$ mA; for $f = 10.0$ kHz, $I_{rms} = 4.30$ mA.
17. $0 \le f < 60$ Hz
19. 37.8 Ω
21. 0.10 kΩ
23. **(a)** 74.6 Hz **(b)** 3.35 W
25. 0.967
27.

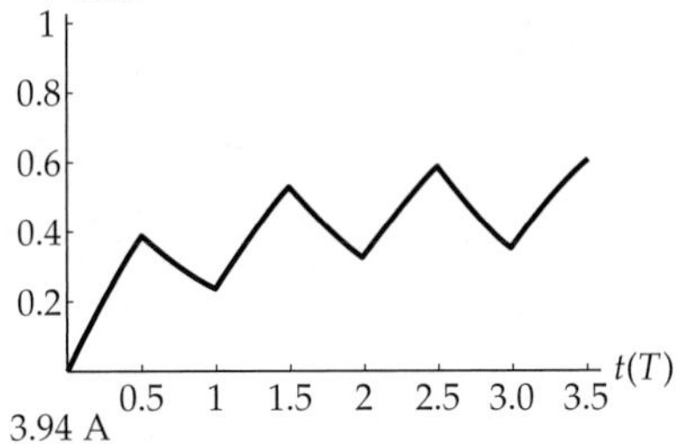

29. 3.94 A
31. 37.5 mA
33. **(a)** 0.602 **(b)** 0.106 W
35. 8.7 GHz $< f < \infty$
37. **(a)** 0.22 kHz **(b)** 0.10 kV **(c)** 63 V
39. increases; 78°
41.

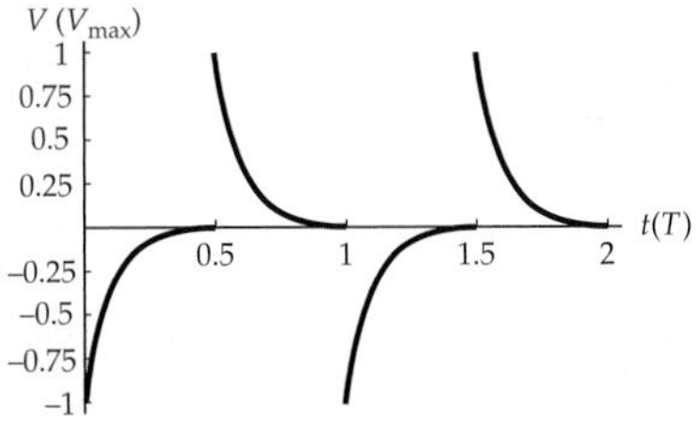

43. 1.51 kΩ
45. **(a)** 8.7 A **(b)** 4.3 A
47. **(a)** 0.962 **(b)** increases **(c)** 0.998
49. **(a)** 0.173 V **(b)** 7.84 V **(c)** 1.84 V **(d)** greater
51. 7.9 kΩ or 5.3 kΩ
53. 94 μH
55. **(a)** 49.6 Hz **(b)** 49.6 Hz
57. **(a)** 0.15 kHz **(b)** 0.15 kHz
59. **(a)** decrease **(b)** 62.5 Hz
61. 572 mA
63. 88.4 Ω
65. **(a)** 5.1 Ω **(b)** 2.1 A **(c)** Reducing R increases P_{av}.
67. 11 Hz
69. **(a)** 3.44 mA **(b)** 10.5 nF
71. **(a)** 292 Ω **(b)** greater
73. **(a)** capacitor **(b)** 0.910 μF
75. **(a)** 23 W **(b)** 0.86 W **(c)** 0.86 W
77. **(a)** 0.60 **(b)** 85 μF or 300 μF
79. **(a)** $\dfrac{V_{max}}{2}$ **(b)** $\dfrac{V_{max}}{\sqrt{2}}$
81. 81 Ω; 608 μH
83. **(a)** increased **(b)** 307 Ω **(c)** 0.354 A

Chapter 25

1. positive x-direction
3. **(a)** z-direction **(b)** x-direction **(c)** positive y-direction
5. **(a)** $+z$ **(b)** $-z$ **(c)** $-x$ **(d)** $-x$
7. **(a)** zero **(b)** $(1.22\ \text{N/C})\hat{\mathbf{x}} + (0.696\ \text{N/C})\hat{\mathbf{y}}$
9. 4.1×10^{16} m
11. 1.14
13. **(a)** away **(b)** 0.15c
15. 3.00×10^8 m/s
17. father
19. 4.945×10^{14} Hz
21. 2.82 kHz
23. **(a)** 8.238×10^{14} Hz **(b)** 8.203×10^{14} Hz
25. 1.0×10^{18} Hz
27. 1.7×10^3 waves
29. 2.40 m
31. **(a)** 9.38×10^{14} Hz to 1.07×10^{15} Hz **(b)** UV-A

33. **(a)** decreases **(b)** $\frac{3}{4}$
35. 97 s
37. 85.2 m
39. **(a)** 0.2 km **(b)** 2 m
41. 1.33×10^{-10} T
43. **(a)** 0.81 kV/m **(b)** 1.7 kW/m^2 **(c)** 0.87 kW/m^2
45. 61.4 V/m
47. **(a)** 70 mW/m^2 **(b)** 0.70 mW/m^2
49. 6.6×10^{-4}
51. 1.7×10^{-11} J
53. 2.3 mN
55. **(a)** NIF **(b)** NIF **(c)** NIF
57. **(a)** 0.2 mJ **(b)** 3.8 kW/cm^2 **(c)** 3.8×10^5
59. **(a)** 0.28 kW/m^2 **(b)** 4.6 cm
61. **(a)** 34 mJ **(b)** 2.1 μPa
63. **(a)** 1.56×10^{12} W **(b)** 4.95×10^{17} W/m^2 **(c)** 1.37×10^{10} V/m
65. 0.213
67. $\cos^{-1}\frac{1}{\sqrt{5}}$
69. **(a)** (c) is the smallest; (a) and (b) tie. **(b)** (c) is the smallest (0); (a) and (b) tie (9.25 W/m^2). The results verify part (a).
71. **(a)** *d*-glutamic acid **(b)** 0.576 mW/m^2 **(c)** 0.732 mW/m^2
73. **(a)** $0.375I_0 \cos^2(\theta - 30.0°)$ **(b)** $I_{max} = 0.375I_0$

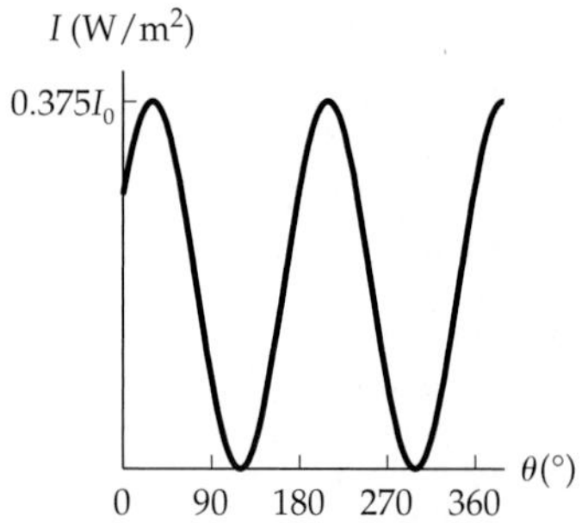

(c) 30.0° or 210.0°
75. 2.00×10^{-11} m
77. 0.40 m
79. 9.0 mm
81. **(a)** increase **(b)** 915 rev/s
83. **(a)** less **(b)** 0.19 kV/m **(c)** four; two
85. 2.10 kHz
87. 31.3 m^2
89. **(a)** 0.540 W/m^2 **(b)** 0.272 W/m^2
91. 19.7 W
93. **(b)** 17.4°
95. 3.85×10^{26} W
97. 49°
99. **(a)** 48.0° **(b)** decrease

Chapter 26

1. 38°
3. 2θ
5. 55 cm
7. **(a)** 35 cm below **(b)** 9.9° **(c)** You will still see the buckle.
9. 7.50 ft
11. 1.13 m
13. **(a)** 61 m **(b)** increase
15. 10 km
17. −7.80 cm
19. 60.0 cm in front of the mirror
21. **(a)** 63 cm in front of the mirror; 16 cm tall **(b)** inverted
23. **(a)** 40 cm behind the mirror; 10 cm tall **(b)** upright
25. 9.6 cm
27. **(a)** 4.2 cm behind the surface of the globe **(b)** 11 cm
29. **(a)** 0.11 km in front of the mirror **(b)** real **(c)** −5.5
31. **(a)** 66 cm **(b)** 17 cm
33. **(a)** upright **(b)** 3.5 ft
35. 1.0 m
37. 1.25
39.

θ_i(°)	10.0	20.0
% Error	6%	4%

41. 42°
43. 40°
45. 22.3 ns
47. 3.6 cm
49. $\sqrt{2}$
51. **(a)** 1.4 **(b)** decreased
53. **(a)** increase **(b)** 23.7°
55. 1.7
57. **(a)** $\frac{1}{3}|f|$ **(b)** upright **(c)** virtual
59. **(a)** just to the left of f **(b)** upright **(c)** virtual
61. **(a)** just to the left of Lens 2. **(b)** inverted **(c)** virtual
63. 26 cm
65. 12 cm
67. **(a)** 45 mm **(b)** −0.0091
69. **(a)** the lens with focal length f_1 **(b)** 1.20 m
71. **(a)** 56 cm **(b)** −0.070
73. **(a)** concave **(b)** −2.2 m
75. **(a)** virtual **(b)** −43 cm **(c)** concave
77. 0.33°
79. 46.2 cm from the lens
81. **(a)** It will decrease the number of reflections. **(b)** five
83. **(a)** smaller **(b)**

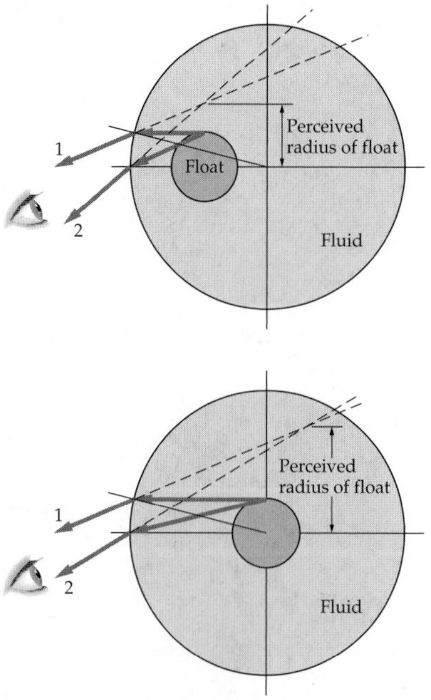

85. **(a)** 56 cm behind the mirror **(b)** upright **(c)** 43 cm
87. $\frac{1}{x}$
89. **(a)** 48.8° **(b)** Neither; it stays the same.
91. **(a)** 61.8°
93. **(a)** −0.67 **(b)** 0.45
95. −30.0 cm
97. $f = \left(\frac{m}{m-1}\right)d_0$
103. **(a)** 0.83 cm **(b)** 4.2 cm
105. **(a)** 2.5 cm **(b)** 5.0 cm to the left of the lens

Chapter 27

1. **(a)** 1.3 cm **(b)** 1.0 cm
3. 14 cm
5. A, C, B, D
7. **(a)** 2.8
(b)

f-number	Diameter (mm)
2.8	20
4	14
8	6.9
11	5.0
16	3.4

9. 8.5 m
11. $\frac{1}{1000}$ s
13. **(a)** 2 **(b)** $\frac{1}{1000}$ s
15.

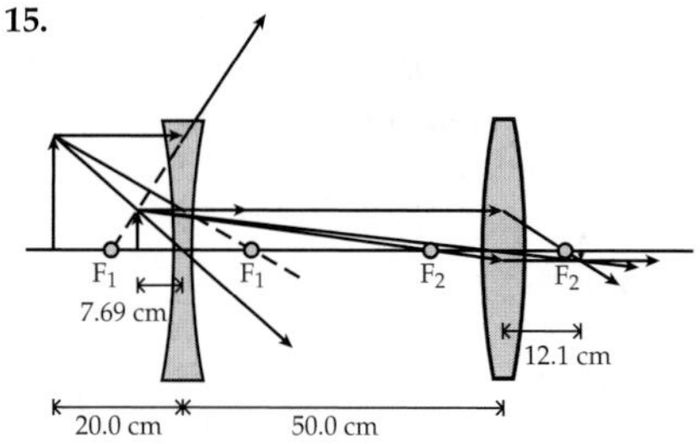

17. 0.80 m
19. 60.7 cm
21. **(a)** The final image is located 5 mm in front of the lens closest to the object. **(b)** 0.15
23. **(a)** increase **(b)** 3.5 mm
25. **(a)** 12 cm to the left of the converging lens **(b)** −0.12
27. −135 cm
29. 8.4 cm
31. 61 cm; 1.6 diopters
33. **(a)** nearsighted **(b)** diverging **(c)** −5.7 diopters
35. **(a)** 45.8 diopters **(b)** decreases
37. 24.2 cm
39. **(a)** diverging **(b)** converging **(c)** distant objects: −0.20 diopters; near objects: +2.3 diopters
41. far: 2.37 m; near: 37.7 cm
43. **(a)** 47 cm to the right of lens 2; 0.38 **(b)** 3.0 cm to the right of lens 2; −0.61 **(c)** 15.8 cm to the right of lens 2; 0.316
45. **(a)** 9.042×10^{-3} rad **(b)** 2.1 m

47. **(a)** f_1 **(b)** $M_1 = 6.0$; $M_2 = 2.9$
49. **(a)** 6.47 cm **(b)** 3.96
51. 3.72
53. −25
55. **(a)** 4.10 mm **(b)** −8.0 cm
57. 0.325 mm
59. **(a)** 0.21 m **(b)** −32
61. 1.8 m
63. **(a)** 6.00 **(b)** 35.0 cm
65. **(a)** 35.0 cm **(b)** 36.9 cm
67. **(a)** right lens **(b)** 2.5
69. 11°
71. **(a)** 18 mm **(b)** 14
73. **(a)** 41.7 diopters **(b)** increases
75. **(a)**

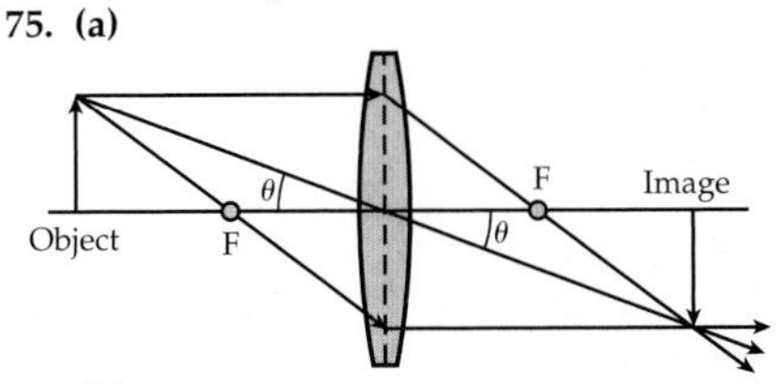

(b)

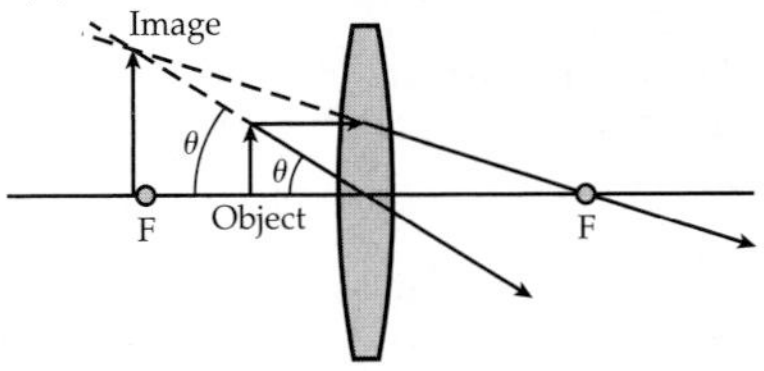

(c)

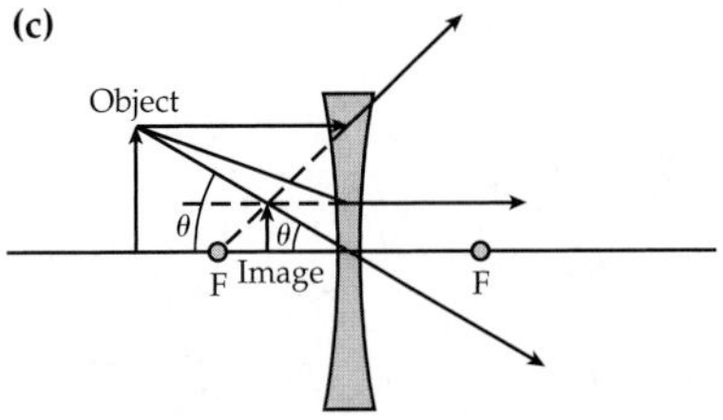

(d) A simple magnifier works by helping the eye view an object that is closer than the near point, thus making it appear larger.
77. **(a)** 2.42 cm **(b)** 2.42 cm
79. 4.6 cm
81. **(a)** positive **(b)** 4.3 diopters
83. **(a)** 73.8 cm **(b)** 42.0
85. 83 cm
89. **(a)** 16 cm in front of the lens **(b)** real **(c)** −0.80 **(d)** inverted
91. **(a)** 39.1 cm to the right of lens 2; −0.0863 **(b)** 93.0 cm to the right of lens 2; −0.600 **(c)** 27.3 cm to the left of lens 2; 0.545
93. **(a)** $x = 0.67$ m **(b)** 0.35; upright
97. **(a)** less **(b)** −351 cm
99. **(a)** greater **(b)** 2.81 diopters

Chapter 28

1. destructively
3. 134 m
5. 0.18 kHz
7. 457 Hz; 915 Hz
9. 0.68 kHz; 2.0 kHz
11. **(a)** 600 m **(b)** maximum **(c)** 200 m
13. 623 nm
15. **(a)** 415 nm **(b)** increase **(c)** 587 nm
17. 154 μm
19. 658 nm
21. 86.9 cm
23. **(a)** 2.24 cm **(b)** increased
25. 482 nm and 619 nm
27. 411 nm, 513 nm, and 685 nm
29. **(a)** destructive **(b)** 3.5 m
31. **(a)** destructive interference **(b)** constructive interference **(c)** constructive interference
33. **(a)** 102 nm **(b)** thinner
35. **(a)** 503.2 nm **(b)** 670.9 nm and 402.6 nm
37. 1.8 μm
39. 69.9 μm
41. 46.9 cm
43. 14
45. 3.20 cm
47. 2.1 m
49. 7.4 mm
51. **(a)** 1.0×10^{-5} rad **(b)** 6.5 μm
53. **(a)** 1.3×10^{-3} rad **(b)** 0.12 km
55. 0.12 nm
57. 456 nm
59. **(a)** 480 nm **(b)** greater
61. 2
63. violet, orange, violet, orange, violet, violet, orange, violet
65. **(a)** For $m > 1$, $\sin\theta = \dfrac{m\lambda}{d} > 1$, which is impossible because $\sin\theta \leq 1$. **(b)** $d < 930$ nm
69. 39 μm
71. **(a)** 93.7 nm **(b)** 93.7 nm
73. 528 nm
75. 81 μm
77. 491 nm
79. **(a)** 474 nm **(b)** greater
81. 401 nm and 515 nm
83. $\dfrac{2d}{\lambda_{\text{vacuum}}} - \dfrac{1}{2} = m$ where $m = 0, 1, 2, \ldots$
85. 6.0 μm
87. **(a)** greater **(b)** 0.63 μm
89. **(a)** greater **(b)** 15.1°

Chapter 29

1. c
3. 0.78c
5. 0.99986c
7. 748 m
9. 0.80c
11. 0.20 rad/s
13. **(a)** less **(b)** 55 beats/min
15. 1.01 ms
17. 4:32 P.M.
19. 0.80c
21. 0.761c
23. **(a)** 0.24 m × 0.50 m × 0.50 m **(b)** 0.059 m^3
25. 4.1 ly
27. **(a)** 120 m **(b)** 150 m **(c)** 0.60c
29. **(a)** 0.411c **(b)** 8.77 m
31. 0.92c
33. 1.00c
35. 0.38c
37. 0.53c
39. **(a)** 0.50c **(b)** more **(c)** 0.73c
41. **(a)** 8.8×10^{14} kg·m/s **(b)** 1.2×10^{15} kg·m/s
43. 0.98c
45. 0.38 m/s
47. **(a)** 0.141c **(b)** 0.312c
49. 0.19 nJ
51. 4.4×10^{-16} kg
53. 1.02 MeV
55. 0.88c
57. 0.35 kg
59. **(a)** 0.75% **(b)** 69%
61. **(a)** 43.5 J **(b)** greater **(c)** 7.00×10^{8} J
63. 2.4 kg
65. 2.96×10^{6} km
67. **(a)** $v \approx 1.00c$ **(b)** 3.6×10^{9} eV $\ll 1.0 \times 10^{20}$ eV
69. 1.20×10^{-18} kg·m/s
71. **(a)** 0.643c **(b)** 2.29×10^{-22} kg·m/s **(c)** -2.29×10^{-22} kg·m/s **(d)** 9.85 km/s
73. **(a)** 9.37×10^{7} m/s **(b)** less **(c)** 9.06×10^{7} m/s
75. **(a)** 16 m **(b)** 0.11 km
77. **(a)** 1.6 m **(b)** 76°
79. 140 MeV
81. **(a)** greater **(b)** 0.393c
83. 199 pulses/s
87. 0.55c
89. **(a)** −0.956c **(b)** −0.371c

Chapter 30

1. 3100 K
3. **(a)** 1.6×10^{11} Hz **(b)** 1.9 mm
5. **(a)** 2 **(b)** 1.035
7. **(a)** halogen **(b)** 1.2 **(c)** halogen
9. **(a)** 4.792×10^{13} Hz **(b)** 3.18×10^{-20} J
11. 4.3×10^{32} photons/s
13. 4.4×10^{18} photons/s
15. 2.08×10^{15} Hz
17. **(a)** 2.5×10^{20} photons/s **(b)** 1.7×10^{12} photons/s
19. **(a)** 1.09×10^{15} Hz; 275 nm **(b)** ultraviolet
21. **(a)** red **(b)** blue **(c)** red: 4.9×10^{20} photons/s; blue: 5.8×10^{19} photons/s
23. **(a)** aluminum **(b)** Al: 1.03×10^{15} Hz; Ca: 6.93×10^{14} Hz
25. **(a)** cadmium **(b)** Zn: 0.19 eV; Cd: 0.30 eV
27. **(a)** 369 eV **(b)** 4.00×10^{14} Hz $\leq f \leq 1.03 \times 10^{15}$ Hz
29. 6.377×10^{-34} J·s
31. 13 cm; much larger
33. 0.52 μm
35. 3.25 m/s
37. **(a)** blue **(b)** 8.8×10^{-28} kg · m/s; 1.7×10^{-27} kg · m/s
39. **(a)** 1.59×10^{16} photons/s **(b)** -1.05×10^{-27} kg·m/s **(c)** 1.67×10^{-11} N
41. 14 keV
43. 112°
45. **(a)** The change in wavelength is the same for both photons. **(b)** X-ray **(c)** visible: 9.3×10^{-4}%; X-ray: 16%
47. **(a)** 0.315 nm **(b)** incident: 3.94 keV; scattered: 3.88 keV **(c)** 60 eV
49. 41.4 pm; 4.80 fJ
51. 13.0 km/s
53. 2.1×10^{-36} m
55. **(a)** 1.58 km/s **(b)** 62.4°
57. **(a)** electron **(b)** 1836
59. $\dfrac{h}{\sqrt{2mqV}}$
61. 1.4×10^{-20} kg·m/s

63. 1.1×10^{-26} J
65. 1.8×10^{-25} J
67. **(a)** 7.0×10^{-25} kg·m/s **(b)** 1.7 eV
69. 62 nm
71. 3.65×10^{-17} W
73. **(a)** 1.6×10^{15} photons/s
(b) 3.3×10^{-19} N
75. 1×10^{19} photons
77. 0.58 nm
79. 9200 K; No
81. **(a)** decreased **(b)** −10.2 eV
83. 4.31 km/s
85. **(a)** 0.0732 nm **(b)** 3.44 nm
87. **(a)** stays the same **(b)** 1.25 eV
89. **(a)** less **(b)** 54°

Chapter 31

1. 8.4×10^{-16}
3. 5.6 pJ
5. 371.2 nm
7. 121.5 nm; 102.6 nm; 97.23 nm
9. **(a)** 121.5 nm **(b)** 820.4 nm
11. 3.64×10^{-3}
13. 0.850 eV
15. **(a)** 6.62×10^{-25} kg·m/s
(b) 3.17×10^{-34} J·s
17. 7
19. **(a)** 0.967 eV **(b)** less than
(c) 0.266 eV
21. **(a)** 2.56×10^{-13} m **(b)** less than
(c) 3.17 nm
23. **(a)** 5.697 nm **(b)** 218 eV
25. **(a)** 4 **(b)** decrease **(c)** 0.544 eV
27. 44.2 km/s
29. 0.332 nm
31. 1.32 nm
33. 0, 1, 2, 3, 4
35. **(a)** 2 **(b)** none **(c)** 5 **(d)** none
37. **(a)** 75 **(b)** 76 **(c)** −2.35 meV
39. **(a)** greater than **(b)** $\sqrt{2}$
41.

	n	ℓ	m_ℓ	m_s
$1s^1$	1	0	0	$-\frac{1}{2}$
$1s^2$	1	0	0	$\frac{1}{2}$
$2s^1$	2	0	0	$-\frac{1}{2}$
$2s^2$	2	0	0	$\frac{1}{2}$
$2p^1$	2	1	−1	$-\frac{1}{2}$
$2p^2$	2	1	−1	$\frac{1}{2}$
$2p^3$	2	1	0	$-\frac{1}{2}$
$2p^4$	2	1	0	$\frac{1}{2}$
$2p^5$	2	1	1	$-\frac{1}{2}$
$2p^6$	2	1	1	$\frac{1}{2}$

43.

n	ℓ	m_ℓ	m_s
3	0	0	$-\frac{1}{2}$
3	0	0	$\frac{1}{2}$

45.

	n	ℓ	m_ℓ	m_s
$1s^1$	1	0	0	$-\frac{1}{2}$
$1s^2$	1	0	0	$\frac{1}{2}$
$2s^1$	2	0	0	$-\frac{1}{2}$
$2s^2$	2	0	0	$\frac{1}{2}$
$2p^1$	2	1	−1	$-\frac{1}{2}$
$2p^2$	2	1	−1	$\frac{1}{2}$
$2p^3$	2	1	0	$-\frac{1}{2}$
$2p^4$	2	1	0	$\frac{1}{2}$
$2p^5$	2	1	1	$-\frac{1}{2}$
$2p^6$	2	1	1	$\frac{1}{2}$
$3s^1$	3	0	0	$-\frac{1}{2}$
$3s^2$	3	0	0	$\frac{1}{2}$

47. **(a)** 8 **(b)** 18 **(c)** 32
49.

	n	ℓ	m_ℓ	m_s
$5d^1$	5	2	−2	$-\frac{1}{2}$
$5d^2$	5	2	−2	$\frac{1}{2}$
$5d^3$	5	2	−1	$-\frac{1}{2}$
$5d^4$	5	2	−1	$\frac{1}{2}$
$5d^5$	5	2	0	$-\frac{1}{2}$
$5d^6$	5	2	0	$\frac{1}{2}$
$5d^7$	5	2	1	$-\frac{1}{2}$
$5d^8$	5	2	1	$\frac{1}{2}$
$5d^9$	5	2	2	$-\frac{1}{2}$
$5d^{10}$	5	2	2	$\frac{1}{2}$

51. 66.9 keV
53. 78
55. **(a)** 6.44 eV **(b)** 1.53×10^5 photons
57. 3.28×10^{15} Hz
59. **(a)** 2.55 eV **(b)** 5.55×10^{-28} J
(c) the same
61. **(a)** 2.74×10^{15} W/m^2 **(b)** 7.09 MeV
63. **(a)** 0.85 eV **(b)** −1.70 eV
65. **(a)** 7 **(b)** −74.0 eV
67. 0.328 nm
69. **(a)** 389 **(b)** 2.67 m **(c)** less than
73. **(a)** less than **(b)** 0.665 nm

Chapter 32

1. **(a)** $Z = 92$; $N = 146$; $A = 238$
(b) $Z = 94$; $N = 145$; $A = 239$
(c) $Z = 60$; $N = 84$; $A = 144$
3. **(a)** 7.0 fm **(b)** 4.7 fm
5. **(a)** 2.3×10^{17} kg/m^3
(b) same as
(c) 2.3×10^{17} kg/m^3
7. **(a)** 6.0×10^6 m/s
(b) 0.30 pm
(c) less than
9. **(a)** 3.7 fm
(b) 240
11. **(a)** more than
(b) 7.4 fm
(c) 5.9 fm
13. ${}^{4}_{2}\mathrm{He}$
15. ${}^{14}_{6}\mathrm{C}$
17. **(a)** ${}^{212}_{84}\mathrm{Po} \longrightarrow {}^{208}_{82}\mathrm{Pb} + {}^{4}_{2}\mathrm{He}$; 8.95 MeV
(b) ${}^{239}_{94}\mathrm{Pu} \longrightarrow {}^{235}_{92}\mathrm{U} + {}^{4}_{2}\mathrm{He}$; 5.244 MeV
19. **(a)** ${}^{18}_{9}\mathrm{F} \longrightarrow {}^{18}_{8}\mathrm{O} + e^+ + v$; 1.146 MeV
(b) ${}^{22}_{11}\mathrm{Na} \longrightarrow {}^{22}_{10}\mathrm{Ne} + e^+ + v$; 2.331 MeV
21. **(a)** ${}^{66}_{29}\mathrm{Cu}$ **(b)** 0.2 MeV
23. 78 s
25. 27.0 min
27. **(a)** 0.115 h^{-1}
(b) 1.7×10^9 nuclei
29. 1.97×10^4 y
31. 1.44×10^4 y
33. 3.90×10^3 y
35. 1559 MeV
37. **(a)** 8.790 MeV/nucleon
(b) 7.570 MeV/nucleon
39. 15.66 MeV
41. 3
43. ${}^{1}_{0}\mathrm{n} + {}^{235}_{92}\mathrm{U} \longrightarrow {}^{88}_{38}\mathrm{Sr} + {}^{136}_{54}\mathrm{Xe} + 12\,{}^{1}_{0}\mathrm{n}$;
126.5 MeV
45. 1.2×10^6 kg
47. 4.033 MeV
49. 5.494 MeV
51. **(a)** 4.33×10^9 kg/s
(b) 0.0307%
53. 1.0×10^3 rad
55. **(a)** 0.38 J **(b)** 0.54 mK
57. **(a)** 6.87×10^{11} electrons
(b) 0.0775 J
(c) 93.0 rem
59. **(a)** ${}^{206}_{80}\mathrm{Hg}$ **(b)** ${}^{239}_{93}\mathrm{Np}$ **(c)** ${}^{11}_{5}\mathrm{B}$
61. 250 rem
63. **(a)** 1.32×10^3 Bq
(b) reduced by a factor of 2
65. **(a)** 57 N **(b)** 5.4 MeV **(c)** 5.4 MeV
67. **(a)** type B **(b)** 1.00 d
69. 10 km
71. **(a)** 2.0×10^{18} reactions **(b)** 0.78 mg
73. 33.5×10^6 rad
75. 7.4 mJ
77. 18.5 mK

Photo Credits

FM—p. ii Cosmo Condina/Photographers Choice/Getty Images, Inc.

Chapter 1—p. 1 Getty Images Inc. **p. 2 (top)** Oliver Meckes/Ottawa/Photo Researchers, Inc. **p. 2 (bottom)**NASA **p. 3** Courtesy of BIPM (Bureau International des Poids et Mesures/International Bureau of Weights and Measures, www.bipm.org) **p. 4 (top)** U.S. Department of Commerce **p. 4 (middle)** © 2000 by Sidney Harris **p. 6** Jeff Greenberg/PhotoEdit **p. 7** AP Wide World Photos **p. 8** David Frazier Photolibrary/Photo Researchers, Inc. **p. 9** CNRI/Science Photo Library/Photo Researchers, Inc. **p. 10** Amaldi Archives, Dipartimento di Fisica, Universita "La Sapienza," Rome, courtesy AIP Emilio Segre Visual Archives **p. 15 (top)** Michael Mancuso/Omni-Photo Communications, Inc. **p. 15 (bottom)** Matt Moran/USDA Forest Service **p. 16** Young-Jin Son, Drexel University College of Medicine

Chapter 2—p. 17 AP Wide World Photos **p. 24** Swerve/Alamy Images **p. 26** NASA Headquarters **p. 28** AP Wide World Photos **p. 38 (top)** James Sugar/Stockphoto.com/Black Star **p. 38 (bottom)** Alaska Stock **p. 41** CORBIS-NY **p. 50 (left)** Worlds of Fun Photo by Dan Feicht **p. 50 (right)** Stephen Dalton/Photo Researchers, Inc. **p. 51** Sidney Harris **p. 52** Frank Labua/Pearson Education/PH College

Chapter 3—p. 55 Mark Downey/Getty Images Inc.-Photodisc **p. 56** Jeff Greenberg/Photo Researchers, Inc. **p. 61** EyeWire Collection/Getty Images-Photodisc **p. 66** George Whiteley/Photo Researchers, Inc. **p. 68** AP Wide World Photos **p. 74** US Department of Defense **p. 75** © National Maritime Museum Picture Library, London, England

Chapter 4—p. 79 Bjorn Larsen/Alamy Images **p. 84 (left)** Loren M. Winters, North Carolina School of Science and Mathematics **p. 84 (right)** NASA Headquarters **p. 87 (left)** Krafft/Photo Researchers, Inc. **p. 87 (right)** Jerry Driendl/Getty Images Inc.-Taxi **p. 90 (left)** Michael Quinton/MINDEN PICTURES **p. 90 (middle)** John Terence Turner/Alamy Images **p. 90 (right)** Galen Rowell/Mountain Light/Explorer/Photo Researchers, Inc. **p. 94** Courtesy of the Archives, California Institute of Technology **p. 96** Stephen Dalton/Photo Researchers, Inc. **p. 100** Gregory G. Dimijian/Photo Researchers, Inc. **p. 103** NASA Headquarters

Chapter 5—p. 107 © Wally McNamee/CORBIS **p. 109** James S. Walker **p. 111 (middle)** David Lawrence/Prime Images, Inc. **p. 111 (bottom)** Bruce Bennett Studios, Inc./Getty Images **p. 116** William J. Hughes Technical Center, Federal Aviation Administration **p. 124** © Pete Saloutos/CORBIS **P. 128** NASA/Johnson Space Center **p. 134** © Norbert Wu/DRK PHOTO **p. 139** AP Wide World Photos **p. 140** Mark Moffett/Minden Pictures

Chapter 6—p. 141 Corbis RF **p. 143 (top)** Ian Aitken © Rough Guides **p. 143 (bottom)** Ernest Braun/Getty Images Inc.-Stone Allstock **p. 147** Jack Dykinga/Getty Images Inc.-Stone Allstock **p. 149 (left)** Thomas Florian/Index Stock Imagery, Inc. **p. 149 (right)** W. Kenneth Hamblin **p. 153 (top)** oote boe/Alamy Images **p. 153 (bottom)** Science Museum/The Image Works **p. 157** © Galen Rowell/CORBIS **p. 163** © Doug Wilson/CORBIS **p. 164 (left)** AP Wide World Photos **p. 164 (middle)** Official U.S. Navy photo by Mark Meyer **p. 164 (right)** Bombardier Transportation Sweden **p. 166** Chris Priest/Science Photo Library/Photo Researchers, Inc. **p. 169** Photofest **p. 174** Color Box/Getty Images Inc.-Taxi **p. 175** The British Library/Heritage Image Partnership/The Image Works

Chapter 7—p. 179 Rich Pilling/MLB Photos/Getty Images **p. 182 (left)** Robert E. Daemmrich **p. 182 (right)** © Mimmo Jodice/Corbis **p. 192** Dr. T. J. McMaster/Wellcome Trust Medical Photographic Library **p. 195** Philippe Achache/Getty Images Inc.-Liaison **p. 203** John E. Bortle

Chapter 8—p. 204 Photolibrary.Com **p. 209 (left)** Grant Heilman/Grant Heilman Photography, Inc. **p. 209 (right)** Jay Brousseau **p. 210** Philip H. Coblentz/World Travel Images, Inc. **p. 212** Barry Bland/Alamy Images **p. 215 (top)** © Bill Nation/Corbis Sygma **p. 215 (bottom)** Salt River Project, Phoenix, Arizona **p. 220** Jeff Greenberg/Omni-Photo Communications, Inc. **p. 227** Federal Motor Carrier Safety Administration, Federal Highway Administration, Colorado Division **p. 231** Stephen Derr/Getty Images Inc.-Image Bank **p. 232** Lester Lefkowitz/Getty Images Inc.-Stone Allstock

Chapter 9—p. 240 INSADCO Photography/Alamy Images **p. 246 (top)** © Harold & Esther Edgerton Foundation, 2003, courtesy of Palm Press, Inc. **p. 246 (bottom)** © Wally McNamee/CORBIS **p. 247** J. Scott Altenbach, University of New Mexico **p. 250** NASA Headquarters **p. 252** NASA Headquarters **p. 253 (left)** J. McIsaac/Bruce Bennett Studios, Inc./Getty Images **p. 253 (right)** © Franck Seguin/CORBIS **p. 261** Richard Megna/Fundamental Photographs, NYC **p. 264** Alexander Calder (1898–1976). *Myxomatose.* 1953. Sheet metal, rod, wire, and paint. 256.5 × 408.9 cm (101 × 161 in) © Copyright ARS, NY. Copyright Art Resource, NY. Private Collection, New York, NY, USA **p. 268** Loren M. Winters **p. 270 (top)** NASA Headquarters **p. 270 (bottom)** Fred Bavendam/Peter Arnold, Inc. **p. 274** © Jose Azel/Aurora/PNI **p. 278** Michael Rinaldi/U.S. Navy News Photo **p. 280** U.S. Geological Survey/U.S. Department of the Interior

Chapter 10—p. 281 James Havey/Creative Eye/MIRA.com **p. 282 (left)** NASA Headquarters **p. 282 (middle)** Novastock/PhotoEdit **p. 282 (right)** Barry Dowsett/Photo Researchers, Inc. **p. 284** Fred Espenak/Photo Researchers, Inc. **p. 291 (top left)** Loren M. Winters **p. 291 (top right)** Karl Weatherly/Getty Images Inc.-Photodisc **p. 291 (bottom left)** Novosti/Sovfoto/Eastfoto **p. 291 (bottom right)** Bill Gallery/Stock Boston **p. 295** Jerry Marshall **p. 299** Mehau Kulyk/Science Photo Library/Photo Researchers, Inc. **p. 308** Jeff Hester/Arizona State University/NASA Headquarters **p. 309 (left)** A. Ramey/Stock Boston **p. 309 (right)** Arthur Tilley/Getty Images Inc.-Taxi **p. 312 (left)** NASA/Ames Research Center/Ligo Project **p. 312 (right)** © Loren M. Winters, North Carolina School of Science and Mathematics **p. 313** Yomega

Chapter 11—p. 315 AP Wide World Photos **p. 316** Kevin M. Law/Alamy Images **p. 318** Benelux Press B.V./Index Stock Imagery, Inc. **p. 322** Richard Megna/Fundamental Photographs, NYC **p. 328** Getty Images Inc.–Photodisc **p. 332 (top)** Murray Close/Paramount Pictures/Photofest **p. 332 (bottom left)** Dana White/PhotoEdit **p. 332 (bottom right)** Allan Price/Omni Photo Communications, Inc. **p. 338** Glyn Kirk/Getty Images Inc.-Stone Allstock **p. 340 (top left)** Hasler, Pierce, Palaniappan, Manyin/NASA Goddard Laboratory for Atmospheres **p. 340 (top right)** Photo by Arthur Harvey, courtesy of the Miami Herald **p. 340 (bottom)** Chandra X-Ray Center/Smithsonian Astrophysical Observatory/NASA/Marshall Space Flight Center **p. 344 (left)** Pearson Education/PH College **p. 344 (right)** Photograph by Erik Gustafson, courtesy AIP Emilio Segre Visual Archives, Margrethe Bohr Collection **p. 345** Gravity Probe B at Stanford University

Chapter 12—p. 358 NASA and The Hubble Heritage Team (STScI/AURA). R. G. French (Wellesley College), J. Cuzzi (NASA/Ames), L. Dones (SwRI), and J. Lissauer (NASA/Ames) **p. 359** The Burndy Library, Dibner Institute for the History of Science and Technology, Cambridge, Massachusetts, and the Grace K. Babson Collection of the Works of Sir Isaac Newton **p. 363** NASA/JPL/University of Texas' Center for Space Research/GeoForschungsZentrum (GFZ) Potsdam **p. 364 (left)** NASA Headquarters **p. 364 (right)** NASA/Jet Propulsion Laboratory **p. 370 (left and right)** NASA Headquarters **p. 372 (left)** NASA Headquarters **p. 372 (right)** NASA/Johnson Space Center **p. 378 (left)** E. R. Degginger/Color-Pic, Inc. **p. 378 (right)** © Mark Pilkington/Geological Survey of Canada/Science Photo Library/Photo Researchers, Inc. **p. 381** David Nunuk/Science Photo Library/Photo Researchers, Inc. **p. 385** W. Kenneth Hamblin **p. 391 (left)** NASA/Johnson Space Center **p. 391 (right)** NASA Headquarters **p. 393 (left and right)** NASA/Jet Propulsion Laboratory

Chapter 13—p. 394 Richard Megna/Fundamental Photographs, NYC **p. 395** BSIP Agency/Index Stock Imagery, Inc. **p. 406** NASA Headquarters **p. 408 (left)** Loren M. Winters **p. 408 (right)** Terje Rakke/Getty Images Inc.-Image Bank **p. 416** GeoSpectra Corporation, Bowling Green, Ohio **p. 417 (left)** © Globe Photos, Inc. 2004 **p. 417 (right)** Thomas D. Mangelsen/Peter Arnold, Inc. **p. 420 (left)** © Phil Schermeister/CORBIS **p. 420 (right)** University of Washington Libraries. Special Collections, UW21422 **p. 424** Harmony On-Line **Fig. 13–23** Jerry Marshall **p. 427** Craighead Research/Cornell University **Fig. 13–27** GONG/NSO/AURA/NSF **p. 428** Scott Camazine/Photo Researchers, Inc.

Chapter 14—p. 430 Andy Sacks/Getty Images Inc.-Stone Allstock **p. 431 (top)** Corbis Digital Stock **p. 431 (bottom)** Jeff Bray/Index Stock Imagery, Inc. **p. 437** Leonard Lessin/Peter Arnold, Inc. **p. 440 (left)** PSF/Photolibrary.Com **p. 440 (right)** © Phillip Colla/Seapics.com **p. 441 (left)** Eurelios/Phototake NYC **p. 441 (right)** Visuals Unlimited **p. 451 (left)** David P. Wojtowicz, Jr., Department of Atmospheric Sciences, University of Illinois at Urbana-Champaign **p. 451 (right)** Michal Heron/Pearson Education/PH College **p. 457** Richard Megna/Fundamental Photographs, NYC **p. 459 (left)** © Neal Preston/CORBIS **p. 459 (right)** Tomas del Amo/Index Stock Imagery, Inc. **p. 462 (left)** Photoedit **p. 462 (right)** Wada Mitsumiro/Getty Images Inc.-Liaison **Fig. 14–44** © Doug Wechsler

Chapter 15—p. 476 Richard Megna/Fundamental Photographs, NYC **p. 479 (top)** Animals Animals © David Hamman **p. 479 (bottom left)** Richard Megna/Fundamental Photographs, NYC **p. 479 (bottom right)** Michal Heron/Pearson Education/PH College **p. 484** Richard Megna/Fundamental Photographs, NYC **p. 489** Life Measurement, Inc. **p. 490 (left)** Carl Purcell/Photo Researchers, Inc. **p. 490 (middle)** E. R. Degginger/Color-Pic, Inc. **p. 490 (right)** J. Cochin/Photo Researchers, Inc. **p. 492** Brock May/Photo Researchers, Inc. **p. 494** Bluestone Productions/Taxi/Getty Images Inc. **p. 499 (left)** Tom Myers/Photo Researchers, Inc. **p. 499 (right)** Tom McHugh/Photo Researchers, Inc. **p. 503** Robert B. Suter, Vassar College **p. 508** Yoav Levy/Phototake NYC **p. 509** Grace Davies Photography **p. 511** Jerry Marshall

Chapter 16—p. 514 LANL/Photo Researchers, Inc. **p. 521** Dallas & John Heaton/Corbis/Bettmann **p. 522 (left)** Bob Daemmrich/Stock Boston **p. 522 (right)** Joe Sohm/The Image Works **p. 526** Phil Degginger/Color-Pic, Inc. **p. 532 (left)** Animals Animals © Michael Fogden **p. 532 (right)** E. R. Degginger/Color-Pic, Inc. **p. 535** Pete Oxford/Nature Picture Library **p. 536 (left)** John Chumack/Photo Researchers, Inc. **p. 536 (middle)** Ron Dahlquist/Getty Images Inc.-Stone Allstock **p. 536 (right)** Thomas Kitchin/Tom Stack & Associates, Inc. **p. 537** Sheila McKinnon/Creative Eye/MIRA.com **p. 543** Jim Ross/NASA/Dryden Flight Research Center

Chapter 17—p. 548 Richard Megna/Fundamental Photographs, NYC **Fig. 17–4** Carey B. Van Loon **p. 554** Tom Bochsler/Pearson Education/PH College **p. 562 (left)** Jon Arnold/Getty Images Inc.-Taxi **p. 562 (right)** Jim Wark/Peter Arnold, Inc. **p. 565** © Woods Hole Oceanographic Institution **p. 568 (left)** Steve McCutcheon/Visuals Unlimited **p. 568 (right)** Malin Space Science Systems/Jet Propulsion Laboratory/NASA Headquarters **p. 569** Gerald & Buff Corsi/Visuals Unlimited **p. 570 (left)** © Philip Gould/CORBIS **p. 570 (right)** Frank Oberle/Getty Images Inc.-Stone Allstock **p. 571** European Space Agency **p. 581** Ralph White/CORBIS-NY

Chapter 18—p. 585 Adam Hart-Davis/Photo Researchers, Inc. **p. 600** Olaf Soot/Getty Images Inc.-Stone Allstock **p. 601 (left)** Sargent-Welch/VWR International **p. 601 (right)** Porterfield/Chickering/Photo Researchers, Inc. **p. 613** Jeff Hunter/Getty Images Inc. **p. 614 (top)** Yorgos Nikas/Biol. Reprod./Getty Images Inc.-Stone Allstock **p. 614 (middle)** Photo Lennart Nilsson/Bonnier Alba AB, A CHILD IS BORN, Dell Publishing Company **p. 614 (bottom)** Tim Brown/Getty Images Inc.-Stone Allstock

Chapter 19—p. 625 Landmann-Benali/Getty Images Inc.-Liaison **p. 628 (left)** Pearson Learning Photo Studio **p. 628 (right)** Charles D. Winters/Photo Researchers, Inc. **p. 630** Jim Pickerell/The Stock Connection **Fig. 19–13** Yoav Levy/Phototake NYC **p. 648 (left)** Michael J. Lutch/Boston Museum of Science **p. 648 (right)** Paul Katz/Index Stock Imagery, Inc. **p. 660** Steve P. Hopkin/Getty Images Inc.-Taxi

Chapter 20—p. 662 Richard Megna/Fundamental Photographs, NYC **p. 677** Craig X. Sotres/Pearson Education/PH College **Fig. 20–14 (left)** Paul Silverman/Fundamental Photographs, NYC **Fig. 20–14 (right)** Frank LaBua/Pearson Education/PH College **p. 683** Robert F. Bukaty/AP Wide World Photos **p. 684 (left)** Tom Pantages **p. 684 (right)** Richard Megna/Fundamental Photographs, NYC **p. 685** J. Reid/Custom Medical Stock Photo, Inc.

Chapter 21—p. 695 Paul Silverman/Fundamental Photographs, NYC **p. 697 (left)** Warren Faidley/DRK Photo **p. 697 (right)** Jean-Louis Dubois/Jacana/Photo Researchers, Inc. **p. 699 (top)** EyeWire Collection/Getty Images-Photodisc **p. 699 (bottom)** LEDtronics, Inc. **p. 702** Yoav Levy/Phototake NYC **p. 703** Tony Freeman/PhotoEdit **p. 704** Frank Labua/Pearson Education/PH College **p. 719** Raph Krubner/The Stock Connection **p. 720 (top)** Department of Clinical Radiology, Salisbury District Hospital/Science Photo Libary/Photo Researchers, Inc. **p. 720 (middle)** Michael Dalton/Fundamental Photographs, NYC

Chapter 22—p. 732 Reprinted by permission from R. J. P. Williams, Biomineralization: Iron and the origin of life. *Nature* 343:213–214 (1990). Copyright © 1990 Macmillan Magazines Limited. Image courtesy of Hojatollah Vali, McGill University. **Fig. 22–3a** Phil Degginger/Color-Pic, Inc. **Fig. 22–3b** Grantpix/Photo Researchers, Inc. **p. 741** James Holmes/Oxford Centre for Molecular Sciences/Science Photo Library/Photo Researchers, Inc. **p. 744 (left)** NASA Transition Region and Coronal Explorer **p. 744 (middle)** Louis A. Frank, University of Iowa **p. 744 (right)** Kennan Ward/Corbis/Stock Market **p. 757** High Field Magnet Laboratory, University of Nijmegen, Nijmegen, The Netherlands **p. 761** Richard Megna/Fundamental Photographs, NYC **p. 765** Sim Laboratoire/Photo Researchers, Inc.

Chapter 23—p. 768 AP Wide World Photos **p. 769** Richard Megna/Fundamental Photographs, NYC **p. 774 (left)** Andrew Rafkind/Getty Images Inc.-Stone Allstock **p. 774 (right)** Paul Silverman/Fundamental Photographs, NYC **p. 782 (left)** Gary Ladd/Photo Researchers, Inc. **p. 782 (middle)** Lincoln Potter/Getty Images Inc.-Liaison **p. 782 (right)** L. S. Stepanowicz/Visuals Unlimited **p. 792 (left)** Bert Krages/Visuals Unlimited **p. 792 (middle)** Philip Condit II/Getty Images Inc.-Stone Allstock **p. 792 (right)** Jack Plekan/Fundamental Photographs, NYC

Chapter 24—p. 803 Simon Frasier/Photo Researchers, Inc. **p. 809 (left)** Frank Labua/Pearson Education/PH College **p. 809 (middle top)** Wernher Krutein/Getty Images Inc.-Liaison **p. 809 (middle bottom)** David Buffington/Getty Images Inc.-Photodisc **p. 809 (right)** Richard Megna/Fundamental Photographs, NYC **p. 827** Jack Hollingsworth/Getty Images Inc.-Photodisc

Chapter 25—p. 837 Leonard Lessin/FBFA/Photo Researchers, Inc. **p. 838** Lester Lefkowitz/Corbis/Stock Market **p. 843 (left)** National Optical Astronomy Observatory/Association of Universities for Research in Astronomy/National Science Foundation **p. 843 (right)** NASA Headquarters **p. 846** Doug Johnson/Photo Researchers, Inc. **p. 847 (top left)** Zac Macaulay/Getty Images Inc.-Image Bank **p. 847 (top middle)** Janie Airey/Getty Images/Digital Vision **p. 847 (top right)** © Martin Harvey/CORBIS All Rights Reserved **p. 847 (bottom left)** Courtesy of SE-IR Corporation, Goleta, California **p. 847 (bottom right)** O. Brown, R. Evans, and M. Carle, University of Miami Rosenstiel School of Marine and Atmospheric Science, Miami, Florida **p. 848 (middle)** NASA Headquarters **p. 848 (bottom)** International Atomic Energy Agency **p. 853** Eric Clifton and Greg Neaveill/Peoria Astronomical Society, Inc. **Fig. 25–17** Fundamental Photographs, NYC **p. 857** Peter Aprahamian/Sharples Stress Engineers Ltd./Photo Researchers, Inc. **p. 859 (middle)** Warren Bolster/Getty Images Inc.-Stone Allstock **p. 859 (bottom left, both)** E. R. Degginger/Color-Pic, Inc. **p. 859 (bottom right)** Carol & Don Spencer/Visuals Unlimited **p. 860 (both)** Tom Pantages **p. 869** David Rose/NIRS-Rose1

Chapter 26—p. 870 © Royalty-Free/CORBIS **p. 872 (top)** James S. Walker **p. 872 (bottom)** Digital Light Processing/Texas Instruments Incorporated **p. 873** Windsor Castle, Royal Library/Archive für Kunst und Geschichte, Berlin **p. 874** © Aviation-images.com **p. 877** Goodrich Corportion Electro-Optical Systems **p. 885** Richard Megna/Fundamental Photographs, NYC **p. 887** © Charles O'Rear/CORBIS All Rights Reserved **p. 890 (left)** Ken Kay/Fundamental Photographs, NYC **p. 890 (right)** Hank Morgan/Photo Researchers, Inc. **p. 891** E. R. Degginger/Color-Pic, Inc. **p. 892** Craig Tuttle/Corbis/Stock Market **p. 897** Roy Bishop **p. 901 (top & bottom)** Richard Megna/Fundamental Photographs, NYC

Chapter 27—p. 909 © Roger Ressmeyer/CORBIS **p. 911 (top)** Diane Hirsch/Fundamental Photographs, NYC **p. 911 (bottom)** Diane Schiumo/Fundamental Photographs, NYC **p. 916** Bonnie Kamin/PhotoEdit **p. 921** Lawrence Migdale/Pix **p. 927 (top)** © Roger Ressmeyer/CORBIS **p. 927 (bottom)** Jeff J. Daly/Fundamental Photographs, NYC **p. 932** Mike Wilkes/Nature Picture Library **p. 935** Custom Medical Stock Photo, Inc. **Fig. 27–24b** © Aviation-images.com

Chapter 28—p. 937 Corbis Digital Stock **p. 939** Richard Megna/Fundamental Photographs, NYC **p. 942** Tom Pantages **Fig. 28–11b** Tom Pantages **p. 947** Michael Ventura/PhotoEdit **p. 949** Alamy Images **p. 950** Jeremy Burgess/Science Photo Library/Photo Researchers, Inc. **p. 952** Richard Megna/Fundamental Photographs, NYC **p. 954** Ken Kay/Fundamental Photographs, NYC **Fig. 28–21** Reproduced by permission from Michel Cagnet, Maurice Franzon, and Jean Claude Thierr, *Atlas of Optical Phenomena*. New York: Springer-Verlag, 1962. © 1962 by Springer-Verlag GmbH & Co. **Fig. 28–22** Reproduced by permission from Michel Cagnet, Maurice Franzon, and Jean Claude Thierr, *Atlas of Optical Phenomena*. New York: Springer-Verlag, 1962. © 1962 by Springer-Verlag GmbH & Co. **p. 955 (bottom)** The Image Finders 2003 **Fig. 28–23** Museum of Fine Arts, Springfield, MA., Robert J. Freedman Collection. © 2005 Artists Rights Society (ARS), New York/ADAGP, Paris. **Fig. 28–28** Photo by M. H. F. Wilkins. Courtesy of Biophysics Department, King's College,

London, England. **p. 960 (left)** Phil Degginger/Color-Pic, Inc. **p. 960 (right)** Charles Krebs/Corbis/Stock Market **p. 963** Richard Megna/Fundamental Photographs, NYC **p. 964** E. R. Degginger/ Color-Pic, Inc.

Chapter 29—p. 970 Eric Reiman, University of Arizona/Science Photo Library/Photo Researchers, Inc. **p. 971** Lotte Jacobi Collection, University of New Hampshire **Fig. 29-11** A. C. Searle and C. M. Savage, The Australian National University. Animations at http://www.anu.edu.au/ Physics/Searle. **p. 987 (left)** Alexander Tsiaras/Stock Boston **p. 987 (right)** Courtesy of SOHO/Extreme-ultraviolet Imaging Telescope consortium. SOHO is a project of international cooperation between the European Space Agency (ESA) and NASA. **Fig. 29–16** Hank Morgan/Rainbow **Fig. 29–25a** Space Telescope Science Institute **Fig. 29–25b** European Space Agency/ NASA Headquarters **p. 994 (left)** Julian Baum/Photo Researchers, Inc. **p. 994 (right)** NASA Headquarters

Chapter 30—p. 1003 Natural History Museum Picture Library, London/ Getty Images Inc.-Stone Allstock **p. 1005** Getty Images Inc.-Photodisc **p. 1012 (left)** Eunice Harris/Photo Researchers, Inc. **p. 1012 (right)** Stock Trek/Getty Images Inc.-Photodisc **Fig. 30–10a** The Burndy Library, Cambridge, Massachusetts **Fig. 30–10b** DESY/HASYLAB, Hamburg, Germany **Fig. 30–11** Dr. N. Niimura/Tom Pantages **Fig. 30–12a** Dr. Everett Anderson/Science Photo Library/Photo Researchers, Inc. **Fig. 30–12b** Scott Camazine/Photo Researchers, Inc. **Fig. 30-14** Hitachi, Ltd., Advanced Research Laboratory **Fig. 30–19a** Courtesy of International Business Machines Corporation. Unauthorized use not permitted. **Fig. 30–19b** R. J. Driscoll, M. G. Youngquist, & J. D. Baldeschwieler, California Institute of Technology/Science Photo Library/Photo Researchers, Inc. **p. 1029** James S. Walker **p. 1030** John Chumack/Photo Researchers, Inc. **p. 1031** Esther Schmidt/VALAN PHOTOS **p. 1033** Paul Zahl/Photo Researchers, Inc.

Chapter 31—p. 1034 Eduardo Kac, GFP Bunny, 2000. Alba, the fluorescent bunny. Photo: Chrystelle Fontaine. Courtesy Julia Friedman Gallery. **p. 1036** Kitt Peak National Observatory **p. 1039** AIP Emilio Segre Visual Archives, Uhlenbeck Collection **p. 1055** China Newphoto/Landov LLC **p. 1058 (left)** GCA/CNRI/Phototake NYC **p. 1058 (right)** Richard Shock/ Getty Images Inc.-Stone Allstock **p. 1060** Yoav Levy/Phototake NYC **p. 1061** Philippe Plailly/SPL/Photo Researchers, Inc. **p. 1063 (top left)** Getty Images, Inc.–Photodisc **p. 1063 (top right)** Dr. Peter Dawson/ Science Photo Library/Photo Researchers, Inc. **p. 1063 (bottom left)** Tom McHugh/Photo Researchers, Inc. **p. 1063 (bottom right)** Bjorn Bolstad/ Photo Researchers, Inc.

Chapter 32—p. 1071 National Cancer Institute/Getty Images Inc.- Photodisc **p. 1080** Michael Dalton/Fundamental Photographs, NYC **p. 1085** Will & Deni McIntyre/Photo Researchers, Inc. **p. 1087** Corbis/ Sygma **p. 1093 (left)** Catherine Pouedras/Science Photo Library/Photo Researchers, Inc. **p. 1093 (middle)** Martin Bond/Science Photo Library/ Photo Researchers, Inc. **p. 1093 (right)** U.S. Department of Energy/ Science Photo Library/Photo Researchers, Inc. **p. 1095** Randy Montoya/ Sandia National Laboratories **p. 1096 (both)** Yoav Levy/Phototake NYC **p. 1098 (top left)** CNRI/Science Photo Library/Photo Researchers, Inc. **p. 1098 (right)** Wellcome Dept. of Cognitive Neurology/Science Photo Library/Photo Researchers, Inc. **p. 1098 (middle)** UHB Trust/Getty Images Inc.-Stone Allstock **p. 1109** Photo courtesy of The Coleman Company, Inc.

Index

Aberration
  lens, 927–928
  spherical, 877, 927
Absolute zero, 518–519, 520, 585
  third law of thermodynamics and, 614
Absorption spectrum, 1037
Accelerated frames of reference, 991
Acceleration, 25–34, 401–405
  angular. *See* Angular acceleration
  average, 25–26, 29
  of car, 28, 29, 67–68, 190, 196
  of center of mass, 266, 268–269
  centripetal, 162–167, 291–294, 370
  constant, 29–34, 37–43, 81–83, 286
  dimensions of, 5
  force and, 110–112
  in free fall, 40–43, 83
  as function of time, 42
  graphical interpretation of, 26–29
  of gravity. *See* Acceleration of gravity
  instantaneous, 26, 27
  mass and, 110–111
  momentum and, 244
  Newton's second law of motion and, 110–118
  nonzero, 28
  problems, 49
  of projectile, 83
  from rest, 188–189, 190
  of rocket, 269–270
  in simple harmonic motion, 401–402, 404
  speed and, 29
  tangential, 167, 293
  time, position, and, 32–33
  time and, 25, 30, 32–33, 42, 402
  typical, 26
  units of, 112
  upward and downward, 40–43
  velocity and, 28–29, 111, 402–403
  x and y components of, 122–124
Acceleration of gravity, 38–39, 83, 94, 362–365, 375, 405, 775, 776
  altitude and, 210, 211–212, 363–364
  latitude and, 38
  of Moon, 364–365
  range and, 94
Acceleration vectors, 67–69
  instantaneous, 68
  problems, 76–77
  velocity vectors vs., 69
Accelerators, 988, 991
AC circuits. *See* Alternating-current circuits
Accommodation, 910
Accommodation ranges, 932
Accretion disk, 994
Achromatic lenses, 928
Acousto-optic modulation (AOM), 959
Actinium-227, 1082
Action-reaction forces, 118–121
Action-reaction pairs, internal forces and, 249
Active Noise Reduction (ANR), 455
Activity, 1083, 1086, 1087
Adair, Robert, 473
Addition
  of physical quantities, 7
  of vectors, 61–63, 65, 69–70
    problems, 76
  of velocities, relativistic, 981–983
  of velocity vectors, 69–70
Adhesive tape, charge on, 657
Adiabatic cooling, 599, 600
Adiabatic heating, 595, 599
Adiabatic processes, 593–595, 598–599, 600
  isotherms vs., 598
Age-related macular degeneration (AMD), 1061
Aging, space travel and, 975–977
Air
  dielectric constant of, 681
  dielectric strength of, 683
  index of refraction of, 885
  speed of sound in, 437, 438
  vibrating columns of, 459–463
Air bags, 50
Airburst over Pennsylvania (2001), 513
Air carts, elastic collisions of, 259, 260–261, 267
Air conditioners, vs. heat engines, 605
  problems, 622
Aircraft carrier, takeoff and landing, 139, 935–936
Airplanes
  airplane EMF, 799
  lift and, 498
  takeoff distance, 34
Air pressure, 552
  boiling temperature and, 567
Air resistance
  free fall and, 38
  projectile motion and, 83, 94
  range and, 94, 96
  work and, 185, 186
Air splint, 479
Air tracks, 108–109, 140
Air wedge, 945–946
Albireo (double star), 1030
Alcator fusion experiment, 800
Alkali metals, 1055
Allen, Bryan, 195
Alpha decay, 1078–1080, 1082, 1083
Alpha particles, 1035–1036, 1074, 1077, 1078
Alpha rays
  penetrating ability, 1078
  relative biological effectiveness for, 1096
Alternating-current circuits, 697, 803–836
  alternating voltages and currents, 804–809
    problems, 832–833
  capacitors in, 809–812
  direct-current circuits vs., 792
  finding average power, 805–807
  high-frequency and low-frequency limits of, 823
  inductors in, 817–820
  properties of, 822
  RC circuits, 812–816
  resonance in electric circuits, 824–829
  RLC circuits, 820–824
  RL circuits, 818–820
Alternating-current generator, 782, 804
Alternating voltages, 804–809
  problems, 832–833
Altitude, acceleration of gravity and, 210, 211–212, 363–364
Aluminum, 700, 1054
Alveoli, 504
Amber, 625
  electric charge and, 626
AMD (age-related macular degeneration), 1061
American National Standards Institute (ANSI), 730
Americium-241, 1080
Ammeter, 720–721, 1009, 1010
Ampère, André-Marie, 696
Ampere (A), defined, 696
Ampère's law, 749–751
  magnetic field in solenoid and, 753–754
Amplified rebound height, 280
Amplitude, 397–398
  motion of mass on a spring and, 405
  of oscillating mass, 418, 420
  wave superposition and, 452, 453
Analyzer, polarizer and, 855–856, 857
Andromeda galaxy, 843
Angel Falls (Venezuela), 15
Angle
  critical, for total internal reflection, 888–889
  direction, 57, 60, 63, 64
  force at angle to displacement, 182–184
  general launch, 89–92
  phase, 815–816, 820–821, 822
  scattering, 1014, 1015, 1016
  work and, 182–184, 185
  zero launch, 84–89

Angle of incidence, 871, 873, 888
  Brewster's angle, 891
Angle of reflection, 871, 873
Angle of refraction, 888
Angle of the velocity, 95
Angstrom (Å), 845
Angular acceleration, 285–286, 289, 315
  average, 285
  instantaneous, 285
  problems, 308–309, 350–351
  sign convention for, 285
  torque and, 319–322
Angular displacement, 283, 343
Angular frequency, defined, 400
Angular kinematic equations, vs. linear, 287
Angular magnification (M), 920–922, 923
  of compound microscope, 923
  of telescope, 925–926
Angular momentum, 241, 315, 335–343
  Bohr orbits and, 1040, 1046–1047
  of circular motion, 335, 336, 337
  conservation of, 315, 338–343, 368
  defined, 335
  magnitude of, 1048
  problems, 353–354
  sign convention for, 337
  spin, 1048
Angular momentum quantum number, 1053
  orbital, 1048
Angular momentum vectors, 344–345
Angular position, 282–283, 400
  defined, 282
  of fringes, 942
  problems, 308–309
  sign convention for, 282
Angular quantity, linear analogs, 286, 299, 320
Angular resolution, 955
Angular size, 920–922, 924, 924–925, 925
Angular speed, 284, 285, 286–288, 290, 335, 336, 339–342, 343, 400
  rolling object and, 294–295
Angular velocity, 283–286
  instantaneous, 284
  problems, 308–309
  right-hand rule for, 344
  sign convention, 284
Angular velocity vector, 344
Animals, infrasonic and ultrasonic communications, 440–441
Annihilation, electron-positron, 988
ANR (Active Noise Reduction), 455
Antenna, 838, 846
  radio, 938, 939
Antielectron, 988
  See also Positron
Antifreeze, biological, 569
Antilock braking systems (ABS), 149
Antimatter, 970, 988–989
Antineutrino, 1081–1082
Antinode, 456, 459, 461
Antiparticle, 1077
Antiquarks, 1101
Antireflection coating, 968
Antiscalding devices, 522
Antitheft devices, magnetic, 781
AOM (acousto-optic modulation), 959
Aperture, 912–913, 926
Apochromatic lenses, 928
***Apollo 11* reflector, 868**
***Apollo 15* landing on Moon, 53**
Apparent depth, 887
Apparent weight, 126–128
Applications
  of Archimedes' principle, 488–493
  of Bernoulli's equation, 498–501
  of equations of motion, 35–37
    problems, 50–51
  of torque, 333–335
Aqueous humor, 910
Archerfish, 96
Arches National Park (Utah), 332
Archimedes' principle, 486–493
  applications of, 488–493
  problems, 509–510
Arc length, 283
Area
  dimensions of, 5
  in heat conduction, 532
  pressure and, 477–478, 479
  shear modulus and, 563–564
  Young's modulus and, 561
Area expansion, 523–524
Argon, 1054, 1055
***Armageddon* (movie), 378, 380**
Arrow of time, 600, 610
Arteriole, blood speed in, 510
Asteroids, 370, 378–380, 380, 390, 392
Astigmatism, 919
Astronaut jet packs, 116
Astronauts
  measuring mass of, 427
  weightlessness training, 128
Astronomical mirrors, 877
Astronomical systems, energy conservation in, 377–384
Atmos clock, 202
Atmosphere
  depth of, 512
  evaporating, 570–571
  of Mars, 580
  of Moon, 382
  planetary, 382
  root mean square speed of, 558–559
Atmospheric pressure, 479–480, 482–483, 500
  defined, 479
Atom(s), 626, 1035
  Bohr, 310, 690
    See also Bohr model
  early models of, 694, 1035–1036
    problems, 1067–1068
  electric current in, 755
  electric forces and, 625
  line spectrum of, 1037
  nuclear model of, 1035–1036
  Rutherford's planetary model of, 694, 1035–1036
  structure of, 627
Atomic clocks, 4, 975
Atomic-force microscopy (AFM), 192
  flexing cantilever used in, 192–193
Atomic mass, 552
  electrons and, 1078
  measuring, 742
Atomic mass unit (u), defined, 1073
Atomic number (Z), 1072
  in alpha decay, 1078–1079
  in beta decay, 1080
  gamma decay and, 1082
  for stable nuclei, 1076
Atomic physics, 1034–1070
  Bohr's model of hydrogen atom, 632–633, 692, 1039–1046
  early models of the atom, 1035–1036
  multielectron atoms, 1050–1055
  periodic table, 1054–1055
  quantum mechanical hydrogen atom, 1047–1050
  spectrum of atomic hydrogen, 1036–1039, 1043–1045
Atomic radiation, 1055–1064
  fluorescence and phosphorescence, 1062–1064
  lasers, 866, 867, 938, 1059–1062
  problems, 1069
  X-rays, 846, 848, 865, 1055–1058, 1109
Atomic systems, role of electric vs. gravitational forces in, 632–633
Atomizer, 499
Attractive forces, 503, 625, 626, 733
  polarization and, 629
Atwood's machine, 160–161, 235, 239, 311, 353
Aurora australis, 744
Aurora borealis, 744
Autoclave, 567
Average acceleration, 29
  for car traveling in circle, 68
  defined, 25
  graphical interpretation, 27
Average acceleration vector, 67–68
  defined, 67
Average angular acceleration, defined, 285
Average angular velocity, defined, 283
Average force, collisions and, 244–246
Average power
  in alternating-current circuits, 822
  defined, 195
Average speed, 19–21
  defined, 19
  problems, 48–49
  sign convention for, 19

Average velocity, 21–23, 31
defined, 21
graphical interpretation of, 22–23, 25, 30
problems, 48–49
See also Velocity(ies)
Average velocity vector, defined, 66
Avogadro, Amedeo, 551
Avogadro's number, 551
a waves, 677
Axis
principal, 875–876
transmission, 854–857
Axis of rotation, 282
angular velocity and, 344
conservation of angular momentum and, 339–340
moment of inertia and, 298, 299
static equilibrium and, 323–325
See also Torque

Back emf, 784, 785
Bacterial motion, 50
Bacterium, weighing, 427
Bailey, Donovan, 7
Balance
center of mass and, 330–332
problems, 353
Ballistic pendulum, 256–257
Ballistocardiograph, 251
Ballistocardiographic (BCG) scale, 251
Balloons
buoyancy of, 490
electric charge and, 628
pressure and, 478
Balls
dropping ball, 85–86
inflating, 549, 551
pressurizing, 480
Balmer, Johann Jakob, 1037
Balmer line, 1036
Balmer series, 1037–1038, 1039, 1044
Banking, of roads, 164–165
Bar, 479
Bar magnets, 732, 733, 734, 752–753, 775
induction and, 772–773
magnetic field of, 737
Barnes, Randy, 105
Barometer, 482
Barrington Meteor Crater, 378
Barton, Otis, 507, 581
Baryons, 1100, 1101
Baseball
catching, 218
force to hold, 350
hearing good hit on, 473
hitting, 244–246, 443–444
pitching, 117–118, 287–288
Bathysphere, 507–508, 581
Bats, 247, 440, 442
Battery(ies), 695, 697–699
charged, 690
in circuit with inductor, 788
electromotive force and, 698–699
internal resistance of, 707
rating car, 727
work done by, 698
Battery check meters, 704
Bay of Fundy, 385
Beat frequency, 463–464
calculating, 464
defined, 464
Beats, 463–465
problems, 473
Beauty (quark), 1101
Becquerel, Antoine Henri, 1078
Becquerel (Bq), 1083
Beebe, Charles William, 507, 581
bel, 444
Bell, Alexander Graham, 444
Benzene, 885
Bernoulli's equation, 495–501
applications of, 498–501
change in height, 496–497
change in speed, 495–496
general case, 497–498
problems, 510–511
Beryllium, 1053, 1054
Beta decay, 1080–1082, 1083
Beta particles, 1077
See also Electron(s)
Beta rays, 1077
penetrating ability, 1078
relative biological effectiveness for, 1096
Betelgeuse, 1005, 1030
Bethe, Hans, 1094
Big Bang, Doppler effect and evidence for, 452
Big Ben clock, 78
Bimetallic strips, 521–522
Binary stars, 391, 967
Binding energy, nuclear, 1089–1090
Binoculars, 890, 949
Biologically equivalent dose, 1096
Bioluminescence, 508
Birds
sound intensity of songs, 443
wading, countercurrent exchange in, 535
Bismuth, 1076
Blackbody, 538, 1004
Blackbody radiation, 1004–1007, 1036
Planck's hypothesis of quantized energy and, 1005–1007
problems, 1030
Black holes, 384, 474, 993–994
Blood flow, 9–10, 501, 502, 503
Blood flow meter, 451
Blood pressure, 503
Blood speed in arteriole, 510
Blowfly flight, 800, 802
Blue light, 858–859, 943
Blue quarks, 1101
Blue whale, 15, 491
BMMD, 406, 427
Boats
acceleration of, 115
speed of, 31
Bobbing for apples, 404–405
Bod Pod, 489
Body density, measuring, 488–489
Body fat, measuring, 489
Body Mass Measurement Device (BMMD), 406, 427
Body temperature, 532, 535
heat transport in human body, 546
Bohr, Niels, 344, 632, 1039
Bohr atom, 310, 690
Bohr model, 632–633, 692, 1039–1046
assumptions of, 1039–1040
magnetic fields in, 767
problems, 1068–1069
Bohr orbits, 1040–1042
de Broglie wavelengths and, 1046–1047
energy of, 1042–1043
energy of K-shell electron and, 1056–1057
Bohr radius, 1041
Boiling points, 517, 520
raising, 567
Bolometer, 701–702
Boltzmann, Ludwig, 550
Boltzmann constant, 550
Bone, stretching, 561–562
Bone loss in space, preventing, 313
Born, Max, 1047
Boron, 1053, 1054
Bottlenose dolphin, 491
Bottles, as musical instruments, 459–460, 464–465
Bottom (quark), 1101
Bow and arrow, 214
Boyle, Robert, 552
Boyle's law, 552–553
Brackett series, 1038
Brahe, Tycho, 367, 368
Brain
brain function and magnetic fields, 765
electroencephalograph and, 677
magnetism and, 756
near-infrared brain scans, 868–869
PET scans of, 970, 988–989, 1098
power consumption of, 203
transcranial magnetic stimulation, 767, 801
Brain growth, 16, 138
Brain power, 546
Braking, magnetic, 777
Breathing, surface tension and, 503–504
Breezes, 536
Bremsstrahlung, 1056
Brewster, David, 891
Brewster's angle, 891

Bridges, resonance and, 420
Bright fringes, 945
conditions for, in two-slit experiment, 940, 941–943
principal maxima, 957–959
in single-slit diffraction, 951, 952
See also Constructive interference; Interference fringes
British engineering system of measurement, 4, 479
British thermal unit (btu), 527
Bromine, 1055
Bronchoscope, 890
Bug zappers, 792
Bulk modulus, 564, 565
Bullet train, 471
Bungee jumping, 408
Buoyancy, 476, 486–487
applications of, 488–493
problems, 509
Buoyant force, 486–487
b waves, 677

Calder, Alexander, 264
Calorie (food), 527
calorie (heat), 526
Calorimeter, 529
Calorimetry, 529–530
Camera, 912–913
lenses, 912–913, 928, 949
problems, 931–932
Canoes, 119, 249, 250–251
Cantilevers, 427
Capacitance (C), 677
computer keyboards and, 682
defined, 678
of parallel-plate capacitor, 678–680
filled with dielectric, 681–682
Capacitive reactance, 809–810, 817
Capacitor phasor, 811
Capacitors, 662, 677–683
charging, 717–719
dielectrics and, 680–683
discharging, 719
electric energy storage and, 683–685
hazards, 685
limiting behavior of, 718
problems, 691–692
See also Parallel-plate capacitor
Capacitors in circuits, 714–717, 719
in alternating-current circuits, 809–812
problems, 833
in parallel, 714–715
phasor diagrams for, 810–812
problems, 729
in series, 715–717
Car batteries, rating, 727
Carbon
electronic configuration, 1054
in periodic table, 1053
Carbon-14, 1072
beta decay of, 1080–1081
Carbon-14 dating, 1087–1089
Carbon dioxide, 582, 885
Carnot, Sadi, 602
Carnot cycle, 602–604, 606
problems, 621–622, 624
Carnot's theorem, 603
Cars
acceleration, 28, 29, 67–68, 190, 196
antilock braking systems, 149
coasting, 185–186
collisions, 257–258
energy recovery technology in, 784
horsepower of, 195, 355
hybrid, 784
performance, 195–196
speeding, 36–37
stopping distance, 35–36, 149
Cartesian diver, 491
Cassegrain astronomical telescope, 936
Catapult launcher, 202
Cataracts, 931
CAT scan, 1058
Cavendish, Henry, 365
Cavendish experiment, 365–366
Cavitation, 474
CD
interference in, 950
linear and angular speeds of, 290
playing, 312–313
reflection from, 960–961
Cell
electric potential in, 689, 690
radiation and, 1095
Cell membrane
dielectrics and, 693–694
electric current through, 726
energy storage in, 692
in nerve cell, electric field within, 660, 661
Celsius, Anders, 516
Celsius scale, 516
absolute zero, 518–519
compared to Kelvin and Fahrenheit scales, 520
conversion to Fahrenheit scale, 517–518
conversion to Kelvin scale, 520
Center of curvature, 875–876
Center-of-curvature ray (C ray), 878, 879
Center of mass, 240, 264–269
acceleration of, 266, 268–269
of arm, 265
balance and, 330–332
problems, 353
location of, 264–265
motion of, 266–269
of person, 351
problems, 276–277
for two objects, 264
velocity of, 266
weight of, 269
x coordinate of, 265
y coordinate of, 265
Center of percussion, 314
***centi*, 5**
Centimeter (cm), 4
Centrifuge, 166–167, 174, 291–293, 312
Centripetal acceleration, 162–167, 370
of rotating object, 291–294
tangential acceleration vs., 293
Centripetal force, 163, 167
on planet in orbit, 369–370
tides and, 385
Ceres, attraction of, 389
Cesium, 1054
cgs system, 112
Chain reactions, 1092–1093
Chameleon tongue, acceleration of, 50
Characteristic x-rays, 1056, 1057
Charge. *See* Electric charge
Charged battery, 690
Charged particles in magnetic field, motion of, 739–744
Charging by induction, 648–649
Charles, Jacques, 554
Charles's law, 554
Charmed (quark), 1101
Checking solution, problem solving and, 12, 39
Chemical properties, in periodic table, 1054–1055
Chemical reactions, temperature and, 559
Chest X-ray, 1109
Chicxulub (Mexico), meteor crater, 378
Chlorine, 1054, 1055
Chromatic aberration, 927–928
Chromatic musical scale, 458
Chromium-51, 1097
Ciliary muscles, 910
Circle, as special case of ellipse, 368
Circuit breakers, 807, 808–809
Circuits. *See* Electric circuits
Circular interference fringes, 946–947
Circular motion, 68, 162–167
angular momentum of, 335, 336, 337
in magnetic field, 741–743
problems, 174–175
simple harmonic motion and, 399–405
See also Rotational motion
Circular opening, first dark fringe for diffraction pattern of, 954
Classical kinetic energies, 989
Classical velocity addition, 981, 983
Clausius, Rudolf, 608
Climate
adiabatic heating and cooling and, 599, 600
water and, 529
Clocks
atmos, 202
atomic, 4, 975
light, 973–974

oscillators and, 394
pendulums and, 394, 413, 415
Clockwise rotation, 284
Closed path, 205–208
Clothesline, tension in, 157–158
Cobalt, 755
Coefficient of kinetic friction, 143, 144
Coefficient of linear expansion, 521, 523, 524
Coefficient of performance (COP), 605–608
for heat pump, 607–608
for refrigerator, 605–606
Coefficient of static friction, 147, 148
Coefficient of viscosity, 501–502
Coefficient of volume expansion, 521, 524
Coefficients of thermal expansion, 521
Coherent light, 938
Coil
induced emf of rotating, 781–782
inductance of, 785
torque on, 747–748
"Cold" reservoir, 601, 602–604, 605
Collector plate, 1009, 1010
Collisions, 240
average force, 244–246
of cars, 257–258
elastic, 253, 259–263
impulse, 244–248
inelastic, 253–258
photon-electron, 1014–1016
of projectiles, 91–92
problem, 105
rotational, 341–343
Colonoscope, 890
Color television, 957
Combination circuits, 710
Comets, 358, 368, 378, 381, 391
Hale-Bopp, 381, 853
Wild 2, 393
Communications satellites, 372
Commutator, 748
Compact disk. ***See*** **CD**
Completely inelastic collisions, 253–254
Complete submersion, 488–489
Compound microscopes, 923–924
problems, 933–934
Compressed spring, 411–412
increase in mass, 988
Compressible flow, equation of continuity for, 494
Compressing the ground, 238
Compton, Arthur Holly, 1014
Compton effect, 1014–1016
problems, 1032
Compton shift formula, 1015
Compton wavelength, 1015, 1033
Computerized axial tomography (CAT) scan, 1058
Computers
cooling, 583
electric potential and, 689
keyboards, 682
periodic motion and, 395
Concave lens(es), 891
image formed by, 892
imaging characteristics of, 893
pair of prisms vs., 892
ray tracing for, 892–893
in two-lens system, 913–914
Concave meniscus lenses, 891
Concave mirrors, 875–876
focal point and focal length for, 876
image formation with, 879
imaging characteristics, 880
mirror equation and, 881
ray tracing for, 878
Conduction, 531–535
area and length in, 532
problems, 545
Conduction electrons, 630
Conductor, 531, 629–630
excess charge on, 646–647
ideal spherical, 675–676
resistivities of, 700
shielding and, 646–648
zero field within, 647
Cones, 910
Conical pendulum, 176
Connected objects, 158–161
problems, 173–174
Consciousness, centripetal acceleration and losing, 312
Conservation of angular momentum, 315, 338–343, 368
problems, 354
Conservation of electric charge, 625, 627–628, 711
Conservation of energy. ***See*** **Energy conservation**
Conservation of linear momentum, 248–252
Conservation of mechanical energy, 214–222
defined, 215
problems, 234–235
Conservation of momentum, 240
energy conservation vs., 255
problems, 275–276
for a system of objects, 250
Conservative forces, 205–208
defined, 206, 207
examples, 207
mechanical energy and, 214–222
problems, 233–234
spring force as, 212
work done by, 208–214
problems, 234
Constancy of speed of light, 971–972
Constant acceleration, 81–83, 286
freely falling objects and, 37–43
motion with, 29–34
problems, 49–50
Constant-acceleration equations of motion, 30, 31, 32, 34, 82
applications, 35–37
Constant electric field, electric potential for, 665–666
Constant force, work done by, 180–186, 190–191
Constant mass, 243–244
Constant pressure, 554–555, 589–591
constant volume vs., 597
Constant pressure specific heats, 596–599
Constant temperature, 553
isothermal processes and, 591–593
Constant velocity, 24, 80–81
on x-versus-t graph, 23
Constant-velocity motion, 741
Constant volume, 589–591
adding heat to system of, 591
constant pressure vs., 597
Constant-volume gas thermometer, 518, 549
Constant volume specific heats, 596–599
Constructive interference, 452, 453–455, 463, 938–939, 942, 945
in diffraction grating, 958
path lengths and, 1018
in thin films, 947–949
when scattering from crystal, 1018
See also Bright fringes
Contact forces, 120–121
Contact lenses, 909, 913, 919, 935
Container, influence on specific heat, 530
Continuity, 493–494
equation of, 494
problems, 510
Contour map, 230
Contracted length, 978
Control rods, 1093
Convection, 535–536
problems, 545
Converging lenses, 891–892, 916, 917–918, 927
Convex lens(es), 891
imaging characteristics of, 893
as magnifying glass, 919
pair of prisms vs., 892
ray tracing for, 892, 893
in two-lens system, 913–914
Convex meniscus lenses, 891
Convex mirrors, 875–876
focal point and focal length for, 876
image formation with, 878–879
imaging characteristics, 880
mirror equation and, 882
ray tracing for, 878
Conveyor belt, 178
Coordinate system, 18, 31
circular motion, 163
connected objects, 159
free-body diagrams and, 113, 114
inclines and, 130–131, 145
momentum and, 241, 242
Newton's law and, 121
one-dimensional, 18
two-dimensional, 56, 57
vector components and, 56
COP (coefficient of performance), 605–608

Copernicus, 367, 368
Copper, 700
Copper-64, 1097
Copper rod, levitation of, 745–746
Coral, fluorescence in, 1063
Cormorants, diopter change in diving, 932
Cornea, 910, 917
Corner reflectors, 875
Corrective lenses, 913–919
problems, 932–933
Cosine
pendulum and, 413
simple harmonic motion as, 397–398
Cosmic background radiation, 1030
Cosmic rays, 1001
carbon-14 dating and, 1087
typical dose, 1097
Cosmology, 1102
Coulomb (C), 626–627, 664
Coulomb's law, 630–637
Bohr orbits and, 1040
nuclear stability and, 1075
problems, 657–658
spherical charge distributions and, 637
superposition and, 634–636
Counterclockwise rotation, 284
Countercurrent exchange, 535
Crab nebula, 308, 340
angular acceleration of, 313
Cracking your knuckles, 474
Credit-card magnetic strips, 765, 774
Credit card readers, 774
Crests, 432, 433
Crick, F. H. C., 959
Cricket thermometer, 16, 428, 546
Critical angle, for total internal reflection, 888–889
Critically damped oscillations, 418, 419
Critical point, 567
Critical reactor, 1093
Critical temperatures, 702
Crossed polarizers, 856–857, 858
Cross product, 317
Cruise, Tom, 332
Crying child, intensity level in decibels of, 445–446
Crystal
diffraction grating in, 959
diffraction of X-rays and particles by, 1018–1020
liquid, 857–858
Cunningham, Tolson, 1058
Curie, Marie, 1083
Curie, Pierre, 1083
Curie (Ci), defined, 1083
Curling stones, 261–262
Current. *See* **Electric current (I)**
Current loop, 752–753
applying Lenz's law to magnet moving to and from, 775
magnetic forces between, 753
magnetic forces on, 746
problems, 764, 765
rectangular, 746–747
Current phasor, 804, 811, 812
Curvature, center of, 875–876
C ray, 878, 879
Cutoff frequency, 1010–1011
Cylinder, Gaussian surface and, 652

Dactyl, 370, 390, 956–957
Damped oscillations, 418–419
Damping, resonance curves, 420
Damping constant, 418
Dark-adapted vision, 1009, 1031
Dark fringes, 945
for diffraction pattern of circular opening, 954
single-slit diffraction and, 951–952
two-slit experiment and, 940, 941–942, 1020, 1021
See also Destructive interference; Interference fringes
Daughter nucleus, 1078, 1079
da Vinci, Leonardo, 175, 381, 873
da Vinci Code, 175
Davisson, C. J., 1018
DC circuits, 696, 792
Dead Sea, 490
Death Valley (California), 147
de Broglie, Louis, 1017
de Broglie hypothesis, 1017–1020, 1021
problems, 1032
de Broglie wavelength, 1017–1018
Bohr orbits and, 1046–1047
problems, 1068–1069
Decay
alpha, 1078–1080, 1082, 1083
beta, 1080–1082, 1083
gamma, 1082
muon, 977
radioactive, 1077–1083
uranium, 1078–1079, 1082–1083
Decay constant, 1084–1085
Decay rate (R), 1083, 1086
of carbon-14, 1087
Deceleration, 29, 51
***deci*, 5**
Decibels (dB), 444–446
scale, 445, 446
de Coulomb, Charles Augustin, 626
***Deep Impact* (movie), 378**
***Deep Space I* spacecraft, 140**
***Deep View 66* (submarine), 509**
Deepwater waves, 470
Defibrillator, 662, 684–685, 685, 692
Deformation, elastic, 560–565
Degrees, 283
Degree terminology, Kelvin scale and, 520
***deka*, 5**
Delay circuits, 719
DeLoach, Joseph, 48
Delta notation, 19
Deltoid muscle, 356
Democritus, 1035
Density
buoyant force and, 486–487
defined, 477
of Earth, 366
electric charge, 675–676
of electric field lines, 643
energy, 685, 789, 802, 849
floatation and, 490–491
of fluids, 477
of gold, 477
measuring body, 488–489
measuring with hydrometer, 507
of Moon, 364, 366
nuclear, 1073–1075
Pascal's principle and, 485–486
problems, 508
surface charge, 637
of water, 477, 525
Dental X-rays, 865
Depth
apparent, 887
of atmosphere, 512
pressure and, 481–486
Destructive interference, 452, 453–455, 463, 938–939, 942, 945
CD operation and, 950
path lengths and, 1018
in thin films, 947–949
two-slit experiment and, 1021
See also Dark fringes
Detached retina, 1069
Detonator ride (Worlds of Fun, Kansas City), 50
Deuterium-deuterium reaction, 1094
Dewar, Sir James, 538
Diamagnetism, 756–757
Diamond, 885
Diatomic ideal gas, internal energy of, 560
Didrikson, Babe, 103
Dielectric breakdown, 683
Dielectric constants, 681
Dielectrics, 680–683
effects of, 681
problems, 691–692
Dielectric strength, 683
Diesel engines, 595
Diffraction, 950–954
of circular opening, 954
particle, by crystals, 1018–1020
problems, 967
resolution and, 954–957
X-ray, 959, 1018
Diffraction gratings, 957–961
problems, 967–968
Diffraction patterns, 1018–1022
Diffuse reflection, 872

Digital Micromirror Device (DMD), 872
Digital movie projection, 872
Dimension, 4
Dimensional analysis, 4–5, 434
problems, 14
Dimensionally consistent, 5
Dinosaur sounds, 474
Diopter, 916, 932
Dipole
electric field lines for, 644–645
equipotential surfaces for, 675
Direct-current circuits, 696
alternating-current circuits vs., 792
Direct-current power, conversion to alternating-current power, 806
Direction, 56, 108
acceleration vectors and, 67–69
circular motion and, 162–165
distance and, 56
of electric current, 698
in alternating-current circuits, 803
in induced current, 775
of electric field, 638–639, 640, 674, 853
electric potential and, 664–666
of electric field lines, 643
electric shielding and, 648
of force, 325–326
in Coulomb's law, 631
electric force, 635, 636
magnetic force, 737–739, 744–745, 752
of magnetic field, 734
momentum and, 241–243
of motion and energy conservation, 668
pressure and, 479–480
of propagation of electromagnetic wave, 839–840
of propagation of light, 885, 886
of relative motion, contracted lengths in, 979–980
of rotation, 284, 318–319
of vector, 60
unit vector, 64–65
velocity vector, 11, 69
weight and, 126–128
Directionality, 585, 600
Direction angle, 63, 64
of vector A, 57, 60
Discharging a capacitor, 719
Disk
kinetic energy of, 300–301
moment of inertia of, 298
Disk drives, 774
Disorder, entropy and, 612–614
Dispersion, 896–897
problems, 906
Displacement, 18–19
angular, 283, 343
average velocity and, 21
buoyant force as weight of displaced fluid, 487
defined, 18
force at an angle to, 182–184
force in direction of, 180–182
problems, 47–48
projectile motion and, 83
sign convention for, 19
weight and, 487, 490
Wien's displacement law, 1004–1005
Displacement vector, 56, 66
defined, 66
problems, 76–77
Distance
angular size and, 920
Coulomb's law and, 632
defined, 18
dimensions of, 5
direction and, 56
displacement vs., 19
electric potential energy and, 669–670
force and, 486
as function of time, 80
gravity and, 360
intensity variation and, 442–443
in interference fringes, 942–943
nonconservative work and, 224–225
plane mirror and, 873–874
probability as function of, 1049
problems, 47–48
sign convention for, 18
speed as function of, 383–384
stopping, 35–36, 136, 149
typical, 3
velocity as function of, 32
See also Object distance
Distribution of speeds, 556–557
Diverging lens(es), 891, 892
chromatic aberration in, 928
in telescope, 925
vision correction and, 915, 916
Diving
in bathysphere, 581
conservation of angular momentum in, 338
direction of forces in, 325–326
nonconservative work in, 224–225
speed in, 86–87
time and velocity in, 39
Diving sea mammals, 491
Division, of physical quantities, 6
DMD (Digital Micromirror Device), 872
Doloriometer/Algorimeter, 511
Dolphin ultrasound, 470
Domains, magnetic, 756
Doppler, Christian, 446
Doppler effect (electromagnetic waves), 843–845
Doppler effect in sound waves vs., 843–844
Nexrad, 844–845
Doppler effect (sound waves), 446–452
Doppler effect in electromagnetic waves vs., 843–844
general case, 450–452
moving observer and, 446–448, 450–452
moving source and, 448–452
problems, 471
sign convention for, 447, 449, 450–451
Doppler radar, 451, 844
Doppler red-shift, 843, 960
Doppler shift, 844
Dot product, work as, 183
Double concave lenses, 891
See also Concave lens(es)
Double convex lenses, 891, 892
See also Convex lens(es)
Double star, 1030
Down (quark), 1101
Drag racing, 33
Driven oscillations, 419–420
Drop tower at Glenn Research Center, 53
Dumbbell-shaped object, moments of inertia and, 296, 299
Dump trucks, static friction and, 147–148
D waves, 677
Dynamic microphone, 773
Dynamics, 107
Dynamo theory of magnetism, 735
Dyne (dyn), 112

Eagle, trajectory of, 81
Ear, human, 460–461, 471, 472
Earth
acceleration of gravity on, 38–39
atmosphere of, 382
evaporating, 570–571
visible range of frequency and, 848
average density of, 366
convection currents and, 535–536
crust, maximum depth of, 512
Earth-Moon distance, 389, 875
electric field of, 661
electric potential of, 689
escape speed of, 381–382
gravitational force between Moon and, 359, 365, 632
gravitational potential energy of, 374–376
gravitational strength of, 362–363
magnetic field of, 734–735, 737
geological record of, 756
mass of, 366
moment of inertia and, 298–299, 341
orbit of, 369
patterned ground on, 568
precession of, 345
rotation of, 83, 309
speed of, 369, 389
tides on, 385
weighing, 365–366
Earth-Moon center of mass, 277
ECG (electrocardiogram), 395, 676–677

Echocardiograms, 441
Echolocation, 440, 442
Eddington, Arthur, 992
Eddy currents, 777–778
EEG (electroencephalograph), 677
Efficiency, 601–602
Carnot cycle and maximum, 602–604
Eiffel, Alexandre, 521
Eiffel Tower, 521
Einstein, Albert, 358, 970, 986, 1007–1008
general theory of relativity, 345, 384, 991–995
special theory of relativity, 971–972
Einstein cross, 993
Einthoven's triangle, 677
EKG (electrocardiogram), 395, 676–677
Elastic collisions, 253, 259–263
defined, 259
in one dimension, 259–261
problems, 276
in two dimensions, 261–263
Elastic deformation, 565
problems, 581
solids and, 560–565
Elastic limit, 565
Electrical safety codes, 730
Electric charge, 625, 626–629, 1073
charge separation, 627–629
conservation of, 625, 627–628, 711
conservation of energy of, 667
density of, 675–676
induced, 647
magnetic force and, 735–739
polarization and, 629
positive and negative, 626, 643, 738
problems, 657
quantization of, 629
sign of, 638
systems, 644
time for *RC* circuit vs., 718
See also Coulomb's law
Electric circuits, 696–697
analyzing, using loops, 713
analyzing simple circuit, 712
with capacitors, 714–717, 719, 809–812
combination, 710
defined, 696
energy and power in, 702–705
finding equivalent capacitance in, 716–717
Kirchhoff's rules for, 711–714
measuring current in, 720–721
measuring voltage in, 720–721
oscillations in, 824–829
primary, 769
RC circuits, 717–720, 812–816
with resistors, 705–710, 719
resonance in, 803, 824–829
safety features in household, 807–809
secondary, 769
solenoids and, 753–755
Electric current (I), 696–699
alternating, 804–809
problems, 832–833
in alternating-current capacitor circuit, 810–812
in alternating-current inductor circuit, 817–818
defined, 696
direction of, 698
in alternating-current circuits, 803
of induced current, 775, 776–777
eddy currents, 777–778
electrons and, 755
induced, 769–770, 775, 776–778
loops of, 746–748
magnetic fields and, 748–752, 768
problems, 764
magnetic force and, 745–746
measuring, 720–721
for phase angle, 816
problems, 726
relation to voltage, 791, 810–812, 827
in *RC* circuit, 812–813
in *RLC* circuit, 820
resistance and, 699–702
RMS, 809–810, 819, 829
time dependence of, in *RL* circuit, 787
time vs., for *RC* circuit, 717
Electric dipole, 644–645
Electric eel, 642–643
Electric energy
conversion of mechanical energy to, 768
mechanical work and, 778–781
problems, 799
storage of, 683–685
problems, 692
Electric field(s), 625, 637–643
at conductor surfaces, 647–648
creation by changing magnetic flux, 779
defined, 638
direction of, 638–639, 640, 664–666, 674, 853
of Earth, 661
electromagnetic waves and, 838–840
energy density of, 685, 849
equipotential surfaces and, 674–676
problems, 691
magnetic field and, 739
motion in, 740
net, 640–641
of point charge, 639–640
problems, 658–659
production by magnetic fields, 768, 769–770
rate of change of electric potential and, 664–666
ratio to magnetic field in electromagnetic waves, 850
superposition and, 640–643
total, 640
work on charged particle, 740
Electric field lines, 643–646
parallel-plate capacitor and, 645–646
problems, 659
rules for drawing, 643–644
Electric fish, 642–643, 693
Electric flux
defined, 649
Gauss's law and, 650–652
problems, 659
sign of, 649, 651
Electric forces, 625, 630–637
determining direction of, 635, 636
magnetic forces vs., 739–741
net force, 634–635
role in atomic systems, 632–633
spherical charge distributions, 637, 646
superposition, 634–636
work done by, 663
See also Coulomb's law
Electric generators, 781–783
Electric guitar pickups, 773–774
Electric motors, 783–784
magnetic torque and, 748, 783
Electric potential (*V*), 663–666
defined, 663–664
electric field and rate of change of, 664–666
human body and, 676–677
loop rule and, 711–712
of point charges, 669–673
problems, 690–691
problems, 689–690
superposition of, 670–672
Electric potential energy (*U*), 663–666
change in, 663
distance and, 669–670
energy conservation and, 667–668
problems, 689–691
superposition and, 672–673
Electric power, 702–705
problems, 727
Electric shielding, 646–648
Electric shock hazard, 807–808, 809
Electric system, energy conservation in, 669
Electric voltage. *See* **Voltage(s)**
Electrocardiogram (ECG or EKG), 395, 676–677
Electrodialysis, 639
Electroencephalograph (EEG), 677
Electrognathography, 800
Electromagnet. *See* **Solenoid**
Electromagnetic flowmeter, 740–741
Electromagnetic force, 1099, 1102
Electromagnetic radiation, 1004
Electromagnetic waves, 536, 837–869
electromagnetic spectrum, 845–848
problems, 865–866
energy and momentum in, 849–853
problems, 866–867
polarization and, 853–860
production of, 838–840
problems, 864

propagation of, 841–845
direction of, 839–840
problems, 864–865
Electromagnetism, 837
Electromotive force (emf), 697–699
back emf, 784, 785
induced emf, 769–770, 772–773, 778, 781–782
motional emf, 776–779
Electron(s), 1035
atomic mass and, 1078
beta particles, 1077
Compton wavelength of, 1015
conduction, 630
diffraction pattern of, 1021
direction of flow of, 698
electric charge and, 626–629, 1073
electric current and, 755
in hydrogen atom, 1040
as lepton, 1100
mass of, 627, 988, 1073, 1081
path of, in wire, 699
photoelectric effect and, 1009–1012
properties of, 627
spin of, 755
wavelength of photon emitted by, 1044–1045
Electron escape speed, 693
Electronic configurations, 1053–1054
Electronic flash, 684
Electron microscopes, 1020
Electron neutrino, 1100
Electron-positron annihilation, 988
Electron probability clouds, 1049–1050
Electron spin quantum number, 1048
Electron volt (eV), 664, 1008
Electrostatic force, 630–631
Electrostatic precipitators, 648
Electrostatic shielding, 647–648
Electroweak theory, 1102
Elementary particles, 1099–1102
Elephants, communication between, 440
Elephant seal, 491
Elevations, fluid pressure and, 496
Elevator, apparent weight and, 126, 128
Eleventh General Conference of Weights and Measures, 2
Ellipse, 368
Elliptical orbit, 381
El Niño event, 526
emf (electromotive force). *See* **Electromotive force (emf)**
Emission nebulas, 1036
Emission spectrum, 1037
Emissivity, 537, 538
Emitter, 1009
Empire State Building, climbing, 201
Endoscopes, 890
Endoscopic surgery, bacterial infection from, 629
Energy, 204, 768
of Bohr orbit, 1042–1043
conservation of. *See* Energy conservation
dimensions of, 5
in electric circuits, 702–705
in electromagnetic waves, 849–853
entropy of universe and, 612
from gasoline, 619
internal, 209, 559–560, 586–588
mass-energy equivalence, 986–989, 1073
problems, 1001
nuclear binding, 1089–1090
problems, 727
quantized, 1005
Planck's hypothesis of, 1005–1007
relativistic, 986–989
problems, 1001
released in nuclear fission reaction, 1090–1091
released in nuclear fusion reaction, 1093
rest, 986, 1100, 1101
stored in magnetic field, 788–790
problems, 800
total, 986
uncertainty in, 1024
waves and, 431
work-energy theorem, 186–190, 214, 222, 343
See also Electric energy; Electric potential energy (U); Heat; Kinetic energy; Mechanical energy
Energy conservation, 299–304, 341, 666–669
of electric charge, 667
in electric system, 669
first law of thermodynamics and, 586–588
gravity and, 377–384
problems, 391
Lenz's law and, 775
loop rule and, 711–712
mechanical energy, 214–222
momentum conservation vs., 255
in oscillatory motion, 409–412
problems, 426–427
phase changes and, 574–576
problems, 583
photon scattering and, 1014
problems, 311–312, 690
specific heats and, 530
in transformers, 791
Energy density, 685, 789, 802, 849
of electric field, 685, 849
of electromagnetic wave, 849
Energy-level diagram, 1043
Energy levels
of hydrogen, 1048, 1049
maximum number of electrons in, 1052
of multielectron atoms, 1050–1051
Energy plot, 228–230
Energy recovery technology in cars, 784
Energy transfer. *See* **Heat flow**
Energy usage, 705
Engines, 281, 600–604
diesel, 595
See also Heat engines
Entoptic halos, 969
Entropy, 608–612
heat and, 608–612
order and disorder and, 612–614
problems, 622
Sun and, 623
in universe, 610–612, 613
Entropy change, 608, 609–610
Equations
of angular velocity, 286
Bernoulli's equation, 495–501
Compton shift, 1015
constant-acceleration equations of motion, 30, 31, 32, 34, 82
of continuity, 494
kinematic, 287
linear vs. angular, 287
magnitude of electron's charge, 626–627
mirror, 878, 880–884
of motion, 29–34, 82, 84–87
applications of, 35–37
problems, 50–51
x equation of motion, 80–81
y equation of motion, 80–81
Poiseuille's equation, 503
of projectile motion, 83–84
Schrödinger's, 1047, 1048, 1049
of state, for ideal gases, 549–552
thin-lens, 894–896, 914
transformer, 791
Equilibrium
determining, 575
of object at rest, 332
oscillations and, 394
phase, 565–569
rotational, 324, 328
static. *See* Static equilibrium
of suspended object, 331
thermal, 515, 516, 529
translational, 154–158, 324, 328
Equilibrium vapor pressure, 566
Equipotentials, 230
electric field and, 674–676
ideal conductors, 675–676
problems, 236, 691
Equivalence
of physical laws, 971
principle of, 992
Equivalent capacitance
for capacitors in parallel, 714–715
for capacitors in series, 716–717
finding for electric circuit, 716–717
Equivalent resistance, 705–710
Kirchhoff's rules vs., 713
for resistors in series, 705–707
Escape ramps, 227
Escape speed, 381–384, 693, 993–994
Estimation, order-of-magnitude calculations, 10–11
Eta, 1100

Eta Carinae, 252
Ethyl alcohol, 885
Europa, 370
Evaporation, 569–571
problems, 581–582
Event, 974, 975
Excess charge, 646–647
Excimer laser, 1060
Excited states, 1043
of hydrogen, probability clouds for, 1050
Expansion joints, 522–523
External force, 113–118
connected objects and, 158–159
momentum and, 249–252
nonzero net, 268–269
zero net, 266–268
External torque, 339
Extrasolar planet, 390
Extraterrestrial volcano, 390
Eye
largest, 935
octopus, 931
Eyeglasses, 909, 913, 918
Eye (human), 910–912
adaptation to range of frequencies, 848
farsightedness and shape of, 917
length contraction and, 980
nearsightedness and shape of, 915
problems, 931–932
resolution of, 969
tracking movement of magnetically, 781
Eyepiece, 923

Fahrenheit, Gabriel, 516
Fahrenheit scale, 516–518
compared to Kelvin and Celsius scales, 520
conversion to Celsius scale, 517–518
Faraday, Michael, 678, 769
Faraday's law of induction, 772–774
calculating magnitude of induced emf and, 778
Lenz's law and, 775–778
problems, 797–798
Farad (F), 678
Far point, 911
Farsightedness, 917–919
FDG (fluorodeoxyglucose), 1097–1098
***femto*, 5**
Femtometer (fm), 1074
Fermi, Enrico, 10, 1074, 1081, 1093
Fermi (fm), defined, 1074
Fermi Problems, 10
Ferris wheel, 174
Ferromagnetism, 755–756
Ferry speed, 29
Fever thermometers, 702
Feynman, Richard, 94
Fields, electric. *See* Electric field(s)
Fifth harmonic, 460
Fireflies, cold light Of, 1033
Fireman's pole, 125
First harmonic, 456–461
First law of thermodynamics, 585, 586–588
problems, 619
First minimum. *See* Dark fringes
Fishes
electric, 642–643, 693
fish scale, 127
swim bladders, 491
Fishing
rotational work and, 343
torque and angular acceleration and, 320–321
Fission reactions, 986, 1090–1093
Fizeau, Armand, 841–842
Flashlight, 697, 698
Flea's jump, 214, 238
Floatation, 490–493
Fluids, 476–513
Archimedes' principle and buoyancy, 486–487
Bernouilli's equation, 495–501
applications of, 498–501
critical point and, 567
defined, 476
density of, 477
fluid flow and continuity in, 493–494
problems, 510
gravitational potential energy in, 484
pressure in, 477–480
applications, 498–499
pressure-speed relationship, 495–496
seeking own level, 483–485
static equilibrium in, 481–486
surface tension in, 503–504
Torricelli's law, 500–501
viscosity in, 501–503
volume flow rate, 502–503
See also Ideal gases
Fluorescence, 1062–1064
Fluorescent light, 938
Fluorescent light bulb, 1062–1063
Fluorine, 1054, 1055
Fluorodeoxyglucose (FDG), 1097–1098
***f*-number, 912–913**
Foamcrete, 116–117
Focal length, 876, 906, 911–912, 915
for concave mirror, 876
of contact lenses, 919
for convex mirror, 876
object distance and, 895
refractive power and, 916
sign convention for, 882, 895
in thin-lens equation, 894
Focal point
for concave and convex mirrors, 876
on convex lens, 892
Focal-point ray (F ray), 878, 879
lenses and, 892
Food energy, converting to mechanical energy, 211
Foot, forces in, 352
Football, 93, 255–256
"lob" pass vs. a "bullet," 105
Foot (ft), 4
Footwear safety, 730–731
Force(s), 108
acceleration and, 110–112
at angle to the displacement, 182–184
buoyant, 486–487
centripetal, 163, 167, 369–370, 385
conservative and nonconservative. *See* Conservative forces; Nonconservative forces
contact, 120–121
deformation of solid and, 563–564
direction of, 325–326
in direction of displacement, 180–182
distance and, 486
electromagnetic, 1099, 1102
electrostatic, 630–631
exerted on the Moon, 138
external. *See* External force
frictional, 108, 142–149
horizontal components of, 327–329
kinetic friction and, 142–146
measurement of, 112
net gravitational, 360–362
normal forces, 114, 128–132, 142, 143, 144, 166
nuclear, stability and, 1075–1076
origin of pressure and, 555–556
on pendulums, 414
radial, 317
repulsive, 626, 733
restoring, 394
of running, 139
shear deformation of solid and, 563
spring, 152–154, 212
static equilibrium and, 323–326
for structural stability, 327–328
tangential, 316–317
torque for, 317–319
typical, 112
units of, 112
variable, 190–194
vector nature of, 108, 121–124
problems, 137–138
vertical components of, 327–329
weight as, 124–125
work and. *See* Work
See also Gravity; Static equilibrium; Torque
Force constant, 152
Force field. *See* Electric field(s)
Force meter, 110
Force vectors, kinetic friction and, 145–146
Forensics, fluorescence in, 1063
Fosbury, Dick, 274

Fosbury flop, 265, 274
Fountain Hills, Arizona, 52
Fountain jets, 87
Frames of reference, 109–110
accelerated, 991
in gravitational field, 991
inertial, 109, 971, 991
velocity relative to, 981
Francium, 1054, 1055
Franklin, Benjamin, 626
F ray, 878, 879, 892
Free-body diagrams, 113–118
Free fall, 37–43, 358, 359
acceleration in, 40–43, 83
defined, 38
gravity and, 37–43
problems, 51–53
from rest, 39–40
weightlessness and, 128
Free space
permeability of, 750
permittivity of, 650
Freezing points, 517, 520
French Academy of Sciences, 2
Frequency(ies)
angular, 400
beat, 463–464
behavior of capacitor as function of, 810
of current in *RL* and *RC* circuits, 818–819
cutoff, 1010–1011
defined, 395
Doppler effect and, 449–450
large and small, 822–824
natural, 419–420, 803, 825, 826
quantum and, 1006
in *RC* circuit, 814
of sound wave, 439–441
temperature, electromagnetic radiation, and, 1004
wavelength and, 845–848
of waves, 433
Frequency bands, 846
Fresnel lenses, 935
Friction, 108–109, 588–589
heat engine and, 603
kinetic, 142–146, 205–208, 225–226
magnetic, 777
as nonconservative force, 205, 209
origin of, 142
static, 146–149
Frictional force, 108, 142–149
problems, 170–171
Fringes. *See* Interference fringes
***From the Earth to the Moon* (Verne), 381, 390**
Frost heaving, 568
Frost wedging, 568
Fundamental forces, 1099
evolution of, 1102
Fundamental mode of vibration, 456–461
Fuses, 807, 808, 809
Fusion
lasers for, 866
latent heat of, 571–572, 575
nuclear, 1093–1095
Fusion curve, 567, 568
Fusion reactions, 987, 1094

Gaits, analyzing, 417
Galaxy
diameter, 2
Doppler effect and light of, 451–452
mass, 3
Galileo (Galileo Galilei), 25, 37–38, 109, 358, 370, 384, 412–413, 413, 841, 865, 925
Galileo's telescope, 934
Galileo thermometer, 907
Gallinule, 478
Gallium, 1055
Galvanometer, 748
Gamma decay, 1082
Gamma rays, 846, 848
penetrating ability, 1078
in positron-emission tomography, 988–989
relative biological effectiveness for, 1096
Gas(es). *See* Fluids; Ideal gases
Gas lanterns, mantles in, 1109
Gasoline, energy from, 619
Gauge pressure, 480
Gaussian surface, 651–652
Gauss's law, 645, 650–652
parallel-plate capacitor and, 678
problems, 659
Gay-Lussac, Joseph, 554
Gecko foot, 140
Geiger, Hans, 1035
Gell-Mann, Murray, 1100–1101
General launch angle, 89–92
problems, 102–103
General loops, 747
General relativity, 384, 991–995
gyroscopes and, 345
problems, 1001
Generators
alternating-current, 782, 804
electric, 781–783
problems, 799
Geomagnetism, 734–735
Geometrical optics, 870–908
dispersion and the rainbow, 896–897
forming images with plane mirror, 872–875
ray tracing and mirror equation, 877–884
ray tracing for lenses, 891–894
reflection of light, 870, 871–872
refraction of light, 884–891
spherical mirrors, 875–880
thin-lens equation, 894–896, 914
Geometric center, 331
Geosynchronous satellites, 372–373
Germer, L. H., 1018
GFI (ground fault interrupter), 808, 809
GFP bunny, 1034, 1064
GFP (green fluorescent protein), 1034, 1063–1064
***giga*, 5**
Glacial speed, 15
Glare, reflected, 859–860
Glashow, Sheldon, 1102
Glass
electric charge and, 626
index of refraction of, 885, 886
light propagating through, 888
light ray passing through slab of, 908
Global Positioning System (GPS), 372, 373, 390
Gold
resistivity and, 700
weight and density of, 477–480
Golden ratio, 279
Golf, 90–91, 225
***Gossamer Albatross*, 195, 201**
***Gossamer Condor*, 195**
GPS, 372, 373, 390
Graduation fling, 216–218
Gram (g), 4, 112
Grandfather clock, 394, 415
See also Clocks
Grand unified theory (GUT), 1102
Grasshopper liftoff, 139
Grating spectroscopes, 959–960
Gravitational force, net, 360–362
Gravitational lensing, 384, 993
Gravitational potential energy, 209, 210–212, 374–377
Bernoulli's equation and, 497
in fluid, 484
height and, 210–212, 227–228
kinetic energy and, 215, 227–228
"Gravitational slingshot" effect, 280
Gravitational strength (g), 38–39, 362–363
See also Acceleration of gravity
Gravitational work, 187
Gravitropism, 170
Gravity, 358–393
acceleration of. *See* Acceleration of gravity
as conservative force, 205–208
Coulomb's law and Newton's law of, 632
energy conservation and, 377–384
problems, 391
fluid flow and, 496
free fall and, 37–43
as fundamental force of nature, 1099, 1102
general relativity and, 991–995
gravitational attraction of spherical bodies, 362–367
problems, 390
gravitational potential energy, 209, 210–212, 215, 227–228, 374–377, 484, 497
problems, 391

Gravity (*Continued*)
hadrons and, 1100
Kepler's laws of orbital motion and, 367–374
light and, 992–993
motional emf and, 776
pendulums and, 414, 415–416
planetary evaporation and, 570
projectile motion and, 79
radiation pressure and, 852
role in atomic systems, 632–633
tides, 384–386
torque and, 331
universal law of gravitation, 358–362, 632
problems, 389–390
weight as gravitational force, 124–128
work and, 186–187, 188
Gravity escape system (GES), 183
Gravity maps, 416
Gravity wave detectors, 995
Gravity waves, 995
Green fluorescent protein (GFP), 1034, 1063–1064
Green quarks, 1101
Grindstones, 297
Grounded plugs, 808, 809
Ground fault circuit interrupter (GFCI), 808, 809
Grounding, 649
Ground spider, polarization and, 863, 868
Ground state, 1043
of hydrogen, probability cloud for, 1049
Pauli exclusion principle and, 1051–1053
Guitars, 458–459, 459, 773–774
Gurneys, 181
GUT (grand unified theory), 1102
Gyroscopes, 345–346

H4 chronometer, 75
Hadrons, 1100
quark composition, 1101
Hafele, J. C., 975
Hahn, Otto, 1090
Hail vs. rain, momentum of, 246–247
Hair, thickness of, 14
Hale reflecting telescope, 904, 909, 932
Half-life (T1/2), 1076, 1083–1089
defined, 1085
problems, 1107–1108
Halley, Edmund, 107, 358
Halley's comet, 391
Halogen light bulbs, 1030
Halogens, 1055
Hanging in air, 90
Harmonics, 456–463
Harmonic wave, 431
Harmonic wave functions, 436–437
problems, 470
Harrison, John, 75
Hawking, Stephen, 358
HDTV, 968
Headlights, distinguishing, 955–956, 956
Heads-up display, 874
Hearing aids, 774
Hearing a pin drop, 473
Hearing loss, measuring, 473
Hearing range, human, 440
Heart
electrocardiographs and, 676–677
heartbeat detectors, 251
pacemakers, 720
power output of, 202, 511
pump rate, 510
Heat, 514
defined, 515
entropy and, 608–612
latent, 571–574
laws of thermodynamics and, 585
measurement units of, 526–527, 529
mechanical equivalent of, 526, 527
mechanical work and, 526–528
problems, 544
resistance heating, 703–704
sign conventions for, 528, 586, 587
specific heats. *See* Specific heat(s)
thermal expansion and, 520–526
work and, 209, 586–588
See also Temperature; Thermodynamic processes
Heat capacity, 528–529
Heat death, 613
Heat engines, 600–604
air conditioners vs., 605
Carnot cycle and, 602–604, 606
efficiency of, 601–602
entropy change and, 609–610
heat pumps vs., 605
living organisms as, 614
maximum efficiency of, 602–604
maximum work by, 603
problems, 621–622
refrigerators vs., 605
Heat flow, 515, 531–538, 592–593
by conduction, 531–535
convection, 535–536
disorder and, 612–613
phase changes and energy conservation, 574–576
radiation, 536–538
second law of thermodynamics and, 600
thermal equilibrium and, 529
Heat pumps, 606–608
heat engine vs., 605
problems, 622
Heat transfer, 515
Heavy ions, relative biological effectiveness for, 1096
***hecto*, 5**
Height
amplified rebound, 280
change in, of fluid, 496–497
determining, 58, 227
gravitational potential energy and, 210–212, 227–228
maximum, 96–97
speed and, 215, 217–219
walking and, 113
Heisenberg, Werner, 1022
Heisenberg uncertainty principle, 1021–1025
energy and time and, 1024
momentum and position and, 1022–1024, 1025
problems, 1032
Helical motion, 743–744
Helicopter thrust, 277–278
Helioseismology, 428
Helium
in blimp, amount of, 580
Bohr model and, 1042
electronic configuration, 1054
in periodic table, 1053
Helium-neon laser, 1059–1060
Hematocrit value, 292, 502
Henry, Joseph, 785
Henry (H), 785
Hero of Alexandria, 601
Hero's engine, 601
Hertz, Heinrich, 395, 838
Hertz (Hz), 395
High-definition television (HDTV), resolving lines on, 968
High-temperature superconductivity, 702
High-voltage electric power transmission, 792
Hohmann transfer, 373
Hole, expansion of, 523
Hollow shapes
kinetic energy of, 301
moments of inertia of, 298
Holograms, 1061–1062
Honeybee
electrostatic charge of, 658, 660
velocity of, 252
Hooke, Robert, 153
Hooke's law, 152–153, 154, 396, 562, 565
Hoop, moment of inertia of, 297, 298
Horizontal components of force, 327–329
Horizontal motion, 79
independence of, 84
Horizontal resistors, 712–713
Horsepower (hp), 195, 355
Horseshoe magnet, 733
"Hot" reservoir, 601, 602–604, 605
Hubble Deep Field, 843
Hubble's law, 452
Hubble Space Telescope (HST), 252, 346, 372, 877, 967, 1012
Human centrifuge, at NASA's Manned Spacecraft Center, 174
Human radiation, 866
Humans
breathing, surface tension in, 503–504
cold hands and feet, 535

countercurrent heat exchange in arm of, 535
ear, 460–461, 471, 472
electric potential and, 676–677
evaporative cooling, 569
hearing range, 440
human polar bears, net radiated power from, 537–538
lung capacity, 550
measuring mass, 406
perception of pitch, 458
power output, 195, 201
resistivity and, 726–727
sound perception, 444
stretching bone, 561–562
vision, 1033
"wave," 431
See also Brain; Eye (human); Heart
Hummingbird acceleration, 82–83
Hurricane Andrew, 340
Hurricanes, 339–340
Huygens Probe, 16
Huygens's principle, 941, 950
"Hybrid" cars, 784
Hydraulic lift, 485–486
Hydroelectric power, 781, 782
Hydrogen, 1053
electronic configuration, 1054
speed distribution for, 570
Hydrogen atom
Bohr's model, 632–633, 1039–1046
assumptions of, 1039–1040
problems, 1068
dissociation energy for, 1030–1031
probability clouds and, 1049–1050
quantum mechanical, 1047–1050
spectral series in, 1037–1039, 1044
spectrum of, 1036–1039, 1043–1045
problems, 1068
states of, 1049–1050
Hydrometer, measuring density with, 507
Hydrostatic Paradox I and II, 512

Ice
change in entropy and, 609
formation of, 574
index of refraction, 885
latent heat and, 571, 572
melting, 568, 575–576
Icebergs, 492–493, 525, 526
Ice block, acceleration of, 129–130
Ice cream maker, latent heat and, 574
Ice man of the Alps, 1087, 1088
Ida, 370, 390, 956–957
Ideal conductors, 675–676
Ideal gases, 548, 549–555
constant pressure of, 554–555
equation of state for, 549–552
internal energy of, 559–560
isotherms and, 552–554, 591–592
kinetic theory of, 555–560
problems, 579–580
specific heats for, 596–599
work done by, 589–595
Ideal gas law, 550
Ideal heat pump, 606, 607
Ideal inductor, 818
Ideal pulley, 150
Ideal reflector, 538
Ideal springs, 153
Illusions, optical, 887–888
Image
finding, for multilens system, 914
real, 879, 882, 893
virtual, 878, 882, 893, 918
Image distance, 873–874, 878, 879, 880–884
sign convention for, 882, 895
Image formation
with concave mirror, 879
with convex mirrors, 878–879
in human eye, 910
with plane mirror, 872–875
problems, 903
ray tracing, 878–880
Image orientation, 873–874, 880
Image size, 874, 880
Immunofluorescence, 1063
Impedance (Z)
in alternating-current circuits, 822
current and, 826–827
in *RC* circuits, 812–815
in *RLC* circuit, 820
in *RL* circuits, 818
Impulse, 244–248
defined, 244
magnitude of, 248
momentum and, 245–246
problems, 275
as vector, 245
Incandescent light, 938
Incandescent light bulb, 703, 853
Incidence, angle of, 871, 873, 888
Brewster's angle, 891
Incident ray, 871
Incline
coordinate system and, 130–131, 145
moment of inertia and, 301–302
object on, 130–131
traveling uphill, 196, 197
weighing block on, 280
Incoherent light, 938
Incompressible fluid, equation of continuity for, 494
Independence of motion, 79, 83, 84, 87
Index of refraction (*n*), 885–890
angular resolution and, 955
for common substances, 885
defined, 885
direction of propagation of light and, 885, 886
human eye and, 910
phase changes due to reflection and, 944–945
Induced charge, 647
Induced current, 769–770
direction of, 775, 776–777
eddy current, 777–778
force and, 778
Induced emf, 769–770, 781–782
magnitude of, 772–773, 778
maximum and minimum values, 782
Inductance, 784–786
defined, 785
problems, 799
of solenoid, 785–786
Induction
charging by, 648–649
Faraday's law of, 772–774
magnetic, 769
Induction loops, 774
Induction stove, 778
Inductive reactance, 817
Inductor circuits, 817–818
Inductor phasor, 820, 821
Inductors, 786
in alternating-current circuits, 817–820
problems, 834
energy stored in, 788–790
in *RL* circuits, 787–788
Inductor voltage, 822
Inelastic collisions, 253–258
defined, 253
in one dimension, 254–257
problems, 276
in two dimensions, 257–258
Inertia
law of, 109
moment of, 295–299, 301–302, 338, 339, 341
oscillations and, 394, 415
pulley and, 333
Inertial confinement, 1095
Inertial frame of reference, 109, 971, 991
Infrared radiation, 536
Infrared receptors, 847
Infrared waves, 846, 847
Infrasonic sounds, 440–441, 441
Ink-jet printers, 646
In phase, 453–455
Insects, tracking movement magnetically, 781
Instantaneous acceleration
defined, 26
graphical interpretation of, 27
Instantaneous acceleration vector, 68
Instantaneous angular acceleration, defined, 285
Instantaneous angular velocity, defined, 284
Instantaneous speed, 23

Instantaneous velocity, 23–25
  calculating, 24
  defined, 23
  graphical interpretation of, 25
  problems, 49
  See also Average velocity; Velocity(ies)
Instantaneous velocity vector, 66–67
Instantaneous voltage, 811–812
Insulated windows, 534–535
Insulators, 531, 629–630
  adiabatic process and, 593–595
  resistivities of, 700
Intact, 917
Intensity (*I*), 850–851
  intensity level vs., 444, 445–446
  polarizer and, 854–855, 856
  of sound, 441–446
  sunlight, 442, 866
Intensity level, 444–446
  calculating, 445
  defined, 444
  intensity vs., 444, 445–446
Interference, 938–940
  constructive, 452, 453–455, 463, 938–939, 942, 945, 947–949, 958, 1018
  destructive, 452, 453–455, 463, 938–939, 942, 945, 947–949, 950, 1018, 1021
  problems, 964–965
  in reflected waves, 944–950
    problems, 966–967
  two-slit experiment, 940–944
  wave, 452, 453–455
    problems, 471–472
Interference fringes
  in air wedge, 945–946
  circular, 946–947
  spacing of, 943–944
  in two-slit experiment, 940–944
  See also Bright fringes; Dark fringes
Interference patterns, 453, 454, 463
  diffraction gratings and, 957–958
  diffraction patterns, 1018–1022
  in two-slit experiment, 1020–1021
  wave properties and, 1018
Intermolecular spring, 560–561
Internal energy
  of ideal gases, 559–560
  of system, 586
  work, heat, and, 209, 586–588
Internal forces, momentum and, 249–252
Internal resistance, 707
Internal torque, 339
International Bureau of Weights and Standards, 3
International Space Station, 373
Interstellar magnetic field, 800
Intracorneal rings, 917, 931
Intraocular lens, 935
Inverse square dependence, 360
Inversion under reflection, 436
Io, 370, 370–371, 385–386
  volcanoes on, 52, 103, 390
Iodine, 1055
Iodine-131, 1097, 1107
Ionization, 1043
Iridescence, 961
Iris, 911
Iron
  ferromagnetism and, 755–756
  resistivity and, 700
Irradiated food, 848
Irreversible process, 588
  entropy change and, 608–612
Isothermal curve, vs. adiabatic curve, 594
Isothermal processes, 591–593
Isotherms, 552–554
  adiabats vs., 598
Isotopes, 1072
  separation of, 742–743

Jack and Jill, 122–123
Jellyfish, fluorescence in, 1063
***Jeopardy!* (TV), "buzzing in" on, 234**
Jernigan, Tamara, 406
Jiffy (unit of time), 15
Jordan, Michael, 52
Joule, James Prescott, 180, 526
Joule (J), 180, 187, 527, 664
  defined, 180
Juggling, 94
Jumping, 247–248
Junction, 711
Junction rule, 711, 712
Jupiter
  atmosphere evaporation and, 570
  mass of, 370–371
  tidal locking on, 385–386

Kamerlingh-Onnes, Heike, 702
Kaon, 1100
Keating, R. E., 975
Keck reflecting telescopes, 927
Kelvin, William Thomson, Lord, 519, 603
Kelvin scale, 519–520
  Carnot cycle and, 603
  Charles's law and, 554
  compared to Celsius and Fahrenheit scales, 520
  conversion to Celsius scale, 520
  degree terminology and, 520
  entropy change and, 608
  ideal gas law and, 550
  kinetic energy of gas and, 557
  special significance of, 586
Kepler, Johannes, 367–368
Kepler's laws of orbital motion, 367–374
  Kepler's first law, 368
  Kepler's second law, 368–369
  Kepler's third law, 369–373
  orbital maneuvers and, 373–374
  problems, 390–391
Keratometer, 919, 935
***kilo*, 4, 5**
Kilocalorie (kcal), defined, 526
Kilogram (kg), 2, 112, 124
  defined, 3
  standard, 3
Kilohertz (kHz), 395
Kilowatt-hours (kWh), 705
Kinematic equations, 287
Kinematics
  average velocity, 21–23, 31
  constant acceleration, 29–34
  distance, 18
  energy conservation solution, 215–217
  equations of motion, 29–34, 82, 84–87
  free fall, 37–43, 358, 359
  general launch angle, 89–92
  instantaneous velocity, 23–25
  one-dimensional, 17–54
  projectile motion, 79, 83–97
  quadratic formula from, 54
  rotational, 286–289
  speed, 19–21
  two-dimensional, 79–106
  zero launch angle, 84–89
Kinetic energy, 179
  alpha decay and, 1079
  beta decay and, 1081
  classical, 989
  collisions and, 253–258
  defined, 187
  in elastic collisions, 259, 261, 262, 263
  of fluid element, 495, 496
  as function of time, 411
  gravitational potential energy and, 215, 227–228
  in inelastic collisions, 253–258
  maximum, 410–411, 1010–1012
  of object falling toward Earth, 379
  in oscillatory motion, 409–412
  of photoelectrons, 1010, 1011, 1012
  potential energy and, 209
  pressure vs., 498
  relativistic, 989–990
  of rolling motion, 300–304
  rotational, 295–299, 300–301, 302, 303
  temperature and, 557–559
  typical kinetic energies, 187
  work-energy theorem and, 186–190, 214, 222, 343
    problems, 200–201
Kinetic friction, 142–146, 225–226
  coefficient of, 143, 144
  as nonconservative force, 205–208
  normal force and, 142, 143, 144
  rules of thumb, 143–144
Kinetic theory of gases, 555–560
  problems, 580–581
Kingfisher, average speed of, 20

Kirchhoff, Gustav, 711
Kirchhoff's rules, 711–714
applications, 712–714
equivalent resistance vs., 713
junction rule, 711, 712
loop rule, 711–712
problems, 728–729
Krypton, 1055
K shell, 1056–1057, 1058

Ladder
balance and, 332
vertical and horizontal components of force and, 329
Lagoon Nebula, 1036
Lakes, ecology of, 526
Lamps, "touch-sensitive," 715
Land breezes, 535, 536
Landing site, 87–89
Laser(s), 938, 1059–1062
for fusion, 866
laser surgery, 867
safety, 867
Laser eye surgery, 1060–1061
problems, 1069, 1070
Laser Interferometer Gravitational Wave Observatory (LIGO), 995
Laser printers, 630, 959
LASIK (laser in situ keratomileusis), 1060, 1061
Latent heats, 571–574
defined, 571
of fusion, 571–572, 575
problems, 582–583
specific heats vs., 573
of sublimation, 571
of vaporization, 571
Launch angle, 94, 96
general, 89–92
zero, 84–89
Launch from the origin, 89–90, 91
Lava, 536
Lava bomb, 40–42, 87
Law(s)
Coulomb's, 630–637, 1040, 1075
Hooke's, 152–153, 154, 396, 562, 565
of inertia, 109
Lenz's, 775–778, 779
of Malus, 854, 856
of nature, 2
Ohm's, 699–702, 703, 712
of reflection, 871–872
Snell's, 885–886, 888, 891, 908
Wien's displacement law, 1004–1005
Laws of thermodynamics, 585–624
first law, 585, 586–588
problems, 619
second law, 585, 600, 610, 613, 614
third law, 585, 614
zeroth law, 515–516, 586
***LC* circuits**
natural frequency of, 825, 826
oscillations in, 824–826
LCD (liquid-crystal display), 857–858, 860
Lead-207, 1082, 1083
Leaf falling to ground, nonconservative work in, 223–224
Leg, as physical pendulum, 417, 418
Length, 1, 2–3
changing length of solid and, 560–563
in heat conduction, 532
of pendulum and period, 413, 415
proper, 978, 979
relativity of, 978–980, 983
problems, 999–1000
of string, pitch and, 458, 459
See also Focal length
Length contraction, 978–980
problems, 999–1000
Lens(es), 891
aberrations in, 927–928
in binoculars, 949
in cameras, 912–913, 928, 949
in combination, 913–919
problems, 932–933
in compound microscopes, 923
converging, 891–892, 916, 917–918, 927
corrective, 913–919
problems, 932–933
diverging, 891, 892, 915, 916, 925, 928
Fresnel, 935
in human eye, 910
ray tracing for, 891–894
problems, 905–906
sign conventions for, 895
in telescopes, 925–926, 927
thin-lens equation, 894–896, 914
See also Concave lens(es); Convex lens(es)
Lenticular cloud, 600
Lenz, Heinrich, 775
Lenz's law, 775–778
determining the direction of the current using, 779
problems, 798–799
Leptons, 1099–1100
Leucippus, 1035
Levitation, magnetic, 757
Lewis, Gilbert Newton, 15
Lift, 498
Lifting, vertical, 188–189
Light
blue, 858–859, 943
coherent, 938
cutoff frequency and frequency of, 1010–1011
direction of propagation of, 885, 886
Doppler effect and, 446, 451–452
fluorescent, 938
gravity and, 992–993
interference of reflected, 945
passing through polarizers, 853–858
photoconductive material and, 630
photon model of, 1008
polarization by scattering and reflection, 858–860
polarized vs. unpolarized, 853, 854
red, 896, 897, 948
reflection of, 858–860, 870, 871–872, 888–890, 944–945
refraction of, 884–891
speed of, 3, 841–843, 865, 885, 886
constancy of, 971–972
in vacuum, 841, 971, 972
ultraviolet, 536, 846, 848, 866
violet, 896, 897
visible, 536, 837, 846, 847–848
wave nature of, 431, 937
See also Physical optics
white, 896, 897, 1011–1012
See also Photon(s)
Light amplification, 1059
Light bulb, 779–781
fluorescent, 1062–1063
halogen, 1030
incandescent, 703, 853
intensity and, 850–851
three-way, 707, 709, 730
Light clock, 973–974
Light dimmers, 819–820
Light fringes, two-slit experiment and, 1020
Lightning, 692, 697
thunder and distance of, 438–439
Lightning rod, 648, 676
Light power, 779–780
work converted to light energy, 780–781
Light sails, 1014
LIGO, 995
Limits, problem solving and, 12
Lindbergh, Charles, 274
Linear distance, from central fringe, 942–943
Linear expansion, 520–523
coefficient of, 521, 523, 524
Linear kinematic equations, vs. angular, 287
Linearly polarized wave, 853
Linear momentum, 241–243, 335
conservation of, 248–252
defined, 241
problems, 275
Linear motion, angular momentum in, 337
Linear position, of fringes, 942
Linear quantities
angular analogs, 286, 299, 320
rotational quantities and, 289–294
problems, 309–310
Linear reasoning, 190
Linear speed, 290
rolling object and, 294–295
Line of sight, electromagnetic waves and, 840
Line spectrum, 1037
Linnaeus, Carolus, 516
Liquid crystal, 857–858
Liquid-crystal display (LCD), 857–858, 860

Liquids, 476
See also Fluids
Lithium
Bohr model and, 1042, 1043
electronic configuration, 1054
in periodic table, 1053
Living systems, entropy and, 613–614
Location, vectors and, 61–62
Lodestone, 732
Longitudinal waves, 432
Long-Range Lineup System (LRLS), 935–936
Loop detectors on roadways, 802
Loop rule, 711–712
application of, 712–714
Loop-the-loop track, 239
Los Alamos National Laboratory, 441
Loudest animal, 473
Loudness, beat frequency and, 464
Loudspeaker
sound intensity of, 442
sound produced by, 432
Luminiferous ether, 972
Lunar gravity, 364–365
Lunar module, 391
Lung capacity, 510, 550
Lyman series, 1038–1039, 1044

M87, 994
McCarthy, Eddie, 1058
MacCready, Paul, 195
Macular degeneration, age-related, 1061
Magnesium, 1054
Magnet(s)
bar, 732, 733, 734, 737, 752–753, 772–773, 775
horseshoe, 733
permanent, 733
refrigerator, 734
wire loop and, problems on, 798
Magnetar, magnetic field, 737, 766
Magnetic antitheft devices, 781
Magnetic braking, 777
Magnetic confinement, 1095
Magnetic dipole field, 645
Magnetic disk drives, 774
Magnetic domains, 756
Magnetic field(s), 732, 733–735, 756
applying Lenz's law to, 776
in Bohr model, 767
brain function and, 765
circular motion in, 741–743
of current loop, 752–753
direction of, 734
of Earth, 734–735, 737, 756
electric currents and, 748–752, 768
problems, 764
electric field and, 739
production of, 768, 769–770
electromagnetic waves and, 838–840
energy stored in, 788–790
problems, 800
helical motion in, 743–744
interstellar, 800
of long, straight wire, 748–749, 750–751
motion of charged particles in, 739–744
problems, 762–763
radioactivity in, 1077–1078
ratio to electric fields in electromagnetic waves, 850
of and in solenoid, 754
typical, 737
Magnetic field lines, 733–734
inside solenoid, 753
Magnetic field right-hand rule, 749
Magnetic flux, 770–771
defined, 770
energy stored in magnetic field and, 788–790
generators and motors and, 781–784
inductance and, 784–786
Lenz's law and, 775–778
mechanical work and electric energy and, 778–781
problems, 797
RL circuits and, 787–788
transformers and, 790–792
units of, 770
See also Faraday's law of induction
Magnetic forces
between current-carrying wire, 751–752
on current-carrying wire, 744–746
problems, 763–764
between current loops, 753
on current loops, 746
electric forces vs., 739–741
magnitude of, 735–737
on moving charges, 735–739
problems, 762
properties of, 736
right-hand rule (RHK) for, 737–739
Magnetic friction, 777
Magnetic induction, 769
Magnetic levitation, 757
Magnetic moment, 747
Magnetic poles
of bar magnet, 733
of Earth, 734–735
Magnetic quantum number, 1048
Magnetic reed switches, 755
Magnetic resonance imaging (MRI), 1098–1099
magnetic field and, 737
solenoid in, 755, 765, 797
Magnetic stars, 777
Magnetic torque, 746–748
electric motors and, 748, 783
problems, 764
Magnetism
current loops, 746, 752–753, 775
dynamo theory of, 735
electric currents, magnetic fields, and Ampere's law, 748–752
loops of current and magnetic torque, 746–748
magnetic field. *See* Magnetic field(s)
magnetic force exerted on current-carrying wire, 744–746
magnetic force on moving charges, 735–739
in matter, 755–757
motion of charged particles in magnetic field, 739–744
solenoids, 753–755, 785–786, 789
temperature and, 756
Magnetite, 756
Magnetoencephalography, 765
Magnification (m), 882, 883–884
angular, 920–922, 923, 925–926
comparing, 922
of compound microscope, 923
defined, 894
of image in thin-lens equation, 894
of magnifying glass, 922
sign convention for, 882, 895
of telescope, 926
Magnifying glass, 919–923
problems, 933
Magnitude(s), 11, 108
of angular momentum, 1048
beat frequency and, 464
capacitance and, 678–679
of charges on capacitor, 714
circular motion and, 162–163
of electric field due to point charge, 639
of electron's charge, 626–627
of electrostatic force, 630–631
of field inside solenoid, 753–754
of induced emf, 772–773, 778
of magnetic field produced by current loop, 753
of magnetic force, 735–737
of momentum, 241–243, 984, 985
Newton's third law of motion and, 118
of vector, 56, 57
weight and, 125
Malus, Etienne-Louis, 854
Malus, law of, 854, 856
Manned Maneuvering Units (MMUs), 116
Marconi, Guglielmo, 838
Marianas Trench, 581
Mars
atmosphere of, 580
landing of exploration rovers on, 104, 864
mass of, 366–367
meteorites from, 391
orbit of, 368
patterned ground, 568
Marsden, Ernest, 1035
Mass(es), 1, 2, 3, 108
acceleration and, 110–111
atomic or molecular, 552, 742, 1078
center of, 240, 264–269, 330–332
constant, 243–244
of Earth, 366
of electron, 627, 988, 1073, 1081
of Jupiter, 370–371
Kepler's third law and, 370

of Mars, 366–367
mass-energy equivalence, 986–989, 1073
problems, 1001
mass-per-length vs., 435
moment of inertia and continuous distribution of, 296, 297–298
momentum of, 241–243
of neutron, 627, 1073
of photon, 1013
problems, 1031–1032
of proton, 627, 1073
radioactive decay and, 1077
relativistic momentum of, 983–985
rest, 985, 989, 1013
of string and speed of wave on string, 434
suspended from pulley, 333
systems with changing, 269–270
torque and, 326
typical, 3, 108
units of, 4, 112
on vertical spring, 408–409
weight vs., 3, 124–125
Mass difference, 1090, 1093
Masseter muscle, 354–355
Mass number (A), 1072
in alpha decay, 1078–1079
in beta decay, 1080, 1081
gamma decay and, 1082
Mass on a spring, 228–229, 988
factors affecting the motion of, 406–408
natural frequency of, 420, 825, 826
period of, 405–409
problems, 426
simple harmonic motion and, 396–399,402–403
Mass on vertical spring, 408–409
Mass-per-length
defined, 434
mass vs., 435
of string, pitch and, 459
Mass spectrometer, 741, 742
Matter
antimatter and, 970, 988–989
magnetism in, 755–757
phases of, 548
See also Ideal gases; Liquids; Solids
Matter-antimatter annihilation, 988
Matter waves, 1046
Maximum acceleration, of simple harmonic motion, 402, 404
Maximum efficiency, Carnot cycle and, 602–604
Maximum height, 96–97
Maximum kinetic energy, 410–411, 1010–1012
Maximum potential energy, 410, 411
Maximum range, 94
Maximum speed, 197
of simple harmonic motion, 401
Maximum work, by heat engine, 603
Maxwell, James Clerk, 556, 838, 842
Maxwell's equations, 1036
Maxwell speed distribution, 556
RMS speed and, 558
in sweat, 569
Measurement
converting units of, 8–10
problems, 15
units of, 1, 2–4
problems, 14
See also Units of measurement
Mechanical advantage, 172
Mechanical energy, 377–378
conservation of, 214–222
problems, 234–235
conversion from food energy, 211
conversion to electric energy, 768
defined, 214
work done by nonconservative force and, 222–225
Mechanical equivalent of heat, 526, 527
Mechanical work
electric energy and, 778–781
problems, 799
heat and, 526–528
problems, 544
Mechanics, 17
quantum, 1047
Medicine
applications of ultrasound in, 440, 441
Doppler shift uses in, 451
X-rays in, 865, 1056, 1058, 1109
***mega*, 5**
Megahertz (MHz), 395
Melting temperature, fusion curve and, 568
Mendeleev, Dmitri, 1055
Mercury
in barometer, 483
in thermometer, 518
Mercury, Sun and, 371–372
Merry-go-round, 290–291
angular momentum and, 337
rotational collision and, 342
torque and, 317
walking on, 354
Mesons, 1100, 1101
Metal detectors, 827
Metals, as conductors, 630
Metastable state, 1059
Meteor, infrasound and, 441
Meteorite ALH84001, 391
Meteorites, 50, 203, 378, 391
Meter (m), 2, 19, 65, 66
defining by speed of light, 3, 842
Mica, 681, 683
Michelson, A. A., 865, 972
***micro*, 5**
Microcoulomb, 627
Microcurie, 1083
Microfarad, 678
Microhematocrit, 292
Microhematocrit centrifuge, 292
Micromirror devices, 872
Microphone, 773
Microscopes, 923
atomic-force (AFM), 192–193
compound, 923–924
problems, 933–934
early, 922–923
electron, 1020
scanning tunneling, 1025–1026
Microwave relay dishes, 838
Microwaves, 846, 847
intensity of, 442
Mid-ocean ridge, 756
Midpoint ray (M), 892
Midshipman fish, "love song" of, 474
***milli*, 4, 5**
Millicurie (mCi), 1083
Millihenry (mH), 785
Millikan, Robert A., 629, 1033
Millimeters of mercury, 483
Minimum speed in projectile motion, 88–89
Mirages, 887–888
Mirror equation, 878, 880–884
problems, 903–904
Mirror image, locating, 873–874
Mirrors
astronomical, 877
concave. *See* Concave mirrors
convex. *See* Convex mirrors
parabolic, 877
plane, 872–875
sign convention for, 876, 882–883
spherical, 875–877
***Mission Impossible*, 332**
mks system (SI units), 2–4, 112
MMUs (Manned Maneuvering Units), 116
Mobile, balance and, 264, 330–331
Moderator, 1093
Molar specific heat, 596–597
Mole(s), 551–552
counting molecules and, 551
defined, 551
ideal gas equation in terms of, 551
Molecular mass, 552, 742, 1078
Molecular oscillations, 425
Molecule(s)
mass of, 552, 742, 1078
moles and, 551
number of, and pressure of ideal gas, 549–555
oscillatory motion between turning points in, 229
speed distribution of, 556–557
Molybdenum, 1056, 1057–1058
Moment arm, 317–318
Moment of inertia, 295–299
conservation of angular momentum and, 338, 339, 341
continuous distribution of mass and, 296, 297–298
defined, 296
discrete masses and, 296
problems, 311
speed on incline and, 301–302

Momentum, 240
acceleration and, 244
angular. *See* Angular momentum
change in, 241–242, 245, 246
conservation of, 240, 250, 255
energy conservation vs., 255
problems, 275–276
direction and, 241–243
in electromagnetic waves, 849–853
problems, 866–867
external force and, 249–252
impulse and, 245–246
internal force and, 249–252
linear, 241–243, 248–252, 335
Newton's second law and, 243–244, 250–251
origin of pressure and, 555–556
of photon, 1013–1014
problems, 1031–1032
photon scattering and, 1014–1016
relativistic, 983–985
problems, 1000–1001
of system of objects, 242–243
uncertainty in, 1021–1024, 1025
as vector, 241, 242–243
wavelength of particle and, 1017
Momentum-impulse theorem, 245
Monoatomic ideal gas
adiabatic processes and, 598, 599
internal energy of, 560
molar specific heat for, 597
Monochromatic light, 938
Montgolfier brothers, 381
Moon
acceleration of gravity and, 364–365
atmosphere evaporation and, 571
atmosphere of, 382
density of, 364, 366
distance to Earth, 389, 875
escape speed of, 382
force exerted on, 138
gravitational force between Earth and, 359, 365, 632
position of, 77
projectile motion on, 94
tides on Earth and, 385
velocity of, 77
Morley, E. W, 972
Mormon Flat Dam (Arizona), 215
Motion
of center of mass, 266–269
circular. *See* Circular motion
with constant acceleration, 29–34
constant-velocity, 741
equations of, 29–34, 82, 84–87
applications, 35–37
problems, 50–51
helical, 743–744
horizontal, 79, 84
independence of, 79, 83, 84, 87
nontangential, 336
in one dimension, 17, 22
orbital, 359, 367–374
oscillatory, 229
periodic, 395–396
projectile. *See* Projectile motion
relativity of, 971
retrograde, 367
rolling, 294–295, 300–304
rotational. *See* Rotational motion
simple harmonic. *See* Simple harmonic motion
spinning, 339, 345–346
tangential, 336
translational, 294–295
in two dimensions, 79, 80–83
problems, 101
See also Projectile motion
vertical, 79, 84
See also Dynamics; Kinematics; Kinetic energy; Newton's laws of motion; Oscillations
Motional emf, 776–779
qualitative, 776–778
quantitative, 778–779
Motors
electric, 783–784
magnetic torque and, 748
problems, 799
Mountain climbing, 88–89, 211–212
Movement
measuring magnetically, 800
tracking magnetically, 781
Movies, physics in, 506
Moving observer, Doppler effect and, 446–448, 450–452
Moving source, Doppler effect and, 448–452
MRI. *See* Magnetic resonance imaging (MRI)
Multielectron atoms, 1050–1055
problems, 1069
Multimeter, 720, 721
Multiplication
of physical quantities, 6
of unit vectors, 64–65
Muon, 1100
decay of, 977
length contraction for, 979, 980
Muonium, 1068
Muon neutrino, 1100
Muscae volitantes, 912
Muscle cells, work done in, 202
Musical scale, chromatic, 458
Mutchkin, 15
Mutual inductance, 784
Mylar, 681
***Mysterious Island* (Verne), 58**
***Myxomatose* (Calder), 264**

***nano*, 5**
Nanometers (nm), 845
Nasal strips, 153–154, 175, 237
National Electric Code, 731
Natural frequency, 419–420, 803
of *LC* circuits, 825, 826
Natural law, 2
Navigation
gyroscopes in, 345, 346
using polarized light, 858
Near-infrared spectroscopy (NIRS), 868–869
Near point, 911–912
Nearsightedness, 915–917
Negative charge, 626
electric field lines and, 643
magnetic force for, 738
Negative displacement, 19
Negative ion, 628
Negative of vector, 63–64
Negative work, 184–186, 187
spring and, 193–194
Neon, 1054, 1055
Neoprene rubber, 681, 683
Neptune, 358
Nerve axons, 16
Nerve cells
electric charge problems, 659, 661
electric field within, 660, 661
Nerve impulses, 48
Net electric field, 640–641
Net electric force, 634–635
Net gravitational force, 360–362
Net radiated power, 537–538
Neurology, tuning forks in, 425
Neutral charge, 626, 633
Neutrality (electric), 626, 633
Neutrino, 1081–1082, 1100
Neutron(s), 1072
charge of, 1073
conversion of, 1080
as hadron, 1100
mass of, 627, 1073
moderating, 276
quark model of, 693
relative biological effectiveness for, 1096
Neutron diffraction, 1019–1020
Neutron number (*N*), 1072
for stable nuclei, 1076
Neutron star, 995
angular speed of, 340
radius of, 1109
Newton, Sir Isaac, 25, 51, 358, 384, 897, 926
Newtonian reflector, 926
Newton (N), 112, 124, 631
Newton's law of universal gravity, 358–362
Coulomb's law and, 632
Newton's laws of motion, 107
applications of. *See* Circular motion; Connected objects; Frictional force; Spring(s); String, tension in; Translational equilibrium

Coulomb's law and, 631–632
first law of motion, 108–110
incompleteness of, 970, 972
second law of motion, 110–118
applications, 333–335
circular motion and, 162
connected objects and, 158–159, 160, 161
kinetic friction and, 142, 145
momentum and, 243–244, 250–251
problems, 136
for rotational motion, 320, 338
for a system of particles, 266, 268
translational equilibrium and, 154
third law of motion, 118–121, 249
problems, 136–137
Newton's rings, 946–947, 966
Nexrad, 844–845
Nickel, 755–756
Night vision (dark-adapted), 1009, 1031
NIRS (near-infrared spectroscopy), 868–869
Nitrogen, 570, 1053, 1054
Noble gases, 1055
Nodes, 456, 457, 459
Noggin, 15
Noise reduction, 455
Noise standards, OSHA, 474
Nonconservative forces, 205–208, 209
examples, 207
problems, 233–234
work done by, 222–227
problems, 235–236
Nonreflective coating, 949
Nontangential motion, angular momentum of, 336
Nonzero acceleration, 28
Nonzero net external force, 268–269
Nonzero velocity, 28
Normal, 871, 886
Normal force(s), 114, 128–132, 166
kinetic friction and, 142, 143, 144
problems, 138–139
North pole
of Earth, 733, 734–735
of magnet, 733
Nuclear binding energy, 1089–1090
problems, 1108
Nuclear density, 1073–1075
Nuclear fission, 986, 1090–1093
problems, 1108
Nuclear forces
stability and, 1075–1076
strong, 1075–1076, 1099, 1100, 1102
weak, 1099–1100, 1102
Nuclear fusion, 1093–1095
problems, 1108
Nuclear physics
constituents and structure of nuclei, 1072–1076
elementary particles, 1099–1102
half-life and radioactive dating, 1076, 1083–1089
nuclear binding energy, 1089–1090
practical applications, 1095–1099
problems, 1108
radioactivity, 1076–1083
unified forces and cosmology, 1102
Nuclear power, 986
Nuclear radiation, 1076–1083
half-life, 1083–1087
Nuclear reactor, 987, 1093
Nuclear size, 1073–1075
Nucleon, 1072, 1089
binding energy per, 1090
Nucleus (nuclei), atomic, 1036, 1071
constituents and structure of, 1072–1076
problems, 1107
daughter, 1078, 1079
as function of time and the decay constant, 1084–1085
parent, 1078
Numbering system, for bright and dark fringes, 942

Object beam, 1061–1062
Object distance, 873–874, 877, 880–884
focal length and, 895
sign convention for, 883, 895
Objective (microscope), 923
Observer
Doppler effect and moving, 446–448
role in motion, 109–110
Occlusion in artery, 511
Ocean wave, 431
Octopi, propulsion of, 270
Octopus eyes, 931
Oersted, Hans Christian, 732, 748, 769
Ohm, 699, 809–810, 813
Ohm, Georg Simon, 699
Ohmmeter, 700
Ohm's law, 703
loop rule and, 712
problems, 726–727
resistance and, 699–702
Oil spills, 524
Old Faithful, 54
Omega, 1100
One-dimensional coordinates, 18
One-dimensional coordinate system, 18
One-dimensional motion, 17
along *x* axis, 22
Open circuit, 697
Opposite phase, 454–455
Optical fibers, 890
Optical illusions, 887–888
Optical instruments, 909–936
camera, 912–913, 928, 949
compound microscope, 923–924
human eye, 781, 848, 910–912, 915, 917, 969, 980
lens aberrations, 927–928
lenses in combination and corrective optics, 913–919
magnifying glass, 919–923
telescopes, 909, 924–927
Optically active compounds/molecules, 858, 867, 869
Optical pyrometer, 536
Optical tunneling, 1025
Optical tweezers, 1014
Optic nerve, 910
Optics. *See* Geometrical optics; Physical optics
***Optics* (Ptolemy), 904**
Orbital angular momentum quantum number, 1048
Orbital maneuvers, 373–374
Orbital motion, 359
Kepler's laws of, 367–374
problems, 390–391
Orbits (atomic)
in Bohr model, 1040–1043, 1046–1047, 1056–1057
calculating time for, 743
Orbits (planetary), 358, 367–372
Order, entropy and, 612–614
Order-of-magnitude calculations, 10–11
problems, 15
Organ pipes, 462
Oscillations, 229, 394–429
beats as, 464
damped, 418–419
driven, 419–420
in electric circuits, 824–829
energy conservation in, 409–412
frequencies of, 803
molecular, 425
pendulum, 412–418
periodic motion, 395–396
period of mass on a spring, 405–409
simple harmonic motion. *See* Simple harmonic motion
uniform circular motion and simple harmonic motion, 399–405
Oscillatory motion. *See* Oscillations
Oscilloscope, 437
OSHA (Occupational Safety and Health Administration), 445
noise standards, 474
Overdamped oscillations, 419
Overhang
maximum, 353
unlimited, 279
Oxygen, 1053, 1054
exciting oxygen molecule, 1030
speed distribution for, 570
Ozone, UV radiation absorbed by, 848

Pacemakers, 720, 730, 755
switches, 764
Pain threshold, measuring, 511
Paleomagnetism, 735
Paper, dielectrics and, 681, 683
Parabolic mirror, 877
Parabolic path, of projectiles, 84, 87
Parallel, capacitors in, 714–715
Parallel, resistors in, 707–709
Parallel arrangements, heat flow and, 533–534
Parallelogram method of vector addition, 62

Parallel-plate capacitor, 645–646
capacitance of, 678–680
when filled with dielectric, 681–682
dielectrics and, 681–682
electric potential and, 665–666
equipotential surfaces and electric fields and, 674–675
Gaussian surface and, 652
See also Capacitors
Parallel ray (*P* ray), 871, 872, 876–877, 878, 879
on concave mirror, 876
on convex mirror, 876
lenses and, 892
Paramagnetism, 756
Pararescue jumpers, 104
Paraxial rays, 877
Parent nucleus, 1078
Particles
diffraction of by crystals, 1018–1020
elementary, 1099–1102
Newton's second law for system of, 266, 268
point, 17, 336
wavelike behavior of, 1017
wave-particle duality, 853, 954, 1020–1021
See also individual particles
Pascal, Blaise, 479
Pascals (Pa), 479
Pascal's principle, 485–486
Paschen series, 1038–1039, 1044
Path, speed and, 217–218
Path lengths
in diffraction grating, 958
interference patterns and, 1018
in two-slit experiment, 941–942
Patterned ground, 568
Paul, Les, 773
Pauli, Wolfgang, 344, 1051, 1081
Pauli exclusion principle, 1051–1053
quarks and, 1101
PDT (photodynamic therapy), 1061
Pendulum, 394, 395, 412–418
ballistic, 256–257
conical, 176
driving, 419
natural frequency of, 419
period of, 414–417, 418
physical, 416–418
problems, 427
simple, 413–416
Percussion, center of, 314
Perfect radiator, 537
Performance, coefficient of (COP), 605–608
Period
defined, 285, 395
of physical pendulum, 416–417, 418
of planet, 369
of simple pendulum, 414–415, 416
time expressed in terms of, 403
of wave, 433
Periodic motion, 395–396
problems, 425
See also Simple harmonic motion
Periodic table, 1054–1055
problems, 1069
Period of mass on spring, 405–409
problems, 426
Period of rotation, 285
Permanent magnets, 733
Permeability of free space, 750
Permittivity of free space, 650
Perseus Black Hole, 474
***peta*, 5**
PET (positron-emission tomography), 970, 988–989, 1097–1098
Pfund series, 1038
Phase, of waves, 938–939
Phase angle, 815–816
in alternating-current circuits, 822
in *RLC* circuit, 820–821, 822
Phase changes, 548
energy conservation and, 574–576
problems, 583
due to reflection, 944–945
temperature and, 549
thin film and, 947–949
Phase diagram, 567, 568
for carbon dioxide, 582
for water, 582
Phase equilibrium, 565–569
problems, 581–582
Phase relation, of voltage and current, 810–812, 827
Phases of matter, 548
See also Ideal gases; Liquids; Solids
Phase transitions, 1102
Phasor diagram
for alternating-current circuits, 822
for alternating-current inductor circuit, 818
for capacitor circuits, 810–812
for inductor circuits, 817–818
for *RC* circuit, 812
at resonance, 828
for *RLC* circuit, 820–822, 826
for *RL* circuit, 818
Phasors, 804, 805, 811, 812
Phoropter, 916
Phosphorescence, 1064
Phosphorus, 1054
Photocells, 1012–1013
Photoconductive material, 630
Photocopiers, 630
Photodynamic therapy (PDT), 1061
Photoelastic stress analysis, 857
Photoelectric effect, 1009–1013
problems, 1030–1031
Photoelectron, 1009, 1010, 1011, 1012
Photon(s), 1003, 1007–1016
change in electron orbit and, 1040, 1044–1045
energy of, 1007
high-energy. *See* Gamma rays
momentum of, 1013–1014
problems, 1030–1031
radioactivity and, 1077
rest mass of, 1013
Photon-electron collision, 1014–1016
Photon scattering, 1014–1016
problems, 1032
Photorefractive keratectomy (PRK), 1060–1061, 1069
Photosynthesis, 15
Photovoltaic cells, 1012
Physical laws, equivalence of, 971
Physical optics, 937–969
diffraction, 950–954
diffraction gratings, 957–961
interference in reflected waves, 944–950
resolution, 954–957
superposition and interference, 938–940
Young's two-slit experiment, 940–944
Physical pendulums, 416–418
Physics
defined, 2
problem solving in, 11–13
Piano, 458
shape of, 430
tuning, 473
Pickering, Edward C., 1070
Pickering series, 1070
***pico*, 5**
Picofarad, 678
Pigeon in flight, power output of, 203
Pion, 1100
Piston under pressure, height of, 553, 554–555
Pitch, 439, 458, 459
Pitcher's mounds, 101
Pit vipers, infrared receptors in, 847
Pivot point, 416, 418
Pixels, 957
Planck, Max, 970, 1005
Planck's constant, 1006
Heisenberg's uncertainty principle and, 1022, 1024–1025
in hydrogen atom, 1040
photoelectric effect and, 1011
Planck's quantum hypothesis, 1005–1007
problems, 1030
Plane mirror, forming images with, 872–875
problems, 903
Planets
atmospheres of, 382
orbits, 358, 367–372
See also individual planets; Kepler's laws of orbital motion
Plane waves, 871
Plano-convex lenses, 891
Plaque in artery, 510
Plasma (nuclear), 1095
Plates, in capacitor, 677–678

Plate tectonics, 535–536
Plimsoll line, 490, 491
Plum-pudding model of the atom, 1035
Point charge
electric field lines for, 643
electric field of, 639–640
electric flux for, 650
electric potential of, 669–673
problems, 690–691
equipotentials for, 674, 675
forces between, 630–631
Pointillism, 957
Point masses, gravitational force and, 359, 362, 365
Point particles, 17
angular momentum for, 336
Point source, intensity with distance from, 443
Poise, 502
Poiseuille, Jean Louis Marie, 502
Poiseuille's equation, 503
Poisson, Simeon D., 954
Poisson's bright spot, 954
Polaris, 284, 345
Polarization, 629, 853–860
problems, 867–868
by scattering and reflection, 858–860
total, 890–891
Polarization charge, 629
Polarized light, vs. unpolarized, 853, 854
Polarized plugs, 808, 809
Polarizer, 853–858
Polarizing filters, 856
Polaroid sunglasses, 859–860
Polaroid vision in spider, 868
Pole vaulting, 246
Population inversion, 1059
Porro prisms, 890
Position, 18
acceleration and, 31–33
angular, 282–283, 400, 942
energy as function of, in simple harmonic motion, 410
as function of time, 31–33, 39, 40, 42, 82–83
gravitational potential energy vs., 228
harmonic wave function and, 436–437
problems, 47–48
of shadow, as function of time, 400
time vs., 33, 397, 398, 400
uncertainty in, 1022–1024, 1025
velocity as function of, 30–31, 32, 34, 36, 39, 40, 42, 82
Position vector, 65–66
problems, 76–77
Positive charge, 626
electric field lines and, 643
magnetic force for, 738
Positive displacement, 19
Positive ion, 628
Positive work, 184, 185, 187
spring and, 193–194
Positron, 988, 1077
Positron-emission tomography (PET), 970, 988–989, 1097–1098
Potassium, 1054
Potential differences (V), 681, 684, 695, 701
Potential energy, 209–214
defined, 209
as function of time, 410
gravitational, 209, 210–212, 215, 227–228, 374–377, 484, 497
problems, 391
in oscillatory motion, 409–412
pendulum and, 413–414
of springs, 212–214
zero level, 210, 375
See also Electric potential energy (U)
Potential energy curves, 227–230
problems, 236
Potential hill, 670, 674
Potential well, 380, 670
Pot of flowers, tension in wires of hanging, 156–157
Pound (lb), 112
Power, 195–197
in alternating-current inductor circuit, 817
average, 195, 822
consumed by capacitor in alternating current circuit, 810, 812
of dental drill, 354
in electric circuit, 702–705
for phase angle, 816
problems, 201
produced by a torque, 343
radiated, 536–538
in *RC* circuit, 816
rotational work and, 343
problems, 354
speed and, 195–197
typical value of, 195
Power factor, 815
deriving phase angle from, 821
Power output of heart, 202, 511
Prairie dog burrows, 499
***P* ray. *See* Parallel ray (*P* ray)**
Precession, 345
Prefixes, of measurement, 4
Premature infants, respiratory distress in, 504
Pressure, 477–480
atmospheric, 479–480, 482–483, 500
constant, 554–555, 589–591, 597
defined, 477
depth and, 481–486
direction and, 479–480
equilibrium vapor pressure, 566
in fluid, 477–480
applications, 498–499
pressure-speed relationship, 495–496
fusion curve and, 568
gauge, 480
of ideal gas, 549–555
in isotherm, 553
in kinetic theory of gases, 557
origin of, 555–556
phase equilibrium and, 566
problems, 508
radiation, 852–853, 1014
weight of fluid and, 481
Pressure cooker, 567
Pressure-volume plot, 592
adiabatic processes on, 594
constant-pressure process on, 589, 590
constant-volume process on, 591
isotherms on, 592
Primary circuit, 769
Principal axis, 875–876
Principal maxima, 957–959
Principal quantum number, 1047
***Principia* (Newton), 359**
Principle of equivalence, 992
Printers
ink-jet, 646
laser, 630, 959
Prism(s)
concave lens vs., 892
convex lens vs., 892
Porro, 890
refraction in, 888
Prismatics, 896–897
PRK (photorefractive keratectomy), 1060–1061, 1069
Probability, as function of distance, 1049
Probability clouds, electron, 1049–1050
Problem solving, 11–13
checking solution, 12, 39
free-body diagrams, 113–118
sketching, 12, 31, 216, 223, 257
strategizing, 12, 36
Projectile, defined, 83
Projectile motion, 79, 83–97
acceleration components of, 83
with air resistance, 83, 94
basic equations, 83–84
collision in, 91–92
problem, 105
defined, 79
launch angle of, 85
arbitrary, 89
general, 89–92
zero, 84–89
maximum height and, 96–97
problems, 103–104, 105
parabolic path, 84, 87
range of, 93–94, 96
problems, 103–104, 105
symmetry in, 94–96
problems, 103–104
trajectory of horizontal launch of, 85
v_x and v_y as functions of displacement, 83
v_x and v_y as functions of time, 83
x and y as functions of time, 83

Proper length (L_0), 978, 979
Proper time, 974, 978
Propulsion, 269–270
problems, 277–278
Proton(s), 1072
electric charge of, 627, 1073
as hadron, 1100
magnetic resonance imaging and, 1098–1099
mass of, 627, 1073
relative biological effectiveness for, 1096
Proton number. *See* **Atomic number (Z)**
Proton-proton cycle, 1094
Ptolemy, 367, 368, 904
Pulley, 150–152, 154–156, 174
angular acceleration and, 289, 320
connected objects and, 159, 160–161
with constant angular acceleration, 286
mass suspended from, 333
torque and, 334–335
Pulling sled, 189–190
Pulsar, 1002
angular speed of, 340–341
Pulsilogium, 413
Pumpkin toss/"Punkin Chunkin," 102, 103
Pupil, 911
Purley, David, 51
PV plot. *See* **Pressure-volume plot**
P wave, 677
Pyrex glass, 681, 683
Pythagorean theorem, 293
vectors and, 57–58, 58

QCD (quantum chromodynamics), 1102
QED (quantum electrodynamics), 1102
QRS complex, 677
Quadratic formula, from kinematics, 54
Quantities, 2
linear and angular, 286, 299, 320
rotational, 289–294
Quantization, 625, 1003
of electric charge, 629
Quantized energy, 1005–1007
Quantum, 1006
Quantum chromodynamics (QCD), 1102
Quantum effects, superconductivity and, 702
Quantum electrodynamics (QED), 1102
Quantum mechanical hydrogen atom, 1047–1050
problems, 1069
Quantum mechanics, 1047
Quantum number(s), 1006–1007
angular momentum, 1053
orbital, 1048
electron spin, 1048
magnetic, 1048
for multielectron atoms, 1050, 1051
for quantum mechanical hydrogen atom, 1047–1048
Quantum physics, 1003–1033, 1034
blackbody radiation, 1004–1007, 1036
de Broglie hypothesis, 1017–1020, 1021
Heisenberg uncertainty principle, 1021–1025
mass and momentum of photon, 1013–1014
photons and photoelectric effect, 1007–1013
photon scattering and Compton effect, 1014–1016
Planck's quantum hypothesis, 1005–1007
quantum tunneling, 1003, 1025–1026
wave-particle duality, 853, 954, 1020–1021
Quantum tunneling, 1003, 1025–1026
Quark confinement, 1101
Quark model of the neutron, 693
Quarks, 1100–1102
Quarter-wave coatings, 949
Quartz, 885
Quasar, 960
Quasi-static processes, 588
q wave, 677

Radar, Doppler, 451, 844
Radar guns, 451
Radial force, 317
Radial keratotomy (RK), 917
Radian (rad), 282, 283, 398, 402, 1096
Radiated power, 536–538
Radiation, 536–538
atomic. *See* Atomic radiation
biological effects of, 1095–1097, 1108
electromagnetic, 1004
human, 866
in hydrogen atom, 1040
nuclear, 1076–1083
half-life, 1076, 1083–1089
problems, 545
typical dose of natural, 1097
See also Spectrum
Radiation dosages, 1097
Radiation pressure, 852–853, 1014
Radioactive dating, 1083–1089
problems, 1107–1108
Radioactive decay, 1077–1083
alpha decay, 1078–1080, 1082, 1083
beta decay, 1080–1082, 1083
gamma decay, 1082
radioactive decay series, 1082–1083
Radioactive pharmaceutical, 1108
Radioactive tag, 1107
Radioactive tracers, 1097
Radioactivity, 1076–1083
in the bones, 1108
problems, 1107
Radio antennas, 938, 939
Radio tuning, 420, 826, 827, 840
Radio waves, 431, 837, 840, 846
Radium-226, 1085
Radius
Bohr, 1041
nuclear, 1073–1075
Radon, 1076, 1085, 1086, 1097
Radon-222, 1085
Rails, 478
Rain
estimating raindrops in storm, 10
largest raindrops, 583
momentum of hail vs., 246–247
Rainbow, 848, 897
problems, 906
Rainbow holograms, 1062
Rain shadow, 599
Raising bucket, 154–156
Range, 93–94, 96
launch angle vs., 96
Ray diagrams
for convex mirror, 876
to derive mirror equation, 880–881
to derive thin-lens equation, 894
Rayleigh's criterion, 954–955
Rays, 871
Ray tracing, 877–880
for lenses, 891–894
problems, 905–906
problems, 903–904
RBE (relative biological effectiveness), 1096
***RC* circuits, 717–720**
alternating-current, 812–816, 819
average power consumed by, 816
behavior as, 717, 719
impedance and, 812–815
maximum current, 812–813
phase angle and power factor and, 815–816
problems, 729–730, 833
Reactance
capacitive, 809–810, 817
inductive, 817
Reaction time, 52
Real image, 879, 882, 893
Rear-view mirrors, 880
Recoil, 251, 270
Rectangular current loops, 746–747
Red giant, 1094
Red light, 896, 897, 948
Red quarks, 1101
Red shift, 451, 843
of quasar, 960
Reference beam, 1061–1062
Reflected ray, 871
Reflection gratings, 960–961
Reflection of light, 858–860, 870, 871–872
angle of, 871, 873
law of, 871–872
phase changes due to, 944–945
problems, 902–903
from rough surface, 872
from smooth surface, 871, 872
total internal, 888–890
***Reflections on the Motive Power of Fire* (Carnot), 602**

Reflections/reflected waves, 436
interference in, 944–950
Reflector
corner, 875
ideal, 538
Newtonian, 926
Refraction, 870, 884
angle of, 888
index of, 885–890, 910, 944–945, 955
qualitative properties of, 886
Refraction of light, 884–891
problems, 904–905
total internal reflection, 888–890
total polarization, 890–891
Refractive power, of lens, 916
Refractors, 925–927
Refrigerator magnet, 734
Refrigerators, 605–606
problems, 622
Relative biological effectiveness (RBE), 1096
Relative velocity, 69–71
problems, 77
vector addition and, 69–70
Relativistic distortion, 980
Relativistic length, 983
Relativity, 970–1002
general relativity, 345, 384, 991–995
of length and length contraction, 978–980, 983
mass-energy equivalence, 986–989, 1073
nuclear binding energy and, 1089–1090
postulates of special relativity, 971–972
relativistic addition of velocities, 981–983
relativistic energy, 986–989
relativistic kinetic energy, 989–990
relativistic momentum, 983–985
problems, 1000–1001
relativistic universe, 991
of time and time dilation, 972–977
Remote controls, 847
Remote-sensing, 847
Rem (roentgen equivalent in man), 1096–1097
Repulsive forces, 626, 733
Resistance, 789–790
in alternating-current *RLC* circuit, 823–824
equivalent, 705–710, 713
internal, 707
Ohm's law and, 699–702
problems, 726–727
in *RC* circuit, 812–815, 816
See also Air resistance
Resistance heating, 703–704
Resistivity, 700–701, 726–727
defined, 700
temperature and, 702
Resistor circuit, 804, 805, 806–807
Resistors
analyzing complex circuit of, 710
in electric circuits, 705–710, 719
electric power dissipation in, 703–704
horizontal and vertical, 712–713
in parallel, 707–709
problems, 727–728
in *RL* circuits, 787–788
in series, 705–707
problems, 727–728
Resolution, 954–957
diffraction-limited, 957
of eye, 969
problems, 967
Resonance, 420, 826–829
bridges and, 420
in electrical circuits, 803, 824–829
problems, 835
in mechanical systems, 826
off-resonance current, finding, 828–829
Resonance curve, 420
Rest energy, 986
for hadrons, 1100
for leptons, 1100
for quarks and antiquarks, 1101
Rest mass, 985, 989
of photon, 1013
Restoring force, 394
Retina, 910
detached, 1069
Retrograde motion, 367
Retroreflector arrays, 875
Reversible engine, 609, 611
Reversible processes, 588–589
defined, 588
entropy change and, 610
idealized, 589
Revolutions (rev), 283
Rigel, 536, 1005
Right-hand rule
for angular velocity, 344
applied to electromagnetic wave, 839, 840
magnetic field, 749
for magnetic forces, 737–739
for torque, 344
Riverbend Maximum Security Institution, 251
River crossing, relative velocity of, 71
RK (radial keratotomy), 917
***RLC* circuits**
alternating-current, 820–824
natural frequency of, 826
problems, 834–835
resonance in, 827
RMS currents in, 829
***RL* circuits, 787–788**
alternating-current, 818–820
problems, 800
RMS currents
in alternating current capacitor circuit, 809–810
in *RL* and *RC* circuits, 819
in *RLC* circuit, 829
RMS speed, 558–559
speed distribution and, 558
RMS values, 805–807, 849–850
maximum vs. minimum values, 806
Roads, banked, 164–165
Roche limit, 386
Rockets
escape speed and, 381–384
exploding, 269
propulsion, 269–270
problems, 277–278
Rods
stretching, 560–561
thermal conductivity of, 532, 533–534
thermal expansion of, 520–521
Rods (retinal), 910
Roentgen, Wilhelm, 1055, 1058
Roentgen equivalent in man (REM), 1096–1097
Roentgen (R), defined, 1095
Roller coasters, 207, 215
Roller pigeons, 313
Rolling motion, 294–295
kinetic energy of, 300–304
problems, 310–311
without slipping, 294
Romer, Ole, 841
Roof, forces on, 498–499, 499
Root mean square (rms) speed, 558–559
Root mean square values, 805–807, 849–850
Rotation
axis of, 282, 298, 299, 323–325, 339–340, 344
direction of, 284, 318–319
period of, 285
Rotational analogies, 320
Rotational collisions, 341–343
Rotational equilibrium, 324, 328
Rotational kinematics, 286–289
problems, 309
Rotational kinetic energy, 295–299, 300–301, 302, 303
defined, 296
problems, 311
Rotational motion, 281, 282
angular acceleration and, 285–286
angular position and, 282–283
angular velocity and, 283–286
conservation of angular momentum and, 338–343
as natural phenomena, 281
Newton's second law for, 320, 338
vector nature of, 344–346
See also Circular motion
Rotational quantities, linear quantities and, 289–294
problems, 309–310
Rotational work, 343
problems, 354
Rotation of Earth, 83, 309
Rotorua (New Zealand), meteor crater, 378
Rounding corner, 163–164
Round-off error, 8

Rubber
as insulator, 630
neoprene, 681, 683
Rubidium, 1054
Rumford, Count (Benjamin Thompson), 526
Runaway chain reaction, 1092
Running, force of, 139
Running shoes, 237
Russell Traction System, 173
Rutherford, Ernest, 694, 1035–1036
Rutherford model of the atom, 694, 1035–1036
Rydberg atoms, 1070
Rydberg constant, 1037, 1044, 1045

Saccadic motion, 781
Safety features in household electrical circuits, 807–809
Saint Elmo's fire, 648, 683
Salam, Abdus, 1102
Salt, raising boiling point with, 567
Samara, flight of, 138
Sand bag, in free fall, 42–43
Satellites
communications, 372
elliptical orbit of, 381
geosynchronous, 372–373
GPS, 390
orbital maneuvers, 373–374
weather, 372
Saturn, rings of, 358, 386
Saturn V rocket, 270
Scalar
defined, 11, 56
distance as, 18
gravitational potential energy as, 376
kinetic energy and, 187
multiplying unit vectors by, 64–65
vectors vs., 11, 56
work as, 180
Scalar components, of vector, 57
Scandium, 1055
Scanning electron micrograph (SEM), 1020
Scanning tunneling microscopy (STM), 1025–1026
Scattering, 858–860
Scattering angle, 1014, 1015, 1016
Schriver, William, 102
Schrödinger, Erwin, 1047
Schrödinger's equation, 1047, 1048, 1049
Schwarzschild radius, 994
Scientific notation, 7
Scorpions, fluorescence in, 1063
Scotopic vision, 1009
Scott, David, 38
Scuba tanks, 583
Sea breezes, 535, 536
Seafloor spreading, 756
Seatbelts, 140
Secondary circuit, 769
determining voltage in, 790–791
Second harmonic, 456, 457, 460, 461
Second law of thermodynamics, 585, 600, 610, 613, 614
Second (s), 2, 4
Security systems, 251
Seismograph, 773
Selenium, as photoconductor, 630
Self-inductance, 784, 786
Semiconductors, 630, 700, 702
SEM (scanning electron micrograph), 1020
Series
capacitors in, 715–717
radioactive decay, 1082–1083
resistors in, 705–707
spectral, 1037–1039, 1044, 1070
Series arrangements, heat flow and, 534
Shadow, diffraction and, 952
Shallow-water waves, 470
Shapes
hollow, 298, 301
solid, 298, 301, 563–564
SharkPOD (Protective Oceanic Device), 643
Sharks, electric shark repellent, 643
Sharp point, intense electric field near, 648
Shear deformation, 563
Shear modulus, 563–564
Shea Stadium, 15
Shell, 1051–1055
Shepard, Alan, 94
Shielding, 646–648
Shock absorbers, 419
Shock hazard, electric, 807–808, 809
Shock wave lithotripsy (SWL), 440, 441
Shoes, flats vs. heels, 356
Short circuit, 708, 719
Shot put record, 105
Shutter speed, camera, 913
Sigma, 1100
Sign
of electric charge and direction of magnetic force, 736
of magnification, 882
of microscope images, 923
of power, 812
of power in inductor, 818
of telescope images, 924
Sign conventions
for angular acceleration, 285
for angular momentum, 337
for angular position, 282
for angular velocity, 284
for average speed, 19
for displacement, 19
for distance, 18
for Doppler effect, 447, 449, 450–451
for electric charge, 638
for electric flux, 649, 651
for focal length, 882, 895
for heat, 528, 586, 587
for image distance, 882, 895
for lenses, 895
for magnification, 882, 895
for mirrors, 876, 882–883
for torque, 318
for work, 587
for x and y components of vector, 59
Significant figures, 5–8
problems, 14–15
Silicon, 1054
Silver, resistivity and, 700
Simple harmonic motion, 396–399
acceleration in, 401–402
acceleration vs. time in, 402
energy as function of position in, 410
energy as function of time in, 411
pendulums and, 413–414
position vs. time in, 397, 398, 400
problems, 425
total energy in, 410, 411
uniform circular motion and, 399–405
velocity in, 401
velocity vs. time in, 401
Simple pendulum, 413–416
Simultaneity, 975
Sine
pendulum and, 413, 414
simple harmonic motion as, 397–398
Single-slit diffraction, 951–954
Siphon for irrigation, 512
Siri's formula, 489
SI (Système Internationale) units, 2–4, 112
See also Units of measurement
Skateboarding, 219–220
Sketching
free-body diagrams, 113–118
problem solving and, 12, 31, 216, 223, 257
Skids, 164–165
Sky
color of, 858–859
rainbow in, 897
Skylab, 200, 389
Sled acceleration, 123–124
Sleeping metabolic rate, 544
Slides, 144, 145–146, 183, 219, 220
Sling, vertical and horizontal components of force and, 328–329
Slinky, wave on, 437–438
Slug, 4, 112
Smearing, diffraction-induced, 954
Smoke detector, 1079–1080, 1108
Snell's law, 885–886, 888, 891, 908
Soap bubbles, 947
Sodium, 1054, 1055
Sodium pump in cells, 690
Sojourner, 364
Solar constant, 546
Solar energy panels, 1012, 1013
Solar system model of the atom, 1035–1036

Solar wing, sailing, 1014
Solenoid, 753–755
energy stored in magnetic field of, 789
inductance of, 785–786
problems, 765, 797, 799
superconducting, 765
Solid-liquid phase boundary, 568
Solids, 548
changing length of, 560–563
changing shape of, 563–564
changing volume of, 564–565
elastic deformation and, 560–565
problems, 581
kinetic energy of, 301
moments of inertia of, 298
speed of sound in, 438
Sonograms, 440, 441
Sonoluminescence, 595
Sound
decibels, 444–446
longitudinal waves and, 432
speed of, 437, 438–439, 441
wave properties of, 438
See also Sound waves
Sound energy, work and, 209
Sound intensity, 441–446
problems, 470–471
Sound perception, of humans, 444
Sound waves, 437–441
Doppler effect, 446–452
frequency of, 439–441
problems, 470
South pole
of Earth, 733, 735
of magnet, 733
Space, gyroscopes in, 346
Spacecraft
gravitational force and, 360–362
maneuvering, 373–374
Voyager I and *Voyager II*, 865
Space shuttles, 372, 373
Discovery, 26
orbit, 393
Space Station, 1013
Space travel, biological aging and, 975–977
Spark plug, 689
Speakers, in phase, 454–455
Special theory of relativity postulates, 971–972
problems, 998
Specific heat(s), 528–531
at atmospheric temperature and pressure, 529
conservation of energy, 530
at constant volume, 595–599
defined, 529
of ice, 572
for ideal gas, 596–599
latent heats vs., 573
molar, 596–597
problems, 544–545, 621
of water, 529, 573
Spectroscopes, grating, 959–960
Spectrum
electromagnetic, 845–848
of hydrogen, 1036–1039, 1043–1045
wavelength of visible lines in, 1037
Specular beam, 960
Specular reflection, 872
Speed, 11
acceleration and, 29
angular, 284, 285, 286–288, 290, 294–295, 335, 336, 339–342, 343, 400
average, 19–21
Doppler effect and, 450–451
of Earth, 369, 389
effect of change in initial, 220–221
elastic collisions and, 259–263
electric potential energy and, 667–668
escape, 381–384, 693, 993–994
of fluid, 495–496, 500–501
of fluid vs. pressure, 495–496
applications, 498–499
as function of distance from center of Earth, 383–384
as function of time, 26–27
glacial, 15
height difference and, 215, 217–219
inelastic collisions and, 254–258
instantaneous, 23
linear, 290, 294–295
magnitude of momentum as function of, 984
maximum, 197, 401
of meteorites, 378
minimum, in projectile motion, 88–89
path and, 217–218
potential energy curves and, 229
power and, 195–197
root mean square (rms), 558–559
of sound, 437, 438–439, 441
speed distribution of molecules, 556–557
symmetry and, 94–95
tangential, 289–291, 293
time dilation as function of, 974
velocity, acceleration, and, 28–29
of wave on a string, 433–436
of waves, 433
work and, 186–190
Speed distribution
of molecule, 556–557
root mean square speed and, 558
for water molecules in sweat, 569
Speed of light (c), 841–843
astronomical perspective on, 843
constancy of, 971–972
index of refraction and, 885, 886
measuring, 865
meter and, 3, 842
in vacuum, 841, 971, 972
Speedometer, 23, 24, 777
Speed settings, on camera, 913
Sphere
Gaussian surface and, 651
ideal spherical conductor, 675–676
spherical charge distributions, 637, 646
Spherical aberration, 877, 927
Spherical bodies, gravitational attraction of, 362–367
problems, 390
Spherical mirrors, 875–877
problems, 903
ray tracing for, 878–880
Spherical waves, 871
Sphygmomanometer, 508
Spider webs, 420, 428, 581
Spin angular momentum, 1048
Spinning motion, 339, 345–346
See also Rotational motion
Spinning wheel, 288–289, 294, 303, 304
Spontaneous emission, 1059
Spray, 494
Spring(s), 221–222
compression of, 411–412, 988
Hooke's law and, 152–153, 154
mass on. *See* Mass on a spring
in parallel, 203
potential energy of, 212–214
problems, 171–172
rules of thumb, 153
in series, 203
turning points, 228–229
vertical, 408–409
work and, 191–194
Spring forces, 152–154, 212
Spring scale, 153
Sprint training, 21–22
Sputnik, 391
Square brackets, 5
SR-71 Blackbird, 543
Stability
binding energy and nuclear, 1089–1090
nuclear forces and, 1075–1076
structural, 327–328
Standard meter, 2
Standing waves, 455–463, 1046
electron probability clouds and three-dimensional, 1049–1050
problems, 472
in vibrating columns of air, 459–463
in column of air closed at one end, 460
in column of air open at both ends, 462
waves on a string and, 455–459
Star of Africa, 15
Stars
binary, 391, 967, 1030
explosions of, 252, 340–341
magnetic, 777
neutron, 340, 995, 1109
nuclear fusion in, 1094
temperatures of, 536, 1005

Star trails, 284
***Starts from rest*, 188**
State, 1049
equation of, 549–552
excited, 1043, 1050
ground, 1043, 1049, 1051–1053
State function, 588
entropy as, 608
State of a system, 588
See also Thermodynamic processes
Static equilibrium, 157
conditions for, 323–324
in fluids, 481–486
problems, 508–509
lamp in, 327
problems, 351–352
zero torque and, 323–329
Static friction, 146–149
coefficient of, 147, 148
rules of thumb, 147
Steam engine, 600–601
Steam heat, 573–574
Steel sheets, making, 545
Stefan-Boltzmann law, 537
Stenosis, 503
Stimulated emission, 1059
STM (scanning tunneling microscopy), 1025–1026
Stopping distance, 35–36, 136, 149
Storm, force of, 278
Strain, 565
Strange (quark), 1101
Strassman, Fritz, 1090
Strategizing, problem solving and, 12, 36
Stress, 565
photoelastic stress analysis, 857
String, tension in, 150–152, 159, 177, 434, 458, 459
problems, 171–172
Strong nuclear force, 1075–1076, 1099, 1100, 1102
Structural stability, 327–328
Styrofoam, 564, 565, 574
Sublimation, latent heat of, 571
Sublimation curve, 567
Submarine
communicating with, 866
tourist, 509
Submerged volume, 492–493
Submersion, complete, 488–489
Subscripts, of velocities, 70
Subshell, 1051–1055
Subtitles for the hearing impaired, movie, 874
Subtraction
of physical quantities, 7
of vectors, 63–64, 65
problems, 76
Sulfur, 1054
Sulfur dioxide, center of mass of, 277
Sun
blackbody radiation and, 1005
convection currents and, 536
distance of planet from and its period, 369
entropy and, 623
eruption of matter on, 744
evaporating, 1108
fusion reactions on, 987
gravitational field of, 992
helioseismology, 428
intensity of light from, 442, 866
Mercury and, 371–372
nuclear fusion in, 1094
orbital motion and, 367, 368–369, 370
proton-proton cycle of, 1094
sunspots on, 427, 536, 737
tides and, 385
Sunsets, 859
Sunspots, 427, 536, 737
Superconducting solenoid, 765
Superconductivity, 702
Superposition, 360–362, 938–940
of electric fields, 640–643
of electric potential, 670–672
electric potential energy and, 672–673
of forces, 634–636
problems, 471–472, 964–965
wave, 452–453
Supersonic erosion, 511
Supersonic waterfall, 53
Surface charge density, 637
Surface tension, 503–504
problems, 511
Surfactant, 504
Suspension bridge, longest, 543
Suspension point, 330, 331
Sweat, 569–570
Swim bladders, 491
Swing, natural frequency of, 420
Switch, opening or closing on *RL* circuit, 787
SWL (shock wave lithotripsy), 440, 441
Symbol
for battery, 697
for capacitors, 714
for resistor, 700
Symmetry
of flight, 41–42
in projectile motion, 94–96
Système International (SI), 2
System of objects, momentum of, 242–243
Systems with changing mass, 269–270
problems, 277–278

Tacoma Narrows Bridge, 420
Talus slope, 149
Tangential acceleration, 167
centripetal acceleration vs., 293
of rotating object, 293
Tangential force, 316–317
Tangential motion, angular momentum of, 336
Tangential speed, 293
of rotating object, 289–291
Tape recording, magnetic, 774
Tau, 1100
Tau neutrino, 1100
T coil (telecoil), 774
Teflon, dielectrics and, 681, 683
Telecommunications, optical fibers and, 890
Telescopes, 909, 924–927
Hubble Space Telescope, 252, 346, 372, 877, 967, 1012
length of, 926
problems, 934, 935, 936
Television accelerator, 1001
Televisions
capacitors in, 685
color, 957
screens, 646
tuning, 827, 840
waves, 846
Temperature, 514, 515
blackbody radiation and, 1004–1005
body, 532, 535, 546
chemical reactions and, 559
coldest place in universe, 545
constant, 553, 591–593
cricket thermometer, 546
distribution of speeds and, 556
excited states and, 1043
ferromagnets and, 756
fusion curve and melting, 568
greatest change in, 543
heat added or removed vs., 571–572
heat required for change in, 572–573
hottest living things, 546
isothermal processes, 591–593
kinetic energy and, 557–559
lowest, on Earth, 543
phase changes and, 549
phase equilibrium and, 566
and pressure of ideal gas, 549–555
of stars, 536, 1005
using color to measure, 536
volume for ideal gas vs., 554–555
zeroth law of thermodynamics and, 515–516
See also Heat; Thermodynamic processes
Temperature change, 524
Temperature dependence, 701–702
Temperature scales, 516–520
absolute zero, 518–519, 520, 585, 614
Celsius scale, 516, 517–519, 520
Fahrenheit scale, 516–518, 520
Kelvin scale. *See* Kelvin scale
problems, 543
TEM (transmission electron micrograph), 1020
Tension
translational equilibrium and, 154–158
work and energy to calculate, 239

Tension, in strings, 150–152, 159, 177
pitch and, 458, 459
problems, 171–172
speed of wave and, 434
***tera*, 5**
Terminals, 697
Termometro Lentos, size of floats in, 907
Tesla, Nikola, 736
Tesla (T), 736–737
Theremin, 682, 683
Thermal conductivity, 532, 533–534
Thermal contact, 515
Thermal equilibrium, 515, 516
heat flow and, 529
Thermal expansion, 520–526
area, 523–524
coefficients of, 521
linear, 520–523, 524
problems, 543–544
relation between linear and volume expansion, 524
special properties of water, 525
volume, 524–525
Thermal expansion joints, 522–523
Thermistors, 702
Thermodynamic processes, 588–595
adiabatic, 593–595
constant pressure, 589–591
constant volume, 589–591
isothermal, 591–593
problems, 619–621
summary of characteristics, 595
Thermodynamics, 515
laws of. *See* Laws of thermodynamics
Thermograms, 847
Thermometers, 702
bimetallic strip as, 522
clinical, 520
constant-volume gas thermometer, 518, 549
Thermonuclear fusion reaction, 1094
Thermos, 538
Thin films, 947–949
Thin-lens equation, 894–896, 914
problems, 906
Third harmonic, 456, 457, 460
Third law of thermodynamics, 585, 614
Thompson, Benjamin (Count Rumford), 526
Thomson, J. J., 1035
Thomson model of the atom, 1035
Thorium, 1109
3-D movies, 863
Three-dimensional standing waves, 1049–1050
Thrust, 270
Thunder, distance of lightning strike and, 438–439
Thundersticks at ball games, 474
Thyroid, treating, 1097, 1107
Tidal bulges, 385
Tidal locking, 385–386
Tides, 384–386
problems, 391–392
Tile, thermal conductivity of, 532
Time(s), 1, 2, 3–4
acceleration and, 25, 30, 32–33, 42, 402
arrow of, 600, 610
charge on capacitor and, 717–718
distance as function of, 80
energy as function of, in simple harmonic motion, 411
expressed in terms of period, 403
harmonic wave function and, 436–437
position as function of, 31–33, 39, 40, 42, 82–83
position of shadow as function of, 400
position vs., 33, 397, 398, 400
potential energy as function of, 410
proper, 974, 978
rate of change of angular momentum and, 338
relativity of, 972–977
problems, 999
speed as function of, 26–27
typical, 4
uncertainty in, 1024
velocity as function of, 30–31, 32, 34, 36, 39, 40, 42, 82
velocity vs., 401
wave speed and, 435
Time constant, 717
Time dependence
in alternating-current circuits, 803
of current in *RL* circuit, 787
root mean square value of quantity with sinusoidal, 805
of voltage and current in alternating-current capacitor circuit, 810–812
Time dilation, 973–977
problems, 999
Tires, static friction and, 149
Titan, gravity on, 390
***Titanic*, 481**
Toboggan, 131–132
Top (quark), 1101
Tops, angular momentum and, 344
Tornadoes, conservation of angular momentum and, 340
Torque, 315, 316–329
angular acceleration and, 319–322
applications of, 333–335
defined, 317
external, 339
gravity and, 331
internal, 339
magnetic, 746–748, 783
mass and, 326
power produced by, 343
problems, 350–353
right-hand rule for, 344
sign convention for, 318
work done by, 343
zero, 323–332, 368
Torque vectors, 344–345
Torricelli, Evangelista, 482, 512
Torricelli's law, 500–501
Torsiversion, correcting, 352
Tossing objects, 216–218
Total electric field, 640
Total energy, 986
Total internal reflection, 888–890
Total polarization, 890–891
Total work, 184–186, 591
Traction, 150, 151, 173
Trains
accelerating, 27–28
colliding cars, 253, 254–255
Trajectory, 90
of eagle, 81
of horizontal launch of projectile, 85
Transcranial magnetic stimulation (TMS), 767, 801
Transformers, 790–792
problems, 800
transformer equations, 791
Transition elements, 1055
Translational analogies, 320
Translational equilibrium, 154–158, 324, 328
problems, 172–173
Translational motion, 294–295
Transmission axis, 854–857
Transmission electron micrograph (TEM), 1020
Transverse waves, 431, 840
Triboelectric charging, 628
Triceps muscle, 356
Triceratops, 351
Trigonometric relations, vectors and, 57–58
Trip distance, 19
Triple point, 548, 567, 568
Tritium, 1089–1090
Troughs, 432, 433
Truth (quark), 1101
Tsunami, 469, 473
Tungsten, 700, 1056, 1057
Tuning forks, use in neurology, 425
Tunneling, 1025–1026
Turbine, world's fastest, 309
Turbulence, 404
Turning points, 228–229
of motion, 401, 410
Turn signals, 719
Turntables
rotational collision and, 341–342
uniform circular motion and simple harmonic motion and, 399–400
T wave, 677
Two-dimensional coordinate system, 56, 57
Two-lens system, 913–914
Two-slit experiment, 940–944
problems, 965–966
wave-particle duality and, 1020–1021
***Tyrannosaurus rex*, 139, 355**

Ultracentrifuge, 167
Ultrasonic sounds, 440, 441
Ultrasound scan, 440, 441

Ultraviolet catastrophe, 1005, 1007
Ultraviolet light (UV rays), 536, 846, 848, 866
Uncertainty, measurement and, 6, 7
Uncertainty principle, Heisenberg, 1021–1025
Underdamped oscillations, 418, 419
Underwater vision, 894
Unified field theory, 1102
Unified forces, 1102
Uniform circular motion, simple harmonic motion and, 399–405
problems, 425–426
Uniform sphere, gravity and, 362–365
United Nations pendulum, 427
United States, geometric center, 331
U.S. Army Corps of Engineers, 291
Units of measurement, 1, 2–4
of acceleration, 112
of capacitance, 678
of capacitive reactance, 809–810
conversion of, 8–10
of electric charge, 626–627
of electric current, 696
of electric potential, 664
of electric power, 702
of energy usage, 705
of force, 112
of heat, 526–527, 529
of impedance, 813
of inductance, 785
of intensity of sound, 442
of magnetic flux, 770
of mass, 4, 112
of radioactive activity, 1083
of refractive power, 916
of resistance, 699
of torque, 316
of wavelengths, 845
See also individual units
Unit vectors, 64–65
problems, 76
Universal gas constant, 551
Universal gravitation constant, 359, 365
Universal law of gravitation, 358–362
Coulomb's law and, 632
problems, 389–390
Universe
entropy in, 610–612, 613
heat death of, 613
relativistic, 991
Unpolarized light, 853–858
polarized vs., 853, 854
Up (quark), 1101
Uranium, 1076
alpha decay, 1078–1079
enriching, 581
separation of isotopes, 742–743
Uranium-235
decay of, 1082–1083
nuclear fission and, 1090–1093
Uranium-238, 1085
decay of, 1079
Uranus, 358
UV rays, 536, 846, 848, 866

Vacuum
dielectric constant of, 681
free fall in, 38
speed of light in, 841, 971, 972
Van de Graaff generator, 628
van Leeuwenhoek, Antoni, 922–923
Vaporization, latent heat of, 571
Vapor pressure curve, 566, 581–582
Variable air capacitor, 680
Variable force, work done by, 190–194
problems, 201
Variable Specific Impulse Magnetoplasma Rocket (VASIMR), 139
Vasodilation, 511
Vector(s), 55–78
addition of, 65, 69–70
graphically, 61–62
problems, 76
using components, 62–63
components of, 56–60, 64–65, 111
problems, 74–75
defined, 11, 56
displacement, 56, 66
force as, 108, 121–124
force vectors, kinetic friction and, 145–146
impulse as, 245
momentum as, 241, 242–243
negative of, 63–64
position, 65–66
relative motion and, 69–71
rotational motion and, 344–346
scalars vs., 11, 56
subtraction of, 63–64, 65
problems, 76
unit, 64–65
See also Acceleration vectors; Displacement vector; Position vector; Torque; Velocity vectors
Vector direction angle, 57, 60
Vector sum, 61
Velocity(ies), 11, 21–25, 401
acceleration and, 28–29, 111, 402–403
angle of, 95
angular, 283–286, 344
average, 21–23, 25, 30, 31
of center of mass, 266
constant, 23, 24, 80–81
dimensions of, 5
in elastic collisions, 259–260
as function of distance, 32
as function of position, 30–31, 32, 34, 36, 39, 40, 42, 82
as function of time, 30–31, 32, 34, 36, 39, 40, 42, 82
instantaneous, 23–25
momentum and, 241
of Moon, 77
Newton's first law of motion and, 109–110
nonzero, 28
relative, 69–71
relativistic addition of, 981–983
problems, 1000
in simple harmonic motion, 401
speed, acceleration, and, 28–29
subscripts of, 70
time vs., 401
See also Acceleration
Velocity curve, 32
Velocity selector, 740, 741
Velocity vectors, 11, 56, 61, 66–67
acceleration vectors vs., 69
addition of, 69–70
average, 66
instantaneous, 66–67
problems, 76–77
for projectile launched at origin, 95
Ventricular fibrillation, 660
Verne, Jules, 50, 53, 58, 381, 382, 390
Vertical components of force, 327–329
Vertical lifting, 188–189
Vertical motion, 79
independence of, 84
Vertical resistors, 712–713
Vertical spring, mass on, 408–409
Very Large Array (VLA), 846
Vibrating columns of air, 459–463
Victoria Falls, 101
Violet light, 896, 897
Violin string, tension in, 177
Virtual image, 878, 882, 893, 918
Virtual objects, 883
Virus, 2
Viscosity, 501–503
problems, 511
Visible light, 536, 837, 846, 847–848
Vision
corrective lenses for, 913–919
dark-adapted, 1009, 1031
human, 1033
underwater, 894
Visualization, problem solving and, 12
Vitreous humor, 910
VLA (Very Large Array), 846
Volta, Alessandro, 664
Voltage(s), 803
alternating, 804–809
problems, 832–833
in alternating-current capacitor circuit, 810–812
in alternating-current inductor circuit, 817–818
of capacitor being charged, 683
determining, in secondary circuit, 790–791
inductor, 822
instantaneous, 811–812
measuring, 720–721
for phase angle, 816

relation to current, 791, 810–812, 827
in *RC* circuit, 812–813
in *RLC* circuit, 820
Voltage phasor, 804, 805, 811, 812
Voltmeters, 720–721
Volt (V), 698
defined, 664
electric potential vs., 664
electron, 664, 1008
Volume
adiabatic process and, 595
Bernoulli's equation and, 497
changing volume of solid, 564–565
constant, 589–591, 597
dimensions of, 5
in isotherm, 553
and pressure of ideal gas, 549–555
submerged, 492–493
temperature for ideal gas vs., 554–555
Volume expansion, 524–525
coefficient of, 521, 524
Volume flow rate, 502–503
***Voyager I* and *Voyager II* spacecraft, communicating with, 865**
***Voyage to the Bottom of the Sea* (movie), 506**
***v*-versus-*t* graph, 26, 27, 29, 30**

Walking speed, 6–7, 417
Warped space, 994, 995
Washington Monument, baseball thrown from top of, 102
Water
climate and, 529
density of, 477, 525
dielectric constant, 681
fusion curves, 568
index of refraction, 885
phase diagram for, 582
seeking own level, 483–485
specific heat of, 529, 573
speed distribution for, 569
thermal behavior of, 520, 525–526
vapor-pressure curve for, 566, 581–582
See also Ice
Water faucets, 522
Water fountain, 500–501
Water molecule, 75
center of mass of, 279
Water pipes, 526
Water purification, 639
Water waves, 432
deepwater and shallow-water, 470
motion of, 433
Watson, J. D., 959
Watt, James, 195
Watts per meter squared, 442
Watt (W), 195, 702
Wave(s)
beats, 463–465
defined, 431
gravity, 995
harmonic, 431
harmonic wave functions, 436–437
intensity and, 442
interference, 452, 453–455
light, 431
longitudinal, 432
matter, 1046
ocean, 431
period of, 433
in phase, 938–939
radio, 431, 837, 840, 846
reflected, 436, 944–950
sound intensity, 441–446
sound waves, 437–441, 446–452
sources in phase, 453–455
speed of, 433
spherical, 871
standing, 455–463, 1046, 1049–1050
superposition, 452–453
television, 846
transverse, 431, 840
types of, 431–433
problems, 469–470
water, 432, 433, 470
wavelength, frequency, and speed of, 432–433
See also Electromagnetic waves
Wave fronts, 871
Wave function, 1049
Wavelength, 432–433
calculating for Balmer series, 1037
Compton, 1015, 1033
Doppler-shifted, 448
frequency and, 845–848
of particle, momentum and, 1017
of photon emitted by electron, 1044–1045
red-shifted, from quasar, 960
Wave-particle duality, 954, 1020–1021
electromagnetic waves and, 853
problems, 1032
Wave properties of light, 937
See also Physical optics
Waves on a string, 431, 433–436
motion of, 432
problems, 470
standing, 455–459
Weak nuclear force, 1099–1100, 1102
Weather satellites, 372
Weber, Wilhelm, 770
Weber (Wb), 770
Weddell seal, 491
Wedgwood, Josiah, 536
Weed pulling, torque required for, 350
Weight, 1, 124–128
apparent, 126–128
buoyant force and, 487
of center of mass, 269
circular motion and, 165
defined, 124
displacement and, 487, 490
mass vs., 3, 124–125
normal force and, 128–132
problems, 138
Weightlessness, 128, 406
on equator, 312
Weightlifter, work done by, 182
Weinberg, Steven, 1102
Whales, 473
communication among, 440–441
Wheel(s)
right-hand rules for angular velocity and torque and, 344
rolling motion, 294–295
spinning, 288–289, 294, 303, 304
static friction and, 149
torque and, 318
Whistle, Doppler effect of, 449–450
White light, 896, 897, 1011–1012
Wien's displacement law, 1004–1005
Wind, 498–499, 499
Windmill, angular momentum of, 338
Windows
double-paned, 547
heat flow through, 533, 534–535
Windshield wipers, 719
Wing of a hawkmoth, 234
Wire, current-carrying
forces between, 751–752
magnetic field of, 748–749, 750
magnetic force exerted on, 744–746
problems, 763–764
potential difference in, 701
resistance of, 700–701
Wishing well, speed of sound and, 439
Woodpecker impact, 15
Work
angle and, 182–184, 185
on charged particle by electric field, 740
constant force and, 180–186, 190–191
problems, 200
defined when angle between force and displacement is θ, 182
defined when force is in direction of displacement, 180
done by conservative forces, 208–214
problems, 234
done by expanding gas, 589–590
done by gas, 589–595
done by nonconservative forces, 222–227
problems, 235–236
done by nonconstant force, 191
done by variable force, 190–194
problems, 201
done on fluid element, 495
as dot product, 183
electric force and, 663
electromotive force and, 698

Work (*Continued*)
entropy and, 611–612
force and, 179
gravity and, 186–187, 188
heat, internal energy, and, 209, 586–588
maximum, by heat engine, 603
mechanical work, 778–781
heat and, 526–528
positive and negative, 184–186, 187, 193–194
power and, 195–197
rotational, 343
sign convention for, 587
speed and, 186–190
total, 184–186, 591
typical values of, 181
zero, 184, 187
See also Mechanical work
Work-energy theorem, 186–190, 214, 222, 343
Bernoulli's equation and, 495–501
problems, 200–201
Work function, 1009–1010
Wrench, torque and, 316

***x* axis, 18, 56, 57**
***x* component of acceleration, 122, 123**
***x* component of vector, 57**
***x* coordinate of center of mass, 265**
Xenon, 1055
***x* equation of motion, 80–81**
X-ray(s), 846, 848, 1055–1058
damage, 1108
medical use of, 865, 1056, 1058, 1109
relative biological effectiveness for, 1096
X-ray diffraction, 959, 1018
by crystals, 1018
X-ray photon scattering, 1014–1016
X-ray tubes, 1056
***x* unit vector, 64**
***x*-versus-*t* graph, 22–23, 33**
***x*-versus-*t* values, 24**

***y* axis, 56, 57**
***y* component of acceleration, 122, 123, 124**
***y* component of vector, 57**
***y* coordinate of center of mass, 265**
***y* equation of motion, 80–81**
Yerkes refractor, 926, 927
Young, Thomas, 561, 940
Young's modulus, 561–562, 563
Young's two-slit experiment, 940–944
problems, 965–966
Yo-yo, 304, 313
Yttrium-90, 1097
***y* unit vector, 64**

***z* axis, 56**
Z dependence, 1042, 1043
Zero, absolute, 518–519, 520, 585, 614
Zero field within a conductor, 647
Zero launch angle, 84–89
problems, 101
Zero level, of potential energy, 210, 375
Zero net external force, 266–268
Zero rest mass, 1013
Zeros, 7
Zeroth law of thermodynamics, 515–516, 586
Zero torque
balance and, 330–332
gravity and, 368
problems, 351–352
static equilibrium and, 323–329
Zero work, 184, 187
Zhuang Yong, 48
Zinc, 1055
Z machine, 1095
Zweig, George, 1100–1101

MULTIPLES AND PREFIXES FOR METRIC UNITS

Multiple	Prefix (Abbreviation)	Pronunciation
10^{24}	yotta- (Y)	yot'ta (*a* as in *a*bout)
10^{21}	zetta- (Z)	zet'ta (*a* as in *a*bout)
10^{18}	exa- (E)	ex'a (*a* as in *a*bout)
10^{15}	peta- (P)	pet'a (as in *peta*l)
10^{12}	tera- (T)	ter'a (as in *terra*ce)
10^{9}	giga- (G)	ji'ga (*ji* as in *ji*ggle, *a* as in *a*bout)
10^{6}	mega- (M)	meg'a (as in *mega*phone)
10^{3}	kilo- (k)	kil'o (as in *kilo*watt)
10^{2}	hecto- (h)	hek'to (*heck-toe*)
10	deka- (da)	dek'a (*deck* plus *a* as in *a*bout)
10^{-1}	deci- (d)	des'i (as in *deci*mal)
10^{-2}	centi- (c)	sen'ti (as in *senti*mental)
10^{-3}	milli- (m)	mil'li (as in *mili*tary)
10^{-6}	micro- (μ)	mi'kro (as in *micro*phone)
10^{-9}	nano- (n)	nan'oh (*an* as in *an*nual)
10^{-12}	pico- (p)	pe'ko (*peek-oh*)
10^{-15}	femto- (f)	fem'toe (*fem* as in *fem*inine)
10^{-18}	atto- (a)	at'toe (as in an*atom*y)
10^{-21}	zepto- (z)	zep'toe (as in *zep*pelin)
10^{-24}	yocto- (y)	yock'toe (as in *sock*)

THE GREEK ALPHABET

Alpha	A	α
Beta	B	β
Gamma	Γ	γ
Delta	Δ	δ
Epsilon	E	ε
Zeta	Z	ζ
Eta	H	η
Theta	Θ	θ
Iota	I	ι
Kappa	K	κ
Lambda	Λ	λ
Mu	M	μ
Nu	N	ν
Xi	Ξ	ξ
Omicron	O	o
Pi	Π	π
Rho	P	ρ
Sigma	Σ	σ
Tau	T	τ
Upsilon	Υ	υ
Phi	Φ	ϕ
Chi	X	χ
Psi	Ψ	ψ
Omega	Ω	ω

SI BASE UNITS

Physical Quantity	Name of Unit	Symbol
Length	meter	m
Mass	kilogram	kg
Time	second	s
Electric current	ampere	A
Temperature	kelvin	K
Amount of substance	mole	mol
Luminous intensity	candela	cd

SOME SI DERIVED UNITS

Physical Quantity	Name of Unit	Symbol	SI Unit
Frequency	hertz	Hz	s^{-1}
Energy	joule	J	$kg \cdot m^2/s^2$
Force	newton	N	$kg \cdot m/s^2$
Pressure	pascal	Pa	$kg/(m \cdot s^2)$
Power	watt	W	$kg \cdot m^2/s^3$
Electric charge	coulomb	C	$A \cdot s$
Electric potential	volt	V	$kg \cdot m^2/(A \cdot s^3)$
Electric resistance	ohm	Ω	$kg \cdot m^2/(A^2 \cdot s^3)$
Capacitance	farad	F	$A^2 \cdot s^4/(kg \cdot m^2)$
Inductance	henry	H	$kg \cdot m^2/(A^2 \cdot s^2)$
Magnetic field	tesla	T	$kg/(A \cdot s^2)$
Magnetic flux	weber	Wb	$kg \cdot m^2/(A \cdot s^2)$

SI UNITS OF SOME OTHER PHYSICAL QUANTITIES

Physical Quantity	SI Unit
Density (ρ)	kg/m^3
Speed (v)	m/s
Acceleration (a)	m/s^2
Momentum, impulse (p)	$kg \cdot m/s$
Angular speed (ω)	rad/s
Angular acceleration (α)	rad/s^2
Torque (τ)	$kg \cdot m^2/s^2$ *or* $N \cdot m$
Specific heat (c)	$J/(kg \cdot K)$
Thermal conductivity (k)	$W/(m \cdot K)$ *or* $J/(s \cdot m \cdot K)$
Entropy (S)	J/K *or* $kg \cdot m^2/(K \cdot s^2)$ *or* $N \cdot m/K$
Electric field (E)	N/C or V/m

APPENDICES IN THE TEXT

Appendix A	Basic Mathematical Tools	p. APP-1
Appendix B	Typical Values	p. APP-11
Appendix C	Planetary Data	p. APP-12
Appendix D	Electrical Circuit Elements	p. APP-13
Appendix E	Periodic Table of the Elements	p. APP-14
Appendix F	Properties of Selected Isotopes	p. APP-15